MEMPHIS LIBRARIES
P9-CDJ-437
STOP
NOT TO BE TAKEN FROM THE LIBRARY!

Comprehensive Natural Products Chemistry

Comprehensive Natural Products Chemistry

Editors-in-Chief

Sir Derek Barton†

Texas A&M University, USA

Koji Nakanishi

Columbia University, USA

Executive Editor

Otto Meth-Cohn

University of Sunderland, UK

Volume 8

MISCELLANEOUS NATURAL PRODUCTS INCLUDING MARINE NATURAL PRODUCTS, PHEROMONES, PLANT HORMONES, AND ASPECTS OF ECOLOGY

Volume Editor

Kenji Mori

Science University of Tokyo, Japan

1999

ELSEVIER

AMSTERDAM – LAUSANNE – NEW YORK – OXFORD – SHANNON – SINGAPORE – TOKYO

Elsevier Science Ltd., The Boulevard, Langford Lane, Kidlington, Oxford, OX5 1GB, UK

First edition 1999

Library of Congress Cataloging-in-Publication Data
Comprehensive natural products chemistry / editors-in-chief, Sir Derek Barton, Koji Nakanishi ; executive editor, Otto Meth-Cohn. -- 1st ed.
p. cm.
Includes index.
Contents: v. 8. Miscellaneous natural products including marine natural products, pheromones, plant hormones, and aspects of ecology / volume editor Kenji Mori
1. Natural products. I. Barton, Derek, Sir, 1918-1998. II. Nakanishi, Koji, 1925- . III. Meth-Cohn, Otto.
QD415.C63 1999
547.7--dc21 98-15249

British Library Cataloguing in Publication Data
Comprehensive natural products chemistry
1. Organic compounds
I. Barton, Sir Derek, 1918-1998 II. Nakanishi Koji III. Meth-Cohn Otto
572.5

ISBN 0-08-042709-X (set : alk. paper)
ISBN 0-08-043160-7 (Volume 8 : alk. paper)

∞™ The paper used in this publication meets the minimum requirements of the American National Standard for Information Sciences—Permanence of Paper for Printed Library Materials, ANSI Z39.48–1984.

Typeset by BPC Digital Data Ltd., Glasgow, UK.
Printed and bound in Great Britain by BPC Wheatons Ltd., Exeter, UK.

Contents

Introduction

For many decades, Natural Products Chemistry has been the principal driving force for progress in Organic Chemistry.

In the past, the determination of structure was arduous and difficult. As soon as computing became easy, the application of X-ray crystallography to structural determination quickly surpassed all other methods. Supplemented by the equally remarkable progress made more recently by Nuclear Magnetic Resonance techniques, determination of structure has become a routine exercise. This is even true for enzymes and other molecules of a similar size. Not to be forgotten remains the progress in mass spectrometry which permits another approach to structure and, in particular, to the precise determination of molecular weight.

There have not been such revolutionary changes in the partial or total synthesis of Natural Products. This still requires effort, imagination and time. But remarkable syntheses have been accomplished and great progress has been made in stereoselective synthesis. However, the one hundred percent yield problem is only solved in certain steps in certain industrial processes. Thus there remains a great divide between the reactions carried out in living organisms and those that synthetic chemists attain in the laboratory. Of course Nature edits the accuracy of DNA, RNA, and protein synthesis in a way that does not apply to a multi-step Organic Synthesis.

Organic Synthesis has already a significant component that uses enzymes to carry out specific reactions. This applies particularly to lipases and to oxidation enzymes. We have therefore, given serious attention to enzymatic reactions.

No longer standing in the wings, but already on-stage, are the wonderful tools of Molecular Biology. It is now clear that multi-step syntheses can be carried out in one vessel using multiple cloned enzymes. Thus, Molecular Biology and Organic Synthesis will come together to make economically important Natural Products.

From these preliminary comments it is clear that Natural Products Chemistry continues to evolve in different directions interacting with physical methods, Biochemistry, and Molecular Biology all at the same time.

This new Comprehensive Series has been conceived with the common theme of "How does Nature make all these molecules of life?" The principal idea was to organize the multitude of facts in terms of Biosynthesis rather than structure. The work is not intended to be a comprehensive listing of natural products, nor is it intended that there should be any detail about biological activity. These kinds of information can be found elsewhere.

The work has been planned for eight volumes with one more volume for Indexes. As far as we are aware, a broad treatment of the whole of Natural Products Chemistry has never been attempted before. We trust that our efforts will be useful and informative to all scientific disciplines where Natural Products play a role.

D. H. R. Barton† K. Nakanishi O. Meth-Cohn

Preface

It is surprising indeed that this work is the first attempt to produce a "comprehensive" overview of Natural Products beyond the student text level. However, the awe-inspiring breadth of the topic, which in many respects is still only developing, is such as to make the job daunting to anyone in the field. Fools rush in where angels fear to tread and the particular fool in this case was myself, a lifelong enthusiast and reader of the subject but with no research base whatever in the field!

Having been involved in several of the *Comprehensive* works produced by Pergamon Press, this omission intrigued me and over a period of gestation I put together a rough outline of how such a work could be written and presented it to Pergamon. To my delight they agreed that the project was worthwhile and in short measure Derek Barton was approached and took on the challenge of fleshing out this framework with alacrity. He also brought his long-standing friend and outstanding contributor to the field, Koji Nakanishi, into the team. With Derek's knowledge of the whole field, the subject was broken down into eight volumes and an outstanding team of internationally recognised Volume Editors was appointed.

We used Derek's 80th birthday as a target for finalising the work. Sadly he died just a few months before reaching this milestone. This work therefore is dedicated to the memory of Sir Derek Barton, Natural Products being the area which he loved best of all.

OTTO METH-COHN
Executive Editor

SIR DEREK BARTON

Sir Derek Barton, who was Distinguished Professor of Chemistry at Texas A&M University and holder of the Dow Chair of Chemical Invention died on March 16, 1998 in College Station, Texas of heart failure. He was 79 years old and had been Chairman of the Executive Board of Editors for Tetrahedron Publications since 1979.

Barton was considered to be one of the greatest organic chemists of the twentieth century whose work continues to have a major influence on contemporary science and will continue to do so for future generations of chemists.

Derek Harold Richard Barton was born on September 8, 1918 in Gravesend, Kent, UK and graduated from Imperial College, London with the degrees of B.Sc. (1940) and Ph.D. (1942). He carried out work on military intelligence during World War II and after a brief period in industry, joined the faculty at Imperial College. It was an early indication of the breadth and depth of his chemical knowledge that his lectureship was in physical chemistry. This research led him into the mechanism of elimination reactions and to the concept of molecular rotation difference to correlate the configurations of steroid isomers. During a sabbatical leave at Harvard in 1949–1950 he published a paper on the "Conformation of the Steroid Nucleus" (*Experientia*, 1950, **6**, 316) which was to bring him the Nobel Prize in Chemistry in 1969, shared with the Norwegian chemist, Odd Hassel. This key paper (only four pages long) altered the way in which chemists thought about the shape and reactivity of molecules, since it showed how the reactivity of functional groups in steroids depends on their axial or equatorial positions in a given conformation. Returning to the UK he held Chairs of Chemistry at Birkbeck College and Glasgow University before returning to Imperial College in 1957, where he developed a remarkable synthesis of the steroid hormone, aldosterone, by a photochemical reaction known as the Barton Reaction (nitrite photolysis). In 1978 he retired from Imperial College and became Director of the Natural Products Institute (CNRS) at Gif-sur-Yvette in France where he studied the invention of new chemical reactions, especially the chemistry of radicals, which opened up a whole new area of organic synthesis involving Gif chemistry. In 1986 he moved to a third career at Texas A&M University as Distinguished Professor of Chemistry and continued to work on novel reactions involving radical chemistry and the oxidation of hydrocarbons, which has become of great industrial importance. In a research career spanning more than five decades, Barton's contributions to organic chemistry included major discoveries which have profoundly altered our way of thinking about chemical structure and reactivity. His chemistry has provided models for the biochemical synthesis of natural products including alkaloids, antibiotics, carbohydrates, and DNA. Most recently his discoveries led to models for enzymes which oxidize hydrocarbons, including methane monooxygenase.

The following are selected highlights from his published work:

The 1950 paper which launched Conformational Analysis was recognized by the Nobel Prize Committee as the key contribution whereby the third dimension was added to chemistry. This work alone transformed our thinking about the connection between stereochemistry and reactivity, and was later adapted from small molecules to macromolecules e.g., DNA, and to inorganic complexes.

Barton's breadth and influence is illustrated in "Biogenetic Aspects of Phenol Oxidation" (*Festschr. Arthur Stoll*, 1957, 117). This theoretical work led to many later experiments on alkaloid biosynthesis and to a set of rules for *ortho-para*-phenolic oxidative coupling which allowed the predication of new natural product systems before they were actually discovered and to the correction of several erroneous structures.

In 1960, his paper on the remarkably short synthesis of the steroid hormone aldosterone (*J. Am. Chem. Soc.*, 1960, **82**, 2641) disclosed the first of many inventions of new reactions—in this case nitrite photolysis—to achieve short, high yielding processes, many of which have been patented and are used worldwide in the pharmaceutical industry.

Moving to 1975, by which time some 500 papers had been published, yet another "Barton reaction" was born—"The Deoxygenation of Secondary Alcohols" (*J. Chem. Soc. Perkin Trans. 1*, 1975, 1574), which has been very widely applied due to its tolerance of quite hostile and complex local environments in carbohydrate and nucleoside chemistry. This reaction is the chemical counterpart to ribonucleotide® deoxyribonucleotide reductase in biochemistry and, until the arrival of the Barton reaction, was virtually impossible to achieve.

In 1985, "Invention of a New Radical Chain Reaction" involved the generation of carbon radicals from carboxylic acids (*Tetrahedron*, 1985, **41**, 3901). The method is of great synthetic utility and has been used many times by others in the burgeoning area of radicals in organic synthesis.

These recent advances in synthetic methodology were remarkable since his chemistry had virtually no precedent′in the work of others. The radical methodology was especially timely in light of the significant recent increase in applications for fine chemical syntheses, and Barton gave the organic community an entrée into what will prove to be one of the most important methods of the twenty-first century. He often said how proud he was, at age 71, to receive the ACS Award for Creativity in Organic Synthesis for work published in the preceding five years.

Much of Barton's more recent work is summarized in the articles "The Invention of Chemical Reactions—The Last 5 Years" (*Tetrahedron*, 1992, **48**, 2529) and "Recent Developments in Gif Chemistry" (*Pure Appl. Chem.*, 1997, **69**, 1941).

Working 12 hours a day, Barton's stamina and creativity remained undiminished to the day of his death. The author of more than 1000 papers in chemical journals, Barton also held many successful patents. In addition to the Nobel Prize he received many honors and awards including the Davy, Copley, and Royal medals of the Royal Society of London, and the Roger Adams and Priestley Medals of the American Chemical Society. He held honorary degrees from 34 universities. He was a Fellow of the Royal Societies of London and Edinburgh, Foreign Associate of the National Academy of Sciences (USA), and Foreign Member of the Russian and Chinese Academies of Sciences. He was knighted by Queen Elizabeth in 1972, received the Légion d'Honneur (Chevalier 1972; Officier 1985) from France, and the Order of the Rising Sun from the Emperor of Japan. In his long career, Sir Derek trained over 300 students and postdoctoral fellows, many of whom now hold major positions throughout the world and include some of today's most distinguished organic chemists.

For those of us who were fortunate to know Sir Derek personally there is no doubt that his genius and work ethic were unique. He gave generously of his time to students and colleagues wherever he traveled and engendered such great respect and loyalty in his students and co-workers, that major symposia accompanied his birthdays every five years beginning with the 60th, ending this year with two celebrations just before his 80th birthday.

With the death of Sir Derek Barton, the world of science has lost a major figure, who together with Sir Robert Robinson and Robert B. Woodward, the cofounders of *Tetrahedron*, changed the face of organic chemistry in the twentieth century.

Professor Barton is survived by his wife, Judy, and by a son, William from his first marriage, and three grandchildren.

A. I. SCOTT
Texas A&M University

Reprinted from *Tetrahedron*, 1998, **54**, 8847

Contributors to Volume 8

Professor W. Francke
Institut für Organische Chemie, Universität Hamburg, Martin Luther King Platz 6, D 20146 Hamburg, Germany

Professor J. B. Harborne
Department of Botany, Plant Science Laboratories, University of Reading, Whiteknights, Reading RG6 6AS, UK

Dr. N. Hirai
Division of Applied Life Sciences, Graduate School of Agriculture, Kyoto University, Kyoto 606-8502, Japan

Dr. H. Imaseki
Graduate Division of Biochemical Regulation, Nagoya University, Obata-Minami 2-4-19, Moriyama, Nagoya, Japan

Professor M. Ishibashi
Faculty of Pharmaceutical Sciences, Hokkaido University, Sapporo 060, Japan

Dr. H. Iwamura
Division of Applied Life Sciences, Graduate School of Agriculture, Kyoto University, Kyoto 606-8502, Japan

Dr. Y. Kamiya
FRP, RIKEN, Hirosawa 2-1, Wako-shi, Saitama 351-01, Japan

Professor J. Kobayashi
Faculty of Pharmaceutical Sciences, Hokkaido University, Sapporo 060, Japan

Professor E. D. Morgan
Department of Chemistry, Keele University, Keele, Staffordshire ST5 5BG, UK

Professor K. Mori
Department of Chemistry, Science University of Tokyo, Kagurazaka 1-3, Shinjuku-ku, Tokyo 162, Japan

Professor N. Murofushi
Department of Applied Biological Chemistry, Graduate School of Agriculture and Agricultural Life Sciences, The University of Tokyo, Bunkyo-ku, Tokyo 113-8657, Japan

Professor T. Nihira
Department of Biotechnology, Faculty of Engineering, Osaka University, Yamadaoka 2-1, Suita-shi, Osaka 565, Japan

Dr. Y. Sakagami
School of Agricultural Sciences, Nagoya University, Chikusa-ku, Nagoya 464-01, Japan

Dr. S. Schulz
Institut für Organische Chemie, Universität Hamburg, Martin-Luther-King-Platz 6, D-20146 Hamburg, Germany

Dr. H. Tsuji
Department of Biology, Kobe Women's University, 2-1 Aoyama, Higashisuma, Suma, Kobe 654, Japan

Dr. J. Ueda
College of Integrated Arts and Sciences, Osaka Prefecture University, 1-1 Gakuen-cho, Sakai, Osaka 593, Japan

Professor I. D. Wilson
Department of Chemistry, Keele University, Keele, Staffordshire ST5 5BG, UK

Dr. Y. Yamada
Department of Biotechnology, Faculty of Engineering, Osaka University, 2-1 Yamada-oka, Suita-shi, Osaka 565, Japan

Dr. H. Yamane
Biotechnology Research Center, The University of Tokyo, Yayoi 1-1-1, Bunkyo-ku, Tokyo 113, Japan

Dr. T. Yokota
Department of Biosciences, Teikyo University, Utsunomiya 320, Japan

Abbreviations

The most commonly used abbreviations in *Comprehensive Natural Products Chemistry* are listed below. Please note that in some instances these may differ from those used in other branches of chemistry

A	adenine
ABA	abscisic acid
Ac	acetyl
ACAC	acetylacetonate
ACTH	adrenocorticotropic hormone
ADP	adenosine 5'-diphosphate
AIBN	2,2'-azobisisobutyronitrile
Ala	alanine
AMP	adenosine 5'-monophosphate
APS	adenosine 5'-phosphosulfate
Ar	aryl
Arg	arginine
ATP	adenosine 5'-triphosphate
B	nucleoside base (adenine, cylosine, guanine, thymine or uracil)
9-BBN	9-borabicyclo[3.3.1]nonane
BOC	*t*-butoxycarbonyl (or carbo-*t*-butoxy)
BSA	*N*,*O*-bis(trimethylsilyl)acetamide
BSTFA	*N*,*O*-bis(trimethylsilyl)trifluoroacetamide
Bu	butyl
Bu^n	*n*-butyl
Bu^i	isobutyl
Bu^s	*s*-butyl
Bu^t	*t*-butyl
Bz	benzoyl
CAN	ceric ammonium nitrate
CD	cyclodextrin
CDP	cytidine 5'-diphosphate
CMP	cytidine 5'-monophosphate
CoA	coenzyme A
COD	cyclooctadiene
COT	cyclooctatetraene
Cp	h^5-cyclopentadiene
Cp*	pentamethylcyclopentadiene
12-Crown-4	1,4,7,10-tetraoxacyclododecane
15-Crown-5	1,4,7,10,13-pentaoxacyclopentadecane
18-Crown-6	1,4,7,10,13,16-hexaoxacyclooctadecane
CSA	camphorsulfonic acid
CSI	chlorosulfonyl isocyanate
CTP	cytidine 5'-triphosphate
cyclic AMP	adenosine 3',5'-cyclic monophosphoric acid
CySH	cysteine
DABCO	1,4-diazabicyclo[2.2.2]octane
DBA	dibenz[*a*,*h*]anthracene
DBN	1,5-diazabicyclo[4.3.0]non-5-ene

DBU	1,8-diazabicyclo[5.4.0]undec-7-ene
DCC	dicyclohexylcarbodiimide
DEAC	diethylaluminum chloride
DEAD	diethyl azodicarboxylate
DET	diethyl tartrate (+ or -)
DHET	dihydroergotoxine
DIBAH	diisobutylaluminum hydride
Diglyme	diethylene glycol dimethyl ether (or bis(2-methoxyethyl)ether)
DiHPhe	2,5-dihydroxyphenylalanine
Dimsyl Na	sodium methylsulfinylmethide
DIOP	2,3-*O*-isopropylidene-2,3-dihydroxy-1,4-bis(diphenylphosphino)butane
dipt	diisopropyl tartrate (+ or -)
DMA	dimethylacetamide
DMAD	dimethyl acetylenedicarboxylate
DMAP	4-dimethylaminopyridine
DME	1,2-dimethoxyethane (glyme)
DMF	dimethylformamide
DMF-DMA	dimethylformamide dimethyl acetal
DMI	1,3-dimethyl-2-imidazalidinone
DMSO	dimethyl sulfoxide
DMTSF	dimethyl(methylthio)sulfonium fluoroborate
DNA	deoxyribonucleic acid
DOCA	deoxycorticosterone acetate
EADC	ethylaluminum dichloride
EDTA	ethylenediaminetetraacetic acid
EEDQ	*N*-ethoxycarbonyl-2-ethoxy-1,2-dihydroquinoline
Et	ethyl
EVK	ethyl vinyl ketone
FAD	flavin adenine dinucleotide
Fl	flavin
FMN	flavin mononucleotide
G	guanine
GABA	4-aminobutyric acid
GDP	guanosine 5'-diphosphate
GLDH	glutamate dehydrogenase
gln	glutamine
Glu	glutamic acid
Gly	glycine
GMP	guanosine 5'-monophosphate
GOD	glucose oxidase
G-6-P	glucose-6-phosphate
GTP	guanosine 5'-triphosphate
Hb	hemoglobin
His	histidine
HMPA	hexamethylphosphoramide (or hexamethylphosphorous triamide)
Ile	isoleucine
INAH	isonicotinic acid hydrazide
IpcBH	isopinocampheylborane
Ipc_2BH	diisopinocampheylborane
KAPA	potassium 3-aminopropylamide
K-Slectride	potassium tri-*s*-butylborohydride

LAH	lithium aluminum hydride
LAP	leucine aminopeptidase
LDA	lithium diisopropylamide
LDH	lactic dehydrogenase
Leu	leucine
LICA	lithium isopropylcyclohexylamide
L-Selectride	lithium tri-*s*-butylborohydride
LTA	lead tetraacetate
Lys	lysine
MCPBA	*m*-chloroperoxybenzoic acid
Me	methyl
MEM	methoxyethoxymethyl
MEM-Cl	ß-methoxyethoxymethyl chloride
Met	methionine
MMA	methyl methacrylate
MMC	methyl magnesium carbonate
MOM	methoxymethyl
Ms	mesyl (or methanesulfonyl)
MSA	methanesulfonic acid
MsCl	methanesulfonyl chloride
MVK	methyl vinyl ketone
NAAD	nicotinic acid adenine dinucleotide
NAD	nicotinamide adenine dinucleotide
NADH	nicotinamide adenine dinucleotide phosphate, reduced
NBS	*N*-bromosuccinimider
NMO	*N*-methylmorpholine *N*-oxide monohydrate
NMP	*N*-methylpyrrolidone
PCBA	*p*-chlorobenzoic acid
PCBC	*p*-chlorobenzyl chloride
PCBN	*p*-chlorobenzonitrile
PCBTF	*p*-chlorobenzotrifluoride
PCC	pyridinium chlorochromate
PDC	pyridinium dichromate
PG	prostaglandin
Ph	phenyl
Phe	phenylalanine
Phth	phthaloyl
PPA	polyphosphoric acid
PPE	polyphosphate ester (or ethyl *m*-phosphate)
Pr	propyl
Pr^i	isopropyl
Pro	proline
Py	pyridine
RNA	ribonucleic acid
Rnase	ribonuclease
Ser	serine
Sia_2BH	disiamylborane
TAS	tris(diethylamino)sulfonium
TBAF	tetra-*n*-butylammonium fluoroborate
TBDMS	*t*-butyldimethylsilyl
TBDMS-Cl	*t*-butyldimethylsilyl chloride
TBDPS	*t*-butyldiphenylsilyl
TCNE	tetracyanoethene

TES	triethylsilyl
TFA	trifluoracetic acid
TFAA	trifluoroacetic anhydride
THF	tetrahydrofuran
THF	tetrahydrofolic acid
THP	tetrahydropyran (or tetrahydropyranyl)
Thr	threonine
TMEDA	*N*,*N*,*N'*,*N'*,tetramethylethylenediamine[1,2-bis(dimethylamino)ethane]
TMS	trimethylsilyl
TMS-Cl	trimethylsilyl chloride
TMS-CN	trimethylsilyl cyanide
Tol	toluene
TosMIC	tosylmethyl isocyanide
TPP	tetraphenylporphyrin
Tr	trityl (or triphenylmethyl)
Trp	tryptophan
Ts	tosyl (or *p*-toluenesulfonyl)
TTFA	thallium trifluoroacetate
TTN	thallium(III) nitrate
Tyr	tyrosine
Tyr-OMe	tyrosine methyl ester
U	uridine
UDP	uridine 5'-diphosphate
UMP	uridine 5'-monophosphate

Contents of All Volumes

An Historical Perspective of Natural Products Chemistry

KOJI NAKANISHI

Columbia University, New York, USA

To give an account of the rich history of natural products chemistry in a short essay is a daunting task. This brief outline begins with a description of ancient folk medicine and continues with an outline of some of the major conceptual and experimental advances that have been made from the early nineteenth century through to about 1960, the start of the modern era of natural products chemistry. Achievements of living chemists are noted only minimally, usually in the context of related topics within the text. More recent developments are reviewed within the individual chapters of the present volumes, written by experts in each field. The subheadings follow, in part, the sequence of topics presented in Volumes 1–8.

1. ETHNOBOTANY AND "NATURAL PRODUCTS CHEMISTRY"

Except for minerals and synthetic materials our surroundings consist entirely of organic natural products, either of prebiotic organic origins or from microbial, plant, or animal sources. These materials include polyketides, terpenoids, amino acids, proteins, carbohydrates, lipids, nucleic acid bases, RNA and DNA, etc. Natural products chemistry can be thought of as originating from mankind's curiosity about odor, taste, color, and cures for diseases. Folk interest in treatments for pain, for food-poisoning and other maladies, and in hallucinogens appears to go back to the dawn of humanity

For centuries China has led the world in the use of natural products for healing. One of the earliest health science anthologies in China is the Nei Ching, whose authorship is attributed to the legendary Yellow Emperor (thirtieth century BC), although it is said that the dates were backdated from the third century by compilers. Excavation of a Han Dynasty (206 BC–AD 220) tomb in Hunan Province in 1974 unearthed decayed books, written on silk, bamboo, and wood, which filled a critical gap between the dawn of medicine up to the classic Nei Ching; Book 5 of these excavated documents lists 151 medical materials of plant origin. Generally regarded as the oldest compilation of Chinese herbs is Shen Nung Pen Ts'ao Ching (Catalog of Herbs by Shen Nung), which is believed to have been revised during the Han Dynasty; it lists 365 materials. Numerous revisions and enlargements of Pen Ts'ao were undertaken by physicians in subsequent dynasties, the ultimate being the Pen Ts'ao Kang Mu (General Catalog of Herbs) written by Li Shih-Chen over a period of 27 years during the Ming Dynasty (1573–1620), which records 1898 herbal drugs and 8160 prescriptions. This was circulated in Japan around 1620 and translated, and has made a huge impact on subsequent herbal studies in Japan; however, it has not been translated into English. The number of medicinal herbs used in 1979 in China numbered 5267. One of the most famous of the Chinese folk herbs is the ginseng root *Panax ginseng*, used for health maintenance and treatment of various diseases. The active principles were thought to be the saponins called ginsenosides but this is now doubtful; the effects could well be synergistic between saponins, flavonoids, etc. Another popular folk drug, the extract of the Ginkgo tree, *Ginkgo biloba* L., the only surviving species of the Paleozoic era (250 million years ago) family which became extinct during the last few million years, is mentioned in the Chinese Materia Medica to have an effect in improving memory and sharpening mental alertness. The main constituents responsible for this are now understood to be ginkgolides and flavonoids, but again not much else is known. Clarifying the active constituents and mode of (synergistic) bioactivity of Chinese herbs is a challenging task that has yet to be fully addressed.

The Assyrians left 660 clay tablets describing 1000 medicinal plants used around 1900–400 BC, but the best insight into ancient pharmacy is provided by the two scripts left by the ancient Egyptians, who were masters of human anatomy and surgery because of their extensive mummification practices. The Edwin Smith Surgical Papyrus purchased by Smith in 1862 in Luxor (now in the New York Academy of Sciences collection), is one of the most important medicinal documents of the ancient Nile Valley, and describes the healer's involvement in surgery, prescription, and healing practices using plants, animals, and minerals. The Ebers Papyrus, also purchased by Edwin Smith in 1862, and then acquired by Egyptologist George Ebers in 1872, describes 800 remedies using plants, animals, minerals, and magic. Indian medicine also has a long history, possibly dating back to the second millennium BC. The Indian materia medica consisted mainly of vegetable drugs prepared from plants but also used animals, bones, and minerals such as sulfur, arsenic, lead, copper sulfate, and gold. Ancient Greece inherited much from Egypt, India, and China, and underwent a gradual transition from magic to science. Pythagoras (580–500 BC) influenced the medical thinkers of his time, including Aristotle (384–322 BC), who in turn affected the medical practices of another influential Greek physician Galen (129–216). The Iranian physician Avicenna (980–1037) is noted for his contributions to Aristotelian philosophy and medicine, while the German-Swiss physician and alchemist Paracelsus (1493–1541) was an early champion who established the role of chemistry in medicine.

The rainforests in Central and South America and Africa are known to be particularly abundant in various organisms of interest to our lives because of their rich biodiversity, intense competition, and the necessity for self-defense. However, since folk-treatments are transmitted verbally to the next generation via shamans who naturally have a tendency to keep their plant and animal sources confidential, the recipes tend to get lost, particularly with destruction of rainforests and the encroachment of "civilization." Studies on folk medicine, hallucinogens, and shamanism of the Central and South American Indians conducted by Richard Schultes (Harvard Botanical Museum, emeritus) have led to renewed activity by ethnobotanists, recording the knowledge of shamans, assembling herbaria, and transmitting the record of learning to the village.

Extracts of toxic plants and animals have been used throughout the world for thousands of years for hunting and murder. These include the various arrow poisons used all over the world. *Strychnos* and *Chondrodendron* (containing strychnine, etc.) were used in South America and called "curare," *Strophanthus* (strophantidine, etc.) was used in Africa, the latex of the upas tree *Antiaris toxicaria* (cardiac glycosides) was used in Java, while *Aconitum napellus*, which appears in Greek mythology (aconitine) was used in medieval Europe and Hokkaido (by the Ainus). The Colombian arrow poison is from frogs (batrachotoxins; 200 toxins have been isolated from frogs by B. Witkop and J. Daly at NIH). Extracts of *Hyoscyamus niger* and *Atropa belladonna* contain the toxic tropane alkaloids, for example hyoscyamine, belladonnine, and atropine. The belladonna berry juice (atropine) which dilates the eye pupils was used during the Renaissance by ladies to produce doe-like eyes (belladona means beautiful woman). The Efik people in Calabar, southeastern Nigeria, used extracts of the calabar bean known as esere (physostigmine) for unmasking witches. The ancient Egyptians and Chinese knew of the toxic effect of the puffer fish, fugu, which contains the neurotoxin tetrodotoxin (Y. Hirata, K. Tsuda, R. B. Woodward).

When rye is infected by the fungus *Claviceps purpurea*, the toxin ergotamine and a number of ergot alkaloids are produced. These cause ergotism or the "devil's curse," "St. Anthony's fire," which leads to convulsions, miscarriages, loss of arms and legs, dry gangrene, and death. Epidemics of ergotism occurred in medieval times in villages throughout Europe, killing tens of thousands of people and livestock; Julius Caesar's legions were destroyed by ergotism during a campaign in Gaul, while in AD 994 an estimated 50,000 people died in an epidemic in France. As recently as 1926, a total of 11,000 cases of ergotism were reported in a region close to the Urals. It has been suggested that the witch hysteria that occurred in Salem, Massachusetts, might have been due to a mild outbreak of ergotism. Lysergic acid diethylamide (LSD) was first prepared by A. Hofmann, Sandoz Laboratories, Basel, in 1943 during efforts to improve the physiological effects of the ergot alkaloids when he accidentally inhaled it. "On Friday afternoon, April 16, 1943," he wrote, "I was seized by a sensation of restlessness... ." He went home from the laboratory and "perceived an uninterrupted stream of fantastic dreams ... ." (*Helvetica Chimica Acta*).

Numerous psychedelic plants have been used since ancient times, producing visions, mystical fantasies (cats and tigers also seem to have fantasies?, see nepetalactone below), sensations of flying, glorious feelings in warriors before battle, etc. The ethnobotanists Wasson and Schultes identified "ololiqui," an important Aztec concoction, as the seeds of the morning glory *Rivea corymbosa* and gave the seeds to Hofmann who found that they contained lysergic acid amides similar to but less potent than LSD. Iboga, a powerful hallucinogen from the root of the African shrub *Tabernanthe iboga*, is used by the Bwiti cult in Central Africa who chew the roots to obtain relief from fatigue and hunger; it contains the alkaloid ibogamine. The powerful hallucinogen used for thousands of years by the American Indians, the peyote cactus, contains mescaline and other alkaloids. The Indian hemp plant, *Cannabis sativa*, has been used for making rope since 3000 BC, but when it is used for its pleasure-giving effects it is called cannabis and has been known in central Asia, China, India, and the Near East since ancient times. Marijuana, hashish (named after the Persian founder of the Assassins of the eleventh century, Hasan-e Sabbah), charas, ghanja, bhang, kef, and dagga are names given to various preparations of the hemp plant. The constituent responsible for the mind-altering effect is 1-tetrahydrocannabinol (also referred to as 9-THC) contained in 1%. R. Mechoulam (1930–, Hebrew University) has been the principal worker in the cannabinoids, including structure determination and synthesis of 9-THC (1964 to present); the Israeli police have also made a contribution by providing Mechoulam with a constant supply of marijuana. Opium (morphine) is another ancient drug used for a variety of pain-relievers and it is documented that the Sumerians used poppy as early as 4000 BC; the narcotic effect is present only in seeds before they are fully formed. The irritating secretion of the blister beetles, for example *Mylabris* and the European species *Lytta vesicatoria*, commonly called Spanish fly, was used medically as a topical skin irritant to remove warts but was also a major ingredient in so-called love potions (constituent is cantharidin, stereospecific synthesis in 1951, G. Stork, 1921–; prep. scale high-pressure Diels–Alder synthesis in 1985, W. G. Dauben, 1919–1996).

Plants have been used for centuries for the treatment of heart problems, the most important being the foxgloves *Digitalis purpurea* and *D. lanata* (digitalin, diginin) and *Strophanthus gratus* (ouabain). The bark of cinchona *Cinchona officinalis* (called quina-quina by the Indians) has been used widely among the Indians in the Andes against malaria, which is still one of the major infectious diseases; its most important alkaloid is quinine. The British protected themselves against malaria during the occupation of India through gin and tonic (quinine!). The stimulant coca, used by the Incas around the tenth century, was introduced into Europe by the conquistadors; coca beans are also commonly chewed in West Africa. Wine making was already practiced in the Middle East 6000–8000 years ago; Moors made date wines, the Japanese rice wine, the Vikings honey mead, the Incas maize chicha. It is said that the Babylonians made beer using yeast 5000–6000 years ago. As shown above in parentheses, alkaloids are the major constituents of the herbal plants and extracts used for centuries, but it was not until the early nineteenth century that the active principles were isolated in pure form, for example morphine (1816), strychnine (1817), atropine (1819), quinine (1820), and colchicine (1820). It was a century later that the structures of these compounds were finally elucidated.

2. DAWN OF ORGANIC CHEMISTRY, EARLY STRUCTURAL STUDIES, MODERN METHODOLOGY

The term "organic compound" to define compounds made by and isolated from living organisms was coined in 1807 by the Swedish chemist Jons Jacob Berzelius (1779–1848), a founder of today's chemistry, who developed the modern system of symbols and formulas in chemistry, made a remarkably accurate table of atomic weights and analyzed many chemicals. At that time it was considered that organic compounds could not be synthesized from inorganic materials *in vitro*. However, Friedrich Wöhler (1800–1882), a medical doctor from Heidelberg who was starting his chemical career at a technical school in Berlin, attempted in 1828 to make "ammonium cyanate," which had been assigned a wrong structure, by heating the two inorganic salts potassium cyanate and ammonium sulfate; this led to the unexpected isolation of white crystals which were identical to the urea from urine, a typical organic compound. This well-known incident marked the beginning of organic chemistry. With the preparation of acetic acid from inorganic material in 1845 by Hermann Kolbe (1818–1884) at Leipzig, the myth surrounding organic compounds, in which they were associated with some vitalism was brought to an end and organic chemistry became the chemistry of carbon compounds. The same Kolbe was involved in the development of aspirin, one of the earliest and most important success stories in natural products chemistry. Salicylic acid from the leaf of the wintergreen plant had long been used as a pain reliever, especially in treating arthritis and gout. The inexpensive synthesis of salicylic acid from sodium phenolate and carbon dioxide by Kolbe in 1859 led to the industrial production in 1893 by the Bayer Company of acetylsalicylic acid "aspirin," still one of the most popular drugs. Aspirin is less acidic than salicylic acid and therefore causes less irritation in the mouth, throat, and stomach. The remarkable mechanism of the anti-inflammatory effect of aspirin was clarified in 1974 by John Vane (1927–) who showed that it inhibits the biosynthesis of prostaglandins by irreversibly acetylating a serine residue in prostaglandin synthase. Vane shared the 1982 Nobel Prize with Bergström and Samuelsson who determined the structure of prostaglandins (see below).

In the early days, natural products chemistry was focused on isolating the more readily available plant and animal constituents and determining their structures. The course of structure determination in the 1940s was a complex, indirect process, combining evidence from many types of experiments. The first effort was to crystallize the unknown compound or make derivatives such as esters or 2,4-dinitrophenylhydrazones, and to repeat recrystallization until the highest and sharp melting point was reached, since prior to the advent of isolation and purification methods now taken for granted, there was no simple criterion for purity. The only chromatography was through special grade alumina (first used by M. Tswett in 1906, then reintroduced by R. Willstätter). Molecular weight estimation by the Rast method which depended on melting point depression of a sample/camphor mixture, coupled with Pregl elemental microanalysis (see below) gave the molecular formula. Functionalities such as hydroxyl, amino, and carbonyl groups were recognized on the basis of specific derivatization and crystallization, followed by redetermination of molecular formula; the change in molecular composition led to identification of the functionality. Thus, sterically hindered carbonyls, for example the 11-keto group of cortisone, or tertiary hydroxyls, were very difficult to pinpoint, and often had to depend on more searching experiments. Therefore, an entire paper describing the recognition of a single hydroxyl group in a complex natural product would occasionally appear in the literature. An oxygen function suggested from the molecular formula but left unaccounted for would usually be assigned to an ether.

Determination of C-methyl groups depended on Kuhn–Roth oxidation which is performed by drastic oxidation with chromic acid/sulfuric acid, reduction of excess oxidant with hydrazine, neutralization with alkali, addition of phosphoric acid, distillation of the acetic acid originating from the C-methyls, and finally its titration with alkali. However, the results were only approximate, since *gem*-dimethyl groups only yield one equivalent of acetic acid, while primary, secondary, and tertiary methyl groups all give different yields of acetic acid. The skeletal structure of polycyclic compounds were frequently deduced on the basis of dehydrogenation reactions. It is therefore not surprising that the original steroid skeleton put forth by Wieland and Windaus in 1928, which depended a great deal on the production of chrysene upon Pd/C dehydrogenation, had to be revised in 1932 after several discrepancies were found (they received the Nobel prizes in 1927 and 1928 for this "extraordinarily difficult structure determination," see below).

In the following are listed some of the Nobel prizes awarded for the development of methodologies which have contributed critically to the progress in isolation protocols and structure determination. The year in which each prize was awarded is preceded by "Np."

Fritz Pregl, 1869–1930, Graz University, Np 1923. Invention of carbon and hydrogen microanalysis. Improvement of Kuhlmann's microbalance enabled weighing at an accuracy of 1 μg over a 20 g range, and refinement of carbon and hydrogen analytical methods made it possible to perform analysis with 3–4 mg of sample. His microbalance and the monograph *Quantitative Organic Microanalysis* (1916) profoundly influenced subsequent developments in practically all fields of chemistry and medicine.

The Svedberg, 1884–1971, Uppsala, Np 1926. Uppsala was a center for quantitative work on colloids for which the prize was awarded. His extensive study on ultracentrifugation, the first paper of which was published in the year of the award, evolved from a spring visit in 1922 to the University of Wisconsin. The ultracentrifuge together with the electrophoresis technique developed by his student Tiselius, have profoundly influenced subsequent progress in molecular biology and biochemistry.

Arne Tiselius, 1902–1971, Ph.D. Uppsala (T. Svedberg), Uppsala, Np 1948. Assisted by a grant from the Rockefeller Foundation, Tiselius was able to use his early electrophoresis instrument to show four bands in horse blood serum, alpha, beta and gamma globulins in addition to albumin; the first paper published in 1937 brought immediate positive responses.

Archer Martin, 1910–, Ph.D. Cambridge; Medical Research Council, Mill Hill, and Richard Synge, 1914–1994, Ph.D. Cambridge; Rowett Research Institute, Food Research Institute, Np 1952. They developed chromatography using two immiscible phases, gas–liquid, liquid–liquid, and paper chromatography, all of which have profoundly influenced all phases of chemistry.

Frederick Sanger, 1918–, Ph.D. Cambridge (A. Neuberger), Medical Research Council, Cambridge, Np 1958 and 1980. His confrontation with challenging structural problems in proteins and nucleic acids led to the development of two general analytical methods, 1,2,4-fluorodinitrobenzene (DNP) for tagging free amino groups (1945) in connection with insulin sequencing studies, and the dideoxynucleotide method for sequencing DNA (1977) in connection with recombinant DNA. For the latter he received his second Np in chemistry in 1980, which was shared with Paul Berg (1926–, Stanford University) and Walter Gilbert (1932–, Harvard University) for their contributions, respectively, in recombinant DNA and chemical sequencing of DNA. The studies of insulin involved usage of DNP for tagging disulfide bonds as cysteic acid residues (1949), and paper chromatography introduced by Martin and Synge 1944. That it was the first elucidation of any protein structure lowered the barrier for future structure studies of proteins.

Stanford Moore, 1913–1982, Ph.D. Wisconsin (K. P. Link), Rockefeller, Np 1972; and William Stein, 1911–1980, Ph.D. Columbia (E. G. Miller); Rockefeller, Np 1972. Moore and Stein cooperatively developed methods for the rapid quantification of protein hydrolysates by combining partition chromatography, ninhydrin coloration, and drop-counting fraction collector, i.e., the basis for commercial amino acid analyzers, and applied them to analysis of the ribonuclease structure.

Bruce Merrifield, 1921–, Ph.D. UCLA (M. Dunn), Rockefeller, Np 1984. The concept of solid-phase peptide synthesis using porous beads, chromatographic columns, and sequential elongation of peptides and other chains revolutionized the synthesis of biopolymers.

High-performance liquid chromatography (HPLC), introduced around the mid-1960s and now coupled on-line to many analytical instruments, for example UV, FTIR, and MS, is an indispensable daily tool found in all natural products chemistry laboratories.

3. STRUCTURES OF ORGANIC COMPOUNDS, NINETEENTH CENTURY

The discoveries made from 1848 to 1874 by Pasteur, Kekulé, van't Hoff, Le Bel, and others led to a revolution in structural organic chemistry. Louis Pasteur (1822–1895) was puzzled about why the potassium salt of tartaric acid (deposited on wine casks during fermentation) was dextrorotatory while the sodium ammonium salt of racemic acid (also deposited on wine casks) was optically inactive although both tartaric acid and "racemic" acid had identical chemical compositions. In 1848, the 25 year old Pasteur examined the racemic acid salt under the microscope and found two kinds of crystals exhibiting a left- and right-hand relation. Upon separation of the left-handed and right-handed crystals, he found that they rotated the plane of polarized light in opposite directions. He had thus performed his famous resolution of a racemic mixture, and had demonstrated the phenomenon of chirality. Pasteur went on to show that the racemic acid formed two kinds of salts with optically active bases such as quinine; this was the first demonstration of diastereomeric resolution. From this work Pasteur concluded that tartaric acid must have an element of asymmetry within the molecule itself. However, a three-dimensional understanding of the enantiomeric pair was only solved 25 years later (see below). Pasteur's own interest shifted to microbiology where he made the crucial discovery of the involvement of "germs" or microorganisms in various processes and proved that yeast induces alcoholic fermentation, while other microorganisms lead to diseases; he thus saved the wine industries of France, originated the process known as "pasteurization," and later developed vaccines for rabies. He was a genius who made many fundamental discoveries in chemistry and in microbiology.

The structures of organic compounds were still totally mysterious. Although Wöhler had synthesized urea, an isomer of ammonium cyanate, in 1828, the structural difference between these isomers was not known. In 1858 August Kekulé (1829–1896; studied with André Dumas and C. A. Wurtz in Paris, taught at Ghent, Heidelberg, and Bonn) published his famous paper in Liebig's *Annalen der Chemie* on the structure of carbon, in which he proposed that carbon atoms could form C–C bonds with hydrogen and other atoms linked to them; his dream on the top deck of a London bus led him to this concept. It was Butlerov who introduced the term "structure theory" in 1861. Further, in 1865 Kekulé conceived the cyclo-hexa-1:3:5-triene structure for benzene (C_6H_6) from a dream of a snake biting its own tail. In 1874, two young chemists, van't Hoff (1852–1911, Np 1901) in Utrecht, and Le Bel (1847–1930) in Paris, who had met in 1874 as students of C. A. Wurtz, published the revolutionary three-dimensional (3D) structure of the tetrahedral carbon Cabcd to explain the enantiomeric behavior of Pasteur's salts. The model was welcomed by J. Wislicenus (1835–1902, Zürich, Würzburg, Leipzig) who in 1863 had demonstrated the enantiomeric nature of the two lactic acids found by Scheele in sour milk (1780) and by Berzelius in muscle tissue (1807). This model, however, was criticized by Hermann Kolbe (1818–1884, Leipzig) as an "ingenious but in reality trivial and senseless natural philosophy." After 10 years of heated controversy, the idea of tetrahedral carbon was fully accepted, Kolbe had died and Wislicenus succeeded him in Leipzig.

Emil Fischer (1852–1919, Np 1902) was the next to make a critical contribution to stereochemistry. From the work of van't Hoff and Le Bel he reasoned that glucose should have 16 stereoisomers. Fischer's doctorate work on hydrazines under Baeyer (1835–1917, Np 1905) at Strasbourg had led to studies of osazones which culminated in the brilliant establishment, including configurations, of the Fischer sugar tree starting from D-(+)-glyceraldehyde all the way up to the aldohexoses, allose, altrose, glucose, mannose, gulose, idose, galactose, and talose (from 1884 to 1890). Unfortunately Fischer suffered from the toxic effects of phenylhydrazine for 12 years. The arbitrarily but luckily chosen absolute configuration of D-(+)-glyceraldehyde was shown to be correct sixty years later in 1951 (Johannes-Martin Bijvoet, 1892–1980). Fischer's brilliant correlation of the sugars comprising the Fischer sugar tree was performed using the Kiliani (1855–1945)–Fischer method via cyanohydrin intermediates for elongating sugars. Fischer also made remarkable contributions to the chemistry of amino acids and to nucleic acid bases (see below).

4. STRUCTURES OF ORGANIC COMPOUNDS, TWENTIETH CENTURY

The early concept of covalent bonds was provided with a sound theoretical basis by Linus Pauling (1901–1994, Np 1954), one of the greatest intellects of the twentieth century. Pauling's totally interdisciplinary research interests, including proteins and DNA is responsible for our present understanding of molecular structures. His books *Introduction to Quantum Mechanics* (with graduate student E. B. Wilson, 1935) and *The Nature of the Chemical Bond* (1939) have had a profound effect on our understanding of all of chemistry.

The actual 3D shapes of organic molecules which were still unclear in the late 1940s were then brilliantly clarified by Odd Hassel (1897–1981, Oslo University, Np 1969) and Derek Barton (1918–1998, Np 1969). Hassel, an X-ray crystallographer and physical chemist, demonstrated by electron diffraction that cyclohexane adopted the chair form in the gas phase and that it had two kinds of bonds, "standing (axial)" and "reclining (equatorial)" (1943). Because of the German occupation of Norway in 1940, instead of publishing the result in German journals, he published it in a Norwegian journal which was not abstracted in English until 1945. During his 1949 stay at Harvard, Barton attended a seminar by Louis Fieser on steric effects in steroids and showed Fieser that interpretations could be simplified if the shapes ("conformations") of cyclohexane rings were taken into consideration; Barton made these comments because he was familiar with Hassel's study on *cis-* and *trans*-decalins. Following Fieser's suggestion Barton published these ideas in a four-page *Experientia* paper (1950). This led to the joint Nobel prize with Hassel (1969), and established the concept of conformational analysis, which has exerted a profound effect in every field involving organic molecules.

Using conformational analysis, Barton determined the structures of many key terpenoids such as ß-amyrin, cycloartenone, and cycloartenol (Birkbeck College). At Glasgow University (from 1955) he collaborated in a number of cases with Monteath Robertson (1900–1989) and established many challenging structures: limonin, glauconic acid, byssochlamic acid, and nonadrides. Barton was also associated with the Research Institute for Medicine and Chemistry (RIMAC), Cambridge, USA founded by the Schering company, where with J. M. Beaton, he produced 60 g of aldosterone at a time when the world supply of this important hormone was in mg quantities. Aldosterone synthesis ("a good problem") was achieved in 1961 by Beaton ("a good experimentalist") through a nitrite photolysis, which came to be known as the Barton reaction ("a good idea") (quotes from his 1991 autobiography published by the American Chemical Society). From Glasgow, Barton went on to Imperial College, and a year before retirement, in 1977 he moved to France to direct the research at ICSN at Gif-sur-Yvette where he explored the oxidation reaction selectivity for unactivated C–H. After retiring from ICSN he made a further move to Texas A&M University in 1986, and continued his energetic activities, including chairman of the *Tetrahedron* publications. He felt weak during work one evening and died soon after, on March 16, 1998. He was fond of the phrase "gap jumping" by which he meant seeking generalizations between facts that do not seem to be related: "In the conformational analysis story, one had to jump the gap between steroids and chemical physics" (from his autobiography). According to Barton, the three most important qualities for a scientist are "intelligence, motivation, and honesty." His routine at Texas A&M was to wake around 4 a.m., read the literature, go to the office at 7 a.m. and stay there until 7 p.m.; when asked in 1997 whether this was still the routine, his response was that he wanted to wake up earlier because sleep was a waste of time—a remark which characterized this active scientist approaching 80!

Robert B. Woodward (1917–1979, Np 1965), who died prematurely, is regarded by many as the preeminent organic chemist of the twentieth century. He made landmark achievements in spectroscopy, synthesis, structure determination, biogenesis, as well as in theory. His solo papers published in 1941–1942 on empirical rules for estimating the absorption maxima of enones and dienes made the general organic chemical community realize that UV could be used for structural studies, thus launching the beginning of the spectroscopic revolution which soon brought on the applications of IR, NMR, MS, etc. He determined the structures of the following compounds: penicillin in 1945 (through joint UK–USA collaboration, see Hodgkin), strychnine in 1948, patulin in 1949, terramycin, aureomycin, and ferrocene (with G. Wilkinson, Np 1973—shared with E. O. Fischer for sandwich compounds) in 1952, cevine in 1954 (with Barton Np 1966, Jeger and Prelog, Np 1975), magnamycin in 1956, gliotoxin in 1958, oleandomycin in 1960, streptonigrin in 1963, and tetrodotoxin in 1964. He synthesized patulin in 1950, cortisone and cholesterol in 1951, lanosterol, lysergic acid (with Eli Lilly), and strychnine in 1954, reserpine in 1956, chlorophyll in 1960, a tetracycline (with Pfizer) in 1962, cephalosporin in 1965, and vitamin B_{12} in 1972 (with A. Eschenmoser, 1925–, ETH Zürich). He derived biogenetic schemes for steroids in 1953 (with K. Bloch, see below), and for macrolides in 1956, while the Woodward–Hoffmann orbital symmetry rules in 1965 brought order to a large class of seemingly random cyclization reactions.

Another central figure in stereochemistry is Vladimir Prelog (1906–1998, Np 1975), who succeeded Leopold Ruzicka at the ETH Zürich, and continued to build this institution into one of the most active and lively research and discussion centers in the world. The core group of intellectual leaders consisted of P. Plattner (1904–1975), O. Jeger, A. Eschenmoser, J. Dunitz, D. Arigoni, and A. Dreiding (from Zürich University). After completing extensive research on alkaloids, Prelog determined the structures of nonactin, boromycin, ferrioxamins, and rifamycins. His seminal studies in the synthesis and properties of 8–12 membered rings led him into unexplored areas of stereochemisty and chirality. Together with Robert Cahn (1899–1981, London Chemical Society) and Christopher Ingold (1893–1970, University College, London; pioneering mechanistic interpretation of organic reactions), he developed the Cahn–Ingold–Prelog (CIP) sequence rules for the unambiguous specification of stereoisomers. Prelog was an excellent story teller, always had jokes to tell, and was respected and loved by all who knew him.

4.1 Polyketides and Fatty Acids

Arthur Birch (1915–1995) from Sydney University, Ph.D. with Robert Robinson (Oxford University), then professor at Manchester University and Australian National University, was one of the earliest chemists to perform biosynthetic studies using radiolabels; starting with polyketides he studied the biosynthesis of a variety of natural products such as the C_6–C_3–C_6 backbone of plant phenolics, polyene macrolides, terpenoids, and alkaloids. He is especially known for the Birch reduction of aromatic rings, metal–ammonia reductions leading to 19-norsteroid hormones and other important products (1942–) which were of industrial importance. Feodor Lynen (1911–1979, Np 1964) performed studies on the intermediary metabolism of the living cell that led him to the demonstration of the first step in a chain of reactions resulting in the biosynthesis of sterols and fatty acids.

Prostaglandins, a family of 20-carbon, lipid-derived acids discovered in seminal fluids and accessory genital glands of man and sheep by von Euler (1934), have attracted great interest because of their extremely diverse biological activities. They were isolated and their structures elucidated from 1963 by S. Bergström (1916–, Np 1982) and B. Samuelsson (1934–, Np 1982) at the Karolinska Institute, Stockholm. Many syntheses of the natural prostaglandins and their nonnatural analogues have been published.

Tetsuo Nozoe (1902–1996) who studied at Tohoku University, Sendai, with Riko Majima (1874–1962, see below) went to Taiwan where he stayed until 1948 before returning to Tohoku University. At National Taiwan University he isolated hinokitiol from the essential oil of *taiwanhinoki*. Remembering the resonance concept put forward by Pauling just before World War II, he arrived at the seven-membered nonbenzenoid aromatic structure for hinokitiol in 1941, the first of the troponoids. This highly original work remained unknown to the rest of the world until 1951. In the meantime, during 1945–1948, nonbenzenoid aromatic structures had been assigned to stipitatic acid (isolated by H. Raistrick) by Michael J. S. Dewar (1918–) and to the thujaplicins by Holger Erdtman (1902–1989); the term tropolones was coined by Dewar in 1945. Nozoe continued to work on and discuss troponoids, up to the night before his death, without knowing that he had cancer. He was a remarkably focused and warm scientist, working unremittingly. Erdtman (Royal Institute of Technology, Stockholm) was the central figure in Swedish natural products chemistry who, with his wife Gunhild Aulin Erdtman (dynamic General Secretary of the Swedish Chemistry Society), worked in the area of plant phenolics.

As mentioned in the following and in the concluding sections, classical biosynthetic studies using radioactive isotopes for determining the distribution of isotopes has now largely been replaced by the use of various stable isotopes coupled with NMR and MS. The main effort has now shifted to the identification and cloning of genes, or where possible the gene clusters, involved in the biosynthesis of the natural product. In the case of polyketides (acyclic, cyclic, and aromatic), the focus is on the polyketide synthases.

4.2 Isoprenoids, Steroids, and Carotenoids

During his time as an assistant to Kekulé at Bonn, Otto Wallach (1847–1931, Np 1910) had to familiarize himself with the essential oils from plants; many of the components of these oils were compounds for which no structure was known. In 1891 he clarified the relations between 12 different monoterpenes related to pinene. This was summarized together with other terpene chemistry in book form in 1909, and led him to propose the "isoprene rule." These achievements laid the foundation for the future development of terpenoid chemistry and brought order from chaos.

The next period up to around 1950 saw phenomenal advances in natural products chemistry centered on isoprenoids. Many of the best natural products chemists in Europe, including Wieland, Windaus, Karrer, Kuhn, Butenandt, and Ruzicka contributed to this breathtaking pace. Heinrich Wieland (1877–1957) worked on the bile acid structure, which had been studied over a period of 100 years and considered to be one of the most difficult to attack; he received the Nobel Prize in 1927 for these studies. His friend Adolph Windaus (1876–1959) worked on the structure of cholesterol for which he also received the Nobel Prize in 1928. Unfortunately, there were chemical discrepancies in the proposed steroidal skeletal structure, which had a five-membered ring B attached to C-7 and C-9. J. D. Bernal, Mineralogical Museums, Cambridge University, who was examining the X-ray patterns of ergosterol (1932) noted that the dimensions were inconsistent with the Wieland–Windaus formula. A reinterpretation of the production of chrysene from sterols by Pd/C dehydrogenation reported by Diels (see below) in 1927 eventually led Rosenheim and King and Wieland and Dane to deduce the correct structure in 1932. Wieland also worked on the structures of morphine/strychnine alkaloids, phalloidin/amanitin cyclopeptides of toxic mushroom *Amanita phalloides*, and pteridines, the important fluorescent pigments of butterfly wings. Windaus determined the structure of ergosterol and continued structural studies of its irradiation product which exhibited antirachitic activity "vitamin D." The mechanistically complex photochemistry of ergosterol leading to the vitamin D group has been investigated in detail by Egbert Havinga (1927–1988, Leiden University), a leading photochemist and excellent tennis player.

Paul Karrer (1889–1971, Np 1937), established the foundations of carotenoid chemistry through structural determinations of lycopene, carotene, vitamin A, etc. and the synthesis of squalene, carotenoids, and others. George Wald (1906–1997, Np 1967) showed that vitamin A was the key compound in vision during his stay in Karrer's laboratory. Vitamin K (K from "Koagulation"), discovered by Henrik Dam (1895–1976, Polytechnic Institute, Copenhagen, Np 1943) and structurally studied by Edward Doisy (1893–1986, St. Louis University, Np 1943), was also synthesized by Karrer. In addition, Karrer synthesized riboflavin (vitamin B_2) and determined the structure and role of nicotinamide adenine dinucleotide phosphate ($NADP^+$) with Otto Warburg. The research on carotenoids and vitamins of Karrer who was at Zürich University overlapped with that of Richard Kuhn (1900–1967, Np 1938) at the ETH Zürich, and the two were frequently rivals. Richard Kuhn, one of the pioneers in using UV-vis spectroscopy for structural studies, introduced the concept of "atropisomerism" in diphenyls, and studied the spectra of a series of diphenyl polyenes. He determined the structures of many natural carotenoids, proved the structure of riboflavin-5-phosphate (flavin-adenine-dinucleotide-5-phosphate) and showed that the combination of NAD-5-phosphate with the carrier protein yielded the yellow oxidation enzyme, thus providing an understanding of the role of a prosthetic group. He also determined the structures of vitamin B complexes, i.e., pyridoxine, *p*-aminobenzoic acid, pantothenic acid. After World War II he went on to structural studies of nitrogen-containing oligosaccharides in human milk that provide immunity for infants, and brain gangliosides. Carotenoid studies in Switzerland were later taken up by Otto Isler (1910–1993), a Ruzicka student at Hoffmann-La Roche, and Conrad Hans Eugster (1921–), a Karrer student at Zürich University.

Adolf Butenandt (1903–1998, Np 1939) initiated and essentially completed isolation and structural studies of the human sex hormones, the insect molting hormone (ecdysone), and the first pheromone, bombykol. With help from industry he was able to obtain large supplies of urine from pregnant women for estrone, sow ovaries for progesterone, and 4,000 gallons of male urine for androsterone (50 mg, crystals). He isolated and determined the structures of two female sex hormones, estrone and progesterone, and the male hormone androsterone all during the period 1934–1939 (!) and was awarded the Nobel prize in 1939. Keen intuition and use of UV data and Pregl's microanalysis all played important roles. He was appointed to a professorship in Danzig at the age of 30. With Peter Karlson he isolated from 500 kg of silkworm larvae 25 mg of a-ecdysone, the prohormone of insect and crustacean molting hormone, and determined its structure as a polyhydroxysteroid (1965); 20-hydroxylation gives the insect and crustacean molting hormone or ß-ecdysone (20-hydroxyecdysteroid). He was also the first to isolate an insect pheromone, bombykol, from female silkworm moths (with E. Hecker). As president of the Max Planck Foundation, he strongly influenced the postwar rebuilding of German science.

The successor to Kuhn, who left ETH Zürich for Heidelberg, was Leopold Ruzicka (1887–1967, Np 1939) who established a close relationship with the Swiss pharmaceutical industry. His synthesis of the 17- and 15-membered macrocyclic ketones, civetone and muscone (the constituents of musk) showed that contrary to Baeyer's prediction, large alicyclic rings could be strainless. He reintroduced and refined the isoprene rule proposed by Wallach (1887) and determined the basic structures of many sesqui-, di-, and triterpenes, as well as the structure of lanosterol, the key intermediate in cholesterol biosynthesis. The "biogenetic isoprene rule" of the ETH group, Albert Eschenmoser, Leopold Ruzicka, Oskar Jeger, and Duilio Arigoni, contributed to a concept of terpenoid cyclization (1955), which was consistent with the mechanistic considerations put forward by Stork as early as 1950. Besides making the ETH group into a center of natural products chemistry, Ruzicka bought many seventeenth century Dutch paintings with royalties accumulated during the war from his Swiss and American patents, and donated them to the Zürich Kunsthaus.

Studies in the isolation, structures, and activities of the antiarthritic hormone, cortisone and related compounds from the adrenal cortex were performed in the mid- to late 1940s during World War II by Edward Kendall (1886–1972, Mayo Clinic, Rochester, Np 1950), Tadeus Reichstein (1897–1996, Basel University, Np 1950), Philip Hench (1896–1965, Mayo Clinic, Rochester, Np 1950), Oskar Wintersteiner (1898–1971, Columbia University, Squibb) and others initiated interest as an adjunct to military medicine as well as to supplement the meager supply from beef adrenal glands by synthesis. Lewis Sarett (1917–, Merck & Co., later president) and co-workers completed the cortisone synthesis in 28 steps, one of the first two totally stereocontrolled syntheses of a natural product; the other was cantharidin (Stork 1951) (see above). The multistep cortisone synthesis was put on the production line by Max Tishler (1906–1989, Merck & Co., later president) who made contributions to the synthesis of a number of drugs, including riboflavin. Besides working on steroid reactions/synthesis and antimalarial agents, Louis F. Fieser (1899–1977) and Mary Fieser (1909–1997) of Harvard University made huge contributions to the chemical community through their outstanding books *Natural Products related to Phenanthrene* (1949), *Steroids* (1959), *Advanced Organic Chemistry* (1961), and *Topics in Organic Chemistry* (1963), as well as their textbooks and an important series of books on Organic Reagents. Carl Djerassi (1923–, Stanford University), a prolific chemist, industrialist, and more recently a novelist, started to work at the Syntex laboratories in Mexico City where he directed the work leading to the first oral contraceptive ("the pill") for women.

Takashi Kubota (1909–, Osaka City University), with Teruo Matsuura (1924–, Kyoto University), determined the structure of the furanoid sesquiterpene, ipomeamarone, from the black rotted portion of spoiled sweet potatoes; this research constitutes the first characterization of a phytoallexin, defense substances produced by plants in response to attack by fungi or physical damage. Damaging a plant and characterizing the defense substances produced may lead to new bioactive compounds. The mechanism of induced biosynthesis of phytoallexins, which is not fully understood, is an interesting biological mechanistic topic that deserves further investigation. Another center of high activity in terpenoids and nucleic acids was headed by Frantisek Sorm (1913–1980, Institute of Organic and Biochemistry, Prague), who determined the structures of many sesquiterpenoids and other natural products; he was not only active scientifically but also was a central figure who helped to guide the careers of many Czech chemists.

The key compound in terpenoid biosynthesis is mevalonic acid (MVA) derived from acetyl-CoA, which was discovered fortuitously in 1957 by the Merck team in Rahway, NJ headed by Karl Folkers (1906–1998). They soon realized and proved that this C_6 acid was the precursor of the C_5 isoprenoid unit isopentenyl diphosphate (IPP) that ultimately leads to the biosynthesis of cholesterol. In 1952 Konrad Bloch (1912–, Harvard, Np 1964) with R. B. Woodward published a paper suggesting a mechanism of the cyclization of squalene to lanosterol and the subsequent steps to cholesterol, which turned out to be essentially correct. This biosynthetic path from MVA to cholesterol was experimentally clarified in stereochemical detail by John Cornforth (1917–, Np 1975) and George Popják. In 1932, Harold Urey (1893–1981, Np 1934) of Columbia University discovered heavy hydrogen. Urey showed, contrary to common expectation, that isotope separation could be achieved with deuterium in the form of deuterium oxide by fractional electrolysis of water. Urey's separation of the stable isotope deuterium led to the isotopic tracer methodology that revolutionized the protocols for elucidating biosynthetic processes and reaction mechanisms, as exemplified beautifully by the cholesterol studies. Using MVA labeled chirally with isotopes, including chiral methyl, i.e., -CHDT, Cornforth and Popják clarified the key steps in the intricate biosynthetic conversion of mevalonate to cholesterol in stereochemical detail. The chiral methyl group was also prepared independently by Duilio Arigoni (1928– , ETH, Zürich). Cornforth has had great difficulty in hearing and speech since childhood but has been helped expertly by his chemist wife Rita; he is an excellent tennis and chess player, and is renowned for his speed in composing occasional witty limericks.

Although MVA has long been assumed to be the only natural precursor for IPP, a non-MVA pathway in which IPP is formed via the glyceraldehyde phosphate-pyruvate pathway has been discovered (1995–1996) in the ancient bacteriohopanoids by Michel Rohmer, who started working on them with Guy Ourisson (1926–, University of Strasbourg, terpenoid studies, including prebiotic), and by Duilio Arigoni in the ginkgolides, which are present in the ancient *Ginkgo biloba* tree. It is possible that many other terpenoids are biosynthesized via the non-MVA route. In classical biosynthetic experiments, ^{14}C-labeled acetic acid was incorporated into the microbial or plant product, and location or distribution of the ^{14}C label was deduced by oxidation or degradation to specific fragments including acetic acid; therefore, it was not possible or extremely difficult to map the distribution of all radioactive carbons. The progress in ^{13}C NMR made it possible to incorporate ^{13}C-labeled acetic acid and locate all labeled carbons. This led to the discovery of the nonmevalonate pathway leading to the IPP units. Similarly, NMR and MS have made it possible to use the stable isotopes, e.g., ^{18}O, ^{2}H, ^{15}N, etc., in biosynthetic studies. The current trend of biosynthesis has now shifted to genomic approaches for cloning the genes of various enzyme synthases involved in the biosynthesis.

4.3 Carbohydrates and Cellulose

The most important advance in carbohydrate structures following those made by Emil Fischer was the change from acyclic to the current cyclic structure introduced by Walter Haworth (1883–1937). He noticed the presence of a- and ß-anomers, and determined the structures of important disaccharides including cellobiose, maltose, and lactose. He also determined the basic structural aspects of starch, cellulose, inulin, and other polysaccharides, and accomplished the structure determination and synthesis of vitamin C, a sample of which he had received from Albert von Szent-Györgyi (1893–1986, Np 1937). This first synthesis of a vitamin was significant since it showed that a vitamin could be synthesized in the same way as any other organic compound. There was strong belief among leading scientists in the 1910s that cellulose, starch, protein, and rubber were colloidal aggregates of small molecules. However, Hermann Staudinger (1881–1965, Np 1953) who succeeded R. Willstätter and H. Wieland at the ETH Zürich and Freiburg, respectively, showed through viscosity measurements and various molecular weight measurements that macromolecules do exist, and developed the principles of macromolecular chemistry.

In more modern times, Raymond Lemieux (1920–, Universities of Ottawa and Alberta) has been a leader in carbohydrate research. He introduced the concept of *endo*- and *exo*-anomeric effects, accomplished the challenging synthesis of sucrose (1953), pioneered in the use of NMR coupling constants in configuration studies, and most importantly, starting with syntheses of oligosaccharides responsible for human blood group determinants, he prepared antibodies and clarified fundamental aspects of the binding of oligosaccharides by lectins and antibodies. The periodate–potassium permanganate cleavage of double bonds at room temperature (1955) is called the Lemieux reaction.

4.4 Amino Acids, Peptides, Porphyrins, and Alkaloids

It is fortunate that we have China's record and practice of herbal medicine over the centuries, which is providing us with an indispensable source of knowledge. China is rapidly catching up in terms of infrastructure and equipment in organic and bioorganic chemistry, and work on isolation, structure determination, and synthesis stemming from these valuable sources has picked up momentum. However, as mentioned above, clarification of the active principles and mode of action of these plant extracts will be quite a challenge since in many cases synergistic action is expected. Wang Yu (1910–1997) who headed the well-equipped Shanghai Institute of Organic Chemistry surprised the world with the total synthesis of bovine insulin performed by his group in 1965; the human insulin was synthesized around the same time by P. G. Katsoyannis, A. Tometsko, and C. Zaut of the Brookhaven National Laboratory (1966).

One of the giants in natural products chemistry during the first half of this century was Robert Robinson (1886–1975, Np 1947) at Oxford University. His synthesis of tropinone, a bicyclic amino ketone related to cocaine, from succindialdehyde, methylamine, and acetone dicarboxylic acid under Mannich reaction conditions was the first biomimetic synthesis (1917). It reduced Willstätter's 1903 13-step synthesis starting with suberone into a single step. This achievement demonstrated Robinson's analytical prowess. He was able to dissect complex molecular structures into simple biosynthetic building blocks, which allowed him to propose the biogenesis of all types of alkaloids and other natural products. His laboratory at Oxford, where he developed the well-known Robinson annulation reaction (1937) in connection with his work on the synthesis of steroids became a world center for natural products study. Robinson was a pioneer in the so-called electronic theory of organic reactions, and introduced the use of curly arrows to show the movements of electrons. His analytical power is exemplified in the structural studies of strychnine and brucine around 1946–1952. Barton clarified the biosynthetic route to the morphine alkaloids, which he saw as an extension of his biomimetic synthesis of usnic acid through a one-electron oxidation; this was later extended to a general phenolate coupling scheme. Morphine total synthesis was brilliantly achieved by Marshall Gates (1915–, University of Rochester) in 1952.

The yield of the Robinson tropinone synthesis was low but Clemens Schöpf (1899–1970) , Ph.D. Munich (Wieland), Universität Darmstadt, improved it to 90% by carrying out the reaction in buffer; he also worked on the stereochemistry of morphine and determined the structure of the steroidal alkaloid salamandarine (1961), the toxin secreted from glands behind the eyes of the salamander.

Roger Adams (1889–1971, University of Illinois), was the central figure in organic chemistry in the USA and is credited with contributing to the rapid development of its chemistry in the late 1930s and 1940s, including training of graduate students for both academe and industry. After earning a Ph.D. in 1912 at Harvard University he did postdoctoral studies with Otto Diels (see below) and Richard Willstätter (see below) in 1913; he once said that around those years in Germany he could cover all *Journal of the American Chemical Society* papers published in a year in a single night. His important work include determination of the structures of tetrahydrocannabinol in marijuana, the toxic gossypol in cottonseed oil, chaulmoogric acid used in treatment of leprosy, and the Senecio alkaloids with Nelson Leonard (1916–, University of Illinois, now at Caltech). He also contributed to many fundamental organic reactions and syntheses. The famous Adams platinum catalyst is not only important for reducing double bonds in industry and in the laboratory, but was central for determining the number of double bonds in a structure. He was also one of the founders of the *Organic Synthesis* (started in 1921) and the *Organic Reactions* series. Nelson Leonard switched interests to bioorganic chemistry and biochemistry, where he has worked with nucleic acid bases and nucleotides, coenzymes, dimensional probes, and fluorescent modifications such as ethenoguanine.

The complicated structures of the medieval plant poisons aconitine (from *Aconitum*) and delphinine (from *Delphinium*) were finally characterized in 1959–1960 by Karel Wiesner (1919–1986, University of New Brunswick), Leo Marion (1899–1979, National Research Council, Ottawa), George Büchi (1921–, mycotoxins, aflatoxin/DNA adduct, synthesis of terpenoids and nitrogen-containing bioactive compounds, photochemistry), and Maria Przybylska (1923–, X-ray).

The complex chlorophyll structure was elucidated by Richard Willstätter (1872–1942, Np 1915). Although he could not join Baeyer's group at Munich because the latter had ceased taking students, a close relation developed between the two. During his chlorophyll studies, Willstätter reintroduced the important technique of column chromatography published in Russian by Michael Tswett (1906). Willstätter further demonstrated that magnesium was an integral part of chlorophyll, clarified the relation between chlorophyll and the blood pigment hemin, and found the wide distribution of carotenoids in tomato, egg yolk, and bovine corpus luteum. Willstätter also synthesized cyclooctatetraene and showed its properties to be wholly unlike benzene but close to those of acyclic polyenes (around 1913). He succeeded Baeyer at Munich in 1915, synthesized the anesthetic cocaine, retired early in protest of anti-Semitism, but remained active until the Hitler era, and in 1938 emigrated to Switzerland.

The hemin structure was determined by another German chemist of the same era, Hans Fischer (1881–1945, Np 1930), who succeeded Windaus at Innsbruck and at Munich. He worked on the structure of hemin from the blood pigment hemoglobin, and completed its synthesis in 1929. He continued Willstätter's structural studies of chlorophyll, and further synthesized bilirubin in 1944. Destruction of his institute at Technische Hochschule München, during World War II led him to take his life in March 1945. The biosynthesis of hemin was elucidated largely by David Shemin (1911–1991).

In the mid 1930s the Department of Biochemistry at Columbia Medical School, which had accepted many refugees from the Third Reich, including Erwin Chargaff, Rudolf Schoenheimer, and others on the faculty, and Konrad Bloch (see above) and David Shemin as graduate students, was a great center of research activity. In 1940, Shemin ingested 66 g of ^{15}N-labeled glycine over a period of 66 hours in order to determine the half-life of erythrocytes. David Rittenberg's analysis of the heme moiety with his home-made mass spectrometer showed all four pyrrole nitrogens came from glycine. Using ^{14}C (that had just become available) as a second isotope (see next paragraph), doubly labeled glycine $^{15}NH_2{}^{14}CH_2COOH$ and other precursors, Shemin showed that glycine and succinic acid condensed to yield δ-aminolevulinate, thus elegantly demonstrating the novel biosynthesis of the porphyrin ring (around 1950). At this time, Bloch was working on the other side of the bench.

Melvin Calvin (1911–1997, Np 1961) at University of California, Berkeley, elucidated the complex photosynthetic pathway in which plants reduce carbon dioxide to carbohydrates. The critical $^{14}CO_2$ had just been made available at Berkeley Lawrence Radiation Laboratory as a result of the pioneering research of Martin Kamen (1913–), while paper chromatography also played crucial roles. Kamen produced ^{14}C with Sam Ruben (1940), used ^{18}O to show that oxygen in photosynthesis comes from water and not from carbon dioxide, participated in the *Manhattan* project, testified before the House UnAmerican Activities Committee (1947), won compensatory damages from the US Department of State, and helped build the University of California, La Jolla (1957). The entire structure of the photosynthetic reaction center (>10 000 atoms) from the purple bacterium *Rhodopseudomonas viridis* has been established by X-ray crystallography in the landmark studies performed by Johann Deisenhofer (1943–), Robert Huber (1937–), and Hartmut Michel (1948–) in 1989; this was the first membrane protein structure determined by X-ray, for which they shared the 1988 Nobel prize. The information gained from the full structure of this first membrane protein has been especially rewarding.

The studies on vitamin B_{12}, the structure of which was established by crystallographic studies performed by Dorothy Hodgkin (1910–1994, Np 1964), are fascinating. Hodgkin also determined the structure of penicillin (in a joint effort between UK and US scientists during World War II) and insulin. The formidable total synthesis of vitamin B_{12} was completed in 1972 through collaborative efforts between Woodward and Eschenmoser, involving 100 postdoctoral fellows and extending over 10 years. The biosynthesis of fascinating complexity is almost completely solved through studies performed by Alan Battersby (1925–, Cambridge University), Duilio Arigoni, and Ian Scott (1928–, Texas A&M University) and collaborators where advanced NMR techniques and synthesis of labeled precursors is elegantly combined with cloning of enzymes controlling each biosynthetic step. This work provides a beautiful demonstration of the power of the combination of bioorganic chemistry, spectroscopy and molecular biology, a future direction which will become increasingly important for the creation of new "unnatural" natural products.

4.5 Enzymes and Proteins

In the early days of natural products chemistry, enzymes and viruses were very poorly understood. Thus, the 1926 paper by James Sumner (1887–1955) at Cornell University on crystalline urease was received with ignorance or skepticism, especially by Willstätter who believed that enzymes were small molecules and not proteins. John Northrop (1891–1987) and co-workers at the Rockefeller Institute went on to crystallize pepsin, trypsin, chymotrypsin, ribonuclease, deoyribonuclease, carboxypeptidase, and other enzymes between 1930 and 1935. Despite this, for many years biochemists did not recognize the significance of these findings, and considered enzymes as being low molecular weight compounds adsorbed onto proteins or colloids. Using Northrop's method for crystalline enzyme preparations, Wendell Stanley (1904–1971) at Princeton obtained tobacco mosaic virus as needles from one ton of tobacco leaves (1935). Sumner, Northrop, and Stanley shared the 1946 Nobel prize in chemistry. All these studies opened a new era for biochemistry.

Meanwhile, Linus Pauling, who in mid-1930 became interested in the magnetic properties of hemoglobin, investigated the configurations of proteins and the effects of hydrogen bonds. In 1949 he showed that sickle cell anemia was due to a mutation of a single amino acid in the hemoglobin molecule, the first correlation of a change in molecular structure with a genetic disease. Starting in 1951 he and colleagues published a series of papers describing the alpha helix structure of proteins; a paper published in the early 1950s with R. B. Corey on the structure of DNA played an important role in leading Francis Crick and James Watson to the double helix structure (Np 1962).

A further important achievement in the peptide field was that of Vincent Du Vigneaud (1901–1978, Np 1955), Cornell Medical School, who isolated and determined the structure of oxytocin, a posterior pituitary gland hormone, for which a structure involving a disulfide bond was proposed. He synthesized oxytocin in 1953, thereby completing the first synthesis of a natural peptide hormone.

Progress in isolation, purification, crystallization methods, computers, and instrumentation, including cyclotrons, have made X-ray crystallography the major tool in structural. Numerous structures including those of ligand/receptor complexes are being published at an extremely rapid rate. Some of the past major achievements in protein structures are the following. Max Perutz (1914, Np 1962) and John Kendrew (1914–1997, Np 1962), both at the Laboratory of Molecular Biology, Cambridge University, determined the structures of hemoglobin and myoglobin, respectively. William Lipscomb (1919–, Np 1976), Harvard University, who has trained many of the world's leaders in protein X-ray crystallography has been involved in the structure determination of many enzymes including carboxypeptidase A (1967); in 1965 he determined the structure of the anticancer bisindole alkaloid, vinblastine. Folding of proteins, an important but still enigmatic phenomenon, is attracting increasing attention. Christian Anfinsen (1916–1995, Np 1972), NIH, one of the pioneers in this area, showed that the amino acid residues in ribonuclease interact in an energetically most favorable manner to produce the unique 3D structure of the protein.

4.6 Nucleic Acid Bases, RNA, and DNA

The "Fischer indole synthesis" was first performed in 1886 by Emil Fischer. During the period 1881–1914, he determined the structures of and synthesized uric acid, caffeine, theobromine, xanthine, guanine, hypoxanthine, adenine, guanine, and made theophylline-D-glucoside phosphoric acid, the first synthetic nucleotide. In 1903, he made 5,5-diethylbarbituric acid or Barbital, Dorminal, Veronal, etc. (sedative), and in 1912, phenobarbital or Barbipil, Luminal, Phenobal, etc. (sedative). Many of his syntheses formed the basis of German industrial production of purine bases. In 1912 he showed that tannins are gallates of sugars such as maltose and glucose. Starting in 1899, he synthesized many of the 13 α-amino acids known at that time, including the L- and D-forms, which were separated through fractional crystallization of their salts with optically active bases. He also developed a method for synthesizing fragments of proteins, namely peptides, and made an 18-amino acid peptide. He lost his two sons in World War I, lost his wealth due to postwar inflation, believed he had terminal cancer (a misdiagnosis), and killed himself in July 1919. Fischer was a skilled experimentalist, so that even today, many of the reactions performed by him and his students are so delicately controlled that they are not easy to reproduce. As a result of his suffering by inhaling diethylmercury, and of the poisonous effect of phenylhydrazine, he was one of the first to design fume hoods. He was a superb teacher and was also influential in establishing the Kaiser Wilhelm Institute, which later became the Max Planck Institute. The number and quality of his accomplishments and contributions are hard to believe; he was truly a genius.

Alexander Todd (1907–1997, Np 1957) made critical contributions to the basic chemistry and synthesis of nucleotides. His early experience consisted of an extremely fruitful stay at Oxford in the Robinson group, where he completed the syntheses of many representative anthocyanins, and then at Edinburgh where he worked on the synthesis of vitamin B_1. He also prepared the hexacarboxylate of vitamin B_{12} (1954), which was used by D. Hodgkin's group for their X-ray elucidation of this vitamin (1956). M. Wiewiorowski (1918–), Institute for Bioorganic Chemistry, in Poznan, has headed a famous group in nucleic acid chemistry, and his colleagues are now distributed worldwide.

4.7 Antibiotics, Pigments, and Marine Natural Products

The concept of one microorganism killing another was introduced by Pasteur who coined the term antibiosis in 1877, but it was much later that this concept was realized in the form of an actual antibiotic. The bacteriologist Alexander Fleming (1881–1955, University of London, Np 1945) noticed that an airborne mold, a *Penicillium* strain, contaminated cultures of *Staphylococci* left on the open bench and formed a transparent circle around its colony due to lysis of *Staphylococci*. He published these results in 1929. The discovery did not attract much interest but the work was continued by Fleming until it was taken up further at Oxford University by pathologist Howard Florey (1898–1968, Np 1945) and biochemist Ernst Chain (1906–1979, Np 1945). The bioactivities of purified "penicillin," the first antibiotic, attracted serious interest in the early 1940s in the midst of World War II. A UK/USA team was formed during the war between academe and industry with Oxford University, Harvard University, ICI, Glaxo, Burroughs Wellcome, Merck, Shell, Squibb, and Pfizer as members. This project resulted in the large scale production of penicillin and determination of its structure (finally by X-ray, D. Hodgkin). John Sheehan (1915–1992) at MIT synthesized 6-aminopenicillanic acid in 1959, which opened the route for the synthesis of a number of analogues. Besides being the first antibiotic to be discovered, penicillin is also the first member of a large number of important antibiotics containing the ß-lactam ring, for example cephalosporins, carbapenems, monobactams, and nocardicins. The strained ß-lactam ring of these antibiotics inactivates the transpeptidase by acylating its serine residue at the active site, thus preventing the enzyme from forming the link between the pentaglycine chain and the D-Ala-D-Ala peptide, the essential link in bacterial cell walls. The overuse of ß-lactam antibiotics, which has given rise to the disturbing appearance of microbial resistant strains, is leading to active research in the design of synthetic ß-lactam analogues to counteract these strains. The complex nature of the important penicillin biosynthesis is being elucidated through efforts combining genetic engineering, expression of biosynthetic genes as well as feeding of synthetic precursors, etc. by Jack Baldwin (1938–, Oxford University), José Luengo (Universidad de León, Spain) and many other groups from industry and academe.

Shortly after the penicillin discovery, Selman Waksman (1888–1973, Rutgers University, Np 1952) discovered streptomycin, the second antibiotic and the first active against the dreaded disease tuberculosis. The discovery and development of new antibiotics continued throughout the world at pharmaceutical companies in Europe, Japan, and the USA from soil and various odd sources: cephalosporin from sewage in Sardinia, cyclosporin from Wisconsin and Norway soil which was carried back to Switzerland, avermectin from the soil near a golf course in Shizuoka Prefecture. People involved in antibiotic discovery used to collect soil samples from various sources during their trips but this has now become severely restricted to protect a country's right to its soil. M. M. Shemyakin (1908–1970, Institute of Chemistry of Natural Products, Moscow) was a grand master of Russian natural products who worked on antibiotics, especially of the tetracycline class; he also worked on cyclic antibiotics composed of alternating sequences of amides and esters and coined the term depsipeptide for these in 1953. He died in 1970 of a sudden heart attack in the midst of the 7th IUPAC Natural Products

Symposium held in Riga, Latvia, which he had organized. The Institute he headed was renamed the Shemyakin Institute.

Indigo, an important vat dye known in ancient Asia, Egypt, Greece, Rome, Britain, and Peru, is probably the oldest known coloring material of plant origin, Indigofera and Isatis. The structure was determined in 1883 and a commercially feasible synthesis was performed in 1883 by Adolf von Baeyer (see above, 1835–1917, Np 1905), who founded the German Chemical Society in 1867 following the precedent of the Chemistry Society of London. In 1872 Baeyer was appointed a professor at Strasbourg where E. Fischer was his student, and in 1875 he succeeded J. Liebig in Munich. Tyrian (or Phoenician) purple, the dibromo derivative of indigo which is obtained from the purple snail Murex bundaris, was used as a royal emblem in connection with religious ceremonies because of its rarity; because of the availability of other cheaper dyes with similar color, it has no commercial value today. K. Venkataraman (1901–1981, University of Bombay then National Chemical Laboratory) who worked with R. Robinson on the synthesis of chromones in his early career, continued to study natural and synthetic coloring matters, including synthetic anthraquinone vat dyes, natural quinonoid pigments, etc. T. R. Seshadri (1900–1975) is another Indian natural products chemist who worked mainly in natural pigments, dyes, drugs, insecticides, and especially in polyphenols. He also studied with Robinson, and with Pregl at Graz, and taught at Delhi University. Seshadri and Venkataraman had a huge impact on Indian chemistry. After a 40 year involvement, Toshio Goto (1929–1990) finally succeeded in solving the mysterious identity of commelinin, the deep-blue flower petal pigment of the Commelina communis isolated by Kozo Hayashi (1958) and protocyanin, isolated from the blue cornflower Centaurea cyanus by E. Bayer (1957). His group elucidated the remarkable structure in its entirety which consisted of six unstable anthocyanins, six flavones and two metals, the molecular weight approaching 10 000; complex stacking and hydrogen bonds were also involved. Thus the pigmentation of petals turned out to be far more complex than the theories put forth by Willstätter (1913) and Robinson (1931). Goto suffered a fatal heart attack while inspecting the first X-ray structure of commelinin; commelinin represents a pinnacle of current natural products isolation and structure determination in terms of subtlety in isolation and complexity of structure.

The study of marine natural products is understandably far behind that of compounds of terrestrial origin due to the difficulty in collection and identification of marine organisms. However, it is an area which has great potentialities for new discoveries from every conceivable source. One pioneer in modern marine chemistry is Paul Scheuer (1915–, University of Hawaii) who started his work with quinones of marine origin and has since characterized a very large number of bioactive compounds from mollusks and other sources. Luigi Minale (1936–1997, Napoli) started a strong group working on marine natural products, concentrating mainly on complex saponins. He was a leading natural products chemist who died prematurely. A. Gonzalez Gonzalez (1917–) who headed the Organic Natural Products Institute at the University of La Laguna, Tenerife, was the first to isolate and study polyhalogenated sesquiterpenoids from marine sources. His group has also carried out extensive studies on terrestrial terpenoids from the Canary Islands and South America. Carotenoids are widely distributed in nature and are of importance as food coloring material and as antioxidants (the detailed mechanisms of which still have to be worked out); new carotenoids continue to be discovered from marine sources, for example by the group of Synnove Liaaen-Jensen, Norwegian Institute of Technology). Yoshimasa Hirata (1915–), who started research at Nagoya University, is a champion in the isolation of nontrivial natural products. He characterized the bioluminescent luciferin from the marine ostracod *Cypridina hilgendorfii* in 1966 (with his students, Toshio Goto, Yoshito Kishi, and Osamu Shimomura); tetrodotoxin from the fugu fish in 1964 (with Goto and Kishi and co-workers), the structure of which was announced simultaneously by the group of Kyosuke Tsuda (1907–, tetrodotoxin, matrine) and Woodward; and the very complex palytoxin, $C_{129}H_{223}N_3O_{54}$ in 1981–1987 (with Daisuke Uemura and Kishi). Richard E. Moore, University of Hawaii, also announced the structure of palytoxin independently. Jon Clardy (1943–, Cornell University) has determined the X-ray structures of many unique marine natural products, including brevetoxin B (1981), the first of the group of toxins with contiguous *trans*-fused ether rings constituting a stiff ladder-like skeleton. Maitotoxin, $C_{164}H_{256}O_{68}S_2Na_2$, MW 3422, produced by the dinoflagellate *Gambierdiscus toxicus* is the largest and most toxic of the nonbiopolymeric toxins known; it has 32 alicyclic 6- to 8-membered ethereal rings and acyclic chains. Its isolation (1994) and complete structure determination was accomplished jointly by the groups of Takeshi Yasumoto (Tohoku University), Kazuo Tachibana and Michio Murata (Tokyo University) in 1996. Kishi, Harvard University, also deduced the full structure in 1996.

The well-known excitatory agent for the cat family contained in the volatile oil of catnip, *Nepeta cataria*, is the monoterpene nepetalactone, isolated by S. M. McElvain (1943) and structure determined by Jerrold Meinwald (1954); cats, tigers, and lions start purring and roll on their backs in response to this lactone. Takeo Sakan (1912–1993) investigated the series of monoterpenes neomatatabiols, etc.

from Actinidia, some of which are male lacewing attractants. As little as 1 fg of neomatatabiol attracts lacewings.

The first insect pheromone to be isolated and characterized was bombykol, the sex attractant for the male silkworm, *Bombyx mori* (by Butenandt and co-workers, see above). Numerous pheromones have been isolated, characterized, synthesized, and are playing central roles in insect control and in chemical ecology. The group at Cornell University have long been active in this field: Tom Eisner (1929–, behavior), Jerrold Meinwald (1927–, chemistry), Wendell Roeloff (1939–, electrophysiology, chemistry). Since the available sample is usually minuscule, full structure determination of a pheromone often requires total synthesis; Kenji Mori (1935–, Tokyo University) has been particularly active in this field. Progress in the techniques for handling volatile compounds, including collection, isolation, GC/MS, etc., has started to disclose the extreme complexity of chemical ecology which plays an important role in the lives of all living organisms. In this context, natural products chemistry will be play an increasingly important role in our grasp of the significance of biodiversity.

5. SYNTHESIS

Synthesis has been mentioned often in the preceding sections of this essay. In the following, synthetic methods of more general nature are described. The Grignard reaction of Victor Grignard (1871–1935, Np 1912) and then the Diels–Alder reaction by Otto Diels (1876–1954, Np 1950) and Kurt Alder (1902–1956, Np 1950) are extremely versatile reactions. The Diels–Alder reaction can account for the biosynthesis of several natural products with complex structures, and now an enzyme, a Diels–Alderase involved in biosynthesis has been isolated by Akitami Ichihara, Hokkaido University (1997).

The hydroboration reactions of Herbert Brown (1912–, Purdue University, Np 1979) and the Wittig reactions of Georg Wittig (1897–1987, Np 1979) are extremely versatile synthetic reactions. William S. Johnson (1913–1995, University of Wisconsin, Stanford University) developed efficient methods for the cyclization of acyclic polyolefinic compounds for the synthesis of corticoid and other steroids, while Gilbert Stork (1921–, Columbia University) introduced enamine alkylation, regiospecific enolate formation from enones and their kinetic trapping (called "three component coupling" in some cases), and radical cyclization in regio- and stereospecific constructions. Elias J. Corey (1928–, Harvard University, Np 1990) introduced the concept of retrosynthetic analysis and developed many key synthetic reactions and reagents during his synthesis of bioactive compounds, including prostaglandins and gingkolides. A recent development is the ever-expanding supramolecular chemistry stemming from 1967 studies on crown ethers by Charles Pedersen (1904–1989), 1968 studies on cryptates by Jean-Marie Lehn (1939–), and 1973 studies on host–guest chemistry by Donald Cram (1919–); they shared the chemistry Nobel prize in 1987.

6. NATURAL PRODUCTS STUDIES IN JAPAN

Since the background of natural products study in Japan is quite different from that in other countries, a brief history is given here. Natural products is one of the strongest areas of chemical research in Japan with probably the world's largest number of chemists pursuing structural studies; these are joined by a healthy number of synthetic and bioorganic chemists. An important Symposium on Natural Products was held in 1957 in Nagoya as a joint event between the faculties of science, pharmacy, and agriculture. This was the beginning of a series of annual symposia held in various cities, which has grown into a three-day event with about 50 talks and numerous papers; practically all achievements in this area are presented at this symposium. Japan adopted the early twentieth century German or European academic system where continuity of research can be assured through a permanent staff in addition to the professor, a system which is suited for natural products research which involves isolation and assay, as well as structure determination, all steps requiring delicate skills and much expertise.

The history of Japanese chemistry is short because the country was closed to the outside world up to 1868. This is when the Tokugawa shogunate which had ruled Japan for 264 years was overthrown and the Meiji era (1868–1912) began. Two of the first Japanese organic chemists sent abroad were Shokei Shibata and Nagayoshi Nagai, who joined the laboratory of A. W. von Hoffmann in Berlin. Upon return to Japan, Shibata (Chinese herbs) started a line of distinguished chemists, Keita and Yuji Shibata (flavones) and Shoji Shibata (1915–, lichens, fungal bisanthraquinonoid pigments, ginsenosides); Nagai returned to Tokyo Science University in 1884, studied ephedrine, and left a big mark in the embryonic era of organic chemistry. Modern natural products chemistry really began when three extraordinary organic chemists returned from Europe in the 1910s and started teaching and research at their respective faculties:

Riko Majima, 1874–1962, C. D. Harries (Kiel University); R. Willstätter (Zürich): Faculty of Science, Tohoku University; studied urushiol, the catecholic mixture of poison ivy irritant.

Yasuhiko Asahina, 1881–1975, R. Willstätter: Faculty of pharmacy, Tokyo University; lichens and Chinese herb.

Umetaro Suzuki, 1874–1943, E. Fischer: Faculty of agriculture, Tokyo University; vitamin B_1(thiamine).

Because these three pioneers started research in three different faculties (i.e., science, pharmacy, and agriculture), and because little interfaculty personnel exchange occurred in subsequent years, natural products chemistry in Japan was pursued independently within these three academic domains; the situation has changed now. The three pioneers started lines of first-class successors, but the establishment of a strong infrastructure takes many years, and it was only after the mid-1960s that the general level of science became comparable to that in the rest of the world; the 3rd IUPAC Symposium on the Chemistry of Natural Products, presided over by Munio Kotake (1894–1976, bufotoxins, see below), held in 1964 in Kyoto, was a clear turning point in Japan's role in this area.

Some of the outstanding Japanese chemists not already quoted are the following. Shibasaburo Kitazato (1852–1931), worked with Robert Koch (Np 1905, tuberculosis) and von Behring, antitoxins of diphtheria and tetanus which opened the new field of serology, isolation of microorganism causing dysentery, founder of Kitazato Institute; Chika Kuroda (1884–1968), first female Ph.D., structure of the complex carthamin, important dye in safflower (1930) which was revised in 1979 by Obara *et al.*, although the absolute configuration is still unknown (1998); Munio Kotake (1894–1976), bufotoxins, tryptophan metabolites, nupharidine; Harusada Suginome (1892–1972), aconite alkaloids; Teijiro Yabuta (1888–1977), kojic acid, gibberrelins; Eiji Ochiai (1898–1974), aconite alkaloids; Toshio Hoshino (1899–1979), abrine and other alkaloids; Yusuke Sumiki (1901–1974), gibberrelins; Sankichi Takei (1896–1982), rotenone; Shiro Akabori (1900–1992), peptides, C-terminal hydrazinolysis of amino acid ; Hamao Umezawa (1914–1986), kanamycin, bleomycin, numerous antibiotics; Shojiro Uyeo (1909–1988), lycorine; Tsunematsu Takemoto (1913–1989), inokosterone, kainic acid, domoic acid, quisqualic acid; Tomihide Shimizu (1889–1958), bile acids; Kenichi Takeda (1907–1991), Chinese herbs, sesquiterpenes; Yoshio Ban (1921–1994), alkaloid synthesis; Wataru Nagata (1922–1993), stereocontrolled hydrocyanation.

7. CURRENT AND FUTURE TRENDS IN NATURAL PRODUCTS CHEMISTRY

Spectroscopy and X-ray crystallography has totally changed the process of structure determination, which used to generate the excitement of solving a mystery. The first introduction of spectroscopy to the general organic community was Woodward's 1942–1943 empirical rules for estimating the UV maxima of dienes, trienes, and enones, which were extended by Fieser (1959). However, Butenandt had used UV for correctly determining the structures of the sex hormones as early as the early 1930s, while Karrer and Kuhn also used UV very early in their structural studies of the carotenoids. The Beckman DU instruments were an important factor which made UV spectroscopy a common tool for organic chemists and biochemists. With the availability of commercial instruments in 1950, IR spectroscopy became the next physical tool, making the 1950 Colthup IR correlation chart and the 1954 Bellamy monograph indispensable. The IR fingerprint region was analyzed in detail in attempts to gain as much structural information as possible from the molecular stretching and bending vibrations. Introduction of NMR spectroscopy into organic chemistry, first for protons and then for carbons, has totally changed the picture of structure determination, so that now IR is used much less frequently; however, in biopolymer studies, the techniques of difference FTIR and resonance Raman spectroscopy are indispensable.

The dramatic and rapid advancements in mass spectrometry are now drastically changing the protocol of biomacromolecular structural studies performed in biochemistry and molecular biology. Herbert Hauptman (mathematician, 1917–, Medical Foundation, Buffalo, Np 1985) and Jerome Karle (1918–, US Naval Research Laboratory, Washington, DC, Np 1985) developed direct methods for the determination of crystal structures devoid of disproportionately heavy atoms. The direct method together with modern computers revolutionized the X-ray analysis of molecular structures, which has become routine for crystalline compounds, large as well as small. Fred McLafferty (1923–, Cornell University) and Klaus Biemann (1926–, MIT) have made important contributions in the development of organic and bioorganic mass spectrometry. The development of cyclotron-based facilities for crystallographic biology studies has led to further dramatic advances enabling some protein structures to be determined in a single day, while cryoscopic electron micrography developed in 1975 by Richard Henderson and Nigel Unwin has also become a powerful tool for 3D structural determinations of membrane proteins such as bacteriorhodopsin (25 kd) and the nicotinic acetylcholine receptor (270 kd).

Circular dichroism (c.d.), which was used by French scientists Jean B. Biot (1774–1862) and Aimé Cotton during the nineteenth century "deteriorated" into monochromatic measurements at 589 nm after R.W. Bunsen (1811–1899, Heidelberg) introduced the Bunsen burner into the laboratory which readily emitted a 589 nm light characteristic of sodium. The 589 nm $[a]_D$ values, remote from most chromophoric maxima, simply represent the summation of the low-intensity readings of the decreasing end of multiple Cotton effects. It is therefore very difficult or impossible to deduce structural information from $[a]_D$ readings. Chiroptical spectroscopy was reintroduced to organic chemistry in the 1950s by C. Djerassi at Wayne State University (and later at Stanford University) as optical rotatory dispersion (ORD) and by L. Velluz and M. Legrand at Roussel-Uclaf as c.d. Günther Snatzke (1928–1992, Bonn then Ruhr University Bochum) was a major force in developing the theory and application of organic chiroptical spectroscopy. He investigated the chiroptical properties of a wide variety of natural products, including constituents of indigenous plants collected throughout the world, and established semiempirical sector rules for absolute configurational studies. He also established close collaborations with scientists of the former Eastern bloc countries and had a major impact in increasing the interest in c.d. there.

Chiroptical spectroscopy, nevertheless, remains one of the most underutilized physical measurements. Most organic chemists regard c.d. (more popular than ORD because interpretation is usually less ambiguous) simply as a tool for assigning absolute configurations, and since there are only two possibilities in absolute configurations, c.d. is apparently regarded as not as crucial compared to other spectroscopic methods. Moreover, many of the c.d. correlations with absolute configuration are empirical. For such reasons, chiroptical spectroscopy, with its immense potentialities, is grossly underused. However, c.d. curves can now be calculated nonempirically. Moreover, through-space coupling between the electric transition moments of two or more chromophores gives rise to intense Cotton effects split into opposite signs, exciton-coupled c.d.; fluorescence-detected c.d. further enhances the sensitivity by 50- to 100-fold. This leads to a highly versatile nonempirical microscale solution method for determining absolute configurations, etc.

With the rapid advances in spectroscopy and isolation techniques, most structure determinations in natural products chemistry have become quite routine, shifting the trend gradually towards activity-monitored isolation and structural studies of biologically active principles available only in microgram or submicrogram quantities. This in turn has made it possible for organic chemists to direct their attention towards clarifying the mechanistic and structural aspects of the ligand/biopolymeric receptor interactions on a more well-defined molecular structural basis. Until the 1990s, it was inconceivable and impossible to perform such studies.

Why does sugar taste sweet? This is an extremely challenging problem which at present cannot be answered even with major multidisciplinary efforts. Structural characterization of sweet compounds and elucidation of the amino acid sequences in the receptors are only the starting point. We are confronted with a long list of problems such as cloning of the receptors to produce them in sufficient quantities to investigate the physical fit between the active factor (sugar) and receptor by biophysical methods, and the time-resolved change in this physical contact and subsequent activation of G-protein and enzymes. This would then be followed by neurophysiological and ultimately physiological and psychological studies of sensation. How do the hundreds of taste receptors differ in their structures and their physical contact with molecules, and how do we differentiate the various taste sensations? The same applies to vision and to olfactory processes. What are the functions of the numerous glutamate receptor subtypes in our brain? We are at the starting point of a new field which is filled with exciting possibilities.

Familiarity with molecular biology is becoming essential for natural products chemists to plan research directed towards an understanding of natural products biosynthesis, mechanisms of bioactivity triggered by ligand–receptor interactions, etc. Numerous genes encoding enzymes have been cloned and expressed by the cDNA and/or genomic DNA-polymerase chain reaction protocols. This then leads to the possible production of new molecules by gene shuffling and recombinant biosynthetic techniques. Monoclonal catalytic antibodies using haptens possessing a structure similar to a high-energy intermediate of a proposed reaction are also contributing to the elucidation of biochemical mechanisms and the design of efficient syntheses. The technique of photoaffinity labeling, brilliantly invented by Frank Westheimer (1912–, Harvard University), assisted especially by advances in mass spectrometry, will clearly be playing an increasingly important role in studies of ligand–receptor interactions including enzyme–substrate reactions. The combined and sophisticated use of various spectroscopic means, including difference spectroscopy and fast time-resolved spectroscopy, will also become increasingly central in future studies of ligand–receptor studies.

Organic chemists, especially those involved in structural studies have the techniques, imagination, and knowledge to use these approaches. But it is difficult for organic chemists to identify an exciting and worthwhile topic. In contrast, the biochemists, biologists, and medical doctors are daily facing exciting life-related phenomena, frequently without realizing that the phenomena could be understood or at least clarified on a chemical basis. Broad individual expertise and knowledge coupled with multidisciplinary research collaboration thus becomes essential to investigate many of the more important future targets successfully. This approach may be termed "dynamic," as opposed to a "static" approach, exemplified by isolation and structure determination of a single natural product. Fortunately for scientists, nature is extremely complex and hence all the more challenging. Natural products chemistry will be playing an absolutely indispensable role for the future. Conservation of the alarming number of disappearing species, utilization of biodiversity, and understanding of the intricacies of biodiversity are further difficult, but urgent, problems confronting us.

That natural medicines are attracting renewed attention is encouraging from both practical and scientific viewpoints; their efficacy has often been proven over the centuries. However, to understand the mode of action of folk herbs and related products from nature is even more complex than mechanistic clarification of a single bioactive factor. This is because unfractionated or partly fractionated extracts are used, often containing mixtures of materials, and in many cases synergism is most likely playing an important role. Clarification of the active constituents and their modes of action will be difficult. This is nevertheless a worthwhile subject for serious investigations.

Dedicated to Sir Derek Barton whose amazing insight helped tremendously in the planning of this series, but who passed away just before its completion. It is a pity that he was unable to write this introduction as originally envisaged, since he would have had a masterful overview of the content he wanted, based on his vast experience. I have tried to fulfill his task, but this introduction cannot do justice to his original intention.

ACKNOWLEDGMENT

I am grateful to current research group members for letting me take quite a time off in order to undertake this difficult writing assignment with hardly any preparation. I am grateful to Drs. Nina Berova, Reimar Bruening, Jerrold Meinwald, Yoko Naya, and Tetsuo Shiba for their many suggestions.

8. BIBLIOGRAPHY

"A 100 Year History of Japanese Chemistry," Chemical Society of Japan, Tokyo Kagaku Dojin, 1978.
K. Bloch, *FASEB J.*, 1996, **10**, 802.
"Britannica Online," 1994–1998.
Bull. Oriental Healing Arts Inst. USA, 1980, **5**(7).
L. F. Fieser and M. Fieser, "Advanced Organic Chemistry," Reinhold, New York, 1961.
L. F. Fieser and M. Fieser, "Natural Products Related to Phenanthrene," Reinhold, New York, 1949.
M. Goodman and F. Morehouse, "Organic Molecules in Action," Gordon & Breach, New York, 1973.
L. K. James (ed.), "Nobel Laureates in Chemistry," American Chemical Society and Chemistry Heritage Foundation, 1994.
J. Mann, "Murder, Magic and Medicine," Oxford University Press, New York, 1992.
R. M. Roberts, "Serendipity, Accidental Discoveries in Science," Wiley, New York, 1989.
D. S. Tarbell and T. Tarbell, "The History of Organic Chemistry in the United States, 1875–1955," Folio, Nashville, TN, 1986.

8.01 Overview

KENJI MORI
Science University of Tokyo, Japan

8.01.1 SCOPE OF VOLUME 8

In volume 8 we consider miscellaneous natural products, including marine natural products, pheromones, plant hormones, and aspects of chemical ecology. In other words, we discuss the chemistry of biofunctional molecules called bioactive natural products. The discussion, however, will be confined to the compounds of relatively low molecular weight in accordance with the traditional coverage of natural products chemistry. We do not describe the detailed physical and chemical properties of individual compounds, because other comprehensive reference volumes are available.[1]

Bioactive natural products with low molecular weights are classified as shown in Table 1. Semiochemicals are extensively treated in this volume in connection with ecology.[2] The term "chemical ecology" has been widely accepted as a new discipline in the study of the chemistry and biology of semiochemicals after the publication of a book entitled *Chemical Ecology* in 1970[3] and also after the launch of the *Journal of Chemical Ecology* in 1974. Plant, insect, and microbial hormones are also treated in depth in the volume. Vitamins and antibiotics will not be covered because their chemistry is now one of the established fields of classical natural products chemistry.

Table 1 Classification of bioactive natural products.

Name	*Definition*
(A) Vitamins	Biofunctional molecules which are taken in as food constituents, being essential to the proper nourishment of the organism. Derived from *vita* (L.) = life+amine
(B) Hormones	Biofunctional molecules which are secreted and pass into the target organ of the same individual. Derived from *horman* (Gk.) = stir up
(C) Antibiotics	Biofunctional molecules mainly of microbial origin which kill other microorganisms. Derived from *anti* (Gk.) = against+*bios* (Gk.) = made of life
(D) Semiochemicals	Biofunctional molecules which spread information between individuals (=signal substances). Derived from *semio* (Gk.) = sign
(a) Pheromones	Biofunctional molecules which are used for communication between individuals within the same species. Derived from *pherein* (Gk.) = to carry+*horman* (Gk.) = stir up
(b) Allelochemicals	Biofunctional molecules which are used for communication between individuals belonging to different species. Derived from *allelon* (Gk.) = of each other
(1) Allomones	Biofunctional molecules which evoke advantageous reactions for their producers. Derived from *allos* (Gk.) = other
(2) Kairomones	Biofunctional molecules which evoke advantageous reactions for their receivers. Derived from *kairo* (Gk.) = opportune
(3) Synomones	Biofunctional molecules which evoke advantageous reactions for both their producers and receivers. Derived from *syn* (Gk.) = together with

8.01.2 GENERAL REMARKS ON THE INVESTIGATION OF BIOACTIVE NATURAL PRODUCTS

8.01.2.1 Developmental Stages of Studies on Bioactive Natural Products

Let us first consider the process by which the investigation of a bioactive natural product develops. As shown in Figure 1, the first step in the discovery of a bioactive natural product is to observe carefully a biological phenomenon and to speculate on the cause of that phenomenon. For example, in gibberellin research, the first step was the observation in Japan in 1898 that the infection of rice seedlings by the fungus *Gibberella fujikuroi* Sawada et Wollenweber causes elongation of the seedlings to bring about the so-called "bakanae" disease, a destructive pest which reduces the yield of rice in Asia.[4] The second stage of the research is to prove the participation of a chemical substance in that specific phenomenon. In gibberellin research, Kurosawa[5] proved that chemical substances with low molecular weights produced by *G. fujikuroi* caused the elongation of rice seedlings.[4]

Then comes the third crucial stage of the isolation and structure elucidation of the bioactive natural product responsible for that phenomenon. In 1938, Yabuta and Sumiki isolated the plant hormone gibberellins as crude crystals, which elongated rice seedlings.[6] The correct gross structure of gibberellin A_3 (**1**) (Figure 2) was proposed by Cross *et al.* in 1959.[7] With the established structure of a bioactive compound, one can proceed with further chemical or biochemical research. Chemists and biochemists begin to clarify the biosynthesis and the mode of action of that bioactive compound. On the other hand, synthetic chemists attempt the synthesis of that compound. In the case of the gibberellins, their synthesis was undertaken by many groups, culminating in total synthesis by Nagata, Corey, Mander[8,9] and others. Mori *et al.*'s relay synthesis of (±)-gibberellin A_4 (**2**) in 1969 was the first success in this area.[10] As to the biosynthesis and the mode of action of the gibberellins, many workers are extensively involved.[4,9]

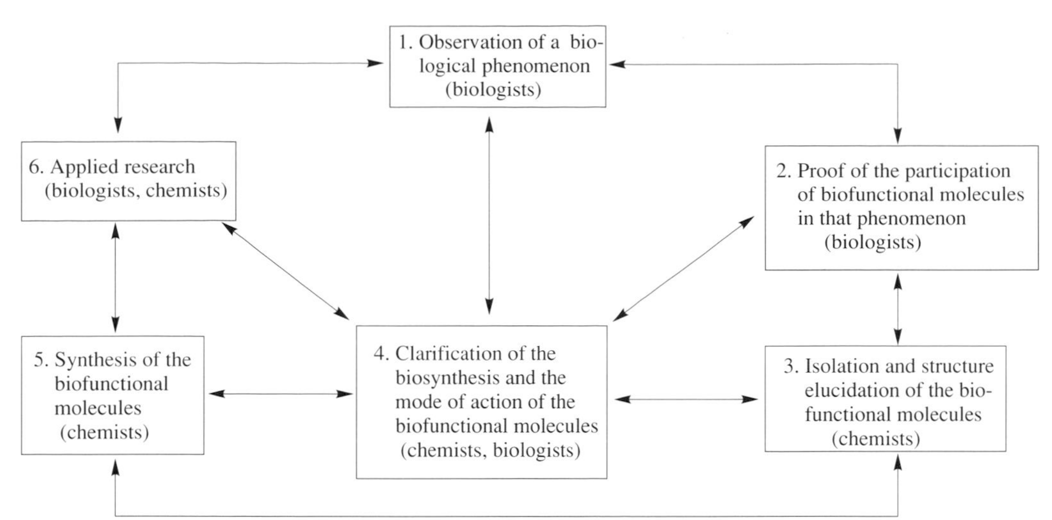

Figure 1 Developmental stage of studies on bioactive natural products: each stage is mutually interrelated with other stages.

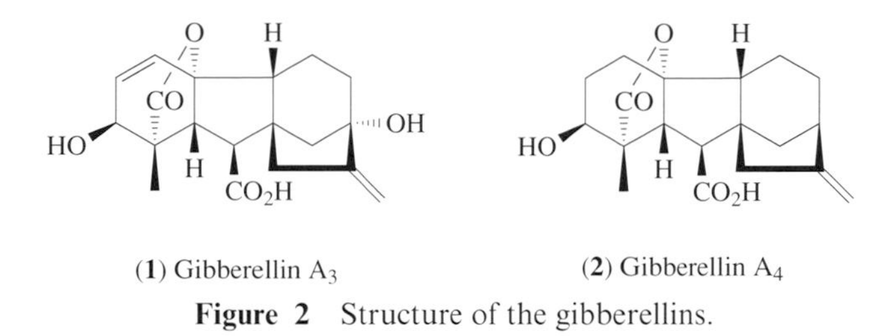

Figure 2 Structure of the gibberellins.

Finally, application of a particular bioactive compound in agriculture or medicine is the practical goal of the research. Chemists will synthesize many analogues and derivatives, and biologists will evaluate their biological effects. If one can find a really useful compound, it will be commercialized for practical application. For example, gibberellin A_3 (**1**) is used in Japan to produce seedless grapes.

8.01.2.2 Modern Trends in the Investigation of Bioactive Natural Products

Thanks to the development of microanalytical techniques and efficient separation methods, it is possible to determine the structure of a bioactive natural product with less than 1 mg of material. In Table 2, examples are given with regard to the amounts of the samples employed for the structure elucidation of insect pheromones and hormone. When Butenandt *et al.*[11] studied bombykol (**3**), the female sex pheromone of the silkworm moth *Bombyx mori* L., in 1961, they isolated 12 mg of the crystalline 4-(*p*-nitrophenylazo)benzoate of bombykol (**3**) from half a million scent glands of the female silkworm moths generated from more than 1 million silk cocoons bought in Germany, Italy, and Japan. A valuable and highly recommended account of the reflections on the study of bombykol (**3**) was presented by Hecker and Butenandt.[9] More recent examples[13–16] in Table 2 show that the structures could be proposed even with microgram quantities.

8.01.2.3 Importance of Synthesis in Studies on Bioactive Natural Products

The proposal of a structure for a bioactive natural product must be supported by synthesis. Synthesis is necessary to establish the absolute configuration of a biofunctional molecule, because it is impossible to do so with microgram quantities of that natural product.

Table 2 Amounts of the samples employed for the structure elucidation of some insect pheromones and hormones.

Researchers and year of the work	*Name of compound*	*Structure*[a]	*Amount of sample* (mg)
Butenandt *et al.*[11] 1961	Bombykol (pheromone of *Bombyx mori*)	 (**3**)	12 (as a derivative)
Röller *et al.*[13] 1967	Juvenile hormone I (from *Hyalophora cecropia*)	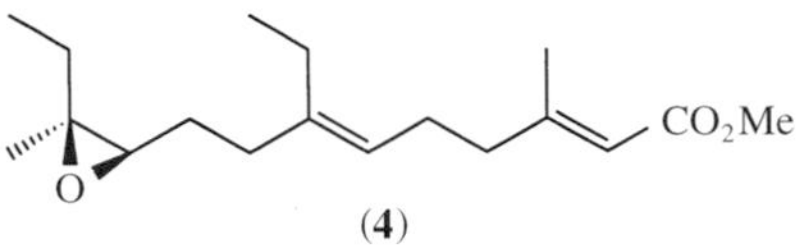 (**4**)	0.3
Silverstein and co-workers[14] 1974	Sulcatol (pheromone of *Gnathotrichus sulcatus*)	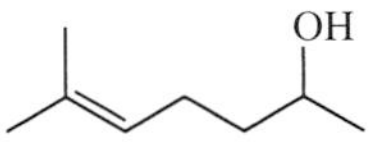 (**5**) *R*:*S* = 35:65	0.5
Persoons *et al.*[15] 1976	Periplanone-B (pheromone of *Periplaneta americana*)	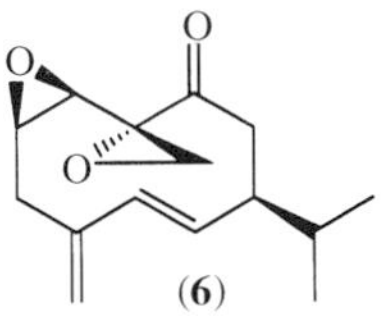 (**6**)	0.2
Oliver *et al.*[16] 1992	Pheromone of *Biprorulus bibax*	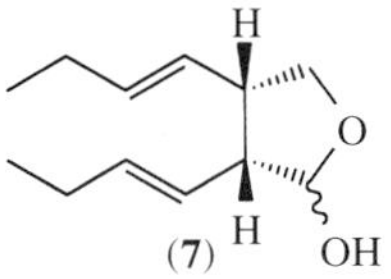 (**7**)	0.075

[a] The stereostructures including *cis*/*trans* isomerism were determined by synthesis.

In Table 3, examples are given to show how often wrong structures were proposed for important bioactive natural products. The false structure could be disproved by the fact that a synthetic compound with the proposed structure was biologically inactive. In the case of the female pheromone of the gypsy moth (*Lymantria dispar*), Jacobson *et al.* proposed it to be gyptol (**8**) in 1960.[17] The synthetic gyptol (**8**) was biologically inactive, and the true pheromone turned out to be disparlure (**9**).[18] In 1963, the female pheromone of the American cockroach (*Periplaneta americana*) was claimed to be a cyclopropane compound (**10**) by Jacobson *et al.*[19] Again, it was incorrect, and the true pheromone was shown to be periplanone-B (**6**) by Persoons *et al.*[15] Subsequently, in 1966, Jacobson and co-workers proposed the female pheromone of the pink bollworm moth (*Pectinophora gossypiella*) to be propylure (**11**).[20] However, the true pheromone was shown to be gossyplure (**12a**+**12b**) by Hummel *et al.*[21] In the case of Kögl and Erxleben's auxin-a (**13**),[22] the proposed structure (**13**) finally turned out to be a product of scientific fraud.[23] These examples are extreme cases of incorrect structural proposals.

Nevertheless, one should not forget that the proposed structures are not always correct. After their brilliant structure elucidation of periplanone-B (**6**),[15] the major female-produced sex pheromone of the American cockroach, Persoons and co-workers proposed the structure of periplanone-A, the minor and very unstable component of the pheromone, as (**14**).[24,25] They also noticed its facile rearrangement to a stable and biologically inactive compound (**15**) as shown in Scheme 1.[24,25] The structure (**15**) was supported by a synthesis of its racemate from (**16**) and 1,3-butadiene by Mori and Igarashi.[26] After more than 10 years of confusion and contradiction, the correct structure of Persoons and co-workers' compound was shown to be not (**14**) but (**17**), which was the thermal decomposition product of the genuine pheromone (**18**).[27] The genuine periplanone-A was (**18**), which was isolated by Hauptmann *et al.* in 1986.[28] Pyrolysis of (**18**) under gas chromatographic conditions caused its isomerization to (**17**). An X-ray analysis of (**19**), the reduction product of (**17**), was the key to solving the problem.[27]

After rigorous purification compound (**17**), generated by pyrolysis of (**18**)[27] and isolated by Persoons and co-workers in 1978, turned out to be biologically inactive and was therefore given the name isoperiplanone-A.[29] A concise review of this structural problem was presented by Mori *et al.*[30]

Table 3 Examples of wrong and correct structures proposed for some bioactive natural products.

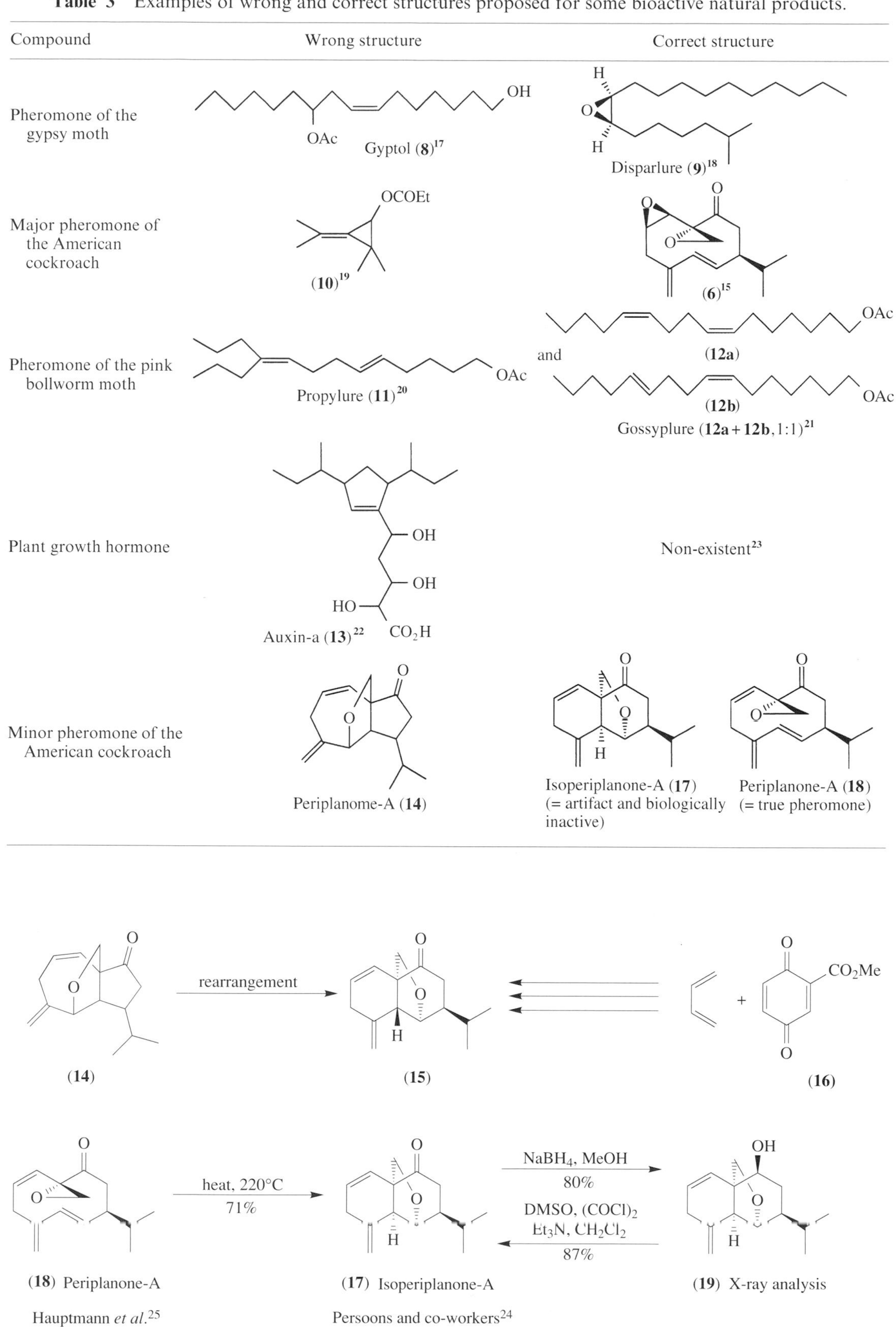

Scheme 1

8.01.3 SIGNIFICANCE OF CHIRALITY

The importance of chirality has been well recognized by organic chemists since the nineteenth century. The first optical resolution was achieved in 1848 by Pasteur, and the concept of a carbon center being asymmetric was proposed in 1874 by van't Hoff and Le Bel, and then in 1883 it was Pasteur who made the profound remark that the universe is dissymmetric. In the latter half of the twentieth century, organic stereochemistry has made such great progress that a much more precise knowledge about the dissymmetric nature of the bioactive natural products has become available. We can investigate the correct enantiomeric composition of bioactive natural products and can also clarify the relationships between absolute configuration and bioactivity. These investigations became possible by the advance in analytical techniques coupled with progress in enantioselective synthesis.

8.01.3.1 Determination of the Absolute Configuration of Bioactive Natural Products in Combination with Synthesis of Their Enantiomers and Stereoisomers

As discussed in Section 8.01.1.2.2, bioactive natural products are usually obtained in milligram to microgram quantities. It is seldom possible to determine the absolute configuration of a bioactive compound by dealing with the natural product itself. Exceptionally, in the case of a crystalline substance, its stereostructure may be solved by X-ray analysis. Often it is necessary to secure a reference sample. One must therefore synthesize the pure enantiomers or at least an enantiomerically enriched sample of the bioactive natural products with known absolute configuration by enantio-selective synthesis.

Once a synthetic sample is secured with known absolute configuration, one should compare its various properties with those of the natural product so as to identify its absolute configuration. In the case of pheromones, this analytical aspect has been reviewed twice.[31,32]

8.01.3.1.1 Determination of absolute configuration by chiroptical methods

The classical and accepted method of assignment of the absolute configuration is by comparison of the specific rotation of an unknown sample with that of the reference. The magnitude of the specific rotation, however, varies very much according to the structure of the compound. The largest $[\alpha]_D$ value observed among insect pheromones was $[\alpha]_D^{24} = -547$ (hexane) in the case of periplanone-A (**18**), the minor component of the American cockroach pheromone,[27] while some epoxy pheromones show $[\alpha]_D$ values of ~ 0 to $+0.6$.[33,34] In the latter cases, an attempt to measure the specific rotation of the natural product does not provide useful information with regard to its absolute configuration.

In general, however, if the sign and the magnitude of the specific rotation of an enantiomerically pure sample are known through synthetic work, and if the natural product is available in an amount large enough to measure its specific rotation, comparison of the sign of the specific rotation will enable the researcher to assign the absolute configuration of the natural product.

Because the magnitude of the specific rotation is usually larger at short wavelengths, optical rotary dispersion (ORD) is more useful in studying the stereochemistry of such scarce materials as hormones and pheromones. If a chiral natural product has a carbonyl chromophore, measurement of its circular dichroism (CD) can be informative.

8.01.3.1.2 Determination of absolute configuration by NMR methods

High-field NMR measurement at 300–600 MHz is often useful if it is done in the presence of a chiral shift reagent or after derivatization with a chiral derivatizing reagent.[31,32]

8.01.3.1.3 Determination of absolute configuration by HPLC or GC with chiral stationary phases

HPLC with cellulose-based chiral stationary phases[35] or GC with cyclodextrin-based chiral stationary phases[36] are two popular methods for the determination of the absolute configuration of bioactive natural products, especially pheromones.[32] Even with achiral stationary phases, HPLC and GC are useful in analyzing chiral natural products after derivatization with a chiral derivatizing reagent.[32] Cyclodextrin-based GC columns are particularly useful for the stereochemical studies of pheromones.[32]

8.01.3.1.4 Determination of the absolute configuration of bioactive natural products by a combination of enantioselective synthesis and comparison of the synthetic samples with the natural products

The absolute configuration of sordidin (**20**), the male-produced aggregation pheromone of the banana weevil *Cosmopolites sordidus*, was determined as 1*S*,3*R*,5*R*,7*S* by its enantioselective synthesis followed by the GC comparison of the synthetic sample with the natural pheromone.[37] Scheme 2 summarizes the synthesis of (±)-, (+)-, and (−)-sordidin (**20**). The commercially available dibromide (**21**) was converted into (**22**) Lithiation of (**22**) was followed by the reaction with (±)-propylene oxide (**23**) to give (**24a**), which was further converted into a mixture of (±)-sordidin (**20**) and its stereoisomer (±)-(**25**). This mixture was separated by preparative GC to give (±)-(**20**). By using (**22**) and (*S*)-propylene oxide (**23**), (1*R*,3*S*,5*S*,7*R*)-(−)-(**20**) was synthesized. Inversion of the configuration at C-2 of (2*S*)-(**24a**) gave (2*R*)-(**24a**), which afforded (1*S*,3*R*,5*R*,7*S*)-(+)-sordidin (**20**). This (+)-isomer was identical with the natural pheromone when compared by GC (Cyclodex B column).

In 1995, Yamamura and co-workers isolated phyllanthurinolactone from the fresh nyctinastic plant *Phyllanthus urinaria* L. as its leaf-closing factor.[38] It was bioactive only for that plant in the daytime at a very low concentration of 1×10^{-7} mol l^{-1}. They proposed the structure (**26**) for phyllanthurinolactone, although the absolute configuration of the aglycone part remained unknown. Audran and Mori synthesized four stereoisomers (**26**)–(**29**) (Figure 3) of phyllanthurinolactone and found only (**26**) to be bioactive.[39] The natural phyllanthurinolactone must therefore be (**26**).

The assignment of the absolute configuration of a bioactive natural product is thus possible by a combination of its synthesis with analytical comparison or bioassay.

8.01.3.2 Natural Products are not Always Enantiomerically Pure

Many of the primary metabolites are found as pure enantiomers such as D-glucose in starch and L-amino acids in proteins. However, it is well known that monoterpenes are not always enantiomerically pure, and they often exist in nature as racemates. Norin studied the enantiomeric compositions of monoterpenes including α-pinene (**30**) (Figure 4) in common spruce *Picea abies*.[40] A variation of the enantiomeric ratio of α-pinene was found, from 90% of the (−)-isomer to almost 80% of the (+)-isomer. The variation was observed among the different parts (wood, bark, root, and needles) of the tree, and also among the individuals. The ecological implication of this variation is being actively investigated.

Figure 4 also shows some other examples of natural products which are enantiomerically impure. Through the synthesis of enantiomerically pure karahana lactone (**31**),[41] dihydroactinidiolide (**32**),[42] and 2,5-epoxy-6,8-megastigmadiene (**33**),[43] and comparing their specific rotations with those of their corresponding natural sources, it was revealed that the natural products were enantiomerically impure. These degraded carotenoids might have been generated by degradation of carotenoids via a nonenzymatic photosensitized oxygenation process devoid of enantioselectivity.

A general observation is that not only monoterpenes but also some diterpenes exist in both enantiomeric forms. For example, (−)-kaur-16-ene (**34a**) is the precursor of the plant hormone gibberellins and is ubiquitous among higher plants. It is also found in *Gibberella fujikuroi*, the fungal producer of the gibberellins (Figure 5). Its antipode, (+)-kaur-16-ene (**34b**), however, is also known as the constituent of the plant *Podocarpus ferrugineus*.[44]

Steroids such as cholesterol (**35**) and triterpenes such as lanosterol (**36**) (Figure 6) are believed to occur naturally as single enantiomers. It is a remarkable feat of nature to produce single enantiomers of these isoprenoid compounds despite their structural complexity. However, there is an exception even among triterpenes. The triterpene limatulone (**37**) has been proved to exist as its racemate (**37a**) and also as the *meso*-isomer (**37b**).[45,46] In 1985, Faulkner and co-workers reported on the isolation of limatulone (**37**), a defensive metabolite of the limpet *Achmeia* (*Collisella*) *limatula*.[47] This triterpene is a potent feeding inhibitor against fish and crabs, and even at the level of 0.05% dry weight of food pellets induces regurgitation in the intertidal fish *Gibbonsia elegans*, a known

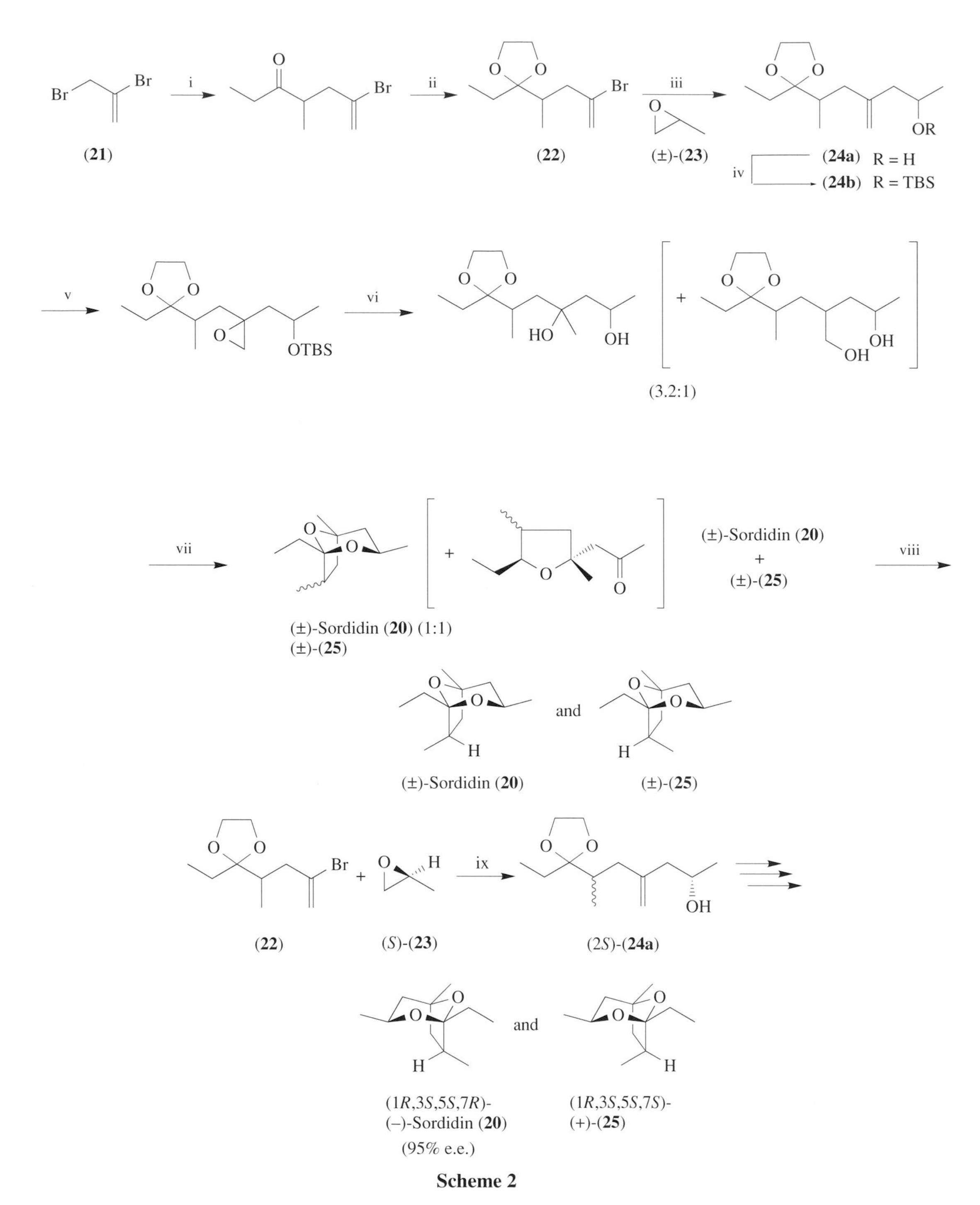

Scheme 2

i, 1.5 equiv. Et_2CO, 1.5 equiv. LDA, THF (69%); ii, 2 equiv. $HO(CH_2)_2OH$, $TsOH \cdot H_2O$, C_6H_6 (92%); iii, (a) 1 eqiv. Bu^sLi, THF; (b) 1.3 equiv. (±)-propylene oxide (**23**); (c) 1 equiv. $BF_3 \cdot OEt_2$ (70%); iv, 1.5 equiv. TBSCl, 3 equiv. imidazole, cat. DMAP, DMF (99%); v, 1.5 equiv. MCPBA, 5 equiv. $NaHCO_3$, CH_2Cl_2 (75%); vi, 8 equiv. $LiAlH_4$, THF (69%); vii, 1.5 equiv. $TsOH \cdot H_2O$, CH_2Cl_2; SiO_2 chromatography (40% of a mixture of (±)-sordidin (**20**) and (±)-(**25**)); viii, preparative GLC (PEG 20M, 25 m × 6 mm i.d.); ix, (a) 1 equiv. Bu^sLi, THF; (b) 1 equiv. $BF_3 \cdot OEt_2$ (70%); x, (a) Ph_3P, $PhCO_2H$, $EtO_2CN = NCO_2Et$, THF; (b) NaOMe, MeOH (82%)

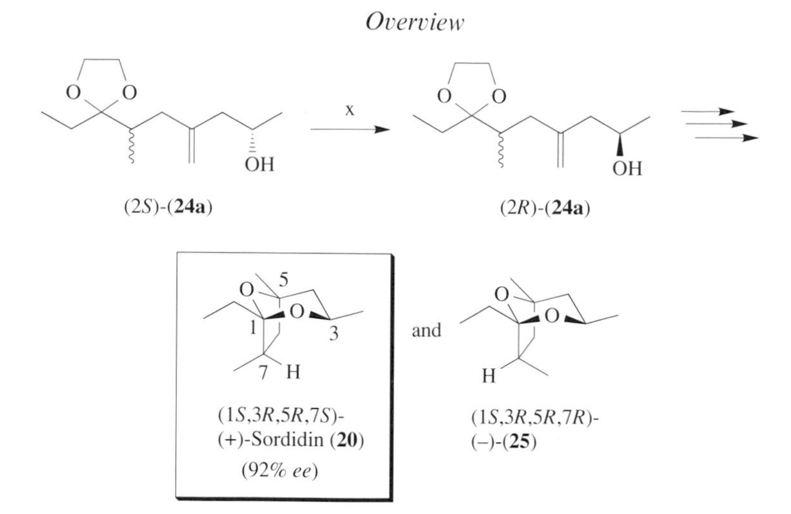

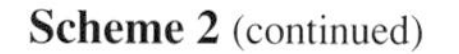

Scheme 2 (continued)

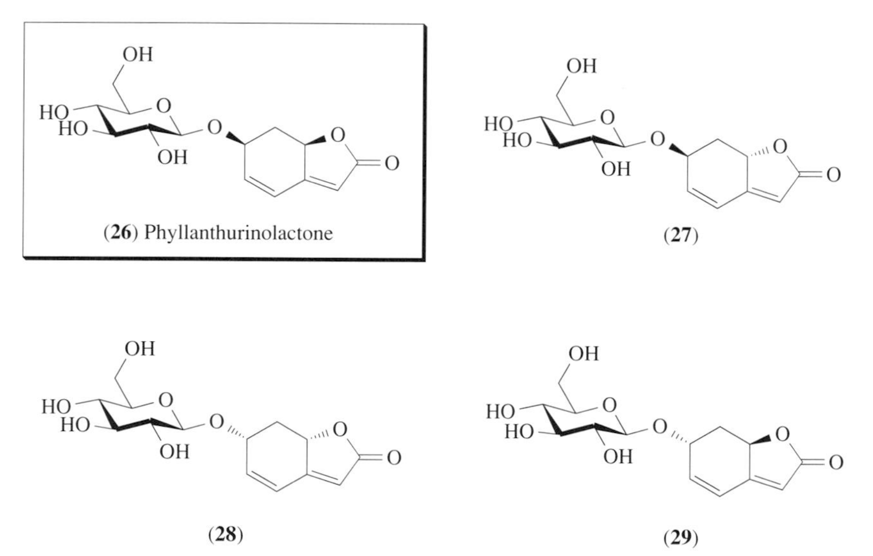

Figure 3 Structure of phyllanthurinolactone (**26**), the leaf-closing factor of *Phyllanthus urinaria* L., and its three stereoisomers (**27**)–(**29**); only (**26**) is bioactive.

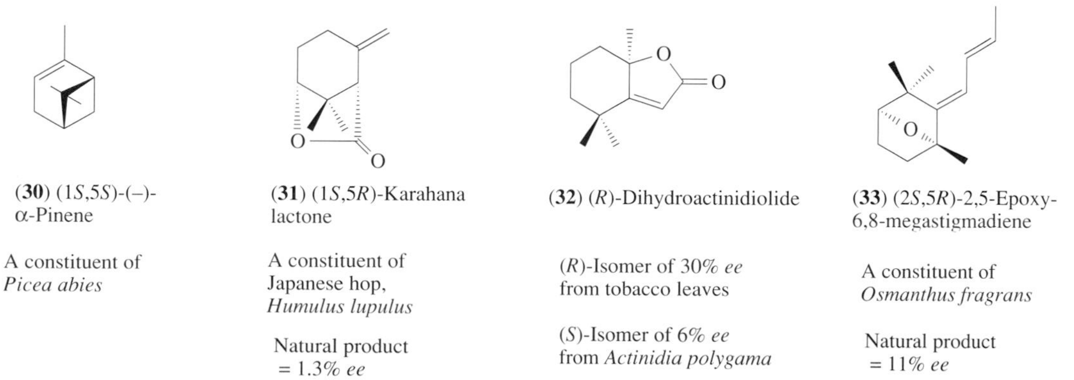

Figure 4 Examples of natural products which occur as enantiomeric mixtures.

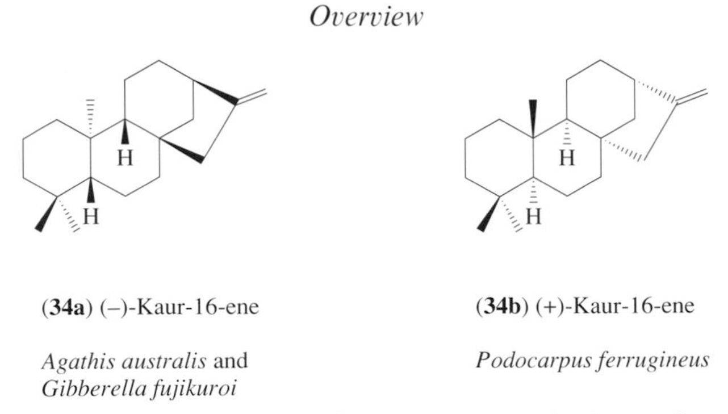

Figure 5 A diterpene kaur-16-ene is found in nature in both enantiomeric forms.

limpet predator. The natural product was reported to be optically inactive,[47] inferring that it must be either (±)-(**37a**) or *meso*-(**37b**). Mori *et al.* synthesized both of them.[45] The differences in the ^{1}H and ^{13}C NMR spectra of the synthetic (**37a**) and (**37b**) were significant, despite overall spectral resemblances between the two. These spectra were then compared with the authentic spectra of natural limatulone. The ^{1}H and ^{13}C NMR spectra of the natural product were identical with those of (±)-limatulone (**37a**). However, the ^{1}H NMR spectrum of a less bioactive fraction from the HPLC separation of *Achmeia limatula* metabolite[46] coincided with that of synthetic *meso*-(**37b**). It is clear that the limpet *Achmeia limatula* produced both (±)-(**37a**) and *meso*-(**37b**). It must be added that under chemical conditions so far examined, no retroaldol–aldol process took place to convert (**37a**) to (**37b**) or vice versa. This indicates the presence of a nonstereoselective biosynthetic pathway leading to limatulone. The mechanism by which the limpet biosynthesizes the three stereoisomers (+)-(**37a**), (−)-(**37a**), and *meso*-(**37b**) simultaneously must await further investigation. The bioactivity of the enantiomers is also not yet established. In short, nature does not always produce enantiomerically pure compounds.

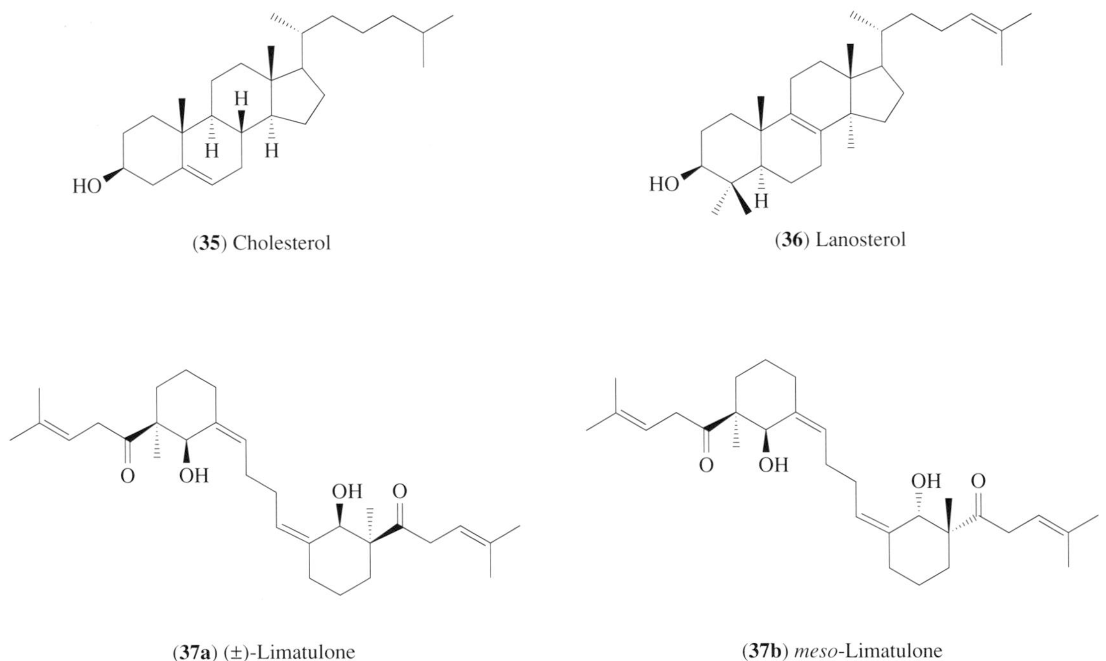

Figure 6 Cholesterol (**35**) and lanosterol (**36**) are enantiomerically pure, but limatulone exists as the racemate (**37a**) and the *meso*-isomer (**37b**)

8.01.3.3 Is a Single Enantiomer Always Responsible for Bioactivity?

In the case of a chiral and bioactive natural product, it is generally believed that only one enantiomer is bioactive and its antipode is inactive. As shown in Figure 7, (*S*)-glutamic acid (**38**) has taste, whereas its antipode is completely devoid of taste. Only the naturally occurring (6*S*,1′*S*)-hernandulcin (**39**) is a sweetener.[48] Mori and co-workers synthesized the pure enantiomers of juvenile hormone (JH) I (**4**)[49,50] and JH III (**40**).[51] Only the naturally occurring (+)-JH I (**4**) and (+)-JH III (**40**) were extremely active as insect juvenile hormones.[52,53]

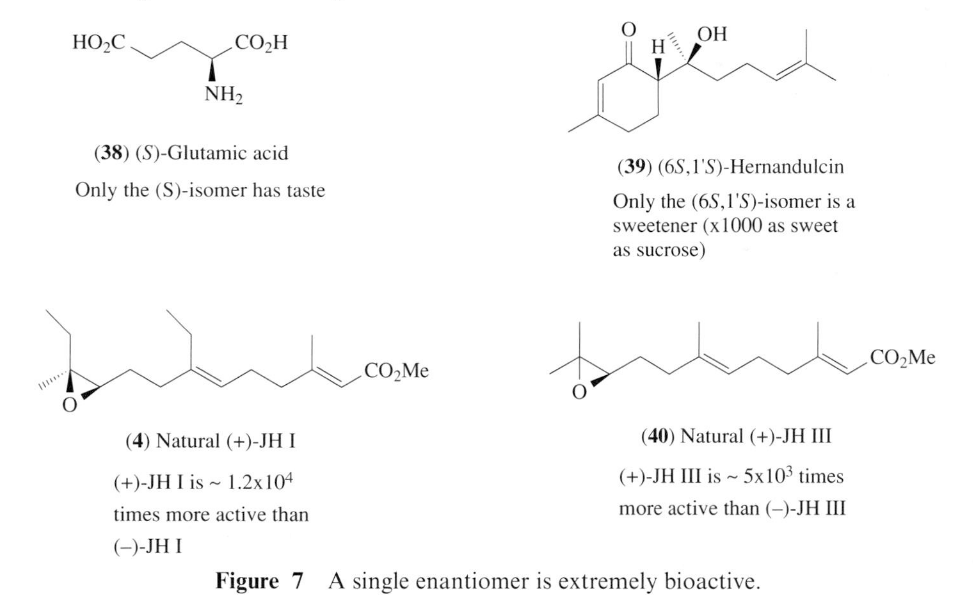

Figure 7 A single enantiomer is extremely bioactive.

Let us examine another case. (+)-Nepetalactone (**41**) (Figure 8) is a monoterpene isolated by McElvain *et al.* in 1941 as the cat attractant in catnip.[54] In 1987, it was reisolated as the sex pheromone of the vetch aphid (*Megoura viciae*) by Pickett and co-workers.[55] They then clarified that only the (+)-isomer is active as the aphid pheromone. It seemed of interest to establish whether both the enantiomers of (**41**) are active as the cat attractant or not. The enantiomers of (**41**) were synthesized and bioassayed on Japanese cats.[56] Both of them were extremely active in the cats at the 0.01 mg dosage level.[56] Pheromone perception by vetch aphids as a means of distinguishing between the (+)- and (−)-isomers of nepetalactone is a more rigorous process than perception by cats.

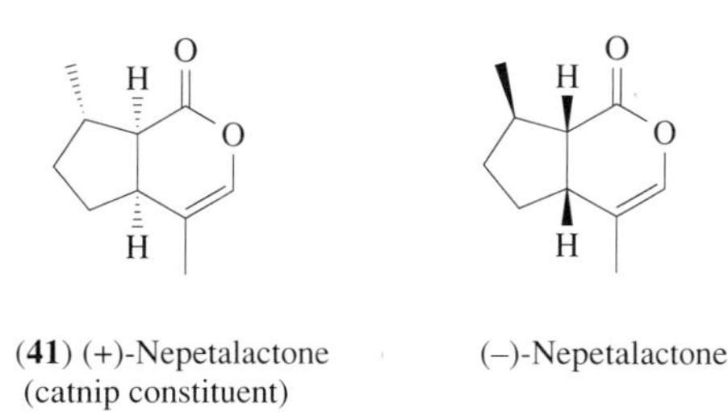

Figure 8 Only (+)-nepetalactone (**41**) is active as the aphid pheromone, whereas both the enantiomers are active as the cat attractant.

8.01.3.4 Stereochemistry–Bioactivity Relationships Among Pheromones

As exemplified in Figures 9–12, the relationship between stereochemistry and pheromone activity is far from straightforward.[57] It was discovered that insects utilize chirality to enrich and diversify their communication system, with work commencing in the early 1970s. The stereochemistry–bioactivity relationship can be divided into 10 categories as detailed below.

8.01.3.4.1 Only a single enantiomer is bioactive and its antipode does not inhibit the action of the active stereoisomer

This is the most common relationship, and ~60% of the chiral pheromones belong to this category. Most other bioregulators such as hormones also belong to this category. Many workers therefore believed that this relationship must be the only one, emphasizing the importance of a single bioactive enantiomer. However, this is merely one of the diverse relationships which were found in the case of pheromones.

The upper part of Figure 9 shows three individual pheromones belonging to this category. In 1974. Silverstein and co-workers synthesized the enantiomers of (**42**), the alarm pheromone of the leaf-cutting ant (*Atta texana*), and found (*S*)-(**42**) to be about 400 times more active than the (*R*)-enantiomer.[58] Also in 1974, Mori synthesized both enantiomers of *exo*-brevicomin (**43**), the pheromone of the western pine beetle (*Dendroctonus brevicomis*).[59] Bioassay of the enantiomers of (**43**) by Wood *et al.* showed only (+)-(**43**) to be bioactive.[60] (1*R*,5*S*,7*R*)-Dehydro-*exo*-brevicomin (**44**) is a chemical signal of the male house mouse (*Mus musculus*) and a potential multipurpose pheromone. This mammalian chemical communication is now known to be an enantioselective process.[61]

(1) Only a single enantiomer is bioactive, and its antipode does not inhibit the action of the active stereoisomer:

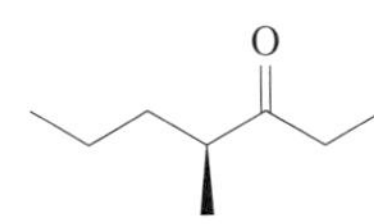

(*S*)-(**42**) *Atta texana* alarm pheromone

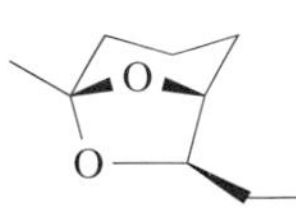

(1*R*,5*S*,7*R*)-(**43**) *exo*-Brevicomin

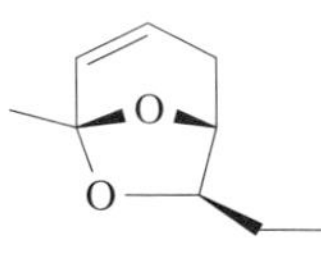

(1*R*,5*S*,7*R*)-(**44**) Dehydro-*exo*-brevicomin

etc.

(2) Only one enantiomer is bioactive, and its antipode inhibits the action of the pheromone:

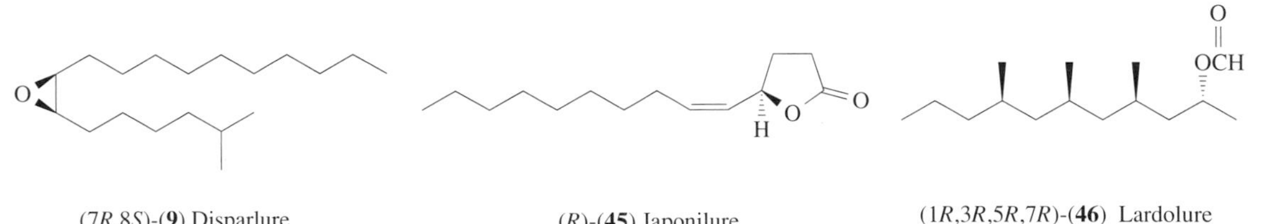

(7*R*,8*S*)-(**9**) Disparlure   (*R*)-(**45**) Japonilure   (1*R*,3*R*,5*R*,7*R*)-(**46**) Lardolure

etc.

Figure 9 Stereochemistry and pheromone activity.

8.01.3.4.2 Only one enantiomer is bioactive and its antipode inhibits the action of the pheromone

The enantiomers of disparlure (**9**) (Figure 9), the pheromone of the gypsy moth (*Lymantria dispar*), was first synthesized in 1974 by Marumo and co-workers[62] and then by Mori and co-workers.[31] Electroantennographic (EAG) and behavioral responses of the gypsy moth to the enantiomers showed that (7*R*,8*S*)-(+)-(**9**) was the most effective, followed by (±)-(**9**), whereas (7*S*,8*R*)-(−)-(**9**) inhibited the activity of the (+)-isomer.[62] Under field conditions, males of the gypsy moth and males of the nun moth (*Lymantria monacha*) responded to (7*R*,8*S*)-(**9**). However, the addition of (7*S*,8*R*)-(−)-(**9**) significantly suppressed the response by *L. dispar*, whereas (7*S*,8*R*)-(**9**) did not have such an effect on the response of *L. monacha*.[63] EAG studies using the differential receptor saturation technique suggested the existence of one receptor type having greater affinity for (7*S*,8*R*)-(**9**) than for (7*R*,8*S*)-(**9**).[64]

Japonilure ((*R*)-(+)-(**45**)) is the female sex pheromone of the Japanese beetle (*Popillia japonica*). Since (±)-(**45**) was inactive, Tumlinson *et al.* carefully studied the relationship between the enantiomeric purity of (**45**) and its bioactivity.[65] The bioactive enantiomer is (*R*)-(**45**), whereas (*S*)-(**45**) strongly inhibits the action of (*R*)-(**45**). Accordingly, (−)-(**45**) of 99% *ee* was about two-thirds as active, that of 90% *ee* was about one-third as active, and that of 80% *ee* was about one-fifth as active as the pure (*R*)-(**45**). Both (−)-(**45**) of 60% *ee* and (±)-(**45**) were inactive.[65] In 1996, Leal found that a scarab beetle, *Anomala osakana*, uses (*S*)-(**45**) as its female pheromone, whereas (*R*)-(**45**) interrupts the attraction caused by (*S*)-(**45**).[66]

Lardolure (1*R*,3*R*,5*R*,7*R*)-(−)-(**46**) is the aggregation pheromone isolated from the acarid mite *Lardoglyphus konoi*. Only (1*R*,3*R*,5*R*,7*R*)-(**46**) is active against another mite, *Carpoglyphus lactis*.[67] The antipode (1*S*,3*S*,5*S*,7*S*)-(**46**) is neither active nor inhibitory against *C. lactis*.[67] Against *L. konoi*, however, (1*R*,3*R*,5*R*,7*R*)-(**46**) is active and (1*S*,3*S*,5*S*,7*S*)-(+)-(**46**) is inhibitory. Therefore, a mixture of all the possible stereoisomers of (**46**) is only marginally active against *L. konoi*.[67]

For the practical use of pheromones belonging to this category, one must synthesize highly pure enantiomers in order to allow trap capture of the insects. Enantiomerically pure commercial products are now available for (7*R*,8*S*)-disparlure (**9**) and (*R*)-japonilure (**45**).

8.01.3.4.3 *Only one enantiomer is bioactive and its diastereomer inhibits the action of the pheromone*

Serricornin ((4*S*,6*S*,7*S*)-(**47**)) (Figure 10) is the female-produced sex pheromone of the cigarette beetle, *Lasioderma serricorne*. Bioactivity of the stereoisomers of (**47**) was studied carefully by Chuman and co-workers in the course of developing practical pheromone traps.[68] Only (4*S*,6*S*,7*S*)-(**47**) was bioactive. Its (4*S*,6*S*,7*R*)-isomer was inhibitory against the action of (4*S*,6*S*,7*S*)-(**47**). Accordingly, the commerical pheromone lure must be manufactured without contamination with the (4*S*,6*S*,7*R*)-isomer.

(3) Only one enantiomer is bioactive, and its diastereomer inhibits the action of the pheromone:

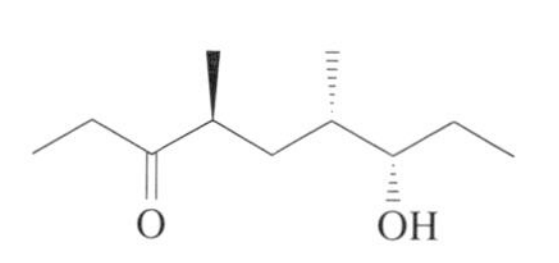

(4*S*,6*S*,7*S*)-(**47**) Serricornin

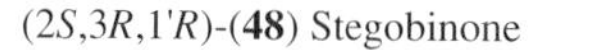

(2*S*,3*R*,1′*R*)-(**48**) Stegobinone

etc.

(4) The natural pheromone is a single enantiomer, and its antipode or diastereomer is also active:

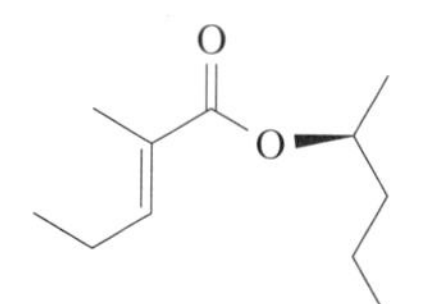

(*S*)-(**49**) Dominicalure 1

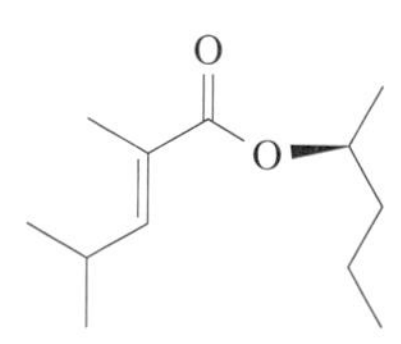

(*S*)-(**50**) Dominicalure 2

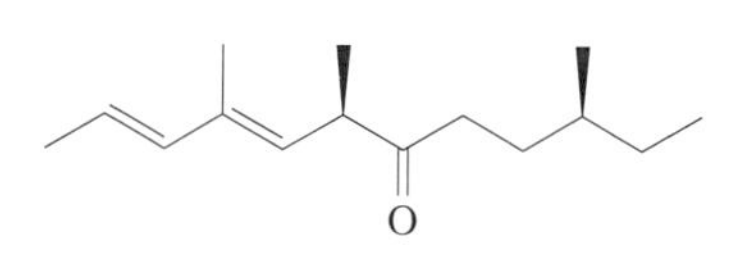

(3*S*,7*R*)-(**51**) *Matsucoccus feytaudi* pheromone

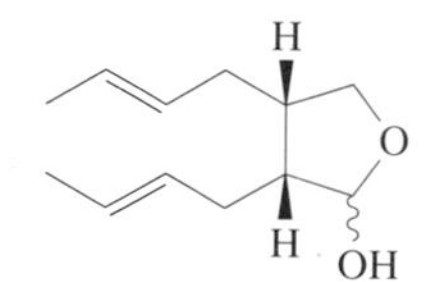

(3*R*,4*S*)-(**7**) *Biprorulus bibax* pheromone

(3*S*,11*S*)-(**52**) *Blattella germanica* pheromone

Figure 10 Stereochemistry and pheromone activity.

Stegobinone ((2*S*,3*R*,1′*R*)-(**48**)) is one of the two pheromone components of the female drugstore beetle, *Stegobium paniceum*. Very low bioactivity of the racemic and diastereomeric mixture of stegobinone indicated the presence of inhibitor(s) in the synthetic products. It was later shown that the addition of (2*S*,3*R*,1′*S*)-epistegobinone to (**48**) significantly reduces the response of male drugstore beetles.[69]

8.01.3.4.4 The natural pheromone is a single enantiomer and its antipode or diastereomer is also active

Dominicalure 1 ((*S*)-(**49**)) (Figure 10) and dominicalure 2 ((*S*)-(**50**)) are the male-produced aggregation pheromone components of the lesser grain borer, *Rhyzopertha dominica*. They are attractive to both sexes of that insect. The natural (*S*)-(**49**) and (*S*)-(**50**) were about twice as active as the unnatural (*R*)-(**49**) and (*R*)-(**50**) as assayed by their field test.[70]

Females of the maritime pine scale (*Matsuccocus feytaudi*) use (3*S*,7*R*)-(**51**) as the sex pheromone. Its (3*R*,7*R*)-isomer also showed bioactivity similar to that of the natural pheromone, while *M. feytaudi* males responded very weakly to the two other stereoisomers.[71] It therefore seems that the stereochemistry at C-3 is not important for the expression of bioactivity.

The male spined citrus bug (*Biprorulus bibax*) produces (3*R*,4*S*)-(**7**) as the aggregation pheromone. The opposite (3*S*,4*R*)-isomer of the pheromone is also bioactive, indicating that *B. bibax* does not discriminate between the enantiomers.[72] The male German cockroach (*Blattella germanica*) does not discriminate among the four stereoisomers of the female-produced sex pheromone, although the natural product is (3*S*,11*S*)-(**52**).[73]

8.01.3.4.5 The natural pheromone is an enantiomeric mixture and both the enantiomers are separately active

Female Douglas fir beetles (*Dendroctonus pseudotsugae*) produce an average of a 55:45 mixture of (*R*)-(**53**) and (*S*)-(**53**) (Figure 11).[74] The combined effect of the enantiomers was additive, rather than synergistic, and both the enantiomers are required for maximum response.

Male southern pine beetles (*Dendroctonus frontalis*) produce an 85:15 mixture of (1*S*,5*R*)-frontalin (**54**) and its (1*R*,5*S*)-isomer. In laboratory and field bioassays, the response of *D. frontalis* was significantly greater to the mixture of (1*S*,5*R*)-(**54**) and α-pinene than to (1*R*,5*S*)-(**54**) and α-pinene.[75] EAG studies showed that antennal olfactory receptor cells were significantly more responsive to (1*S*,5*R*)-(**54**) than to (1*R*,5*S*)-(**54**).[75] Both the enantiomers stimulated the same olfactory cells, suggesting that each cell has at least two types of enantioselective acceptors.

8.01.3.4.6 The different enantiomers or diastereomers are employed by the different species

(*S*)-Ipsdienol (**55**) was first isolated as the pheromone component of the California five-spined ips (*Ips paraconfusus*). Both the bark beetles *I. calligraphus* and *I. avulsus* respond to (*R*)-(**55**).[76] The pine engraver *I. pini* in the USA responds to a mixture of the enantiomers of (**55**). Detailed study on variation of the enantiomeric purity of (**55**) in *I. pini* was reported by Seybold *et al.*[77]

Complicated relationships between stereochemistry and bioactivity of the pine sawfly pheromones have been extensively studied.[40] For example, the white pine sawfly (*Neodiprion pinetum*) uses (1*S*,2*S*,6*S*)-(**56**) as its sex pheromone,[78,79] whereas (1*S*,2*R*,6*R*)-(**57**) is used as the pheromone by the introduced pine sawfly (*Diprion similis*) in the USA.[79]

Chirality of pheromones is important for discriminating between two species of the winter-flying geometrid moths in Middle Europe. Thus, (6*R*,7*S*)-(**58**) is the pheromone of *Colotois pennaria*, whereas *Erannis defoliaria* uses (6*S*,7*R*)-(**58**) as its pheromone.[80]

8.01.3.4.7 Both enantiomers are necessary for bioactivity

Sulcatol (**5**) (Figure 12) is the male-produced aggregation pheromone of the ambrosia beetle, *Gnathotrichus sulcatus*. Neither (*R*)-(**5**) nor (*S*)-(**5**) was bioactive, but when combined to give a racemic mixture, the synthetic (**5**) was more active than the natural pheromone, which was a mixture of (*R*)-(**5**) and (*S*)-(**5**) in the ratio 35:65.[81] In the case of the grain beetle *Cryptolestes turcicus*, neither (*R*)-(**59**) nor (*S*)-(**59**) was bioactive as the aggregation pheromone. However, a mixture of them ((*R*)-(**59**):(*S*)-(**59**) = 85:15) was bioactive.[82]

(5) The natural pheromone is an enantiomeric mixture, and both the enantiomers are separately active:

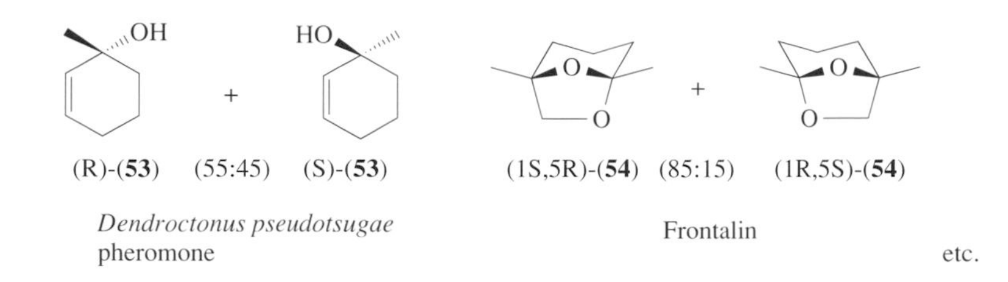

(6) The different enantiomers or diastereomers are employed by the different species:

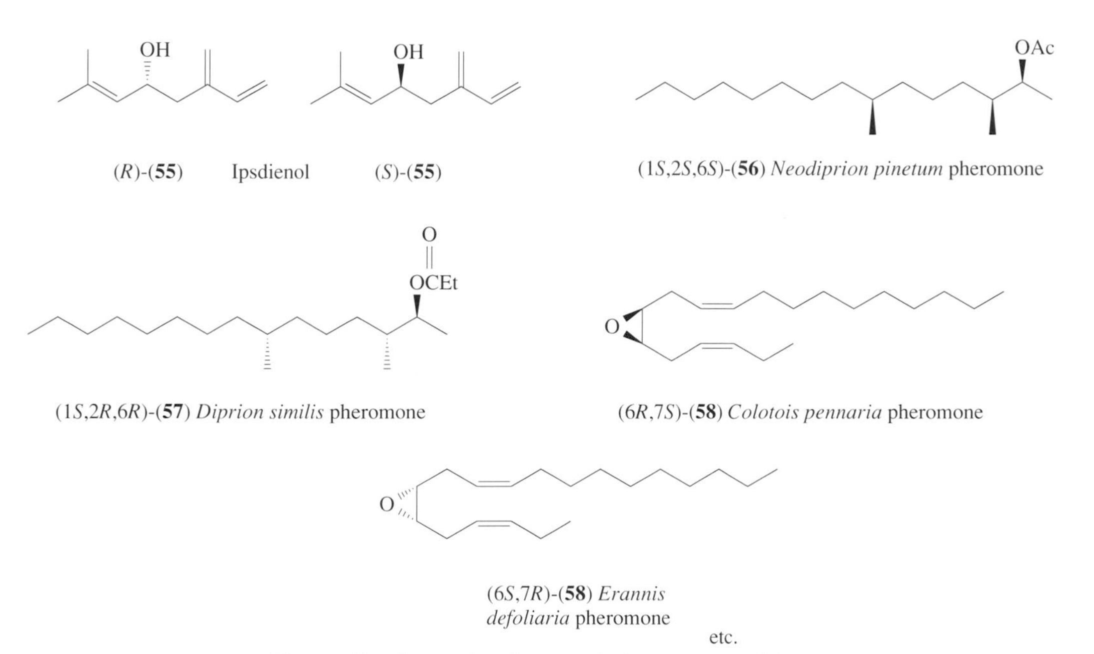

Figure 11 Stereochemistry and pheromone activity.

8.01.3.4.8 One enantiomer is more active than the other stereoisomer(s), but an enantiomeric or a diastereomeric mixture is more active than the enantiomer alone

The smaller tea tortrix moth (*Adoxophyes* sp.) uses (**60**) as a minor component of its pheromone bouquet, and (*R*)-(**60**) was found to be slightly more active then (*S*)-(**60**). Further field tests suggested that there is an optimum *R*:*S* ratio of 95:5 for trapping of males.[83]

In the case of the ant *Myrmica scabrinodis*, the naturally occurring mixture of (*R*)-(**61**) and (*S*)-(**61**) (*R*:*S* = 9:1) was more attractive than the pure (*R*)-(**61**) or (±)-(**61**), while (*S*)-(**61**) was inactive.[84]

Tribolure ((4*R*,8*R*)-(**62**)) is the male-produced aggregation pheromone of the red-flour beetle, *Tribolium castaneum*. Suzuki *et al.* found that (4*R*,8*R*)-(**62**) was as active as the natural pheromone, while a mixture of (4*R*,8*R*)-(**62**) and its (4*R*,8*S*)-isomer in a ratio of 8:2 was about 10 times more active than (4*R*,8*R*)-(**62**) alone.[85]

8.01.3.4.9 One enantiomer is active on male insects while the other is active on females

Mori *et al.*'s synthetic enantiomers of olean (**63**),[86] the female olive fruit fly (*Bactrocera oleae*) pheromone, were bioassayed in Greece.[87] The (*R*)-isomer was active against the males, whereas the (*S*)-isomer was active against the females. The natural olean was racemic.[87]

(7) Both the enantiomers are necessary for bioactivity:

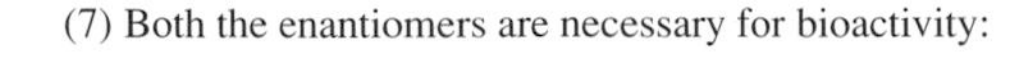

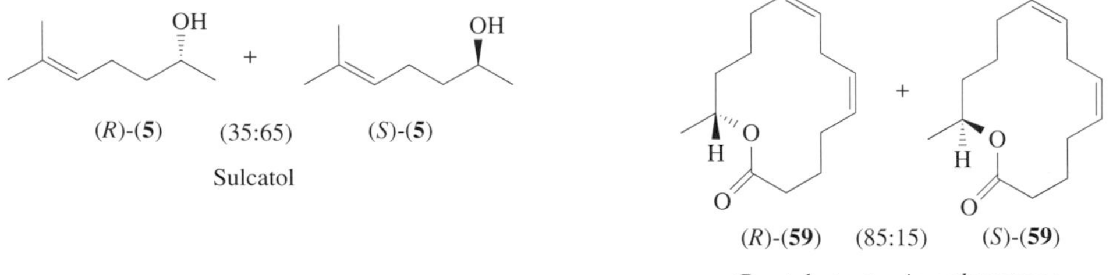

(8) One enantiomer is more active than the other stereoisomer(s), but an enantiomeric or
a diastereomeric mixture is more active than that enantiomer alone:

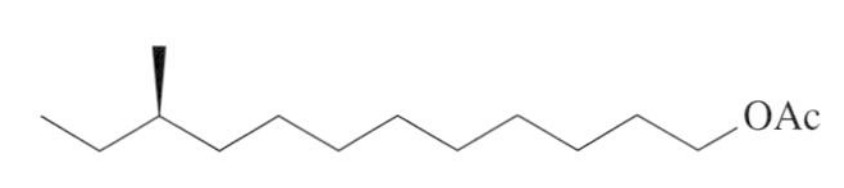

(*R*)-(**60**) Pheromone of *Adoxophyes* sp. (*R*)-(**61**) *Myrmica scabrinodis* pheromone (4*R*,8*R*)-(**62**) Tribolure

(9) One enantiomer is active on male insects, while the other is active on females:

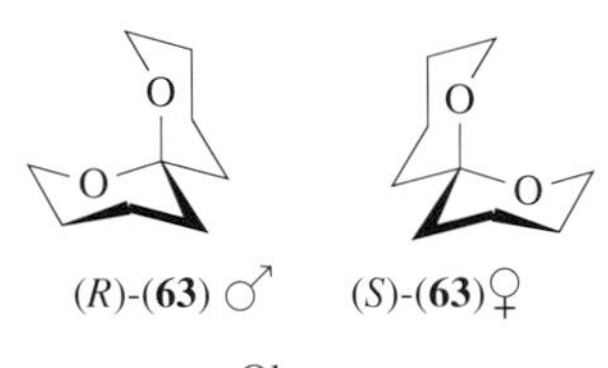

(10) Only the *meso*-isomer is active:

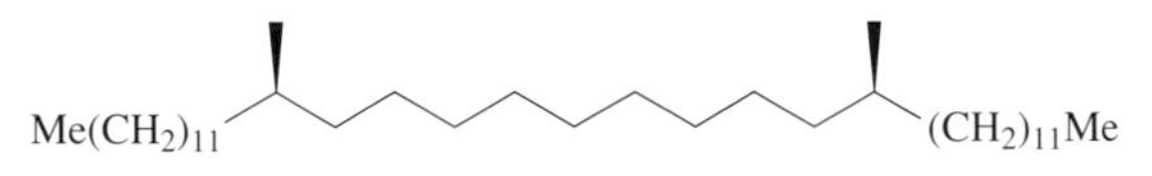

(13*R*,23*S*)-(**64**) *Glossina pallidipes* pheromone

Figure 12 Stereochemistry and pheromone activity.

8.01.3.4.10 Only the meso-*isomer is active*

In the cases of the tsetse fly sex pheromones, *meso*-alkanes seem to be bioactive. Thus, (13*R*,23*S*)-(**64**) was active as the sex-stimulant pheromone of the female tsetse fly, *Glossina pallidipes*. Neither the (13*R*,23*R*)-nor the (13*S*,23*S*)-isomer was bioactive.[88]

8.01.4 FUTURE PERSPECTIVES IN THE INVESTIGATION OF BIOACTIVE NATURAL PRODUCTS

The urgent need of society to preserve the global ecological system makes it important to understand more about the roles of bioactive natural products in the environment. The more we understand the roles, the better we may be able to protect our ecological system. Our chemical knowledge of terrestrial and marine organisms will enable us to increase our food supply. Further understanding of insect physiology and behavior will make it easier to manage the insect pests in agriculture and forestry. Undoubtedly, the new structures resulting from marine research will provide further prototypes to develop new drugs. Above all, our greatest enjoyment is if we can understand more about the chemistry of our fellow creatures. The following chapters in this volume will certainly help us to build up our own ideas as to the new research trends in the twenty-first century.

ACKNOWLEDGMENT

I wish to thank Dr. H. Takikawa for his help.

8.01.5 REFERENCES

1. J. Buckingham (ed.), "Dictionary of Natural Products," Chapman & Hall, London, 1994, vols 1–7.
2. J. Harborne, "Introduction to Ecological Biochemistry," 3rd edn., Academic Press, New York, 1988.
3. E. Sondheimer and J. B. Simeone (eds.), "Chemical Ecology," Academic Press, New York, 1970.
4. S. Tamura, in "Gibberellins," ed. N. Takahashi, B. O. Phinney, and J. MacMillan, Springer, New York, 1991, p. 1.
5. E. Kurosawa, *Taiwan Hakubutsugakkai Kaiho* (*Bull. Nat. Hist. Soc. Taiwan*), 1926, No. 87, 213.
6. T. Yabuta and Y. Sumiki, *Nippon Nogeikagaku Kaishi* (*J. Agric. Chem. Soc. Jpn.*), 1938, **14**, 1526.
7. B. E. Cross, J. F. Grove, J. MacMillan, J. S. Moffatt, T. P. C. Mulholland, and J. C. Seaton, *Proc. Chem. Soc.*, 1959, 302.
8. L. N. Mander, *Nat. Prod. Rep.*, 1988, 541.
9. L. N. Mander, *Chem. Rev.*, 1992, **92**, 573.
10. K. Mori, M. Shiozaki, N. Itaya, M. Matsui, and Y. Sumiki, *Tetrahedron*, 1969, **25**, 1293.
11. A. Butenandt, R. Beckmann, and E. Hecker, *Hoppe-Seyler's Z. Physiol. Chem.*, 1961, **324**, 71.
12. E. Hecker and A. Butenandt, in "Techniques in Pheromone Research," ed. H. E. Hummel and T. A. Miller, Springer, New York, 1984, 1.
13. H. Röller, K. H. Dahm, C. C. Sweeley, and B. M. Trost, *Angew. Chem. Int., Ed. Engl.*, 1967, **6**, 179.
14. K. J. Byrne, A. A. Swigar, R. M. Silverstein, J. H. Borden, and E. Stokkink, *J. Insect Physiol.*, 1974, **20**, 1895.
15. C. J. Persoons, P. E. J. Verwiel, F. J. Ritter, E. Talman, P. J. F. Nooijen, and W. J. Nooijen, *Tetrahedron Lett.*, 1976, 2055.
16. J. E. Oliver, J. R. Aldrich, W. R. Lusby, R. M. Waters, and D. G. James, *Tetrahedron Lett.*, 1992, **33**, 891.
17. M. Jacobson, M. Beroza, and W. A. Jones, *J. Am. Chem. Soc.*, 1961, **83**, 4819.
18. B. A. Bierl, M. Beroza, and C. W. Collier, *Science*, 1970, **170**, 87.
19. M. Jacobson, M. Beroza, and R. Yamamoto, *Science*, 1963, **139**, 48.
20. W. A. Jones, M. Jacobson, and D. F. Martin, *Science*, 1966, **152**, 1516.
21. H. E. Hummel, L. K. Gaston, H. H. Shorey, R. S. Kaae, K. J. Byrne, and R. M. Silverstein, *Science*, 1973, **181**, 873.
22. F. Kögl and H. Erxleben, *Hoppe-Seyler's Z. Physiol. Chem.*, 1934, **227**, 51.
23. K. Mori, A. Kamada, and M. Kido, *Liebigs Ann. Chem.*, 1991, 775.
24. E. Talman, P. E. J. Verwiel, F. J. Ritter, and C. J. Persoons, *Isr. J. Chem.*, 1978, **17**, 227.
25. C. J. Persoons, P. E. J. Verwiel, F. J. Ritter, and W. J. Nooyen, *J. Chem. Ecol.*, 1982, **8**, 439.
26. K. Mori and Y. Igarashi, *Tetrahedron*, 1990, **46**, 5101.
27. S. Kuwahara and K. Mori, *Tetrahedron*, 1990, **46**, 8083.
28. H. Hauptmann, G. Mühlbauer, and H. Saas, *Tetrahedron Lett.*, 1986, **27**, 6189.
29. C. J. Persoons, F. J. Ritter, P. E. J. Verwiel, H. Hauptmann, and K. Mori, *Tetrahedron Lett.*, 1990, **31**, 1747.
30. K. Mori, S. Kuwahara, and Y. Igarashi, *Pure Appl. Chem.*, 1990, **62**, 1307.
31. K. Mori, in "Techniques in Pheromone Research," eds. H. E. Hummel and T. A. Miller, Springer, New York, 1984, p. 323.
32. K. Mori, in "Methods in Chemical Ecology", eds. J. Millar and K. Haynes, Chapman & Hall, New York, in press.
33. K. Mori and N. P. Argade, *Liebigs Ann. Chem.*, 1994, 695.
34. K. Mori, T. Takigawa, and M. Matsui, *Tetrahedron*, 1979, **35**, 833.
35. E. Yashima and Y. Okamoto, *Bull. Chem. Soc. Jpn.*, 1995, **68**, 3289.
36. W. A. König, "Gas Chromagtographic Enantiomer Separation with Modified Cyclodextrins," Hüthig, Heidelberg, 1992.
37. T. Nakayama and K. Mori, *Liebigs Ann.*, 1997, 1075.
38. M. Ueda, T. Shigemori-Suzuki, and S. Yamamura, *Tetrahedron Lett.*, 1995, **36**, 6267.
39. G. Audran and K. Mori, *Eur. J. Org. Chem.*, 1998, 57.
40. T. Norin, *Pure Appl. Chem.*, 1996, **68**, 2043.
41. K. Mori and H. Mori, *Tetrahedron*, 1985, **41**, 5487.
42. K. Mori and Y. Nakazono, *Tetrahedron*, 1986, **42**, 283.
43. K. Mori and H. Tamura, *Tetrahedron*, 1986, **42**, 2643.
44. J. Buckingham (ed.), "Dictionary of Natural Products," Chapman & Hall, London, 1994, vol. 3. p. 3489.
45. K. Mori, H. Takikawa, and M. Kido, *J. Chem. Soc., Perkin Trans. 1*, 1993, 169.
46. D. J. Faulkner, *Nat. Prod. Rep.*, 1994, **11**, 384.
47. K. F. Albizati, J. R. Pawlik, and D. J. Faulkner, *J. Org. Chem.*, 1985, **50**, 3428.
48. K. Mori and M. Kato, *Tetrahedron*, 1986, **42**, 5895.
49. K. Mori and M. Fujiwhara, *Tetrahedron*, 1988, **44**, 343.
50. K. Mori and M. Fujiwhara, *Liebigs Ann. Chem.*, 1990, 369.
51. K. Mori and H. Mori, *Tetrahedron*, 1987, **43**, 4097.
52. S. Sakurai, T. Ohtaki, H. Mori, M. Fujiwhara, and K. Mori, *Experimenta*, 1990, **46**, 220.

53. H. Kindle, M. Winistörfer, B. Lanzrein, and K. Mori, *Experientia*, 1989, **45**, 356.
54. S. M. McElvain, R. D. Bright, and P. R. Johnson, *J. Am. Chem. Soc.*, 1941, **63**, 1558.
55. G. W. Dawson, D. C. Griffiths, N. F. Janes, A. Mudd, J. A. Pickett, L. J. Wadhams, and C. M. Woodcock, *Nature (London)*, 1987, **325**, 614.
56. K. Sakurai, K. Ikeda, and K. Mori, *Agric. Biol. Chem.*, 1988, **52**, 2369.
57. K. Mori, *Biosci. Biotechnol. Biochem.*, 1996, **60**, 1925.
58. R. G. Riley, R. M. Silverstein, and J. C. Moser, *Science*, 1974, **183**, 760.
59. K. Mori, *Tetrahedron*, 1974, **30**, 4223.
60. D. L. Wood, L. E. Browne, B. Ewing, K. Lindahl, W. D. Bedard, P. E. Tilden, K. Mori, G. B. Pitman, and P. R. Hughes, *Science*, 1976, **192**, 896.
61. M. V. Novotny, T. M. Xie, S. Harvey, D. Wiesler, B. Jemiolo, and M. Carmack, *Experientia*, 1995, **51**, 738.
62. S. Iwaki, S. Marumo, T. Saito, M. Yamada, and K. Katagiri, *J. Am. Chem. Soc.*, 1974, **96**, 7842.
63. J. P. Vité, D. Klimetzek, G. Loskant, R. Hedden, and K. Mori, *Naturwissenschaften*, 1976, **63**, 582.
64. J. R. Miller, K. Mori, and W. L. Roelofs, *J. Insect Physiol.*, 1977, **23**, 1447.
65. J. H. Tumlinson, M. G. Klein, R. E. Doolittle, T. E. Ladd, and A. T. Proveaux, *Science*, 1977, **197**, 789.
66. W. S. Leal, *Proc. Natl. Acad. Sci. USA*, 1996, **93**, 12 112.
67. Y. Kuwahara, M. Matsumoto, Y. Wada, and T. Suzuki, *Appl. Entomol. Zool.*, 1991, **26**, 85.
68. M. Mori, K. Mochizuki, M. Kohno, T. Chuman, A. Ohnish, H. Watanabe, and K. Mori, *J. Chem. Ecol.*, 1986, **12**, 83.
69. H. Kodama, K. Mochizuki, M. Kohno, A. Ohnishi, and Y. Kuwahara, *J. Chem. Ecol.*, 1987, **13**, 1859.
70. H. J. Williams, R. M. Silverstein, W. E. Burkholder, and A. Khorramshani, *J. Chem. Ecol.*, 1981, **7**, 759.
71. H. Jactel, P. Menassieu, M. Lettere, K. Mori, and J. Einhorn, *J. Chem. Ecol.*, 1994, **20**, 2159.
72. D. G. James and K. Mori, *J. Chem. Ecol.*, 1995, **21**, 403.
73. R. Nishida and H. Fukami, *Mem. Coll. Agric. Kyoto Univ.*, 1983, No. 122, 1.
74. B. S. Lindgren, G. Gries, H. D. Pierce, Jr., and K. Mori, *J. Chem. Ecol.*, 1992, **18**, 1201.
75. T. L. Payne, J. V. Richerson, J. C. Dickens, J. R. West, K. Mori, C. W. Brisford, R. L. Hedden, J. P. Vité, and M. S. Blum. *J. Chem. Ecol.*, 1982, **8**, 873.
76. J. P. Vité, G. Ohloff, and R. F. Billings, *Nature (London)*, 1978, **272**, 817.
77. S. J. Seybold, T. Ohtsuka, D. L. Wood, and I. Kubo, *J. Chem. Ecol.*, 1995, **21**, 995.
78. M. Kraemer, H. C. Coppel, F. Matsumura, T. Kikukawa, and K. Mori, *Environ. Entomol.*, 1979, **8**, 519.
79. J. I. Olaifa, F. Matsumura, T. Kikukawa, and H. C. Coppel, *J. Chem. Ecol.*, 1988, **14**, 1131.
80. G. Szöcs, M. Tóth, W. Francke, F. Schmidt, P. Philipp, W. A. König, K. Mori, B. S. Hansson, and C. Löfstedt, *J. Chem. Ecol.*, 1993, **19**, 2721.
81. J. H. Borden, L. Chong, J. A. McLean, K. N. Slessor, and K. Mori, *Science*, 1976, **192**, 894.
82. J. G. Millar, H. D. Pierce, Jr., A. M. Pierce, A. C. Oehlschlager, and J. H. Borden, *J. Chem. Ecol.*, 1985, **11**, 1071.
83. Y. Tamaki, H. Noguchi, H. Sugie, A. Kariya, S. Arai, M. Ohba, T. Terada, T. Suguro, and K. Mori, *Jpn. J. Appl. Entomol. Zool.*, 1980, **24**, 221.
84. M.-C. Cammaerts and K. Mori, *Physiol. Entomol.*, 1987, **12**, 381.
85. T. Suzuki, J. Kozaki, R. Sugawara, and K. Mori, *Appl. Entomol. Zool.*, 1984, **19**, 15.
86. K. Mori, T. Uematsu, K. Yanagi, and M. Minobe, *Tetrahedron*, 1985, **41**, 2751.
87. G. Haniotakis, W. Francke, K. Mori, H. Redlich, and V. Schurig, *J. Chem. Ecol.*, 1986, **12**, 1559.
88. P. G. McDowell, A. Hassanali, and R. Dransfield, *Physiol, Entomol.*, 1985, **10**, 183.

8.02
Plant Hormones

NOBORU MUROFUSHI and HISAKAZU YAMANE
The University of Tokyo, Japan

YOUJI SAKAGAMI and HIDEMASA IMASEKI
Nagoya University, Japan

YUJI KAMIYA
RIKEN, Saitama, Japan

HAJIME IWAMURA and NOBUHIRO HIRAI
Kyoto University, Japan

HIDEO TSUJI
Kobe Women's University, Japan

TAKAO YOKOTA
Teikyo University, Japan

and

JUNICHI UEDA
Osaka Prefecture University, Japan

8.02.1 INTRODUCTION

Plant hormones have not been as extensively studied, nor are they as well known, even to biological scientists, as ordinary "hormones" of animals. However, research on plant hormones began at the beginning of the nineteenth century. Auxin was the first plant hormone to be identified, and evidence of its occurrence was obtained from 1910 to the late 1920s. Although the recognition of gibberellins as a plant hormone (group) occurred much later than the initial isolation of the first gibberellin (gibberellin A_1), their discovery occurred as early as auxin's, in the 1920s. Also ethylene, as a substance which can exert hormonal effects, has a history as long as that of auxin and the gibberellins. Auxins, gibberellins, and ethylene, together with other plant hormones, i.e., cytokinins (1950s), abscisic acid (1960s), and brassinolide (1970s) were all discovered using a wide range of quite interesting approaches, for example, the discovery of auxin came from quite basic research in plant physiology, gibberellins were found by plant pathology researchers, and research on abscisic acid originated from the agricultural problem of premature fruit and leaf abscission.

A brief survey of the processes influenced by plant hormones should provide some important insight into future research on the isolation and characterization of other bioactive substances.

"Hormones," generally speaking, means hormones of animals, and the classic definition is:

> Organic compounds produced in special organs, glands, tissues, or cells, which are translocated through veins (or like tissues) to specific organs or tissues and which act to stimulate growth, differentiation, or development.

Although this almost defines plant hormones, there are some ambiguous points. The relationship between where plant hormones are produced, how they are translocated, and their mode of action is not the same as for "hormones" (hormones of animals). Additionally, there are other very significant differences between plant hormones and "hormones." For example:

(i) The number of known plant hormones (or plant hormone groups) is fewer than seen for "hormones," only five to seven, depending on how rigorous a definition is used.

(ii) Plant hormones appear to be ubiquitous, i.e., present in all higher plants (and many are also present in lower plants, and even fungi and bacteria). This universality of plant hormones is in marked contrast to the wide diversity of "hormones."

(iii) Each "hormone" appears to have an individual and quite specific role. In contrast, the effects of plant hormones are quite complex, often one plant hormone has a wide range of effects. However, the various phenomena observed when plant hormones are applied exogenously to the plant do not always mean that the plant hormone is directly causing the effect. Some effects of the plant hormone may be direct, but others are indirect, with the site of action being quite remote from location of the end effect. Additionally, there is evidence of rather complicated interactions among various plant hormones.

These unique characteristics of plant hormones have made it quite difficult to investigate their mode of action, and it is partly because of this fact that our understanding of how plant hormones act lags behind that of "hormones."

While plants do not have a nervous system, they do grow and develop, according to each species' individual life cycle. On the basis of this, plant hormones can be defined as minute factors acting as mediators in signal transduction, from the reception of external factors (light, temperature, water, etc.) to internal factors (e.g., aging), to gene expression.

As noted above, five hormone structures/classes are recognized, the auxins, gibberellins, cytokinins, abscisic acid, and ethylene. The designation of a new plant hormone/plant hormone class is rather strict. Thus, it is likely that there are more than five, and even scientists who are reluctant to accept new plant hormones do not claim that all physiology of higher plants can be explained by the actions of these five plant hormone classes.

Studies have afforded two new structural classes with hormone-like characteristics, namely, brassinosteroids and the jasmonic acid-related compounds. In the view of many researchers the brassinosteroids and jasmonic acid-related compounds also satisfy the definition of a plant hormone, and in this chapter we have chosen to accept them as such.

Besides the above five or seven plant hormone groups, phenolic compounds and polyamines have often been described as having plant hormone-like effects. That is, they appear to influence plant growth regulation, even though their effects are generally not so dramatic as those of the "accepted" plant hormones. Due to space considerations we will not cover phenolic substances and polyamines in this chapter.

Antheridiogens are defined as substances which are secreted from prothallia of ferns to induce antheridia in protonemata or immature prothallia. On the basis of this definition, antheridiogens are thought by many researchers to be a kind of "primer pheromone." Interestingly, all known antheridiogens are closely related to gibberellins, or are gibberellin derivatives, although there are still some antheridiogens which have not been characterized. Furthermore, compounds with the characteristic antheridiogen structure have been isolated from higher plants, though their function in higher plants is not understood.

Because of the close structural similarity between gibberellins and antheridiogens, they are included in the gibberellin section (see Section 8.02.3.7).

It is also likely that there are other plant hormones, as yet uncharacterized. The case of florigen, though, may be an exception. This mysterious substance was reported in the 1930s as a hypothetical, translocatable plant hormone that could induce flower bud formation. Yet despite intensive research, florigen remains unidentified, like a mountain not yet climbed.

Many scientists in the field of natural products chemistry possess an intense interest in searching for new plant hormones. While one could argue that this chapter should deal mainly with isolation, structure determination, and organic synthesis of plant hormones, it will mainly focus on the biological activities of plant hormones and on their biosynthesis.

When auxin was discovered as the principal factor that promoted elongation of seedlings, it is likely that no one imagined that it had other functions, for example, callus induction. Similarly, when abscisic acid was isolated, no one could have guessed that it would eventually be found also to control stomatal closure. As previously pointed out, the very diverse effects induced by plant hormones have attracted the interests of many researchers, and this research has often led to practical use in agriculture, horticulture, and other applied fields.

While the physiological activities of each plant hormone/hormone group are discussed extensively in this chapter, the following points should be considered as well. First, some effects caused by application of a plant hormone should not always be assumed to be a direct effect of the plant hormone *per se*. For example, some effects which were initially thought to be due to auxins were later found to be caused by ethylene, once it was confirmed that application of auxins can induce an ethylene-synthesizing enzyme. Therefore, it is necessary to consider interactions among plant hormones when the effects of the various plant hormones are discussed. Second, it is determined within each plant hormone structural group which are the active forms among many homologues. For instance, there are many gibberellins now characterized (112 at present), and some plant species may have 40 or more different gibberellins present. However, only a few of these gibberellins appear to be active forms ("active gibberellins") in elongation growth of the shoot, others being biosynthetic precursors, or inactive catabolites. Possible roles for gibberellins in other growth and development processes, such as flowering, apical dominance, senescence, etc., currently remain at the speculative stage. In the period when a large number of new gibberellins were being isolated, many studies regarding the biological activities of gibberellins were also carried out. However, at that time the biosynthetic relationship between each gibberellin was not well known and often was not considered. It is now generally accepted that there are likely to be only a few specific "active gibberellins" that play causal roles in plant growth and development. Thus, precursor gibberellins do not function in the action mechanism, but rather are converted to "active gibberellins." Similarly, other gibberellins will be inactive forms, produced at various stages along the biosynthetic pathway. Surprisingly, many inactive gibberellin catabolites are produced prior to formation of an "active gibberellin," and these inactivation steps may be one of the primary ways by which plants control the effective level of "active gibberellins." Generally, inactivation of plant hormones is a very rapid event, and this suggests that careful analysis of these metabolic steps is needed when researching the kinetics of plant hormone action.

Thus, a complete understanding of the biosynthesis and metabolism of plant hormones is essential in any study of their activities and effects on plant growth and development events. Studies on biosynthesis and the various biosynthetic pathways will quickly extend into fields such as isolation and characterization of enzymes, the regulation mechanism of enzyme induction, enzyme action and so on.

Because there is a wide diversity in the biosynthetic routes of plant hormones, depending on plant species, organ type, growth stage, etc., it will often be difficult to initially establish biosynthetic pathways. For example, despite decades of intensive research, the biosynthetic routes of auxin biosynthesis have not been completely established.

Furthermore, even though well-established pathways are known to exist for some hormone classes (i.e., the gibberellins), one must keep an open mind that other pathways, major or minor, may coexist. For example, while the mevalonic pathway in terpene biosynthesis is now accepted for gibberellins, the nonmevalonic pathway proposed for the biosynthesis of isoprenoids suggests the possibility that gibberellins could also be synthesized through this route (see Section 8.02.3.5.1).

Research on hormone biosynthesis now extends to the fields of molecular biology and genetic engineering, and knowledge gained from these studies should quickly lead to a more complete and useful integration of chemistry and biology.

The rapid development in our knowledge of plant hormones is very much due to progress in analytical methodology. Two of the most important methodologies have been GC–MS and immunoassay. GC–MS has been especially useful as a highly reliable analytical method for organic chemists for minute levels of plant components, making it ideally suited for plant hormone identification and quantification. However, immunoassay, which is regarded by many researchers as a very important analytical method, cannot, unfortunately, always be safely used for the study of plant hormones. This is because immunoreactions occur primarily with high molecular weight compounds and all the known plant hormones are of low molecular weight. Thus, plant hormones can become effective antigens only by coupling them with proteins, a process that requires skills in organic chemistry, for carrying out the coupling reactions, designing the structure of the spacer, and deciding where to attach the plant hormones to the protein. In fact, sophisticated methods and an extensive knowledge of chemical structure are essential to allow for the discrimination of minute substances consisting of many homologues with only slight structural differences. This type of problem is especially important for gibberellins, which have over 100 homologues.

The practical application of plant hormones is well established in agriculture and horticulture, and synthetic mimetics are also frequently used. Often, the synthetic compounds are more potent than the natural plant hormones, a classic case being 2,4-dichlorophenoxyacetic acid (2,4-D), which at low doses mimics the auxin, indoleacetic acid, and at high doses acts as a herbicide on many broadleaved plants. 2,4-D is more active than indoleacetic acid because it is not easily inactivated, unlike indoleacetic acid. Thus, some synthetic compounds which act as modified hormones are described in this chapter. Results obtained from the use of synthetic mimetics of natural plant hormones are important not only from the viewpoint of practical uses, but also for the investigation of structure–activity relationships of the natural plant hormones.

A major focus of plant hormone research is on molecular mechanisms of biosynthesis and regulation of hormone action, especially at the gene expression level. This requires not only a background in plant hormone physiology and biochemistry, but also knowledge of molecular biology techniques.

Finally, the reader is referred to other useful books concerning plant hormones.[1–8]

8.02.2 AUXINS

8.02.2.1 Brief History

The first plant physiological study on auxins was reported by Charles Darwin and his son Francis in 1880.[9] They described in their book, "Power of Movement in Plants," that, when seedlings of a monocotyledon (*Phalaris canariensis*) are freely exposed to a lateral light, some influence is transmitted from the upper to the lower point, causing the latter to bend. This hypothesis was further developed through studies of phototropism using *Avena* coleoptiles by Rothert, Fitting, Boysen Jensen, Paal, and Söding.[10] It was proposed from these studies that the transmitted influence (stimulus) was a growth substance. Went showed that a growth substance diffused from the coleoptile tips into an agar block was influenced by light.[11] This agar block could bend the *Avena* coleoptile when it was put on one side of the coleoptile after removal of the tip. Thus, the substance which diffused into the agar block can replace the light-influenced *Avena* coleoptile. Since then, the *Avena* curvature test was established and this has been the most general bioassay system to qualify and quantify auxins because the curve angle is proportional to the amount of the growth substance in the agar. Using this bioassay, purification of this growth substance was begun. Because the amount of the growth substance in higher plants is very small, other rich sources were investigated. Kögl and his co-workers isolated three active compounds and named them auxins (from the Greek auxin, to increase).[12] Auxin a and auxin a lactone were purified from human urine and auxin b from malt. These three compounds were chemically characterized, but subsequent attempts to repeat these experiments were not successful. The original samples left in Kögl's laboratory were analyzed by modern methods and revealed completely different chemical structures from those proposed.[13] Auxin b lactone was synthesized according to the proposed structure but it had no auxin activity.[14] Kögl *et al.* isolated another active compound from urine and named it hetero auxin.[15] The structure of hetero auxin was determined to be indole-3-acetic acid (IAA) which was then isolated from the yeast, *Rhizopus*, and alkaline hydrolysate of corn meal as an auxin.[16,17] Subsequently, the first isolation of free IAA from a higher plant, immature kernels of *Zea mays*, was reported by Haagen-Smit *et al.*[18] Since then, IAA has been isolated from many plants (see Table 1). IAA is now recognized to be an active principle of auxin.

Table 1 Isolation of IAA from plants.

Plant material	*Ref.*
Zea mays	
Kernels	18
Coleoptiles	19
Roots	20
Avena sativa	
Seedlings	21
Grains	22
Phaseolus mungo, etiolated seedlings	23
Brassica pekinensis, clubroots	24
Undaria pinnatifida, thalli	25
Fagales castanea, insect gall	26
Phaseolus vulgarus, shoots	27
Gossypium hirsutum, ovules	28
Nicotiana tabacum, callus	29
Tropaeolum majus, embryo	30
Oryza sativa, bran	31
Pseudotsuga menziesii, shoots	32
Ricinus communis, xylem sap	33
Triticum aestivum, grain	34
Lathyrus maritimus, seeds	35
Vitis vinifera, fruit	36
Chamaecyparis lawsoniana, leaves	37
Lygodium flexuosum, leaves	38
Euphorbia escula, roots	39
Pisum sativum, flowers	40
Phaseolus vulgaris, seeds	41

8.02.2.2 Chemistry

8.02.2.2.1 Purification and analysis

(i) Purification

Most of the auxins in plants are IAA and its derivatives and are easily oxidized and degraded by oxygen and enzymes, because they possess an indole skeleton. Therefore, the extraction from plant materials and the purification process should be carried out rapidly in the dark at low temperature.[42]

A general purification procedure of auxins, especially IAA, is as follows: (i) solvent extraction with MeOH or acetone using a homogenizer or a blender recommended for large materials, (ii) fractionation with silica gel column and/or TLC, and (iii) isolation with HPLC using an ODS column. For the extraction, an antioxidant such as ascorbic acid, diethyldithiocarbamate, or BHT (2,6-di-*t*-butyl-4-methylphenol) is often added to the solvent to avoid oxidation.[43] A phosphate buffer is as effective as an organic solvent, but extraction should be done at low temperature to avoid degradation by enzymes (oxidase and esterase).[44] Paper chromatography was previously the most popular method for auxin purification,[45] but now it is rarely used. HPLC is now the most effective method for the final purification of auxins and a reverse-phase column such as ODS is usually used.[46] For HPLC, UV 280 nm absorption is usually used to detect IAA or its derivatives, which is the maximum absorption of an indole chromophore. Fluorescent detection can be used and is more sensitive than UV; irradiation wavelength is 280 nm and emission wavelength is 350 nm.[47] The amount of auxins in plant material is usually small; therefore, large amounts of the starting material must be treated to determine the chemical structure of auxin with physicochemical methods (NMR, IR, MS, etc.). GC–MS, however, is now the most common and convenient method to identify and quantify auxins when authentic samples are available.

(ii) Analysis

Bioassay has been an important method for analyzing auxin from the beginning of auxin research. Although the historical role of bioassay in analyzing auxin appears to have finished, the bioassay still plays a role in some cases. The first auxin bioassay, the *Avena* curvature test, is still valuable in establishing whether a certain compound has auxin activity or not.[48] This method needs some skill to detect the auxin activity and measure the amount of auxins. Many bioassay systems simpler than the *Avena* curvature test have been developed.[49] Bioassay methods are still effective for investigating a new auxin and examining auxins in a crude sample, but now other methods such as immunological methods are usually used for quantification of auxins.

An antibody for IAA was first reported by Fuchs and Fuchs in 1969[50] and the monoclonal antibody was also developed.[51] Now, commercially available antibodies of IAA can be used for radioimmunoassay and enzyme immunoassay.[52] Partial purification of samples is recommended before the auxin immunoassay. It should be noted, since the carboxy function of IAA is generally used for bonding to a macromolecule to make the immunogen, ester, and amide derivatives of IAA show high cross-activities to the antibody. The immunoassay is a simple, quick, and inexpensive method, but it can not essentially analyze the exact amounts and chemical structures of auxins.

GC is a sensitive and convenient analytical method for volatile compounds. IAA should be converted to volatile derivatives for the GC analysis. Many derivatives of IAA and column packing materials for each derivative have been reported (Table 2). An FID (hydrogen flame ionization detector) is generally used and the minimum detection limit is 10 ng. An ECD (electron capture detector) can be used for the halogen-containing derivatives and the detection limit is approximately 1000 times lower than that of FID.[53] GC–MS is a system where a mass spectrometer is used as the detector of GC and is the most sensitive and reliable method for the analysis of a small molecular weight compound. By comparing with the retention time and mass spectrum of the standard sample, identification and quantification are possible. The detection limit of IAA with GC–MS is about 10 ng. The single ion monitoring (SIM) method is the most sensitive method by which only the typical ions of a compound, for example, molecular and strong fragment ions, are analyzed by mass spectrometry and data are stored in a computer. The retention times and ion strengths of each ion are plotted on a mass chromatogram and 1–10 pg IAA can be quantified.[54]

Table 2 GC analysis of auxin derivatives.

Derivatization	*Reagent*	*GC packing material*	*Ref.*
Methylesterification	CH_2N_2: diazomethane	SE-54 DB-1	54 55
Trimethylsilylation	BSTFA: *N,O*-bis(trimethylsilyl)trifluoroacetamide	Me silicon SE-30	56 44
Trifluoroacetylation	TFAA: trifluoroacetic anhydride	SE-54	54
Heptafluorobutyrylation	HFBI: heptafluorobutyrylimidazole	OV-101	57

GC is a sensitive and reliable system, but derivatization is necessary for the acidic auxins such as IAA. However, sample derivatization is not needed for HPLC analysis. LC–MS in which the compounds separated by HPLC are introduced into a mass spectrometer has become available, although the LC–MS instrument is still expensive. The detection limit of auxins is 0.1–1.0 ng using atmospheric pressure chemical ionization (APCI) in the authors' laboratory. Although this method is not completely established, it will be an effective method to analyze auxins in the future.

8.02.2.2.2 Structures

(i) Natural auxins

IAA is the most abundant natural auxin of higher plants (Table 1). Other natural auxins having indole skeletons are IAA derivatives or have structures which are easily converted to IAA in plants. The isolation of indole-3-acetonitrile, indole-3-acetoamide, indole-3-acetaldehyde, indole-3-acetaldoxime, and indole-3-ethanol were reported from plants and they are putative intermediates of the biosynthetic pathway of IAA. Chlorinated IAA was isolated from immature seeds of pea by Marumo *et al.*[79] This compound seemed to be a typical auxin of Leguminosae, but the immature seeds of the pine tree also contain 4-chloro-IAA. Structures, origins, and references of natural auxins are listed in Table 3. Indoleglucosinolates were isolated from Brassicacae and non-enzymatically convert to IAA (Table 4). Conjugated IAAs were also isolated from many kinds of plants and they are thought to be metabolites and/or storage forms of IAA (Table 5). Three compounds, *p*-hydroxyphenylacetic acid,[58] phenylacetic acid,[59] and phenylacetonitrile[60] which do not have an indole skeleton but show auxin activities were also isolated from plants.

(ii) Other natural compounds related to auxin

Many auxin antagonists or antiauxins are known to exist in plants. Here, physiologically interesting compounds, except abscisic acid, are described.

Using a physicochemical method, Bruinsma reported that IAA equivalently distributed to the light-irradiated side and the shadow side, then he showed a growth inhibitor, xanthoxin, accumulated on the light side of sunflower coleoptile.[100] This result suggested a hypothesis contrary to the Cholodony–Went theory of phototropism in which the auxin moved to the shadow side, causing bending of the coleoptile. According to Bruinsma's hypothesis, Hasegawa and co-workers[101–105] identified several growth inhibitors (Table 6a).[101–108] The structures of these compounds are unique to plant species; therefore, it seems that Bruinsma's hypothesis does not completely contradict the Cholodony–Went theory. Yamamura and co-workers isolated the active principles of leaf-closing factors from several plants and they showed antagonistic activities to auxins (Table 6b).[106–108] Oligosaccharides were identified as antiauxins from the xyloglucan hydrolysate with cellulase.[109] These compounds are interesting, because oligosaccharides may be liberated from cell walls when the auxin loosens the cell walls, resulting in cell expansion.

8.02.2.2.3 Biological activities

(i) Plant and organ level

(a) Stem growth. Generally speaking, exogenously applied auxin does not promote stem growth of intact plants.[110] But some exceptional results have also been reported, for such cases where the stem is under low growth conditions, for example dwarf cultivars,[111] ethylene treatments,[112] apical removal,[113] flower removal[114] and leaf removal,[115] and the elongation was observed in a short time[116] or a small part of the stem.[117] The exogenously applied auxin is more effectively taken up via a stem or leaf with a lanolin paste or a spray method than via roots from a water solution.

Auxin promotes the elongation of explants from the stems of many plants.[118] Under usual conditions, the elongation effects of auxin are observed in two stages. The first one is similar to acid growth and the elongation can be observed after 10–20 min of auxin treatment. In both stages, auxin induces the loosening of the cell wall and promotes the cell elongation. In contrast, auxin inhibits the growth of the hook explants from young buds.[119] A high dose of auxin usually inhibits

Table 3 Naturally occurring auxins.

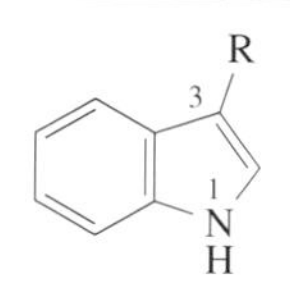

R	*Name*	*Plant material*	*Ref.*
CH_2CO_2H	indole-3-acetic acid	see Table 1	
$CH_2CH_2CO_2H$	indole-3-propionic acid	*Cucurbita pepo* *Pisum sativum*	61 62
$CH_2CH_2CH_2CO_2H$	indole-3-butyric acid	*Nicotiana tabacum* Several plants	63 64
CH_2CHO	indole-3-acetoaldehyde	*Helianthus annuus* *Emblica officinalis*	65 66
CH_2CH_2OH	indole-3-ethanol	*Cucumis sativus* *Desmanthus illinoensis*	67 68
CH_2CONH_2	indole-3-acetoamide	*Phaseolus mungo* *Euphorbia escula*	69 39
CH_2CN	indole-3-acetonitrile	*Brassica oleracea* *Cruciferae*	70 71
$CH_2CH{=}NOH$	indole-3-acetoaldoxime	*Brassica pekinensis* *Nicotiana tabacum*, crown gall	72 73
$CH_2CH{=}CHCO_2H$	indole-3-acrylic acid	*Lens culinaris*	74
CO_2H	indole-3-carboxylic acid	*Pisum sativum* *Nicotiana tabacum*, crown gall	75 76
CHO	indole-3-carboxaldehyde	*Brassica oleracea* *Pinus sylvestris*	77 78

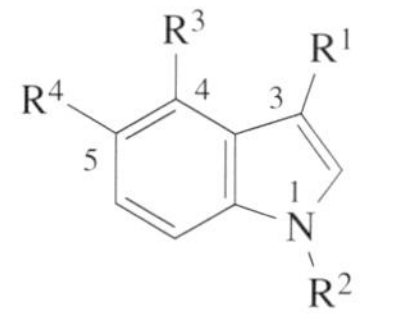

R^1, R^2, R^3, R^4	*Name*	*Plant material*	*Ref.*
$R^1 = CH_2CO_2H$, R^2, $R^4 = H$, $R^3 = Cl$	4-chloroindole-3-acetic acid	*Pisum sativum* *Vicia amurensis*	79 80
$R^1 = CH_2CO_2H$, R^2, $R^3 = H$, $R^4 = OH$	5-hydroxyindole-3-acetic acid	tomato	81
$R^1 = CH_2CN$, $R^2 = OMe$, R^3, $R^4 = H$	1-methoxyindole-3-acetonitrile	*Brassica pekinensis*	24
$R^1 = CH_2CN$, $R^3 = OMe$, R^2, $R^4 = H$	4-methoxyindole-3-acetonitrile	*Brassica pekinensis*	82

the growth of explants.[120] The promotion and inhibition observed after the treatment with a high concentration of auxin can partly be explained by the effect of ethylene which is induced by the auxin.[121]

The amount of endogenous auxin in stems is usually high under the good growing conditions, but there are many reports that the growth rate of stems is independent of the amount of endogenous auxin.[122] When the amount of auxin is measured along the stems, the good growing part does not contain much auxin, but the highest amount of auxin is detected at the top of the stem because auxin is activated or synthesized there and moves to the lower part.[123]

Table 4 Indoleglucosinolates.

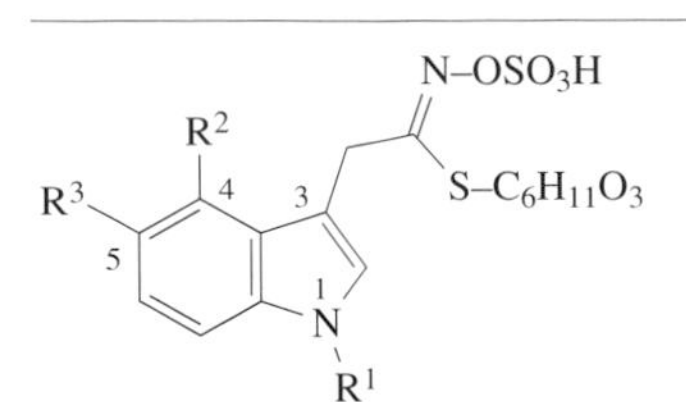

R^1, R^2, R^3	*Name*	*Plant material*	*Ref.*
R^1, R^2, R^3 = H	glucobrassin	*Brassica oleracea* *Brassica napus*	83 84
R^1 = OMe, R^2, R^3 = H	neoglucobrassin	*Brassica oleracea* *Brassica napus*	85 86
R^2 = OH, R^1, R^3 = H	4-hydroxyglucobrassin	*Brassica napobrassica* *Brassica napus*	87 88
R^2 = OMe, R^1, R^3 = H	4-methoxyglucobrassin	*Brassica juncea* *Capparis spinosa*	89 90
R^1 = COMe, R^2, R^3 = H	1-acetylglucobrassin	*Tovaria pendula*	91
R^3 = OH, R^1, R^2 = H	5-hydroxyglucobrassin	*Brassica oleracea*	92
R^3 = Me, R^1, R^2 = H	5-methylglucobrassin	*Brassica oleracea*	92
R^1 = SO_3H, R^2, R^3 = H	1-sulfoglucobrassin	*Isatis tinctoria*	93

Table 5 Conjugated auxins.

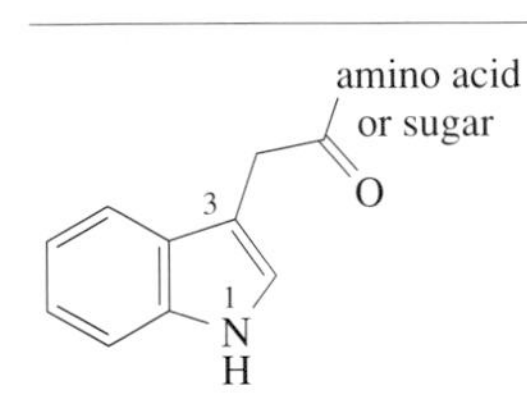

Name	*Plant material*	*Ref.*
Indole-3-acetyl-L-aspartic acid	*Gleditschia tricanthos*	94
Monomethyl 4-chloroindole-3-acetyl-L-aspartate	*Pisum sativum*	95
2-*O*-(Indole-3-acetyl)-*myo*-inositol 1-DL-(Indole-3-acetyl)-*myo*-inositol	*Zea mays*	96
5-*O*-β-L-Arabinopyranosyl-2-*O*-(indole-3-acetyl)-*myo*-inositol	*Zea mays*	97
5-*O*-L-Galactosyl-2-*O*-(indole-3-acetyl)-*myo*-inositol	*Zea mays*	98
di-*O*-(Indole-3-acetyl)-*myo*-inositol tri-*O*-(Indole-3-acetyl)-*myo*-inositol	*Zea mays*	99
4-*O*-(Indole-3-acetyl)-D-glucopyranose 6-*O*-(Indole-3-acetyl)-D-glucopyranose	*Zea mays*	98

(b) Leaves. The production of auxin is very high in the developing young leaves and the growth rate of leaves is related to the amount of auxin produced.[124] The exogenously applied auxin does not show the promoting activity of the whole leaf growth. The main vein growth is promoted by the auxin treatment and this induces the morphological changes in leaves.[125] The epinasty of petioles was observed by auxin treatment, but this is thought to be the effect of ethylene induced by the

Table 6 Naturally occurring antiauxins.

Compound	*Plant material*	*Ref.*
a		
Cinnamic acid derivatives (raphanusol A, B)	*Raphanus sativus*	101, 102
3-Methylthiomethylene-2-pyrrolidinethione (raphanusanin)	*Raphanus sativus*	103, 104
6-Methoxy-2-benzoxazolinone (MBOA)	*Zea mays*	105
b		
Potassium chelidonate	*Cassia mimosoides*	106
	Cassia occidentalis	106
Trigonelline	*Aeschynomene indica*	107
Pyllanthurinolactone	*Phyllanthus urinaria*	108

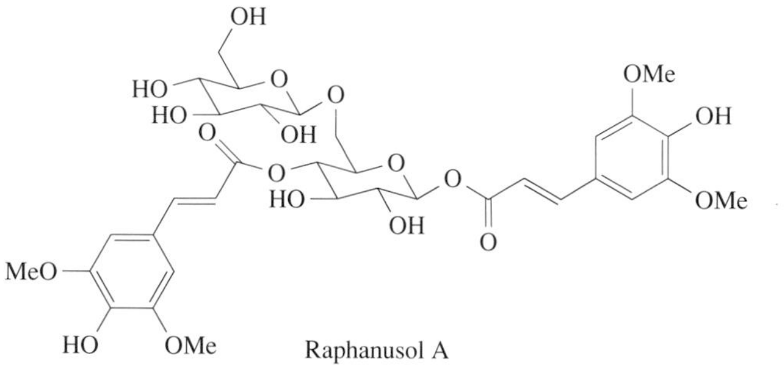

Raphanusol A

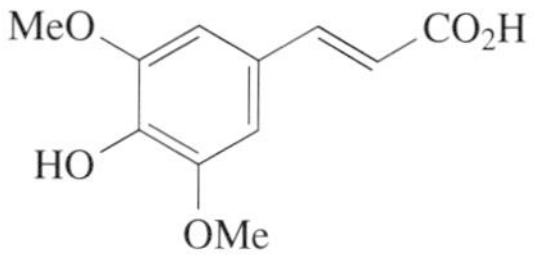

Raphanusol B

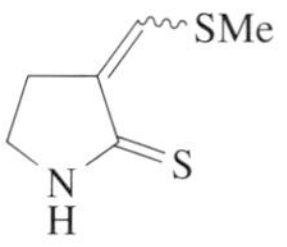

Raphanusanin

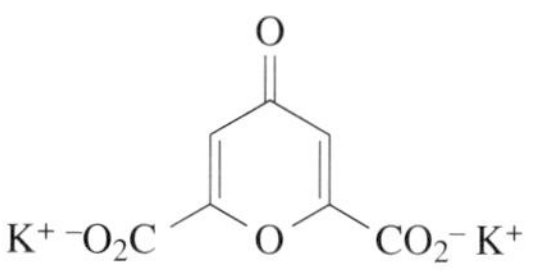

Potassium chelidonate

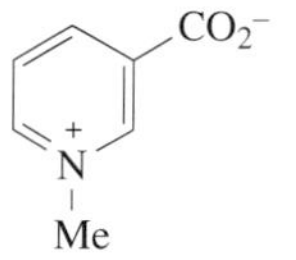

Trigonelline

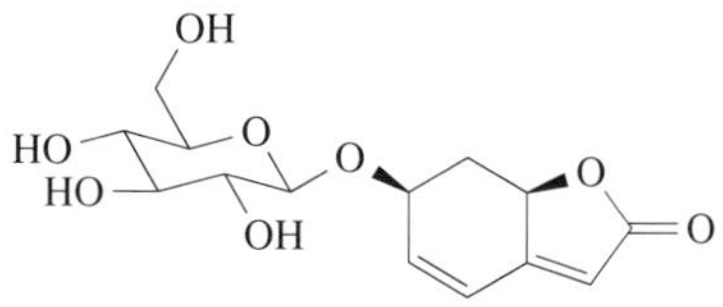

Phyllanthurinolactone

auxin.[126] The auxin treatment is effective in opening the closed leaves when leaves close under dark conditions in some plants.[127]

(c) Roots. Auxin inhibits root growth generally, because the optimum concentration of auxin to roots is much lower than that of stems.[128,129] The usual endogenous auxin level is inhibitory to root growth. A negative relationship was observed between the amount of endogenous auxin and the growth rate of roots. However, the promotion effect of auxin to root growth was also reported. This growth promotion was increased with ethylene inhibitors. Therefore, the inhibitory effect of auxin on roots growth is considered to be the effect of ethylene induced by the auxin treatment, although there are reports which are contradictory to the above results.

The time course of the growth rate of roots is complex after the treatment of auxin. When the auxin dose is low, the growth is promoted in the beginning but the total growth is inhibited because the growing time is shortened. When the auxin dose is high, the growth rate decreases steeply in the beginning and then returns gradually.

The tip of roots contains the highest amount of auxin and the amount decreases to the base of the roots. At least a part of auxin contained in the tip of roots is synthesized and transported from the top of the terrestrial part.

(d) Seeds and buds. Auxin rarely affects the induction and break of seed dormancy and germination.[130] The increase of auxin with germination was reported, which was not considered to break the seed dormancy but relate to the growth of other parts after germination. Chlorinated auxin, 4-chloro-IAA is active in stimulating pod growth in deseeded pea, where other auxins are weak or inactive.[131]

Apical dormancy is one of the most typical physiological phenomena controlled by auxin.[132] The removal of the stem tip, which includes a very young leaf in dicots, stops the auxin supply and activates the dormant lateral buds. Since exogenous auxin can replace the stem tip and auxin transport inhibitor, such as 2,3,5-triiodobenzoic acid (TIBA), can break apical dormancy, auxin from the tip was considered to inhibit directly the growth of lateral buds.[133] But the amount of auxin increases in the activated lateral buds after the removal of the stem tip and the buds begin growth. It is now believed that auxin may inhibit indirectly growth of lateral buds. Although there are many hypotheses explaining the role of auxin in the apical dormancy, the following three hypotheses seem to be important: (i) auxin maintains the high level of abscisic acid (ABA) in the lateral buds and the acid keeps them dormant,[134] (ii) auxin induces ethylene biosynthesis in the lateral buds and lateral buds do not grow in a high level of ethylene,[135,136] and (iii) auxin accumulates nutritional substances in the apical bud and the lateral buds do not receive enough nutrition to grow.[137,138] In general, rich nutritious conditions weaken the apical dominance.

(e) Differentiation. Auxin promotes the differentiation of the vascular bundle.[139] In general, the concentration of auxin necessary to induce phloem differentiation is lower than that of the vessel. The amount of endogenous or polar transport auxin is proportional to the differentiation level of the vascular bundle.[140] The differentiation of the lateral root primordium is promoted by auxins.[141] Auxins, especially indolebutyric acid, promote the rooting of plant cuttings. Callus formation and growth are stimulated by auxin. The root formation on the callus is promoted by the relatively high ratio of auxin to cytokinin. Embryogenic cell formation on the callus also needs auxin, but somatic embryo formation from embryogenic cells is usually promoted in non- or poor-auxin conditions.[142]

(f) Sexual reproduction.[143,144] Auxin promotes or inhibits flower bud formation depending on plant species, timing, concentration of auxin, and type of auxin. The inhibition of a flower bud formation is observed in the short-day plants. The formation of female flowers is promoted or their ratio is increased by auxin treatment. In contrast, the promotive effect of auxin on male flower formation was also reported. In some species, the decrease of endogenous auxin is related to flower bud formation, although the opposite results were also reported. Generally, auxin promotes fruiting and fruit growth. The parthenocarpy of many plants is promoted by auxin treatment.

(g) Others. Auxin inhibits the formation of the abscission layer and the abscission of various organs. However, auxin sometimes promotes abscission, especially under highly concentrated exogenous auxin treatment. In general, the decrease of endogenous auxin accelerates the abscission of organs.[145]

Auxin attracts photoassimilatory products—accumulation of products was observed in the tissues near the auxin-treated point.[146] This product-attracting effect of auxin may be related to the sink function maintained in the auxin. The translocation of assimilatory products occurs before the growth induced by auxin treatment; therefore, it is not a passive flow to the parts consuming products.[147] When auxin promotes the translocation of the assimilatory product, the loading to phloem is also promoted by auxin.[148] Auxin also maintains a high level of invertase activity which is related to unloading from phloem. Auxin has also been reported to attract inorganic nutrition.[149]

(ii) Cell level

Plant cell expansion is limited by the rigidity of the cell wall and auxin brings about cell wall loosening.[150] Since cell wall loosening gives rise to the reduction of cell turgor, water flows into the cells, resulting in cell expansion. The epidermal cell layers are most sensitive to auxin treatment, because tissues elongate very little with auxin treatment when the outer epidermal cell layers are removed by peeling.[151] Several studies have shown that dynamic properties of the cell wall are changed after auxin treatment. It was suggested that the mechanisms of cell wall loosening in monocots and dicots are different. In dicots, xyloglucan breakdown was accelerated by auxin.[152] Auxin induced the soluble xyloglucan from the cell wall in a very short time after auxin treatment. In monocots, antibodies of β-D-glucans and β-D-glucanases inhibit auxin-induced cell elongation, although their effects are partial.[153] However, the effect of hydrolytic enzymes on wall loosening has been brought into question by the study of maize coleoptile.[154] Another candidate enzyme in cell wall loosening is xyloglucan endotransglycosidase which transfers small xyloglucan chains to other polymer chains.[155] The rise of this enzyme activity is proportional to the cell elongation, although this enzyme itself cannot induce cell elongation. Two proteins, expansins which were isolated from cucumber cell walls, could induce wall extension and showed no polysaccharide hydrolytic activities.[156] A protein with similar properties was also obtained from the cell walls of oat coleoptile.[157]

Auxin promotes synthesis and incorporation of new cell wall material. The activity of β-glucan synthase was observed within 10–15 min of auxin application.[158] Cell wall polymers are synthesized in Golgi and auxin induces an increase in Golgi size.[159] The synthesis of cell wall proteins is also promoted by auxin treatment. It was suggested that most of the cell wall proteins are not synthesized rapidly enough to initiate the rise in growth rate.[160]

Auxin induces a rise in H^+-ATPase activity at the plasma membrane a short time after auxin application. The role of H^+-ATPase in auxin activity is described in the next section.

(iii) Molecular level

(a) Metabolism of nucleic acids and proteins. From the beginning of 1950, the effect of auxin was investigated with regard to the change in the total amount of DNA, RNA, and protein.[161] Since then, the incorporation of isotope labeled compounds in DNA or protein has been investigated. These studies showed that auxin promoted the transcription and translation processes.[162] The results explained the slow reaction of auxin, for example, several hours to days after auxin treatment. Since the typical auxin effect is observed 10–30 min after auxin treatment, this rapid response was studied using an RNA and protein biosynthesis inhibitors. These inhibitors affected the cell elongation induced by auxin; therefore, a hypothesis (Gene Theory) was proposed that the cell elongation caused by auxin occurred by gene expression.

Since the "Gene Theory" could not explain clearly the fast response of plant cell to auxin, the "Acid Growth Theory" was proposed. In 1971, Hager[163] proposed that the rapid cell elongation was caused by auxin decreasing the pH outside cells, which increases the efflux of H^+ into the cell wall.[164] This theoretical mechanism was thought not to be related to gene expression because of the very rapid reaction. The H^+ release by auxin, however, was shown to need protein synthesis because inhibitors of protein synthesis prevented H^+ release. This theory, which, however, remains controversial, has promoted studies on auxin's effect on structures, ion permeability, and H^+-ATPase in membranes.[164]

(b) H^+-ATPase and H^+ efflux. The H^+ efflux induced by auxin seems to be caused by the increasing current through the plasma membrane H^+-ATPase.[165] Many workers have measured the H^+ efflux and correlated it with the elongation induced by auxin.[166] In most of these studies, the increased H^+ efflux has been reported to start with the similar timing of the elongation rise. Moreover, H^+ efflux and cell elongation were correlated to plasma membrane hyperpolarization in oat coleoptiles.[165] Carboxylic acids were found to induce membrane hyperpolarization in millimolar concentrations, but it was concluded that this was nonspecific.[167] The study of structurally auxin-related active and inactive compounds showed that H^+-efflux abilities of compounds are correlated to elongation activities. However, a study published in 1991 suggested that the hyperpolarization was not auxin specific.[168] The auxin-mediated H^+ efflux is inhibited by the protein synthesis inhibitors, such as cycloheximide. Although we do not have direct evidence, the synthesis of H^+-ATPase itself and/or other regulatory proteins are perhaps accelerated by auxin treatment. We can say so far that H^+-efflux-mediated H^+-ATPase is necessary for cell wall loosening and elongation, but is not sufficient to explain the growth induced by auxin.

Protoplasts have been used for the electrophysiological assays of auxin.[169] The hyperpolarization of plasma membrane was observed within 2 min of auxin treatment of tobacco protoplasts. In these experiments, the biological activities of auxin and auxin analogues are well correlated to the hyperpolarization induced by them. Root tissues are more sensitive to auxin than terrestrial tissues and protoplasts prepared from roots also showed lower optimum concentration for the membrane hyperpolarization induced by auxin. In auxin–resistant mutants, maximum hyperpolarizaiton of their protoplasts was observed at a much higher concentration of auxin than the wild type.[170] This methodology measuring hyperpolarization using protoplasts seems to be a powerful tool to investigate the receptor and signal transduction of auxins.

(c) Gene expression. Research into gene expression by the rapid response to auxin has developed due to advances in molecular biology. Many researchers have reported that the special peptides or

proteins were expressed by auxin treatment in the very early stage of auxin response and their corresponding cDNA were cloned, but most of their functions have not been elucidated. Several of these peptides and proteins are known to belong to a glutathione *S*-transferase (GST) superfamily. The function of GST in plants has not yet been elucidated although it has been studied extensively in animals. The key enzyme of ethylene biosynthesis, 1-aminocyclopropan-1-carboxylate (ACC) synthase, is activated by auxin and its mRNA is detected within 30 min of auxin addition. Arg1 was obtained from mung bean hypocotyls and showed homology to fatty acid desaturase. In genes induced by auxin of the promoter regions, PS-IAA4/5 of pea, GH3 and SAUR of kidney bean, and par B of tobacco are well studied. For the details of auxin-regulated genes, the reader is referred to a review by Takahashi *et al.*[171]

(d) Receptors and signal transduction. Since 1942 there have been many reports which suggested the existence of auxin-binding proteins, and the theory of two-point attachment was proposed based on the structure and molecular properties of auxins.[172]

Several auxin-binding proteins have been reported, which have been purified by affinity chromatography or photoaffinity labeling reagents. An azidoauxin was used for the photoaffinity labeling reagent and several proteins were purified. From the membrane protein fraction of *Cucurbita pepo*, 40 kDa and 42 kDa proteins were obtained and it was suggested that they were auxin transport proteins.[173] The 31 kDa and 24 kDa proteins were purified from the soluble fraction of *Hyoscyamus muticus* and they were recognized by the antibody of β-1,3-glucanase.[174] The 60 kDa soluble protein among four purified proteins from *Z. mays* has a homology to a β-glucosidase.[175] The 25 kDa protein from *H. muticus* and 24 kDa protein from *Arabidopsis thaliana* were reported to be a GST.[176,177] It is interesting that auxin-regulated proteins are reported to belong to a GST superfamily and auxin-binding property is also observed. These proteins are not considered to be receptors or binding proteins of auxin, but their physiological role in plant cells should be important.

The soluble protein purified by affinity chromatography from *Phaseolus aureus* showed that it promoted RNA synthesis and auxin-dependent protein synthesis. This soluble auxin-binding protein was identified to be a glutathione-dependent formaldehyde dehydrogenase.[178] From the microsome fraction of *Z. mays*, auxin-binding protein 1 (ABP1) was purified.[179,180] Its binding ability is very high (K_d (NAA)=0.1 μM) and its K_d values for synthetic auxins are proportional to their physiological activities. The amino acid sequence of ABP1 was determined from cDNA, and ABP1 genes were discovered in *Arabidopsis*, tobacco, and strawberry. ABP1 seems the most plausible candidate to be an auxin-binding protein. The following experiments also supported ABP1 as one of the auxin receptors. The C-terminal peptide of ABP1 was applied to the stoma guard cell of *Vicia faba* and reversibly inhibited the K^+ uptake, which was a similar response to the high dose auxin treatment. The depolarization sensitivity to auxin was increased 1000-fold by the addition of purified ABP1 to tobacco protoplast and this depolarization was completely inhibited by the ABP1 antibody. The monoclonal antibody recognizing the estimated auxin binding domain of ABP1 showed the auxin activity itself. For details of ABP1, the reader is referred to other reviews.[181–183]

There are many reports supporting the existence of signal transduction systems in which Ca^{2+}, phosphoinositides, protein kinases, and phospholipases are related. Auxin induces a change of Ca^{2+} concentration in cytoplasm.[184] The components of a phosphoinositide pathway necessary for the induction of intracellular Ca^{2+} release have been reported in plants, although this pathway in plants remains unclear.[185] Several reports showed that auxin affected the phosphorylation of proteins.[186] There are reports of auxin inducing an increase in phospholipase C activity[187] and phospholipase A_2 activity.[188] We now have some limited knowledge of the signal transduction system related to auxin. However, it will become more systematic as in the near future as receptor studies of auxin develop.

8.02.2.2.4 Biosynthesis

The biosynthetic pathway of auxin, IAA, is not completely clarified in higher plants. Since tryptophan is chemically related to IAA and is an amino acid distributed in all plants, the precursor of IAA is thought to be tryptophan. Several reports, however, revealed that the biosynthetic pathways do not proceed via tryptophan. In this chapter, the L-Trp pathway is described first and the other pathway second.

(i) Triptophan pathway

The biosynthetic pathway of IAA from Trp as a precursor is classified into five routes as shown in Scheme 1.

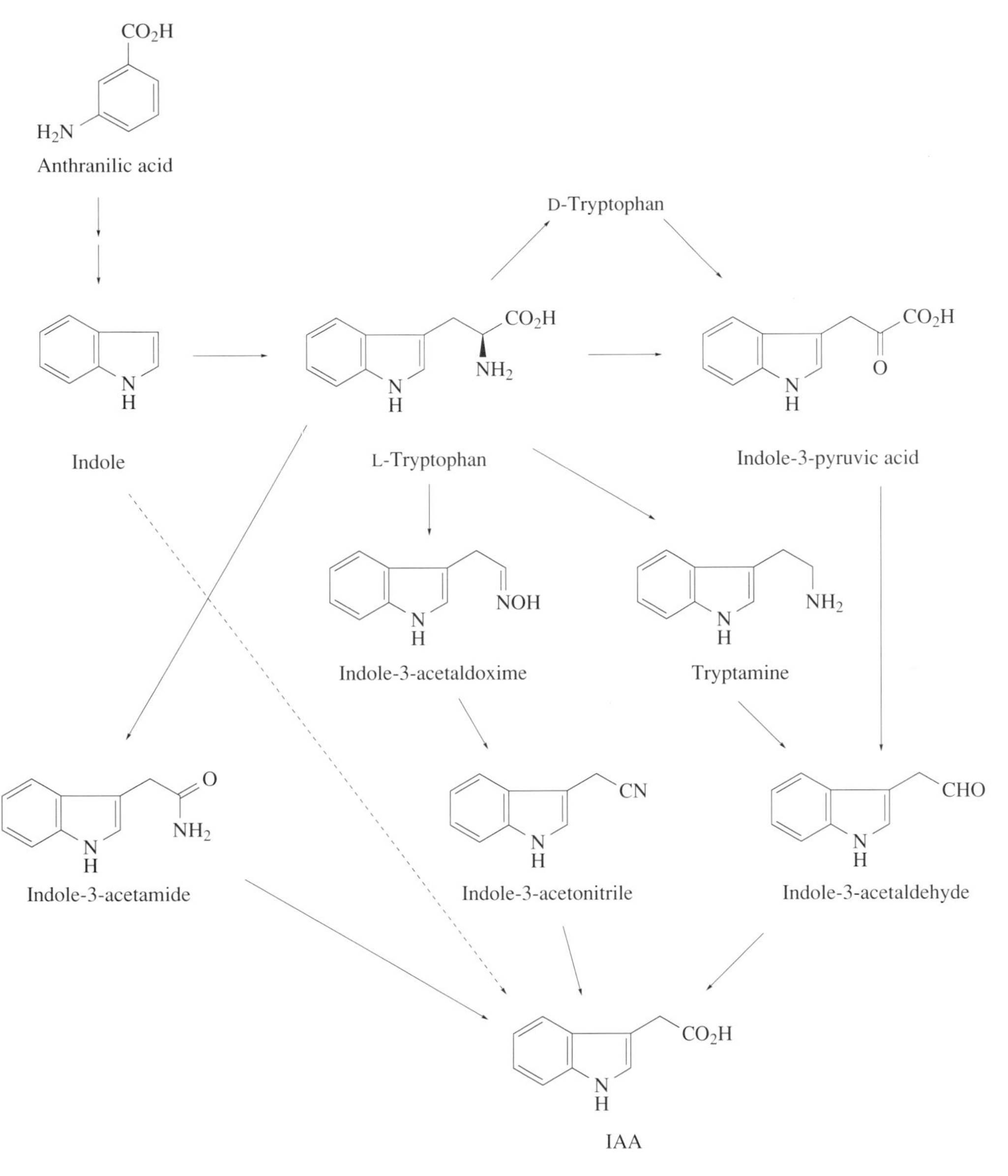

Scheme 1

(a) D-Trp. L-Trp is converted to D-Trp by racemase and then metabolized to indolepyruvic acid.[189] The amount of D-Trp is reported to be as small as IAA and the racemization step is the rate-limiting step of this pass. The malonylated Trp was believed to contain only the D form of Trp,[190] but contrary evidence was subsequently reported.[191,192] This pathway is now recognized not to be the main biosynthetic pathway of IAA.

(b) Indolepyruvic acid. L-Trp is directly converted to indolepyruvic acid by amino transferase and then IAA is synthesized via an indole aldehyde.[193] But the key enzyme of this pathway, amino transferase obtained from plants, showed a very high K_m value for Trp and the genes encoding the enzymes of this pathway have not been isolated from plants. Another problem is that the key intermediate, indolepyruvic acid, is a chemically very unstable compound; therefore, the analysis of the endogenous level of this compound and incorporation experiments are very difficult.

(c) Tryptamine. L-Trp is metabolized to tryptamine by tryptophan decarboxylase and then IAA is synthesized via indoleacetaldehyde. The gene encoding tryptophan decarboxylase was isolated from *Catharanthus roseus*. The transgenic tobacco introduced tryptophan decarboxylase accumulated tryptamine but did not overproduce IAA.[194] Since tryptamine is not a general compound in plants, this pathway may not be widespread.

(d) Indoleacetonitrile. L-Trp is converted to indoleacetaldoxime and then to indoleacetonitrile.[195] The nitrilase which converts indoleacetonitrile to IAA was cloned in *Arabidopsis*.[196] This pathway was found first in the *Brassicacae*, because indoleacetonitrile and its related compounds, such as indoleglucosinolates, are typical indole compounds in *Brassicacae*. Therefore, this route was not considered to be generally important, but now indoleacetonitrile itself seems to be an important precursor of IAA in the tryptophan-independent pathway.

(e) Indoleacetamide. This pathway is well-established in bacteria but not in plants.[197] *Agrobacterium tumefaciens* are phytopasogenic bacteria and induce a tumor in the roots of dicots. *A. tumefaciens* possesses Ti plasmid coding two enzymes necessary for the IAA synthesis from indoleacetoamide. These enzymes are tryptophan monooxygenase, converting Trp to indoleacetamide and indoleacetamide hydrolase, converting indoleacetamide to IAA. Although this pathway seems to be widespread in bacteria, the existence of these two enzymes in plants has been reported.[198]

(ii) Tryptophan-independent pathway

Tryptophan auxotroph plants showed that IAA was synthesized without tryptophan.[199] Maize mutants fed with ^{15}N-labeled anthranilate did not synthesize tryptophan but did synthesize IAA to levels 50-fold above those of the wild type. The trp2 and trp3 mutants of *Arabidopsis* accumulated a considerable amount of IAA when fed anthranilate. The trp2 mutant also accumulates indole and indoleacetonitrile indicating that the biosynthetic pathway of IAA separates from that of tryptophan at the point of formation of indole or a very similar intermediate. This pathway is very interesting but many questions need to be answered: what kind of compounds are intermediate from indole to indoleacetonitrile, what kind of enzymes catalyze their metabolism, and does this pathway exist in a wild type plant?

8.02.2.2.5 Synthetic auxins

(i) Auxin agonists

After the structure of auxin was determined, the auxin activities of many synthetic compounds were examined. Among them, Zimmerman and Hitchcock discovered that 2,4-D showed strong auxin-like activity.[200] In plant tissue culture experiments, 2,4-D is the most abundantly used auxin, replacing IAA which is not stable enough for long-term cultures. Moreover, they found that high concentrations of 2,4-D killed plants, especially dicots. Thus, 2,4-D was the first synthetic herbicide and since then, phenoxyacetic acid derivatives have been widely synthesized to develop new herbicides. Phenylacetic acid was reported as a natural auxin and its synthetic derivatives were applied to herbicides. Naphthaleneacetic acid is used not only for tissue cultures but also for a plant growth regulation as a form of acetamide, for example, as a rooting promoter. Naturally occurring indolebutyric acid is also used as a rooting promoter. All monochloro- and dichloroindoleacetic acids were synthesized as derivatives of 4-Cl-IAA, and among them, 5,6-Cl_2-IAA showed the highest auxin activity. These compounds and appropriate references[200–214] are shown in Figure 1.

(ii) Auxin antagonists

In the late 1940s, Thimann and Bonner[215] showed that TIBA was antiauxin and Bonner[216] reported that 2,4-dichloroanisole inhibited the elongation of *Avena* coleoptile induced by IAA or 2,4-D. TIBA is now known to be an auxin transport inhibitor. Many antiauxins have been synthesized and their biological activities examined since 1950 (Figure 2). These are classified into phenoxyisobutyric acid derivatives, naphthoxypropionic or isobutyric acid, halogenated benzoic acid, indolyl derivatives, and others. Pyrazoloisoindole and pyrazolyl benzoic acid derivatives were reported to be inhibitors of auxin transport. The salts of maleic hydrazide are used as plant growth regulators in agriculture. A fluorinated indolyl compound (TFIBA) seems to be a useful compound for practical use. These compounds and appropriate references[215–226] are shown in Figure 2.

(a) Phenoxyacetic acid and phenoxypropionic acid derivatives

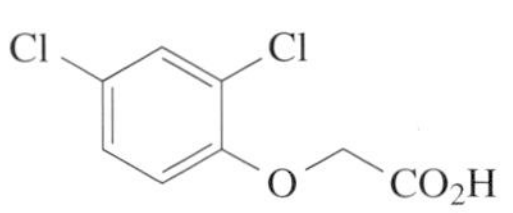

2,4-Dichlorophenoxyacetic acid[200]

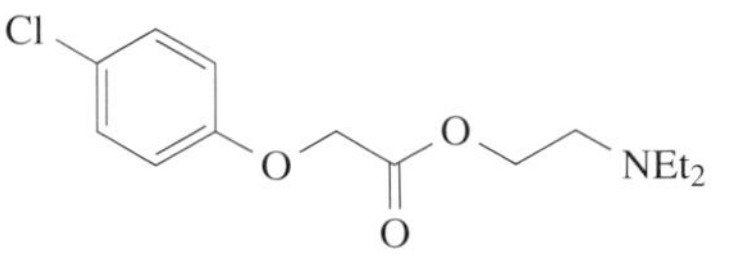

2-*N*,*N*-Dimethylaminoethyl 4-chlorophenoxyacetate[201]

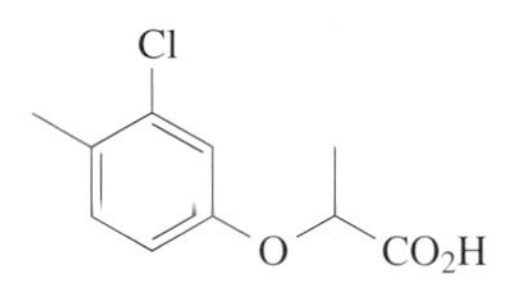

2-(3-Chloro-4-methylphenoxy)propionic acid[202]

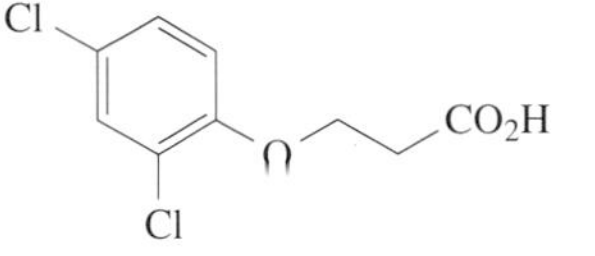

3-(2,4-Dichlorophenoxy)propionic acid[203]

(b) Indole derivatives

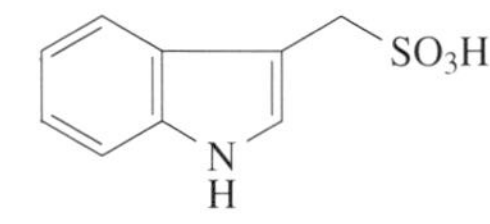

Indole-3-methanesulfonic acid[204]

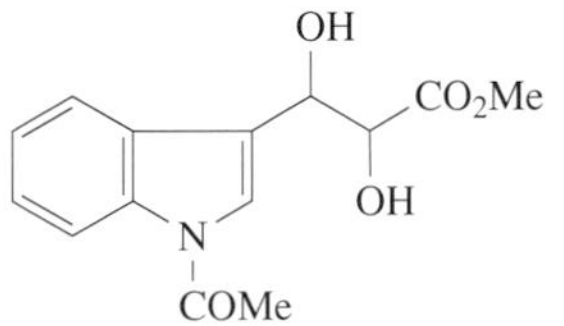

Methyl 1-acetylindole-3-glycerate[205]

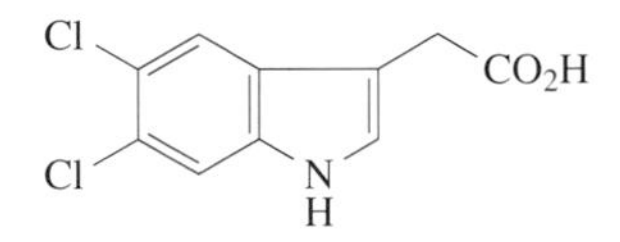

5,6-Dichloroindole-3-acetic acid[206]

(c) Benzoisothiazole and benzoisothiazoline derivatives

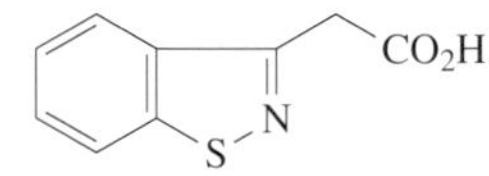

1,2-Benzisothiazol-3-ylacetic acid[207]

Ethyl 3-oxo-1,2-benzisothiazolin-2-yl-carboxylate[208]

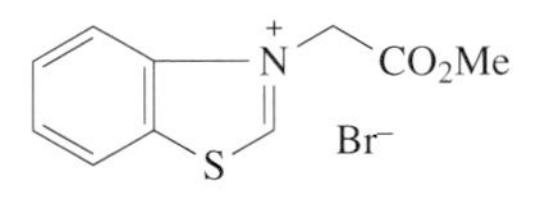

3-Carbomethoxymethylbenzothiazolium bromide[209]

(d) Others

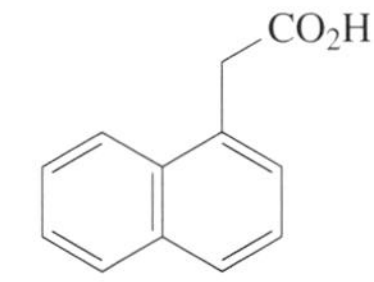

1-Naphthaleneacetic acid[210]

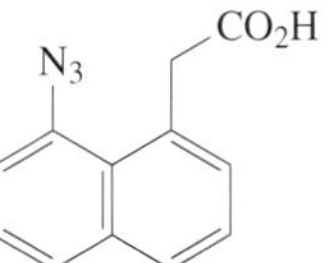

5-Azido-1-naphthaleneacetic acid[211]

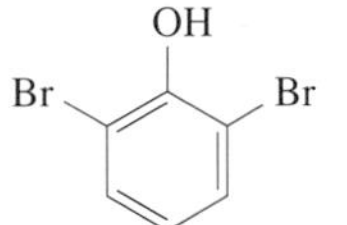

2,6-Dibromophenol[212]

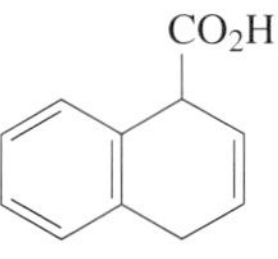

1,4-Dihydro-1-naphthoic acid[213]

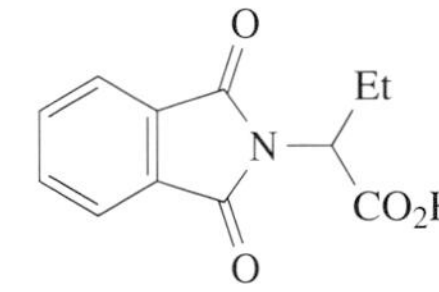

N-(1-carboxypropyl)phthalimide[214]

Figure 1 Synthetic auxins. Relevant references are shown next to each structure.[200–214]

8.02.3 GIBBERELLINS

8.02.3.1 Introduction

The gibberellins (GAs) are a large family of tetracyclic diterpenoid compounds, some of which are bioactive growth regulators, controlling such diverse processes as germination, cell elongation and division, and flower and fruit development.[227] There are two different types of GAs, the C_{20}-GAs (**1**) which have 20 carbon atoms, and the C_{19}-GAs (**2**) in which the C-20 carbon has been lost as carbon dioxide during metabolism. For most of the C_{19}-GAs, the carboxylic acid at C-19 bonds

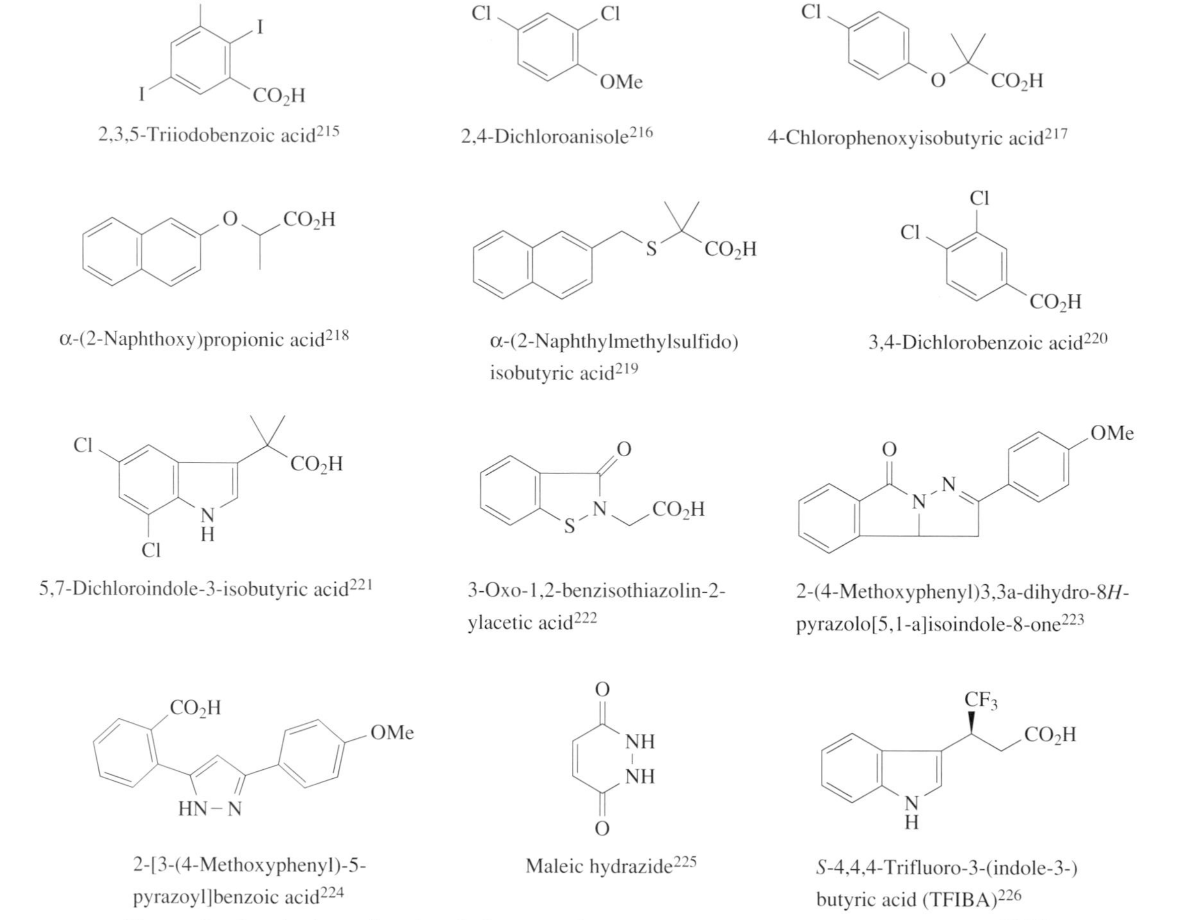

Figure 2 Synthetic antiauxins. Relevant references are shown next to each structure.[215–226]

to C-10 to give a lactone ring.[228] Many structural modifications can be made to the *ent*-gibberellane skeleton. Each different GA which is found to be naturally occurring and whose structure has been chemically characterized is allocated a number.[229] In 1997, 112 GAs were known and they are numbered from GA_1 to GA_{112} (Figure 3). Variations in GA structure are due to different oxidative states of C-20, namely methyl, hydroxymethyl, aldehyde, or carboxylic acid. Additional functional groups can also be added to the *ent*-gibberellane skeleton, especially to the C_{19}-GAs. Introduction into the ring system of hydroxy groups occurs frequently, and less common substituents are double bond and epoxide groups. The position and/or stereochemistry of these functional groups is very important. For example, the presence of the 3-hydroxy group in α and β stereochemistry can result in quite different physiological activity.[230] GAs were first isolated from the fungus *Gibberella fujikuroi* in which they occur in large quantities as secondary metabolites. They are now known to be present in several other species of fungus, in some ferns,[231] and in many higher plants.[228] Of the 112 known GAs, 87 have been identified only in higher plants, 12 are present only in *G. fujikuroi*, and 13 are present in both. Many different GAs can be present in one plant. GAs can also form conjugates;[232] the most common occurring naturally are glucoside or glucosyl esters. The role of the conjugates is not clear, however they may be formed to metabolize active GAs. Not all GAs have high biological activity. Many of the GAs within a plant are precursors or deactivation products of the active GA.[233] A knowledge of GA biosynthetic and metabolic pathways is fundamental to determine which GAs have biological activity *per se*. By using single gene dwarf mutants[234] and specific chemical growth retardants (e.g., prohexadione[235]) which inhibit specific metabolic steps, it is possible to determine which GA(s) in a plant is (are) the active hormone(s) for a particular growth or developmental event. In this chapter the authors focus on the biosynthesis of gibberellins, because rapid progress in molecular cloning of GA biosynthesis enzymes has thrown new light on GA biosynthesis and its regulation. There are several books and review chapters on the chemistry, biological activity, and biosynthesis of gibberellins.[227,228,230–234,236,237]

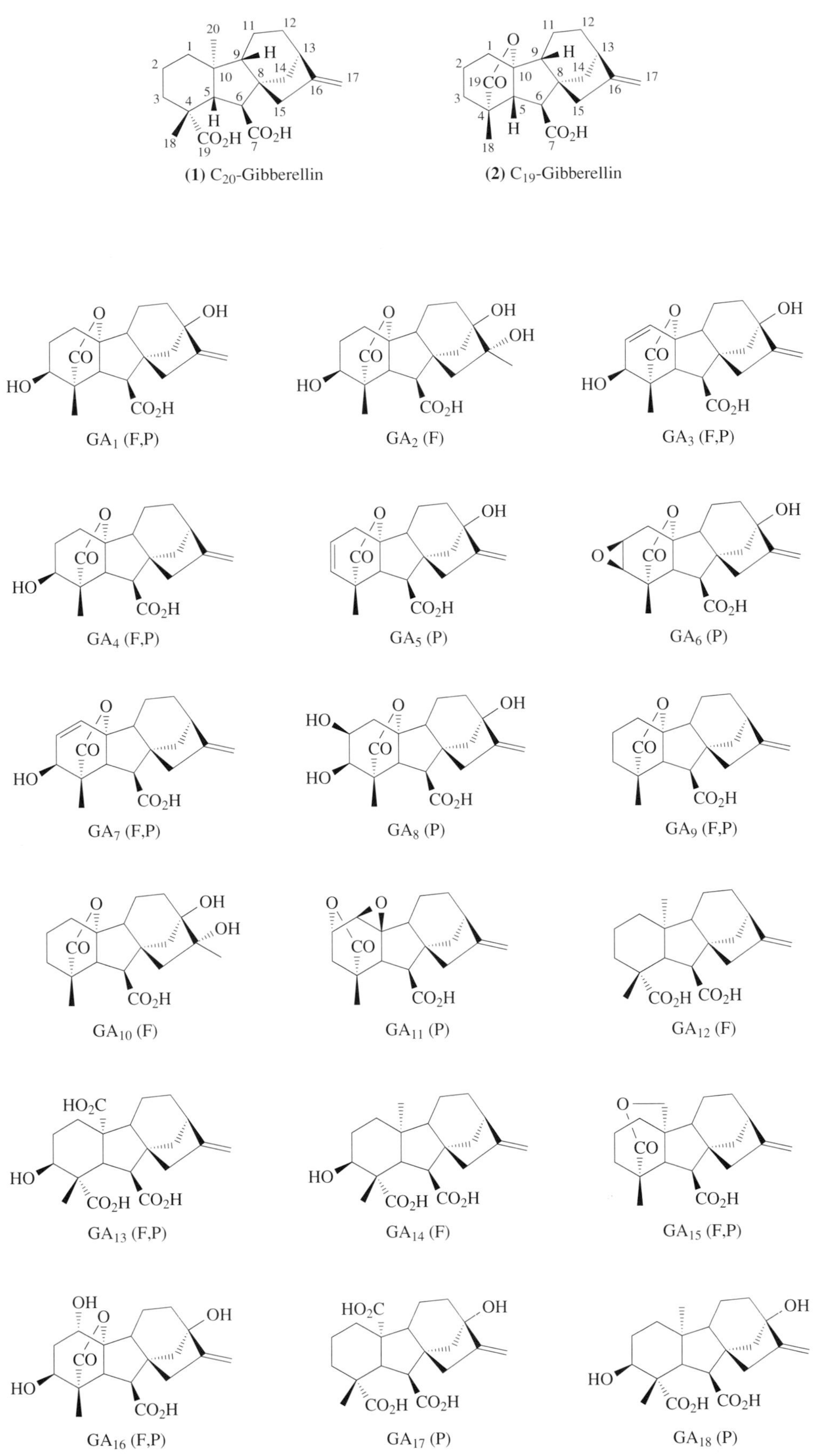

Figure 3 Chemical structures of free GAs. P (plant) and F (fungus) in parentheses show origin.

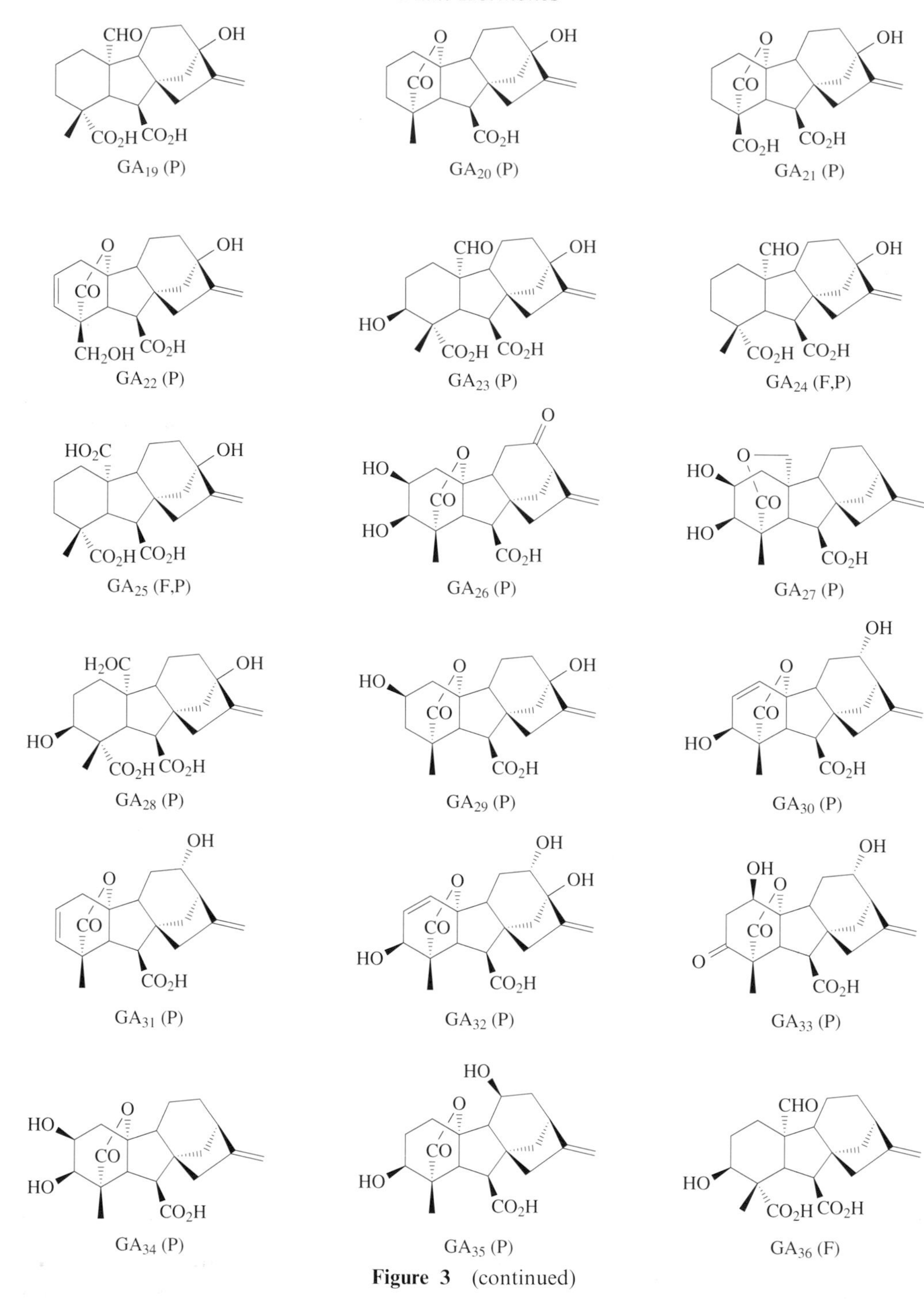

Figure 3 (continued)

8.02.3.2 History

8.02.3.2.1 Fungal gibberellins

The plant hormone, gibberellin, was first isolated from secondary metabolites of a rice pathogen called "bakanae-byokin" in Japanese, which means "foolish seedling pathogen." In this disease the infected rice plants are taller than non-infected plants and the color of leaves becomes yellowish-green and they are often withered. The disease has been shown to be caused by the fungal infection of a plant pathogen, *Fusarium moniliforme* and the perfect stage was later named *G. fujikuroi* (Saw.). In 1926 Eiichi Kurosawa showed that a sterile culture filtrate from a culture of *G. fujikuroi* would

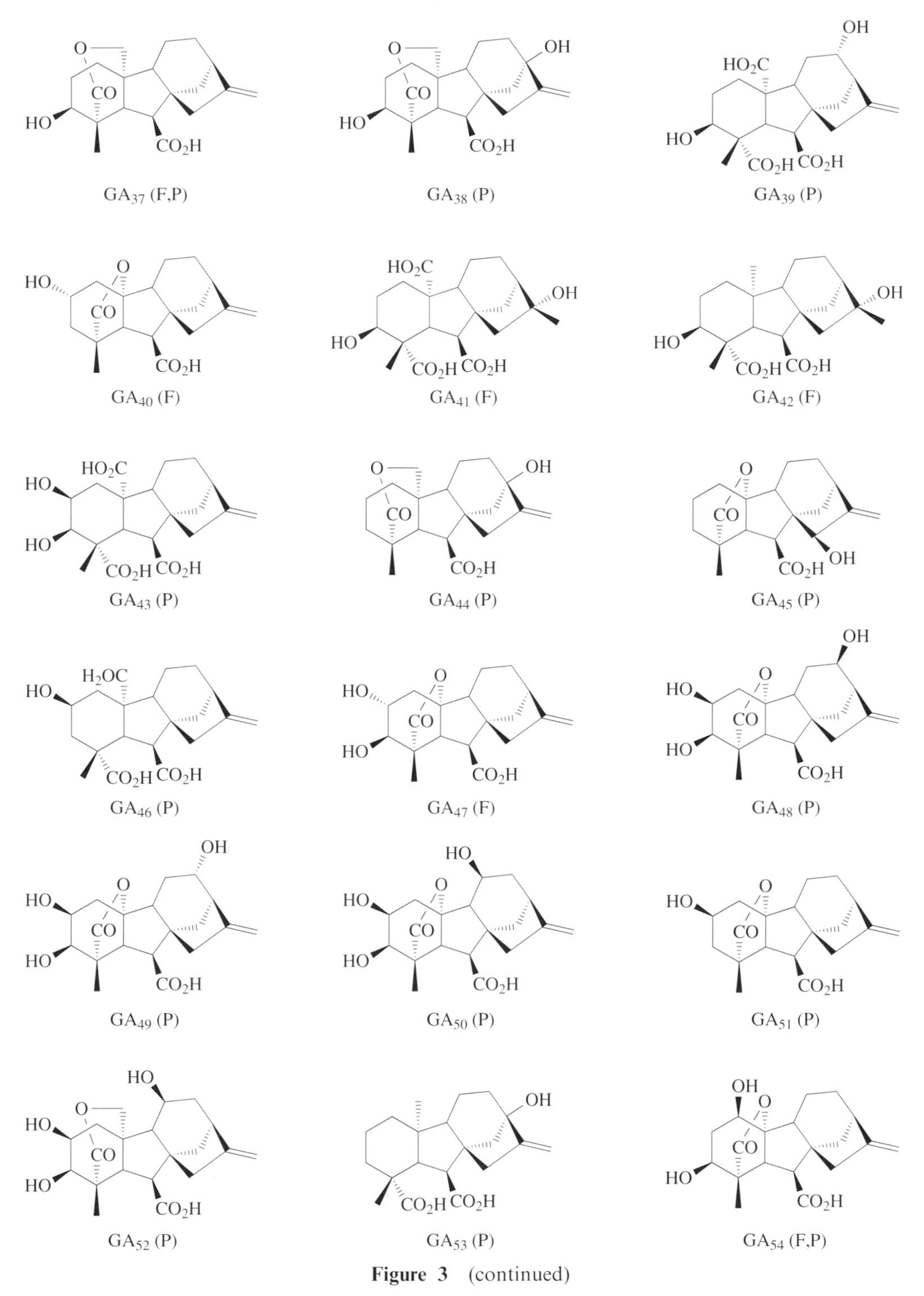

Figure 3 (continued)

mimic the "bakanae" symptoms produced by the fungus itself.[238] He concluded that the activity of his sterile extract was not due to enzyme action but rather to some kind of chemical or a toxin, which is responsible for the symptoms of the disease. This landmark research led to studies on the isolation of the active principle from culture filtrate of *G. fujikuroi* by Yabuta and Sumiki, at the University of Tokyo. In 1938 they isolated two active components which were named gibberellin A and gibberellin B.[239] GA research in Japan was interrupted during World War II. After the war, the research was resumed and the early reports on GAs by Japanese scientists attracted many researchers throughout the world. In 1954 the ICI group reported the isolation of an active principle from *G.*

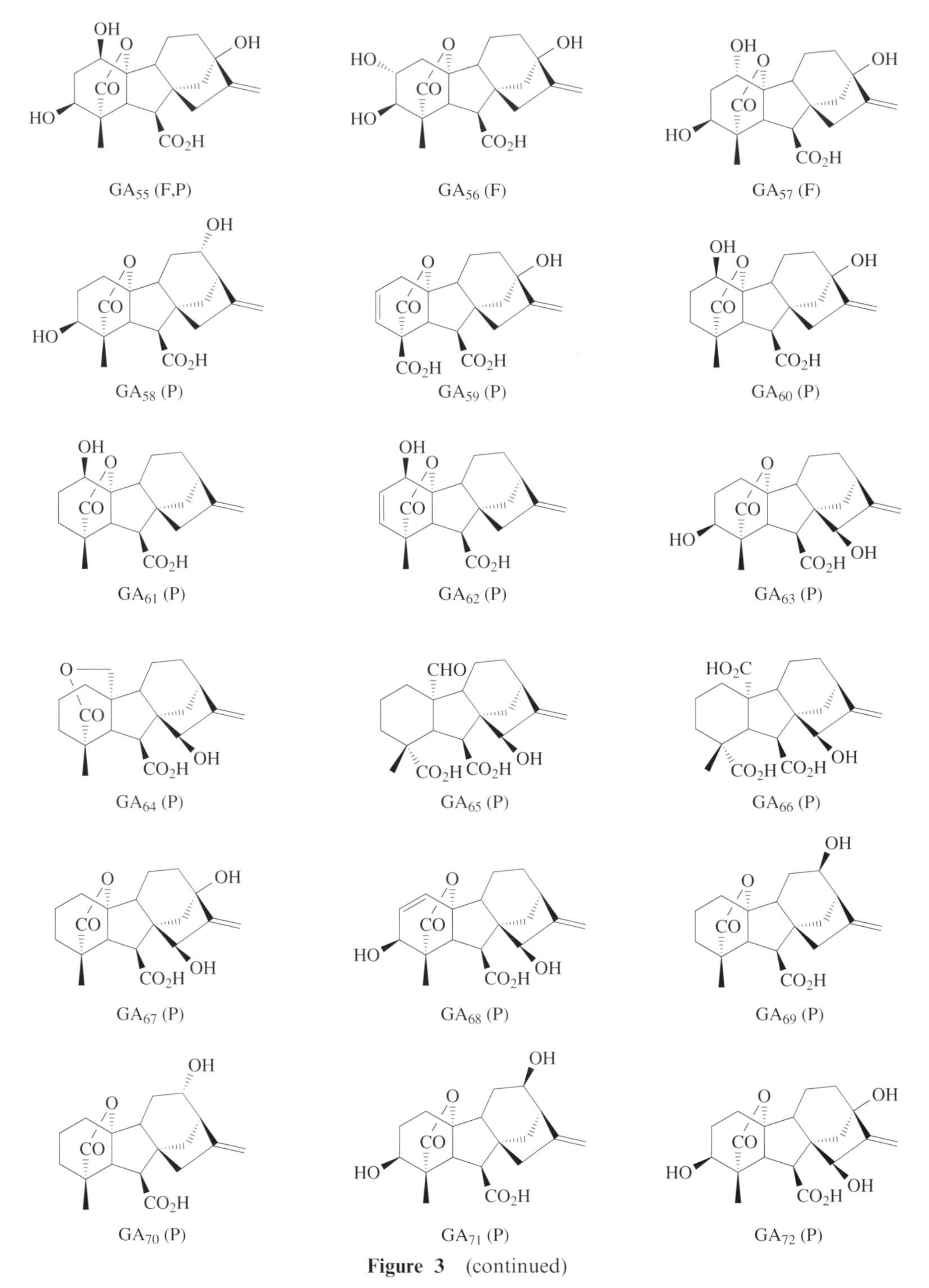

Figure 3 (continued)

fujikuroi that had plant growth-promoting activity.[240] The biological activity was similar to gibberellin A and B, but its chemical nature was clearly different. They called it gibberellic acid. The Northern Regional Research Laboratory (NRRL) group in the USA reported the separation of their gibberellin A into two components,[241] gibberellin X and gibberellin A. At the same time, the University of Tokyo group reinvestigated the purity of their gibberellin A and separated it into three gibberellins, A_1, A_2, and A_3. Direct comparison among various GAs isolated by the three groups disclosed that gibberellin A_1 (University of Tokyo) and gibberellin A (NRRL) were identical and gibberellic acid (ICI), gibberellin X, and gibberellin A_3 (University of Tokyo) were also identical. From 1957 to 1968 more than 10 different GAs were isolated and later GAs were isolated as

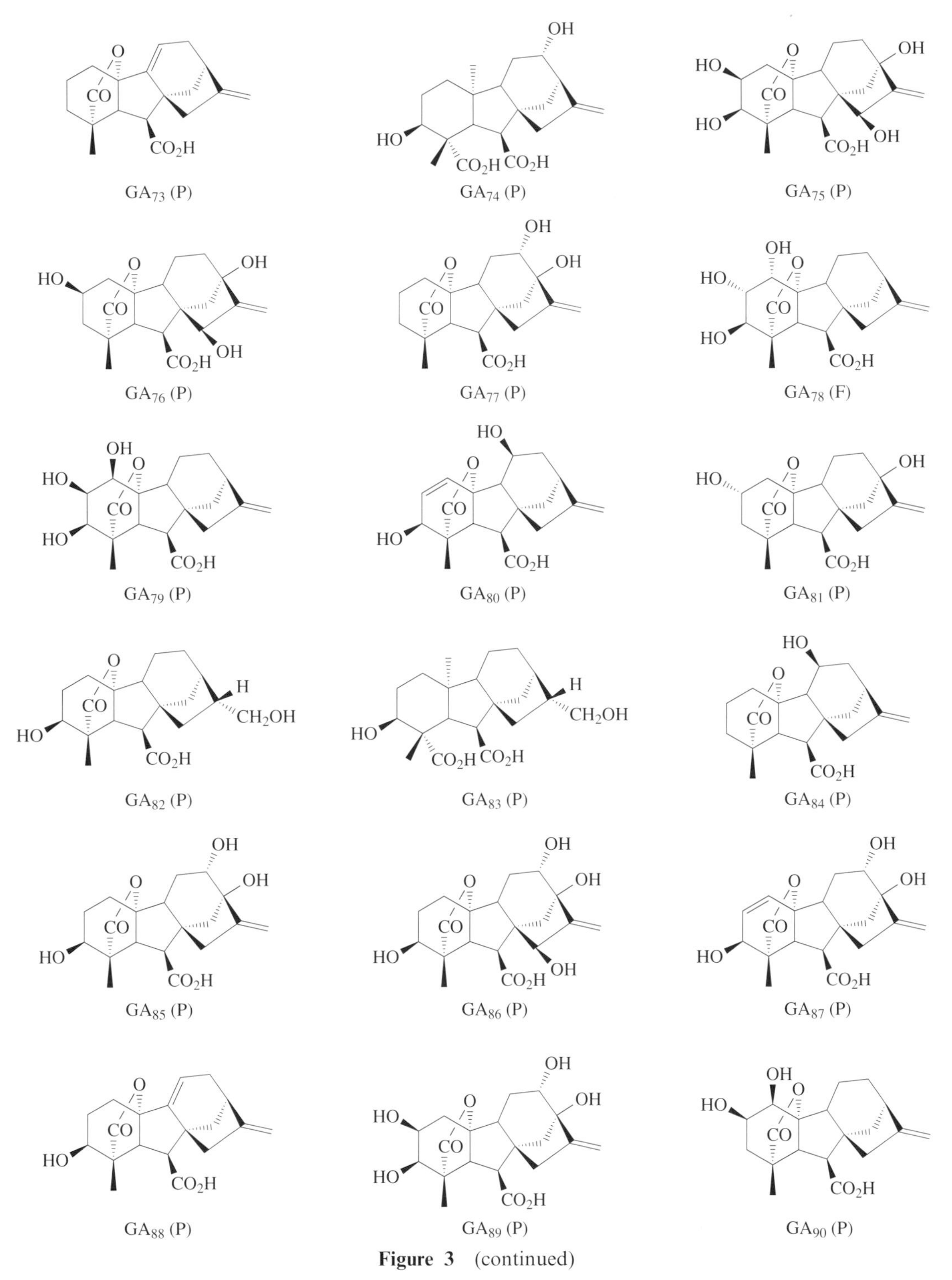

Figure 3 (continued)

endogenous plant growth regulators, and naming the GAs became a problem. In 1968 MacMillan and Takahashi proposed the naming of GAs by the use of a chronological numbering system.[229] They are responsible for assigning numbers to new gibberellins as they are identified, however, when the system was initiated it was not envisaged that so many GAs would be discovered. So far, over 110 GAs are known and this causes problems for plant physiologists and chemists, as there is no correlation of structure with number. In 1979 Rademacher and Graebe reported the isolation of GA_4 from culture filtrate of *Sphaceloma manihoticola*, a plant pathogen of cassava.[242] In 1995 a new fungus, *Phaeosphaeria* sp. L487, which produces GA_1 and GA_4, but not GA_3, was isolated. The

GA_{91} (P) GA_{92} (P) GA_{93} (P)

GA_{94} (P) GA_{95} (P) GA_{96} (P)

GA_{97} (P) GA_{98} (P) GA_{99} (P)

GA_{100} (P) GA_{101} (P) GA_{102} (P)

GA_{103} (P) GA_{104} (P) GA_{105} (P)

GA_{106} (P) GA_{107} (P) GA_{108} (P)

Figure 3 (continued)

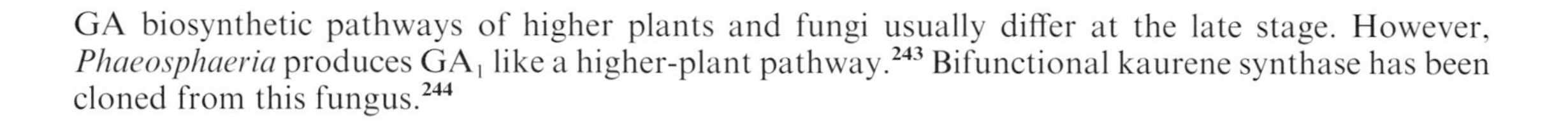

GA biosynthetic pathways of higher plants and fungi usually differ at the late stage. However, *Phaeosphaeria* produces GA_1 like a higher-plant pathway.[243] Bifunctional kaurene synthase has been cloned from this fungus.[244]

8.02.3.2.2 Plant gibberellins

In 1951 Mitchell *et al.* reported the occurrence of growth-stimulating substances, later interpreted as GA-like substances, from an extract of immature bean seeds.[245] Immature seeds usually contain

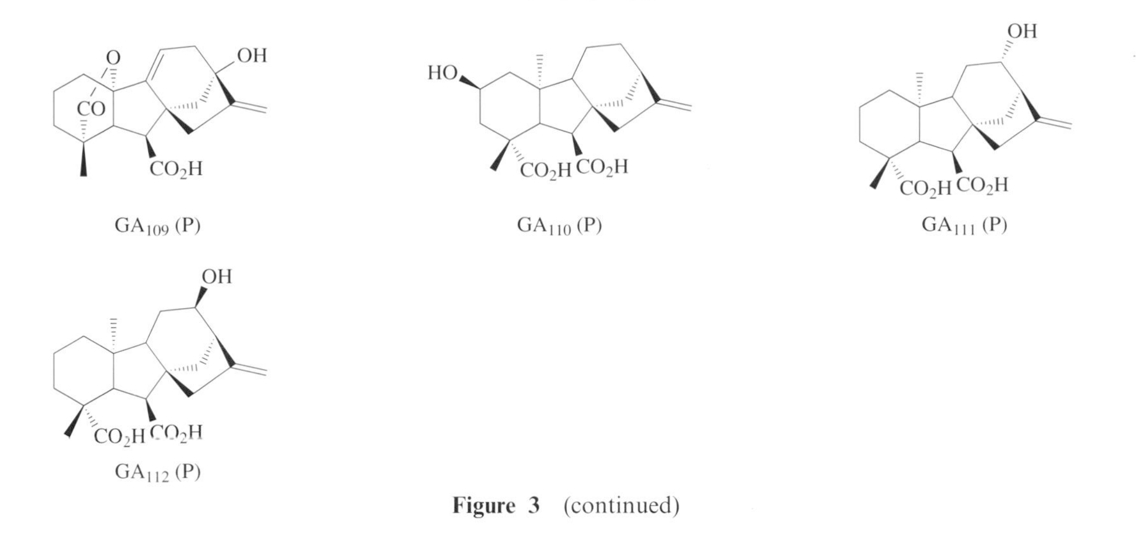

Figure 3 (continued)

high levels of GAs compared with vegetative tissue. Most of the earlier work in plants focused on immature seeds. MacMillan and Suter[246] reported the isolation of GA_1 from immature seeds of scarlet runner bean, *Phaseolus multiflorus* (also called *P. coccineus*) and later they found GA_5, GA_6, and GA_8. West and Phinney[247] reported the isolation of bean factor-I and -II from immature seeds of *Phaseolus vulgaris* and they identified them later as GA_1 and GA_5, respectively. In 1959 Kawarada and Sumiki[248] reported the identification of GA_1 in water sprouts of mandarin orange (*Citrus unshu*). This was the first isolation of GA_1 from vegetative tissue of plants. These pioneering studies stimulated research by many plant physiologists and much data appeared in the literature showing the occurrence of GA-like compounds in higher plants. In 1966, Murofushi *et al.*[249] succeeded in identifying a new gibberellin, GA_{19}, from young shoots of bamboo (*Phyllostachys edulis*). They isolated 14 mg from the water extract of 4.4×10^4 kg of bamboo shoots. After this, more than 90 GAs were isolated from higher plants. Some of them were identified without isolation by comparing their GC–MS data with those of authentic GA analogues prepared by chemical synthesis from structurally known GAs. Before the invention of GC–MS, relatively large amounts of plant materials were necessary for isolation and identification of GAs (kg base). It subsequently became possible to identify picogram levels of GAs unambiguously using sensitive GC-Selected Ion Monitoring (GC-SIM). There are several good reviews and books detailing the historical background of GAs.[250–252]

8.02.3.3 Chemistry

8.02.3.3.1 Isolation and characterization

As of 1997, over 112 GAs and about 20 conjugated GAs are known. The isolation and characterization of each GA reflects the development of technology in organic chemistry of natural products.[253] Originally, large quantities of GAs were required for structural studies. Isolation procedures involved extremely large amounts of materials and extensive purification steps.[249] Several GA bioassay systems[230] have been used for the detection of active fractions from chromatographic purification. The development of techniques such as HPLC greatly reduced the amounts of materials required for structure determination.[254] Identification of known GAs requires only picogram quantities. Vegetative tissues usually contain several nanograms or picograms of GAs per gram fresh weight. For this purpose 10–50 g fresh weight of plant materials are needed for GC-SIM identification. Endogenous GAs can be identified from immature seeds, which contain higher levels of GAs, using less than 1 g fresh weight. There are several good books and reviews concerning the isolation and identification of GAs.[255–257] Solvent or solid extraction from plants and HPLC purification followed by GC–MS identification can be used to identify GAs.

(i) Solvent fractionation

Plant materials are usually extracted by homogenization in methanol or 80% aqueous acetone, followed by filtration. This extraction procedure is repeated 3–4 times. The organic solvent in the filtrate is removed *in vacuo* and the residual aqueous phase is submitted to solvent fractionation to give an acidic ethyl acetate-soluble (AE) fraction, a neutral ethyl acetate-soluble fraction (NE), an acidic 1-butanol soluble fraction (AB), and a neutral 1-butanol fraction (NB). Most free GAs partition into the AE fraction. However, GA_9, GA_{12}, and GA_{25} sometimes appear in the NE fraction; GA_{29} is not completely extracted with ethyl acetate, and GA_{32}, which is almost nonextractable with ethyl acetate, is extractable with 1-butanol and partitions into the AB fraction. All free GAs are unstable in acid conditions and therefore the pH should be kept between 2.5 and 3.0 during the acidic solvent fractionation. Conjugated GAs show different behavior from free GAs in a solvent fractionation procedure. Gibberellin glycosyl ether (GA-GEt) cannot be extracted with ethyl acetate but can be extracted with 1-butanol and is partitioned into the AB fraction. Some of nonpolar gibberellin glycosyl ester (GA-GEs) can be extracted with ethyl acetate, but more polar GA-GEs can be extracted with 1-butanol into the NB fraction, because the acid group is esterified to glucose. After solvent extraction and solvent fractionation, GAs were purified by countercurrent distribution, charcoal adsorption chromatography, silisic acid adsorption chromatography, and partition chromatography. These methods sometimes require large amounts of plant materials, because of low recovery of GAs during purification. Details of large-scale purification can be found elsewhere.[237,253]

(ii) Solid-phase extraction

Improvements in the convenience and speed of plant hormone purification prior to physicochemical analyses have resulted from replacing solvent–solvent partition steps with solid phases packed into a small disposable column or cartridge, allowing rapid sample preparation. Generally, recoveries of GAs are good and this procedure requires small solvent volumes. These procedures usually involve combinations of reverse phase (silica bonded C_{18}, ODS) and ion-exchange and silica gel matrices.[254] Purification of GAs by one single HPLC is difficult due to the broad range of polarities of GAs. Small columns of C_{18} have been used to separate GAs and their glucosyl conjugates from less polar kaurenoid precursors.[258] Group separation of GAs by adsorption on silica gel cartridges gives substantial purification and in combination with anion-exchange and C_{18} cartridges, can form the basis of a quick and convenient purification scheme.

8.02.3.3.2 HPLC purification and GC–MS analysis

Preparative and analytical HPLC is used in all physicochemical assays for GAs and immunoassays. Diversity of structural difference of GAs makes it difficult to use immunoassays for GA analysis. GC–MS analysis is recommended rather than immunoassay, if a sufficient quantity of the sample is available. Reverse-phase chromatography on C_{18} bonded to microparticulate silica (ODS) has become a standard technique for GA analysis. When sample purification is difficult, normal phase HPLC, ion-exchange, and gel-permeation chromatography have been used as alternative or supplementary steps of purification. All the basic information is available in the review by Rivier and Crozier.[259]

Analysis of GAs by GC–MS has been discussed in numerous review articles.[255,256] A comprehensive GC–MS data book of GA and related mass spectra has been published[257] either as hard copy or as computer software. Full scan GC–MS spectra and Kovats Retention Indices make it possible to identify GAs without having authentic samples.[257] Measurement of isotope dilution by mass spectrometry is well established and suitable isotope-labeled GAs can be purchased.[254] For high sensitivity in quantitative analyses the mass spectrometer is normally operated in the SIM mode, for which quadruple instruments are particularly well suited.

8.02.3.4 Biological Activities

As already described, 112 free GAs have been isolated and characterized from higher plants and/or fungi. Of these free GAs, only a limited number of GAs are physiologically active, in other words, bioactive *per se*. Other GAs are biosynthetic intermediates to the physiologically active GAs or deactivation products synthesized according to the branch from the main biosynthetic pathways. The physiologically active GAs generally show high biological activities on plant growth, although the biological activities of GAs vary depending on the assay plants.[237] Of a variety of biological effects of GAs on plants, this section mainly focuses on two distinct effects, the shoot elongation and synthesis of hydrolytic enzymes within cereal aleurone, which have been employed extensively as model systems for studies of mode of action of GAs.

8.02.3.4.1 Effects on shoot elongation

The most dramatic effect of GAs is on shoot elongation in intact plants. This effect, which is caused mainly by cell elongation and in part by cell division, is particularly prominent when GAs are applied to plants at the seedling stage. The response of maturing plants to GAs is generally quite low or negligible.

Most bioassay systems for GAs have been developed based on their growth-promoting effect on intact plants. GA-deficient single gene mutants such as dwarf maize (*Z. mays* L.) *d*1 and *d*5 and dwarf rice (*Oryza sativa* L.) cv. Tan-ginbozu and cv. Waito-C have been widely used as assay plants to estimate biological activities of GAs[260,261] (the endogenous, physiologically active GA which mainly controls vegetative growth in maize and rice is GA_1[262]). Tan-ginbozu and *d*5 are GA-deficient mutants which are shown to be blocked for steps early in the GA biosynthetic pathway.[263,264] These mutants are responsive not only to physiologically active GAs, such as GA_1, GA_3, GA_4, and GA_7, but also to their precursors which are located after the block. The dwarf maize *d*1 and Waito-C are blocked for the final step leading to the physiologically active GAs (the 3β-hydroxylating step; see Section 8.02.3.5),[263,265] and therefore physiologically active GAs with 3β-hydroxy are highly active in the assays using these dwarf mutants, but the activity of their precursors is very low or negligible.

Cucumber and lettuce hypocotyl assays were also used as assay plants. The activity spectrum of GAs in the lettuce hypocotyl assay is similar to those in *d*5 and Tan-ginbozu assays, but different from that in the cucumber assay[235] (the endogenous, physiologically active GA which mainly controls vegetative growth in cucumber is GA_4, a non-13-hydroxylated GA[261]). In cucumber assay, non-13-hydroxy-GAs such as GA_4, GA_7, and GA_9 show high activity, although GA_9 was shown to be active as a result of its conversion to GA_4.[266]

The activity spectra of 20 selected GAs in the dwarf rice, dwarf maize, and cucumber hypocotyl assays are shown in Table 7. On the basis of information on growth-promoting activity of GAs,[260,261,267–270] the structure–activity relationships are summarized as follows: (i) GAs with a hydroxy at C-3β such as GA_1, GA_3, GA_4, GA_7, and GA_{32} (especially GA_3, GA_7, and GA_{32} which possess the 1,2-didehydro-3β-hydroxy structure) are highly active in dwarf maize and dwarf rice assays (other highly active GAs are probably converted into the corresponding physiologically active GAs in the respective assay plants). (ii) The activity of GAs with a hydroxy at C-2 such as GA_8, GA_{29}, and GA_{34} is quite low (GAs are inactivated by hydroxylation at C-2). (iii) In general, the C-11β and C-12α hydroxys have little influence on growth-promoting activity, although the C-12α hydroxy reduces activity in cucumber hypocotyl assay. (iv) The C-13 hydroxy greatly reduces activity in cucumber hypocotyl assay. (v) GAs with carboxyls at C-4, C-6, and C-10 such as GA_{13}, GA_{17}, and GA_{25} are almost inactive. (vi) Methyl esters of GAs are almost inactive except bioassays on formation of sexual organs in the fern *Lygodium japonicum* (see Section 8.02.3.7.).

GA glucosides generally showed very weak activity compared with their aglycons.[257,265] The GA glucosides are probably not hydrolyzed in the assay plant. However, in the dwarf rice immersion assay under nonaseptic conditions, some GA glucosides showed high activity. This was because the corresponding aglycons were produced by microbial degradation. In fact, in aseptic conditions such high activity was not observed.[257] However, some GA glucosyl esters showed high activity even in the dwarf rice micro-drop method. It was indicated that the exogenously applied GA glucosyl esters were rapidly hydrolyzed in the rice plant.[270]

8.02.3.4.2 Effects on enzyme activities

Exogenously applied GA was found to increase greatly the activity of α-amylase in the barley endosperm.[271] Since then this phenomenon has been employed as a GA bioassay system. In this assay, physiologically active GAs are highly active, while the activity of their precursors is very low or negligible.[237] It is probable that the precursors are not metabolized into the corresponding physiologically active GAs in the endosperm.

In the barley aleurone cells, the activity of protease as well as that of α-amylase is enhanced by GA application,[272] and the increase of activities of α-amylase and protease was based on *de novo*

Table 7 Relative activities of GAs.

GA	*Dwarf rice*		*Dwarf maize*		
	Tan-ginbozu	*Waito-C*	d*1*	d5	*Cucumber*
GA_1	+ + +	+ + +[b]	+ + +	+ + +	+[d]
GA_3	+ + + +[a]	+ + + +	+ + + +	+ + + +	+ +[c]
GA_4	+ +	+ +	+ + +	+ + +	+ + +
GA_7	+ + +	+ + +	+ + + +	+ + + +	+ + + +
GA_8	+	+	0[e]	0	0
GA_9	+ +	+	+	+ +	+ + +
GA_{13}	0	n.t.[f]	+	0	0
GA_{17}	+	n.t.	+	0	0
GA_{19}	+ + +	+	+	+ + +	0
GA_{20}	+ + +	+	+	+ + +	0
GA_{24}	+ + +	+	+	+	+ + +
GA_{25}	0	n.t.	0	0	+
GA_{29}	+	n.t.	+	+	0
GA_{30}	+ + +	+ +	+ + +	+ + +	+
GA_{32}	+ + + +	+ + + +	+ + + +	+ + +	+ + +
GA_{34}	0	0	0	0	0
GA_{35}	+ +	+ +	+ +	+ +	+ + +
GA_{37}	+ +	+ +	+ +	+ +	+ +
GA_{38}	+ +	+ +	+ + +	+ +	+
GA_{51}	0	n.t.	n.t.	n.t.	0

[a]+ + + +, very high. [b]+ + +, high. [c]+ +, moderate. [d]+, low. [e]0, very low/inactive. [f]n.t., not tested.

synthesis in the aleurone cells. Such increase of activities of the hydrolytic enzymes by GA application has been observed in common with other cereals. In the cereal aleurone cells, the activities of other enzymes such as β-glucanase, ribonuclease, esterase, catalase, peroxidase, citrate synthase, isocitrate lyase, and so on were also increased by GA application.[273,274]

Based on the above evidence, the physiological role of GAs in the cereal grains is speculated to be as follows:[274] GAs secreted from the developing seed embryo move to the aleurone layer, and stimulate the *de novo* synthesis of α-amylase and protease. These enzymes move out of the aleurone and into the endosperm to hydrolyze starch and storage proteins, respectively, resulting in production of sugars and amino acids which are used for further development of the seed embryo. Other enzymes, whose activities are increased by GAs, perform hydrolytic functions in the endosperm or maintenance functions within the aleurone cells.

8.02.3.4.3 Other effects

GAs are also active in breaking dormancy of tubers or seeds.[275] In light- or cold-requiring seeds of *Lactuca*, *Nicotiana*, *Perilla*, and so on, GAs can replace the stimuli in the induction of seed germination. Sprouting of dormant potato tubers is also induced by treatment with GA_3 (2–3 ppm solution).

GAs show effects not only in the vegetative growth stage but in the reproductive stages of plants. Though, in general, GA treatment does not induce flower formation in short-day plants, it induces flower formation in some species of long-day plants under short-day conditions.[276,277] However, low temperature is required for induction of flowering in some plants, which are called cold-requiring plants. GA treatment can also replace cold-treatment in the flower formation of many species of cold-requiring plants, most of which are long-day plants.

In some species of plants producing unisexual flowers such as cucumber and hemp, GA treatment increased the number of male flowers.[275,277] This suggests that GA promotes male sex expression in these plants. It is quite interesting that GAs and related compounds have been isolated as antheridiogens in Schizaeaceous ferns that induce formation of the male sexual organ (see Section 8.02.3.7).

Parthenocarpic fruits can be produced by GA treatment in various plant species such as almond, apple, peach, pear, tomato, grapes, and so on.[274,278] In grapes such as the Delaware, muscat, berry A, and Thompson varieties, GA application induces parthenocarpic fruits and increases the size of the resultant fruits and so GA has been of practical use in production of seedless grapes.[237]

8.02.3.5 The Gibberellin Biosynthesis Pathway

GA biosynthesis can be divided into three stages:[233] (i) the biosynthesis of *ent*-kaurene, (ii) conversion of *ent*-kaurene to GA_{12}-aldehyde (Scheme 2), and (iii) the biosynthesis after GA_{12}-aldehyde (Scheme 3). We separate these three stages because they are catalyzed by different types of enzymes and their localization in cells is also different. The first stage processes take place in plastids and are catalyzed by soluble enzyme. The second stage processes are catalyzed by membrane-bound cytochrome P450 enzymes. The reactions of the last stage are catalyzed by 2-oxoglutarate-dependent dioxygenases. There are two parallel GA metabolic pathways in most higher plants,[233,236,279] one from GA_{12} to GA_4 and the second leading to the 13-hydroxylated GA, GA_1 from GA_{53}. They are called the early nonhydroxylation and the early 13-hydroxylation pathways, respectively (Scheme 3).

8.02.3.5.1 *The biosynthesis of* ent-*kaurene*

(i) Cyclization of geranylgeranyl diphosphate to copalyl diphosphate

ent-Kaurene is synthesized by the two-step cyclization of geranylgeranyl diphosphate (GGDP) via the intermediate, copalyl diphosphate (CDP). The enzymes that catalyze these reactions are referred to as the A and B activities, respectively, of *ent*-kaurene synthase. However, a more logical nomenclature was proposed by MacMillan.[280] Thus, the conversion of GGDP to CDP is catalyzed by copalyl diphosphate synthase (CPS) and CDP to *ent*-kaurene by *ent*-kaurene synthase (KS). The mechanism of these cyclization reactions has been discussed.[281] The biosynthesis of GGDP from mevalonic acid is common to many terpenoid pathways.[282] A nonmevalonate pathway to isoprenoids, involving pyruvate and glyceraldehyde-3-phosphate, has been proposed in green algae.[283] Such a pathway may operate in plastids of higher plants, given the difficulty in demonstrating the incorporation of mevalonate into isoprenoids in these organelles. CPS and KS were first separated by anion-exchange chromatography on extracts of *Marah macrocarpus* endosperm.[284] There were also indications for the involvement of two enzymes from studies on GA-deficient mutants; work with cell-free extracts of young fruits suggests that the dwarf tomato mutants, *gib-1* and *gib-3*, have lesions at CPS and KS, respectively.[285]

Koornneef and van der Veen isolated a GA-deficient mutant from *A. thaliana*.[286] Among the fast-neutron-generated mutants, *ga1-3* contains a large deletion (5 kb) and failed to recombine with the ethane methanesulfonate (EMS)-treated mutants.[287] Sun *et al.* utilized *ga1-3* to clone the *GA1* locus by genomic subtraction. Cosmid clones containing wild-type DNA inserts spanning the deletion in *ga1-3* complemented the dwarf phenotype when integrated into the *ga1-3* genome by T-DNA transformation.[288] *GA1* cDNA contains a 2.4 kb open reading frame, which was shown by functional analysis to encode CPS.[289] Although *ga1-3* contains a large deletion and genomic Southern analysis indicated that *GA1* is a single-copy gene, the mutant produces low amounts of GAs, suggesting that there are *GA1* homologues in *Arabidopsis*. The *Ls* locus of the pea was shown to encode CPS. The *ls-1* dwarf mutant possessed reduced CPS activity in a cell-free system from immature seeds. Confirmation was obtained after cloning CPS from pea by RT-PCR, its identity being confirmed by expression in *E. coli* of a glutathione *S*-transferase fusion protein with CPS activity.[290] The maize *An-1* gene was cloned by transposon tagging and the gene probably encodes CPS.[291]

(ii) CDP to ent-*kaurene*

KS was purified from endosperm of pumpkin (*Cucurbita maxima*), which is a rich source of GA-biosynthetic enzymes. The enzyme, which had a predicted M_r of 81 000, required divalent cations, such as Mg^{2+}, Mn^{2+}, and Co^{2+}, for activity.[292] The K_m for CDP was 0.35 μM. Purification of KS was quickly followed by its molecular cloning.[293] The isolated full-length cDNA was expressed in *E. coli* as a fusion protein, with maltose-binding protein, which converted [^{3}H]CDP to *ent*-[^{3}H]kaurene. The KS transcript is abundant in growing tissues, such as apices and developing cotyledons, and is present in every organ in pumpkin seedlings. Although it is difficult to compare mRNA abundance across species, it appears that KS is expressed at much higher levels than is CPS. Transcripts of CPS are undetectable in leaves of *Arabidopsis*[289] and pea[290] by Northern blot analysis.

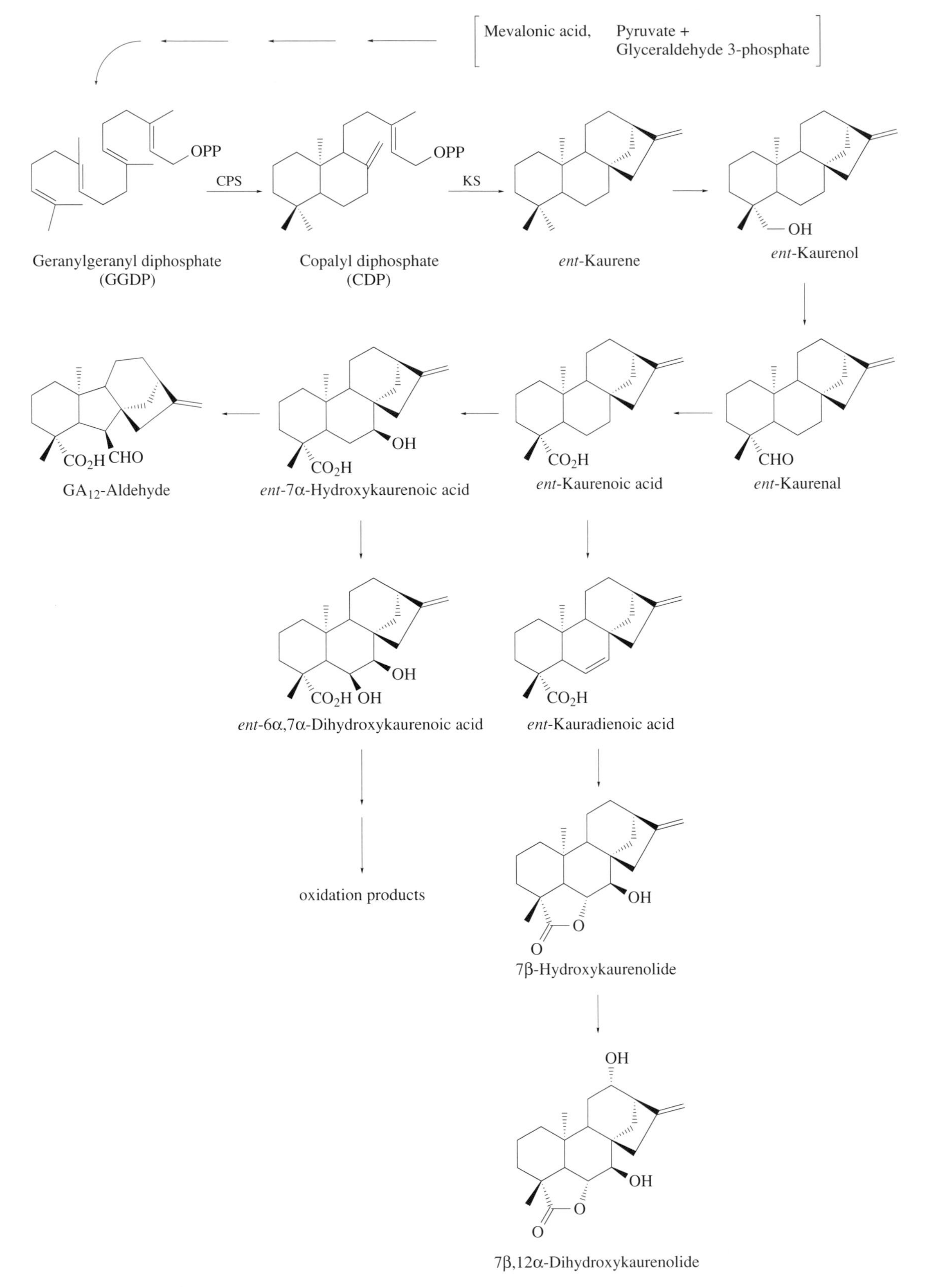
Mevalonic acid,
Pyruvate +
Glyceraldehyde 3-phosphate
OPP
CPS
OPP
KS
OH
Geranylgeranyl diphosphate
(GGDP)
Copalyl diphosphate
(CDP)
ent-Kaurene
ent-Kaurenol
CO2H CHO
GA12-Aldehyde
OH
CO2H
ent-7α-Hydroxykaurenoic acid
CO2H
ent-Kaurenoic acid
CHO
ent-Kaurenal
OH
CO2H OH
ent-6α,7α-Dihydroxykaurenoic acid
CO2H
ent-Kauradienoic acid
oxidation products
OH
O
O
7β-Hydroxykaurenolide
OH
OH
O
O
7β,12α-Dihydroxykaurenolide

Scheme 2

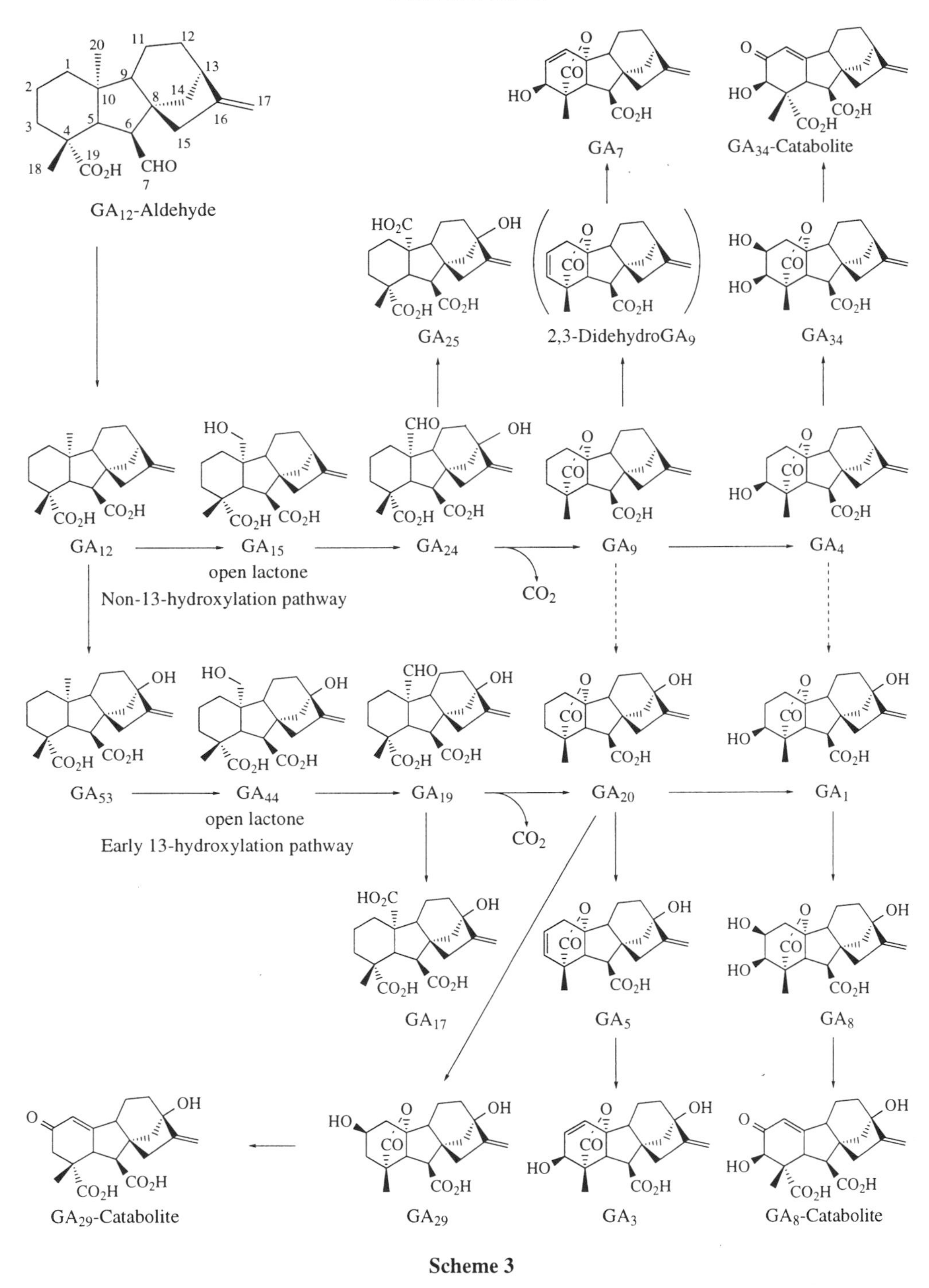

Scheme 3

This is consistent with strict regulation of the first step of *ent*-kaurene synthesis from the abundant GGDP.

The deduced amino acid sequence of KS shares significant homology with other terpene cyclases, with highest homology (51% amino acid similarity) with CPS from *Arabidopsis*[289] and maize.[291] It contains the DDXXD motif, which is conserved in casbene synthase[294] 5-*epi*-aristolochene synthase,[295] and limonene synthase[296] and is proposed to function as a binding site for the divalent metal ion–diphosphate complex. CPS lacks the DDXXD motif, consistent with its catalytic activity not involving cleavage of the diphosphate group.

(iii) Subcellular localization of ent*-kaurene synthesis*

Although there has been evidence for *ent*-kaurene synthesis in plastids for over 20 years, unequivocal confirmation of this was only provided in 1995. By precise use of marker enzymes to assess plastid purity and GC–MS to identify enzyme products, Aach *et al.* clearly demonstrated that CPS/KS activity (GGDP to *ent*-kaurene) is localized in developing chloroplasts from wheat seedlings and leucoplasts from pumpkin endosperm.[297] Mature chloroplasts contained little activity. These results were supported by the recent cDNA cloning of CPS[289–291] and KS.[293] The first 50 *N*-terminal amino acids of the GA1 protein (*Arabidopsis* CPS)[270] are rich in serine and threonine with an estimated pI of 10.2. Such properties are common features of precursors of many chloroplast-localized proteins, such as the small subunit of Rubisco.[298] The transit is cleaved on entry into the plastid to produce a functional mature protein. Incubation of a ^{35}S-labeled *Arabidopsis* pre-CPS of 86 kDa with isolated pea chloroplasts resulted in transport into the chloroplasts and processing to a 76 kDa protein.[289]

Plastids are the major site of production of GGDP and most of the GGDP synthases cloned in plants have transit peptides for plastid transport.[299] Localization of GGDP synthase from *Capsicum annum* in plastids has been demonstrated immunocytochemically.[300] GGDP is a common precursor for many plastid-localized terpenoids, including carotenoids and the phytol side chain of chlorophyll. Overexpression of phytoene synthase, which converts GGDP to phytoene, in transgenic tomato resulted in a lower chlorophyll content than in wild-type plants and a dwarf phenotype that was partially reversed by applying GA_3.[301] The endogenous GA concentrations in apical shoots of the transgenic plants were reduced to about 3% of that in wild-type shoots. If, as suggested, over-production of phytoene has depleted GGDP content, resulting in reduced synthesis of GA and chlorophyll, the three pathways must share the same pool of GGDP and, thus, be interdependent.

8.02.3.5.2 ent-*Kaurene to GA_{12}-aldehyde*

The highly hydrophobic *ent*-kaurene is oxidized by membrane-bound monooxygenases to GA_{12}. The enzymes require NADPH and oxygen and, on the basis of the early demonstration that *ent*-kaurene and *ent*-kaurenal oxidation are inhibited by carbon monoxide with reversibility by light at 450 nm, are all assumed to involve cytochrome P450.[302] The involvement of cytochrome P450 in *ent*-kaurenoic acid 7β-hydroxylase in the fungus, *G. fujikuroi*, has also been shown.[303] At least one of the enzymes (GA_{12}-aldehyde synthase) is associated with the endoplasmic reticulum in pea embryos and pumpkin endosperm, requiring transport of *ent*-kaurene, or perhaps a later intermediate, from the plastid.

Several GA-deficient dwarf mutants are defective in *ent*-kaurene oxidase activity. In pea, the lh^i (*lh-2*) mutation affects stem elongation and seed development.[299] Cell-free extracts from immature *lh-2* seeds were deficient in *ent*-kaurene oxidase activity relative to wild-type seeds; the three steps from *ent*-kaurene to *ent*-kaurenoic acid were affected, suggesting that a single enzyme might catalyze these reactions.[304] However, unequivocal verification of this must await the availability of pure enzyme because a regulatory function for *LH* cannot be excluded. The Tan-ginbozu mutant of rice (*dx*) is probably also deficient in *ent*-kaurene oxidase activity. Application of uniconazole, an *ent*-kaurene oxidase inhibitor, mimics the phenotype of Tan-ginbozu and produces endogenous GA concentrations similar to those in the mutant.[264] Based on growth responses to applied *ent*-kaurene, Tan-ginbozu and *lh* were thought to have lesions prior to *ent*-kaurene synthesis. The subsequent findings indicate that such application experiments may give misleading results.

Although less progress has been made with the monooxygenases than the other enzymes, the cloning of the *Dwarf*3 (*D3*) gene of maize should enable rapid progress in characterizing this group of enzymes. The *D3* gene, obtained by transposon tagging, was found, on the basis of its deduced amino acid sequence, to encode a member of a new class of cytochrome P450 monooxygenases[305] with closest homology to sterol hydroxylases. Unfortunately, there is uncertainty about the step catalyzed by the D3 protein.

8.02.3.5.3 Pathways from GA_{12}-aldehyde

The enzymes involved in the third stage of the pathway are soluble oxidases that utilize 2-oxoglutarate as a cosubstrate. These 2-oxoglutarate-dependent dioxygenases belong to a family of nonheme Fe-containing enzymes that have been the subject of several reviews.[306,307] The enzymes show considerable diversity of function, but are clearly related on the basis of conserved amino acid sequences.

The reactions known to be catalyzed by 2-oxoglutarate-dependent dioxygenases are shown as a network of pathways in Scheme 3. Although they were originally delineated in developing seeds,[233] both early 13-hydroxylation and early nonhydroxylation pathways have now been demonstrated in vegetative tissues. The individual steps between GA_{12}-aldehyde and GA_3 and GA_8 were demonstrated in intact maize shoots by applying each isotopically labeled intermediate and identifying its immediate metabolite by GC–MS.[308] The dioxygenases in GA biosynthesis will be discussed in detail.

(i) 7-Oxidation

Oxidation at C-7 from an aldehyde to a carboxylic acid may be catalyzed by either dioxygenases or monooxygenases. Pumpkin endosperm contains both 7-oxidase activities. The dioxygenase activity from pumpkin has been partially purified and shown to have a very low pH optimum, whereas the monooxygenase is most active above pH 7. The two types of activity also differ in their substrate specificities; the monooxygenase is specific for GA_{12}-aldehyde, while the soluble activity oxidizes several hydroxylated GA_{12}-aldehyde derivatives.[309] The presence of both types of enzyme in a single tissue may indicate subcellular compartmentation of GA-biosynthetic pathways.

(ii) 13- and 12α-Hydroxylation

As for GA 7-oxidase, both dioxygenase and monooxygenase forms of these hydroxylases have been described. The soluble, 2-oxoglutarate-dependent 13-hydroxylase detected in cell-free extracts from spinach leaves is still the only example of a dioxygenase with this activity.[310] 13-Hydroxylases in pumpkin endosperm,[309,311] developing pea embryos,[312] and barley embryos/scutella[313] are of the monooxygenase type. The preferred substrate for the 13-hydroxylases is probably GA_{12}, although other GAs are hydroxylated to some extent. GA_{12}-aldehyde is 13-hydroxylated in embryo cell-free systems from *Phaseolus coccineus*[314] and *P. vulgaris*[315] indicating that GA_{53}-aldehyde is an intermediate in the 13-hydroxylation pathway in these tissues. In peas, GA_{15} and GA_9 were hydroxylated only when their lactones were opened by hydrolysis.[312] Although "late" 13-hydroxylation (on GA_9 or GA_4) can often be demonstrated, it may be relatively inefficient and accompanied by hydroxylation at other positions on the C and D rings. However, in some species, such as *Picea abies*, it would appear to be the major pathway.[316]

Both forms of the 12α-hydroxylases are present in pumpkin seed, the monooxygenase hydroxylating GA_{12}-aldehyde (GA_{12} is not a substrate),[309] while the dioxygenase utilizes a variety of GA tricarboxylic acid substrates.[311] The monooxygenase has a low pH optimum and may thus catalyze part of the same pathway as the soluble 7-oxidase, for which 12α-hydroxy GA_{12}-aldehyde is a substrate.

(iii) 20-Oxidation

Formation of the C_{19}-GA skeleton requires successive oxidation of C-20 from a methyl group through the alcohol and aldehyde, from which this carbon atom is lost as CO_2 (Scheme 3). A single enzyme (GA 20-oxidase) can catalyze this reaction sequence, although the number of enzymes that are actually involved *in vivo* is unknown. Clear evidence that GA 20-oxidases are multifunctional was obtained after purification of the enzyme to homogeneity from pumpkin endosperm.[317] The enzyme converted GA_{12} to GA_{15}, GA_{24}, and GA_{25}; and GA_{53} to GA_{44}, GA_{19}, and GA_{17}, with a small amount of putative GA_{20} produced at high protein concentrations. The purification of the GA 20-oxidase from pumpkin was quickly followed by the cloning of a cDNA that encoded this enzyme.[318] The expressed fusion protein of the cDNA was functionally active and catalyzed the same reactions as the native enzyme. The aldehyde intermediate, GA_{24}, was converted to both GA_{25}

and GA_9, although the latter was obtained in less than 1% yield. The derived amino acid sequence corresponds to a protein of 43.3 kDa, which is close to that estimated for the native enzyme from gel filtration (44 kDa), and contains the conserved regions found in other plant dioxygenases.

The cloning of the GA 20-oxidase cDNA from pumpkin seeds enabled the isolation of homologous clones from other species. Three GA 20-oxidase cDNAs were cloned from *Arabidopsis*.[319,320] Confirmation that all three cDNAs encoded GA 20-oxidases was obtained by heterologous expression in *E. coli*. In contrast to the pumpkin enzyme, the C_{19}-GAs were the major products. Expression of the GA 20-oxidase genes is tissue specific. Gibberellin 20-oxidase cDNAs have been cloned from at least seven species with multiple genes being found in several of them.[236] Their encoded amino acid sequences share a relatively low degree of sequence conservation, with amino acid identities ranging from 50 to 75%. With the exception of the enzyme from pumpkin seed, the proteins have very similar functions, converting 20-methyl GAs to the corresponding C_{19} lactones. The structural differences that determine whether the 20-oxo intermediates are oxidized to C_{19}-GAs or to tricarboxylic acids are likely to be subtle. The 20-oxidases from pumpkin, *Marah*, and *Arabidopsis* prefer nonhydroxylated substrates to the 13-hydroxylated analogues. This is consistent with the types of GAs found in the tissues in which these enzymes are present. For example, GA_4, which is not 13-hydroxylated, is the major GA in *Arabidopsis* shoots.[321] In contrast, a GA 20-oxidase cloned from shoots of rice, in which 13-hydroxy C_{20}-GAs are the predominant forms, oxidizes GA_{53} more efficiently than it does GA_{12}.[322]

GA_{44} oxidase activity from spinach leaves was separated by anion-exchange chromatography from GA_{53} oxidase and GA_{19} oxidase activities, which coeluted.[292] These last two activities are induced by transfer of plants to long days, whereas GA_{44} oxidase activity is not photoperiod-sensitive.[323] The spinach GA_{44} oxidase converts the lactone form of this GA, as do cell-free systems from pea shoots[233] and germinating barley embryos,[313] whereas GA 20-oxidases from immature seeds require a free alcohol at C-20 for oxidation to occur.[312,324] Detailed studies with recombinant GA 20-oxidase, produced by expression of one of the *Arabidopsis* cDNAs in *E. coli*, revealed that this enzyme also required a free alcohol function and that oxidation of the alcohol was much slower than that of the methyl and aldehyde substrates. An unexpected difference between the spinach GA_{44} oxidase and the recombinant *Arabidopsis* GA 20-oxidase is in the stereospecific removal of a hydrogen atom during oxidation of the C-20 alcohol intermediates. It was shown using GA_{15} or GA_{44} labeled stereospecifically with deuterium, that the *Arabidopsis* enzyme removes the *pro-R* H atom on conversion of the free alcohol to the aldehyde.[325] In contrast, GA_{44}, as the lactone, is oxidized with loss of the *pro-S* H, by cell-free extracts of spinach leaves. This observation provides further evidence for the existence of a distinct lactone oxidase; the different stereochemistry of the reactions is presumably due to the fixed orientation of C-20 in the lactone as opposed to it assuming an energetically more favored conformation as the free alcohol.

(iv) 3β-Hydroxylation and related reactions

3β-Hydroxylation results in the conversion of the C_{19}-GAs GA_{20} and GA_9 to GA_1 and GA_4, respectively, in the final step in the formation of physiologically active GAs. There is now increasing evidence that, in common with GA 20-oxidases, certain GA 3β-hydroxylases may be multifunctional. An enzyme purified from developing embryos of *P. vulgaris* catalyzed 2,3-desaturation and 2β-hydroxylation reactions, in addition to 3β-hydroxylation.[326] GA_{20} and GA_9 were about equally reactive as substrates. A 3β-hydroxylase from the same source also epoxidized GA_5 to GA_6.[327] The enzyme could utilize non-2β-hydroxylated C_{19} (γ-lactone) GAs or 19-20 δ-lactone C_{20}-GAs as substrates, the latter presumably acting as structural analogues of the former, which are the natural substrates. An enzyme that 3β-hydroxylated GA_{15} to give GA_{37} was partially purified from pumpkin endosperm.[328] It did not possess desaturase or 2β-hydroxylase activities.

The pumpkin GA 3β-hydroxylase has the typical properties of a 2-oxoglutarate-dependent dioxygenase. In particular, it was possible to demonstrate a 1 : 1 stoichiometry between the formation of hydroxy GA and succinate, once uncoupled oxidation of 2-oxoglutarate was subtracted. In contrast, although the *P. vulgaris* enzyme requires 2-oxoglutarate for activity, Smith *et al.*[326] could find no evidence that this compound functioned as a substrate. They suggested that ascorbate may serve as the cosubstrate, as it does in the related enzyme, 1-aminocyclopropane-1-carboxylic acid (ACC) oxidase. However, 2-oxoglutarate is essential for full 3β-hydroxylase activity, whereas it serves no function for ACC oxidase activity. The nature of the *P. vulgaris* 3β-hydroxylase is unresolved.

The desaturase activity of 3β-hydroxylases provides the first step in the production of GA_3. Unequivocal evidence for conversion of GA_{20} to GA_3 via GA_5 was obtained in shoots of *Z. mays*, so establishing a new biosynthetic pathway.[329] There was no indication that GA_5 was converted to GA_1, a reduction that is without precedent in GA biosynthesis. The equivalent pathway for non-13-hydroxylated GAs was demonstrated in cell-free systems from immature seeds of *Marah* and apple.[330] Enzyme activity present in both endosperm and developing embryos of *Marah* results in the conversion of GA_9 to GA_4 and 2,3-dehydro GA_9, which is further oxidized to GA_7. The *Marah* and apple systems have a marked preference for non-13-hydroxylated substrates; although both systems converted GA_5 to GA_3, GA_{20} was metabolized to GA_1 (3β-hydroxylation), GA_{29} (2β-hydroxylation), and GA_{60} (1β-hydroxylation), but not to GA_5, by the *Marah* system and was unmetabolized by the apple preparation. The branch pathway from GA_{20} to GA_3 occurs also in barley embryos and may be common, although not ubiquitous, in higher plants.

The conversions of GA_5 to GA_3, and of 2,3-didehydro GA_9 to GA_7 are unusual reactions that are initiated by loss of the 1β-H.[330] Hydrogen abstraction is accompanied by rearrangement of the 2,3 double bond to the 1,2 position and hydroxylation on C-3β. This enzymatic activity may also result in the 1β-hydroxylation of GA_{20} and GA_5, also observed in *Marah*. The enzyme that converts GA_5 to GA_3 requires 2-oxoglutarate, but not added Fe^{2+}, for activity and, in contrast to most other related dioxygenases, it is not inhibited by iron chelators.[308] Its activity, however, is reduced by Mn^{2+} and other metal ions, the inhibition by Mn^{2+} being reversed by Fe^{2+}. It would appear that Fe is bound very tightly at the active site.

GA_3 and GA_1 formation in maize is apparently catalyzed by one enzyme.[331] As well as affecting the conversion of GA_{20} to GA_1, the *dwarf-1* mutation reduces formation of GA_5 and the conversion of GA_5 to GA_3. If *DWARF-1* is a structural gene, a single enzyme must catalyze all three reactions. In contrast, the *le* (3β-hydroxylation) mutation of pea was reported not to affect the conversion of GA_5 to GA_3.

The *GA4* gene (3β-hydroxylase) of *Arabidopsis* has been cloned by T-DNA insertion.[332] The gene encodes a dioxygenase that has relatively low amino acid sequence identity with the GA 20-oxidases; it has 30% identity (50% similarity) with the *Arabidopsis* stem-specific 20-oxidase (GA_5). Expression of the *ga4* cDNA in *E. coli* has confirmed that it encodes a 3β-hydroxylase.[333] The preferred substrate for the recombinant enzyme is GA_9, for which the K_m is 10-fold lower than that for GA_{20}. Thus, as with the *Arabidopsis* GA 20-oxidases, the presence of a 13-hydroxy group reduces substrate affinity for the enzyme. The enzyme also epoxidizes the 2,3-double bond in GA_5 and 2,3-didehydro GA_9, and hydroxylates certain C_{20}-GAs, albeit with low efficiency. This activity could account for the presence of 3β-hydroxy C_{20}-GAs in *Arabidopsis*. The mutant with the T-DNA insertion is a semi-dwarf with very little likelihood of the mutant gene encoding an active 3β-hydroxylase. Residual growth in this mutant must, therefore, result from the action of other enzymes. Another putative 3β-hydroxylase has been cloned in the laboratory of one of the authors (Y. Kamiya).

(v) 2β-Hydroxylation and related reactions

Hydroxylation on C-2β results in the formation of inactive products and is, therefore, important for turnover of the physiologically active GAs. The natural substrates for these enzymes are normally C_{19}-GAs, although 2β-hydroxy C_{20}-GAs are also found in plant tissues, particularly where the concentration of C_{20}-GAs is high. 2β-Hydroxylases have been partially purified from cotyledons of *P. sativum*[334] and *P. vulgaris*.[335] There is evidence that, for both sources, at least two enzymes with different substrate specificities are present. Two activities from cotyledons of imbibed *P. vulgaris* seeds were separable by cation-exchange chromatography and gel filtration. The major activity, corresponding to an enzyme of M_r 2.6×10^4 by size-exclusion HPLC, hydroxylated GA_1 and GA_4 in preference to GA_9 and GA_{20}, while GA_9 was the preferred substrate for the second enzyme (M_r 4.2×10^4).

Formation of 2-keto derivatives (GA catabolites) by further oxidation of 2β-hydroxy GAs (Scheme 3) occurs in several species, but is particularly prevalent in developing seeds[336] and roots of pea.[337] The conversion of GA_{29} to GA_{29}-catabolite in pea seeds was inhibited by prohexadione-calcium, an inhibitor of 2-oxoglutarate-dependent dioxygenase, indicating that the reaction is catalyzed by an enzyme of this type.[337] Formation of the GA catabolites could be initiated by oxidation either at C-1 or C-2α.

8.02.3.6 Regulation of GA Biosynthesis

The role of GAs as mediators of environmental stimuli is well established. Factors such as photoperiod and temperature can modify GA metabolism by changing the flux through specific steps in the pathway. Work has shown that GA biosynthesis is modified by the action of GA itself in a type of feedback regulation. The mechanisms underlying these regulatory processes can be investigated as a result of advances in the molecular biology of GA biosynthesis.

8.02.3.6.1 Feedback regulation

The presence of abnormally high concentrations of C_{19}-GAs in certain GA-insensitive dwarf mutants, such as *Rht3* wheat,[338] *Dwarf-8* maize,[339] and *gai Arabidopsis*,[340] indicates a link between GA action and biosynthesis. In maize, there is a gene-dosage effect, with a 60-fold increase in the GA_1 content of homozygous *Dwarf-8* shoots, compared with wild-type and a 33-fold increase in the heterozygote. It has been proposed that GA 20-oxidase is a primary target for feedback regulation.[338] In addition to an elevated C_{19}-GA content, the GA-insensitive dwarfs often contain lower amounts of C_{20}-GAs than their corresponding wild-types, suggesting increased GA 20-oxidase activity. Conversely, overgrowth mutants, such as slender (*sln*) barley and *la cry^s* pea, that grow as if saturated with GA, even in its absence, contain reduced amounts of C_{19}GAs and elevated amounts of C_{20}-GAs.[341] It appears that the slender mutation activates the GA signal transduction pathway, even in the absence of GA, and may thereby cause constitutive down-regulation of GA 20-oxidase activity.

With the availability of GA 20-oxidase cDNA clones, it has been possible to begin a molecular analysis of the feedback mechanism. Transcript levels for each of the three *Arabidopsis* GA 20-oxidase genes are much higher in the *ga1-3* (CPS-deficient) mutant than in wild type and are very substantially reduced by treating the mutant with GA_3.[319] The reduction occurs within 1–3 h, long before a growth response is discernible, confirming that it is not related to growth rate and indicating that the message is turned over rapidly.[342] Strong down-regulation of GA 20-oxidase transcript levels by GA has also been observed in pea[341] and rice.[322] Low endogenous GA concentration, as in mutants or after treatment with a biosynthesis inhibitor, consistently resulted in increased mRNA levels. Conversely, these levels were substantially reduced by application of GA. Other enzymes in the pathway, particularly the GA 3β-hydroxylase, may also be subject to feedback regulation.

8.02.3.6.2 Regulation by light

The involvement of GAs in the photoperiod-induced bolting of long-day rosette plants is well documented.[231] In spinach (*Spinacia oleracea*), changes in GA concentrations and enzyme activity in cell-free systems on transfer from short days (SD) to long days (LD) are consistent with enhanced oxidation of GA_{53} and GA_{19} in LD.[323] The activities of GA_{53} and GA_{19} 20-oxidases, now known to be the same enzyme, increase in the light and decrease in the dark. Furthermore, there are higher amounts of GA 20-oxidase mRNA in plants grown in LD than those in SD or in total darkness.[343] It has been suggested that, in LD, there is sufficient GA 20-oxidase activity to raise the GA_1 concentration above the threshold required for stem extension. In fact, light appears to increase the total flux through the pathway, since *ent*-kaurene synthesis is also enhanced in LD in spinach and in *Agrostemma githago*. Although GA_{53} 20-oxidase activity is regulated by light, oxidation of GA_{44} in the lactone form remains at high constant levels irrespective of light or dark treatment. As discussed earlier, this latter activity is probably due to another enzyme, which is not under light regulation.

Despite many attempts to implicate GA metabolism in phytochrome-mediated changes in growth rate, supporting evidence is sparse. Enhancement of GA_{20} 3β-hydroxylation by far-red light has been observed in lettuce[344,345] and cowpea epicotyls.[346] In the latter case, higher GA_1 concentrations in plants grown in far-red light were due also to reduced 2β-hydroxylation and were accompanied by heightened tissue responsiveness to GA. However, in peas, enhanced shoot elongation by treatment with far-red-rich light was not associated with increased GA_1 content.[347] There is also no evidence to suggest that dark-grown peas or sweet peas contain more GA_1 than light-grown plants. In fact, work with phytochrome- and GA-deficient mutants of pea indicates that growth inhibition by red light, which is mediated by phytochrome B, is due to altered responsiveness to GA, rather than to changes in the concentration of GA_1.[348] Several phytochrome-deficient mutants, such as the *ein* mutant of *Brassica rapa*[349] and the *ma3^R* mutant of *Sorghum*,[350] have an overgrowth phenotype and were originally thought to contain abnormally high GA levels. Although the GA_1 content of these plants may be elevated, this is apparently not the cause of their altered phenotype. Phytochrome B-deficient mutants of cucumber[351] and pea[348] contain comparable amounts of active GAs to the wild-types, but show an enhanced response to GAs. In *Sorghum*, altered GA content is due to a shift in the phase of a diurnal fluctuation in GA concentrations. It was proposed that phytochrome deficiency disrupted diurnal regulation of the conversion of GA_{19} to GA_{20}.

Gibberellin metabolism is sensitive to light quantity. When pea seedlings were grown in low irradiance, GA_{20} concentration increased sevenfold compared with plants grown in high irradiance, whereas, in plants grown in the dark, the GA_{20} content was reduced to 25% of that in high irradiance.[352] Moreover, the response of the seedling to exogenous GA_1 was heightened in the dark. These results indicate that the rate of GA 20-oxidation is sensitive to light intensity. It has been shown in the laboratory of one of the authors (Y. Kamiya) that the transcript levels of GA 20-oxidase in pea seedlings are regulated by phytochromes. When dark-grown pea seedlings were transferred to light, a red light pulse increased the transcript levels of GA 20-oxidase in apical buds, but a far-red light pulse reduced it. Both phytochrome A and B regulate this reaction.

8.02.3.6.3 Regulation by temperature

Induction of seed germination (stratification) or of flowering (vernalization) by exposure to low temperatures, are processes in which GAs have been implicated. There are, however, few examples in which GAs have been shown unequivocally to mediate the temperature stimulus. The most extensively studied system is *Thlaspi arvense*, in which stem extension and flowering are induced by exposure to low temperatures followed by a return to higher temperatures.[353,354] The same effect can be obtained without cold-induction by application of GAs, the most active of those tested being GA_9. In noninduced plants, *ent*-kaurenoic acid accumulates to high concentrations in the shoot tip, the site of perception of the cold stimulus, whereas, after vernalization and return to high temperatures, the level of this intermediate falls within days to relatively low values. Metabolism of labeled *ent*-kaurenoic acid to GA_9 could be demonstrated in thermoinduced shoot tips, but not in noninduced material. Furthermore, microsomes from induced shoots metabolized *ent*-kaurenoic acid and *ent*-kaurene, but microsomes from noninduced shoots were much less active for both activities. Leaves, or microsomes extracted from leaves, from thermoinduced and noninduced plants metabolized *ent*-kaurenoic acid to the same extent. These results are consistent with regulation of *ent*-kaurenoic acid 7β-hydroxylase and, to a lesser degree, *ent*-kaurene oxidase by cold treatment in shoot tips of *Thlaspi*.

Although the mechanism for thermoinduction of GA biosynthesis is not yet known, it has been suggested that cold treatment may allow increased rates of gene expression, possibly via demethylation of the promoters.

The sections above focus mainly on GA biosynthesis and its regulation because basic GA chemistry is well established and is the subject of many books and reviews. Some of the topics written about by the authors consider further molecular biological subjects and readers may have an impression of too much biochemistry. However, all the DNA- and RNA-related reactions can also be thought of as pure chemical reactions and with a strong chemistry background, the reactions can be fully understood.

8.02.3.7 Antheridiogens

Since Döpp[355,356] demonstrated that gametophytes of the bracken fern, *Pteridium aquilinum*, produced hormonal substances inducing antheridial formation, it has been reported that developing gametophytes of numerous species of ferns produce antheridium-inducing substances,[357] which have been named antheridiogens.[358] Based on cross-testing of biological activity, it has been found that antheridiogens can be classified into several types of compounds.[359] However, all the antheridiogens characterized so far are derived from Schizaeaceous ferns, and are all GA-related compounds.[360–369] Isolation and characterization, biological activities, and biosynthesis of antheridiogens in Schizaeaceous ferns are now described briefly.

8.02.3.7.1 Isolation and characterization

In 1971, Nakanishi *et al.*[360] isolated a major antheridiogen in *Anemia phyllitidis*. It was tentatively named A_{An}, and its chemical structure was assigned to **Ia** (Figure 4) based on spectroscopic analyses.

However, the structure of A_{An} was later refined to structure **Ib** by total synthesis of **Ia** and **Ib**,[370] and the tentative name A_{An} was replaced by antheridic acid.[344] Since natural antheridic acid was synthesized starting from GA_7, it was shown to possess the same stereochemistry as that of gibberellin.[371] Antheridic acid was also shown to be a major antheridiogen in the other three *Anemia* species, *A. hirsuta*, *A. rotundifolia*, and *A. flexuosa*.[361,364] 3α-Hydroxy-9,15-cyclo-GA_9 (**IIe**) and 3-*epi*-GA_{63} (**IVd**) were also identified as minor antheridiogens in *A. phyllitidis*.[367,368] In another *Anemia* species, *A. mexicana*, a novel GA-like compound was isolated as a major antheridiogen.[372] The chemical structure of this compound was shown to be 1β-hydroxy-9,15-cyclo-GA_9 (**IIb**) based on synthetic studies.[366]

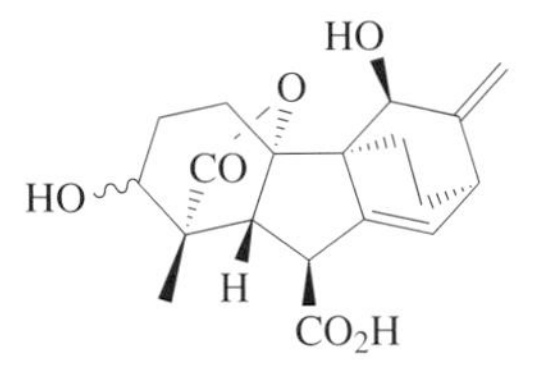

20-Norantherid-8(14),16-diene derivatives
Ia (β-OH): 3-*epi*-Antheridic acid
Ib (α-OH): Antheridic acid*

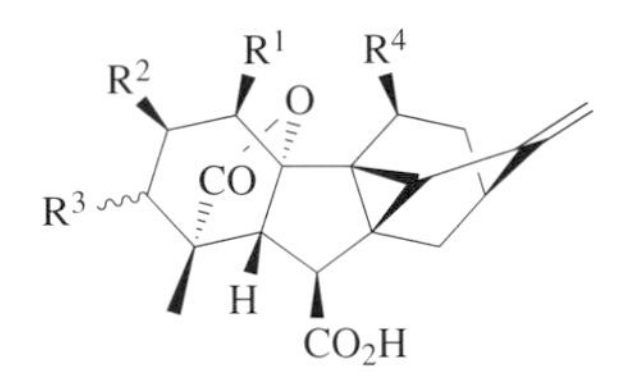

ent-9,15-Cyclo-20-norgibberell-16-diene derivatives
IIa ($R^1 = R^2 = R^3 = R^4$ = H): GA_{103}
IIb (R^1 = OH, $R^2 = R^3 = R^4$ = H): GA_{104}*
IIc ($R^1 = R^3 = R^4$ = H, R^2 = OH): GA_{105}
IId ($R^1 = R^2 = R^4$ = H, R^3 = β-OH): GA_{106}
IIe ($R^1 = R^2 = R^4$ = H, R^3 = α-OH): GA_{107}*
IIf ($R^1 = R^2 = R^3$ = H, R^4 = OH): GA_{108}

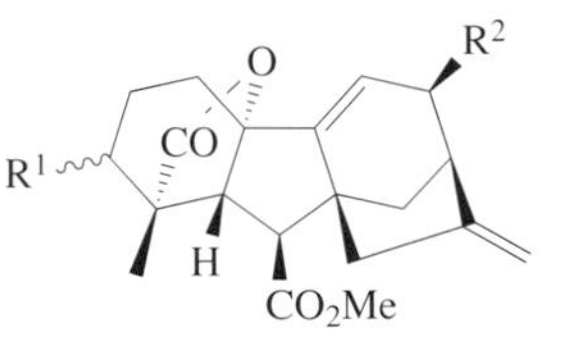

ent-20-Norgibberell-9(11),16-diene derivatives
IIIa ($R^1 = R^2$ = H): GA_{73}-Me*
IIIb (R^1 = β-OH, R^2 = H): GA_{88}-Me*
IIIc (R^1 = α-OH, R^2 = H): 3-*epi*-GA_{88}-Me*
IIId (R^1 = H, R^2 = OH): GA_{96}-Me*

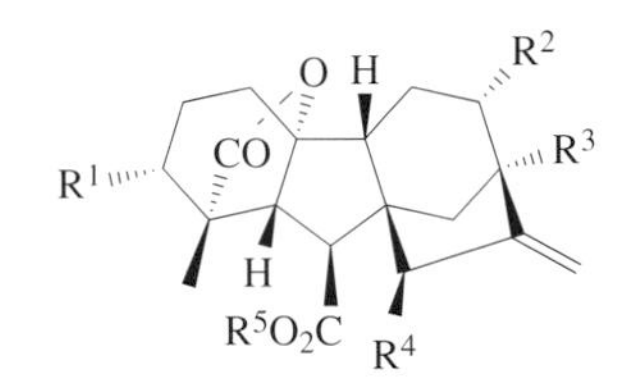

ent-20-Norgibberell-16-ene derivatives
IVa ($R^1 = R^2 = R^3 = R^4$ = H, R^5 = Me): GA_9-Me*
IVb ($R^1 = R^3 = R^4$ = H, R^2 = OH, R^5 = Me): GA_{70}-Me*
IVc ($R^1 = R^2 = R^4$ = H, R^3 = OH, R^5 = Me): GA_{20}-Me*
IVd ($R^1 = R^4$ = OH, $R^2 = R^3 = R^5$ = H): 3-*epi*-GA_{63}*

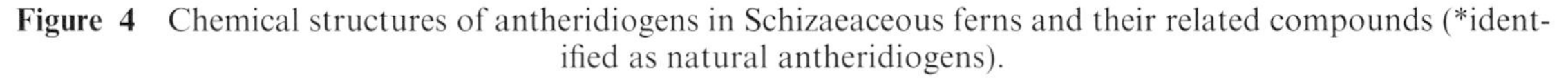

Figure 4 Chemical structures of antheridiogens in Schizaeaceous ferns and their related compounds (*identified as natural antheridiogens).

The antheridiogens from *Lygodium* species have all been found to be GA methyl esters. GA_{73} methyl ester (**IIIa**; GA_{73}-Me) was identified as the principal biologically active antheridiogen in three species of ferns, *Lygodium japonicum*, *Lygodium circinnatum*, and *Lygodium flexuosum*,[347,351] and GA_9-Me (**IVa**), GA_{20}-Me (**IVc**), GA_{70}-Me (**IVb**), GA_{88}-Me (**IIIb**), 3-*epi*-GA_{88}-Me (**IIIc**), and GA_{96}-Me (**IIId**) were identified as minor antheridiogens in *L. circinnatum*.[363,369]

The above results indicate that the antheridiogens from the Schizaeaceous ferns can be classified into four basic structures, that is, 20-norantherid-8(14),16-diene, *ent*-9,15-cyclo-20-norgibberell-16-ene, *ent*-20-norgibberell-9(11),16-diene, and *ent*-20-norgibberell-16-ene as shown in Figure 4, and that their biosynthetic relationships are closely related.

It should also be noted that some of the antheridiogens and related compounds such as 3-*epi*-GA_{63}, GA_{73}, GA_{88}, 3-*epi*-GA_{88}, 9,15-cyclo-GA_9 (GA_{103}, **IIa**), 1β-hydroxy-9,15-cyclo-GA_9, (GA_{104}, **IIb**), 2β-hydroxy-9,15-cyclo-GA_9 (GA_{105}, **IIc**), 3β-hydroxy-9,15-cyclo-GA_9 (GA_{106}, **IId**), 3α-hydroxy-9,15-cyclo-GA_9 (GA_{107}, **IIe**), and 11β-hydroxy-9,15-cyclo-GA_9 (GA_{108}, **IIf**) were also identified from immature apple seeds, and new GA numbers from GA_{103} to GA_{108} were allocated to these 9,15-cyclo-GAs.[373]

8.02.3.7.2 Biological activities

In *A. phyllitidis*, the principal antheridiogen, antheridic acid, induced antheridial formation in light-grown prothallia at 8.6×10^{-9} M and induced dark spore germination at 8.6×10^{-10} M.[374] The levels of activity of minor antheridiogens in *A. phyllitidis*, GA_{107} and 3-*epi*-GA_{63}, were slightly less than antheridic acid in both antheridial formation and dark spore germination assays.[367,368] The level of activity of synthetic (±)-antheridic acid was about half that of natural antheridic acid, the unnatural enantiomer is likely to be inactive, while synthetic (±)-3-*epi*-antheridic acid was about one order of magnitude less active than (±)-antheridic acid in both assays.[374] It should be noted that antheridic acid possessing a 3α-hydroxy was more active than 3-*epi*-antheridic acid possessing a 3β-hydroxy in antheridial formation and dark spore germination assays in *A. phyllitidis*, because in general 3β-hydroxy-GAs such as GA_1, GA_3, GA_4, and GA_7 are highly active in most GA-bioassay systems and the activities of gibberellin derivatives with a 3α-hydroxy are quite low.[267]

In *L. japonicum*, the principal antheridiogen, GA_{73}-Me, induced antheridial formation in dark-grown protonemata at 10^{-14} M, induced dark spore germination at 10^{-11} M, and inhibited archegonial formation in light-grown prothallia at 10^{-11} M.[375] A minor antheridiogen in *L. japonicum*, GA_9-Me, was two to four orders of magnitude less active than GA_{73}-Me in the three bioassays.[363,375] It is noteworthy that antheridiogens in *L. japonicum* showed inhibitory activity on archegonial formation as well as inducing activity on antheridial formation and dark spore germination, because antheridiogens had been considered not to affect archegonial formation in fern gametophytes since Döpp[356] reported that antheridiogens in *P. aquilinum* had no effect on archegonial formation. The effects of antheridiogens on archegonial formation require further investigation.

As already described, it was found that GA_{73}-Me is a principal antheridiogen in *L. circinnatum* and *L. flexuosum* as well as *L. japonicum*. The production of GA_{73}-Me in the first two species was more than 1000 times larger than that in *L. japonicum*, while response to GA_{73}-Me of dark-grown protonemata of the former two *Lygodium* ferns in the antheridial formation assay was more than 100 times less sensitive than that of *L. japonicum*.[369] These results suggest that the capacity of antheridiogen production in the gametophytes of these *Lygodium* ferns is closely related to the response to the antheridiogen.

The antheridiogens and their derivatives also showed plant growth-promoting activity. In the dwarf rice assay using cv. Tan-ginbozu, (±)-antheridic acid only showed significant activity at 1000 ng per plant, and the (±)-3-epimer was one order more active than (±)-antheridic acid. Gibberellin A_9, GA_{73}, and GA_{103} were all active to show significant activity at a dosage of 1 ng per plant.[375] However, in the cucumber hypocotyl assay GA_{73} and GA_9 were highly active, showing significant activity at a dosage of 100 ng per plant, GA_{103} being slightly active at a dosage of 1000 ng per plant.[375] This suggests that cucumber recognized the C/D ring structure strictly and/or did not convert GA_{103} into the active form.

8.02.3.7.3 Biosynthesis

As already described, close relationships could be speculated between the four basic structures of the antheridiogens from Schizaeaceous ferns. The 9,15-cyclo-GA structure could be a precursor of antheridic acid and/or GA_{73}-Me. Alternatively, these four basic structures could be biosynthesized according to branches from known GA biosynthetic pathways. Based on the above speculation, feeding experiments using isotope-labeled substrates have been carried out to clarify biosynthetic pathways of the antheridiogens.

In prothallia of *A. phyllitidis*, antheridic acid was shown to be biosynthesized from GA_{103} via GA_{107} in *A. phyllitidis* (Scheme 4; *, identified as native compounds in the culture filtrate of *A. phyllitidis* prothallia),[367] although GA_{103} has not been identified as a native antheridiogen in *A. phyllitidis*.[367] Gibberellin A_{73}-Me was found to be biosynthesized from GA_{24} via GA_{73} in prothallia of *L. circinnatum*, and subsequently GA_{73}-Me was hydroxylated to give monohydroxy derivatives such as GA_{88}-Me, 3-*epi*-GA_{88}-Me, and GA_{96}-Me (Scheme 5; *, identified as native compounds in *L. circinnatum* prothallia and/or the culture filtrate).[369,376] It was also shown that GA_9-Me is biosynthesized from GA_{24} via GA_9, and that GA_9-Me is hydroxylated to give monohydroxy derivatives such as GA_{20}-Me and GA_{70}-Me (Scheme 5).[376] Conversion of GA_{103} possessing a 9,15-cyclo-GA structure into GA_{73}-Me was not observed in *L. circinnatum*.[376]

8.02.4 CYTOKININS

8.02.4.1 History

Cytokinin was discovered in 1955 by Skoog's group as a factor required for the cell division of callus tissues from the pith of *Nicotiana tabacum* cv. Wisconsin No. 38[377] after a long effort to isolate

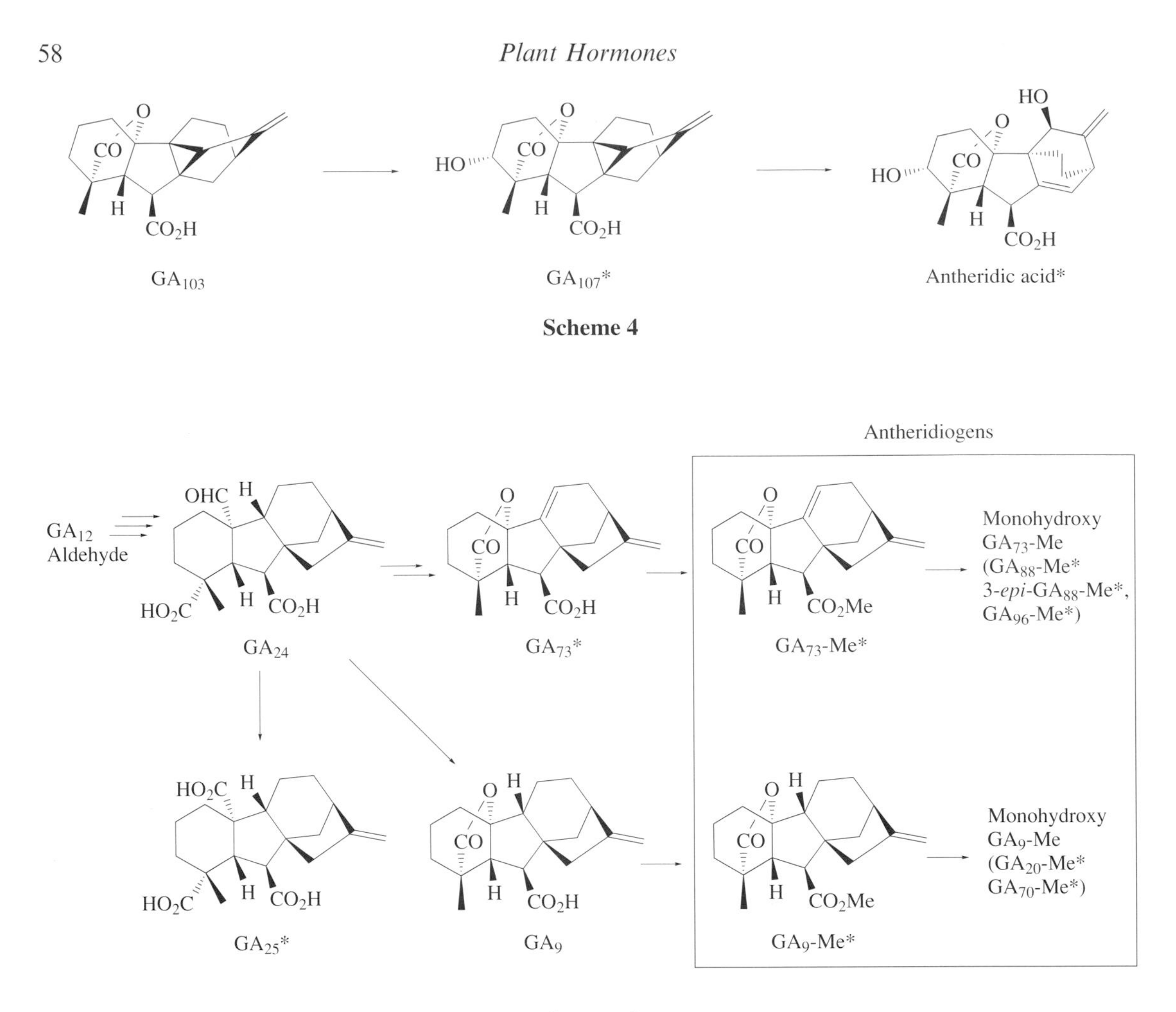

Scheme 4

Scheme 5

such substances from natural sources. From autoclaved herring sperm DNA, kinetin was first isolated as crystals by Miller *et al.*[378] Shortly after the discovery of kinetin, a large number of compounds were synthesized and tested for cytokinin activity. A number of N^6-substituted adenines and phenylurea derivatives were found to have the activity.[379] In 1961, Miller purified an N^6-substituted adenine with high cytokinin activity from immature corn grains.[380] This work was followed by Letham who determined its structure and named it zeatin, the first natural cytokinin to be discovered.[381,382] The term "cytokinin", which is now widely used, was proposed by Letham[383] in 1963 to designate such compounds that have kinetin-like activity. Although the stimulation of cell division in plant tissue cultures is the most characteristically associated property of cytokinins, the mechanism of this action is not well understood at the time of writing.

Not long after the first observations of kinetin-stimulated cell division, three groups of workers reported nearly simultaneously that cell enlargement in disks of etiolated leaves was markedly increased in the presence of cytokinins.[384–386] This effect was observed even in light-grown, fully mature leaves which had ceased expanding.[384] Kinetin treatment of hypocotyls of dicots was found to cause an increase in fresh and dry weight more than double that of the controls, although no elongation occurred.[387]

Within a few months of the discovery of kinetin, it was found that it promoted seed germination. In lettuce seeds, cytokinins can substitute for the red light requirement in breaking seed dormancy[385,388] as well as synergize with light in promoting germination.[389] This particular activity of cytokinins is probably only another aspect of their stimulatory effect on cell enlargement, as indicated by the work of Haber and Luippold.[390]

In 1957, Richmond and Lang[391] demonstrated that kinetin retarded senescence of green leaves. In detached leaves, kinetin is able to postpone for a number of days the disappearance of chlorophyll (Chl) and degradation of proteins, which normally accompanies the senescence process of leaves placed in the dark. Since then, there have been a large number of papers describing the effects of cytokinins on the metabolism of cellular constituents of high molecular weight such as DNA, RNA, and proteins.[392–395] In the early 1960s, Osborne[396] and other groups observed that treatment with cytokinins of leaf disks incubated in the dark maintained the ratio of RNA or protein to DNA, which was lowered in untreated controls. This suggests that a critical action of this hormone in senescence retardation might be the maintenance of protein-synthesizing machinery, perhaps by regulating RNA synthesis. Cytokinins also increase aminoacyl-tRNA synthetase activity in the leaf disks.[397] An alternative hypothesis to explain cytokinin-induced delay of senescence is based on the observations that the activities of proteinase, RNase, and DNase increased in senescing leaf disks, and that cytokinins prevented rises in the activities of these hydrolyzing enzymes so that the levels of protein, RNA, and DNA were maintained.[393,395,398]

Möthes found that kinetin acted as a mobilizing agent.[399] When tobacco leaves were sprayed in one spot with kinetin, and a ^{14}C-labeled amino acid was supplied either through the petiole or on the surface of the leaf at a site remote from the kinetin-treated area, autoradiographs showed that radioactive material congregated in precisely the area sprayed with kinetin.[400] Cytokinins have the ability to attract (forming a so-called attraction center) a large number of substances in this manner, and to prevent the movement of leaf components out of the treated area. The correlated action of cytokinins may account for the senescence-inducing effect of fruits and stem apices on leaves.[379,401]

In the late 1960s, cytokinins were found to stimulate stomatal opening[402] and hence transpiration.[403–405] Kuraishi claimed that cytokinin effects such as Chl retention,[406] Chl synthesis,[407] and amino acid accumulation[408] could result from cytokinin-stimulated transpiration. However, Müller and Leopold[409] noted that kinetin-induced transport took place in the phloem independently of water flow in the xylem.

Application of a cytokinin to lateral buds induces their growth by releasing them from apical dominance.[410] An interesting relationship between the cytokinin effect on apical dominance and fasciation of plants caused by the microorganism *Corynebacterium fascians* was described by Samuels.[411] The symptom of this disease is the loss of apical dominance with subsequent appearance of a "witches broom" of growing shoots. N^6-Methylaminopurine and N^6-(γ,γ-dimethylallylamino)purine (N^6-isopentenyladenine, ipA) were identified as active substances in the extract from the bacteria.[412]

One of the more interesting activities of the cytokinins is their partial prevention of the toxic effects of certain pathogens or adverse effects of environmental stresses. *Pseudomonas tabaci* which causes "wildfire" disease of tobacco produces a toxin, a structural analogue of methionine, which damages RNA metabolism of the host, resulting in a breakdown in protein synthesis.[413] The protection provided by cytokinins in this system may well be related to their senescence-retarding capacity which involves a maintenance of protein synthesis.

One of the truly dramatic responses to cytokinins is the formation of organs which takes place under appropriate conditions in a variety of tissue cultures. At certain balanced levels of IAA and kinetin in the tobacco pith tissue cultures, the tissue grows as an amorphous, undifferentiated callus. However, increasing the ratio of kinetin with respect to auxin results in the formation of buds which may grow into shoots and ultimately into complete tobacco plants under the right conditions.[414] In the reverse conditions, where the kinetin to auxin ratio is lowered, roots appear on the pith.[414]

In the 1960s, Fox[415–417] reported that plant tissues requiring cytokinins for growth incorporated benzyladenine and 6-methylpurine into their RNA. The incorporation in soybean tissue cultures was almost exclusively into the tRNA fraction. The pioneering work of Zachau *et al.*[418] identified ipA as a constituent present adjacent to the 3′ end of the anticodon in tRNAs from yeast. Armstrong *et al.*[419,420] were the first to notice that ipA and its corresponding analogues occur only in those tRNA species responding to codons beginning with uridine. A number of cytokinins were found in tRNA from a variety of organisms.[421] Since then, there has been the issue of whether cytokinins exert their biological effects through their incorporation into tRNA. At the time of writing, this is unresolved. Both theoretical consideration and the evidence available so far indicate that the cytokinin moieties in tRNA are formed by modification of preformed polynucleotide chains.[393,394]

In 1971, Fletcher and McCullagh[422] found that benzyladenine stimulated Chl accumulation during greening of etiolated cucumber cotyledons. Thus it has been shown that cytokinins not only inhibit the loss of Chl in senescing leaves, but also enhance the formation of Chl in greening tissues. Naito and co-workers[423–427] have shown that benzyladenine induces an increase in nuclear DNA, leading to endopolyploidy in growing bean leaves. Cytokinins also stimulated chloroplast DNA synthesis and chloroplast replication.[425,428–431] How cytokinins reinitiate DNA synthesis in the nucleus and chloroplast where it has already been completed remains to be elucidated.

Among a variety of activities of cytokinins, stimulation of cell division in cytokinin-requiring tissue cultures,[432,433] retardation of senescence of green tissues in darkness,[434] and promotion of Chl synthesis in greening of etiolated tissues[435] have been used for the bioassay of this hormone. The latter two are often used as convenient rapid methods, because Chl is easily quantified.

In accordance with progress in plant physiology and biochemistry, cytokinin effects on a wide spectrum of metabolic reactions and enzymes have been reported.[393,395,436] Such diverse manifestations of cytokinin activity are no doubt secondary phenomena, appearing in response to some, as yet unknown, more basic role of this hormone. Among them, however, the stimulation of replication, development, and function of chloroplasts is a unique activity of this hormone.

Since the 1980s, reports on the identification and determination of natural cytokinins and their metabolism have been increasing, supported by the development of analytical techniques.[437–439] There have been also an increasing number of papers on the mechanism of the action of cytokinins, including cytokinin-binding proteins, cytokinin signal transduction pathways, and the control of gene expression during the early period of cytokinin treatments.[439] In these works, a number of new approaches have been developed, such as: studying the effects of increased levels of cytokinins using plants transformed with genes for cytokinin-synthesizing enzymes; and screening cytokinin-resistant mutants and analyzing their genes to elucidate the pathways of cytokinin signal transduction.

8.02.4.2 Chemistry

8.02.4.2.1 Isolation

Cytokinins in plant tissues are usually extracted by aqueous methanol or ethanol.[440] Extraction with hot water is less effective. Cytokinin nucleotides are easily hydrolyzed by nonspecific phosphatases in plant tissues to corresponding nucleosides and bases, and the enzymes are not completely inactivated by the conventional extraction solvents. To minimize the effects of phosphatases, extraction by perchloric acid and a solvent system containing formic acid is recommended.[438] Even extraction by hot water or hot alcohols is not sufficient to avoid the hydrolysis. For soft tissues like callus, a method to add the tissues without grinding in ethanol so that the final concentration of the mixture is 80% has been recommended, attaining more than 80% recovery of cytokinins.[440]

Solvent partitioning is useful for preliminary clean-up of the crude aqueous extracts.[441] *n*-Butyl alcohol, and less often ethyl acetate, is used to extract bases, leaving polar nucleosides and nucleotides in the aqueous layer. To obtain a good recovery the partition coefficient of each cytokinin species should be considered. Table 8 summarizes the partition data of representative cytokinins.[440,442,443]

Table 8 Partition coefficients ($C_{n\text{-BuOH}}/C_{aq.}$) of naturally occurring cytokinins.

Cytokinin	*Partition coefficient*	*Ref.*	*Partition coefficient*	*Ref.*	*Partition coefficient*	*Ref.*
Zeatin	1.0 [pH 2.5]	440	1.59 [pH 3.0]	441	6.26 [pH 7.0]	441
	6.7 [pH 8.0]	443				
Dihydrozeatin	0.6 [pH 2.5]	440	8.6 [pH 8.2]	440		
9-Ribosylzeatin	1.1 [pH 2.5]	440	2.5 [pH 8.0]	443		
O-Glucosylzeatin	0.64 [pH 8.0]	443				
O-Glucosyl-9-ribosylzeatin	0.27 [pH 8.0]	443				
ipA	5.7 [pH 2.5]	440	10.7 [pH 3.0]	441	40.4 [pH 7.0]	441
	>20 [pH 8.2]	440				
2-Methylthio-ipA	>20 [pH 2.5]	440	>20 [pH 8.0]	443		
9-Ribosyl ipA	5.7 [pH 2.5]	440	11 [pH 8.2]	440		

Ion exchange chromatography has been often used for the first step of purification. The recovery percentage of zeatin and analogous cytokinins has been reported to be more than 90% when cellulose phosphate is used. For final or subfinal purification, Sephadex LH-20 has been widely used, which should give an excellent recovery of applied cytokinins. The elution profiles of some cytokinins have been examined and summarized.[440] The most commonly used technique is HPLC. It has a marked advantage over GC in that the derivatization steps to a volatile analogue are not necessary, and over PC and TLC in its resolution capability and high recovery efficacy. Octadecyl silane (ODS) is the most commonly used packing material. It is used not only in the final purification step but also in initial steps to concentrate cytokinins from crude extracts.[438]

Zeatin, ipA and their derivatives including ribosides, ribotides, and glucosides have been isolated and identified from dozens of plant species, and the reviews have been given by Letham and Palni[437] and Koshimizu and Iwamura.[438]

8.02.4.2.2 Structure

Numerous cytokinins have been isolated, not only from plant tissues but also from fungi, bacteria, algae, and marine organisms. These naturally occurring cytokinins are all N^6-substituted adenine and its derivatives. The N^6-side chain is mostly isoprenoid, and in a few cases it is aralkyl, as mentioned below. Figure 5 shows the structures of representatives of naturally occurring cytokinins; details are documented in the review by Koshimizu and Iwamura.[438]

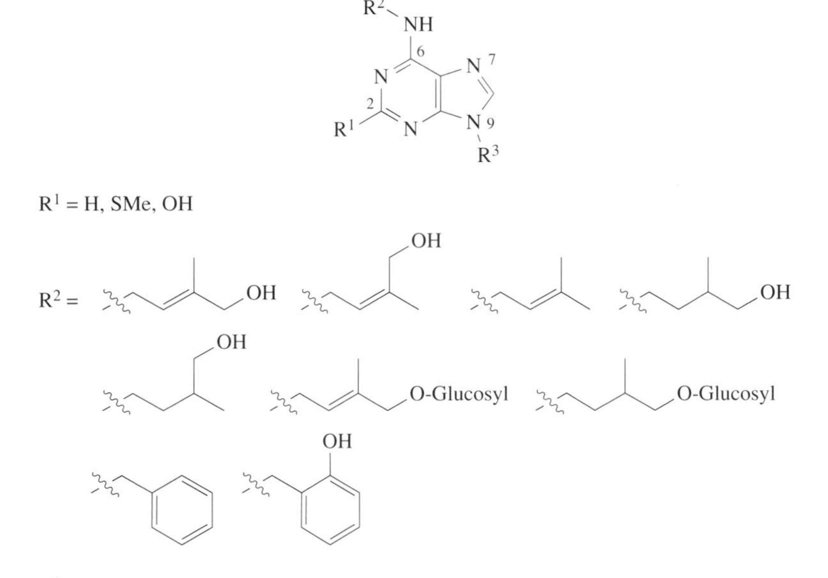

R³ = H, ribosyl, 5'-phosphoribosyl, glucosyl

Figure 5 Structures of representative naturally occurring cytokinins.

(*E*)-Zeatin, the cytokinin first identified to occur in plants (*Z. mays*), has an isoprenoidal (*E*)-4-hydroxy-3-methylbut-2-enyl side chain. Following the discovery of (*E*)-zeatin, a number of its derivatives have been found to occur naturally; they are 9-ribosylzeatin, 9-ribosylzeatin 5′-phosphate, 9-glucosylzeatin, 9-glucosylzeatin 6′-phosphate, 7-glucosylzeatin (raphanatin), 2-methylthio counterparts of these, and 2-hydroxyzeatin.[438] The ribosides and ribotides, especially the latter, are the structures that occur abundantly. (*E*)-Zeatin-2′-deoxyriboside has been isolated from plant pathogenic *Pseudomonas amygdali*.[444]

The glucosides are not only widely distributed but also identified as dominant metabolites of cytokinins fed to plant tissues.[438,440] The derivatives in which the 4-hydroxy group of the side chain moiety of zeatin and its riboside is substituted by glucopyranosyloxy, *O*-glucosylzeatin and its riboside, have been isolated from a number of plant species.

(*Z*)-Zeatin (the geometrical isomer of (*E*)-zeatin), dihydrozeatin (N^6-(4-hydroxy-3-methylbutyl)adenine), 9-ribosyl and *O*-glucosyl-9-ribosyl derivatives of these, and ipA, its 9-ribosyl, 9-(5′-phosphoribosyl), 2-methylthio-9-ribosyl and 9-glucosyl derivatives are also found in plants. 9-Ribosyl-(*Z*)-zeatin is a common cytokinin riboside present in plant tRNA hydrolysate, and its 2-methylthio derivative also occurs in tRNA.[438] Thus, the possibility that cytokinins isolated from plant sources may include those cytokinins formed by enzymatic cleavage of tRNA during extraction processes has not been excluded.

Diphenylurea, a compound that is structurally different from N^6-substituted adenines but shows cytokinin activity, was once isolated from coconut milk as its cell division-promotive principle, but it has been considered to be an artifact that had been contaminated in the extraction solvent.[445] Yet, diphenylurea and its analogues constitute an important class of cytokinin-active compounds.

L-2-(6-((*E*)-4-hydroxy-3-methylbut-2-enylamino)purin-9-yl)alanine (lupinic acid) is an alanine conjugate of (*E*)-zeatin found as a novel metabolite of (*E*)-zeatin in *Lupinus angustifolia* seedlings, and the structure was confirmed by synthesis. An analogous alanine conjugate is also produced in *P. vulgaris* seedlings when fed benzyladenine. Another amino acid conjugate, 3-(3-amino-3-carboxypropyl)-6-(3-methyl-2-butenylamino)purine (discadenine), has been isolated as a spore germination inhibitor from a cellular slime mold *Dictyostelium discoideum*. A detailed summary of naturally occurring cytokinins including these conjugates has been given by Letham and Palni,[437] and also by Koshimizu and Iwamura[438] including those found in tRNA.

The cytokinins with nonisoprenoid side chains, benzyladenine, its 9-riboside and 9-ribosyl-N^6-(*o*-hydroxybenzyl)adenine occur naturally,[446–448] and may constitute the second structural class of natural cytokinins.

8.02.4.2.3 Synthesis

The first synthesis of cytokinins has been achieved by reacting 6-methylthiopurine with an appropriate amine. The reaction is normally carried out in a sealed tube. The 6-methylthio group is more susceptible to nucleophilic attack than the 2-methylthio group. Thus, the reaction of 2,6-dimethylthiopurine with 4-hydroxy-3-methylbut-2-enylamine or isopentenylamine affords 2-methylthiozeatin or 2-methylthio-ipA. 6-Methylthio analogues are less reactive than the corresponding 6-chloro derivatives, especially at a riboside level, and 9-ribosylzeatin has been prepared from 6-chloro-9-β-D-ribofuranosylpurine, without recourse to a sealed tube reaction.[449] The coupling of 6-chloropurine derivatives with an amine is most frequently used for the preparation of N^6-substituted adenine derivatives. This method was, however, inadequate for producing 2-methylthio-9-ribosylzeatin from 6-chloro-2-methylthio-9-β-D-ribofuranosylpurine. An alternative route to ribosides is to couple an appropriate amine with 2,6-dichloro-9-(2,3,5-triacetyl-β-D-ribofuranosyl)purine that is the product of a fusion reaction between 2,6-dichloropurine and fully acetylated ribofuranose and, after deblocking, treat the product with sodium methylmercaptide.[450] Zeatin ribotide has also been prepared by this method from 6-chloropurine ribotide, in addition to a conventional phosphorylation of the precursor isopropylidene riboside.[451] The rearrangement reaction of 1-alkyladenine derivatives to N^6-alkyladenines has been utilized for the preparation of 9-ribosyl-ipA 5′-phosphates.[452]

9-β-D-Glucopyranosylzeatin, identified as a metabolite of zeatin, has been prepared from 9-β-D-glucopyranosyl-6-chloropurine by the coupling reaction. Direct glycosylation reaction of adenine base gives a mixture of positional isomers, and thus 7-β-D-glucosylzeatin, another metabolite, has been prepared by ring closure of *N*-glucosylimidazole precursor so as to give unequivocally the 7-glucoside.[453] Direct coupling of acetobromoglucose with zeatin to obtain *O*-β-D-glucopyranosylzeatin was reportedly unsuccessful, probably because of the low solubility of zeatin in solvents and the ease of formation of *N*-glucoside. The compound was synthesized by the coupling of 6-chloropurine with *O*-glucosylated 4-hydroxy-3-methylbut-2-enylamine.[454]

The synthesis of a novel alanine conjugate of zeatin, lupinic acid, involved Michael addition of 2-trifluoroacetylaminoprop-2-enoate, which was prepared by *N*-trifluoroacetylation of 2-chloroaniline and following dehydrochlorination, to 6-chloropurine, followed by selective hydrolysis of the methyl ester, and the coupling of 6-chloropurin-9-ylalanine with the side chain amine.[454] The synthesis of discadenine, another amino acid conjugate of the cytokinin ipA involves direct alkylation of the cytokinin by α-phthalimido-α-bromobutyrate followed by deblocking.[455]

Shaw[456] has reviewed the synthesis of naturally occurring cytokinins and Koshimizu and Iwamura[438] have reviewed synthetic cytokinins and anticytokinins.

8.02.4.3 Biological Activity

8.02.4.3.1 Mechanisms of stimulation of greening

The marked effect of cytokinins on greening of etiolated tissues has been well documented. Pretreatment of etiolated cytoledons with cytokinins not only eliminates the lag period of Chl formation, but also stimulates the rate of accumulation of Chl after exposure of the cotyledons to light.[422] Synthesis of 5-aminolevulinic acid (ALA) is the rate-limiting step in the biosynthesis of Chl. Fletcher *et al.*[457] showed that when cucumber cotyledons were treated with benzyladenine, the rate of synthesis of ALA doubled both in light and darkness. Advances in understanding the biochemistry of Chl synthesis have greatly enhanced the study of the action of cytokinins on Chl synthesis. ALA synthesis by the C_5 pathway consists of the following three steps[458,459] (Scheme 6, after Kannangara *et al.*)[459] (i) Glutamate is converted to glutamyl-tRNA by the same reaction as the first step of protein synthesis. (ii) The glutamate bound to tRNA is reduced to glutamate 1-semialdehyde (GSA) with NADPH by glutamyl-tRNA reductase. (iii) GSA is transaminated to yield ALA. Which of the three reactions is stimulated by cytokinins? With cucumber cotyledons, Masuda *et al.*[460] showed that the activity of GSA aminotransferase in the third reaction was unchanged by benzyladenine treatment. Benzyladenine stimulates the accumulation of GSA in the presence of gabaculine, a potent inhibitor of the third reaction, to the same degree as that of ALA in the absence of the inhibitor.[461] These observations indicate that benzyladenine should act prior to the third reaction. Benzyladenine stimulates the first reaction by increasing the levels of $tRNA^{Glu}$, without any increase in the activity of glutamyl-tRNA synthetase nor the levels of glutamate.[462] However, the effect on the $tRNA^{Glu}$ level is not specific. The same extent of stimulatory effect on the levels of $tRNA^{Phe}$ was observed.[463] Benzyladenine doubles the glutamyl-tRNA reductase activity in the second reaction by increasing the levels of mRNA for this enzyme (Figure 6). Either in benzyladenine-treated cotyledons or untreated controls, the activity of this enzyme has been shown to be stoichiometrically equivalent to that of the overall ALA synthesis,[460] suggesting that this step is rate-limiting. Thus benzyladenine promotes ALA synthesis through the stimulation of the second reaction, which is the first committed step of porphyrin biosynthesis.

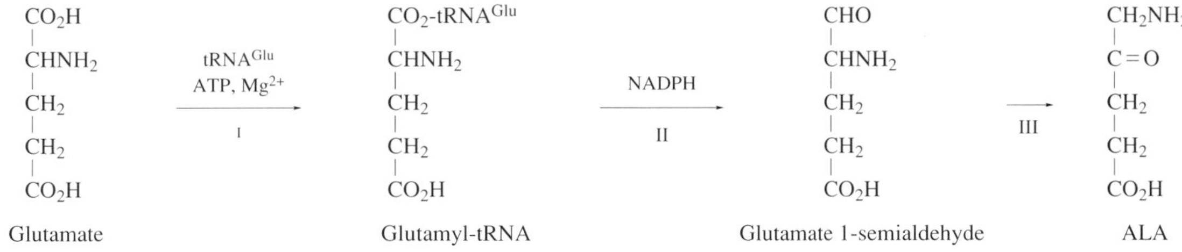

$$\underset{\text{Glutamate}}{HO_2C{-}CH(NH_2){-}CH_2{-}CH_2{-}CO_2H} \xrightarrow[\text{I}]{tRNA^{Glu},\ ATP,\ Mg^{2+}} \underset{\text{Glutamyl-tRNA}}{tRNA^{Glu}{-}O_2C{-}CH(NH_2){-}CH_2{-}CH_2{-}CO_2H} \xrightarrow[\text{II}]{NADPH} \underset{\text{Glutamate 1-semialdehyde}}{OHC{-}CH(NH_2){-}CH_2{-}CH_2{-}CO_2H} \xrightarrow[\text{III}]{} \underset{\text{ALA}}{H_2NCH_2{-}C(=O){-}CH_2{-}CH_2{-}CO_2H}$$

Scheme 6

8.02.4.3.2 *Transformation with* ipt *gene*

A number of workers have generated transgenic plants with increased levels of cytokinins, in an attempt to examine the physiological changes induced by these conditions.[439] The *ipt* (isopentenyl transferase gene) sequence of *A. tumefaciens* was inserted in plant cells, either under the control of its own promoter or behind other types of plant promoters. However, these transgenic plants do not always show the same response as normal plants, presumably due to the changes in sensitivity to cytokinins of the cells in transgenic plants which have grown under high levels of endogenous cytokinins. In an attempt to analyze the effect of a localized increase in cytokinins, the *ipt* coding sequence has been fused to promoters of heat shock genes (*hsp*).[464–467] Estruch *et al.*[470] placed a transposon *Ac* between a 35S cauliflower mosaic virus promoter and the *ipt* coding region. The results indicate that transposition of *Ac* away from the 35S-*Ac*-*ipt* construct causes expression of *ipt* and cytokinin production, which could then induce adventitious bud formation in cells that had not been activated by *Ac* transposition. Li *et al.*[469] fused the *ipt* coding sequence to one end of the bidirectional small-auxin-up-regulated (SAUR) promoter and the β-glucuronidase (GUS) gene to the other end, and introduced it into tobacco plants. The transformed plants showed a variety of phenotypes typically associated with increased cytokinin levels throughout the plants, though certain phenotypes indicated a local response to localized increase in cytokinins such as numerous adventitious buds forming over the petioles and leaf veins.

8.02.4.3.3 *Cytokinin-binding proteins*

Although a number of reports on cytokinin-binding proteins have appeared, characterization of cytokinin receptors is in its infancy. Among the earliest reports on cytokinin-binding proteins from wheat germ,[470,471] the best characterized was CBF-1, which has a similar amino acid sequence to vicilin-type seed storage protein.[472] This protein has a reasonably high affinity ($K_d = 10^{-7}$ M) for

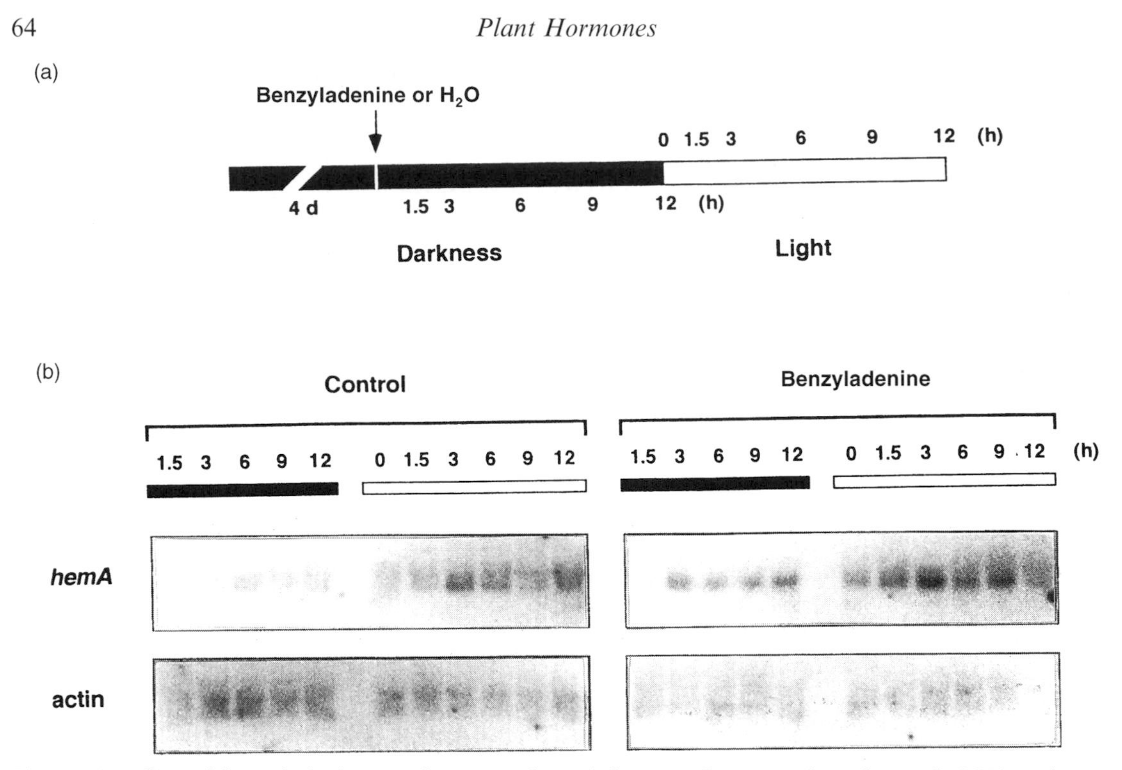

Figure 6 Effect of benzyladenine on the expression of the gene for cucumber glutamyl-tRNA reductase, *hemA*.[460] (a) A schematic illustration of the various treatments. Etiolated cucumber cotyledons were incubated with 50 mM benzyladenine or water (control) in darkness and then illuminated. Samples were harvested after the indicated times. (b) Autoradiogram showing results of RNA gel blot analysis. Hydbridization was carried out using *hemA* cDNA as the probe, with pea actin cDNA as a reference.

active cytokinins, but much lower affinity for nonactive analogues. Photoaffinity labeling using benzyladenine derivatives showed that a single His residue of CBF-1 was labeled, indicating a very specific interaction of cytokinin with the protein. The physiological significance of storage proteins as cytokinin-binding proteins is not clear.

A cytokinin-binding protein having a molecular mass of 40–45 kDa and a K_d of 1.5×10^{-8} M for zeatin was purified from barley leaves.[473] Similar binding proteins were later found in wheat, maize, and rye.[474] Momotani and Tsuji[475] purified a water soluble cytokinin-binding protein from tobacco leaves by successive affinity chromatography. The predominant protein in this fraction has an apparent molecular mass of 31 kDa. Using equilibrium dialysis, K_d for benzyladenine was found to be 10^{-7} M, and binding of benzyladenine to this protein was blocked by other cytokinins, but not by adenine or ATP.

Nagata *et al.*[476] used tritiated versions of diphenyl urea derivatives to study cytokinin-binding proteins from etiolated mung bean seedlings. A partially purified protein that was isolated had a molecular mass of about 21 kDa with a K_d of 10^{-8}–10^{-9} M for zeatin or benzyladenine. Either free diphenylurea-type molecule or the N^6-substituted adenyl cytokinin bases competed efficiently in binding assays, but the cytokinin ribosides did not.

Not many of the reported cytokinin-binding proteins are functionally well characterized. However, there are a number of reports describing the effects of barley leaf cytokinin-binding proteins on transcriptional regulation.[477–485] Kulaeva *et al.*[484] purified a protein with cytokinin-binding properties from cytosol of barley leaves by anti-idiotype antibodies which were isolated from antibenzyladenine serum by chromatography on antibenzyladenine antibody-Sepharose. They showed that this protein activated *in vitro* RNA synthesis in the presence of benzyladenine.

In 1993, Mitsui *et al.*[486] isolated cDNAs encoding a 57 kDa cytokinin-binding protein from a tobacco cDNA library. Sequence analysis revealed significant homology between this protein and *S*-adenosyl-L-homocysteine hydrolase from other organisms. The enzyme catalyzes the reversible hydrolysis of *S*-adenosyl-L-homocysteine which is one of the products of methyl-transfer reactions from *S*-adenosyl-L-methionine (SAM) and is a competitive inhibitor of all SAM-dependent reactions.

The effects of exogenously applied cytokinins on plant respiration have been studied in great detail.[487] Some cytokinins, especially benzyladenine, are known inhibitors of alternative (cyanide-resistant) electron transport in isolated plant mitochondria[488,489] and intact tissues.[490] Brinegar *et al.*[491] detected two polypeptides with molecular masses of 57 kDa and 32 kDa by photoaffinity labeling of purified mung bean mitochondria using tritiated 2-azido-N^6-benzyladenine. The former is assumed to be the α and/or β subunit of F_1ATPase, and the latter appears to be an integral membrane protein.

Although many cytokinin-binding proteins have been isolated, it has been difficult to establish which of these are true receptor proteins, because they are not fully characterized in their function. Also, it has not been determined what orders of magnitude of K_d are required for cytokinin-binding proteins, because there is no information about the actual concentration at the site of cytokinin action in the cell, though endenogenous hormone levels usually are presented on a fresh weight basis. Furthermore, immunocytochemical localization has shown that levels of cytokinin can vary dramatically between cell and tissue types.[492,493]

8.02.4.3.4 *Analysis of signal transduction pathway*

Alteration in the levels of cytokinins is not the only type of mutation that could show phenotypes indicative of changes in the cytokinin physiology of the plant. Another type of mutants are those that are insensitive to exogenously supplied cytokinins, or that act as though cytokinins are always present at high levels. Such mutants are expected to carry alterations in some components of the signal transduction pathway, including the hormone receptors. Therefore, a successful genetic approach to studying cytokinin signal transduction is to use these mutants. For the analysis of signal transduction, it is interesting to note that mRNA accumulation from cytokinin-inducible genes can also be caused by a variety of stresses.[494]

Most cytokinin-habituated clones of *N. tabacum* cv. Havana can be readily reversed by forming complete plants. Cultures of pith tissue from such regenerated plants required cytokinin for continuous growth.[495,496] Therefore, the habituation was not mutation, but represented an epigenetic state. However, some habituated clones have been reported to be genetic mutations. The physiological basis of habituation is not known. Habituated clones consist of two classes: (i) cytokinin-overproducing ones; and (ii) cytokinin-response ones. The latter can be used for cytokinin signal transduction analysis. However, the cytokinin-insensitive mutants have two possibilities: (i) Steps in a cytokinin-mediated cell cycle control system may be constitutively activated, while other activities under cytokinin control are unaffected. (ii) Cytokinin-mediated controls necessary to advance the cell cycle may be replaced by other factors which would be produced in the cytokinin habituation process.

Cytokinin-resistant mutants have been isolated in *Nicotiana* spp., *Arabidopsis* spp., and *Physcomitrella* spp. (moss).[497] *Nicotiana plumbaginifolia* mutants were selected by their ability to survive and exhibit a somewhat normal phenotype in the presence of 20 mM benzyladenine.[498] Ethyl methanesulfonate-mutagenized M2 seeds of *A. thaliana* were germinated in the presence of a dose of cytokinin (2.5 mM) that would express a cytokinin root syndrome, i.e., inhibition of primary root elongation and stimulation of root hair elongation, in the wild type. Seedlings with roots that appeared normal were selected. Thus cytokinin-resistant mutants *ckr1* were isolated.[499] However, later studies have shown that these mutants were not resistant to cytokinins, but to ethylene which is produced by cytokinin treatment.[500] Similarly, *ckr1* mutants from tobacco were not cytokinin-resistant, but were those defective in abscisic acid biosynthesis.[498,501] In 1995 a cytokinin-resistant mutant *cyr1* of *Arabidopsis* with abbreviated shoot development was isolated.[502] Cytokinin-resistant mutants isolated from *Physcomitrella patens* exhibit phenotypes of aberrant morphology and little gametophore production in the absence of exogenous cytokinin.[503] Kakimoto[504] has isolated, by activation T-DNA (transferred DNA) tagging, *Arabidopsis* mutants that exhibit typical cytokinin responses including rapid cell division and shoot formation in tissue culture in the absence of exogenous cytokinin. The tagged gene, *CKI1* (cytokinin-independent) encodes a protein similar to the two-component regulators, suggesting that *CKI1* is involved in cytokinin signal transduction, possibly as a cytokinin receptor.

Cytokinins have long been known to replace the effects of light in a number of plant developmental processes. The overlapping role of cytokinin and light raises the interesting question of whether cytokinins and light act independently to affect developmental processes or whether cytokinins are involved in the sequence of events initiated by physiologically active photoreceptors. Sano and Youssefian[505] have shown that the level of mRNA for the wheat *wpk4* gene encoding a putative protein kinase increases in response to light, nutrient deprivation, and cytokinins. In the presence of a cytokinin antagonist, however, the *wpk4* mRNA level is not increased by either light or nutrient deprivation, indicating that the light and nutrient signals that induce *wpk4* mRNA accumulation may be mediated through cytokinins. Chory *et al.*[506] isolated a series of *det* (de-etiolated) mutants of *A. thaliana*, which show many characteristics of light-grown plants even when grown in complete darkness. Therefore, wild-type *DET* genes may block some steps in the greening process. Chory *et al.*[507] presented a model in which cytokinins act to decrease the activity of DET1 or DET2, which then permit the elaboration of the downstream light-regulated process of greening. This model could be related to cytokinin-repressed genes which will be described below. *DET2* has been identified as the gene encoding a reductase involved in the brassinosteroid biosynthetic pathway.[508]

8.02.4.3.5 Regulation of gene expression

Modulation of gene expression by cytokinins has been reported by many authors. For example, cytokinins induce the accumulation of mRNA for the large and small subunits of RuBisCO (ribulose-1,5-biphosphate carboxylase/oxygenase)[509–511] and the light harvesting Chl *a*/*b* protein[510–512] in etiolated cotyledons of cucumber and pumpkin or tobacco cell cultures, as well as in *Lemna gibba* cultured in the dark.[513] Benzyladenine stimulates light induction of nitrate reductase in excised barley leaves in the presence of nitrate.[514] Nuclear run-on studies have shown that enhancement of the accumulation of mRNAs for hydroxypyruvate reductase gene and the *Sesbania rostrata* early nodulin gene *SrEnod2* by cytokinins are at the levels of transcription[515] and posttranscription,[516] respectively.

Various reports describe changes in gene expression after 12 h to several days of treatment with cytokinins,[509–513] or under conditions of constant high levels of a cytokinin after transformation with the *ipt* gene from *A. tumefaciens*.[517] To understand the primary and immediate action of the hormone, however, information is needed about changes in gene expression which occur shortly after application of the hormone. A few reports have dealt with fairly rapid changes that occur within several hours of application of a cytokinin. Crowell *et al.*[518] and Crowell and Amasino[519] isolated cDNAs for mRNAs that showed a rapid increase or decrease in abundance within 4 h after addition of zeatin to cultured soybean cells starved of cytokinin, though these genes responded to auxin as well. Dominov *et al.*[520] isolated a cDNA for a mRNA which accumulated in cultured tobacco cells within 5 h of benzyladenine treatment. Benzyladenine rapidly increases levels of mRNA for the above-mentioned *wpk4* in wheat seedlings.[505] Zeatin also causes a large and rapid increase in phosphoenolpyruvate carboxylase and carbonic anhydrase mRNA levels.[521] Toyama *et al.*[522] have shown that mRNA for *psaL* which encodes a subunit of photosystem I is increased within 2 h of treating etiolated cucumber cotyledons with benzyladenine. Crowell and Amasino[519] isolated the cDNA for a cytokinin-repressed mRNA which exhibited homology to a gene for iron superoxide dismutase. The level of the mRNA for phytochrome in cucumber cotyledons is decreased by 2 h of benzyladenine treatment, as well as by red light.[523] Chen *et al.*[524] showed that the abundance of some translatable mRNAs in excised pumpkin cotyledons changed 60 min after application of benzyladenine. Nuclear run-on assays have shown that the rate of transcription of the hydroxypyruvate reductase gene was enhanced within 60 min of treatment of etiolated pumpkin cotyledons with benzyladenine.[515] Kinetin also stimulates the rate of rRNA transcription in *A. thaliana* within 60 min of treatment.[525]

Teramoto *et al.*[526] showed that benzyladenine caused changes in the levels of 16 mRNAs within 6 h in etiolated cotyledons of cucumber, three of the mRNAs being markedly repressed. They isolated by differential screening two cDNAs for cytokinin-repressed genes, *CR9*[527] and *CR20*[528], that are repressed within 2–4 h after application of benzyladenine to etiolated cucumber cotyledons (Figure 7). Expression of these two genes could be closely related to the action of endogenous cytokinins.[528] However, the pattern of diurnal changes in expression of the two genes are opposite to each other; the transcripts of *CR9* increase in the light period, while those of *CR20* increase in the dark period.[528] *CR9* shows a high homology with *lir1*, a light-induced gene from rice. *CR20* has no open reading frame long enough to encode a polypeptide, but contains a unique sequence which could form branched stems, suggesting that it may work as a noncoding functional RNA.[529] Cytokinins would initiate cotyledon growth and its related processes by repressing the expression of these genes which may block the above processes.[530] Using large-scale differential hybridization, Toyama *et al.*[531] isolated 86 cDNAs for mRNAs which decreased within 2 h of benzyladenine

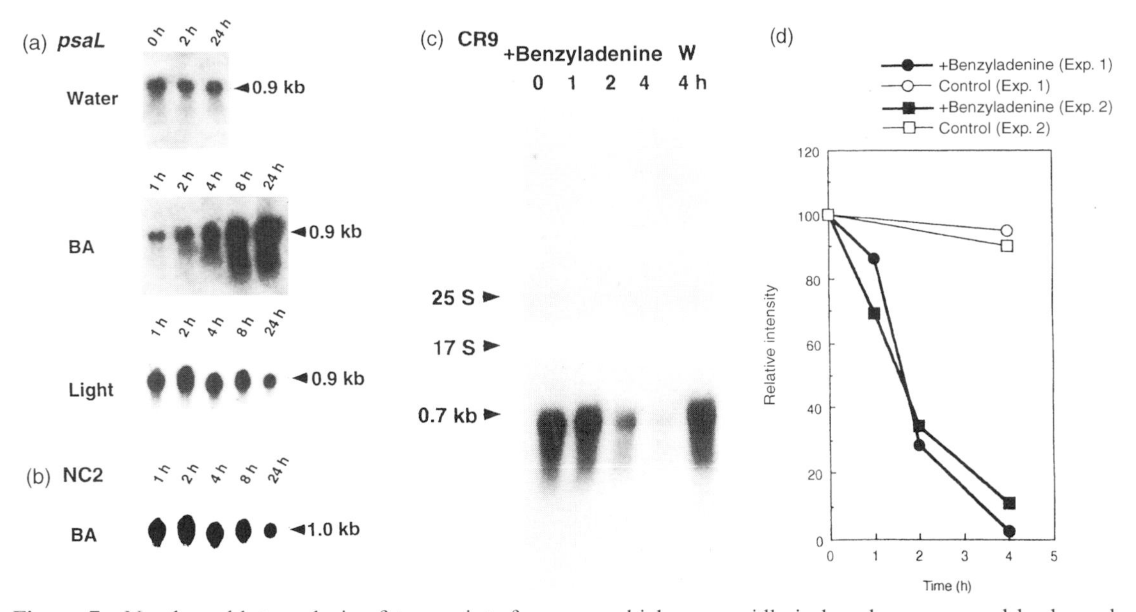

Figure 7 Northern blot analysis of transcripts for genes which are rapidly induced or repressed by benzyladenine. (a) Changes in the levels of mRNA for *psaL*, a rapidly cytokinin-induced gene, in cucumber cotyledons during treatment with benzyladenine or water.[522] Etiolated cotyledons were incubated with 10 mM benzyladenine in darkness (BA) or with water in the light (light) for the indicated times. Controls were incubated with water in the dark (water). Northern hydridization was carried out with *psaL* cDNA as a probe. (b) Changes in *NC2* mRNA levels under the same conditions as benzyladenine in (a). cDNA for *NC2*, whose mRNA level is unchanged by benzyladenine treatment, was used as a reference in Northern blot analysis. (c) Changes in the levels of mRNA for *CR9*, a rapidly cytokinin-repressed gene.[527] Etiolated cucumber cotyledons were incubated with 10 mM benzyladenine for the indicated times in darkness. Controls (W) were incubated with water for 4 h. Total RNA was analyzed by Northern blotting with *CR9* cDNA as a probe. (d) Band intensities were quantified by densitometry.

treatment. Some of them are homologous to those for catalase, 3-hydroxy-3-methylglutaryl CoA reductase and lectin.

Determining the position of these genes in the pathway from cytokinin signal perception to the physiological activities reported so far will be a crucial area for research.

8.02.4.3.6 Anticytokinins

Anticytokinins have been developed based on the activity assayed with tobacco callus derived from cytokinin-dependent *N. tabacum* L. cv Wisconsin No. 38. They are divided into two classes, adenylate and nonadenylate anticytokinins.[532] The former class of compounds has an immediate structural analogy to N^6-substituted adenine cytokinins; they are: N^7-substituted 7-amino-3-methylpyrazolo[4,3-*d*]pyrimidines; N^4-substituted 4-amino-7-(β-D-ribofuranosyl)-, N^4-substituted 4-amino-2-methylthio-, and N^4-substituted 4-amino-2-methylpyrrolo[2,3-*d*]pyrimidines; and N^4-substituted 4-amino-2-methylthiopyrido[2,3-*d*]pyrimidines (Figure 8). Among these, 2-methylpyrrolo[2,3-*d*] pyrimidines and 2-methylthiopyrido[2,3-*d*]pyrimidines are the compounds whose antagonistic natures have been established kinetically by the method of Lineweaver and Burk.[533] Structurally analogous 4-alkoxy-2-methylthiopyrrolo[2,3-*d*]pyrimidines have been prepared and reported to retard the action of cytokinins in tobacco callus assay, but their antagonistic nature has not been explicitly examined. *s*-Triazine anticytokinins and carbamate anticytokinins are the compounds of nonadenylate structures, and their antagonistic natures have been established kinetically.[532,534]

Antagonists are of potential value in studies of the mechanism of action of biologically active compounds, being used to block the function of the agonists and thus extending the studies of their action. A pyrazolo[4,3-*d*]pyrimidine anticytokinin was tested in cytokinin-autonomous tobacco callus.[535] The compound inhibited the growth of the callus, but the inhibition was reversed by added cytokinin. The result was interpreted to show that cytokinin autonomous tissue uses endogenous cytokinins. Cytokinins, as well as cAMP and traumatic acid, promote meristematic divisions of

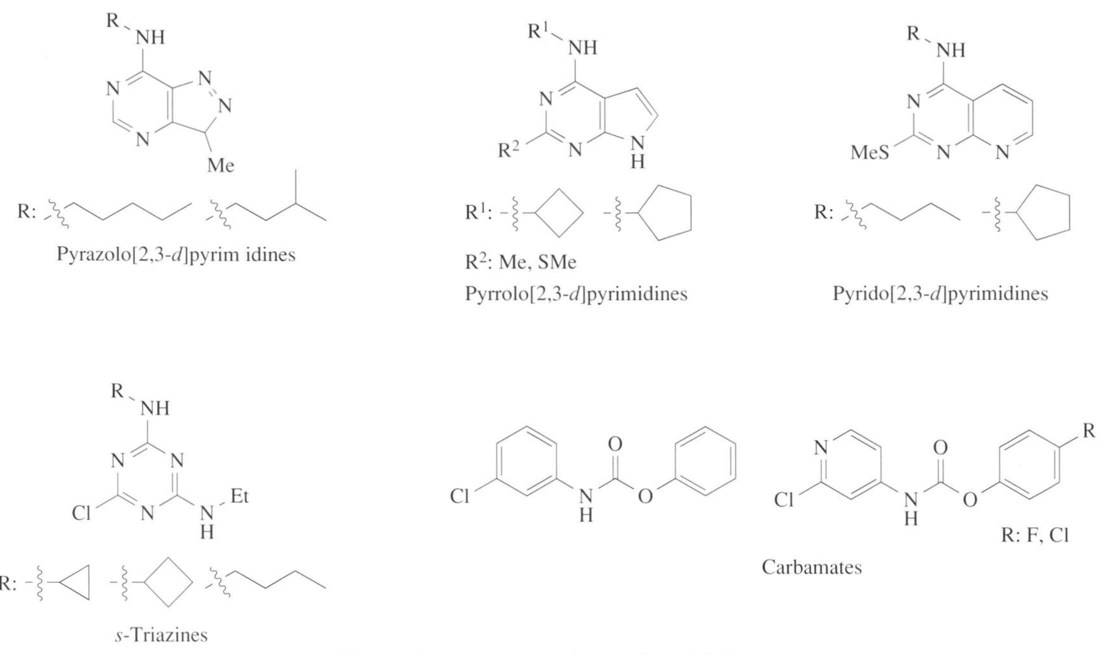

Figure 8 Representative anticytokinins.

epidermis of *Torenia* stem segments to form meristematic zones which develop into adventitious buds without further addition of growth regulators. 6-Cyclobutylamino-2-methylpyrrolo[2,3-*d*]pyrimidine, an anticytokinin, caused 50–80% inhibition of the meristematic zone formation when applied in the presence of benzyladenine, but did not inhibit the zone formation caused by cAMP and traumatic acid.[536,537] This led the researchers to propose that the mechanisms that promote meristematic zone formation by cAMP and traumatic acid are separate from that of cytokinins. Cytokinins induce adventitious shoots in horseradish (*Armoracia rusticana*) hairy roots in the dark, and enhance the number of shoots in the presence of light. A series of 2-methylpyrrolo[2,3-*d*]pyrimidine anticytokinins partially inhibited the shoot formation in the light, and this inhibition was overcome by the application of benzyladenine, suggesting that light irradiation may cause an increase of endogenous level of cytokinins.[538] *s*-Triazine anticytokinin effectively inhibited the increase in *wpk4* mRNA levels induced by light deprivation or nutrient deprivation, indicating that these signals are mediated by cytokinins.[505] When tobacco plants were wounded in the presence of benzyladenine, jasmonic and salicylic acids were accumulated, and this effect was erased by the application of a *s*-triazine anticytokinin.[539] Based on this experiment and those with transgenic plants, it was concluded that cytokinins are indispensable for control of endogenous levels of jasmonic and salicylic acids, putative signaling molecules of wounding and pathogen infection.

N^6-Isopentenyladenine and 2,3,5-triiodobenzoic acid, an antiauxin, promoted tillering in barley seedlings, and 4-chloro-2-cyclobutylamino-6-ethylamino-*s*-triazine, an anticytokinin, and α-naphthaleneacetic acid, an auxin, inhibited it.[540] The results suggest that the auxin/cytokinin interaction observed for the lateral bud formation in dicotyledonous plants operates similarly. Both adenylate and urea cytokinins stimulated growth and polyamine content of green and etiolated radish (*Raphanus sativa*) cotyledons, and this promotion was partially removed by subsequent application of 2-methylthiopyrrolo[2,3-*d*]pyrimidine or carbamate anticytokinin.[541] Lateral bud growth stimulation of zeatin in *Lycopersicon esculentum* seedlings was not overcome by the application of 3-methylpyrazolo[4,3-*d*]pyrimidine and 2-methylthiopyrrolo[2,3-*d*]pyrimidine anticytokinins 24 h after cytokinin application. However, a simultaneous application of anticytokinin and cytokinin or anticytokinin treatment followed by cytokinin application caused a significant blocking of bud growth.[542] These results were thought to indicate that the timing of application is important in determining the nature of responses and the involvement of cytokinin in cotyledonal bud growth. Tracheal element differentiation in isolated mesophyll cells of *Zinnia elegans* in a medium containing auxin and cytokinins was inhibited or delayed if the medium was supplemented with an anticytokinin, antiauxin, or auxin transport inhibitor.[543] The agonist/antagonist interaction appears to operate in this plant system as well.

Cytokinins induce the formation of betacyanins in the cotyledons and hypocotyls of derooted seedlings of *Amaranthus caudatus*. A pyrrolo[2,3-*d*]pyrimidine anticytokinin, 6-cyclopentylamino-2-methylpyrrolo[2,3-*d*]pyrimidine, was also found to induce the pigments, but 6-cyclohexylamino and 6-isobutylamino analogues did not.[544] The anticytokinin-active cyclohexylamino compound caused germination of positive-photoblastic seeds of *Lactuca sativa* in the dark as well as N^6-adenylate cytokinins.[542] These cases show that anticytokinins developed based on the assay with tobacco callus do not necessarily act in the same way in other systems, and also indicate species differences and complexity of the cytokinin receptors.

s-Triazine and carbamate anticytokinins have been found to cause flowering in seedlings of *Asparagus officinalis*.[534] 2-Chloro-4-cyclohexylamino-6-ethylamino-*s*-triazine, the strongest anticytokinin in the *s*-triazine series in the tobacco assay, induced the highest degree of flowering, but the highly flower-inducing members in the carbamate series were not necessarily strong anticytokinins. Antagonism by benzyladenine could not be observed clearly by virtue of the toxicity of the cytokinins for the growth of seedlings,[545] and thus it is not clear if the anticytokinin nature of the compounds is responsible for the novel flower-inducing activity.

8.02.4.4 Biosynthesis and Metabolism

8.02.4.4.1 Synthesis of ipA ribotide

The biosynthetic pathway of cytokinins has not been shown clearly except for *A. tumefaciens*-induced crown gall tumors. This lack of knowledge stems from the difficult conditions in tracer experiments: the endogenous cytokinins are present at extremely low levels in plant tissues, and the most likely precursor is purine which, however, plays the central role in cellular metabolism.[437] The favored model for cytokinin biosynthesis predicts the addition of dimethylallylpyrophosphate (DMAPP) to the N^6 position of 5′-AMP by an isopentenyl transferase (IPT), yielding ipA ribotide[439] (Scheme 7, pathway modified from Morris[546]). The 5′-AMP is an obligatory substrate, neither adenine nor adenosine are suitable.[547,548] Nucleotides of cytokinins are of special significance in cytokinin metabolism:[437] they appear to be the initial products of cytokinin biosynthesis,[549] and the first metabolites formed in appreciable amounts from exogenous cytokinin bases, suggesting that the ribotide formation associates with cytokinin uptake and transport across the cell membrane.[437] The precursor of DMAPP appears to be mevalonic acid.[437] Some authors[439] define cytokinin metabolism as the conversion of ipA ribotide to any other N^6-substituted adenine cytokinins. N^6-Isopentenyladenine ribotide can be converted to other types of cytokinin ribotides, ribosides, and free bases, including zeatin and dihydrozeatin.

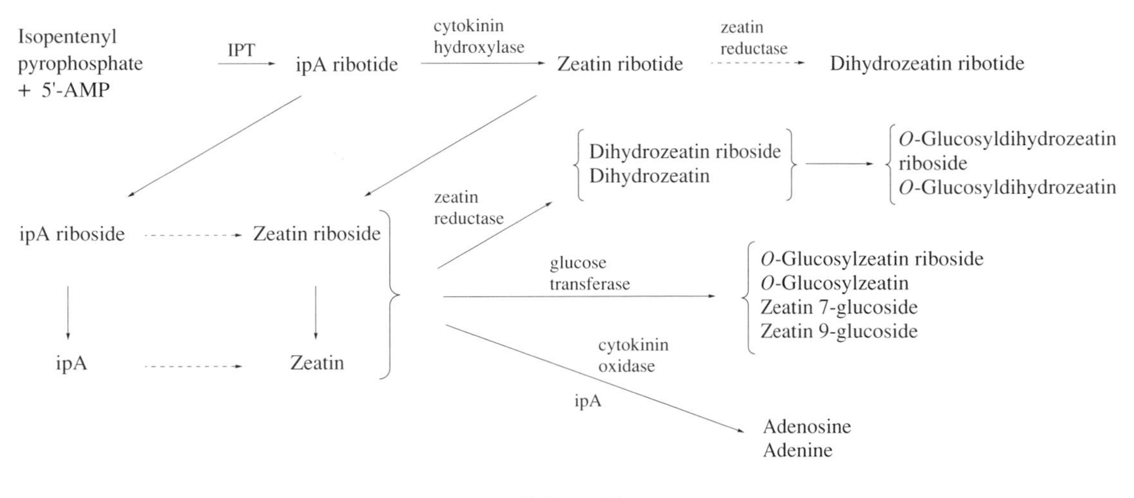

Scheme 7

8.02.4.4.2 tRNA-derived cytokinin molecules

Beside this *de novo* route, cytokinin molecules have been found in certain tRNAs as the base adjacent to the 3′ end of the anticodon,[393,395,550] as described above. Zeatin is associated with plant and bacterial tRNA mainly in the *cis* form.[395,437,551] However, tRNA from pea shoots contains both the *cis* and *trans* forms.[552] Furthermore, substantial amounts of *trans*-zeatin have been found in tRNA from *Vinca rosea* crown gall tumors, and *trans*-zeatin is also the major free cytokinin in this tissue.[553] *cis–trans*-Zeatin isomerase has been partially purified.[554] The presence of this enzyme suggests that tRNA-derived molecules could be brought into the pool of active cytokinins. Even in the absence of the Ti plasmid-encoded cytokinin biosynthetic genes, *ipt* and *tzs*, *A. tumefaciens* strains still release significant amounts of ipA into the culture medium. However, a deletion/insertion mutant of tRNA : isopentenyl transferase gene, *miaA*, no longer releases ipA into the medium.[555] Transfer RNA, therefore, makes a small but significant contribution to the total amount of cytokinin normally secreted by *Agrobacterium* strains. From these results, it cannot be excluded that cytokinin molecules in tRNA could be precursors of free cytokinins. However, the following observations suggest that plants normally use the *de novo* pathway. (i) The levels of free cytokinins in tissues are much higher than those contained in the tRNA, for example, free cytokinins present in pea root tips exceed by a factor of 27 the amount present in the tRNA.[556] (ii) A tRNA turnover rate high enough to accumulate free cytokinins has not been observed.[395,557] (iii) In some organisms, particularly higher plants, certain cytokinins in tRNA do not occur as free compounds and vice versa.[437]

8.02.4.4.3 The ipt *gene*

Chen and Melitz[548] extracted and purified IPT from tobacco cell lines capable of greening in the absence of exogenous cytokinin. In 1974 Miller[558] provided unequivocal evidence that crown gall tumors contained elevated levels of cytokinins. Later studies showed that this increase in cytokinin content is due to activity of the *ipt* gene encoded in the *tmr* locus of the T-DNA from *A. tumefaciens*.[559–564] However, characterization of the enzyme has been achieved mainly through examination of the protein overexpressed in bacteria. In spite of the importance of IPT, more detailed biochemical analysis of this T-DNA-encoded protein has not been carried out. A variety of different bacteria have been shown to contain *ipt* or *ipt*-like type genes.[546,565,566] The characterization of regions of homology between these enzymes, the determination of their active site, and their interaction with substrates could help the development of strategies for gene isolation or the synthesis of inhibitors for cytokinin biosynthesis.

8.02.4.4.4 Interconversion of ribotides, ribosides, and free bases

An important way to regulate the availablility of active cytokinins in plant tissues is through metabolic interconversions. The major conversions are (i) the dephosphorylation and deribosylation of ipA ribotide, (ii) the hydroxylation of the side chain, (iii) the reduction of the side chain, and (iv) the glycosylation of the adenine ring and side chain.

N^6-Isopentenyladenine ribotide undergoes dephosphorylation and deribosylation to yield the riboside and free base, respectively. The interconversions may be caused by such enzymes as adenosine phosphorylase, adenosine kinase, 5′-nucleotidase (5′-ribonucleotide phosphohydrolase), adenosine nucleosidase (adenosine ribohydrolase), and adenine phosphoribosyl transferase (AMP : pyrophosphate phosphoribosyl transferase).[567] A number of observations suggest that ribosides are translocation forms of cytokinins in the xylem,[568] and the free base is one of the active forms of cytokinins. Thus the regulation of the interconversion of cytokinin ribotides, ribosides, and free bases may play a significant role in maintaining adequate levels of active cytokinins in plant cells.[569]

8.02.4.4.5 Hydroxylation of ipA

The N^6-side chain is hydroxylated to yield *trans*-zeatin derivatives (zeatin ribotide, riboside, and *trans*-zeatin) by cytokinin hydroxylase. Inhibitor studies indicate that the reaction is cytochrome P450 dependent.[570] Experiments with labeled precursors showed that *trans*-hydroxylation of isopentenyl-type cytokinins to yield zeatin-type cytokinins occurred principally at the nucleotide level.[571,572]

8.02.4.4.6 Reduction of the side chain

The side chain of zeatin-type cytokinins is reduced to yield dihydrozeatin derivatives (dihydrozeatin ribotide, riboside, and dihydrozeatin). The zeatin reductase was purified from bean embryos, and shown to use NADPH as the cofactor.[573] The major cytokinin conjugate in the stem and petiole of bean seedlings has been identified as dihydrozeatin riboside. Reduction of the side chain of zeatin-type cytokinins appears to serve as a mechanism to protect against excess cytokinin destruction.[574]

8.02.4.4.7 Conjugation with sugars

Cytokinin molecules can be further modified, often by *O*-glycosylation of the side chain of zeatin or dihydrozeatin, and *N*-glycosylation at the 7 and 9 positions of the adenine base. Cytokinin-7-glucosyl transferase uses uridine 5′-diphosphate-glucose (UDPG) as glucose donor.[575] N^3-Glucosides have not yet been found as metabolites of naturally occurring cytokinins but of benzyladenine.[551] Auer and Cohen[576] have identified a new cytokinin conjugate as a benzyladenine disaccharide, 6-benzylamino-9-[*O*-glucopyranosyl-(1→3)-ribofuranosyl]purine.

N-glycosides, which are shown to be very stable and have low biological activity *in vitro*, appear to withdraw cytokinin from the potential pool.[437] As cytokinin oxidase degrades zeatin-N^7-glucoside and zeatin-N^9-glucoside,[577] the metabolic stability of these glucosides in plant tissues may be due to compartmentation or to low levels of the oxidase. Cytokinin glucosides, which are assumed to be storage forms, occur in coconut milk.[551] In contrast to the stability of *N*-glucosides, the *O*-glycosides are deglycosylated readily by β-glucosidase to release free cytokinins. *N*-glucosides are insensitive to this hydrolase.[551] Interconversion between *O*-glycosides and free cytokinins may be involved in homeostatic control of active cytokinin levels.[437,557] Zeatin *O*-glucosyl and *O*-xylosyl transferases were partially purified from embryos of *P. vulgaris*.[578,579] Lightfoot[580] reported that the cDNA clones XZT82 and XZT205, which had been considered candidates for those of zeatin *O*-xylosyl transferase gene, did not encode this enzyme.

8.02.4.4.8 Degradation of cytokinins

Cytokinins can be degraded through the activity of a cytokinin oxidase which removes the isopentenyl side chain, yielding the adenine or adenosine from the cytokinin base or nucleoside, respectively.[437,581] This reaction requires molecular oxygen.[582] Cytokinin oxidase was purified from wheat, maize kernels,[582] and callus tissues of tobacco[583] and bean,[581] and proved to be a glycoprotein. Synthetic cytokinins with an aromatic substituent at position N^6 of the purine ring, such as kinetin and benzyladenine, are resistant to this oxidase.[582] Cytokinin oxidase activity increases 1 h after application of a cytokinin to *Phaseolus* callus tissues,[581] and this induction is inhibited by a urea derivative, thidiazuron.[584] The activity of this enzyme is also increased as a consequence of elevated cytokinin levels which are caused by the derepression of the tetracycline-dependent *ipt* gene transcription in tobacco calli and plants transformed with *ipt*.[585] Auxin enhances the activity of cytokinin oxidase which is purified from tobacco callus tissues transformed with *ipt* from the T-DNA.[572] This enhancement could contribute to the reduction in cytokinin levels which have been elevated in the transgenic tissues.

8.02.4.4.9 Regulation of cytokinin metabolism

Environmental factors and chemical stimuli appear to alter biosynthesis, metabolism, and translocation of cytokinins in plant tissues[568] and thus influence development.[538,586] There have been numerous reports showing a correlation between various steady state levels of specific cytokinin metabolites and specific responses of the plant or plant tissue. Leaves of derooted and debudded bean seedlings rapidly transform [8-^{14}C]zeatin into zeatin-*O*-glucoside which is subsequently degraded. However, rooted debudded seedlings yield mainly zeatin riboside phosphates.[587] The regulation of hydroxylation of ipA to yield *trans*-zeatin is critical in sex determination of *Mercuralis annua*. Treatment of male plants with *trans*-zeatin resulted in female flowers.[588–590]

8.02.4.4.10 Enzymes involved in cytokinin metabolism

The active cytokinins have not been identified clearly, mainly because mutants or inhibitors that block particular metabolic steps are not available. Although there are reports on crude activities for virtually all of the major steps in cytokinin metabolism[437,557] few of these enzymes have been purified in sufficient quantities to be used in antibody production or protein sequencing. A monoclonal antibody raised against the purified zeatin-*O*-xylosyl transferase[591] has been used to identify a λgt22 cDNA clone containing an insert proposed to encode the enzyme.[592]

The *rolC* (locus for root induction) gene of the T-DNA from *Agrobacterium rhizogenes* involved in the evocation of hairy root disease has been used to alter cytokinin metabolism. Transgenic plants overexpressing *rolC* exhibited a variety of phenotypes suggestive of cytokinin effects.[593,594] The *rolC* protein produced by overexpression of *rolC* in *E. coli* showed a β-glucosidase activity that could cleave 7- and 9-linked glucose residues from zeatin and other cytokinin glucosides.[595] However, inconsistent results[593,594] have been reported as to whether active cytokinin pools in the transgenic plants are increased as a result of the expected activity of the enzyme.

8.02.5 ABSCISIC ACID

8.02.5.1 Brief History

Abscisic acid (ABA, **3**), was the fifth plant hormone to be discovered, after auxin gibberellin, cytokinin, and ethylene. ABA was discovered by three independent lines of research.[596] Immature cotton fruits easily abscise from shoots several days after anthesis. This physiological phenomenon known as abortion allows the plant to produce excellent seeds but is a serious problem for cotton cultivators. Carns found that an auxin inhibitor was involved in the abscission.[597] It was purified from hexane extracts of cotton burs using the abscission test of cotyledonal nodes in cotton explants in 1961, and was called "abscisin" after the abscission of cotton fruits.[598] However, Ohkuma and other members of Addicott's group failed to isolate the inhibitor from the hexane extracts for structure elucidation.[599] Later, they isolated a new inhibitor "abscisin II" which showed a high activity in the abscission test from acetone extracts of immature cotton fruits. The basic structure of "abscisin II" was reported by Ohkuma *et al.* in 1965,[600] and Cornforth *et al.* soon confirmed its structure by synthesizing it.[601] It is not clear whether the first inhibitor "abscisin" was the same compound as "abscisin II." The apical flowers of inflorescences of yellow lupins also abort like cotton fruits. Van Steveninck suggested that an auxin inhibitor which was produced in the basal immature seeds promoted the abortion.[602] This inhibitor was identified as the same compound as "abscisin II" by Koshimizu *et al.* in 1966.[603] Phillips and Wareing suggested that the dormancy of terminal buds of sycamore maple was regulated by a growth inhibitor which was formed in the leaves.[604] This inhibitor was named "dormin," and purified from the sycamore maple leaves by Wareing *et al.* in 1964.[605] The structure of "dormin" was found to be the same as "abscisin II" by Cornforth *et al.* in 1965.[606] The two names, "abscisin II" and "dormin," were unified to "abscisic acid" and the abbreviation "ABA" was agreed at the Sixth International Congress on the Plant Growth Substances in 1967.[607]

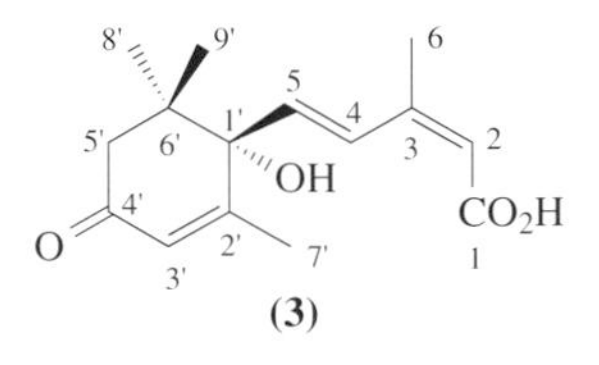

(3)

ABA has an asymmetric carbon at the C-1′, and natural ABA is a (+)-isomer. Cornforth *et al.* applied Mills's rule to 1′,4′-*cis*-diol (**4**) and 1′,4′-*trans*-diol (**5**) of ABA converted from (+)-ABA, and proposed that the absolute configuration of (+)-ABA is *R*.[608] (The absolute configuration proposed by Cornforth *et al.* in 1967 was *S* according to the old definition of *R*/*S*, but since the definition of *R*/*S* was changed in 1966,[609] the new definition is used to avoid confusion). Doubt was cast on the absolute configuration by Taylor and Burden in 1972 who showed a conversion of violaxanthin (**6**) to (+)-ABA via xanthoxin (**7**) and ABA-aldehyde (**8**) as shown in Scheme 8.[610] They suggested that the configuration of the C-1′ of (+)-ABA is *S* since the stereochemistry of C-6 of violaxanthin had been elucidated by X-ray analysis to be *S*. The degradation of (+)-ABA

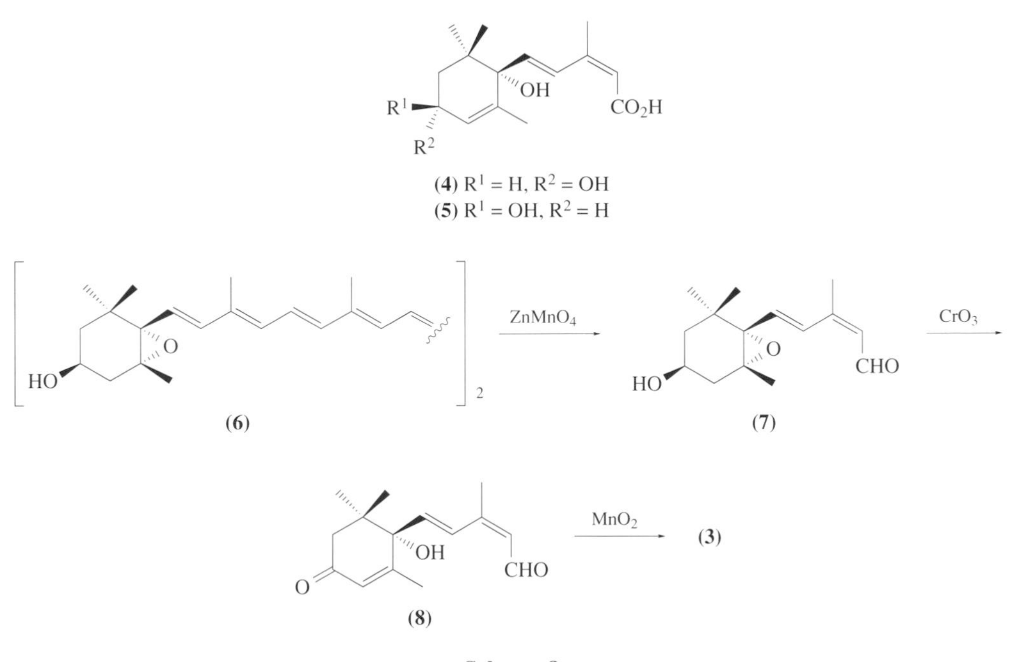

Scheme 8

to a derivative of dimethylmalic acid,[611] the exciton chirality method,[612] and the chiral synthesis of (+)-ABA,[613,614] supported Taylor and Burden's proposal,[610] and the absolute configuration of the C-1′ of (+)-ABA was finally concluded to be *S*.

ABA occurs in all higher plants, ferns, mosses, liverworts, and algae. In early studies, ABA was not detected in liverworts, and Pryce had proposed that lunularic acid (**9**), which has a structure similar to that of ABA, plays an ABA-like physiological role in liverworts.[615] However, subsequent work using gas chromatography–selected ion monitoring (GC–SIM) analysis showed that two species of liverworts contained ABA.[616] As a result of the universal distribution and many physiological studies ABA has been classified as a plant hormone.[596,617] More than 20 species of phytopathogenic fungi including *Cercospora* and *Botrytis* are known to produce ABA in culture media.[618,619] Since a strain of *Botrytis cinerea* which does not produce ABA is virulent to tomato plants,[620] ABA would not be involved in the infection ability of the phytopathogenic fungi. The presence of ABA in mammalian brains was reported in 1986,[621] and it was suggested that ABA may also be a hormone in mammals. However, further studies on its presence have not been reported. Exogenous ABA has no effects on rats or insects.[622]

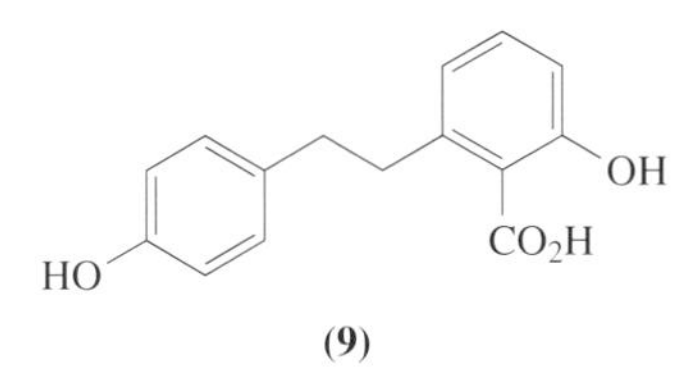

(9)

8.02.5.2 Chemistry

8.02.5.2.1 Isolation and analysis

ABA is usually extracted from plants and other sources using methanol. Methanolysis of the glucosyl ester of ABA sometimes occurs to give its methyl ester as an artifact,[623] and acetone is used for extraction to avoid the methanolysis. The methyl ester of ABA is not always an artifact. Its occurrence in stigma of tobacco has been confirmed by extraction using acetone.[624] Ethyl acetate is used to extract ABA and its natural metabolites, phaseic acid (**10**) and dihydrophaseic acid (**11**), and its epimer (**12**), from an aqueous solution after concentration of a methanol solution. ABA is a weak acid with a pK_a of 4.7, and the partition coefficient between ethyl acetate and water at pH

2.5 is 10.[625] Conjugated forms of ABA and the metabolites are extracted from the aqueous solution by *n*-butyl alcohol saturated with water.

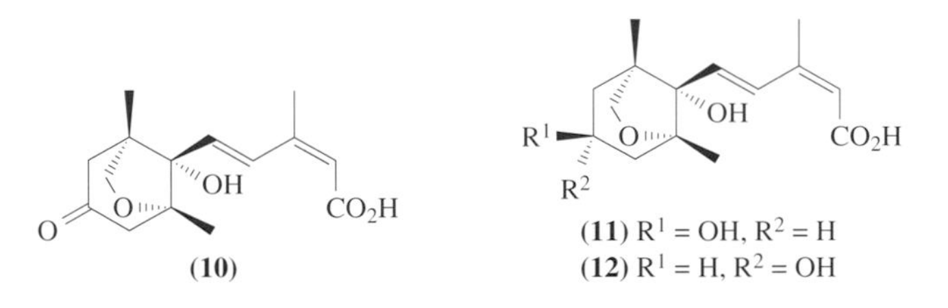

(10)

(11) R^1 = OH, R^2 = H
(12) R^1 = H, R^2 = OH

Isolation and chromatographic analysis of ABA are according to a method for a weak acid. Column packings suitable for a rough separation of ABA are active charcoal, silica gel, and others. ABA is observed as a dark spot on a thin layer plate of silica gel containing fluorescent dye under UV irradiation at 254 nm. The colorless spot of ABA on the plate becomes bright yellow after spraying with a 5% solution of sulfuric acid in ethanol followed by heating, and the spot shows yellowish green fluorescence under UV irradiation at 365 nm.[626] Octadecyl silica gel columns are often used in HPLC analysis of ABA.[625] The detection limit of ABA at 260 nm is about 1 ng. The methyl ester of ABA is detected by GC using OV-1, OV-17, SE-30, and XE-60 columns.[627] A GC analysis with an electron capture detector (ECD) is one of the most convenient methods to detect ABA with high selectivity and sensitivity. The detection limit of the methyl ester of ABA by ECD is 1–10 pg.[625]

ABA is detected and quantified by enzyme immunoassay with a monoclonal antibody.[628] The antibody against ABA bound to protein through the C-4′ carbonyl group does not show cross-reactivity against the glucosyl ester of ABA (**13**) and its metabolites, while the antibody against ABA bound to protein through the C-1 carboxy group recognizes not only ABA but also its glucosyl and methyl esters. The difference between contents assayed by the two antibodies corresponds usually to the content of glucosyl ester of ABA. The detection limit of immunoassay with a monoclonal antibody is about 0.02 pmol.

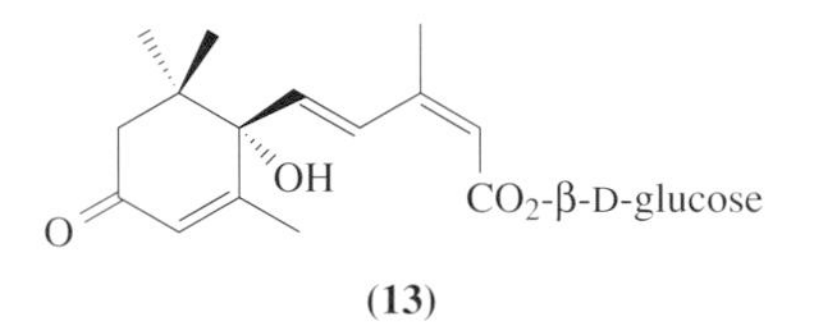

(13)

The optical resolution of (±)-ABA, (±)-phaseic acid, and the methyl ester of (±)-dihydrophaseic acid is achieved by HPLC with columns packed with chiral materials.[629–631] Affinity gel which has a bound anti-(+)-ABA antibody can be used for the optical resolution.[632] The methyl ester of (±)-ABA is chemically resolved after conversion to its (±)-1′,4′-*trans* and *cis*-diols followed by the Sharpless asymmetric epoxidation separately.[633] The (+)-enantiomers of each 1′,4′-diol are not reactive to the epoxidation; the (+)-enantiomers remaining after the epoxidation are converted to (+)-ABA.

8.02.5.2.2 Physiochemical properties

The melting points of (+)-ABA and its racemate are 160–161 °C and 188–190 °C, respectively.[601,634] ABA is converted to a γ-lactone (**14**) under strong acidic conditions (formic acid–hydrochloric acid) as shown in Scheme 9, and with an alkaline treatment the γ-lactone changes to a short-lived ion (**15**) with a resonance structure which shows a purple color.[635] The 1′-hydroxy group is resistant to acetylation, and the six-membered ring aromatizes with migration of the methyl groups by treatment with acetic anhydride and *p*-toluenesulfonic acid to give an aromatic derivative (**16**).[636] Hydrogens and an oxygen of ABA can be exchanged with hydrogens and an oxygen from a medium. Hydrogens at C-3′,-5′, and -7′ are easily exchanged with deuterium of deuterated water under alkaline conditions via an enol form to give a hexadeuterated ABA (**17**).[637] The order of ease of exchange is 5′-$H_{pro\text{-}S}$, 5′-$H_{pro\text{-}R}$, and 3′-H and 7′-H. Hydrogens at the C-2 and C-6 are also exchangeable in an alkaline solution although it takes longer than two months.[638] 4′-Carbonyl oxygen is exchanged with an oxygen of water under alkaline conditions.[639] ABA has two reduction voltages, E_p −1.3 V and −1.75 V, because conjugation of the side chain and the enone groups is precluded by the 1′-hydroxy group. The side chain of ABA, 3-methyl-2,4-pentadienoic acid, is more electrophilic than the enone. The methyl ester of ABA is easily reduced at a mercury cathode to give a bicyclic compound (**18**), as shown in Scheme 10.[640]

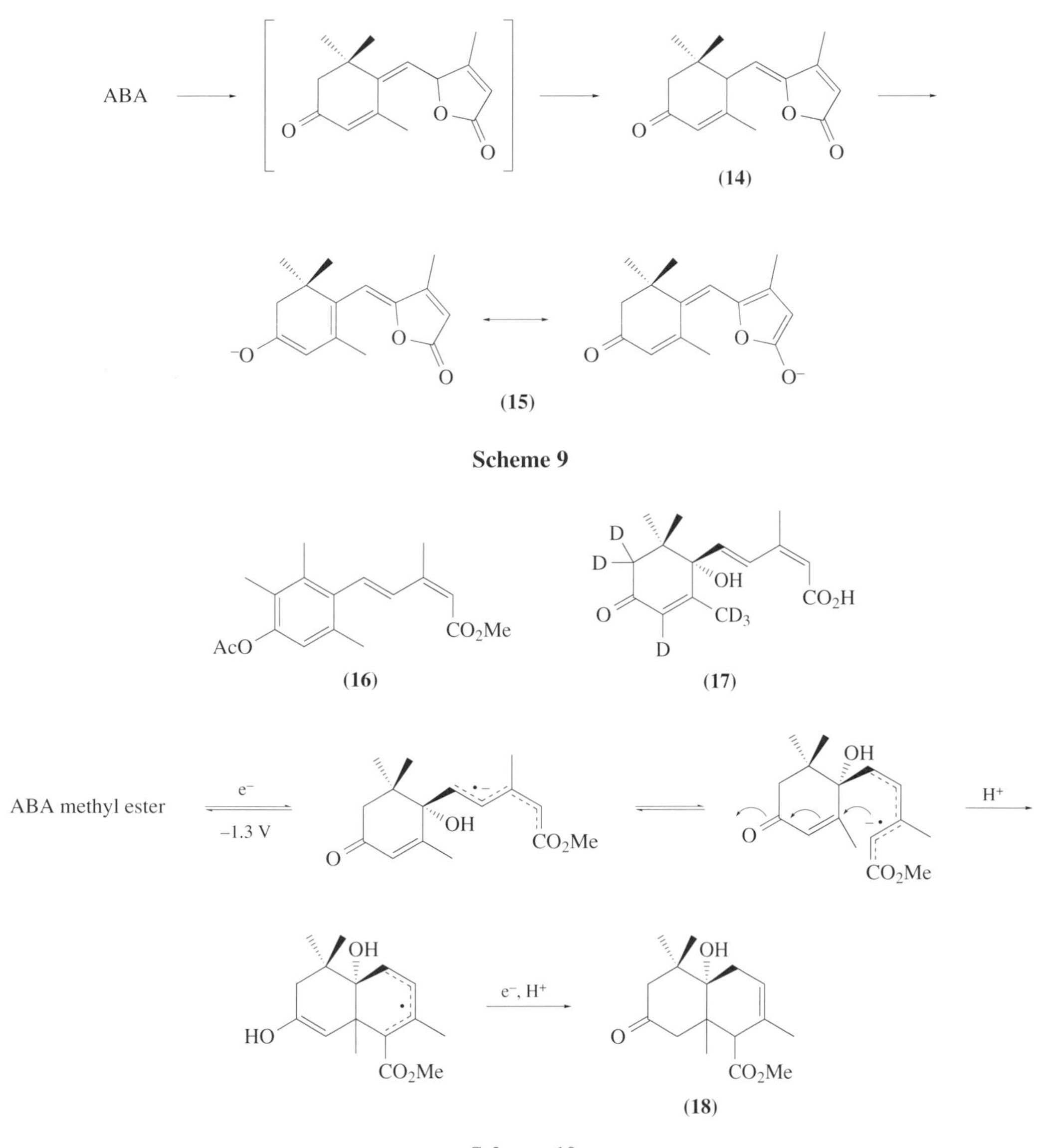

Scheme 9

Scheme 10

In the IR spectrum, the methyl ester of ABA shows absorption bands at 3610 cm^{-1} (the 1′-hydroxy group), 1667 cm^{-1} (the 4′-carbonyl group), and 1710 cm^{-1}, 1637 cm^{-1} and 1604 cm^{-1} (the side chain).[641] The methyl ester of phaseic acid shows an absorption band at 1723 cm^{-1} for the 4′-carbonyl group. The methyl ester of dihydrophaseic acid shows an absorption band at 3552 cm^{-1} which is assigned to the 4′α-hydroxy group since its strength does not decrease even in a dilute solution due to hydrogen bonding between the 4′-alcoholic hydrogen and the 8′-ethereal oxygen.[642] This assignment means that the C-4′ of dihydrophaseic acid is *S* and that the C-4′ of *epi*-dihydrophaseic acid is *R*.

ABA absorbs UV with maxima at 240 nm (ε 2.1×10^4, a shoulder peak), 260 nm (ε 2.6×10^4) and 320 nm (ε 50) in an acidic methanol solution.[643] The first and third maxima are assigned to the π–π^* and n–π^* transitions of the enone group in the ring, respectively. The second maximum is assigned to the π–π^* transition of the doubly conjugated carboxylic acid group in the side chain. ABA is moderately photosensitive due to these groups. Irradiation with UV with a wavelength shorter than 305 nm isomerizes the 2-(*Z*)-double bond to (*E*) to give an equilibrium mixture of ABA and its 2-(*E*)-isomer (**19**) with a ratio of 1 : 1, and also causes decomposition of ABA to unidentified compounds by the excitation of the π–π^* transition of the side chain and the enone groups. Since the excitation of the n–π^* transition of the enone group forms a diradical-like structure at C-2′, C-3′, and C-4′, UV at a wavelength longer than 320 nm can be used to label ABA-binding sites by cross-linking to other components like protein.[643]

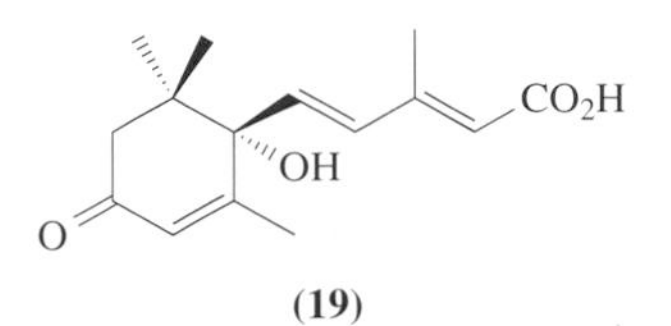

(19)

ABA has a strong optical activity, and its specific optical rotation is +430° in an acidic methanol solution.[644] In the optical rotatory dispersion (ORD) and the c.d. spectra, ABA shows a positive Cotton effect from 300 nm to 200 nm.[612,644] Phaseic acid, dihydrophaseic acid and its epimer, which did not have the enone group, show a small specific optical rotation with a minus value and also a negative plain curve in the ORD.

The ^{1}H NMR signals of ABA are summarized in Table 9. In the ^{1}H NMR spectrum of ABA, the chemical shift δ 7.77 ppm of the 4-H is lower than that of usual protons bound to double bonds. This down-field shift is explained by the deshielding effect of the 1-carboxyl group. The same proton of 2-(*E*)-ABA is observed at δ 6.5 ppm. A W-coupling is observed between the 5′-H at a lower field than the other 5′-H and the 8′-H, and between the 5′-H at a higher field than the other 5′-H and the 3′-H. Since a half-chair conformation with the pseudoaxial side chain can allow such a relationship (Figure 9), the signal can be assigned at a higher field to the 5′-$H_{pro\text{-}R}$ having an equatorial orientation, and at a lower field to the 5′-$H_{pro\text{-}S}$ having an axial orientation.[637] However, strictly speaking, the NMR signals of ABA at room temperature are the average of the signals of different conformers, as described later. Signals of the 5′-H of (+)- and (−)-ABAs in the ^{1}H NMR spectrum are separated in the presence of γ-cyclodextrin.[645] This method can be applied to determination of the enantiomeric ratio. The ^{1}H NMR spectrum of phaseic acid shows two W-couplings between the 5′-$H_{pro\text{-}S}$ and the 8′-$H_{pro\text{-}R}$, and between the 5′-$H_{pro\text{-}R}$ and the 3′-$H_{pro\text{-}S}$ because the conformation of the ring is fixed to a chair with the side chain axial (Figure 9).[646] Eu(hfc)$_3$ separates the signals of the 7′- or 9′-H of methyl esters of (+)- and (−)-phaseic acids.[647] In the ^{1}H NMR spectrum of *epi*-dihydrophaseic acid, the signal of the 4′-hydroxy proton is observed as a doublet by coupling with the 4′-H, meaning that the 4′-hydroxy group is relatively fixed by hydrogen bonding with the 8′-oxygen.[642] The 8′-$H_{pro\text{-}S}$ is shifted to a lower field by the 4′-hydroxy group close to the 8′-$H_{pro\text{-}S}$. The 4′-hydroxy proton of dihydrophaseic acid does not show the deshielding effect on the 8′-$H_{pro\text{-}S}$. The ^{13}C NMR signals of ABA are assigned as shown in Table 9.[637,648] An INEPT (Insensitive Nuclei Enhanced by Polarization Transfer) method showed that one of the methyl groups at the C-6′ at a higher field than the other methyl group is bound to the 8′-H, indicating that a signal at δ 23.1 ppm is the C-8′ and a signal at δ 24.3 ppm is the C-9′.[637]

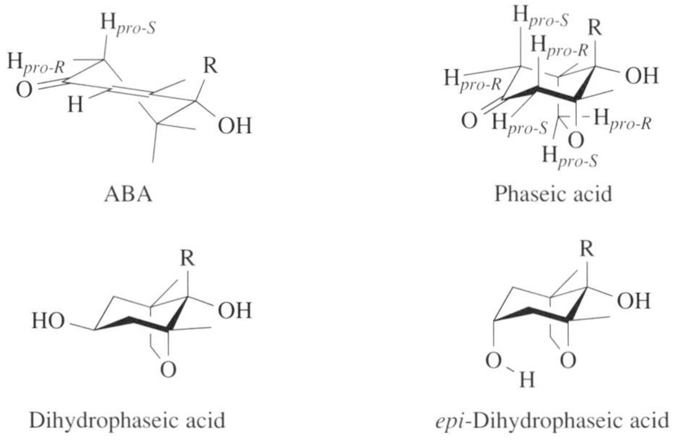

Figure 9 Steric structures of ABA with a preferred conformation and its metabolites (R = the side chain).

In the electron impact (EI) mass spectrum, the methyl ester of ABA gives a molecular ion at *m*/*z* 278, a dehydrated ion at *m*/*z* 260, and major fragment ions at *m*/*z* 246 [M-methanol]$^+$, 190, 162, 134, and 125. The major fragmentation pathway of the methyl ester of ABA, as shown in Scheme

Table 9 Assignment of ^1H and ^{13}C NMR signals of ABA.

^1H NMR (300 MHz, CD_3OD)				^{13}C NMR (24 MHz, $CDCl_3$)		
δ (ppm)	*Multiplicity*	J (Hz)	*Assignment*	δ (ppm)	*Multiplicity*	*Assignment*
1.03	s		9′-H	19.1	q	C-7′
1.06	s	[a]	8′-H	21.4	q	C-6
1.93	d	1.3	7′-H	23.1	q	C-8′
2.03	d	1.3	6-H	24.3	q	C-9′
2.18	dd	16.9 and 0.9	5′-$H_{pro\text{-}R}$	41.7	s	C-6′
2.53	d	16.9[a]	5′-$H_{pro\text{-}S}$	49.7	t	C-5′
5.74	br s		2-H	79.9	s	C-1′
5.92	m		3′-H	118.1	d	C-2
6.23	dd	16.2 and 0.4	5-H	127.0	d	C-3′
7.77	dd	16.2 and 0.6	4-H	128.3	d	C-4
				136.8	d	C-5
				151.4	s	C-3
				163.0	s	C-2′
				170.9	s	C-1
				198.3	s	C-4′

[a]The coupling constant between 8′-H and 5′-H is very small.

11, was proposed by Gray *et al.* with labeled methyl esters of ABA.[639] The cleavage of the bond between C-4′ and C-5′ of an ion radical (**20**) followed by the elimination of isobutylene through a retro Diels–Alder reaction gives an intermediate ion radical (**21**). The ion radical is cyclized, and a methanol is eliminated to give a bicyclic ion radical (**22**) at *m/z* 190 which is a base peak in the spectrum. By further elimination of the carbon monoxide, hydrogen radical, and methyl radical, other fragment ions are formed. An ion (**23**) at *m/z* 125 with a second intensity is derived from the side chain. The strong ions at *m/z* 190 and 125 are selected as characteristic ions of the methyl ester of ABA for detection and quantitation by GC–SIM. Negative chemical ionization (NCI) mass spectrometry detects ABA with a high sensitivity since the negative molecular ion $[M]^-$ of methyl ester of ABA is more stable than the positive molecular ion $[M]^+$ due to the high electrophilicity of ABA. The NCI mass spectrum shows $[M]^-$ at *m/z* 278 as a base peak, and other fragment ions at *m/z* 310, 260, 245, 141, and 152.[649] It has been suggested that fragment ions at *m/z* 141 and 152 are derived from the dioxygenated ion radical (**24**) at *m/z* 310 which is not observed without contamination of air in carrier gas.[649] Combination of SIM with NCI gives highly selective and sensitive detection of ABA, the lowest detection limit is 0.3 pg which is 200 times lower than that of EI–SIM.[650] The detection limit can be decreased to 20 fg by using the pentafluorobenzyl ester of ABA which has higher electrophilicity than the methyl ester.[651] This is the lowest detection limit among the several detection methods of ABA including the immunoassay.

8.02.5.2.3 Structure–activity relationship

(i) Functional groups essential for the activities

The 2-(*Z*)-pentadienoic acid moiety seems to be essential for activity since modification of the side chain decreases the activity of ABA. 2-(*E*)-ABA (**19**) is inactive,[652] and the elongation and the shortening of the length of the side chain decrease the activity.[653,654] An analogue (**25**) where the geometry of the C-2 is fixed to (*Z*) by introduction of a benzene ring shows low activity probably due to the bulkiness of the benzene ring.[655] An acetylenic aldehyde (**26**) having a triple bond at C-4 shows high activity,[656] suggesting that the recognition of this part by receptors is not strict. Another analogue (**27**) having a phenyl group at the C-1′ shows weak activity.[657] 1′-Aldehyde and 1′-alcohol of ABA are active derivatives which may be converted to ABA by oxidation in plants.[656,658] The methyl ester of ABA is active in long-term assays and is not active in the stomatal assay,[603,659] suggesting that the activity of the ester in long-term assays is expressed after hydrolysis.

The 1′-hydroxy group is not essential for the activity of ABA. (+)-1′-Deoxy-ABA retains activity although it is lower than that of ABA.[653] The biological activities of (+)-1′-deoxy-1′-fluoro-ABA are 0.1 to 0.05 those of (+)-ABA, and almost equal to those of (+)-1′-deoxy-ABA.[660] The property

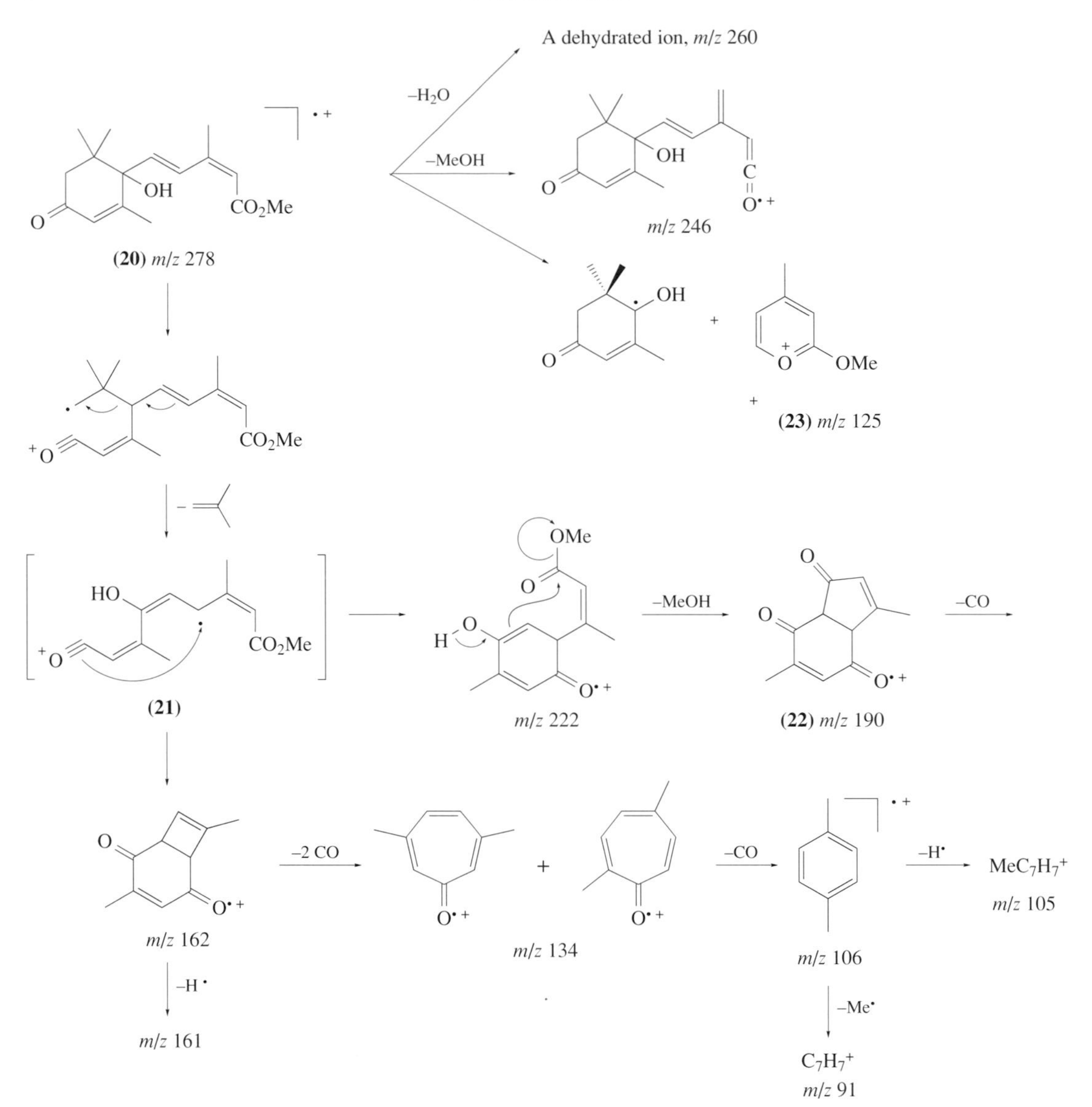
A dehydrated ion, m/z 260
–H2O
–MeOH
OH
CO2Me
O
(20) m/z 278
OH
C
O• +
m/z 246
OH
+
OMe
(23) m/z 125
CO2Me
HO
CO2Me
(21)
OMe
O
H
O
O• +
m/z 222
–MeOH
O
O
O• +
(22) m/z 190
–CO
O
O• +
m/z 162
–H •
m/z 161
–2 CO
+
O• +
O• +
m/z 134
–CO
m/z 106
–Me•
C7H7+
m/z 91
–H•
MeC7H7+
m/z 105

Scheme 11

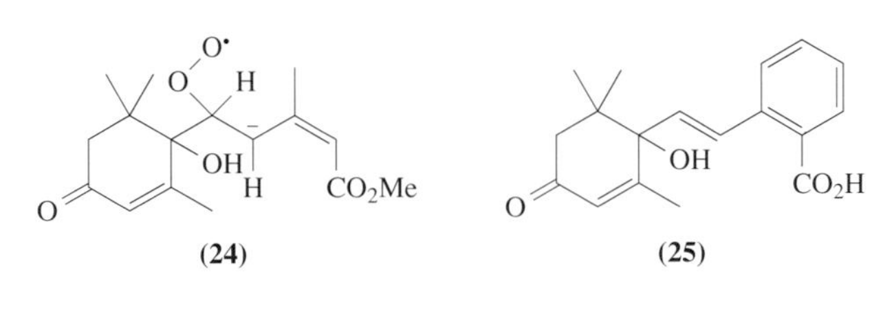
O•
O
H
OH
H
CO2Me
O
(24)
OH
CO2H
O
(25)

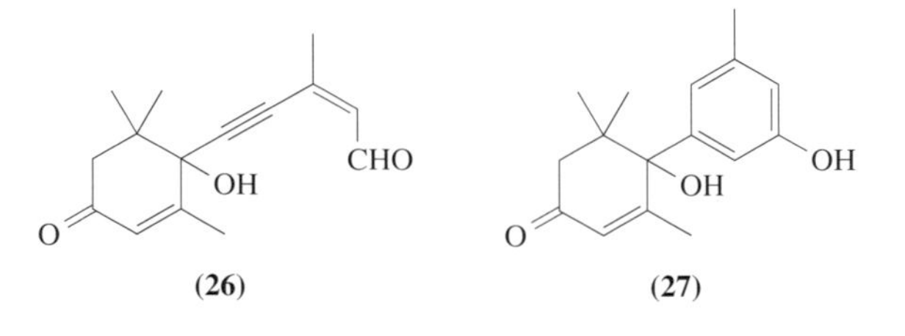
OH
CHO
O
(26)
OH
OH
O
(27)

of a monofluoro group is similar to that of oxygen of a hydroxy group except for lack of the ability to donate hydrogen. Thus the 1′-hydroxy group of (+)-ABA interacts with binding proteins as an uncharged hydrogen-bonding donor. 1′-*O*-Methyl-ABA shows weak activities, indicating that the hydroxy group should be free.[661] This is coincident with the result with (+)-1′-deoxy-1′-fluoro-ABA.

The activities of (+)-7′- and 8′- nor-ABAs and of (+)- 8′,9′-dinor-ABA have shown that the C-7′ is critical for the activity while C-8′ and C-9′ are not.[662] The C-7′ must interact with binding sites. The activities of the analogues alkylated at C-7′, C-8′, and C-9′ have shown that the decrease in activity by alkylating the C-8′ of (+)-ABA is smaller than that by alkylating the C-7′ and C-9′ of (+)-ABA, indicating that the activity is relatively unaffected by the bulky group at the C-8′ of (+)-ABA.[663] The space around the C-8′ in (+)-ABA seems not to have been involved in the interaction with binding sites for exhibiting activity. The C-8′ might be essential only for metabolic inactivation by hydroxylation. 7′-Difluoro-ABA is as active as ABA.[664]

The 4′-carbonyl group is probably not essential since 4′-deoxy-ABA and 1′,4′-diols of ABA (**4**) and (**5**) are active.[665,666] However, the receptors for ABA cannot perceive a bulky group at the C-4′ since 4′-*p*-aminobenzoyl hydrazone of ABA lacks the activity.[667] 3′-Thioether of ABA for affinity chromatography retains an activity, showing that ABA is tolerant of the 3′-modification.[668] The activities of other analogues have been summarized by Walton.[669]

(ii) Active conformation

Since the ring of ABA is not constrained, ABA can adopt many conformations. The possible conformations of the ring in ABA are represented using the torsion angle notation in Table 10.[670] The X-ray analysis of a crystal of ABA shows that the ring adopts a slightly distorted sofa, similar to the sofa $\mathbf{S_1}$ which has the nondistorted enone and the pseudoaxial side chain.[671,672] The ^{1}H NMR spectrum has revealed that the preferred ring conformation of ABA in solution is the half-chair $\mathbf{HC_1}$ with a pseudoaxial side chain.[637,673] This conformation is coincident with the negative Cotton effect with $\Delta\varepsilon$ −2.34 derived from the n–π^* transition of the enone at 320 nm in the c.d. spectrum which indicates that the torsion angle of C-3′—C-4′ is about 10–20°.[612,613,674–676] The computer-aided conformational analysis has suggested that the energy difference between the most stable half-chair $\mathbf{HC_1}$ and its inverted form $\mathbf{HC_2}$ with the pseudoequatorial side chain is 3 kcal mol^{-1},[673] meaning that the $\mathbf{HC_1}$/$\mathbf{HC_2}$ ratio in conformational equilibrium at 300 K is about 99.4:0.6 from the Gibbs equation, $\Delta G_0 = -RT\ln K$. The energy barrier to the ring inversion between two half-chairs has not been examined. Since the ^{1}H NMR spectrum of ABA at 368 K is the same as that at 300 K,[673] the ^{1}H signals at 300 K must already be those averaging $\mathbf{HC_1}$ and $\mathbf{HC_2}$ owing to the low energy barrier to interconversion. A nuclear Overhauser effect is observed not only between 5-H and 5′-$H_{pro\text{-}S}$/7′-H/9′-H but also between 5-H and 8′-H, showing that the half-chair conformation $\mathbf{HC_2}$ with a pseudoequatorial side chain also exists. ABA in solutions would exist as an equilibrium mixture of the two conformers, $\mathbf{HC_1}$ and $\mathbf{HC_2}$. Other forms, the sofas $\mathbf{S_{1-4}}$, 1,3-diplanars $\mathbf{DP_{1-4}}$, and boats $\mathbf{B_{1-2}}$, are probably transient, short-lived conformations in the course of inversion between $\mathbf{HC_1}$ and $\mathbf{HC_2}$.

Considering the low barrier to interconversion and the thermodynamic stabilization in binding to the active site of the receptor, in addition to the half-chairs, the short-lived forms can be the active conformation of ABA. The active conformation has been investigated by the activities of the analogues which prefer different conformations. Analogues with a conformation resembling the active conformation of ABA should at least be as active as ABA. The allenic analogue (**28**) with the equatorial side chain is inactive although it shows activity after conversion to ABA.[677] The (1′*S*,2′*S*)-2′,3′-dihydro-ABA (**29**) is active, whereas (1′*S*,2′*R*)-2′,3′-dihydro-ABA (**30**) is inactive.[678] The two dihydro-ABAs may adopt a chair form with axial and equatorial side chains, respectively, due to the steric repulsion between the 1,3-diaxial methyl groups, 7′ and 9′, and 7′ and 8′, respectively. These three examples suggest that the active conformation has an axial side chain, rather than an equatorial one.

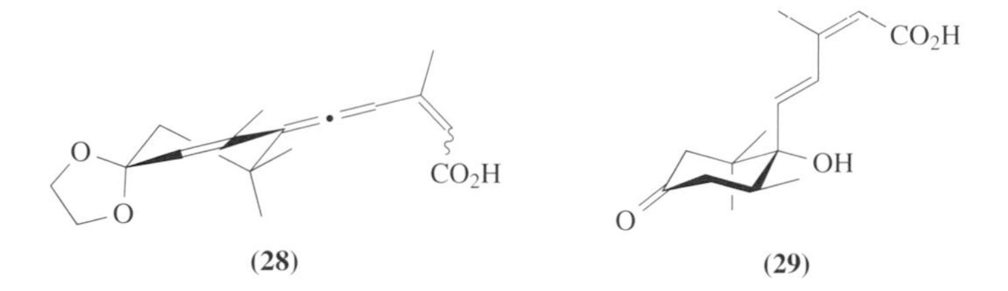

Table 10 Representation of idealized conformations of the cyclohexenone ring in ABA using the torsion angle notation and orientations of the side chain and C-9′ (R = the side chain).

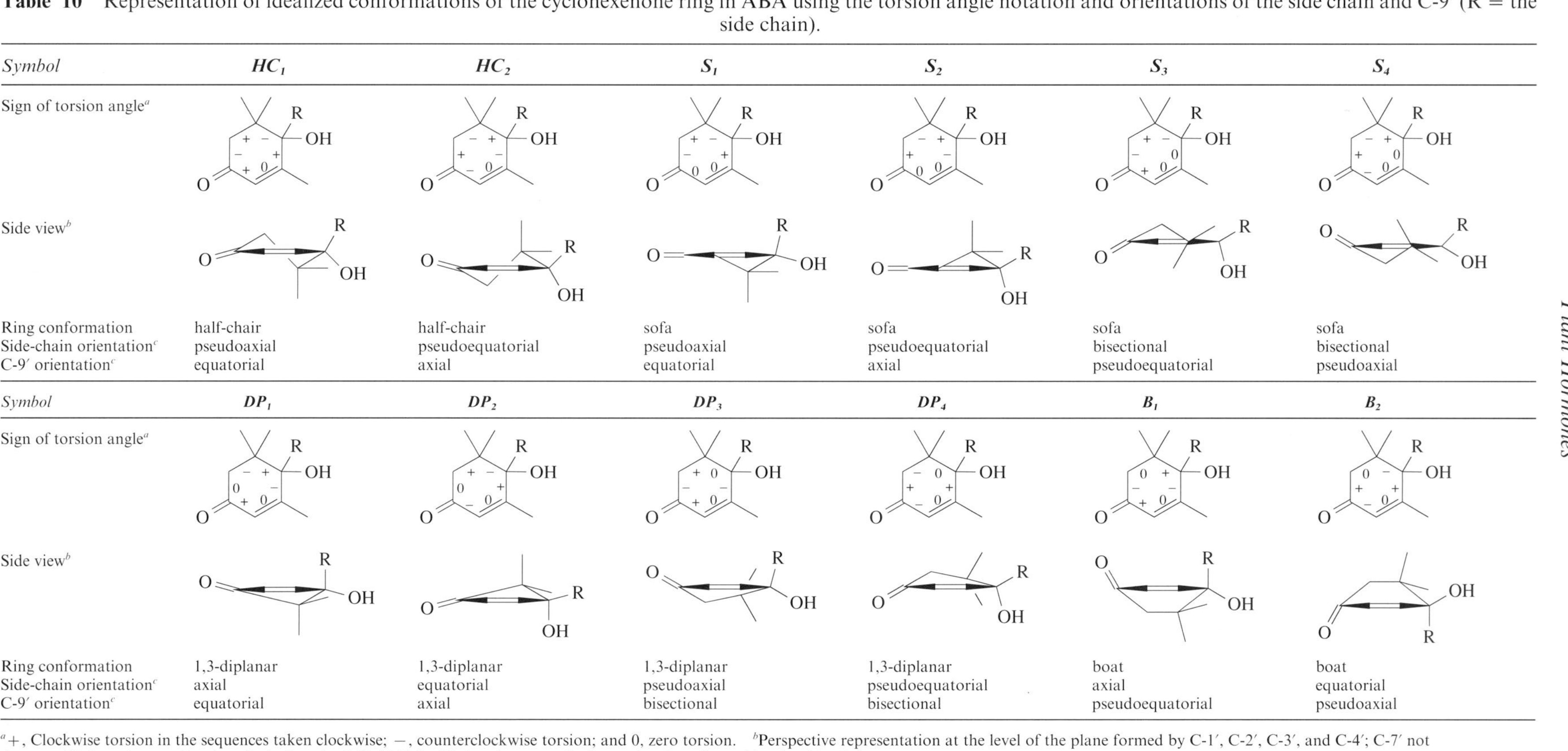

Symbol	***HC₁***	***HC₂***	***S₁***	***S₂***	***S₃***	***S₄***
Sign of torsion angle[a]						
Side view[b]						
Ring conformation	half-chair	half-chair	sofa	sofa	sofa	sofa
Side-chain orientation[c]	pseudoaxial	pseudoequatorial	pseudoaxial	pseudoequatorial	bisectional	bisectional
C-9′ orientation[c]	equatorial	axial	equatorial	axial	pseudoequatorial	pseudoaxial
Symbol	***DP₁***	***DP₂***	***DP₃***	***DP₄***	***B₁***	***B₂***
Sign of torsion angle[a]						
Side view[b]						
Ring conformation	1,3-diplanar	1,3-diplanar	1,3-diplanar	1,3-diplanar	boat	boat
Side-chain orientation[c]	axial	equatorial	pseudoaxial	pseudoequatorial	axial	equatorial
C-9′ orientation[c]	equatorial	axial	bisectional	bisectional	pseudoequatorial	pseudoaxial

[a] +, Clockwise torsion in the sequences taken clockwise; −, counterclockwise torsion; and 0, zero torsion. [b] Perspective representation at the level of the plane formed by C-1′, C-2′, C-3′, and C-4′; C-7′ not shown. [c] Orientations to the plane including the C-2′–C-3′ double bond, perpendicular to the plane formed by the side chain.

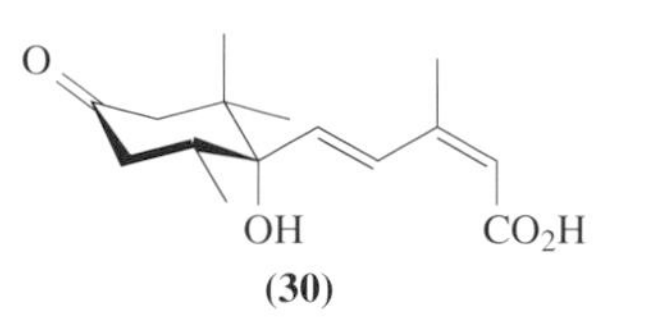

This suggestion has been further examined for the cyclopropane analogues synthesized by Todoroki *et al.*[670] Chemical and physical properties of cyclopropanes are similar to those of alkenes. Introduction of the cyclopropane into C-5′–C-6′ can constrain the orientation of the subsituents on the ring, regardless of the cyclohexadienone-like planar or boat conformation of the ring, as shown in Figure 10. 5′β,9′-Cyclo-ABA (**31**) in which the 6′β-substituent is constrained essentially to the axial-like orientation between axial and bisectional orientation shows no activity, while 5′α,8′-cyclo-ABA (**32**) and an achiral analogue (**33**) with no axial-like substituent at C-6′β exhibits activity equivalent to ABA. Most conformational changes of ABA are represented by the orientations of the side chain and the 6′-methyl groups, the C-8′, and C-9′. Therefore, the activities of the analogues suggest that the active conformation of ABA is not the half-chair $\mathbf{HC_2}$ with the pseudoequatorial side chain and the pseudoaxial methyl at C-6′β, but is close to the other half-chair $\mathbf{HC_1}$ with the pseudoaxial side chain and the pseudoequatorial methyl at C-6′β. Phaseic acid is inactive in most assays, although its cyclohexanone ring is constrained to the chair form where the side chain is fixed in an axial position owing to the bridged bicyclic system (Figure 9). Therefore, the active conformation of ABA would be a conformation where the C-9′ is equatorial and the side chain is between pseudoaxial and bisectional, that is close to the favored half-chair $\mathbf{HC_1}$ with the side chain pseudoaxial. In binding to the active site on the receptor, ABA probably tilts the side chain to outside of the ring, that is, to the bisectional orientation with the C-9′ equatorial. The computer-aided and ^{1}H NMR analyses indicate that the bonding between C-1 and C-2 can rotate freely, but the rotation angles of the bondings between C-3 and C-4 and between C-5 and C-1′ are 20° and 180°, respectively.[673] However, the active conformation of the side chain is unclear.

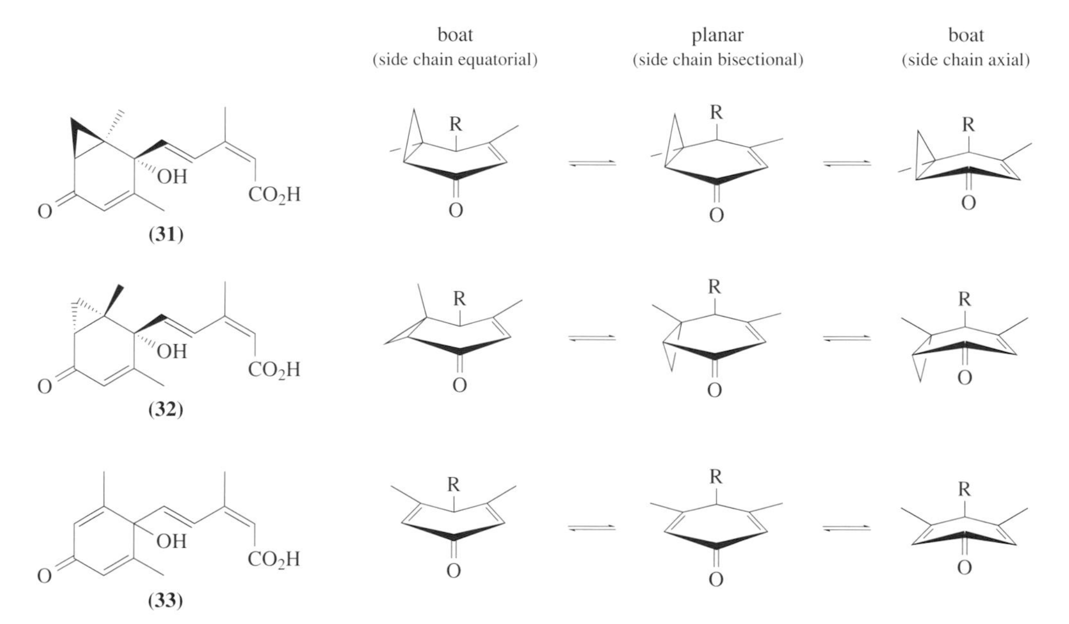

Figure 10 Ring conformational change of the cyclopropane analogues and the achiral analogue (R = the side chain).

Unnatural (−)-ABA shows one-half to one-third of the activity of (+)-ABA in many bioassys,[679] and this small difference in activity between the enantiomers has been explained by the pseudo-symmetry of the molecule which is derived from the 2,6,6-trimethyl-cyclohex-2-en-4-one.[680] Figure 11 shows the steric structures of (+)- and (−)-ABAs with the preferable conformation, a half-chair with the pseudoaxial side chain, view from the carbonyl group at C-4′. In the (−)-ABA molecule, the C-7′ corresponds to the C-9′ of (+)-ABA, the C-9′ corresponds to the C-7′ of (+)-ABA, and the C-8′ occupies the space facing the *re*-face of the C-2′ in (+)-ABA, whereas a methyl group corresponding to the C-8′ of (+)-ABA is absent. This hypothesis has been tested by the analogues alkylated at the C-7′ and C-9′ of (+)- and (−)-ABA, and the achiral analogue (**33**).[667] The decrease in the activity of (−)-7′- and 9′-alkyl-ABAs correlates with that of (+)-9′- and 7′-alkyl-ABAs, respectively, and the achiral analogue shows activity intermediate between (+)- and (−)-ABAs. Thus the pseudosymmetry would be the reason why (−)-ABA shows high activity. The activity of (−)-ABA is low in the assay of stomatal closure,[663] which suggests that the receptors in stomatal cells are more specific to (+)-ABA than to (−)-ABA.

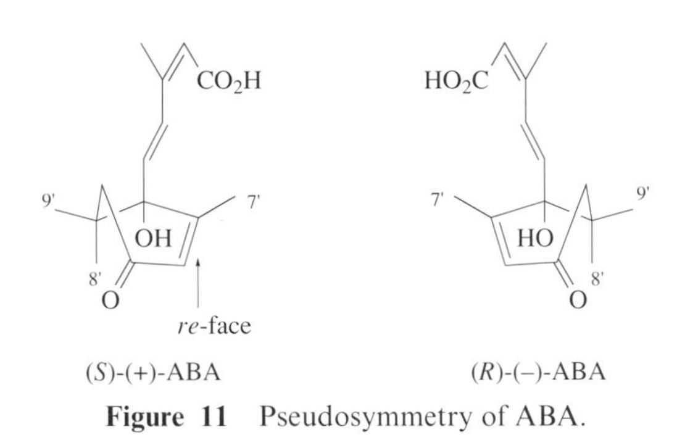

Figure 11 Pseudosymmetry of ABA.

8.02.5.3 Biological Activities

ABA regulates many physiological processes, including germination, growth, flowering, seed maturation, senescence, and adaptation to various environmental stresses such as water deficiency, low temperature, and freezing.[617,681] Exogenous ABA shows unique activities on plants which reflect its physiological role as a hormone.

8.02.5.3.1 Stomatal closure

One characteristic activity of ABA is its effect on stomata which protects plants from water stress. At 10^{-7} M, ABA given through a transpiration stream from cut ends of shoots causes stomatal closure.[682] The activity is more effective in epidermal strips floated on a buffer solution than in shoots. At 10^{-10} M, ABA closes stomata in epidermal strips at pH 5.5.[683] The activity at pH 6.8 is 10^3 times lower than that at pH 5.5, suggesting that the active form of ABA is not a dissociated acid but an undissociated acid. The stimulation study on the redistribution of ABA in plant tissues after pH changes of apoplasts and synplasts has suggested that ABA accumulates in guard cell walls according to the anion-trap mechanism for weak acids. Hartung and Slovik have proposed that ABA may be an ideal "stress messenger" for stomata.[684]

The effect on stomata in epidermal strips strongly suggests that ABA acts on guard cells directly by binding to the receptors. An experiment with photoaffinity labeling has suggested that ABA-binding sites are located in plasma membranes of guard cells.[652] The iontophoretic microinjection of ABA into guard cells does not inhibit opening of closed stomata,[685] which suggests also that the binding sites are located outside of the guard cells, probably in the plasma membranes of guard cells. However, there is still controversy on the location of the ABA-binding sites in guard cells. Another microinjection experiment with the caged ABA (**34**) which releases ABA by irradiation with UV has suggested two transduction pathways and an intracellularly located ABA receptor.[686] Guard cells may have two binding sites, one in the plasma membranes and another in the cytosol, which may be concerned with rapid responses for closing stomata and slow responses for induction of adaptive reactions, respectively. ABA induces a rapid increase of cytosolic Ca^{2+} ions in guard cells preceding stomatal closure.[687] Since a Ca^{2+}-modulated protein phosphatase has been shown to be involved in signal transduction of ABA in the regulation of stomatal aperture in *A. thaliana*,[688,689] ABA may express its effect via dephosphorylation of phosphorylated proteins. ABA causes disruption of cortical microtubules in guard cells of *V. faba* L.,[690] but the disruption of microtubules seems not to be directly involved in stomatal closure.

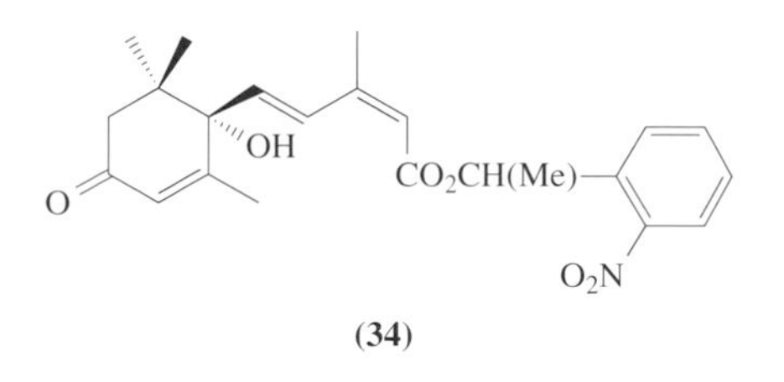

(34)

8.02.5.3.2 Inhibition of seed germination

ABA inhibits germination of many seeds at 10^{-5}–10^{-7} M. The suppression of α-amylase induction would be a cause of the inhibition by ABA in starch seeds. The induction of α-amylase by gibberellin in barley half-seeds without embryos is inhibited by ABA at 10^{-7} M.[691] Barley seeds have two isozymes of α-amylase, and ABA suppresses biosynthesis of the isozyme with a higher isoelectric point than that of the other isozyme.[692] The suppression is antagonized by gibberellin. ABA induces new proteins in barley aleurone layers, and the induction is inhibited by gibberellin.[693] The proteins might suppress the induction of α-amylase. In lipid seeds the target of ABA seems to be different from that in starch seeds since ABA inhibits the increase of isocitrate lyase activity which catalyzes β-oxidation of fatty acids during germination of bean seeds.[694] Gibberellin antagonizes the inhibitory effect of ABA on the increase of isocitrate lyase activity.

The receptors for ABA in the aleurone cells may face outwards. ABA microinjected into a single barley aleurone cell has no effect on the α-amylase expression and secretion,[695] and monoclonal antibodies generated against guard cell protoplasts of pea suppress the expression of ABA-induced gene in barley aleurone protoplasts.[696] It has been suggested that the inhibitory effect of ABA on α-amylase biosynthesis is mediated by Ca^{2+} ions since ABA suppresses the increase of the cytosolic levels of Ca^{2+} by gibberellin.[697] Cutler *et al.* screened mutants of *A. thaliana* that showed an enhanced response to exogenous ABA in the seed germination tests, and found that the responsive gene encodes the β subunit of a protein farnesyl transferase.[698] Farnesyl transferase may act as a negative regulator of ABA in the signal transduction.

8.02.5.3.3 Other activities

ABA inhibits the growth of seedlings and hypocotyls of many plants at 10^{-5}–10^{-7} M. In contrast to the inhibitory effect, ABA promotes the growth of tomato and bean shoots at low concentrations,[699] and increases the volume of wheat protoplasts in the presence of Ca^{2+}.[700] ABA also has both effects on root growth. The growth of root tips of pea is promoted at a low concentration of ABA, and is inhibited at a high concentration of ABA.[701] Such properties of ABA suggest that growth at certain stages is regulated by endogenous levels of ABA, that is, ABA is not only a brake but also an accelerator.

Diverse effects of exogenous ABA on organs and tissues have been shown.[702] ABA suppressed germination of buds and is thought to be involved in dormancy. ABA fed to one side of root tips caused root curvature, suggesting involvement of ABA in gravitropism of the root. Since ABA induces the morphogenesis of leaves of water plants from a submersed type to a floating type, ABA induced by water stress might act as a morphogen in some water plants. ABA generally inhibits induction of flower buds, although flowering of some short-day plants is promoted by ABA. The sink effect of ABA on accumulation of sucrose in immature seeds is controversial, but ABA may be involved in storage of starch and lipids during development of embryos. Application of ABA before exposure to low temperature confers cold or freezing tolerance on plants.

ABA has been found to induce or inhibit more than 70 proteins and genes.[703] A protein, dehydrin is expressed in a period of late embryogenesis of corn seeds before beginning of desiccation. Changes in the endogenous levels of dehydrin are correlated with that of ABA, and ABA can induce the expression of dehydrin, suggesting that dehydrin induced by endogenous ABA protects embryos from desiccation. Dehydrin has a domain which is rich in lysine and serine and may retain water molecules in the domain. Lea proteins involved in the desiccation of cotton seeds are induced by ABA. The *Rab*16 gene of rice is induced by ABA and osmotic stress. Its expression is regulated by a cis element, and the binding protein of the element was found to be a bZip protein. These properties of proteins induced by ABA suggest that the proteins are involved in adaptation to the stresses.

8.02.5.4 Biosynthesis and Metabolism

8.02.5.4.1 Biosynthesis

ABA has a sesquiterpene skeleton consisting of three isoprene units, which suggests that ABA is directly synthesized from mevalonic acid via a biosynthetic pathway of sesquiterpenes. Labeled mevalonic acid is incorporated into ABA by leaves, mesocarps, immature seeds, and cultured cells,[704–706] thus supporting the direct synthesis of ABA. However, the chemical conversion of violaxanthin to (+)-ABA by Taylor and Burden has revealed that ABA may be indirectly biosynthesized via carotenoids.[610] Some studies indicate that the carotenoid pathway may be more important although there is still controversy.[707] Scheme 12 illustrates the two pathways.

(i) Origin of the atoms

Feeding experiments with labeled mevalonic acids revealed the origin of hydrogens and carbons of ABA which may discriminate between the two pathways.[708] The hydrogen at C-2 is derived from the 4-$H_{pro\text{-}R}$ of mevalonic acid.[709] This suggests that the 2-(*Z*)-double bond of ABA is formed in(*E*) at the first step, since the retention of 2-$H_{pro\text{-}S}$ of isopentenyl pyrophosphate which corresponds to 4-$H_{pro\text{-}R}$ of mevalonic acid gives an (*E*) double bond as in the biosynthesis of farnesol and phytoene, according to the biosynthetic rule of isoprenoids. The (*E*) double bond must isomerize to (*Z*) in a later step of the biosynthesis. The origin of the hydrogens at C-4 and C-5 is 2-$H_{pro\text{-}R}$ and 5-$H_{pro\text{-}S}$ of mevalonic acid,[710] respectively, meaning that the formation of the (*E*) double bond is the same as that of conjugated double bonds of carotenoids. The hydrogen at C-3′ of ABA retains 2-$H_{pro\text{-}R}$ of mevalonic acid as well as the hydrogen on the double bond in the ε-ring of *β*,ε-carotenoids.[710] One of the two hydrogens at C-5′ of ABA retains 4-$H_{pro\text{-}R}$ of mevalonic acid.[710] ABA which was biosynthesized in deuterated water has a deuterium at *pro-R* of the C-5′, showing that the other proton is derived from a proton in a medium.[711] Such retention of the hydrogens is also coincident with the origin of hydrogens in the *β* and ε rings of carotenoids. Therefore, the ring formation of ABA probably occurs as shown in Scheme 13 (reproduced by permission of The Biochemical Society and Portland Press from *Biochem. J.*, 1984, **220**, 325.) The double bond at C-1 is located in the upper face (*re*-face) and the C-6 attacks the C-1 from the lower face. Subsequently the proton at the C-6 or the *β*-proton at C-4 eliminates to give the *β*- or α-type of six-membered ring, respectively.[637] Two methyl groups, C-6 and C-7′, are derived from the methyl group, C-6, of mevalonic acid.[712] C-6′ and C-9′ of ABA have been shown to be derived from carbons of a single acetic acid unit,[637,713] revealing that the C-9′ of ABA retains the C-6 of mevalonic acid since a single acetic acid unit forms a bond between the C-3 and C-6 of mevalonic acid, as shown in Scheme 14. This means that the C-8′ of ABA is derived from the C-2 of mevalonic acid. The origin of the methyl groups of ABA can be also understood by the ring formation of carotenoids.

The origin of the oxygens has been investigated with ABA which was biosynthesized in plants grown in $^{18}O_2$ or $H_2{}^{18}O$. The mass spectral analysis of the ABA showed that one oxygen of the 1-carboxy group and the 1′- and 4′-oxygens are derived from a molecular oxygen, and that the residual oxygen of the 1-carboxy group is derived from water.[714,715] The major positions labeled with $^{18}O_2$ depend on plant species, organs, maturity of fruits, and water conditions, and are classified into three types. First, leaves of cocklebur, immature fruits of avocado, and immature embryos of maize incorporate ^{18}O from $^{18}O_2$ mainly into the carboxylic oxygen.[716,717] The labeling of the carboxylic oxygen in these plant tissues which is actively synthesizing ABA suggests that the biosynthetic precursor of ABA is an oxidized carotenoid which can give a C_{15} aldehyde by oxidative cleavage at C-9. Second, the 1′-hydroxy oxygen is labeled in roots of cocklebur under water-stressed conditions, and in mature fruits of apple.[715,716] Third, mature fruits of avocado incorporate $^{18}O_2$ into both oxygens at the C-1 and C-1′.[716] The incorporation of $^{18}O_2$ into the 4′-hydroxy group is relatively low in all cases. Tissues in which the biosynthesis of ABA is not active would incorporate $^{18}O_2$ into nonoxidized carotenoids to form a C_{15} aldehyde with high contents of ^{18}O at the C-1, C-1′, and C-4′, but the labels at the C-1 and C-4′ might be lost by exchange with oxygen of water.

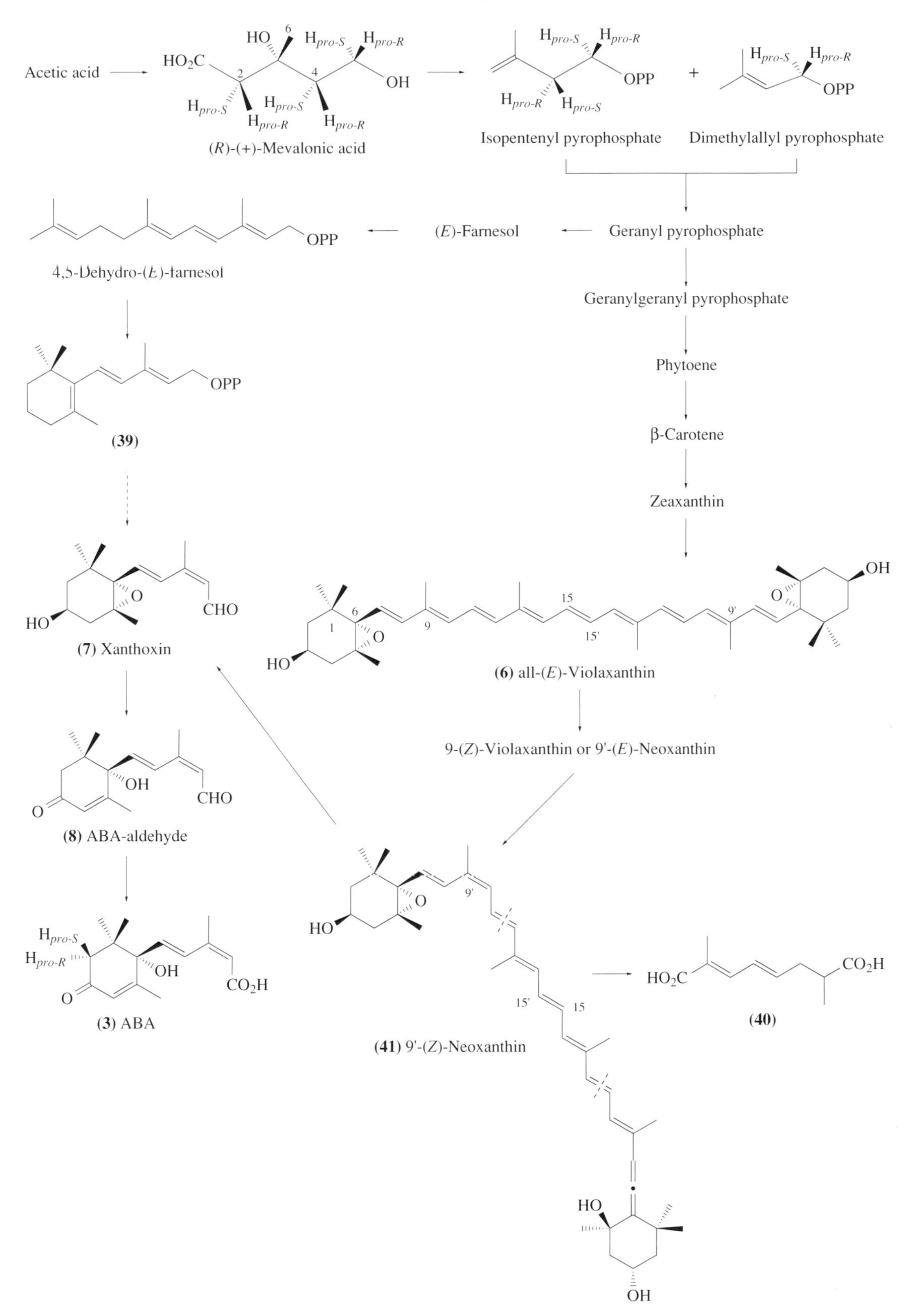
Acetic acid
HO
6
HO2C
2
4
Hpro-S
Hpro-R
OH
Hpro-S
Hpro-R
(R)-(+)-Mevalonic acid
OPP
Isopentenyl pyrophosphate
+
Dimethylallyl pyrophosphate
OPP
(E)-Farnesol
Geranyl pyrophosphate
4,5-Dehydro-(E)-farnesol
Geranylgeranyl pyrophosphate
Phytoene
(39)
β-Carotene
Zeaxanthin
CHO
(7) Xanthoxin
OH
15
9'
9
15'
1
6
(6) all-(E)-Violaxanthin
9-(Z)-Violaxanthin or 9'-(E)-Neoxanthin
(8) ABA-aldehyde
CO2H
(3) ABA
HO2C
(40)
(41) 9'-(Z)-Neoxanthin

Scheme 12

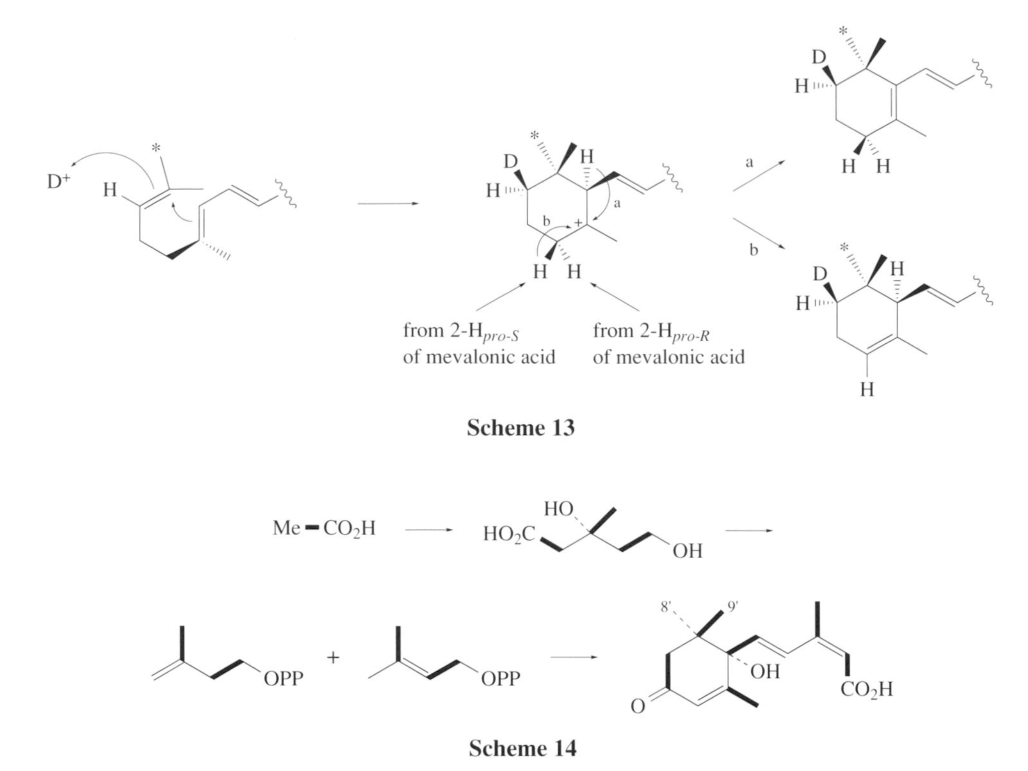

Scheme 13

Scheme 14

(ii) Immediate precursors

There are several C_{15} compounds which have been proposed as immediate precursors for ABA. 1′-Deoxy-ABA is metabolized to ABA in a low yield,[718] but it may not be a precursor of ABA since its endogenous occurrence has not been elucidated. 1′,4′-*trans*-Diol (**5**) of ABA is contained in plant tissues and organs, and it is converted to ABA as well as 1′,4′-*cis*-diol (**4**).[719,720] These diols might be metabolites of ABA formed by reduction of the 4′-carbonyl group rather than a precursor for ABA.[721] ABA-aldehyde (**8**) is easily oxidized to give ABA in plants.[721] The aldehyde can also be oxidized nonenzymatically. However, the experiments with ABA-deficient mutants has suggested that the oxidation of ABA-aldehyde is a critical step for synthesizing ABA. The ABA-deficient tomato mutants, *flacca* and *sitiens*, cannot convert ABA-aldehyde to ABA but convert it to 2-(*E*)-ABA-alcohol (**35**).[722] Other ABA-deficient mutants, *droopy* of potato, Az 34 of bean, and CKR 1 of *Nicotiana*, cannot convert ABA-aldehyde to ABA either.[723–725] These studies on the mutants, the endogenous occurrence of ABA-aldehyde, and the ^{18}O labeling experiment strongly suggest that ABA-aldehyde is an immediate precursor of ABA.[721]

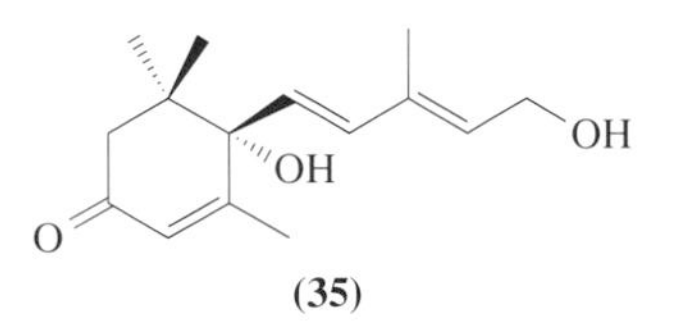

(35)

Xanthoxin (**7**) is endogenous in many plants,[726] and it is metabolized to ABA in plant tissues and cell-free systems.[727,728] The ABA-deficient mutants, *flacca* and *sitiens*, cannot convert xanthoxin to ABA.[727] It has been suggested that xanthoxin is converted to ABA via ABA-aldehyde. These findings suggest that xanthoxin is a precursor for ABA-aldehyde.[724] However, contradictory results have been reported. Carbon-14-mevalonic acid is incorporated into ABA in avocado fruits, but the label is not incorporated into xanthoxin which was added to the fruits as a cold trap.[729] Incorporation of $^{18}O_2$ into the C-1′ of ABA has been suggested to occur after cleavage of a carotenoid.[716] Plants contain more 2-(*E*)-xanthoxin than xanthoxin,[726] and 2-(*E*)-xanthoxin is not converted to ABA.[727] Formation of xanthoxin *in vivo* from carotenoids has not been elucidated although it is yielded *in vitro* by photooxidation and enzymatic degradation of violaxanthin. Xanthoxin might be a precursor in the direct pathway.[730] 4′-Hydroxy-β-ionylideneacetic acid (**36**), 1′,2′-epoxy-β-ionylideneacetic acid (**37**), and xanthoxin acid (**38**) are metabolized to ABA in plants,[729,731,732] suggesting that the ring in β-ionylideneethanol pyrophosphate (**39**), a cyclized form of 4,5-dehydro-(*E*)-farnesol, is oxidized in a similar way to that of oxidation of carotenoids to give ABA. Further investigation may be needed to confirm whether xanthoxin is a precursor of ABA in the carotenoid pathway.

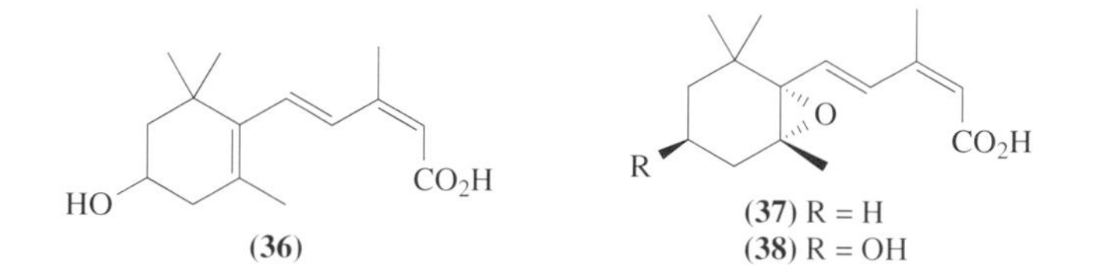

(iii) Potent carotenoids as precursors

The analysis of ABA-deficient mutants has suggested that oxidized carotenoids are the precursors for ABA. *Flacca* and *sitiens* contain 2,7-dimethyl-octa-2,4-dienedioic acid (**40**) at a higher level than the wild type.[733] The structure of (**40**) resembles the central moiety of carotenoids, suggesting that (**40**) is a by-product of ABA biosynthesis via the carotenoid pathway. If bonds between C-11 and C-12, and between C-11′ and C-12′ of 9′-(*Z*)-neoxanthin (**41**) are cleaved, the carotenoid will give one fragment of C_{10} and two fragments of C_{15}. The C_{10} fragment and one of the C_{15} fragments may be converted to (**40**) and 2-(*E*)-ABA-alcohol,[734] respectively, although Milborrow *et al.* have suggested that (**40**) is a decomposed compound of carotenoids and not a by-product of ABA biosynthesis.[735] Another ABA-deficient mutant, *aba2*, of *N. plumbaginifolia* is impaired in the epoxidation of zeaxanthin.[736] The gene, *ABA2*, encodes a chloroplast-imported protein sharing similarities with different monooxygenases and oxidases of bacterial origin.

Several carotenoids have been proposed as potent precursors of ABA from the feeding experiments and the quantitative analyses. Bean leaves under water-deficient conditions incorporate a label from $^{14}CO_2$ into ABA with a ratio of 30–70% of lutein and violaxanthin, and a label from ^{18}O-violaxanthin is also incorporated into the 1′-hydroxy group or the 4′-carbonyl group of ABA.[737] This suggests that lutein and violaxanthin are precursors of ABA. A cell-free system prepared from orange pericarps can metabolize ^{14}C-labeled β-carotene to violaxanthin and ABA.[738] Bean seedlings grown in the dark can accumulate ABA in response to water deficiency although the endogenous levels of carotenoids are low. The increase in ABA and its metabolites, phaseic acid and dihydrophaseic acid, is accompanied by the decrease in violaxanthin, 9-(*Z*)-violaxanthin, and 9′-(*Z*)-neoxanthin.[739] The total increase in ABA and its metabolites is coincident with the decrease in the carotenoids. ABA and xanthophylls including 9′-(*Z*)-neoxanthin of bean seedlings grown in deuterated water are labeled in a similar ratio.[740] It has been suggested that 9′-(*Z*)-neoxanthin is converted from violaxanthin.[741]

Based on these findings, the biosynthetic pathway of ABA via carotenoids has been proposed to be β-carotene→zeaxanthin→all-(*E*)-violaxanthin (**6**)→9-(*Z*)-violaxanthin or 9′-(*E*)-neoxanthin → 9′-(*Z*)-neoxanthin (**41**)→xanthoxin (**7**)→ABA-aldehyde (**8**)→ABA (**1**) (Scheme 12). Nonhebel and Milborrow have shown that tomato plants grown with deuterated water gives ABA labeled with 14 deuteriums and xanthoxin labeled with one deuterium, and suggested that the pool size of the carotenoid precursor may be 35 times larger than ABA.[742]

(iv) Fungi

The biosynthetic pathways of ABA in the phytopathogenic fungi, *Cercospora rosicola*, *Cercospora cruenta*, *Cercospora pini-densiflorae*, and *Botrytis cinerea* have been investigated. *C. rosicola* produces α-ionylideneethanol, α-ionylideneacetic acid, 4′-hydroxy-α-ionyldeneacetic acid, and 1′-deoxy-ABA, and these compounds are converted to ABA by this fungus.[743] The hydroxylation at C-1′ seems to occur at the last step before ABA. This suggests that *C. rosicola* synthesizes ABA via a direct pathway not via the carotenoid pathway.[648] *C. cruenta* produces 4′-hydroxy-γ-ionylideneacetic acid (**42**) and 1′,4′-dihydroxy-γ-ionylideneacetic acid (**43**) in addition to ABA.[744] These γ-acids are characteristic compounds of this fungus and *Stemphylium* sp.[745] *C. cruenta* converts the γ-acids to ABA.[746] 1′,4′-*trans*-Diol (**5**) of ABA was found in culture broth of *C. pini-densiflorae* and *B. cinerea*, and it has been suggested that it is a precursor of ABA in these fungi.[620,747,748]

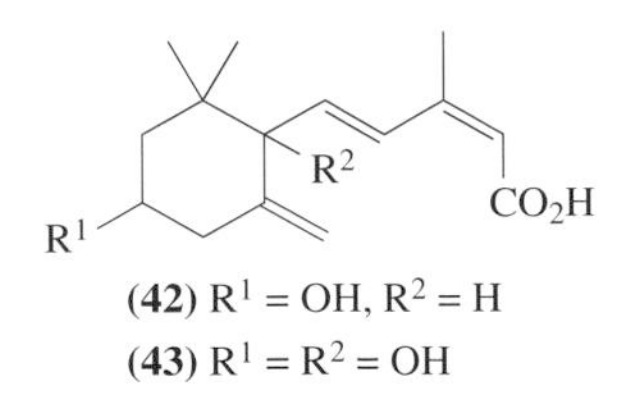

The plant genes for ABA biosynthesis may have been transferred into phytopathogenic fungi in a process of coevolution between host plants and parasites. However, these fungi biosynthesize ABA via direct pathways, and there is no evidence that the carotenoid pathway operates in these fungi.

8.02.5.4.2 Metabolism

The metabolic inactivation of (+)-ABA in plants is classified into the modification of the ring moiety and the conjugation with hydrophilic compounds as summarized in Scheme 15. Dehydrovomifoliol (**44**), vomifoliol (**45**), and drummondol (**46**) are known as natural compounds related to ABA by a short side chain,[749,750] but the conversion of ABA into these compounds by plants has not been confirmed.

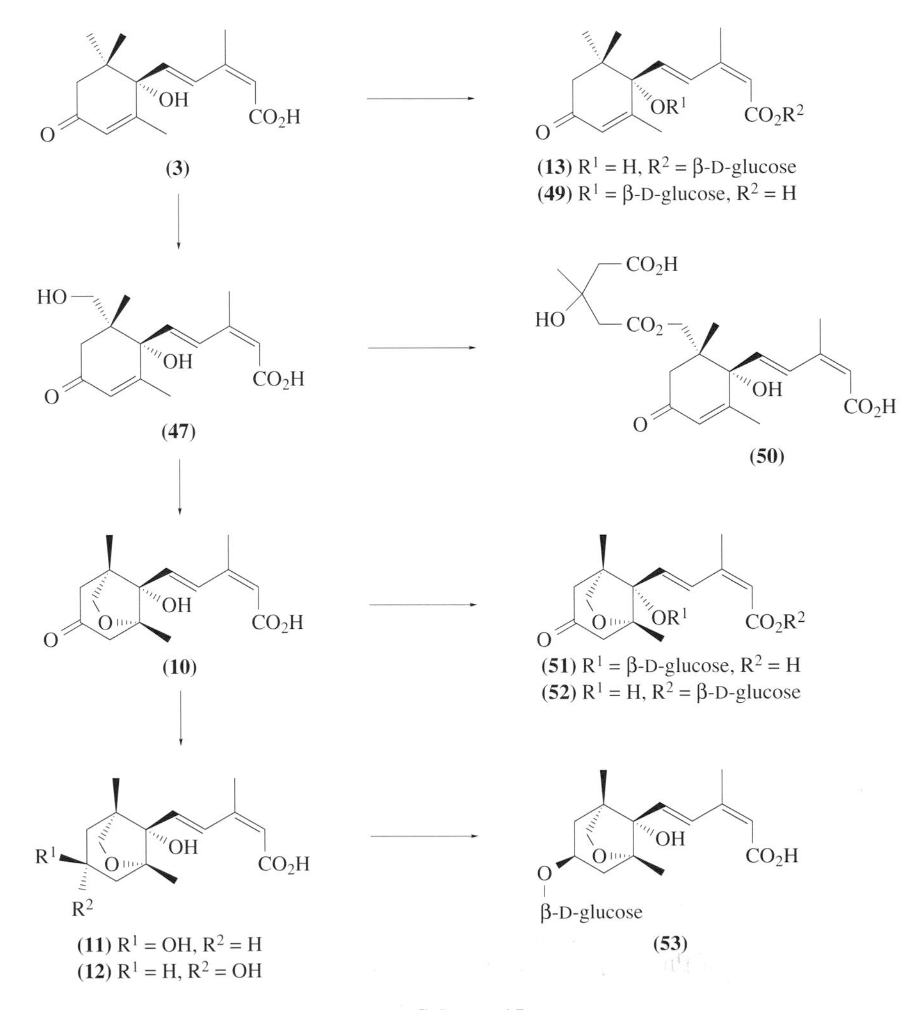

Scheme 15

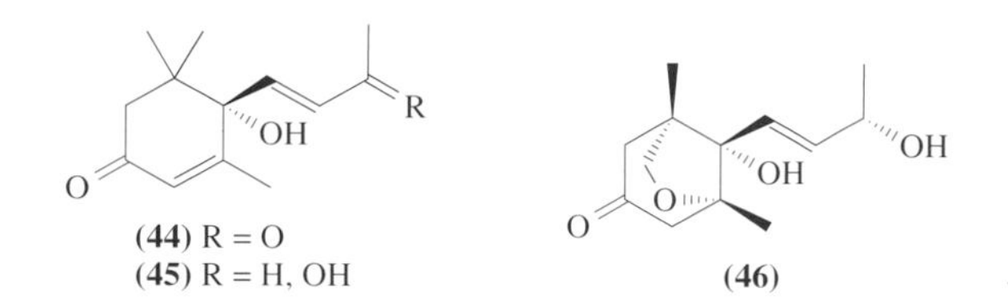

(**44**) R = O
(**45**) R = H, OH
(**46**)

(i) Modification of the ring moiety

ABA is first hydroxylated at C-8′ to give 8′-hydroxy-ABA (**47**). The 8′-hydroxylating activity has been detected in a microsomal fraction from liquid sperm of Eastern wild cucumber.[751] Oxygen and NADPH are essential factors for the enzyme, and the reaction is inhibited by carbon monoxide, suggesting that the enzyme is monooxygenase, a cytochrome P450. The 8′-hydroxylase seems to be induced by ABA itself.[752] 8′-Hydroxy-ABA in plant tissues is detected after derivatization to the acetate or the trimethylsilyl ether.[751,753] However, 8′-hydroxy-ABA is extremely labile, isomerizing spontaneously to phaseic acid (**10**); it has been isolated only once, by Milborrow in 1969.[754]

The isomerization of 8′-hydroxy-ABA to phaseic acid is an intramolecular Michael addition: the 8′-hydroxy group attacks C-2′ from the α-face to form an enol intermediate (**48**), which is converted to phaseic acid. This isomerization is probably regulated by an enzyme. The NMR analysis of dihydrophaseic acid converted from the hexadeuterated ABA (**17**) by tomato seedlings has shown that the deuterium at C-3′ of ABA is retained at 3′-*pro-R* of phaseic acid.[645] The 3′-$H_{pro\text{-}R}$ of phaseic acid exchanges with a proton from water faster than the 3′-$H_{pro\text{-}S}$ via an enol in the presence of base. This means that the more labile proton is retained in the phaseic acid formed in tomato seedlings, suggesting that the addition of a proton to the enol intermediate (**48**) from the α-face of the ring must be catalyzed by an isomerase. The composition of the metabolites of 3′-fluoro-ABA (as described later) also points to the involvement of the isomerase in the reaction.[755] Phaseic acid is more stable than 8′-hydroxy-ABA, but in a solution it exists along with a small amount of 8′-hydroxy-ABA which can be trapped by boric acid.[756]

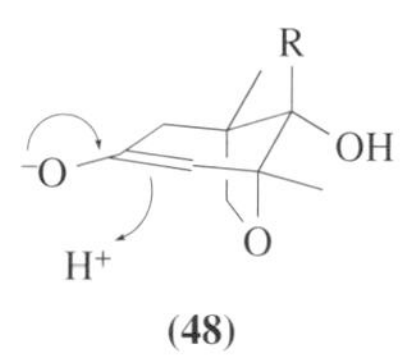

(**48**)

Phaseic acid is further converted to dihydrophaseic acid (**11**) and *epi*-dihydrophaseic acid (**12**) in plants. The endogenous level of *epi*-dihydrophaseic acid in plants is about 10% of the level of dihydrophaseic acid. *epi*-Dihydrophaseic acid is not an artifact formed during extraction since *epi*-dihydrophaseic acid converted from [4′-^{18}O]-ABA by bean seedlings retains the label at C-4′, as does dihydrophaseic acid.[627] Enzymatic activity reducing the 4′-carbonyl group of phaseic acid has been found in a soluble fraction from liquid sperm of Eastern wild cucumber.[751] It is unknown whether the same enzyme catalyzes the reduction of phaseic acid to give dihydrophaseic acid and its epimer. Dihydrophaseic acid might be further metabolized in plants since the presence of 5′-hydroxydihydrophaseic acid has been suggested in ferns and maize.[757,758] However, the metabolism of the ring moiety after dihydrophaseic acid has not been confirmed.

The biological activity of 8′-hydroxy-ABA cannot be tested due to its instability, but several studies have suggested that 8′-hydroxy-ABA is active. The borate of 8′-hydroxy-ABA induces biosynthesis of lipid and oil body protein in embryos of *Brassica napus*.[756] 8′-Monfluoro-ABA, which mimics 8′-hydroxy-ABA, is as active as ABA,[660] and 8′-methoxy-ABA (**57**) is more active than ABA.[759] The activity of phaseic acid is low in many biological assays,[669] but, like ABA, it shows high activity in the inhibition of α-amylase induction.[760] 8′-Hydroxy-ABA which is partly isomerized from phaseic acid might show this activity. Dihydrophaseic acid and its epimer are almost inactive in the assays tested.[669] The inactivation of ABA would be achieved partly in the step involving isomerization of 8′-hydroxy-ABA to phaseic acid, and completed in the step reducing phaseic acid to dihydrophaseic acid and its epimer.

(ii) Conjugation

ABA is conjugated with β-D-glucose at the 1-carboxy and the 1′-hydroxy groups to form the glucosyl ester (**13**) and the glucoside (**49**), respectively. The glucosyl ester is a major conjugate of ABA, and found in many plants.[761] Glucose transferase of ABA which synthesizes the glucosyl ester from ABA and UDP-glucose has been partly purified from cultured cells of *Macleaya microcarpa*.[762] The optimum pH of this enzyme is 5.0, suggesting that the glucosylation occurs in the vacuole. The major role of conjugation would be excretion of ABA by compartmentation into a vacuole. It has been suggested that this ABA conjugate does not subsequently release ABA.[763] Thus plants exposed to water deficiency supply ABA by *de novo* synthesis. The 1′-*O*-glucoside exists in tomato plants and apple seeds.[764] The 1′-*O*-glucoside not only decomposes to the methyl ester of ABA and glucose by treatment with diazomethane but also isomerizes spontaneously to the glucosyl ester in an acidic methanol solution, probably due to the steric repulsion. A bound form of ABA which releases ABA and its methyl ester upon mild acid hydrolysis has been isolated from peas and barley.[765] This "adduct" might be a conjugate of the 4′-en-4′-ol form of ABA.

Labile 8′-hydroxy-ABA is conjugated with β-hydroxy-β-methylglutaric acid to form a stable ester (**50**).[766] This conjugate is a characteristic metabolite of *Robinia pseudacacia*. The 1′-*O*-β-D-glucoside (**51**) of phaseic acid has been found in apple seeds and tomato leaves fed with (±)-[2-^{14}C]-ABA.[767] Exogenous phaseic acid is converted to its 1-*O*-β-D-glucosyl ester (**52**).[768] The glucoside is an endogenous metabolite in plants. The 4′-*O*-β-D-glucoside (**53**) of dihydrophaseic acid occurs in avocado fruits and has been isolated from tomato seedlings fed with (±)-[2-^{14}C]-ABA.[769,770] *epi*-Dihydrophaseic acid is converted to a polar metabolite by tomato seedlings, the metabolite is probably its 4′-*O*-glucoside.[771]

(iii) (−)-ABA

The metabolism of unnatural (−)-ABA is different from that of (+)-ABA. Exogenous (+)-ABA is mainly metabolized to phaseic acid, dihydrophaseic acid, and its 4′-*O*-glucoside, and the formation of glucosyl ester and glucoside of ABA is minor. However, the glucosyl ester is the major metabolite of (−)-ABA.[771,772] This suggests that the major part of the glucosyl ester of ABA formed in plants fed with (±)-ABA is that of (−)-ABA.

The ABA-hydroxylating enzyme in avocado fruits and cultured maize cells does not distinguish strictly (−)-ABA from (+)-ABA.[631,773] These plants metabolize (−)-ABA to unnatural phaseic and dihydrophaseic acids and a small amount is metabolized to (*R*)-(−)-7′-hydroxy-ABA (**54**).[772,773] The hydroxylation of C-7′ of (−)-ABA may be explained by the pseudosymmetry of ABA. In the half-chair conformation with the pseudoequatorial side chain (**HC_2** in Table 10), C-7′ of (−)-ABA corresponds spatially to C-8′ of (+)-ABA. The 8′-hydroxylase of (+)-ABA might not distinguish C-7′ of (−)-ABA from C-8′ of (+)-ABA. However, it has not been proved that 8′-hydroxylase of (+)-ABA catalyzes the 7′-hydroxylation of (−)-ABA. The natural occurrence of 7′-hydroxy-ABA and the formation of 7′-hydroxy-ABA from (+)-ABA fed to cultured cells suggest the occurrence of 7′-hydroxylase of (+)-ABA which may be concerned with the 7′-hydroxylation of (−)-ABA.[774,775]

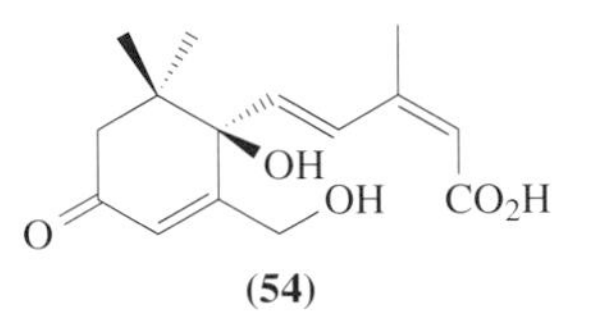

(54)

(iv) Metabolism-resistant analogues

ABA could make a great contribution to agriculture because it can protect plants from various environmental stresses. However, its commercial application remains minimal, mainly because of the rapid metabolism of applied ABA.[776] In order to solve this problem, highly active and long-lasting analogues of ABA that resist metabolic inactivation by modification of the ring moiety have been developed by Todoroki *et al.*[755,759,777] and other groups.[778,779] The analogues are classified into two types: type 1, resistant to the hydroxylation of C-8′, and type 2, resistant to the isomerization to phaseic acid.

(+)-8′,8′,8′-Trifluoro and (+)-8′,8′-difluoro-ABA (**55** and **56**, respectively) are analogues of type 1.[777,778] The replacement of hydrogen by fluorine at C-8′ of ABA should block hydroxylation at C-8′ since the energy of the C—F bond is higher than that of the C—H bond. The fluorinated analogues show stable activity, and are 20–30 times more active than ABA in a long-term assay.[777] (+)-8′,8′,8′-Trifluoro-ABA is the most active analogue of ABA. Deuterium labeling is also expected to confer metabolic stability on ABA by the primary isotope effect. (+)-7′,7′,7′,8′,8′,8′,9′,9′,9′-Nonadeutero-ABA synthesized by Lamb *et al.* is metabolized to octadeuterophaseic acid at a slower rate than (+)-ABA.[779]

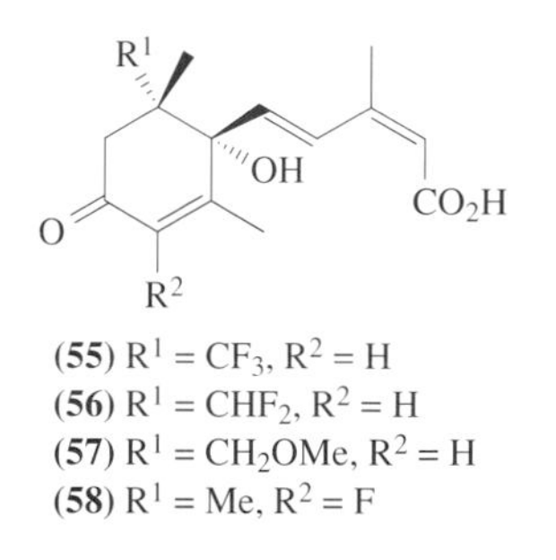

(**55**) $R^1 = CF_3$, $R^2 = H$
(**56**) $R^1 = CHF_2$, $R^2 = H$
(**57**) $R^1 = CH_2OMe$, $R^2 = H$
(**58**) $R^1 = Me$, $R^2 = F$

Type 2 analogue, (+)-8′-methoxy-ABA (**57**), which does not have the nucleophilicity of the 8′-oxygen, also shows long-lasting activity.[759] (+)-3′-Fluoro-ABA (**58**) is another analogue of type 2, and resists the cyclization by means of another mechanism.[755] Substitution of the 3′-hydrogen by fluorine increases the electron density of C-2′ by pushing the π electron at C-3′ toward C-2′, so this analogue can acquire resistance to the nucleophilic addition of the 8′-oxygen. The 3′-fluoro-8′-hydroxy-ABA (**59**) converted from (+)-3′-fluoro-ABA in bean shoots is stable although it converts gradually to 3′β- and 3′α-fluorophaseic acids (**60** and **61**, respectively) and these three metabolites coexist at equilibrium in the ratio of 51 : 7 : 42 (Scheme 16, percentages are the compositions of methyl esters, R is the side chain of ABA).[755] This indicates that the isomerization of 8′-hydroxy-ABA to phaseic acid can be controlled by regulation of the electron density of C-2′. The equal activity of 3′-fluoro-ABA to ABA could be caused by the conversion of 3′-fluorophaseic acids to 3′-fluorodihydrophaseic acids (**62** and **63**), as shown in Scheme 16. The discrepancy between the ratio of (**60**) and (**61**), 1 : 6, and the ratio of (**62**) and (**63**), 1 : 1, has suggested the participation of the enzyme in the isomerization of (**59**) to (**60**) via the enol intermediate (**64**).

8.02.6 ETHYLENE

8.02.6.1 Biological Activity and Physiological Roles

Ethylene is the simplest molecule which shows biological activity in higher plants. It has no obvious physiological effect on mammals, insects, or single-cell organisms, although it has anesthetic effects on mammals. There are several microorganisms which produce large amounts of ethylene, but its physiological role is not known. As with the physiological activity of the other plant hormones, the biological effects of ethylene vary during the developmental stages of higher plants. This is exemplified by the fact that the apparent effects on different organs at different stages are not related to each other and that corresponding organs of different species (or cells at different locations in the same species) often respond in an opposite way.[780,781]

The earliest recognized effect of gaseous substances on plants was probably stimulation of leaf abscission.[780] Precocious abscission of street trees had been observed near broken illuminating gas pipelines in Germany and the USA from the late nineteenth to early twentieth centuries, though the active component was not identified. Careful studies made by a Russian scientist who has observed unusual growth behavior of etiolated pea plants grown in a chemical laboratory revealed that the most active component was ethylene contained in leaked illuminating gas.[782] The biological activity of ethylene is summarized in Table 11.

Endogenous ethylene has many physiological effects. When dicot seeds germinate underground, etiolated seedlings emerge. Etiolated seedlings of dicot plants have unique morphology; a pair of plumules are closed tightly, and a short stem portion directly below the plumules is bent making a hook. When such seedlings grow above the ground and are exposed to light, the hook becomes

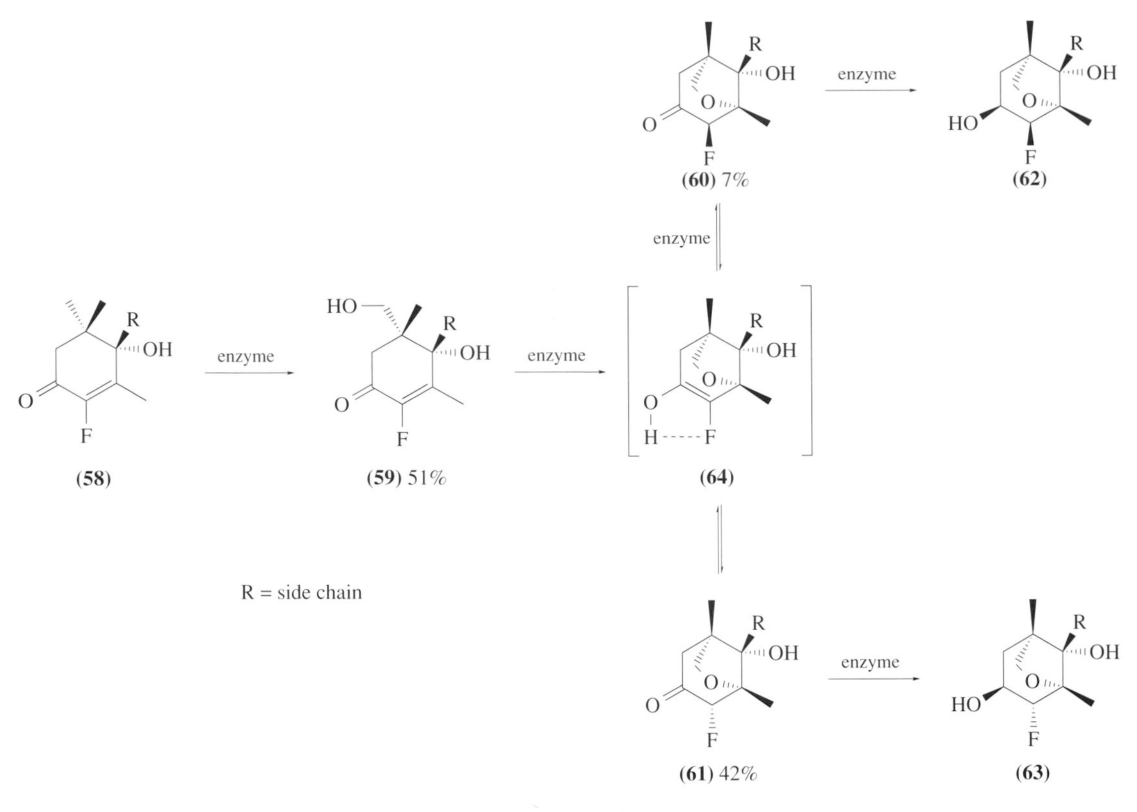

Scheme 16

straight and the closed plumules start to open. This unique morphology of etiolated seedlings is entirely maintained by ethylene that is produced by seedlings. The effect of light is to reduce the rate of ethylene production and increase evolution of carbon dioxide, which is also a competitive inhibitor of ethylene action.[783,784] The unique morphology has been thought to prevent the apical meristem from physical damage that might be imposed by soil particles during growth underground.

Many semiaquatic plants have root systems below the water surface, and leaves and flowers above the water surface, with both portions connected by rachis, petiole, or stem. When such plants are submerged by flooding, elongation growth of rachis, petiole, or stem is greatly stimulated and pushes up the terrestrial portion above the water surface. Ethylene stimulates the elongation growth of the rachis, petiole, or stem of such plants, and when submerged, the plants increase ethylene production.[785] The ecophysiological interpretation of the phenomenon is that the response is to avoid drowning.

Leaf abscission is a programmed organ death. Unlike animals, plants grow and form organs at apical meristems located at the terminal ends of stems, branches, and roots, adding more new organs to the existing shoots. Senescence and abscission of old leaves are mechanisms that degrade cellular components in old leaves and supply nutrients to younger, actively growing organs, followed by removal of "empty" leaves. Shortly before abscission occurs at the abscission zone in the pulvinus, a burst of ethylene production takes place at the abscission zone.[785] Ethylene induces several hydrolytic enzymes including a specific isoform of cellulase necessary for cell separation. If the effect of ethylene is suppressed, leaf abscission never occurs.[786] When part of a plant organ is physically damaged, for example cut, several cell layers adjacent to the cut surface are activated to produce and accumulate protective materials such as suberin to seal off the cut surface and antimicrobial substances to protect invasion of pathogenic microorganisms. Such cell activation is mimicked by ethylene, and tissue wounding indeed stimulates ethylene production from the wound-affected cells. In this case ethylene clearly serves as a mediator of cell activation to prevent the deterioration of the entire organism.[787]

Thus, ethylene plays important roles in many stages of plant development and environmental responses. As described earlier, the effect of ethylene can be beneficial to a plant at a certain period and in a certain tissue, but harmful at other periods or in other tissues. For this to be the case, ethylene production must be precisely regulated by many factors, both endogenous and exogenous.

Table 11 Biological effects of ethylene.

Tissue/organ level
Promotion of fruit ripening
Promotion of senescence of leaves and petals
Stimulation of adventitious root formation
Flower formation in Bromelidiaceae
Increase in flowering rate or break of flower bud dormancy
Change to female flowers of developing flowers in Cucurbitaceae, Euphorbiaceae, and Cannabinaceae
Stimulation of root hair formation
Inhibition of flower formation in many long and short day plants
Diageotropic growth in etiolated *Pisum sativum* and *Vicia faba* seedlings
Breaking of dormancy
—seeds of clover, peanuts, and lettuce
—bulbs of gladiolus and iris
Abscission of leaves, flowers, and fruits
Inhibition of auxin polar transport
Cell level
Inhibition of cell elongation and stimulation of cell lateral expansion
—stem cells of many dicot plants
Stimulation of cell elongation
—outer cells of hook of etiolated dicot seedlings
—cells of upper side of leaf petioles of dicot seedlings (causes petiole epinasty)
—internodes and mesocotyl cells of rice seedlings and deep-water rice
—rachis, stems, or petioles of some of semiaquatic plants, such as *Callitriche platycaepa*, *Ranunculus sceleratus*, *Sagittaria pygmaea*, *Potamogeton distinctus*, and *Regenllidium diphyllum*
Stimulation of water uptake in pulvinus
—causes bending of lamina or petioles
Inhibition of cell division
—stem and root apical meristem of many plants
Biochemical level
Induction of many enzymes, such as peroxidase, phenylalanine-ammonia lyase, chitinase, glucanase, cinnamic-4-hydroxylase, and chalcone synthase
Swelling of mitochondria
Change in membrane permeability
Increased secretion of rubber latex and resins

8.02.6.2 Structural Requirement for Biological Activity

Burg and Burg[788] examined various analogues of ethylene for biological activity to suppress elongation of pea stems (Table 12). Ethane did not affect growth and acetylene showed significant activity, but less than ethylene; an unsaturated bond is essential for activity, a double bond being more active than a triple bond. Replacement of one of the hydrogen atoms of ethylene by a large group (propylene and 1-butene) reduces the biological activity according to the molecular size of the analogue, indicating that the extra bulk hinders binding to the receptor site. Analogues that lack a terminal carbon atom with an unsaturated bond (carbon dioxide and 2-butene) have no biological activity. The introduction of halogen atoms that change the distribution of electrons around the molecule (e.g., chloroethylene and bromoethylene) also reduces biological activity.

Other small molecules which have a terminal carbon with an unsaturated bond (e.g., carbon monoxide) show low but definite biological activity. These results indicate that the molecular requirement for biological activity is a small molecule with a terminal carbon connected by multiple bonding. The same degree of biological activity among ethylene analogues has been demonstrated for stimulation of abscission.

8.02.6.3 Biosynthesis

8.02.6.3.1 Microbial system

A wide variety of microorganisms produce ethylene. Although the physiological significance of ethylene in those microorganisms is not known, many of them are either pathogenic or epiphytic to plants, and may affect local cell growth and differentiation of the plant as a result of ethylene

Table 12 Relative biological activity of ethylene and its derivatives. Elongation growth of excised sections of etiolated pea seedlings was measured at various concentrations of each gas and the concentration in the gas phase for half-maximum activity was determined (modified from Burg and Burg[788]).

Compound	*Concentration in gas phase* ($\mu L\ L^{-1}$)
Ethylene	0.1
Propylene	10
Chloroethylene	140
Carbon monoxide	270
Acetylene	280
Fluoroethylene	430
Methylacetylene	800
Bromoethylene	1 600
Ethylacetylene	1.1×10^4
1-Butene	2.7×10^4
1,3-Butadiene	5.0×10^5
Ethane	inactive

production. Microorganisms produce ethylene by two different pathways, depending on species. *Penicillium digitatum* and *Pseudomonas syringae* use 2-oxoglutarate and *Cryptococcus albidus* uses L-methionine as substrate.[791]

(i) The 2-oxoglutarate-dependent pathway

By feeding experiments of radioactive substrates, Chou and Yang[789] demonstrated that either 2-oxoglutarate or glutamate was a direct precursor of ethylene in *P. digitatum*. Goto and Hyodo[790] established a cell-free ethylene-forming system from *P. syringae* pv. *phaseolicola* that required 2-oxoglutarate as substrate, and histidine and Fe^{2+} as cofactors. The ethylene-forming enzymes were later purified from *P. digitatum*[791] and *P. syringae*[792] and found to catalyze oxidation of 2-oxoglutarate to 1 mol of ethylene and 3 mol of carbon dioxide in the presence of arginine, Fe^{2+}, and oxygen. The enzyme from *P. syringae* is bifunctional and ethylene is formed when the enzyme dioxygenates 2-oxoglutarate, but when both arginine and 2-oxoglutarate are monooxygenated, the two substrates are degraded. While 2 mol of 2-oxoglutarate are degraded to 2 mol of ethylene, 1 mol each of 2-oxoglutarate and arginine are degraded. The ethylene-forming enzyme of *P. syringae* is encoded by an endogenous plasmid (pPSP1) and its gene has been isolated.[793]

(ii) The 2-oxo-4-methylthiobutyrate-dependent pathway

Some microorganisms in culture show methionine-dependent ethylene formation. In studies with *E. coli*, 2-oxo-4-methylthiobutyrate (KMB) produced from methionine by transamination was suggested as a precursor of ethylene, and later a cell-free system which produced ethylene from KMB in the presence of NAD(P)H, EDTA-Fe^{3+} and oxygen was established.[794] The proposed mechanism for a similar enzyme purified from *C. albidus* involves reduction of EDTA-Fe^{3+} to EDTA-Fe^{2+} by the enzyme, reduction of oxygen to superoxide by EDTA-Fe^{2+}, dismutation of superoxide to hydrogen peroxide, reduction by EDTA-Fe^{2+} of hydrogen peroxide to hydroxy radical, and oxidation of KMB by the hydroxy radical to ethylene.[795]

8.02.6.3.2 Higher plant system

(i) ACC pathway

The major pathway of ethylene biosynthesis in higher plants includes L-methionine, *S*-adenosylmethionine (AdoMet), and 1-aminocyclopropane-1-carboxylic acid (ACC) as the intermediates, and this pathway is commonly called the ACC pathway (Scheme 17; after Miyazaki and Yang).[796]

The determination of ACC as an immediate precursor greatly speeded the progress of investigation on ethylene biosynthesis.[797] Some lower plants such as the semiaquatic fern *Regnellidium diphyllum* and the liverwort *Riella helicophylla* produce ethylene but do not use ACC as a precursor, and there is convincing evidence for the presence of a nonACC pathway.[798]

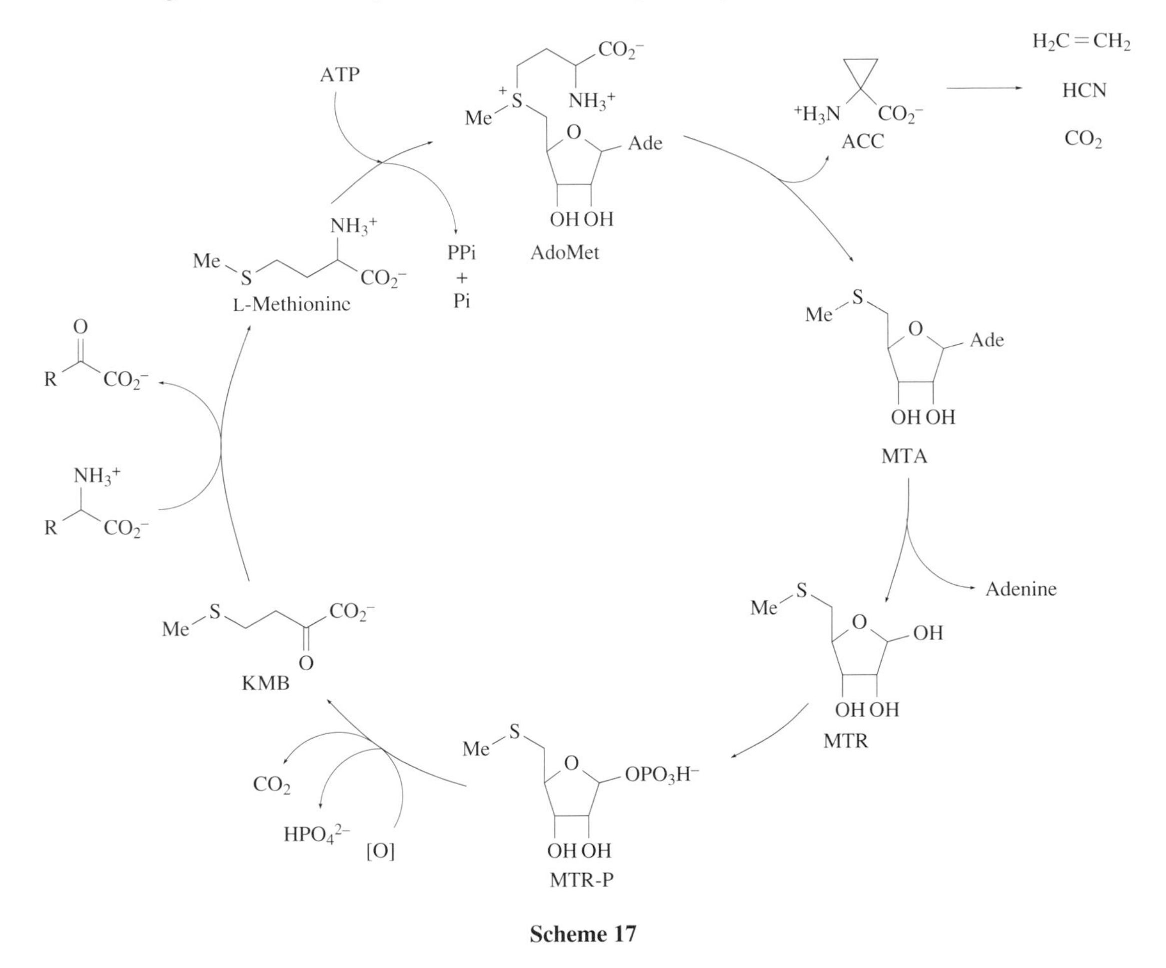

Scheme 17

In an induced ethylene-producing system such as auxin-treated, wounded tissues or ripening fruits, comparison of the rates of ethylene production and the endogenous levels of ACC content under the various conditions reveals that the endogenous levels of ACC is a primary determinant of the rate of ethylene production from tissues:[799,800] the ACC-forming reaction is the rate-limiting step of ethylene biosynthesis.

ACC synthase activity which catalyzes conversion of AdoMet to ACC in the presence of 5′-pyridoxal phosphate as a cofactor was detected in extracts of ripe tomato fruits[801,802] and auxin-treated stems,[800,803] and later in many other ethylene-producing tissues. In tissues that produce little ethylene, ACC synthase activity is always low, but increases concomitant with the rise of ethylene production in tissues. However, many low ethylene-producing tissues efficiently produce ethylene from exogenously supplied ACC. Thus, ACC-dependent ethylene-forming enzyme (EFE, now identified as ACC oxidase) is thought to be a constitutive enzyme or induced earlier than ACC synthase. These observations clearly indicate that ACC synthase is the rate-limiting enzyme of ethylene biosynthesis.

(ii) Methionine cycle

The ACC pathway is coupled to a cyclic process which regenerates L-methionine[786] (Scheme 17). A unique feature of this cyclic pathway (the methionine cycle) is that methylthioadenosine produced together with ACC from AdoMet is recycled to produce methionine. Adams and Yang[804] found that when ^{35}S-methionine was supplied to tissue plugs from climacteric apple fruits (ethylene producing) the plugs produced both radioactive 5′-methylthioadenosine (MTA) and 5′-methylthioribose (MTR), but plugs from preclimacteric fruits (nonethylene producing) did not.

The methionine cycle operates in animals and microorganisms in relation to polyamine synthesis. Thus, enzymes which catalyze all of these reactions are present in all organisms. However, two enzymes in the ACC pathway, ACC synthase (AdoMet methylthioadenosine-lyase) and ACC oxidase, are unique to higher plants. ACC is also malonylated to form *N*-malonyl ACC, which is inactive as a precursor of ethylene.[805,806]

The biological significance of the methionine cycle is of great importance for synthesis of ethylene and polyamines, both biologically active compounds. A small amount of methionine serves as a catalyst of the cyclic process while ethylene or polyamines are continuously produced. Thus, even in tissues in which supply of free methionine through its *de novo* synthesis or degradation of proteins is limited, ethylene and polyamines are produced in large amounts if a sufficient supply of ATP is available. It can be said that the real source of ethylene and polyamines is ATP.

8.02.6.4 Regulation of Biosynthesis

The rate of ethylene production by plant tissues varies considerably with a number of factors, and its regulatory system is highly complex. The physiological aspects of the regulation of ethylene production are detailed in other reviews.[807–810]

There are several characteristics of the regulatory systems of ethylene production. First, the rate of ethylene production is regulated developmentally. The basic program of ethylene biosynthesis during normal development is determined genetically. As described in an earlier section, endogenous ethylene plays various roles in the normal epigenetic development of plants. An increased level of endogenous ethylene plays a key role in maintenance of the morphology of dicot seedlings, senescence of leaves and flower petals, abscission of leaves, flowers, petals, and young fruits, and ripening of fruits. However, these increases in ethylene production are transient and soon dissipate during the epigenetic program. Thus, ethylene biosynthesis increases or decreases frequently in specific tissues or organs depending upon the developmental stages.

Second, the rate of ethylene production is affected greatly by changes in environmental factors. Plants growing in the wild receive two kinds of environmental stimuli. One is the regular, daily alternate changes of light and dark, and of high and low temperature. These changes are also coupled with the annual photoperiodic cycle. Another is irregular change by a number of transient events in nature, and include: unusual drought or flooding; unusually high or low temperatures; cutting, wounding, or stress caused by the physical force of wind, insects, or mammals; touch or pressure exerted by other objects; infection by pathogenic microorganisms, and so on. Although regular environmental changes are necessary factors for normal plant development, the irregular changes are not, and may kill the plant. In many cases where plants receive irregular stimuli, ethylene serves as a mediator which modulates multiple metabolic processes that may establish tolerance to harsh environmental changes or that heal wounds. This characteristic indicates that ethylene synthesis is regulated by a set of genes specific to different physical and chemical stimuli.

Third, ethylene production is regulated by other plant hormones. Physiological observations clearly indicate that plant hormones interact with one another, and ethylene is no exception.[811] However, the plant hormone interaction in physiological processes has been difficult to dissect biochemically. Ethylene biosynthesis, a series of defined biochemical reactions, is induced by auxin in vegetative tissues, and the auxin action is synergistically enhanced by cytokinin and antagonized by ABA. Ethylene itself regulates its own synthesis in both ways; autocatalytic and autoinhibitory depending upon tissues. Therefore, any factors that affect endogenous levels of auxin, cytokinin, and ABA will regulate ethylene synthesis.

The major regulatory steps in ethylene biosynthesis are endogenous levels of ACC synthase and ACC oxidase. However, the major factor that determines the rate of ethylene synthesis is the endogenous level of free ACC and there is a good correlation between the two.[800,812] As ACC is rapidly metabolized to *N*-malonylACC (MACC),[806,813] or transported into vacuoles,[814] a dynamic balance of ACC synthesis and ACC metabolism (or vascular transport) is an important factor, and high endogenous ACC synthase activity is required to maintain active ethylene synthesis. Cessation of ACC synthase formation quickly reduces ethylene production because ACC synthase is rapidly inactivated.[815,816] ACC synthase, ACC oxidase, and ACC *N*-malonyltransferase are now recognized as inducible enzymes, but, in many cases, an increase of ACC oxidase activity precedes an increase of ACC synthase activity, and *N*-malonytransferase is induced by ethylene. These observations indicate that the induction mechanism of ACC synthase governs the regulation of ethylene production.

8.02.6.4.1 Regulation of ACC synthase induction

Increases in ACC synthase activity in response to auxin[800,803] and wounding[817] are inhibited by inhibitors of protein synthesis and RNA synthesis, suggesting that synthesis of ACC synthase is transcriptionally regulated. Evidence for *de novo* synthesis of ACC synthase was obtained by the density labeling method for the wound-induced tomato enzyme[818] and the elicitor-induced parsley cell enzyme.[819] However, in cultured parsley[820] and tomato cells,[819] elicitor-induced increases of ACC synthase activity are not suppressed by cordycepin, an inhibitor of RNA synthesis. These observations suggest that in some cases, increases in ACC synthase are regulated at the post-transcriptional level. It should be noted that although the primary translation product of ACC synthase mRNA is enzymatically active,[821,822] deletion of some 60 amino acids at the C-terminus increases the specific activity several folds[823,824] and the presence of the C-terminal deleted enzyme in cells is observed together with the primary translation product.[825] Thus, there is a possibility that posttranslational modification further increases the overall endogenous activity of ACC synthase.

Immunochemical and fluorographic analysis of *in vitro* translation products of mRNAs from fresh and wounded tissues of winter squash mesocarp[825] and tomato pericarp[826,827] clearly showed the transcriptional control of ACC synthase. Translatable mRNA for ACC synthase is not detectable in fresh tissue, but dramatically increases after wounding with a lag period of a few hours. After wounding, there is a good correlation between the time course profiles of ethylene production, ACC synthase activity, and the relative abundance of the mRNA.

The cloning of cDNAs for ACC synthase has enabled the direct comparison of steady-state levels of mRNA under various conditions. In all tissues examined, induction of ACC synthase is transcriptionally regulated; mRNA is not present at a detectable level in tissues producing no ethylene. RNA blots analyzed with cloned cDNAs as probes show that the genes of ACC synthase isozymes are differentially regulated by different stimuli. Out of two genes of winter squash (*C. maxima*), CM-*ACS1* and CM-*ACS2*, CM-*ACS1* is expressed only when tissues are mechanically wounded, whereas CM-*ACS2* is expressed when vegetative tissues are treated with auxin.[821,828] In addition, the wound-induced expression of CM-*ACS1* is observed in both fruit mesocarp (reproductive tissue) and hypocotyls (vegetative tissue) but the auxin-induced expression of CM-*ACS2* is limited to vegetative tissues. CM-*ACS1* expression is never induced by auxin and CM-*ACS2* never by wounding. Auxin treatment to excised (wounded) hypocotyls causes a small but significant accumulation of CM-*ACS1* mRNA together with a large amount of CM-*ACS2* mRNA, but intact seedlings sprayed with auxin do not accumulate CM-*ACS1* mRNA. This result indicates that the wound-induced expression of CM-*ACS1* is further enhanced by auxin. The wound-induced expression of CM-*ACS1* is not affected by cytokinin, but stimulated by abscisic acid, whereas the auxin-induced expression of CM-*ACS2* is stimulated by cytokinin but suppressed by ABA.[829] Exogenous ethylene suppresses expression of both CM-*ACS*1[821,828] and CM-*ACS2*.[830] Treatments with aminoethoxyvinylglycine (AVG), (aminooxy)$_2$ acetic acid (AOA), or 2,5-norbornadiene (NBD) of wounded or auxin-treated tissues stimulate the wound-induced or auxin-induced expression of the respective genes, indicating that endogenously produced ethylene also is repressing the expression of ACC synthase genes.

In the mung bean (*Vigna radiata*) two genes (VR-*ACS1* and VR-*ACS6*) out of a family of six were shown to be expressed.[831,832] Neither VR-*ACS1* nor VR-*ACS6* mRNA are detectable in intact hypocotyls. The expression of the VR-*ACS6* gene is specifically induced by auxin treatment, whereas expression of VR-*ACS1* is induced by cycloheximide.[833] Like auxin-specific CM-*ACS2*, the auxin-induced expression of VR-*ACS6* is stimulated by cytokinin and suppressed by ethylene and ABA. Cycloheximide has often been observed to induce the expression of some plant genes, for example, a zucchini ACS isogene, CP-*ACS1A*. It is speculated that the expression of CP-*ACS1A* is under the control of a short-lived repressor protein.[834]

Among five genes of tomato, at least four genes (LE-*ACS2*, LE-*ACS3*, LE-*ACS4*, and LE-*ACS5*) were reported to be expressed.[835] LE-*ACS2* and LE-*ACS4* are not expressed in intact mature green fruits, but are expressed in ripening fruits. Accumulation of LE-*ACS2* mRNA starts earlier than LE-*ACS4* during ripening, and a steady-state level of LE-*ACS2* mRNA is more abundant than that of LE-*ACS4*.[834] However, the response of the two genes to tissue wounding is different. The expression of LE-*ACS2* is greatly stimulated, while that of LE-*ACS4* is repressed after wounding of tissue.[836,837] Expression of LE-*ACS2* is induced by both wounding and ripening, but that of LE-*ACS4* is expressed only during ripening. Fruit ripening is accompanied by the loss of cell-to-cell adhesion which eventually leads to softening of fruits. The loss of cell-to-cell adhesion will generate a wound signal, and the wound signal induces expression of LE-*ACS2*. While LE-*ACS4* is expressed in response to the ripening signal, its expression is repressed by the wound signal, resulting in less accumulation of mRNA than LE-*ACS4* mRNA in red ripe fruits. Expression of LE-*ACS3* is induced in hypocotyl sections by auxin.[838] Studies also reveal that a low level of LE-*ACS3* mRNA is detected in intact mature green fruits but not in seedlings. Wounding of mature green fruits does not appreciably change mRNA levels, but when excised green fruits are treated with auxin, expression of LE-*ACS3* is greatly stimulated. Moreover, expression of LE-*ACS3* is rapidly induced in intact seedlings sprayed with auxin, indicating that expression of this gene is induced specifically by auxin.[839] Thus, in tomato, at least three genes (LE-*ACS2*, LE-*ACS3*, and LE-*ACS4*) are expressed in a stimulus-specific manner.

ACC synthase isozyme genes also show tissue-specific or developmentally regulated expression. LE-*ACS2* and LE-*ACS4* are expressed in fruits but not in etiolated seedlings. LE-*ACS2* is not expressed in young and mature leaves, roots, petals, or pistils, but is expressed in stamen, mature and senescent anthers, and senescent petals.[837] Similar organ specificity of expression of ACC synthase genes is observed in carnation flowers.[840]

8.02.6.4.2 Regulation of ACC oxidase induction

EFE is found in a number of plant tissues,[841] and has been thought to be a constitutive enzyme. However, ACC oxidase was originally identified as the product of a mRNA (pTOM13) whose abundance increased during ripening and after wounding of tomato fruits.[842,843] ACC oxidase gene was also identified as a senescence-related gene of flowers.[844–846] These results indicate that although ACC oxidase is present constitutively in many vegetative tissues, it is also inducible in fruits during ripening, in flowers during senescence, and by wounding. In unripe fruit,[847–849] and in newly opened flowers,[850–853] ACC oxidase activity is at a very low level and mRNAs for the enzyme are not detected, but both enzyme activity and mRNA levels increase dramatically during fruit ripening and flower senescence.

The expression of ACC oxidase isozyme genes is regulated differentially and in a tissue-specific manner. In petunia, each of three isogenes (PH-*ACO1*, PH-*ACO3*, and PH-*ACO4*) is expressed in different parts of the flower.[854] Tissue-specific expression of ACC oxidase gene is also found in the flower of an orchid, *Phalaenopsis*.[851,852]

There is a good evidence that ethylene synthesis (probably expression of either ACC synthase or ACC oxidase) is also under the control of another gene. One gene of tomato, designated *E8*, is transcriptionally activated at the onset of ripening when ethylene production starts to increase.[854] Its expression is induced in unripe tomato fruits by ethylene treatment. Fruits harvested from tomato plants transformed with an *E8* antisense gene showed greatly reduced levels of *E8* protein in fruits throughout the ripening period. However, the ethylene production rates of antisense fruits are several-fold higher than untransformed fruits.[855,856] The mechanism of this overproduction of ethylene in the antisense fruits is not known, but it is likely that the *E8* gene product suppresses ethylene biosynthesis.

8.02.6.5 Genetic Engineering

Transformation of tomato plants with the antisense gene of ACC oxidase results in substantially reduced levels of ethylene production in fruits and wounded leaves.[857,858] Similarly, tomato plants transformed with the antisense gene of ACC synthase also bear fruits which produce little ethylene and do not fully ripen even after 70 days of pollination.[859] These fruits, however, ripen normally when treated with ethylene or propylene for 15 days. The accumulation of mRNAs for ACC oxidase or ACC synthase does not occur in the fruits of plants transformed with antisense ACC oxidase or antisense ACC synthase, respectively, whereas mRNAs of other ripening-related genes accumulate normally. The results indicate that introduction of genes which repress transcription or translation of either ACC oxidase or ACC synthase or which reduce ACC levels can be a useful tool to extend longevity of crops deteriorated by ethylene action. Using this idea, a bacterial ACC deaminase gene has been used to obtain tomato plants whose fruits show low production of ethylene and delayed ripening.[860,861]

Application of sense and antisense gene engineering has been successful in controlling ethylene production from tomato fruits, and the technology may be used for other fruits and vegetables. However, the phenomenon of gene silencing of the endogenous homologous genes has become a problem in transgenic plants, and extensive examination of the mechanism of gene silencing is needed for further application of the technology.

8.02.7 BRASSINOSTEROIDS

8.02.7.1 Brief History

Brassinosteroids (BRs) are a group of steroidal plant hormones. In 1970, Mitchell *et al.* partially purified growth hormones termed brassins from rape pollen, which markedly promoted growth of pinto bean (*P. vulgaris* cv. Pinto) seedlings.[862] They found that the biological activities of brassins are distinct from those of auxin and GAs. However, little attention was paid to this finding. In 1979 the active principle named brassinolide was isolated as the first BR and was determined to be a unique steroidal compound.[863] Prior to these discoveries, in 1968 Marumo *et al.* partially purified three biologically active compounds from leaves of *Distylium racemosum*, which exhibit strong activity in the rice lamina inclination bioassay: these compounds were later found to be BRs.[864] In 1982, Yokota *et al.* isolated castasterone from insect galls of chestnut and postulated that it was a biosynthetic precursor of brassinolide.[865] Since then, most new BRs have been successfully isolated in Japan from various plant sources. Since BRs show a broad spectrum of biological activity, it has been difficult to define their physiological function. However, it has been found that BR biosynthesis and sensitivity mutants of *A. thaliana* and *P. sativum* (garden pea) show dwarfism. These findings, coupled with biosynthetic studies, established that BRs are important hormones for the normal growth of plants. There are many reviews dealing with natural product chemistry, physiology, molecular biology, and application of BRs.[866–873]

8.02.7.2 Chemistry

8.02.7.2.1 Distribution

BRs are distributed in a wide range of lower and higher plants. Plants in which BRs have been found include monocots (*Erythronium japonicum*, *Lilium elegans*, *Lilium longiflorum*, *Oryza sativa*, *Secale cereale*, *Triticum aestivum*, *Tulipa gesneriana*, *Typha latifolia*, and *Z. mays*), dicots (*Alnus glutinosa*, *Apium graveolens*, *A. thaliana*, *Beta vulgaris*, *Brassica campestris*, *Cassia tora*, *Castanea crenata*, *Catharanthus roseus*, *Citrus unshiu*, *D. racemosum*, *Dolichos lablab*, *Fagopyrum esculentum*, *Helianthus annuus*, *Lycopersicon esculentum*, *Ornithopus sativus*, *Pharbitis purpurea*, *P. vulgaris*, *P. sativum*, *Raphanus sativum*, *R. pseudacacia*, *Thea sinensis*, *V. faba*, and *V. radiata*), gymnosperms (*Cryptomeria japonica*, *Cupressus arizonica*, *Picea sitchensis*, *Pinus silvestris*, *Pinus thunbergii*), algae (*Hydrodictyon reticulatum*), and pteridophytes (*Equisetum arvense*).[870] As yet, no fungus has been found to produce BRs.

BRs have been detected in various plant organs including stem, leaf, pollen, and seed.[870] Pollen, anthers, and seeds are especially rich sources of BRs. Localization of BRs in amyloplats in pollen has been demonstrated.[874]

8.02.7.2.2 Structure

The structure determination of natural BRs has relied largely on ^{1}H NMR and electron impact-based GC–MS as methaneboronate derivatives. Over 40 BRs have been shown to occur naturally and their structures are shown in Fig. 12. They have structural differences in either the A ring, B ring, or side chain, arising from modifications through biosynthesis and metabolism.[869,870,872]

BRs are classified as C_{27}, C_{28}, and C_{29} steroids according to the difference in the side chain carbon skeleton. The C_{27} BRs have no alkyl substituent at C-24. The C_{28} BRs have either a methylene, α-methyl, or β-methyl at C-24 (note that in the side chain stereochemistry of steroids, groups in front of the plane are defined to be α-oriented). The C_{29} BRs have either an ethylidene or α-ethyl group at C-24, and include a group of BRs having a methylene at C-24 and an additional methyl at C-25.

BRs that carry $2\alpha,3\alpha$-vicinal diol in the A ring are the most biologically active. BRs having $2\alpha,3\beta$-, $2\beta,3\alpha$-, or $2\beta,3\beta$-vicinal hydroxys may be metabolites from, and are less active than, BRs with $2\alpha,3\alpha$-vicinal hydroxys. BRs having either an α-hydroxy, β-hydroxy, or ketone at C-3 in the A ring are biosynthetic precursors of $2\alpha,3\alpha$-diol BRs. The biosynthetic roles of 2,3-epoxybrassinosteroids are not yet known. The substituent in the B ring is either 6-deoxo, 6-oxo (ketone), or 6-oxo-7-oxa (lactone), which are biosynthetically interrelated. α-Oriented vicinal hydroxys at C-22 and C-23 in the side chain are characteristic of natural BRs. Cathasterone is the exception: it has a single hydroxy at C-22 and is a precursor of BRs having a 22, 23-vicinal diol.

8.02.7.2.3 Analysis

BRs are present only at low concentrations in plant vegetative tissues (submicrogram to a few micrograms per kilogram wet weight), so their extraction and purification are important steps before instrumental analysis. Typical procedures for analysis of BRs have been described by Yokota *et al.*[875] BRs are separated from lipids by partitioning between hexane and 80% methanol. BRs partitioned into the 80% methanol fraction are further purified by conventional column chromatography using silica gel, Sephadex LH-20, or charcoal. Column chromatography using Sephadex LH-20 and charcoal is an effective purification procedure in that BRs can be purified as a group. Respective BRs are usually separated by HPLC using a reversed-phase support (octadesylsilica).

The rice lamina inclination bioassay has been frequently used to guide purification and isolation of BRs from various plant extracts and also to evaluate biological potency of natural and synthetic BRs.[868] This bioassay is very sensitive to BRs and relatively insensitive to impurities contained in the plant extracts.[876] However, it should be noted that plant extracts often contain compounds synergistically enhancing BR activity.[873] Immunoassay for castasterone and brassinolide has been developed.[877] This immunoassay system also recognizes other naturally occurring BRs although generally to a lesser extent, and has been used to detect naturally occurring BRs in the partially purified plant extracts of stems and seeds of *P. vulgaris*.[877]

The most rigorous identification of BRs in plant extracts relies on GC–MS or selected ion monitoring (SIM).[868] For GC, a couple of vicinal hydroxys in BRs are reacted with methaneboronic acid in pyridine to produce bismethaneboronates. In cases of 2-deoxy-BRs, an unreacted hydroxy at C-3 should then be trapped by trimethylsilylation. Because BRs having epimeric substituents at C-2, C-3, or C-24 occur naturally and exhibit analogous retention times and mass spectra, use of Kovats indices is recommended.[875] Mass spectrometric data on methaneboronate derivatives of BRs have been reviewed by Ikekawa and Takatsuto.[878] For quantitation of the endogenous levels of BRs, deuterated BRs should be added to plant extracts as internal standards before purification starts.[868]

HPLC has been used to detect BRs in plant extracts, although it is less informative than GC–MS because retention times are the only criteria for identification. Several boronic acid reagents have been used to obtain fluorescent boronates of BRs.[879] Fluorescent boronates such as dansylaminophenylboronates and phenanthreneboronate are detected by a fluorimetric detector, while ferroceneboronates are monitored by an electrochemical detector. These derivatives can be detected at subnanogram levels. A microanalytical LC–MS method for the determination of BRs as their boronates has been developed using atmospheric pressure chemical ionization and electrospray ionization methods.[880] An LC–MS system using fast atom bombardment was developed to successfully identify acyl conjugated teasterones from lily pollen.[881] Fast atom bombardment mass spectra of several BRs have been reported.[882] ^{1}H NMR and ^{13}C NMR data also have been published.[883,884]

8.02.7.2.4 Synthesis

Most natural BRs have been synthesized and studies have been reviewed by Adam *et al.*[869] and Marquardt and Adam.[885] The syntheses of brassinolide and castasterone have been investigated by

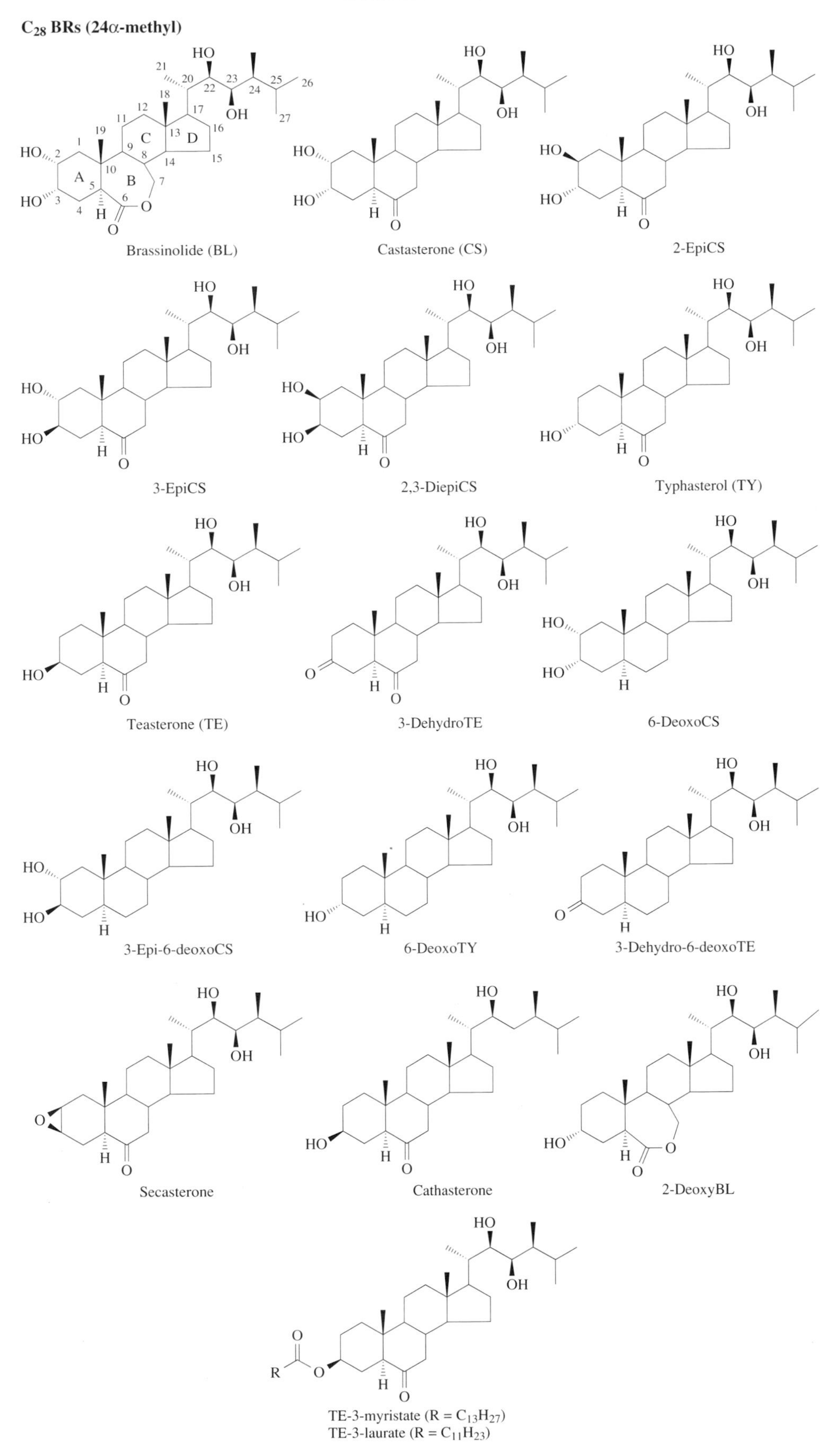

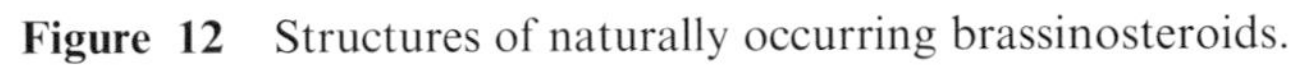
TE-3-myristate (R = $C_{13}H_{27}$)
TE-3-laurate (R = $C_{11}H_{23}$)

Figure 12 Structures of naturally occurring brassinosteroids.

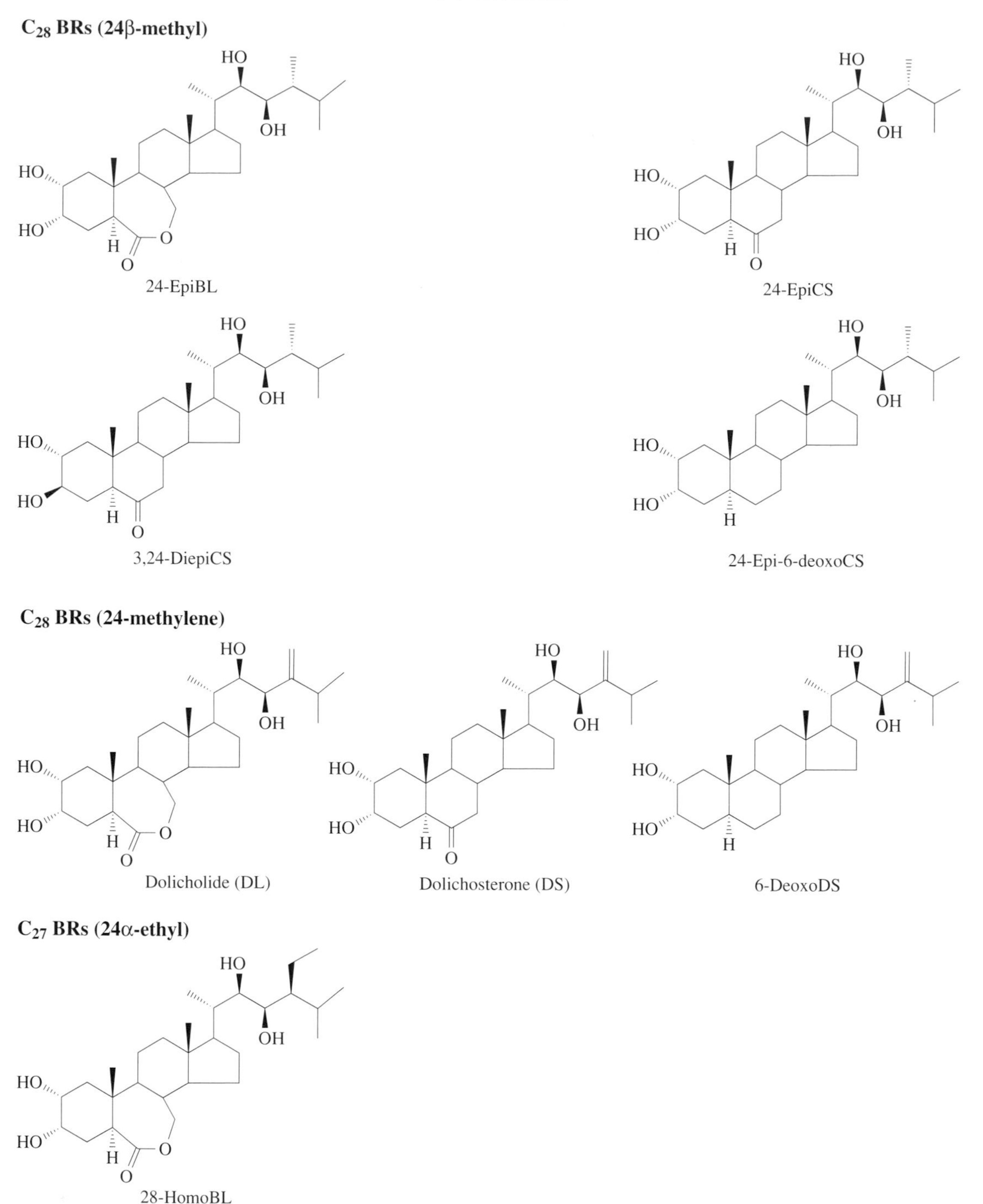

Figure 12 (continued)

many workers. Methods of introduction of α-oriented vicinal glycol in the A ring and a lactone or ketone group in the B ring have been established. In many cases, steroid rings were synthesized using a C_{22} aldehyde obtained from stigmasterol. Construction of the side chain containing four contiguous chiral centers has been extensively studied and work is ongoing. Efficient syntheses of brassinolide have been reported.[886,887] Along with brassinolide, 24-epibrassinolide and 28-homobrassinolide are often used for large-scale application studies and these now can be effectively synthesized from plant sterols such as ergosterol, brassicasterol, and stigmasterol (see McMorris *et al.*[887]).

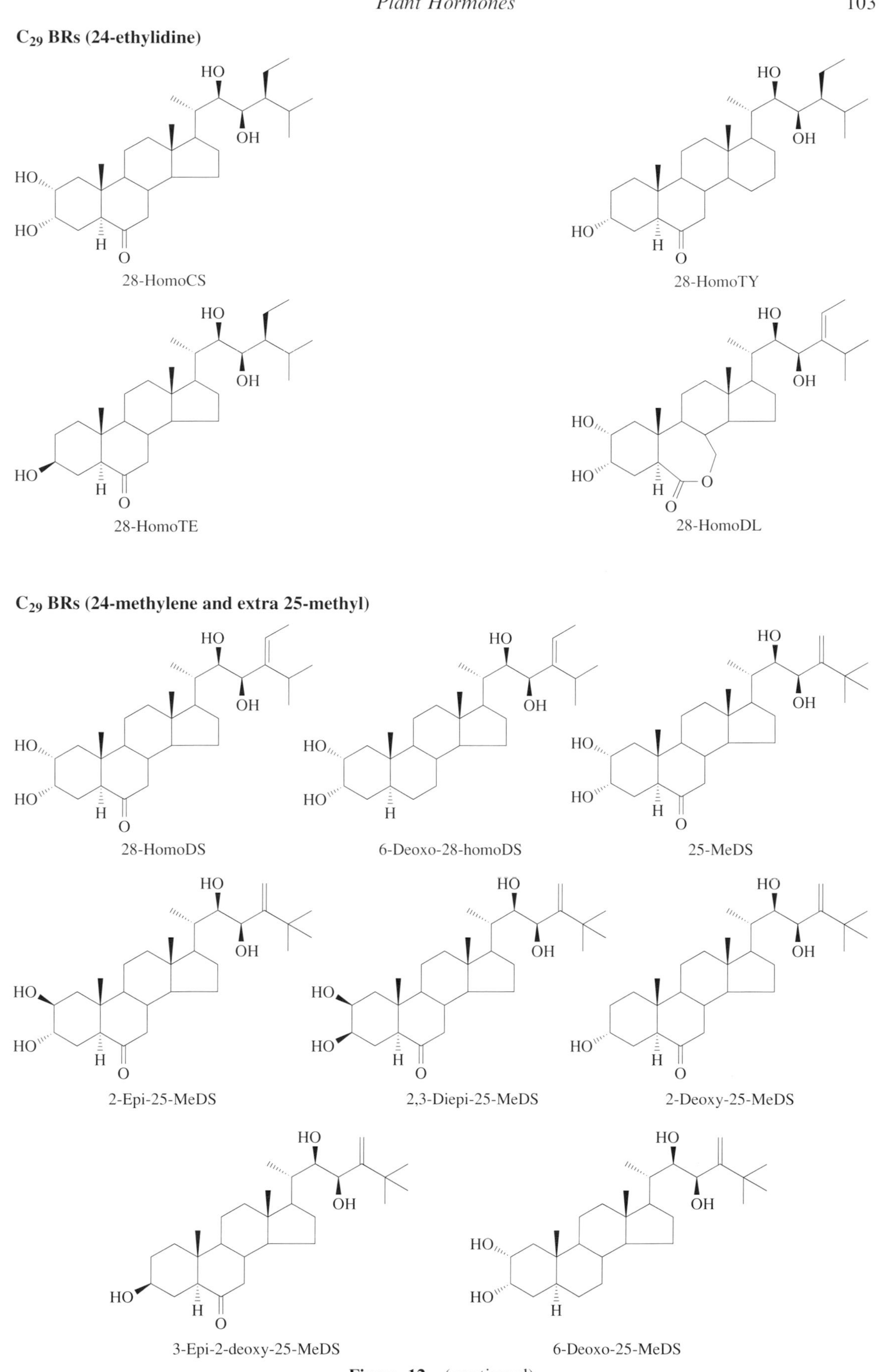

Figure 12 (continued)

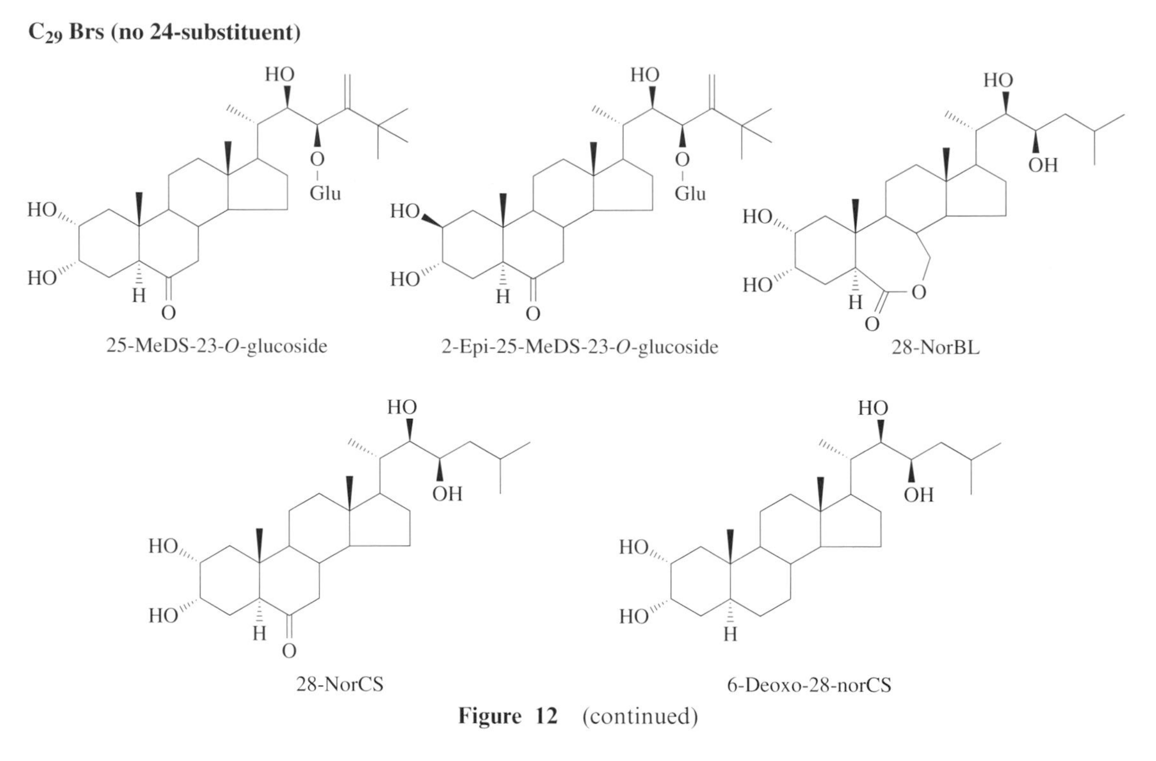

Figure 12 (continued)

8.02.7.3 Biological Activities

8.02.7.3.1 Structure–activity relationships

The structure–activity relationships of the natural and synthetic BRs have been examined using various systems including the bean second internode bioassay, the bean first internode curvature bioassay, the rice lamina inclination assay, and bioassays using tomato and radish seedlings.[888] The structure–activity relationships of BRs vary depending on the bioassay systems, although the overall features are not markedly different.

In the bean second internode bioassay, the biological activity of BRs is markedly influenced by slight structural changes.[889] For example, loss of the methyl at C-24 nearly abolishes the biological activity. Castasterone is 100-fold less active than brassinolide, indicating that the B ring lactone rather than the 6-oxo group is quite important in eliciting biological activity. The overall structural requirements of BRs with the *trans*-A/B ring system for biological activity are 2α,3α-vicinal diol in the A ring, 7-oxa-6-one (lactone) in the B ring, and 22*R*,23*R*- or 22*S*,23*S*-diol and 24-methyl in the side chain.

In the rice lamina inclination bioassay, extensive studies on structure–activity relationship of BRs were conducted.[890,891] In this bioassay, structural requirements are less stringent. For example, castasterone is only four times less active than brassinolide. Favorable functionalities are 2α,3α-vicinal diol or 3α,4α-vicinal diol in the A ring, 7-oxa-6-one or 6-ketone in the B ring, as well as methyl, ethyl, methylene, or ethylidene at C-24, or an additional methyl at C-25 (terminal tertiary butyl group) in the side chain.[888] It is noteworthy that 25-methylbrassinolide is more biologically active than brassinolide, a very potent naturally occurring BR.[892] Introduction of a hydroxy at C-25 also enhances the biological activity of 24-epibrassinolide.[893] It has been reported that the activity of BRs depends on the spatial arrangement of oxygen atoms.[894]

8.02.7.3.2 Physiology

Physiological and molecular studies have indicated that BRs work independently of auxins and gibberellins.[871–873] As summarized in several reviews,[871–873] the biological activities of BRs include stem elongation, pollen tube growth, leaf bending, leaf unrolling, inhibition of root growth and lateral root formation, proton pump activation, acceleration of 1-aminocyclopropane-1-carboxylic acid production, increase of transverse-oriented microtubles, and xylogenesis.

BR biosynthesis and response mutants of *A. thaliana*[895–898] and pea[899] have been isolated and these mutants are all dwarf. The short stature is due to shortened cell lengths and the phenotype of BR biosynthesis mutants is rescued by exogenous BRs, indicating that BRs cause elongation of cells, thereby promoting the growth of plants.[899] The dwarf mutants of *Arabidopsis* show, when grown in the dark, many of the characteristics of light-grown plants, including cotyledon expansion, primary leaf initiation, anthocyanin accumulation, and depression of light-regulated gene expression. Therefore, BRs seem to be implicated in the suppression of photomorphogenesis in the dark and BR deficiency may cause deetiolation phenotypes.[895–898]

BRs have been known to increase the yield of crops, although their effects vary with environmental conditions.[867] BRs can alleviate or protect injuries caused by various stresses including chilling, heat, salt, nutrition, and disease infection.[867–873] The molecular and biochemical basis of these antistress effects are not fully understood.

8.02.7.4 Biosynthesis and Metabolism

8.02.7.4.1 Biosynthesis

BRs are likely to be derived from sterols with the same alkyl substituents in the side chain as theirs.[872] Thus, 28-Nor-BRs (C_{27}-BRs) may come from cholesterol. C_{28}-BRs have an α-methyl, β-methyl, or methylene at C-24 and these may be derived from campesterol, 24-epicampesterol, and 24-methylenecholesterol, respectively. C_{29}-BRs have a 24α-ethyl, 24-ethylidene, or 24-methylene-25-methyl moiety and these may be synthesized from sitosterol, isofucosterol, and 24-methylene-25-methylcholesterol, respectively.

Biosynthetic pathways of brassinolide (C_{28}-BR with 24α-methyl) from campesterol were established by using normal and transformed cells of *Catharanthus roseus* (Scheme 18; asterisks indicate tentative pathways).[870,872,900–908] Several experiments using intact plants suggest that this biosynthetic pathway is also present in intact plants.[907,909] The first reaction in the pathway is the reduction of the Δ^5 double bond of campesterol to form campestanol.[905] The gene encoding this enzyme was isolated by using the dwarf mutant *det-2* of *Arabidopsis* which lacks this enzyme.[895] Campestanol is further converted to castasterone through bifurcation of the early or late C-6 oxidation pathway.

In the early C-6 oxidation pathway, campesterol is hydroxylated at C-6 to give 6α-hydroxycampestanol.[905] Oxidation of the hydroxy at C-6 gives 6-oxocampestanol.[905] The pathway from 6-oxocampesterol to cathasterone through C-22 hydroxylation has not yet been established, although both are endogenous in *C. roseus* cells. This may be because the endogenous pool of cathasterone is 500-fold lower than that of the putative precursor, 6-oxocampestanol.[906] Then, hydroxylation occurs at C-23 in the side chain to yield teasterone having 22,23-vicinal diol.[903] This reaction is carried out by a cytochrome P450 called CYP90, the gene encoding this protein being clarified by analyzing the dwarf mutant *cpd* of *Arabidopsis*.[896] Teasterone is converted to typhasterol.[902] This reaction converting 3β-hydroxy to 3α-hydroxy is mediated through 3-dehydroteasterone.[904] Although 3-dehydroteasterone has not been detected as the metabolite of teasterone, it is converted to typhasterol. This conversion was also detected in the cells of the lily shoot apex.[910] Such epimerization of the 3-hydroxy has also been observed in the metabolism of bile acids, ecdysteroids, and cardenolide. It should be noted that typhasterol is converted back to teasterone through the same mechanism. Typhasterol is hydroxylated at C-2α to give rise to castasterone.[902] Castasterone is finally converted to brassinolide, the most potent BR.[900,901] The pathway of teasterone to castasterone was confirmed in the intact plants of *C. roseus*, tobacco, and rice, although conversion of castasterone to brassinolide has not been observed except in *C. roseus*.[909] Altogether, it is concluded that the early C-6 oxidation pathway is quite an important synthetic pathway of BRs in plants.

An alternative pathway leading to brassinolide via 6-deoxo-BRs is designated the late C-6 oxidation pathway where a vicinal diol was introduced prior to the oxidation at C-6.[908] This pathway is also most likely to start from campestanol because levels of 6-deoxo-BRs were extremely low in the dwarf pea mutant *lkb* which has a lesion in the synthesis of campesterol.[899] Thus, it is hypothesized that campestanol is converted to 6-deoxoteasterone through double hydroxylation in the side chain. Further modifications of the structure leading to castasterone were experimentally established using *C. roseus* cells.[908] 6-Deoxoteasterone is isomerized to 6-deoxotyphasterol via 3-dehydro-6-

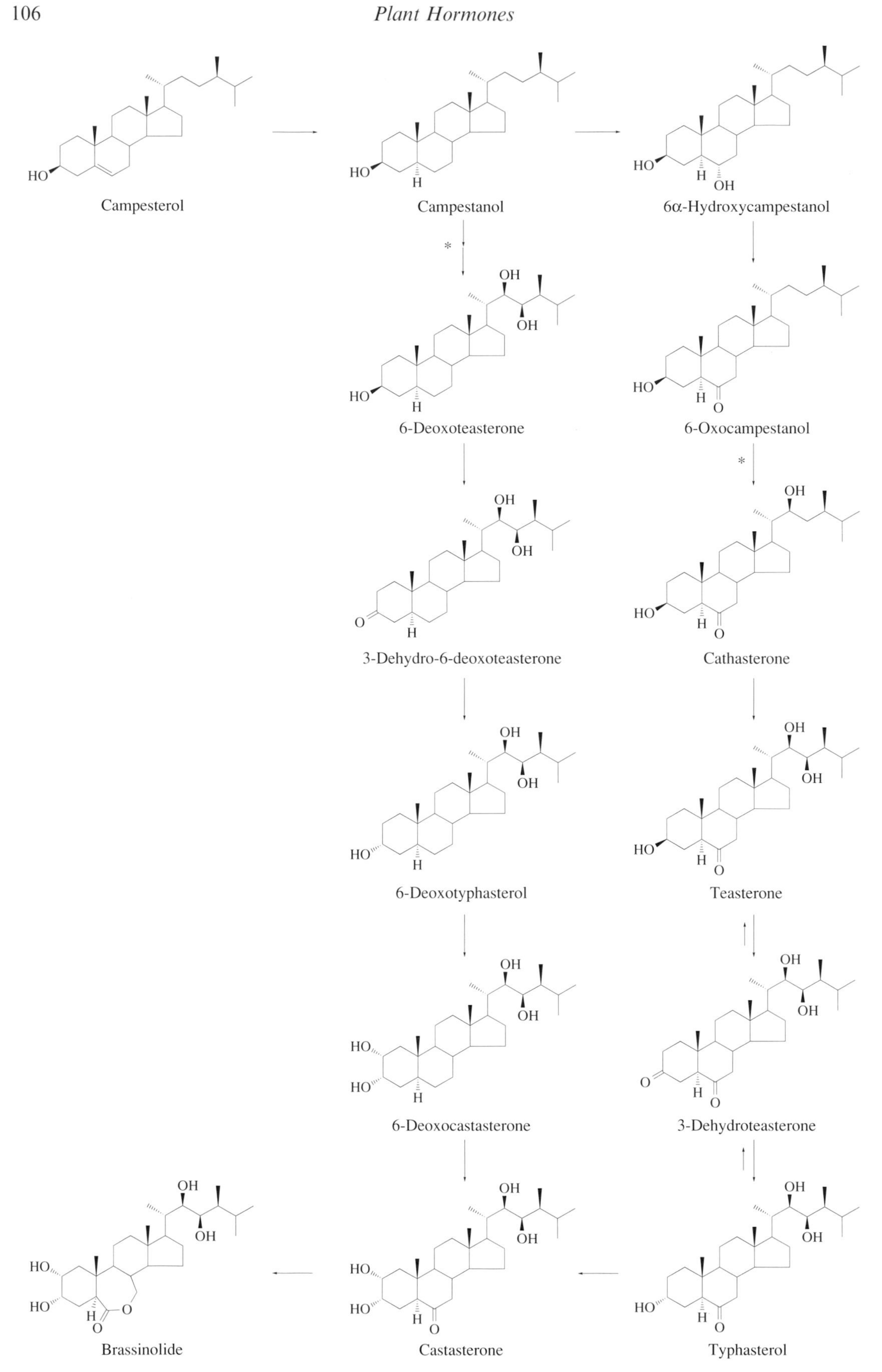
HO
Campesterol
HO
H
Campestanol
HO
H
OH
6α-Hydroxycampestanol
*
OH
OH
HO
H
6-Deoxoteasterone
HO
H
O
6-Oxocampestanol
*
OH
OH
O
H
3-Dehydro-6-deoxoteasterone
OH
HO
H
O
Cathasterone
OH
OH
HO
H
6-Deoxotyphasterol
OH
OH
HO
H
O
Teasterone
OH
OH
HO
HO
H
6-Deoxocastasterone
OH
OH
O
H
O
3-Dehydroteasterone
OH
OH
HO
HO
H
O
O
Brassinolide
OH
OH
HO
HO
H
O
Castasterone
OH
OH
HO
H
O
Typhasterol

Scheme 18

deoxoteasterone. 6-Deoxotyphasterol is further oxidized at C-2α to 6-deoxocastasterone which, in turn, is converted to castasterone. Because of the very low biological activity of 6-deoxo-BRs in the rice lamina inclination assay, they had been considered to be biosynthetic cul-de-sacs. However, the presence of the late C-6 oxidation pathway indicates that 6-deoxo-BRs have a role in plant growth.

A triazole-retardant uniconazole which has been known to inhibit gibberellin biosynthesis seems to be an inhibitor of BR biosynthesis as well, because application of uniconazole to pea plants reduces the content of castasterone, a major BR.[911] To support this, xylogenesis inhibition caused by uniconazole is recovered by BRs,[912] while stunted lateral root formation caused by uniconazole is inhibited by brassinolide.[913] Biological activity of BRs is suppressed by a metabolite (KM-01) isolated from a fungus *Drechslera avenae*.[914]

8.02.7.4.2 Metabolism

Metabolism of brassinolide, castasterone, and their 24-epimers has been reported.

Although brassinolide and castasterone are major BRs, very little has been known about their metabolism. Brassinolide was converted to 23-*O*-glucoside in explants of mung bean seedlings.[915] Because 23-*O*-glucosides of 25-methyldolichosterone and its 2-epimer have been isolated from seeds of *P. vulgaris*,[916] 23-*O*-glucosylation seems to be a major deactivation process in beans. Castasterone is a precursor of brassinolide but was not converted to brassinolide in mung bean. The major metabolites seem to be a mixture of glycosidic and nonglycosidic compounds although their structures have not been determined.[915] Castasterone was reported to be converted to 3-epicastasterone as a minor metabolite in the seedlings of tobacco and rice.[909] 3-Epicastasterone along with 2-epicastasterone and 2,3-diepicastasterone are endogenous in the immature seeds of *P. vulgaris*,[917] suggesting that epimerization of 2- and 3-hydroxys may be the deactivation process of castasterone. In seedlings and leaf explants of rice, brassinolide and castasterone were largely converted to water-soluble metabolites including sulfate ester-like compounds although their structures have not been elucidated.[918]

The metabolism of 24-epicastasterone and 24-epibrassinolide has been extensively investigated using cultured cells of *Ornithopus sativus*[893,919,920] and tomato (*Lycopersicon esculentum*).[921–923] These BRs suffered 3-epimerization in both plant cells. However, when 24-epibrassinolide was fed to *Cucumis sativus* seedlings, epimerization of the 2-hydroxy group occurred.[924] Furthermore, epimerization of 2- and 3-hydroxys also occurs in castasterone and brassinolide as described above, and, therefore, does not seem to be affected by the structural difference in the B ring and side chain. It seems likely that 3-epimerization occurs through 24-epi-3-dehydrocastasterone because it was isolated in the hydrolyzate of 24-epicastasterone metabolites from tomato.[920]

The metabolites of 24-epicastasterone and 24-epibrassinolide obtained in *O. sativus* cells contained 3,24-diepicastasterone and 3,24-diepibrassinolide with the 3β-hydroxys esterified with lauric acid, myristic acid, and palmitic acid.[922] Lily pollen contains teasterone with the 3β-hydroxy esterified with lauric acid and myristic acid.[925] Thus, the 3β-hydroxy rather than the 3α-hydroxy is susceptible to esterification. In tomato, 2-*O*-glucoside and 3-*O*-glucoside of 3,24-diepicastasterone were detected.[920] Thus, the formation of 3-β-hydroxy is an important process for further conjugation.

Introduction of hydroxys occurs at C-20, C-25, and C-26. In *O. sativus* cells, conversion of 24-epicastasterone to 20*R*-hydroxy-3,24-diepicastasterone was observed.[923] Further, 2α,3α-dihydroxy-5α-pregnane-6,20-dione and 2α,3α,6β-trihydroxy-5α-pregnane-20-one, which are considered to be derived from 20*R*-hydroxy-3,24-diepicastasterone, were detected in the metabolite. 24-Epibrassinolide was similarly metabolized to afford 20*R*-hydroxy-3,24-diepibrassinolide and 2α,3β-dihydroxy-B-homo-7-oxa-5α-pregnane-6,20-dione.[923] In tomato cells, hydroxylation at C-25 and C-26 was prevalent. 24-Epicastasterone and 24-epibrassinolide were converted to their 25-*O*-glucoside and 26-*O*-glucoside, respectively, although their aglycones were not detected as the metabolites.[903,919] 25-Hydroxylation also seems to occur after 3-epimerization because of the formation of 3,24-diepi-25-hydroxycastasterone from 24-epicastasterone in tomato cells,[920] and of 3,24-diepi-25-hydroxybrassinolide from 24-epibrassinolide in *O. sativus* cells.[923] The biological activity of 24-epibrassinolide is increased about 10-fold by 25-hydroxylation, but is decreased by 26-hydroxylation.[893] This is reminiscent of the promoting effect of 25-methylation on the biological activity of brassinolide and dolicholide.[888] 12β-Hydroxylation of both 24-epicastasterone and 24-epibrassinolide was observed in the fungus *Cunninghamella echinulata*.[926]

8.02.8 JASMONIC ACID AND RELATED COMPOUNDS

8.02.8.1 Introduction

Jasmonic acid and its related compounds belong to the group of sesquiterpenes, some of which are well known as odoriferous components of essential oils of plants. *cis*-Jasmone was first isolated as a fragrant constituent of the essential oil of jasmine (*Jasminum grandiflorum*) and its chemical structure elucidated.[927] Then, in 1962, methyl jasmonate was isolated from the same source.[928] These compounds are very useful in the cosmetics and perfume industries because of their aromatic odor.

In 1971, jasmonic acid was isolated from the culture filtrate of a fungus of *Botryodiplodia theobromae* (*Lasiodiplodia theobromae*) as a plant growth inhibitor and its chemical structure elucidated.[929] This was the first report describing the physiological effects of jasmonic acid and its related compounds on plant growth and development. Ten years later, methyl jasmonate and jasmonic acid were isolated and identified as senescence-promoting substances from different plant species.[930,931] These reports showing the potent physiological activities of jasmonic acid and methyl jasmonate stimulated much research into jasmonic acid and its related compounds. As a result, an additional specific activity of the compound was discovered: the induction or promotion of tuber formation. A glucoside of hydroxylated jasmonic acid named tuberonic acid glucoside was isolated as a tuber-promoting substance, and its chemical structure was determined in 1989.[932] The distribution of jasmonic acid and methyl jasmonate in the plant kingdom was intensively investigated using physicochemical assays (e.g., GLC, HPLC, MS). Radioimmunoassays for jasmonic acid and its derivatives indicated that they are widely distributed in Angiospermae, Gymnospermae, and Pteridophyta.[933] Green algae such as *Euglena*,[934] *Chlorella*,[935] and *Dunaliella*,[936] as well as the red algae *Gelidium*[937] also contain jasmonic acid and its related compounds. Up to 1997, various free acid forms as well as conjugate forms of jasmonic acid and its related compounds have been found in fungi, and lower and higher plants. However, no glucosyl esters of these compounds have been found. Jasmonic acid and its related compounds identified in the plant kingdom are summarized in Table 13.

However, the distribution of jasmonic acid and its related compounds is not restricted to plants. Methyl jasmonate and its epimer have also been reported in an insect, *Grapholitha molesta*.[970,971]

Multiple physiological effects caused by exogenous application of jasmonic acid and its related compounds suggest their involvement in signal transduction pathways and gene expression. Indeed, novel proteins (jasmonic acid-related compounds-induced proteins: JIPs) induced by the application of these compounds were detected in 1986.[972,973] These observations led to the study of the mode of action of jasmonic acid-related compounds at the molecular level, resulting in the successful isolation of the methyl jasmonate-inducible gene, *cyp93A1*, from soybean suspension culture cells by using the differential display method. This gene has been found to encode a novel cytochrome P450.[974]

According to the results of intensive studies on the identification of jasmonic acid and its related compounds described above, it is concluded that these compounds are widely distributed throughout the plant kingdom. This observation together with their physiological activities suggest that jasmonic acid and its related compounds are a new class of plant growth regulators. The status of jasmonic acid-related compounds as candidates for plant hormones has been discussed, as well as that of brassinosteroids. This chapter considers the chemical, biochemical, physiological, and molecular biological aspects of jasmonic acid and its related compounds. These topics have all been the subject of reviews.[975–979] The term "jasmonates" has generally been used to describe jasmonic acid and its related compounds according to Parthier's consideration.[972] However, this term includes the esters of jasmonic acid and related compounds. In this chapter the term "jasmonic acid-related compounds" is used instead of "jasmonates" in order to avoid confusion.

8.02.8.2 Chemistry

8.02.8.2.1 Chemical structure

Jasmonic acid-related compounds form a relatively large group of sesquiterpene carboxylic acids based on the cyclopentane ring. Over 20 structurally different jasmonic acid-related compounds have been isolated from the culture filtrates of fungi, algae, and lower and higher plants. Conjugate forms of these compounds as well as their free forms have been found. The chemical structures of these compounds are given in Figure 13.

Table 13 Jasmonic acid-related compounds in the plant kingdom.

	Source	*Compound*	*Ref.*
FUNGI			
Botryodiploida theobromae Pat.	culture filtrate	jasmonic acid	929, 938, 939
(*Lasiodiplodia theobromae* Griff.		(+)-7-isojasmonic acid	
& Maubl)		(±)-9,10-dihydro-7-isojasmonic acid	
		(+)-4,5-didehydro-7-isojasmonic acid	
		(+)-11,12-didehydro-7-isojasmonic acid	
		(+)-cucurbic acid	
		ethyl (+)-7-isojasmonate	
		(+)-3-oxo-2-(2*Z*-pentenyl)cyclopentylpropionic acid	
		(+)-3-oxo-2-(2*Z*-pentenyl)cyclopentylbutyric acid	
Gibberella fujikuroi	culture filtrate	(+)-7-isojasmonic acid	940
		(−)-jasmonic acid	
		4,5-didehydro-9,10-dihydrojasmonic acid	
		N-jasmonoyl-(*S*)-isoleucine	
		N-dihydrojasmonoyl-(*S*)-isoleucine	
RHODOPHYTA			
Gelidium latifolium		jasmonic acid	937
CHLOROPHYTA			
Euglena gracilis	cells	jasmonic acid	934
Chlorella sp.	cells	jasmonic acid	935
		methyl jasmonate	
Dunaliella tertiolecta	cells	jasmonic acid	936
Dunaliella salina	cells	jasmonic acid	936
PTERIDOPHYTA			
Equisetum arvense	leaf	jasmonic acid	941
		6-*epi*-7-isocucurbic acid	
Equisetum sylvaticum	leaf	4,5-didehydrojasmonic acid	941
Anemia phyllitidis	spore	cucurbic acid	942
		6-*epi*-cucurbic acid	
		6-*epi*-7-isocucurbic acid	
GYMNOSPERMAE			
Pinaceae			
Pinus mugo	pollen	*N*-[(−)-jasmonoyl]-(*S*)-isoleucine	943
		N-[7-isocucurbinoyl]-(*S*)-isoleucine	
ANGIOSPERMAE			
Juglandaceae			
Juglans regia	male flower	cucurbic acid	942
		6-*epi*-cucurbic acid	
		6-*epi*-7-isocucurbic acid	
Fagaceae			
Castanea crenata Sieb. et Zucc.	leaf, gall	jasmonic acid	944
Fagus sylvatica L.	leaf	jasmonic acid	933
Quercus robur L.	leaf	jasmonic acid	933
Moraceae			
Ficus superba Miq. var. *japonica*	leaf	jasmonic acid	945
		methyl jasmonate	
Papaveraceae			
Eschscholtzia californica	culture cells	jasmonic acid	946
		methyl jasmonate	
Cruciferae			
Brassica napus L.	immature seeds	jasmonic acid	947
Rosaceae			
Malus sylvestris Mill.	immature seeds	jasmonic acid	933
		methyl jasmonate	
Leguminosae			
Phaseolus vulgaris L.	immature seeds	(−)-jasmonic acid	944
	immature pods	jasmonic acid	933
Vicia faba L.	pods	(−)-jasmonic acid	948
	immature seeds	(−)-jasmonic acid	938
		(+)-7-isojasmonic acid	
		(−)-9,10-dihydrojasmonic acid	
		(+)-6-*epi*-7-isojasmonic acid	
		3,7-didehydrojasmonic acid	
	apical bud	*N*[(−)-jasmonoyl]-(*S*)-tryptophan	949
	flower	*N*[(−)-jasmonoyl]-(*S*)-tryptophan	950–952
		N[(−)-jasmonoyl]-(*S*)-tyrosine	
		N-[(+)-cucurbinoyl]-(*S*)-tryptophan	

Table 13 (continued)

	Source	Compound	Ref.
Vicia narbonensis L.	immature pods	jasmonic acid	933
Phaseolus coccineus L.	pods	jasmonic acid	933
Pisum sativum L.	immature pods	jasmonic acid	933
Glycine max Merrill	pods	jasmonic acid	933
Dolichos lablab L.	immature seeds	jasmonic acid	944
	pods	jasmonic acid	933
Calliandra haematocephala Hassk.	pods	jasmonic acid	933
Lupinus albus L.	immature pods	jasmonic acid	933
Mimosa pudica L.	leaf, stem	jasmonic acid	953
Rutaceae			
Citrus aurantifoia Swingle	immature fruit	jasmonic acid	933
Citrus sinensis Osbeck	fruit	jasmonic acid	933
Citrus limon Brum.	pericarp	methyl 7-isojasmonate	954
Linaceae			
Linum usitatissimum L.	immature seeds	jasmonic acid	947
Theaceae			
Camellia japonica L.	anther, pollen	jasmonic acid	955
		methyl jasmonate	
Camellia sasanqua L.	anther, pollen	jasmonic acid	955
		methyl jasmonate	
Camellia sinensis L.	anther, pollen	jasmonic acid	955
		methyl jasmonate	
Cleyera ochnacea DC.	mature leaf	jasmonic acid	931
Oleaceae			
Jasminum grandiflorum L.	essential oil	(−)-methyl jasmonate	928
		cis-jasmone	927
		12-hydroxyjasmonic acid lactone	956
Labiatae			
Rosmarinus officinalis L.	essential oil	(−)-methyl jasmonate	957
Perilla frutescens	fresh leaf	5′-β-D-glucopyranosyloxyjasmonic acid	958
		3-β-D-glucopyranosyl-3-*epi*-2-isocucurbic acid	
Solanaceae			
Solanum tuberosum L.	etiolated seedlings	jasmonic acid	933
	leaf	tuberonic acid glucoside	932
Solanum demissum	leaf (short day)	jasmonic acid	959
	leaf (long day)	jasmonic acid	959
		11-hydroxyjasmonic acid	
		12-hydroxyjasmonic acid	
Cucurbitaceae			
Cucurbita maxima L.	immature seeds	jasmonic acid	933
Cucurbita pepo L.	seeds	(+)-cucurbic acid	960, 961
		cucurbic acid glucoside	
		methyl cucurbate glucoside	
Compositae			
Helianthus annuus L.	immature seeds	jasmonic acid	933
Helianthus tuberosus L.	leaf	jasmonic acid	962, 963
		methyl tuberonate glucoside	962
Praxelis clematidea	dry plant	*N*-(12-acetoxyjasmonoyl)-phenylalanine methyl ester	964
Artemisia absinthium L.	leaf	(−)-methyl jasmonate	930
Artemisia tridentata Nutt.	leaf	methyl jasmonate	965
Gramineae			
Oryza officinalis	leaf	(−)-jasmonic acid	966
Secale cereale	immature caryopses	jasmonic acid	942
		cucurbic acid	
		6-*epi*-cucurbic acid	
		6-*epi*-7-isocucurbic acid	
Triticum aestivum L.	leaf	jasmonic acid	967
	coleoptile	jasmonic acid	967
	root	jasmonic acid	967
Liliaceae			
Allium cepa L.	leaf, root	jasmonic acid	968
Orchidaceae			
Cymbidium foberi	volatile components	methyl jasmonate	969
		methyl 7-isojasmonate	
Cymbidium virescence	volatile components	methyl jasmonate	969
		methyl 7-isojasmonate	

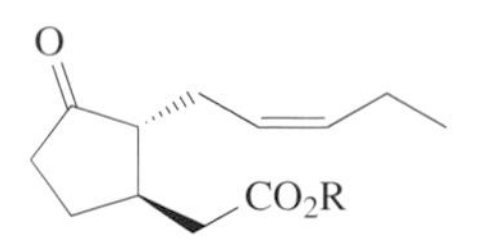

R = H Jasmonic acid
R= Me Methyl jasmonate

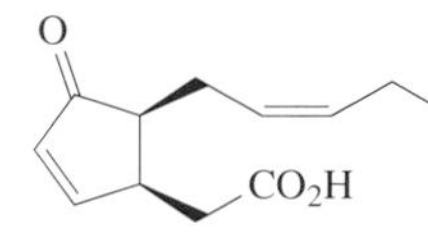

(+)-4,5-Didehydrojasmonic acid

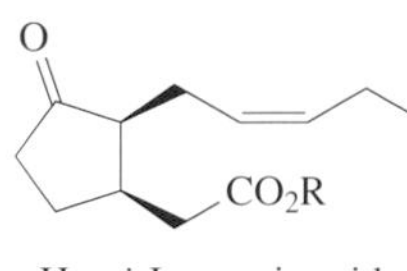

R = H *epi*-Jasmonic acid
R = Me *epi*-Methyl jasmonate

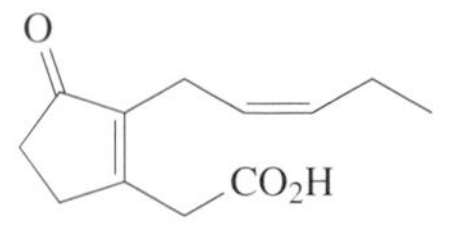

3,7-Didehydrojasmonic acid

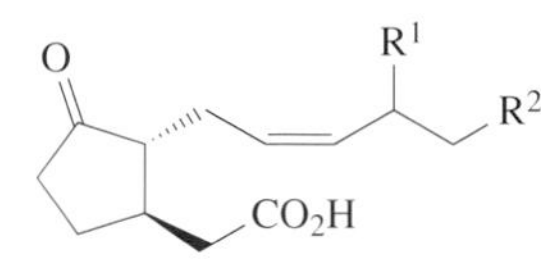

R^1 = H, R^2 = OH
11-Hydroxyjasmonic acid
R^1 = OH, R^2 = H
12-Hydroxyjasmonic acid

R = CO_2H 3-Oxo-2-(2*Z*-pentenyl) cyclopentenylpropionic acid
R = CH_2CO_2H 3-Oxo-2-(2*Z*-pentenyl) cyclopentenylbutyric acid

R = H Tuberonic acid
R = β-D-Glucopyranosyl Tuberonic acid glucoside

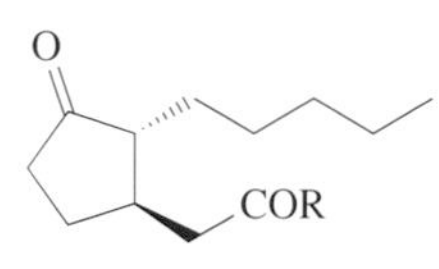

R = OH (–)-Dihydrojasmonic acid
R = isoleucine
N-[(–)-dihydrojasmonyl]-isoleucine

R = phenylalanine methyl ester
N-(12-acetoxyjasmononyl)-(*S*)-phenylalanine methyl ester

(+)-9,10-Dihydro-7-isojasmonic acid

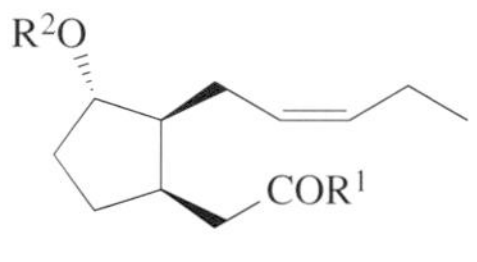

R^1 = OH, R^2 = H
(+)-Cucurbic acid
R^1 = tryptophan, R^2 = H
[(+)-Cucurbinoyl]-(*S*)-tryptophan
R^1 = OH, R^2 = β-D-glucopyranosyl
Cucurbic acid gluc oside
R^1 = OMe, R^2 = β-D-glucopyranosyl
Methyl cucurbate glucoside

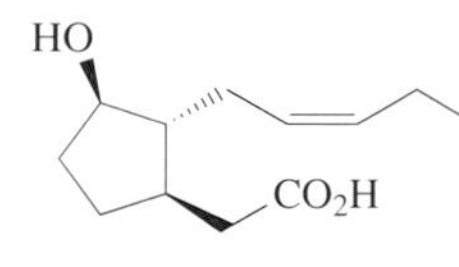

6-*epi*-7-Isocucurbic acid

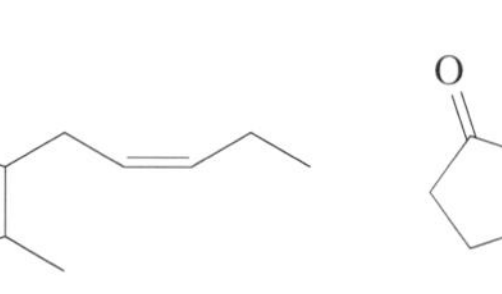

cis-Jasmone

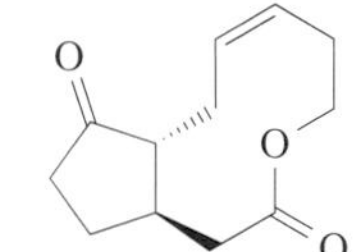

12-Hydroxyjasmonic acid lactone

Figure 13 Structural formulae of jasmonic acid and its related compounds.

Numbering the carbon atom in jasmonic acid is done in two ways: one based on the IUPAC rule and the other based on that of linolenic acid as a start substrate in its biosynthetic pathway (Figure 14). The variety of chemical structures in jasmonic acid-related compounds is based on functional groups substituted at positions C-3 (C-1), C-6 (C-3), and C-7 (C-2). Because of the presence of the chiral centers at the C-3 (C-1) and C-7 (C-2) positions, there are two diastereoisomers and respective enantiomers.

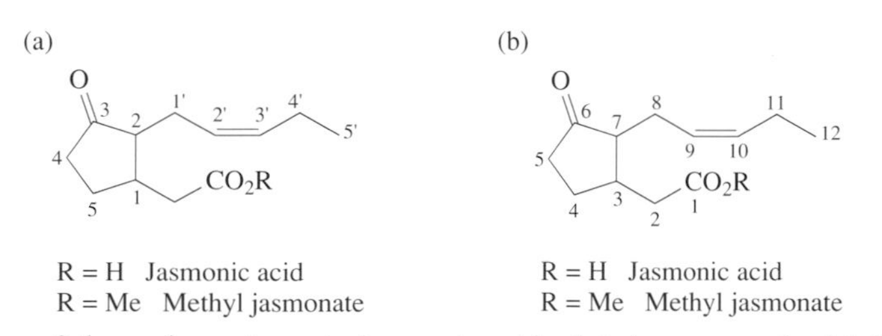

(a) R = H Jasmonic acid
R = Me Methyl jasmonate

(b) R = H Jasmonic acid
R = Me Methyl jasmonate

Figure 14 Numbering of the carbon atoms in jasmonic acid-related compounds. (a) Based on the IUPAC rule, (b) based on that of linolenic acid.

There are numerous reports of the occurrence of the (–)-form of jasmonic acid-related compounds [(3*R*,7*R*)] as the native form. However, its enantiomeric forms of (+)-jasmonic acid [(3*S*,7*S*)] and (+)-methyl jasmonate [3*S*,7*S*] are really the naturally occurring forms based on the results of their biosynthetic pathway.[980–984] The (+)-form of these compounds transforms into the (−)-form by isomerization during the isolation procedure, resulting in a molecular equilibrium of about 9:1((−)-form:(+)-form).[985] In *B. theobromae*, only (+)-jasmonic acid was found as the native biosynthetic product.[938,986] There is no evidence, however, that the (−)-forms of jasmonic acid-related compounds do not exist in organisms as the native forms. Indeed, the possible presence of (−)-jasmonoid glucoside in plant tissues has been reported.[958] In this case, the 7*S*-isomer of this compound was undetectable. As well as plant hormones, other conjugate forms of jasmonic acid-related compounds have also been found, usually conjugated with amino acid molecules.[949–952] There is no evidence of the naturally occurring glucosyl ester form of jasmonic acid. As shown in Table 14, the physicochemical properties of jasmonic acid-related compounds have been reported in relation to their uptake by mesophyll protoplasts of barley.[987]

Table 14 Physicochemical properties of jasmonic acid-related compounds.

	Compounds[c]					
Parameter	*JA*	*JA-Me*	*7-iso-CA*	*6-epi-7-iso-CA*	*(+)-JA-Leu*	*(−)-JA-Leu*
Molecular weight	210	224	212	212	323	323
Dissociation (20 °C) pK_a	4.50		4.50	4.50	4.15	4.10
Distribution coefficients[a]						
$K_{D(mol)}$	57.8	6.8	24.7	9.0	206	230
$K_{D(ion)}$	0.077					
Relative conductance[b] p_{rel} (from K_D)	1.33×10^{-3}					
Collander terms $K_{D(mol)} M_r^{-1.5}$	0.019	0.002	0.008	0.029	0.036	0.040
Permeability coefficients calculated from uptake experiments at pH 6.0 P_s ($\times 10^{-6}$ ms^{-1})	3.24	0.05	2.67	0.98	0.23	0.30

Source: Dathe *et al.*[987]
[a]K_D denotes the distribution coefficient (octanol:water). K_D is calculated from experimental data assuming that only the neutral molecule species is soluble in octanol. $K_{D(mol)}$ denotes the K_D for the undissociated molecule species and $K_{D(ion)}$ for the corresponding phytohormone anion. [b]p_{rel} is the anion permeability of a membrane relative to its permeability to the neutral hormone molecule species, calculated as $K_{D(ion)}/K_{D(mol)}$. [c]JA, jasmonic acid; JA-ME, methyl jasmonate; CA, cucurbic acid; JA-Leu, *N*-jasmonoyl-(*S*)-leucine.

8.02.8.2.2 Chemical synthesis

A total synthesis of (±)-methyl jasmonate was first reported in 1962 by Demole and Stoll[988] at almost the same time as its discovery. Numerous reports of its chemical synthesis have since been reported.[989] For instance, a highly efficient procedure for producing (±)-methyl jasmonate starting from cheaply available adipate has been investigated, resulting in the successful achievement of a higher than 60% total yield overall the six steps from adipic acid.[990]

Since molecules of jasmonic acid-related compounds have two chiral centers, stereocontrolled synthesis of methyl epijasmonate is required. Attempts at a stereoselective organic synthesis of the pure enantiomers of methyl jasmonate have been made, resulting in an efficient synthesis of methyl epijasmonate from 2-oxabicyclo[3,3,0]-oct-6-en-3-one in a 20% yield through 11 steps.[991] Based on this method, chiral synthesis of tuberonic acid ([3*R*,7*S*]-12-hydroxy-epijasmonic acid) has also been carried out.[992] The short and simple route starting from norbornene to epijasmonoids which is completely stereocontrolled has also been reported.[993]

(±)-2-[2,3-$^2H_2$]Jasmonic acid and its methyl ester which are used as the internal standards in a quantitative analysis of natural jasmonic acid and methyl jasmonate have been prepared from 2-(2-pentynyl)cyclopent-2-enone through catalytic semideuteration of acetylenic intermediates with deuterium gas in pyridine.[994]

Chemical syntheses of conjugates of jasmonic acid with amino acids[995,996] and glucose[939] have also been reported. (±)-Jasmonic acid obtained from alkaline hydrolysis of synthetic (±)-methyl jasmonate was activated with isobutyl chloroformate and triethylamine in tetrahydrofuran at low temperature. The aminolysis of the resulting mixed anhydride with the lithium salt of each (*S*)-amino acid afforded a mixture of the diastereoisomeric conjugates. The synthesis of glucosyl ester of (±)-jasmonic acid was performed in two steps. The first step was the synthesis of (±)-jasmonic acid-*β*-D-2,3,4,6-tetraacetylglucopyranosyl ester by the reaction of (±)-jasmonic acid with α-acetobromoglucose and triethylamine in acetone. The second step was an enzymatic removal of four acetyl groups, resulting in (±)-jasmonic acid glucosyl ester giving a 22% yield. This glucosyl ester is very unstable under acidic or alkaline conditions.

8.02.8.2.3 Isolation and identification

(i) Extraction, separation, and identification

In general, the concentrations of jasmonic acid-related compounds in plant tissues are relatively low, like those of plant hormones. Jasmonic acid-related compounds should be extracted from plant tissues using a water-miscible organic solvent such as ethanol, methanol, or acetone in the usual manner.[997] After the evaporation of the organic solvent, a residual aqueous phase should be partitioned with an appropriate organic solvent. As the acidic form of these compounds is a short-chain alkylcyclopentanone or alkylcyclopentane carboxylic acid of $pK_a = 4.5036$, the aqueous solution is adjusted to pH 2–3 with HCl and partitioned with ethyl acetate, diethyl ether, benzene, or chloroform. Chloroform is recommended to isolate jasmonic acid-related compounds conjugated with amino acid moieties.

Crude extract containing jasmonic acid-related compounds is further purified using several chromatographies. Partition or adsorption column chromatography, TLC, ion-exchange or gel permeation chromatography are recommended for the isolation of jasmonic acid-related compounds. Detailed procedures for the purification of jasmonic acid-related compounds have been reviewed by Ueda *et al.*[977] For the quantitative analysis of jasmonic acid-related compounds from relatively small amounts of source materials, ion-exchange column chromatography and preparative HPLC are recommended rather than absorption chromatographies.

HPLC is suitable for the analysis and separation of jasmonic acid-related compounds. As jasmonic acid-related compounds isolated from natural sources or synthesized chemically consist of an equilibrium mixture of the 7-epimer (2-epimer), attempts to separate the enantiomers have been made using HPLC after converting them to their diastereomeric ketal with (−)-2,3-butanediol[998] or esters of (−)-borneol.[999] Direct resolution using a column with a chiral stationary phase such as Chiralpak AS is a convenient method.[968]

Methyl esters of jasmonic acid-related compounds can be easily separated and analyzed by GLC, and the identification and the quantitation of jasmonic acid-related compounds can be successfully performed by using GLC and mass spectrometry (GC–MS). GC–SIM is also recommended not only for identification but also for quantitation. For the quantitative analysis, an internal standard such as a deuterium-labeled jasmonic acid-related compound should be used. Fragmentation of the methyl esters or the ethyl ester of jasmonic acid-related compounds have been reported.[977]

(ii) Physicochemical detection

Detection of jasmonic acid-related compounds has been based on their chemical properties.[977] One can detect jasmonic acid-related compounds on thin-layer plates by spraying with anisaldehyde reagent, which consists of acetic acid–sulfuric acid–anisaldehyde (100:2:1, v/v/v), with a 5% solution of vanillin in concentrated sulfuric acid or with 1% potassium permanganate solution, or by exposure to iodine vapor. Jasmonic acid-related compounds are also detected under UV radiation as fluorescent spots after spraying with 5% H_2SO_4–EtOH followed by heating at 130 °C for 5 min.

(iii) Immunoassay

Two immunological techniques, a radioimmunoassay and a nonradioactive enzyme-linked immunosorbent assay, have been introduced in order to achieve sensitive analysis of jasmonic acid-related compounds without the extensive purification necessary for bioassay and physicochemical determination.[994,1000,1001]

In the radioimmunoassay, the antiserum was raised in rabbits against (±)-jasmonic acid linked to bovine serum albumin. The tracer was tritium-labeled (±)-jasmonic acid. Cross-reactivity studies using several compounds structurally related to jasmonic acid demonstrated the antiserum to be specific for jasmonic acid. The detection limit was ca. 2 ng of jasmonic acid in its methyl ester form.

An enzyme-linked immunoassay was also demonstrated using monoclonal antibodies of sufficient selectivity and affinity. An IgG1(kappa) immunoglobulin of a monoclonal antibody for the analysis of (3*R*,7*R*)-jasmonic acid and its methyl ester was used to set up a competitive enzyme-linked immunoassay employing (3*R*,7*R*)-jasmonic acid-related compounds coupled to alkaline phosphatase as a tracer. The assay was linear between ca. 10 pg and 10 ng. Except for (3*R*,7*S*)- and (3*RS*,7*RS*)-jasmonic acid methyl esters, almost no cross-reactions of jasmonic acid-related compounds, synthetic derivatives, and biosynthetic precursors were found.

(iv) Bioassay

As jasmonic acid-related compounds exert similar physiological effects to ABA, bioassay systems for the detection of ABA can be used to detect jasmonic acid-related compounds. A seed germination test on lettuce seeds,[955] a growth inhibition test on rice[955] or lettuce seedlings,[1002] a chlorophyll degradation test on oat leaf segments,[930] or an abscission test on bean petiole explants[1003–1005] are recommended as bioassays for the detection of jasmonic acid-related compounds in extracts of plant tissues. Like ABA, jasmonic acid-related compounds have potent activities on these assay systems. As described below, the presence of sugar molecules in assay systems partially or completely nullifies the activities of jasmonic acid.[1006]

8.02.8.3 Structure–Activity Relationships

Structure–activity relationships of jasmonic acid-related compounds have intensively been studied in the bioassay systems of oat leaf senescence,[930] rice seedling growth,[955] and potato tuber formation.[1007]

As described above, there are four isomeric forms of jasmonic acid (two enantiomers and a diastereomer of each). In the senescence-promoting effect on oat leaf segments and the growth inhibition of rice seedlings, (−) forms have been shown to be more active than (+) or (±) form,[931] nevertheless they are rapidly isomerized. Our knowledge of the physiological effects of naturally occurring jasmonic acid or methyl jasmonate is very limited, because of its probable rapid epimerization. The biological activities of four isomeric forms of methyl jasmonate in several bioassay systems have been studied.[1008] According to the results obtained from this study, the absolute configurations of the two side chains with respect to the plane of the cyclopentanone ring for each activity are different, suggesting that there are different receptors which trigger reactions leading to the individual activity.

All the stereoisomers of cucurbic acid and its various analogues were stereoselectively synthesized in a racemic form, and their inhibitory activities on the root growth of rice seedlings and plant height of young corn plants were investigated.[1009] An essential factor for inhibitory activity was a *cis* configuration of the C-3 (C-1) and the C-7 (C-2) substituents.

Jasmonic acid was much less active than methyl jasmonate in promoting oat leaf senescence, and this relationship was true for the other compounds tested as well as for ABA. The reduction of the unsaturated bond in the substituent at the C-7 (C-2) position and the keto group at the C-6 (C-3) position greatly reduced the activity. The length of the *n*-alkyl substituents at the C-7 (C-2) position also had a significant effect on the activity. From these results, it is concluded that the important functional groups for the senescence-promoting activity of jasmonic acid-related compounds are the methyl acetate at the C-3 (C-1) position, the 2′-*cis*-pentenyl or *n*-pentyl group at the C-7 (C-2) position, and the keto group at the C-6 (C-3) position.[931]

Several types of jasmonic acid-related compounds were synthesized, and their inhibitory activities on rice seedling growth were investigated. The structure–activity relationship on the inhibitory activity of these compounds was slightly different from that of the promoting activity on leaf senescence. The free acid or methyl ester form at the C-3 (C-1) position, 2′-*cis*-pentenyl or *n*-pentyl group at the C-7 (C-2) position, and keto or hydroxy group at the C-6 (C-3) position were essential for inhibition activity. Furthermore, 4-acetylnonanoic acid showed an inhibitory effect similar to jasmonic acid, suggesting that this is the fundamental structure for the inhibitory activity of jasmonic acid-related compounds on rice seedling growth.[1002] Similar relationships of (±)-cucurbic acid analogues have been found in the root growth of rice seedlings and the growth of maize seedlings.[1009]

Twelve compounds related to jasmonic acid were synthesized and their potato tuber-inducing activities were investigated. Jasmonic acid and its methyl ester showed strong activity. Comparison of the activities suggests that those parts of the structure that are essential for the activity are the carboxy group or its esters at the C-3 (C-1) position, the pentenyl side chain at the C-7 (C-2), and the oxygen atom at the C-6 (C-3).[1007]

An attempt to determine which of *cis* or *trans* stereoisomers of methyl jasmonate is responsible for their biological activity has been studied using stereochemically locked *cis*- and *trans*-7-methyl derivatives. The results showed that the induction of the locking methyl group at the C-7 (C-2) position considerably decreased activity, suggesting that the presence of the methyl group lowers affinity for the methyl jasmonate receptor, presumably owing to a steric effect.[1010]

The structure–activity relationship of jasmonic acid-related compounds for the accumulation of cathepsin D inhibitor and proteinase inhibitor II mRNA was significantly different from that for the physiological response of plant tissues to these compounds.[1011,1012] An acetyl side chain, whether methylated or not, at the C-3 (C-1) position, an *n*-pentenyl side chain at the C-7 (C-2) position, and a keto group at the C-6 (C-3) position of the cyclopentane ring of jasmonic acid-related compounds were all important for its activity. These structural requirements seem to be different from those for the induction of potato tuber formation, the promotion of senescence, and the inhibition of seedling growth.

8.02.8.4 Biosynthesis and Metabolism

8.02.8.4.1 Biosynthetic pathway

Because the chemical structure of jasmonic acid-related compounds resembles that of prostaglandin, when jasmonic acid was first discovered it was speculated that it was biosynthesized from arachidonate. However, the biosynthetic pathway of jasmonic acid was subsequently found to start from linolenic acid.[980–984]

As shown in Scheme 19, α-linolenic acid is converted to 13(*S*)-hydroperoxylinolenic acid by lipoxygenase and then to an unstable allene oxide compound, 12,13(*S*)-epoxylinolenic acid, by hydroperoxide dehydrase. Nonenzymatically, 13(*S*)-hydroperoxylinolenic acid is spontaneously hydrolyzed and converted into a stable compound, 12-oxo-13-hydroxy-9(*Z*),15(*Z*)-octadecadienoic acid. However, 12-oxo-*cis*-10,15-phytodienoic acid, an important key intermediate compound, is produced from 12,13(*S*)-epoxylinolenic acid by allene oxide cyclase (AOC). The double bond in the cyclopentanone ring of 12-oxo-*cis*-10,15-phytodienoic acid is saturated in the presence of NADPH by reductase, resulting in the formation of 3-oxo-2-(2′-*cis*-pentenyl)-cyclopentaneoctanoic acid. This oxidized product is further oxidized in the β-oxidation system and then (+)-jasmonic acid (*cis*-jasmonic acid) is yielded as the final product.

The characterization and functions of the first and the second enzymes of the jasmonic acid pathway, lipoxygenase[1013–1015] and hydroperoxide cyclase,[1016–1019] respectively, have been well known in many plant species and in cell-free extracts. The third enzyme, AOC, was partially purified, indicating that the enzyme is a soluble protein of molecular weight 4.5×10^4.[1020] The fourth enzyme, 12-oxophytodienoic acid reductase, has been characterized from the kernel and seedlings of maize.[981] The molecular weight of the enzyme is estimated to be 5.4×10^4 by gel filtration. The optimum pH of this enzyme is relatively broad, 6.8–9.0.

Evidence obtained following purification of the enzyme allene oxidase synthase (AOS) from flax seed has allowed the identification of this enzyme as having catalytic activity for a cytochrome P450 family of hemoproteins (molecular weight 5.5×10^4) and illustrates the cooperation of different oxygenases in the jasmonic acid biosynthetic pathway.[1021] Its relationship to the P450 gene family was established from the proposed primary structure as deduced from the cDNA. The full-length transcript encodes a protein of 536 amino acids containing a 58 amino acid leader sequence that has the features of a mitochondrial or chloroplast transit peptide.[1022] These results suggest that the site of jasmonic acid biosynthesis is not in protoplasm but in organelles, especially in chloroplasts.

It has been shown that transgenic potato plants overexpressing flax AOS contain six- to 12-fold higher levels of jasmonic acid than the nontransformed plants. In this case, *pin* 2 (proteinase inhibitor II) genes were not expressed in transgenic potato leaves.[1022] These results suggest that an increase in the endogenous levels of jasmonic acid alone is not able to trigger a constitutive expression of wound- or water stress-responsive genes (see below).

8.02.8.4.2 Metabolism

There are only a few reports dealing with the metabolism of jasmonic acid-related compounds. The metabolism of jasmonic acid-related compounds applied exogenously in excised shoots of

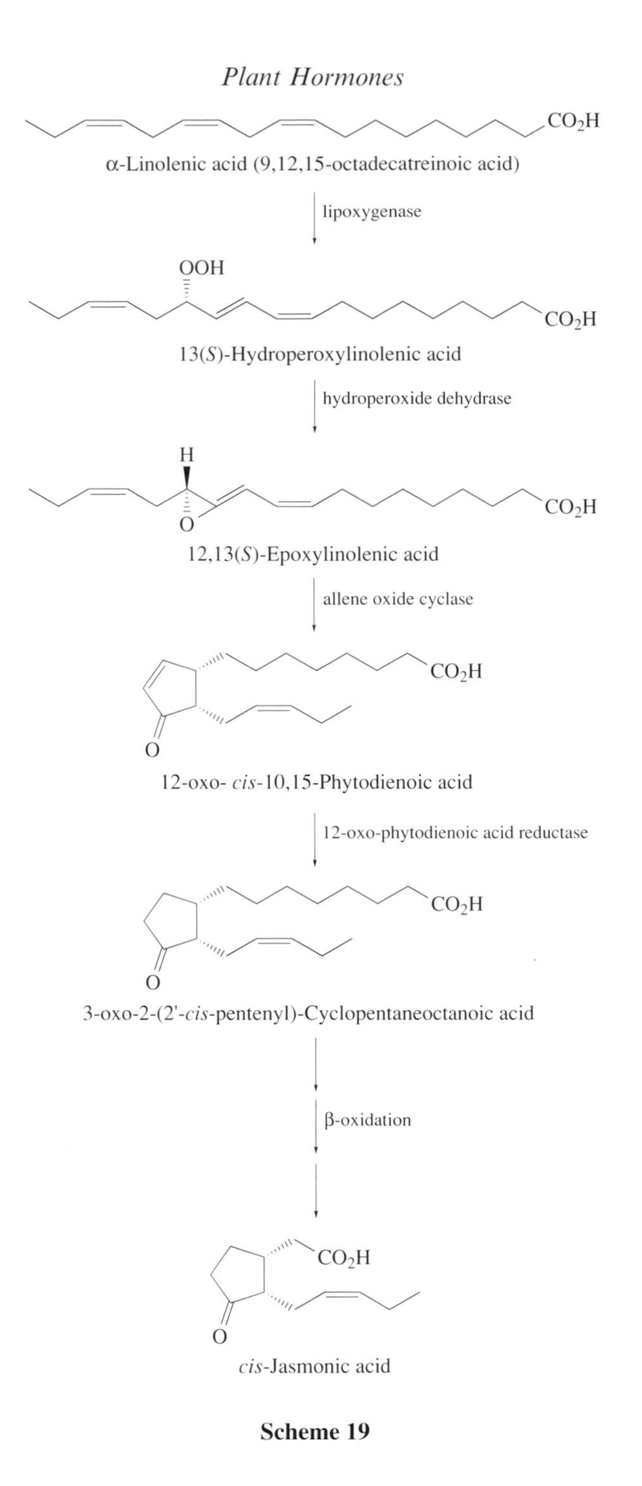

Scheme 19

barley seedlings, tomato and potato in intact plants, or in cell suspension culture has been extensively studied.

When the jasmonic acid-related compound, [2-^{14}C] and uniform labeled [^{3}H]($\pm$)-9,10-dihydrojasmonic acid were applied to detached barley seedlings, this compound was converted into two major and some minor metabolites through hydroxylation, glucosylation, and conjugation with amino acids. Based on spectroscopic investigation, the major metabolites were identified as (−)-9,10-dihydro-11-hydroxyjasmonic acid and its *O*(11)-*β*-D-glucopyranoside. To a lesser extent, (−)-9,10-dihydro-12-hydroxyjasmonic acid was also detected[1023–1025] (Scheme 20; major routes indicated by bold arrows, minor ones by standard arrows). Under these experimental conditions, glucosylation apparently took place only with the C-11 hydroxylated dihydrojasmonic acid.

In a suspension culture of tomato (*Lycopersicon peruvianum*) cells, uniformly labeled [^{3}H]dihydrojasmonic acid was enzymatically metabolized into dihydrojasmonic acid glucosyl ester by dihydrojasmonic acid glucosyl transferase in the presence of UDP-glucose. In this case, no glucosyl ether metabolite was detected.[939]

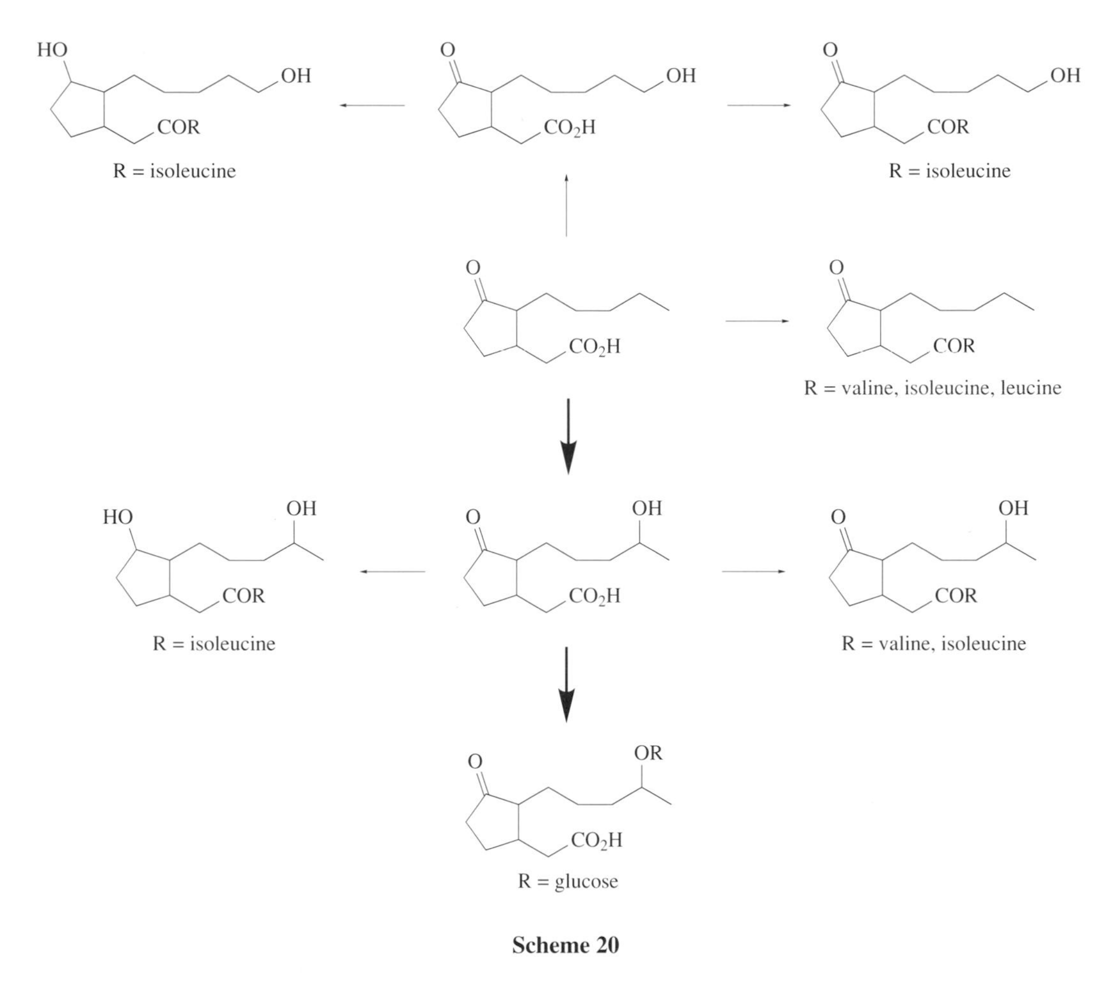

Scheme 20

Tuberonic acid glucoside isolated from the leaves of potato plants as a potato tuber-forming substance is thought to be formed from jasmonic acid by its hydroxylation and then glucosylation. In relation to tuberization in potato plants, the metabolism of tuberonic acid has been studied.[1026] [2-^{14}C](+)-Jasmonic acid incorporated into intact potato plants was metabolized to tuberonic acid glucoside within 2 weeks. More than half of the total radioactivity, however, was still associated with the treated leaves up to 25 days after the application of [2-^{14}C](+)-jasmonic acid.

Endogenous levels not only of jasmonic acid but also the conjugates of *N*-((−)-jasmonoyl)-(*S*)-valine, *N*-((−)-jasmonoyl)-(*S*)-leucine and *N*((−)-jasmonoyl)-(*S*)-isoleucine have been shown to increase in barley leaf tissues subjected to the condition of osmotic stress. These results strongly suggest that the conversion of jasmonic acid into the conjugation forms is easily carried out in stress conditions.[1027]

8.02.8.5 Multiple Physiological Effects

Jasmonic acid-related compounds applied exogenously to many kinds of plant systems exert multiple effects, either inhibition or promotion.[972,973,975,979] Some of the effects are thought to be similar to those of ABA. Almost all of the results were obtained using (±)-jasmonic acid-related compounds synthesized commercially.

One of characteristic effects of jasmonic acid-related compounds is a senescence-promoting activity. Almost all studies of the role of jasmonic acid-related compounds on leaf senescence reported the marked loss of chlorophyll in green leaf tissues, the leaves becoming yellow.[930,931,1028] This has led to the discussion of whether jasmonic acid-related compounds are true endogenous plant hormones or stress factors in promoting loss of chlorophyll.[972] Similar arguments might be raised in other physiological activities, such as the promotion of senescence.

The role of jasmonic acid-related compounds in the life cycles of plants has also been studied, because of the many similarities in the effectiveness of jasmonic acid-related compounds and ABA in physiological responses. In dormant seeds of *Acer tataricum*, jasmonic acid was able to substitute for cold stratification and to induce germination though the endogenous level of jasmonic acid in dormant seeds was equivalent to that of stratified ones.[945] Seasonal changes in jasmonic acid concentration and the total amounts of jasmonic acid involved in the growth of seeds and pericarp of soybean fruit have been investigated.[1029] The concentration of jasmonic acid in the seed increased up to stage II and decreased thereafter during intensive seed growth, whereas in the pericarp the amount of jasmonic acid increased up to stage IV. Based on these results, the possible role of jasmonic acid in assimilate translocation and pericarp senescence was demonstrated. Seasonal changes in the total amounts of jasmonic acid and methyl jasmonate as well as ABA in *Ficus superba* var. *japonica* leaves which corresponded to the vernal abscission were investigated. The level of jasmonic acid was low and remained almost constant during the plant's life cycle. However, methyl jasmonate increased gradually as leaves aged to maximum in early spring, again corresponding to leaf abscission.[1030]

The mode of physiological action of jasmonic acid-related compounds is not clear. Evidence for physiological action of these compounds could be obtained in the study on the inhibitory effect of jasmonic acid on IAA-induced elongation of oat coleoptile segments.[1031] Jasmonic acid inhibits IAA-induced elongation of oat coleoptile segments within 2 h of incubation in the dark. Some parameters of cell elongation, oxygen consumption, the osmolarity of the cell sap, or mechanical property of the cell wall, were not affected by the simultaneous addition of jasmonic acid in the presence or absence of IAA, but an increase in the amount of the cell wall polysaccharides was prevented. The presence of sugar molecules such as glucose, fructose, and sucrose could substantially nullify jasmonic acid-induced growth inhibition in the presence of IAA. From the results of this study together with the results of a further study using [^{14}C]glucose, jasmonic acid has been shown to inhibit IAA-induced elongation of oat coleoptile segments by interfering with some aspects of sugar metabolism of cell wall polysaccharides, especially their biosynthesis.[1006,1031]

The formation of potato tubers[1032] and onion bulbs[1033] seems to be caused by disruption of cortical microtubules. Jasmonic acid and its methyl ester were found to disrupt cortical microtubules in cultured tobacco cells. They exerted a microtubule-disrupting effect only in cells at the S phase of the cell cycle.[1034] Neither microtubules in preprophase bands, spindles, and phragmoplasts nor cortical microtubules at stages of the cell cycle other than the S phase were disrupted by the application of these compounds. Considering these facts, the growth-promoting effects of jasmonic acid-related compounds on potato tubers and onion bulbs seem to depend on their disruption of cortical microtubules.

8.02.8.6 Mode of Action at the Molecular Level

8.02.8.6.1 JIPs and signal transduction

The multiple physiological effects caused by the application of jasmonic acid-related compounds suggest that they are involved in many metabolic and signal-transduction processes, including direct effects on DNA. Parthier[972,973] and Reinbothe *et al.*[978] reviewed the possible function of jasmonic acid and methyl jasmonate as signaling molecules. Extensive work on the effects of jasmonic acid-related compounds on these processes resulted in the identification of jasmonic acid-related compounds-induced specific polypeptides in both intact plants and explants of grasses[1035] and legumes.[1036] Proteins induced by jasmonic acid-related compounds are summarized in Table 15. JIPs have been shown to be vegetative storage proteins, proteinase inhibitors, lipoxygenase and some other proteins involved in the defense against pathogens, and physical and chemical stresses.

Tomato leaves exposed to gaseous methyl jasmonates accumulated high levels of soluble protein that inhibited papain in continuous light conditions, but did so at a significantly lower rate in the dark.[1037] A study using methyl jasmonate-insensitive mutants of *A. thaliana* ecotype Columbia also shows altered expression by methyl jasmonate of AtVsp (vegetative storage protein of *A. thaliana*) which is homologous to the soybean VspA and VspB by methyl jasmonate.[1038]

Two polypeptides of molecular weights 2.3×10^4 and 1.5×10^4 were induced by methyl jasmonates not only in leaf segments but also in the intact seedlings of barley (*Hordeum vulgare*). The protein

Table 15 Jasmonic acid-related compounds-induced proteins (JIPs).

Protein	*Plant source*	*Ref.*
Proteinase inhibitors I and II	*Lycopersicon esculentum* L. (leaves)	1039–1042
Proteinase inhibitor II	*Solanum tuberosum* L. (leaves)	1043–1046
Trypsin inhibitor	alfalfa leaves	1039
Lipoxygenase	*Glycine max* Merrill, *Pisum sativum* L.	1047
Thionin	*Hordeum vulgare* L. (leaves)	1048
Phenylalanine ammonia lyase	*Glycine max* Merrill (cell cultures)	1049
Chalcon synthetase	*Glycine max* Merrill (cell cultures)	1050
Vegetative storage proteins	*Glycine max* Merrill (seedlings, leaves, cell cultures)	1036, 1050–1057
Napin, cruciferin	rape seed embryo	947
Others (functions unknown)	*Hordeum vulgare* L. (leaf segments)	1048, 1058–1063
	cotton cotyledons	1064
	Lycopersicon esculentum L. (leaves)	1039

of molecular weight 1.5×10^4 was found to correspond to the high-molecular-weight precursor of a leaf thionin.[1048] However, the function of the protein of molecular weight 2.3×10^4 is not clear. Tissue-specific expression of the gene coding this protein has been analyzed immunocytochemically and by *in situ* hybridization in barley, suggesting that the function of this protein might be involved in osmotic stress during development of the seedlings.[1065] The fact that jasmonic acid-related compounds induce accumulation of novel abundant proteins in plant tissues suggests that these compounds concomitantly repress synthesis of most pre-existing ("control") proteins. From the results shown by *in vitro* translation and northern hybridization, the impairment of "control" proteins occurs at the level of translation initiation.[1066]

The involvement of jasmonic acid-related compounds in the wound signal transduction pathway has been studied intensively. Wounding by leaf-chewing insects or by mechanical damage induces the synthesis of defensive proteinase inhibitor proteins in plant tissues. Several chemical compounds including oligosaccharides,[1067] ABA,[1068] systemin,[1069] and linolenic acid and jasmonic acid[1070] have been identified as signal molecules in the signal transduction pathway. Endogenous levels of linoleic acid and linolenic acid in tomato (*L. esculentum*) leaves increased within a short period after wounding, correlating with the accumulation of jasmonic acid.[1071] However, salicylic acid inhibits the conversion of 13-(*S*)-hydroperoxylinolenic acid to 12-oxophytodienoic acid, thereby inhibiting the signaling pathway by blocking synthesis of jasmonic acid.[1072] Furthermore, it has been reported that the inhibition of synthesis of proteinase inhibitor proteins and mRNAs by salicylic acid occurs at a step after the synthesis of jasmonic acid but preceding transcription of the inhibitor genes.[1072] Endogenous levels of jasmonic acid and salicylic acid related to signal transduction in wounded tobacco plants are regulated by benzylaminopurine.[1073] These results strongly suggest that jasmonic acid-related compounds are indeed involved in the signaling pathway for the induction of specific proteins.

Judging from the results described here, several types of specific receptor for jasmonic acid-related compounds in an individual physiological phenomenon are functioning in plant cells. The regulation mechanisms for these receptors have not been found. Detailed studies of the interaction of jasmonic acid-related compounds with other plant hormones in signal transduction pathways and molecular-level studies of the receptors of jasmonic acid-related compounds are required.

8.02.9 REFERENCES

1. J. MacMillan (ed.), "Hormonal Regulation of Development I. Molecular Aspects of Plant Hormones," Encyclopedia of Plant Physiology (New Series) Springer-Verlag, Berlin, 1980, vol. 9.
2. L. G. Nickell, "Plant Growth Regulating Chemicals," CRC Press, Boca Raton, FL, 1983, vol. 1, 2.
3. T. K. Scott (ed.), "Hormonal Regulation of Development II. The Functions of Hormones from the Level of the Cell to the Whole Plant," Encyclopedia of Plant Physiology (New Series), Springer-Verlag, Berlin, 1984, vol. 10.
4. N. Takahashi (ed.), "Chemistry of Plant Hormones," CRC Press, Boca Raton, FL, 1986.
5. A. Crozier and J. R. Hillman (eds.), "The Biosynthesis and Metabolism of Plant Hormones, Society of Experimental Biology, Seminar Series 23," Cambridge University Press, Cambridge, 1984.
6. P. J. Davies (ed.), "Plant Hormones and Their Role in Plant Growth and Development," Martinus Nijhoff, Dordrecht, 1987.
7. S. S. Purohit (ed.), "Hormonal Regulation of Plant Growth and Development," Kluwer, Dordrecht, 1988, vol. 2.
8. T. C. Moore, "Biochemistry and Physiology of Plant Hormones," 2nd edn., Springer-Verlag, New York, 1989.
9. C. Darwin and F. Darwin, "The Power of Movement in Plants," John Murray, London, 1880.

10. H. Söding, in "Encyclopedia of Plant Physiology," ed. W. Rhuland, Springer, Berlin, 1961, vol. XIV, p. 450.
11. F. W. Went, *Recl. Trav. Bot. Neerl.*, 1928, **25**, 1.
12. F. Kögl and D. G. F. R. Kostermans, *Z. Physiol. Chem.*, 1934, **228**, 113.
13. J. A. Vliegenthart and J. F. G. Vliegenthart, *Recl. Trav. Chim.*, 1966, **85**, 1266.
14. Y.-S. Hwang and M. Matsui, *Agric. Biol. Chem.*, 1968, **32**, 81.
15. F. Kögl, A. J. Haagen-Smit, and H. Erxleben, *Z. Physiol. Chem.*, 1934, **228**, 104.
16. K. V. Thimann, *J. Biol. Chem.*, 1935, **109**, 279.
17. A. J. Haagen-Smit, W. D. Leech, and W. R. Bergren, *Am. J. Bot.*, 1942, **29**, 500.
18. A. J. Haagen-Smit, W. B. Dandliker, S. H. Wittwer, and A. E. Murneek, *Am. J. Bot.*, 1946, **33**, 118.
19. M. S. Greenwood, S. Shaw, J. R. Hillman, A. Ritchie, and M. B. Wilkins, *Planta*, 1972, **108**, 179.
20. M. C. Elliott and M. S. Greenwood, *Phytochemistry*, 1974, **13**, 239.
21. R. S. Bandurski and A. Schulze, *Plant Physiol.*, 1974, **54**, 257.
22. F. W. Percival and R. S. Bandurski, *Plant Physiol.*, 1976, **58**, 60.
23. T. Okamoto, Y. Isogai, and T. Koizumi, *Chem. Pharm. Bull.*, 1967, **15**, 159.
24. M. Nomoto and S. Tamura, *Agric. Biol. Chem.*, 1970, **34**, 1590.
25. H. Abe, M. Uchiyama, and R. Sato, *Agric. Biol. Chem.*, 1972, **36**, 2259.
26. T. Yokota, M. Okabayashi, N. Takahasi, I. Shimura, and K. Umeya, "Plant Growth Substances 1973," Hirokawa, Tokyo, 1974, p. 28.
27. J. C. White, G. C. Medlow, J. R. Hillman, and M. B. Wilkins, *J. Exp. Bot.*, 1975, **26**, 419.
28. W. W. Shindy and O. E. Smith, *Plant Physiol.*, 1975, **55**, 550.
29. M. Nishio, S. Zuhi, T. Ishii, T. Furuya, and K. Syono, *Chem. Pharm. Bull.*, 1976, **24**, 2038.
30. T. Przbyllok and W. Nagl, *Z. Pflanzenphysiol.*, 1977, **84**, 463.
31. Y. Suzuki, H. Kinashi, S. Takeuchi, and A. Kawarada, *Phytochemistry*, 1977, **16**, 635.
32. J. L. Caruso, R. G. Smith, L. M. Smith, T.-T. Cheng, and J. D. Daves Jr., *Plant Physiol.*, 1978, **62**, 841.
33. J. R. F. Allen, A. M. Greenway, and D. A. Baker, *Planta*, 1979, **144**, 299.
34. P. Gaskin, P. S. Kirkwood, J. R. Lenton, J. MacMillan, and M. E. Radley, *Agric. Biol. Chem.*, 1980, **44**, 1589.
35. K. C. Engvild, H. Egsgaard, and E. Larsen, *Physiol. Plant*, 1981, **53**, 79.
36. R. Budini, S. Girroti, A. M. Pierpaoli, and D. Tonelli, *Microchem. J.*, 1982, **27**, 365.
37. D. Blakesley, J. F. Hall, G. D. Weston, and M. C. Elliott, *J. Chromatogr.*, 1983, **258**, 155.
38. S. Kaur and N. Punetha, *Kalikasan*, 1983, **12**, 158.
39. S. J. Nissen and M. E. Foley, *Plant Physiol.*, 1987, **84**, 287.
40. M. Katayama, S. V. Thiruvikraman, and S. Marumo, *Plant Cell Physiol.*, 1988, **29**, 889.
41. K. Bialek and J. D. Cohen, *Plant Physiol.*, 1989, **90**, 398.
42. T. Yokota, N. Murofushi, and N. Takahashi, in "Encyclopedia of Plant Physiology," ed. J. MacMillan, Springer, Berlin, 1980, vol. 9, p. 113.
43. G. Guin, D. L. Brummet, and R. C. Beier, *Plant Physiol.*, 1986, **81**, 997.
44. B. Sundberg, *Physiol. Plant*, 1990, **78**, 293.
45. B. B. Stowe and K. V. Thimann, *Arch. Biochem. Biophys.*, 1954, **51**, 499.
46. T. T. Lee, A. N. Starratt, and J. J. Jevnikar, *J. Chromatogr.*, 1985, **325**, 340.
47. A. Crozier, J. B. Zaerr, and R. O. Morris, *J. Chromatogr.*, 1980, **198**, 57.
48. H. Kaldewey, J. L. Wakhloo, A. Weis, and H. Jung, *Planta*, 1969, **84**, 1.
49. D. R. Reeve and A. Crozier, in "Encyclopedia of Plant Physiology," ed. J. MacMillan, Springer, Berlin, 1980, vol. 9, p. 203.
50. S. Fuchs and Y. Fuchs, *Biochim. Biophys. Acta*, 1969, **192**, 528.
51. R. Mertens, J. Eberle, A. Arnscheidt, A. Ledebur, and E. W. Weiler, *Planta*, 1985, **166**, 389.
52. R. Atzorn, U. Geier, and G. Sandberg, *J. Plant Physiol.*, 1989, **135**, 522.
53. M. Hofinger, *Phytochemistry*, 1980, **19**, 219.
54. L. Rivier and M. Saugy, *J. Plant Growth Regul.*, 1986, **5**, 1.
55. B. H. Brown, A. Crozier, G. Sandberg, and E. Jensen, *Phytochemistry*, 1986, **25**, 299.
56. A. M. Monteiro, G. Sandberg, and A. Crozier, *Phytochemistry*, 1987, **26**, 327.
57. J. R. F. Allen, L. Rivier, and P.-E. Pilet, *Phytochemistry*, 1982, **21**, 525.
58. Y. Isogai, S. Nomoto, T. Noma, and T. Okamoto, in "Plant Growth Substances 1973," Hirokawa, Tokyo, 1974, p. 9.
59. F. Wightman, in "Plant Growth Regulation," ed. P. E. Pilet, Springer, Berlin, 1977, p. 75.
60. A. W. Wheeler, *Ann. Bot. (London)*, 1977, **41**, 867.
61. L. M. Segal and F. Wightman, *Physiol. Plant*, 1982, **56**, 367.
62. E. A. Schneider, C. W. Kazakoff, and F. Wrightman, *Planta*, 1985, **165**, 232.
63. M. H. Bayer, *Plant Physiol.*, 1969, **44**, 267.
64. E. Epstein and J. Ludwig-Müller, *Physiol. Plant*, 1993, **88**, 382.
65. R. Rajagopal, *Physiol. Plant*, 1967, **20**, 655.
66. S. Ram and T. R. Rao, *New Phytol.*, 1981, **88**, 53.
67. D. L. Rayle and W. K. Purves, *Plant Physiol.*, 1967, **42**, 520.
68. A. C. Thompson, G. F. Nicollier, and D. F. Pope, *J. Agric. Food Chem.*, 1987, **35**, 361.
69. Y. Isogai, T. Okamoto, and T. Koizumi, *Chem. Pharm. Bull.*, 1967, **15**, 151.
70. E. R. H. Jones, H. B. Henbest, G. F. Smith, and J. A. Bentley, *Nature*, 1952, **169**, 485.
71. M. E. Wall, H. Taylor, P. Perera, and M. C. Wani, *J. Nat. Prod.*, 1988, **51**, 129.
72. J. Ludwig-Müller and W. Hilgenberg, *Physiol. Plant*, 1988, **74**, 240.
73. T. Rausch, S. C. Minocha, W. Hilgenberg, and G. Kahl, *Physiol. Plant*, 1985, **63**, 335.
74. M. Hofinger, X. Monseur, M. Pais, and F. X. Jarreau, *Phytochemistry*, 1975, **14**, 475.
75. E. R. H. Jones and W. C. Taylor, *Nature*, 1957, **179**, 1138.
76. S. Nakagawa, D. S. Tjokrokusumo, A. Sakurai, I. Yamaguchi, N. Takahashi, and K. Syono, *Plant Cell Physiol.*, 1987, **28**, 485.
77. C. A. Bradfield and L. F. Bjeldanes, *J. Agric. Food Chem.*, 1987, **35**, 46.

78. A. Ernsten and G. Sandberg, *Physiol. Plant*, 1986, **68**, 511.
79. S. Marumo, H. Hattori, H. Abe, and K. Munakata, *Nature*, 1968, **219**, 959.
80. M. Katayama, S. V. Thiruvikraman, and S. Marumo, *Plant Cell Physiol.*, 1987, **28**, 383.
81. G. B. West, *J. Pharmacol.*, Suppl., 1959, **11**, 275T.
82. K. Wakabayasi, M. Nagao, T. Tahira, Z. Yamaizumi, M. Katayama, S. Marumo, and T. Sugimura, *Mutagenesis*, 1986, **1**, 423.
83. R. Gmelin, M. Saarlvirta, and A. I. Virtanen, *Suom. Kemistil.*, 1960, **3B**, 172.
84. F. Shahidi and J. E. Gabon, *J. Food Qual.*, 1989, **11**, 421.
85. R. Gmelin and A. I. Virtanen, *Suom. Kemistil.*, 1962, **35B**, 34.
86. S. G. Dungey, J. P. Sang, N. E. Rothnie, M. V. Palmer, D. G. Burke, R. B. Knox, E. G. Williams, E. P. Hilliard, and P. A. Salisbury, *Phytochemistry*, 1988, **27**, 815.
87. R. J. W. Truscott, P. K. Johnstone, I. R. Minchinton, and J. P. Sang, *J. Agric. Food Chem.*, 1983, **31**, 863.
88. D. I. McGregor, *Can. J. Plant Sci.*, 1988, **68**, 367.
89. M. V. Palmer, S. P. Yeung, and J. P. Sang, *J. Agric. Food Chem.*, 1987, **35**, 262.
90. H. Schraudolf, *Phytochemistry*, 1989, **28**, 259.
91. H. Schraudolf and R. Baeuerle, *Z. Naturforsch., C: Biosci*, 1986, **41**, 526.
92. J. K. Goetz and H. Schraudolf, *Phytochemistry*, 1983, **22**, 905.
93. M. C. Elliot and B. B. Stowe, *Phytochemistry*, 1970, **9**, 1629.
94. H.-D. Klämbt, *Naturwissenschaften*, 1960, **47**, 398.
95. H. Hattori and S. Marumo, *Planta*, 1972, **102**, 85.
96. C. Labarca, P. S. Nicholls, and R. S. Bandursky, *Biochem. Biophys. Res. Commun.*, 1965, **20**, 641.
97. M. Ueda and R. S. Bandurski, *Plant Physiol.*, 1969, **44**, 1175.
98. R. S. Bandarsky, in "Plant Growth Substances," ed. N. B. Mandava, American Chemical Society, Washington DC, 1979, p. 1.
99. A. Ehman and R. S. Bandarski, *Carbohydr. Res.*, 1974, **36**, 1.
100. J. Bruinsma, C. M. Karssen, M. Benschop, and J. B. van Dort, *J. Exp. Bot.*, 1975, **26**, 411.
101. K. Hasegawa and K. Miyamoto, *Plant Cell Physiol.*, 1980, **21**, 363.
102. K. Hasegawa and T. Hase, *Plant Cell Physiol.*, 1981, **22**, 303.
103. N. Harada, H. Ono, H. Hagiwara, H. Uda, K. Hasegawa, and M. Sakoda, *Tetrahedron Lett.*, 1991, **32**, 6761.
104. M. Sakoda, K. Hasegawa, and K. Ishizuka, *Phytochemistry*, 1991, **30**, 57.
105. K. Hasegawa, S. Togo, M. Urashima, J. Mizutani, S. Kosemura, and S. Yamamura, *Phytochemistry*, 1992, **31**, 3673.
106. E. M. Iyoshi, Y. Shizuri, and S. Yamamura, *Chem. Lett.*, 1987, 511.
107. M. Ueda, M. Niwa, and S. Yamamura, *Phytochemistry*, 1995, **39**, 817.
108. M. Ueda, T. Shigemori-Suzuki, and S. Yamamura, *Tetrahedron Lett.*, 1995, **36**, 6267.
109. G. J. MacDougall and S. C. Fry, *Plant Physiol.*, 1989, **89**, 883.
110. P. B. Goodwin, in "A Comprehensive Treatise," eds. D. S. Lemam, P. B. Goodwin, and T. J. V. Higgins, Elsevier/North-Holland Biomedical Press, Amsterdam, 1978, vol. II, p. 31.
111. S. Kuraishi and R. M. Muir, *Plant Cell Physiol.*, 1964, **5**, 61.
112. A. Apelbaum and S. P. Burg, *Plant Physiol.*, 1972, **50**, 125.
113. D. D. Shushu and E. G. Cutter, *Can. J. Bot.*, 1990, **68**, 965.
114. T. J. Gianfagna, G. J. Wulster, and G. S. Teiger, *HortScience*, 1986, **21**, 461.
115. A. Yahalom, B. L. Epel, and Z. Glinka, *Physiol. Plant*, 1988, **72**, 428.
116. J. L. Hall, D. A. Brummell, and J. Gillespie, *New Phytol.*, 1985, **100**, 341.
117. C. M. S. Carrington and J. Esnard, *J. Exp. Bot.*, 1988, **39**, 441.
118. D. A. Brummell and J. L. Hall, *Plant Cell Environ.*, 1987, **10**, 523.
119. H. Kazama and M. Katsumi, *Plant Cell Physiol.*, 1976, **17**, 467.
120. R. Colombo and P. Ferrari-Bravo, *Plant Sci. Lett.*, 1982, **25**, 247.
121. K. Miyamoto and S. Kamisaka, *Physiol. Plant*, 1988, **74**, 457.
122. M. J. Vesper and C. L. Kuss, *Planta*, 1990, **182**, 486.
123. S. Horemans, H. A. Van Onckelen, and J. A. De Greef, *J. Exp. Bot.*, 1986, **37**, 1525.
124. A. Rodriguez, M. J. Canal, and R. S. Tames, *Physiol. Plant*, 1988, **73**, 92.
125. A. B. Hayes and J. A. Lippincott, *Am. J. Bot.*, 1981, **68**, 305.
126. J. H. Palmer, in "Encyclopedia of Plant Physiology," eds. R. P. Pharis and D. M. Reid, Springer, Berlin, 1985, vol. 11, p. 139.
127. N. Morimoto, C. Shichijo, S. Watanabe, S. Suda, and T. Hashimoto, *Physiol. Plant*, 1986, **68**, 196.
128. M. L. Evans, in "Encyclopedia of Plant Physiology," ed. T. K. Scott, Springer, Berlin, 1984, vol. 10, p. 23.
129. L. J. Feldman, *Ann. Rev. Plant Physiol.*, 1984, **35**, 223.
130. L. E. Powell, in "Plant Hormones and their Role in Plant Growth and Development," ed. P. J. Davies, Martinus Nijhoff, Dordrecht, 1987, p. 539.
131. D. M. Reinecke, J. A. Ozga, and V. Magnus, *Phytochemistry*, 1995, **40**, 1361.
132. M. G. Cline, *Physiol. Plant*, 1994, **90**, 230.
133. M. H. M. Goldsmith, *Annu. Rev. Plant Physiol.*, 1977, **28**, 439.
134. A. Everat-Bourbouloux, *Physiol. Plant*, 1987, **70**, 648.
135. W. Russell and K. V. Thimann, in "Plant Growth Substances 1988," eds. R. P. Pharis and S. B. Rood, Springer, Berlin, 1990, p. 419.
136. T. K. Prasad and M. G. Cline, *Plant Physiol.*, 1987, **83**, 505.
137. T. Sachs, *Isr. J. Bot.*, 1970, **19**, 484.
138. G. F. W. Gocal, R. P. Pharis, E. C. Yeung, and D. Pearce, *Plant Physiol.*, 1991, **95**, 344.
139. R. Aloni, *Annu. Rev. Plant Physiol.*, 1987, **38**, 179.
140. B. Sundberg and C. H. A. Little, *Plant Physiol.*, 1990, **94**, 1721.
141. M. A. W. Hinchee and T. L. Rost, *Bot. Gaz.*, 1986, **147**, 137.
142. C. Borkird, J. H. Choi, and Z. R. Sung, *Plant Physiol.*, 1986, **81**, 1143.
143. G. Bernier, *Annu. Rev. Plant Physiol. Plant Mol. Biol.*, 1988, **39**, 175.

144. C. J. Brady, *Annu. Rev. Plant Physiol.*, 1987, **38**, 155.
145. R. Sexton and J. A. Roberts, *Annu. Rev. Plant Physiol.*, 1982, **33**, 133.
146. J. W. Patrick and K. H. Steains, *J. Exp. Bot.*, 1987, **38**, 203.
147. C. H. A. Little, B. Sundberg, and A. Ericsson, *Tree Physiol.*, 1990, **6**, 177.
148. P. M. Hayes and J. W. Patrick, *Planta*, 1985, **166**, 371.
149. D. K. Small and D. A. Morris, *Plant Growth Regul.*, 1990, **9**, 329.
150. D. J. Cosgrove, *New Phytol.*, 1993, **124**, 1.
151. D. G. Pope, *Ann. Bot.*, 1982, **49**, 493.
152. T. Hoson, Y. Sone, A. Misaki, and Y. Masuda, *Physiol. Plant*, 1993, **87**, 142.
153. M. Inoue and D. J. Nevin, *Plant Physiol.*, 1991, **96**, 426.
154. H. Hohl, Y. N. Hong, and P. Schopfer, *Plant Physiol.*, 1991, **95**, 1012.
155. S. J. McQueen-Mason, S. C. Fry, D. M. Durachko, and D. J. Cosgrove, *Planta*, 1993, **190**, 327.
156. S. J. McQueen-Mason, D. M. Durachko, and D. J. Cosgrove, *Plant Cell*, 1992, **4**, 1425.
157. Z.-C. Li, D. M. Durachko, and D. J. Cosgrove, *Planta*, 1993, **191**, 349.
158. U. Kutschera and W. R. Briggs, *Plant Physiol.*, 1987, **84**, 1361.
159. M. E. Cunninghame and J. L. Hall, *Protoplasma*, 1985, **125**, 230.
160. A. Dietz, U. Kutschera, and P. M. Ray, *Plant Physiol.*, 1990, **93**, 432.
161. K. V. Thimann, in "The Physiology of Plant Growth and Development," ed. M. B. Wilkins, McGraw-Hill, London, 1969, p. 1.
162. D. A. Brummell and J. L. Hall, *Plant Cell Environ.*, 1987, **10**, 523.
163. A. Hager, H. Menzel, and A. Kraus, *Planta*, 1971, **100**, 47.
164. D. L. Rayle and R. Cleland, *Curr. Topics Dev. Biol.*, 1977, **11**, 187.
165. A. P. Senn and M. H. M. Goldsmith, *Plant Physiol.*, 1988, **88**, 131.
166. W. S. Peters and H. Felle, *J. Plant Physiol.*, 1991, **137**, 655.
167. G. W. Bates and M. H. M. Goldsmith, *Planta*, 1983, **159**, 231.
168. H. Felle, W. Peters, and K. Palme, *Biochem. Biophys. Acta*, 1991, **1064**, 199.
169. A. Tretyn, G. Wagner, and H. H. Felle, *J. Plant Physiol.*, 1991, **139**, 187.
170. W. H. Shen, E. Daviond, C. David, H. Barbier-Brygoo, and J. Tempe, *Plant Physiol.*, 1990, **94**, 554.
171. Y. Takahashi, S. Ishida, and T. Nagata, *Plant Cell Physiol.*, 1995, **36**, 383.
172. W. L. Porter and K. V. Thimann, *Phytochemistry*, 1965, **4**, 229.
173. G. R. Hicks, D. L. Rayle, and T. L. Lomax, *Science*, 1989, **245**, 52.
174. H. Macdonald, A. M. Jones, and P. J. King, *J. Biol. Chem.*, 1991, **266**, 7393.
175. J. Feldwisch, A. Vente, R. Zettl, L. Bakon, N. Campos, and K. Palme, *Biochem. J.*, 1994, **302**, 15.
176. J. Bilang and A. Sturm, *Plant Physiol.*, 1995, **109**, 253.
177. R. Zettl, J. Schell, and K. Palme, *Proc. Natl. Acad. Sci. USA*, 1994, **91**, 689.
178. S. Sugaya and S. Sakai, *Plant Sci.*, 1996, **114**, 1.
179. M. Löbler and D. Klämbt, *J. Biol. Chem.*, 1985, **260**, 9848.
180. S. Shimomura, T. Sotobayashi, M. Futai, and T. Fukui, *J. Biochem.*, 1986, **99**, 1513.
181. A. M. Jones, *Annu. Rev. Plant Physiol. Plant Mol. Biol.*, 1994, **45**, 393.
182. R. M. Napier, *J. Exp. Bot.*, 1995, **46**, 1787.
183. R. M. Napier and M. A. Venis, *New Phytol.*, 1995, **129**, 167.
184. H. R. Irving, C. A. Gehring, and R. W. Parish, *Proc. Natl. Acad. Sci. USA*, 1992, **89**, 1790.
185. B. K. Drøbak, *Plant Physiol.*, 1993, **102**, 705.
186. B. W. Poovaiah, A. S. N. Reddy, and J. J. McFadden, *Physiol. Plant*, 1987, **69**, 569.
187. B. Zbell and C. Walter-Back, *J. Plant Physiol.*, 1988, **133**, 353.
188. B. Andre and G. F. E. Scherer, *Planta*, 1991, **185**, 209.
189. D. M. Law, *Physiol. Plant*, 1987, **70**, 626.
190. M. H. Zenk and H. Scherf, *Planta*, 1964, **62**, 350.
191. Y. Sakagami, H. Tan, K. Manabe, M. Higashi, and S. Marumo, *Biosci. Biotech. Biochem.*, 1995, **59**, 1362.
192. Y. Liu, A. L. Silverstone, Y. M. Wu, and S. F. Yang, *Phytochemistry*, 1995, **40**, 691.
193. T. P. Cooney and H. M. Nonhebel, *Planta*, 1991, **184**, 368.
194. D. D. Songstad, V. De Luca, N. Brisson, W. G. W. Kurz, and C. L. Neesler, *Plant Physiol.*, 1990, **94**, 1410.
195. R. Rajagopal, K. Tsurusaki, S. Kuraishi, and N. Sakurai, *Plant Physiol.* (*Life Sci. Adv.*), 1993, **12**, 17.
196. B. Bartel and G. R. Fink, *Proc. Natl. Acad. Sci. USA*, 1994, **91**, 6649.
197. E. Kemper, S. Waffenschmidt, E. W. Weiler, T. Rausch, and J. Schröder, *Planta*, 1985, **163**, 257.
198. M. Kawaguchi, S. Fujioka, A. Sakurai, Y. T. Tamaki, and K. Syono, *Plant Cell Physiol.*, 1993, **34**, 121.
199. J. Normanly, J. P. Slovin, and J. D. Cohen, *Plant Physiol.*, 1995, **107**, 323.
200. P. W. Zimmerman and A. E. Hitchcock, *Contrib. Boyce Thompson Inst.*, 1942, **12**, 321.
201. L. Conti, *Boll. Chem. Farm.*, 1968, **107**, 325.
202. B. Aberg, *Swed. J. Agric. Res.*, 1973, **3**, 49.
203. J. Beraud, *C. R. Acad. Sci. Ser. D*, 1972, **274**, 866.
204. J. D. Cohen, B. G. Baldi, and K. Bialek, *Plant Physiol*, 1985, **77**, 195.
205. N. Frdmann, E. Libbert, and U. Schiewer, *Flora* (*Jena*), *Abt. A*, 1969, **160**, 350.
206. T. Hatano, M. Katayama, and S. Marumo, *Experientia*, 1987, **43**, 1237.
207. C. Branca and E. Gaetani, *Atti. Accad. Naz. Lincei., Cl. Sci. Fis., Mat. Natur., Rend.*, 1973, **54**, 275.
208. C. A. Maggiali, M. R. Mingiarfi, C. Branca, and D. Ricci, *Farmaco. Ed. Sci.*, 1983, **38**, 935.
209. V. Sutoris, P. Bajci, V. Sekerka, and J. Halgas, *Chem. Pap.*, 1988, **42**, 249.
210. F. E. Gardner, P. C. Marth, and L. P. Batjer, *Science*, 1939, **90**, 208.
211. M. A. Venis and E. W. Thomas, *Phytochemistry*, 1990, **29**, 381.
212. D. B. Harper and R. L. Wain, *Ann. Appl. Biol.*, 1969, **64**, 395.
213. T. Fujita, K. Kawazu, T. Mitsui, and M. Katsumi, *Phytochemistry*, 1967, **6**, 889.
214. H. Koch, *Sci. Pharm.*, 1971, **39**, 209.
215. K. V. Thimann and W. D. Bonner, Jr., *Plant Physiol.*, 1948, **23**, 158.

216. J. Bonner, *Am. J. Bot.*, 1949, **36**, 429.
217. M. M. Moloney and P. E. Pilet, *Planta*, 1981, **153**, 447.
218. M. S. Smith, R. L. Wain, and F. Wightman, *Nature*, 1952, **169**, 883.
219. B. Åberg, *Physiol. Plant*, 1951, **4**, 627.
220. B. Åberg, *Swed. J. Agric. Res.*, 1981, **11**, 107.
221. T. Hatano, Y. Kato, M. Katayama, and S. Marumo, *Experientia*, 1989, **45**, 400.
222. C. Branca, C. A. Maggiali, M. R. Mingiardi, and D. Ricci, *J. Plant Growth Regul.*, 1982, **1**, 243.
223. E. M. Beyer, Jr., *Plant Physiol.*, 1972, **50**, 322.
224. E. M. Beyer, Jr., A. L. Johnson, and P. B. Sweetser, *Plant Physiol.*, 1976, **57**, 839.
225. A. C. Leopold and W. H. Klein, *Science*, 1951, **114**, 9.
226. M. Katayama and R. K. Gautam, *Biosci. Biotech. Biochem.*, 1996, **60**, 755.
227. P. J. Davies, "Plant Hormones," ed. P. J. Davies, 2nd edn., Kluwer, Dordrecht, 1995, p. 1.
228. J. R. Bearder, in "Hormonal Regulation of Development," Encyclopedia of Plant Physiology (New Series), ed. J. MacMillan, Springer, Berlin, 1980, vol. 9, p. 22.
229. J. MacMillan and N. Takahashi, *Nature*, 1968, **217**, 170.
230. G. V. Hoad, in "The Biochemistry and Physiology of Gibberellins," ed. A. Crozier, Praeger, New York, 1983, vol. 2, p. 57.
231. H. Yamane, in "Gibberellins," eds. N. Takahasi, B. O. Phinney, and J. MacMillan, Springer, New York, 1991, p. 378.
232. G. Sembdner, D. Gross, H.-W. Liebisch, and G. Schneider, in "Hormonal Regulation of Development," Encyclopedia of Plant Physiology (New Series), ed. J. MacMillan, Springer, Berlin, 1980, vol. 9, p. 281.
233. J. E. Graebe, *Annu. Rev. Plant Physiol.*, 1987, **38**, 419.
234. J. J. Ross, *Plant Growth Regul.*, 1994, **15**, 193.
235. I. Nakayama, Y. Kamiya, M. Kobayashi, H. Abe, and A. Sakurai, *Plant Cell Physiol.*, 1990, **31**, 1183.
236. P. Hedden and Y. Kamiya, *Annu. Rev. Plant Physiol. Plant Mol. Biol.*, 1997, **48**, 431.
237. N. Takahashi, I. Yamaguchi, and H. Yamane, in "Chemistry of Plant Hormones," ed. N. Takahashi, CRC Press, Florida, 1986, p. 57.
238. E. Kurosawa, *Trans. Natl. Hist. Soc. Formosa*, 1926, **16**, 213.
239. T. Yabuta and Y. Sumiki, *J. Agric. Chem. Soc. Jpn*, 1938, **14**, 1526.
240. P. J. Curtis and B. E. Cross, *Chem. Ind.*, 1954, 1066.
241. F. H. Stodola, K. B. Raper, D. I. Fennell, H. F. Conway, V. E. Sohns, C. T. Langford, and R. W. Jackson, *Arch. Biochem. Biophys.*, 1955, **54**, 240.
242. W. Rademacher and J. E. Graebe, *Biochem. Biophys. Res. Commun.*, 1979, **91**, 35.
243. H. Kawaide, T. Sassa, and Y. Kamiya, *Phytochemistry*, 1995, **39**, 305.
244. H. Kawaide, R. Imai, T. Sassa, and Y. Kamiya, *J. Biol. Chem.*, 1997, **272**, 21 706.
245. J. W. Mitchell, D. P. Skaggs, and W. P. Anderson, *Science*, 1951, **114**, 159.
246. J. MacMillan and P. J. Suter, *Naturwissenschaften*, 1958, **45**, 46.
247. C. A. West and B. O. Phinney, *J. Am. Chem. Soc.*, 1959, **81**, 2424.
248. A. Kawarada and Y. Sumiki, *Bull. Agric. Chem. Soc. Jpn.*, 1959, **23**, 343.
249. N. Murofushi, S. Iriuchijima, N. Takahashi, S. Tamura, J. Kato, Y. Wada, E. Watanabe, and T. Aoyama, *Agric. Biol. Chem.*, 1966, **30**, 917.
250. S. Tamura, in "Gibberellins," eds. N. Takahashi, B. O. Phinney, and J. MacMillan, Springer, New York, 1991, p. 1.
251. B. O. Phinney, in "The Biochemistry and Physiology of Gibberellins," ed. A. Crozier, Praeger, New York, 1983, p. 19.
252. J. MacMillan, *Annu. Rev. Plant Physiol. Plant Mol. Biol.*, 1996, **47**, 1.
253. T. Yokota, N. Murofushi, and N. Takahashi, in "Hormonal Regulation of Development," Encyclopedia of Plant Physiology (New Series), ed. J. MacMillan, Springer, Berlin, 1980, vol. 9, p. 113.
254. P. Hedden, *Annu. Rev. Plant Physiol. Plant Mol. Biol.*, 1993, **44**, 107.
255. M. H. Beale and C. L. Willis, in "Methods in Plant Biochemistry," eds. B. V. Charlwood and D. V. Banthorpe, 1991, vol. 7, p. 289.
256. P. Hedden, in "Gas Chromatography/Mass Spectrometry: Modern Methods of Plant Analysis," new series, eds. H. F. Linskens and J. F. Jackson, Springer, Berlin, 1986, vol. 3, p. 1.
257. P. Gaskin and J. MacMillan, "GC–MS of the Gibberellins and Related Compounds: Methodology and a Library of Spectra," University of Bristol, 1991.
258. M. Koshioka, K. Takeno, F. D. Beall, and R. P. Pharis, *Plant Physiol.*, 1983, **73**, 398.
259. L. Rivier and A. Crozier, "Principles and Practice of Plant Hormone Analysis," Academic Press, London, 1987, vols. 1 and 2.
260. T. Yokota, N. Murofushi, N. Takahashi, and M. Katsumi, *Phytochemistry*, 1971, **10**, 2943.
261. H. Yamane, I. Yamaguchi, T. Yokota, N. Murofushi, N. Takahashi, and M. Katsumi, *Phytochemistry*, 1973, **12**, 255.
262. B. O. Phinney and C. R. Spray, in "Plant Growth Substances 1982," ed. P. F. Waring, Academic Press, London, 1982, p. 101.
263. S. Fujioka, H. Yamane, C. R. Spray, P. Gaskin, J. MacMillan, B. O. Phinney, and N. Takahashi, *Plant Physiol.*, 1988, **88**, 1367.
264. S. Ogawa, T. Toyomasu, H. Yamane, N. Murofushi, R. Ikeda, Y. Morimoto, Y. Nishimura, and T. Omori, *Plant Cell Physiol.*, 1996, **37**, 363.
265. Y. Murakami, in "Plant Growth Substances 1970," ed. D. J. Carr, Springer, Berlin, 1972, p. 166.
266. M. Nakayama, H. Yamane, N. Murofushi, N. Takahashi, L. N. Mander, and H. Seto, *J. Plant Growth Regul.*, 1991, **10**, 115.
267. P. W. Brian, J. F. Grove, and T. P. C. Mulholland, *Phytochemistry*, 1967, **6**, 1475.
268. A. Crozier, C. C. Kuo, R. C. Durley, and R. P. Pharis, *Can. J. Bot.*, 1970, **48**, 867.
269. D. R. Reeve and A. Crozier, *J. Exp. Bot.*, 1974, **25**, 431.
270. K. Hiraga, H. Yamane, and N. Takahashi, *Phytochemistry*, 1974, **13**, 2371.
271. J. V. Jacobsen, P. M. Chandler, T. J. V. Higgins, and J. A. Zwar, in "Plant Growth Substances 1982," ed. P. F. Waring, Academic Press, New York, 1982, p. 111.

272. J. V. Jacobsen and J. E. Varner, *Plant Physiol.*, 1968, **42**, 1596.
273. T. G. Brock and P. B. Kaufman, in "Plant Physiology: A Treatise," ed. F. C. Steward, vol. X: "Growth and Development," ed. R. G. S. Bidwell, Academic Press, San Diego, CA, 1991, p. 277.
274. L. G. Paleg and C. A. West, in "Plant Physiology: A Treatise," ed. F. C. Steward, vol. VIB: "Physiology of Development: The Hormones," Academic Press, New York, 1972, p. 146.
275. P. Saunders, in "Phytohormones and Related Compounds: A Comprehensive Treatise, Vol. II, Phytohormones and the Development of Higher Plants," eds. D. S. Letham, P. B. Goodwin, and T. J. V. Higgins, Elsevier/North-Holland, Amsterdam, 1978, p. 423.
276. J. A. D. Zeevaart, *Annu. Rev. Plant Physiol.*, 1976, **27**, 321.
277. J. A. D. Zeevaart, in "The Biochemistry and Physiology of Gibberellins," ed. A. Crozier, Praeger, New York, 1983, vol. 1, p. 99.
278. S. H. Witter, in "Phytohormones and Related Compounds: A Comprehensive Treatise, Vol. II, Phytohormones and the Development of Higher Plants," eds. D. S. Letham, P. B. Goodwin, and T. J. V. Higgins, Elsevier/North-Holland, Amsterdam, 1978, p. 599.
279. V. M. Sponsel, in "Plant Hormones," ed. P. J. Davies, 2nd edn., Kluwer, Dordrecht, 1995, p. 66.
280. J. MacMillan, *Nat. Prod. Rep.*, 1997, **14**, 221.
281. R. C. Coolbaugh, in "The Biochemistry and Physiology of Gibberellins," ed. A. Crozier, Praeger, New York, 1983, vol. 1, p. 53.
282. J. Chappell, *Annu. Rev. Plant Physiol. Plant Mol. Biol.*, 1995, **46**, 521.
283. J. Schwender, M. Seemann, H. K. Lichtenthaler, and M. Rohmer, *Biochem. J.*, 1996, **316**, 73.
284. J. D. Duncan and C. A. West, *Plant Physiol.*, 1981, **68**, 1128.
285. R. J. Bensen and J. A. D. Zeevaart, *J. Plant Growth Regul.*, 1990, **9**, 237.
286. M. Koornneef and J. H. van der Veen, *Theor. Appl. Genet.*, 1980, **58**, 257.
287. M. Koornneef, J. van Eden, C. J. Hanhart, and A. M. M. de Jongh, *Genet. Res.*, 1983, **41**, 57.
288. T. P. Sun, H. M. Goodman, and F. M. Ausubel, *Plant Cell*, 1992, **4**, 119.
289. T. P. Sun and Y. Kamiya, *Plant Cell*, 1994, **6**, 1509.
290. T. Ait-Ali, S. M. Swain, J. B. Reid, T. P. Sun, and Y. Kamiya, *Plant J.*, 1997, **11**, 443.
291. R. J. Bensen, G. S. Johal, V. C. Crane, J. T. Tossberg, P. S. Schnable, R. B. Meeley, and S. P. Briggs, *Plant Cell*, 1995, **7**, 75.
292. T. Saito, H. Abe, H. Yamane, A. Sakurai, N. Murofushi, K. Takio, N. Takahashi, and Y. Kamiya, *Plant Physiol.*, 1995, **109**, 1239.
293. S. Yamaguchi, T. Saito, H. Abe, H. Yamane, N. Murofushi, and Y. Kamiya, *Plant J.*, 1996, **10**, 203.
294. C. J. D. Mau and C. A. West, *Proc. Natl. Acad. Sci. USA*, 1994, **91**, 8497.
295. P. J. Facchini and J. Chappell, *Proc. Natl. Acad. Sci. USA*, 1992, **89**, 11 088.
296. S. M. Colby, W. R. Alonso, E. V. Katahira, D. J. McGarvey, and R. Croteau, *J. Biol. Chem.*, 1993, **268**, 23 016.
297. H. Aach, G. Böse, and J. E. Graebe, *Planta*, 1995, **197**, 333.
298. K. Keegstra, L. D. Olsen, and S. M. Theg, *Annu. Rev. Plant Physiol. Plant Mol. Biol.*, 1989, **40**, 471.
299. G. E. Bartley, P. A. Scolnik, and G. Giuliano, *Annu. Rev. Plant Physiol. Plant Mol. Biol.*, 1994, **45**, 287.
300. M. Kuntz, S. Römer, C. Suire, P. Hugueney, J. H. Weil, R. Schantz, and B. Camara, *Plant J.*, 1992, **2**, 25.
301. R. G. Fray, A. Wallace, P. D. Fraser, D. Valero, P. Hedden, P. M. Bramley, and D. Grierson, *Plant J.*, 1995, **8**, 693.
302. P. J. Murphy and C. A. West, *Arch. Biochem. Biophys.*, 1969, **133**, 395.
303. J. C. Jennings, R. C. Coolbaugh, D. A. Nakata, and C. A. West, *Plant Physiol.*, 1993, **101**, 925.
304. S. M. Swain, J. B. Reid, and Y. Kamiya, *Plant J.*, 1997, **12**, 1329.
305. R. G. Winkler and T. Helentjaris, *Plant Cell*, 1995, **7**, 1307.
306. E. De Carolis and V. De Luca, *Phytochemistry*, 1994, **36**, 1093.
307. A. G. Prescott, *Annu. Rev. Plant Physiol. Plant Mol. Biol.*, 1996, **47**, 245.
308. M. Kobayashi, C. R. Spray, B. O. Phinney, P. Gaskin, and J. MacMillan, *Plant Physiol.*, 1996, **110**, 413.
309. P. Hedden, J. E. Graebe, M. H. Beale, P. Gaskin, and J. MacMillan, *Phytochemistry*, 1984, **23**, 569.
310. S. J. Gilmour, A. B. Bleecker, and J. A. D. Zeevaart, *Plant Physiol.*, 1987, **85**, 87.
311. T. Lange, P. Hedden, and J. E. Graebe, *Planta*, 1993, **189**, 340.
312. Y. Kamiya and J. E. Graebe, *Phytochemistry*, 1983, **22**, 681.
313. E. Großelindemann, M. J. Lewis, P. Hedden, and J. E. Graebe, *Planta*, 1992, **188**, 252.
314. C. G. N. Turnbull, A. Crozier, L. Schwenen, and J. E. Graebe, *Planta*, 1985, **165**, 108.
315. M. Takahashi, Y. Kamiya, N. Takahashi, and J. E. Graebe, *Planta*, 1986, **168**, 190.
316. T. Moritz, *Physiol. Plant*, 1995, **95**, 67.
317. T. Lange, *Planta*, 1994, **195**, 108.
318. T. Lange, P. Hedden, and J. E. Graebe, *Proc. Natl. Acad. Sci. USA*, 1994, **91**, 8552.
319. A. L. Phillips, D. A. Ward, S. Uknes, N. E. J. Appleford, T. Lange, A. K. Huttly, P. Gaskin, J. E. Graebe, and P. Hedden, *Plant Physiol.*, 1995, **108**, 1049.
320. Y. L. Xu, L. Li, K. Wu, A. J. M. Peeters, D. A. Gage, and J. A. D. Zeevaart, *Proc. Natl. Acad. Sci. USA*, 1995, **92**, 6640.
321. M. Talon, M. Koornneef, and J. A. D. Zeevaart, *Proc. Natl. Acad. Sci. USA*, 1990, **87**, 7983.
322. T. Toyomasu, H. Kawaide, H. Sekimoto, C. von Numers, A. L. Phillips, P. Hedden, and Y. Kamiya, *Physiol. Plant*, 1997, **99**, 111.
323. S. J. Gilmour, J. A. D. Zeevaart, L. Schwenen, and J. E. Graebe, *Plant Physiol.*, 1986, **82**, 190.
324. P. Hedden and J. E. Graebe, *J. Plant Growth Regul.*, 1982, **1**, 105.
325. J. L. Ward, G. S. Jackson, M. H. Beale, P. Gaskin, and P. Hedden, *J. Chem. Soc., Chem. Commun.*, 1997, 13.
326. V. A. Smith, P. Gaskin, and J. MacMillan, *Plant Physiol.*, 1990, **94**, 1390.
327. S. S. Kwak, Y. Kamiya, A. Sakurai, N. Takahashi, and J. E. Graebe, *Plant Cell Physiol.*, 1988, **29**, 935.
328. T. Lange, A. Schweimer, D. A. Ward, P. Hedden, and J. E. Graebe, *Planta*, 1994, **195**, 98.
329. S. Fujioka, H. Yamane, C. R. Spray, B. O. Phinney, P. Gaskin, J. MacMillan, and N. Takahashi, *Plant Physiol.*, 1990, **94**, 127.
330. K. S. Albone, P. Gaskin, J. MacMillan, B. O. Phinney, and C. L. Willis, *Plant Physiol.*, 1990, **94**, 132.

331. C. R. Spray, M. Kobayashi, Y. Suzuki, B. O. Phinney, P. Gaskin, and J. MacMillan, *Proc. Natl. Acad. Sci. USA*, 1996, **93**, 10 515.
332. H. H. Chiang, I. Hwang, and H. M. Goodman, *Plant Cell*, 1995, **7**, 195.
333. P. Hedden, personal communication.
334. V. A. Smith and J. MacMillan, *Planta*, 1986, **167**, 9.
335. D. L. Griggs, P. Hedden, and C. M. Lazarus, *Phytochemistry*, 1991, **30**, 2507.
336. V. M. Sponsel, *Planta*, 1983, **159**, 454.
337. J. J. Ross, J. B. Reid, S. M. Swain, O. Hasan, A. T. Poole, P. Hedden, and C. L. Willis, *Plant J.*, 1995, **7**, 513.
338. N. E. J. Appleford and J. R. Lenton, *Planta*, 1991, **183**, 229.
339. S. Fujioka, H. Yamane, C. R. Spray, M. Katsumi, B. O. Phinney, P. Gaskin, J. MacMillan, and N. Takahashi, *Proc. Natl. Acad. Sci. USA*, 1988, **85**, 9031.
340. M. Talon, M. Koornneef, and J. A. D. Zeevaart, *Planta*, 1990, **182**, 501.
341. D. N. Martin, W. M. Proebsting, T. D. Parks, W. G. Dougherty, T. Lange, M. J. Lewis, P. Gaskin, and P. Hedden, *Planta*, 1996, **200**, 159.
342. A. L. Phillips, personal communication.
343. K. Wu, L. Li, D. A. Gage, and J. A. D. Zeevaart, *Plant Physiol.*, 1996, **110**, 547.
344. T. Toyomasu, H. Tsuji, H. Yamane, M. Nakayama, I. Yamaguchi, N. Murofushi, N. Takahashi, and Y. Inoue, *J. Plant Growth Regul.*, 1993, **12**, 85.
345. T. Toyomasu, H. Yamane, M. I. Yamaguchi, N. Murofushi, N. Takahashi, and Y. Inoue, *Plant Cell Physiol.*, 1992, **33**, 695.
346. J. F. Martínez-García and J. L. García-Martínez, *Physiol. Plant*, 1992, **86**, 236.
347. J. B. Reid, O. Hasan, and J. J. Ross, *J. Plant Physiol.*, 1990, **137**, 46.
348. J. L. Weller, J. J. Ross, and J. B. Reid, *Planta*, 1994, **192**, 489.
349. P. F. Devlin, S. B. Rood, D. E. Somers, P. H. Quail, and G. C. Whitelam, *Plant Physiol.*, 1992, **100**, 1442.
350. K. L. Childs, M. M. Cordonnier-Pratt, L. H. Pratt, and P. W. Morgan, *Plant Physiol.*, 1992, **99**, 765.
351. E. López-Juez, M. Kobayashi, A. Sakurai, Y. Kamiya, and R. E. Kendrick, *Plant Physiol.*, 1995, **107**, 131.
352. H. Gawronska, Y. Y. Yang, K. Furukawa, R. E. Kendrick, N. Takahashi, and Y. Kamiya, *Plant Cell Physiol.*, 1995, **36**, 1361.
353. J. D. Metzger, *Plant Physiol.*, 1985, **78**, 8.
354. J. D. Metzger, *Plant Physiol.*, 1990, **94**, 151.
355. W. Döpp, *Ber. Dtsch. Bot. Ges.*, 1950, **63**, 139.
356. W. Döpp, *Ber. Dtsch. Bot. Ges.*, 1959, **72**, 11.
357. U. Näf, K. Nakanishi, and M. Endo, *Bot. Rev.*, 1975, **41**, 315.
358. U. Näf, *Physiol. Plant*, 1966, **19**, 1079.
359. U. Näf, *Plant Cell Physiol.*, 1968, **9**, 27.
360. K. Nakanishi, M. Endo, U. Näf, and L. F. Johnson, *J. Am. Chem. Soc.*, 1971, **93**, 5579.
361. P. R. Zanno, M. Endo, K. Nakanishi, U. Näf, and C. Stein, *Naturwissenschaften*, 1972, **59**, 512.
362. E. J. Corey, A. G. Myers, N. Takahashi, H. Yamane, and H. Schraudolf, *Tretrahedron Lett.*, 1986, **27**, 5083.
363. H. Yamane, N. Takahashi, K. Takeno, and M. Furuya, *Planta*, 1979, **147**, 251.
364. H. Yamane, K. Nohara, N. Takahashi, and H. Schraudolf, *Plant Cell Physiol.*, 1987, **28**, 1203.
365. H. Yamane, Y. Satoh, K. Nohara, M. Nakayama, N. Murofushi, H. Takahashi, K. Takeno, M. Furuya, M. Furber, and L. N. Mander, *Tetrahedron Lett.*, 1988, **29**, 3959.
366. M. Furber, L. N. Mander, J. E. Nester, N. Takahashi, and H. Yamane, *Phytochemistry*, 1989, **28**, 63.
367. T. Yamauchi, N. Oyama, H. Yamane, N. Murofushi, N. Takahashi, H. Schraudolf, M. Furber, L. N. Mander, G. L. Patrick, and B. Twitchin, *Phytochemistry*, 1991, **30**, 3247.
368. T. Yamauchi, N. Oyama, H. Yamane, N. Murofushi, H. Schraudolf, D. Owen, and L. N. Mander, *Phytochemistry*, 1995, **38**, 1345.
369. T. Yamauchi, N. Oyama, H. Yamane, N. Murofushi, H. Schraudolf, M. Pour, M. Furber, and L. N. Mander, *Plant Physiol.*, 1996, **111**, 741.
370. E. J. Corey and A. G. Myers, *J. Am. Chem. Soc.*, 1985, **107**, 5574.
371. M. Furber and L. N. Mander, *J. Am. Chem. Soc.*, 1987, **109**, 6389.
372. J. E. Nester, F. Veysey, and R. C. Coolbaugh, *Planta*, 1987, **170**, 26.
373. N. Oyama, T. Yamauchi, H. Yamane, N. Murofushi, M. Agatsuma, M. Pour, and L. N. Mander, *Biosci. Biotech. Biochem.*, 1996, **60**, 305.
374. K. Takeno, H. Yamane, K. Nohara, N. Takahashi, E. J. Corey, A. G. Myers, and H. Schraudolf, *Phytochemistry*, 1987, **26**, 1855.
375. K. Takeno, H. Yamane, T. Yamauchi, N. Takahashi, M. Furber, and L. N. Mander, *Plant Cell Physiol.*, 1989, **30**, 201.
376. T. Yamauchi, N. Oyama, H. Yamane, N. Murofushi, H. Schraudolf, M. Pour, H. Seto, and L. N. Mander, *Plant Physiol.*, 1997, **113**, 773.
377. C. O. Miller, F. Skoog, M. H. Von Staltza, and F. M. Strong, *J. Am. Chem. Soc.*, 1955, **77**, 1392.
378. C. O. Miller, F. Skoog, F. S. Okumura, M. H. Von Staltza, and F. M. Strong, *J. Am. Chem. Soc.*, 1956, **78**, 1375.
379. J. E. Fox, in "Physiology of Plant Growth and Development," ed. M. B. Wilkins, McGraw-Hill, London, 1969, p. 85.
380. C. O. Miller, *Proc. Natl. Acad. Sci. USA*, 1961, **47**, 170.
381. D. S. Letham, *Life Sci.*, 1963, **8**, 569.
382. D. S. Letham, J. S. Shannon, and I. R. McDonald, *Proc. Chem. Soc.* (*London*), 1964, 230.
383. D. S. Letham, *N. Zeal. J. Bot.*, 1963, **1**, 336.
384. S. Kuraishi and F. Okumura, *Bot. Mag. Tokyo*, 1956, **69**, 817.
385. C. O. Miller, *Plant Physiol.*, 1956, **31**, 318.
386. R. A. Scott, Jr. and J. L. Liverman, *Plant Physiol.*, 1956, **31**, 321.
387. R. S. deRopp, *Plant Physiol.*, 1956, **31**, 253.
388. A. H. Haber and N. E. Tolbert, *Plant Physiol.*, 1957, **32** (suppl.), xlix.

389. C. O. Miller, *Plant Physiol.*, 1958, **33**, 115.
390. A. H. Haber and H. J. Luippold, *Plant Physiol.*, 1960, **35**, 168.
391. A. Richmond and A. Lang, *Science*, 1957, **125**, 650.
392. B. I. S. Srivastava, *Int. Rev. Cytol.*, 1967, **22**, 349.
393. F. Skoog and D. J. Armstrong, *Annu. Rev. Plant Physiol.*, 1970, **21**, 359.
394. H. Kende, *Int. Rev. Cytol.*, 1971, **31**, 301.
395. R. H. Hall, *Annu. Rev. Plant Physiol.*, 1973, **24**, 415.
396. D. J. Osborne, *Plant Physiol.*, 1962, **37**, 595.
397. C. Jayabaskaran, M. Kuntz, P. Guillemaut, and J.-H. Weil, *Plant Physiol.*, 1990, **92**, 136.
398. R. K. Atkin and B. I. S. Srivastava, *Physiol. Plant*, 1969, **22**, 742.
399. K. Möthes, *Sov. Plant Physiol.*, 1972, **19**, 863.
400. K. Möthes, *Naturwissenschaften*, 1960, **47**, 337.
401. A. C. Leopold and M. Kawase, *Am. J. Bot.*, 1964, **51**, 294.
402. H. Meidner, *J. Exp. Bot.*, 1967, **18**, 556.
403. A. Livné and Y. Vaadia, *Physiol. Plant*, 1965, **18**, 658.
404. J. E. Pallas and J. E. Box, *Nature*, 1970, **227**, 87.
405. M. Zeroni and M. A. Hall, in "Encyclopedia of Plant Physiology," ed. J. MacMillan, Springer, Berlin, 1980, vol. 9, p. 511.
406. S. Kuraishi, *Plant Cell Physiol.*, 1976, **17**, 875.
407. E. Uheda and S. Kuraishi, *Plant Cell Physiol.*, 1978, **19**, 825.
408. S. Kuraishi and F. Ishikawa, *Plant Cell Physiol.*, 1977, **18**, 1273.
409. K. Müller and A. C. Leopold, *Planta*, 1966, **68**, 167.
410. T. Sachs and K. V. Thimann, *Nature*, 1964, **201**, 939.
411. R. M. Samuels, Ph.D. Thesis, Indiana University, 1961.
412. J. P. Helgeson and N. J. Leonard, *Proc. Natl. Acad. Sci. USA*, 1966, **56**, 60.
413. L. Lovrekovich and G. L. Farkes, *Nature*, 1963, **198**, 710.
414. F. Skoog and C. O. Miller, in "Biological Action of Growth Substances," Symposium of the Society of Experimental Biologists, ed. H. K. Porter, Cambridge University Press, Cambridge, 1957, no. 11, p. 118.
415. J. E. Fox, *Plant Physiol.*, 1964, **39** (suppl.), xxxi.
416. J. E. Fox, *Plant Physiol.*, 1966, **41**, 75.
417. J. E. Fox and C.-M. Chen, *J. Biol. Chem.*, 1967, **242**, 4490.
418. H. Zachau, G. Dütting, and H. Feldmann, *Angew. Chem.*, 1966, **78**, 392.
419. D. J. Armstrong, F. Skoog, L. H. Kirkegaard, A. E. Hampel, R. M. Bock, I. Gillam, and G. M. Tener, *Proc. Natl. Acad. Sci. USA*, 1969, **63**, 504.
420. D. J. Armstrong, W. J. Burrows, F. Skoog, K. L. Roy, and D. Söll, *Proc. Natl. Acad. Sci. USA*, 1969, **63**, 834.
421. J. L. Stoddart and M. A. Venis, in "Encyclopedia of Plant Physiology," ed. J. MacMillan, Springer, Berlin, 1980, vol. 9, p. 445.
422. R. A. Fletcher and D. McCullagh, *Planta*, 1971, **101**, 88.
423. K. Naito, H. Tsuji, and I. Hatakeyama, *Physiol. Plant*, 1978, **43**, 367.
424. Y. Soeda, I. Kinoshita, and H. Tsuji, *Plant Growth Regul.*, 1990, **9**, 165.
425. K. Naito, H. Tsuji, I. Hatakeyama, and K. Ueda, *J. Exp. Bot.*, 1979, **30**, 1145.
426. E. Momotani, I. Kinoshita, E. Yokomura, and H. Tsuji, *Plant Cell Physiol.*, 1990, **31**, 621.
427. I. Kinoshita, E. Yokomura, and H. Tsuji, *Biochem. Physiol. Pflanzen*, 1991, **187**, 167.
428. W. M. Laetsch and P. Boasson, in "Hormonal Regulation in Plant Growth and Development," eds. H. Kaldewey and Y. Varder, Verlag Chemie, Weinheim, 1972, p. 453.
429. K. Naito, K. Ueda, and H. Tsuji, *Protoplasma*, 1981, **105**, 293.
430. I. Kinosita and H. Tsuji, *Plant Physiol.*, 1984, **76**, 575.
431. E. Momotani, K. Aoki, and H. Tsuji, *J. Exp. Bot.*, 1991, **42**, 1287.
432. T. Murashige and F. Skoog, *Physiol. Plant*, 1962, **15**, 473.
433. C. O. Miller, in "Modern Methods of Plant Analysis," Springer, Berlin, 1963, vol. 6, p. 194.
434. D. J. Osborne and D. R. McCalla, *Plant Physiol.*, 1961, **36**, 219.
435. R. A. Fletcher, V. Kallidumbil, and P. Steele, *Plant Physiol.*, 1982, **69**, 675.
436. J. Guern and C. Péaud-Lenoël (eds.), "Metabolism and Molecular Activities of Cytokinins," Springer, Berlin, 1981.
437. D. S. Letham and L. M. S. Palni, *Annu. Rev. Plant Physiol.*, 1983, **34**, 163.
438. K. Koshimizu and H. Iwamura, in "Chemistry of Plant Hormones," ed. N. Takahashi, CRC, Boca Raton, FL, 1988, p. 154.
439. A. N. Binns, *Annu. Rev. Plant Physiol. Plant Mol. Biol.*, 1994, **45**, 173.
440. R. Horgan, in "Isolation of Plant Growth Substances," ed. J. Hillman, Cambridge University Press, Cambridge, 1978, p. 97.
441. R. Horgan, in "Progress in Phytochemistry," eds. L. Reinhold, J. B. Harbone, and T. Swain, Pergamon, Oxford, 1981, vol. 7, p. 137.
442. D. S. Letham, *Planta*, 1974, **118**, 361.
443. I. M. Scott, G. C. Martin, R. Horgan, and J. K. Heald, *Planta*, 1982, **154**, 273.
444. A. Evidente, N. S. Iacobellis, R. Vellone, A. Sisto, and G. Surico, *Phytochemistry*, 1989, **28**, 2603.
445. E. M. Shantz and F. C. Steward, *J. Am. Chem. Soc.*, 1955, **77**, 6351.
446. S. K. Nandi, L. M. S. Palni, D. S. Letham, and O. C. Wong, *Plant Cell Environ.*, 1989, **12**, 273.
447. D. Ernst, W. Schäfer, and D. Oesterhelt, *Planta*, 1983, **159**, 222.
448. H. J. Chaves das Neves and M. S. S. Pais, *Biochem. Biophys. Res. Commun.*, 1980, **95**, 1387.
449. G. Shaw, B. M. Smallwood, and D. V. Wilson, *J. Chem. Soc. C*, 1966, 921.
450. H. J. Vreman, R. Y. Schmitz, F. Skoog, A. J. Playtis, C. R. Frihart, and N. J. Leonard, *Phytochemistry*, 1974, **13**, 31.
451. G. Shaw, B. M. Smallwood, and D. V. Wilson, *J. Chem. Soc. C*, 1968, 1516.
452. W. A. H. Grimm and N. J. Leonard, *Biochemistry*, 1967, **6**, 3625.
453. D. E. Cowley, C. C. Duke, A. J. Liepa, J. K. MacLeod, and D. S. Letham, *Aust. J. Chem.*, 1978, **31**, 1095.

454. C. C. Duke, J. K. MacLeod, R. E. Summons, D. S. Letham, and C. W. Parker, *Aust. J. Chem.*, 1978, **31**, 1291.
455. M. Uchiyama and H. Abe, *Agric. Biol. Chem.*, 1977, **41**, 1549.
456. G. Shaw, in "Cytokins: Chemistry, Activity, and Function," eds. D. W. S. Mok and M. C. Mok, CRC, Boca Raton, FL, 1994, p. 15.
457. R. A. Fletcher, C. Teo, and A. Ali, *Can. J. Bot.*, 1973, **51**, 937.
458. A. Schön, G. Krupp, S. Gough, S. Berry-Lowe, C. G. Kannangara, and D. Söll, *Nature*, 1986, **322**, 281.
459. C. G. Kannangara, S. P. Gough, P. Bruyant, J. K. Hoober, A. Kahn, and D. von Wettstein, *Trends Biochem. Sci.*, 1988, **13**, 139.
460. T. Masuda, H. Ohta, Y. Shioi, H. Tsuji, and K. Takamiya, *Plant Cell Physiol.*, 1995, **36**, 1237.
461. T. Masuda, C. G. Kannangara, K. Takamiya, and H. Tsuji, in "Research in Photosynthesis," ed. N. Murata, Kluwer, Dordrecht, 1992, vol. 3, p. 43.
462. T. Masuda, R. Tanaka, Y. Shioi, K. Takamiya, C. G. Kannangara, and H. Tsuji, *Plant Cell Physiol.*, 1994, **35**, 183.
463. T. Masuda, Y. Komine, H. Inokuchi, C. G. Kannangara, and H. Tsuji, *Plant Physiol. Biochem.*, 1992, **30**, 235.
464. J. I. Medford, R. Horgan, Z. El-Sawi, and H. J. Klee, *Plant Cell*, 1989, **1**, 403.
465. T. Schmülling, S. Beinsberger, J. De Greef, J. Schell, H. Van Onckelen, and A. Spena, *FEBS Lett.*, 1989, **249**, 401.
466. A. C. Smigocki, *Plant Mol. Biol.*, 1991, **16**, 105.
467. C. M. Smart, S. R. Scofield, M. W. Bevan, and T. A. Dyer, *Plant Cell*, 1991, **3**, 647.
468. J. J. Estruch, E. Prinsen, H. V. Onckelen, J. Schell, and A. Spena, *Science*, 1991, **254**, 1364.
469. Y. Li, G. Hagen, and T. J. Guilfoyle, *Dev. Biol.*, 1992, **153**, 386.
470. G. M. Polya and A. W. Davis, *Planta*, 1978, **139**, 139.
471. J. L. Erion and J. E. Fox, *Plant Physiol.*, 1981, **67**, 156.
472. A. C. Brinegar, G. Cooper, A. Stevens, C. R. Hauer, J. Shabanowitz, D. F. Hunt, and J. E. Fox, *Proc. Natl. Acad. Sci. USA*, 1988, **85**, 5927.
473. G. A. Romanov, V. Y. Taran, L. Chvojka, and O. N. Kulaeva, *J. Plant Growth Regul.*, 1988, **7**, 1.
474. G. A. Romanov, in "Physiology and Biochemistry of Cytokinins in Plants," eds. M. Kamínek, D. W. S. Mok, and E. Zazímalová, SPB Academic, The Hague, 1992, p. 133.
475. E. Momotani and H. Tsuji, *Plant Cell Physiol.*, 1992, **33**, 407.
476. R. Nagata, E. Kawachi, Y. Hashimoto, and K. Shudo, *Biochem. Biophys. Res. Commun.*, 1993, **191**, 543.
477. A. C. Brinegar, in "Cytokinins: Chemistry, Activity, and Function," eds. D. W. S. Mok and M. C. Mok, CRC, Boca Raton, FL, 1994, p. 217.
478. E. G. Romanko, S. Y. Selivankina, and A. K. Ovcharov, *Sov. Plant Physiol.*, 1982, **29**, 408.
479. E. G. Romanko, S. Y. Selivankina, I. E. Moshkov, and G. V. Novikova, *Sov. Plant Physiol.*, 1986, **33**, 823.
480. S. Y. Selivankina, E. G. Romanko, A. K. Ovcharov, and E. I. Kharchenko, *Sov. Plant Physiol.*, 1982, **29**, 208.
481. S. Y. Selivankina, E. G. Romanko, E. A. Burkhanova, N. N. Karavaiko, V. V. Zayakin, and O. N. Kulaeva, *Biokhimiya*, 1985, **50**, 47.
482. A. B. Fedina, É. A. Burkhanova, and V. I. Kharchenko, *Sov. Plant Physiol.*, 1987, **34**, 263.
483. S. Y. Selivankina, E. G. Romanko, N. N. Karavaiko, I. E. Moshkov, G. V. Novikova, and O. N. Kulaeva, *Dokl. Akad. Nauk. SSSR*, 1988, **299**, 254.
484. O. N. Kulaeva, N. N. Karavaiko, I. E. Moshkov, S. Y. Selivankina, and G. V. Novikova, *FEBS Lett.*, 1990, **261**, 410.
485. S. Y. Selivankina, E. G. Romanko, G. V. Novikova, D. G. Muromtseva, and O. N. Kulaeva, *Sov. Plant Physiol.*, 1988, **35**, 205.
486. S. Mitsui, T. Wakasugi, and M. Sugiura, *Plant Cell Physiol.*, 1993, **34**, 1089.
487. M. E. Musgrave, in "Cytokinins: Chemistry, Activity, and Function," eds. D. W. S. Mok and M. C. Mok, CRC, Boca Raton, FL, 1994, p. 167.
488. C. O. Miller, *Plant Physiol.*, 1982, **69**, 1274.
489. P. Dizengremel, M. Chauveau, and J. Roussaux, *Plant Physiol.*, 1982, **70**, 585.
490. M. E. Musgrave and J. N. Siedow, *Physiol. Plant*, 1985, **64**, 161.
491. C. Brinegar, G. Shah, and G. Cooper, in "Plant Hormone Signal Perception and Transduction," eds. A. R. Smith, A. W. Berry, N. V. J. Harpham, I. E. Moshkov, G. V. Novikova, O. N. Kulaeva, and M. A. Hall, Kluwer, Dordrecht, 1996, p. 83.
492. J. Eberle, T. L. Wang, S. Cook, B. Wells, and E. W. Weiler, *Planta*, 1987, **172**, 289.
493. L. Sossountzov, R. Maldiney, B. Sotta, I. Sabbagh, Y. Habricot, M. Bonnet, and E. Miginiac, *Planta*, 1988, **175**, 291.
494. J. C. Thomas, E. F. McElwain, and H. J. Bohnert, *Plant Physiol.*, 1992, **100**, 416.
495. A. Binns and F. Mein, Jr., *Proc. Natl. Acad. Sci. USA*, 1973, **70**, 2660.
496. F. Meins, Jr. and C. E. Hansen, in "Physiology and Biochemistry of Cytokinins in Plants," eds. M. Kamínek, D. W. S. Mok, and E. Zazímalová, SPB Academic, The Hague, 1992, p. 83.
497. T. L. Wang, in "Cytokinins: Chemistry, Activity, and Function," eds. D. W. S. Mok and M. C. Mok, CRC, Boca Raton, FL, 1994, p. 255.
498. A. D. Blonstein, A. D. Parry, R. H. Horgan, and P. J. King, *Planta*, 1991, **183**, 244.
499. W. Su and S. H. Howell, *Plant Physiol.*, 1992, **99**, 1569.
500. A. J. Cary, W. Liu, and S. H. Howell, *Plant Physiol.*, 1995, **107**, 1075.
501. P. Rousselin, Y. Kraepiel, R. Maldiney, E. Miginiac, and M. Caboche, *Theor. Appl. Genet.*, 1992, **85**, 213.
502. J. Deikmann and M. Ulrich, *Planta*, 1995, **195**, 440.
503. N. W. Ashton, D. J. Cove, and D. R. Featherstone, *Planta*, 1979, **144**, 437.
504. T. Kakimoto, *Science*, 1996, **274**, 982.
505. H. Sano and S. Youssefian, *Proc. Natl. Acad. Sci. USA*, 1994, **91**, 2582.
506. J. Chory, C. Peto, R. Feinbaum, L. Pratt, and F. Ausubel, *Cell*, 1989, **58**, 991.
507. J. Chory, D. Reinecke, S. Sim, T. Washburn, and M. Brenner, *Plant Physiol.*, 1994, **104**, 339.
508. J. Li, P. Nagpal, V. Vitart, A. Chao, S. Fujioka, Y.-H. Choi, M. Guha-Biswas, T. C. McMorris, S. Takatsuto, T. Yokota, D. W. Russell, A. Sakurai, and J. Chory, in "Proceedings of the 23rd Annual Meeting of the Plant Growth Regulation Society of America," University of Calgary, Calgary, 1996, p. 11.
509. S. Lerbs, W. Lerbs, N. L. Klyachko, E. G. Romanko, O. N. Kulaeva, R. Wollgiehn, and B. Parthier, *Planta*, 1984, **162**, 289.

510. M. O. Abdelghani, L. Suty, J. N. Chen, J.-P. Renaudin, and B. Teyssendier de la Serve, *Plant Sci.*, 1991, **77**, 29.
511. T. Ohya and H. Suzuki, *Plant Cell Physiol.*, 1991, **32**, 577.
512. B. Teyssendier de la Serve, M. Axelos, and C. Péaud-Lenoël, *Plant Mol. Biol.*, 1985, **5**, 155.
513. S. Flores and E. M. Tobin, *Plant Mol. Biol.*, 1988, **11**, 409.
514. J.-L. Lu, J. R. Ertl, and C.-M. Chen, *Plant Mol. Biol.*, 1990, **14**, 585.
515. B. R. Andersen, G. Jin, R. Chen, J. R. Ertl, and C.-M. Chen, *Planta*, 1996, **198**, 1.
516. D. L. Silver, A. Pinaev, R. Chen, and F. J. de Bruijn, *Plant Physiol.*, 1996, **112**, 559.
517. J. Memelink, J. H. C. Hoge, and R. A. Schilperoort, *EMBO J.*, 1987, **6**, 3579.
518. D. N. Crowell, A. T. Kadlecek, M. C. John, and R. M. Amasino, *Proc. Natl. Acad. Sci. USA*, 1990, **87**, 8815.
519. D. N. Crowell and R. M. Amasino, *Plant Physiol.*, 1991, **95**, 711.
520. J. A. Dominov, L. Stenzler, S. Lee, J. J. Schwarz, S. Leisner, and S. H. Howell, *Plant Cell*, 1992, **4**, 451.
521. B. Sugiharto, J. N. Burnell, and T. Sugiyama, *Plant Physiol.*, 1992, **100**, 153.
522. T. Toyama, H. Teramoto, and G. Takeba, *Plant Cell Physiol.*, 1996, **37**, 1038.
523. J. L. S. Cotton, C. W. Ross, D. H. Byrne, and J. T. Colbert, *Plant Mol. Biol.*, 1990, **14**, 707.
524. C.-M. Chen, J. Ertl, M.-S. Yang, and C.-C. Chang, *Plant Sci.*, 1987, **52**, 169.
525. R. J. Gaudino and C. S. Pikaard, *J. Biol. Chem.*, 1997, **272**, 6799.
526. H. Teramoto, E. Momotani, and H. Tsuji, *Physiol. Plant*, 1993, **87**, 584.
527. H. Teramoto, E. Momotani, G. Takeba, and H. Tsuji, *Planta*, 1994, **193**, 573.
528. H. Teramoto, T. Toyama, G. Takeba, and H. Tsuji, *Planta*, 1995, **196**, 387.
529. H. Teramoto, T. Toyama, G. Takeba, and H. Tsuji, *Plant Mol. Biol.*, 1996, **32**, 797.
530. H. Teramoto, E. Momotani, G. Takeba, and H. Tsuji, *Plant Growth Regul.*, 1996, **18**, 59.
531. T. Toyama, H. Teramoto, G. Takeba, and H. Tsuji, *Plant Cell Physiol.*, 1995, **36**, 1349.
532. H. Iwamura, in "Cytokinins: Chemistry, Activity, and Function," eds. D. W. S. Mok and M. C. Mok, CRC, Boca Raton, FL, 1994, p. 43.
533. H. Lineweaver and D. Burk, *J. Am. Chem. Soc.*, 1934, **56**, 658.
534. H. Iwamura, in "Plant Growth Substances 1988," eds. R. P. Pharis and S. B. Rood, Springer, Berlin, 1990, p. 179.
535. H. Iwamura, S. Murakami, J. Koga, S. Matsubara, and K. Koshimizu, *Phytochemistry*, 1979, **18**, 1265.
536. S. Tanimoto and H. Harada, *Plant Cell Physiol.*, 1982, **23**, 1371.
537. S. Tanimoto and H. Harada, *Biol. Plant.*, 1984, **26**, 337.
538. T. Saitou, H. Kamada, and H. Harada, *Plant Sci.*, 1992, **86**, 161.
539. H. Sano, S. Seo, N. Koizumi, T. Niki, H. Iwamura, and Y. Ohashi, *Plant Cell Physiol.*, 1996, **37**, 762.
540. H. Suge and H. Iwamura, *Jpn. J. Crop. Sci.*, 1993, **62**, 595.
541. I. G. Sergiev, V. S. Alexieva, and E. N. Karanov, *J. Plant Physiol.*, 1994, **145**, 266.
542. L. H. Aung, *Biol. Plant.*, 1986, **28**, 407.
543. D. L. Church and A. W. Galston, *Phytochemistry*, 1988, **27**, 2435.
544. H. Iwamura, N. Masuda, K. Koshimizu, and S. Matsubara, *Phytochemistry*, 1979, **18**, 217.
545. K. Yanosaka, H. Iwamura, M. Shinozaki, and K. Yoshida, *Plant Cell Physiol.*, 1991, **32**, 447.
546. R. O. Morris, *Annu. Rev. Plant Physiol.*, 1986, **37**, 509.
547. Y. Taya, Y. Tanaka, and S. Nishimura, *Nature*, 1978, **271**, 545.
548. C. M. Chen and D. K. Melitz, *FEBS Lett.*, 1979, **107**, 15.
549. L. M. S. Palni, R. Horgan, N. M. Darrall, T. Stuchbury, and P. F. Wareing, *Planta*, 1983, **159**, 50.
550. L. Maréchal-Drouard, J. H. Weil, and A. Dietrich, *Annu. Rev. Plant Physiol. Plant Mol. Biol.*, 1993, **44**, 13.
551. G. Sembdner, D. Gross, H.-W. Liebisch, and G. Schneider, in "Encyclopedia of Plant Physiology," ed. J. MacMillan, Springer, Berlin, 1980, vol. 9, p. 281.
552. H. J. Verman, F. Skoog, C. R. Frihart, and N. J. Leonard, *Plant Physiol.*, 1972, **49**, 848.
553. L. M. S. Palni and R. Horgan, *Planta*, 1983, **159**, 178.
554. N. V. Bassil, D. W. S. Mok, and M. C. Mok, *Plant Physiol.*, 1993, **102**, 867.
555. J. Gray, S. B. Gelvin, R. Meilan, and R. O. Morris, *Plant Physiol.*, 1996, **110**, 431.
556. K. C. Short and J. G. Torrey, *Plant Physiol.*, 1972, **49**, 155.
557. B. A. McGaw, in "Plant Hormones and their Role in Plant Growth and Development," Nijhoff, Dordrecht, 1987, p. 76.
558. C. O. Miller, *Proc. Natl. Acad. Sci. USA*, 1974, **71**, 334.
559. F. Heidekamp, W. G. Dirkse, J. Hille, and H. van Ormondt, *Nucleic Acids Res.*, 1983, **11**, 6211.
560. C. F. Barry, S. G. Rogers, R. T. Fraley, and L. Brand, *Proc. Natl. Acad. Sci. USA*, 1984, **81**, 4776.
561. D. E. Akiyoshi, D. A. Regier, G. Jen, and M. P. Gordon, *Nucleic Acids Res.*, 1985, **13**, 2773.
562. I. Buchmann, F. J. Marner, G. Schröder, S. Waffenschmidt, and J. Schröder, *EMBO J.*, 1985, **4**, 853.
563. A. N. Binns and M. F. Thomashow, *Annu. Rev. Microbiol.*, 1988, **42**, 575.
564. P. C. Zambryski, *Annu. Rev. Plant Physiol. Plant Mol. Biol.*, 1992, **43**, 465.
565. G. K. Powell and R. O. Morris, *Nucleic Acids Res.*, 1986, **14**, 2555.
566. M. Crespi, E. Messens, A. B. Caplan, M. Van Montagu, and J. Desomer, *EMBO J.*, 1992, **11**, 795.
567. C.-N. Chen, D. K. Melitz, and F. W. Clough, *Arch. Biochem. Biophys.*, 1982, **214**, 634.
568. D. S. Letham, in "Phytohormones and Related Compounds: A Comprehensive Treatise," eds. D. S. Letham, P. G. Goodwin, and T. J. V. Higgins, Elsevier/North-Holland, Amsterdam, 1978, vol. 1, p. 205.
569. M. Laloue and C. Pethe, in "Proceedings of the 11th International Conference on Plant Growth Substances," ed. P. F. Wareing, Academic Press, London, 1982, p. 185.
570. C.-M. Chen and S. M. Leisner, *Plant Physiol.*, 1984, **75**, 442.
571. T. Stuchbury, L. M. Palni, R. Horgan, and P. F. Wareing, *Planta*, 1979, **147**, 97.
572. R. Zhang, X. Zhang, J. Wang, D. S. Letham, S. A. McKinney, and T. J. V. Higgins, *Planta*, 1995, **196**, 84.
573. R. C. Martin, M. C. Mok, G. Shaw, and D. W. S. Mok, *Plant Physiol.*, 1989, **90**, 1630.
574. D. J. Armstrong, S. G. Kim, M. C. Mok, and D. W. S. Mok, in "Metabolism and Molecular Activities of Cytokinins," eds. J. Guern and C. Péaud-Lenoël, Springer, Berlin, 1981, p. 97.
575. B. Entsch and D. S. Letham, *Plant Sci. Lett.*, 1979, **14**, 205.
576. C. A. Auer and J. D. Cohen, *Plant Physiol.*, 1993, **102**, 541.

577. B. A. McGaw and R. Horgan, *Planta*, 1983, **159**, 30.
578. J. E. Turner, D. W. S. Mok, M. C. Mok, and G. Shaw, *Proc. Natl. Acad. Sci. USA*, 1987, **84**, 3714.
579. S. C. Dixon, R. C. Martin, M. C. Mok, G. Shaw, and D. W. S. Mok, *Plant Physiol*, 1989, **90**, 1316.
580. D. A. Lightfoot, *Plant Physiol.*, 1994, **105** (suppl.), 70.
581. J. M. Chatfield and D. J. Armstrong, *Plant Physiol.*, 1986, **80**, 493.
582. C. D. Whitty and R. H. Hall, *Can. J. Biochem.*, 1974, **52**, 789.
583. J. Wang and D. S. Letham, *Plant Sci.*, 1995, **112**, 161.
584. S. C. Capelle, D. W. S. Mok, S. C. Kirchner, and M. C. Mok, *Plant Physiol.*, 1983, **73**, 796.
585. V. Motyka, M. Faiss, M. Strnad, M. Kamínek, and T. Schmülling, *Plant Physiol.*, 1996, **112**, 1035.
586. B. Brzobohatý, I. Moore, and K. Palme, *Plant Mol. Biol.*, 1994, **26**, 1483.
587. P. F. Wareing, R. Horgan, J. E. Henson, and W. Davis, in "Plant Growth Regulation," ed. P. E. Pilet, Springer, Berlin, 1977, p. 147.
588. B. Durand and R. Durand, *Plant Sci.*, 1991, **80**, 49.
589. B. Durand and R. Durand, *Plant Sci.*, 1991, **80**, 107.
590. J.-P. Louis, C. Augur, and G. Teller, *Plant Physiol.*, 1990, **94**, 1535.
591. R. C. Martin, R. R. Martin, M. C. Mok, and D. W. S. Mok, *Plant Physiol.*, 1990, **94**, 1290.
592. R. C. Martin, M. C. Mok, and D. W. S. Mok, *Plant Physiol.*, 1993, **102** (suppl.), 24.
593. O. Nilsson, T. Moritz, N. Imbault, G. Sandberg, and O. Olsson, *Plant Physiol.*, 1993, **102**, 363.
594. T. Schmülling, M. Fladung, K. Grossmann, and J. Schell, *Plant J.*, 1993, **3**, 371.
595. J. J. Estruch, D. Chriqui, K. Grossmann, J. Schell, and A. Spena, *EMBO J.*, 1991, **10**, 2889.
596. F. T. Addicott (ed.), "Abscisic Acid," Praeger, New York, 1983.
597. H. R. Carns, *Annu. Rev. Plant Physiol.*, 1966, **17**, 295.
598. W.-C. Liu and H. R. Carns, *Science*, 1961, **134**, 384.
599. K. Ohkuma, N. Hirai, and N. Kondo, in "Handbook of Plant Hormones," eds. N. Takahashi and Y. Masuda, Baifukan, Tokyo, vol. 2, p. 1.
600. K. Ohkuma, F. T. Addicott, O. E. Smith, and W. E. Thiessen, *Tetrahedron Lett.*, 1965, 2529.
601. J. W. Cornforth, B. V. Milborrow, and G. Ryback, *Nature*, 1965, **206**, 715.
602. R. F. M. Van Steveninck, *Nature*, 1959, **183**, 1246.
603. K. Koshimizu, H. Fukui, T. Mitsui, and Y. Ogawa, *Agric. Biol. Chem.*, 1966, **30**, 941.
604. I. D. J. Phillips and P. F. Wareing, *J. Exp. Bot.*, 1958, **9**, 350.
605. P. F. Wareing, C. F. Eagles, and P. M. Robinson, in "Régulateurs Naturels de la Croissance Végétale," ed. J. P. Nitsch, Centre National de la Recherche Scientifique, Paris, 1964, p. 377.
606. J. W. Cornforth, B. V. Milborrow, G. Ryback, and P. F. Wareing, *Nature*, 1965, **205**, 1269.
607. F. T. Addicott, J. L. Lyon, K. Ohkuma, W. E. Thiessen, H. R. Carns, O. E. Smith, J. W. Cornforth, B. V. Milborrow, G. Ryback, and P. F. Wareing, *Science*, 1968, **159**, 1493.
608. J. W. Cornforth, W. Draber, B. V. Milborrow, and G. Ryback, *Chem. Commun.*, 1967, 114.
609. R. S. Cahn, C. Inglod, and V. Prelog, *Angew. Chem. Int. Ed., Engl.* 1966, **5**, 385.
610. H. F. Taylor and R. S. Burden, *Proc. R. Soc. London B*, 1972, **180**, 317.
611. G. Ryback, *J. Chem. Soc., Chem. Commun.*, 1972, 1190.
612. N. Harada. *J. Am. Chem. Soc.*, 1973, **95**, 240.
613. M. Koreeda, G. Weiss, and K. Nakanishi, *J. Am. Chem. Soc.*, 1973, **95**, 239.
614. K. Mori, *Tetrahedron*, 1974, **30**, 1065.
615. R. J. Pryce, *Phytochemistry*, 1972, **11**, 1759.
616. M. Nakayama, T. Takase, and T. Yokota, in "Abstracts of The XV International Botanical Congress, Yokohama, 1993," 1993, p. 388.
617. W. J. Davies and H. G. Jones (eds.), "Abscisic Acid—Physiology and Biochemistry," Bios, Oxford, 1991.
618. G. Assante, L. Merlini, and G. Nasini, *Experientia*, 1977, **33**, 1556.
619. K. Dörffling and W. Petersen, *Z. Naturforschung*, 1984, **39**, 683.
620. J. Kettner and K. Dörffling, *Planta*, 1995, **196**, 627.
621. M. T. Le Page-Degrivry, J. N. Bidard, E. Rouvier, C. Bulard, and M. Lazdunski, *Proc. Natl. Acad. Sci. USA*, 1986, **83**, 1155.
622. S. N. Visscher, in "Abscisic Acid," ed. F. T. Addicott, Praeger, New York, 1983, p. 553.
623. B. V. Milborrow and R. Mallaby, *J. Exp. Bot.*, 1975, **26**, 741.
624. T. Matsuzaki and A. Koiwai, *Agric. Biol. Chem.*, 1986, **50**, 2193.
625. A. J. Ciha, L. Brenner, and W. A. Brun, *Plant Physiol.*, 1977, **59**, 821.
626. R. Antoszewski and R. Rudnicki, *Anal. Biochem.*, 1969, **32**, 233.
627. J. A. D. Zeevaart and B. V. Milborrow, *Phytochemistry*, 1976, **15**, 493.
628. P. Hedden, *Annu. Rev. Plant Physiol. Plant Mol. Biol.*, 1993, **44**, 107.
629. Y. Okamoto, R. Aburatani, and K. Hatada, *J. Chromatogr.*, 1988, **448**, 454.
630. M. Okamoto and H. Nakazawa, *J. Chromatogr.*, 1990, **508**, 217.
631. M. Okamoto and H. Nakazawa, *Biosci. Biotech. Biochem.*, 1993, **57**, 1768.
632. J. P. Knox and G. Galfre, *Anal. Biochem.*, 1986, **155**, 92.
633. H. Yamamoto and T. Oritani, *Biosci. Biotech. Biochem.*, 1994, **58**, 992.
634. K. Ohkuma, J. L. Lyon, F. T. Addicott, and O. E. Smith, *Science*, 1963, **142**, 1592.
635. R. Mallaby and G. Ryback, *J. Chem. Soc., Perkin Trans. II*, 1972, 919.
636. M. H. Beale, A. Chen, P. A. Harrison, and C. L. Willis, *J. Chem. Soc., Perkin Trans. I*, 1993, 3061.
637. B. V. Milborrow, *Biochem. J.*, 1984, **220**, 325.
638. R. D. Willows, A. G. Netting, and B. V. Milborrow, *Phytochemistry*, 1991, **30**, 1483.
639. R. T. Gray, R. Mallaby, G. Ryback, and V. P. Williams, *J. Chem. Soc., Perkin Trans. II*, 1974, 919.
640. B. Terem and J. H. P. Utley, *Electrochim. Acta*, 1979, **24**, 1081.
641. J. MacMillan and R. J. Pryce, *Tetrahedron*, 1969, **25**, 5893.
642. B. V. Milborrow, *Phytochemistry*, 1975, **14**, 1045.
643. M. H. M. Cornelussen, C. M. Karssen, and L. C. van Loon, *Phytochemistry*, 1995, **39**, 959.

644. B. V. Milborrow, *Planta*, 1967, **76**, 93.
645. S. R. Abrams, M. J. T. Reaney, G. D. Abrams, T. Mazurek, A. C. Shaw, and L. V. Gusta, *Phytochemistry*, 1989, **28**, 2885.
646. B. V. Milborrow, N. J. Carrington, and G. T. Vaughan, *Phytochemistry*, 1988, **27**, 757.
647. T. Kitahara, K. Touhara, H. Watanabe, and K. Mori, *Tetrahedron*, 1989, **45**, 6387.
648. R. D. Bennett, S. M. Norman, and V. P. Maier, *Phytochemistry*, 1990, **29**, 3473.
649. T. G. Heath, D. A. Gage, J. A. D. Zeevaart, and J. T. Watson, *Org. Mass Spectrom.*, 1990, **25**, 655.
650. L. Rivier and M. Saugy, *J. Plant Growth Regul.*, 1986, **5**, 1.
651. A. G. Netting and B. V. Milborrow, *Biomed. Environ. Mass Spectrom.*, 1988, **17**, 281.
652. C. Hornberg and E. W. Weiler, *Nature*, 1984, **310**, 321.
653. T. Oritani and K. Yamashita, *Agric. Biol. Chem.*, 1970, **34**, 830.
654. T. Oritani and K. Yamashita, *Agric. Biol. Chem.*, 1970, **34**, 198.
655. S. C. Chen and J. M. MacTaggart, *Agric. Biol. Chem.*, 1986, **50**, 1097.
656. P. J. Orton and T. A. Mansfield, *Planta*, 1974, **121**, 263.
657. B.-T. Kim, T. Asami, K. Morita, C.-H. Soh, N. Murofushi, and S. Yoshida, *Biosci. Biotech. Biochem.*, 1992, **56**, 624.
658. K. Raschke, R. D. Firn, and M. Pierce, *Planta*, 1975, **125**, 149.
659. P. E. Kriedemann, B. R. Loveys, G. L. Fuller, and A. C. Leopold, *Plant Physiol.*, 1972, **49**, 842.
660. Y. Todoroki, N. Hirai, and K. Koshimizu, *Phytochemistry*, 1995, **40**, 633.
661. P. A. Rose, B. Lei, A. C. Shaw, D. L. Barton, M. K. Walker-Simmons, and S. R. Abrams, *Phytochemistry*, 1996, **41**, 1251.
662. M. K. Walker-Simmons, P. A. Rose, A. C. Shaw, and S. R. Abrams, *Plant Physiol.*, 1994, **106**, 1279.
663. S. Nakano, Y. Todoroki, N. Hirai, and H. Ohigashi, *Biosci. Biotech. Biochem.*, 1995, **59**, 1699.
664. P. A. Rose, S. R. Abrams, and L. V. Gusta, *Phytochemistry*, 1992, **31**, 1105.
665. T. Oritani and K. Yamashita, *Agric. Biol. Chem.*, 1974, **38**, 801.
666. D. C. Walton and E. Sondheimer, *Plant Physiol.*, 1972, **49**, 290.
667. D. R. C. Hite, W. H. Outlaw, Jr., and M. A. Seavy, *Physiol. Plant*, 1994, **92**, 79.
668. J. J. Balsevich, G. G. Bishop, and G. M. Banowetz, *Phytochemistry*, 1996, **44**, 215.
669. D. C. Walton, in "Abscisic Acid," ed. F. T. Addicott, Praeger, New York, 1983, p. 113.
670. Y. Todoroki, S. Nakano, N. Hirai, and H. Ohigashi, *Tetrahedron*, 1996, **52**, 8081.
671. H. Ueda and J. Tanaka, *Bull. Chem. Soc. Jpn.*, 1977, **50**, 1506.
672. V. H. W. Schmalle, K. H. Klaska, and O. Jarchow, *Acta Crystallogr.*, 1977, **33**, 2218.
673. R. D. Willows and B. V. Milborrow, *Phytochemistry*, 1993, **34**, 233.
674. G. Ohloff, E. Otto, V. Rautenstrauch, and G. Snatzke, *Helv. Chim. Acta*, 1973, **56**, 1874.
675. R. J. Abraham and M. S. Lucas, *J. Chem. Soc., Perkin Trans. II*, 1988, 669.
676. G. Snatzke, *Tetrahedron*, 1965, **21**, 413.
677. B. V. Milborrow and S. R. Abrams, *Phytochemistry*, 1993, **32**, 827.
678. M. K. Walker-Simmons, R. J. Anderberg, P. A. Rose, and S. R. Abrams, *Plant Physiol.*, 1992, **99**, 501.
679. E. Sondheimer, E. C. Galson, Y. P. Chang, and D. C. Walton, *Science*, 1971, **174**, 829.
680. B. V. Milborrow, in "Plant Growth Substances 1985," ed. M. Bopp, Springer, Berlin, 1986, p. 108.
681. A. M. Hetherington and R. S. Quatrano, *New Phytol.*, 1991, **119**, 9.
682. A. K. Dhawan and D. M. Paton, *Annu. Bot.*, 1980, **45**, 493.
683. A. B. Ogunkanmi, D. J. Tucker, and T. A. Mansfield, *New Phytol.*, 1973, **72**, 277.
684. W. Hartung and S. Slovik, *New Phytol.*, 1991, **119**, 361.
685. B. E. Anderson, J. M. Ward and J. I. Schroeder, *Plant Physiol*, 1994, **104**, 1177.
686. A. C. Allan, M. D. Fricker, J. L. Ward, M. H. Beale, and A. J. Trewavas, *Plant Cell*, 1994, **6**, 1319.
687. S. M. Assmann, *Annu. Rev. Cell Biol.*, 1993, **9**, 345.
688. J. Leung, M. Bouvier-Durand, P.-C. Morris, D. Guerrier, F. Chefdor, and J. Giraudat, *Science*, 1994, **264**, 1448.
689. K. Meyer, M. P. Leube, and E. Grill, *Science*, 1994, **264**, 1452.
690. C.-J. Jiang, N. Nakajima, and N. Kondo, *Plant Cell Physiol.*, 1996, **37**, 697.
691. E. M. Sivori, V. Sonvico, and N. O. Fernandez, *Plant Cell Physiol.*, 1971, **12**, 993.
692. R. C. Nolan, L.-S. Lin, and T.-H. D. Ho, *Plant Mol. Biol.*, 1987, **8**, 13.
693. L.-S.Lin and T.-H. D. Ho, *Plant Physiol.*, 1986, **82**, 289.
694. J. Dommes and D. H. Northcote, *Planta*, 1985, **165**, 513.
695. S. Gilroy and R. L. Jones, *Plant Physiol.*, 1994, **104**, 1185.
696. M. Wang, S. Heimovaara-Dijkstra, R. M. Van der Meulen, J. P. Knox, and S. J. Neill, *Planta*, 1995, **196**, 271.
697. S. Gilroy and R. L. Jones, *Proc. Natl. Acad. Sci. USA*, 1992, **89**, 3591.
698. S. Cutler, M. Ghassemian, D. Bonetta, S. Cooney, and P. McCourt, *Science*, 1996, **273**, 1239.
699. A. A. Abou-Mandour and W. Hartung, *Z. Pflanzenphysiol.*, 1980, **100**, 25.
700. M. E. Bossen, A. Tretyn, R. E. Kendrick, and W. J. Vredenberg, *J. Plant Physiol.*, 1991, **137**, 706.
701. D. H. Gaither, D. H. Lutz, and L. E. Forrence, *Plant Physiol.*, 1975, **55**, 948.
702. J. A. D. Zeevaart and R. A. Creelman, *Annu. Rev. Plant Physiol. Plant Mol. Biol.*, 1988, **39**, 439.
703. P. M. Chandler and M. Robertson, *Annu. Rev. Plant Physiol. Plant Mol. Biol.*, 1994, **45**, 113.
704. B. V. Milborrow and D. R. Robinson, *J. Exp. Bot.*, 1973, **24**, 537.
705. R. C. Noddle and D. R. Robinson, *Biochem. J.*, 1969, **112**, 547.
706. A. K. Cowan and I. D. Railton, *J. Plant Physiol.*, 1987, **131**, 423.
707. A. D. Parry and R. Horgan, *Physiol. Plant.*, 1991, **82**, 320.
708. B. V. Milborrow, in "Abscisic Acid," ed. D. R. Robinson and G. Ryback, *Biochem. J.*, 1969, **113**, 895.
710. B. V. Milborrow, *Biochem. J.*, 1972, **128**, 1135.
711. R. D. Willows and B. V. Milborrow, *Phytochemistry*, 1989, **28**, 2641.
712. B. V. Milborrow, *Phytochemistry*, 1975, **14**, 2403.
713. R. D. Bennett, S. M. Norman, and V. P. Maier, *Phytochemistry*, 1981, **20**, 2343.
714. R. A. Creelman and J. A. D. Zeevaart, *Plant Physiol.*, 1984, **75**, 166.
715. R. A. Creelman, D. A. Gage, J. T. Stults, and J. A. D. Zeevaart, *Plant Physiol.*, 1987, **85**, 726.

716. J. A. D. Zeevaart, T. G. Heath, and D. A. Gage, *Plant Physiol.*, 1989, **91**, 1594.
717. D. A. Gage, F. Fong, and J. A. D. Zeevaart, *Plant Physiol.*, 1989, **89**, 1039.
718. T. L. Wang, S. K. Cook, R. J. Francis, M. J. Ambrose, and C. L. Hedley, *J. Exp. Bot.*, 1987, **38**, 1921.
719. M. Okamoto, N. Hirai, and K. Koshimizu, *Phytochemistry*, 1987, **26**, 1269.
720. G. T. Vaughan and B. V. Milborrow, *Aust. J. Plant Physiol.*, 1987, **14**, 593.
721. C. D. Rock and J. A. D. Zeevaart, *Plant Physiol.*, 1990, **93**, 915.
722. I. B. Taylor, R. S. T. Linforth, R. J. Al-Naieb, W. R. Bowman, and B. A. Marples, *Plant Cell Environ.*, 1988, **11**, 739.
723. S. C. Duckham, I. B. Taylor, R. S. T. Linforth, R. J. Al-Naieb, B. A. Marples, and W. R. Bowman, *J. Exp. Bot.*, 1989, **40**, 901.
724. R. K. Sindhu, D. H. Griffin, and D. C. Walton, *Plant Physiol.*, 1990, **93**, 689.
725. A. D. Parry, A. D. Blonstein, M. J. Babiano, P. T. King, and R. Horgan, *Planta*, 1991, **183**, 237.
726. R. D. Firn, R. S. Burden, and H. F. Taylor, *Planta*, 1972, **102**, 115.
727. A. D. Parry, S. J. Neill, and R. Horgan, *Planta*, 1988, **173**, 397.
728. R. K. Sindhu and D. C. Walton, *Plant Physiol.*, 1987, **85**, 916.
729. B. V. Milborrow and M. Garmston, *Phytochemistry*, 1973, **12**, 1597.
730. R. A. Creelman, *Physiol. Plant.*, 1989, **75**, 131.
731. T. Oritani and K. Yamashita, *Agric. Biol. Chem.*, 1979, **43**, 1613.
732. B. V. Milborrow and R. C. Noddle, *Biochem. J.*, 1970, **119**, 727.
733. R. S. T. Linforth, W. R. Bowman, D. A. Griffin, P. Hedden, B. A. Marples, and I. B. Taylor, *Phytochemistry*, 1987, **26**, 1631.
734. R. S. T. Linforth, W. R. Bowman, D. A. Griffin, B. A. Marples, and I. B. Taylor, *Plant Cell Environ.*, 1987, **10**, 599.
735. B. V. Milborrow, H. M. Nonhebel, and R. D. Willows, *Plant Sci.*, 1988, **56**, 49.
736. E. Marin, L. Nussaume, A. Quesada, M. Gonneau, B. Sotta, P. Hugueney, A. Frey, and A. Marion-Poll, *EMBO J.*, 1996, **15**, 2331.
737. Y. Li and D. C. Walton, *Plant Physiol.*, 1987, **85**, 910.
738. A. K. Cowan and G. R. Richardson, *Plant Cell Physiol.*, 1993, **34**, 969.
739. Y. Li and D. C. Walton, *Plant Physiol.*, 1990, **92**, 551.
740. A. D. Parry, M. J. Babiano, and R. Horgan, *Planta*, 1990, **182**, 118.
741. A. D. Parry and R. Horgan, *Phytochemistry*, 1991, **30**, 815.
742. H. M. Nonhebel and B. V. Milborrow, *J. Exp. Bot.*, 1986, **37**, 1533.
743. S. J. Neill, R. Horgan, D. C. Walton, and C. A. M. Mercer, *Phytochemistry*, 1987, **26**, 2515.
744. T. Oritani and K. Yamashita, *Agric. Biol. Chem.*, 1987, **51**, 275.
745. T. Sassa, H. Mitobe, and E. Haruki, *Agric. Biol. Chem.*, 1988, **52**, 1625.
746. T. Oritani and K. Yamashita, *Agric. Biol. Chem.*, 1985, **49**, 245.
747. M. Okamoto, N. Hirai, and K. Koshimizu, *Phytochemistry*, 1988, **27**, 2099.
748. N. Hirai, M. Okamoto, and K. Koshimizu, *Phytochemistry*, 1986, **25**, 1865.
749. M. Takasugi, M. Anetai, N. Katsui, and T. Masamune, *Chem. Lett.*, 1973, 245.
750. R. G. Powell and C. R. Smith, *J. Nat. Prod.*, 1981, **44**, 86.
751. D. F. Gillard and D. C. Walton, *Plant Physiol.*, 1976, **58**, 790.
752. S. J. Uknes and T.-H. D. Ho, *Plant Physiol.*, 1984, **75**, 1126.
753. A. A. Adesomoju, J. I. Okogun, D. E. U. Ekong, and P. Gaskin, *Phytochemistry*, 1980, **19**, 223.
754. B. V. Milborrow, *Chem. Commun.*, 1969, 966.
755. Y. Todoroki, N. Hirai, and H. Ohigashi, *Tetrahedron*, 1995, **51**, 6911.
756. J. Zou, G. D. Abrams, D. L. Barton, D. C. Taylor, M. K. Pomeroy, and S. R. Abrams, *Plant Physiol*, 1995, **108**, 563.
757. H. Yamane, S. Fujioka, C. R. Spray, B. O. Phinney, J. MacMillan, P. Gaskin, and N. Takahasi, *Plant Physiol.*, 1988, **86**, 857.
758. S. Fujioka, H. Yamane, C. R. Spray, P. Gaskin, J. MacMillan, B. O. Phinney, and N. Takahashi, *Plant Physiol.*, 1988, **88**, 1367.
759. Y. Todoroki, N. Hirai, and K. Koshimizu, *Biosci. Biotech. Biochem.*, 1994, **58**, 707.
760. R. D. Hill, J.-H. Liu, D. Durnin, N. Lamb, A. Shaw, and S. R. Abrams, *Plant Physiol.*, 1995, **108**, 573.
761. S. J. Neill, R. Horgan, and J. K. Heald, *Planta*, 1983, **57**, 371.
762. H. Lehmann and H. R. Schütte, *Z. Pflanzenphysiol.*, 1980, **96**, 277.
763. B. V. Milborrow, *J. Exp. Bot.*, 1978, **29**, 1059.
764. B. R. Loveys and B. V. Milborrow, *Aust. J. Plant Physiol.*, 1981, **8**, 571.
765. A. G. Netting, R. D. Willows, and B. V. Milborrow, *Plant Growth Regul.*, 1992, **11**, 327.
766. N. Hirai, H. Fukui, and K. Koshimizu, *Phytochemistry*, 1978, **17**, 1625.
767. B. V. Milborrow, in "Plant Growth Substances 1979," ed. F. Skoog, Springer, Berlin, 1980, p. 262.
768. J. A. D. Zeevaart and G. L. Boyer, in "Plant Growth Substances 1982," ed. P. F. Wareing, Academic Press, London, 1982, p. 335.
769. N. Hirai and K. Koshimizu, *Agric. Biol. Chem.*, 1983, **47**, 365.
770. B. V. Milborrow and G. T. Vaughan, *Aust. J. Plant Physiol.*, 1982, **9**, 361.
771. G. T. Vaughan and B. V. Milborrow, *J. Exp. Bot.*, 1984, **35**, 110.
772. G. L. Boyer and J. A. D. Zeevaart, *Phytochemistry*, 1986, **25**, 1103.
773. J. J. Balsevich, A. J. Cutler, N. Lamb, L. J. Friesen, E. U. Kurz, M. R. Perras, and S. R. Abrams, *Plant Phsyiol.*, 1994, **106**, 135.
774. H. Lehmann and L. Schwenen, *Phytochemistry*, 1988, **27**, 677.
775. C. R. Hampson, M. J. T. Reaney, G. D. Abrams, S. R. Abrams, and L. V. Gusta, *Phytochemistry*, 1992, **31**, 2645.
776. B. R. Loveys, in "Abscisic Acid—Physiology and Biochemistry," eds. W. J. Davies and H. G. Jones, Bios, Oxford, 1991, p. 245.
777. Y. Todoroki, N. Hirai, and K. Koshimizu, *Phytochemistry*, 1995, **38**, 561.
778. B. T. Kim, Y. K. Min, T. Asami, N. K. Park, I. H. Jeong, K. Y. Cho, and S. Yoshida, *Bioorg. Med. Chem. Lett.*, 1995, **5**, 275.
779. N. Lamb, N. Wahab, P. A. Rose, A. C. Shaw, S. R. Abrams, A. J. Cutler, P. J. Smith, L. V. Gusta, and B. Ewan, *Phytochemistry*, 1996, **41**, 23.

780. F. B. Abeles, "Ethylene in Plant Biology," Academic Press, New York, 1973.
781. M. Lieberman, *Annu. Rev. Plant Physiol.*, 1979, **30**, 533.
782. D. Neljubow, *Ber. Deut. Botan. Ges.*, 1911, **29**, 97.
783. B. G. Kang, C. S. Yocum, S. P. Burg, and P. M. Ray, *Science*, 1967, **156**, 958.
784. J. D. Goeschl, H. K. Pratt, and B. A. Bonner, *Plant Physiol.*, 1967, **42**, 1077.
785. M. B. Jackson and D. J. Osborne, *Nature*, 1970, **225**, 1019.
786. R. Sexton, L. N. Lewis, A. J. Trewavas and P. L. Kelly, in "Ethylene and Plant Development," eds. J. A. Roberts and G. A. Tucker, Butterworths, London, 1985, p. 173.
787. H. Imaseki, in "Hormonal Regulation of Development III," eds. R. P. Pharis and D. M. Reid, Springer, Berlin, 1985, p. 485.
788. S. P. Burg and E. A. Burg, *Plant Physiol.*, 1967, **42**, 144.
789. T. W. Chou and S. F. Yang, *Arch. Biochem. Biophys.*, 1973, **157**, 73.
790. M. Goto and H. Hyodo, *Plant Cell Physiol.*, 1987, **28**, 405.
791. H. Fukuda, H. Kitajima, T. Fujii, M. Tazaki, and T. Ogawa, *FEMS Microbiol. Lett.*, 1989, **59**, 1.
792. K. Nagahama, T. Ogawa, T. Fujii, M. Tazaki, S. Tanase, Y. Morino, and H. Fukuda, *J. Gen. Microbiol.*, 1991, **137**, 2281.
793. H. Fukuda, T. Ogawa, K. Ishihara, T. Fujii, K. Nagahama, T. Omata, Y. Inoue, S. Tanase, and Y. Morino, *Biochem. Biophys. Res. Commun.*, 1992, **188**, 826.
794. J. E. Ince and C. J. Knowles, *Arch. Microbiol.*, 1986, **146**, 151.
795. H. Fukuda, M. Takahashi, T. Fujii, M. Tazaki, and T. Ogawa, *FEMS Microbiol. Lett.*, 1989, **60**, 107.
796. J. H. Miyazaki and S. F. Yang, *Physiol. Plant*, 1987, **69**, 366.
797. D. O. Adams and S. F. Yang, *Proc. Natl. Acad. Sci. USA*, 1979, **76**, 170.
798. D. J. Osborne, in "The Plant Hormone Ethylene," eds. A. K. Mattoo and J. C. Suttle, CRC, Boca Raton, FL, 1991, p. 193.
799. G. Bufler, Y. Mor, M. S. Reid, and S. F. Yang, *Planta*, 1980, **150**, 439.
800. H. Yoshii and H. Imaseki, *Plant Cell Physiol.*, 1981, **22**, 369.
801. Y.-B. Yu, D. O. Adams, and S. F. Yang, *Arch. Biochem. Biophys.*, 1979, **198**, 280.
802. T. Boller, R. C. Herner, and H. Kende, *Planta*, 1979, **145**, 293.
803. Y. B. Yu and S. F. Yang, *Plant Physiol.*, 1979, **64**, 1074.
804. D. O. Adams and S. F. Yang, *Plant Physiol.*, 1977, **60**, 892.
805. N. E. Hoffman, S. F. Yang, and T. McKeon, *Biochem. Biophys. Res. Commun.*, 1982, **104**, 765.
806. C. Kionka and N. Amrhein, *Planta*, 1984, **162**, 226.
807. A. K. Mattoo and W. B. White, in "The Plant Hormone Ethylene," eds. A. K. Mattoo and J. C. Suttle, CRC, Boca Raton, FL, 1991, p. 21.
808. F. B. Abeles, P. W. Morgan, and M. E. Saltveit, Jr., "Ethylene in Plant Biology," 2nd edn., Academic Press, San Diego, CA, 1992.
809. H. Hyodo, in "The Plant Hormone Ethylene," eds. A. K. Mattoo and J. C. Suttle, CRC, Boca Raton, FL, 1991, p. 43.
810. S. F. Yang and N. E. Hoffman, *Annu. Rev. Plant Physiol.*, 1984, **35**, 155.
811. J. C. Suttle, in "The Plant Hormone Ethylene," eds. A. K. Mattoo and J. C. Suttle, CRC, Boca Raton, FL, 1991, p. 115.
812. N. E. Hoffman and S. F. Yang, *J. Am. Soc. Hort. Sci.*, 1980, **105**, 492.
813. N. E. Hoffman, Y. Liu, and S. F. Yang, *Planta*, 1983, **157**, 518.
814. M. Guy and H. Kende, *Planta*, 1984, **160**, 281.
815. H. Kende and T. Boller, *Planta*, 1981, **151**, 476.
816. H. Yoshii and H. Imaseki, *Plant Cell Physiol.*, 1982, **23**, 639.
817. H. Hyodo, K. Tanaka, and J. Yoshisaka, *Plant Cell Physiol.*, 1985, **26**, 161.
818. M. A. Acaster and H. Kende, *Plant Physiol.*, 1983, **72**, 139.
819. J. Chappell, K. Hahlbrock, and T. Boller, *Planta*, 1984, **161**, 475.
820. G. Felix, D. G. Grosskopf, M. Regenass, C. W. Basse, and T. Boller, *Plant Physiol.*, 1991, **97**, 19.
821. N. Nakajima, H. Mori, K. Yamazaki, and H. Imaseki, *Plant Cell Physiol.*, 1990, **31**, 1021.
822. T. Sato and A. Theologis, *Proc. Natl. Acad. Sci. USA*, 1989, **86**, 6621.
823. H. Mori, N. Nakagawa, T. Ono, N. Yamagishi, and H. Imaseki, in "Cellular and Molecular Aspects of the Plant Hormone Ethylene," eds. J. C. Pech, A. Latche, and C. Balague, Kluwer, Dordrecht, 1993, p. 1.
824. N. Li and K. Mattoo, *J. Biol. Chem.*, 1994, **269**, 6908.
825. N. Nakajima, N. Nakagawa, and H. Imaseki, *Plant Cell Physiol.*, 1988, **29**, 989.
826. A. B. Bleecker, G. Robinson, and H. Kende, *Planta*, 1988, **173**, 385.
827. L. Edelman and H. Kende, *Planta*, 1990, **182**, 635.
828. N. Nakagawa, H. Mori, K. Yamazaki, and H. Imaseki, *Plant Cell Physiol.*, 1991, **32**, 1153.
829. N. Yamagishi *et al.*, unpublished.
830. I. S. Yoon *et al.*, unpublished.
831. J. R. Botella, J. M. Arteca, C. D. Schlagnhaufer, R. N. Arteca, and A. T. Phillips, *Plant Mol. Biol.*, 1992, **20**, 425.
832. W. T. Kim, A. Silverstone, W. K. Yip, J. G. Dong, and S. F. Yang, *Plant Physiol.*, 1992, **98**, 465.
833. I. S. Yoon, H. Mori, J. H. Kim, B. G. Kang, and H. Imaseki, *Plant Cell Physiol.*, 1997, **38**, 217.
834. J. E. Lincoln, A. D. Campbell, J. Oetiker, W. H. Rottmann, P. W. Oeller, N. F. Shen, and A. Theologis, *J. Biol. Chem.*, 1993, **268**, 19422.
835. W.-K. Yip, T. Moore, and S. F. Yang, *Proc. Natl. Acad. Sci. USA*, 1992, **89**, 2475.
836. D. C. Olson, J. A. White, L. Edelman, R. N. Harkins, and H. Kende, *Proc. Natl. Acad. Sci. USA*, 1991, **88**, 5340.
837. W. H. Rottmann, G. F. Peter, P. W. Oeller, J. A. Keller, N. F. Shen, B. P. Nagy, L. P. Taylor, A. D. Campbell, and A. Theologis, *J. Mol. Biol.*, 1991, **222**, 937.
838. W. K. Yip, J. G. Dong, J. W. Kenny, G. A. Thompson, and S. F. Yang, *Proc. Natl. Acad. Sci. USA*, 1990, **87**, 7930.
839. H. Tatsuki, I. S. Yoon, H. Mori, and H. Imaseki, "Proceedings of International Symposium on Plant Molecular Biology, Lucknow," ed. P. Nath, in press.

840. A. Drory, S. Mayak, and W. R. Woodson, *J. Plant Physiol.*, 1993, **141**, 663.
841. A. C. Cameron, C. A. L. Fenton, Y. Yu, D. O. Adams, and S. F. Yang, *HortScience*, 1979, **14**, 178.
842. C. J. S. Smith, A. Slater, and D. Grierson, *Planta*, 1986, **168**, 94.
843. A. J. Hamilton, M. Bouzayan, and D. Grierson, *Proc. Natl. Acad. Sci. USA*, 1991, **88**, 7434.
844. H. Wang and W. R. Woodson, *Plant Physiol.*, 1991, **96**, 1000.
845. K. A. Lawton, K. G. Raghothama, P. B. Goldsbrough, and W. R. Woodson, *Plant Physiol.*, 1990, **93**, 1370.
846. H. Wang and W. R. Woodson, *Plant Physiol.*, 1992, **100**, 535.
847. M. J . Holdsworth, C. R. Bird, R. Jay, W. Schuch, and D. Grierson, *Nucleic Acid Res.*, 1987, **15**, 731.
848. D. J. McGarvey, R. Sirevag, and R. E. Christoffersen, *Plant Physiol.*, 1992, **98**, 554.
849. J. G. Dong, J. C. Fernanndez-Maculet, and S. F. Yang, *Proc. Natl. Acad. Sci. USA*, 1992, **89**, 9789.
850. W. R. Woodson, K. Y. Park, A. Drory, P. B. Larsen, and H. Wang, *Plant Physiol.*, 1992, **99**, 526.
851. J. A. Nadeau, X. S. Zhang, H. Nair, and S. D. O'Neill, *Plant Physiol.*, 1993, **103**, 31.
852. S. D. O'Neill, J. A. Nadeau, X. S. Zhang, A. Q. Bui and A. H. Halevy, *Plant Cell*, 1993, **5**, 419.
853. X. Tang, A. M. T. R. Gomes, A. Bhatia and W. R. Woodson, *Plant Cell*, 1994, **6**, 1227.
854. J. E. Lincoln and R. L. Fisher, *Mol. Gen. Genet.*, 1988, **212**, 71.
855. L. Penarrubia, M. Aguilar, L. Margossian, and R. L. Fischer, *Plant Cell*, 1992, **4**, 681.
856. L. Penarrubia, M. Aguilar, L. Margossian, and R. L. Fischer, in "Cellular and Molecular Aspects of the Plant Hormone Ethylene," eds. J. C. Pech, A. Latech, and C. Balague, Kluwer, Dordrecht, 1993, p. 100.
857. A. J. Hamilton, G. W. Lycett, and D. Grierson, *Nature*, 1990, **346**, 284.
858. S. Picton, S. L. Barton, M. Bouzayen, A. J. Hamilton, and D. Grierson, *Plant J.*, 1993, **3**, 469.
859. P. W. Oeller, L. Min-Wong, L. P. Taylor, D. A. Pike, and A. Theologis, *Science*, 1991, **254**, 437.
860. H. J. Klee, M. B. Hayford, K. A. Kretzmer, G. F. Barry, and G. M. Kishore, *Plant Cell*, 1991, **3**, 1187.
861. R. E. Sheehy, V. Ursin, S. Vanderpan, and W. R. Hiatt, in "Cellular and Molecular Aspects of the Plant Hormone Ethylene," eds. J. C. Pech, A. Latech, and C. Balague, Kluwer, Dordrecht, 1993, p. 106.
862. J. W. Mitchell, N. Mandava, J. F. Worley, and J. R. Plimmer, *Nature*, 1970, **225**, 1065.
863. M. D. Grove, G. F. Spencer, W. K. Rohwedder, N. Mandava, J. F. Worley, J. D. Warthen, G. L. Steffens, J. Flippen-Anderson, and J. C. Cook, Jr., *Nature*, 1979, **281**, 216.
864. S. Marumo, H. Hattori, H. Abe, Y. Nonoyama, and K. Munakata, *Agric. Biol. Chem.*, 1968, **32**, 528.
865. T. Yokota, M. Arima, and N. Takahashi, *Tetrahedron Lett.*, 1982, **23**, 1275.
866. N. B. Mandava, *Annu. Rev. Plant Physiol. Plant Mol. Biol.*, 1988, **39**, 23.
867. H. G. Cutler, T. Yokota, and G. Adam (eds.), "Brassinosteroids—Chemistry, Bioactivity and Applications," ACS Symposium Series No. 474, American Chemical Society, Washington, DC, 1991.
868. S. Takatsuto, *J. Chromatogr.*, 1994, **658**, 3.
869. G. Adam, A. Porzel, J. Schmidt, B. Schneider, and B. Voigt, in "Studies in Natural Products Chemistry," ed. A. Rahman, Elsevier, Oxford, 1996, vol. 18, p. 495.
870. S. Fujioka and A. Sakurai, *Nat. Prod. Rep.*, 1997, **14**, 1.
871. S. D. Clouse, *Plant J.*, 1996, **10**, 1.
872. T. Yokota, *Trends Plant Sci.*, 1997, **2**, 137.
873. J. M. Sasse, *Physiol. Plant.*, 1997, **100**, 696.
874. P. E. Taylor, K. Spuck, P. M. Smith, J. M. Sasse, T. Yokota, P. G. Griffiths, and D. W. Cameron, *Planta*, 1993, **189**, 91.
875. T. Yokota, T. Matsuoka, K. Takahashi, and M. Nakayama, *Phytochemistry*, 1996, **42**, 509.
876. K. Wada, S. Marumo, H. Abe, T. Morishita, K. Nakamura, M. Uchiyama, and K. Mori, *Agric. Biol. Chem.*, 1984, **48**, 719.
877. T. Yokota, S. Watanabe, Y. Ogino, I. Yamaguchi, and N. Takahashi, *J. Plant Growth Regul.*, 1990, **9**, 151.
878. N. Ikekawa and S. Takatsuto, *Mass Spectrosc.*, 1984, **32**, 55.
879. K. Gamoh and S. Takatsuto, *J. Chromatogr.*, 1994, **658**, 17.
880. K. Gamoh, M. C. Prescott, L. J. Goad, and S. Takatsuto, *Bunseki Kagaku*, 1996, **45**, 523.
881. S. Asakawa, H. Abe, Y. Kyokawa, S. Nakamura, and M. Natsume, *Biosci. Biotech. Biochem.*, 1994, **58**, 219.
882. G. M. Caballero, O. T. Centurion, L. Galagovsky, and E. G. Gros, in "Proceedings of the 23rd Annual Meeting of the Plant Growth Regulation Society of America, Calgary, 1996," ed. T. D. Davis, Plant Growth Regulation Society of America, 1996, p. 50.
883. A. Porzel, V. Marquardt, G. Adam, G. Massiot, and D. Zeigen, *Magn. Reson. Chem.*, 1992, **30**, 651.
884. T. Ando, M. Aburatani, N. Koseki, S. Asakawa, T. Mouri, and H. Abe, *Magn. Reson. Chem.*, 1993, **31**, 94.
885. V. Marquardt and G. Adam, in "Chemistry of Plant Protection," ed. W. Ebing, Springer, Heidelberg, 1991, vol. 7, p. 103.
886. M. Aburatani, T. Takeuchi, and K. Mori, *Agric. Biol. Chem.*, 1987, **51**, 1909.
887. T. C. McMorris, R. G. Charez, and P. A. Patil, *J. Chem. Soc., Perkin Trans. 1*, 1996, 295.
888. T. Yokota and K. Mori, in "Molecular Structure and Biological Activity of Steroids," eds. M. Bohl and W. L. Duax, CRC, Boca Raton, FL, 1992, p. 317.
889. M. J. Thompson, W. J. Meudt, N. B. Mandava, S. R. Dutky, W. R. Lusby, and D. W. Spaulding, *Steroids*, 1982, **39**, 89.
890. K. Wada and S. Marumo, *Agric. Biol. Chem.*, 1981, **45**, 2579.
891. S. Takatsuto, N. Ikekawa, T. Morishita, and H. Abe, *Chem. Pharm. Bull.*, 1987, **35**, 211.
892. K. Mori and T. Takeuchi, *Liebigs. Ann. Chem.*, 1988, 815.
893. T. Hai, B. Schneider, and G. Adam, *Phytochemistry*, 1995, **40**, 443.
894. C. Brosa, J. M. Capdevila, and J. Zamora, *Tetrahedron*, 1996, **52**, 2435.
895. J. Li, P. Nagpal, V. Vitart, T. C. McMorris, and J. Chory, *Science*, 1996, **272**, 398.
896. M. Szekeres, K. Németh, Z. Koncz-Kélmán, J. Mathur, A. Kauschmann, T. Altmann, G. P. Rédei, F. Nagy, J. Schell, and C. Koncz, *Cell*, 1996, **85**, 171.
897. S. D. Clouse, M. Langford, and T. C. McMorris, *Plant Physiol.*, 1996, **111**, 671.
898. A. Kauschmann, A. Jessop, C. Koncz, M. Szekeres, L. Willmitzer, and T. Altmann, *Plant J.*, 1996, **9**, 701.
899. T. Nomura, M. Nakayama, J. B. Reid, Y. Takeuchi, and T. Yokota, *Plant Physiol.*, 1997, **113**, 31.

900. T. Yokota, Y. Ogino, N. Takahashi, H. Saimoto, S. Fujioka and A. Sakurai, *Agric. Biol. Chem.*, 1990, **54**, 1107.
901. H. Suzuki, S. Fujioka, S. Takatsuto, T. Yokota, N. Murofushi, and A. Sakurai, *J. Plant Growth Regul.*, 1993, **12**, 101.
902. H. Suzuki, S. Fujioka, S. Takatsuto, T. Yokota, N. Murofushi, and A. Sakurai, *J. Plant Growth Regul.*, 1994, **13**, 21.
903. S. Fujioka, T. Inoue, S. Takatsuto, S. Yanagisawa, T. Yokota, and A. Sakurai, *Biosci. Biotech. Biochem.*, 1995, **59**, 1543.
904. H. Suzuki, T. Inoue, S. Fujioka, S. Takatsuto, T. Yanagisawa, T. Yokota, N. Murofushi, and A. Sakurai, *Biosci. Biotech. Biochem.*, 1994, **58**, 1186.
905. H. Suzuki, T. Inoue, S. Fujioka, T. Saito, S. Takatsuto, T. Yokota, N. Murofushi, T. Yanagisawa, and A. Sakurai, *Phytochemistry*, 1995, **40**, 1391.
906. S. Fujioka, T. Inoue, S. Takatsuto, T. Yanagisawa, T. Yokota, and A. Sakurai, *Biosci. Biotech. Biochem.*, 1995, **59**, 1973.
907. Y.-H. Choi, S. Fujioka, A. Harada, T. Yokota, S. Takatsuto and A. Sakurai, *Phytochemistry*, 1996, **43**, 593.
908. Y.-H. Choi, S. Fujioka, T. Nomura, A. Harada, T. Yokota, S. Takatsuto, and A. Sakurai, *Phytochemistry*, 1997, **44**, 609.
909. H. Suzuki, S. Fujioka, S. Takatsuto, T. Yokota, N. Murofushi, and A. Sakurai, *Biosci. Biotech. Biochem.*, 1995, **59**, 168.
910. H. Abe, C. Honjo, Y. Kyokawa, S. Asakawa, M. Natsume, and M. Narushima, *Biosci. Biotech. Biochem.*, 1994, **58**, 986.
911. T. Yokota, Y. Nakamura, N. Takahashi, M. Nonaka, H. Sekimoto, H. Oshio, and S. Takatsuto, in "Gibberellins," eds. N. Takahashi, B. O. Phinney, and J. MacMillan, Springer, New York, 1991, p. 339.
912. T. Iwasaki and H. Shibaoka, *Plant Cell Physiol.*, 1991, **32**, 1007.
913. M. Kawaguchi, H. Imaizumi-Anraku, S. Fukai, and K. Syôno, *Plant Cell Physiol.*, 1996, **37**, 461.
914. S.-K. Kim, T. Asano, and S. Marumo, *Biosci. Biotech. Biochem.*, 1995, **59**, 1934.
915. H. Suzuki, S.-K. Kim, N. Takahashi, and T. Yokota, *Phytochemistry*, 1993, **33**, 1361.
916. T. Yokota, S.-K. Kim, Y. Kosaka, Y. Ogino, and N. Takahashi, in "Conjugated Plant Hormones—Structure, Metabolism and Function," eds. K. Schreiber, H. E. Schütte, and G. Sembdner, Institute of Plant Biochemistry, Halle, Academy of Sciences GDR, 1987, p. 288.
917. S.-K. Kim, in "Brassinosteroids—Chemistry, Bioactivity and Applications," eds. H. G. Cutler, T. Yokota, and G. Adam, American Chemical Society, Washington, 1991, p. 26.
918. T. Yokota, K. Higuchi, Y. Kosaka, and N. Takahashi, in "Progress in Plant Growth Regulation," eds. C. M. Karssen, L. C. van Loon, and D. Vrengdenhil, Kluwer, Dordrecht, 1992, p. 298.
919. B. Schneider, A. Kolbe, A. Porzel, and G. Adam, *Phytochemistry*, 1994, **36**, 319.
920. T. Hai, B. Schneider, A. Porzel, and G. Adam, *Phytochemistry*, 1996, **41**, 197.
921. A. Kolbe, B. Schneider, A. Porzel, B. Voigt, G. Krauss, and G. Adam, *Phytochemistry*, 1994, **36**, 671.
922. A. Kolbe, B. Schneider, A. Porzel, J. Schmidt, and G. Adam, *Phytochemistry*, 1995, **38**, 633.
923. A. Kolbe, B. Schneider, A. Porzel, and G. Adam, *Phytochemistry*, 1996, **41**, 163.
924. N. Nishikawa, H. Abe, M. Natsume, A. Shida, and S. Toyama, *J. Plant Physiol.*, 1995, **147**, 294.
925. S. Asakawa, H. Abe, N. Nishikawa, M. Natsume, and M. Koshioka, *Biosci. Biotech. Biochem.*, 1996, **60**, 1416.
926. B. Voigt, A. Porzel, H. Naumann, C. Hörhold-Schubert, and G. Adam, *Steroids*, 1993, **58**, 320.
927. L. Ruzicka and M. Pfeiffer, *Helv. Chim. Acta*, 1933, **16**, 1208.
928. E. Demole, E. Lederer, and D. Mercier, *Helv. Chim. Acta*, 1962, **45**, 675.
929. D. C. Aldridge, S. Galt, D. Giles, and W. B. Turner, *J. Chem. Soc. (C)*, 1971, 1623.
930. J. Ueda and J. Kato, *Plant Physiol.*, 1980, **66**, 246.
931. J. Ueda and J. Kato, *Agric. Biol. Chem.*, 1982, **46**, 1975.
932. T. Yoshihara, El-S. A. Omer, H. Koshino, S. Sakamura, Y. Kikuta, and Y. Koda, *Agric. Biol. Chem.*, 1989, **53**, 2835.
933. A. Meyer, O. Miersch, C. Büttner, W. Dathe, and G. Sembdner, *J. Plant Growth Regul.*, 1984, **3**, 1.
934. J. Ueda, K. Miyamoto, T. Sato, and Y. Momotani, *Agric. Biol. Chem.*, 1991, **55**, 275.
935. J. Ueda, K. Miyamoto, M. Aoki, T. Hirata, T. Sato, and Y. Momotani, *Bull Univ. Osaka Pref., Ser. B*, 1991, **43**, 103.
936. S. Fujii, R. Yamamoto, K. Miyamoto, and J. Ueda, *Phycol. Res.*, 1997, **45**, 223.
937. M. V. Krupina and W. Dathe, *Z. Naturforsch. Teil C*, 1991, **46**, 1127.
938. O. Miersch, G. Sembdner, and K. Schreiber, *Phytochemistry*, 1989, **28**, 339.
939. O. Miersch, B. Wrobel, and G. Sembdner, in "Proceedings of the International Symposium, Conjugated Plant Hormones," eds. K. Schreiber, H. R. Schutte, and G. Sembdner, VEB Deutscher Verlag der Wissenschaften, Berlin, 1986, p. 333.
940. B. E. Cross and G. R. B. Webster, *J. Chem. Soc. (C)*, 1970, 1839.
941. W. Dathe, O. Miersch, and J. Schmidt, *Biochem. Physiol. Pflanz.*, 1989, **185**, 83.
942. W. Dathe, C. Schindler, G. Schneider, J. Schmidt, A. Porzel, E. Jensen, and I. Yamaguchi, *Phytochemistry*, 1991, **30**, 1909.
943. H.-D. Knöfel and G. Sembdner, *Phytochemistry*, 1995, **38**, 569.
944. H. Yamane, H. Takagi, H. Abe, T. Yokota, and N. Takahashi, *Plant Cell Physiol.*, 1981, **22**, 689.
945. V. Berestetzky, W. Dathe, T. Daletskaya, L. Musatenko, and G. Sembdner, *Biochem. Physiol. Pflanz.*, 1991, **187**, 13.
946. H. Gundlach, M. J. Müller, T. M. Kuchan, and M. Zenk, *Proc. Natl. Acad. Sci. USA*, 1992, **89**, 2389.
947. R. W. Wilen, G. J. H. van Rooijen, D. W. Pearce, R. P. Pharis, L. A. Holbrook, and M. N. Moloney, *Plant Physiol.*, 1991, **95**, 399.
948. W. Dathe, H. Rönsch, A. Preiss, W. Schade, and G. Sembdner, *Planta*, 1981, **153**, 530.
949. G. Schneider, R. Kramall, and C. Brückner, *J. Chromatogr.*, 1989, **438**, 459.
950. C. Brückner, R. Kramell, G. Schneider, H.-D. Knöfel, G. Sembdner, and K. Schreiber, *Phytochemistry*, 1986, **25**, 2236.
951. C. Brückner, R. Kramell, G. Schneider, and G. Sembdner, in "Proceedings of the International Symposium, Conjugated Plant Hormones," eds. K. Schreiber, H. R. Schutte, and G. Sembdner, VEB Deutscher Verlag der Wissenschaften, Berlin, 1986, p. 308.
952. C. Brückner, R. Kramell, G. Schneider, J. Schmidt, A. Preiss, G. Sembdner, and K. Schreiber, *Phytochemistry*, 1988, **27**, 275.

953. S. Tsurumi and Y. Asahi, *Physiol. Plant.*, 1985, **64**, 207.
954. R. Nishida and T. E. Acree, *J. Agric. Food Chem.*, 1984, **32**, 1001.
955. H. Yamane, H. Abe, and N. Takahashi, *Plant Cell Physiol.*, 1982, **23**, 1125.
956. E. Demole, G. Willhalm, and M. Stoll, *Helv. Chim. Acta*, 1964, **47**, 1152.
957. L. Crabalona, *Comp. Rend. Acad. Sci., Paris, Ser. C*, 1967, **264**, 2074.
958. T. Fujita, K. Terato, and M. Nakayama, *Biosci. Biotech. Biochem.*, 1996, **60**, 432.
959. H. Helder, O. Miersch, D. Vreugdenhil, and G. Sembdner, *Physiol. Plant.*, 1993, **88**, 647.
960. K. Koshimizu, H. Fukui, S. Usuda, and T. Mitsui, "Plant Growth Substances 1973," Hirokawa Publishing, Tokyo, 1974, p. 86.
961. S. Fukui, K. Koshimizu, Y. Yamazaki, and S. Usuda, *Agric. Biol. Chem.*, 1977, **41**, 189.
962. T. Yoshihara, H. Matsuura, A. Ichihara, Y. Kikuta, and Y. Koda, in "Progress in Plant Growth Regulation," eds. C. M. Karssen, L. C. van Loon, and D. Veugdenhil, Kluwer, Dordrecht, 1992, p. 286.
963. H. Matsuura, T. Yoshihara, A. Ichihara, Y. Kikuta, and Y. Koda, *Biosci. Biotech. Biochem.*, 1993, **57**, 1253.
964. F. Bohlmann, P. Wegner, J. Jakupovic, and R. M. King, *Tetrahedron*, 1984, **40**, 2537.
965. E. E. Farmer and C. A. Ryan, *Proc. Natl. Acad. Sci. USA*, 1990, **87**, 7713.
966. G. C. Neto, Y. Kono, H. Hyakutake, M. Watanabe, Y. Suzuki, and A. Sakurai, *Agric. Biol. Chem.*, 1991, **55**, 3097.
967. W. Dathe, A. D. Parry, J. K. Heald, I. M. Scott, O. Miersch, and R. Horgan, *J. Plant Growth Regul.*, 1994, **13**, 59.
968. H. Yamane, N. Takahashi, J. Ueda, and J. Kato, *Agric. Biol. Chem.*, 1981, **45**, 1709.
969. A. Omata, S. Nakamura, K. Yomogida, K. Moriaki, and I. Watanabe, *Agric. Biol. Chem.*, 1990, **54**, 1029.
970. T. C. Baker, R. Nishida, and W. L. Roelofs, *Science*, 1981, **214**, 1359.
971. R. Nishida, T. C. Baker, and W. L. Roelofs, *J. Chem. Ecol.*, 1982, **8**, 947.
972. B. Parthier, *J. Plant Growth Regul.*, 1990, **9**, 1.
973. B. Parthier, *Bot. Acta*, 1990, **104**, 446.
974. G. Suzuki, H. Ohta, T. Kato, T. Igarashi, F. Sakai, D. Shibata, A. Takano, T. Masuda, Y. Shioi, and K. Takamiya, *FEBS Lett.*, 1996, **383**, 83.
975. G. Sembdner and B. Parthier, *Annu. Rev. Plant Physiol. Plant Mol. Biol.*, 1993, **44**, 569.
976. M. Hamberg and H. Gardner, *Biochim. Biophys. Acta*, 1992, **1165**, 1.
977. J. Ueda, K. Miyamoto, and S. Kamisaka, *J. Chromatogr. A*, 1994, **658**, 129.
978. S. Reinbothe, B. Mollenhauer, and C. Reinbothe, *Plant Cell*, 1994, **6**, 1197.
979. K. Miyamoto, M. Oka, and J. Ueda, *Physiol. Plant.*, 1997, **100**, 631.
980. B. A. Vick and D. C. Zimmerman, *Plant Physiol.*, 1984, **75**, 458.
981. B. A. Vick and D. C. Zimmerman, *Plant Physiol.*, 1986, **80**, 202.
982. B. A. Vick and D. C. Zimmerman, *Plant Physiol.*, 1987, **85**, 1073.
983. M. Hamberg, O. Miersch, and G. Sembdner, *Lipids*, 1988, **23**, 521.
984. B. A. Vick and D. C. Zimmerman, *Plant Physiol.*, 1989, **90**, 125.
985. H. von Gerlach and P. Künzler, *Helv. Chim. Acta*, 1978, **61**, 2503.
986. O. Miersch, J. Schmidt, G. Sembdner, and K. Schreiber, *Phytochemistry*, 1989, **28**, 1303.
987. W. Dathe, H.-M. Kramell, W. Daeter, R. Kramell, S. Slovik, and W. Hartung, *J. Plant Growth Regul.*, 1993, **12**, 133.
988. E. Demole and M. Stoll, *Helv. Chim. Acta*, 1962, **45**, 692.
989. T. L. Ho, *Synth. Commun.*, 1974, **4**, 265.
990. H. Kataoka, K. Yamada, J. Goto, and J. Tsujii, *Tetrahedron*, 1987, **43**, 4107.
991. T. Kitahara, T. Nishi, and K. Mori, *Tetrahedron*, 1991, **47**, 6999.
992. T. Kitahara and T. Nishi, *Proc. Jpn Acad. Ser. B*, 1995, **71**, 20.
993. H . Seto and H. Yoshioka, *Chem. Lett.*, 1990, 1797.
994. H. Nojiri, H. Yamane, H. Seto, I. Yamaguchi, N. Murofushi, T. Yoshihara, and H. Shibaoka, *Plant Cell Physiol.*, 1992, **33**, 1225.
995. R. Kramell, J. Schmidt, A. Preiss, G. Schneider, and G. Sembdner, in "Proceedings of the International Symposium, Conjugated Plant Hormones," eds. K. Schreiber, H. R. Schutte, and G. Sembdner, VEB Deutscher Verlag der Wissenschaften, Berlin, 1986, p. 323.
996. R. Kramell, J. Schmidt, G. Schneider, G. Sembdner, and K. Schreiber, *Tetrahedron*, 1988, **44**, 5791.
997. T. Yokota, N. Murofushi, and N. Takahashi, in "Hormonal Regulation of Development I," ed. J. MacMillan, Springer, Berlin, 1980, p. 113.
998. R. Nishida, T. E. Acree, and H. Fukami, *Agric. Biol. Chem.*, 1985, **49**, 769.
999. M. Okamoto and H. Nakazawa, *Biosc. Biotech. Biochem.*, 1992, **56**, 1172.
1000. H.-D. Knöfel, C. Brückner, R. Kramell, G. Sembdner, and K. Schreiber, *Biochem. Physiol. Pflanz.*, 1984, **179**, 317.
1001. T. Albrecht, A. Kehlen, K. Stahl, H.-D. Knöfel, G. Sembdner, and E. W. Weiler, *Planta*, 1993, **191**, 86.
1002. H. Yamane, J. Sugawara, Y. Suzuki, E. Shimamura, and N. Takahashi, *Agric. Biol. Chem.*, 1980, **44**, 2857.
1003. J. Ueda, K. Miyamoto, Y. Momotani, J. Kato, and S. Kamisaka, in "Plant Cell Walls as Biopolymers with Physiological Functions," ed. Y. Masuda, Yamada Science Foundation, Osaka, 1992, p. 307.
1004. J. Ueda, K. Miyamoto, and M. Hashimoto, *J. Plant Growth Regul.*, 1996, **15**, 189.
1005. J. Ueda, Y. Morita, and J. Kato, *Plant Cell Physiol.*, 1991, **32**, 983.
1006. J. Ueda, K. Miyamoto, and S. Kamisaka, *J. Plant Growth Regul.*, 1995, **14**, 69.
1007. Y. Koda, Y. Kikuta, H. Tazaki, Y. Tsujino, S. Sakamura, and T. Yoshihara, *Phytochemistry*, 1991, **30**, 1435.
1008 Y. Koda, Y. Kikuta, T. Kitahara, T. Nishi, and K. Mori, *Phytochemistry*, 1992, **31**, 1111.
1009. H. Seto, Y. Kamuro, Z. Qian, and T. Shimizu, *J. Pesticide Sci.*, 1992, **17**, 61.
1010. Y. Koda, J. L. Ward, and M. H Beale, *Phytochemistry*, 1995, **38**, 821.
1011. A. Ishikawa, T. Yoshihara, and K. Nakamura, *Plant Mol. Biol.*, 1994, **26**, 403.
1012. A. Ishikawa, T. Yoshihara, and K. Nakamura, *Biosci. Biotech. Biochem.*, 1994, **58**, 544.
1013. B. Axelrod, *Adv. Chem. Ser.*, 1974, **136**, 324.
1014. T. Galliard and H. W.-S. Chan, in "The Biochemistry of Plants. A Comprehensive Treatise," eds. P. K. Stumpf and E. E. Conn, Academic Press, New York, 1980, p. 131.
1015. G. A. Veldink, J. F. G. Vliegenthart, and J. Boldingh, *Prog. Chem. Fats Other Lipids*, 1977, **15**, 131.
1016. B. A. Vick and D. C. Zimmerman, *Plant Physiol.*, 1979, **64**, 203.

1017. B. A. Vick and D. C. Zimmerman, *Plant Physiol.*, 1981, **67**, 92.
1018. B. A. Vick and D. C. Zimmerman, *Plant Physiol.*, 1982, **69**, 1103.
1019. B. A. Vick, P. Feng, and D. C. Zimmerman, *Lipids*, 1980, **15**, 468.
1020. M. Hamberg, *Biochem. Biophys. Res. Commun.*, 1988, **156**, 543.
1021. W.-C. Song and A. R. Brash, *Science*, 1991, **253**, 781.
1022. K. Harms, R. Atzon, A. Brash, H. Kühn, C. Wasternack, L. Willmitzer, and H. Peña-Cortés, *Plant Cell*, 1995, **7**, 1645.
1023. A. Meyer, D. Gross, S. Vorkefeld, M. Kummer, J. Schmidt, G. Sembdner, and K. Schreiber, *Phytochemistry*, 1989, **28**, 1007.
1024. A. Meyer, J. Schmidt, D. Gross, E. Jensen, A. Rudolph, S. Vorkefeld, and G. Sembdner, *J. Plant Growth Regul.*, 1991, **10**, 17.
1025. G. Sembdner, A. Meyer, O. Miersch, and C. Brückner, in "Plant Growth Substances 1988," eds. R. P. Pharis and S. B. Rood, Springer Verlag, Berlin, 1990, p. 374.
1026. T. Yoshihara, M. Amanuma, T. Tsutsumi, Y. Okuma, H. Matsuura, and A. Ichihara, *Plant Cell Physiol.*, 1996, **37**, 586.
1027. R. Kramell, R. Atzorn, G. Schneider, O. Miersch, C. Brückner, J. Schmidt, G. Sembdner, and B. Parthier, *J. Plant Growth Regul.*, 1995, **14**, 29.
1028. J. Ueda and J. Kato, *Z. Pflanzenphysiol.*, 1981, **103**, 357.
1029. R. Lopez, W. Dathe, C. Brückner, O. Miersch, and G. Sembdner, *Biochem. Physiol. Pflanz.*, 1987, **182**, 195.
1030. J. Ueda, T. Mizumoto, and J. Kato, *Biochem. Physiol. Pflanz.*, 1991, **187**, 203.
1031. J. Ueda, K. Miyamoto, and M. Aoki, *Plant Cell Physiol.*, 1994, **35**, 1065.
1032. T. Matsuki, H. Tazaki, T. Fujimori, and T. Hogetsu, *Biosci. Biotech. Biochem.*, 1992, **56**, 1329.
1033. T. Mita and H. Shibaoka, *Plant Cell Physiol.*, 1983, **24**, 109.
1034. M. Abe, H. Shibaoko, H. Yamane, and N. Takahashi, *Protoplasma*, 1990, **156**, 1.
1035. R. A. Weidhase, H.-M. Kramell, J. Lehmann, H.-W. Liebisch, W. Lerbs, and B. Parthier, *Physiol. Sci.*, 1987, **51**, 177.
1036. J. M. Anderson, *J. Plant Growth Regul.*, 1988, **7**, 203.
1037. C. J. Bolter, *Plant Physiol.*, 1993, **103**, 1347.
1038. S. Berger, E. Bell, and J. E. Mullet, *Plant Physiol.*, 1996, **111**, 525.
1039. E. E. Farmer, R. R. Johnson, and C. A. Ryan, *Plant Physiol.*, 1992, **98**, 995.
1040. E. E. Farmer and C. A. Ryan, *Proc. Natl. Acad. Sci. USA*, 1990, **87**, 7713.
1041. E. E. Farmer and C. A. Ryan, *Plant Cell*, 1992, **4**, 129.
1042. C. A. Ryan, *Plant Mol. Biol.*, 1992, **19**, 123.
1043. H. Peña-Cortés, J. J. Sánchez-Serrano, R. Mertens, L. Willmitzer, and S. Prat, *Proc. Natl. Acad. Sci. USA*, 1989, **86**, 9851.
1044. R. Lorberth, C. Dammann, M. Ebneth, S. Amati, and J. J. Sánchez-Serrano, *Plant J.*, 1992, **2**, 477.
1045. H. Peña-Cortés, X. Liu, J. J. Sánchez-Serrano, R. Schmid, and L. Willmitzer, *Planta*, 1992, **186**, 495.
1046. H. Peña-Cortés, L. Willmitzer, and J. J. Sánchez-Serrano, *Plant Cell*, 1991, **3**, 963.
1047. D. Bartels, K. Engelhardt, R. Roncarati, K. Schneider, M. Rotter, and F. Salamini, *EMBO J.*, 1991, **10**, 1037.
1048. L. Andersen, W. Becker, K. Schlüter, K. Burges, B. Parthier, and K. Apel, *Plant Mol. Biol.*, 1992, **19**, 193.
1049. H. Gundlach, M. J. Müller, T. M. Kutchan, and M. H. Zenk, *Proc. Natl. Acad. Sci. USA*, 1992, **89**, 2389.
1050. R. A. Creelman, M. L. Tierney, and J. E. Mullet, *Proc. Natl. Acad. Sci. USA*, 1992, **89**, 4938.
1051. J. M. Anderson, *J. Plant Growth Regul.*, 1991, **10**, 5.
1052. J. M. Anderson, S. R. Spilatro, S. F. Klauer, and V. R. Franceschi, *Plant Sci.*, 1989, **62**, 45.
1053. V. R. Franceschi and H. D. Grimes, *Proc. Natl. Acad. Sci. USA*, 1991, **88**, 6745.
1054. J.-F. Huang, D. J. Bantroch, J. S. Greenwood, and P. E. Staswick, *Plant Physiol.*, 1991, **97**, 1512.
1055. H. S. Mason, D. B. DeWald, R. A. Creelman, and J. E. Mullet, *Plant Physiol.*, 1992, **98**, 859.
1056. H. S. Mason, and J. E. Mullet, *Plant Cell*, 1990, **2**, 569.
1057. P. E. Staswick, J.-F. Huang, and Y. Rhee, *Plant Physiol.*, 1991, **96**, 130.
1058. B. Hause, U. zur Nieden, J. Lehmann, C. Wasternack, and B. Parthier, *Bot. Acta*, 1994, **107**, 333.
1059. G. Herrmann, J. Lehmann, A. Peterson, G. Sembdner, R. A. Weidhase, and B. Parthier, *J. Plant Physiol.*, 1989, **134**, 703.
1060. F. Mueller-Uri, B. Parthier, and L. Nover, *Planta*, 1988, **176**, 241.
1061. B. Parthier, C. Brückner, W. Dathe, B. Hause, G. Herrmann, H.-D. Knöfel, H.-M. Kramell, R. Kramell, J. Lehmann, O. Miersch, *et al.*, "Progress in Plant Growth Regulation," eds. C. M. Karssen, L. C. van Loon, and D. Vreugdenhil, Kluwer, Dordrecht, 1992, p. 276.
1062. S. Reinbothe, C. Reinbothe, J. Lehmann, and B. Parthier, *Physiol. Plant.*, 1992, **86**, 49.
1063. R. A. Weidhase, H.-M. Krammell, J. Lehmann, W. Liebisch, W. Lerbs, and B. Parthier, *Plant Sci.*, 1987, **51**, 177.
1064. S. Reinbothe, A. Machmudova, C. Wasternack, C. Reinbothe, and B. Parthier, *J. Plant Growth Regul.*, 1992, **11**, 7.
1065. B. Hause, U. Demus, C. Teichmann, B. Parthier, and C. Wasternack, *Plant Cell Physiol.*, 1996, **37**, 641.
1066. S. Reinbothe, C. Reinbothe, and B. Parthier, *Plant J.*, 1993, **4**, 459.
1067. P. D. Bishop, G. Pearce, J. E. Bryant, and C. A. Ryan, *J. Biol. Chem.*, 1984, **259**, 13 172.
1068. H. Peña-Cortés, T. Albrecht, S. Prat, E. W. Weiler, and L. Willmitzer, *Plant*, 1993, **191**, 123.
1069. A. Kernan and R. W. Thornburg, *Plant Physiol.*, 1989, **91**, 73.
1070. B. A. Vick and D. C. Zimmerman, *Biochem. Biophys. Res. Commun.*, 1983, **111**, 470.
1071. A. Conconi, M. Miquel, J. Browse, and C. A. Ryan, *Plant Phsyiol.*, 1996, **111**, 797.
1072. S. H. Doares, J. Narváez-Vásquez, A. Conconi, and C. A. Ryan, *Plant Physiol.*, 1995, **108**, 1741.
1073. H. Sano, S. Seo, N. Koizumi, T. Niki, H. Iwamura, and Y. Ohashi, *Plant Cell Physiol.*, 1996, **37**, 762.

8.03
Plant Chemical Ecology

JEFFREY B. HARBORNE
University of Reading, UK

8.03.1 CONSTITUTIVE CHEMICAL DEFENCE

8.03.1.1 Introduction

The subject of plant chemical ecology may be considered to have developed to answer a question raised in the 1950s by the more and more frequent discovery in higher plants of complex and varied natural products: what is the purpose of this structural profusion? Up to 1950, only about 8000 natural molecules had been recorded in plant tissues,[1] but during the next decade a wealth of new structures were uncovered. Fraenkel in 1959[2] argued that the *raison d'être* of these so-called secondary metabolites could not be accommodated by the idea that they are simply "waste products" of primary metabolism, accumulating in the plant cell because of the absence of an efficient excretory system. Instead, he described these metabolites as "trigger" substances which induce or prevent the uptake of nutrients by the herbivore.

In 1964, Ehrlich and Raven[3] were among the first to propose a defined ecological role for plant products as defense agents against insect herbivory. They proposed that, through the process of co-evolution, insects are able to detoxify certain defensive agents so that, eventually, they may become feeding attractants. Such a hypothesis helps to explain the relatively restricted feeding preferences of many types of insect, but especially the Lepidoptera. Research was therefore initiated in many laboratories to test this co-evolutionary theory and to attempt to establish a defense role for secondary compounds. In addition, modifications to this basic theory were proposed to Feeny,[4] Rhoades and Cates[5] and Bryant *et al.*[6] to take into account the facts that some plants are ecologically "apparent" (e.g., trees) and others are not and that secondary metabolites are costly to produce by the plant. In this section, some of the key findings which have subsequently established that this coevolutionary theory is a reasonable one will be summarized.

There are a variety of factors that have to be considered when proposing a defensive role for secondary metabolites. It is important to know which of all the major and minor constituents in a given host plant are active and furthermore how their concentrations fluctuate during the life of that plant. Such basic information is still rarely available for even some of the best known crop plants. The localization of these metabolites is important, since compounds at the surface (in trichomes) or in epidermal cells will be more rapidly detected by phytophagous insects than those hidden away in the mesophyll layer. Of course, leaf miners are exceptional and are only deterred from feeding by substances within the leaf.

The distribution of secondary substances within the different parts of the plant is also a significant factor, since chemical defense may be limited to the more vulnerable tissues (e.g., young leaves, ripening fruits). Also, variation in chemistry can occur within the same leaf, between leaves on the same branch, or between plants in the same population. Environmental parameters can dramatically alter the concentrations of secondary compounds, especially in leaf tissue. Other features of importance are the relative solubilities of the compounds present and hence their dietary impact on the herbivore, the relative stability within the plant, and the rate of turnover.

Equally important in the determination of a defensive role is to establish the effects of a target compound on representative herbivores. Compounds toxic to insects may have little effect on mammalian herbivores, while some insects may be relatively impervious to the effects of mammalian toxins. With insects, feeding deterrence needs to be established preferably against an insect species capable of feeding on the host plant in question. The behavior of larvae of adapted specialist insects and unadapted generalists can be quite different (see Chapter 8.05). Again, secondary chemistry can have different effects on phytophagous insects, from antifeedance and feeding deterrence to harmful effects on reproductive performance. Finally, laboratory results have to be tested in the field to confirm that the right herbivore has been recognized. Laboratory experiments, for example, with the *Vernonia* leaf sesquiterpene lactones showed insect antifeedant activity, whereas field observations showed that the compounds only deterred mammalian grazing.[7]

The role of secondary metabolites as defensive agents in plants is thus a major topic in this review. This defensive role can be enhanced by induction through animal grazing and this will be discussed in the following section. Rather convincing evidence that plant toxins are effective against animals is provided by the fact that certain toxins are "borrowed" dietarily by aposematic insects and then used in their defense against bird and other animal predation. It should also not be forgotten that some secondary compounds are produced by plants to attract animals to carry out pollination and to distribute the seed. These subjects will be considered in later sections. Some attention will be given to the biochemistry of plant–plant interactions, since secondary constituents can mediate in both hostile and beneficial interactions between plants growing in natural ecosystems. The final sections will be devoted to plant–microbial interactions and the way that micro-organisms produce secondary compounds to damage their higher plant hosts and that higher plants produce such constituents to defend themselves from microbial infection.

The primary literature on plant chemical ecology has been reviewed in a text book[8] and a series of four review articles published in *Natural Product Reports*.[9–12] Other notable reviews that deserve mention are by Haslam,[13] Hartley and Jones,[14] and Herms and Mattson.[15]

8.03.1.2 The Phenolic Barrier

More attention has been given to phenolic constituents as defensive barriers in plants than to any other class of secondary metabolite. This is partly because of their universal distribution in green plants and partly because their concentrations are easily measured in plant tissues by means of simple color reactions. Also, plant phenolics are known for their generally negative effects on insects and hence they have played an important role in early theories of plant–herbivore interactions.[4,5] The original idea of all phenolics having a similar nonspecific effect on herbivores because of their ability to bind with protein, however, has now been discarded. Different phenolics vary in their biological activity and in their effects on herbivores. Some have a direct toxic effect. Others may require release from a bound form (as a glycoside) to interact and yet others may require oxidation via phenolase for their toxic properties to become apparent.

As has been argued elsewhere,[16] measurement of total leaf phenolic content via the Folin–Ciocalteau or other reagents is relatively meaningless in indicating what particular phenolics are present in a plant. Far more rewarding results can be obtained by monitoring individual phenolics or the presence of particular subclasses (e.g., condensed tannins) and in this review attention will be concentrated on experiments where such measurements have been made. Phenolics are conveniently classified into four subclasses of increasing molecular weight: (i) low molecular weight, such as salicin; (ii) phenylpropanoids, such as chlorogenic acid; (iii) flavonoids, such as rutin; and (iv) tannins. They will be considered in this sequence in this review.

Some examples of simple phenolics being feeding barriers to insect and mammalian herbivores are shown in Table 1. There has been some concentration of effort on the salicylic acid-based phenols of beech and willow trees and the herbivores that feed on them.[17,18] Undoubtedly, examination of simple phenolic constituents of other woody angiosperms would reveal further examples of harmful toxins. There is some evidence in the case of *Betula platyphylla* that a metabolite of the naturally occurring phenol is the harmful agent, interfering with the digestion in the animal of the normal dietary nutrients. Thus, platyphylloside undergoes stepwise reduction *in vivo* to centrolobol and its is the latter compound which builds up in the gut of the hare to cause feeding inhibition.[19] Similarly, metabolism also takes place with salicortin which occurs in the leaf of *Populus* (Figure 1). It is hydrolyzed *in vivo* to 6-hydroxycyclohexenone, which is the active agent.[20]

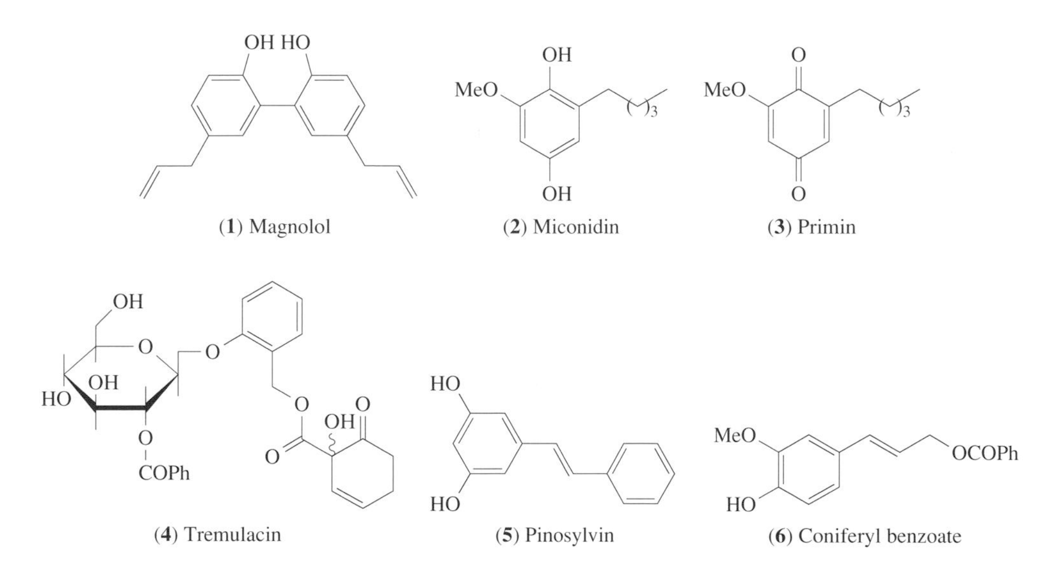

The concentration of the phenolic in the plant is a key factor in deterrence and it is the accumulation of phenols in particular parts of the plant which represents a feeding barrier. Such concentrations of toxin can sometimes be circumvented. Thus, the meadow vole, *Microtus pennsylvanicus*, is able to reduce the phenolic toxicity of the gymnosperm tree *Picea glauca* by cutting branches off and leaving them to stand on the winter snow for 2–3 days before eating. During this time, phenolic levels drop from 2.8% to 1.5% dry weight. Additionally, the terpenoid content (see the next section) may also affect feeding and, in the case of this meadow vole, it completely avoids feeding on leaves of *Pinus strobus* irrespective of the phenolic content because of the high content of myrcene and bornyl acetate.[21]

Table 1 Examples of simple phenolics that have been implicated as defensive agents against herbivores.

Phenolic[a]	*Occurrence*	*Effect on herbivore*
o-Pentadecenylsalicylic acid	Leaf trichomes of *Pelargonium* × *hortorum*	Toxic to two-spotted spider mite *Tetranychus urticae*
Magnolol (**1**)	Leaf of *Magnolia virginiana*	Toxic to larvae of the moth *Callosomia promethea*
Miconidin (**2**) and primin (**3**)	Leaf trichomes of *Primula obconica*	Antifeedant to larvae of *Heliothis armigera*
Salicortin and tremulacin (**4**)	Leaves of *Populus* spp.	Toxic to large willow beetle *Phratora vulgatissima*
Salicylaldehyde	Leaf of *Populus balsamifera*	Antifeedant to snowshoe hare *Lepus americanus*
Pinosylvin (**5**) and its methyl ether	Buds of *Alnus crispa*	Antifeedant to snowshoe hare *Lepus americanus*
Platyphylloside	Buds and internodes of *Betula platyphylla*	Feeding inhibitor to mountain hare, moose and goat
Coniferyl benzoate (**6**)	Flower buds and catkins of *Populus tremuloides*	Feeding deterrent to ruffed grouse *Bonasa umbellus*

[a] For references, see text and references 9–12

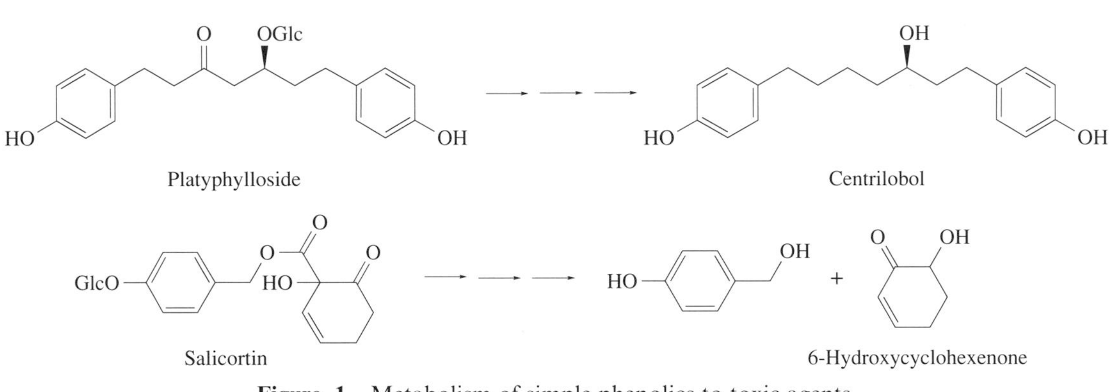

Figure 1 Metabolism of simple phenolics to toxic agents.

Insects can overcome phenolic barriers to their feeding by tolerating or otherwise metabolizing a particular toxin. This happens, for example, with the silkworm *Callosamia securifera*, which is monophagous on the magnolia tree, *M. virginiana*. Two related generalist silkworms, *C. angustifera* and *C. promethea*, do not survive, because of the toxic lignans present, magnolol (**1**) and a related biphenyl ether. When an acceptable host leaf is painted with magnolol at the same concentration as in the magnolia leaf, the specialist *C. securifera* survives but the two generalists lose out.[22]

Some phenylpropanoids, caffeic acid esters, and related structures are of widespread occurrence in the plant kingdom (Figure 2). They are less likely to be useful defensive agents in that many insects may be adapted to them and hence would be expected to tolerate their dietary presence. Nevertheless, feeding inhibition has been observed for caffeic acid, the 3′-methyl ether ferulic acid, and various derivatives. For example, ferulic acid is released from a bound form in maize seed and is antifeedant at a concentration of 0.05 mg g^{-1} to the maize weevil, *Sitophilus zeasmais*. Similarly, chlorogenic acid (**7**) is a feeding deterrent to the leaf beetle, *Lochmaea capreae cribrata*, feeding on the Salicaceae.[23] Chlorogenic acid also occurs in leaf trichomes of tomato and reduces growth of early instars of the cotton bollworm, *Helicoverpa zea*. Again, the chlorogenic acid analogue, 1-caffeoyl-4-deoxyquinic acid, present in leaves of the wild groundnut *Arachis paraguaensis*, inhibits growth of the tobacco armyworm *Spodoptera litura*.[24]

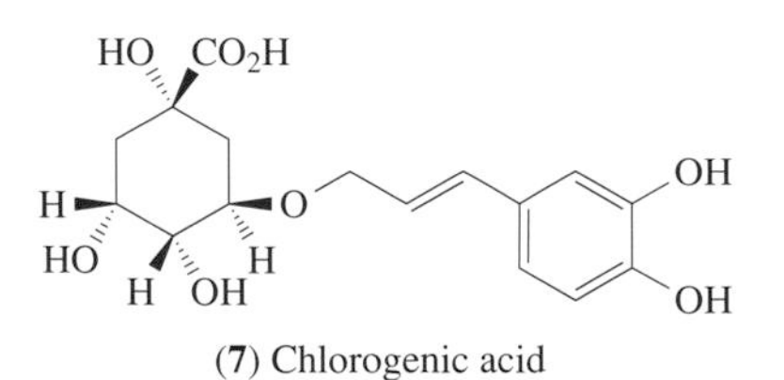

(**7**) Chlorogenic acid

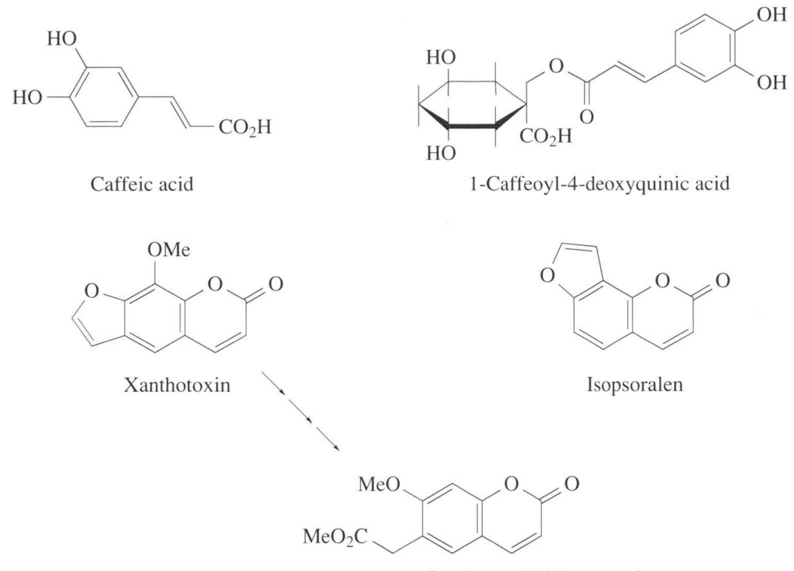

Figure 2 Phenylpropanoids as feeding inhibitors to insects.

Dietary caffeic acid may not necessarily be harmful and indeed may on occasion be a beneficial component. This appears to be true of the tree locust *Anacridium melanorhoda*. This phenolic is retained from the diet by the insect. It is bound to the cuticle, where it stabilizes the protein. It is used as an alternative to dietary tyrosine and has a sparing effect on nitrogen uptake.[25]

One group of masked phenolics, the furanocoumarins, have been implicated as defensive agents in the Umbelliferae, a family where they occur regularly. They are notable in being phototoxic, namely that their toxicity to animal life is enhanced in the presence of sunlight. Furanocoumarins usually occur in plants as mixtures of related structures. Berenbaum *et al.*[26] have demonstrated that such mixtures act synergistically in the interaction between *Helicoverpa zea* and the fruits of the parsnip *Pastinaca sativa*. They are more protective against this herbivore than when a single-structure xanthotoxin is applied at the same concentration as the furanocoumarin mixture.

The toxicity of xanthotoxin to insects is related to the relative rate of detoxification. The black swallowtail *Papilio polyxenes* tolerates it, because it can metabolize 95% of a dietary dose within 1.5 h to an open-chain compound. By contrast, the armyworm *Spodoptera frugiperda* is sensitive to its toxic effects because it detoxifies it much more slowly.[27] Xanthotoxin is a linear coumarin, whereas isopsoralen is an angular coumarin and even umbellifer specialist insects find it difficult to metabolize these structural analogues. These furanocoumarins are also toxic to mammals. The rock hyrax *Procaria capensis syriaca* dies within 20 h if fed with shoots of *Pituranthos triradiata*, which contain between 0.6% and 1.7% dry weight of furanocoumarin.

Most flavonoids are water soluble, occurring in the vacuoles of leaves and flowers. Their defensive role seems to be limited, since for most phytophagous insects they are regular dietary components. Indeed, larvae of ~10% butterfly species sequester and store them in their tissues.[28] Nevertheless, there are some Lepidopteran species sensitive to dietary flavonoids, notably *Helicoverpa zea* and *H. virescens*. Typically, cyanidin 3-glucoside (**8**), a common anthocyanin, added to the diet at a 0.07% concentration causes 50% inhibition of larval growth during 5 days of feeding. The reason for this is not clear, but it may be related to the inability of the insect to absorb its nutritional requirements from such a diet.[28]

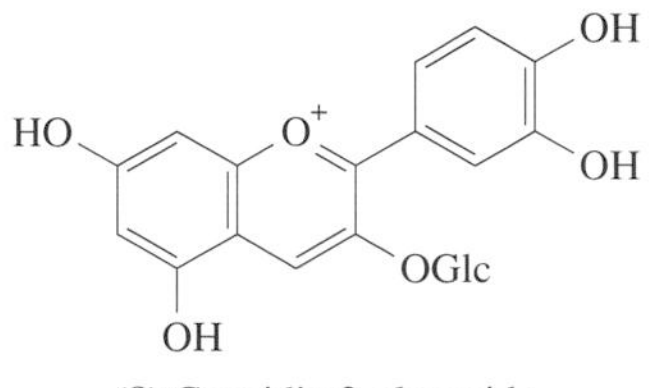

(**8**) Cyanidin 3-glucoside

There are additionally lipophilic flavonoids in plants with a more restricted distribution and here there is evidence of toxicity and feeding deterrence. The rotenoids which occur in the roots of legumes such as *Derris elliptica*, are well known to be insecticidal and toxic to fish. Again, the prenylated flavanones present in *Lonchocarpus* seed are toxic to the predating mouse, *Liomys salvini*. In captivity, this mouse prefers to starve rather than feed on the seeds of this plant.[29]

Much attention has been devoted to the plant tannins as feeding barriers, because of the very widespread occurrence of one class, the condensed tannins (or proanthocyanidins), in most woody plants. That they are biologically active is evident from their well-known ability to bind to proteins. There have been considerable difficulties in measuring their quantities in leaf and other plant tissue, but these have been overcome and several methods of accurate determination are now available.[30] The majority of experiments carried out since the late 1980s confirm the view that the condensed tannins at least are significant and powerful feeding barriers to both phytophagous insects and grazing animals. Some of the more important supporting evidence follows.[31]

(i) The toxic effects of condensed tannins in unadapted animals are well established. Typically, weaning hamsters that are treated with diet containing 4% dry weight of sorghum tannin suffer weight loss and then perish within 3–21 days.

(ii) Coevolutionary adaptation in mammals to high tannin diets is well established. This involves the increased synthesis of a series of unique proline-rich proteins in the parotid glands. These salivary proteins have a high affinity for condensed tannins and remove them by binding them at an early stage in the digestive process.

(iii) Adapted animals may still avoid feeding on plants, when certain types of tannin are present. Thus, snowshoe hares in Alaska show a threefold preference for leaves of *Purshia tridentata* over those of *Coleogne ramossissima*. This difference in feeding behavior is due to chemical variations in the procyanidins present. In *Coleogne*, the polymers are based on epicatechin units, whereas in *Purshia*, they are based on both catechin and epicatechin units in a 1:1 ratio.

(iv) In moths and butterflies, dietary tannins lower the growth rate, although it is only plants with more than average tannin concentrations which deter feeding. Experiments with *Aphis craccivora*, a pest of the groundnut *Arachis hypogaea*, show that it is deterred from feeding when the procyanidin content in the phloem of the petiole reaches more than 0.3% fresh weight. Aphids forced to feed on cultivars with a high tannin content show a twofold decline in reproductive rates. Interestingly, tannin production in the groundnut is channeled towards the phloem to provide aphid resistance, since other parts of the plant are essentially tannin free.[31]

8.03.1.3 Toxic Terpenoids

Terpenoids, from volatile monoterpenoids to involatile triterpenoids, are broadly defensive against herbivory on plants (Figure 3). Not only have individual compounds been implicated as being toxic or antifeedant (Table 2), but also mixtures of related structures often synergize to produce a deterrent or toxic effect. Most groups of herbivores can be inhibited from feeding, including insects, molluscs, birds, and especially geese, and many browsing and grazing animals. The defensive role of terpenoids has been reviewed in detail elsewhere,[32] only the salient points with reference to insect and mammalian feeding will be mentioned here.

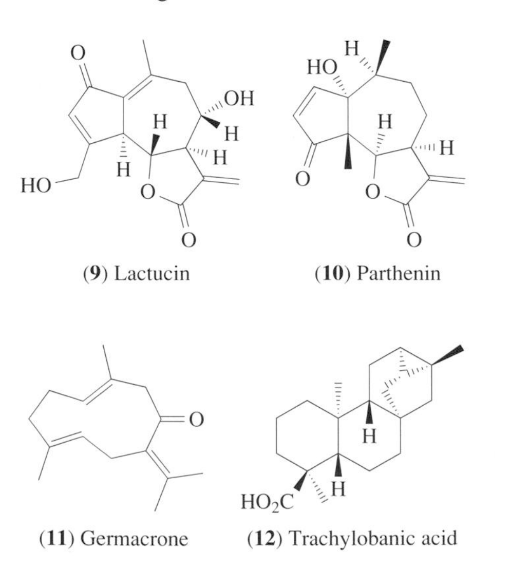

(**9**) Lactucin (**10**) Parthenin

(**11**) Germacrone (**12**) Trachylobanic acid

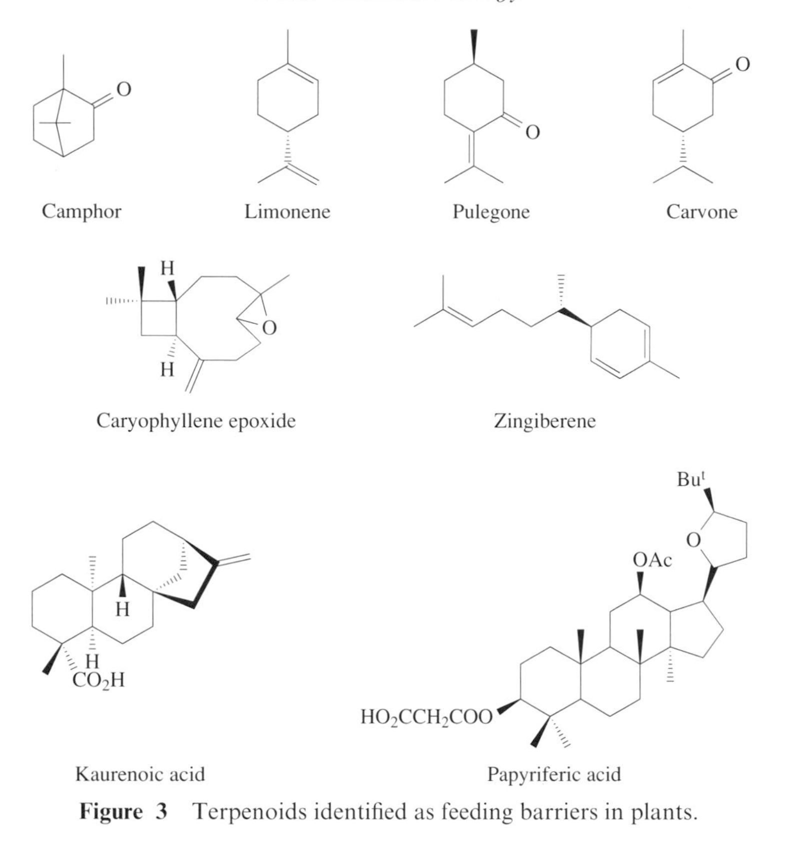

Figure 3 Terpenoids identified as feeding barriers in plants.

Table 2 Terpenoids identified as barriers to herbivore feeding.

Terpenoid[a]	*Occurence*	*Effect on herbivore*
Camphor	White spruce leaf	Antifeedant to snowshoe hare
Limonene	Bark of *Pinus ponderosa*	Feeding deterrent to pine bark beetle, *Dendroctonus*
Pulegone and carvone	*Satureja douglasii* leaf	Feeding deterrent to slug, *Ariolimus dolichophallus*
Lactucin (**9**) and 8-deoxylactucin	*Cichorium intybus* leaf	Antifeedant to locust, *Schistocerca gregaria*
Caryophyllene epoxide	*Melampodium divaricatum* leaf	Arrests leaf-cutting ant, *Atta cephalotes*
Zingiberene	Leaf trichome of *Lycopersicon hirsutum*	Toxic to the Colorado beetle
Parthenin (**10**)	*Parthenium hysterophorus* leaf	Toxic to flour beetle *Tribolium confusum*
Germacrone (**11**)	*Ledum groenlandicum* leaf	Grazing deterrent to snowshoe hare
Kaurenoic and trachylobanoic acids (**12**)	Floret of sunflower *Helianthus annuus*	Toxic to larvae of moth *Homeosoma electellum*
Papyriferic acid	Paper birch, *Betula resinifolia*	Antifeedant to snowshoe hare

[a] Listed as monoterpenoids, sesquiterpenoids, diterpenoids, and triterpenoids, respectively. For references, see text and references 9–12.

Monoterpenoids are generally toxic to unadapted insects. This has been demonstrated with locusts, which reject a range of monoterpenes which have been tested by applying them to artificial diets at 0.01% dry weight. Adapted insects may use monoterpene mixtures as feeding cues, but they can become susceptible to high concentrations or to non-host plant compounds. The pine bark beetle, *Dendroctonus brevicomis*, for example, is adapted to pine trees high in α- and β-pinene, myrcene, and 3-carene, but avoids feeding on trees which are high in limonene. What is true of monoterpenoids is true of the higher terpenoids (see Table 2). Sesquiterpene lactones, in particular, appear to discourage insects feeding on the plants where they are present. Several diterpenoids are antifeedants and the triterpenoid azadirachtin (**13**) is well known to be a potent insecticidal agent.

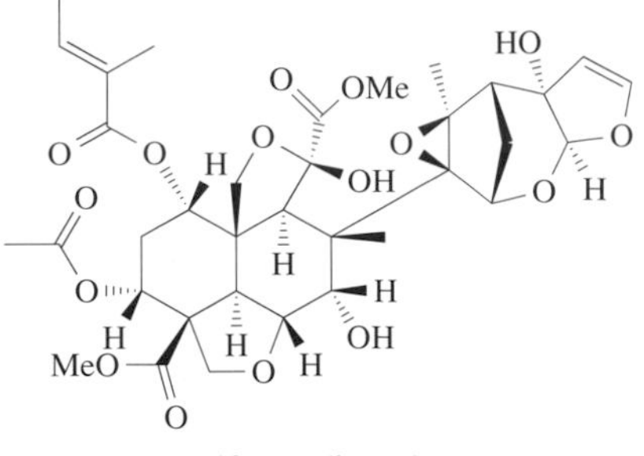

(**13**) Azadirachtin

Seasonal variation or distribution within the plant can determine the effectiveness of terpenoid toxins. For example, juvenile leaves of the holly, *Ilex opaca*, contain 135 mg g^{-1} dry weight of saponins, which arrests feeding by the Southern red mite *Oligonychus ilicis*. Mature leaves have much lower levels (30 mg g^{-1} dry weight) but these are then protected by physical structures. In the leek, *Allium porrum*, saponins are concentrated in the flowers, driving the leek moth larva *Acrolepiopsis assectella* to feed exclusively on the more expendable leaves. The relative amounts of saponin are 0.03% dry weight in leaves and 0.2–0.4% dry weight in the flowers.

The leaves of most gymnosperms and of many angiosperm trees and shrubs are rich in monoterpenoid and sesquiterpenoid mixtures and there is increasing evidence that they are defensive against many mammalian feeders, including deer, hares, and voles. The concentration is a major factor in defense and this may vary seasonally, as happens in the shrub *Chrysothamnus nauseasus*. Leaves are not eaten in the summer, when the sesquiterpenes such as (*E*)-β-farnesene (**14**), β-humulene (**15**), and (γ)-muurolene (**16**) reach a total concentration of 80 μg g^{-1} dry weight. This level drops to 18 μg g^{-1} dry weight in the winter, when the leaves are browsed by the mule deer, *Odocoileus nemionus*.[33]

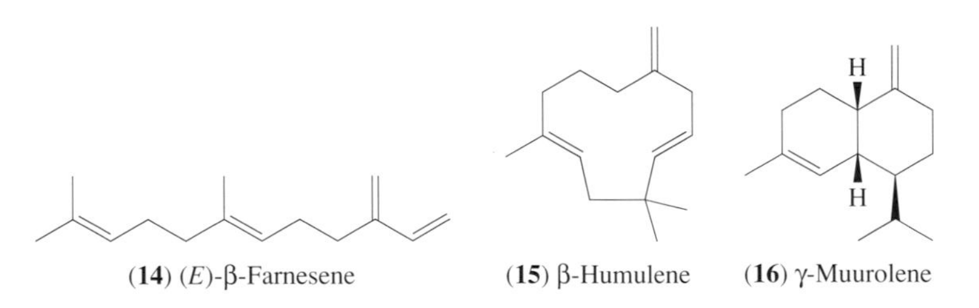

(**14**) (*E*)-β-Farnesene (**15**) β-Humulene (**16**) γ-Muurolene

Adaptation to a terpene-rich diet has been observed in possums and gliders feeding on *Eucalyptus* leaves. This occurs by efficient detoxification (as in ringtail possum *Pseudocheirus peregrinus*) or by avoiding the inhibitory effect of terpenes on the microorganisms of the rumen. Thus the greater glider *Petauroides volans* avoids the deleterious effect on the microbial population in their hindguts by absorbing the terpenes through the stomach and small intestine and detoxifying them via the liver.[34]

8.03.1.4 Nitrogen-based Toxins

Many plant alkaloids are both bitter tasting and acutely toxic and therefore appear to be obvious barriers to animal feeding. Yet ironically, the defensive role of alkaloids has not been explored to the same extent as that of other nitrogen-containing plant toxins. Thus, more attention has been devoted by ecologists to cyanogenic glycosides, glucosinolates, and non-protein amino acids. These compounds will be considered here first before passing on to the alkaloids.

Cyanogenic glycosides have been likened to a two-edged sword in that on enzymic breakdown they yield two different classes of toxin, namely cyanide, which is a respiratory inhibitor, and an aldehyde or ketone, which is directly toxic. Typically, the bound glucoside linamarin is broken down in a two-stage process to yield cyanide and acetone (Figure 4).

The defensive role of linamarin, lotaustralin, and other commonly occurring cyanogens has been reviewed[35–37] and they have been shown to be effective variously in deterring feeding by molluscs, Lepidoptera, deer, sheep, rabbits, and voles. Detoxification of cyanide is possible, either by the enzyme rhodanese, which converts it into thiocyanate, or by the enzyme β-cyanoalanine synthase

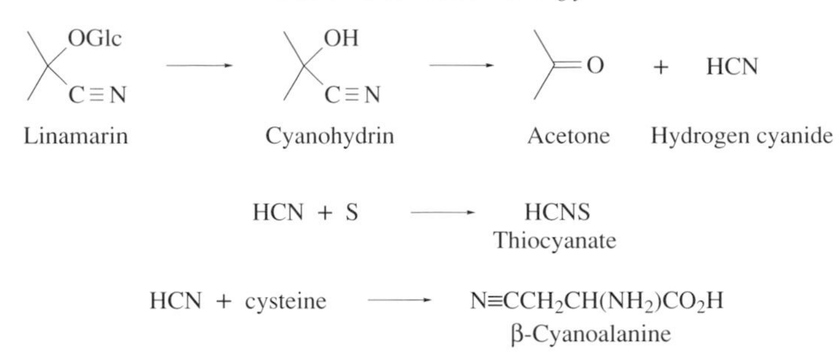

Figure 4 Release of cyanide and ketone from the cyanogenic glucoside linamarin and detoxification of cyanide to thiocyanate or β-cyanoalanine.

to produce β-cyanoalanine (Figure 4). Unfortunately, both of these detoxification products can have harmful effects: thiocyanate, a metabolite of cassava cyanogen in humans, has goitrogenic effects; β-cyanoalanine, a metabolite of cyanogen in insects, is a neurotoxin.

A special feature of cyanogenesis in many plants where it occurs is its variable nature. This is especially pronounced in both clover and birdsfoot trefoil, where populations can vary from having about 5% cyanogenic to those with 100% cyanogenic. Much argument has been spent over this feature, but most ecologists recognize the advantages to a plant of maintaining a variable toxic defense in keeping the herbivore "guessing" about the palatability of a particular plant and having to adapt via induced detoxification enzymes to sampling cyanogenic forms. One final point may be made about the protective value of cyanogenesis in young seedlings of clover. Measurements of cyanogenic content within the plant indicate larger amounts in the growing stem and cotyledons than in the leaves. Thus, garden slugs will restrict their feeding to the leaves, leaving the more vital organs to regenerate following grazing.[38]

Like cyanogenic glycosides, the glucosinolates or mustard oil glycosides are bound toxins which yield the free toxin, an isothiocyanate, following enzymic hydrolysis (Figure 5). Their distribution in nature is limited to $\sim$15 plant families, but in many cases, as in the Cruciferae, they occur universally and in considerable abundance. Overall, the evidence is strong that glucosinolates and their products function in plant defense against generalized consumers, including mammals, birds, aphids, grasshoppers, beetles, flies, and mites.[39] Even more telling is the fact that crucifer specialists are restricted in their ability to feed on their chosen food plants by the high concentration that are often present in young tissues. The toxicity of sinigrin, a common glucosinolate of crucifers (Figure 5), to unadapted insects has been demonstrated by feeding it to larvae of the black swallowtail butterfly *Papilio polyxenes* by infiltrating a 0.1% solution into a normal food plant, namely celery. This was sufficient to cause 100% mortality to the larvae.[40]

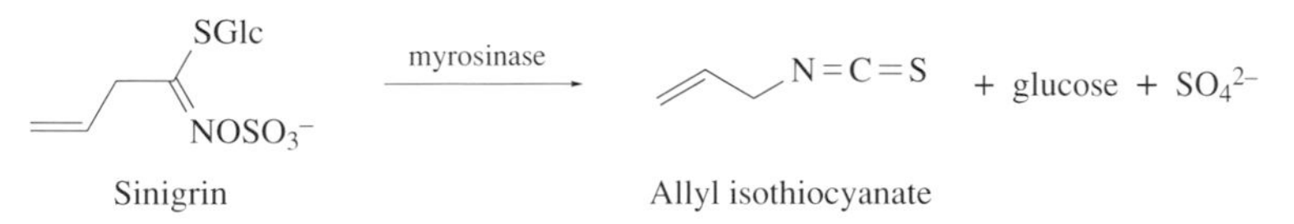

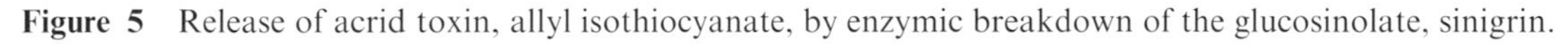
Figure 5 Release of acrid toxin, allyl isothiocyanate, by enzymic breakdown of the glucosinolate, sinigrin.

The ecological strategy in defense of the mustard plant *Sinapis alba* is to protect the vulnerable young tissues with *p*-hydroxybenzylglucosinolate (sinalbin) (**17**). The high concentrations in young cotyledons (20 mmol dm^{-3}) and young leaves (up to 10 mmol dm^{-3}) effectively deter feeding by both specialist insect, e.g., the flea beetle *Phyllotetra cruciferae* and a generalist insect such as an armyworm. As the plant grows, the concentration drops so that older leaves have between 2 mmol dm^{-3} and 3 mmol dm^{-3} sinalbin. At these levels, there will be some stimulation for the flea beetle to feed, but the more generalist insect will still be deterred from feeding.[41] A similar strategy operates in the wild crucifer *Schouwia pupurea*, which is fed upon by the locust *Schistocerca gregaria* so that the locust is forced to feed on the dried senescing rather than the fresh green leaves.[42]

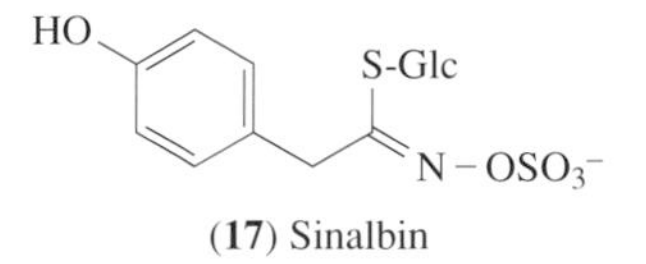

(**17**) Sinalbin

The next group of nitrogen-containing toxins to be considered are the nonprotein amino acids. They accumulate especially in the seeds of the Leguminosae but are also widely distributed in other plant families. These compounds are largely structural analogues of one or other of the protein amino acids and are antimetabolites in their mode of action. They are liable to be incorporated into protein synthesis, when imbibed dietetically, and this usually has disastrous consequences. A number of non-protein amino acids of legume seeds are notably neurotoxic (see Table 3) and cause neurolathyrism in humans and domestic animals. Canavanine and L-Dopa, both of which are regularly found in some quantity in seeds of tropical legume trees, have an ecological role in protecting these seeds from bruchid beetle attack. Two legume tree species may be growing adjacent to each other; the seeds of one, protected by L-Dopa at a concentration of 6–9% dry weight, will be free of bruchid infestation, whereas the seeds of the second species lacking a protective chemical are riddled with bruchid borings.

Table 3 Effects of nonprotein amino acids on herbivores.

Nonprotein amino acid	*Plant source*	*Effect on animals*
Albizziine (**18**)	Seeds of *Albizia* spp.	Insecticidal
α-Amino-β-oxalylaminopropionic acid	Seeds of *Lathyrus* spp.	Produces neurolathyrism in humans, horses, and cattle
Azetidine-2-carboxylic acid	Rhizome and leaf of *Polygonatum multiflorum*	Larvicidal
Canavanine	Seeds of *Canavalia ensiformis*	Toxic to bruchid beetles
β-Cyanoalanine	Seeds of *Vicia* spp.	Neurotoxic
α,γ-Diaminobutyric acid	Seeds of *Acacia* spp.	Toxic to mammals
Dopa	Seeds of *Mucuna* spp.	Toxic to bruchid beetles
Hypoglycin (**19**)	Fruits of *Blighia sapida*	Toxic and hypoglycemic in humans
Indospicine (**20**)	Leaf of *Indigofera* spp.	Hepatotoxic in mammals
Se-methylselenocysteine	Seeds of *Astragalus* spp.	Causes "blind staggers" in grazing animals
Mimosine (**21**)	Seeds and leaf of *Mimosa pudica*	Causes hair loss and is goitrogenic in calves

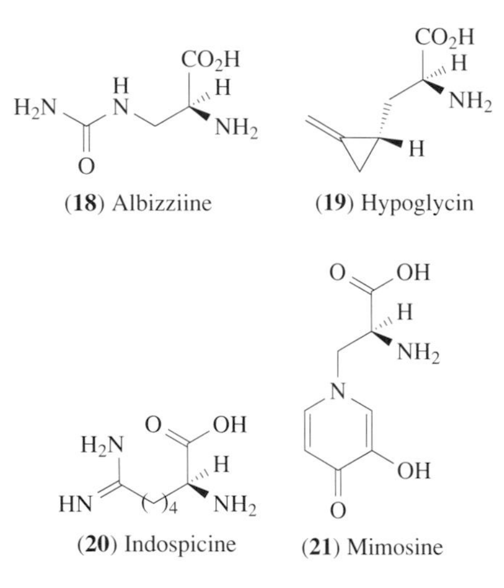

(**18**) Albizziine (**19**) Hypoglycin

(**20**) Indospicine (**21**) Mimosine

While the presence of nonprotein amino acids may provide a general defense against insect predation on seeds, some individual species may coevolve to overcome the barrier and survive the harmful effects of the toxin. This happens with larvae of the bruchid beetle *Caryedes brasiliensis*, which in Costa Rica feeds exclusively on seeds of *Dioclea megacarpa* that contain large amounts of canavanine. The beetle is able to detoxify the canavanine with the enzyme arginase, which produces canaline and urea (Figure 6). The canaline is further broken down, providing more useful nitrogen, together with the urea, for the further growth of the larva. In spite of this success, it is still true that canavanine and other related structures (see Table 3) are highly damaging dietary constituents to the majority of insects. Rosenthal[43] has shown, for example, that feeding larvae of the tobacco hornworm, *Manduca sexta*, with an agar-based diet containing 0.05% canavanine, creates dramatic growth aberrations in the pupae and adults and renders them infertile.

Considering that over 10 000 plant alkaloids are known and that these may be present in over 20% of angiosperm families, it is remarkable how little is known of their ecological significance.

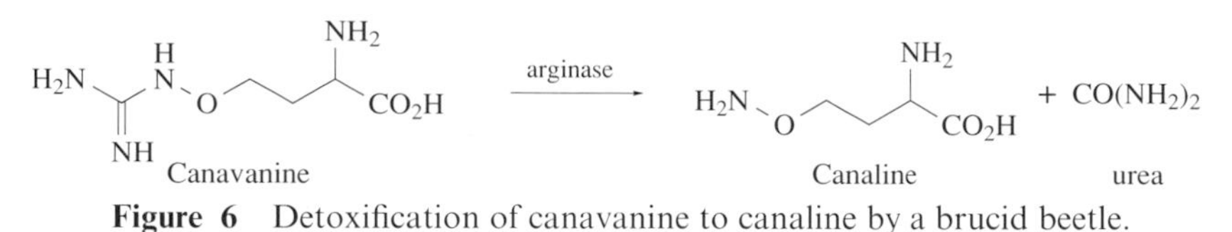

Figure 6 Detoxification of canavanine to canaline by a brucid beetle.

Most attention has been devoted to the accidental poisoning of domestic animals, especially cows, which graze on alkaloid-containing plants in the Leguminosae (e.g., *Lupinus* spp.) and in the Compositae (e.g., *Senecio* spp.). Wild animals such as deer and rabbits avoid feeding on them in general. The toxicity of the pyrrolizidine alkaloids (PAs) of ragwort *Senecio jacobeae* and other *Senecio* plants is notorious and ~50% of all domestic cattle deaths worldwide are due to poisoning by these alkaloids. They are dangerous to life because they are metabolized *in vivo* to produce a more toxic agent which has thc ability to bind to macromolecules such as DNA in the liver. Typically, a *Senecio* alkaloid such as senecionine undergoes hydrolysis to retronecine and this undergoes dehydrogenation to the related pyrrole. This, and a further breakdown product, (*E*)-4-hydroxyhex-2-enal, are responsible for liver damage (Figure 7). Hence, PAs are known to be hepatotoxic.[44]

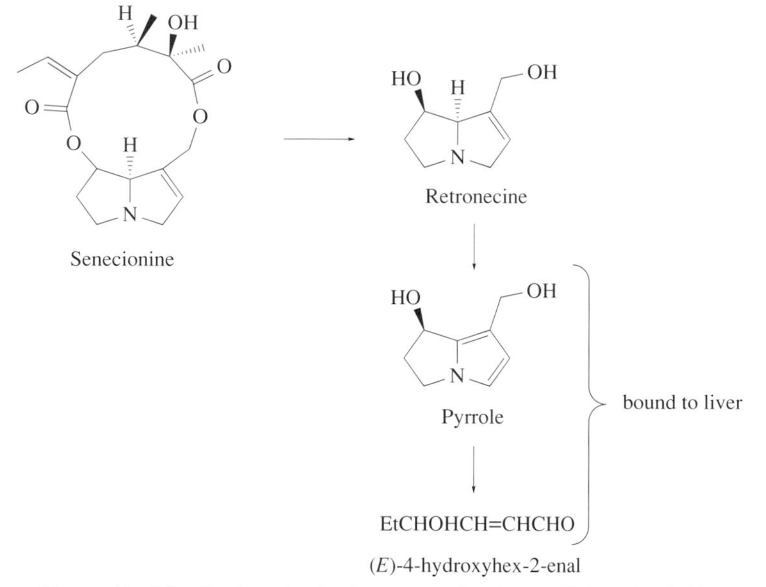

Figure 7 Metabolism *in vivo* in mammals of pyrrolizine alkaloids.

Another group of toxic alkaloids are the quinolizidine alkaloids (QAs), e.g., anagyrine (**22**) and cytisine (**23**), of lupin plants. They are directly toxic to sheep and have also been implicated as teratogenic agents. Their protective role in plants has been demonstrated by offering rabbits and hares the choice of feeding on low-QA "sweet" lupins or high-QA "bitter" lupins, when the latter are largely avoided in favor of the sweet-tasting plants.[45]

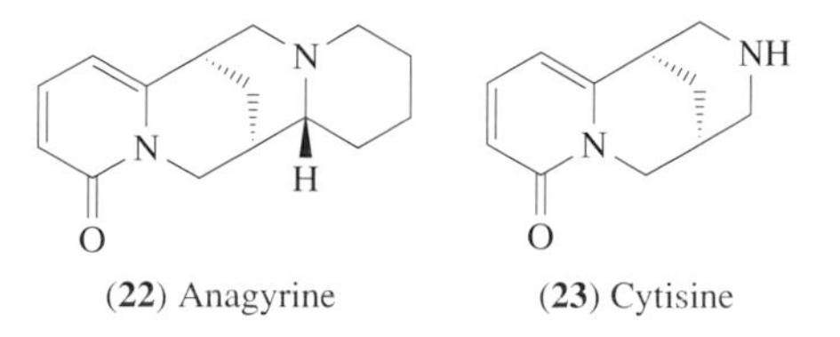

There is increasing evidence that in certain plants, alkaloids are formed in high concentration to protect juvenile tissues and that these concentrations drop dramatically as the tissues mature and physical structure takes over to defend from herbivore feeding. The pattern of accumulation of caffeine (**24**) and theobromine (**25**) in the coffee plant *Coffea arabica* is closely correlated with such a defense strategy. During leaf development, the alkaloid content reaches 4% dry weight, whereas in the maturing leaf, the rate of biosynthesis decreases exponentially from 17 mg d^{-1} to

0.016 mg d^{-1} per gram of leaf. Soft young fruit (bean) tissue is similarly protected, and there is a sharp drop in alkaloid content as the coffee bean ripens.[9]

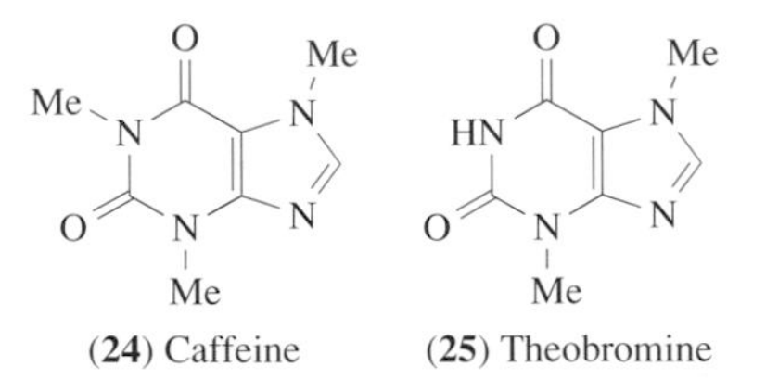

(**24**) Caffeine (**25**) Theobromine

A similar study of ergot alkaloid synthesis in *Ipomoea parasitica* showed a high concentration in young seedlings and also during flowering, but low amounts of alkaloid in between. Feeding experiments with the larvae of the moth *Heliothis virescens* showed that these ergot alkaloids, such as lysergol (**26**), were deterrent to feeding and also reduced fertility. Other classes of alkaloid, e.g., harman, lupin, and indole, have been shown to be antifeedant or harmful to phytophagous insects, so that there is circumstantial evidence for a useful defense role for these organic bases.[11] Other herbivores such as slugs seem to be able to consume alkaloid-containing plants with impunity, so that the alkaloids still appear to be defensive to only a limited number of potential herbivores.

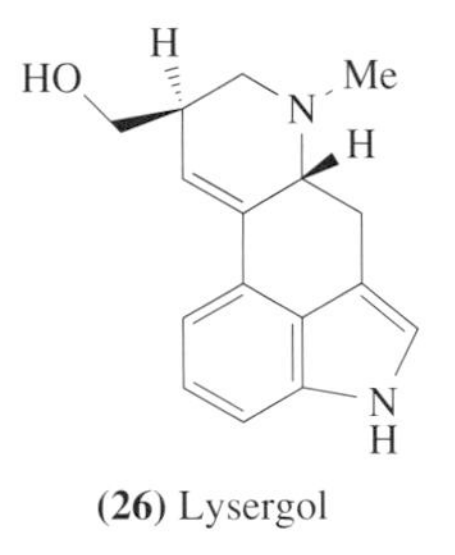

(**26**) Lysergol

8.03.1.5 Other Aspects of Chemical Defense

Many other plant compounds, which do not fall into the three main categories listed above, are also potentially defensive against herbivory. Fluoroacetate and oxalate ions must at least be mentioned because of their considerable toxicity. Fluoroacetate, $CH_2FCO_2^-$, occurs in *Dichapetalum cymosum* (Dichapetalaceae) in species of *Gastrolobium* and *Oxylobium* (Leguminosae), and in a variety of other sources. It is highly toxic to mammals, since it is incorporated into the Krebs cycle and then blocks it at the fluorocitrate stage. Cattle poisoning is well known in South Africa and Australia after animals have grazed on these plants. Interestingly, some native fauna in Australia have co-evolved with these poisonous plants and can feed on them without harm. This is true of the grey kangaroo, *Macropus fuliginosus*, although it is still not entirely clear how it is able to cope with the fluoroacetate poison.[46] The potential threat of the organic acid anion oxalate depends on which cation it is associated with in the plant. Thus plants with calcium oxalate, which is insoluble, are relatively safe to eat, whereas plants with the soluble potassium oxalate (e.g., *Setaria sphacelata*) may be toxic.[47]

If secondary compounds do have a protective function against herbivory, they are most likely to be located where they are most readily perceived by animals, namely at the leaf surface. Hence, a first line of defense, especially against insect feeders, are secondary metabolites localized in glandular hairs or trichomes principally on the upper surface. Some examples of phenolics and terpenoids that are so located have been given in earlier sections. Further examples are quoted in Juniper and Southwood.[48] A second line of defense in some plants is provided by the leaf wax, which itself may be a barrier to feeding. Additionally $\sim$50% of angiosperm species contain "extra" lipophilic secondary constituents mixed in the wax. It is likely that many of these constituents are repellent to insects. These is certainly evidence in the case of some varieties of *Sorghum* that the leaf wax alkanes themselves are distasteful to *Locusta migratoria*.[49]

A third line of defense in plants from insect grazing is latex production. Latex has been reported in over 12 000 plant species and one of its main functions appears to be to protect those plants which contain if from herbivory. The effectiveness of latex, a viscous liquid consisting of a suspension of rubber particles, as a feeding deterrent is often reinforced by the presence of terpenoid toxins (e.g. sesquiterpenes or diterpenes) or of alkaloids. Experimental evidence for a defensive role has been mainly confined to studies with ants. However, Dussourd and Eisner[50] have established that many mandibular insects, in order to feed on latex-bearing plants, have to overcome the latex defense by vein-cutting behavior. Thus, larvae of the Monarch butterfly, *Danaus plexippus*, feeding on milkweed plants cut the leaf veins before feeding distal to the cuts. Vein cutting blocks the flow of latex to the feeding sites and represents a counter-adaptation by the insect to the plant's defense. Unadapted insects such as the armyworm, *Spodoptera eridania*, have not learned this behavior and are repelled from feeding by latex droplets.

Other barriers to both insect feeding and mammalian grazing involve the physical make-up of plant tissues and especially the extent of lignified cell walls in the leaves. Monkeys, for example, tend to concentrate their feeding on young flush leaves of trees to avoid the toughness and rigidity of the mature tissues.

Finally, it is important to point out that the effectiveness of secondary metabolites as defense agents may be strengthened by the presence of inorganic compounds. Many grasses contain crystalline occlusions, known as raphides, composed of calcium carbonate, and these probably restrict mammalian grazing on such plants. Calcium chloride may also contribute to plant resistance. Harada *et al.*[51] reported that calcium chloride interacts with the leaf diacylglycerols in providing the resistance in *Nicotiana benthamiana* to aphid feeding. Resistant forms have 10–100 times more calcium chloride in the leaf than susceptible cultivars.

Some plants, growing on particular soils, have the ability to accumulate toxic metal ions in their tissues, usually by chelating them with either organic acids (citrate, oxalate) or with small peptides, called phytochelatins. Such plants are therefore toxic to grazing animals and hence will be protected from herbivory. This has been demonstrated in the case of the nickel accumulator *Thlaspi montanum*, which may contain up to 3000 ppm nickel in its tissues. Several Lepidopteran and grasshopper larvae, when fed on these leaves or on an artificial diet containing nickel, showed acute toxicity to nickel at 1000 ppm.[52]

8.03.2 INDUCED CHEMICAL DEFENSE

8.03.2.1 Proteinase Inhibitor Synthesis

One way in which a higher plant might reduce the "metabolic costs" of synthesizing and storing toxins is to produce the defensive agent only when it is actually needed, i.e., in direct response to insect feeding. There is evidence from the work of Ryan[53] that some plants may sometimes be able to respond rapidly to insect attack by the production of specific proteins which are proteinase inhibitors and are able ultimately to deter further feeding. Thus, a Colorado beetle feeding on potato or tomato leaves can cause the rapid accumulation of proteinase inhibitors, even in parts of the plant distant from the site of attack. The process is mediated by a proteinase inhibitor inducing factor known as PIIF, which is released into the vascular system. Within 48 h of leaf damage, the leaves may contain up to 2% of the soluble protein as a mixture of two proteinase inhibitors. Subsequently, the presence of the proteinase inhibitors in the leaf is detected by the beetle, which avoids further feeding and moves on to another plant (see Figure 8).

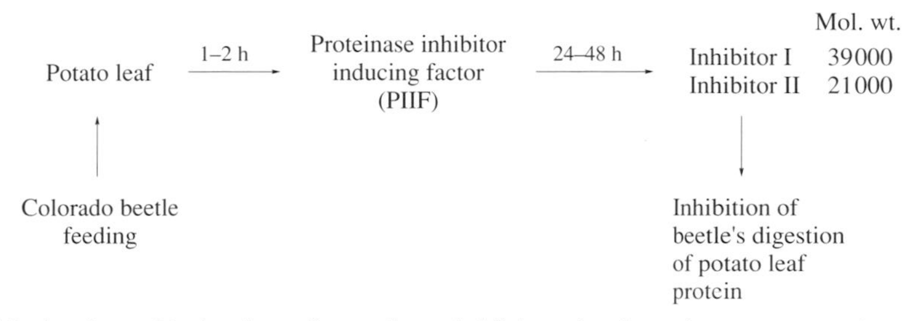

Figure 8 Mechanism of induction of proteinase inhibitors in plants in response to insect herbivory.

In theory, the inhibitors, if taken in the diet, will have an adverse effect on the insect's ability to digest and utilize the plant protein, since they inhibit the protein-hydrolyzing enzymes trypsin and chymotrypsin. In fact proteinase inhibitors are well known as constitutive constituents of many plant seeds, where they have a similar protective role in deterring insect feeding.

PIIF induction is also brought about by mechanical wounding of plant tissue so that it is not yet entirely clear how specific this effect is to herbivore grazing. The nature of chemical signal PIIF has been explored in tomato plants and it is a small peptide, called systemin. This peptide is 10 000 times more active than oligosaccharides, which also have the ability to trigger this defense system.[53] A volatile chemical, the fatty acid metabolite methyl jasmonate, may also be involved in the signaling process.

The PIIF-like activities have been detected in extracts of 37 plant species representing 20 families, so this mechanism may well be a general one. The general effectiveness of trypsin inhibitors in deterring herbivory has been elegantly demonstrated in unrelated genetic engineering experiments. A gene encoding a cowpea trypsin inhibitor was transferred to tobacco. As a result, transformed leaves were more resistant to budworm (*Heliothis virescens*) feeding than the original plants. Ecological experiments have also shown that PIIF induction in the tomato reduces the grazing by larvae of the armyworm, *Spodoptera littoralis*, within 48 h, with avoidance being most pronounced on the young leaves.[54]

While proteinase inhibitors are formed *de novo* as a result of PIIF induction, it is possible that there may be an increase in the synthesis of enzymic protein as a response to mechanical or insect-feeding leaf damage. In tomato plants, for example, increases in phenolase, peroxidase, and lipoxygenase have been observed (Table 4). The tomato plant, however, responds differently according to which insect is feeding. Thus, lepidopteran larvae induced the production of more phenolase and lipoxygenase, whereas leaf miners produced only an increase in peroxidase. Increasing the phenolase content of the leaf probably makes the leaf more unpalatable, since the phenolase will oxidize the internal phenolics to toxic oxidation products during the feeding process.[55]

Table 4 Examples of induced plant defense.

Chemicals induced (status)[a]	*Plant species*	*Inducing insect or mollusc*
Proteins		
Proteinase inhibitors (*de novo*)	*Lycopersicon esculentum*	*Helicoverpa zea*
Phenolases (increases)		
Peroxidases (increases)		
Lipoxygenases (increases)		
Secondary compounds		
Nicotine alkaloids (220%)	*Nicotiana sylvestris*	*Manduca sexta*
Tropane alkaloids (186%)	*Atropa acuminata*	*Arian ater*
Furanocoumarins (215%)	*Pastinaca sativa*	*Trichoplusia ni*
Indolic glucosinolates (+ve)	*Brassica napus*	*Psylliodes chrysocephala*
Aliphatic glucosinolates (−ve)		
Volatiles		
4,8-Dimethyl-1,3,7-nonatriene, and β-ocimene (**27**)	*Cucumis sativa*	Spider mite
4,8-Dimethyl-1,3,7-nonatriene, β-ocimene, linalool, and methyl salicylate	*Phaseons lunatus*	Spider mite
Linalool, guaiacol, and 3-octanone	*Glycine max*	*Pseudoplusia includens*
4,8-Dimethyl-1,3,7-nonatriene and linalool (**28**)	*Zea mays*	*Spodoptera exigua*

[a] For references, see references 9–12.

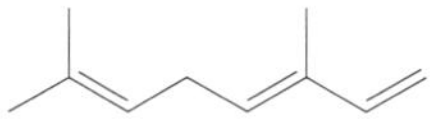

HO

(**27**) β-Ocimene 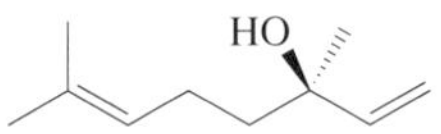 (**28**) Linalool

8.03.2.2 Phytochemical Induction

A related form of induced defense, apparently quite distinct from the PIIF system, has been observed in a variety of plants. The effect is relatively rapid and the leaves become unpalatable to animals within a matter of hours or a few days. It may be short term, disappearing after the insect has stopped feeding, or long term, extending over to the following season in trees. The chemical changes involve an increase in the concentration of existing toxins, sufficient to lead to herbivore avoidance. The effect is separate from a localized "wound response," which only takes place immediately around the site of damage, since it can often be discerned throughout the plant.

Such increases in toxin synthesis have been observed in two alkaloid-containing plants. One is the wild tobacco species *Nicotiana sylvestris*, which contains nicotine (**29**) and nornicotine (**30**) as the major alkaloids. Larval feeding induced a 220% increase in alkaloid content throughout the plant over a period of 5–10 days. Mechanical damage which avoids cutting the secondary veins produced a smaller response (170%). In fact, the tobacco hornworm *Manduca sexta*, when feeding on the tobacco leaf, avoids cutting through the secondary veins. It thus avoids triggering off the fullest response in the leaf, which can be as much as 400% of the control if the simulated damage includes damaging the vein. The nicotine alkaloids are synthesized in the roots and transported up into the leaf, and this was apparent in experiments in which pot-bound plants with confined roots failed to show any significant alkaloid increase after mechanical damage. Similar experiments with the tropane alkaloids in leaves of *Atropa acuminata* showed a maximum increase of 164% over the control 8 days after mechanical damage or slug feeding. Repeated mechanical damage at 11 day intervals increased the response to 186% of the control, but this effect fell off with time. Further experiments showed that only 9% of the leaf area needed to be removed mechanically or by animal feeding to produce the maximum response (Table 4).

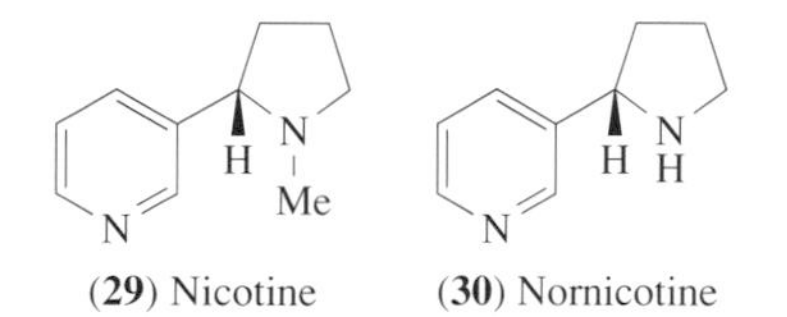

(**29**) Nicotine (**30**) Nornicotine

Another well-investigated example of induced chemical defense is the wild parsnip, *Pastinaca sativa*, which produces five furanocoumarins in the leaves. Artificial damage increased furanocoumarin synthesis to 162% of the control, while feeding by the generalist insect *Trichoplusia ni* increased it to 215%. Furthermore, larvae of *T. ni* grew very slowly on induced leaves, and larvae on artificial diet supplemented with furanocoumarins were similarly affected. The response of oil seed rape, *Brassica napus*, to insect infestation or leaf damage is quite distinctive and involves the massive accumulation of indole glucosinolates, which are barely detectable in the control. There is a corresponding reduction in the amounts of the aliphatic glucosinolates of the plant, so that the total titer of glucosinolate increases only slightly.

Other examples where induced changes in protective chemistry have been recorded were given by Tallamy and Raupp.[56] For every plant that shows a positive response, there is another plant where no detectable change in palatability occurs. Environmental factors also determine the magnitude of the response. Again, the response may disappear as the plant grows older and concomitantly more resistant to grazing. For example, two-year-old trees of *Pinus contorta* respond to defoliation by increasing the concentration of both terpenes and tannins in the needles, whereas 10-year-old trees fail to show any increases. The nature of the elicitor of this increase in secondary metabolism is still under investigation, but it is already apparent that treating plants with jasmonic acid (**31**) or methyl jasmonate can sometimes trigger off the increase in synthesis.[57]

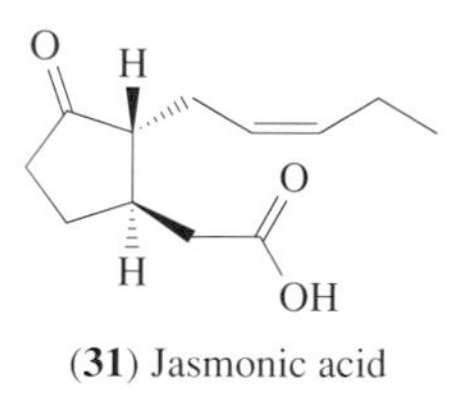

(**31**) Jasmonic acid

8.03.2.3 Plants Cry for Help via Volatiles

An even more interesting and remarkable plant–animal interaction involving induced chemical changes has been observed by Dicke *et al.*[58] In response to herbivory, some plants have developed the means to release volatile chemicals, which are particularly attractive to parasitoids of their herbivores, which then visit the plant and destroy the herbivores. As Dicke *et al.* put it, plants may "cry for help" when attacked by spider mites and predatory mites come to the rescue. Much research has been conducted on the spider mite *Tetranychus urticae*, the predatory mite *Phytoseiulus persimilis*, and the host plants. The chemicals released seem to be plant species specific. Cucumber plants infested by the spider mite release β-ocimene (**27**) and 4,8-dimethyl-1,3,7-nonatriene and are only moderately attractive to the predatory mites, whereas Lima beans release a cocktail of linalool (**28**), β-ocimene, the nonatriene, and methyl salicylate, which is highly attractive. A further advantage to the plant world is that the volatile released may alert uninfested neighboring plants so that they become better protected from spider mite attack. Thus cotton seedlings, when infected by these mites, release volatile cues which both attract predatory mites and also alert neighboring plants to withstand herbivore attack.

The systemic release of volatile chemicals which mediate in plant–herbivore–predator interactions has been observed in other plant systems. Corn (*Zea mays*) seedlings respond to beet armyworm (*Spodoptera exigua*) attack by releasing volatiles, which attract parasitic wasps, *Cotesia marginiventris*, to attack the herbivore. The response occurs throughout the plant and not only at the site of damage. The chemicals involved include linalool, which is released at the rate of 1 ng h^{-1} before damage and at the rate of 110 ng h^{-1} 6 h after armyworm attack (Table 4).

A similar tritrophic system exists in the case of the soya bean plant, the soya bean looper *Pseudoplusia ineludens*, and its parasitoid *Microplitis demolitor*. Here the volatiles again include linalool, but the more important attractants are guaiacol (**32**) and 3-octanone. These last two compounds do not appear to be released in appreciable amounts from the plant, but are released from the insect frass and are formed within the larvae from dietary sources. In other plants such as cotton and cowpea, the release of green leaf volatiles (e.g., (*E*)-2-hexenal and (*E*)-2-hexen-1-ol) appears to be sufficient to attract parasitic wasps to attack leaf-feeding caterpillars.

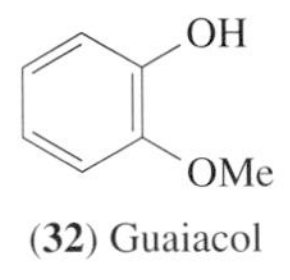

(**32**) Guaiacol

The cotton plant, when fed upon by the beet armyworm, produces 4,8-dimethyl-1,3,7-nonatriene, which is the most common new volatile formed in these experiments (Table 4). Biosynthetic experiments with $^{13}CO_2$ have shown that it is produced *de novo* on insect infestation and does not occur in the intact leaf in bound form.[59]

8.03.3 SEQUESTRATION OF PLANT TOXINS BY INSECTS

One of the most remarkable features of plant–animal interactions in the natural world is the ability of certain insects to sequester plant toxins from their food plants in the larval stage. They then move these toxins into the adult imago and both larva and adult thus gain protection from predation. The best known and most widely studied example is the Monarch butterfly, the larvae of which absorb cardiac glycosides from the food plant *Asclepias curassavica* so that the adult insect is protected from bird and mice predation. This phenomenon, which has been recognized for 19 classes of plant toxin (Table 5) does provide valuable evidence for the defensive role of these, particular chemicals (**33–49**) in the plants where they occur. Such compounds would hardly be "borrowed" in this way unless they were effective repellents to the range of animals which would predate upon these otherwise defenseless insects. There is also good evidence that insects selecting this life style have to adapt to the toxicity of these agents, often conjugating them *in vivo* and storing them safely in cuticular tissues. Furthermore, such insects regularly advertise their toxicity by adopting a warning coloration or being aposematic.[60]

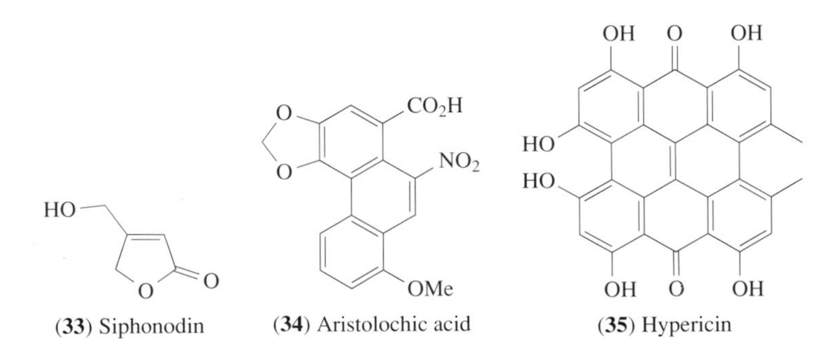

(**33**) Siphonodin (**34**) Aristolochic acid (**35**) Hypericin

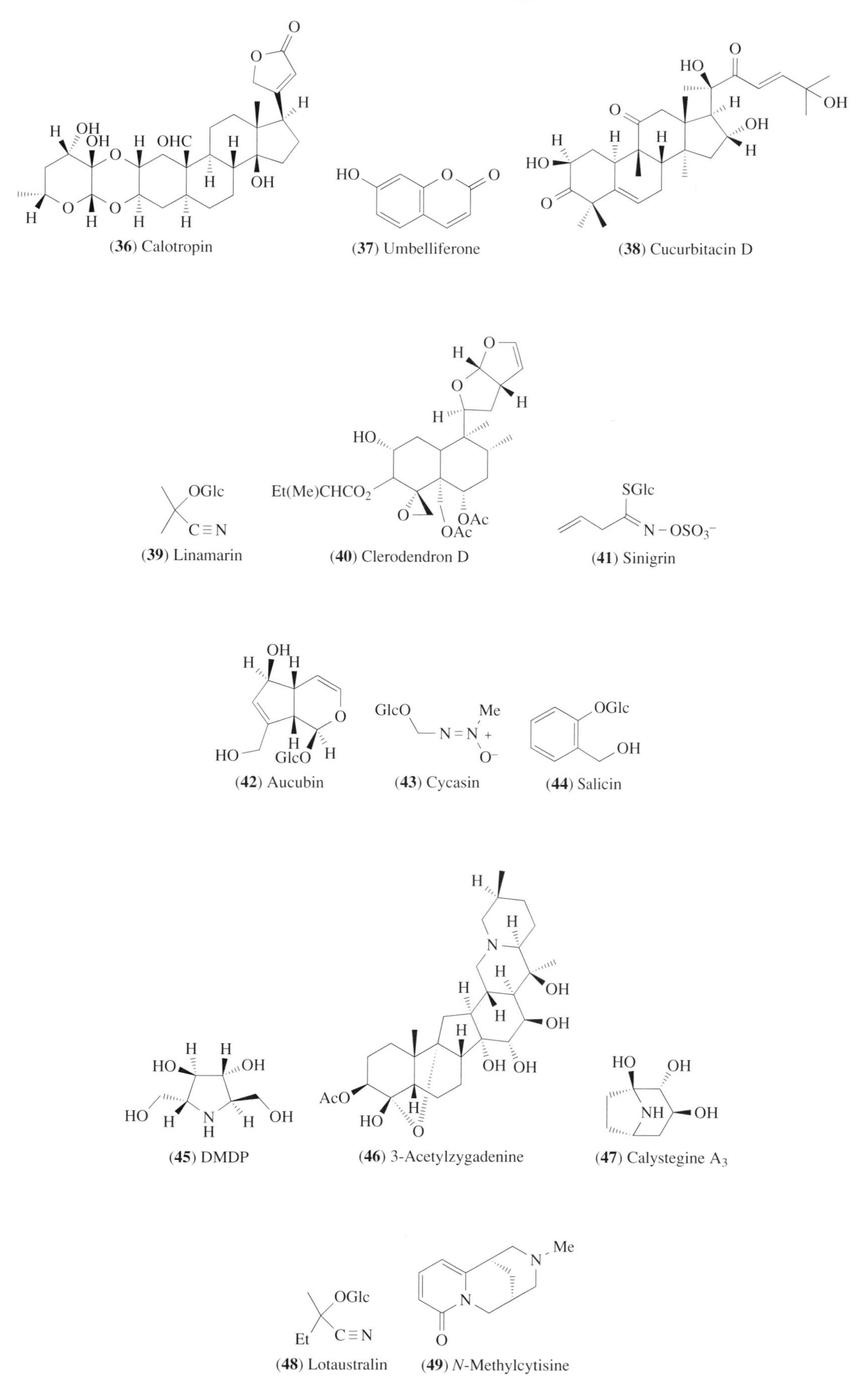

(36) Calotropin

(37) Umbelliferone

(38) Cucurbitacin D

(39) Linamarin

(40) Clerodendron D

(41) Sinigrin

(42) Aucubin

(43) Cycasin

(44) Salicin

(45) DMDP

(46) 3-Acetylzygadenine

(47) Calystegine A_3

(48) Lotaustralin

(49) *N*-Methylcytisine

Table 5 Classes of plant toxin sequestered by insects and stored for defense.

Class of chemical	*Typical structure*	*Plant source*	*Insect storing it*
Aliphatic lactones	Siphonidin (**33**)	*Euonymus europaeus*	Moth, *Yponomeuta cagnagellus*
Aristolochic acids	Aristolochic acid (**34**)	*Aristolochia* spp.	Butterfly, *Battusa archidamus*
Bianthraquinones	Hypericin (**35**)	*Hypericum hirsutum*	Beetle, *Chrysolina brunsvicensis*
Cardiac glycosides	Calotropin (**36**)	*Asclepias* spp.	Butterfly, *Danaus plexippus*
Coumarins	Umbelliferone (**37**)	*Prunus mahaleb*	Moth, *Yponomenta mahalebellus*
Cucurbitacins	Cucurbitacin D (**38**)	*Cucurbita* spp.	Beetle, *Diabrotica balteata*
Cyanogenic glycosides	Linamarin (**39**)	*Lotus corniculatus*	Moth, *Zygaena trifolii*
Diterpenoids	Clerodendron D (**40**)	*Clerodendrum trichotomum*	Sawfly
Fluoroacetate	Fluoroacetate	*Dichapetalum cymosum*	Moth, *Sindrisa albimaculata*
Glucosinolates	Sinigrin (**41**)	*Brassica oleracea*	Butterfly, *Pieris brassicae*
Iridoids	Aucubin (**42**)	*Plantago lanceolata*	Butterfly, *Euphydryas cynthia*
Methylazoxymethanols	Cycasin (**43**)	*Zamia floridana*	Butterfly, *Eumaeus atala*
Phenols	Salicin (**44**)	*Salix* spp.	Beetle, *Chrysomela aenicollis*
Polyhydroxy alkaloids	2,5-DM-3,4-DP (**45**)	*Omphalea* spp.	Moth, *Urania fulgens*
Pyrazines	3-Isopropyl-2-methoxypyrazine	*Asclepias curassavica*	Butterfly, *Danaus plexippus*
Pyrrolizidine alkaloids	Retronecine	*Senecio* spp.	Moth, *Arctia caja*
Quinolizidine alkaloids	Cytisine	*Cytisus scoparius*	Aphid, *Aphis cytisorum*
Steroidal alkaloids	3-Acetylzygadenine (**46**)	*Veratrum album*	Sawfly, *Rhadinocerea nodicornis*
Tropane alkaloids	Calystegine A_3 (**47**)	*Datura wrightii*	Hawkmoth, *Acherantia atropus*

More than one class of toxin may be utilized by a given insect. The Monarch butterfly, *Danaus plexippus*, obtains cardiac glycosides and pyrazines from the food plants (Table 5) and also borrows pyrrolizidine alkaloids in the adult stage from borage leaves or *Eupatorium* floral nectars. These alkaloids are incidentally both protective and also utilized as a source of male pheromone. The dietary pyrazines are present in the warning odor secretions of the adult, providing protection again from bird predation.[61]

The same toxins may occasionally be sequestered in the diet *and* synthesized *de novo* by the insect. This is true of the cyanogenic glycosides linamarin (**39**) and lotaustralin (**48**), which are synthesized by the burnet moth *Zygaena trifolii* from valine and isoleucine. They can also be readily acquired from the food plant, *Lotus corniculatus*. As a result of this duplication, the adult moth contains exceptionally high concentrations of cyanogenic glycoside.[62]

Some metabolism and/or conjugation of plant toxins commonly occur within the insect. This happens with both the cardiac glycosides and the pyrrolizidine alkaloids in the various Lepidoptera listed in Table 5. In the case of the cardiac glycosides of *Asclepias*, each species used as a food plant has a specific pattern of glycosides which is modified *in vivo* to a different mixture of constituents. By means of these patterns, as revealed by one-dimensional TLC, it is possible to recognize from the compounds found in the adult which food plant the larva fed upon.

The phenolic salicin (**44**), which is sequestered from the food plant *Salix* by a chrysomelid beetle (Table 5), is hydrolyzed *in vivo* and then oxidized to the corresponding aldehyde, salicylaldehyde. It is the latter compound which becomes the principal component of the defensive secretion in the beetle.

In general, insects only borrow some of the toxins that are potentially available to them in the food plant. The broom aphid, *Aphis cytisorum*, feeding on *Petteria ramentacea*, only takes in the alkaloid cytisine (**23**), rejecting at the same time anagyrine and *N*-methylcytisine (**49**). Such borrowed toxins may occasionally be passed up the food chain. For example, the ragwort aphid, *Aphis jacobaeae*, borrows pyrrolizidine alkaloids, as *N*-oxides, from the food plant. The ladybird, *Coccinella septempunctata*, feeding on these aphids, supplements its own alkaloid defensive secretion with the aphid alkaloids.[9–12]

While all the classes of chemical now reported to be sequestered by insects are defensive (Table 5), it is possible for these insects to sequester and store other classes of secondary metabolite. This applies to both the carotenoid and flavonoid pigments of plants. The purpose of such sequestration is still not entirely clear, and much remains to be known about the fate of dietary secondary metabolites in insects.

8.03.4 PLANT COMPOUNDS INVOLVED IN INSECT OVIPOSITION

8.03.4.1 Oviposition Stimulants

Plant chemicals play an important role, together with visual cues, in attracting phytophagous insects to their chosen host plants for both feeding and oviposition. In fact, in most cases, the association depends on the female adult butterfly or moth locating the right plant species, laying her eggs on the leaves, and for the eggs to hatch out and the larvae to feed. We know now in a reasonable number of examples that particular secondary constituents characteristic of these host plants (i.e., compounds **50–61**) are the major attractants to such oviposition (Table 6). What is unexpected is that the chemicals are largely involatile components on the surface or within the leaf and are not volatile in nature. Recognition therefore depends on direct contact with the chemistry of the leaf.

Table 6 Plant-derived oviposition stimulants of insects.

Oviposition stimulants	*Plant source*	*Female adult insect*
Methyleugenol (**50**), asarone (**51**), xanthotoxin (**52**), falcarindiol	*Daucus carota*	Carrot fly, *Psila rosa*
Luteolin 7-malonylglucoside and chlorogenic acid	*Daucus carota*	Black swallowtail, *Papilio polyxenes*
Vicenin-2 (**53**), narirutin (**54**), adenosine, and bufotenine (**55**)	*Citrus unshui*	Swallowtail, *Papilio xuthus*
Sterols, steryl ferulate	*Oryza sativa*	Rice grain weevil, *Sitophilus zeamais*
Sinigrin, glucobrassicin (**56**)	*Brassica oleracea*	Cabbage white, *Pieris brassicae*
Allyl isothiocyanate	*Brassica oleracea*	Diamond back moth, *Plutella maculipennis*
Aristolochic acids and sequoyitol (**57**)	*Aristolochia debilis*	Piperine swallowtail, *Atrophaneura alcinane*
Aucubin (**42**) and catalpol (**58**)	*Plantago lanceolata*	Buckeye butterfly, *Junonia coenia*
Rutin and other flavonol glycosides	*Ascelpias curassavica*	Monarch, *Danaus plexippus*
Cardenolides	*Asclepias humistrata*	Monarch, *Danaus plexippus*
Moracin C (**59**)	*Morus alba*	Mulberry pyralid
(+)-Isocotylocrebine (**60**), 7-demethyltylopharine (**61**)	*Tylophora tanakae*	Danaid butterfly, *Ideopsis similis*
2′-Acetylsalicin	*Salix pentandra*	Shoot gall sawfly, *Enura amarinae*

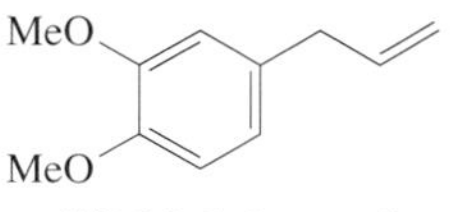

(**50**) Methyleugenol

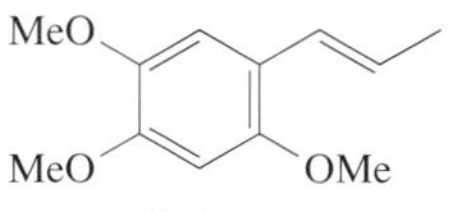

(**51**) Asarone

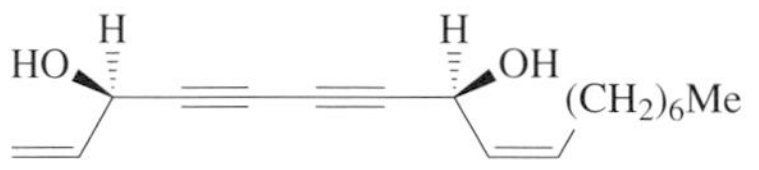

(**52**) Falcarindiol

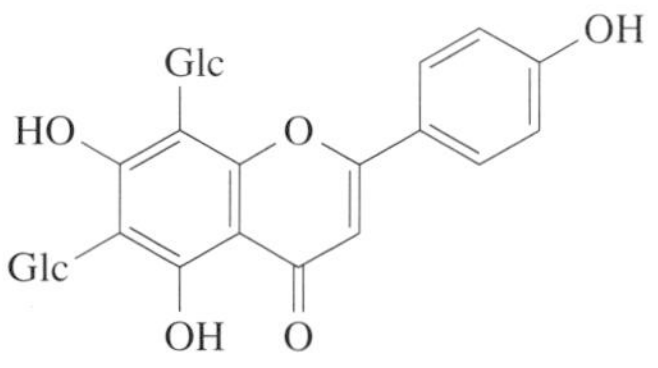

(**53**) Vicenin-2

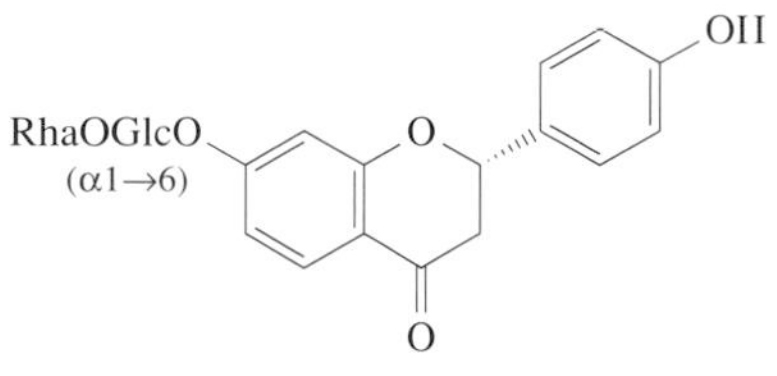

(**54**) Narirutin

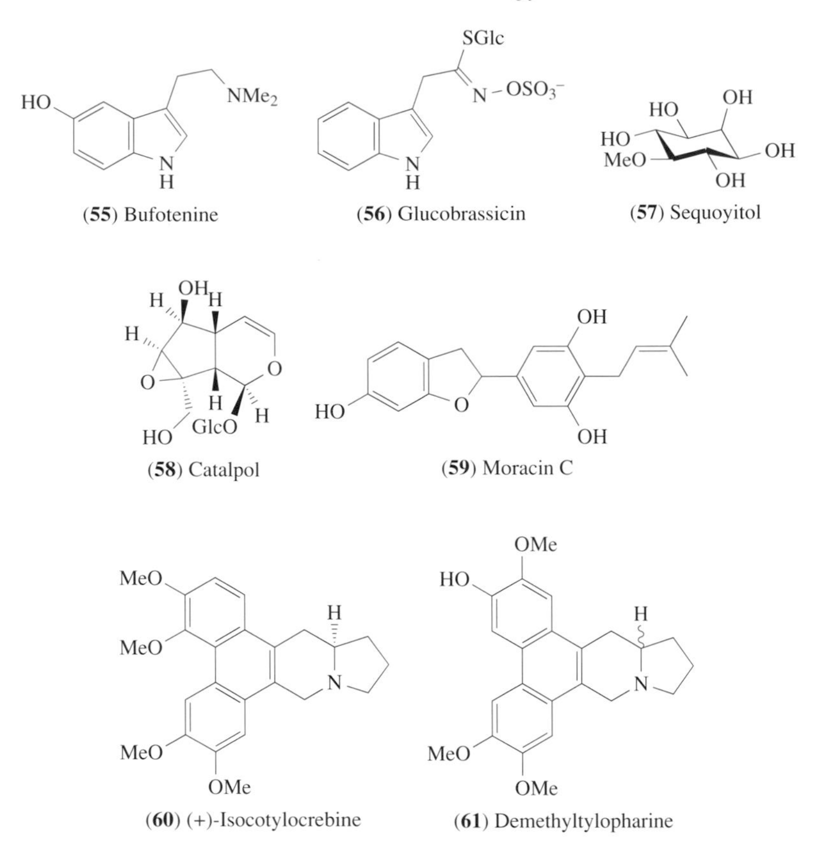

(55) Bufotenine (56) Glucobrassicin (57) Sequoyitol

(58) Catalpol (59) Moracin C

(60) (+)-Isocotylocrebine (61) Demethyltylopharine

Much is known about the choice of plant species for oviposition by the female Monarch butterfly. An important feature is the concentration of cardiac glycoside, since this determines the survival of the subsequent generation. Plants containing 200–500 μg g^{-1} wet weight are preferred. Those with lower levels are avoided, since the larvae subsequently would not be able to absorb enough cardenolide to be properly defended. Equally, higher levels of cardenolide are disadvantageous since larvae are liable to suffer physiological strain trying to absorb excessive amounts of toxin. Flavonol glycosides also seem to be important oviposition attractants in the case of *Asclepias curassavica* and a mixture of four such glycosides have been characterized as oviposition stimulants to the monarch.[63]

Ovipositing females of the Monarch clearly distinguish by odor between young and old leaves of *A. curassavica*, always laying on young leaves. Headspace analysis has failed to reveal any one volatile as a cue for the young leaves. However, there are quantitative differences, which may be operative. Thus, *trans*-α-farnesene and linalool accumulate in young leaf odor, whereas α-thujene and methyl salicylate are dominant in the odor of older leaves and may be responsible for the rejection of old leaves.[64]

Chemical specificity of the leaf attractants has been established in the use of black swallowtail butterfly ovipositing on carrot leaves. A mixture of luteolin 7-(6″-malonylglucoside) and chlorogenic acid is most effective (Figure 9). The related luteolin 7-glucoside, also a component of the carrot leaf, is quite inactive. Whereas this swallowtail depends on two phenolic constituents for recognizing the carrot leaf, the carrot fly relies on a mixture of phenylpropenes, furanocoumarins, and a polyalkyne in the leaf wax (Table 6) to find its food plant.[65]

It is apparent in some cases that mixtures of unrelated structures may synergize to provide oviposition stimulants (Table 6). This is particularly apparent in *Papilio xuthus* and *P. protenor*, two related swallowtails living on rutaceous plants, where mixtures of flavanone glycosides, glycosylflavones, and organic bases provide the attractive cocktail for egg laying.

8.03.4.2 Oviposition Deterrents

Little can be said about the chemical factors which guide ovipositing female Lepidoptera away from unsuitable host plants, since only a few deterrents have so far been identified (Table 7). The

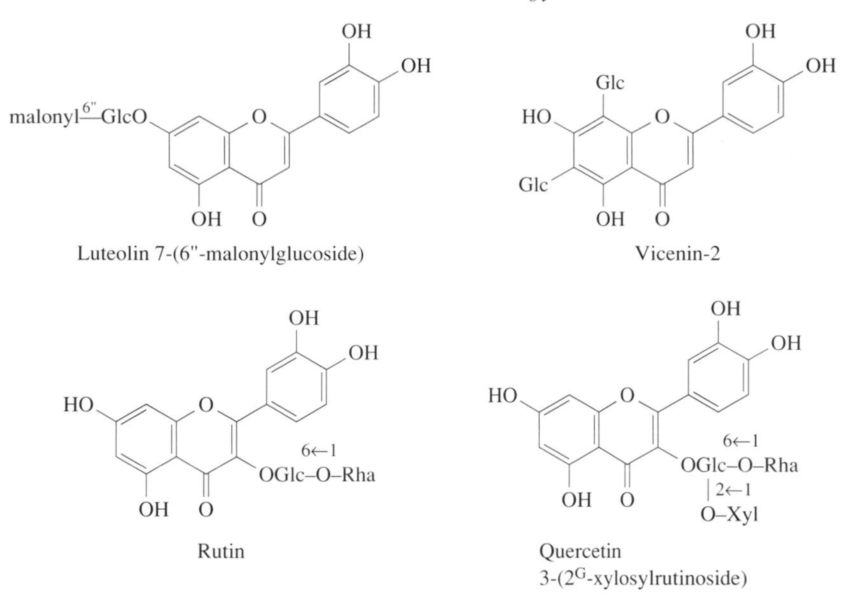

Figure 9 Some flavonoids involved in insect oviposition.

role of cardiac glycosides in *Cheiranthus* and *Erysimum* in discouraging egg laying on these two crucifer plants is fairly clear. These compounds are toxic to the larvae and would be fatal as soon as the eggs hatched out.[66] These plants are avoided by the females in spite of the fact that they both contain glucosinolates, which under other circumstances would be attractive (see Table 6).

Table 7 Plant-derived oviposition deterrents in butterfly.

Oviposition deterrent	*Plant source*	*Butterfly*
Two strophanthidin glycosides	Wallflower, *Cheiranthus* × *allianii*	*Pieris brassicae*, *P. rapae*
Four cardiac glycosides	*Erysimum cheiranthoides*	*Pieris rapae*
Quercetin 3-(2^G-xylosylrutinoside)	*Orixa japonica*	*Papilio xuthus*
8-Prenyldihydrokaempferol 7-glucoside (**62**)	*Phellodendron amurense*	*Papilio xuthus*, *P. protenor*

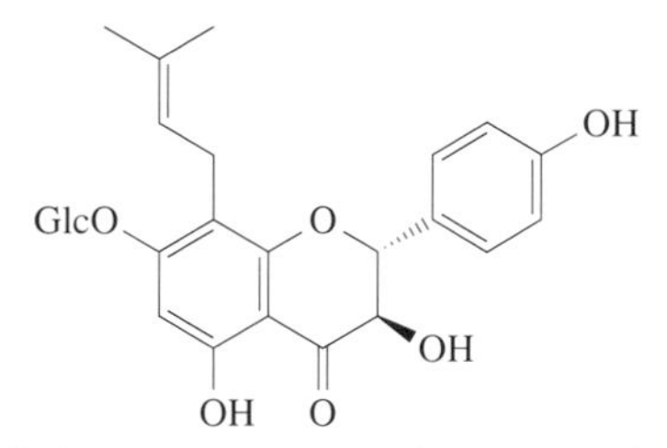

(**62**) 8-prenyldihydrokaempferol-7-glucoside

Oviposition deterrence in the swallowtail, *Papilio xuthus*, can be produced by a small change in the structure of the flavonol glycoside present in the leaves of potential host plants. Thus the insect is stimulated to oviposit on *Citrus* plants by the presence of quercetin 3-rutinoside (rutin) among others. It is, however, deterred from oviposition on the non-host plant *Orixa japonica* because the leaves contain quercetin 3-(2^G-xylosylrutinoside) (Figure 9). Thus, the simple addition of the extra sugar xylose apparently turns rutin from an attractant into a repellent.[67]

8.03.5 PLANT CHEMISTRY AND POLLINATION

8.03.5.1 Flower Pigments

The view that plant pollinators show particular color preferences when visiting flowers for purposes of pollination and nectar gathering is still broadly accepted by plant ecologists.[68] Thus, bees are known to prefer blue flowers, although they will also be attracted to yellow but not red. By contrast, pollinating birds show a preference for scarlet colors or red–yellow bicolors. Lepidoptera are more variable in their choice of flower colors but are attracted to pink. An increasing number of plant species have been recognized which have more than one type of pollinator. For example, the plant *Ipomopsis aggregata* (Polemoniaceae) with mainly red (pelargonidin-based) flowers is pollinated by hummingbirds for much of the season. Later, however, hawkmoths take over this role and the proportion of plants with white instead of red flowers increases.[69]

A correlation between flower color, pollinator, and anthocyanidin type was originally established by general surveys throughout the plant kingdom. More recently, such correlations have been confirmed by detailed surveys in two families, the Labiatae and Polemoniaceae, where information on pollinators is readily available. They show that pelargonidin or pelargonidin–cyanidin mixtures predominate in bird-pollinated plants, whereas delphinidin or delphinidin–cyanidin mixtures occur in bee-pollinated species (Table 8).[70]

Table 8 Correlation between anthocyanidin type, flower color, and pollinator in the Labiatae and Polemoniaceae.

Number of plants investigated	*Flower color range*	*Anthocyanidin type (numbers)*[a]	*Pollinators*
Labiatae			
6	Scarlet	Pg (2) Pg/Cy (4)	Bird
18	Red–purple to purple	Pg (2) Cy (11) Cy/Pn (5)	Lepidoptera and bees
26	Violet to blue	Cy/Dp (13) Dp (5) Dp/Mv (8)	Bees and beeflies
Polemoniaceae			
6	Scarlet–red	Pg (3) Pg/Cy (1) Cy (2)	Bird
6	Pink to violet	Cy (1) Cy/Dp (4) Dp (1)	Lepidoptera
6	Violet to blue	Cy/Dp (3) Dp (3)	Bee

[a] Pg, pelargonidin (**63**); Cy, cyanidin (**64**); Pn, peonidin; Dp, delphinidin (**65**); Mv, malvidin (**66**) (numbers of species with this anthocyanidin type are given in parentheses).

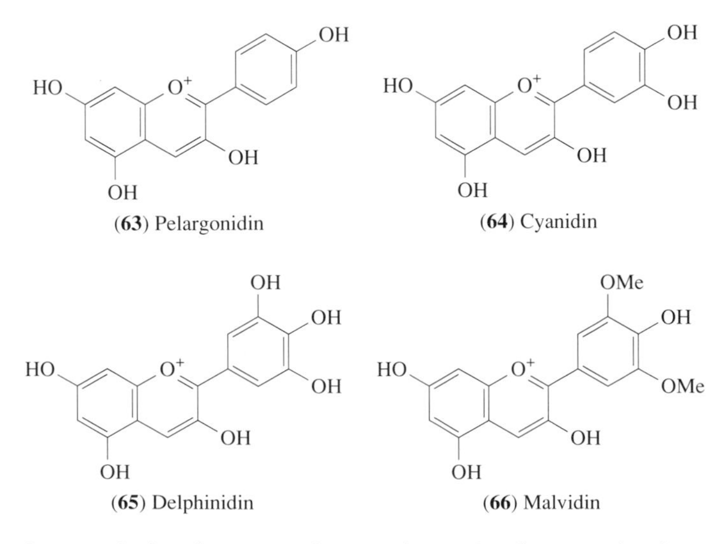

It follows that plant evolution has moved towards scarlet flower color in tropical New World climates where bird pollinators are common. By contrast, evolution has favored blue flower color in temperate climates, where bees are major pollinators. Blue flower color is largely due to delphinidin, petunidin, or malvidin glycosides, although occasionally cyanidin glycosides can be involved (Table 9). In order to produce a blue flower color *in vivo* anthocyanins have either to contain aromatic acylation within the structure or also to have specific copigments occurring with them in the cell vacuoles to shift their purple colors towards the blue region. Anthocyanins are linked to their flavone copigments by hydrogen bonding, but there may also be present a metal ion to stabilize the complex. Two metals, iron and magnesium, have so far been characterized in these pigment complexes.[71] Some idea of the range of anthocyanins identified in blue flowered species can be gained from Table 10. The majority of these anthocyanins are polyacyl glycosides (Figure 10).

Table 9 Anthocyanins identified in blue-flowered plant species.

Plant species	*Anthocyanin*	*Factors providing shift to blue color*
Commelinaceae		
Commelina communis	Delphinidin 3-(*p*-coumaryl-glucoside)-5-malonyl-glucoside	Flavone copigment, Fe–Mg complex
Compositae		
Centaurea cyanus	Cyanidin 3-succinylglucoside-5-glucoside	Flavone copigment, Fe–Mg complex
Cichorium intybus	Dimalonyldelphin	Flavone copigment
Senecio cruentus	Delphinidin 3-malonylglucoside-7-di(caffeyl-glucoside)-3′-caffeylglucoside	Intramolecular aromatic acylation
Convolvulaceae		
Evolvulus pilosus	Delphin with malonic acid and two caffeylglucoses	Intramolecular aromatic acylation
Pharbitis nil	Peonidin 3-sophoroside-5-glucoside acylated with three caffeylglucoses	Intramolecular aromatic acylation
Leguminosae		
Lupinus cv.	Delphinidin 3-malonylglucoside	Flavone copigment
Clitoria ternatea	Delphinidin 3-malonylglucoside-3′,5′-di(*p*-coumarylglucosyl)glucoside	Flavone copigment, intramolecular aromatic acylation
Ranunculaceae		
Aconitum chinense	Delphinidin 3-rutinoside-7-di(*p*-coumarylglucoside)	Intramolecular aromatic acylation
Delphinium hybridum	Delphinidin 3-rutinoside-7-tetra(4-hydroxybenzoyl-glucosyl)glucoside	Intramolecular aromatic acylation

Table 10 Flavonoid pigments identified as providing ultraviolet patterning in yellow flowered species.

Yellow or colorless flavonoids[a]	*Species*
Chalcones: marein (**67**)[b] and coreopsin (**68**)[b]	*Coreopsis bigelovii*
Quercetagetin[b] and patuletin[b] 7-glucosides (**69**)	*Eriophyllum* spp.
Quercetin 3- and 7-glucoside	*Helianthus annuus*
Chalcone: coreopsin[b]; aurone: sulfurein (**70**)[a]	*H. gracilentus*
Quercetagetin[b] and patuletin[b] 7-glucosides	*Rudbeckia hirta*
Chalcone: isosalipurposide (**71**)[b]	*Oenothera* spp.
Chalcone: isosalipurposide[b] or quercetin glycosides	*Potentilla* spp.
Gossypetin 3′-methyl ether[b] 3-rutinoside	*Coronilla valentina*

[a] For references, see reference 72. [b] Yellow pigments.

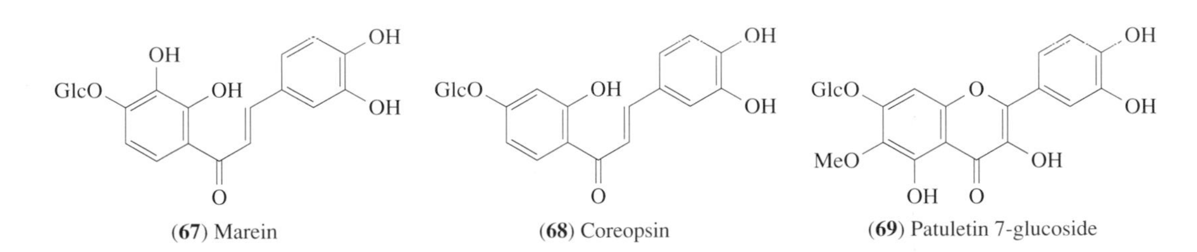

(**67**) Marein (**68**) Coreopsin (**69**) Patuletin 7-glucoside

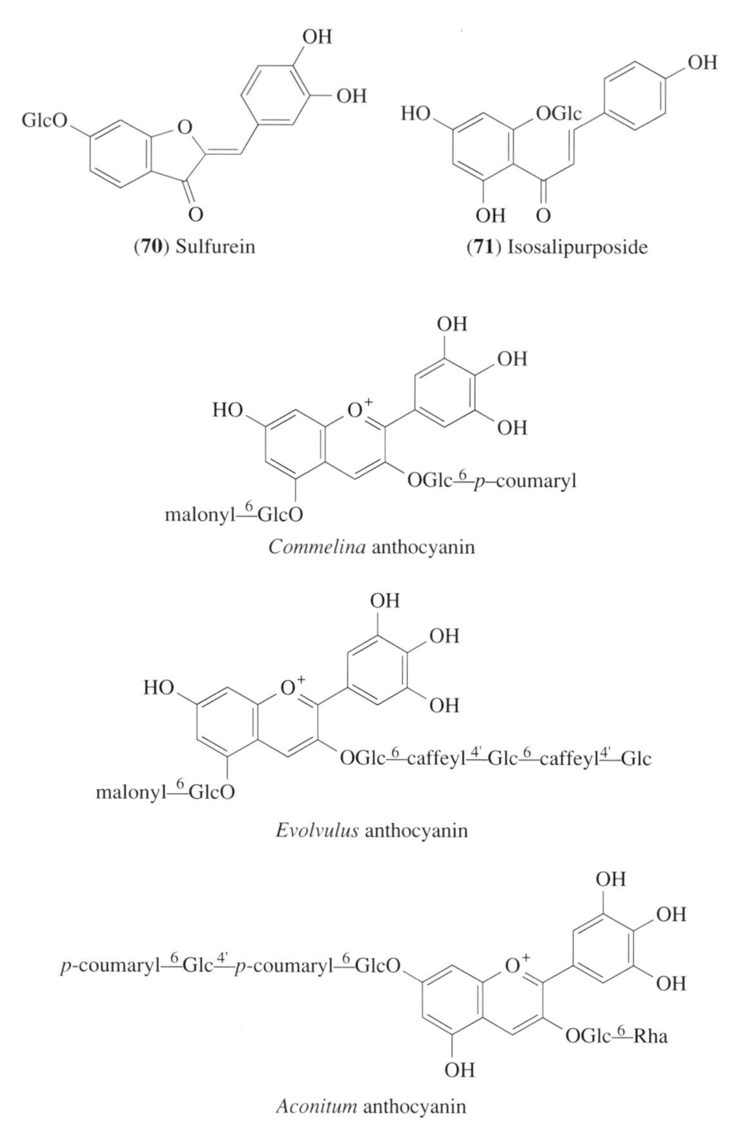

Figure 10 Anthocyanins identified in blue-flowered plant species.

Yellow is an important bee flower color and the presence of carotenoids in the chromoplasts provides such visible color. However, a number of such flowers also have ultraviolet patterning, which is detected by bees, which means that they are more efficient at locating the nectar and collecting the pollen. Such ultraviolet patterning is due to the occurrence of water-soluble flavonoids located specifically in the inner parts of the flower. They quench the bright reflectance of the carotenoid so that the inner parts are dark and absorbing. A range of chalcones or yellow flavonols, notably some gossypetin (**72**) derivatives, have been shown to be responsible for this UV patterning (Table 10). Colorless flavonoids, e.g., quercetin glycosides, have also been detected in a few instances. From the functional point of view, the presence of yellow flavonoids is more efficient, since these compounds are able to contribute to visible yellow color at the same time as they provide ultraviolet patterning.[72]

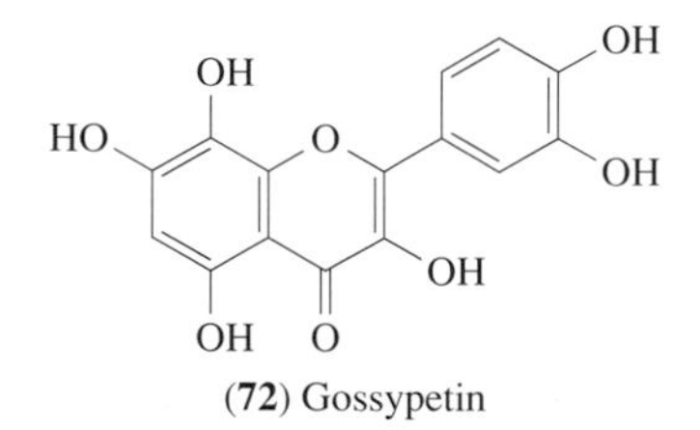

(**72**) Gossypetin

8.03.5.2 Floral Volatiles

The floral volatiles play an important role in attracting pollinators to the plant. They may attract a pollinating bee or wasp from a distance of several meters. Fruity or aminoid odors are attractive to beetles, sweet smells to bees, moths, and butterflies, musty or fruity odors to bats, and fecal odors to dung flies. Research using headspace analysis has indicated the major floral volatiles in a representative sample of flowering plants (Table 11). General reviews of floral volatiles and pollination biology have been provided by Knudsen *et al.*[73] and Harrewijn *et al.*[74] Some of the main findings in this area of plant ecology will now be mentioned.

Table 11 Floral volatiles of bat-, bee-, beetle-, butterfly-, moth-, and fly-pollinated plants.

Floral volatiles[a]	*Plant species*	*Pollinator*
Dimethyl trisulfide (24.3%), dimethyl disulfide, dimethyl tetrasulfide, etc.	*Crescentia cujeta*	Bat
Squalene (26.5%) nerol, geraniol, hydrocarbons	*Dactylanthus taylorii*	Bat
Geraniol, citral, farnesol, etc.	*Ophrys* spp.	*Andrena* male bee
Carvone oxide	*Catasetum maculatum*	*Eulaema* male bee
Linalool (95%) + its oxides	*Daphne mezereum*	*Colletes* bee
Indole, 1,2,4-trimethoxybenzene, cinnamaldehyde	*Cucurbita* spp.	Diabroticite beetle
Methyl anthranilate and isoeugenol (**73**)	*Cimifuga simplex*	Butterfly
Methyl benzoate (25%), linalool (50%), geraniol (12%)	*Platanthera chlorantha*	Moth
Ethyl acetate, monoterpenes, and aliphatics	*Zygogynum* spp.	Moth
trans-Ocimene (**74**) (46%), 1,8-cineole (12%)[b]	*Brugmansia × candida*	Hawkmoth
2-Heptanone (16%), indole (16%), germacrene B (**75**) (18%), *p*-cresol (3%)	*Arum maculatum*	Dungfly

[a] Only major components are listed; values in parentheses are average percentages of total floral odor. [b] Tropane alkaloids, thought to be present, could not be detected.

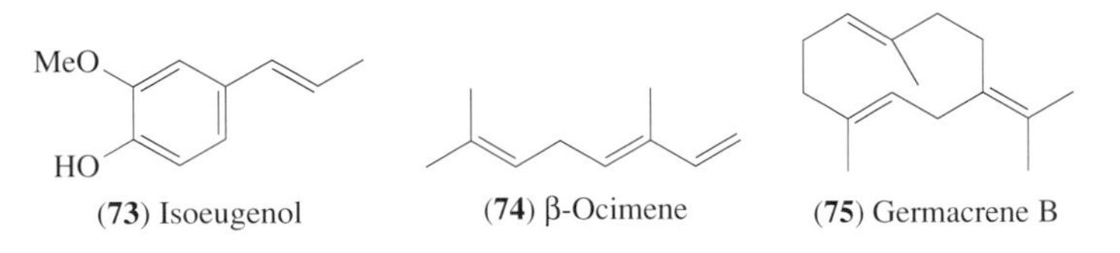

(**73**) Isoeugenol (**74**) β-Ocimene (**75**) Germacrene B

The chemistry of aroid odors has been somewhat controversial in that simple amines such as hexylamine were earlier reported from *Arum maculatum* and several related species. A reinvestigation of *A. maculatum* failed to indicate any amines in the headspace. Instead, indole, *p*-cresol, germacrene B (**75**), and 2-heptanone were detected as major constituents. The plant is pollinated by females of the owl midge, *Psychoda phalaenoides*, which otherwise feeds on cow dung. Both indole and *p*-cresol were detected in the headspace of the dung, so these two compounds appear to the most important attractants.[75] Incidentally, indole and skatole are the major "distasteful" odors of another aroid plant, the voodoo lily *Sauromatum guttatum*. Odors unpleasant to the human nose are also dominant in bat-pollinated flowers, and a series of methyl sulfides (Table 11) have been identified in *Crescentia cujeta* and several other bat-pollinated plants.[76]

In sweet-smelling plants, individual constituents may dominate (e.g., linalool in *Daphne mezereum*), but more usually there are several components, which act synergistically to attract the pollinator (e.g., as in the moth-pollinated *Platanthera*) (see Table 11). The floral scent is usually released at the right time of day for the particular pollinator, e.g., during the day for bee-pollinated flowers. For moth-pollinated species, it may be at dusk or even later in the night. Thus, ocimene (**74**) is released from flowers of *Mirabilis jalaba* as night between 6.00 p.m. and 8.00 p.m. Different parts of the flower may have slightly different odors. This is true in *Rosa rugosa* and *R. canina*, where bees can select out pollen for collection from the rest of the flower. The compound geranylacetone (**76**), for example, is specific to the pollen and is not found in the floral odor.[77]

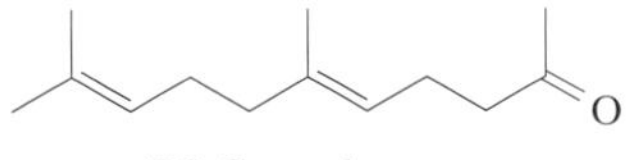

(**76**) Geranylacetone

In the distinctive pollination of orchid flowers of the genus *Ophrys* by male bees of the genus *Andrena*, a large number of scent constituents are involved. Pseudocopulation of the flower by the male bee depends on the flower having the same shape, same scent, and same color as the female bee. In *Ophrys lutea*, for example, 1-octanol, decyl acetate, and linalool are common to the floral volatiles and to the pheromonal odor of the female bee. Other studies on the *Ophrys–Andrena* volatiles were described by Borg-Karlson *et al.*[78] and other examples of orchid pollination requiring specific floral volatiles have been given.[9–12]

8.03.5.3 Nectar and Pollen Constituents

Nectar is an important source of food for most animal pollinators. Nectar chemistry does vary within certain limits and it is possible to suggest that many plant species modify the nectar components, through natural selection, to suit the needs of particular pollinators. The major components of nectars are simple sugars in solution, the sugar content varying from 15% to 75% by weight. The three common sugars are glucose, fructose, and sucrose, but traces of various oligosaccharides (e.g., raffinose, **77**) are sometimes present. There are distinct quantitative differences in the proportions of the three common sugars and angiosperm species can be divided into three groups, according to whether sucrose is dominant, glucose and fructose are dominant, or all three sugars occur in equal amounts. There is thus an evolutionary trend from nectar, which is mainly sucrose, to nectar, which is mainly glucose and fructose. Such a trend would correspond to some extent to the sugar preferences of the particular pollinators (see Table 12).

Gal(α1→6)Glc(α1→2)Fru

(**77**) Raffinose

Table 12 Amounts of amino acids and sugar types of nectars attractive to different pollinators.

Amount of amino acid on histidine scale[a]	*Dominant sugar*	*Pollinator*
9.0	Glucose/fructose	Carrion and dung flies
5.4	Sucrose	Butterflies
5.4	Sucrose	Settling moths
4.6	Sucrose	Bees
4.4	Sucrose	Hawkmoths
3.9	Sucrose	Hummingbirds
3.9	Glucose/fructose	Passerine birds
3.6	Glucose/fructose	Bats

[a] Based on ninhydrin colours of single drops of nectar, compared with known concentrations of histidine solutions. A score of 8 corresponds to about 1 mg ml^{-1} amino acid.

Lipid is an alternative source of energy to sugar, and lipid bodies replace nectar sugar in some 49 genera of the Scrophulariaceae, Iridaceae, Krameriaceae, Malpighiaceae, and Orchidaceae. These are all bee pollinated and the oil is mainly used by the bees for feeding their young. These lipids appear to be chemically distinct from the triglyceride seed oils. Indeed, in species of *Krameria*, free fatty acids have been characterized. These are all saturated acids with chain lengths between C_{16} and C_{22} and all have an acetate substituent in the β-position.[79]

Small amounts of protein amino acids are also present in nearly all nectars. The 10 amino acids essential for insect nutrition are often present and there is no doubt that nectars are a useful source of nitrogen, especially to insects such as butterflies, which have few other ways of acquiring amino acids at the adult stage. It is much less important for bird pollinators and there are indications that amino acid concentrations are related to the needs of the different pollinating vectors (Table 12).[80]

Plant nectars may contain toxins, which are presumably derived from their synthesis in other plant parts. Alkaloids have been most frequently detected, but several other classes have also been noted (Table 13). The alkaloid content may vary from the traces (0.106 μg g^{-1} fresh weight) in the tobacco plant nectar to as much as 273 μg g^{-1} fresh weight of tropane alkaloids in the deadly nightshade, *Atropa belladonna*.[81] The purpose of toxin accumulation in nectars is still uncertain, although a defensive role against herbivores or an undesirable animal visitor is certainly possible.

The formation of iridoids in the nectar of the plant *Catalpa speciosa* is apparently to protect the plant from ants, which are nectar thieves.[82]

Table 13 Toxins of plant nectars.

Class	*Compound*	*Plant nectar*
Alkaloid	Hyoscyamine (**78**)	*Atropa belladonna*
Phenolic	Arbutin	*Arbutus unedo*
Alkaloid	Hyoscyamine	*Brugmansia aurea*
Alkaloid	Pyrrolizidines	*Eupatorium* spp.
Alkaloid	Quinolizidines	*Lupinus polyphyllus*
Alkaloid	Nicotine	*Nicotiana tabacum*
Iridoids	Catalpol (**79**)	*Catalpa speciosa*
Diterpenoid	Acetylandromedol[a] (**80**)	*Rhododendron ponticum*
Alkaloid	Pyrrolizidines[a]	*Senecio jacobaea* and other spp.
Alkaloid	Quinolizidines	*Sophora microphylla*
Sugar	Mannose[b]	*Tilia cordata*

[a] These toxins are carried through from nectar to the honey stored by bees in their hives. [b] Toxic to bees, since they are unable to metabolize it.

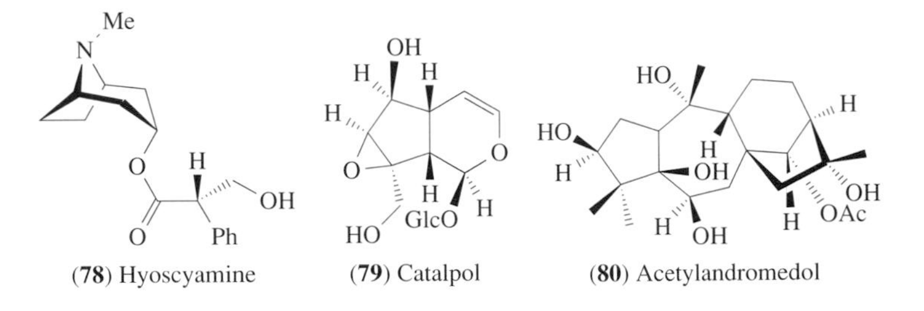

(**78**) Hyoscyamine (**79**) Catalpol (**80**) Acetylandromedol

Occasionally, the toxins in the nectar may be collected during the process of pollination by certain butterflies. This is true of adult Ithomiines and Danaids, which have a requirement for pyrrolizidine alkaloids both for defence and for pheromone production. These alkaloids are obtained from nectar of *Eupatorium* and *Senecio* species, which grown in their respective habitats (see Section 8.03.2).

Pollen, like nectar, is largely nutritional and is collected and eaten by bees and beetles. Carotenoids are present in many pollens, providing yellow color, and function in improving pollen detection by the pollinator. All pollens also contain small amounts of flavonol glycosides, particularly such compounds as kaempferol and 3-sophoroside (**81**) isorhamnetin 3-sophoroside (**82**). Until recently, the occurrence of these flavonol glycosides was obscure. However, there is now evidence in the *Petunia* flower that the pollen flavonol kaempferol 3-sophoroside (**81**) has an essential role in assisting the germination of the pollen when it lands on the stigma. During the process, a specific β-glycosidase removes the protecting sugars to release the free aglycone. The kaempferol formed is probably a growth promoter and at the same time prevents the introduction of pathogens into the pistil.[83]

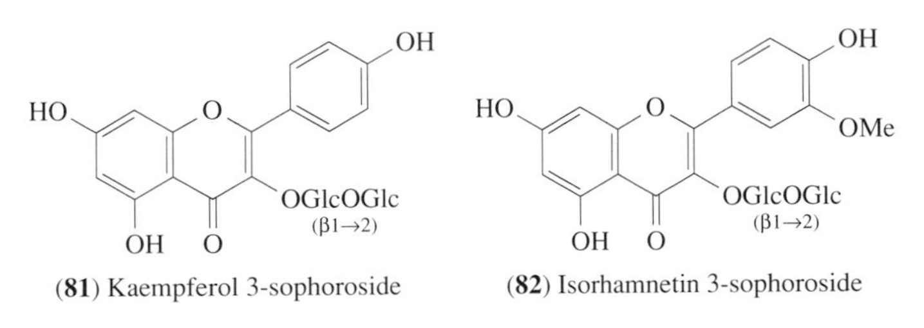

(**81**) Kaempferol 3-sophoroside (**82**) Isorhamnetin 3-sophoroside

8.03.6 PLANT CHEMISTRY AND SEED DISPERSAL

The ripe fruit is the one part of the plant which is likely to be undefended chemically, since it is provided for animals in return for the widespread dispersal of the seed that lies within the fruit. By contrast, the seed and the seedcoat usually possess some chemical toxins, although they are often also well protected by physical structures. This is to ensure that the seed is not consumed along with the fruit.

The unripe fruit will, however, differ from the ripe fruit in being protected to some degree from herbivory, since the seed within is not yet ready for distribution. The green chlorophylls of the leaf may, for example, camouflage the unripe fruit so that it is not seen by a herbivore. There may be alkaloids present, as in green tomato fruits, which discourage animal feeding. Other chemical traits, such as acidity, bitterness, or astringency, may deter the majority of herbivores. Chemical changes during ripening will reduce or eliminate these barriers. Attractive colors, odors, and flavors will develop during ripening and advertise the readiness of the fruit for eating.

Although our knowledge of the chemistry of cultivated fruits is considerable, we know much less about that of wild species. The chemical ecology of fruits and their seeds has not been as intensively investigated as that of plant leaves, so that it is sometimes necessary to extrapolate from what we know of the cultivated species. Here, it is intended to review briefly the chemical attractants of fruits—the colors, the odor principles, and the flavors—and then to consider the chemical defense of plant seeds.

The ripe fruit is usually exposed to herbivores by its attractive and distinctive color, which may be provided largely by carotenoids and anthocyanins. Green fruits will contain chlorophyll, but most orange and red fruits are colored by a range of carotenoids. β-Carotene, which is yellow, is often abundant in yellow-colored fruits and the red lycopene (**83**) is a common pigment of red fruits. Red to purple black fruits generally have anthocyanins present. The red range of colors, which overlaps with that of carotenoids, is usually due to the presence of cyanidin-derived structures, while most blue to purple–black fruits are based on delphinidin (e.g., **84**). Much information is available on fruit pigments of cultivated plants[84] but less is known about fruit color in wild plants.

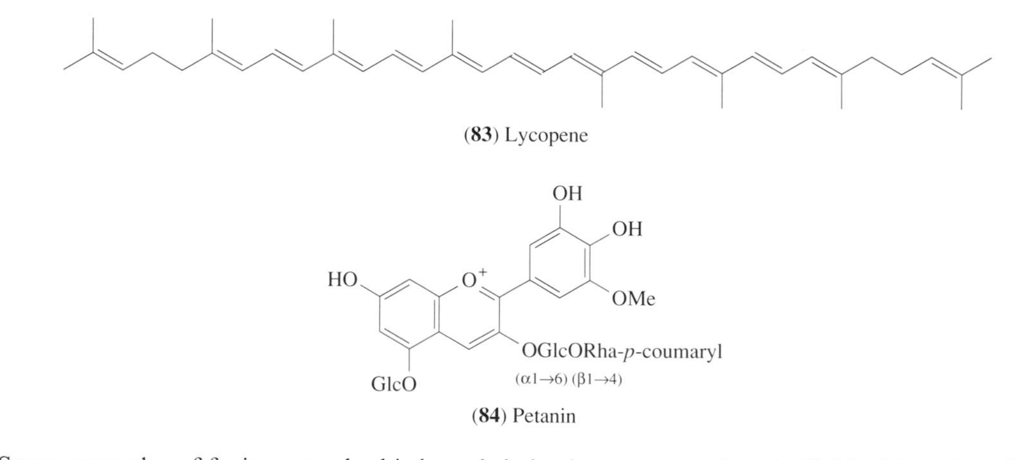

(**83**) Lycopene

(**84**) Petanin

Some examples of fruits eaten by birds and their pigments are given in Table 14. Red and black colors are much preferred by birds, with blue coming a poor third (5–7% of wild fruits). Most blue fruits are colored by anthocyanins, although some additional factor (copigmentation or metal for chelation) must be present to shift the usual purple anthocyanin towards the blue region. The presence of anthocyanins in blue fruited species has been confirmed by a survey of 26 species in 18 genera in Costa Rica, India, Florida, and Malaysia.[85] The absence of anthocyanin from the blue fruits of *Elaeocarpus angustifolius* is striking. Microscopic investigation showed that color is produced by the presence of an iridosome structure beneath the outer cell wall of the adaxial epidermis. This unique iridescent blue appears to be of selective advantage in ensuring that the fruit is readily

Table 14 Pigments of some fruits eaten by frugivorous birds.

Color class	*Plant species*	*Pigments*
Red	*Empetrum rubrum*	Anthocyanins: cyanidin 3-glycosides
	Rosa canina	Carotenoids, e.g., lycopene
	Solanum dulcamara	Carotenoids, e.g., lycopene
	Taxus baccata	Carotenoids, e.g., lycopene
Blue	*Elaeocarpus angustifolius*	None (structural color)
	Vaccinium spp.	Anthocyanins: delphinidin glycosides
Black	*Atropa belladonna*	Anthocyanin: petunidin triglycoside
	Empetrum nigrum	Anthocyanins: delphinidin 3-glycosides
	Vitis vinifera	Anthocyanins: malvidin 3-glucoside

apparent to the cassowary bird, *Casuarius casuarius*, and to fruit-eating pigeons. The brilliant blue color persists even when the mesocarp is almost completely senescent or has been consumed by beetles. This plant species produces a structural color in the fruit which is presumably superior to and more stable than the usual anthocyanin pigmentation.[85]

Many fruits have attractive odors, which are pleasant to humans (Table 15).[86] This is not surprising considering that humans and other primates have an ecological role in dispersing the seed of such plants. Chimpanzees living in African rainforest areas depend on fruits of trees such as figs for much of their diet. Through their feces, they distribute the seeds along the forest floor and their food choice will determine those tree species which are able to regenerate through seed and those which cannot.

Table 15 Chemical principles of fruit odors.

Fruit	*Components identified as aroma principles*[a]
Almond	Benzaldehyde
Apple	Ethyl 2-methylbutanoate
Banana	Amyl acetate, amyl propionate, and eugenol
Coconut	γ-Nonalactone
Cucumber	$CH_3CH_2CH{=}CHCH_2CH_2CH{=}CHCHO$
Ginkgo	Butanoic and hexanoic acid
Grapefruit	(+)-Nootkatone (**85**), 1-*p*-menthene-8-thiol (**86**)
Lemon	Citral
Mandarin orange	Methyl *N*-methylanthranilate and thymol (**87**)
Mango	Car-3-ene, dimethylstyrene
Quince	Ethyl 2-methylbutanoate
Passion fruit	Methyl salicylate, eugenol, and isoeugenol
Peach	γ-Undecalactone
Pear	Ethyl *trans*-2,*cis*-4-decadienoate
Pepper	2-Isobutyl-3-methoxypyrazine
Pineapple	Furaneol (**88**) and mesifurane (**89**)
Raspberry	1-(*p*-Hydroxyphenyl)-3-butanone
Vanilla	Vanillin

[a] Many fruits have several other minor aroma principles in addition to those listed. For further details, see reference 86.

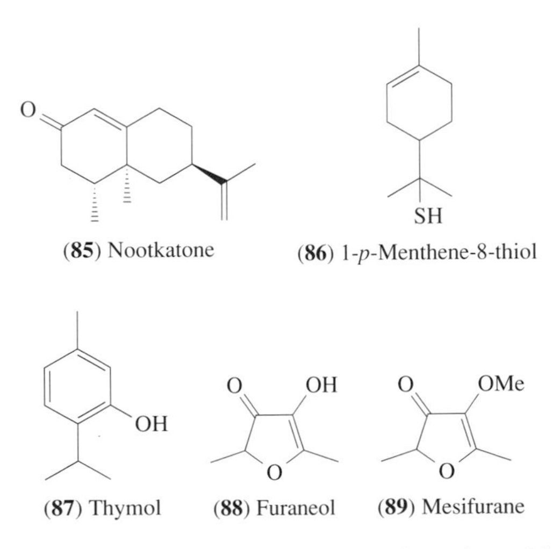

(**85**) Nootkatone (**86**) 1-*p*-Menthene-8-thiol

(**87**) Thymol (**88**) Furaneol (**89**) Mesifurane

The odor principles of most fruits consumed by humans have been identified and much is known of the chemistry involved (Figure 11). Some fruits have single aroma principles (e.g., citral in lemons), others have mixtures of several components (e.g., banana, Table 15), and yet others have complex mixtures (e.g., 10 or more terpenes in apricots). Many of these odor components occur in glycosidic form in the unripe fruit and are released by enzymic hydrolysis during ripening.

Some fruits have odors repellent to humans, reflecting the fact that other animals besides primates may be responsible for eating the fruit and dispersing the seed. This is true of the durian fruit, *Duria zibethium*, which is rejected by some humans because of its offensive sewage-like smell. Many animals consume the fruit in the Malaysian rainforest, but elephants are particularly attracted to

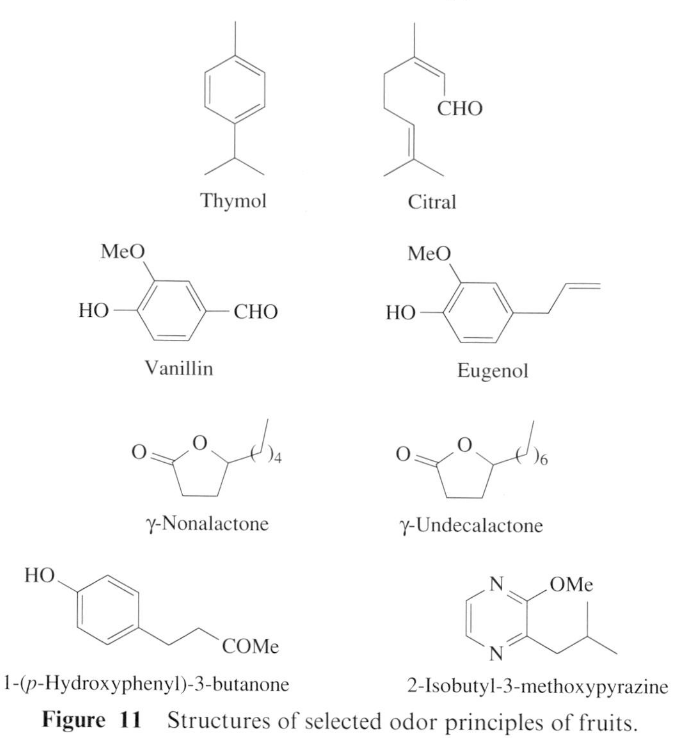

Figure 11 Structures of selected odor principles of fruits.

the foul odor and are also good seed dispersers. The reason why the fruit of that most ancient of gymnosperms, *Ginkgo biloba*, smells of rancid butter, due to a mixture of organic acids, is more obscure but it was presumably attractive to the seed disperser (a reptile?) of a past age.

In addition to the aromas, fruits contain many nonvolatile constituents which contribute to taste and flavor. The most important and universal is sugar, providing sweetness of taste. Sugar is in the form of sucrose, glucose, and fructose, with either sucrose or glucose and fructose being dominant. Sucrose-rich fruits can provide problems of digestion to some birds which lack the enzyme sucrase, which hydrolyzes sucrose to glucose and fructose. This is true of starlings, which are liable to develop osmotic diarrhea if they eat such fruits, and this can be fatal. Sucrose-rich fruits include the peach and the apricot. Sweetness can occasionally be provided in fruit more efficiently by the presence of sweet proteins. Such proteins may be 3000 times sweeter then sucrose. Examples of fruits with sweet protein are the plants *Dioscoreophyllum cumminsii* and *Thaumatococcus daniellii*. A considerable number of secondary compounds with sweet tastes have also been encountered in plants, but most of these structures are not specifically produced to provide sweetness in plant fruits.[87] Thus, the sweet diterpenoid stevioside (**90**) is synthesized in leaves of *Stevia rebaudiana*, while the triterpenoid glycyrrhizin (**91**) of *Glycyrrhiza glabra* is confined to the roots of that plant.

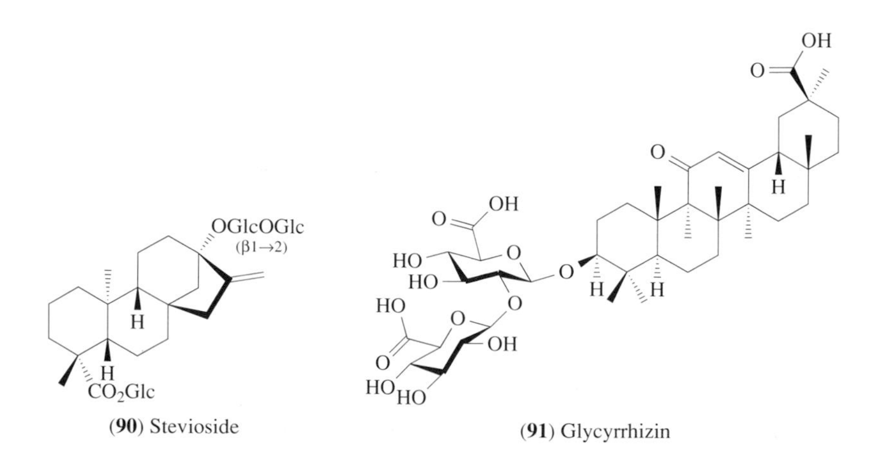

(**90**) Stevioside

(**91**) Glycyrrhizin

Acidity or sourness is another characteristic of fruit flavor. This is due to the accumulation of simple organic acids, such as citrate, malate, tartrate, and oxaloacetate. The concentrations of such acids is largely reduced during the ripening process and any remaining acidity is often counter-balanced by the sugar that is present in the ripe fruit.

Astringency in fruits is largely due to the presence of condensed tannins or flavolans (see Section 8.03.1.2). In small amounts, tannins may provide an attractive feature of ripe fruits, providing a counterbalance to the blandness of sweetness. In larger amounts (e.g., 1.6% wet weight), the astringency may become pronounced and hence the fruit is unpalatable. The best known cultivated fruit with high levels of tannin is the persimmon, *Diospyros kaki*, and here efforts have been made to reduce the astringency by chemical treatment.

There is some evidence that tannin levels change during fruit ripening, so that tannins become inactivated either by polymerization or by complexing with pectin. As a result, they become less astringent and hence more attractive for eating. While this may be true for cultivated fruits,[88] it is not clear how far it is true for wild species. A study of tannin levels in fruits eaten by chimpanzees showed some decreases in concentration with ripening. There was, however, little correlation between high tannin levels and whether the fruit was eaten or not.[89] In the author's own unpublished work on wild figs eaten by chimpanzees, little evidence was found that high tannin levels were responsible for fruit rejection. A complication in the case of wild figs is the fact that tannin is also present in the seeds as well as the pulp, so that it is not always clear what is being rejected within the fruit.[90]

It has been proposed by Janzen[91] that ripe fruits may still contain some chemical constituents which may have a harmful effect on the herbivore because of the need to deter the "wrong" seed dispersers. This would apply to animals which either damage the seed in eating the fruit or which fail to distribute the seed away from the parent plant. The example quoted by Janzen[91] is of ripe *Andira inermis* fruits, which have a potent antibiotic in the juicy pulp. This has no effect on the Costa Rican fruit bats, which avidly eat the fruit and properly disperse the seed. However, such fruits are rejected by cattle and pigs, because the antibiotic inhibits the gut flora required for proper digestion. Such animals would feed under the parent tree and not move away to disperse the seed.

How far protective chemistry is involved generally in fruits in selecting out the "right" seed dispersal agents is still undetermined. However, Barnea *et al.*[92] have evidence that fruits eaten by frugivorous birds are mildly toxic, even when ripe. The purpose here would seem to be to prevent the consumption of too many fruits in any one foraging bout and hence to regulate seed retention time. This in turn ensures better seed dispersal by the bird, since only a few seeds will be deposited at any one site at one time. Chemical analysis of ripe fruits of ivy, holly, and hawthorn confirmed that saponins, flavonoids, and cyanogens were present in the pulp and were mildly distasteful to foraging blackbirds, starlings, and redwings. Likewise, ornithological observations showed that feeding bouts were limited to 1.3–5.3 min and numbers of seeds eaten per bout varied between four and six.[92]

Turning finally in this section to the defensive chemistry of seeds, we have the situation where much is known about the occurrence of toxins in seeds but little is known about the defensive role. Many toxic constituents from alkaloids to cyanogens and monoterpenoids to diterpenoids have been encountered in seeds.[93] A few examples are given in Table 16. Unfortunately, in many cases, there is only circumstantial evidence that these toxins protect the seed from herbivory. Those examples where something is known of defensive chemistry will be described.

Table 16 Some toxins of seeds.

Compound	*Class*[a]	*Source*
Alkaloid	Caffeine	Guarana, *Paullinia cupana*
	Ajaconine (**92**)	Delphinium, *Delphinium ajacis*
	Atropine (**93**)	Deadly nightshade, *Atropa belladonna*
	Cytisine (**23**)	*Laburnum anagyroides*
Cyanogen	Amygdalin	Bitter almond, *Prunus amygdalus*
Furanocoumarin	Xanthotoxin	Wild parsnip, *Pastinaca sativa*
Monoterpene	α-Thujone	White cedar, *Thuja occidentalis*
Diterpene	Columbin (**94**)	Serendipity berry, *Dioscoreophyllum cumminsii*
Sesquiterpene lactone	Anisatin (**95**)	Japanese star anise, *Illicium anisatum*
Protein	Abrin	Jequirity, *Abrus precatorius*

[a] For references, see reference 93.

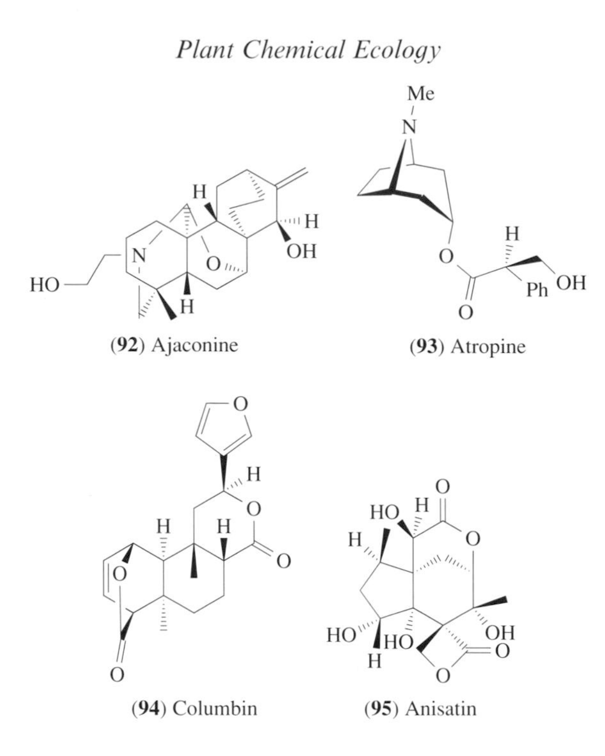

(**92**) Ajaconine (**93**) Atropine

(**94**) Columbin (**95**) Anisatin

The protective role of α-thujone (**96**) in the cone of the white cedar, *Thuja occidentalis*, is fairly clear. This substance makes up to 26% of an oil mixture which is secreted in oval blisters of the seed coat. This toxic seed coat slows down feeding by the red squirrel seed predator, since it carefully removes the seed coat every time before eating the rest of the tissue.[91]

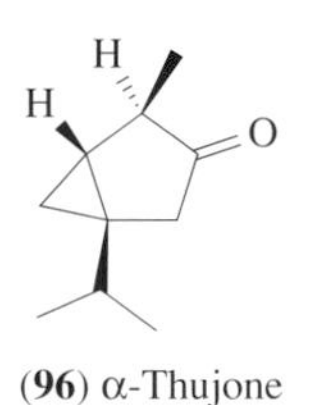

(**96**) α-Thujone

The protective role of the purine alkaloid caffeine (**97**) in the seed coat (at a concentration of 1.64% dry weight) of the guarana fruit has been established by Baumann *et al.*[94] Their experiments show that caffeine is not released from the seed while the fruit is being digested by toucans and guans, because there is a powerful diffusion barrier preventing the birds' intoxication. The level of caffeine in the seed is 3–5 times that of the coffee bean and this would be toxic to the bird if the seed coat were damaged during eating the fruit.

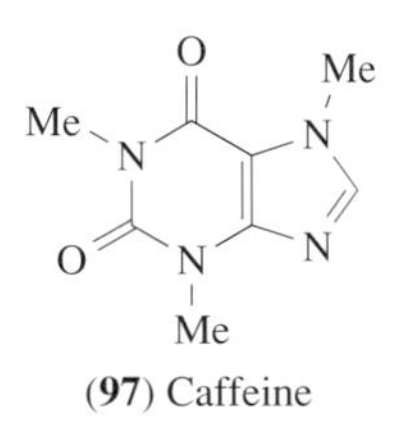

(**97**) Caffeine

One further example where seed toxins have been examined in an ecological context is the furanocoumarins which occur in seed of wild parsnip, *Pastinaca sativa*, and herbivory by the parsnip webworm, *Depressaria pastinacella*. While the furanocoumarins do not provide complete protection, they can limit predation by this insect. The mixture of compounds present slows down the metabolism within the insect to detoxify them. There appears to be competition for binding sites between rapidly metabolized furanocoumarins (e.g., xanthotoxin) and those that are turned over more slowly (e.g., bergapten, **98**).[95] Such furanocoumarin protection may operate in other plants of the family Umbelliferae that parsnip belongs to, since these substances are regular constituents of the seeds. Other toxins also occur in umbellifer seeds such as mono- and sesquiterpenoid mixtures.[96]

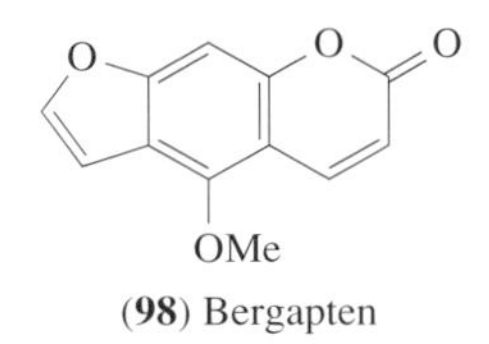

(98) Bergapten

A deliberate attempt by Hendry *et al.*[97] to determine chemical factors responsible for seed longevity in the soil in 80 species within the British flora revealed that *o*-dihydroxyphenols were universally present. Seed longevity was directly related to *o*-dihydroxyphenol concentration. Furthermore, some 75% of the seeds also contained cyanogens. There is evidence here that seed phenolics have a protective role against herbivory and also probably against microbial decomposition.

8.03.7 ALLELOPATHY

It has long been recognized that non-volatile secondary substances may leak out of the plants that synthesize them and they may be excreted from leaves, stems, or roots into the environment. Lipophilic constituents in the leaf wax on the surface may be leached either by rainfall or by the steady drip from leaves under misty conditions. Similarly, volatile compounds in leaf hairs or in floral scents may be slowly lost into the environment by evaporation. In both cases, potentially toxic constituents may accumulate in the soil surrounding the producer plant and, if not destroyed by microbial activity, they may exert a deleterious effect on the growth of either the producer plant or on neighboring plant species. In the Californian chaparral, bare patches in the soil may arise around shrubs which are inhibiting the germination of seeds of competing annual species. Such interactions are termed "autotoxicity" in the case of self-inhibition of plant growth and "allelopathy" in the more general circumstances of one plant species inhibiting the growth of another.

There are clearly a number of experimental problems in fully establishing a case history of allelopathy and some ecologists would argue that all the necessary evidence has rarely accumulated in any one particular interaction.[98] The release of the supposed toxin from the plant and its subsequent presence in the soil in sufficient concentration to exert a deleterious effect have to be firmly demonstrated. Again, it is important to show that the toxin is not readily metabolized by the common soil microbial agents. Furthermore, the toxin has to be tested against plant species that are likely to compete with the producer plant in the natural ecosystem. The commonly used bioassay against germination of lettuce seeds has little relevance to most natural allelopathic interactions.

Much has been written about the somewhat controversial subject of allelopathy. Three books have reviewed the evidence,[99–101] but see also other publications.[8–12,102] Here it is intended to discuss the range of secondary compounds involved and to consider how far the different metabolites are likely to exert an allelopathic effect. The majority of natural substances implicated in allelopathy are either phenolic in nature (Table 17) or else terpenoid or fatty acid based (Table 18). The phenolics are generally released from leaves, though occasionally (e.g., cinnamic acid) exudation from roots occurs. By contrast, the terpenoids are usually volatilized from the leaf surface in a hot Mediterranean-type climate.

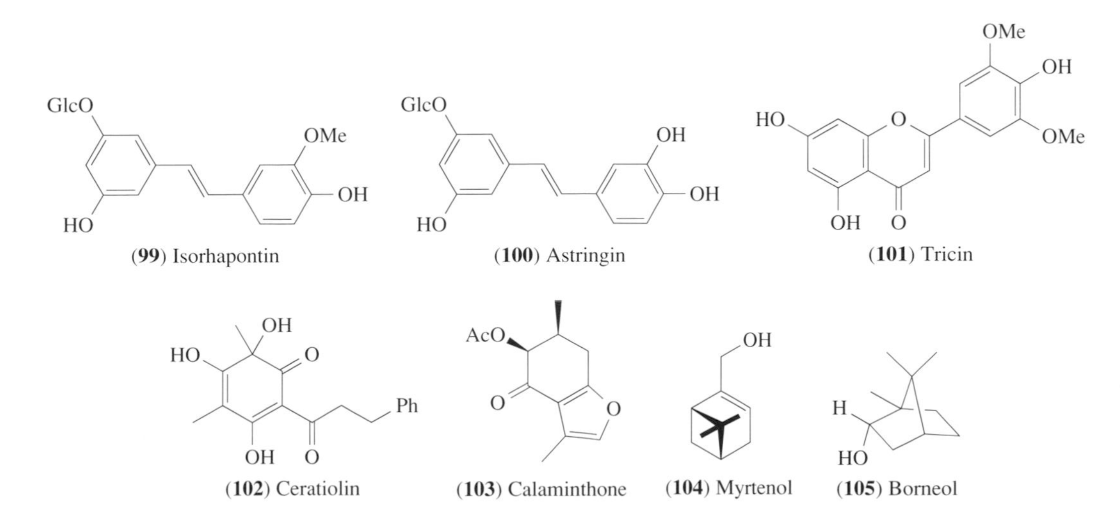

(99) Isorhapontin (100) Astringin (101) Tricin

(102) Ceratiolin (103) Calaminthone (104) Myrtenol (105) Borneol

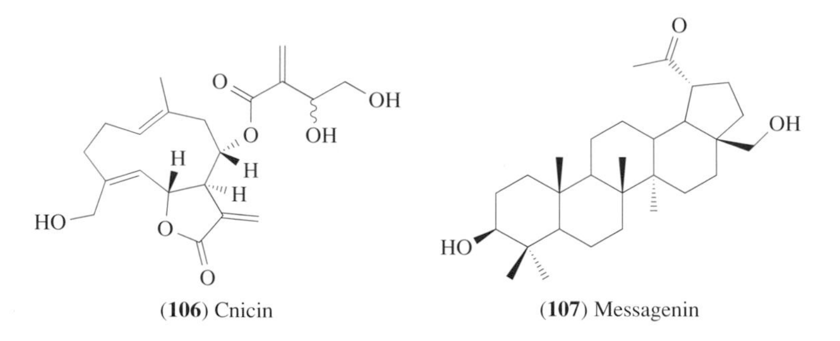

Table 17 Phenolic constituents implicated as allelopathic agents in plants.

Phenolic	*Producer plant*	*Receiver plant(s)*
Phenols		
Hydroquinone, vanillic acid, ferulic acid, *p*-hydroxybenzoic acid	Chamise, *Adenostoma fasciculatum*, and *Arctostaphylos glandulosa*	Inhibits seed germination in annual competitors
Phenolic acid		
Salicylic acid	*Quercus falcata*	Inhibits competing undergrowth
Benzaldehyde		
3-Acetyl-6-methoxybenzaldehyde	Californian shrub, *Encelia farinosa*	Inhibits growth of *Malacothrix* spp.
Cinnamic		
(*E*)-Cinnamic acid	Rubber plant, *Parthenium argentatum*	Autotoxic
Quinone		
Juglone	Walnut tree, *Juglans regia*	*Alnus glutinosa* and other spp.
Lignan		
Nordihydroguaiaretic acid	Creosote bush, *Larrea tridentata*	Seedlings of eight herbs
Stilbenes		
Isorhapontin (**99**), astringin (**100**)	Bark of *Picea engelmannii*	Seed inhibition of other conifers
Flavone		
Tricin (**101**)	Grass, *Agropyron repens*	Inhibits root growth in competing spp.
Cinnamic		
3,4-Methylenedioxycinnamate	*Asparagus officinalis*	Inhibits seedling growth in cress
Dihydrochalcone		
Ceratiolin (**102**)[a]	*Ceriatola ericoides*	Inhibits seeds of competitor species

[a] During leaching from leaf decomposes to dihydrocinnamic acid, the active agent.

One of the best studied allelopathic agents is juglone, released from the walnut tree, *Juglans regia*, where it occurs in bound form as the related quinhydrone glucoside (Figure 12). Some 16 species of herbs and trees have been shown to be sensitive to juglone, with inhibitory effects on growth at 1×10^{-6} mol dm^{-3} and wilting and death at a concentration of 1×10^{-3} mol dm^{-3}. Furthermore, juglone has been shown to persist in the soil under the walnut tree. High concentrations were found in the top layers (0–8 cm) but significant amounts (1 μg g^{-1} of soil) were detectable at a depth of 1.8 m. The amounts present in soil under mixed stands of walnut and alder are sufficient to explain the mortality that alder suffers in such mixed plantations.[103] Finally, microbial turnover of juglone has been explored, and while a bacterium capable of detoxifying juglone has been isolated, it is a relatively rare organism in most types of soil.[104]

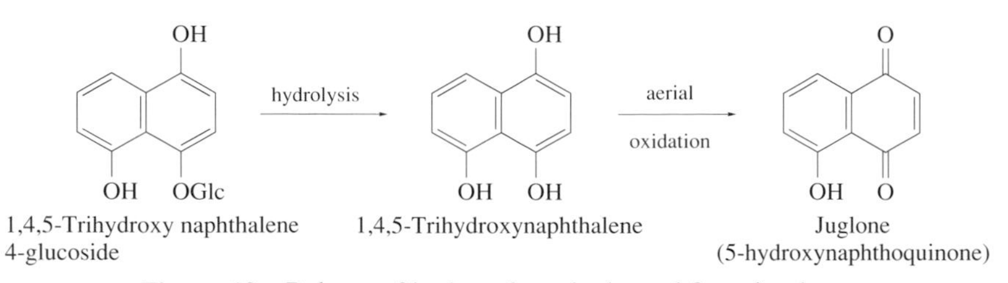

Figure 12 Release of juglone from its bound form in plants.

Table 18 Terpenoid and other constituents implicated as allelopathic agents in plants.

Secondary constituents	*Producer plant*	*Receiver plants(s)*
Terpenoids 1,8-Cineole, camphor, α- and β-pinene, camphene	Sagebrush, *Salvia leucophylla*, and *Artemisia californica*	Inhibits germination of seeds of competing annual species
Terpenoids Menthofuran, epievodone, calaminthone (**103**), ursolic acid	*Calamintha ashei*	Inhibits seeds of competing species
Terpenoids Camphor, myrtenol (**104**), borneol (**105**) and carvone	*Conradina canescens*	Inhibits seeds of competing species
Sesquiterpene lactone Cnicin (**106**)	*Centaurea maculosa*	Inhibits competing annual species
Lupane triterpene Messagenin (**107**)	*Melilotus messanensis*	Inhibits other dicot species
Hydrocarbons 2-Heptan-one and 2-heptanol	*Amaranthus palmeri*	Germination of onion and carrot seeds
Fatty acids C_{14}–C_{22} acids	*Polygonum aviculare*	Inhibits bermuda grass, *Cynodon dactylon*
Fatty acid derived α-Terthienyl and phenylheptatriyne	*Tagetes erecta*	Seedlings of test species
Alkaloid Caffeine	*Coffea arabica*	Autotoxic
Anthranilic acids Benzoxazinones	*Secale cereale*	Inhibits root growth in competing species

One phenolic that has been most frequently invoked as an allelopathic constituent is ferulic acid (3-methoxy-4-hydroxycinnamic acid) (Table 17), but there is still some question as to whether such a common plant acid is likely to be an effective inhibitor of germination or growth. It is unlikely to persist in the soil, since its ready microbial degradation to vanillic and protocatechuic acids can even occur during the short period when it is undergoing seed germination bioassays. Another reason for doubting the potency of ferulic acid as an allelopathic agent is the experience with *Asparagus* roots. Dried roots yielded ferulic acid and several related structures, all with only moderate inhibitory activity. By contrast, extraction of fresh root yielded 3,4-methylene-dioxycinnamic acid (Figure 13), not detectable previously, which turned out to be 10 times more toxic to test species than ferulic acid.[105]

Ferulic acid and other cinnamic and benzoic acids, which occur very widely in the leaves of many plants, are present in bound form, usually as esters, and hydrolysis must occur at some stage, since model experiments of leaching from grass leaves indicate the detection of the free acids rather than of the bound form.[106] Of the various simple phenolics implicated in allelopathy, the more likely candidates are compounds of restricted occurrence such as hydroquinone and salicylic acid (see Figure 13) rather than the almost universal hydroxyl cinnamic and benzoic acids (Table 17).

Evidence that volatile leaf terpenoids present in perennial shrubs can exert an allelopathic effect on competing annual species was earlier obtained by Muller and Chou[107] from studies on *Salvia leucophylla* and *Artemisia californica* growing in the Californian chapparal. Terpenoids were also implicated in several other earlier studies of allelopathy exerted by *Eucalyptus globosus*, *E. camuldensis*, *Artemisia absinthium* and *Sassafras albidum*.[99] However, the best data supporting the view that terpenes can produce inhibitory zones of growth on grasses were obtained by Fischer[102] for the labiate shrubs *Calamintha ashei* and *Conradina canescens*. A variety of monoterpenoids have been implicated, including well-known common terpenes and several relatively specific constituents such as calaminthone, menthofuran, and epievodone (Table 18 and Figure 14).

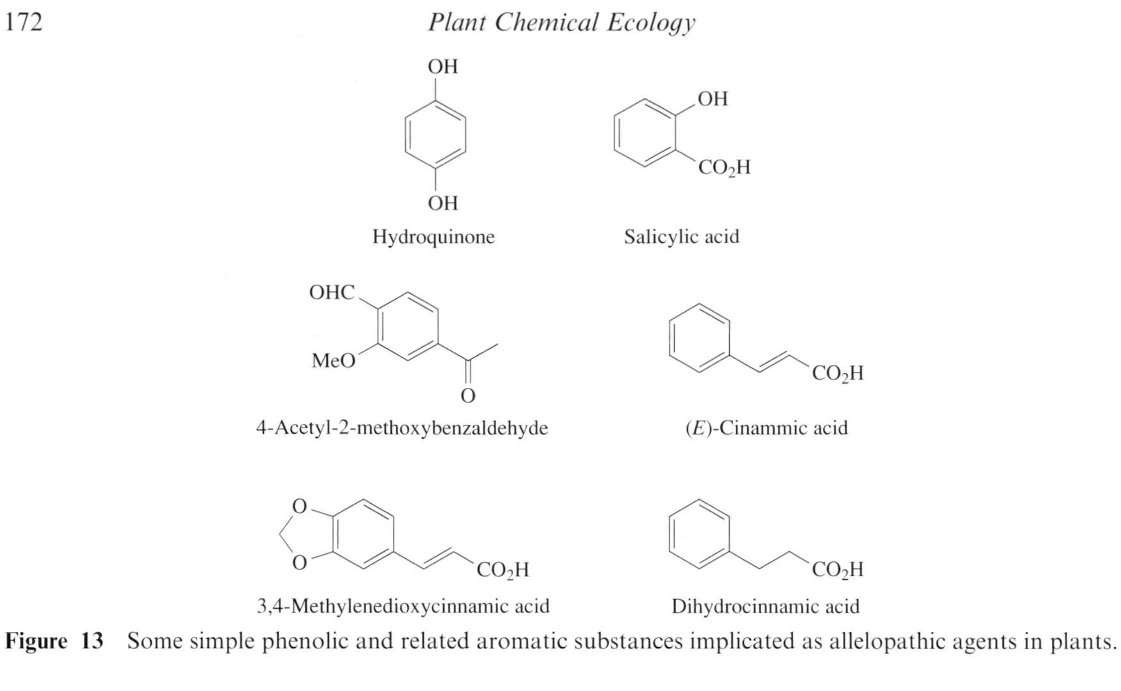

Figure 13 Some simple phenolic and related aromatic substances implicated as allelopathic agents in plants.

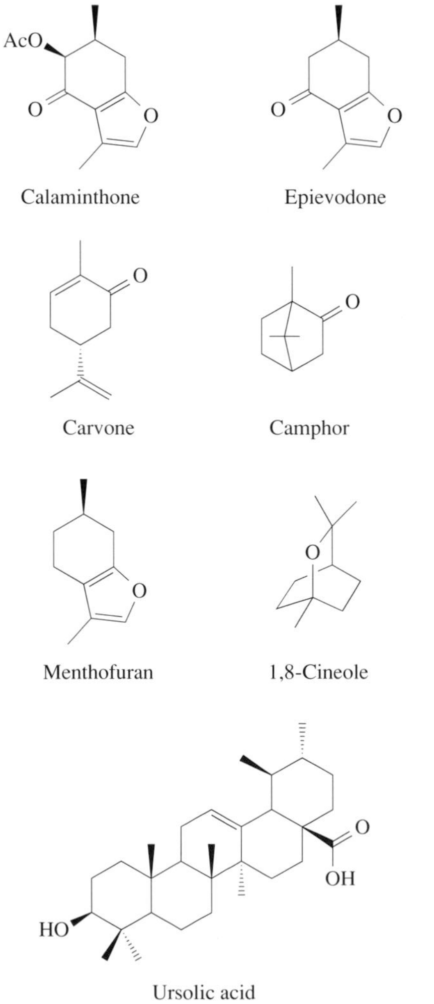

Figure 14 Some terpenoids implicated as allelopathic agents in plants.

A significant component of leaf washings of the above two shrubs, which grow in the scrub community of sandhills in Florida, is the triterpenoid ursolic acid. This compound is a natural detergent and Fischer[102] hypothesized that it improves the penetration of the monoterpenes into the target grass seeds, causing them to fail to germinate.

One question often asked regarding the possible allelopathic effects of monoterpenes is their presumed low water solubility and hence their inability to interact in the soil with seeds of other plant species. This has been answered by Fischer and co-workers,[108] who found that this low solubility is a myth and that oxygenated monoterpenes in particular have water solubilities sufficient for them to produce toxic effects. Thus, monoterpene ketones (e.g., menthone) are soluble in the range 155–6990 ppm, while alcohols such as geraniol are soluble in the range 183–1360 ppm. When tested against seed germination, the monoterpenes of labiate shrubs can exhibit phytotoxic damage in water at concentrations under 100 ppm.[108]

The allelopathic properties of both phenolic and terpenoid agents must depend to a considerable extent on their persistence over time in the upper layers of soil. Few studies have been carried out on their survival rates, but there is some information on their rate of degradation in leaf litter deposited in the soil under certain trees. Fresh leaves from the litter under the tree *Umbellularia californica* contain about 16% dry weight of terpene, but this amount dropped by between 63 and 86% 1 year after leaf fall. Loss of monoterpene from leaf litter was correlated with water solubility so that terpene ketones disappeared first and also losses were greatest in moist sites after rainfall.[109]

The loss of phenolics from leaf litter (and from soil) is probably due to degradation by aerial or enzymic oxidation, but some leaching by rainfall must presumably also take place. Significant concentrations, however, remain in leaf litter for several years after leaf fall. Thus, phenolic contents of up to 1.8 g kg^{-1} (measured as tannic acid equivalents) have been detected in pine leaf litter from northern California.[110]

That phenolics in the soil can have an adverse effect on seed germination in spruce, *Picea abies*, has been demonstrated by Pellissier.[111] He found that phenolics in the humus layer of soil prevented the natural regeneration of spruce, by inhibiting both seed germination and seedling growth. 3,4-Dihydroxybenzoic acid, *p*-hydroxybenzoic acid, catechol, and *p*-hydroxyacetophenone, identified in the humus, were implicated as a synergistic mixture causing this allelopathy.

One final question about secondary constituents and allelopathy may be raised: do substances other than phenolics or terpenoids have a role in allelopathic interactions? The answer would seem to be in the negative, although autotoxic effects of the alkaloid caffeine have been reported in coffee plantations.[100] Deliberate experiments to implicate the alkaloids of the *Cinchona* plants in allelopathic effects on seeds failed[112] and similarly no allelopathic effects could be detected for the mustard oil, allyl isothiocyanate, which is released by *Brassica napus* and other crucifer crops into the surrounding soil.[113]

8.03.8 BIOCHEMISTRY OF SYMBIOTIC ASSOCIATIONS

8.03.8.1 Flavonoids, Nitrogen Fixation, and Mycorrhizal Associations

As has already been mentioned (Section 8.03.7), it is well established that secondary plant constituents may be exported from the plant into the soil via the root. The concentrations thus exuded may be quite low, but nevertheless they may have significant effects on bacteria and fungi that inhabit the soil around the plant roots. In particular, flavonoids have been identified as signalling agents in the legume root–*Rhizobium* association and also as having inhibitory or promoting effects on some mycorrhizal fungi which infect higher plant roots.

In the case of the legume–*Rhizobium* symbiosis, it has been recognized that certain flavonoids exuded from the legume roots in minute quantities are essential for nitrogen fixation to take place, since they are responsible for switching on the nodulating genes in the infecting *Rhizobium* species. There is significant specificity in the process and in the three symbioses mainly studied (Table 19), different flavone or flavanone aglycones are involved. This ties in with the fact that symbiosis between *Rhizobium* species and legume hosts occurs in a highly species-specific fashion.[114] The flavonoids involved (Figure 15) are either common flavonoid structures (e.g., the flavone luteolin) or are 5-deoxyflavones, which occur characteristically in plants of the Leguminosae.

The interaction that has been most closely studied is that of *Rhizobium leguminosarum* with pea *Pisum sativum*.[115] Several other flavonoids are capable of inducing node gene expression in addition

Table 19 Variation in the flavonoid nodulating signal in different legume–rhizobial symbioses.

Symbiosis	*Inducing flavonoids*
Rhizobium trifolii on clover	7,4′-Dihydroxyflavone, 7-methoxy-4′-hydroxyflavone, 7,4′-dihydroxy-3′-methoxy-flavone
Rhizobium meliloti on lucerne	Luteolin (5,7,3′,4′-tetrahydroxyflavone)
Rhizobium leguminosarum on pea	Eriodictyol (5,7,3′,4′-tetrahydroxyflavanone), apigenin 7-glucoside

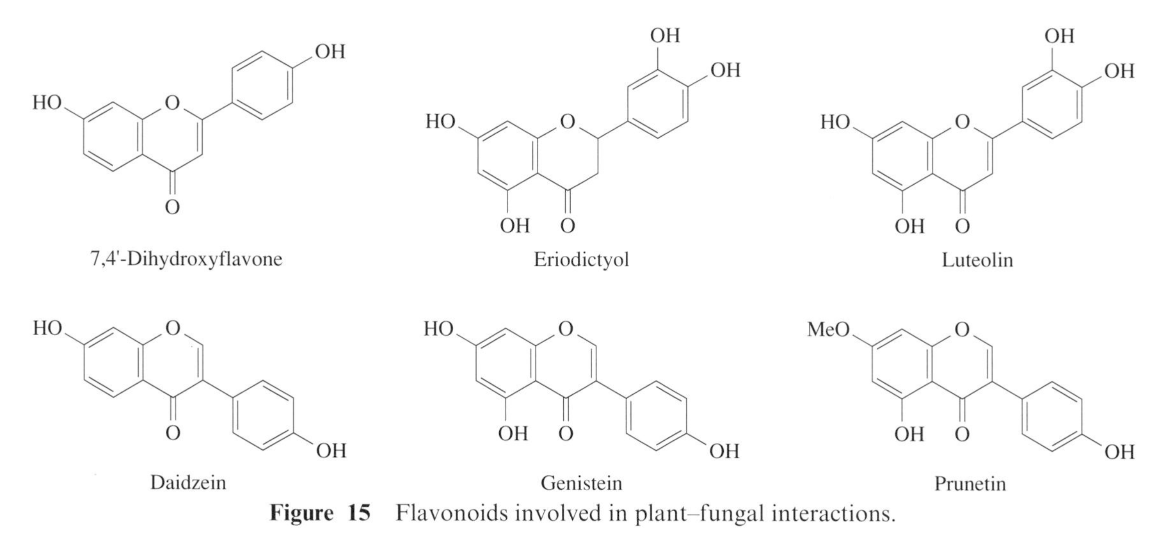

Figure 15 Flavonoids involved in plant–fungal interactions.

to the two substances, eriodictyol and apigenin 7-glucoside, which appear to be present naturally in pea root exudates (Table 20). These flavonoids are all active at a low concentration (e.g., 10 nmol dm^{-3}). It is notable that in almost all cases the flavonoids are only active in the aglycone form; the one exception seems to be apigenin 7-glucoside.

Table 20 Inducing and inhibiting flavonoids of nod gene expression in *Rhizobium leguminosarum* on pea.

Inducing flavonoid	*Relative activity at* 10 *nmol* dm^{-3}
Apigenin 7-glucoside	55
Apigenin	59
Luteolin	73
Naringenin	58
Eriodictyol	75
Hesperitin	250
Inhibiting flavonoid	Inhibition at 5 μmol dm^{-3} (%)
Kaempferol (flavonol)	60
Daidzein (isoflavone)	75
Genistein (isoflavone)	81

Equally important is the finding that other flavonoids, especially the isoflavones daidzein and genistein, are inhibitors of nodulating gene expression (Table 20). The concentrations required for inhibition, however, are two orders of magnitude higher than that required for gene induction.[115] Hence flavonoids have the capacity of regulating the expression of these nodulating genes in the bacteria, by being able to switch the process both on and off. This is linked with the fact that host plant legumes are known to limit the number of nodules present in the root system. In this economically important process of nitrogen fixation, flavonoids play a crucial role in determining the extent of nodulation in a variety of legume species.

Turning to the looser symbiotic association that exists in the soil between plants and fungi, it is apparent that isoflavones similar to those involved in nitrogen fixation are the main active agents.

Thus, prunetin (7-*O*-methylgenistein) (Figure 15), excreted from the roots of pea, is an attractant to zoospores of the fungus *Aphanomyces euteiches*, a causal agent of root rot in this plant. Similarly, genistein and its methyl ethers are attractants towards *Phytophthora sojae*.[116] By contrast, in the attraction of zoospores of *Aphanomyces cochlioides*, a causal fungus of spinach root rot, the host-specific 5-hydroxy-6,7-methylenedioxyflavone is involved rather than an isoflavone.[117] Finally, it should be mentioned that genistein and its 4′-methyl ether, biochanin A, have been reported to inhibit growth in the mycorrhizal fungus *Gigaspora margarita*,[118] in spite of the fact that ordinary flavonoids are known to promote the development of these mycorrhizal fungi more generally.[119]

8.03.8.2 Chemical Communication Between Host Plants and Their Parasites

There are more than 3000 higher plant species which have the ability to form specialized intrusive organs called haustoria, by which they attach themselves to other higher plants and feed on them. Biochemically, such plant parasites are of interest since the ability of their seeds to germinate and establish a haustorial connection depends entirely on chemical triggers, which exude from the roots of the unsuspecting host plant. Much effort has been expended in determining what chemicals pass from host to parasite and it is now clear that the compounds involved are typical secondary constituents.

The chemistry of the stimulants exuded from host plant roots to trigger off germination of the seeds of the parasitic witchweed *Striga asiatica* has recently been resolved through the identification of three closely related lactones. The first is strigol (Figure 16), characterized from maize, *Zea mays*, and millet, *Panicum miliaceum*.[120] Small amounts of strigol are also released from roots of *Sorghum*, another host plant, but the major stimulant of *Sorghum* is sorgolactone.[121] A third related compound has been obtained from cowpea roots, *Vigna unguiculata*, which stimulates germination of the closely related parasite *Striga gesneroides*.[122] The three substances (Figure 16) are incredibly active in that concentrations in the range 10^{-11}–10^{-15} mol dm^{-3} have been found to set off *Striga* seed germination. For this group of sesquiterpene lactones active as *Striga* germination stimulants, the collective name "strigolactone" has been proposed.[123] *Orobanche* is another genus of parasites depending on seed germination stimulants and a similar strigolactone has been characterized in root exudates from one of its host plants.

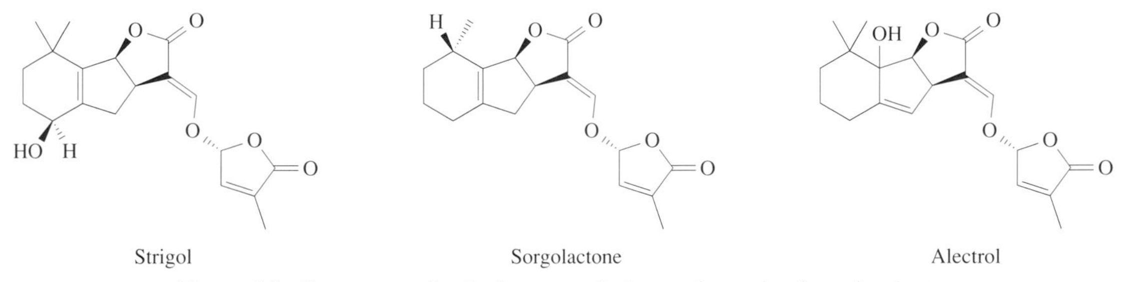

Figure 16 Structures of strigolactones, *Striga* seed germination stimulants.

The second message of the *Striga–Sorghum* interaction—the haustorial inducer—has not yet been fully identified, but it may be closely related to 2,6-dimethoxy-*p*-benzoquinone. This compound is produced in surface-abraded sorghum roots exposed to a *Striga* enzyme, and is able to initiate haustorial formation; however, it is not formed by undamaged roots.[123] Haustoria-inducing compounds have been described from legumes parasitized by *Agalinis purpurea*. The simple dihydrostilbene xenognisin A and the isoflavone xenognisin B have been isolated from root exudates of tragacanth, a species of *Astragalus*. Again, the triterpene sayasapogenol A (Figure 17) with haustoria-inducing activity was obtained from roots of *Lespedesa sericea*, also a host of *Agalinis*. Hence, more than one class of secondary constituent have the ability to act as second messenger in the host–parasite interactions of *Agalinis*.[124]

8.03.8.3 Chemical Transfer from Host to Parasite

While higher plant parasites are usually dependent on their host for primary metabolites, this is not so in terms of secondary chemistry and most such parasites develop their own range of terpenoid and phenolic constituents, which are different from those of the host. Species of the *Orobanche*

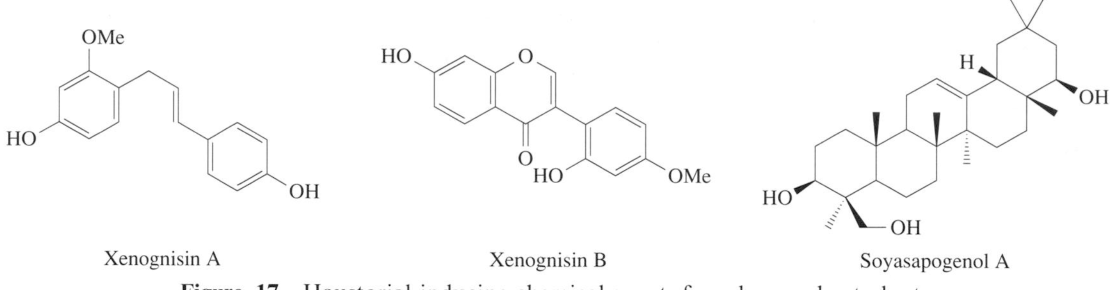

Figure 17 Haustorial-inducing chemical agents from legume host plants.

parasites, for example, produce a range of caffeic acid esters, including orobanchoside, which are characteristic of these plants. This does not mean, however, that secondary constituents cannot on occasion pass from one plant to another, through the haustorial connection. In fact, a number of examples have been found where host alkaloids are taken up by semiparasites belonging to the genera *Castilleja*, *Pedicularis*, and *Cuscuta* (Table 21).[125–127]

Table 21 Transfer of alkaloids from host plants to parasitic plants.

Host plant	*Alkaloid*	*Semiparasite*
Senecio atratus	Senecionine (Figure 7)	
Lupinus spp.	*N*-Methylcytisine	*Castilleja* spp.
Thermopsis spp.	*N*-Methylcytisine	
Senecio trangularis	Senecionine	*Pedicularis bracteosa*
Thermopsis montana	Anagyrine (**22**)	*P. crenulata*
Lupinus argenteus	Lupanine (**108**)	*P. racemosa*
Picea engelmanni	Pinidinol (**109**)	*P. bracteosa*
Genista acanthaclada	Lupanine	
	Cytisine (**23**)	*Cuscuta palaestina*
	Anagyrine	

(**108**) Lupanine (**109**) Pinidinol

It may be significant that the transfers observed all involve alkaloids, since it is known that these substances can be moved around plants via the phloem as *N*-oxides. Furthermore, the uptake of alkaloids by the semiparasite is probably beneficial and will supplement the defensive armory already in place in these plants. Thus *Castilleja* and *Pedicularis* spp. are known to synthesize their own iridoid toxins.

It may be noted that uptake of alkaloids from the host plants listed in Table 21 is selective, and only one or more of the alkaloids of the host are found in the parasite. For example, only three of the 20 alkaloids in *Genista* are found in the dodder *Cuscuta palaestina*. *Pedicularis bracteosa* is exceptional in being able to borrow the pyrrolizidine senecionine (see Figure 7) when it is attached to an annual weed, a *Senecio* species, and the pyrrolidine pinidinol (**109**) when it parasitizes the gymnosperm tree *Picea engelmanni*. Such a catholic borrowing from two such different host plants must be quite rare. One other feature of this phenomenon is that the amount of alkaloid borrowed can be quite variable and studies of semiparasite populations show some individuals with high levels and others with trivial quantities of borrowed toxin.

8.03.9 MYCOTOXINS

Mycotoxins are secondary metabolites of fungal origin which contaminate certain plants, e.g., ground nuts and cereals, as a result of infection. They pose a threat to animals eating such plant tissues since many mycotoxins are highly poisonous. The best known group of mycotoxins are the aflatoxins, which are formed in ground nuts infected by *Aspergillus flavus*. The aflatoxins are lethal coumarins, readily detected in contaminated plant material by their intense fluorescences. Their toxic effects in birds and mammals are well documented.[128] Some mycotoxins such as cercosporin and rugulosin are known to be phototoxic and the same is true of the dibenzopyrone mycotoxin of *Alternaria* species, alternariol.

Whether the occurrence of mycotoxins in plants is purely fortuitous or whether there is an underlying ecological rationale is still debatable. Plant–microbe associations are not necessarily harmful to the plant, and some such associations may well be beneficial to both partners. Thus, there is evidence that ryegrasses grow better if they are infected by an endophytic fungus and one capable of synthesizing mycotoxins. In such cases, and several will be reviewed below, the plant may derive an extra benefit from the production of mycotoxins, since the plant would be better able to deter herbivores in the natural ecosystem.

The first example refers to a bacterial rather than a fungal infection and is concerned with the production of the corynotoxins. These are poisonous principles of the annual ryegrass, *Lolium rigidum*, that has been infected by *Corynebacterium rathayi*. Toxicity here is due to the interaction of three different organisms, since the bacterial infection is introduced into the plant via a nematode, *Anguina agrostis*. This nematode, carrying the bacterium, produces a gall in the seedhead of the grass and then the bacterium multiplies in the gall, at the same time producing the toxins. Infected grasses are very poisonous to grazing sheep. The corynetoxins are complex glycolipids that contain *N*-acetylglucosaminyltunicaminyluracil in amide linkages with different fatty acids.[129]

A second example of mycotoxins benefiting the infected plant is the case of *Lolium perenne* infected by the fungus *Acremonium loliae*. A major mycotoxin produced is lolitrem B (Figure 18), a tremorgenic agent producing "ryegrass staggers" in sheep and cattle which feed on the infected pasture grass. Eradication of this fungus from ryegrass would therefore seem to be desirable agriculturally, since it would remove a potential threat to livestock. However, it is clear that the presence of the fungus not only increases the vigor of the grass but also protects it against insect pests, such as the Argentinian stem weevil, *Listronotus bonariensis*. Rowan *et al.*[130] have found that insect resistance of infected ryegrass is due to the presence of a novel alkaloid, peramine, and not due to lolitrem B. Peramine has been detected in the mycelium of *Acremonium loliae*, so that it is clearly of fungal origin. It is probably formed biosynthetically by the cyclization of proline with arginine, whereas lolitrem B is presumably derived from an indole isoprenoid precursor (Figure 18). The defensive chemicals formed in these grass–fungal endophyte associations have been reviewed in depth.[131]

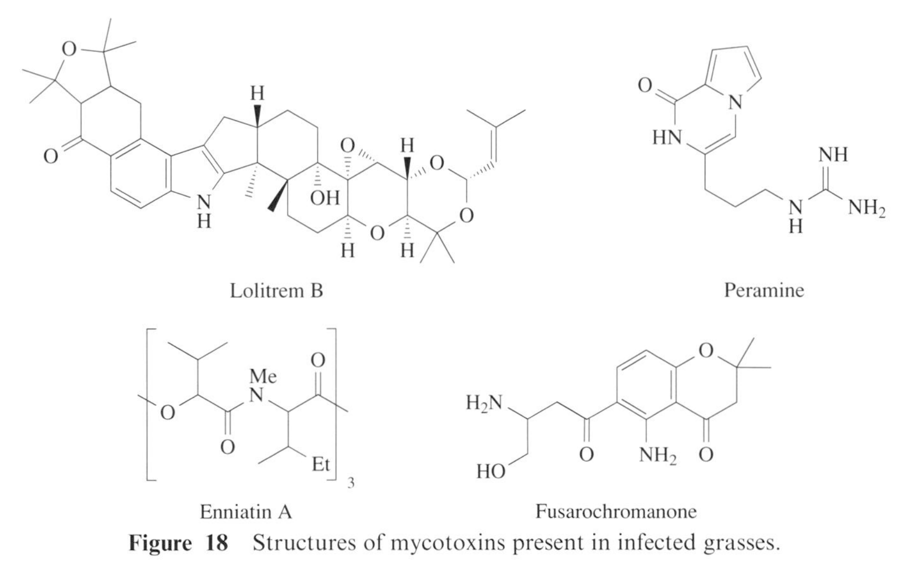

Figure 18 Structures of mycotoxins present in infected grasses.

Another example of a mycotoxin-producing fungus conferring resistance to insects of a host plant is that of *Fusarium avenaceum* infecting the foliage of balsam fir, *Abies balsamea*. This would explain the unexpected collapse of infestations of the spruce budworm, *Choristoneura fumiferana*, which happens from time to time on this tree.[132] The toxin enniatin A, a cyclohexadepsipeptide, has been identified in the foliage of balsam fir and in feeding experiments to the budworm larvae it is toxic when applied at a concentration of 0.04%. The spread of this cereal pathogen to a gymnosperm is a remarkable feature of this complex interaction. *Fusarium* species which infect cereals produce several other mycotoxins in addition to enniatin A. One is fusarochromanone, produced by *F. roseum* infecting this oat crop in Alaska. This compound is toxic to poultry and reduces the hatchability of fertile hens' eggs.[133]

It is appropriate to close this section on mycotoxins with a reference to the ergot alkaloids, produced by the ergot fungus *Claviceps purpurea* infecting rye, since these are the oldest known groups of mycotoxins. They have recently been found for the first time in tall fescue grass, *Festuca arundinacea*. The fungal endophyte is *Acremonium coeniphialum* and its infection on tall fescue causes toxic symptoms in grazing cattle, including "fescue foot," a gangrene of the hooves. Ergovaline was the principle ergot alkaloid detected, with the total alkaloid content varying from between 1.5 mg and 14 mg per kilogram of plant tissue. The fungal origin of the ergot alkaloids was established by their production in pure cultures by this fungus.[134]

8.03.10 PHYTOTOXINS

The term phytotoxin is applied by plant pathologists to microbial substances which are responsible for producing the symptoms of disease in higher plants. Other terms used for these fungal or bacterial toxins include "pathotoxin" to indicate their pathogenic effects, and "host-selective toxin" to indicate that it is specifically related to susceptibility in the host plant. In many but not all cases, phytotoxins are produced by the invading microorganism once the higher plant is infected and they may be responsible for causing the disease symptoms, which vary from chlorosis, wilting, or necrosis to growth abnormalities or death. Such symptoms can often be reproduced by inoculating healthy plants with trace amounts of the pure toxin. Although phytotoxins are occasionally formed in liquid culture by the fungus or bacterium, the emphasis has been to isolate the phytotoxin from infected plant tissue. In almost all cases, phytotoxins are active at very low concentrations and application of solutions of the toxin at 1×10^{-3} mol dm^{-3} will generally produce a necrotic lesion in a test plant.

Once, relatively few toxins were fully characterized, but with advances in experimental techniques many more structures have been uncovered in recent decades. A bewildering range of secondary metabolites have been described from simple low-molecular-weight constituents, through complex diterpenoids, lactones, macrolides, peptides, and phenolics to proteins and polysaccharides.

The basis of action of these toxins on the higher plant infected is still largely unknown. Nevertheless, much work has been devoted to fusicoccin, the toxin of *Fusicoccum amygdali* on almond trees. Its toxic effects have been traced to an interference with the function of the leaf plasmalemma.[135] Other toxins are thought to scavenge essential metal ions from the plant (e.g., fusaric acid (**110**)) or to interfere with ion transport across the chloroplast membrane.[136] Here, attention is mainly devoted to outlining the structural variation that has been encountered to date among the known phytotoxins.

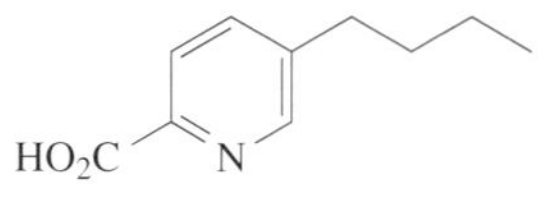

(**110**) Fusaric acid

Although most phytotoxins have relatively complex structures, there are a few examples where simple molecules have been recognized as toxic agents (Figure 19). The simplest is 3-(methylthio)panoic acid (see Figure 19), which is produced by the cassava pathogen *Xanthomonas campestris* pv. *manihotis* on cassava. It is effective at levels of 6 μg g^{-1} fresh weight in producing characteristic symptoms of bacterial blight in cassava leaves.[137] A second simple toxin is tryptophol, produced by *Drechslera nodulosum*, which is a pathogen of goose grass, *Eleusine indica*.[138] A third simple toxin is *S*-dihydrophenylalanine, formed by the fire-blight pathogen *Erwinia amylovora*, which attacks pear and apple trees. It is able to upset the redox potential of the plant cell and may be a trigger in infected plants for the suppression of the hypersensitive response.[130]

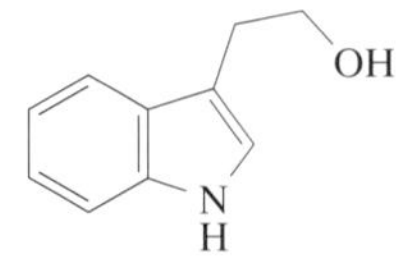

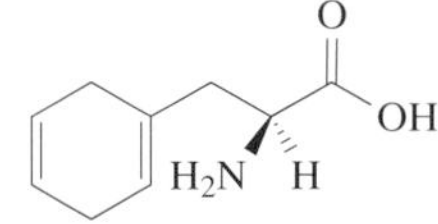

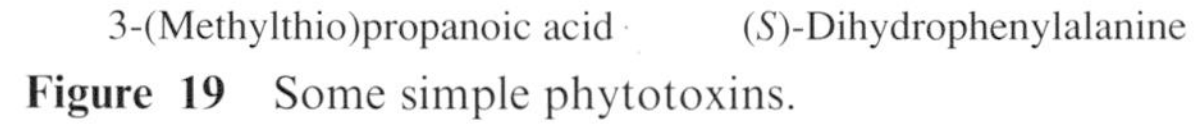

Tryptophol   3-(Methylthio)propanoic acid   (*S*)-Dihydrophenylalanine

Figure 19 Some simple phytotoxins.

The best known terpenoid phytotoxins are undoubtedly the diterpenoid fusicoccin, from *Fusicoccum amygdali*, which has already been mentioned above, and the sesquiterpenoid helminthosporoside. The latter is actually a mixture of three closely related structures, illustrated here with the compound helminthosporoside A (Figure 20). The notable feature of all three isomers is the presence of digalactosyl residues at both ends of the molecule, with galactose taking up the rare furanose configuration. These sugar attachments are essential for phytotoxic activity. Helminthosporoside is responsible for the red "eye spot" symptom of sugar cane and is produced by the infecting fungus, *Helminthosporium sacchari*.[136]

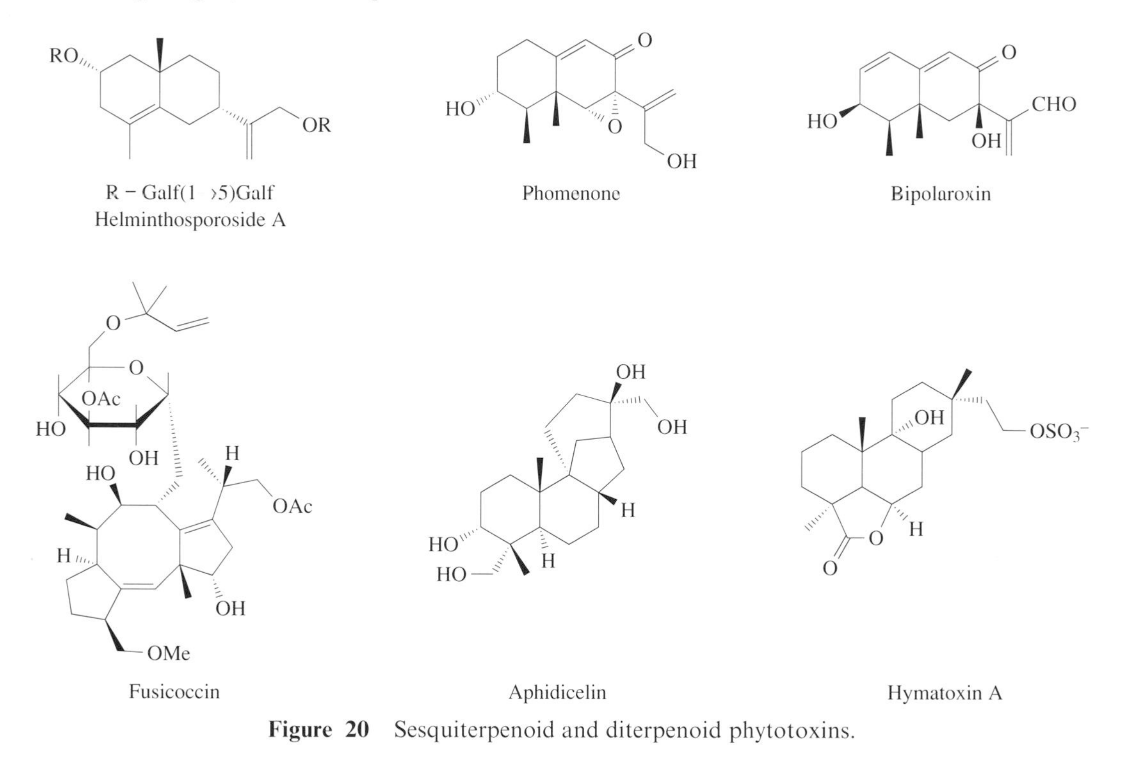

Figure 20 Sesquiterpenoid and diterpenoid phytotoxins.

Other examples of sesquiterpenoid phytotoxins are the eremophilane epoxide, phomenone, causing wilt in tomatoes infected by *Phoma destructiva* and bipolaroxin, from *Bipolaris cynodontis*, a fungal pathogen of bermuda grass, *Cynodon dactylon*. Other examples of diterpenoid phytotoxins are aphidicolin, from *Phoma betae*, a pathogen of beet, *Beta vulgaris*, and hymatoxin A, from *Hypoxylon mammatum*, which is a parasite on aspen, *Populus tremuloides*. It is unusual in having a sulfate substituent (Figure 20).[9–11]

Polyketide-based lactone toxins have been recognized in a number of fungal pathogens, but especially in the genus *Alternaria*. Mixtures of toxins are common. Thus, both *Alternaria helianthi* and *A. chrysanthemi*, pathogens of sunflower and chrysanthemum, respectively, produce the two toxins radicinin and deoxyradicinin (Figure 21). Similarly, *Alternaria solani* on potato produces three related pyrones as well as the lactone zinnolide. A more complex lactone fijiensin is one of the six toxins produced by *Mycosphoerella fijiensis*, the pathogen causing "black sigatoka" disease in bananas. The other toxins are 2,4,8-trihydroxytetralone, juglone, 2-carboxy-3-hydroxycinnamic acid, isoochracinic acid (**111**) (a phthalide), and 3,4,6,8-tetrahydroxytetralone (**112**).[140]

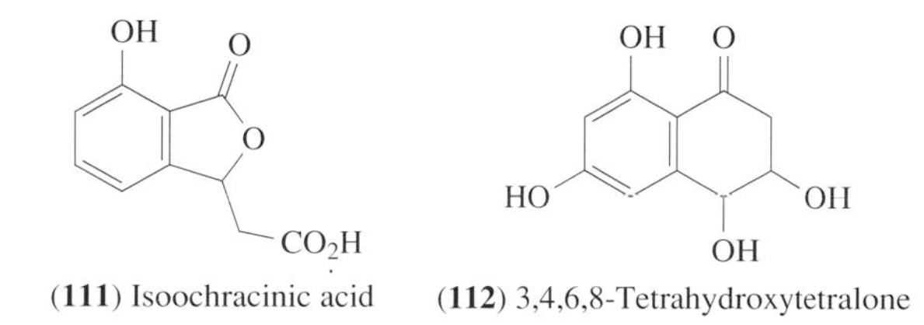

(**111**) Isoochracinic acid (**112**) 3,4,6,8-Tetrahydroxytetralone

A variety of polyketide-based phenolic phytotoxins have been characterized from time to time but it is not clear whether the phenolic group is essential for phytotoxicity. A small selection of phenolic toxins are shown in Figure 22. The toxicity of phomazin is unexpected, since this is a

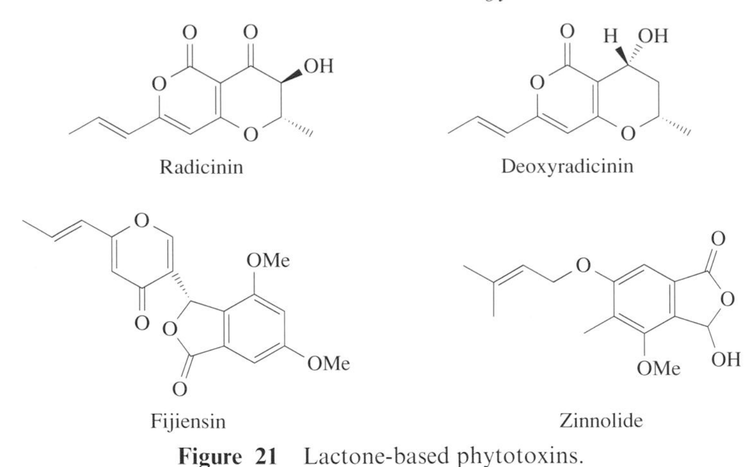

Figure 21 Lactone-based phytotoxins.

simple derivative of 2,4-dihydroxybenzoic acid. It is the causal agent of stem canker, *Phomopsis helianthi*, in sunflowers and, in fact, is highly phytotoxic. Application of 5 μg produces a brown lesion on sunflower leaf within 24 h.[141] The toxicity of eutypine, the causal agent of the "dying arm" disease in the grapevine, is more acceptable, since it has an alkynic substitution in its structure.[142] The third phenolic illustrated, cyperine, which is a diphenyl ether, is synthesized by the fungal pathogen *Ascochyta cypericola* when it infects the purple nutsedge *Cyperus rotundus*.[143] The fourth phenolic, de-*O*-methyldiaporthin, is the isocoumarin phytotoxin of *Drechslera succans*, a pathogen of oats and rye grass. It is relatively nonselective in its action and will produce necrotic lesions in nonhost plants at the 1 μg level.[144]

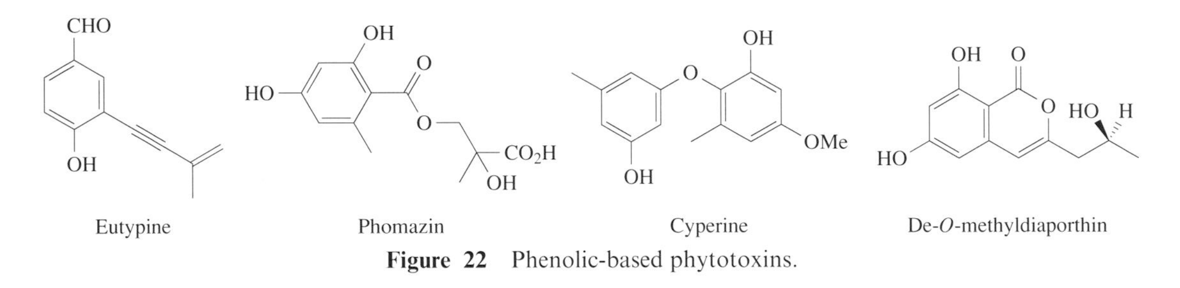

Figure 22 Phenolic-based phytotoxins.

A variety of nitrogen-based phytotoxins have been described, namely some amino acid-based structures and various cyclic or linear peptides. Those based on tryptophan, which include the cyclochalasins from *Helminthosporium*, have been reviewed by Cutler[145] and further details are therefore available. It is appropriate to illustrate this group of toxins with the structure of victorin C, a pathotoxin of the oat disease *Cochliobolus victoriae* (syn. *Helminthosporium victoriae*), which was isolated in crude form ~20 years before its structure was determined (Figure 23). It required a Herculean effort in chemical analysis to determine the position of the unusual amino acid residues in this complex cyclic peptide, and in the several closely related analogues which accompany it in *Cochliobolus*-infected plants.[146]

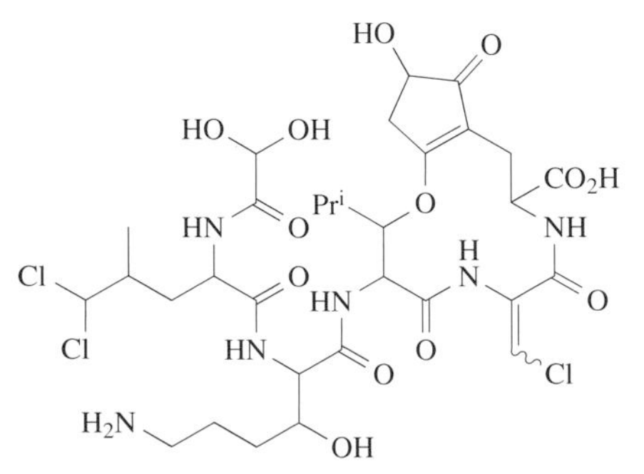

Figure 23 Structure of victorin C of *Cochliobolus victoriae*.

One further group of natural phytotoxins are those which are monocyclic. Some examples of compounds with nine- or ten-membered rings are illustrated in Figure 24. Putaminoxin, together with putaminoxin B (the related structure with a pentyl side chain) and putaminoxin C, are the phytotoxins of *Phoma putaminum*. This is the causal pathogen of leaf necrosis in *Erigeron annuus*.[147] Pinolidoxin is the toxic agent of *Ascochyta pinodes*, a fungus which causes the pea anthracnose disease.[148] Cornexistin, accompanied by the 14-hydroxyl derivative, is a phytotoxin of the fungus *Paecilomyces variotti*. Its high activity against broadleaf weeds and its small effect on cereal crops suggest that it might be a useful herbicide.[149]

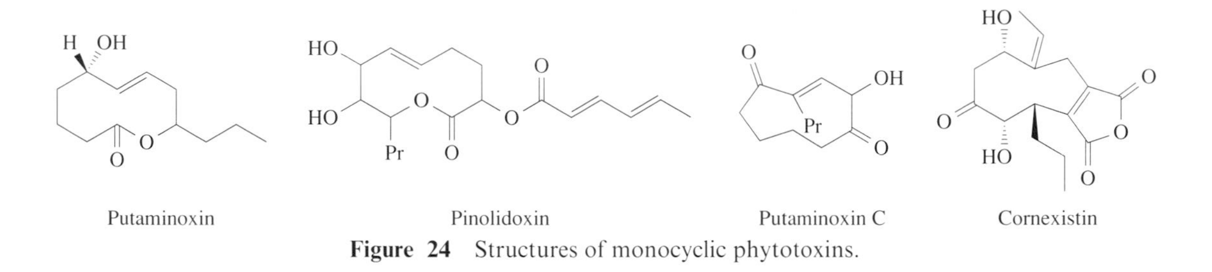

Figure 24 Structures of monocyclic phytotoxins.

This survey of some natural phytotoxins that have been characterized would not be complete without a mention of macromolecular toxins. Various fungal proteins, glycoproteins, and polysaccharides have been implicated as phytotoxic agents, particularly being responsible for wilting symptoms in higher plants. The best known are ceratoulmin, a small protein from *Ceratocystis ulmi*, the fungus responsible for Dutch elm disease, and malseccin, a glycoprotein producing disease symptoms in leaves of lemon infected by *Phoma tracheiophila*. Malseccin contains mannose, galactose, and glucose residues attached to a peptide backbone. There are a number of other macromolecules implicated in the symptoms of plant disease or in the elicitation of a response to infection and, in the case of proteins, sequencing data are sometimes available.[150]

8.03.11 CONSTITUTIVE ANTIMICROBIAL DEFENSE

8.03.11.1 Toxins at the Plant Surface

Just as the first line of chemical defense in a plant against herbivory is at the leaf surface (see Section 8.03.1.1), so it is against microbial infection. Lipophilic substances, mainly terpenoids and phenolics (see Table 22), have been recognized in leaf hairs, or in the leaf wax, which are antimicrobial and which occur in sufficient concentration to represent a barrier to the germination or growth of fungal spores. Theoretically, the same substance, if it is located at the surface, might provide protection against both herbivory and microbial invasion. In practice, however, compounds that have been reported to be antifeedants (see Chapter 8.05) are rarely tested for antimicrobial activity and vice versa, so it is not possible to assume that they are necessarily bifunctional. Again, several antifungal substances may co-occur in the leaf wax and, while synergism may well operate, it has rarely been examined in plant–microbial interactions.

Three of the best known antifungal agents at the leaf surface are sclareol, episclareol, and 2-ketoepimanool (Figure 25), which occur in the wax of *Nicotiana glutinosa* leaves. Sclareol and episclareol effectively inhibit fungal growth *in vitro*, possibly because they are structural analogues of biosynthetic precursors of gibberellic acid and interfere with normal hormonal development in the fungus. Likewise, when 2-ketoepimanool is applied to leaf surfaces of susceptible cultivars of *N. tabacum*, it suppresses the emergence of conidial germ tubes of *Erysiphe cichoracearum* and hence prevents powdery mildew developing. Another series of diterpenoids have been encountered in the leaf wax of the tobacco plant *N. tabacum* and appear to provide resistance here to blue mold disease.[151]

Several flavonoids that exhibit antifungal activity have been reported to be present on leaf surfaces (Table 22). Pinocembrin, for example, is secreted in leaf glands of *Populus deltoides* and is particularly active against the fungus *Melampsora medusae*. Sakuranetin (Figure 25) occurs in the leaf glands of blackcurrant bushes, but only on the upper surface. It successfully inhibits germination of conidia of *Botrytis cinerea*. Likewise, the flavanone 6-isopentenylnaringenin (**117**) is an antifungal agent present in the resin of the hop plant *Humulus lupulus*.[152]

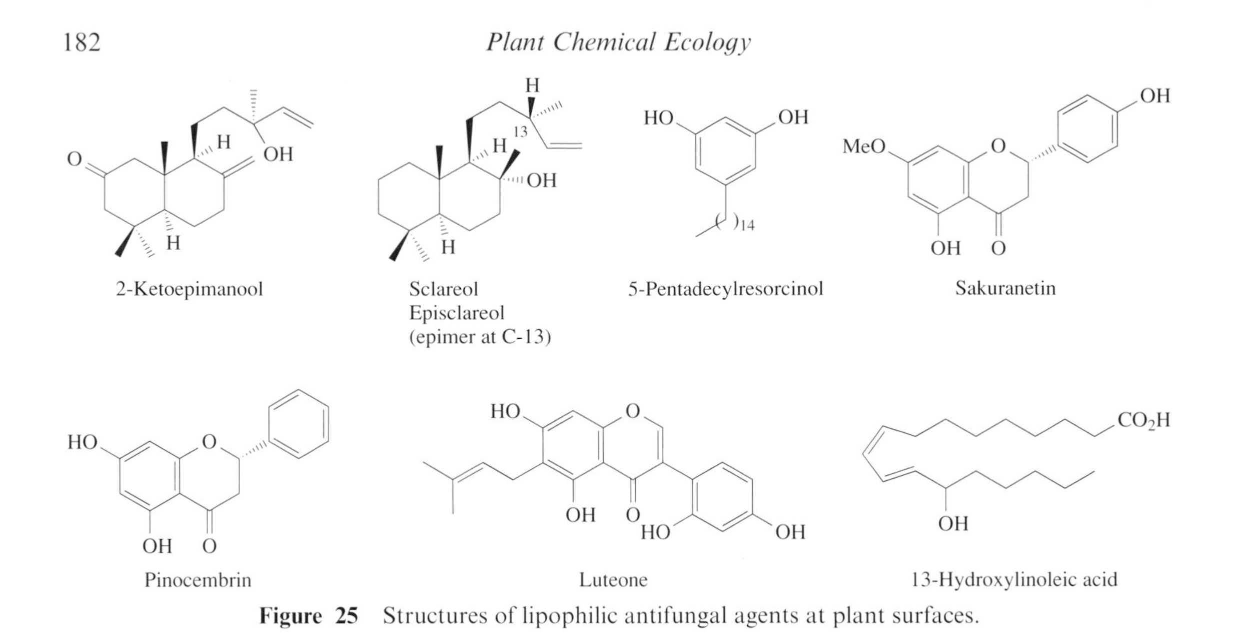

Figure 25 Structures of lipophilic antifungal agents at plant surfaces.

Table 22 Some examples of lipophilic antifungal constituents in plants.

Secondary metabolite	*Source*[a]
Phenolics	
Chrysin dimethyl ether (**113**)	Leaf surface, *Helichrysum nitens*
2,6-Dimethoxybenzoquinone	Root, *Croton lacciferus*
6-Isopentenylnaringenin	Resin, *Humulus lupulus*
Luteone	Leaf surface, *Lupinus albus*
5-Pentadecylresorcinol	Fruit peel, *Mangifera indica*
Pinocembrin	Leaf gland, *Populus deltoides*
Quercetin 7,3′-dimethyl ether	Leaf surface, *Wedelia biflora*
Sakuranetin	Leaf glands, *Ribes nigrum*
Terpenoids	
Cucurbitacin I (**114**)	Fruit, *Ecbollium elaterium*
α-4,8,13-Duvatriene-1,3-diol (**115**)	Leaf surface, *Nicotiana tabacum*
2-Ketoepimanool	Leaf wax, *Nicotiana glutinosa*
Pisiferic acid (**116**)	Leaf wax, *Chamaecyparis pisifera*
Sclareol/isosclareol	Leaf wax, *Nicotiana glutinosa*
Fatty acid derivatives	
2-Decanone	Stem bark, *Commiphora rostrata*
13-Hydroxylinoleic acid	Rice leaf, *Oryza sativa*
Stearic acid	Needle surface, *Pinus radiata*
1,2,4-Trihydroxyheptadec-16-yne	Fruit peel, *Persea americana*

[a] For references, see references. **151**–**156**.

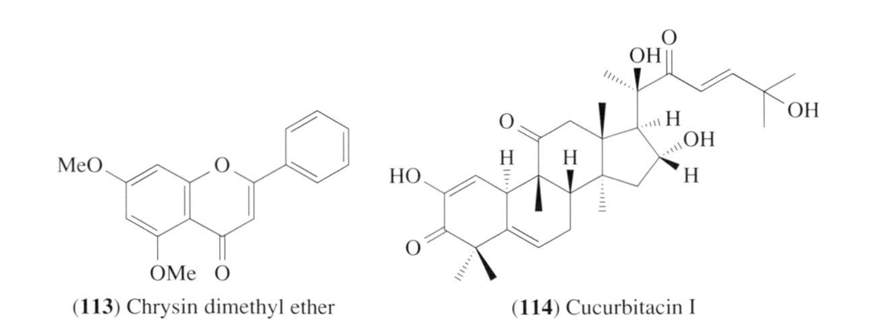

(**113**) Chrysin dimethyl ether (**114**) Cucurbitacin I

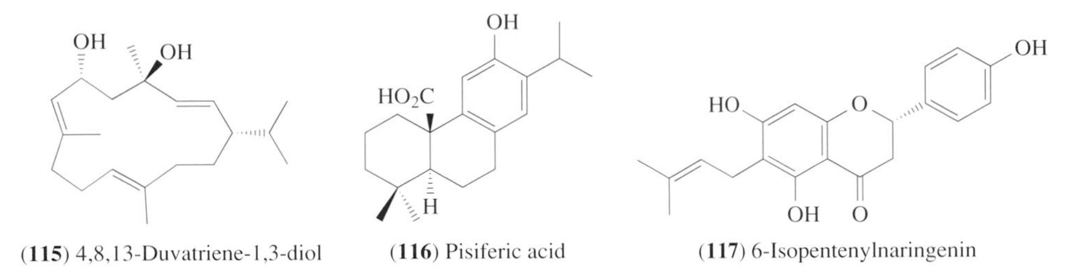

(**115**) 4,8,13-Duvatriene-1,3-diol (**116**) Pisiferic acid (**117**) 6-Isopentenylnaringenin

Isoflavonoids, particularly when they are isoprenylated, are lipophilic in character and may be encountered on plant surfaces. This is true of luteone (Figure 25) and the related wighteone (**118**), which occur on leaf surfaces of *Lupinus albus* and other *Lupinus* species. Antifungal activity is pronounced against *Helminthosporium carbonum* and there is good circumstantial evidence that these isoflavones are protective antifungal agents. Many similar antifungal isoflavonoids have also been encountered on the surface of lupin roots.[153] Pathogenic fungi may be able to overcome such defense at the surface by metabolizing and detoxifying the substances in question. Indeed, luteone is metabolized by cultures of *Botrytis cinerea*. The isoprenyl side chain is hydrated and ring closes on the 2-hydroxyl group to yield a much less fungitoxic metabolite.[154]

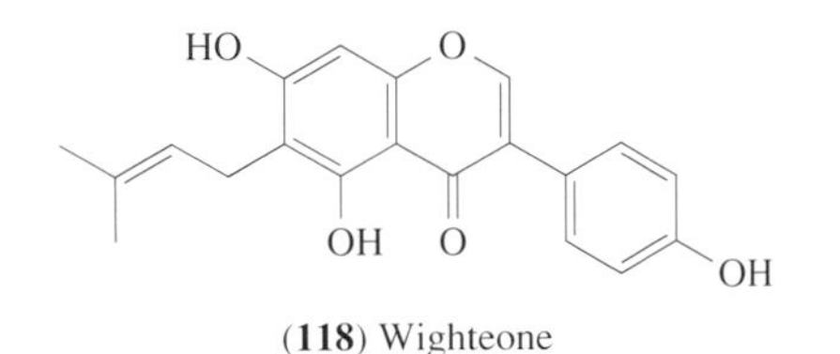

(**118**) Wighteone

The surfaces of unripe fruits may contain antifungal agents that protect them from infection until ripening takes place. This happens in the case of the mango, *Mangifera indica*, where 5-pentadecylresorcinol and related structures accumulate in the peel and prevent infection by the black spot disease *Alternaria alternata*. After ripening, those resorcinols disappear from the peel.[151] A similar system operates in the case of the fruit of the avocado, *Persea americana*, but the antifungal agents in the unripe peel are long-chain alcohols, such as 1,2,4-trihydroxyheptadec-16-yne.[151]

Perhaps the most unexpected class of surface antifungal agents are long-chain fatty acids and their hydroxyl derivatives. Such compounds, including stearic acid, have been detected in the needle surface of *Pinus radiata* and appear to protect the plant from the pine disease organism *Dothistroma pini*. Likewise, a number of epoxy and hydroxyl linoleic and linolenic acids (Figure 25) are reported to accumulate on the leaf surface of rice leaves and help to prevent infection by the blast disease *Pyricularia oryzae*.[155] Antifungal activity in hydroxyl fatty acids has also been observed in fungal–fungal interactions. Thus, Bowers *et al.* found 8-hydroxylinoleic acid to be produced by *Laetisaria arvalis* in order to attack the competing soil fungus *Pythium ultimum*.[156]

8.03.11.2 Bound and Other Toxins

A familiar concept in ecological biochemistry is the storage within plants of toxins needed for protection in a safe, inactive, bound form. The cyanogenic glycosides, for example, are a bound form of the animal toxin cyanide, which is released from the glycoside following enzymic hydrolysis (see Section 8.03.1.4). Such a mechanism also operates against disease attack in plants. There is evidence, for example, that the cyanogen linamarin in *Lotus corniculatus* is protective against some fungal organisms. Nevertheless, the leaf-spot pathogen *Stemphylium loti* can overcome the release of cyanide by detoxifying it via the enzyme formamidehydrolyase with the production of formamide, $HCONH_2$.[157]

The most frequently encountered bound toxins with antifungal properties in plants are saponins (Table 23). The interaction between plant and micro-organism may, however, be relatively complex. For example, the ivy leaf produces two related saponins, hederasaponin B and C (Figure 26), which are stored in the cell vacuoles. When the ivy leaf is damaged by penetration of an invading fungus, the two saponins undergo partial hydrolysis, with loss of the sugars attached to the 26-carboxylic acid group. The α- and β-hederins thus formed are highly toxic to a range of fungi. The *in vivo* concentration in the plant can reach 150 mg ml^{-1} plant juice. This is a powerful barrier, considering the most fungi are completely inhibited by α-hederin at 50–250 μg ml^{-1}. Significantly, if there is further loss of sugar at the 3-positions of α- and β-hederin, the resulting sapogenins (Figure 26) are completely inactive. The toxicity of α- and β-hederin to most fungi is thought to be due to their interference with fungal cell membranes at particular sites where sterols are attached.[157] Other examples of saponins which have been implicated as bound antifungal agents are given in Table 23. The oat plant is particularly well protected in this way, and two different sets of active saponins have been isolated from leaf and root, respectively.[151]

Table 23 Saponins identified as antifungal agents in plants.

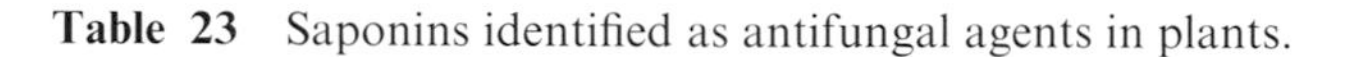

Saponin	*Source*[a]
Avenacin A and B (**119**)	*Avena sativa* root
Avenacoside A and B (**120**)	*Avena sativa* leaf
Camellidin I and II (**121**)	*Camellia japonica* leaf
Cyclamin (**122**)	*Cyclamen* cv. leaf
Hederagenin 3-glucoside	*Dolichos kilimandschariens* leaf
Hederasaponin B and C	*Hedera helix* leaf
Sakurasosaponin (**123**)	*Rapanea melanophloeos*

[a] For references, see reference 151.

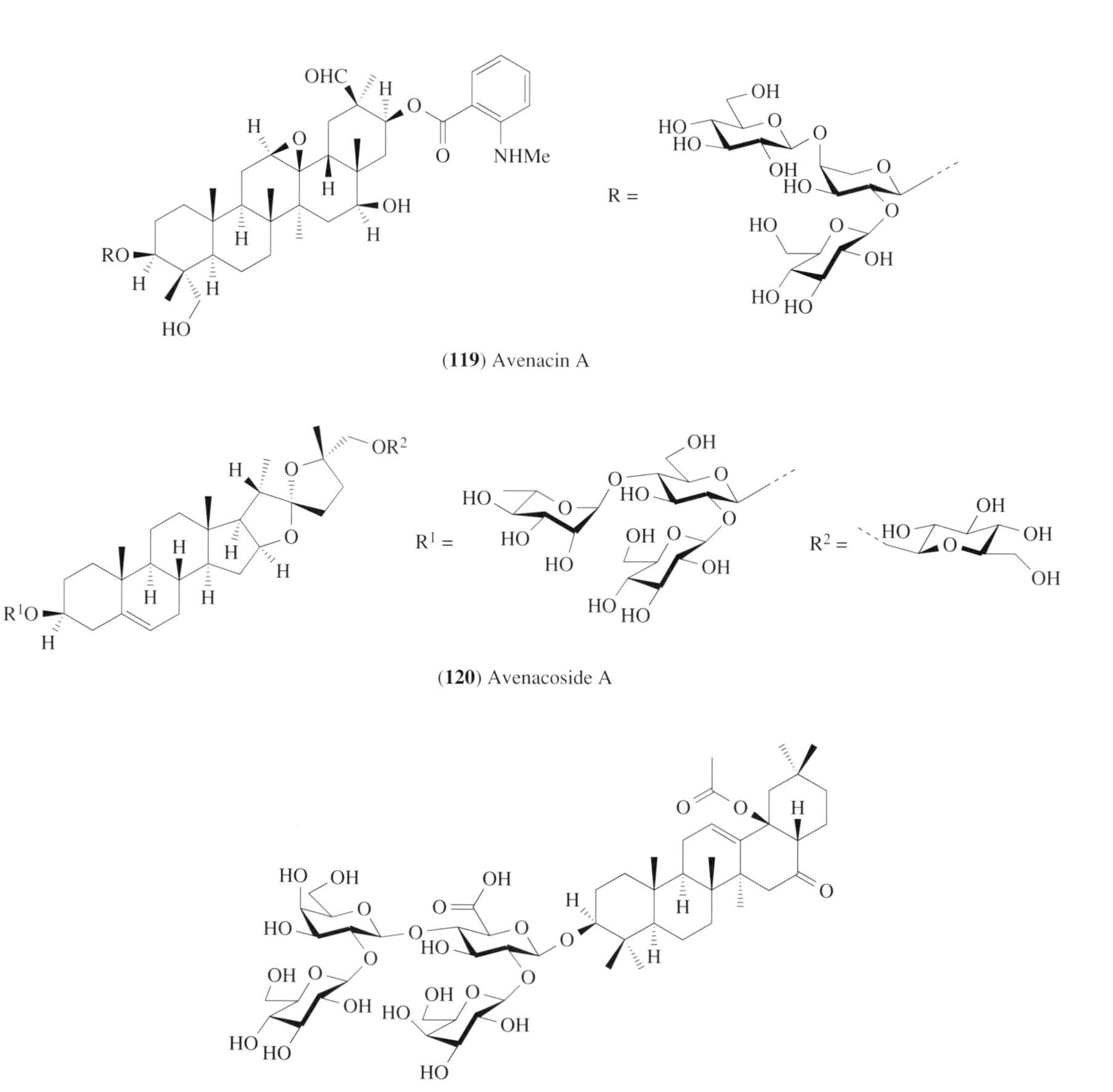

(**119**) Avenacin A

(**120**) Avenacoside A

(**121**) Camellidin I

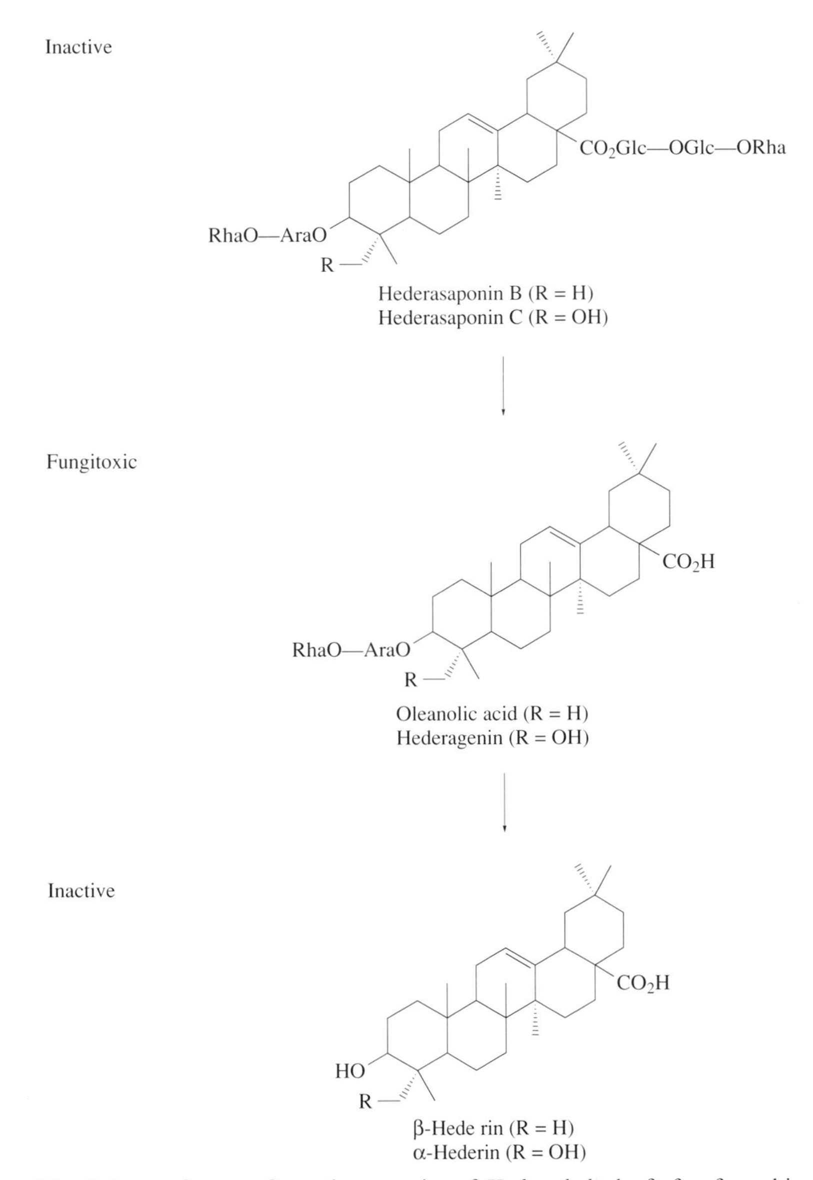

Figure 26 Release of sugars from the saponins of *Hedera helix* leaf after fungal invasion.

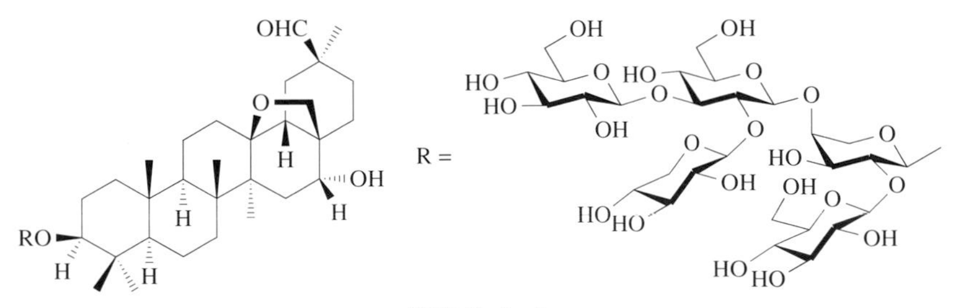

(**122**) Cyclamin

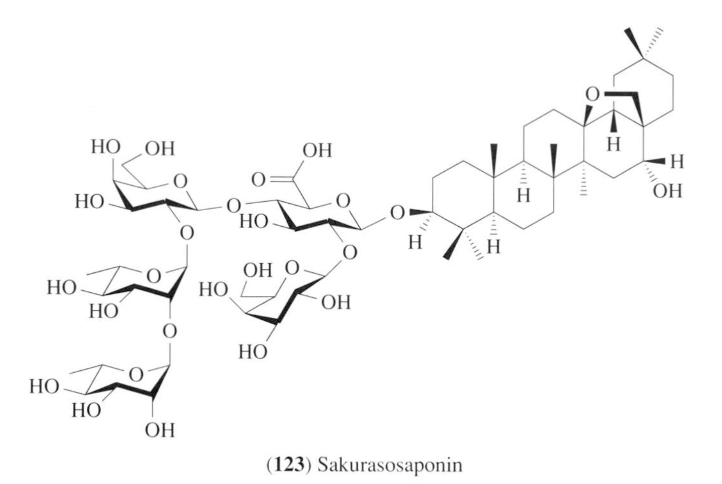

(**123**) Sakurasosaponin

Studies of bound toxins active against microbial invasion have largely been confined to single species, but in at least one family, the Rosaceae, it appears that such a defense system operates regularly. A survey of Rosaceae leaves for a phytoalexin response (Section 8.03.12) showed that the great majority of plants investigated (~130 species) were devoid of such a response. Instead, many of them contained catechin-like phenolics, which on inoculation with fungal spores underwent enzymic hydrolysis or oxidation to give products toxic to the invading organism. In the case of species of *Pyrus*, a very simple reaction took place on fungal inoculation, with the natural leaf phenolic glucoside arbutin undergoing enzymic hydrolysis to release the fungitoxic hydroquinone.[158]

While low-molecular-weight constituents are regularly uncovered as constitutive defense agents against microbial attack, there is also good evidence that proteins are also present, or are induced, to provide protection. Various chitinases and "pathogenesis-related" proteins have been described.[159] More recently, a number of cysteine-rich antimicrobial peptides have been found, especially in seeds, and have been named "plant defensins." Five have been isolated, for example from seeds of horse chestnut, *Clitoria ternata*, *Dahlia merckii*, and *Heuchera sanguinea*.[160] All produced 50% inhibition of fungal growth at concentrations of between 1 and 25 μg ml^{-1}. In radish seed, there are two defensins located in outer cell layers of the seed coat that are preferentially released on germination to create a microenvironment around the seed in which fungal growth is suppressed.[161]

8.03.12 PHYTOALEXINS

As a consequence of the coevolution of plants with their microbial parasites, plants have elaborated a complex series of defensive barriers, capable of providing them with resistance to disease. In addition to the many low-molecular-weight antimicrobial agents detected at the leaf surface or in bound form within the leaf (Section 8.03.11), there is the possibility of an induced defense system which only operates when the plant is infected by or inoculated with a microorganism. Such a defense system involves the *de novo* synthesis of new secondary metabolites which are not present normally in the healthy plant. These substances are called phytoalexins and are, by definition, formed specifically by microbial (usually fungal) induction and are also significantly antimicrobial in their properties.

The first phytoalexin to be characterized was the pterocarpan pisatin, from *Pisum sativum*. This work was carried out in 1960 and, since that time,[162] a considerable amount of effort has been expended on characterizing these phytoalexins from many different higher plants. Books have been devoted to phytoalexins[163,164] and a number of reviews have appeared.[151,165] Here, attention will be centered on the comparative biochemistry of the phytoalexin response and on the more recent data published since the early 1980s.

The first point to make is that a significant number of phytoalexins reported are uniquely formed in this response and have not been described previously as natural products, at least within the higher plants. This is true, for example, of the sulfur-containing indole phytoalexins of the family Cruciferae (Table 24). Although only a few species have been examined, if appears that the family characteristically produce mixtures of novel indole-based constituents, some with distinctive additional ring systems, and nearly all having one or more sulfur substituents (Figure 27). Most of these substances show antifungal activity at a concentration of 100 ppm and a sufficient amount is induced in infected tissue to ward off infection.[166] Another family with a very distinctive phytoalexin response is the Rosaceae. Here the response is largely lacking in leaf tissue,[158] but is relatively pronounced in the sapwood of the woody members of the family (e.g., cotoneaster, pear, and apple). Two series of phytoalexins are formed, biphenyls and dibenzofurans (Table 25). In spite of the close biogenetic link between these two classes of metabolite (Figure 28), there is no evidence of their occurring together in the same phytoalexin response and plants tend to produce *either* biphenyls *or* dibenzofurans as phytoalexins. All the phytoalexins of the Rosaceae are uniquely reported from the family, and such structures are rarely encountered elsewhere in the plant kingdom.[167]

Table 24 Phytoalexins induced in the Cruciferae.

Phytoalexins	*Plant species*[a]
Spirobrassinin, cyclobrassinin, oxymethoxybrassinin, methoxybrassinin, brassinin, dioxybrassinin, brassicanal A–C	Leaf, *Brassica campestris*
Cyclobrassinin sulfoxide, brassilexin	Leaf, *Brassica juncea*
Methoxybrassinin, cyclobrassinin	Leaf, *Brassica napus*
Camalexin, methoxycamalexin	Leaf, *Camelina sativa*
Spirobrassinin, methoxybrassinin, brassinin, oxymethoxybrassinin	Leaf, *Raphanus sativus*

[a] For references, see reference 166.

Table 25 Biphenyls and dibenzofurans as phytoalexins in the family Rosaceae.

Phytoalexin	*Plant source*[a]
Biphenyls	
Aucuparin	Leaf, *Sorbus aucuparia*
2′-Hydroxyaucuparin	Sapwood, *Sorbus aucuparia*
2′-Methoxyaucuparin	Sapwood, *Malus domestica*
4′-Methoxyaucuparin	Sapwood, *Chaenomeles japonica*
Isoaucuparin (**124**)	Sapwood, *Sorbus aucuparia*
Rhaphiolepsin	Leaf, *Rhaphiolepsis umbellata*
Dibenzofurans	
α-Cotonefuran	Sapwood, *Cotoneaster acutifolius*
β-Cotonefuran	
γ-Cotonefuran	
δ-Cotonefuran	
ε-Cotonefuran	
Eriobofuran	Sapwood, *Photinia davidiana*
7-Methoxyeriobofuran	
9-Hydroxyeriobofuran	
α-Pyrufuran	Sapwood, *Pyrus communis*
β-Pyrufuran	
γ-Pyrufuran	
6-Hydroxy-α-pyrufuran	Sapwood, *Mespilus germanicus*
6-Methoxy-α-pyrufuran	
7-Hydroxy-6-methoxy-α-pyrufuran	
2,8-Dihydroxy-3,4,7-trimethoxydibenzofuran	Sapwood, *Cydonia oblonga*

[a] For references, see reference 167.

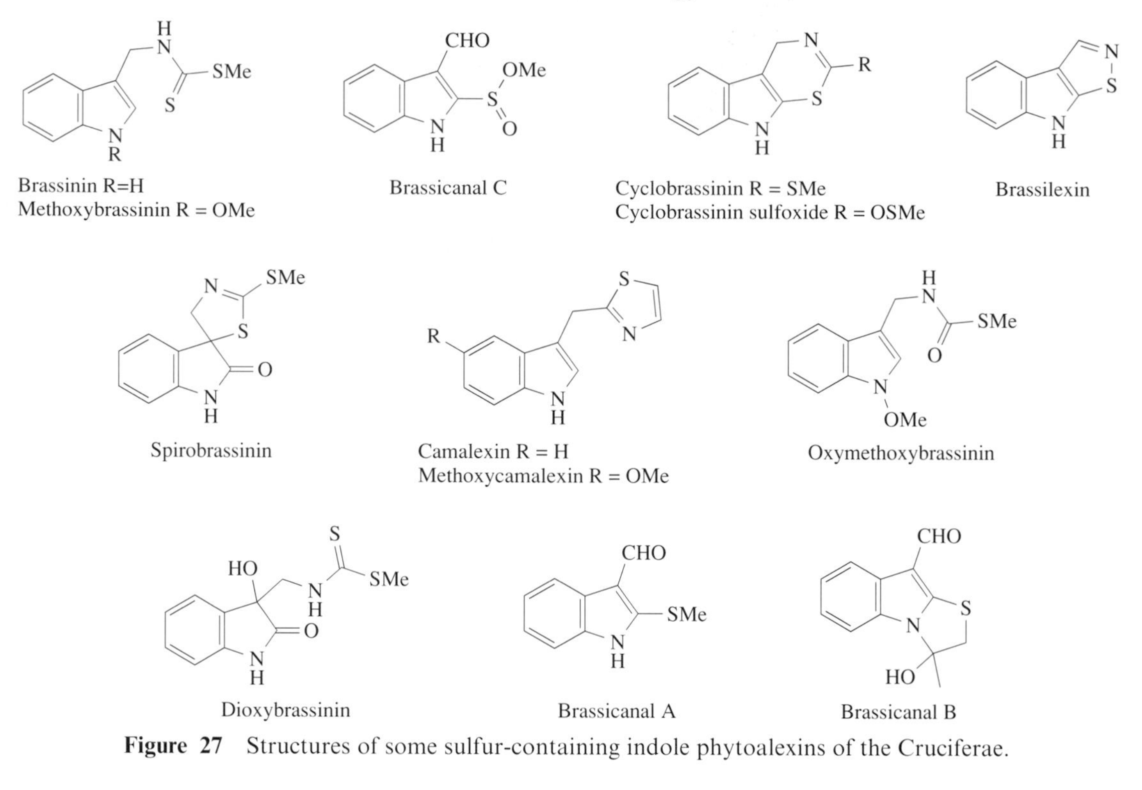

Figure 27 Structures of some sulfur-containing indole phytoalexins of the Cruciferae.

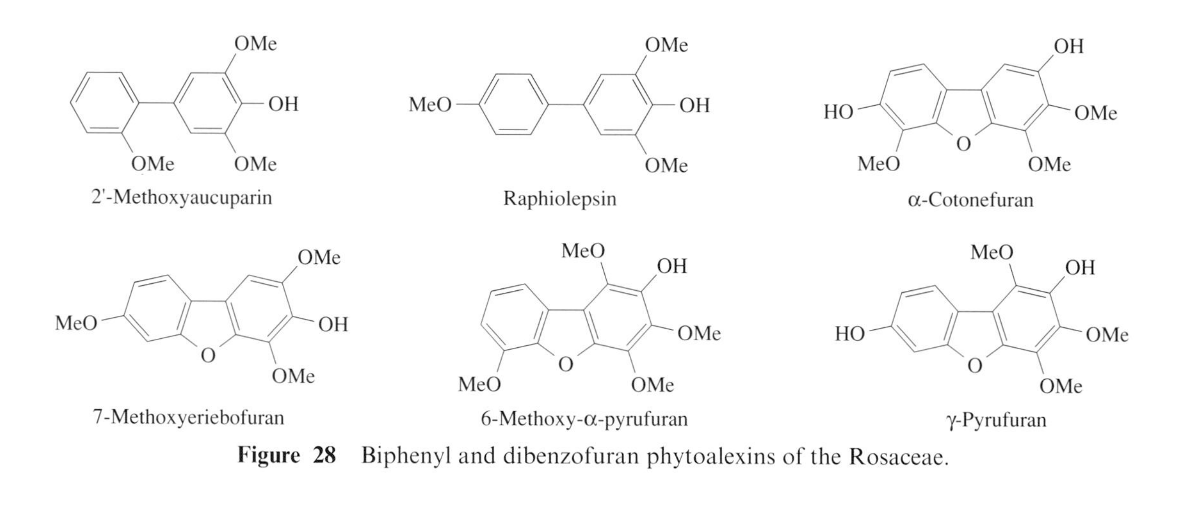

Figure 28 Biphenyl and dibenzofuran phytoalexins of the Rosaceae.

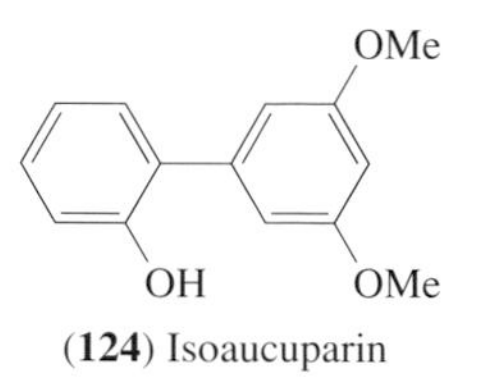

(**124**) Isoaucuparin

In the majority of families surveyed for phytoalexin induction, the type of antifungal agent produced tends to be similar to the constitutive defense chemistry of that family. This is particularly pronounced in Leguminosae, where over 500 species have been studied.[168] The phytoalexins are mainly isoflavonoids, a class of compound typical of the family, and are further subdivided into isoflavones, isoflavanones, pterocarpans, and isoflavans. Such phytoalexins have been recognized mainly in the leaves of herbaceous members of the Leguminosae, whereas the same isoflavonoid subclasses are present constitutively in woody members, and are notably heartwood constituents. In spite of this chemical overlap with the constitutive chemistry of the Leguminosae, there are still a considerable number of individual isoflavonoids which are uniquely produced in the phytoalexin response. This is true of pisatin, the phytoalexin of the pea, kievitone, an isoflavanone phytoalexin of *Phaseolus vulgaris*, and glyceollin I, one of at least five phytoalexins formed in the leaf of the soya bean, *Glycine max*. Most of the minor phytoalexins of Leguminosae are also only known as induced antifungal agents. This applies to lathodoratin, an unusual chromone from infected *Lathyrus odoratus* leaves, and wyerone, an alkynic from infected leaves of the broad bean, *Vicia faba* (Figure 29).[168]

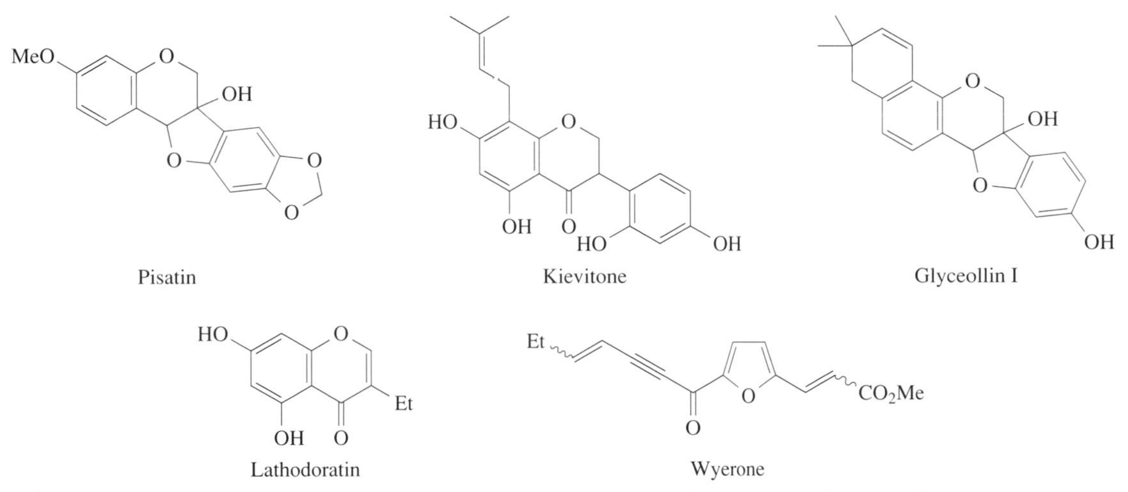

Figure 29 Some typical phytoalexins formed as antifungal agents in Leguminosae.

Whereas some plant families are relatively uniform in their phytoalexin response, e.g., the Cruciferae, Leguminosae, Rosaceae, and Solanaceae, others are capable of synthesizing several different classes of phytoalexin. In the Compositae, for example, alkynics, coumarins, sesquiterpene lactones, and acetophenones have been reported, although only about nine species have been examined (Table 26 and Figure 30).[169] Likewise, in the grass family, the Gramineae, five different classes of phytoalexin have been reported (Table 27). The rice plant is particularly rich in phytoalexin structures and three subclasses of diterpenoid in addition to the flavonoid, sakuranetin, may be formed (Table 27 and Figure 31).[151]

Table 26 Phytoalexins of the Compositae.

Class and phytoalexin	*Source*[a]
Acetylenic	
12,13-Epoxy-2,4,6,8,10-tridecapentayne	} *Arctium lappa* leaf
1-Tridecan-3,5,7,9,11-pentayne	
Safynol, dehydrosafynol	*Carthamus tinctorius*
(*E*)-Mycosinol, (*Z*)-mycosinol	*Coleostephus myconis*
Coumarins	
Scopoletin, ayapin	*Helianthus annuus*
Sesquiterpene lactone	
Cichoralexin	*Cichorium intybus* leaf
Costunolide, lettucenin A	*Lactuca sativa* leaf
Sonchifolin	*Smallanthus sonchifolius* leaf
Lettucenin A	*Taraxacum officinale*
Acetophenone	
4′-Hydroxy-3′-(3-methylbutanoyl)acetophenone	} *Polymnia sonchifolia*
4′-Hydroxy-3′-prenylacetophenone	

[a] For references see 12 and 169.

Table 27 Phytoalexins of the Gramineae.

Class and phytoalexin	*Plant source*[a]
Anthocyanidins	
Luteolinidin	*Saccharum officinarum* leaf
Luteolinidin, apigeninidin 5-caffeylarabinoside	*Sorghum bicolor* leaf
Anthranilic acids	
Avenalumin I, II, and III	*Avena sativa* leaf
HDIBOA glucoside	*Triticum aestivum* leaf
Diterpenoids	
Momilactone A and B	*Oryza sativa* leaf
Oryzalexin A–E, oryzalexin S	*Oryza sativa* leaf
Phytocassane A–D	*Oryza sativa* leaf
Flavanones	
Sakuranetin	*Oryza sativa* leaf
Stilbenes	
Piceatannol	*Saccharum officinarum*
Resveratrol	*Festuca versuta*

[a] For references, see reference 151.

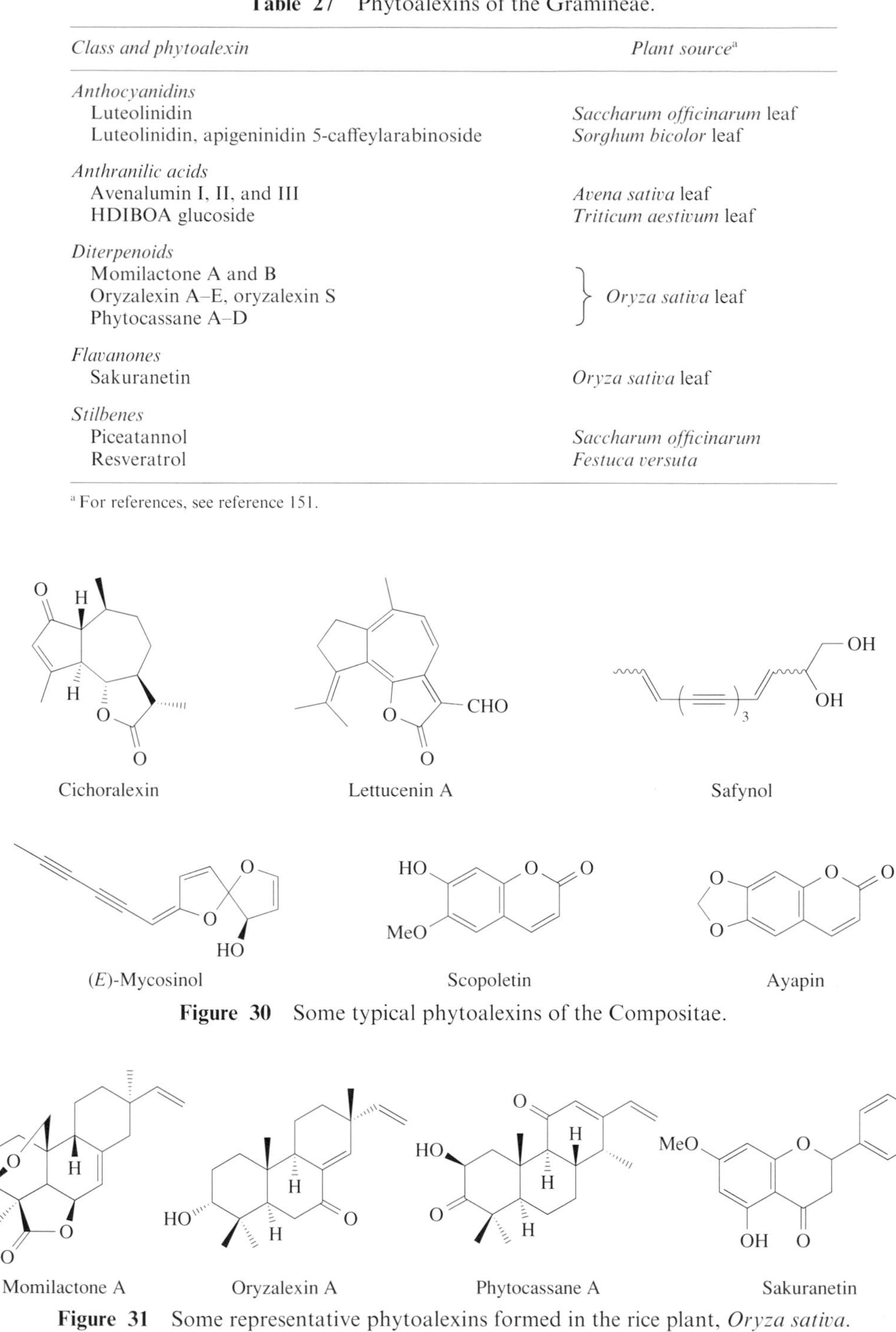

Figure 30 Some typical phytoalexins of the Compositae.

Figure 31 Some representative phytoalexins formed in the rice plant, *Oryza sativa*.

Some classes of phytoalexin have been reported in several unrelated family sources. Thus hydroxystilbenes, such as resveratrol, have been found in peanut, *Arachis hypogaea* (Leguminosae), in the vine, *Vitis vinifera* (Vitacea), in *Veratrum grandiflorum* (Liliaceae), and in *Festuca versuta* (Gramineae). Additionally, the dihydro derivative of resveratrol is a phytoalexin in tubers of *Dioscorea dumentorum* (Dioscoreaceae).[151] Again, acetophenone-based phytoalexins have been encountered in as many as four unrelated plant families (Table 28).[12]

In some plant species, one or only a few phytoalexins are induced. For example, in *Coleostephis myconis*, the major phytoalexin is (*E*)-mycosinol and the isomer (*Z*)-mycosinol may also be detected (Figure 30).[170] In other plant species, there may be considerable complexity. In the cocoa plant,

Table 28 Acetophenones as phytoalexins in plants.

Acetophenone	*Plant source*[a]	*Family*
2′,6′-Dihydroxy-4′methoxy-	*Sanguisorba minor* root	Rosaceae
3′,5′-Dimethoxy-4′-hydroxy-(2-hydroxy)-	*Carica papaya* fruit	Caricaceae
4′-Hydroxy- 3′,4′-Dihydroxy-	} *Theobroma cacao* leaf	Sterculiaceae
4′-Hydroxy-3′-(3-methylbutanoyl)- 4′-Hydroxy-3′-prenyl-	} *Polymnia sonchifolia* leaf	Compositae
2′-Hydroxy-4′,6′-Dimethoxy	*Citrus limon* leaf	Rutaceae

[a] For references, see text.

Theobroma cacao, 4-hydroxyacetophenone, 3,4-dihydroxyacetophenone, and the triterpenoid arjunolic acid are produced on elicitation, in addition to elemental sulfur.[171] In the rice plant, 12 diterpenoids and one flavanone can be formed (Table 27). The most complex reaction, however, is in the carnation, *Dianthus caryophyllus*, where three dianthalexins and 27 anthranilamides have variously been detected. Although the amides can be formed *in vitro* from the dianthalexins, there is good evidence that both alexin and amide are present in the phytoalexin mixture (Figure 32).

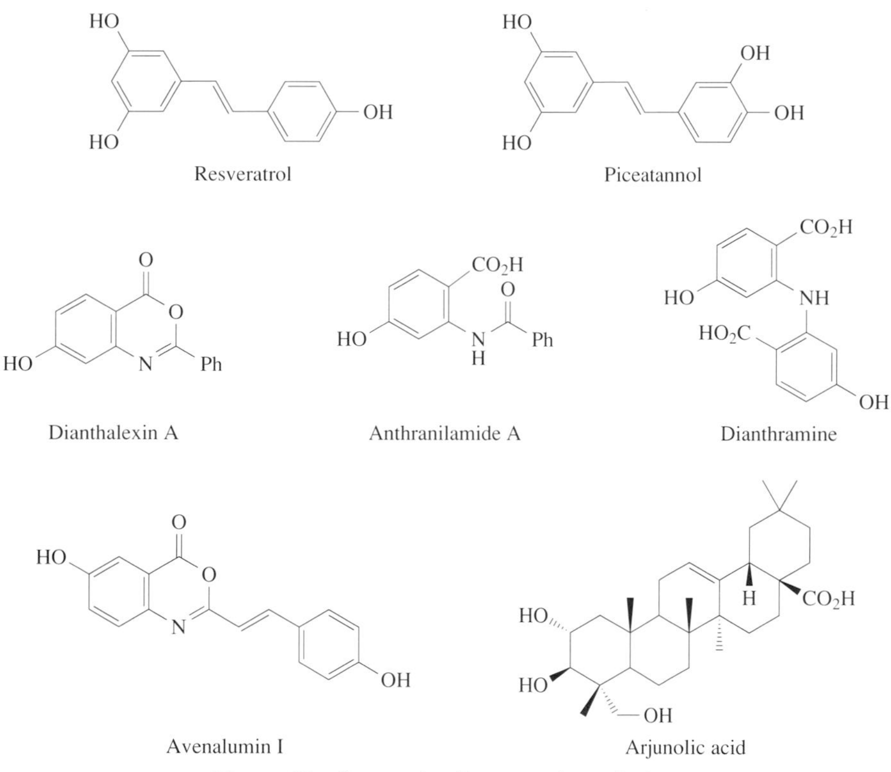

Figure 32 Some miscellaneous phytoalexins.

A novel feature of the carnation system is the presence in infected plants of an "antiphytoalexin," which switches off the phytoalexin response. The compound dianthramine (Figure 32), an anthranilic acid derivative, accumulates in susceptible varieties and reduces the plant's ability to resist fungal invasion.[172]

Phytoalexins are regarded as multisite toxicants and are able to interfere in a variety of ways with fungal cells and with fungal growth. The great majority are lipophilic, with a limited degree of water solubility. Almost all phytoalexins are thus free of glycosidic conjugation, so that the presence of sugar linkages is exceptional. Two rare examples are a tropolone glucoside from bark of *Cupressus sempervirens* infected by *Diplodia pinea* and betavulgarin 2′-glucoside from roots of sugar beet, *Beta vulgaris*, infected with *Rhizoctonia solani* (Figure 33).[12]

To summarize the known distribution and variation in phytoalexin response, it is estimated that up to 750 plant species have exhibited a positive reaction, with the synthesis of one or more identified

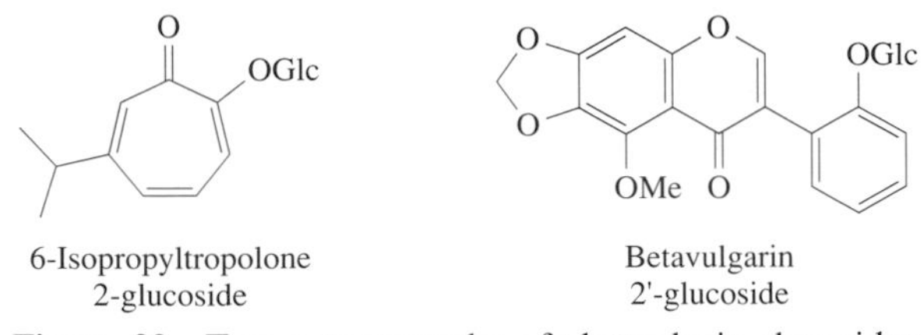

Figure 33 Two rare examples of phytoalexin glycosides.

antifungal constituents. These plants are representative of over 32 families, so that the response is a fairly general one within the flowering plants. Over 200 structures have been found, the majority of them being unique to the phytoalexin elicitation process.

It is also apparent that the ability to produce phytoalexins as a disease-resistance mechanism is absent from some plants. This is hardly surprising, considering the number of other chemical and structural barriers that have been uncovered to prevent fungal infection entering the plant. At least one plant species, *Lupinus albus*, in ~500 studied in the Leguminosae, failed to produce any phytoalexin.[173] A later survey of the Rosaceae in leaf tissue showed only four species responding out of 130 species tested.[158] Here constitutive phenolic barriers seem to be well developed in the leaf. In other Rosaceae tissues, notably the sapwood, the response to phytoalexin induction was much more fully operative.[167] One whole family, the Cucurbitaceae, has been recognized for some time to lack the phytoalexin response. This family is notable for responding to infection by systemic acquired resistance, an immunological reaction initiated by injection of a single leaf of a susceptible plant by the pathogen. The resistance is provided by the synthesis of pathogenesis-related and other proteins.[174]

The coevolutionary interaction between plant and fungus by which phytoalexins are synthesized to ward off the disease also requires that the pathogenic organism should have the ability to further metabolize the plant-produced phytoalexin to inactive products. The metabolism of a representative sample of known phytoalexins has been extensively studied and the results confirm the ability of pathogens to detoxify the antifungal agents placed in their path. The fate of the pea phytoalexins pisatin and maackiain is typical (Figure 34). Thus pisatin is detoxified in a single step via a demethylase enzyme to yield the corresponding free phenol, 6a-hydroxymaackiain, which is essentially nontoxic. The enzyme pisatin demethylase has been surveyed in a number of pea pathogens

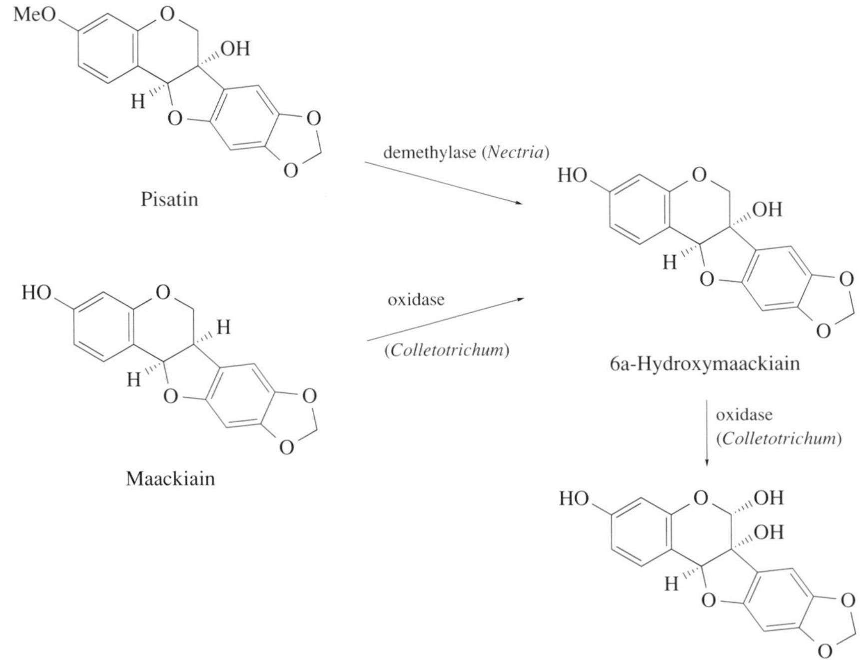

Figure 34 Further metabolism of pea phytoalexins by *Nectria haematococca* and *Colletotrichum gloeosporioides*.

and it is apparent that virulence is dependent on the presence of a very active enzyme system. In the pea pathogen, *Nectria haematococca*, different isolates vary 100-fold in the amount of enzyme present and it is only the highly active strains which are able to infect the pea plant and break through the pisatin barrier.

What is even more telling is that the gene for pisatin demethylase has been transferred by cloning to a nonpathogen on pea, *Cochliobolus heterostrophus*, converting it into a quasipathogen as a result. The conclusion from this work is that unless a particular pathogen strain does not elicit or alternatively degrade the host phytoalexin, then that phytoalexin constitutes an effective defense to fungal infection.[175]

More generally in the further metabolism of isoflavonoid phytoalexins, oxidations regularly take place. Thus, maackiain, a minor pea phytoalexin, is first oxidized by *Colletotrichum gloeosporioides* to 6a-hydroxymaackiain and then further to the dihydroxy derivative. This is then probably cleaved between the adjacent hydroxyl groups to produce several nonaromatic products.[176]

The phytoalexin system would appear, in summary, to be a particularly efficient mechanism of plant resistance, since it only comes into effect *de novo* at the point of infection and specifically produces antimicrobial metabolites at that site. During evolution, plants might evolve more than one pathway of phytoalexin synthesis and this is apparent in a small number of plants (e.g., rice *Oryza sativa*). Furthermore, the process may be turned off if other equally efficient mechanisms arise and there is experimental evidence for this too. Finally, phytoalexins are a group of structurally diverse secondary constituents belonging to many different classes, which nevertheless have a well-defined function as antimicrobial agents. This diversity in structure makes it more understandable why so many constitutive plant molecules are required to provide the plant with equally important defense against insect herbivory and mammalian grazing.

8.03.13 REFERENCES

1. W. Karrer, "Konstitution und Vorkommen der organischen Pflanzenstoffe," Birkhäuser, Basel, 1958.
2. G. S. Fraenkel, *Science*, 1959, **129**, 1466.
3. P. R. Ehrlich and P. H. Raven, *Evolution*, 1964, **18**, 586.
4. P. Feeny, in "Biochemical Interactions between Plants and Insects," eds. J. W. Wallace and R. L. Mansell, Plenum, New York, 1976, p. 1.
5. D. F. Rhoades and R. G. Cates, *Recent Adv. Phytochem*, 1976, **10**, 168.
6. J. P. Bryant, P. J. Kuropat, S. M. Cooper, and N. Owen-Smith, *Nature (London)*, 1989, **340**, 227.
7. W. C. Burnett, S. B. Jones, and T. J. Mabry, in "Biochemical Aspects of Plant and Animal Coevolution," ed. J. B. Harborne, Academic Press, London, 1978, p. 233.
8. J. B. Harborne, "Introduction to Ecological Biochemistry," 4th edn., Academic Press, London, 1993.
9. J. B. Harborne, *Nat. Prod. Rep.*, 1986, **3**, 323.
10. J. B. Harborne, *Nat. Prod. Rep.*, 1989, **6**, 85.
11. J. B. Harborne, *Nat. Prod. Rep.*, 1993, **10**, 327.
12. J. B. Harborne, *Nat. Prod. Rep.*, 1997, **14**, 83.
13. E. Haslam, *J. Chem. Ecol.*, 1988, **14**, 1780.
14. S. E. Hartley and C. G. Jones, in "Plant Ecology," ed. M. J. Crawley, Blackwell, Oxford, 1997, p. 284.
15. D. A Herms and W. J. Mattson, *Q. Rev. Biol.*, 1992, **67**, 283.
16. J. B. Harborne, in "Methods in Plant Biochemistry, Volume 1, Plant Phenolics," ed. J. B. Harborne, Academic Press, London, 1989, p. 1.
17. R. L. Lindroth and A. V. Weisbrod, *Biochem. Syst. Ecol.*, 1991, **19**, 97.
18. R. T. Palto, S. K. Anderson, and G. Iason, *Chemoecology*, 1993, **4**, 153.
19. K. Sunnerheim-Sjöberg and P.-G. Knutsson, *J. Chem. Ecol.*, 1995, **21**, 1339.
20. P. B. Reichardt, J. P. Bryant, B. R. Mattes, T. P. Clausen, F. S. Chapin, III, and M. Meyer, *J. Chem. Ecol.*, 1990, **16**, 1941.
21. J. Roy and J.-M. Bergeron, *J. Chem. Ecol.*, 1990, **16**, 735.
22. K. S. Johnson, J. M. Scriber, and M. Nair, *J. Chem. Ecol.*, 1996, **22**, 1955.
23. K. Matsuda and S. Senbo, *Appl. Entomol. Zool.*, 1986, **21**, 411.
24. P. C. Stevenson, J. C. Anderson, W. M. Blaney, and M. S. J. Simmonds, *J. Chem. Ecol.*, 1993, **19**, 2917.
25. E. A. Bernays and S. Woodhead, *Science*, 1982, **216**, 201.
26. M. R. Berenbaum, J. K. Nitao, and A. R. Zangerl, *J. Chem. Ecol.*, 1991, **17**, 207.
27. D. L. Bull, G. W. Ivie, R. C. Beier, N. W. Pryor, and E. H. Oertli, *J. Chem. Ecol.*, 1984, **10**, 893.
28. R. J. Grayer, F. M. Kimmins, D. E. Padgham, J. B. Harborne, and D. V. Ranga Rao, *Phytochemistry*, 1992, **31**, 3795.
29. D. H. Janzen, L. E. Fellows, and P. G. Waterman, *Biotropica*, 1990, **22**, 272.
30. P. G. Waterman and S. Mole, "Analysis of Phenolic Plant Metabolites," Blackwell, Oxford, 1994.
31. J. B. Harborne, in "Polyphenols 94, Plama de Mallorca," INRA, Paris, 1995, p. 19.
32. J. B. Harborne, in "Ecological Chemistry and Biochemistry of Plant Terpenoids," eds. J. B. Harborne and F. A. Tomas-Barberan, Clarendon, Oxford, 1991, p. 399.
33. S. C. Halls, D. R. Gang, and D. J. Weber, *J. Chem. Ecol.*, 1994, **20**, 2055.
34. W. J. Foley, E. V. Lassak, and J. Brophy, *J. Chem. Ecol.*, 1987, **13**, 2115.

35. D. S. Seigler, in "Herbivores, Their Interactions with Secondary Plant Metabolites," eds. G. A. Rosenthal and M. R. Berenbaum, 2nd edn., Academic Press, San Diego, 1991, vol. 1, p. 35.
36. D. A. Jones, in "Cyanide Compounds in Biology," eds. D. Evered and S. Harnett, Wiley, Chichester, 1988, p. 151.
37. A. Nahrstedt, in "Cyanide Compounds in Biology," eds. D. Evered and S. Harnett, Wiley, Chichester, 1988, p. 131.
38. J. C. Horrill and A. J. Richards, *Heredity*, 1986, **56**, 277.
39. S. Lauda and S. Mole, in "Herbivores, Their Interactions with Secondary Metabolites," eds. G. A. Rosenthal and M. R. Berenbaum, 2nd edn., Academic Press, San Diego, 1991, vol. 1, p. 124.
40. J. M. Erickson and P. Feeny, *Ecology*, 1974, **55**, 103.
41. R. P. Bodnaryk, *J. Chem. Ecol.*, 1991, **17**, 1543.
42. S. Ghaout, A. Louveaux, A. M. Manguet, M. Deschamps, and Y. Rahal, *J. Chem. Ecol.*, 1991, **17**, 1499.
43. G. A Rosenthal, in "Herbivores, Their Interactions with Secondary Metabolites," eds. G. A. Rosenthal and M. R. Berenbaum, 2nd edn., Academic Press, San Diego, 1991, vol. 1, p. 1.
44. A. R. Mattocks, "Chemistry and Toxicology of Pyrrolizidine Alkaloids," Academic Press, London, 1986.
45. M. Wink, *Theor. Appl. Genet.*, 1988, **75**, 225.
46. R. J. Mead, A. J. Oliver, D. R. King, and P. H. Hubach, *Oikos*, 1985, **44**, 55.
47. L. F. James, in "Effects of Poisonous Plants on Livestock," eds. R. F. Keeler, K. R. van Kampen, and L. F. James, Academic Press, New York, 1978, p. 139.
48. B. Juniper and T. R. E. Southwood (eds.), "Insects and the Plant Surface," Edward Arnold, London, 1986.
49. S. Woodhead and R. F. Chapman, in "Insects and the Plant Surface," eds. B. Juniper and T. R. E. Southwood, Edward Arnold, London, 1986, p. 123.
50. D. E Dussourd and T. Eisner, *Science*, 1987, **237**, 898.
51. H. Harada, H. Takahashi, T. Matsuzaki, and M. Hagimori, *J. Chem. Ecol.*, 1996, **22**, 1579.
52. S. N. Martens and R. S. Boyd, *Oecologia*, 1994, **98**, 379.
53. C. A. Ryan, *Plant Mol. Biol.*, 1992, **19**, 123.
54. P. J. Edwards, S. D. Wratten, and E. A. Parker, *Oecologia*, 1992, **91**, 266.
55. M. J. Stout, J. Workman, and S. S. Duffey, *J. Chem. Ecol.*, 1994, **20**, 2575.
56. D. W. Tallamy and M. J. Raupp (eds.), "Phytochemical Induction by Herbivores," Wiley, New York, 1991.
57. R. P. Bodnaryk, *Phytochemistry*, 1994, **35**, 301.
58. M. Dicke, M. W. Sabelis, and J. Takabayashi, *Symp. Biol. Hung.*, 1990, **39**, 127.
59. P. M. Paré and J. H. Tumlinson, *Nature (London)*, 1997, **385**, 30.
60. M. Rothschild, in "Insect–Plant Interactions," ed. H. van Emden, Oxford University Press, Oxford, 1973, p. 59.
61. M. Rothschild, B. P. Moore, and W. V. Brown, *Biol. J. Linn. Soc.*, 1984, **23**, 375.
62. A Nahrstedt, *Recent Adv. Phytochem.*, 1996, **30**, 217.
63. M. Haribal and J. A. A. Renwick, *Phytochemistry*, 1996, **41**, 139.
64. G. Bergström, M. Rothschild, I. Groth, and C. Crighton, *Chemoecology*, 1995, **6**, 147.
65. E. Städler and H. R. Buser, *Experientia*, 1984, **40**, 1157.
66. M. Rothschild, H. Alborn, G. Stenhagen, and L. M. Schoonhoven, *Phytochemistry*, 1988, **27**, 101.
67. R. Nishida, T. Ohsugi, H. Fukami, and S. Nakajima, *Agric. Biol. Chem.*, 1990, **54**, 1265.
68. M. J. Crawley (ed.), "Plant Ecology," 2nd edn., Blackwell, Oxford, 1997, p. 262.
69. K. N. Paige and T. G. Whitham, *Science*, 1985, **227**, 315.
70. J. B. Harborne, in "Advances in Labiate Science," eds. R. M. Harley and T. Reynolds, Royal Botanic Gardens, Kew, 1992, p. 307.
71. J. B. Harborne and C. A. Williams, *Nat. Prod. Rep.*, 1995, **12**, 639.
72. J. B. Harborne and R. J. Grayer, in "The Flavonoids: Advances in Research Since 1986," ed. J. B. Harborne, Chapman & Hall, London, 1994, p. 589.
73. J. T. Knudsen, L. Tollsten, and G. Bergström, *Phytochemistry*, 1993, **33**, 253.
74. P. Harrewijn, A. K. Minks, and C. Mollema, *Chemoecology*, 1995, **6**, 55.
75. G. C. Kite, *Biochem. Syst. Ecol.*, 1995, **23**, 343.
76. J. T. Knudsen and L. Tollsten, *Bot. J. Linn. Soc.*, 1995, **119**, 45.
77. H. E. M. Dobson, J. Bergström, G. Bergström, and I. Groth, *Phytochemistry*, 1987, **26**, 3171.
78. A. K. Borg-Karlson, I. Groth, L. Agren, and B. Kullenberg, *Chemoecology*, 1993, **4**, 39.
79. B. B. Simpson, J. L. Neff, and D. Seigler, *Nature (London)*, 1977, **267**, 150.
80. H. G. Baker and I. Baker, *Plant Syst. Evol.*, 1986, **151**, 175.
81. A. Detzel and M. Wink, *Chemoecology*, 1993, **4**, 8.
82. A. G. Stephenson, *J. Chem. Ecol.*, 1982, **8**, 1025.
83. T. Vogt, P. Pollak, N. Tarlyn, and L. P. Taylor, *Plant Cell*, 1994, **6**, 11.
84. J. Gross, "Pigments in Fruits," Academic Press, London, 1987.
85. D. W. Lee, *Nature (London)*, 1991, **349**, 260.
86. H. Kameoka, in "Modern Methods of Plant Analysis," New Series, eds. H. F. Linskens and J. F. Jackson, Springer, Berlin, 1986, vol. 3, p. 254.
87. A. D. Kinghorn, R. Suttisri, and I. S. Lee, *Proc. Phytochem. Soc. Eur.*, 1995, **37**, 165.
88. J. L. Goldstein and T. Swain, *Phytochemistry*, 1963, **2**, 371.
89. R. W. Wrangham and P. G. Waterman, *Biotropica*, 1983, **15**, 217.
90. J. B. Harborne, *Proc. Phytochem. Soc. Eur.*, 1997, **41**, 353.
91. D. H. Janzen, in "Biochemical Aspects of Plant and Animal Coevolution," ed. J. B. Harborne, Academic Press, London, 1978, p. 163.
92. A. Barnea, J. B. Harborne, and C. Pannell, *Biochem. Syst. Ecol.*, 1993, **21**, 421.
93. J. B. Harborne and H. Baxter, "Dictionary of Plant Toxins," Wiley, Chichester, 1996.
94. T. W. Baumann, B. H., Schulthess, and K. Hänni, *Phytochemistry*, 1995, **39**, 1063.
95. M. R. Berenbaum and A. R. Zangerl, *Recent Adv. Phytochem.*, 1996, **30**, 1.
96. C. A. Williams and J. B. Harborne, *Phytochemistry*, 1972, **11**, 1981.
97. G. A. F. Hendry, K. Thompson, C. J. Moss, E. Edwards, and P. C. Thorpe, *Funct. Ecol.*, 1994, **8**, 658.
98. J. D. Weidenhamer, *Agron. J.*, 1996, **88**, 866.

99. E. L. Rice, "Allelopathy," 2nd edn., Academic Press, New York, 1984.
100. A. R. Putnam, and C. S. Tang, "The Science of Allelopathy," Wiley, Chichester, 1986.
101. Inderjit, K. M. M. Dakshini, and F. A. Einhellig (eds.) "Allelopathy, Organisms, Processes and Applications," American Chemical Society, Washington, DC, 1995.
102. N. H. Fischer, in "Ecological Chemistry and Biochemistry of Plant Terpenoids," eds. J. B. Harborne and F. Tomas-Barberan, Clarendon, Oxford, 1991, p. 377.
103. F. Ponder and S. H. Tadros, *J. Chem. Ecol.*, 1985, **11**, 937.
104. G. B. Williamson and J. D. Weidenhamer, *J. Chem. Ecol.*, 1990, **16**, 1739.
105. A. C. Hartung, M. G. Nair, and A. R. Putnam, *J. Chem. Ecol.*, 1990, **16**, 1707.
106. C. H. Chou, *J. Chem. Ecol.*, 1989, **15**, 2149.
107. C. H. Muller and C. H. Chou, in "Phytochemical Ecology," ed. J. B. Harborne, Academic Press. London, 1972, p. 201.
108. J. D. Weidenhamer, F. A. Macias, N. H. Fischer, and G. B. Williamson, *J. Chem. Ecol.*, 1993, **19**, 1799.
109. S. E. Wood, J. F. Gaskin, and J. H. Langenheim, *Biochem. Syst. Ecol.*, 1995, **23**, 581.
110. P. R. Northup, Z. Yu, R. A. Dahlgren, and K. A. Vogt, *Nature* (*London*), 1995, **377**, 227.
111. F. Pellissier, *Phytochemistry*, 1994, **36**, 865.
112. R. J. Aerts, W. Snoeijer, E. van der Meijden, and R. Verpoorte, *Phytochemistry*, 1991, **30**, 2947.
113. D. N. Choesin and R. E. J. Boerner, *Am. J. Bot.*, 1991, **78**, 1083.
114. H. P. Spaink, C. A. Wijffelman, E. Pees, R. J. H. Okker, and B. J. J. Lugtenberg, *Nature* (*London*), 1987, **328**, 337.
115. J. L. Firmin, K. E. Wilson, L. Rossen, and A. W. B. Johnston, *Nature* (*London*), 1986, **324**, 90.
116. P. F. Morris and E. W. B. Ward, *Physiol. Mol. Plant Pathol.*, 1992, **40**, 17.
117. T. Horio, Y. Kawabata, T. Takayama, S. Tahara, J. Kawabata, Y. Fukushi, H. Nishimura, and J. Mizutani, *Experientia*, 1992, **48**, 410.
118. L. I. R. Vargas, A. F. Schmittenner, and T. L. Graham, *Phytochemistry*, 1993, **32**, 851.
119. S. Chabot, R. Bel Rhlid, R. Chénevert, and Y. Piché, *New Phytol.*, 1992, **122**, 461.
120. B. A. Siame, Y. Weerasuriya, K. Wood, G. Ejeta, and L. G. Butler, *J. Agric. Food Chem.*, 1993, **41**, 1486.
121. C. Hauck, S. Müller, and H. Schildknecht, *J. Plant Physiol.*, 1992, **139**, 474.
122. S. Muller, C. Hauck, and H. Schildknecht, *J. Plant Growth Regul.*, 1992, **11**, 77.
123. L. G. Butler, in "Allelopathy, Organisms, Processes and Applications," eds. Inderjit, K. M. M. Dakshini, and F. R. Einhellig, American Chemical Society, Washington, DC, 1995, p. 158.
124. M. Chang and D. G. Lynn, *J. Chem. Ecol.*, 1986, **12**, 561.
125. F. R. Stermitz and G. H. Harris, *J. Chem. Ecol.*, 1987, **13**, 1917.
126. M. J. Schneider and F. R. Stermitz, *Phytochemistry*, 1990, **29**, 1811.
127. M. Wink and L. Witte, *J. Chem. Ecol.*, 1993, **19**, 441.
128. J. E. Smith and M. O. Moss, "Mycotoxins: Formation, Analyses and Significance," Wiley, Chichester, 1985.
129. J. L. Frahn, J. A. Edgar, A. J. Jones, P. A. Cockrum, N. Anderton, and C. C. J. Culvenor, *Aust. J. Chem.*, 1984, **37**, 165.
130. D. D. Rowan, M. B. Hunt, and D. L. Gaynor, *J. Chem. Soc., Chem. Commun.*, 1986, 935.
131. M. R. Siegel and L. P. Bush, *Recent Adv. Phytochem.*, 1996, **30**, 81.
132. D. B. Strongman, G. M. Strunz, P. Giguére, C. M. Yu, and L. Calhoun, *J. Chem. Ecol.*, 1988, **14**, 753.
133. S. V. Pathra, W. B. Gleason, Y. W. Lee, and C. J. Mirocha, *Can. J. Chem.*, 1986, **634**, 1308.
134. P. C. Lyons, R. D. Plattner, and C. W. Bacon, *Science*, 1986, **232**, 487.
135. E. Marré, *Prog. Phytochem.*, 1980, **6**, 253.
136. G. A. Strobel, *Annu. Rev. Plant Physiol.*, 1974, **25**, 541.
137. D. Perreaux, H. Maraite, and J. A. Meyer, *Physiol. Mol. Plant Pathol.*, 1986, **28**, 323.
138. F. Sugawara and G. A. Strobel, *Phytochemistry*, 1987, **26**, 1349.
139. G. J. Feistner, *Phytochemistry*, 1988, **27**, 3417.
140. A. A. Stierle, R. Upadhyay, J. Hershenhorn, G. A. Strobel, and G. Molina, *Experientia*, 1991, **47**, 853.
141. C. Mazars, M. Rossignol, P. Auriol, and A. Klaebe, *Phytochemistry*, 1990, **29**, 3441.
142. P. Tey-Rulh, I. Phillipe, J. M. Renaud, G. Tsoupras, P. de Angelis, J. Fallot, and R. Tabacchi, *Phytochemistry*, 1991, **30**, 471.
143. A Stierle, R. Upadhyay, and G. Strobel, *Phytochemistry*, 1991, **30**, 2191.
144. Y. F. Hallock, J. Clardy, D. S. Kenfield, and G. A. Strobel, *Phytochemistry*, 1988, **27**, 3123.
145. H. G. Cutler, in "Toxic Action of Marine and Terrestrial Alkaloids," ed. M. Blum, Alakan, Fort Collins, CO, 1995, p. 125.
146. V. Macko, T. J. Wolpert, W. Acklin, B. Jaun, J. Seibl, J. Meili, and D. Arigoni, *Experientia*, 1985, **41**, 1366.
147. A. Evidente, R. Lanzetta, R. Capasso, A. Andolfi, M. Vurro, and M. C. Zonno, *Phytochemistry*, 1997, **44**, 1041.
148. A. Evidente, R. Lanzetta, R. Capasso, M. Vurro, and A. Bottalico, *Phytochemistry*, 1993, **34**, 999.
149. S. C. Fields, L. Mireles-Lo, and B. C. Gerwick, *J. Nat. Prod.*, 1996, **59**, 698.
150. I. A. Dubery, D. Meyer, and C. Bothma, *Phytochemistry*, 1994, **35**, 307.
151. R. J. Grayer and J. B. Harborne, *Phytochemistry*, 1994, **37**, 19.
152. S. Mizobuchi and Y. Sato, *Agric. Biol. Chem.*, 1984, **48**, 2771.
153. J. L. Ingham, S. Tahara, and J. B. Harborne, *Z. Naturforsch., Teil C*, 1983, **38**, 194.
154. S. Tahara, J. L. Ingham, and J. Mizutani, *Agric. Biol. Chem.*, 1985, **49**, 1775.
155. T. Kato, Y. Yamaguchi, T. Namai, and T. Hirukawa, *Biosci. Biotech. Biochem.*, 1993, **57**, 283.
156. W. S. Bowers, H. C. Hoch, P. H. Evans, and M. Katayama, *Science*, 1986, **232**, 105.
157. J. B. Harborne and J. L. Ingham, in "Biochemical Aspects of Plant and Animal Coevolution," ed. J. B. Harborne, Academic Press, London, 1978, p. 343.
158. T. Kokubun and J. B. Harborne, *Z. Naturforsch., Teil C*, 1994, **49**, 628.
159. A Schlumbaum, F. Mauch, U. Vögeli, and T. Boller, *Nature* (*London*), 1986, **324**, 365.
160. R. W. Osborn, G. W. de Samblanx, K. Thevissen, I. Goderis, S. Torrekens, F. van Leuven, S. Attenborough, S. B. Rees, and W. F. Broekaert, *FEBS Lett.*, 1995, **368**, 257.
161. F. R. G. Terras, K. Eggermont, V. Kovaleva, N. V. Raikhal, R. W. Osborn, A. Kester, S. B. Rees, S. Torrekens, F. van Leuven, J. Van der Heyden, B. P. A. Cammue, and W. F. Broekaert, *Plant Cell*, 1995, **7**, 573.

162. I. A. M. Cruickshank and D. R. Perrin, *Nature (London)*, 1960, **187**, 799.
163. J. A. Bailey and J. W. Mansfield (eds.) "Phytoalexins," Blackie, Glasgow, 1982.
164. J. A. Callow (ed.) "Biochemical Plant Pathology," Wiley, Chichester, 1983.
165. D. Gottstein and D. Gross, *Trees*, 1992, **6**, 55.
166. D. Gross, *Z. Pflanzenkr. Pflanzenschutz*, 1993, **100**, 433.
167. T. Kokubun and J. B. Harborne, *Phytochemistry*, 1995, **40**, 1649.
168. J. L. Ingham, in "Advances in Legume Systematics," eds R. M. Polhill and P. H. Raven, Royal Botanic Gardens, Kew, 1981, p. 599.
169. J. B. Harborne, in "Compositae in Systematics," eds. D. J. N. Hind and H. J. Beentje, Royal Botanic Gardens, Kew, 1996, p. 207.
170. P. S. Marshall, J. B. Harborne, and G. S. King, *Phytochemistry*, 1987, **26**, 2493.
171. M. L. V. Resende, J. Flood, J. D. Ramsden, M. G. Rowan, M. H. Beale, and R. M. Cooper, *Physiol. Mol. Plant Pathol.*, 1996, **48**, 347.
172. G. J. Niemann, *Phytochemistry*, 1993, **34**, 319.
173. J. B. Harborne, J. L. Ingham, L. King, and M. Payne, *Phytochemistry*, 1976, **15**, 1485.
174. J. Kúc, *Annu. Rev. Phytopathol.*, 1995, **33**, 275.
175. N. T. Keen, *Plant Mol. Biol.*, 1992, **19**, 109.
176. S. Soby, R. Bates, and H. Van Etten, *Phytochemistry*, in press.

8.04
Pheromones

WITTKO FRANCKE and STEFAN SCHULZ
Universität Hamburg, Germany

8.04.1 INTRODUCTION

Apart from optical, acoustic, and tactile stimuli, living organisms use chemical cues for the transmission of information. To transfer a message, all these channels may be used simultaneously or in distinct sequences, according to a specific hierarchy. Chemotaxis is a general archaic principle, and a wide range of living beings, from microorganisms to primates are capable of releasing chemical signals ("semiochemicals") which are used in intra- and interspecific communication.

The importance of the "chemical channel" and its role in communication depend on the development and use of alternative mechanisms. Humans appear to base communication primarily on sound and vision. In contrast, odor communication seems to be particularly important in insects; these relatively small living beings, who have to take their bearings in a spacious environment, have developed a multitude of glands for the production and release of chemical signals.[1]

Chemical signaling obviously evolved several times and for different reasons. The development of cell-aggregating compounds and the establishment of recognition substances appear instrumental to the formation of multicellular organisms; compounds used for communication within such cell aggregations may be regarded as antecedents of hormones and neurotransmitters, etc.

Inter- and intraspecific chemical defense in host–parasite interactions or in the struggle for living space among animals, plants, and microorganisms includes a broad spectrum of compounds which range from behavior-mediating deterrents or irritants to physiologically active compounds such as growth inhibitors, toxins, and antibiotics. Defense compounds are mostly used against several species and are often produced in large amounts.

Coevolution of plants and herbivores may form a special basis for the origin of systems of chemical communication.[2–4] Food constituents which attract both sexes will cause the discovery of sexual partners at places where food is found. Mimicking such signals through the storage and release of compounds by animals with the aim of attracting or deterring others will lead to the experience that metabolites of food and/or host compounds (or appropriate imitations thereof) may be advantageously used to specify a chemical message and thus facilitate its decoding by the receiver. In fact, striking similarities between volatiles from animals and plants point to coevolutionary processes in the relation between herbivores and their host plants.[5]

While the secretion of compounds which were simply sequestered from the environment or which represent products of primary transformations provides information about the quality of a food source or a potential habitat, endogenously produced semiochemicals merely describe the physiological state of the emitter.

According to their mode of action, semiochemicals which are biosynthesized by the emitter have been classified as intraspecifically active "pheromones"[6]—sub-groups "releaser" (causing a change of behavior in the receiver) and "primer" (causing a physiological impact on the receiver)—and as interspecifically active "allelochemicals"—sub-groups "kairomone" (beneficial for the receiver), "allomone" (beneficial for the emitter), and "synomone" (beneficial for the emitter and receiver)—see Figure 1. A review of this terminology was given by Nordlund.[7]

"Sex pheromones" of moths, produced by the females to attract males, are typical releasers. The "aggregation pheromones" of bark beetles represent another type of releaser: they are produced by one sex but attract both sexes and thus induce mass attack to overcome the host tree's resistance during colonization. "Trail pheromones" of ants, marking compounds, and alarm substances are further examples. Typical primers are compounds which are transmitted to suppress the development of ovaries in the receiver, to accelerate puberty, block pregnancy, or synchronize estrus. Whereas kairomones are predominantly used in host selection and prey location by interspecifics, allomones represent the whole spectrum of compounds which are used for camouflage and for defense in its widest sense, while synomones play a major role in symbiosis.

Although the above terminology basically represents a useful system, the fact that the classification of semiochemicals is predominantly carried out according to their biological function may sometimes cause confusion. Predators of several bark beetle species locate their hosts by orienting towards their prey's pheromones:[8] concerning the above definitions, in this example the same compound is produced as an intraspecifically acting pheromone, and it is interspecifically "illegitimately" used as a kairomone. An intraspecific attractant (pheromone) may interspecifically act as a repellent

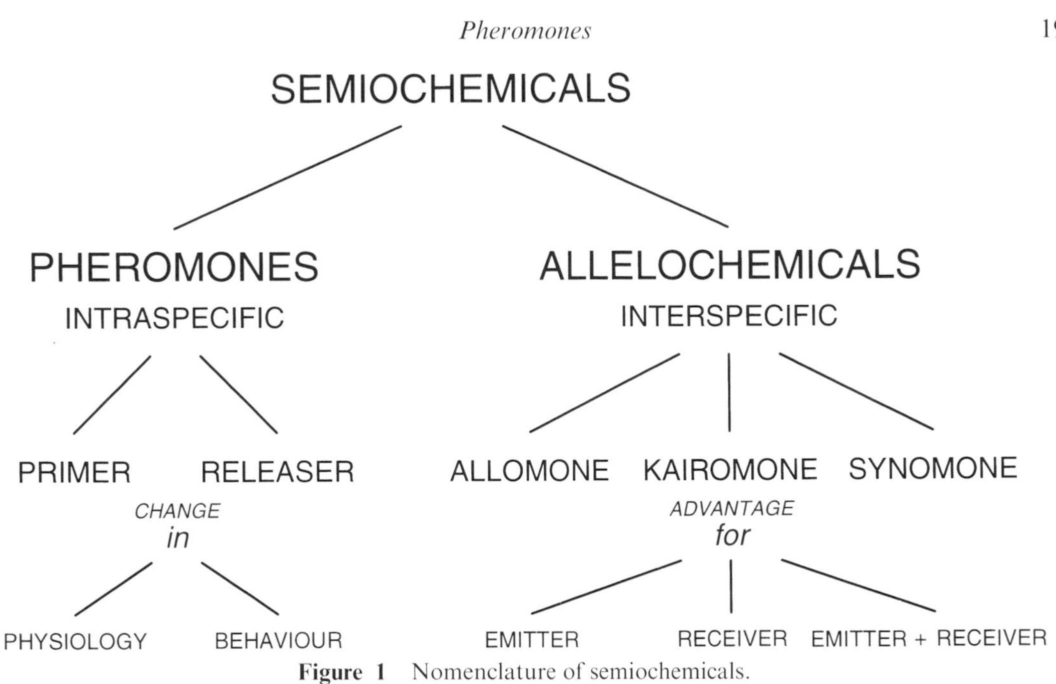

Figure 1 Nomenclature of semiochemicals.

(allomone) for a closely related species. A pheromone of a certain prey species may also be produced by a predator who deceitfully uses it as an allomone:[9] bolas spiders, *Mastophora cornigera*, attract their prey, male moths, by the release of (*Z*)-9-tetradecenyl acetate,[10] which is one of the most widespread sex pheromone components of female moths.[11] Moreover, a behavior-mediating releaser may at the same time act at a physiological level as a primer: 9-oxo-(*E*)-2-decenoic acid, which is secreted by the honey bee queen,[12] is a component of her pheromone blend, which forms the chemical basis of retinue response and suppresses the development of ovaries in the workers. The terminology for those semiochemicals which are simply sequestered from the environment or which are produced by associated microorganisms and thus not synthesized by the emitter is not well defined; the term "infochemicals" has been used.

Chemical communication over long distances requires specific signals which can be clearly distinguished as true "chemical messages," which are different from the "background noise" caused by odor compounds of the environment (see Figure 2). Generally, this is achieved on the basis of unique multicomponent mixtures of less specific compounds which are preferred over the use of species-specific substances, the production of which would require the (costly) development of species-specific enzymes. In many cases, several constituents of the bouquet act synergistically. The aggregation pheromone of the bark beetle, *Pityogenes chalcographus*, is represented by a mixture of 2-ethyl-1,6-dioxaspiro[4.4]nonane[13] and methyl (2*E*,4*Z*)-2,4-decadienoate.[14] The pure spiroacetal is only slightly active whereas the pure ester is not attractive at all, but a 95:5 mixture is highly attractive to both sexes. The spiroacetal is not species-specific as a bark beetle pheromone since it has also been identified as a plant volatile.[15] The above methyl ester is also a component of the secretion of the stink bug, *Euschistus*,[16] and, together with the corresponding ethyl ester, accounts for the aroma of Bartlett pears.[17] Whether the biosynthesis of spiroacetal and the ester proceeds via a common precursor remains to be investigated.

Enantiomeric composition of chiral compounds plays an important role in semiochemistry. Bark beetles which use chiral oxygenated monoterpenes in chemical communication may produce species-specific enantiomeric compositions and react with great sensitivity to non-natural proportions.[18] The female sex attractant of the Japanese beetle, *Popillia japonica*, is a chiral γ-lactone, and male response in these scarab beetles is significantly reduced by the presence of only 1% of the antipode.[19] In contrast, the bouquet which honey bee queens release to induce retinue response contains both enantiomers of 9-hydroxy-(*E*)-2-decenoic acid, and among other essential compounds the racemate is necessary to elicit this specific behavior.[20] Relations between enantiomeric composition and biological activity have been extensively discussed by Mori.[21–23]

Closely related species may produce bouquets which are chemically very similar. In such cases, species specificity of a signal may be based on quantitative differences (including enantiomeric proportions) in the secreted bouquet.

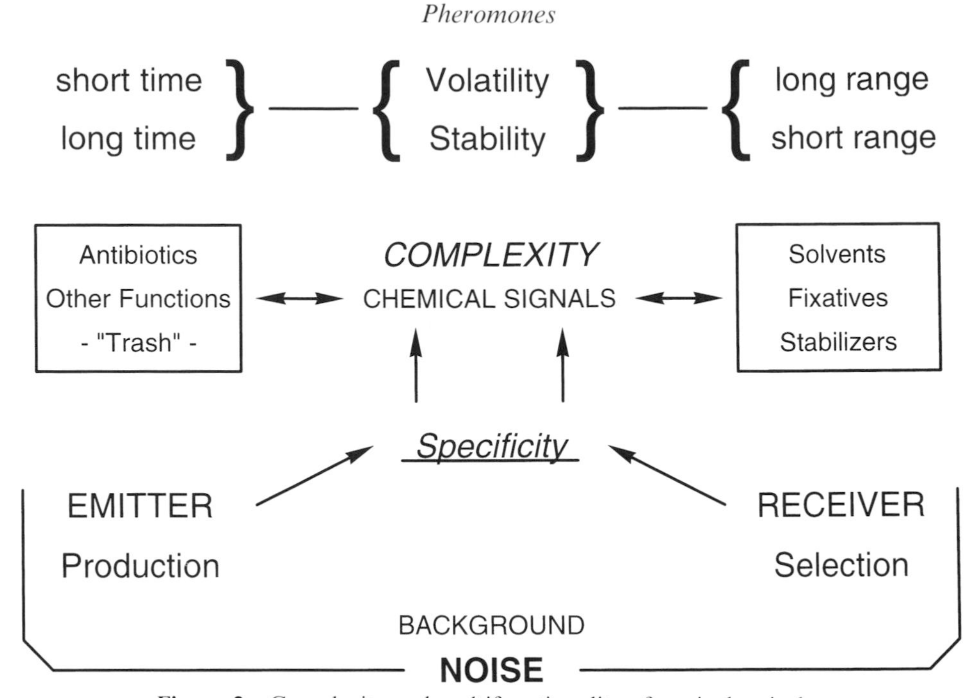

Figure 2 Complexity and multifunctionality of semiochemicals.

Differences in the sensitivity and selectivity at the receptor site and differences in the transformation of a perceived signal into a distinct behavioral pattern at the receiver's central nervous system are important elements in the perception of chemical messages. Males of the bark beetle, *Ips typographus*, release the terpenoid 2-methyl-3-buten-2-ol to attract both sexes of the species,[24] while workers of the hornet, *Vespa crabo*, secrete this compound to induce alarm among conspecific workers.[25] The same alcohol has also been found among the volatiles of citrus fruits[26] and hops[27] and has a strong narcotic effect on mice.[27] As another example, in addition to many other hydrocarbons, (*Z*)-9-tricosene is particularly widespread among insect volatiles: in most cases it does not seem to show any behavior-mediating activity. However, houseflies produce and obviously use it as a sex attractant and mating stimulant.[28]

During biosyntheses of active compounds, the action of less specific enzymes may lead to the formation of "by-products," which cause a "fine tuning" of the bouquet (see Figure 2). When two different species use the same compound as a major attractant, the minor components need not necessarily be the same and, although intraspecifically inactive, they may well be used as interspecifically active repellents which facilitate species discrimination. Females of the cabbage moth, *Mamestra brassicae*, and the closely related *M. suasa* produce (*Z*)-11-hexadecenyl acetate as the dominant sex pheromone component;[29] however, males of the former species are strongly repelled by small amounts of the corresponding aldehyde, which is released by the females of the latter species.

Many compounds present in multicomponent secretions may show more than one function: the same compound may act intraspecifically as an attractant and interspecifically as a repellent. For example, the above-mentioned "queen substance" of honey bees is intraspecifically active at both the behavioral and physiological levels. Only little is known about antibiotic functions of the secretions; it appears likely that methyl 4-hydroxybenzoate, which is part of the retinue behavior-inducing secretion of the honey bee queen,[20] plays such a role, since it is well known for its antibiotic properties. It would be worthwhile to pay more attention to screening semiochemicals and related compounds for additional activities, particularly from insect secretions.[30] Methyl 4-hydroxybenzoate has also been described as a component of the sex pheromone of female dogs[31] and has proved to be behavior mediating in butterflies.[32]

Differences in vapor pressures and thermodynamic stabilities of the constituents of a multicomponent mixture may reveal information on the age of the signal and/or on the distance of the emitter (see Figure 2). However, not all of the secreted compounds must necessarily show any behavioral activity. Many components—and often those which represent major quantities—just act as surfactants, spreading factors, or solvents. They may also represent protecting agents or diluents for thermodynamically unstable semiochemicals or fixatives for highly volatile components. Polar defensive compounds secreted by attacked animals are often embedded (dissolved) in nonpolar media, which after topical application of the mixture carry the active substances through the aggressor's cuticle.[33,34] In short, semiochemicals and accompanying compounds are frequently mixed in a kind of "cosmetic formulation," including stabilizers and preservation agents. The frass of bark beetles containing their aggregation pheromones acts as a true slow-release system.

Some compounds (loosely referred to as "trash" in Figure 2) seem to be "just there," with no (or an unknown) biological function. Such compounds may have had a specific function during earlier evolutionary stages or, as products of less specific enzymes; they may facilitate evolution through introduction of disposable variation.

Generally, the primary goal of most of the behavior-mediating odoriferous compounds seems to be the induction of the receiver's attention in a high state of alertness, which might be one of the reasons for striking structural relationships among semiochemicals.

This review deals predominantly with pheromones, i.e., with compounds which have been definitively proven to mediate intraspecific behavior. In addition, it attempts to give a cross-section on semiochemicals and related compounds, aiming at the comparison of chemical structures of compounds which occur in the context of chemical communication. Other compilations of pheromones, allelochemicals, and other compounds identified from insects have appeared earlier.[33–40] The evolution of chemical defense in arthropods has been treated several times.[33,34,41] Excellent reviews on the synthesis of semiochemicals have been published by Mori.[42–45]

8.04.2 ISOLATION AND IDENTIFICATION TECHNIQUES

Most of the active compounds used as semiochemicals are produced in particularly small amounts, and the separation and identification of those components in complex multicomponent mixtures represents a challenge even to laboratories with the highest analytical standards.

Biologically active material from insects may be obtained by total extraction or from surface washings with organic solvents. The drawback of this rather crude method lies in the tremendous amount of extracted body lipids which usually cause serious difficulties during further separation steps. Extraction of dissected glands may greatly facilitate the whole procedure. Using on-column GC techniques, solid-sample injection of glands or body tissues gives a true picture of volatile compounds present in the investigated sample. However, these methods represent "snapshot photos" of the state of affairs at a particular moment and do not necessarily reflect the composition of the "active principle," which is actually released: some pheromones are generated from precursors just as they "leave" the gland. In contrast, headspace collections of volatiles give much better pictures concerning released compounds with regard to both qualitative and quantitative compositions of the blends; an optimal sampling technique is, of course, most important. Comprehensive compilations of techniques used in pheromone have appeared.[46,47] Solid-phase microextraction (SPME)[48] seems to be a promising alternative for collecting semiochemicals.[49–51] Since not all of the secreted substances are biologically active, on-line combinations of GC separations with electroantennographic methods greatly facilitate the detection of compounds which are actually perceived by the receiver.[52]

Detection of a target compound is usually followed by identification. NMR techniques give the most significant data; however, their application is limited because of the relatively low sensitivity, even of modern instruments (~ 1 μg for ^{1}H-NMR and 10–15 μg for a C–H correlation analysis), and because the isolation procedures necessary for NMR analysis are difficult and time consuming. In addition, isolation procedures often involve large numbers of individuals. In contrast, on-line methods such as GC/MS, GC/FT-IR and two-dimensional gas chromatography coupled to tandem mass spectrometry (GC–GC/MS–MS) combine high selectivity on the separation side and high sensitivity (picogram to nanogram scale) on the detector side. Unfortunately mass spectral data are often less specific, and postulated structures need to be scrutinized through independent syntheses.

The more significant spectroscopic or chromatographic data are available for a target compound, the smaller will be the amount needed for structure elucidation. In contrast, the more specific the data necessary for structure elucidation (e.g., X-ray analysis), the higher will be the amount of the target compound required.

Besides the classical derivatization reactions used in GC, some special microreactions and mass spectrometric investigations of the derivatives produced are particularly useful in structure elucidation:[53] catalytic hydrogenation removes double bonds; reaction with various complex hydrides reduces esters and/or carbonyl compounds to alcohols; addition of dimethyl disulfide to unsaturated compounds allows the determination of double-bond positions. In many cases these reactions may be carried out with minute amounts of crude mixtures. Microtechniques have been reviewed.[54]

Once the gross structure of an active chiral compound has been identified, determination of the absolute configuration of the natural product is essential because the target organism often reacts differently to non-natural enantiomers or diastereomers or needs defined mixtures for maximum response. Determination of enantiomeric compositions by GC or HPLC may be achieved either by separation of diastereomeric derivatives on achiral stationary phases or by direct separation on chiral stationary phases. The use of modified cyclodextrins (Lipodex) has opened up a new dimension in GC determinations of enantiomeric compositions of both natural and synthetic samples of chiral compounds.[55] "Chiral amplification" has been introduced as a strategy for determining the absolute configuration of natural products when the target compound has a functional group in a chiral environment and when the racemate is accessible.[56]

Considerations of biogenetic relationships of compounds which belong to the same odor bouquet or which occur in a related species may suggest which compounds might be expected in a certain context. Based on reasonable assumptions including possible biosynthetic pathways of the target compounds, such "biogenetic analysis" may be particularly useful in identification procedures.

In any case, only the *experimentum crucis*, i.e., generation of the investigated phenomenon in a successful bioassay by using the synthetic equivalent of the natural signal, will provide unambiguous proof of the correctness of any identification procedure for a semiochemical.

8.04.3 GENERAL REMARKS

Semiochemicals may not only be ranked in terms of their origin or mode of action, but may also be classified with respect to their chemical structures. According to the general processes of primary metabolism, the variability in chemical structures of compounds used for communication appears not to be unlimited, even with respect to secondary transformations: communication systems seem to be established according to archaic concepts which follow distinct rules. Apart from coevolutionary processes, convergences, and coincidences, it appears that elementary principles exist in the "chemical languages" of living beings. In addition to examples mentioned above, striking demonstrations are found among volatile constituents of elephants: (*Z*)-7-dodecenyl acetate, the sex pheromone of female Asian elephants, *Elephas maximus*,[57] is known to be one of the most widespread sex pheromones of female moths.[11] In addition, the musth temporal gland secretion of male elephants contains 1,5-dimethyl-6,8-dioxabicyclo[3.2.1]octane[58] (frontalin), a well known aggregation pheromone of *Dendroctonus* bark beetles.[59]

No distinct structure–activity relationships can be deduced from known semiochemicals. Suitable pheromones, however, seem to form complexes with receptors through non-bonding close-range interactions, which facilitates processing and dissociation mechanisms to avoid permanent signals at the receiver's site. In contrast, many allomones which are used for defense and which are designed to disrupt the aggressor's orientation over a longer period often form covalent bonds with the receptors. Hence strong Michael acceptors or Michael donors, 1-alken-3-ones, and 1,3-dicarbonyl compounds, are exceptions among pheromones but are used for defense, like quinones. A survey of the structures of known pheromones reveals that the majority are represented by compounds of medium polarity with boiling points between 150 and 250 °C, e.g., alcohols, esters, and oxygen heterocycles with unbranched or branched chains, possibly including one or more double bonds.

In this chapter, structures of behavior-mediating compounds are classified and discussed according to their carbon skeletons based on principal biosynthetic mechanisms. Although in most cases detailed biosyntheses are unknown, the compounds are grouped into three sections: unbranched "acetogenins," terpenoid "isoprenoids," and "propanogenins and related compounds." The last group includes polyketids formally formed from propanoate units (or methylmalonate) and compounds which obviously originate from a mixed biosynthesis involving propanoate and acetate units (or malonate), since they show a distinct branching pattern. These three "types" cover the vast majority of behavior-mediating compounds. As already indicated in the Introduction, the present compilation is strongly insect-oriented, since chemical communication in these animals has been most successfully investigated. Although incomplete, related areas are touched upon selectively, to provide a picture of general chemical structures that play a major role in the rapidly developing field called "chemical ecology."

8.04.4 ACETOGENINS: PHEROMONES WITH UNBRANCHED CARBON SKELETONS

Acetogenins are natural products formed by condensation of acetate units. They represent long-chain, unbranched compounds, which often contain few functional groups. The ultimate building block, acetyl-coenzyme A (acetyl-CoA), is produced in the primary metabolic process by β-oxidation and thiolytic cleavage of fatty acids, oxidative decarboxylation of pyruvate, or activation of acetate by reaction with ATP and CoA. Common primary metabolites of almost all living cells are fatty acids, typical acetogenins (Scheme 1). The biosynthesis of fatty acids starts with acetyl-CoA, which reacts with hydrogen carbonate, catalyzed by the enzyme acetyl-CoA carboxylase, to form malonyl-CoA, thus enhancing the nucleophilicity of the α-carbon of the acetate unit. Malonyl-CoA is transferred to the acyl carrier protein (ACP) and through catalysis by fatty acid synthetase (FAS) reacts with acetyl-CoA to form 3-oxobutanoyl-ACP. The keto group is then reduced by a 3-oxoacyl-ACP reductase to form (*R*)-3-hydroxybutanoyl-ACP, which in turn is transformed by a 3-hydroxyacyl-ACP dehydratase to 2-butenoyl-ACP. The final reaction, catalyzed by an enoyl-ACP reductase, leads to the saturated butanoyl-ACP. By condensation with additional malonyl-CoA building blocks (acetate units), long-chain, unbranched fatty acids are synthesized.[60]

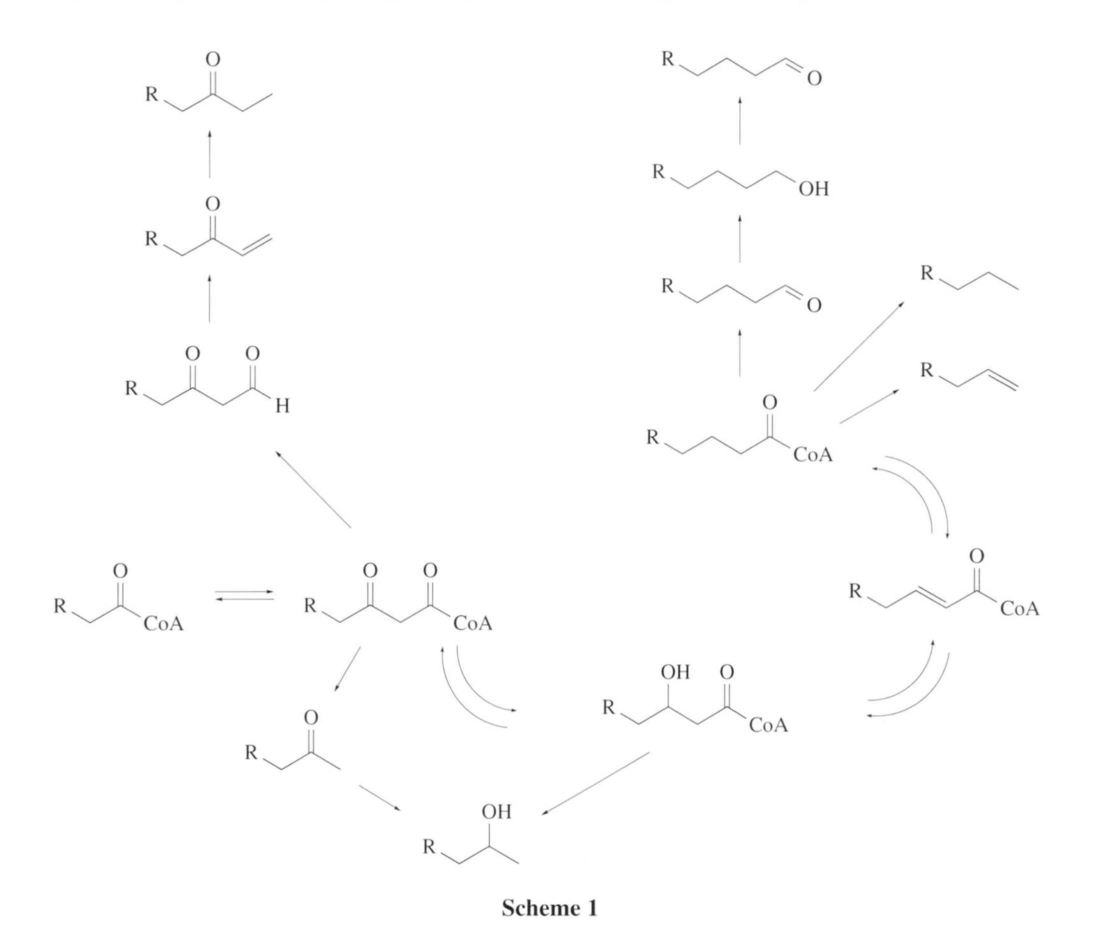

Scheme 1

Fatty acids are the origin of a large number of pheromone structures and are modified in many ways to form a multitude of aliphatic compounds. These compounds may be formed by transformation of the final products of the fatty acid synthesis such as palmitic or stearic acid or through modifications during their biosynthesis. Fatty acids usually occur as even-numbered bishomologues, because of their origin from acetate units. When propanoyl-CoA forms the starter instead of acetyl-CoA, compounds are produced which show a chain elongated by one additional carbon atom.

During the incorporation of an acetate unit, a reductive removal of the carbonyl oxygen is necessary to form saturated alkyl chains. When the oxygen is retained, so-called polyketides are formed. They cyclize easily and often form aromatic compounds or contain multiple functional groups because of their polyoxoacyl structures.[60]

The following section will focus only on acetogenins and polyketides with unbranched carbon skeletons. In many cases the exact biosynthetic pathway leading to such compounds is unknown, despite the considerable knowledge accumulated especially on the biosynthesis of lepidopteran sex pheromones (see below). Nevertheless, several major routes to similar compounds occurring in other organisms have been elucidated and similar mechanisms may take place in insects to form most of the described components:

(i) Condensation of acetate units or degradation of fatty acids by β-oxidation yields compounds with an even number of carbon atoms. Modifications of the acyl group and secondary reaction processes such as oxygenation or insertion of double bonds can alter the product. As an example, hexadecanol is formed from hexadecanoyl-CoA by reduction.

(ii) Decarboxylation of the acyl precursor leads to compounds with an odd number of carbon atoms in the chain, e.g., undecane from dodecanoyl-CoA.

(iii) Other products, mainly C_6, C_8, C_9, and C_{12} compounds, can be formed by oxidative cleavage of unsaturated fatty acids, e.g., linolenic acid.[61,62] Lipoxygenases oxidize fatty acids to hydroperoxy acids, which upon chain cleavage yield aldehydes, which may be further metabolized. These processes have been intensively studied in plants,[62] but rarely verified to take place in insects. Nevertheless, similar components have been identified as pheromones and plant volatiles.

(iv) Starting the fatty acid biosynthesis with propanoyl-CoA instead of acetyl-CoA leads to odd-numbered acyl precursors, which may be modified by the processes described above. Odd-numbered chains may also be formed by α-oxidation of the acyl group, a process not well investigated yet.[63]

(v) Carbon chains generated by the processes mentioned above may be further modified by oxidation of activated or nonactivated carbon atoms and desaturation. This leads to the wide variety of compound classes encountered as insect semiochemicals.

The intermediates of the fatty acid metabolism and catabolism form an important pool for the biosynthesis of low molecular weight semiochemicals. Some routes to important compound classes are depicted in Scheme 1. Generally, transformations retaining the carboxy carbon lead to compounds with an even number of carbon atoms along the chain whereas decarboxylative processes yield compounds with an odd number of carbon atoms. In contrast, the opposite is true when the biosynthesis of the backbone starts with propanoyl-CoA instead of acetyl-CoA. Odd-numbered 2-alkanones are formed by decarboxylation of 3-oxoacyl-CoA precursors.[64,65] Reduction of the ketones leads to chiral 2-alkanols, which may also be formed by decarboxylation of 3-hydroxyacyl-CoA precursors.[66] The acyl-CoA intermediates formed by hydrogenation of alkenoyl-CoA are the precursors for alkanes with odd numbers of carbon atoms.[67] Corresponding 1-alkenes are obtained by oxidative decarboxylation of acyl-CoA precursors, as has been shown in the formation of the 1-alkenes produced by the beetle *Tribolium castaneum*.[68] In addition to hydrolytic cleavage to free fatty acids, the acyl-CoA precursors can also be reduced to aldehydes and further to 1-alkanols, which may be reoxidized to the aldehydes. These pathways are followed in the biosynthesis of many moth pheromones. In most cases the pheromonal aldehyde is formed from an alcohol precursor. Nevertheless, in some cases direct reduction of an acyl-CoA precursor to the aldehyde seems to take place.[69,70]

Additional structural diversity is obtained by combining acids and 1- and 2-alkanols to form esters. For example, from only three acids and the corresponding primary alcohols nine different esters can be formed. The set of chiral 2-alkanols would yield another 18 esters. Including the corresponding three aldehydes and three methyl ketones a bouquet of 42 different compounds may be generated and used as a chemical signal. Blends consisting of such compounds which also include quantitative variation are often found in bees and some other insects as constituents of exocrine glands.[33,38]

The less widespread 3-alkanones may be formed through reduction of 3-oxoacyl-CoA. First 3-ketoaldehydes are synthesized. These compounds have been identified in the defensive secretion of termite soldiers.[71] Further reduction and elimination of water lead to vinyl ketones, which have also been isolated from termites.[71] The 3-alkanones are subsequently formed by hydrogenation of the double bond and may be further reduced to 3-alkanols. Products of these processes have frequently been identified in the exocrine secretion of insects.[33,37] They, and also their acyl precursors, may be further modified by oxygenation along the chain, dehydration to form double bonds, and other processes. A common character of these substances is that the enzymes involved are less specific with respect to chain length of the target compound, so that often mixtures of bishomologues occur. Some exceptions are represented by sex pheromones, which are found more or less as unique compounds in specific glands.

Although many volatiles have been identified from insects, the true function of these compounds, be it as pheromones or allomones, has been established in far fewer cases. For example, 1- and 2-

alkanols, aldehydes, carboxylic acids, and especially esters have frequently been found in the secretions of many Hymenoptera, but only in a few cases has the biological function been established.[33,38] The cuticle of insects is frequently covered with complex mixtures of long-chain alkanes and alkenes, sometimes accompanied by oxidized derivatives such as alcohols, aldehydes, and esters. The primary function of these lipids is to prevent desiccation of the organism; however, often species-specific mixtures of cuticular constituents are also used for communication. By pattern recognition of such mixtures, which in social insects may show pronounced caste specificity and ontogenetic changes, species identification, establishment of colonies, differences in races, kin recognition, or host–parasite relations are mediated. Such phenomena will not be discussed here; a review provides an excellent overview.[67] Only in rare cases has true pheromonal activity been shown for distinct components of surface lipids.[67] In the following section, compounds will be discussed for which pheromonal activity has been proved unambiguously. Acetogenic components of exocrine secretions from insects, particularly ants, bees, and termites, for which no function could unambiguously be established, have been reviewed elsewhere.[33,38,67] The chemistry of pheromones of social insects has also been reviewed.[72–74]

8.04.4.1 Lepidopteran Sex Pheromones

The sex pheromones of female moths, which attract males over long distances, are the best investigated group of insect pheromones. In part this is due to the insects' economic importance as pests but also because of their stereotypical structures. A complete list of identified pheromones and attractants (compounds attracting males, but not established to originate from females) for more than 1500 species investigated has been published.[11]

Lepidopteran sex pheromones can be divided into several chemical classes. Most of them are long-chain unbranched alcohols, aldehydes, or acetates with up to three double bonds. The chain length varies between 10 and 18 carbon atoms. A typical compound is (*Z*)-11-tetradecenyl acetate, which is a component of the sex pheromone of more than 50 species and attractive to more than 100. Species specificity is very often achieved by specific mixtures of these compounds including geometrical and positional isomers. For example, the commercially important pheromone of the pink bollworm moth, *Pectinophora gossypiella*, consists of a 65:35 mixture of the geometrical isomers (7*Z*,11*Z*)- and (7*Z*,11*E*)-7,11-hexadecadienyl acetate.[75] The corn earworm, *Heliothis zea*, uses a mixture of (*Z*)-11-hexadecenal as the major and (*Z*)-9-hexadecenal as a minor component.[76] In many insect species pheromone biosynthesis is induced by a pheromone biosynthesis activating neuropeptide (PBAN).[1,77]

Pheromones of Lepidoptera are frequently formed from palmitic or stearic acid (Scheme 2). Moths possess a unique enzyme, Δ^{11}-desaturase,[78] which plays an important role in the formation of the corresponding compounds. The redbanded leafroller, *Argyrotaenia velutinana*, uses a mixture of (*Z*)- and (*E*)-11-tetradecenyl acetate as pheromone. Minor components are (*E*)- and (*Z*)-9-dodecenyl acetate and 11-dodecenyl acetate. The first step in their biosynthesis (Scheme 2) is chain shortening of the acyl derivatives of palmitic acid or stearic acid to myristic acid through β-oxidation. Subsequently, specific Δ^{11}-desaturases form the corresponding (*E*)- and (*Z*)-11-tetradecenoic acids. These are reduced to the alcohols and esterified with acetyl-CoA to form the tetradecenyl acetates. Chain shortening of the 11-tetradecenoic acids furnishes 9-dodecenoic acids, while chain shortening of myristic acid yields dodecanoic acid. Again, the Δ^{11}-desaturase produces 11-dodecenoic acid, which is subsequently converted into 11-dodecenyl acetate.[79–81] The example shows how very specific compounds and mixtures can be produced by differential use of chain shortening and desaturation. The stereochemical course of Δ^{11}-desaturase has been investigated by Camps[81] and Boland.[82] In addition to the common Δ^{11}-desaturase, some other desaturases have been described from moths for pheromone production: Δ^{9}-,[83] Δ^{10}-,[84] Δ^{13}-,[85] and Δ^{14}-desaturase.[86] Furthermore, general Δ^{5}-, Δ^{6}-, Δ^{9}-, Δ^{12}-, and Δ^{15}-desaturases have been found in insects,[87] which may in some cases also be involved in pheromone production.

Conjugated diene pheromones, such as the first pheromone identified, bombykol, (10*E*,12*Z*)-10,12-hexadecadien-1-ol, from the silkworm, *Bombyx mori*,[88] can be biosynthesized by two different routes. In *Bombyx mori*, palmitic acid is transformed first into (*Z*)-11-hexadecenoic acid, which is then converted into (10*E*,12*Z*)-10,12-hexadecadienoic acid by a unique 10,12-desaturase.[89] This acid is then reduced to form bombykol. In contrast, the apple moth, *Epiphyas postvittana*, forms (*E*)-11-hexadecenoic acid first, which is then chain shortened and again desaturated by a Δ^{11}-desaturase to yield (9*E*,11*E*)-9,11-tetradecadienoic acid, the precursor of the pheromone (9*E*,11*E*)-9,11-tetra-

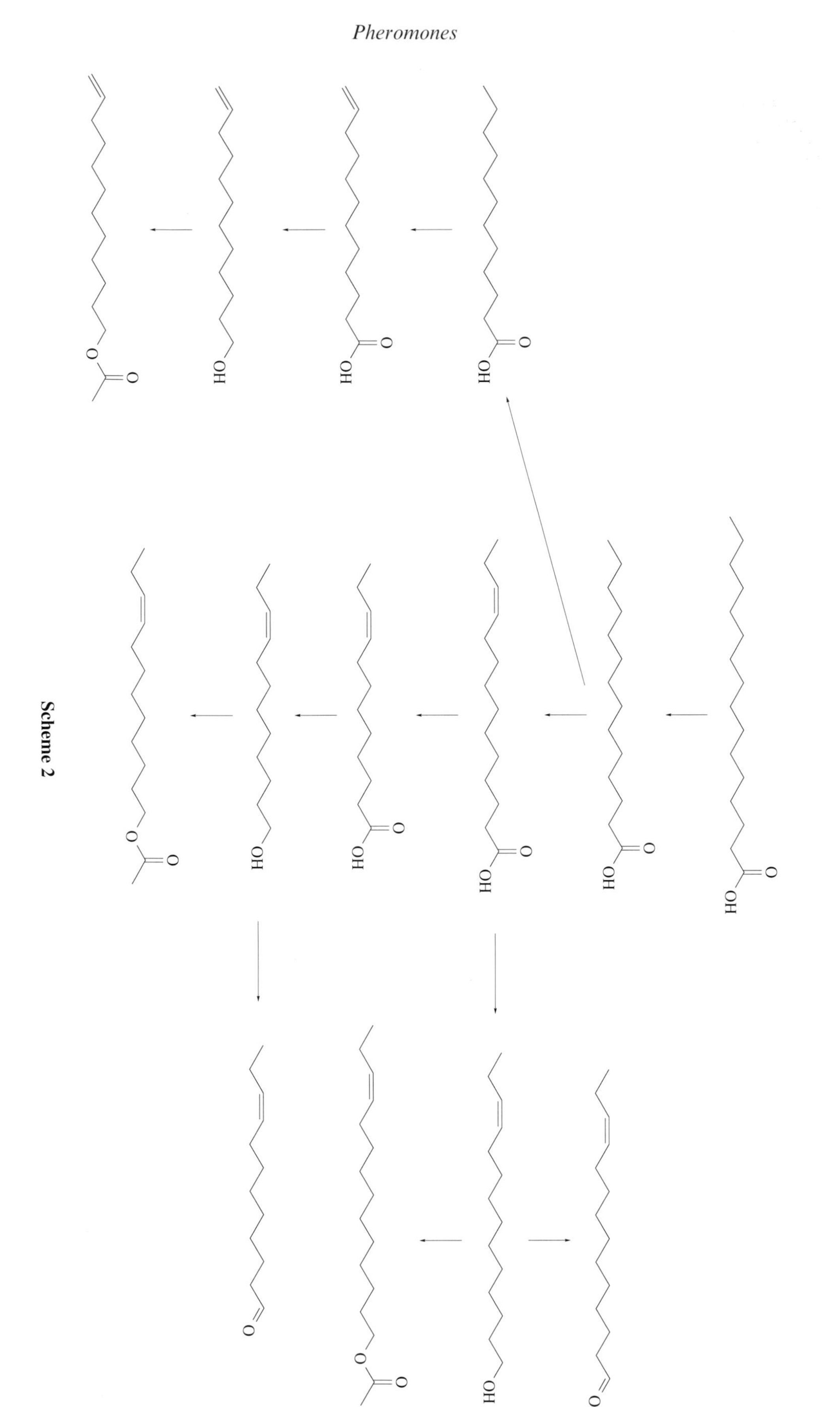

Scheme 2

decadienyl acetate.[90] The lepidopteran sex pheromones usually contain a terminal acetate, alcohol, or aldehyde group. As stated above, the carboxyl group of the acyl precursor is reduced to the alcohol, which may be esterified by an unspecific esterase to form the acetates.[91] Whether aldehydes are produced directly from the acyl precursor or from the alcohols via reoxidation remains to be clarified.[92] Acyl precursors of the pheromones can be stored as components of triacylglycerols.[92]

Owing to their biosynthesis, most of these pheromones contain an even number of carbon atoms in the chain. Nevertheless, some pheromones with odd-numbered carbon chains are known. The cranberry fruitworm, *Acrobasis vaccinii*, produces a sex pheromone consisting of (8*E*,10*Z*)-8,10-pentadecadienyl acetate and (*E*)-9-pentadecenyl acetate.[93] Another pyralid moth, *Chilo auricilius*, uses a mixture of (*Z*)-8-tridecenyl acetate, (*Z*)-9-tetradecenyl acetate, and (*Z*)-10-pentadecenyl acetate as a sex pheromone.[94] This pattern may point to the involvement of α-oxidation in the pheromone biosynthesis. By a sequence involving α-oxidation, the common pheromone precursor (*Z*)-11-hexadecenoic acid may lose one carbon after another and the resulting 10-pentadecenoic, 9-tetradecenoic, and 8-tridecenoic acids may be the precursors of the pheromone components. Finally, the tomato pinworm, *Keiferia lycopersicella*, uses (*E*)-4-tetradecenyl acetate as the sex pheromone, which shows an unusual position of the double bond.[95]

Apart from monoenes and conjugated dienes, nonconjugated dienes and trienes have been identified as moth sex pheromones. Examples are (3*E*,13*Z*)-3,13-octadecadien-1-ol, a component of some pheromones of sessiid moths, and (10*E*,12*E*,14*Z*)-10,12,14-hexadecatrienyl acetate of *Glyphodes pyloalis*.[11]

Some less common structures of lepidopteran pheromones are shown in Figure 3. Processionary moths, *Thaumetopoea* spp., use (*Z*)-13-hexadecen-11-yn-1-yl acetate (**1**) and related compounds as pheromones, which contain a triple bond,[11] formed by a specific desaturase.[85] Rarely other than acetic acid is used as acid component in pheromone esters. However, (*Z*)-7-dodecenyl propanoate (**2**) and butyrate are minor components of the pheromone of the noctuid moth, *Pseudoplusia includens*.[11] The lymantriid moth, *Euproctis chrysorrhoea*, uses the atypical (7*Z*,13*Z*,16*Z*,19*Z*)-7,13,16,19-docosatetraenyl 2-methylpropanoate (**3**) as pheromone.[11] In the unique pheromone structure of *Bucculatrix thurberiella*, the common terminal functional group is replaced by a nitrate group. The active principle is a mixture of (*Z*)-8- and (*Z*)-9-tetradecenyl nitrate (**4**).[96]

The compounds showing the general features discussed so far make up more than 80% of those identified as female lepidopteran sex pheromones. The second largest group are skipped conjugated polyenes with up to four double bonds and corresponding oxidation products. Often they contain an odd number of carbon atoms ranging in chain length from C_{17} to C_{23}. The active hydrocarbons are mostly 6,9-alkadienes and 3,6,9-alkatrienes showing the *Z*-configuration. The most common compound of this type is (3*Z*,6*Z*,9*Z*)-3,6,9-henicosatriene (**5**). In the arctiid moth *Estigmene acrea* and *Phragmatobia fuliginosa* it has been shown to be formed from linolenic acid by chain elongation and decarboxylation.[97] Tetraenes occur less often. (3*Z*,6*Z*,9*Z*)-1,3,6,9-Nonadecatetraene is a pheromone of *Operophtera* spp. and the bishomologue (3*Z*,6*Z*,9*Z*)-1,3,6,9-henicosatetraene is the pheromone of the geometrid moth *Epirrita autumnata* and the arctiid moth *Tyria jacobaeae*.[11] A mixture of (3*Z*,6*Z*,9*Z*,11*Z*)- and (3*Z*,6*Z*,9*Z*,11*E*)-1,3,6,9-nonadecatetraene and (3*Z*,6*Z*,9*Z*)-1,3,6,9-nonadecatriene make up the pheromone of the geometrid *Alsophila pometaria*.[98] This type of structure is so far only known from geometrid, arctiid, and noctuid moths. In many species, the alkenes are precursors for the corresponding chiral epoxides, which are formed by selective epoxidation of one double bond. More than 40 such epoxyalkenes have been identified.[11]

Each of the double bonds of the dienes or trienes may be epoxidized. The regioselectivity and stereoselectivity of the epoxidation may differ between species. For example, the female sex pheromone of *Agriopis aurantiaria* is (6*Z*,9*Z*)-(3*S*,4*R*)-3,4-epoxy-6,9-nonadecadiene, while the related species, *Agriopis marginaria*, uses a 10:3 mixture of (3*Z*,9*Z*)-(6*S*,7*R*)-6,7-epoxy-3,9-nonadecadiene and (3*Z*,6*Z*,9*Z*)-3,6,9-nonadecatriene.[99] Several geometrid species are attracted by (3*Z*,6*Z*)-9,10-epoxy-3,6-nonadecadiene.[100] The absolute configuration of the epoxide plays an important role for the biological activity. The two related geometrids *Erannis defoliaria* and *Colotois pennaria* are sympatric and active during the same time. Whereas *Erannis defoliaria* uses (3*Z*,9*Z*)-(6*S*,7*R*)-6,7-epoxy-3,9-nonadecadiene (**6**) as the major pheromone component, the other species uses its enantiomer. Both species respond only to their own enantiomer and are repelled by the other.[99] The only epoxytrienes known so far are (3*Z*,6*Z*)-9,10-epoxy-1,3,6-henicosatriene and its C_{20} homologue from the pheromone glands of the arctiid *Hyphantria cunea*.[101] The active pheromone components of this species are (3*Z*,6*Z*)-(9*S*,10*R*)-9,10-epoxy-3,6-henicosatriene, (9*Z*,12*Z*,15*Z*)-9,12,15-octadecatrienal (**7**), and (9*Z*,12*Z*)-9,12-octadecadienal.[102] The same mixture is used by another arctiid, *Estigmene acrea*. Studies on pheromone biosynthesis in this species showed that all three compounds are synthesized from linolenic acid.[97] Thus, despite the fact that (**7**) shows the

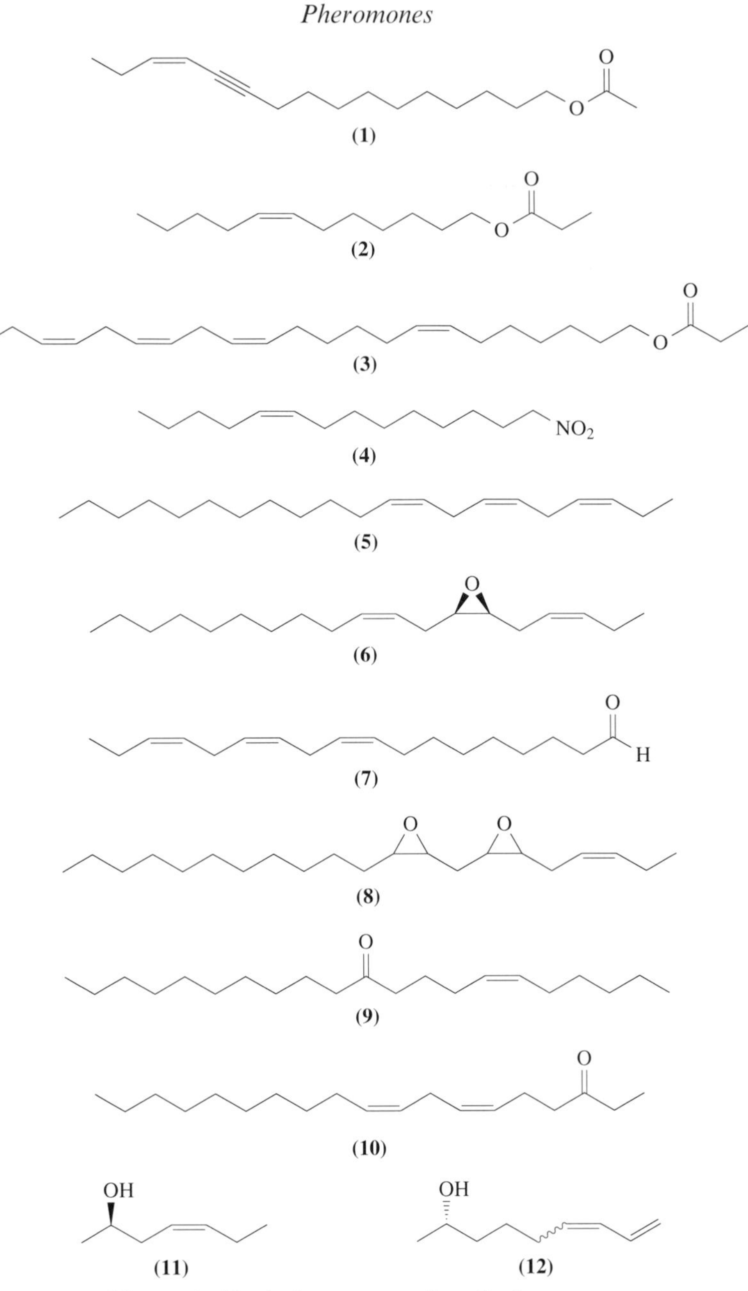

Figure 3 Typical structures of moth pheromones.

structural characters of the typical moth pheromone described above, it is biosynthetically more closely related to the polyene pheromones. Another unique epoxide is (7*R*,8*S*)-7,8-epoxyoctadecane (monachalure), a component of the nun moth pheromone which also contains the corresponding branched 2-methyl epoxide, (+)-disparlure.[103] A single diepoxide has been identified as sex pheromone: the lymantriid moth *Leucoma salicis* uses (3*Z*)-*cis*-6,7-*cis*-9,10-diepoxyhenicosadiene (**8**) as a sex pheromone.[104]

Alkenones derived from skipped conjugated alkenes occur more rarely. The ketones (*Z*)-7-nonadecen-11-one and (*Z*)-7-icosen-11-one are pheromones of *Carposina* species. The related (*Z*)-6-henicosen-11-one (**9**) is active in several *Orgyia* species. The pheromone of the douglas fir tussock moth, *Orgyia pseudotsugata*, consists of (**9**) and the very unstable (6*Z*,8*E*)-6,8-henicosadien-11-one.[105] A mixture of (6*Z*,9*Z*)-6,9-nonadecadien-3-one (**10**) and (3*Z*,6*Z*,9*Z*)-3,6,9-nonadecatriene comprises the pheromone of *Peribatodes rhomboidaria*.[11]

A third group of unbranched sex pheromone components is made up either by secondary alcohols or their esters with medium chain length fatty acids. The female sex pheromone of the moth *Eriocrania cicatricella* consists of three related C_7 components, (*R*)-2-heptanol, (*Z*)-4-hepten-2-

one, and (4*Z*,2*R*)-4-hepten-2-ol (**11**).[106] Similar nonenols have been identified as the female sex pheromones of the leaf miner moth, *Stigmella malella*. A 10:3 *E–Z* mixture of the pure (*S*)-enantiomers of 6,8-nonadien-2-ol (**12**) proved to be highly attractive to males in the field.[107] In zygaenids of the genus *Harrisina*, (*S*)-2-butyl (*Z*)-7-tetradecenoate acts as pheromone.[11] The psychid *Oiketicus kirbyi* uses a sex pheromone consisting of the esters of (*R*)-2-pentanol and octanoic, nonanoic, decanoic, and dodecanoic acid. Another bag worm moth, *Megalophanes viciella*, uses 1-methylethyl octanoate as female sex pheromone.[108] In these cases the typical pheromone ester composition of a long-chain alcohol and a short-chain acid is reversed.

Whereas the sex pheromones of the Lepidoptera are chemically relatively similar, other insects use more structurally diverse compounds. Therefore, in the following section acetogenic pheromones will be discussed according to increasing chain length.

8.04.4.2 C_1 Units

Compounds containing only one carbon are normally not believed to be of acetogenic origin, but will be included here for completeness. The first compound isolated from insects was formic acid, obtained by Markgraf as early as 1749 by distillation of ants. Many ants contain this acid in their poison glands. It functions as an alarm pheromone in some species,[109] and its biosynthesis has been studied in detail.[69] It is formed from amino acids, such as serine, glycine, or histidine, which can transfer one carbon unit to tetrahydrofolate. The resulting 5,10-methylenetetrahydrofolate is oxidized to 10-formyltetrahydrofolate, which can liberate formic acid. Formates have occasionally been identified as pheromones: the branched compounds neryl formate and lardolure which are active in mites will be discussed in Sections 8.04.5.2 and 8.04.6.3, respectively. The described biosynthetic pathway is obviously not acetogenic, but other C_1 compounds may principally be formed by degradation of acetate. Methanol occurs in some pheromone esters (see the following sections). Its biosynthetic origin in insects is unknown. Many insects have receptors for CO_2, but this compound serves mainly as an environmental cue, indicating living organisms which are searched for as hosts.[110]

8.04.4.3 C_2 Units

Hydrolysis of acetyl-CoA produces free acetic acid, which is used as an alarm pheromone by the ant *Crematogaster scutellaris*.[111] Ethyl acetate is a component of the male aggregation pheromone of the Mediterranean fruit fly, *Ceratitis capitata*.[112] Acetates of long-chain alcohols very often represent sex pheromones of Lepidoptera, as has been discussed earlier. Ethanol, the reduction product of acetyl-CoA, is also often found esterified with fatty acids. The attractivity of some bark beetle pheromones is enhanced by ethanol emanating from rotting wood, where it is formed by degradation of carbohydrates.[113]

8.04.4.4 C_3 Units

Propanoates, generated from propanoyl-CoA, are rarely identified as substructures of pheromone esters. Examples are the already mentioned moth sex pheromone (**2**) and tetradecyl propanoate (see Section 8.04.4.15). Additional examples such as monoterpenoid propanoates used by scale insects and methyl branched alkylpropanoates used by *Diabrotica* beetles and several sawfly species will be discussed in Sections 8.04.6.3 and 8.04.6.4, respectively. 2-Methylethyl carboxylates of long-chain fatty acids comprise the pheromone bouquet of the hide beetle *Dermestes maculatus*.[114] 2-Propanol may be formed via acetoacetate.

8.04.4.5 C_4 Chains

The first condensation in the fatty acid biosynthesis is the reaction between acetyl-CoA and malonyl-CoA to form acetoacetyl-CoA, which is reduced to (*R*)-3-hydroxybutanoyl-CoA. The corresponding (3*R*)-3-hydroxybutanoic acid (**13**) acts together with its estolide, (3*R*,3*R*)-3-(3-hydroxybutanoyloxy)butanoic acid (**14**), as the web reduction pheromone of female *Linyphia tri-*

angularis and *Linyphia tenuipalpis* spiders.[115] These compounds represent the only known pheromone from spiders. In the stink bug, *Campyloma verbasci*, two C_4 units are joined together via an ester linkage. The female pheromone is a mixture of butyl butanoate and (*E*)-2-butenyl butanoate (**15**).[116] A different oxidation pattern occurs in (2*RS*,3*RS*)-2,3-butanediol (**16**), a component of the male sex pheromone of the cockroach *Eurycotis floridana*.[117] Another component of the *Ceratitis capitata* pheromone is 3,4-dihydro-2*H*-pyrrole (**17**), which may be regarded as a condensation product of 4-hydroxybutanal with ammonia;[112] however, it may also be formed by degradation of the amino acid proline.

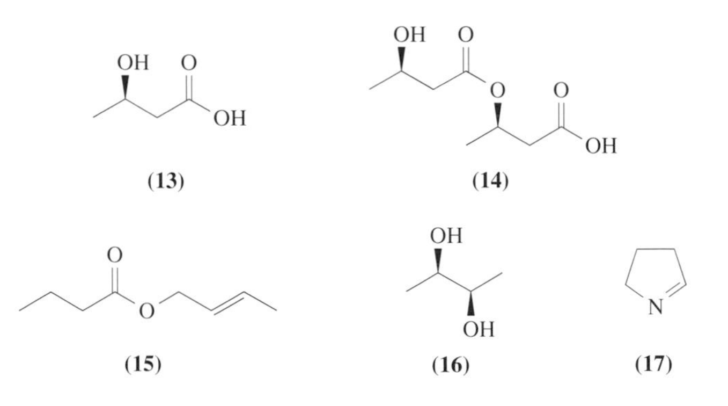

8.04.4.6 C_5 Chains

Compounds with five carbon atoms are rarely found as pheromones. The beetle *Limonius californicus* uses pentanoic acid as a female sex pheromone.[118] The male aggregation pheromone of the maize and rice weevils (*Sitophilus* spp.) contains 3-pentanone. It might be biogenetically related to the main component of the pheromone 5-hydroxy-4-methyl-3-heptanone;[119] see Section 8.04.6.3 and Figure 18.

8.04.4.7 C_6 Chains

Compounds containing six carbon atoms in the chain may be formed either by condensation of three acetyl-CoA units or oxidative degradation of polyunsaturated fatty acids.[62] Hexanoic acid is the female sex pheromone in the beetle *Limonius canus*[120] and synergizes the action of the female sex pheromones in *Trogoderma* spp.[121] The trail pheromone of the ant *Lasius fuliginosus* was reported to consist of hexanoic acid together with the corresponding C_7–C_{10} and C_{12} acids,[109] but a reinvestigation showed that these acids are not active.[122] Instead, the ants use mellein and derivatives thereof as trail pheromone (see Sections 8.04.4.11 and 8.04.6.1). Methyl and ethyl hexanoate are part of the pheromone blend of the fruit fly *Drosophila virilis*.[123] The stink bug, *Leptoglossus zonatus*, uses an alarm pheromone which consists of hexanoic acid and a variety of derivatives, (*E*)-2-hexenal, hexyl acetate, 1-hexanol, and hexanal.[124] (*E*)-2-Hexenal is a typical defensive compound of many stink bugs, but functions also as a component of the male aggregation pheromone in several true bugs of the genus *Podisus*.[116,125,126] The reactive (*E*)-4-oxo-2-hexenal is a common constituent of bug defensive secretions.[116] In the bug *Riptortus clavatus*, the attractant pheromone of males consists of the related (*E*)-2-hexenyl (*E*)-2-hexenoate (**18**) and (*E*)-2-hexenyl (*Z*)-3-hexenoate (**19**) and of myristyl 2-methylpropanoate.[127] These compounds also act as kairomone for its parasitoid, *Ooencyrtus nezarae*. 1-Hexanol has been shown to be a male aggregation pheromone component of the ambrosia beetle, *Platypus flavicornis*,[128] and is a component of the attractive pheromone of the ant *Oecophylla longinoda*, which uses hexanal as an alarm pheromone.[109] Hexanal is also a female attractant of the fly *Sarcophaga bullata*, where both sexes emit this compound.[129] Myristic, palmitic, and stearic acid esters of 1-hexanol along with long-chain hydrocarbons are used by males of the pierid butterfly, *Colias eurytheme*, for releasing an acceptance behavior of the females.[130] A second oxidation in the chain occurs in 4-hexanolide (**20**), reported as a pheromone of some *Trogoderma* beetles.[131]

All these compounds contain a terminally oxidized carbon. In contrast, the diols and ketols of some of cerambycid beetle pheromones show oxidation along the chain, only. The male sex pheromones of longhorn beetles *Hylotrupes bajulus* and *Pyrrhidium sanguineum* consist of (*R*)-3-hydroxy-

2-hexanone (**21**), (2*R*,3*R*)-2,3-hexanediol (**22**), and (2*S*,3*R*)-2,3-hexanediol,[132] whereas in *Anaglyptus subfasciatus* only (**21**) is active.[133]

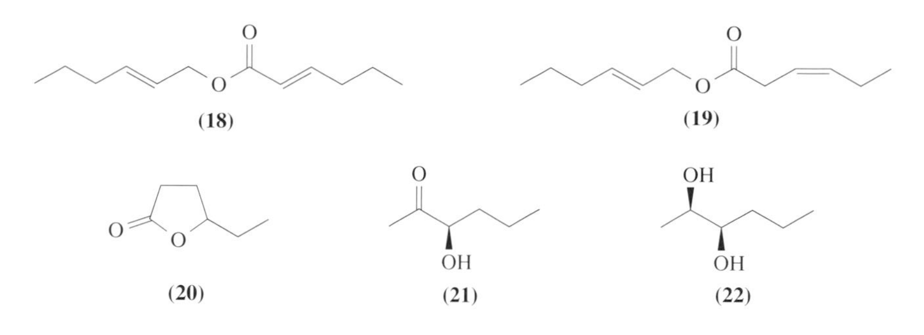

8.04.4.8 C_7 Chains

The simple alcohols 1- and 2-heptanol are components of the aggregation pheromones of female *Dendroctonus jeffreyi*[134] and *Dendroctonus vitei* bark beetles.[135] Whereas the primary alcohol is more active in the former species, the secondary one is more active in the latter species. (*R*)-2-Heptanol, together with the corresponding ketone, 2-heptanone, is a pheromone of female caddisflies, *Rhyacophila fasciata*.[136] 2-Heptanone is also suggested to act as a forage-marking pheromone by honeybees.[137] The C_7 alcohols and ketones of the primitive moth *Eriocrania cicatricella* have been discussed above (see Section 8.04.4.1). Midchain oxidation is found in 4-heptanone, an aggregation pheromone component of the cockroach species *Blatella craniifer* and *Eublaberus distanti*, which also includes 1-octanol.[138] Cyclohexene derivatives, pheromone components of the bark beetle *Dendroctonus pseudotsugae*, may be formed from acyclic precursors: Intramolecular condensation of 2,6-heptandione would yield 3-methylcyclohexenone, seudenone,[139] and give rise to the corresponding alcohol, seudenol,[140] and also to 1-methyl-2-cyclohexen-1-ol,[141] the product of a simple allylic rearrangement.

8.04.4.9 C_8 Chains

Males of the rice bug, *Leptocorisa chinensis*, use a mixture of 1-octanol and (*E*)-2-octenyl acetate as attractant pheromone.[142] In the related *L. oratorius*, (*E*)-2-octenal and octyl acetate act as alarm pheromone.[143] Ethyl (*E*)-3-octenoate is a component of the male aggregation pheromone of the Mediterranean fruit fly, *Ceratitis capitata*.[112] One of the most volatile sex pheromones of moths, 1-methylethyl octanoate, is produced by females of the bag worm moth, *Megalophanes viciella*.[108] Optically active 3-octanol has been identified as a pheromone of the mandibular glands of the ants *Myrmica scabrinodis* and *M. rubra*. The first species uses a 9:1 (3*R*)/(3*S*) mixture, which is more effective than the pure (*R*)-enantiomer.[144] *Crematogaster* spp. seem to produce predominantly (*S*)-3-octanol,[145] and in addition contain 3-octanone as major alarm pheromone component.[146] The unsaturated analogue (*R*)-1-octen-3-ol (**23**) is part of the pheromone blend of some *Oryzaephilus* beetles.[147] This typical aroma component of mushrooms is formed from linoleic acid by the action of a lipoxygenase.[148] Cerambycid beetles use not only C_6 but also C_8 compounds, showing a distinct 1,2-oxidation pattern. The male sex pheromones of *Xylotrechus* spp. consist of (2*S*,3*S*)-2,3-octanediol (**24**) and (*S*)-2-hydroxy-3-octanone (**25**),[149,150] while *Anaglyptus subfasciatus* uses (*R*)-3-hydroxy-2-octanone (**26**) together with (**21**).[133] The absolute configuration of (**24**) is different from that of the corresponding C_6 pheromones of other cerambycids (see Section 8.04.4.7). In contrast to most other moth species, male hepialid moths attract their females with long-range sex pheromones. Whereas *Hepialus californicus* uses the dihydropyrone (*R*)-2-ethyl-6-methyl-4*H*-2,3-dihydropyran-4-one (hepialone) (**27**),[151] the opposite substitution pattern, but corresponding stereochemistry,[152] is found in (2*R*)-6-ethyl-2-methyl-4*H*-2,3-dihydropyran-4-one (**28**), the pheromone of *Hepialus hecta*. The pheromone bouquet of the latter species also shows some unique bicyclic acetals which will be discussed in Section 8.04.6.3. Related 2,3-dihydropyrans have been identified as exocrine gland constituents of some bee species.[153] The polyketide methyl 6-methylsalicylate (**29**), a pheromone of some ants,[154] is formed by condensation and folding of four acetate units. The biosynthetic pathway leading to this compound has been elucidated in detail in the fungus *Penicillium patulum*.[60] Its decarboxylation leads to *m*-cresol, a defensive compound of some beetles.[35] The related 6-methylsalicylic aldehyde (**30**) is a pheromone of some mite species.[155]

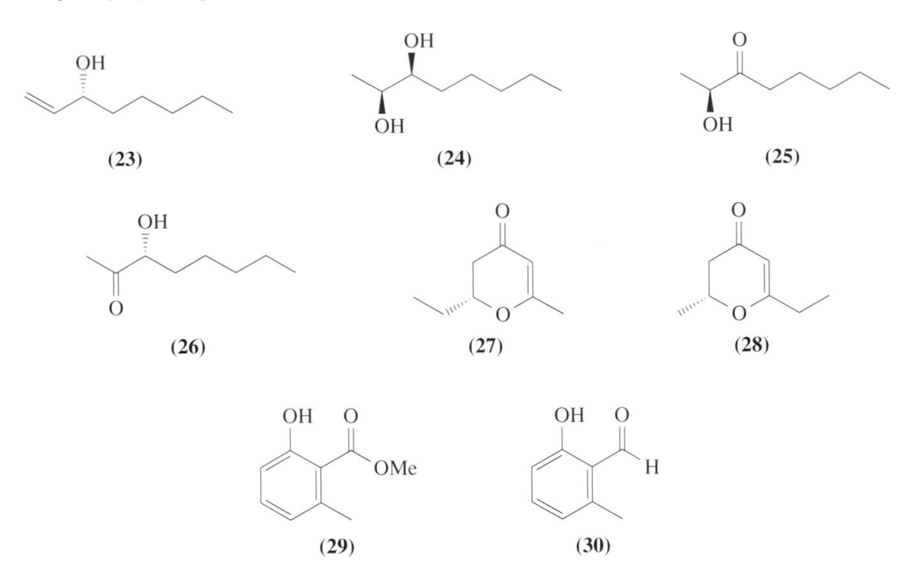

8.04.4.10 C_9 Chains

Pheromones with nine carbon atoms in the chain are particularly abundant, which may be due to the fact that the common unsaturated C_{18} acids can be enzymatically cleaved to such compounds. In studies on the biosynthesis of the pheromone of the wax moth, *Galleria melonella*, such cleavage has been shown to take place. The males use nonanal (together with undecanal) as sex pheromones. Experiments with [^{14}C]oleate showed that the insects synthesize nonanal (but not undecanal) *de novo* by cleavage of oleate at the double bond.[156] 1-Nonanol is the male sex pheromone of another pyralid moth, *Achroia innotata*.[157] Nonanal has also been identified as a component of the aggregation pheromone of nymphs from the gregarious locust *Schistocerca gregaria*, which also contains nonanoic acid and the corresponding C_6, C_8, and C_{10} acids.[158] The male aggregation attachment pheromone of the tick *Amblyomma variegatum* is a mixture of nonanoic acid, nitrophenol, and methyl salicylate.[159] Males of the fruit fly, *Anastrepha suspensa*, use (*Z*)-3-nonen-1-ol and (3*Z*,6*Z*)-3,6-nonadien-1-ol (**31**) together with two lactones as male sex pheromone.[160] The scarab beetle, *Anomala albopilosa*, and other *Anomala* spp. produce (*E*)-2-nonen-1-ol and (*E*)-2-nonenal (**32**) as sex pheromone.[161] The lactone, (2*Z*,6*Z*)-2,6-nonadien-4-olide (**33**) is presumed to be an aphrodisiac of male *Aphomia gularis* moth.[162] The female pheromone of the caddisfly, *Rhyacophila fasciata*, is made up by a mixture of 2-nonanone and 2-nonanol, whereas *Hydropsyche angustipennis* uses 2-nonanone, (*Z*)-6-nonen-2-one, 2-decanone, and 2-octanone.[136] As mentioned above, a 6*E*/6*Z* mixture of (2*S*)-6,8-nonadien-2-ol (**12**) forms the female sex pheromone of the moth *Stigmella malella*.[107]

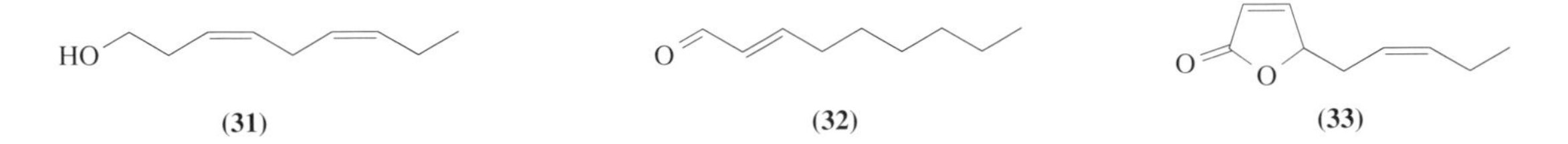

Midchain oxidation is found in 5-nonanol, a component of the curculionid beetle, *Metamasius hemipterus*,[163] and also characteristic for the formation of spiroacetals, a class of compounds known as pheromones from bark beetles and flies. Thus, (5*R*,7*S*)-7-methyl-1,6-dioxaspiro[4.5]decane (**34**) has been identified as pheromone component of the bark beetles *Leperisinus varius* (together with 2-nonanone),[164] *Pityophthorus* spp., and *Conophthorus* spp.,[165] where it acts as a male repellent. In *Pityogenes chalcographus*, (2*S*,5*R*)-2-ethyl-1,6-dioxaspiro[4.4]nonane (**35**) is a component of the male aggregation pheromone, whereas the (2*S*,5*S*)-stereoisomer is inactive.[166] With respect to the reaction of the female-produced pheromone, the olive fly, *Dacus oleae*, shows a striking difference between the sexes: the natural pheromone is racemic, but males are attracted by (*R*)-1,7-dioxaspiro[5.5]decane (**36**) whereas females are attracted by the (*S*)-enantiomer.[167,168] Biosynthetic studies indicated that (**36**) is formed from malonate units. In males, diethyl 5-oxo-1,9-nonanedioate has been identified, a possible precursor of (**36**).[169]

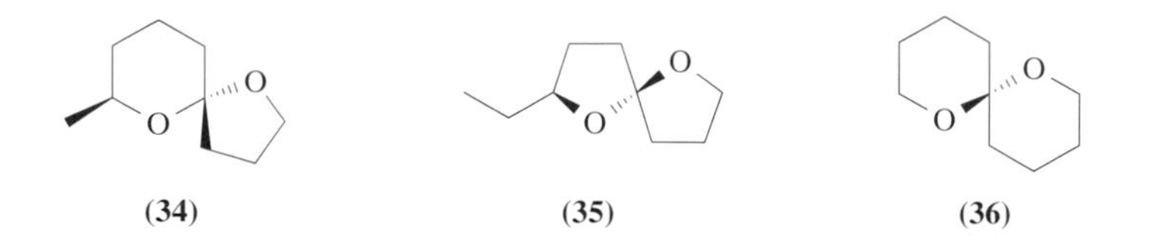

(**34**) (**35**) (**36**)

The pheromone systems of bark beetles can be fairly complex, and apart from terpenes (see Section 8.04.5) they may include bicyclic acetals such as brevicomin. One of the most widespread pheromones in *Dendroctonus* species is (+)-*exo*-brevicomin (**37**).[170] The biosynthesis of this compound obviously involves (*Z*)-6-nonen-2-one as an intermediate.[171] Interestingly, this ketone and its reduction product, (2*S*,6*Z*)-6-nonen-2-ol (**38**), play a role in the communication system of some caddisfly species.[172] (*E*)-6-Nonen-2-ol and the corresponding ketone, which upon epoxidation and ring closure would yield *endo*-brevicomin, were also found in caddisflies.[172] The bark beetle *Dryocoetes affaber* uses (+)-*endo*-brevicomin as the main male aggregation pheromone component whereas the (−)-enantiomer has an inhibiting effect and (+)-*exo*-brevicomin shows synergistic properties. The (+)-*exo*- to (+)-*endo* ratio is 1:2, whereas the related species *D. confusus* shows a 9:1 ratio.[173] Interestingly, 3,4-dehydro-*exo*-brevicomin (**39**) is a component of the aggression pheromone of male mice[174] which again exemplifies the striking similarities between semiochemicals of very different organisms (see Section 8.04.1). Many other related spiroacetals and bicyclic acetals, also containing additional oxygen functionalities, have been identified in bark beetles,[175] flies,[176] bees,[177] and wasps,[164] but their biological functions remain unknown.

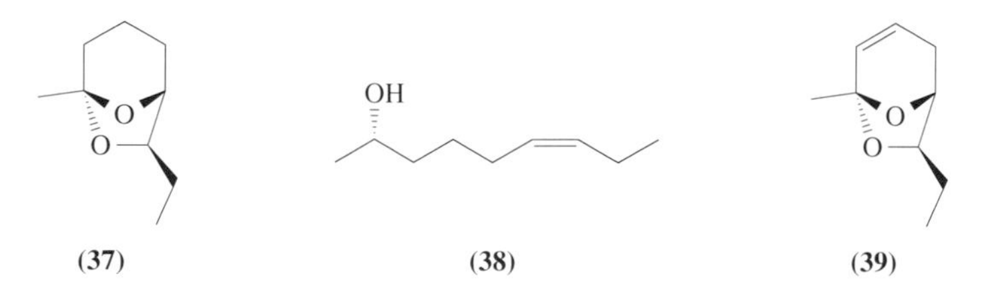

(**37**) (**38**) (**39**)

8.04.4.11 C_{10} Chains

Decyl acetate is an alarm pheromone component of the thrips *Frankliniella occidentalis*.[178] From the bug *Nezara viridula*, (*E*)-4-oxo-2-decenal (**40**), a typical compound of the defensive secretion of stink bugs, is reported to be an attractant and arrestant of second instar larvae.[179] The other known C_{10} pheromones are derivatives of decanoic acid: (*Z*)-3-decenoic acid is the female sex pheromone of the furniture carpet beetle, *Anthrenus flavipes*.[180] The stink bug *Plautia stali* uses methyl (2*E*,4*E*,6*Z*)-2,4,6-decatrienoate (**41**) as an aggregation pheromone.[181] The male pheromone of the bark beetle *Pityogenes chalcographus* is composed of the related methyl (2*E*,4*Z*)-2,4-decadienoate (**42**) and the spiroacetal (**35**).[182] The ester (**42**) is also used by stink bugs of the genus *Euschistus* as an aggregation pheromone.[116] Further oxidation in the chain leads to (6*E*)-[(1*E*)-pentenyl]-pyran-2-one (**43**), a queen pheromone component of the fire ant *Solenopsis invicta*.[183] The well known queen substance of honey bees, 9-oxo-(*E*)-2-decenoic acid (**44**), attracts drones and prohibits the building of queen cells by workers, while the racemate of (*E*)-9-hydroxy-2-decenoic acid (**45**) induces retinue behavior.[184,185] These acids are synthesized by the bees starting from stearic acid. After hydroxylation at the $\omega - 1$ position, the chain is shortened by β-oxidation. Subsequent oxidation of the hydroxy acid (**45**) furnishes (**44**).[186] By ring closure of (**45**), 9-decanolide, phoracantholide I, can be formed, which is part of the defensive secretion of the cerambycid beetle, *Phoracantha synonyma*.[187] The bees, *Apis dorsata* and *Apis florea* use (2*E*)-2-decenyl acetate as sting alarm pheromone.[188] The pentaketide mellein (**46**) is a trail pheromone of some ants including *Lasius fuliginosus* (see Section 8.04.4.7),[122,189] which also use branched chain derivatives (see Section 8.04.6.1). Mellein is also present in male pheromone glands of the oriental fruit moth, *Grapholita molesta*, where it acts as a male–male repellent.[190] The *R*-enantiomer is used as a pheromone by the male wax moth, *Aphomia sociella*.[191] This microbial metabolite may be produced by the fungus *Aspergillus ochraceus*, which has been detected on the moth.

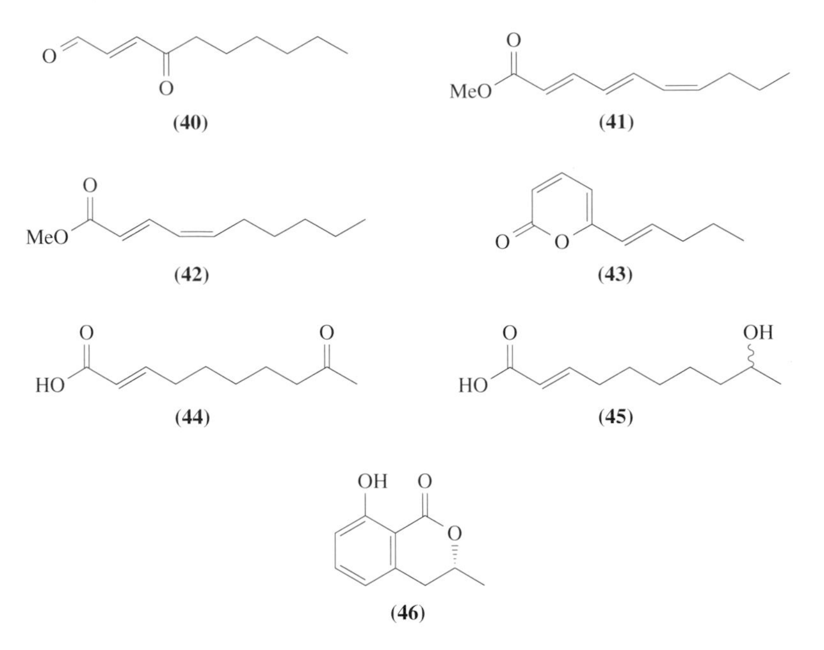

8.04.4.12 C_{11} Chains

Undecane is the sex pheromone of the ant *Formica lugubris*.[192] The related tridecane and (*Z*)-4-tridecene are also active, but to a lesser degree. Several other ant species use undecane as part of the alarm pheromone.[109,192] This shows that a functional group is not a prerequisite for pheromonal activity. Even very simple compounds may be active when used in an appropriate biological context. Undecane has also been found to be the sex pheromone of the mite *Caloglyphus rodriguezi*[193] and a component of aggregation pheromones of the ant *Camponotus pennsylvanicus*,[109] in addition to the cockroaches *Blatella craniifer* and *Eublaberus distanti*.[138] Other ants use oxidized C_{11} compounds: in *Bothroponera soror*, attraction is released by 2-undecanol, whereas 2-undecanone induces alarm.[109] 3-Undecanone is a component of the attraction pheromone of *Oecophylla longinoda*.[109] Spiroacetals showing 11 carbon atoms have also been identified: females of the bee *Andrena wilkella* contain (2*S*,6*R*,8*S*)-2,8-dimethyl-1,7-dioxaspiro[5.5]undecane (**47**) in their cephalic secretion. The compound functions as an aggregation pheromone for males[194] and also occurs in several fruit fly species and bugs; however, its function in these insects is unknown. The female sex pheromone of the varied carpet beetle *Anthrenus verbasci* consists of a 85:15 mixture of (*Z*)- and (*E*)-5-undecenoic acid.[195]

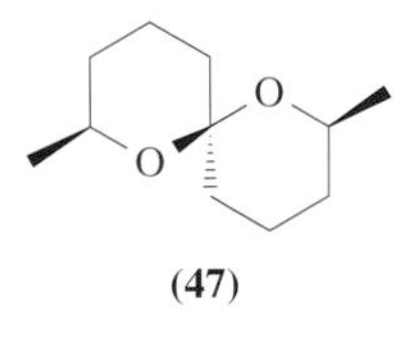

(47)

8.04.4.13 C_{12} Chains

In the tsetse fly, *Glossina morsitans*, dodecane is a component of its larviposition pheromone. It attracts gravid females to the larvae.[196] 1-Dodecanol is a component of the male sex pheromone of some cockroaches.[117] The corresponding acetate functions as alarm pheromone in the thrips, *Frankliniella occidentalis*,[178] while the butanoate is a female sex pheromone in the carpet beetle, *Anthrenus sarnicus*.[197] Another beetle, *Cyclas formicarius*, uses (*Z*)-3-dodecenyl (*E*)-2-butenoate as female sex pheromone.[198] In *Reticulitermes* termites, (3*Z*,6*Z*,8*E*)-3,6,8-dodecatrien-1-ol (**48**) functions as a trail pheromone,[199,200] but another termite, *Pseudacanthotermes spiniger*, uses (**48**) as a sex pheromone.[201] The 2-methylethyl esters of (5*Z*)-, (7*Z*)-, and (9*Z*)-dodecenoic and dodecanoic acids are components of the female sex pheromone bouquet of the hide beetle *Dermestes maculatus*, which also contains related C_{14}, C_{16}, and C_{18} esters (see below).[114] In this case the combination of a long-chain alcohol and a short fatty acid usually found in lepidopteran sex pheromones is reversed, but it strongly resembles the bouquet of some Zygaenid and Psychid moths. Lactones are commonly used as pheromones, especially in beetles: the aggregation pheromone of the scarab beetle, *Anomala cuprea*, is (*R*)-4-dodecanolide,[202] while the unsaturated analogue (4*R*)-(5*Z*)-5-dodecen-4-olide (**49**) is the sex pheromone of *Anomala albopilosa* and other *Anomala* spp.[161,203] Formal addition of water to the double bond leads to (4*S*,6*S*)-6-hydroxy-4-dodecanolide (**50**), which enhances the activity of the male courtship pheromone of the day flying butterfly *Idea leuconoe*, which also contains homologues of (**50**) in the pheromone glands.[204]

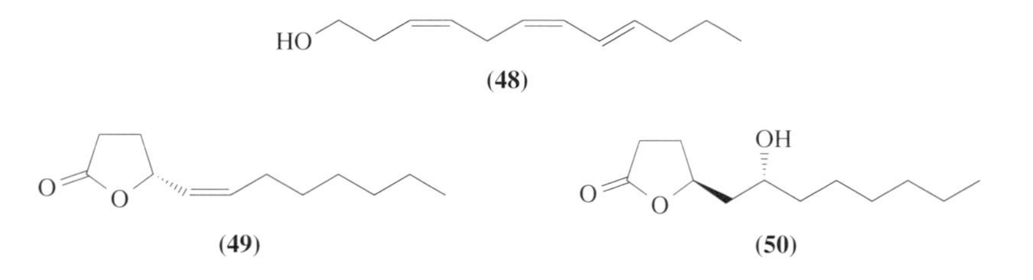

Macrocyclic lactones are typical constituents of the male aggregation pheromones of cucujiid beetles. In *Cryptolestes pusillus*, (*Z*)-3-dodecen-12-olide (**51**) is the main pheromone component. The related (11*S*)-(3*Z*)-3-dodecen-11-olide (**52**) is the pheromone of the rusty grain beetle, *C. ferrugineus*, while the *R*-enantiomer along with (3*Z*,6*Z*)-(11*R*)-3,6-dodecadien-11-olide (**53**) acts as an attractant in *Oryzaephilus mercator*. A mixture of (**53**) and (3*Z*,6*Z*)-3,6-dodecadien-12-olide (**54**) is used by *O. surinamensis*.[205] These macrolides are formed from oleic or linoleic acid, which are chain shortened and oxidized at the ω or $\omega - 1$ position.[206] Methyl jasmonate (**55**) and *epi*-methyl jasmonate (**56**) are components of the male pheromone of the moth *Grapholita molesta*.[207,208] Feeding experiments showed that neither compounds is sequestered from plants. In plants, the well known elicitor jasmonic acid and its epimer are formed from linolenic acid when the leaf is damaged.[62] An unusual compound, (1′*E*)-(3*R*,4*S*)-3,4-bis(1′-butenyl)tetrahydro-2-furanol (**57**), has been identified as the male aggregation pheromone of the bug *Biprorulus bibax*.[116,209] It is presumably formed by coupling of two C_6 units such as 2-hexenal.

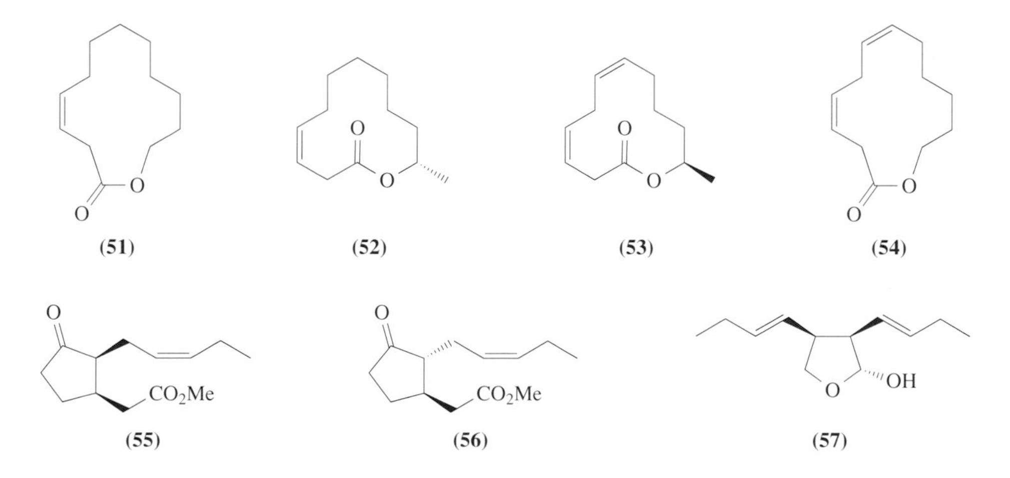

8.04.4.14 C_{13} Chains

Several fruit flies use C_{13} compounds as male aggregation pheromones. In *Drosophila hydei*, 2-tridecanone and some esters are used,[210] whereas (*S*)-1-methyldodecyl acetate is the pheromone of *Drosophila mulleri*.[211] The unsaturated analogue (9*E*)-(1*S*)-1-methyl-9-dodecenyl acetate (**58**) is a component of the pheromone of the gall midge *Mayetiola destructor*.[212] In several ant species C_{13} alkaloids have been found, where they are predominantly used as defensive compounds. Nevertheless, monomorine I (**59**) and monomorine III (**60**) are reported to show a weak trail pheromone activity in the pharaoh ant *Monomorium pharaonis*.[109]

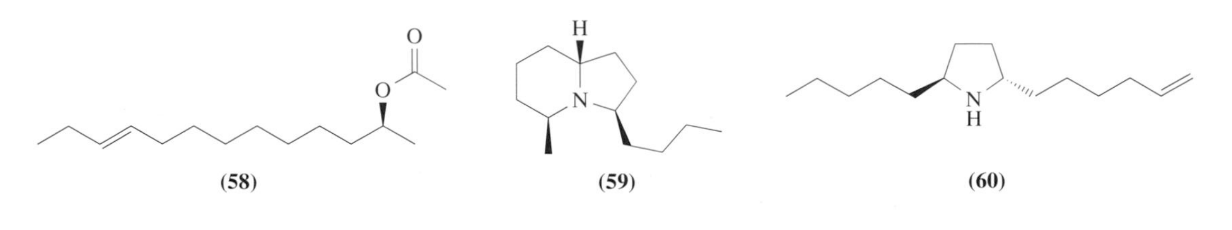

(**58**) (**59**) (**60**)

8.04.4.15 C_{14} Chains

Tetradecane is an aggregation pheromone component of the cockroach *Blattella craniifer*.[138] Females of the subterranean termite *Reticulitermes flavipes* use tetradecyl propanoate as sex pheromone.[213] The female sex pheromone of the scarab beetles *Anomala orientalis* and *Blitopertha orientalis* is a 9:1 mixture of (*Z*)- and (*E*)-7-tetradecen-2-one.[214] A similar *Z*/*E* ratio is found in (*R*)-5-tetradecen-4-olide (**61**), the sex pheromone of the Japanese beetle, *Popillia japonica*, which also contains small amounts of tetradecan-4-olide. As little as 1% of the wrong *S*-enantiomer of (**61**) inhibits the attractiveness of the pheromone.[215] This enantiomer is the pheromone of the related Osaka beetle, *Anomala osakana*.[216] As in *P. japonica*, small amounts of the opposite enantiomer inhibit response. The lactone (**61**) is also a component of the pheromone of *Anomala octiescostata*.[203] The soybean beetle, *Anomala rufocuprea*, uses methyl (*Z*)-5-tetradecenoate as female sex pheromone.[217] Free acids occur in carpet beetles: *Attagenus elongatulus* uses (*Z*,*Z*)-3,5-tetradecadienoic acid (**62**) as female sex pheromone[218] whereas the *E*,*Z*-isomer is active in *A. megatoma*.[219] The 1-methylethyl esters of (5*Z*)-, (7*Z*)-, and (9*Z*)-tetradecenoic and tetradecanoic acid are components of the pheromone bouquet of *Dermestes maculatus* (see also C_{12}, C_{16}, and C_{18}, chains).[114] The male sex pheromone of the dried bean beetle, *Acanthoscelides obtectus*, shows an unusual allenic structure, methyl (*R*,2*E*)-2,4,5-tetradecatrienoate (**63**).[220] In addition to the above mentioned C_{12} macrolides, cucujiid beetles also contain corresponding C_{14} compounds. Thus, an 85:15 mixture of (13*R*)- and (13*S*)-(5*Z*,8*Z*)-5,8-tetradecadien-13-olide (**64**) is the pheromone of *Cryptolestes turcicus*. In addition, the *R*-enantiomer is a pheromone synergist in *Oryzaephilus surinamensis*. Such a synergistic function has also been reported for (*Z*)-5-tetradecen-13-olide in other *Cryptolestes* spp.[205]

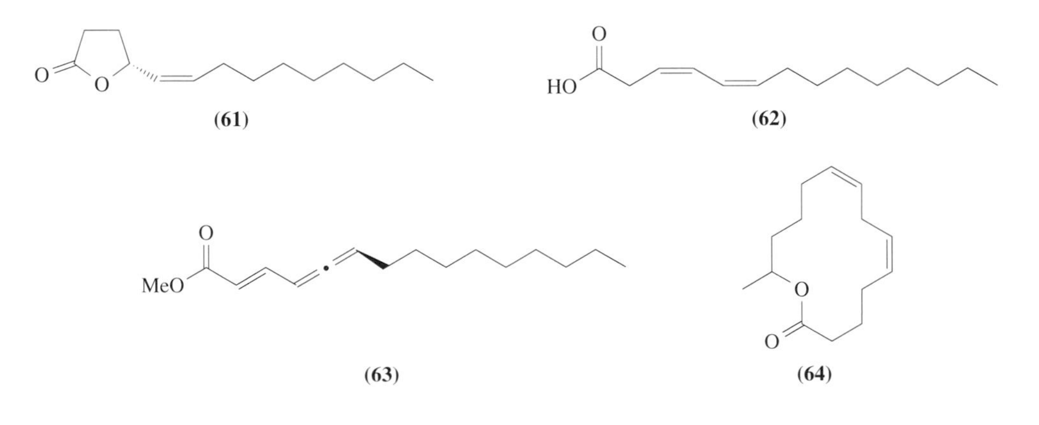

(**61**) (**62**) (**63**) (**64**)

8.04.4.16 C_{15} Chains

Pentadecane is a component of the larviposition pheromone of the tsetse fly, *Glossina morsitans*,[196] and 1-pentadecene has been identified from *Tribolium* beetles.[67] A mixture of (6*Z*)- and (7*Z*)-pentadecene, accompanied by small amounts of the corresponding tetradecenes and (*Z*)-5-tridecene, compose the alarm pheromone of the mite *Tyrophagus neiswanderi*.[221] (*S*)-1-Methyltetradecyl acetate and 2-pentadecanone are male aggregation pheromone components of *Drosophila buschkii*.[222] The latter is also a synergist of the pheromone of *Drosophila hydei*.[223]

8.04.4.17 C_{16} Chains

Hexadecane is a component of the pheromone of the moth *Acrolepiopsis assectella*.[224] Methyl palmitate together with common C_{18} methyl esters acts as brood pheromone in honey bees (see below).[225] The esters 1-methylethyl (*Z*)-9-hexadecenoate and the corresponding hexadecanoate are components of the pheromone bouquet of *Dermestes maculatus* (see also C_{12}, C_{14}, and C_{18} chains).[114] The parasitoid wasp *Ascogaster reticulatus* uses the typical lepidopteran pheromone (*Z*)-9-hexadecenal as sex pheromone,[226] while this compound functions as trail pheromone in the ant *Iridomyrmex humilis*.[109] The lactone (*R*)-5-hexadecanolide (**65**) of the wasp *Vespa orientalis* stimulates the construction of queen cells by workers.[227] The doubly unsaturated lactone (4*R*,7*Z*)-7,15-hexadecadien-4-olide (**66**) has been identified as the sex pheromone of the scarab beetle, *Heptophylla picea*.[228] An unusual lactone, (5*R*,6*S*)-6-acetoxy-5-hexadecanolide (**67**), is the oviposition attractant pheromone of the mosquito *Culex pipiens*.[229] It was isolated from eggs and stimulates other females to oviposit at the same site. A compound of mixed biogenetic origin is the oviposition deterring pheromone of the cherry fly, *Rhagoletis cerasi*. In contrast to the *Culex* compound (**67**), this glucoside (**68**) inhibits oviposition on fruits where eggs of this species have already been deposited.[230] The pheromone is a mixture of the (8*R*,15*R*)- and (8*S*,15*R*)-stereoisomers.

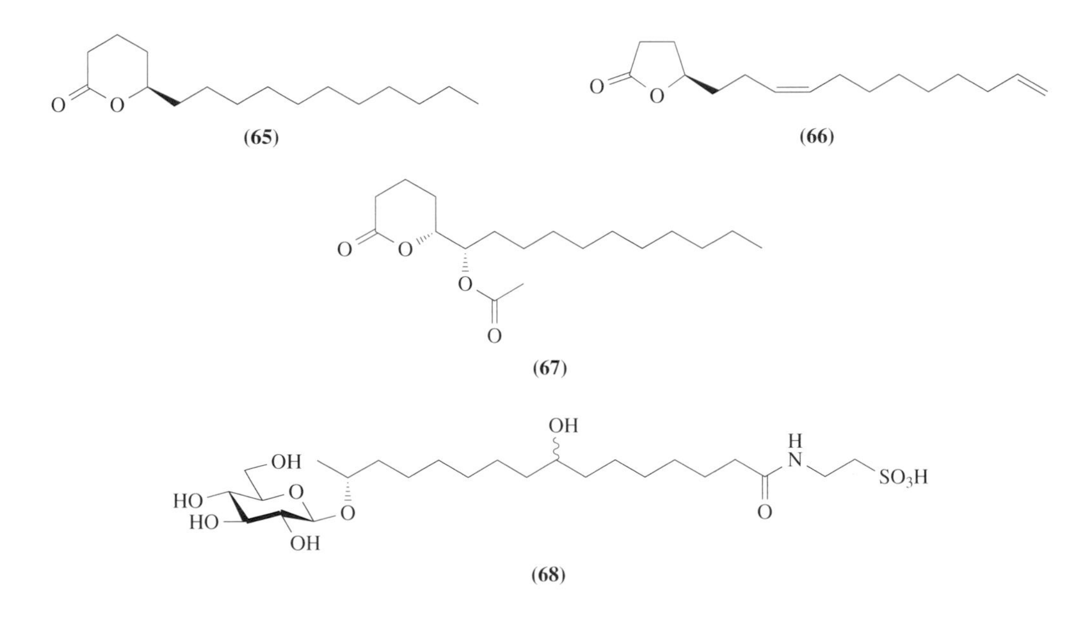

8.04.4.18 C_{17} Chains

The fly *Lycoriella mali* uses heptadecane as a female sex pheromone component,[231] whereas males of the moth *Acrolepiopsis assectella* release the same compound from their hairpencils together with other less abundant alkanes as courtship pheromone.[224] The major aggregation pheromone component of several *Drosophila* spp. is (*Z*)-10-heptadecen-2-one.[232] The flour beetle *Tribolium confusum* has been reported to use a mixture of 1-heptadecene, 1-hexadecene, and 1-pentadecene as components of the sex pheromone;[67] however, 4,8-dimethyldecanal seems to be the correct structure of the major *Tribolium* pheromone (see Section 8.04.5.6). The decarboxylation product of linoleic acid, (6*Z*,9*Z*)-6,9-heptadecadiene, induces alarm in mites of the genus *Tortonia*[233] and it has been identified as a component of sex pheromones of geometrid moths (see above).

8.04.4.19 C_{18} Chains

Methyl esters and free stearic, oleic, linoleic, and linolenic acid act as egg dispersion pheromones of the moth *Lobesia botrana*[234] and the mosquito *Culex quinquefasciatus*.[235] The corresponding methyl esters are produced by honey bee larvae and induce capping of drone and worker brood cells.[225] The glyceride 1,2-dioleyl-3-palmitylglycerol (**69**) also functions as brood pheromone in this bee.[236] The 1,3-dioleyl isomer is inactive. Several *Drosophila* spp. use the "lepidopteran-like" (*Z*)-11-octadecenyl acetate (**70**) as an aggregation pheromone and antiaphrodisiac.[237–239] The large pine weevil, *Hylobius abietis*, contains a blend of hexadecyl, octadecyl, (*E*)- and (*Z*)-9-octadecenyl, and icosyl acetate, which functions as part of the female sex pheromone.[240] 1-Methylethyl oleate is part of the pheromone bouquet of *Dermestes maculatus*.[114] The macrolide (9*Z*,11*E*)-(13*R*)-9,11-octadecadien-13-olide (**71**), possibly derived from linoleic acid by oxidation at C-13 and ring closure, has been identified as the main constituent of the pheromone glands of male *Heliconius pachinus* butterflies; however, the biological function has not been tested.[241]

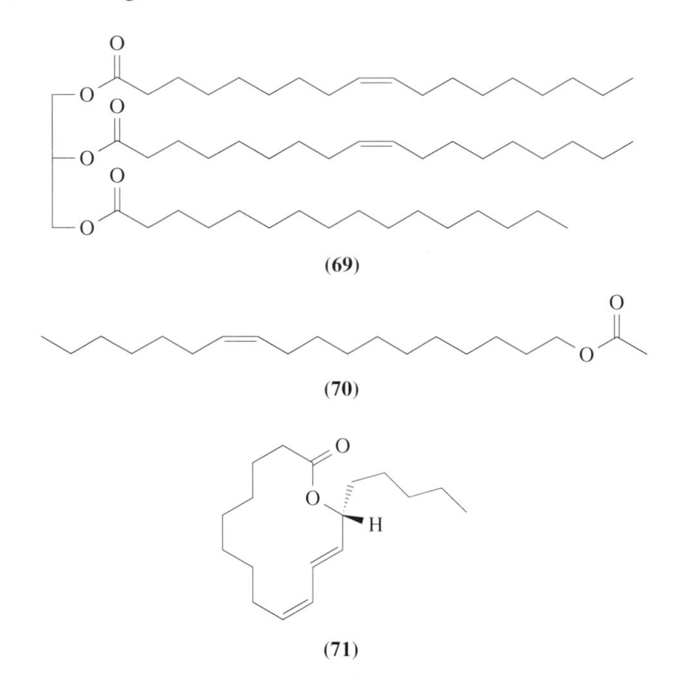

8.04.4.20 C_{19} Chains

Nonadecane is part of the courtship pheromone of *Acrolepiopsis assectella*.[224] Polyenic compounds identified from moths have been discussed in Section 8.04.4.1.

8.04.4.21 C_{20} Chains

In the primitive eusocial bee, *Lasioglossum zephyrum*, a blend of macrolides is used for chemical communication (see also Section 8.04.5.1). The most abundant, icosanolide, is accompanied by octadecanolide, docosanolide, tetracosanolide, docosenolide, and octadecenolide.[242] The alcohol (*Z*)-11-icosen-1-ol acts as an alarm pheromone and foraging cue of the honey bee, *Apis mellifera*, which contains it in the sting gland.[243] The same compound is used by the beewolf, *Philanthus triangulum*, as marking pheromone for mating territories.[244] The corresponding acetate is an aggregation pheromone of some fruit flies of the genus *Drosophila*.[245]

8.04.4.22 C_{21} Chains

Henicosane is part of the courtship pheromone of *Acrolepiopsis assectella*,[224] whereas the alkene (*Z*)-7-henicosene acts together with (*Z*)-7-tricosene as the female sex pheromone of the beetle *Aleochara curtula* and is also used by young males to avoid aggression from older males.[246] (*Z*)-9-Henicosene is a male aggregation pheromone component in the fruit flies *Drosophila americana*,

D. lummei, and *D. novamexicana*.[67] The corresponding *Z*-10-isomer has this function in *D. virilis*. In addition, the complete blend contains branched butenoates and hexanoates.[123] For C_{21} compounds in moths, see Section 8.04.4.1.

8.04.4.23 C_{23} Chains

The female sex pheromone of the common housefly, *Musca domestica*, consists of (*Z*)-9-tricosene,[67] which is also the precursor for the pheromone components (*Z*)-14-tricosen-10-one and 9,10-epoxytricosane.[247] In *Drosophila melanogaster*, (*Z*)-7-tricosene functions as an antiaphrodisiac.[67] Such compounds are transferred from the males to the females during mating and inhibit further mating attempts by other males.

8.04.4.24 C_{24} Chains

Males of the cockroach *Periplaneta japonica* are able to determine the sex of conspecifics by the absence or presence of octadecyl (*Z*)-9-tetracosenoate.[138]

8.04.4.25 C_{25} and Longer Chains

In the fly *Fannia canicularis*, (*Z*)-9-pentacosene functions as female mating stimulant pheromone.[67] Other long-chain alkenes identified in flies which function either as female mating stimulants or sex recognition components include (*Z*)-5-tricosene, (*Z*)-11-pentacosene, (*Z*)-9-heptacosene, (*Z*)-13-heptacosene, (7*Z*,11*Z*)-7,11-heptacosadiene, (*Z*)-13-nonacosene, (*Z*)-14-nonacosene, (*Z*)-9-hentriacontene, (*Z*)-11-hentriacontene, (*Z*)-9-tritriacontene, (*Z*)-11-tritriacontene, (*Z*)-13-tritriacontene, and (5*Z*,27*Z*)-5,27-tritriacontadiene.[67,248,249] The female courtship pheromone of the braconid wasp, *Cardiochiles nigriceps*, consists of (7*Z*,13*Z*)-7,13-heptacosadiene.[250] In the sawfly, *Pikonema alaskensis*, (9*Z*,19*Z*)-9,19-nonacosadiene, (9*Z*,19*Z*)-9,19-hentriacontadiene, (9*Z*,19*Z*)-9,19-tritriacontadiene, (9*Z*,19*Z*)-9,19-pentatriacontadiene, and (9*Z*,19*Z*)-9,19-heptatriacontadiene are active female sex pheromone constituents.[251] Nevertheless, they may only be precursors of the truly active components (*Z*)-5-tetradecen-1-ol and (*Z*)-10-nonadecenal, which they form upon oxidative cleavage.[252] The same may be true for another parasitoid wasp, *Macrocentrus grandii*, which uses a mixture of (9*Z*,13*Z*)-9,13-alkadienes as pheromone. In this species, 4-tridecenal is also active.[67] Nymphs of the cockroach *Nauphoeta cinerea* use a mixture of long-chain epoxides, (9*Z*,43*Z*)-, (11*Z*,41*Z*)-, (11*Z*,43*Z*)-, and (11*Z*,45*Z*)-25,26-epoxyhenpentacontadienes, as recognition pheromone, which inhibits wing raising of adults.[253] The (25*S*,26*R*)-enantiomer of the (9*Z*,43*Z*)-isomer is more active than its antipode. These compounds resemble the lepidopteran pheromones of the epoxide type (see Section 8.04.4.1).

8.04.5 ISOPRENOIDS IN SYSTEMS OF CHEMICAL COMMUNICATION

The biosynthesis of isoprenoids has been thoroughly investigated, particularly in plants.[254] The "classical" synthesis involves mevalonate, (*R*)-3,5-dihydroxy-3-methylpentanoate, which is formed from three acetate units. Elimination of water and carbon dioxide yields 3-methyl-3-butenyl diphosphate, which forms an equilibrium with 3-methyl-2-butenyl diphosphate by the action of an isopentenyl diphosphate isomerase. Coupling of these two C_5 units yields geranyl diphosphate, the parent compound of monoterpenes (Scheme 3). Apart from this "mevalonate" pathway, amino acids such as valine or leucine may serve as starters for the formation of 3-methyl-2-pentenyl diphosphate (for a review see reference 255). Another "non-mevalonate" pathway leading to monoterpenes (at least in bacteria) has been described by Rohmer *et al.*[256,257] Glyceraldehyde-3-phosphate and a C_2 unit derived from pyruvate decarboxylation are the precursors of 3-methyl-3-butenyl diphosphate via a deoxy-D-xylulose.[258–261]

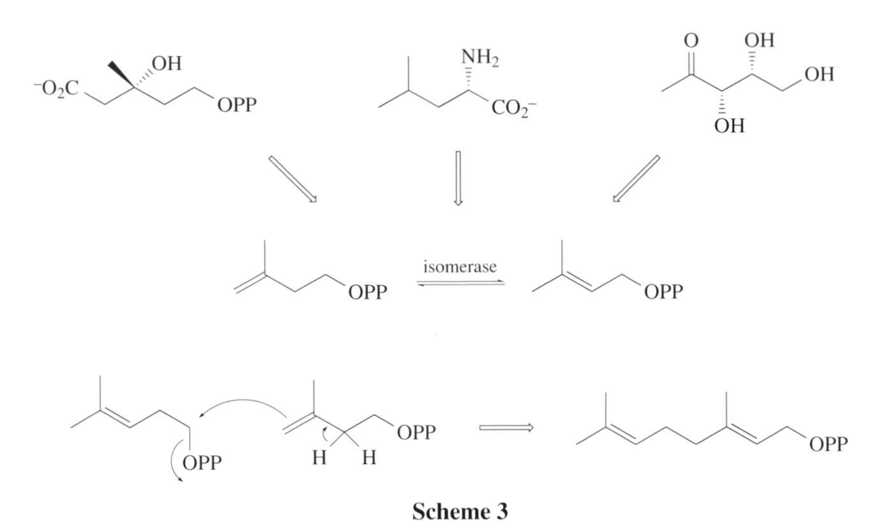

Scheme 3

Almost all types of signals, from sex pheromones to highly potent defense substances, are found among the isoprenoids. Many of these compounds seem to be directly sequestered from plants or represent simple transformation products thereof. It has been shown, however, that *de novo* syntheses may also take place.[262–266] Sometimes associated microorganisms play an important role in the production of terpenoids; they may be involved in *de novo* synthesis and also in secondary transformations of plant compounds.[267,268]

8.04.5.1 Isoprene-derived Structures/Hemiterpenes and Related Compounds

"Oxygenated isoprene" has been identified from many species. However, it is not always clear whether such iso-C_5 units represent metabolites of higher terpenes or compounds of *de novo* syntheses via mevalonate. On the other hand, widespread insect volatiles such as 3-methylbutanol (or 2-methylbutanol) and the corresponding aldehydes and carboxylic acids including unsaturated derivatives (which are frequently used as repellents)[34,269] may certainly be produced from leucine or isoleucine. Isovaleric and isobutyric acid were identified in the interdigital secretions of the reindeer, *Rangifer tarandus tarandus*, and were found to elicit a response.[270]

2-Methyl-3-buten-2-ol (**72**) (Figure 4) has already been mentioned as an aggregation pheromone of bark beetles[24] and as an alarm pheromone of hornets.[25] In the bark beetle *Ips typographus*, 2-methyl-3-buten-2-ol proved to be synthesized *de novo*.[271] 3-Methyl-3-buten-1-ol (**73**) is a pheromone of the larch bark beetle, *Ips cembrae*,[272] and, together with its methyl ether and the isomeric 3-methyl-2-buten-1-ol (**74**), a component of the secretion from the preorbital gland of an antelope species.[273] Esters of these alcohols with a series of long-chain fatty acids are constituents of Dufour's gland secretions of several bee species,[274,275] and in the primitively eusocial bee, *Lasioglossum malachurum*, they may be involved in mate finding.[73] Hemiterpenoid esters such as 3-methyl-2-butenyl 3-methyl-2-butenoate (**75**) were found in the poison glands of workers of the European hornet, *Vespa crabo*;[276] however, nothing has been reported on the biological activity of these esters. In contrast, 3-methyl-1-butyl acetate (**76**) is one of the most potent alarm pheromones in honey bees,[277] while the corresponding acetamide is an alarm pheromone of several species of yellowjackets.[278]

A higher oxygenated isoprenoid, 3-hydroxy-3-methylbutan-2-one (**77**), is a volatile constituent of ambrosia beetles[279] and induces an extremely high response in electroantennograms (EAG signals) in *Trypodendron lineatum*.[280] The same compound was also found in the mandibular glands of the parasitic wasp *Rhyssa persuasoria*.[281]

Sulfur-containing components account for the characteristic smell of odor marks of many mammals. Several of these compounds showing hemiterpene structures could be identified in the anal gland secretions of mustelids:[282–285] 3-methyl-3-butenyl methyl sulfide (**78**), 2,2-dimethylthietane

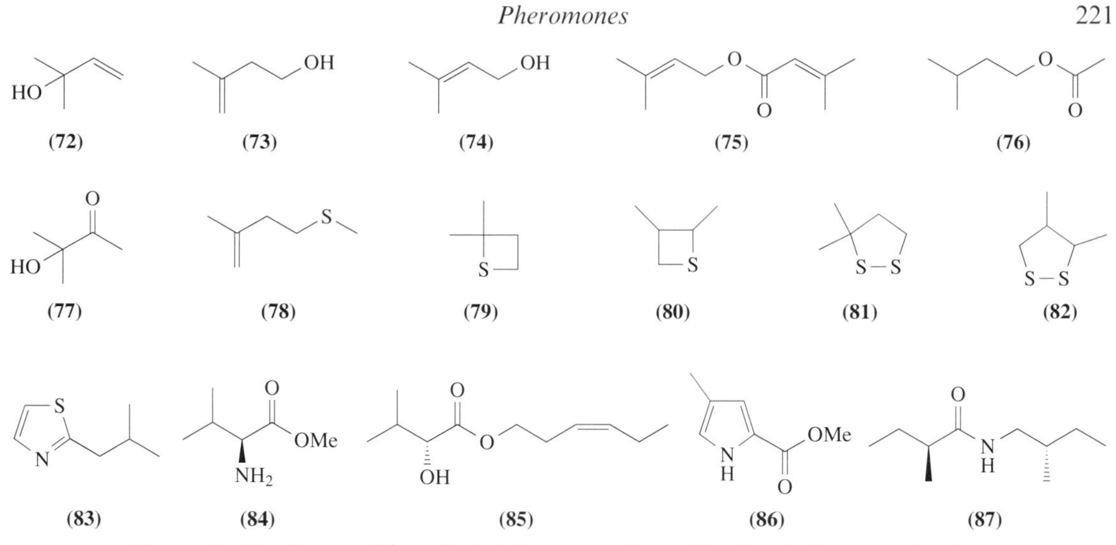

Figure 4 Hemiterpenoids and related compounds used in chemical communication.

(**79**), 2,3-dimethylthietane (**80**), 3,3-dimethyl-1,2-dithiolane (**81**), 3,4-dimethyl-1,2-dithiolane (**82**), and bis(3-methylbutyl) disulfide. The compounds are part of species-specific blends, which are used in odor marking and during defense. The methyl sulfide (**78**), the complement of the corresponding ether,[273] was also found in the urine of red fox.[286] Cryptic isovaleric acid is represented by 2-isobutyl-1,3-thiazole (**83**) and its 4,5-dihydro product, which are constituents of the preorbital gland secretion of antelopes[287] and which seem to play a role in territorial marking.[288] The heterocycle (**83**) was previously known as a natural flavor component of tomatoes. A homologue, 2-(2′-methylbutyl)-4,5-dihydrothiazole, is a volatile compound from the urine of mice, which, along with a second component, is synergistically active in inducing male aggressive behavior.[289]

It remains to be investigated whether all of these compounds are true isoprene derivatives or whether they originate from amino acids. It is obvious that valine provides the source for the corresponding methyl ester (**84**), a pheromone component of the scarab beetle, *Phyllophaga anxia*[290] (which in addition uses L-isoleucine methyl ester as a pheromone component, previously known from another scarab[291]). Also, the acid moiety in the male-produced aggregation pheromone in the assassin bug, *Pristhesancus plagipennis*, (*Z*)-3-hexenyl (*R*)-2-hydroxy-3-methylbutanoate (**85**), points to valine as the parent compound.[292] The structure of methyl 4-methylpyrrole-2-carboxylate, a widespread trail pheromone of leafcutting ant species,[293] strongly suggests its origin from leucine. The sex pheromone of the long horn beetle, *Migdolus fryans*, (**87**) indicates L-isoleucine to be the parent compound.[294] Esters of (7*Z*)- and (9*Z*)-octadecenol with isovaleric acid and 2-methylbutanoic acid (S:R = 7:2), possibly derived from leucine and isoleucine, were identified as components of the female sex pheromone of the brown tailed moth, *Euproctis similis*.[295] 2-Methylpropanoic and 2-methylbutanoic acid form aliphatic, terpenoid, and aromatic esters in the defensive secretion of the leaf beetle, *Chrysomela lapponica*.[296] Pyrrole-2-carboxylic acid, possibly derived from proline, is the main esterifying moiety in the buprestins, β-D-glucose 1,2,6-triesters, the bitter principles in jewel beetles, Buprestidae.[297]

8.04.5.2 Monoterpenes

Monoterpenes are particularly important in systems of bark beetle communication,[18,298,299] but they have also been identified in many other insect species (Figure 5). As typical plant constituents, monoterpene hydrocarbons play a role in host selection: the pine beauty moth, *Panolis flammea*, can distinguish between different ratios of α-pinene (**88**) and β-pinene (**89**) in selecting its oviposition site.[300] On the other hand, α- and β-pinene are part of the alarm pheromones of certain aphid species.[301] Because of their habitat, almost all bark beetles attacking coniferous trees contain monoterpene hydrocarbons; however, pronounced "pheromonal" activity of these compounds could be shown for only a few species;[302] details will be discussed later.

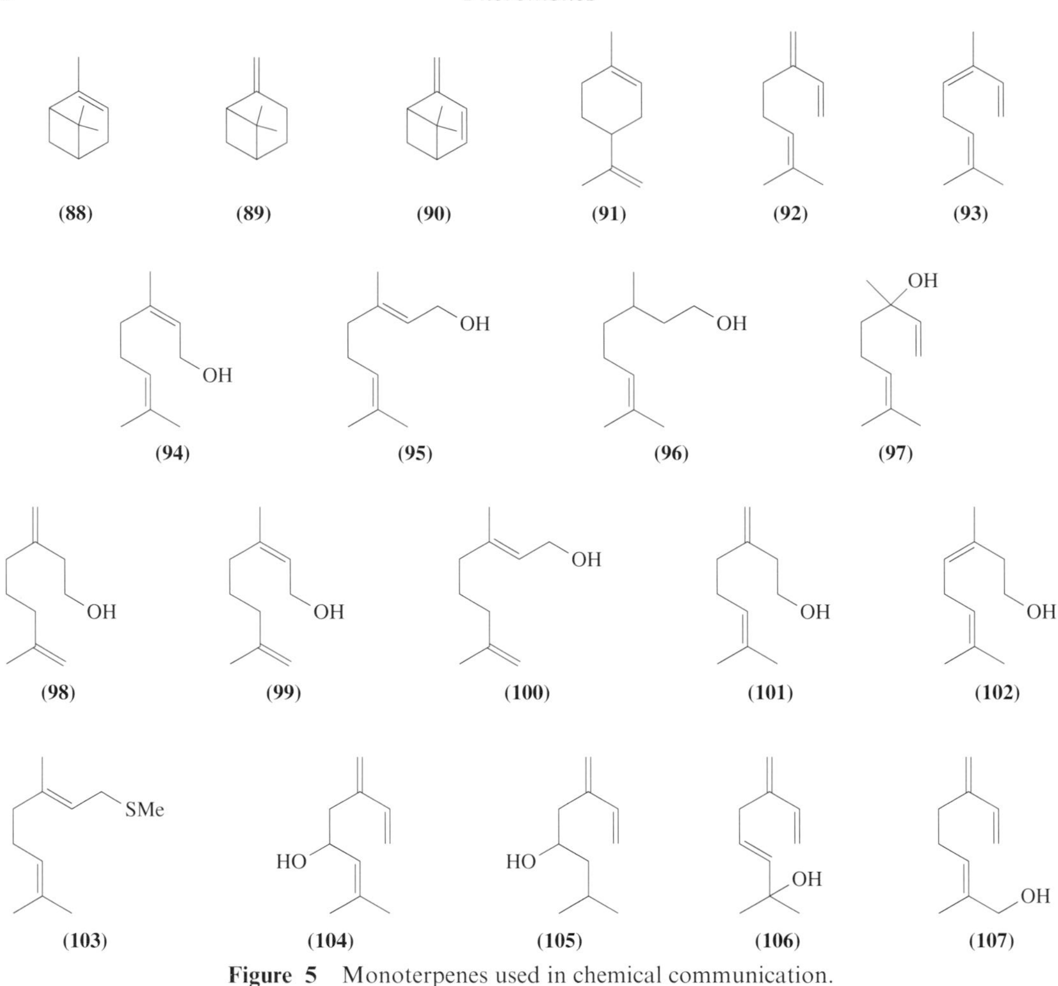

Figure 5 Monoterpenes used in chemical communication.

The soldiers of many termite species contain various series of monoterpene hydrocarbons in their frontal glands, α-pinene, β-pinene, limonene (**91**), mycrene (**92**), and ocimene (**93**) being most abundant.[35,303] Workers of some ant species, too, secrete these components from their poison gland[33,303] and *Tenebrionid* flour beetle species are also capable of producing and using them as part of their defensive secretions.[304] The role of such compounds in male *Pieris* butterflies is unclear.[305] Generally, bees, ants, and particularly termites[303,306,307] seem to use monoterpene hydrocarbons for defense or as solvents for more toxic defense compounds. In termites, the enantiomeric composition of the components may be highly significant; however, the reason for this is not clear.[308,309] The defensive secretion of the termite *Nasutitermes princeps* contains large amounts of enantiomerically pure (+)-α-pinene,[310] while the limonene secreted by workers of the ant *Myrmicaria eumenoides* proved to be the almost optically pure (+)-enantiomer.[311] In this ant species, the monoterpene has at least a dual function, since it acts as a solvent and spreading factor for a highly poisonous alkaloid which the ants use to attack their prey, and in addition it is a long-range intraspecific recruitment pheromone. Also, bugs may use optically active monoterpenes as (biologically active?) "solvents" for their defense compounds, α,β-unsaturated aliphatics.[312]

Monooxygenated acyclic terpenes and terpene esters are abundant in bees,[38] certainly at least in part because of their close contact with flowers during pollination. Typical flower volatiles such as nerol (**94**) and geraniol (**95**) and also citral (the mixture of the corresponding aldehydes) and nerolic and geranic acid are particularly widespread. Other common open-chain monoterpenes include citronellol (**96**) and linalool (**97**). Citronellol is a male sex attractant of the two-spotted spider mite, *Tetranychus urticae*.[313] Neryl formate is an alarm pheromone of the cheese mite, *Tyrophagus putrescentiae*.[314] Neryl myristate was isolated from the acarid mite *Aleuroglyphus ovatus*, but its function is unknown.[315] A mixture of propanoates of the geraniol isomers, 7-methyl-3-methylene-7-octen-1-ol (**98**), and (*Z*)-3,7-dimethyl-2,7-octadien-1-ol (**99**) and its (*E*)-isomer (**100**) comprise the female sex pheromone of the San José scale, *Quadraspidiotus perniciosus*.[316] A further geraniol isomer, γ-geraniol (**101**), is part of the aggregation pheromone of the bark beetle *Pityogenes quadridens*[317] while *cis*-isogeraniol (**102**) is a recruitment pheromone of the ant *Leptogenys diminuta*.[318] The interesting thioterpene geranyl methylsulfide (**103**) is a component of the pheromone system of the male danaine butterfly, *Idea leuconoe*.[319]

Series of esters of geraniol with acetic, butanoic, hexanoic, octanoic, and decanoic acid were found in Dufour's gland secretions of many *Andrenid* bees; also "true" terpenoid esters such as geranyl geranoate and several partially hydrogenated esters of the same type have been identified from bees, wasps, and hornets.[303,320] No biological function of these compounds has been demonstrated for Hymenoptera; however, geranyl acetate is a sex pheromone component of the Mediterranean fruit fly, *Ceratitis capitata*.[321] Geranyl hexanoate and geranyl octanoate could be identified as female-specific compounds in click beetles, *Agriotes* spp.[322] These compounds are sex pheromones and can be used to determine the distribution of click beetles through monitoring with baited traps.[323]

Neral, geranial, and linalool form part of the aggregation pheromone or mate-seeking cue of *Colletes* bees[324,325] and represent sex pheromone components of a parasitic wasp.[326] (*R*)-(−)-Linalool, a minor component of the sex pheromone of the scarab beetle, *Holotrichia parallela*, enhances the attractiveness of the major component, L-isoleucine methyl ester, and shows an interesting periodicity in its production.[327] Stingless bees, *Lestrimelitta limao*, which do not forage from plants, secrete large amounts of citral during their raids, to disorient prey bee species while plundering their nests.[328] *Centris* bees use a mixture of acyclic monoterpenes to set up territories by marking grass stems.[329] The concentration of citral in the mandibular glands of certain leaf-cutting ants[330] can reach such concentrations (possibly used for defense) that a lemon-tasting drink may be produced from the ethanolic extracts of worker heads. Although the biological function of citral in the scent of male *Pierid* butterflies is not entirely clear,[305,317,331] the mixture acts as part of the alarm pheromone of certain mite species.[332] Some ant species[303] and *Anthophorid* bees contain nerolic acid and geranic acid in their cephalic secretions. Nothing is known about their biological activity in these insects; however, in honey bee workers the acids have been identified along with nerol and citral as components of the Nasonov pheromone,[333] which plays a role in worker attraction, nest finding, foraging, and marking of food sources.

Monoterpenes play a particularly important role in host selection and mass aggregation of bark beetles. Insects attacking conifers have to overcome both physical and chemical obstacles during host colonization. Defense through sticky and toxic oleoresin is involved in the most effective mechanisms in tree resistance against bark beetle invasion. However, bark beetles have developed a series of strategies to survive and successfully colonize selected host trees. Oxygenated monoterpenes found in bark beetles may be conversion products of host terpenes, may be simply sequestered from the host, or may be synthesized *de novo*. Under natural conditions, oxygenated monoterpenes may result from all three sources simultaneously.

Derivatization of host plant monoterpenes by bark beetles may originally have represented a detoxification mechnism, but has obviously developed into an important principle in the formation of volatile signals which bark beetles use for intra- and interspecific communication. Based on investigations on the effects of the exposure of bark beetles to vapors of monoterpene hydrocarbons,[334–336] it is generally accepted that in most cases oxygenated monoterpenes present in bark beetles are generated from resinous compounds of the host trees. Allylic oxidation or hydration of unsaturated hydrocarbons followed by secondary reactions such as further oxidation, hydrogenation, or rearrangements seem to be important mechanisms in the production of bark beetle pheromones.[298,337]

Whereas some species oxidize host terpenes more randomly and thus generate an array of rather unspecific volatiles[338] which transmit little information, others use highly selective enzyme systems for the production of unique olfactory signals. None of the oxygenated monoterpenes identified represents a species-specific compound *per se*, and species specificity of pheromonal signals is accomplished by differences in the combination of compounds or their enantiomers and/or different modes in perception. The problem has been extensively discussed.[18,170,298,299,339–342]

The myrcene derivatives ipsdienol (**104**) and ipsenol (**105**), which were first identified by Silverstein *et al.*,[343] are typical male-specific aggregation pheromones of many *Ips* spp., but they also play a role in host colonization in other bark beetles such as *Pityokteines*[344,345] and *Xylocleptes*.[346] Enantiomeric composition of these terpenols is instrumental with respect to the behavior-mediating capacity of the signal[347] (even in different populations of the same species[348,349]) and may be used in chemotaxonomy.[350] The corresponding ketones, ipsdienone and ipsenone, were found to be present in males of *Ips sexdentatus* and *I. cribricollis*, respectively[341] (redetermination of the species originally regarded as *I. lecontei* showed that it was actually *I. cribricollis*).

The biological function of the ketones in the beetles is unknown; they may be involved in interconversion reactions which lead from ipsdienol to ipsenol.[342] Ipsdienone was detected early in the essential oil of *Lippia asperifolia*[351] and was also found in other *Lippia* spp.[352] Ipsdienol also proved to be a major component in the floral fragrance of several species of orchids that are pollinated by male *Euglossine* bees; the racemate was shown to be attractive in field tests.[353] Whereas ipsenol may be produced from ipsdienol by a hydrogenation cascade (possibly via ipsdienone and ipsenone), the tertiary alcohol amitinol (**106**) represents a product of an allylic rearrangement of ipsdienol. Amitinol, first identified in an American *Ips* bark beetle,[354] is also a constituent of the essential oil of *Ledum palustre*.[355] Another product of an allylic oxidation of myrcene, myrcenol (**107**), was also identified as a bark beetle pheromone.[356] However, again the compound had been previously known from an essential oil.[357] It has been shown that *Ips* bark beetles can synthesize ipsenol, ipsdienol, amitinol, and myrcenol *de novo*, which adds an exciting facet to the chemical ecology of bark beetles.[358]

Oxygenated monoterpenes, which are found in almost every bark beetle species attacking coniferous trees, include *cis*-verbenol (**108**), *trans*-verbenol, and myrtenol (**109**), which represent primary products of allylic oxidation of the host terpene, α-pinene (Figure 6). Further oxidation leads to the corresponding carbonyl compounds, verbenone (**110**) and myrtenal, which, too, are common bark beetle volatiles. 1,4-Elimination of water from verbenol yields 4-methylene-6,6-dimethylbicyclo-[3.3.1]hept-2-ene (verbenene) (**90**), which was found as a behavior-mediating volatile emitted by females of the bark beetle *Dendroctonus rufipennis*.[359] Oxygenation of β-pinene produces *trans*-pinocarveol (**111**) and the corresponding ketone, pinocarvone. Among these bicyclic terpenes, *cis*-verbenol is a particularly important aggregation pheromone in *Ips* spp., whereas *trans*-verbenol is used by *Dendroctonus* spp. Both sexes of *Ips* species oxidize the chiral α-pinene enantioselectively:[335,360] (4*S*)-*cis*-verbenol is produced from (−)-α-pinene, whereas (+)-α-pinene yields (4*S*)-*trans*-verbenol. Verbenone, which in bark beetles appears to be formed largely from verbenols due to the action of associated microorganisms,[267,268] seems to act as a general inhibitor, which the beetles use to avoid overpopulation and which induces shifting of the attack to another tree.[343,361] In contrast, verbenone in the scent of *Knautia* flowers is attractive to *Zygaenid* moths.[362] Simple rearrangements of verbenone yield the well known flower constituent chrysanthenone (**112**), which was also found in bark beetles;[352] however, its biological significance is unknown. The predatory beetle *Rhizophagus grandis* is linked to its specific prey, larvae of the bark beetle *Dendroctonus micans* by volatile signals which account for long-range orientation and which regulate oviposition. The active blend is made up by simple monoterpenes present in the larval frass, (−)-fenchone (**113**) being the most important component.[363]

The cyclohexane derivatives (**114**), (**116**), and (**117**), (*Z*)-ochtodenol and the two ochtodenals, and also (+)-*cis*-1-(2-hydroxyethyl)-1-methyl-2-isopropenylcyclobutane (grandisol) (**118**), which are probably derived from myrcene (**93**) or γ-geraniol (**101**), are male sex pheromones of the boll weevil.[364] In bioassays with grandisol enantiomers of the highest optical purity, it was found that only the (+)-enantiomer is behavior mediating.[365] The same mixture as in the boll weevil also represents the pheromone of *Curculio caryae*, but in different relative proportions. In contrast to the boll weevil, the pecan weevil produces almost racemic grandisol.[366] (+)-Grandiosic acid, (1*R*,2*S*)-1-Methyl-2-(1-methylethenyl)cyclobutaneacetic acid, is the major sex pheromone of males of the plum curculionid, *Conotrachelus nenuphar*.[367] The ochtodenals and the corresponding carboxylic acids were also found as volatile constituents of the Lady beetle, *Coleomegilla maculata*.[368] While (*E*)-ochtodenol (**115**) and grandisol are part of the aggregation pheromones of some *Pityogenes* bark beetles,[317] grandisol—almost the pure (1*R*,2*S*)-enantiomer—and the corresponding aldehyde, grandisal, are aggregation pheromones of several *Pissodes* weevils.[369] Interestingly, the enantiomeric composition of grandisal is different in different species: whereas *Pissodes nemorensis* produces almost pure (1*S*,2*R*)-grandisal, *Pissodes strobi* contains a mixture of the (1*S*,2*R*)-enantiomer and its antipode in a ratio of 4:6. The *trans*-isomer of grandisol was found as a plant constituent,[370] which was called fragranol. The tricyclic acetal lineatin (**119**), a higher oxygenated derivative of grandisol, is an aggregation pheromone of several *Trypodendron* spp., ambrosia beetles.[371] The natural product was shown to be the (1*S*,4*R*,5*S*,8*S*)-(+)-enantiomer.[372,373] Another terpene containing a cyclobutane system is (1*R*-*cis*)-(+)-1-acetoxymethyl-2,2-dimethyl-3-(1-methylvinyl)cyclobutane (**120**), the sex pheromone of the mealy bug, *Planococcus citri*,[374] which shows an interesting linkage between two isoprene units.

The *p*-menthane derivative α-terpineol is a relatively widespread oxygenated monoterpene. Its (+)-enantiomer (**121**) has been described as a component of the pheromone of the predatory spined soldier bug, *Podisus maculiventris*, which is also attractive to yellowjackets, *Vespula maculifrons*.[375] Its isomer, terpinen-4-ol, which is also contained in the *Podisus* pheromone, is also fairly common.

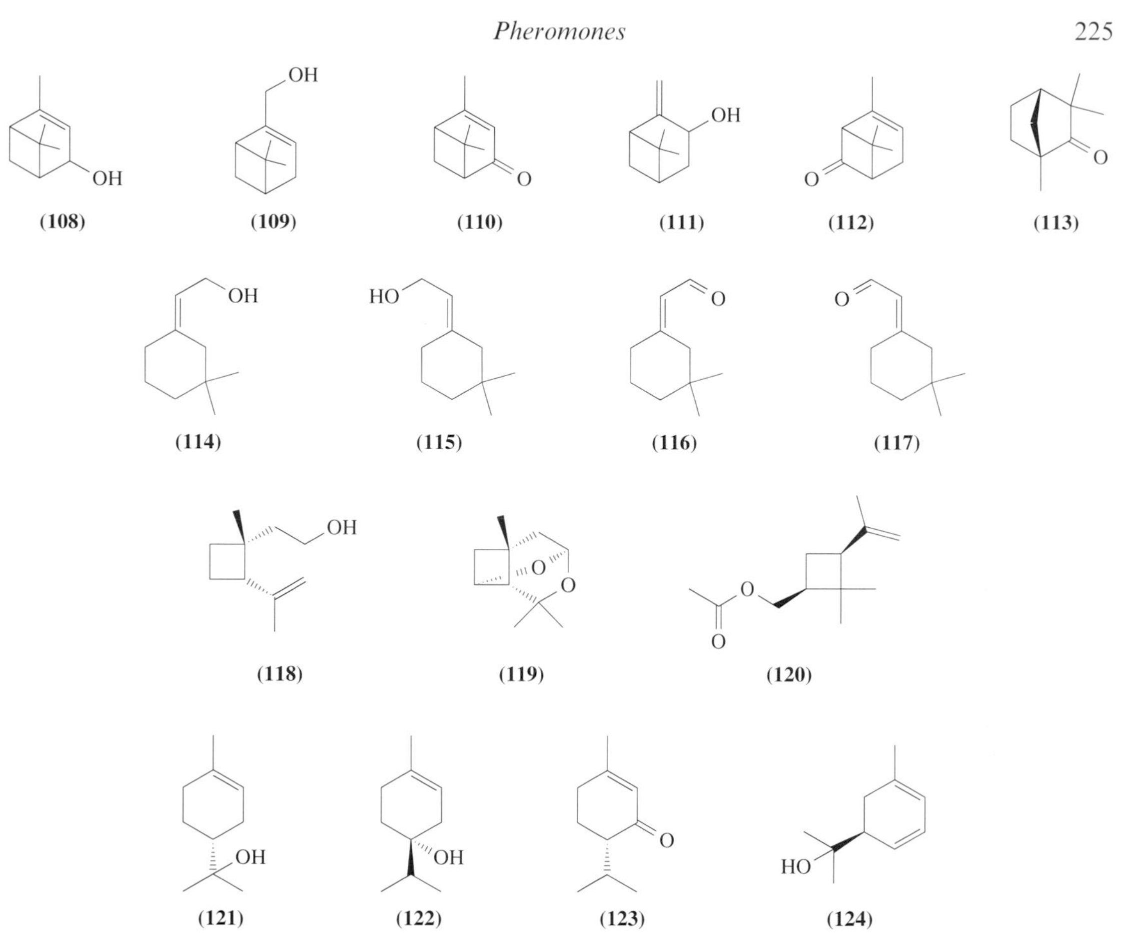

Figure 6 Oxygenated monoterpenes used in chemical communication.

The (−)-enantiomer (**122**) is an aggregation pheromone of the bark beetle *Polygraphus poligraphus*.[376] Another *p*-menthene, (+)-isopiperitenone (**123**), is an alarm pheromone of the acarid mite *Tyrophagus similis*.[377] The interesting *m*-menthadienol (3*S*)-(−)-1-methyl-5(1-hydroxy-1-methylethyl)-1,3-cyclohexadiene (**124**) is produced by the bark beetle *Ips sexdentatus*, boring under stress in 3-carene-rich highly resinous pine trees and possibly released as a repellent ("warning signal"?)[352]

Higher oxygenated monoterpenes, showing various biological functions have been identified in several insect species (Figure 7). The diol (**125**) is believed to act as a kind of solvent for pheromones of male monarch butterflies,[378] whereas the corresponding diacetate is a presumed pheromone of the Australasian predaceous bug *Oechalia schellenbergii*.[379] The dicarboxylic acid callosobruchusic acid (**126**), showing the *R*-configuration, is part of the sex pheromone of the azuki bean weevil, *Callosobruchus chinensis*.[380] (*E*)-8-Oxocitronellyl acetate (**127**) is a constituent of the defensive secretion in rove beetles[381] and is also present in the mandibular gland secretion of *Panurginus* bees.[382] 8-Hydroxylinalool (**128**) is part of the defense secretions of leaf beetles, *Chrysomela* spp.[383] 2,3-Epoxyneral (**129**),[384] which shows a (2*R*,3*R*)-configuration in acarid mites, *Caloglyphus* spp.,[385] the two dialdehydes *β*-acaridial (**130**), and its conjugated isomer, α-acaridial,[386,387] and also dehydrocineol (**131**),[388] have been isolated from mites. (2*S*,3*S*)-Epoxyneral was found in the mite *Tyrophagus perniciosus*, but its function is not fully understood. *β*-Acaridial is the sex pheromone of *Caloglyphus polyphyllae*.[389] 1,8-Cineole (**132**), along with α-pinene, limonene, and straight-chain compounds, was found in bugs.[390] Among other components, ant lions use (racemic) nerol oxide (**133**) as a volatile signal,[391] while digging wasps *Bembix* spp. emit rose oxide (**134**) from their mandibular glands.[392] The latter had already been previously identified from longhorn beetles.[393] Further oxygen heterocycles which are known as volatile constituents of flowers include the pyranoid (**135**) and furanoid (**136**) forms of linalool oxide, which are components of the androconial secretion of *Amauris* butterflies.[394]

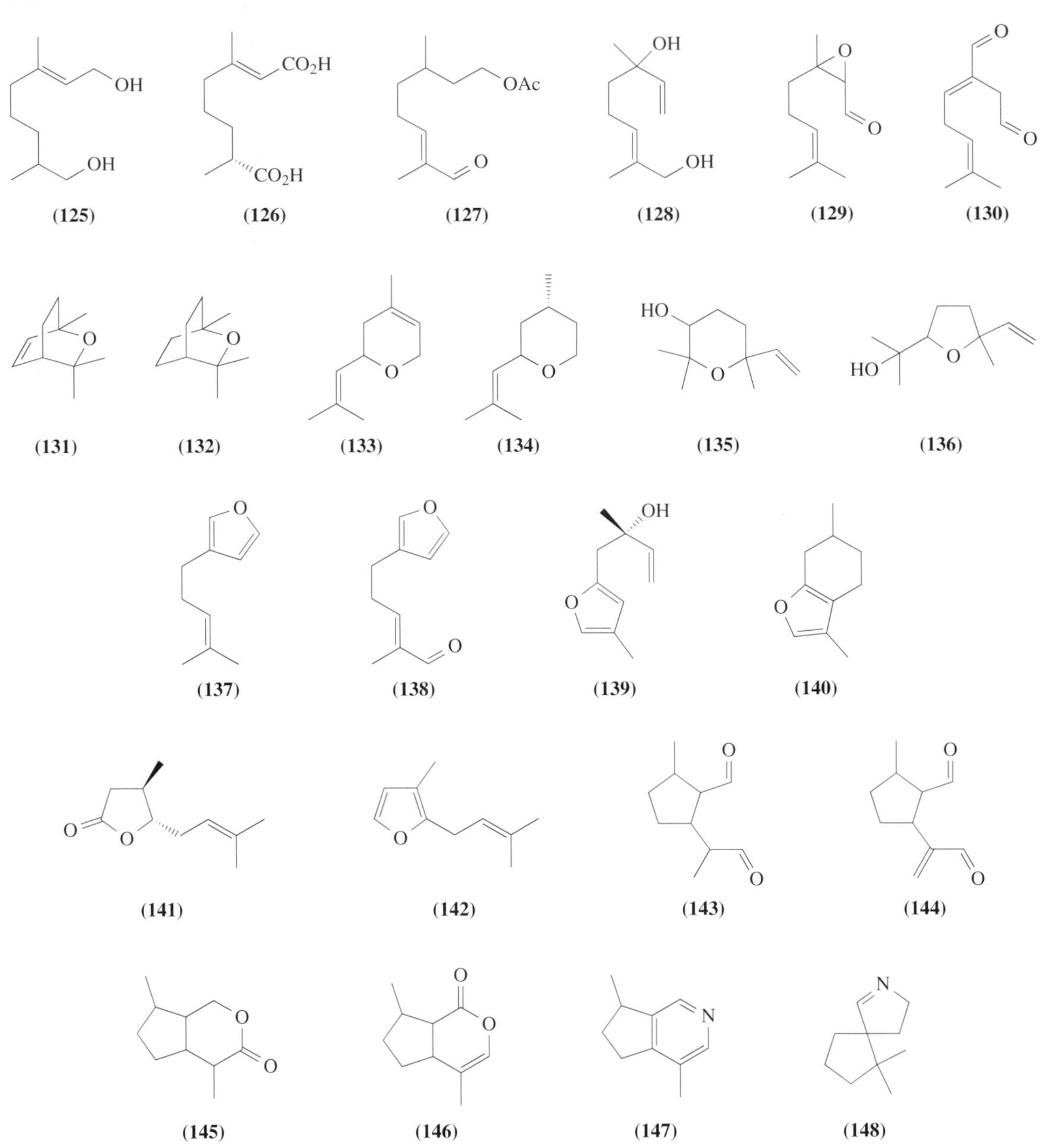

Figure 7 Further oxygenated monoterpenes used in chemical communication.

Furan derivatives have also been described from insects: perillene (**137**) was found in ant species,[395] whereas the aldehyde (perillenal) (**138**) was identified in a saw fly species.[396] The linalool derivative elymnia furan (**139**) was isolated from the abdominal secretion of males of *Elymnias thryallis*, a tropical butterfly.[397] It shows the (*S*)-configuration and has an oxygenation pattern similar to that of the well-known plant constituent menthofuran (**140**). Structural relationships between (3*S*,4*R*)-3,7-dimethyl-6-octen-4-olide, eldanolide (**141**), which represents the sex pheromone of the male sugar-cane borer, *Eldana saccharina*,[398–400] and a plant constituent such as rose furan (**72**)[401] are also obvious and again point to the close relationship between insect and plant volatiles.

Various iridoids have been reported from a wide range of insect and plant sources. Isomers of iridodial (**143**) and dolichodial (**144**) and also lactones such as iridomyrmecin (**145**), nepetalactone (**146**), and the corresponding doubly unsaturated lactones are widespread defensive compounds among Formicine ants, Staphylinid and Cerambycid beetles, and larvae of many leaf beetles.[33,34,381,402–404] Dolichodial was also isolated from a plant source, *Teucrium marum*.[405] Nepetalactone, which has long been known from catnip,[406] accounts for the characteristic response of cats to catnip. The compound has also been identified as a sex pheromone of the aphid *Megoura viciae*.[407] The aphid pheromone shows (4a*S*,7*S*,7a*R*)-configuration and acts synergistically together with the corresponding lactol. Nepetalactone and its lactol seem to be fairly widespread as sex pheromones of aphids.[408,409] The biological background of this phenomenon and the chemistry of aphid cyclopentanoids have been thoroughly reviewed.[410,411] Also in aphids, stereochemistry of chiral semiochemicals seems to play an important role.[412] Actinidin (**147**), an alkaloid derivative of iridoids, is a defense compound of several rove beetle species.[381,413] A second interesting terpene alkaloid is the spirocyclic azomethine polyzonimine (**148**), a defense substance of the millipede *Polyzonium rosalbum*.[414]

8.04.5.3 Sesquiterpenes

The majority of sesquiterpenes identified in insects are represented by open-chain molecules (Figure 8). Among other compounds, (*E*,*E*)-α-farnesene (**149**) is a component of the sex pheromone of the Mediterranean fruit fly, *Ceratitis capitata*.[415] Acting as an allomone, a mixture of (*E*,*E*)- and (*Z*,*E*)-α-farnesene, produced by apples during ripening, is an attractant to larvae of the codling moth, *Laspeyresia pomonella*, and acts as an oviposition stimulant for gravid female adults.[416] In the ant *Solenopsis invicta*, (*Z*,*E*)-α-farnesene is a constituent of the Dufour's gland secretion which acts as a trail pheromone.[417] This compound may also be a pheromone component in a beetle.[418] From aphids, several isomers of α- and β-farnesene were identified,[301] among which (*E*)-β-farnesene (**150**),[419] shows the strongest effects as an alarm pheromone. Both α- and β-farnesene were found in the tarsal glands of the springbok antelope;[420] (*E*,*E*)-α-farnesene was also identified in male Hepialid moths, *Hepialus humuli*.[421] (*E*,*E*)-α-Farnesene and (*E*)-β-farnesene are present in elevated amounts in the urine of dominant male mice and may play an important role in the territorial marking behavior and thus influence the reproductive potential.[422]

(*E*,*E*)-Farnesol (**151**) has frequently been reported as a component of the secretions of Dufour's glands of Andrenid bees,[423,424] of the Nasonov glands of honey bee workers, of the labial glands of bumble bees,[425,426] and of the mandibular glands of leaf-cutting ants.[330] The same range of carboxylic acids which form geranyl esters in several bee species (see above) are also present as farnesyl esters.[423,424,427,428] Farnesyl hexanoate, the most common farnesyl ester, has also been identified in the tarsal glands of the European roe deer.[429] Nerolidol (**152**) has only seldom been found in insects; it is a constituent of the sex-attracting secretion produced by male dung beetles, *Kheper* spp.,[430] it was earlier identified as part of the alarm pheromone of the mite *Tetranychus urticae*,[431] and was also detected in Danaid butterflies.[394] (*S*)-2,3-Dihydrofarnesol (terrestrol) is used as a territorial marking substance by several bumble bee species;[425,432,433] the acetate of this alcohol and the corresponding aldehyde are also present in several ant species.[434] In addition to (*E*)-farnesol, the two hydrated farnesols 3,7,11-trimethyl-2,10-dodecadien-1,7-diol (**153**) and 3,7,11-trimethyl-2-dodecen-1,7,11-triol are constituents of the temporal gland secretion of the African elephant,[435] and it has been suggested that these compounds are used in scent marking. The stereochemistry of the compounds is, however, unknown. Similar derivatives of nerolidol such as (**154**) are known from caparrapi oil.[436] (*E*,*E*)-Farnesal and the (*Z*,*E*)-isomer were identified from the frontal gland of workers of a termite species and are believed to be involved in chemical defense,[437] while males of the rice moth use the same compounds to attract females.[438] In the stored product mite *Suidasia medanensis*, the two isomers are part of the alarm pheromone.[439] The interesting acyclic sesquiterpene nomadone (**155**) is a constituent of the mandibular gland secretion of *Nomada* spp.[317,440] The biological function of this ketone and several very closely related acyclic sesquiterpenes and norsesquiterpene ketones which form species-specific blends in the cephalic secretions of *Nomada* bees is not clear; however, they seem to play a role in intra- and interspecific territorial behavior.

The monocyclic sesquiterpene (−)-germacrene A (**156**) is an alarm pheromone of the alfalfa aphid *Therioaphis maculata*.[441] This very thermolabile compound was first isolated from a gorgonian coral[442] and, among other terpenes, is also present in the frontal gland of a termite species[443] and in the defensive osmeterial secretion of swallowtail butterflies (Papilionids), which contain various additional sesquiterpenes and monoterpenes.[444] The two germacrane derivatives periplanone A (**157**) and periplanone B (**158**) are the sex pheromones produced by females of the American cockroach, *Periplaneta americana*. While periplanone B had already been known for some time,[445] and its absolute configuration determined,[446] structure elucidation of periplanone A took some time[447] and caused a lot of confusion which finally was clarified through unambiguous synthesis.[448] Some additional germacrane derivatives could be identified.[449] Finally, the nomencalture of the periplanones was adjusted.[450] The sex pheromone of the smoky brown cockroach, *Periplaneta fuliginosa*,

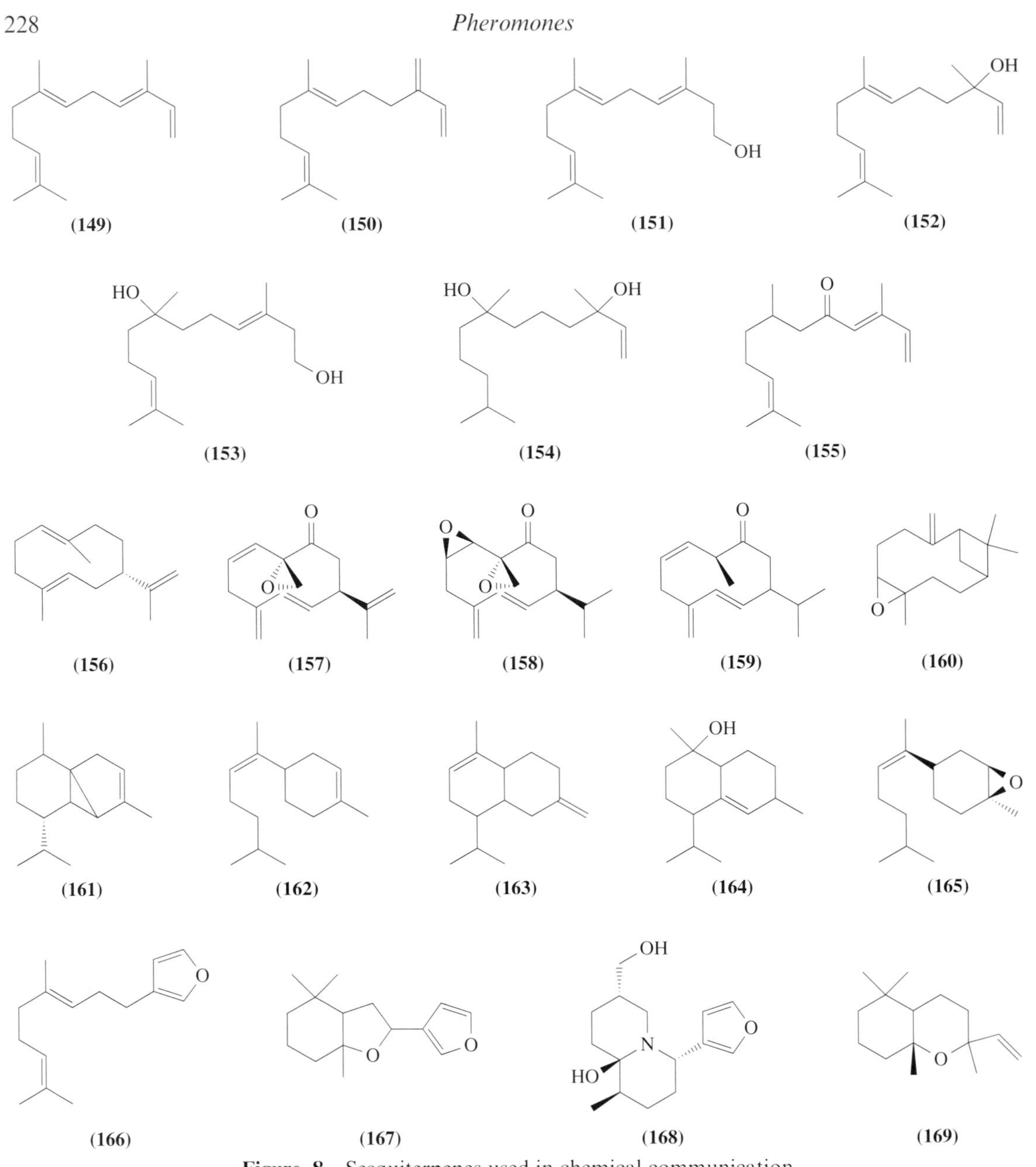

Figure 8 Sesquiterpenes used in chemical communication.

was found to be periplanone-D (**159**),[451] while the pheromone of *Periplaneta brunnea* was reported to be the hydrocarbon (2*Z*,6*E*,8*S*)-1(14),2,5(15),6-germacratetraene, called periplanene-Br.[452]

A bicyclic oxygenated sesquiterpene, which shows some structural relationships to the germacrane system, is caryophyllene oxide (**160**). It forms a major volatile component of the osmeterial secretion of larvae of the swallowtail, *Papilio memnon*,[444] and it was also identified as a constituent of the essential oil of tropical plants, where it acts as a toxic deterrent to leaf-cutting ants.[453] In bark beetle communication, sesquiterpenes are far less important than monoterpenes: the host-produced hydrocarbon α-cubebene (**161**), which is an attractant for elm bark beetles,[454] appears to represent an exception. However, further examples are likely to be discovered. Other sesquiterpene hydrocarbons such as δ-cadinene (**162**) and α-bisabolene (**163**) could be identified from the defensive secretions of termites; their oxygenated derivatives were also found.[303] The cadinol (**164**) is a constituent of the frontal gland secretion of *Subulitermes* soldiers,[455] and it is also a component of the wing scents of male Lycaenid butterflies, where it is believed to play a role in courtship behavior.[456] (*Z*)-2-(3′,4′-Epoxy-4′-methylcyclohexyl)-6-methyl-2,5-heptadiene, (*Z*)-α-bisabolene oxide (**165**), is a component of the sex pheromone blends of stink bugs, *Nezara* spp.[457,458] The occurrence of different mixtures of diastereomers of the epoxide in different species and subspecies has been discussed in the context of evolutionary pressure caused by predators which use bisabolene oxide as a kairomone.[459,460] In *Nezara viridula* the pheromone is a mixture of the *trans*- and *cis*-isomers which varies in different strains. Mixtures of the (−)-enantiomers proved to be biologically active; however, the absolute configuration of the natural products remains to be determined.[461] Similarly to monoterpenes, appropriate structural elements of sesquiterpenes may form furan subunits: dendrolasin (**166**), which plays an important role in the defensive chemistry of *Lasius* ants,[462] is the sesquiterpene analogue of perillene (**137**). Further oxygenation of dendrolasin would lead to the structure of ancistrofuran (**167**), which is a defensive compound of *Ancistrotermes* termites.[463] Exemplified by (−)-castoramin (**168**), an interesting range of furan sesquiterpenes occurs in the scent gland of the Canadian beaver, *Castor fiber*,[464] but not much is known about their actual function. The sesquiterpene analogue of the pyranoid form of linalool oxide (**135**), caparrapioxide (**169**), previously known from plant sources, was identified from *Amitermes* termite defensive secretions;[465] its function is unknown.

8.04.5.4 Diterpenes

Because of their relatively low volatility, diterpenes (Figure 9) seem to be less suitable for odor communication. The acyclic geranyllinalool (**170**), possibly the (*R*)-enantiomer, the analogue of nerolidol, is part of the defensive secretion of a termite,[466] while geranylgeraniol (**171**), which is complementary to farnesol, was identified in Dufour's glands of workers of *Formica* ants[467] and in the labial glands of males of several bumble bees, *Bombus* spp.[432] Almost nothing is known about the biological function of these compounds. Geranylcitronellol, which is the prenyl homologue of the above-mentioned bumble-bee-marking pheromone, 2,3-dihydrofarnesol (terrestrol), was also found in the labial glands of *Bombus*,[468] *Psithyrus*,[432] and *Pyrobombus* bumble bees.[426] Oxidation products of these alcohols, geranylgeranial and geranylcitronellal, are components of the mandibular gland secretions of *Lasius* ants.[434]

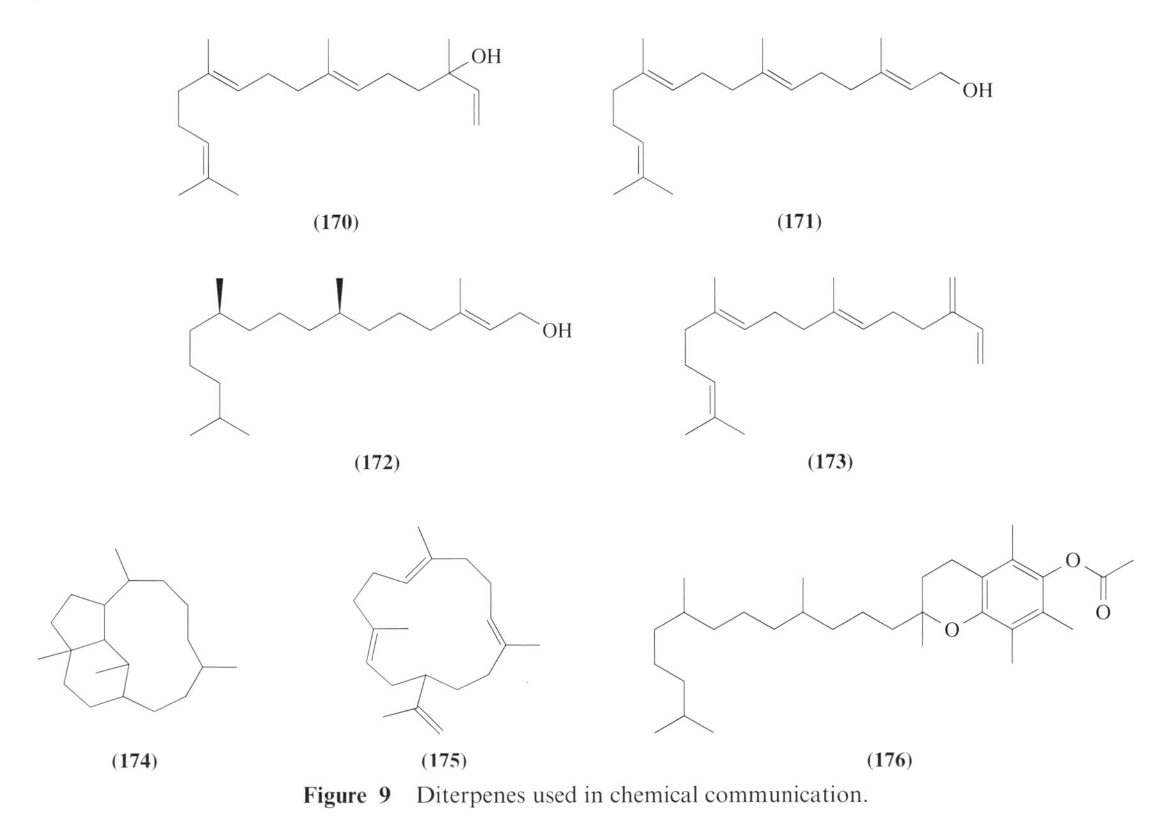

Figure 9 Diterpenes used in chemical communication.

Complementary to the farnesyl esters, esters of geranylgeraniol are also known from *Hymenoptera*.[432,434,467] All-*trans*-geranylgeranyl acetate and geranylgeraniol are recruitment pheromones in the Dufour's gland of the ponerine ant *Ectatomma ruidum*.[469] Geranylgeranyl octanoate is a Dufour's gland product in several North American *Andrena* bees.[428] Corresponding esters such as geranylgeranyl docosanoate and phytyl dodecanoate were identified from the surface wax of cotton buds and anthers, where they represent phagostimulants for adult boll weevils, *Anthonomus grandis*.[470] A whole range of geranylcitronellyl esters of fatty acids were found in *Bombus hypnorum* and *B. terrestris*; the absolute configuration of the compounds and their biological function are unknown.[471,472] (*E*)-Phytol (**172**), one of the most typical plant constituents, is a component of the wing gland pheromone of males of the Pyralid moth, *Ephestia elutella*, and plays a role in courtship success.[473] Springene (**173**), the prenyl homologue of (*E*)-β-farnesene, is a component of the complex mixture identified from the dorsal gland of the springbok antelope.[420] The dorsal gland secretion of both sexes of the collared peccary, *Tayassu tajacu*, contains geranylgeraniol, β-springene, and two prenyl homologues of α-farnesene, (3*E*,6*E*,10*E*)-α-springene, and its (3*Z*)-isomer.[474] The female secretion contains sex-specific esters of fatty alcohols with straight- and branched-chain carboxylic acids.[474] A number of functions for the dorsal secretions have been discussed, including territorial scent marking, herd recognition, and alarm and reproductive information. Most diterpenes have been identified from the frontal gland secretions of termites,[38,303] which seem to use them for defense when smearing them on aggressors. Most of these components are derivatives of the tricyclic trinervitane system (**174**).[475–477] However, monocyclic diterpenes are also known: (*E*,*E*,*E*)-neocembrene (cembrene A) (**175**) is a trail pheromone of the termite *Trinervitermes bettonianus*.[478] The compound could be isolated directly from the trail and from workers and alate females and proved to be absent in soldiers; however, in another termite species, *Cubitermes umbratus*, the same compound and other monocyclic diterpenes are reported to be part of the defense secretion of soldiers.[479] Neocembrene was also found in the secretions of *Nasutitermes* termites[480] and in *Monomorium* ants.[481] Cembrene was originally identified as a constituent of pine resin.

Tocopheryl acetates such as (**176**), which show a perhydroditerpene as a substructure, are constituents of the pupal exocrine secretion of the squash beetle, *Epilachna borealis*.[482]

8.04.5.5 Steroids

Only a few steroids are known to act as pheromone (Figure 10). Although cholesterol is a widespread constituent of the surface lipids of many insects, no behavior-mediating capacity has been reported. However, cholesteryl oleate (**177**) proved to be the mounting sex pheromone of the hard tick, *Dermacentor variabilis*.[483] This compound is secreted by sexually mature females during feeding and accumulates on the surface of the body cuticle, indicating successful parasitism. Mate-seeking males, attracted to the females by the pheromone 2,6-dichlorophenol (for a review on tick pheromones, see reference 484), recognize and mount the feeding females. Cholesteryl esters forming specific blends were also found in other tick species and are discussed as important factors in minimizing heterospecific matings in nature.[485] 5β-Cholestan-3-one (**178**) and the corresponding 3,24-diketone are components of the larval trail following pheromone of tent caterpillars, *Malacosoma* spp.[486,487] In contrast to earlier findings, the actual active principle is (**178**) and not the diketone. The aggregation pheromone of the German cockroach, *Blattella germanica*, comprises interesting chlorinated sitosterol-derived glucosides, blattellastanoside A (**179**) and B (**180**),[488] the structures of which could be unambiguously determined by total synthesis.[489] Extreme olfactory sensitivity of mature and gonadally regressed goldfish has been reported for a potent steroidal pheromone, 17α,20β-dihydroxy-4-pregnen-3-one (**181**).[490] Other steroids such as 21-hydroxy-(**181**), 4-androstene-3,17-dione, and its 17β-hydroxy derivative, testosterone, are also involved in the complex of fish pheromones.[491] Waterbeetles (Dytiscidae) store large amounts of testosterone and the interesting cross-conjugated 1,2-dehydrotestosterone (**182**), cortexone, and other C-21 steroids such as cybisterone (**183**)[492–497] in their pygidial glands as defense compounds.

Particularly exciting semiochemicals are the odorous C_{19}-steroids 5α-androst-16-en-3-one, the "boar-taint steroid" (**184**), and the corresponding alcohol, which have attracted much attention because of their action as sex pheromones in pigs and their possible importance in human social interactions. The steroidal alcohol was isolated as early as 1944 from pig testes[498] and later in human male axillary sweat,[499] while its glucuronide was known from human urine.[500] The boar-taint pheromone is synthesized in the testes of boars and released to the bloodstream; it is delivered to the salivary glands where it is reduced to the alcohol, which represents the actual boar pheromone, stimulating the sow's standing reflex during copulation.

The ketone (**184**) was also found in humans,[501,502] and there are several contradictory papers concerning the pheromonal effect of the two steroids on human behavior.[503] Some authors found a positive response[504–506] whereas others reported no influence.[507] Anyway, the compounds seem to

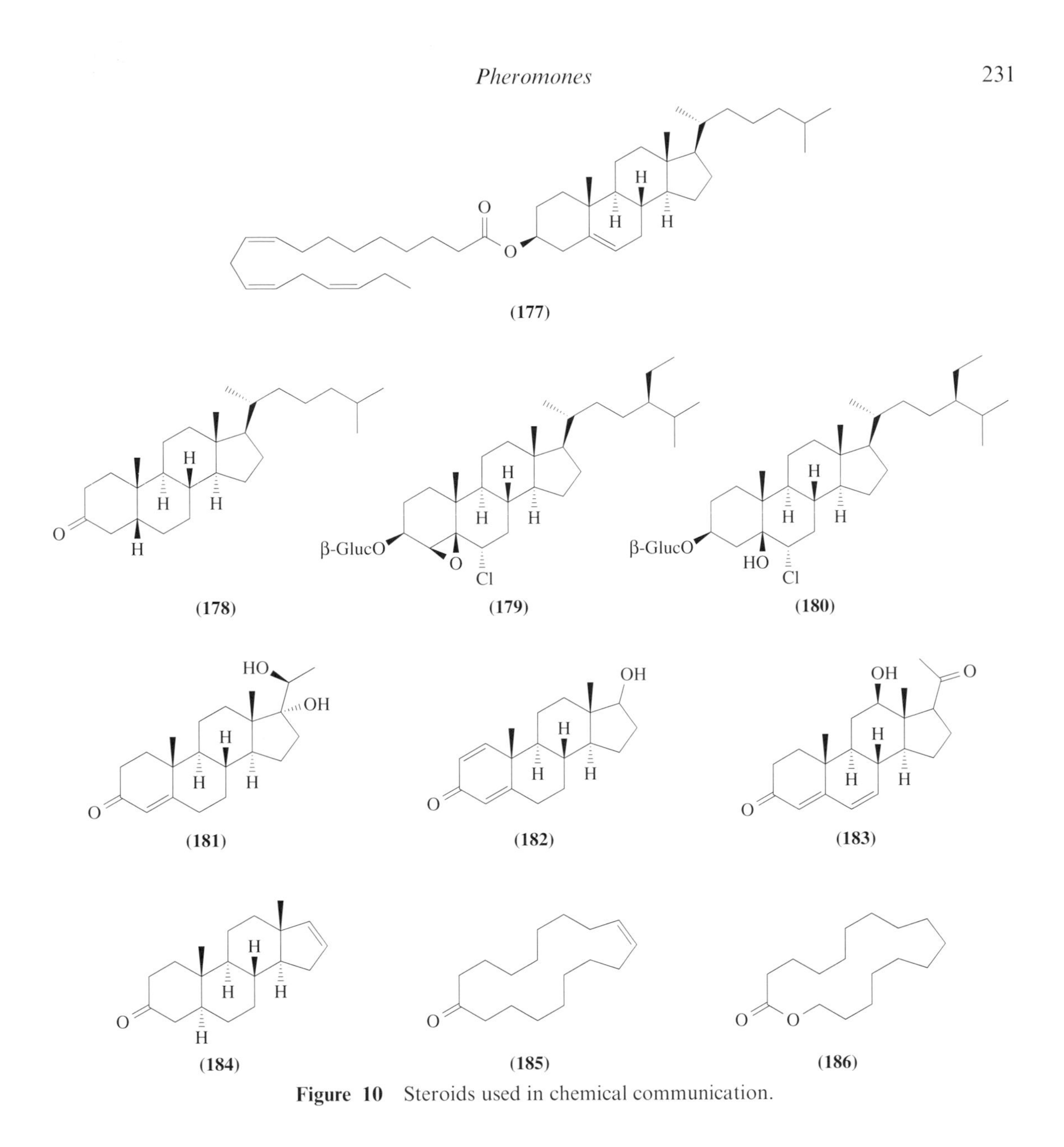

Figure 10 Steroids used in chemical communication.

contribute to the profiles of axillary odor.[508,509] Striking similarities between the molecular structures of the musk-smelling androstenone and exciting-smelling macrocyclic compounds such as civetone (**185**) and hexadecanolide (**186**) or exaltolide are obvious. Interestingly, (**184**) was also identified in the occipital gland secretion of the Bactrian camel[510] and also in vegetables such as parsnip and celery.[511]

8.04.5.6 Degradation Products and Rearranged Terpenes

Many natural products show a typical branching pattern of terpenes without strictly fulfilling the isoprene rule. These compounds may represent either results of rearrangement reactions or degradation products of higher terpenes, or they may reflect a biosynthesis which involves homomevalonate and/or additional acetate or propanoate units (Scheme 4).

A formal retroaldol reaction (Scheme 4(A)) would split off two carbon atoms from the "head part" of a terpene, leaving a "bis-norterpenoid." In contrast, oxygenation at the double bond carrying the geminal dimethyl group would remove three carbon atoms, yielding a "bis-homoterpenoid" (Scheme 4(B)). The same skeleton may, however, also be formed from two propanoate and three acetate units. There are mechanisms which cause fragmentation to remove four carbon atoms through oxidative cleavage forming, e.g., (*E*)-4,8-dimethyl-1,3,7-nonatriene from nerolidol.[512]

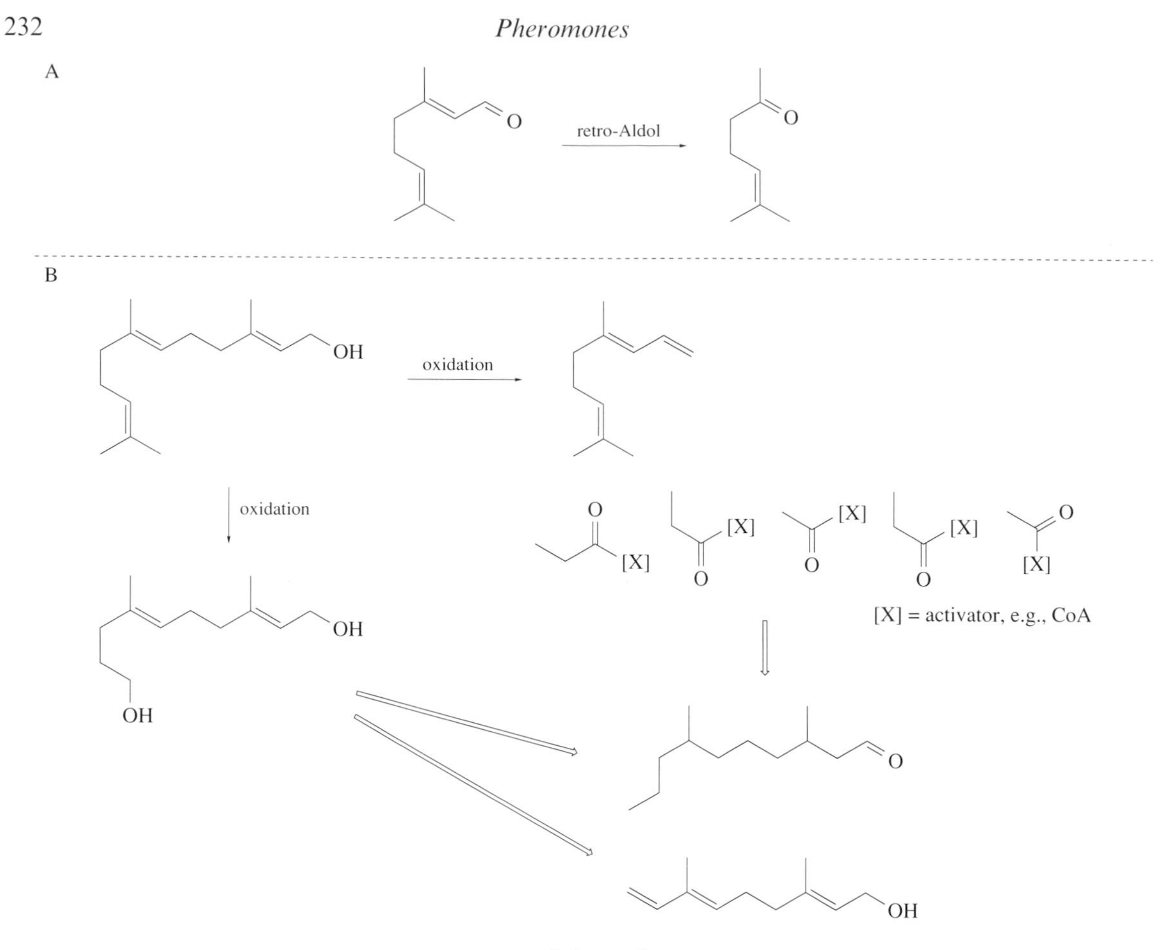

Scheme 4

Similar reactions starting at the "tail part" of a terpene may produce homoterpenes (e.g., homoocimene (**229**), see below); however, such structures would generally suggest the participation of homomevalonate.

An almost ubiquitous terpenoid is 6-methyl-5-hepten-2-one (**187**) (Figure 11), possibly produced from a geranyl precursor via route A in Scheme 4. It is a constituent of the essential oils of many plants and has been found in the resin of coniferous trees and in the culture media of fungi. It is widespread as a volatile constituent of cephalic and abdominal secretions of ants and bees[38] and has been also identified in beetles,[513–515] male butterflies,[331,516] and mammals.[517] Little is known about the biological activity of this ketone. Some ant species obviously use it for defense or as an alarm pheromone. The cursorial spider *Habronestes bradleyi*, a specialist predator of the highly territorial and aggressive meat ant, *Iridomyrmex purpureus*, locates his prey by using its alarm pheromone, 6-methyl-5-hepten-2-one, as a kairomone.[518] In the aphid hyperparasitoid, *Alloxysta victrix*, the same ketone is discussed as a spacing pheromone,[519] and it has the same effect in the cherry-oat aphid, *Rhopalosiphum padi*.[520] During interactions between the robber bee, *Lestrimelitta limao*, and the stingless bee, *Tetragonisca angustula*, nest guards of the latter species are alarmed by cephalic secretions containing geranial, neral, and 6-methyl-5-hepten-2-one. The cleptoparasites emit this mixture during their raid, and *Tetragonisca* workers become particularly aggressive when they smell this ketone, since it indicates the presence of the robbers.[521] Oxygenation of 6-methyl-5-hepten-2-one by an oxidase system attacking the double bond followed by ring closure may yield 1,3,3-trimethyl-2,7-dioxabicyclo[2.2.1]heptane (**188**), which was found in Dufour's gland of the ant *Aenictus rotundatus*.[522] The prenylated bicyclic acetal which may originate from geranylacetone and also the corresponding product formed from the bisnorditerpenoid 6,10,14-trimethyl-5,9,13-pentadecan-2-one were also present along with a wide range of diterpenes.

6-Methyl-5-hepten-2-ol (sulcatol) (**189**), which is frequently associated with the ketone, is an aggregation pheromone of bark beetles, *Gnathotrichus* spp.[523,524] These ambrosia beetles produce species-specific enantiomeric mixtures of the alcohol and the natural proportions are essential for maximum response. Cyclic derivatives of 6-methyl-5-hepten-2-ol could be identified as bark beetle

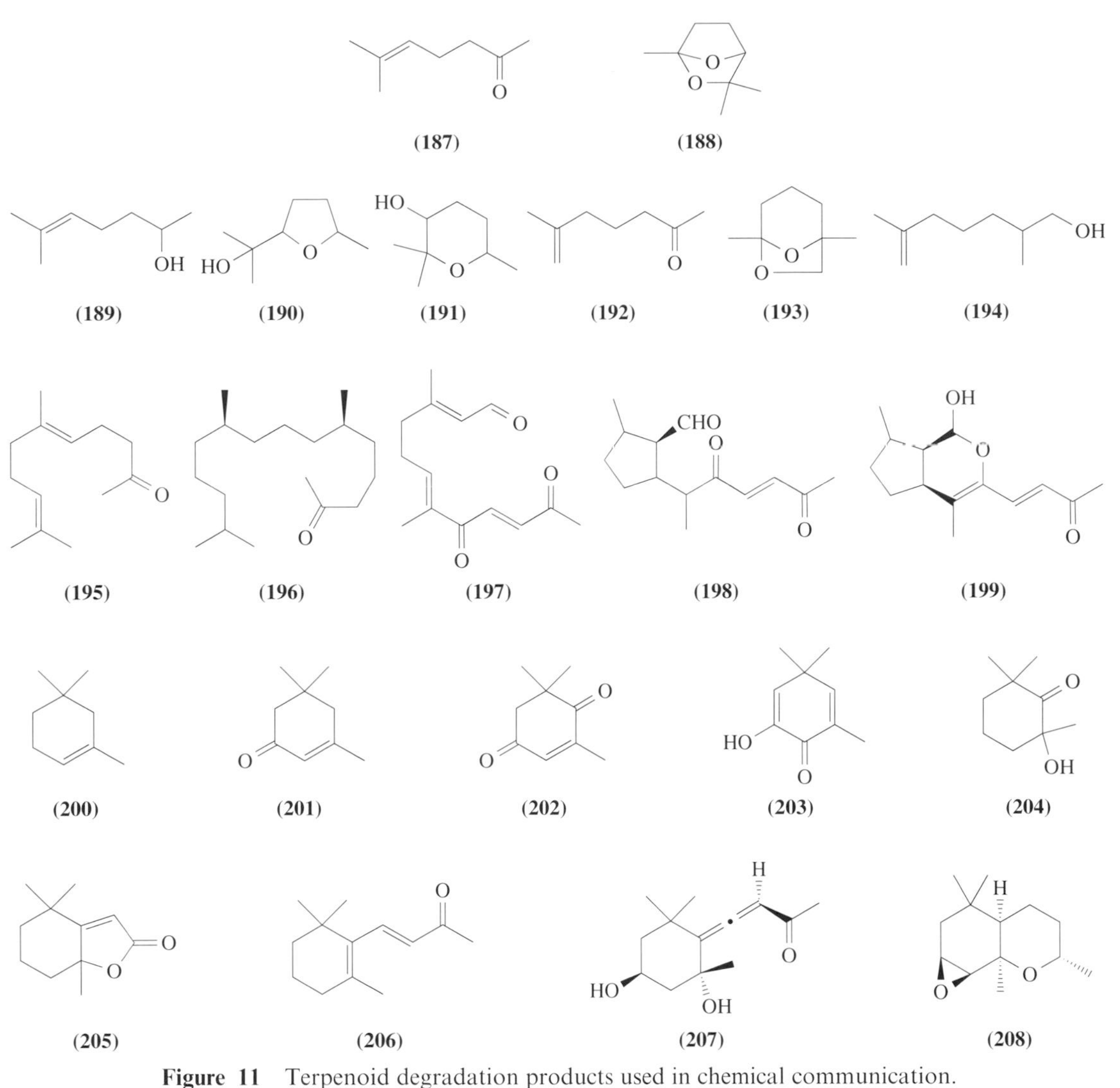

Figure 11 Terpenoid degradation products used in chemical communication.

pheromones; apparently, oxidation at the double bond as described above for the parent ketone followed by ring closure yields either 2-(1-hydroxy-1-methylethyl)-5-methyltetrahydrofuran (pityol) (**190**) or 2,2,6-trimethyl-3-hydroxytetrahydropyran (vittatol) (**191**). Whereas males of *Pityophthorus pityographus* release *trans*-pityol showing (2*R*,5*S*)-configuration,[525] the elm bark beetle, *Pteleobius vittatus*, uses *cis*-pityol and *cis*-vittatol of as yet unknown absolute configurations.[513] (2*R*,5*S*)-Pityol is also an aggregation pheromone of cone beetles, *Conophtorus* spp.[526,527] The structural relations between sulcatol, pityol, and vittatol are essentially the same as those between linalool (**97**) and the pyranoid and furanoid forms of linalool oxides (**135**) and (**136**). Two other bisnorterpenes, 6-methyl-3,5-heptadien-2-one, which is also present in the tarsal glands of the springbok,[420] and 6-methyl-6-hepten-2-one (**192**), could be identified as volatile constituents of the bark beetle, *Dendroctonus simplex*.[352] The latter ketone might serve as a precursor of the bicyclic acetal 1,5-dimethyl-6,8-dioxobicyclo[3.2.1]octane (frontalin) (**193**), which under physiological conditions would be easily produced from 6,7-epoxy-6-methylheptan-2-one.[528] Apart from brevicomin (**38**) (see Section 8.04.4.10), frontalin is an important aggregation pheromone in many *Dendroctonus* bark beetles;[18,59] in *Dendroctonus simplex* it occurs as the pure (−)-enantiomer.[352] It has also been found as a volatile constituent of the temporal gland of male African elephants.[58] The norterpene 2,6-dimethyl-6-hepten-1-ol (**194**) is a constituent of the mandibular glands of males of the ant *Lasius niger* (cited in ref. 529), while its isomer, 2,6-dimethyl-5-hepten-1-ol, seems to be family widespread in bees and ants.[38] The compound may be produced by α-oxidation of a monoterpene precursor followed by loss of one carbon atom. The sulfate of dihydro-(**194**) has been identified from marine organisms where it shows (*R*)-configuration.[530] Compounds (**192**) and (**194**) show the same 2-propenyl tail as the scale pheromones (**98**)–(**100**).

Similar degradation processes which yield 6-methyl-5-hepten-2-one (**187**) may furnish geranylacetone (**195**) from a sesquiterpene precursor. Both (**187**) and (**195**) are present in the urine of the red fox[286] and in the dorsal gland of the springbok[420] and represent the two main components in the mandibular gland secretions of the cleptoparasitic bee, *Holcopasites calliopsidis*.[531] While no behavioral activity of these compounds has been reported for bees, they are believed to possess a communicative function in the fox. Degradation of phytol (**102**) would produce 6,10,14-trimethylpentadecan-2-one (hexahydrofarnesylacetone) (**196**), a well known plant constituent which is also found in the hairpencil secretion of male Danaid butterflies.[532] The corresponding alcohol was first identified in the male African sugar-cane borer, *Eldana saccharina*,[533] and was reported as a courtship pheromone of female rice moths, *Corcyra cephalonica*.[534] A careful study involving synthetic samples of all stereoisomers strongly suggested the natural bioactive compound to show (2*R*,6*R*,10*R*)-configuration.[535] The methyl branchings thus show the same stereochemistry as in phytol.

The acyclic norsesquiterpene gyrinal (gyrinidal) (**197**) is a defensive compound of water beetles, *Gyrinus* and *Dineutes* spp.[536–538] Cyclic compounds such as gyrinidione (**198**) and the bicyclic hemiacetal gyrinidone (**199**)[538,539] were also identified from the secretion of whirligig beetles. Gyrinidone shows some structural relations to the lactol produced from nepetalactone (**146**), which belongs to the set of aphid alarm pheromones discussed above.[407] The three polyoxygenated norsesquiterpenes are highly toxic to fish. By analogy with the formation of iridoids from citral, a biosynthesis starting from farnesal has been suggested;[539] however, higher terpenoids may certainly also serve as precursors.

A second group of degradation products shows a trimethylcyclohexyl moiety and may represent degradation products of carotenoids. β-Cyclogeraniolene (**200**) and its isomers represent minor constituents in the frontal gland secretion of the termite *Ancistrotermes cavithorax*,[463] whereas isophorone (**201**), oxoisophorone (**202**), and some other cyclohexanone derivatives were found in the species-specific odor bouquets from male scent organs of Danaine and Ithomiine butterflies.[394] Oxoisophorone and isophorone have also been identified as trace constituents of the defensive secretion of a grasshopper species,[540] while isophorone and the isomeric γ-phorone are volatile constituents of females of the bark beetle, *Ips typographus*.[541] 2-Hydroxy-4,4,6-trimethyl-2,5-cyclohexadien-1-one (lanierone) (**203**) is a component of the complex aggregation signal of male *Ips pini* bark beetles.[349,542] Such terpenoid cyclohexenone derivatives represent widespread plant constituents and are also known to be produced by microorganisms. 2-Hydroxy-2,6,6-trimethylcyclohexanone (**204**) is a component of the supracaudal gland secretion of the red fox.[543] It is part of a mixture which contains dihydroactinidiolide (**205**) and β-ionone (**206**). Dihydroactinidiolide, a constituent of tobacco and other plant sources, was identified in queens of the fire ant *Solenopsis invicta*[544] and in male scent organs of the butterfly *Idea iasonia*.[317] The close relationship to the termite sesquiterpene ancistrofurane (**167**) is obvious. The particularly widespread β-ionone, representing a cyclic analogue of geranylacetone (**195**), belongs to a group of bisnorterpenoids which are frequently found in plants. Some ionone derivatives also occur in insects. The allenic ketone (**207**) was identified from the defensive froth of a grasshopper, *Romalea* sp., and it may originate from allenic carotenoids.[545] An edulane derivative, the epoxide (**208**), was found in males of Danaid butterflies, *Euploea* spp.[546] The compound shows the same structural relationship to caparrapi oxide (**169**), as does vittatol (**191**), to the pyranoid form of linalool oxide (**135**) (loss of a vinyl group). Additionally, the secretion of *Euploea* contained some less oxygenated edulanes, which were already known from plants.[547] Also, males of the Danaid *Danaus plexippus* secrete edulane derivatives;[317] however, the biological significance of these compounds is unclear.

The interesting tricyclic rearranged terpene cantharidin (**209**) (Figure 12) is relatively widespread in insects.[33] Males of the meloid beetle *Lytta vesicatoria* (Spanish fly) are able to produce it from mevalonate and from farnesol and transfer it to the females (who do not produce this defensive compound) during copulation.[548] Cantharidin acts as an intraspecific attractant and appears to play an essential role in female choice.[549,550]

A unique class of compounds showing a pentamethyl cyclopentane skeleton is represented by α-(**210**) and β-necrodol (**211**). Along with lavandulol (**212**) they are defense compounds of the carrion beetle, *Necrodes surinamensis*.[551,552] Lasiol, 2,3,6-trimethyl-5-hepten-1-ol (**213**), a male-specific volatile identified in the ant *Lasius meridionalis*, also represents a lavandulol skeleton. The natural compound shows (2*R**,3*R**)-configuration;[553] however, despite the availability of pure enantiomers,[554] neither the absolute configuration nor its biological function are known. Whether (1*S*,2*R*,5*R*)-2-ethyl-1,5-dimethyl-6,8-dioxabicyclo[3.2.1]octane (bicolorin) (**214**), the male-produced aggregation pheromone of the beech bark beetle, *Taphrorychus bicolor*,[555] represents a rearranged or degraded terpene or whether it is produced from other sources remains to be investigated.

The terpenoid structures of several sex pheromones of female scale insects share an unusual

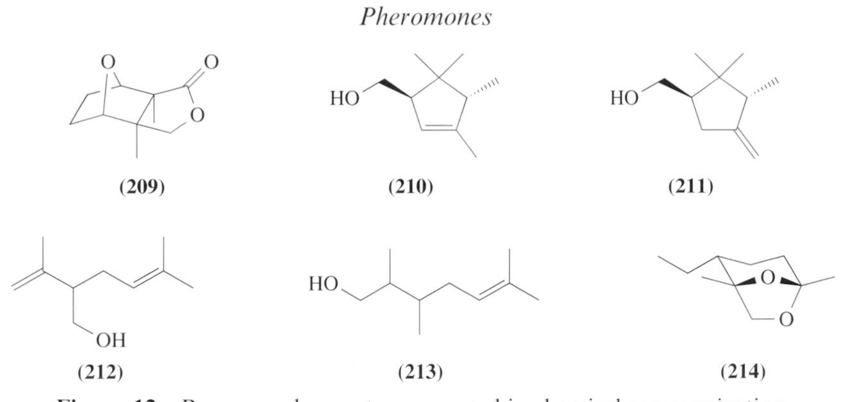

Figure 12 Rearranged monoterpenes used in chemical communication.

substitution pattern. They represent esters of typical acyclic monoterpene alcohols (see (**98**)–(**100**), pheromone components of the San José scale), which carry an additional side-chain at C-6. The propanoate (**215**) is the pheromone of the white peach scale[556] and the acetate (**216**) that of the yellow scale, *Aonidiella citrina* (Figure 13).[557] Interestingly, the carbon skeleton of the sex pheromone of the Comstock mealy bug, *Pseudococcus comstocki*, (*R*)-3-acetoxy-2,6-dimethyl-1,5-heptadiene (**217**),[558,559] would be generated upon splitting the bonds between C-5 and C-6 in (**215**). The two norsesquiterpenes (**218**) and (**219**) make up the sex pheromone of the California red scale, *Aonidiella aurantii*.[560,561] The close relationships in the chemistry of scale pheromones are striking and point to similar origins. Although nothing is known about the biosyntheses of the compounds, their carbon skeletons indicate that they may be derived from higher terpenes. Actually, cembrene (**175**) or an oxygenated derivative would yield such degradation products. Similar processes are known from plants. 5-(1-Methylethyl)-8-methyl-(*E*)-6,8-nonadien-2-one (solanone) (**220**), a bisnorsesquiterpene of Burley tobacco, is assumed to be produced through degradation of diterpene alcohols, which are structurally related to cembrene. Several representatives of such compounds, e.g., 2,7,11-cembratriene-4,6-diol (duvatriendiol) (**221**), are well known from tobacco.[562] Structural relations between solanone and the scale pheromones (**215**)–(**219**) are obvious.

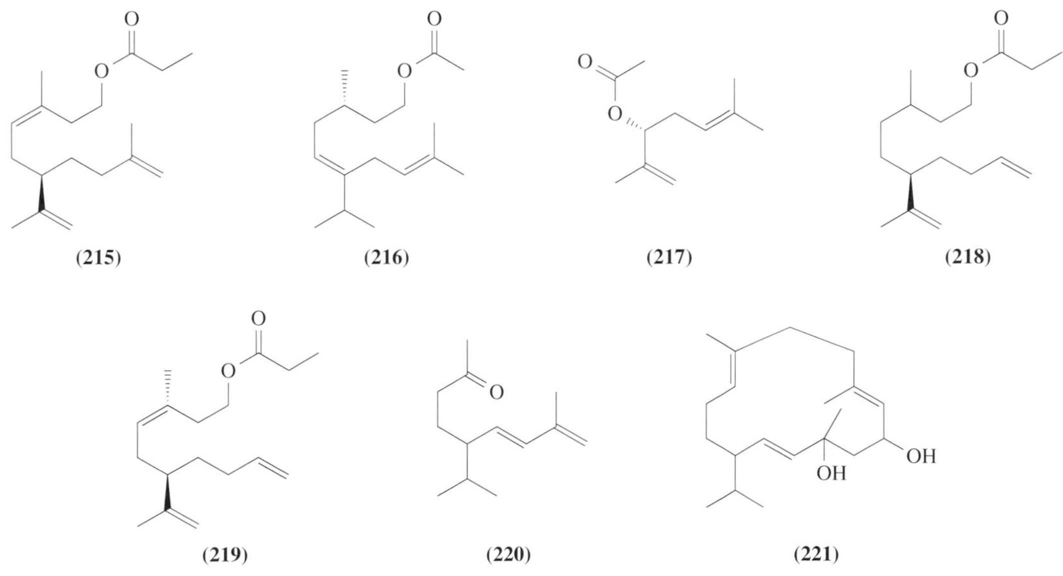

Figure 13 Pheromones of scale insects and structurally related compounds.

The terpenoid (**222**) (Figure 14) and the corresponding dicarboxylic acid, identified from the hairpencil secretions of *Danaus* sp.,[563] have been interpreted as fixatives for volatiles, particularly important in promoting adherence of pheromone-transfer particles. While (**222**) is related to geraniol, complementary linalool derivatives (**223**) are components of the hairpencil secretion of another danaid butterfly, *Euploea sylvester*.[394] The compounds may be produced from sesquiterpene precursors.

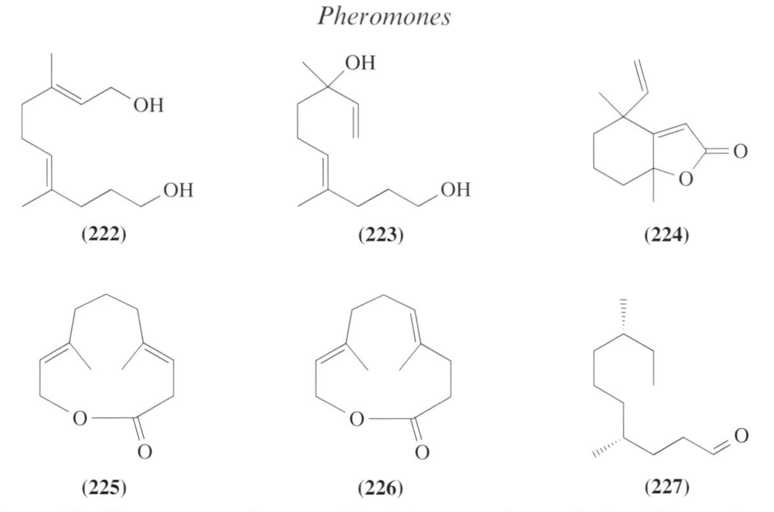

Figure 14 Pheromones of terpenoid origin or products of mixed biosyntheses.

The carbon skeleton of (**222**)–(**226**) is also present in (4*R*,8*R*)-4,8-dimethyldecanal (**227**), the male-produced sex pheromone of the red flour beetle *Tribolium*,[564,565] which may be produced by degradation of farnesol (see Scheme 4(B)). However, the formation according to a mixed polyketide biosynthesis involving a sequence such as acetate–propanoate–acetate–propanoate–acetate followed by simple secondary transformations (see Scheme 4) would yield the same carbon skeleton. Thus, in fact, the compound does not necessarily need to be a terpenoid at all (see Section 8.04.6.1).

Homologues of dihydroactinidiolide (**205**) are represented by the bicyclic lactone 2,6-dimethyl-2-vinyl-7-oxabicyclo[4.3.0]non-9-en-8-one (anastrephin) (**224**) and its 2-epi-isomer, which are components of the sex pheromone blend of male Mexican and Caribbean fruit flies.[566] The compounds seem to be products of an intramolecular cyclization rearrangement of the monocyclic lactone (3*E*,8*E*)-4,8-dimethyldecadien-10-olide (suspensolide) (**225**), which is another volatile component of the Caribbean fruit fly, *Anastrepha suspensa*.[567] In addition to the cucujolides, pheromones of cucujid beetles, which clearly represent acetogenins (see above), ferrulalactone I (**226**), an isomer of suspensolide, was identified as a species-specific component of the aggregation pheromone of the rusty grain beetle, *Cryptolestes ferrugineus*.[568] Similarly to anastrephin (**224**), the biogenesis of which was postulated to involve a terpene precursor, the other compounds shown in Figure 14 may well represent terpenoids.

Whether 4-methyl-3,5-hexadienoic acid, which may have a spacer function in territorial behavior of conspecific males of the fly *Urophora stylata*,[569] represents a degradation product of a terpenoid or whether it is generated from other sources (isoleucine?) remains to be investigated.

8.04.5.7 Homoterpenes

It has been shown that 3-methylenepentyl diphosphate, which is directly related to homomevalonate, can be isomerized both to 3-methyl-2-pentenyl diphosphate and to 3-methyl-3-pentenyl diphosphate through the action of an isopentenyl diphosphate isomerase (see also Scheme 3).[570] Scheme 5 shows that a terpene biosynthesis involving either of the three homoisoprenoids **X**–**Z** will lead to various types of homoterpenes: **X** will furnish ethyl branchings along the chain, **Y** will yield vicinal dimethyl branchings and **Z** will produce an ante-iso branching at the end of the chain. All these principles are verified in the insect juvenile hormones: the incorporation of homomevalonate into juvenile hormones has been shown.[571] JH I[572]—biological activity rests with (10*R*,11*S*)-configuration at the epoxide moiety, as has been unambiguously proven in bioassays with enantiomerically pure samples[573,574]—shows a sequence of **Z**+**X**+a conventional isopentenyl diphosphate. Whereas JH 0[575] is made up of a sequence **Z**+**X**+**X** and hence shows two ethyl groups along the chain, the juvenile hormone 4-methyl-JH I[576] shows a vicinal dimethyl grouping which points to a combination of **Z** + **X** + **Y**.

(*E*)-3-Methyl-7-methylene-1,3,8-nonatriene (**228**) (Figure 15) is produced by the bark beetle, *Ips typographus*, during certain stress situations and has a repellent effect.[352] The homoocimene (**229**), apparently showing (*E*,*E*)-configuration,[577] is a minor compound in the Dufour's glands of the ant *Labidus praedator*, which contain major amounts of (*E*)-β-ocimene.[578]

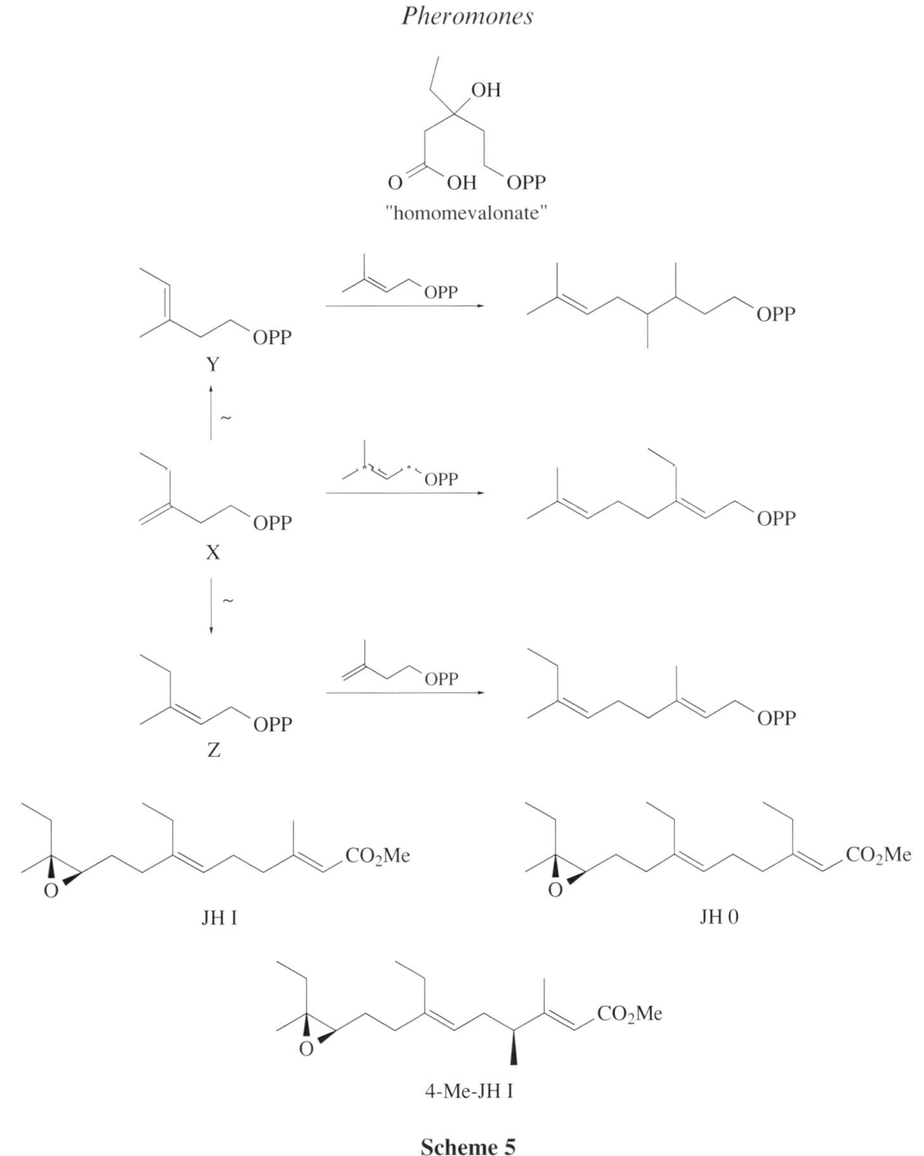

Scheme 5

A monocyclic sesquiterpene analogue of (**216**) is (4*Z*,4′*E*)-4-(1,5′-dimethyl-4-heptenylidene)-1-methylcyclohexane (homo-γ-bisabolene) (**230**), the major component of the male sex pheromone of the cereal pest *Eurygaster integriceps*.[579]

The racemate of the homoterpene 10-homonerol oxide (**231**) forms a main component among the volatile compounds of the thoracic gland of the ant lion, *Grocus bore*.[403]

Homofarnesenes and (*Z*,*E*)-α-bishomofarnesene (**232**) are constituents of the Dufour's gland secretion of *Myrmica* ants,[580–583] where it forms a mixture with straight-chain hydrocarbons and other homofarnesenes. Faranal (**233**), the trail pheromone of the Pharaoh's ant, *Monomorium pharaonis*,[584] shows a structure which points to incorporation of homomevalonate in a way similar to the biosynthesis of 4-methyl-JH I (see Scheme 5). The same branching pattern is also present in the homofarnesene (**234**), one of the trail pheromone components of the fire ant, *Solenopsis invicta*,[585] and in 9-methylgermacrene-B (**235**), the proposed structure for the sex pheromone of the sand fly, *Lutzomyia longipalpes*, an important vector for leishmaniasis.[586]

8.04.6 PROPANOGENINS AND RELATED COMPOUNDS

8.04.6.1 Structural Considerations/Biosynthesis

Branched chains (Figure 16) may be formed in various ways. Formal steps are outlined in Scheme 6.

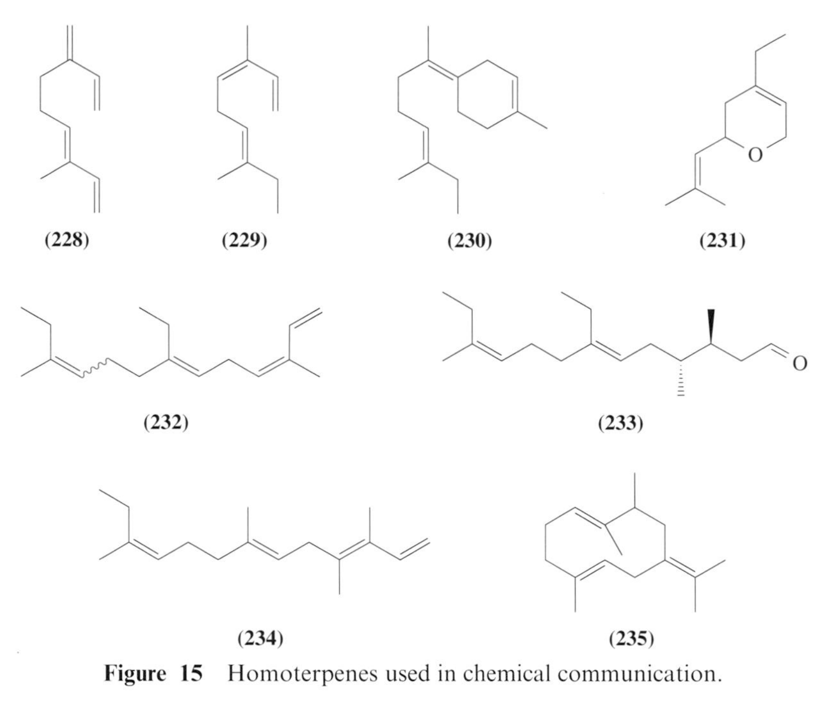

Figure 15 Homoterpenes used in chemical communication.

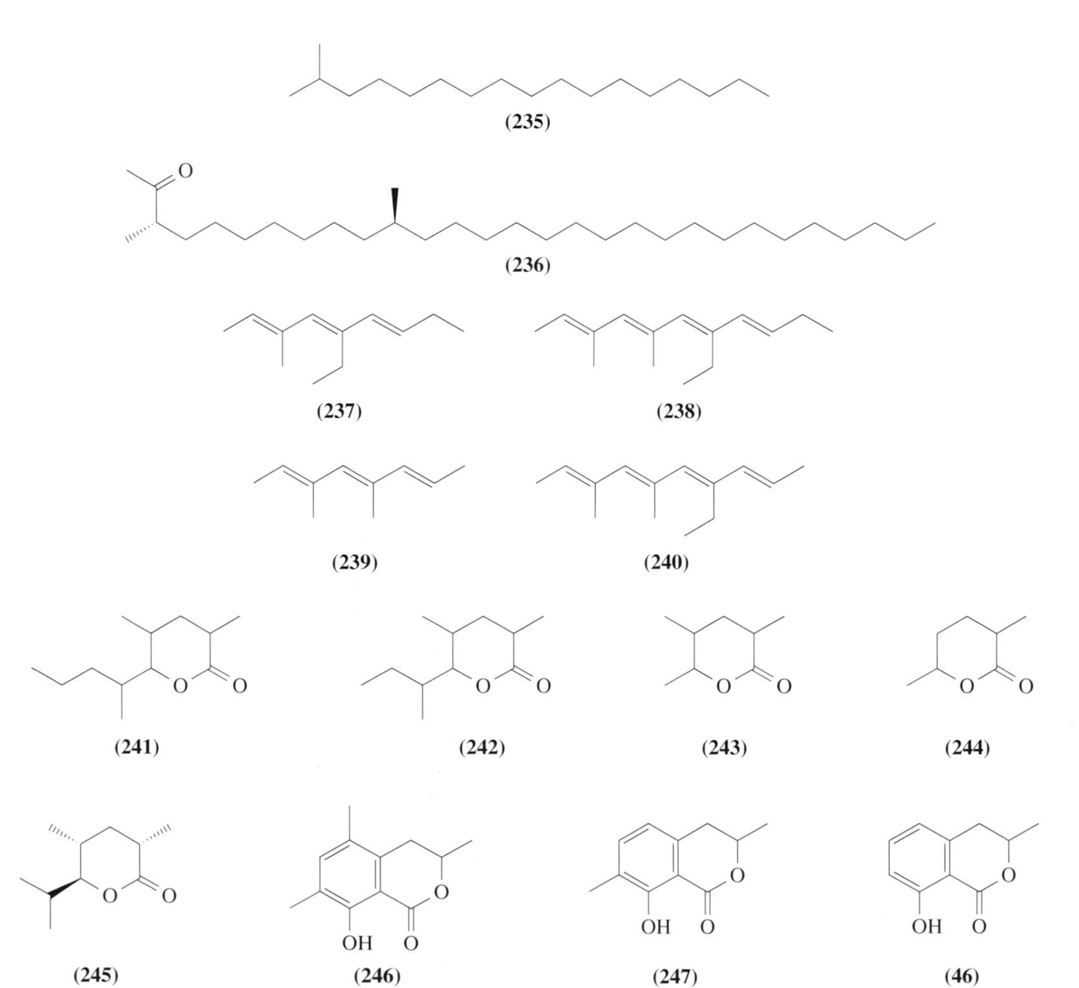

Figure 16 Branched pheromone structures of nonterpenoid origin.

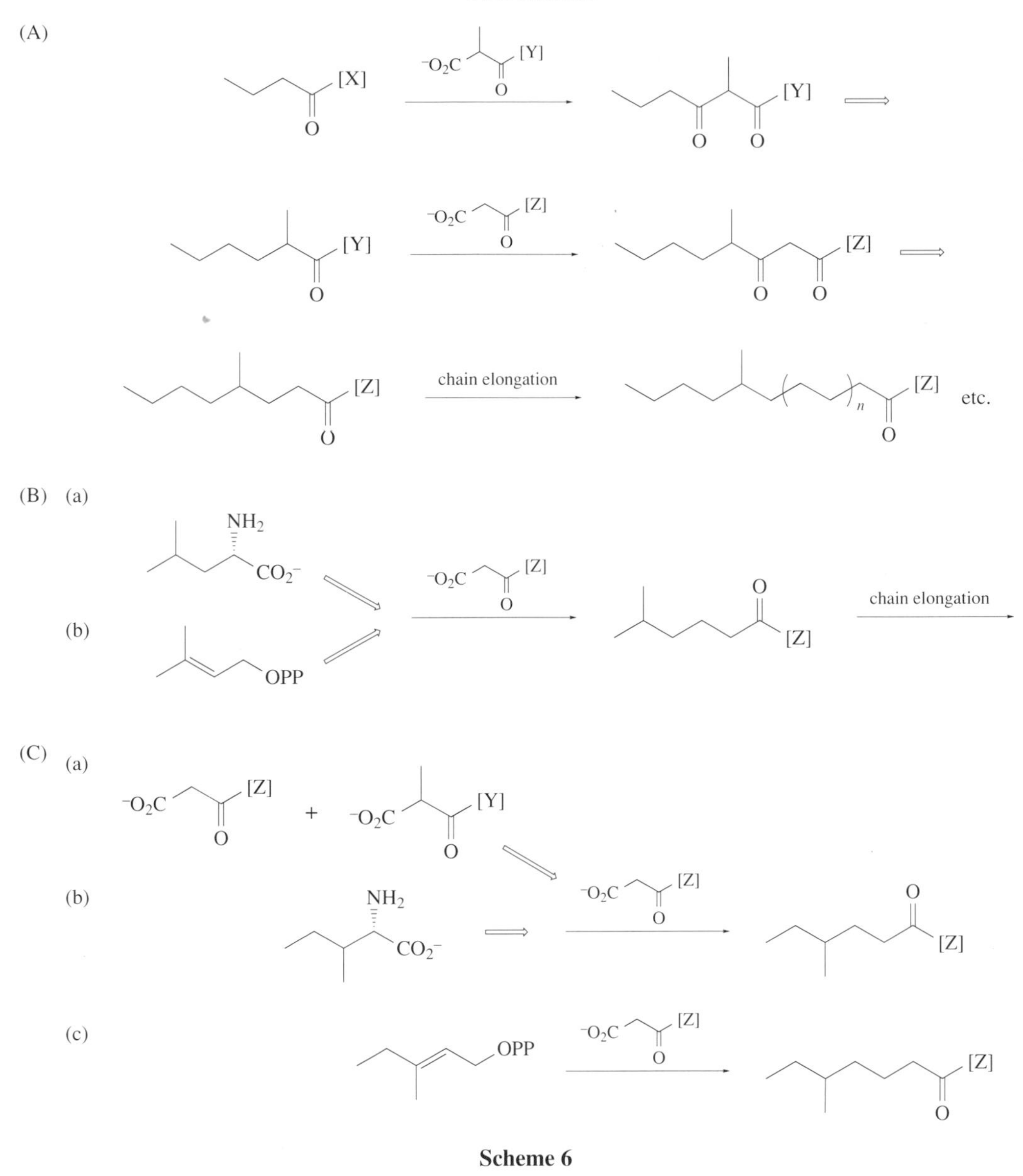

Scheme 6

In Scheme 6(A), when during chain elongation along an acetogenic route an acetate unit (or malonate) is replaced by propanoate, methyl branching in the final compound will be the result. The same structure is produced when methylmalonate (or succinate) is incorporated instead of propanoate and when the additional carbon atom is lost through decarboxylation. Accordingly, butanoate will yield (the very rare) ethyl branching. Consequently, several branchings in the same compound will be separated by an uneven number of methylene groups.

In Scheme 6(B), terminal branching ("iso branching") occurs when an "isoprene unit" such as 3-methyl-2-butenyl diphosphate or leucine (which will lose one carbon atom during the sequence) is involved in the starting phase (see also Sections 8.04.5.1 and 8.04.5.6).

In Scheme 6(C), ante-iso branching may be produced from either a "homo-isoprene unit" (see also Scheme 5) or isoleucine as the starter or when the first condensation step involves acetate (malonate) and a propanoate (methylmalonate) unit (see also Section 8.04.5.6). Consequently, when chain elongation of branched starters proceeds via Scheme 6(B) or 6(C)(c), the incorporation of a propanoate unit at a later stage will result in a compound showing an even number of methylene groups between the first branching points. In contrast, in corresponding compounds produced from the starters in Scheme 6(C)(a) or 6(C)(c) or when valine is the starter instead of leucine in Scheme 6(B)(a), the two branching points are separated by an uneven number of carbon atoms.

Addition of methyl groups during chain elongation ("chain methylation") with C_1 units such as methionine or fragmented acetate[587] or chain modification with other alkyl-transferring agents may occur at random.

An entirely different branching is produced upon aldol condensation of carbonyl compounds, eventually followed by hydrogenation. Some examples have been reported from bugs and ants.[33,38]

As outlined in Scheme 6, the general scope of the biosynthesis of many branched compounds seems to be obvious; however, only a few detailed investigations have been reported. In 2-methylheptadecane (**235**), the sex pheromone of Arctiid moths, *Holomelina* spp.,[11] leucine serves as the methyl branch donor. The hydrocarbon is produced through decarboxylation of 17-methyloctadecanoic acid.[588] Similarly, methyl-branched fatty acids serve as precursors to branched hydrocarbons identified in the German cockroach, *Blatella germanica*.[589] In these insects, labeling experiments with radioactive valine, isoleucine, and methionine indicated that the carbon skeleton of all three amino acids may serve as methyl branch donors. It was also shown that succinate and propanoate serve to produce methylmalonyl-CoA. The methyl branching is therefore produced early during chain elongation.[590] In the German cockroach, no biological activities could be attributed to the branched hydrocarbons found on the surface of the cuticula; however, oxygenated compounds such as 3,11-dimethylnonacosan-2-one (**236**), a component of the female-specific wing-raising pheromone, is produced from the parent hydrocarbon via the alcohol. Only females can oxidize the alkane.[591] The ketone (**236**) and 3,11-dimethyl-29-hydroxynonacosan-2-one show (3*S*,11*S*)-configuration.[592–594] Another component of the contact pheromone 3,11-dimethylheptacosan-2-one has been identified from the German cockroach.[595] Compared with the main component (**236**), it lacks one acetate unit. A study on the production site of these ketones showed a relation to corpora allata activity and suggested that the internal synthesis and accumulation are followed by transport to the epicuticular surface.[596]

The biosynthesis of the unique branched polyenes which make up the male-produced aggregation pheromones of nitidulid beetles, *Carpophilus* spp., has been carefully studied by Bartelt and coworkers. In *Carpophilus davidsoni* and *C. freemani*, (2*E*,4*E*,6*E*)-5-ethyl-3-methyl-2,4,6-nonatriene (**237**) is built up from one acetate unit and one propanoate unit followed by the incorporation of two butanoate units, accompanied by the loss of the carboxyl carbon from the second butanoate.[597,598] Similarly, the biosynthesis of (2*E*,4*E*,6*E*)-7-ethyl-3,5-dimethyl-2,4,6,8-undecatetraene (**238**), another major component, involves an additional propanoate unit which accounts for carbon atoms 3 and 4 and the additional methyl branching. The biosyntheses of (2*E*,4*E*,6*E*,8*E*)-3,5,7-trimethyl-2,4,6,8-undecatetraene and (2*E*,4*E*,6*E*,8*E*)-5,7-diethyl-3-methyl-2,4,6,8-undecatetraene, identified in *Carpophilus davidsoni*, in addition to (3*E*,5*E*,7*E*)-5-ethyl-7-methyl-3,5,7-undecatriene, the pheromone of *Carpophilus mutilatus*, follow the same principal pattern.[597] The biosynthesis of the latter compound starts with butanoate and does not involve acetate.[597] The general biosynthetic scheme of these compounds seems to follow conventional chain elongation; however, instead of the stereotypical acetate, other acyl units are also involved.

The termination of the chain is particularly interesting, since it obviously includes "head-to-head" coupling of two acyl units and loss of one carbon atom. In (**237**) and (**238**) this terminating group is butanoate.[598] Consequently, in (2*E*,4*E*,6*E*)-3,5-dimethyl-2,4,6-octatriene (**239**) from *Carpophilus davidsoni* this should be propanoate. According to their structures, the biosyntheses of (2*E*,4*E*,6*E*,8*E*)-3,5,7-trimethyl-2,4,6,8-decatetraene and its 7-ethyl homologue (**240**), pheromone components of *Carpophilus brachypterus*,[599] should either start with an acetate unit and finish with propanoate or, less likely, proceed in the opposite direction. The same compounds or similar blends as those mentioned above make up the pheromones of other Nitidulid species also[600–605] and have been successfully tested in the field.[606]

Another important study on the biosynthesis of branched pheromones has been carried out by Bestmann *et al.*,[607] who investigated the incorporation of marked propanoate into lactones, known from *Camponotus* ants.[608] They found that 3,5-dimethyl-6-(1′-methylbutyl)tetrahydro-2*H*-pyran-2-one (**241**) is formed from four propanoate (methylmalonyl) units. In *Camponotus herculaneus* (**241**) induced trail following.[609] The compound had been identified earlier as part of the queen recognition pheromone of the fire ant, *Solenopsis invicta*,[610,611] and was named invictolide. Bioassays with optically pure enantiomers synthesized by Mori and Nakazono[612] revealed that the biological activity rests with (3*R*,5*R*,6*R*,1′*R*)-configuration.[613] The lactones (**242**) and (**243**), which also act as trail pheromones in some *Camponotus* species, are obviously produced from an acetate starter followed by incorporation of three propanoate units or from acetate and two propanoates, respectively.[607] Although its biosynthesis is unknown, *cis*-2-methylhexanolide (**244**), a major volatile emitted by the carpenter bee, *Xylocopa hirsutissima*,[614] may be produced from an acetate starter, a second acetate unit, and a propanoate unit. Similarly, (3*S*,5*R*,6*S*)-3,5-dimethyl-6-(1-methylethyl)-3,4,5,6,-tetrahydro-2*H*-pyran-2-one (**245**), which together with (*Z*)-4-tridecenal acts as a pheromone of the parasitic wasp *Macrocentrus grandii*,[615] may be formed from an amino acid precursor such as valine as the starter, which accounts for the isobranching of the side chain, and which is elongated by two propanoate units.

Mixed biosynthesis involving acetate and propanoate units was also reported for the formation of the 3,4-dihydroisocoumarin (**246**), which had been identified as a trail pheromone of *Lasius* ants and *Camponotus* ants.[189,616] The 3,4-dihydroisocoumarin (**247**) is a trail pheromone of *Camponotus rufipes*, whereas mellein (**46**), which is made up of five acetate units, is the trail pheromone of *Lasius fuliginosus* and *Formica rufa*. Mellein (**46**), which is relatively widespread in nature, and dihydroisocoumarins, other than (**246**) and (**247**), were found in the rectal glands of *Formica* and *Camponotus* species.[607] Behavior tests with *Lasius niger* and electrophysiological investigations including *Formica* species showed the (*R*)-enantiomer of mellein to be much more active than its antipode.[617] According to bioassays, the stereogenic center in the active homologues seems to show (*R*)-configuration also.[122,189]

8.04.6.2 Pheromones from Propanoate and Mixed Structures Showing a Terminal Functional Group

In a manner similar to that of the polyenic structures of Nitidulid pheromones (see Figure 15), the linkage between the subunits of branched pheromones may be evident as a double bond. In some pheromones of curculionid beetles, even the oxygen is kept at the connection site, yielding β-hydroxy acids or α,β-unsaturated acids.

Propanoic acid has never been reported as a pheromone. It was found in the osmeterial secretion of a papilionid butterfly and in a myrmicine ant[618] and is considered to be a component of the defensive secretions. Some pheromones of scales (see Section 8.04.5.6 and Figure 13) and root worms and also pine sawflies (see Section 8.04.6.4) are esters of propanoic acid.

Two propanoate units are coupled in 1-ethylpropyl (2*S*,3*R*)-2-methyl-3-hydroxypentanoate (**248**) (Figure 17), the male-produced aggregation pheromone of the granary weevil, *Sitophilus granarius*.[619–622] Elimination of water in the acid moiety of (**248**) and esterification with (*S*)-2-pentanol will yield dominicalure 1 (**249**), the aggregation pheromone of the lesser grain borer beetle, *Rhizoperta dominica*.[623] The 1-methylethyl ester of the same acid proved to be the pheromone of the Bostrychid beetle, *Prostephanus truncatus*.[624] A minor component of the *Prostephanus* pheromone proved to be 1-methylethyl (2*E*,4*E*)-2,4-dimethyl-2,4-heptadienoate (**250**), the acid moiety of which is obviously made up of three propanoate units. A second component in *Rhizoperta* was identified as (*S*)-1-methylbutyl (2*E*)-2,4-dimethyl-2-pentenoate, dominicalure 2.[623] In this case, the starting unit, contributing four carbon atoms and producing iso-branching, may well originate from an amino acid. Invictolide (**241**), made up of four propanoate units, has already been mentioned (see Figure 15). Five propanoate units are represented by supellapyrone (**251**), the female-produced sex pheromone of the brown-banded cockroach, *Supella longipalpa*,[625] which proved to show a (2′*R*,4′*R*)-configuration in the side chain.[626,627]

Ethyl 4-methyloctanoate (**252**), the most important aggregation pheromone of males of the coconut rhinoceros beetle, *Oryctes rhinoceros*,[628] shows a mixed structure involving three acetate units and one propanoate unit. It is accompanied by ethyl 4-methylheptanoate which, instead of two acetate units (or one butanoate), shows propanoate as the starter. The ester (**252**) is also a pheromone of the related *Oryctes monocerus*.[629] Another mixed structure, possibly built up by a sequence of propanoate–acetate–propanoate–acetate, is represented by (*R*)-4-methyl-1-nonanol, the pheromone of the meal worm, *Tenebrio molitor*.[630] Methyl 2,6,10-trimethyltridecanoate (**253**), a major component in the pheromone blend of stink bugs, *Euschistus* spp.,[16] possibly involves four propanoate and two acetate units. The biosynthesis of the male sex pheromone of the stink bug *Stiretrus anchorago*, 6,10,13-trimethyltetradecan-1-ol (**254**),[631] which was identified in the secretion of the sternal gland, may involve two propanoate and three acetate units. Mori and Wu[632] prepared all four stereoisomers for bioassays with pure enantiomers. The structure of the alcohol part of the pheromone of the tussock moth *Euproctis pseudoconspera*, (*R*)-10,14-dimethylpentadecyl 2-methylpropanoate (**255**),[633] although looking like an elongated monoterpene, seems to be biogenetically close to (**254**) and to (*Z*)-16-methyl-9-heptadecenyl 2-methylpropanoate, the sex pheromone of *Euproctis taiwana*.[634] The blend of 2-methylpropanoates which make up the pheromone of *Euproctis similis*[295] and the highly unsaturated 2-methylpropanoate of *Euproctis chrysorrhoea*[11] have already been mentioned (see Section 8.04.5.1). Similarly to the hydrocarbon (**235**), the iso-branched pheromones mentioned above may be produced from a starter originating from an amino

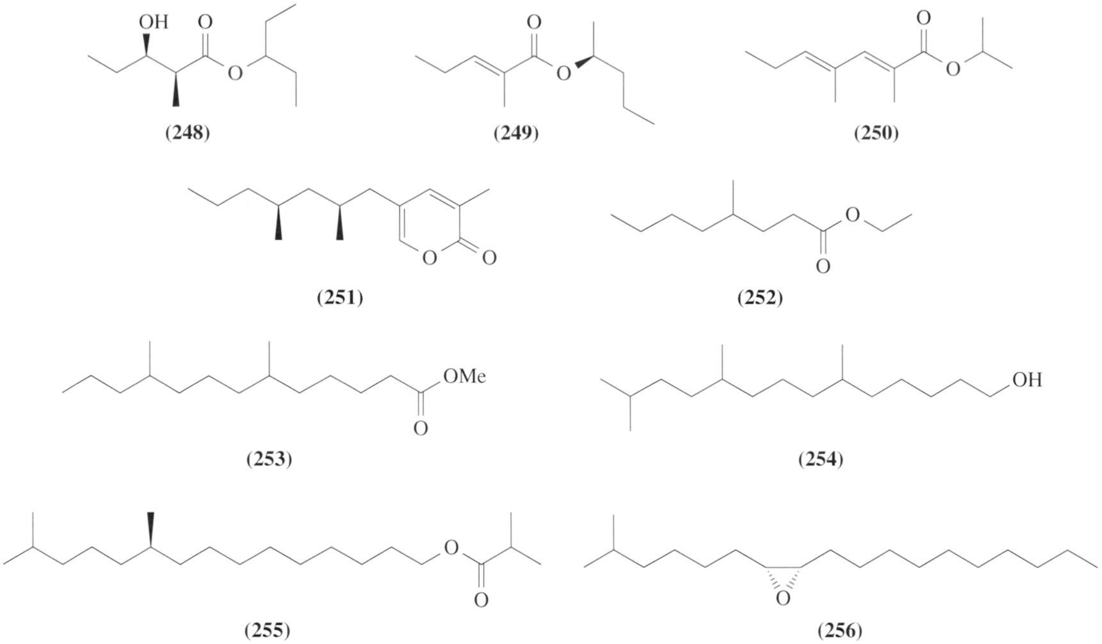

Figure 17 Pheromones from propanoate and mixed structures showing a terminal functional group.

acid. This may also be true for (*Z*)-2-methyl-7-octadecene, a minor component of the gypsy moth pheromone,[103] which uses the corresponding epoxide, (7*R*,8*S*)-7,8-epoxy-2-methyloctadecane (**256**), as the most important female sex attractant.[635]

8.04.6.3 Branched Ketones, Corresponding Alcohols and Related Compounds

As mentioned above (see Section 8.04.4 and Scheme 1), methyl ketones may be formed from 3-oxoacyl precursors after decarboxylation while ethyl ketones may be produced via vinyl ketones. Of course, $\omega-2$ oxygenation of longer chains may also take place; however, Bartelt and co-workers' investigations on the biosyntheses of Nitidulid pheromones[597,598] opened up a particularly interesting alternative. They found that chain elongation may be terminated by a "head-to-head" coupling of two acyl chains accompanied by the loss of one carbon atom. As exemplified in Scheme 7, which shows the possible formation of pheromone components of *Sitophilus* weevils and its general extension to biogeneses of other branched pheromones and where [X] and [Y] are activators and R^2 may be a proton or a methyl or ethyl group, reaction of the acyl precursor of the granary weevil pheromone (**248**) with a propanoate unit, accompanied by the loss of one carbon atom, would yield (4*S*,5*R*)-5-hydroxy-4-methyl-3-heptanone (sitophilure) (**257**) (Figure 18), the aggregation pheromone of the rice weevil, *Sitophilus oryzae*.[636,637] In Scheme 7 the longer chain is shown to retain the carbonyl group. According to Bartelt and co-workers' results, the reverse coupling—although possible—appears less likely. Following this hypothetical mechanism, close biogenetic relations between several pheromones become obvious (see Figure 18), and 3-pentanone found in *Sitophilus* spp.[119] should be produced from two propanoate units.

The diketone (**258**), the pheromone of the pea weevil, *Sitona lineatus*,[638,639] still retains the oxygen of the first two propanoate units whereas (*S*)-4-methyl-3-hepanone (**259**) retains only the former acyl group of the parent acid precursor. The ketone (**259**), the alarm pheromone of the leaf cutting ant, *Atta texana*,[640] was the first chiral insect volatile for which it was shown that only one enantiomer was biologically active. Meanwhile, (**259**) has been identified in many insects including other ant species,[641–643] in "daddy-long-legs," opilionid spiders,[644] in mutillid wasps,[645] and in caddisflies.[172] The corresponding alcohol, 4-methyl-3-heptanol, is often associated with the ketone,[641] be it as a pure enantiomer or as a mixture of diastereomers. Interestingly, the stereogenic center carrying the methyl group is conserved with (*S*)-configuration. In *Scolytus* bark beetles, (3*S*,4*S*)-4-methyl-3-heptanol is an important component of the aggregation pheromone[646] whereas its diastereomer, (3*R*,4*S*)-4-methyl-3-heptanol, acts as a trail pheromone in the ant *Leptogenys diminuta*.[647] 4-Methyl-3-hexanol, the biosynthesis of which may start with acetate instead of propanoate, has also been

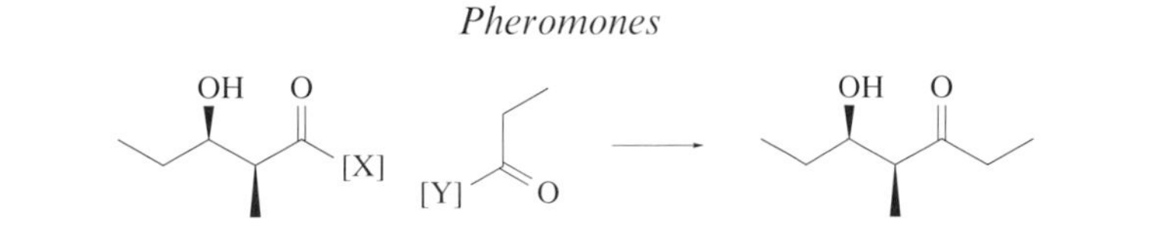

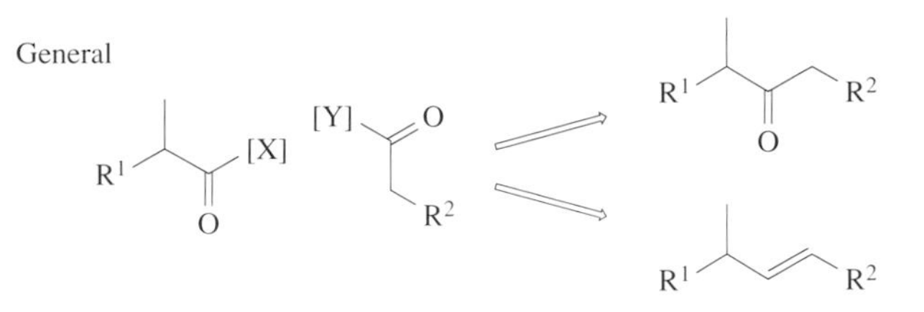

Scheme 7

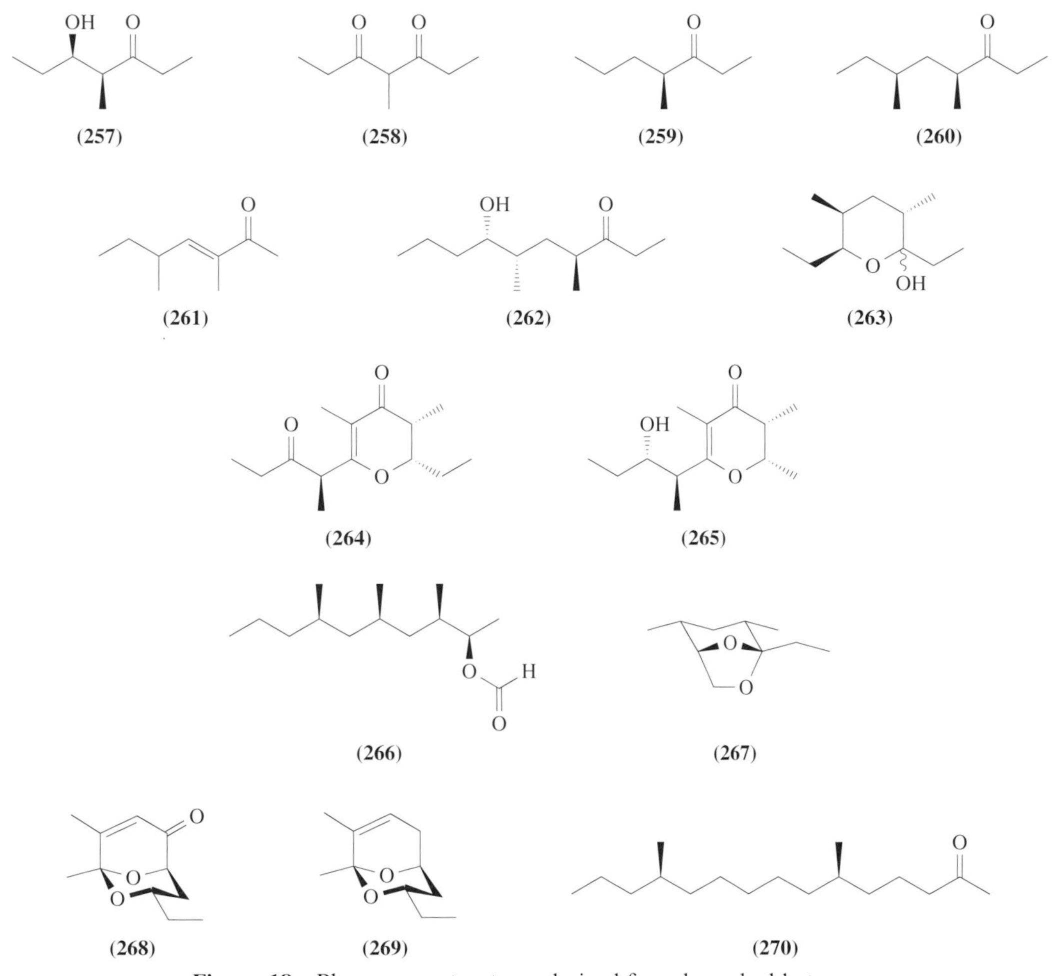

Figure 18 Pheromone structures derived from branched ketones.

found in several insect species.[648] An extended chain incorporating an additional propanoate unit is represented by (6*E*)-4,6-dimethyl-6-octen-3-one (leiobunone),[643] which was identified in "daddy-long-legs," *Leiobunum* sp.,[644] (4*S*,6*S*)-4,6-Dimethyloctan-3-one (**260**), a presumed pheromone of caddisflies,[172] and (4*E*)-4,6-dimethyl-4-octen-3-one, (manicone),[649] which is contained in the mandibular gland secretion of *Manica* ants and which acts as an alarm pheromone. In *Manica rubida*, the stereogenic center shows (*S*)-configuration.[650] Interesting minor components are represented by homomanicone and bis-homomanicone, (*E*)-4,6-dimethyl-4-nonen-3-one and (*E*)-4,6-dimethyl-4-decen-3-one, which suggests propanoate as the starter for the nonanone and two acetate units or a butanoate for the decanone. Normanicone, (*E*)-3,5-dimethyl-3-hepten-2-one (**261**), is a methyl ketone which suggests that this time the terminating group is an acetate unit instead of propanoate (see Scheme 7). In the caddisfly, *Hesperophylax occidentalis*, 6-methylnonan-3-one was identified as a sex pheromone.[651] Although its absolute configuration is unknown, it is postulated here, following again the above-mentioned conservative principles in chain formation, that it should show (*S*)-configuration, which is also represented in (4*S*,6*S*)-4,6-dimethylnonan-3-one, another caddisfly ketone.[172] This ketone and 4,6-dimethyl-3-octanone (dihydromanicone) have been identified earlier in the Mutillid, *Dasymutilla occidentalis*.[645] Although the stereochemistry of the ketones in these wasps is unknown, according to the ideas outlined above they may show (4*S*,6*S*)-configuration. 6-Methyl-3-octanone which together with straight-chain 3-ketones accounts for alarm/attraction in some *Crematogaster* ants,[652] also belongs to the set of compounds discussed here.

A 3-nonanone skeleton which still retains the oxygen of the starting unit is (4*S*,6*S*,7*S*)-7-hydroxy-4,6-dimethyl-3-nonanone (serricornin) (**262**), the sex pheromone of the cigarette beetle, *Lasioderma serricorne*.[653] The compound forms a 1:3 equilibrium with its cyclic hemiacetal (**263**),[654] and its activity is strongly inhibited by the nonnatural 4*S*,6*S*,7*R*-stereoisomer.[655] Elongation of serricornin by an additional propanoate unit will form the dihydropyran (2*S*,3*R*,1′*R*)-2,3-dihydro-2-ethyl-3,5-dimethyl-6-(1′-methyl-2′-oxobutyl)-4*H*-pyran-4-one (β-serricorone) (**264**). The compound is accompanied by its (1′*S*)-epimer, α-serricorone, and its reduction product, serricorole, which shows (1′*S*,2′*S*)-configuration.[656–658] These compounds show a bifunctional role in the communication system of *Lasioderma*. It should be mentioned that Chuman *et al.*,[653] during their investigations on the cigarette beetle, were the first to point out close relations between pheromone structures that involve propanoate units. The relations between invictolide (**241**) and serricorone (**264**) are obvious; the latter structure is possibly terminated by propanoate according to Scheme 7 whereas the former is finished by lactonization. In (2*S*,3*R*,1′*S*,2′*S*)-2,3-dihydro-2,3,5-trimethyl-6-(2′-hydroxy-1′-methylbutyl)-4*H*-pyran-4-one, stegobiol (**265**), the pheromone of males of the drugstore beetle, *Stegobium paniceum*,[659,660] the methyl terminus may originate from acetate. Stegobinone, showing a carbonyl group at C-2′ in the side chain of stegobiol and which is another pheromone component of the drugstore beetle, was first identified by Kuwahara *et al.*[661] Its absolute configuration was established by independent syntheses.[662] See also reference 663. The furniture beetle, *Anobium punctatum*, an Anobiid like the cigarette beetle and the drugstore beetle, seems to use the same system as the drugstore beetle.[664] In the drugstore beetle, 1′-epistegobinone strongly inhibits response.[665]

(1*R*,3*R*,5*R*,7*R*)-1,3,5,7-Tetramethyldecyl formate (lardolure) (**266**), the aggregation pheromone of the acarid mite *Lardoglyphus konoi*, appears to be made up by four propanoate units and an acetate stopper (with one carbon atom excised; see Scheme 7). The compound was identified by Kuwahara *et al.*,[666] and its absolute configuration was established by Mori and Kuwahara[667,668] through independent syntheses.

A cryptic ethyl ketone with a mixed structure is represented by (1*S*,2*R*,4*S*,5*R*)-2,4-dimethyl-5-ethyl-6,8-dioxabicyclo[3.2.1]octane (α-multistriatin) (**267**), an important component of the aggregation pheromone of elm bark beetles, *Scolytus* spp. The compound was first identified in *Scolytus multistriatus*[669] and its absolute configuration was established by independent syntheses.[670,671] Structural relations between the caddisfly compound (4*S*,6*S*)-4,6-dimethyl-3-nonanone[172] and α-multistriatin are obvious. Even the stereochemical position of the methyl groups is the same. Two other bicyclic acetals, (2*R*,3*S*,5*S*)-1,8-dimethyl-3-ethyl-2,9-dioxabicyclo[3.3.1]non-7-ene (**268**) and (1*R*,3*S*,5*S*)-1,8-dimethyl-3-ethyl-2,9-dioxabicyclo[3.3.1]non-7-en-6-one (**269**), are components of the male-produced pheromone of the swift moth, *Hepialus hecta*.[152,672] The compounds are accompanied by small amounts of the 1-ethyl homologues and the 3-nor compounds which show only a methyl group in this position.

The structures of the pheromones of the southern corn rootworm, *Diabrotica undecimpunctata*, (*R*)-10-methyl-2-tridecanone,[673,674] and that of *Diabrotica balteata*, (6*R*,10*R*)-6,12-dimethyl-2-pentadecanone[675,676] (**270**), suggest propanoate as the starter, incorporation of a second or third propanoate unit, respectively, and an acetate terminus, according to Scheme 7.

A group of ketones representing the structures of male-produced aggregation pheromones of several scales, *Matsucoccus* spp., carry the carbonyl group more towards the middle of the chain. They show distinct structural relations, and their biosyntheses should be similar.[677] The terminating unit should be propanoate (see Scheme 7).

The pheromone of Israeli pine scale was identified to be (2*E*,6*E*,8*E*)-(5*R*)-5,7-dimethyl-2,6,8-decatrien-4-one (**271**) (Figure 19). The insects produce a (6*E*)/(6*Z*)-mixture of ~75:25;[678] however, (**271**) is much more active than its 6*Z*-isomer. Field tests showed that the compound was highly attractive also to a predator and that nonnatural stereoisomers did not influence trap catches.[679] The pheromone of the maritime pine scale, *Matsucoccus feytaudi*, is (8*E*,10*E*)-(3*S*,7*R*)-3,7,9-trimethyl-8,10-dodecadien-6-one (**272**);[680,681] field tests were successful.[682] The first example of this type of scale pheromone was that of (2*E*,4*E*)-(6*R*,10*R*)-4,6,10,12-tetramethyl-2,4-tridecadien-7-one (matsuone) (**273**), the pheromone of three *Matsucoccus* pine bast scales.[683–685]

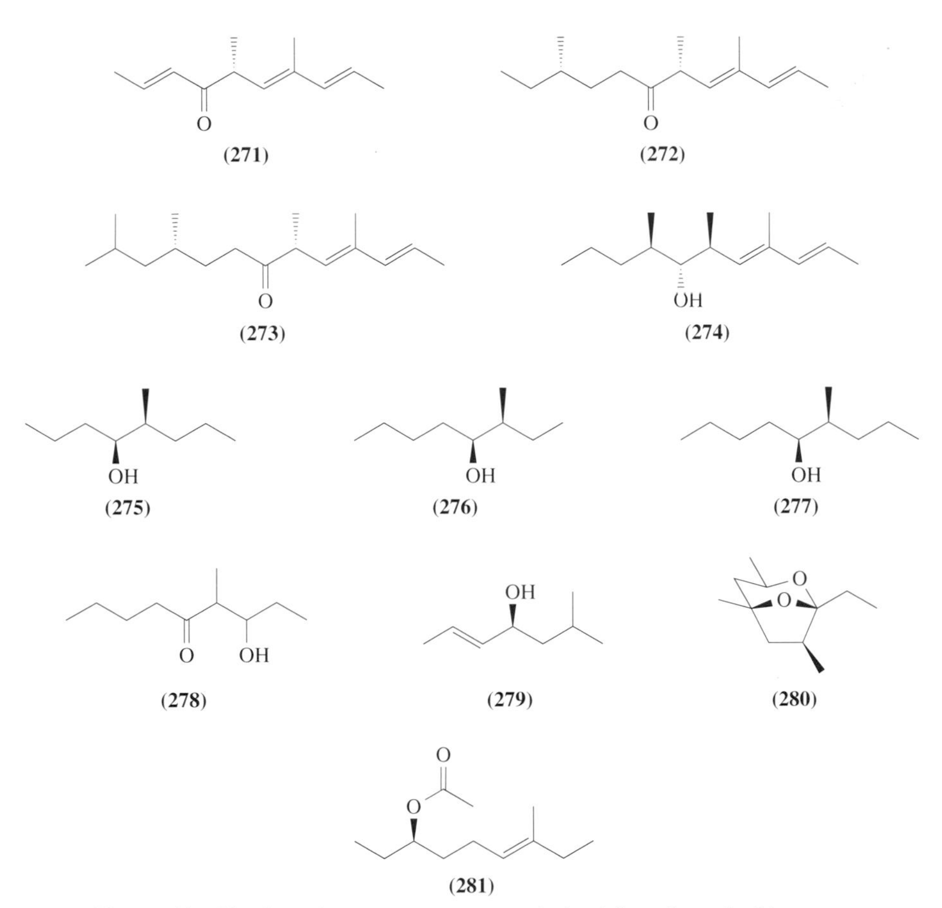

Figure 19 Further pheromone structures derived from branched ketones.

The structure of the pheromone of the wood roach, *Cryptocercus punctulatus*, (7*E*,9*E*)-(4*R*,5*R*,6*S*)-4,6,8-trimethyl-7,9-decadien-5-ol (**274**), closely resembles that of the *Matsucoccus* pheromones.[686,687] The biosyntheses of these compounds may involve acetate and propanoate units (that of (**273**) may start with a valine derivative) and the final steps may be similar to those of the Nitidulid pheromones.[598] The terminating unit should be propanoate and the final coupling should be accompanied by the loss of one carbon, reduction and elimination of water (see Scheme 7).

A series of simple branched secondary alcohols have been identified as male-produced aggregation pheromones of some weevils, *Rhynchophorus* and related species. (4*S*,5*S*)-5-Methyl-4-octanol (cruentol) (**275**) is the pheromone of the palmetto weevil, *Rhynchophorus cruentatus*,[688,689] whereas its isomer, (3*S*,4*S*)-3-methyl-4-octanol (phoenicol) (**276**) is the pheromone of the African palm weevil, *Rhynchophorus phoenicis*.[688,690,691] The homologue of (**276**), (4*S*,5*S*)-4-methyl-5-nonanol (ferruginol) (**277**), was identified in the African palm weevil, *Rhynchophorus ferrugineus*, and some related species.[692–694] Ferruginol is also the pheromone of the sugar cane weevil, *Metamasius hemipterus*, where it is accompanied by 2-methyl-4-heptanol, 2-methyl-4-octanol, 5-nonanol, and 3-hydroxy-4-methyl-5-nonanone (**278**).[695–697] No bioassays with these minor components have been reported. (2*E*)-(4*S*)-Methyl-2-hepten-4-ol (rhynchophorol) (**279**) is the pheromone of the American palm weevil, *Rhynchophorus palmarum*.[698–700] The corresponding epoxide was also found to be present,[701] but no bioassays have been reported.

Similarly to the scales discussed above, these weevils do not seem to be very sensitive to the presence of nonnatural stereomers, since racemic mixtures proved to be active in the field. This greatly facilitates their use in large-scale integrated pest management. Some species also contain ketones, corresponding to the pheromone alcohols; however, they do not show behavioral activity.

A higher degree of oxygenation along the chain is represented in (1*S*,3*R*,5*R*,7*S*)-1-ethyl-3,5,7-trimethyl-2,8-dioxabicyclo[3.2.1]octane, (sordidin) (**280**), the aggregation pheromone of the banana weevil, *Cosmopolites sordidus*.[702–704]

The structure of rhynchophorol is closely related to that of 2-methyl-4-heptanol, a minor component in *Metamasius hemipterus*.[701] The corresponding ketone was already known as an alarm pheromone of *Tapinoma* ants.[705]

An unusual structure is represented by (2*R*,6*E*)-3-acetoxy-7-methyl-6-nonene (quadrilure), (**281**), the pheromone of the square-necked grain beetle, *Cathartus quadricollis*.[706,707] The beetles also contain the alcohol and the ketone, which do not seem to be attractive.

8.04.6.4 The "ante-iso" Type

As mentioned above (see Section 8.04.5) ante-iso branching may be generated in various ways: it may involve isoleucine or homomevalonate or an acetate–propanoate sequence at the start. It certainly may also be generated through degradation. A survey of higher molecular pheromones that play a role in chemical communication reveals a distinct group of compounds which show the same relative configuration at the stereogenic center of the ante-iso branching, some of which have already been discussed earlier in this section and in Section 8.04.5.6 (see structures (**222**)–(**227**) in Figure 14), and it awaits detailed investigations as to whether these compounds represent degraded isoprenoids or polyketids. A typical example that fits both ways is tribolure (**227**), (Figure 20), the sex pheromone of the red flour beetle, *Tribolium castaneum*.[564,565] During bioassays a mixture of the (4*R*,8*R*)- and (4*R*,8*S*)-stereoisomers proved to be more active than the pure (4*R*,8*R*)-enantiomer;[708] the enantiomeric composition of the natural product is unknown, however.

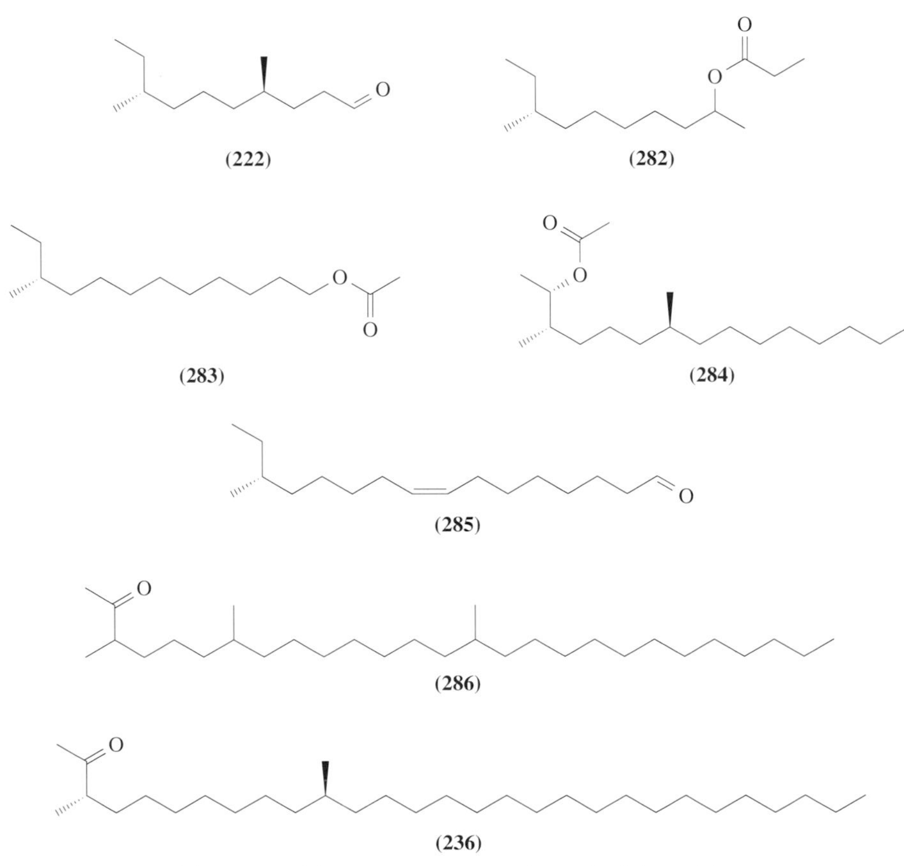

Figure 20 Pheromones showing "ante-iso" branching.

1,7-Dimethylnonyl propanoate (**282**), a female-produced sex pheromone of several corn rootworms, conservatively keeps (*R*)-configuration at the methyl branching whereas the stereochemistry at the oxygen function may vary with species (including mixtures).[709,710] As an exception among Lepidoptera, tea tortrix moths, *Adoxophyes* spp., use a branched acetate, 10-methyldodecyl acetate

(**283**), as the female sex pheromone.[711] It is accompanied by the bishomologue 12-methyltetradecyl acetate[712,713] and showed optimum activity at an (*R*)/(*S*) ratio of 95:5.[714]

Females of pine sawflies (Diprionidae) use the acetate or propanoate of 1,2,6-trimethyldecanol (diprionol) as pheromones, (1*S*,2*S*,6*S*)-diprionyl acetate (**284**) being most widespread.[715] Species recognition is based on stereoisomeric composition, and small amounts of diastereomers of the main component seem to act synergistically or inhibitorily. The European species *Neodiprion sertifer* produces (**284**) along with very small amounts of the (1*S*,2*S*,6*R*)- and (1*S*,2*R*,6*R*)-isomers.[716] A homologue showing an extended chain at the lipophilic side and a chain-shortened component and even a bis-nor structure were also identified in *Neodiprion sertifer*.[717] A stereoisomer of the latter compound, (1*S*,2*R*,6*R*)-1,2,6-trimethyldodecanol, proved to be most important in *Diprion pini*. Esterified as the acetate or propanoate, it proved to be attractive in the field, the propanoate showing higher activity.[718] Also in this species stereoisomers, a nor-product, and higher homologues were detected. In the USA, the related *Diprion similis* uses a stereoisomer of **284**, (1*S*,2*R*,6*R*)-diprionyl acetate, as the pheromone.[719] The complex relationships in the chemistry of sawfly pheromones have been discussed by Norin.[720] Pheromone-related compounds in pupal and adult females of *Neodiprion sertifer* of different ages and in different parts of the body have been investigated.[721] Trimethyltridecan-2-ols have also been identified as pheromones of sawfly species.[722]

14-Methyl-8-hexadecenal (trogodermal) (**285**), the female sex pheromone of khapra beetles, *Trogoderma* spp., constantly shows (*R*)-configuration whereas the geometry of the double bond may vary with species (including mixtures).[723–726]

Among a series of ketones identified from the locust *Schistocerca gregaria*,[727] 3,7-dimethylheptacosan-2-one and 3,7,15-trimethylheptacosan-2-one (**286**) are most closely related to the pheromones of the German cockroach, e.g., (**236**) and 3,11-dimethyl-29-hydroxynonacosan-2-one. The stereochemistry and the biological function of the locust ketones are unknown; however, it may be pointed out that locusts and cockroaches are relatively closely related.

The configuration at the ante-iso branching in the compounds shown in Figure 18 is different from that of isoleucine, which speaks against a biosynthesis involving this amino acid at an early stage. A starting sequence with an acetate–propanoate coupling appears likely. The fact that the structures of some *Diprion* pheromones would fit the configuration of isoleucine and the biogenesis of the cockroach ketones (see above) may suggest, however, incorporation of leucine followed by oxidation and inversion of the α-methyl group.

8.04.6.5 Hydrocarbons

Almost all surfaces of living beings are covered with layers of lipids, among which hydrocarbons play an important role (Figure 21). These fractions may form highly complex mixtures of predominantly straight-chain and methyl- and dimethylalkanes and -alkenes. The majority of the compounds carry 19–39 carbon atoms. Modern analytical techniques greatly facilitate the separation and structural elucidation even of trace amounts. While the branching pattern of methyl-, dimethyl-, or trimethylalkanes can be determined by GC/MS even in nonseparable peaks,[728] assignment of absolute configuration of minute amounts of chiral hydrocarbons remains an unsolved challenge.

Especially in social insects, ants, bees, wasps, and termites, patterns of hydrocarbons are believed to play a major role in nestmate recognition and also in predator–prey interactions and host–parasite relationships. Profiles of cuticular hydrocarbons have been used in studies on kinship and in chemotaxonomy. Whereas biological activities including functions in the context of chemical communication have frequently been described for complex mixtures, relatively few examples have been reported where single compounds or very simple mixtures are used as pheromones. According to their volatility, low-boiling hydrocarbons may act as signals over long distances (sexual attraction) whereas high-boiling hydrocarbons account for close-range recognition including contact pheromones.

Apart from Arctiid[11] and Lymantriid moths,[11,103] some Geometrids and leaf miner moths use branched pheromones as female sex pheromones. Females of the leaf miner, *Lyonetia clerkella*, produce 14-methyl-1-octadecene.[729] Although the racemate was active, biological activity rests with the *S*-enantiomer.[730] A homologue, 10,14-dimethyl-1-octadecene (**287**), has been identified as a pheromone of the related species *Lyonetia prunifoliella*.[731] The pheromone of the coffee leaf miner, *Perileucoptera coffeella*, 5,9-dimethylpentadecane (**288**), shows a similar branching pattern.[732] Its bishomologue, 5,9-dimethylheptadecane, is the pheromone of *Leucoptera malifoliella* (*scitella*).[733] During field tests in Hungary, only the (5*S*,9*S*)-stereoisomer proved to be attractive.[734] In the

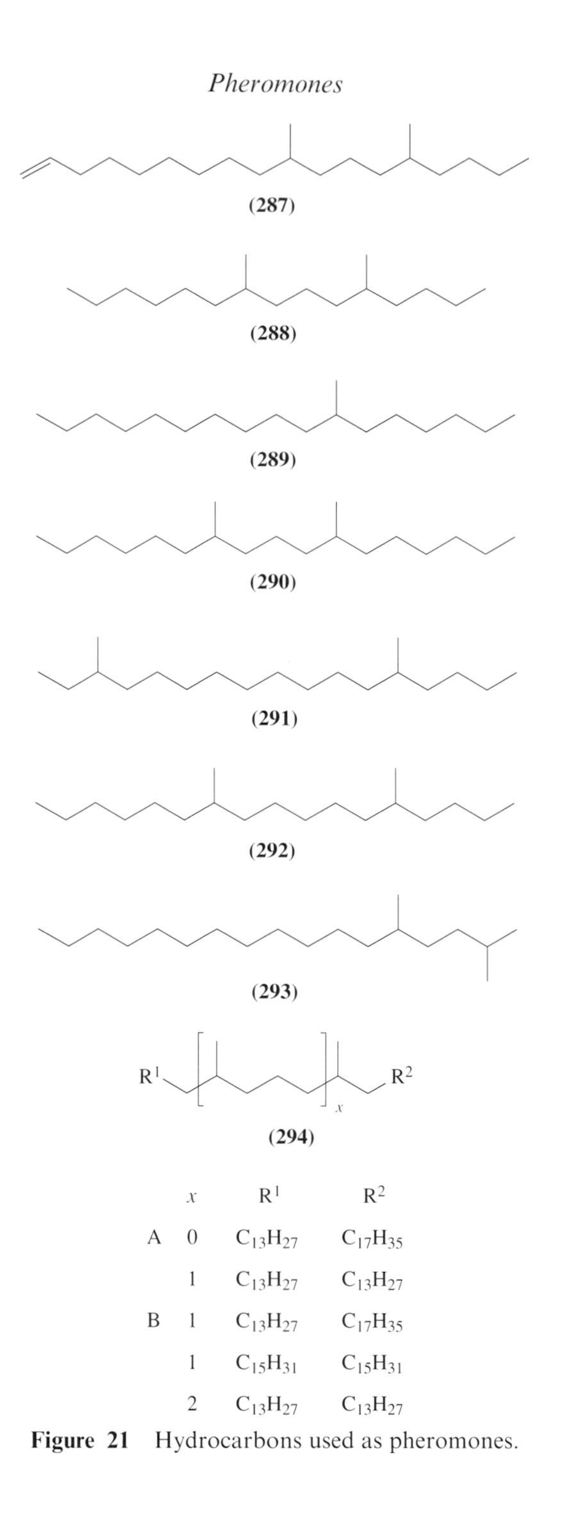

	x	R^1	R^2
A	0	$C_{13}H_{27}$	$C_{17}H_{35}$
	1	$C_{13}H_{27}$	$C_{13}H_{27}$
B	1	$C_{13}H_{27}$	$C_{17}H_{35}$
	1	$C_{15}H_{31}$	$C_{15}H_{31}$
	2	$C_{13}H_{27}$	$C_{13}H_{27}$

Figure 21 Hydrocarbons used as pheromones.

Spanish population of *Leucoptera malifoliella*, 5,9-dimethyloctadecane was identified as a minor component,[735] which is structurally even closer to **(287)**. Some Geometrid moths use similar hydrocarbons as sex pheromones. In the spring hemlock looper, *Lambdina athasaria*, it is a blend of 7-methylheptadecane (**289**) and 7,11-dimethylheptadecane (**290**).[736] The same compounds (enantiomeric composition unknown) are used by the pitch pine looper, *Lambdina pellucidaria*.[737] The western false hemlock looper, *Nepytia freemani*, uses 3,13-dimethylheptadecane (**291**), which according to field tests shows (3*S*,13*R*)-configuration.[738] Although (5*R*,11*S*)-5,11-dimethylheptadecane (**292**) is the major pheromone component of *Lambdina fiscellaria*,[739] its activity is strongly synergized by the minor compounds (*S*)-7-methylheptadecane (**289**) and (*S*)-2,5-dimethylheptadecane (**293**).[740]

As indicated above, stereochemical assignments of all these low-boiling hydrocarbons are exclusively based on behavior assays; whether the natural compounds are indeed represented by pure stereoisomers awaits analytical proof.

Although nothing is known about the biosyntheses of these branched hydrocarbons, it seems likely—specifically with regard to their typical branching pattern—that propanoate units are incorporated. The biosynthesis of (**293**) may involve leucine as a starter, similarly to valine in the case of 2-methylheptadecane, the pheromone of Arctiid moths.[588] According to structural immanence and "conservative branching," compounds (**287**)–(**293**) should show the same absolute configuration at corresponding stereogenic centers (see Figure 20).

In the communication system of *Agromyza frontella*, another dimethylalkane, 3,7-dimethylnonadecane, showing three methylene groups between the branching points, plays a major role.[741]

The same branching pattern occurs in 13,17-dimethylnonatriacontane, a host-seeking kairomone for the wasp, *Trichogramma nubilale*, which is an egg parasite on the corn borer moth, *Ostrinia nubilalis*.[742] It is obvious that such compounds bearing more than 40 carbon atoms act only at close range; they may, however, be transported over longer distances adsorbed on scales. The hydrocarbons known as female contact pheromones of tsetse flies, *Glossina* spp., and stable flies, *Stomoxys* spp., also share a typical 1,5-dimethyl branching. Most important compounds are compiled in Figure 21. The gross structure (**294**), which clearly shows that the central part of the compounds is made up of a "propanoate–acetate sequence," provides an overview which is specified in the accompanying table. Sex stimulants of the stable fly, *Stomoxys calcitrans* are 15-methyltritriacontane and 15,19-dimethyltritriacontane (**294A**).[743,744] The hydrocarbons (**294B**) are sex stimulants of *Glossina morsitants morsitans*, the major vector of the sleeping sickness in East Africa.[745] During field tests with 17,21-dimethyltriacontane, only the *meso*-form was bioactive.[746] Similarly, in 15,23-dimethylpentatriacontane, the sex stimulant of *Glossina pallidipes*,[747] only the *meso*-form proved to be active.[748] Other tsetse fly species show more distant branchings. In both 11,23- and 13,25-dimethylheptatriacontane methyl branching is interrupted by 11 methylene groups,[749] i.e., the propanoate groups are separated by five acetate units.

The occurrence of high-boiling branched hydrocarbons as principal contact pheromones is by no means restricted to flies: (*Z*)-21-methyl-8-pentatriacontene is a contact sex pheromone on the female body surface of the yellow-spotted longhorn beetle, *Psacothea hilaris*.[750]

ACKNOWLEDGMENTS

The authors are indebted to Professor Kenji Mori, Science University of Tokyo, for kindly revising the manuscript, to Mr. Takahiro Onoda, University of Tokyo, for his careful checking of the references, to Mr. Michael Specht, Hamburg University, for the drawing of figures, schemes, and formulas, and to Mrs. Anja Raffay, Hamburg University, for typing the manuscript.

8.04.7 REFERENCES

1. R. T. Cardé and A. K. Minks (eds.), "Insect Pheromone Research. New Directions," Chapman & Hall, New York, 1997.
2. K. C. Spencer (ed.), "Chemical Mediation of Coevolution," Academic Press, San Diego, CA, 1991.
3. G. A. Rosenthal and M. R. Berenbaum (eds.), "Herbivores: Their Interactions with Secondary Metabolites," Academic Press, San Diego, CA, 1991.
4. R. Karban and I. T. Baldwin, "Induced Responses to Herbivory," University of Chicago Press, Chicago, IL, 1997.
5. S. Schulz, *Eur. J. Org. Chem.*, 1998, 13.
6. P. Karlson and M. Lüscher, *Nature (London)*, 1959, **183**, 55.
7. D. A. Nordlund, in "Semiochemicals: Their Role in Pest Control," eds. D. A. Nordlund, R. L. Jones, and W. J. Lewis, Wiley, New York, 1981, pp. 13–28.
8. U. Kohnle and J. P. Vité, *Z. Angew. Entomol*, 1984, **98**, 504.
9. M. K. Stowe, T. C. J. Turlings, J. H. Loughrin, W. J. Lewis, and J. H. Tumlinson, *Proc. Natl. Acad. Sci. USA*, 1995, **92**, 23.
10. M. K. Stowe, J. H. Tumlinson, and R. R. Heath, *Science*, 1987, **236**, 964.
11. H. Arn, M. Tóth, and E. Priesner, "List of Sex Pheromones of Lepidoptera and Related Attractants," 2nd edn., International Organization for Biological Control, West Palearctic Regional Section, Wädenswil, 1992. The Pherolist is available at three Web sites: Cornell University, Geneva, NY, USA, http://www.nysaes.cornell.edu/pheronet/; Institut National de Recherches Agronomiques (INRA), Versailles, France, http://quasimodo. versailles.inra.fr/pherolist/pherolist.htm; and Max-Planck-Institut fur Verhaltensphysiologie, Seewiesen, Germany, http://www.mpi-seewiesen. mpg.de/~kaisslin/pheronet/pherolist.htm.
12. C. G. Butler, R. K. Callow, F. R. S., and N. C. Johnston, *Proc. R. Soc. London, Ser. B*, 1961, **155**, 417.
13. W. Francke, V. Heemann, B. Gerken, J. A. A. Renwick, and J. P. Vité, *Naturwissenschaften*, 1977, **64**, 590.
14. J. A. Byers, G. Birgersson, J. Löfqvist, and G. Bergström, *Naturwissenschaften*, 1988, **75**, 153.
15. R. Kaiser, "The Scent of Orchids," Elsevier, Amsterdam, 1993.
16. J. R. Aldrich, J. E. Oliver, W. R. Lusby, J. P. Kochansky, and M. Borges, *J. Chem. Ecol.*, 1994, **20**, 1103.

17. D. E. Heinz and W. G. Jennings, *J. Food Sci.*, 1966, **31**, 69.
18. D. L. Wood, *Annu. Rev. Entomol.*, 1982, **27**, 411.
19. J. H. Tumlinson, M. G. Klein, R. E. Doolittle, T. L. Ladd, and A. T. Proveaux, *Science*, 1977, **197**, 789.
20. K. N. Slessor, L.-A. Kaminski, G. G. S. King, J. H. Borden, and M. L. Winston, *Nature*, (*London*), 1988, **332**, 354.
21. K. Mori, *Tetrahedron*, 1985, **45**, 3233.
22. K. Mori, *J. Chem. Soc., Chem. Commun.*, 1997, 1153.
23. K. Mori, *Eur. J. Org. Chem.*, 1998, 1479.
24. A. Bakke, P. Frøyen, and L. Skattebøl, *Naturwissenschaften*, 1977, **64**, 98.
25. H. J. Veith, N. Koeniger, and U. Maschwitz, *Naturwissenschaften*, 1984, **71**, 328.
26. M. G. Moshonas, and P. E. Shaw, *J. Agric. Food Chem.*, 1972, **20**, 1029.
27. R. Hänsel, R. Wohlfart, and H. Coper, *Z. Naturforsch., Teil C*, 1980, **35**, 1096.
28. D. A. Carlson, M. S. Mayer, D. L. Silhacek, J. D. James, M. Beroza, and B. A. Bierl, *Science*, 1971, **174**, 76.
29. M. Tóth, G. Szöcz, C. Löfstedt, B. S. Hansson, and M. Subchev, *Entomol. Exp. Appl.*, 1986, **42**, 291.
30. J. M. Brand, H. M. Page, W. A. Lindner, and A. J. Markovetz, *Naturwissenschaften*, 1989, **76**, 277.
31. M. Goodwin, K. M. Gooding, and F. Regnier, *Science*, 1979, **203**, 559.
32. R. Nishida, C. S. Kim, K. Kawai, and H. Fukami, *Chem. Express*, 1990, **5**, 497.
33. M. S. Blum, "Chemical Defenses in Arthropods," Academic Press, New York, 1981.
34. K. Dettner, *Annu. Rev. Entomol.*, 1987, **32**, 17.
35. W. J. Bell and R. T. Cardé (eds.), "Chemical Ecology of Insects," Chapman & Hall, New York, 1984.
36. M. N. Inscoe, in "Insect Suppression with Controlled Release Pheromone Systems," eds. A. F. Kydonieus and M. Beroza, CRC Press, Boca Raton, FL, 1982, vol. 2, p. 201.
37. E. D. Morgan and N. B. Mandava (eds.), "Handbook of Natural Pesticides: Pheromones," CRC Press, Boca Raton, FL, 1985, vol. IV, part A.
38. E. D. Morgan and N. B. Mandava (eds.), "Handbook of Natural Pesticides: Pheromones," CRC Press, Boca Raton, FL, 1985, vol. IV, part B.
39. M. S. Mayer and J. R. McLaughlin, "Handbook of Insect Pheromones and Sex Attractants," CRC Press, Boca Raton, FL, 1991.
40. L. F. Alves, *Progr. Chem. Org. Nat. Prod.*, 1988, **53**, 1.
41. J. M. Pasteels and J.-C. Grégoire, *Annu. Rev. Entomol.*, 1983, **28**, 263.
42. K. Mori, in "The Total Synthesis of Natural Products," ed. J. ApSimon, Wiley, New York, 1981, vol. 4, p. 1.
43. K. Mori, in "The Total Synthesis of Natural Products," ed. J. ApSimon, Wiley, New York, 1992, vol. 9, p. 1.
44. K. Mori, Tetrahedron Report No. 252, *Tetrahedron*, 1989, **45**, 3233.
45. K. Mori, in "Stereocontrolled Organic Syntheses," ed. B. M. Trost, Blackwell, Oxford, 1994.
46. H. E. Hummel and T. A. Miller (eds.), "Techniques in Pheromone Research," Springer, New York, 1984.
47. T. E. Acree and D. M. Soderlund (eds.), "Semiochemistry, Flavours and Pheromones," de Gruyter, Berlin, 1985.
48. J. Pawliszyn, "Solid Phase Micro Extraction. Theory and Practice," Wiley–VCH, New York, 1997.
49. C. Malosse, P. Ramirez-Lucas, D. Rochat, and J. Morin, *J. High Resolut. Chromatogr.*, 1995, **18**, 669.
50. I. Maier, G. Pohnert, S. Pantke-Boecker, and W. Boland, *Naturwissenschaften*, 1996, **83**, 378.
51. A.-K. Borg-Karlson and R. Mozuraitis, *Z. Naturforsch., Teil C*, 1996, **51**, 599.
52. H. Arn, E. Städler, S. Rauscher, *Z. Naturforsch., Teil C*, 1975, **30** 722.
53. A. B. Attygalle and E. D. Morgan, *Angew. Chem.*, 1988, **100**, 475; *Angew. Chem., Int. Ed. Engl.*, 1988, **27**, 460.
54. J. G. Millar and K. Haynes (eds.), "Methods in Chemical Ecology," Chapman & Hall, New York, 1988.
55. W. A. König, "Gas Chromatographic Enantiomer Separation with Modified Cyclodextrins," Hüthig, Heidelberg, 1992.
56. X. Shi, W. S. Leal, and J. Meinwald, *Bioorg Med. Chem.*, 1996, **4**, 297.
57. L. E. L. Rasmussen, T. D. Lee, W. L. Roelofs, A. Zhang, and G. D. Daves, Jr., *Nature* (*London*), 1996, **379**, 684.
58. T. E. Perrin, L. E. L. Rasmussen, R. Gunawardena, and R. A. Rasmussen, *J. Chem. Ecol.*, 1996, **22**, 207.
59. G. W. Kinzer, A. F. Fentiman, Jr., T. F. Page, Jr., R. L. Foltz, J. P. Vité, and G. B. Pitman, *Nature* (*London*), 1969, **221**, 477.
60. D. O'Hagan, "The Polyketide Metabolites," Horwood, Chichester, 1991.
61. L. Crombie, D. O. Morgan, and E. H. Smith, *J. Chem. Soc., Perkin Trans. 1*, 1991, 567.
62. B. A. Vick and D. C. Zimmerman, in "The Biochemistry of Plants," eds. P. K. Stumpf and E. E. Conn, Academic Press, Orlando, FL, 1987, p. 53.
63. M. Luckner, "Secondary Metabolism in Microorganisms, Plants, and Animals," Springer, Berlin, 1990.
64. W. Franke, A. Platzeck, and G. Eichhorn, *Arch. Mikrobiol.*, 1961, **40**, 73.
65. R. C. Lawrence, *J. Gen. Microbiol.*, 1966, **44**, 393.
66. R. Tressl and F. Drawert, *Z. Naturforsch., Teil B*, 1971, **26**, 774.
67. D. R. Nelson and R. J. Blomquist, in "Waxes: Chemistry, Molecular Biology and Functions," ed. R. J. Hamilton, Oily Press, Dundee, 1995, p. 1.
68. C. Frössl and W. Boland, *J. Chem. Soc., Chem. Commun.*, 1991, 1731.
69. M. S. Blum, *Annu. Rev. Entomol.*, 1987, **32**, 381.
70. G. D. Prestwich and G. J. Blomquist, "Pheromone Biochemistry," Academic Press, Orlando, FL, 1987.
71. G. D. Prestwich and M. S. Collins, *J. Chem. Ecol.*, 1982, **8**, 147.
72. E. Engels, W. Engels, G. Lübke, W. Schröder, and W. Francke, *Apidologie*, 1993, **24**, 539.
73. M. Ayasse, W. Engels, A. Hefetz, J. Tengö, G. Lübke, and W. Francke, *Insectes Soc.*, 1993, **40**, 41.
74. R. K. Vander Meer, M. D. Breed, K. E. Espelie, and M. L. Winston, "Pheromone Communication in Social Insects," Westview Press, Boulder, CO, 1998.
75. C. E. Linn, Jr., and W. L. Roelofs, *J. Chem. Ecol.*, 1985, **11**, 1583.
76. J. A. Klun, J. R. Plimmer, B. A. Bierl-Leonhardt, A. N. Sparks, M. Primiani, O. L. Chapman, G. H. Lee, and G. Lepone, *J. Chem. Ecol.*, 1980, **6**, 165.
77. A. K. Raina, and J. A. Kleen, *Science*, 1984, **225**, 531.
78. L. B. Bjostad and W. E. Roelofs, *Science*, 1983, **220**, 1387.
79. W. L. Roelofs and R. A. Jurenka, *Bioorg. Med. Chem.*, 1996, **4**, 461.

80. N. Fang, P. E. A. Teal, and J. H. Tumlinson, *Arch. Insect Biochem. Physiol.*, 1995, **29**, 35.
81. I. Navarro, I. Font, G. Fabriàs, and F. Camps, *J. Am. Chem. Soc.*, 1997, **119**, 11335.
82. W. Boland, C. Frössl, M. Schöttler, and M. Tóth, *J. Chem. Soc., Chem. Commun.*, 1993, 1155.
83. C. Löfstedt and M. Bengtsson, *J. Chem. Ecol.*, 1988, **14**, 903.
84. S. P. Foster and W. L. Roelofs, *Arch. Insect Biochem. Physiol.*, 1988, **8**, 1.
85. G. Arsequell, G. Fabriàs, and F. Camps, *Arch. Insect Biochem. Physiol.*, 1990, **14**, 47.
86. C. Zhao, C. Löfstedt, and X. Wang, *Arch. Insect Biochem. Physiol.*, 1990, **15**, 57.
87. M. de Renobales, C. Cripps, D. W. Stanley-Samuelson, R. A. Jurenka, and G. J. Blomquist, *Trends Biol. Sci.*, 1987, **12**, 364.
88. A. Butenandt, R. Beckmann, D. Stamm, and E. Hecker, *Z. Naturforsch., Teil B*, 1959, **14**, 283.
89. T. Ando, T. Hase, R. Arima, and M. Uchiyama, *Agric. Biol. Chem.*, 1988, **52**, 473.
90. S. P. Foster and W. L. Roelofs, *Experientia*, 1990, **46**, 269.
91. H. J. Bestmann, M. Herrig, and A. B. Attygalle, *Experientia*, 1987, **43**, 1033.
92. J. H. Tumlinson, P. E. A. Teal, and N. Fang, *Bioorg. Med. Chem.*, 1996, **4**, 451.
93. L. M. McDonough, A. L. Averill, H. G. Davis, C. L. Smithhisler, D. A. Murray, P. S. Chapman, S. Voerman, L. J. Dapsis, and M. M. Averill, *J. Chem. Ecol.*, 1994, **20**, 3269.
94. B. F. Nesbitt, P. S. Beevor, A. Cork, D. R. Hall, H. David, and V. Nandagopal, *J. Chem. Ecol.*, 1986, **12**, 1377.
95. R. E. Charlton, J. A. Wyman, J. R. McLaughlin, J.-W. Du, and W. L. Roelofs, *J. Chem. Ecol.*, 1991, **17**, 175.
96. D. R. Hall, P. S. Beevor, D. G. Campion, D. J. Chamberlain, A. Cork, R. D. White, A. Almestar, and T. J. Henneberry, *Tetrahedron Lett.*, 1992, **33**, 4811.
97. G. S. Rule and W. L. Roelofs, *Arch. Insect Biochem. Physiol.*, 1989, **12**, 89.
98. J. W. Wong, P. Palaniswamy, E. W. Underhill, W. F. Steck, and M. D. Chisholm, *J. Chem. Ecol.*, 1984, **10**, 463.
99. G. Szöcs, M. Tóth, W. Francke, F. Schmidt, P. Philipp, W. A. König, K. Mori, B. S. Hansson, and C. Löfstedt, *J. Chem. Ecol.*, 1993, **19**, 2721.
100. J. W. Wong, E. W. Underhill, S. L. MacKenzie, and M. D. Chisholm, *J. Chem. Ecol.*, 1985, **11**, 727.
101. M. Tóth, H. R. Buser, A. Peña, H. Arn, K. Mori, T. Takeuchi, L. N. Nikolaeva, and B. G. Kovalev, *Tetrahedron Lett.*, 1989, **30**, 3405.
102. A. S. Hill, B. G. Kovalev, L. N. Nikolaeva, and W. L. Roelofs, *J. Chem. Ecol.*, 1982, **8**, 383.
103. G. Gries, R. Gries, G. Khaskin, K. N. Slessor, G. G. Grant, J. Liška, and P. Kapitola, *Naturwissenschaften*, 1996, **83**, 382.
104. R. Gries, D. Holden, G. Gries, P. D. C. Wimalaratne, K. N. Slessor, and C. Saunders, *Naturwissenschaften*, 1997, **84**, 219.
105. G. Gries, K. N. Slessor, R. Gries, G. Khaskin, P. D. C. Wimalaratne, T. G. Gray, G. G. Grant, A. S. Tracey, and M. Hulme, *J. Chem. Ecol.*, 1997, **23**, 19.
106. J. Zhu, M. V. Kozlov, P. Philipp, W. Francke, and C. Löfstedt, *J. Chem. Ecol.*, 1995, **21**, 29.
107. M. Tóth, G. Szöcs, E. J. van Nieukerken, P. Philipp, F. Schmidt, and W. Francke, *J. Chem. Ecol.*, 1995, **21**, 13.
108. W. Francke and E. Priesner, unpublished results.
109. B. D. Jackson and E. D. Morgan, *Chemoecology*, 1993, **4**, 125.
110. D. K. Jewett and L. B. Bjostad, *J. Chem. Ecol.*, 1996, **22**, 1331.
111. J. M. Pasteels, D. Daloze, and J. L. Boevé, *J. Chem. Ecol.*, 1989, **15**, 1501.
112. E. B. Jang, D. M. Light, R. G. Binder, R. A. Flath, and L. A. Carvalho, *J. Chem. Ecol.*, 1994, **20**, 9.
113. J. A. Byers, *Experientia*, 1989, **45**, 271.
114. W. Francke, A. R. Levinson, T.-L. Jen, and H. Z. Levinson, *Angew. Chem., Int. Ed. Engl.*, 1979, **18**, 796.
115. S. Schulz and S. Toft, *Science*, 1993, **260**, 1635.
116. J. R. Aldrich, in "Chemical Ecology of Insects II," eds. R. T. Cardé and W. J. Bell, Chapman & Hall, New York, 1995, p. 318.
117. J.-P. Farine, J.-L. Le Queré, J. Duffy, C. Everaerts, and R. Brossut, *J. Chem. Ecol.*, 1994, **20**, 2291.
118. M. Jacobson, C. E. Lilly, and C. Harding, *Science*, 1968, **159**, 208.
119. N. R. Schmuff, J. K. Phillips, W. E. Burkholder, H. M. Fales, C.-W. Chen, P. P. Roller, and M. Ma, *Tetrahedron Lett.*, 1984, **25**, 1533.
120. L. I. Butler, L. M. McDonough, J. A. Onsager, and B. J. Landis, *Environ. Entomol.*, 1975, **4**, 229.
121. R. E. Greenblatt, W. E. Burkholder, J. H. Cross, R. F. Cassidy, Jr., R. M. Silverstein, A. R. Levinson, and H. Z. Levinson, *J. Chem. Ecol.*, 1977, **3**, 337.
122. F. Kern, R. W. Klein, E. Janssen, H.-J. Bestmann, A. B. Attygalle, D. Schafer, and U. Maschwitz, *J. Chem. Ecol.*, 1997, **23**, 779.
123. R. J. Bartelt, A. M. Schaner, and L. L. Jackson, *Physiol. Entomol.*, 1986, **11**, 367.
124. W. S. Leal, A. R. Panizzi, and C. C. Niva, *J. Chem. Ecol.*, 1994, **20**, 1209.
125. J. R. Aldrich, W. R. Lusby, and J. P. Kochansky, *Experientia*, 1986, **42**, 583.
126. J. R. Aldrich, in "Semiochemistry: Flavors and Pheromones," eds. T. E. Acree and D. M. Soderlund, de Gruyter, Berlin, 1985, p. 95.
127. W. S. Leal, H. Higuchi, N. Mizutani, H. Nakamori, T. Kadosawa, and M. Ono, *J. Chem. Ecol.*, 1995, **21**, 973.
128. J. A. A. Renwick, J. P. Vité, and R. F. Billings, *Naturwissenschaften*, 1977, **64**, 226.
129. J. E. Girard, F. J. Germino, J. P. Budris, R. A. Vita, and M. P. Garrity, *J. Chem. Ecol.*, 1979, **5**, 125.
130. J. W. Grula, J. D. McChesney, and O. R. Taylor, Jr., *J. Chem. Ecol.*, 1980, **6**, 241.
131. U. Ravid, R. M. Silverstein, and L. R. Smith, *Tetrahedron*, 1978, **34**, 1449.
132. F. Schröder, R. Fettköther, U. Noldt, K. Dettner, W. A. König, and W. Francke, *Liebigs Ann. Chem.*, 1994, 1211.
133. W. S. Leal, X. Shi, K. Nakamuta, M. Ono, and J. Meinwald, *Proc. Natl. Acad. Sci. USA*, 1995, **92**, 1038.
134. R. A. A. Renwick and G. B. Pitman, *Environ. Entomol.*, 1979, **8**, 40.
135. J. A. A. Renwick, P. R. Hughes, and J. P. Vité, *J. Insect Physiol.*, 1975, **21**, 1097.
136. C. Löfstedt, B. S. Hansson, E. Petersson, P. Valeur, and A. Richards, *J. Chem. Ecol.*, 1994, **20**, 153.
137. A. Vallet, P. Cassier, and Y. Lensky, *J. Insect Physiol.*, 1991, **37**, 789.
138. B. S. Fletcher and T. E. Bellas, in "Handbook of Natural Pesticides: Pheromones," eds. E. D. Morgan and N. B. Mandava, CRC Press, Boca Raton, 1985, vol. IV, part B, p. 207.

139. J. P. Vité, G. B. Pitman, A. F. Fentiman, Jr., and G. W. Kinzer, *Naturwissenschaften*, 1972, **59**, 469.
140. E. L. Plummer, T. E. Stewart, K. Byrne, G. T. Pearce, and R. M. Selverstein, *J. Chem. Ecol.*, 1977, **2**, 307.
141. L. M. Libbey, A. C. Oehlschlager, and L. C. Ryker, *J. Chem. Ecol.*, 1983, **9**, 1533.
142. W. S. Leal, Y. Ueda, and M. Ono, *J. Chem. Ecol.*, 1996, **22**, 1429.
143. N. E. Gunawardena and M. K. Bandumathie, *J. Chem. Ecol.*, 1993, **19**, 851.
144. M.-C. Cammaerts and K. Mori, *Physiol. Entomol.*, 1987, **12**, 381.
145. J. M. Brand, *J. Chem. Ecol.*, 1985, **11**, 177.
146. J. M. Brand and V. Pretorius, *Biochem. Syst. Ecol.*, 1986, **14**, 341.
147. A. C. Oehlschlager, A. M. Pierce, H. D. Pierce, Jr., and J. H. Borden, *J. Chem. Ecol.*, 1988, **14**, 2071.
148. M. Wurzenberger and W. Grosch, *Biochim. Biophys. Acta*, 1984, **795**, 163.
149. K. Iwabuchi, J. Takahashi, and T. Sakai, *Appl. Entomol. Zool.*, 1987, 22, 110.
150. T. Sakai, Y. Nakagawa, J. Takahashi, K. Iwabuchi, and K. Ishii, *Chem. Lett.*, 1984, 263.
151. I. Kubo, T. Matsumoto, D. L. Wagner, and J. N. Shoolery, *Tetrahedron Lett.*, 1985, **26**, 563.
152. S. Schulz, W. Francke, W. A. König, V. Schurig, K. Mori, R. Kittmann, and D. Schneider, *J. Chem. Ecol.*, 1990, **16**, 3511.
153. W. Francke, W. Mackenroth, W. Schröder, S. Schulz, J. Tengö, E. Engels, W. Engels, R. Kittmann, and D. Schneider, *Z. Naturforsch., Teil C*, 1985, **40**, 145.
154. J. M. Brand, R. M. Duffield, J. G. MacConnell, M. S. Blum, and H. M. Fales, *Science*, 1973, **179**, 388.
155. Y. Kuwahara, in "Modern Acarology," eds. F. Dusbabek and V. Bukva, Academia, Prague, 1991, p. 43.
156. S. P. Schmidt and R. E. Monroe, *Insect Biochem.*, 1976, **6**, 377.
157. T. Nemoto, Y. Kuwahara, and T. Suzuki, *Appl. Entomol. Zool.*, 1990, **25**, 261.
158. B. Torto, P. G. N. Njagi, A. Hassanali, and H. Amiani, *J. Chem. Ecol.*, 1996, **22**, 2273.
159. R. Schöni, E. Hess, W. Blum, and K. Ramstein, *J. Insect Physiol.*, 1984, **30**, 613.
160. J. L. Nation, *Environ. Entomol.*, 1975, **4**, 27.
161. W. S. Leal, M. Hasegawa, M. Sawada, M. Ono, and S. Tada, *J. Chem. Ecol.*, 1996, **22**, 2001.
162. Y. Kuwahara, *Appl. Entomol. Zool.*, 1980, **15**, 478.
163. P. Ramirez-Lucas, C. Malosse, P.-H. Ducrot, M. Lettere, and P. Zagatti, *Bioorg. Med. Chem.*, 1996, **4**, 323.
164. W. Francke, G. Hindorf, and W. Reith, *Naturwissenschaften*, 1979, **66**, 618.
165. W. Francke, J. Bartels, H. Meyer, F. Schröder, U. Kohnle, E. Baader, and J. P. Vité, *J. Chem. Ecol.*, 1995, **21**, 1043.
166. W. Francke, V. Heemann, B. Gerken, J. A. A. Renwick, and J. P. Vité, *Naturwissenschaften*, 1977, **64**, 590.
167. R. Baker, R. Herbert, P. E. Howse, O. T. Jones, W. Francke, and W. Reith, *J. Chem. Soc., Chem. Commun.*, 1980, 52.
168. G. Haniotakis, W. Francke, K. Mori, H. Redlich, and V. Schurig, *J. Chem. Ecol.*, 1986, **12**, 1559.
169. J. G. Pomonis and B. E. Mazomenos, *Int. J. Invertebr. Reprod. Dev.*, 1986, **10**, 169.
170. J. H. Borden, in "Comprehensive Insect Physiology, Biochemistry, and Pharmacology," eds. G. A. Kerkut and L. I. Gilbert, Pergamon, New York, 1985, vol. 6, p. 257.
171. D. Vanderwel and A. C. Oehlschlager, *J. Am. Chem. Soc.*, 1992, **114**, 5081.
172. W. Francke, J. Bergmann, and C. Löfstedt, unpublished results.
173. A. D. Camacho, H. D. Pierce, Jr., and J. H. Borden, *J. Chem. Ecol.*, 1994, **20**, 111.
174. M. V. Novotny, T. M. Xie, S. Harvey, D. Wiesler, B. Jemiolo, and M. Carmack, *Experientia*, 1995, **51**, 738.
175. W. Francke, F. Schröder, P. Philipp, H. Meyer, V. Sinnwell, and G. Gries, *Bioorg. Med. Chem.*, 1996, **4**, 363.
176. M. T. Fletcher, and W. Kitching, *Chem. Rev.*, 1995, **95**, 789.
177. W. Francke, W. Reith, G. Bergström, and J. Tengö, *Z. Naturforsch., Teil C*, 1981, **36c**, 928.
178. C. R. Teerling, H. D. Pierce, Jr., J. H. Borden, and D. R. Gillespie, *J. Chem. Ecol.*, 1993, **19**, 681.
179. C. Pavis, C. Malosse, P. H. Ducrot, and C. Descoins, *J. Chem. Ecol.*, 1994, **20**, 2213.
180. H. Fukui, F. Matsumura, M. C. Ma, and W. E. Burkholder, *Tetrahedron Lett.*, 1974, 3563.
181. H. Sugie, M. Yoshida, K. Kawasaki, H. Noguchi, S. Moriya, K. Takagi, H. Fukuda, A. Fujiie, M. Yamanaka, Y. Ohira, T. Tsutsumi, K. Tsuda, K. Fukumoto, M. Yamashita, and H. Suzuki, *Appl. Entomol. Zool.*, 1996, **31**, 427.
182. J. A. Byers, G. Birgersson, J. Löfqvist, and G. Bergström, *Naturwissenschaften*, 1988, **75**, 153.
183. J. R. Rocca, J. H. Tumlinson, B. M. Glancey, and C. S. Lofgren, *Tetrahedron Lett.*, 1983, **24**, 1889.
184. D. A. Shearer, R. Boch, R. A. Morse, and F. M. Laigo, *J. Insect Physiol.*, 1970, **16**, 1437.
185. K. N. Slessor, L.-A. Kaminski, G. G. S. King, and M. L. Winston, *J. Chem. Ecol.*, 1990, **16**, 851.
186. E. Plettner, K. N. Slessor, M. L. Winston, and J. E. Oliver, *Science*, 1996, **271**, 1851.
187. B. P. Moore and W. V. Brown, *Aust. J. Chem.*, 1976, **29**, 1365.
188. H. J. Veith, J. Weiss, and N. Koeniger, *Experientia*, 1978, **34**, 423.
189. H. J. Bestmann, F. Kern, D. Schäfer, and M. C. Witschel, *Angew. Chem., Int. Ed. Engl.*, 1992, **31**, 795.
190. R. Nishida, H. Fukami, T. C. Baker, W. L. Roelofs, and T. E. Acree, in "Proceedings of the Amercian Chemical Society Symposium, Semiochemistry: Flavors and Pheromones, Washington, DC, 1983," ed. T. E. Acree and D. M. Soderlund, de Gruyter, Berlin, 1985, p. 47.
191. G. Kunesch, P. Zagatti, A. Pouvreau, and R. Cassini, *Z. Naturforsch., Teil C*, 1987, **42**, 657.
192. F. Walter, D. J. C. Fletcher, D. Chautems, D. Cherix, L. Keller, W. Francke, W. Fortelius, R. Rosengren, and E. L. Vargo, *Naturwissenschaften*, 1993, **80**, 30.
193. N. Mori, Y. Kuwahara, K. Kurosa, R. Nishida, and T. Fukushima, *Appl. Entomol. Zool.*, 1995, **30**, 415.
194. W. Francke, W. Reith, G. Bergström, and J. Tengö, *Naturwissenschaften*, 1980, **67**, 149.
195. Y. Kuwahara and S. Nakamura, *Appl. Entomol. Zool.*, 1985, **20**, 354.
196. R. K. Saini, A. Hassanali, J. Andoke, P. Ahuya, and W. P. Ouma, *J. Chem. Ecol.*, 1996, **22**, 1211.
197. D. E. Finnegan and J. Chambers, *J. Chem. Ecol.*, 1993, **19**, 971.
198. R. R. Heath, J. A. Coffelt, P. E. Sonnet, F. I. Proshold, B. Dueben, and J. H. Tumlinson, *J. Chem. Ecol.*, 1986, **12**, 1489.
199. A. Tai, F. Matsumura, and H. C. Coppel, *J. Org. Chem.*, 1969, **34**, 2180.
200. M. Tokoro, M. Takahashi, K. Tsunoda, R. Yamaoka, and K. Hayashiya, *Wood Res.*, 1991, **78**, 1.
201. C. Bordereau, A. Robert, O. Bonnard, and J. L. Le Quere, *J. Chem. Ecol.*, 1991, **17**, 2177.
202. W. S. Leal, *Naturwissenschaften*, 1991, **78**, 521.
203. W. S. Leal, M. Hasegawa, M. Sawada, M. Ono, and Y. Ueda, *J. Chem. Ecol.*, 1994, **20**, 1643.

204. R. Nishida, S. Schulz, C. S. Kim, H. Fukami, Y. Kuwahara, K. Honda, and N. Hayashi, *J. Chem. Ecol.*, 1996, **22**, 949.
205. A. C. Oehlschlager, G. G. S. King, H. D. Pierce, Jr., A. M. Pierce, K. N. Slessor, J. G. Millar, and J. H. Borden, *J. Chem. Ecol.*, 1987, **13**, 1543.
206. D. Vanderwel, H. D. Pierce, Jr., A. C. Oehlschlager, J. H. Borden, and A. M. Pierce, *Insect Biochem.*, 1990, **20**, 567.
207. T. C. Baker, R. Nishida, and W. L. Roelofs, *Science*, 1981, **214**, 1359.
208. R. Nishida, T. C. Baker, and W. L. Roelofs, *J. Chem. Ecol.*, 1982, **8**, 947.
209. K. Mori, M. Amaike, and H. Watanabe, *Liebigs Ann. Chem.*, 1993, 1287.
210. C. Antony, T. L. Davis, D. A. Carlson, J.-M. Pechine, and J.-M. Jallon, *J. Chem. Ecol.*, 1985, **11**, 1617.
211. R. J. Bartelt, A. M. Schaner, and L. L. Jackson, *J. Chem. Ecol.*, 1989, **15**, 399.
212. S. P. Foster, M. O. Harris, and J. G. Millar, *Naturwissenschaften*, 1991, **78**, 130.
213. J. L. Clément, H. Lloyd, P. Nagnan, and M. S. Blum, *Sociobiology*, 1989, **15**, 19.
214. A. Zhang, H. T. Facundo, P. S. Robbins, C. E. Linn, Jr., J. L. Hanula, M. G. Villani, and W. L. Roelofs, *J. Chem. Ecol.*, 1994, **20**, 2415.
215. J. H. Tumlinson, M. G. Klein, R. E. Doolittle, T. L. Ladd, and A. T. Proveaux, *Science*, 1977, **197**, 789.
216. W. S. Leal, *Proc. Natl. Acad. Sci. USA*, 1996, **93**, 12112.
217. Y. Tamaki, H. Sugie, and H. Noguchi, *Appl. Entomol. Zool.*, 1985, **20**, 359.
218. H. Fukui, F. Matsumura, A. V. Barak, and W. E. Burkholder, *J. Chem. Ecol.*, 1977, **3**, 539.
219. R. M. Silverstein, J. O. Rodin, W. E. Burkholder, and J. E. Gorman, *Science*, 1967, **157**, 85.
220. D. F. Horler, *J. Chem. Soc. C*, 1970, 859.
221. Y. Kuwahara, W. S. Leal, Y. Nakano, Y. Kaneko, H. Nakao, and T. Suzuki, *Appl. Entomol. Zool.*, 1989, **24**, 424.
222. A. M. Schaner, L. D. Tanico-Hogan, and L. L. Jackson, *J. Chem. Ecol.*, 1989, **15**, 2577.
223. R. A. Moats, R. J. Bartelt, L. L. Jackson, and A. M. Schaner, *J. Chem. Ecol.*, 1987, **13**, 451.
224. E. Thibout, S. Ferary, and J. Auger, *J. Chem. Ecol.*, 1994, **20**, 1571.
225. Y. Le Conte, G. Arnold, J. Trouiller, and C. Masson, *Naturwissenschaften*, 1990, **77**, 334.
226. Y. Kainoh, T. Nemoto, K. Shimizu, S. Tstsuki, T. Kusano, and Y. Kuwahara, *Appl. Entomol. Zool.*, 1991, **26**, 543.
227. T. Ikan, R. Gottlieb, E. D. Bergmann, and J. Ishay, *J. Insect Physiol.*, 1969, **15**, 1709.
228. W. S. Leal, S. Kuwahara, M. Ono, and S. Kubota, *Bioorg. Med. Chem.*, 1996, **4**, 315.
229. B. R. Laurence, K. Mori, T. Otsuka, J. A. Pickett, and L. J. Wadhams, *J. Chem. Ecol.*, 1985, **11**, 643.
230. B. Ernst and B. Wagner, *Helv. Chim. Acta*, 1989, **72**, 165.
231. J. G. Kostelc, J. E. Girard, and L. B. Hendry, *J. Chem. Ecol.*, 1980, **6**, 1.
232. A. M. Schaner and L. L. Jackson, *J. Chem. Ecol.*, 1992, **18**, 53.
233. Y. Kuwahara, M. Ohshima, M. Sato, K. Kurosa, S. Matsuyama, and T. Suzuki, *Appl. Entomol. Zool.*, 1995, **30**, 177.
234. D. Thiéry, B. Gabel, P. Farkas, and M. Jarry, *J. Chem. Ecol.*, 1995, **21**, 2015.
235. Y.-S. Hwang, G. W. Schultz, and M. S. Mulla, *J. Chem. Ecol.*, 1984, **10**, 145.
236. N. Koeniger, and H. J. Veith, *Experientia*, 1983, **39**, 1051.
237. L. L. Jackson, M. T. Arnold, and G. J. Blomquist, *Insect Biochem.*, 1981, **11**, 87.
238. A. M. Schaner, R. J. Bartelt, and L. L. Jackson, *J. Chem. Ecol.*, 1987, **13**, 1777.
239. K. Hedlund, R. J. Bartelt, M. Dicke, and L. E. M. Vet, *J. Chem. Ecol.*, 1996, **22**, 1835.
240. P. Kalo, *J. Chromatogr.*, 1985, **323**, 343.
241. M. Miyakado, J. Meinwald, and L. E. Gilbert, *Experientia*, 1989, **45**, 1006.
242. B. H. Smith, R. G. Carlson, and J. Frazier, *J. Chem. Ecol.*, 1985, **11**, 1447.
243. J. A. Pickett, I. H. Williams, and A. P. Martin, *J. Chem. Ecol.*, 1982, **8**, 163.
244. J. O. Schmidt, C. A. McDaniel, and R. T. S. Thomas, *J. Chem. Ecol.*, 1990, **16**, 2135.
245. A. M. Schaner, K. J. Graham, and L. L. Jackson, *J. Chem. Ecol.*, 1989, **15**, 1045.
246. K. Peschke, *J. Chem. Ecol.*, 1987, **13**, 1993.
247. G. J. Blomquist, J. W. Dillwith, and J. G. Pomonis, *Insect Biochem.*, 1984, **14**, 279.
248. H. T. Bolton, J. F. Butler, and D. A. Carlson, *J. Chem. Ecol.*, 1980, **6**, 951.
249. T. Nemoto, M. Doi, K. Oshio, H. Matsubayashi, Y. Oguma, T. Suzuki, and Y. Kuwahara, *J. Chem. Ecol.*, 1994, **20**, 3029.
250. T. C. Syvertsen, L. L. Jackson, G. J. Blomquist, and S. B. Vinson, *J. Chem. Ecol.*, 1995, **21**, 1971.
251. R. J. Bartelt, R. L. Jones, and H. M. Kulman, *J. Chem. Ecol.*, 1982, **8**, 95.
252. R. J. Bartelt, R. L. Jones, and T. P. Krick, *J. Chem. Ecol.*, 1983, **9**, 1343.
253. K. Mori and N. P. Argade, *Liebigs Ann. Chem.*, 1994, 695.
254. J. W. Porter and S. L. Spurgeon (eds.), "Biosynthesis of Isoprenoid Compounds," Wiley, New York, 1981, vol. 1.
255. T. Bach, *Lipids*, 1995, **30**, 191.
256. M. Rohmer, M. Seemann, S. Horbach, S. Bringer-Meyer, and H. Sahm, *J. Am. Chem. Soc.*, 1996, **118**, 2564.
257. J. G. Zeidler, H. K. Lichtenthaler, H. U. May, and F. W. Lichtenthaler, *Z. Naturforsch., Teil C*, 1997, **52**, 15, and references cited therein.
258. W. Eisenreich, S. Sagner, M. H. Zenk, and A. Bacher, *Tetrahedron Lett.*, 1997, **38**, 3889.
259. G. Flesch and M. Rohmer, *FEBS Lett.*, 1998, **405**, 175.
260. H. K. Lichtenthaler, J. Schwender, A. Disch, and M. Rohmer, *FEBS Lett.*, 1997, **400**, 271.
261. J. Piel, J. Donath, K. Bandemer, and W. Boland, *Angew. Chem., Int. Ed. Engl.*, 1998, **37**.
262. G. M. Happ and J. Meinwald, *J. Am. Chem. Soc.*, 1965, **87**, 2507.
263. J. Meinwald, G. M. Happ, J. Labows, and T. Eisner, *Science*, 1966, **151**, 79.
264. N. Mitlin and P. A. Hedin, *J. Insect Physiol.*, 1974, **20**, 1825.
265. H. J. Bestmann, J. Erler, O. Vostrowsky, and L. Wasserthal, *Z. Naturforsch., Teil C*, 1993, **48**, 510.
266. S. J. Seybold, D. R. Quilici, J. A. Tillman, D. Vanderwel, D. L. Wood, and G. J. Blomquist, *Proc. Natl. Acad. Sci. USA*, 1995, **92**, 8393.
267. J. M. Brand, J. W. Bracke, L. N. Britton, A. J. Markovetz, and S. J. Barras, *J. Chem. Ecol.*, 1976, **2**, 195.
268. A. Leufvén, G. Bergström, and E. Falsen, *J. Chem. Ecol.*, 1984, **10**, 1349.
269. B. S. Davidson, T. Eisner, B. Witz, and J. Meinwald, *J. Chem. Ecol.*, 1989, **15**, 1689.
270. A. Brundin, G. Andersson, K. Andersson, T. Mossing, and L. Källquist, *J. Chem. Ecol.*, 1978, **4**, 613.
271. B. S. Lanne, P. Ivarsson, P. Johnsson, G. Bergström, and A.-B. Wassgren, *Insect Biochem.* 1989, **19**, 163.

272. J. T. Stoakley, A. Bakke, J. A. A. Renwick, and J. P. Vité, *J. Appl. Entomol.*, 1978, **86**, 174.
273. B. V. Burger and P. J. Pretorius, *Z. Naturforsch., Teil C*, 1988, **42**, 1355.
274. R. M. Duffield, W. E. LaBerge, J. H. Cane, and J. H. Wheeler, *J. Chem. Ecol.*, 1982, **8**, 535.
275. M. Ayasse, W. Engels, A. Hefetz, G. Lübke, and W. Francke, *Z. Naturforsch., Teil C*, 1990, **45**, 709.
276. J. W. Wheeler, M. T. Shamin, P. Brown, and R. M. Duffield, *Tetrahedron Lett.*, 1983, **24**, 5811.
277. N. Koeniger, J. Weiss, and U. Maschwitz, *J. Insect Physiol.* 1979, **25**, 467.
278. P. J. Landolt, R. R. Heath, H. C. Reed, and K. Manning, *Fla. Entomol.*, 1995, **78**, 101.
279. W. Francke, V. Heemann, and K. Heyns, *Z. Naturforsch., Teil C*, 1974, **29**, 243.
280. T. Payne, personal communication.
281. N. W. Davies and J. L. Madden, *J. Chem. Ecol.*, 1985, **11**, 1115.
282. H. Schildknecht, I. Wilz, F. Enzmann, N. Grund, and M. Ziegler, *Angew. Chem.*, 1976, **88**, 228; *Angew. Chem., Int. Ed. Engl.*, 1976, **15**, 242.
283. D. R. Crump, *J. Chem. Ecol.*, 1980, **6**, 837.
284. H. Schildknecht and C. Birkner, *Chem. Ztg.*, 1983, **107**, 267.
285. W. F. Wood, C. O. Fisher, and G. A. Graham, *J. Chem. Ecol.*, 1993, **19**, 837.
286. J. W. Jorgenson, M. Novotny, M. Carmack, G. B. Copland, S. R. Wilson, S. Katona, and W. K. Whitten, *Science*, 1978, **199**, 796.
287. B. V. Burger, P. J. Pretorius, J. Stander, and G. R. Grierson, *Z. Naturforsch., Teil C*, 1988, **43**, 731.
288. B. V. Burger, P. J. Pretorius, H. S. C. Spies, R. C. Bigalke, and G. R. Grierson, *J. Chem. Ecol.*, 1990, **16** 397.
289. M. Novotny, S. Harvey, B. Jemiolo, and J. Alberts, *Proc. Natl. Acad. Sci. USA*, 1985, **82**, 2059.
290. A. Zhang, P. S. Robbins, W. S. Leal, C. E. Linn, Jr., M. G. Villani, and W. L. Roelofs, *J. Chem. Ecol.*, 1997, **23**, 231.
291. W. S. Leal, S. Matsuyama, Y. Kuwahara, S. Wakamura, and M. Hasegawa, *Naturwissenschaften*, 1992, **79**, 184.
292. J. G. James, C. J. Moore, J. R. Aldrich, *J. Chem. Ecol.*, 1994, **20**, 3281.
293. J. H. Tumlinson, R. M. Silverstein, J. C. Moser, R. G. Brownlee, and J. M. Ruth, *Nature (London)*, 1971, **234**, 348.
294. W. S. Leal, J. M. S. Bento, E. F. Vilela, and T. M. C. Della Lucia, *Experientia*, 1994, **50**, 853.
295. T. Yasuda, S. Yoshii, and S. Wakamura, *Appl. Entomol. Zool.*, 1994, **29**, 21.
296. M. Hilker and S. Schulz, *J. Chem. Ecol.*, 1994, **20**, 1075.
297. B. P. Moore, W. V. Brown, *J. Aust. Entomol. Soc.*, 1985, **24**, 81.
298. W. Francke and J. P. Vité, *J. Appl. Entomol.*, 1983, **96**, 146.
299. J. A. Byers, *Experientia*, 1989, **45**, 271.
300. S. R. Leather, *Entomol. Exp. Appl.*, 1987, **43**, 295.
301. J. A. Pickett and D. C. Griffiths, *J. Chem. Ecol.*, 1980, **6**, 349.
302. B. S. Lanne, F. Schlyter, J. A. Byers, J. Löfqvist, A. Leufvén, G. Bergström, J. N. C. van der Pers, R. Unelius, P. Baeckström, and T. Norin, *J. Chem. Ecol.*, 1987, **13**, 1045.
303. J. W. Wheeler and R. M. Duffield, in "CRC Handbook of Natural Pesticides," eds. E. D. Morgan and N. B. Mandava, CRC Press, Boca Raton, FL, 1988, part B, p. 59.
304. C. Gnanasunderam, H. Young, and R. F. N. Hutchins, *J. Chem. Ecol.*, 1981, **7**, 889.
305. N. Hayashi, Y. Kuwahara, and H. Komae, *Experientia*, 1978, **34**, 684.
306. R. Baker and S. Walmsley, *Tetrahedron*, 1982, **38**, 1899.
307. G. D. Prestwich, *Tetrahedron*, 1982, **38**, 1911.
308. M. Lindström, T. Norin, I. Valterova, and J. Vrkoc, *Naturwissenschaften*, 1990, **77**, 134.
309. I. Valterova, J. Vrkoc, M. Lindström, and T. Norin, *Naturwissenschaften*, 1992, **79**, 416.
310. C. Everaerts, O. Bonnard, J. M. Pasteels, Y. Roisin, and W. A. König, *Experientia*, 1990, **46**, 227.
311. M. Kaib and H. Dittebrand, *Chemoecology*, 1990, **1**, 3.
312. B. S. Krall, B. W. Zilkowski, S. L. Knight, R. J. Bartelt, and D. W. Whitman, *J. Chem. Ecol.*, 1997, **23**, 1951.
313. S. Regev and W. W. Cone, *Environ. Entomol.*, 1980, **9**, 50.
314. Y. Kuwahara, S. Ishii, and H. Fukami, *Experientia*, 1975, **31**, 1115.
315. W. S. Leal, Y. Kuwahara, and T. Suzuki, *Agric. Biol. Chem.*, 1988, **52**, 1299.
316. R. J. Anderson, M. J. Gieselmann, H. R. Chinn, K. G. Adams, C. A. Henrick, R. E. Rice, and W. L. Roelofs, *J. Chem. Ecol.*, 1981, **7**, 695.
317. W. Francke, J. Bartels, S. Krohn, S. Schulz, E. Baader, J. Tengö, and D. Schneider, *Pure Appl. Chem.*, 1989, **61**, 539.
318. A. B. Attygalle, S. Steghaus-Kovac, V. U. Ahmad, U. Maschwitz, O. Vostrowsky, and H. J. Bestmann, *Naturwissenschaften*, 1991, **78**, 90.
319. S. Schulz and R. Nishida, *Bioorg. Med. Chem.*, 1996, **4**, 341.
320. W. Francke, W. Schröder, A.-K. Borg-Karlsson, G. Bergström, and J. Tengö, *Z. Naturforsch., Teil C*, 1987, **42**, 169.
321. E. B. Jang, D. M. Light, R. G. Binder, R. A. Flath, and L. A. Carvalho, *J. Chem. Ecol.*, 1994, **20**, 9.
322. A. K. Borg-Karlson, L. Ågren, H. Dobson, and G. Bergström, *Experientia*, 1988, **44**, 531.
323. I. Kudryavtsev, K. Siirde, K. Läätz, V. Ismailov, and V. Pristavko, *J. Chem. Ecol.*, 1993, **19**, 1607.
324. A. Hefetz, S. W. T. Batra, and M. S. Blum, *Experientia*, 1979, **35**, 319.
325. J. H. Cane and J. O. Tengö, *J. Chem. Ecol.*, 1981, **7**, 427.
326. D. C. Robacker and L. B. Hendry, *J. Chem. Ecol.*, 1977, **3**, 563.
327. W. S. Leal, M. Sawada, S. Matsuyama, Y. Kuwahara, and M. Hasegawa, *J. Chem. Ecol.*, 1993, **19**, 1381.
328. M. S. Blum, R. M. Crewe, W. E. Kerr, L. H. Keith, A. W. Garrison, and M. M. Walker, *J. Insect Physiol.*, 1970, **16**, 1637.
329. S. B. Vinson, H. J. Williams, G. W. Frankie, J. W. Wheeler, M. S. Blum, and R. E. Coville, *J. Chem. Ecol.*, 1982, **8**, 319.
330. H. Schildknecht, *Angew. Chem.*, 1976, **88**, 235; *Angew. Chem., Int. Ed. Engl.*, 1976, **15**, 214.
331. G. Bergström and L. Lundgren, *Zoon, Suppl.*, 1973, **1**, 67.
332. Y. Kuwahara, T. Sato, and T. Suzuki, *Appl. Entomol. Zool.*, 1991, **26**, 501.
333. J. A. Pickett, I. H. Williams, A. P. Martin, and M. C. Smith, *J. Chem. Ecol.*, 1980, **6**, 425.
334. J. A. A. Renwick, P. R. Hughes, and T. DeJ TY., *J. Insect Physiol.*, 1973, **19**, 1735.
335. J. A. A. Renwick, P. R. Hughes, and I. S. Krull, *Science*, 1976, **191**, 199.
336. P. R. Hughes, *J. Insect Physiol.*, 1974, **20**, 1271.

337. H. D. Pierce, Jr., J. E. Conn., A. C. Oehlschlager, and J. H. Borden, *J. Chem. Ecol.*, 1987, **13**, 1455.
338. W. Francke and V. Heemann, *Z. Angew. Entomol.*, 1976, **82**, 117.
339. J. P. Vité, G. Ohloff, and R. F. Billings, *Nature*, 1978, **272**, 817.
340. U. Kohnle, J. P. Vité, C. Erbacher, J. Bartels, and W. Francke, *Entomol. Exp. Appl.*, 1988, **49**, 43.
341. W. Francke, M.-L. Pan, J. Bartels, W. A. König, J. P. Vité, S. Krawielitzki, and U. Kohnle, *J. Appl. Entomol.*, 1986, **101**, 453.
342. R. H. Fish, L. E. Browne, and B. J. Bergot, *J. Chem. Ecol.*, 1984, **10**, 1057.
343. R. M. Silverstein, J. O. Rodin, and D. L. Wood, *Science*, 1966, **154**, 509.
344. C. M. Harring, *Z. Angew. Entomol.*, 1978, **85**, 281.
345. J. E. Macías-Sámano, J. H. Borden, H. D. Pierce, Jr., R. Gries, and G. Gries, *J. Chem. Ecol.*, 1997, **23**, 1333.
346. D. Klimetzek, J. Köhler, S. Krohn, and W. Francke, *J. Appl. Entomol.*, 1989, **107**, 304.
347. S. J. Seybold, *J. Chem. Ecol.*, 1993, **19**, 1809.
348. D. R. Miller, J. H. Borden, and K. N. Slessor, *J. Chem. Ecol.*, 1996, **22**, 2157.
349. D. R. Miller, J. H. Borden, and K. N. Slessor, *J. Chem. Ecol.*, 1989, **15**, 233.
350. S. J. Seybold, T. Ohtsuka, D. L. Wood, and I. Kubo, *J. Chem. Ecol.*, 1995, **21**, 995.
351. Y. R. Naves, *Helv. Chim. Acta*, 1948, **31**, 29.
352. W. Francke, J. Bartels, H. Meyer, F. Schröder, U. Kohnle, E. Baader, and J. P. Vité, *J. Chem. Ecol.*, 1995, **21**, 1043.
353. W. M. Whitten, H. G. Hills, and N. H. Williams, *Phytochemistry*, 1988, **27**, 2759.
354. R. M. Silverstein, J. O. Rodin, D. L. Wood, and L. E. Browne, *Tetrahedron*, 1966, **22**, 1929.
355. M. v. Schantz, K. G. Widén, and R. Hiltunen, *Acta Chem. Scand.*, 1973, **27**, 551.
356. J. A. Byers, F. Schlyter, G. Birgersson, and W. Francke, *Experientia*, 1990, **46**, 1209.
357. R. Granger, J. Parset, and J. P. Girard, *Phytochemistry*, 1972, **11**, 2301.
358. P. Ivarsson and G. Birgersson, *J. Insect Physiol.*, 1995, **41**, 843.
359. G. Gries, J. H. Borden, R. Gries, J. P. Lafontaine, E. A. Dixon, H. Wieser, and A. T. Whitehead, *Naturwissenschaften*, 1992, **79**, 367.
360. D. Klimetzek and W. Francke, *Experientia*, 1980, **36**, 1343.
361. J. A. A. Renwick and J. P. Vité, *Nature (London)*, 1969, **224**, 1222.
362. C. M. Naumann, P. Ockenfels, J. Schmitz, F. Schmidt, and W. Francke, *Entomol. Gen.*, 1991, **15**, 255.
363. J. C. Grégoire, D. Couillin, R. Krebber, W. A. Koenig, H. Meyer, and W. Francke, *Chemoecology*, 1992, **3**, 14.
364. J. H. Tumlinson, R. C. Gueldner, D. D. Hardee, A. C. Thompson, P. A. Hedin, and J. P. Minyard, *J. Org. Chem.*, 1971, **36**, 2616.
365. J. C. Dickens and K. Mori, *J. Chem. Ecol.*, 1989, **15**, 517.
366. P. A. Hedin, D. A. Dollar, J. K. Collins, J. G. Dubois, P. G. Mulder, G. H. Hedger, M. W. Smith, and R. D. Eikenbary, *J. Chem. Ecol.*, 1997, **23**, 965.
367. F. J. Eller and R. J. Bartelt, *J. Nat. Prod.*, 1996, **59**, 451.
368. P. A. Hedin, V. A. Phillips, and R. J. Dysart, *J. Miss. Acad. Sci.*, 1988, **33**, 59.
369. B. E. Hibbard and F. X. Webster, *J. Chem. Ecol.*, 1993, **19**, 2129.
370. F. Bohlmann, C. Zdero, and U. Faass, *Chem. Ber.*, 1973, **106**, 2904.
371. J. H. Borden, J. R. Handley, B. D. Johnston, J. G. MacConnell, R. M. Silverstein, K. N. Slessor, A. A. Swigar, and D. T. W. Wong, *J. Chem. Ecol.*, 1979, **5**, 681.
372. V. Schurig, R. Weber, D. Klimetzek, U. Kohnle, and K. Mori, *Naturwissenschaften*, 1982, **69**, 602.
373. D. Klimetzek, J. P. Vité, and K. Mori, *Z. Angew. Entomol.*, 1980, **89**, 57.
374. B. A. Bierl-Leonhardt, D. S. Moreno, M. Schwarz, J. Fargerlund, and J. A. Plimmer, *Tetrahedron Lett.*, 1981, **22**, 389.
375. J. R. Aldrich, W. R. Lusby, and J. P. Kochansky, *Experientia*, 1986, **42**, 583.
376. U. Kohnle, W. Francke, and A. Bakke, *Z. Angew. Entomol.* 1985, **100**, 5.
377. Y. Kuwahara, A. Akimoto, W. S. Leal, H. Nakao, and T. Suzuki, *Agric. Biol. Chem.*, 1987, **51**, 3441.
378. J. Meinwald, W. T. Thompson, and T. Eisner, *Tetrahedron Lett.*, 1971, 3485.
379. J. R. Aldrich, J. E. Oliver, G. K. Waite, C. Moore, and R. M. Waters, *J. Chem. Ecol.*, 1996, **22**, 729.
380. K. Tanaka, K. Ohsawa, H. Honda, and I. Yamamoto, *J. Pestic. Sci.*, 1981, **6**, 75.
381. M. Jefson, J. Meinwald, S. Nowicki, K. Hicks, and T. Eisner, *J. Chem. Ecol.*, 1983, **9**, 159.
382. J. W. Wheeler, J. Avery, F. Birmingham, and R. M. Duffield, *Insect. Biochem.*, 1984, **14**, 391.
383. S. Schulz, J. Gross, and M. Hilker, *Tetrahedron*, 1997, **53**, 9203.
384. W. S. Leal, Y. Kuwahara, T. Suzuki, Y. Nakano, and H. Nakao, *Agric. Biol. Chem.*, 1989, **53**, 295.
385. N. Mori, Y. Kuwahara, and K. Kurosa, *Bioorg. Med. Chem.*, 1996, **4**, 289.
386. W. S. Leal, Y. Kuwahara, and T. Suzuki, *Agric. Biol. Chem.*, 1989, **53**, 875.
387. W. S. Leal, Y. Kuwahara, Y. Nakano, H. Nakao, and T. Suzuki, *Agric. Biol. Chem.*, 1989, **53**, 1193.
388. F. O. Ayorinde, J. W. Wheeler, and R. M. Duffield, *Tetrahedron Lett.*, 1984, **25**, 3525.
389. W. S. Leal, Y. Kuwahara, T. Suzuki, and K. Kurosa, *Naturwissenschaften*, 1989, **76**, 332.
390. T. O. Olagbemiro and B. W. Staddon, *J. Chem. Ecol.*, 1983, **9**, 1397.
391. P. Baeckström, G. Bergström, F. Björkling, H. Hui-Zhu, H.-E. Högberg, U. Jacobsson, G.-Q. Lin, J. Löfqvist, T. Norin, and A.-B. Wassgren, *J. Chem. Ecol.*, 1989, **15**, 61.
392. W. Francke, G. Bergström, and J. Tengö, unpublished results.
393. G. Vidari, M. de Bernardi, M. Pavan, and L. Ragozzino, *Tetrahedron Lett.*, 1973, 4065.
394. S. Schulz, W. Francke, J. Edgar, and D. Schneider, *Z. Naturforsch., Teil C*, 1988, **43**, 99.
395. R. Bernardi, C. Cardani, D. Ghiringhelli, A. Selva, M. Baggini, and A. Pavan, *Tetrahedron Lett.*, 1967, 3893.
396. G. Ahlgren, G. Bergström, J. Löfqvist, A. Jansson, and T. Norin, *J. Chem. Ecol.*, 1979, **5**, 309.
397. S. Schulz, M. Steffensky, and Y. Rosin, *Liebigs Ann. Chem.*, 1996, 941.
398. G. Kunesch, P. Zagatti, J. Y. Lallemand, A. Debal, and J. P. Vigneron, *Tetrahedron Lett.*, 1981, **22**, 5271.
399. J. P. Vigneron, R. Méric, M. Larchevêque, A. Debal, J. Y. Lallemand, G. Kunesch, P. Zagatti, and M. Gallois, *Tetrahedron*, 1984, **40**, 3521.
400. T. Uematsu, T. Umemura, and K. Mori, *Agric. Biol. Chem.*, 1983, **47**, 597.
401. G. Büchi, E. st. Kovats, P. Enggist, and G. Uhde, *J. Org. Chem.*, 1968, **33**, 1227.
402. G. W. K. Cavill, in "Cyclopentanoid Terpene Derivatives," eds. W. I. Taylor and A. R. Battersby, Dekker, New York, 1969, p. 203.

403. R. M. Smith, J. J. Brophy, G. W. K. Cavill, and N. W. Davies, *J. Chem. Ecol.*, 1979, **5**, 727.
404. M. Rowell-Rahier and J. M. Pasteels, *J. Chem. Ecol.*, 1986, **12**, 1189.
405. U. M. Pagnoni, A. Pinetti, R. Trave, and L. Garanti, *Aust. J. Chem.*, 1976, **29**, 1375.
406. S. M. McElvain, P. M. Walters, and R. D. Bright, *J. Am. Chem. Soc.*, 1942, **64**, 1828.
407. G. W. Dawson, D. C. Griffiths, N. F. Jones, A. Mudd, J. A. Pickett, L. J. Wadhams, and C. M. Woodcock, *Nature (London)*, 1987, **325**, 614.
408. R. Lilley, J. Hardie, L. A. Merritt, J. A. Pickett, L. J. Wadhams, and C. M. Woocock, *Chemoecology*, 1995, **5/6**, 43.
409. B. J. Gabrýs, H. J. Gadomski, Z. Klukowski, J. A. Pickett, G. T. Sobota, L. J. Wadhams, and C. M. Woodcock, *J. Chem. Ecol.*, 1997, **23**, 1881.
410. J. A. Pickett, L. J. Wadhams, C. M. Woodcock, and J. Hardie, *Annu. Rev. Entomol.*, 1992, **37**, 67.
411. G. W. Dawson, J. A. Pickett, and D. W. M. Smiley, *Bioorg. Med. Chem.*, 1996, **4**, 351.
412. J. Hardie, L. Peace, J. A. Pickett, D. W. M. Smiley, J. R. Storer, and L. J. Wadhams, *J. Chem. Ecol.*, 1997, **23**, 2547.
413. T. E. Bellas, W. V. Brown, and B. P. Moore, *J. Insect Physiol.*, 1974, **20**, 277.
414. J. Smolanoff, A. F. Kluge, J. Meinwald, A. McPhail, R. W. Miller, K. Hicks, and T. Eisner, *Science*, 1975, **188**, 734.
415. R. Baker, R. H. Herbert, and G. G. Grant, *J. Chem. Soc., Chem. Commun.*, 1985, 824.
416. O. R. W. Sutherland, C. H. Wearing, and R. F. N. Hutchins, *J. Chem. Ecol.*, 1977, **3**, 625.
417. R. K. Vander Meer, F. Alvarez, and C. S. Lofgren, *J. Chem. Ecol.*, 1988, **14**, 825.
418. G. Yarden, A. Shani, and W. S. Leal, *Bioorg. Med. Chem.*, 1996, **4**, 283.
419. W. S. Bowers, L. R. Nault, R. E. Webb, and S. R. Dutky, *Science*, 1972, **177**, 1121.
420. B. V. Burger, M. le Roux, H. S. C. Spies, V. Truter, and R. C. Bigalke, *Z. Naturforsch., Teil C*, 1980, **36**, 340.
421. S. Schulz, W. Francke, W. A. König, V. Schurig, K. Mori, R. Kittmann, and D. Schneider, *J. Chem. Ecol.*, 1990, **16**, 3511.
422. M. Novotny, S. Harvey, and B. Jemiolo, *Experientia*, 1990, **46**, 109.
423. G. Bergström and J. Tengö, *Chem. Scr.*, 1974, **5**, 28.
424. J. Tengö and G. Bergström, *J. Kans. Entomol. Soc.*, 1978, **51**, 521.
425. D. H. Calam, *Nature (London)*, 1969, **221**, 856.
426. B. G. Svensson and G. Bergström, *Insectes Soc.*, 1977, **24**, 213.
427. J. Tengö and G. Bergström, *J. Chem. Ecol.*, 1975, **1**, 253.
428. A. Fernandes, R. M. Duffield, J. W. Wheeler, and W. E. LaBerge, *J. Chem. Ecol.*, 1981, **7**, 453.
429. H. Schildknecht, personal communication.
430. B. V. Burger, Z. Munro, M. Röth, H. S. C. Spies, V. Truter, G. D. Tribe, and R. M. Crewe, *Z. Naturforsch., Teil C*, 1983, **38**, 848.
431. S. Regev and W. W. Cone, *Environ. Entomol.*, 1975, **4**, 307.
432. B. Kullenberg, G. Bergström, and S. Ställberg-Stenhagen, *Acta Chem. Scand.*, 1970, **24**, 1481.
433. G. Bergström and B. Svensson, *Zoon, Suppl.*, 1973, **1**, 61.
434. G. Bergström and J. Löfqvist, *J. Insect Physiol.*, 1970, **16**, 2353.
435. J. W. Wheeler, L. E. Rasmussen, F. Ayorinde, I. O. Buss, and G. L. Smuts, *J. Chem. Ecol.*, 1982, **8**, 821.
436. J. Borges del Castillo, C. J. W. Brooks, and M. M. Campbell, *Tetrahedron Lett.*, 1966, 3731.
437. R. Baker, M. Edwards, D. A. Evans, and S. Walmsley, *J. Chem. Ecol.*, 1981, **7**, 127.
438. P. Zagatti, G. Kunesch, F. Ramiandrasoa, C. Malosse, D. R. Hall, R. Lester, and B. F. Nesbitt, *J. Chem. Ecol.*, 1987, **13**, 1561.
439. W. S. Leal, Y. Kuwahara, T. Suzuki, and K. Kurosa, *Agric. Biol. Chem.*, 1989, **53**, 2703.
440. W. Francke, S. Krohn, and J. Tengö, *J. Chem. Ecol.*, 1991, **17**, 557.
441. C. Nishino, W. Bowers, M. E. Montgomery, L. R. Nault, and M. W. Nielson, *J. Chem. Ecol.*, 1977, **3**, 349.
442. A. J. Weinheimer, W. W, Youngblood, P. H. Washecheck, T. K. B. Karns, and L. S. Ciereszko, *Tetrahedron Lett.*, 1970, 497.
443. R. Baker, H. R. Coles, M. Edwards, D. A. Evans, P. E. Howse, and S. Walmsley, *J. Chem. Ecol.*, 1981, **7**, 135.
444. K. Honda, *J. Chem. Ecol.*, 1981, **7**, 1089.
445. C. J. Persoons, P. E. J. Verwiel, F. J. Ritter, E. Talman, P. J. F. Nooijen, and W. J. Nooijen, *Tetrahedron Lett.*, 1976, 2055.
446. M. A. Adams, K. Nakanishi, W. C. Still, E. V. Arnold, J. Clardy, and C. J. Persoons, *J. Am. Chem. Soc.*, 1979, **101**, 2495.
447. H. Hauptmann, G. Mühlbauer, and H. Sass, *Tetrahedron Lett.*, 1986, **27**, 6189.
448. S. Kuwahara and K. Mori, *Tetrahedron*, 1990, **46**, 8083.
449. M. Biendl, H. Hauptmann, and H. Sass, *Tetrahedron Lett.*, 1989, **30**, 2367.
450. C. J. Persoons, F. J. Ritter, P. E. J. Verwiel, H. Hauptmann, and K. Mori, *Tetrahedron Lett.*, 1990, **31**, 1747.
451. S. Takahashi, K. Watanabe, S. Saito, and Y. Nomura, *Appl. Entomol. Zool.*, 1995, **30**, 357.
452. H. Y. Ho, H. T. Young, R. Kou, and Y. S. Chow, *Bull. Inst. Zool. Acad. Sin.*, 1992, **31**, 225 (*Chem. Abstr.*, 118-07-056530).
453. J. J. Howard, T. P. Green, and D. F. Wiemer, *J. Chem. Ecol.*, 1989, **15**, 2279.
454. G. T. Pearce, W. E. Gore, R. M. Silverstein, J. W. Peacock, R. A. Cuthbert, G. N. Lanier, and J. B. Simeone, *J. Chem. Ecol.*, 1975, **1**, 115.
455. G. D. Prestwich, *Biochem. Syst. Ecol.*, 1979, **7**, 211.
456. L. Lundgren and G. Bergström, *J. Chem. Ecol.*, 1975, **1**, 399.
457. R. Baker, M. Borges, N. G. Cooke, and R. H. Herbert, *J. Chem. Soc., Chem. Commun.*, 1987, 414.
458. J. R. Aldrich, W. R. Lusby, B. E. Marron, K. C. Nicolaou, M. P. Hoffmann, and L. T. Wilson, *Naturwissenschaften*, 1989, **76**, 173.
459. G. K. Waite and W. R. Lusby, *Z. Naturforsch., Teil C*, 1993, **48**, 73.
460. M. A. Ryan, C. J. Moore, and G. H. Walter, *Comp. Biochem. Physiol. B*, 1995, **111**, 189.
461. P. Brézot, C. Malosse, K. Mori, and M. Renou, *J. Chem. Ecol.*, 1994, **20**, 3133.
462. A. Quilico, F. Piozzi, and M. Pavan, *Tetrahedron*, 1957, **1**, 177.
463. R. Baker, P. H. Briner, and D. A. Evan, *J. Chem. Soc., Chem. Commun.*, 1978, 981.
464. B. Maurer and G. Ohloff, *Helv. Chim. Acta*, 1976, **59**, 1169.

465. D. A. Evans, R. Baker, and P. E. Howse, in "Chemical Ecology: Odor Communication in Animals," ed. F. Ritter, Elsevier/North-Holland, Amsterdam, 1979, p. 213.
466. R. Baker, A. H. Parton, and P. E. Howse, *Experientia*, 1982, **38**, 297.
467. G. Bergström and J. Löfqvist, *J. Insect Physiol.*, 1973, **19**, 877.
468. G. Bergström, P. Bergman, M. Appelgren, and J. O. Schmidt, *Bioorg. Med. Chem.*, 1996, **4**, 515.
469. H. J. Bestmann, E. Janssen, F. Kern, and B. Liepold, *Naturwissenschaften*, 1995, **82**, 334.
470. G. H. McKibben, M. J. Thompson, W. L. Parrot, A. C. Thompson, and W. R. Lusby, *J. Chem. Ecol.*, 1985, **11**, 1229.
471. M. Ayasse, T. Marlovits, J. Tengö, T. Taghizadeh, and W. Francke, *Apidologie*, 1995, **26**, 163.
472. A. Hefetz, T. Taghizadeh, and W. Francke, *Z. Naturforsch., Teil C*, 1996, **51**, 409.
473. P. L. Phelan, P. J. Silk, C. J. Northcott, S. H. Tan, and T. C. Baker, *J. Chem. Ecol.*, 1986, **12**, 135.
474. J. S. Waterhouse, J. Ke, J. A. Pickett, and P. J. Weldon, *J. Chem. Ecol.*, 1996, **22**, 1307.
475. S. Cerrini, D. Lamba, I. Valterová, M. Budesinský, J. Vrkoč, and F. Tureček, *Collect. Czech Chem. Commun.*, 1987, **52**, 707.
476. I. Valterová, M. Budesinky, J. Vrkoč, and G. D. Prestwich, *Collect. Czech Chem. Commun.*, 1990, **55**, 1580.
477. S. H. Goh, C. H. Chuah, J. Vadiveloo, and Y. P. Tho, *J. Chem. Ecol.*, 1990, **16**, 619.
478. P. G. McDowell and G. W. Oloo, *J. Chem. Ecol.*, 1984, **10**, 835.
479. G. D. Prestwich, D. F. Wiemer, J. Meinwald, and J. Clardy, *J. Am. Chem. Soc.*, 1978, **100**, 2560.
480. A. J. Birch, W. V. Brown, J. E. T. Corrie, and B. P. Moore, *J. Chem. Soc., Perkin Trans. 1*, 1972, 2653.
481. J. P. Edwards and J. Chambers, *J. Chem. Ecol.* 1984, **10**, 1731.
482. A. B. Attygalle, S. R. Smedley, T. Eisner, and J. Meinwald, *Experientia*, 1996, **52**, 616.
483. J. G. C. Hamilton, D. E. Sonenshine, and W. R. Lusby, *J. Insect Physiol.*, 1989, **35**, 873.
484. D. E. Sonenshine, *Annu. Rev. Entomol.*, 1985, **30**, 1.
485. H. Sobbhy, M. G. Aggour, D. E. Sonenshine, and M. J. Burridge, *Exp. Appl. Acarol.*, 1994, **18**, 265.
486. T. D. Fitzgerald and F. X. Webster, *J. Chem. Ecol.*, 1993, **71**, 1511.
487. T. D. Fitzgerald, *J. Chem. Ecol.*, 1993, **19**, 449.
488. M. Sakuma and H. Fukami, *J. Chem. Ecol.*, 1993, **19**, 2521.
489. K. Mori, K. Fukamatsu, and M. Kido, *Liebigs Ann. Chem.*, 1993, 665.
490. P. W. Sorensen, T. J. Hara, and N. E. Stacey, *J. Comp. Physiol. A*, 1987, **160**, 305.
491. M. V. Novotny, T.-M. Xie, S. Harvey, D. Wiesler, B. Jemiolo, and M. Carmak, *Experientia*, 1995, **51**, 738.
492. H. Schildknecht, H. Birringer, and U. Maschwitz, *Angew. Chem., Int. Ed. Engl.*, 1967, **6**, 558.
493. H. Schildknecht and D. Holz, *Angew. Chem., Int. Ed. Engl.*, 1967, **6**, 881.
494. H. Schildknecht, D. Holz, and U. Maschwitz, *Z. Naturforsch., Teil B*, 1967, **22**, 938.
495. H. Schildknecht, R. Siewerdt, and U. Maschwitz, *Liebigs Ann. Chem.*, 1967, **703**, 182.
496. H. Schildknecht and W. Körnig, *Angew. Chem., Int. Ed. Engl.*, 1968, **7**, 62.
497. H. Schildknecht, H. Birringer, and D. Krauss, *Z. Naturforsch., Teil B*, 1969, **24**, 38.
498. V. Prelog and L. Ruzicka, *Helv. Chim. Acta*, 1944, **27**, 61.
499. B. W. L. Brooksbank, R. Brown, and J.-A. Gustafsson, *Experientia*, 1974, **30**, 864.
500. B. W. L. Brooksbank and G. A. D. Haslewood, *Biochem. J.*, 1961, **80**, 488.
501. D. B. Gower, *J. Steroid Biochem.*, 1972, **3**, 45.
502. T. K. Kwan, D. J. H. Trafford, H. L. J. Makin, and D. B. Gower, *Biochem. Soc. Trans.*, 1989, **17**, 749.
503. R. E. Maiworm and W. U. Langthaler, in "Chemical Signals in Vertebrates 6," eds. R. L. Doty and D. Mueller-Schwarze, Plenum, New York, 1992, p. 575.
504. J. J. Cowley, A. L. Johnson, and B. W. L. Brooksbank, *Psychoneuroendocrinology*, 1977, **2**, 159.
505. M. Kirk-Smith, D. A. Booth, D. Carroll, and P. Davies, *Res. Commun. Psychol. Psychiatr. Behav.*, 1978, **3**, 379.
506. D. Benton, *Biol. Psychol.*, 1982, **15**, 249.
507. S. L. Black and C. Biron, *Behav. Neural Biol.*, 1982, **34**, 326.
508. J. N. Labows, K. J. McGinley, and A. M. Kligman, *J. Soc. Cosmet. Chem.*, 1982, **33**, 193.
509. P. Rennie, K. T. Holland, A. J. Mallet, W. J. Watkins, and D. B. Gower, in "Chemical Signals in Vertebrates 5," eds. D. W. McDonald, D. Mueller-Schwarze, and S. E. Natynczuk, Oxford University Press, Oxford, 1990, p. 55.
510. F. Ayorinde, J. W. Wheeler, C. Wemmer, and J. Murtaugh, *J. Chem. Ecol.*, 1982, **8**, 177.
511. R. Claus and H. O. Hoppen, *Experientia*, 1979, **35**, 1674.
512. A. Gäbler, W. Boland, U. Preiss, and H. Simon, *Helv. Chim. Acta*, 1991, **74**, 1773.
513. D. Klimetzek, J. Bartels, and W. Francke, *J. Appl. Entomol.*, 1989, **107**, 518.
514. K. Dettner and G. Schwinger, *Experientia*, 1987, **43**, 458.
515. P. Ivarsson, B.-I. Henrikson, and A. E. Stenson, *Chemoecology*, 1996, **7**, 191.
516. M. C. Birch and A. Hefetz, *Bull. Entomol. Soc. Am.*, 1987, **33**, 222.
517. B. S. Goodrich, E. R. Hesterman, K. E. Murray, R. Mykytowycz, G. Stanley, and G. Sugowdz, *J. Chem. Ecol.*, 1978, **4**, 581.
518. R. A. Allan, M. A. Elgar, and R. J. Capon, *Proc. R. Soc. London, Ser. B*, 1996, **263**, 69.
519. S. G. Micha, J. Stammel, and C. Hoeller, *Eur. J. Entomol.*, 1993, **90**, 439.
520. A. Quiroz, J. Pettersson, J. A. Pickett, L. J. Wadhams, and H. M. Niemeyer, *J. Chem. Ecol.*, 1997, **23**, 2599.
521. D. Wittmann, R. Radtke, J. Zeil, G. Lübke, and W. Francke, *J. Chem. Ecol.*, 1990, **16**, 631.
522. N. J. Oldham, E. D. Morgan, B. Gobin, E. Schoeters, and J. Billen, *J. Chem. Ecol.*, 1994, **20**, 3297.
523. J. H. Borden, L. Chong, J. A. McLean, K. N. Slessor, and K. Mori, *Science*, 1976, **192**, 894.
524. J. H. Borden and J. A. McLean, *J. Chem. Ecol.*, 1979, **5**, 79.
525. W. Francke, M. L. Pan, W. A. König, K. Mori, P. Puapoomchareon, H. Heuer, and J. P. Vité, *Naturwissenschaften*, 1987, **74**, 343.
526. G. Birgersson, G. L. Debarr, P. De Groot, M. J. Dalusky, H. D. Pierce, Jr., J. H. Borden, H. Meyer, W. Francke, L. E. Espelie, and C. W. Berisford, *J. Chem. Ecol.*, 1995, **21**, 143.
527. H. D. Pierce, Jr., P. De Groot, J. H. Borden, S. Ramaswamy, and A. C. Oehlschlager, *J. Chem. Ecol.*, 1995, **21**, 169.
528. A. L. Perez, R. Gries, G. Gries, and A. C. Oehlschlager, *Bioorg. Med. Chem.*, 1996, **4**, 445.
529. J. Wu and K. Mori, *Agric. Biol. Chem.*, 1991, **55**, 2667.
530. S. De Rosa, A. Milone, A. Crispino, A. Jaklin, and A. De Giulio, *J. Nat. Prod.*, 1997, **60**, 462.

531. A. Hefetz, G. C. Eickwort, M. S. Blum, J. Cane, and G. E. Bohart, *J. Chem. Ecol.*, 1982, **8**, 1389.
532. S. Schulz, M. Boppré, and R. I Vane-Wright, *Philos. Trans. R. Soc. London, Ser. B*, 1993, **342**, 161.
533. B. V. Burger, W. M. Mackenroth, D. Smith, H. S. C. Spies, and P. R. Atkinson, *Z. Naturforsch., Teil C*, 1985, **40**, 847.
534. D. R. Hall, A. Cork, R. Lester, B. F. Nesbitt, and P. Zagatti, *J. Chem. Ecol.*, 1987, **13**, 1575.
535. K. Mori, H. Harada, P. Zagatti, A. Cork, and D. R. Hall, *Liebigs Ann. Chem.*, 1991, 259.
536. H. Schildknecht, H. Neumaier, and B. Tauscher, *Liebigs Ann. Chem.*, 1972, **756**, 155.
537. J. Meinwald, K. Opheim, and T. Eisner, *Proc. Natl. Acad. Sci. USA*, 1972, **69**, 1208.
538. J. Wheeler, S. Oh, E. F. Benfield, and S. E. Neff, *J. Am. Chem. Soc.*, 1972, **94**, 7589.
539. J. R. Miller, L. B. Hendry, and R. O. Mumma, *J. Chem. Ecol.*, 1975, **1**, 59.
540. T. Eisner, L. B. Hendry, D. B. Peakall, and J. Meinwald, *Science*, 1971, **172**, 277.
541. G. Birgersson, F. Schlyter, J. Löfqvist, and G. Bergström, *J. Chem. Ecol.*, 1984, **10**, 1029.
542. S. A. Teale, F. X. Webster, A. Zhang, and G. N. Lanier, *J. Chem. Ecol.*, 1991, **17**, 1159.
543. E. S. Albone, *Nature (London)*, 1975, **256**, 575.
544. J. R. Rocca, J. H. Tumlinson, B. M. Glancey, and C. S. Lofgren, *Tetrahedron Lett.*, 1983, **24**, 1889.
545. J. Meinwald, K. Erickson, M. Hartshorn, Y. C. Meinwald, and T. Eisner, *Tetrahedron Lett.*, 1968, 2959.
546. W. Francke, S. Schulz, V. Sinnwell, W. A. König, and Y. Roisin, *Liebigs Ann. Chem.* 1989, 1195.
547. M. Winter, K.-H. Schulte-Elte, A. Velluz, J. Limacher, W. Pickenhagen, and G. Ohloff, *Helv. Chim. Acta*, 1979, **62**, 131.
548. J. R. Sierra, W. D. Woggon, and H. Schmid, *Experientia*, 1976, **32**, 142.
549. J. Meinwald, paper presented at the 32nd IUPAC Congress, Stockholm, 2–7 August 1989.
550. M. Frenzel, K. Dettner, D. Wirth, J. Waibel, and W. Boland, *Experientia*, 1992, **48**, 106.
551. B. Roach, T. Eisner, and J. Meinwald, *J. Org. Chem.*, 1990, **55**, 4047.
552. R. T. Jacobs, G. I. Feutrill, and J. Meinwald, *J. Org. Chem.*, 1990, **55**, 4051.
553. H. A. Lloyd, T. H. Jones, A. Hefetz, and J. Tengö, *Tetrahedron Lett.*, 1990, **31**, 5559.
554. T. Kasai, H. Watanabe, and K. Mori, *Bioorg. Med. Chem.*, 1993, **1**, 670.
555. W. Francke, F. Schröder, U. Kohnle, and M. Simon, *Liebigs Ann. Chem.*, 1996, 1523.
556. R. R. Heath, J. R. McLaughlin, J. H. Tumlinson, T. R. Ashley, and R. E. Doolittle, *J. Chem. Ecol.*, 1979, **5**, 941.
557. W. L. Roelofs, M. J. Gieselmann, K. Mori, and D. S. Moreno, *Naturwissenschaften*, 1982, **69**, 348.
558. T. Negishi, M. Uchida, Y. Tamaki, K. Mori, T. Ishiwatari, S. Asano, and K. Nakagawa, *Appl. Entomol. Zool.*, 1980, **15**, 328.
559. K. Mori and H. Ueda, *Tetrahedron*, 1981, **37**, 2581.
560. W. Roelofs, M. Gieselmann, A. Cardé, H. Tashiro, D. S. Moreno, C. A. Henrick, and R. J. Anderson, *J. Chem. Ecol.*, 1978, **4**, 211.
561. R. J. Anderson, K. G. Adams, H. R. Chinn, and C. A. Henrick, *J. Org. Chem.*, 1980, **45**, 2229.
562. E. Demole and P. Enggist, *Helv. Chim. Acta*, 1975, **58**, 1602.
563. J. Meinwald, W. R. Thompson, T. Eisner, and D. F. Owen, *Tetrahedron Lett.*, 1971, 3485.
564. T. Suzuki and K. Mori, *Appl. Entomol. Zool.*, 1983, **18**, 134.
565. H. Z. Levinson and K. Mori, *Naturwissenschaften*, 1983, **70**, 190.
566. M. A. Battiste, L. Strekowski, D. P. Vanderbilt, M. Visnick, R. W. King, and J. Nation, *Tetrahedron Lett.*, 1983, **24**, 2611.
567. T. Chuman, J. Sivinski, R. R. Heath, C. O. Calkins, J. H. Tumlinson, M. A. Battiste, R. L. Wydra, L. Strekowski, and J. L. Nation, *Tetrahedron Lett.*, 1988, **29**, 6561.
568. J. W. Wong, V. Verigin, A. C. Oehlschlager, J. H. Borden, H. D. Pierce, Jr., A. M. Pierce, and L. Chong, *J. Chem. Ecol.*, 1983, **9**, 451.
569. M. Frenzel, K. Dettner, W. Boland, and P. Erbes, *Experientia*, 1990, **46**, 542.
570. T. Koyama, K. Ogura, and S. Seto, *J. Biol. Chem.*, 1973, **248**, 8043.
571. R. C. Jennings, J. J. Kenneth, and D. A. Schooley, *J. Chem. Soc., Chem. Commun.*, 1975, 21.
572. H. Röller, K. H. Dahm, C. C. Sweely, and B. M. Trost, *Angew. Chem., Int. Ed. Engl.*, 1967, **6**, 179.
573. K. Mori and M. Fujiwhara, *Tetrahedron*, 1988, **44**, 343.
574. S. Sakurai, T. Ohtaki, H. Mori, M. Fujiwhara, and K. Mori, *Experientia*, 1990, **46**, 220.
575. B. J. Bergot, G. C. Jamieson, M. A. Ratcliff, and D. A. Schooley, *Science*, 1980, **210**, 336.
576. T. Koyama, K. Ogura, F. C. Baker, G. C. Jamieson, and D. A. Schooley, *J. Am. Chem. Soc.*, 1987, **109**, 2853.
577. W. Francke and E. D. Morgan, unpublished results.
578. S. J. Keegans, J. Billen, E. D. Morgan, and O. A. Gökcen, *J. Chem. Ecol.*, 1993, **19**, 2705.
579. B. W. Staddon, A. Abdollahi, J. Parry, M. Rossiter, and D. W. Knight, *J. Chem. Ecol.*, 1994, **20**, 2721.
580. A. B. Attygalle, M. C. Cammaerts, and E. D. Morgan, *J. Insect Physiol.*, 1983, **29**, 27.
581. E. D. Morgan, in "Insect Communication," ed. T. Lewis, Academic Press, London, 1984, p. 169.
582. B. D. Jackson, M.-C. Cammaerts, E. D. Morgan, and A. B. Attygalle, *J. Chem. Ecol.*, 1990, **16**, 82
583. B. D. Jackson and E. D. Morgan, *Chemoecology*, 1993, **4**, 125.
584. M. Kobayashi, T. Koyama, K. Ogura, S. Seto, F. J. Ritter, and I. E. M. Brüggemann-Rotgans, *J. Am. Chem. Soc.*, 1980, **102**, 6602.
585. F. M. Alvarez, R. K. Vander Meer, and C. S. Lofgren, *Tetrahedron*, 1987, **43**, 2897.
586. J. G. C. Hamilton, G. W. Dawson, and J. A. Pickett, *J. Chem. Ecol.*, 1996, **22**, 1477.
587. J. L. C. Wright, T. Hu, J. L. Lachlan, J. Needham, and J. A. Walter, *J. Am. Chem. Soc.*, 1996, **118**, 8757.
588. R. E. Charlton and W. L. Roelofs, *Arch. Insect Biochem. Physiol.*, 1991, **18**, 81.
589. P. Juárez, J. Chase, and G. J. Blomquist, *Arch. Biochem. Biophys.*, 1992, **293**, 333.
590. J. Chase, R. A. Jurenka, C. Schal, P. P. Halarnkar, and G. J. Blomquist, *Insect Biochem.*, 1990, **20**, 149.
591. J. Chase, K. Touhara, G. D. Prestwich, C. Schal, and G. J. Blomquist, *Proc. Natl. Acad. Sci. USA*, 1992, **89**, 6050.
592. R. Nishida, Y. Kuwahara, H. Fukami, and S. Ishii, *J. Chem. Ecol.*, 1979, **5**, 289.
593. K. Mori, T. Suguro, and S. Masuda, *Tetrahedron Lett.*, 1979, 3447
594. K. Mori, T. Suguro, and S. Masuda, *Tetrahedron*, 1981, **37**, 1329.
595. C. Schal, E. L. Burns, R. A. Jurenka, and G. J. Blomquist, *J. Chem. Ecol.*, 1990, **16**, 1997.
596. C. Schal, X. Gu, E. L. Burns, and G. J. Blomquist, *Arch. Insect Biochem. Physiol.*, 1994, **25**, 375.

597. R. J. Petroski, R. J. Bartelt, and D. Wiesleder, *Insect Biochem. Mol. Biol.*, 1994, **24**, 69.
598. R. J. Bartelt and D. Wiesleder, *Bioorg. Med. Chem.*, 1996, **4**, 429.
599. R. N. Williams, M. S. Ellis, and R. J. Bartelt, *Entomol. Exp. Appl.*, 1995, **77**, 141.
600. R. J. Bartelt, D. Wiesleder, P. F. Dowd, and R. D. Plattner, *J. Chem. Ecol.*, 1992, **18**, 379.
601. R. J. Bartelt, D. G. Carlson, R. S. Vetter, and T. C. Baker, *J. Chem. Ecol.*, 1993, **19**, 107.
602. R. J.. Bartelt, K. L. Seaton, and P. F. Dowd, *J. Chem. Ecol.*, 1993, **19**, 2203.
603. R. J. Petroski, R. J. Bartelt, and R. S. Vetter, *J. Chem. Ecol.*, 1994, **20**, 1483.
604. R. J. Bartelt and D. G. James, *J. Chem. Ecol.*, 1994, **20**, 3207.
605. R. J. Bartelt, D. K. Weaver, and R. T. Arbogast, *J. Chem. Ecol.*, 1995, **21**, 1763.
606. D. G. James, R. J. Bartelt, and C. J. Moore, *J. Chem. Ecol.*, 1996, **22**, 1541.
607. H. J. Bestmann, E. Übler, and B. Hölldobler, *Angew. Chem., Int. Ed. Engl.*, 1997, **36**, 395.
608. T. H. Jones and H. M. Fales, *Tetrahedron Lett.*, 1983, **24**, 5439.
609. H. J. Bestmann, U. Haak, F. Kern, and B. Hölldobler, *Naturwissenschaften*, 1995, **82**, 142.
610. J. R. Rocca, J. H. Tumlinson, B. M. Glancey, and, C. S. Lofgren, *Tetrahedron Lett.*, 1983, **24**, 1889.
611. J. R. Rocca, J. H. Tumlinson, B. M. Glancey, and C. S. Lofgren, *Tetrahedron Lett.*, 1983, **24**, 1893.
612. K. Mori and Y. Nakazono, *Tetrahedron*, 1986, **42**, 6459.
613. M. S. Mayer and J. R. McLaughlin, "Handbook of Insect Pheromones and Sex Attractants," CRC Press, Boca Raton, FL, 1991, p. 622.
614. J. W. Wheeler, S. L. Evans, M. S. Blum, H. H. V. Velthuis, and J. M. F. de Camargo, *Tetrahedron Lett.*, 1976, 4029.
615. P. D. Swedenborg, R. L. Jones, H.-Q. Zhou, I. Shin, and H.-W. Liu, *J. Chem. Ecol.*, 1994, **20**, 3373.
616. E. Übler, F. Kern, H. J. Bestmann, B. Hölldobler, and A. B. Attygalle, *Naturwissenschaften*, 1995, **82**, 523.
617. F. Kern and H. J. Bestmann, *Z. Naturforsch., Teil C*, 1994, **49**, 865.
618. Cited in ref. 31, p. 163.
619. J. K. Phillipps, S. F. P. Miller, J. F. Andersen, H. M. Fales, and W. E. Burkholder, *Tetrahedron Lett.*, 1987, **28**, 6145.
620. J. K. Phillips, J. M. Chong, J. F. Andersen, and W. E. Burkholder, *Entomol. Exp. Appl.*, 1989, **51**, 149.
621. K. Mori and M. Ishikura, *Liebigs Ann. Chem.*, 1989, 1263.
622. H. Z. Levinson, A. Levinson, Z. Ren, and K. Mori, *J. Appl. Entomol.*, 1990, **110**, 203.
623. H. J. Williams, R. M. Silverstein, W. E. Burkholder, and A. Khorramshahi, *J. Chem. Ecol.*, 1981, **7**, 759.
624. A. Cork, D. R. Hall, R. J. Hodges, and J. A. Pickett, *J. Chem. Ecol.*, 1991, **17**, 789.
625. R. E. Charlton, F. X. Webster, A. Zhang, C. Schal, D. Liang, L. Sreng, and W. L. Roelofs, *Proc. Natl. Acad. Sci. USA*, 1993, **90**, 10202.
626. K. Mori and Y. Takeuchi, *Proc. Jpn. Acad., Ser. B*, 1994, **70**, 143.
627. W. S. Leal, X. Shi, D. Liang, C. Schal, and J. Meinwald, *Proc. Natl. Acad. Sci. USA*, 1995, **92**, 1033.
628. R. H. Hallett, A. L. Perez, G. Gries, R. Gries, H. D. Pierce, Jr., J. Yue, A. C. Oehlschlager, L. M. Gonzales, and J. H. Borden, *J. Chem. Ecol.*, 1995, **21**, 1549.
629. J. P. Morin, D. Rochat, C. Malosse, M. Lettéré, R. Desmier de Chenin, H. Wibwo, and C. Descoins, *C. R. Acad. Sci., Ser. III*, 1996, **319**, 595.
630. Y. Tanaka, H. Honda, K. Ohsawa, and J. Yamamoto, *Nippon Noyaku Gakkaishi*, 1989, **14**, 197 (*Chem. Abstr.*, 111-21-189524).
631. J. Kochansky, J. R. Aldrich, and W. R. Lusby, *J. Chem. Ecol.*, 1989, **15**, 1717.
632. K. Mori and J. Wu, *Liebigs Ann. Chem.*, 1991, 783.
633. A. Ichikawa, T. Yasuda, and S. Wakamura, *J. Chem. Ecol.*, 1995, **21**, 627.
634. T. Yasuda, S. Wakamura, and N. Arakaki, *J. Chem. Ecol.*, 1995, **21**, 1813.
635. B. A. Bierl, M. Beroza, and W. Collier, *J. Econ. Entomol.*, 1972, **65**, 659.
636. J. K. Phillips, C. A. Walgenbach, J. A. Klein, W. E. Burkholder, N. R. Schmuff, and H. M. Fales, *J. Chem. Ecol.*, 1985, **11**, 1263.
637. C. A. Walgenbach, J. K. Phillips, W. E. Burkholder, G. G. S. King, K. N. Slessor, and K. Mori, *J. Chem. Ecol.*, 1987, **13**, 2159.
638. M. M. Blight, J. A. Pickett, M. C. Smith, and L. J. Wadhams, *Naturwissenschaften*, 1984, **71**, 480.
639. M. M. Blight and L. J. Wadhams, *J. Chem. Ecol.*, 1987, **13**, 733.
640. R. G. Riley, R. M. Silverstein, and J. C. Moser, *Science*, 1974, **183**, 760.
641. M. S. Blum, "Chemical Defenses in Arthropods," Academic Press, New York, 1981, p. 152.
642. R. R. do Nascimento, E. D. Morgan, W. A. König, and T. M. C. Della Lucia, *J. Chem. Ecol.*, 1997, **23**, 1569.
643. B. Hölldobler, N. J. Oldham, E. D. Morgan, and W. A. König, *J. Insect Physiol.*, 1995, **41**, 739.
644. J. Meinwald, A. F. Kluge, J. E. Carrel, and T. Eisner, *Proc. Natl. Acad. Sci. USA*, 1971, **68**, 1467.
645. H. M. Fales, T. M. Jaouni, J. O. Schmidt, and M. S. Blum, *J. Chem. Ecol.*, 1980, **6**, 895.
646. L. J. Wadhams, M. E. Angst, and M. M. Blight, *J. Chem. Ecol.*, 1982, **8**, 477.
647. S. Steghaus-Kovac, U. Maschwitz, A. B. Attygalle, R. T. S. Frighetto, N. Frighetto, O. Vostrowsky, and H. J. Bestmann, *Experientia*, 1992, **48**, 690.
648. W. Francke, unpublished results.
649. H. M. Fales, M. S. Blum, R. W. Crewe, and J. M. Brand, *J. Insect Physiol.*, 1972, **18**, 1077.
650. H. J. Bestmann, A. B. Attygalle, J. Glasbrenner, R. Riemer, O. Vostrowsky, M. G. Constantino, G. Melikyan, and E. D. Morgan, *Liebigs Ann. Chem.*, 1988, 55.
651. L. B. Bjostad, D. K. Jewett, and D. L. Bringham, *J. Chem. Ecol.*, 1996, **22**, 103.
652. R. H. Schaffrahn and M. K. Rust, *Southwest. Entomol.*, 1989, **14**, 49.
653. T. Chuman, K. Mochizuki, M. Mori, M. Kohno, K. Kato, and M. Noguchi, *J. Chem. Ecol.*, 1985, **11**, 417.
654. K. Mori and H. Watanabe, *Tetrahedron*, 1985, **41**, 3423.
655. M. Mori, K. Mochizuki, M. Kohno, T. Chuman, A. Ohnishi, H. Watanabe, and K. Mori, *J. Chem. Ecol.*, 1986, **12**, 83.
656. T. Chuman, K. Mochizuki, K. Kato, M. Ono, and A. Okubo, *Agric. Biol. Chem.*, 1983, **47**, 1413.
657. T. Ebata and K. Mori, *Agric. Biol. Chem.*, 1987, **51**, 2925.
658. T. Imai, H. Kodama, T. Chuman, and M. Kohno, *J. Chem. Ecol.*, 1990, **16**, 1237.
659. H. Kodama, M. Ono, M. Kohno, and A. Ohnishi, *J. Chem. Ecol.*, 1987, **13**, 1871.

660. K. Mori and T. Ebata, *Tetrahedron*, 1986, **42**, 4685.
661. Y. Kuwahara, H. Fukami, R. Howard, S. Ishii, F. Matsumura, and W. E. Burkholder, *Tetrahedron*, 1978, **34**, 1769.
662. R. W. Hoffmann, W. Ladner, K. Steinbach, W. Massa, R. Schmidt, and G. Snatzke, *Chem. Ber.*, 1981, **114**, 2786.
663. K. Mori, S. Sano, Y. Yokoyama, M. Bando, and M. Kido, *Eur. J. Org. Chem.*, 1998, 1135.
664. P. R. White and M. C. Birch, *J. Chem. Ecol.*, 1987, **13**, 1695.
665. H. Kodama, K. Mochizuki, M. Kohno, A. Ohnishi, and Y. Kuwahara, *J. Chem. Ecol.*, 1987, **13**, 1859.
666. Y. Kuwahara, L. T. M. Yen, Y. Tominaga, K. Matsumoto, and Y. Wada, *Agric. Biol. Chem.*, 1982, **46**, 2283.
667. K. Mori and S. Kuwahara, *Tetrahedron*, 1986, **42** 5539.
668. K. Mori and S. Kuwahara, *Tetrahedron*, 1986, **42** 5545.
669. G. T. Pearce, W. E. Gore, R. M. Silverstein, J. W. Peacock, R. A. Cuthbert, G. N. Lanier, and J. B. Simeone, *J. Chem. Ecol.*, 1975, **1**, 115.
670. G. T. Pearce, W. E. Gore, and R. M. Silverstein, *J. Org. Chem.*, 1976, **41**, 2797.
671. K. Mori, *Tetrahedron*, 1976, **32**, 1979.
672. V. Sinnwell, W. Schulz, W. Francke, R. Kittmann, and D. Schneider, *Tetrahedron Lett.*, 1986, **27**, 2091.
673. P. L. Guss, J. H. Tumlinson, P. E. Sonnett, and J. R. McLaughlin, *J. Chem. Ecol.*, 1983, **9**, 1363.
674. S. Senda and K. Mori, *Agric. Biol. Chem.*, 1983, **47**, 795.
675. T. Chuman, P. L. Guss, R. E. Doolittle, J. R. McLaughlin, J. L. Krysan, J. M. Schalk, and J. H. Tumlinson, *J. Chem. Ecol.*, 1987, **13**, 1601.
676. J. R. McLaughlin, J. H. Tumlinson, and K. Mori, *J. Econ. Entomol.*, 1991, **84**, 99.
677. E. Dunkelblum, R. Gries, G. Gries, K. Mori, and Z. Mendel, *J. Chem. Ecol.*, 1995, **21**, 849.
678. E. Dunkelblum, Z. Mendel, F. Assael, M. Harel, L. Kerhoas, and J. Einhorn, *Tetrahedron Lett.*, 1993, **34**, 2805.
679. E. Dunkelblum, Z. Mendel, G. Gries, R. Gries, L. Zegelman, A. Hassner, and K. Mori, *Bioorg. Med. Chem.*, 1996, **4**, 489.
680. J. Einhorn, P. Menassieu, C. Malosse, and P.-H. Ducrot, *Tetrahedron Lett.*, 1990, **31**, 6633.
681. K. Mori, T. Furuuchi, and H. Kiyota, *Liebigs Ann. Chem.*, 1994, 971.
682. H. Jactel, P. Menassieu, M. Letteré, K. Mori, and J. Einhorn, *J. Chem. Ecol.*, 1994, **20**, 2159.
683. B. E. Hibbard, G. N. Lanier, S. C. Parks, Y. T. Qi, F. X. Webster, and R. M. Silverstein, *J. Chem. Ecol.*, 1991, **17**, 89.
684. K. Mori, T. Furuuchi, and K. Matsuyama, *Liebigs Ann. Chem.*, 1995, 2093.
685. X. Shi, F. X. Webster, and J. Meinwald, *Tetrahedron Lett.*, 1995, **36**, 7201.
686. J. L. Le Queré, R. Brossut, C. A. Nalepa, and O. Bonnard, *J. Chem. Ecol.*, 1991, **17**, 811.
687. K. Mori and M. Itou, *Liebigs Ann. Chem.*, 1992, 87.
688. A. L. Perez, G. Gries, R. Gries, R. M. Giblin-Davis, and A. C. Oehlschlager, *J. Chem. Ecol.*, 1994, **20**, 2653.
689. K. Mori and N. Murata, *Liebigs Ann. Chem.*, 1995, 697.
690. K. Mori, H. Kiyota, and D. Rochat, *Liebigs Ann. Chem.*, 1993, 865.
691. D. Rochat, F. Akamou, A. Sangare, D. Mariau, and K. Mori, *C. R. Acad. Sci., Ser. III*, 1995, **318**, 183.
692. A. L. Perez, R. H. Hallet, R. Gries, G. Gries, A. C. Oehlschlager, and J. H. Borden, *J. Chem. Ecol.*, 1996, **22**, 357.
693. A. C. Oehlschlager, R. N. B. Prior, A. L. Perez, R. Gries, G. Gries, H. D. Pierce, Jr., and S. Laup, *J. Chem. Ecol.*, 1995, **21**, 1619.
694. R. M. Giblin-Davis, R. Gries, G. Gries, E. Peña-Rojas, J. Pinzón, J. E. Peña, A. L. Perez, H. D. Pierce, Jr., and A. C. Oehlschlager, *J. Chem. Ecol.*, 1997, **23**, 2287.
695. K. Mori, H. Kiyota, C. Malosse, and D. Rochat, *Liebigs Ann. Chem.*, 1993, 1201.
696. C. Malosse, P. Ramirez-Lucas, and D. Rochat, *J. High. Resolut. Chromatogr.*, 1995, **18**, 669.
697. A. L. Perez, Y. Campos, C. M. Chinchilla, A. C. Oehlschlager, G. Gries, R. Gries, R. M. Giblin-Davis, G. Castrillo, J. E. Peña, R. E. Duncan, L. M. Gonzales, H. D. Pierce, Jr., R. C. McDonald, and R. Andradé, *J. Chem. Ecol.*, 1997, **23**, 869.
698. A. C. Oehlschlager, H. D. Pierce, Jr., B. Morgan, P. D. C. Wimalaratne, K. N. Slessor, G. G. S. King, G. Gries, R. Gries, J. Borden, C. M. Chinchilla, and R. G. Mexzan, *Naturwissenschaften*, 1992, **79**, 134.
699. K. Mori and K. Ishigami, *Liebigs Ann. Chem.*, 1992, 1195.
700. D. Rochat, C. Malosse, M. Letteré, P.-H. Ducrot, P. Zagatti, M. Renou, and C. Descoins, *J. Chem. Ecol.*, 1991, **17**, 2127.
701. D. Rochat, C. Malosse, M. Letteré, P. Ramirez-Lucas, J. Einhorn, and P. Zagatti, *C. R. Acad. Sci., Ser. II*, 1993, **316**, 1737.
702. J. Beauhaire, P.-H. Ducrot, C. Malosse, D. Rochat, I. O. Ndiege, and D. O. Otieno, *Tetrahedron Lett.*, 1995, **36**, 1043.
703. K. Mori, T. Nakayama, and H. Takikawa, *Tetrahedron Lett.*, 1996, **37**, 3741.
704. M. T. Fletcher, C. Moore, and W. Kitching, *Tetrahedron Lett.*, 1997, **38**, 3475.
705. A. Hefetz and H. A. Lloyd, *J. Chem. Ecol.*, 1983, **9**, 607.
706. H. D. Pierce, Jr., A. M. Pierce, B. D. Johnston, A. C. Oehlschlager, and J. H. Borden, *J. Chem. Ecol.*, 1988, **14**, 2169.
707. D. S. Dodd, H. D. Pierce, Jr., and A. C. Oehlschlager, *J. Org. Chem.*, 1992, **57**, 5250.
708. T. Suzuki, J. Kozaki, R. Sugawara, and K. Mori, *Appl. Entomol. Zool.*, 1984, **19**, 15.
709. P. L. Guss, P. E. Sonnet, R. L. Carney, T. F. Branson, and J. H. Tumlinson, *J. Chem. Ecol.*, 1984, **10**, 1123.
710. P. L. Guss, P. E. Sonnet, R. L. Carney, J. H. Tumlinson, and P. J. Wilkin, *J. Chem. Ecol.*, 1985, **11**, 21.
711. Y. Tamaki, H. Noguchi, H. Sugie, R. Sato, and A. Kariya, *Appl. Entomol. Zool.*, 1979, **14**, 101.
712. H. Noguchi, H. Sugie, Y. Tamaki, and Y. Oomasu, *Jpn. J. Appl. Entomol. Zool.*, 1985, **29**, 278.
713. H. J. Bestmann, R. T. S. Frighetto, N. Frighetto, and O. Vostrowsky, *Liebigs Ann. Chem.*, 1990, 829.
714. Y. Tamaki, H. Noguchi, H. Sugie, A. Kariya, S. Arai, M. Ohba, T. Terada, T. Suguro, and K. Mori, *Jpn. J. Appl. Entomol. Zool.*, 1980, **24**, 221.
715. D. M. Jewett, F. Matsumura, and H. C. Coppel, *Science*, 1976, **192**, 51.
716. A.-B. Wassgren and G. Bergström, *J. Chem. Ecol.*, 1995, **21**, 987.
717. E. Hedenström, H.-E. Höögberg, A.-B. Wassgren, G. Bergström, J. Löfqvist, B. Hansson, and O. Anderbrant, *Tetrahedron*, 1992, **48**, 3139.
718. G. Bergström, A.-B. Wassgren, O. Anderbrant, J. Fägerhag, H. Edlund, E. Heckenström, H.-E. Högberg, C. Geri, M. A. Anger, M. Varama, B. S. Hansson, and J. Löfqvist, *Experientia*, 1995, **51**, 370.
719. J. I. Olaifa, F. Matsumura, T. Kikukawa, and H. C. Coppel, *J. Chem. Ecol.*, 1988, **14**, 1131.

720. T. Norin, *Pure Appl. Chem.*, 1996, **68**, 2043.
721. A.-B. Wassgren, O. Anderbrant, J. Löfqvist, B. Hansson, G. Bergström, E. Hedenström, and H.-E. Högberg, *J. Insect Physiol.*, 1992, **38**, 885.
722. G. Bergström, A.-B. Wassren, O. Anderbrand, S. Ochieng, F. Östrand, B. Hansson, E. Hedenström, and H.-E. Högberg, *Naturwissenschaften*, 1998, **85**, 244.
723. R. E. Greenblatt, W. E. Burkholder, J. C. Cross, R. C. Byler, and R. M. Silverstein, *J. Chem. Ecol.*, 1976, **2**, 285.
724. J. H. Cross, R. C. Byler, R. F. Cassidy, Jr., R. M. Silverstein, R. E. Greenblatt, W. E. Burkholder, A. R. Levinson, and H. Z. Levinson, *J. Chem. Ecol.*, 1976, **2**, 457.
725. R. M. Silverstein, R. F. Cassidy, W. E. Burkholder, T. J. Shapas, H. Z. Levinson, A. R. Levinson, and K. Mori, *J. Chem. Ecol.*, 1980, **6**, 911.
726. H. Z. Levinson, A. R. Levinson, and K. Mori, *Naturwissenschaften*, 1981, **68**, 480.
727. W. Francke, U. Brunnemann, and G. Schmidt, unpublished results.
728. R. E. Doolittle, A. T. Proveaux, H. T. Alborn, and R. R. Heath, *J. Chem. Ecol.*, 1995, **21**, 1677.
729. H. Sugie, Y. Tamaki, R. Sato, and M. Kumakura, *Appl. Entomol. Zool.*, 1984, **19**, 323.
730. R. Sato, N. Abe, H. Sugie, M. Kato, K. Mori, and Y. Tamaki, *Appl. Entomol. Zool.*, 1986, **21**, 478.
731. R. Gries, G. Gries, G. G. S. King, and C. T. Maier, *J. Chem. Ecol.*, 1997, **23**, 1119.
732. W. Francke, M. Tóth, G. Szöcs, W. Krieg, H. Ernst, and E. Buschmann, *Z. Naturforsch., Teil C*, 1988, **43**, 787.
733. W. Francke, S. Franke, M. Tóth, G. Szöcs, P. Guerin, and H. Arn, *Naturwissenschaften*, 1987, **74**, 143.
734. M. Tóth, G. Helmchen, U. Leikauf, Gy. Sziráki, and G. Szöcs, *J. Chem. Ecol.*, 1989, **15**, 1535.
735. M. Riba, J. A. Rosell, M. Eizaguirre, R. Canela, and A. Guerrero, *J. Chem. Ecol.*, 1990, **16**, 1471.
736. R. Gries, G. Gries, J. Li, C. T. Maier, C. R. Lemmon, and K. N. Slessor, *J. Chem. Ecol.*, 1994, **20**, 2501.
737. C. T. Maier, R. Gries, and G. Gries, *J. Chem. Ecol.*, 1998, **24**, 491.
738. G. G. S. King, R. Gries, G. Gries, and K. N. Slessor, *J. Chem. Ecol.*, 1995, **21**, 2027.
739. J. Li, R. Gries, G. Gries, K. N. Slessor, G. G. S. King, W. W. Bowers, and R. J. West, *J. Chem. Ecol.*, 1993, **19**, 1057.
740. J. Li, G. Gries, R. Gries, J. Bikič, and K. N. Slessor, *J. Chem. Ecol.*, 1993, **19**, 2547.
741. Y. Carrière, J. G. Millar, J. N. McNeil, D. Miller, and E. W. Underhill, *J. Chem. Ecol.*, 1988, **14**, 947.
742. S. Shu, P. D. Swedenborg, and R. L. Jones, *J. Chem. Ecol.*, 1990, **16**, 521.
743. P. E. Sonnet, *J. Chem. Ecol.*, 1984, **10**, 771.
744. Y. Naoshima and H. Mukaidani, *J. Chem. Ecol.*, 1987, **13**, 325.
745. D. A. Carlson, P. A. Langley, and P. Huyton, *Science*, 1978, **201**, 750.
746. T. W. Coates and P. A. Langley, *Overseas Development Administration and University of Bristol, Tse Tse Research Laboratory Annual Report*, 1980, **44**, 3007.
747. D. A. Carlson, D. R. Nelson, P. A. Langley, T. W. Coates, T. L. Davis, and M. E. Leegwater-van der Linden, *J. Chem. Ecol.*, 1984, **10**, 429.
748. P. G. McDowell, A. Hassanali, and R. Dransfield, *Physiol. Entomol.*, 1985, **10**, 183.
749. K. Matsuyama and K. Mori, *Biosci. Biotechnol. Biochem.*, 1994, **58**, 539.
750. M. Fukaya, T. Yasuda, S. Wakamura, and H. Honda, *J. Chem. Ecol.*, 1996, **22**, 259.

8.05
Insect Hormones and Insect Chemical Ecology

E. DAVID MORGAN and IAN D. WILSON
Keele University, UK

8.05.1 INTRODUCTION

Over half of all species of living things are insects. Since they first evolved from other arthropods 300 million years ago in the Devonian period, they have had plenty of time to evolve into a great number of variations on their primitive form. Insect chemistry, on the other hand, is a subject of the last three or four decades. Born of natural curiosity and a desire to find new ways to control insect pests, encouraged by increasing resolution of chromatographic methods and increasing sensitivity of spectroscopy, it has revealed a whole new world of chemistry, both substances made by insects and substances made by plants for the benefit or detriment of insects. The subject is now too large to cover in entirety. That great part of it called pheromones is treated in Chapter 8.04. Insect hormones and insect chemical ecology are covered here. It would be a brave person who would attempt to say exactly what insect chemical ecology is, but at least the title will induce the curious reader to enquire further. In this chapter it includes those natural products closely connected with insects that have not been already considered under plant chemical ecology in Chapter 8.03.

Although it took a long time to discover the fact, we now know that insect growth and development is controlled by hormones. A brief description should place that subject in context and permit a discussion of substances that interfere with normal development. All insects grow through alternate periods of feeding, followed by casting off their rather rigid exterior cuticle, and expanding the body before the new cuticle hardens. The process of shedding the old skin and producing a new one is called molting or ecdysis. The intermediate stages are called instars. Most insects pass through several larval or nymphal instars before becoming mature adults. One large group of insects, called the Hemimetabola, hatch from the egg into nymphs that are very similar in appearance to the adult form (Figure 1), but only become winged and sexually mature as adults (e.g., locusts, cockroaches, bugs, and termites). The other large group, the Holometabola (including moths, flies, beetles, bees, and ants) go through a process of complete metamorphosis. The larvae have an appearance quite different from the adults. The larvae molt into a resting stage, the pupa, in which the tissues are almost completely rearranged to produce the adult (Figure 1). Primitive insects, the Ametabola, do not have wings, and adults differ from larvae only in the former having mature sexual organs.

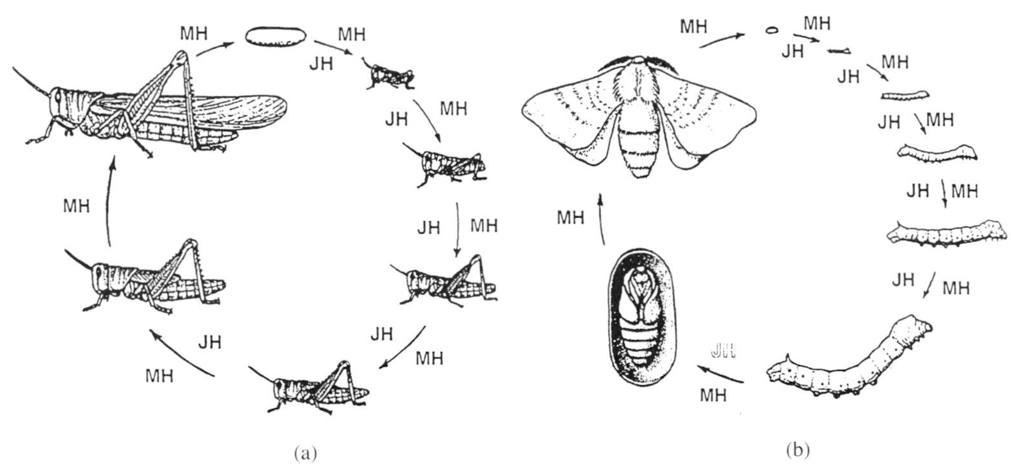

Figure 1 Simplified life cycle of (a) a hemimetabolous insect and (b) a holometabolous insect, showing the points of intervention of molting hormone (MH) and juvenile hormone (JH).

Each of the steps from embryo through the larval or nymphal instars (and, for holometabolous insects, the pupa) to adult are controlled by hormones. The process of molting or ecdysis is controlled by the molting hormone, while the presence or absence of the juvenile hormone determines whether the next stage is to an immature or mature individual. Low levels of juvenile hormone in the holometabolous larva produces a pupa as the next stage (Figure 1). In turn the prothoracic glands which produce the moulting hormone are stimulated to do so by a prothoracicotropic hormone (PTTH) from the brain, and other neurosecretory cells of the brain stimulate the corpus allatum to produce the juvenile hormone. A further step in the understanding of this system of stimulation and control is indicated in Figure 2. It must, however, be emphasized that this is a simplification of the full story, and that other processes besides molting (e.g., metabolism, water balance, pheromone production, reproduction, caste determination, and migration) are all controlled by hormones. For a more accurate and detailed discussion, the reader is recommended to read Nijhout.[1]

Insect hormones fall into two distinct groups, the well-defined small molecules produced by glandular tissues, the molting and metamorphosis hormones, and a much larger group of peptides and proteins produced from neurosecretory cells.

8.05.2 INSECT JUVENILE HORMONE

Insect juvenile hormone (JH) is something of a misnomer, because the hormone as well as operating in juvenile insects reappears in adult insects, where it has a function (not yet clearly established) in reproduction. The name, however, is now firmly established. In nymphs and larvae its presence ensures that each molt is to another immature stage. Its absence from the last larval instar of Hemimetabola and the holometabolous pupa ensures that the next molt is to the adult form. Metamorphosis is then not controlled by a metamorphosis hormone but rather by the absence of a hormone. Further research has shown it to have many other functions, including control of

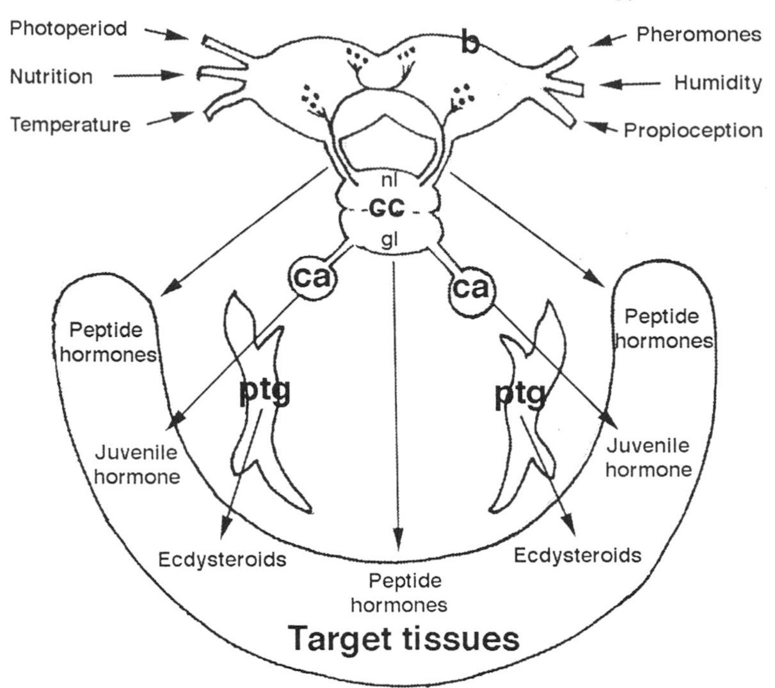

Figure 2 An idealized representation of the insect brain (b), the corpus cardiacum (cc), corpora allata (ca), and prothoracic glands (ptg) and peripheral tissues, showing some of the stimuli and secretions considered here. The black dots represent neurosecretory cells, nl is the neurosecretory lobe, and gl is the glandular lobe of the corpus cardiacum.

diapause (arrested development of larvae or pupae during winter months), synthesis of egg protein (vitellin), development of ovaries, phase development in locusts and aphids, determination of caste between queens and workers in honeybees, and control of production of and response to pheromones. A survey of literature references on the effect of JH in different species is available,[2] and Nijhout[1] gives a discussion.

The juvenile hormone is produced in the corpus allatum, a group of secretory cells connected to and close behind the brain (Figure 2). In many insects it appears as two groups of cells near the esophagus, but in Diptera and some Hemiptera it is fused with other structures into a "ring gland" around the esophagus.

The existence and nature of the juvenile hormone was made clear from the experimental work of Wigglesworth with the bug *Rhodnius prolixus*. By joining individuals at different stages of development so that they shared their hemolymph, or by implanting or removing corpora allata, he was able to see the effects of the hormone on subsequent molts.[3] Juvenile insects could be made to molt to precocious pupae or adults, or development could be arrested with formation of extra larval stages. The hormone could even reverse development, for example producing a stage with some larval character from a pupa. Wigglesworth also showed that vitellogenesis, the production of egg proteins in adult females, was controlled by active corpora allata.[4]

8.05.2.1 Bioassays

All isolations of biologically active compounds require a good bioassay. For JH activity, Wigglesworth first developed the *Tenebrio* assay,[5] but later Gilbert and Schneidermann developed a *Galleria* wax assay, which was generally adopted because it was very specific and sensitive, and it is still in use.[6] A small amount of cuticle is removed from a young *Galleria* pupa, and the wound is sealed with wax and a paraffin solution of the material to be assayed. The wound tissue that forms is very sensitive to JH activity, and in its presence a patch of pupal cuticle is formed at that spot in the emerging adult. The method can detect approximately 5 pg of JH. The subject of JH bioassays was considered in detail by Gilbert and Schneidermann,[6] and later by Staal.[7] The bioassays were centrally important in the period of isolation and identification of JH and in the early work on JH analogs (see Sections 8.05.2.8 and 8.05.2.9). The wax test is still the most sensitive method of detection, but less specific than chemical methods (see Section 8.05.2.5).

8.05.2.2 Isolation

The extremely low titer of JH in insect larvae frustrated attempts at its chemical isolation for some time. Various attempts to isolate active substances gave, among others, farnesol (**1**) and farnesal (**2**), both of which showed moderate JH activity. That led to the testing of various terpene derivatives, among which farnesyl methyl ether (**3**) and diethyl farnesylamine (**4**) were very active, but they were still much less active than the natural hormone. When methyl farnesoate was treated with hydrogen chloride in ethanol, a mixture of more than 20 compounds was formed, which showed activity, and the most active compound, (**5**), was isolated and identified. It had an activity approaching that of JH. Finally, the discovery that methyl 10,11-epoxyfarnesoate (**6**) was very active led Bowers to predict that JH would have a structure close to that of (**6**).[8]

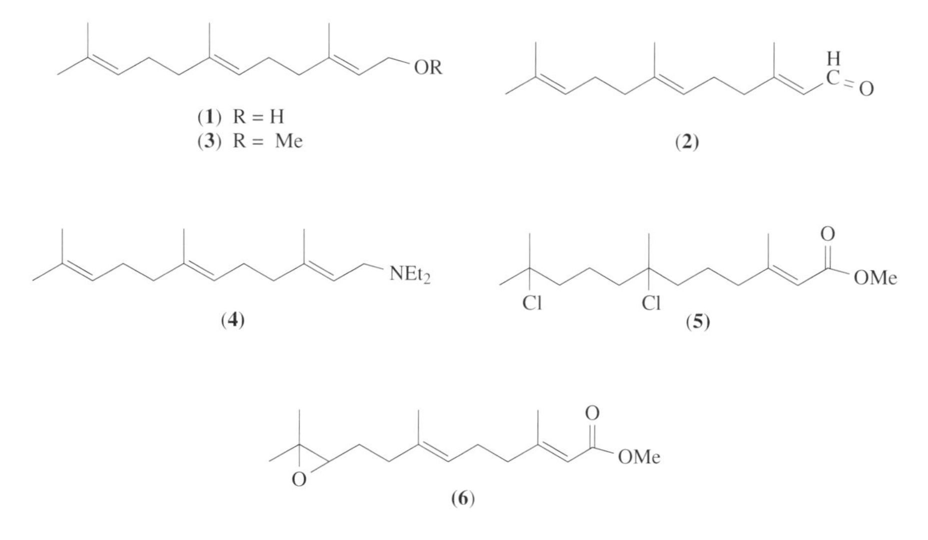

When it was discovered that adult males of the giant silk moth *Hyalophora cecropia* apparently stored the compound in their abdomens,[9] and contained relatively massive amounts of JH (subsequently estimated at 1–2 μg per insect, or about 50 times as much as other insects), various groups began the isolation of the substance, leading to its obtention in a pure form in 1965.[10,11] A chemical structure followed two years later.[12]

8.05.2.3 Structure

The structure of the substance from isolated male *H. cecropia* abdomens was based upon mass, IR, and NMR spectra. The mass spectrum established the molecular formula as $C_{18}H_{30}O_3$. Three moles of hydrogen were taken up in catalytic hydrogenation of 20 μg, and an oxygen atom was lost. Cleavage of 15 μg with OsO_4 and HIO_4 gave levulinic aldehyde, identified by GC. The presence of ethyl branches and an epoxide was deduced from the NMR spectrum, obtained on 200 μg. With other data the structure (**7**) was deduced.[12] The geometry of the 2,3 double bond was secure from spectroscopy, but the stereochemistry of the remaining centers was unknown. Synthetic studies of a mixture of geometric isomers established the correctness of the structure (**7**).[13] Methyl (*E,E,Z*)-10,11-epoxy-3,11-dimethyl-7-ethyl-2,6-tridecadienoate had both identical spectra to that of the natural JH and the highest biological activity of all the isomers. This left two chiral centers at C-10 and C-11 undetermined. Measurement of the optical rotatory dispersion (ORD) showed that the compound had a plane positive dispersion curve, so it was not racemic. The absolute configuration at C-10 and C-11 was determined by three groups at about the same time, both by stereospecific synthesis of the two possible enantiomers, starting from materials of known absolute configuration,[14] and by acid-catalyzed opening of the epoxide to a diol, and determination of the chirality of the doubly asymmetric 1,2-diol produced.[15,16] These final elements of structure enabled the compound to be assigned the (10*R*, 11*S*) configuration (**8**). This enantiomer had more than six times the biological activity of its optical antipode. This intense activity on the structure and synthesis of the JH compound was in progress in several laboratories from 1965 to 1971 (for a more detailed description, see Trost[17]).

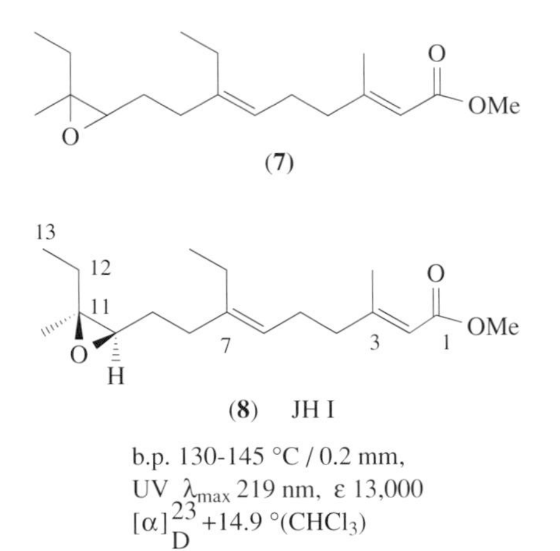

Shortly after this first compound was isolated, smaller amounts (10–13%) of a second compound of closely similar structure, but of slightly lower bioassay activity, was isolated from the *H. cecropia* abdomen extracts.[18] A lengthy paper describes many of the problems encountered, for example the small amounts of hormone available were found to be very sensitive to acids and metal surfaces.[19] The elucidation of the structure (**9**) followed closely on the first, and it was found to be structurally very similar to the initial compound, except it had a methyl rather than an ethyl branch at C-7. Henceforward the compounds were called JH I and JH II. Within a few years a third compound, JH III (**10**), was isolated from tissue cultures of the corpora allata of the tobacco hornworm *Manduca sexta* using radiolabeled methionine, when another closely related labeled compound was found during separation by HPLC.[20] With only 7 μg of the pure compound, it was possible to show by spectroscopic and microchemical methods that it was a homolog of the two earlier compounds, and the same structure as the methyl epoxyfarnesoate prepared by Bowers *et al.*[8] Some NMR shifts of the JH compounds are given in Table 1.

Table 1 NMR shifts for some JH compounds as δ (ppm from TMS).

Carbon atom	^1H *JH I*, 400 *MHz*	^1H *JH II*, 250 *MHz*	^1H *JH III*, 100 *MHz*	^{13}C *JH I*, *not specified*	^{13}C *JH II*, *not specified*	^{13}C *JH III*, 25 *MHz*	^{13}C *JHB*$_3$, *not specified*
O-Me	3.61 (s)	3.68	3.70 (s)	50.75	50.72	50.75	50.87
1	—	—	—	167.14	167.20	167.20	167.01
2	5.57 (br)	5.67 (m)	5.68 (m)	115.31	115.31	115.34	115.64
3	—	—	—	159.83	159.89	159.86	158.84
4	—	—	—	41.19	40.78	40.83	37.57[a]
5	—	—	—	33.29	36.51	36.33	35.54[a]
6	5.05 (br t, *J* 6.4)	5.14 (m)	5.22 (m)	122.91	123.44	123.50	63.95[b]
7	—	—	—	141.28	135.38	135.32	58.31[c]
8	—	—	—	27.26[a]	27.03[a]	27.47[a]	26.65[a]
9	—	—	—	23.17	16.03	16.03	24.71[a]
10	2.50 (dd *J* 7.3, 5.1)	2.70 (dd *J* 6.7, 5.7)	2.70 (t, *J* 6)	61.78	61.72	58.27	62.83[b]
11	—	—	—	64.65	64.53	64.09	60.56[c]
12	—	—	1.27 (s)	21.59	21.59	18.75	18.76[d]
13	0.98 (t *J* 7.8)	1.00 (t *J* 7.5)	xxx	9.65	9.50	xxx	xxx
3-Me	2.13 (d *J* 1.2)	2.17 (d *J* 1.3)	2.18 (d *J* 1.3)	25.80[a]	25.80[a]	24.86[a]	24.79
7-Et CH_3—	0.98 (t *J* 7.8)	xxx	xxx	13.10	xxx	xxx	xxx
7-Et —CH_2—	—	—	—	25.57[a]	xxx	xxx	
7-Me	xxx	1.62 (s)	1.63 (s)	—	25.92[a]	25.95[a]	18.63[d]
11-Me	1.16 (s)	1.27 (s)	1.32 (s)	18.84	18.78	18.81	16.42[d]

Coupling constants *J* in hertz; xxx indicates atom not present.
[a,b,c,d]Atoms with the same superscript letter in the same compound have not been distinguished and values are interchangeable.

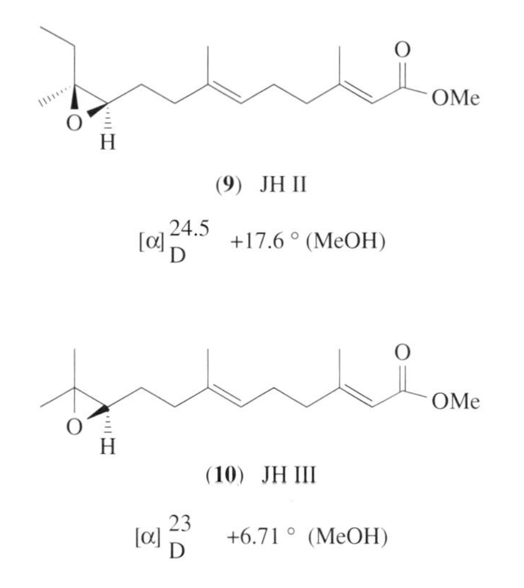

(**9**) JH II

$[\alpha]_D^{24.5}$ +17.6 ° (MeOH)

(**10**) JH III

$[\alpha]_D^{23}$ +6.71 ° (MeOH)

In the authors' work on ant secretions, four homologous sesquiterpenes with structures based on farnesene (**11**) had already been found in *Myrmica* ants.[21] It was therefore very satisfying when a fourth JH was isolated from extracts of *M. sexta* embryos, called JH 0 (**15**).[22] With only 200 ng of material, structure was determined chiefly by GC-MS, in both EI and CI modes. Then the symmetry was broken by the discovery of yet another from *M. sexta* eggs called 4-MeJH I or iso-JH 0 (**16**).[23] The absolute configurations of (**15**) and (**16**) are still not confirmed.

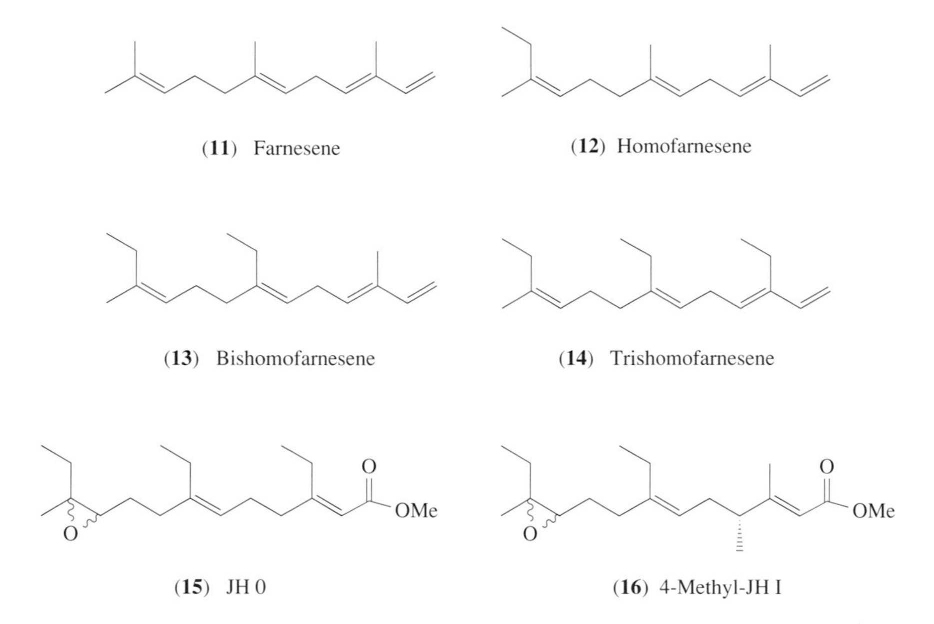

(**11**) Farnesene

(**12**) Homofarnesene

(**13**) Bishomofarnesene

(**14**) Trishomofarnesene

(**15**) JH 0

(**16**) 4-Methyl-JH I

Much more recently, JH III bisepoxide (JHB_3) was discovered, first as an unknown JH derivative in the blowfly *Phormia regina*.[24] Another group, having found only traces of JH III synthesized by the ring glands of *Drosophila melanogaster* larvae isolated with the help of L-[^{3}H-methyl]methionine, a more polar JH compound. By GC–MS they found a molecular mass of 282 and base peak of m/z 43, and they obtained a ^{1}H NMR spectrum. They deduced it was a bisepoxide, and confirmed it by synthesizing a racemate from methyl farnesoate.[25] They also deduced its chirality as (2*E*,6*S*,7*S*,10*R*)-JH III bisepoxide (**17**) by the opening of the epoxides with thiophenol to give methyl 6,10-bis(thiophenyl)-7,11-dihydroxyfarnesoate (**18**). The absolute configuration was confirmed by comparison of biosynthetic JHB_3 and synthetic stereoisomers by HPLC and cyclodextrin-modified capillary electrophoresis.[26] There was no direct evidence at this stage that JHB_3 was the natural hormone, because there is no dipteran larval bioassay, but the evidence is accumulating that JH III bisepoxide is the only JH of the larvae of higher Diptera. Another JH compound, similar to JH III but with different chromatographic properties, has been tentatively identified in the stink bug *Plautia stali*. Confirmation will have to await its purification and structure determination.

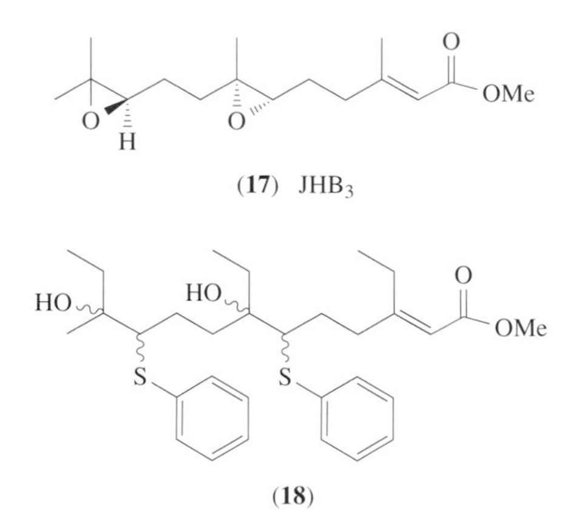

(17) JHB_3

(18)

It appears that JH III (**10**) is the most widely distributed of all the compounds, and is found in all the insect orders so far examined. It seems to be only the Lepidoptera that produce the full range of compounds, and JH 0 and 4-MeJH I have only been found in lepidopteran eggs and embryos.

Although the various JHs have different activities when injected into different insects (e.g., in the *Tenebrio* assay, the relative potency of JH I:JH II:JH III is 16 000:600:1, although JH III is the natural hormone in this species), there seems to be no difference in the function of the different compounds.

8.05.2.4 Synthesis

The great burst of synthetic activity from 1965 to 1971 gave first nonstereospecific syntheses, then with the correct geometry, and finally enantioselective syntheses. The challenge of JH was a great stimulus to finding ways to create trisubstituted double bonds of defined geometry. A comprehensive review of syntheses of JH compounds up to 1979 covers the methods then available.[27]

It must seem surprising for such relatively simple structures that enantiomerically pure JHs were not available until the mid-1980s. Mori and Mori synthesized enantiomerically pure JH III starting from dimethylcyclohexanedione,[28] which was first reduced to (*S*)-3-hydroxy-2,3-dimethyl-cyclohexanone (**19**) by baker's yeast, and by a further 19 steps gave the natural (*R*)-(+)-JH III (**10**) in an overall yield of 3.6%. The unnatural (*S*)-(−)-JH III was obtained in 2.3% yield in 20 steps from (**19**). There was no satisfactory method available to determine the optical purity of the products. The only successful method found was use of the chiral solvating reagent (*R*)-(−)-2,2,2-trifluoro-1-(9-anthryl)ethanol (**20**). Addition of this to (*R*)- and (*S*)-JH III gave sufficiently large differences in their NMR spectra at 400 MHz in $CDCl_3$ to distinguish them clearly. They were able to show that their products were ~100% *e.e.*, and the optical rotation they obtained for (**10**) was greater than that previously accepted.

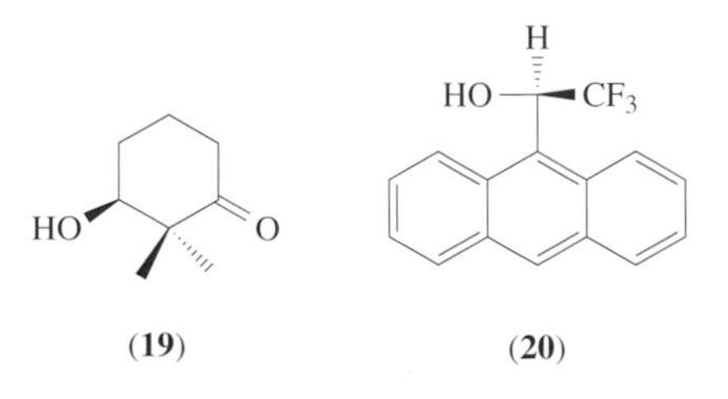

(19) **(20)**

Mori and Fujiwhara then made the natural (+)-JH I and (+)-JH II from methyl ethylcyclo-hexandione (**21**), which was reduced to (**22**) only by the yeast *Pichia terricola* (Scheme 1). Then using much the same strategy as above, they synthesized (+)-JH I (**8**) in 2.75% yield from the hydroxyketone in 19 steps and (+)-JH II (**9**) in 1.2% overall yield.[29] More recently, another synthesis of enantiomerically pure JH II has appeared which depends upon the stereoselective alkylation and carbonyl reduction of an optically pure β-ketosulfoxide.[30] They also found the only satisfactory way to determine the enantiomeric purity of the product was to use the chiral solvating agent (**20**).

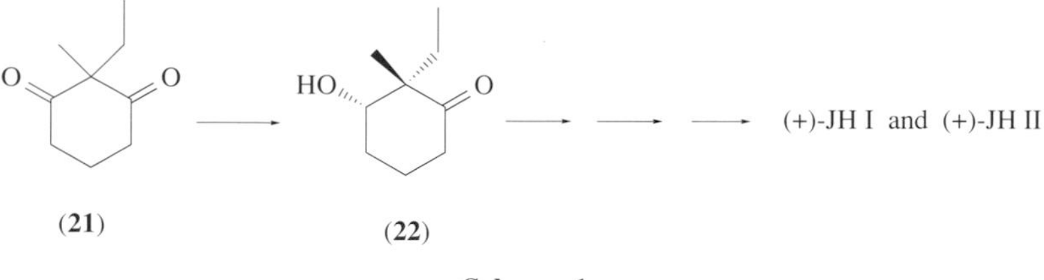

Scheme 1

JHB_3 has been synthesized from geraniol (Scheme 2) via Sharpless epoxidation, oxidation to the aldehyde, and chain lengthening by a Wittig reaction. Sharpless dihydroxylation gave (**23**), which was ring closed via its mesylate, and the product (**24**) was reduced with Wilkinson's catalyst to natural JHB_3 (**17**), of greater than 99.5% stereochemical purity. HPLC was used to remove small amounts of (10*S*) and (7*R*) enantiomers.[31]

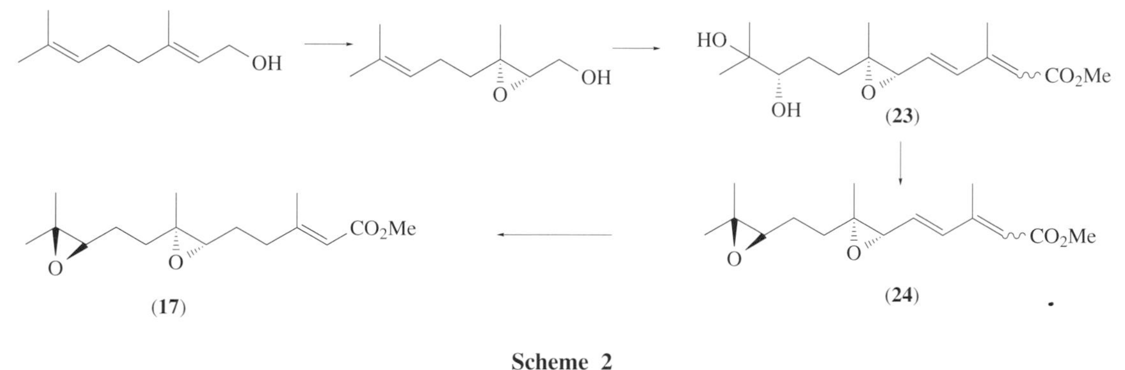

Scheme 2

8.05.2.5 Chemical Analysis

The minute amounts of JH compounds present in insects makes tough demands on chemical methods for their detection and quantitation. Their low masses and low polarity make them suitable for GC, but unless considerable purification is performed first, they can be lost among the many other substances present in a chromatogram when using flame ionization detection. Various alternatives have been offered, for example conversion to a 10-(2,4-dichlorobenzoyloxy)-11-methoxy derivative, which is suitable for electron capture detection, and gave a limit of detection of 3 pg ml^{-1} of insect blood.[32] A more reliable method is GC–MS, either underivatized[33] or as a derivative.[34] The latter method with selective ion monitoring can be used for still greater selectivity and sensitivity. The procedure is outlined in Scheme 3. Acid-catalyzed addition of [$^2H_3$]methanol to the epoxide after a simplified clean-up procedure gave recoveries of 75–80% and a detection limit of 10–40 pg g^{-1} of tissue.[34] The derivatized JHs yield fewer fragment ions and the diagnostic ions appear in parts of the spectrum uncluttered with background ions. There is a comprehensive review of techniques for identification and quantification of JH by Baker.[35]

The production of JH in insects can be followed by injection of [^{14}C]methylmethionine, or by the culture of corpus allata in a medium containing it. The labeled methyl group is incorporated into the methyl ester of the JH (see Section 8.05.2.6).

Radioimmunoassay (RIA) methods and antisera are available for JH determinations, and this method has been used largely by biologists for identifying JH in insects.[36] There is a summary available on reports of measurement of JH titer in a large number of species.[2] A recent improvement in RIA claims to make it significantly more sensitive: one antiserum, equally sensitive for JH I, II, and III, has a limit of 100–130 pg per sample, and another, specific for JH II, is useful down to 35 pg per sample.[37] RIA methods must be used with caution because many other lipid substances give false-positive reactions.[38]

Insect tissue containing JHs (e. g. 3-6 ml of haemolymph, or 5-15 g whole insects)

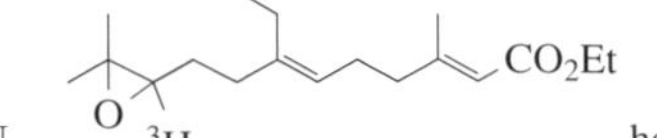

1. Add CH_3CN, , homogenize, filter
2. Partition filtrate between pentane-water, collect and concentrate pentane

Crude lipid extract (glycerides, sterols, fatty acids and JHs)

3. C_{18} RP Sep-pak, elute with acetonitrile
4. Pasteur pipette alumina column, elute with ether
5. Add d_4-MeOH, TFA,

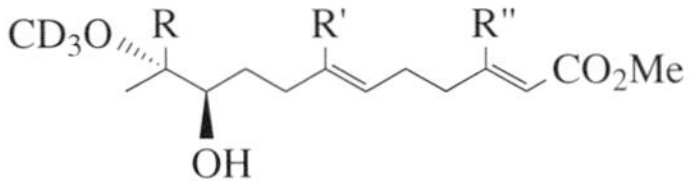

CD_3O R R' R'' CO_2Me

OH

6. HPLC silica, elute ether-CH_2Cl_2

70-80% yield of [^{3}H]-standard, 10^7-10^8 purification factor

Scheme 3

8.05.2.6 Biosynthesis

The biosynthesis of JH III was presumed to proceed via the usual route for acetate through acetoacetate, mevalonate, isopentenyl pyrophosphate, and farnesyl pyrophosphate, and initially it was thought that the extra methyl groups were derived from *S*-adenosyl methionine, by analogy with the formation of ethyl groups in the side chains of plant sterols. Experiments with labeled methionine, however, showed that the label was incorporated only into the ester methyl group, itself an important discovery that was useful in tracking the formation and metabolism of JH compounds. It was shown that labeled acetate and propionate were incorporated into JH II and JH III by incubation with corpora allata of *Manduca sexta*. Either [1-^{14}C]propionate or [2-^{14}C]propionate and [2-^{14}C]acetate were incorporated into JH II in the ratio 1:3, while propionate was not incorporated into JH III.[39] [2-^{14}C]Mevalonic acid was similarly incorporated into both JH II and JH III. Exactly where the propionate, replacing acetate, was incorporated was found by administering [1-^{14}C]propionate to *Hyalophora cecropia* pupae, and isolating and degrading the labeled JH I.[40] Only atoms C-7 and C-11 were labeled. [5-^{3}H]Homomevalonic acid (**25**) was also synthesized and shown to be incorporated via homoisopentenyl pyrophosphate (**26**) and isopentenyl pyrophosphate (**27**) into JH II, but not JH III, in the predicted manner (Scheme 4).[41] Presumably in Lepidoptera the enzymes that assemble the JH compounds are highly specific, since we do not find a random mixture of all the possible acetate plus propionate compounds. From farnesyl pyrophosphate, or its homo- or bishomo-homolog (**28**), the pathway proceeds to farnesoic acid, or its homologs (**29**). The sequence is completed by epoxidation and esterification. The epoxidation is by an enzyme linked to cytochrome P450; methylation uses *S*-adenosylmethionine. It appears that different insects perform the last two steps in different orders.[42] In *M. sexta* the free acid is epoxidized and then methylated to the ester, but there is evidence that in other insects (locusts and cockroaches, which are hemimetabolous insects), methylation to the ester is followed by epoxidation (Scheme 5). In the higher Diptera the JH III bisepoxide has been shown to be formed from farnesoic acid via epoxidation and then methylation, because methyl farnesoate is not epoxidized to JH III or the bisepoxide, and JH III is not epoxidized to the bisepoxide, but farnesoic acid is converted to the bisepoxide.[43] Biosynthetic studies in many species have been discussed by Schooley and Baker.[42]

Insects seem to be versatile employers of biosynthetic propionate and acetate together or alternatively to produce whole families of compounds, in ways that are unknown among higher animals but which have their parallels among microorganisms. The JH compounds are a good example. There are a number of other examples among pheromones and secretions. It has been shown that pig liver prenyltransferase (which assembles isoprenyl units into farnesyl pyrophosphate) will accept homomevalonate and incorporate it into homofarnesenes, so it has been suggested that insects have a special ability to construct the homomevalonate units, not found in higher animals.

In spite of their lipophilic character, the JH compounds are sufficiently water-soluble (up to 5×10^{-5} M) to be carried free in the insect hemolymph (concentration in insect larvae $< 10^{-7}$ M), yet to be transported in the body they seem to be bound to carrier proteins (JH-binding proteins), which protect them from degradation by hemolymph enzymes, since JH compounds are relatively unstable and are readily inactivated by esterases.

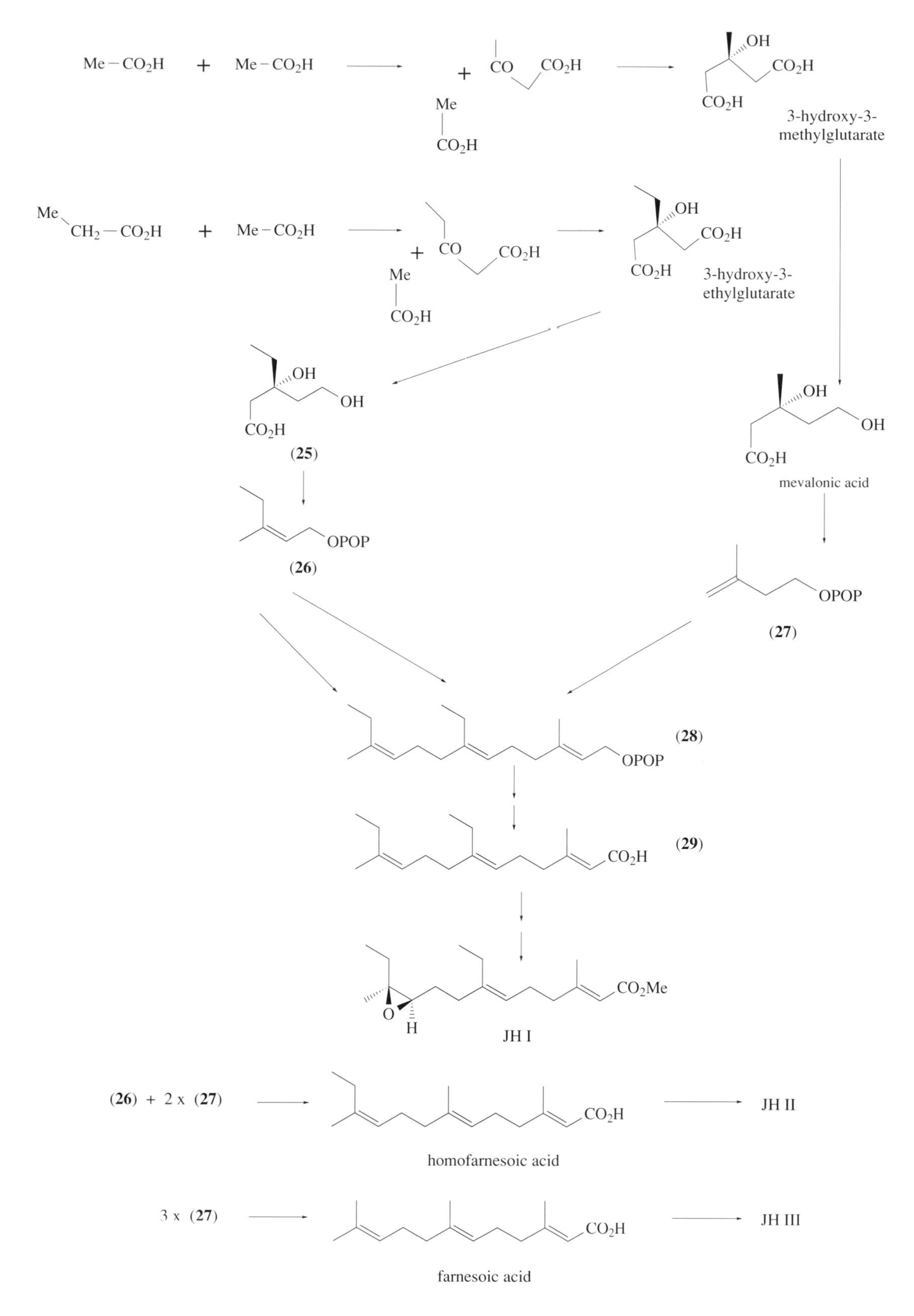
Me – CO2H
+
Me – CO2H
CO
CO2H
Me
CO2H
OH
CO2H
CO2H
3-hydroxy-3-methylglutarate
Me
CH2 – CO2H
+
Me – CO2H
CO
CO2H
Me
CO2H
OH
CO2H
CO2H
3-hydroxy-3-ethylglutarate
OH
OH
CO2H
(25)
OPOP
(26)
OH
OH
CO2H
mevalonic acid
OPOP
(27)
(28)
OPOP
(29)
CO2H
CO2Me
O
H
JH I
(26) + 2 x (27)
CO2H
JH II
homofarnesoic acid
3 x (27)
CO2H
JH III
farnesoic acid

Scheme 4

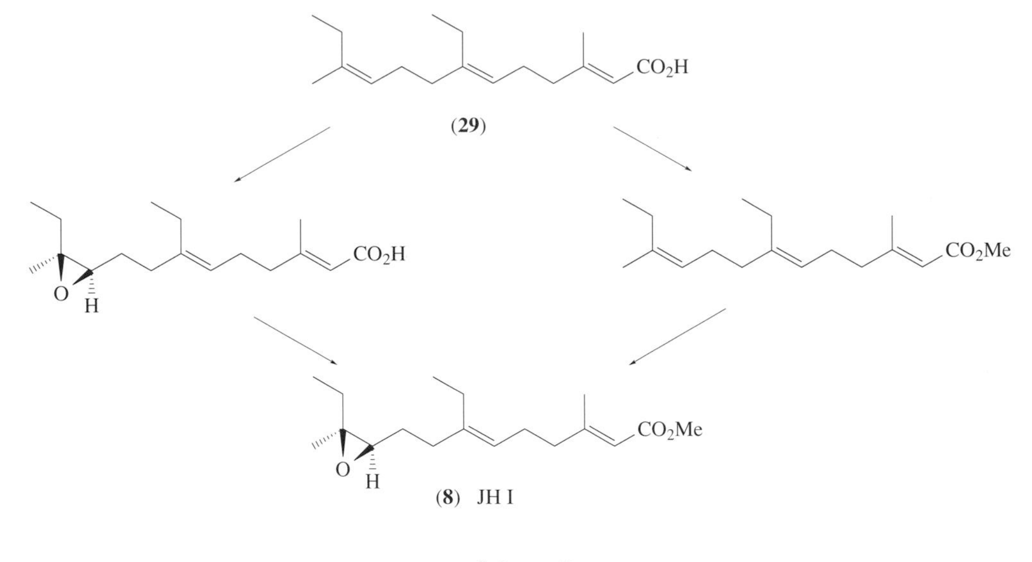

Scheme 5

8.05.2.7 Inactivation

The essential steps of JH inactivation are hydrolysis of the ester and opening of the epoxide to a diol, to give a dihydroxyfarnesoic acid (**30**) as the final product. Again the experimental evidence depends upon the species. In some the free acid is produced first, in others epoxide opening occurs first (Scheme 6).[44] When the metabolism of (10*R*)-[^{3}H]JH III and (10*R*, 11*S*)-[^{3}H]JH I was studied by injection into larvae of *Manduca sexta*, it was found that JH III was metabolized much more rapidly than JH I. Both compounds were converted to the corresponding diol acid, but also to a phosphate conjugate, while for JH I the conjugate was the major product.[45]

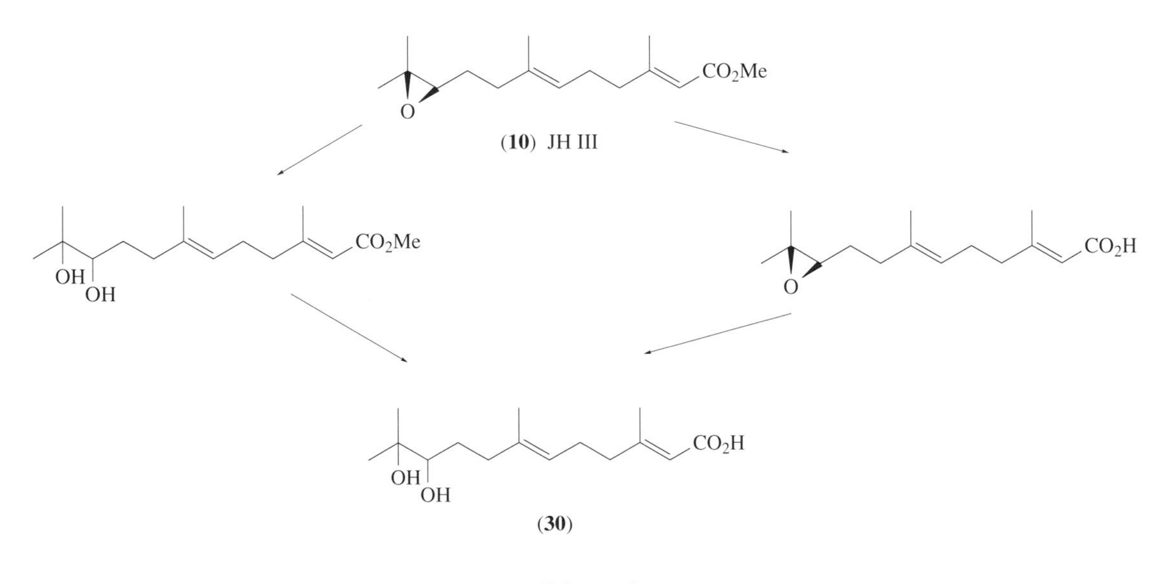

Scheme 6

The available evidence up to the mid-1990s indicated that insects do not store JH, so production is dependent upon rate of synthesis. What stimulates production seems to vary with the species, and is too complex to consider here.[46] However, more recent work suggests there is another product, probably an acylglycerol derivative of JH, biosynthesized by the corpora allata in at least three species of Lepidoptera, and stored in that gland.[47] The maintenance of the physiological levels of

JH depend upon interacting effects of activity of the corpora allata in synthesis, carrier proteins to transport the JH, and degradative enzymes to remove it. Insects are a large, diverse, and ancient group of organisms, and seem to have developed many independent pathways. The relative importance of these effects depends very much upon the particular insect species.

8.05.2.8 JH Mimics

The discovery that JH had a profound influence on insect development and the finding of highly active JH mimics such as (**5**) early in the investigation of JH generated hopes of finding substances for insect pest control that would be effective through upsetting insect development. JH exposure at the wrong time in the development cycle is almost always lethal, and JH is readily absorbed through the cuticle. Mimics should disrupt metamorphosis and egg production. The JH compounds themselves soon proved to be too unstable to light and oxygen, too expensive to synthesize, and too easily hydrolyzed by a number of enzymes to be useful. There was a period of intense activity in a search for and testing of more stable compounds which could be used as JH mimics. Titles of papers such as "Cecropia juvenile hormone: harbinger of a new age in pest control" indicate the mood of the period.[48] Several thousand compounds were synthesized with JH activity (cf. Jacobson *et al.*[49] and Pallas and Menn[50]). The optimistic assumptions that JH mimics would be widely used were, however, not sustained, as several disadvantages became apparent. They only act on certain stages of the insect life cycle, most effectively at the end of the last larval instar, but larvae continue to feed. Compounds found very active in bioassay proved ineffective in use. Some species, for example leaf miners or those inside fruit or seeds, could not be contacted with the substance. It was thought, too, that insects would not develop resistance to mimics of such essential substances; that was found to be untrue, and some species developed resistance to the synthetic mimics as rapidly as they did to conventional pesticides.[51]

Today there is a small number of compounds that have reached commercial production and find use in some special situations. Twenty of them are listed in a table by Sehnal.[52] The first of these discovered was hydroprene (**31**) (Altozar, Gencor, Gentrol, Mator, or ZR 512), which though slow acting is a very effective agent for the control of cockroaches. The best known, and the first registered for commercial use, is methoprene (**32**) (trade names Altocid, Precor, Apex, and ZR 515), used to control mosquito larvae, at concentrations as low as 0.025 lb acre^{-1} (4.6 g Ha^{-1}). It is also used in cattle salt blocks at a concentration of 0.02% for control of pests infesting livestock feces. As Apex it is used to control sciarid flies in mushroom culture, and as Kabat against pests in stored tobacco. In Great Britain it is very useful (as Pharorid) for the control of *Monomorium pharaonis* (pharaoh's ant) in hospitals. It is also reported to be used (as Manta) to increase natural silk production: it is sprayed on last-instar silkworms to give increased formation of silk fiber. Both hydroprene and methoprene are very active in insects and are nontoxic to mammals (LD_{50} > 34 000 mg kg^{-1} acute, orally in rats, for both).

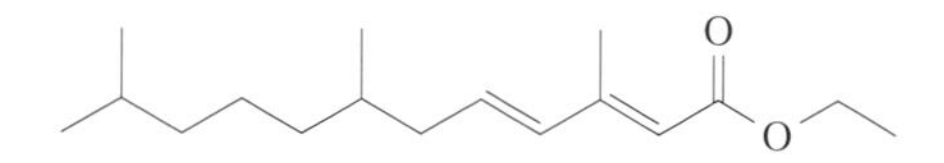

(**31**) Hydroprene

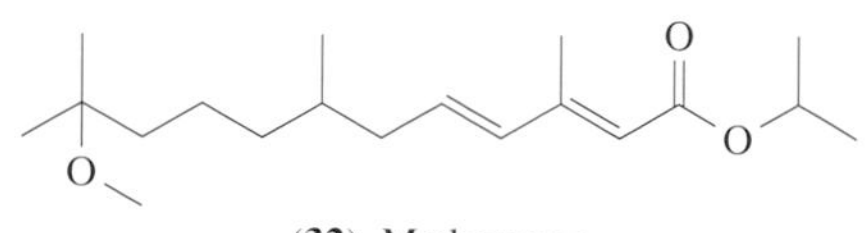

(**32**) Methoprene

Others are kinoprene (**33**) (trade names Enstar, ENF-70531, and ZR 777) used to control aphids and white fly on ornamental plants in greenhouses, epofenonane (**34**) based on the structure of a natural plant compound (see Chapter 8.02.10.2), fenoxycarb (**35**) (trade names Award, Insegar, Logic, Varikill, ABG-6215, and others) used in horticulture, and pyriproxyfen (**36**), which has been shown to inhibit emergence of 50% of adult *Chironomus fusciceps* at a concentration of 0.0018 ppm and 90% inhibition at 0.0537 ppm.[53] A survey on advances in juvenoids is presented by Henrick.[54]

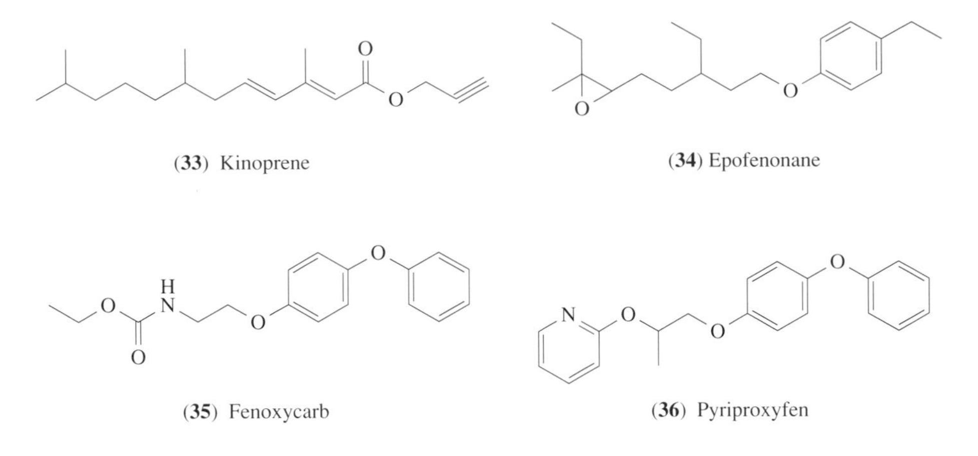

(**33**) Kinoprene

(**34**) Epofenonane

(**35**) Fenoxycarb

(**36**) Pyriproxyfen

8.05.2.9 Phytojuvenoids

Once the bioassay for juvenile hormone was established (see Section 8.05.2.1) it was possible to look at many materials for JH activity. Six years before the isolation of the first JH it was shown that there was weak JH activity in microorganisms and plants.[55] Even farnesol (**1**), widely distributed in nature, particularly in plants, had weak JH properties. After the isolation of the first JH compounds there began an intensive search through the plant kingdom for JH compounds or JH activity, which by analogy with the phytoecdysteroids (see Section 8.05.3.5) were termed phytojuvenoids. Unlike the ecdysteroids, which are found widely distributed in plants, and are 100–1000 times more concentrated in some plants than in insects, the JHs have only once been found in plants. Relatively large amounts of natural JH III and its precursor methyl farnesoate were discovered in a Malaysian sedge, *Cyperus iria* (Cyperaceae), some years ago, but the authors are not aware of any further reports of that kind,[56] although many plant substances of diverse structure have been found with JH activity. These substances and some of the thousands of synthetic juvenoids can have biological activity more than a thousand time greater than the JHs themselves, which suggests a rather low specificity of the JH binding site or activity at a number of relevant sites. Their overt biological activity is discussed by Sehnal.[52] Whether these plant substances are produced to protect the plant from insect predators cannot be answered. It has been reported that many isoprenoid juvenoids are much more effective when ingested by phytophagous insects than when applied topically.[57] There is little or no direct evidence that in nature they function effectively in this way.

8.05.2.9.1 Juvabiones

Just about the time the isolation and identification of juvenile hormone was proceeding, a curious incident led to the discovery of a group of plant compounds with a JH effect. The bug *Pyrrhocoris apterus*, raised in the laboratory for experimental purposes, was found to develop abnormally when cultured in the USA, while the same strain behaved normally when raised in Europe.[58] The cause was quickly traced to the paper which was used to line their cages. Something in the American paper made from the Canadian balsam fir (*Abies balsamea*) was affecting their maturation. It was called the paper factor, and soon shown to be a sesquiterpene ester, and called juvabione (**37**),[59] in reality the methyl ester of todomatuic acid, isolated much earlier from the turpentine oil of the Todo fir (*Abies sachalinensis*).[60] Mori quickly produced a synthesis of racemate (**37**) from *p*-methoxyacetophenone,[61] and this was followed by a synthesis of natural (+)-juvabione from (*R*)-(+)-limonene.[62] Subsequently (+)-dehydrojuvabione (**38**) was isolated in larger quantities from *A. balsamea* wood[63] but having only 1/10 the activity in bioassay, as were two further closely related juvenoids (**39**), and (**40**), and others from the wood of *Abies lisiocarpa* (**39**), (**41**), (**42**), (**43**), and (**44**).[64]

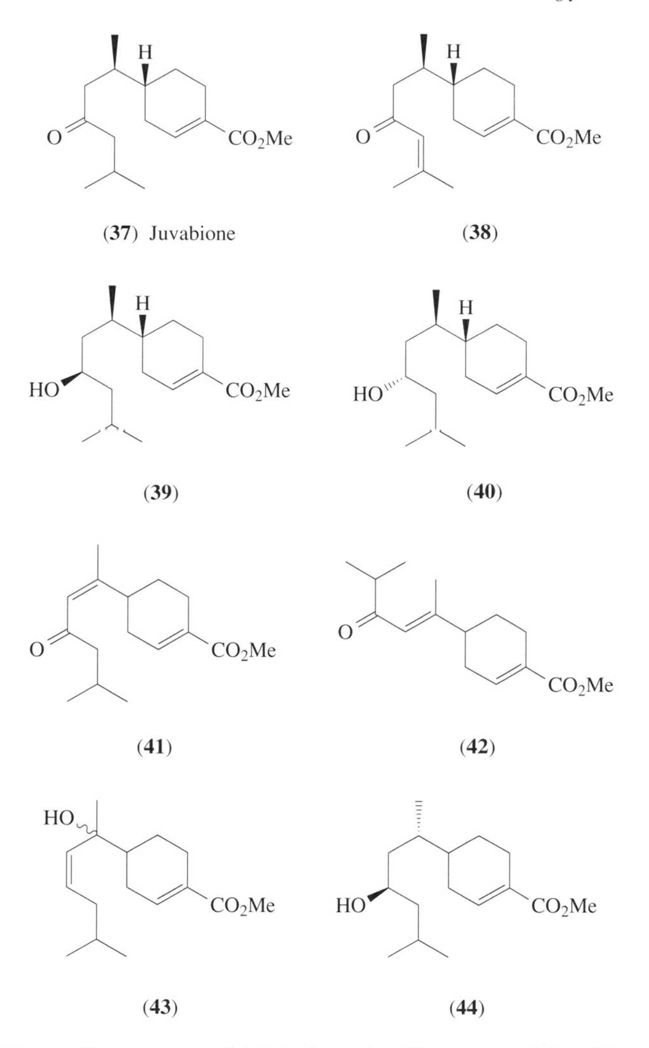

(37) Juvabione (38) (39) (40) (41) (42) (43) (44)

By a fortunate accident, *P. apterus*, which led to the discovery of juvabiones, is extraordinarily sensitive to these JH analogs, but the juvabiones are essentially only active on pyrrhocorid bugs, and themselves are worthless as insect control agents.

8.05.2.9.2 Other phytojuvenoids

In one survey of 343 species of higher plants for JH activity using two species of insect (the mealworm *Tenebrio molitor* and the milkweed bug *Oncopeltus fasciatus*) in bioassays, one particularly active compound was isolated in 0.0003% yield from the roots of a common weed known as the American coneflower (*Echinacea angustifolia*).[65] The compound, called echinolone (**45**), has a superficial resemblance to JH. It is noteworthy that the same plant contains the powerful insecticide echinacein, an unsaturated isobutylamide (see Section 8.05.5.1).

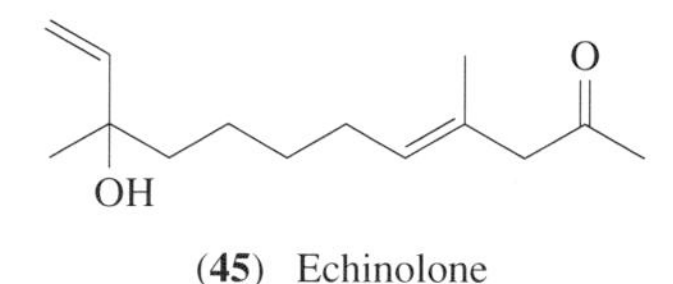

(**45**) Echinolone

The oil of sweet basil (*Ocimum basilicum*) gave two exceptionally active compounds, juvocimene-1 (**46**) and juvocimeme-2 (**47**),[66] which, although having structures close to JH themselves, are both several orders of magnitude more powerful in bioassays on the milkweed bug (*O. fasciatus*) than JH III, and juvocimene-2 is 10 times more active than juvocimene-1.[66] Sesamin (**48**), known since 1928 to be present in sesame (*Sesamum indicum*) oil, and sesamolin (**49**), known since 1907, already recognized as having insecticidal synergist action, were also found to have some JH activity.[67] It is reported that JH activity is confined to gymnosperms and angiosperms. This may be so or it may simply be that in spite of the search that was made, insufficient numbers of other plants were examined to discover activity.

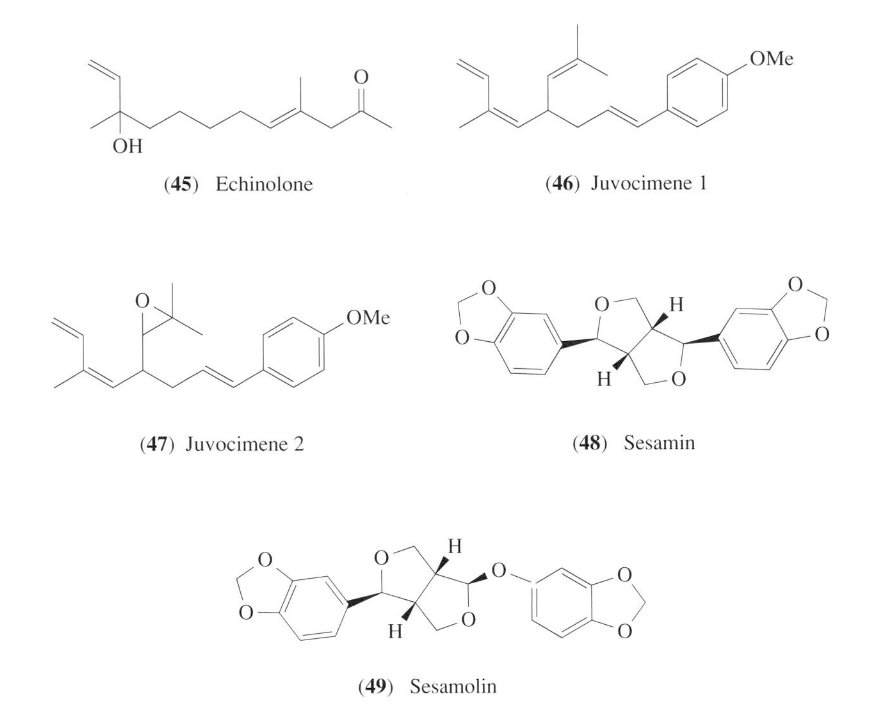

(45) Echinolone (46) Juvocimene 1

(47) Juvocimene 2 (48) Sesamin

(49) Sesamolin

8.05.2.10 Precocenes

In the course of the intensive screening of plants for JH mimics and anti-JH activity, Bowers discovered that an extract of the common ornamental garden plant *Ageratum houstonianum* caused the second-instar larvae of the milkweed bug *Oncopeltus fasciatus* to undergo two normal molts and then metamorphose prematurely to miniature adultoids. This observation led to the isolation of two anti-JH substances, which he called precocenes I (**50**) and II (**51**),[68] and which are present in a number of *Ageratum* species. Precocene I had been isolated earlier as a methoxychromene[69] and synthesized from 6-methoxycoumarin,[70] but the precocene name has stuck.

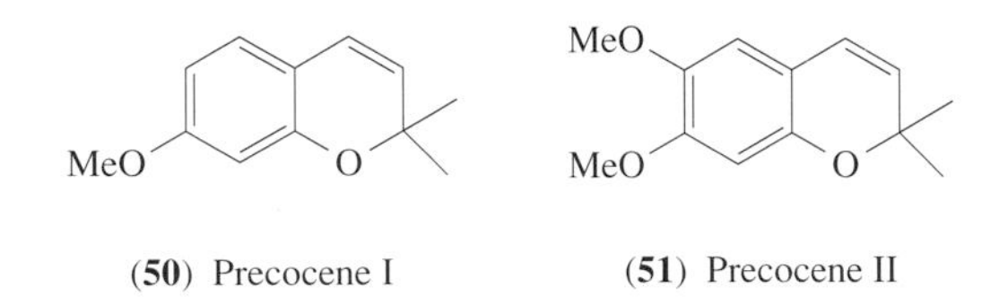

(**50**) Precocene I (**51**) Precocene II

The precocenes have been shown to have a toxic effect on the corpora allata, and prevent these glands from making JH. The double bond of the chroman ring is activated by the *p*-methoxy group, and is epoxidized by a cytochrome P450 mixed-function oxidase involved in the late stages of JH production (Scheme 5), and becomes permanently bound to it.[71] They are effective in causing premature metamorphosis only in some Hemiptera, Homoptera, and Orthoptera, and they have little or no effect on the holometabolous insects.[72] They do not inactivate the corpora allata of mosquitos and honeybees, but do have a toxic effect. They are also hepatotoxic to rats. They are therefore not interesting as materials for insect control, but have been valuable tools in the study of insect physiology and behavior. Bowers considers them as "suicide substrates," substances accumulated in the plant, themselves nontoxic but converted to toxic products by the insect's own metabolism. A study of their biosynthesis in plants with [2-^{14}C]acetate and [2-^{14}C]mevalonate has shown that the aromatic ring is acetate derived and the other five carbon atoms of the rings are from a terpenoid.[73] It should be noted that the group of acylchromenes of type (**52**), which are insecticidal, are not epoxidized in inactivation and do not function as anti-JH substances.

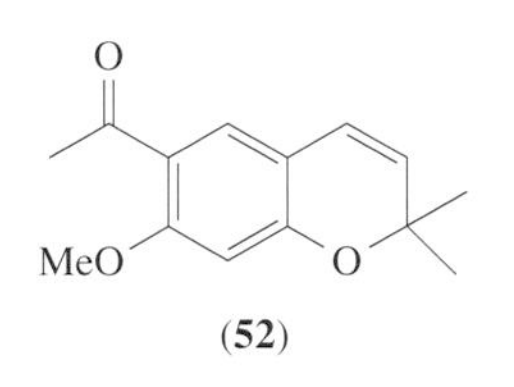

(**52**)

8.05.3 ECDYSTEROIDS

The introduction to this chapter shows how the process of arthropod development is under the control of hormones. Molting, or ecdysis, is mediated by a group of hormones known collectively as ecdysteroids. The first example of this class of hormones, ecdysone (**53**), was first isolated in a pure form from pupae of the silk moth *Bombyx mori* in 1954.[74] The ecdysteroids form a group of polyhydroxylated steroids that have been found to be widespread in nature. Zooecdysteroids have been found in arthropods such as insects and crustaceans and some nonarthropod invertebrates, while large numbers of structurally diverse phytoecdysteroids have been isolated from plants.

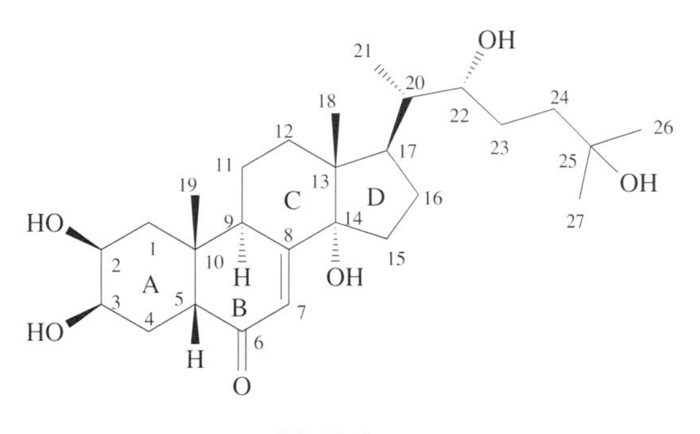

(**53**) Ecdysone

Molting in insects appears to be initiated by neurosecretory cells in the *corpus cardiacum* which secrete a prothoracicotropic hormone (see Section 8.05.4.1). This hormone stimulates the prothoracic glands (or their equivalent) to synthesize and release ecdysone, or 3-dehydroecdysone (**54**), into the hemolymph (where 3-dehydroecdysone is rapidly converted to ecdysone by the 3β-reductases present in hemolymph and various other tissues). Ecdysone is transported by the hemolymph to various tissues, principally fat body and malpighian tubules, which convert it into 20-hydroxyecdysone (**55**). 20-Hydroxyecdysone is generally considered to be the more potent and thus more physiologically active hormone. In some insect species (*Oncopeltus*, *Melanogaster*, *Apis mellifera*) a structurally related C-20-hydroxylated ecdysteroid, makisterone A (**56**) is observed rather than 20-hydroxyecdysone, while in some crustaceans the main C-20-hydroxylated molting hormone seems to be ponasterone A (a 25-deoxy-20-hydroxyecdysteroid) (**57**).

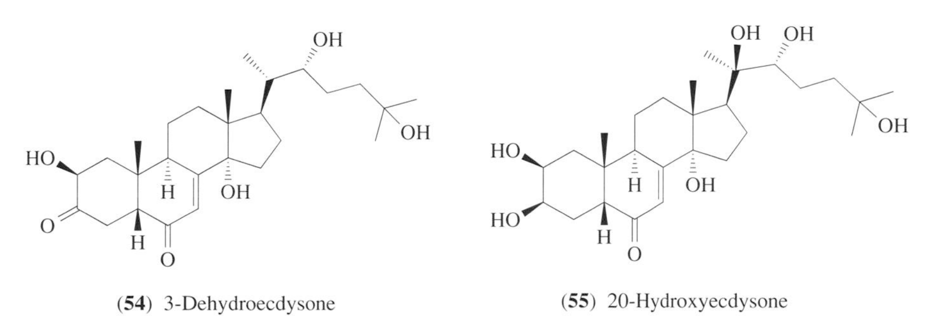

(**54**) 3-Dehydroecdysone (**55**) 20-Hydroxyecdysone

The very widespread occurrence of this type of compound in plants, including ecdysone and 20-hydroxyecdysone together with many structural variants on the basic ecdysteroid pattern, has been

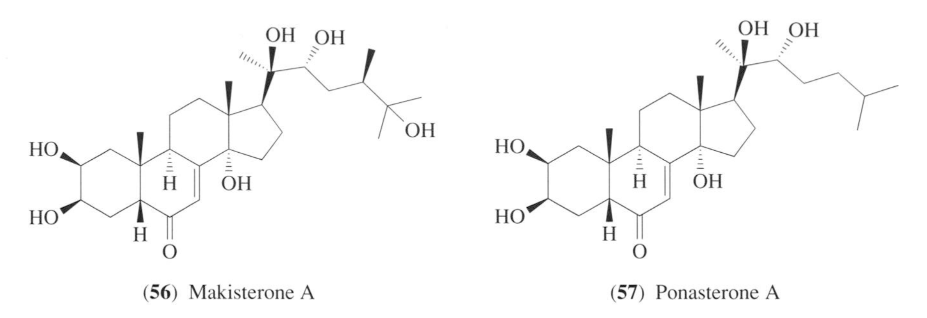

(**56**) Makisterone A (**57**) Ponasterone A

the cause of much speculation as to their function. Although a role as a plant hormone has been suggested, and cannot be ruled out, it seems more likely that these phytoecdysteroids act as defensive substances. They are thought to protect the plants that contain them against the depredations of phytophagous insects.

8.05.3.1 Bioassays

The initial bioassay systems developed for the detection of the ecdysteroids were based on the larval-to-pupal molt of a number of dipteran species (e.g., *Calliphora*,[75,76] *Sarcophaga*,[77] and *Musca*.[78] In these assays, last-instar larvae are ligated, and those in which the abdomen remains in the larval form are taken for use in the assay. Extracts, or pure compounds, are then injected into the abdomen, following which signs of development are sought (e.g., tanning of the cuticle). An alternative *in vivo* assay, which does not require the injection of the test material, is based on the use of ligated larvae of *Chilo suppressalis*, which are simply dipped into methanolic extracts or solutions of the test material.[79] These tests provide sensitivities in the range 5–50 ng of 20-hydroxyecdysone per test. Bioassays of this type were important in early work to monitor the purification of active molting hormones from large-scale extraction and isolation programs and for screening purposes. They were also employed to some extent for quantitative analysis in the determination of the titers of the hormones in insects during development, and in attempts to measure the relative potency of different ecdysteroids. However, the use of these *in vivo* systems for the establishment of quantitative structure–activity relationships (QSAR) has been more problematic. This is because of the potential for less active compounds to be converted, via metabolism in the test system, into physiologically more active ones (e.g., the conversion of ecdysone into the more active 20-hydroxyecdysone). Further, the potential for different rates of deactivation or penetration to the sites of action can also result in difficulties in interpreting data.

For QSAR work, and for relatively high-throughput screening purposes, *in vitro* systems are somewhat better suited. A variety of such *in vitro* systems have been designed, including the puffing of polytene chromosomes,[80] morphogenesis in imaginal disks[81] (including a system that allows for metabolic "activation" as a result of the presence of fat body[82]), and systems based on Kc cell lines derived from *Drosophila melanogaster*.[83] The use of such systems has been reviewed.[84] More recently, an *in vitro* assay based on the use of an ecdysteroid-responsive B_{11} cell line, also derived from *D. melanogaster*, has been described.[85] In this assay the cells are grown in 96-well plates, and their densities measured turbimetrically via a microplate reader. Ecdysteroids, and ecdysteroid agonists, cause the cells to clump together, producing a reduction in absorbance. The system can also be used to search for antagonists by assessing the ability of compounds to prevent or limit the expected reduction in absorbance, resulting from co-incubation of the test material with 20-hydroxyecdysone.

8.05.3.2 Occurrence

The ecdysteroids and related substances have now been shown to be widespread in both arthropods and many species of plant. Indeed, ecdysteroids have been found in algae and fungi. From the alga *Laurencia pinnata* a series of pinnasterols has been isolated (pinnasterol (**58**), 2-acetylpinnasterol (**59**), 2-acetyl-14α-hydroxypinnasterol (**60**), 3-acetyl-14α-hydroxypinnasterol (**61**), and 2-acetyl-14α-hydroxy-22-*epi*-pinnasterol (**62**))[86,87] and from the fungus *Polyporus umbellatus* a series of polyporusterones A to G (**63**–**69**).[88] Although these are from a lower plant and a fungus, they illustrate the way in which families of similar ecdysteroids are often found in one species. The phytoecdysteroids most commonly encountered are 20-hydroxyecdysone and polypodine B (**70**), often present together with a "cocktail" of minor phytoecdysteroids. There is an interesting hypothesis of the adaptive advantage of such a mixture.[89] The most commonly encountered zooecdysteroids are ecdysone and 20-hydroxyecdysone. Thus far, in excess of 250 ecdysteroids, or ecdysteroid-like compounds, have been isolated from arthropod or plant sources.[90] At first it was thought that plant ecdysteroids were confined to the ferns and gymnosperms, the older part of the plant kingdom, but later it was found that they are much more widely distributed. Of the 180 or more families of mono- and dicotyledons, gymnosperms and ferns and their allies, which have been examined for ecdysteroids, only the Urticaceae, Cannabaceae, Cruciferae, the monospecific Gingkoaceae and the grasses, Graminae (with the possible exception of the genus *Briza*) have not yielded ecdysteroids from some species. A study of 1056 species of Japanese plants from 738 genera and 186 families using the *Chilo* dipping test for ecdysteroids (see Section 8.05.3.2) found that 5.1% of species gave a positive reaction for their presence.[91] Dinan is conducting a massive survey of plant seeds, with the intention of covering 5000 species, or approximately 1% of the estimated world flora, using two different antisera in a radioimmunoassay (see Section 8.05.3.8.5).[92] With more than 4500 plant species examined, he has found that about 6% of them have given an ecdysteroid-positive reaction. Highest levels by radioimmunoassay in seeds have been in *Gomphrena affinis* Amaranthaceae (0.85%), *Serratula tinctoria* Compositae (1.17%), and *Silene burchellii* Caryophyllaceae (0.73%). In the Chenopodiaceae, which he has studied intensively, there is a strong correlation between presence of ecdysteroids in the seeds and other parts of the plant during later stages of development.[92]

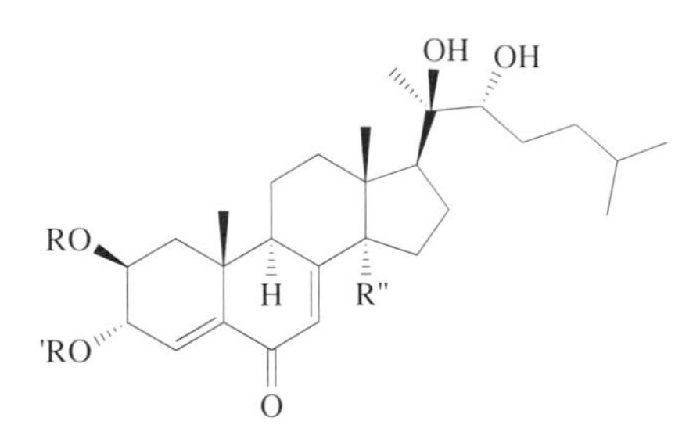

(**58**) R = R' = R" = H Pinasterol
(**59**) R = Ac, R' = R" = H 2-Acetylpinnasterol
(**60**) R = Ac, R' = H, R" = OH 2-Acetyl-14a-hydroxypinnasterol
(**61**) R = H, R' = Ac, R" = OH 3-Acetyl-14a-hydroxypinnasterol

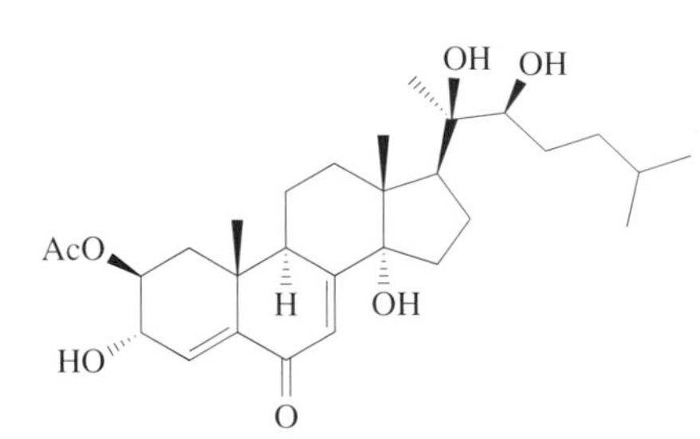

(**62**) 2-Acetyl-14a-hydroxy-22-*epi*-pinnasterol

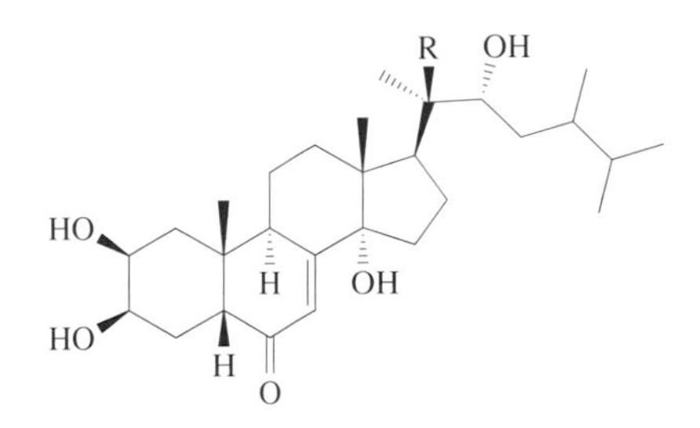

(**63**) R = OH Polyporusterone A
(**68**) R = H Polyporusterone F

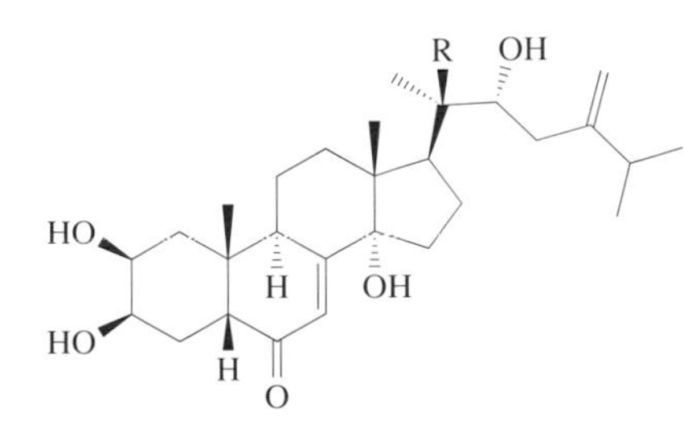

(**64**) R = OH Polyporusterone B
(**69**) R = H Polyporusterone G

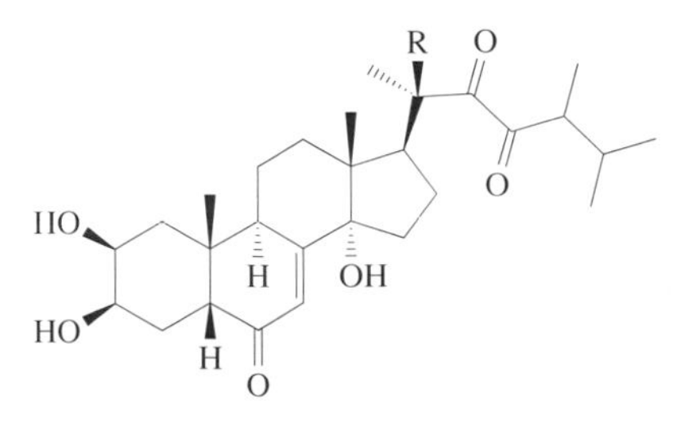

(**65**) R = OH Polyporusterone C
(**67**) R = H Polyporusterone E

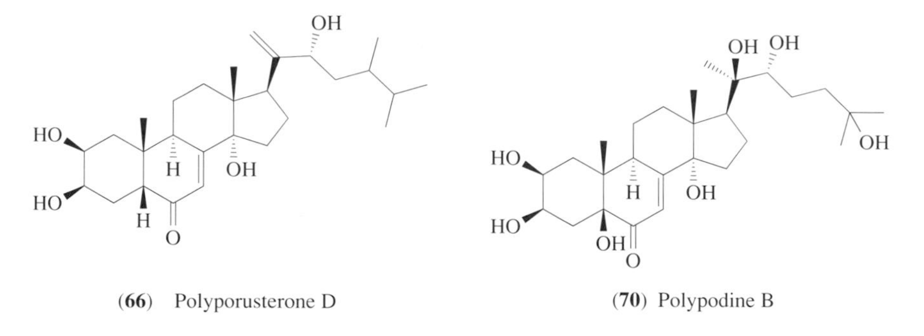

(**66**) Polyporusterone D (**70**) Polypodine B

An exact definition of the structural features required to say what constitutes an ecdysteroid has proved to be rather difficult to achieve. Strictly, any such definition probably requires a steroid to be in possession of a *cis*-fused A,B ring combined with a 7-en-6-one and a 14α-hydroxyl group, as suggested by Lafont and Horn.[93] A somewhat broader classification has been used here, which uses as its basis the compounds collected together in *The Ecdysone Handbook*.[90]

The variations in structure that are seen in the phyto- and zooecdysteroids includes the total number of hydroxyl groups present on the nucleus and side chain and the number of carbon atoms. The number of carbon atoms in ecdysteroids depends on the structure of the side chain and the phytosterol from which the compound was derived and the extent to which side chain cleavage occurs. The side chain can show very great structural variety, incorporating a γ-lactone, as in cyasterone (**71**), or δ-lactone, as in ajugalactone (**72**), or a carboxylic acid. In addition, conjugation is known to occur at the hydroxyl groups with inorganic species (e.g., sulfate and phosphate), organic acids (acetates, benzoates, cinnamates, coumarates, and long-chain fatty acids), and sugars (galactosides, xylosides, glucosides, etc.). Acetonides and methyl ethers have also been observed.

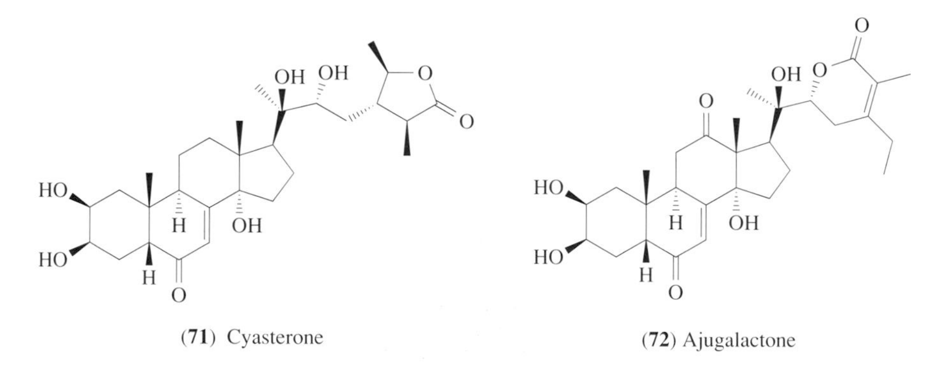

(**71**) Cyasterone (**72**) Ajugalactone

8.05.3.3 Structure and Isolation

The initial undertaking of the isolation of ecdysone from insect material was a heroic task that ranks as a classic piece of natural product chemistry. Initial work by Plagge and Becker, made possible by the development of the *Calliphora* bioassay, concentrated on extracts of *Calliphora* pupae. These studies were interrupted by Becker's death in the Second World War. However, further studies by Butenandt and Karlson employing pupae of the silk moth (*Bombyx mori*) began in 1943. This organism provided a better source of the hormone, and from 500 kg of pupae some 25 mg of pure ecdysone (**53**) was finally isolated.[74] Further work led to the isolation of a second, more polar, hormone.[94] This was named β-ecdysone, or ecdysterone, in order to differentiate it from the first hormone (which also became known as α-ecdysone), and was identified as 20-hydroxyecdysone (**55**).

The determination of the structure of ecdysone, reviewed by Horn,[95] was beyond the immediate capabilities of the time. Initial studies suggested a molecular formulae of $C_{18}H_{30}O_4$ or $C_{18}H_{32}O_4$, with IR and UV data suggesting an α,β-unsaturated ketone. X-ray crystallography and MS gave a more reliable molecular mass of 464, leading to the revision of the molecular formulae to $C_{27}H_{44}O_6$, and to the suggestion that the hormone was a steroid. However, the elucidation of the complete structure was a long and complex task, only brought to fruition 11 years after the original isolation of the hormone in a pure form with the X-ray crystallographic studies of Huber and Hoppe, which finally yielded a complete structure and stereochemistry.[96] As a result of this work the structure was shown to be 2β,3β,14α-(22*R*,25*S*)-pentahydroxy-5β-cholest-7-en-6-one (**53**), m.p. 238–239 °C, $[\alpha]_D^{20}$ + 64.7° (MeOH), soluble in polar solvents.

As well as studies on the hormones controlling insect molting, work was also being undertaken to investigate this process in crustaceans. Studies on extracts of the marine crayfish (*Jasus lalandei*) resulted in the isolation of 2 mg of a crustacean molting hormone (from 1 t of material), which was called "crustecdysone."[97] This hormone was identified as 20-hydroxyecdysone (**55**), m.p. 241–242.5 °C, $[\alpha]_D$ + 61.8° (MeOH), and shown to be identical with the substance called ecdysterone or β-ecdysone previously isolated from *B. mori*.

At the same time that these studies on the arthropod molting hormones were being described, work was also being undertaken on plants. In 1966, Nakanishi was able to report the discovery of a polyhydroxy steroid ponasterone A (**57**), isolated from the leaves of *Podocarpus nakaii* Hay, that was the 25-deoxy analog of 20-hydroxyecdysone.[98] This was rapidly followed by the isolation of 20-hydroxyecdysone itself from the Australian pine (*Podocarpus elatus*).[99]

The finding of the ecdysteroids in plants, often in much larger amounts than in insects and crustaceans, proved to be of considerable benefit to those studying the mode of action of the hormones in arthropods. In addition, the structural variety of the plant-derived phytoecdysteroids enabled structure–activity relationships to be investigated. Levels of total ecdysteroids of 0.001–0.1% of the dry weight of plants are common, but extreme examples are in the stems of *Diploclisia glauclescens*[96] (Menispermaceae) containing 3.2%, the roots of *Cyanotis arachnoidae*[96] (Commelinaceae), which contain 2.9%, and the seeds of *Leuzea carthamoides*[96] (Compositae) which contain 1.9%

The methods used for the isolation of the ecdysteroids in the early studies (reviewed in the literature[95,103,104]) involved large quantities of samples and lengthy and tedious work-up procedures. These approaches have been largely superseded by modern, high-efficiency chromatographic methods. In addition, the comparatively modest amounts of pure material required by modern spectroscopic instruments for structural characterization has greatly reduced the amounts of material that need to be isolated, further simplifying the task of the ecdysone chemist. Indeed, the continuing advances that have been seen in the in-line coupling of techniques such as NMR, and particularly MS, to liquid chromatography mean that isolation in a pure form may no longer be necessary for the characterization and identification of novel ecdysteroids.[105] Isolation methods for plant and arthropod-derived compounds are described below (see Sections 8.05.3.4 and 8.05.3.5).

8.05.3.4 Zooecdysteroids

Concentrations of ecdysteroids in arthropods are generally in the range 10^{-9}–10^{-6} g g^{-1} of tissue. The types of compounds to be found in arthropods range from relatively nonpolar acyl esters, such as 20-hydroxyecdysone 22-palmitate (**73**) from the lepidopteran *Helicoverpa virescens*, through mid-polar hormones such as ecdysone and 20-hydroxyecdysone, to polar and ionic substances such as ecdysone 22-phosphate (**74**) from *Schistocerca gregaria*, *Locusta migratoria*, and *Bombyx mori*, and the side chain-oxidized ecdysonoic acids. The relatively low concentrations of these substances present at most stages of the life cycle means that purification can require a number of stages. In most schemes the first step is the extraction of the insect material into a polar solvent such as aqueous methanol or ethanol. The defatting of this extract, required to remove co-extracted lipids, is readily accomplished by partitioning the extract against a nonpolar solvent such as petroleum ether or hexane. This may then be followed by further solvent partition steps, or normal-phase column chromatography on silica gel. Depending upon their polarity the ecdysteroids can be eluted

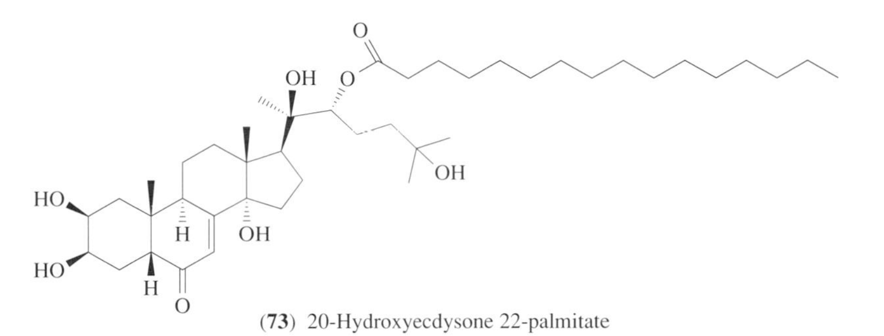

(**73**) 20-Hydroxyecdysone 22-palmitate

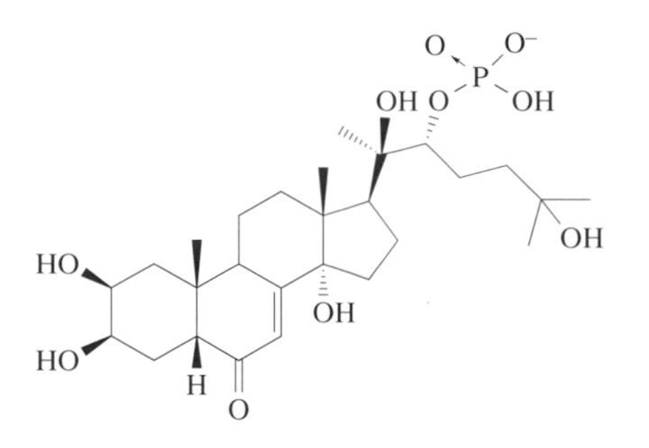

(**74**) Ecdysone 22-phosphate

with solvents of increasing eluotropic strength to give partially purified fractions. These fractions may then require further purification, which may be effected by reversed-phase chromatography, preparative-scale TLC, etc. A suitable method for the isolation of a range of ecdysteroids from insects has been described by Rees and Isaac.[106]

8.05.3.5 Phytoecdysteroids

The phytoecdysteroids are, for the most part, generally present in much higher concentrations than found in insects, and consequently usually require less effort to isolate in a pure form. It should be noted that the titer of ecdysteroids in a particular species can vary with the season, and also the location within the plant (i.e., roots versus leaves, etc.). Plants can be screened by RIA for ecdysone-like immunoreactive compounds or for molting hormone activity by an appropriate bioassay. Plants showing the desired positive result can then be taken through an appropriate extraction and isolation scheme monitored by the assay used for screening the whole plant until such time as a particular chromatographic peak (or peaks) can be identified as being associated with the activity.

As with arthropods, the initial extraction from plant material is usually best performed using a polar solvent such as ethanol or methanol, or acetone, acetonitrile, or methanol plus water mixtures. Following extraction, the removal of lipids and other nonpolar materials is then best achieved by the use of solvent partition with a nonpolar solvent (e.g., hexane–aqueous methanol 7:3). The ecdysteroids remain in the polar, aqueous phase. More polar contaminants can also be removed by partition with an apolar solvent such as ethyl acetate or 1-butanol. Chromatography can then be used to fractionate the partially purified concentrate into individual components. Generally, reversed-phase HPLC is employed,[107] but other techniques including TLC and droplet counter-current chromatography have been used (see Section 8.05.3.8).[108]

8.05.3.6 Structure Determination

The determination of the structure of new ecdysteroids has been greatly simplified by the data that have been generated on the large number of structures that have already been identified. Collations of data on the ecdysteroids are to be found in a number of reviews.[90,109]

8.05.3.6.1 UV spectroscopy

The presence of the 7-en-6-one group in ecdysteroids results in a relatively strong UV absorption, λ_{max} 242 nm, with a molar absorptivity ε of 10 000–16 000 $M^{-1}\,cm^{-1}$. For ecdysone, ε is 12 400 $M^{-1}\,cm^{-1}$ at 242 nm in ethanol.[109] The absence of a 14α-hydroxyl group causes a hypsochromic shift to 248 nm. The UV spectra of compounds such as ecdysone are clearly too unspecific for use in identification, but do facilitate the isolation of these materials by providing a suitable means for detection following chromatography. The ready loss of the 14α-hydroxyl group in the presence of acid results in the production of compounds with absorptions at 244 nm and 293 nm. While the absorption at 242 nm is characteristic of most ecdysteroids, the UV spectra can be helpful where unusual double bonds or pendent benzylic or other groups are present.

8.05.3.6.2 Fluorescence

The ecdysteroids do not possess native fluorescence, but in the presence of sulfuric acid or aqueous ammonia can be made to fluoresce. Excitation wavelengths range from 376 nm to 410 nm with emission between 400 nm and 460 nm. The induction of fluorescence in ecdysteroids has been used as the basis of several quantitative assays with limits of detection of approximately 5 ng.[110,111]

8.05.3.6.3 Optical activity

Because the ecdysteroids possess chiral centers they are optically active. Measurements in methanolic solution reveal a modest specific rotation $[\alpha]_D$ of between 60° and 80°.[109]

8.05.3.6.4 Optical rotatory dispersion and circular dichroism

Because of the unsaturated carbonyl present in the B ring of the ecdysteroids the optical rotatory dispersion (ORD) spectrum shows two Cotton effects. When measured in dioxan the amplitude of the positive Cotton effect is typically between 40° and 80°. Such data are useful because they can provide information on the type of A/B ring fusion. Values of 40–80° are thus typical of *cis* fusion, while 5α-compounds show much larger effects. For example, 5αH-20-hydroxyecdysone shows a negative Cotton effect (α_{240}) of −475°. Loss of the C-7 conjugated double bond (e.g., in cheilanthones A (**75**) and B (**76**)) also causes a large negative Cotton effect between −160° and −190°. However, these examples are exceptions, and the ORD curves for most ecdysteroids are essentially superimposable.

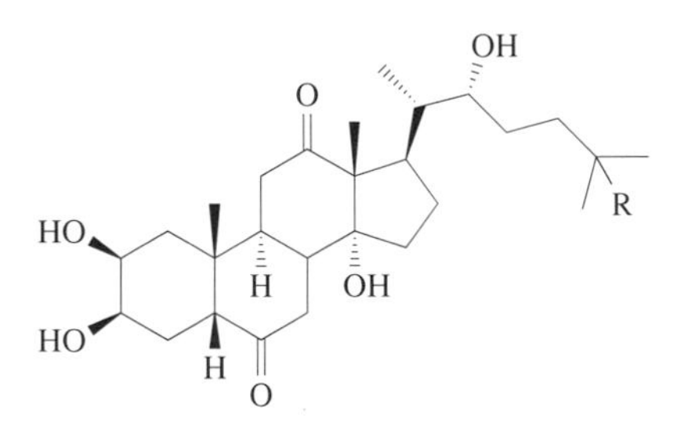

(**75**) R = OH Cheilanthone A
(**76**) R = H Cheilanthone B

The circular dichroism (c.d.) spectra of the ecdysteroids have also been shown to be useful for obtaining information pertaining to the A/B ring junction. The amplitude of the Cotton effect for *trans* A/B ring-fused compounds is greater than for *cis* compounds.[109] The c.d. spectra of dibenzoate derivatives have been used to assign the absolute configuration of the *vicinal* diols on the A ring using the "dibenzoate chirality rule."[112,113]

8.05.3.6.5 Infrared spectroscopy

The presence of a number of hydroxyl groups on the ecdysteroids ensures a strong absorption between 3340 cm^{-1} and 3500 cm^{-1} in the IR spectrum.[109] In addition the α,β-unsaturated ketone provides a characteristic absorption in the region 1640–1670 cm^{-1} (shifted to 1690 cm^{-1} in the 5β-hydroxyecdysteroids such as polypodine B (**70**)), with a weaker alkene stretch at ~1612 cm^{-1}. Loss of the 7-ene in compounds such as cheilanthones A (**75**) and B (**76**) also results in a shift to ~1684 cm^{-1}. Some of the more unusual phytoecdysteroids contain lactones (absorbing between 1700 cm^{-1} and 1800 cm^{-1}) or additional double bonds giving, in the case of ecdysteroids such as kaladasterone (**77**), absorptions at 1650 cm^{-1} and 1605 cm^{-1}. Other phytoecdysteroids, containing aromatic substituents (e.g., 20-hydroxyecdysterone 2-cinnamate or 3-*p*-coumarate, both from *Dacrydium intermedium*), show an ester carbonyl absorption (1720 cm^{-1}). Certain zooecdysteroids, for example the 3-dehydro compounds, also show a further carbonyl absorption in addition to that due to the

7-en-6-one. In 3-dehydroecdysone (**54**) this is observed at 1720 cm^{-1}, and in 3-dehydro-20-hydroxy-ecdysone at 1700 cm^{-1}.

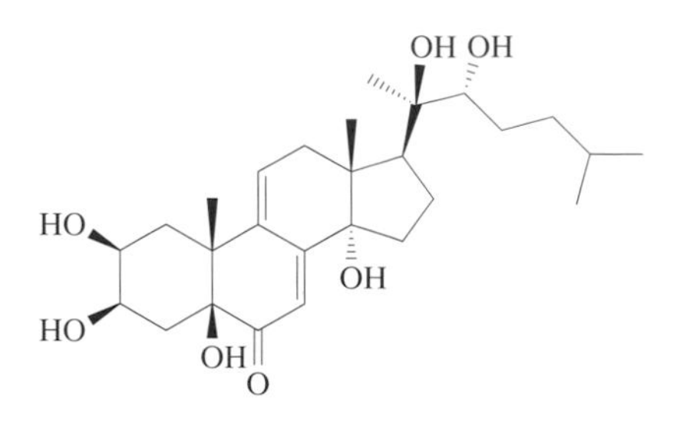

(**77**) Kaladasterone

8.05.3.6.6 *Mass spectrometry*

The mass spectra of the ecdysteroids are very valuable for characterization and identification, as it is possible to derive diagnostic ions for the steroid nucleus and side chain. These data, in conjunction with appropriate NMR spectra (see below) can provide the necessary structural information for unequivocal identification.

In the electron impact (EI) spectra obtained from the ecdysteroids the molecular ions are usually weak (~1%), if present at all, because of the ready dehydration of the parent to give more abundant [M-18] ions.[95,109] Further losses of 18 are usually observed in the high-mass region of the spectrum caused by the sequential loss of further molecules of water. Where a 20,22-*vicinal* diol is present in the side chain, for example in 20-hydroxyecdysone, ions due to the facile cleavage of this group are usually readily apparent. For 20-hydroxyecdysone this cleavage gives rise to an ion at *m/z* 99 (Scheme 7), frequently forming the base peak of the spectrum. Changes in the side chain give rise to diagnostic changes in the mass of this ion. As well as the ion at *m/z* 99 there is often an ion at *m/z* 81 (**78**), resulting from side chain cleavage. In the case of fragments due to the steroid nucleus, all compounds containing the full 2β,3β,14α-cholest-7-en-6-one structure, and the 20,22-*vicinal* diol, will show similar fragmentation patterns with a weak ion at *m/z* 363 (**79**) for the full nucleus and a stronger ion at *m/z* 345 as a result of the loss of the 14α-hydroxyl group. Where an additional hydroxyl group is present on the ecdysteroid nucleus the mass of this ion is modified to *m/z* 361, with one less, to *m/z* 329, and so on.

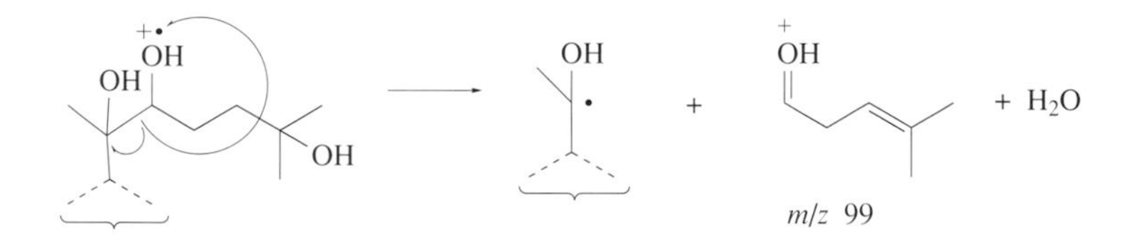

Scheme 7

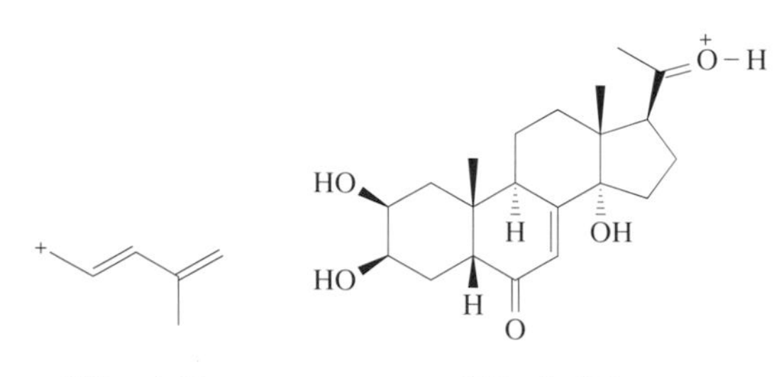

(**78**) m/z 81 (**79**) *m/z* 363

Given the masses of the nucleus and side chain it is possible to decide on the location and the number of hydroxyl groups present, and to determine whether the side chain contains additional groups such as a methyl or a lactone. For the determination of the molecular mass of the parent ion a "softer" ionization technique such as fast-atom, or fast-ion, bombardment (FAB, FIB) is required. This technique has been applied to a range of ecdysteroids including phosphate and acetyl phosphate conjugates.[114] Using tandem MS in combination with FAB–MS enables both molecular mass and fragmentation data to be obtained. Such data can also be obtained via HPLC–MS and HPLC–MS–MS.[107]

8.05.3.6.7 NMR spectroscopy

The ecdysteroids are poorly soluble in chloroform, so most of the early NMR data generated on the ecdysteroids was obtained in [^{2}H]pyridine.[109] Deuterated methanol and dimethyl sulfoxide have also been used. The increasing field strength of NMR spectrometers has reduced the sample requirements, and therefore makes solubility less of a problem, and the NMR spectra of ecdysteroids can now be obtained in most common solvents.

The ^{1}H NMR spectra of the ecdysteroids are complex, with the mass of CH and CH_2 signals unresolved. The methyl signals, however, appear as sharp peaks, most of which are singlets. Much information can be obtained from the position of these methyl resonances and whether they show coupling to other protons.[93,109] Hydroxylation at C-20 produces a marked downfield shift of the signals of both the C-18 and C-21 methyl resonances. In ecdysone the C-8 and C-21 methyl groups give rise to signals at 0.7 ppm and 1.25 ppm (d, $J = 6$ Hz) (pyridine), respectively, while in 20-hydroxyecdysone these signals appear at 1.20 ppm and 1.56 ppm for the C-18 and C-21 methyl groups. Removing the proton at C-20 and replacing it with a hydroxyl group results in the loss of the coupling seen for the C-21 methyl group of ecdysone. Changes in the A ring of these compounds (e.g., 2-deoxyecdysteroids or epimerization at C-3, etc.) has relatively little effect on the position of the methyl resonances. The introduction of the 14α-hydroxyl group is responsible for a small downfield chemical shift in the signal for the C-18 methyl group of approximately 0.06–0.08 ppm, while at the same time the signal for the C-7 proton is simplified to a doublet.

Other useful signals, well separated from the complex envelope of overlapped CH and CH_2 resonances, include those for the 2-H_a and 3-H_e protons, which for ecdysone in pyridine are observed at 4.15 ppm (as a complex multiplet, $w_{1/2}$ 22 Hz) and 4.14 ppm (multiplet, $w_{1/2}$ 9 Hz), respectively. In addition, the protons at C-5 (3.01 ppm, doublet of doublets, $J = 13$ Hz and 4 Hz), C-7 (6.25 ppm, doublet, $J = 2.5$ Hz), and C-9 H_a (3.54 ppm, multiplet, $w_{1/2} = 22$ Hz) can also provide useful diagnostic data. The complete ^{1}H NMR assignments for ecdysone and 20-hydroxyecdysone have been determined,[115] and the values in CD_3OD are given in Table 2.

In addition to ^{1}H NMR spectra, a considerable body of information has been amassed concerning the ^{13}C NMR spectra of ecdysteroids. The ^{13}C NMR spectroscopy of these compounds has advantages over ^{1}H NMR in that the signals are spread out over $\sim$100 ppm compared to a few parts per million for the latter. The assignments for the two important ecdysteroids, ecdysone and 20-hydroxyecdysone, in deuteropyridine are given in Table 2.

The signal for the carbon of the carbonyl group at C-6 is detected between 201 ppm and 204 ppm, with the unsaturated C-7 and C-8 carbons appearing near 121 ppm and 165 ppm, respectively. The C-5 carbon is found at $\sim$51 ppm in 5β-H ecdysteroids, but at 79–80 ppm in 5β-OH compounds such as polypodine B. Other changes that result from 5β-hydroxylation include changes to the signals at C-6 ($\sim$2 ppm upfield), C-3 ($\sim$1.5–2 ppm downfield), and C-19 (5–6 ppm upfield). There are also changes in the positions of the C-4 and C-10 signals.

8.05.3.7 Synthesis

The synthesis of an ecdysteroid is a complex undertaking, nevertheless two routes to ecdysone were reported within a remarkably short time of the elucidation of the structure by two separate industrial research groups (Syntex[116] and a joint Shering–Hoffmann–La Roche collaboration[117]). Both groups started with stigmasterol, which was both readily available and allowed relatively easy

Table 2 ^{1}H and ^{13}C NMR spectral data for ecdysone and 20-hydroxyecdysone as δ (ppm from TMS).

	^{1}H spectra in CD_3OD at 250 MHz			*^{13}C spectra in C_5D_5N at 62.9 MHz*	
Atom	*Ecdysone*	*20-Hydroxyecdysone*	*Atom*	*Ecdysone*	*20-Hydroxyecdysone*
1-H_a	1.43	1.43	C-1	38.08 (t)	38.09 (t)
1-H_e	1.78	1.78	C-2	68.10 (d)	68.33 (d)
2-H_a	3.83 (ddd, 12, 3, 3)	3.83 (m, $w_{1/2}$ 22) 3)	C-3	68.10 (d)	68.23 (d)
3-H_e	3.94 (ddd, 3, 3, 3)	3.94 (m, $w_{1/2}$ 8) 3)	C-4	32.45 (t)	32.53 (t)
4-H_a	1.65	1.65	C-5	51.41 (d)	51.48 (d)
4-H_e	1.75	1.75	C-6	203.36 (s)	203.56 (s)
5-H	2.38 (dd, 12, 5)	2.38 (dd, 13, 4)	C-7	121.61 (d)	121.79 (d)
7-H	5.81 (d, 2.5)	5.85 (d, 2.5)	C-8	165.53 (s)	166.11 (s)
9-H_a	3.14 (m, $w_{1/2}$ 22)	3.09 (m, $w_{1/2}$ 21)	C-9	34.63 (d)	34.67 (d)
11-H_a	1.65	1.65	C-10	38.76 (s)	38.80 (s)
11-H_e	1.78	1.78	C-11	21.20 (t)	21.29 (t)
12-H_a	2.10 (ddd, 13,13,5)	2.13 (ddd, 13,13,5)	C-12	31.49 (t)	32.19 (t)
12-H_e	1.7-1.8	1.85	C-13	47.70 (s)	48.27 (s)
15α-H	2.00	2.00	C-14	83.97 (s)	84.42 (s)
15β-H	1.53	1.55	C-15	32.03 (t)	31.88 (t)
16α-H	1.98[a]	1.95	C-16	26.74 (t)	21.61 (t)
16β-H	1.48[a]	1.75	C-17	48.28 (d)	50.28 (d)
17-H	2.01	2.39 (m)	C-18	15.89 (q)	17.99 (q)
22-H	3.59 (m, $w_{1/2}$ 16)	3.33 (dd, 11, 2)	C-19	24.55 (q)	24.55 (q)
23α-H	1.30	1.30	C-20	43.04 (d)	77.09 (s)
23β-H	1.60	1.65	C-21	13.74 (q)	21.77 (q)
24α-H	1.75	1.75	C-22	74.07 (d)	77.75 (d)
24β-H	1.45	1.45	C-23	25.69 (t)	27.59 (t)
18-Me	0.73 (s)	0.89 (s)	C-24	42.55 (t)	42.64 (t)
19-Me	0.97 (s)	0.96 (s)	C-25	69.80 (s)	69.86 (s)
21-Me	0.95 (d, 7.5)	1.18 (s)	C-26	30.09 (q)	30.10 (q)
26-Me	1.19 (s)	1.19 (s)	C-27	30.01 (q)	30.15 (q)
27-Me	1.20 (s)	1.20 (s)			

Source: Girault and Lafont.[115]
$w_{1/2}$ in hertz.
[a]Assignments are interchangeable.

access to the side chain. Since that time a further half dozen synthetic routes to ecdysone (via stigmasterol, ergosterol, and diosgenin) have been described.[118–123] In addition to the synthesis of ecdysone, routes to 20-hydroxyecdysone, from the ecdysone synthetic intermediate 2b,3b-acetoxy-5a-hydroxypregnan-7-en-6-one 20-methyl carboxylate,[124] pregnenolone,[125] 20β-benzoyloxy-5-pregnan-3β-ol,[126] and a D-ring aromatic steroid,[127] have also been developed. Routes have also been elaborated for the synthesis of a range of complex ecdysteroids including 2-deoxyecdysone,[128,129] 2,25-deoxyecdysone,[128] 2-deoxy-3-epi-20-hydroxyecdysone,[130] 22-deoxy-20-hydroxyecdysone,[131] ponasterone A (**57**),[124] rubrosterone (**80**), and others.[132–135]

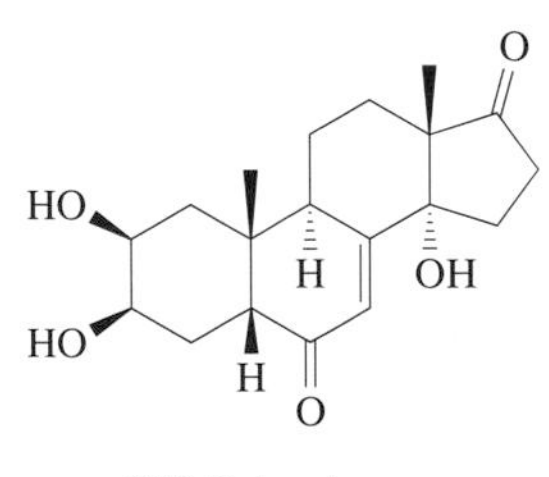

(**80**) Rubrosterone

Most of the recent work on the synthesis of the ecdysteroids has concentrated on the fabrication and elaboration of the side chain, or the use of a common ecdysteroid for conversion into other, less readily available, compounds such as sulfates, acetates, 3-epimers, and dehydrocompounds.[136,137] The synthesis of the ecdysteroids has been reviewed.[138]

8.05.3.8 Chemical Analysis

In insects the ecdysteroids are generally present in quite low concentrations (10^{-9}–10^{-6} g g^{-1}), and quantitative analysis usually requires extensive sample clean up and sensitive methods of detection. Sample preparation methods for chromatographic analysis have been reviewed.[139] In the case of plants, much higher concentrations may be achieved, whereby the ecdysteroids can account for up to 15% dry weight, depending upon the species, season, and other conditions. Sample preparation in such circumstances is often less demanding than for arthropod samples. A variety of qualitative and quantitative methods that have been used to determine the ecdysteroids are summarized below.

8.05.3.8.1 Thin-layer chromatography

Thin-layer chromatography (TLC) or planar chromatography has been widely exploited for the separation of the ecdysteroids, using both normal- and reversed-phase systems.[140] A suitable general normal-phase solvent system for use with silica gel TLC plates is chloroform–ethanol 4:1 (v/v).[140] For reversed-phase TLC on C_{18} bonded plates, methanol–water mixtures (1:1, v/v) provide good separations.[140] The identification of ecdysteroids by TLC is usually by comparison with the R_f values of authentic standards, but direct TLC–MS and TLC–MS/MS have been described.[141,142] Where quantitative high-performance TLC (HPTLC) methods have been sought, using scanning densitometry,[143] a sensitivity of the order of 10–100 ng per spot has been obtained, which is more than adequate for most plant-derived samples. Detection in TLC is usually by fluorescence quenching, but spray reagents have also been used to provide color reactions of varying degrees of specificity. The most widely used is the vanillin–sulfuric acid spray, which when warmed produces a green color for ecdysone.

8.05.3.8.2 Gas–liquid chromatography

Gas–liquid chromatography (GLC), particularly with electron capture detection (ECD), was important as a quantitative method for the ecdysteroids before reliable HPLC methods were available. The exquisite sensitivity and selectivity of the ECD for these compounds, due to the presence of the unsaturated ketone group,[100] still make this the method of choice for difficult samples where recourse to HPLC–MS is not practicable. However, the successful operation of GLC methods for the ecdysteroids requires care as the analytes are involatile and thermally labile, and can only be analyzed by GLC following derivatization of the hydroxyl groups to form TMS ethers. Unless carefully controlled, silylation can lead to the formation of a mixture of TMS derivatives, and multiple peaks on chromatography.[144] Most of the work on the GLC of the ecdysteroids was performed using packed columns, but later capillary columns were used.[145]

8.05.3.8.3 Column liquid chromatography

Although less sensitive than GLC–ECD, HPLC is the most widely used chromatographic technique for the quantification of ecdysteroids. This is largely due to the strong UV absorption at 254 nm, which provides a suitable chromophore for detection down to $\sim$10 ng on column. Increased sensitivity can be obtained, if required, by utilizing MS, or MS–MS detection. The lack of any requirement for derivatization greatly adds to the attraction of HPLC for trace ecdysteroid analysis compared to GLC. Both normal- and reversed-phase systems have been developed for the separation and quantification of zoo- and phytoecdysteroids.[140] Typical reversed-phase methods employ C_{18}-bonded phases and acetonitrile–methanol or water–methanol solvent systems ($\sim$25:80 or 50:50

v/v, respectively). A suitable normal-phase system for the chromatography of ecdysone and 20-hydroxyecdysone on silica gel is based on dichloromethane–2-propanol–water 125:30:2 (v/v/v). The HPLC of phytoecdysteroids has been reviewed.[107,140] In addition to HPLC, supercritical fluid chromatography (SFC)[146] and droplet counter-current chromatography (DCCC)[147] have been employed for the separation and, in the case of the latter, isolation of the ecdysteroids.

8.05.3.8.4 Capillary electrophoresis

Capillary electrophoresis is a relatively recently introduced technique offering the possibility of very high separation efficiencies with small sample requirements. The bulk of the ecdysteroids are nonionic, and separations based on capillary electrophoretic methods have therefore been performed using micellar electrokinetic chromatography (MEKC). Extracts of the plants *Silene nutans* and *Silene otites*, together with the ecdysteroids present in extracts of the eggs of the desert locust *Schistocerca gregaria*, were analyzed by MEKC using a phosphate:borate:sodium dodecylsulfate buffer system at pH 9.4 modified with 5% methanol. With UV detection at 254 nm, 175 pg of ecdysone on column could be observed,[148] although, because of the small volumes ($\sim$4 nl) that can be applied, this corresponded to $\sim$35 μg ml^{-1}.

8.05.3.8.5 Radioimmunoassays

Radioimmunoassay (RIA) methods for ecdysteroids were, like GLC–ECD, an early development for the detection and analysis of ecdysteroids. While capable of good sensitivity and high throughput, they suffer from the problem of poor specificity. Antisera raised against one particular ecdysteroid may show significant cross-reactivity with a range of related compounds such as metabolites, often of widely differing biological activity. This has had the consequence that where a mixture of ecdysteroids is present, the result obtained by the RIA alone may reflect neither the true total of the ecdysteroids present nor the molting hormone activity of the sample. The use of the technique to detect which fractions of a chromatographic eluent contain immunoreactive material does, however, provide a means for monitoring the isolation of RIA-positive compounds. The development and use of immunoassays for ecdysteroids has been reviewed.[149]

8.05.3.9 Biosynthesis

The biosynthesis of ecdysteroids in both arthropods and plants has been the subject of investigation over a period of many years, but all of the steps in the process cannot yet be said to have been fully elucidated.[150,151]

8.05.3.9.1 Arthropods

The biosynthesis of ecdysteroids in arthropods during the larval stages has been shown to take place in specific "molting" glands such as the prothoracic gland in locusts or the ring gland in Diptera, or their equivalents in crustaceans, the "Y" organs.[152] The activity of the prothoracic gland seems to be controlled by the brain via the proteinaceous prothoracicotropic hormone (see Section 8.05.4.1). Certain other tissues have been shown to be capable of ecdysteroid biosynthesis as well as these ecdysial glands. In adult females, biosynthesis has been shown to occur in the ovaries, and the enocytes and testes of some insects have also been proposed as sources of these hormones.

In general, insects cannot synthesize cholesterol *de novo*, and rely on dietary sources. Phytophagous insects may be able to undertake the dealkylation of suitable plant sterols to form cholesterol, while carnivorous species will obtain it directly from their prey. The ability of insects to dealkylate plant sterols for this purpose is by no means universal, so that some species have the C-28 ecdysteroid makisterone A as their molting hormone.

Numerous studies have now been performed, both *in vivo* and *in vitro*, using a variety of radiolabeled precursors in order to elucidate the biosynthetic pathway. An abbreviated summary of the main stages is given in Scheme 8. It is generally accepted that the first few steps involve the production of 7-dehydrocholesterol (**81**).[150,151,153] The next steps in the pathway appear to involve the introduction of the 7-en-6-one and 14α-hydroxyl groups together with the A,B-*cis* ring junction. It is likely that the subsequent introduction of the hydroxylation on the side chain proceeds via hydroxylation at C-25, followed by that at C-22 to produce 2-deoxyecdysone (**82**). Hydroxylation at C-2 would then follow to form ecdysone. However, much remains unclear. There is evidence that more than one route to ecdysone is possible. Some *in vitro* studies have demonstrated that 25-hydroxycholesterol can be efficiently converted into 3-dehydroecdysone, probably via 7-dehydro-25-hydroxycholesterol.[154] In some crustaceans, where the molting hormone is ponasterone A (**57**), the glands produce 25-deoxyecdysone. In addition to ecdysone, the biosynthetic glands that produce these molting hormones have been reported to produce 3-dehydroecdysone (**54**) as a major secretory product.[152,154] The 3-dehydroecdysone is rapidly converted into ecdysone by enzymes present in the hemolymph, and this ecdysone is then rapidly converted to the major active compound 20-hydroxyecdysone by a variety of tissues.[152,154] The biosynthesis of the ecdysteroids has been reviewed.[150–152,154]

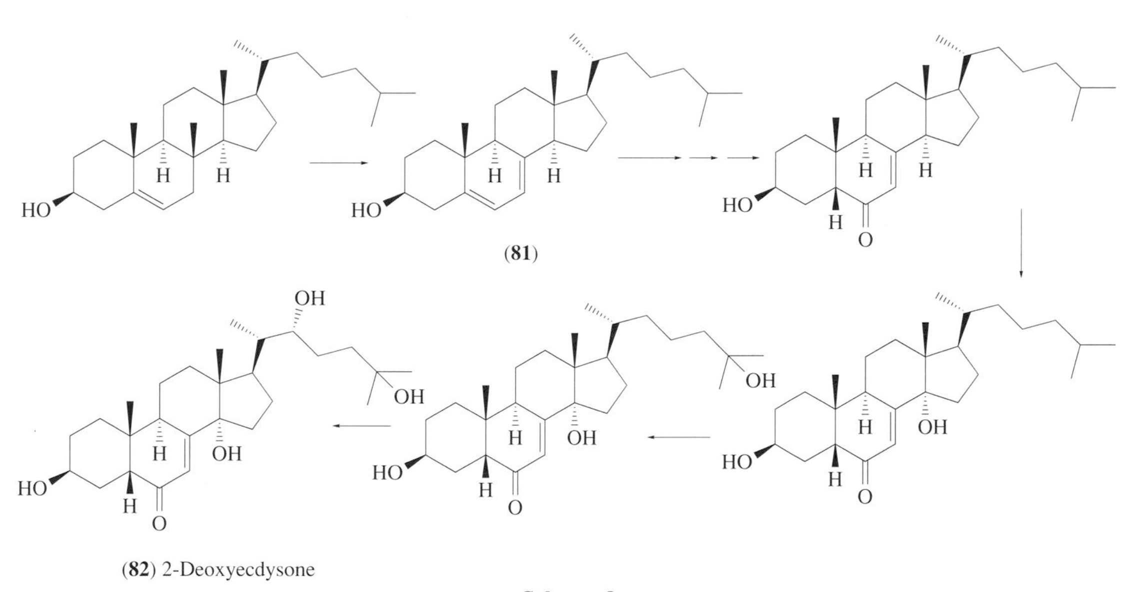

Scheme 8

8.05.3.9.2 Plants

The situation regarding the biosynthesis of ecdysteroids in plants is, like that for arthropods, still not fully elucidated.[150,155] However, as in the case of insects, early studies demonstrated that radiolabeled cholesterol, or suitable precursors of it such as acetate or mevalonic acid, could be converted into 20-hydroxyecysone, and ecdysone into 20-hydroxyecdysone. Both whole-plant systems and *in vitro* experiments employing cell suspensions or tissue culture have been investigated. Given that ecdysteroids are not uniformly distributed within a given plant, it is quite possible that not all of the tissues produce ecdysteroids, and indeed the sites of production and storage may well be different.

It seems likely that an early step in ecdysteroid biosynthesis in plants is the formation of 7-dehydrocholesterol, as in insects, and this conversion has been observed in *Polypodium vulgare*.[156] Functionalization of the B ring is probably the next step, followed by the introduction of the 14α-hydroxyl group. Hydroxylation of the side chain, and introduction of the C-2 hydroxyl group, then follow, although the order of introduction is not clear. Hydroxylation of the 5-β position is a common feature of phytoecdysteroids (e.g., polypodine B (**70**)), but the point at which this occurs is not known. In spinach, ^{3}H-labeled ecdysone gives rise to both 20-hydroxyecdysone and polypodine B.[157] It seems reasonable to assume that a similar biosynthetic route will be followed for the production of the C_{28} and C_{29} ecdysteroids. Once produced, compounds such as ecdysone or 20-hydroxyecdysone can then be converted into the side chain cleavage compounds such as rubrosterone (**80**) or poststerone (**83**), or to the various esters, glucosides, phosphates, etc., that have been detected in plant extracts.

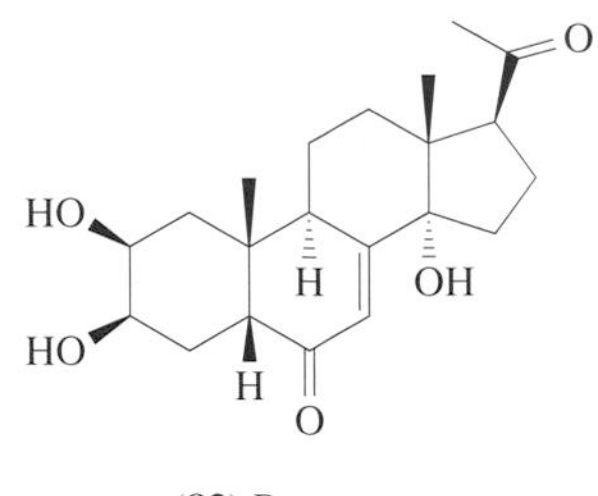

(**83**) Poststerone

8.05.3.10 Metabolism

The metabolism of ecdysteroids, particularly ecdysone, in insects and crustaceans has been the subject of many investigations,[153,158–163] generally following the fate of injected or ingested ^{3}H-labeled ecdysone. These studies have shown that metabolic pathways may vary with species, developmental stages, and tissue. Metabolism can occur on either the side chain or steroid nucleus to produce metabolites of varying degrees of polarity. In general, ecdysone is rapidly converted to 20-hydroxy-ecdysone. Further metabolism of the side chain can include hydroxylation at C-26, from which ecdysonoic acids, for example (**84**), can be generated (Scheme 9). Alternatively, the new hydroxyl group at C-26 can be conjugated to phosphate. Metabolism can also occur at C-22 to give conjugates with long-chain fatty acids, phosphate, and, less commonly, acetate, glucose, or glycolic acid. Side chain cleavage between C-20 and C-22 has also been demonstrated. In the case of the ecdysteroid nucleus, common metabolic reactions, some of which may be reversible, include the formation of 3-dehydro compounds, and from them the corresponding 3-epimers. Conjugates to phosphate or acetate at C-2 or C-3 have been identified. Combinations of metabolism on both the side chain and nucleus are possible to form a wide range of products, for example 3-acetyl-26-hydroxyecdysone.

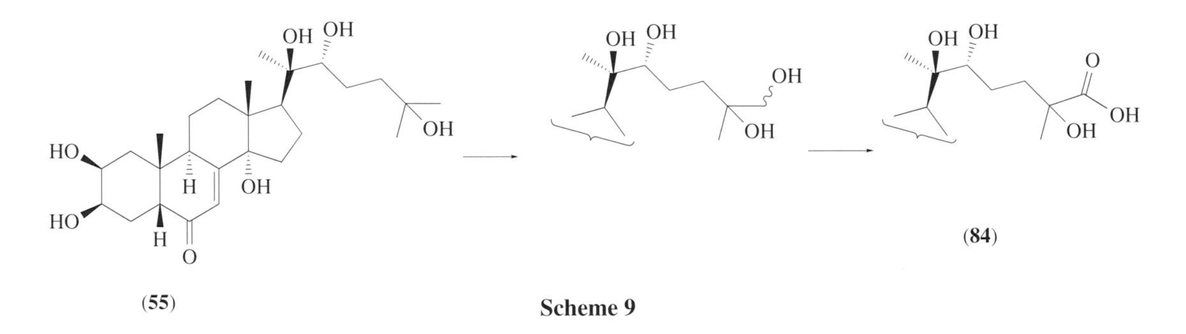

Scheme 9

Some of these conjugated ecdysteroids may in fact represent storage forms of the hormone, as suggested by the presence of large amounts of these compounds in newly laid eggs, rather than simply deactivation or excretory products.

Very little is known about the metabolic fate of ecdysteroids in plants.

8.05.3.11 Structure and Activity

The relationship between structure and biological activity within the ecdysteroid family has been of considerable interest to those trying to understand the complexity revealed by the isolation of so many compounds with similar structures. Biological activity in the context of the ecdysteroids means molting hormone activity as no convincing evidence for a hormonal role in plants currently exists. The observed activity in insects depends upon a number of features, and also to some extent upon the test system used. *In vitro* systems depending upon specific receptors (see Section 8.05.3.1) have advantages in that they exclude factors such as metabolic activation or biological half-life. However, when trying to understand the role of a compound isolated from a plant, metabolic activation within the "target" organism may well be an important factor. For example, phytoecdysteroids such as 20-hydroxyecdysone bearing a 3-*p*-coumarate have high molting hormone activity, which has been attributed to rapid hydrolysis to the free hormone *in vivo*.[164] Similarly, 2-deoxyecdysone and 2-deoxy-20-hydroxyecdysone are also as active as 20-hydroxyecdysone *in vivo*,[165] and this is probably due to conversion to 20-hydroxyecdysone.

In hormonally active ecdysteroids a *cis*-fused A,B ring is present, with a 5β-hydrogen or 5β-hydroxyl group.[164] The presence of a 5β-hydroxyl group can in fact confer higher biological activity than a 5β-hydrogen, for example polypodine B (**70**) is ~500 times more active than 20-hydroxyecdysone *in vitro*.[81] The corresponding 5α analogs are all much less active.

A 3β-hydroxyl group is also important for good activity as the 3-dehydro- and 3-epiecdysteroids are significantly less potent than ecdysone in bioassays[136,166] (3-epi-20-hydroxyecdysone has only 7–10% of the activity of 20-hydroxyecdysone in the *Musca* assay[166]). In addition, the presence of both the 7-en-6-one and the 14α-hydroxyl groups appears to be essential for biological activity (the cheilanthones (**75**) and (**76**), which lack the 7,8 double bond, are inactive[167]). The ecdysteroids, unlike mammalian hormones, also require the presence of the side chain for activity, and compounds lacking this feature, for example rubrosterone (**80**) and poststerone (**83**), are inactive.[81] Certain features are required in the side chain, in particular a 22-hydroxyl group, of the correct configuration, appears to be a requirement for high molting hormone activity (both 22-deoxy-20-hydroxyecdysone[168] and 22-*epi*-20-hydroxyecdysone[164] are essentially devoid of activity). The most active compounds contain a 20,22-diol, and *in vitro* 20-hydroxyecdysone is ~500 times more potent than ecdysone.[81] Further hydroxylation of the side chain to give 20,26-dihydroxyecdysone results in a lowering of activity,[169] while oxidation to ecdysonoic acids presumably leads to a further inactivation of the hormone. The C-25 hydroxyl group does not appear to be essential for high activity and, indeed, in *in vitro* tests, ponasterone A is considerably more potent than 20-hydroxyecdysone.[81] However, it is possible to have considerable structural variation in the side chain and retain molting hormone activity, as evidenced by the cases of makisterone A (**56**)[170] and podecdysone B[171] (makisterone C) (**85**) with methyl and ethyl substituents at C-24, respectively, and lactone-containing compounds such as cyasterone (**71**).[172]

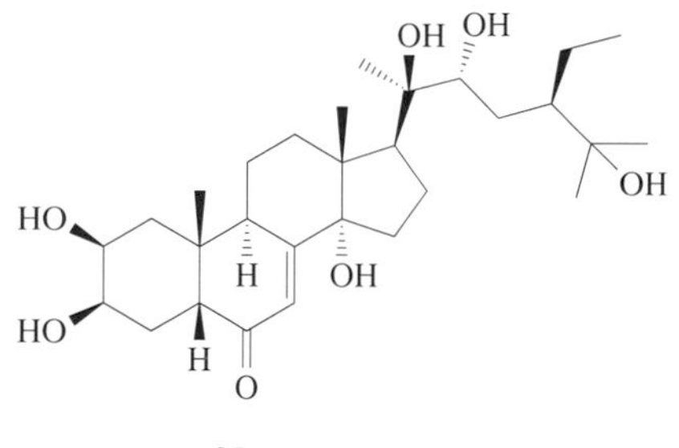

(**85**) Podecdysone B

8.05.3.12 Ecdysteroids as Insect Control Agents

Given the postulated function of the ecdysteroids as defensive compounds, protecting the plants that produce them from attack by phytophagous insects, it is perhaps no surprise that the highly potent 20-hydroxyecdysone and polypodine B are so frequently encountered in plants scattered throughout the Kingdom. For larvae of the silkworm *Bombyx mori*, 20-hydroxyecdysone is lethal, while ecdysone (which is rarely found in plants) produces supernumerary moults.[173] For other species (*Pieris brassicae*, *Chilo partellus*) ecdysteroids are feeding deterrents,[102] while some species are very tolerant to large amounts. The pink bollworm *Pectinophora gossypiella* and the leek moth

Acrolepiopsis assectella undergo extra and incomplete moults, so that they die of starvation.[174,175] Cotton does not contain ecdysteroids. *Helicoverpa armigera* and *Locusta migratoria* showed no detrimental effects when fed with 400–1000 ppm of 20-hydroxyecdysone. In *H. armigera* it was rapidly converted in the gut to an ester of a mixture of long-chain fatty acids at C-22, and then excreted.[176] The subject is discussed by Dinan, who points out that a group of plant substances that are nontoxic to mammals (ecdysteroids are present in some food crops like spinach, *Spinacia oleracea*, and are not destroyed by cooking[92]), and have the potential to disrupt the endocrine system of phytophagous insects, are attractive for the development of biorational control and for the genetic engineering of crops to produce more resistant varieties.[92] As yet, there is little progress in this direction, but very few pest insects have been examined for the effects of dietary ecdysteroids.

8.05.4 INSECT NEUROPEPTIDES

The third great division of insect hormones is the neuropeptides, which control some well-known, and some less well-understood, functions. Most of the chemical messengers that function between the central nervous system and the subordinate organs appear to be oligopeptides or small proteins. Early work by Kopec in 1919 and 1922[177] on the then-called "brain hormone" of insects (see Section 8.05.4.1) was the first demonstration that the nervous system of any animal had an endocrine function. That neurons, the cellular part of the nervous system, produced substances that controlled processes in peripheral parts of the body was a completely novel idea then. It was only after the significance of neurosecretory cells in the vertebrate brain was demonstrated in 1928 that the experiments of Kopec received attention and the study of neurosecretion in insects began. The distribution of neurosecretory cells and the position of neurohemal organs in the insect body are illustrated in Figure 3.

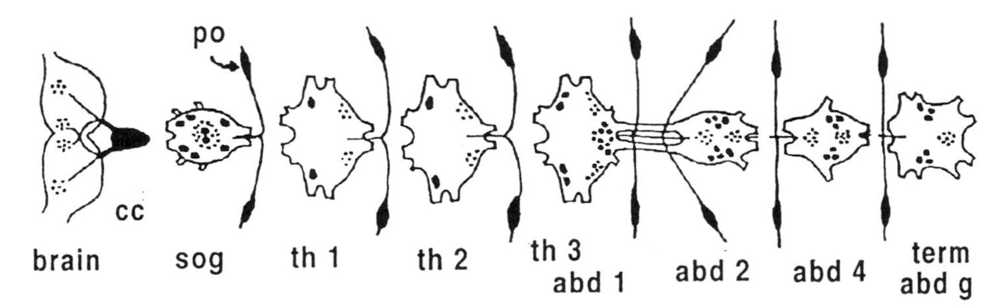

Figure 3 Schematic representation of the central and peripheral nervous system of an insect showing the distribution of neurosecretory cells (black dots) and neurohemal organs (black swellings); abd, abdominal ganglion; cc, corpus cardiacum; po, perisympathetic organ; sog, subesophageal ganglion; th, thoracic ganglion. (Reproduced by permission of Elsevier from G. A. Kerkut and L. I. Gilbert (eds.), "Comprehensive Insect Physiology, Biochemistry and Pharmacology," 1985, vol. 7, p. 67.)

Progress in the isolation and identification of neuropeptides has been limited by several factors: the very small amounts of these substances available in insects for isolation, their instability under many of the conditions required for their isolation, the small number of cells that produce them or the tiny size of the organ, and the lack of clear and sensitive bioassays in many cases. There is little chance of progress if most of the material has to be used at each stage of purification to locate the activity. Early methods for their isolation were reviewed by Stone and Mordue.[178] The quantities normally present in organs or hemolymph are of the order of femtomoles or picograms. The work has been considerably aided in recent years by the use of reverse-phase HPLC and the development of sensitive MS techniques for the determination of amino acid sequences in peptides.[179] The first of these was fast atom bombardment (FAB), which was soon followed by laser and plasma desorption (including MALDI or matrix-assisted laser desorption ionization), and then electrospray (ES) methods, which allow direct coupling of the separation of the peptide by HPLC to its sequencing by MS. With LC–ES–MS the molecular masses of proteins above 130 kDa can be determined with a detection limit of about 1 pmol. Another technique that is speeding the identification of neuropeptides is recombinant DNA technology for genes coding for neuropeptides, allowing larger quantities of them to be produced for structure determination.

At least 10 processes controlled by neurohormones are considered here, and another five processes are recognized as hormone controlled, for which no substances have yet been isolated and identified. Only two insect neuropeptide hormones, proctolin (see Section 8.05.4.2) and locust adipokinetic hormone (see Section 8.05.4.3), were known by the mid-1970s; now over 60 have been identified. Considering the number and diversity of insect species, it is surprising that there are so few variations on the structures of the small-molecule hormones. The situation is different with the neuropeptides. Many of them appear to be characteristic of the species only. At a conference in 1984, it was estimated that there must be well over 100 insect neuropeptides, but three years later, when it began to be known that they can occur in groups with similar structures and properties, it was realized the number must be much larger. They will never all be isolated from all insect species, so any consideration of them will be necessarily fragmentary. Table 3 lists some of the important neuropeptides described here where the structure is known. Since the time of writing, a review of 128 pages by Gäde has appeared listing all the known neuropeptides.[250]

8.05.4.1 Prothoracicotropic Hormone

According to Bollenbacher and Bowen,[251] PTTH was the first neurohormone to be described in any animal and the first insect hormone discovered. The early experiments of Kopec, supported later by those of Wigglesworth,[179] demonstrated the existence in the brain of a factor necessary for molting, called the "brain hormone." He traced this hemolymph-soluble factor to the anterior central part of the brain. Later, Williams showed that molting and development required factors from both the brain and the prothoracic glands (see Figure 2), and that the "brain hormone" was required to activate the prothoracic glands to produce the true molting hormone. When later it was realized that the brain produced a number of neuropeptides, the name "brain hormone" was replaced by prothoracicotropic hormone (PTTH). The earlier methods used for its isolation have been reviewed and its known chemical characteristics tabulated.[252] It was partially characterized from four species of Lepidoptera, but chiefly from *Manduca sexta* and *Bombyx mori*. Progress was slowed by its apparent inhomogeneity, since *Bombyx* PTTH gave multiple peaks during HPLC, but it is now concluded that the first few residues were missing from some of the proteins, although they still showed activity. All that is known of *Manduca* PTTH is that it is an asymmetric acidic homodimeric peptide with intra- and intermolecular disulfide bonds.[253] Larval prothoracic glands can be equally stimulated by large or small PTTHs, but pupal gland responds more to the large PTTH. In *B. mori* it is a larger molecule of mass 22 kDa, a dimer of two identical units of 109 amino acids each, held together by one disulfide bond, and with a pentasaccharide chain attached to asparagine at position 41 (Table 3).[180,181] Active PTTH was isolated from the brains of pupae. For the initial work, 5.4 μg of pure PTTH was obtained from brains of 5×10^5 heads (3.75 kg), and for the complete structure a further 3×10^6 heads were used. Following their isolation of *Bombyx* PTTH, the Japanese group have isolated the gene from *Bombyx* and incorporated it into *Escherichia coli*, and by gene expression produced more of the hormone.[181] By cloning of the *Samia cynthia ricini* gene they have also produced *Samia* PTTH (Table 3).[181] There is relatively low homology between the *Bombyx* and *Samia* PTTH (50%), and *Samia* PTTH has an extra 15 residues at the carboxyl terminus. Each PTTH showed activity in its own species only. A smaller molecule of mass 5 kDa from the brains of adult moths is now called bombyxin (see Section 8.05.4.9), and is not PTTH. By immunological methods one group said that *Bombyx* PTTH is also present in the brain of *M. sexta*,[253] but another says there is no cross-activity between *Bombyx* and *Manduca*.[254] The American group says that the PTTHs show no sequence similarity to any other protein but do share a common ancestral gene with vertebrate growth factors.[180] The PTTH of *Drosophila melanogaster* has been isolated from $\sim 4 \times 10^5$ larvae, via delipidation, salt extraction, heat treatment, column chromatography, and HPLC. The 50 μg of pure hormone was shown to be a glycosylated protein of mass 66 kDa with a 45 kDa peptide portion. Its amino acid sequence is not yet known and there is no indication of homology with any known protein.[183] A discussion of the function of PTTH can be found in Nijhout.[177]

8.05.4.2 Proctolin

Proctolin has the distinction of being the first insect neuropeptide to be chemically identified[255] (Arg-Tyr-Leu-Pro-Thr, Table 3), although its exact function is still unclear. The original isolation

Table 3 Neuropeptide hormones with a known function identified in insects.

Substance	*Insect source*	*Structure*	*Ref.*
PTTH	*Bombyx mori*	Homo dimer of: Gly-Asn-Ile-Gln-Val-Glu-Asn-Gln-Ala-Ile-Pro-Asp-Pro-Pro-Cys-Thr-Cys-Lys-Tyr-Lys-Lys-Glu-Ile-Glu-Asp-Leu-Gly-Glu-Asn-Ser-Val-Pro-Arg-Phe-Ile-Glu-Thr-Arg-Asn-Cys-Asn-Lys-Thr-Gln-Gln-Pro-Thr-Cys-Arg-Pro-Pro-Tyr-Ile-Cys-Lys-Glu-Ser-Leu-Tyr-Ser-Ile-Thr-Ile-Leu-Lys-Arg-Arg-Glu-Thr-Lys-Ser-Gln-Glu-Ser-Leu-Glu-Ile-Pro-Asn-Glu-Leu-Lys-Tyr-Arg-Trp-Val-Ala-Glu-Ser-His-Pro-Val-Ser-Val-Ala-Cys-Leu-Cys-Thr-Arg-Asp-Tyr-Gln-Leu-Arg-Tyr-Asn-Asn-Asn	180
		(Disulfide bridges at Cys17-Cys54, Cys40-Cys96, and Cys48-Cys98, interchain bridge at Cys15-Cys15′, carbohydrate chain Manα1-6Manβ1-4GlcNAcβ1-4(Fucα1-6)GlcNAc- is attached to Asn41)	181 182
PTTH	*Samia cynthia ricini*	Homo dimer of: Gly-Asp-Leu-Arg-Arg-Glu-Lys-His-Asn-Gln-Ala-Ile-Gln-Asp-Pro-Pro-Cys-Ser-Cys-Gly-Tyr-Thr-Gln-Thr-Leu-Leu-Asp-Phe-Gly-Lys-Asn-Ala-Phe-Pro-Arg-His-Val-Val-Thr-Arg-Asn-Cys-Ser-Asp-Gln-Gln-Gln-Ser-Cys-Leu-Phe-Pro-Tyr-Val-Cys-Lys-Glu-Thr-Leu-Tyr-Asp-Val-Asn-Ile-Leu-Lys-Arg-Arg-Glu-Thr-Ser-Thr-Gln-Ile-Ser-Glu-Glu-Val-Pro-Arg-Glu-Leu-Lys-Phe-Arg-Trp-Ile-Gly-Glu-Lys-Trp-Gln-Ile-Ser-Val-Gly-Cys-Met-Cys-Thr-Arg-Asp-Tyr-Arg-Asn-Ser-Thr-Glu-Asp-Tyr-Gln-Pro-Arg-Leu-Leu-Thr-Lys-Ile-Ile-Gln-Gln-Arg-Asp-Leu-Ser-Tyr-Asn-Asn-Asn (Disulfide bridges at Cys19-Cys55, Cys42-Cys97, and Cys49-Cys99, interchain bridge at Cys15-Cys15′, carbohydrate chain Manα1-6Manβ1-4GlcNAcβ1-4(Fucα1-6)GlcNAc- is attached to Asn41)	181
Proctolin	*Periplaneta americana* and others	Arg-Tyr-Leu-Pro-Thr	183
Adiokinetic hormones			
Lia-AKH	*Libellus auripennis*	pGlu-Val-Asn-Phe-Thr-Pro-Ser-TrpNH$_2$	184
Ani-AKH	*Anax imperator*	pGlu-Val-Asn-Phe-Ser-Pro-Ser-TrpNH$_2$	185
See Pea-CAH-I	*Mastotermes darwiniensis*	pGlu-Val-Asn-Phe-Ser-Pro-Asn-TrpNH$_2$	186
	Trinervitermes trinervoides		186
Emp-AKH	*Empusa pennata*	pGlu-Val-Asn-Phe-Thr-Pro-Asn-TrpNH$_2$	187
	Sphodromantis sp.		187
Grb-AKH	*Gryllus bimaculatus*	pGlu-Val-Asn-Phe-Ser-Thr-Gly-TrpNH$_2$	188
	Achaeta domestica		189
	Grylloides sigillatus		190
	Romalea microptera		191
Scg-AKH-II	*Schistocerca gregaria*	pGlu-Leu-Asn-Phe-Ser-Thr-Gly-TrpNH$_2$	192
	Schistocerca nitans		193
	Heterodes namaqua		190
	Acanthoproctus cervinus		190
	Libanasidus vittatus		194
	Anabrus simplex		194
Lom-AKH-II	*Locusta migratoria*	pGlu-Val-Asn-Phe-Ser-Ala-Gly-TrpNH$_2$	192
Taa-AKH	*Tabanus atratus*	pGlu-Leu-Thr-Phe-Thr-Pro-Gly-TrpNH$_2$	195
Lom-AKH-III	*Locusta migratoria*	pGlu-Leu-Asn-Phe-Thr-Pro-Trp-TrpNH$_2$	196
Psi-AKH	*Pseudagrion inconspicuum*	pGlu-Val-Asn-Phe-Thr-Pro-Gly-TrpNH$_2$	197
Mas-AKH	*Manduca sexta*	pGlu-Leu-Thr-Phe-Thr-Ser-Ser-Trp-GlyNH$_2$	198
	Helicoverpa zea		199
	Bombyx mori		200
Lom-AKH-I	*Locusta migratoria*	pGlu-Leu-Asn-Phe-Thr-Pro-Asn-Trp-Gly-ThrNH$_2$	192
	Schistocerca gregaria		201
Mcsp-AKH	*Magicicada* sp.	pGlu-Val-Asn-Phe-Ser-Pro-Ser-Trp-Gly-AsnNH$_2$	186
	Cacama valavata		202
Hypertrehaloemic hormones			
Tem-HrTH	*Tenebrio molitor*	pGlu-Leu-Asn-Phe-Ser-Pro-Asn-TrpNH$_2$	203
	Zophobas rugipes		203
	Onymacris plana		204
	Onymacris rugatipennis		204
	Physadesmia globosa		204
	Polyphaga aegyptiaca		205

Table 3 (continued)

Substance	*Insect source*	*Structure*	*Ref.*
Pht-HrTH	*Phormia terranovae*	pGlu-Leu-Thr-Phe-Ser-Pro-Asp-TrpNH_2	206
	Drosophila melanogaster		207
Poa-HrTH	*Polyphaga aegyptiaca*	pGlu-Ile-Thr-Phe-Thr-Pro-Asn-TrpNH_2	205
Hez-HrTH	*Helicoverpa zea*	pGlu-Leu-Thr-Phe-Ser-Ser-Gly-Thr-Gly-AsnNH_2	208
Cam-HrTH-II	*Carausius morosus*	pGlu-Leu-Thr-Phe-Thr-Pro-Asn-Trp-Gly-ThrNH_2	188
	Sipyloidea sipylus		209
	Extatosoma tiaratum		210
Cam-HrTH-I	*Carausius morosus*	pGlu-Leu-Thr-Phe-Thr-Pro-Asn-Trp(hexose?)-Gly-ThrNH_2	211
Plc-HrTH-II	*Platyneura capensis*	pGlu-Val-Asn-Phe-Ser-Pro-Ser-Trp-Gly-ThrNH_2	212
Bld-HrTH	*Blaberus discoidalis*	pGlu-Val-Asn-Phe-Ser-Pro-Gly-Trp-Gly-ThrNH_2	213
	Nauphoeta cinerea		214
	Leucophaea maderae		210
	Gromphadorhina portentosa		210
	Blatella germanica		215
Hypotrehalosemic hormone (?)			
Taa-HoTH	*Tabanus atratus*	pGlu-Leu-Thr-Phe-Thr-Pro-Gly-Trp-Gly-TyrNH_2	195
Cardioaccelerators			
Periplanetin	*Periplaneta americana*	pGlu-Val-Asn-Phe-Ser-Pro-Asn-TrpNH_2	216
Pea-CAH-I			210
	Blatta orientalis		217
	Leptinotarsa decemlineata		
	Trinervitermes trinervoides		187
Periplanetin	*Periplaneta americana*	pGlu-Leu-Thr-Phe-Thr-Pro-Asn-TrpNH_2	216
Pea-CAH-II			210
	Blatta orientalis		
	Leptinotarsa decemlineata		217
Corazonin	*Periplaneta americana*	pGlu-Thr-Phe-Gln-Tyr-Ser-Arg-Gly-Trp-Thr-AsnNH_2	218
Leucosulfakinin LSK-II	*Leucophaea maderae*	pGlu-Ser-Asp-Asp-Tyr(SO_3H)-Gly-His-Met-Arg-PheNH_2	219
Leucosulfakinin LSK-I	*Leucophaea maderae*	Glu-Glu-Phe-Glu-Asp-Tyr(SO_3H)-Gly-His-Met-Arg-PheNH_2	220
Diuretic hormones			
Diuretic hormone	*Locusta migratoria*	Cys-Leu-Ile-Thr-Asn-Cys-Pro-Arg-GlyNH_2 | | NH_2Gly-Arg-Pro-Cys-Asn-Thr-Ile-Leu-Cys	221
Diuretic hormone	*Locusta migratoria*	Met-Gly-Met-Gly-Pro-Ser-Leu-Ser-Ile-Val-Asn-Pro-Met-Asp-Val-Leu-Arg-Gln-Arg-Leu-Leu-Leu-Glu-Ile-Ala-Arg-Arg-Arg-Leu-Arg-Asp-Ala-Glu-Glu-Gln-Ile-Lys-Ala-Asn-Lys-Asp-Phe-Leu-Gln-Gln-Ile	222,223
Diuretic hormone DH-2	*Manduca sexta*	Ser-Phe-Ser-Val-Asn-Pro-Ala-Val-Asp-Ile-Leu-Gln-His-Arg-Tyr-Met-Glu-Lys-Val-Ala-Gln-Asn-Asn-Arg-Asn-Phe-Leu-Asn-Arg-Val	224
Diuretic hormone DH-1	*Manduca sexta*	HArg-Met-Pro-Ser-Leu-Ser-Ile-Asp-Leu-Pro-Met-Ser-Val-Leu-Arg-Gln-Lys-Leu-Ser-Leu-Glu-Lys-Glu-Arg-Lys-Val-Lys-Ala-Leu-Arg-Ala-Ala-Ala-Asn-Arg-Asn-Phe-Leu-Asn-Asp-IleNH_2	225
Diuretic hormone	*Tenebrio molitor*	Ser-Pro-Thr-Ile-Ser-Ile-Thr-Ala-Pro-Ile-Asp-Val-Leu-Arg-Lys-Thr-Trp-Glu-Gln-Glu-Arg-Ala-Arg-Lys-Gln-Glu-Arg-Ala-Arg-Lys-Gln-Met-Val-Lys-Asn-Arg-Glu-Phe-Leu-Asn-Ser-Leu-Asn	226
Diuretic hormone	*Musca domestica*	Asn-Lys-Pro-Ser-Leu-Ser-Ile-Val-Asn-Pro-Leu-Asp-Val-Leu-Arg-Gln-Arg-Leu-Leu-Leu-Glu-Ile-Ala-Arg-Arg-Gln-Met-Lys-Glu-Asn-Thr-Arg-Gln-Val-Glu-Leu-Asn-Arg-Ala-Ile-Leu-Lys-Asn-Val	227
	Stomoxys calcitrans		227
Diuretic hormone	*Periplaneta americana*	Thr-Gly-Ser-Gly-Pro-Ser-Leu-Ser-Ile-Val-Asn-Pro-Leu-Asp-Val-Leu-Arg-Gln-Arg-Leu-Leu-Leu-Glu-Ile-Ala-Arg-Arg-Arg-Met-Arg-Gln-Ser-Gln-Asp-Gln-Ile-Gln-Ala-Asn-Arg-Glu-Ile-Leu-Gln-Thr-Ile	228
Diuretic hormone	*Acheta domestica*	Thr-Gly-Ala-Gln-Ser-Leu-Ser-Ile-Val-Ala-Pro-Leu-Asp-Val-Leu-Arg-Gln-Arg-Leu-Met-Asn-Glu-Leu-Asn-Arg-Arg-Arg-Met-Arg-Glu-Leu-Gln-Gly-Ser-Arg-Ile-Gln-Gln-Asn-Arg-Gln-Leu-Leu-Thr-Ser-Ile	229

Table 3 (continued)

Substance	*Insect source*	*Structure*	*Ref.*
Pheromone biosynthesis-activating neuropeptides			
PBAN	*Helicoverpa zea*	Leu-Ser-Asp-Asp-Met-Pro-Ala-Pro-Ala-Thr-Pro-Ala-Asp-Gln-Glu-Met-Tyr-Arg-Gln-Asp-Pro-Glu-Gln-Ile-Asp-Asp-Ser-Arg-Thr-Lys-Tyr-Phe-Ser-Pro-Arg-LeuNH_2	230
PBAN	*Bombyx mori*	Leu-Ser-Glu-Asp-Met-Pro-Ala-Thr-Pro-Ala-Asp-Gly-Glu-Met-Tyr-Gln-Pro-Asp-Pro-Glu-Glu-Met-Glu-Ser-Arg-Thr-Arg-Tyr-Phe-Ser-Pro-Arg-LeuNH_2	231
PBAN	*Lymantria dispar*	Leu-Ala-Asp-Asp-Met-Pro-Ala-Thr-Met-Ala-Asp-Gln-Glu-Val-Tyr-Arg-Pro-Glu-Pro-Glu-Gln-Ile-Asp-Ser-Arg-Asn-Lys-Tyr-Phe-Ser-Pro-Arg-LeuNH_2	232
PBAN	*Leucania separata*	Lys-Leu-Ser-Tyr-Asp-Asp-Lys-Val-Phe-Glu-Asn-Val-Glu-Phe-Thr-Pro-Arg-LeuNH_2	233
Diapause hormone			
Egg diapause	*Bombyx mori*	Thr-Asp-Met-Lys-Asp-Glu-Ser-Asp-Arg-Gly-Ala-His-Ser-Glu-Arg-Gly-Ala-Leu-Cys-Phe-Gly-Pro-Arg-LeuNH_2	234
Allatotropins			
AT	*Manduca sexta*	Gly-Phe-Lys-Asn-Val-Glu-Met-Met-Thr-Ala-Arg-Gly-PheNH_2	235
Allatostatins			
Dip-AST-5	*Diploptera punctata*	Asp-Arg-Leu-Tyr-Ser-Phe-Gly-Leu	236
Dip-AST-4	*Diploptera punctata*	Asp-Arg-Leu-Tyr-Ser-Phe-Gly-LeuNH_2	237
Dip-AST-2	*Diploptera punctata*	Gly-Gly-Ser-Leu-Tyr-Ser-Phe-Gly-LeuNH_2	237
Dip-AST-8	*Diploptera punctata*	Asp-Gly-Arg-Met-Tyr-Ser-Phe-Gly-LeuNH_2	238
Dip-AST-9	*Diploptera punctata*	Gly-Asp-Gly-Arg-Leu-Tyr-Ala-Phe-Gly-LeuNH_2	236
Dip-AST-3	*Diploptera punctata*	Gly-Asp-Gly-Arg-Leu-Tyr-Ser-Phe-Gly-LeuNH_2	237
Dip-AST-6	*Diploptera punctata*	Tyr-Pro-Gln-Glu-His-Arg-Phe-Ser-Phe-Gly-LeuNH_2	238
Dip-AST-7	*Diploptera punctata*	Ala-Pro-Ser-Gly-Ala-Gln-Arg-Leu-Tyr-Gly-Phe-Gly-Leu	228
Dip-AST-1	*Diploptera punctata*	Ala-Pro-Ser-Gly-Ala-Gln-Arg-Leu-Tyr-Gly-Phe-Gly-LeuNH_2	237
Dip-AST-B	*Diploptera punctata*	Ala-Tyr-Ser-Try-Val-Ser-Glu-Tyr-Lys-Arg-Leu-Pro-Val-Tyr-Asn-Phe-Gly-LeuNH_2	239
Bl-AST-1	*Blatella germanica*	Leu-Tyr-Asp-Phe-Gly-LeuNH_2	240
Bl-AST-2	*Blatella germanica*	Asp-Arg-Leu-Tyr-Ser-Phe-Gly-LeuNH_2	240
Bl-AST-3	*Blatella germanica*	Ala-Gly-Ser-Asp-Gly-Arg-Leu-Tyr-Ser-Phe-Gly-LeuNH_2	240
Bl-AST-4	*Blatella germanica*	Ala-Pro-Ser-Ser-Ala-Gln-Arg-Leu-Tyr-Gly-Phe-Gly-LeuNH_2	240
Mas-AS	*Manduca sexta*	Gln-Val-Arg-Phe-Arg-Gln-Cys-Tyr-Phe-Asn-Pro-Ile-Ser-Cys-Phe	241
Grb-AST-A1	*Gryllus bimaculatus*	Ala-Gln-His-Gln-Tyr-Ser-Phe-Gly-Leu NH_2	242
Grb-AST-B1	*Gryllus bimaculatus*	Gly-Try-Glu-Asp-Leu-Asn-Gly-Gly-TrpNH_2	243
Grb-AST-B2	*Gryllus bimaculatus*	Gly-Try-Arg-Asp-Leu-Asn-Gly-Gly-TrpNH_2	243
Grb-AST-B3	*Gryllus bimaculatus*	Ala-Trp-Arg-Asp-Leu-Ser-Gly-Gly-TrpNH_2	243
Grb-AST-B4	*Gryllus bimaculatus*	Ala-Trp-Glu-Arg-Phe-His-Gly-Ser-TrpNH_2	243
Grb-AST-A2	*Gryllus bimaculatus*	Ala-Gly-Gly-Arg-Gln-Tyr-Glu-Phe-Gly-LeuNH_2	242
Callatostatins			
Cal-3	*Calliphora vomitoria*	Ala-Asn-Arg-Tyr-Gly-Phe-Gly-LeuNH_2	244
Cal-4	*Calliphora vomitoria*	Asp(or Asn?)-Arg-Pro-Tyr-Ser-Phe-Gly-LeuNH_2	244
Cal-5	*Calliphora vomitoria*	Gly-Pro-Pro-Tyr-Asp-Phe-Gly-MetNH_2	244
Hyp-Met-Cal	*Calliphora vomitoria*	Gly-Pro-Hyp-Tyr-Asp-Phe-Gly-MetNH_2	242
Cal-2	*Calliphora vomitoria*	Leu-Asn-Glu-Glu-Arg-Arg-Ala-Asn-Arg-Tyr-Gly-Phe-Gly-LeuNH_2	244
Cal-1	*Calliphora vomitoria*	Asp-Pro-Leu-Asn-Glu-Glu-Arg-Arg-Ala-Asn-Arg-Tyr-Gly-Phe-Gly-LeuNH_2	244
Bombyxin	*Bombyx mori*	A chain: H-Gly-Ile-Val-Glu-Asp-Glu-Cys-Cys-Leu-Arg-Pro-Cys-Ser-Val-Asp-Val-Leu-Leu-Ser-Tyr-CysH B chain: pGlu-Gln-Pro-Gln-Ala-Val-His-Thr-Tyr-Cys-Gly-Arg-His-Leu-Aln-Arg-Thr-Leu-Ala-Asp-Leu-Cys-Trp-Glu-Ala-Gly-Val-AspH (A and B chain linked by disulfides)	245
Eclosion hormone			
EH	*Manduca sexta* *Bombyx mori*	HAsn-Pro-Ala-Ile-Ala-Thr-Gly-Tyr-Asp-Pro-Met-Glu-Ile-Cys-Ile-Glu-Asn-Cys-Ala-Gln-Cys-Lys-Lys-Met-Leu-Gly-Ala-Trp-Phe-Glu-Gly-Pro-Leu-Cys-Ala-Glu-Ser-Cys-Ile-Lys-Phe-Lys-Gly-Lys-Leu-Ile-Pro-Glu-Cys-Glu-Asp-Phe-Ala-Ser-Ile-Ala-Pro-Phe-Leu-Asn-Lys-LeuH	246 247

Table 3 (continued)

Substance	*Insect source*	*Structure*	*Ref.*
Ecdysis-triggering hormone			
Mas-ETH	*Manduca sexta*	Ser-Asn-Glu-Ala-Ile-Ser-Pro-Phe-Asp-Gln-Gly-Met-Met-Gly-Tyr-Val-Ile-Lys-Thr-Asn-Lys-Asn-Ile-Pro-Arg-$MetNH_2$	248
Others			
Salivation-stimulating	*Locusta migratoria*	Glu-Val-Gly-Asp-Leu-Phe-Lys-Glu-Trp-Leu-Gln-Gly-Asn-Met-Asn	249

procedure used kilogram quantities of whole cockroaches, countercurrent distribution, paper electrophoresis, and other techniques, more recent methods include ion-paired reverse-phase HPLC.[256] There are several bioassays for proctolin, including the cockroach heart, hyperneural muscle, or hindgut, and the locust tibia muscle. Along with other substances such as acetylcholine, 5-hydroxytryptamine, and octopamine, it accelerates the heartbeat in insect hearts, and it causes contraction of insect muscle at nanomolar concentrations. It is debatable, therefore, whether it should be classed as a hormone or a neurotransmitter. Little is known about the residence time of neuropeptides generally, but it has been shown that [3,5-$^3H_2$-Tyr]proctolin is destroyed within minutes by proteolytic enzymes when in contact with insect tissues that respond to proctolin.[257]

8.05.4.3 Adipokinetic Hormone

The first true neuropeptide with a clearly defined function to be isolated, structurally identified, and synthesized is the locust adipokinetic hormone.[201] The insect flight muscle is one of the most active metabolizing tissues known, consuming energy at a very high rate. The desert locust and the migratory locust undertake long sustained flights, which require the mobilization of lipids to provide energy for the flight muscle. In a pioneering study it was shown that a neural factor was released from the glandular lobes of the corpus cardiaca (see Figure 2) soon after flight began, to mobilize diglycerides from the fat body. The isolation was followed by measuring the amount of lipid in the hemolymph colorimetrically, and the structure was determined by a combination of manual sequencing and MS. The first adipokinetic hormone (AKH) isolated, from the African migratory locust (*Locusta migratoria*) is a blocked decapeptide with pyroglutamic acid (**86**) at the N terminus and an amide at the C terminus.[201] The structure was confirmed by synthesis.[258] This is now called Lom-AKH-I (Table 3), according to the system proposed by Raina and Gäde.[259] Soon after the first, a second substance, an octapeptide called Lom-AKH-II (Table 3), was isolated from *L. migratoria*.[192] More than 10 such AKH peptides have been identified. All are structurally related and all contain pyroglumatic acid at the N-terminal end (Table 3). One substance previously known as a cardioaccelerating substance (see later in this section) in other insects has been identified as an AKH in the termites *Mastotermes darwiniensis* and *Trinervitermes trinertoides*.[186] The extraction, purification, and structure determination of these peptides has been reviewed by Gäde.[260] The most active is Mcsp-AKH from the cicada *Magicada* sp., which causes an increase of over 100% in the lipids in the hemolymph at 0.625 pmol (and of carbohydrates at 2.5 pmol).[261] They do not appear to have species-specific action, for example locust AKH is active in the tobacco hornworm *Manduca sexta*, but a complication arises because the *M. sexta* substance controls lipid release in the adult and carbohydrate release in the larva. Structurally related are some hypertrehalosemic hormones (HrTH), which raise the level of trehalose (**87**), the characteristic sugar circulating in the hemolymph of insects. The cockroach *Periplaneta americana* is very sensitive to hypertrehalosemic factors from other insects, making it a good bioassay subject. However, it was found that extracts of *Locusta* corpora cardiaca, strongly active on *Periplaneta*, had no effect on its own trehalose level. This led to the realization that there may be a *hypo*trehalosemic factor too, and one of these was claimed to have been isolated, but was later shown to be strongly hyperlipemic and hypertrehalosemic.[262] These are included in Table 3: all are isolated from corpora cardiaca and show a high degree of homology. They are all blocked at both ends, the N terminus with pyroglutamic acid, the C terminus with an amide, and they all have an aromatic amino acid (Phe or Tyr) at position 4 and a nonpolar acid at position 2 (Val or Leu, or in one case Ile). There are polar amino acids at positions 3 and 5. They are arranged in Table 3 by increasing chain length.

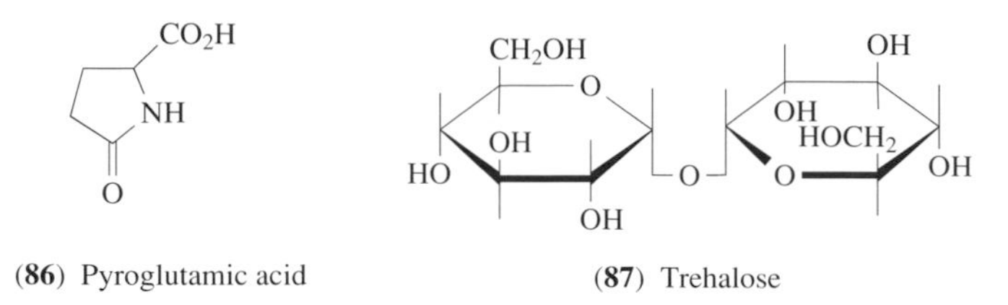

(**86**) Pyroglutamic acid (**87**) Trehalose

Some investigators have used antibodies to known vertebrate peptides in the search for new insect neuropeptides, and in this way have discovered some peptides that, like proctolin, do not clearly fall into the category of hormones but are closely allied in structure to known hormones. Some so-called cardioaccelerators, such as periplanetins, corazonin, and leukosulfakinins, which raise the insect heart rate closely resemble the AKHs and are frequently grouped with them. Two from termites have AKH activity.[262] In the interests of comparison they are included in Table 3.

8.05.4.4 Diuretic Hormone

Insects are able to control their water balance over a severe range of environmental conditions. Since a simple and quantitative bioassay was described by Maddrel for measuring water expulsion from the malpighian tubules, a number of insects have been examined for the existence of a diuretic hormone. In some (*Glossinia*, *Anopheles*, and *Rhodnius*) it is produced by 12 neurosecretory cells of the mesothoracic ganglion (Figure 3), in others (*Locusta*, *Schistocerca*, *Pieris*, *Papilio*, and *Carausius*) it comes from the brain. In those insects so far studied, it appears to form a heterogeneous family of small polypeptides. Six diuretic hormones have been identified. Two have been identified in *Locusta*: the smaller is an antiparallel dimer of two identical nonapeptides, held together by disulfide bridges[221] (Table 3). The first coleopteran diuretic hormone was identified in *Tenebrio molitor*,[226] and is a 37 amino acid peptide of 4371 Da with a C-terminal end that is quite different from others of this class (Table 3). Another from *Manduca* is a polypeptide of 41 amino acids without disulfide linkages.[225]

8.05.4.5 PBAN

The enormous advances in the chemistry and biosynthesis of insect pheromones, especially among the lepidoptera, have expanded enquiry and understanding further into the initiation and release of sexual pheromones. It is now known that a pheromone biosynthesis-activating neuropeptide (PBAN) exists which initiates the process. The first such substances identified were isolated from the subesophageal ganglia of *Helicoverpa* and *Bombyx*. In the former it is a 34 amino acid peptide;[230] in the latter it is very similar but with 33 amino acids.[231] The *Bombyx* PBAN has an amino acid sequence identical to that of a neuropeptide isolated from the armyworm *Leucania separata* as a melanizing and reddish-coloring hormone.[233] Armyworm larvae change their pigmentation from reddish to black when crowded, and this color change has been shown to be hormone controlled. The above two PBANs also share some similarities with other hormones, such as the diapause neuropeptide of *Bombyx* (see Section 8.05.4.6) and vertebrate insulin-like growth factors. The true PBAN of *Leucania* is a much smaller peptide of 18 amino acids (Table 3).[233]

8.05.4.6 Diapause Hormone

Arrested development can occur at any stage of the insect life cycle, egg, larva, pupa, or adult, to avoid adverse conditions. Only one substance regulating this, an egg diapause hormone, from the heads of adult *Bombyx mori* has been identified. To illustrate the scale of quantities used in this work, from 2.4 million heads ($\sim$4.4 kg) was obtained 65 μg of a substance called EDH-A, a peptide of 3.3 kDa, and 5.8 μg of EDH-B of 2.0 kDa.[234] Later, a pure product was obtained and identified as a peptide of 24 amino acids[263] (Table 3), with a sequence partially resembling *Bombyx* PBAN (see Section 8.05.4.5). The egg diapause hormone is produced in just two cells of the subesophageal ganglion, and is apparently much more abundant in adult males.

8.05.4.7 Allatotropins and Allatostatins

The production of JH by the corpora allata (see Section 8.05.2) is controlled either by the brain (*Manduca*) or by a factor in the hemolymph (*Leptinotarsa decemlineata*). The factors responsible for stimulating JH production are called allatotropins. That from *Manduca*, a tridecapeptide, has been identified (Table 3).[235] It has no structural relationship to other known peptides.

In two cockroaches and a bug, severing the nervous connection between the brain and corpus allatum causes the continuous secretion of JH, so that the action of an allatostatin or allatohibin from the brain was postulated. Five small peptides, allatostatins A1 to A5, that inhibit the formation of JH in the adult cockroach *Diploptera punctata* were first isolated and identified (Table 3).[237] Further allatostatins have followed. There are 13 from *Diploptera*. Those from the cricket *Gryllus bimaculatus* are different from all the others so far known.[242,243] A number of peptides from *Calliphora vomitoria*, called callatostatins, are active allatostatins in *Diploptera* but were inactive in *Calliphora*.[244,264]

8.05.4.8 Bursicon

Two simultaneous studies in 1962 demonstrated the existence of a hormone required for tanning of the cuticle in adult blowflies. It soon became evident that this hormone, given the name bursicon,[265] was widely distributed in insects (locust, wax moth, flour beetle, and milkweed bug). The properties indicated it was none of the then known hormones (JH, ecdysone or PTTH), and none of these could induce tanning.[266] In Diptera it is produced from the brain neurosecretory cells, in other orders it is thought to be from the terminal abdominal ganglion. Bursicon is a large peptide of still unknown structure. It is unlikely to be species-specific. A number of unsuccessful attempts have been made to isolate it, but it is apparently unstable in the pure state. The most recent estimate of its mass is a single-chain peptide of 30 kDa, and it is suggested that it may have the same structure throughout most insects.[267]

8.05.4.9 Bombyxin

In the course of isolation of PTTH of *Bombyx mori*, a much smaller protein was obtained from the brain, shown to be produced in a group of medial neurosecretory cells, and called bombyxin.[245] Its normal physiological role is still unclear but it may have a part in carbohydrate metabolism (see Section 8.05.4.3). It also stimulates cell division. It is a 5 kDa heterodimer, with a high similarity to vertebrate insulin both in its amino acid sequence and its tertiary structure. A number of insects have been shown to produce peptides similar to insulin, but there is no cross-activity between bombyxin and insulin, and the injection of insulin into insects has not produced any agreed effect. It is still not clear if the hypoglycemic factors of insects and the insulin-like materials are the same.

There is also a factor in the corpus cardiacum with a hyperglycemic action similar to that of glucagon in the vertebrate liver.[268] It has been suggested that glucagon could be the insect hyperglycemic hormone. A number of investigations have not yet resolved this question, but peptides approaching the mass of bovine glucagon (3.5 kDa) have been isolated. These have no similarity to the HrTH described earlier (see Section 8.05.4.3).

8.05.4.10 Eclosion and Ecdysis-triggering Hormone

The emergence of the adult silkworm from its pupa is shown to be dependent upon a neurohormone from the head, acting directly upon the abdominal nerve cord, which in turn releases an eclosion hormone (EH) that governs the sequence of muscular actions required in emergence from the pupal case. The eclosion hormone has been extracted from the heads of *Manduca sexta* adults just before their emergence (eclosion), and identified as a 62 amino acid protein (Table 3).[246,247] A very similar protein has been isolated from *Bombyx* along with three others, closely related.[269] It is thought that this hormone is not species-specific.

Another step in the process of eclosion has been clarified by the isolation of an ecdysis-triggering hormone (ETH) from the pupae of *M. sexta*.[248] The hormone from epitracheal cells distributed through each pupal segment release a peptide of 2940 Da, Mas-ETH (Table 3). Its structure, which is novel and unrelated to the EH, has been confirmed by synthesis. Injection of the hormone into molting animals rapidly induces the behavior of pre-ecdysis and ecdysis.

8.05.4.11 Other Neurohormones

Other neuropeptides have been isolated that might claim a place here, but there are difficulties in the definition of a neurohormone evident from the discussion of various authors.[249] There is, for example, a salivation-stimulating peptide obtained from 2500 salivary glands of *Locusta*[270] of 1779 Da (Table 3). The description of peptides isolated from one insect and tested for hormone action in another species increases the difficulties. There are a number of myotropic peptides that affect muscle contraction known in insects, but they are outside the scope of this chapter. A review lists 56 peptides of known sequence, isolated from *Locusta* and *Schistocerca*, most of them being myotropic.[271] We have attempted to draw a middle line, but will have omitted some substances that fall into this uncertain category.

8.05.4.12 Neuropeptides as Insect Control Agents

The stimulus to isolate and identify more neuropeptides is at least threefold: to understand more of the basic physiology of insects, to understand through comparison with vertebrate peptides how new peptides evolved, and to find new ways to control insect pests. Synthetic insecticides from all the major groups have been shown to release various neuropeptides, including diuretic hormone, hyperglycemic factors, and AKH.[272] Most chemical messengers that function between the central nervous system and subordinate organs appear to be peptides or small proteins. The vulnerability of the neuroendocrine system has been emphasized by Richards and Hoffmann.[273] As neuropeptides are so vital to insect development and reproduction, it is logical to think of the neuroendocrine system of insects as a target for methods of pest control,[274] and these substances have received a lot of attention in this respect.[252] Studies of human neuropeptide pharmacology have led to the discovery of a number of useful drugs, which either inhibit or stimulate some neuropeptide action. As yet, the strategy has been disappointing with insects. Proctolin, because of its simple structure, has served as a model for testing synthetic peptides. Miller lists 32 analog peptides which have been tested on cockroach rectum or locust tibia, of which only one (*p*-methoxyphenylalanyl-proctolin) showed substantial activity.[275]

8.05.5 PLANT SUBSTANCES TOXIC AND DETERRENT TO INSECTS

In the 300 million years that insects and plants have coexisted on the earth, they have had many opportunities to adapt to each other. While some insects feed on other animals or on dead material, plant and animal, the great majority live on plants. In response the plants have developed many strategies to protect themselves, prominent among them being chemical defense. Insects in their turn have developed ways to overcome the plant toxins designed to repel them. The result is not a victory but a stalemate (unless man intervenes with a monoculture). It is therefore unreasonable to expect to find in plants substances that will be universally deterrent or toxic to insects.

One man's meat...

The tobacco hornworm (*Manduca sexta*) has adapted to live on *Nicotiana* plants, which are avoided by other insects, and the stored-product beetle (*Lasioderma serricorne*) can feed on the processed leaves, so gaining the name of cigarette beetle.

...is another man's poison

On the other hand, some wild species of *Nicotiana* are very toxic to *M. sexta*, apparently because they contain *N*-acylnornicotines, while the tobacco budworm (*Helicoverpa virescens*), another nicotine specialist, is unaffected by these compounds.[276] The nicotine in *Nicotiana* plants is a very powerful poison to most phytophagous insects. It is such an effective toxicant that crude nicotine (as an extract of the leaves of *Nicotiana glauca*) was used as a commercial insecticide before 1945 (see Section 8.05.5.6.1).

What we might expect to find in plants are substances that are avoided by a large group of species. The idea of a universal insecticide is as unreasonable as a panacea in human medicine. That idea was abandoned long ago, and we now look to specific remedies for specific diseases. Interest should turn towards what groups of insects avoid certain plants and why. In this way we shall learn more about the chemical ecology that links plants and insects, and the industrial chemist will learn what is most effective to use against insect pests.

It has been pointed out by Duffey and Stout in a review of plant defenses against insects that it is not correct to attribute plant defenses to certain chemicals in the plant in isolation.[277] Rather, plant defenses are composed of an array of enzymes, including proteinase inhibitors, amylase inhibitors, phytohemaglutinins, and others, as well as so-called toxic phytochemicals which act in concert to cause a greater or lesser effect on the insect predator. They give examples from their studies of lepidopteran larvae feeding on tomato plants. It is meaningless to think of phenolics, proteinase inhibitors, proteins, saponins, and lipids as examples of classes of defensive or nutritive chemicals with inherent and constant properties, independent of the other substances in the plant. While the correctness of what they say is not challenged, there are many cases of natural products with interesting effects on insects, such as the JH mimics (see Section 8.05.2.8) or anti-JH compounds (see Section 8.05.2.10) described earlier. The natural products chemist is interested in the discovery of these materials and to learn why they have their special properties toward insects. If in pursuit of this knowledge one discovers something useful in controlling pests, one is doubly rewarded. Substances that inhibit insects from feeding on a plant or disrupt their growth and development when ingested are particularly interesting, because satisfactory bioassays exist to guide their isolation. Observation of plants that do not appear to be much attacked by insects can be a helpful clue, but that has its limitations. *Ginkgo biloba* L. (Ginkgoaceae) is a tree that is subject very little to insect damage, but when extracts of various parts of the tree were examined by feeding them to insects and noting the amount of feces produced, no compounds of specially interesting antifeedant effect were discovered. The immunity of the tree seems to consist of the combined effects of a number of moderately effective substances, including ginkgolides (highly functionalized diterpenes) (see Section 8.05.5.3.3) and anacardic acids (isoprenyl-substituted salicylic acids).[278] What is known of the fate of secondary metabolism plant substances ingested by insects has been comprehensively reviewed by Brattsten,[279] and the enzymes involved in detoxification by Ahmad *et al.*[280]

Plant products can be direct-acting toxins such as the classical insecticides nicotine, pyrethrins, rotenone, quassin, and sabidalla, or they can be indirect-acting compounds, which include feeding deterrents, oviposition deterrents, repellents, growth and development regulators, antihormones, and hormone mimics.[281] An antifeedant is a substance that inhibits feeding but does not kill the insect directly, though the insect may remain on the plant until it dies of starvation. It is not quite the same as a substance that has an odor which causes an insect to avoid it, that is, a repellant.

The following sections are arranged as far as possible according to their biosynthetic origins, that is compounds derived from fatty acids, acetogenins, terpenes, mixed origins, phenylpropanoids, and alkaloids.

8.05.5.1 Isobutylamides

An extract of black pepper (*Piper nigrum*, Piperaceae) has strong insecticidal action against several insects. Long-chain polyunsaturated isobutylamides from species of Piperaceae, Asteraceae, and Rutaceae have long been recognized as having insect toxicity. Pellitorine (*N*-isobutyl-2*E*,4*E*-decadienylamide) (**88**) from *Anacyclus pyrethrum* (Asteraceae) was the first to be isolated, in 1895 by Dunstan and Garnett, but because of its instability through oxidation it was not obtained truly pure and its structure determined until much later.[282] Spilanthol (affinin or *N*-isobutyl-2*E*,6*Z*,8*E*-decatrienylamide) (**89**) from *Heliopsis longipes* is active against *Musca domestica* and *Tenebrio molitor*. Spilanthol from the aerial parts of paracress *Spilanthes oleraceae* was shown to be identical with affinin from the Mexican plant *H. longipes*, which in turn was confused with *Erigon affinis*, hence the misleading name. Natural spilanthol has toxicity to houseflies comparable to pyrethrin, but all-*trans*-spilanthol is nontoxic to houseflies while retaining its knock-down effect. Anacyclin (**90**), also from *A. pyrethrum*, contains a conjugated diene and a conjugated diyne.[282] Herculin (**91**) from the bark of *Zanthoxylum clava-herculis* lacks a conjugated diene but is as toxic as pyrethrins to *M. domestica*, ticks, and mosquito larvae. Echinacein (neoherculin or α-sanshool) (**92**) from *Echinacea angustifolia*, *E. pallida*, *Zanthoxylum piperitum*, and *Z. clava-herculis* (all Rutaceae) is also comparable to pyrethrin with houseflies. A number of isobutylamides have been isolated from *Spilanthes mauritiana* and tested on mosquito larvae.[283]

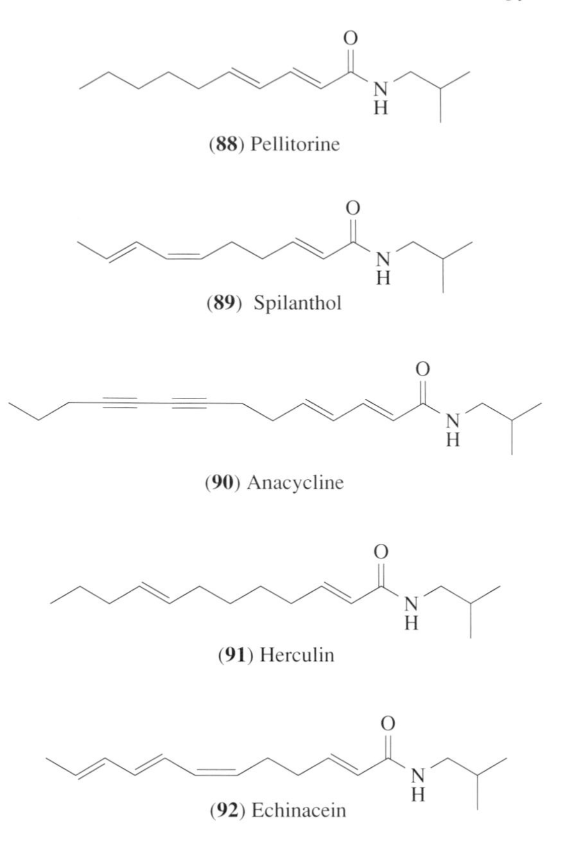

(**88**) Pellitorine

(**89**) Spilanthol

(**90**) Anacycline

(**91**) Herculin

(**92**) Echinacein

Essential features for activity are one or two double bonds conjugated to the amide carbonyl and more double bonds further away. The unsaturated isobutylamides are soft, waxy, low-melting solids that are soluble in nonpolar solvents. They are present in the plants in relatively low concentrations. They share with pyrethrin a pungent taste and rapid knock-down effect on insects. When pure, they are all very pungent for humans, and cause numbness of the mouth and profuse salivation. Their toxicity to mammals is high. In mammals and insects they act as voltage-dependent blockers of sodium channels in nerve cells. There are no direct studies of their biosynthesis, but are presumably formed from acetate in the normal way.

Their instability in air precluded their use as insecticides, but the realization that insects were often able to acquire resistance to ester insecticides such as pyrethrum by using esterases to detoxify them led to renewed interest in the isobutylamides, which should be more resistant to hydrolysis. Work on natural and synthetic isobutylamides including their effects on insects, structure elucidation, and synthesis are discussed by Miyakado *et al.*[284]

Related to these unsaturated fatty acid amides are the medium-chain isobutylamides and other amides terminating in a methylenedioxybenzene from the Piperaceae, such as piperine (**93**) and pipercide (**94**) from *P. nigrum*, and guineensine (**95**) from *P. guineense.*[285] Even the simple fagaramide (**96**) from the root bark of *Zanthoxylum* (= *Fagara*) *zanthoxyloides* and *Z. macrophylla* (Rutaceae) is toxic and inhibits insect growth. For some insects these compounds have an activity approaching pyrethrins. More data on insect toxicity are given by Holyoke and Reese.[286] There has been some attempt to correlate structure and activity.[287]

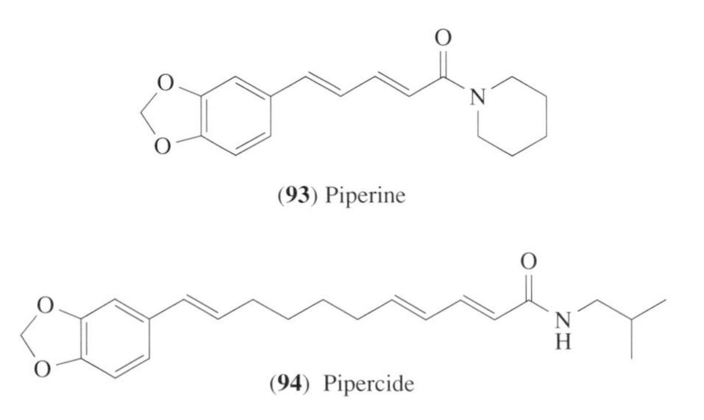

(**93**) Piperine

(**94**) Pipercide

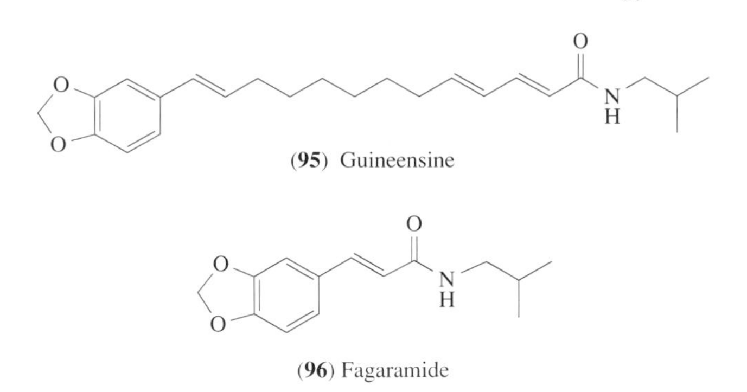

(**95**) Guineensine

(**96**) Fagaramide

8.05.5.2 Annonins

The toxic and insecticidal action of the seeds of a number of species of *Annona*, including *A. squamosa* or sweetsop and *A. muricata* or soursop (Annonaceae), have long been recognized. For a time the action was attributed to annonaines, phenylalanine-derived alkaloids, but since the discovery of the annonins in 1982 it has been recognized that they, exemplified by squamosin (annonin I, squamosin A) (**97**) from several species of *Annona* and *Rollinia*, and asimicin (**98**), from *Asimina triloba* (the North American pawpaw tree), *Rollinia sylvatica*, and *Xylopia aromatica*, are the active agents. They represent the one important group of plant acetogenins which are toxic and antifeedant to insects, but have been found only in the Annonaceae. They also have antitumor, immunosuppressive, antiprotozoal, anthelmintic, and antimicrobial properties, and have therefore received a great deal of attention. Well over 200 examples have now been isolated. Their chemistry has been reviewed in detail with physical and spectroscopic data for new compounds.[288–290]

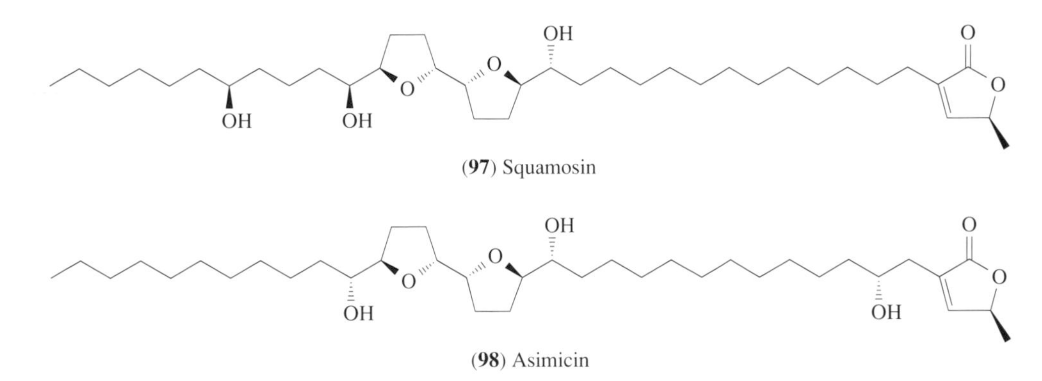

(**97**) Squamosin

(**98**) Asimicin

They are slow-acting toxins, particularly towards Lepidoptera and leafhoppers,[291,292] but asimicin is active against the melon or cotton aphid (*Aphis gossypii*) and mosquito larvae also. They inhibit mitochondrial respiration rather as rotenone does, but through specific inhibition of complex I (NADH–ubiquinone oxidoreductase), with a net depletion of ATP.[293] They also have high mammalian toxicity.

Biosynthetically, they are derived from C_{32} or C_{34} long-chain acids combined with a 2-propanol group at C-2 to form a γ-lactone, the fatty acid portion having one, two, or three tetrahydrofuranyl groups with flanking hydroxyls.

8.05.5.3 Terpenes

Terpenes form the largest class of secondary plant products that interact with insects. The subject is vast and the information about effects on insects is so fragmentary that some omissions are inevitable. There is a review that considers plant–herbivore aspects,[294] and Reese and Holyoke[295] list some terpenes with their effects on insects.

For 40 years it was accepted that isoprenes and isopentyl pyrophosphate were synthesized in plants from acetate via mevalonate. However, the work of Rohmer has shown that terpenes in plants may not only be biosynthesized from acetate but also from desoxyxylulose (from the breakdown of sugars), glyceraldehyde 3-phosphate, and pyruvate.[296] The acetate–mevalonate path is followed in the cytoplasm, but the alternative desoxyxylulose route operates in chloroplast-bound tissues. This discovery may change considerably our views on terpene biosynthesis.

8.05.5.3.1 Monoterpenes

Monoterpenes are very widely distributed in plants. It is generally accepted now that they are produced specifically for the purpose of the plant, since they can be shown to be synthesized even in very young tissues.[297] Large amounts of volatile terpenes released into the air can act as fumigants to deter insects. Monoterpenes can also act as volatile signaling compounds, either attracting insects to their host plants or repelling them, and there is much evidence that insects produce them themselves as pheromones and defensive secretions, but these aspects are beyond the present discussion.

Monoterpenes are generally regarded as nontoxic to mammals. They are used as natural and artificial flavorings in food, cosmetics, and pharmaceuticals, and even as pharmaceuticals themselves, for example camphor and eucalyptol. There are, on the other hand, various reports of insecticidal monoterpenes. Limonene (**99**) is sufficiently toxic to the cat flea (*Ctenocephalides felis*) to be used to control it on cats and dogs.[298] Linalool (**100**) is also used against pet fleas and in insecticidal sprays for house plants, while menthol (**101**) is used against tracheal mites of honeybees.[299] Simmonds and Blaney tested 11 monoterpenes for insecticidal activity against *Spodoptera littoralis*. Pulegone (**102**), menthone (**103**), isomenthone (**104**), and isopulegone (**105**) were the most toxic (with LD_{50} topically applied, of 1.2 mg g^{-1}, 2.2 mg g^{-1}, 3.1 mg g^{-1}, and 3.4 mg g^{-1}, respectively), and myrcene (**106**), carvone (**107**), and linalool (**100**) had the greatest antifeedancy, though these were not outstanding compared to some clerodane diterpenes.[300] Rice and Coats have similarly examined a large number of monoterpenes, seeking the structural elements required for activity against insects, using the housefly with topical, fumigant, and ovicidal tests.[301] Thymyl trifluoroacetate (**108**) was the most effective fumigant, followed by menthol (**101**), and fenchone (**109**), thymol (**110**), and geranyl acetate (**111**) were the most effective in the topical test, and geraniol (**112**), geranyl propionate (**113**), terpineol (**114**), carvacrol (**115**), and menthone (**103**) were as ovicidal as a pyrethrin standard. Ketones were generally more effective than alcohols in the topical and ovicidal tests.[301] In 48 h bioassays with the western corn rootworm (*Diabrotica vergifera*) *d*-limonene was only 10 times less active than commercially used organophosphates or carbamates for this pest.[302]

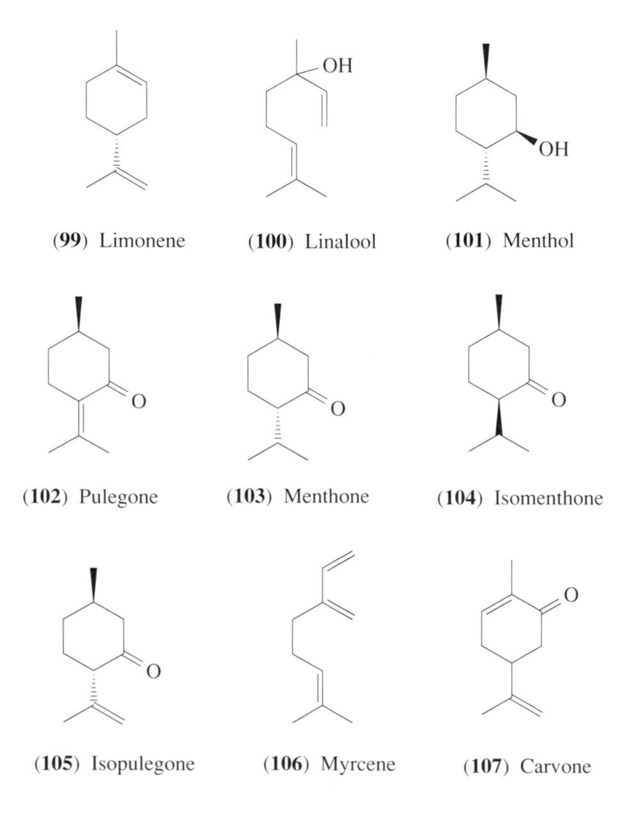

(**99**) Limonene (**100**) Linalool (**101**) Menthol

(**102**) Pulegone (**103**) Menthone (**104**) Isomenthone

(**105**) Isopulegone (**106**) Myrcene (**107**) Carvone

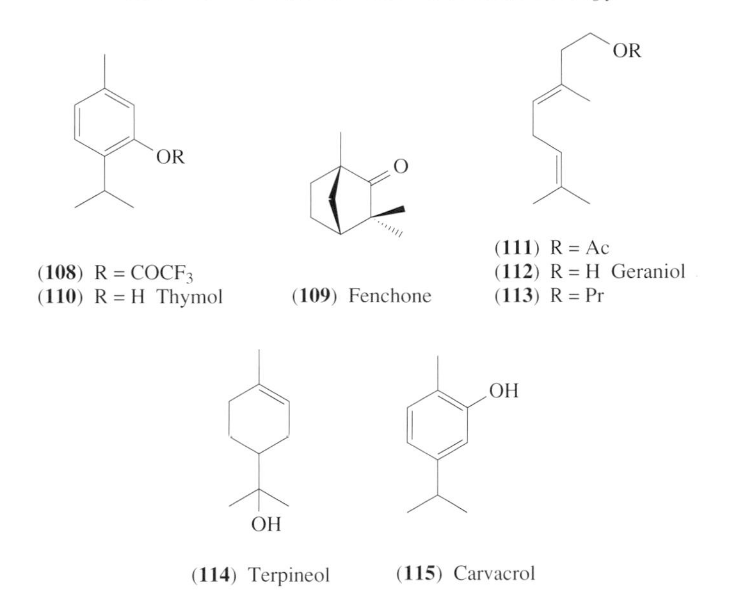

An interesting group are the iridoid glycosides, sugar conjugates of monoterpenes. The first iridoid, iridomyrmecine, was discovered in the pygidial gland secretion of the ant *Iridomyrmex humilis* (see Section 8.05.6.4) The glycosides were later discovered in plants and are now recognized as widely distributed and protect the plant from insect herbivores.[303] They are extremely bitter compounds, soluble in water and crystallizable from polar solvents. According to the review by Bowers,[304] there are about 600 iridoid glycosides known from at least 57 plant families. Aucubin (**116**) is found in all parts of *Aucuba japonica*, cornin or verbenalin (**117**) is produced by *Verbena officinalis* and *Cornus florida*, and loganin (**118**) from *Strychnos nux-vomica* is a precursor of the indole alkaloides in that plant. Catalpol (**119**) occurs in *Plantago lanceolata*, *Buddlea globosa*, and *B. variabilis*, and as various esters of the 6-hydroxyl group, for example catalposide (catalpin), the *p*-hydroxybenzoyl ester in the fruit of *Catalpa* spp., globularin, the (*E*)-cinnamoyl ester, and globularicisin (picroside I) the (*Z*)-cinnamoyl ester. Some, such as plumieride (**120**) from *Plumeria acutifolia*, *P. lancifolia*, and *P. rubra* (Apocynaceae), have been extended by further biosynthesis with acetoacetate. Iridoid glycosides can represent up to 25% of the dry weight of leaves, but are found in all parts of some plants, for example antirrhinoside is found in all parts of the garden shrub *A. japonica*. Some adapted insects sequester the iridoid glycosides from plants to use for their own defense.[305] There is apparently nothing known about the structure–activity relationship of these compounds or how they are metabolized.

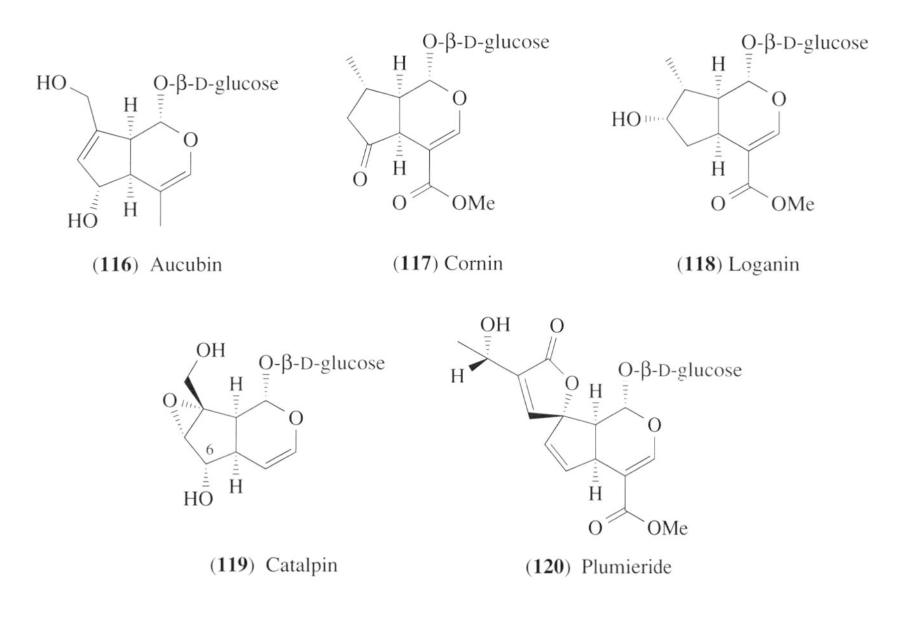

8.05.5.3.2 Sesquiterpenes

The sesquiterpenes present a number of compounds of interesting activity against insects. They fall mostly into four types, with an assortment of others, chiefly from the Compositae. The most active compounds are listed in Table 4 with their plant sources. Generally these sesquiterpenes are crystalline compounds, readily extracted with dichloromethane and isolated by column chromatography. The drimane dialdehydes, represented by warburganal, are a good example. Warburganal (**121**) from *Warburgia salutaris* (= *ugandensis*) and other *W. spp.* (Canellaceae) is a broad-spectrum antifeedant.[306] It is a strong deterrent for the African armyworm (*Spodoptera exempta*), the bollworm *S. littoralis*, the Colorado potato beetle (*Leptinotarsa decemlineata*), and the aphid *Mysus persicae*, but ineffective against several American insects, for example *S. eridania*, *Manduca sexta*, and *Schistocerca vaga*. Polygodial (**122**) from the water pepper (*Polygonum hydropiper*, Polygonaceae), *Drymis lanceolata* (= *aromatica*) (Winteraceae), and the bark of *Warburgia stuhlmannii*, the first of this group isolated,[307] is very similar in activity, and has also been isolated from marine nudibranchs. Others of the series with strong antifeedant activity include cinnamodial (or ugandensidial) (**123**) from the bark of *Cinnamosa fragrans* and the heartwood of the muziga tree (*Warburgia salutaris*), muzigadial (**124**), a rearranged drimedane from *W. salutaris*, and drimenol (**125**) from *Drymis winteri*.[308] All these compounds share with other plant antifeedants and toxicants the effect of producing a hot taste in the human mouth. All have been synthesized, and the syntheses have been reviewed.[309,310] Their chemistry and antifeedant properties have also been reviewed.[311] It is said that the requirement for activity in the group is the unsaturated aldehyde with the second aldehyde group at a fixed distance from it.[312] The compounds act directly on taste receptors; in some lepidopteran larvae they block the stimulatory effect of sugars on the chemoreceptors of the mouthparts.[313] As aphid antifeedants, natural (−)-polygodial and the synthetic (+) isomer were equally active, and had similar phytotoxicity and fish toxicity and the same hot taste to humans.[314]

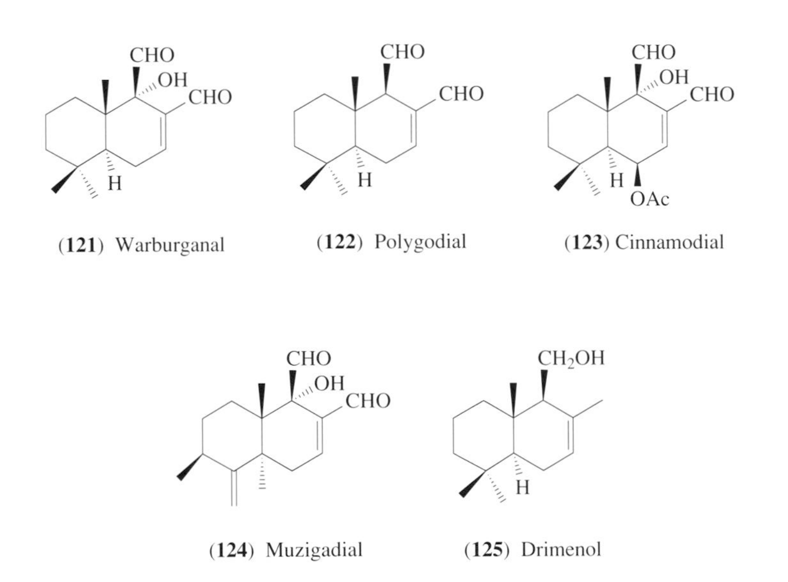

(**121**) Warburganal (**122**) Polygodial (**123**) Cinnamodial

(**124**) Muzigadial (**125**) Drimenol

As the largest group of plant sesquiterpenes known, it is not surprising that many of the sesquiterpene lactones are antifeedant to insects. The reason for the lactone toxicity is unknown. They tend also to combine antifeedancy with cytotoxicity. A review of over 50 sesquiterpene lactones which show activity on insect feeding and development has pointed to the importance of an α-cyclopentenone and γ-lactone fragments for this activity.[315] Such a clear correlation is not possible to see in the examples given here. Because of their number, they are gathered into Table 4, with data on their source and a registry number where possible for further information. The lactones comprise bicyclic gemacranolide types (**126**)–(**136**), tricyclic guaianolide types (**137**)–(**145**), pseudoguaianolides (**146**)–(**149**), a secoguaianolide (**150**), and a few others (**151**)–(**155**) (Table 4). The vast majority are said to be found in the Asteraceae,[316] where they are located in glandular hairs or latex ducts.[317] It has been noted that those from the Umbelliferae form a distinct stereochemical group at the ring fusion positions. These examples do not exhaust the list of similar structures. There are many compounds which are either not active or have not been tested for activity with insects. The leaves of *Gingko biloba* (Gingkoaceae) are well known for their resistance to many insects. It would

Table 4 Sesquiterpenes which are antifeedant and toxic to insects.

Structure	*Molecular formula*	*Reference no.*[a]	*Name*	*Plant source and family*
(121)	$C_{15}H_{22}O_3$	62994-47-2 2116628	Warburganal	*Warburgia salutaris* (= *ugandensis*) and other *Warburgia* spp., Canellaceae
(122)	$C_{15}H_{22}O_2$	6754-20-7 1874733	Polygodial	*Polygonum hydropiper*, Polygonaceae *Drymis lanceolata* (= *aromatica*), Winteraceae *Warburgia stuhlmannii*, *W. ugandensis*, Canellaceae *Dendrodius krebsii*, *D. lumbata*, *D. nigra*, *D. tuberculosa* (nudibranchs)
(123)	$C_{17}H_{24}O_5$	23599-45-3 2594744	Cinnamodial (ugandensidial)	*Cinnamosma fragrans*, *Warburgia salutaris* (= *ugandensis*), Canellaceae
(124)	$C_{15}H_{20}O_3$	66550-09-2 2980962	Muzigadial (canellal)	*Warburgia salutaris*, *Canella winteriana*, Canellaceae
(125)	$C_{15}H_{26}O$	468-68-8 2329053	Drimenol	*Drymis winteri*, Winteraceae
(126)	$C_{20}H_{26}O_6$	6856-01-5 3595346	Eupatoriopicrin	*Eupatorium cannabinum*, *Chaenactis carpoclinia*, *C. douglasii*, *Eriophyllum stachaedifolium*, Compositae
(127)	$C_{23}H_{28}O_{10}$	11091-29-5 5322631	Glaucolide A	*Vernonia glauca* and many other *Vernonia* spp., Compositae
(128)	$C_{22}H_{30}O_6$	26560-24-7 4717426	Laserolide	*Laser trilobium*, Umbelliferae
(129)	$C_{17}H_{22}O_5$	41059-80-7 6573183	Lipiferolide	*Liriodendron tulipifera*, Magnoliaceae *Michelia fuscata*
(130)	$C_{21}H_{24}O_9$	35852-26-7 6546061	Melampodin A	*Melampodium heterophyllum*, *M. leucanthum*, *M. pilosum*, Compositae
(131)	$C_{25}H_{30}O_{12}$	60295-53-6 1278037	Melampodinin	*Melampodium americanum*, *M. diffusum*, *M. longipes*, Compositae
(132)	$C_{19}H_{24}O_6$	19889-00-0 6739641	Onopordopicrin	*Arctium minus*, *Berkheya speciosa*, *Brachylaena elliptica*, *B. discolor*, *B. rotundata*, *Onopordum acanthium* and other *Onpordum* spp., Compositae
(133)	$C_{15}H_{20}O_4$	26931-94-2 4235037	Salonitenolide	*Centaurea salonitana*, *Cnicus benedictus*, *Berkleya speciosa*, *Jurinea maxima*, Compositae
(134)	$C_{17}H_{22}O_6$	39815-40-2 1437034	*epi*-Tulipinolide diepoxide	*Liriodendron tulpifera*, Magnoliaceae
(135)	$C_{22}H_{28}O_8$	38458-58-1 1301775	Eucannabinolide (schkuhrin I, hyodorilactone A)	*Schkuhria pinnata*, *Eupatorium cannabinum*, *E. sachalinense*, Compositae
(136)	$C_{25}H_{34}O_9$	70434-09-2 1276264	Schkuhrin II	*Schkuhria pinnata*, Compositae
(137)	$C_{29}H_{40}O_{10}$	41929-11-7 4283480	Archangelolide	*Laserpitium archangelica*, Umbelliferae
(138)	$C_{15}H_{18}O_5$	29431-83-2 4488932	Chrysartemin A (canin)	*Artemisia cana* and other *Artemisia* spp., *Tanacetum parthenium*, *Handelia trichophylla*, Compositae
(139)	$C_{17}H_{28}O_5$	29431-84-3 6574349	Chrysartemin B (artecanin)	*Artemisia cana*, *Chrysanthemum macrophyllum*, Compositae
(140)	$C_{25}H_{34}O_7$	68852-48-2 1359117	Gradolide	*Laserpitium siler*, Umbelliferae
(141)	$C_{15}H_{18}O_4$	22489-66-3 1292298	Grossheimin (grosshemin)	*Grossheimia macrocephala*, *Amberboa lipii* (= *Centaurea lipii*), *Cynara scolymus*, *Chartolepis intermedia*, *Venidium decurens*, Compositae
(142)	$C_{22}H_{30}O_7$	38114-47-5 1357890	Isomontanolide	*Laserpitium siler*, Umbelliferae
(143)	$C_{23}H_{32}O_3$	68799-88-2 1359254	Polhovalide	*Laserpitium siler*, Umbelliferae
(144)	$C_{27}H_{38}O_{10}$	50657-07-3 4731117	Trilobalide	*Laser trilobium*, Umbelliferae
(145)	$C_{15}H_{18}O_3$	65017-97-2 4694498	Xerantholide	*Xeranthemum cylindraceum*, Compositae
(146)	$C_{15}H_{20}O_4$	2571-81-5 1290096	Coronopilin	Many *Ambrosia*, *Iva*, *Parthenium* spp., *Hymenoclea salsoda*, Compositae
(147)	$C_{17}H_{22}O_5$	31299-06-6 1262044	Ligulatin B (incanin)	*Parthenium ligulatum*, *P. incanum*, *P. schotti*, *P. tomentosum*, Compositae
(148)	$C_{17}H_{22}O_6$	22621-72-3 1353887	Tetraneurin A	*Parthenium alpinum*, *P. cineracum*, *P. confertum*, *P. fruticosum*, *P. hysterophorus*, Compositae
(149)	$C_{17}H_{22}O_5$	19202-92-7 962645	Tenulin	*Helenium tenuifolium*, *H. elegans*, Compositae
(150)	$C_{17}H_{22}O_5$	26791-72-0 1294633	Xanthumin	*Xantium chasei*, *X. chinensis*, *X. occidentale*, *X. strumarium*, Compositae
(151)	$C_{15}H_{22}O_2$	19906-72-0 5743465	Bakkenolide A (fukinanolide)	*Petasites* spp., *Senecio pyrimidalis*, *Cacalia hastata*, *Ligularia calthaefolia*, *Homogyne alpina*, Compositae

Table 4 (continued)

Structure	*Molecular formula*	*Reference no.*[a]	*Name*	*Plant source and family*
(**152**)	$C_{15}H_{20}O_2$	546-43-0 6568980	Alantolactone (helenin)	*Inula helenium*, Compositae
(**153**)	$C_{15}H_{18}O_3$	481-06-1 1291318	α-Santonin	*Artemisia maritima*, and widely in other *Artemisia* spp., Compositae
(**154**)	$C_{19}H_{18}O_7$	21871-10-3 1630244	Vernodalin	*Vernonia amygdalina*, *V. guineensis*, Compositae
(**155**)	$C_{15}H_{18}O_7$	33570-04-6 6875294	Bilobalide	*Gingko biloba*, Gingkoaceae
(**156**)	$C_{15}H_{20}O_2$	22489-40-3 5274069	Pinguisone	*Aneura pinguis*, Hepatica
(**157**)	$C_{17}H_{28}O_4$	20071-59-4 1319928	Shirimodiol 8-acetate	*Parabenzoin trilobium*, Lauraceae
(**158**)	$C_{19}H_{30}O_5$	20071-58-3 1688972	Shirimodiol diacetate	*Parabenzoin trilobium*, Lauraceae
(**159**)	$C_{34}H_{46}O_{13}$	139979-81-0 5370316	Angulatin A	*Celastrus angulatus*, Celastraceae
(**160**)	$C_{32}H_{40}O_{14}$	6361860	Celangulin	*Celastrus angulatus*
(**161**)	$C_{20}H_{24}O_8$	65388-17-2 1407953	Vernodalol	*Vernonia anthelmintica*, Compositae
(**162**)	$C_{30}H_{30}O_8$	303-45-7 1917878	Gossypol	*Gossypium hirsutum*, *Thespesia populnea*, Malvaceae
(**163**)	$C_{15}H_{24}O$	6137324	Caryophyllene α-oxide	*Gossypium hirsutum*, Malvaceae

[a] Broken numbers are *Chemical Abstracts* registry numbers while unbroken numbers are Beilstein registry numbers.

be more convenient for this plant to consider all its known compounds together. Having decided on a chemical classification, the lactone bilobalide (**155**) (0.0096% yield from leaves) should be mentioned here. It is much more toxic to insects than the diterpene gingkolides (see Section 8.05.5.3.3), and comparable to synthetic insecticides, with low mammalian toxicity (LD_{50} greater than 500 mg kg^{-1} in mice) and no mutagenicity or phytotoxicity. It is a fast-acting contact poison, causing tremor and paralysis, but is relatively nontoxic to the tobacco cutworm, housefly, mosquito, cockroach, and spider mite.[318]

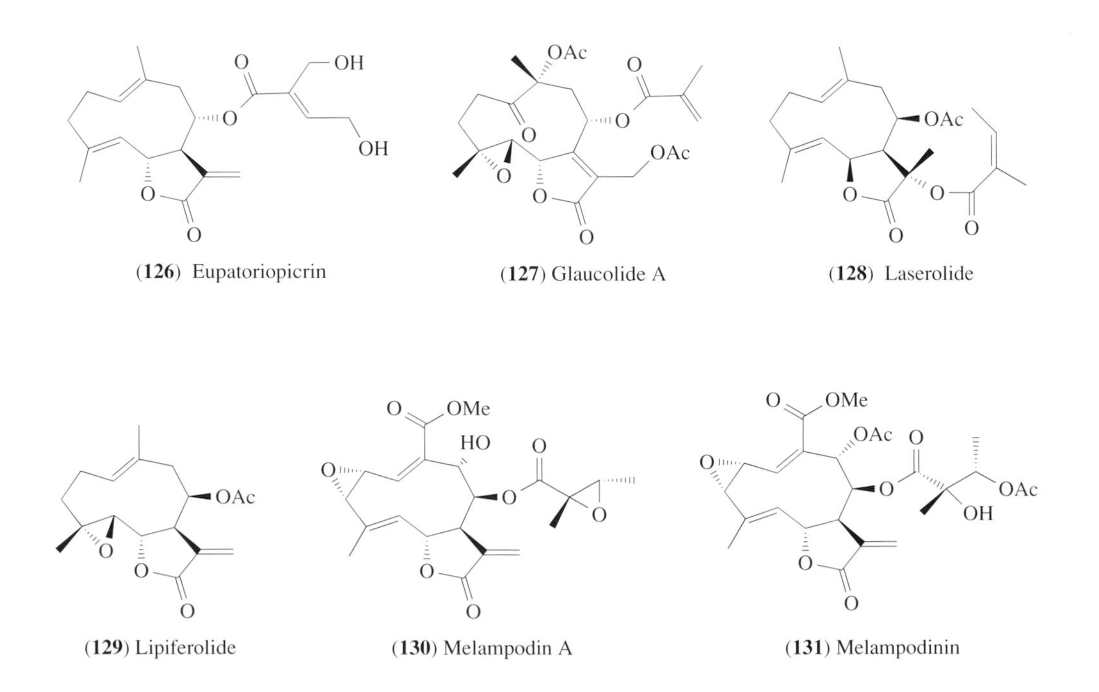

(**126**) Eupatoriopicrin
(**127**) Glaucolide A
(**128**) Laserolide
(**129**) Lipiferolide
(**130**) Melampodin A
(**131**) Melampodinin

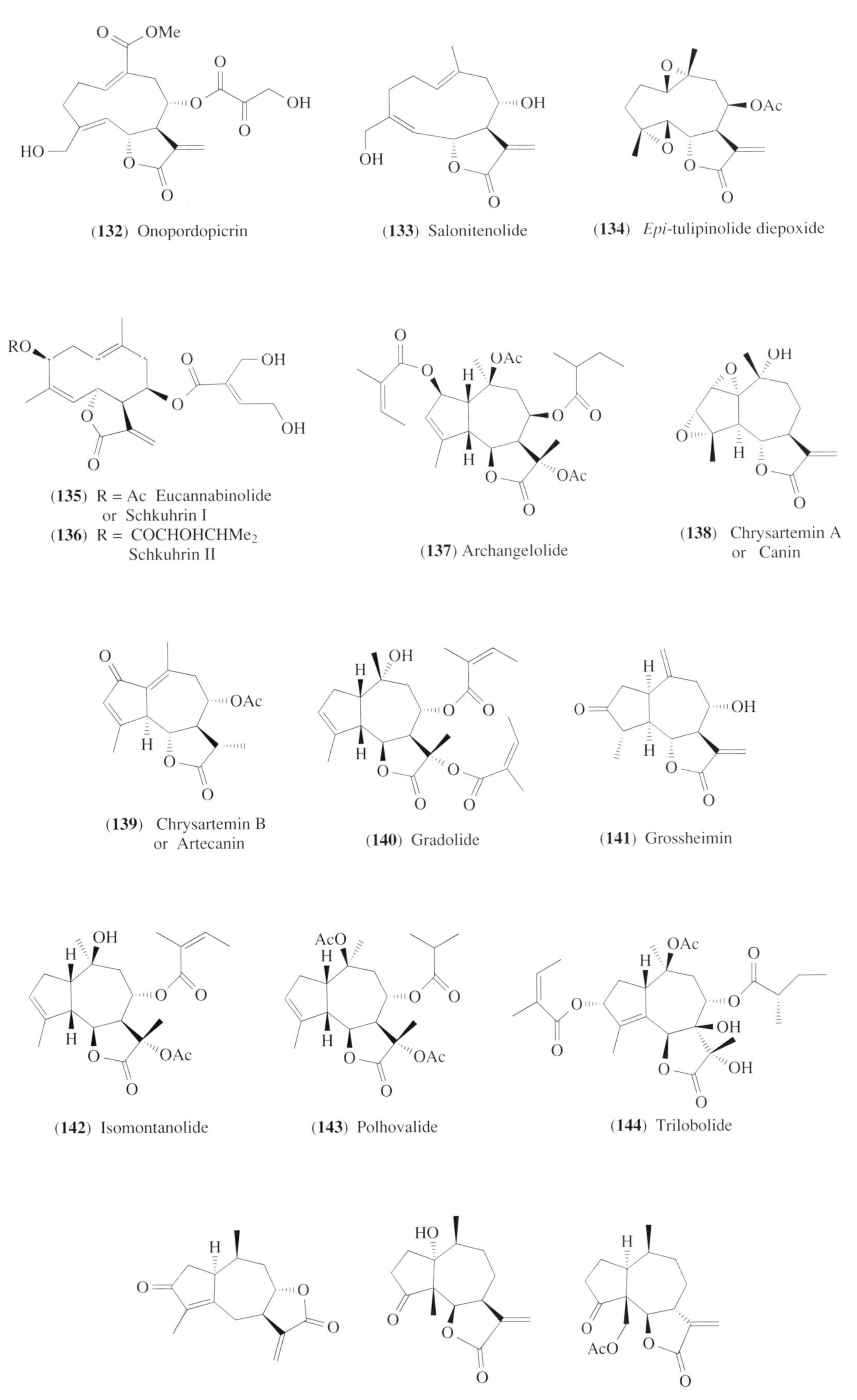

(**132**) Onopordopicrin

(**133**) Salonitenolide

(**134**) *Epi*-tulipinolide diepoxide

(**135**) R = Ac Eucannabinolide or Schkuhrin I

(**136**) R = $COCHOHCHMe_2$ Schkuhrin II

(**137**) Archangelolide

(**138**) Chrysartemin A or Canin

(**139**) Chrysartemin B or Artecanin

(**140**) Gradolide

(**141**) Grossheimin

(**142**) Isomontanolide

(**143**) Polhovalide

(**144**) Trilobolide

(**145**) Xerantholide

(**146**) Coronopilin

(**147**) Ligulatin B

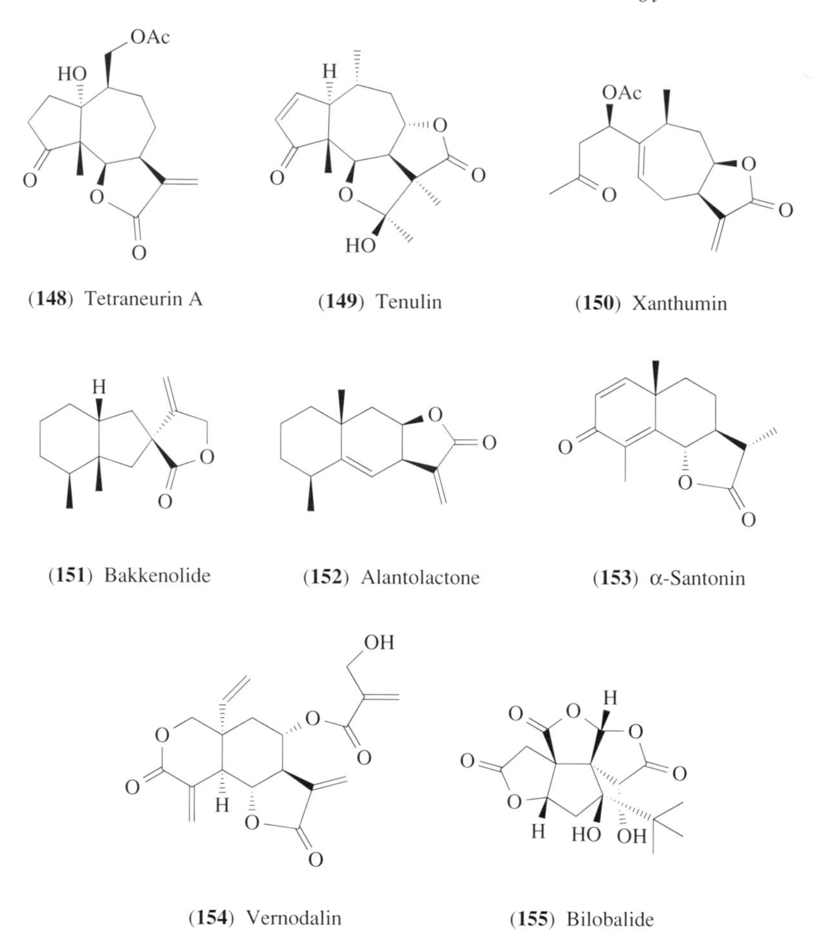

(**148**) Tetraneurin A (**149**) Tenulin (**150**) Xanthumin

(**151**) Bakkenolide (**152**) Alantolactone (**153**) α-Santonin

(**154**) Vernodalin (**155**) Bilobalide

The remaining sesquiterpene examples (**156**)–(**163**) form no clear chemical group. Angulatin A (**159**) and celangulin (**160**) from the root bark of *Angulatus celastris* are antifeedant for a variety of insects, c.f. the structure of wilforine (see Section 8.05.5.6).[319] An extract of *A. celastris* is used as insecticide in China. Gossypol (**162**), from cotton (*Gossypium hirsutum*), though different in appearance from the others, is a sesquiterpene dimer. The extensive research on it and related compounds from cotton has been reviewed.[298] It inhibits the growth of a number of lepidopterans at rather high concentrations.[295,320] It accounts for the resistance of cultivars of cotton to the bollworm *Spodoptera littoralis*. Not much is known about how resistant species detoxify it, but with so many hydroxyl groups it is probably conjugated. Caryophyllene oxide (**163**), present in the same glands as gossypol, is also toxic to *Helicoverpa virescens* and synergizes the growth-inhibiting effect of gossypol.[321]

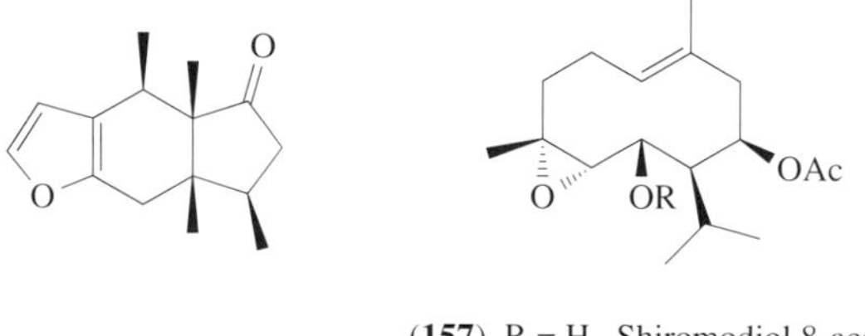

(**156**) Pinguisone

(**157**) R = H Shiromodiol 8-acetate
(**158**) R = Ac Shiromodiol diacetate

8.05.5.3.3 Diterpenes

The variety of plant diterpenes deterrent or toxic to insects is more restricted than the sesquiterpenes. The largest and most important group are the clerodanes (see Section 8.05.5.3.4). Other

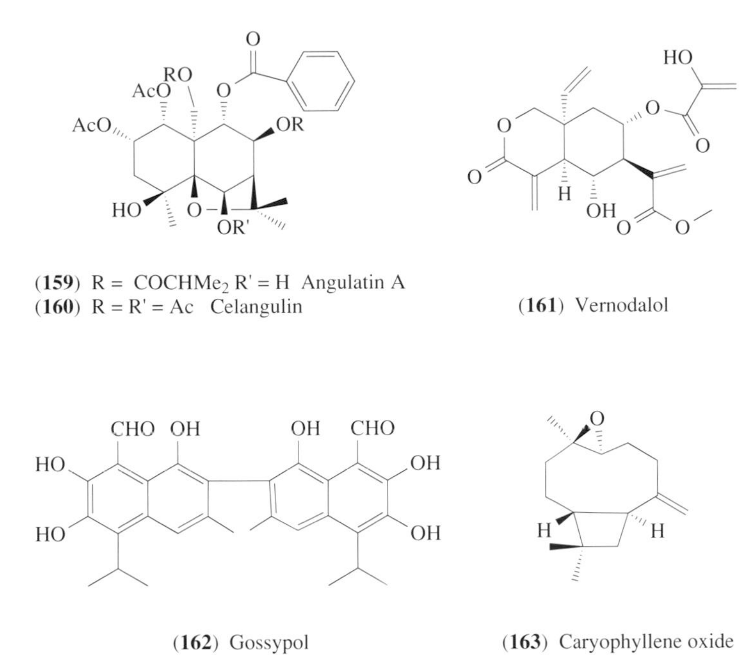

(**159**) R = $COCHMe_2$ R' = H Angulatin A
(**160**) R = R' = Ac Celangulin

(**161**) Vernodalol

(**162**) Gossypol

(**163**) Caryophyllene oxide

types include the intact kaurene diterpenes such as inflexin (**164**) from the leaves of *Isodon inflexus* and isodomedin (**165**) from *I. shikokianus* (Labiatae), the B ring-opened shikodonin (**166**), also from *I. shikokianus*, and the bisnorditerpenes of the nagilactones, strong feeding deterrents from *Podocarpus* species, of which nagilactone D (**167**) from *Podocarpus nagi* and *P. gracilior* is the most active. The closely related podolactone C (**168**), from *Podocarpus neriifolius* containing a sulfoxide group is one of several diterpenes toxic to the larvae of *Musca domestica*. More details are given by Holyoke and Reese.[286] The grayanolide diterpene rhodojaponin III (**169**) from the flowers of *Rhododendron molle* is a very powerful toxin to *Spodoptera frugiperda* (LD_{50} 8.8 ppm) but surprisingly is nontoxic to *Helicoverpa virescens*.[322] The norditerpenes gingkolides A (**170**) to C (**172**) from the root bark and leaves of *Gingko biloba* are toxic to insects by topical application and antifeedant for leaf-cutting ants. Gingkolide C is most toxic, and A the least.[323] They are active against the brown plant hopper *Nilaparvata lugens*, an important pest of rice in Asia.

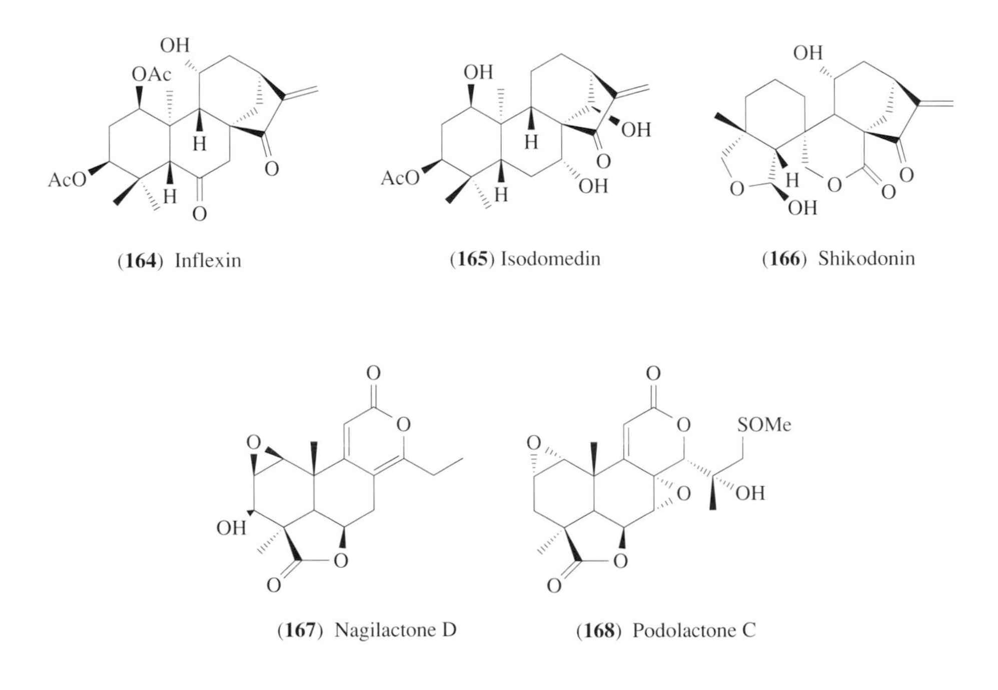

(**164**) Inflexin

(**165**) Isodomedin

(**166**) Shikodonin

(**167**) Nagilactone D

(**168**) Podolactone C

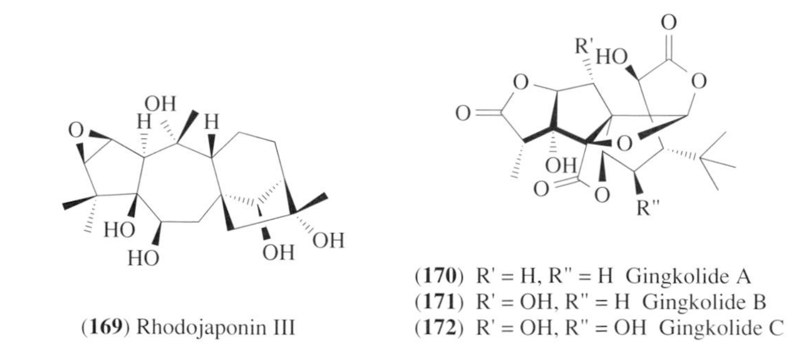

(**169**) Rhodojaponin III

(**170**) R' = H, R'' = H Gingkolide A
(**171**) R' = OH, R'' = H Gingkolide B
(**172**) R' = OH, R'' = OH Gingkolide C

8.05.5.3.4 Clerodanes

Clerodin was first isolated from the leaves of bhat (*Clerodendron infortunatum*, Verbenaceae), an ayurvedic medicinal shrub, by extraction with petrol. Clerodin was obtained in 0.12% yield from dry leaves, and shown to be toxic to worms, tadpoles, and others.[324,325] After several attempts to determine its structure, it was finally solved by a combination of chemical studies and an X-ray crystal structure of a bromolactone, which included correction of the molecular formula to $C_{24}H_{34}O_7$ (**173**).[326,327] Clerodin is a crystalline solid, m.p. 161–162 °C, $[\alpha]_D^{30}$ −37.6° (EtOH). The absolute configuration was reversed in 1979 by re-examining the crystals of the bromolactone after the isolation of other clerodin types with the opposite configuration.[328] This caused some confusion, so those with the same configuration as clerodin are called *neo*-clerodanes (**174**), and those of the opposite configuration are *ent-neo*-clerodins. The commonly accepted numbering is shown on structure (**174**). Clerodin has some resemblances to azadirachtin (see Section 8.05.5.3.6), though it is a smaller, simpler molecule. Both have a very bitter taste, and a substituted decalin portion attached through a quaternary carbon atom to an oxygenated ring, fused through an acetal group to a dihydrofuran ring. Whether these compounds function in the same way is not known.

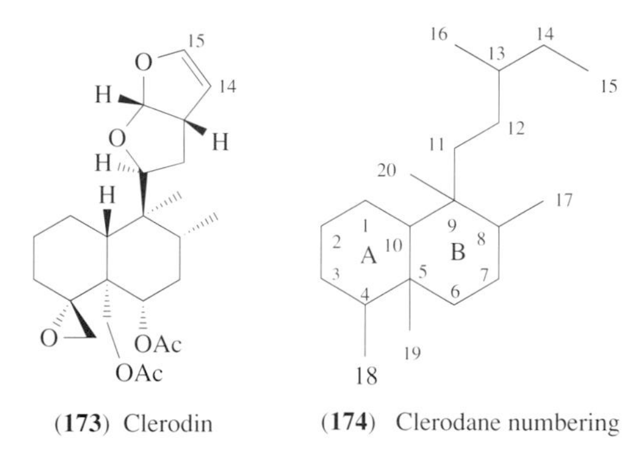

(**173**) Clerodin

(**174**) Clerodane numbering

Except for the demonstration by Banergee at the time of isolation that clerodin was toxic to some lower animals, attention was focused on the structure of clerodin and the isolation of clerodane diterpenes, which are found chiefly in the Labiatae, Compositae, and Euphorbiacae, until in 1972 the group of Munakata isolated clerodendrin A (**175**) (0.04%) from the leaves of the shrub *Clerodendron tricotomum* Thunb., using an antifeedant test with the tobacco cutworm (*Spodoptera litura*) on sweet potato leaves.[329] They then isolated caryoptin (**176**), 14,15-dihydrocaryoptin, and caryoptin hemiacetal (**177**) as well as clerodin, 14,15-dihydroclerodin, and clerodin hemiacetal (**178**) from *Caryoptera divaricata* (Verbenaceae), showing that they all had antifeedant properties against some insects, generally polyphagous ones.[330] They went on to screen 23 species from 13 plant families from Japan and 13 species of Verbenaceae from Taiwan. Among them they found 13 species of Verbenaceae with antifeedant properties against *S. litura*.[331] Following on from their discoveries many other diterpenes with a clerodane skeleton have been isolated, so that over 700 clerodane compounds are now known. A list of all the clerodanes by family (including 19 families of higher plants) up to 1991 includes 793 structures.[332] There is also a review of those of the Labiatae,[333] and literature surveys.[334,335]

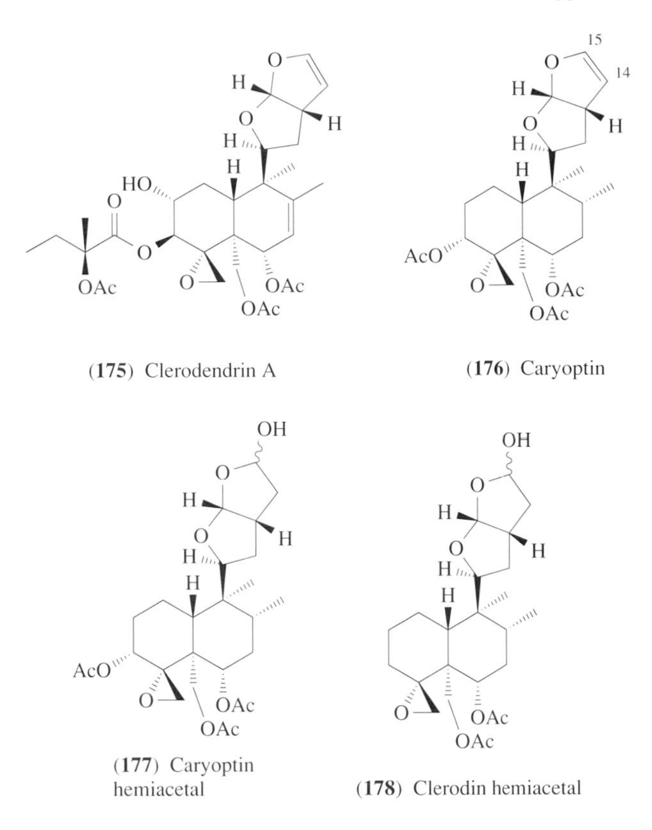

(**175**) Clerodendrin A

(**176**) Caryoptin

(**177**) Caryoptin hemiacetal

(**178**) Clerodin hemiacetal

(i) Biological activity

Biological activity has been reported for only a small number of clerodanes, but it has been pointed out[332] that few of those tested have been found inactive. Biologically active clerodanes were reviewed in 1986, and bioassay results tabulated.[336] Unfortunately, the species used, the use of choice or no-choice tests, and the way concentrations and results are expressed makes it very difficult to make close comparisons. Most of the tests are with *Spodoptera* species. In feeding bioassays with *Spodoptera litura*, 3.6 ppm of clerodin causes 50% inhibition of feeding, and with *Spodoptera littoralis* 50 ppm causes complete inhibition,[337] but the greatest single species activity listed is for the compound ajugarin I (**179**),[338] with 80–100% feeding inhibition for *Schistocerca gregaria* at 0.06 ppm. The same compound was much less active against *S. littoralis* (300 ppm on leaf disks of *Ricinus communis*) or *S. exempta* (100 ppm on leaf disks of *Zea mays*).[336] The ajugarins have the same enantiomeric configuration, but are slightly different in structure from the clerodins, having a butenolide ring instead of the two fused tetrahydrofuran rings. They were obtained from the leaves of *Ajuga remota* (Labiatae), with ajugarin I in 0.05% yield. Ajugarins II (**180**) (0.002%) and III (**181**) (0.001%) were less active, while ajugarins IV (**182**) and V (**183**) were inactive. Other compounds showing exceptional activity were ajugapitin (**184**) from *Ajuga chamaepitys* (Labiatae) and 4,15-dihydroajugapitin from *Ajuga pseudoiva*, both of which were active at 0.3 ppm against *S. littoralis*.[339] Clerodin and dihydroclerodin are both active at 50 ppm against *S. littoralis*. Ley's group have isolated from *Scutellaria woronowii* (Juz.) (Labiatae) two compounds, jodrellins A (**185**) (0.005% of dry weight of plant) and B (**186**) (0.023% by weight, also in *S. galericulata*), which are the most active clerodane antifeedants isolated so far.[340] The extra C-2 to C-19 oxygen link seems to increase the antifeedant activity of the jodrellins relative to clerodin. Using the antifeedant index (AI $= [(C - T)/(C + T)]\%$ where C and T are the weight of the control and test disks eaten, respectively), at 100 ppm Jodrellin B had an AI of 100, Jodrellin A had an AI of 92, and clerodin had an AI of 74. The same ratio of activities held at lower concentrations. Jodrellin B had an AI of 83 at 50 ppm, but this fell to 54 at 1 ppm.[340] Structures were determined largely by NMR spectroscopy, using nuclear Overhauser effects and COSY techniques. Tables of ^{1}H and ^{13}C NMR data on a large number of clerodanes are recorded, together with mass spectral data,[341] nuclear Overhauser effects, and crystallographic data.[342,343]

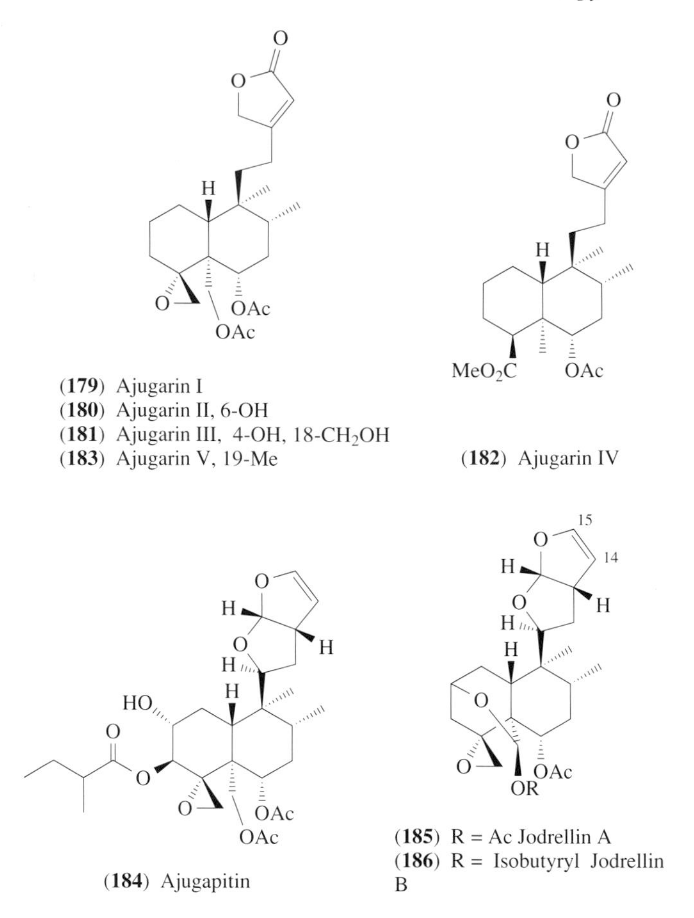

(**179**) Ajugarin I
(**180**) Ajugarin II, 6-OH
(**181**) Ajugarin III, 4-OH, 18-CH_2OH
(**183**) Ajugarin V, 19-Me

(**182**) Ajugarin IV

(**184**) Ajugapitin

(**185**) R = Ac Jodrellin A
(**186**) R = Isobutyryl Jodrellin B

The chlorinated clerodanes tafricanins A (**187**) and B (**188**) were isolated from *Teucrium africanum* (Labiatae), and their structures determined by X-ray methods.[344] Tafricanin B is as active as clerodin against *Locusta*, but tafricanin A is only weakly active. This lower activity of compounds with C=O compared with CH_2OH seems to be fairly common among the diterpenes. There are only four other chlorine-containing diterpenes known from terrestrial sources. The biological activity of clerodanes from a range of species of *Baccharis* (Compositae), *Teucrium* (Labiatae), and *Salvia* (Labiatae) has been examined but without finding any exceptionally active compounds.[345]

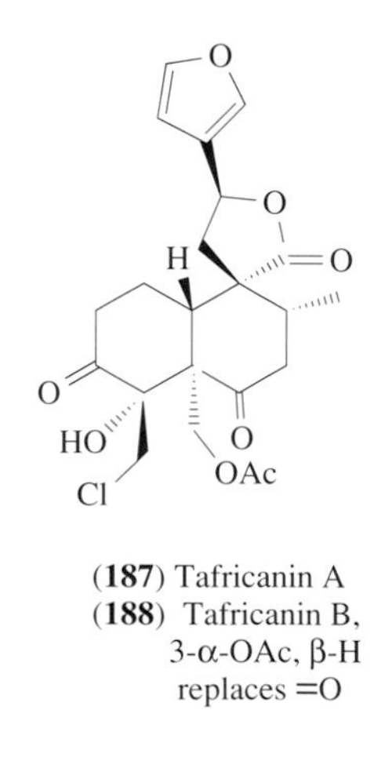

(**187**) Tafricanin A
(**188**) Tafricanin B, 3-α-OAc, β-H replaces =O

In a systematic search of the Labiatae using feeding bioassays on *S. littoralis* and *S. frugiperda*, of 24 species of *Teucrium* (out of 100 known species), 25% showed activity against *S. littoralis*, and all nine species of *Ajuga* were active. The most active extract was from *Ajuga iva*.[300] In the same study, 35 *neo*-clerodanes were tested, and 16 were significantly active with *S. littoralis* and 13 with *S. frugiperda*, but there was no correlation between plant activity and their content of known clerodanes.[300] This study also showed how antifeedant activity changes from month to month, with maxima in April and October for *Teucrium polium* and in May and October for *Ajuga reptans*.[300]

Clerodanes are antifeedants for a large number of species, especially among the Lepidoptera, but they are feeding stimulants for the turnip sawfly (*Athalia rosae*, Hymenoptera). The larvae feed on crucifers, but the adults avoid them and feed on the leaves of *Clerodendron trichotomum* (Verbenaceae), thereby absorbing and sequestering 8 mg per fly of clerodendrin A (**175**), which makes their surface very bitter and apparently acts to protect them from predators such as birds and lizards (Section 8.05.6.8).[344]

The bioassays used for clerodane antifeedancy are very similar to the filter paper disk tests described in Section 8.05.5.3.6(ii), and results are usually expressed as the concentration of a solution which, when disks are immersed in it, causes 80–100% protection of the disk from being eaten. An interestingly different study has concentrated on substances deterrent to leaf-cutting ants. Plants observed not to be taken by raids of the leaf-cutter *Acromyrmex octospinosus* in Panama were selected. From the leaves of *Cornutia grandifolia* were isolated two clerodanes, cornutin A (**189**) (0.083%) and cornutin B (**190**) (0.144%).[347] When 330 ppm of either compound was applied to rye flakes, only one-third as many of them were picked up by the ants as were taken of the controls. Cornutin A was slightly more active.[347] Kolavenol (**191**), originally from *Melompodium divaricatum*, but found in many other sources, is very weakly active against feeding (5500 ppm gave a reduction to one-third pick-up in the test above) by another leaf-cutting ant, *Atta cepholotes*, but is not toxic to them or their mutualistic fungus.[348,349] The four compounds (**192**)–(**195**) from *Detarium microcarpum* are claimed to show strong antifeedant activity at 1% against the termite *Reticulitermes speratus* in a paper disk assay.[350] Activity may not be high, but antifeedancy against termites is interesting.

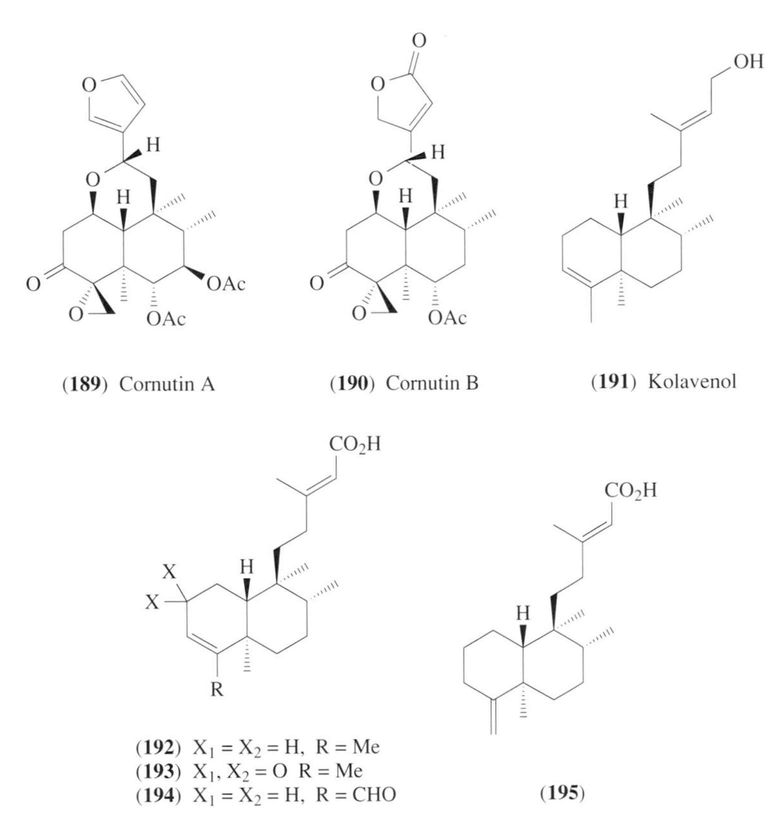

(**189**) Cornutin A (**190**) Cornutin B (**191**) Kolavenol

(**192**) $X_1 = X_2 = H$, R = Me
(**193**) $X_1, X_2 = O$ R = Me
(**194**) $X_1 = X_2 = H$, R = CHO

(**195**)

(ii) Synthesis and biosynthesis

The interest generated in clerodins in the 1970s stimulated the search for synthetic routes to produce more of these compounds. The first complete synthesis of a natural compound was of annonene (**196**),[351] and the first of a biologically active compound was of (±)-ajugarin I (**179**).[352] The final stage of forming the 4,18-epoxide was the most difficult step. The synthetic methods up to 1986 have been fully reviewed.[309] Many of the natural clerodanes have been converted to derivatives, and have been reconstructed through partial synthesis, but clerodin itself has not yet been synthesized.

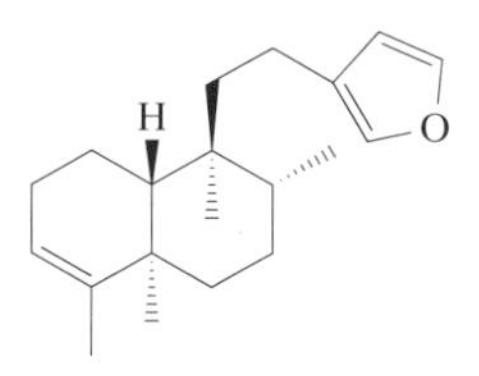

(**196**) Annonene

The biosynthesis of clerodane diterpenes follows the general route proposed by Ruzicka. It has not been studied directly in higher plants, but work with a liverwort, using ^{13}C-labeled acetate and mevalonate, producing clerodane types, has confirmed what might be predicted for bicyclic diterpene formation generally (Scheme 10).[353] The route shown gives A,B *trans* products, but if the pathway "pauses" at (**197**), then by migration of either of the C-4 methyl groups, both *cis* and *trans* products can arise, and both are found in nature. The rearrangement to give the final clerodane structure is different from that originally proposed by Barton.[354]

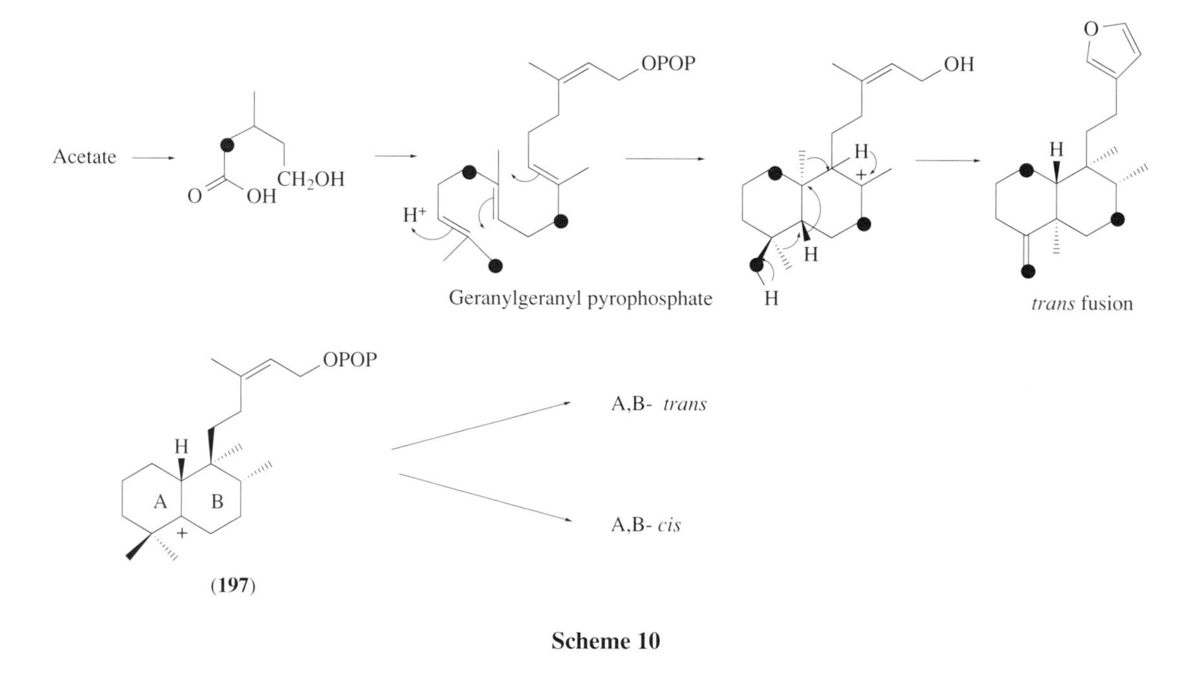

Scheme 10

(iii) Structure and activity

It is evident from physiological studies that clerodanes are not toxic but induce an avoidance response through the sensilla around the mouth part of the insect. The group of Saito have conducted a number of experiments with injected clerodin in the tobacco cutworm (*Spodoptera litura*) without finding any noticeable toxic effects. However, when they applied clerodin solution to various mouthparts, feeding was reduced and the most sensitive part was the maxillary palps.[355] No effect was found when it was applied to the hypopharynx or antennae.

Simmonds and Blaney[300] have done much in exploring the antifeedant and neurophysiological properties of clerodanes, but in spite of the vast number of natural, synthetic, and semisynthetic clerodanes available now, it is still not possible to make any strong conclusions about structure–activity relationships. In a study of the activity against *S. littoralis* and *S. frugiperda* of 35 compounds of which 16 were active, the most active compound was 12-*epi*-teucvin (**198**), but that was not much greater than teucvin itself, showing that the disposition of the furan ring had little effect on activity.[300]

It is interesting to compare clerodendrin A (**175**), which is active as an antifeedant, with the derivative (**199**), which is inactive,[330] and teumarin (**200**), teumassilin (**201**), and 16,19-diacetylteumassilin (**202**), which all have similar low activities (AI of 24–32 at 100 ppm), with ajugarin (**179**).[300] One of the compounds they tested, montanin B (**203**) (from *Teucrium montanum*) showed respectable activity (AI of 42 at 100 ppm), but its derivative with an acetyl and a glucose moiety attached, teuflavoside (**204**) (from *Teucrium flavum*), was a feeding stimulant, particularly to *S. frugiperda* in the same test, and the extract of *T. flavum* was inactive to *S. frugiperda*, even though it contained other active compounds.

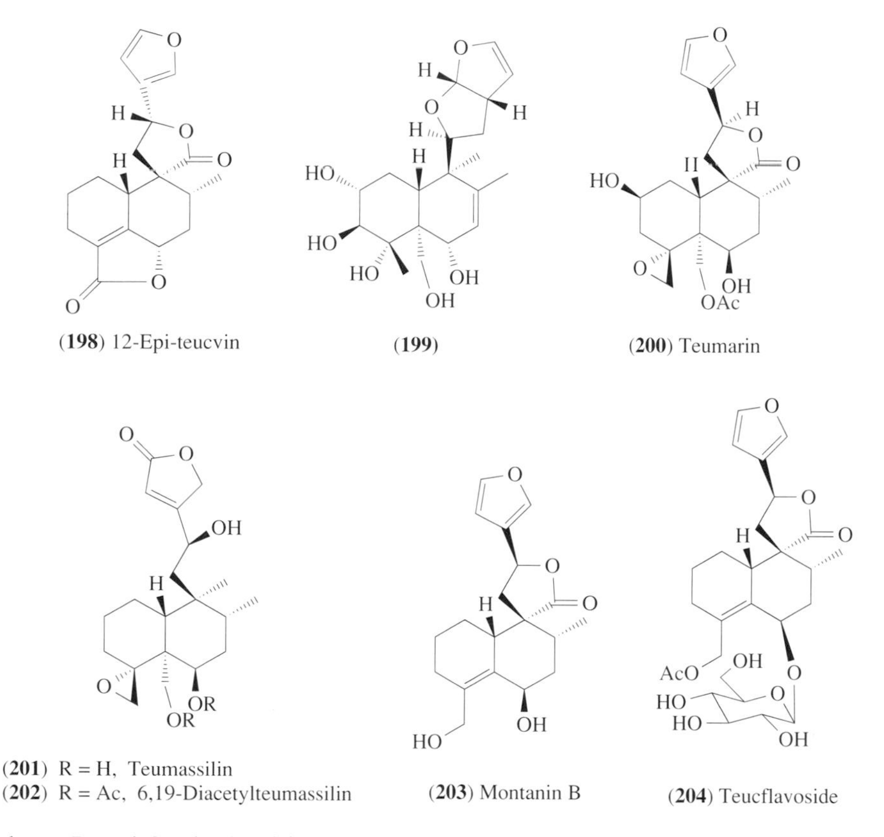

(**198**) 12-Epi-teucvin (**199**) (**200**) Teumarin

(**201**) R = H, Teumassilin
(**202**) R = Ac, 6,19-Diacetylteumassilin (**203**) Montanin B (**204**) Teucflavoside

Clerodendrons B and C, obtained from *Clerodendron inerme*, and which differ from clerodendron A (**175**) only in the group esterified at C-3, are claimed to have growth-inhibitory as well as antifeedant activity.[356] This seems to deserve confirmation.

The large number of clerodanes available from many species, and the relative ease by which they can be isolated, might make them good candidates for commercial exploitation, but perhaps the limited range of species for which they are active has discouraged their possible commercial use.

8.05.5.3.5 Ryanodine

Through an industrial–academic collaboration of Merck and Rutgers University in the 1940s a systematic search among plants led to the discovery of another group of insect toxicants, represented by ryanodine.[357] *Ryania* are shrubs and trees of the family Flacourtiaceae found in northern South America. *Ryania speciosa*, *R. tomentosa*, *R. acuminata*, *R. sagotiana*, and *R. subuliflora* all have insecticidal properties in their roots, stem, and leaves. The powdered plant material, containing 0.2% of ryanodine, was used for a time as an insecticide with very low mammalian toxicity (LD_{50} orally in rats 750 mg kg^{-1}). Solvent extraction gave a concentrate 700 times more effective than the powdered wood.

Ryanodine (**205**) was obtained by water extraction of root or stem, followed by crystallization from ether in 0.1% yield. It is a stable crystalline compound, m.p. 219–220 °C, $[\alpha]_D^{25}$ +26° (MeOH). It is soluble in polar organic solvents.[358] Ryanodine is not an alkaloid but a pyrrolecarboxylic ester of ryanodol ($C_{20}H_{32}O_8$), a diterpene that has been folded in a complex way. Its X-ray crystallographic analysis[359] did not immediately end the discussion of its structure.[360] The ^{1}H and ^{13}C NMR spectra are recorded,[361] and it has been synthesized.[362] 9,21-Didehydroryanodine is an active minor constituent of ryania. Other esters of ryanodol have been found in *R. speciosa*[363] and other plants subsequently.[286]

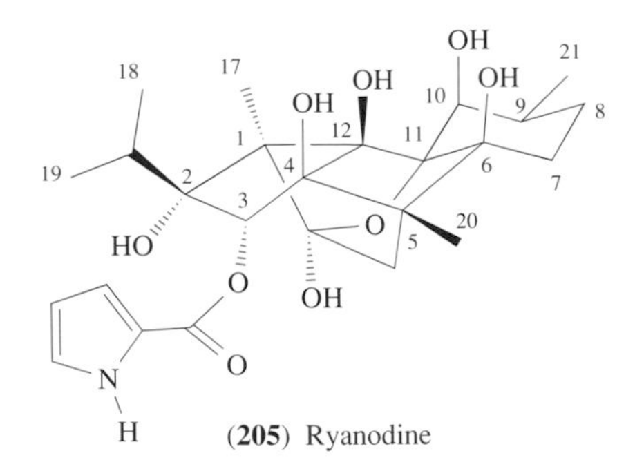

(**205**) Ryanodine

As a toxin, it is slow-acting, being absorbed through the integument and through the gut. Ryania is more stable than pyrethrin and nicotine, but not as potent as these. Data on the activity against insects of crude ryania[364] and of ryanodine[365] are available. An effort has been made to learn something of the structure–activity relationships from testing a large number of derivatives.[366] Ryanodol and dehydroryanodol are also active. Ryanodine acts on the Ca^{2+} release channels of the sarcoplasmic reticulum of muscles and the brain.[366,367] The channels are held open so that calcium ions flow into the cells. It is the most potent inhibitor of control of these channels known, and has become a popular neurophysiological experimental tool.[368] Mammalian toxicity is moderately low (LD_{50} orally in rats 750 mg kg^{-1}),

The biosynthesis of ryanodol has not been studied, but one can see how suitable folding of geranylgeranyl pyrophosphate (**206**), cyclization, and oxidation can give rise to it. Details of its metabolism and environmental fate are unknown.

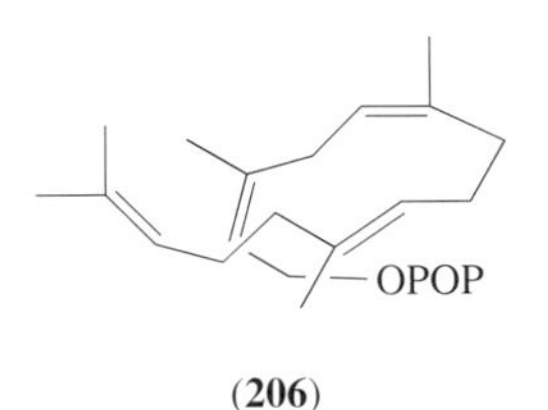

(**206**)

8.05.5.3.6 Triterpenenes

At least four groups of triterpenoids are of interest here, the C_{30} cucurbitacins, the C_{26} skeleton limonoids, the quassinoids with C_{20} (there are a few with 18-, 19-, or 25-carbon skeletons), and the saponins, which have either triterpenoid or steroid skeletons.

(i) Cucurbitacins

The cucurbitacins are a still-expanding group of triterpenoids, first isolated[369] in 1954, chiefly found in the Cucurbitaceae, but also occurring in at least five other plant families. They are generally known as the bitter principles of the Cucurbitaceae. They have an intact triterpenoid skeleton, as illustrated by cucurbitacin A (**207**), and differ in their pattern of oxygenation. They are toxic to insects and mammals, and have antitumor and antigibberellin properties as well. They are potent feeding and oviposition deterrents to insects not adapted to living on the cucurbit plants.[370] Some chrysomelid beetles are immune to their toxicity, and feed on the plants, either storing the cucurbitacins in hemolymph, eggs, and body tissues against predation, or metabolizing and excreting them (see Section 8.05.6.8). Once adapted to cucurbit plants, the insects begin to use the cucurbitacins to locate their hosts. Some luperine beetles can detect 1–3 ng of pure cucurbitacin B (**208**). It has been shown that the cucurbitacins act as antagonists of the insect molting hormones (ecdysteroids) at their receptors, and that the α,β-unsaturated ketone of the side chain is important for this activity.[371]

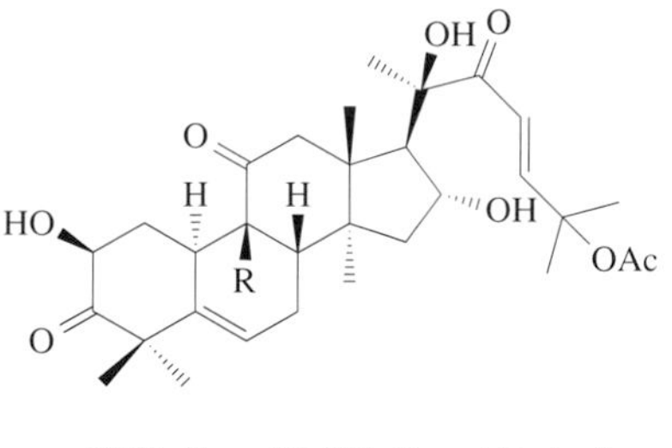

(**207**) R = CH_2OH Cucurbitacin A
(**208**) R = Me Cucurbitacin B

(ii) Limonoids

Of particular interest are the limonoids, in which four carbon atoms of the parent triterpene have been lost from the side chain and the remainder of the side chain has been converted to a furan ring. They are found chiefly in the Meliaceae and Rutaceae. The parent of the group is limonin (**209**), a bitter-tasting, crystalline tetranortriterpenoid from citrus fruits, first isolated in 1841, it has m.p. 298 °C, $[a]_D$ $-128°$ (c = 1.21 in acetone). Its chemistry was reviewed by Maier *et al.*[372] It is strongly antifeedant for the Colorado potato beetle (*Leptinotarsa decemlineata*).[373] Nomilin (**210**), from the same source,[374] is antifeedant for *Spodoptera frugiperda* and *Trichoplusia ni*.[375] Limonin is active against *L. decemlineata*, but 10 times less active than nomilin for *Helicoverpa zea*. Limonin and nomilin have their A ring opened by oxidation, and another antifeedant of this type is harrisonin (**211**), from the roots of *Harrisonia abyssinica* (Simaroubaceae). Toonacilin (**212**) and 6-acetyltoonacilin (**213**) are antifeedants from *Toona ciliata* (= *Cedrela toona*) (Simaroubaceae), with the B ring open, and are toxic to *Hypsipyla grandella* and *Epilachnis varivestis*. *Trichilia roka* (Meliaceae) contains trichilin A (**214**), and other trichilins. It is closely related to sedanin, a diacetate of toosendanin (see Section 8.05.5.3.7(vi)), and a growth inhibitor of *Helicoverpa virescens*, *H. zea*, *S. frugiperda*, *S. eridania*, and *Epilachnis varivestis*. The limonoids of Meliaceae and their insecticidal properties are treated in more detail by Isman *et al.*[376] Gedunin (**215**), with the D ring opened, from *Cedrela odorata* from Costa Rica, and also found in *Azadirachta indica* with close relatives, is active against a number of insect species.[377] The important C ring-opened limonoids of neem (*A. indica*) require a separate section (see Section 8.05.5.3.7).

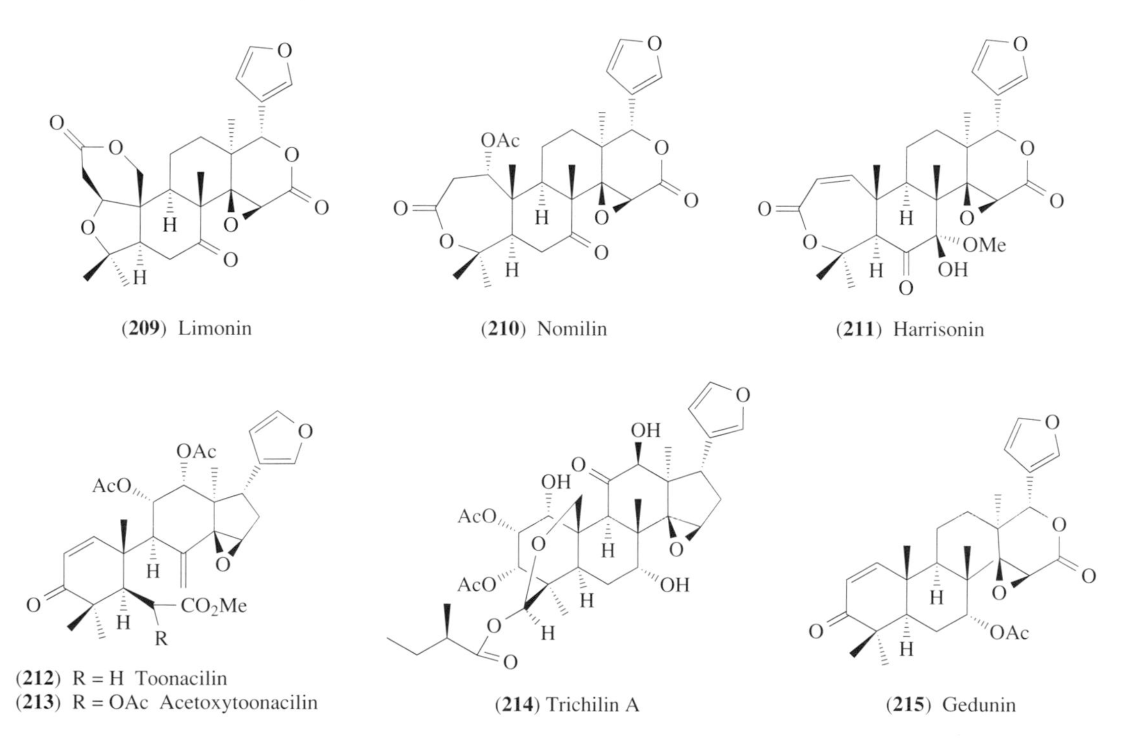

(**209**) Limonin

(**210**) Nomilin

(**211**) Harrisonin

(**212**) R = H Toonacilin
(**213**) R = OAc Acetoxytoonacilin

(**214**) Trichilin A

(**215**) Gedunin

(iii) Quassinoids

Quassia wood (*Quassia amara*, Simaroubaceae) from north-eastern South America was for a century the source of a natural insecticide Surinam quassia, the active compounds being quassin (**216**) and neoquassin, the corresponding 16-alcohol. It was largely replaced later by Jamaica quassia, from *Picrasma excelsa*, which contains quassin and picrasins A (**217**) to G.[378] The picrasins were first identified from *Picrasma ailanthoides*.[379] There is confusion over the names because those isolated from *P. ailanthoides* were called nigakilactones (A to N). Picrasin A is the same as nigakilactone G. Picrasin B (**218**), also found in *Picrasma quassioides*, *Soulamea pacheri*, and *Quassia africana*, is nigakilactone I and also simalikilactone B, and picrasins C to G are close relatives. All the quassinoids possess a single methyl group at C-4. Quassin and picrasins are obtained from the powdered woods by extraction with hot water and, after further purification, are obtained in 0.1–0.2% yields.[380] Quassin and neoquassin have a long chemical history.[381] Their structures were settled only by modern spectroscopic methods.[382]

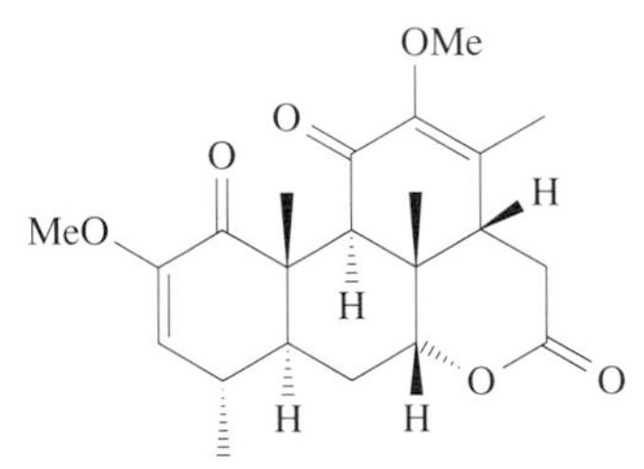

(**216**) Quassin

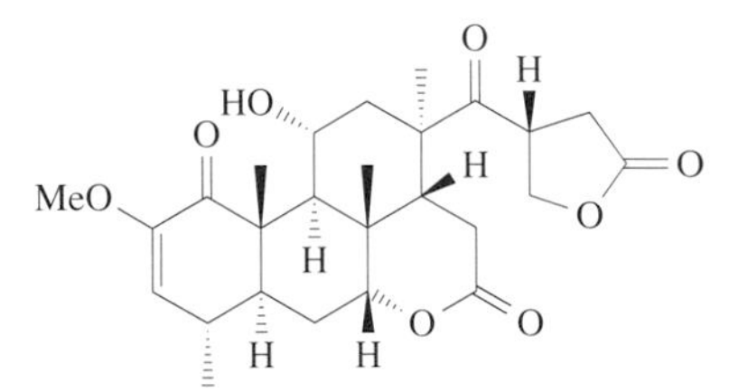

(**217**) Picrasin A

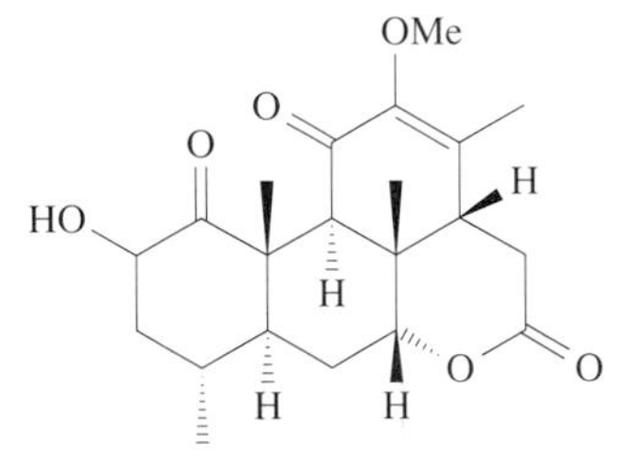

(**218**) Picrasin B

Quassin is a very bitter-tasting crystalline compound, m.p. 221–222 °C, $[a]_D^{20}$ +34.5°. It is soluble in polar organic solvents. Racemic quassin has been synthesized,[383,384] and the mass spectra of the picrasins are recorded.[385] There are no data on the insect toxicity of pure quassin.

The biological activity of quassia extracts on over 100 insect species are recorded, but biological activity is rather limited. The activity is greatest with sawflies and aphids.[379] In both insects and mammals the action seems to be against the nervous system, and the toxicity to the silkworm and the rabbit are similar on a weight basis. One group of quassinoids from *Simaba multiflora* and *Soulamea soulameoides* were tested against *Helicoverpa virescens* and *Spodoptera frugiderda*, and their insect growth inhibitory effects paralleled their antileukemic and cytotoxic activity. The most effective one was 6α-senecioyloxychaparrinone (**219**), which was comparable to azadirachtin as a growth inhibitor but was less toxic.[386] In another study the growth-disrupting and antifeedant properties of 46 natural and semisynthetic quassinoids were examined on the tobacco budworm (*H. virescens*) and the black cutworm (*Agrostis ipsilon*).[387] Bruceine B (**220**) from seeds of *Brucea amarissima* is toxic to *Locusta migratoria*, and isobruceine A (**221**) from *Soulamea tomentosum* is antifeedant to *Spodoptera eridania*. Simalikilactone D (**222**) from *Q. africana* is also a very potent feeding inhibitor for *Epilachna varivestis* and *S. eridania* at 50 ppm, and is also reported to be amebicidal, antimalarial, and antiviral. Samaderin A (**223**) from the bark and seeds of *Samaderia indica* is an example of a quassinoid degraded to a C_{18} skeleton, but it is still active insecticidally. This compound and many others are also cytotoxic.

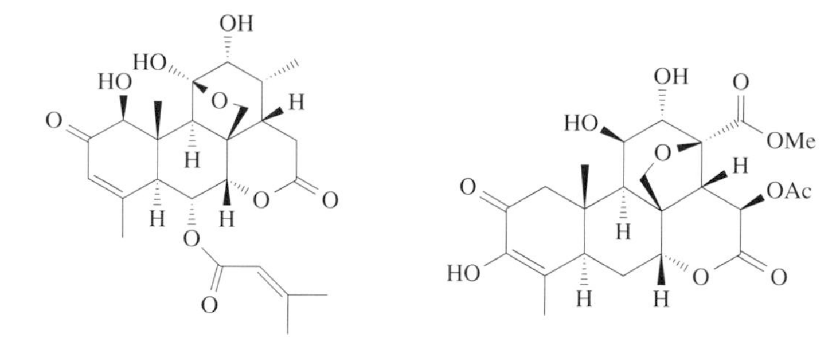

(**219**) 6α-Senecioyloxychapparinone (**220**) Bruceine B

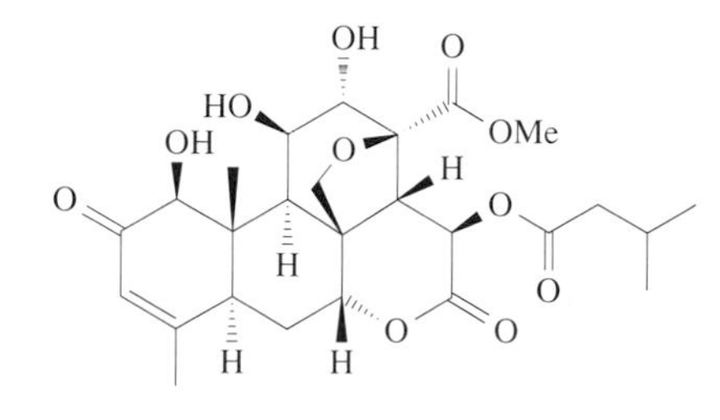

(**221**) Isobruceine A

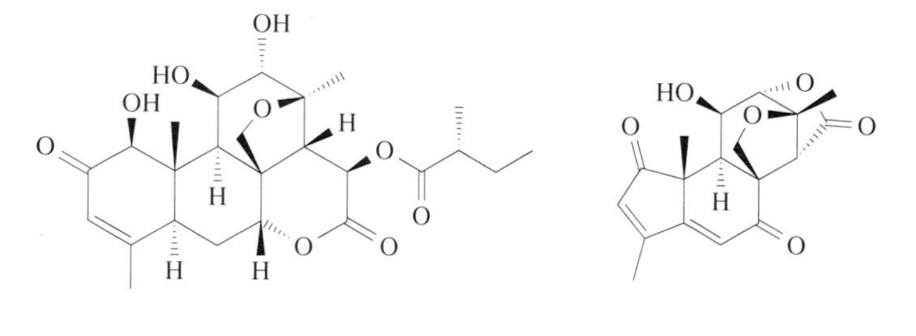

(**222**) Simalikilactone D (**223**) Samaderin A

The chemistry and biosynthesis of the quassinoids has been reviewed.[388,389] Thirty-three compounds were listed in 1970, and there are now over 100 known, all from the Simaroubaceae. They have been shown to be derived biosynthetically from the terpenoid *apo*-euphol.[390]

(iv) Saponins

Steroidal saponins such as digitalin (**224**) from *Digitalis purpurea* (Scrophulariaceae) are toxic to many but not all insects. Triterpenoid saponins such as α-hederin (**225**) from ivy (*Hedera helix*, Araliaceae) leaves are also said to be toxic, though it is difficult to find suitable data recorded. Since saponins form complexes with sterols, and cholesterol or an equivalent phytosterol is a vitamin for

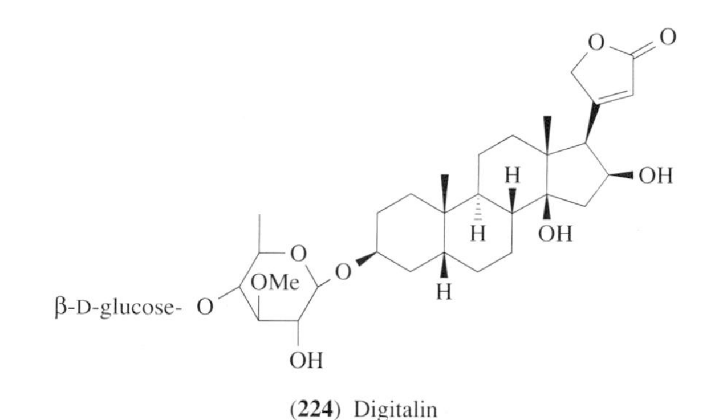

(**224**) Digitalin

insects (see Section 8.05.3.9.1), it is possible that saponins act in the gut to reduce available sterols. Some experiments have demonstrated that the toxic effects of saponins can be overcome by supplementary feeding with cholesterol. The subject is discussed by Gershenzon and Croteau.[291]

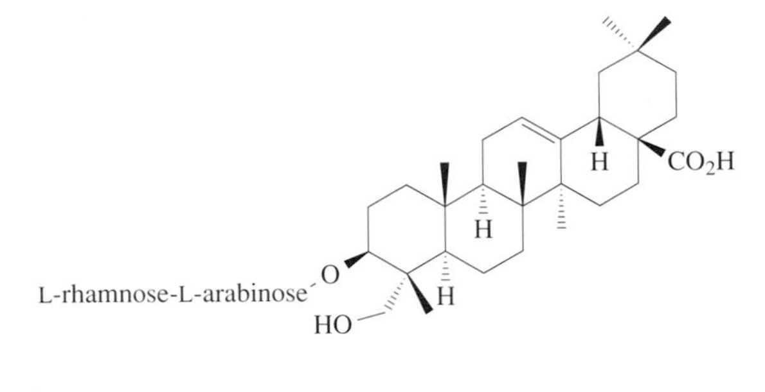

(**225**) α-Hederin

8.05.5.3.7 Neem triterpenoids

One of the most interesting groups of anti-insect phytochemicals and the most promising as new natural pesticides are the neem triterpenoids. The neem tree of South Asia (*Azadirachta indica* A. Jussieu 1830, synonyms *Melia azadirachta* L. and *Antelaea azadirachta* L., Meliaceae) has a long tradition in Indian folklore and ayurvedic medicine for valuable properties, including repelling insects. It was well known from observation that swarming locusts did not eat the leaves of the neem tree, but it is surprising that this was not followed up by a scientific investigation sooner. Many chemical investigations of neem oil and seeds were made, and some triterpenoids isolated and identified, but these were not related to biological properties. In 1950 it was reported that neem leaves showed insect repellency in stored grain, and in 1963 it was reported that an ethanol extract of neem seeds, after the oil was expelled, was repellent to locusts. In 1966 a systematic study of the Neem seeds using a desert locust (*Schistocerca gregaria*) bioassay to find the source of the deterrency[391] resulted in the isolation of the compound azadirachtin (**226**), a highly oxidized triterpenoid derivative, and the most powerful of a group of insect feeding deterrents and growth and development disruptants.[392] These substances have aroused so much interest that the neem tree is the subject of a book which includes detailed coverage of the substances and their biological properties.[393]

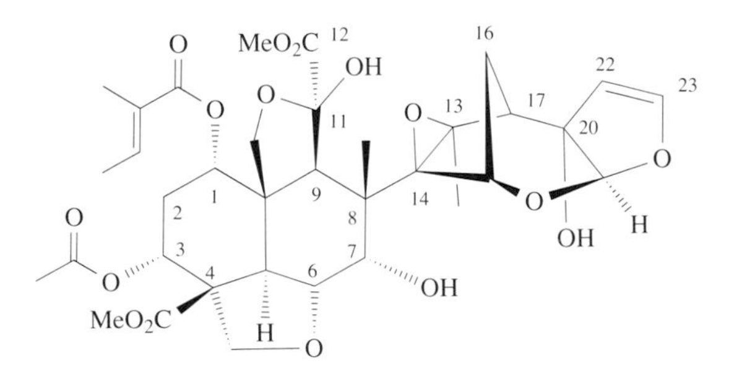

(**226**) Azadirachtin

(i) Isolation

Azadirachtin accounts for 0.2–0.8% by weight of neem seed kernels, and 30–50% of the mass of seeds is triglyceride oil. The original isolation procedure required a lengthy process of grinding the seeds, solvent partitions to remove the oil and water-soluble material, and several column chromatographic treatments, following the process with a bioassay for locust feeding deterrence.[393] Various shorter procedures have been published. A full list of references to isolation methods is given by Kraus.[377] A method that has been used and refined over some time requires the use of columns of Florex RVM, an attapulgite clay which is less polar than silica or alumina, from which the polar triterpenoids can more easily be recovered.[394] A method has been recommended for the isolation of 12 neem triterpenoids[377] that is probably quicker and less costly in solvents than the lengthy HPLC method originally used for these minor compounds. More recently, Biotage flash chromatography for up to 5 g of seed extract has been shown to give higher recoveries of azadirachtin.[395] However, very pure azadirachtin can best be obtained by repeated preparative HPLC; indeed, crystalline azadirachtin was first obtained in this way.[396] A crude leaf extract has antifeedant properties, but the authors were unable to demonstrate that azadirachtin was present; Sundaram has since found azadirachtin in five parts of the tree collected in south India.[397] He recorded azadirachtin values of 0.025% for neem seed kernels, 0.0006% in leaves, 0.0004% in bark, 0.00024% in roots, and 0.00015% in the stem, based on HPLC analysis. An azadirachtin type in which tigloyl is replaced by cinnamoyl has also been identified in leaves.[377] The compound therefore seems to be present in all parts of the tree, but is much more concentrated in the seeds.

Isolation of each of the neem triterpenoids is difficult because of the number of compounds present with very similar chromatographic properties. In the original isolation of azadirachtin it was recognized that there was another compound of slightly lower polarity present in smaller amount in the extracts. It was not pursued then, and attention was concentrated on the structure and properties of the major compound present. Later Kraus's group isolated and identified this compound as 3-tigloylazadirachtol[398] (**227**), m.p. 204–206 °C, $[\alpha]_D^{20}$ −69°. It is the second most abundant of the group, and can be up to 20% of the amount of azadirachtin. Still more recently, Govindachari's group have isolated several other minor products from the seeds, related to azadirachtin, to which they have given names azadirachtin D (**228**), the third most abundant of the group, and azadirachtins H (**229**), I (**230**), and K (**231**).[399] They have also isolated 13-desepoxyazadirachtin (**232**) from seeds,[400] and another group has identified (**233**), an azadirachtin that resembles the azadirachtol (**227**).[401] A further compound, 1-tigloyl-3-acetylazadirachtinin (**234**), first isolated from seeds,[402] has subsequently been found as an acid-catalyzed rearrangement product of azadirachtin. When first formed from azadirachtin, it is found as the (11*S*)-epimer, which slowly changes to an equilibrium mixture. Rembold claimed to have separated azadirachtin into a number of isomers which he called azadirachtins A to D, and later isolated three other compounds azadirachtins E to G.[403] The only isomer of (**226**) is (**231**). Azadirachtin B had already been identified as (**227**), and the structural evidence for azadirachtin D was correctly given later. No spectral evidence was presented for the other structures proposed. Meliantriol (**235**), a simpler triterpenoid, was originally described as a locust antifeedant,[404] but its isolation is not believed to have been repeated, and its antifeedant properties have not been confirmed. Isolation of further new neem triterpenoids now depends on HPLC with UV detection. Any compounds not exhibiting UV absorption are presently lost.

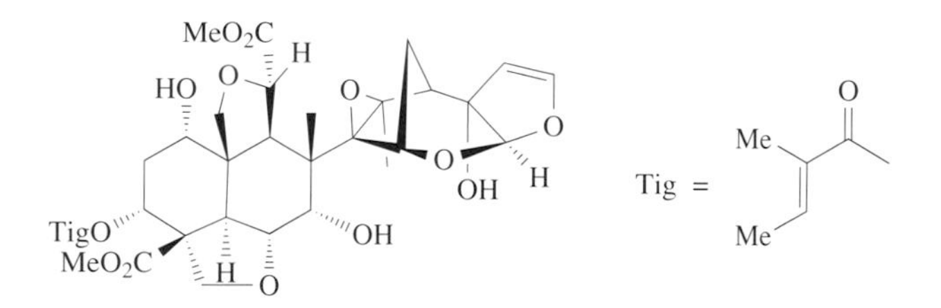

(**227**) 3-Tigloylazadirachtol

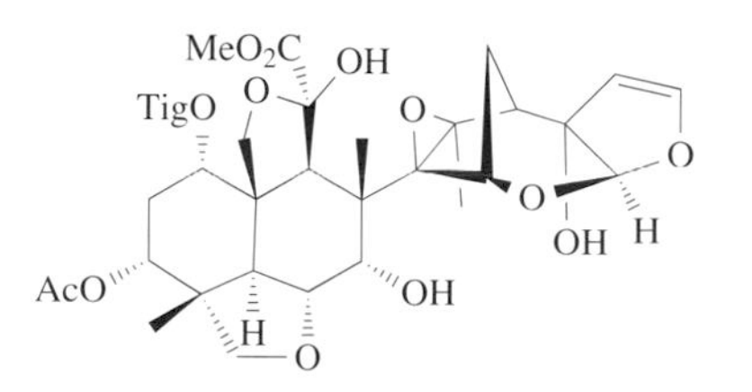

(**228**) 3-Acetyl-1-tigloyl-11-hydroxy-meliacarpin
(azadirachtin D)

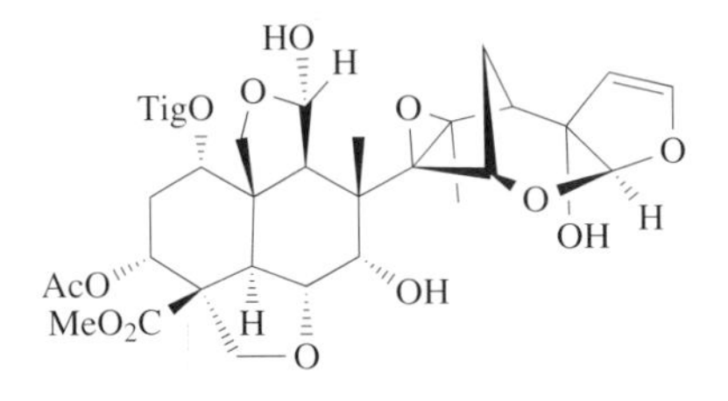

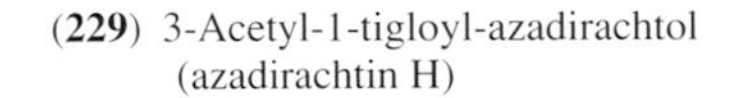

(**229**) 3-Acetyl-1-tigloyl-azadirachtol (azadirachtin H)

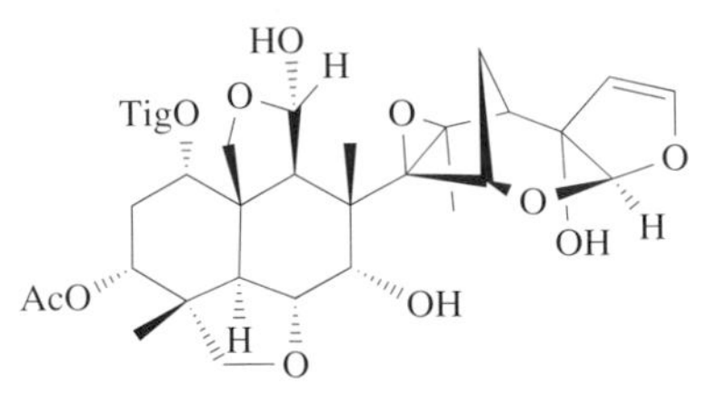

(**230**) 3-Acetyl-1-tigloylmeliacarpin (azadirachtin I)

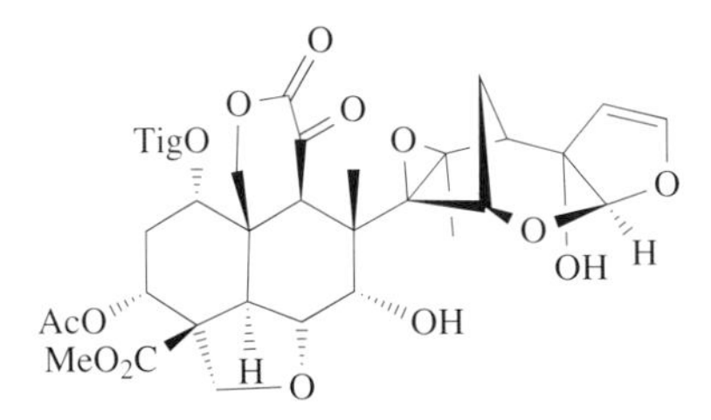

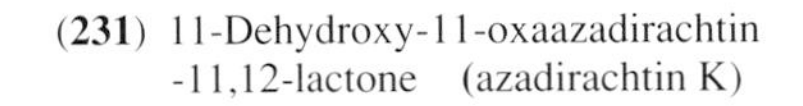

(**231**) 11-Dehydroxy-11-oxaazadirachtin -11,12-lactone (azadirachtin K)

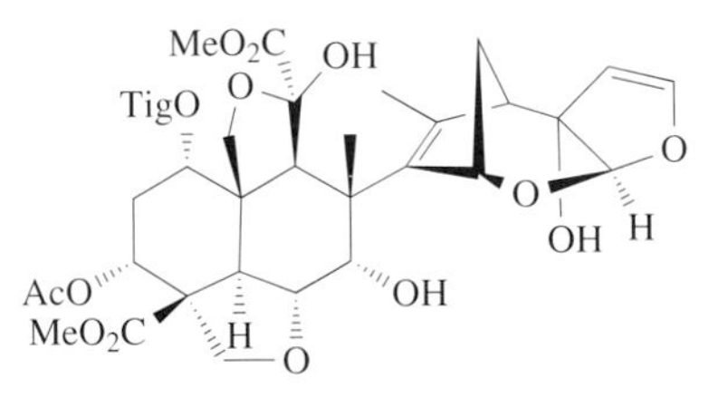

(**232**) 13-Desepoxyazadirachtin

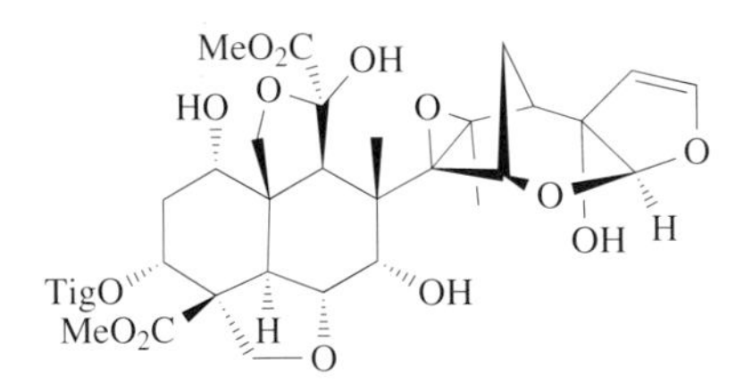

(**233**)

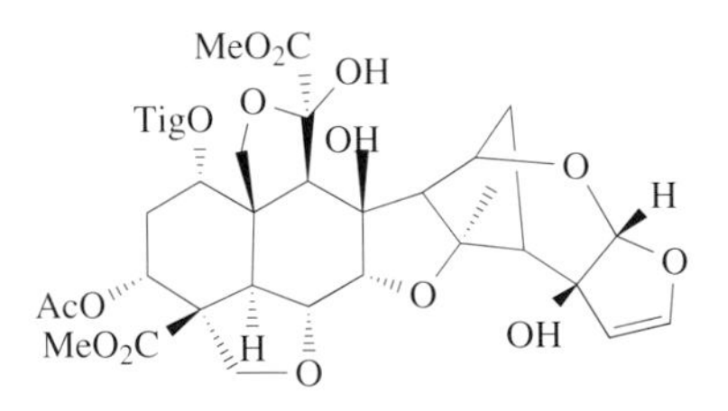

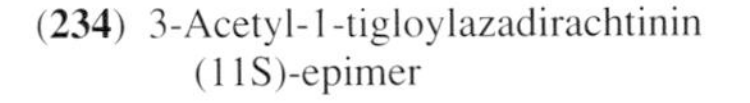

(**234**) 3-Acetyl-1-tigloylazadirachtinin (11S)-epimer

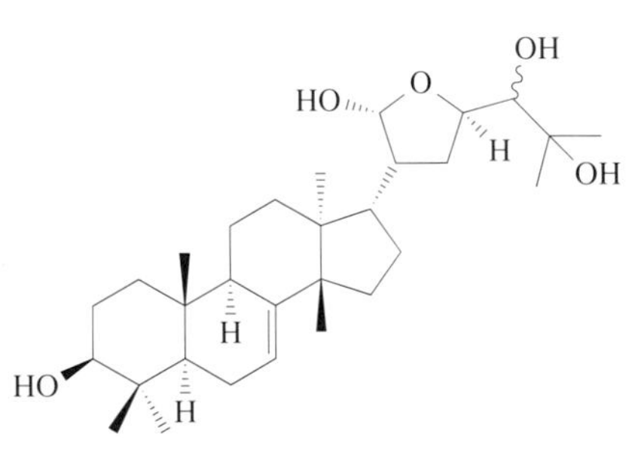

(**235**) Meliantriol

(ii) Bioassays

The locust feeding test was used for the original isolation of azadirachtin, with filter paper disks impregnated with sucrose (controls) or sucrose plus fractions of neem extracts (experimental).[377] The amount of the filter paper eaten was assessed after 8 h. That procedure is still used, with glass fiber disks replacing filter paper, and the method is also used with several lepidopteran larvae. Feeding behavior assays have been reviewed.[405] The first demonstration of an effect upon growth and development of insects was described by Ruscoe.[406] From this has been developed the assay for disruption of development, as used with the Mexican bean beetle (*Epilachna varivestis*).[407] *E. varivestis* is rather insensitive to the feeding inhibitory effect but highly sensitive to the growth-disrupting effect. Other effects on insects, such as on reproduction, mobility, and egg-laying, are covered in Schmutterer's book.[393] Over 150 compounds have been isolated from the seeds, leaves, bark, and twigs of the neem tree, the great majority of them triterpenoids, but relatively few of them have been biologically tested, and only a handful tested on more than one species.

(iii) Chemical structure

Azadirachtin, a white microcrystalline powder, m.p. 156–160 °C, $[\alpha]_D^{24.8}$ −60°, from carbon tetrachloride containing solvent of crystallization or m.p. 174 °C, $[\alpha]_D^{24.8}$ −71.4°, from ethyl acetate–hexane, has 11 functional groups, which presented a considerable problem in structure determination. Studies by the methods available at the time began immediately after its isolation in the 1960s. The molecular formula $C_{35}H_{44}O_{16}$ followed from high-resolution MS of a bistrimethylsilyl derivative, and some structural elements from NMR studies and some simple derivatives.[408] The presence of a number of closely related compounds in seeds, and its sensitivity to strong acid and alkali, made it difficult to isolate in a pure state in large quantities for its structure elucidation. It was only after the development of higher field strength and new techniques in NMR spectroscopy, and with the help of an X-ray structure on a simple crystalline derivative, that its structure was finally solved, 18 years after its first isolation.[402,409] The absolute configuration was determined to be (1*S*, 3*R*) by NMR and X-ray methods on MTPA esters of a simpler degradation product.[410] Azadirachtin is therefore a C-*seco*-tetranortriterpenoid, with a large number of alterations from the parent triterpene structure. It has 16 asymmetric centers, seven of them at quaternary carbon atoms.

The structures of the other triterpenoids closely related to azadirachtin have been solved subsequently, in each case by MS and NMR techniques, with some help from the formation of simple derivatives. The names of the neem triterpenoids are confused. Kraus has proposed three basic structures, the azadirachtol (**236**), azadirachtin (**237**), and meliacarpin (**238**) groups, in an attempt to provide some system,[6] although these do not cover all possibilities, for example structures (**229**), (**230**), and (**231**). The azadirachtols have a tetrahydrofuran ring between carbon atoms 11 and 19, the azadirachtins have a hemiacetal at the same place (free hydroxyl group on C-11), and the meliacarpins are like azadirachtins except that they have an unaltered methyl group attached to C-4 instead of the CO_2CH_3 group. Some other minor products have been isolated. Azadirachtols are known from neem seeds, where tigloyl is replaced by dihydrotigloyl or isobutyryl groups. 3-Desacetylazadirachtin, and others where dihydrotigloyl, isobutyryl, and isovaleryl replace tigloyl, and one with isocaproyl at C-1 and epoxymethacryloyl at C-3, have all been isolated as minor products from seeds.[377] Two epimers, vepaol (**239**) and isovepaol (**240**), are 22,23-dihydro-23-methoxyazadirachtins. Meliacarpins are more common in *Melia azedarach* (see Section 8.05.5.3.7(vi)). An 11-oxo derivative of azadirachtin I (**241**) is also a minor product from neem seed extract. Most of these minor compounds have been isolated by Kraus's group.

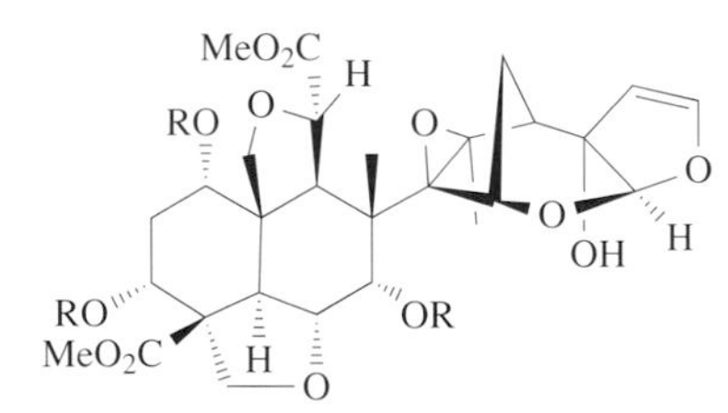

(**236**) an azadirachtol

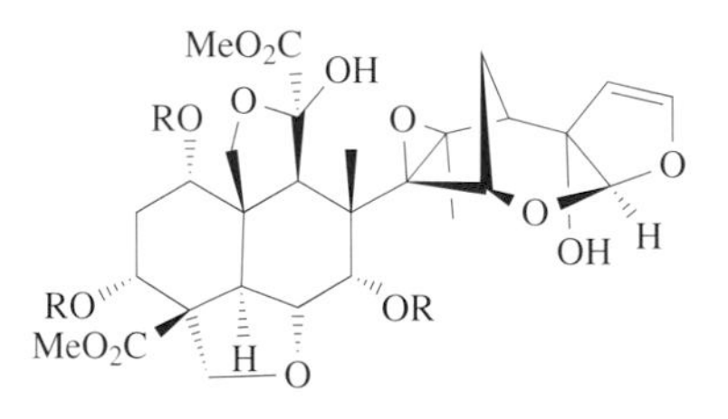

(**237**) an azadirachtin

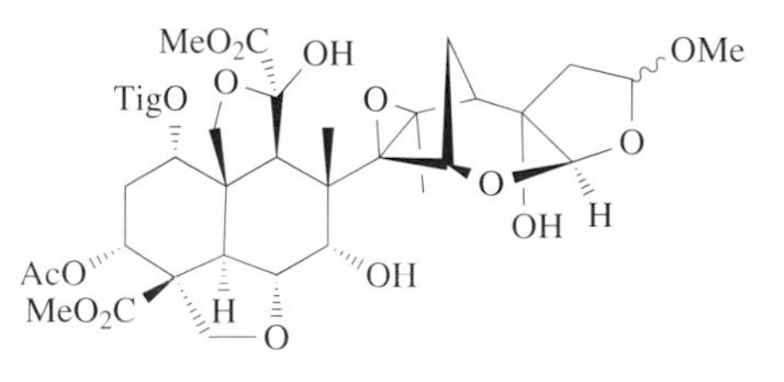

(**239**) β-OMe 22,23-Dihydro-23β-methoxy-azadirachtin (vepaol)

(**240**) α-OMe 22,23-Dihydro-23α-me thoxy-azadirachtin (isovepaol)

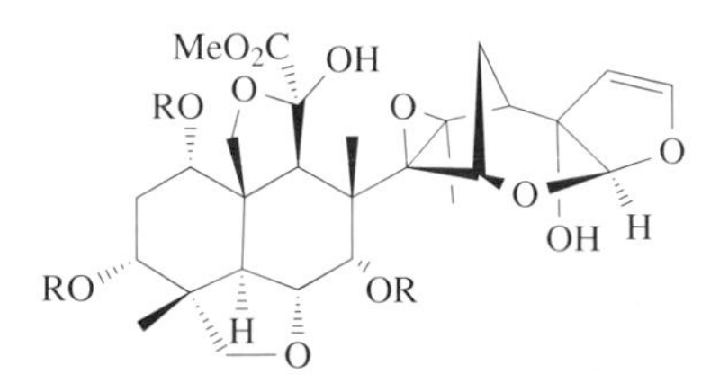

(**238**) a meliacarpin

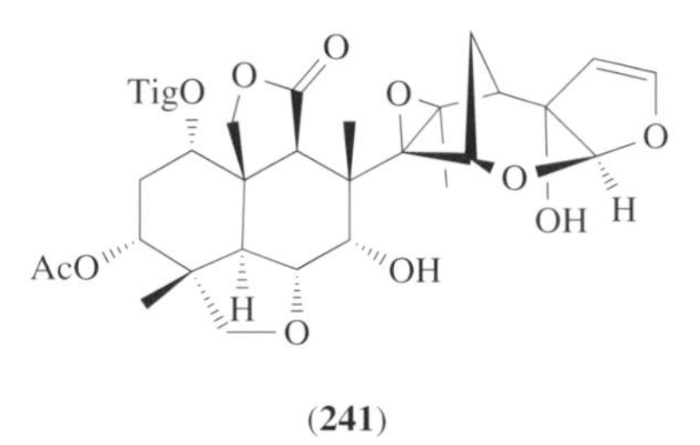

(241)

(iv) Chemical properties

Azadirachtin and its close relatives are colorless solids, crystallizable only when in a very pure state. They are soluble in lower alcohols, acetone, ether, dichloromethane, and chloroform, but are only slightly soluble in water, and difficult to dissolve directly, although once dissolved in methanol the solution can readily be diluted with water. They have an extremely bitter taste. They are rather unstable even under mildly alkaline conditions, are more stable at lower pH values but are decomposed by strongly acid conditions. Mild alkaline conditions first remove the 3-acetate. Acetylation gives an 11-acetate and some 20-acetate. Trimethylsilylation under increasingly severe conditions gives first an 11-TMS ether, then an 11,20-bis-TMS ether, and only under forcing conditions is the 7-hydroxyl group also silylated. The 22,23 double bond of the dihydrofuran ring is very reactive, but mild hydrogenation can saturate it while another part of the molecule is transformed. The chemical reactions of azadirachtin itself have been reviewed, together with a review of work on its synthesis.[411] Simpler models of both the left-hand **(242)**, **(243)**, and **(244)**, and right-hand **(245)**, **(246)**, and **(247)**, parts of azadirachtin (as drawn) have been synthesized by Ley[411] and Mori,[412] but the difficult task of joining these two parts has not yet been achieved. None of the other neem triterpenoids has yet been synthesized.

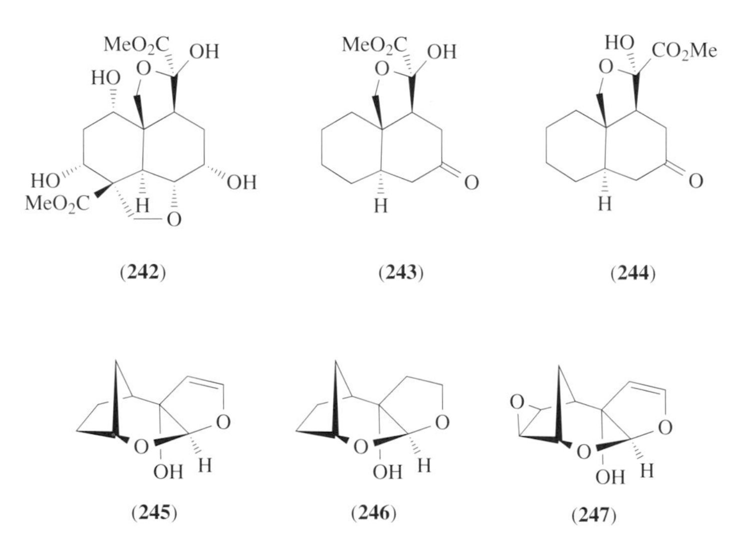

(242) **(243)** **(244)**

(245) **(246)** **(247)**

The mass spectra of this group of compounds all show successive losses of water, acetic acid, tiglic acid, and methanol or methoxyl groups. While the spectra are valuable for fixing the molecular mass, they are of little use for identification. That depends heavily on NMR spectra and chromatographic retention times, provided standards are available. Two tabulations of selected ^{1}H and ^{13}C NMR data are available.[394,413]

(v) Biosynthesis

There have been some experiments on feeding labeled acetic and mevalonic acids to neem leaves and fruit but no attempts yet on the biosynthesis of **(226)**. There are, however, so many triterpenoids known from the seeds, of varying degrees of oxidation and rearrangement from the parent triterpenol, that it is possible to propose a very likely plan (Scheme 11) for its formation, although the exact sequence of reactions cannot be predicted.[377,413] The first group of compounds, with the side chain intact, are called protolimonoids, and loss of four carbon atoms from the side chain and conversion of the remainder to a furan ring gives the limonoids, which are characteristic triterpenoids of the Meliaceae family, which includes the mahogany trees. Further reactions inside the plant can lead to opening of one of the rings, in this case the C-ring giving a series of C-*seco*-limonoids, which are common in *Azadirachta indica*. Several changes are shown in some steps of Scheme 11.

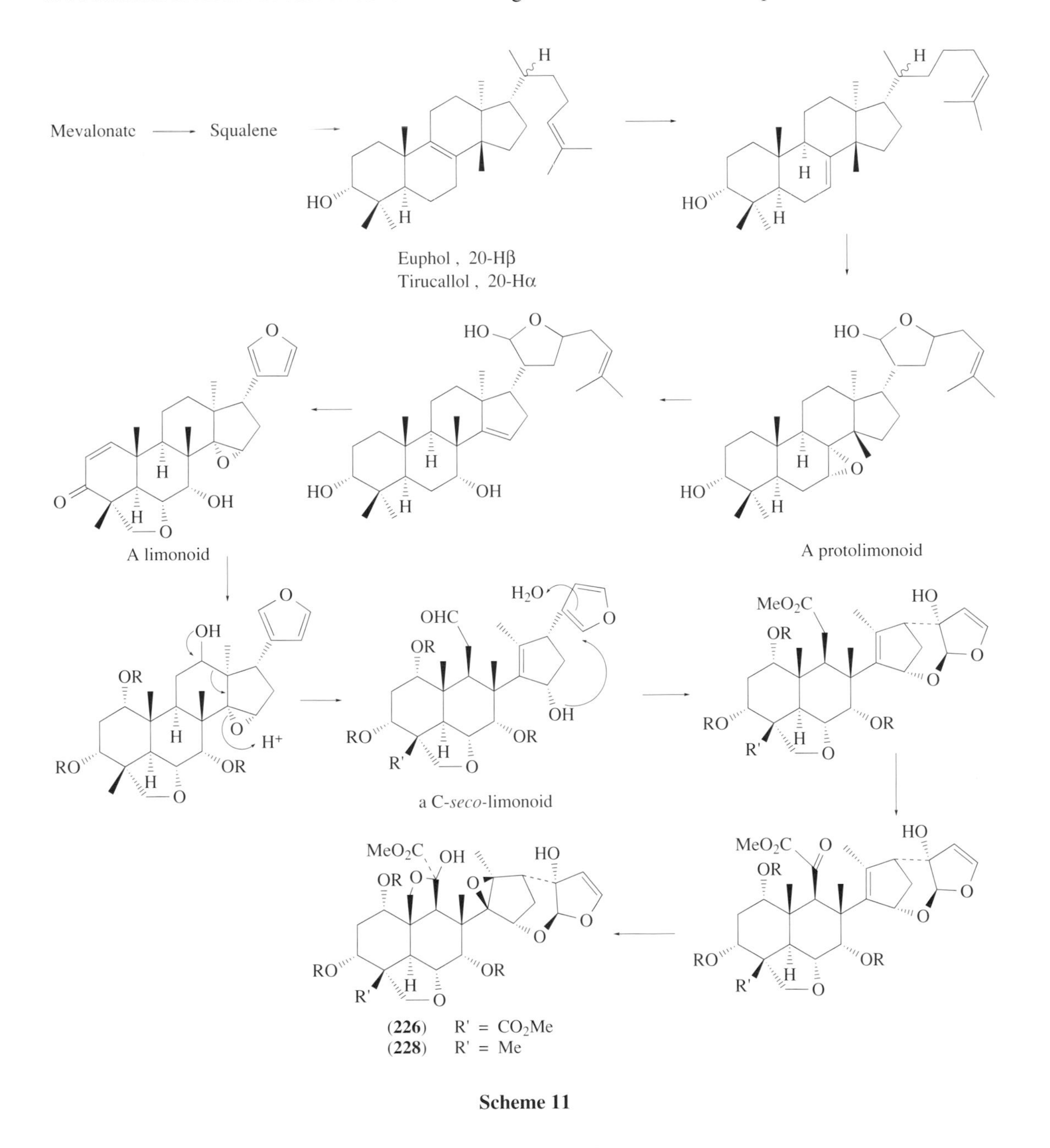

Scheme 11

There has been interest in the tissue culture of neem, to obtain plant strains that produce more azadirachtin, and to produce azadirachtin directly by cell culture. Both have been shown to be achievable in principle. Azadirachtin and azadirachtin I have been produced in cell culture in concentrations of 0.9% per gram of dry tissue, comparable with that obtainable from seeds.[414] Nimbin (**248**), salannin (**249**), their desacetyl derivatives (**250**) and (**251**), and (**234**) have also been isolated and identified from the dispersed cells.[415]

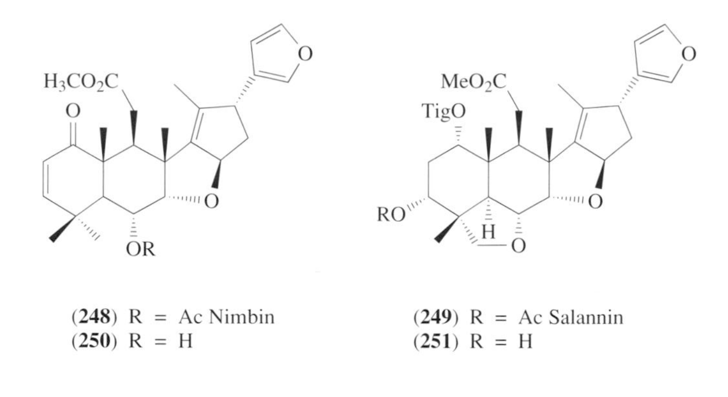

(**248**) R = Ac Nimbin
(**250**) R = H

(**249**) R = Ac Salannin
(**251**) R = H

(vi) Related compounds and related species

The major triterpenoids in neem seed kernels are usually nimbin (**248**) and salannin (**249**) and their desacetyl derivatives (**250**) and (**251**), although in some phenotypes there is much more salannin than nimbin.[394] Nimbin has no significant effect on insects but salannin has moderate antifeedant and growth-disrupting properties.[377] Only two other species of *Azadirachta* are known. *Azadirachta excelsa* (Jack) (= *A. integrifoliola*) is a tree of the wetter areas of South Asia, known as marrango in the Philippine Islands.[416] This plant, once almost extinct, is now being replanted in Asia and is being introduced to other tropical areas with sufficient rainfall. The seeds contain azadirachtin, 1-tigloyl-3-acetylazadirachtol, and marrangin, or 1-acetylazadirachtin H (**252**), which has an activity greater than or equal to azadirachtin in some insect tests. The Thai neem (*Azadirachta siamensis*, Val.) was formerly regarded as a variety of *A. indica*, but is now given species status. Its seeds contain azadirachtin, marrangin, nimbin, and salannin, but much less of the last two than in Indian neem.[417]

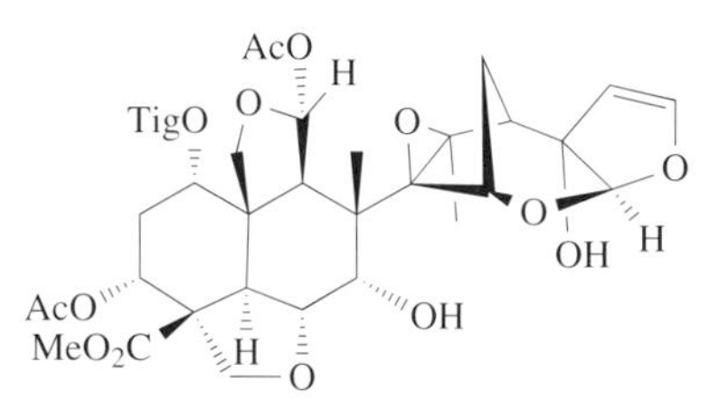

(**252**) Marrangin

In the related *Melia* genus many former species from south Asia and Africa are now considered to be only one species, *Melia azedarach* L., the Persian lilac or Chinaberry tree.[418] *M. azedarach* is often confused with *A. indica*, but there are many differences, most of all the seeds of *M. azedarach* are extremely hard, and the kernels are smaller and contain meliatoxins, poisonous to man, many animals, and fish. The leaves were also known not to be eaten by locust swarms. The seeds contain a number of triterpenoids including meliacarpins (**238**), but not azadirachtin. *Melia toosendan* from south China and Japan may or may not be a separate species. The bark contains a limonoid, toosendanin (**253**), which has insect feeding-deterrent and growth-disrupting properties.[419] It is about 100 times less toxic to Lepidoptera than azadirachtin,[420] but its vertebrate toxicity (LD_{50} orally in mice >10 g kg^{-1}) is very low. Its mode of action in insects is unknown, but thought to be a presynaptic blocking agent in the neuromuscular system. *Melia volkensii* Gürke is a distinct species of the drier parts of east Africa. Its seeds do not contain azadirachtin but a C-*seco*-limonoid, volkensin (**254**), an insect antifeedant.[421] There has been no intensive search for anti-insect compounds among the 500–550 species of Meliaceae, but many limonoids have been isolated from the more accessible species, and no compounds as active as those of neem have been found.

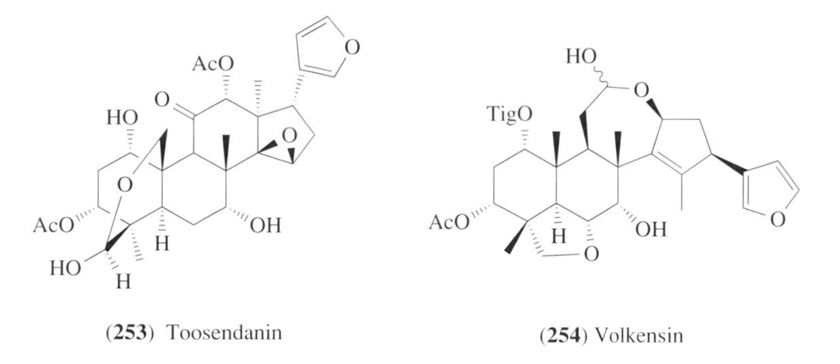

(**253**) Toosendanin

(**254**) Volkensin

(vii) Physiological properties

There is a large literature on the feeding-deterrent, growth-regulating, and reproductive effects of azadirachtin and its analogues for a number of insect species. Azadirachtin is an extremely effective antifeedant against the desert locust (*Schistocerca gregaria*) (effective at 0.04 ppm), a polyphagous insect. It is less active with the migratory locust (*Locusta migratoria*), which feeds mostly on grasses, and is essentially inactive against *S. americana* and eight North American grasshoppers which do not meet the neem tree in their natural habitat. This accords with the hypothesis that only insects in contact with these plants develop a capacity to recognize and avoid them. It was discovered that inhibition in locusts was associated with the sensilla on the mandibular palps, because cauterizing these sensilla caused the locusts to resume feeding on the food treated with azadirachtin. Unlike some antifeedants to which insects become habituated and then begin to feed again, azadirachtin retains its effectiveness. As indicated above, antifeedancy varies greatly between species, but azadirachtin and its analogues are particularly effective deterrents to a number of lepidopteran insect pests. The effects of azadirachtin on feeding have been studied in detail by Schoonhoven[422] and Blaney,[423] and Mordue *et al.* have made an interesing comparison of feeding deterrence and toxicity between the locust and an aphid.[424] Mordue and Blackwell, in a comprehensive review of the physiological effects of azadirachtin, discuss in detail what is known of its chemoreception, endocrine, and tissue effects.[425] They give tables of antifeedant (40 species) and growth-regulating (12 species) effects for known concentrations of azadirachtin. Another compilation of data on effects on insects covers 413 species from 15 insect orders.[426] Unfortunately, some of the data are from experiments with ill-defined materials such as "neem oil" which may contain little or no azadirachtin. The effect of azadirachtin and neem extracts on the activity of insects is also reviewed.[427]

If locusts are forced to feed on material containing azadirachtin, or if it is injected, then they display growth-disrupted effects. The growth-disrupting effects are more important because they affect a much larger number of insect species, and the potency is more consistent between species. The growth-disrupting effects of azadirachtin are noted first in prolongation of larval instars or sometimes early death, a lower rate of feeding, and an inability to molt, to complete molting, to escape from the old cuticular case, or to molt to deformed individuals, especially adults with deformed or crumpled wings. The effective dose for mortality by injection into the hemolymph generally varies between 1 $\mu g\ g^{-1}$ and 4 $\mu g\ g^{-1}$ body weight (see Table 2 in Mordue and Blackwell[425]). In the *Epilachna* bioassay (see Section 8.05.5.3.7(ii)), 0.25 ppm of azadirachtin causes 50% mortality, and the surviving pupae and adults are severely deformed. It causes no antifeeding effect and no acute toxicity at this concentration.

Effects on insect reproduction are often noticed. Ovarian development, fecundity, vitellogenin synthesis, öogenesis, ovarian ecdysteroid synthesis, oviposition, spermiogenesis, and male potency may all be affected by azadirachtin. Through a misunderstanding, it was suggested at an early stage that azadirachtin was an ecdysteroid antagonist.[406] Much attention was subsequently given to examining the effects of azadirachtin on apolysis, ecdysis, and hemolymph titers of ecdysteroids. Certainly many adverse effects are observed, but whether these are due to a direct effect on the prothoracic glands (see Figure 2), upon PTTH (see Section 8.05.4.1), or the brain is unproven. It has been shown clearly *in vitro* that azadirachtin does not affect the prothoracic glands, or the stimulation of ecdysone formation by PTTH. The corpora cardiaca (source of PTTH) of locusts treated with [^{3}H]azadirachtin were heavily labeled but the brain was not. Effects on juvenile hormone (see Section 8.05.2) were equally confusing.

It is suggested that azadirachtin works on some more fundamental site of insect metabolism. Certainly, cytochrome P450 levels are significantly reduced in treated locusts, and the ecdysone 20-monooxygenase half-life is considerably reduced. The evidence is discussed in detail in Mordue and Blackwell's review.[425] They conclude that the toxic effects of azadirachtin cannot be explained by the effects on the endocrine system alone. The compound has direct effects on a range of tissues and organs, which suggests either a number of different activities or a specific toxicity to all cells which becomes obvious in some more than others.[425] It is surprising that when [^{3}H]azadirachtin is fed to locusts 75% of it is excreted unchanged within 24 h, while the remaining 25% remains bound to various tissues (chiefly the malpighian tubules) but is not quickly metabolized.[428]

There is no indication of mammalian toxicity in azadirachtin. Much of the toxicity data on neem refer to the seed oil, which is not relevant. The oil contains large quantities of ricinoleic acid as glycerides, which is likely to cause problems with many mammals. Azadirachtin is nontoxic to rats at doses as high as 8500 mg kg^{-1} orally, and no dermal toxicity was noted.[429] Short-term toxicity data were provided by Larson for the licensing of Margosan-O, a neem insecticide containing 20–25% neem oil and 3000 ppm of azadirachtin. No unfavorable results of these tests were reported.[430] Summaries of toxicity tests on a number of neem products, leaves, oil, extracts, and residual seed cake have been collected by Jacobson.[431,432]

(viii) Structure–activity relationships

Feeding-deterrent tests can be done much more quickly than growth-disruption tests; as a result we know much more about the effect of differences in structure on antifeedancy than on developmental effects, which is unfortunate since ultimately more insects are susceptible to the developmental effects. Structure–activity relationships are also complicated by a wide variation in effect depending on the test insect chosen.

No simple change in the azadirachtin molecule destroys its feeding deterrent properties. Azadirachtin can be hydrogenated to 22,23-dihydroazadirachtin, which is more stable than **(226)**, but is equally active with **(226)** in both feeding deterrency and developmental effect. Using tritium instead of hydrogen therefore gives a valuable radiolabeled compound for metabolism studies. Further hydrogenation of the tigloyl double bond produces a slightly less active compound. Acetylation (at C-11) or deacetylation gives slightly less activity with locusts.[433] Addition of one, two, or three trimethylsilyl groups successively lowers activity. Ley's group have made and tested with *Spodoptera littoralis* a number of derivatives and synthetic products.[411] Some of these have a group added at C-23. Activity can again be summarized by the observation that increasing size of the added group gives lower activity. No derivatives have been found, with the possible exception of 11-methoxyazadirachtin or marrangin **(252)**, that are more active than azadirachtin. Initially it seemed that the model of the right-hand portion of the azadirachtin molecule **(226)** had all the antifeedant effects of azadirachtin with *S. littoralis*,[434] but further experiments with *Spodoptera frugiperda*, *Helicoverpa armigera*, and *H. virescens* have not supported this,[435] and Mori reports **(247)** to be inactive and **(245)** and **(246)** to have only 1/100 the activity of **(226)** in locusts.[412] More recently, Ley has found that a model of the right-hand side of azadirachtin **(243)** is exceptionally active among these model compounds with an AI_{50} (antifeeding index of 50%) of 9.4 ppm for *S. littoralis* (azadirachtin has an AI_{50} of 0.06 ppm under the same conditions).[436] Also interesting is that the epimer **(244)** is many times less active.

Less is known about the structure–activity relationships for growth and development disruption, except to note that substances such as salannin **(249)** (EC_{50} in the *Epilachna* test $>$ 100 ppm) which lack the right-hand portion of azadirachtin also lack the growth-disrupting action, and that the rearrangement product of azadirachtin **(234)** retains its antifeedant properties but not its growth-disrupting effect while the meliacarpins **(238)** retain growth disruption but show little antifeedancy. According to Rembold and Puhlmann, biological activity in the *Epilachna* test depends upon the polarity of ring A: increasing polarity by removing the esterified groups increases activity;[437] attaching groups at C-7 or C-11 reduced activity. Blaney *et al.* have studied structure–activity relationship effects for reduced growth and increased mortality in four Lepidoptera, *S. littoralis*, *S. frugiperda*, *H. armigera*, and *H. virescens*.[434]

(ix) Insecticidal prospects

Azadirachtin has been known for many years, but its commercial exploitation has been slow. It is available in the USA, but the pesticide registration laws in most countries make it almost impossible to register a neem product for use on food crops. Pure azadirachtin would be prohibitively expensive to produce, and the nature of many of the minor components in a neem seed extract are not known, far less their toxicity. The anecdotal evidence from around the world of the lack of harmful effects of neem fruit and seeds is not enough. Until the usefulness of neem products is clearly demonstrated in the USA, and public pressure causes the change of laws in other countries, the market is severely restricted.

The systemic action of azadirachtin has been long known.[438] Translocation is a valuable property in a pesticide. Sundaram has demonstrated the systemic action with young aspen and spruce trees. Azadirachtin applied as a solution to the roots of young trees was translocated through the plants, and especially to areas of new growth.[439] In aspens it was taken into the roots in 3 h and into the foliage by 3 d.

One might expect resistance to evolve to a single product much faster than to a mixture, to the advantage of a neem product, since it must be used as a mixture of compounds. The peach aphid (*Myzus persicae*) raised in contact with pure azadirachtin for 40 generations developed a ninefold resistance to the compound compared with unexposed controls, while aphids that had been exposed to the same concentration of azadirachtin, but in a mixture as neem seed extract, developed no resistance.[440] Similarly, *Spodoptera litura* became desensitized to azadirachtin in its pure form on cabbage by repeated exposure but not as a mixture of products from neem.[441]

Early neem products had not overcome the problem of stability in storage and proper formulation for field use. Suppliers are now alert to the storage problem and can control it.[430] Instability in the field, particularly to UV light, has been reported. The need for correct formulation was not recognized. Sundaram and Curry have shown that formulation with a photostabilizer such as benzophenone or lecithin prolongs the field life of azadirachtin and neem extracts.[442] In one report, azadirachtin as 1% Margosan-O was as effective in the field 3 weeks after application as an organophosphate and was more effective than a *Bacillus thuringiensis* product.[443]

The German government overseas aid organization GTZ has done a lot to inform farmers in the developing world about the value and use of neem seed extract. A simple booklet explains how to harvest, dry, store, and grind the seeds, and to produce a milky suspension which can be splashed onto crops or filtered through cloth and sprayed. It claims good control of beetle, moth, and butterfly larvae, good control of grasshoppers, leaf-miners, and leaf- and plant-hoppers and fair control of adult beetles, aphids, and white flies, but does not recommend it for mealybugs, scale insects, fruit maggots, or spider mites.[444]

If a world market for neem products is established, there should be no difficulty in satisfying demand. It is estimated there are 14 million trees in India alone, producing 400 000 t of seed per year, and unknown numbers in Africa, with new plantations being started in many countries, particularly in Saudi Arabia, West Africa, and Central and South America. The neem tree is a fast-growing evergreen with relatively low needs for water. A 15-year-old tree can yield up to 20 kg of fruit or 2 kg of seed kernels. An Indian company is prepared to process 10 000 t of neem seed per year. Its broad spectrum of activity, low application dose (10–25 g of azadirachtin per hectare), systemic action, and low prospects of a build up of resistance make it a very promising product when laws permit its use.

8.05.5.4 Pyrethrins

Pyrethrum powder, as used since classical times, is the dried flowerheads of *Chrysanthemum cinerariaefolium* Vis. (Compositae) or sometimes *Chrysanthemum coccineum* Willd. or a few other species of that genus.[445] The crude pyrethrins are extracted from the dried flowers with a nonpolar solvent, usually light petroleum, and then partitioned from the light petroleum into nitromethane. The relatively pure mixture of compounds (**255**) was first obtained by Staudinger and Ruzicka as semicarbazones for structural studies reported in the 1920s. They found two compounds, pyrethrins I (**256**) and II (**257**), but in the 1940s two more, cinerins I (**258**) and II (**259**), were isolated, and finally in 1965 with the help of GC two more minor products, jasmolins I (**260**) and II (**261**), were found. The determination of structure proceeded by classical methods, confirmed by the synthesis of cinerin I by Crombie and Harper in 1950.[446] Crombie's group synthesized jasmolin I (**259**) before it was found in nature and also performed the total synthesis of pyrethrin I. The chemistry has been reviewed in detail.[445,447–449] The absolute stereochemistry was not correctly established until 1972 by an X-ray structure of the 6-bromo-2,4-dinitrophenylhydrazone of the entire pyrethrin I.[450] The pyrethrins are all viscous, colorless liquids, distillable under high vacuum, soluble in a range of organic solvents, insoluble in water, rapidly oxidized in air, and unstable to light, moisture, and alkali. Their optical rotation properties are as follows: pyrethrin I, $[\alpha]_D^{20}$ −14°; pyrethrin II, $[\alpha]_D^{19}$ +14.7°; cinerin I, $[\alpha]_D^{20}$ −22°; and cinerin II, $[\alpha]_D^{16}$ +16° (jasmolins I and II, rotations not recorded). The stereochemistry of the cyclopropane portion is (1*R*,3*S*) and the rethrolone is (*S*) with *Z* double bonds. The ^{1}H NMR spectra[451] and the ^{13}C NMR spectra[452] have all been studied and tabulated. All the possible isomers and enantiomers have been synthesized, and it is known that the natural isomers are the most active.

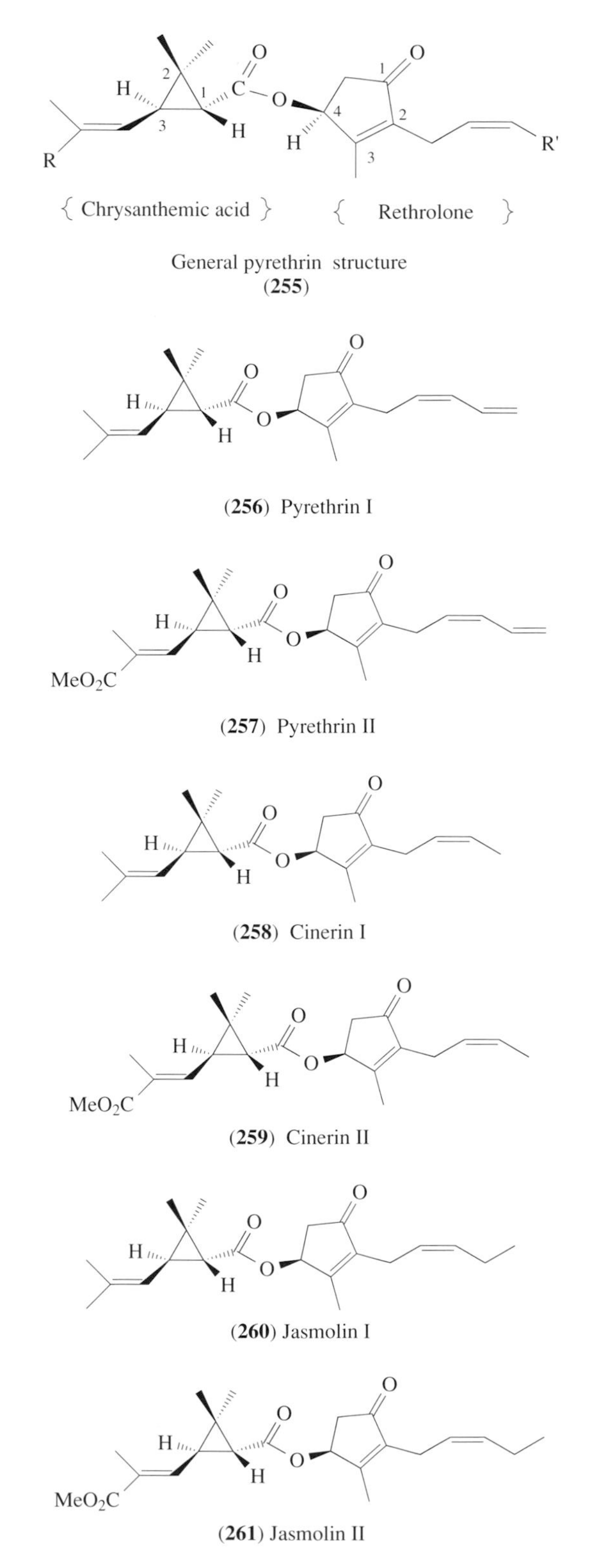

General pyrethrin structure
(**255**)

(**256**) Pyrethrin I

(**257**) Pyrethrin II

(**258**) Cinerin I

(**259**) Cinerin II

(**260**) Jasmolin I

(**261**) Jasmolin II

8.05.5.4.1 Biological activity

The activity of various isomers for the housefly is tabulated.[448] Different research groups have produced conflicting results on the relative insecticidal activity among the pyrethrins, but it is generally accepted that pyrethrin I is the most toxic of them, and pyrethrin II has the greatest knock-down effect. Comparisons of activity for the housefly (*Musca domestica*) and the mustard beetle (*Phaedon cochleariae*) are tabulated.[448] The distinctive action of pyrethrins is to cause rapid paralysis of insects. The "knock-down" occurs almost instantly even with a sublethal dose, but the insect can recover. Much higher doses are required to kill. Pyrethrins readily penetrate the insect cuticle, but they are only weakly toxic when fed to insects. They act by binding to the cell membrane of nerve cells so that the sodium channels are held open (DDT functions in the same way), so that the nerves go on firing continuously to exhaustion.[453] Several compounds, themselves nontoxic and inactive, synergize the pyrethrin action (see Section 8.05.5.4.3), and change the relative order of toxicity among the pyrethrins. Pyrethrins can cause severe contact reaction in sensitive persons.

8.05.5.4.2 Biosynthesis

At the time when only "pyrethrin I" and "pyrethrin II" (actually mixtures of pyrethrins and cinerins) were known, (±)-2-[^{14}C]mevalonic acid was fed to the ovules of the flowers, and these were later extracted with acetone and diluted with cold pyrethrin.[454] The radiolabeled pyrethrin mixture was separated via the dinitrophenylhydrazones into pyrethrins I and II. Hydrolysis of pyrethrin I phenylhydrazone showed that all the radioactivity was in the chrysanthemic acid. Ozonolysis of the acid gave equally labeled acetone and caronic acid (Scheme 12). The acetone was converted to iodoform, and Kuhn–Roth oxidation of the caronic acid gave acetic acid with half the activity of the caronic acid, showing the label was in the methyl group. Hydrolysis and ozonolysis of pyrethrin II gave caronic acid and pyruvic acid, equally labeled.[455] The pyruvic acid gave unlabeled iodoform. Further labeling experiments with 1-[^{14}C]mevalonic acid gave no labeled pyrethrins.[454] Therefore no mevalonic acid was incorporated into the rethrolones. When 2-[^{14}C]acetic acid was used, both the acid and rethrolone portions were labeled in the ratio 0.33:0.67. With L-[^{14}C]*S*-methylmethionine, 20 times as much label was found in pyrethrin II as in pyrethrin I.[455] A series of labeling experiments are summarized in a table in Matsui and Yamamoto.[448] In insects and mammals, pyrethrin I is oxidized to a carboxylic acid at a terminal methyl group of the chrysanthemic acid, and the same may happen in plants. The biosynthesis of the rethrolone portion has never been studied in detail, but it is evidently produced via a polyketide. What is known of the biosynthetic route is summarized in Scheme 13. The involvement of an enzyme sulfhydryl group has been demonstrated in two other examples of irregular monoterpene formations, and is presumed to operate here in the same way.[456]

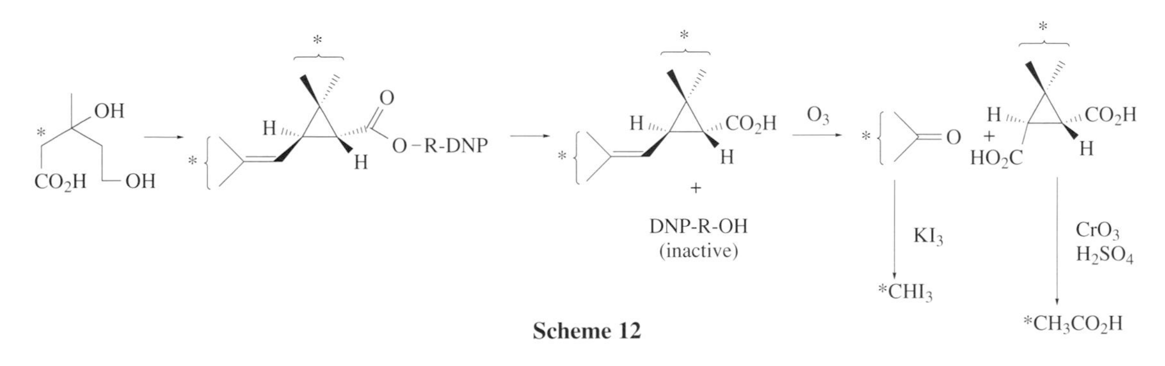

Scheme 12

8.05.5.4.3 Metabolism

The pyrethrins are detoxified by oxidation by cytochrome P450 and by hydrolysis by carboxylesterases. The compounds are relatively safe to mammals (LD_{50} orally in rats ~1500 mg kg^{-1}), because the rate of oxidation and particularly the ester hydrolysis for them is much faster, probably 10 times faster than in insects. In the housefly, the major metabolic products are conjugated to glucose, while glucuronates are produced in the rat.[457] The detoxifying enzymes are inhibited by a number of compounds, not themselves particularly toxic such as methylenedioxyphenyl compounds, for example piperonyl butoxide. These "synergists" can increase the potency of pyrethrin I about 30 times. Commercial sprays usually contain 0.03–0.1% pyrethrins with 10 times more synergist.

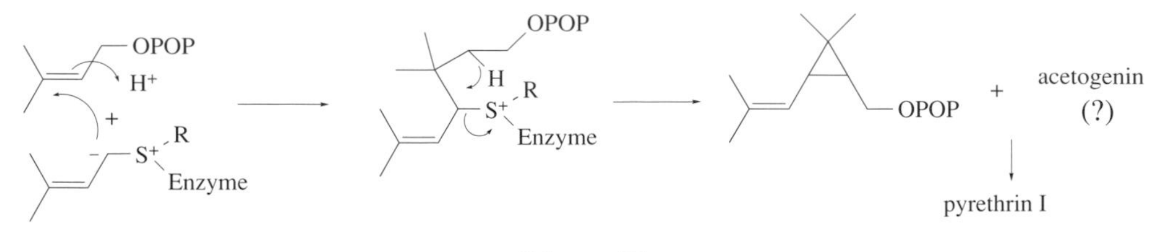

Scheme 13

From the time of Staudinger in the 1920s, model compounds adding a synthetic acid to the rethrolone or another alcohol to chrysanthemic acid have been made. Only a few of them showed activity comparable to natural pyrethrins. Benzyl and furanomethyl chrysanthemates have relatively good activity. Since they are the naturally occurring compounds with the broadest range of toxicity to insects, they were the first, and so far only, group to provide the basis of a new class of synthetic pesticide. More recently, a great number of compounds have been made by varying the acid or rethrolone portion or both, to control the hydrolysis and oxidation, giving effective synthetic pyrethrins, with final structures which bear little resemblance to the original pyrethrin. These are all now referred to as pyrethroids. Natural pyrethrins are still extracted and used commercially and compete with synthetic products, especially in the household market.

8.05.5.5 Phenylpropanoids

Although plants make prolific use of the shikimic acid pathway to aromatic compounds, we do not find the wealth of substances with strong toxicity among aromatic compounds that we find among the terpenoids or alkaloids. The qualification *strong toxicity* must be made because many of the common coumarins, flavonoids, and tannins are moderately deterrent to insects. The outstanding examples here are the rotenones (see Section 8.05.5.5.1).

Some simple phenylpropanoids show insect toxicity. Apiole (**262**) from parsley (*Petroselenium crispum*) and dillapiole (**263**) from dill (*Anethum graveolens*), *Ligusticum scotinum*, *Orthodon formosanus*, and *Crithmum maritimum* (all Umbelliferae) and the leaves of *Piper aduneum* and *P. novae-hollandae* (Piperaceae) are insecticidal and show some synergism for other toxins. Apiole is also found in camphor wood (*Cinnamomum camphora*, Lauraceae) and in the leaves of *Piper angustifolia*. The monomethoxy compound carpacin (**264**) from the bark of the carpano tree (*Cinnamomum sp.*) and *Justicia prostrata* (Acanthaceae) is also toxic.

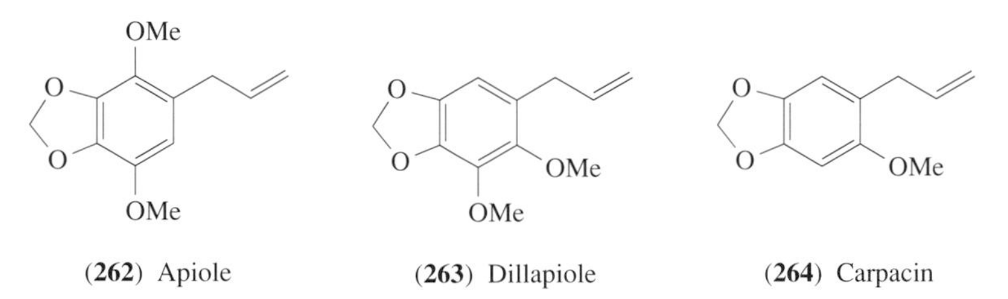

(**262**) Apiole (**263**) Dillapiole (**264**) Carpacin

Hordenine (**265**) is antifeedant for a number of sap-feeding insects at 10–100 ppm.[458] Gramine (**266**) is antifeedant for the bark beetle *Scolytus multistriatus* at 2 mM.[459] Both are found in barley (*Hordeum vulgare*, Graminae) and a host of other plants, hordenine in the ornamental grass *Phalaris arundinacea*, and the cacti *Ariocarpus scapharostrus* and *A. kolschoubeyanus* and gramine in the giant reed (*Arundo donax*), *Phalaris arundinacea*, the sugar maple (*Acer saccharinum*), and the Canadian maple (*Acer rubrum*). Myricoside (**267**) is a more complex antifeedant for *Spodoptera exempta* at 10 ppm from the roots of *Clerodendron myricoides* (Verbenaceae).[460]

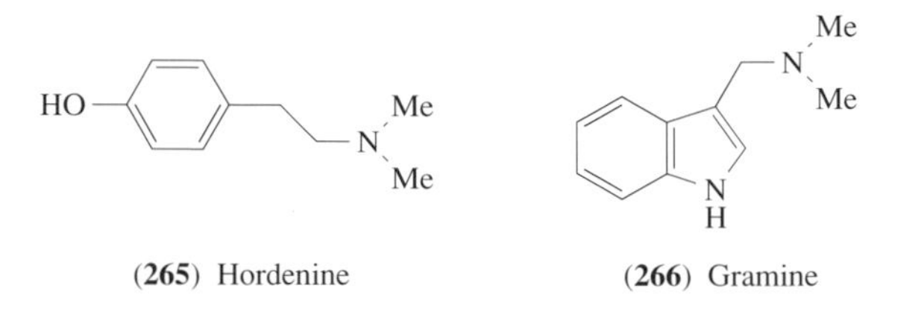

(**265**) Hordenine (**266**) Gramine

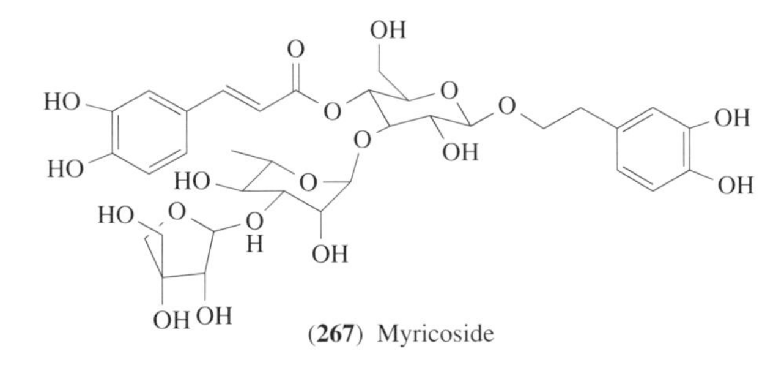

(**267**) Myricoside

Eugenol (**268**), *o*-, *m*-, and *p*-coumaric acids (**269**), (*Z*)-caffeic acid (**270**), and sinapic acid (**271**) all inhibit growth of the greenbug (*Schizaphis graminum*), as do some other phenolic compounds.[461] These compounds should be compared with piperine (**93**) and fagaramide (**96**) (see Section 8.05.5.1). Chlorogenic acid (3-caffeoylquinic acid) (**272**), a product of two important intermediates in the biosynthesis of phenylpropanoids, occurs widely in dicotyledonous plants, particularly in willows (*Salix* spp.), yet it is an antifeedant for a number of leaf-eating beetles.[462] Salicin (**273**), widely distributed and not only found in *Salix* spp., is strongly deterrent to *Bombyx mori* at 2×10^{-7} M.[463]

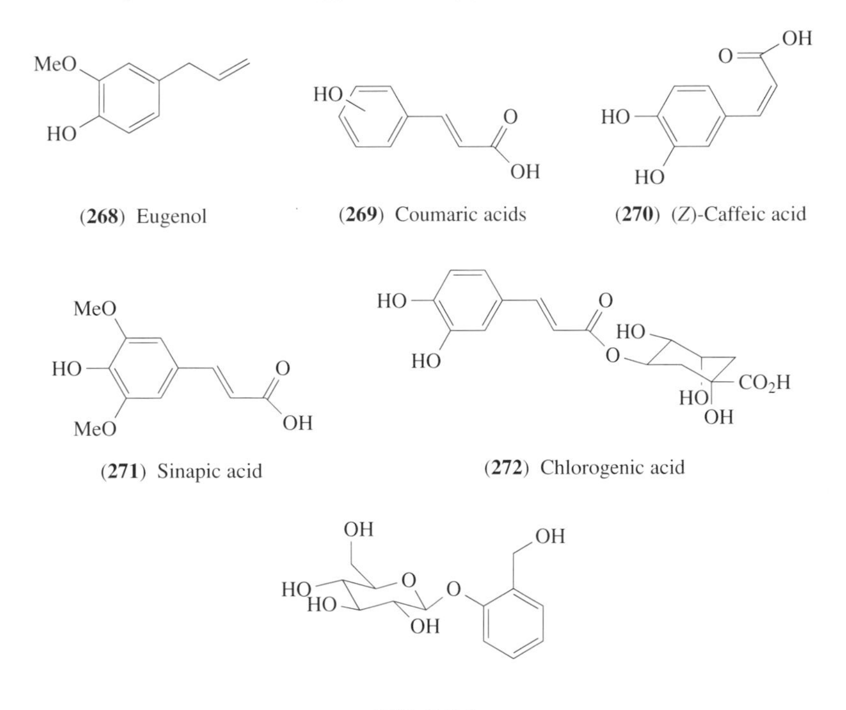

(**268**) Eugenol

(**269**) Coumaric acids

(**270**) (*Z*)-Caffeic acid

(**271**) Sinapic acid

(**272**) Chlorogenic acid

(**273**) Salicin

Among the chromenes and coumarins, the acetylchromene encecalin (**274**) from the desert sunflower (*Encelia californica*) and *Hemizonia congesta*, *Eupatorium glandulosum*, *Ageratum scoridonoides*, and *Lagascea rigida* (all Compositae) has been studied in detail,[464] and is toxic by contact and antifeedant to a wide variety of insects. The precocenes, also chromenes, are discussed in Section 8.05.2.10. The coumarin glycoside cichoriin (6,7-dihydroxycoumaran 7-glucoside) (**275**) from the flowers of commercial chicory (*Cichorium intybus*) and other Compositae including species of *Artemisia*, *Centaurea*, *Kuelpinia*, *Laurea*, and *Sonchus* is an antifeedant for locusts.

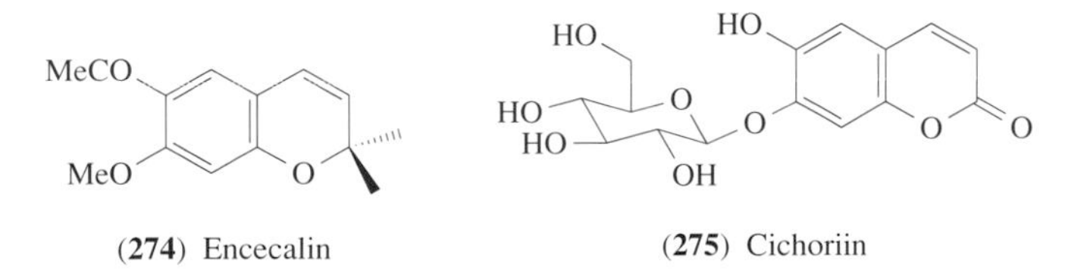

(**274**) Encecalin

(**275**) Cichoriin

Crombie *et al.* isolated the contact insecticide of *Mammea americana* and *M. africana* (Guttiferaceae) though, presumably because it was not isolated in a pure state, it was not named, but referred to as Mammea insecticide X. It is an isoprenylated coumarin (**276**). The main compound has an isobutyl chain attached to the ketone at C-8, but there were minor components with *s*-butyl, propyl, and isopropyl chains as impurities.[465] Surangin B from *Mammea longifolia* roots with a geranyl group at C-6 was more active. The compounds were toxic to *Musca domestica* through uncoupling oxidative phosphorylation.[465]

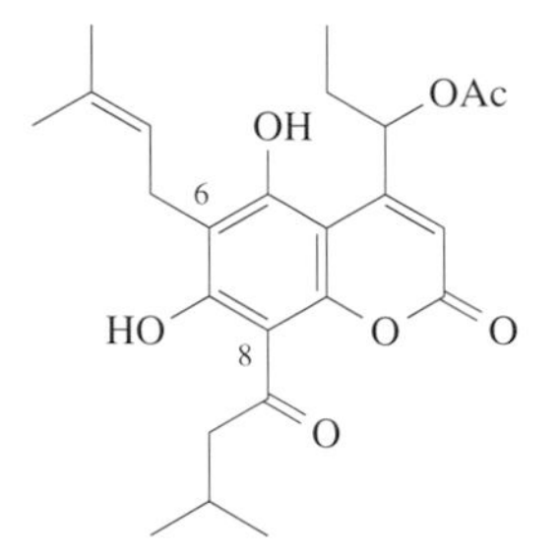

(**276**) Mammea insecticide X

Reese and Holyoke list a number of flavonoids which have moderate toxicity to insects.[295] Notable among them are phloretin (**277**) from the root bark of the apple tree, and widely found in Rosaceae, Ericaceae, and Symplococeae, and its glucoside phloridzin (**278**) from the bark of apple, pear, plum and cherry, and the leaves and skin of apples and many Rosaceae. Both compounds are feeding deterrents for the greenbug (*Schizaphis graminum*) and the peach aphid (*Mysus persicae*). Hesperetin or eriodictyol 4′-methyl ether (**279**) from Rutaceae has the same action. Its isomer, eriodictyol 3′-methyl ether, from *Artemisia campestris* and *Tanacetum sibiricum* (Compositae) and *Eriodictyon* spp. (Hydrophyllaceae) is also a feeding deterrent for *S. graminum* and *M. persicae*. Hesperetin 7-rutinoside is the principal flavonoid of oranges and lemons, but information on its antifeedant properties was not found, but rutin (quercitin-3-rutinoside) (**280**), found in many plants and comprising up to 3% of the dry weight of buckwheat (*Fagopyron esculentum*, Polygonaceae), is toxic to *Helicoverpa zea* and *H. virescens*. Rutin at 0.25% of the diet halves larval growth.[466] The effects of a number of flavonoids on the growth of *H. zea* have been recorded.[467]

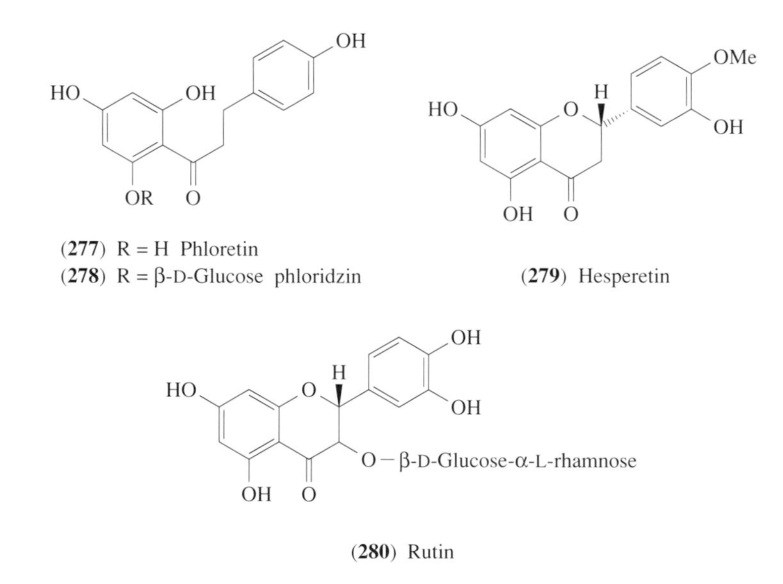

(**277**) R = H Phloretin
(**278**) R = β-D-Glucose phloridzin

(**279**) Hesperetin

(**280**) Rutin

Some furanocoumarins typical of Umbelliferae and Rutaceae are toxic and mutagenic. Psoralen (**281**), xanthotoxin (**282**), and bergapten (**283**) from *Thamnosma montana* are phototoxic. They intercalate into DNA and react with it when irradiated with long UV wavelengths. They are lethal to generalist insect herbivores, but adapted species tolerate them by rapid excretion.[468] Xanthotoxin is metabolized in the gut of the black swallowtail butterfly (*Papilio polyxenes*) to two products (**284**) and (**285**) (Equation (1)). Some susceptible species produce the same metabolites, but at a rate 33 times slower, so that a 60-fold higher concentration accumulates.[469] Isopimpinellin (5,8-dimethoxypsoralen) (**286**) is toxic in the dark or light.[470] Isopimpinellin from *Orixa japonica* is reported to be used to insect-proof books in Japan.[471] Xanthotoxin also obtained from *O. japonica* is a feeding deterrent for the cockroaches *Blattella germanica*, *Periplaneta americana*, and *Neostylopyga rhombifolia* at 67 ppm.[472] The subject of light-activated pesticides has been reviewed by Berenbaum.[473]

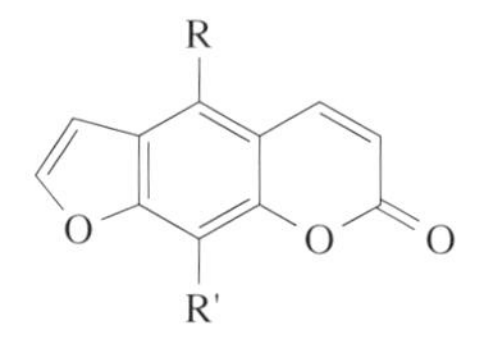

(**281**) R = R' = H, Psoralen
(**283**) R = OMe, R' = H Bergapten
(**286**) R = R' = OMe Isopimpinellin

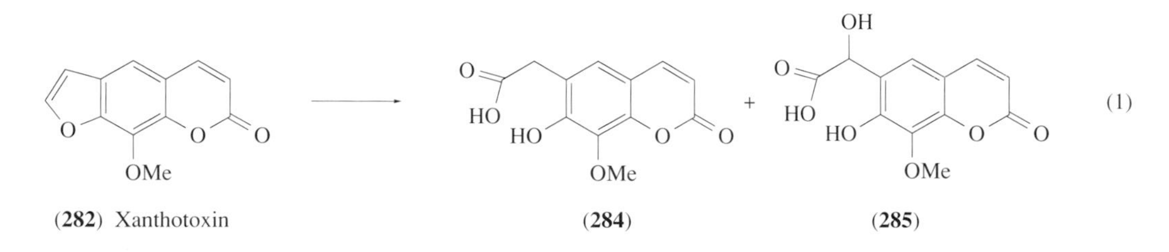

(**282**) Xanthotoxin (**284**) (**285**) (1)

Lignans and lignins are generally nontoxic to insects, but Gang *et al.*[474] list 13 from nine plants that are toxic to a variety of insects. Two lignans from *Justica procumbens*, justicidins A (**287**) and B (**288**), were tested only on *B. mori*, but were lethal at 20 ppm.[475] According to Berenbaum, the reason for the small number of examples here is because only a few studies of lignans have been made for insecticidal activity.[476]

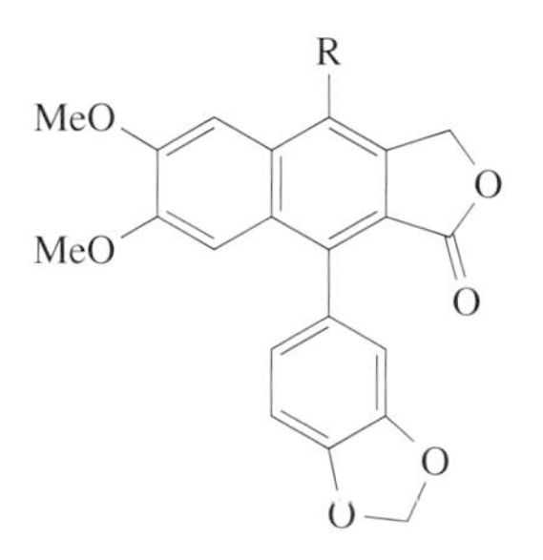

(**287**) R = OMe Justicidin A
(**288**) R = H Justicidin B

Many quinones show low activity against insects, and a few show moderate activity. Juglone (**289**), from the bark and shells of the walnuts *Juglans regia* and *J. nigra*, and *Carya ovata* and *C. illinoensis* (all Juglandaceae), is often quoted as a feeding deterrent, but 1600 ppm are required for complete inhibition of feeding by the bark beetle *Scolytus multistriatus*. Tectoquinone (**290**), deoxylapachol (**291**), and 1,4-dihydroxy-2-methylnaphthoquinone (**292**), all from teak wood, (*Tectona grandis*, Verbenaceae), were reported repellent to termites, but have attracted no further interest.

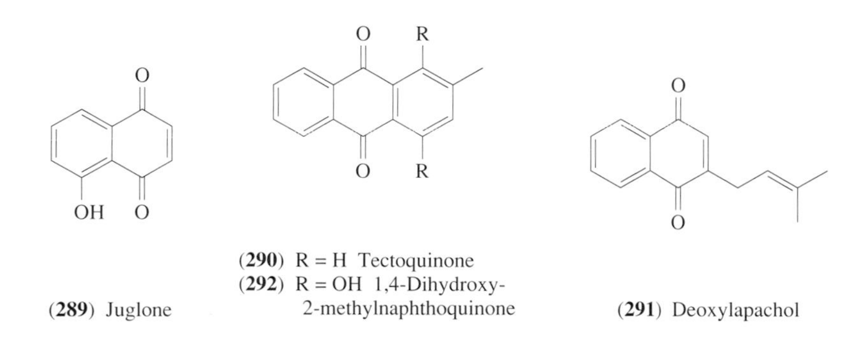

(**289**) Juglone

(**290**) R = H Tectoquinone
(**292**) R = OH 1,4-Dihydroxy-2-methylnaphthoquinone

(**291**) Deoxylapachol

A small group of aromatic compounds not fitting into the usual groups are the rocaglamides. Rocaglamide (**293**) is a highly substituted benzofuran from *Aglaia odorata* and *A. elliptica*, ornamental shrubs of the family Meliaceae from India and Malaysia.[477] It inhibits larval growth and is a slow-acting toxin to the variegated cutworm (*Peridroma saucia*) and Asian armyworm, although feeding ceases almost immediately.[478] There are other related compounds in the plants.

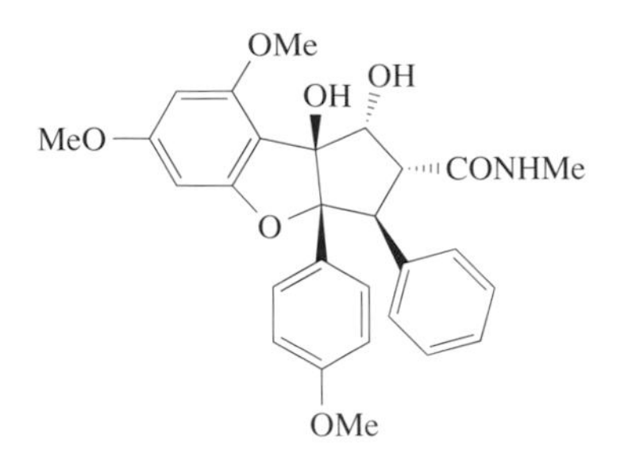

(**293**) Rocaglamide

8.05.5.5.1 Rotenone

A chemical interest in rotenone grew from observation that people in South-East Asia use *Derris* plant material as a fish poison. *Lonchocarpus*, *Tephrosia*, *Mundulea*, and other plants (all Leguminosae) were used similarly in Africa and South America. The earliest recorded use of *Derris* root as an insecticide is in 1848, against leaf-eating caterpillars. Isolation studies began in 1892 with the work of Geoffroy, who isolated what we know as rotenone from *Lonchocarpus nicou*.[479] This was followed by the isolation of the same compound in Japan in 1912 from *Derris chinensis*, called roten in Taiwan. Because the compound was recognized as a ketone, it was called rotenone. In 1916 the same compound was again isolated from *Derris elliptica*, but called tubotoxin. Subsequently a number of related rotenoids were isolated from various leguminous plants. In 1923, tubotoxin and rotenone were shown to be one compound, and in 1929 the correct molecular formula of $C_{23}H_{22}O_6$ was assigned, with more of its structure solved. The complete structure (**294**) was solved simultaneously in 1932 in Britain, France, Japan, and the USA. The structural work was reviewed[480] in 1933.

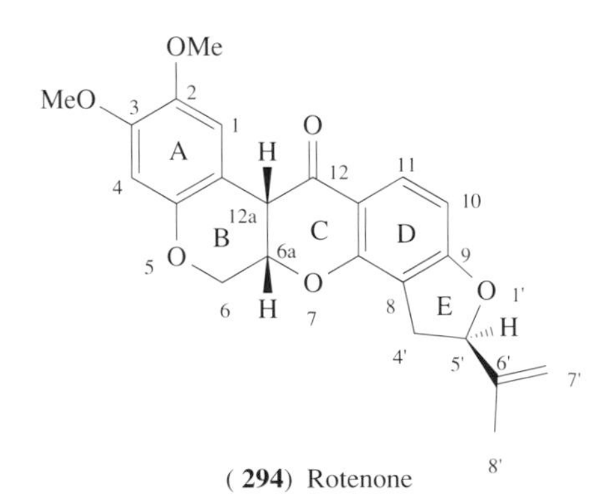

(**294**) Rotenone

Rotenone is a crystalline compound, m.p. 165–166 °C, $[\alpha]_D^{20}$ −288° (benzene). It is practically insoluble in water but readily soluble in many organic solvents. It decomposes when exposed to light and air, it is hydroxylated at C-7, then dehydrated, and is also decomposed by alkali. The chemistry of rotenones was extensively studied by Crombie's group.[481] The absolute configuration was established[482] in 1960 (the three asymmetric centers are 5′*R*,6a*S*,12a*S*), and the crystal structure[483] in 1975. The molecule is shaped like butterfly wings, with two flat sections, rings A and B lying at an angle to rings C, D, and E, folded at the 6a–12a bond. The first synthesis was recorded[484] in 1961 (see also Miyano[485]). The ^{1}H and ^{13}C NMR spectra of rotenone have been studied by Crombie,[486,487] and further studies of the ^{1}H spectra of 12 common rotenoids[488] have been made. The mass spectrum has been published.[489] Rotenoids are defined as chromanochromanones, that is, with four fused rings. Crombie's systematic name for rotenone[482] is 16αβ,12αβ,4′,5′-tetrahydro-2,3-dimethoxy-5′β-isopropenylfurano-(3′,2′,8′,9)-6*H*-rotoxen-12-one.

A number of compounds closely related to rotenone were subsequently isolated from other species of Leguminosae, principally from the roots, but also from the aerial parts. A comprehensive review of their chemistry to 1970 lists 10 compounds.[490] By 1980, there were 38, of which rotenone itself is the most widely distributed. It has been isolated from at least 61 species of Leguminosae, particularly species of *Derris*, *Lonchocarpus*, *Millettia*, and *Tephrosia* and one of Scrophulariaceae (leaves of *Verbascum thapsus*).[491] The next most common are deguelin (**295**) and tephrosin (**296**), both isolated from 21 species, 12a-hydroxyrotenone (**297**) from 13 species, elliptone (**298**) from eight species, and toxicarol (**299**) and sumatrol (**300**), both from five species.[491] The simplest is munduserone (**301**) from *Mundulea serica* root bark.[492] They all have the same absolute configuration as rotenone at carbon atoms 6a and 12a.[493] At least 68 species of plants, including 21 species of *Tephrosia*, 12 of *Derris*, and 10 of *Milletia*, have been reported to contain rotenone or rotenoids. The principal sources are *D. elliptica* and *D. malaccensis*, which also contain malaccol (**302**), from Malaysia and Indonesia, and *Lonchocarpus utilis* and *L. uruca* from South America. Roots of *D. elliptica* contain 5–9% rotenone, with a maximum of 13%. Seven rotenoids are also found in plants as glycosides.

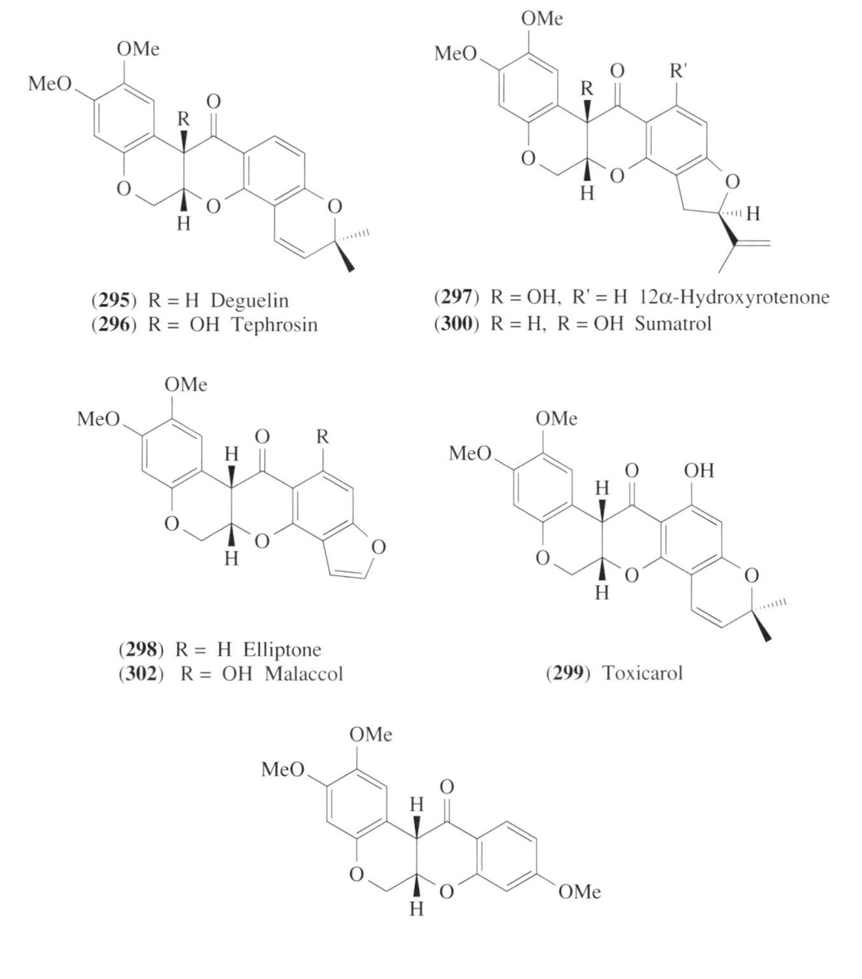

(**295**) R = H Deguelin
(**296**) R = OH Tephrosin

(**297**) R = OH, R′ = H 12α-Hydroxyrotenone
(**300**) R = H, R = OH Sumatrol

(**298**) R = H Elliptone
(**302**) R = OH Malaccol

(**299**) Toxicarol

(**301**) Munduserone

(i) Biological activity

The rotenoids are the only group of flavonoids known to be strongly toxic to insects. They are especially active against leaf-eating Coleoptera and Lepidoptera, but rotenone is reported to have no effect on *Heliothis virescens*.[494] Aphids are killed by 3 ppm of rotenone in solution, and the LD_{50}

for the silkworm *Bombyx mori* is 3 μg g^{-1} body weight, but rotenone is 5–10 times more active than any of the others.[495] Rotenoids are considered to be only moderately toxic to mammals (LD_{50} orally in rats 132 mg kg^{-1}), although more harmful by inhalation than by ingestion, and harmless to plants. Other rotenoids have higher mammalian toxicity, and deguelin is a strong irritant, toxic by inhalation. Their specific action is to depress cellular respiration by interfering with electron transport in the oxidation of NADPH to $NADP^+$ by cytochrome *b*, which then inhibits the mitochondrial oxidation of Krebs cycle intermediates.[496] There is a dramatic fall in oxygen uptake on applying rotenone to insect cells. One review gives a table comparing the toxicity of derris dust, pyrethrum, and nicotine for seven pest insect species.[490] It demonstrates that the toxicity of rotenone varies widely with species, as is found for other compounds. There are also 6a,12a-dehydrorotenoids known in plants. These have essentially planar molecules and are generally very weakly active or inactive. A table of rotenoids affecting insect mortality is given by Holyoke and Reese.[295]

(ii) Biosynthesis

Structural similarities and distribution in nature connect the rotenoids with isoflavones, which in turn arise via the shikimic acid pathway and a phenyl-C_3 precursor which is chain-lengthened with acetate followed by cyclization with aryl migration to an isoflavonoid.[497] The conversion of [2-^{14}C]7-hydroxy-2′,4,5′-trimethoxyisoflavone (**303**) into [6-^{14}C]amorphigenin (**304**) by *Amorpha fruticosa* seedlings was the first direct demonstration of the biosynthetic origins of rotenone.[498] Labeling experiments by Crombie's group using *Derris elliptica* plants and radiolabeled [1-^{14}C]phenylalanine gave [12-^{14}C]rotenone. Similarly, [2-^{14}C]phenylalanine gave [12a-^{14}C]rotenone, [3-^{14}C]phenylalanine gave [6a-^{14}C]rotenone, and [CH_3-^{14}C]methionine gave [6-^{14}C]rotenone.[499] Later, Crombie's group demonstrated the isoprenoid origin of the E ring; the (*E*)-methyl group of rotenoic acid becomes the methylene of rotenone.[500] Mevalonic acid was only rather poorly incorporated. The biosynthesis can then be summarized in Scheme 14.

At least eight oxidative metabolites of rotenone have been identified,[501] including 12a-hydroxyrotenone (**297**), and other oxidations occur at the isopropenyl group.

Partially because of their complex structures and no doubt because of the specificity of structure required for activity against insects, the rotenoids have not been subjected to the same intense interest by synthetic chemists that the pyrethrins received; however, there have been some advances that show how stability and activity can be increased.[502]

8.05.5.6 Alkaloids

Of the myriad of plant alkaloids, the great majority show toxicity to animals in some way, yet relatively few are recorded as being toxic to insects. That small number of records is probably more of a reflection of the investigator's interests in looking at alkaloids than of the true proportion of insect-toxic ones. It may indicate a greater expectation of finding substances with selective toxicity to insects among CHO compounds than among alkaloids. This section surveys the alkaloid group with representative examples of toxic or antifeedant compounds and the insects they affect. The special cases of nicotine (see Section 8.05.5.6.1) and ceveratrine alkaloids (see Section 8.05.5.6.2) are treated separately. The definition of alkaloids as being basic nitrogen-containing plant natural products is used here, so, for example, nonbasic ryanodine and colchicine are not considered as alkaloids.

Among simple plant alkaloids, vasicinol (7-hydroxyvasicine or 7-hydroxypeganine) (**305**) from the roots of *Adhatoda vasica* (Acanthaceae) and vasicinone (**306**) from the aerial parts of *Peganum harmala* (Zygophyllaceae) and the roots and leaves of plants from other families are both moderate antifeedants for *Aulacophora foveicollis* and *Epilachna vijintioctopunctata*.[503] The leaves of *Derris elliptica* (see Section 8.05.5.5.1) and the seeds of *Lonchocarpus sericeus* (Leguminosae) contain 2,5-dihydroxymethyl-3,4-dihydroxypyrrolidine (**307**), which is a potent inhibitor of glucosidase and an antifeedant and toxin for the locusts *Schistocerca gregaria* and *Locusta migratoria*, and the armyworms *Spodoptera exempta*, *S. littoralis*, and *Prodenia litura*.[504] Castanospermine (**308**) from the seeds of *Castanospermum australe* (Leguminosae) is also known to be toxic to the pea aphid (*Acyrthosiphon pisum*) and other insects as well as to mammals.[505] Pilocarpine (**309**) from *Pilocarpus jaborandi* (Rutaceae) is toxic to *Bombyx mori* and *Pieris brassicae* at very low doses as well as having mammalian toxicity. It is hardly surprising that strychnine (*Strychnos* spp.) is equally toxic to these

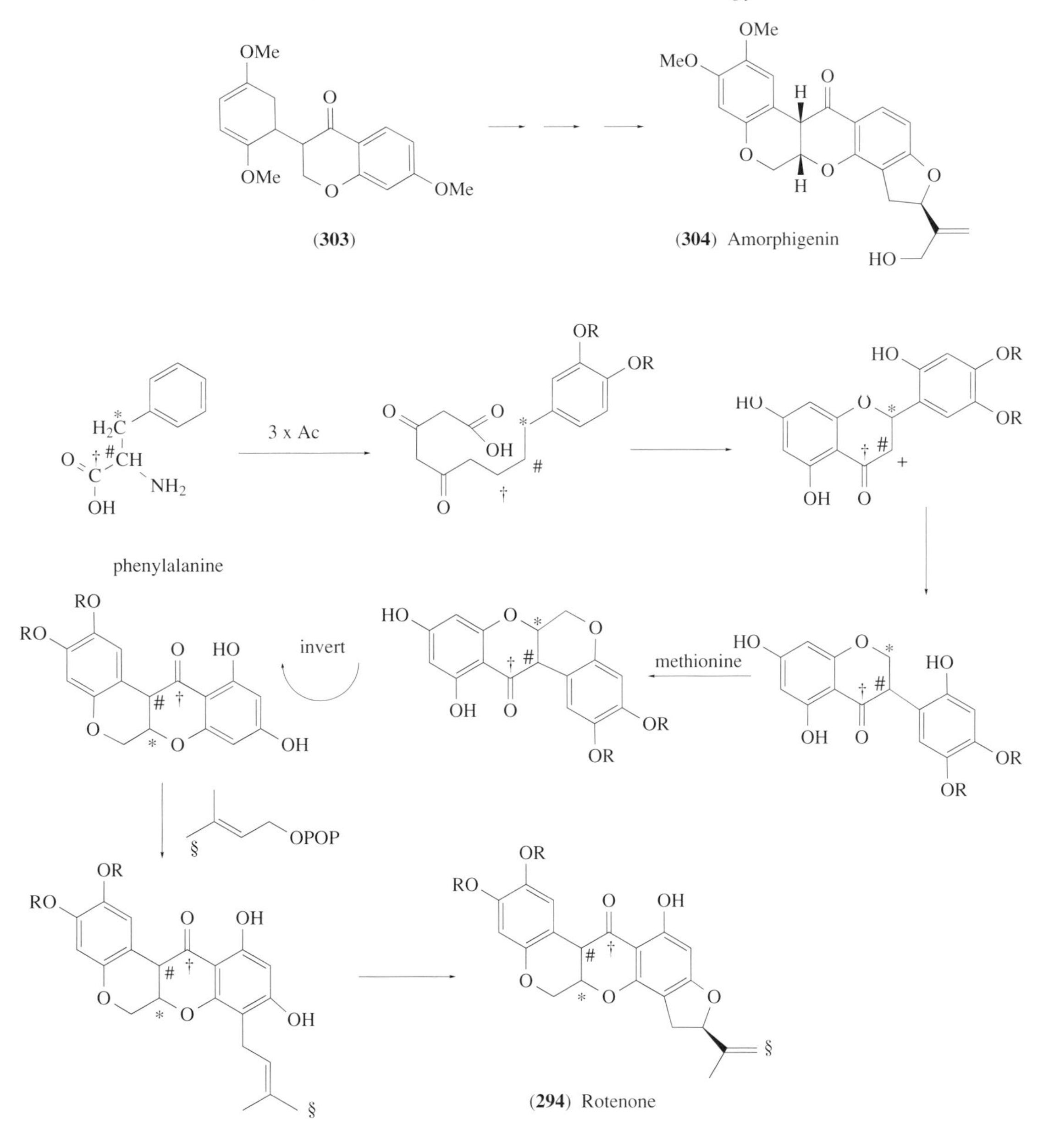

Scheme 14

species. Schoonhoven has studied the feeding deterrence of a number of familiar alkaloids, including quinine, strychnine, solanine, and tomatine (see later in this section). Quinine was active on *B. mori* at 5×10^{-4} M and on *P. brassicae* at 2×10^{-6} M. Strychnine was active on *B. mori* at 10^{-7} M, and on *P. brassicae* at 10^{-5} M.[506]

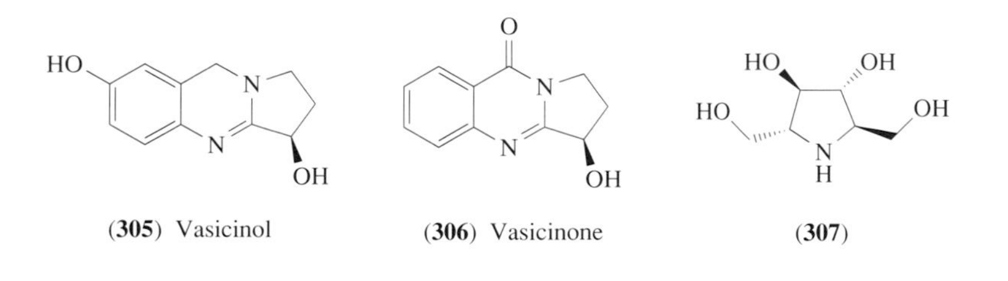

(**305**) Vasicinol (**306**) Vasicinone (**307**)

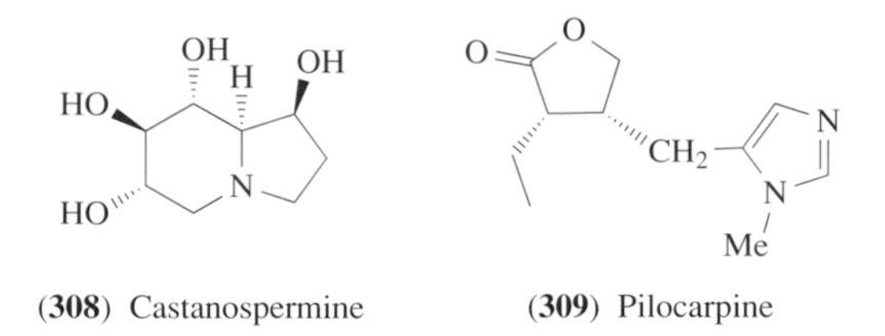

(**308**) Castanospermine (**309**) Pilocarpine

It is interesting to compare the structure of ammodendrine (**310**), from *Ammodendron conollyii* and several *Sophora* spp. (Leguminosae), with that of nicotine. Ammodendrine is reported to be toxic to many insects. The quinolizidine alkaloids lupinine (**311**) from *Lupinus luteus* and other *Lupinus* spp. (Leguminosae) and *Anabasis aphylla* (Chenopodiaceae) and aphylline (**312**), also from *A. aphylla*, have been found to be toxic for insects. Two examples of insect-toxic alkaloids unearthed by bioassay-guided separation are of one new and one well-known alkaloid. The first is (−)-7*epi*-deoxynupharidine (**313**) from *Nuphar japonica* (Nymphaceae), which has an LC_{50} of ~4 μmol ml^{-1} of the diet for *Drosophila melanogaster* larvae.[507] The same investigative group found that the familiar (−)-canadine (tetrahydroberberine) (**314**) from tubers of *Corydalis bulbosa* (Papaveraceae)[508] killed 99.7% of *D. melanogaster* larvae at 1.40 μmol ml^{-1} of the diet, and the LC_{50} was 0.91 μmol ml^{-1}.

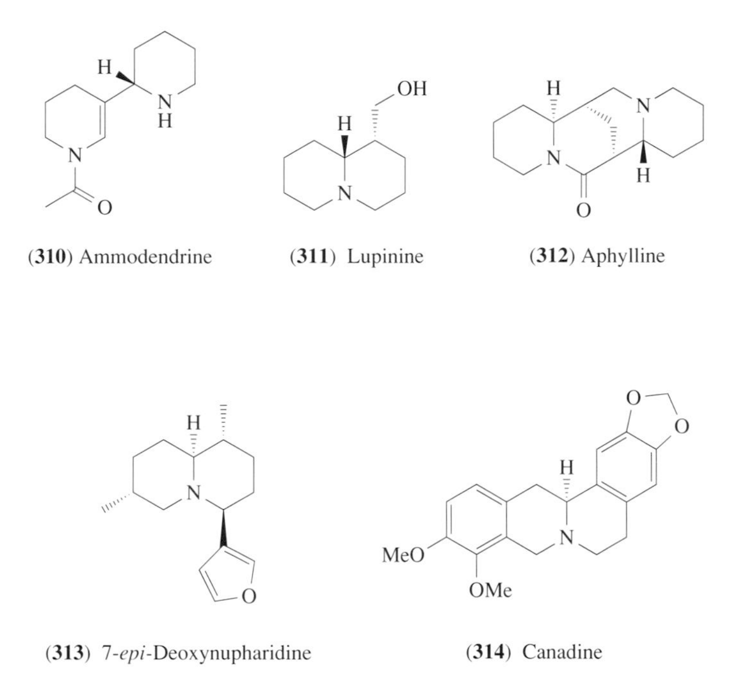

(**310**) Ammodendrine (**311**) Lupinine (**312**) Aphylline

(**313**) 7-*epi*-Deoxynupharidine (**314**) Canadine

Cocaine (**315**) from *Erythroxylon coca* leaves (Erythroxylaceae) at 0.5% on tomato leaves killed *Manduca sexta* larvae. It was estimated that the amount of cocaine in coca leaves would inhibit the growth of about 80% of *Manduca* larvae. The toxic effect was attributed to the blockage of reuptake of the neurotransmitter octopamine.[509] A number of species of *Zanthoxylum* (Rutaceae) are reported to be notably toxic. *Zanthoxylum monophyllum* contains zanthophylline (**316**), which is moderately toxic to the alfalfa weevil (*Hypera postica*), the range caterpillar (*Hemileuca oliviae*) and the grasshopper *Melanoplus sanguinipes*.[510] A number of complex alkaloids of the wilfordine type, which are composed of a pyridine dicarboxylic acid esterified with a sesquiterpene polyol, have been found to have insect toxicity. For example, the thunder-god vine (*Tripterygium wilfordii*, Celastraceae) is used in China as an insecticide.[320] It contains a number of wilfordine alkaloids including wilforine (**317**) (compare with angulatin (**159**) and celangulin (**160**), see Section 8.05.5.3.2), which is strongly active against the diamondback moth (*Plutella xylostella*). The insect toxin celabenzene (**318**), of quite different structure, is also present in *T. wilfordii* and *Maytensus mossambicensis* (Celastraceae).

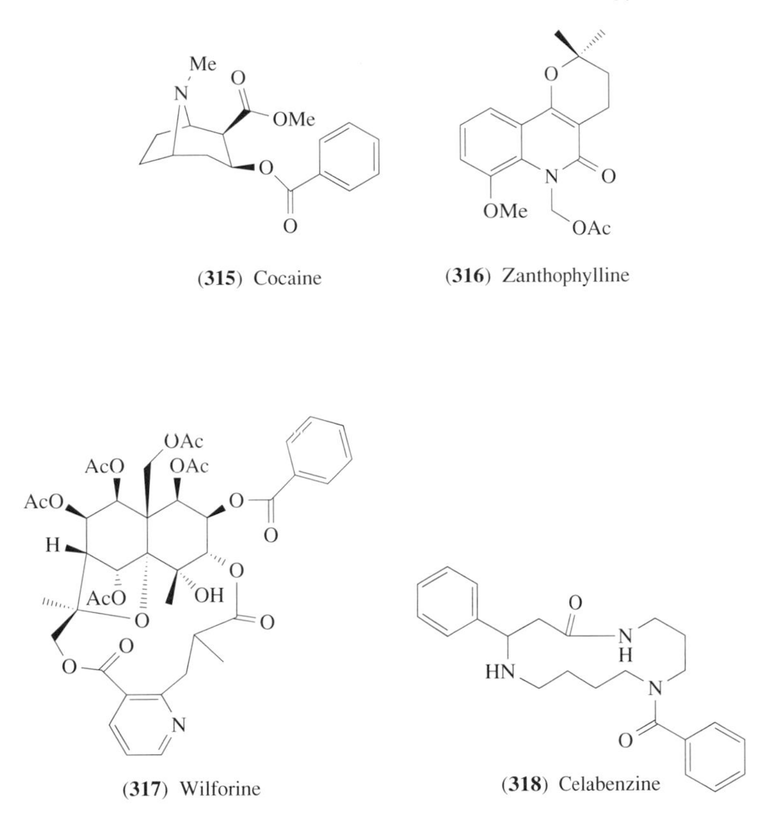

(**315**) Cocaine (**316**) Zanthophylline

(**317**) Wilforine (**318**) Celabenzine

The pyrrolizidine alkaloids are said to be toxic, and there is a great deal of literature on their sequestration from host plants by adapted insects to protect them from predators (see Section 8.05.6.8), but it is very difficult to find data on the insect toxicity of pyrrolizidines. A review on their chemistry does not mention insect toxicity.[511] Examples are senecionine (**319**) from *Senecio vulgaris*, an antifeedant to *L. migratoria* at 10 ppm,[512] and senkirkine (**320**), isolated from a number of *Senecio* spp. and other Compositae, and tested on the spruce budworm (*Choristoneura fumiferana*) and found to be active[513] at 1 mM.

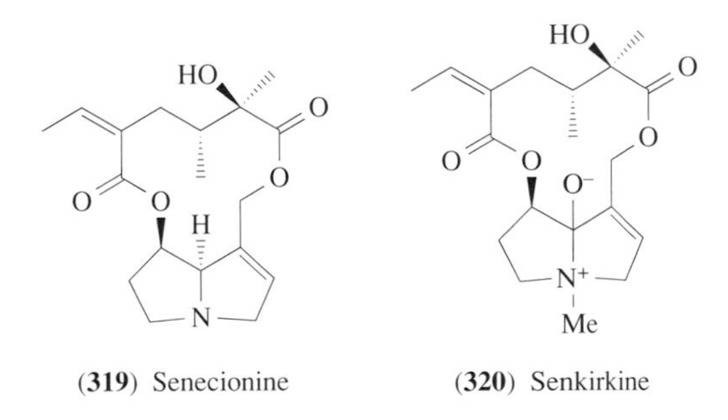

(**319**) Senecionine (**320**) Senkirkine

The isoquinoline alkaloid isoboldine (**321**), found in plants among 13 families, is a feeding inhibitor of *Spodoptera litura* and *Calospilos miranda* at 200 ppm.[514] The indole alkaloid cimicidine (0.03% yield from the plant) and its more complex and more active relation haplophytine (0.007–0.03%) (**322**) from the Mexican cockroach plant (*Haplophyton cimicidum*, Apocynaceae) are toxic to a range of insects.[515] The LD_{50} of haplophytine is 18 μg g^{-1} for the German cockroach (*Blattella germanica*).

Solanine and tomatine would not deserve consideration in such a brief review of insecticidal alkaloids on the strength of their toxicity,[506] but since they come from such everyday plants and

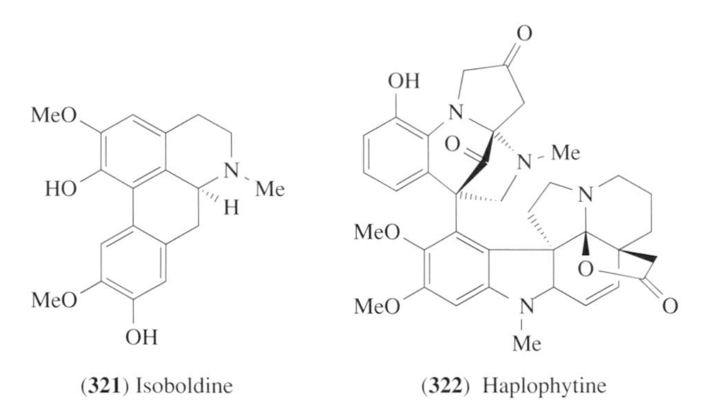

(**321**) Isoboldine

(**322**) Haplophytine

have been used as insecticides, they can be mentioned as representatives of steroidal alkaloids. Solanine (α-solanine or solatunine) (**323**) occurs with closely related glycosides of the aglycone solanidine in the potato (*Solanum tuberosum*), woody nightshade (*S. nigrum*), and the tomato (*Lycopersicon esculentum*) (Solanaceae). It is present in "sunburnt" potato tubers (green parts) and fresh potato sprouts (0.04%). A crude extract as the hydrochloride has been used as a commercial insecticide in South America in spite of its high mammalian toxicity (LD_{50} orally in rats 590 mg kg^{-1}, LDLo in humans 2.8 mg kg^{-1}). Tomatine (α-tomatine or lycopericin) (**324**) from the leaves (and fruit of some strains) of the tomato *L. esculentum* and other species of *Lycopersicon* and *Solanum* has also been used as an insecticide, especially for the Colorado potato beetle (*Leptinotarsa decemlineata*). It completely inhibits feeding of *P. brassicae* at 40 ppm, it inhibits growth of *Helicoverpa zea*[516] at 0.4 mg kg^{-1}, and has high mammalian toxicity, (LD_{50} orally in rats 900 mg kg^{-1}).

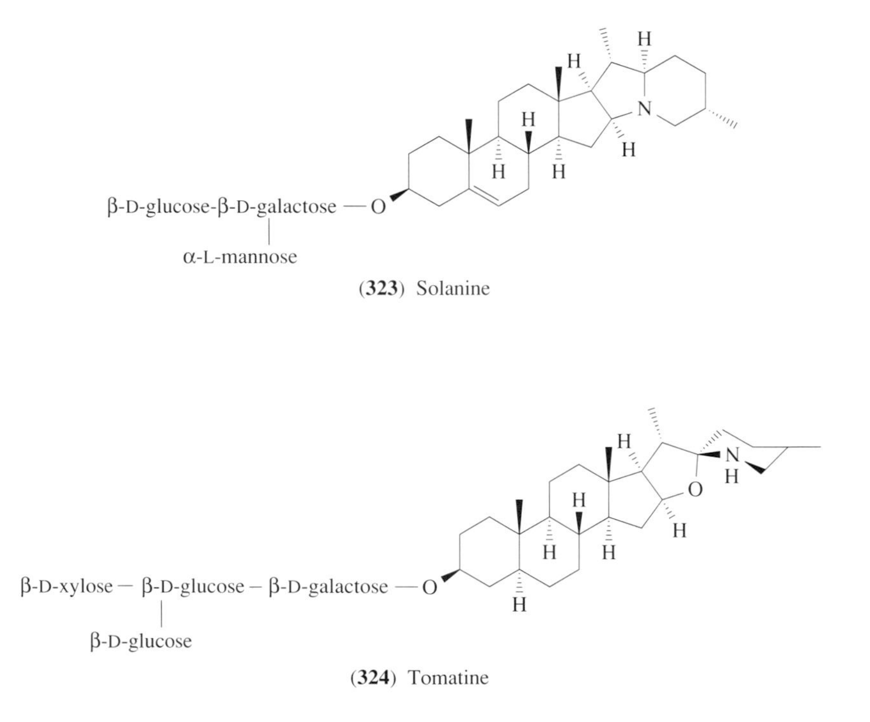

(**323**) Solanine

(**324**) Tomatine

Canavanine (**325**) is a basic amino acid that can be included here. It is obtained from jack beans (*Canavalia ensiformis*, Leguminosae), but is more widely distributed. It is toxic because it resembles arginine and is "mistakenly" incorporated into proteins, so it is a potent growth inhibitor for many organisms. Larvae of *Caryedes* beetles live in the seeds and are able to metabolize the canavanine for their nitrogen requirements. They convert it to canaline (also toxic) and urea, and then from canaline to homoserine and ammonia (Scheme 15).[517] Some other unusual amino acids from plants are listed by Reese and Holyoke.[295]

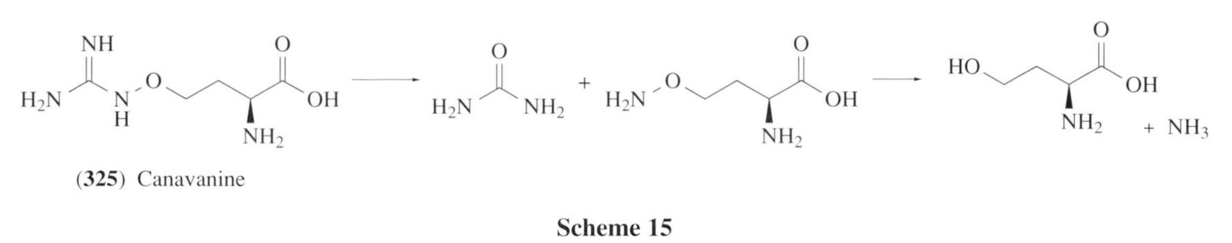

(**325**) Canavanine

Scheme 15

8.05.5.6.1 Tobacco alkaloids

Tobacco has the oldest record in chemical ecology for toxicity to insects. Tobacco extracts were recorded as used as a spray on crops in 1690, and used as a dust or spray throughout the eighteenth century, with numerous references to its use.[518] More concentrated forms of nicotine became available in the twentieth century until nicotine sulfate (40% solution) became a much-used insecticide, only slowly replaced by synthetic insecticides for some purposes. Nicotine (**326**) is found in at least 18 species of *Nicotiana*, and altogether in 24 genera of 12 families and even in different classes and orders. It is found in *Duboisia hopwoodi* from Australia, deadly nightshade (*Atropa belladonna*), both Solanaceae, common milkweed (*Asclepias syriaca*, Asclepiadaceae), clubmoss (*Lycopodium clavatum*), and horsetails (*Equisetum arvense*). Anabasine (**327**) is found in *Anabasis aphylla* (Chenopodiaceae) and the garden flower *Zinnia elegans* (Compositae) as well as in *Nicotiana* spp. Anabasine and anabaseine (**328**) have been found in the venom glands of several species of *Aphaenogaster* and *Messor* ants (see Section 8.05.5.6.4(ii)), and nicotine together with pyrrolidone and phenylacetaldehyde form the osmeterial gland secretion of larvae of the gypsy moth (*Lymantria dispar*).[519] Nicotine itself is present in *Nicotiana tabacum* leaves from 2% to 14%, and 18% in *N. rustica*, which was cultivated especially for nicotine extraction. It is accompanied by varying amounts of related alkaloids, which include anabasine, anabaseine, nornicotine (**329**), nicotyrine (**330**), and anatabine (**331**). Leete lists 45 tobacco alkaloids.[520]

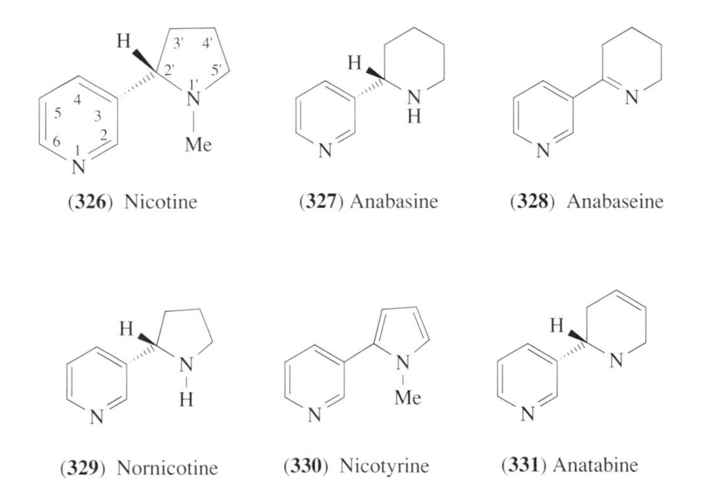

(**326**) Nicotine (**327**) Anabasine (**328**) Anabaseine

(**329**) Nornicotine (**330**) Nicotyrine (**331**) Anatabine

Formerly the stalks and leaf ribs of smoking tobacco were used for nicotine extraction, but when processes were discovered for incorporating these into smoking materials, *N. tabacum* became too valuable for insecticidal use, and *N. rustica* was cultivated for this purpose, and still is in China. Since it exists largely as salts in the plant, for isolation, alkali is added and the mixture steam distilled.

Nicotine has a long chemical history of isolation as an impure oil, then pure, and finally in 1893 the correct molecular structure (**326**) proposed by Pinner, confirmed by synthesis by Pictet.[521] Nicotine is a colorless liquid, b.p. 246 °C, $[\alpha]_D^{20}$ $-169°$, basic (pK_a 3.22 and 8.11), miscible with water below 60 °C, and miscible with several organic solvents. It decomposes in air, turning brown, becoming more viscous and changing from odorless to unpleasant. It decomposes to a mixture containing nicotinic acid, oxynicotine, nicotyrine, cotinine, myosmine, methylamine, and ammonia.[522] The chemistry of nicotine and its close relatives has been covered in various places including the multivolume series *The Alkaloids*.[523]

(i) Biological activity

Nicotine is a contact toxicant, regarded as very poisonous to mammals and insects (LD_{50} orally in mice 230 mg kg^{-1}, half that of rotenone, but 1000 times more toxic intravenously). It is a very effective antagonist of acetylcholine, and it affects ganglia of the central nervous system, facilitating trans-synaptic conduction at low concentrations and blocking it at high concentrations. Studies of the molecular mechanism of the toxicity of nicotinoids explain why nicotine is much more toxic to mammals than to insects.[524] It does not have a wide spectrum of activity against insects, and is most effective against small, soft-bodied insects, since it is most effectively absorbed through the cuticle.[525] Its vapor penetrates the cuticle, and human skin. Negherbon lists the effect of nicotine as a fumigant on 57 species.[526] In insects it causes death slowly over an hour, preceeded by tremor, convulsions, and paralysis. It is still very effective for aphids used as a 0.05–0.06% spray. Anabasine, nornicotine, and other nicotinoids possess the same molecular dimensions to act at acetylcholine receptor sites, so we would expect them to be similarly toxic. In some cases they are more toxic than nicotine; in others less toxic. Anabasine sulfate is 5–10 times more toxic to the aphid *Aphis rumicis* than nicotine sulfate.[527] The volatility and autoxidation of nicotine means there is little problem of persistence. It is relatively nontoxic to plants, but will affect the growth and flowering of some plants. It is expensive to produce, unpleasant to handle, extremely toxic to man and mammals, and not a particularly effective insecticide. Holyoke and Reese give a table of chemical data on the nicotine alkaloids and a table of references to the effects of nicotine on a long list of insect species.[496] The use of nicotine as an insecticide has been comprehensively reviewed.[528]

(ii) Biosynthesis and metabolism

Biosynthesis of the alkaloids occurs chiefly in the roots, and is translocated to the stem and leaves.[520,529] Nicotinic acid arises from glyceraldehyde phosphate and aspartic acid and, because the hydrogen atom at C-6 is exchanged, goes through a dihydro intermediate. Ornithine is the source of the pyrrolidine ring. It is known that the C-5 amino group and the C-2 hydrogen are preserved.[530] The process is outlined in Scheme 16. Contrary to expectations, other nicotinoids are not made in exactly the same way. Anabasine is formed from nicotinic acid and lysine, but the lysine is converted first to cadaverine.

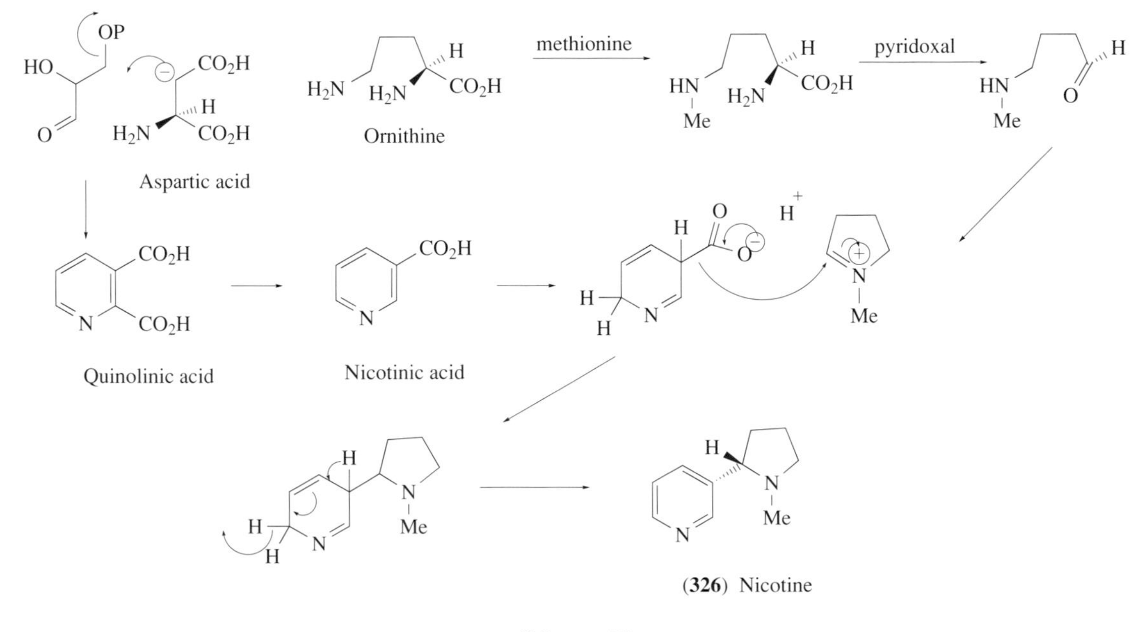

Scheme 16

The metabolism of nicotine in plants and mammals was the subject of long and somewhat inconclusive investigation, but cotinine (**332**) is the major metabolite.[520] Some tobacco-feeding insects such as the tobacco wireworm (*Conoderus vespertinus*) and the cigarette beetle (*Lasioderma serricorne*) metabolize nicotine to several other alkaloids, principally cotinine.[531] The aphid *Mysus* *persicae* avoids the xylem-translocated nicotine by selectively feeding on the phloem.[531] Others, such as the tobacco hornworm (*Manduca sexta*) and the tobacco budworm (*Heliothis virescens*), were said to excrete it unchanged in their feces, but more recent work on *M. sexta* has shown that after feeding there are significant levels of nicotine in the hemolymph and the midgut and hindgut and their contents, and that it is not secreted unchanged but almost exclusively as metabolites, the major one being cotinine *N*-oxide.[532] Their nerves are said to be relatively insensitive to nicotine compared to other insects.

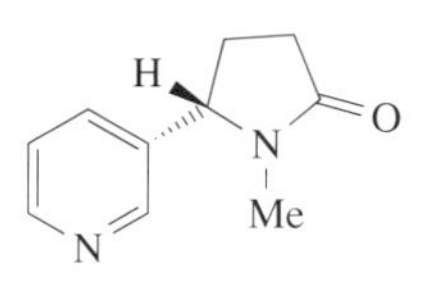

(**332**) Cotinine

During the 1940s there was considerable activity in the USA on synthesizing and testing of new nicotine derivatives. That was put aside with the arrival of DDT, but there has been more recent work.[528]

8.05.5.6.2 Sabadilla

Sabadilla, the powdered seeds of *Schoenocaulon officinale* A. Gray (= *Sabadilla officinarum* Brant and *Veratrum sabadilla* Retz, Liliaceae) from South America was used as an insecticide from the sixteenth century, and especially from 1850 to 1950. Its main active compounds are veratridine (**333**) and cevadine (**334**), esters of the nontoxic veracevine (**335**). Together they represent 0.8% of the commercial powder. Cevadine is the most toxic to insects (LD_{50} 0.5 μg g^{-1} for milkweed bugs) but it has a high mammalian toxicity (LD_{50} 5.8 mg kg^{-1} peritoneally in mice). Thrips are insects which are particularly difficult to control, so it is interesting that cevadine and veratridine are highly toxic to citrus thrips (*Scirtothrips citri*), with LC_{50} values of 18.25 ng cm^{-2} and 29.91 ng cm^{-2}, respectively.[533] Because the ester group confers toxicity to veracevine, variation of the ester is a promising way to greater activity. The 3,5-dimethoxybenzoyl ester of veracevine (**335**) has approximately twice the activity against insects but no greater mammalian toxicity than (**333**).[533] The closely related protoveratrine A (**336**) from the rhizomes of *Veratrum album* is also insecticidal (LD_{50} 2.2 μg g^{-1} for the milkweed bug) but with a high mammalian toxicity.[534] These alkaloids disappear from the leaf surface quite quickly when used as an insecticide. Crosby lists 20 compounds from *Schoenocaulon* and *Veratrum* with a wide spectrum of activity against insects, especially Hemiptera and Homoptera, but ineffective toward aphids.[535] The pharmacology of this group of alkaloids has been reviewed.[536] They operate by affecting sodium channels in membranes.[534]

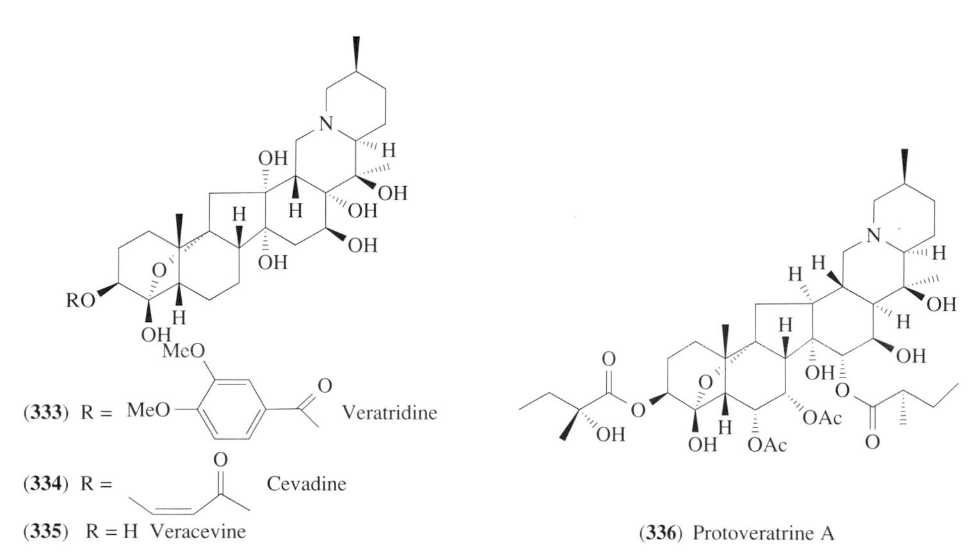

(**333**) R = MeO–C₆H₃(MeO)–CO Veratridine

(**334**) R = (Z)-2-methylbut-2-enoyl Cevadine

(**335**) R = H Veracevine

(**336**) Protoveratrine A

8.05.6 INSECT TOXINS

Insects have evolved many ways of repelling predators, and the use of chemicals is but one of them. Many of the chemicals used at close range are of only moderate toxicity, such as benzoquinones or salicylaldehyde, or they are substances with a physical effect (gummy or slimy), or have an unpleasant odor (to humans) such as quinoline or isobutyric acid. These are covered in the encyclopedic review of arthropod chemical defenses by Blum.[537] Another very good early review of the subject is by Eisner.[538] Insect alkaloids were reviewed in 1986.[539] Chemical toxins from insects can be divided into those stored and released from specialized glands, for example venoms, and those contained in the blood or gut or body generally, for example cantharidin or pederine, which are released by so-called reflex bleeding.

8.05.6.1 Cantharidin

Cantharidin (**337**) is a defensive secretion of meloid beetles. It forms about 0.25–0.5% of the body weight, and is stored in the hemolymph and male genitalia. The hemolymph of *Meloe proscarabeus* contains 26% cantharidin. When disturbed, the insect bleeds as a reflex from the leg joints. Cantharidin is also found in their eggs and larvae. It was obtained crystalline in 1810. Its structure has been known since 1904, and there are many syntheses available. Its structure is symmetrical, and it is not optically active. It is crystalline, m.p. 218 °C, and soluble in organic solvents and oils.

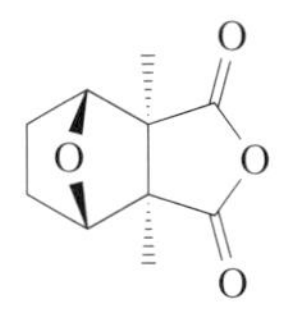

(**337**) Cantharidin

The effect of cantharidin on many insects is recorded, but concentrations of 0.5% or greater were required for lethality.[540] Cantharidin is highly toxic in humans (LDLo orally 428 mg). It is a vesicant and extreme irritant to all tissues. Since it is best known from *Lytta* (= *Cantharis*) *vesicatoria*, found frequently in Spain, crude cantharidin has been known as Spanish fly, and has an unwarranted reputation as an aphrodisiac. The cantharidin content of Spanish fly varies from zero to high values. There are many cases of death and severe injury through its use. It inhibits protein phosphatase 2A, which is important in signal transduction.[541]

Cantharidin is synthesized only by the males. Females acquire it from males through copulation, and it passes thence to eggs and larvae. A number of methods for its quantification by GC and HPLC have been described.[542] The biosynthesis of the compound has received much attention. Through use of labeled acetate, mevalonate, and multiply labeled farnesol, it has been shown to be formed through farnesol with loss of carbon atoms 1 and 5–7 (Scheme 17).[542]

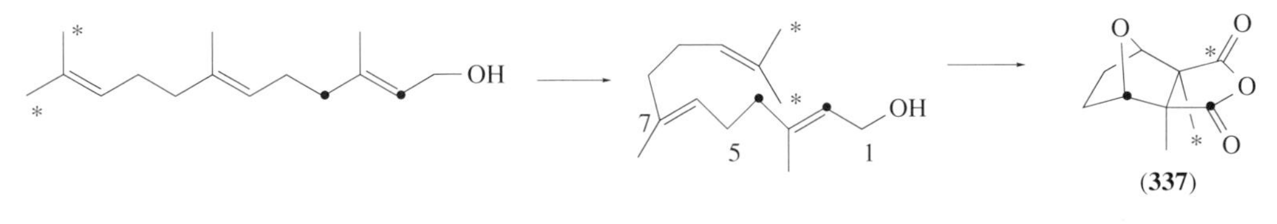

Scheme 17

Although many insects are deterred by the substance (as little as 10^{-5} M cantharidin in a sugar solution will stop ants from drinking it), some insects are attracted to cantharides beetles or the pure compound.[543] So-called canthariphilous beetles (Coleoptera) and flies (Diptera) take up and store cantharidin rather like some Lepidoptera sequester pyrrolizidine alkaloids from their plant hosts (see Section 8.05.6.8).[543]

The cantharid beetle (*Chauliognathus lecontei*) emits drops of secretion from glands on each side of its abdomen containing (8*Z*)-dihydromatricaria acid (**338**).[544]

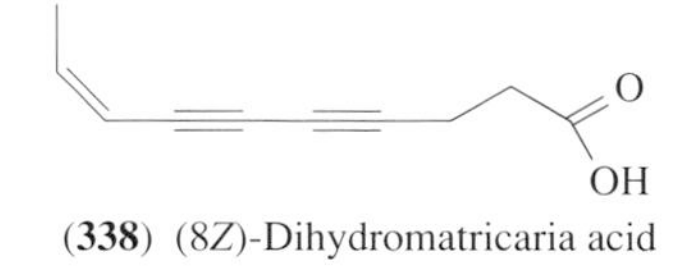

(**338**) (8Z)-Dihydromatricaria acid

8.05.6.2 Pederine

Pederine (**339**) is a powerful vesicant produced by a small number of Coleoptera, of the staphylinid family.[545] *Paederus fuscipes* of southern Europe, *P. sabalus* in Africa, and a few other *Paederus* spp. release hemolymph containing pederine when disturbed or crushed. Individual insects contain about 1 μg of pederine. The first proposed structure[546] was corrected[547] and confirmed by the crystal structure of a di-*p*-bromobenzoate. It is accompanied by much smaller quantities of pseudopederine (**340**) and pederone (**341**). It is soluble in polar organic solvents and crystallized from nonpolar solvents, m.p. 112–112.5 °C. It has been synthesized several times, including a total synthesis of (+)-pederine,[548] and NMR and mass spectral data are available. It is extremely toxic, with an LD_{50} of 2 mg kg^{-1} or less for humans. It inhibits mitosis and blocks protein synthesis in cell cultures at 1–10 ng ml^{-1}. Dettner has studied the ecology of cantharidin and pederine.[549] Studies of its biosynthesis demonstrate the incorporation of acetate and propionate, suggesting a polyketide intermediate.[550] The defensive secretions of other staphylinid beetles so far investigated are relatively poor toxins. Tables of these, and secretions of other beetle families up to about 1980, are given by Weatherston and Percy.[551]

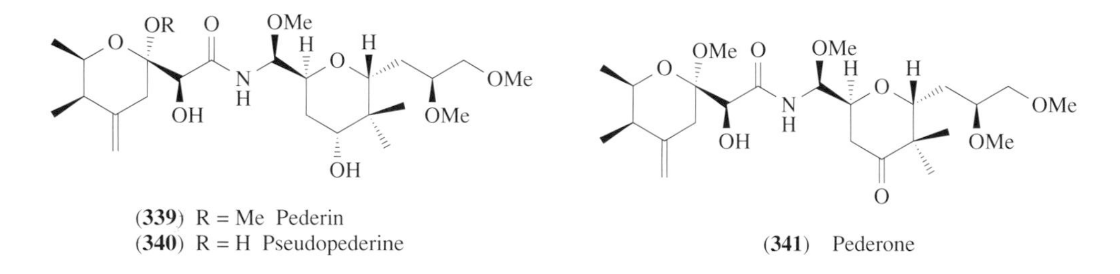

(**339**) R = Me Pederin
(**340**) R = H Pseudopederine

(**341**) Pederone

8.05.6.3 Coccinellines

Ladybirds (Coleoptera: Coccinellidae) also practice reflex bleeding, using a small number of alkaloids called coccinellines that have been studied by Pasteels and Daloze.[537,552,553] They have been comprehensively reviewed.[554] There are about 85 μg of coccinelline (**342**) in a beetle of the *Coccinella* spp., which gives it a bitter taste. It is accompanied by about 7 μg of precoccinelline (**343**). Convergine (**344**) and hippodamine (**345**) are found in *Hypodamia* spp. and others, myrrhine (**346**) in *Myrrha*, and propyleine (**347**) in *Propylaea* spp. Through feeding with [1-^{14}C]acetate and [2-^{14}C]acetate, they have been shown to be synthesized through a polyketide (Scheme 17). Adaline (**348**) is from the same polyketide but with a different linkage. More complex coccinellines have also been discovered. *Exochomus quadripustulus* produces (**349**), a dimer between a coccinelline and a novel azanaphthylene,[555] and *Chilorcorus cacti* produces two, (**350**) and (**351**), of very similar structure.[554] The Mexican bean beetle (*Epilachna varivestis*) is a coccinellid beetle that produces a number of alkaloids,[554] including the azamacrolide epilachnine (**352**). This has been shown to be biosynthesized from oleic acid and serine (Equation (2)).[556] Some coccinellids contain alkaloids of plant origin, sequestered by phytophagous insects and obtained from them by the labybirds preying on them. While all these compounds are repellent to many other arthropods and higher animals, detailed information about their toxicity is largely lacking.[557]

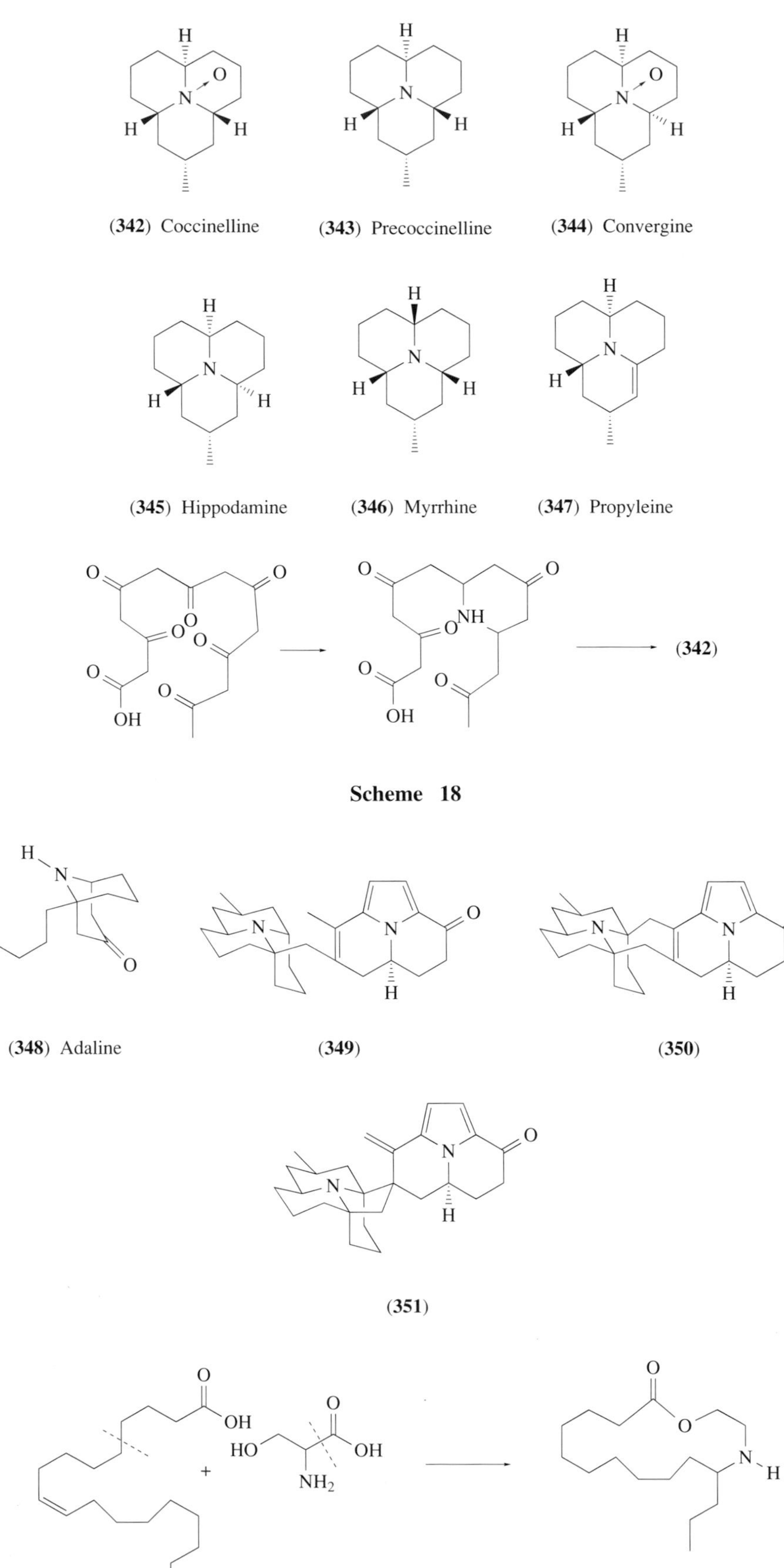

H
N
O
H
H
(342) Coccinelline
(343) Precoccinelline
(344) Convergine
(345) Hippodamine
(346) Myrrhine
(347) Propyleine
OH
NH
(342)
Scheme 18
(348) Adaline
(349)
(350)
(351)
HO
NH₂
(352) Epilachnine
(2)

8.05.6.4 Iridomyrmecin

Pavan isolated iridomyrmecin (**353**) from the pygidial glands of the ant *Iridomyrmex humilis*.[558] He claimed that it was insecticidal,[559] but later work showed its effect is very weak.[560] The monoterpenes isoiridomyrmecin (**354**), iridodial (**355**), dolichodial (**356**), and isodihydronepetalactone (**357**) and others were subsequently discovered in related ants; collectively these are called iridoids. They are the characteristic defensive compounds of the pygidial glands of dolichoderine ants, but they have been found elsewhere, for example in cerambycid and staphylinid beetles,[537,547] and the iridoid glycosides are found widely in plants (see Section 8.05.5.3.1). The structures of the iridoids have been reviewed by Weatherston[561] and Blum,[537] and more recently.[562] The biosynthesis from geraniol, suggested by Robinson and co-workers,[563] has been largely confirmed by the studies of Boland and co-workers.[564]

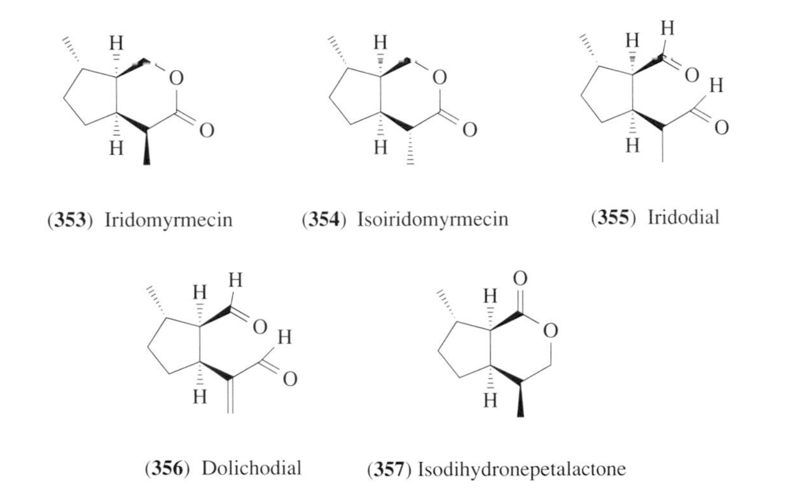

(**353**) Iridomyrmecin (**354**) Isoiridomyrmecin (**355**) Iridodial

(**356**) Dolichodial (**357**) Isodihydronepetalactone

Actinidine (**358**) is an alkaloid found with the iridoids in insects. It was for a time thought to be an artefact, since it can be formed from iridoids and amino acids by heating in a gas chromatograph, but it is a genuine product, occurring, for example, in the defensive glands of rove beetles (Staphylinidae).[565] No record of its pharmacological properties was found.

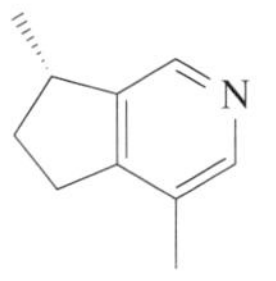

(**358**) Actinidine

8.05.6.5 Others

The blood of at least two species of Lepidoptera, *Arctia caja* and *Utethesia bella*, contains a choline ester, 3,3-dimethylacryloylcholine (**359**), which they excrete through defensive glands.[566] There is the equivalent of 1–2 mg of acetylcholine per insect. There are several examples of grasshoppers that excrete a mixture of hemolymph and air or glandular secretion when disturbed.[537,538]

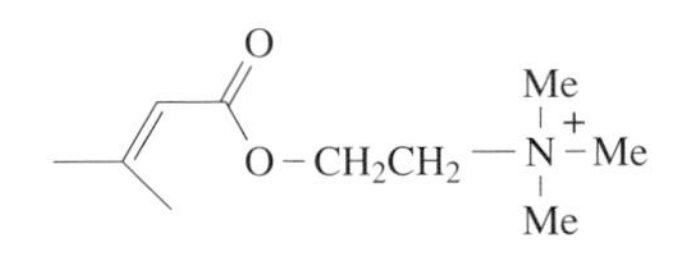

(**359**)

Eisner *et al.* found that fireflies (Coleoptera: Lampyridae) produce novel steroid pyrones called lucibufagins, of type (**360**), in which R can be H, acetate, propionate, or isobutyrate. The lucibufagins are released by reflex bleeding, and are responsible at least in part for the unpalatability of these insects for birds.[567] They bear a close resemblance to the bufadienolides of toads and plants.

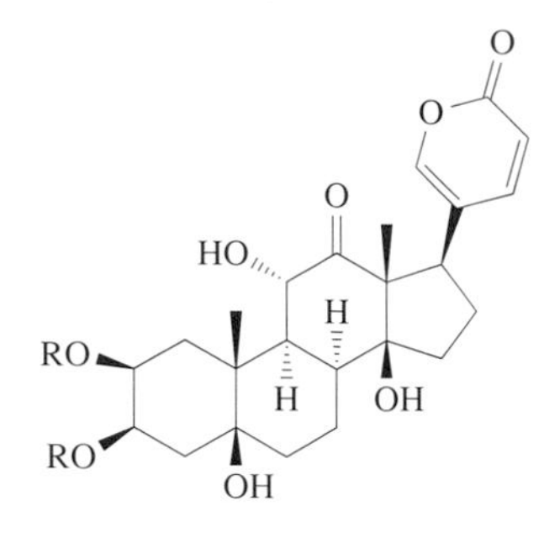

(**360**) Lucibufagins

8.05.6.6 Venoms

Venoms injected by an abdominal stinging apparatus occur only in the Hymenoptera. Some Coleoptera with biting or sucking mouthparts, and a few Neuroptera also produce venoms. The venoms of Hymenoptera have been thoroughly covered by Piek's book, including the chemistry and pharmacology.[568] By far the majority of hymenopteran venoms are proteinaceous. In the past, the quantities available for study were so small that progress in the identification of pure components was slow. Advances in isolation and analytical methods has made them more accessible to identification and testing. Table 5 shows a very interesting comparison between the lethality of hymenopteran venoms, reproduced from Schmidt.[569] Histamine is frequently present in the venom of bees and wasps. 5-Hydroxytryptamine is found in the venom of social bees and wasps, but not in that of ants or solitary wasps.

Table 5 Lethal capacity of venoms of ants and other Hymenoptera.

Species	*Venon* LD_{50} (mg kg^{-1})	*Amount of venom per individual* (μg)	*Colony size*	*Lethal capacity (number of stings for LD_{50} dose)* 25 g mammal	2 kg mammal
Mutillidae					
Dasymutilla klugii	71	420	Solitary	4.2	—[a]
Anthophoridae					
Diadasia rinconis rinconis	76	32	Solitary	59	—[a]
Pompilidae					
Pepsis formosa pattoni	65	2500	Solitary	0.65	—[a]
Apidae					
Apis mellifera	3.5	50	Many thousands	1.8	140
Vespidae					
Vespula (Paravespula) pennsylvanica	10.7	70	Thousands	3.8	310
Formicidae					
Pseudomyrmex mexicanus	8.0	16	Tens to hundreds	12.5	—[a]
Dinoponera grandis	38	550	Tens to hundreds	1.7	—[a]
Paraponera clavata	6.0	180	Hundreds	0.8	67
Eciton burchelli	10	60	Many thousands	4.2	330
Ectatomma quadridens	6.5	160	Hundreds	1.0	81
Pogonomyrmex maricopa	0.15	25	Many thousands	0.15	12

Source: Schmidt.[569]
[a]Lethal capacity is so low that lethality is not a likely threat from this species.

8.05.6.6.1 Honeybee venom

Honeybee (*Apis mellifera*) venom is now more thoroughly characterized (Table 6) than any other animal venom, although little is known about the effects of individual components on insects. It contains histamine, dopamine, and norepinephrine, which is fewer small-molecule substances than in wasp venom, but more than in ant venom. Roughly half the mass of the venom in *A. mellifera* and *A. cerana* is melittin, a 26-residue peptide (**361**), chiefly responsible for the lysis of red blood cells. The compound has been synthesized. It has detergent-like properties, lowers the surface tension of water, and tends to form a tetramer in solution.[570] There are small variations in structure of melittin from *A. florea* (**362**) and *A. dorsata* (**363**).[570] Melittin F is a fragment of melittin containing only residues 1–7 (Gly to Lys in (**361**)). Phosphalipase A_2 is the major allergen of venom, and its primary amino acid sequence is known,[568] but it has a carbohydrate portion attached of variable and still uncertain structure. The hyaluronidase catalyzes the hydrolysis of hyaluronic acid, the mucopolysaccharide of connective tissue. The other enzymes (Table 6) are less well identified. Apamine (**364**) is a neurotoxic octadecapeptide causing convulsions in mice at 0.5 mg kg^{-1} It has two disulfide bridges which reform on mild oxidation of the reduced form. It has been synthesized together with many analogues for intense pharmacological study, partly because it is the smallest known peptide neurotoxin. Peptide 401 (**365**), though very similar in structure to apamin, has very different biological properties, and its most notable effect is to release histamine from mast cells. Secapin is a 25-residue peptide with no apparent physiological effect. Tertiapin is a very minor component, a 21-residue peptide with a structure very similar to peptide 401, and similar but weaker activity. The functions of the minor components remain to be established. Young queens have venom half as lethal as that of workers, and queens 1–2 years old have inactive venom. The venom of *A. cerana* contains large amounts of oily (11*Z*)-eicosen-1-ol[571] (see Section 8.05.6.4(i)).

Table 6 Composition of the venom of the honeybee *Apis mellifera*.

Type of compound	*Compound*	*Percentage of venom by weight*	*Amount per insect* (nmol)
Small molecules	Histamine	0.66–1.6	5–10
	Dopamine	0.13–1.0	2.7–5.5
	Noradrenaline	0.1–0.7	0.9–4.5
Peptides	Mellitin	40–50	10–12
	Mellitin-F	0.01	0.0003
	Apamine	3	0.75
	Peptide 401	2	0.6
	Secapin	0.5	0.13
	Tertiapin	0.1	0.03
	Procamine A, B	1.4	2.0
Enzymes	Phospholipase A_2	10–12	0.23
	Hyaluronidase	1–2	0.03
	Acid phosphomonoesterase	1.0	—
	α-D-Glucosidase	0.6	—
	Lysophospholipase	1.0	0.03

Source: Banks and Shipolini.[570]

Gly-Ile-Gly-Ala-Val-Leu-Lys-Val-Leu-Thr-Thr-Gly-Leu-Pro-Ala-Leu-Ile-Ser-Trp-Ile-Lys-Arg-Lys-Arg--Gln-GlnNH$_2$ (**361**)
Ile Ala Thr Asn Lys (**362**)
Ile Ser Glu (**363**)

Thr-Glu-Pro-Ala-Lys
Cys-Asn-Cys
Ala
S-S S-S
Leu-Cys-Ala-Arg-Arg-Cys-Gln-Gln-His-NH$_2$

(**364**) Apamin

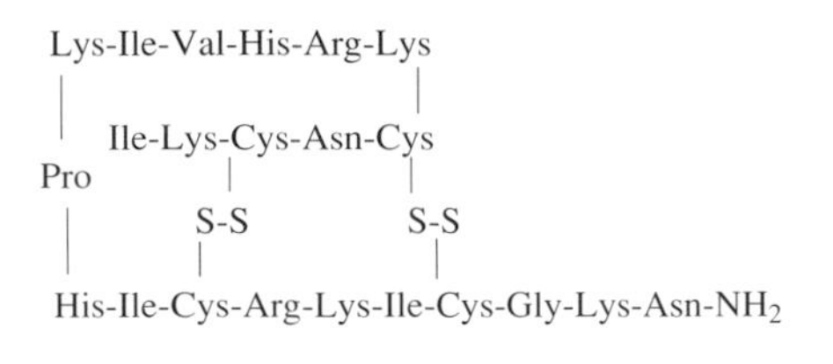

(**365**) Peptide 401

8.05.6.6.2 Other bees

Our knowledge of the venom of other bees is still poor.[572] Bumblebee venom has very similar phosphalipase A_2 and hyaluronidase to that of honeybee venom.

8.05.6.6.3 Wasp venoms

Social wasp venoms contain amines (histamine, tyramine, serotonin, and catecholamines), peptides, and proteins, including many hydrolases,[573] but quite different in nature from those in bee venom. The hornet *Vespa crabro* produces 19 mg of hydroxytryptamine per gram of venom sac.[574] Polyamines include putrescine, spermidine, and spermine. Pain-causing kinins, varying with species, but all related in structure to bradykinin, and mastoparans (mast cell degranulating peptides) which release histamine, and peptides that attract macrophages have been purified and identified.[573]

The great majority of known species of Hymenoptera are solitary wasps. Their venom is applied much more against other insects than that of the honeybee and social wasps.[575] *Philanthus triangulatus*, the bee wolf, paralyzes worker bees to provide live food for its larvae. The paralyzing substance was found not to be a peptide but a butyryltyrosine derivative of a polyamine, philanthotoxin-433 (**366**).[576] It and analogues have been synthesized for pharmacological study, and shown to be noncompetitive inhibitors of the nicotinic acetylcholine receptors.

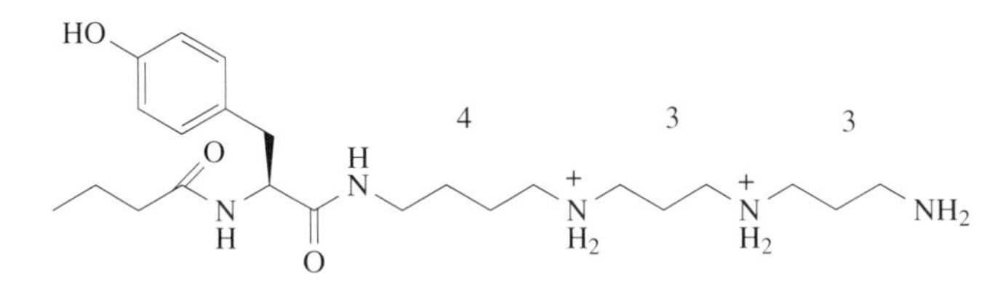

(**366**) Philanthotoxin-433

8.05.6.6.4 Ant venoms

The venom glands of ants have a greater diversity of function and chemistry than those of other Hymenoptera. Both Schmidt[569] and Blum[577] have reviewed the subject.

(i) Formic acid

The first identified of insect secretions was formic acid.[578] It is produced in the venom glands of formicine ants at up to 60% aqueous solution and sprayed, since formicine ants do not possess a sting, during defense or attack. It has been shown to be biosynthesized chiefly from serine, using both the α- and β-carbon atoms, or the α-carbon atom of glycine, via N^5,N^{10}-methylentetrahydrofolate and 10-formyltetrahydrofolate.[579]

Formic acid is also used by a long list of carabid beetles:[580] it is, for example, sprayed by the carabid beetle *Galerita janus* from two abdominal glands near the anus. Frequently, formic acid or acetic acid is accompanied by a linear hydrocarbon mixture to aid spreading of the acid on the nonpolar cuticular surface. In ants it is presumed that the hydrocarbons of the Dufour gland are discharged with the formic acid. The defensive secretion of many larvae of Lepidoptera of the family Notodontidae contain formic acid, for example *Heterocampa manteo* discharges 27% aqueous formic acid with 2-undecanone and 2-tridecanone as droplets suspended in it, while *Schizura concinna* uses formic acid containing 2-tridecanone, decyl, and dodecyl acetates.[581] Such two-phase secretions seem to be common in insect offense and defense.

(ii) Alkaloids

A few genera of ants produce alkaloidal venoms. The alkylpyrrolidines, -pyrrolines, -piperidines, and -tetrahydropyridines of *Solenopsis* ants have been reviewed.[569,582] From a chemical point of view the genus *Solenopsis* is conveniently divided into two subgenera. *Solenopsis* (*Solenopsis*) spp., accidently imported into the USA and known there as imported fire ants because of their fierce sting, produce 2-alkyl-6-methylpiperidines (**367**), a few tetrahydropyridines (**368**), and *N*-methylpiperidines (**369**), where *n* varies from 6 to 14 and the side chain may contain a double bond in the middle of the chain. *Trans* isomers are produced chiefly by *S. invicta* and *cis* isomers by *S. xyloni* and *S. geminata*. About 95% of the venom consists of these alkaloids, with about 19 µg in a single worker ant.[583] The venom produces necrotic, hemolytic, antibiotic, and toxic properties.[584] There are droplets of aqueous protein in the oily alkaloids which may be responsible for the allergenic properties of the venom.

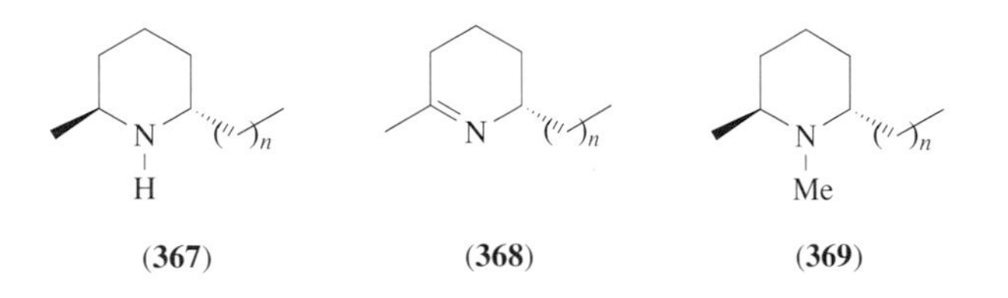

The subgenus *Solenopsis* (*Diplorhoptrum*), known as thief ants, produce dialkyl-pyrrolines and -pyrrolidines (**370**)–(**373**), where *n* can vary between 4 and 8 and *m* can vary from 1 to 5, with one alkyl group always having an even number of carbon atoms and one having an odd number, and only the *trans* arangement of substituents. The *Diplorhoptrum* ants are not particularly troublesome for stinging, and their venom contains relatively small amounts of the alkaloids. Pyrroline and pyrrolidine alkaloids have also been found in *Megalomyrmex foreli* (Myrmicinae).[585] A few simple pyrrolizidine (**374**) and indolizidine (**375**) alkaloids have been reported from *Solenopsis* and *Monomorium* spp.[586] The compounds have been synthesized, but their pharmacological action has not been reported.

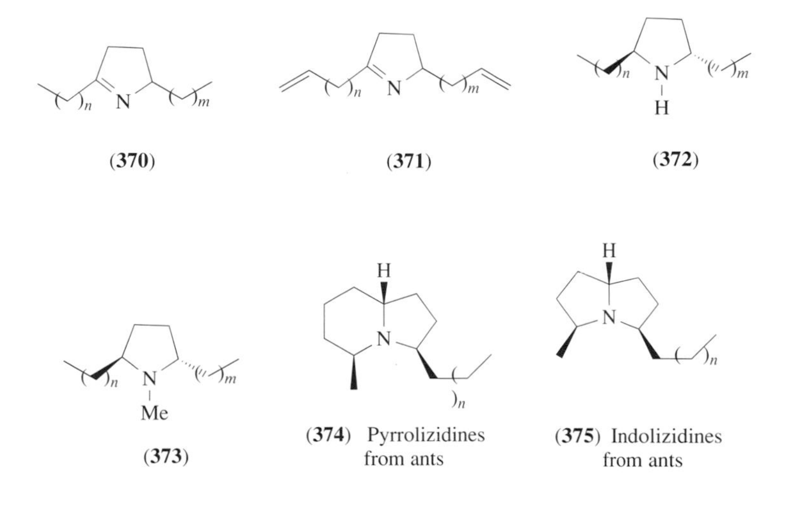

More recently, tricyclic alkaloids, called tetraponerines (**376**) have been found in *Tetraponera* sp. (Pseudomyrmecinae), which have a spatulate sting, so the venom is quite clearly spread on the surface of prey and not injected.[587] Both epimers at positions 5 and 10a are known.

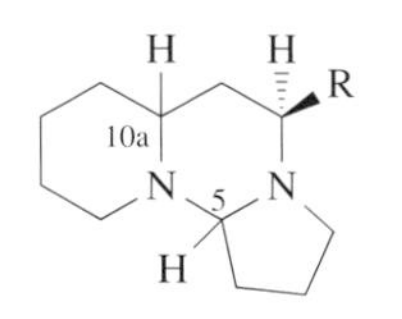

(**376**) Tetraponerines
R = C_3H_7 , C_4H_9 or C_5H_{11}

Francke *et al.* have identified another series of oxygenated indolizidine alkaloids of type (**377**) in the very large poison glands of species of *Myrmicaria*. The alkaloids, when applied to their cuticle, immobilize antagonists. In the African ant *Myrmicaria eumenoides* both epimers at position 5 were found, but epimerization at this position is rapid.[588] In some colonies of *M. opaciventris* they identified pyrroloindolizidines of type (**378**) and dimers of these, and in other colonies they found tiny amounts of the monomers and dimers and large amounts of trimer (**379**).[589] In *M. striata* the same decacyclic alkaloid of molecular mass 663 (**379**) was found.[590] Their structures were deduced from NMR spectral studies. The substance (**379**) is poisonous to termites, applied topically, but the presence of some monoterpenes is crucial for spreading and lowering the viscosity of the venom. The proposed biosynthetic route to this series is illustrated in Scheme 19.

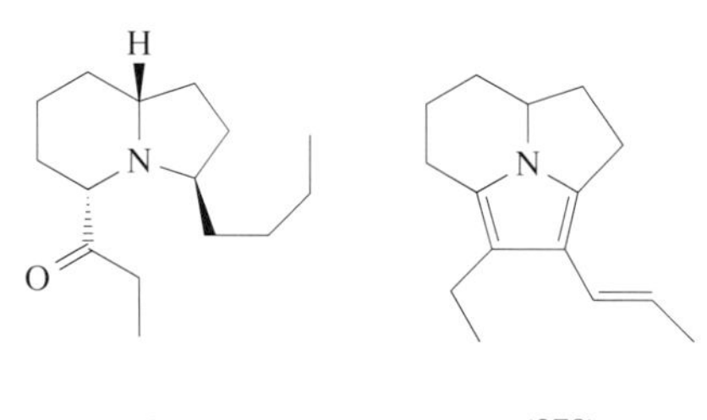

(**377**) (**378**)

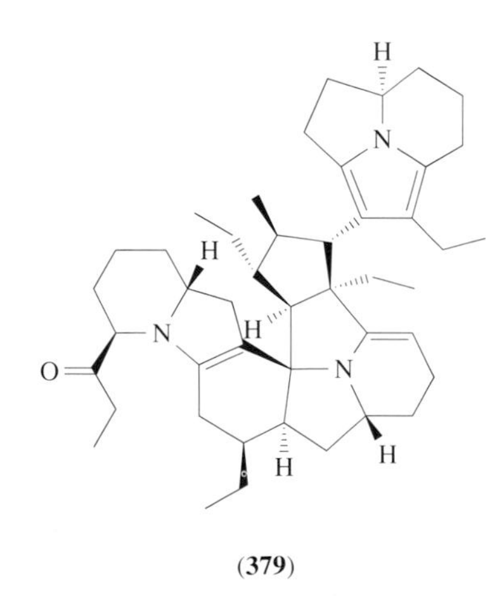

(**379**)

A few species of *Messor* and *Aphenogaster*, scattered in different parts of the world, have venoms of tobacco alkaloids. In *Messor ebeninus*, *Aphenogaster fulva*, and *A. tennesseensis* it is chiefly anabaseine (**380**) with some anabasine,[582] in *Messor capitatus* it is anabasine, and in *M. mediorubra* it is anabasine with very small amounts of anabaseine and *N*-methylanabasine (E. D. Morgan *et al.*, unpublished findings). Nothing is yet known about the biosynthesis of these nicotine alkaloids in insects.

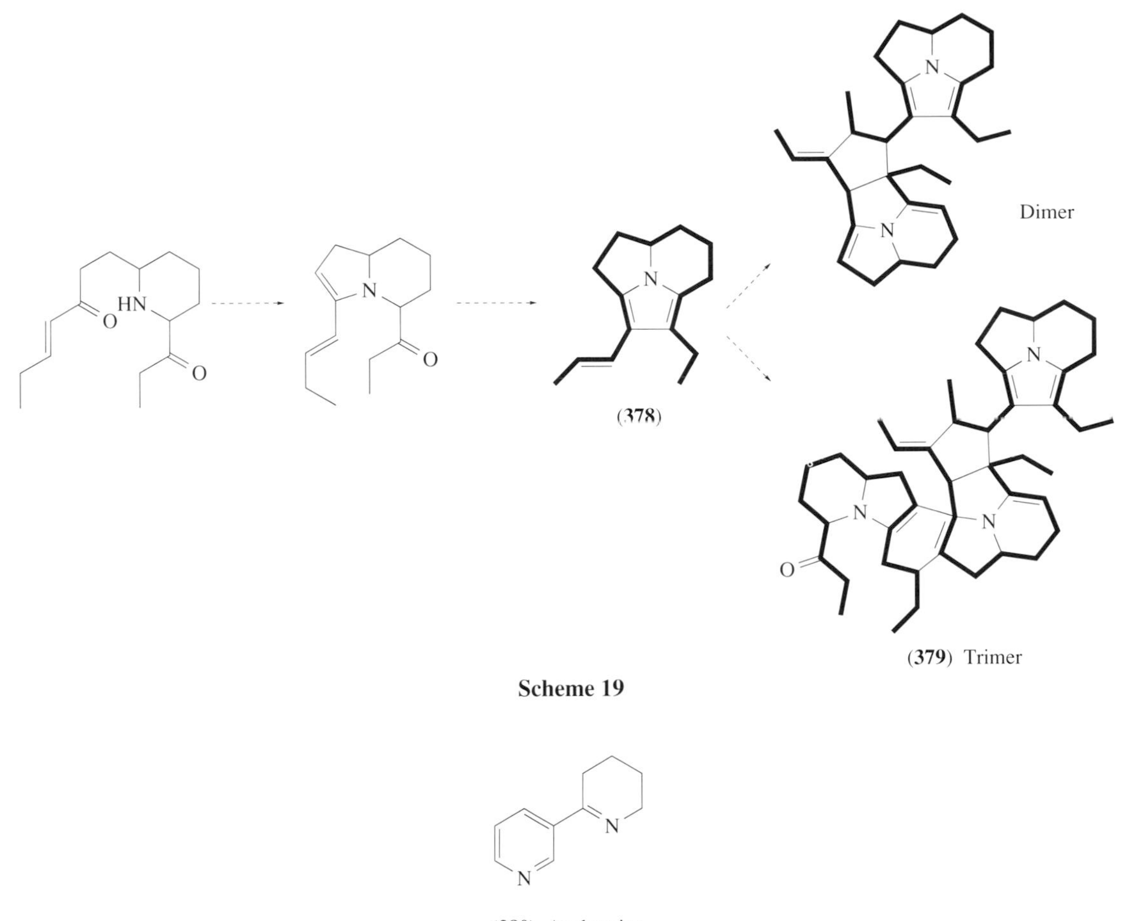

Scheme 19

(**380**) Anabaseine

(iii) Protein venom

The great majority of venomous ant species nevertheless have proteinaceous venoms. Little is yet known of their composition. Of the small-molecule substances found in bees and wasps, only histamine is found in ant venom.

A 25 amino acid peptide called poneratoxin (**381**) has been identified as the most neurotoxic fraction in the venom of *Paraponera clavata* (Ponerinae).[597]

Phe-Leu- Pro-Leu-Leu-Ile-Leu-Gly-Ser-Leu-Leu-Met-Thr-Pro-Pro-Val-Ile-Glu-Ala-Ile- His-Asp-Ala-Gln-Arg-NH_2

(**381**) Poneratoxin

(iv) Others

Crematogaster ants (Myrmicinae) produce in their Dufour glands, and expose on their sting lance, a complex mixture of cross-conjugated polyenones (**382**), which have been called long-chain electrophilic contact poisons.[592] The long-chain portion has one, two, or no double bonds. They are stored as acetates, and on release are hydrolyzed to the alcohols and oxidized to the aldehydes. The mixture is applied to the surface of attackers with the spatulate sting, which the *Crematogaster* ants are able to bring forward over their bodies. The aldehydes have been shown to be highly toxic by topical application on *Myrmica* ants.[593]

Similar cross-conjugated dienones, gyrinidal (**383**) (48% of the mixture in two species), gyrinidione (**384**) (36% of the contents), and others, were found much earlier in gyrinid water beetles.[594,595] They

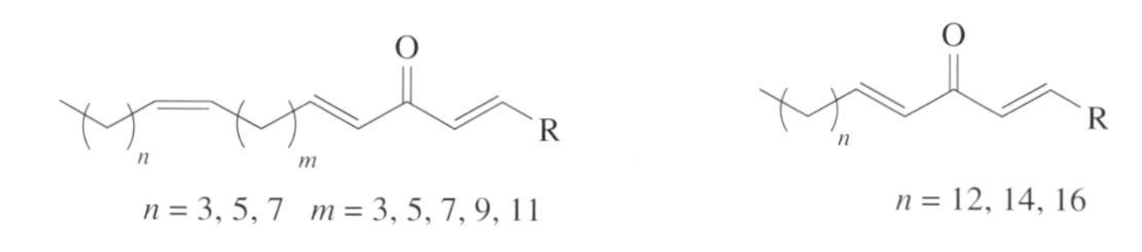

(382) R = CH_2OAc, CH_2OH, CHO

were as toxic to fish as cortexone (See Section 8.05.6.7). The resemblance in structure to the iridoids (See Section 8.05.6.4) is noteworthy. The cephalic secretion of the stingless bee *Trigona* (*Oxytrigona*) *mellicolor* (the fire bee), which plunders nests of *Apis*, contains formic acid and **(385)** and **(386)**.[596] The secretion causes severe blistering on human skin, but this has not been attributed to any particular component.

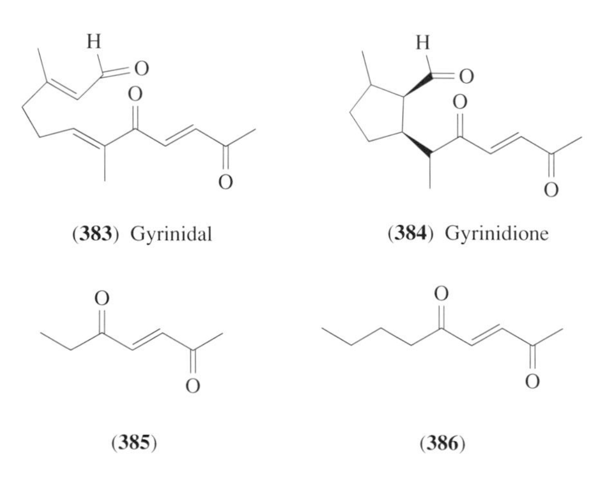

(383) Gyrinidal

(384) Gyrinidione

(385)

(386)

8.05.6.6.5 Lepidopteran venoms

The larvae of at least 13 families of Lepidoptera possess hairs filled with venom that easily become detached when the larva is disturbed. They cause a stinging sensation and eruptions on animal skin (urtication). Histamine is certainly present in many of these urticating hairs. In *Dirphia* spp. the histamine concentration ranges from 75 mg g^{-1} to 200 mg g^{-1}. Little is yet known of their chemistry, but the hairs are probably filled with protein venom. The subject has been reviewed.[597]

8.05.6.6.6 Hemipteran venoms

Among the Hemiptera, defensive glands are common, but the substances are chiefly volatile alcohols, aldehydes, acids, and esters.[598] Members of one group of the family Lygaedae contain cardiac glycosides sequestered from their food.[599] About 2500 species of Hemiptera have adopted predatory behavior and can have venom in their salivary glands. Most notable among them are the Reduviidae or assassin bugs. They can inflict a painful sting, but the content of their saliva has not been extensively studied.

8.05.6.6.7 Cockroaches

Cockroaches have a variety of defensive gland secretions, and because they come into contact with human food can be allergenic, but the defensive compounds, such as 2-hexenal or quinones, are not notably toxic.[600]

8.05.6.7 Toxins from Defensive Glands

Defensive sterols are known to be produced only in beetles, and then only in the three families Dytiscidae, Chrysomelidae, and Lampyridae, with the great majority in the Dytiscidae. It is suggested that the evolutionary production of these sterols enable the Dytiscidae to adopt an aquatic environment in the presence of many vertebrate predators. Blum lists 19 sterols, all from the Dytiscidae.[601] They frequently secrete C_{21} corticosteroids from neck glands, and in one species produce both estrone (**387**) and testosterone (**388**). The most common steroid encountered is cortexone (11-deoxycorticosterone or 4-pregen-21-ol-3,20-dione) (**389**). Water beetles produce from 3 mg to 1 mg each. It is reported to be a powerful narcotic for freshwater fish. In humans it causes excessive fluid and salt retention and hypertension.

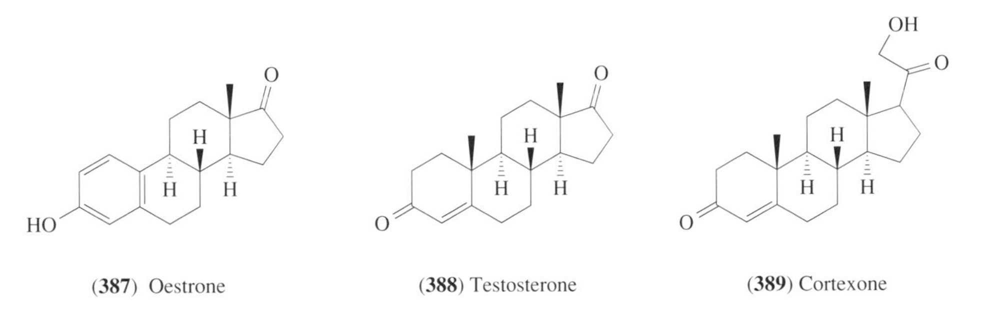

(**387**) Oestrone (**388**) Testosterone (**389**) Cortexone

Chrysomelid beetles of the genera *Chrysolina*, *Chrysochloa*, and *Dlochrysa* contain mixtures of cardenolides, which they appear to make themselves.[602] For example, the glands of *Chrysolina coerulans* contain periplogenin (**390**), sarmentiogenin (**391**), and bipindogenin (**392**), while *C. herbacea* contains both these and their 3-xylosides.

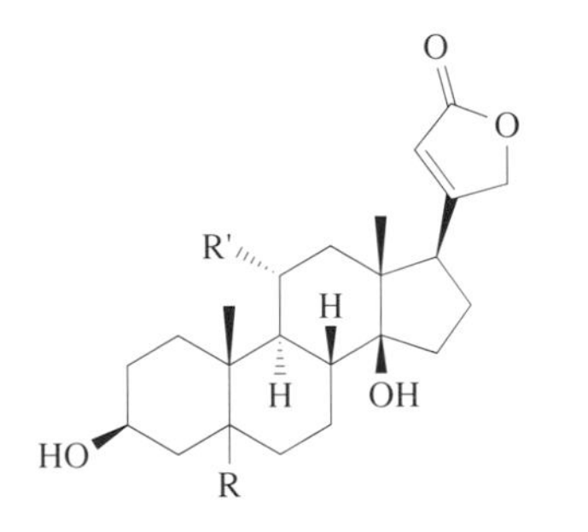

(**390**) R = α-H, R' = OH Sarmentiogenin
(**391**) R = β-OH, R' = H Periplogenin
(**392**) R = β-OH, R' = OH Bipindogenin

Springtails (Collembola) are primitive wingless insects, formerly presumed to be devoid of chemical defenses, but a giant springtail (*Tetrodontophora bielanensis*) has been shown to secrete droplets of sticky defensive fluid, containing pyridopyrazines, chiefly (**393**) and (**394**), from tiny pores distributed over its body.[603] The compounds are not present in its food, and are therefore presumed to be synthesized by the insect. The synthesized compounds, applied to the cuticle of a beetle (*Nebria brevicollis*) induced disorientation and cleaning action in the beetle. Compound (**394**) was the more active.[604]

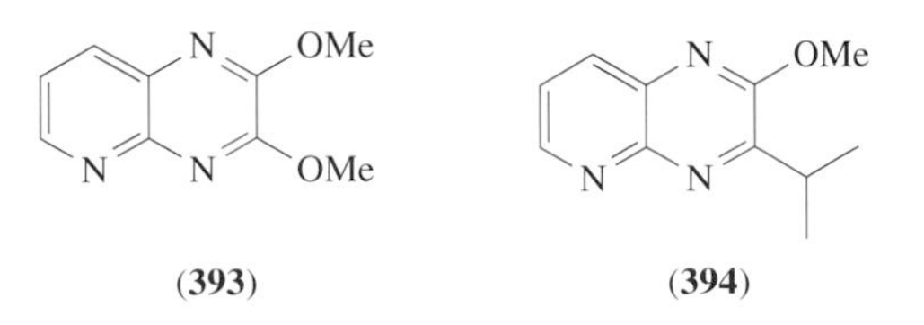

(**393**) (**394**)

8.05.6.8 Sequestration of Plant Substances

There are many examples of groups of insects which feed on toxic plants and are able to escape the effects of the toxins but retain them in their bodies, sometimes modifying them slightly, to make the insect in turn toxic to predators.[605] Rothschild lists at least six orders of insects in which some species can do this.[605] The species are highly selective in what they sequester, and the range is wide, but cardenolides (e.g., digitalin (**224**) and ouabain (**395**) from *Strophanthus* and *Acokanthera* spp. of Apocynaceae) and pyrrolizidine alkaloids (e.g., senicionine (**319**), senkirkine (**320**), and seneciphylline (**396**) from *Senecio* spp. and monocrotaline (**397**) from *Crotolaria* spp.) are the most commonly encountered. There have been extensive studies of butterflies and moths that feed on groundsel (*Senecio vulgaris*), which contains principally senecionine (**319**), and ragwort (*S. jacobaea*), with chiefly seneciphylline (**396**).[606] Eisner and Meinwald have shown how the alkaloids are sequestered, how the female checks the male suitor for alkaloids and how they are passed to the eggs via the sperm.[607] Studies of a sawfly (*Rhadinoceraea nodicornis*) which feeds on *Veratrum album* found that the ceveratrine alkaloids (see Section 8.05.5.6.2) can be treated in one of four ways: they can be sequestered, they can be partially metabolized and then sequestered, they can be secreted intact, or they can be secreted degraded.[608] Frequently the sequestered compounds are slightly modified in structure. The chrysomelid beetle *Diabrotica speciosa*, which normally feeds on curcubit plants and accumulates the bitter cucurbitacins, when fed on [^{14}C]cucurbitacin B (**398**) converted it to 23,24-dihydrocucurbitacin D, that is, removes the C-25 acetate and reduces the side chain double bond.[346]

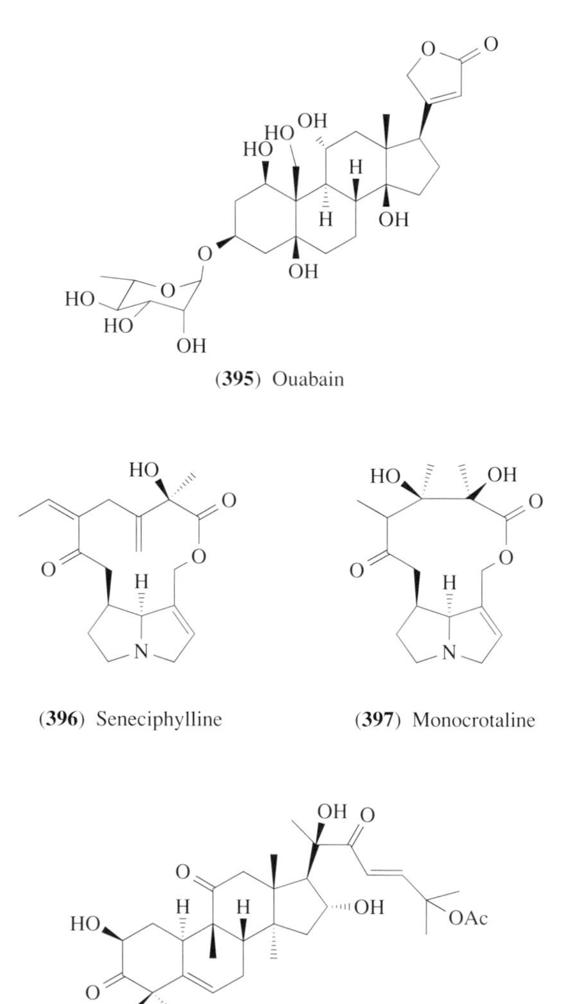

(**395**) Ouabain

(**396**) Seneciphylline

(**397**) Monocrotaline

(**398**) Cucurbitacin B

Rothschild has studied the monarch butterfly (*Danaus plexippus*), which feeds on milkweed (*Asclepias curassivaca*), containing cardiac glycosides, and stores them essentially unchanged, and apparently at no cost to the development of the insect.[609] The butterfly *Parnassius phoebus* feeding on *Sedum stenopetalum* accumulates sarmentosin (**399**), a bitter-tasting cyanoglucoside of unknown toxicity, as much as 500 μg per insect, stored chiefly in the wings and eggs.[610] Arctiid moths, unlike the sequestering butterflies, feed on a wide range of plants containing pyrrolizidine alkaloids, cardenolides, or others,[611] for example *Seiractia echo* feeds on *Cycas* spp. (Cycadaceae), which contains cycasin (**400**), which with its aglycone methylazoxymethanol is a potent toxin, carcinogen, and mutagen for vertebrates.[612] On the other hand, some species of *Oreana* leaf beetles, which do not live on alkaloid-containing plants, make protective cardenolides from dietary sterols. Other species are adaptable, and can make cardenolides and feed on senecionid plants, sequester alkaloids, and use both for protection, while still others are only able to sequester the pyrrolizidine alkaloids.[613] Sinigrin (**401**), found widely in Cruciferae species, is itself not particularly toxic, but allyl isothiocyanate (**402**), released from it by enzymes when the leaves are eaten, is very toxic and mutagenic. Some Lepidoptera are immune to the effects of (**402**), and use it to locate their host plants, but even they experience toxicity if the levels of sinigrin are raised above what is normally encountered. Aristolochic acid (**403**), a nitrophenanthrene from *Aristolochia* spp. which causes cardiac and respiratory arrest, is sequestered by the papilionid butterfly (*Pachlioptera aristolochiae*).[614] The sequestration of clerodendrin by the turnip sawfly has already been noted (see Section 8.05.5.3.4(i)).

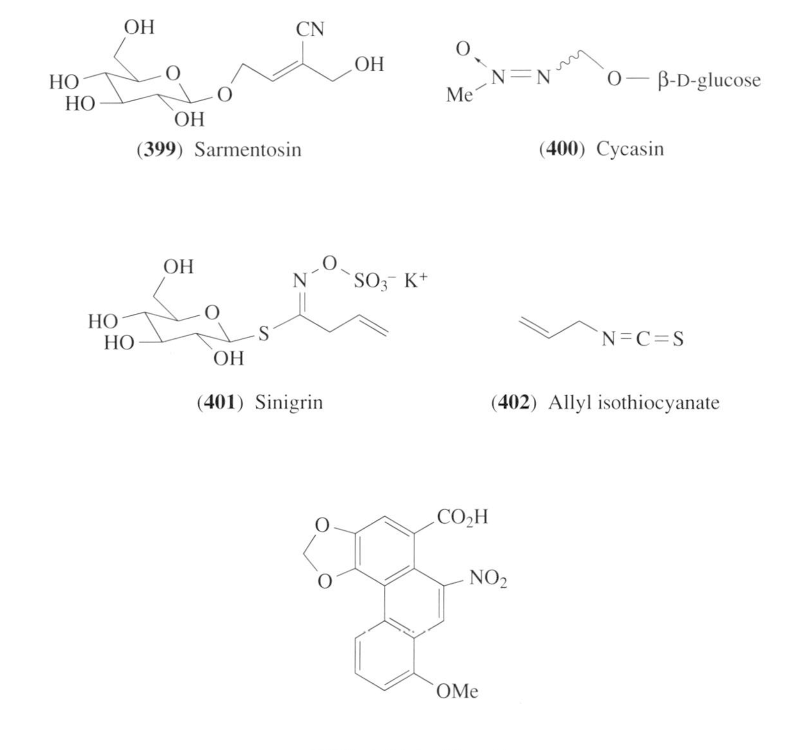

(**399**) Sarmentosin

(**400**) Cycasin

(**401**) Sinigrin

(**402**) Allyl isothiocyanate

(**403**) Aristolochic acid

Some insects produce regurgitates or defecate when disturbed. The case of the grasshoppers *Poekiloceris* or *Romalea*, which probably disgorge cardenolides from their host plants in this way, are frequently quoted. The unusual C_{13} allenic ketone romallenone, secreted in an odorous foam by *Romalea microptera* along with quinones and phenols, does not appear to have notable toxicity, but the secretion is simply repellent to possible predators.

While oleandrin (**404**) is the principal cardenolide of *Nerium oleander*, the bug *Aspidiotus nerii* feeding on the plant sequesters adynerin (**405**), a minor constituent.[615] The aphid *Aphis nerii* (Homoptera) feeding on *N. oleander* sequesters three of its cardiac glycosides.[616] Then the ladybird *Coccinella unidecimpunctata* sequesters cardenolides from the aphids, but another ladybird, *C. septempunctata*, feeding on the aphids did not contain cardenolides.[617] The male of the beetle *Neopyrochroa flabellata* (Pyrochroidae) eats other beetles producing cantharidin and offers a little cantharidin to the female before copulation. Females prefer mates containing cantharidin.[618] It is feasible that one sample of alkaloids can protect up to four groups of organisms in turn. A plant may produce pyrrolizidine alkaloids, which are taken up and stored in their bodies by insects feeding on the plant. These insects are in turn preyed upon by coccinellid beetles, which devour the insects but retain the alkaloids for protection from some predators. But they may in turn be eaten by frogs, which again retain the alkaloids in their skin, still for protection from predators.[557] The subject of sequestered plant substances has been reviewed by Blum,[619] and by Rosenthal and Berenbaum.[620]

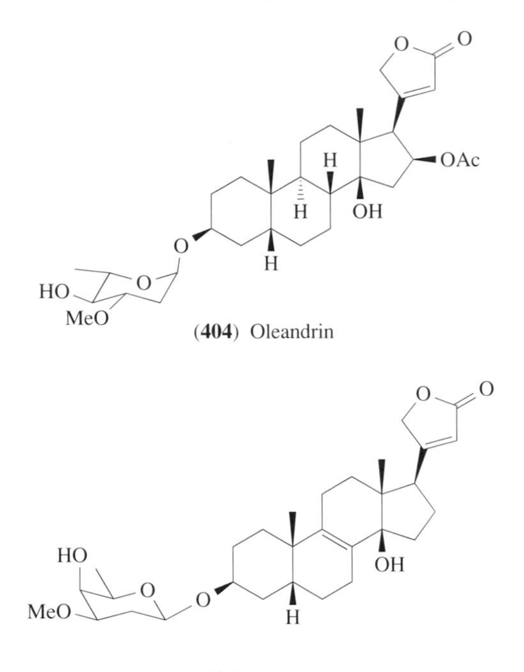

(**404**) Oleandrin

(**405**) Adynerin

8.05.6.9 A Parting Shot

We must finally mention the special case of bombardier beetles (*Brachynus* spp.), which, when attacked, react with a little on-the-spot chemistry. They hold in an abdominal chamber a mixture of hydroquinone and 25% aqueous hydrogen peroxide.[621] On disturbance, a muscular contraction mixes this with some catalase and peroxidase enzymes in an outer chamber, producing quinone, oxygen, and water almost instantly and with sufficient heat to raise the temperature of the sudden discharge to 100 °C. There is an audible "pop." Moreover the insect can direct it anywhere through 360°.

8.05.7 REFERENCES

1. H. F. Nijhout, "Insect Hormones," Princeton University Press, Princeton, NJ, 1994.
2. N. Wakabayashi and R. M. Waters, in "Handbook of Natural Pesticides," eds. E. D. Morgan and N. B. Mandava, CRC Press, Boca Raton, FL, 1985, vol. III, part A, p. 87.
3. V. B. Wigglesworth, "The Physiology of Insect Metamorphosis," Cambridge University Press, Cambridge, 1954.
4. V. B. Wigglesworth, *Q. J. Microbiol. Sci.*, 1936, **79**, 91.
5. V. B. Wigglesworth, *J. Insect Physiol.*, 1958, **2**, 73.
6. L. I. Gilbert and H. A. Schneidermann, *Trans. Am. Microsc. Soc.* 1960, **79**, 38.
7. G. B. Staal, in "Insect Juvenile Hormones: Chemistry and Action," eds. J. J. Menn and M. Beroza, Academic Press, New York, 1972, p. 69.
8. W. S. Bowers, M. J. Thompson, and E. C. Uebel, *Life Sci.*, 1965, **4**, 2323.
9. C. M. Williams, *Nature*, 1956, **178**, 212.
10. A. S. Meyer, H. A. Schneidermann, and L. I. Gilbert, *Nature*, 1965, **206**, 272.
11. H. Röller and J. S. Bjerke, *Life Sci.*, 1965, **4**, 1617.
12. H. Röller, K. H. Dahm, C. C. Sweely, and B. M. Trost, *Angew. Chem., Int. Ed. Engl.*, 1967, **6**, 179.
13. K. H. Dahm, B. M. Trost, and H. Röller, *J. Am. Chem. Soc.*, 1967, **89**, 5292.
14. D. J. Faulkner and M. R. Petersen, *J. Am. Chem. Soc.*, 1971, **93**, 3766.
15. K. Nakanishi, D. A. Schooley, M. Koreeda, and J. Dillon, *J. Chem. Soc., Chem. Commun.*, 1971, 1235.
16. A. S. Meyer, E. Hanzmann, and R. C. Murphy, *Proc. Natl. Acad. Sci. USA*, 1971, **68**, 2312.
17. B. M. Trost, in "Insect Juvenile Hormones: Chemistry and Action," eds. J. J. Menn and M. Beroza, Academic Press, New York, 1972, p. 217.
18. A. S. Meyer, H. A. Schneidermann, E. Hanzmann, and J. H. Ko, *Proc. Natl. Acad. Sci. USA*, 1968, **60**, 853.

19. A. S. Meyer, E. Hanzmann, H. A. Schneidermann, L. I. Gilbert, and M. Boyette, *Arch. Biochem. Biophys.*, 1970, **137**, 190.
20. K. J. Judy, D. A. Schooley, L. L. Dunham, M. S. Hall, B. J. Bergot, and J. B. Siddall, *Proc. Natl. Acad. Sci. USA*, 1973, **70**, 1509.
21. E. D. Morgan, K. Parry, and R. C. Tyler, *Insect Biochem.*, 1979, **9**, 117.
22. B. J. Bergot, G. C. Jamieson, M. A. Ratcliff, and D. A. Schooley, *Science*, 1980, **210**, 336.
23. B. J. Bergot, F. C. Baker, D. C. Cerf, G. C. Jamieson, and D. A. Schooley, "Juvenile Hormone Biochemistry: Action, Agonism and Antagonism," Elsevier-North Holland, Amsterdam, 1981.
24. M. A. Liu, G. L. Jones, J. G. Stoffolano, and C. M. Yin, *Physiol. Entomol.*, 1988, **13**, 69.
25. D. S. Richard, S. W. Applebaum, T. J. Sliter, F. C. Baker, D. A. Schooley, C. C. Reuter, V. C. Henrich, and L. I. Gilbert, *Proc. Natl. Acad. Sci. USA*, 1989, **86**, 1421.
26. A. J. Herlt, R. W. Rickards, R. D. Thomas, and P. D. East, *J. Chem. Soc., Chem. Commun.*, 1993, 1497.
27. K. Mori, *Recent Dev. Chem. Nat. Carbon Compounds*, 1979, **9**, 9.
28. K. Mori and H. Mori, *Tetrahedron*, 1987, **43**, 4097.
29. K. Mori and M. Fujiwhara, *Tetrahedron*, 1988, **44**, 343.
30. H. Kosugi, O. Kanno, and H. Uda, *Tetrahedron: Asymmetry*, 1994, **5**, 1139.
31. R. W. Rickards and R. D. Thomas, *Tetrahedron Lett.*, 1993, **34**, 8369.
32. M. G. Peter, K. H. Dahm, and H. Röller, *Z. Naturforsch.*, 1975, **31c**, 129.
33. H. Rembold, H. Hagenguth, and J. Rascher, *Anal. Biochem.*, 1980, **101**, 356.
34. B. J. Bergot, M. A. Ratcliff, and D. A. Schooley, *J. Chromatogr.*, 1981, **204**, 231.
35. F. C. Baker, in "Morphogenetic Hormones of Arthropods," ed. A. P. Gupta, Rutgers University Press, New Brunswick, NJ, 1990, part 1, p. 389.
36. Z.-Y. Huang, G. E. Robinson, and D. W. Borst, *J. Comp. Physiol.*, 1994, **174**, 731.
37. W. G. Goodman, A. P. Orth, Y. C. Toong, R. Ebersohl, K. Hiruma, and N. A. Granger, *Arch. Insect Biochem. Physiol.*, 1995, **30**, 295.
38. N. A. Granger and W. G. Goodman, in "Immunological Techniques in Insect Biology," eds. L. I. Gilbert and T. A. Miller, Springer-Verlag, New York, 1988, p. 215.
39. D. A. Schooley, K. J. Judy, B. J. Bergot, M. S. Hall, and J. B. Siddall, *Proc. Natl. Acad. Sci. USA*, 1973, **70**, 2921.
40. M. G. Peter and K. H. Dahm, *Helv. Chim. Acta*, 1975, **58**, 1037.
41. R. C. Jennings, K. J. Judy, and D. A. Schooley, *J. Chem. Soc., Chem. Commun.*, 1975, 21.
42. D. A. Schooley and F. C. Baker, in "Comprehensive Insect Physiology, Biochemistry and Pharmacology," eds. G. A. Kerkut and L. I. Gilbert, Pergamon, Oxford, 1985, vol. 7, p. 363.
43. B. D. Hammock and G. B. Quisstad, in "Progress in Pesticide Biochemistry," eds. D. H. Hutson and T. R. Roberts, Wiley, New York, 1981, vol. 1, p. 1.
44. P. Moshitzky and S. W. Applebaum, *Arch. Insect Biochem. Physiol.*, 1995, **30**, 225.
45. P. P. Halarnkar, G. P. Jackson, K. M. Straub, and D. A. Schooley, *Experientia*, 1993, **49**, 988.
46. H. F. Nijhout, "Insect Hormones," Princeton University Press, Princeton, NJ, 1994, p. 115.
47. N. A. Granger, W. P. Janzen, and R. Ebersohl, *Insect Biochem. Mol. Biol.*, 1995, **25**, 427.
48. A. S. Meyer, in "Insect Juvenile Hormones: Chemistry and Action," eds. J. J. Menn and M. Beroza, Academic Press, New York, 1972, p. 317.
49. M. Jacobson, M. Beroza, D. L. Bull, H. R. Bullock, W. F. Chamberlain, T. P. McGovern, R. E. Redfern, R. Sarmiento, M. Schwarz, P. E. Sonnet, N. Wakabayashi, R. M. Waters, and J. E. Wright, in "Insect Juvenile Hormones: Chemistry and Action," eds. J. J. Menn and M. Beroza, Academic Press, New York, 1972, p. 249.
50. F. M. Pallos and J. J. Menn, in "Insect Juvenile Hormones: Chemistry and Action," eds. J. J. Menn and M. Beroza, Academic Press, New York, 1972, p. 303.
51. B. D. Hammock, in "Comprehensive Insect Physiology, Biochemistry and Pharmacology," eds. G. A. Kerkut and L. I. Gilbert, Pergamon, Oxford, 1985, vol. 7, p. 431.
52. F. Sehnal, in "Endocrinology of Insects," eds. R. G. H. Downer and H. Laufer, Liss, New York, 1983, p. 657.
53. M. Takagi, Y. Tsuda, and Y. Wada, *J. Am. Mosquito Control Assoc.*, 1995, **11**, 474.
54. C. A. Henrick, in "Insect Chemical Ecology, Proceedings of a Conference Held at Tábor, 1990," ed. I. Hrdy, Academia, Prague, 1991, p. 429.
55. H. A. Schneidermann, L. I. Gilbert, and M. J. Weinstein, *Nature*, 1960, **188**, 1041.
56. Y. C. Toong, D. A. Schooley, and F. C. Baker, *Nature*, 1988, **333**, 170.
57. K. Sláma, in "Herbivores: Their Interaction with Secondary Plant Metabolites," eds. G. A. Rosenthal and D. H. Janzen, Academic Press, New York, 1979, p. 683.
58. K. Slama and C. M. Williams, *Proc. Natl. Acad. Sci. USA*, 1965, **54**, 411.
59. W. S. Bowers, H. M. Fales, M. J. Thompson, and E. C. Uebel, *Science*, 1966, **154**, 1020.
60. R. Tutihasi and T. Hanazawa, *J. Chem. Soc. Jpn.*, 1940, **61**, 1041.
61. K. Mori and M. Matsui, *Tetrahedron Lett.*, 1967, 2515.
62. B. A. Pauson, H.-C. Cheung, S. Gurbaxani, and G. Saucy, *J. Chem. Soc., Chem. Commun.*, 1968, 1057.
63. V. Cerny, L. Dolejs, L. Labler, F. Sôrm, and K. Slama, *Tetrahedron Lett.*, 1967, 1053.
64. J. F. Manville, L. Greguss, K. Slama, and E. von Rudloff, *Coll. Czech Chem. Commun.*, 1977, **42**, 3658.
65. M. Jacobson, R. E. Redfern, and G. D. Mills, *Lloydia*, 1975, **38**, 473.
66. W. S. Bowers and R. Nishida, *Science*, 1980, **209**, 1030.
67. W. S. Bowers, *Science*, 1968, **161**, 895.
68. W. S. Bowers, T. Ohta, J. S. Cleere, and P. A. Marsella, *Science*, 1976, **193**, 542.
69. A. R. Alertsen, *Acta Chem. Scand.*, 1955, **9**, 1725.
70. R. Livingstone and R. B. Watson, *J. Chem. Soc.*, 1957, 1509.
71. G. T. Brooks, G. E. Pratt, and R. C. Jennings, *Nature*, 1979, **281**, 570.
72. W. S. Bowers in "Endocrinology of Insects" eds. R. G. H. Downer and H. Laufer, Liss, New York, 1983, p. 517.
73. A. V. Vyas and N. B. Mulchandani, *Phytochemistry*, 1980, **19**, 2597.
74. A. Butenandt and P. Karlson, *Z. Naturforsch.*, 1954, **9b**, 389.
75. G. Fraenkel, *Proc. R. Soc. London, Ser. B.* 1922, **118**, 1.

76. G. Fraenkel and J. Zdarek, *Biol. Bull.*, 1976, **139**, 138.
77. T. Ohtaki, R. D. Milkman, and C. M. Williams, *Proc. Natl. Acad. Sci. USA*, 1967, **58**, 981.
78. J. N. Kaplanis, L. A. Tabor, M. J. Thompson, W. E. Robbins, and T. J. Shortino, *Steroids*, 1966, **8**, 625.
79. Y. Sato, M. Sakai, S. Imai, and S. Fujioka. *Appl. Entomol. Zool.*, 1968, **3**, 49.
80. G. Richards, *J. Insect Physiol.*, 1978, **24**. 329.
81. J. W. Fristrom and M. A. Yund, in "Invertebrate Tissue Culture, Research Applications," ed. K. Maramorosch, Academic Press, New York, 1976, p. 161.
82. H. Oberlander, *Experientia*, 1974, **30**, 1409.
83. L. Cherbas, C. D. Yonge, P. Cherbas, and C. M. Williams, *Wilhelm Roux's Arch.*, 1980, **189**, 1.
84. B. Cymborowski, "Ecdysone, From Chemistry to Mode of Action," ed J. Koolman, Georg Thieme, Stuttgart, 1989, p. 144.
85. C. Y. Clement, D. A. Bradbrook, R. Lafont, and L. Dinan, *Insect Biochem. Mol. Biol.*, 1993, **23**, 187.
86. A. Fukuzawa, Y. Kumagai, T. Masamune, A. Furusaki, C. Katayama, and T. Matsumoto, *Tetrahedron Lett.*, 1981, **22**, 4085.
87. A. Fukuzawa, M. Miyamoto, Y. Kumagai, and T. Masamune, *Phytochemistry*, 1986, **25**, 1305.
88. T. Ohsawa, M. Yukawa, C. Takao, M. Murayama, and H. Bando, *Chem. Pharm. Bull.*, 1992, **40**, 143.
89. C. G. Jones and R. D. Firn, *Trans. Roy. Soc. B*, 1991, **333**, 273.
90. R. Lafont and I. D. Wilson, "The Ecdysone Handbook," 2nd edn., The Chromatographic Society, Nottingham, 1997.
91. S. Imai, T. Toyosato, M. Sakai, Y. Sato, S. Fujioka, E. Murata, and M. Goto, *Chem. Pharm. Bull.*, 1969, **17**, 335.
92. L. Dinan, *Russ. J. Plant Physiol.*, 1998, **45**, 296.
93. R. Lafont and D. H. S. Horn, in "Ecdysone, From Chemistry to Mode of Action," ed. J. Koolman, Georg Thieme Stuttgart, 1989, p. 39.
94. H. Hoffmeister, *Angew. Chem., Int. Ed. Engl.*, 1966, **5**, 248.
95. D. H. S. Horn, in "Naturally Occurring Pesticides," eds. M. Jacobson and D. G. Crosby, Dekker, New York, 1971, p. 333.
96. R. Huber and W. Hoppe, *Chem. Ber.*, 1965, **98**. 2403.
97. F. Hampshire and D. H. S. Horn, *J. Chem. Soc., Chem. Commun.*, 1966, 37.
98. K. Nakanishi, M. Kooreda, S. Sasaki, M. L. Chang, and H. Y. Hsu, *J. Chem. Soc., Chem. Commun.*, 1966, 915
99. M. N. Galbraith and D. H. S. Horn, *J. Chem. Soc., Chem. Commun.*, 1966, 905.
100. B. M. R. Bandara, L. Jayasinghe, V. Karunaratne, G. P. Wannigamma, M. Bokel, W. Kraus, and S. Sotheeswaran, *Phytochemistry*, 1989, **28**, 1073.
101. W.-S. Chou and H.-S. Lu, in "Developments in Endocrinology, Vol. 7: Progress in Ecdysone Research," ed. J. A. Hoffmann, Elsevier-North Holland, Amsterdam, 1980, p. 281.
102. R. Lafont, A. Bouthier, and I. D. Wilson, in "Insect Chemical Ecology," ed. I. Hrdy, Academia, Prague, 1991, p. 197.
103. E. D. Morgan and C. F. Poole, *Adv. Insect Physiol.*, 1976, **12**, 17.
104. E. D. Morgan and I. D. Wilson, in "Handbook of Natural Pesticides: Methods," ed. N. B. Mandava, CRC Press, Boca Raton, FL, 1985, vol. 2, p. 3.
105. R. Lafont, I. D. Wilson, E. D. Morgan, and B. Wright, *J. Chromatogr. A*, 1998, **799**, 333.
106. H. H. Rees and R. E. Isaac, *Methods Enzymol.*, 1985, **111**, 377.
107. R. Lafont, N. Kaouadji, E. D. Morgan, and I. D. Wilson, *J. Chromatogr. A*, 1994, **658**, 55.
108. I. Kubo, J. A. Klocke, I. Ganjian, N. Ichikawa, and T. Matsumoto, *J. Chromatogr. A*, 1983, **257**, 157.
109. I. D. Wilson, in "Handbook of Natural Pesticides," ed. E. D. Morgan and N. B. Mandava, 1987, CRC Press, Boca Raton, FL, 1987, vol. 3, part A, p. 15.
110. M. W. Gilgan and M. E. Zinck, *Steroids*, 1972, **20**, 95.
111. J. Koolman, *Insect Biochem.*, 1986, **10**, 381.
112. N. Harada and K. Nakanishi, *J. Am. Chem. Soc.*, 1969, **91**, 3989.
113. M. Kooreda, N. Harada, and K. Nakanishi, *J. Chem. Soc., Chem. Commun.*, 1969, 548.
114. R. E. Issac, M. E. Rose, H. H. Rees, and T. W. Goodwin, *J. Chem. Soc., Chem. Commun.*, 1982, 249.
115. J.-P. Girault and R. Lafont, *J. Insect Physiol.*, 1988, **34**, 701.
116. J. B. Siddall, A. D. Cross, and J. H. Fried, *J. Am. Chem. Soc.*, 1966, **88**, 862.
117. A. Furlenmeier, A. Furst, A. Langemann, G. Waldvogel, P. Hocks, U. Kerb, and P. Wiechert, *Experientia*, 1966, **22**, 573
118. I. T. Harrison, J. B. Siddall, and J. H. Fried, *Tetrahedron Lett.*, 1966, 3457.
119. D. H. R. Barton, P. G. Feakins, J. P. Poyser, and P. G. Sammes, *J. Chem. Soc. C*, 1970, 1584.
120. E. Lee, Y.-T. Liu, P. H. Solomon, and K. Nakanishi, *J. Am. Chem. Soc.*, 1976, **98**, 1634
121. Y.-W. Lee, E. Lee, and K. Nakanishi, *Tetrahedron Lett.*, 1980, **21**, 4323.
122. H. Mori, K. Shibata, K. Tsuneda, and M. Sawai, *Chem. Pharm. Bull.*, 1968, **16**, 563.
123. H. Mori, K. Shibata, and M. Sawai, *Tetrahedron*, 1971, **27**, 1157.
124. G. Huppi and J. B. Siddall, *J. Am. Chem. Soc.*, 1967, **89**, 6790.
125. U. Kerb, R. Wiechert, R. Furlenmeier, and A. Fürst, *Tetrahedron Lett.*, 1968, 4277.
126. H. Mori and K. Shibata, *Chem. Pharm. Bull.*, 1969, **17**, 1970.
127. T. Kametani, M. Tsubuki, and H. Nemoto, *Tetrahedron Lett.*, 1980, **21**, 4855.
128. M. N. Galbraith, D. H S. Horn, and J. A. Thomson, *Experientia*, 1975, **31**, 873.
129. C. Hetru, Y. Nakatani, B. Luu, and J. A. Hoffmann, *Nouv. J. Chim.*, 1983, **7**, 587.
130. M. N. Galbraith, D. H. S. Horn, E. J. Middleton, and R. J. Hackney, *Aust. J. Chem.*, 1969, **22**, 1059.
131. M. N. Galbraith, D. H. S. Horn, E. J. Middleton, and R. J. Hackney, *Aust. J. Chem.*, 1969, **22**, 1517.
132. T. Takemoto, Y. Hikino, H. Hikino, S. Ogawa, and N. Nishimoto, *Tetrahedron Lett.*, 1968, 3053.
133. K. Shibata and H. Mori, *Tetrahedron*, 1971, **27**, 1149.
134. P. Hocks, U. Kerb, R. Wiechert, A. Furlenmeier, and A. Furst, *Tetrahedron Lett.*, 1968, 4281.
135. W.-M. Tom, Y. J. Abdul-Hajj, and M. Kooreda, *J. Chem. Soc., Chem. Commun.*, 1975, 24.
136. K.-D. Spindler, J. Koolman, F. Mosora, and H. Emmerich, *J. Insect Physiol.*, 1977, **23**, 441.
137. J. Koolman and K.-D. Spindler, *Hoppe-Seyler's Z. Physiol. Chem.*, 1977, **358**, 1339.
138. T. Kametani and M. Tsubuki, in "Ecdysone, From Chemistry to Mode of Action," ed. J. Koolman, Georg Thieme, Stuttgart, 1989, p. 74.

139. I. D. Wilson, E. D. Morgan, and S. Murphy, *Anal. Chim. Acta*, 1990, **236**, 145.
140. R. Lafont, E. D. Morgan, and I. D. Wilson, *J. Chromatogr. A*, 1994, **658**, 31.
141. I. D. Wilson, R. Lafont, and P. Wall, *J. Planar Chromatogr.*, 1988, **1**, 357
142. I. D. Wilson, R. Lafont, R. G. Kingston, and C. J. Porter, *J. Planar Chromatogr.*, 1990, **3**, 359.
143. M. Bathori, I. Mathe, L. Praszna, K. Rischak, and H. J. Kalasz, *J. Planar Chromatogr.*, 1996, **9**, 264.
144. C. R. Bielby, E. D. Morgan, and I. D. Wilson, *J. Chromatogr.*, 1986, **351**, 57.
145. R. P. Evershed, J. G. Mercer, and H. H. Rees, *J. Chromatogr.*, 1987, **390**, 357.
146. M. W. Raynor, J. P. Kithinji, K. D. Bartle, D. E. Games, I. M. Mylchreest, R. Lafont, E. D. Morgan, and I. D. Wilson, *J. Chromatogr.*, 1989, **467**, 292.
147. M. Zhang, M. J. Stout, and I. Kubo, *Phytochemistry*, 1992, **31**, 247.
148. P. Davies, R. Lafont, T. Large, E. D. Morgan, and I. D. Wilson, *Chromatographia*, 1993, **37**, 37.
149. L. Reum and J. Koolman, in "Ecdysone, From Chemistry to Mode of Action," ed. J. Koolman, Georg Thieme, Stuttgart, 1989, p. 131.
150. H. H. Rees, in "Comprehensive Insect Physiology, Biochemistry and Pharmacology," eds. G. A. Kerkut and L. I. Gilbert, Pergamon, Oxford, 1985, vol. 7, p. 249.
151. H. H. Rees, in "Ecdysone, From Chemistry to Mode of Action", ed. J. Koolman, Georg Thieme, Stuttgart, 1989, p. 152.
152. M. L. Grieneisen, *Insect Biochem. Mol. Biol.*, 1994, **24**, 115.
153. R. Lafont, J.-L. Connat, J.-P. Delbeque, P. Porcheron, C. Dauphin-Villemant, and M. Garcia, in "Recent Advances in Insect Biochemistry and Molecular Biology," eds. E. Ohnishi, H. Sonobe, and S. Y. Takahashi, University of Nagoya Press, Nagoya, 1995, p. 45.
154. D. Bocking, C. Dauphin-Villemant, J.-Y. Toullec, C. Blais, and R. Lafont, *C. R. Acad. Sci. Paris*, 1994, **317**, 891.
155. J. H. Adler and R. J. Grebenok, *Lipids*, 1995, **30**, 257.
156. N. J. De Souza, E. L. Ghisalberti, H. H. Rees, and T. W. Goodwin, *Phytochemistry*, 1970, **9**, 1247.
157. R. J. Grebenok and J. H. Adler, *Phytochemistry*, 1993, **33**, 341.
158. C. Blais and R. Lafont, *Hoppe-Seyler's Z. Physiol. Chem*, 1984, **365**, 809.
159. G. F. Wierich, in "Ecdysone, From Chemistry to Mode of Action," ed. J. Koolman, Georg Thieme, Stuttgart, 1989, p. 174.
160. M. F. Meister, J.-L. Dimarcq, C. Keppler, C. Hetru, M. Lagueux, R. Lafont, B. Luu, and J. A. Hoffmann, *Mol. Cell. Endocrinol.*, 1985, **41**, 27.
161. C. Blais, J.-F. Modde, P. Beydon, and R. Lafont, in "Invertebrate Cell Systems in Applications," ed. J. Mitsuhashi, CRC Press, Boca Raton, FL, 1989, vol. 1, p. 99.
162. R. Lafont and J.-L. Connat, in "Ecdysone, From Chemistry to Mode of Action," ed. J. Koolman, Georg Thieme, Stuttgart, 1989, p. 167.
163. D. Bocking, C. Dauphin-Villemant, and R. Lafont, *Eur. J. Entomol.* 1995, **92**, 63.
164. R. Bergamasco and D. H. S. Horn, in "Progress in Ecdysone Research", ed. J. A. Hoffmann, Elsevier, Amsterdam, 1980, p. 229.
165. Y. K. Chong, M. N. Galbraith, and D. H. S. Horn, *J. Chem. Soc., Chem. Commun.*, 1970, 1217.
166. M. J. Thomson, J. N. Kaplanis, W. E. Robbins, S. R. Dutky, and H. N. Nigg, *Steroids*, 1974, **24**, 359.
167. A. Faux, M. N. Galbraith, D. H. S. Horn, E. J. Middleton, and J. A. Thomson, *J. Chem. Soc., Chem. Commun.*, 1970, 243.
168. M. N. Galbraith, D. H. S. Horn, E. J. Middleton, and J. A. Thompson, *Experientia*, 1973, **29**, 19.
169. M. J. Thompson, J. N. Kaplanis, W. E. Robbins, and R. T. Yamamoto, *J. Chem. Soc., Chem. Commun.*, 1967, 650.
170. S. Imai, M. Hori, S. Fujioka, E. Murata, M. Goto, and K. Nakanishi, *Tetrahedron Lett.*, 1968, 3883.
171. M. N. Galbraith, D. H. S. Horn, E. J. Middleton, and R. J. Hackey, *J. Chem. Soc., Chem. Commun.*, 1969, 402.
172. H. Hikino, Y. Hikino, K. Nomoto, and T. Takemoto, *Tetrahedron*, 1968, **24**, 4895.
173. Y. Tanaka, *Eur. J. Entomol.*, 1995, **92**, 155.
174. I. Kubo, J. A. Klocke, and S. Asano, *J. Insect Physiol.*, 1983, **29**, 307.
175. C. Arnault and K. Sláma, *J. Chem. Ecol.*, 1986, **12**, 1979.
176. P. D. Robinson, E. D. Morgan, I. D. Wilson, and R. Lafont, *Physiol. Entomol.*, 1987, **12**, 321.
177. S. Kopec, *Biol. Bull.*, 1922, **42**, 323.
178. J. V. Stone and W. Mordue, *Insect Biochem.*, 1980, **10**, 229.
179. E. de Hoffmann, J. Charette, and V. Stroobant, "Mass Spectrometry," Wiley, Chichester, 1996.
180. D. P. Muehleisen, R. S. Gray, E. J. Katahira, M. K. Thomas, and W. E. Bollenbacher, *Peptides*, 1993, **14**, 531.
181. A. Suzuki, S. Nagata, and H. Kataoka, *ACS Symp. Ser.* 1996, **658**, 268.
182. J. Ishibashi, H. Kataoka, A. Isogai, A. Kawakami, H. Saegusa, Y. Yagi, A. Mizoguchi, H. Ishizaki, and A. Suzuki, *Biochemistry*, 1994, **33**, 5912.
183. A. J. Kim, G. H. Cha, K. Kim, L. I. Gilbert, and C. C. Lee, *Proc. Natl. Acad. Sci. USA*, 1997, **94**, 1130.
184. G. Gäde, *Z. Physiol. Chem.*, 1990, **371**, 475.
185. G. Gäde, M. P. E. Janssens, and R. Kellner, *Peptides*, 1994, **15**, 1.
186. W. Liebrich, R. Kellner, and G. Gäde, *Peptides*, 1995, **16**, 559.
187. G. Gäde, *Hoppe-Seyler's Z. Physiol. Chem.*, 1991, **372**, 193.
188. G. Gäde and K. L. Rinehart, *Biochem. Biophys. Res. Commun.*, 1987, **149**, 908.
189. J. P. Woodring, H. W. Fescemeyer, J. A. Lockwood, A. B. Hammond, and G. Gäde, *Comp. Biochem. Physiol. A*, 1989 **92**, 65.
190. G. Gäde, *Z. Physiol. Chem.*, 1992, **373**, 1169.
191. G. Gäde, C. Hilbich, K. Beyreuther, and K. L. Rinehart, *Peptides*, 1988, **9**, 681.
192. K. J. Siegert, P. J. Morgan, and W. Mordue, *Z. Physiol. Chem.*, 1985, **366**, 723.
193. G. Gäde, G. J. Goldsworthy, M. H. Schaffer, J. C. Cook, and K. L. Rinehart, *Biochem. Biophys. Res. Commun.*, 1986, **134**, 723.
194. G. Gäde, S. E. Reynolds, and J. R. Beeching, in "Perspectives in Comparative Endocrinology," eds. K. G. Davey, R. E. Peter, and S. S. Tobe, National Research Council, Ottawa, 1994, p. 119.
195. H. Jaffe, A. K. Raina, C. T. Riley, B. A. Fraser, R. J. Nachman, V. W. Vogel, Y.-S. Zhang, and D. K. Hayes, *Proc. Natl. Acad. Sci. USA*, 1989, **86**, 8161.

196. R. C. H. M. Oudejans, F. P. Kooiman, W. Heerma, C. Versluis, A. J. Slotboom, and A. M. Th. Beenakkers, *Eur. J. Biochem.*, 1991, **195**, 351.
197. M. P. E. Janssens, R. Kellner, and G. Gäde, *Biochem. J.*, 1994, **302**, 539.
198. R. Ziegler, K. Eckart, H. Schwarz, and R. Keller, *Biochem. Biophys. Res. Commun.*, 1985, **133**, 337.
199. H. Jaffe, A. K. Raina, C. T. Riley, B. A. Fraser, G. M. Holman, R. M. Wagner, R. L. Ridgway, and D. K. Hayes, *Biochem. Biophys. Res. Commun.*, 1986, **135**, 622.
200. J. Ishibashi, H. Kataoka, H. Nagasawa, A. Isogai, and A. Suzuki, *Biosci. Biotechnol. Biochem.*, 1992, **56**, 66.
201. J. V. Stone, W. Mordue, K. E. Batley, and H. R. Morris, *Nature*, 1976, **263**, 207.
202. J. A. Veenstra and H. H. Hagedorn, *Arch. Insect Biochem. Physiol.*, 1995, **29**, 391.
203. G. Gäde and G. Rosinski, *Peptides*, 1990, **11**, 455.
204. G. Gäde, *S. Afr. J. Zool.*, 1994, **29**, 11.
205. G. Gäde and R. Kellner, *Gen. Comp. Endocrinol.*, 1992, **86**, 119.
206. G. Gäde, H. Wilps, and R. Kellner, *Biochem. J.*, 1990, **269**, 309.
207. M. H. Schaffer, B. E. Noyes, C. A. Slaughter, G. C. Thorne, and S. J. Gaskell, *Biochem. J.*, 1990, **269**, 315.
208. H. Jaffe, A. K. Raina, C. T. Riley, B. A. Fraser, T. G. Bird, C.-M. Tseng, W. Vogel, Y.-S. Zhang, and D. K. Hayes, *Biochem. Biophys. Res. Commun.*, 1988, **155**, 344.
209. G. Gäde, *Physiol. Entomol.*, 1989, **14**, 405.
210. G. Gäde and K. L. Rinehart, *Z. Physiol. Chem.*, 1990, **371**, 345.
211. G. Gäde, R. Kellner, K. L. Rinehart, and M. L. Proefke, *Biochem. Biophys. Res. Commun.*, 1992, **189**, 1303.
212. G. Gäde and M. P. E. Janssens, *Z. Physiol. Chem.*, 1994, **375**, 803.
213. T. K. Hayes, L. L. Keeley, and D. W. Knight, *Biochem. Biophys. Res. Commun.*, 1986, **140**, 674.
214. G. Gäde and K. L. Rinehart, *Biochem. Biophys. Res. Commun.*, 1986, **141**, 774.
215. K. J. Siegert and W. Mordue, *Physiol. Entomol.*, 1986, **11**, 205.
216. J. L. Witten, M. H. Schaffer, M. O'Shea, J. C. Cook, M. E. Hemling, and K. L. Rinehart, *Biochem. Biophys. Res. Commun.*, 1984, **124**, 350.
217. G. Gäde and R. Kellner, *Peptides*, 1989, **10**, 1287.
218. J. A. Veenstra, *FEBS Lett.*, 1989, **250**, 231.
219. R. J. Nachman, G. M. Holman, W. F. Haddon, and N. Ling, *Science*, 1986, **234**, 71.
220. R. J. Nachman, G. M. Holman, B. J. Cook, W. F. Haddon, and N. Ling, *Biochem. Biophys. Res. Commun.*, 1986, **140**, 357.
221. J. P. Proux, C. A. Miller, J. P. Li, R. L. Carney, A. Girardie, M. Delaage, and D. A. Schooley, *Biochem. Biophys. Res. Commun.*, 1987, **149**, 180.
222. E. Lehmberg, R. B. Ota, K. Furuya, D. S. King, S. W. Applebaum, H.-J. Ferenz, and D. A. Schooley, *Biochem. Biophys. Res. Commun.*, 1991, **179**, 1036.
223. I. Kay, C. H. Wheeler, G. M. Coast, N. F. Totty, O. Cusinato, M. Patel, and G. J. Goldsworthy, *Z. Physiol. Chem.*, 1991, **372**, 929.
224. M. B. Blackburn, T. G. Kingan, W. Bodnar, J. Shabanowitz, D. F. Hunt, T. Kempe, R. M. Wagner, A. K. Raina, M. E. Schnee, and M. C. Ma, *Biochem. Biophys. Res. Commun.*, 1991, **181**, 927.
225. H. Kataoka, R. G. Troetschler, J. P. Li, S. J. Kramer, R. L. Carney, and D. A. Schooley, *Proc. Natl. Acad. Sci. USA*, 1989, **86**, 2976.
226. K. Furuya, K. M. Schegg, H. Wang, D. S. King, and D. A. Schooley, *Proc. Natl. Acad. Sci. USA*, 1995, **92**, 12 323.
227. F. L. Clottens, G. M. Holman, G. M. Coast, N. F. Totty, T. K. Hayes, I. Kay, A. I. Mallet, M. S. Wright, J.-S. Chung, O. Truong, and D. L. Bull, *Peptides*, 1994, **15**, 971.
228. I. Kay, M. Patel, G. M. Coast, N. F. Totty, A. I. Mallet, and G. J. Goldsworthy, *Regul. Pept.*, 1992, **42**, 111.
229. I. Kay, G. M. Coast, O. Cusinato, C. H. Wheeler, N. F. Totty, and G. J. Goldsworthy, *Z. Physiol. Chem.*, 1991, **372**, 505.
230. A. K. Raina, H. Jaffe, T. G. Kempe, P. Keim, R. W. Blacher, H. M. Fales, C. T. Riley, J. A. Klun, R. L. Ridgway, and D. K. Hayes, *Science*, 1989, **244**, 796.
231. A. Kitamura, N. Nagasawa, H. Kataoka, T. Ando, and A. Suzuki, *Agric. Biol. Chem.*, 1990, **54**, 2495.
232. E. P. Masler, A. K. Raina, R. M. Wagner, and J. P. Kochansky, *Insect Biochem. Mol. Biol.*, 1994, **24**, 829.
233. S. Matsumoto, A. Fonagy, M. Kurihara, K. Uchiumi, T. Nagamine, M. Chijimatsu, and T. Mitsui, *Biochem. Biophys. Res. Comm.*, 1992, **182**, 534.
234. M. Isobe, K. Hasegawa, J. Kubota, and T. Goto, *Agric. Biol. Chem.*, 1976, **40**, 1189.
235. H. Kataoka, R. G. Toschi, J. P. Li, L. G. Carney, D. A. Schooley, and S. J. Kramer, *Science*, 1989, **243**, 1481.
236. C. S. Garside, T. K. Hayes, and S. S. Tobe, *Peptides*, 1997, **18**, 17.
237. A. P. Woodhead, B. Stay, S. L. Seidel, M. A. Khan, and S. S. Tobe, *Proc. Natl. Acad. Sci. USA*, 1989, **86**, 5997.
238. A. P. Woodhead, M. A. Khan, B. Stay, and S. S. Tobe, *Insect Biochem. Mol. Biol.*, 1994, **24**, 257.
239. X. Belles, J. L. Maestro, M. D. Piulachs, A. H. Johnsen, H. Duve, and A. Thorpe, *Regul. Pept.*, 1994, **53**, 237.
240. G. E. Pratt, D. E. Farnsworth, K. F. Fok, N. R. Siegel, A. L. McCormack, J. Shabanowitz, D. F. Hunt, and R. Feyereisen, *Proc. Natl. Acad. Sci. USA*, 1991, **88**, 2412.
241. S. J. Kramer, A. Tochi, C. A. Miller, H. Kataoka, G. B. Quistad, J. P. Li, R. L. Carney, and D. A. Schooley, *Proc. Natl. Acad. Sci. USA*, 1991, **88**, 9458.
242. M. W. Lorenz, R. Kellner, and K. H. Hoffmann, *Regul. Pept.*, 1995, **57**, 227.
243. M. W. Lorenz, R. Kellner, and K. H. Hoffmann, *J. Biol. Chem.*, 1995, **270**, 21103.
244. H. Duve, A. H. Johnsen, A. G. Scott, C. G. Yu, K. J. Yagi, S. S. Tobe, and A. Thorpe, *Proc. Natl. Acad. Sci. USA*, 1993, **90**, 2456.
245. H. Nagasawa, H. Kataoka, A. Isogai, S. Tamura, A. Suzuki, A. Mizoguchi, Y. Fujiwara, S. Y. Takahashi, and H. Ishizaki, *Proc. Natl. Acad. Sci. USA*, 1986, **83**, 5840.
246. H. Kataoka, R. G. Troetschler, S. J. Kramer, B. J. Cesarin, and D. A. Schooley, *Biochem. Biophys. Res. Commun.*, 1987, **146**, 746.
247. T. Marti, K. Takio, K. A. Walsh, G. Terzi, and J. W. Truman, *FEBS Lett.*, 1987, **219**, 415.
248. D. Zitnan, T. G. Kingan, J. L. Hermesman, and M. E. Adams, *Science*, 1996, **271**, 88.
249. I. Orchard and B. G. Loughton, in "Comprehensive Insect Physiology, Biochemistry and Pharmacology," eds. G. A. Kerkut and L. I. Gilbert, Pergamon, Oxford, 1985, vol. 7, p. 61.

250. G. Gäde, *Prog. Chem. Org. Nat. Prod.* 1997, **71**, 1.
251. W. E. Bollenbacher and M. F. Bowen, in "Endocrinology of Insects," eds. R. G. H. Downer and H. Laufer, Liss, New York, 1983, vol. 1, p. 89.
252. W. Mordue and P. J. Morgan, in "Handbook of Natural Pesticides," eds. E. D. Morgan and N. B. Mandava, CRC Press, Boca Raton, FL, 1987, vol. 3, part A, p. 153.
253. R. S. Gray, D. P. Muehleisen, E. J. Katahira, and W. E. Bollenbacher, *Peptides*, 1994, **15**, 777.
254. I. Sauman and S. M. Reppert, *Dev. Biol.*, 1996, **178**, 418.
255. A. N. Starratt and B. E. Brown, *Life Sci.*, 1975, **17**, 1253.
256. A. N. Starratt and M. E. Stevens, *J. Chromatogr.*, 1980, **194**, 421.
257. G. B. Quistad, M. E. Adams, R. M. Scarborough, R. L. Carney, and D. A. Schooley, *Life Sci.*, 1984, **34**, 569.
258. C. E. Broomfield and P. M. Hardy, *Tetrahedron Lett.*, 1977, 2201.
259. A. K. Raina and G. Gäde, *Insect Biochem.*, 1988, **18**, 785.
260. G. Gäde, in "Chromatography and Isolation of Insect Hormones and Pheromones," eds. A. R. McCaffery and I. D. Wilson, Plenum, New York, 1980, p. 165.
261. A. Raina, L. Pannel, J. Kochansky, and H. Jaffe, *Insect Biochem. Mol. Biol.*, 1995, **25**, 929.
262. J. P. Woodring and D. J. Leprince, *J. Insect Physiol.*, 1992, **38**, 775.
263. K. Imai, T. Konno, Y. Nakazawa, T. Komiya, M. Isobe, K. Koga, T. Goto, T. Yagunima, K. Sakakibara, K. Hasegawa, and O. Yamashita, *Proc. Jpn. Acad.* 1991, **67B**, 98.
264. H. Duve, A. H. Johnsen, A. G. Scott, P. East, and A. Thorpe, *J. Biol. Chem.*, 1994, **269**, 21 059.
265. G. Fraenkel and C. Hsiao, *J. Insect Physiol.*, 1965, **11**, 513.
266. S. E. Reynolds, in "Comprehensive Insect Physiology, Biochemistry and Pharmacology," eds. G. A. Kerkut and L. I. Gilbert, Pergamon, Oxford, 1985, vol. 8, p. 335.
267. U. K. Kaltenhauser, J. Kellermann, K. Andersson, F. Lottspeich, and H. W. Honegger, *Insect Biochem. Mol. Biol.*, 1995, **25**, 525.
268. J. E. Steele, *Nature*, 1961, **192**, 680.
269. T. Kono, H. Nagasawa, A. Isogai, H. Fugo, and A. Suzuki, *Insect Biochem.*, 1991, **21**, 185.
270. D. Veelaert, L. Schoofs, P. Proost, J. van Damme, B. Devresse, J. van Beeuman, and A. De Loof, *Regul. Pept.*, 1995, **57**, 221.
271. L. Schoofs, D. Veelaert, J. Vanden Broek, and A. De Loof, *Peptides*, 1997, **18**, 145.
272. S. H. P. Maddrell and S. E. Reynolds, *Nature*, 1972, **236**, 404.
273. G. Richards and J. A. Hoffmann, in "Handbook of Natural Pesticides," eds. E. D. Morgan and N. B. Mandava, CRC Press, Boca Raton, FL, vol. 3, part A, 1987, p. 1.
274. J. J. Menn and A. B. Borkovec, *J. Agric. Food Chem.*, 1989, **37**, 271.
275. T. A. Miller, in "Endocrinology of Insects," eds. R. G. H. Downer and H. Laufer, Liss, New York, 1983, p. 101.
276. J. E. Huesing and D. Jones, *Phytochemistry*, 1987, **26**, 1381.
277. S. S. Duffey and M. J. Stout, *Arch. Insect Biochem. Physiol.*, 1996, **32**, 3.
278. Y. Fu-Shun, K. A. Evans, L. H. Stevens, T. A. van Beek, and L. M. Schoonhoven, *Entomol. Exper. Appl.* 1990, **54**, 57.
279. L. B. Brattsten, in "Molecular Aspects of Insect–Plant Associations," eds. L. B. Brattsten and S. Ahmad, Plenum, New York, 1986, p. 221.
280. S. Ahmad, L. B. Brattsten, C. A. Mullin, and S. J. Yu, in "Molecular Aspects of Insect–Plant Associations," eds. L. B. Brattsten and S. Ahmad, Plenum, New York, 1986, p. 73.
281. E. D. Morgan and N. B. Mandava, (eds.), "Handbook of Natural Pesticides," CRC Press, Boca Raton, FL, 1987, vol. 2, 1990, vol. 4.
282. L. Crombie, *J. Chem. Soc.*, 1955, 999.
283. A. Hassanali and W. Lwande, *ACS Symp. Ser.*, 1989, **387**, 78.
284. M. Miyakado, I. Nakayama, and N. Ohno, *ACS Symp. Ser.*, 1989, **387**, 173.
285. M. Miyakado, I. Nakayama, H. Yoshioka, and N. Nakatani, *Agric. Biol. Chem.*, 1979, **43**, 1609.
286. C. W. Holyoke and J. C. Reese, in "Handbook of Natural Pesticides," eds. E. D. Morgan and N. B. Mandava, CRC Press, Boca Raton, FL, 1987, vol. 3B, p. 67.
287. W. S. K. Gbewonyo, D. J. Candy, and M. Anderson, *Pestic. Sci.*, 1993, **37**, 57.
288. J. K. Ruprecht, Y.-H. Hui, and J. L. McLaughlin, *J. Nat. Prod.*, 1990, **53**, 237.
289. L. Zeng, Q. Ye, N. H. Oberlies, G. Shi, Z.-M. Gu, K. He, and J. L. McLaughlin, *Nat. Prod. Rep.*, 1996, **13**, 275.
290. A. Cavé, B. Figadère, A. Laurens, and D. Cortes, *Prog. Chem. Org. Nat. Prod.*, 1997, **70**, 81.
291. J.-P. Girault, M. Bathori, E. Varga, K. Szendrei, and R. Lafont, *J. Nat. Prod.*, 1990, **53**, 279.
292. J. L. McLaughlin, L. Zeng, N. H. Oberlies, D. Alfonso, H. A. Johnson, and B. A. Cummings, *ACS Symp. Ser.*, 1997, **658**, 117.
293. Y. J. Ahn, M. Kwon, H. M. Park, and C. K. Han, *ACS Symp. Ser.*, 1997, **658**, 90.
294. J. Gershenzon and R. Croteau, in "Herbivores: their Interactions with Secondary Plant Metabolites," 2nd ed., eds. G. A. Rosenthal and D. H. Jansen, Academic Press, New York, 1991, vol. 1, p. 165.
295. J. C. Reese and C. W. Holyoke, in "Handbook of Natural Pesticides," eds. E. D. Morgan and N. B. Mandava, CRC Press, Boca Raton, FL, 1987, vol. 3B, p. 21.
296. H. K. Lichtentahler, M. Rohmer, and J. Schwender, *Physiol. Plant*, 1997, **101**, 643.
297. D. V. Banthorpe and B. V. Charlwood, in "Chemistry of Terpenes and Terpenoids," ed. A. A. Newman, Academic Press, New York, 1972, p. 337.
298. M. G. Collart and W. F. Hink, *Entomol. Exper. Appl.* 1986, **42**, 225.
299. J. O. Moffett, R. L. Cox, M. Ellis, R. Rivera, W. T. Wilson, T. D. Cardoso, and C. Vargas, *J. Southwest. Entomol.*, 1989, **14**, 57.
300. M. S. J. Simmonds and W. M. Blaney, in "Advances in Labiate Science," eds. R. M. Hartley and T. Reynolds, Royal Botanic Gardens, Kew, 1992, p. 375.
301. P. J. Rice and J. R. Coats, *ACS Symp. Ser.*, 1994, **557**, 92.
302. J. R. Coats, L. L. Karr, and C. D. Drewes, *ACS Symp. Ser.*, 1991, **449**, p. 305.
303. E. Gowan, B. A. Lewis, and R. Turgeon, *J. Chem. Ecol.*, 1995, **21**, 1781.
304. M. D. Bowers, in "Herbivores. Their Interactions with Secondary Plant Metabolites," eds. G. A. Rosenthal and M. R. Berenbaum, Academic Press, San Diego, CA, 1991, vol. 1, p. 297.

305. M. D. Bowers, *J. Chem. Ecol.*, 1984, **10**, 1567.
306. I. Kubo, Y-W. Lee, M. J. Pettei, F. Pillewicz, and K. Nakanishi, *J. Chem. Soc., Chem Commun.*, 1976, 1013.
307. C. S. Barnes and J. W. Loder, *Aust. J. Chem.*, 1962, **15**, 322.
308. H. H. Appel, C. J. W. Brooks, and K. H. Overton, *J. Chem. Soc.* 1959, 3322.
309. Ae. de Groot and T. A. van Beek, *Rec. Trav. Chim. Pays-Bas*, 1987, **106**, 1.
310. S. V. Ley, in "Pesticide Science and Biotechnology," eds. R. Greenhalgh and T. R. Roberts, Blackwell, Oxford, 1987, p. 25.
311. B. J. M. Jansen and Ae. de Groot, *Nat. Prod. Rep.*, 1991, **8**, 309.
312. G. L. Fritz, G. D. Mills, J. D. Warthen, and R. M. Waters, *J. Chem. Ecol.*, 1989, **15**, 2607.
313. J. L. Frazier, in "Molecular Aspects of Insect–plant Associations," eds. L. B. Bratsten and S. Ahmed, Plenum Press, New York, 1986, p. 1.
314. Y. Asakawa, G. W. Dawson, D. C. Griffiths, J.-Y. Lallemand, S. V. Ley, K. Mori, A. Mudd, M. Pezechk-Leclaire, J. A. Pickett, H. Watanabe, C. M. Woodcock, and Z. Zhong-Ning, *J. Chem. Ecol.*, 1988, **14**, 1845.
315. A. K. Picman, *Biochem. System. Ecol.*, 1986, **14**, 255.
316. F. C. Seaman, *Bot. Rev.*, 1982, **48**, 121.
317. S. B. Rees and J. B. Harborne, *Phytochemistry*, 1985, **24**, 2225.
318. Y. J. Ahn, M. Kwon, H. M. Park, and C. K. Han, *ACS Symp. Ser.*, 1997, **658**, 90.
319. M. Wang, H. Qin, M. Kong, and Y.-Z. Li, *Phytochemistry*, 1991, **30**, 3931.
320. R. D. Stipanovic, A. A. Bell, and M. J. Lukefahr, *ACS Symp. Ser.*, 1977, **62**, 197.
321. G. H. Gunasena, S. B. Vinson, H. J. Williams, and R. D. Stipanovic, *J. Econ. Entomol.*, 1988, **81**, 93.
322. J. A. Klocke, M.-Y. Hu, S.-F. Chiu, and I. Kubo, *Phytochemistry*, 1991, **30**, 1797.
323. Y. J. Ahn, M. Kwon, H. M. Park, and C. K. Han, *ACS Symp. Ser.*, 1997, **658**, 90.
324. H. N. Banergee, *Sci. Cult.*, 1936, **2**, 163 (*Chem. Abstr.*, 1937, **31**, 209).
325. H. N. Banergee, *J. Indian Chem. Soc.*, 1937, **14**, 51.
326. G. A. Sim, T. A. Hamor, I. C. Paul, and J. M. Robertson, *Proc. Chem. Soc.*, 1961, 75.
327. D. H. R. Barton, H. T. Cheung, A. D. Cross, L. M. Jackman, and M. Martin-Smith, *Proc. Chem. Soc.*, 1961, 76.
328. D. Rogers, G. G. Ünal, D. J. Williams, S. V. Ley, B. S. Joshi, and K. R. Ravindranath, *J. Chem. Soc., Chem. Commun.*, 1979, 97.
329. N. Kato, M. Takahashi, M. Shibayama, and K. Munakata, *Agric. Biol. Chem.*, 1972, **36**, 2579.
330. N. Kato, M. Takahashi, M. Shibayama, and K. Munakata, *Phytochemistry*, 1973, **12**, 1833.
331. N. Kato, M. Takahashi, M. Shibayama, K. Munakata, and Y.-L. Chen, *Agric. Biol. Chem.*, 1974, **38**, 1045.
332. A. T. Merrit and S. V. Ley, *Nat. Prod. Rep.*, 1992, **9**, 243.
333. L. Rodriguez-Hahn, B. Esquivel, and J. Cárdenas, *Prog. Chem. Org. Nat. Prod.*, 1994, **63**, 107.
334. J. R. Hanson, *Nat. Prod. Rep.*, 1995, **12**, 207.
335. J. R. Hanson, *Nat. Prod. Rep.*, 1996, **13**, 59.
336. T. A. van Beek and Ae. de Groot, *Rec. Trav. Chim. Pays-Bas*, 1986, **105**, 513.
337. A. G. Antonious and T. Saito, *Appl. Entomol. Zool.*, 1983, **18**, 40.
338. I. Kubo, Y.-W. Lee, V. Balogh-Nair, K. Nakanishi, and A. Chapya, *J. Chem. Soc., Chem. Commun.*, 1976, 949.
339. X. Belles, F. Camps, J. Coll, and M. D. Piulachs, *J. Chem. Ecol.*, 1985, **11**, 1439.
340. J. C. Anderson, W. M. Blaney, M. D. Cole, L. L. Fellows, S. V. Ley, R. N. Sheppard, and M. S. J. Simmonds, *Tetrahedron Lett.*, 1989, **30**, 4737.
341. P. Y. Malakov, G. Y. Papanov, B. Rodríguez, M. C. de la Torre, M. S. J. Simmonds, W. M. Blaney, and I. M. Boneva, *Phytochemistry*, 1994, **37**, 147.
342. B. Rodríguez, M. C. de la Torre, A. Perales, P. Y. Malakov, G. Y. Papanov, M. S. J. Simmonds, and W. M. Blaney, *Tetrahedron*, 1994, **50**, 5451.
343. M. C. de la Torre, G. Domínguez, B. Rodríguez, A. Perales, M. S. J. Simmonds, and W. M. Blaney, *Tetrahedron*, 1994, **50**, 13 553.
344. J. R. Hanson, D. E. A. Rivett, S. V. Ley, and D. J. Williams, *J. Chem. Soc., Perkin Trans. 1*, 1982, 1005.
345. M. E. Sosa, C. E. Tonn, and O. S. Giordano, *J. Nat. Prod.*, 1994, **57**, 1262.
346. R. Nishida and H. Fukami, *J. Chem. Ecol.*, 1990, **16**, 151.
347. T.-B. Chen, D. L. Galinis, and D. F. Wiemer, *J. Org. Chem.*, 1992, **57**, 862.
348. T. D. Hubert and D. F. Wiemer, *Phytochemistry*, 1985, **24**, 1197.
349. J. J. Howard, J. Cazin, and D. F. Wiemer, *J. Chem. Ecol.*, 1988, **14**, 59.
350. L. Lajide, P. Escoubas, and J. Mizutani, *Phytochemistry*, 1995, **40**, 1101.
351. S. Takahashi, T. Kusumi, and H. Kakizawa, *Chem. Lett.*, 1979, 515.
352. S. V. Ley, N. S. Simpkins, and A. J. Whittle, *J. Chem. Soc., Chem. Commun.*, 1983, 503.
353. K. Nabeta, T. Ishikawa, and H. Okuyama, *J. Chem. Soc., Perkin Trans. 1*, 1995, 3111.
354. D. H. R. Barton and D. Elad, *J. Chem. Soc.*, 1956, 2090.
355. A. G. Antonious, T. Saito, and K. Nakamuta, *J. Pestic. Sci.*, 1984, **9**, 143.
356. L. J. M. Rao, J. Pereira, and K. N. Gurudutt, *Phytochemistry*, 1993, **34**, 572.
357. R. E. Heal, E. F. Rogers, R. T. Wallace, and O. Starnes, *Lloydia*, 1950, **13**, 89.
358. E. F. Rogers, F. R. Koniuszy, J. Shavel, and K. Folkers, *J. Am. Chem. Soc.*, 1948, **70**, 3086.
359. S. N. Srivastava and M. Przybylska, *Can. J. Chem.*, 1968, **46**, 795.
360. K. Wiesner, *Adv. Org. Chem.*, 1972, **8**, 295.
361. A. L. Waterhouse, I. Holden, and J. E. Casida, *J. Chem. Soc., Perkin Trans. 2*, 1985, 1011.
362. P. Deslongchamps, *Pure Appl. Chem.*, 1977, **49**, 1329.
363. P. R. Jefferies, R. F. Toia, B. Brannigan, I. Pessah, and J. E. Casida, *J. Agric. Food. Chem.*, 1992, **40**, 142.
364. A. González-Coloma, D. Terrero, A. Perales, P. Escoubas, and B. M. Fraga, *J. Agric. Food. Chem.*, 1996, **44**, 296.
365. D. G. Crosby, in "Naturally Occurring Insecticides," eds. M. Jacobson and D. G. Crosby, Dekker, New York, 1971, p. 198.
366. P. R. Jefferies and J. E. Casida, *ACS Symp. Ser.*, 1994, **551**, 130.
367. M. Schmitt, A. Turberg, M. Londershausen, and A. Dorn, *Pestic. Sci.*, 1996, **48**, 375.
368. P. R. Jefferies, P. Yu, and J. E. Casida, *Pestic. Sci.*, 1997, **51**, 33.

369. P. R. Enslin, *J. Sci. Food Agric.*, 1954, **5**, 410.
370. D. W. Tallamy, J. Still, N. P. Ehresman, P. M. Gorski, and C. E. Mason, *Environ. Entomol.*, 1997, **26**, 678.
371. L. Dinan, P. Whiting, J.-P. Girault, R. Lafont, T. S. Dhadialla, D. E. Cress, B. Mugat, C. Antoniewski, and J.-A. Lepesant, *Biochem. J.*, 1997, **327**, 643.
372. V. P. Maier, S. Hasegawa, R. D. Bennett, and L. C. Echols, *ACS Symp. Ser.*, 1980, **143**, 63.
373. K. D. Murray, A. R. Alford, E. Groden, F. A. Drummond, R. H. Storch, M. D. Bentley, and P. M. Sugathapala, *J. Econ. Entomol.*, 1993, **86**, 1793.
374. O. H. Emerson, *J. Am. Chem. Soc.*, 1948, **70**, 545.
375. M. A. Altieri, M. Lippmann, L. L. Schmidt, and I. Kubo, *Protect. Ecol.*, 1984, **6**, 91.
376. M. B. Isman, J. T. Arnason, and G. H. N. Towers, in "The Neem Tree," ed. H. Schmutterer, VCH, Weinheim, 1995, p. 652.
377. W. Kraus, in "The Neem Tree," ed. H. Schmutterer, VCH, Weinheim, 1995, p. 35.
378. H. Hikino, T. Ohta, and T. Takemoto, *Phytochemistry*, 1975, **14**, 2473.
379. T. Murae, T. Tsuyuki, T. Ikeda, T. Nishihama, S. Masuda, and T. Takahashi, *Tetrahedron*, 1971, **27**, 5147.
380. E. London, A. Robertson, and H. Worthington, *J. Chem. Soc.*, 1950, 3431.
381. D. G. Crosby, in "Naturally Occurring Insecticides," eds. M. Jacobson and D. G. Crosby, Dekker, New York, 1971, p. 178.
382. Z. Valenta, S. Papadopoulos, and C. Podesva, *Tetrahedron*, 1961, **15**, 100.
383. G. Vidari, S. Ferriño, and P. A. Grieco, *J. Am. Chem. Soc.*, 1984, **106**, 3539.
384. N. Stojanac and Z. Valenta, *Can. J. Chem.*, 1991, **69**, 853.
385. T. Murae, A. Sugie, Y. Moriyama, T. Tsuyuki, and T. Takahashi, *Org. Mass Spectrom.* 1974, **8**, 297.
386. J. A. Klocke, M. Arisawa, S. S. Handa, A. D. Kinghorn, G. A. Cordell, and N. R. Farnsworth, *Experientia*, 1985, **41**, 379.
387. Z. Lidert, K. Wing, J. Polonsky, Y. Imakura, M. Okano, S. Tani, Y.-M. Lin, H. Kiyokawa, and K-H. Lee, *J. Nat. Prod.*, 1987, **50**, 442.
388. J. D. Connolly, K. H. Overton, and J. Polonsky, *Prog. Phytochem.*, 1970, **2**, 385.
389. J. Polonsky, *Prog. Chem. Org. Nat. Prod.*, 1985, **47**, 221.
390. J. Moron and J. Polonsky, *Tetrahedron Lett.*, 1968, 385.
391. J. H. Butterworth and E. D. Morgan, *J. Chem Soc., Chem. Commun.*, 1968, 23.
392. J. H. Butterworth and E. D. Morgan, *J. Insect Physiol.*, 1971, **17**, 969.
393. H. Schmutterer (ed.), "The Neem Tree, *Azadirachta indica* A. Juss and Other Meliaceous Plants," VCH, Weinheim, 1995, pp. 696.
394. S. Johnson and E. D. Morgan, *J. Chromatogr. A*, 1997, **761**, 53.
395. A. P. Jarvis, E. D. Morgan, and C. Edwards, *Phytochem. Anal.*, in press.
396. T. R. Govindachari, G. Gopalakrishnan, R. Raghunathan, and S. S. Rajan, *Curr. Sci.* (*India*), 1994, **66**, 295.
397. K. M. S. Sundaram, *J. Environ. Sci. Health*, 1996, **B31**, 913.
398. A. Klenk, M. Bokel, and W. Kraus, *J. Chem Soc., Chem. Commun.*, 1986, 523.
399. T. R. Govindachari, G. Sandhya, and S. P. Ganesh Raj, *Indian J. Chem.*, 1992, **31B**, 295.
400. T. R. Govindachari and G. Gopalakrishnan, *Phytochemistry*, 1997, **45**, 397.
401. C. S. S. R. Kumar, M. Srinivas, and S. Yakkundi, *Phytochemistry*, 1996, **43**, 451.
402. W. Kraus, M. Bokel, A. Bruhn, R. Cramer, I. Klaiber, A. Klenk, G. Nagl, H. Pöhnl, H. Saldo, and B. Vogler, *Tetrahedron*, 1987, **43**, 2817.
403. H. Rembold, in "The Neem Tree," ed. M. Jacobson, CRC Press, Boca Raton, FL, 1989, p. 47.
404. D. Lavie, M. K. Jain, and S. R. Shpan-Gabrielith, *J. Chem Soc., Chem. Commun.*, 1967, 910.
405. W. M. Blaney and M. S. J. Simmonds, in "The Neem Tree," ed. H. Schmutterer, VCH, Weinheim, 1995, p. 171.
406. C. N. E. Ruscoe, *Nature New Biol.*, 1972, **236**, 159.
407. H. Schmutterer and H. Rembold, *Z. Angew. Entomol.*, 1980, **89**, 179.
408. J. H. Butterworth, E. D. Morgan, and G. R. Percy, *J. Chem. Soc., Perkin Trans. 1*, 1972, 2445.
409. J. N. Bilton, H. B. Broughton, P. S. Jones, S. V. Ley, Z. Lidert, E. D. Morgan, H. S. Rzepa, R. N. Sheppard, A. M. Z. Slawin, and D. J. Williams, *Tetrahedron*, 1987, **43**, 2805.
410. S. V. Ley, H. Lovell, and D. J. Williams, *J. Chem Soc., Chem. Commun.*, 1992, 1304.
411. S. V. Ley, A. A. Denholm, and A. Wood, *Nat. Prod. Rep.*, 1993, 109.
412. H. Watanabe, T. Watanabe, and K. Mori, *Tetrahedron*, 1996, **52**, 13939.
413. P. S. Jones, S. V. Ley, E. D. Morgan, and D. Santafianos, in "The Neem Tree," ed. M. Jacobson, CRC Press, Boca Raton, FL, 1989, p. 19.
414. A. P. Jarvis, E. D. Morgan, S. A. van der Esch, F. Vitali, S. V. Ley, and A. Pape, *Nat. Prod. Lett.*, 1997, **10**, 95.
415. A. P. Jarvis, E. D. Morgan, S. A. van der Esch, F. Vitali, S. V. Ley, and A. Pape, unpublished data.
416. H. Schmutterer and K. Ermel, in "The Neem Tree," ed. H. Schmutterer, VCH, Weinheim, 1995, p. 598.
417. K. Sombatsiri, K. Ermel, and H. Schmutterer, in "The Neem Tree," ed. H. Schmutterer, VCH, Weinheim, 1995, p. 585.
418. K. R. S. Ascher, H. Schmutterer, C. P. W. Zebitz, and S. N. H. Naqvi, in "The Neem Tree," ed. H. Schmutterer, VCH, Weinheim, 1995, p. 605.
419. S.-F. Chiu, in "The Neem Tree," ed. H. Schmutterer, VCH, Weinheim, 1995, p. 642.
420. S.-F. Chiu, *J. Appl. Entomol.*, 1989, **107**, 185.
421. H. Rembold and R. W. Mwangi, in "The Neem Tree," ed. H. Schmutterer, VCH, Weinheim, 1995, p. 647.
422. L. M. Schoonhoven, *Entomol. Exper. Appl.* 1982, **31**, 57.
423. W. M. Blaney and M. S. J. Simmonds, in "Biology of Grasshoppers," eds. R. F. Chapman and A. Joern, Wiley, New York, 1990, p. 1.
424. A. J. Mordue, A. J. Nisbet, M. Nasiruddin, and E. Walker, *Entomol. Exp. Appl.* 1996, **80**, 69.
425. A. J. Mordue and A. Blackwell, *J. Insect Physiol.*, 1993, **39**, 903.
426. H. Schmutterer and R. P. Singh, in "The Neem Tree," ed. H. Schmutterer, VCH, Weinheim, 1995, p. 326.
427. H. Schmutterer and H. Wilps, in "The Neem Tree," ed. H. Schmutterer, VCH, Weinheim, 1995, p. 204
428. E. S. Garcia, B. Subrahmanyam, T. Müller, and H. Rembold, *J. Insect Physiol.*, 1989, **35**, 743.

429. H. Schmutterer, in "Natural Pesticides from the Neem Tree and Other Tropical Plants," eds. H. Schmutterer and K. R. S. Ascher, GTZ, Eschborn, 1984, p. 31.
430. R. O. Larson, in "The Neem Tree," ed. M. Jacobson, CRC Press, Boca Raton, FL, 1989, p. 155.
431. M. Jacobson, in "The Neem Tree," ed. M. Jacobson, CRC Press, Boca Raton, FL, 1989, p. 133.
432. M. Jacobson, in "The Neem Tree," ed. H. Schmutterer, VCH, Weinheim, 1995, p. 484.
433. E. D. Morgan, in "Natural Pesticides from the Neem Tree," eds. H. Schmutterer, K. R. S. Ascher, and H. Rembold, GTZ, Eschborn, 1981, p. 43.
434. W. M. Blaney, M. S. J. Simmonds, S. V. Ley, J. C. Anderson, and P. L. Toogood, *Entomol. Exp. Appl.* 1990, **55**, 149.
435. W. M. Blaney, M. S. J. Simmonds, S. V. Ley, J. C. Anderson, S. C. Smith, and A. Wood, *Pestic. Sci.* 1994, **40**, 169.
436. M. L. de la Puente, S. V. Ley, M. S. J. Simmonds, and W. M. Blaney, *J. Chem. Soc., Perkin Trans. 1*, 1996, 1523.
437. H. Rembold and I. Puhlmann, in "The Neem Tree," ed. H. Schmutterer, VCH, Weinheim, 1995, p. 222.
438. J. S. Gill and C. T. Lewis, *Nature*, 1971, **232**, 402.
439. K. M. S. Sundaram, *J. Environ. Sci. Health, Part B*, 1996, **31**, 1289.
440. R. Feng and M. B. Isman, *Experientia*, 1995, **51**, 831.
441. M. K. Bomford and M. B. Isman, *Entomol. Exp. Appl.* 1996, **81**, 307.
442. K. M. S. Sundaram and J. Curry, *J. Environ. Sci. Health, Part B*, 1996, **31**, 1041.
443. M. T. AliNiazee, A. Alhumeyri, and M. Saeed, *Can. Entomol.*, 1997, **129**, 27.
444. Anon., Deutsche Gesellschaft für Technische Zusammenarbeit (GTZ), Eschborn, no date.
445. J. E. Casida, "Pyrethrum, the Natural Insecticide," Academic Press, New York, 1973.
446. L. Crombie and S. H. Harper, *J. Chem. Soc.*, 1950, 1152.
447. L. Crombie and M. Elliott, in *Prog. Chem. Org. Nat. Prod.*, 1961, **19**, 120.
448. M. Matsui and I. Yamamoto, in "Naturally Occurring Insecticides," eds. M. Jacobson and D. G. Crosby, Dekker, New York, 1971, p. 4.
449. M. Elliott and N. F. Janes, in 'Pyrethrum, the Natural Insecticide," ed. J. E. Casida, Academic Press, New York, 1973, p. 55.
450. M. J. Begley, L. Crombie, D. J. Simmonds, and D. A. Whiting, *J. Chem. Soc., Chem. Commun.*, 1972, 1276.
451. A. F. Bramwell, L. Crombie, P. Hemesley, G. Pattenden, M. Elliott, and N. F. Janes, *Tetrahedron*, 1969, **25**, 1727.
452. L. Crombie, G. Pattenden, and D. J. Simmonds, *J. Chem. Soc., Perkin Trans. 1*, 1975, 1500.
453. T. Narahashi, in "The Physiology of Insect Central Nervous Systems," eds. J. E. Treherne and J. W. L. Beament, Academic Press, New York, 1965, p. 1.
454. M. P. Crowley, P. J. Godin, H. S. Inglis, M. Snarey, and E. M. Thain, *Biochem. Biophys. Acta*, 1962, **60**, 312.
455. P. J. Godin, H. S. Inglis, M. Snarey, and E. M. Thain, *J. Chem. Soc.*, 1963, 5878.
456. W. W. Epstein and C. D. Poulter, *Phytochemistry*, 1973, **12**, 737.
457. L. B. Bratsten, in "Molecular Aspects of Insect–plant Associations," eds. L. B. Bratsten and S. Ahmad, Plenum, New York, 1986, p. 211.
458. S. Kurata and K. Sogawa, *Appl. Entomol. Zool.*, 1976, **11**, 89.
459. D. M. Norris, *ACS Symp. Ser.*, 1977, **62**, 215.
460. R. Cooper, P. H. Solomon, I. Kubo, K. Nakanishi, J. N. Shoolery, and J. L. Occolowitz, *J. Am. Chem. Soc.*, 1980, **102**, 7953.
461. G. W. Todd, A. Getahun, and D. C. Cress, *Ann. Entomol. Soc. Am.*, 1971, **64**, 718.
462. K. Matsuda and S. Senbo, *Appl. Entomol. Zool.*, 1986, **21**, 411.
463. L. M. Schoonhoven, *Rec. Adv. Phytochem.*, 1972, **5**, 197.
464. M. B. Isman, *ACS Symp. Ser.*, 1989, **387**, 44.
465. L. Crombie, D. E. Games, N. J. Haskins, and G. F. Reed, *J. Chem. Soc., Perkin Trans. 1*, 1972, 2255.
466. S. S. Duffey and M. B. Isman, *Experientia*, 1981, **37**, 574.
467. C. A. Elliger, B. C. Chan, and A. C. Waiss, *Naturwissenschaften*, 1980, **67**, 358.
468. M. R. Berenbaum, *Econ. Entomol.*, 1981, **6**, 345.
469. G. W. Ivie, D. L. Bull, R. C. Beier, N. W. Pryor, and E. H. Oertli, *Science*, 1983, **221**, 374.
470. J. A. Klocke, M. F. Balandrin, M. A. Barnby, and R. B. Yamasaki, *ACS Symp. Ser.*, 1989, **387**, 136.
471. T. Yajima, N. Kato, and K. Munakata, *Agric. Biol. Chem.*, 1977, **41**, 1263.
472. T. Yajima and K. Munakata, *Agric. Biol. Chem.*, 1979, **43**, 1701.
473. M. R. Berenbaum, *ACS Symp. Ser.*, 1987, **339**, 206.
474. D. R. Gang, A. T. Dinkova-Kostova, L. B. Davin, and N. G. Lewis, *ACS Symp. Ser.*, 1997, **658**, 58.
475. S. Murakoshi, T. Kamikado, C.-F. Chang, A. Sakurai, and S. Tamura, *Jpn. J. Appl. Entomol. Zool.*, 1976, **20**, 26.
476. M. R. Berenbaum, *ACS Symp. Ser.*, 1989, **387**, 11.
477. J. Janprasert, C. Satasook, P. Sukumalanand, D. E. Champagne, M. B. Isman, P. Wiriyachitra, and G. H. N. Towers, *Phytochemistry*, 1993, **32**, 67.
478. C. Satasook, M. B. Isman, and P. Wiriyachitra, *Pestic. Sci.*, 1992, **36**, 53.
479. E. Geoffroy, *Ann. Inst. Colon. Marseilles*, 1896, **2**, 1.
480. F. B. LaFarge, H. L. Haller, and L. E. Smith, *Chem. Rev.*, 1993, **12**, 182.
481. L. Crombie, *Prog. Chem. Org. Nat. Prod.*, 1963, **21**, 275.
482. G. Büchi, J. S. Kaltenbronn, L. Crombie, P. J. Godin, and D. A. Whiting, *Proc. Chem. Soc.* 1960, 274.
483. M. J. Begley, L. Crombie, and D. A. Whiting, *J. Chem. Soc., Chem. Commun.*, 1975, 850.
484. M. Miyano, A. Kobayashi, and M. Matsui, *Agric. Biol. Chem.*, 1960, **24**, 540.
485. M. Miyano, *J. Am. Chem. Soc.*, 1965, **87**, 3958.
486. D. L. Adam, L. Crombie, and D. A. Whiting, *J. Chem. Soc.* 1966, 542.
487. L. Crombie, G. W. Kilbee, and D. A. Whiting, *J. Chem. Soc., Perkin Trans. 1*, 1975, 1497.
488. D. G. Carlson, D. Weisleder, and W. H. Tallent, *Tetrahedron*, 1973, **29**, 2731.
489. R. I. Reed and J. M. Wilson, *J. Chem. Soc.* 1966, 5949.
490. H. Fukami and M. Nakajima, in "Naturally Occurring Insecticides," eds. M. Jacobson and D. G. Crosby, Dekker, New York, 1971, p. 71.
491. J. L. Ingham, *Prog. Chem. Org. Nat. Prod.*, 1983, **43**, 1.
492. N. Finch and W. D. Ollis, *Proc. Chem. Soc.* 1960, 176.

493. C. Djerassi, W. D. Ollis, and R. C. Russell, *J. Chem. Soc.* 1961, 1448.
494. H. A. Yoshida and N. C. Toscano, *J. Econ. Entomol.*, 1994, **87**, 305.
495. R. D. O'Brien, *Ann. Rev. Entomol.*, 1966, **17**, 369.
496. C. W. Holyoke and J. C. Reese, in "Handbook of Naturally Occurring Pesticides," eds. E. D. Morgan and N. B. Mandava, CRC Press, Boca Raton, FL, 1987, vol. 3B, p. 67.
497. L. Crombie, *Nat. Prod. Rep.* 1984, **1**, 3.
498. L. Crombie, P. M. Dewick, and D. A. Whiting, *J. Chem. Soc., Perkin Trans. 1*, 1973, 1285.
499. L. Crombie, I. Holden, G. W. Kilbee, and D. A. Whiting, *J. Chem. Soc., Perkin Trans. 1*, 1982, 789.
500. P. Bhandari, L. Crombie, M. Sanders, and D. A. Whiting, *J. Chem. Soc., Chem. Commun.*, 1988, 1085.
501. J. Fukami, I. Yamamoto, and J. E. Casida, *Science*, 1967, **155**, 713.
502. J. L. Josephs and J. E. Casida, *Bioorg. Med. Chem. Lett.*, 1992, **2**, 593.
503. B. P. Saxena, K. Tikku, C. K. Atal, and O. Koul, *Insect Sci. Appl.*, 1986, **7**, 489.
504. W. M. Blaney, M. S. J. Simmonds, S. V. Evans, and L. E. Fellows, *Entomol. Exp. Appl.*, 1984, **36**, 209.
505. D. L. Dreyer, K. C. Jones, and R. J. Molyneux, *J. Chem. Ecol.*, 1985, **11**, 1045.
506. L. M. Schoonhoven, *Rec. Adv. Phytochem.*, 1972, **5**, 197.
507. M. Miyazawa, K. Yoshio, Y. Ishikawa, and H. Kameoka, *Nat. Prod. Lett.*, 1996, **8**, 307.
508. M. Miyazawa, K. Yoshio, Y. Ishikawa, and H. Kameoka, *Nat. Prod. Lett.*, 1996, **8**, 299.
509. J. A. Nathanson, E. J. Hunnicutt, L. Kantham, and C. Scavone, *Proc. Natl. Acad. Sci. USA*, 1993, **90**, 9645.
510. J. L. Capinera and F. R. Stermitz, *J. Chem. Ecol.*, 1979, **5**, 767.
511. J. D. Robins, *Prog. Chem. Org. Nat. Prod.*, 1982, **41**, 115.
512. E. A. Bernays and R. F. Chapman, *Ecol. Entomol.*, 1977, **2**, 1.
513. M. D. Bentley, D. E. Leonard, W. F. Stoddart, and L. H. Zalkow, *Ann. Entomol. Soc. Am.*, 1984, **77**, 393.
514. K. Munakata, in "Chemical Control of Insect Behaviour," eds. H. H. Shorey and J. J. McKelvey, Wiley, New York, 1977, chap. 6, p. 93.
515. D. G. Crosby, in "Naturally Occurring Insecticides," eds. M. Jacobson and D. G. Crosby, Dekker, New York, 1971, p. 213.
516. C. A. Elliger, Y. Wong, B. G. Chan, and A. C. Waiss, *J. Chem. Ecol.*, 1981, **7**, 753.
517. G. A. Rosenthal, *J. Chem. Ecol.*, 1986, **12**, 1145.
518. I. Schmeltz, in "Naturally Occurring Insecticides," eds. M. Jacobson and D. G. Crosby, Dekker, New York, 1971, p. 99.
519. R. Deml and K. Dettner, *Entomol. Generalis*, 1995, **19**, 239.
520. E. Leete, in "Alkaloids, Chemical and Biological Perspectives," ed. S. W. Pelletier, Wiley-Interscience, New York, 1983, vol. 1, p. 85.
521. A. Pictet and A. Rotschy, *Ber. Dtsch. Chem. Ges.*, 1904, **37**, 1225.
522. E. Wada, T. Kisaki, and K. Saito, *Arch. Biochem. Biophys.*, 1959, **79**, 124.
523. L. Marion, in "The Alkaloids, Chemistry and Physiology," eds. R. H. F. Manske and H. L. Holmes, Academic Press, New York, 1950, vol. 1, p. 167, 1960, vol. 6, p. 123.
524. I. Yamamoto, G. Yabuta, M. Tomizawa, T. Saito, T. Miyamoto, and S. Kagabu, *J. Pestic. Sci.*, 1995, **20**, 33.
525. R. D. O'Brien, "Insecticides, Action and Metabolism," Academic Press, New York, 1967.
526. W. O. Negherbon, in "Handbook of Toxicology," ed. W. O. Negherbon, Saunders, Philadelphia, vol. 3, 1959.
527. C. H. Richardson, L. C. Craig, and R. T. Hansberry, *J. Econ. Entomol.*, 1936, **29**, 850.
528. I. Yamamoto, *Adv. Pest Control Res.*, 1965, **6**, 231.
529. C. R. Enzell, J. Wahlberg, and A. J. Aasen, *Prog. Chem. Org. Nat. Prod.* 1977, **34**, 1.
530. K. Nakanishi, T. Goto, S. Ito, S. Natori, and S. Nozoe, "Natural Products Chemistry," Oxford University Press, Oxford, 1983, vol. 3, p. 544.
531. L. S. Self, F. E. Guthrie, and E. Hodgson, *Nature*, 1964, **204**, 300.
532. M. J. Snyder, J. K. Walding, and R. Feyereisen, *Insect Biochem. Mol. Biol.*, 1994, **24**, 837.
533. J. D. Hare and J. G. Morse, *J. Econom. Entomol.*, 1997, **90**, 326.
534. I. Ujváry, B. K. Eya, R. L. Grendell, R. F. Toia, and J. E. Casida, *J. Agric. Food Chem.*, 1991, **39**, 1875.
535. D. G. Crosby, in "Naturally Occurring Insecticides," eds. M. Jacobson and D. G. Crosby, Dekker, New York, 1971, p. 186.
536. W. Fahrig, *Pharmazie*, 1953, **8**, 83.
537. M. S. Blum, "Chemical Defenses of Arthropods," Academic Press, New York, 1981.
538. T. Eisner, in "Chemical Ecology," eds. E. Sondheimer and J. B. Simeone, Academic Press, New York, 1970, p. 157.
539. A. Numata and T. Ibuka, in "The Alkaloids," ed. A. Brossi, Academic Press, London, 1986, vol. 31, p. 193.
540. K. Görnitz, *Arb. Physiol. Angew. Entomol.*, 1937, **4**, 116.
541. C. W. Laidley, E. Cohen, and J. E. Casida, *J. Pharmacol. Exp. Ther.*, 1997, **280**, 1152.
542. J. P. McCormick and J. E. Carrel, in "Pheromone Biochemistry," eds. G. D. Prestwich and G. J. Blomquist, Academic Press, New York, 1987, p. 307.
543. M. Frenzel and K. Dettner, *J. Chem. Ecol.*, 1994, **20**, 1795.
544. J. Meinwald, Y. C. Meinwald, A. M. Chalmers, and T. Eisner, *Science*, 1968, **160**, 890.
545. M. Pavan and G. Bo, *Physiol. Comp. Oecol.*, 1953, **3**, 307.
546. C. Cardani, D. Ghiringhelli, R. Mondelli, and A. Quilico, *Tetrahedron Lett.*, 1965, 2537.
547. T. Matsumoto, M. Yanagiya, S. Maeno, and S. Yasuda, *Tetrahedron Lett.*, 1968, 6297.
548. M. Yanagiya and T. Matsumoto, *Tetrahedron Lett.*, 1982, **23**, 4043.
549. R. L. L. Kellner and K. Dettner, *J. Chem. Ecol.*, 1995, **21**, 1719.
550. C. Cardani, C. Fuganti, D. Ghiringhelli, P. Grasselli, M. Pavan, and M. D. Valcurone, *Tetrahedron Lett.*, 1973, 2815.
551. J. Weatherstone and J. E. Percy, in "Arthropod Venoms," ed. S. Bettini, Springer-Verlag, Berlin, 1978, p. 511.
552. B. Tursch, D. Daloze, J.-C. Braekman, C. Hootele, and J. M. Pasteels, *Tetrahedron*, 1975, **31**, 1541.
553. B. Lebrun, J.-C. Braekman, D. Daloze, and J. M. Pasteels, *J. Nat. Prod.*, 1997, **60**, 1148.
554. A. G. King and J. Meinwald, *Chem. Rev.*, 1996, **96**, 1105.
555. M. Timmermans, J.-C. Braekman, D. Daloze, J. M. Pasteels, J. Merlin, and J.-P Declercq, *Tetrahedron Lett.*, 1992, **33**, 1281.

556. A. B. Attygalle, C. L. Blankespoor, T. Eisner, and J. Meinwald, *Proc. Natl. Acad. Sci. USA*, 1994, **91**, 12 790.
557. J. W. Daly, *Brazil. J. Med. Biol. Res.*, 1995, **28**, 1033.
558. M. Pavan, *Ric. Sci.*, 1949, **19**, 1011.
559. M. Pavan, *Ric. Sci.*, 1950, **20**, 1853.
560. G. W. K. Cavill, D. L. Ford, H. Hinterberger, and D. H. Solomon, *Aust. J. Chem.*, 1961, **14**, 276.
561. J. Weatherstone, *Q. Rev.* 1967, **21**, 287.
562. A. B. Attygalle and E. D. Morgan, *Chem. Soc. Rev.*, 1984, **13**, 245.
563. K. J. Clark, G. I. Fray, R. H. Jaeger, and R. Robinson, *Tetrahedron*, 1959, **6**, 201.
564. N. J. Oldham, M. Veith, W. Boland, and K. Dettner, *Naturwissenschaften*, 1996, **83**, 470.
565. K. Kanehisa, H. Tsumuki, and K. Kawazu, *Appl. Entomol. Zool.*, 1994, **29**, 245.
566. M. Rothschild, T. Reichstein, J. von Euw, R. Aplin, and R. R. M. Harman, *Toxicon*, 1970, **8**, 293.
567. T. Eisner, D. F. Wiemer, L. W. Haynes, and J. Meinwald, *Proc. Natl. Acad. Sci. USA*, 1978, **75**, 905.
568. T. Piek (ed.), "Venoms of the Hymenoptera, Biochemical, Pharmacological and Behavioural Aspects," Academic Press, London, 1986.
569. J. O. Schmidt, in "Venoms of the Hymenoptera, Biochemical, Pharmacological and Behavioural Aspects," ed. T. Piek, Academic Press, London, 1986, p. 425.
570. B. E. C. Banks and R. A. Shipolini, in "Venoms of the Hymenoptera, Biochemical, Pharmacological and Behavioural Aspects," ed. T. Piek, Academic Press, London, 1986, p. 329.
571. J. O. Schmidt, E. D. Morgan, N. J. Oldham, R. R. do Nascimento, and F. R. Dani, *J. Chem. Ecol.*, 1997, **23**, 1929.
572. T. Piek, in "Venoms of the Hymenoptera, Biochemical, Pharmacological and Behavioural Aspects," ed. T. Piek, Academic Press, London, 1986, p. 417.
573. T. Nakajima, in "Venoms of the Hymenoptera, Biochemical, Pharmacological and Behavioural Aspects," ed. T. Piek, Academic Press, London, 1986, p. 309.
574. K. D. Bhoola, J. D. Calle, and M. Schachter, *J. Physiol.*, 1961, **159**, 167.
575. T. Piek and W. Spanjer, in "Venoms of the Hymenoptera, Biochemical, Pharmacological and Behavioural Aspects," ed. T. Piek, Academic Press, London, 1986, p. 161.
576. K. Nakanishi, D. W. Huang, K. Monde, Y. Tokiwa, K. Fang, Y. Liu, H. Jiang, X. F. Huang, S. Matile, P. N. R. Usherwood, and N. Berova, *ACS Symp. Ser.*, 1997, **658**, 339.
577. M. S. Blum, *J. Toxicol.*, 1992, **11**, 115.
578. J. Wray, *Phil. Trans. R. Soc. London*, 1670, **5**, 2063.
579. A. Hefetz and M. S. Blum, *Biochim. Biophys. Acta*, 1978, **543**, 484.
580. J. Weatherston and J. E. Percy, in "Arthropod Venoms," ed. S. Bettini, Springer-Verlag, Berlin, 1981, p. 511.
581. J. Weatherston, J. E. Percy, L. M. MacDonald, and J. A. MacDonald, *J. Chem. Ecol.*, 1979, **5**, 165.
582. A. B. Attygalle and E. D. Morgan, *Chem. Soc. Rev.* 1984, **13**, 245.
583. J. G. MacConnell, M. S. Blum, and H. M. Fales, *Tetrahedron*, 1971, **26**, 1129.
584. J. M. Brand, M. S. Blum, H. M. Fales, and J. G. MacConnell, *Toxicon*, 1972, **10**, 259.
585. T. H. Jones, P. J. DeVries, and P. Escoubas, *J. Chem. Ecol.*, 1991, **17**, 2507.
586. T. H. Jones, J. A. Torres, T. F. Spande, H. M. Garraffo, M. S. Blum, and R. R. Snelling, *J. Chem. Ecol.*, 1996, **22**, 1221.
587. P. Merlin, J. C. Braekman, D. Daloze, and J. M. Pasteels, *J. Chem. Ecol.*, 1988, **14**, 517.
588. W. Francke, F. Schröder, F. Walter, V. Sinnwell, H. Baumann, and M. Kaib, *Annalen*, 1995, 965.
589. F. Schröder, S. Franke, W. Francke, H. Baumann, M. Kaib, J. M. Pasteels, and D. Daloze, *Tetrahedron*, 1996, **52**, 13 539.
590. F. Schröder, V. Sinnwell, H. Baumann, M. Kaib, and W. Francke, *Angew. Chem., Int. Ed. Engl.*, 1997, **36**, 77.
591. T. Piek, B. Hue, P. Mantel, T. Nakajima, and J. O. Schmidt, *Comp. Biochem. Physiol.*, 1991, **99C**, 481.
592. D. Daloze, M. Kaisin, C. Detrain, and J. M. Pasteels, *Experientia*, 1991, **47**, 1082.
593. J. M. Pasteels, D. Daloze, and J. L. Boeve, *J. Chem. Ecol.*, 1989, **15**, 1501.
594. J. Meinwald, K. Opheim, and T. Eisner, *Proc. Natl. Acad. Sci. USA*, 1972, **69**, 1208.
595. J. W. Wheeler, S. K. Oh, E. F. Benfield, and S. E. Neff, *J. Am. Chem. Soc.*, 1972, **94**, 7589.
596. T. E. Rinderer, M. S. Blum, H. M. Fales, Z. Bian, T. H. Jones, S. M. Buco, V. A. Lancaster, R. G. Danka, and D. F. Howard, *J. Chem. Ecol.*, 1988, **14**, 495.
597. A. Delgado Quiroz, in "Venoms of the Hymenoptera, Biochemical, Pharmacological and Behavioural Aspects," ed. T. Piek, Academic Press, London, 1986, p. 555.
598. J. Weatherston and J. E. Percy, in "Arthropod Venoms," ed. S. Bettini, Springer-Verlag, Berlin, 1978, p. 489.
599. J. von Euw, T. Reichstein, and M. Rothschild, *Insect Biochem.*, 1971, **1**, 373.
600. L. M. Roth and D. W. Alsop, in "Arthropod Venoms," ed. S. Bettini, Springer-Verlag, Berlin, 1978, p. 465.
601. M. S. Blum, "Chemical Defenses of Arthropods," Academic Press, New York, 1981, p. 254.
602. J. M. Pasteels and D. Daloze, *Science*, 1977, **197**, 70.
603. M. Budesinský, A. Trka, K. Stránský, and M. Sreibl, *Collect. Czech. Chem. Commun.*, 1986, **51**, 956.
604. K. Dettner, A. Scheuerlein, P. Fabian, S. Schulz, and W. Francke, *J. Chem. Ecol.*, 1996, **22**, 1051.
605. M. Rothschild, *Symp. R. Entomol. Soc. London*, 1972, **6**, 59.
606. R. T. Aplin, M. H. Benn, and M. Rothschild, *Nature*, 1968, **219**, 747.
607. T. Eisner and J. Meinwald, in "Pheromone Biochemistry," eds. G. D. Prestwich and G. J. Blomquist, Academic Press, New York, 1987, p. 251.
608. U. Schaffner, J.-L. Boeve, H. Gefeller, and U. P. Schlunegger, *J. Chem. Ecol.*, 1994, **20**, 3233.
609. C. A. Dixon, J. M. Erickson, D. N. Kellett, and M. Rothschild, *J. Zool.*, 1978, **185**, 437.
610. R. Nishida and M. Rothschild, *Experientia*, 1995, **51**, 267.
611. M. Rothschild and R. T. Aplin, in "Chemical Releasers in Insects," ed. A. S. Tahori, Gordon and Breach, New York, 1971, p. 177.
612. H. J. Teas, *Biochem. Biophys. Res. Commun.*, 1967, **26**, 686.
613. J. M. Pasteels, S. Dobler, M. Rowell-Rahier, A. Ehmke, and T. Hartmann, *J. Chem. Ecol.*, 1995, **21**, 1163.
614. J. von Euw, T. Reichstein, and M. Rothschild, *Israel J. Chem.*, 1969, **6**, 659.
615. M. Rothschild, J. von Euw, and T. Reichstein, *Proc. R. Soc. London, Ser. B*, 1973, **183**, 227.

616. M. Rothschild and T. Reichstein, *Nova Acta Leopold. Suppl.*, 1976, **7**, 507.
617. M. Rothschild, J. von Euw, and T. Reichstein, *J. Insect Physiol.*, 1970, **16**, 1141.
618. T. Eisner, S. R. Smedley, D. K. Young, M. Eisner, B. Roach, and J. Meinwald, *Proc. Natl. Acad. Sci. USA*, 1996, **93**, 6499.
619. M. S. Blum, "Chemical Defenses of Arthropods," Academic Press, New York, 1981, p. 411.
620. G. A. Rosenthal and M. R. Berenbaum, "Herbivores: Their Interaction with Secondary Plant Metabolites," 2nd edn., Academic Press, New York, 1991.
621. H. Schildknecht, E. Machwitz, and U. Machwitz, *Z. Naturforsch.*, 1968, **23B**, 1213.

8.06
Microbial Hormones and Microbial Chemical Ecology

YASUHIRO YAMADA and TAKUYA NIHIRA
Osaka University, Japan

8.06.1 INTRODUCTION

A new field of research has emerged that has shown that microorganisms are capable of communication between individual cells utilizing minute amounts of compounds that behave as "signal substances." Produced by some of the cells, these compounds diffuse to other cells and thereby cause morphological or physiological changes to occur or lead to synchronization of the cell's life cycle. The trigger to generate these substances is the experience of stress by the organism. Stress includes environmental factors such as the shortage of nutrient sources, signals from their hosts, the cell density of the organism, and some previously programmed indication by certain genes. These key compounds have been termed "microbial hormones" or "pheromones." Often they are common to one genus, in some cases with small variations. In others they are unique to one species.

The first group of substances reviewed are the autoregulating butyrolactones produced by the gram-positive *Streptomyces*, followed by the homoserine lactone autoinducers that are used by gram-negative bacteria. These small molecules have a common skeleton and work has shown their widespread distribution in these genera. Next to be discussed are the compounds that regulate the morphological differentiation of prokaryotic organisms such as gram-negative *Bacillus subtilis* bacteria, referred to as the "competence pheromones" and "spore-inducing factors." These are peptides acting together with other nonpeptide compounds of as yet unknown structure. A group of five polypeptide signal substances, which mediate the complex life cycle and morphogenesis of the gram-negative *Mixococcus xanthus* bacterium are then considered and the peptides utilized by the gram-negative *Enterococcus faecalis* as sex pheromones, controlling the conjugative plasmid transfer, are described. Finally, microbial signal substances of as yet unknown mode-of-action are examined.

The isolation and structure elucidation of these compounds has proved very time consuming due to their extremely low concentration and difficult assay and purification, often requiring tons of culture medium to produce, at best, a few milligrams of compound. The combined efforts of numerous chemists and microbiologists were required to unravel the complex problems involved. Reference to the genes involved in the biosynthesis of the signal substances uses italic script, while the same letters in roman script refer to the proteins derived from these genes.

8.06.2 AUTOREGULATORS OF STREPTOMYCETES

8.06.2.1 General Aspects of *Streptomyces*-derived Microbial Hormone-like Compounds (Autoregulators) with a Butyrolactone Skeleton

Streptomyces is the major genus in Actinomycete which is known as a great producer of secondary metabolites including antibiotics, pigments, enzyme inhibitors, and other versatile physiologically active compounds. In 1992 the number of microbial secondary metabolites reported was ~16 500 and the number is increasing by about 500 per year. Sixty-seven percent of them were isolated from Actinomycete, 20% from other fungi, and 18% from bacteria. Therefore, Actinomycete is a most important source of useful drugs. The most remarkable feature of microorganisms belonging to this family is their diversity in species and also in products. For example, one species, *S. griseus*, has produced 187 physiologically active compounds (1990). As for the diversity of their products, they cover β-lactams, peptides, aminoglycosides, macrolides, polyenes, tetracyclines, quinones, polyethers, nucleosides, etc., including polyketides, substances generated from shikimate, and terpenoids from the viewpoint of their biosynthesis.

Another significant aspect of Actinomycete is their rather complex life cycle, although they belong to the prokaryote, gram-positive group. Many Actinomycete have their life cycle accompanied with cytodifferentiation, e.g., fungi which belong to the eukaryote class. Morphological changes in Actinomycete involve first substrate mycelium formation, then aerial mycelium formation, arthrospore formation, and germination of spore to substrate mycelium, as shown in Figure 1.

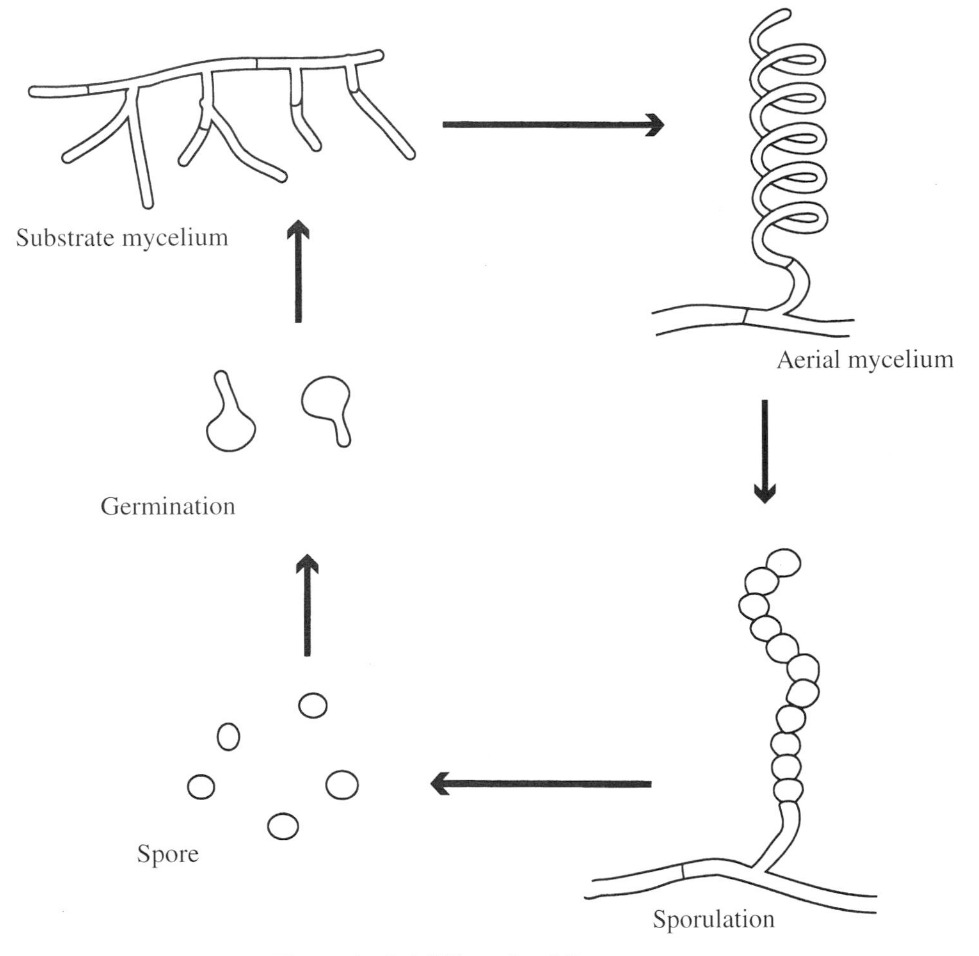

Figure 1 Brief life cycle of *Streptomyces*.

Since Actinomycete is a very important family of microorganisms in the pharmaceutical and fermentation industries, it has a long history as the target of many researches into its taxonomy and the structure and chemistry of its secondary metabolites. The fermentation technology of Actinomycete has also been widely investigated. However, the history of their molecular biology or molecular genetics is still in its infancy compared with other prokaryotes such as *Escherichia coli*, or *B. subtilis*, due to the extreme diversity of the species as mentioned previously, their GC-rich DNA and their difficulty of transformation. Consequently, metabolic engineering of the secondary metabolites is, for example, still an unexplored field of research. Nevertheless, in the 1990s, the molecular genetics and molecular biology of *Streptomyces* is progressing steadily, especially in *S. coelicolor* and *S. lividans* by the pioneering efforts of Hopwood *et al.* at the John Innes Centre, UK. As a result, new aspects of the genetics, molecular biology, and metabolic engineering of *Streptomyces* have been developing fast.

In the mid-1960s, Khokhlov *et al.* at the Shemyakin Institute of Bioorganic Chemistry (at that time part of the USSR Academy of Science) discovered a factor in *S. griseus*, a streptomycin producer. This factor, produced in the culture broth of wild-type *S. griseus*, complemented the streptomycin production and spore formation minus mutant at very low concentration. They named this factor the A-factor (autoregulating factor). In the mid-1970s its structure was determined as

2-isocapryloyl-3-hydroxymethol butyrolactone. Several types of butyrolactone factors which induce secondary metabolite production and morphological differentiation in *Streptomyces* have since been discovered and their common distribution in *Streptomyces* has been confirmed by Gräfe in Germany and Yamada and co-workers in Japan. These signal substances are produced in some cells and diffuse to other cells to cause synchronized morphological differentiation, such as aerial mycelium and spore formation or physiological differentiation leading to the production of secondary metabolites. These factors might be called microbial pheromones or hormones. Since their structures can be classified into a few types with a common butyrolactone skeleton, these factors may be regarded as a kind of microbial hormone in a broad sense, with the term "pheromone" restricted to species-specific signal substances.

As Khokhlov, the first discoverer of this factor, proposed the general name "autoregulator" for this unique signal substance, this term is used in this review, although autoregulator is also used to imply an intracellular proteinic regulator in molecular biology.

In this section, butyrolactone autoregulators of *S. griseus*, *S. virginiae* and *S.* sp. FRI-5 are reviewed.

8.06.2.2 A-factor of *Streptomyces griseus*

8.06.2.2.1 Introduction

The first discoverer of this signal substance, Khokhlov, published about 10 articles, mainly in Russian, during 1967–1978 which opened the research field of autoregulators in *Streptomyces*. In 1980, he published a review[1] in English about the idea which led his group to the discovery of this unique factor together with the history of their research. Readers who are interested in the details of their results should refer to this article and the references therein.

Khokhlov's group obtained a number of *S. griseus* streptomycin minus mutants and found that one pair of the mutants, *S. griseus* 1439 and 751, complemented the production of streptomycin. When strains 1439 and 751 were cultured together, the yield of the antibiotic was the same as that of a control wild type. They found a very small amount of a factor in the culture broth of strain 751 which completely restored the production of streptomycin in strain 1439. They named this the A-factor, meaning "autoregulating factor." They disclosed that A-factor is commonly distributed among streptomycin-producing *S. griseus* and many of their streptomycin minus mutants were restored to antibiotic production by A-factor addition. They investigated the production of the intermediates of streptomycin biosynthesis in its minus mutants and did not find any of the possible intermediates. Consequently, they concluded that A-factor is a pleiotropic regulator which triggers the production of streptomycin. Khokhlov *et al.* also revealed that A-factor restored formation of spores in streptomycin minus mutants. Accordingly, A-factor turned out to be a very potent pleiotropic regulator of morphological and physiological differentiation of *S. griseus*.

8.06.2.2.2 Structure and synthesis

Khokhlov and co-workers carried out the isolation of A-factor and determined its structure in the mid-1970s[2,3] and they describe the very difficult purification procedures due to low concentrations of the target factor produced with much solid residue. They estimated that 20 mg of oily crude A-factor was obtained together with other homologues from 0.5 ton of culture broth.

They repeated this procedure[2] and finally obtained a crystalline A-factor diphenylcarbanilate derivative) (**1**) with the empirical formula $C_{27}H_{32}N_2O_6$, from which they deduced the molecular formula of A-factor as $C_{13}H_{22}O_4$. On the basis of spectroscopy of the diphenylcarbamate derivative and of A-factor itself and from the information obtained by chemical degradation, as shown in Scheme 1, Khokhlov and co-workers proposed the structure of A-factor as 2*S*-isocapryloyl-4*S*-hydroxymethylbutyrolactone (**2**). The absolute configurations at C-2 and C-3 were deduced from the circular dichroism spectrum of A-factor.[1]

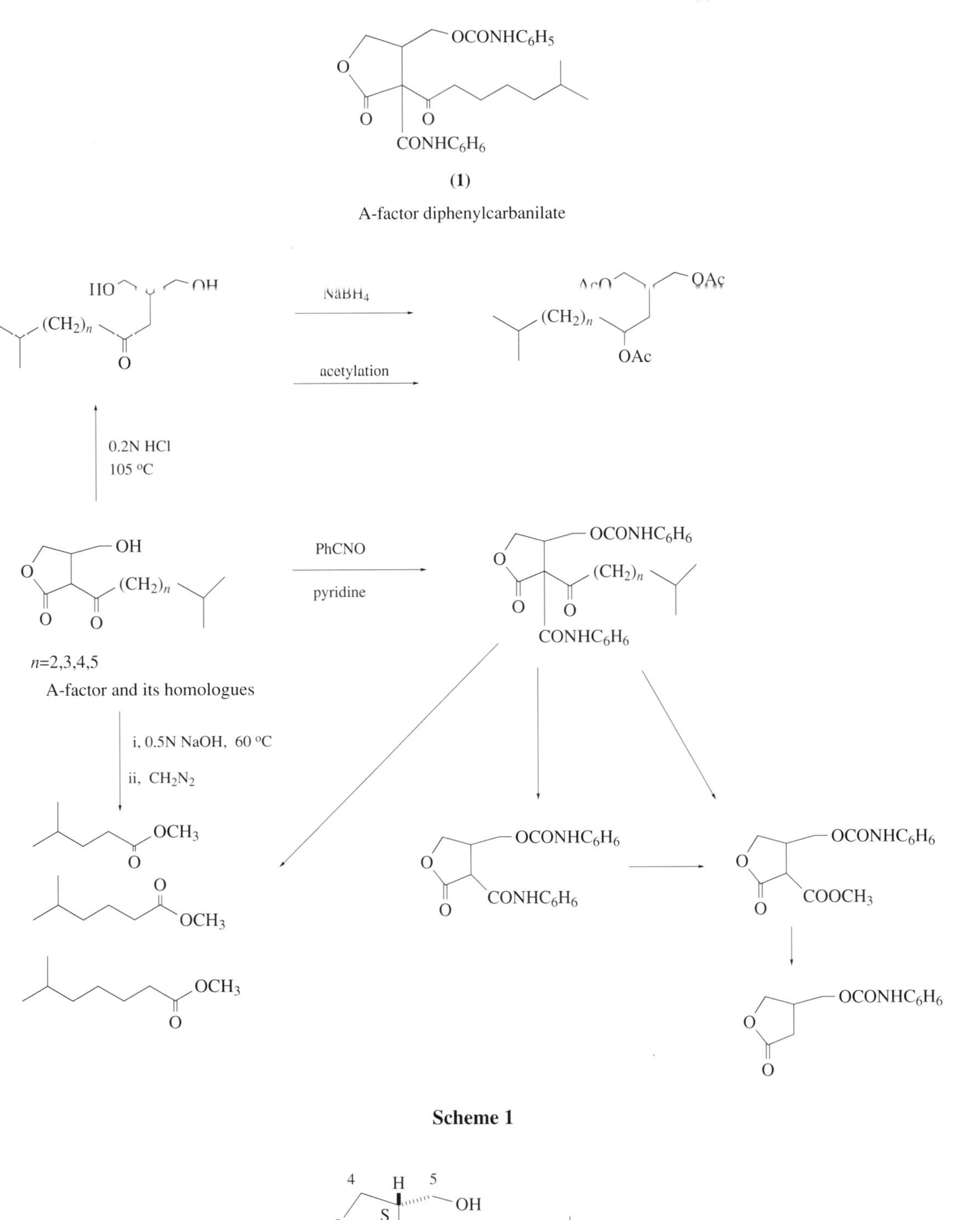

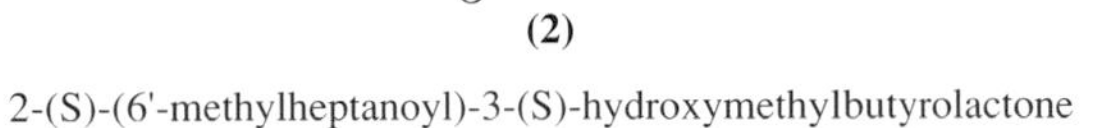
2-(S)-(6'-methylheptanoyl)-3-(S)-hydroxymethylbutyrolactone

Khokhlov and co-workers also synthesized A-factor and its analogues,[4] as shown in Scheme 2. They studied the structure and activity relationship using synthesized A-factor analogues and showed that the hydroxymethyl group at C-3 is essential for its activity and also the length of the alkyl side chain at C-2 is critical for induction of streptomycin biosynthesis.

Mori *et al.*[5,6] confirmed Khokhlov's absolute configuration as 2*R*-isocapryloyl-4*R*-hydroxymethylbutyrolactone (Scheme 3) by synthesizing natural A-factor from *S*-(−)-paraconic acid.

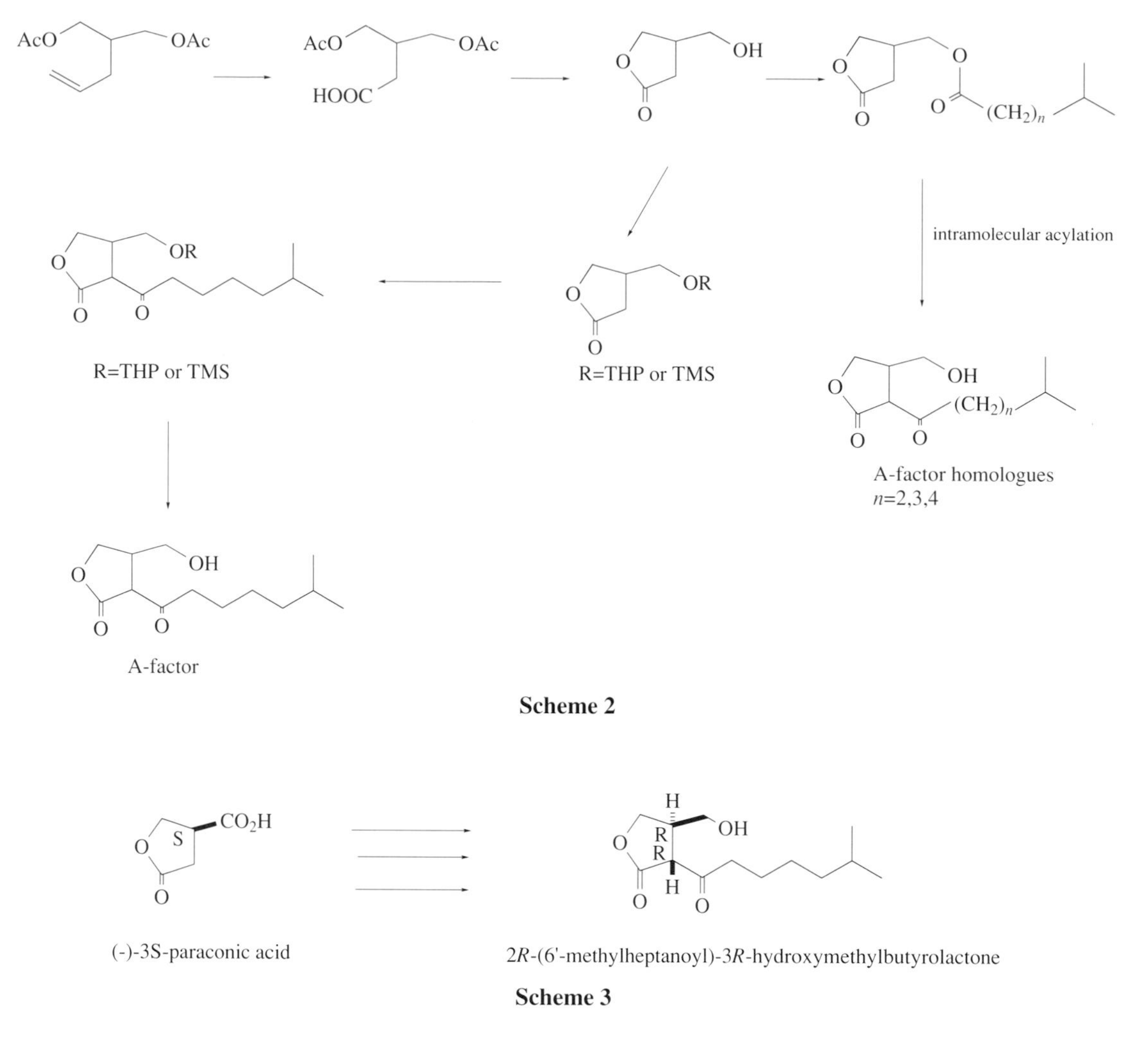

Scheme 2

Scheme 3

8.06.2.2.3 Mode-of-action

In the early 1980s, Beppu and co-workers at Tokyo University started molecular biological studies of A-factor. They confirmed the pioneering work of the Russian group.[1] Using stereoselectively synthesized A-factor they found that it induces streptomycin resistance in its producer via expression of Sm-6-phosphotransferase.[7,8]

They discovered one gene in *S. griseus*, named *afsA*, which conferred the ability to produce A-factor. The *afsA* gene codes for a protein of 301 amino acid residues.[9] The homologous genes were found in other Streptomycete, *S. bikiniensis* and *S. coelicolor*, which produce autoregulators of A-factor activities.[10,11] In *S. griseus*, *afsA* is located on a giant plasmid and it is easily cured with acridine orange or other curing reagents.[12] The protein AfsA is deeply involved in the biosynthesis of A-factor and it is possible that this protein is an enzyme which catalyzes some important step in the biosynthetic pathway of this butyrolactone autoregulator. However, there is no biochemical evidence *in vitro* or *in vivo* of its action.

The mode of signal transduction with A-factor is initiated by its binding to the A-factor receptor protein, ArpA. This A-factor receptor protein was actually isolated from cells of *S. griseus* using tritium-labeled A-factor prepared as shown in Scheme 4.[13] ArpA is a cytoplasmic protein of MW 29 kDa with 272 amino acid residues and there are 30 to 40 molecules of ArpA in each cell.[14] The K_d value of A-factor with ArpA is a few nM. ArpA has a dimeric tertiary structure and a helix-turn-helix motif which characterizes a DNA-binding domain in its N-terminus part.[14] As *S. griseus* ArpA deficient mutants produce streptomycin and show morphogenesis without the effect of A-factor, ArpA is supposed to be a repressor of an unidentified gene *X* which may trigger the production of antibiotics or cytodifferentiation.[15]

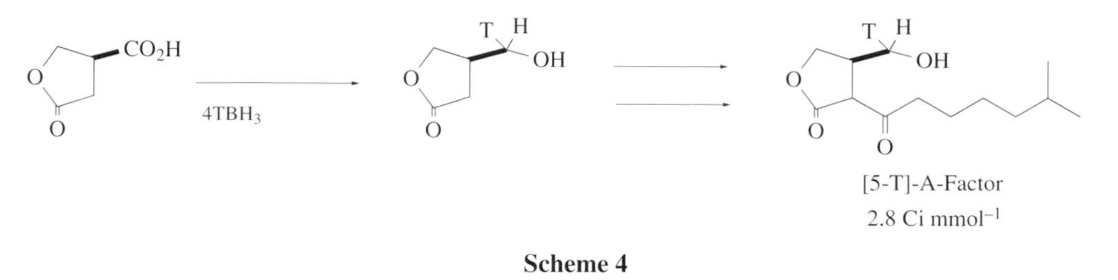

Scheme 4

How does the signal on A-factor, transduced by its binding with ArpA, lead to streptomycin production, antibiotic resistance, and morphogenesis? The Tokyo University group proposed a chart of the A-factor regulatory network, as shown in Figure 2. The signal transduction initiated by the A-factor–ArpA complex finally triggers the expression of A-factor-dependent-binding protein (Adp)[16] and it consequently induces a regulator, StrR.[17] StrR is the main pleiotropic up-regulator of streptomycin production and it releases the production of AphD which causes streptomycin resistance. StrR is also presumed to prime the transcription of a gene cluster which contains nearly 30 genes for streptomycin biosynthesis.[18] As for the mechanism of initiation in morphogenesis such as sporulation, there is no data (1997).

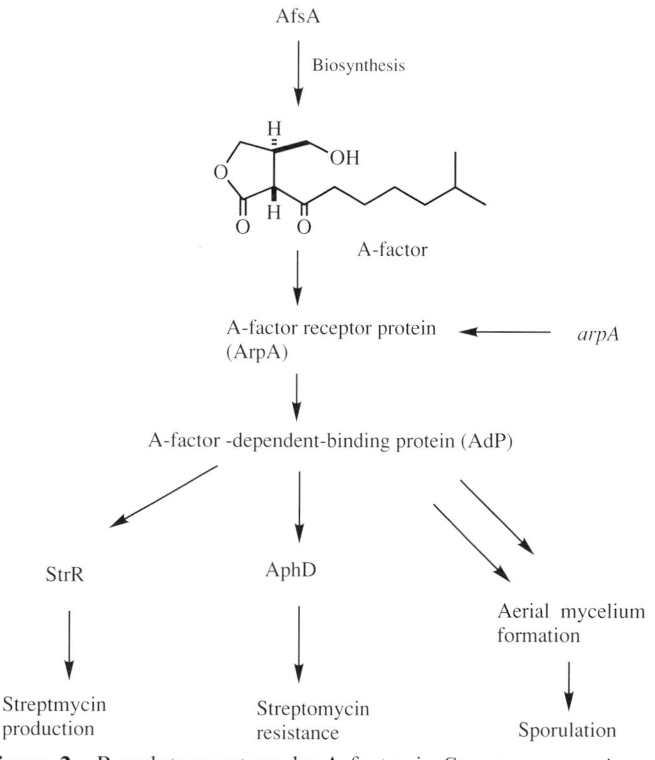

Figure 2 Regulatory sytems by A-factor in *Streptomyces griseus*.

A speculative model for regulation of streptomycin biosynthesis in *S. griseus*, including the gene transcription chart, is shown in Figure 3. In this model,[12] the A-factor receptor protein, ArpA, represses the transcription of unidentified gene *X* in the absence of A-factor. At a rather early stage of culture, AfsA expression induces onset of A-factor appearance and it diffuses to other cells to bind with ArpA. Then ArpA releases from the promoter site of gene *X* leading to its transcription and protein X expression. Protein X may activate the *adp* gene and its product, A-factor-dependent-binding protein, next up-regulates promoter PA for gene *strR*. The StrR protein triggers expression of *aphD* and other genes for streptomycin biosynthesis.

Generally speaking, since regulatory systems of living organisms are very complex even in rather primitive microorganisms, these systems often have multiple circuits to compensate each other and this model may be more complex.

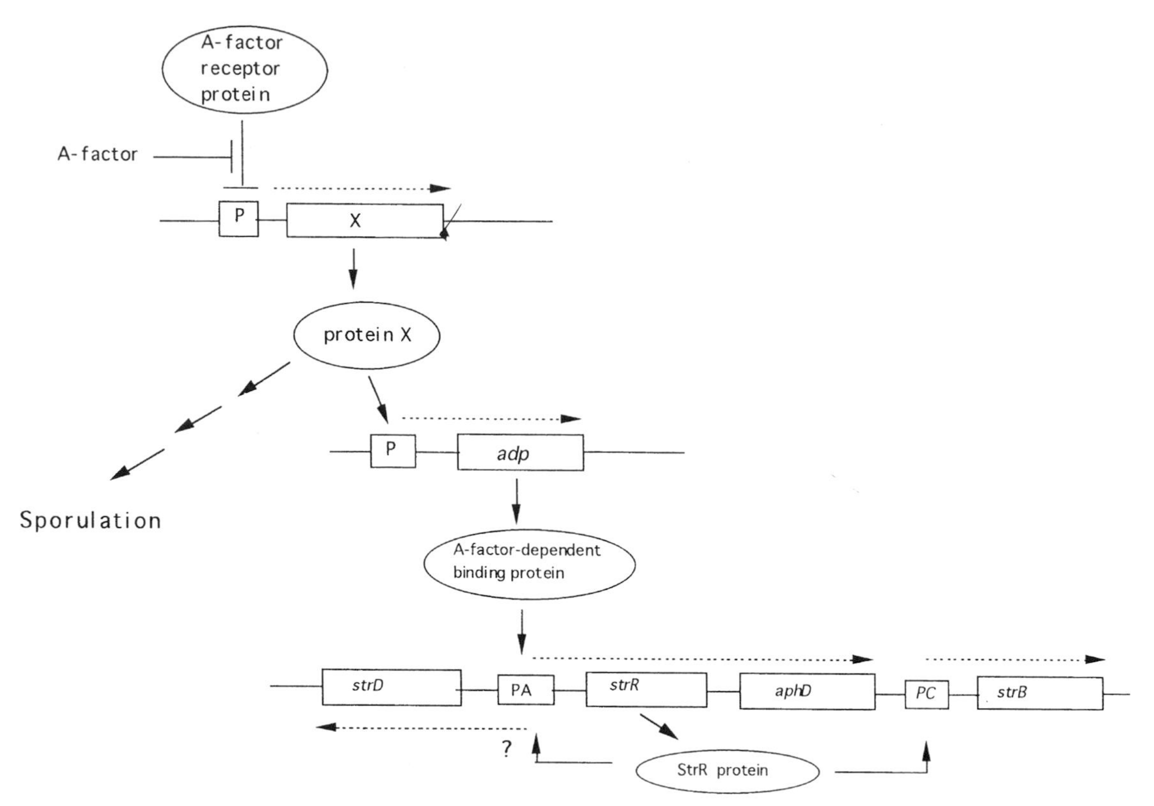

Figure 3 Possible pathway of A-factor receptor mediated signal transmission.

8.06.2.3 Virginiae Butanolides of *Streptomyces virginiae*

8.06.2.3.1 Introduction

The history of the autoregulator in *S. virginiae* goes back to the end of the 1960s. At that time, Yanagimoto and Terui at Osaka University were working on the production of virginiamycin using *S. virginiae* and they found that some inducer of antibiotic production is present in the culture broth of this actinomycete.[19] They named this compound IM (inducing material) and studied it from the standpoint of fermentation technology. They did not derive any mutant but took the time course of the growth, IM production, and virginiamycin production. IM is produced at 12 h into cultivation and it induces the production of virginiamycin after 4 h (i.e., at 16 h into culture). Since the assay system could not depend upon the antibiotic-deficient mutants as in the case of A-factor, a new analytical method was elaborated to estimate the quantity of IM produced in the culture broth. In this analysis, a sample of the culture broth or its extract was added to an 8 h culture of *S. virginiae* and, if it had activity, after 4 h additional cultivation it led to the production of virginiamycin, which was determined by bioassay on a plate culture.

Yanagimoto *et al.* tried to isolate this signal substance, IM, and partially purified it. They found that it was a low molecular weight, ethyl acetate extractable oily substance with a lactone moiety and some hydroxy groups.[20] Later, in the mid-1980s, Yamada and co-workers purified IM and revealed that IM is a mixture of five homologous butyrolactone derivatives[21,22] having the same skeleton as A-factor and the already-known Gräfe's factors[23,24] which were isolated from other *Streptomyces* species. These five autoregulators were named virginiae butanolide A–E (VB A–E). In this section, the structures of VB, their chemical synthesis, biosynthesis, and mode-of-action are described.

8.06.2.3.2 Structures of VB and their chemical synthesis

As the concentration of VB is less than 100 $\mu g\ l^{-1}$ in a culture broth of *S. virginiae*, only about 1 mg or less of each VB A–E was isolated from the ethyl acetate extract of tons of culture broth, after tedious purification processes and repeated time-consuming assay procedures.[21,22] At first, the structures of VB A, B, C, D, and E were assigned as shown in Figure 4 on the basis of their NMR and mass spectroscopy, and the chemical synthesis of VB C. At this stage the configuration of the 1′-hydroxyalkyl group on C-2 and hydroxymethyl group on C-3 was presumed to be *cis* (i.e., the same as Gräfe's factors) on the basis of coupling constants between C2-H and C3-H. Furthermore, the configuration of the hydroxy group on C-6 (C-1′ in the alkyl side chain) was not assigned. Later, the nOe of every proton on this butyrolactone skeleton was analyzed in detail, which supported the first proposed *cis* configuration. Many attempts to obtain crystalline VB A derivatives suitable for X-ray crystallography were unsuccessful.

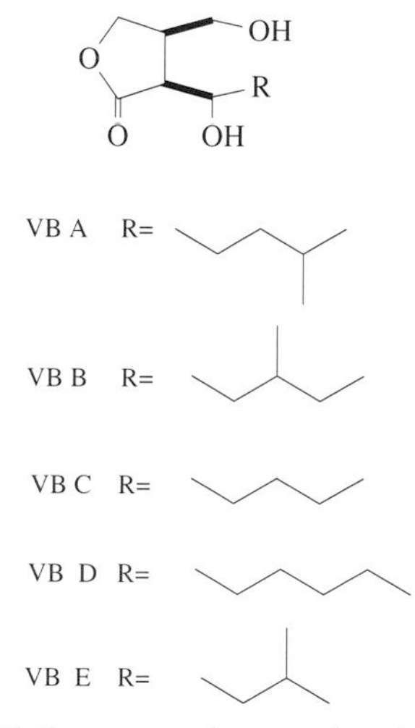

Figure 4 Initially proposed stereochemistry of VB A–E.

Synthetic racemic *cis* and *trans* VB A were transformed into their corresponding bicycloacetals, (**6a**) and (**6b**) (Scheme 5), and the coupling constants and NOE of protons on these fixed-ring systems were determined in detail.[25]

These results clearly showed that the former *cis* configurational assignment in natural VB A was wrong and it is actually *trans*. The configuration of the hydroxy group on C-6 was determined to be α. Since Mori and Chiba had already determined the absolute configuration of VB A, B, and C on C-3 as *R*,[26] the structures of the VBs were confirmed as 2*R*, 3*R*, 6*S*, as shown in Figure 5. The former *trans* isomers, which included one of Gräfe's factors,[23] IM-2,[27] and another synthetic factor,[21] were actually *trans* and the configuration of the hydroxy group on C-6 is β, as shown in Figure 6.

8.06.2.3.3 Biosynthesis of VB A

Since the concentration of VBs in a culture broth of *S. virginiae* is at nM level, it is very difficult to conduct incorporation experiments to clarify their biosynthetic pathway using stable isotope-labeled precursors. However, during studies of the distribution of VBs in *Streptomyces*, Yamada and co-workers found one *S. antibioticus* strain which produced a small amount of VB A. Therefore, biosynthetic studies of VB A were conducted with *S. antibioticus*.[28] The building blocks of the VB A molecule turned out to be two acetic acids, one isovaleric acid, and glycerol, as shown in Scheme 6. The presence of 3-oxo-7-methyloctanoyl CoA as a key intermediate was confirmed by intact incorporation of *N*-acetylcysteamine thioester of [2,3-^{13}C]-3-oxo-7-methyloctanoic acid into VB A.[29] The fate of protons on glycerol during incorporation into the VB A molecule was also studied and it was found that the C-2 proton is lost and one of the C-1 protons is also stereoselectively replaced.[30] On the basis of these results, two plausible biosynthetic pathways from 3-oxo-7-methyloctanoyl CoA were considered, as shown in Scheme 7. Yamada and co-workers prepared the possible intermediates (**7**)–(**11**) and tested their conversion into VB A with a crude cell-free enzyme system.[31] The dihydroxyacetone ester of 3-oxo-7-methyloctanoic acid (**10**) was the best substrate and this reaction was NADH-dependent. The phosphorylated counterparts (**7**)–(**9**) in pathway A turned

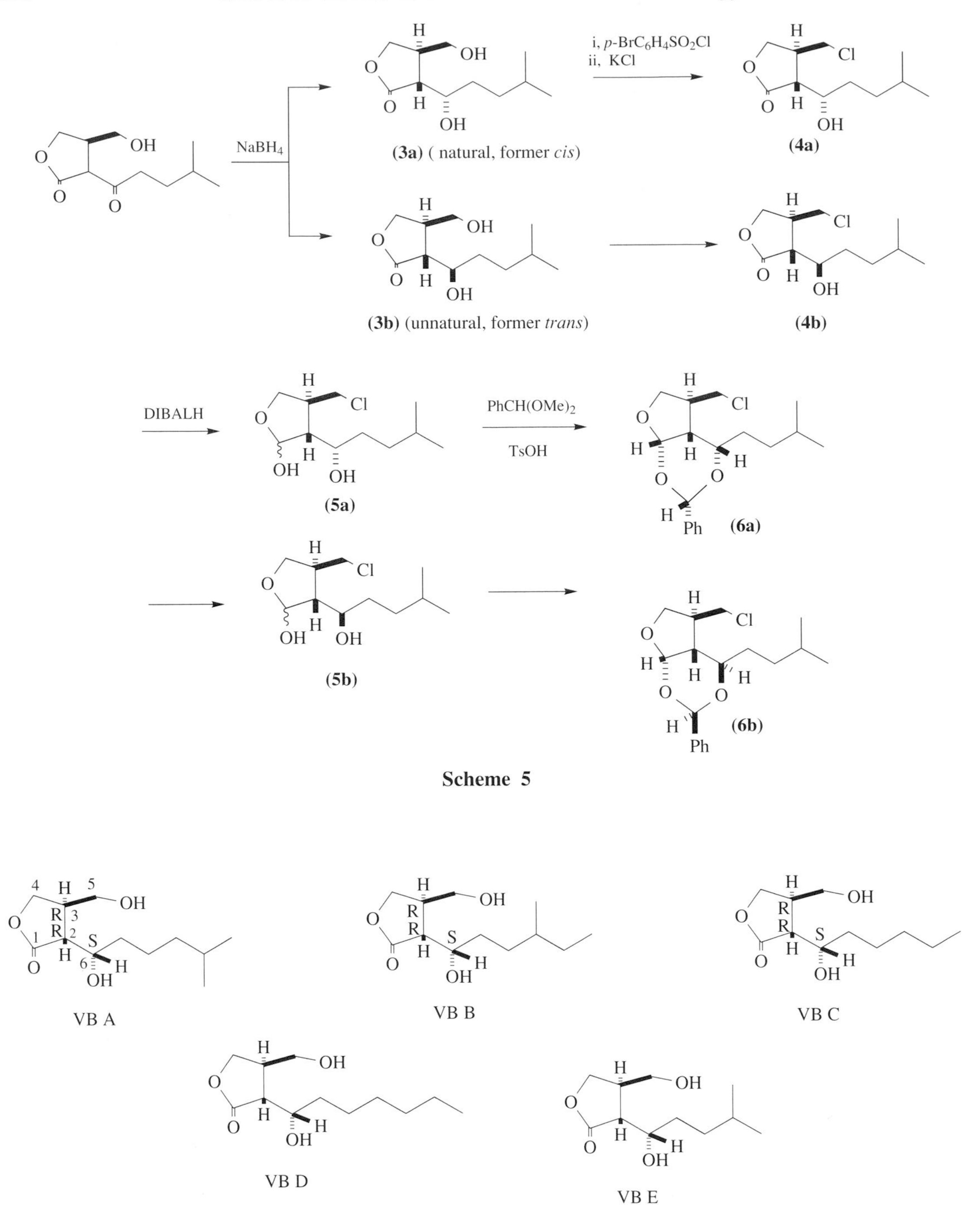

Scheme 5

Figure 5 Structures of virginiae butanolides A–E.

out to be poor substrates. 6-Dehydro VB A (**11**) was also efficiently transformed to VB in the presence of NADPH. Consequently, pathway B was regarded as preferable to pathway A. In Scheme 8 the final two steps of VB A biosynthesis is illustrated in detail, taking into consideration the stereochemistry of NADH hydride delivery to form a butyrolactone ring. The biosynthetic pathway of VB A in *S. antibioticus* might be common to other A-factor or IM-2 type autoregulators.

The starter or number of extenders in the key intermediate, 3-oxo-acyl CoA, determines the chain length of the C-2 alkyl group. The redox state of the C-6 carbon [oxo (A-factor type) or hydroxy (VB and IM-2 types)] and the orientation of the hydroxy group (α: VB type or β: IM-2 type) depends upon the character of the final NADPH-dependent hydrogenases.

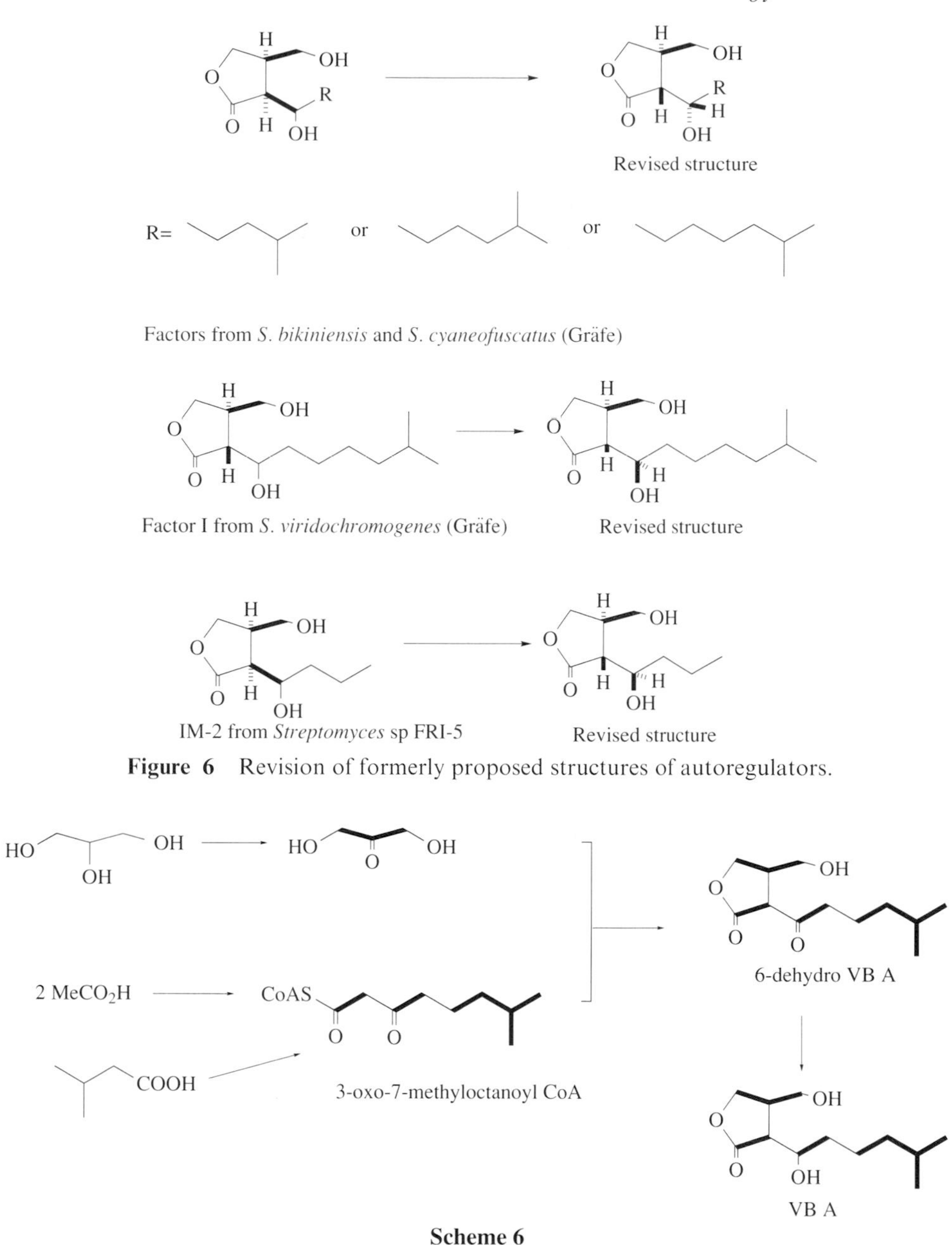

Figure 6 Revision of formerly proposed structures of autoregulators.

Scheme 6

8.06.2.3.4 Mode of action of VBs

The VB receptor protein in *S. virginiae* was actually characterized, isolated, and cloned by Yamada and co-workers as the first example among the butyrolactone-type autoregulator receptors in streptomycete. First, the structure and activity relationships of this unique signal molecule were investigated.[32] It was found that every functional group, the lactone ring, the two hydroxy groups, the length of the C-2 alkyl group, and the stereochemistry in this compact molecule were essential to express its physiological activity. On the basis of this result, the tritium-labeled ligand [11,12-$^3H_2$]VB C_7(= VB D) was prepared, as shown in Scheme 9, and assay systems for purification of VB receptor protein were developed with this ligand.[33] The VB receptor was identified in the cytoplasm of *S. virginiae* cells as a very small section of protein that contains only 30 to 40 molecules per genome. Its Kd value with this ligand was estimated to be about 1.1 nM in the crude state.[34] A protein named VbrA (VB receptor A) shadowed the real VB receptor protein at every purification step and at first it was mistaken as the receptor.[35] VbrA was eliminated finally and the real VB receptor protein, BarA (butyrolactone autoregulator receptor A) was purified.[36] The molecular weight of BarA was estimated to be 26 kDa and BarA forms a dimer in the native state. The *barA*

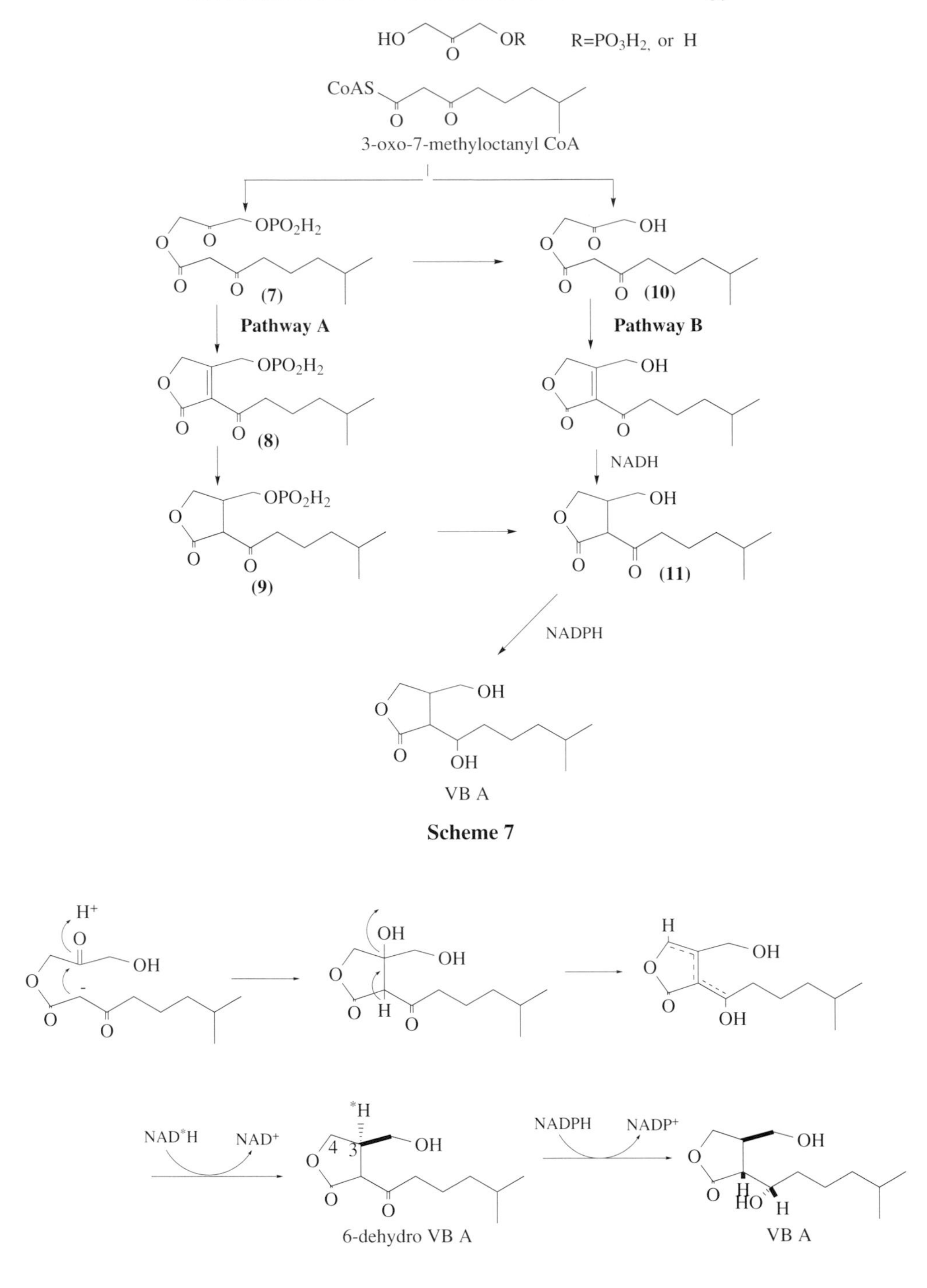

Scheme 7

Scheme 8

gene codes 232 amino acids and at its N-terminus, there is a helix-turn-helix motif which strongly suggests the feature of a DNA binding site. Evidence obtained *in vivo* and *in vitro* indicates that BarA binds to specific sites of *S. virginiae* DNA and acts as a repressor. When VB binds with BarA on DNA, it is released from DNA leading to transcription of the downstream genes. As most of the enzymes or their genes and loci on the chromosome for the biosynthesis of virginiamycin M and S are not yet determined, the direct regulation systems between the initiation by VB-bound BarA and the final activation mechanism of the biosynthetic genes are problems. However, the pleiotropic role and triggering mechanisms of VB and BarA at the initiation stage will soon be clarified. The regulation systems of differentiation in living organisms are generally complex and there are many loops of signal transmission compensating each other. These systems also depend on the stage of the cell cycle or the age of the living organism.

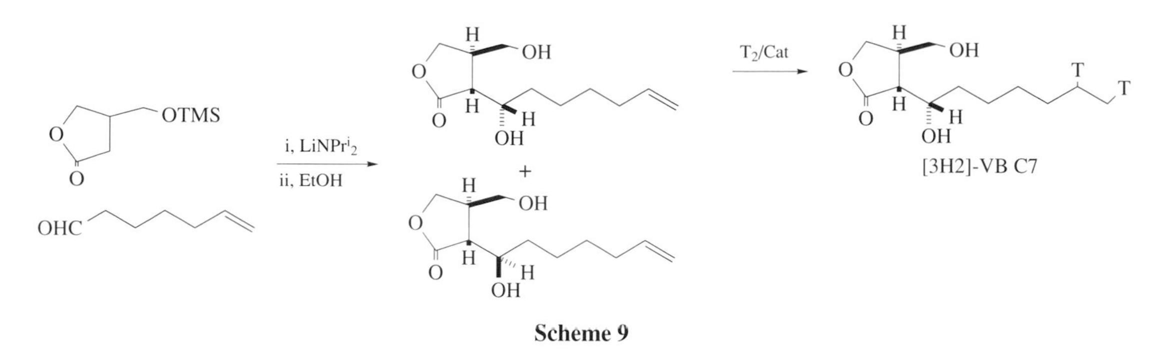

Scheme 9

In *S. virginiae*, this general rule is not an exception, even though it belongs to a rather primitive prokaryote compared with yeasts or fungi, which are eukaryotic. For example, addition of VB at an early stage of culture, after 0–2 h of inoculation, suppresses virginiamycin production and this suppression continues for more than 24 h of culture. In this case, cell growth and morphological differentiation is normal. On the basis of these data and results of *barA* gene disruption experiments, early onset of the next regulator by VB-bound BarA leads to down-regulation of antibiotic production, indicating the importance of VB appearance timing and the complexity of this regulation system.

8.06.2.4 IM-2 of *Streptomyces* sp. FRI-5

8.06.2.4.1 Introduction

Streptomyces sp. FRI-5 is intrinsically a producer of D-cycloserine and it produces a blue pigment in the presence of quantities of nonalactone (10 μg ml^{-1}) or when it grows on rather poor medium. The presence of an endogenous autoregulator which induces blue pigment formation was proved by Yanagimoto and Enatsu.[37] Yamada and co-workers have purified it from the isobutyl alcohol extract of more than 1 ton of culture broth and obtained less than 1 mg of an autoregulator named IM-2. The work involved 2.5 years of purification and development of assay procedures.[27] IM-2 is more hydrophilic than VB and the salient point of its activity is the induction of nucleoside-type antibiotics which are quite different from D-cycloserine.

8.06.2.4.2 Structure

Since IM-2 is more hydrophilic than VB, 1150 l of culture broth was extracted with isobutyl alcohol. This extract gave ~715 g residue which was finally purified to 580 μg of pure IM-2 after repeated reverse-phase chromatography. The structure of IM-2 was determined by comparison of its ^{1}H-NMR with that of synthetic 2-(1′-hydroxyalkyl)-3-hydroxymethyl-butyrolactone, as shown in Figure 6. The β orientation of the hydroxy group at C-6[25] and the absolute configuration, 2*R*, 3*R*, 6*R*,[38] (**12**) were determined later. The effect of the C-2 alkyl chain length on activity was investigated with synthesized IM-2 analogues (**13**).

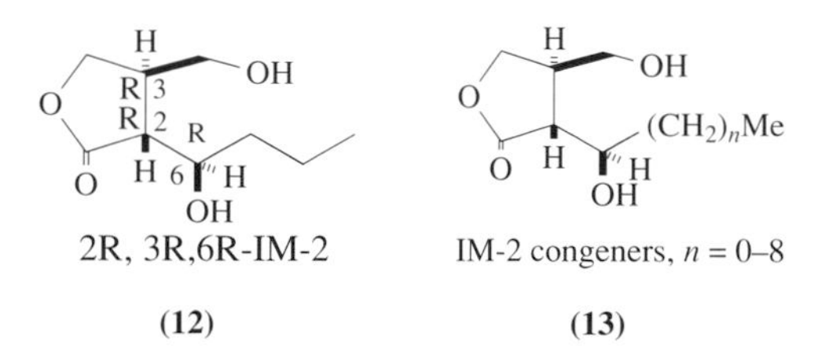

The most active chain lengths are butyl ($n = 2$) or valeryl ($n = 3$) and their minimum effective concentration is less than 1 ng ml^{-1}. On the other hand, analogues with shorter or longer alkyl side chains showed remarkably poor activities, one hundredth or one ten thousandth, compared with those of the nature butyl or synthetic valeryl derivatives.[27]

8.06.2.4.3 Effects of IM-2 in Streptomyces *sp. FRI-5*

When IM-2 was added at 5 h cultivation, cell growth was suppressed in a couple of hours and blue pigment was rapidly induced. After 36 h cultivation, nucleoside-type antibiotics including showdomycin, minimycin, and their stereoisomers were identified in the culture broth, and D-cycloserine was not detected at all[39] (Figure 7). *Streptomyces* sp. FRI-5 is isolated as a D-cycloserine producer at first and it produces only D-cycloserine without addition of IM-2 under these culture conditions. Consequently, when environmental conditions such as nutrient and aeration states deteriorate, IM-2 production might be induced and this autoregulator subsequently triggers the alteration of the secondary metabolite pattern from D-cycloserine to nucleoside-type antibiotics and blue pigments.

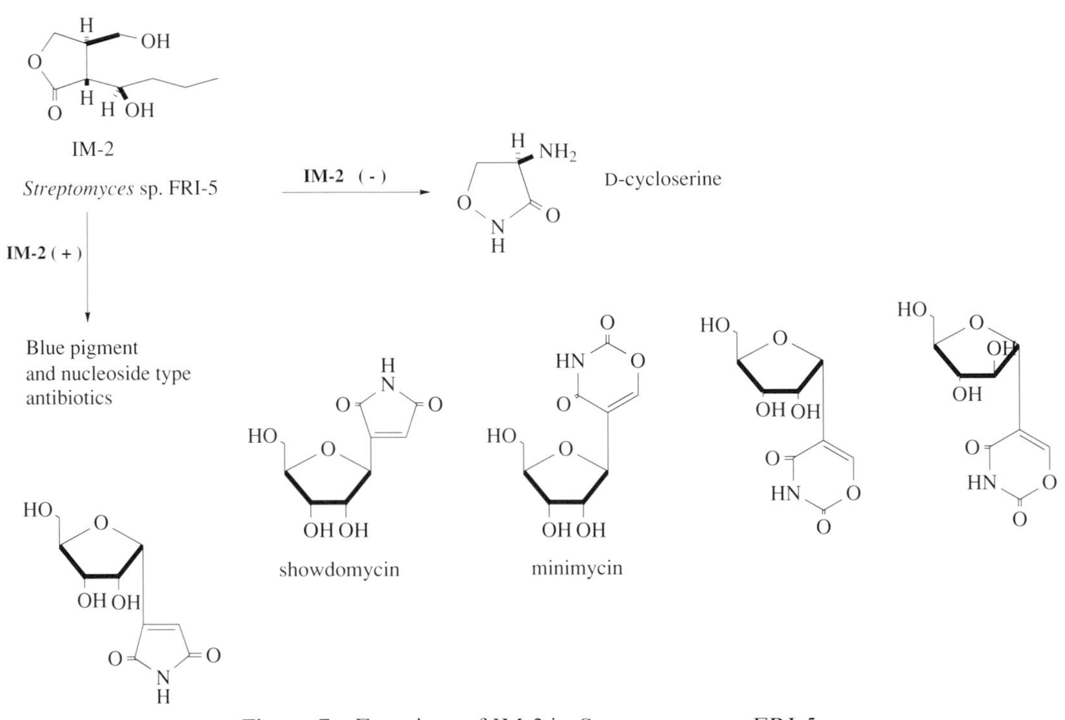

Figure 7 Functions of IM-2 in *Streptomyces* sp. FRI-5.

8.06.2.4.4 Mode-of-action

IM-2 receptor protein was isolated and identified from *Streptomyces* sp. FRI-5. Yamada and co-workers prepared the tritium-labeled ligand, [9,10-$^3H_2$]-2-(1′-β-hydroxypentyl)-3-hydroxymethylbutyrolactone (**14**) ([^{3}H]-IM-2-C_5), and purified IM-2 receptor protein (FarA) with it.[40] FarA is a cytoplasmic protein and its molecular weight was estimated at ~27 kDa by SDS-PAGE. FarA exists in a dimer form of 54 kDa in the native state. The Kd value of FarA with [^{3}H]-IM2-C_5 is 1.3 nM and its binding activity is specific to IM-2.[40] The *farA* gene encodes 221 amino acids and also has a helix-turn-helix motif on its N-terminus, indicating its DNA binding property as a regulator of transcription. Therefore, the IM-2 receptor, FarA, is also presumed to play a pleiotropic role in co-operation with IM-2 to initiate physiological differentiation.

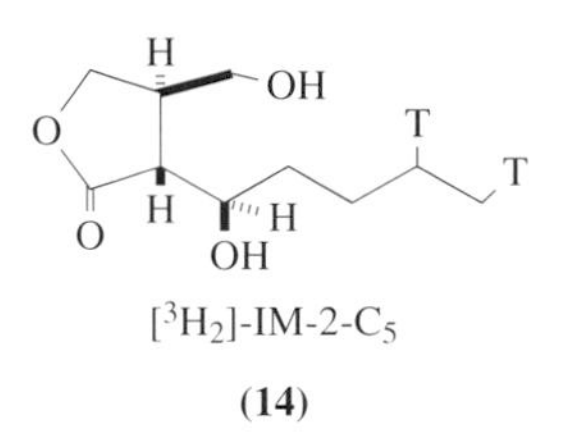

[$^{3}H_{2}$]-IM-2-C_5

(14)

8.06.2.5 Gräfe's Factors

8.06.2.5.1 Structures and functions

Grafe *et al.* screened many Streptomycete for the autoregulators which revive the anthracycline production and aerial mycelium formation of *S. griseus* ZIMET 43682, an anthracycline and aerial mycelium minus mutant. They isolated new autoregulators from *S. viridochromogenes*,[23] *S. bikiniensis*, and *S. cyaneofuscatus*[24] using this mutant as an indicator of anthracycline production and mycelium formation. Their isolates have a 3-hydroxymethylbutyrolactone skeleton in common with A-factor and have a 1′-hydroxy alkyl substituent on C-2, as shown in Figure 6. The stereochemistry of the two substituents on C-2 and C-3 of the butyrolactone ring was revised and the configuration of the hydroxy group on C-6 was determined later.[25] However, their absolute configurations have not yet been determined (1997). Gräfe's autoregulators are not endogenous and tens of nanograms are required to express their activities, which is 10-fold higher compared with those of A-factor, VB, and IM-2. Gräfe's factors from *S. bikiniensis* and *S. cyaneofuscatus* were obtained in mixture form[24] and consequently, their specific activities cannot exactly be estimated. However, the values are still enough to classify Gräfe's factors as autoregulators. These results showed that *Streptomyces* autoregulators are not strictly species specific, as are insect pheromones, but they share common features among *Streptomyces*-like hormones in eukaryotes.

8.06.2.6 Conclusion

The three known types of streptomycete autoregulators (**15**)–(**17**) have a common skeleton, 2-(1′-oxo or hydroxy-alkyl)-3-hydroxymethol butyrolactone, and the configuration of the two substituents on this butyrolactone ring is *trans*. They differ by the redox state on C-6 (oxo in A-factor, hydroxy in VB and IM-2), the orientation of the hydroxy groups on C-6, and the length and branching of the C-2 alkyl groups. Their absolute configurations are 3*R* for A-factor; 2*R*, 3*R*, 6*S* for VBs; and 2*R*, 3*R*, 6*R* for IM-2; and the presence of their enantiomers or *cis*-type diasteroisomers in Streptomycete could not be ruled out.

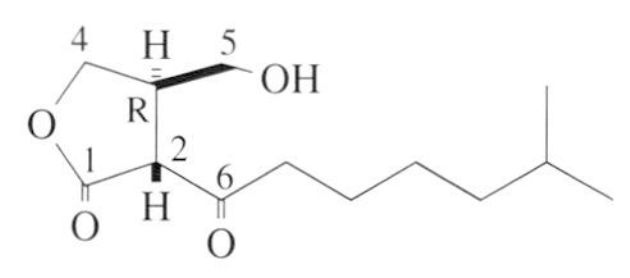

(**15**) A-factor *S. griseus*

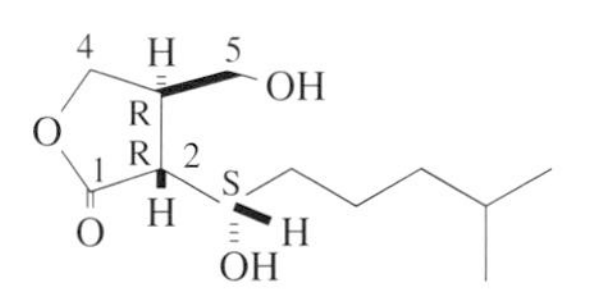

(**16**) Virginiae Butanolide A *S. virginiae*

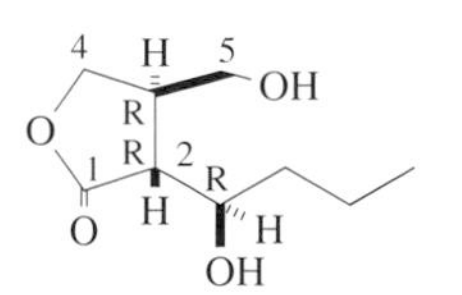

(**17**) IM-2 *Streptomyces* sp FRI-5

Each of these three autoregulators has its specific receptor protein and these receptors also have common features such as molecular size, dimer formation, their location in cytoplasm, a helix-turn-helix motif of the DNA binding site on the N-terminus, and their repressor-like conduct to regulate the transcription of the pleiotropic genes which trigger morphological and physiological differentiation of *Streptomyces*. Therefore, the modes-of-action of these three types of autoregulators may fundamentally bear a resemblance to each other.

According to the surveys of Yamada and co-workers,[28,41] Gräfe's group,[42] and Beppu's group,[7] $\sim 60\%$ of *Streptomyces* have these autoregulators. Although the roles of these signal substances in each of the surveyed *Streptomyces* species remains unknown (1997), it may be suggested that, in some *Streptomyces*, autoregulators play a very important role for their survival strategy, especially in nutrient-rich environments. From an ecological point of view, the function of autoregulators may help to clarify the long-discussed role of secondary metabolite production, especially that of antibiotics. No one has ever detected antibiotics in the natural environment of poor nutrient states such as in soils or river water. However, in the nutrient-rich state in an artificial culture broth or in nature, such as in a rhizosphere, on a fallen fruit, or on dead animal bodies, many species of microorganisms propagate and compete for survival. When they sense the deficiency of a carbon or nitrogen source, or at a period when their already programmed cell growth cycle determines, autoregulators are produced from some cells and they diffuse to other cells to mediate synchronized cytodifferentiation for spore formation or production or antibiotics to eliminate other microorganisms.

As the organic synthesis of these low molecular weight unique Streptomycete autoregulators is rather easy and a very small amount is enough for use due to their high activities, their application is promising. For example, VB could be utilized to increase the yield of virginiamycin in *S. virginiae*. When VB was added at the appropriate time during the fermentation process to produce virginiamycin, the yield of the antibiotic doubled compared with that without VB addition.[43,44]

Knowledge about these Streptomycete hormone-like compounds is still very limited in many aspects, for example their structure versatility, the enzymes involved in their biosynthesis, and above all their mode-of-action. When we consider the very important position of *Streptomyces* and other Actinomycete in nature from an ecological point of view and their great ability for secondary metabolite production, the development of research on Streptomycete hormones is very promising, not only for the basic science of microbiology, but also for application to induce still-unknown abilities of Streptomycete and new secondary metabolite production. One set of *Streptomyces* species together with an autoregulator could have numerous applications, such as microbial pesticides.

8.06.3 AUTOINDUCERS WITH *N*-ACYL HOMOSERINE LACTONE SKELETON IN GRAM-NEGATIVE BACTERIA

8.06.3.1 Introduction

Since the 1970s, some gram-negative bacteria which are mainly symbiotic or pathogenic to animals or plants, have turned out to have signal substances named "autoinducers." When these gram-negative bacteria grow as symbionts or on infected states, they sense the environmental information in their hosts, or the host's own signals that relate to a defence mechanism. On the other hand, bacteria communicate with each other using small molecule signal substances named autoinducers, to respond to the environmental changes, mainly to increase cell densities. Autoinducers are also called "quorum sensing factors," because the expression of their activities is dependent upon the cell densities of bacteria which secrete them.

Autoinducers are unique low molecular weight compounds with a common skeleton of *N*-acyl homoserine lactone and are effective at very low concentration ($\sim$ 10s nM). Already known gram-negative bacteria that have autoinducers are *Vibrio fischeri*, *V. harveyi*, *Pseudomonas aeruginosa*, *Agrobacterium tumefaciens*, *Erwinia carotovora*, and *Rhizobium leguminosarum*. *V. fischeri* and *V. harveyi* are symbionts of fish and squids, living in their light organs where they cause luminescence. *P. aeruginosa* is an opportunistic infectant of wounds or of some organs and becomes virulent depending upon the state of the host. *A. tumefaciens* is a pathogen of plants and causes the crown gall, a cancer in plants. *E. carotovora* is also a plant pathogen. *R. leguminosarum* is a very important leguminous plant symbiont which causes nodules in plant roots for nitrogen fixation. Although the cell densities of these bacteria in the free-living state are low, those of the infected state are very high, such as 10^{9-11} ml^{-1}, due to enriched nutrient sources of carbon and nitrogen. At these high cell densities, the concentration of secreted autoinducers also reaches some critical value and triggers the expression of several effects such as luminescence, excretion of lytic enzymes, production of antibiotics, induction of conjugal transfer of plasmids, etc., which are advantageous to the symbionts or pathogens and in some cases to the hosts. In this section, the structures, functions and modes-of-action of these autoinducers are described.

8.06.3.2 Autoinducers of *Vibrio fischeri* (VAI-1)

8.06.3.2.1 Introduction

V. fischeri, a gram-negative bacterium, is a symbiont in the light organ of a marine monocentrid fish and is the cause of its luminescence. The mechanism of the luminescence is as follows.[45] The luminescence takes place in the cells of *V. fischeri*. In their cells, an enzyme, luciferase, catalyzes oxidation of its substrate, the alkyl aldehyde (mainly tetradecanal) and $FMNH_2$ in the presence of oxygen to the carboxylic acid, FMN, and H_2O as follows.

$$O_2 + RCHO + FMNH_2 \longrightarrow H_2O + RCO_2H + FMN + h\nu \quad (1)$$

In this oxidation reaction, the emission of light at 490 nm occurs leading to the luminescence of the cells and the light organ. At the end of the 1960s, Kempner and Hanson,[46] and Nealson *et al.*[47] found that *V. fischeri* growing in shaken flasks suddenly produces luciferase under certain medium conditions.

In freshly inoculated cultures, the luciferase operon is inactive and during the exponential period of growth its activation occurs. This phenomenon was referred to as the conditioning of the medium and the idea of "autoinduction" was introduced. This autoinduction system is regulated by an autoinducer which accumulates in the medium at a constant rate.[48] When its concentration reaches a critical value, it stimulates the transcription of the luciferase operon which results in the synthesis of the luminous system. In 1981 Eberhard *et al.* isolated the autoinducer and identified its structure as *N*-acyl homoserine lactone.[49] Also, in 1983 Engebrecht *et al.* cloned the bioluminescence genes of *V. fischeri* in *E. coli*[50] and later identified each gene and gene product.[51] Thereafter, the self-regulatory mechanism involving the *V. fischeri* autoinducer, later abbreviated as VAI-1, has been investigated in detail mainly using the transformed *E. coli* bioluminescent system.[52,53]

8.06.3.2.2 Structure and properties of VAI-1

The purification of *V. fischeri* autoinducer, VAI-1, was as difficult as that of the autoregulators of *Streptomyces*, due to its low concentration in the culture broth. Eberhard *et al.* finally succeeded in isolating VAI-1 from *V. fischeri* strain MJ-1 using the dim mutant *V. fischeri* strain B-61 which responds to the addition of the exogenous autoinducer as the assay system.[49] From the ethyl acetate extract of 6 l of culture broth, 2.7 mg of VAI-1 was obtained in 50% yield, after repeated HPLC. Its structure was determined as *N*-(3-oxo-hexanoyl)-homoserine lactone (**18**) by spectroscopy, including the NMR, IR, and high-resolution mass spectrum. VAI-1 was synthesized from 3-oxohexanoic acid and racemic homoserine lactone and it was identical to the natural material. As regards the absolute configuration of C-2 on the lactone ring, there is no information reported, although three reports of the synthesis of VAI-1 analogues have since appeared.[54–56] In some reviews, the absolute configuration of VAI-1 is represented as C-2 (*S*) and this is reasonable from the viewpoint of biosynthesis. In the purified state, the autoinducer is active at 10 ng ml^{-1} concentration and maximum inducing activity was obtained at about 1 $\mu g\ ml^{-1}$. At greater than 10 $\mu g\ ml^{-1}$ concentration, it inhibits the luminescence.[49] Kaplan and Greenberg showed that VAI-1 diffuses freely through the cell membrane using tritium-labeled VAI-1[57] and this property is involved in the regulation system of luminescence. Many analogues of VAI-1 were prepared and their activities or antagonistic activities for the induction of bioluminescence[54] and receptor-binding activities[56] were investigated. Consequently, every feature of this compact, but unique, structure is essential for its activity and it is very specific to the species.

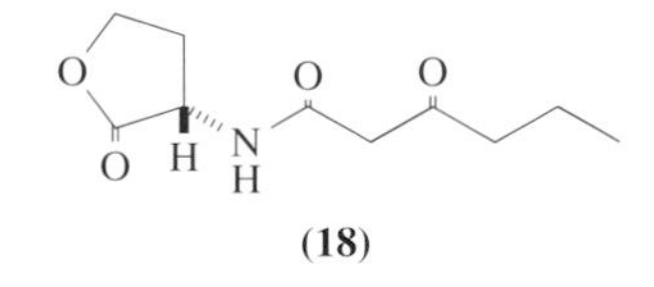

(18)

N-3-oxohexanoyl-L-homoserine lactone
VAI-1

8.06.3.2.3 *Biosynthesis of VAI-1*

The *luxI* gene, one of the seven genes cloned from *V. fischeri* which code peptides for light emission in the original strain and also in *E. coli*, was suggested to be VAI-1 synthase.[50,51] *E. coli* harboring the plasmids which contain the *luxI* gene has the ability to produce VAI-1 and this fact indicates that *E. coli* has one set of substrates for the biosynthesis of VAI-1 molecules. Eberhard *et al.* showed that the precursors of VAI-1 are *S*-adenosyl methionine and 3-oxo-hexanoyl CoA, using *V. fischeri* cells and a crude cell-free system.[58] From the studies of kinetics of the incorporation reactions, the presence of 3-oxo-hexanoyl ACP (acyl-carrier protein) as an intermediate was suggested. It was also proved that homoserine lactone was not the precursor of VAI-1. The possible biosynthetic pathway of VAI-1[53] is shown in Scheme 10. The synthesis of VAI-1 is positively autoregulated by itself in *V. fischeri* and this phenomenon is one portion of the regulatory mechanism in bioluminescence.[58]

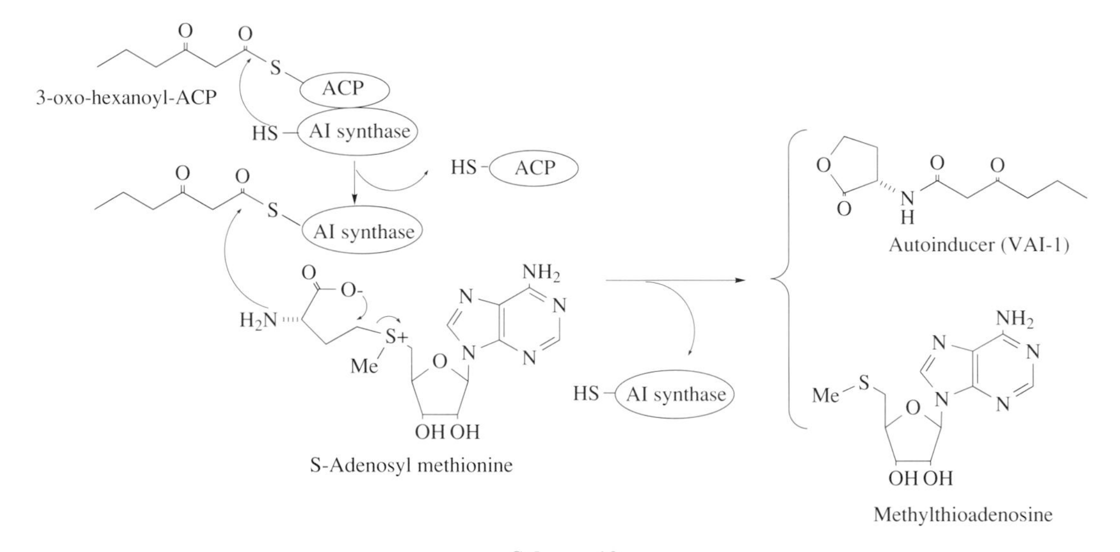

Scheme 10

8.06.3.2.4 *Mode-of-action of VAI-1*

In *V. fischeri*, there is a gene cluster, named the *lux* gene, which encodes all proteins for the expression of bioluminescence in a ~9 kb DNA fragment.[50,51] The *lux* gene is able to function in *E. coli* to produce luminescence. Their arrangement is shown in Figure 8. *luxA* and *luxB* encode proteins corresponding to the α-subunit and β-subunit of luciferase, respectively. *luxC*, *luxD*, and *luxE* encode fatty acyl reductase, acyl transferase, and acyl protein synthase, respectively, forming the fatty acid reductase complex which supplies alkyl aldehydes to the bioluminescence system and recycles the fatty acids produced to the alkyl aldehydes. Another gene, named *luxG*, is located downstream of *luxE* and its function is not yet clear.[59] On the other hand, *luxI* and *luxR* are genes for the regulation of *luxC,D,A,B,E*.[60–62] *luxI* might encode the autoinducer synthase as previously described. *luxR* is transcribed divergently to other *lux* genes and it encodes the regulator protein which binds autoinducer VAI-1. LuxI and LuxR are two key proteins for the regulation of bioluminescence in combination with the autoinducer VAI-1 in *V. fischeri*. LuxR is a polypeptide consisting of 250 amino acids residues. It is a membrane-associated protein[63] numbering about 500 per cell. Genetic evidence suggests that LuxR may exist as the multimer form.[64] LuxR consists of two domains,[65–67] the regulator domain and the activator domain, as shown in Figure 9.[53] In the activator domain, LuxR has a helix-turn-helix motif of a DNA binding site (no. 4 region) at the C-terminus portion and no. 5 region which activates the transcription of *luxC,D,A,B,E* by binding the *lux* box[62] which locates on the promoter site of the *lux* genes operon in combination with RNA polymerase (RPO).[68] The regulator domain prevents the binding of activator domain to the *lux* box in the absence of an autoinducer. When autoinducer VAI-1 binds to the binding region[69,70] (no. 2), the intramolecular regulation is released and the activator domain is able to bind to the *lux* box to trigger the transcription of *lux* genes leading to bioluminescence. LuxR was overproduced in *E. coli* and it showed VAI-1 binding activity *in vivo*, but after purification it showed neither VAI-1 binding activity nor DNA binding activity in the presence of VAI-1 *in vitro*.[71] Therefore, the functions of the five regions of LuxR in Figure 9 were mainly determined by the methodology of molecular genetics *in vivo*. Thus, LuxR in combination with VAI-1 activates *lux* genes including *luxI* which catalyzes the synthesis of more VAI-1 and this positive regulation system triggers the burst of luminescence, although there is some negative regulation of *luxR* transcription by LuxR itself. A second autoinducer of *V. fischeri* (VAI-2) has been isolated and its structure was identified as *N*-octanoyl-L-homoserine lactone (**19**).[72] The synthesis of VAI-2 is not dependent on *luxI* but on another gene, named *ainS*, which is distinct from *luxI* by restriction mapping and Southern hybridization. VAI-2 can activate *lux* operon transcription via LuxR in the absence of VAI-1 and equal activation of luminescence required 25- to 45-fold more of it than of VAI-1. VAI-2 is supposed to competitively inhibit the association of VAI-1 with LuxR and controls early onset of the luminescence burst at lower cell density,[73] preventing the consumption of energy. The discovery of this autoinduction system, VAI-2 and *ainS*, indicates that the regulatory system of genes even in prokaryotes is not so simple, but involves multiple controlling circuits.

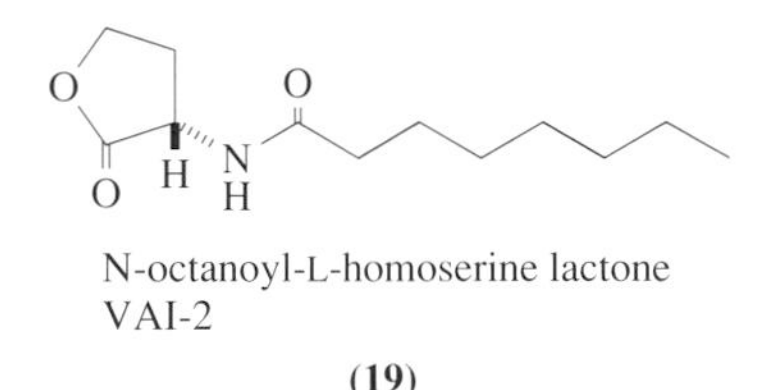

N-octanoyl-L-homoserine lactone
VAI-2

(19)

8.06.3.3 Autoinducer of *Vibrio harveyi* (HAI-1)

8.06.3.3.1 Structure and properties of HAI-1

V. harveyi is also a gram-negative bacterium that emits bioluminescence depending upon its cell densities as described in the section on *V. fischeri*. *V. harveyi* is found as a free-living bacterium in the marine environment or on the surface of various marine animals, whereas *V. fischeri* MJ-1 grows as a symbiont in the pure state in the light organ of the fish *Monosentris japonicus*.

V. harveyi produces very small amounts of autoinducer (HAI-1), which Cao and Meighen[74] have purified. Starting from 6 l of "conditioned medium" of *V. harveyi*, repeated HPLC purification gave partially purified HAI-1. The NMR spectrum at this stage indicated that there was a β-hydroxybutyryl moiety in the molecule. The mass spectrum of the further-purified HAI-1 strongly suggested that HAI-1 is *N*-3-hydroxybutyryl homoserine lactone and this structure was confirmed by chemical synthesis. Later the absolute configuration of the C-3 hydroxy group was determined as *R* by chemical synthesis of both enantiomers and the 3(*R*)-hydroxybutyryl moiety was proved to be the direct off-shoot from the fatty acid biosynthesis and not from the fatty acid degradation pathway by the inhibition of HAI-1 biosynthesis with cerulenin.[75] Consequently, the structure of HAI-1 was assigned as *N*-(3-*R*-hydroxybutyryl)-L-homoserine lactone (**20**). This compound fully induced bioluminescence in *V. harveyi* at 1 μg ml^{-1} and it did not show any inhibition at 10 μg ml^{-1} as does VIA-1 in *V. fischeri*. HAI-1 is specific to *V. harveyi* and is not active in *V. fischeri* and vice versa for VAI-1.

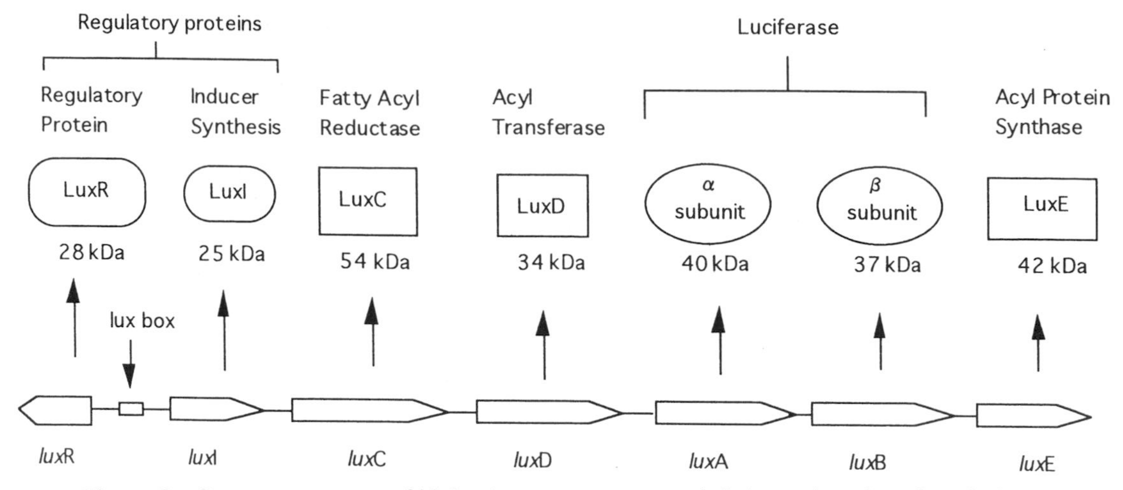

Figure 8 Gene arrangement of bioluminescence operon and their products in *Vibrio fischeri*.

Regulator domain Activator domain

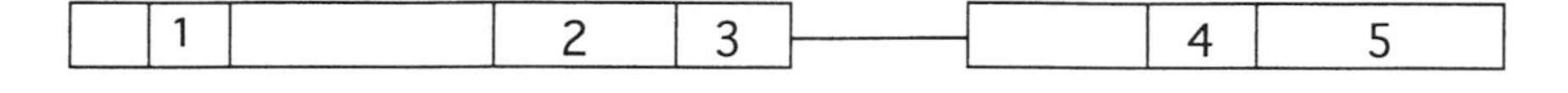

1. *lux*R Autoregulation region.
2. Autoinducer binding region.
3. Multimerization region.
4. HTH, DNA-binding region.
5. C-terminal region required for transcriptional activation.

Figure 9 Schematic diagram of LuxR regulator protein.

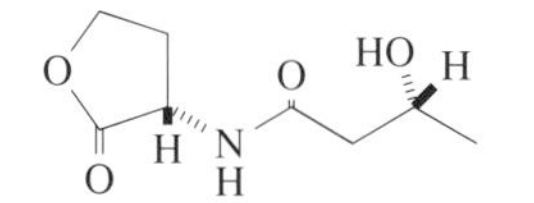

N-(3-*R*-hydroxybutyryl)-L-homoserine lactone, HVI-1

(20)

8.06.3.3.2 Mode-of-action of HAI-1

The structural genes of luciferases in *V. harveyi*, i.e., *lux* genes, were initially cloned in *E. coli* in the early 1980s[76,77] and later in the 1980s other enzymes involved in bioluminescence were identified[78–81] as operons of *luxC,D,A,B,E,G,H*, similar to *V. fischeri*. However, their regulatory genes are not physically linked with these *lux* genes, unlike those of *V. fischeri*. Therefore, it has taken rather a long time to clarify the autoinduction system of *V. harveyi*. It has no genes belonging to the *luxI* and *luxR* family and its signal transmission pathway is quite different from that of *V. fischeri*. The *luxR* gene was cloned first as a regulatory gene from *V. harveyi*[82] and its product LuxR turned out to be a transcriptional activator of the *lux* promoter.[83] *luxR* encodes 205 amino acid residues and it has the possible DNA-binding sequence in its N-terminus domain. Although it was named LuxR, its amino acid sequence has no homology with the LuxR of *V. fischeri* and its function is also quite different.

In the early 1990s, other regulatory genes, *luxL*, *luxM*, *luxN*,[84] *luxO*,[85] *luxP*, and *luxQ*,[86] were identified. *luxL* and *luxM* encode proteins of 168 and 216 amino acid residues, respectively and are required for the production of HAI-1. Therefore, their function can be supposed to be analogous to that of *luxI*, although there is no homology of amino acid sequences among them. Mutation of *luxL* and *luxM* did not completely eliminate the autoinducer activity, which indicated the presence of another autoinducer, HAI-2, in *V. harveyi*. *luxN* encodes a polypeptide of 849 amino acid residues and it is required for response to autoinducer HAI-1 as a mediator. The amino acid sequence of LuxN indicated that it belongs to the subclass of two-component signal transduction proteins which function as sensors of environmental factors and adaptive responsive regulators in gram-negative bacteria. Usually, two-component systems consist of one set of two proteins, one sensor kinase protein and another response regulator protein, which communicate by transphosphorylation. In LuxN, both the sensor kinase domain and the response regulator domain occur in the same peptide. *luxL, M, N* form one operon and they seem to work as one set with HAI-1. *luxO* encodes a protein of 453 amino acid residues and has a domain of the response regulator module of a two-component system. *luxO* also has a DNA-binding motif. Destruction of gene *luxO* causes the constitutive expression of bioluminescence in *V. harveyi* independent of cell densities and this fact demonstrates that LuxO is a repressor of *lux* enzyme genes. Consequently, LuxO is thought to be the key regulatory protein which directly switches on or off the transcription of *lux* enzyme genes in combination with LuxR according to the signal from LuxN, as shown in Figure 10. Thus, LuxM and L synthesize the signal HAI-1 and LuxN, its sensor, transmits this signal to LuxO to derepress the transcription of *lux* genes of enzymes forming the signaling system 1 loop in Figure 10. *luxP* and *luxQ* encode proteins of 245 and 851 amino acid residues, respectively. The amino acid sequence of LuxP has a homology with that of the periplasmic ribose-binding proteins of *E. coli* and *Salmonella*. On the other hand, LuxQ is homologous with LuxN and it also seems to belong to the subclass of two-component family proteins. *luxP* and *luxQ* are supposed to form signaling system 2 in combination with unknown autoinducer, HAI-2, and its synthase, signal 2 in Figure 10.

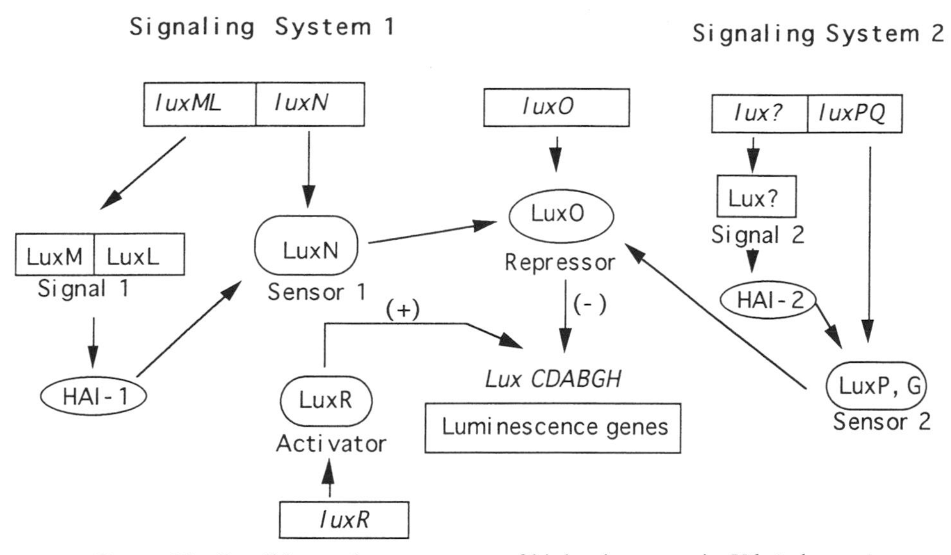

Figure 10 Possible regulatory systems of bioluminescence in *Vibrio harveyi*.

In *V. harveyi*, there are autoinduction systems of bioluminescence depending upon cell densities and an autoinducer which has the same skeleton of *N*-acyl homoserine lactone as in *V. fischeri*, but its regulation system is remarkably different from that of *V. fischeri*.

8.06.3.4 Autoinducers of the LuxR–LuxI Family

8.06.3.4.1 Introduction

Previously, two autoinduction systems of *V. fischeri* and *V. harveyi* in which the target enzymes for the bioluminescence are the same were described. Their autoinducers also have the same *N*-acyl homoserine lactone skeleton, but the regulatory systems with these autoinducers are remarkably different from each other. Other cell-density-dependent autoinduction systems in some gram-negative bacteria have been discovered and most of them were found to belong to the *V. fischeri* type, which is classified as the LuxR–LuxI family. Autoinduction systems of *Pseudomonas aeruginosa*, *Erwinia carotovora*, *Agrobacterium tumefaciens*, and *Rhizobium leguminosarum* are introduced briefly in this section as the representatives of opportunistic infectious pathogens, plant pathogens, and plant symbionts.

8.06.3.4.2 Autoinducers of **Pseudomonas aeruginosa**

P. aeruginosa is a common bacterium in the environment and is well known as an opportunistic pathogen which often infects immuno-compromised individuals causing fatal syndromes. Their infection involves some virulence factors including exotoxin A,[87,88] two elastolytic proteases named LasA and LasB, alkaline protease (ArpA),[89,90] and rhamnolipids as biosurfactants.[91] The extracellular production of these virulence factors also depends upon cell densities[92] and there are two autoinducers identified in *P. aeruginosa*. The first is *N*-(3-oxododecanoyl)-L-homoserine lactone (**21**) (PAI-1)[93] and the second is *N*-butyryl-L-homoserine lactone (**22**) (factor 2).[94] PAI-1 was purified as the autoinducer of elastase LasB induction and is active in combination with LasR, which is the product of gene *lasR*.[92,95] LasR corresponds to the LuxR of *V. fischeri*, being homologous along its entire length with the LuxR protein.[52] LasR regulates *lasA*, *aprA*, and *toxA* genes. Another regulatory gene, *lasI* was found immediately downstream of *lasR*. *lasI* is transcribed in the same direction with *lasR* and the *lasI* product LasI corresponds to LuxI in *V. fischeri*, having a homologous amino acids sequence with it.[92] As LasI is involved in the synthesis of PAI-1, its function is also similar to that of LuxI. Therefore, LasR and LasI form one set of regulatory proteins and they belong to the LuxR–LuxI family.

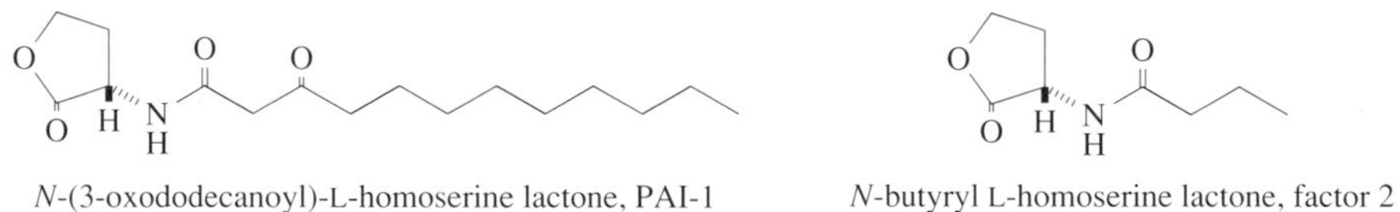

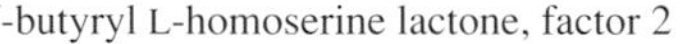

N-(3-oxododecanoyl)-L-homoserine lactone, PAI-1

(21)

N-butyryl L-homoserine lactone, factor 2

(22)

N-Butyryl-L-homoserine lactone, factor 2, was isolated as an autoinducer of rhamnolipid production in *P. aeruginosa*.[94] Factor 2 functions with RhlR to induce the production of the enzymes, rhamnosyl transferase, RhlA and RhlB.[96] Factor 2 is synthesized by RhlI and the transcription of *rhlI* itself is induced by RhlR in conjugation with factor 2. Thus, in *P. aeruginosa* there are two sets of the LuxR–LuxI family regulatory systems with each autoinducer.[96] Cross-regulation between RhlR and LasR regulators was observed. This phenomenon and the two sets of regulatory systems in *V. fischeri* involving VAI-1 and VAI-2 are the similar autoinducer-dependent complex systems in prokaryote gram-negative bacteria.

8.06.3.4.3 Autoinducers of **Erwinia carotovora**

E. carotovora is one of the broad range of plant pathogens which cause soft-rot disease colonizing in the host plant tissue. *E. carotovora* secretes cell-wall-degrading enzymes such as pectate lyase, pectin lyase, polygalacturonase, cellulase, and protease for the maceration of the tissue. These lytic exoenzymes enhance the proliferation of the pathogen in plants. The production of these virulence factors is dependent on cell densities of the pathogen.[97] Pirhonen *et al.* discovered a gene named *expI* in *E. carotovora* subsp. *carotovora* SCC3193 as a regulatory gene of these virulence factors.[98] *expI* encodes the polypeptide, ExpI, of 217 amino acid residues and it is similar to LuxI in size and is homologous in its amino acid sequence. *expI* mutants did not produce the lytic enzymes and exogenous addition of *N*-(3-oxohexanoyl)-homoserine lactone, VAI-1, complemented this phenotype. Therefore, ExpI is thought to be involved in the synthesis of the autoinducer which is presumed to be the same as VAI-1, functioning similarly to LuxI in *V. fischeri*. Another regulatory gene, *expR*, which corresponds to LuxR, was found close downstream of *expI* and they are transcribed with convergently overlapping 3′-ends.[97]

E. carotovora ATCC39048 produces a carbapenem antibiotic, (5*R*)-carbapen-2-em-3-carboxylic acid (**23**) and the induction of carbapenem is regulated also by *N*-(3-oxohexanoyl)-L-homoserine lactone. Bainton *et al.* prepared two types of carbapenem minus mutants (Car^-), group 1 and group

2, and found that the group 1 mutant produced a low molecular diffusible factor which restored carbapenem synthesis in group 2 mutants, but not vice versa.[99] The factor was purified and identified with *N*-(3-oxohexanoyl)-L-homoserine lactone which is the same as that of *V. fischeri* (VAI-1). The absolute configuration was confirmed by chemical synthesis. A number of VAI-1 analogues were synthesized and their structure–activity relationships were investigated.[100] Once again, the structure is very specific and small changes, such as the length of acyl groups, remarkably decreased the inducing activities. Group 2 mutants were presumed to be a defect of gene *carI*, corresponding to *luxI*, *expI*, which is involved in VAI-1 biosynthesis.[101] Another part of the regulatory gene *carR* was isolated and its product CarR turned out to be a homologue of LuxR.[102] CarR seems to be specific for the activation of the carbapenem synthesis gene and does not activate the exoenzyme system as does ExpR.

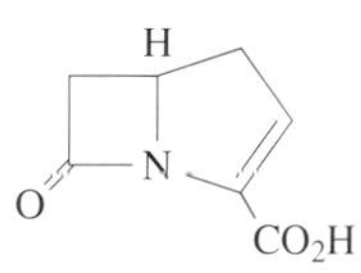

(5*R*)-carbapen-2-em-3-carboxylic acid

(23)

8.06.3.4.4 Autoinducers of Agrobacterium tumefaciens

A. tumefaciens is a phytopathogenic bacterium which causes crown gall tumors in infected plants. *A. tumefaciens* has the Ti plasmids (pTi) and there are two DNA transfer processes associated with these Ti plasmids. One is the T-DNA transfer from *A. tumefaciens* to a host plant and the other is pTi transfer from *Agrobacterium* donor cells to recipient cells. These two DNA transfers are thought to be accompanied by a process of conjugal transfer of single-stranded DNA from a donor cell to a recipient cell. T-DNA on pTi after being transferred to the host cells causes their neoplasias and crown gall tumors. In crown galls, the production of opine amino acids derivatives (**24**) and (**25**), takes place and the opines become the nutrient source for *Agrobacterium*. Therefore, for *A. tumefaciens*, pTi is a very important virulence factor that provides it with an advantageous state in the host plant. Consequently, the transfer of pTi from a donor cell to a recipient cell to increase the number of plasmid-harboring cells in the host plant is also advantageous for *A. tumefaciens* from the viewpoint of their ecology and the opines induce the conjugal transfer of pTi between them. There are two types of opines, the octopine type (**24**) and the nopaline type (**25**). Octopine-type pTi induces production of octopine in crown gall and nopaline-type pTi induces nopaline similarly. In 1991, Zhang and Kerr found a diffusible compound that is excreted from a donor cell and enhances the transconjugal transfer of the Ti plasmid in *A. tumefaciens*.[103] This factor was named the conjugation factor (CF) and it was characterized. CF is a low molecular compound and its production is dependent on the presence of pTi but appears to be common for both octopine- and nopaline-type plasmids. CF is very biologically active, affecting donor but not recipient bacterial cells, but it does not promote the aggregation of cells. Later, the same group purified CF from nopaline-type *A. tumefaciens* and determined its structure as *N*-(3-oxooctanoyl)-L-homoserine lactone (**26**) by spectroscopy and this structure was confirmed by chemical synthesis.[104] At the same time the regulatory gene, *traR*, was identified on nopaline-type pTi whose product, TraR induces conjugal transfer of pTi in conjunction with CF.[105] As TraR is a homologue of LuxR and CF is a cognate of VAI-1, CF is called an *Agrobacterium* autoinducer, AAI. Another regulatory gene for conjugal transfer genes, *traI*, was also found on nopaline-type pTi.[106] TraI is a homologue of the corresponding LuxI and it is involved in AAI synthesis. Similarly, the corresponding regulatory genes, *traR* and *traI*, were identified on octopine-type pTi.[107] The nopaline-type TraR–TraI set bears a close resemblance to octopine-type TraR–TraI.

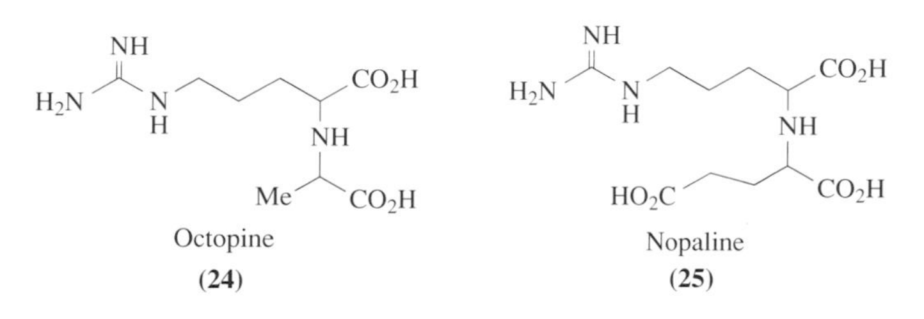

Octopine **(24)**

Nopaline **(25)**

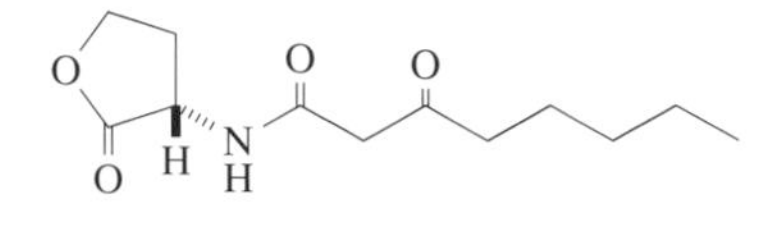

N-(3-oxooctanoyl)-L-homoserine lactone,
AAI,CF

(26)

8.06.3.4.5 Autoinducers of Rhizobium leguminosarum

R. leguminosarum is one of the symboints of plants belonging to the *Leguminoseae*. *R. leguminosarum* infects the host plant and forms nodules in which bacterial cells become a differentiated form, the bacteroid, and they fix nitrogen to produce ammonia. Many bacterial genes are required for the symbiosis, such as nodulation (*nod*) genes, nitrogen fixation (*nif*) genes, and rhizosphere-expressed (*rhi*) genes. In *R. leguminosarum* bv. *viciae*, these genes for the symbiosis are mainly located on the symbiotic plasmid (sym plasmid), namely pRL1JI. In *rhi* genes, there is an operon *rhiABC* and downstream of it, *rhiR*, which is transcribed convergently to *rhiABC*.[108] All of these *rhi* genes have been cloned and sequenced, revealing that RhiR shows significant similarities to LuxR.[108] The biochemical roles of the *rhi* genes have not been determined, but they are supposed to play a role in the plant–symbiont interaction being expressed in the rhizosphere. In 1996, a Netherlands group isolated and determined the structure of bacteriocin *small* of *R. leguminosarum* bv. *viciae* RBL1390.[109] Bacteriosin *small* might be thought to be a very specific antibiotic which inhibits the growth of some *R. leguminosarum* strains, such as strain 248, which is also harboring sym plasmid pRL1JI. From 10 l of culture broth ~0.5 mg of *small* was obtained. The structure of *small* was assigned as *N*-(3*R*-hydroxy-7*Z*-tetradecenoyl)-L-homoserine lactone (**27**) by spectroscopy, including H-NMR and ^{13}C-NMR. Therefore, this compound belongs to the family of autoinducers of LuxI–LuxR cognates. At the same time, a University of Iowa group also published the isolation and structure of an autoinducer from *R. leguminosarum* using the assay system *rhiA* gene activation with reporter gene *lacZ*.[110] This autoinducer was named *R. leguminosarum* autoinducer RLAI and its structure was identical with that of bacteriocin *small*, although the position of the double bond and the absolute configuration were not determined. It was discovered that RLAI is required to activate the rhizosphere-expressed *rhiABC* operon together with RhiR and a growth-inhibition function encoded by sym plasmid pRL1JI as bacteriocin *small*. The ecological role of growth inhibition in the host–symbiont relationship will eventually be revealed and it may be one of the most important aspects of the cell-density-dependent autoinducers. The genes required for the synthesis of the autoinducer are not on the sym plasmid and they should encode the polypeptides corresponding to LuxI.

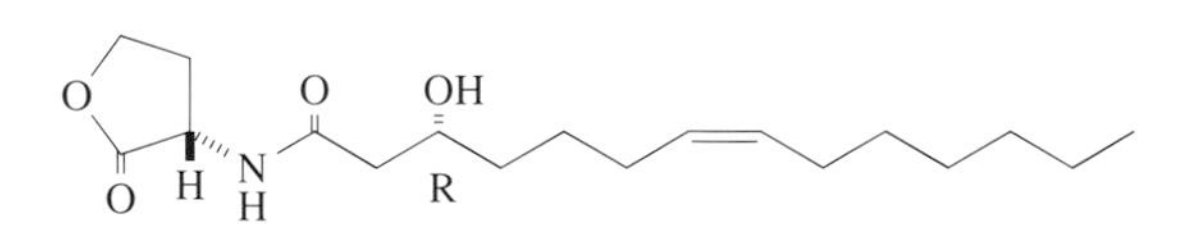

N-(3*R*-hydroxy-7Z-tetradecenoyl)-L-homoserine lactone, bacteriocin small, RLAI

(27)

8.06.4 COMPOUNDS REGULATING MORPHOLOGICAL DIFFERENTIATION IN PROKARYOTES

8.06.4.1 Introduction

Intercellular signaling is a key factor determining the initiation and coordination of cellular differentiation. In eukaryotic cells, differentiation into a specialized form of cells is a critical and primary aspect of their life cycle, while in prokaryotic cells differentiation has been considered to be a rather minor or exceptional event. However, even a unicellular microorganism, such as *E. coli*,[111,112] has been found to differentiate when a stationary phase is reached. To understand a signaling pathway fully, it is important to define the chemical and biochemical nature of the signaling molecules and the mechanism by which cells produce and respond to the molecules.

8.06.4.2 *Bacillus subtilis*

8.06.4.2.1 *Physical and morphological differentiation*

After *E. coli*, the gram-positive soil bacterium *B. subtilis* is genetically the most-characterized bacterium among prokaryotes.[113] The genus *Bacillus* belongs to the family of endospore-forming, gram-positive rods and cocci, to which the genera *Sporolactobacillus*, *Clostridium*, *Desulfotomaculum*, *Sporosarcina*, and *Oscillospira* also belong.[114] *B. subtilis* is the representative strain of the genus *Bacillus*, and, in addition to the typical exponential phase under the rich nutrient conditions growing by binary fission, has developed several strategies, called late-growth responses, to survive under unfavorable environmental conditions. The late-growth responses are synthesis and secretion of degradative enzymes (proteinases and α-amylase) to degrade other bacteria and to supply additional nutrients, production of antibiotics such as gramicidins to suppress the growth of other microorganisms, development of motility and competence, and finally formation of dormant and resistant endospores.

Competence in *B. subtilis* is the natural ability of cells to take up exogenous DNA. Under appropriate growth conditions, usually in minimum medium supplemented with glucose, during the transition from the exponential to the stationary growth phase, a subpopulation of cells in a culture differentiates to become competent. Competent cells are metabolically less active and have a different buoyant density than noncompetent cells. This differentiation requires several regulatory genes and results in the production of specialized proteins that bind and take up DNA independently of nucleotide sequence.[115]

Endospores are the dormant and resistant life forms of *B. subtilis*, which will germinate and start a new round of vegetative growth when exposed to adequate nutrient conditions. Spore formation requires the expression of a large number of genes (probably more than 100), and the whole sporulation event is divided into eight stages, stage 0 to stage VII. During stage 0, vegetative cells somehow sense the nutrient depletion and start to prepare for endospore development. At stage I, nucleoids of the vegetative cells have rearranged to form axial filaments of chromatin. At stage II, the cell divides asymmetrically to produce two daughter cells with different developmental fates. The smaller cell, known as the forespore, is destined to become the endospore, and the larger cell, known as the mother cell, will engulf the forespore and later lyse to release the spore. Engulfment of the forespore protoplast by the mother cell is completed in stage III. During stage IV, a layer of peptidoglycan known as the cortex is laid down between the membranes of the engulfed prespore. Subsequent maturation of the spores during the late stage V (spore protein coats are synthesized), VI (development of refractility and heat resistance), and finally stage VII, results in the lysis of the mother cell and release of the mature spores. The expression of genes necessary for each stage has been well studied, and is controlled primarily by the production of six sigma factors (σ^A, σ^E, σ^F, σ^G, σ^H, and σ^K).[116] σ^A is the primary sigma factor necessary for vegetative growth, and the other five sigma factors will be produced sequentially in the order σ^H, σ^F, σ^E, σ^G, σ^K. Each sigma factor binds with core RNA polymerase forming a holoenzyme with a different promotor specificity, and activates transcription of a different set of genes. These sigma factors control not only the temporal pattern of gene expression, but also the spatial distribution of gene expression.

Although much genetic knowledge has been accumulated on the regulatory cascade leading to the late-growth responses, such as production of degradative enzymes, competence development and spore formation, knowledge of the factors which initiate such responses is very limited.

8.06.4.2.2 *Competence pheromones*

The competence of *B. subtilis* has been suggested to be influenced by extracellular signaling factors.[117–119] However, only in 1994 was the biochemical nature of these factors uncovered.[120] Grossman's group used the expression of the *lacZ* structural gene fused to the promotor of the late-competence gene (*comG*) as an indicator, and from the spent medium purified two peptide factors called competence pheromones. One was Ala-Asp-Pro-Ile-Thr-Arg-Gln-X-Gly-Asp and the other was Asp-Pro-Ile-Thr-Arg-Gln-X-Gly-Asp with $K_{1/2}$ = 3–4 nM. Because the two factors differ only by Ala at the N-terminus and a sample containing different ratios of the peptides showed similar activity, both factors have been suggested as similarly active. A computer search of the databases revealed that the sequence matched to nine of the last 10 amino acids of a small open reading frame (called *comX*, encoding 45 amino acids) situated between *comQ* and *comP*, both of which are known to participate in competence development. The DNA sequence of the *comX* gene predicted that the unknown residue X in the competence pheromones was originally tryptophan. However, the typical UV absorption arising from the indole moiety was missing in the native competence pheromones, indicating that the tryptophan moiety was modified somehow. Mass measurements revealed that the natural factor is 206 Da larger than the expected peptide moiety. The modifying group was found to be essential in inducing the competence because the synthetic nine or 10 amino acid peptides had no pheromone activity. The structure of the modifying group has not been identified. A gene *comQ* immediately upstream of *comX* seemed to participate in producing the competence pheromone, probably either by proteolytic cleavage of the propeptide or modification of the tryptophan residue.

Grossman and co-workers also revealed the presence of a second signaling factor, called the competence-stimulating factor (CSF).[121] Although not purified completely, CSF seemed also to be a small peptide of between 520 and 720 Da. Competence pheromones and CSF are distinct not only biochemically but also on the response pathways. While competence induction by CSF requires the oligopeptide permease encoded by *spo0K*, the competence pheromone requires the membrane-bound histidine protein kinase encoded by *comP*. Genetic analysis using double mutants indicated that CSF is on the same signaling pathway as Spo0K oligopeptide permease, and the competence pheromone is on the same signaling pathway as ComP histidine protein kinase. The surface location of both Spo0K oligopeptide permease and ComP histidine kinase may allow the assumption of their receptor-like function (Figure 11), although more direct biochemical evidence is awaited to clarify the initial event following the formation of the signaling factors. Nothing is known about how the production of these signal compounds, such as CSF and competence pheromone, is triggered.

8.06.4.2.3 Spore-inducing factors

Much knowledge has been accumulated on the signal-transducing pathway leading to sporulation initiation.[122] Genes participating in the sporulation initiation are referred to as *spo0* genes or locuses, and so far nine such loci, i.e. *spo0A*, *0B*, *0E*, *0F*, *0H*, *0J*, *0K*, *kinA*, and *kinB* genes, have been identified (Figure 12). Spo0A protein encoded by *spo0A* gene is central in the signal transduction. Spo0A is a kind of transcriptional regulator, which acts either as a repressor or an activator of the transcription of target genes when it is phosphorylated. When an environmental signal indicating less favored conditions for growth is transmitted to one of the two protein kinases KinA or KinB encoded by *kinA* or *kinB* genes, KinA or KinB kinase autophosphorylates its aspartic residue in the presence of ATP. This phosphate group will be transferred to Spo0F protein by KinA or KinB protein kinases, then to Spo0A protein by the phosphoprotein phosphotransferase activity of Spo0B protein. The phosphorylated form of Spo0A protein (Spo0A $\sim$ P) will then repress the transcription of the inhibitory genes contained in its promotor region, the 7-bp so-called "0A box" TGNCGAA (where N can be any base, but is usually A or T[123]), or activate the transcription of the necessary genes such as *spoIIA*, *spoIIE*, or *spoIIG*. The oligopeptide transport system, which is encoded by the *spo0K* locus, is essential for sporulation initiation, thus suggesting the presence of peptide signal molecules for activating the KinA or KinB protein kinases. Actually, there is some evidence that sporulation is triggered by a kind of peptide. Grossman and Losick[124] have reported the existence of extracellular differentiation factor A (EDF-A) in the spent medium of *B. subtilis*. EDF-A is a pronase-sensitive and dialyzable compound which can induce spore formation to low-density cells. The production of EDF-A was impaired in *spo0A* or *spo0B* mutants, but restored in a *spo0AabrB* double mutant, suggesting that EDF-A production is under the negative control of AbrB repressor protein. In addition to the compounds of peptidyl nature, there is also some evidence that non-peptidyl compounds trigger sporulation. Srinivasan and Halvorson reported the presence of an endogenous factor, called sporogen, which induces spore formation of *B. cereus*,[125,126] and Waldburger *et al.*[127] reported the presence of pronase-resistant sporulation factor (SF) in *B. subtilis*, which can induce spore formation at low-density cells. However, because these factors have not been purified, the similarity or identity between sporogen, EDF-A, or SF is unknown.

8.06.4.2.4 Future aspects

Major late-growth responses, such as the production of degradative enzymes, competence development, and sporulation are not independent events. There is substantial evidence that each event

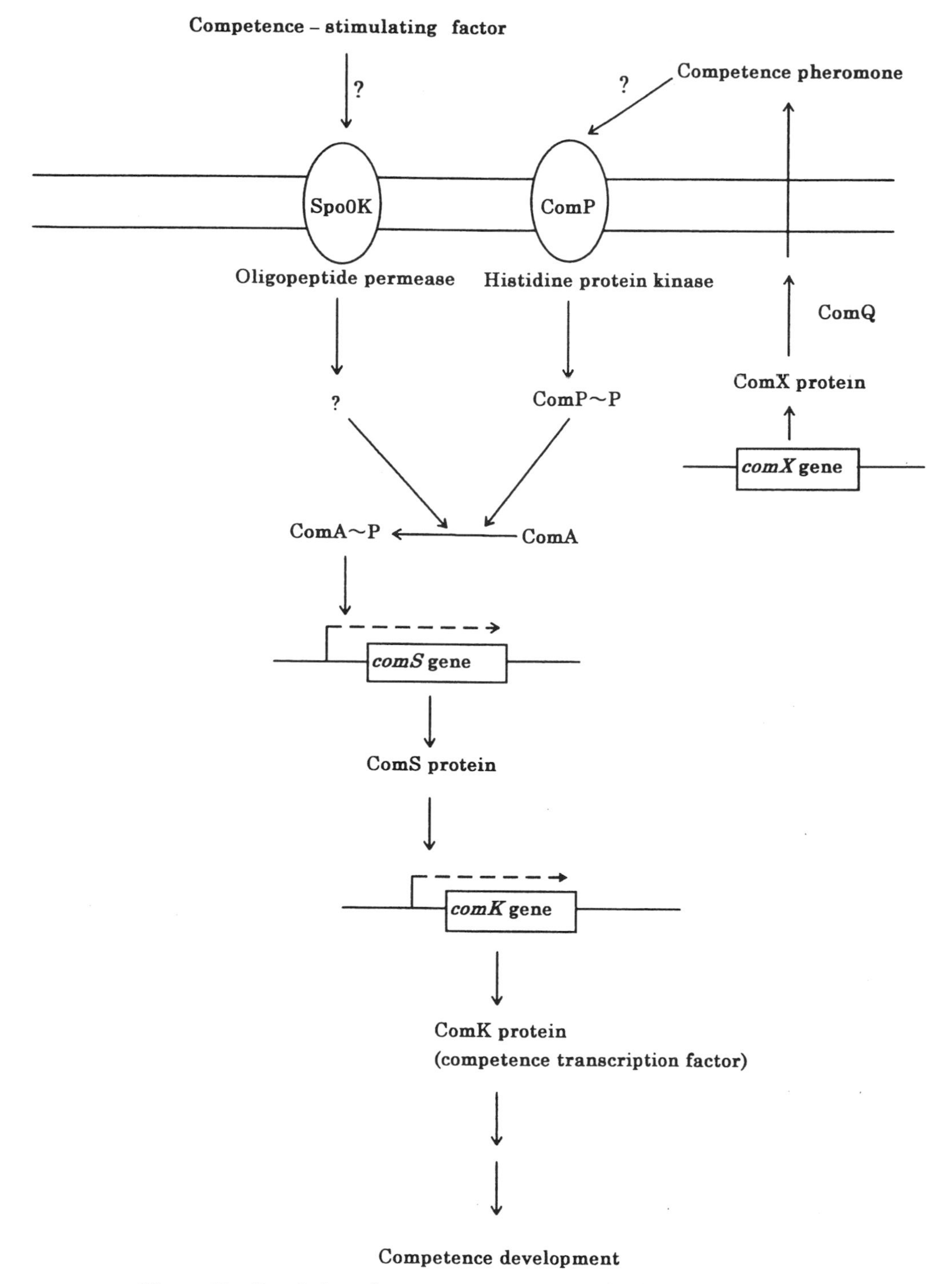

Figure 11 Regulation of competence gene expression in *Bacillus subtilis*.

is correlated. For example, the production of CSF, affecting competence development, requires *spo0A*, *0H*, *0K*, and *abrB* genes, which are in turn known to function in sporulation initiation.[128] The unphosphorylated form of response regulator DegU (which constitutes the two-component system DegS–DegU[129] controlling the synthesis of the degradative enzyme) is necessary for competence development,[130] while the phosphorylated form of DegU is known to be required for spore formation. Mutants lacking the competence pheromone are naturally devoid of competence development, but still undergo sporulation on the usual sporulation medium. However, on a

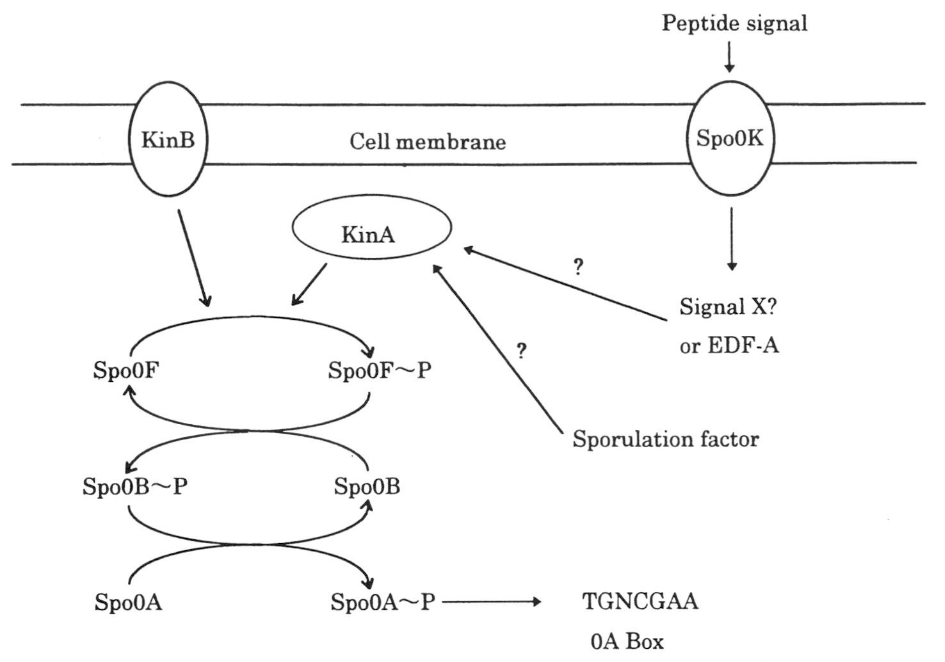

Figure 12 Regulation of sporulation initiation in *Bacillus subtilis*.

minimum medium, sporulation of the mutant was impaired and the addition of the competence pheromone restored the sporulation, indicating that under some nutrient conditions, the competence pheromone also plays a role in the sporulation event.

8.06.4.3 *Myxococcus xanthus*

8.06.4.3.1 Morphological differentiation of Myxococcus xanthus

M. xanthus is a member of the so-called myxobacteria. The myxobacteria are a group of fruiting body-forming and gliding gram-negative soil bacteria constituted from genera *Myxococcus*, *Archangium*, *Cystobacter*, *Melittangium*, *Stigmatella*, *Polyangium*, *Nannocystis*, and *Chondromyces*,[131] characterized by their social behavior and complex developmental cycle. Although at least 32 species are known in the myxobacteria, most research has focused on *M. xanthus*,[132] which made it possible to examine its social and developmental events at the cellular and molecular level. When starved on a solid surface and at high cell density, growth of *M. xanthus* slows, and after 4 h the cells begin to congregate. Initially, the aggregates are asymmetric, but by 12 h thousands of cells have accumulated and the aggregates have become elliptical or circular mounds, called fruiting bodies, about 0.2 mm high, composed of $\sim 10^5$ cells. Early in aggregation, ridge-like accumulations of cells move co-ordinately and rhythmically over the fruiting surface, a phenomenon called ripping. At about 20 h, cells inside the fruiting body progressively differentiate from $5\ \mu m \times 0.5\ \mu m$ rods to dormant spherical cells, called myxospores, that have several coats and are heat and desiccation resistant.[133,134] When nutrients are restored, myxospores germinate and the vegetative cells grow and multiply (Figure 13).

In 1978, Hagen *et al.*[136] isolated a large number of developmentally defective mutants of *M. xanthus* by many independent treatments of UV, *N*-methyl-*N'*-nitro-*N*-nitrosoguanidine (NTG), or ethyl methanesulfonate-mutagenesis, and showed that the developmental defect of some mutants could be restored by mixing with wild-type cells. Such conditional mutants were classified into four groups. Each group was distinguished from the others by the fact that the developmental defect by members of each group was complemented by mixing either with wild-type cells or with a member of another group. It was concluded that wild-type cells produce at least four different intercellular signals required for normal development, and that each group of conditional mutants was defective in its ability to produce one of the four signals, but apparently could respond to signals from signal-producing cells. Later, in 1993, Downard *et al.*[137] found one more mutant group, and now it is

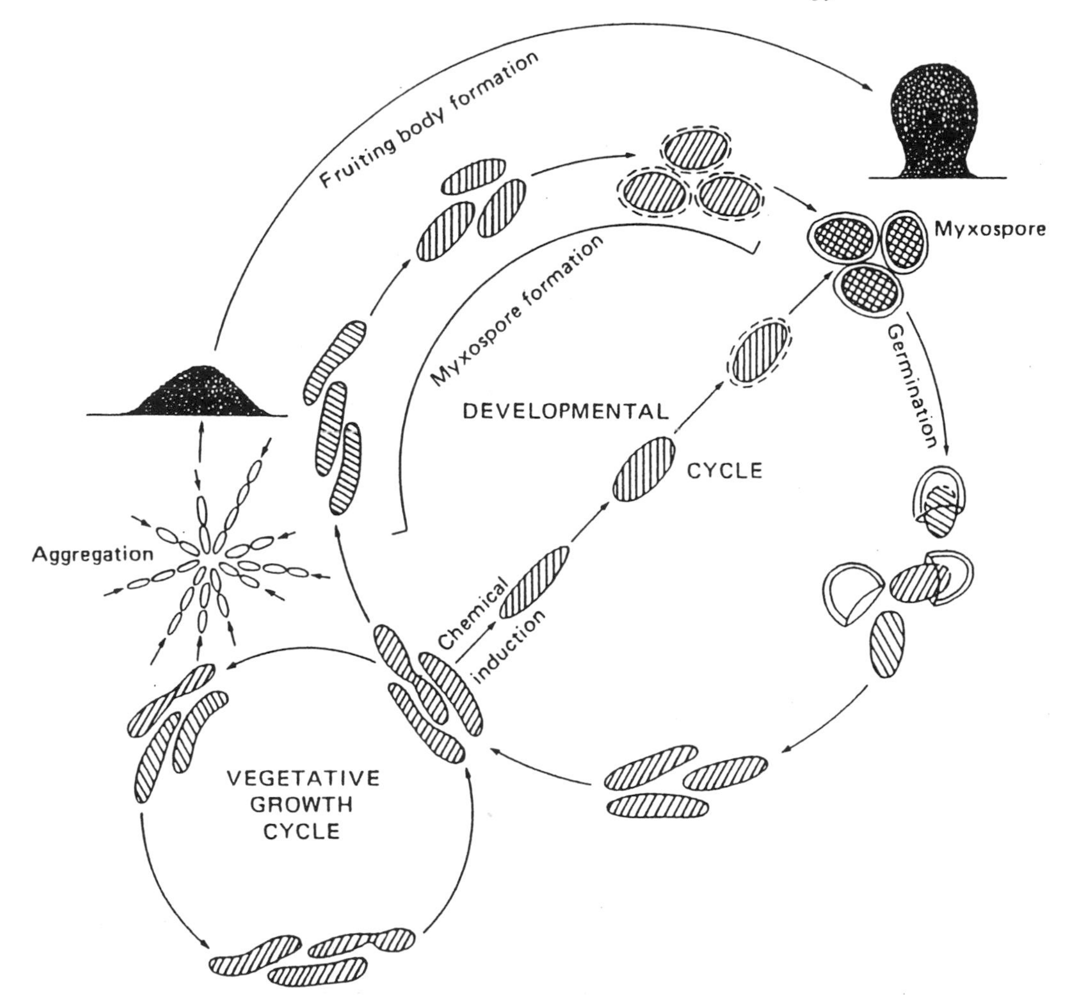

Figure 13 Diagram of the life cycle of *Myxococcus xanthus*. The fruiting body is not drawn to scale but is a few hundredths of a millimeter in diameter. The vegetative cells are about 5 to 7 μm by 0.7 μm. Reprinted from Ref. 135 with permission of the publisher.

known that there are at least five signals regulating the developmental processes of *M. xanthus*. The five signals are referred to as A-, B-, C-, D-, and E-signals, and those defective in the production of each signal are designated as *asg* (A-signal-generating), *bsg*, *csg*, *dsg*, and *esg* mutants, respectively. The chemical natures of A- and E-signal have been elucidated, but those of B-, C-, and D-signals remain unknown (Table 1).

Table 1 The five developmental signals in *Myxococcus xanthus*.

Signaling group	*Signal*	*Mutation*	*Gene product*	*Ref.*
A	Mixture of amino acids and peptides	*asgA*	Transmitter domain of His protein kinase Receiver domain of response regulator	138–140
		asgB	DNA-binding transcriptional regulator	141
		asgC	Sigma factor	142
B	Unknown	*bsgA*	ATP-dependent protease	143–145
C	Unknown	*csgA*	Short-chain alcohol dehydrogenase	146,147
D	Fatty acids?	*dsg*	Translation initiation factor IF3	148,149
E	Fatty acid	*esg*	E1 decarboxylase of the branched-chain keto-acid dehydrogenase	150,151

Source: Dworkin.[132]

8.06.4.3.2 Signals and genes regulating M. xanthus *morphological differentiation*

(i) A-signal

During the 24 h developmental cycle of *M. xanthus*, A-signal is released by wild-type cells (asg^+) after 1–2 h of development, and 18 of the 38 conditional mutants isolated by Hagen *et al.*[136] were A-signal deficient (*asg*) mutants. To detect A-signal activity efficiently, one indicator strain was created from an *asg* mutant by transposon Tn5 *lac* insertion. The indicator strain possesses *lacZ* (encoding β-galactosidase) as transcriptional fusion with an early developmental gene and shows the A-signal dependent expression of β-galactosidase activity. With this indicator strain, two kinds of A-signal active compounds were isolated from the spent medium of wild-type cells.[138,152,153] One was heat labile and was identified as two proteinases of 10 and 27 kDa. The 27 kDa proteinase has a trypsin-like substrate specificity, and commercial proteinases were found to have A-signal activity. The other A-signal active compounds from the spent culture were heat-stable low molecular weight compounds, and were identified as a mixture of amino acids and peptides. The most active amino acids were Tyr, Pro, Phe, Trp, Leu, and Ile with threshold concentrations ranging from 8 to 14 μM. Because the A-signal activity of many authentic peptides were equal to or less than the sum of activities of their constituent amino acids, Kuspa *et al.*[138] concluded that there seemed to be no specialized A-signal peptides. A-signal activities of proteinases, peptides, and free amino acids can be connected by a pathway in which proteinases degrade some unknown proteins to peptides and then to free amino acids. The primary A-signal seems to be a subset of active amino acids, and proteinases and peptides have A-signal activity only because they can yield active amino acids. The substrates for the proteinases have not been identified.

So far, the defect in A-signal production was mapped to three unlinked loci (*asgA*, *asgB*, and *asgC*). The *asgA* encodes a histidine protein kinase composed of two domains that are highly homologous to members of a two-component regulatory system.[139,140] The two-component system is a prokaryotic regulatory system usually composed of two different proteins; one is a sensor kinase which detects environmental conditions and autophosphorylates its own histidine residue, and the other is a response regulator which accepts the phosphate group from the phosphorylated sensor kinase and regulates gene expression.[154] AsgA protein is unique in that on one protein it has a transmitter domain of the sensor kinase and a receiver domain of the response regulator. It has been proposed that AsgA functions within a phosphorelay that senses starvation and responds by altering the expression of genes required for production of A-signal. *asgB* gene,[141] on the other hand, codes for a putative transcriptional factor containing on its C-terminus a helix-turn-helix DNA-binding motif, and *asgC* gene[142] was found to encode the major sigma factor of *M. xanthus*. Therefore, the simplest regulation cascade for A-signal production is as follows: when a starvation signal is sensed by AsgA protein histidine kinase, the response-regulator part of AsgA will act on the *asgB* gene so as to allow its expression by an RNA polymerase holoenzyme containing a major sigma factor encoded by *asgC*; the resulting AsgB protein, in turn, will activate transcription of genes for the proteinases, and the proteinases will be secreted and degrade unknown protein(s) to produce A-signal active free amino acids.

(ii) B-signal

Some of the *bsg* mutants were located in a single locus, *bsgA*,[143] and *bsgA* mutants seem to be blocked earlier than other signal mutants. Of 30 or so developmentally induced genes tested, including those induced immediately after starvation, all were found to be dependent on *bsgA* for normal levels of transcription.[155] The *bsgA* gene was independently cloned and sequenced by two groups,[144,145] revealing that BsgA is a 90.4 kDa protein homologous to an ATP-dependent protease La of *E. coli* encoded by *lon*. Because protease La functions in *E. coli* in the degradation of unnecessary or abnormal proteins, Gill *et al.*[144] postulated that the BsgA protein may function normally to destroy a repressor of development. The identity of B-signal is unknown (1997) and the relation of *bsgA* with the ability to produce B-signal remains to be solved.

(iii) C-signal

C-signal acts later than B- or A-signals, being required for normal expression of nearly all genes that begin to be expressed after 6 h into development.[155,156] All the C-signal deficient (*csg*) mutations were mapped on a single locus, named *csgA*.[157] The *csgA* gene has been cloned, sequenced, and estimated to encode a 17.7 kDa protein[146] based on the DNA sequence of the gene. Kim and Kaiser[158,159] purified a C-signal active protein from starved wild-type cells by solubilization with a detergent, and have shown that the internal amino acid sequence obtained from the purified C-signal protein matched exactly to that encoded on the *csgA* gene. Furthermore, their C-signal protein reacted with an antibody raised against recombinant CsgA protein. The C-signal protein is a dimer of 17 kDa subunits, and is tightly associated with the extracellular matrix of developing *M. xanthus* cells, which explains the reason why C-signaling requires close contact between the cells. CsgA has been referred to as C-signal itself. However, Lee *et al.*[147] have shown that CsgA protein has a $NAD(P)^+$ binding pocket and that that binding pocket is necessary for CsgA activity. They have further shown that strains whose csgA alleles contain amino acid substitutions in the $NAD(P)^+$ binding pocket fail to develop. CsgA protein and its suppressor protein SocA[160] have been reported to have good homology with short-chain alcohol dehydrogenases whose catalytic activity require $NAD(P)^+$. Therefore, the activity of CsgA as a developmental signal may be related to its proposed properties as a member of short-chain alcohol dehydrogenases and function to produce the real C-signal. The identity of the real C-signal or the substrate for CsgA remains unknown.

(iv) D-signal

D-signal was first required at about 4 h[161] in development. The *dsg* locus was cloned by the ability to rescue a *dsg* mutant,[161] and DNA sequencing revealed that the *dsg* product is 50% and 51% identical to the amino acid sequence of translation initiation factor IF3 of *E. coli* and *B. stearothermophilus*, respectively.[148] Although some of the point mutations in the *dsg* locus resulting in *dsg* phenotype allowed cells to grow, loss-of-function mutation was found to be lethal,[162,163] meaning that *dsg* encoding translational initiation factor IF3 of *M. xanthus*[163] was essential for its viability. It is hard to imagine that the translational initiation factor plays a role as an extracellular signal. Therefore, *dsg* should function in producing the real D-signal or in translation of proteins transmitting the signal. As a candidate for D-signal, Rosenbluh and Rosenburg[149] reported that autocide AMI, a mixture of saturated and unsaturated free fatty acids produced by wild-type cells, could rescue the developmental defect of *dsg* mutants.

(v) E-signal

It has been shown that E-signal is required early in development (at about 3 to 5 h), as demonstrated by the lack of expression of an early developmental gene (*tps*) in an E-signal deficient mutant.[137,151] This implies that E-signal probably acts before C-signal. Both aggregation and sporulation were greatly reduced in the *esg* mutant with sporulation occurring at 10^{-4} to 10^{-5} of the wild-type level. The *esg* locus, mutations on which caused a defect in E-signal production, was originally located and cloned using Tn5 insertion, and DNA sequencing[150] revealed that the *esg* locus encodes α and β subunits that constitute the E1 component of a branched-chain keto-acid dehydrogenase complex involved in branched-chain amino acid metabolism. These dehydrogenases are composed of the E1, E2, and E3 enzymes and they act to convert the branched-chain keto acids produced from branched-chain amino acids to CoA derivatives of short branched-chain fatty acids isovaleric acid, isobutyric acid, and methylbutyric acid. Furthermore, the developmental defect of some *esg* mutants were rescued by short branched-chain fatty acids, especially by isovalerate, suggesting that these short fatty acids or longer branched-chain fatty acids such as i-15:0[150,164] may be the E-signal itself.

8.06.4.4 *Enterococcus faecalis*

8.06.4.4.1 Sex pheromones controlling conjugative plasmid transfer

In gram-positive cocci, conjugation via cell mating is a major mechanism of horizontal genetic transfer, and the genes transferred often include antibiotic resistance and virulence determinants.

E. faecalis (previously known as *S. faecalis*), a typical gram-positive coccus, is a resident of the intestinal tracts of humans and most animals,[165] and is frequently involved in urinary tract and other infections. *E. faecalis* often possesses plasmid(s) which can transfer by conjugation very efficiently even in liquid culture, and this transfer plays a key role in the acquisition of not only drug resistance but also certain virulence traits, such as hemolysin determinants by the organism. Although DNA exchange by conjugation such as that mediated by the sex factor F in *E. coli* is more famous in gram-negative bacteria, conjugal plasmid transfer in gram-positive, non-streptomycete bacteria has also been known since the mid-1970s, and *E. faecalis* was one of the first of such phenomena to be reported.[166] Since then, many plasmids and transposable elements have been identified in *E. faecalis*, and it became evident that some plasmids transfer very efficiently in liquid culture (e.g., 10^{-2} per donor), while others transfer poorly but could transfer reasonably well on solid surfaces such as filter membranes. It was eventually found that those plasmids that transferred in the liquid culture encoded a response system to an extracellular peptide called sex pheromones that induced a series of biosynthetic events resulting in efficient conjugation.[167]

Plasmid-free (recipient) strains of *E. faecalis* secrete a family of heat-stable, protease-sensitive sex pheromones with specificities for donors carrying various conjugative plasmids (Figure 14). Bacteria carrying a particular plasmid respond by synthesizing a surface material that facilitates formation of mating aggregates with nearby recipients. The induced surface material on donors is referred to as the aggregation substance, and its receptor on the recipient has been designated as the binding substance (BS). BS is also present on the donor surface, the reason for inducible self-clumping of donor cells. There is evidence that BS is in part made up of lipoteichoic acid, because very low concentrations of purified lipoteichoic acid could inhibit aggregation.[168] After receiving a copy of the particular plasmid, the recipient shuts down production of the pheromone but continues to produce pheromones specific for other plasmids. As many as five different pheromones are known to be produced by a single recipient cell, and it is likely that this number is actually much larger. The specificities of pheromone response generally correspond to the incompatibility groups of pheromone-inducible conjugative plasmids.[169]

In contrast to the production of pheromones by recipient cells, each plasmid in donor cells in turn produces a peptide which acts as a specific competitive inhibitor of the corresponding pheromone.[170] The purpose of the inhibitory peptide would be to prevent premature induction by a pheromone too low for actual mating, namely when recipient cells are too far away or too diluted to encounter, or to prevent self-induction by a small amount of the corresponding pheromone in the donor cells which escaped from the shutdown process in the donor cells. The best-characterized plasmid systems in *E. faecalis* are pAD1 and pCF10, and readers are referred to earlier reviews.[167,171,172] for detailed information on the genes and corresponding proteins encoded on each plasmid participating in the conjugal transfer.

8.06.4.4.2 Structures of sex pheromones and pheromone inhibitors

Several sex pheromones and related inhibitors have been isolated and characterized by Suzuki and co-workers,[170,173–177] and their structures are shown in Table 2. All the pheromones and the inhibitory peptides are short, hydrophobic octa- or heptapeptides. All the pheromones are active at concentrations as low as 50 pM, and in the case of cCF10, a donor cell harboring pCF10 could detect as few as one or two molecules. The amino terminus is especially important for recognition.

8.06.5 OTHER MICROBIAL HORMONE-LIKE SIGNAL SUBSTANCES

Finally, it is emphasized that there are some other well-known microbial hormone-like compounds, such as trisporic acid (**28**), the mating pheromone of plus and minus mycelia of *Blakeslea trispora* for sexual spore formation;[178–180] sclerosporin (**29**), the main sporogenic substance from *Sclerotinia fruticola*;[181,182] AO 1 (**30**), the sporogenic substance of *Aspergillus oryzae*;[183,184] and fruiting-inducing cerebroside (**31**) in a basidiomycete, *Schizophyllum commune*.[185,186] These low

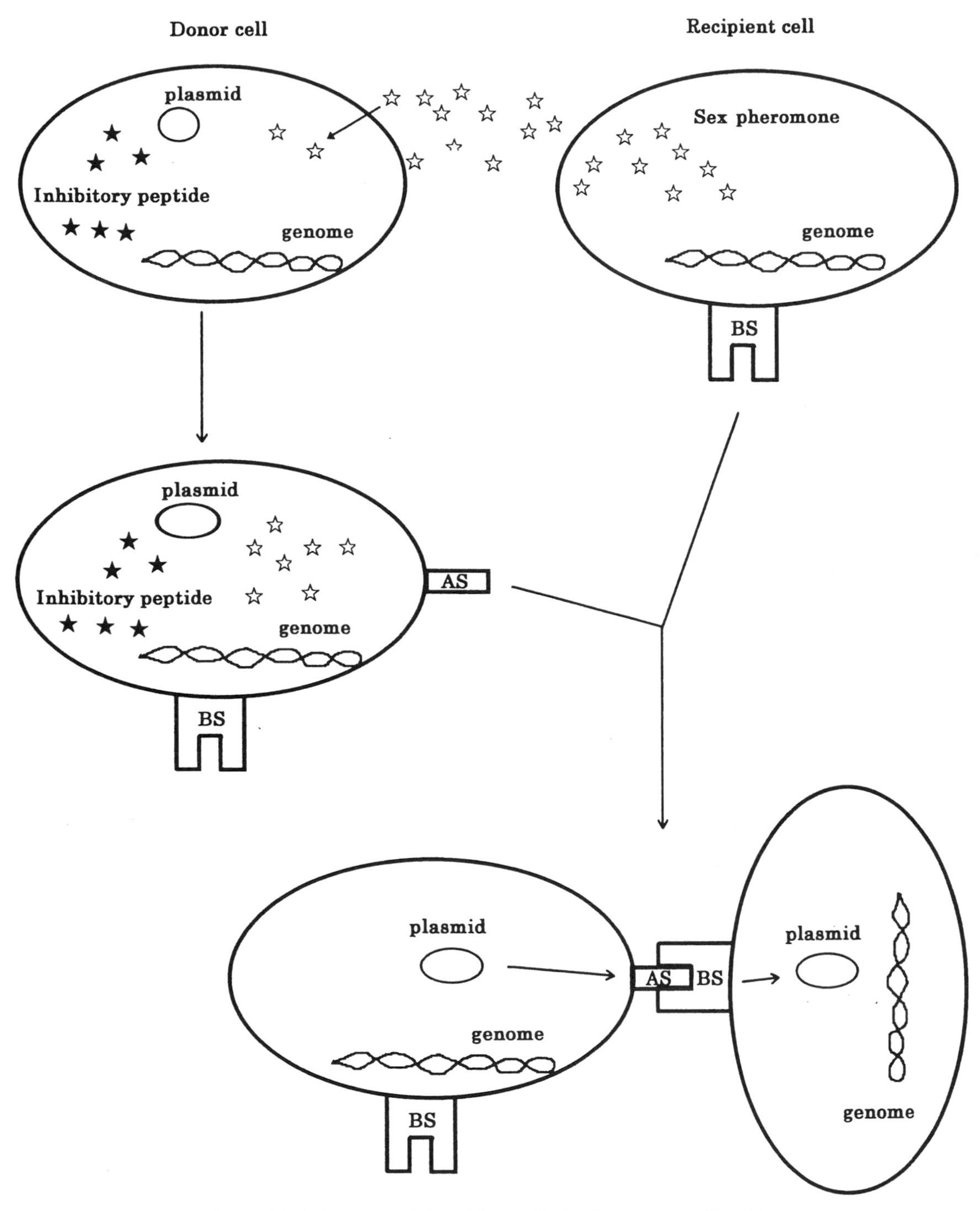

Figure 14 Diagram of plasmid transfer in *Enterococcus faecalis*.

molecular weight endogenous signal substances induce cytodifferentiation in each fungus at very low concentration and may be considered as microbial hormones or pheromones. The absolute configurations of sclerosporin (**29**),[187,188] sporogen-AO 1 (**30**),[189,190] and fruiting-body inducing cerebroside (**31**)[191–193] were determined in detail by chemical synthesis. However, according to editorial policy, they are referred to only briefly in this book due to their lack of mode-of-action. As for the well-studied peptidic mating factors, α- and a-factors in *Saccharomyces cerevisiae*, there is an established review[194] written by experts on yeast and readers are advised to refer to it.

Table 2 Structures of enterococcal sex pheromones and inhibitors.

Pheromone or inhibitor	Plasmid encoding		Peptide structure	Molecular weight	Ref.
	Pheromone	Inhibitor			
Pheromone					
cPD1	pPD1		H-Phe-Leu-Val-Met-Phe-Leu-Ser-Gly-OH	912	173
cAD1	pAD1		H-Leu-Phe-Ser-Leu-Val-Leu-Ala-Gly-OH	818	174
cAM373	pAM373		H-Ala-Ile-Phe-Ile-Leu-Ala-Ser-OH	733	175
cCF10	pCF10		H-Leu-Val-Thr-Leu-Val-Phe-Val-OH	789	176
Inhibitor					
iPD1		pPD1	H-Ala-Leu-Ile-Leu-Thr-Leu-Val-Ser-OH	828	177
iAD1		pAD1	H-Leu-Phe-Val-Val-Thr-Leu-Val-Gly-OH	846	170

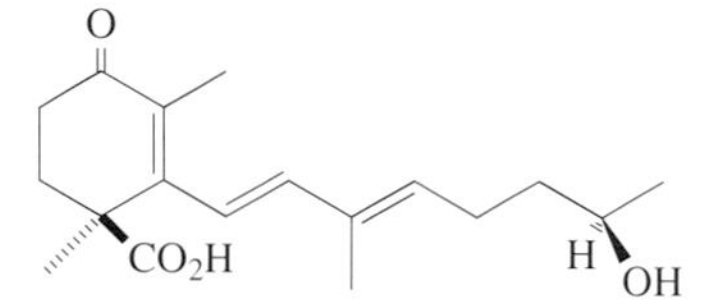

(28) Trisporic acid C *Blakeslea trispora*

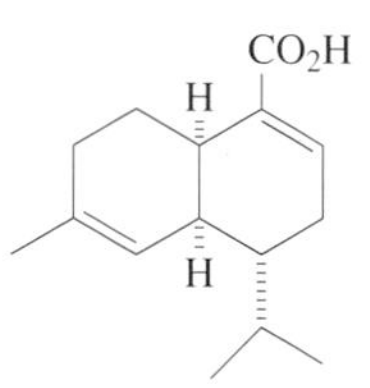

(29) Sclerosporin *Sclerotinia fructicola*

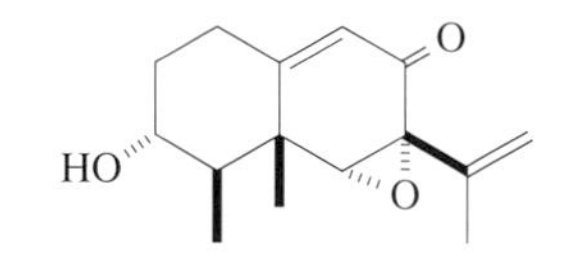

(30) Sporogen-AO1 *Aspergillus oryzae*

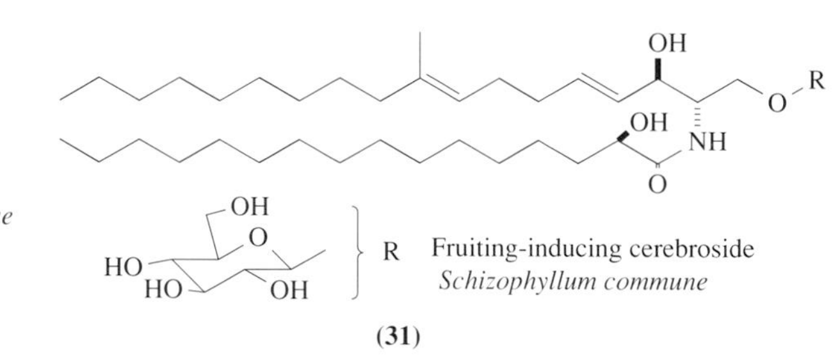

Fruiting-inducing cerebroside *Schizophyllum commune*

(31)

8.06.6 REFERENCES

1. A. S. Khokhlov (IUPAC), "Frontiers of Bioorganic Chemistry and Molecular Biology," Pergamon Press, Oxford, 1980, p. 201.
2. S. A. Pliner, E. M. Kleiner, E. Y. Kornitskaya, I. I. Tovarova, B. V. Rosynov, G. M. Smirnova, and A. S. Khokhlov, *Bioorganicheskaya Chimica*, 1975, **1**, 70.
3. E. M. Kleiner, S. A. Pliner, V. B. Soifer, V. V. Onoprienko, T. A. Balashova, B. V. Rosynov, and A. S. Khokhlov, *Bioorganocheskaya Chimica*, 1976, **2**, 1142.
4. E. M. Kleiner, V. V. Onoprienko, S. A. Pliner, V. S. Soifer, and A. S. Khokhlov, *Bioorganicheskaya Chimica*, 1977, **3**, 424.
5. K. Mori and K. Yamane, *Tetrahedron*, 1982, **38**, 2919.
6. K. Mori, *Tetrahedron*, 1983, **39**, 3107.
7. O. Hara and T. Beppu, *J. Antibiot.*, 1982, **35**, 349.
8. O. Hara and T. Beppu, *J. Antibiot.*, 1982, **35**, 1208.
9. S. Horinouchi, H. Suzuki, M. Nishiyama, and T. Beppu, *J. Bacteriol.*, 1989, **171**, 1206.
10. S. Horinouchi, Y. Kumada, and T. Beppu, *J. Bacteriol.*, 1984, **158**, 481.
11. O. Hara, S. Horinouchi, T. Uozumi, and T. Beppu, *J. Gen. Microbiol.*, 1983, **129**, 2939.
12. S. Horinouchi and T. Beppu, *Mol. Microbiol.*, 1994, **12**, 859.
13. K. Miyake, S. Horinouchi, M. Yoshida, N. Chiba, K. Mori, N. Nogawa, N. Morikawa, and T. Beppu, *J. Bacteriol.*, 1989, **171**, 4298.
14. H. Onaka, N. Ando, T. Nihira, Y. Yamada, T. Beppu, and S. Horinouchi, *J. Bacteriol.*, 1995, **177**, 6083.
15. K. Miyake, T. Kuzuyama, S. Horinouchi, and T. Beppu, *J. Bacteriol.*, 1990, **172**, 3003.
16. D. Vujaklija, S. Horinouchi, and T. Beppu, *J. Bacteriol.*, 1993, **175**, 2652.
17. D. Vujaklija, K. Ueda, S.-K. Hong, T. Beppu, and S. Horinouchi, *Mol. Gen. Genet.*, 1991, **229**, 119.
18. J. Distler, K. Mansouri, G. Mayer, M. Stockmann, and W. Piepersberg, *Gene*, 1992, **115**, 105.
19. M. Yanagimoto and G. Terui, *J. Ferment. Technol.*, 1971, **49**, 611.
20. M. Yanagimoto, Y. Yamada, and G. Terui, *Hakkokogaku*, 1979, **57**, 6.
21. Y. Yamada, K. Sugamura, K. Kondo, M. Yanagimoto, and H. Okada, *J. Antibiot.*, 1987, **40**, 496.

22. K. Kondo, Y. Higuchi, S. Sakuda, T. Nihira, and Y. Yamada, *J. Antibiot.*, 1989, **42**, 1873.
23. U. Gräfe, W. Schade, I. Eritt, W. F. Fleck, and L. Radics, *J. Antibiot.*, 1982, **35**, 1722.
24. U. Gräfe, G. Reinhardt, W. Schade, I. Eritt, W. F. Fleck, and L. Radics, *Biotechnol. Lett.*, 1983, **5**, 591.
25. S. Sakuda and Y. Yamada, *Tetrehedron Lett.*, 1991, **32**, 1817.
26. K. Mori and N. Chiba, *Liebigs Ann. Chem.*, 1990, 31.
27. K. Sato, T. Nihira, S. Sakuda, M. Yanagimoto, and Y. Yamada, *J. Ferment. Bioeng.*, 1989, **68**, 170.
28. H. Ohashi, Y. H. Zheng, T. Nihira, and Y. Yamada, *J. Antibiot.*, 1989, **42**, 1191.
29. S. Sakuda, A. Higashi, T. Nihira, and Y. Yamada, *J. Am. Chem. Soc.*, 1990, **112**, 898.
30. S. Sakuda, A. Higashi, S. Tanaka, T. Nihira, and Y. Yamada, *J. Am. Chem. Soc.*, 1992, **114**, 663.
31. S. Sakuda, S. Tanaka, K. Mizuno, O. Sukcharoen, T. Nihira, and Y. Yamada, *J. Chem. Soc., Perkin Trans. 1*, 1993, 2309.
32. T. Nihira, Y. Shimizu, H. S. Kim, and Y. Yamada, *J. Antibiot.*, 1988, **41**, 1828.
33. H. S. Kim, T. Nihira, H. Tada, M. Yanagimoto, and Y. Yamada, *J. Antibiot.*, 1989, **42**, 769.
34. H. S. Kim, H. Tada, T. Nihira, and Y. Yamada, *J. Antibiot.*, 1990, **43**, 692.
35. S. Okamoto, T. Nihira, H. Kataoka, A. Suzuki, and Y. Yamada, *J. Biol. Chem.*, 1992, **267**, 1093.
36. S. Okamoto, K. Nakamura, T. Nihira, and Y. Yamada, *J. Biol. Chem.*, 1995, **270**, 12 319.
37. M. Yanagimoto and T. Entasu, *J. Ferment. Technol.*, 1983, **61**, 545.
38. K. Mizuno, S. Sakuda, T. Nihira, and Y. Yamada, *Tetrahedron*, 1994, **50**, 10 849.
39. K. Hashimoto, T. Nihira, S. Sakuda, and Y. Yamada, *J. Ferment. Bioeng.*, 1992, **73**, 449.
40. M. Ruengjitchatchawalya, T. Nihira, and Y. Yamada, *J. Bacteriol.*, 1995, **170**, 551.
41. K. Hashimoto, T. Nihira, and Y. Yamada, *J. Ferment. Bioeng.*, 1992, **73**, 61.
42. I. Eritt, U. Gräfe, and W. F. Fleck, *Z. Allg. Mikrobiol.*, 1984, **24**, 3.
43. Y. K. Yang, H. Shimizu, S. Shioya, K. Suga, T. Nihira, and Y. Yamada, *Biotechnol. Bioeng.*, 1995, **46**, 437.
44. Y. K. Yang, M. Morikawa, H. Shimizu, S. Shioya, K. Suga, T. Nihira, and Y. Yamada, *Biotechnol. Bioeng.*, 1996, **49**, 437.
45. E. A. Meighen, *Microbiol. Rev.*, 1991, **55**, 123.
46. E. S. Kempner and F. E. Hanson, *J. Bacteriol.*, 1968, **95**, 975.
47. K. H. Nealson, T. Platt, and J. W. Hastings, *J. Bacteriol.*, 1970, **104**, 313.
48. K. H. Nealson, *Arch. Microbiol.*, 1977, **112**, 73.
49. A. Eberhard, A. L. Burlingame, C. Eberhard, G. L. Kenyon, K. H. Nealson, and N. J. Oppenheimer, *Biochemistry*, 1981, **20**, 2444.
50. J. Engebrecht, K. Nealson, and M. Silverman, *Cell*, 1983, **32**, 773.
51. J. Engebrecht and M. Silverman, *Proc. Natl. Acad. Sci. USA*, 1984, **81**, 4154.
52. W. C. Fuqua, S. C. Winans, and E. P. Greenberg, *J. Bacteriol.*, 1994, **176**, 269.
53. C. Fuqua, S. C. Winans, and E. P. Greenberg, *Annu. Rev. Microbiol.*, 1996, **50**, 727.
54. A. Eberhard, C. A. Widrig, P. McBath, and J. B. Schneller, *Arch. Microbiol.*, 1986, **146**, 35.
55. H. B. Kaplan, A. Eberhard, C. Widrig, and E. P. Greenberg, *J. Labelled Compd. Radiopharm.*, 1985, **22**, 387.
56. A. L. Schaefer, B. L. Hanzelka, A. Eberhard, and E. P. Greenberg, *J. Bacteriol.*, 1996, **178**, 2897.
57. H. B. Kaplan and E. P. Greenberg, *J. Bacteriol.*, 1985, **163**, 1210.
58. A. Eberhard, T. Longin, C. A. Widrig, and S. J. Stranick, *Arch. Microbiol.*, 1991, **155**, 294.
59. E. Swartzman, S. Kapoor, A. F. Graham, and E. A. Meighen, *J. Bacteriol.*, 1990, **172**, 6797.
60. J. Engebrecht and M. Silverman, *Nucleic Acids Res.*, 1987, **15**, 10 455.
61. J. H. Devine, G. S. Shadel, and T. O. Baldwin, *Proc. Natl. Acad. Sci. USA*, 1989, **86**, 5688.
62. J. H. Devine, C. Countryman, and T. O. Baldwin, *Biochemistry*, 1988, **27**, 837.
63. D. Kolibachuk and E. P. Greenberg, *J. Bacteriol.*, 1993, **175**, 7307.
64. S. H. Choi and E. P. Greenberg, *Mol. Marine Biol. Biotechnol.*, 1992, **1**, 408.
65. J. Slock, D. VanRiet, D. Kolibachuk, and E. P. Greenberg, *J. Bacteriol.*, 1990, **172**, 3974.
66. S. H. Choi and E. P. Greenberg, *Proc. Natl. Acad. Sci. USA*, 1991, **88**, 11 115.
67. S. H. Choi and E. P. Greenberg, *J. Bacteriol.*, 1992, **174**, 4064.
68. A. M. Stevens, K. M. Dolan, and E. P. Greenberg, *Proc. Natl. Acad. Sci. USA*, 1994, **91**, 12 619.
69. K. A. Poellinger, J. P. Lee, J. V. Parales, Jr., and E. P. Greeenberg, *FEMS Microbiol. Lett.*, 1995, **129**, 97.
70. B. L. Hanzelka and E. P. Greenberg, *J. Bacteriol.*, 1995, **177**, 815.
71. H. B. Kaplan and E. P. Greenberg, *Proc. Natl. Acad. Sci. USA*, 1987, **84**, 6639.
72. A. Kuo, N. V. Blough, and P. V. Dunlap, *J. Bacteriol.*, 1994, **176**, 7558.
73. A. Kuo, S. M. Callahan, and P. V. Dunlap, *J. Bacteriol.*, 1996, **178**, 971.
74. J. G. Cao and E. A. Meighen, *J. Biol. Chem.*, 1989, **264**, 21 670.
75. J. G. Cao and E. A. Meighen, *J. Bacteriol.*, 1993, **175**, 3856.
76. R. Belas, A. Mileham, D. Cohn, M. Hilman, M. Simon, and M. Silverman, *Science*, 1982, **218**, 791.
77. D. H. Cohn, R. C. Ogden, N. J. Abelson, T. O. Baldwin, K. H. Nealson, M. I. Simon, and A. J. Mileham, *Proc. Natl. Acad. Sci. USA*, 1983, **80**, 120.
78. C. M. Miyamoto, A. D. Graham, M. Boylan, J. F. Evans, K. W. Hasel, E. A. Meighen, and A. F. Graham, *J. Bacteriol.*, 1985, **161**, 995.
79. C. Miyamoto, D. Byers, A. F. Graham, and E. A. Meighen, *J. Bacteriol.*, 1987, **169**, 247.
80. C. M. Miyamoto, M. Boylan, A. F. Graham, and E. A. Meighen, *J. Biol. Chem.*, 1988, **263**, 13 393.
81. C. M. Miyamoto, A. E. Graham, and E. A. Meighen, *Nucleic Acids Res.*, 1988, **16**, 1551.
82. R. E. Showalter, M. O. Martin, and M. R. Silverman, *J. Bacteriol.*, 1990, **172**, 2946.
83. E. Swartzman, M. Silverman, and E. A. Meighen, *J. Bacteriol.*, 1992, **174**, 7490.
84. B. L. Bassler, M. Wright, R. E. Showalter, and M. R. Silverman, *Mol. Microbiol.*, 1993, **9**, 773.
85. B. L. Bassler, M. Wright, and M. R. Silverman, *Mol. Microbiol.*, 1994, **12**, 403.
86. B. L. Bassler, M. Wright, and M. R. Silverman, *Mol. Microbiol.*, 1994, **13**, 273.
87. B. H. Iglewski and D. Kabat, *Proc. Natl. Acad. Sci. USA*, 1975, **72**, 2284.
88. B. H. Iglewski, J. Sadoff, M. J. Bjorn, and E. S. Maxwell, *Proc. Natl. Acad. Sci. USA*, 1978, **75**, 3211.
89. R. A. Bever and B. H. Iglewski, *J. Bacteriol.*, 1988, **170**, 4309.

90. E. Kessler and M. Safrin, *J. Bacteriol.*, 1988, **170**, 5241.
91. U. A. Ochsner, A. K. Koch, A. Fiechter, and J. Reiser, *J. Bacteriol.*, 1994, **176**, 2044.
92. L. Passador, J. M. Cook, M. J. Gambello, L. Rust, and B. H. Iglewski, *Science*, 1993, **260**, 1127.
93. J. P. Pearson, K. M. Gray, L. Passador, K. D. Tucker, A. Eberhard, B. H. Iglewski, and E. P. Greenberg, *Proc. Natl. Acad. Sci. USA*, 1994, **91**, 197.
94. J. P. Pearson, L. Passador, B. H. Iglewski, and E. P. Greenberg, *Proc. Natl. Acad. Sci. USA*, 1995, **92**, 1490.
95. P. C. Seed, L. Passador, and B. H. Iglewski, *J. Bacteriol.*, 1995, **177**, 654.
96. J. M. Brint and D. E. Ohman, *J. Bacteriol.*, 1995, **177**, 7155.
97. G. P. C. Salmond, B. W. Bycroft, G. S. A. B. Stewart, and P. Williams, *Mol. Microbiol.*, 1995, **16**, 615.
98. M. Pirhonen, D. Flego, R. Heikinheimo, and E. T. Palva, *EMBO J.*, 1993, **12**, 2467.
99. N. J. Bainton, P. Stead, S. R. Chhabra, B. W. Bycroft, G. P. C. Salmond, G. S. A. B. Stewart, and P. Williams, *Biochem. J.*, 1992, **288**, 997.
100. S. R. Chhabra, P. Stead, N. J. Bainton, G. P. C. Salmond, G. S. A. B. Stewart, P. Williams, and B. W. Bycroft, *J. Antibiot.*, 1993, **46**, 441.
101. S. Jones, B. Yu, N. J. Bainton, M. Birdsall, B. W. Bycroft, S. R. Chhabra, A. J. R. Cox, P. Golby, P. J. Reeves, S. Stephens, M. K. Winson, G. P. C. Salmond, G. S. A. B. Stewart, and P. Williams, *EMBO J.*, 1993, **12**, 2477.
102. S. McGowan, M. Sebaihia, S. Jones, B. Yu, N. Bainton, P. F. Chan, B. Bycroft, G. S. A. B. Stewart, P. Williams, and G. P. C. Salmond, *Microbiology*, 1995, **141**, 541.
103. L. Zhang and A. Kerr, *J. Bacteriol.*, 1991, **173**, 1867.
104. L. Zhang, P. J. Murphy, A. Kerr, and M. E. Tate, *Nature*, 1993, **362**, 446.
105. K. R. Piper, S. B. Von Bodman, and S. K. Farrand, *Nature*, 1993, **362**, 448.
106. I. Hwang, P. L. Li, L. Zhang, K. R. Piper, D. M. Cook, and M. E. Tate, *Proc. Natl. Acad. Sci. USA*, 1994, **91**, 4639.
107. W. C. Fuqua and S. C. Winans, *J. Bacteriol.*, 1994, **176**, 2796.
108. M. T. Cubo, A. Economou, G. Murphy, A. W. B. Johnston, and J. A. Downie, *J. Bacteriol.*, 1992, **174**, 4026.
109. J. Schripsema, K. E. E. De Rudder, T. B. Van Vliet, P. P. Lankhorst, E. De Vroom, J. W. Kijne, and A. A. N. Von Brussel, *J. Bacteriol.*, 1996, **178**, 366.
110. K. M. Gray, J. P. Pearson, J. A. Downie, B. E. A. Boboye, and E.P. Greenberg, *J. Bacteriol.*, 1996, **178**, 372.
111. D. A. Siegele and R. Kolter, *J. Bacteriol.*, 1992, **174**, 345.
112. G. W. Huisman and R. Kolter, *Science*, 1994, **265**, 537.
113. A. L. Sonenshein, J. A. Hoch, and R. Losick (eds.), "*Bacillus subtilis* and Other Gram-positive Bacteria: Biochemistry, Physiology, and Molecular Genetics," American Society for Microbiology, Washington, DC, 1993.
114. P. H. A. Sneath, in "Bergey's Manual of Systematic Bacteriology," ed. P. H. A. Sneath, Williams & Wilkins, Baltimore, MD, 1986, vol. 2, p. 1104.
115. D. Dubnau, *Microbiol. Rev.*, 1991, **91**, 9397.
116. C. P. Moran, Jr., in "*Bacillus subtilis* and Other Gram-positive Bacteria: Biochemistry, Physiology, and Molecular Genetics," eds. A. L. Sonenshein, J. A. Hoch, and R. Losick, American Society for Microbiology, Washington, DC, 1993, p. 653.
117. A. Akrigg, S. R. Ayad, and G. R. Barker, *Biochem. Biophys. Res. Commun.*, 1967, **28**, 1062.
118. A. Akrigg and S. R. Ayad, *Biochem. J.*, 1970, **117**, 397.
119. H. Joenje, M. Gruber, and G. Venema, *Biochim. Biophys. Acta*, 1972, **262**, 189.
120. R. Magnuson, J. Solomon, and A. D. Grossman, *Cell*, 1994, **77**, 207.
121. J. M. Solomon, R. Magnuson, A. Srivastava, and A. D. Grossman, *Genes Dev.*, 1995, **9**, 547.
122. J. A. Hoch, in "*Bacillus subtilis* and Other Gram-positive Bacteria: Biochemistry, Physiology, and Molecular Genetics," eds. A. L. Sonenshein, J. A. Hoch, and R. Losick, American Society for Microbiology, Washington, DC, 1993, p. 747.
123. M. Strauch, V. Webb, G. Spiegelman, and J. A. Hoch, *Proc. Natl. Acad. Sci. USA*, 1990, **87**, 1801.
124. A. D. Grossman and R. Losick, *Proc. Natl. Acad. Sci. USA*, 1988, **85**, 4369.
125. V. R. Srinivasan and H. O. Halvorson, *Nature*, 1963, **197**, 100.
126. V. R. Srinivasan, *Nature*, 1966, **209**, 537.
127. C. Waldburger, D. Gonzalez, and G. H. Chambliss, *J. Bacteriol.*, 1993, **175**, 6321.
128. D. Dubnau, in "*Bacillus subtilis* and Other Gram-positive Bacteria: Biochemistry, Physiology, and Molecular Genetics," eds. A. L. Sonenshein, J. A. Hoch, and R. Losick, American Society for Microbiology, Washington, DC, 1993, p. 555.
129. T. Msadek, F. Kunst, and G. Rapoport, in "*Bacillus subtilis* and Other Gram-positive Bacteria: Biochemistry, Physiology, and Molecular Genetics," eds. A. L. Sonenshein, J. A. Hoch, and R. Losick, American Society for Microbiology, Washington, DC, 1993, p. 729.
130. I. Smith, in "*Bacillus subtilis* and Other Gram-positive Bacteria: Biochemistry, Physiology, and Molecular Genetics," eds. A. L. Sonenshein, J. A. Hoch, and R. Losick, American Society for Microbiology, Washington, DC, 1993, p. 785.
131. H. D. McCurdy, in "Bergey's Manual of Systematic Bacteriology," ed. P. H. A. Sneath, Williams & Wilkins, Baltimore, MD, 1986, vol. 3, p. 2139.
132. M. Dworkin, *Microbiol. Rev.*, 1996, **60**, 70.
133. M. Inouye, S. Inouye, and D. R. Zusman, *Proc. Natl. Acad. Sci. USA*, 1979, **76**, 209.
134. D. White, *Int. Rev. Cytol.*, 1981, **72**, 203.
135. M. Dworkin, "Developmental Biology of the Bacteria," Benjamin/Cummings, Menlo Park, CA, 1986.
136. D. C. Hagen, A. P. Bretscher, and D. Kaiser, *Dev. Biol.*, 1978, **64**, 284.
137. J. Downard, S. V. Ramaswamy, and K.-S. Kil, *J. Bacteriol.*, 1993, **175**, 7762.
138. A. Kuspa, L. Plamann, and D. Kaiser, *J. Bacteriol.*, 1992, **174**, 3319.
139. L. Plamann, Y. Li, B. Cantwell, and J. Mayor, *J. Bacteriol.*, 1995, **177**, 2014.
140. Y. Li and L. Plamann, *J. Bacteriol.*, 1996, **178**, 289.
141. L. Plamann, J. M. Davis, B. Cantwell, and J. Mayor, *J. Bacteriol.*, 1994, **176**, 2013.
142. J. M. Davis, J. Mayor, and L. Plamann, *Mol. Microbiol.*, 1995, **18**, 943.
143. R. E. Gill and M. G. Cull, *J. Bacteriol.*, 1986, **168**, 341.
144. R. E. Gill, M. Karlok, and D. Benton, *J. Bacteriol.*, 1993, **175**, 4538.
145. N. Tojo, S. Inouye, and T. Komano, *J. Bacteriol.*, 1993, **175**, 4545.
146. T. J. Hagen and L. J. Shimkets, *J. Bacteriol.*, 1990, **172**, 15.

147. B.-U. Lee, K. Lee, J. Mendez, and L. J. Shimkets, *Genes Dev.*, 1995, **9**, 2964.
148. Y. L. Cheng, L. V. Kalman, and D. Kaiser, *J. Bacteriol.*, 1994, **176**, 1427.
149. A. Rosenbluh and E. Rosenberg, *J. Bacteriol.*, 1989, **171**, 1513.
150. D. R. Toal, S. W. Clifton, B. A. Roe, and J. Downard, *Mol. Microbiol.*, 1995, **16**, 177.
151. J. Downard and D. Toal, *Mol. Microbiol.*, 1995, **16**, 171.
152. L. Plamann, A. Kuspa, and D. Kaiser, *J. Bacteriol.*, 1992, **174**, 3311.
153. A. Kuspa, L. Plamann, and D. Kaiser, *J. Bacteriol.*, 1992, **174**, 7360.
154. J. A. Hoch and T. J. Shilhavy (eds.), "Two-component Signal Transduction," ASM Press, Washington, DC, 1995.
155. L. Kroos and D. Kaiser, *Genes Dev.*, 1987, **1**, 840.
156. S.-F. Li and L. J. Shimkets, *J. Bacteriol.*, 1993, **175**, 3648.
157. L. J. Shimkets and S. J. Asher, *Mol. Gen. Genet.*, 1988, **211**, 63.
158. S. K. Kim and D. Kaiser, *Proc. Natl. Acad. Sci. USA*, 1990, **87**, 3635.
159. S. K. Kim and D. Kaiser, *Cell*, 1990, **61**, 19.
160. K. Lee and L. J. Shimkets, *J. Bacteriol.*, 1996, **178**, 977.
161. Y. Cheng and D. Kaiser, *J. Bacteriol.*, 1989, **171**, 3719.
162. Y. Cheng and D. Kaiser, *J. Bacteriol.*, 1989, **171**, 3727.
163. L. V. Kalman, Y. L. Cheng, and D. Kaiser, *J. Bacteriol.*, 1994, **176**, 1434.
164. G. Bartholomeusz and J. Downard, in "Abstracts of the 22nd International Conference in Biol. Myxobacteria," 1995, p. 31.
165. J. O. Mundt, in "Bergey's Manual of Systematic Bacteriology," ed. P. H. A. Sneath, Williams & Wilkins, Baltimore, MD, 1986, vol. 2, p. 1063.
166. A. E. Jacob and S. J. Hobbs, *J. Bacteriol.*, 1974, **117**, 360.
167. D. B. Clewell, in "Bacterial Conjugation," ed. D. B. Clewell, Plenum, New York, NY, 1993, p. 349.
168. E. E. Ehrenfeld, R. E. Kessler, and D. B. Clewell, *J. Bacteriol.*, 1986, **168**, 6.
169. R. Wirth, A. Friesenegger, and T. Horaud, *Mol. Gen. Genet.*, 1992, **233**, 157.
170. M. Mori, A. Isogai, Y. Sakagami, M. Fujino, C. Kitada, D. B. Clewell, and A. Suzuki, *Agric. Biol. Chem.*, 1986, **50**, 539.
171. G. M. Dunny, B. A. B. Leonard, and P. J. Hedberg, *J. Bacteriol.*, 1995, **177**, 871.
172. D. B. Clewell, *Cell*, 1993, **73**, 9.
173. A. Suzuki, M. Mori, Y. Sakagami, A. Isogai, M. Fujino, C. Kitada, R. A. Craig, and D. B. Clewell, *Science*, 1984, **226**, 849.
174. M. Mori, Y. Sakagami, M. Narita, A. Isogai, M. Fujino, C. Kitada, R. A. Craig, D. B. Clewell, and A. Suzuki, *FEBS Lett.*, 1984, **178**, 97.
175. M. Mori, H. Tanaka, Y. Sakagami, A. Isogai, M. Fujino, C. Kitada, B. A. White, F. Y. An, D. B. Clewell, and A. Suzuki, *FEBS Lett.*, 1986, **206**, 69.
176. M. Mori, Y. Sakagami, Y. Ishii, A. Isogai, C. Kitada, M. Fujino, J. C. Adsit, G. M. Dunny, and A. Suzuki, *J. Biol. Chem.*, 1988, **263**, 14 574.
177. M. Mori, H. Tanaka, Y. Sakagami, A. Isogai, M. Fujino, C. Kitada, D. B. Clewell, and A. Suzuki, *J. Bacteriol.*, 1987, **169**, 1747.
178. H. Van den Ende, *J. Bacteriol.*, 1986, **96**, 1289.
179. D. J. Austin, J. D. Bu'Lock, and G. W. Gooday, *Nature*, 1969, **223**, 1178.
180. T. Reschke *Tetrahedron Lett.*, 1969, 3435.
181. M. Katayama and S. Marumo, *Agric. Biol. Chem.*, 1978, **42**, 505.
182. M. Katayama and S. Marumo, *Tetrahedron Lett.*, 1979, 1773.
183. S. Tanaka, K. Wada, M. Katayama, and S. Marumo, *Agric. Biol. Chem.*, 1984, **48**, 3189.
184. S. Tanaka, K. Wada, S. Marumo, and H. Hattori, *Tetrahedron Lett.*, 1984, **25**, 5907.
185. G. Kawai and Y. Ikeda, *Biochim. Biophys. Acta*, 1982, **719**, 612.
186. G. Kawai and Y. Ikeda, *Biochim. Biophys. Acta*, 1983, **754**, 243.
187. T. Kitahara, T. Matsuoka, M. Katayama, S. Marumo, and K. Mori, *Tetrahedron Lett.* 1984, **25**, 4685.
188. T. Kitahara, H. Kurata, T. Matsuoka, and K. Mori, *Tetrahedron*, 1985, **41**, 5475.
189. K. Mori and H. Tamura, *Liebigs Ann. Chem.*, 1988, 97.
190. T. Kitahara, H. Kurata, and K. Mori, *Tetrahedron*, 1988, **44**, 4339.
191. K. Mori and Y. Funaki, *Tetrahedron Lett.*, 1984, **25**, 5291.
192. K. Mori and Y. Funaki, *Tetrahedron*, 1985, **41**, 2369.
193. K. Mori and Y. Funaki, *Tetrahedron*, 1985, **41**, 2379.
194. G. F. Sprague, Jr. and J. W. Thorner, in "Molecular and Cellular Biology of the Yeast *Saccharomyces*, Gene Expression," eds. E. W. Jones, J. R. Pringle, and J. R. Broach, Cold Spring Harbor Laboratory Press, NY, 1992, p. 657.

8.07
Marine Natural Products and Marine Chemical Ecology

JUN'ICHI KOBAYASHI and MASAMI ISHIBASHI
Hokkaido University, Sapporo, Japan

8.07.1 INTRODUCTION

The oceans cover nearly 70% of the whole surface area of the earth and more than 30 phylums and 500 000 species of marine organisms live in them. In the oceans the circumstances are quite different from those on the land. The undersea environment is a closed system with high salinity, high pressure, and relatively constant temperature. Animals, plants, and microorganisms living in the ocean therefore are expected to produce quite different secondary metabolites from those produced by terrestrial organisms. Since the 1970s, a great number of new marine natural products have been isolated from various marine organisms. Most of these marine natural products possess a variety of unique chemical structures that have never been encountered among natural products of terrestrial origins, whereas these marine natural products frequently exhibit interesting biological activity which may be of great importance in many fields of biological sciences. The aims of research projects concerned with marine natural products may be (i) to find novel compounds that are useful as leads for drug development, (ii) to provide good tools for basic studies of life science, and (iii) to study the roles and biological functions of secondary metabolites in the life of marine organisms. The third subject, which is called "marine chemical ecology," has, since the mid 1980s, become a field of study.

This chapter describes studies on marine natural products, particularly those of interest from the viewpoint of marine chemical ecology, and consists mainly of two parts. The first part deals with marine natural products related to marine chemical ecology. Classification of the phenomena associated with marine chemical ecology is arbitrary, and here we describe them in seven sections (Sections 8.07.2–8). In the latter part of this chapter (Section 8.07.9), topics in marine natural products chemistry are described from various viewpoints irrespective of the relationships to ecological subjects. As the dividing line between the sections may sometimes be obscure, the selection of compounds and topics is arbitrary and not necessarily comprehensive.

A series of excellent reviews on marine natural products chemistry, published by Faulkner,[1–12] cover all the literature describing marine natural products, organized phylogenetically. A special issue of *Chemical Reviews* appeared in 1993,[13] providing broad aspects of contributions on marine natural products chemistry. The present authors wrote a review in 1992[14] covering the nitrogen-containing secondary metabolites isolated from marine organisms, mainly reported in the late 1980s.

Several good books or reviews dealing with general or specialized subjects in marine natural products research have been published,[15–21] in particular, those describing the role of marine natural products in chemical ecology.[22–25]

8.07.2 FEEDING ATTRACTANTS AND STIMULANTS

Feeding is one of the most fundamental behaviors of all living organisms. It was suggested by many biological studies that chemical substances or chemical changes in the environment may initiate the feeding behavior of marine organisms and promote the ingestion of foods, and those chemicals may be designated as "attractants" and "stimulants," respectively. This sections deals with studies of feeding attractants and stimulants of marine animals. Excellent reviews have been published on this subject by Sakata.[26–28] This section contains descriptions of feeding attractants of stimulants of particularly fish and mollusks.[26–29]

8.07.2.1 Fish

It is well known that chemical substances participate in feeding behaviors of fish and this phenomenon is called "chemoreception." Fish possess a sense of smell and taste and their sensitivity is much higher than that of man (Table 1).[28] The gustatory organs (taste buds) of fish are distributed not only in the mouth but also on the palp, lip, and skin. The senses of taste and smell cannot be clearly categorized for fish because signal communications by chemical substances are mediated with water. It has been suggested that the gustatory organs of fish may play a role in receiving signals of chemical substances over long distances.

Table 1 Comparison of taste sensitivity of man and fish.[28,30]

	Lowest concentration of taste (mol l^{-1})		
Substance	*Man*	*Fish*[a]	*Ratio*
Raffinose		1/245 760	
Sucrose	1/91	1/81 921	900:1
Lactose	1/16	1/2560	160:1
Glucose	1/13	1/20 480	1575:1
Galactose	1/9	1/5120	569:1
Fructose	1/24	1/61 440	2560:1
Arabinose	1/13	1/15 360	1182:1
Saccharin	1/9091	1/1 536 000	169:1
Quinine hydrochloride	1/1 030 928	1/24 576 000	24:1
Sodium chloride	1/100	1/20 480	205:1
Acetic acid	1/1250	1/204 800	164:1

[a] Cypriniformes family.

Hashimoto *et al.*[31] studied feeding attractants and stimulants of the eel *Anguilla japonica* in extracts of mussels *Tapes japonicus* by watching the behavior of eels using samples mixed with gelatin. Seven amino acids were isolated in the active fraction, and among them the effective concentrations of glycine (Gly), L-alanine (Ala), and L-arginine (Arg) were revealed to be 2×10^{-5} mM, 1×10^{-7} mM, and 5×10^{-8} mM, respectively. Mixtures of these amino acids were more effective than each pure amino acid, suggesting that the synergetic effect of amino acids may be important. Quaternary ammonium salts, nucleic acid derivatives, and organic acids proved to be inactive.

Amino acids were generally identified as feeding attractants and stimulants of fish of other kinds (Table 2). They were mostly normal amino acids such as Gly, Ala, and Arg, while unusual amino acids were also reported such as arcamine (**1**) and strombine (**2**).[32] In addition to amino acids, inosine, inosine-5′-monophosphate (5′-IMP (**3**)), adenosine-5′-monophosphate (5′-AMP (**4**)), betaine [Bet, $(CH_3)_3N^+CH_2CO_2^-$], and trimethylamine *N*-oxide [TMAO, $(CH_3)_3N^+O^-$] were also reported as active substances of turbot, plaice, or Dover sole (Table 2).

Table 2 Feeding attractants and stimulants of fish.

Predator fish	*Prey animals*	*Active substances*	*Ref.*
Bathystoma rimator	*Arca zevra*	arcamine (**1**)	32
Bathystoma rimator	*Strombus gigas*	strombine (**2**)	32
Merlangius merlangus	*Arenicola marina*	Gly, Ala, Ser, Thr, Leu, Glu, Val	27, 33
Chrysophrys major	*Perinereis vancaurica tetradentata*	Gly, Ala, Val, amphoteric fluorescent substance	34
Lagodon rhomboides	*Penacus duorarum*	Gly, Asp, Ile, Phe, Bet	27
Salmo gairdnerii		Gly, Ala, α-Aba,[a] Val	35
Scophthalmus maximus	squid extract	5′-IMP, inosine	36
Pleuronectes platessa	squid extract	amino acids, AMP, TMAO	37
Limanda limanda	squid extract	amino acids	37
Soleo solea	*Mytilus edulis*	Bet	27

[a] α-Aba: α-aminobutyric acid.

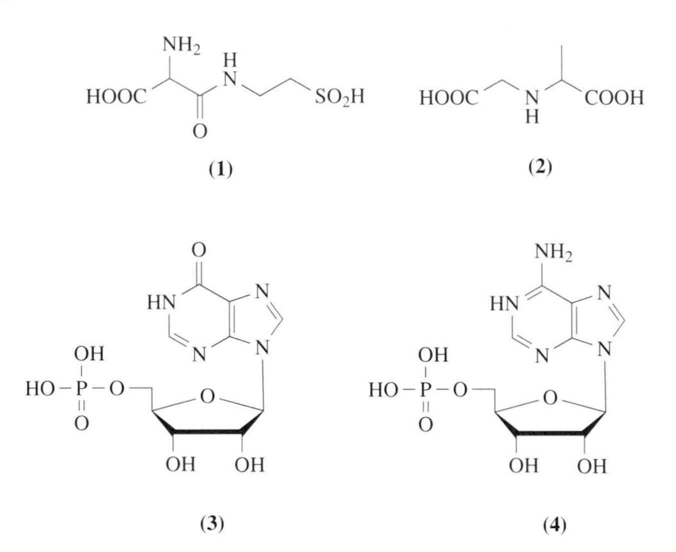

Ina and Matsui[34] studied the relationship between the feeding-stimulating activity and concentration of three amino acids, Gly, L-Ala, and L-Val; the results are shown in Figure 1, in which biting counts represented how many times the fish (sea bream *Chrysophyrys major*) bit the sample kneaded with starch and corresponded to the feeding-stimulating activity. Ina and co-workers[38] also examined the structure-activity relationships of neutral amino acids as feeding stimulants for sea bream *C. major* and revealed the following important structural factors. (i) It was essential for feeding-stimulating activity that amino acids contain α-hydrogen, and free α-amino and α-carboxyl groups. (ii) Active amino acids possess a carbon chain with less than five carbons. More than six carbons reduced the effectiveness. Those with an aromatic ring (e.g., Phe) were inactive. (iii) The methyl group of the terminal position of the aliphatic chain was necessary; the activity was almost lost for those whose terminal methyl was substituted with other functionalities (e.g., Cys, Ser, Orn, and cysteic acid). (iv) Di- and tripeptides were inactive, even those consisting of active amino acids. (v) Only L-amino acids have activity. (vi) L-Cys, L-Pro, and L-Trp exhibited strong synergistic responses to active amino acids such as Gly, L-Ala, and L-Val, whereas the responses of α-Aba, Ser, Thr, Asp, Glu, Asn, and Met were weak.

8.07.2.2 Mollusks

The phylum Mollusca is represented by classes of Gastropoda and Bibalvia, and the former contains commercially important seafood products such as abalones and turbots. It is known[26–29] that gastropods search about for foods and "chemoreception" may be participating in their feeding behaviors, and some amino acids, propionic acid, and trimethylamine were shown to play important roles in the feeding preferences or chemoreception of particular species of gastropods. Abalones *Haliotis* sp. were reported to feed on most algae, and their feeding preference on brown algae were

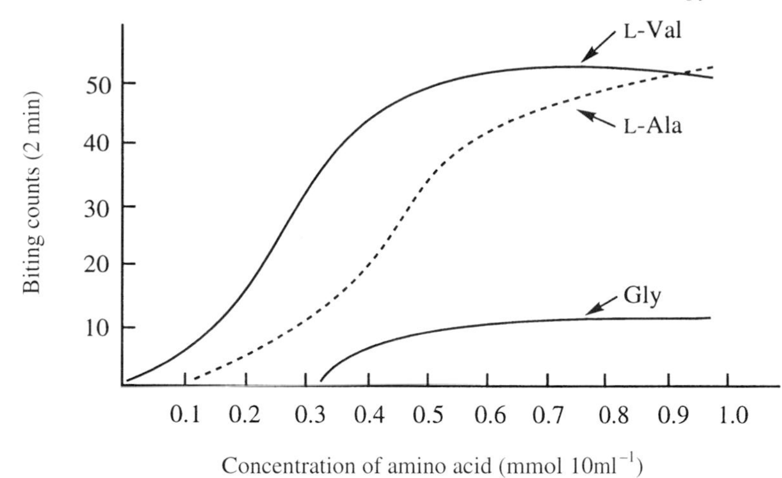

Figure 1 Relationship between the feeding-stimulating activity and concentration of Gly, L-Ala, and L-Val.

shown by examination of the stomach contents. Very few studies, however, have been described on the chemical investigation of the feeding behaviors of abalones.

Sakata *et al.*[39] studied the chemical substances involved in feeding behaviors of the herbivorous gastropod *Haliotis* sp. and developed an excellent biological test of feeding preference of the young abalone *Haliotis discus* using Avicel plates. Abalones, about one year old and starved for at least a day before the test, moved around the test plate, and when they found a spot where their phagostimulant was absorbed, they bit off the Avicel from the sample zone. This test therefore showed not the feeding-attracting activity but the feeding-stimulating activity. Using this method, the methanol extracts of several algae were examined to reveal that extracts of brown alga *Undaria pinnatifida* were the most active. Isolation and identification of the feeding stimulant of the young abalones contained in the extracts of *U. pinnatifida* were studied using the Avicel plate tests to obtain two kinds of active substances, digalactosyldiacylglycerols (DGDG (**5**)) and phosphatidylcholines (PC (**6**)).[40]

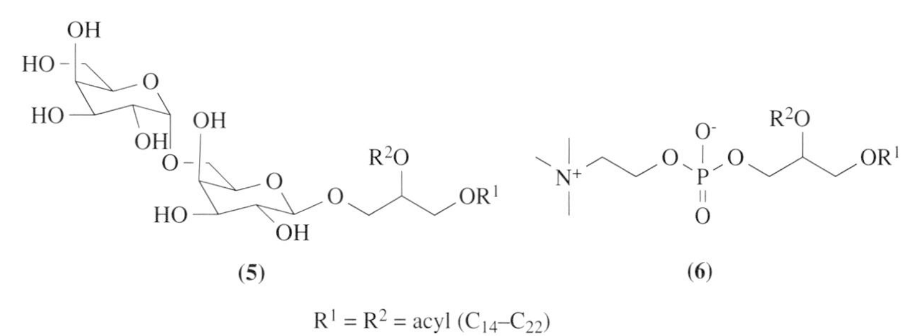

Feeding stimulants of a seahare *Aplysia juliana* were investigated from extracts of a green alga *Ulva pertusa* to isolate two kinds of genetic glycerolipids, DGDG (**5**) and 1,2-diacylglyceryl-4′-*O*-(*N*,*N*,*N*-trimethyl)-homoserine (DGTH (**7**)).[41] The composition of fatty acid methyl esters (C_{14}–C_{22}) liberated from DGTH (**7**) was analyzed by GLC after hydrolysis with methanolic KOH, and FD and FAB mass spectral data provided evidence for identification of DGTH (**7**). From the less polar fractions of the silica gel column chromatography of the $CHCl_3$/MeOH (1:1) extract of *U. pertusa*, monogalactosyldiacylglycerols (MGDG (**8**)) was obtained and revealed to be less active. The *Aplysia* clearly responded to filter paper on which only 100 μg of DGTH (**7**) was absorbed while DGDG (**5**) was as active as DGTH (**7**).

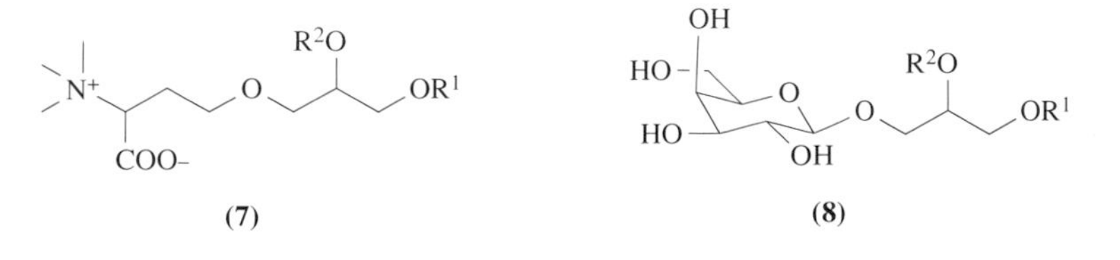

Sakata and co-workers also reported that the Avicel plate method was applicable to investigation of the feeding stimulants for other gastropods such as the turban shell *Turbo cornutus* and the top shell *Omphalius pfeifferi*.[28–30] The turban shell *T. cornutus* moved around for food at night and often fed on a brown alga *Eisenia bicyclis*. Feeding stimulants of *T. cornutus* isolated from the methanol extract of *E. bicyclis* were revealed to be DGDG (**5**), PC (**6**), and 6-sulfoquinovosyldiacylglycerol (SQDG (**9**)). The methanol extract of the green alga *Ulva pertusa* was also studied to isolate feeding-stimulating substances of the top shell *O. pfeifferi* to identify the active glycerolipids such as DGDG (**5**), DGTH (**7**), and SQDG (**9**).[42] Comparisons of the feeding-stimulating activity of glycerolipids isolated from brown and green algae are summarized in Table 3. It is interesting that SQDG (**9**) was very active for the turban shell *T. cornutus* and the top shell *O. pfeifferi* but inactive for the abalone *Haliotis discus*. DGDG (**5**), PC (**6**), and DGTH (**7**) were active for all four gastropods, and among them, PC (**6**) was the most active substance. In addition, the glycerolipids, DGDG (**5**), PC (**6**), and SQDG (**9**), were revealed to be feeding stimulants not only of mollusks but also of echinoderms such as the sea urchin *Strongylocentrotus intermedius*.[27,28]

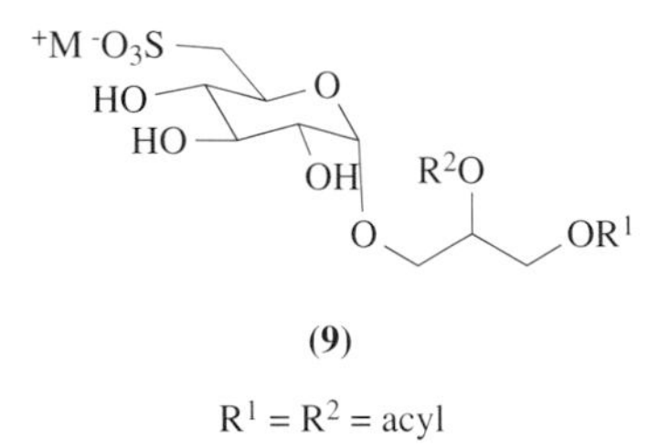

(9)

$R^1 = R^2 = $ acyl

Table 3 Feeding-stimulating activity of glycerolipids for gastropods.[a]

	Haliotis discus	*Turbo cornutus*	*Omphalius pfeifferi*	*Aplysia juliana*
DGDG (**5**)	14–28	15–25	18–23	800
PC (**6**)	10–20	15	< 15	< 50
DGTH (**7**)	10–20	24	< 10	100
SQDG (**9**)	> 300	20–40	< 20	> 1000

[a] Minimum amount of sample for each sample zone of the Avicel plate test or for each filter paper of the test for seahare (μg).

8.07.3 PHEROMONES

8.07.3.1 Sex Attractants of Algae

Brown algae or seaweeds are widely distributed throughout the oceans. Female gametes of marine brown algae release and attract males by chemical signals. In the early 1970s, Müller *et al.*[43] studied the sex attractant of a cosmopolitan brown alga *Ectocarpus siliculosus* and observed that culture dishes of the mature gynogametes of *Ectocarpus* release a faint aromatic fragrance, which was not perceived with the male culture. The scented substance was lipophilic and can be adsorbed on charcoal. When a cotton yarn or filter paper which assimilated the scented substance released from the mature gynogametes of *Ectocarpus* was presented to androgametes, the male gametes congregated around the allurement. By using this phenomenon as a biological test, the attractant substance of *Ectocarpus* was investigated from as many as 14 900 culture dishes of mature gynogametes (ca. 1 kg) to isolate 92 mg of the active substance, named as ectocarpene, which was identified to be (+)-(6*S*)-(1*Z*-butenyl)-1,4-cycloheptadiene (**10**) from mass and NMR spectroscopic data as well as degradation experiments. Diimide reduction of the side-chain of (**10**) followed by ozonolysis of the resulting 6-butyl-1,4-cycloheptadiene afforded (*R*)-2-butylsuccinic acid. Ectocarpene (**10**) was the first sex-attractant of algae whose chemical structure was established. In following years, a series of 11 hydrocarbons (**10**)–(**20**) was isolated and identified as pheromones that were involved as chemical signals in the sexual reproduction of more than 40 species of brown algae (Table 4).[44–47]

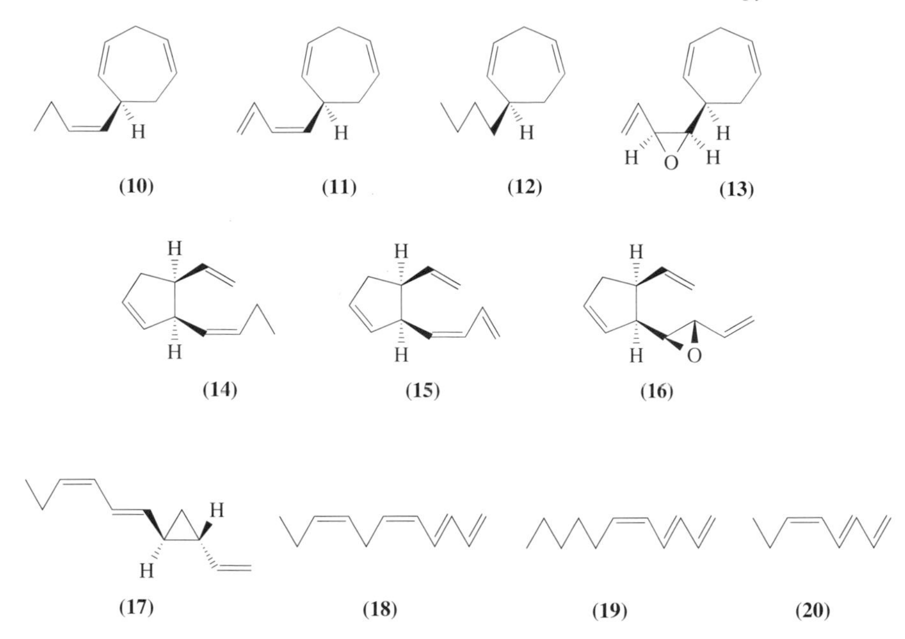

Table 4 Male gametes-attracting substances of brown algae.

Pheromone	*Order*	*Algal species*
Ectocarpene (**10**)	Ectocarpales	*Ectocarpus siliculosus*
	Sphacelariales	*Sphacelaria rigidula*
	Dictyosiphonales	*Adenocystis utricularis*
Desmarestene (**11**)	Desmarestiales	*Desmarestia aculeata*
Dictyotene (**12**)	Dictyotales	*Dictyota dichotoma*
Lamoxirene (**13**)	Laminariales	*Laminaria digita*
Multifidene (**14**)	Cutleriales	*Cutleria multifida*
	Laminariales	*Chorda tomemtosa*
Viridiene (**15**)	Syringodermatales	*Syringoderma phinneyi*
Caudoxirene (**16**)		*Perithalia caudata*
Hormosirene (**17**)	Fucales	*Hormosira banksii*
	Fucales	*Xiphophora chondrophylla*
	Durvillaeales	*Durvillaea potatorum*
	Scytosiphonales	*Scytosiphon lomentaria*
Finavarrene (**18**)	Fucales	*Ascophyllum nodosum*
	Dictyosiphonales	*Dictyosiphon foeniculaceus*
Cystophorene (**19**)	Fucales	*Cystophora siliquosa*
Fucoserratene (**20**)	Fucales	*Fucus serratus*

These sex pheromones were all lipophilic and volatile compounds consisting of C_8 or C_{11} linear or monocyclic hydrocarbons or their epoxides; monocyclic ones contained a cyclopropane, cyclopentene, or cycloheptadiene structure. Identification of these compounds was based on a combination of GC and MS, and comparison with synthetic authentic compounds. For instance, the presence of spermatozoid-releasing and -attracting substances released from the eggs was first suggested for the *Fucus* species in the first half of the twentieth century,[48] while the structure of the active compound, fucoserratene (**20**),[49] was only disclosed two years later than that of ectocarpene (**10**) because of the difficulty of cultivation of algae of the *Fucus* species. The structural elucidation of fucoserratene (**20**) was established by means of UV, mass, and NMR spectroscopic studies of (**20**) and its perhydrogenation product, viz., *n*-octane. The positions and geometrics of alkenes were revealed by comparison of gas chromatographic behaviors with all isomeric conjugated 1,3,5- and 2,4,6-octatrienes.

As shown in Table 4, the relationships between chemical structures of pheromones and taxonomical classifications of algae appeared unclear. Mature female gametophytes of brown algae of the order Laminariales, which includes the large kelps used as food such as tangle, secrete a highly volatile material which induces an explosive discharge of antheridia and spermatozoids. The active substance was investigated by mass cultures of female gametophytes of *Laminaria digitata*. Suspensions of cultures of mature female gametophytes with developed oogonia and eggs were extracted by circulating a stream of air in a closed-loop system,[50] and volatile compounds were collected on a filter of activated carbon. After desorption with dichloromethane, the volatile substances were analyzed by glass capillary gas chromatography. When a SiO_2 particle adsorbing the active substance was added to mature male gametophytes, mass spermatozoids were released and large haloes of spermatozoids were successively formed around the particle.[51] After six years of research, the active substance was identified to be lamoxirene (**13**), as only minute quantities could be produced and isolated in the laboratory. Lamoxirene (**13**) induced the mass release of male gametes of *Laminaria digitata* within 8–12 s at a threshold of ca. 50 pmol. Lamoxirene (**13**) also induces the spermatozoid releasing and attracting of five species of brown algae of the Laminariales order.[52] Kajiwara observed that each of the culture suspensions of mature female gametophytes of eight species of Japanese brown algae (*Laminaria japonica*, *L. angustata*, *L. angustata f. longissima*, *L. coriacea*, *L. diabolica*, *Cymathaese japonica*, *Kjellmaniella crassifolia*, and *Undaria pinnatifida*) of the Laminariales order induced release of spermatozoids from antheridia of all these eight species of brown algae. Among the 64 combinations, no differences were observed in the spermatozoid-releasing activity.[44] Thus, lamoxirene (**13**) proved to be a common spermatozoid-releasing and -attracting substance to brown algae belonging to the Laminariales order.

The sexual reproduction of brown algae of orders Scytosiphonales and Chordariales is an example of isogamy. A pleasant fragrance emanated from the motile female gametes of brown algae of these orders when settled on a surface to attract androgametes. The secretions from the female algal cells of *Colpomenia bullosa* (Scytosiphonales) were collected on charcoal. The volatile constituents were desorbed with dichloromethane and analyzed by GC, GC–MS, and HPLC to compare with those of the secretions from the male cells. A characteristic peak of the female secretions was collected by preparative HPLC (Zorbax ODS) to obtain ca. 100 μg of active substance, which was revealed to be homosirene (**17**) on the basis of GC–MS and ^{13}C NMR spectral data.[53] The absolute configuration of (**17**) obtained from the female secretions of *C. bullosa* was established as (1*R*,2*R*) by chiral HPLC analysis. The female secretions of *Colpomenia bullosa*, *Scytosiphon lomentaria* (Scytosiphonales), and *Analipus japonicus* (Chordariales) were examined by GC and GC–MS under low-temperature conditions to avoid thermal isomerization of homosirene (**17**) into ectocarpene (**10**). It was revealed that the secretions of *C. bullosa* and *S. lomentaria* contained 5% of ectocarpene (**10**) in addition to homosirene (**17**) (95%). The female secretions of *A. japonicus* was shown to consist of 88% of ectocarpene (**10**) and 12% of homosirene (**17**). The optical purity of homosirene (**17**) from gynogametes of thalli of brown algae of different geographic origins was analyzed to reveal that the ratio of enantiomers, (−)-(1*R*,2*R*)-(**17**) and (+)-(1*S*,2*S*)-(**17**), varied with the species and sometimes depended on the locality (Table 5).[47] Brown algae appeared to produce characteristic enantiomeric mixtures of pheromones, which has been presumed, though without any experimental substantiation, to be a simple means for identification of the chemical signals of their own species. The separation of the enantiomers of the alkenic hydrocarbons from the pheromone blend of seaweeds was achieved by GC on modified cyclodextrins as chiral stationary phases.[54]

Table 5 Enantiomer compositions of hormosirene (**17**) from secretions of female gametes or thalli of brown algae.

Genus and species	*Origin*	*Major enantiomer*	*ee (%)*
Dictyopteris acrostichoides	Sorrento, Australia	(−)-(1*R*, 2*R*)	74.2
Dictyopteris membranaceae	Villefranche, France	(−)-(1*R*, 2*R*)	71.2
Dictyopteris prolifera	Hikoshima, Japan	(−)-(1*R*, 2*R*)	90.0
Dictyopteris undulata	Hikoshima, Japan	(−)-(1*R*, 2*R*)	92.0
Analipus japonicus	Muroran, Japan	(+)-(1*S*, 2*S*)	66.0
Analipus japonicus	Akkeshi, Japan	(+)-(1*S*, 2*S*)	90.0
Durvillaea potatorum	Sorrento, Australia	(−)-(1*R*, 2*R*)	51.7
Haplospora globosa	Halifax, Nova Scotia	(+)-(1*S*, 2*S*)	83.3
Hormosira banksii	Flinders, Australia	(−)-(1*R*, 2*R*)	82.8
Xiphophora gladiata	Hobart, Tasmania	(−)-(1*R*, 2*R*)	72.3
Xiphophora chondrophylla	Flinders, Australia	(−)-(1*R*, 2*R*)	82.0

The biosynthesis of these marine algal pheromones is of great interest, and extensive investigations have been carried out.[47] The C_{11} hydrocarbons such as (*S*)-ectocarpene (**10**), dictyotene (**12**), and finavarrene (**18**) have also been isolated from terrestrial higher plants. Model experiments using

Senecio isatideus (Asteraceae) showed that these C_{11} hydrocarbons were derived from unsaturated C_{12} carboxylic acids (**21**) or (**22**) through oxidative decarboxylation, as shown in Scheme 1. In marine algae, however, the C_{11} hydrocarbons were revealed not to arise from the unsaturated C_{12} carboxylic acids precursors (**21**) or (**22**), but were derived from the polyunsaturated C_{20} fatty acids.[55] Externally supplied [$^2H_8$]-arachidonic acid (**23**) was very effectively converted into labeled dictyotene (**12**) in high yield by female gametes of *Ectocarpus siliculosus* (Scheme 1). The deuteration pattern of [$^2H_4$]-dictyotene (**12**) indicated that the $C_{11}H_{18}$ hydrocarbon (**12**) was formed from the C-10–C-20 positions of the arachidonic acid ([$^2H_8$]-(**23**)). From this result, the $C_{11}H_{16}$ hydrocarbons with one more degree of unsaturation such as (*S*)-ectocarpene (**10**), homosirene (**17**), and finavarrene (**18**) were implied to be generated from the aliphatic terminus of the more highly unsaturated eicosapentaenoic acid (**24**).[56]

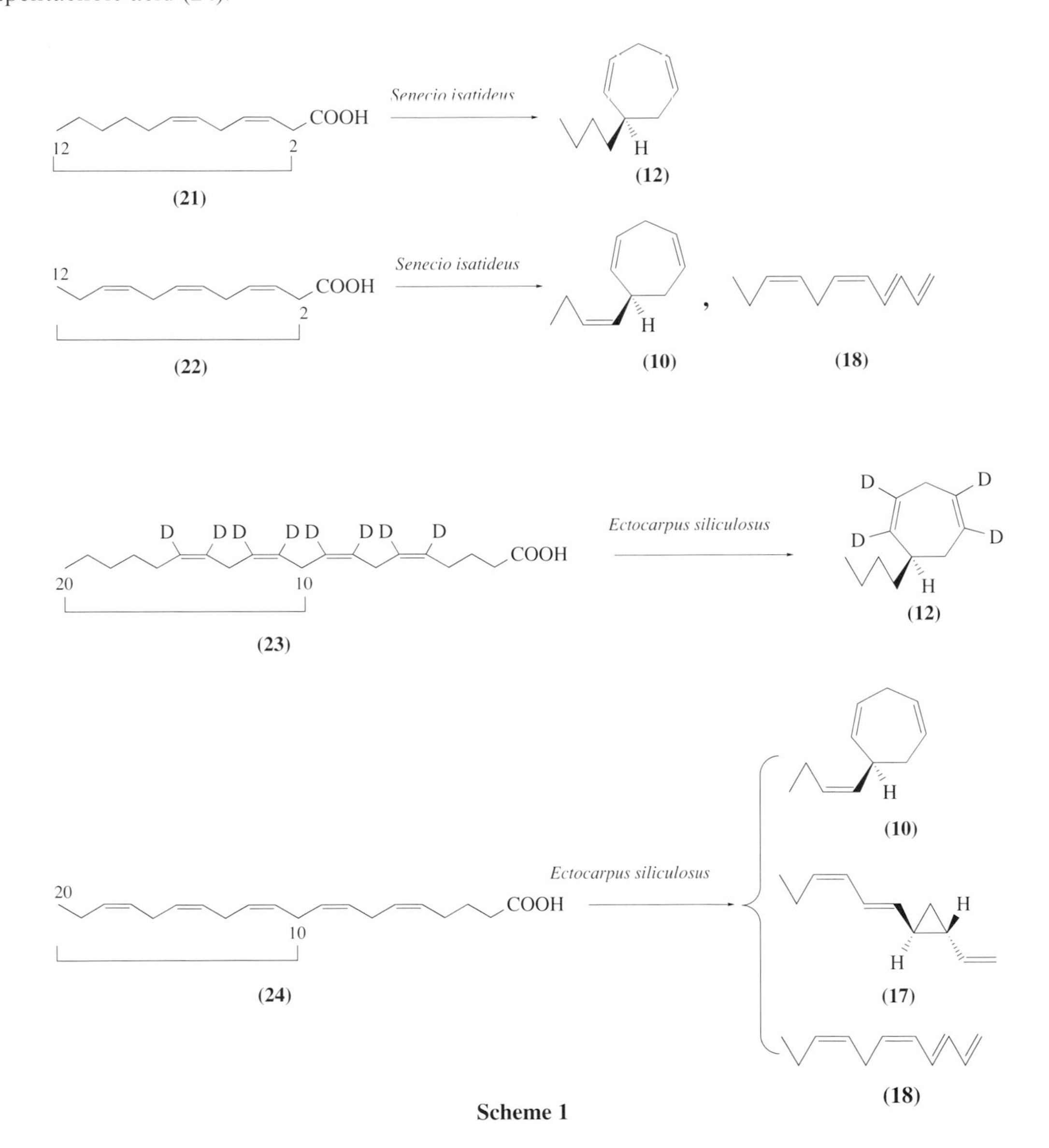

Scheme 1

The first functionalization of the eicosanoid (**24**) was assumed to involve 9-lipoxygenase to yield 9-hydroperoxyicosa-(5*Z*,7*E*,11*Z*,14*Z*,17*Z*)-pentaenoic acid (9-HPEPE (**25**), Scheme 2), which was postulated to release the *cis*-disubstituted cyclopropane (**26**) with (1*R*,2*S*)-configuration and the C_9-dicarbonyl fragment (**27**) by a homolytic cleavage with hydroperoxide lyase.[56]

An uncommon C_{11}-hydrocarbon, 2*Z*,4*Z*,6*E*,8*Z*-undecatriene (giffordene (**28**)) is the major product of the fertile gametophytes of a brown alga *Giffordia mitchellae*.[57] The unusual stereochemistry of giffordene (**28**) may be accounted for by involvement of a spontaneous antarafacial [1.7]-hydrogen shift of a thermolabile intermediate (**29**) with a terminally unsaturated (1,3*Z*,5*Z*,8*Z*)-tetraene

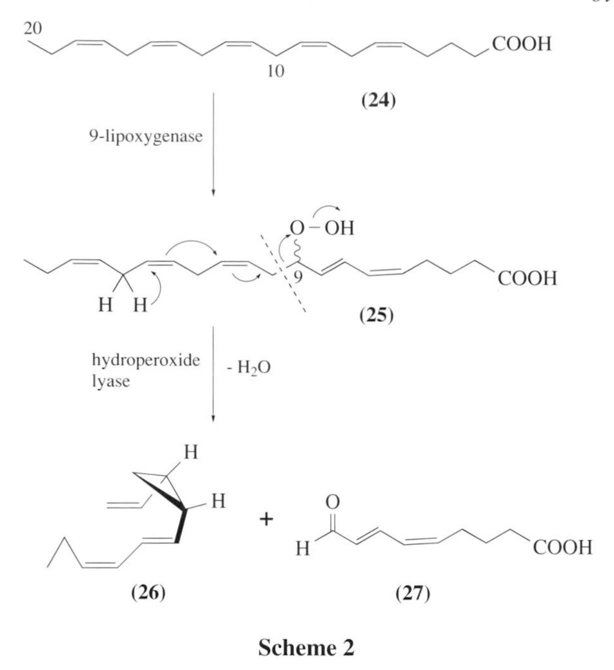

Scheme 2

structure (Scheme 3), which could be supposed to be generated from the 9-HPEPE precursor (**25**) with an appropriate cisoid conformation. The thermolabile precursor (**29**) was synthesized under low-temperature conditions ($< -30\,^{\circ}C$),[58] and kinetic data showed that a half-life of compound (**29**) under the natural environment (18 °C) was ~2.5 h. The activation energies of the [1.7]-hydrogen shift ($E_a = 67.4$ kJ mol^{-1}; $\Delta S_{298} = -91.9$ J mol^{-1} K^{-1}) was considerably lower than that of the well-known [1.7]-hydrogen shift in the conversion of previtamin D_3 into vitamin D_3. The biosynthesis of 7-methylcyclooctatriene (**30**), isolated from the Mediterranean brown alga *Cutleria multifida*,[59] may be accounted for by another pericyclic reaction, viz., an 8πe electrocyclization of (1,3*Z*,5*Z*,7*E*)-nonatriene (**31**), as shown in Scheme 3. This thermolabile precursor (**31**) was also prepared under low-temperature conditions ($< -30\,^{\circ}C$) to determine kinetic data.[58] The half-life of acyclic precursor (**31**) was limited to a few minutes at ambient temperature, and the activation energies of the 8πe electrocyclization reaction ($E_a = 59.4$ kJ mol^{-1}; $\Delta S_{298} = -89.7$ J mol^{-1} K^{-1}) were the lowest values known for natural pericyclic reactions.

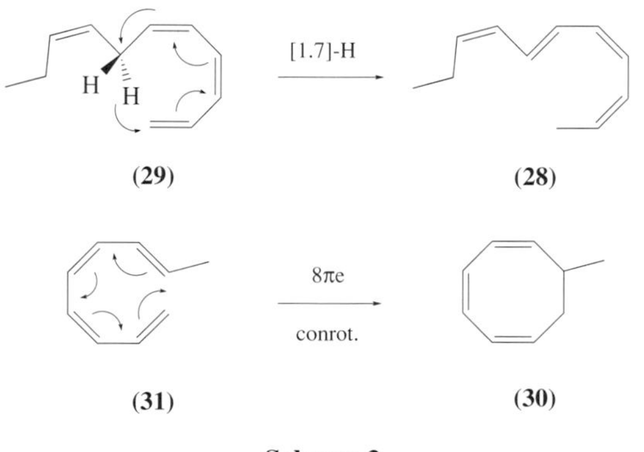

Scheme 3

The generation of (*S*)-ectocarpene (**10**) and other 6-substituted cyclohepta-1,4-dienes such as desmarestene (**11**), dictyotene (**12**), and lamoxirene (**13**) may be assumed to proceed by rearrangement of the thermally labile divinylcyclopropane (**26**) (Scheme 2) through a spontaneous [3,3] sigmatropic reaction (Cope rearrangement) as shown in Equation (1). To verify this hypothesis, the thermally labile divinylcyclopropane (**26**) and its analogues (**32**) and (**33**) were prepared and the activation energies of the Cope rearrangements yielding (**10**), (**11**), and bisnor-derivative (**34**), respectively, were determined, as shown in Table 6.[60] The two temperatures in Table 6 were typical for the surroundings of marine algae in spring in Arctic and Mediterranean regions, and the half-lives for the transformation proved to be relatively long (18–77 min).

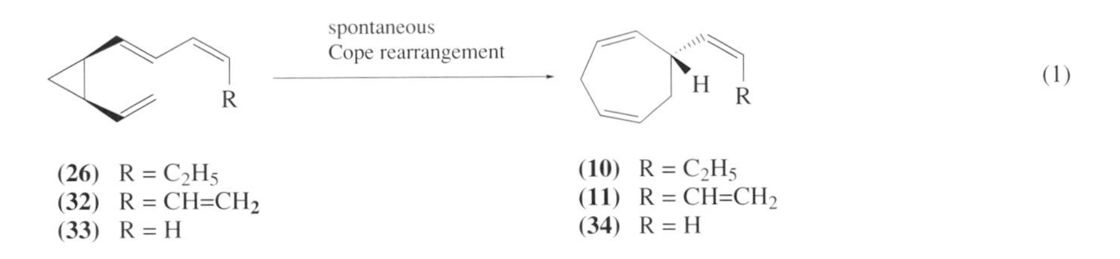

Table 6 Activation parameters and half-lives for the cope rearrangement of cyclopropanes.

Cyclo-propane	*Cyclo-heptadiene*	E_a (kJ mol^{-1})	ln *A*	$t_{1/2}$ (8 °C) (min)	$t_{1/2}$ (18 °C) (min)
(26)	**(10)**	63.8 ± 1.2	18.9 ± 0.4	56	21
(32)	**(11)**	62.6 ± 1.5	18.5 ± 0.5	45	18
(33)	**(34)**	64.9 ± 1.2	19.0 ± 0.4	77	30

The cyclohepta-1,4-dienes **(10)**–**(13)** were considered as sex pheromones of brown algae, but a question arose as to whether thermolabile cyclopropane precursors like **(26)** were actually the active pheromones. Comparative biological tests of **(10)** and **(26)** using male gametes of *Ectocarpus siliculosus* were carried out and revealed that the unstable cyclopropane **(26)** was much more active than the stable cycloheptadiene **(10)**. A threshold concentration of **(10)** was determined as approximately 10 nmol L^{-1} in seawater, while that of **(26)** was found to be significantly lower, around 5 pmol L^{-1}. The release of the pheromone by female gametes, chemotactic orientation of male gametes, and fusion of sexual cells all occurs within only a few minutes. The half-life for the rearrangement of **(26)** was significantly longer than the time required for a sexual encounter in algae. Thus, it may be suggested that the degradation of the cyclopropanes by Cope rearrangement inactivates algal pheromones. Marine brown algae have developed the fastest and most effective signal system for spontaneous deactivation of attractant and/or release factors. These results suggested that systems in which cycloheptadienes were identified as active factors should be reexamined since thermally labile cyclopropyl precursors may be the actual pheromones and have even lower threshold concentrations.

The biosynthetic sequence involving the functionalization at C-9 of unsaturated C_{20} fatty acid, as depicted in Scheme 2, was further supported by a study on the algal pheromones produced by a freshwater diatom *Gomphonema parvulum*,[61] which was revealed to produce significant amounts of C_{11} hydrocarbons; its major volatile component was identified as homosirene **(17)**. Approximately 10^7–10^8 cells of *G. parvulum* were sonicated, centrifugated, and the supernatant crude preparation was treated with deuterium-labeled arachidonic acid **(23)** to give deuterium-labeled dihydrohomosirene **(35)** together with 9-oxo-nona-5*Z*,7*E*-dienoic acid **(27)**, as shown in Equation (2). The polar fragment **(27)** was identified by comparison with a synthetic authentic specimen. When the experiment was conducted in the presence of $^{18}O_2$ gas, the oxygen isotope was incorporated into the aldehyde function of **(27)**. These observations further supported the current concept of the algal pheromone biosynthesis (Scheme 2).

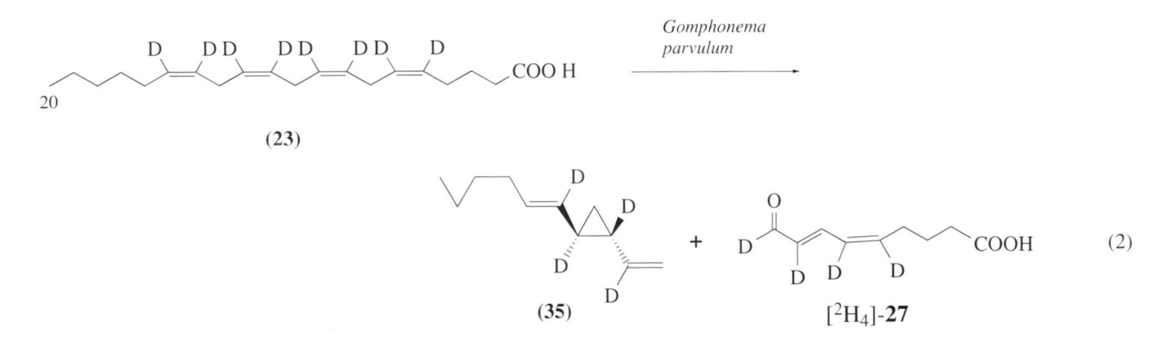

8.07.3.2 Others

Some secondary metabolites such as steroids and prostaglandins are identified as pheromonal attractants of teleosts.[62] 17α,20β-Dihydroxy-4-pregnen-3-one (**36**) was revealed to be a potent female sex pheromone of goldfish *Carassius auratus*,[63] Ovulating goldfish release (**36**), and milt (sperm and seminal fluid) volume was increased by exposure to (**36**). The goldfish olfactory epithelium was extremely sensitive to (**36**), and the increase in milt volume induced by (**36**) was abolished by sectioning the medial olfactory tracts, which was implicated in the control of sex behavior in male goldfish.

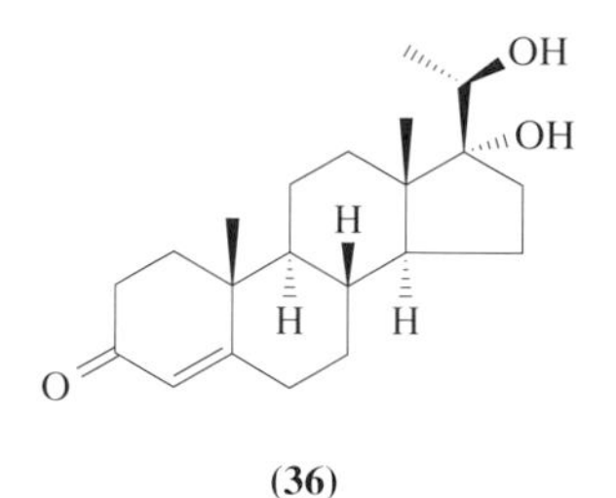

(36)

8.07.4 SYMBIOSIS

8.07.4.1 Invertebrates and Microalgae

Symbiotic association (living together; the wide sense of "symbiosis") of animals, plants, or microorganisms with different species is frequently found and there are three styles of symbiosis, (i) parasitism, (ii) commensalism, and (iii) mutualism. The first case implies that by living together only one of a pair (the symbiont) is advantaged but the partner (the host) is damaged. In the second case, only the symbiont is advantaged while the host is neither advantaged nor damaged. The third one is the narrow sense of "symbiosis," in which both the symbiont and host are advantaged by living together.

Symbiotic association between marine invertebrates and microalgae is a well-known phenomenon classified as mutualism (the third case), and Table 7 shows a list of symbiotic marine microalgae and their host invertebrates.[64] These symbiotic microalgae live inside the cells of host invertebrates. Flatworms take microalgae from their food and the algae go into the parenchyma under the epidermis of the hosts. Symbionts of bivalves live in the edges of the mantle, the hematocoel, or the renal tissues. Sometimes symbionts of bivalves are found inside the gill cells or in the ameboid hemocytes. Nudibranchs have their symbiotic algae reside in the branched hepatopancreas on their back processes or in the alimentary canals.

The morphology and life history of some species of symbiotic microalgae are almost the same as those of free-living species, but those for other species are quite different. Dinoflagellates of the genus *Symbiodinium* living inside invertebrates are always found as vegetative cells which have very thin and faint cell walls and no flagella, while those of free-swimming ones have very bold cell walls. The symbiotic dinoflagellates *Amphidinium* sp. and *Prorocentrum* sp. found inside flatworms have two flagella and cell walls, but the cell walls of *Amphidinium* sp. appear relatively thin. Prasinophytes of free-swimming species are found in the periods of vegetable cells or moving spores; the latter have four flagella with equal length. In both periods, free-swimming prasinophytes possess cell walls and stigma, while those symbiotically living inside invertebrates lose both cell walls and stigma and their cells are wrapped by plasma-membrane.

The way in which the symbiotic algae are handed down to the next generation of the host invertebrates also depends on the combinations of algae and invertebrates. Protozoans give their symbiont algal cells almost equally to the divided daughter cells during cell fission. For flatworms, bivalves, and nudibranchs, the symbiotic associations with microalgae are reconstructed by the next generation. On their oviposition, however, it is frequently found that microalgae of the parent host cells are adhered to the egg capsule, and hatched larvae of the host invertebrates can easily take the algae into their own cells as symbionts.

These symbiotic associations of microalgae and invertebrates are maintained by quite dexterous mechanisms and succeed for generations. Sponges recognize cyanophytes by their choanocyte or digestive cells as symbiotic algae and have a mechanism not to digest algae. It is now recognized

Table 7 Symbiotic marine microalgae and their host invertebrates.[63]

Host invertebrates	*Genus*	*Symbiotic microalgae*	*Genus*
Sponges	*Cliona*	dinoflagellate	*Symbiodinium*
	Aplysina	cyanophyte	
	Jaspis	cyanophyte	
Coelenterates	*Clavularia*	dinoflagellate	*Symbiodinium*
(Octocorallia)	*Heliopora*	dinoflagellate	*Symbiodinium*
	Lobophylum	dinoflagellate	*Symbiodinium*
	Jubipora	dinoflagellate	*Symbiodinium*
	Xenia	dinoflagellate	*Symbiodinium*
	Zoantles	dinoflagellate	*Symbiodinium*
Tunicates	*Trididemnum*	prochlorophyte	*Prochloron*
	Didemnum	prochlorophyte	*Prochloron*
	Diplosoma	prochlorophyte	*Prochloron*
	Lissoclinum	prochlorophyte	*Prochloron*
Flatworms	*Convoluta*	prasinophyte	*Tetraselmis*
	Convoluta	diatom	*Lichmorphora*
	Convoluta	dinoflagellate	*Symbiodinium*
	Amphiscolops	prasinophyte	*Tetraselmis*
	Amphiscolops	dinoflagellate	*Symbiodinium*
	Amphiscolops	dinoflagellate	*Amphidinium*
	Amphiscolops	dinoflagellate	*Prorocentrum*
	Pseudaphanostoma	dinoflagellate	*Symbiodinium*
	Haplodiscus	dinoflagellate	*Amphidinium*
Bivalves	*Hippopus*	dinoflagellate	*Symbiodinium*
	Corculum	dinoflagellate	*Symbiodinium*
	Fragum	dinoflagellate	*Symbiodinium*
	Tridaena	dinoflagellate	*Symbiodinium*
Nudibranchs	*Aeolidiella*	dinoflagellate	*Symbiodinium*
	Melibe	dinoflagellate	*Symbiodinium*
	Elysia	chloroplast of Codiales	
	Plakobrunchus	chloroplast of Codiales	

that the microalgae are fundamental to the biology of their hosts, and their carbon and nitrogen metabolisms are linked in important ways. Table 8 shows substances given to host invertebrates by symbiotic algae. Glycerol is the main chemical substance released by symbiotic dinoflagellate *Symbiodinium* sp. For coelenterates and bivalves, more than 20–40% of the carbon metabolites produced by photosynthesis of *Symbiodinium* sp. was released into the host cells as glycerol. Galactose, glycolic acid, alanine, glutamic acid, and lipid particles are also reported as released substances from symbiotic algae to the host invertebrates.

Table 8 Substances given by symbiotic algae to host invertebrates.[64]

Symbiont	*Host*	*Carbon metabolites*
Dinoflagellate	coelenterates	
Symbiodinium sp.	*Pocillopora damicornis*	glycerol
Symbiodinium sp.	*Porites poriles*	glycerol
Symbiodinium sp.	*Condylactis gigantea*	glycerol, lipid particles
	bivalve	
Symbiodinium sp.	*Tridaena crocea*	glycerol
	nudibranch	
Symbiodinium sp.	*Melibe pilosa*	lipid particles?
Symbiodinium sp.	*Pteraeolidia ianthina*	glycerol
Cyanophyte	*Placobranchus ianthobapsus*	galactose
Chloroplast of Codiales	*Tridachia crispata*	galactose
Prochlorophyte	tunicate	
	Diplosoma virens	glycolic acid
Prasinophyte	flatworm	
Tetraselmis convolutae	*Convoluta roscoffensis*	alanine, glutamic acid

Nitrogen fixation in coral reef sponges with symbiotic cyanobacteria has been reported.[65] The presence of unicellular cyanobacteria in sponges was confirmed by electron microscopy. Cyanobacteria were observed in large vacuolated cells (termed bacteriocytes) throughout the tissue of a sponge *Siphonochalina tabernacula*. The cyanobacteria observed in the thin brown pigmented ectosome of a sponge *Theonella swinhoei* occupied 10–15% of the ectosome volume. Cyanobacteria numbers decrease away from the ectosome and were not evident in the deeper part of the endosome, while symbiotic bacteria were rare in the ectosome and particularly numerous in the endosome. Nitrogenase activity was determined using the acetylene reduction technique immediately after collection of the sponges on reef-based platforms. Nitrogenase activity was detected in the two sponges (*S. tabernacula* and *T. swinhoei*) which harbor symbiotic cyanobacteria, and was absent in the third sponge *Inodes erecta* which lacked cyanobacteria and a symbiont. The activity of the ectosome of *T. swinhoei* was higher than the endosome; the former contained cyanobacteria while the latter was almost free of cyanobacteria. In addition, when sponge tissues were incubated in light or dark considerably more ethylene was produced in the illuminated sample. The nitrogenase activity was therefore concluded to be due to symbiotic cyanobacteria.

What is the purpose of symbiotic association with invertebrates for algae? One family of colonial didemnid ascidians (Urodhordata) harbor the symbiotic unicellular prokaryotic alga *Prochloron* sp., which synthesizes chlorophylls a and b but no phycobilin pigments. During investigation of conditions for laboratory cultivation of this unusual alga, it was found that *Prochloron* sp. is a naturally occurring tryptophan auxotroph that survives in nature by close association with the host.[66] Cell division in *Prochloron* cultures, freed from the host ascidian *Diplosoma similis*, occurred under acidic conditions (pH 5.5) in the presence of tryptophan. Among 21 common levorotatory amino acids tested as a supplement to the basal medium, only L-tryptophan affected *Prochloron* growth. The metabolic dysfunction that rendered *Prochloron* auxotrophic may involve only the initial step of the tryptophan biosynthetic pathway from chorismic acid (the final intermediate common to both tyrosine and tryptophan synthesis) since shikimic acid alone did not support the algal growth but anthranilic acid did. The growth rates of *Prochloron* culture were slow compared with those of many other algal cultures. Although the growth rate of *Prochloron* in the acidian *D. similis* was not known, it may be imagined that symbiotic association with didemnid ascidians is advantageous to the alga. It was also reported[64] that recycling of nitrogen metabolites such as uric acid or ammonia was carried out by symbiotic microalgae and was also beneficial to algal physiology. Thus, it may be profitable for algae to live in the host invertebrates as symbionts, compared with free living in the oligotrophic environments of tropical or subtropical oceans lacking in inorganic or other nitrogenous nutrients.

Investigations by Muscatine and co-workers on the "host factor" that participates in symbiotic association between dinoflagellates and reef-building corals or other tropical anthozoans have revealed that free amino acids exhibited "host-factor" activity and induced the release of photosynthate from symbiotic dinoflagellate *in vitro*.[67] Symbiotic dinoflagellates incubated with a crude homogenate of their own host tissue fixed $^{14}CO_2$ in the light and released a substantial fraction of the fixed carbon to the incubation medium, principally as [^{14}C]glycerol, [^{14}C]alanine, and [^{14}C]glucose. In contrast, the release of labeled organic carbon compounds by dinoflagellates incubated without homogenate of the host tissue was significantly less. Thus, it was shown that a factor (a chemical substance, i.e., a "host factor") present in the homogenate of the host tissue evoked the release of fixed carbon as low molecular weight compounds from symbiotic dinoflagellates. The host factor of the Hawaiian reef coral *Pocillopora damicornis* was investigated from the crude aqueous extract of this host using size-exclusion chromatography and HPLC. The host-factor active fraction showed absorbance features characteristic of the hydrophilic mycosporine-like amino acids abundant in marine cnidarians. Further investigation on isolation of individual mycosporine-like amino acids from lyophilized methanolic extracts of *P. damicornis* tissues led to identification of the host factor as a set of free amino acids. Synthetic amino acid mixtures, based on the measured free amino acid pools of *P. damicornis* tissues, not only induced the selective release of ^{14}C-labeled photosynthetic products from isolated symbiotic dinoflagellates but also enhanced total $^{14}CO_2$ fixation. Comparison of the radiochromatograms of the ^{14}C-labeled photosynthetic products released and fixed by symbiotic dinoflagellates of *P. damicornis* incubated with a synthetic free amino acid pool showed that the release of ^{14}C-labeled products was selective and not the result of dinoflagellate lysis. In addition, the released products were qualitatively identical to those released by dinoflagellates incubated in crude aqueous extracts of their own host. The products released by dinoflagellates isolated from *P. damicornis* differed from those released by dinoflagellates isolated from the sea anemone *Aiptasia pulchella*. These differences may be a reflection of the diversity of dinoflagellate species found in symbiosis with cnidarians, rather than host-factor specificity.

Symbiotic dinoflagellates of flatworms produce a variety of secondary metabolites with unique bioactivity such as strong cytotoxicity or vasoconstrictive activity, which is described in Section 8.07.9.4 The biological significance of these unique metabolites in the lives of dinoflagellates and/or flatworms is unknown.[68]

8.07.4.2 Others

Embryos of the shrimp *Palaemon macrodactylus* are known to resist infection by the pathogenic fungus *Lagenidium callinectes*. Fenical and co-workers found that a penicillin-sensitive bacterial strain (*Alteromonas* sp.) that was consistently isolated from healthy embryos effectively inhibited the growth of the fungus *L. callinectes in vitro*.[69] This bacterial strain was shown to produce and release an antifungal compound, which was identified as 2,3-indolinedione (**37**) (known as isatin). Bacteria-free embryos which were exposed to the fungus quickly died, whereas embryos reinoculated with the bacteria or treated only with 2,3-indolinedione (**37**) thrived. Thus it was clearly revealed that the symbiotic *Alteromonas* sp. bacteria protect shrimp embryos from fungal infection by producing and liberating the antifungal metabolite (**37**).

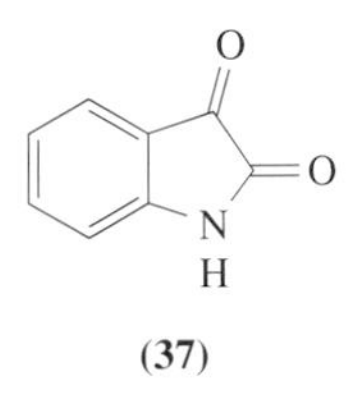

(37)

The species-specific partnership between anemone fish and the giant sea anemone is a well-known phenomenon of symbiosis in marine environments. The anemone fish recognize their specific partner anemone on the basis of chemical substances secreted by the anemone. Naya and co-workers studied the chemical substances secreted by the sea anemone which elicit symbiotic behavior of the fish and identified several nitrogenous compounds that induce characteristic symbiotic movements.[70,71] The sea anemone *Radianthus kuekenthali*, which was a specific host of the fish *Amphiprion perideraion*, was homogenized, and fractionation of the aqueous layer of the extract yielded 48 μg of a cationic compound as the sole active factor, which was named amphikuemin (**38**). The structure of (**38**) was established by hydrolysis, spectroscopic data, and synthesis. The synthetic L-enantiomer of (**38**) was identical with natural material in all respects including biological activity. Amphikuemin (**38**) was active in the attracted swimming assay using *A. perideraion* at a concentration of 10^{-10} M. Analogues of amphikuemin (**38**) were also prepared; *N*-demethyl-L-amphikuemin was totally devoid of activity, while the activity of the synthetic quaternary salt 4′-demethyl-L-amphikuemin was greatly reduced to 2.0×10^{-8} M.

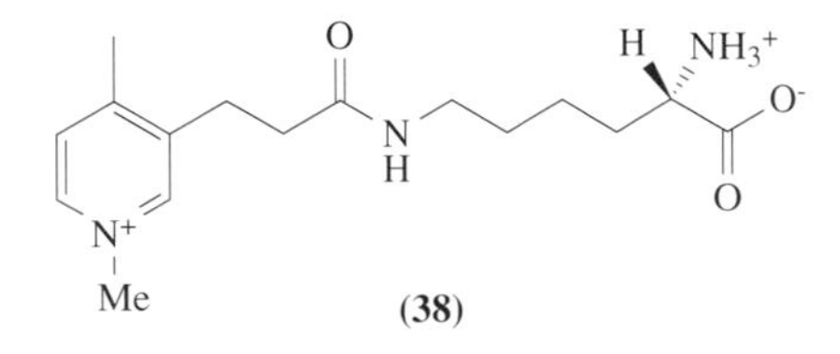

(38)

The organic layer of the extract of *R. kuekenthali* was also examined to give 16 tryptophan-derived alkaloids, dihydroaplysinopsins (**39**)–(**44**) and aplysinopsins (**45**)–(**54**), which are known as sponge-metabolites.[72] Dihydroaplysinopsins (**39**)–(**44**) induced attracted swimming of *A. perideraion* at the effective dose of 10^{-6} M, more weakly than amphikuemin (**38**). Aplysinopsins (**45**)–(**54**) with structural variations in ring C, double-bond geometry at C-8 and C-1′, and the presence and absence of bromine caused the fish *A. perideraion* to perform an up-and-down movement of the head, like a seesaw. *A. clarkii*, another guest fish of the host animal *R. kuekenthali*, did not respond to either amphikuemin (**38**) or the aplysinopsins and dihydroaplysinopsins. It is thus likely that these two different species of symbiotic fish recognize their common host through different chemicals.

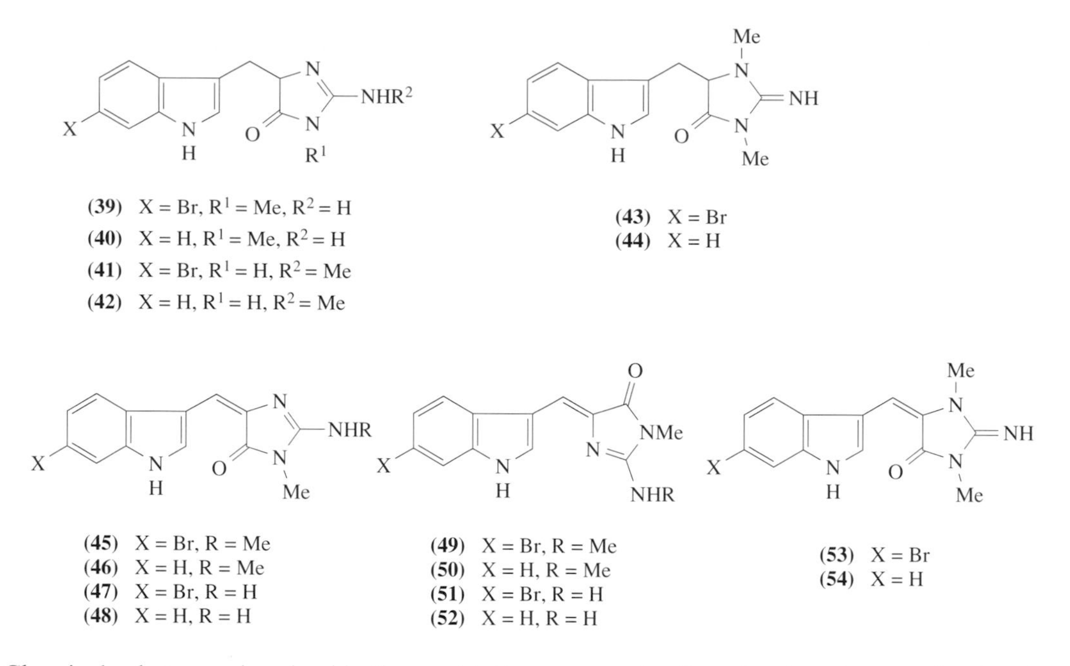

Chemical substances involved in the symbiotic movements of another host/guest pair, *Stoichactis kenti* (sea anemone) and *A. ocellaris* (anemone fish) were also studied, and tyramine (**55**) and tryptamine (**56**) were identified as active substances that induced attracted swimming with tail wagging and active searching behavior, respectively, both at a dose of 10^{-6} M. It was observed that addition of tyramine to a partially purified fraction of the secretion of *S. kenti* increased the activity while further purification sometimes led to reduction in the activity. The symbiosis-inducing activity in the pair of *S. kenti–N. ocellaris*, therefore, may depend on the synergistic effect of multiple unidentified chemicals.

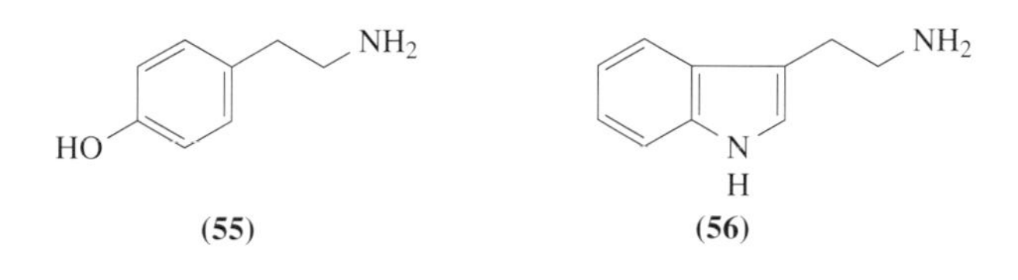

8.07.5 BIOFOULING

Many species of sessile organisms live in the sea. Barnacles, mussels, hydroids, bryozoans, and tunicates are representative. In their life cycles, the larvae of sessile organisms swim and float until they find a suitable place to settle and grow. The search for an appropriate place where they undergo metamorphosis and grow into adults is quite important for them, because it influences their entire life and the conservation of the species. It is believed that they have elaborate chemical control mechanisms to induce the larval settlement at the most satisfactory location for the particular species. They also possess chemical defense systems against larval settlement of other competing sessile organisms.[73–76]

8.07.5.1 Microorganisms

A biofilm formation by marine bacteria is observed in a few hours when a plastic plate is put into the sea. This biofilm formation is followed by attachment of microalgae such as diatoms to give a "slime," which is called "microfouling." The microalgal slime triggers larval settlements and metamorphosis of various marine sessile organisms such as barnacles and blue mussels to lead to "macrofouling." Although it is believed that chemical signals are involved in the attachment of marine bacteria and microalgae, the process is not well understood. Organisms already growing at some ecological site have chemical control systems (allelopathy) for preventing the settling of other attaching organisms such as diatoms.

Research on chemical substances that regulate biofilm formation has been carried out.[77] "Macrofouling" causes serious problems on ship hulls, fishing facilities, and other artificial objects in the sea such as the cooling systems of power plants. Organotin compounds like TBTO [bis-(*n*-tributyltin)-oxide] have been used as antifouling agents, but they exhibit toxic effects on the marine environment including fish and shellfish. Antifouling substances with less toxicity, therefore, are required. On the other hand, substances that induce or promote biofilm formation may be of use in fish farming. A bioassay system was developed using a marine bacterium possessing attaching properties (*Rhodospirillum salexigens* SCRC 113), and evaluation of the biofilm formation was monitored by measuring the quantity of polysaccharide produced by the bacteria as well as the absorbance of the culture medium by the attaching bacteria. As a result, one biofilm formation accelerator, hateramine (**57**), was isolated from a marine sponge, and 10 compounds (**58**)–(**67**) were identified as biofilm formation inhibitors from natural and synthetic compounds. Bisdeacetylsolenolide D (**58**), aeroplysinin-I (**59**), and new nitroalkanes (**60**) and (**61**) were natural products isolated from marine sponges. Ethyl *N*-(2-phenethyl)carbamate (**62**) was isolated from the marine bacterium SCRC3P79 (*Cytophaga* sp.) and its inhibition activity was remarkable. A synthetic analogue (**63**) also exhibited significant inhibition activity of biofilm formation and showed antibacterial activity. Compounds (**64**)–(**67**) were prepared by synthesis and significantly inhibited the biofilm formation not only in the laboratory bioassay but also in the field experiment.

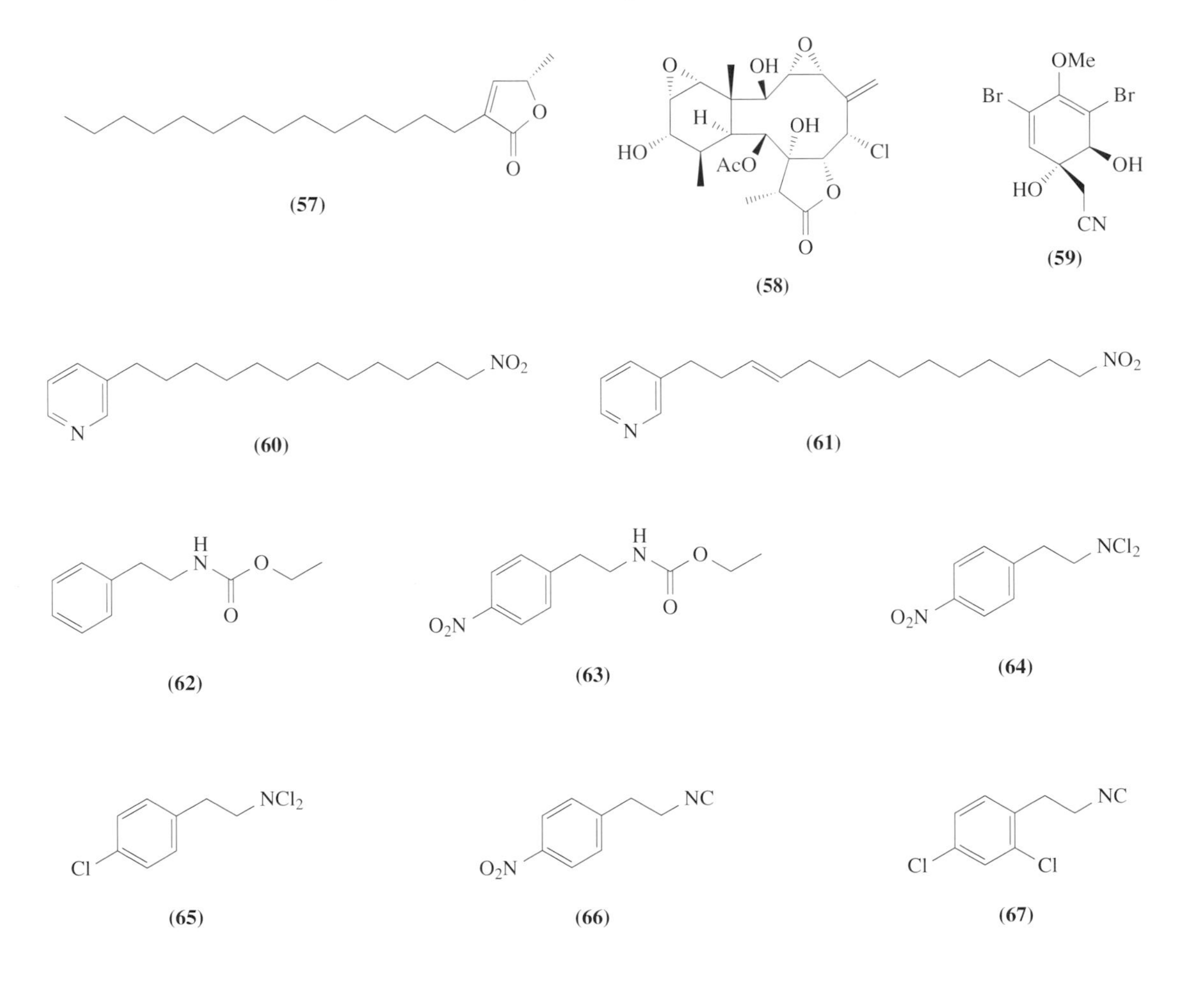

8.07.5.2 Hydrozoa

The larvae (pulanula) of the Japanese hydrozoan *Coryne uchidai* search for their attaching site by swimming and crawling about, and frequently settle on seaweeds. From an extract of the seaweed

Sargassum tortile, δ-tocotrienol epoxides (**68**) and (**69**) were identified as larval settlement/metamorphosis-inducing substances.[73,74] The inducing activity was further confirmed by the use of synthetic (**68**) and (**69**). More interestingly, the epoxide (**68**) also exhibited inhibition of metamorphosis, depending upon the concentration. When the epoxide (**68**) was presented to the larvae for too long, the metamorphosis induced by (**68**) was interrupted at an early stage of polyp formation, but was restarted by diluting with fresh seawater.

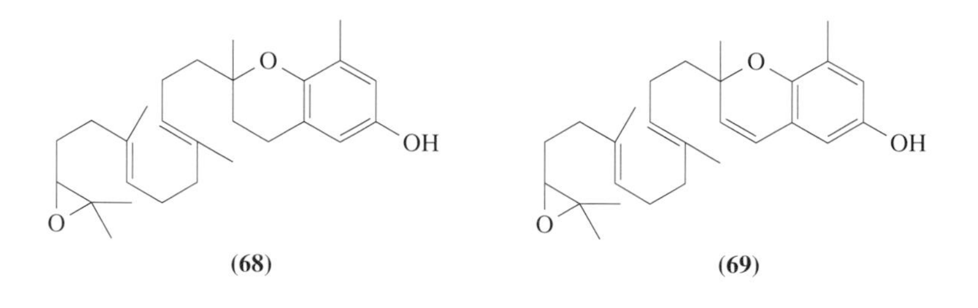

The larval settlement and metamorphosis of a hydroid *Hydractinia echinata* were induced by a biofilm of Gram-negative bacteria. The active substance released from the biofilm is unknown but assumed to be an unstable hydrophobic substance. Metal ions such as Li^+, Rb^+, and Cs^+ are revealed to induce larval metamorphosis, and tumor-promoting phorbol esters are also reported to be active.[78] Bacterial biofilms also induced larval settlement of a hydroid *Halcordyle disticha*. Catecholamines were active in the induction of larval metamorphosis of *H. disticha* but inactive in that of *H. echinata*.[79]

8.07.5.3 Polychaetes

Some lugworms live in colonies of tubes made from sea sand. The larval settlement and metamorphosis of a reef-building tube worm *Phramatopoma californica* were induced by extracts of their tubes, and the active substance was identified as a mixture of fatty acids (C_{14}–C_{22}).[80] Tests using individual fatty acids revealed that $C_{16:1}$, $C_{18:2}$, $C_{20:4}$, and $C_{20:5}$ fatty acids were active with the concentration of 100 μg g^{-1}; only the *Z* isomer of $C_{16:1}$ fatty acid was active, while its *E* isomer was inactive. Too large a dose (1 mg g^{-1}) of $C_{20:4}$ and $C_{20:5}$ fatty acids induced unusual metamorphosis. In addition, the terminal carboxyl group proved to be essential for the metamorphic-inducing activity. Marine bacterial biofilms also induced larval settlement of a polychaetes *Janua brasiliensis*, and the active compound was believed to be biopolymers like glycoproteins.[81]

8.07.5.4 Mollusks

The larvae (veliger) of oysters settle down around their parents as with many other marine invertebrates, and extracts of parent oysters induced the larval settlement. Larval settlements of the eastern oyster *Crassostrea virginica* and the pacific oyster *C. gigas* were induced by the biofilms of Gram-negative bacterium. From the culture broth of this bacterium, L-3,4-dihydroxyphenylalanine (L-DOPA (**70**)) and its polymerized product, a black melanin pigment (molecular weight, 12 000–120 000), were identified as active substances.[82] L-DOPA (**70**) is a biosynthetic precursor of the catecholamines, well-known neurotransmitters, such as (−)-adrenalin (**71**) and (−)-noradrenalin (**72**). L-DOPA (**70**) induced larval settlement and metamorphosis and the most effective concentration was 2.5×10^{-5} M; at this concentration settlement and metamorphosis of approximately 50% of the larvae were induced. However, when (**70**) was presented to the larvae for too long at concentrations higher than 5×10^{-6} M, (**70**) proved to be toxic.[83] (−)-Adrenalin (**71**) and (−)-noradrenalin (**72**) were also strongly active but, interestingly, they induced larval metamorphosis without attaching to suitable substrates. D-DOPA (**73**), dopamine (**74**), acetylcholine (**75**), and GABA (**76**) were revealed to be inactive in larval settlement and metamorphosis of oysters.

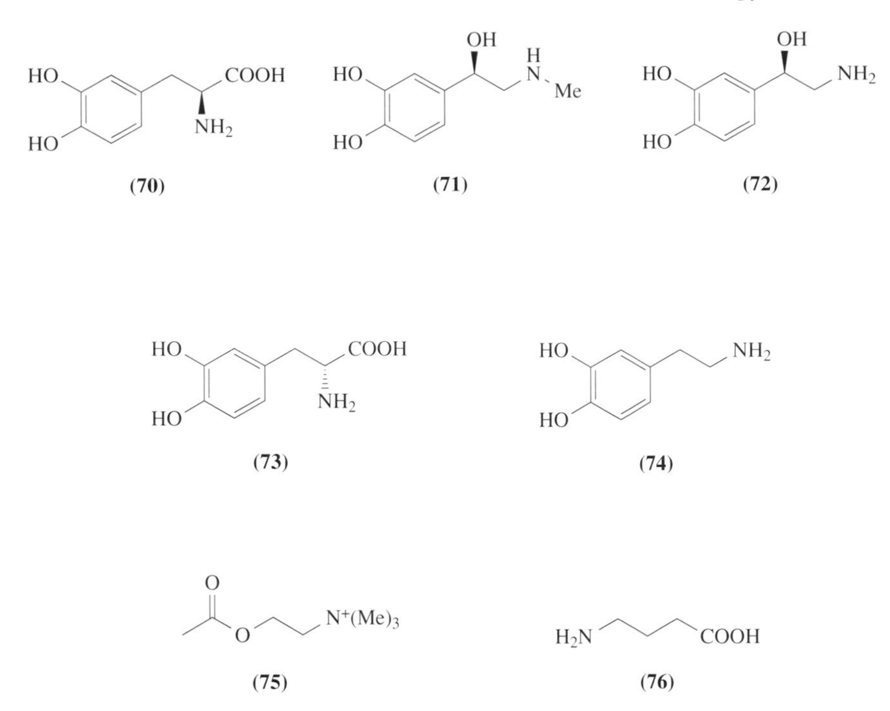

The planktonic larvae of the gastropod mollusk, *Haliotis rufescens*, the large red abalone of the eastern Pacific, are recruited from the plankton to crustose red algae including species of *Lithothamnium*, *Lithophyllum*, and *Hildenbrandia*.[84] This recruitment to the intact crustose red algae is normally dependent upon larval contact with the recruiting algal surface, and the crustose red algae produce or contain molecules that induce substratum-specific settlement, attachment, and metamorphosis of the *Haliotis* larvae. Such molecules were associated with macromolecules and were demonstrated to be produced by a number of red algae, but were not produced by the green or brown algae. The substratum specificity of larval settlement and metamorphosis was shown to result from the unique availability of these chemical inducers at the surfaces of the crustose red algae. By purification based on size-separation by gel-filtration, followed by ion-exchange chromatography over a diethylaminoethyl (DEAE)-acrylamide matrix, the principal inducer was resolved from the red algal phycobiliprotein.[85] Phycoerythrobilin (**77**) (a linear tetrapyrrole bile pigment) and its specific protein conjugate, phycoerythrin, were the accessory photosynthetic pigments responsible for the coloration of the coralline red algae. Both free and conjugated phycoerythrobilin (**77**) exhibited half maximal induction of settling of the *Haliotis* larvae at $\sim 10^{-6}$ M and were toxic to the larvae at concentrations higher than 10^{-5} M. Other linear and cyclic tetrapyrroles and other proteins including hemoglobin, which contains the cyclic tetrapyrrole heme group, were inactive over comparable concentrations.

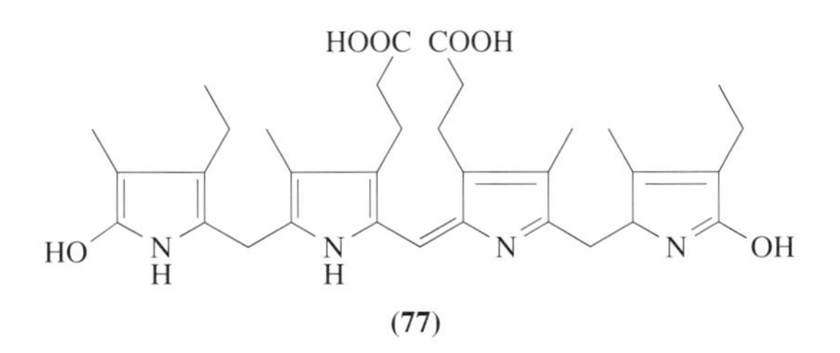

(77)

Table 9 summarizes the induction activity of several extracts of algae and a series of amino acids for larval settlement and metamorphosis of *Haliotis rufescens*.

GABA (γ-aminobutyric acid (**78**)), a neurotransmitter in higher animals, was found to be a potent inducer of rapid settling of the *Haliotis* larvae contained in the crustose red algae. GABA (**78**) was active in inducing behavioral metamorphosis, viz., inducing to settle and begin their characteristically snail-like plantigrade attachment, gliding locomotion, and grazing behavior on the clean glass of the test vials, at concentrations as low as $\sim 10^{-7}$ M, with half-maximal effectiveness (50% settling) at $< 10^{-6}$ M. Prolonged exposure of larvae to higher concentrations of the free inducer (**78**) was

Table 9 Induction activity of larval settlement and metamorphosis of *Haliotis rufescens*.[84]

Inducer	*Settled* (%)
None	0
Lithothamnium sp. and *Lithophyllum* sp.	82
Bossiella sp.	4
Diatoms	0
Bacteria and microalgae	0
Macrocystis pyrifera	0
Lithothamnium extract, 2 μg of protein per ml	2
The same, boiled	6
The same, protease digested; boiled	23
GABA (γ-aminobutyric acid)	⩾ 99
α-Aminobutyric acid (D- and L-)	0
β-Aminobutyric acid (D- and L-)	0
n-Butylamine	0
n-Butyric acid	0
n-Pentanoic acid	0
Succinic acid	0
γ-Guanidinobutyric acid	0
γ-Hydroxybutyric acid	58
δ-Amino-*n*-valeric acid	89
ε-Amino-*n*-caproic acid	74
L-Glutamic acid	12
D-Glutamic acid	0
L-Glutamine	0
L-Aspartic acid	0
Other neurotransmitters and effectors	0

toxic. GABA (**78**) is present in high concentrations ($\sim 10^{-2}$ M) in the crustose coralline red algal extracts. Structural analogues of GABA (**78**) such as γ-hydroxybutyrin acid (**79**), δ-amino-*n*-valeric acid (**80**), and ε-amino-*n*-caproic acid (**81**) showed significant activity in the induction of settling. The absolute requirements for the inducing activity may be the presence of the primary carboxyl group of GABA (**78**) and that of specific substitution at the γ-position. Increasing and decreasing the chain length of GABA homologues progressively decreased their activity. In addition, L-α,ω-diamino acids, such L-α,β-diaminopropionic acid (**82**) and L-lysine (**83**), facilitated, at a concentration of 10^{-5} M, the induction of larval metamorphosis in the presence of GABA (2.5×10^{-7} M).[86] On the other hand, other neurotransmitters and effectors were inactive; tested substances were L-epinephrine, L-norepinephrine, serotonin, histamine, acetylcholine, choline, and indole-3-butyric, -acetic, -propionic, and -acrylic acids.

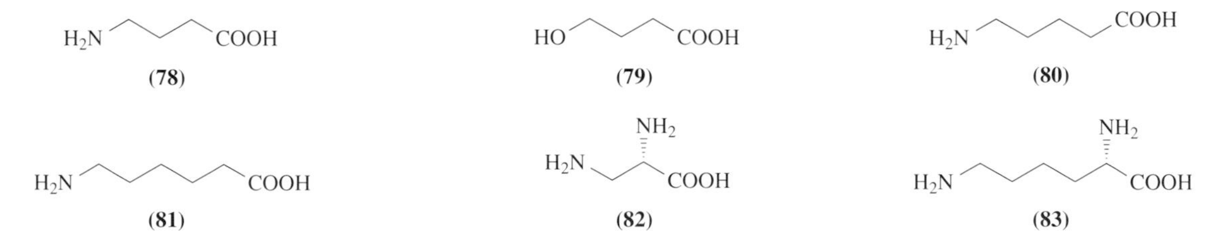

The larval settlement and metamorphosis of *Haliotis rufescens* as well as those of the tube worm *Phramatopoma californica* and the nudibranch *Phestilla sibogae* were induced by increase in the concentration of potassium ion (K^+) in the seawater. Addition of the K^+-channel blocker, tetraethylammonium chloride, inhibited the larval settlement and metamorphosis of *H. rufescens* but affected no changes for *P. californica* and *P. sibogae*.[87]

The neutral methanol extract of the brown alga *Dilophus okamurai* was found to inhibit the settlement and metamorphosis of the swimming larvae (veliger) of the abalone *Haliotis discus hannai*. Two active components were isolated and identified as spartane-type diterpenes (**84**) and (**85**).[88] There herbivorous abalones can scarcely be seen in the community of the brown alga *D. okamurai* which grows in a depth of water very similar to that of the brown alga *Eisenia bicyclis*. Thus it was suggested that *D. okamurai* contains an antifeeding substance against the abalone. The two diterpenes (**84**) and (**85**) inhibited the larval settlement and metamorphosis of the abalone *H. discus hannai* at a concentration of 5 ng ml^{-1}. It is unknown whether these diterpenes are true antifeedants.

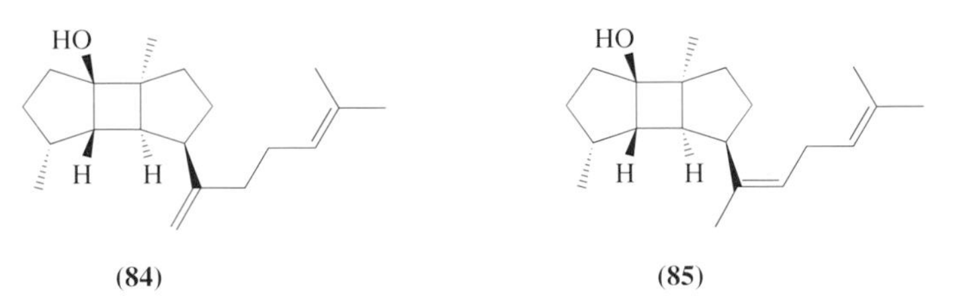

(**84**) (**85**)

The larval settlement and metamorphosis of the scallop *Pecten maximus* were induced by jacaranone (**86**), which was isolated from extracts of the alga *Delesseria sanguinea*, at a concentration of 0.5 mg l^{-1}.[89] Among the compounds isolated from *D. sanguinea* only (**86**) displayed this property.

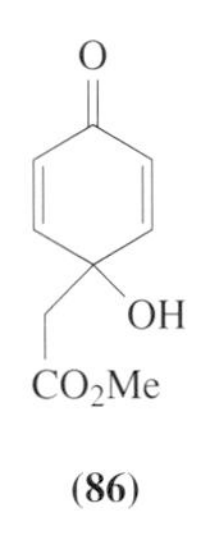

(**86**)

8.07.5.5 Barnacles

Barnacles are one of the representative marine fouling invertebrates which cause serious problems on ship hulls and other man-made objects. Larvae (cypris) of barnacles settle around their parents. Extracts of parents of the Atlantic barnacle *Balanus balanoides* contained a substance that induced larval settling and metamorphosis. This substance was suggested to be a chemically stable protein bridged with yellow pigments (quinones?), which kept their activity after heating but became inactive by treatment with sodium perchlorate. This protein was named "arthropodin" and proved to consist of subunits with molecular weights of 18 000 and 5000–6000.[73,74]

From the adult barnacle *B. amphitrite* a protein with molecular weight 3000–5000 was isolated as a larval-settlement inducing pheromone.[90] This active pheromone was a substance released from the adult barnacle into the water, and the lowest effective concentration was 10^{-10} M. It was revealed that di- or tripeptides also induced larval settlement (Table 10). It seems to be a requirement that active dipeptides have a basic amino acid residue at the carboxyl terminal and the other amino acid is neutral or basic; other combinations of amino acid residues proved to be inactive. The most active peptides were L-His-L-Lys (basic–basic) and L-Leu-Gly-L-Arg (neutral–neutral–basic) and their lowest effective concentration was 2.0×10^{-10} M.

Table 10 Induction activity of dipeptides in larval settlement of the barnacle *B. amphitrite*.[89]

Peptides	*Concentrations at which larval settlement was induced* (M)
Leu–Arg (neutral–basic)	2.0×10^{-8}–2.0×10^{-10}
His–Lys (basic–basic)	2.0×10^{-8}–2.0×10^{-10}
Glu–Lys (acidic–basic)	inactive
Lys–Gly (basic–neutral)	inactive
Glu–Gly (acidic–neutral)	inactive
Gly–Ser (neutral–neutral)	inactive
Arg–Glu (basic–acidic)	inactive
Gly–Asp (neutral–acidic)	inactive
Glu–Glu (acidic–acidic)	inactive

Inducers of larval settlement for barnacles were well discriminated from those of larval metamorphosis. The proteins ("arthropodin") or peptides described above corresponded to the former, while juvenile hormone (**87**) and its synthetic analogue ZR-512 (**88**) were reported as examples of the latter.[73,74,91] ZR-512 (**88**) caused larval metamorphosis of cypris of *B. galeatus* at concentrations as low as 10 ppb without settling to the surface, and was 1000 times as strongly active as compound (**87**), while another analogue ZR-515 (**89**) with methoxy and isopropyl ester groups was inactive. Compound (**88**), however, was ineffective when juvenile hormone (**87**) (0.5–10 ppb) was introduced to the larvae before addition of (**88**). Thus, many things remain to be studied for a fuller understanding of the mechanism and roles of chemical signals of larval settlement and metamorphosis.

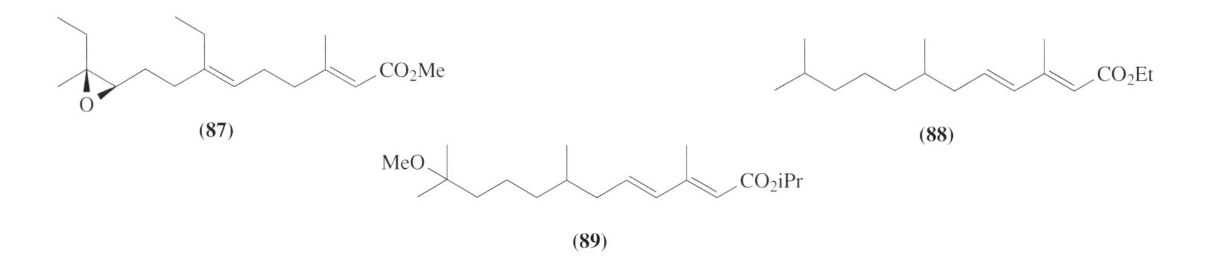

On the other hand, the search for settlement inhibitors has been extensive because of the serious problems caused by marine fouling invertebrates. To clarify the influence of marine bacteria on the settling of fouling invertebrate larvae, marine bacterial products that inhibit settling by cyprids of the barnacle *Balanus amphitrite* were screened. The culture broth of a bacterial strain belonging to *Alteromonas* sp., which was associated with the marine sponge *Halichondria okadai*, effectively inhibited settling of the cyprids.[92] The active substance was isolated from the extract of the mycelium and was identified as ubiquinone-8 (**90**). Compound (**90**) and its related compounds, ubiquinone-0 (**91**), ubiquinone-9 (**92**), ubiquinone-10 (**93**), vitamin K_1 (**94**), and vitamin K_3 (**95**) were examined for inhibitory activities and toxicities. As a result, compound (**91**), which had no polyprenyl group, showed higher inhibitory activity and higher toxicity than other polyprenylated ubiquinones, suggesting that the length of the polyprenyl side-chain was an important factor influencing both inhibitory activity and toxicity of ubiquinones. Similarly, vitamin K_3 (**95**) with no polyprenyl group was more active and more toxic than vitamin K_1 (**94**), which has a polyprenyl group. Bacterial films on solid substrates were reported to play important roles as repellents or attractants for the settlement of marine sessile organism larvae, as discussed in Section 8.07.5.1. As ubiquinones and related compounds are commonly found in bacteria, they may retard the settling of invertebrate larvae.

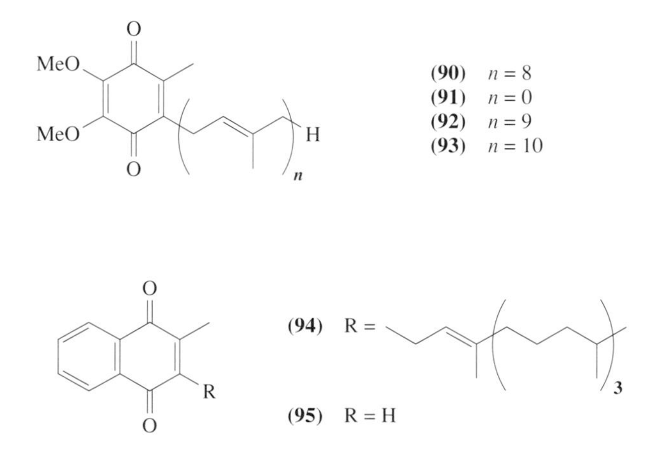

Studies on antifouling substances against barnacle larvae have been reported and some new active secondary metabolites were isolated from marine invertebrates such as the sea pansy, sponges, and nudibranchs. From extracts of the Atlantic sea pansy *Renilla reniformis*, three new briarein-type diterpenes, renillafoulins A (**96**), B (**97**), and C (**98**) were isolated. All three compounds inhibited larval settlement of the barnacle *B. amphitrite* with EC_{50} (50% effective concentration) values ranging from 0.02 μg ml^{-1} to 0.2 μg ml^{-1}.[93]

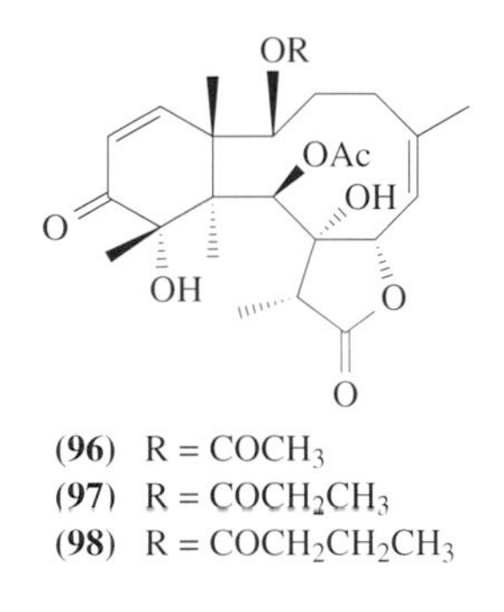

(**96**) R = $COCH_3$
(**97**) R = $COCH_2CH_3$
(**98**) R = $COCH_2CH_2CH_3$

Dimeric bromopyrroles (**99**)–(**101**) isolated from the Caribbean sponge *Agelas conifera* were found to be active in barnacle settlement; the assay using *B. amphitrite* gave settlement-inhibition EC_{50} values of 44 μg ml^{-1} and 21 μg ml^{-1} for compounds (**99**) and (**101**), respectively, and settlement-facilitation EC_{50} values of 29 ng ml^{-1} and 17 ng ml^{-1} for compounds (**99**) and (**100**), respectively.[94]

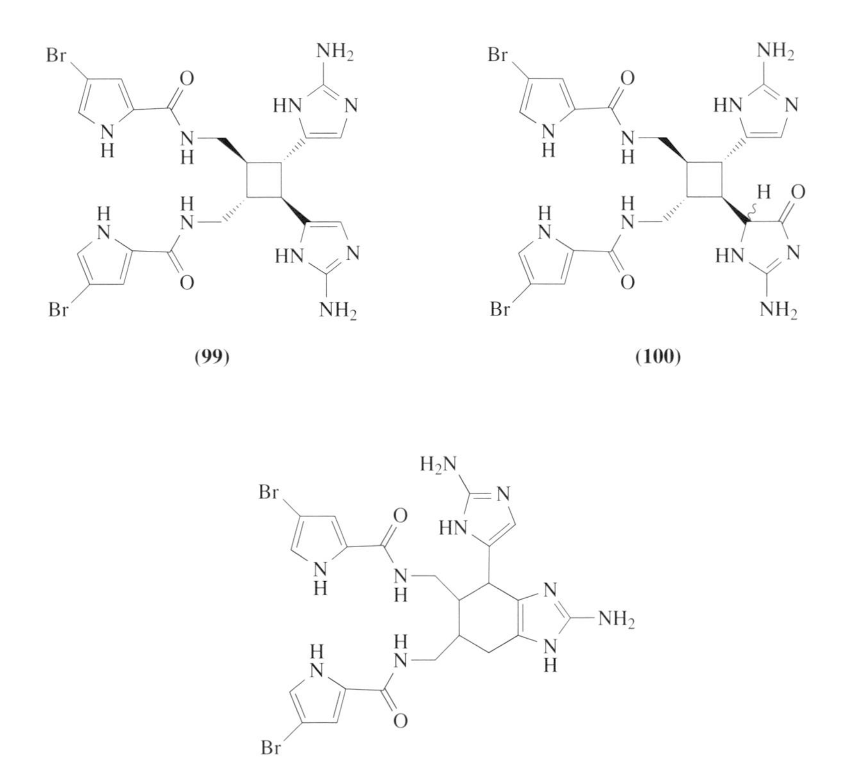

(**99**) (**100**) (**101**)

Antifouling substances from Japanese marine invertebrates have been investigated extensively. From the marine sponge *Acanthella carvernosa*, collected off Yakushima Island, 1000 km southwest of Tokyo, three new diterpene formamides, kalihinenes X (**102**), Y (**103**), and Z (**104**) along with known kalihinol A (**105**) and 10-formamidokalihinene (**106**) were isolated.[95] The new kalihinenes (**102**)–(**104**) inhibited attachment and metamorphosis of cyprid larvae of the barnacle *B. amphitrite* with EC_{50} values of 0.49, 0.45, and 1.1 μg ml^{-1}, respectively, while no toxicity was found at these concentrations. Kalihinol A (**105**) (EC_{50} 0.087 μg ml^{-1}) and 10-formamidikalihinene (**106**) (EC_{50} 0.095 μg ml^{-1}) also showed potent antifouling activity, more active than $CuSO_4$ (EC_{50} 0.15 μg ml^{-1}).

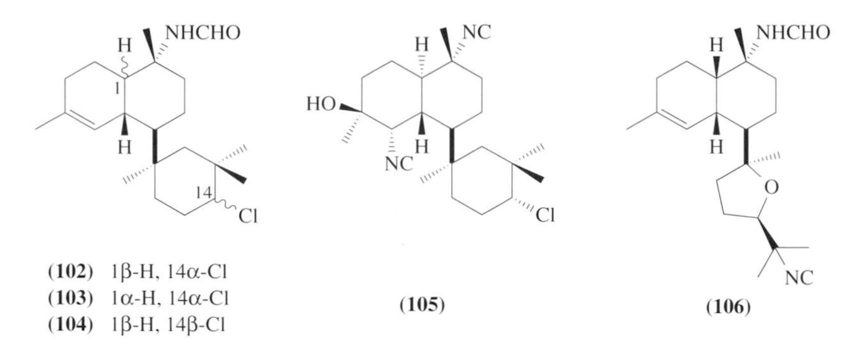

Antifoulants were also obtained from the sponge of the same species *Acanthella carvernosa*, collected off Hachijo-jima Island. The hexane-soluble fraction of both EtOH and MeOH extracts of this sponge were fractionated with the guidance of antifouling activity assay to afford 14 active terpenoids (**107**)–(**120**), seven of which ((**109**)–(**111**), (**111**), (**116**), (**118**)–(**120**)) were new natural products.[96] Most of them contained either isocyanate, isothiocyanate, isocyanide, or formamide groups. Antifouling activities of these terpenoids against cyprid larvae of the barnacle *B. amphitrite* are shown in Tables 11 and 12. Table 11 summarizes percentages of metamorphosed and dead larvae when the barnacle was exposed to terpenoids (**107**)–(**120**) at a concentration of 5 μg ml^{-1} for 48 h. Larval metamorphosis was completely inhibited by compounds (**115**)–(**120**). Table 12 shows percentages of inhibition at three concentrations for kalihinanes (**116**)–(**120**). Isocyanate and isothiocyanate derivatives (**119**) and (**120**) were highly antifouling (EC_{50} ca. 0.05 μg ml^{-1}).

Table 11 Percentages of metamorphosed and dead larvae of the barnacle *B. amphitrite* exposed to compounds (**107**)–(**120**) at a concentration of 5 μg ml^{-1} for 48 h.

Larvae (%)	(**107**)	(**108**)	(**109**)	(**110**)	(**111**)	(**112**)	(**113**)	(**114**)	(**115**)	(**116**)	(**117**)	(**118**)	(**119**)	(**120**)
Metamorphosed	8	37	29	38	4	8	33	42	0	0	0	0	0	0
Dead	0	13	0	0	0	0	0	0	0	0	0	0	0	0

Table 12 Percentages of inhibition of larval settlement and metamorphosis of the barnacle *B. amphitrite*.

Concentration (μg ml^{-1})	(**116**)	(**117**)	(**118**)	(**119**)	(**120**)
5	100	100	100	100	100
0.5	69	61	49	44	67
0.05	24	24	17	50	42

Nudibranchs are well known to sequester secondary metabolites from their sponge diets to protect their soft bodies from predators, and their chemical defense could be used for development of antifouling strategy. In fact, antifouling substances that inhibited larval settlement and metamorphosis of the barnacle *B. amphitrite* were isolated from extracts of four nudibranchs of the family Phyllidiidae from five different collection sites (Table 13). The antifouling substances were

Table 13 Isolation yields and antifouling activity of sesquiterpenes (**121**)–(**130**) from nudibranchs of the family Phyllidiidae (mg/animal).

Species	*Collection site*	(**121**)	(**122**)	(**123**)	(**124**)	(**125**)	(**126**)	(**127**)	(**128**)	(**129**)	(**130**)
Phyllidia pustulosa	Yakushima	1.2	0.4	0.7							
Phyllidia pustulosa	Kuchinoerabu	0.1	0.4		0.2						
Phyllidia pustulosa	Tanegashima		0.3		0.4	0.8	1.6				
Phyllidia pustulosa	Kamikoshiki							13.0			
Phyllidia ocelata	Kamikoshiki								1.0		
Phyllidia varicosa	Shimokoshiki							2.1		2.5	
Phillidiopsis krempfi	Shimokoshiki							1.8	0.8		1.3
Antifouling activity IC_{50} (μg ml^{-1})		10	3.2	7.2	4.6	2.3	0.13	0.14	0.70	0.33	>50

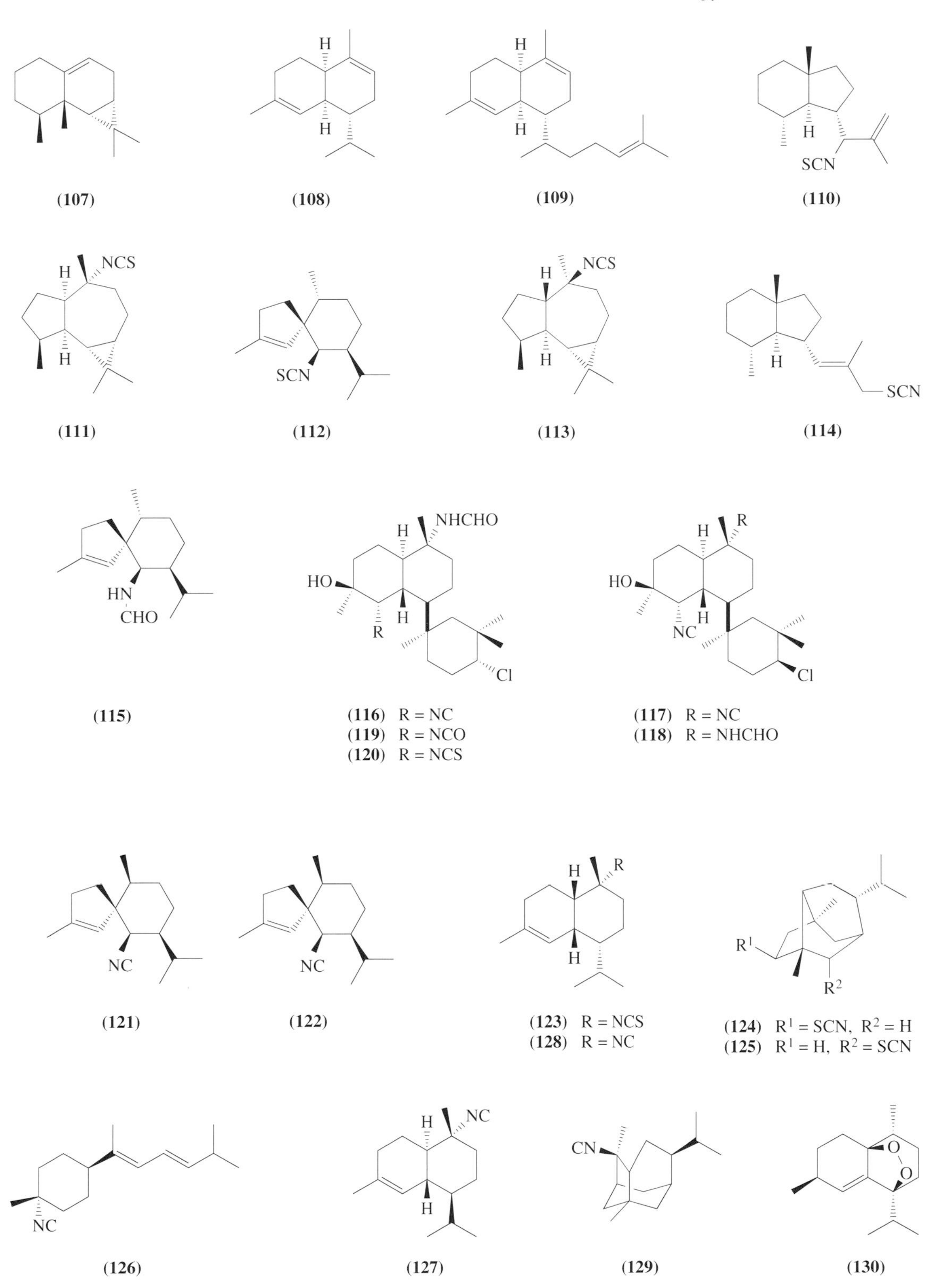

(107) (108) (109) (110)

(111) (112) (113) (114)

(115)

(116) R = NC
(119) R = NCO
(120) R = NCS

(117) R = NC
(118) R = NHCHO

(121) (122)

(123) R = NCS
(128) R = NC

(124) R^1 = SCN, R^2 = H
(125) R^1 = H, R^2 = SCN

(126) (127) (129) (130)

identified as three new sesquiterpene isocyanides **(121)**, **(127)**, and **(129)**, a new sesquiterpene peroxide **(130)**, and six known sesquiterpenes **(122)**–**(126)** and **(128)**.[97] These sesquiterpenes had structural similarities to those isolated from the sponge *Acanthella carvernosa* described above. Table 13 shows the isolation yields and antifouling activities of these sesquiterpenes. Compounds **(126)** and **(127)**, especially, showed potent antifouling activity with IC_{50} values of 0.13 μg ml^{-1} and

0.14 μg ml^{-1}, respectively, while no toxicity was found at these concentrations. Their activities were comparable to that of $CuSO_4$ (IC_{50} 0.15 μg ml^{-1}). It should be noted that (**130**) showed no activity at 50 μg ml^{-1}, though it contained a peroxide moiety.

Different classes of natural products containing a bromopyrrole or bromotyramine moiety, isolated from marine sponges, were revealed to have antifouling activity. The MeOH extract of the sponge *Pseudoceratina purpurea* collected off Hachijo-jima Island, 300 km south of Tokyo, was found to be active against cyprids of the barnacle *B. amphitrite*. Bioassay-guided fractionation of the ether-soluble portion of the MeOH extract using silica gel chromatography, gel-filtration on Toyopearl HW-40, and reverse phase (C_{18}) HPLC afforded eight bromopyrrole or bromotyramine derivatives, ceratinamides A (**131**) and B (**132**), psammaplysins A (**133**) and E (**134**), ceratinamine (**135**), moloka'iamine (**136**), pseudoceratidine (**137**), and 4,5-dibromopyrrole-2-carbamide (**138**). Ceratinamides A (**131**) and B (**132**) were new bromotyrosine derivatives with a spiro[4,6]dioxazundecane skeleton.[98] Ceratinamine (**135**) was a new compound containing an unprecedented cyanoformamide functionality.[99] Pseudoceratidine (**137**) was also a new compound consisting of a spermidine moiety and two 4,5-dibromopyrrole-2-carbonyl units.[100] These compounds exhibited antifouling activity (settlement and metamorphosis inhibitory activity) against cyprid larvae of the barnacle *B. amphitrite* with ED_{50} values ranging from 0.10 μg ml^{-1} to 8.0 μg ml^{-1}, as summarized in Table 14. Ceratinamide A (**131**) and psammaplysin A (**133**) were particularly potent. Compounds (**132**), (**133**), (**135**), and (**137**) were lethal to the larvae at a concentration of 30 μg ml^{-1}, while others were not toxic at this concentration. Thus, (**131**) was considered to be a promising antifouling agent. It was interesting that psammaplysin A (**133**) and 4,5-dibromopyrrole-2-carbamide (**138**) induced larval metamorphosis of the ascidian *Halocynthia roretzi*; this subject is discussed further in Section 8.07.5.6. Psammaplysin E (**134**), ceratinamine (**135**), and moloka'iamine (**136**) also exhibited potent cytotoxicity against P388 murine leukemia cells, while psammaplysin A (**133**) and pseudoceratidine (**137**) were antibacterial against *Flavobacterium marinotypicum*. In addition, compounds (**131**)–(**138**) were inactive against *Alteromonus macleodii*, *Pseudomonas nautica*, *Vibrio alginolyticum*, *Bacillus marinus*, *Penicillium chrysogenum*, *Candida albicans*, and *Mortierella ramanniana* at 10 μg/disk, suggesting that antifouling activity was not associated with antibacterial or antifungal activity.

Table 14 Biological activities of compounds (**131**)–(**138**).

Compounds	*Metamorphosis inducing activity on acidian H. roretzi* ED_{100} (μg ml^{-1})	*Antifouling activity against barnacle B. amphitrite* ED_{50} (μg ml^{-1})	*Antibacterial activity against F. marinotypicum* (mm, 10 μg/disc)	*Cytotoxic activity against P388 cell* IC_{50} (μg ml^{-1})
(**131**)		0.10		> 10
(**132**)		2.4		> 10
(**133**)	1.2	0.27	10	> 10
(**134**)		4.8		2.1
(**135**)		5.0		3.4
(**136**)		4.3		2.1
(**137**)		8.0	15	> 10
(**138**)	25	> 30		> 10

8.07.5.6 Tunicates

Inducers or promoters of larval settlement and metamorphosis of tunicates have also been investigated. As promoters of settlement and metamorphosis of the tunicate *Ciona savignyi* larvae, two pteridine-containing bromophysostigmine alkaloids, urochordamines A (**139**) and B (**140**) were isolated from the tunic (outer body) of the adult tunicate *C. savignyi*, collected off Asamushi, 600 km northeast of Tokyo, in June 1992.[101] They were stereoisomers at the C-9 position and urochordamine A (**139**) promoted larval settlement and metamorphosis in *C. savignyi* at a concentration of 2 ng ml^{-1}; all larvae treated with (**139**) completed settlement and metamorphosis by the time 50% of the larvae in a control group had settled. Urochordamine B (**140**), however, had no activity at the same concentration, suggesting the importance of the stereochemistry of C-9. It

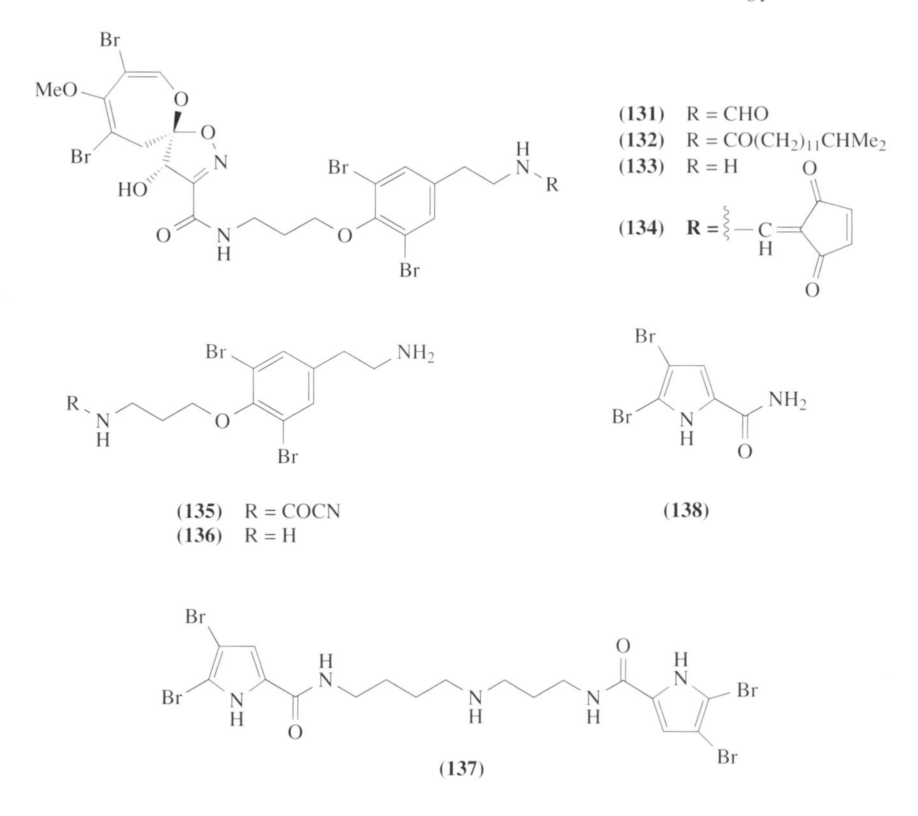

should be noted that compounds (**139**) and (**140**) were not detected in *C. savignyi* collected at the same place in October. Therefore, compounds (**139**) and (**140**) may be produced only during the spawning season. Moreover, the colonial tunicate *Botrylloides* sp. collected in the Gulf of Sagami in June contained both compounds (**139**) and (**140**). It was found, however, that compounds (**139**) and (**140**) were not obtained from the tunic of *C. savignyi*, collected in 1993 or later, thus raising doubts that compounds (**139**) and (**140**) were produced by the tunicates.[76] Urochordamines A (**139**) and B (**140**) were converted into more polar compounds, urochordamines A′ (**141**) and B′ (**142**), respectively, when left standing in protic solvents.[102] Structure–activity relationships of urochordamines were tested on the promoting activity of larval metamorphosis of *H. rorentzi*. The percentage of metamorphosed larvae at a concentration of 25 μg ml^{-1} were similar whether induced by (**139**) or (**141**). The promoting activity of (**140**) was lower than that of (**139**) or (**141**). Interestingly, (**142**) showed no promoting activity at a concentration of 25 μg ml^{-1}. Compound (**141**) induced larval metamorphosis even at a concentration of 0.25 μg ml^{-1}, while (**139**) was ineffective at a concentration of 2.5 μg ml^{-1}. Therefore, the order of promoting activity was (**141**) > (**139**) > (**140**) > (**142**), indicating that both the structure of the pteridine moiety and the stereochemistry at C-9 were important for larval metamorphosis-promoting activity. The effect of exposure time on metamorphosis was also examined. About 20% of 6 h old larvae metamorphosed when exposed to 25 μg ml^{-1} of (**141**) for 15 min, which was a much lower proportion than at 30 min and 60 min exposures. However, exposure to 25 μg ml^{-1} of (**139**) for 60 min was necessary to induce metamorphosis, while 30 min exposure was ineffective. These results may indicate that urochordamines promote larval metamorphosis in ascidians without a chemoreception pathway. In addition, urochordamines induced metamorphosis of pediveliger larvae of the mussel *Mytilus edulis galloprovincialis*: 87%, 73%, 26%, and 50% of larvae metamorphosed in 6 days after treatment with 2.5 μg ml^{-1} of (**139**)–(**142**), respectively. Urochordamines are the first compounds to be identified that induce metamorphosis of pediveliger larvae, implying again that urochordamines do not act via chemoreception, but via an internal mechanism.

Promoters of larval metamorphosis in two kinds of ascidians, *H. roretzi* and *C. savignyi*, were also isolated from the brown alga *Sargassum thunbergii*, and they were identified as diphlorethol (**143**), a mixture of phlorotannins, and sulfoquinovosyl diacylglycerols (SQDG, (**144**) = (**9**)) composed of $C_{16:0}$ (82%) and $C_{18:1}$ (18%) as analyzed by GC.[103] These compounds induced larval metamorphosis of both ascidians at a concentration of 25 μg ml^{-1} and phlorotannin such as (**143**) appeared to act faster than SQDG (**144**). A large number of phlorotannins from brown algae are

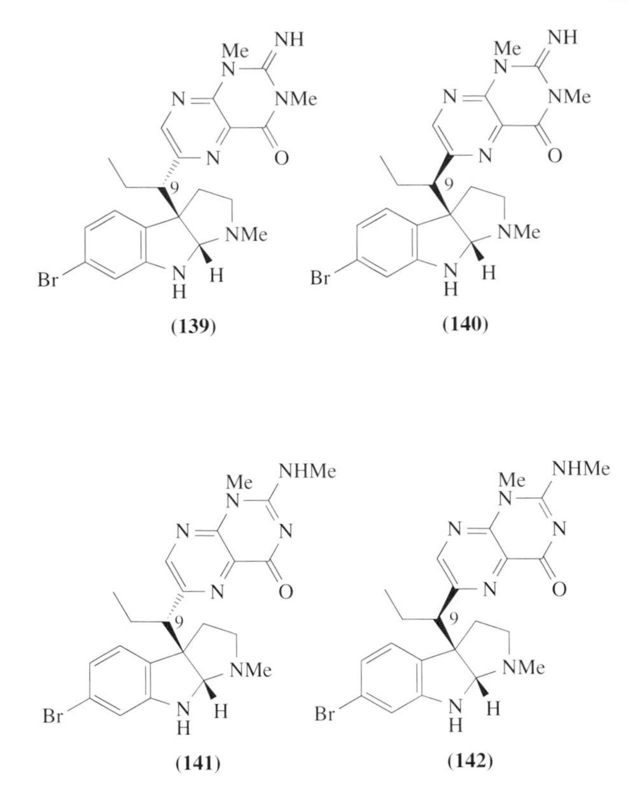

known to be antibiotics, and SQDG (**144**) is a common lipid of chloroplast membrane, and has been known to exhibit various biological activities such as a feeding-stimulating activity for herbivorous gastropods (Section 8.07.2.2). Furthermore, the fact that δ-tocotrienol epoxides (**68**) and (**69**) isolated from the brown alga *S. tortile* were the promoters of larval settlement/metamorphosis in the hydroid *Coryne uchidai* (Section 8.07.5.2) together with the results above led to the idea that brown alga may play important roles in the larval settlement/metamorphosis of marine animals.

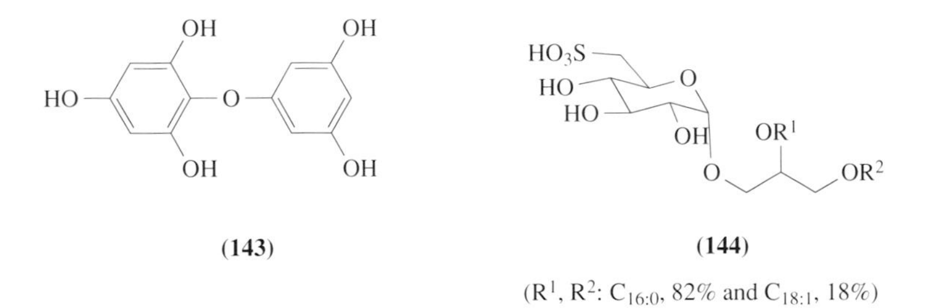

(R^1, R^2: $C_{16:0}$, 82% and $C_{18:1}$, 18%)

N,*N*-Dimethylguanidinium styryl sulfates, (*E*)- and (*Z*)-narains (**145**) and (**146**),[104] 3,4-dihydroxylstyrene dimers (**147**)–(**149**), trimethoxystyrene derivative (**150**), and a hypoxanthin base (**151**) were isolated from the MeOH extract of a marine sponge *Jaspis* sp. as metamorphosis-inducers of ascidian *Halocynthia roretzi* larvae.[105] The EtOH and acetone extracts afforded a triethoxy derivative (**152**) and a naphthyl derivative (**153**), respectively, instead of (**150**). (*Z*)-Narain (**146**), as well as its anion alone, induced larval metamorphosis on *H. roretzi* at a concentration of 5 μM, while *N*,*N*-dimethylguanidine sulfate was inactive at a concentration of 50 μM. (*E*)-Narain (**145**), its anion, (**147**), (**148**), and (**150**) were active at a concentration of 5 μM; (**149**) and (**151**) were much less active, whereas (**153**) did not show any effect. Therefore, the anion moiety, especially the sulfate group, may play an important role as the metamorphosis-inducer. The stereochemistry of the double bond was also important for activity. 4-Carboxy-1,3-dimethylimidazolium ion (**154**), which may be biogenetically related to (**151**), was isolated from the marine sponge *Cacospongia scalaris* as a metamorphosis-inducing compound, showed the same degree of activity as (**151**).

Two new pipecolate derivatives, anthosamines A (**155**) and B (**156**), were isolated from the marine sponge *Anthosigmella* aff. *raromicrosclera* along with a new diketopiperazine (**157**), *cyclo*-(L-Arg-

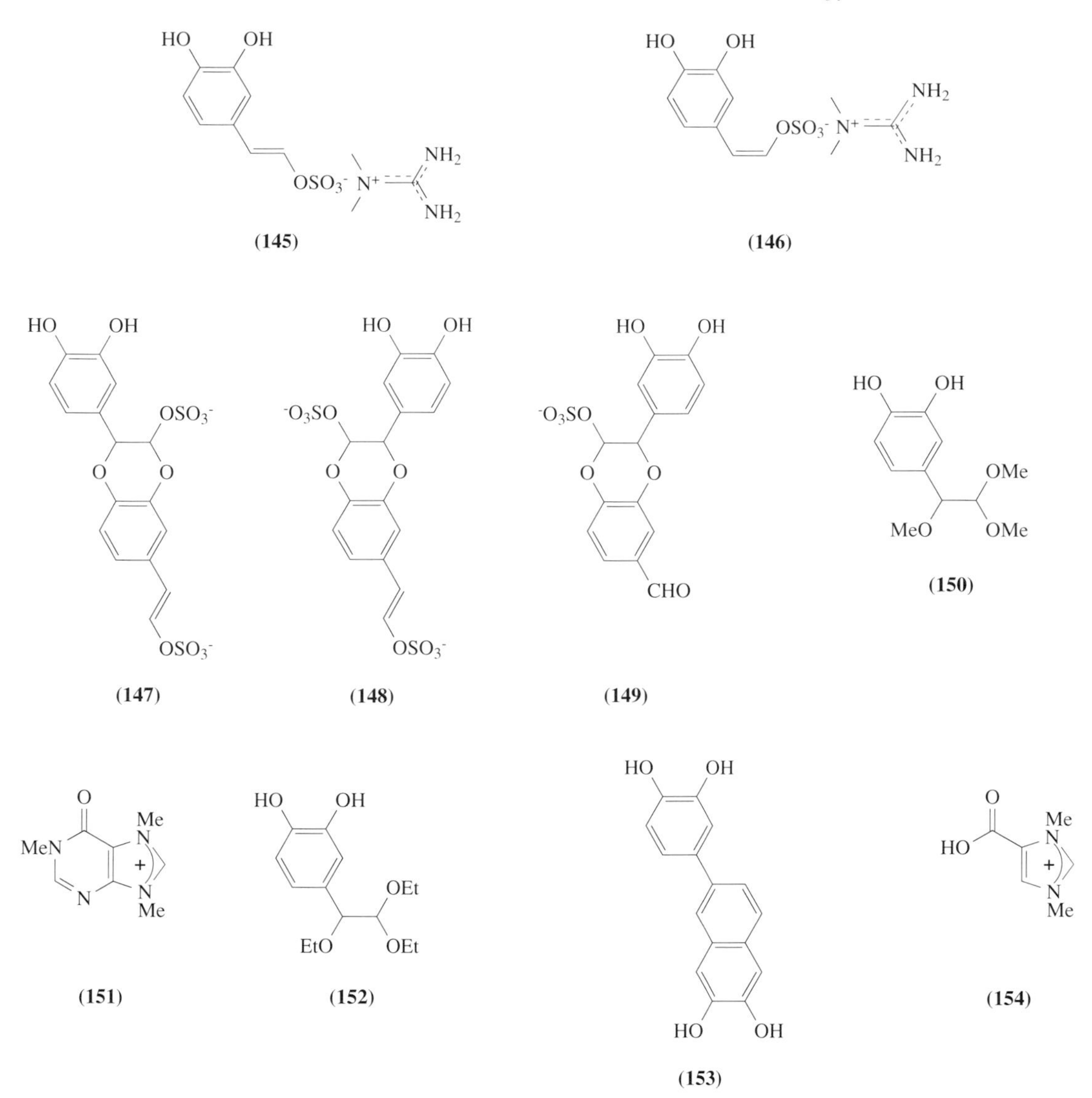

dehydrotyrosine) as metamorphosis-inducers of ascidian *H. roretzi* larvae. Reversible isomerization between amine and iminium ion forms was observed for anthosamines A (**155**) and B (**156**); iminium ion forms (**155b**) and (**156b**) predominate in protic solvent, and amine forms (**155a**) and (**156a**) in aprotic solvent (Equation (3)). Anthosamines A (**155**) and B (**156**) induced larval metamorphosis on the ascidian *H. roretzi* at a concentration of 50 μM completely, while the diketopiperazine (**157**) (50 μM) induced metamorphosis of 50% of larvae. From the Japanese marine sponge *Stelletta* sp., a new bisguanidinium alkaloid, stellettadine A (**158**), containing a norsesquiterpene unit was isolated as a metamorphosis-inducing compound. The absolute configuration of C-6′ was determined as *S* by treatment of (**158**) with $NaIO_4$ in the presence of $RuCl_3$ to afford (*S*)-2-methylglutaric acid. Stelettadine A (**158**) showed metamorphosis-inducing activity with an ED_{100} (100% effective dose) value of 50 μM on ascidian *H. roretzi* larvae.

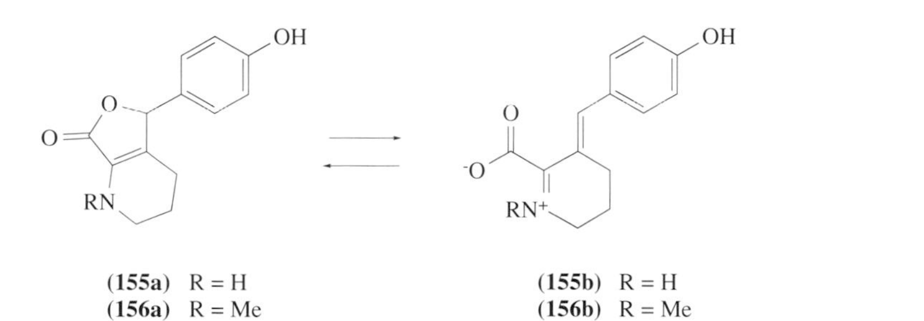

(3)

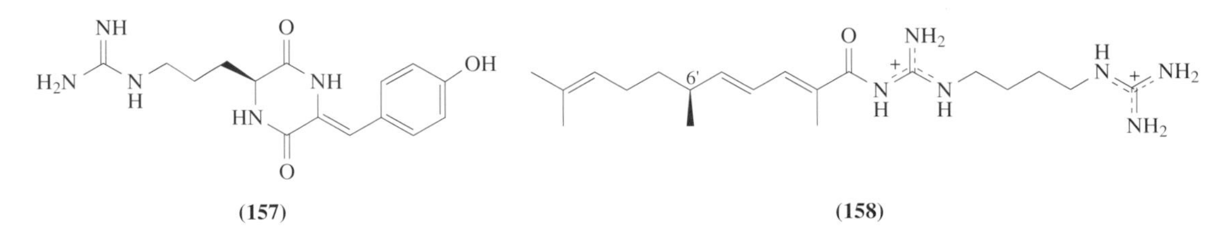

(157) **(158)**

8.07.6 BIOLUMINESCENCE

Bioluminescence of marine organisms is a widely recognized phenomenon, and bioluminescent animals found in the sea are widely distributed over almost all phyla of the animal kingdom from protozoans to vertebrates.[106] Among several bioluminescence mechanisms that have been reported, the luciferin–luciferase (L-L) reaction is probably the most well-known luminous system. In this system, light is emitted as a result of a reaction of a heat-stable substrate (general name: luciferin) with oxygen catalyzed by an enzyme (general name: luciferase), which was initially discovered in the bioluminescence of the luminous beetle *Pyrophorus noctiluca* in 1885. Other luminous systems involve the participation of a luminescent protein (general name: photoprotein) or symbiosis with luminous bacteria.[106]

8.07.6.1 Sea Firefly

Among many bioluminescent marine organisms, the ostracod crustacean *Vargula hilgendorfii* (formerly *Crypridina hilgendorfii*) has played a central role in contributing to an understanding of the chemical and physical bases of bioluminescence. *V. hilgendorfii* is commonly referred to as "umi botaru" in Japan or the "sea firefly." Until the 1960s this animal occurred in great abundance along the south coastal waters of Japan, but their numbers have since been reduced significantly. *V. hilgendorfii* is a small animal (ca. 3 mm long) which lives in sand at the bottom of shallow waters and becomes an active feeder at night. It emits a strongly luminescent secretion into the seawater when it is disturbed. The secretion consists of luciferin and luciferase, thus producing a bright luminous cloud. The light results from the oxidation of luciferin by molecular oxygen, catalyzed by luciferase, and the mechanism of the reaction has been studied extensively. The structure of the *Vargula* luciferin was determined as an imidazopyrazine compound (**159**) in the 1960s,[106] and the products of the reaction were known to be oxyluciferin (**160**) and carbon dioxide. The mechanism of the *Vargula* bioluminescence can be summarized as shown in Scheme 4.[106] The *Vargula* luciferin (**159**) dissociates to its anion (pK_a 8.3 in water), which is then oxidized by molecular oxygen to give a hydroperoxide anion. The hydroperoxide anion is postulated to be hydroperoxide form (a) or peracid form (b). The anion decomposes through a dioxetane intermediate to form oxyluciferin anion in a singlet excited state. In neutral solutions protonation occurs on the excited anion to give the excited neutral oxyluciferin molecule, which subsequently emits light. This bioluminescence system via decomposition of the dioxetanone intermediate is an example of the CIEEL mechanism[107,108] and the unstable dioxetanone intermediate has not been isolated.[109]

Oxidation of *Vargula* luciferin (**159**) in the presence of luciferase with light production gives oxyluciferin (**160**) which cannot be reduced (Scheme 4). On the other hand, luciferin (**159**) is very susceptible to oxygen even in the absence of the enzyme to form "reversibly oxidized luciferin."[106] Thus, the orange-colored aqueous solution of (**159**) turns red in air, and then becomes colorless owing to further autooxidation. The red substance is also formed by chemical oxidation of (**159**) with ferricyanide, lead dioxide, or diphenylpicrylhydrazyl radical (DPPH). It emits light only very slowly in the presence of the enzyme, but the luminescent activity similar to that of the luciferin (**159**) can be restored by chemical reduction with sodium hydrosulfite ($Na_2S_2O_4$). Oxidation of (**159**) with lead dioxide or DPPH affords a product without light production. This product can be reduced to luciferin (**159**) with sodium hydrosulfite or sodium borohydride ($NaBH_4$). Thus it can be recognized as a reversibly oxidized luciferin. The product, named luciferinol, was isolated and its structure was determined to be (**161**), although it is fairly unstable and difficult to handle.[110] Luciferinol (**161**) is a two-electron oxidation product and has no bioluminescent activity with luciferase. By further HPLC analysis, another oxidation product besides (**161**) was found, particularly when ferricyanide was used as the oxidant. Although it was extremely labile, it was named

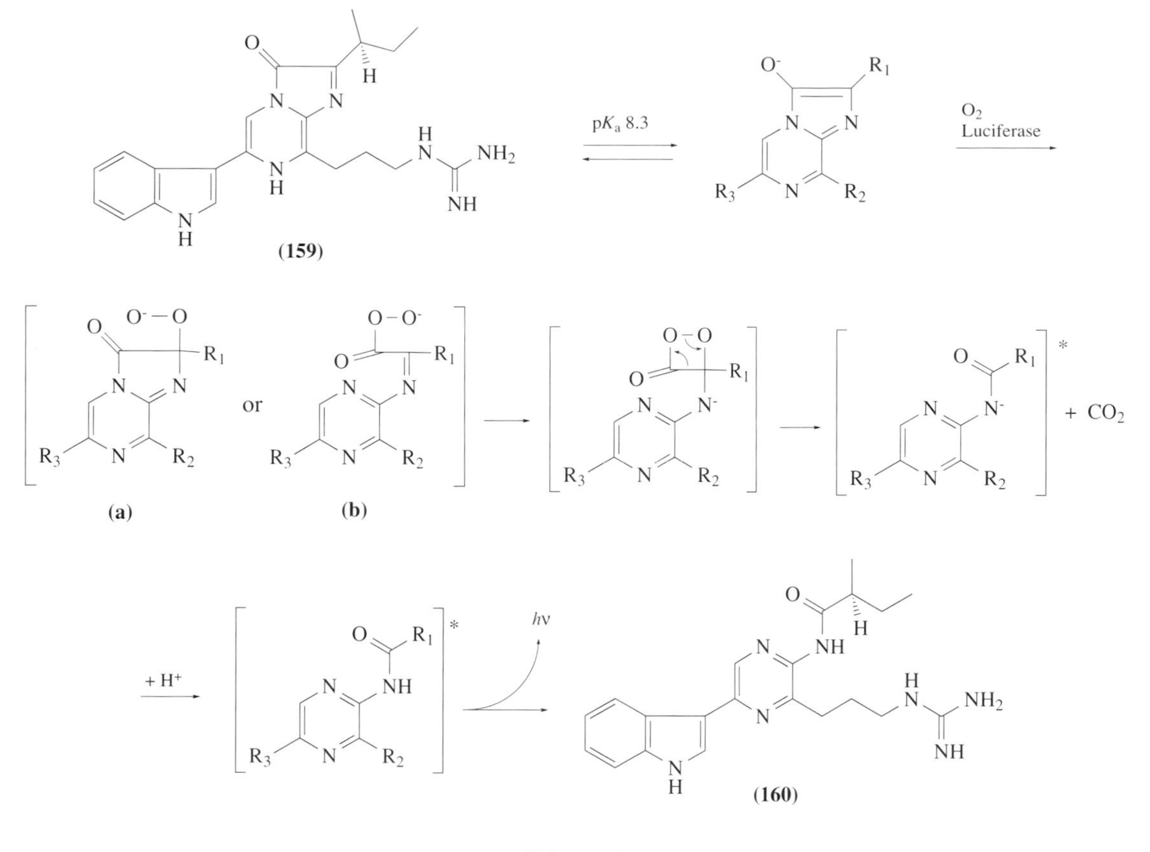

Scheme 4

biluciferyl and its structure was determined as (**162**).[111] It was suggested that a one-electron oxidation of (**159**) gives the luciferyl radical, which dimerizes to its dimer (**162**). Biluciferyl (**162**) was oxidized with DPPH into luciferinol (**161**) and reduced with sodium hydrosulfite or acid into luciferin (**159**). Whereas luciferinol (**161**) produces no light with *Vargula* luciferase, very slow bioluminescence was observed with biluciferyl (**162**) in the presence of the enzyme, indicating that biluciferyl (**162**) is an intermediate in the oxidation of (**159**) into (**161**) and that the slow luminescent reaction observed in the reversibly oxidized luciferin could be due to biluciferyl (**162**).

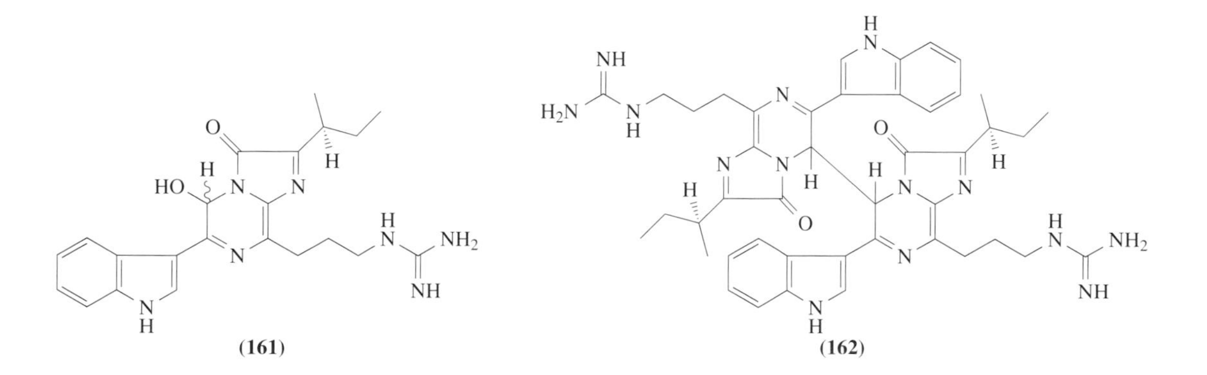

The cloning and nucleotide sequence analysis of the cDNA for the *Vargula* luciferase and the expression of the cDNA in a mammalian cell system has been reported.[112] The *Vargula* luciferase was purified to homogeneity by using tryptamine and *p*-aminobenzamidine affinity column chromatography. Digestion of the luciferase with endopeptides afforded several peptides, and a portion of one peptide fragment was used to design the complementary oligodeoxynucleotide probe. The mRNA was extracted from the *V. hilgendorfii* ground to find powder in liquid nitrogen. The cDNA library was constructed and the nucleotide sequence suggested that the primary structure of the *Vargula* luciferase consisted of 555 amino acid residues in a single polypeptide chain with a calculated molecular weight of 62 171. To establish that the cloned cDNA actually encoded *Vargula* luciferase, a mammalian cell system (monkey COS cells) was used to express cDNA. As a result, luciferase activity was clearly detected in the culture medium of transfected COS cells with some luciferase activity also present in the cell extract.

The enzyme, *Vargula* luciferase, showed significant amino acid sequence homologies in two regions with an N-terminal segment of aequorin, a calcium-binding photoprotein which emits light in the bioluminescent reaction of the jellyfish *Aequorea victoria* (Section 8.07.6.2). In spite of the close similarities in substrate structure and mechanisms of the bioluminescent systems in *Vargula* and *Aequorea*, the enzymes and substrates of the two systems showed only slight light-emitting cross-reactions.[112]

8.07.6.2 Jellyfish

Many bioluminescent organisms are included in Coelenterata (phylum: Cnidaria). The hydrozoan *Aequorea aequorea* (jellyfish) possesses a photoprotein, aequorin, which emits light by the action of Ca^{2+} without molecular oxygen. On the other hand, the anthozoan *Renilla reniformis* (sea pansy) has a luciferin–luciferase reaction system and requires molecular oxygen for light production.[106]

The photoprotein, aequorin (molecular weight 21 400), isolated from the outer margin of the umbrella of the jellyfish *Aequorea victoria* from Friday Harbor, WA, USA, emits blue light in aqueous solution when Ca^{2+} or Sr^{2+} is added in either the presence or absence of molecular oxygen.[113,114] Aequorin consists of two components: an apoprotein (apopaequorin) and a chromophore. The chromophore is made up of coelenterazine (an imidazopyrazine compound (**163**)) and molecular oxygen. The chromophore is attached noncovalently to apoaequorin. The binding of Ca^{2+} to aequorin presumably causes the protein to change its conformation, converting it into an oxygenase, which then catalyzes the oxidation of coelenterazine (**163**) by the bound oxygen, yielding as products light (λ_{max} 470 nm), CO_2, and a blue fluorescent protein (BFP). The BFP is composed of apoaequorin and the oxidized product of (**163**), coelenteramide (**164**). The excited-state coelenteramide (**164**) is the emitter in the reaction. Aequorin may be regenerated from apoaequorin by incubation with coelenterazine (**163**), dissolved oxygen, EDTA, and 2-thioethanol (Scheme 5).

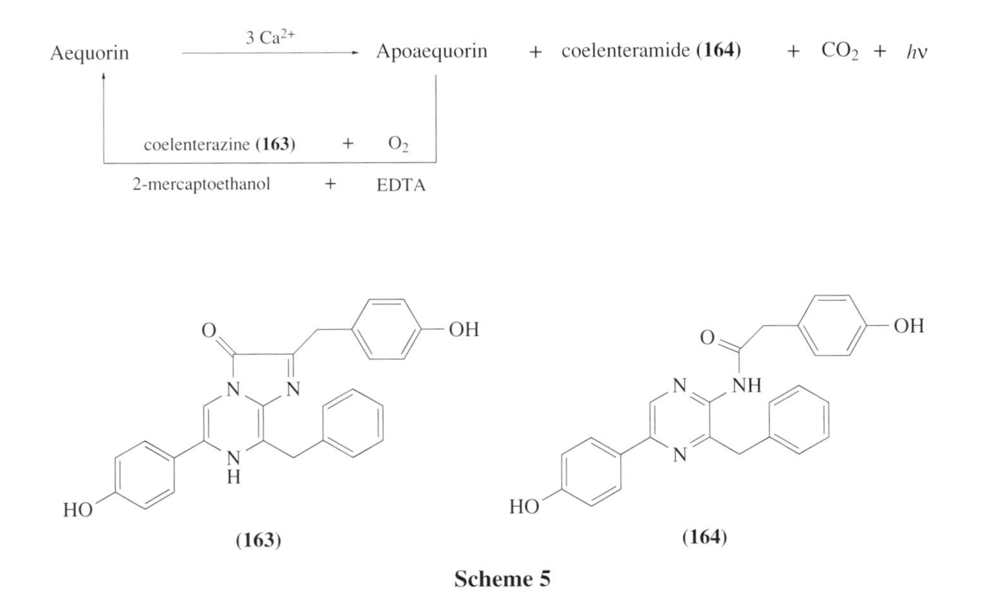

Scheme 5

A cDNA library of *Aequorea* was constructed and clones carrying the cDNA for the Ca^{2+}-dependent photoprotein were isolated by the method of colony hybridization using synthetic oligonucleotide probes. The primary structure of the apoprotein (apoaequorin) deduced from the nucleotide sequence showed that the protein was composed of 189 amino acid residues.[115–117] Intracellular high-affinity Ca^{2+}-binding proteins such as calmodulin, parvalbumin, troponin C, and myosin light chains belong to a superfamily of proteins known a EF-hand proteins. These proteins have either two or four Ca^{2+}-binding regions which show sequence homology with the well-defined helix–loop–helix structures denoted as EF hands in the X-ray structure of carp parvalbumin. The Ca^{2+} dependency of aequorin luminescence indicates the presence of three high-affinity Ca^{2+} sites, making aequorin a candidate for membership in the EF-hand superfamily. Comparison of the amino acid sequence with those of the Ca^{2+}-binding proteins calmodulin and troponin C revealed that three segments of aequorin had properties expected for EF-hand domains. The EF-hand regions had relatively high hydrophilicity, suggesting that the regions are located near the surface of the protein, while the sequence also suggested that the protein had three hydrophobic regions at which the protein may interact with its functional chromophore coelenterazine (**163**).

Various expression plasmids for apoaequorin cDNA were constructed and expressed in *Escherichia coli*. Aequorin was regenerated from the expressed apoaequorin by incubating with coelenterazine, 2-mercaptoethanol, and EDTA.[118] Thus, provided coelenterazine is available, the recombinant DNA technique may outweigh the enormous task associated with collecting and processing large numbers of *Aequorea* and purifying the highly unstable aequorin; it was estimated that 5300 specimens of *Aequorea* were required to collect 70–100 mg of aequorin. The initial procedure of the recombinant DNA method for obtaining a highly active preparation of aequorin[118] was, however, time-consuming and the yield was relatively low. These objections were overcome by an improved procedure which consisted of fusing the apoaequorin cDNA to the signal peptide coding sequence of the outer membrane protein A (*ompA*) of *Escherichia coli*.[119] Subsequent expression of the cDNA in *E. coli* results in high-level production of apoaequorin and the release of the protein into the culture medium. The purification of the protein was then easily achieved by acid precipitation and DEAE-cellulose chromatography. The procedure yielded 7.4 mg of recombinant apoaequorin with a purity greater than 95% from 200 ml of culture medium and, on regeneration with coelenterazine, the recombinant aequorin was fully active with Ca^{2+}.

To understand the mechanism of the bioluminescence reaction, a structure–function relationship was studied with respect to modifying certain of its amino acid residues by carrying out oligonucleotide-directed site-specific mutagenesis of apoaequorin cDNA and expressing the mutagenized cDNA in *Escherichia coli*. Amino acid substitution was made at the three Ca^{2+}-binding sites, the three cysteines, and a histidine in one of the hydrophobic regions. Subsequent assay of the modified aequorin showed that the Ca^{2+}-binding sites, the cysteines, and probably the histidine all play a role in the bioluminescence reaction of aequorin.[120] Further study of the modified aequorin with the three cysteine residues replaced by serine revealed that six of seven modified aequorins had reduced luminescence, whereas the seventh with all three cysteines replaced by serine had luminescence activity equal to or greater than that of the wild-type aequorin. The time required for the regeneration of the triply substituted aequorin was substantially increased compared with the time required for the regeneration of the wild-type aequorin. The results suggested that cysteine plays an important role in the regeneration of aequorin but not in its catalytic activity.[121]

The photoprotein aequorin purified by the standard isolation procedure was not homogeneous and consisted of a mixture of closely related photoproteins, i.e., isoaequorins, which have isoelectric points ranging from pH 4.2 to 4.9. Aequorin samples, extracted from 50 000 specimens of *Aequorea*, were separated with sufficient resolution into various molecular forms of aequorin (isoaequorins) by high-performance liquid chromatography on a TSK DEAE-5PW anion-exchange column by using buffers containing sodium acetate to obtain eight different kinds of the photoprotein, i.e., aequorins A–H. The M_r values ranged from 20 100 (aequorin F) to 22 800 (aequorin A), the luminescence activities ranged from 4.35×10^{15} photons mg^{-1} (aequorin A) to 5.16×10^{15} photons mg^{-1} (aequorin F), and the first-order reaction rate constants of luminescence ranged from $0.95\ s^{-1}$ (aequorin A) to $1.33\ s^{-1}$ (aequorin F). As regards sensitivity to Ca^{2+}, aequorin D was the most sensitive, having a sensitivity about 0.4–0.5 pCa units above that of the least sensitive kind (aequorin A).[122]

The light-emitting reaction of aequorin was selectively triggered by Ca^{2+}. Because of its high sensitivity to Ca^{2+}, aequorin is widely used as a Ca^{2+} indicator in various biological systems. To provide useful semisynthetic Ca^{2+}-sensitive photoproteins, chemical modification of the functional part of aequorin was studied by replacing the coelenterazine moiety in the protein with several synthetic coelenterazine analogues. One of the semisynthetic photoproteins, derived from coelenterazine analogue (**165**) with an extra ethano group, showed highly promising properties for the measurement of Ca^{2+} for two reasons: (i) the rise time of luminescence in response to Ca^{2+} was shortened approximately fourfold compared with native aequorin and (ii) the luminescence spectrum showed two peaks at 405 nm and 465 nm and the ratio of the peak heights was dependent on Ca^{2+} concentration in the range of pCa 5–7, allowing the determination of Ca^{2+} concentration directly from the ratio of two peak intensities (I_{400}/I_{465}). This photoprotein containing the analogue (**165**) was designated *e*-aequorin. The coelenterazine analogue (**166**) with an amino group instead of a hydroxyl group was also incorporated into apoaequorin, yielding a Ca^{2+}-sensitive photoprotein, which indicated that an electrostatic interaction between the phenolate group in the coelenterazine moiety and some cationic center in apoaequorin is not important in native aequorin.[123] Furthermore, 37 coelenterazine analogues were synthesized and incorporated into apoaequorin, yielding 30 semi-synthetic aequorins that had the capacity to emit a significant amount of light in the presence of Ca^{2+} (over 10% of that of natural aequorin). Among them, the photoprotein derived from coelenterazine analogue (**167**) showed by far the highest level of the relative intensity of Ca^{2+}-triggered luminescence of 190 when compared with natural aequorin (relative intensity 1.0). However, since the yield of the incorporation reaction of this photoprotein was low, the overall property of this photoprotein appeared to be not favorable. On the other hand, the photoproteins of *e*-aequorin-type, prepared from the coelenterazine analogues (**168**) and (**169**) containing an ethano bridge, showed fast luminescence reactions, high relative luminescence intensities, and, like *e*-aequorin, the dependency of the spectral distribution of luminescence on the concentration of Ca^{2+}. With the two photoproteins derived from (**168**) and (**169**), the degree of dependence of the luminescence intensity ratio I_{400}/I_{465} on pCa was greater than that with *e*-aequorin, suggesting that these two photoproteins were possibly superior to *e*-aequorin in measuring Ca^{2+} concentration by the ratio method.[124]

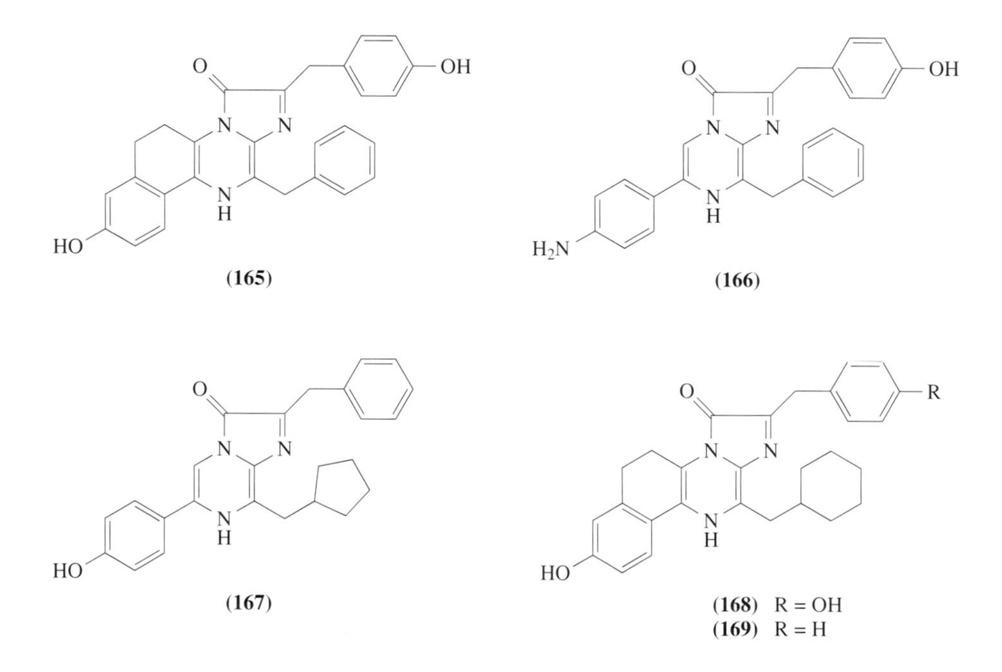

It was suggested that the conformational rigidity of the *p*-hydroxyphenyl group in coelenterazine analogues like (**165**) had some enhancement effects on the light yield, while decreasing light yield was observed for coelenterazine analogues with an alcoholic side-chain due to hydrogen bond formation with one of the nitrogen atoms in the emitter.[125]

The comparison of properties of a recombinant aequorin with those of natural aequorin was studied.[126] Recombinant apoaequorin was obtained by fusing the apoaequorin cDNA to the signal-peptide coding sequence of the outer-membrane Protein A of *Escherichia coli*, subsequent expression of the cDNA in *E. coli*, and purification of the recombinant apoaequorin released into the culture medium by acid precipitation and DEAE-cellular chromatography.[119] In chromatographic behavior the recombinant apoaequorin did not match any of 10 isoaequorins tested, although it was very similar to aequorin J; aequorins I and J[125] were additional isoaequorins isolated subsequent to the initial purification of eight isoaequorins, aequorins A–H.[122] The Ca^{2+}-sensitivity of recombinant aequorin was higher than that of any isoaequorin except aequorin D. The recombinant aequorin exhibited no toxicity when tested in various kinds of cells, even where samples of natural aequorin had been found to be toxic. Properties of four recombinant semisynthetic aequorins (designated *fch*-, *hcp*-, *e*-, and *n*-aequorin, respectively), prepared from the recombinant apoaequorin and synthetic analogues of coelenterazine ((**170**), (**167**), (**165**), and (**171**), respectively), were approximately parallel with those of corresponding semisynthetic aequorins prepared from natural apoaequorin as shown in Table 15. Both recombinant *e*-aequorin and natural *e*-aequorin J luminesced with high values of the luminescence intensity ratio I_{400}/I_{465}, although the ratios were not pCa-dependent.

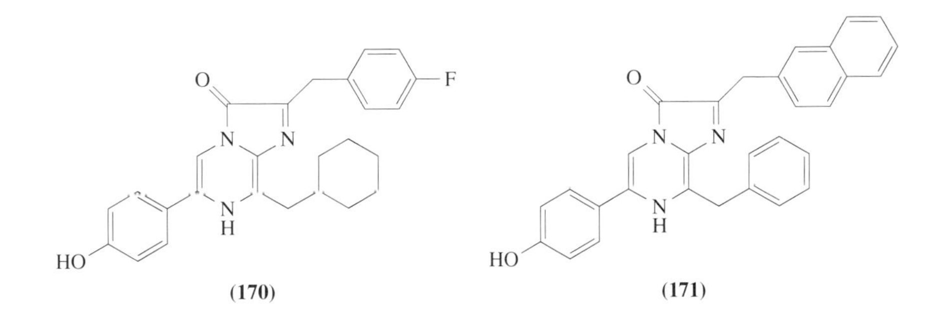

Table 15 Properties of semisynthetic aequorins derived from recombinant aequorin and natural aequorin J.

Semisynthetic aequorin	*Coelenterazine part*	*Light-emitting capacity* (10^{12} quanta μg^{-1})	*Light intensity* (10^8 quanta μg^{-1}) pCa 7	pCa 6	*Without added* Ca^{2+}
Recombinant aequorin (control)	(**163**)	4.80	2.32	240.0	0.04
Recombinant *fch*-aequorin	(**170**)	3.05	210.0		0.5
Recombinant *hcp*-aequorin	(**167**)	2.23	445.0		0.7
Recombinant *n*-aequorin	(**171**)	1.34	0.60	3.8	0.01
Recombinant *e*-aequorin	(**165**)	2.23	4.1		0.2
Aequorin J (control)	(**163**)	4.50	0.56	50.0	0.03
fch-Aequorin J	(**170**)	3.52	52.0		0.15
hcp-Aequorin J	(**167**)	2.88	120.0		0.24
n-Aequorin J	(**171**)	1.21	0.10	0.6	0.01
e-Aequorin J	(**165**)	2.08	2.1		0.16

A group of semisynthetic *e*-type aequorins showed bimodal luminescence, with peaks at 400–405 nm and 440–475 nm, while all other aequorins luminesced with only one peak in the range 440–475 nm. The cause of the spectral variation was studied by various experiments, and it was suggested that the spectrum of Ca^{2+}-triggered luminescence is strongly affected by the ionic charge on the amide N atom of the coelenteramide that is bound to apoaequorin. When the amine N atom is negatively charged, light is emitted with a 440–475 nm peak. In the case of *e*-type aequorins, the negative charge on the amide N atom is less, resulting in the emission of a 400–405 nm peak from the unchanged form of coelenteramide; the intensity ratio of 400–405 nm peak to 440–475 nm peak is determined by the amount of negative charge resting on the amide N atom of *e*-coelenteramide at the time of light emission. Most of the spectral variations in luminescence and fluorescence can be explained on the basis of ionic and hydrophobic interaction between a coelenteramide and apoaequorin.[127]

Bioluminescence activities of semisynthetic aequorins containing coelenterazine analogues with different substituents at the C-2 position of the imidazopyrazinone ring coelenterazine ((**163**) and (**172**)–(**176**)) were studied and it was evident that a hydroxybenzyl group at the C-2 position was essential for efficient luminescence activity; the photoprotein containing coelenterazine analogues (**173**)–(**176**) gave lower bioluminescence activity than aequorin containing either coelenterazine (**163**) or its close structure analogue (**172**).[128] Coelenterazine analogues possessing either the 2- or 8-adamantylmethyl group (**177**) and (**178**) were also prepared and it was found that the bioluminescence intensity of semisynthetic aequorin and the modified aequorin with three cysteine residues replaced by serine containing an 8-adamantylmethyl coelenterazine analogue (**178**) was stronger than that of natural coelenterazine.[129]

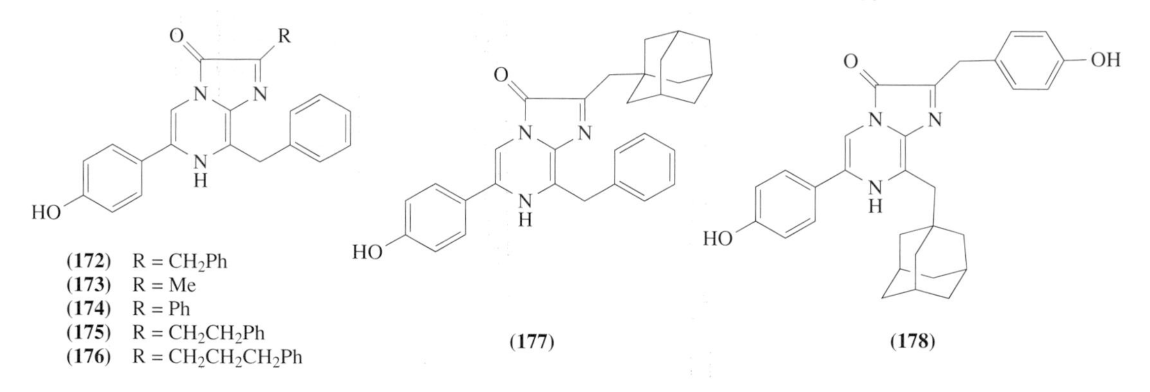

Shimomura and Johnson[130] reported that reduction of aequorin with $NaHSO_3$ afforded a yellow compound, assigned as the 2-hydroxyl derivative (**179**). Based on this result, Shimomura and Johnson suggested the chromophore coelenterazine as the light-emitting species linking to the protein through a peroxidic bond, as illustrated in Figure 2(a), while Cormier and co-workers[131] proposed the noncovalently bound structure for aequorin (Figure 2(b)).

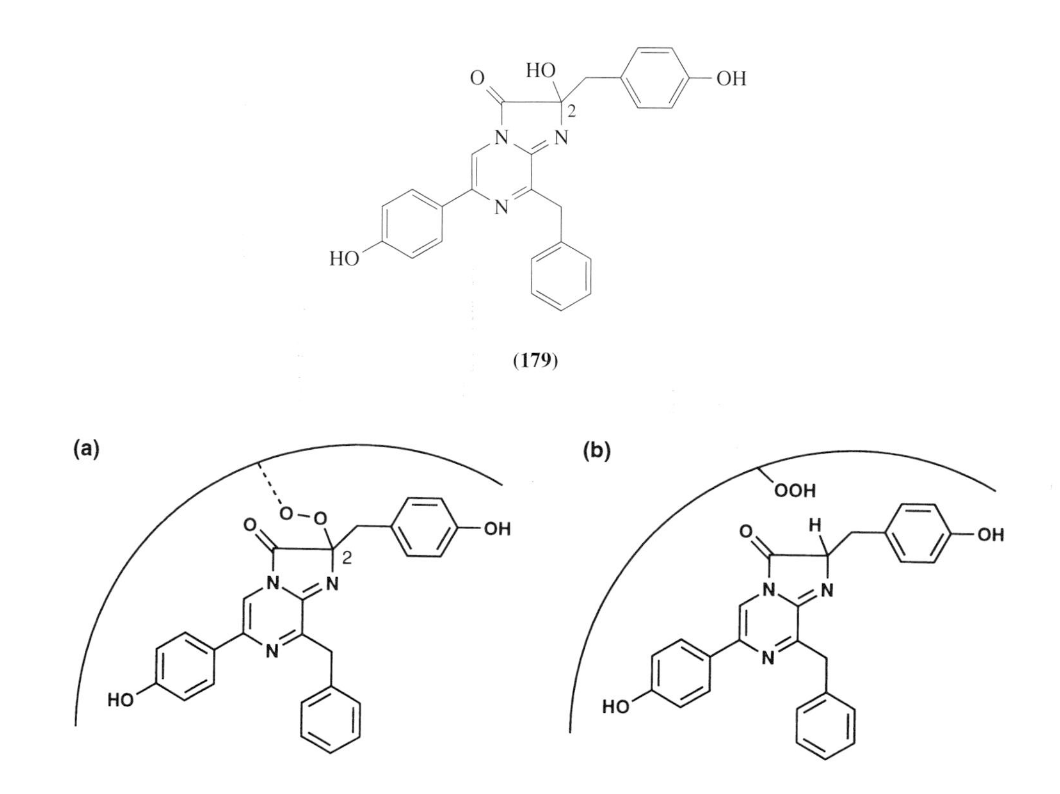

Figure 2 The functional part structure of aequorin proposed by (a) Shimomura and Johnson and (b) Cormier and co-workers.

Based on the ^{13}C NMR spectra of ^{13}C-enriched aequorins, the proposal represented in Figure 2(a) was supported by Kishi and co-workers.[132] Three ^{13}C-enriched coelenterazines (**180a–c**) were prepared and then incorporated into apoaequorin to yield three ^{13}C-enriched aequorins. During the incubation of coelenterazine (**180a**) into apoaequorin to form aequorin, the ^{13}C NMR showed that the C-2 carbon atom underwent an upfield shift (from 129 ppm to 98 ppm) due to a hybridizational change from sp^2 to sp^3. The chemical shift of -CO*C*H(NH-)R usually occurs in the range δ_C 40–60 and that of -CO*C*(OR^1)(NH-)R in the range 80–100 ppm. The observed chemical shift (98 ppm) for C-2 supported the partial structure (a). On the other hand, the chemical shift of the C-9 carbon of coelenterazine was shifted downfield during incubation with apoaequorin (from 130 ppm to 148 ppm). An experiment in the presence of $^{18}O_2$ was also carried out. The C-2 signal of the aequorin derived from (**180b**) incubated in the presence of a mixture of $^{18}O_2$ and $^{16}O_2$ was apparently broader than that of (**180b**)-containing aequorin incubated only with $^{16}O_2$ and accompanied by a shoulder which was observed approximately 0.07 ppm upfield from the peak at δ_C 98. The shoulder was attributable to the ^{18}O isotope effect.

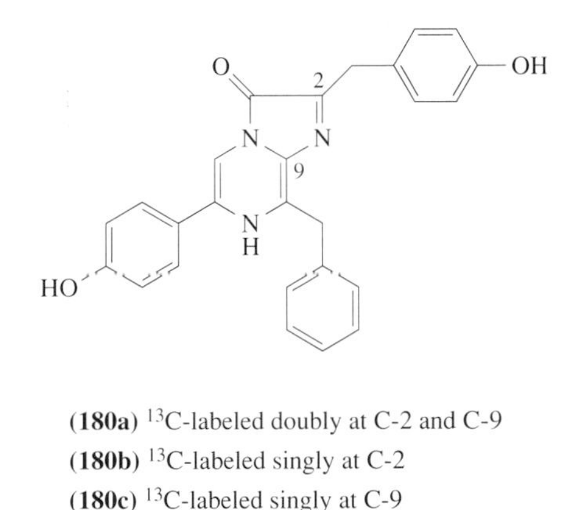

(**180a**) ^{13}C-labeled doubly at C-2 and C-9
(**180b**) ^{13}C-labeled singly at C-2
(**180c**) ^{13}C-labeled singly at C-9

To provide a model for a key intermediate in the bioluminescence of aequorin suggested by Kishi and co-workers,[132] unstable 2-*t*-butyl peroxide (**181**) and stable 2,2-di-*t*-butyl derivative (**182**) were prepared.[133] The chemical shifts of the C-2 and C-9 carbons of these model compounds were δ_C 104.85 and 150.23, respectively, for (**181**) and δ_C 84.83 and 147.88, respectively, for (**182**). The chemical shift of the C-2 carbon of (**181**) appeared 6.85 ppm downfield from the peak of aequorin derived from coelenterazine (**180a**) at δ_C 98,[132] which may have been due to the solvent effect and to the β effect by three methyl carbons in the *t*-butyl group attached to C-2. The chemical shift difference of the C-9 carbon between (**181**) (δ_C 150.23) and (**180a**)-containing aequorin (δ_C 148) was 2.23 ppm. These ^{13}C NMR chemical shift data were consistent with Kishi's proposal that coelenterazine is transformed into the 2-alkyl-2-peroxy-2,3-dihydroimidazo[1,2-*a*]pyrazin-3-one structure in aequorin.

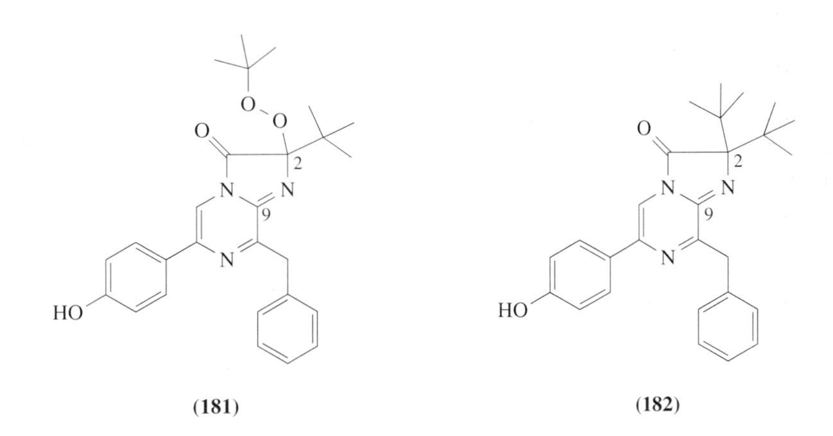

(**181**) (**182**)

Concerning the structure of the yellow compound (YC), Teranishi *et al.* reported that YC should have the 5-oxo-structure (**183**) instead of having hydroxyl group at C-2 as reported previously on the basis of preparing model compounds (**184**) and (**185**) oxygenated at C-2 and C-5 positions, respectively, and comparing their absorption spectra.[134] The three absorption spectra of YC,[130] (**184**), and (**185**), similarly showed two maxima at around 430–450 nm and 290–310 nm, while the relative intensity ratio ($\varepsilon_{430}/\varepsilon_{290}$) of YC (ca. 3) corresponded better to that of the 5-oxy compound (**185**) ($\varepsilon_{430}/\varepsilon_{290} = 3.1$) than to that of the 2-oxy compound (**184**) ($\varepsilon_{430}/\varepsilon_{290} = 0.26$). Thus, the 5-oxo structure (**185**) was suggested as a revised structure of YC, being closely related to luciferinol (**161**) derived by oxidation of *Vargula* luciferin (**159**) (Section 8.07.6.1).

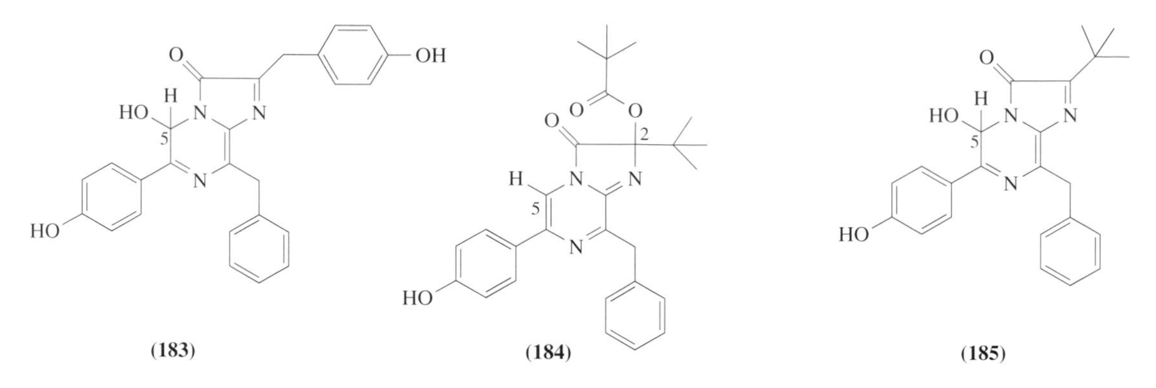

To provide further demonstrations of involvement of 5-oxy coelenterazine analogues in luminescence reactions, synthetic approaches to 5-hydroperoxide were examined. As a result, an unstable *t*-butyldimethylsilyl peroxide of coelenterazine analogue (**186**) was prepared, and this peroxide (**186**) in DMF under either aerobic or anaerobic conditions emitted weak light for 2 days.[135] Furthermore, 5-hydroxyperoxide (**187**) was successfully prepared, which thermally decomposed to give only amide (**188**). From all these results, the structure of YC was suggested as 2-peroxide (**179**).[136]

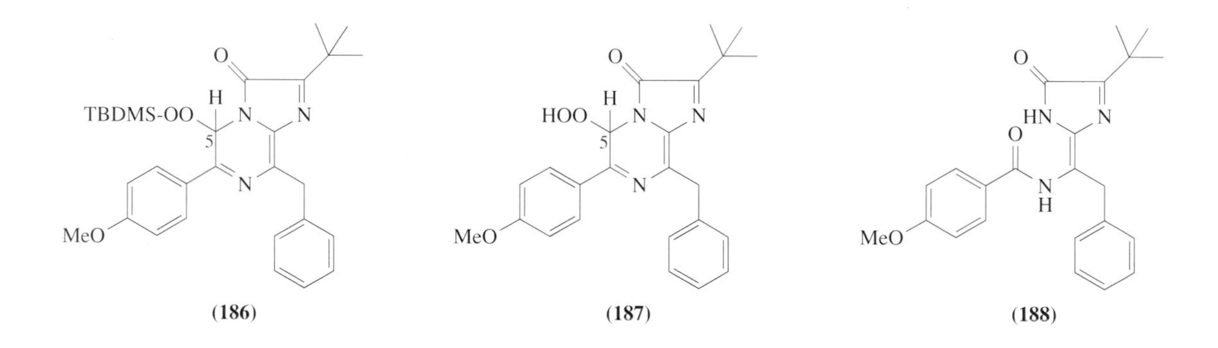

Chemiluminescence and fluorescence studies of the imidazopyrazine derivatives of coelenterazine in basic aprotic solvents indicated that the excited light-emitter was the amide anion (Figure 3(a), see also Scheme 4),[125] whereas on the basis of luminescence studies of regenerated blue fluorescent proteins (BFP) with coelenterazine analogues including *N*-methyl derivatives, the emitting species was assigned to a pheonolate monoanion (Figure 3(b)).[137] The *O*-methyl and *N*-methyl coelenteramide derivatives (**189**) and (**190**), respectively, were prepared, and the fluorescence emission spectrum of (**189**) in BFP could not be observed while incubation of (**190**) with recombinant apoaequorin yielded a BFP with a fluorescence emission spectrum possessing a peak at 480 nm. This emission maximum was similar to the bioluminescence maximum of aequorin, suggesting that the excited-state emitter in aequorin could not be the amide anion (Figure 3(a)) but must be the phenolate anion (Figure 3(b)).[136]

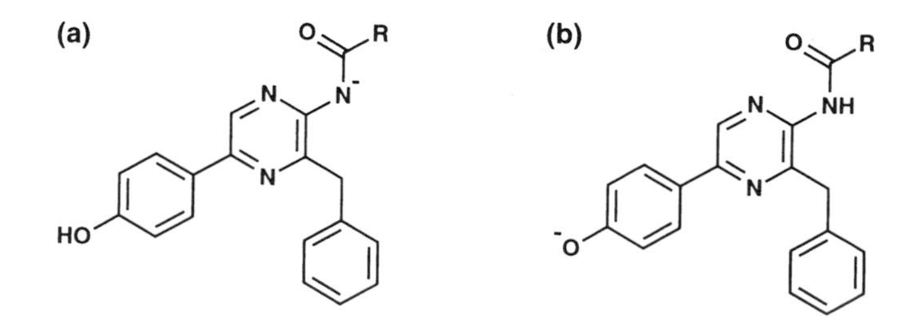

Figure 3 Possible ionized structures of coelenteramide: (a) amide anion, (b) phenolate anion.

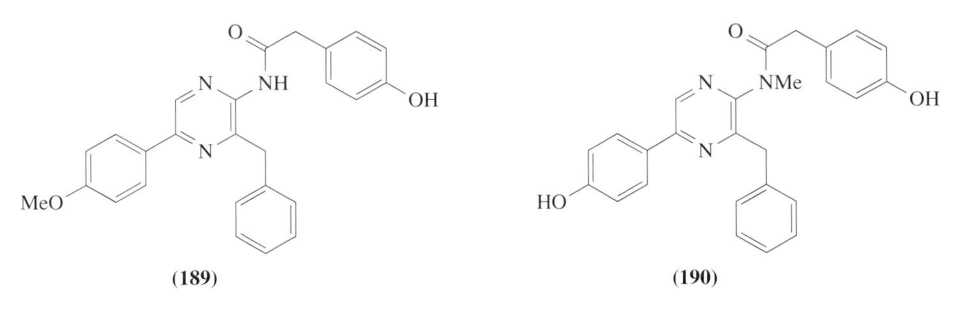

(189) **(190)**

Photoaffinity labeling is an attractive methodology to characterize and identify functional domains and active sites of biomacromolecules. To examine the bioluminescence mechanism and the nature of the active site and the binding mode between coelenterazine and apoaequorin, a photolabile analogue of coelenterazine with a trifluoromethyl diazirine group (**191**) was prepared and the chemi- and bioluminescence of the analogue (**191**) showed that its behavior was almost identical with that of natural coelenterazine in terms of luminescence characteristics, suggesting that the analogue (**191**) occupied the same active site of aequorin and should be a useful photoaffinity label for probing the detailed structure of aequorin.[138–140]

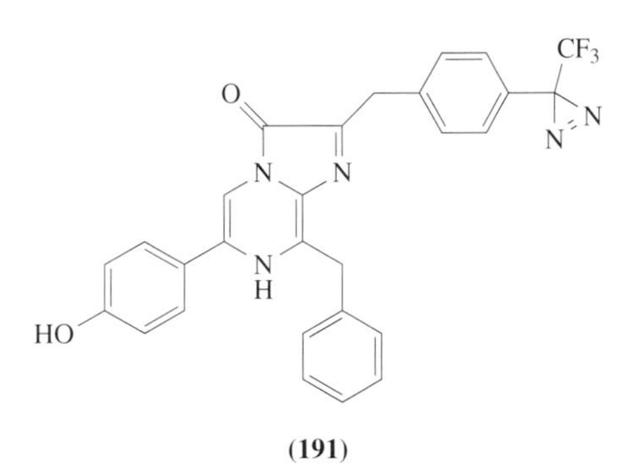

(191)

The jellyfish *Aequorea victoria* emits a bluish-green light from the margin of its umbrella. The light is due to not only the photoprotein aequorin described above but also to a green fluorescent protein (GFP).[136] The GFP was known as a single polypeptide chain consisting of 238 amino acid residues and having a M_r of 27 kD, and containing a modified hexapeptide as a chromophore which emits a green light (λ_{max} 508 nm). The GFP chromophore was proposed to be composed of modified amino acid residues within the polypeptide.[141] The structure of the GFP chromophore was elucidated by releasing it as a hexapeptide upon digestion of the protein with papain with the aid of synthesis of model chromophores and modern two-dimensional (2D) NMR spectroscopic studies.[142] As a result, the GFP hexapeptide chromophore was proposed as (**192**) which was formed by a post-translational modification of the tripeptide Ser^{65}-Tyr^{66}-Gly^{67} in the primary structure.

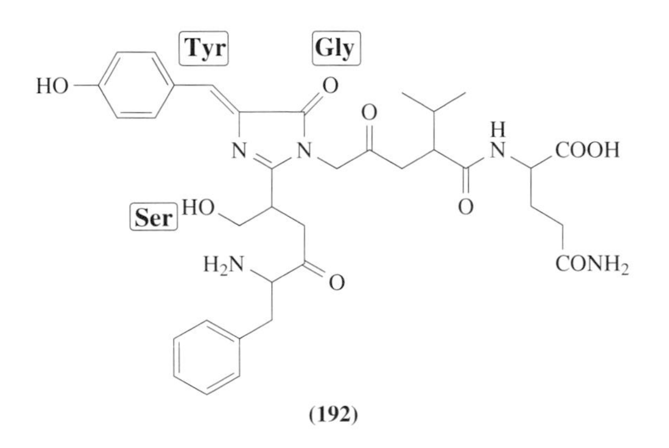

(192)

The chromophore formation through the dehydration–dehydrogenation mechanism in the tripeptide Ser^{65}-Tyr^{66}-Gly^{67} (Scheme 6) was supported by several types of modern mass spectroscopic analyses [ESI (electrospray ionization), MALDI-TOF (matrix-assisted laser desorption/ionization-time of flight), and MALDI-PSD (post source decay) spectra] of the chromophore-containing peptide obtained from the protease digest of GFP, which suggested the loss of 20 a.m.u. corresponding to the dehydration–dehydrogenation process.[143]

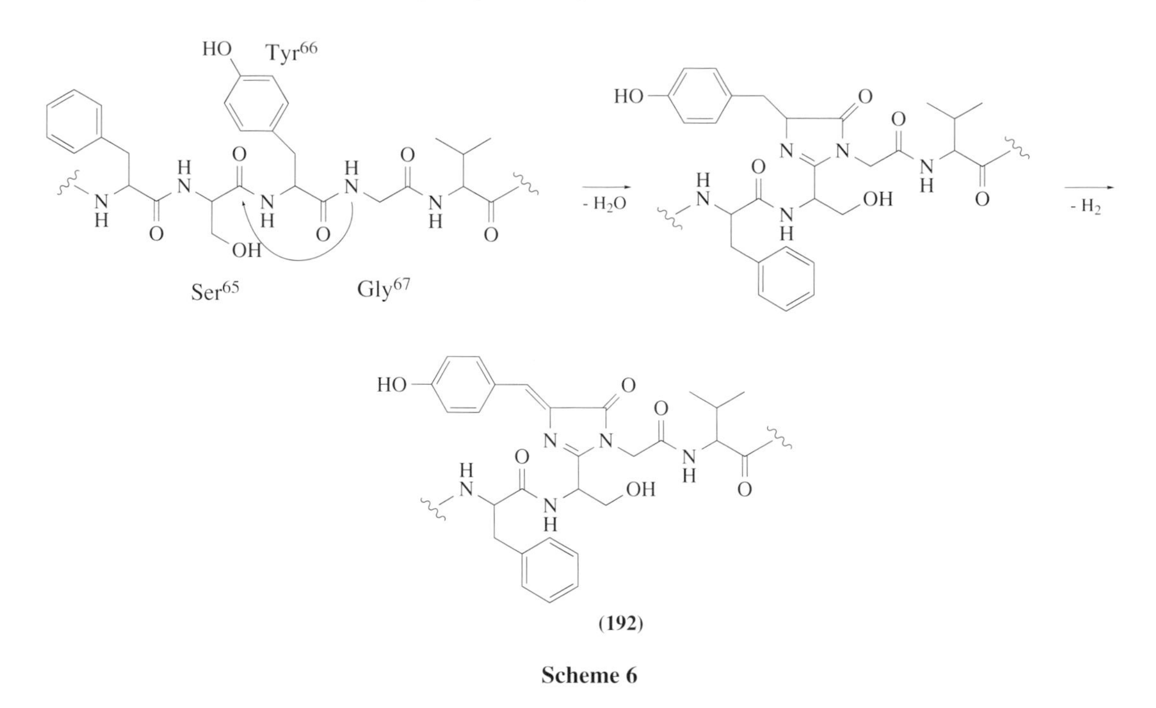

Scheme 6

Expression of the cDNA for *Aequorea* GFP in *E. coli* yielded a fused protein with fluorescence excitation and emission spectra virtually identical to those of native GFP, suggesting that the primary structure of the protein underwent modification to form a chromophore during expression.[144] Because of the marked intrinsic fluorescence emitted by GFP, it may also serve as a reporter or marker in gene expression studies.[145]

8.07.6.3 Squid

A luminous squid, *Watasenia scintillan* (Japanese name: hotaru-ika), is caught in large numbers in Toyama Bay, Japan, in a certain period of the year. It has three tiny black spots, which are luminous organs, located at the tip of each of the ventral pair of arms. The chemistry of the chromophore (luciferin) of *Watasenia* squid was studied in the mid-1970s. *Watasenia* oxyluciferin (**193**) and luciferin (**194**) was extracted from the arm photophores, and they were revealed as the disulfates of coelenterazine (**163**) and coelenteramide (**164**), respectively.[146,147]

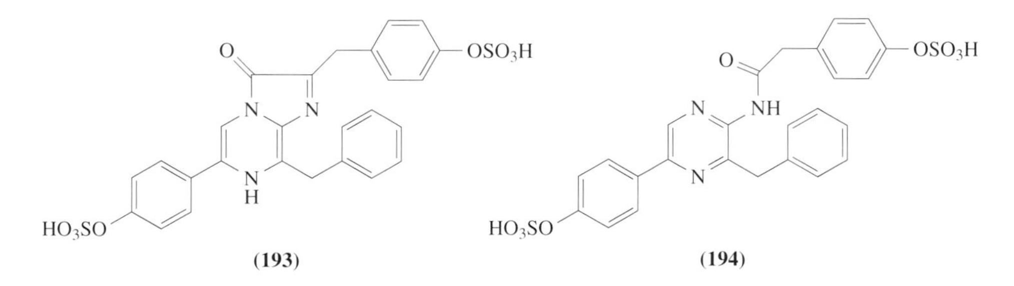

The bioluminescence system of a different luminous squid, *Symplectoteuthis oualaniensis* (Japanese name: tobi-ika), has been studied extensively.[148] *S. oualaniensis* is an oceanic squid common to the western Pacific and Indian Oceans. The squid possesses a large oval organ (major axis, ca. 2 cm) on the anterodorsal surface of the mantle that emits a bright flash of blue light when stimulated. The organism is yellow and consists of numerous ovate [0.4–1.1 mm (major axis) $\times$ 0.3–0.5 mm (minor axis)] photogenic granules. Studies of the biochemical aspects of the luminescence of *S. oualaniensis* revealed that sodium or potassium ions as well as molecular oxygen were necessary for the luminescence,[149] suggesting involvement of a photoprotein different from aequorin (Section 8.07.6.2). The chromophores of this bioluminescence system were isolated from the acetone-powder of the photogenic organs of *S. oualaniensis* and identified as dehydrocoelenterazine (**195**) and its acetone adduct (**196**). Dehydrocoelenterazine (**195**) was postulated to be equivalent to luciferin in this luminescent system and assumed to be stored in a form of conjugated adduct (**197**).[150,151] Furthermore, the photoprotein in *S. oualaniensis* was extracted in high salt solutions at low temperature.[152] It was found that only the high molecular weight fraction from gel filtration chromatography had luminescence activity, suggesting that the luminescent substance is a photoprotein. No dehydrocoelenterazine (**195**) was, however, detected after the luminescence (470 nm) of the homogenate by addition of KCl, indicating that the bioluminescence of *S. oualaniensis* consumed dehydrocoelenterazine (**195**). On the other hand, three solutions containing dehydrocoelenterazine (**195**), its dithiothreitol (DDT) adduct (**198**), and glutathione (GSH) adduct (**199**) were added, respectively, to the protein fraction after the luminescence, and light emission was immediately observed in all three solutions. These findings recorded the reconstruction of the luminous system in this squid.

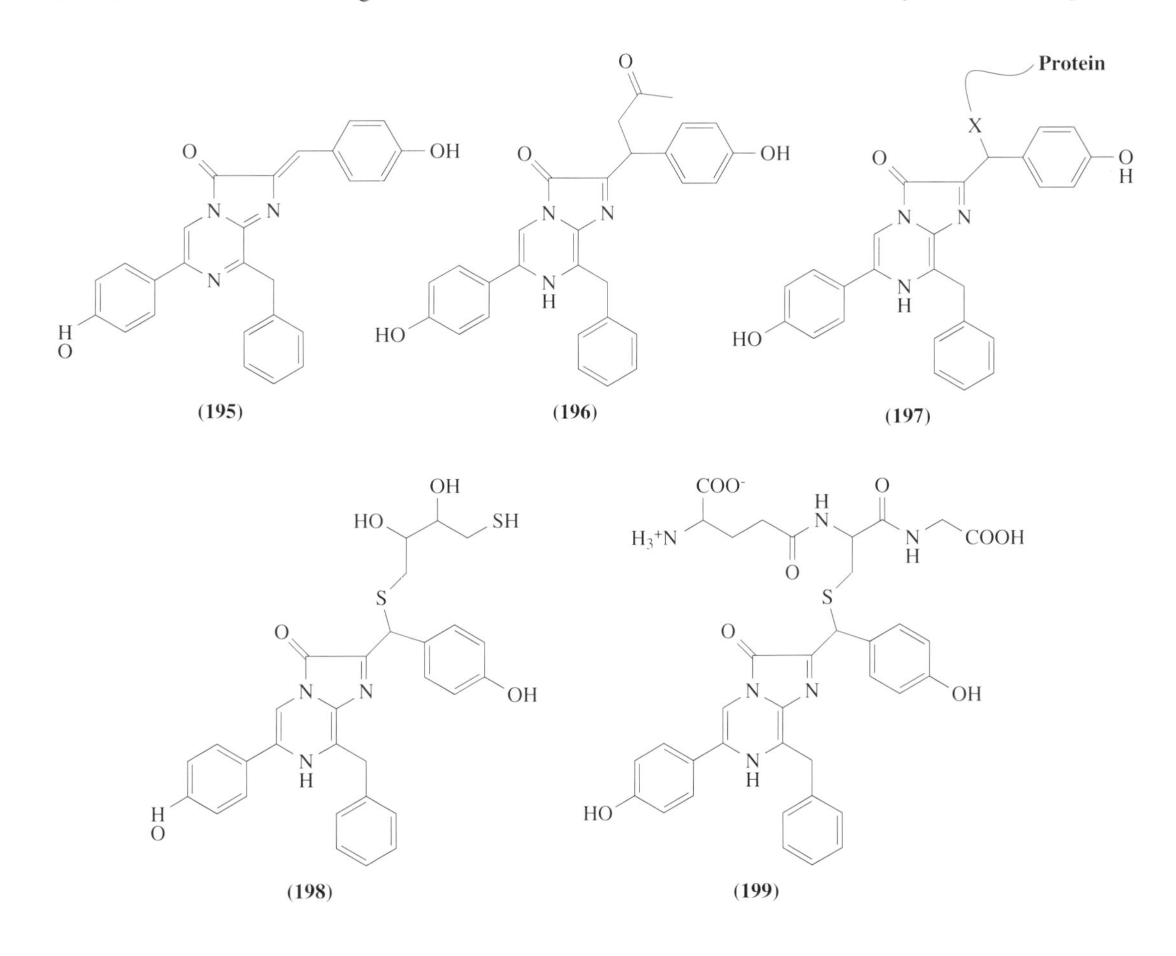

The involvement of a common structure of 3,7-dihydroimidazo[1,2-*a*]pyrazin-3-one (**200**) is known in many marine luminescent systems such as *Vargura*, *Watasenia*, and *Aequorea* bioluminescence, and coelenterazine is the most common luciferin among these systems. Although the luminescence mechanism with these luciferins has been well investigated, no direct evidence of the peroxide intermediate (e.g., Scheme 4) was demonstrated until 1995 due to its unstable nature. In 1995, Usami and Isobe succeeded in collecting evidence of the intermediates of luminescence by synthesizing the peroxide at lower temperatures.[153] They prepared a model compound (**201**) with a *t*-butyl group at the C-2 position enriched with ^{13}C atoms at C-2, C-3, and C-5 positions ((**201a**), (**201b**), and (**201c**), respectively), which were photooxygenated in an NMR tube at $-78\,^{\circ}C$ in CF_3CD_2OD-CD_3OD solution and its ^{13}C NMR spectrum was recorded at that temperature. The ^{13}C NMR spectrum revealed that the signal of the starting coelenterazine analogue (**201**) diminished to half intensity after 3 min photo-irradiation, and to 1/5 after a further 3 min irradiation. The first 3 min irradiation produced signals for two newly formed compounds (**202**) and (**203**), while 6 min irradiation resulted in the higher intensity of the signals for (**202**). When the mixture was warmed, its ^{13}C NMR spectrum showed signals due to the stable photoproduct (**204**) corresponding to coelenteramide (**164**). Assignment of the ^{13}C NMR signals of these compounds are summarized in Table 16. These results provide direct evidence for the generation of the peroxide (**202**) and afforded the first proof of 1,2-dioxetanone (**203**) as a luminescence intermediate.[154] Teranishi and co-workers also reported photooxygenation of (**201**) at $-95\,^{\circ}C$ sensitized with polymer-bound Rose Bengal in CD_2Cl_2 solution, in which the peroxide (**202**) was detected as a major product from the ^1H NMR spectrum at low temperature (below $-50\,^{\circ}C$), while dioxetanone (**203**) was not detected and on warming to $-20\,^{\circ}C$ the amide (**204**) was present almost exclusively.[155]

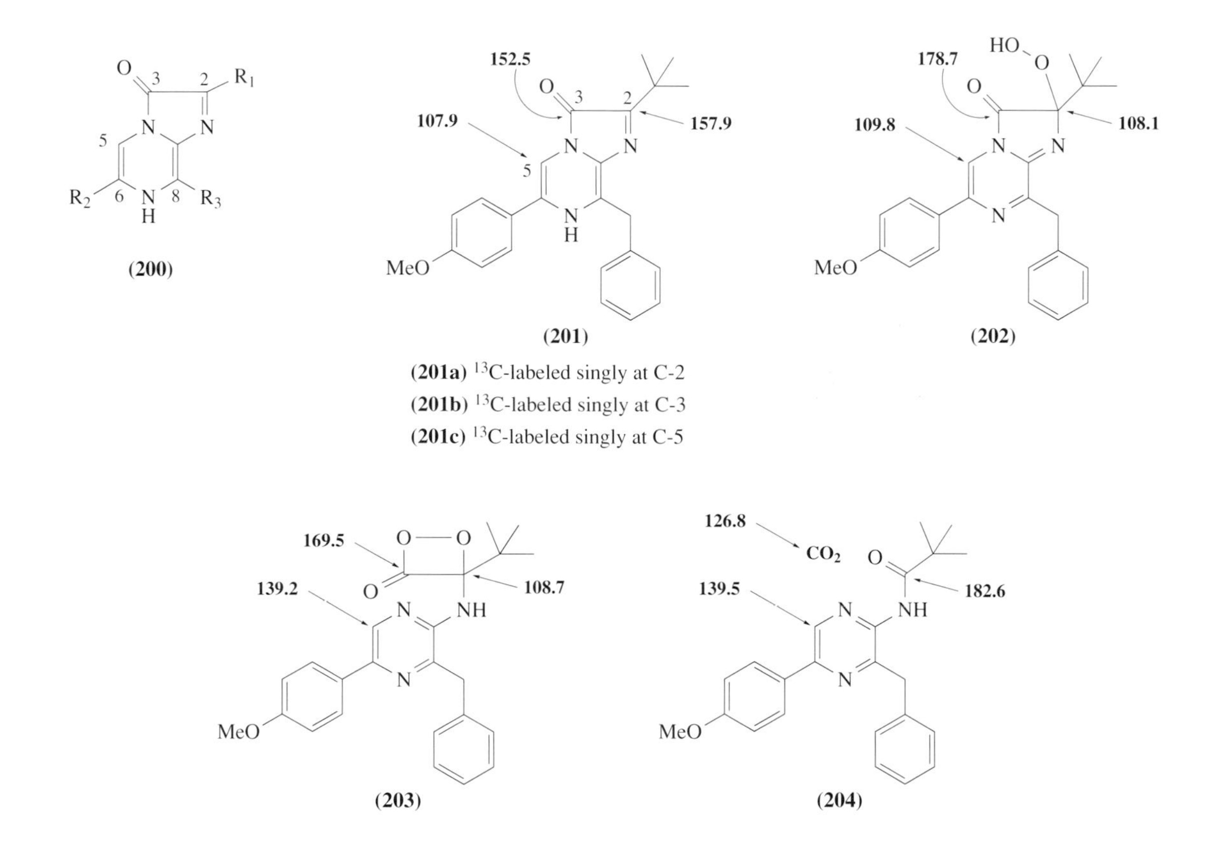

Table 16 ^{13}C NMR data of the coelenterazine analogue (**201**) under photoirradiation experiments.

	C-2	*C-3*	*C-5*	*Assigned structure*
Before irradiation	157.9	152.5	107.9	(**201**)
After photoirradiation	108.1 108.7	178.7 169.5	109.8 139.2	(**202**) (**203**)
After warming	182.6	126.8	139.5	(**204**)

Imidazopyrazinones represented as (**200**) are widely utilized by bioluminescent organisms in the marine environment. Imidazopyrazinones with conjugated substituents at the 6,8-positions were synthesized from the corresponding 3,5-disubstituted-2-aminopyrazines which were prepared from commercially available 2-aminopyrazine by a sequence of reactions including Pd-mediated Stille coupling as a key step.[156] The synthesized imidazopyrazinones showed chemiluminescence. By applying the same method, a benzofuran analogue (**205**) of the *Vargula* luciferin (**159**) was prepared, which reacted with luciferase to give a luminescence but the light yield was about 1/25, which may be due to its poor chemiluminescence.[157]

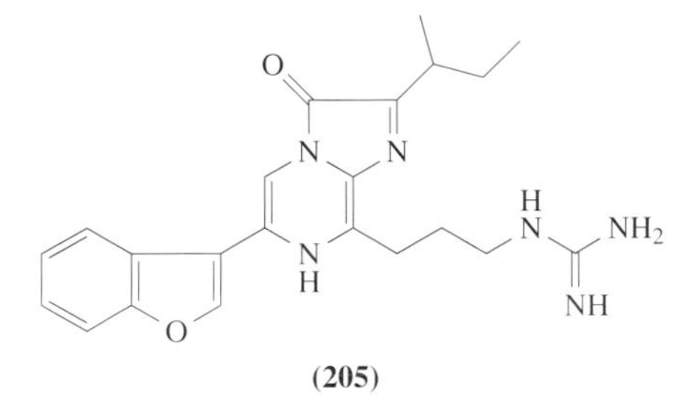

(**205**)

8.07.6.4 Microalgae

Dinoflagellates are ubiquitous microalgae in the oceans and they are responsible for much of the sparkling luminescence elicited at night by disturbing surface waters. The chemical process of light emission involves air oxidation of dinoflagellate luciferin catalyzed by dinoflagellate luciferase. The dinoflagellate bioluminescence system was shown to cross-react with the krill (euphausid shrimp) bioluminescent system,[158] composed of a photoprotein and krill fluorescent substance F. Dinoflagellate luciferin and krill fluorescent substance F share common chemical and spectroscopic properties, and they are both unstable in the presence of oxygen. Krill fluorescent substance F (**206**) was successfully isolated from the krill *Euphausia pacifica* by careful operations using alumina and ion-exchange chromatography at low temperature under an inert atmosphere, and the structures of (**206**) and its air-oxidation product oxy-F (**207**) were elucidated on the basis of chemical degradations and spectroscopic data.[159] Dinoflagellate luciferin (**208**) was isolated from cultured *Pyrocystis lunula*, and its structure was elucidated by comparing the spectroscopic data with those of krill fluorescent substance F (**206**). From the crude extract of luciferin, the air-oxidation product (**209**) with a characteristic blue color was isolated. The nonenzymatic oxidation of dinoflagellate luciferin (**208**) in methanol afforded (**210**) without emission of light, whereas air oxidation in the presence of luciferase proceeded with the emission of light at 474 nm to yield (**211**).[160,161] The absolute stereochemistry of three chiral centers on the D-ring of krill fluorescent substance F (**206**) was established as 17*S*, 18*S*, and 19*S*, respectively, on the basis of chiral HPLC comparison of the ozonolysis product with synthetic compounds, and this result suggested that chlorophylls are the biogenetic origin of krill fluorescent substance F (**206**).[162]

Bioluminescent organisms have thus been investigated extensively and some answers to the question of how they emit light have been obtained, whereas it is still not clear why they emit light.[163] In connection with this subject, the circadian rhythms of microalgae have been studied.[164] In the unicellular marine dinoflagellate *Gonyaulax polyedra*, bioluminescence as well as other biological phenomena including cell division, motility, and photosynthesis exhibits circadian rhythmicity which may persist under constant conditions with a precise period of about 24 h. Using the circadian rhythm of bioluminescence as a bioassay, the period of free-running circadian rhythms in *Gonyaulax* was revealed to be shortened by extracts from mammalian cells. The effect was dose-dependent, accelerating the circadian clock by as much as 4 hours per day. The substance responsible for this effect was isolated from bovine muscle and identified as creatine (**212**). Authentic creatine (**212**) had identical biological effects at micromolar concentrations.[165] A period-shortening substance with similar chemical properties was found also to be present in extracts of *Gonyaulax* itself. The endogenous active substance, termed gonyauline (**213**), was isolated and characterized as a low molecular weight cyclopropanecarboxylic acid.[166] Synthetic ($\pm$)-gonyauline (**213**) also had a similar accelerating effect on the period of the circadian clock.[167] The optically active natural (+)-gonyauline (**213**) as well as its enantiomer, ($-$)-gonyauline, was synthesized by procedures including optical resolution with the aid of brucine, and the absolute configuration of two chiral centers of natural (+)-gonyauline (**213**) was determined as 1*R*, 2*R* by applying the exciton chirality method to the CD spectrum of the bis-α,β-unsaturated ester (**214**), which was prepared from a chirally resolved synthetic intermediate.[168]

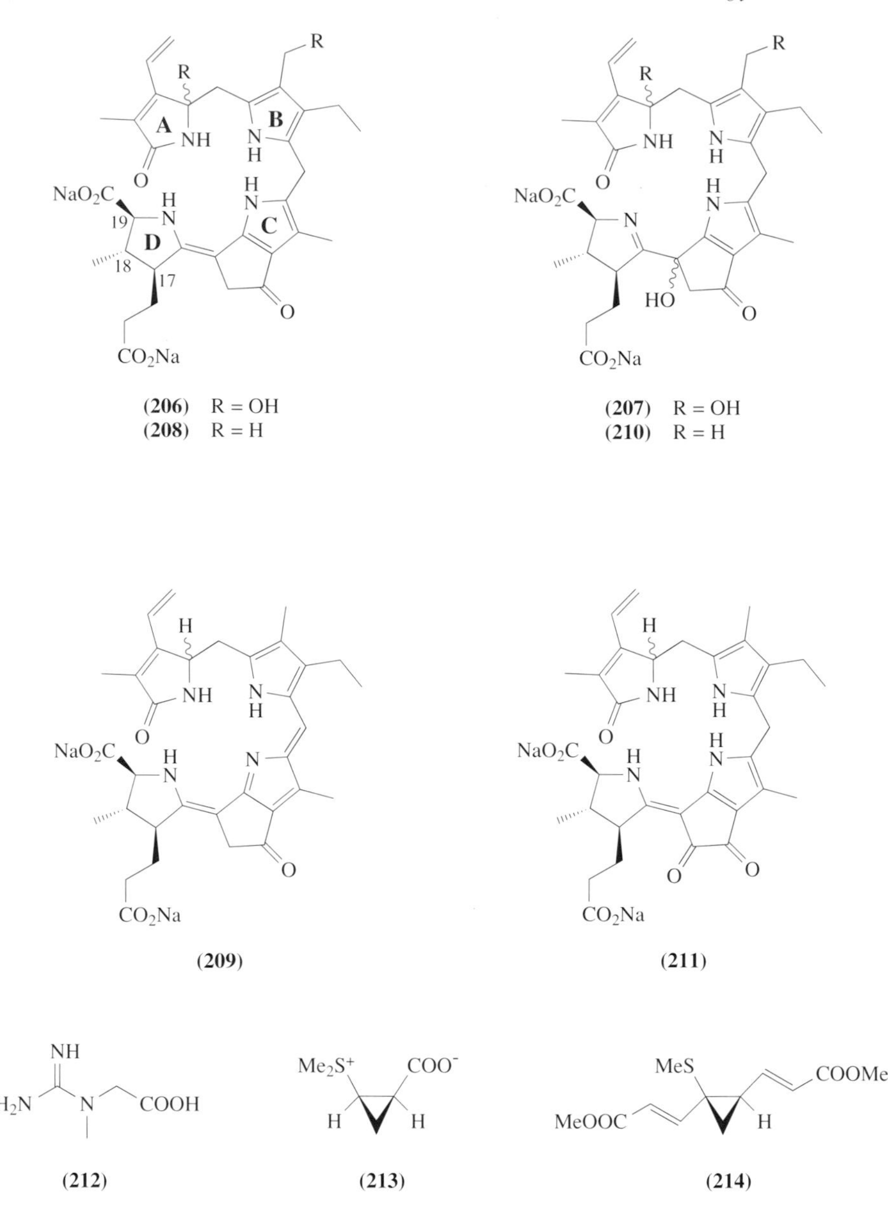

From the structural similarity between gonyauline (**213**) and methionine (**215**) it was estimated that gonyauline (**213**) might be derived directly from methionine (**215**) by methylation and deamination–cyclopropanation reactions accompanied by an inversion of configuration at the C-2 position. Thus, feeding experiments of labeled methionine or its analogues to dinoflagellate *G. polyedra* were carried out. As a result, methyl groups of methionine or *S*-methylmethionine were efficiently incorporated into a sulfonium methyl group of gonyauline (**213**). However, carboxyl carbons of methionine or *S*-methylmethionine at the C-1 position were not incorporated into gonyauline (**213**) even at a high concentration. Thus, it was revealed that gonyauline (**213**) was not derived through direct methylation and deamination–cyclopropanation reactions of methionine (**215**).[169,170] Feeding [2,3-$^{13}C_2$]-methionine, which was synthesized from 2-methylpentane-2,4-diol and [1,2-$^{13}C_2$]-acetonitrile, to *G. polyedra* afforded ^{13}C-labeled gonyauline (**213**) at C-1 and C-3 positions, suggesting that gonyauline (**213**) were derived biogenetically from methionine (**215**) through decarboxylation and elongation of C_1 carbon (Scheme 7).[170] During these biogenetical studies of gonyauline (**213**), another sulfonium compound, gonyol (**216**), was found to be accumulated in *G. polyedra*, and its structure was elucidated by spectroscopic methods and confirmed by chemical synthesis.[169] Feeding experiments showed that gonyol (**216**) was also biogenetically derived from methionine (**215**), but the carboxyl carbon of methionine (**215**) was not incorporated into gonyol (**216**). However, the C-1 and C-2 of gonyol (**216**) were labeled with [1,2-$^{13}C_2$]-sodium acetate in the presence of methionine (**215**). Gonyol (**216**) obtained by the feeding experiment with [2,3-$^{13}C_2$]-methionine was enriched with ^{13}C at C-3 and C-4 positions. These results implied that gonyol (**216**) might be biogenetically derived from methionine (**215**) and acetate through dimethyl-*β*-propiothetin (**217**) or its analogous intermediates (Scheme 7). Although gonyol (**216**) was found as a minor component in *G. polyedra* under normal culture conditions, other dinoflagellates such as *Amphidinium* sp. and *Symbiodinium* sp. contained gonyol (**216**) at various levels from a trace to one of the major components. It seemed likely that dinoflagellates contain various sulfonium compounds and dimethyl-*β*-propiothetin (**217**) might be a common precursor for a methionine cascade to sulfonium compounds including gonyauline (**213**) and gonyol (**216**).

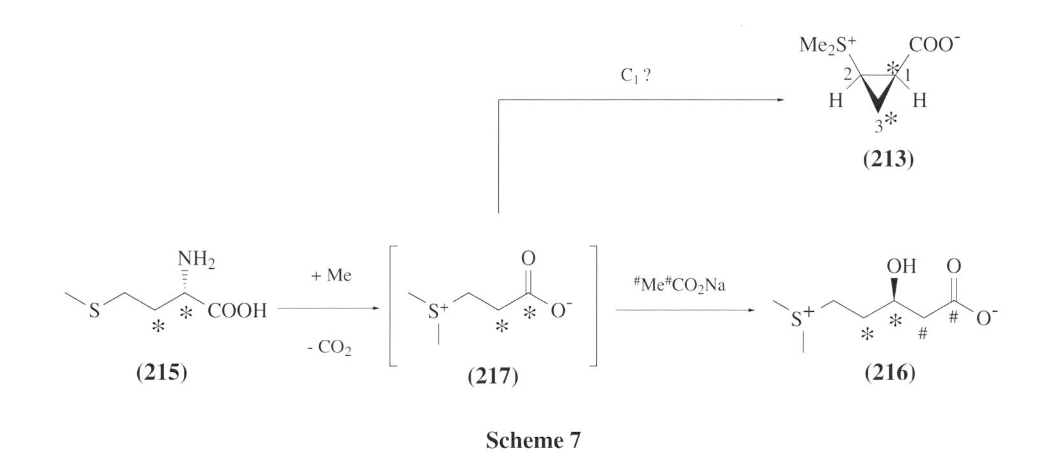

Scheme 7

8.07.7 CHEMICAL DEFENSE INCLUDING ANTIFEEDANT ACTIVITY

There are many soft-bodied and benthic organisms in the marine environment. These organisms appear not to have physical defense mechanisms, while most of them contain unusual secondary metabolites and these secondary metabolites are assumed to have a defensive function. These defensive functions are termed "chemical defense." A number of good reviews have been published on this subject (e.g., on chemical defense of tropical marine algae,[171] nudibranchs,[172] alcyonarian corals,[173] fish,[174] and others[175–178]). In this section, we survey several reports on this subject.

8.07.7.1 Algae

The major function of chemical defense by marine natural products may be feeding deterrence. From the neutral methanol extract of the brown alga *Dilophus okamurai*, two spatane-type diterpenes (**84**) and (**85**) were isolated as inhibitors of the settlement and metamorphosis of the swimming larvae (veliger) of the abalone *Haliotis discus hannai*.[88] The diterpenes (**84**) and (**85**) were also found to be strongly active feeding deterrents for the young abalone. For this bioassay, the Avicel plate method developed by Sakata and co-workers was used (see Section 8.07.2.2). The ethanol solutions (25 μl) of a standard phosphatidylcholine (PC) (10 μg) and the samples, which were prepared by mixing 100 μg of each of the fractions and pure compounds with PC (10 μg), were applied with a microsyringe onto the sample zone (25 mm in diameter) on an Avicel plate. The feeding-deterrent activity of each sample was evaluated by comparing the number of biting traces left on the plates with that of the standard PC. Further studies on feeding-deterrent substances of this alga led to the isolation of seven active compounds related to spatane-type diterpenoids (**218**)–(**224**).[179–181] The feeding-deterrent activity of these compounds was evaluated by the Avicel plate method to reveal the following observations. Compounds (**84**), (**85**), (**219**), and (**222**) exhibited comparably strong feeding-deterrent activity, and compounds (**220**) and (**224**) exhibited moderate activity. Compounds

(**218**) and (**223**) showed weak activity, while compound (**221**) showed very weak activity. It was interesting that although compounds (**218**) and (**223**) exhibited weak feeding-deterrent activity against the young abalone *Haliotis discus hannai*, these metabolites were found to be the strongest feeding deterrents against the sea urchin *Strongylocentrotus nudus*.[181]

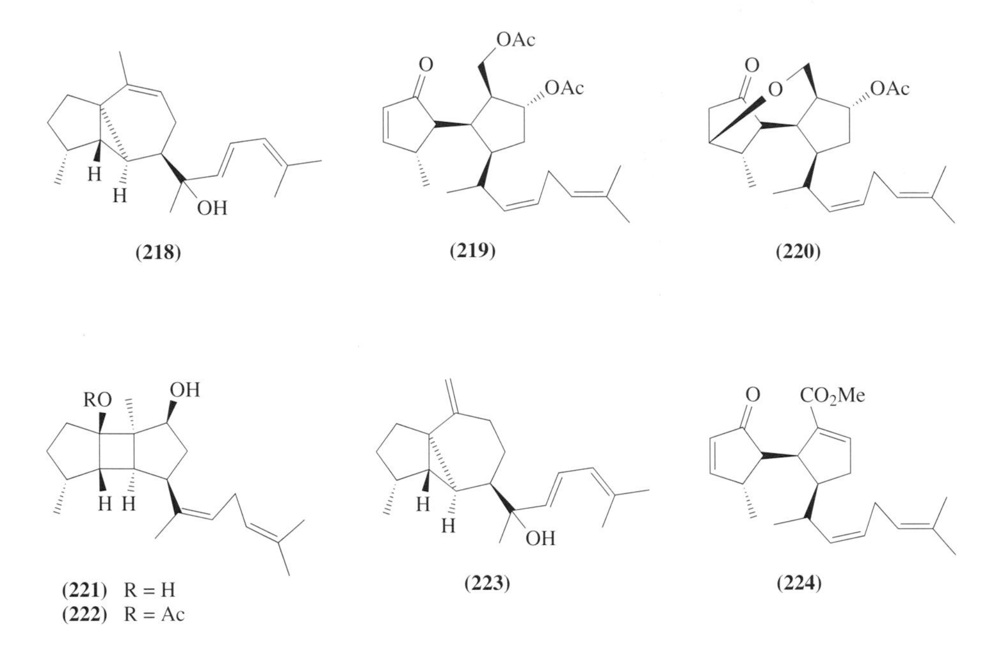

From the brown alga of the genus *Dictyopteris*, collected at Akita Prefecture, Japan, five sesquiterpenoid derivatives, zonarol (**225**), isozonarol (**226**), zonarone (**227**), isozonarone (**228**), and chromazonarol (**229**) were isolated as feeding deterrents against abalones, and the feeding-deterrent activity of these sesquiterpenoids was considerably stronger than that of diterpenes obtained from *Dilophus okamurai*.[182]

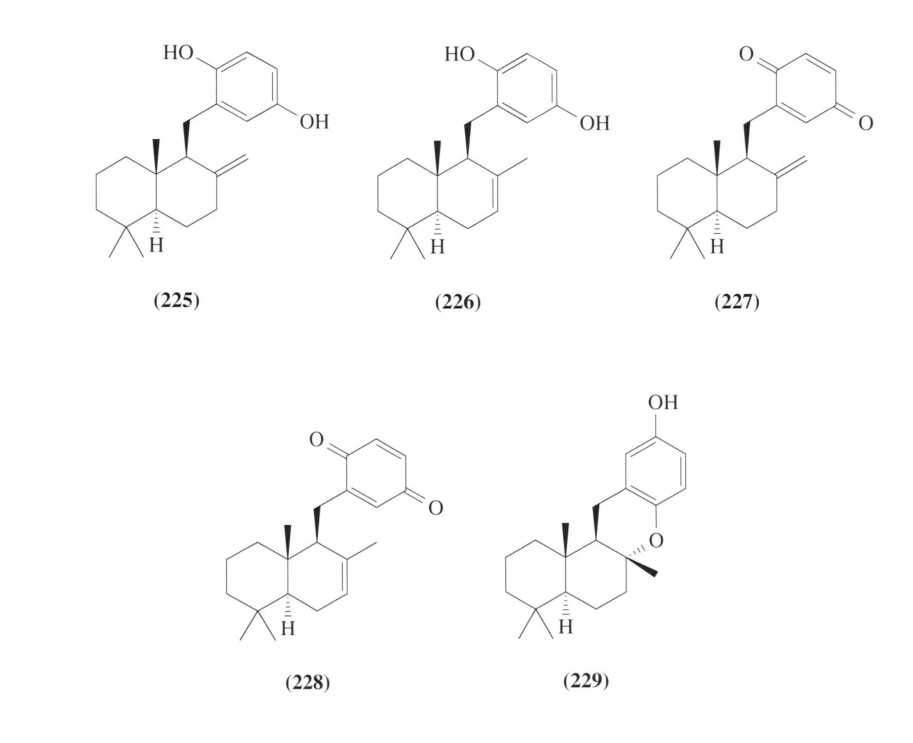

Feeding deterrents of brown alga of the order Laminariales against abalones or sea urchins were identified as water-soluble phenols such as phloroglucinol (**230**) and its oligomers (**231**) and (**232**).[182]

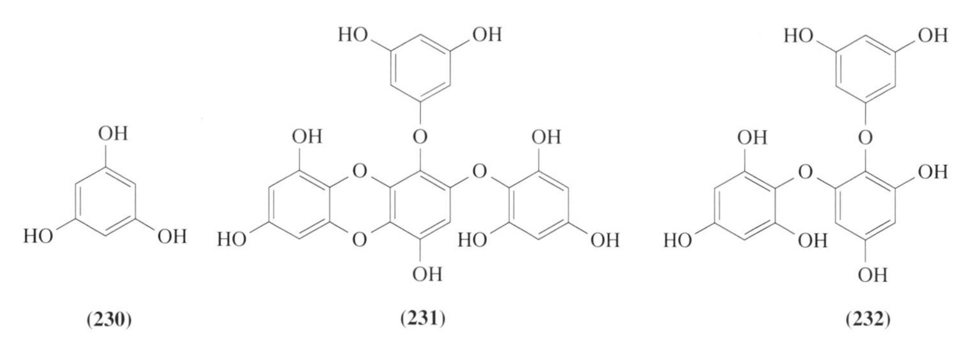

The red alga *Rhodomela* sp. contained bromophenols **(233)**–**(235)** as feeding deterrents against sea urchins, but a co-isolated sulfate (**237**) was inactive, while feeding deterrents of red alga of the genus *Laurencia* against sea urchins was identified as bromine-containing diterpenes **(238)**–**(243)**.[182]

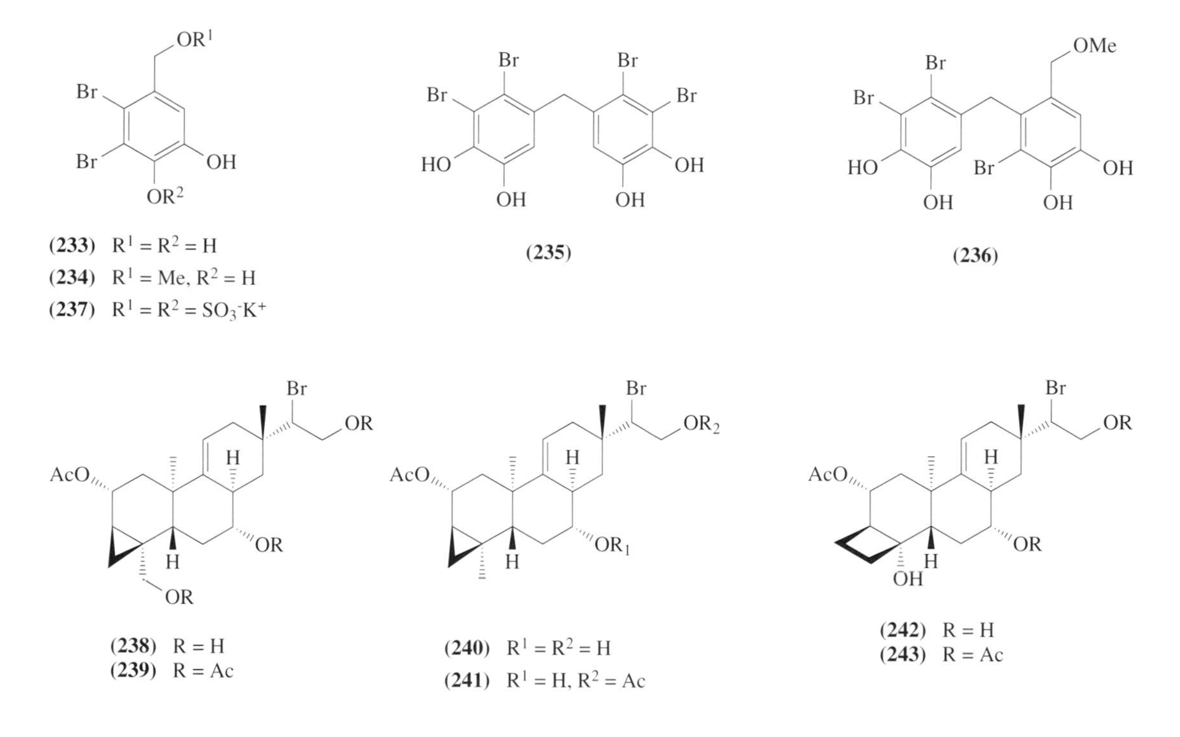

Three prenylated phenols, sporochnols A–C (**244**)–(**246**), were isolated from the Caribbean marine alga *Sporochnus bolleanus*.[183] The major metabolite (**244**) showed significant feeding deterrence toward herbivorous fishes. When sporochnol A (**244**) was incorporated (at 80% of its natural yield) into agar strips containing palatable, freeze-dried algae, it reduced parrotfish consumption of treatment strips by a significant 27% compared with otherwise equivalent control strips. This compound, however, had no effect on feeding by the sea urchin *Diadema antillarum* or the amphipod *Cymadusa filosa*. Because herbivory by reef fishes appears to be the major factor selecting for herbivore defense in tropical seaweeds, it is not unusual to find seaweed defensive metabolites that are effective primarily against fish. Thus, from this observation, the sporochnols appear to function as defensive agents in *Sporochnus bolleanus*.

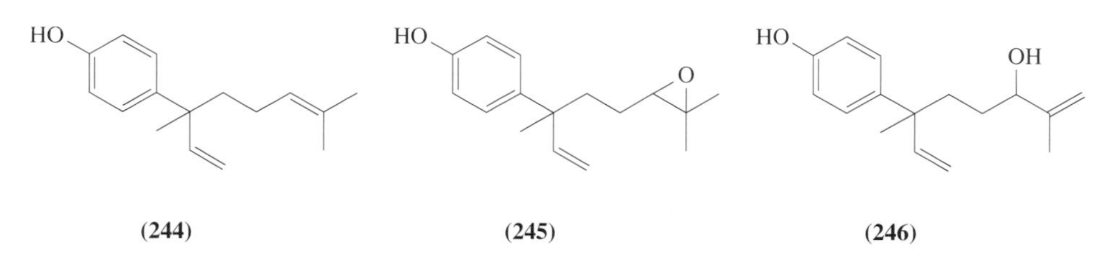

One type of facultative defense of marine organism that has been recognized is the production of predator-induced defenses. This type of defense operates when an attack by a predator acts as a cue for stimulating the synthesis of new or additional defensive compounds. Another type of related facultative or inducible chemical defense is the rapid conversion of one secondary metabolite to another more potent defensive compound upon injury to the organism. The latter type of defense mechanism eliminates the need for an organism to maintain high levels of very deterrent and biologically active substances which may be toxic to the organism itself. Thus, the risk of autotoxicity is minimized. As an example of this type of chemical defense, activation of chemical defense in the tropical green algae *Halimeda* spp. was described.[184] *Halimeda* spp. are among the most common seaweeds on tropical reefs and these seaweeds produce diterpenoid feeding deterrents; the major metabolites were halimedatetraacetate (**247**) and halimedatrial (**248**). It was observed that most species of *Halimeda* on Guam immediately convert the less-deterrent secondary metabolite halimedatetraacetate (**247**) to the more potent feeding deterrent halimedatrial (**248**) when plants are injured by grinding or crushing. This conversion would therefore occur when fish bite or chew *Halimeda* plants. Extracts from injured plants contained higher amounts of halimedatrial (**248**) and were more deterrent toward herbivorous fish than extracts from control plants.

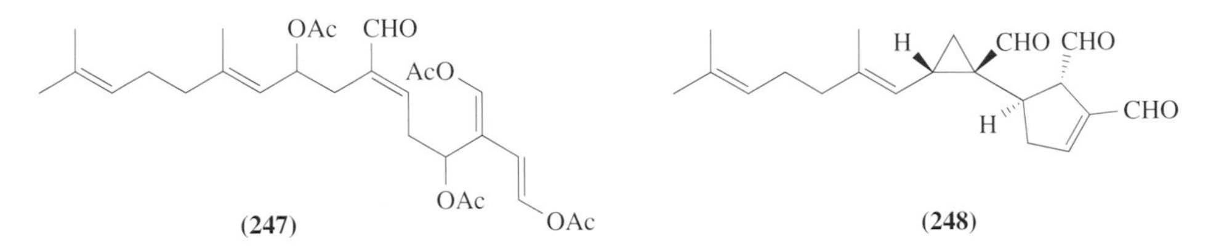

8.07.7.2 Mollusks

Naked nudibranch mollusks exhibit a series of defensive strategies against potential predators, which include the use of chemicals obtained either from the diet or by *de novo* biosynthesis.[172] Two Mediterranean species of *Dendrodoris* nudibranchs have elaborated a very sophisticated defensive strategy against predators. Polygodial (**249**), which was initially known as an antifeedant for insects of the African plant *Warburgia stuhlmanni*,[185] was identified as a defensive allomone of two Mediterranean nudibranchs, *Dendrodoris limbata* and *D. grandiflora*.[186] Polygodial (**249**) was revealed to be an antifeedant to fish, anorectic to insects, and hot-tasting to humans, and these biological activities are most probably due to the simultaneous interaction of both the aldehydic groups with primary amine moieties.[187,188] Polygodial (**249**) was found to be toxic to *D. limbata*,[189] and it was suggested that a related sesquiterpenoid olepupuane (**250**) was the masked form of the allomone present in the animal. The anatomical distribution of the drimane sesquiterpenoids such as (**249**) in different sections and egg masses of the mollusk was investigated. As a result, drimane sesquiterpenoids were found only in the yellow mantle border, in the gills, in the hermaphrodite glands, and in the egg ribbons.[190] TLC analysis of the mantle border revealed the presence of polygodial (**249**), clearly deriving from olepupuane (**250**). Compounds (**249**) and (**250**) were completely absent in other organs. Drimane esters (**251**) with C_{18}–C_{20} fatty acids and euryfuran (**252**) were localized in hermaphrodite glands and egg masses, while the gills possessed 7-deacetoxy-olepupuane (**253**). No drimane sesquiterpenoids were detected in the mucous secretion of *D. limbata*. Analogous studies on *D. grandiflora* showed almost identical results. It seems likely that *Dendrodoris* mollusks have elaborated a very effective mechanism to secure their own survival. Related compounds are localized in different organs of the mollusk, and perform different biological roles. 7-Deacetoxy-olepupuane (**253**) could be the precursor of the defensive allomones, olepupuane (**250**) and polygodial (**249**), whereas the esters (**251**) might play a role during the reproductive cycle.[190]

The ascoglossan mollusk *Cyerce cristallina* has a typical defensive behavior known as autotomy. This animal possesses aposematically colored dorsal appendices called "cerata." When the mollusk is attacked by predators, the cerata are detached from the mantle and exhibit prolonged contractions while secreting large amounts of a toxic mucous secretion. After the autotomic process, the animal provides a striking example of regeneration by completely reproducing the cerata within only 7–10 days. In connection with this phenomenon, seven pyrones, cyercenes A, B, and 1–5 (**254**)–(**260**), were isolated from the cerata of the mollusk *C. cristallina*,[191] and the possible correlation between these pyrone metabolites and the process of chemical defense and regeneration was studied by investigation of the tissue distribution, biological activity, and biogenesis of cyercenes.[192]

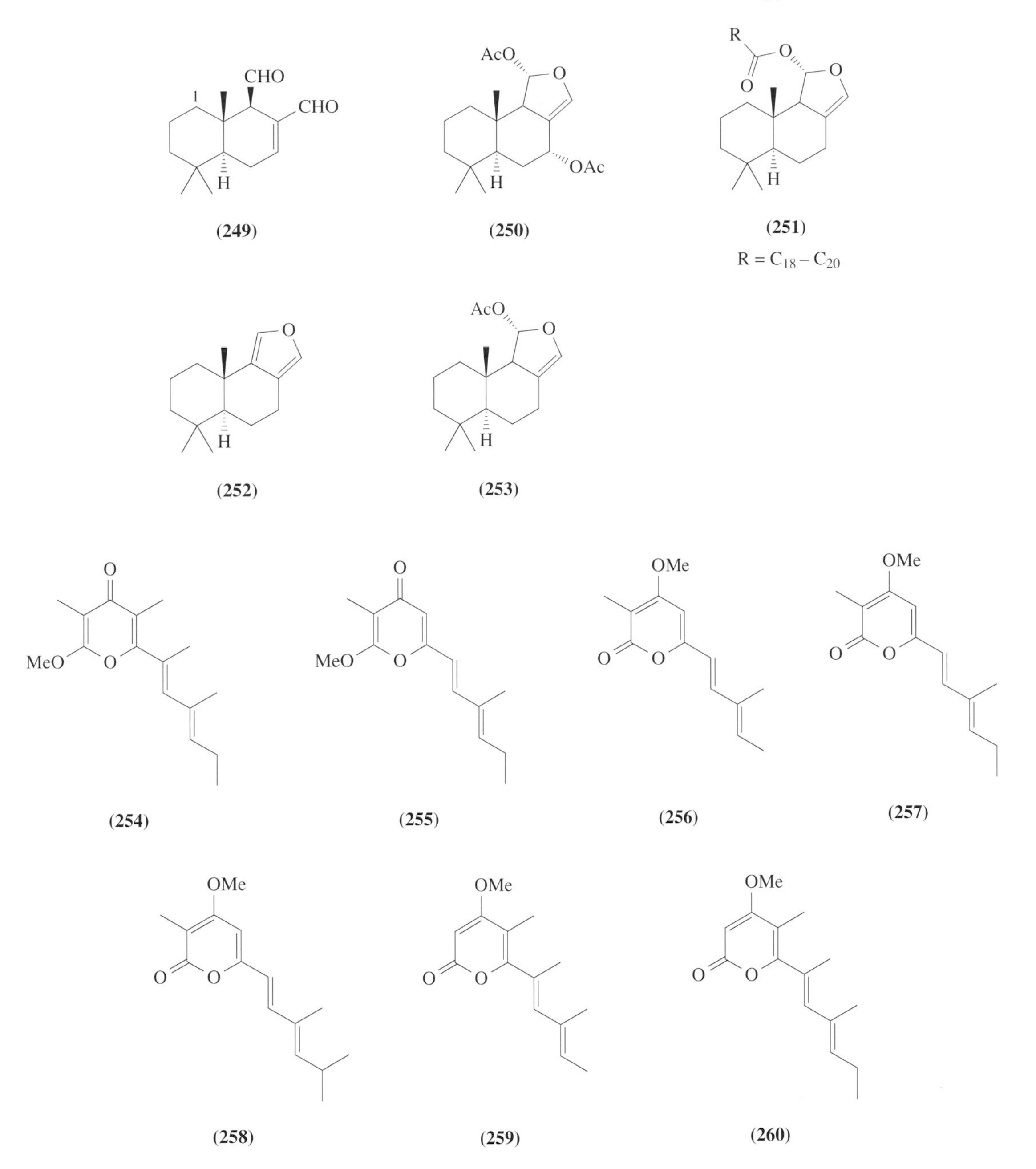

HPLC analysis showed differences in chemical composition between the *C. cristallina* mantle, cerata, digestive gland, and mucous secretion. No cyercene was contained in the digestive gland, while only cyercenes 1 (**256**), 2 (**257**), and 3 (**258**) were found in the mantle. The mucus contained all cyercenes except cyercene A (**254**), and cyercene A (**254**) was only present in the cerata, which contained all seven compounds. The presence of cyercene A (**254**) only in the regenerating tissue, the cerata of *C. cristallina*, may suggest the possible involvement of this compound in the quick regenerative mechanisms. The regeneration-stimulating activity of cyercenes was assayed by a model test of cell growth and differentiation factors, using the *Hydra vulgaris*, which had a comparable speed (normally 8 days) in the regeneration of its head and tentacle to that of *C. cristallina* cerata (7–10 days). Cyercene A (**254**) exhibited 69% enhancement of the *H. vulgaris* average tentacle number at 15 $\mu g\ ml^{-1}$, suggesting that compound (**254**) plays a biological role as a pivotal growth-inducing factor. It was, however, reported that other molecules that act as the main inducing factors of the rapid regeneration of *C. cristallina* may be present in hydrophilic extracts of the mollusk.[191] The cyercenes were contained in the toxic mucus secretion of *C. cristallina*, thus suggesting that these compounds may be involved in the chemical deterrence. Two pyrones (**261**) and (**262**) structurally similar to cyercenes were isolated from the Australian ascoglossan *Cyerce nigricans*, but they were reported to lack the potent ichthyodeterrent properties.[193]

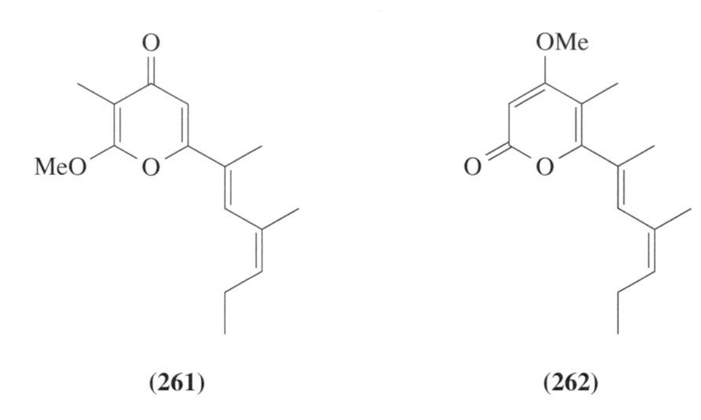

The ichthyotoxicity of the mucus secretion of *C. cristallina* as well as the purified cyercenes was examined by using the mosquito fish *Gambusia affinis* to reveal that all five cyercenes tested (compounds (**254**), (**255**), and (**257**)–(**259**)) and the mucus crude extract were toxic at a concentration of 10 μg ml^{-1}. Thus, cyercenes may be involved in chemical defense in *C. cristallina*. The complete absence of cyercenes and any other related metabolites in the extract of the digestive gland suggested that these compounds were not derived from dietary sources but were biosynthesized *de novo* by *C. cristallina*, and this suggestion was confirmed by an isotope incorporation experiment. The cerata of the mollusks were detached and the "naked" mollusks were incubated in seawater containing [^{14}C]-sodium propionate. Incorporation of almost 60% of the radioactivity was observed in the fractions containing the seven cyercenes isolated by the feeding experiment.[192]

The opisthobranch mollusk *Tethys fimbria* also detaches the dorsal appendices (cerata) during the behavioral defense mechanism known as autotomy. From this nudibranch mollusk prostaglandin(PG)-1,15-lactones of the E series, PGE_3-1,15-lactone 11-acetate (**263**), PGE_2-1,15-lactone (**264**), and PGE_3-1,15-lactone (**265**), were isolated, and these were the first naturally occurring prostaglandin-1,15-lactones.[194] These lactones were contained in the mantle and the cerata of this mollusk, and prostaglandin-1,15-lactones of the F series (**266**) and (**267**) together with PGE_2-1,15-lactone 11-acetate (**268**) were also obtained from the mantle and cerata of the same animal.[195] HPLC analysis of the extracts of both the mantle and the cerata showed the presence of free acids, PGE_2 (**269**) and PGE_3 (**270**).

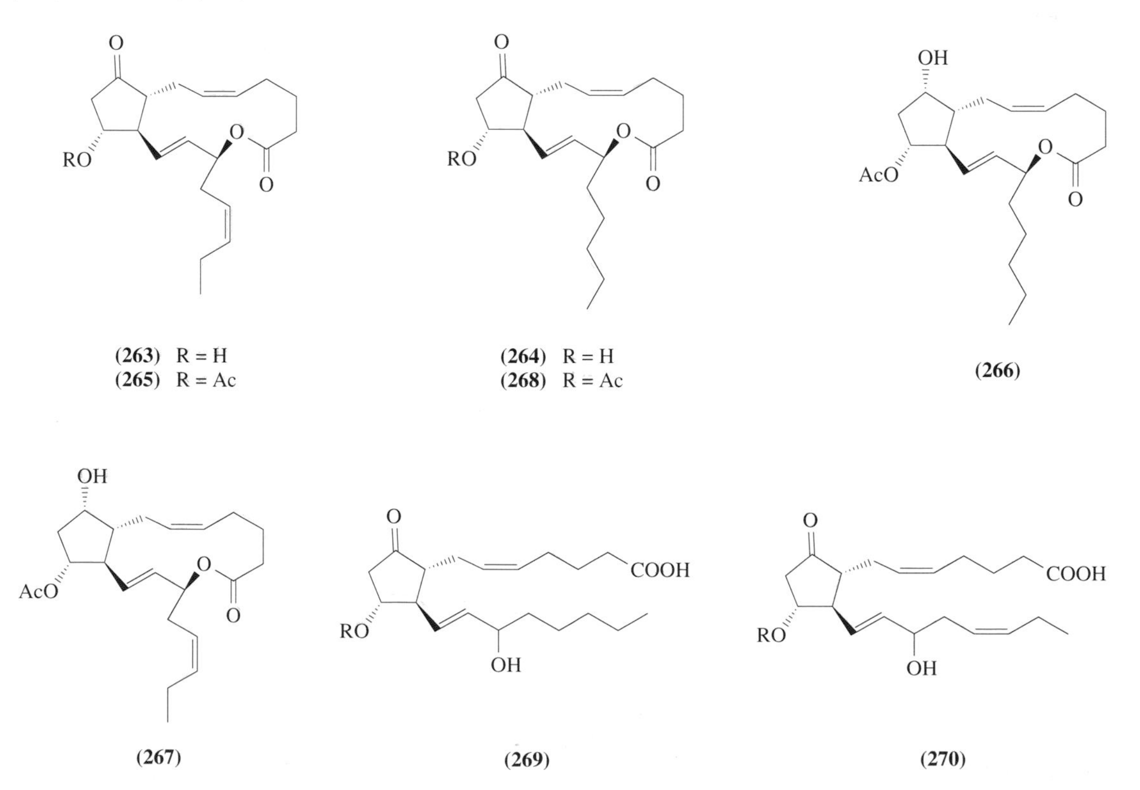

Thus, the biogenetic relationship between PG-1,15-lactones and PG-free acids and their biological roles in *T. fimbria* were investigated.[196] Incorporation experiments on [^{3}H]-PG-free acids revealed that PG-1,15-lactones are (i) synthesized from PG-free acids mainly in the mantle of this animal; (ii) mostly transferred into the cerata; and (iii) converted back into PG-free acids mainly upon detachment of the cerata from the molested mollusk and during their spontaneous contractions. This appears to be a unique mechanism by which PGs, in the form of structurally related compounds, are stored *in vivo* ready to be released in response to a mechanical stimulus to effect their biological action. HPLC analysis showed the absence of PG-free acids and the presence of PG-1,15-lactones in the defensive mucus secretion of *T. fimbria*. Ichthyotoxic activity assay against the mosquito fish *Gambusia affinis* revealed that PG-free acids did not exhibit any toxicity while the 1,15-lactones were toxic at concentrations of 1–10 μg ml^{-1}. From these findings, the nudibranch *T. fimbria* has developed a very economical way of exploiting PG lactones for more than one purpose: (i) as defense allomones in the mollusk defensive secretion; (ii) as inactive precursors of bioactive PG-free acids within the cerata; and (iii) as bioactive lactones in tissues other than the cerata and the mantle, as might be the case for PGF lactones, which are likely to play a role in the control of oocyte production and/or fertilization.[197]

From the Mediterranean cephalaspidean mollusk *Haminoea navicula* two 3-alkylpyridines, haminols A (**271**) and B (**272**) were isolated,[198] and these compounds had structural analogy with the alarm pheromones, in particular with navenone A (**273**), isolated from the Pacific Algajidae *Navanax inermis*.[199] Haminols A (**271**) and B (**272**) also induced alarm response at a concentration of 0.3 mg for (**271**) and 0.1 mg for (**272**).[198]

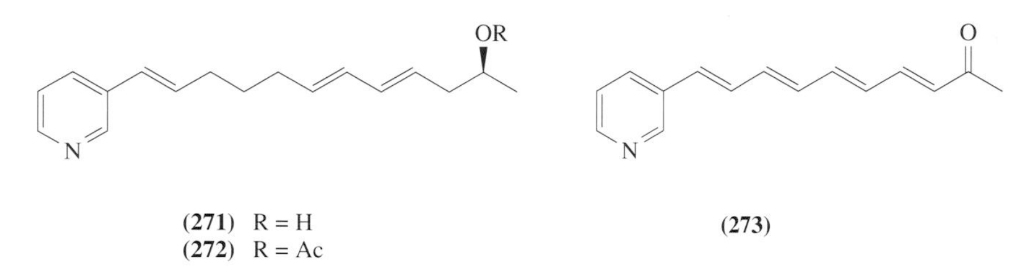

(**271**) R = H
(**272**) R = Ac

(**273**)

The common bright yellow Antarctic lamellarian gastropod *Marseniopsis mollis* was examined for the presence of chemical defense mechanism, and homarine (**274**) was isolated as a major component of ethanolic extracts of this animal.[200] Further HPLC analysis of the mantle, foot, and viscera verified the presence of homarine (**274**) in all body tissues at concentrations ranging from 6 to 24 mg g^{-1} dry tissue.

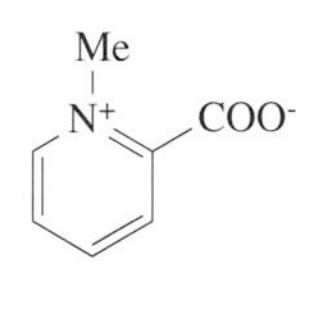

(**274**)

Filter paper disks treated with homarine (**274**) at 0.4 and 4 mg/disk were rejected by the sea star *Odontasier validus*, which was a conspicuous macroinvertebrate predator of the shallow Antarctic benthos. The highest concentration of homarine (**274**) tested not only caused feeding deterrence, but in several sea stars a flight response was noted. Homarine (**274**) was not detected in the Antarctic ascidian *Cnemidocarpa verrucosa*, a presumed primary prey of *M. mollis*. Homarine (**274**) was, however, contained in the epizooites that foul the tunic (primarily the bryozoans and hydroids), suggesting *M. mollis* may ingest and derive its chemicals from them. Since the vestigial internalized shell of *M. mollis* is considered as a primitive condition, the identification of homarine (**274**) as a feeding deterrent may lend support to the hypothesis that chemical defense evolved prior to shell loss in shell-less gastropods.

8.07.7.3 Sponges

Among various marine organisms, marine natural product chemists have found that sponges (Porifera) constitute one of the most interesting sources of bioactive substances. The high frequency of bioactive secondary metabolites in these primitive filter-feeders may be interpreted as the chemical defense of sponges against environmental stress factors such as predation, overgrowth by fouling organisms, or competition for space. Toxic or deterrent sponge metabolites are, consequently, frequently found in habitats such as coral reefs, and sponges growing exposed are usually more toxic than those growing unexposed.[201]

The palatability of crude organic extracts of 71 species of Caribbean sponges from reef, mangrove, and grassbed habitats was examined by laboratory feeding assays employing the common Caribbean wrasse *Thalassoma bifasciatum*. The majority of sponge species (69%) yielded deterrent extracts, but there was considerable inter- and intraspecific variability in deterrency. Reef sponges generally yielded more deterrent extracts than sponges from mangrove or grassbed habitats. There was no relationship between sponge color and deterrency, suggesting that sponges are not aposematic.[202] Sponge species with high concentrations of spicules in their tissues, and with variable spicule morphologies, were examined to assess the palatability to predatory reef fish, revealing that the presence of spicules did not alter food palatability relative to controls for any of the sponges tested. Analysis of the ash content, tensile strength, protein, carbohydrate, lipid content, and total energy content showed that the tissues of palatable sponges were not different from those of chemically deterrent species with regard to the mean ash content, mean tensile strength, protein content, carbohydrate content, and total energy content, but the tissues of chemically defended species did have a higher mean lipid content than those of palatable species.[203]

A striking example of the chemical defense of sponges through accumulation of toxic or deterrent secondary metabolites against fish was provided by the vividly red-colored sponge *Latrunculia magnifica* that was found in the Gulf of Aquaba (Red Sea). This sponge was avoided by fish that readily accepted other cryptic sponges when exposed. When *L. magnifica* was squeezed, the sponge exuded a reddish fluid that caused fish to flee immediately from the vicinity. It was observed that the sponge toxins contained in this juice led to excitation of fish (*Gambusia affinis*) after 5–10 min, jumping, partial paralysis, turning over onto their backs, hemorrhage of the gills, and finally death. Chemical analysis of the sponge afforded the macrocyclic lactones, latrunculins A (**275**) and B (**276**), that were responsible for the strong ichthyotoxicity of the sponge.[204] The toxicity of (**275**) and (**276**) was shown to be at least due to inhibition of acetylcholinesterase.

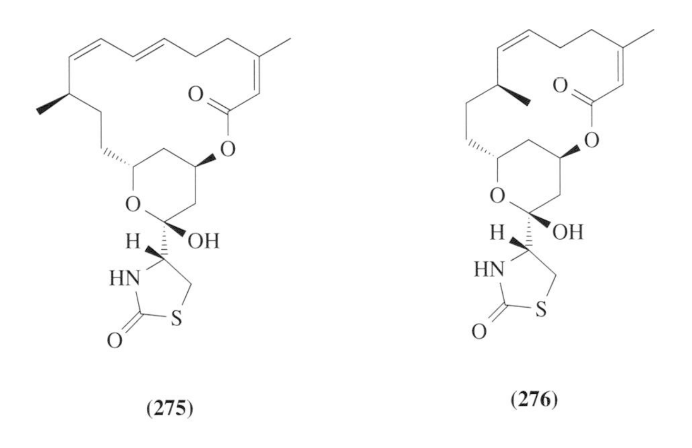

(275) **(276)**

From a Southern Australian sponge *Dysidea* sp. a sesquiterpene dialdehyde (**277**) was isolated together with related sesquiterpenes (**278**)–(**280**). The dialdehyde (**277**) is a 7,8-dihydro derivative of the sesquiterpene polygodial (**249**), which is known as a defensive allomone of nudibranchs (see Section 8.07.7.2). The antifeedant properties of polygodial (**249**) were thought to arise from the double condensation of the dialdehyde functionality with primary amines (e.g., lysine) to form pyrroles (e.g., (**281**)). Although the dialdehyde (**277**) does not incorporate an α,β-unsaturated aldehyde moiety such as in polygodial (**249**), it seemed feasible that (**277**) could condense with primary amines in much the same way as (**249**). Thus, it was proposed that the dialdehyde (**277**) may serve as an antifeedant for the *Dysidea* sp. and other sesquiterpene co-metabolites (**278**)–(**280**) may serve as "inactive" reserves or metabolite by-products, with (**277**) acting as the primary defensive allomone.[205] No biological experimental data, however, were described to support this proposal.

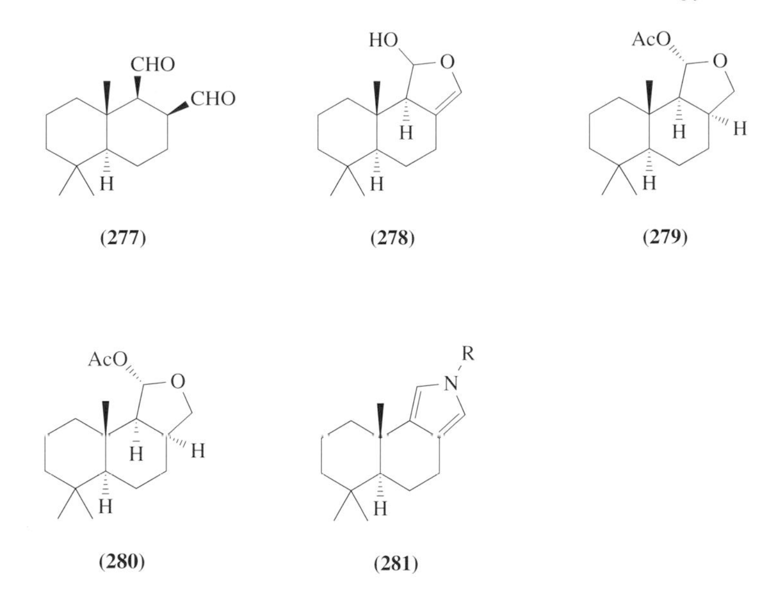

(277) **(278)** **(279)**

(280) **(281)**

Two dialdehyde sesterterpenoids, 12-deacetoxyscalaradial (**282**) and scalaradial (**283**), were isolated from a marine sponge *Cacospongia mollior* collected near Naples.[206] These two dialdehydes showed interesting features in their bioactivities. Both compounds possess the same dialdehyde functionality as polygodial (**249**).

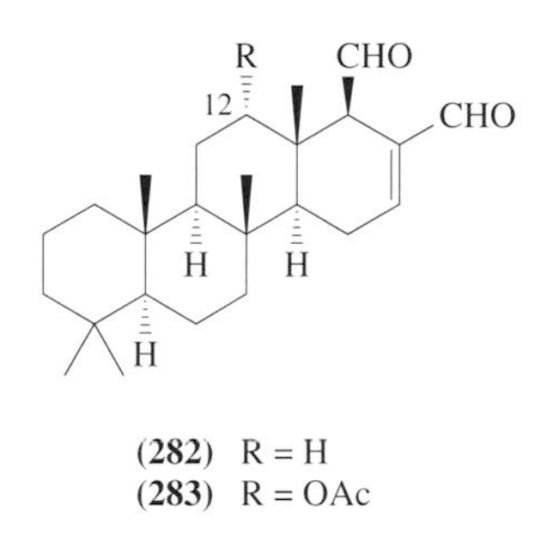

(**282**) R = H
(**283**) R = OAc

Polygodial (**249**) is known to have suitable distances between the two aldehyde groups and an adjacent intracycle unsaturation, and to exhibit antifeedant activity and a hot taste for humans (see Section 8.07.7.2). The scalaradial molecule (**283**), which embodies all these structural features, was tasteless and displayed the same antifeeding effects on fish as polygodial (**249**), but at twice the concentration of (**249**). On the other hand, 12-deacetoxyscalaradial (**282**) showed a similar biological response to polygodial (**249**). On comparing the activities of (**282**) and (**283**), the two compounds were active in the fish feeding inhibition bioassay at concentrations of 30 μg cm^{-2} and 60 μg cm^{-2}, respectively, while (**282**) only was hot to the taste. These results showed that molecular size was not a restrictive factor in these activities, but pointed out the specific importance of the substituent at C-12 in (**282**) and (**283**), or in the equivalent C-1 position of a supposed polygodial derivative. The presence of a bulky substituent in this position, such as the acetoxy group in (**283**), may inhibit the biological activity of the metabolite, either by altering the conformation of the nearby aldehyde group, or altering the surface complementarity between the molecule and the binding site of the receptor.

Chemical studies of the Caribbean sponge *Cacospongia* cf. *linteiformis* (order Dictyoceratida, family Thorectidae) led to the isolation and characterization of a number of bioactive cyclic sesterterpenes without dialdehyde functionality, namely lintenolides A–E (**284**)–(**288**). Antifeedant tests of the *Carassius auratus* fish indicated for lintenolides A–E (**284**)–(**288**) a high activity at a concentration of 30 μg cm^{-2} of food pellets, suggesting their potential role as feeding deterrents.[207,208]

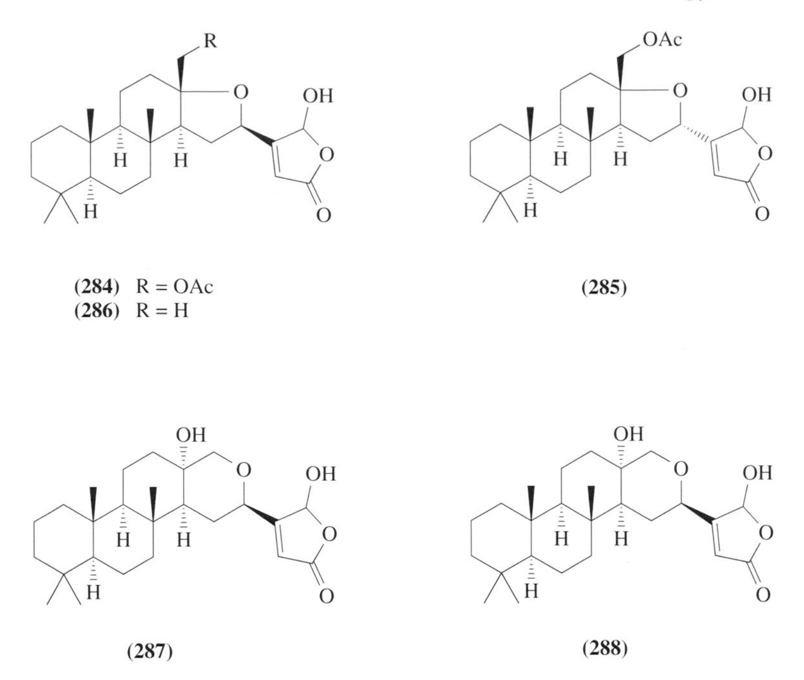

Sponges of the order Verongida are rich sources of brominated secondary compounds, which are biogenetically thought to be derived from bromotyrosine. The Mediterranean sponge *Verongia aerophoba*, found around the Canary Islands, was exceptionally rich in brominated compounds, and it was revealed that the bioactive sponge constituents aeroplysinin-1 (**289**) and the dienone (**290**) are biotransformation products which originate from the biologically inactive or weakly active precursors isofistularin-3 (**291**) and aerophobin-2 (**292**) by enzymatically catalyzed conversions following breakdown of the cellular compartmentation (Scheme 8).[209]

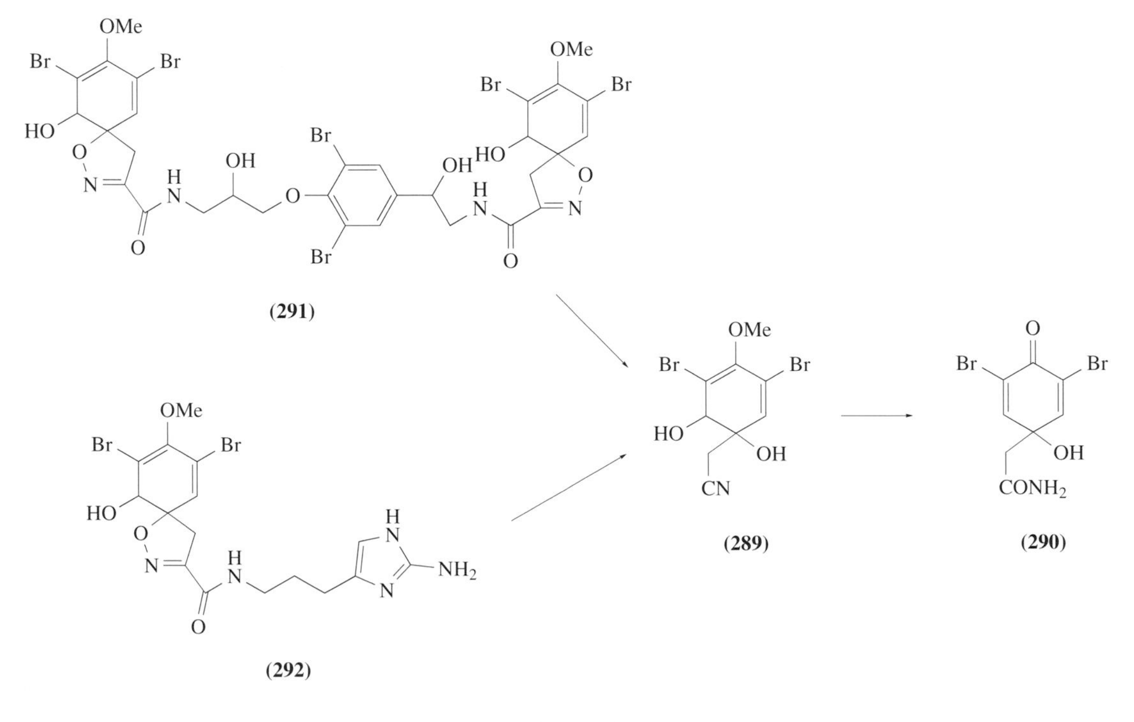

Scheme 8

The enzymatically catalyzed biotransformation processes of (**291**) and (**292**) into (**289**) and (**290**) were suggested by the following observations. (i) HPLC analysis of the 100% MeOH, 50% aqueous MeOH, or 100% H_2O extracts of the *V. aerophoba* sponge showed that (**291**) and (**292**) were the dominating brominated constituents in the 100% MeOH extract in which (**289**) and (**290**) were completely missing, whereas in the 50% aqueous MeOH extract the amounts of (**289**) and (**290**) increased and both (**291**) and (**292**) decreased, and in the aqueous extract (**290**) was dominant. (ii) In the absence of freeze-dried sponge tissue (**291**) as well as (**292**) proved stable when incubated in the presence of MeOH, 50% aqueous MeOH, or 100% H_2O. When exogenous (**291**) or (**292**) was added to lyophilized sponge tissue, the amount of (**290**) formed increased by 20–30% compared with controls lacking exogenously supplied substrate. (iii) No conversion of (**291**) or (**292**) was observed when the assay was carried out using heated sponge at 90 °C for 5 min prior to incubation. This enzymatic conversion was of special interest with regard to the biological activities of these bromotyrosine metabolites. The products of biotransformation, (**289**) and (**290**), were antibiotically active against several Gram-positive and Gram-negative bacteria, whereas the substrates, (**291**) and (**292**), proved to be completely inactive in the same experiments. In addition, (**289**) and (**290**) were found to exhibit pronounced cytotoxicity toward human carcinoma cells (Hela cells) with similar IC_{50} (50% inhibitory concentration) values (3.0–3.2 μM) for both compounds, while compounds (**291**) and (**292**) were revealed to be less cytotoxic (IC_{50} values: 8.5 μM and 99 μM, respectively). The antibiotic activity and the cytotoxicity of (**289**) and (**290**) could be relevant for the chemical defense of *V. aerophoba*, suppressing, for example, an overgrowth of fouling organisms. These compounds are an example of a stress-induced defense mechanism by the enzymatic conversion of preformed biologically inactive storage compounds into highly active defense metabolites. On the other hand, samples of *V. aerophoba* from different islands were collected and HPLC analysis of their extracts revealed that the pattern of brominated compounds was almost superimposable, indicating *de novo* biosynthesis by the sponge or by endosymbiotic microorganisms rather than uptake by filter feeding. The only differences observed between the different samples analyzed were with regard to the total concentrations of brominated compounds which varied from 7.2% to 12.3% of the dry weight, depending on the collection site.[210]

During examination of Porifera collected along the coast of the Bahamas, strong antifeedant activity was discovered in a crude extract of the sponge *Amphimedon compressa*. The MeOH-toluene (3:1) extract of this sponge was extracted successively with EtOAc and Bu^nOH, and the Bu^nOH-soluble material showed strong antifeedant activity. Chromatographic purification of the *n*-BuOH-soluble fraction yielded a polymeric pyridinium alkaloid named amphitoxin (**293**).[211] In laboratory feeding experiments, purified amphitoxin (**293**) deterred feeding of the fish *Thalassoma bifasciatum* at a concentration of 1 mg ml^{-1}, which was one-sixth of its natural concentration; this strong bioactivity suggested a role in the chemical defense of *A. compressa*.

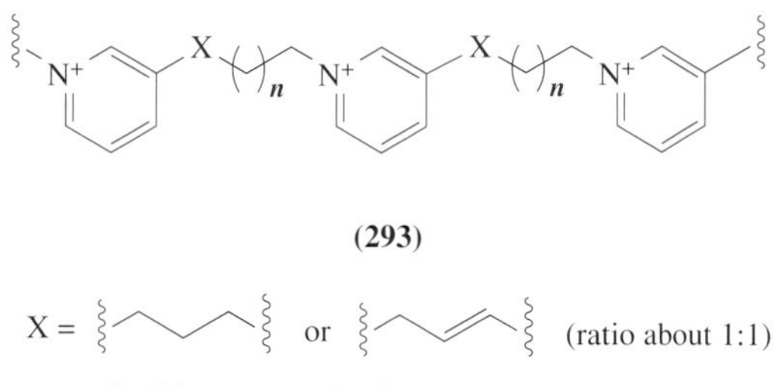

n = 5 (5 is a mean value in saturated chains)

As part of a study of the origins of biologically active substances in marine sponges, zeaxanthin [(3*R*,3′*R*)-dihydroxy-*β*,*β*-carotene (**294**)], which is one of the widely distributed carotenoids in marine organisms, was found to be produced by two species of marine bacteria *Flexibacter* sp., associated with the marine sponge *Reniera japonica*.[212] This carotenoid was also detected in the host sponge, suggesting the transport of zeaxanthin (**294**) from the microorganisms to the host. As zeaxanthin (**294**) plays the role of a quencher and scavenger for active species of oxygen, it was presumed that the sponge accumulates the bacterial product as a defense substance against the active oxygen species produced under irradiation by strong sunlight. It was thought that the bacteria are symbionts of the host sponge and act by obtaining the solid substrate and medium needed for settlement and growth from the host, and by producing and transmitting the biologically active substance to the host.

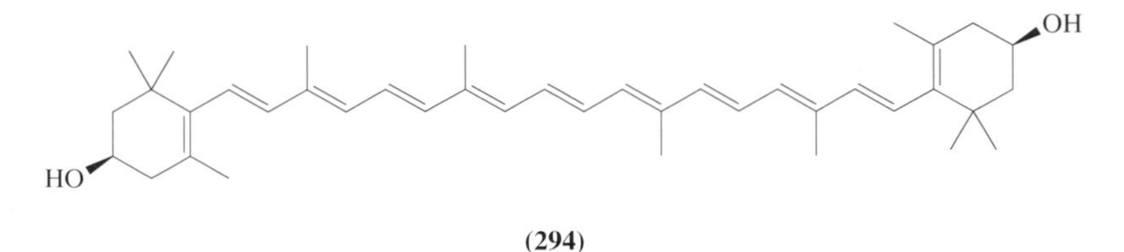

(294)

Scientific dredging operations in Prydz Bay, Antarctica, over the summer of 1990/1991, yielded many collections of marine sponges, including two specimens of the Antarctic marine sponge *Tedania charcoti* from geographically distinct locations. One specimen was collected from trawling around 67 °S, 71 °E at a depth of 439 m and at a water temperature of −2.1 °C, while a second *T. charcoti* was obtained from trawling around 67 °S, 78 °E at a depth of 251–266 m and at a water temperature of −1.6 °C. The aqueous ethanol extract of both these sponges proved to exhibit potent antibacterial properties, inhibiting the growth of strains of the bacteria *Staphylococcus aureus*, a *Micrococcus* sp., a *Serratia* sp., and *Escherichia coli*, as well as the ability to modulate protein phosphorylation in chicken forebrain. Detailed chemical investigations revealed the biological activity to be due to extraordinarily high levels of both cadmium and zinc.[213] It is noteworthy that *Tedania charcoti* possesses the ability to sequester both cadmium and zinc from seawater, where they are present in only trace amounts, as well as to tolerate the accumulation of extraordinarily high levels of these toxic heavy metals. It may be speculated that the high concentration of cadmium and zinc accumulated by *T. charcoti* serves as a natural antibiotic and/or an antifouling agent and/or a toxic defense against predation. It was therefore suggested that there is a possibility that at least some biologically active responses of extracts of marine organisms may be due to inorganic rather than organic substances.

8.07.7.4 Other Invertebrates

The Caribbean encrusting gorgonian octocoral *Erythropodium caribaeorum* deters predation by reef fish. The crude lipid-soluble extracts of *E. caribaeorum* incorporated into carrageenan food strips at the same volumetric concentration as it occurred at gorgonian tissues deterred feeding of a natural assemblage of fish on the reef from which the gorgonian had been collected. Fractionation of the crude extract revealed that the feeding-deterrent effects were present in a fraction containing chlorinated diterpenoids such as erythrolides A (**295**), B (**296**), and D (**297**), while a fraction contained sesquiterpene hydrocarbons in which erythrodiene (**298**) was found as a major compound. Of the three diterpenes, only (**296**) and (**297**), when assayed independently, inhibited feeding of reef

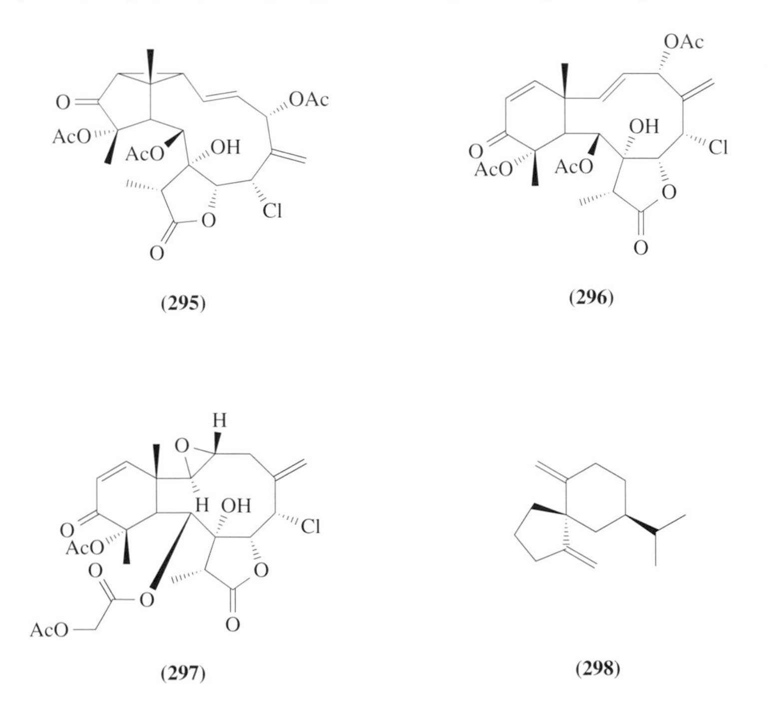

(295) (296)

(297) (298)

fish at the natural concentrations found in *E. caribaeorum*. Thus, the erythrolides appear to defend *E. caribaeorum* from reef predators, while the function of sesquiterpene hydrocarbons like (**298**), which occur in high concentrations in the lipid-soluble extracts of many octocorals, remained unclear.[214]

The starfish *Luidia clathrata* of the northern Gulf of Mexico showed feeding-deterrent properties against marine fish. The body-wall tissues of *L. clathrata* were rejected by the pinfish *Lagodon rhomboides* significantly more frequently than control tissues. Pellets containing EtOH body-wall extract (3 mg ml^{-1} agar) and krill were rejected significantly more often than control pellets containing only krill. Pellets containing body-wall extracts at 0.75 mg ml^{-1} agar were consumed with equal frequency to control pellets. The fish antifeedant activity noted for pellets containing natural tissue concentrations of body-wall extract of *L. rhomboides* indicated the presence of compounds that deter fish predators. From the EtOH extract of the starfish *L. rhomboides* 10 new polyhydroxysteroids (**299**)–(**308**) were isolated, and these polyhydroxysteroids were likely to be responsible for fish antifeedant activity, but the comparatively large amount of compound required

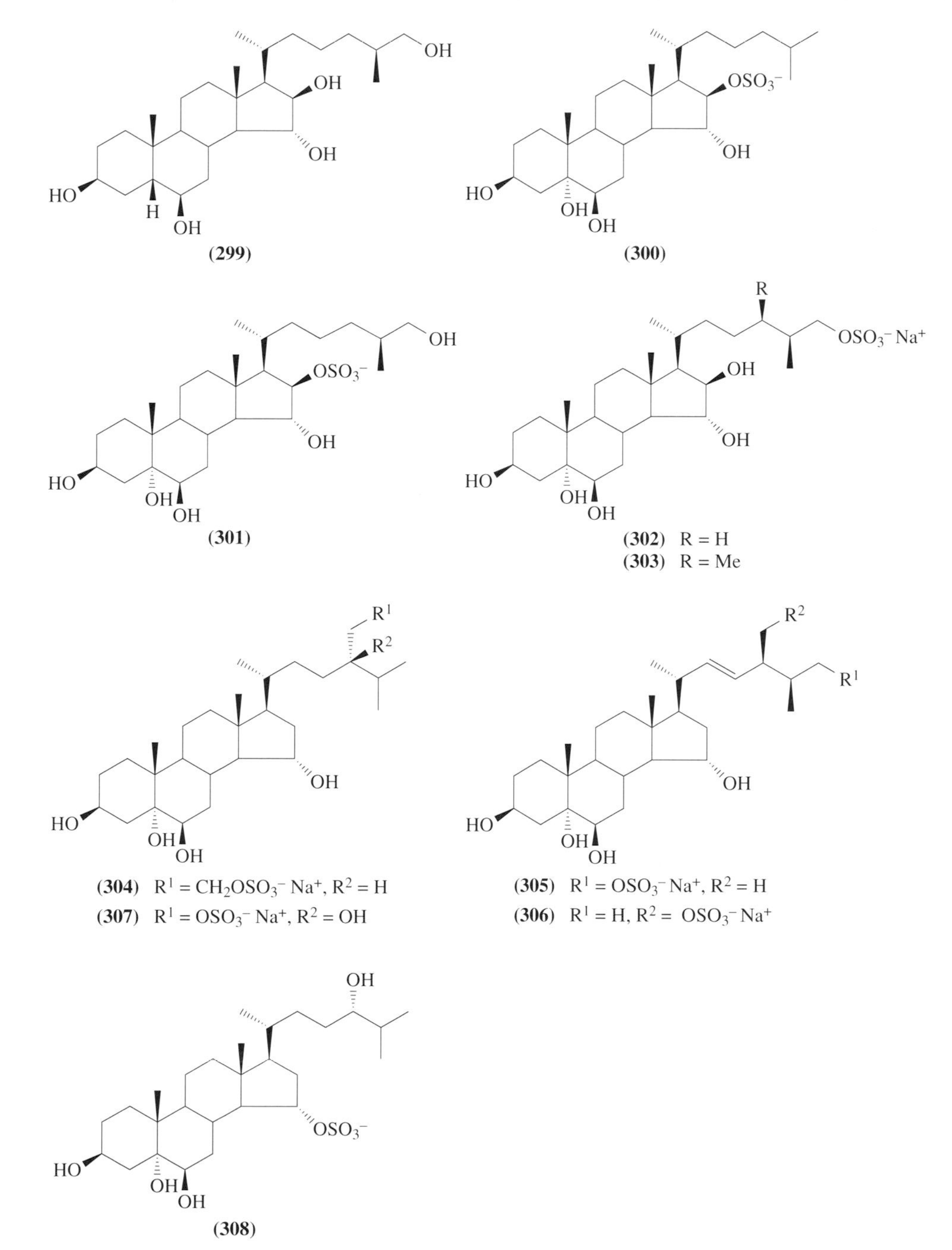

in the fish antifeedant assay precluded this determination. In addition, the EtOH body-wall extract of the starfish *L. rhomboides* significantly inhibited attachment of competent cyprid larvae of the barnacle *Balanus amphitrite* and larvae of the bryozoan *Bugula neritina* at concentrations of 3.0, 0.6, and 0.12 mg ml^{-1} of seawater. The EtOH body-wall extract of *L. rhomboides* also inhibited the growth of two species of Gram-positive bacteria, *Bacillus subtilis* and *Staphylococcus aureus*, at a concentration of 0.75 mg per disk. Purified polyhydroxysteroid compounds tested were all active in inhibition assay of the settlement of larvae, while the antibacterial activity was not necessarily observed in all purified polyhydroxysteroid compounds tested, thus indicating that some polyhydroxysteroids may display a broad spectrum of bioactivity, while others may have specific functional roles.[215]

It has been proposed that hemocytes play important roles in the defense mechanisms of ascidians. Two antibacterial substances, halocyamines A (**309**) and B (**310**) were isolated from the hemocytes of the solitary ascidian *Halocynthia roretzi*,[216] and they were tetrapeptide-like substances containing one bromine atom and were present only in the hemocytes. Both possess antimicrobial activity against several kind of Gram-positive bacteria and yeasts, and against a highly antibiotic-sensitive strain of Gram-negative bacterium. They also showed cytotoxic activity against some cultured mammalian cells. Halocyamine A (**309**) inhibited *in vitro* the growth of two fish RNA viruses, infectious hematopoietic necrosis virus and infectious pancreatic necrosis virus. Pretreatment of RNA virus with halocyamine A (**309**) reduced the infectivity of the virus toward host cells. The growth of marine bacteria *Achromobacter aquamarinus* and *Pseudomonas perfectomarinus* was also inhibited by halocyamine A (**309**) but that of *Alteromonas putrefaciens* and *Vibrio anguillarum* was not. These results suggest that halocyamines may have a role in the defense mechanisms of *H. roretzi* against marine viruses and bacteria.[217]

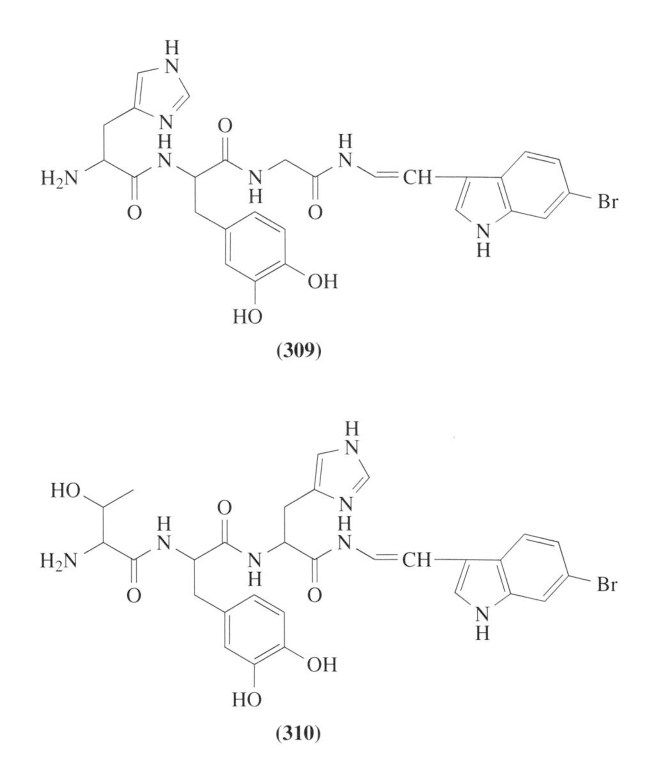

Research with the marine ascidian *Atapozoa* sp. and its nudibranch predators showed the direct chemical link in predator–prey associations involving ascidians and physically vulnerable mollusks and has demonstrated the *in situ* ichthyodeterrent properties of the *Atapozoa* secondary metabolites.[218] Bipyrrole metabolites such as tambjamines A (**311**), C (**312**), E (**313**), and F (**314**) were found in the organic extract of *Atapozoa* sp. and its nudibranch predators of the genus *Nembrotha*.

The unpalatability of *Atapozoa* larvae to coral reef fish and the presence of deterrent quantities of tambjamine C (**312**) in these larvae, along with *in situ* feeding deterrent properties of tambjamine class alkaloids, provides convincing evidence for the chemical protection of *Atapozoa* larvae. Microscopic examination of *Atapozoa* revealed that its intense pigmentation is confined to the granular

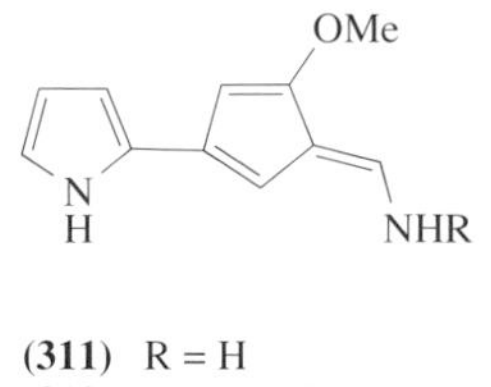

(311) R = H
(312) R = $CH_2CH(Me)_2$
(313) R = CH_2CH_3
(314) R = CH_2CH_2Ph

amebocyte blood cells. Thus, the large quantity (0.5–1.7% dry weight) of the brightly yellow-colored tambjamines should therefore reside within these blood cells. A transmission electron microscope investigation of *Atapozoa* larvae, performed expressly for the purpose of identifying bacterial symbionts, failed to find any significant quantities of bacteria within the granular amebocytes or in any other part of this ascidian. Thus, *Atapozoa* seem to be capable of the *de novo* biosynthesis of the tambjamines.[219]

A sulfated polyhydroxy benzaldehyde, polyclinal (**315**), was isolated from extracts of the temperate colonial ascidian *Polyclinum planum* and the structure was solved by an X-ray crystallographic study. The distribution of polyclinal (**315**) within three distinct regions of the ascidian colonies was investigated. The colonies were dissected into the stalk, the pulpy inner mesenchyme, and the zooid-rich surface layer. The concentration of polyclinal (**315**) in these different colony parts, based on wet weights, were determined to be 5.8×10^{-5} g g^{-1}, 7.8×10^{-4} g g^{-1}, and 2.5×10^{-3} g g^{-1} in the stolons, the cortex, and the zooid-rich outer layers of the colonies, respectively. The higher concentration of polyclinal (**315**) in the zooid-rich surface layer of the colonies suggested that polyclinal (**315**) may function as a chemical defense against predators, which would be consistent with the observation that predator-deterrent gorgonian secondary metabolites were also distributed in the outer more accessible portions of the colonies.[220]

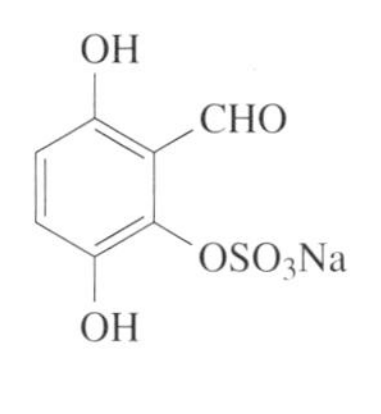

(315)

8.07.7.5 Fish

Soles of the genus *Pardachirus* are characterized by their chemical defense against predation with a copious and ichthyotoxic secretion being discharged upon disturbance. Two chemical classes of bioactive entities, namely, steroid monoglycosides and amphiphilic peptides, were isolated from each secretion of two different biological species.[174] Toxic material contained in the mucosal secretion of the Pacific sole *Pardachirus pavoninus* was investigated to isolate three ichthyotoxic and strongly surfactant peptides, which were named pardaxin P-1 (**316**), P-2 (**317**), and P-3 (**318**).[221]

5 10 15 20 25
(316) Gly-Phe-Phe-Ala-Leu-Ile-Pro-Lys-Ile-Ile-Ser-Ser-Pro-Leu-Phe-Lys-Thr-Leu-Leu-Ser-Ala-Val-Gly-Ser-Ala-
30
Leu-Ser-Ser-Ser-Gly-Glu-Gln-Glu

5 10 15 20 25
(317) Gly-Phe-Phe-Ala-Leu-Ile-Pro-Lys-Ile-Ile-Ser-Ser-Pro-Ile-Phe-Lys-Thr-Leu-Leu-Ser-Ala-Val-Gly-Ser-Ala-
30
Leu-Ser-Ser-Ser-Gly-Gly-Gln-Glu

5 10 15 20 25
(318) Gly-Phe-Phe-Ala-Phe-Ile-Pro-Lys-Ile-Ile-Ser-Ser-Pro-Leu-Phe-Lys-Thr-Leu-Leu-Ser-Ala-Val-Gly-Ser-Ala-
30
Leu-Ser-Ser-Ser-Gly-Glu-Gln-Glu

These peptides have nearly identical sequences and contain both an unusually larger number of serine residues (including a Ser-27–Ser-28–Ser-29 segment) and a large number of hydrophobic residues such as leucines and isoleucines. These 33 amino acid polypeptides fold into ordered structures in trifluoroethanol–water solution and in micelles but adopt a random-coiled structure in aqueous solution. The complete proton NMR spectrum of pardaxin P-2 (**317**) was assigned in CF_3CD_2OD/H_2O (1:1) solution, and the 3D structure was elucidated with distance-restrained molecular dynamics calculations. It was demonstrated that peptide segments within the 7–11 and 14–26 residue stretches were helical while residues at the C- and N-terminus exist predominantly in extended conformations in solution. The dipeptide 12–13 segment connecting the two helices exists as a bend or a hinge allowing the two helices to be oriented in an L-shaped configuration. From these studies pardaxin P-2 (**317**) was shown to adopt a novel amphiphilic helix (7–11)-bend (12–13)-helix (14–26) motif with Pro-13 forming the focal point of the turn or bend between the two helices.[222] The structure and activity of pardaxin and analogues were studied by assaying 13 synthetic pardaxin analogues for their ability to interact with model membranes of phosphatidylcholine. As a result, it was found that (i) an amphipahtic α-helix from isoleucine-14 to leucine-26 was responsible for most of the membrane-perturbing properties of pardaxin; (ii) a hydrophobic N-terminal region enhanced the activity of the isoleucine-14 to leucine-26 α-helix by binding the pardaxin molecule to the lipid bilayer; (iii) a bend centered around ^{12}Ser–^{13}Pro appeared to create overall amphipathicity for the two different helical regions of pardaxin, but this contributed only slightly to potency; and (iv) the C-terminal amino acids were unimportant for membrane-perturbing activity and may be present only to enhance transportation in an aqueous environment prior to membrane binding in the native system.[223] The mechanism of action of pardaxin as well as steroid glycosides such as mosesin-1 (**319**) was investigated based on the permeabilization assay method of a phospholipid bilayer to suggest that it may well be attributed to nonspecific derangement of animal cell membrane without binding to any particular biomolecule. The N-terminal region of pardaxins was revealed not only to function as the strong binding region, but also to enhance the deranging activity (putatively of the α-helical region at the middle of the 33 amino acid sequence) by a factor of 6 when bound. On the other hand, defensive steroid glycosides like (**319**) were shown to be indifferent to the cholesterol content in the lipid bilayer in their permeabilizing action, being unlike the conventional hemolytic saponins. In addition, different kinetics between the steroid glycosides and pardaxins implied that the action of the former is rather transient while that of the latter is persistent, resulting in all-or-none rupture of liposomes. An extent of synergism, possibly with such differently allotted functions in the defensive action, was also shown between the two chemical entities.[224] The conformations of synthetic peptides of different length corresponding to the amino-terminal, central, and carboxyl-terminal regions of pardaxin were studied by circular dichroism spectroscopy.[225] The peptide segments that adopt an ordered conformation showed a similar conformation when present in the entire toxin as suggested by proton magnetic resonance data,[222] and the amino-terminal and central regions of the toxin was indicated to play a role in initiating and maintaining an ordered conformation of pardaxin.

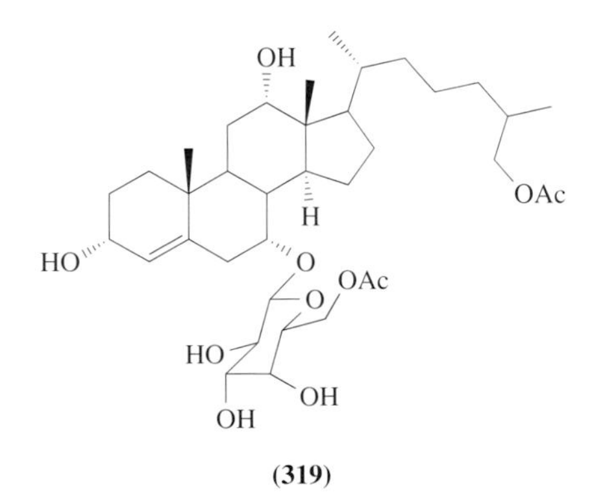

(**319**)

A water-soluble cationic steroidal antibiotic, named squalamine (**320**), was isolated from the dogfish shark *Squalus acanthias*. This aminosterol antibiotic (**320**) exhibited potent bactericidal activity against both Gram-positive and Gram-negative bacteria. In addition, squalamine (**320**) was fungicidal and induced osmotic lysis of protozoa. Squalamine (**320**) is an aminosterol characterized by condensation of an anionic bile salt intermediate with spermidine, which is without precedent in vertebrates. The discovery of squalamine (**320**) implicated a steroid as a potential host-defense agent in vertebrates.[226]

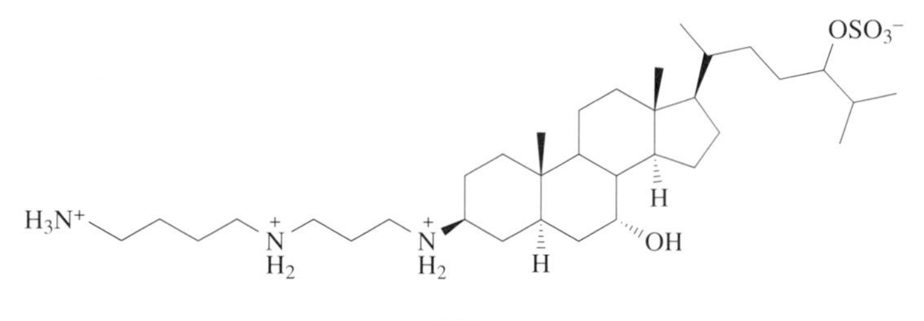

(320)

Squalamine (**320**) was distributed in many tissues of the shark. The liver and gall bladder, the organs in which bile salts are synthesized and stored for secretion into the gastrointestinal tract, were the richest sources identified (4–7 μg g^{-1} of tissue). However, both the spleen and the testes of this animal were also relatively rich sources of squalamine (**320**), each containing ~2 μg g^{-1}. The stomach (1 μg g^{-1}), the gills (0.5 μg g^{-1}), and the intestine (0.02 μg g^{-1}) yielded smaller amounts. It was clear that squalamine (**320**) was also present in organs that were not engaged in the synthesis of bile salts for digestive functions. It was not certain whether a single organ such as the liver was the principal site of synthesis of squalamine (**320**), and it was also possible that squalamine (**320**) was not synthesized by the shark, but rather, derived from an exogenous source present in the shark's food chain. This unusual steroid (**320**) would appear to be the product of an unknown biochemical pathway involving the condensation of spermidine with a steroid. It may be speculated that squalamine (**320**) serves as a systemic antimicrobial agent in this animal, but not direct evidence of this speculation was provided. The biosynthetic pathway of squalamine (**320**), its role in host defense, and its expression after injury and infection have yet to be investigated.

Syntheses of squalamine (**320**) and its mimics were investigated.[227] The mimic (**321**), which possessed a pendant sulfate and a spermidine group at the opposite placement to those of (**320**), exhibited potent antibiotic properties against a broad spectrum of microorganisms.[228] In addition, the mimic (**321**) possessed unusual ionophoric properties and recognized negatively charged phospholipid membranes; (**321**) favored the transport of ions across negatively charged bilayers over ones that are electrically neutral.[229]

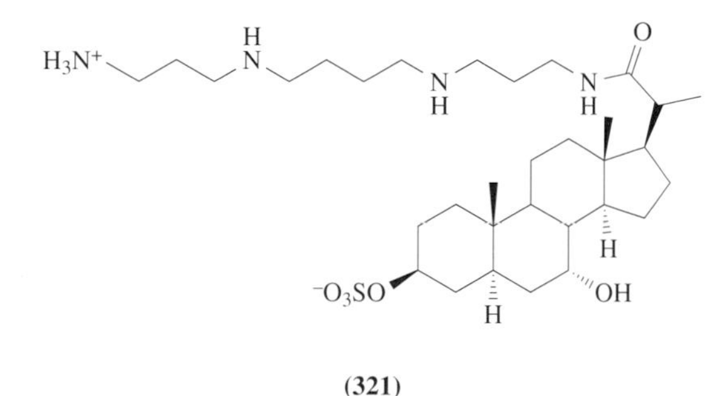

(321)

A lipophilic ichthyotoxin from the defensive mucous skin secretion of soapfish *Diploprion bifasciatum* and *Aulacocephalus temmincki*, named lipogrammistin-A (**322**), was isolated and chemically characterized. Lipogrammistin-A (**322**) not only differed in its chemical type from all other known defense substances of fish, but also was the first macrocyclic polyamine lactam found in the animal kingdom. It showed toxicity to medaka fish and to mice, and hemolytic activity to rabbit erythrocytes. A synthetic study of (**322**) was also performed for further chemical and biological investigation of this compound.[230]

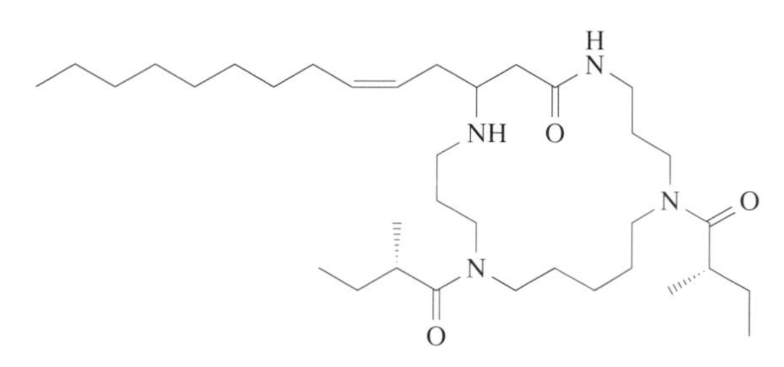

(322)

8.07.8 MARINE TOXINS

8.07.8.1 Cone Shells

8.07.8.1.1 Conus geographus

The cone shells that inhabit tropical and subtropical seas are carnivorous gastropods which catch prey organisms such as fish and shellfish by shooting a venom-containing harpoon-like radular tooth. The venom of one species of cone shell (anboina, *Conus geographus*) is so potent that occasionally it is responsible for human fatalities. Thus, the anboina is called "habu-gai" (shellfish that is dangerous like a venomous snake) and is feared in the Okinawan Island of Japan. On the other hand, cone shells are loved by shell collectors all over the world because of their beautiful colors and shapes. The family Conidae includes approximately 400 species, most of which inhabit the Okinawan Islands.

Cone shells are divided into three groups according to their major prey organisms. A majority feed on various marine worms such as polychaetes (vermivorous); a smaller number prey on shellfish (molluscivorous); and a few feed on small bottom fish (piscivorous). Each species elaborates a venom that reflects the dominant target animals. The venoms of molluscivorous cone shells are highly venomous to gastropods, but are almost ineffective on mice. On the other hand, the venoms of piscivorous Conidae are fatal to both fish and mice. Further, the symptoms in mice differ depending on the species of cone shell involved. These observations suggest a variety of bioactive principles of cone shell venoms among various species of Conidae.

In the early 1950s, Kohn started ecological studies on cone shells,[231] and Endean *et al.* expanded these ecological studies to the toxicology, pharmacology, and biochemistry of Conidae venoms using crude venoms.[232] In 1980, Kobayashi *et al.* started the screening of venoms from about 30 species of cone shells, collected in Okinawa, Japan, by monitoring the pharmacological action on isolated mammalian muscle preparations including skeletal, cardiac, and smooth muscles.[233] In the course of this project, Kobayashi *et al.* have isolated numerous new bioactive compounds with a variety of chemical and pharmacological properties from representative species of worm-, shellfish-, and fish-eating cone shells, and have shown that most of these new compounds are peptide toxins acting on ion channels of the cell membrane.

Among various carnivorous gastropods, cone shells are in a unique position because of the highly specialized radular teeth used as their feeding apparatus. Herbivorous gastropods have hundreds of tiny chitinous radular teeth arranged as in a grater; they shave their algal food with these radular teeth. In the case of cone shells, on the other hand, one central radular tooth has been developed to become a harpoon. The venom apparatus of cone shells consists of a muscular venom bulb, a long coiled venom duct, a radular sheath, radular teeth, and a proboscis. The venom produced in the venom duct is transferred to the radular sheath, where hollow radular teeth are charged with the venom. The bulb contains no venom and is hence thought to provide a pump that transfers the venom from the duct into the radular sheath, which has tens of harpoon-like chitinous radular teeth. The teeth are set one by one along the pharynx to the proboscis, where the foremost tooth is held with its pointed tip forward like a spear. Stinging is accomplished through the thrust of a radular tooth by the proboscis into the prey organism. A tooth is used only once, and if it fails to shoot the prey, that tooth is discarded and a new one from the radular sheath is charged with the venom. The length of radular teeth ranges from several millimeters to 1 cm. The radular tooth has a barb near the tip which prevents it from slipping out once it has taken hold.

Conus geographus is called "anboina" as the animal's habitat is the Gulf of Anbon, Indonesia. The shells are about 10 cm in length. *C. geographus* is highly toxic and responsible for human injury. Fatal cases have been reported from Okinawa, Japan. Collectors are usually stung while swimming with their catch in a net bag on their side or while attempting to examine the shell markings. Stings by *C. geographus* immediately produce severe pain and numbness at the injured area. Then numbness spreads around the mouth and to the extremities. In most cases, this is accompanied by vomiting, dizziness, lachrymation, hypersalivation, and chest pain. In severe cases dyspnea, dysphagia, aphonia, blurring of vision, motor incoordination, generalized pruritus, respiratory paralysis, and death follow.

It has been reported that the crude venom of *C. geographus* causes an inhibitory effect on the contractile response of stimulated skeletal muscle. Kobayashi and co-workers isolated two peptide toxins composed of 22 amino acid residues including 3-hydroxyprolines by monitoring the pharmacological activity.[234] The primary structures of the two peptides named geographutoxins I and II (GTX I and II, (**323**) and (**324**), respectively) are very similar to each other.[235] They differ by four amino acid residues, at positions 8, 14, 18, and 19 from the C terminal. The substitution of the amino acids was explained by single-base substitution in the triplet codons except for the substitution at residue 18 of Gln in GTX I by Met in GTX II. Distinctive features of the primary structures are a basic amino acid cluster in the middle of the sequence, and a Cys–Cys sequence in both N and C terminal regions. The LD_{50} (50% lethal dose) values (i.p.) of GTX I and II in mice are 340 μg kg^{-1} and 110 μg kg^{-1}, respectively.

(**323**) Arg-Asp-Cys-Cys-Thr-Hyp-Hyp-Lys-Lys-Cys-Lys-Asp-Arg-Gln-Cys-Lys-Hyp-Gln-Arg-Cys-Cys-Ala-NH_2 (residues numbered 5, 10, 15, 20, 22)

(**324**) Arg-Asp-Cys-Cys-Thr-Hyp-Hyp-Arg-Lys-Cys-Lys-Asp-Arg-Arg-Cys-Lys-Hyp-Met-Lys-Cys-Cys-Ala-NH_2 (residues numbered 5, 10, 15, 20, 22)

GTX II (**324**) at very low concentrations inhibits twitch responses of skeletal muscle.[236] The IC_{50} value of GTX II (3×10^{-8} M) is somewhat higher than that of tetrodotoxin (1×10^{-8} M). GTX II inhibited action potentials of skeletal muscle without change in the resting potential. Surprisingly, action potentials of the crayfish giant axon, neuroblastoma N1E-115 cells, and cardiac muscle were not affected by GTX II even at high concentrations.[237]

Toxin-binding studies have shown that tetrodotoxin and saxitoxin bind at a common receptor site on sodium channels, which has been designated neurotoxin receptor site 1. GTX II inhibits [^{3}H]saxitoxin binding to sodium channels in skeletal muscle homogenates and T-tubular membranes in close agreement with concentrations that block muscle contraction.[238] However, [^{3}H]saxitoxin binding to membranes of the superior cervical ganglion is not affected by GTX II even at high concentrations. Scatchard analysis of [^{3}H]saxitoxin binding to T-tubular membranes revealed a primarily competitive mode of inhibition of saxitoxin binding by GTX II. These results indicate that GTX II interacts competitively with saxitoxin in binding at neurotoxin receptor site 1 of the sodium channel in a highly tissue-specific manner. Previously no differences had been found experimentally between muscle and nerve sodium channels because of their electrophysiological and pharmacological similarity. Thus, GTX II is the first reagent that discriminates at this site between nerve and muscle sodium channels. GTX II is used to identify these tissue-specific regions of the sodium channel structure at neurotoxin receptor site 1.[239]

Olivera and co-workers isolated from *C. geographus* peptide toxins of three classes, named α-, μ-, and ω-conotoxins.[240,241] α-Conotoxins (GI, GIA, GII, and MI) are composed of 13–15 amino acid residues including two S-S bridges per molecule and are homologous with one another (Table 17). These peptides cause postsynaptic inhibition at the neuromuscular junction and competitively inhibit α-bungarotoxin and curare alkaloid binding to nicotinic acetylcholine receptors. Many pharmacological studies have revealed that α-conotoxins are nicotinic receptor blockers. μ-Conotoxins GVIIIA, GVIIIB, and GVIIIC with 22 amino acid residues are rich in hydroxyproline. These peptide toxins block sodium channels in muscle much more effectively than those in nerve. Geographutoxins I and II (GTX I and II) may be identical with GVIIIA and GVIIIB, respectively. ω-Conotoxins GVIA, GVIB, GVIC, GVIIA, and GVIIB, called "shaker peptides," were also isolated from *C. geographus*. These peptides, consisting of 25–29 amino acid residues, induce a persistent tremor when injected intracerebrally in mice. The transmitter release at the neuromuscular junction of skeletal muscle was blocked by ω-conotoxins GVIA, GVIB, and GVIC, whereas it was not affected by other ω-conotoxins (GVIIA and GVIIB). Electrophysiological experiments indicate that ω-conotoxin GVIIA acts by blocking calcium channels without affecting the propagation of action potentials in the nerve terminal.[242] ω-Conotoxins may provide a valuable tool for defining structural differences among the calcium channel subtypes in muscle and nerve.

The *Conus* peptides are used in hundreds of research laboratories for a wide variety of physiological and pharmacological investigations in both vertebrate and invertebrate nervous systems. Some *Conus* peptides such as ω-conotoxin GVIA have become well-established neurobiological tools, and syntheses of these peptides have also been investigated. For example, μ-conotoxin GIIIA, a 22-amino acid peptide paralytic toxin inhibiting the muscle voltage-activated sodium channels, was synthesized by a solid-phase method. No purification of intermediates was necessary for the synthesis, and a simple air oxidation of the deprotected crude peptide gave the desired toxin.[243] Three photoreactive derivatives of μ-conotoxin GIIIA were also prepared as photoaffinity labeling reagents for muscle-type sodium channels. The reagents competitively inhibited the binding of saxitoxin to the eel sodium channel with K_i values of 11–18 nM. The introduced chromogenic phenyldiazirine group on the toxin was photolyzed efficiently, and spectroscopic properties of the

Table 17 Amino acid sequences of conotoxins.

α	GI	E C C N P A C G R H Y S C[a]
	GIA	E C C N P A C G R H Y S C G K
	GII	E C C H P A C G K H F S C[a]
	MI	G R C C H P A C G K N Y S C[a]
μ	GVIIIA	R D C C T Hy Hy K K C K D R Q C K Hy Q R C C A[a]
	GVIIIB	R D C C T Hy Hy R K C K D R R C K Hy M K C C A[a]
	GVIIIC	R D C C T Hy Hy K K C K D R R C K Hy L K C C A
ω-"Shape-Blocker"	GVIA	C K S Hy G S S C S Hy T S Y N C C R – S C N Hy Y T K R C – – Y[a]
	GVIB	C K S Hy G S S C S Hy T S Y N C C R – S C N Hy Y T K R C – – Y G
	GVIC	C K S Hy G S S C S Hy T S Y N C C R – S C N Hy Y T K R C
ω-"Shaker"	GVIIA	C K S Hy G T Hy C S R G M R D C C T – S C L L Y S N K C R R Y
	GVIIB	C K S Hy G T Hy C S R G M R D C C T – S C L S Y S N K C R R Y
	MVIIA	C K G K G A K C S R L M Y D C C T G S C R – – S G K C[a]
"Sleeper"	GV	G E γ γ L Q γ N Q γ L I R γ K S N[a]

[a] The α-carboxyl is known to be amidated; absence of [a] indicates that no assignment has been made. A (Ala), C (Cys), D (Asp), E (Glu), F (Phe), G (Gly), H (His), K (Lys), L (Leu), N (Asn), P (Pro), Q (Glu), R (Arg), S (Ser), T (Thr), Y (Tyr), γ (γ-carboxylglutamate), Hy (*trans*-4-hydroxyproline).

reagents demonstrated that irradiation and detection can be performed in a spectral region where the absorptions due to most biological macromolecules are negligible.[244]

Structural studies of multiple disulfide cross-linked *Conus* peptides have been conducted. The three intramolecular SS linkages of geographutoxin I (GTX I, μ-conotoxin GIIIA) were examined to show that the SS bridges were between cysteines Cys3 and Cys15, Cys4 and Cys20, and Cys10 and Cys21, indicating that GTX I has a rigid conformation consisting of three loops stabilized by these three SS linkages.[245] The 3D structure of conotoxin GIIIA was determined in aqueous solution by 2D proton NMR and simulated annealing based methods. On the basis of 162 assigned nuclear Overhauser effect (NOE) connectivities obtained at the medium field strength frequency of 400 MHz, conotoxin GIIIA is characterized by a particular folding of the 22-amino acid peptide chain, which is stabilized by three disulfide bridges arranged in a cage at the center of a discoidal structure of approximately 20 Å diameter. The seven cationic side-chains of lysine and arginine residues project radially into the solvent and form potential sites of interaction with the skeletal muscle sodium channel for which the toxin is a strong inhibitor. These results may provide a molecular basis to elucidate the remarkable physiological properties of this neurotoxin.[246]

The amino acid sequence of μ-conotoxin GIIIA (geographutoxin I) was modified by replacing each residue with Ala or Lys to elucidate its active center for blocking sodium channels of skeletal muscle. NMR and CD spectra were virtually identical between native and modified toxins, indicating the similarity of their conformation including disulfide bridges. The inhibitory effect of these modified peptides on twitch contractions of the rat diaphragm showed that Arg at the 13th position and the basicity of the molecule are crucial for the biological action. The segment Lys11–Asp12–Arg13 is flexible, and this may represent a clue to the subtle fit of Arg13 to the specific site of sodium channels. Since known ligands to sodium channels, such as tetrodotoxin, anthopleulin-A, etc., contain guanidino groups as a putative binding moiety, Arg may be a general residue for peptide toxins to interact with the receptor site on sodium channels.[247] Studies of the solution structures of α-conotoxin GI as well as μ-conotoxin GVIA have been carried out using 2D NMR spectroscopy and various structural calculations.[248–250]

Olivera and co-workers also isolated sleeper peptides called conatokins from *C. geographus* (Table 17), which causes a sleep-like state in mice when injected intracerebrally. The conatokins contain 4–5 residues of γ-carboxyglutamate (Gla),[241] and the most well-characterized one, conatokin-G or GV, has five Gla residues[251] and blocks glutamate receptors of NMDA (*N*-methyl-D-aspartate) subtype.[252]

8.07.8.1.2 Other Conus *toxins*

Conus striatus, a piscivorous member of the Conidia, is not sufficiently venomous to be of potential danger to humans, but several severe cases of poisoning by *C. striatus* have been reported.

Kobayashi *et al.* have isolated a glycoprotein (molecular weight 25 000) named striatoxin (StTX) from this venom as a cardiotonic component.[253] Application of StTX to the isolated guinea-pig left atria caused a concentration-dependent inotropic effect; the effect of StTX was reversed in the presence of tetrodotoxin. In the atria, StTX provoked action potentials with a plateau phase of long duration without affecting the maximum rate of rise and amplitude of the action potential and the resting membrane potential. This prolongation was also reversed by tetrodotoxin. These results suggest that StTX causes prolongation of the action-potential duration probably due to slowed inactivation of sodium inward currents and that this may result in an increase in Ca^{2+} availability in cardiac muscle cells. This could explain the cardiotonic action of StTX. Furthermore, in mouse neuroblastoma cells StTX slowed sodium channel inactivation without affecting the time course of activation of the channels, and the voltage dependence of activation was shifted to more negative membrane potentials. The binding of saxitoxin or α-scorpion toxin to sodium channels was not affected by StTX, while that of batrachotoxin was slightly enhanced by the toxin. It may be concluded that StTX interacts with a new receptor site on sodium channels at which specific effects on channel inactivation can occur.[254] In the isolated guinea-pig ileum, StTX caused rhythmic transient contraction followed by relaxation, which was reversed by tetrodotoxin and a low-Na^+ medium. Pharmacological studies on StTX suggested that the contraction of the ileum induced by the toxin is due to the excitation of cholinergic nerves, while the relaxation is mediated through nonadrenergic inhibitory nerves.[255] The minimum lethal dose of StTX in a fish, *Rhodeus ocellatus smithi*, is 1 $\mu g\ g^{-1}$ body weight.

The venom of *Conus magus* is lethal to fish and mice, as are those of *C. geographus* and *C. striatus*, although the average size of *C. magus* is only about 5 cm long and smaller than *C. geographus* and *C. striatus*. A myotoxin (MgTX), a peptide toxin with an estimated molecular weight of 1500, was isolated from the venom of *C. magus* by gel filtration.[256] MgTX elicits an increase in contractile force of the isolated guinea-pig ileum. The mode of action of MgTX is similar to that of the cardiotonic glycoprotein StTX isolated from *C. striatus*. Another active component (MAC) in the venom of *C. magus* was also purified and characterized as a protein of molecular weight between 45 000 and 65 000.[257] MAC exhibits rhythmic transient contractions of the guinea-pig ileum followed by relaxations and a marked increase in the contractile force of the guinea-pig left atria, as in the case of StTX. On the other hand, two peptide toxins, MI and MVIIA, were isolated from the Philippine *C. magus* by Olivera and co-workers.[241] The chemical and pharmacological properties of MI and MVIIA are different from those of MAC and MgTX isolated from the Okinawan *C. magus*, but resemble those of GII and GVIIA isolated from the Philippine *C. geographus*. MI and MVIIA are classified as α- and ω-conotoxins, respectively. MI showed no rhythmic contractions of the guinea-pig ileum, suggesting that the pharmacological properties of MgTX and MI differ from each other. Olivera and co-workers described the purification and first biochemical characterization of an enzymatic activity in venom from *C. magus*, and this enzyme, named conodipine-M, was a novel phospholipase A_2 with a molecular mass of 13.6 kDa, being comprised of two polypeptide chains linked by one or more disulfide bonds.[258]

Conus eburneus is a vermivorous member of the Conidae, which feeds on polychaete worms. The average size of the shell is about 6 cm long and they inhabit sandy seashores. From the venom of the Okinawan *C. eburneus* was purified a vascoactive protein (molecular weight 28 000) named eburnetoxin (ETX).[259] ETX induces contractions of isolated rabbit aorta, resulting from a remarkable increase in Ca^{2+} permeability in the cell membrane.

Conus tessulatus is a vermivorous member of the Conidae as is *C. eburneus*; the average size of the shell is approximately 5 cm long. A vasoactive protein (molecular weight 26 000) designated tessulatoxin (TsTX), was isolated from the venom of the Okinawan *C. tessulatus*.[260] The pharmacological action of TsTX is quite similar to that of ETX, the vasoactive substance from *C. eburneus*. Lazdunski and co-workers isolated a protein toxin consisting of two subunits (molecular weights 26 000 and 29 000) from the venom of the Madagascan *C. tessulatus*.[261] This protein toxin activates the Na^+–Ca^{2+} exchange system by increasing Na^+ permeability of the cell membrane to result in increased Ca^{2+} permeability of cardiac muscle cells, or to induce contraction of smooth muscle.

Conus textile is a molluscivorous member of the Conidae and its prey organisms are closely related molluscivorous Conidae. The average size of the shell is comparable to that of *C. geographus*. Kobayashi and co-workers examined the toxicity of the Okinawan *C. textile* venom against mice, but found that no mouse died even after intraperitoneal injection of a large amount of the venom. The human fatalities recorded in the literature, therefore, may result from the venomous sting of *C. geographus* rather than from that of *C. textile*. Arachidonic acid was isolated from the venom of *C. textile* as a principle of the powerful transient contraction of the isolated guinea-pig ileum. The amount of arachidonic acid in the contents of the venom duct was estimated to be as much as 6 mg g^{-1} wet weight,[262,263] but the role of arachidonic acid in the venom remains uncertain.

From the snail *Conus imperialis*, which preys on polychaete worms, a peptide ligand for nicotinic acetylcholine receptors was isolated by Olivera and co-workers.[264] This peptide, named α-conotoxin ImI, was highly active against the neuromuscular receptor in frogs but not in mice. α-Conotoxin ImI had a sequence that shows striking sequence differences from all α-conotoxins of fish-hunting *Conus*, but its disulfide-bridging was similar. It was suggested that cone venoms may provide an array of ligands with selectivity for various neuronal nicotinic acetylcholine receptor subtypes.

The venom of *Conus nigropunctatus*, a piscivorous member of the Conidae, contained a new peptide toxin, named conotoxin NgVIA,[265] affecting sodium current inactivation, that competes on binding with δ-conotoxin TxVIA purified from *C. textile* venom.[266] The primary structure of conotoxin NgVIA had an identical cysteine framework and similar hydrophobicity as δTxVIA but differed in its net charge. These toxins may enable analysis of the functional significance of the receptor sites in gating mechanisms of sodium channels.

From the venom of the snail-hunting species *Conus marmoreus*, two new peptides, μO-conotoxins MrVIA and MrVIB, were isolated and characterized.[267] These toxins potently blocked voltage-sensitive sodium currents. μO-Conotoxin MrVIA was chemically synthesized and proved indistinguishable from the natural product. The μO-conotoxins showed no sequence similarity to the μ-conotoxins. μO-Conotoxins are as potentially important to sodium channel subtype differentiation as the ω-conotoxins are to calcium channels.

From the milked venom of the eastern Pacific piscivorous species, *Conus purpurascens* (the purple cone), three peptides that cause paralysis in fish were isolated, and the sequence and disulfide bonding pattern of one of these, αA-conotoxin PIVA, was disclosed. The peptide was chemically synthesized in a biologically active form. Electrophysiological experiments and competition binding with α-bungarotoxin demonstrated that αA-conotoxin PIVA acts as an antagonist of the nicotinic acetylcholine receptor at the postsynaptic membrane.[268] *C. purpurascens* uses two parallel physiological mechanisms requiring multiple neurotoxins to immobilize fish rapidly: a neuromuscular block and an excitotoxic shock. Among them, κ-conotoxin PVIIA inhibits the "shaker" potassium channel, while δ-conotoxin PVIA delays sodium-channel inactivation. In general, the evolution of excitotoxins may be influenced by how well the prey is initially tethered. A sea snake that has securely impaled a small fish with venom fangs only uses a neuromuscular block, whereas sea anemones, which initially contact their prey with a single tentacle, have convergently evolved Na^+ and K^+ channel-targeted excitotoxins analogous to those used by *C. purpurascens*.[269]

8.07.8.2 Tetrodotoxin and Saxitoxin

8.07.8.2.1 Tetrodotoxin

Tetrodotoxin (TTX (**325**)) is one of the best known marine natural products, being a toxin responsible for the fatal food poisoning caused by puffer fish.[270,271] TTX (**325**) possesses a unique chemical structure and exhibits potent neurotoxicity by specifically blocking the sodium channels of excitable cell membranes. The structure of TTX was elucidated simultaneously by an American and two Japanese groups in 1964.[272–274] The total synthesis of (±)-TTX was accomplished by Kishi *et al.* in 1972.[275,276] The etiology of TTX (**325**) is of interest because of the wide distribution of the toxin among genetically unrelated animals (Table 18).

Table 18 Distribution of TTX in animals.

	Organisms
Amphibians	newts, frogs
Fish	puffers (*Fugu*), breams (*Ypsiscarus*, *Scarus*), gobies (*Pomacanthus*)
Molluscs	gastropods (*Babylonia*), Octopoda
Echinoderms	starfish (*Astropecten*)
Anthropods	crabs (*Zosimus*)
Annelids	lugworms (*Pseudoptamilla*)
Nemerteans	Anopla (*Lineus*)
Flatworms	Polyclada
Red algae	Cryptonemiales (*Jania*)

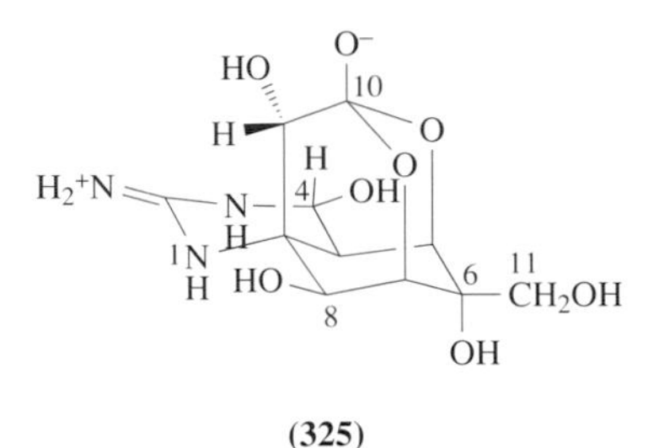

(325)

The marked variations in toxin concentrations in TTX-containing animals according to individual, region, and season suggested the exogenous origin of the toxin in those animals. This was further supported by the observation that puffers raised with artificial baits are unable to develop toxicity; TTX concentrations in cultured puffer fish, *Fugu rubripes*, were determined by indirect competitive enzyme immunoassay with a monoclonal antibody against the toxin, and the average concentrations in skin, muscle, liver, and vescera were 48.9, 11.6, 2.6, and 6.4 ng g^{-1}. These levels were very low and well within accepted tolerances.[277] The primary source of TTX was therefore investigated and demonstrated to be a marine bacterium.[278] The bacterium isolated from a red alga was first assigned to *Pseudomonas* sp., then to *Alteromonas* sp., and finally placed in a new species, *Shewanella alga*. The toxin was identified by Fast Atom Bombardment Spectrometry (FABMS), fluorometric HPLC, dose–survival responses in mice, and degradation into 2-amino-6-(hydroxymethyl)-8-hydroxyquinazoline. Subsequently, a broad spectrum of bacteria were reported to produce the toxin, such as *Vibrio* sp. from a xanthid crab *Atergatis floridus*[279] and *Pseudomonas* sp. from the skin of a puffer fish *Fugu poecilontus*.[280] The amount of toxin in the bacteria was, however, so small that chemists were not able to conduct biosynthetic studies.

In spite of the long history of TTX research, little is known about the biosynthetic origin of this unique molecule. The feeding of radioactive general metabolic precursors to toxic newts and their symbiotic bacteria failed to effect the incorporation of radioactivity into TTX.[281] As an alternative approach, the search for molecules which may provide clues for the biosynthetic pathway of TTX has been extensive. A highly sensitive HPLC method for analyzing TTX has been developed. This analyzer separated analogues on a reversed-phase column and detected fluorescent products produced by heating in sodium hydroxide solution.[282]

A number of new TTX analogues were isolated from puffers, newts, and a frog: tetrodonic acid (**326**), 4-epiTTX (**327**), and 4,9-anhydroTTX (**328**) from Japanese *Takifugu pardalis* and *T. poecilonotus*,[283] 11-norTTX-6(*R*)-ol (**329**) from *Fugu niphobles*,[284] and 11-oxoTTX (**330**) from Micronesian *Arothron nigropunctatus*.[285] From the Okinawan newt *Cynops ensicanda* 6-epi-TTX (**331**) and 11-deoxyTTX (**332**) were isolated,[286] and in this study the structural determination of

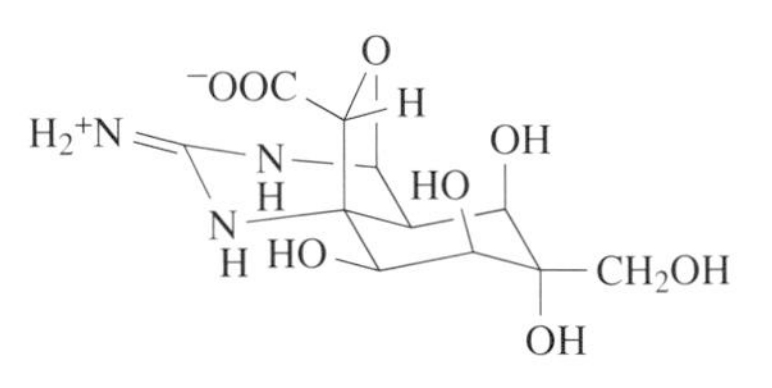

(326)

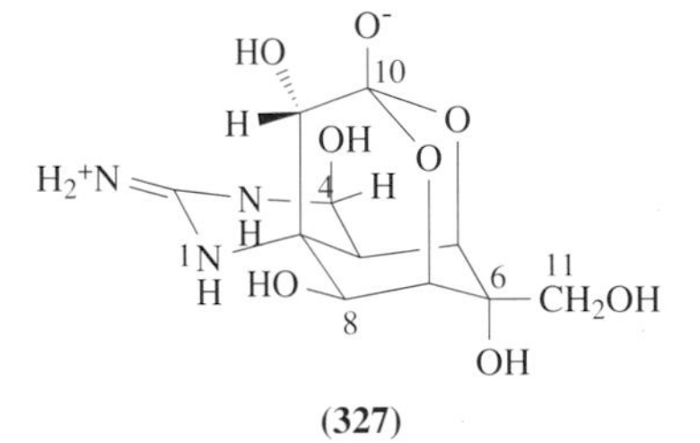

(327)

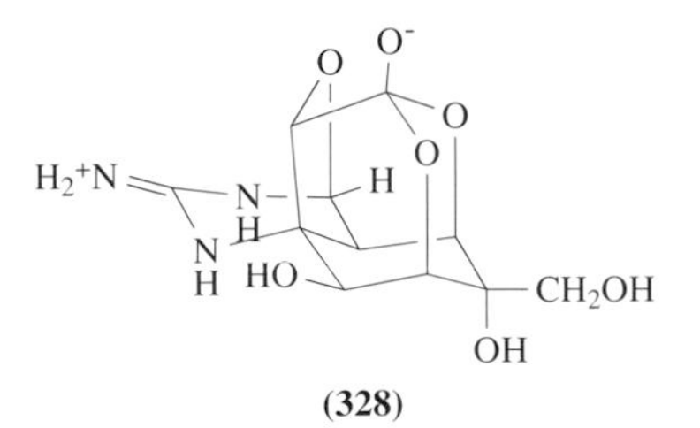

(328)

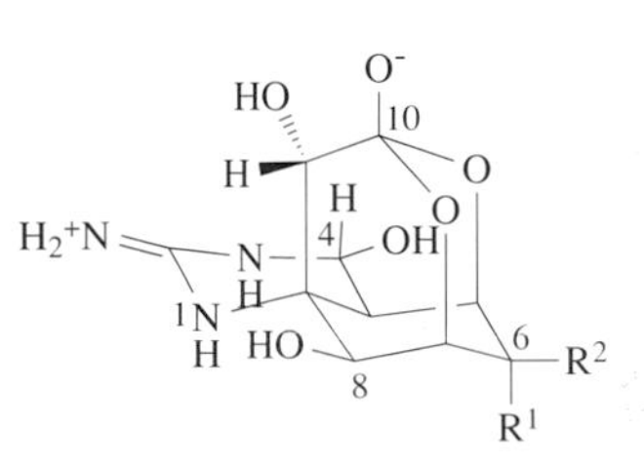

(329) $R^1 = H$, $R^2 = OH$
(330) $R^1 = OH$, $R^2 = CH(OH)_2$
(331) $R^1 = CH_2OH$, $R^2 = OH$
(332) $R^1 = OH$, $R^2 = Me$
(333) $R^1 = OH$, $R^2 = H$

these TTX analogues was achieved mainly through NMR measurements. The poor resolution of ^1H and ^{13}C NMR spectra of TTX caused by hemiacetal-lactone tautomerism (Equation (4)) was markedly improved by the addition of CF_3COOD (1%) to the solvent (4% CD_3COOD/D_2O) to allow the ^1H and ^{13}C signals to be firmly assigned by ^1H-^1H and ^{13}C-^1H COSY measurements.

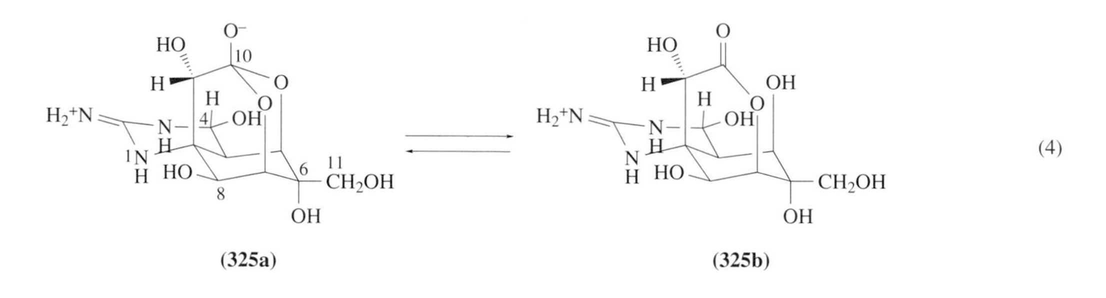

A Costa Rican frog, *Atelops chiriquiensis*, contained TTX (**325**) and chiriquitoxin (**334**). Chiriquitoxin (**334**) was first isolated in 1975 and was as potent as TTX (**335**) in lethality to mice and in blocking the voltage-gated sodium channel, whereas other analogues such as (**326**)–(**328**) have markedly reduced biological activities.[287] It was found that analogues modified at C-6 or C-11 such as (**329**)–(**332**) and (**334**) still retained significant potency to block sodium channels. The structure of chiriquitoxin (**334**) was determined to be a TTX analogue with a glycine molecule attached to C-11. The high potency of chiriquitoxin (**334**) suggested the sodium channel protein had specific binding sites for the C-12 NH_2 and/or the C-13 CO_2H. In addition, functionalities within the channel which interact with the guanidinium and the C-9 and C-10 hydroxyls of TTX (**325**) may be operative. ^1H and ^{13}C NMR data indicated that, under acidic conditions, H-12 of chiriquitoxin (**334**) and its 6,13-lactone derivative (**335**) slowly exchanged within one month with deuterium of solvents. However, the configuration at C-12 did not change.

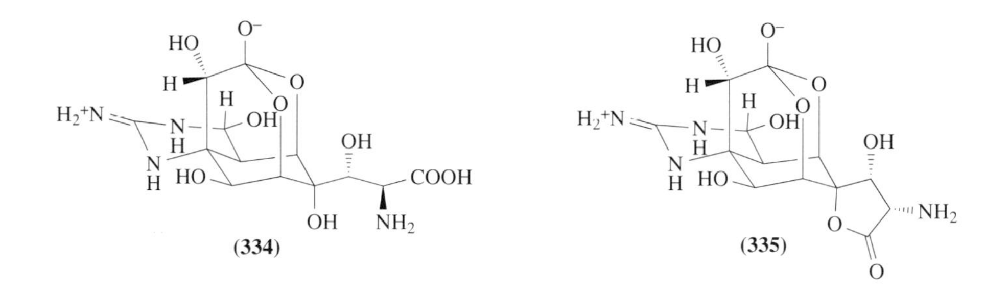

From the southern puffer *Arothron nigropunctatus*, collected in Okinawa, 11-nor-TTX-6(*S*)-ol (**333**) was isolated, and the structure was assigned by NMR data and confirmed by preparation of (**333**) from (**325**).[288] Oxidation of TTX (**325**) with H_5IO_6 followed by reduction with $NaBH_3CN$ afforded two products in the ratio of 1:6, the minor product being assigned as 11-norTTX-6(*R*)-ol (**329**), while the major one was concluded to be its epimer at C-6, 11-norTTX-6(*S*)-ol (**333**).

The *N*-hydroxyl and ring-deoxy derivative of TTX, 1-hydroxy-5,11-dideoxyTTX (**336**) was isolated from the Californian newt *Taricha torosa*.[289] Compound (**336**) was the first 10,7-lactone type analogue of TTX. In addition, 5,6,11-trideoxyTTX (**337**) and its 4-epimer (**338**) were isolated from the puffer fish *Fugu poecilonotus* collected at Shimonoseki, Japan.[290] The discovery of these TTX analogues may be highly significant with respect to the biosynthesis of TTX. Biosynthesis of TTX supposedly involves arginine, analogous with saxitoxin, and a C-5 unit derived from either amino acid, isoprenoids, shikimates, or branched sugars. The occurrence of 6-epiTTX (**331**) and 11-deoxyTTX (**332**) may render an isoprenoid unit a favorable candidate, because it possesses both sp^2 carbons available for Diels–Alder-type condensation and a methyl that remains in (**332**). 11-DeoxyTTX (**332**), 11-norTTX-6-ols (**329**) and (**333**), and 11-carboxyaminomethyl derivative (**334**) were considered to be compounds at different stages of progressive oxidation or their derivatives as in the case of chiriquitoxin (**334**). While structural variations found in (**329**) and (**332**)–(**334**) are limited to the 6,11-positions, i.e., the branching portion, compounds (**336**)–(**338**) were TTX derivatives which lack a ring oxygen function. It seemed, therefore, likely that TTX was formed by the stepwise oxidation of an alicyclic system, and one of the speculations that TTX was formed by the condensation of arginine and a branched sugar was excluded. The discovery of 5,6,11-trideoxyTTX (**337**), which possesses a C-5-methylene and C-6-methine in the cyclohexane ring, also supported the isoprenoid origin (Scheme 9, route A). Alternatively, the 2-aminoperhydropyrimidine portion of TTX (**325**) could be constructed from guanidine and 2-deoxy-3-oxo-D-pentose. Condensation of the product with an isoprenoid would also lead to (**337**) (Scheme 9, route B). Co-occurrence of (**337**) and its 4-epimer (**338**) as an equimolar mixture seemed to favor route B, which is not stereoselective with regard to C-4-OH. If oxidation of C-4 occurred enzymatically after ring formation as in route A, the resulting product would be generated more stereospecifically. However, epimerization at C-4 occurs rather easily.

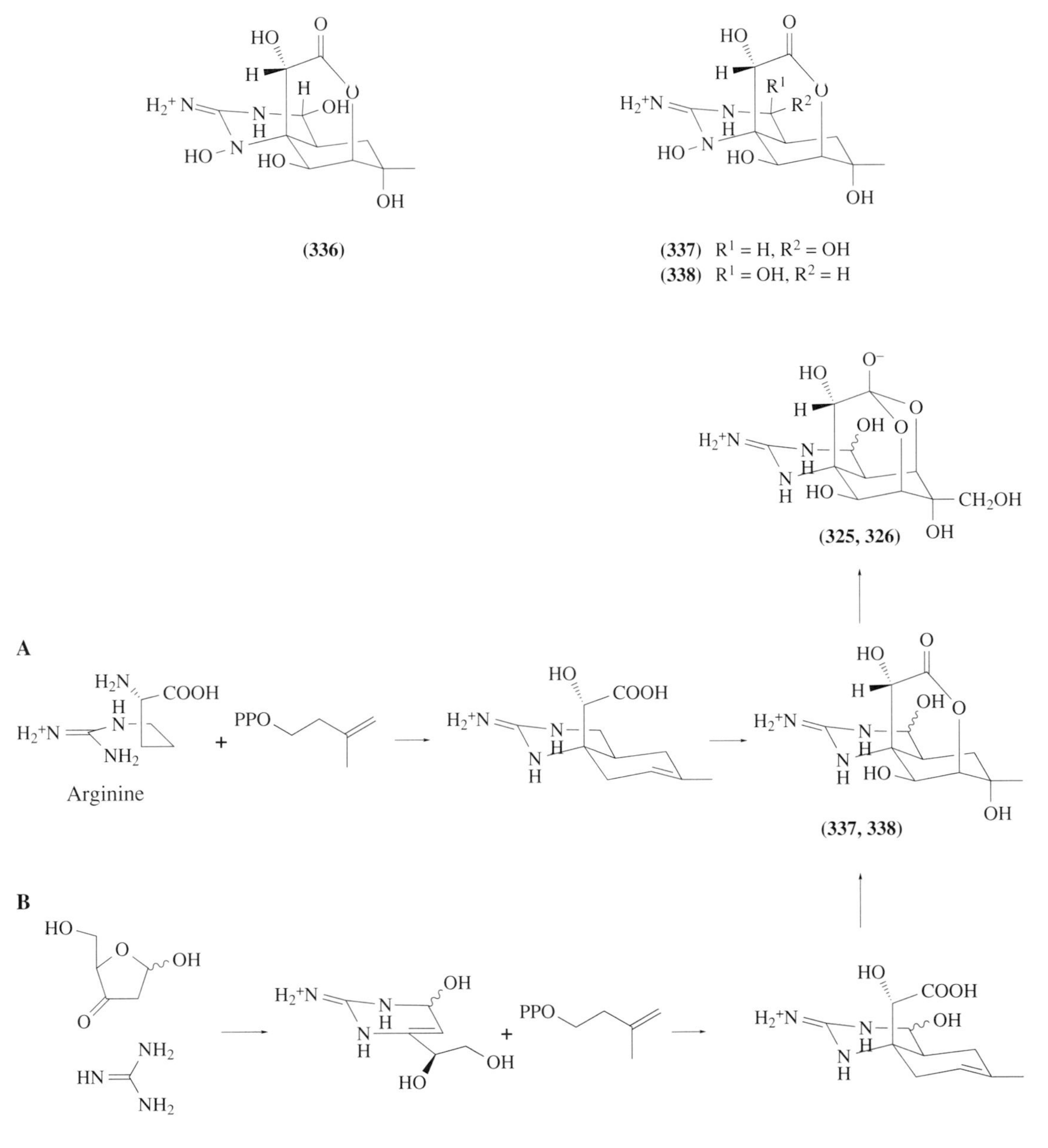

Scheme 9

Several derivatives of TTX analogues were prepared for determination of their binding potencies to sodium channels.[291] 4-EpiTTX (**327**), 4,9-anhydroTTX (**328**), 6-epiTTX (**331**), 11-deoxyTTX (**332**), and chiriquitoxin (**334**) were obtained from natural sources, while tetrodonic acid (**326**),

11-norTTX-6(*S*)-ol (**329**), 11-oxoTTX (**330**), 11-norTTX-6(*S*)-ol (**333**), 11-norTTX-6,6-diol (**339**), TTX-11-carboxylic acid (**340**), and TTX-8-*O*-hemisuccinate (**341**) were prepared chemically (Scheme 10).

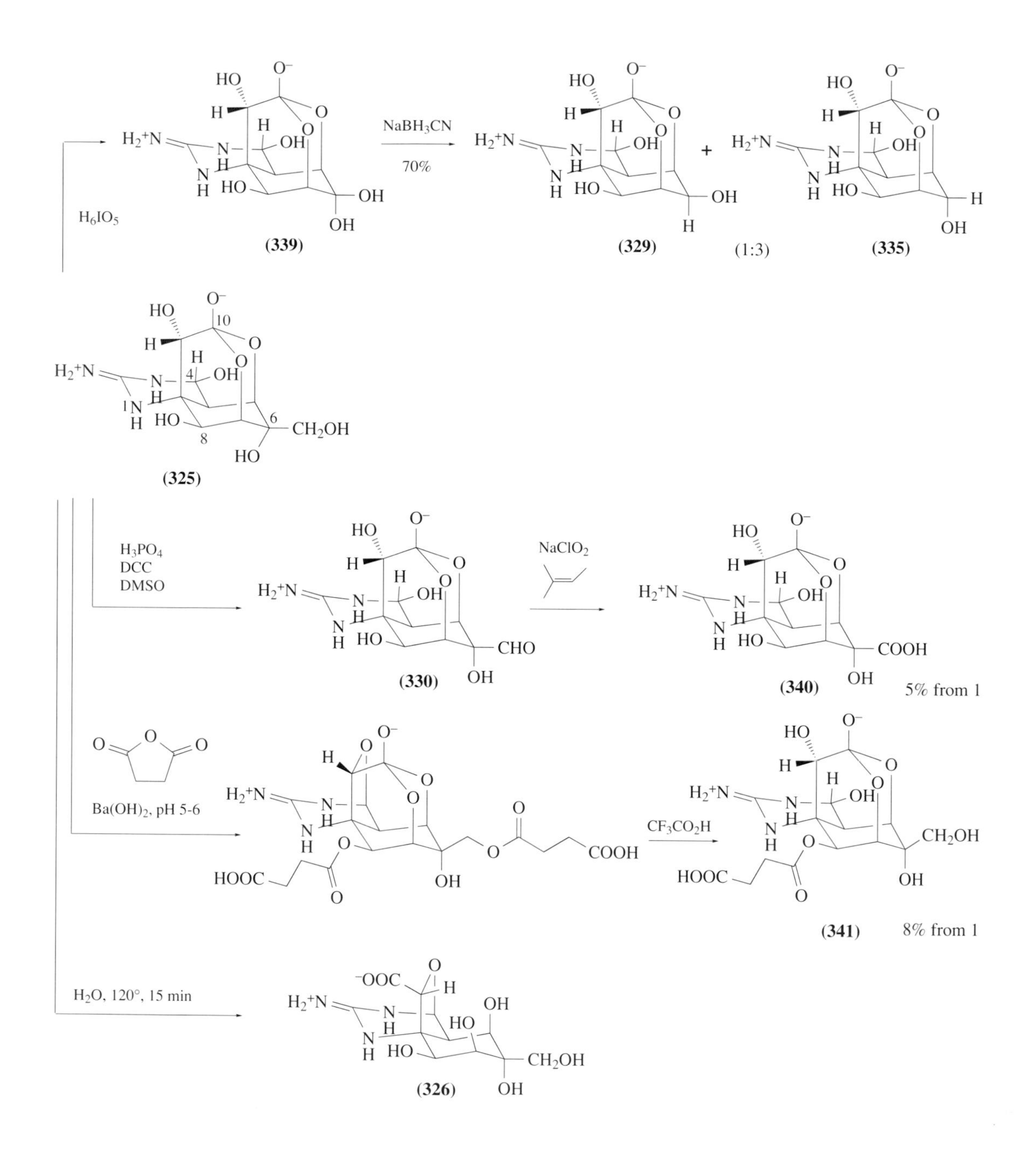

Scheme 10

Isolation, chemical reaction, and purities of the final products were monitored by fluorometric HPLC as mentioned above. The quantity of each derivative was determined by ^{1}H NMR spectroscopy by comparison of the signal intensity of H-4 with that of Bu^{t}OH added to the solution as an internal standard (0.000 38% Bu^{t}OH in 3.96% CD_3COOD-D_2O). Purities of the derivatives proved to be more than 99.3% by HPLC analyses. Their binding potencies to the sodium channel protein were examined by competitive binding assays against ^{3}H-saxitoxin (STX), which shares the same binding site with TTX (**325**) (Table 19).

Table 19 ^{3}H-STX binding inhibition activities of TTX analogues.

Analogues	IC_{50} (nM)	K_i (nM)
(325)	13.7±1.5	1.89±0.21
(326)	> 3000	> 340
(327)	384±45	43.3±5.1
(328)	1142±256	129±20
(329)	186±25	25.7±3.5
(330)	11.6±1.8	1.60±0.25
(331)	295±35	40.7±4.8
(332)	279±38	38.5±5.2
(333)	161±25	22.2±3.5
(334)	6.90±1.9	0.774±0.22
(339)	13.7±1.1	1.54±0.13
(340)	> 3000	> 340
(341)	2553	287

Compounds (**330**), (**334**), and (**339**) retained the high potency of (**325**). The potencies of compounds (**327**), (**329**), and (**331**)–(**333**) were decreased to 1/20–1/40 of (**325**). Moreover, the channel binding potency was almost lost in compounds (**326**), (**328**), (**340**), and (**341**). These results were almost consistent with the stereostructural locations of charged amino acid groups around the binding site of (**325**) deduced from site-directed mutagenesis[292] and the photoaffinity labeling experiments of the sodium channel[293] (see Section 8.07.8.2.3).

8.07.8.2.2 Saxitoxin

Saxitoxin (STX (**342**)) and its analogues are well-known paralytic shellfish poisons (PSP),[294,295] and are the most serious threat to public health among all seafood poisonings. The economic damage caused by accumulation of the toxins in shellfish is immeasurable. STX (**342**) and its analogues are produced by a variety of organisms, including dinoflagellates of the genus *Alexandrium* (formerly *Gonyaulax* or *Protogonyaulax*), *Gymnodinium*, and *Pyrodinium*, a freshwater blue-green alga *Aphanizomenon flos-aquae*, and the red alga *Jania* sp. The presence of STX (**342**) was found in the common Atlantic mackerel *Scomber scombrus*.[296] The occurrence of this toxin in taxonomically varied species as well as quantity differences within the same species suggests the presence of common vectors. STX (**342**) was identified as the toxin in the culture broth of a bacterium isolated from the dinoflagellate *Protogonyaulax tamarensis*.[297] Further confirmation was expected to substantiate PSP-producing bacteria.[298]

(342) $R^1 = R^2 = H$
(344) $R^1 = H, R^2 = OSO_3^-$
(346) $R^1 = OSO_3^-, R^2 = H$

Detection of STX analogues was facilitated by the use of a fluorometric HPLC analyzer, which measures fluorescence derived from oxidation products.[299] The toxins can be divided into two groups, viz., those with and without a hydroxyl group at the N-1 position: saxitoxin (**342**) and neosaxitoxin (**343**). Both groups were further diversified by the presence of 11-*O*-sulfate (**344**)–(**347**) or *N*-sulfate (**348**)–(**353**), the absence of carbamoyl groups (**354**)–(**356**) and the absence of oxygen at C-13 (**357**)–(**359**).[294,295]

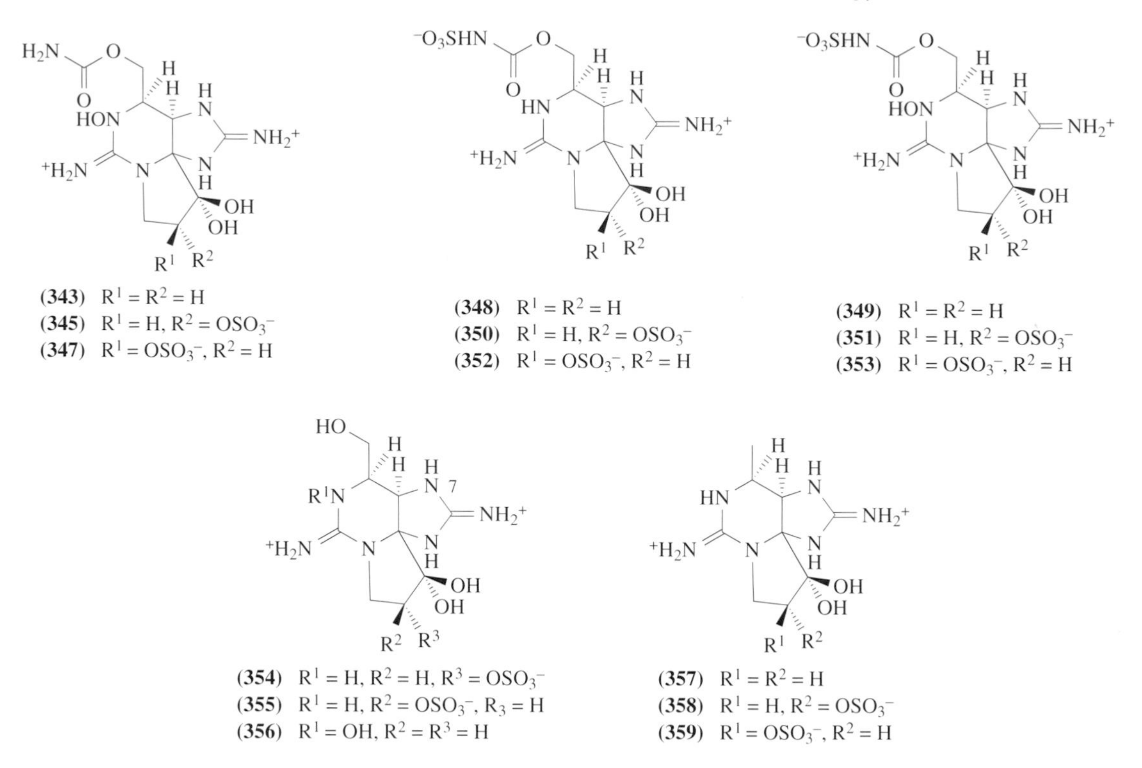

The biosynthesis of STX analogues was extensively investigated by feeding experiments of labeled precursors into *Alexandrium tamarense* and *Aphanizomenon flosaquae*.[295] The carbamoyl group and two guanidinium moieties were shown to be derived from the guanido group of arginine by the feeding experiment of [*guanido*-^{14}C]arginine. The feeding of [1,2-$^{13}C_2$]acetate showed that C-5 and C-6 came from one acetate unit. When [2-^{13}C-2-^{15}N]arginine was fed to the organism, the connectivity of ^{13}C–^{15}N was incorporated intact in neosaxitoxin (**343**). Thus, it was suggested that the skeleton of the tricyclic ring system was formed by a Claisen-type condensation of an acetate unit or its equivalent onto the amino-bearing α carbon of arginine followed by decarboxylation, introduction of a guanidine moiety, and cyclization (Scheme 11). Incorporation of ^{13}C atom was observed at the C-13 position by feeding with [*methyl*-^{13}C]methionine, and feeding experiments with [2-^{13}C-2-$^{2}H_3$]acetate established that the hydrogen found at C-6 was not from the acetate methyl group and that a deuterium atom on the acetate methyl group was shown to have rearranged to the C-5 position. Moreover, feeding [*methyl*-^{13}C,$^{2}H_3$]methionine demonstrated that only one of the methionine methyl hydrogens was left on the side-chain methylene group. From these observations, it was proposed (Scheme 11) that the side-chain carbon (C-13) was derived from a methionine methyl group by the electrophilic attack of *S*-adenosylmethionine (SAM) on a dehydro intermediate followed by hydride migration and elimination of a proton. The resulting exomethylene group was likely to be first converted into an epoxide followed by opening to an aldehyde and subsequent reduction into a hydroxymethyl group; these processes explain the retention of only one deuterium atom. The carbamoyl group may then be introduced to the C-13 carbinol position by use of the guanido group of arginine to build up the STX-type toxins.

Thus, three molecules of arginine may be required to construct one STX molecule. In some organisms, the toxin concentrations in the cell reach 60 pg/cell, which represents an enormous portion of the total organic content of the organisms. The biological significance of these compounds is often discussed and ascribed to self-defense, but the toxigenicity of the organism varies from strain to strain even within the same species with no apparent effect on their chance of survival.[295]

8.07.8.2.3 Sodium channels and TTX/STX

Tetrodotoxin (TTX (**325**)) and saxitoxin (STX (**342**)) are specific and potent blockers of the sodium channel, which is essential for the generation of action potentials in excitable cells. Studies[300] have suggested that these positively charged toxins are bound close to the extracellular orifice of the channel in a region containing negatively charged groups. The complete primary structure of

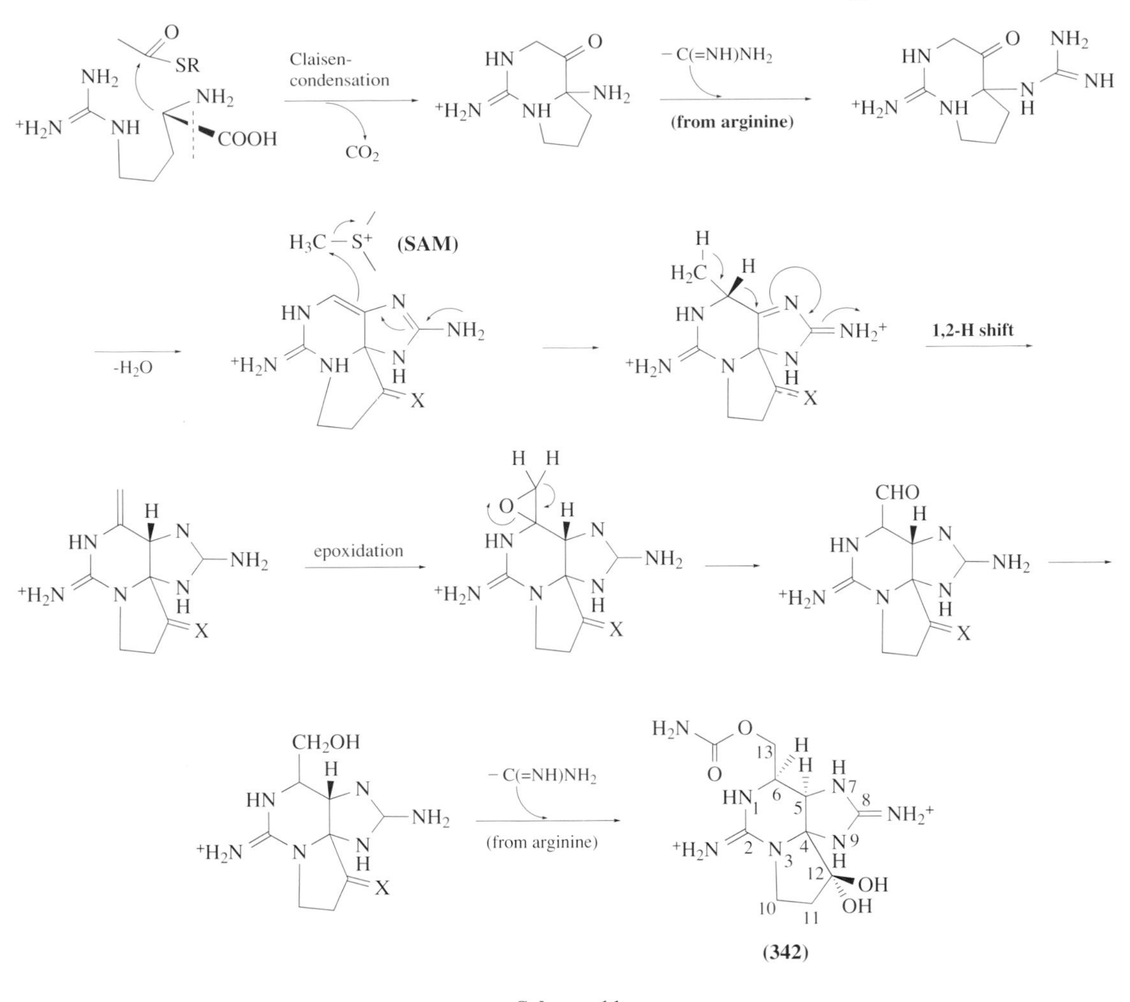

Scheme 11

the sodium channel protein from eel electroplax was first deduced from cDNA in 1984,[301] followed by those from rat brain, skeletal muscle, and heart. These proteins are highly homologous with each other, and each contains four internally homologous domains (repeats I–IV), each of which has six proposed transmembrane helices (segments S1–S6).[300] Single-point mutation of the rat sodium channel that replaced glutamic acid residue 387 (E387) with a glutamine reduced its sensitivity to TTX and STX by more than three orders of magnitude and reduced its single-channel conductance, whereas its macroscopic current properties were only slightly affected.[302] Another mutation that neutralized the aspartic acid residue 384 (D384) to asparagine almost completely eliminated ionic currents. Thus, residues D384 and E387 are suggested to be located at the extracellular mouth or inside the ion-conducting pore of the channel. D384 and E387 belong to the short segment SS2 in the region between the hydrophobic segments S5 and S6 in repeat I. In each of four repeats the S5–S6 region was thought to contain two short segments, SS1 and SS2, that may partly span the membrane as a hairpin and the SS2 segments were postulated as forming part of the channel lining. The effects on toxin sensitivity and single-channel conductance of site-directed mutations in the region encompassing the SS2 segment of each of the four repeats were investigated, focusing mainly on charged residues. As a result, the sensitivity to TTX and STX of the sodium channel was revealed to be strongly reduced by mutations of specific amino acid residues in the SS2 segment of each of the four internal repeats. These residues were found in two clusters of predominantly negatively charged residues and were located at equivalent positions in the SS2 segment of the four repeats. On the other hand, mutations of other amino acid residues in the SS2 and adjacent regions caused minor or insignificant changes in toxin sensitivity. These observations suggested that these two clusters of predominantly negatively charged residues, probably forming ring structures, take part in the extracellular mouth and/or the pore wall of the sodium channel (Figure 4).[292]

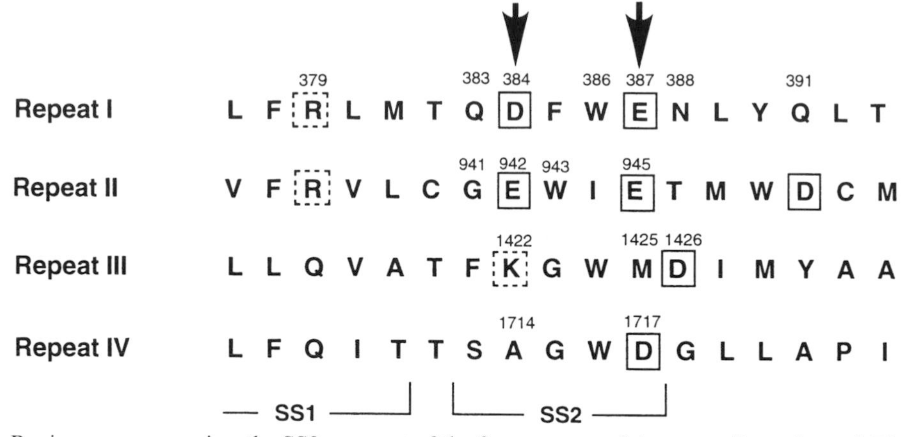

Figure 4 Regions encompassing the SS2 segment of the four repeats of the rat sodium channel. The positions of the SS1 and SS2 segments and the numbers of the amino acid residues mutated are indicated. Negatively charged residues are boxed with solid lines, and positively charged residues with broken lines. The positions of the clusters of residues that were identified as major determinants of toxin sensitivity are indicated by arrows.

It was also proposed that TTX and STX bind to the sodium channels with three points, at least, including the negatively charged groups. Photoaffinity labeling may play an important role in identification of toxin/ligand binding regions within the primary structures of ion channels.[293] Several photoactive TTX derivatives including (diazirino)trifluoroethylbenzoyl TTX (**360**)[303] were prepared and specifically labeled the electroplax sodium channel protein of 250 kDa.

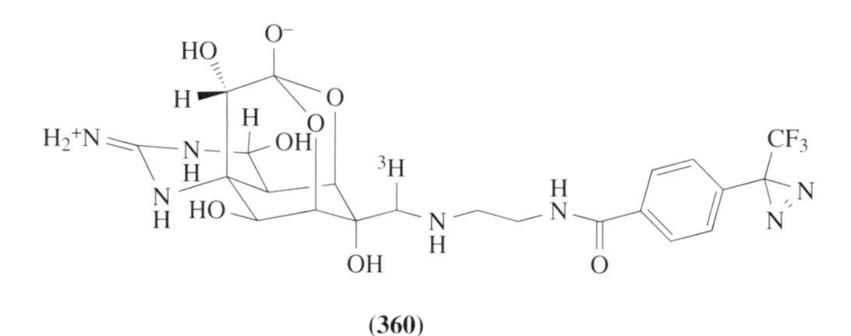

(360)

At least 43% of the labeled sites with a photoactive TTX derivative (**360**) were identified by probing protease-digested labeled fragments with several sequence-directed antibodies. They were located in the loop between segments S5 and S6 of repeat IV, as well as the region containing transmembrane segment S6 and adjacent extracellular and cytoplasmic sequences in repeat III. No photolabeled fragments were detected in the corresponding region of repeat I. These results suggested that C-11 of TTX where the photoreactive moiety was attached oriented to the region between S5 and S6 in repeats III and IV, and the guanidinium group of TTX which was apparently directed opposite to C-11 oriented most likely to the negative charge cluster of repeats I and II (Figure 5).[293]

As discussed in Section 8.07.8.2.1, TTX and its analogues, either isolated from natural sources or prepared by chemical modification of TTX, were tested for binding potencies to sodium channels with the use of [^{3}H]-STX as a competitor (Table 19). Inhibition of ionic bonds between the guanidium group of TTX and acidic clusters in repeats I and II resulted in complete loss of the channel-blocking activity. Oxidation of 11-CH_2OH to CO_2H also caused complete loss of the activity, probably due to repulsion between 11-CO_2H and Asp-1717 in repeat IV. The hemilactal structure was important for ionic interaction with Lys-1422 in repeat III. The effect of epimerization at C-6 was less significant. Reduction of hydrophilic 11-CH_2OH to hydrophobic 11-Me reduced hydrogen bond formation with Asp-1717 and thus resulted in reduced activity. These results of structure–activity relationship studies support the stereostructural locations of charged amino acid groups around the binding site deduced from the site-directed mutagenesis[292,302] and the photoaffinity labeling experiments[293] (Figure 4).

Why are puffer fish not intoxicated by their carrying tetrodotoxin? There are two possible answers to this question. Mutations appear to have occurred in sodium channel molecules of puffer fish to differentiate them from those of TTX-sensitive sodium channels of other animals. Sodium channels of puffer fish showed the lowest affinity to STX, while those of bonito showed higher affinity, and

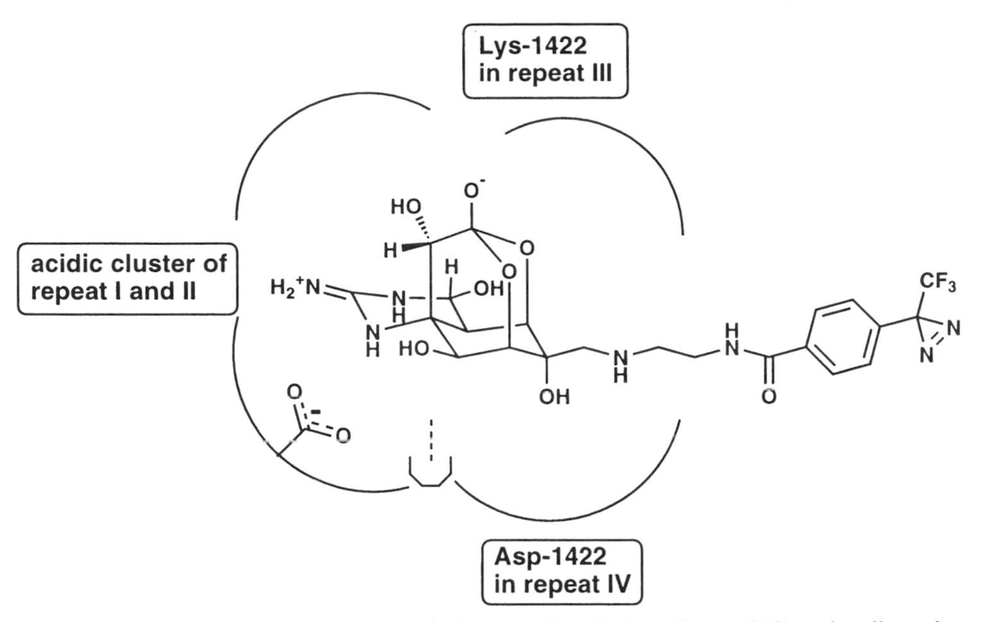

Figure 5 A model of the relative positions of photoreactive TTX analogue (**360**) and sodium channels.

those of rat brain showed the highest. The resistance of puffers to STX (=TTX) would thus be explained. Sequential analysis of cDNA of the sodium channel of puffer fish muscle revealed that the amino acid sequences of the binding site (site-1) differ at three points from those of the rat brain sodium channel. The mutation presumably is responsible for the self-resistant system of puffer fish against their own TTX.[304] The second possible explanation of the resistance mechanism against TTX in puffer fish may be the possible existence of other stronger TTX-binding proteins. The number of TTX-binding proteins in the muscle of puffer fish was estimated to be much greater than that of sodium channel proteins estimated from the binding of brevetoxin (BTX), while in the brain of puffer fish the numbers of TTX-binding proteins and sodium channel proteins (estimated from BTX-binding) were equal.[305] This finding suggested the presence of TTX-binding protein different from sodium channels in the muscle of puffer fish, and it may be speculated that the protein functions in a detoxification pathway for TTX of puffer fish. In connection with this observation, a bullfrog, *Rana catesbeiana*, having low STX-sensitivity, contained in its plasma a soluble component of unknown function that specifically binds to [^{3}H]-STX.[306] The protein responsible for this activity was named saxiphilin and was purified and shown to have a molecular weight of 89 kDa. The function of this protein is unknown but it may possibly be part of a species-specific detoxification mechanism for STX; when STX is administered to the frog, saxiphilin traps the toxin and, consequently, the sodium channels of the frog are not affected.

8.07.8.3 Diarrhetic Shellfish Poisoning

In 1976, mussel poisoning of *Mytilus edulis* broke out in Miyagi Prefecture, Japan. This shellfish poisoning disease was called diarrhetic shellfish poisoning (DSP) since the prominent human symptoms were gastrointestinal disorders such as diarrhea, nausea, vomiting, and abdominal pain. DSP was associated with ingestion of bivalves such as mussels, scallops, or clams which fed on toxic dinoflagellates. DSP was widely distributed in various places in Japan and Europe. The number of known DSP cases in Japan since 1976 exceeds 1300 despite the existence of extensive surveillance; 5000 cases were reported in the single year of 1981 in Spain and there were 400 cases in 1983 in France. Thus, it is a serious problem for public health as well as for the shellfish industry. Although no red tide was sighted during the infestation period, the regional and seasonal variation of shellfish toxicity strongly suggested a planktonic origin of the toxins. The causative organisms were identified as several dinoflagellates of the genus *Dinophysis*, and three kinds of polyether toxins, i.e., okadaic acid and its analogues, pectenotoxins, and yessotoxin, were identified as DSP toxins.[271]

8.07.8.3.1 Okadaic acid and dinophysistoxin

Okadaic acid (OA (**361**)) was first isolated from the sponge *Halichondria okadai*[307] and subsequently it was revealed that the acid was also detected in dinoflagellates *Prorocentrum lima* and *Dinophysis* sp. The toxin responsible for DSP caused by the mussel *Mytilus edulis* was identified as 35-methylokadaic acid, which was named dinophysistoxin 1 (DTX-1 (**362**)).[308] Detection of okadaic acid derivatives was carried out with high sensitivity by a fluorometric HPLC analysis of 9-anthryldiazomethane ester derivatives.[309,310] Okadaic acid (**361**) was toxic (LD_{50} 192 μg kg^{-1}, mouse, i.p.) and showed potent cytotoxicity with ED_{50} values of 1.7 ng ml^{-1} and 17 ng ml^{-1} against P388 and L1210 cells, respectively. In addition, (**361**) was discovered to act as an inhibitor of protein phosphatases 1 and 2A, and widely utilized as a tool for biochemical and pharmacological studies.[311] A series of OA/DTX-related compounds were thus far known. Among them, DTX-3 (**363**)[312] was an unsaturated fatty acid ester attached on 7-OH of DTX-1, and was identified as a main toxic constituent of Japanese scallops, while a saturated acid ester, 7-*O*-palmitoyl DTX-1 (**364**),[313] also obtained from the mussel, was not toxic. The unsaturation in the acyloxy group seemed to play a substantial role in the bioactivity of the toxin. From the dinoflagellate *Prorocentrum lima*, a diol ester (**365**) of okadaic acid was isolated[313] and found to be more toxic (125 μg kg^{-1}, mouse, i.p.) than okadaic acid (**361**). From a mussel digestive gland collected at Bantry Bay, Ireland, where *Dinophysis* sp. occurred and DSP-activity was observed, an isomer of okadaic acid, named dinophysistoxin 2 (DTX-2 (**366**)),[314] was isolated together with okadaic acid (**361**). An underivatized extract of mussel was examined by an ion spray LC–MS method.[315] Extensive spectroscopic studies on (**366**) revealed that this compound bore no methyl group on C-31 but a methyl group was present on C-35. Compound (**366**) had comparable activity to (**361**) and (**362**) in the rat toxicity bioassay.

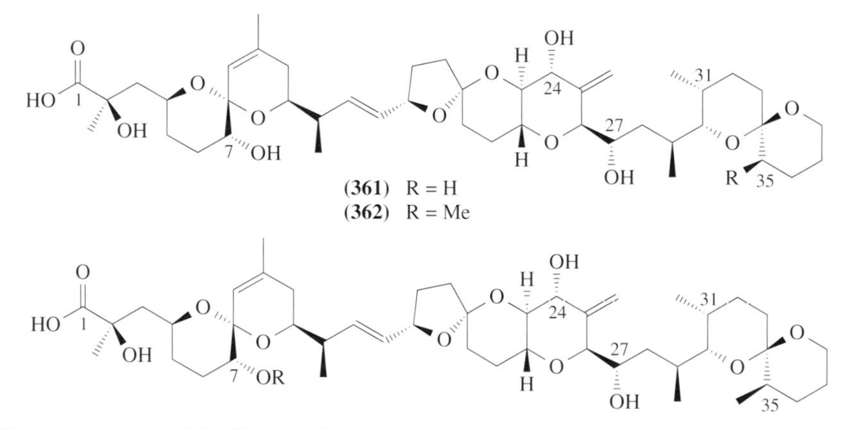

(**361**) R = H
(**362**) R = Me

(**363**) R = acyl [$C_{14:0}$ (13%), $C_{16:0}$ (29%), $C_{18:3}$ (3%), $C_{18:4\omega3}$ (9%), $C_{20:5\omega3}$ (23%), and $C_{22:6\omega3}$ (29%)]
(**364**) R = palmitoyl

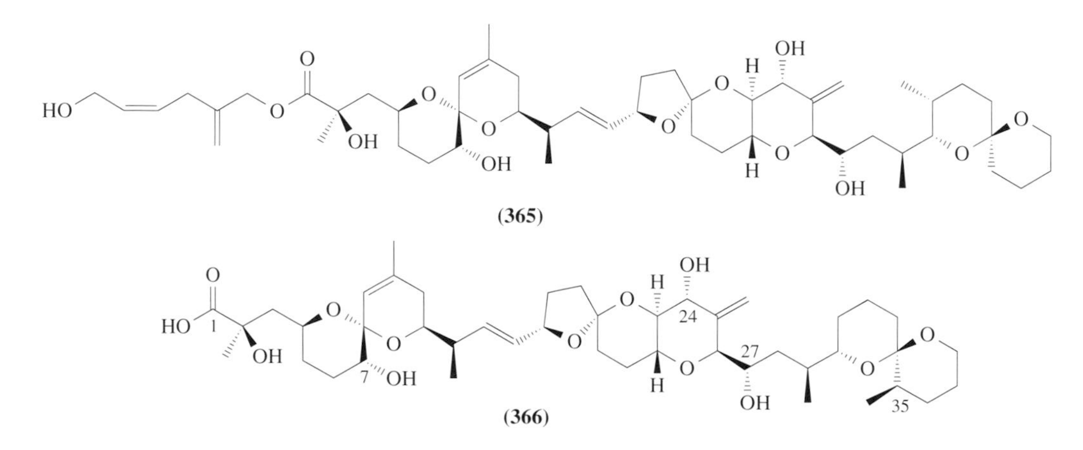

(**365**)

(**366**)

From the cultures of the dinoflagellates *Prorocentrum lima* and *Prorocentrum maculosum*, diol esters (**367**)–(**369**) were obtained.[316] It was observed that the cellular concentration of okadaic acid (**361**) decreased with time in *Prorocentrum* sp. while the concentration of the esters increased, suggesting that the esterified products may be shunt metabolites, produced at a certain stage in the growth cycle, or when the concentration of okadaic acid (**361**) in the cell reached a critical level. It was found that the methyl ester of okadaic acid (**361**) did not show inhibitory activity of protein phosphatases 1 and 2A, suggesting that a free carboxylic acid was essential for the activity. The diol esters (**367**) and (**368**) were also found to be inactive in the inhibition of phosphatases, further supporting the idea that a free carboxyl group is essential for the activity. Water-soluble sulfated DSP toxin derivatives have been isolated; DTX-4 (**370**)[317] from *Prorocentrum lima* and DTX-5a (**371**) and -5b (**372**)[318] from *Prorocentrum muculosum*. The sulfated ester (**370**) was toxic in the mouse bioassay,[317] but its phosphatase inhibition activity was weaker by almost an order of magnitude.[318] The occurrence of these sulfate esters may support the idea that the less active sulfate esters of DSP toxins were the form in which the toxin was stored within, and excreted from, the dinoflagellate cell. Okadaic acid (**361**) was observed in the medium of *P. lima* cultures, suggesting that after exiting the cell the sulfate form was hydrolyzed to the biologically active form, which could not reenter the *Prorocentrum* cell.

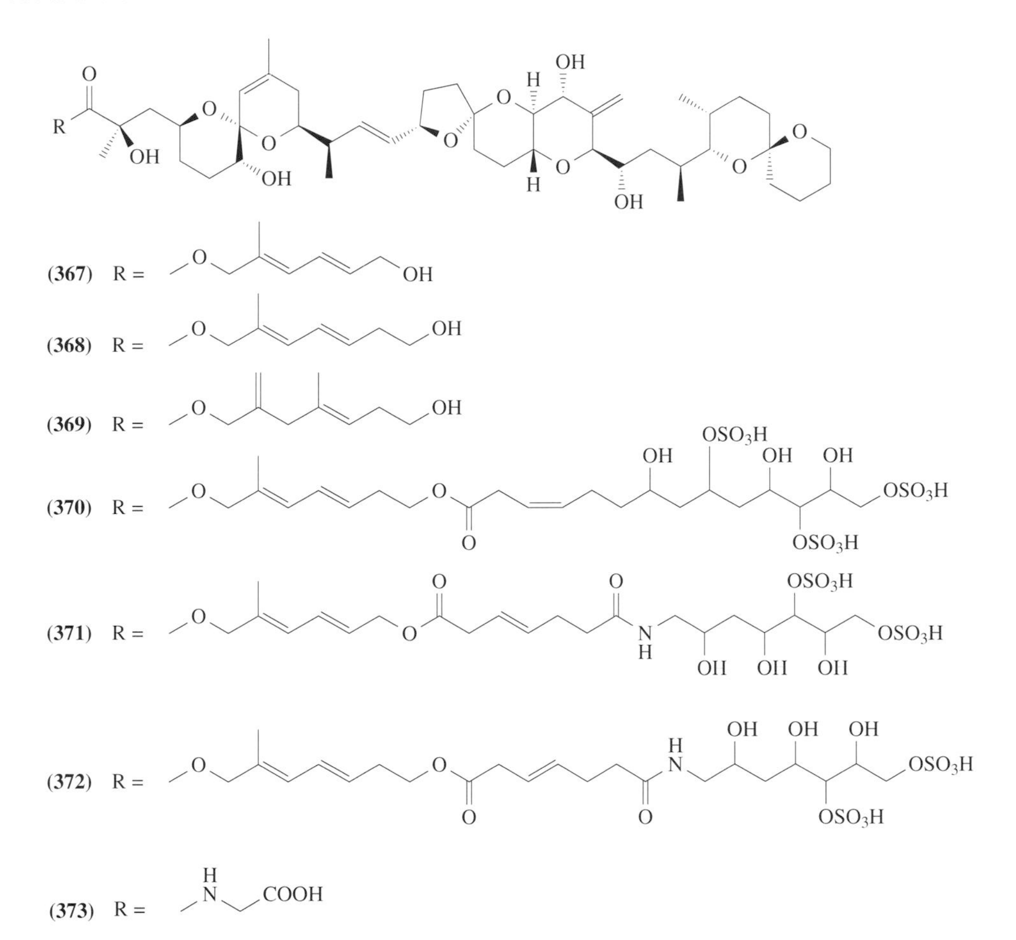

Okadaic acid (**361**) was revealed to be a potent tumor promoter of a new class, different from those belonging to the class of TPA (phorbol-12-tetradecanoate-13-acetate)-type tumor promoters.[319] TPA-type promoters were known to activate protein kinase C, while okadaic acid (**361**) inhibited dephosphorylation of proteins, predominantly serine/threonine residues. Both types of tumor promoters eventually cause the accumulation of essentially the same phosphorylated proteins, which may be involved in tumor promotion. The structure–activity relationships (SAR) within a series of okadaic acid derivatives were investigated by evaluating 17 okadaic acid derivatives as possible tumor promoters by means of three biochemical tests: inhibition of specific [^{3}H]-okadaic acid binding to a particulate fraction of mouse skin containing protein phosphatases, inhibition of protein phosphatase activity, and induction of ornithine decarboxylase in mouse skin.[320] The results indicated that the carboxyl group as well as the four hydroxyl groups at C-2, C-7, C-24, and C-27 of okadaic acid were important for activity. Acanthifolicin (**374**), an episulfide-containing polyether isolated from the sponge *Pandaros acanthifolium*,[321] also gave a positive response in the three biochemical tests as strong as those of okadaic acid (**361**). Glycookadaic acid (**373**), a conjugated form of (**361**) with glycine, proved to be an anticachexia substance.[322] Elucidation of the interrelationship between the fine structure and the activity of (**361**) was studied by NMR, and the formation of a flexible cavity between the C-1 carboxyl group and the C-24 hydroxyl group (Figure 6) was proposed.

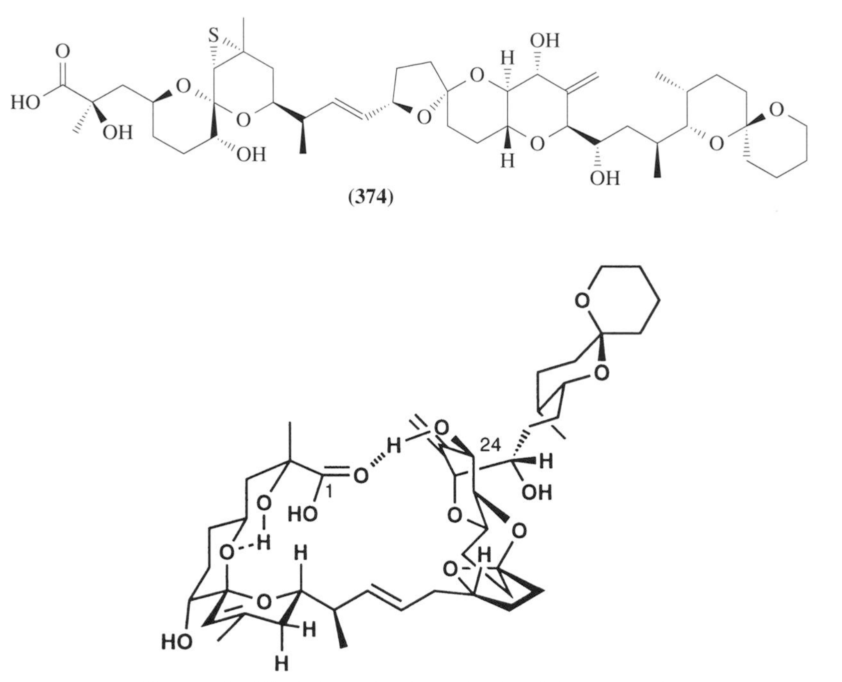

Figure 6 Flexible cavity hypothesis of okadaic acid.

Based on this proposal, a derivative (**375**), the C-24 hydroxyl group of which was epimerized, was prepared (Scheme 12), and was revealed to be less cytotoxic than (**361**) by two orders of magnitude.

The proposed structure of (**361**) including the flexible cavity (Figure 6) was supported by lactonization of okadaic acid (**361**) to give a 3-*O*-methylthiomethyl-1,24-olide (**376**) (Scheme 13). The results of SAR studies on okadaic acid derivatives are summarized in Figure 7,[323] which suggests that (i) the presence of the double bond at C-9/C-10 and the methyl group on C-35 is not important for activity, and (ii) the C-1 carboxyl group and four hydroxyl groups play certain roles in showing the activity. These studies showed that an intramolecular hydrogen bond between the C-1 carboxyl group and the C-24 hydroxyl group was very strong. The triacetate of okadaic acid was chromatographically more polar than (**361**), providing a supportive observation for the strong hydrogen bond network at the C-1/C-24 positions. The tetra-*O*-methyl derivative and C-24 epimer of (**361**), both of which cannot construct the C-1/C-24 hydrogen bond, had considerably reduced activities.

The crystal structures of the *o*-bromobenzyl ester of okadaic acid (**361**)[307] and acanthifolicin (**374**)[321] corresponded well to each other. Three-dimensional structures of okadaic acid (**361**) in organic and aqueous solutions (CD_3OD-$CDCl_3$ or NaOD/D_2O) were analyzed on the basis of long-range carbon-proton coupling constants which were measured by using hetero half-filtered TOCSY and phase-sensitive HMBC techniques,[324] revealing that the conformations in organic and aqueous solutions (Figure 8) resembled each other and that obtained for the crystalline *o*-bromobenzyl ester of (**361**).[307] The methodology using the $^{2,3}J_{C,H}$ and $^{3}J_{H,H}$ values proved to be powerful in elucidation of conformations and configurations of acyclic or macrocyclic portions in natural products of this size.

Several studies on the biosynthesis of okadaic acid and dinophysistoxins reported on the basis of feeding experiments with labeled precursors to the cultures of the dinoflagellates *Prorocentrum* sp. First, the ^{13}C NMR signals of okadaic acid (**361**) were assigned by analyses of 2D NMR spectra, and [1-^{13}C], [2-^{13}C], or [1,2-$^{13}C_2$] sodium acetate was added to cultures of *Prorocentrum lima*.[325] The ^{13}C NMR spectra of labeled compounds showed a 15-fold enhancement of the signal intensities of the labeled carbons. Out of the 44 carbons of (**361**), signals for 39 carbons (carbons other than C-1, C-2, C-37, C-38, and C-44) were enhanced, according to Torigoe and Yasumoto's report.[325] Investigation of the biosynthetic origin of dinophysistoxin-1 (**362**) by addition of [1-^{13}C], [2-^{13}C], or [1,2-$^{13}C_2$] sodium acetate to artificial cultures of *P. lima* revealed enhancement of all carbons except

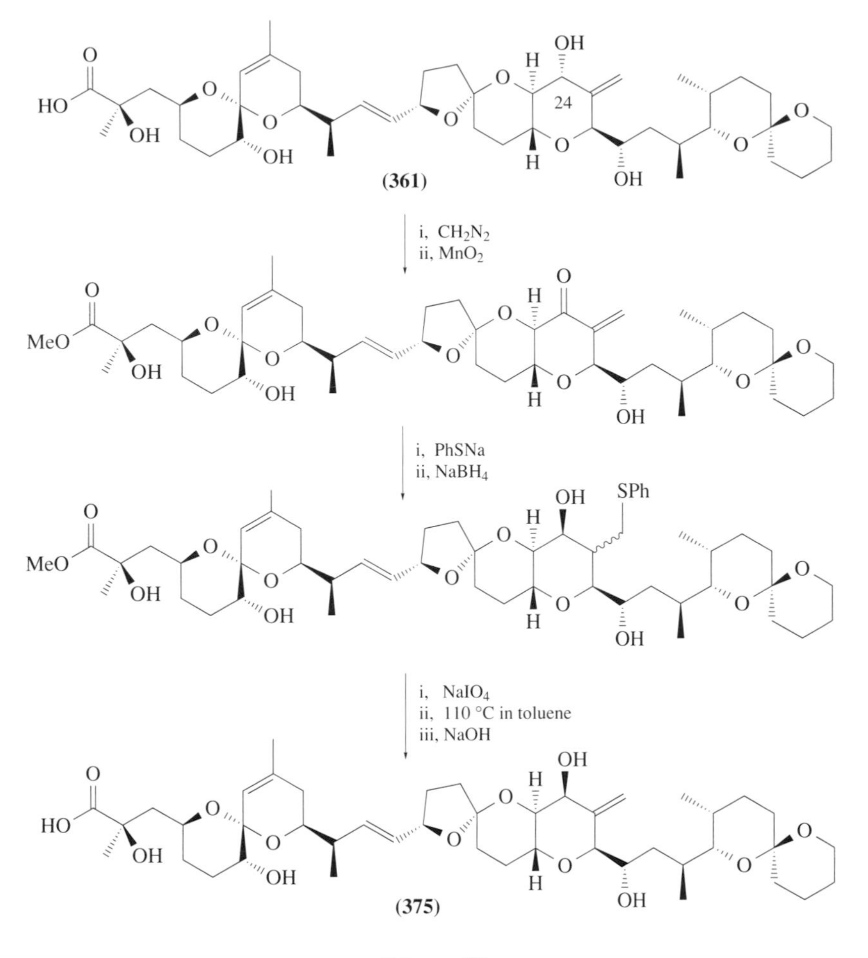

Scheme 12

for the two at C-37 and C-38.[326] Incorporation patterns of intact acetate units to (**361**) or (**362**) are shown in Figure 9. This labeling pattern suggested a type of mixed polyketide involving building blocks formed through dicarboxylic acids such as malonate, succinate, and 3-hydroxyl-3-methylglutarate (Figure 10). In Figure 10, possible biosynthetic precursors of okadaic acid (**361**) and DTX-1 (**362**) proposed by Torigoe and Yasumoto[325] (Figure 10(a)) and Norte *et al.*[326] (Figure 10(b)) are shown. Both Figures 10(a) and 10(b) contain fragments proposed to be derived from classical polyketides. Those in Figure 10(b) contain several methyl branches derived from the C-2 carbon of an acetate attached to carbons derived from the C-1 carbon of an acetate in a linear carbon chain, a feature rarely encountered in polyketide biosynthesis. The participation of dicarboxylic acid precursors had been proposed in the biosynthesis of brevetoxins, the polyether marine toxins isolated from red tide dinoflagellates (see Section 8.07.8.5.2), and it reasonably explained the formation of the "c-m-m-m" moiety (e.g., C-8/C-9/C-10/C-40 in Figures 9 and 10) from acetate-derived precursors (e.g., succinate) coming after two rounds through the tricarboxylic acid (TCA) cycle.

The C-37 and C-38 carbons of okadaic acid (**361**) or DTX-1 (**362**) were not derived from an intact acetate unit, and it has been suggested that these carbons come from an alternative C_2 source. Possible sources of C_2 units include alanine, glycerate, glycine, glycolate, lactate, pyruvate, and succinate. A stable isotope incorporation study was conducted to determine the biosynthetic origin of the C-37/C-38 position in (**361**) and (**362**).[327] While [1,2-$^{13}C_2$]glycine showed small enrichment on feeding to cultures of *P. lima*, [1,2-$^{13}C_2$]calcium glycolate (synthesized from [1,2-$^{13}C_2$]bromoacetic acid) showed enhanced resonances corresponding to C-37 and C-38, both appearing as doublets arising from ^{13}C–^{13}C coupling ($J_{C\text{-}37/C\text{-}38}$ = 35 Hz). Thus, these findings implied that the carbons C-37 and C-38 of okadaic acid (**361**) and DTX-1 (**362**) were derived from glycolate (HO_2CCH_2OH).

The biosynthesis of the water-soluble toxin, DTX-4 (**370**), possessing a sulfated acyl ester chain and the diol ester (**368**) was also investigated by feeding experiments with ^{13}C- and ^{18}O-labeled

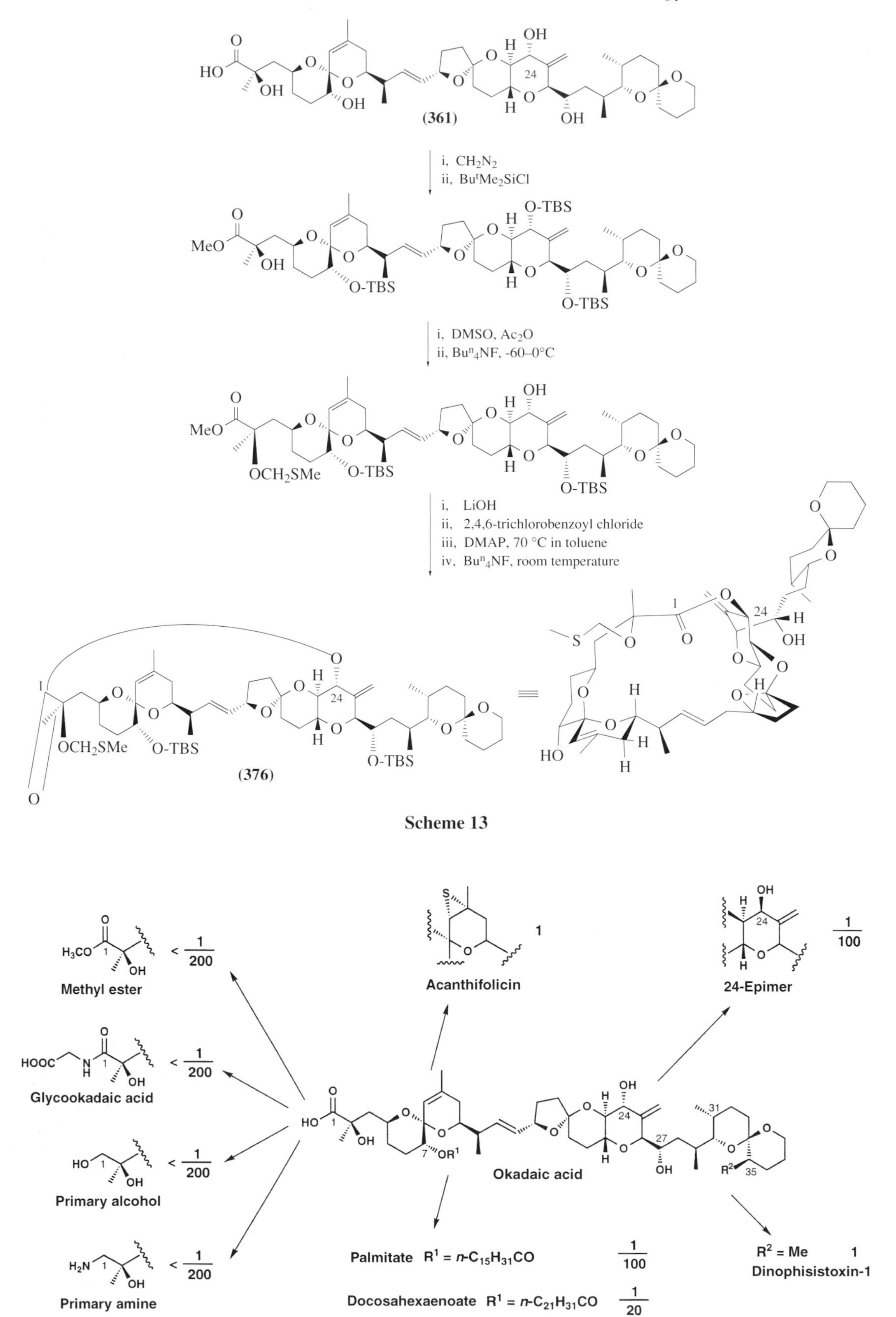

Figure 7 Structure and activity relationships of okadaic acid derivatives. (The activity of okadaic acid was normalized as 1.)

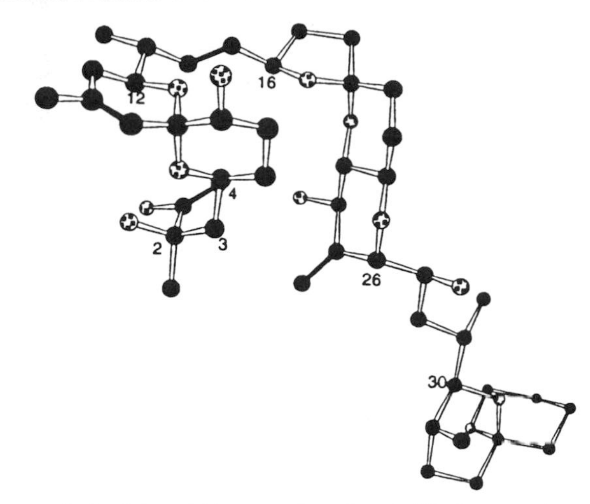

Figure 8 Stereostructure of okadaic acid in organic solvent. (Hydrogen atoms are omitted for clarity.)

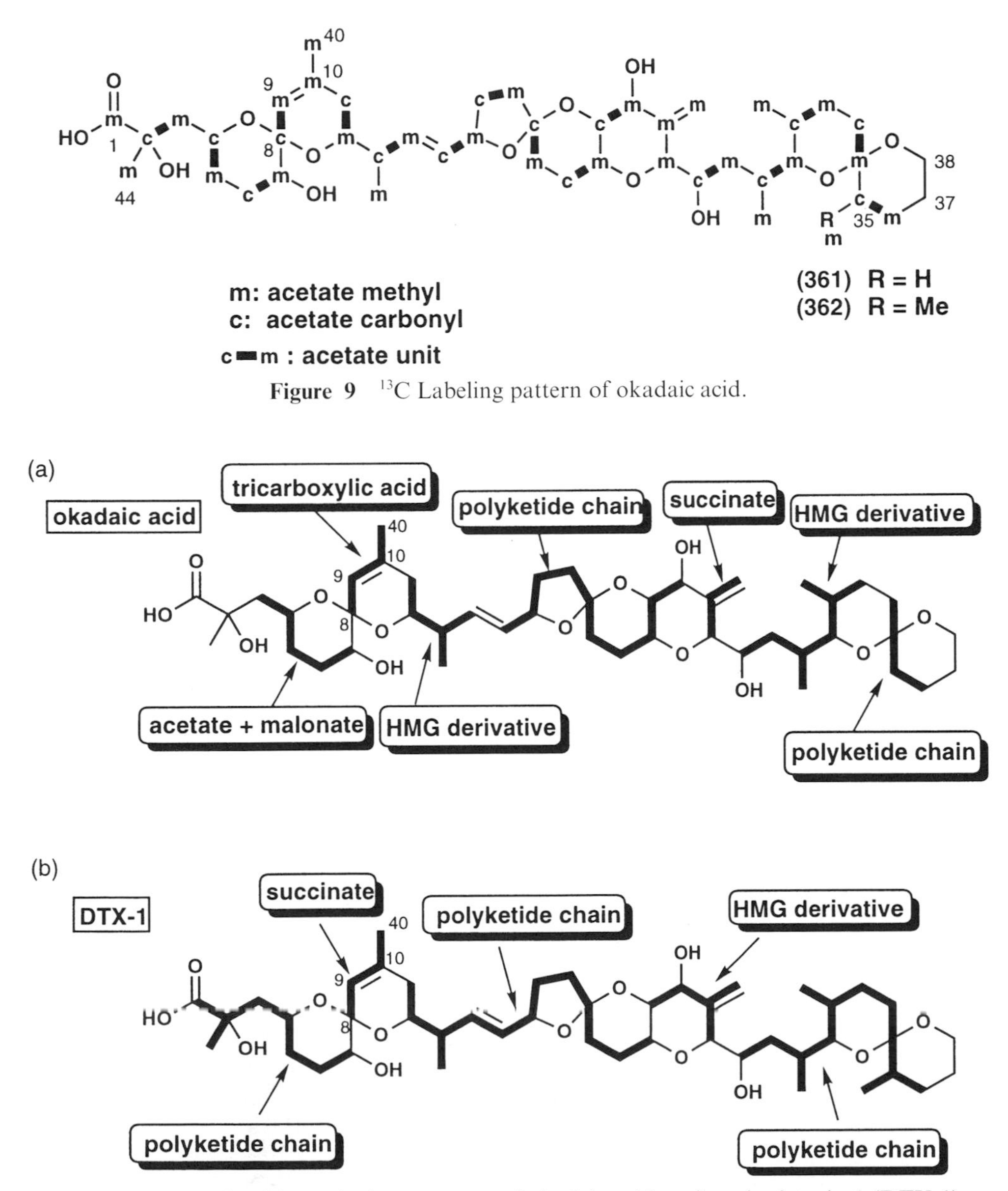

Figure 9 ^{13}C Labeling pattern of okadaic acid.

Figure 10 Possible biosynthetic precursors of okadaic acid or dinophysistoxin-1 (DTX-1).

precursors to *P. lima*.[328] Feeding experiment with [1,2-$^{13}C_2$]glycolate revealed that the C-45/C-46 starter unit of the ester chain of (**368**) or (**370**) was derived from glycolate. The ^{13}C NMR spectrum of the diol ester (**368**) labeled from [2-^{13}C, ^{18}O]glycolate displayed ^{18}O isotopically shifted peaks for C-38 $\Delta\delta = -0.023$ ppm) and C-45 $\Delta\delta = -0.031$ ppm), implying that the ^{13}C—^{18}O bonds in the glycolate precursor were retained in the C-38 and C-45 positions of (**368**). Feeding experiments with [1-^{13}C, $^{18}O_2$]acetate showed that in the okadaic acid skeleton moiety of (**368**) and (**370**) only two positions, C-4 and C-27, retained labeled oxygen while the ^{13}C NMR signals for C-2, C-8, C-19, or C-23, which could in principle bear ^{18}O from ^{18}O-labeled acetate, did not show isotope shifts. In the C_{14} sulfated side-chain of (**370**), ^{18}O isotopically shifted peaks were observed for all five of the oxygen-bearing carbon atoms derived from C-1 of the acetate (Figure 11).

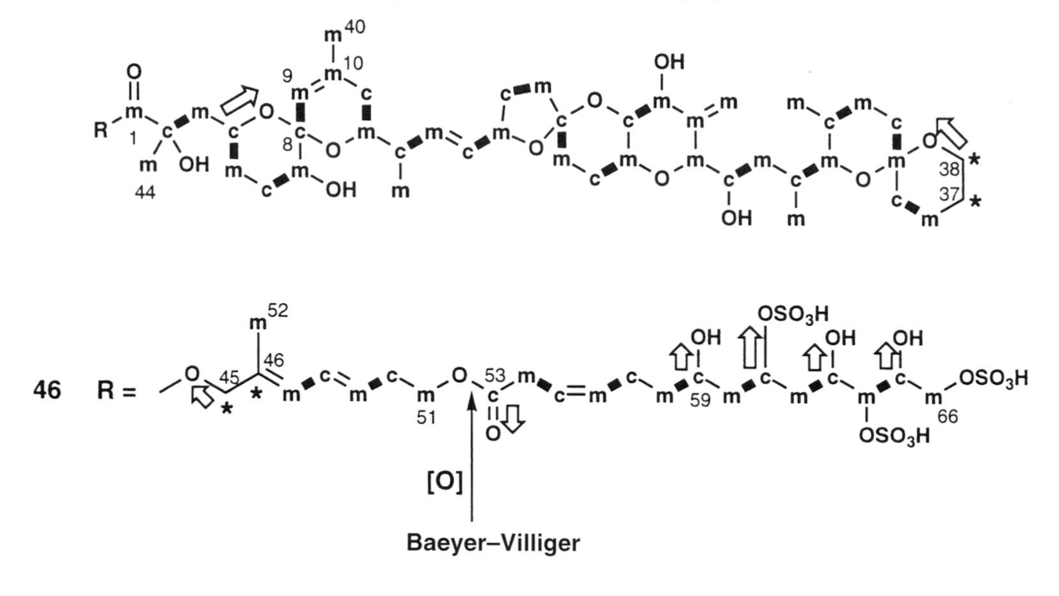

Figure 11 ^{13}C and ^{18}O Labeling patterns of DTX-3 (**370**).

^{18}O was retained at the ester carbonyl C-53 and not at the diol oxygen (C-51), and feeding experiments with [2-$^{13}CD_3$]acetate revealed that the retention of deuterium at C-51 was substantially lower than that of C-66 derived from the methyl of an intact acetate unit in the sulfated ester chain. Although both C-51 and C-66 arise through cleavage of an acetate unit, it was suggested that the cleavage mechanism in each case is different and C-51 and C-53 were once part of an intact acetate unit that is cleaved in a Baeyer–Villiger oxidation step (Figure 11).[329] This hypothesis was confirmed by resolution enhancement of the ^{13}C NMR spectrum of (**370**) labeled from [1,2-$^{13}C_2$]acetate, which revealed narrow satellites corresponding to a $^2J_{COC}$ coupling of 2.6 Hz around the central resonances of both C-51 and C-53. This observation can only be explained if the two carbons arose from the same acetate unit. Baeyer–Villiger oxidation of a polyketide chain is rare and such reactions may be mediated by flavin monooxygenase. In addition, a similar monooxygenase-mediated reaction could be invoked in which a single carbon is ultimately eliminated from the polyketide chain as carbon dioxide. Thus, the generation of isolated carbons derived from the methyl group of an acetate in the polyketide chains of the okadaic acid skeleton was proposed, as shown in Scheme 14; oxidation of a methyl-derived carbon to yield an α diketide followed by a Favorski-type rearrangement, peroxide attack, and collapse of the cyclopropanone would yield a shortened polyketide chain containing an oxidized methyl-derived carbon. Such a process explains how the backbone carbons C-10, C-25, and C-26 in the polyketide chain of okadaic acid each arise from the methyl group of a cleaved acetate unit. This carbon-deletion step may account for the interrupted pattern of acetate units in the chain instead of the participation of dicarboxylic acid precursors (Figures 10(a) and 10(b)).

From the dinoflagellate *Prorocentrum lima* from which okadaic acid and its esters were isolated, prorocentrolide (**377**), a toxic macrocycle, formed from a C_{49} fatty acid incorporating a C_{27} macrolide

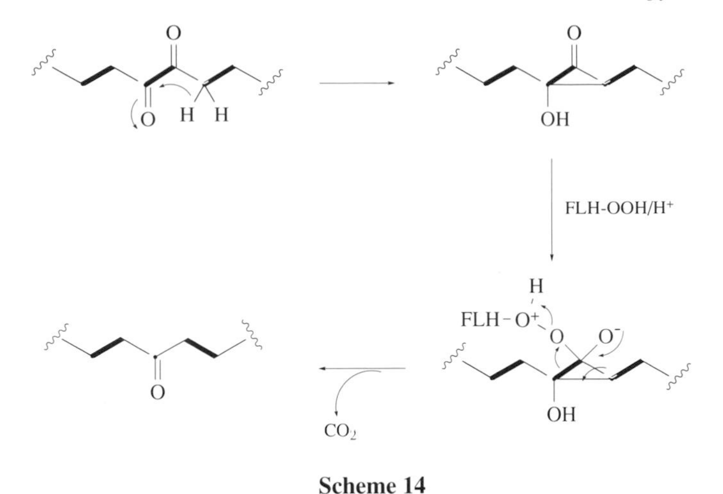

Scheme 14

and a hexahydroisoquinoline in its unique structure.[330] Prorocentrolide (**377**) had a mouse lethality of 0.4 mg kg^{-1} (i.p.), cytotoxicity against L1210 cell (IC_{50} 20 μg ml^{-1}), and negative antimicrobial activities against *Aspergillus niger*, *Candida rugosa*, and *Staphylococcus aureus* at a dose of 80 μg/disk. Prorocentrolide B (**378**) has been obtained from the dinoflagellate *Prorocentrum maculosum* as a fast-acting toxin, showing a type of activity not accounted for by other diarrhetic shellfish poisoning toxins.[331]

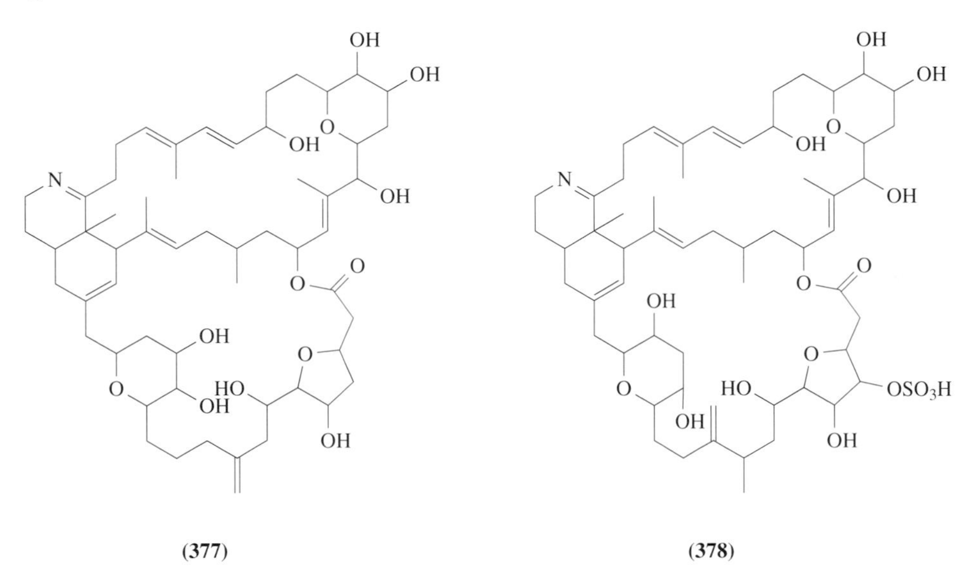

The co-occurrence of prorocentrolide with okadaic acid (**361**) in the dinoflagellate *P. lima* indicated that dinoflagellates are capable of producing polyethers of entirely different skeletons. Biosynthetic studies on prorocentrolide (**377**) were also carried out[325] and revealed ^{13}C labeling patterns as shown in Figure 12. Out of 56 carbons, 44 carbons were labeled by experiments with single ^{13}C-labeled acetates while 49 carbons were revealed to be derived from acetate from feeding experiments with doubly ^{13}C-labeled acetate. Two carbons at C-43 and C-50 were labeled with both methyl and carboxyl carbons of acetates. Three carbons constituting the isoquinoline ring could not be labeled. The probable biosynthetic precursors were proposed from the labeling patterns as shown in Figure 13.

8.07.8.3.2 Pectenotoxin and yessotoxin

Other than okadaic acid, dinophysistoxins, and their analogues, Yasumoto *et al.* identified two kinds of polyether toxins, pectenotoxin and yessotoxin, as DSP toxins. Pectenotoxins (PTX) were

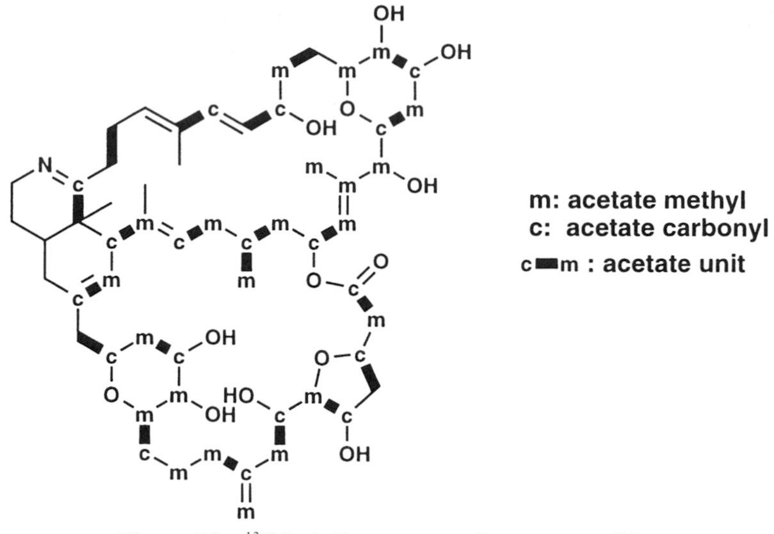

Figure 12 ^{13}C Labeling pattern of prorocentrolide.

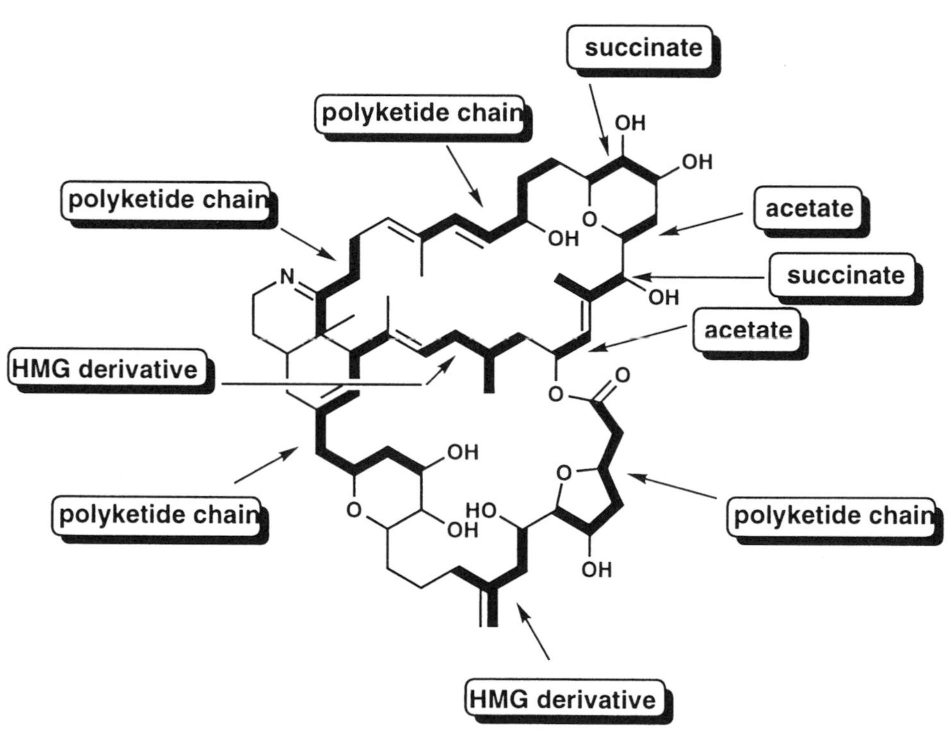

Figure 13 Probable biosynthetic precursors of prorocentrolide.

first isolated from toxic scallops along with DTXs.[312] The digestive glands (200 kg) of the scallop *Patinopecten yessoensis* collected in July 1982, at Mutsu Bay, Japan, were extracted with acetone, and the ether-soluble materials of the acetone extract were subjected to chromatographic separations using columns of silicic acid, Sephadex LH-20, reverse-phase C_8 (LiChroprep RP-8), alumina, Toyopearl HW-40, and HPLC (Develosil silicic acid) to afford five pure toxins, PTX1 (20 mg) as crystals, PTX2 (40 mg), PTX3 (10 mg), PTX4 (7 mg), and PTX5 (0.5 mg). The structure of PTX1 (**379**) was determined by X-ray diffraction methods, and PTX2 (**380**) was identified as 43-deoxy PTX1 on the basis of spectral data.[312] Spectral studies of PTX3 (**381**) showed PTX3 was a

43-aldehyde derivative of PTX2.[332] From a chemical point of view, it was notable that pectenotoxins were substantially different from other described dinoflagellate toxins. Specifically, they differed from others in having a long carbon backbone (C_{40}), a C_{33} lactone ring, and a unique dioxabicyclo moiety. The large oxygen-rich internal cavity was largely similar to the cavities found in the polyether ionophores from terrestrial microorganisms. Coexistence of DTXs and PTXs having such different skeletons is an interesting subject in the biosynthesis of polyether compounds in dinoflagellates.

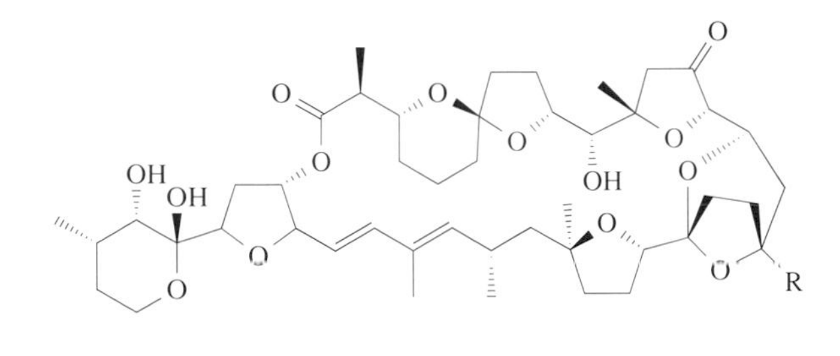

(379) R = CH_2OH
(380) R = Me
(381) R = CHO

Investigation of the cytotoxic constituents of a two-sponge association (*Poecillastra* sp. and *Jaspis* sp.) collected off Cheju and Komun Islands, Korea, led to the isolation of PTX2 (**380**) as an active compound. In an *in vitro* assay, (**380**) displayed very potent cytotoxic activities against human lung (A-549), colon (HT-29), and breast (MCF-7) cancer cell lines. PTX2 (**380**) also exhibited selective cytotoxicity against several cell lines representing ovarian, renal, lung, colon, CNS, melanoma, and breast cancer, with differences in LC_{50} values between sensitive and resistant cell lines of 100-fold or more.[333]

The effects of PTX1 (**379**) on liver cells were investigated by fluorescence microscopy. The *in vitro* application of PTX1 to these cells induced reduction in number and loss of the radial arrangement of microtubules. It also induced disruption of stress fibers and accumulation of actin at the cellular peripheries. Further studies were undertaken to explore the feasibility of using PTX1 as a tool for analyzing the properties of cytoskeletal proteins.[334] Sequential ultrastructural changes were studied in mouse digestive organs after i.p. injections of DSP toxins, DTX-1 (**362**) (see Section 8.07.8.3.1) and PTX1 (**379**). DTX-1 (**362**) produced severe mucosal injuries in the small intestine within 1 h after the administration of the toxin. The injuries were divided into three consecutive stages: extravasation of villi vessels, degeneration of absorptive epithelium, and desquamation of the degenerated epithelium from the lamina propria. In contrast to (**362**), PTX1 (**379**) resulted in no abnormalities in the small intestine, but did cause characteristic liver injuries. Within 1 h after the injection of PTX1 (**379**) numerous nonfatty vacuoles appeared in the hepatocytes around the periportal regions of the hepatic lobules. Electron microscopic observations with colloidal iron demonstrated that these vacuoles originated from invaginated plasma membranes of the hepatocytes.[335]

Yessotoxin is another class of DSP toxins, isolated from digestive glands of the scallop *Patinopecten yessoensis* (84 kg) collected in Mutsu Bay, Japan, in 1986. Their acetone extracts were partitioned between ether and aqueous MeOH, and the methanolic layer was suspended in water and extracted with 1-butanol. The toxic extract was separated by chromatographies using aluminum oxide, two kinds of reversed phase octadecyl silane (ODS) columns, and gel permeation on Toyopearl HW-40 to yield a purified toxin (60 mg), named yessotoxin (YTX (**382**)).[336] The toxin killed mice at a dose of 100 $\mu g\ kg^{-1}$ (i.p.) but caused no fluid accumulation in suckling mice intestines even at a fatal dose. The planar structure of (**382**) was elucidated by detailed analyses of several kinds of 2D NMR techniques. The presence of sulfate ester was suggested by ion chromatography of sulfate ions liberated by solvolysis of (**382**) in pyridine–dioxane (1:1) at 120 °C for 3 h, which also afforded desulfated YTX. YTX (**382**) proved to be a polyether compound having a ladder-shape polycyclic skeleton with an unsaturated side-chain of nine carbons, two sulfate esters, and no carbonyl groups. The negative ion fast atom bombardment tandem mass spectrum of (**382**) showed a series of ions due to fragmentation at positions characteristic of cyclic ethers to demonstrate the applicability of tandem mass spectrometry to the structural elucidation of polyether compounds of this size.[337] The mass spectrometric data alone allowed assignment of the number and location of 11 ether rings of (**382**), which agreed with those previously deduced by NMR studies. The relative configuration of YTX (**382**) and the ring conformation were determined by NMR experiments.[338] Additionally, two other new analogues, 45-hydroxyYTX (**383**) and 45,46,47-trinorYTX (**384**) were isolated from toxic scallops and their relative stereochemistries were also determined.

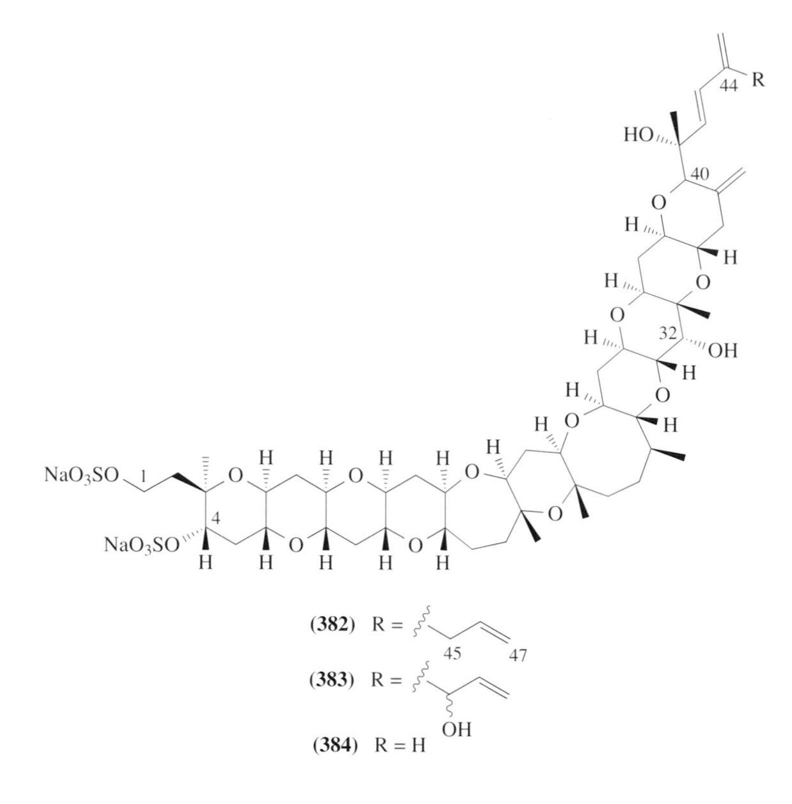

Furthermore, the absolute configuration of YTX (**382**) was studied by applying a modified Mosher's method using the new chiral anisotropic reagent, 2NMA [methoxy-(2-naphthyl)acetic acid (**385**)].[339] Esterification of 32-OH of YTX (**382**) was initially attempted, but was unsuccessful because of its axial orientation. YTX (**382**) was then subjected to solvolysis as described above (pyridine–dioxane (1:1) at 120 °C for 3 h)[336] to give bisdesulfated YTX, which in turn was treated with an excess of acetic anhydride in pyridine at 0 °C for 45 min to give a mixture of recovered bisdesulfated YTX, 1-*O*-acetyl YTX, and 1,4-di-*O*-acetyl YTX, separable by HPLC. 1-*O*-Acetyl YTX (600 μg) thus obtained having free OH at C-4 was divided into two portions, and each portion was treated with (*R*)- and (*S*)-2NMA, EDC, DMAP, and Et_3N in chloroform for 43.5 h at room temperature to give, after HPLC purification, (*R*)- (**386**) and (*S*)-2NMA esters (**387**), respectively. By analyzing the COSY and HOHAHA spectra of (**386**) and (**387**) and the $\Delta\delta$ values $[\delta_R - \delta_S]$ of proton chemical shifts, the absolute configuration of the hydroxyl group at C-4 was determined to be *S*, and, since the relative stereochemistry of the other asymmetric centers was established,[338] the absolute configuration of YTX was determined, as shown in (**382**).[340]

YTX (**382**) had a structural resemblance with brevetoxins (see Section 8.07.8.5.2) and ciguatoxin (see Section 8.07.8.4.1), which are known as potent activators of voltage-gated sodium channels but are not cytotoxins. However, YTX (**382**) did not potentiate those channels and showed cytotoxicity and different toxicological properties. The histopathological response of male mice to YTX (**382**) was compared with that to desulfated YTX. The target organ of the former was the heart and those of the latter were the liver and pancreas. Using electron microscopy, marked intracytoplasmic edema in cardiac muscle cells was seen within 3 h after the i.p. injection of over 300 μg kg^{-1} of YTX (**382**). In contrast, desulfated YTX at the same dose caused severe fatty degeneration and intracellular necrosis in the liver and pancreas, but not in the heart, within 24 h of i.p. injection. Biochemically, the content of triglycerides in the liver of mice treated with desulfated YTX increased about 60 times, and phospholipids twofold more than the control levels of those of mice treated with YTX (**382**).[341]

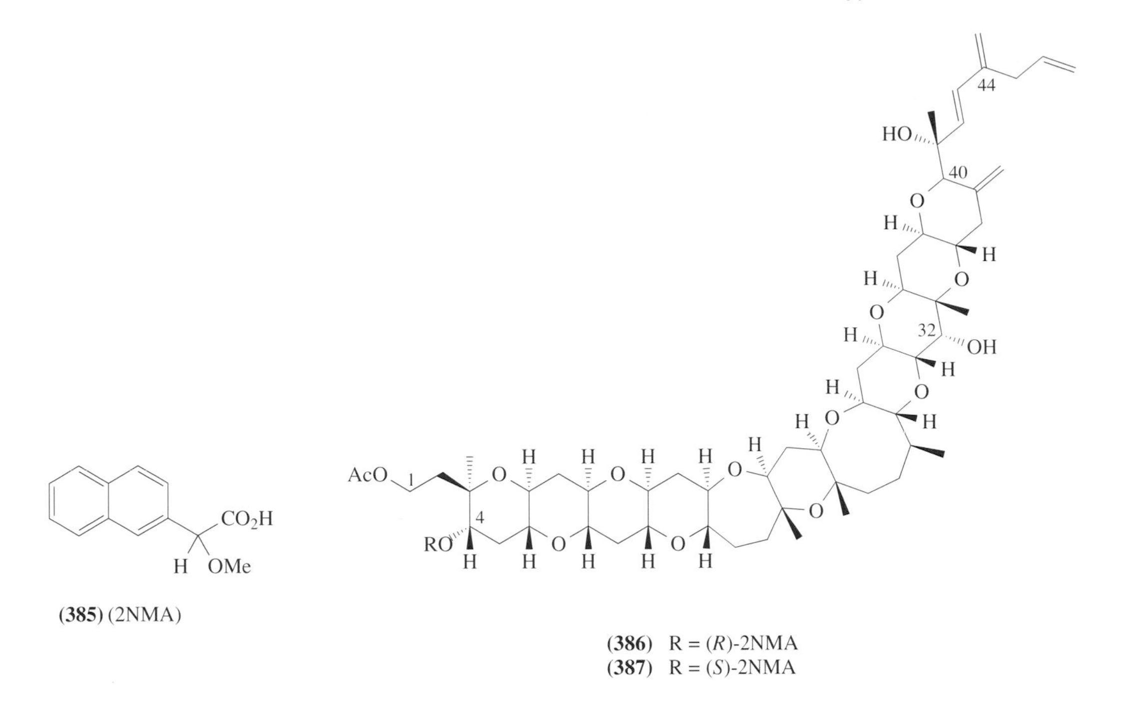

(385) (2NMA)

(386) R = (*R*)-2NMA
(387) R = (*S*)-2NMA

8.07.8.4 Ciguatera

Ciguatera refers to a type of human intoxication resulting from ingestion of coral reef fish.[342] Globally over 20 000 people are estimated to suffer annually from the poisoning, making it one of the largest-scale food poisonings of nonbacterial origin. Though rarely fatal, the poisoning poses a serious threat to public health and tropical fisheries because of its unpredictable occurrence and the implication of numerous fish species. There are two groups of compounds implicated in the poisoning; the main responsible toxins are ciguatoxin (CTX) and its congeners, and the other is maitotoxin (MTX). The clinical symptoms of this poisoning are diverse. In addition to gastrointestinal malaise, vomiting, and diarrhea, which are common in food poisoning, there is a neurological component. Dizziness, tingling of the extremities, and the sensation of temperature reversal are among the salient characteristics of ciguatera poisoning. Other symptoms include joint pain, miosis, erethism, cyanosis, and prostration. Cardiovascular disturbances are low blood pressure and bradycardia. Virtually all victims recover, albeit painfully and slowly. The low fatality rate is due solely to the minute concentration of the toxin in fish flesh.

8.07.8.4.1 Ciguatoxin

Research efforts on ciguatera have been undertaken by institutions in Pacific countries. The historical story of ciguatoxin research at the University of Hawaii was reviewed by Scheuer.[343] His research group initiated the study in 1957, and by 1980 they had isolated 1.3 mg of HPLC-pure ciguatoxin from approximately 1100 kg of moray eels. The toxin was successfully crystallized in MeOH-d_4 solution and the molecular weight was determined to be 1111 Da.[344,345] Although the crystal was not suitable for single-crystal X-ray analysis, the toxin was characterized to be a polyether compound by NMR studies using an inadequate amount of material.

Meanwhile, Yasumoto *et al.* had identified an epiphytic dinoflagellate, *Gambierdiscus toxicus*, as a causative organism in 1977.[346] It was, however, extremely difficult to obtain enough ciguatoxin for structural elucidation. Although *G. toxicus* produced maitotoxin, a second important toxin in ciguatera, no ciguatoxin was isolated when it was cultured. The structure of ciguatoxin was finally elucidated by Yasumoto's group and the resident French public health workers in 1989.[347–349] For that study moray eels (*Gymnothorax javanicus*, ca. 4000 kg) collected in French Polynesian waters were extracted to yield no more than 0.35 mg of ciguatoxin (CTX (**388**)), and 0.75 mg of a congener, CTX4B (**389**), was isolated from the ciguatera causative dinoflagellates, *Gambierdiscus toxicus*, collected in the Gambier Islands in French Polynesia. Using those samples, the structures of CTX (**388**) and CTX4B (**389**) including the relative stereochemistries were successfully elucidated as polycyclic ethers by NMR methods.

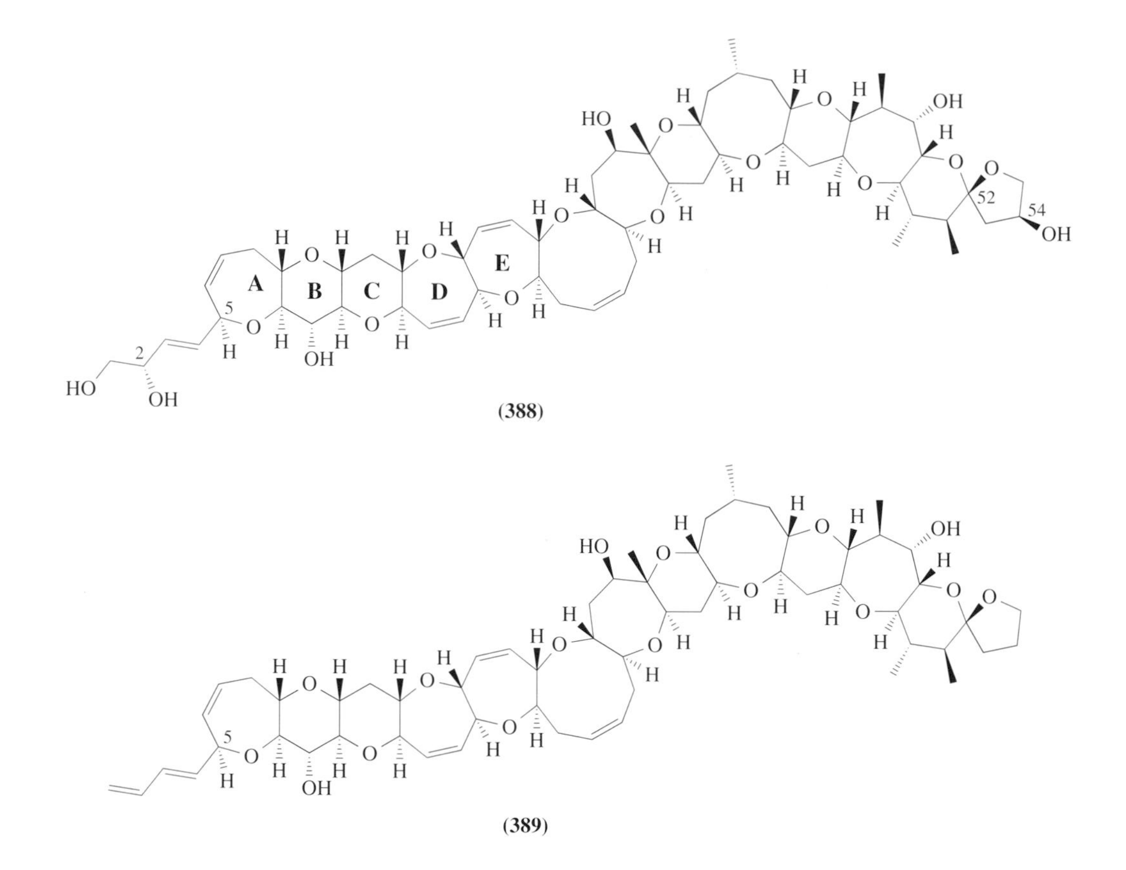

(**388**)

(**389**)

The molecular formulas of CTX (**388**) and CTX4B (**389**) were determined as $C_{60}H_{86}O_{19}$ and $C_{60}H_{84}O_{16}$, respectively, by high-resolution FABMS. The skeletal structure was established mainly on the basis of ^{1}H-^{1}H 2D NMR data. Since some parts of the molecules were unassignable due to extreme broadening or even disappearance of signals, measurement of ^{1}H NMR at low temperature was attempted so as to make the perturbation of the ring slow enough to detect these resonances. The stereochemistry of hydroxyls and methyls substituted on the ether rings was clarified by NOE difference experiments and by combined use of MM2 energy calculations and spectral simulations. The structural difference between CTX (**388**) and CTX4B (**389**) was elucidated by COSY analyses, showing that the latter was a less oxidized entity and had a *trans*-butadiene moiety at one terminus of the molecule and lacked a hydroxyl group at C-54 at the other end. The moray eel, which is placed near the top of the coral ecosystem, tends to contain more polar congeners, while the dinoflagellate produces less polar ones. It was thus suggested that less polar congeners produced by *G. toxicus* are precursors to the more polar toxins found in fish, the latter formed by oxidative enzyme systems in the fish. Interestingly, toxicity of oxidized metabolites was often increased; the lethal potency against mice (i.p.) of CTX (**388**) and CTX4B (**389**) was 0.35 $\mu g\ kg^{-1}$ and ca. 4 $\mu g\ kg^{-1}$, respectively, the former being 11 times more toxic than the latter.

The absolute stereochemistry of the C-5 position of CTX4B (**389**) was suggested to be *R* by the CD spectral data of compound (**390**), which was synthesized stereoselectively and had the AB ring framework of (**389**).[350] The stereochemical assignment of C-54 of CTX (**388**) was confirmed by synthesis of compounds (**391**) and (**392**) corresponding to the KLM ring model of CTX (**388**) and its epimer, and comparison of their ^{1}H NMR data.[351] The ^{13}C NMR assignments of CTX (**388**) were achieved with 1.1 mg of the sample (1 μmol) obtained from carnivorous fish collected in French Polynesia, Kiribati, Micronesia, and Fiji.[352] ^{13}C-decoupled HMQC and HSQC spectra in CD_3CN/D_2O (2:1) and C_5D_5N/D_2O (20:1) revealed $^{1}J_{CH}$ arising from the flexible part of the molecule which yielded no sharp peaks in the 1D ^{13}C NMR spectrum.

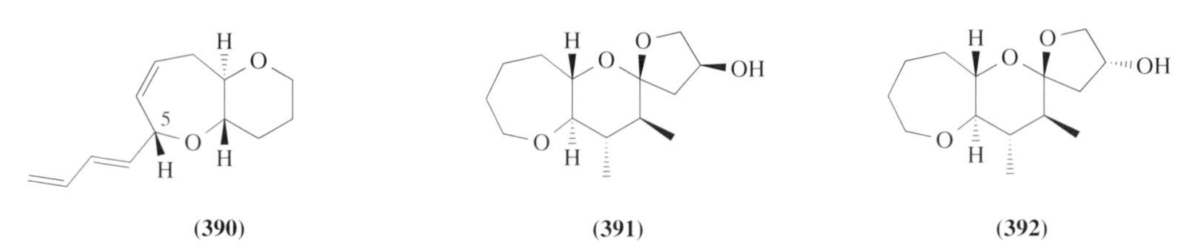

In order to determine the absolute configuration of the C-2 position of CTX (**388**), the 1,2-diol of (**388**) was protected as benzyloxymethyl ethers, and then the 3,4-double bond was oxidatively cleaved by treatment with OsO_4-$NaIO_4$ followed by $NaBH_4$ reduction. The resultant alcohol was esterified with a chiral fluorometric reagent, (*S*)-(+)-2-*t*-butyl-2-methyl-1,3-benzodioxole-4-carboxylic acid [(*S*)-TBMB acid] to give (*S*)-TBMB (**393**), which compared with synthetic references on two HPLC systems (Scheme 15). Starting with only 5 μg of CTX (**388**) the absolute configuration at C-2 was successfully determined to be *S*.[353]

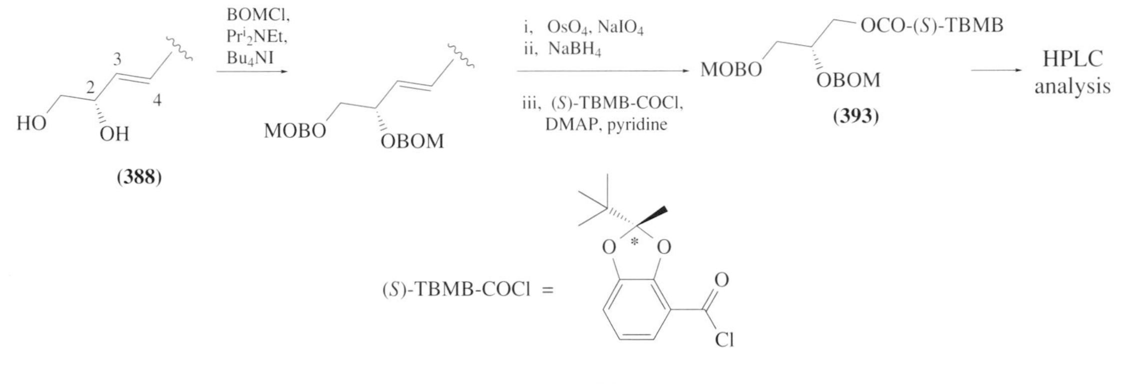

Scheme 15

A number of ciguatoxin analogues have also been structurally elucidated. One strain of the dinoflagellate *Gambierdiscus toxicus* collected at Rangiroa Atoll in French Polynesia was cultured, and from the harvested cells of 1100-L cultures three polyether toxic compounds, gambierol[354] (**394**) (1.2 mg, LD_{50} in mouse 50 μg kg^{-1} (i.p.), $C_{43}H_{64}O_{11}$), CTX3C[355] (**395**) (0.7 mg, 1.3 μg kg^{-1} (i.p.), $C_{57}H_{82}O_{16}$), and CTX4A[356] (**396**) (0.45 mg, 2 μg kg^{-1} (i.p.), $C_{60}H_{84}O_{16}$) were isolated. The ring system of gambierol (**394**) differed from that of CTX (**388**). Ring E in CTX3C (**395**) was an oxocene instead of the oxopene in CTX (**388**) and CTX3C (**395**) had no side-chain attached at C-5 of CTX (**388**). CTX4A (**396**) had the same molecular formula as CTX4B (**389**) and the difference in structure of (**396**) from (**389**) was found in the stereochemistry at the C-52 *spiro* carbon. Thus, these findings demonstrated the diversity of ciguatera toxins and the complex biosynthesis of polyether compounds in *G. toxicus*. Another ciguatoxin analogue, CTX2A1 (**397**) ($C_{57}H_{84}O_{18}$), was isolated from moray eels and its structure was elucidated to be a 2,3-diol derivative of CTX3C (**395**).[353] This was the first CTX analogue to be oxidized in the A ring and was an additional example showing the diverse metabolic modification of CTXs occurring in fish.

To confirm the absolute configuration of the C-5 position of CTX analogues, 100 μg of CTX4A (**396**) was reacted with *p*-bromobenzoyl chloride to yield the tris-*p*-bromobenzoyl derivative. The CD spectrum of CTX4A-tris-*p*-bromobenzoate was comparable with that of the *p*-bromobenzoate derived from a synthetic (5*R*)-AB-ring model compound, thus indicating that the absolute configuration at C-5 was *R*. Consequently, the whole stereostructure of CTX (**388**) was able to be assigned with (2*S*,5*R*)-configuration.[353]

The Caribbean ciguatoxin (< 100 μg) was isolated from a collection of barracuda and horse-eye jack (545 kg) from a ciguatera-endemic region of the Caribbean. Caribbean ciguatoxin was shown to have a molecular weight of 1123 by electrospray mass spectrum. In contrast, CTX (**388**) had a mass of 1111 Da. Structural characterization of the Caribbean ciguatoxin was carried out by the use of micro inverse-detection probes and micro cells for the HMQC and HMBC spectra to result in partial assignment of its structure.[357] Micro inverse-detection probes employed in conjunction with Shigemi micro NMR cells were shown to afford significant improvement in sensitivity. Caribbean ciguatoxin had only six alkenic protons while CTX (**388**) possessed ten. The segment of Caribbean ciguatoxin molecule from C-34 to C-42 positions was identical to (**388**), while Caribbean ciguatoxin lacks the C-54 hydroxyl group, appearing instead to have a hydroxyl at the 53-position.

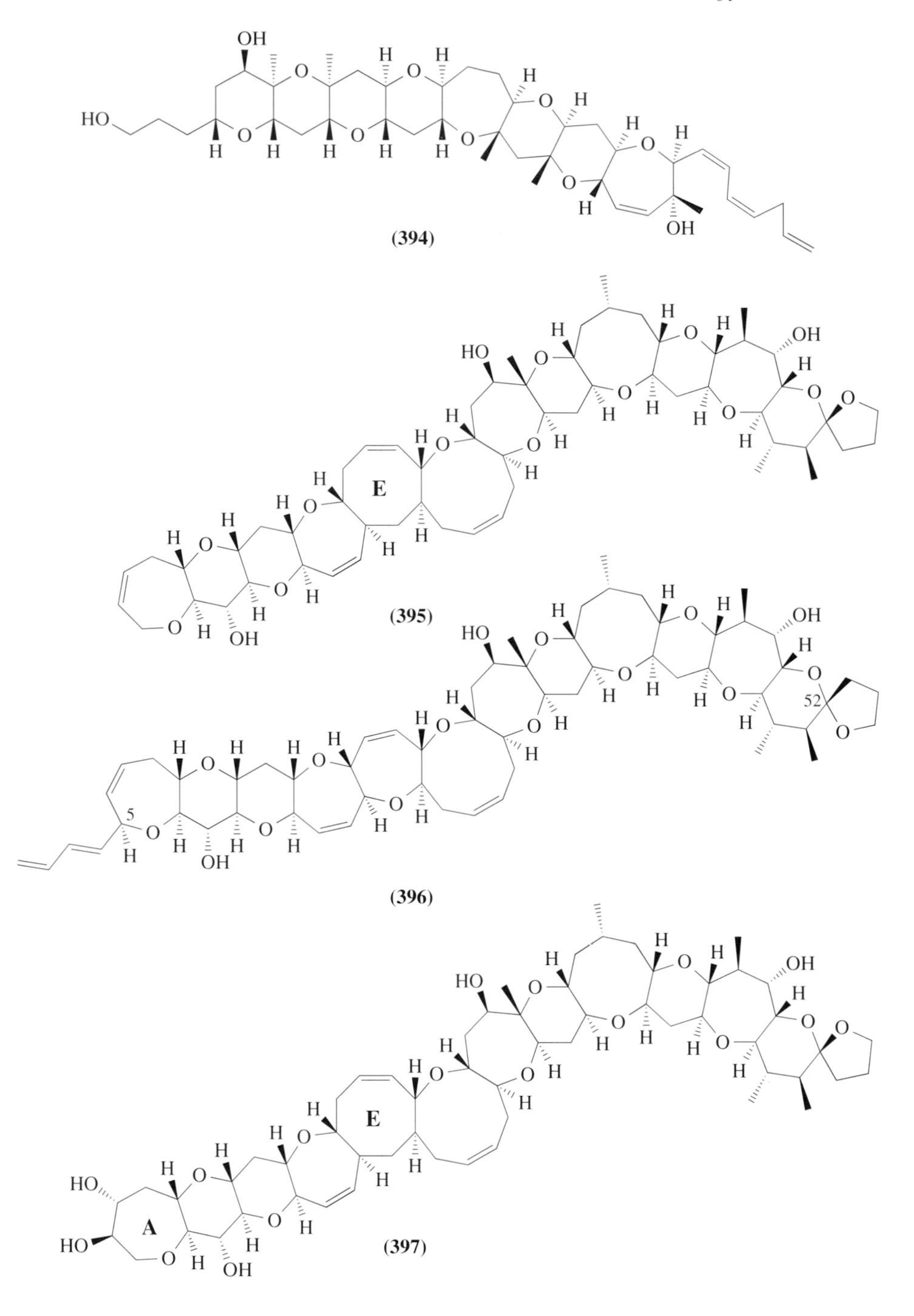

(394)

(395)

(396)

(397)

8.07.8.4.2 Maitotoxin

Maitotoxin (MTX (**398**)) was first discovered by Yasumoto and Satake from the surgeonfish *Ctenochaetus striatus* as one of the causative toxins of ciguatera and was named after the Tahitian fish "maito." The toxins accumulated in herbivorous fish are diverse, and MTX (**398**) was later confirmed to originate from the dinoflagellate *Gambierdiscus toxicus*.[342] MTX (**398**) is one of the most complex natural products having molecular weight 3422 Da (nominal mass) and molecular formula $C_{164}H_{256}O_{68}S_2Na_2$, and its chemical structure has been successfully determined. The lethality of maitotoxin to mice (0.05 μg kg^{-1}, i.p.) is only exceeded by a few proteins. The mechanism underlying its high potency is now explainable by the activation of voltage-independent calcium channels, thus producing a highly enhanced calcium influx of extracellular Ca^{2+} ion across the cell membrane.[358]

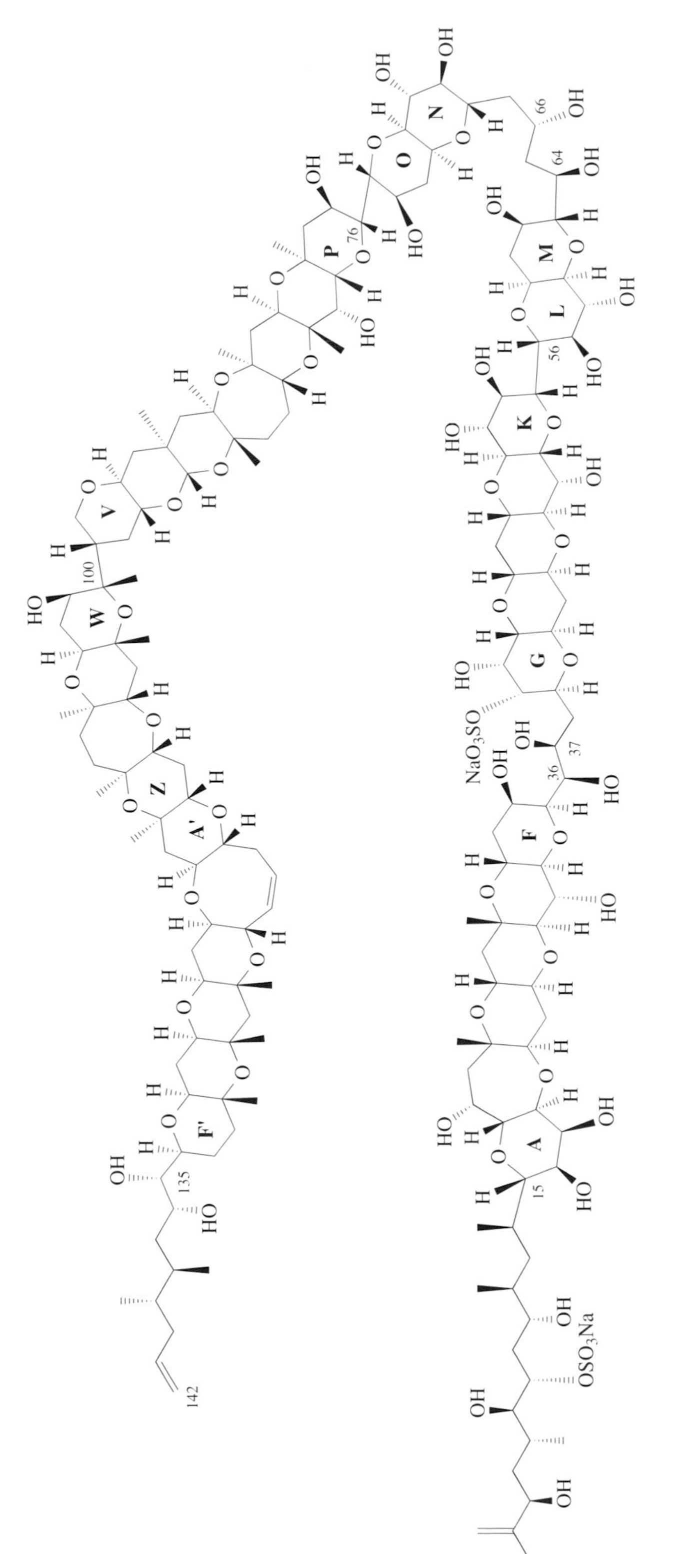

(398)

Isolation of MTX (**398**) was performed by Yasumoto's group by culturing the dinoflagellate *G. toxicus* from the Gambier Islands, French Polynesia in a nutrient-enriched seawater medium.[359] Fernback flasks, each containing 2 l of the medium, were inoculated with seed cultures of the dinoflagellate and maintained at 25 °C for 38 days under illumination of 4000–8000 lux at an 18/6 light:dark photoperiod. After reaching a cell density of about 1000 cells ml^{-1}, the dinoflagellates were collected by filtration and extracted twice with MeOH and twice with MeOH/H_2O (1:1) under reflux. After concentration of the combined extracts by evaporation, the residual suspension was partitioned between CH_2Cl_2 and aqueous MeOH, and the latter was extracted with butanol. The butanol layer was toxic and was subjected to separation by silica gel column, C18 column, Develosil C8, and Develosil TMS-5 column to afford MTX (**398**). The eluates were monitored by the use of mouse lethality (intraperitoneal injection) tests and by a UV flow detector at a wavelength of 225 nm. From 4000 l of culture of the dinoflagellate, 20 mg of MTX (**398**) was obtained as a colorless solid.

The presence of two sulfate ester groups in the molecule was apparent from the IR and mass spectra, and from analyses of solvolysis products (pyridine/dioxane (1:1), 120 °C, 4 h; HPLC analysis on a C18 column with an aqueous solution containing 1.0 mM tetrabutylammonium hydroxide and 0.8 mM 1,3,5-benzenetricarboxylic acid, monitored at 280 nm).[359] Despite its large size, MTX (**398**) had no known repeating units, such as amino acids and sugars, no carbonyl groups, no side-chains (apart from those at the termini) other than methyls or an exomethylene, and no carbocycles. Structural elucidation of MTX (**398**) was carried out by extensive 2D NMR measurements. A large portion of the structure was shown to be a brevetoxin-type polyether. Chiefly due to extensive overlapping of ^1H and ^{13}C NMR signals even in the 2D spectra, MTX (**398**) was treated with $NaIO_4$ followed by reduction with $NaBH_4$, resulting in cleavage of the molecule into two major parts, fragments A (C-1–C-36) and B (C-37–C-135) together with a small C_9 fragment (C-136–C-142).[360] Fragments A and B had molecular weights 964 and 2328, respectively. Their structures were elucidated by extensive 2D NMR and collisionally activated dissociation (CAD) MS/MS experiments. Reassembling the three fragments led to the complete structure of MTX ($C_{164}H_{256}O_{68}S_2Na_2$), encompassing two sulfate esters as described above, 28 hydroxyls, and 32 ether rings.[361] The relative stereostructure was partly assigned by spectral data and was revealed to comprise *trans*-fused polycyclic ethers except for rings L/M and N/O which are *cis*-fused. The stereochemical correlations for the acyclic linkages such as rings K/L, O/P, and V/W were revealed on the basis of NMR data with the aid of molecular mechanics calculations. The stereochemistry of the acyclic parts, C-5–C-14, C-36–C-37, C-64–C-66, and C-135–C-139 remained to be determined in 1994.[362] The structure of MTX (**398**) was confirmed by the complete ^{13}C NMR assignments and NOEs data obtained by a 3D pulsed field gradient NOESY-HMQC spectrum. The ^{13}C-enriched sample for the 3D spectrum was prepared by culturing the dinoflagellate *Gambierdiscus toxicus* in media containing ^{13}C-labeled carbonate and appeared to contain 4% of ^{13}C. The ether linkages and stereochemistry for cyclic ethers were verified on the basis of NOEs derived from the 3D NMR data. The complete ^{13}C NMR assignments of MTX (**398**) further supported the present structure.[363]

Structural studies of MTX (**398**) based on the combination of spectral studies and synthesis of the corresponding compounds were performed by Tachibana's research group collaborating with Yasumoto's group. To confirm the *cis*-fused L/M and N/O ring systems of MTX (**398**), a *cis*-fused 1,6-dioxadecalin system (**399**) was synthesized in a stereocontrolled manner, and the comparison of its ^1H and ^{13}C NMR data with those of natural toxin established the earlier stereochemical assignments.[364] The relative configuration of carbons within the C-63–C-68 acyclic linkage of MTX (**398**) was assigned by synthesis of stereodefined model compounds (**400**)–(**403**) and their comparison with MTX (**398**) in ^1H and ^{13}C NMR spectra. All four diastereomers (**400**)–(**403**) showed different NMR profiles and only (**400**) among these gave virtually identical ^1H and ^{13}C NMR data to those for the C-63–C-68 region of MTX (**398**).[365] The relative configuration of carbons within the C-35–C-39 acyclic linkage of MTX (**398**) was suggested from the NOESY and E.COSY spectra of MTX (**398**), and confirmation of the assignment was achieved by the synthesis of a stereodefined model compound (**404**) and its comparison with MTX (**398**) in ^1H and ^{13}C NMR spectra.[366] From these results stereochemical assignments of the whole molecule of MTX (**398**) except for side chains were completed. The model compounds (**399**)–(**404**) were all optically active, all being prepared starting from D- or L-glucose, and eventually all these model compounds were antipodes of the corresponding portion of MTX (**398**).

Long-range carbon–proton coupling constants ($^{2,3}J_{C,H}$) were measured for MTX (**398**) by hetero-half filter TOCSY (HETLOC) experiments and phase-sensitive HMBC with use of 9 mg of a 4% ^{13}C-enriched sample. The necessary coupling constants within the terminal acyclic portions of MTX (**398**), where NOE analysis was not successful owing to the presumed coexistence of multiple

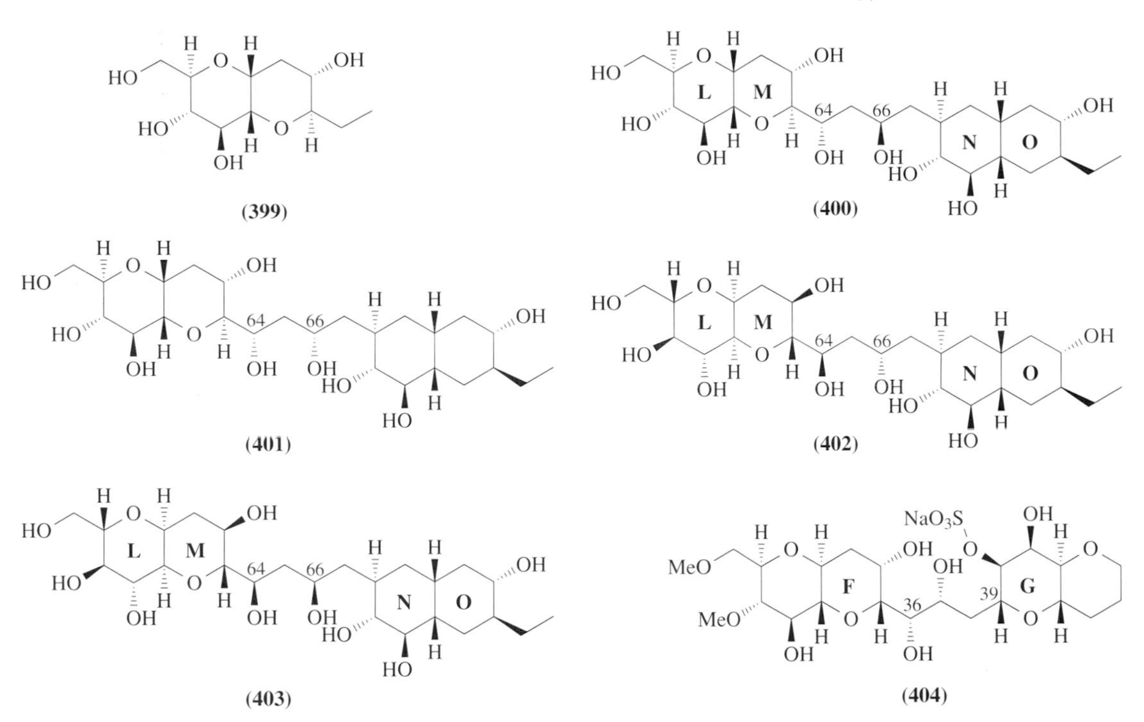

conformers, were thus obtained for the resultant elucidation of relative configurations for the acyclic C-5–C-9 stereogenic centers to be 5*R**, 7*R**, 8*R**, and 9*S**.[367] For the C-9–C-15 portion, from the analyses of conventional NOE, $^3J_{H,H}$-, and $^{2,3}J_{C,H}$-values determined by HETLOC and phase-sensitive HMBC experiments, the relative stereochemical relationships for H-9/H-12, H-12/H-13, and H-14/H-15 were established, leaving only the relationship between H-13 and H-14 unassigned. These findings suggested the configuration of the side chain (C-5–C-15) to be either (**405**) or (**406**) in which the diastereomeric relationship was inverted at the C-13–C-14 bond. Thus, in order to assign this diastereomeric relationship, and also to confirm the relative configurations at C-9, C-12, and C-13 as deduced from the NMR data, the diastereomeric fragments (**405**) and (**406**) were synthesized for comparison of their NMR data with those of MTX (**398**). These fragments were designed also to incorporate the asymmetric centers whose configuration had already been determined (C-5–C-8 and C-15–C-17) so as to reproduce the ^{1}H and ^{13}C chemical shifts of the natural product. In the comparison of the ^{1}H and ^{13}C data of diastereomers (**405**) and (**406**), which were prepared in a stereocontrolled manner, the observed chemical shifts of (**405**) clearly matched those of MTX (**398**) more closely. These results unambiguously established the relative configuration of the C-1–C-15 fragment of MTX (**398**) to be 5*R**, 7*R**, 8*R**, 9*S**, 12*R**, 13*R**, 14*R**, and 15*S**.[368]

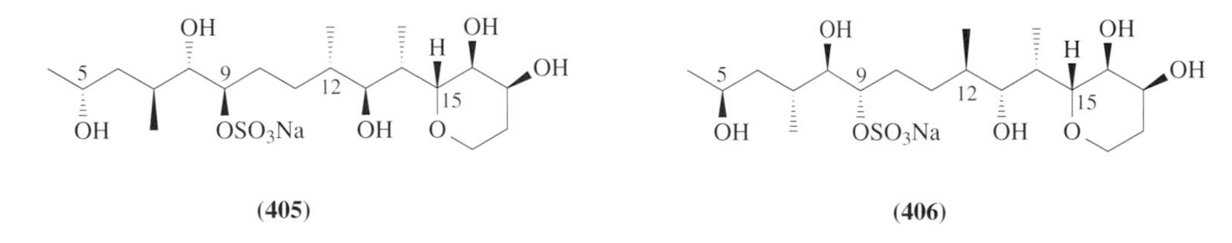

The relative stereochemistry between C-136 and C-138 in the other terminal acyclic chain was unambiguously assigned on the basis of the $^3J_{H,H}$ and $^{2,3}J_{C,H}$ data. The stereochemical relationships among C-134, C-135, and C-136 were also deduced from similar NMR analysis and confirmed by the comparison of the synthesized model compound (**407**) with MTX (**398**) in their ^{1}H NMR. In order to determine the absolute configurations along with the relative configuration between C-138 and C-139, which was not deduced by NMR due to signal broadening, the four stereoisomers (**408**)–(**411**) possible for 3,4-dimethyl-6-hepten-1-ol were synthesized and compared with a degraded fragment of the natural product by GC with a chiral column to identify the diastereomer (**408**) as the natural diastereo- and optical isomer, thus establishing the 138*R*, 139*S* configuration of MTX (**398**). From all of these results, the Tachibana–Yasumoto group established the complete absolute stereochemistry of MTX to be represented by structure (**398**).[369]

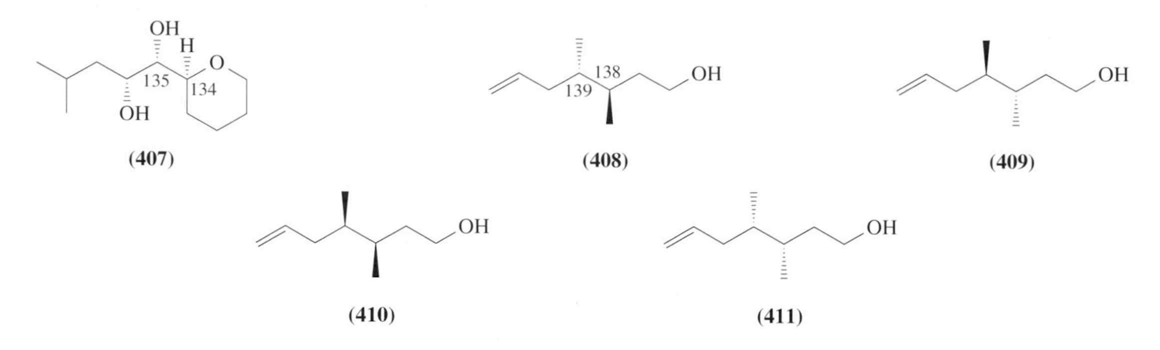

The relative stereochemistry of MTX (**398**) was also studied independently by Kishi and co-workers via organic synthesis.[370] The relative stereochemistry of the C-1–C-15 portion was elucidated via the synthesis of the eight diastereomers possible for the C-1–C-11 portion and the eight diastereomers possible for the C-11–C-15 portion and the comparison of their proton and carbon NMR characteristics with those of MTX (**398**), and the synthesis of two remote diastereomers and comparison of their proton and carbon NMR characteristics with those of MTX (**398**). The relative stereochemistry of the C-35–C-39, C-63–C-68, and C-134–C-142 acyclic portions was established via the synthesis of the diastereomers possible for models of these fragments and comparison of their proton and carbon NMR characteristics with those of MTX (**398**). By these studies, the complete relative stereochemistry of MTX (**398**) was established. Some biogenetic considerations were given to speculate on the absolute configuration of MTX (**398**). The preferred solution conformations of model compounds for the four acyclic portions of MTX (**398**) were elucidated on the basis of vicinal proton coupling constants, and by assembling them, the approximate global conformation of MTX (**398**) was suggested to be represented by the shape of a hook, with the C-35–C-39 portion being its curvature.

In addition to its lethal potency against mice, MTX (**398**) possesses multiple activities, such as hormone stimulation, neurotransmitter release, activation of phosphoinositide degradation, and potentiation of protein kinase, all of which appear to be linked to elevation of the intracellular Ca^{2+} concentration.[371] Thus the toxin, which is commercially available, serves as a tool for studies on cellular events associated with Ca^{2+} flux. As an example of these studies, Daly *et al.* reported that MTX (**398**) elicited a marked influx of $^{45}Ca^{2+}$ into NIH 3T3 fibroblast cells.[372] The influx was blocked by imidazoles (econazole, miconazole, SKF 96365, clotrimazole, calmidazolium) with IC_{50} values from 0.56 μM to 3 μM. The pattern of inhibition of MTX-elicited calcium influx did not correspond to the ability of the agents to block elevation of calcium that ensues through calcium-release activated calcium (CRAC) channels after activation of phosphoinositide breakdown by ATP in HL-60 cells. Tachibana and co-workers studied the blockade of MTX-induced Ca^{2+} influx in rat glioma C6 cells by alkylamines.[373] A series of alkylamines, namely, mono-, di-, and trihexylamine and tetrahexylammonium iodide, were tested for inhibition against Ca^{2+} influx by MTX (**398**) in rat glioma C6 cells. The Ca^{2+} entry was monitored by $^{45}Ca^{2+}$ influx and Ca^{2+}-sensitive fluorescent dye (fura-2) assays. While hexylamine showed no significant inhibition, the other amines exhibited potent inhibition upon $^{45}Ca^{2+}$ influx assays along the increasing number of alkylation; dihexylamine with IC_{50} of approximately 50 μM, trihexylamine with approximately 20 μM, and tetrahexylammonium iodide with approximately 5 μM. C6 cells reportedly lack ion channels other than two types of K^+ channels (delayed rectifier and Ca^{2+}-activated channels), both of which are blocked by charybdotoxin. The toxin did not affect the MTX action, hence suggesting that the inhibition by the amines was not via K^+ channels. Implication of membrane potential was also ruled out by experiments using a voltage-sensitive fluorescent dye, DiSC2(5), where these amines did not alter membrane potential in the presence of MTX (**398**). One of the possible explanations is that MTX (**398**) interacts with a non-voltage-gated and usually inactive channel presumably belonging to the ion channel superfamily, through which the alkylamines block ion entry by a mechanism similar to that proposed for K^+ channels.

8.07.8.4.3 Gambieric acid

Potent antifungal substances were found in one strain of the dinoflagellate *Gambierdiscus toxicus*, an epiphytic species implicated in ciguatera as the source of maitotoxin and ciguatoxins. While these toxins were retained in the algal cells during culture, the antifungals were released into the medium. Activity-guided purification led to the discovery of four polyethers, designated gambieric acids A, B, C, and D (**412**)–(**415**).[374,375] Their property of inhibiting the growth of *Aspergillus niger* was of unprecedented potency, exceeding that of amphotericin B by a factor of 2000. Gambieric acids A (**412**) and B (**413**) and a mixture of gambieric acids C (**414**) and D (**415**) inhibited the growth of *A. niger* at 10, 20, and 10 ng/disk, respectively, by the paper disk method, while amphotericin B and okadaic acid were inhibitory at doses of 20 μg/disk and 10 μg/disk, respectively. Gambieric acid A (**412**) at a dose of 1 mg kg^{-1} showed no toxicity against mice upon an intraperitoneal injection. The cytotoxicity (IC_{50}) of the mixture of gambieric acids C (**414**) and D (**415**) against mouse lymphoma L5178Y cell was 1.1 μg ml^{-1} when monitored by [^{3}H]thymidine incorporation.

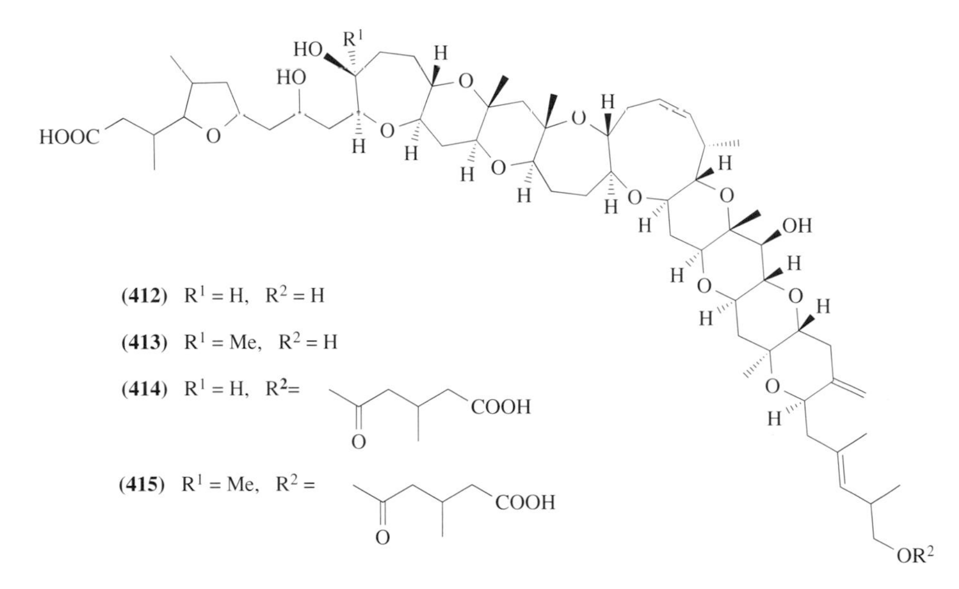

The dinoflagellate *G. toxicus* (GII1 strain) was isolated in the Gambier Islands, French Polynesia, and was cultured in a seawater medium (see Section 8.07.8.4.2). The medium (5000 l), free of algal cells, was passed through a column of Amberlite XAD-2. Antifungal compounds were retained on the column and eluted with methanol. Purification of the eluate by solvent partition and column chromatography afforded gambieric acids A (**412**) (0.6 mg) and B (**413**) (0.15 mg) and a mixture (5.8 mg) of gambieric acids C (**414**) and D (**415**). The structures of gambieric acids were elucidated by NMR and negative FABMS data as well as by hydrolysis to furnish 3-methylglutaric acid. They had novel brevetoxin-type structures consisting of nine contiguous ether rings (7/6/6/7/9/6/6/6/6) and one isolated tetrahydrofuran. Gambieric acids C (**414**) and D (**415**) were 3-methylhemiglutarates of gambieric acids A (**412**) and B (**413**), respectively. The strain GII1 did not produce ciguatoxins, indicating the biosynthetic versatility of this organism. From an ecological point of view, it is interesting to note that epiphytic dinoflagellates release antifungals from the cells, while retaining maitotoxin, which has no antimicrobial activity. Because of their solubility, gambieric acids may stay on the surface of the substrate near the dinoflagellates and exert an allelopathic function against other epiphytic organisms. Maitotoxin, on the other hand, may act as an antifeedant with its extreme toxicity toward higher animals.

8.07.8.5 Other Toxins

8.07.8.5.1 Palytoxin

Palytoxin (**416**) ($C_{129}H_{223}N_3O_{54}$) is one of the most potent and complex marine toxins, associated with marine coelenterates (zoanthids) of the genus *Palythoa*.[376] Its intravenous lethality (LD_{50}) ranged from 0.025 μg kg^{-1} in the rabbit to 0.45 μg kg^{-1} in the mouse. The structural elucidation of this complex toxin was achieved early in the 1980s by three chemistry laboratories, those of Moore and Bartolini,[377] Uemura *et al.*,[378] and Kishi and co-workers.[379] All aspects of stereochemistry of (**416**) were rigorously determined by comparison of synthetic fragments with the natural product.

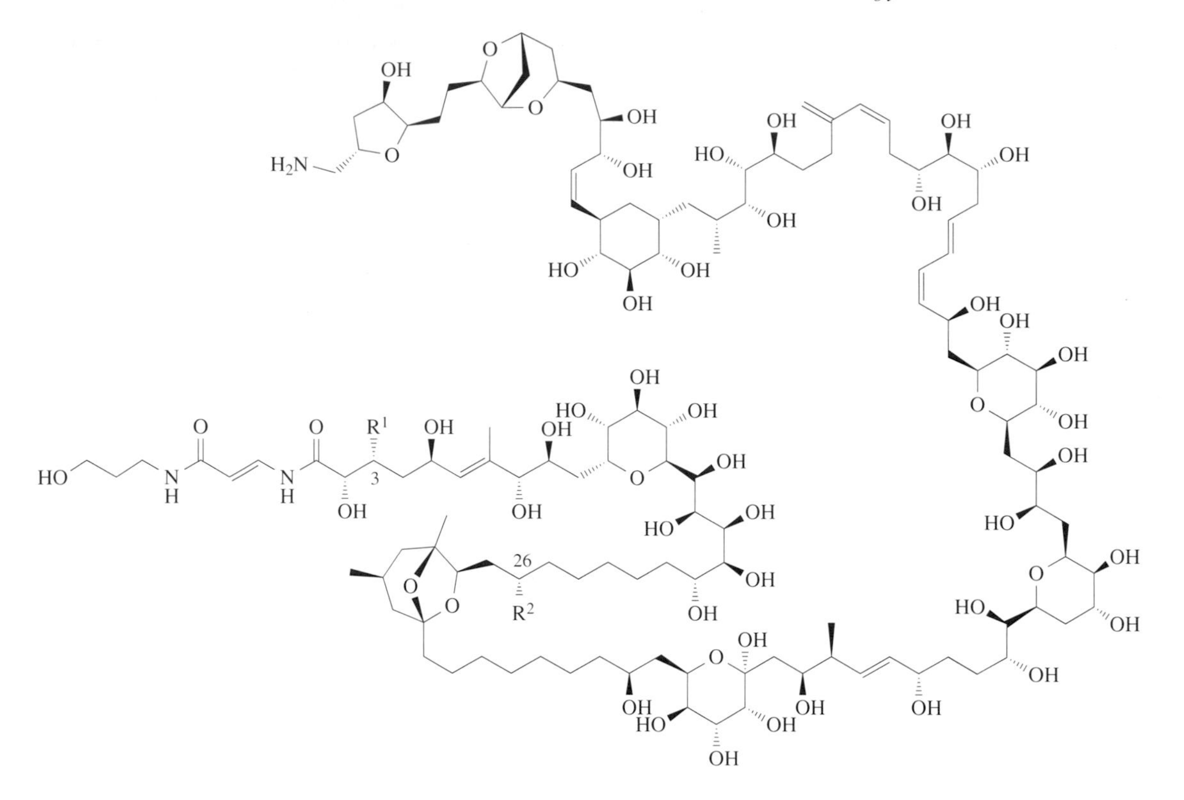

(416) $R^1 = R^2 = Me$

(417) $R^1 = R^2 = H$; OH → H (position not located)

Though first isolated from the zoanthid *Palythoa toxica*, the biogenetic origin of palytoxin (**416**) was questioned because of marked seasonal and regional variations. Sporadic occurrence of palytoxin in an alga, crabs, and a herbivorous fish also suggested that the toxin might be produced by a microorganism and transmitted to the marine food chain, as is the case with ciguatera (see Section 8.07.8.4). A dinoflagellate *Ostereopsis siamensis* is taxonomically closely related to *Gambierdiscus toxicus* (see Section 8.07.8.4), and contains potent toxins, named ostereocins. Yasumoto and co-workers studied the toxins in *O. siamensis*, and identified the major ostereocin as a palytoxin analogue.[380] They collected *O. siamensis* at Aka Island, Okinawa, Japan, and the dinoflagellate was grown for 30 days under the same culture conditions as *G. toxicus*. The extracts of the harvested algal cells were separated by solvent partition and column chromatographies. Purification was monitored by mouse assay following characteristic UV maxima. From 9361 of culture, 3.8 mg of ostereocin D (**417**) ($C_{127}H_{220}N_3O_{53}$) was isolated as a major toxin constituent accounting for 70% of total toxicity. The lethality (LD_{50} value) of (**417**) in mice (i.p.) was 0.75 μg kg^{-1}. Extensive spectral studies revealed that ostereocin D (**417**) was a 3,26-bisdesmethyldeoxy analogue of palytoxin. Determination of the position of the missing hydroxyl group in (**417**) was hampered by heavily overlapping signals of methine protons and oxycarbons. The remainder of ostereocin D (**417**) was believed to be identical regiochemically to palytoxin (**416**), and was deduced to have the same stereochemistry as palytoxin (**416**) on the basis of the analysis of $^3J_{H,H}$ values. From the dinoflagellate *O. siamensis*, palytoxin (**416**) itself was not identified and minor ostreocin analogues were detected by electrospray ionization mass spectrometry. The small structural changes barely affected the mouse lethality; interestingly, the minor analogues had reduced cytotoxicity and hemolytic potency. From this study, though it was not clear whether *O. siamensis* was a symbiont of *Palythoa* spp., the dinoflagellate was strongly indicated to be a source of palytoxin (**416**).

In coral reefs of the Caribbean Sea (Colombia) palytoxin (**416**) was detected in zoanthid species of the genera *Palythoa* and *Zoanthus* by HPLC or assaying the delayed hemolysis in human erythrocytes produced by aqueous extracts (which was inhibited by ouabain pretreatment). The toxin content of the polyps and colonies was highly variable and was not related to their reproductive cycle or to the amount of symbiotic algae. Sequestration of palytoxin (**416**) was observed in crustaceans (*Platypodiella* sp.) living in close association with *Palythoa* colonies and in polychaete worms (*Hermodice carunculata*) feeding on the zoanthids. Resistance of marine animals to the toxin may enable palytoxin to enter food chains.[381]

Extensive pharmacological and biochemical researches have been carried out on palytoxin (**416**). Palytoxin (**416**) was found to exhibit unique biochemical and pharmacological effects such as membrane depolarization, stimulation of arachidonic acid release, stimulation of neurotransmitter release, inhibition of Na^+/K^+-ATPase, induction of contraction of smooth muscle, and tumor-promoting activity.[382] For example, it has been reported that palytoxin (**416**), which at nanomolar concentrations enhances the permeability of mammalian cell membranes to both Na^+ and Ca^{2+}, modulated cytosolic pH in human osteoblast-like Saos-2 cells via an interaction with Na^+/K^+-ATPase.[383] The detailed mechanism of its action accounting for its variable biological effects is still largely unknown.

8.07.8.5.2 Brevetoxin

Brevetoxins are the major toxins in the Florida red tide organism, *Gymnodinium breve*.[384] A dense growth or bloom of dinoflagellates under certain favorable conditions has caused a phenomenon known as "red tide" and the blooms of the dinoflagellate *Gymnodinium breve* have led to massive fish kills, mollusk poisoning, and human food-poisoning along the Florida coast, in the Gulf of Mexico, and in many other regions of the world. In addition, many mass poisonings of humans have also been caused by consumption of affected seafood. The symptoms in human victims included diarrhea, dizziness, respiratory problems, eye irritation, and others.

Lin *et al.* demonstrated the structure of brevetoxin B (**418**), the major toxin responsible for the red-tide poisonings, by X-ray crystallography in 1981.[385] Brevetoxin B (**418**) was isolated from the cultures of *G. breve*, incubated at 25 °C for 21 days under constant illumination with standard fluorescent light. From a medium containing 5×10^8 cells, 5.0 mg of brevetoxin B (**418**) was isolated. Brevetoxin B (**418**) was made up of a single carbon chain locked into a rigid ladder-like structure consisting of 11 contiguous *trans*-fused ether rings, and there had been no precedent for this class of linear-condensed polycyclic ether compounds. Other marine toxins such as ciguatoxin, maitotoxin, and others (see Sections 8.07.8.3 and 8.07.8.4) proved to be included in this class of ladder polyether compounds. The structure of brevetoxin A (**419**), the most toxic component (LC_{100} 4 ng ml^{-1} to guppies), was elucidated by X-ray analysis of the crystals of a dimethyl acetal derivative in 1986 by Shimizu *et al.*[386] In the NMR spectrum of (**419**) certain signals were not observed and some peaks were unusually broadened, which was explained by the speculation that (**419**) undergoes a rather slow conformational change between the boat-chair and crown form in solution. Nakanishi and co-workers also described the structure of (**419**) based on NMR and MS data.[387] Several other toxins were also isolated from the organism *G. breve*, and they were divided into the brevetoxin A and B series by their skeletons.[295] Hemibrevetoxin B (**420**) with a skeleton about half the size of those of brevetoxins was also isolated from *G. breve*.[388]

The biosynthesis of these unprecedented polycyclic ethers was of considerable interest and it was postulated that the all-*trans* cyclic structure could be formed by a cascade of opening of all-*trans* epoxides, which are probably formed by epoxidation of *trans* double bonds. The biosynthetic origins and assignments of ^{13}C NMR signals of brevetoxin B (**418**) were studied by feeding experiments with [1-^{13}C]-, [2-^{13}C]-, and [1,2-$^{13}C_2$]acetates and methyl-^{13}C-methionine.[389] A 2D INADEQUATE experiment was carried out on a sample incorporating [1,2-$^{13}C_2$]acetate to reveal the contiguous carbon pairs derived from one acetate unit. The labeling patterns thus obtained (Figure 14), however, could not be interpreted by the simple acetogenin pathway, and it was proposed that labeled acetate was metabolized through the TCA cycle and incorporated into dicarboxylic acids before being utilized for the toxin biosynthesis (Figure 14).[390,391]

Pharmacological studies have revealed that brevetoxins (BTX) exhibit their damaging effects by acting on the voltage-sensitive domains of sodium channels[392] (specific receptor site 5), which are situated in the membranes of the cells. Nicolaou, who achieved the total synthesis of BTX B (**418**),[393] examined several synthetically prepared structural analogues of BTX B (**418**) in synaptosome receptor binding assays and by functional electrophysiological measurements. A truncated analogue of BTX was not ichthyotoxic at micromolar concentrations, showed decreased receptor-binding affinity, and caused only a shift of activation potential without affecting mean open times or channel inactivation. An analogue with the A-ring carbonyl removed bound to the receptor with nanomolar affinity, produced a shift of activation potential and inhibited inactivation, but did not induce longer mean open times. An analogue in which the A-ring diol was reduced showed low binding affinity,

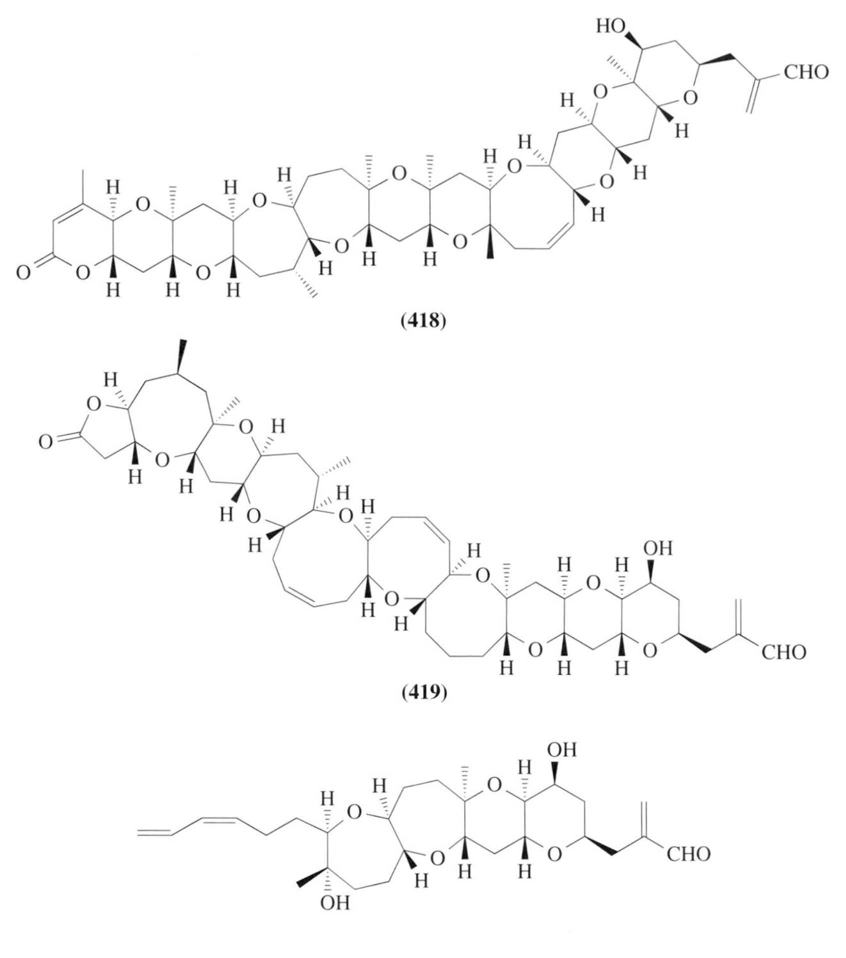

(418)

(419)

(420)

yet populated five subconductance states. These data were consistent with the hypothesis that binding to sodium channels requires an elongated cigar-shaped molecule, approximately 30 Å long. A detailed model was proposed for the binding of brevetoxins to the channel which explained the differences in the effects of the BTX analogues.[394] Nakanishi and co-workers found that BTX also induces selective ion movements across lipid bilayers through transmembrane BTX self-assemblies. They examined the self-assembly of several BTX derivatives in the presence and absence of cations and lipid bilayers using powerful porphyrin chromophores as CD labels. BTX derivatives self-assemble into tubes, which can bind to metals when inserted into the bilayer to form transmembrane pores. Depending on the tendency of the BTX derivative to self-aggregate, it may aggregate in solution before membrane insertion, or may insert itself into the membrane as a monomer before assembling the pore. The active BTX B complex in lipid bilayers is a cyclic, transmembrane self-assembly consisting of antiparallel aligned BTX molecules that can mediate selective ion movement through membranes. The differences in pore formation mechanisms between BTX derivatives may be reflected in differences in pore formation by natural BTX variants, perhaps explaining their varying levels of toxicity.[395] Nakanishi's group also investigated ion movement across large unilamellar egg phosphatidylcholine (PC) vesicles containing 0.3% BTX B (**418**). The transport rates of the ions were found to be sensitive to temperature of vesicle formation, cholesterol concentration, and the ion size. The most dramatic rate increase was achieved with vesicles containing 50% cholesterol. The BTX B skeleton was shown to be oriented perpendicular to the lipid surface. The sensitivity of BTX B channel formation to temperature, presence or absence of cholesterol, etc., suggested that *in vivo* penetration into specific cell membranes and self-assembly stabilization may also be governed by the presence or absence of membrane proteins.[396] BTX B (**418**) was utilized as a model for the study of the scope and limitations of porphyrin chromophores for structural studies

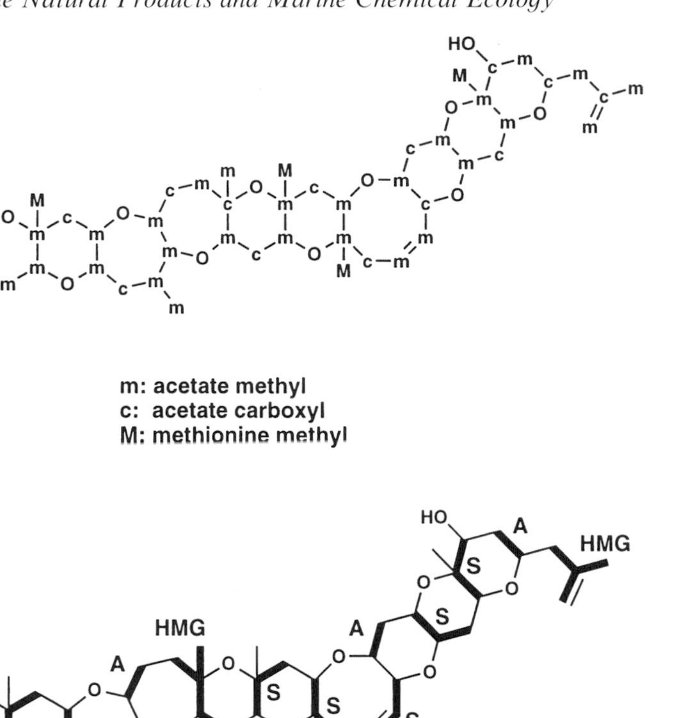

S: succinate or equivalent
HMG: hydroxylmethylglutarate
P: propionate or equivalent
A: acetate
KG: α-keto glutarate or equivalent

Figure 14 Incorporation patterns of labeled acetate into brevetoxin B and hypothetical building blocks of the molecule.

by the exciton coupled circular dichroic (CD) method. Porphyrins at the termini of dimeric steroids and BTX B (**418**) exhibited exciton coupling over interchromophoric distances up to 50 Å (Figure 15). As a result, porphyrin chromophores were found to be promising reporter chromophores for extending the exciton coupled CD method to structural studies of biopolymers.[397]

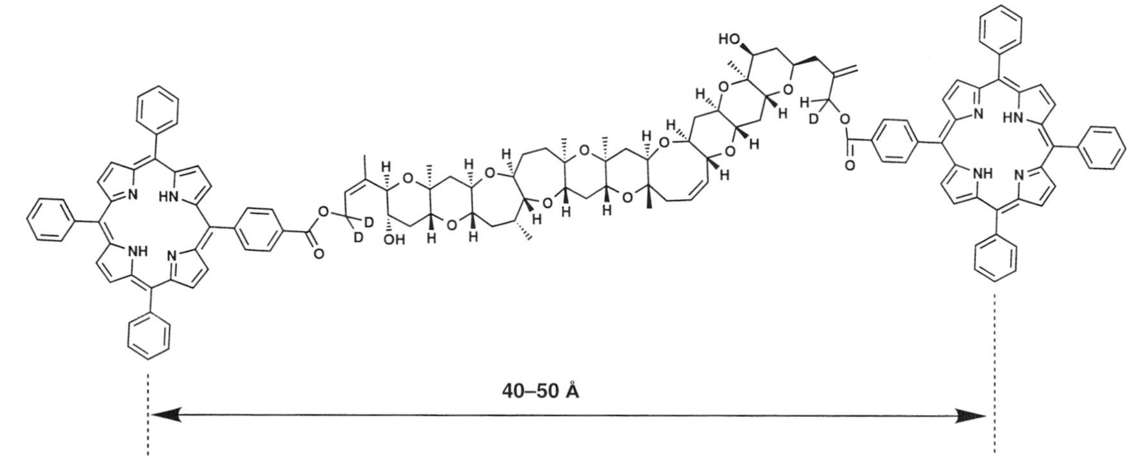

Figure 15 A brevetoxin B-bridged porphyrin dimer for the study of extending the exciton coupled CD method.

Other BTX-related compounds have been isolated and their structures reported. In December 1992 toxicated bivalves were noted for the first time in New Zealand. The toxicated shellfish *Austrovenus stutchburyi* (80 kg) collected at the Bay of Plenty, New Zealand in January 1993, were extracted with 80% MeOH under reflux. The extract was separated by solvent partition and chromatography on columns of SiO_2, ODS, and Sephadex LH-20, followed by reverse-phase HPLC to give 5.4 mg of brevetoxin B_1 (BTX-B_1 (**421**))).[398] The structure was elucidated by comparison of its spectral data with those of BTX B (**418**). NMR techniques revealed that the structures of BTX-B_1 (**421**) and BTX B (**418**) differed only in the functional group of the ring K side-chain; (**421**) had a carboxyl group at C-42 in place of the aldehyde group in (**418**), and (**421**) bore a taurine attached to the C-42 carboxyl group via an amide linkage. The minimum lethal dose of BTX-B_1 (**421**) was 0.05 mg kg^{-1} (i.p.) in mice. The animals exhibited irritability immediately after injection, followed by hind and/or hind-quarter paralysis, severe dyspnea and convolutions prior to death due to respiratory paralysis; these symptoms were very similar to those caused by brevetoxins. However, mice which did not develop respiratory difficulty recovered slowly. Interestingly, BTX-B_1 (**421**) at 100 ng ml^{-1} was not toxic to freshwater "zebra fish" (1.0–1.5 g) in 1 h, unlike BTX-A (**419**), BTX-B (**418**), and other brevetoxin analogues. Although it is well known that several ichthyotoxic brevetoxins, such as BTX-A (**419**) and BTX-B (**418**), are produced by the dinoflagellate *G. breve*, this was the first isolation of a BTX derivative from shellfish. Another brevetoxin analogue, brevetoxin B_3 (BTXB_3 (**422**)), was isolated from the greenshell mussel *Perna canaliculus* collected in January 1993 at the Coromandel Peninsula, North Island of New Zealand.[399] From 30 kg of hepatopancreas 1.2 mg of BTXB_3 (**422**) was obtained as a mixture of two homologues differing in their acyl moieties; the acyl groups were identified as palmitoyl and myristoyl groups by fluorometric HPLC analysis after hydrolysis with 0.1N NaOH in 95% MeOH. Detailed analysis of NMR spectral data showed that in BTXB_3 (**422**) the brevetoxin B skeleton was modified by cleavage of ring D, esterification of the resulting alcohol, and oxidation of the aldehydic terminus. The proposed structure was further supported by negative ion FAB MS/MS experiments. Interestingly, BTXB_3 (**422**) did not kill mice

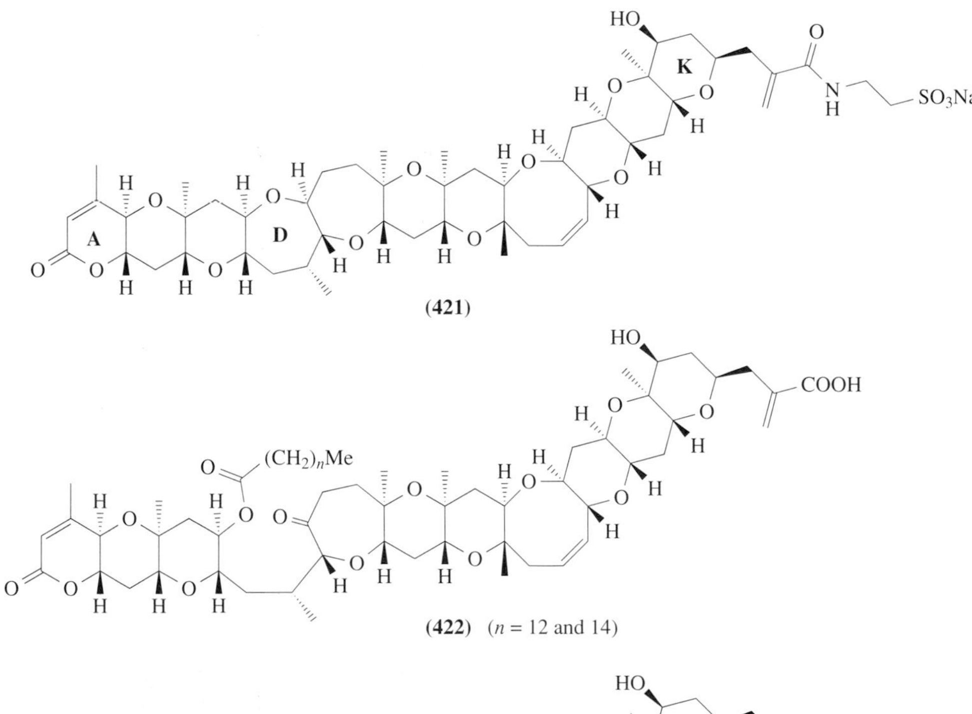

(421)

(422) (n = 12 and 14)

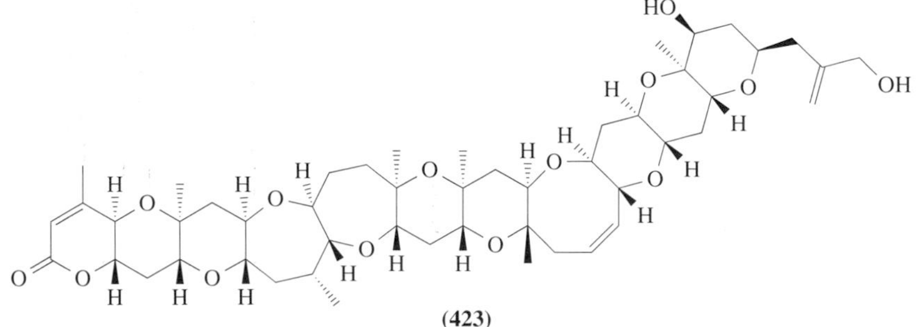

(423)

by i.p. injection at a dose of 300 μg kg^{-1}. It was highly likely that mussels detoxified BTX-B (**418**) to $BTXB_3$ (**422**), as $BTXB_3$ (**422**) had never been detected in dinoflagellates. The ^1H and ^{13}C NMR spectra of another polyether toxin, brevetoxin-3 (**423**), were totally assigned using a series of 2D NMR experiments, which included TOCSY, ROESY, HMQC, HMBC, and IDR (Inverted Direct Response)–HMQC–TOCSY. All work was performed on a sample consisting of 800 μg (0.95 μmol) at 500 MHz.[400]

A different type of marine toxin, gymnodimine (**424**),[401] was isolated from New Zealand oysters *Tiostrea chilensis* collected at Foveaux Strait, South Island. From 3 kg of oysters, 2.0 mg of (**424**) was obtained. Its unprecedented structure was elucidated by spectroscopic methods. When the oysters were found to be toxic at high levels, concurrent blooms of a dinoflagellate *Gymnodinium* cf. *mikimotoi* was also observed. It was unambiguously established that gymnodimine (**424**) was produced by the dinoflagellate *G.* cf. *mikimotoi*, on the basis of large-scale culture of the dinoflagellate, also collected at Foveaux Strait, and HPLC, ^1H NMR, and LC/MS analyses. Mouse lethality (i.p.) of gymnodimine (**424**) was 0.45 mg kg^{-1}. Gymnodimine (**424**) also showed potent ichthyotoxicity against a small freshwater fish *Tanichthys albonubes* at 0.1 ppm at pH 8. Gymnodimine (**424**) was isolated not only from cells but also from the culture medium, suggesting the possibility that it may cause massive fish kills like its related species *Gymnodinium mikimotoi* (formally *G. nagasakiense*) during seasonal blooms of the dinoflagellate.

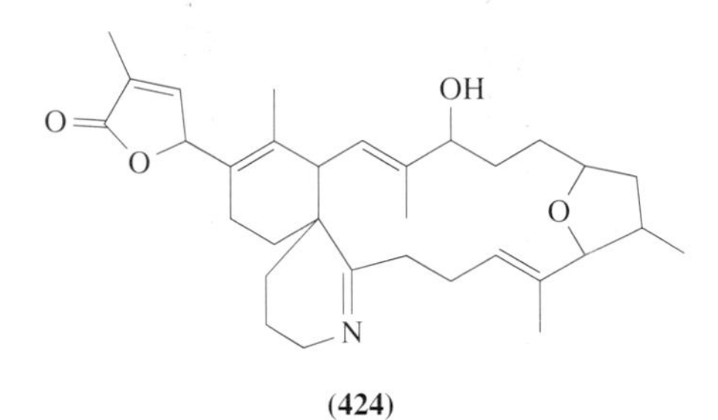

(424)

8.07.8.5.3 Surugatoxin

Surugatoxin (**425**) was first isolated from the toxic ivory shell *Babylonia japonica* by Kosuge *et al.* as a causative toxin of a food-poisoning outbreak in 1965 at Suruga Bay, Japan. Its structure was determined by X-ray analysis in 1972,[402] corrected to (**425**) in 1981,[403] and confirmed by its total synthesis.[404] However, surugatoxin (**425**) was later found to be nontoxic and the real causative agents of intoxication resulting from ingestion of the toxic shell proved to be neosurugatoxin (**426**) and prosurugatoxin (**427**).[405] Pharmacological studies suggested that the toxins had a selective affinity for ganglionic nicotinic receptors, its affinity constant for these receptors being more than three orders of magnitude greater than that of hexamethonium.[406] A Gram-positive bacterium (Aalxll strain), which was associated with the Japanese ivory shell *B. japonica*, was shown to be responsible for the production of surugatoxins.[407]

8.07.8.5.4 Polycavernoside

Fatal human intoxication occurred in Guam in late April 1991, due to ingestion of the red alga *Polycavernosa tsudai* (formerly *Gracilaria edulis*). The responsible toxins were investigated by Yasumoto and co-workers resulting in isolation of two toxins, polycavernoside A (**428**) and B (**429**).[408] The alga *P. tsudai* (2.6 kg) was collected in June 1991, at Tanguisson Beach, Guam, and was extracted with acetone. Solvent partition and column chromatography of the extract, guided by mouse bioassays, yielded polycavernosides A (**428**) (400 μg) and B (**429**) (200 μg). LD_{99} in mice (i.p.) was 200–400 μg kg^{-1} for both. The structure of polycavernoside A (**428**) was elucidated by spectral data, consisting of a 13-membered lactone with a triene side-chain and *O*-α-2,3-di-*O*-methylfucopyranosyl-(1″-3′)-*O*-β-2,4-di-*O*-methylxylopyranoside. The relative stereostructure was deduced mainly from NOE data and the stable conformations were suggested by a force-field calculation (program "Dreiding") for structural confirmation.[409]

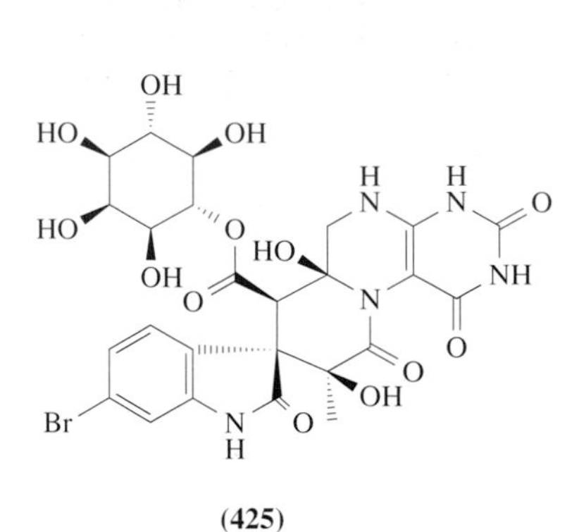

(425)

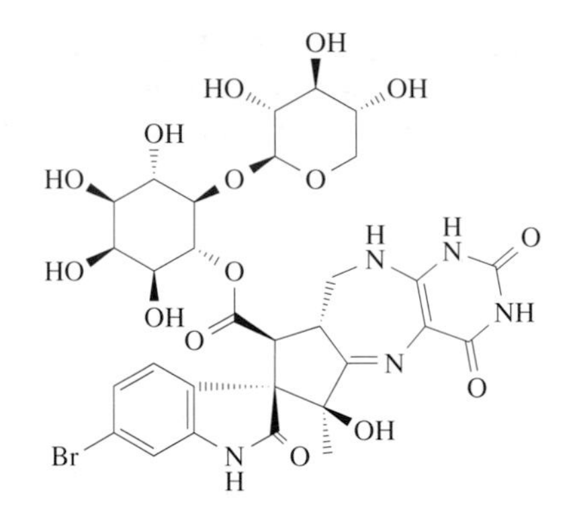

(426)

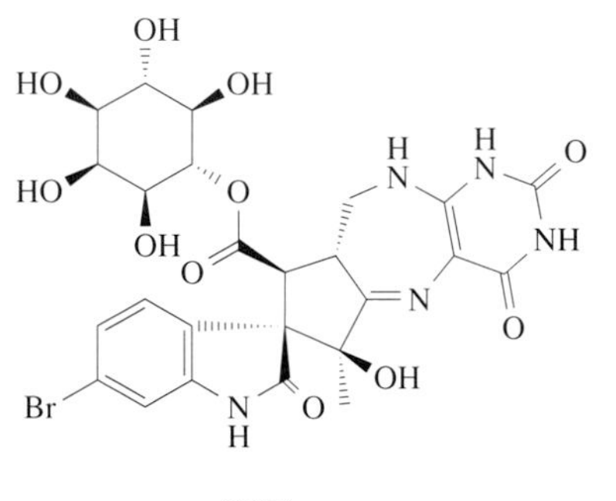

(427)

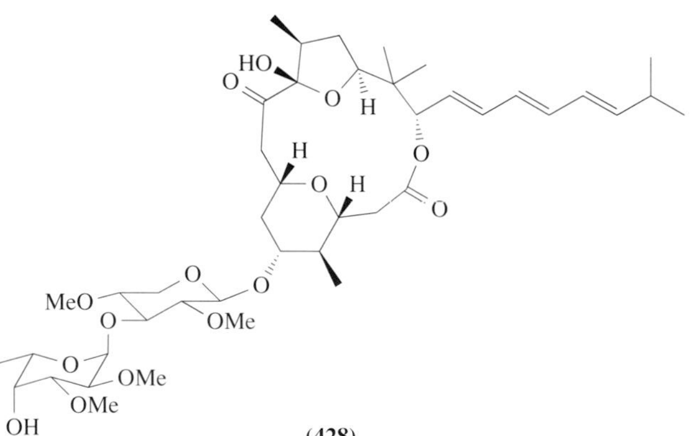

(428)

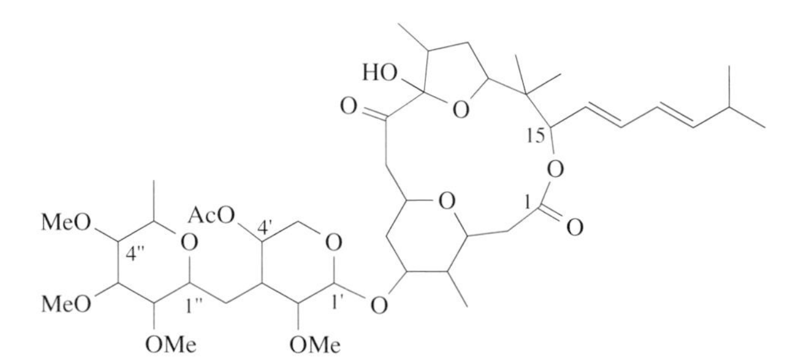

(429)

The algal toxicity rapidly decreased after the incident but rose again, though at lower levels, in the same season of the following year. Three other analogues, polycavernosides A2 (**430**), A3 (**431**), and B2 (**432**), were isolated together with (**428**) from the alga collected on the same beach in June 1992.[410] NMR spectral data of polycavernosides B (**429**) and B2 (**432**) showed that the conjugated triene in the side-chain portion of (**428**) was replaced by a conjugated diene in (**429**) and (**432**). The fucopyranosyl-xylopyranosyl backbone was deduced as a common structure of the glycosidic residue of polycavernosides. Compound (**430**) was the 4′-*O*-demethyl analogue of (**428**), and (**431**) was the *O*-methylated analogue of (**428**) at the 4″-OH. The OMe group on C-4′ in (**428**) was replaced by an OAc group in (**429**), and (**432**) was the 4′-*O*-deacetyl analogue of (**429**). These structures were further supported by FAB/MS/MS by showing prominent ions corresponding to sequential loss of each residue.

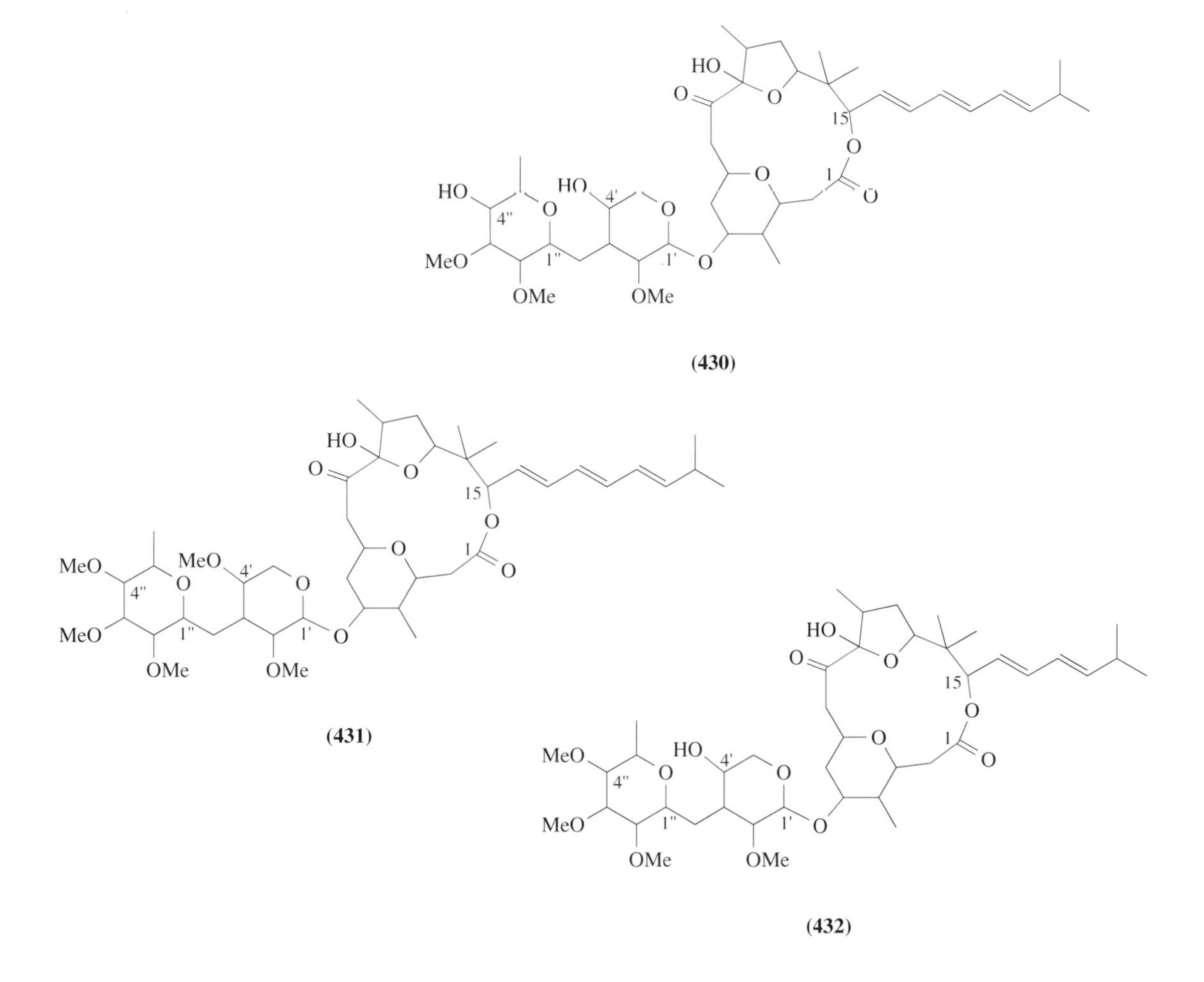

The synthesis of some partial structures of polycavernoside A (**428**) was studied by Murai and co-workers, and they revised the whole relative stereochemistry of polycavernoside A, as shown in Figure 16, by combination of the synthesized sugar and tetrahydropyran parts.[411]

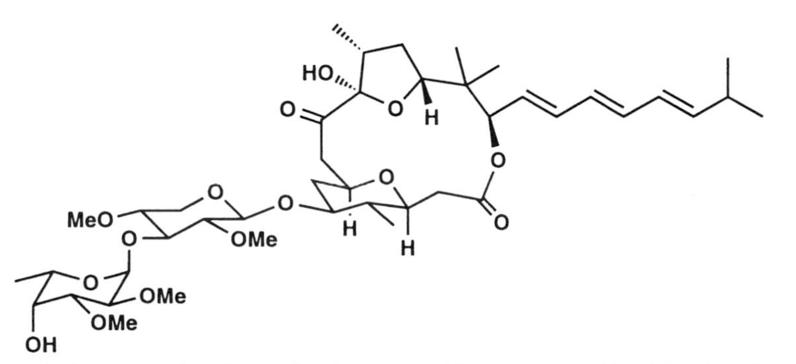

Figure 16 Relative stereochemistry of polycavernoside A, proposed by Murai and co-workers.

A smaller macrocycle, a trioxatridecane, of polycavernosides was reminiscent of the aplysiatoxins, which contained trioxadodecane. The sudden and transient occurrence of the toxins in the alga remained unexplained, but may provide a clue to previous outbreaks of fatal food poisoning caused by algae of the genus *Gracilaria*.

8.07.8.5.5 ***Prymnesin***

Prymnesium parvum was a notorious red-tide organism belonging to Haptophyceae. The flagellate causes serious damage to aquaculture and marine ecology in many parts of the world, the greatest threat being to salmon farmed in Norway. Despite efforts by many research groups, the causative ichthyotoxin named prymnesin evaded purification for a long time and hence chemical and toxicological studies of the toxin have been hampered. In 1994 Yasumoto and co-workers succeeded for the first time in isolating two hemolytic-ichthyotoxic substances, named prymnesin-1 (PRM1) and prymnesin-2 (PRM2 (**433**)).[412] They obtained 10 mg of PRM1 and 15 mg of PRM2 from 400 l of cultures of *P. parvum*. Their hemolytic potencies exceeded that of plant saponins by more than 1000 times. The minimum concentration to cause hemolysis of a 1% mouse blood cell suspension and to kill freshwater fish *Tanichthys albonubes* was ca. 3 nM. FAB MS data on molecular ion species and their isotope distribution pattern containing ^{35}Cl and ^{37}Cl suggested a molecular formula $C_{107}H_{154}O_{44}NCl_3$ for PRM1 and $C_{96}H_{136}O_{35}NCl_3$ for PRM2 (**433**). The presence of chlorine atoms in the molecule was indicated by energy dispersive X-ray analysis. ^{13}C NMR measurement on ^{13}C enriched PRM2 *N*-acetate indicated 96 carbon signals (2 methyls, 24 methylenes, 10 alkenic methines, 53 other methines, and 7 quaternary carbons). Further detailed analyses of both ^{1}H NMR (COSY, HOHAHA, and NOESY) and ^{1}H observed ^{1}H–^{13}C correlation NMR (HSQC and HMBC) data disclosed the unprecedentedly unique structure of PRM2 (**433**).[413] The molecule was characterized by the 14 ether rings (6/6/6/7/6, 6/6, 6/6, 6/6, 6/6, and 6), polyene–polyyne bonds, three chlorine and one nitrogen atoms, and one pentose. Structural confirmation was enhanced by measuring NMR spectra of PRM2 (**433**), the *N*-acetate, and peracetate in two different solvents. The sugar moiety was analyzed by chiral GC after hydrolysis of (**433**) to confirm that the sugar was L-xylose. Perhydro-PRM2 was prepared by hydrogenation of (**433**) (Pd/C, H_2) to confirm the degree of unsaturation. In the ESIMS spectrum of perhydro-PRM2, molecular-related ions were observed at m/z 1963, revealing that the chlorine atom on C-1 had been substituted by a hydrogen atom during hydrogenation. The nature of unsaturation including five double and four triple bonds in (**433**) was thus verified. PRM1 probably had one pentose and one hexose added to the PRM2 skeleton.

The relative stereochemistry of the polyether moiety, C-20–C-74, of (**433**) was determined from the NOE data. The molecular structure of PRM2 (**433**) is conformationally rich with a number of key torsional bonds which were probed with contemporary conformational searching techniques in order to describe the three-dimensional structure of the prymnesin molecule. Important topological features include a backbone twist of 60° and an elongation length of 39.5 Å from ring "A" to ring "N" for the PRM2-molecule in the NOE preferred conformation. These and other molecular structural and energetic features determined by computational techniques were examined with the aim of understanding how PRM2 (**433**) exhibits its potent biological activity in relation to its inherent structural traits.[414]

8.07.8.5.6 ***Pinnatoxin***

Shellfish of the genus *Pinna*, living mainly in shallow areas of the temperate and tropical zones of the Indian and Pacific Oceans, are commonly eaten in China and Japan, and food poisoning resulting from their ingestion occurs frequently. In Japan, six outbreaks involving 2766 people were reported between 1975 and 1991. In 1990 a water-soluble nonexogenous substance which showed a strong nerve toxicity, was isolated from *Pinna attenuate* collected in the South China Sea by Chinese investigators, and the toxin was named pinnatoxin.[415] Chinese investigators conjectured that this toxin was responsible for the human intoxication resulting from *P. attenuata*, which occurred at Guangdong, China, in 1980 and 1989, based on its predominant symptoms, i.e., diarrhea, paralysis, and convulsions. Pinnatoxin significantly increased the contractility of the aortic strips of rabbit and the ileums of guinea pig. The preliminary tests suggested that the actions of pinnatoxin were related to the α-receptor and the M-receptor, and that pinnatoxin is probably a Ca^{2+} channel

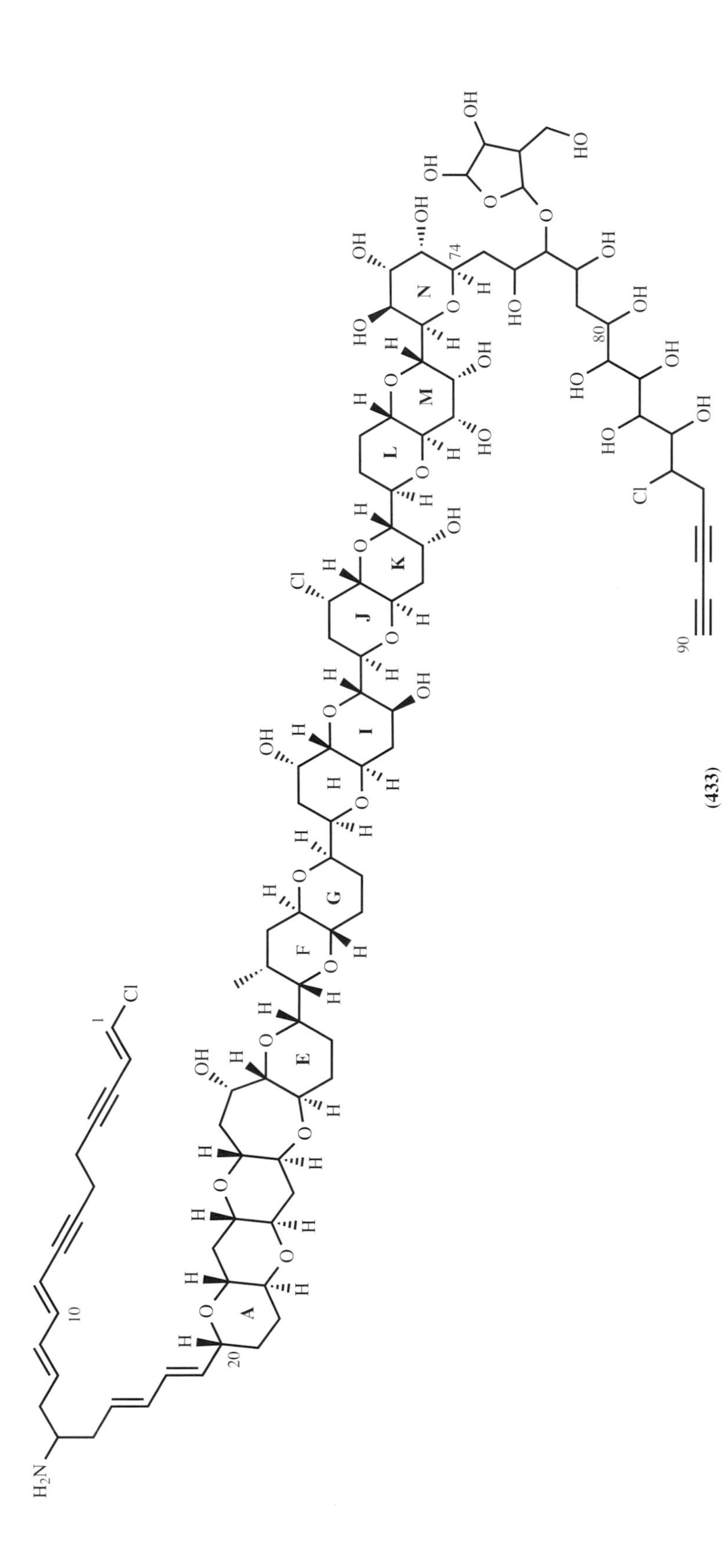

(433)

activator. Thereafter, identification of the specific toxins and clarification of their physiological activity proved to be challenging.

Uemura and co-workers succeeded in isolating four toxins, pinnatoxins A–D (**434**)–(**437**)[416] from *Pinna muricata* and elucidated their structures as an unprecedented class of polyether macrocycle with carboxylate and iminium functionalities.[417] From 45 kg of viscera of *P. muricata* collected in Okinawa, Japan, a bioassay-guided fractionation based on intraperitoneal mouse lethality tests afforded 3.5 mg of pinnatoxin A (**434**) (LD_{99} 180 $\mu g\ kg^{-1}$), 1.2 mg of a mixture of pinnatoxins B (**435**) and C (**436**), and 2 mg of pinnatoxin D (**437**). The mixture of (**435**) and (**436**) was the most toxic (LD_{99} 22 $\mu g\ kg^{-1}$), while pinnatoxin D (**437**) was found to be less toxic (LD_{99} 0.4 $mg\ kg^{-1}$) but exhibits cytotoxicity against P388 leukemia cells (IC_{50} 2.5 $\mu g\ ml^{-1}$). Their planar structures were determined mainly by interpretation of 2D NMR (DQF–COSY, HOHAHA, HSQC–HOHAHA, and HMBC) spectra. Pinnatoxin A (**434**) was revealed to be an amphoteric macrocyclic compound composed of a 6,7-spiro ring, a 5,6-bicyclo ring, and a 6,5,6-trispiroketal ring involving 14 chiral centers. Pinnatoxins B (**435**) and C (**436**) were obtained as a 1:1 mixture and deduced to be diastereomers at the C-34 position, both possessing one more carbon than pinnatoxin A (**434**).[416] Pinnatoxin D (**437**) had three more carbon units (C-34–C-36) than (**434**) and different functionalities at C-21, C-22, and C-28.[418] The biogenesis of the unique structure containing the 6,7-spiro ring (A and G rings) was proposed to be generated through a Diels–Alder reaction, and a relationship to the carbon skeleton of prorocentrolide (**377**) (see Section 8.07.8.3.1) was suggested. The relative stereochemistry of pinnatoxin A (**434**) was deduced, based on detailed analysis of NOESY and ROESY data, and $^3J_{H\text{-}H}$ coupling constants.[419]

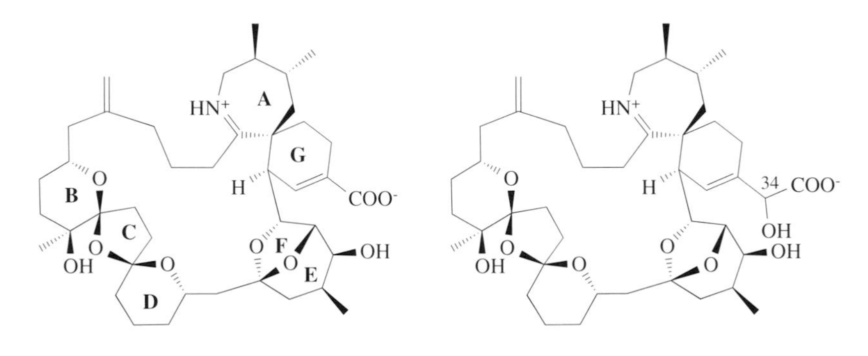

(**434**) (**435, 436**): Diastereomers at C-34

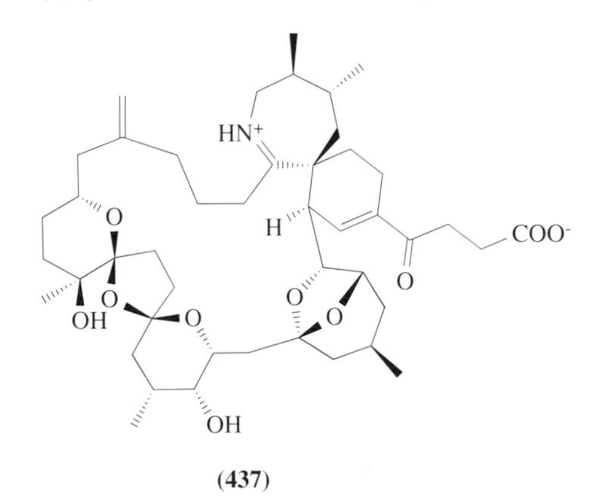

(**437**)

Wright and co-workers studied the digestive glands of mussels *Mytilus edulis* and scallops *Placopecten magellanicus* collected from sites along the eastern shore of Nova Scotia, and isolated a family of macrocylic compounds named spirolides A–D.[420] The structures of two major components of them, spirolides B (**438**) and D (**439**), were elucidated by spectral studies as those containing a spiro-linked tricyclic ether ring system and an unusual seven-membered spiro-linked cyclic iminium moiety. The macrocycle skeleton of spirolides resembled that of pinnatoxins; in contrast with pinnatoxins, spirolides had four less carbons in the macrocycle moiety with no dioxabicyclo-[3.2.1]octane ring moiety (E and F rings). The fact that the spirolides were found in the digestive glands of shellfish, coupled with the observation that they occur on a seasonal basis, usually in

June–July, indicated a microalgal origin of these compounds and their spiro-linked polyether structure might suggest a dinoflagellate source. The spirolides caused potent and characteristic symptoms in the mouse bioassay (LD_{100} 250 μg kg^{-1}, i.p.).

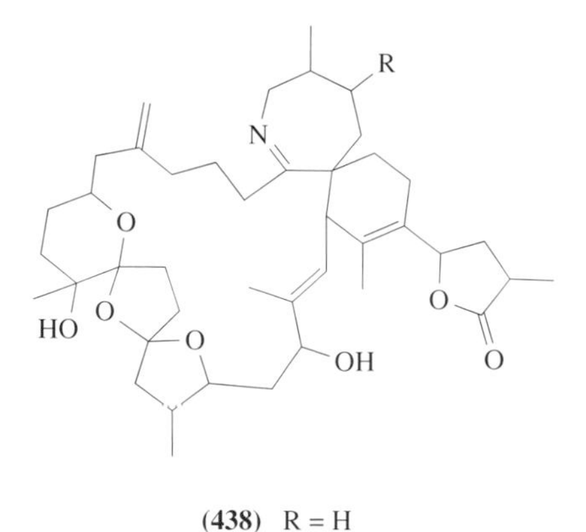

(438) R = H
(439) R = Me

8.07.9 BIOACTIVE MARINE NATURAL PRODUCTS

8.07.9.1 Drug Candidates

Marine natural products are regarded as potentially useful therapeutic agents. Didemnin B (**440**),[421] a cyclic peptide isolated from tunicate *Trididemnum solidum*, has been in clinical trial for some time as a potential anti-cancer drug; some complete or partial responses were observed in CNS-astrocytoma/glioblastoma and non-Hodgkin's lymphoma. Bryostatin 1 (**441**) is a marine-derived compound with anticancer potential, produced by the common fouling marine bryozoan (sea-mat) *Bugula neritina*.[422] Bryostatin 1 (**441**) showed promising Phase I clinical results, and was tested in Phase II human clinical trials by the National Cancer Institute. A potent antitumor compound, aplyronine A (**442**), was isolated from the sea hare *Aplysia kurodai*.[423] Aplyronine A (**442**) exhibited exceedingly potent antitumor activities *in vivo* against P388 murine leukemia (T/C = 545%, 0.08 mg kg^{-1}), Lewis lung carcinoma (T/C = 556%, 0.04 mg kg^{-1}), Ehrlich carcinoma (T/C = 398%, 0.04 mg kg^{-1}), colon 26 carcinoma (T/C = 255%, 0.08 mg kg^{-1}), and B16 melanoma (T/C = 201%, 0.04 mg kg^{-1}). The enantioselective total synthesis of aplyronine A (**442**) was also achieved.[424–426] Other promising antitumor compounds are the ecteinasicidins (see Section 8.07.9.2.3), which showed *in vivo* activity against leukemia, melanoma, ovarian sarcoma, and lung and mammary tumors, including human tumor xenografts. Manoalide (**443**), which was isolated from a marine sponge *Luffariella variabilis*,[427] was found to strongly inactivate phospholipase A_2 by irreversible modification and to show anti-inflammatory properties *in vivo*.[428] Manoalide (**443**) is under clinical trials in the United States as an anti-inflammatory agent. The potent anti-inflammatory diterpene glycoside, pseudopterosin C (**444**),[429] was isolated from the extract of sea whips *Pseudopterogorgia elisabethae*. The extract is a key ingredient in a skin care product that has been commercialized by a cosmetic company.

8.07.9.2 Topics on Tunicates

8.07.9.2.1 Vanadium accumulation by tunicates[430]

Henze discovered high levels of vanadium in blood cells of ascidians, known as sea squirts.[431] After Henze's discovery, many researchers tried to characterize the vanadium in ascidians, with interest in its presence in ascidians or in its possible involvement as an oxygen-carrier other than

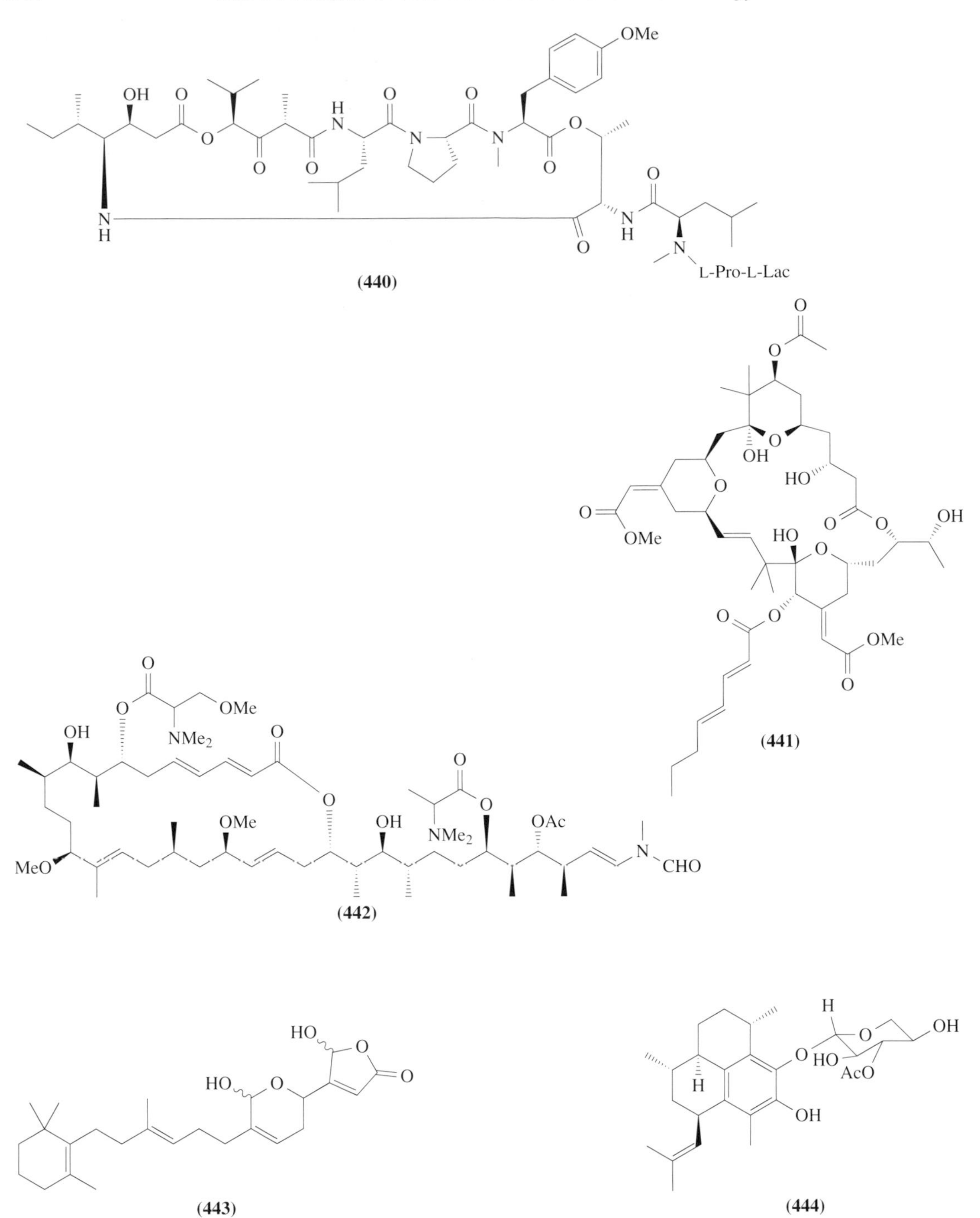

(440)

(441)

(442)

(443)

(444)

iron and copper. However, the mechanism of accumulation and the physiological role of vanadium have not been clarified.

The vanadium in ascidians has been analyzed by a variety of analytical methods, such as colorimetry, emission spectrometry, and atomic absorption spectrometry. These techniques vary widely in sensitivity and precision. Moreover, data were reported in terms of dry weight, ash weight, inorganic dry weight, or amount of protein. Thus, data could not be easily compared.

Michibata *et al.* reexamined the contents of vanadium in several tissues of ascidians by neutron-activation analysis, which is an extremely sensitive method for quantification of vanadium.[432,433] They collected many species of ascidians belonging to two of three suborders, Phlebobranchia and

Stolidobranchia, mainly from the waters around Japan and the Mediterranean. Specimens were dissected into eight samples for analysis, namely, blood cells, plasma, tunic, mantle (muscle), branchial basket, stomach, hepatopancreas, and gonad. The samples were dried and weighed and then they were mineralized at 500 °C and submitted to neutron-activation analysis in a nuclear reactor. Some of the stable vanadium in each sample was converted into the radioactive nucleide, ^{52}V, which emits γ-rays at 1434 keV. Since the frequency of γ-rays emitted depends on the level of stable vanadium in the sample, the original amount of vanadium in the tissues could be calculated. As a result (Table 20), the ascidian species belonging to Phlebobranchia apparently contained higher levels of vanadium than those belonging to Stolidobranchia. Furthermore, blood cells especially contained the highest amounts of vanadium among the tissues examined. Levels of iron and manganese determined simultaneously did not vary much among the members of the two suborders. The highest concentration of 350 mM vanadium was found in the blood cells of *Ascidia gemmata* belonging to Phlebobranchia,[434] 10^7 times higher than in seawater.[435]

Table 20 Concentrations of vanadium in tissues of several ascidians (nM).[430]

	Tunic	*Mantle*	*Branchial basket*	*Serum*	*Blood cells*
Phlebobranchia					
Ascidia gemmata	N.D.	N.D.	N.D.	N.D.	347.2
A. ahodori	2.4	11.2	12.9	1.0	59.9
A. sydneiensis	0.06	0.7	1.4	0.05	12.8
Phallusia mammillata	0.03	0.9	2.9	N.D.	19.3
Ciona intestinalis	0.003	0.7	0.7	0.008	0.6
Stolidobranchia					
Styela plicata	0.005	0.001	0.001	0.003	0.007
Halocynthia roretzi	0.01	0.001	0.004	0.001	0.007
H. aurantium	0.002	0.002	0.002	N.D.	0.004

N.D.: not determined.

Generally, ascidians have 9 to 11 different types of blood cells, which can be grouped into six categories on the basis of their morphology: hemoblasts, lymphocytes, leucocytes, vacuolated cells, pigment cells, and nephrocytes.[436] The vacuolated cells can be further divided into at least four different types: morula cells, signet ring cells, compartment cells, and small compartment cells. The morula cells have been regarded as vanadocytes for many years.[437] An Italian group first demonstrated that the characteristic X ray due to vanadium was not detected from morula cells but from granular amoebocytes, signet ring cells, and compartment cells and, moreover, that vanadium was selectively concentrated in the vacuolar membranes of these cells.[438]

Hirata and Michibata employed density gradient centrifugation for isolation of specific types of blood cells and neutron-activation analysis for quantification of vanadium contents. Vanadium-rich ascidians, *Ascidia ahodori*, *A. sydneiensis samea*, and *A. gemmata*, were used for this analysis. The pattern of distribution of vanadium coincided with that of signet ring cells but not that of morula cells or compartment cells. These results suggested that the signet ring cells were the true vanadocytes.[439]

Vanadium is a multivalent transition metal. Vanadium ions under ordinary aqueous conditions are limited to the oxidation states, +2, +3, +4, and +5, and to only the +3, +4, and +5 oxidation states under physiological conditions.[440] Vanadium ions in the +3 oxidation state [V^{III}] are usually unstable toward air oxidation, and V^{III} ions are hydrolyzed to $V(OH)^{2+}$ which tends to dimerize to $(VOV)^{4+}$ over pH 2.2. Under neutral and alkaline conditions, simple soluble V^{III} compounds without any strong ligands have never been reported. As described below, vanadium ions in the blood cells of ascidians are stable under strongly acidic conditions. Vanadium ions in the +4 oxidation state [V^{IV}] are paramagnetic and give a blue solution of oxo-ions, VO^{2+} (vanadyl ions), under moderately acidic conditions.

It is believed that vanadium in the +5 oxidation state [V^{V}] is dissolved in seawater but this possibility remains to be confirmed. With regard to the oxidation state of vanadium in ascidian blood cells, Henze[431] first suggested the existence of V^{V}, while later many groups reported V^{III}. Noninvasive physical methods including ESR, NMR, and superconducting quantum interference device (SQUID), have since been used to determine the intracellular oxidation state of vanadium.

Such studies indicated that the vanadium ions in ascidian blood cells were predominantly +3, with a small amount of V^{IV}.[441] After separation of the various types of blood cells of *A. gemmata*, Hirata and Michibata made noninvasive ESR measurements of the oxidation state of vanadium in the fractionated blood cells under a reducing atmosphere.[439] Generally only V^{IV} is detectable by ESR spectrometry. Weak signals due to V^{IV} were recorded when the subpopulation of vanadocytes was submitted to ESR spectrometry at 77 K under nitrogen and subsequent bubbling of oxygen gas into the lysate dramatically increased peak height. These results indicate that the oxidation state of vanadium in vanadocytes is predominantly +3, with a small amount of V^{IV}. The ratio of vanadium in the two states is 97.6:2.4. A marine polychaeta *Pseudopotamilla occelata* is reported to contain high levels of V^{III}.[442]

Henze[431] first reported the coexistence of sulfate with vanadium in ascidian blood cells. Vanadium was considered to bind to nitrogenous compound and sulfate to form a complex, designated heamovanadin, which acted as a respiration pigment in ascidian blood cells.[443] Bielig *et al.* suggested that heamovanadin could reduce vanadium.[444] However, the chemical structure of heamovanadin was not determined, even though model compounds were proposed. Kustin and co-workers proposed that heamovanadin was actually an artifact formed by air-oxidation and disputed the possibility that it might be involved in respiration.[445] They isolated a low-molecular compound consisting of three pyrogallol moieties, named tunichrome, from the blood cells of *Ascidia nigra* and *Ciona intestinalis*, which was considered to be involved in the reduction of vanadium.[446] However, tunichrome was absent in the vanadium-containing signet ring cells.[447] Ryan *et al.* observed the reduction of vanadium in the +5 and +4 oxidation states by a tunichrome, designated Mm-1, in buffer solution at pH 7 *in vitro*.[448] It is, however, unclear whether such a reaction could occur in ascidian blood cells.

A considerable amount of sulfate has always been found in ascidian blood cells,[449] suggesting that sulfate might be involved in the biological function and/or the accumulation and reduction of vanadium. However, Frank *et al.* suggested the existence of a nonsulfate sulfur compound such as an aliphatic sulfonic acid in ascidian blood cells.[450] Kanamori and Michibata found that the ratio of the level of sulfate to that of vanadium in the blood cells of the ascidian *Ascidia gemmata* was ca. 1.5 by using Raman spectroscopy, suggesting that sulfate ions were present as counter ions against V^{III}. They also found evidence that an aliphatic sulfonic acid was present in the blood cells.[451]

Henze[431] first reported that a homogenate of ascidian blood cells is extremely acidic and almost all subsequent investigations have supported his observation. However, Kustin and co-workers disputed the earlier reports and reported that the intracellular pH was neutral from measurements of transmembrane equilibrium of ^{14}C-labeled methylamine.[452] Hawkins *et al.*[453] and Brand *et al.*[454] also reported nearly neutral values for the pH of the interior of ascidian blood cells, from chemical shifts in ^{31}P-NMR. However, Frank *et al.*[455] demonstrated that the interior of the blood cells of *Ascidia ceratodes* has a pH of 1.8, based on the finding that the ESR line width accurately reflects the intracellular pH. Michibata *et al.*[434] measured the pH with a microelectrode under anaerobic conditions, and ESR spectrometry for each type of blood cells from vanadium-rich ascidians, *Ascidia gemmata*, *A. ahodori*, and *A. sydeneiensis samea*. Blood cells drawn from each species were fractionated by density-gradient centrifugation. The pH values after conversion to concentrations of protons ($[H^+]$) and levels of vanadium in each layer of cells are compared, showing that the patterns of distribution of protons and vanadium were similar. The signet ring cells contained high levels of vanadium and showed a low intracellular pH in all three species.

Next, Michibata and co-workers examined the presence of H^+-ATPase in the signet ring cells of the ascidian *Ascidia sydneiensis samea*.[456] The vacuolar-type H^+-ATPase is composed of subunits of 72 kDa and 57 kDa. Antibodies prepared against the 72 kDa and 57 kDa subunits of a vacuolar-type H^+-ATPase from bovine chromaffin granules indeed reacted with the vacuolar membranes of signet ring cells. Immunoblotting analysis confirmed that the antibodies reacted with specific antigens in ascidian blood cells. Furthermore, addition of bafilomycin A_1, a specific inhibitor of vacuolar-type H^+-ATPase, inhibited the pumping function of the vacuoles of signet ring cells, and resulted in neutralization of the contents of the vacuoles.

Using neutron-activation analysis and an immunofluorescence method, Michibata *et al.* found that the amount of vanadium per individual increased dramatically two weeks after fertilization. Within two months, the amount accumulated in larvae was about 600 000 times higher than that in the unfertilized eggs of *A. sydneiensis samea*.[457] A vanadocyte-specific antigen, recognized by a monoclonal antibody specific to the signet ring cells, first appeared in the body wall at the same time as the first significant accumulation of vanadium.[458]

In general, heavy metal ions incorporated into the tissues of living organisms are known to bind to macromolecules such as proteins. Wuchiyama and Michibata searched for vanadium-binding proteins in the blood cells of ascidians. Using a combination of SDS–PAGE and flameless atomic absorption spectrometry, they succeeded in isolating at least four different types of vanadium-binding proteins.[459]

Although the unusual phenomenon that some ascidians accumulate vanadium to levels 10 million times higher than that in seawater has attracted researchers in various fields, the physiological roles of vanadium remain to be resolved. Endean[460] and Smith[461] proposed that the cellulose of the tunic might be produced by vanadocytes, while Carlisle[462] suggested that vanadium-containing vanadocytes might reversibly trap oxygen under conditions of low oxygen tension. The hypothesis has also been proposed that vanadium in ascidians acts to protect them against fouling or as an antimicrobial agent.[463] However, these proposals are still conjectural.

8.07.9.2.2 Eudistomins and related alkaloids

Several halogenated *β*-carbolines have been found in marine organisms, especially tunicates (ascidians or sea squirts). Eudistomin refers to a series of *β*-carboline alkaloids isolated from marine tunicates. Eudistomins A–Q (**445**)–(**461**) were extracted from the Caribbean colonial tunicate *Eudistoma olivaceum*.[464–466] Four groups of eudistomins were isolated, including simple *β*-carbolines (eudistomins D (**448**), J (**454**), N (**458**), and O (**459**)), pyrrolyl-*β*-carbolines (eudistomins A (**445**), B (**446**), and M (**457**)), pyrrolinyl-*β*-carbolines (eudistomins G (**451**), H (**452**), I (**453**), P (**460**), and Q (**461**)), and tetrahydro-*β*-carbolines with an oxathiazepine ring (eudistomins C (**447**), E (**449**), F (**450**), K (**455**), and L (**456**)). The structures were elucidated by spectroscopic data including high-resolution FABMS and EIMS and ^{1}H and ^{13}C NMR, and confirmed by syntheses. The stereochemistry of the eudistomins was studied by ^{1}H- and ^{13}C-homonuclear and heteronuclear correlation NMR spectra and analysis of the proton–proton spatial relationships (NOE). The isolated eudistomins were assayed against *Herpes simplex* virus type-1 (HSV-1) and showed antiviral activity. The most active compounds by far are ones containing the oxathiazepine ring (eudistomins C (**447**), E (**449**), K (**455**), and L (**456**)). The eudistomins also exhibited antimicrobial activity to widely different degrees, with the oxathiazepines being generally the most active. Interestingly, a mixture of eudistomins N (**458**) and O (**459**) displayed a remarkable degree of synergism. Synthetic eudistomin N (**458**) or O (**459**) alone is inactive, but a mixture exhibited antimicrobial activity.

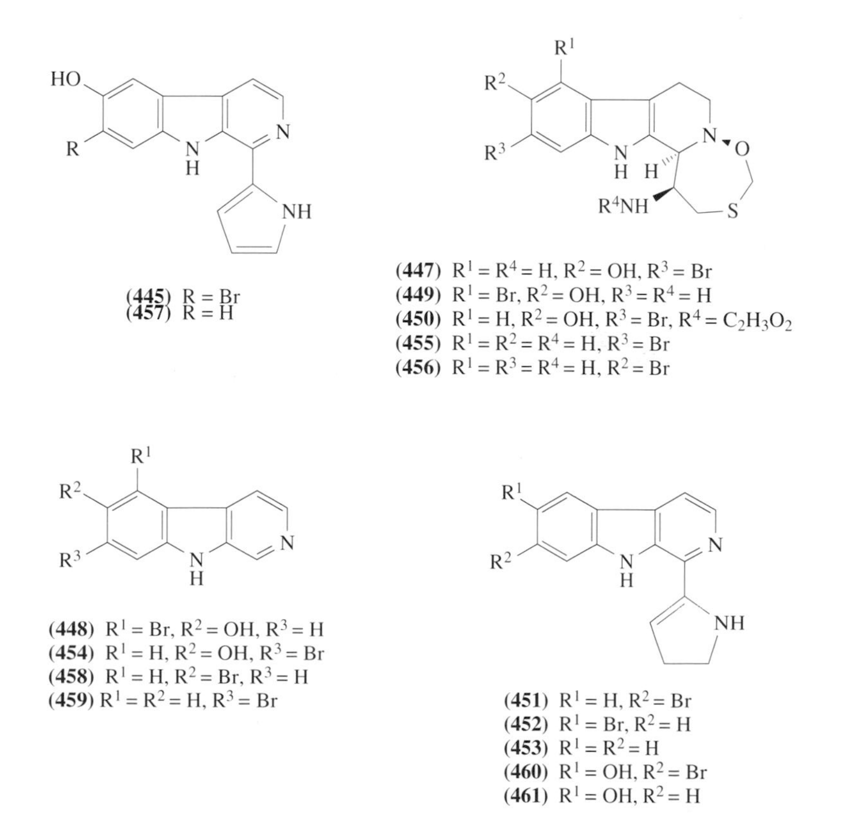

Several eudistomins have been proved to induce calcium release from sarcoplasmic reticulum (SR) in mammalian cells. The application of specific drugs which affect the Ca^{2+}-releasing action from SR is an effective approach to the resolution of an important problem in muscle biology concerning the mechanism in the excitation–contraction coupling between nerve and muscle. The Ca^{2+}-releasing effect is especially pronounced with 7-bromoeudistomin D[467] (BED (**462**)) and 9-methyl-7-bromoeudistomin D[468] (MBED (**463**)). MBED (**463**) was synthesized based on the structure–activity relationship between BED (**462**) and caffeine, a well-known inducer of Ca^{2+} release from SR. MBED (**463**) was found to be ca. 1000 times more potent than caffeine in its Ca^{2+}-releasing action. Ca^{2+} release induced by MBED (**463**) or caffeine was blocked by ruthenium red and high concentration of Mg^{2+}. This result suggests that in this pharmacological property, MBED is similar to caffeine. Thus, MBED can be a valuable tool for elucidating the molecular mechanism of Ca^{2+} release from SR.

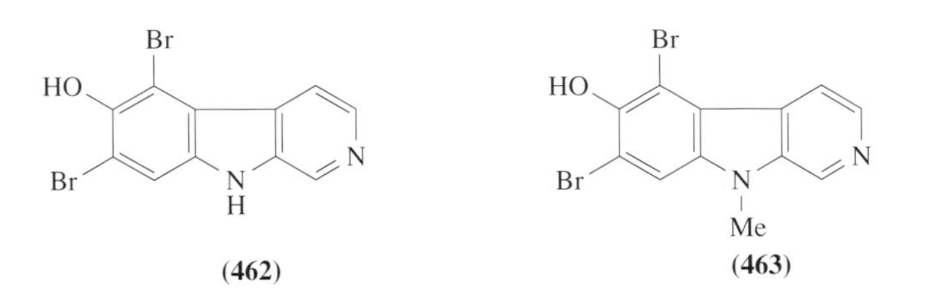

^{3}H-Labeled 9-methyl-7-bromoeudistomin D ([^{3}H]-MBED) was prepared, which binds to the caffeine binding site of terminal cisternae of the skeletal muscle sarcoplasmic reticulum and activates Ca^{2+}-induced Ca^{2+} release (CICR).[469–471] The hepatic microsomal [^{3}H]-MBED binding site distinguishes it from that of skeletal muscle SR. Properties of the binding site of [^{3}H]-MBED were investigated in aortic smooth muscle.[472] The specific activity was higher in microsomes than in other fractions. [^{3}H]-MBED binding sites in smooth muscle microsomes were of a single class with a high affinity, comparable with that in skeletal muscle SR. Caffeine competitively inhibited [^{3}H]-MBED binding, indicating MBED shares the same binding site with caffeine. Solubilization and fractionation of the microsomes gave two fractions of [^{3}H]-MBED binding activities. These results suggest that, in smooth muscle, there are multiple binding sites of [^{3}H]-MBED and caffeinc, which might correspond to different pharmacological actions of caffeine on smooth muscle.

The tissue and subcellular distribution of the binding site of [^{3}H]-MBED were investigated in rabbits.[473] The order of specific activities of total homogenates was liver > brain > other tissues. All binding was completely suppressed by 10 mM caffeine, indicating that all [^{3}H]-MBED binding sites are modulated by caffeine. [^{3}H]-MBED binding sites distributed mainly in membrane fractions rather than soluble fractions in most tissues. In lung and liver, [^{3}H]-MBED binding was enriched in microsomes. [^{3}H]-MBED may be useful as a probe to investigate the actions of caffeine at the molecular level not only in muscles but also in a variety of tissues including liver, kidney, and lung.

Among bromoeudistomin D analogues, the Ca^{2+}-releasing activities of carboline derivatives were higher than those of carbazole derivatives, suggesting that a carboline skeleton is significantly important for the manifestation of Ca^{2+}-releasing activity and Ca^{2+} sensitivity of Ca^{2+}-induced Ca^{2+} release.[474] On the contrary, the analogues which have a carbazole skeleton and bromine at C-6 inhibit both Ca^{2+}- and caffeine-induced Ca^{2+} release. 9-Methyl-substitution of the analogue elevated its Ca^{2+}-releasing activity. Moreover, there is a close correlation between the enhancement of [^{3}H]-ryanodine binding to SR by the analogues and the activation of Ca^{2+} release by them.

4,6-Dibromo-3-hydroxycarbazole (DBHC (**464**)) was synthesized as an analogue of BED (**462**) and its pharmacological properties were examined.[475] In Ca^{2+} electrode experiments, DBHC (**464**) markedly inhibited Ca^{2+} release from SR, induced by caffeine and BED (**462**). [^{3}H]-Ryanodine binding to SR was suppressed by ruthenium red, Mg^{2+} and procaine, but was not affected by DBHC (**464**). [^{3}H]-Ryanodine binding to SR was enhanced by caffeine and BED (**462**). DBHC (**464**) antagonized the enhancement in a concentration-dependent manner. These results suggest that DBHC (**464**) binds to the caffeine-binding site to block Ca^{2+} release from SR. This drug is a novel type of inhibitor for the CICR channels in SR and may provide a useful tool for clarifying the Ca^{2+}-releasing mechanisms in SR.

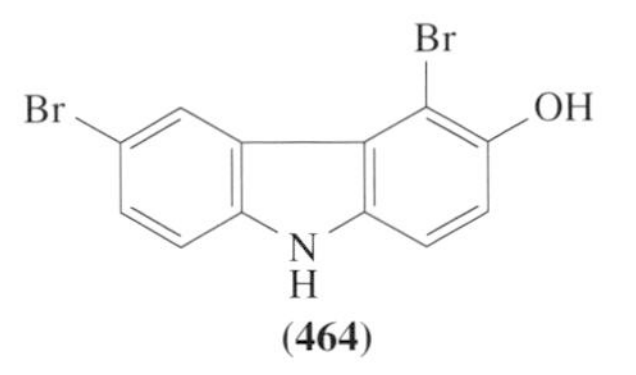

(464)

Eudistomins D (**448**), N (**458**), and O (**459**), and several synthetic β-carbolines with halogeno (Br, Cl, or I) and alkyloxy (RO-, with R = H, Me, or Ac) groups on the benzenoid ring proved to be novel inhibitors of cAMP phosphodiesterase.[476] Moreover, several other synthetic eudistomin D analogues are more active than eudistomin D (**448**) as phosphodiesterase inhibitors.

Eudistomidin A (**465**) was isolated from the Okinawan tunicate *Eudistoma glaucus* and exhibited strong calmodulin-antagonistic activity.[477] Eudistomidin A (**465**) was the first calmodulin antagonist from marine origin and is about 15 times more potent than W-7, a well-known calmodulin antagonist. Eudistomidins B (**466**), C (**467**), and D (**468**) were obtained form the same tunicate.[478] The absolute stereostructure of eudistomidin B (**466**) was elucidated from NMR and CD data, whereas that of eudistomidin C (**467**) was established by synthesis of 6-*O*-methyl-10(*R*)-eudistomidin C. Eudistomidins B, C, and D (**466**)–(**468**) showed potent cytotoxicity against murine leukemia L1210 (IC_{50} = 3.4, 0.36, and 2.4 μg ml^{-1}) and L5178Y (IC_{50} = 3.1, 0.42, and 1.8 μg ml^{-1}) cells, respectively. In addition, eudistomidin B (**466**) activated rabbit heart muscle actomyosin ATPase by 93% at 3×10^{-5} M, while eudistomidin C (**467**) exhibited calmodulin-antagonistic activity ($IC_{50} = 3 \times 10^{-5}$ M). Eudistomidin D (**468**) induced Ca^{2+} release from SR, about 10 times more potent than caffeine. Two new β-carboline alkaloids, eudistomidins E (**469**) and F (**470**), have also been isolated from the same tunicate.[479] Eudistomidins E (**469**) and F (**470**) are structurally unique with the tetrahydropyrimidine ring fused to the β-carboline ring. Since these new β-carboline alkaloids appear to be biogenetically related to eudistomin E (**449**) or eudistomidin C (**467**), the absolute configuration of C-10 in eudistomidins E (**469**) and F (**470**) is assumed to be *S*, the same as that of eudistomin E (**449**) or eudistomidin C (**467**).

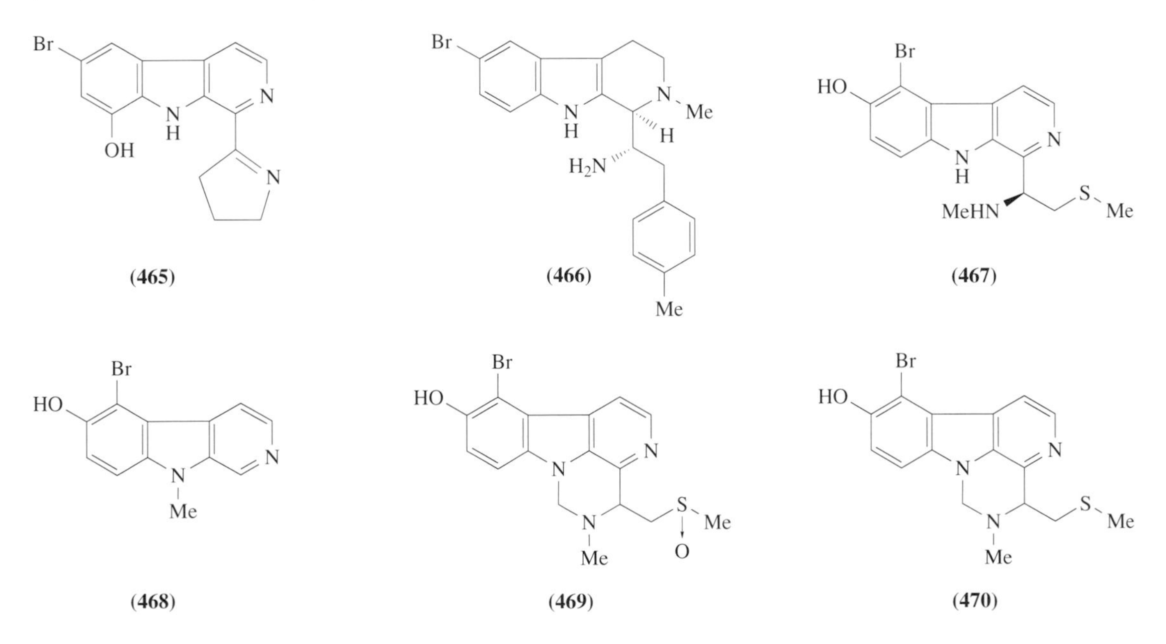

Eudistomin K (**455**) and its sulfoxide (**471**) were isolated from the New Zealand ascidian *Ritterella sigillinoides*.[480,481] The sulfoxide (**471**) also shows antiviral activity against *Polio* and *Herpes simplex*. The structure of eudistomin K (**455**) was determined by X-ray analysis,[482] in which the stereochemistry of the favored invertomer was established, and that of eudistomin K sulfoxide (**471**) by semisynthesis from eudistomin K (**455**). Three other β-carbolines, named eudistomins R, S, and T (**472**)–(**474**), were obtained from a Bermudan tunicate *Eudistoma olivaceum* by using an amino-bonded HPLC column.[483] A 2-methyl-1,2,3,4-tetrahydro-β-carboline with an *N*-methylpyrrolidine at C-1, named woodinine (**475**), was isolated from a New Caledonian ascidian *Eudistoma fragum*,[484] extracts of which exhibited antimicrobial activity.

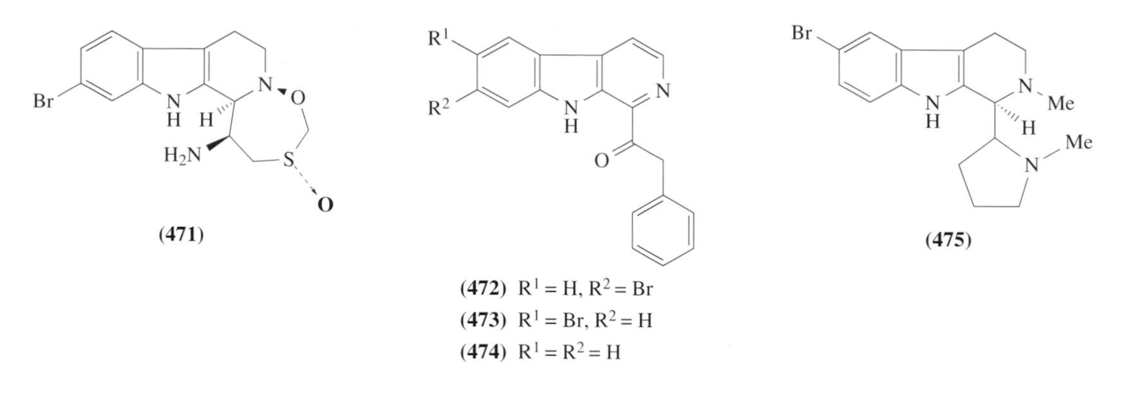

(471)

(472) R¹ = H, R² = Br
(473) R¹ = Br, R² = H
(474) R¹ = R² = H

(475)

The New Caledonian tunicate *Eudistoma album* contained the new eudistalbins[485] A (**476**) and B (**477**) in addition to eudistomin E (**449**). Eudistalbin A (**476**) showed some cytotoxicity against KB cells (IC_{50} = 3.2 μg ml^{-1}), whereas eudistalbin B (**477**) was inactive. The New Caledonian tunicate *Pseudodistoma arborescens* has yielded arborescidines A, B, C, and D (**478**)–(**481**), of which only arborescidine D (**481**) exhibited cytotoxicity against KB cells (IC_{50} = 3 μg ml^{-1}).[486]

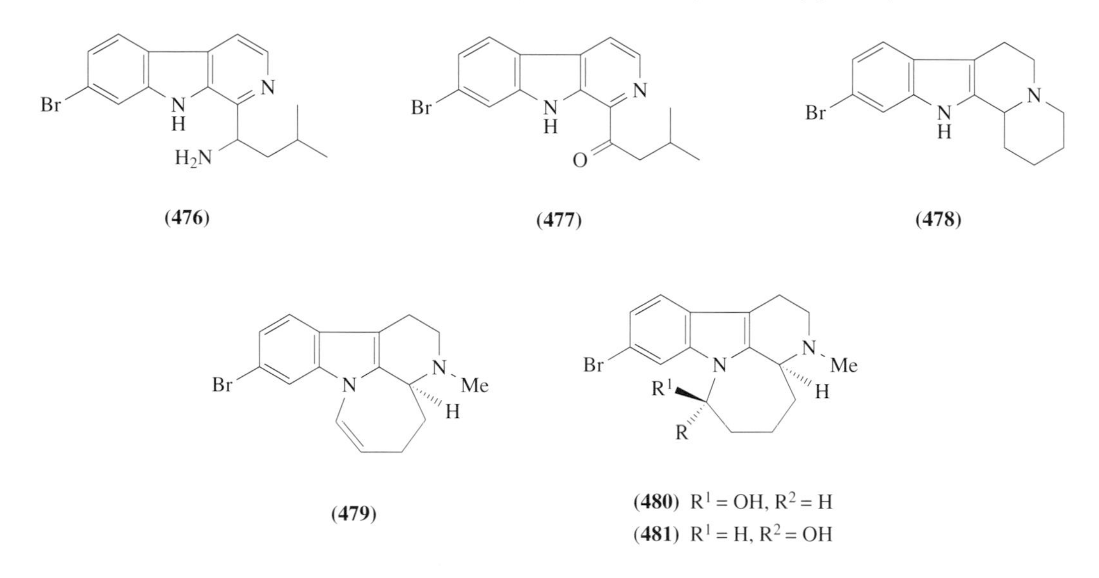

(476) (477) (478)

(479)

(480) R¹ = OH, R² = H
(481) R¹ = H, R² = OH

Eudistomin U (**482**), isolated from the Caribbean ascidian *Lissoclinum fragile*, is the first bisindole among these *β*-carbolines.[487] Spectral data, especially ¹H- and ¹³C NMR data, were in good accordance with those reported for *β*-carboline and indole, thus supporting the proposed structure. Isoeudistomin U (**483**), from the same ascidian, comprises indole and dihydro-α-carboline moieties.[487] These compounds showed a strong antibacterial activity. The structure of eudistomin U (**482**) was confirmed by its syntheses by two groups.

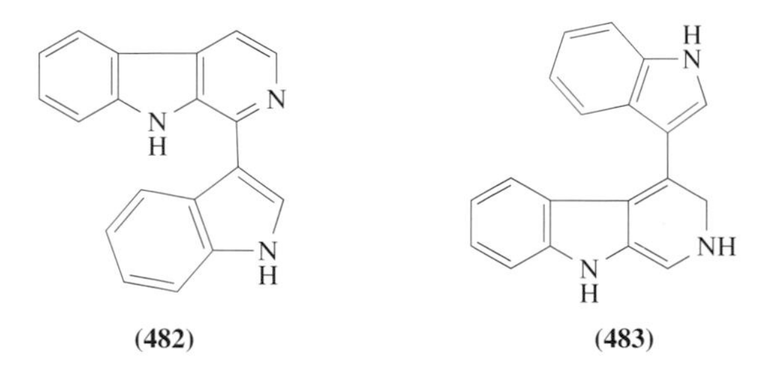

(482) (483)

Biosynthetic studies of eudistomin H (**452**) in the Floridan tunicate *Eudistoma olivaceum* indicate that both 6-bromotryptamine and 6-bromotryptophan are incorporated into eudistomin H (**452**) and both are better than the nonbrominated precursors (Scheme 16).[488] The origin of eudistomin I

(**453**) in the same tunicate has been also investigated by *in vivo* techniques.[488] Tryptophan and proline are the primary precursors to eudistomin I (**453**), while tryptamine serves as an intermediate (Scheme 16).

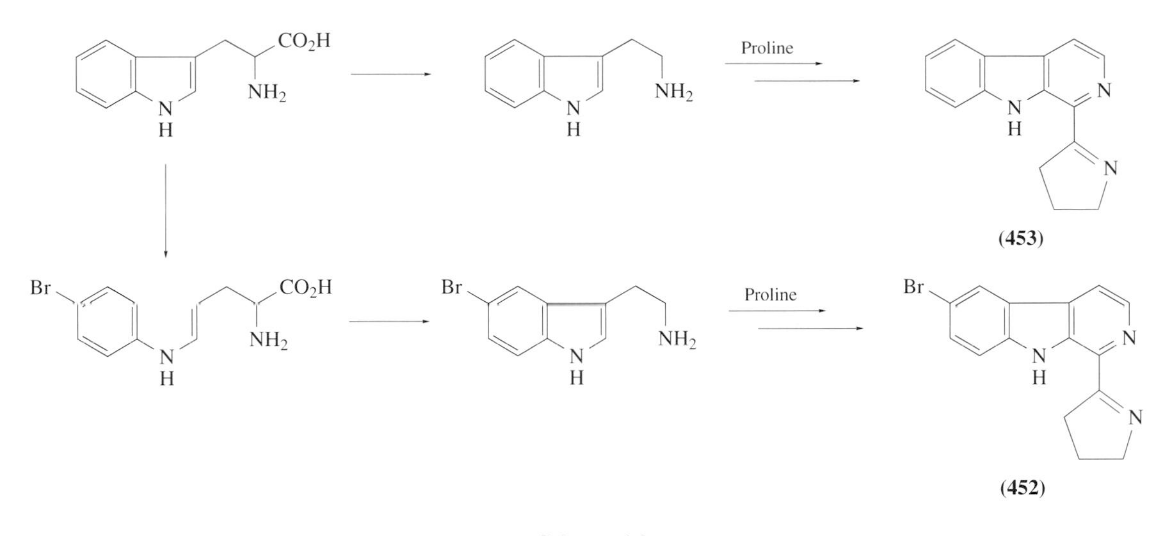

Scheme 16

Marine bryozoans and hydroids also contain β-carboline alkaloids. (*S*)-1-(1′-Hydroxyethyl)-β-carboline (**484**) was obtained from the Tasmanian bryozoan *Costaticella hastata* together with some known β-carbolines previously reported from terrestrial plants.[489] Three new brominated β-carbolines (**485**)–(**487**) were isolated from the Mediterranean hydroid *Aglaophenia pluma*,[490] and the structures were firmly established by synthesis. The terrestrial blue-green alga *Dichothrix baueriana*, which was collected from the Na Pali coast of Kauai, Hawaii, produces bauerines A–C (**488**)–(**490**), which are active against *Herpes simplex* virus.[491]

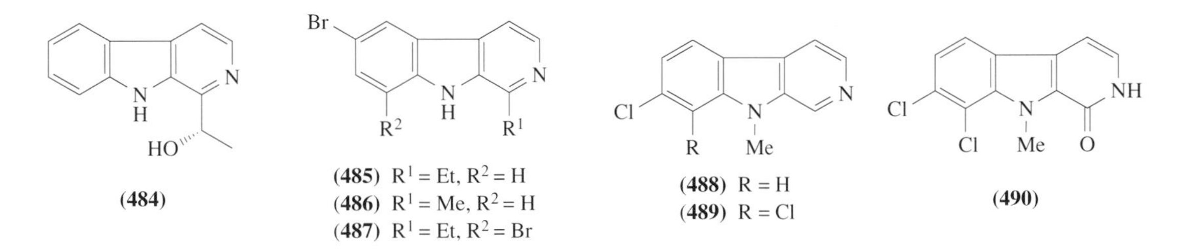

8.07.9.2.3 *New tunicate metabolites*

Tunicates have been recognized as a source of novel bioactive secondary metabolites with unique chemical structures, most of them being nitrogenous metabolites,[492] represented by didemnins (see Section 8.07.9.1)[493] and ecteinascidins.[494,495] This section deals with several natural products isolated from tunicates, particularly those published after 1994.

Seven additional didemnins—didemnins M (**491**), N (**492**), X (**493**), and Y (**494**), nordidemnin N (**495**), epididemnin A_1 (**496**), and acylclodidemnin A (**497**)—were isolated by Rinehart and co-workers from a large amount of extract of the Caribbean tunicate *Trididemnum solidum* (170 kg),[496] which was obtained during preparation of didemnin B for phase I and phase II studies. A polar fraction after a silica gel column chromatography of the extract was subjected to on-line LC/FAB mass analysis employing the moving belt technique and was separated by using high-speed centrifugal countercurrent chromatography (HSCCC), polystyrene-divinylbenzene copolymer gel, reversed- and normal-phase HPLC to give new didemnins. The structures of these compounds were assigned, based on spectral data and chemical degradation studies.

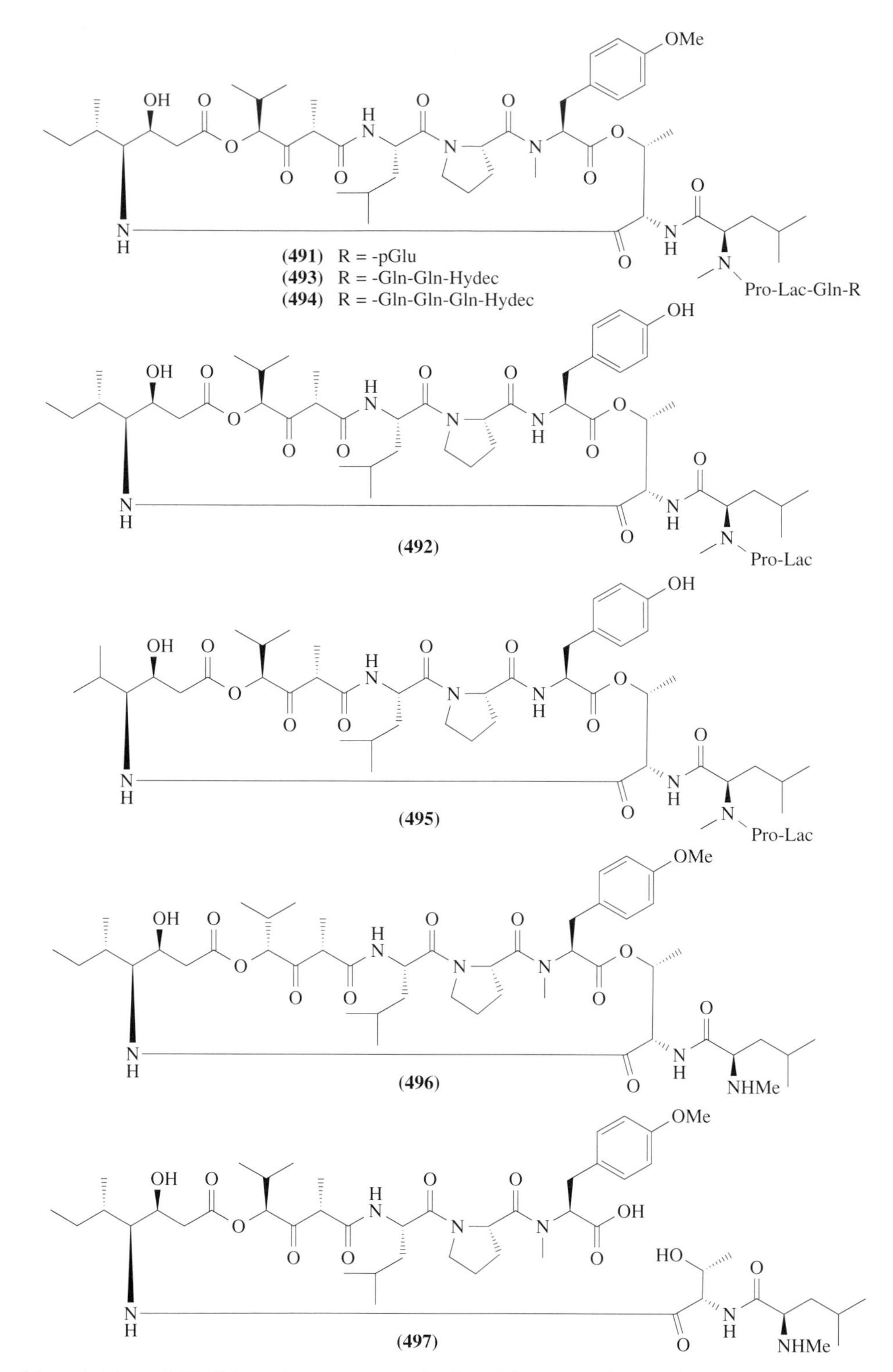

The bioactivities of 42 didemnin congeners, isolated from marine tunicates *Trididemnum solidum* and *Aplidium albicans* or prepared semisynthetically, were compared by using the growth inhibition of various murine and human tumor cells and plaque reduction of HSV-1 and VSV grown on cultured mammalian cells to assess cytotoxicity and antiviral activity. Biochemical assays for macromolecular synthesis (protein, DNA, and RNA) and enzyme inhibition (dihydrofolate reductase, thymidylate synthase, DNA polymerase, RNA polymerase, and topoisomerases I and II) were also performed to specify the mechanisms of action of each analogue. The immunosuppressive activity of the didemnins was determined using a mixed lymphocyte reaction (MLR) assay. These assays revealed that the native cyclic depsipeptide core was an essential structural requirement for most of the bioactivities of the didemnins, especially for cytotoxicity and antiviral activities. The linear side-chain portion of the peptide could be altered with a gain, in some cases, of bioactivities. In particular, dehydrodidemnin B, tested against several types of tumor cells and in *in vivo* studies in mice, as well as didemnin M, tested for the mixed lymphocyte reaction and graft vs. host reaction in murine systems, showed remarkable gains in their *in vitro* and *in vivo* activities compared with didemnin B.[497]

Ecteinascidins 729 (**498**) and 743 (**499**) were reported in 1990 to be isolated from the Caribbean tunicate *Ecteinascidia turbinata* independently by researchers at Harbor Branch Oceanographic Institution[494] and Rinehart's group.[495] Ecteinascidins protect mice *in vivo* against P388 lymphoma, B16 melanoma, M5076 ovarian sarcoma, Lewis lung carcinoma, and the LX-1 human lung and MX-1 human mammary carcinoma xenografts. The crystal structures of two derivatives, 21-*O*-methyl-N^{12}-formyl derivative (**500**) of (**498**) and naturally occurring ecteinascidin 743 N^{12}-oxide (**501**), were reported.[498]

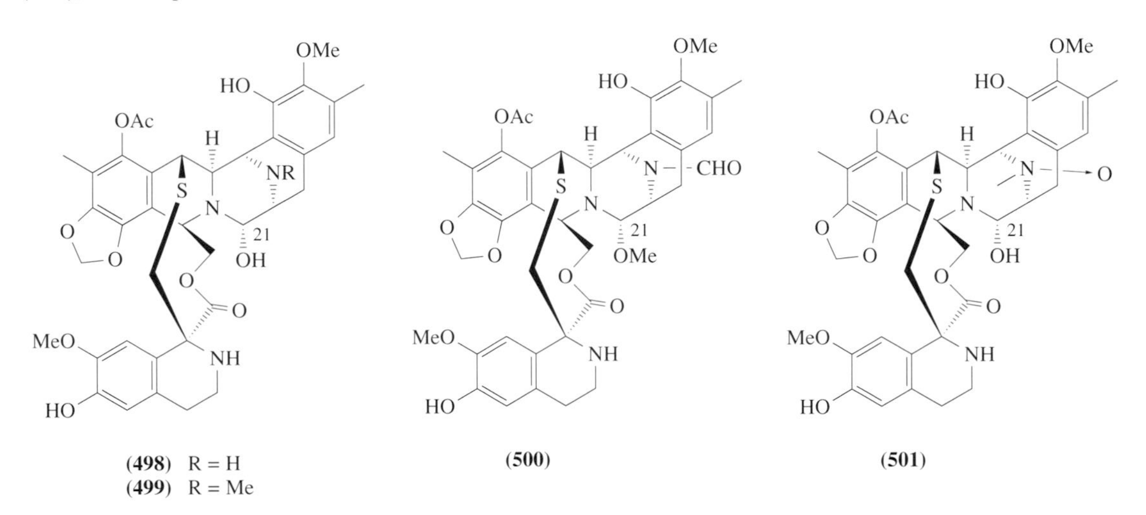

(**498**) R = H
(**499**) R = Me

(**500**)

(**501**)

Ecteinascidins belong to a class of antibiotics that contain tetrahydroisoquinoline units including saframycins,[499] which was known to form a covalent adduct with DNA, most likely with the formation of a bond between the carbinolamine carbon (C-21) and N-2 of guanine in the minor groove of the DNA double helix. Ecteinascidins also bind strongly with DNA, but they do not form permanent covalent adducts. The three-dimensional structure of adducts of ecteinascidins and DNA was analyzed by use of a DNA heptamer model, d(TTGGGAA), and the program QUANTA; it was shown that the ring B platform stacked on the backbone of DNA on one side of the minor groove, following the right-handed curvature of the B-DNA helix, and ring C stacked on the backbone of the other side of the minor groove.[498] Further bioactive ecteinascidins such as (**502**), putative biosynthetic precursors of previously described ecteinascidins, were isolated from the Caribbean tunicate *Ecteinascidia turbinata*. The absolute configuration of the L-cysteine unit of (**502**) was assigned by chiral GC, while a 2D ROESY (rotating-frame Overhauser enhancement spectroscopy) spectrum of its acetate completed the assignment of the stereochemistry of (**502**) as 1*R*,2*R*,3*R*,4*R*,11*R*,13*S*,21*S*,1′*R*.[500]

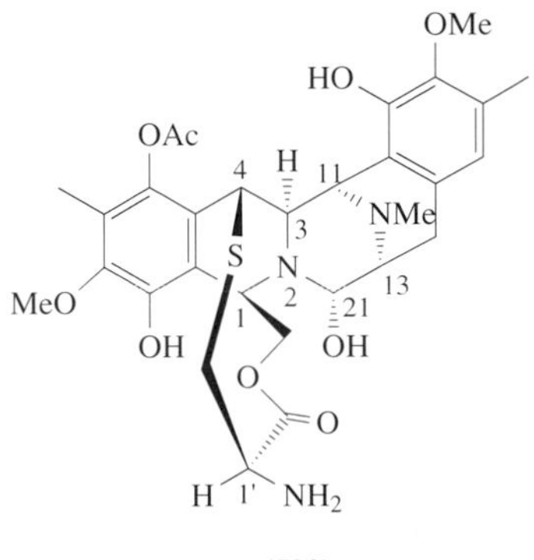

(**502**)

From the Okinawan marine tunicate *Aplidium multiplicatum*, shimofuridin A (**503**),[501] a nucleoside derivative embracing an acylfucopyranoside unit, was isolated and its structure including all absolute configurations was determined by spectral and chemical means. Compound (**503**) exhibited cytotoxicity against murine lymphoma L1210 cells with an IC_{50} value of 9.5 μg ml^{-1} *in vitro*, and antimicrobial activity against fungus *Trichophyton mentagrophytes* (MIC value, 133 μg ml^{-1}) and Gram-positive bacterium *Sarcina lutea* (MIC, 66 μg ml^{-1}). Compound (**503**) also showed endothelin converting enzyme (ECE) inhibition activity (31.2% at 100 μg ml^{-1}). Minor analogues, shimofuridins B–G (**504**)–(**509**)[502] were also isolated from the Okinawan marine tunicate *A. multiplicatum* by HPLC separation and they were stereoisomers of alkenes in the acyl side-chains ((**504**)–(**507**)), or homologues with two more carbons in the second acyl chain ((**508**) and (**509**)).

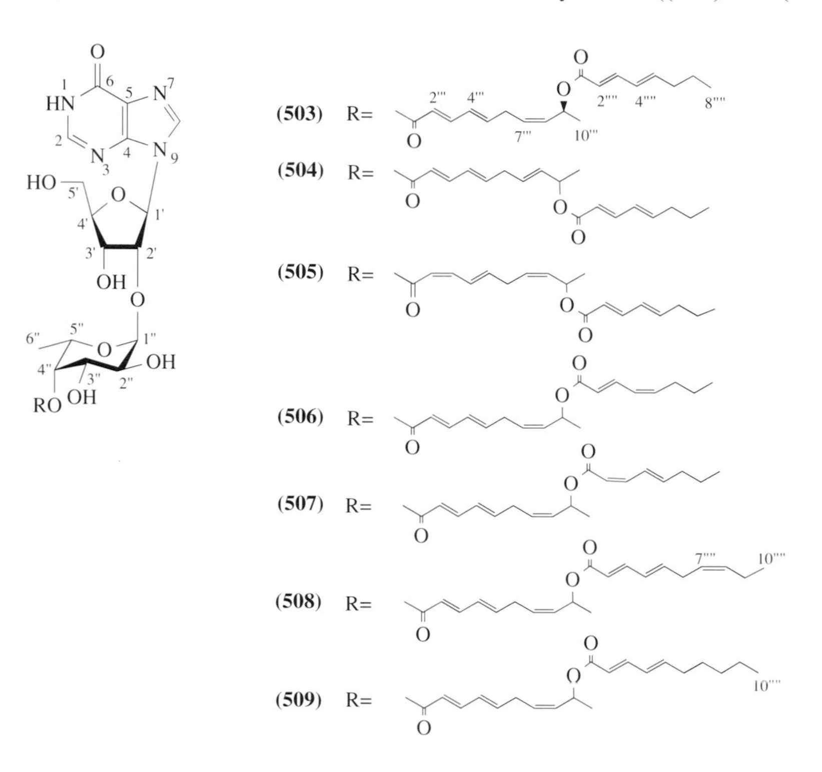

It seems likely that tunicates rarely contain terpenoids and steroids. However, a diterpenoid dimer and steroidal dimers have been obtained from extracts of tunicate. An unprecedented dimeric farnesylated benzoquinone, $C_{42}H_{46}O_5$, named longithorone A (**510**) was isolated from the tunicate *Aplidium longithorax* (Monniot) collected in Palau. The structure was established by single crystal X-ray diffraction. Longithorone A (**510**) was comprised of two subunits each derived from a farnesyl moiety bridging the 2,5-positions of a benzoquinone. Diels–Alder reactions may account for the union of the two subunits and additional ring formations. Longithorone A (**510**) contained five six-membered rings plus bridging 10- and 16-membered rings. Mild cytotoxicity to murine leukemia cells was observed for longithorone A (**510**).[503]

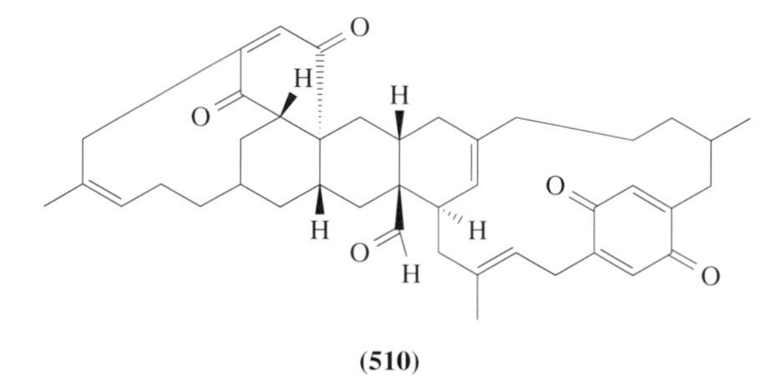

(510)

A cytotoxic dimeric steroidal alkaloid, ritterazine A (**511**),[504] was isolated from the tunicate *Ritterella tokioka*, and its structure elucidated by extensive spectroscopic analyses. It exhibited cytotoxicity against P388 murine leukemia cells with an IC_{50} value of 3.8×10^{-3} μg ml^{-1}. Further examination of this tunicate resulted in isolation of a total of 26 analogues, ritterazines A (**511**)–Z (**536**).[505–507] The structures including absolute stereochemistries were determined by extensive spectroscopic analyses and chemical means. Their structural features were reminiscent of the cephalostatins isolated from the Indian Ocean hemichordate *Cephalodiscus gilchristi*.[508] The most difficult aspect of the structural elucidation of this class of compounds lay in determining the orientation of the steroidal units with respect to the pyrazine ring. Fusetani and co-workers, who isolated the ritterazines, succeeded in preparing *N*-methyl derivatives of ritterazine B (**512**) in order to use the *N*-methyl groups for NOESY experiments.[505] This study revealed that the orientation of the steroidal units in (**512**) was identical with that in cephalostatin 1 (**537**),[508] which was established by X-ray analysis. Furthermore, Fusetani's group succeeded in applying ^{15}N HMBC spectroscopy to determine the two steroidal orientations in ritterazine A (**511**).[509]

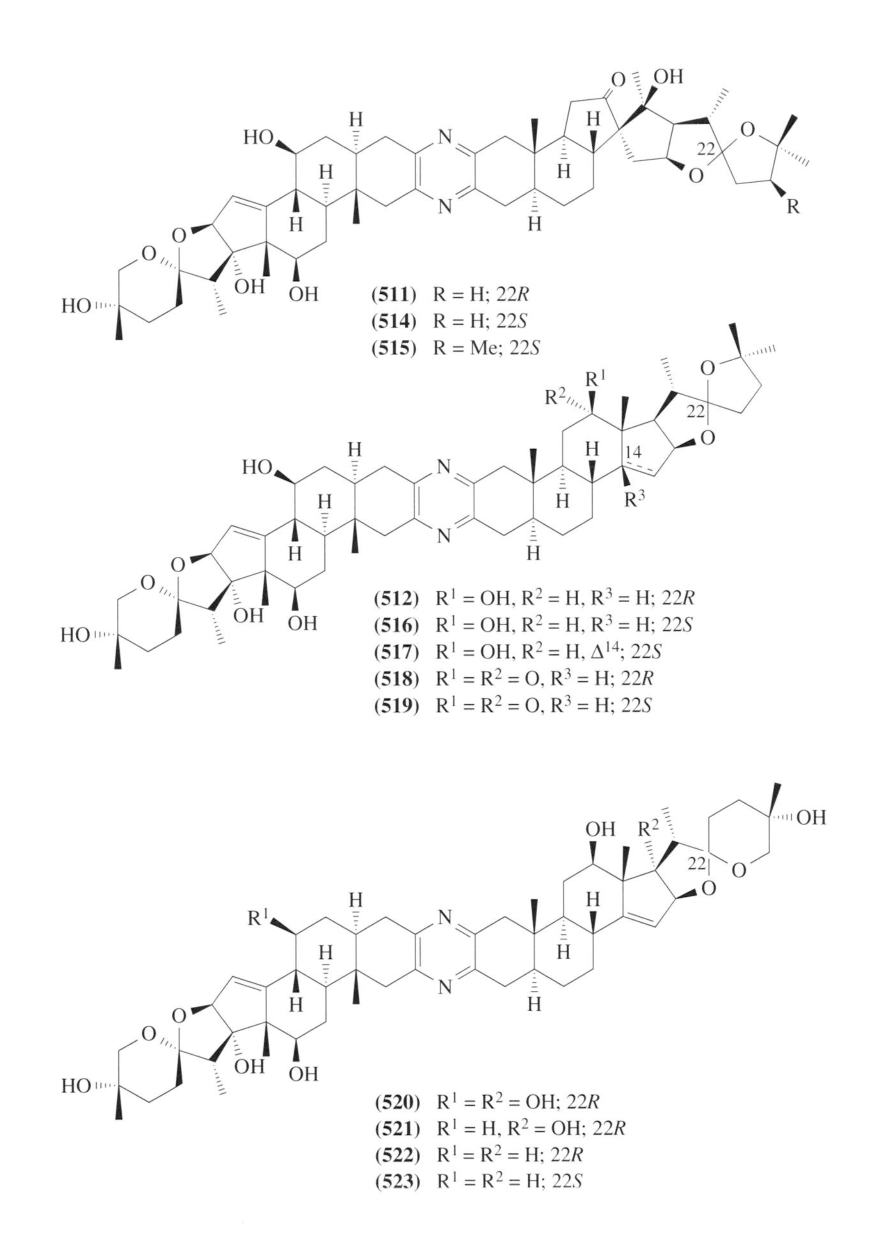

The structure–activity relationship of the 26 natural ritterazines together with several derivatives (**538**)–(**544**) prepared from the most active compound, ritterazine B (**512**), was investigated, and the

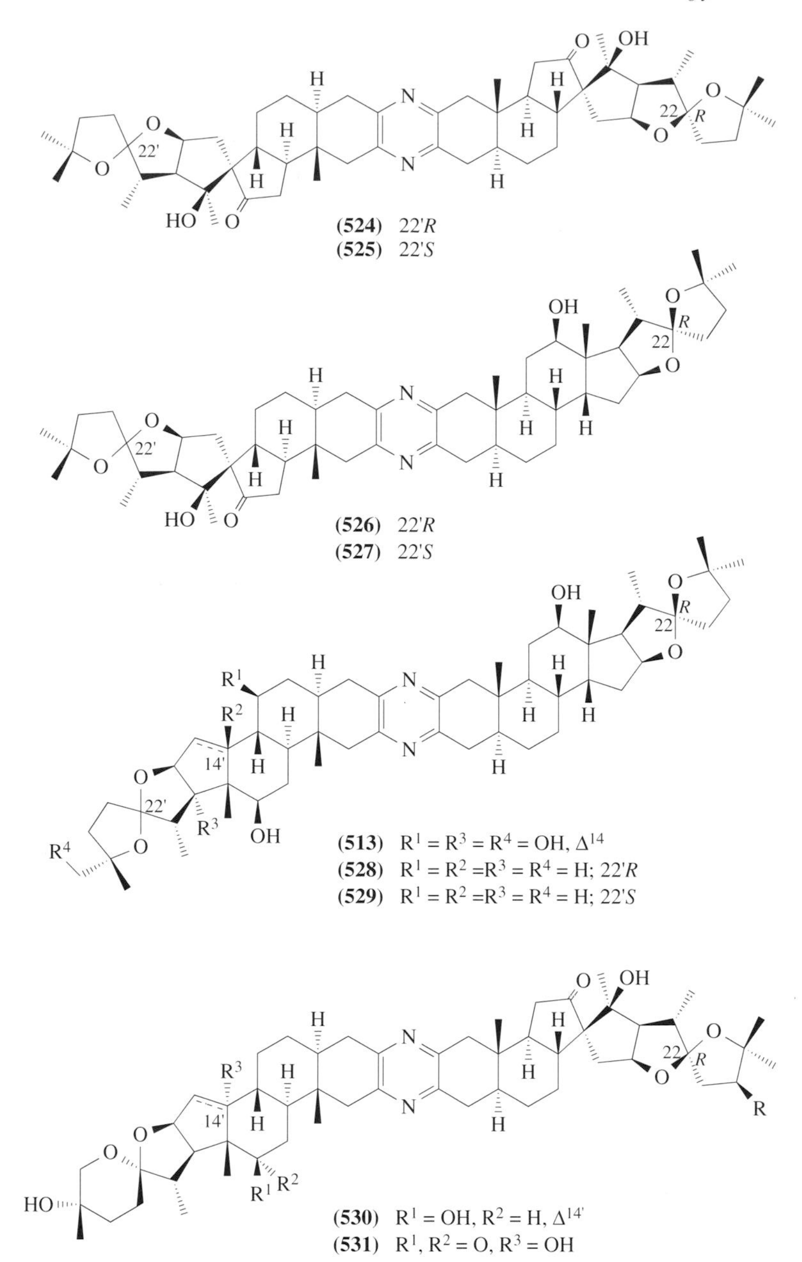

cytotoxicity data are shown in Table 21.[507] Ritterazine B (**512**) which was dissymmetric was the most potent cytotoxin; ritterazines A (**511**), D (**514**), E (**515**), F (**516**), G (**517**), H (**518**), I (**519**), J (**520**), K (**521**), L (**522**), M (**523**), and Y (**535**) were also highly cytotoxic. These compounds had the 5/6 spiro ring and 12′,25′-diol functionalities common in the western hemisphere. Ritterazines N (**524**), O (**525**), P (**526**), Q (**527**), R (**528**), S (**529**), W (**533**), X (**534**), and Z (**536**) having no 5/6 spiroketal were marginally active. Therefore, the 5/6 spiroketal was supposed to possess an important function for the expression of cytotoxicity, because the cleavage of the 5/6 spiroketal decreased their activity. Furthermore, ritterazine T (**530**) which had no conventional steroidal skeletal with OH-12 in the eastern hemisphere, but the 5/6 spiro ring in the western hemisphere, showed weak cytotoxicity. Therefore, the presence of a 5/6 spiro ring in one end, a 5/5 spiro ring in the other end, and the hydroxyl group on C-12 were shown to be important for cytotoxic activity. However, acetylation of secondary hydroxyl groups gave interesting information. Compared with oxidation, acetylation

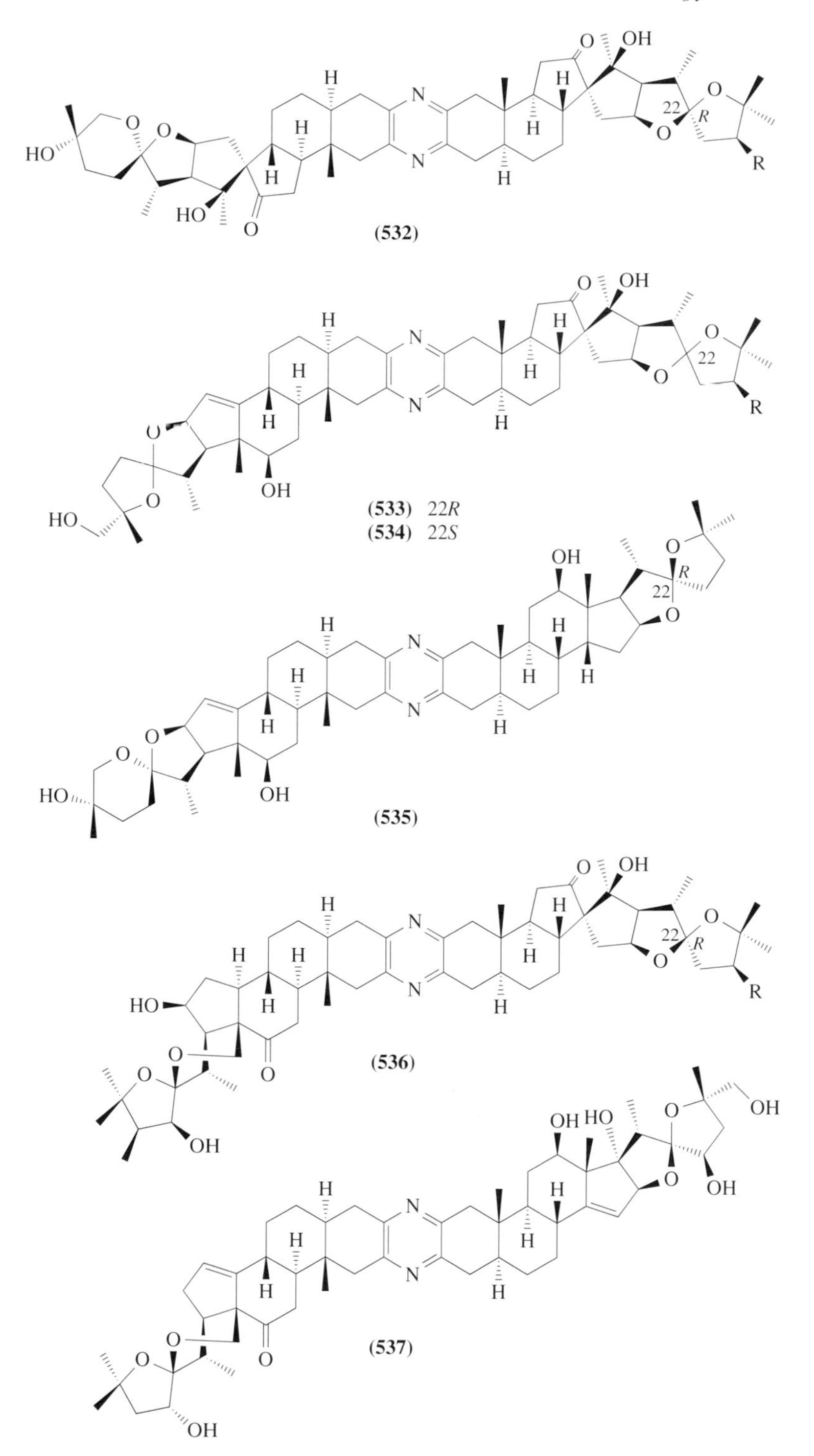

of OH-12 did not decrease cytotoxicity, but induction of acetyl groups in the western hemisphere showed weaker activity. It was possible to consider that the steric hindrance by acetyl groups in the western hemisphere influenced the expression of the cytotoxicity.

It may be generally recognized that peptides are frequently contained in tunicates, represented by didemnins described above. Many of these peptides also show high levels of pharmacological activity.[492,510] Tunicates of the genus *Lissoclinum* are well known as prolific producers of cytotoxic

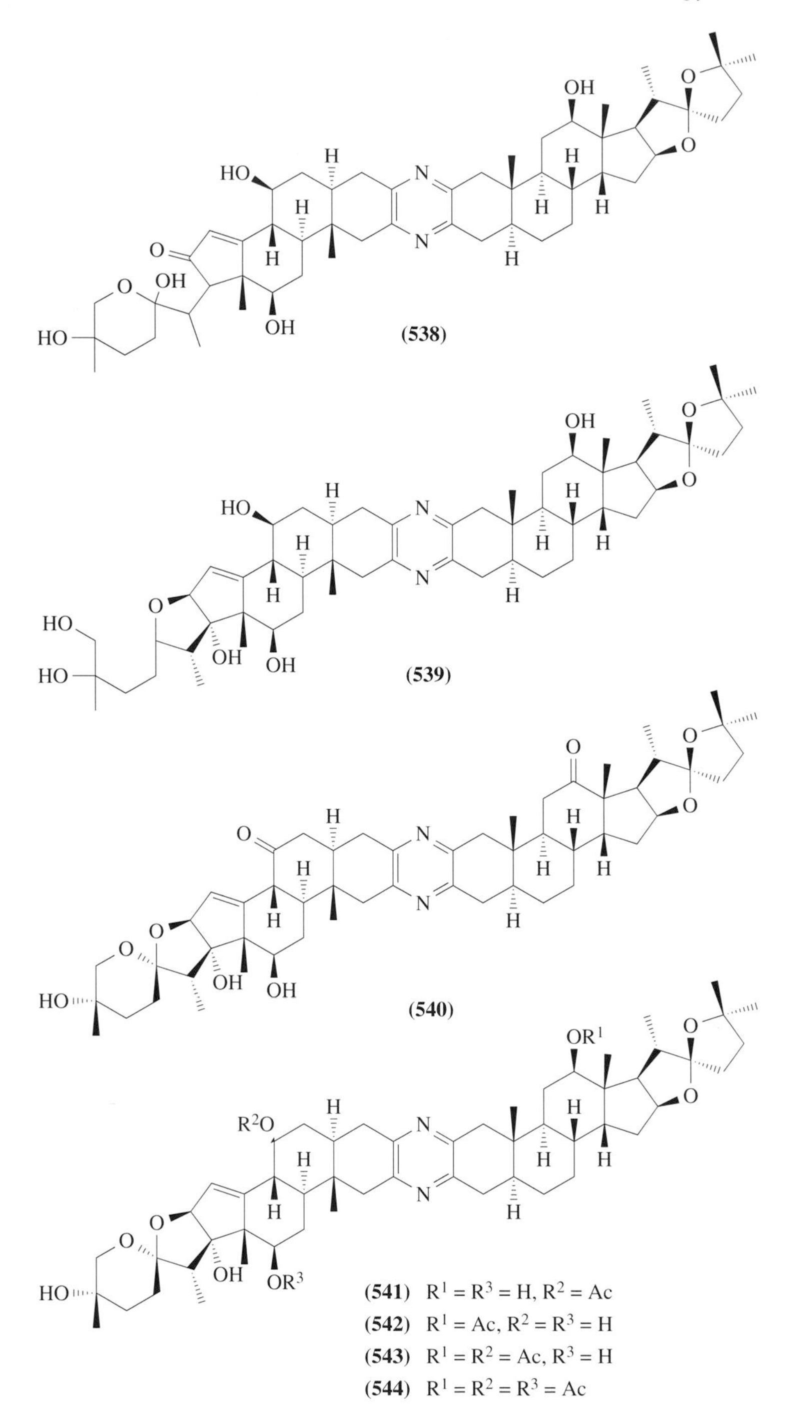

cyclic peptides with highly modified amino acids such as those containing oxazole, oxazoline, thiazole, and thiazoline groups.[511] Two cyclic heptapeptides, nairaiamide A (**545**) and B (**546**), were isolated from the ascidian *Lissoclinum bistratum* collected at Nairai Island, Fiji, and their structures were determined by a combination of spectroscopic techniques, including a natural abundance two-dimensional ^{1}H[^{13}C,^{1}H]-HMQC–TOCSY experiment which was pivotal to assigning and identifying the amino acid residues.[512] ^{1}H[^{13}C]-HMBC and [^{1}H-^{1}H]-ROESY experiments provided the correlations necessary to sequence the peptides. In both peptides, the Ile–Pro amide bonds adopted *cis* configurations.

Table 21 Cytotoxic activities of 34 ritterazine derivatives (**511**)–(**544**) (IC_{50} values against P388 leukemia cells, ng ml^{-1}).

Compound	IC_{50}
(**511**)	3.5
(**512**)	0.15
(**513**)	92
(**514**)	16
(**515**)	3.5
(**516**)	0.73
(**517**)	0.73
(**518**)	16
(**519**)	14
(**520**)	13
(**521**)	9.5
(**522**)	10
(**523**)	15
(**524**)	460
(**525**)	2100
(**526**)	710
(**527**)	570
(**528**)	2100
(**529**)	460
(**530**)	460
(**531**)	2100
(**532**)	2100
(**533**)	3200
(**534**)	3000
(**535**)	3.5
(**536**)	2000
(**537**)	0.001–0.000001
(**538**)	2100
(**539**)	240
(**540**)	18
(**541**)	92
(**542**)	3.5
(**543**)	800
(**544**)	7600

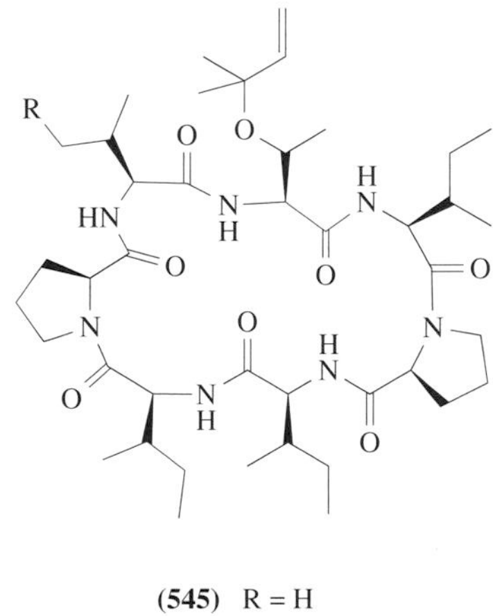

(**545**) R = H
(**546**) R = Me

Cyclodidemnamide (**547**), a cyclic heptapeptide, was isolated from the marine ascidian *Didemnum molle* collected in the Cuyo Islands in the Philippines.[513] From 21.4 g of freeze-dried animal a total of 11.5 mg of (**547**) was isolated. The structure of the new, weakly cytotoxic, heptapeptide was unambiguously assigned using comprehensive NMR analysis in combination with hydrolysis and chiral GC–MS analysis of its component amino acids. The results clearly showed that the phenylalanine unit had the unnatural D configuration, while other amino acid residues obtained from the hydrolysis contained natural L configurations. Computer analysis led to the assignment of a solution conformation that fully rationalized the NOE correlations and shielding effects observed by ^{1}H NMR. One methyl proton (C-32) on a valine residue was observed at unusually high field shift (δ_H 0.15), indicating that this methyl group was directed within the ring interacting with the π-cloud of the thiazole unit. Consistently, NOE correlations were observed from the C-32 methyl protons to the thiazole proton and the NH proton of phenylalanine residue.

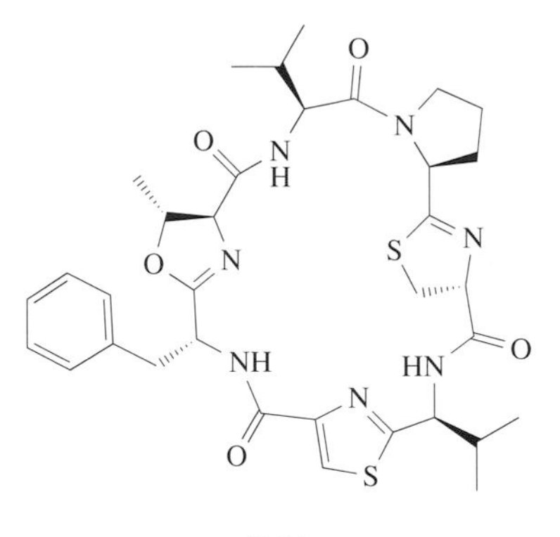

(547)

Three linear cytotoxic tripeptides, virenamides A–C (**548**)–(**550**), were isolated from extracts of the didemnid ascidian *Diplosoma virens* collected on the Great Barrier Reef, Australia.[514] Their structures were deduced from 1D and 2D NMR spectral data and confirmed by HPLC analysis of the constituent amino acids after hydrolysis of the peptides and derivatization with 1-fluoro-2,4-dinitrophen-5-yl-L-alanine amide using Marfey's procedure.[515] The virenamides showed modest cytotoxicity toward a panel of cultured cells: virenamide A (**548**) gave IC_{50} of 2.5 μg ml^{-1} against P388, and 10 μg ml^{-1} against A549, HT29, and CV1 cells. It exhibited topoisomerase II inhibitory activity (IC_{50} 2.5 μg ml^{-1}). Virenamides B (**549**) and C (**550**) both gave IC_{50} of 5 μg ml^{-1} against P388, A549, HT29, and CV1 cells.

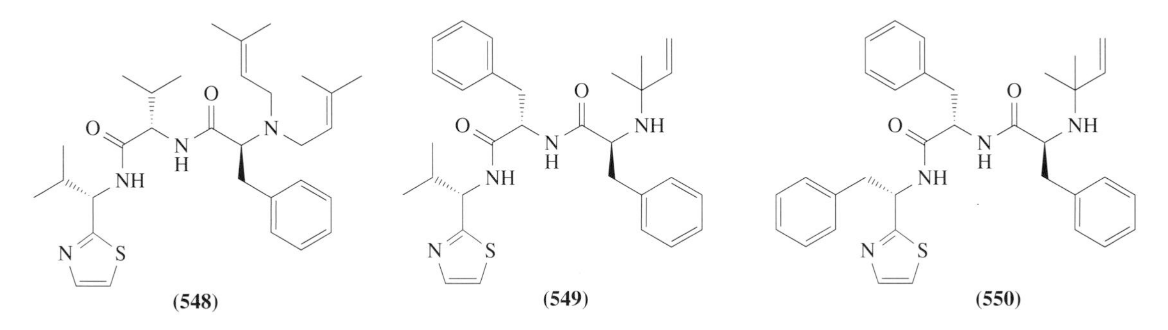

(548) **(549)** **(550)**

Caledonin (**551**), a modified peptide, was isolated from the marine tunicate *Didemnum rodriguesi*.[516] Caledonin (**551**) comprised a central L-phenylalanine residue connected via its amino group to (*S*)-3-amino-5-mercaptopentanoic acid (a new sulfur-containing β-amino acid), and by its carboxyl group to a six-membered cycloguanidine moiety, bearing an *n*-octyl chain. Tunicates are known to produce metal complexing metabolites, such as the tunichromes.[517] Some features of caledonin (**551**), and in particular the hydrophobic chain at one end and the penicillamine-like β-amino acid at the other, suggested that this metabolite was a natural bolaphile which strongly bound Zn^{II} and Cu^{I} ions and may be involved in ion transport across the membranes. A solution of (**551**) in CD_3OD was titrated with $ZnCl_2$ and monitored by ^{1}H NMR spectroscopy. The 2:1 caledonin/$ZnCl_2$ adduct was isolated and characterized by FABMS and NMR spectroscopy.

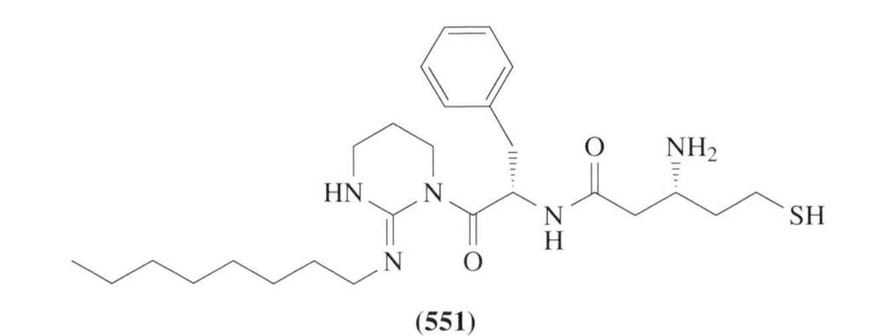

(551)

Four bromotyrosine derivatives, botryllamides A–D (**552**)–(**555**), having striking structural similarities to tunichromes,[517] were isolated from the brilliantly colored Styelid ascidian *Botryllus* sp. from Siquijor Island, the Philippines, and from *B. schlosseri* from the Great Barrier Reef, Australia.[518] Their structures were deduced from 1- and 2-dimensional NMR spectral data. Botryllamide D (**555**) showed marginal cytotoxicity, after 72 h exposure, against the human colon cancer cell line HCT116 (IC_{50} = 17 μg ml^{-1}) but was inactive *in vivo*.

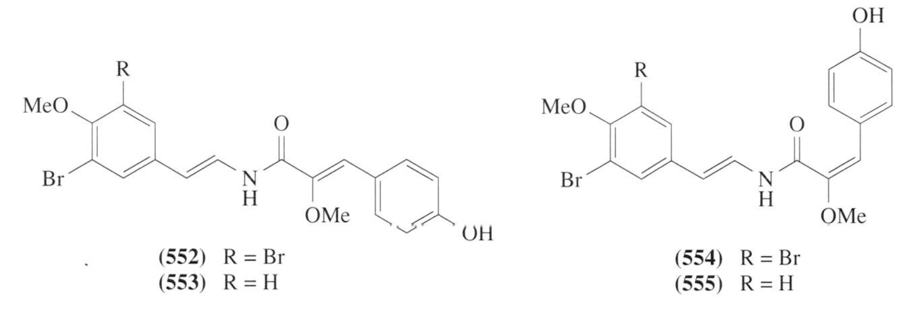

(**552**) R = Br
(**553**) R = H

(**554**) R = Br
(**555**) R = H

Three bromotyrosine-related compounds, polycitone A (**556**) and polycitrins A (**557**) and B (**558**), were isolated from the marine ascidian *Polycitor* sp.[519] The structures of these compounds were established mainly on the basis of NMR spectroscopic data and, in the case of (**556**), also by single-crystal X-ray diffraction analysis using the crystal of its penta-*O*-methyl derivative. Polycitone A (**556**) and polycitrins A (**557**) and B (**558**) represent the first examples of two new classes of marine products which might biogenetically be close to the lamellarins (lamerarin A (**559**)).[520] The penta-*O*-methyl derivative of (**556**) was found to inhibit the growth of SV40 transformed fibroblast cells in a concentration of 10 μg ml^{-1}.

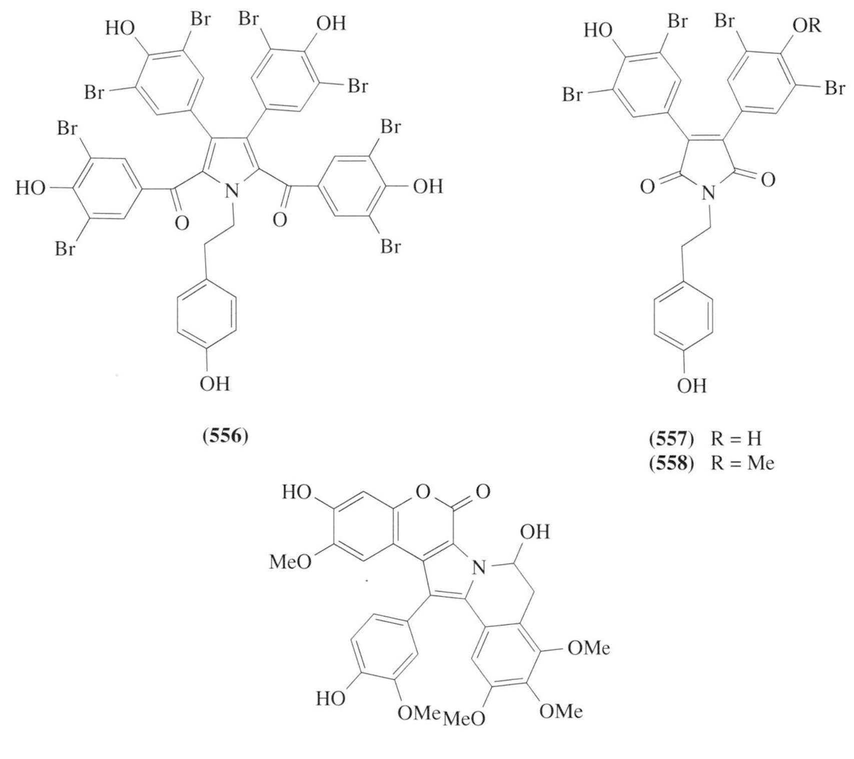

(**556**)

(**557**) R = H
(**558**) R = Me

(**559**)

A fused pentacyclic aromatic alkaloid, cystodamine (**560**), was isolated from a Mediterranean ascidian *Cystodytes dellechiajei* (Polycitoridae).[521] Compound (**560**) was the first example of a marine product displaying a ^{1}H–^{14}N coupling during ^{1}H NMR analysis, which revealed in DMSO-d_6 solution the existence of an aminopyridine moiety (Equation (5)) as the pyridine ring was changed into a pyridone-imine one (**560b**) and one ammonium group was also found (2H, triplet system, J = 52 Hz). This triplet system was due to the ^{1}H–^{14}N coupling, the NH absorption band as an ammonium ion resolving itself into a triplet from spin–spin interaction with the ^{14}N

nucleus ($I = 1$). Cystodamine (**560**) showed cytotoxicity against CEM human leukemic lymphoblasts (IC_{50} 1.0 μg ml^{-1}).

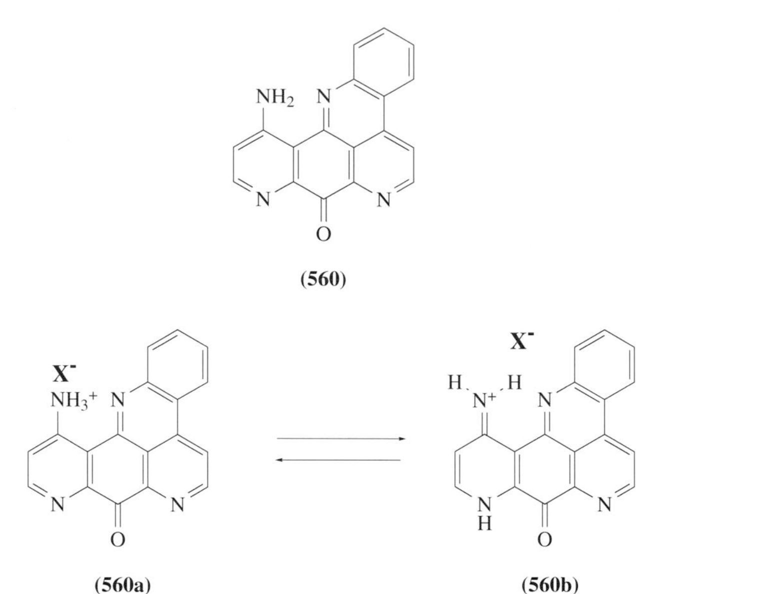

Several additional pyridoacridine alkaloids, dehydrokuanoniamine B (**561**), shermilamine C (**562**), and cystodytin J (**563**),[522] in addition to the known compounds cystodytin A (**564**),[523] kuanoniamine D (**565**),[524] shermilamine B (**566**),[525] and eilatin (**567**),[526] were isolated from a Fijian *Cystodytes* sp. ascidian. These compounds along with a previously reported pyridoacridine, diplamine (**568**),[527] showed dose-dependent inhibition of proliferation in human colon tumor (HCT) cells *in vitro*. All compounds inhibited the topoisomerase (TOPO) II-mediated decatenation of kinetoplast DNA (kDNA) in a dose-dependent manner. The pyridoacridines' ability to inhibit TOPO II-mediated decatenation of kDNA correlated with their cytotoxic potencies and their ability to intercalate into calf thymus DNA. These results suggested that disruption of the function of TOPO II, subsequent to intercalation, is a probable mechanism by which pyridoacridines inhibit the proliferation of HCT cells. Incorporation studies showed that pyridoacridines disrupt DNA and RNA synthesis with little effect on protein synthesis. It appeared that DNA is the primary cellular target of the pyridoacridine alkaloids.

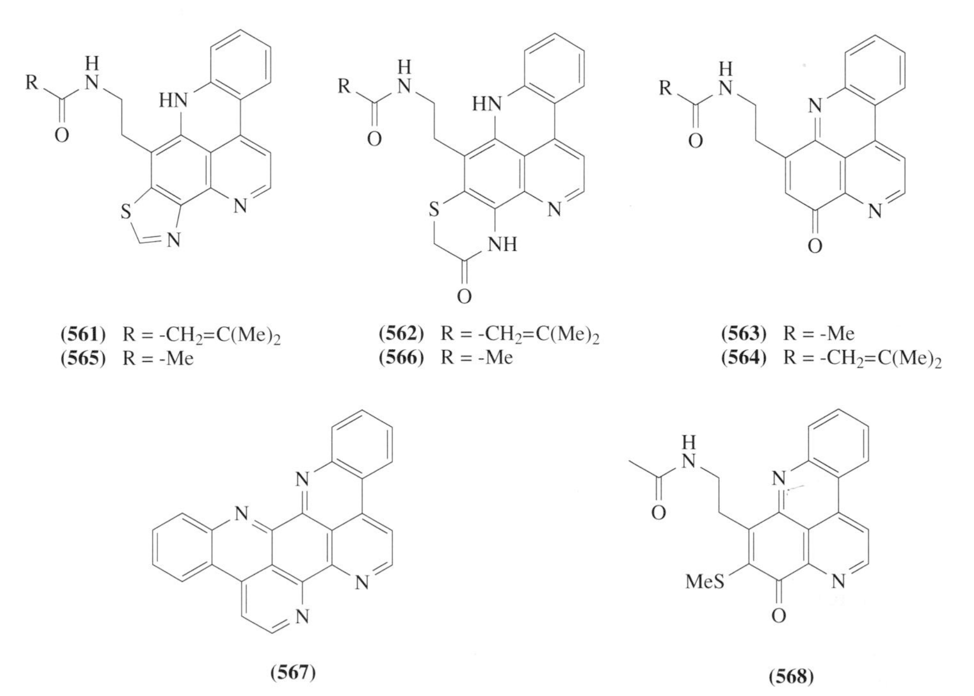

Further pyridoacridine alkaloids, lissoclin A (**569**) and B (**570**), were isolated from *Lissoclinum* sp. collected from the Great Barrier Reef, Australia, along with a tetrahydro-β-carboline, lissoclin C (**571**). Lissoclin A (**569**) underwent photorearrangement to a benzo-1,3-oxathiazoline (**572**).[528]

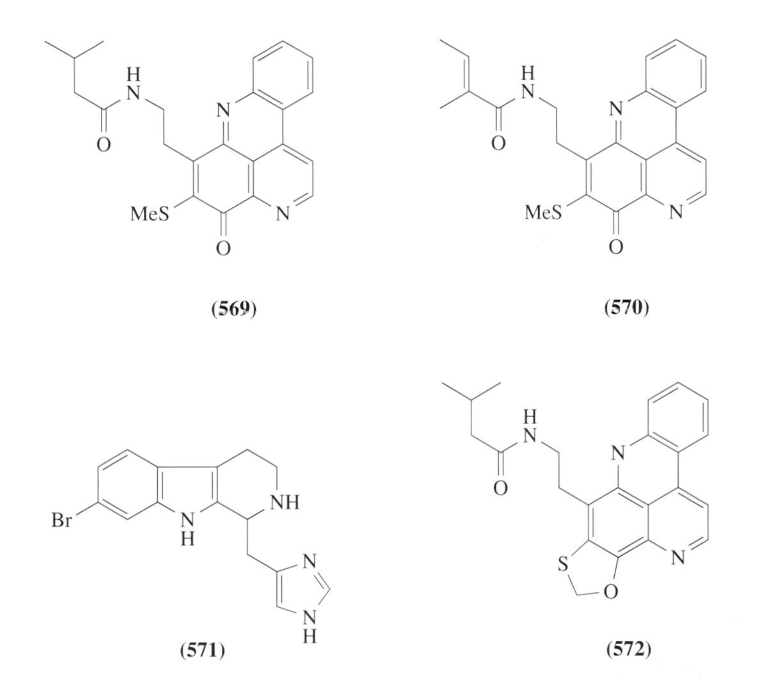

Several unique polysulfides biogenetically related to dopamine were isolated from the tunicate of the genus *Lissoclinum*. Varacin (**573**) was the first benzopentathiepin natural product and the first naturally occurring polysulfide amino acid isolated from *Lissoclinum vareau*, a lavender-colored encrusting species collected in the Fiji Islands. Varacin (**573**) exhibited potent antifungal activity against *Candida albicans* (14 mm zone of inhibition of 2 μg of varacin/disk) and cytotoxicity toward the human colon cancer HCT 116 with an IC_{90} of 0.05 μg ml^{-1}, 100 times the activity of 5-fluorouracil. Varacin (**573**) also exhibited a 1.5 differential toxicity toward the CHO cell line EM9 (chlorodeoxyuridine sensitive) versus BR1 (BCNU resistant), providing preliminary evidence that varacin (**573**) damages DNA.[529] The structure of varacin (**573**) was unambiguously confirmed by total synthesis.[530–532] From *Lissoclinum performatum* collected in Dinard, France, lissoclinotoxin A (**574**) was isolated as a potent antimicrobial and antifungal metabolite with modest cytotoxicity.[533] Its initial structure (**575**) containing a 1,2,3-trithiane ring was later revised to that containing a pentathiepin ring (**574**).[534] Additionally, another pentathepin derivative, lissoclinotoxin B (**576**), was isolated from the same tunicate *L. performatum*.

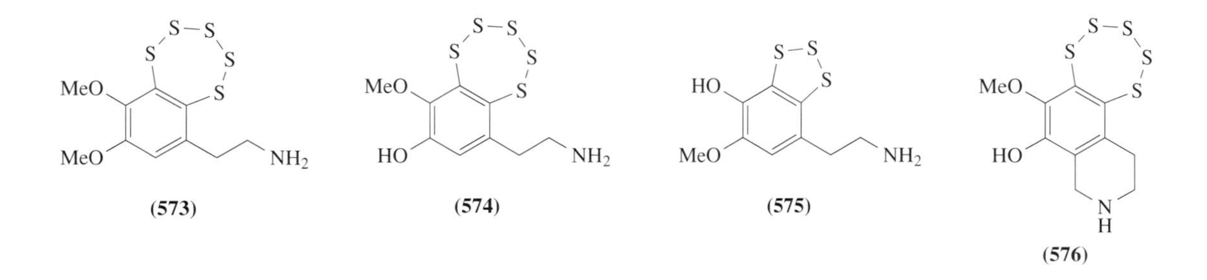

Varacin (**573**) and lissoclinotoxin A (**574**) were found to be chiral, and evidence was provided from unusual stereoisomerism due to restricted inversion about the benzopentathiepin ring, which induced asymmetry into the molecule and caused the protons to become diastereotopic.[528,535] Another polysulfide, lissoclinotoxin C (**577**), and a dibenzotetrathiepin compound, lissoclinotoxin D (**578**), were isolated from the tunicate *Lissoclinum* sp. collected from the Great Barrier Reef,

Australia.[528] The dimeric compound (**578**) exhibited antifungal activity against *Candida albicans*. The ascidian *Lissoclinum japonicum* from Palau contained the antimicrobial and antifungal metabolites, *N,N*-dimethyl-5-(methylthio)varacin (**579**) and 3,4-dimethoxy-6-(2′-*N,N*-dimethylaminoethyl)-5-(methylthio)benzotrithiane (**580**).[536] An inseparable 2:3 mixture of 5-(methylthio)varacin (**581**) and the corresponding trithiane (**582**) was isolated from a different *Lissoclinum* species from Pohnpei, Micronesia and 3,4-desmethylvaracin (**583**) was obtained from a species of *Eudistoma* from Pohnpei. These pentathiepins and trithianes selectively inhibited protein kinase C.[536]

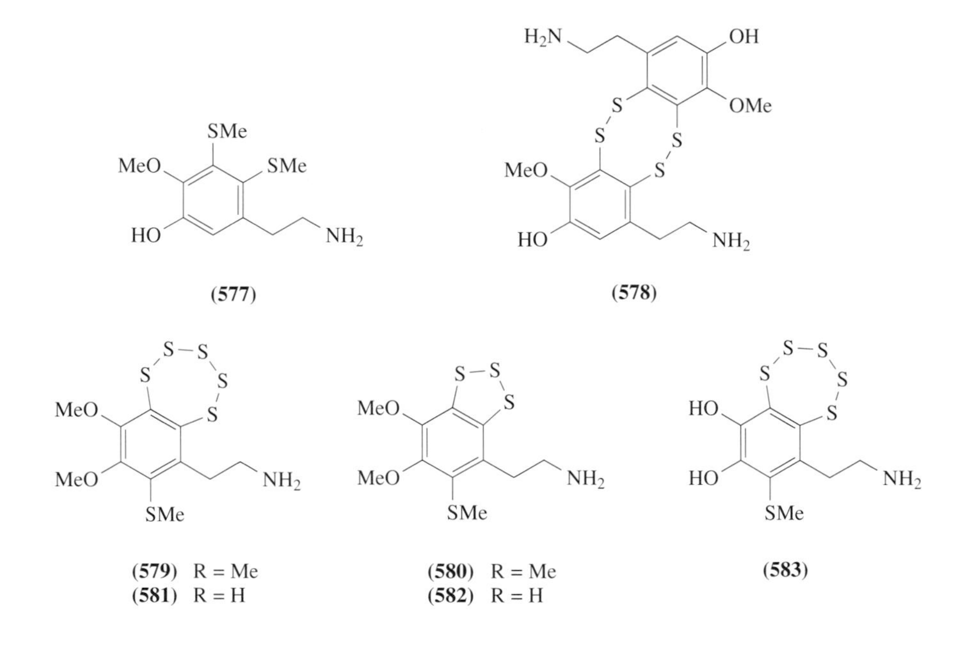

A dimeric disulfide alkaloid, polycarpine (**584**), was isolated almost at the same time independently by two research groups from tunicate *Polycarpa clavata* collected in Western Australia[537] and *Polycarpa aurata* collected in Chuuk, Micronesia.[538] The disulfide (**584**) inhibited the enzyme inosine monophosphate dehydrogenase, and the inhibition could be reversed by addition of excess dithiothreitol.[538] The dihydrochloride of (**584**) was cytotoxic against the human colon tumor cell line HCT-116 at 0.9 μg ml^{-1}.[537]

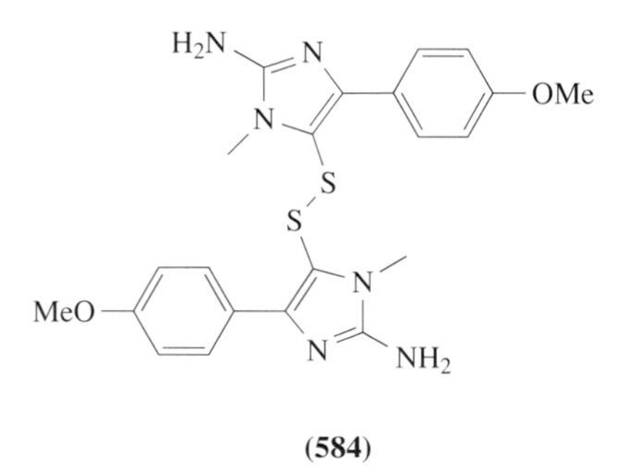

(584)

Four *β*-carboline-based metabolites, didemnolines A–D (**585**)–(**588**) were isolated from an ascidian of the genus *Didemnum*, collected near the island of Rota, Northern Mariana Islands.[539] These *β*-carboline-based metabolites differed from most previously isolated compounds in that they are substituted at the N-9 position of the *β*-carboline ring, rather than at the C-1 position. Didemnolines A–C (**585**)–(**587**) were moderately cytotoxic toward human epidermoid carcinoma KB cells, with sulfoxide-containing (**587**) exhibiting the greatest activity. Compounds (**585**) and (**587**) also exhibited antimicrobial activity.

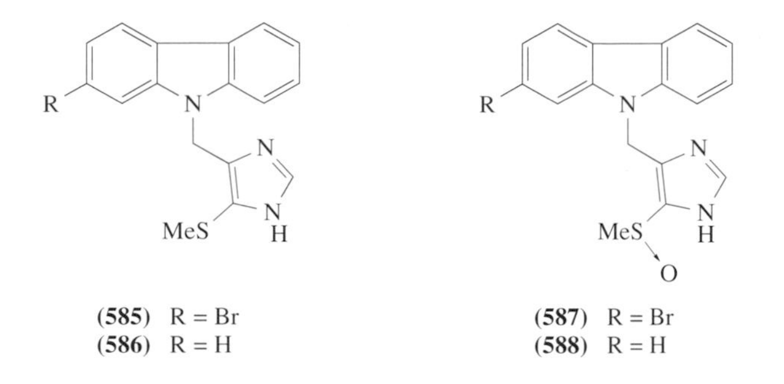

Three fatty acid metabolites didemnilactone A (**589**) and B (**590**) and neodidemnilactone (**591**) were isolated from the tunicate *Didemnum moseleyi* (Herdman). Their structures, including absolute stereochemistry, were established on the basis of spectral studies and chemical synthesis. Didemnilactones exhibited inhibitory activity against lipoxygenase and weak binding activity to leukotriene B4 receptors.[540]

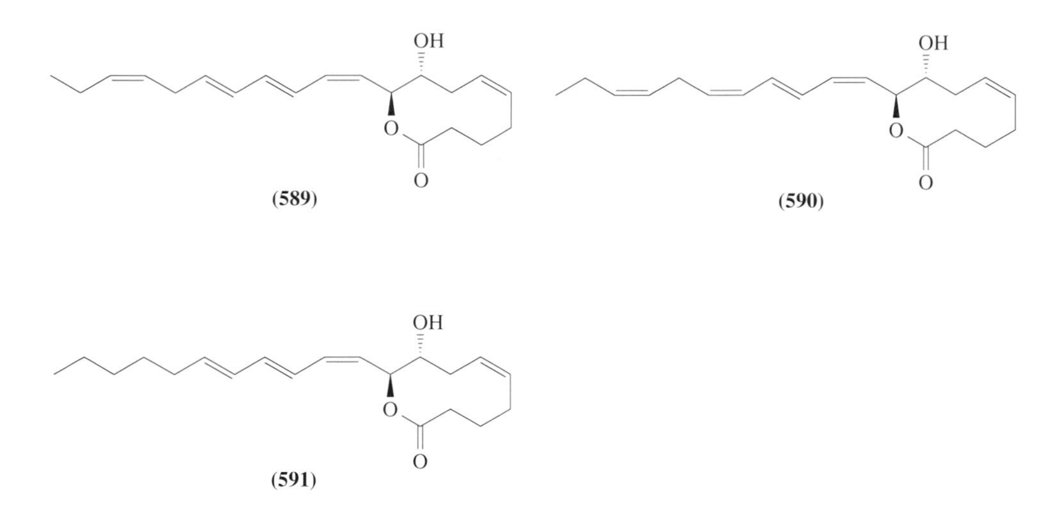

A known microbial antibiotic, enterocin (**592**), and its derivatives (**593**)–(**595**) were isolated from the brown, encrusting ascidian, *Didemnum* sp. collected in Western Australia.[541] Enterosin (**592**) and 5-deoxyenterocin (**593**) were originally isolated from three strains of soil-derived *Streptomyces* species.[542] This was the first observation of enterosins being isolated from nonmicrobial sources, adding further support for the concept that bacteria may produce some of the interesting molecules isolated from marine invertebrates. The limited production of enterocin antibiotics by various *Streptomyces*, and the uniqueness of this structural class, created significant questions about their true origins in biologically complex marine invertebrates. Microscopic examination of a thin section of the alcohol-fixed ascidian tunic of this *Didemnum* sp. showed the presence of large amounts of V-shaped bacteria similar in size and shape to *Arthrobacter* species. Although considerable effort was expended to collect fresh samples and cultivate these endobiotic microorganisms, none of the 20 strains obtained could be confirmed, in culture, to produce the enterocin-based metabolites isolated from the whole animal.[541]

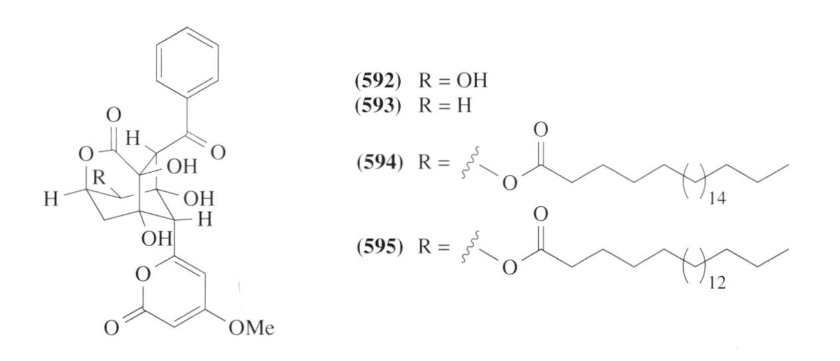

An unusual sulfated mannose homopolysaccharide, kakelokelose (**596**), was isolated from the mucous secretion of the Pacific tunicate *Didemnum molle* collected in Pohnpei, Micronesia, and in Manado, Indonesia, by sequential ultrafiltration, guided by anti-HIV tests. From 1.5 kg of *D. molle* (wet weight), 193 mg of a white solid was obtained, which showed *in vitro* anti-HIV activity. It slowly dissolved in water to give a very viscous solution. Analysis of the NMR data revealed that it consisted of a sequence of 2,3,4-trisulfated mannose units joined through β (1,6) glycosidic linkages.[543]

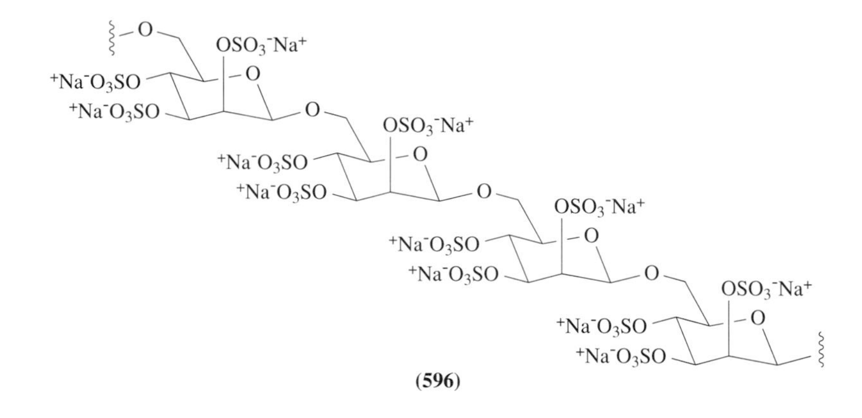

(**596**)

An enediyne antitumor antibiotic, namenamicin (**597**), was isolated from a methanolic extract of the thin encrusting orange ascidian *Polysyncraton lithostrotum* (order: Aplousobranchia, family: Didemnidae) collected at Namenalala Island, Fiji in 10^{-4}% yield (1 mg from 1 kg of frozen tissue), guided by bioautography-directed fractionation of the extract using the biochemical induction assay (BIA).[544] Namenamicin (**597**) contained the same "enediyne warhead" as the calicheamicins; however, the attached carbohydrate moiety differed in replacement of the N—O sugar linkage between the A and B sugars with a C—O, an *S*-methyl substituent at the 4 position of the A sugar, and the absence of a benzoate ring appended to the B sugar. Namenamicin (**597**) exhibited potent *in vitro* cytotoxicity with a mean IC_{50} of 3.5 ng ml^{-1} and *in vivo* antitumor activity in a P388 leukemia model in mice (ILS 40% at 3 μg kg^{-1}). Namenamicin (**597**) also showed potent antimicrobial activity and DNA cleavage experiments indicated that namenamicin (**597**) cleaved DNA with a slightly different recognition pattern than calicheamicin $\gamma_1{}^1$. Bacteria form highly specific symbiotic relationships with marine plants and animals which leads one to speculate on the true biosynthetic origin of namenamicin (**597**). The fact that all of the enediyne antitumor antibiotics previously isolated were products of antinomycetes and namenamicin's extremely low and variable yield from the ascidian lend support to the hypothesis of a microbial origin for this natural product. In order to address the question of compound origin, isolation experiments were carried out in search of a possible producing microorganisms. As a result, 16 micromonospora were isolated from the tissue of *Polysyncraton lithostrotum* and three of these micromonospora were found to produce potent DNA-damaging compounds.

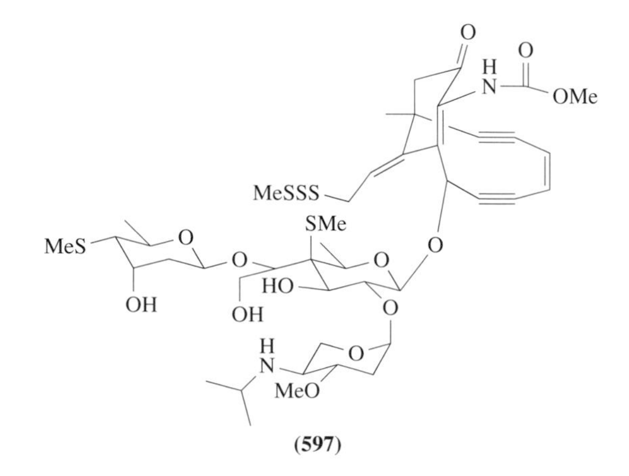

(**597**)

Long-chain amines (**598**)–(**601**) were isolated from New Zealand tunicate *Pseudodistoma novaezelandae*, exhibiting cytotoxic activity against P388 murine leukemia cells and moderate antifungal activity vs. *Candida albicans*.[545] The amines were isolated as a racemate, and (**598**) and (**600**) were obtained as a mixture of (3*E*,5*Z*)-amine and (3*E*,5*E*)-amine in a ratio of 3:2. The synthesis of 2*S*-(**600**) was achieved with high enantiomeric purity in excellent overall yield.[546]

Crucigasterins 277 (**602**), 275 (**603**), and 225 (**604**), three polyunsaturated amino alcohols, were isolated from the Mediterranean tunicate *Pseudodistoma crucigaster*.[547] The structures of these compounds were assigned based on NMR and FABMS data. The absolute stereochemistry of the amino alcohol portion in (**602**) was assigned to be 2*R*,3*S* based on chiral GC comparison of 3-hydroxy-4-aminopentanoic acid, a chemical degradation product of (**602**), with a synthetic sample prepared from L-alanine. Compounds (**602**)–(**604**) exhibited moderate cytotoxicity and antimicrobial activity. These amino alcohols were long-chain 2-amino-3-ols from an ascidian. The 2*R* stereochemistry suggested that these compounds were biosynthesized from D-alanine, unlike the usual plant and mammalian sphingosines, which are derived from L-serine.

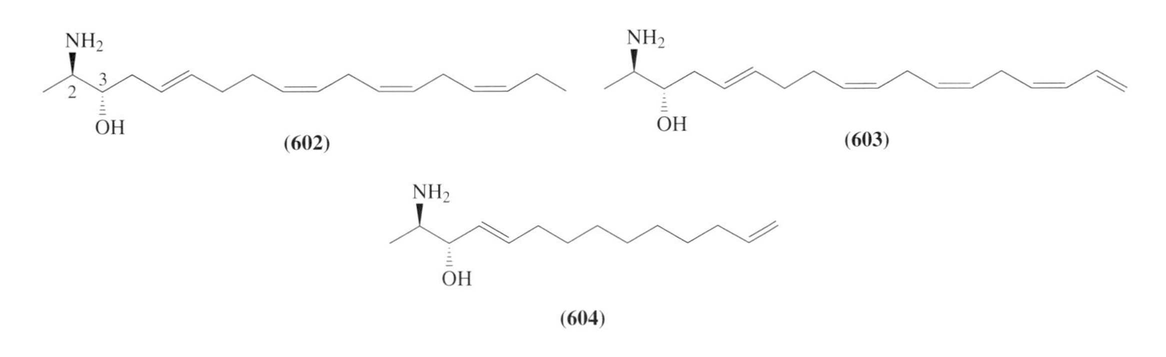

From the ethanol extract of an ascidian *Didemnum* sp., collected from the Great Barrier Reef, Australia, an antifungal amino alcohol, (*R*)-(*E*)-1-aminotridec-5-en-2-ol (**605**), together with (**606**) and (**607**), which were characterized as the *N*-Boc derivatives, were isolated and their absolute stereochemistry was determined by CD spectrum of the dibenzoyl derivative of (**605**) based on exciton coupling theory.[548] In the agar plate disk diffusion assay, amino alcohol (**605**) trifluoroacetate showed moderate activity against *Candida albicans*. The amino alcohols (**606**) and (**607**) (trifluoroacetate salts) both showed activity comparable to that of (**605**), whereas the free base of (**605**), formed upon treatment of the trifluoroacetate salt with K_2CO_3, showed slightly enhanced activity. The structures of compounds (**605**)–(**607**) join an expanding family of modified marine sphinganoids. Sphingosine itself derives from palmitoyl CoA and (*S*)-serine, but the implied biosynthesis of (**605**) appears to require a C_{12} fatty acid and glycine rather than (*S*)-serine.

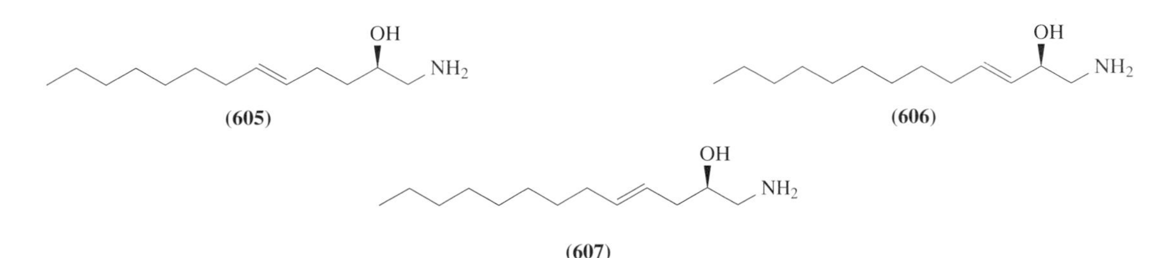

The ascidians of the genus *Clavelina* were found to contain a number of cyclic nitrogenous compounds with unsaturated alkyl side-chains. The ascidian *Clavelina cylindrica* obtained in Bermuda gave the quinolizidine alkaloids clavepictins A (**608**) and B (**609**)[549] together with the indolizidine alkaloids piclavines A_1–A_4 (**610**)–(**613**), B (**614**), and C (**615**).[550] The same ascidian obtained from Venezuela gave pictamine (**616**).[551] Clavepictine A (**608**) and pictamine (**616**) differed only by the length of the side-chain. The ascidian *Clavelina lepadiformis* yielded a decahydroquinoline alkaloid lepadin A (**617**).[552]

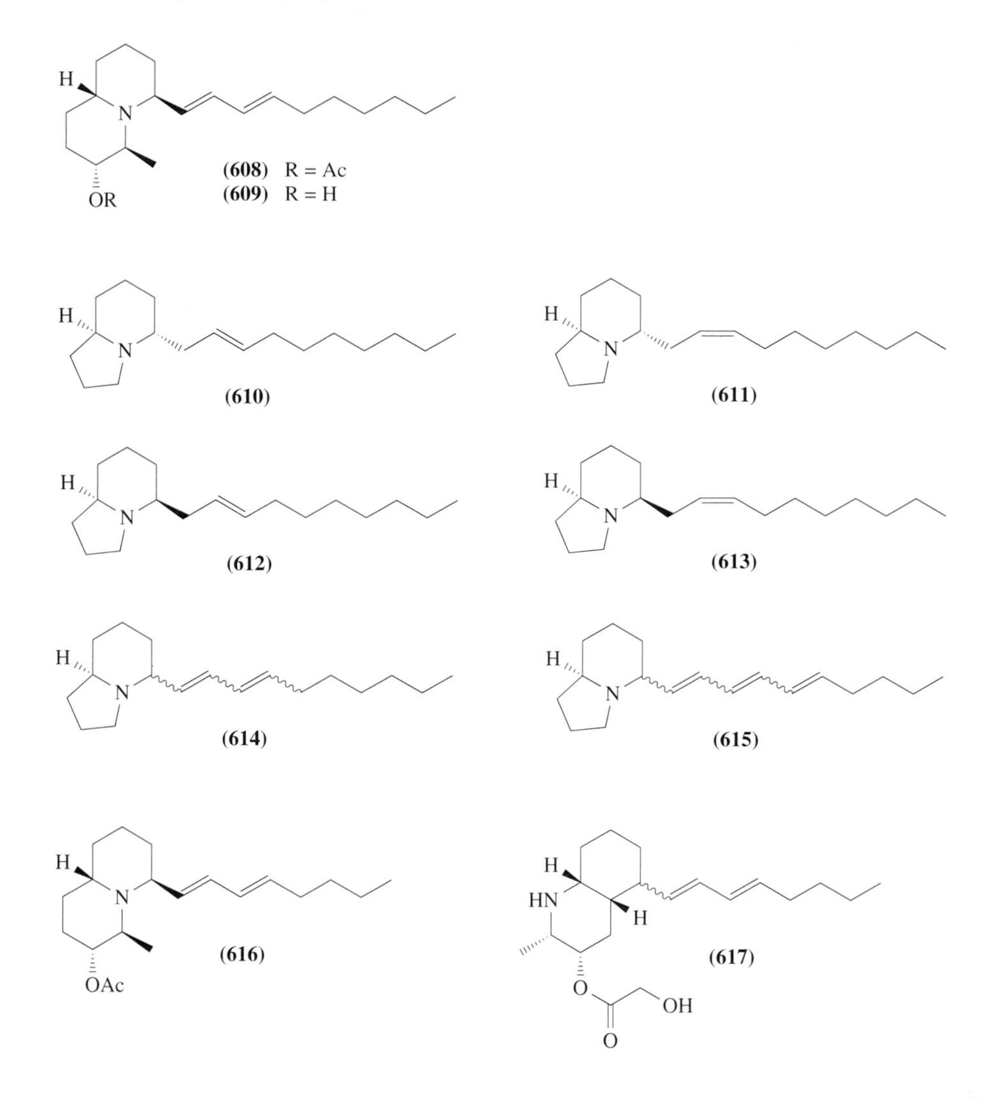

The ascidian *Clavelina cylindrica*, which was collected in Tasmania, yielded two alkaloids, cylindricine A (**618**) and B (**619**).[553] Cylindricine B (**619**) was the first example of the new pyrido-[2,1-*j*]quinoline ring system while cylindricine A (**618**) was the first pyrrolo[2,1-*j*]quinoline known from nature. Single-crystal X-ray studies of both compounds supported the assignment of the structures of the two alkaloids. The absolute configurations were not assigned.

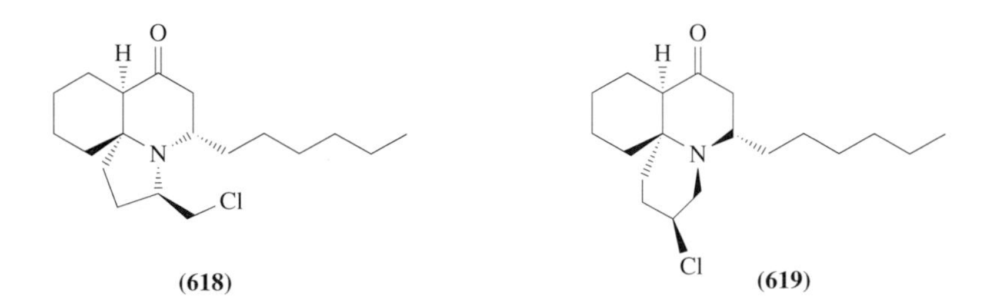

Solutions of either cylindricine A (**618**) or cylindricine B (**619**) both gave, after 6 days, the same equilibrium mixture of 3:2 of (**618**) and (**619**). This process could be followed easily by NMR spectroscopy. The interconversion between (**618**) and (**619**) only occurred when the compounds were present as free bases. Their salts, for example, the picrates, were quite stable in solution at room temperature for several weeks. The interconversion involves the chlorine and the nitrogen participating in a stereospecific ring-opening ring-closing reaction. The reaction, which involves inversion of the nitrogen, may be concerted or involve an aziridinium ion intermediate. The equilibrium mixture of alkaloids (**618**) and (**619**) exhibited bioactivity; it was active in the brine shrimp bioassay causing significant mortality at a level of 3×10^{-2} mmol.

A cytotoxic alkaloid, lepadiformine (**620**), with the same heterocyclic skeleton as (**618**) was isolated from the ascidian *Clavelina lepadiformis*.[554] The structure of (**620**) was established on the basis of chemical properties and by spectroscopic means, including a unique zwitterionic-like moiety. Lepadiformine (**620**) had moderate cytotoxic activity against cell lines of KB, HT29, P388, doxorubicin-resistant P388, and NSCLC-N6 (non-small-cell lung carcinoma). The alkaloid (**620**) also had cycle-dependent and phase-dependent properties. A partial, dose-dependent, G1 phase blockade of the cells was noted after 72 h of growth in a continuous drug exposure experiment. Due to the moderate cytotoxicity and effects on the cell cycle, lepadiformine (**620**) seemed to be of little interest with regard to antitumoral activity. However, the special zwitterionic structure may be of strong interest in terms of biological "proton-transfer mechanisms" since it seemed to be a natural example of models described as "*cis*-decalin amino acid" derivatives having such important properties.

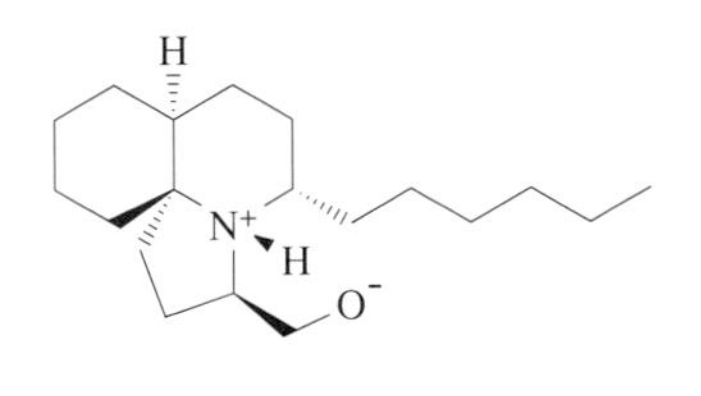

(620)

8.07.9.2.4 Pseudodistomins

This section describes the work of the authors on the unique piperidine alkaloids, pseudodistomins, from an Okinawan marine tunicate.[555]

In 1986 the authors investigated the bioactive substances from an Okinawan tunicate *Pseudodistoma kanoko*, which is an orange-colored compound tunicate and looks like a strawberry (Japanese name, "ichigo-boya"). The material was collected off Ie Island, Okinawa, by SCUBA (−5 m to −10 m). The methanol–toluene (3:1) extract of *P. kanoko* was partitioned between toluene and water. The aqueous layer was successfully extracted with chloroform, ethyl acetate, and 1-butanol. By preliminary screening using mammalian muscle preparations, the chloroform-soluble fraction was found to exhibit marked antispasmodic activity on the isolated guinea-pig ileum; the contractile responses to carbachol and histamine were abolished by this fraction. The chloroform-soluble fraction was therefore subjected to bioassay-guided fractionations using silica gel flash column chromatography eluted with $CHCl_3$/*n*-BuOH/H_2O/AcOH (1.5:6:1:1) followed by reversed-phase HPLC separation (Develosil ODS-5, 50% MeCN with 0.1% TFA) to afford an active fraction, which was positive on the ninhydrin-test on TLC. This active fraction was shown to be a mixture of two components (pseudodistomins A (**621**) and B (**622**)), the separation of which was first carried out after converting them into acetates (**623**) and (**624**), respectively, by ODS-HPLC (YMC-Pack, AM) with 88% MeOH. Acetates (**623**) and (**624**) were used for characterization and structural studies. A small amount of (**621**) and (**622**) (before acetylation) was obtained by careful HPLC (Develosil ODS-5) eluting with 37% MeCN with 0.2% TFA for bioassay purposes. In addition to antispasmodic activity, pseudodistomins A (**621**) and B (**622**) exhibited cytotoxic activity against murine leukemia cells, L1210 and L5178Y, *in vitro* (IC_{50} values: 2.5 μg ml^{-1} and 0.4 μg ml^{-1} against L1210, respectively; 2.4 μg ml^{-1} and 0.7 μg ml^{-1} against L5178Y, respectively). Both compounds (**621**) and (**622**) also exhibited calmodulin antagonistic activity; they both inhibited calmodulin-activated brain phosphodiesterase with IC_{50} values of 3×10^{-5} M, being approximately 3 times more potent than W-7, a well-known synthetic calmodulin antagonist.[556]

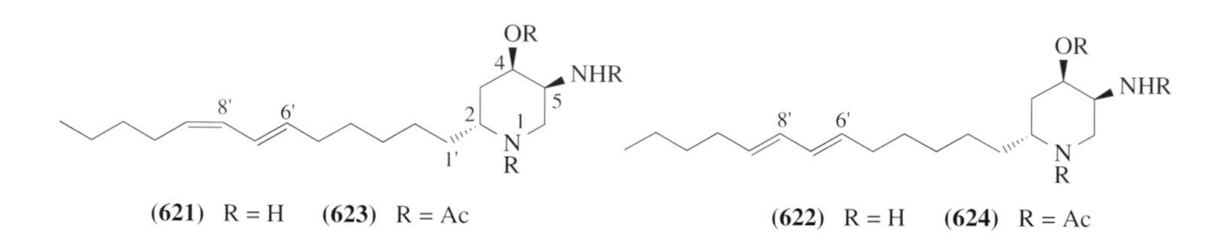

Structural studies were mostly carried out using pseudodistomin B acetate (**624**). In the ^{1}H NMR spectrum of the acetate (**624**), several signals appeared broad and split in an approximately 4:1 ratio; this phenomenon might be ascribed to the presence of two slowly interconverting conformations due to the rotation of the secondary amide group. Although the elucidation of the 1D NMR data was not helped by this phenomenon, 2D NMR data of (**624**) with good quality was recorded on a 400 MHz spectrometer, and analysis of the ^{1}H-^{1}H COSY and ^{13}C-^{1}H COSY spectral data led to a structure for (**624**) containing a piperidine nucleus and an alkyl side-chain with two *E* double bonds. Spectral data for pseudodistomin A acetate (**623**) suggested that pseudodistomin A possesses the same type of structure with different alkene geometry (one *E* and one *Z*). Catalytic hydrogenation of each of the acetates (**623**) and (**624**) afforded the identical tetrahydro derivative (**625**). The relative stereochemistry of the chiral centers at C-2, C-4, and C-5 was determined by the coupling constants and NOE data using the tetrahydroacetate (**625**) (Figure 17). Although it seemed unfavorable that the alkyl side-chain at C-2 is axially oriented, it was proposed that the conformation shown in Figure 17 may be stabilized by an intramolecular hydrogen bond between the N(5)-H and N(1), which was inferred from the FT IR spectrum of a dilute solution of (**625**). Knapp and Hale, who achieved the synthesis of the tetrahydroacetate (**625**) in 1993,[557] indicated that this hydrogen bond would be highly strained, but they also agreed that the conformation in Figure 17 was preferred to the alternative one because of a steric interaction between the N(1)-acetyl group and the alkyl side-chain, which was also supported by the Macromodel calculation.[557]

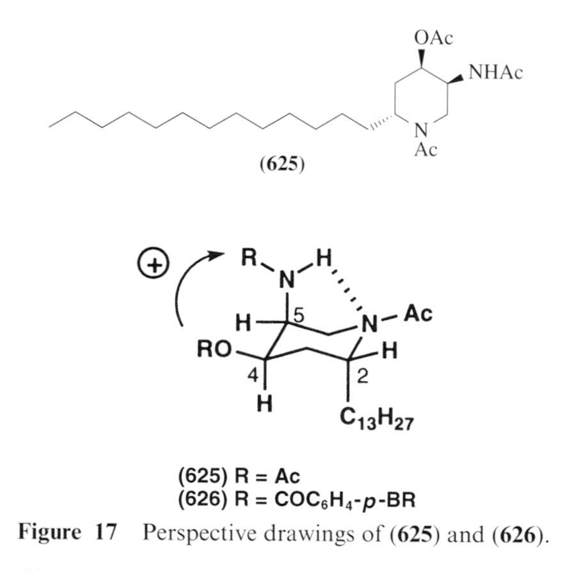

Figure 17 Perspective drawings of (**625**) and (**626**).

The absolute stereochemistry of the C-2, C-4, and C-5 positions was deduced on the basis of the exciton chirality method. The 1-acetyl-4,5-bis(*p*-bromobenzoyl) derivative (**626**) was prepared by partial hydrolysis of (**625**), followed by *p*-bromobenzoylation. The CD spectrum of (**626**) showed a positive Cotton effect, implying the 2*R*, 4*R*, and 5*S*-configurations (Figure 17).

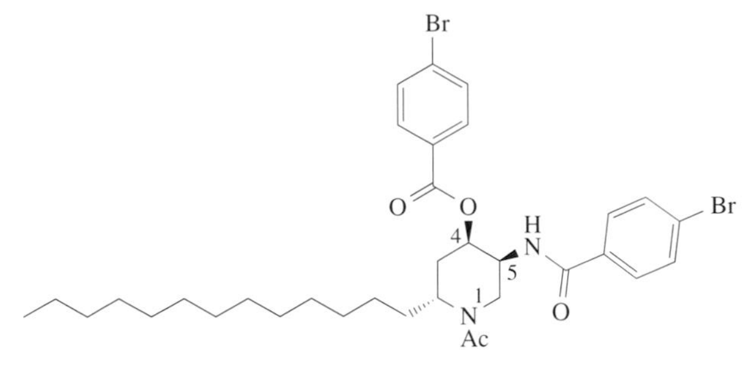

(626)

The authors initially proposed the conjugated diene position in the side-chain of pseudodistomins A and B to be at the 3′,5′ position ((**627**) and (**628**), respectively);[556] this position was, however, later revised to be at the 6′,8′-position ((**621**) and (**622**), respectively) by subsequent studies.

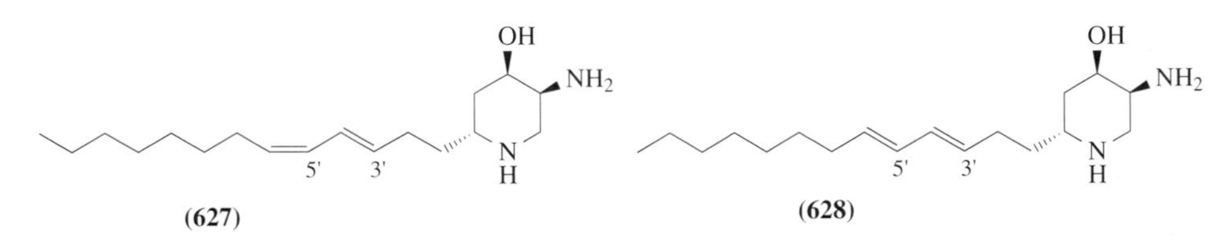

(627) **(628)**

Following publication of the authors' work on the isolation and structure elucidation of pseudodistomins, these compounds were chosen as a target of organic synthesis by several groups, probably because of their unique bioactivities as well as their interesting structure apparently related to sphinganoids. In 1990, Nakagawa *et al.* reported the synthesis of compound (**629**) as a model for (**627**) (initial structure), and they also prepared an optically active compound (**630**) from L-aspartic acid as a key intermediate.[558]

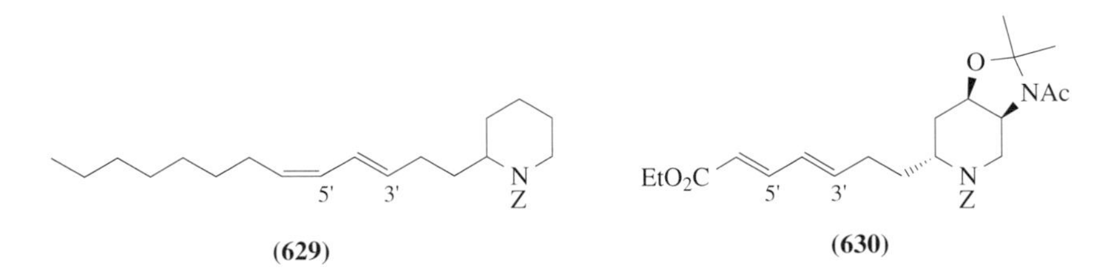

(629) **(630)**

Total synthesis of the tetrahydroacetate (**625**) was achieved by three groups. Natsume and co-workers prepared (±)-(**625**) from a 1,2-dihydropyridine derivative through the singlet oxygen addition reaction.[559] Naito *et al.* described the synthesis of (±)-(**625**) via a route involving the reductive photocyclization of enamide and α-acylamino radical allylation,[560] and they afterwards synthesized optically active (+)-(**625**) through the cycloaddition of a nitrone to (+)-2-aminobut-3-en-1-ol.[561] Knapp and Hale, *vide supra*, prepared (+)-(**625**) as the optically active form from D-serine,[557] and the optical rotation of the synthetic sample of (+)-(**625**) corresponded to that of the natural one. From these studies the piperidine ring absolute stereochemistry of pseudodistomins A (**621**) and B (**622**) was unambiguously confirmed as that described initially by the authors.[556]

Naito and co-workers, however, questioned the side-chain diene position during their synthesis of (±)-(**625**); they prepared the triacetates of (**628**) (3′,5′-diene; the authors' initial structure for pseudodistomin B), and found that the spectral data of the synthetic triacetate were not superimposable onto those of the natural specimen.[562] The side-chain diene position of pseudodistomin B was therefore reinvestigated by chemical degradation experiments. The triacetate (**624**) prepared from a natural specimen of pseudodistomin B (**622**) was treated with ozone followed by reduction with $NaBH_4$ and acetylation with Ac_2O and pyridine afforded the tetraacetate (**631**), the FABMS of which clearly afforded the $(M+H)^+$ ion at m/z 385, implying the side-chain diene position of pseudodistomin B (**622**) to be at the 6′,8′-position.[562] After being made aware of this result, Naito *et al.* prepared the triacetate with the 6′,8′-diene (**624**), whose spectral data proved to be identical with those of the authentic sample of (**624**).[563]

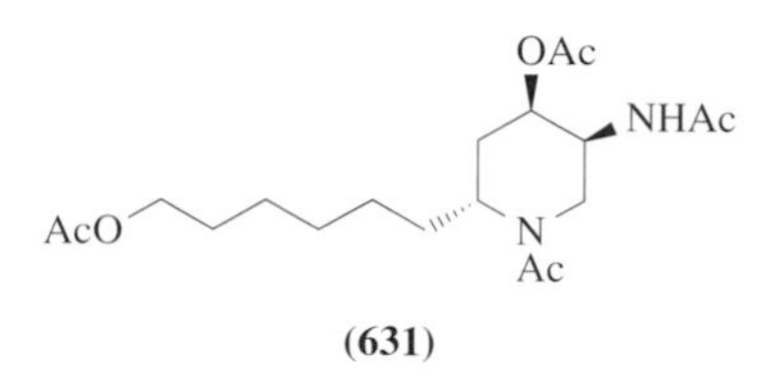

(631)

As the result of the above work, the structure of pseudodistomin A also had to be reexamined. Since a natural specimen of pseudodistomin A was unavailable at that time, the extracts of the tunicate *P. kanoko* were reinvestigated. Pseudodistomins A and B were first isolated from the chloroform-soluble fraction obtained by partition experiments. The toluene-soluble fraction, which had been shown to be less active on the antispasmodic activity assay, was reexamined to detect a ninhydrin-positive spot on TLC. The toluene-soluble fraction was therefore separated by the same procedures as above to succeed in reisolating pseudodistomin A as its acetate (**623**). The pseudodistomin A acetate (**623**) thus obtained was subjected to ozonolysis by the same procedures used for pseudodistomin B acetate (**624**) to give the identical product (**631**) on the basis of TLC, ^{1}H NMR, and EIMS (m/z 325, (M—CH_3CONH_2)+) data. Thus, pseudodistomin A was also revealed to have the 6′,8′-diene (**621**), and the 6′*E*,8′*Z*-configuration was deduced from the combination of the HOHAHA spectrum and coupling constant data of the acetate (**623**).[564]

In the course of the reisolation of pseudodistomin A (**621**), the authors also aimed at isolating other new alkaloids since marine aliphatic amino alcohols related to sphinganoids are of interest to many scientists working in a broad range of biological sciences. The ninhydrin-positive fraction was, after acetylation, carefully examined by HPLC (Develosil ODS-5) eluting with 85% MeOH to afford a new piperidine alkaloid, named pseudodistomin C (**632**), as its acetate (**633**). Before acetylation, (**632**) was also obtained by preparative silica gel TLC ($CHCl_3$/MeOH/H_2O, 6:4:0.7), and was revealed to exhibit cytotoxicity against murine lymphoma L1210 and human epidermoid carcinoma KB cells *in vitro* (IC_{50} values, 2.3 μg ml^{-1} and 2.6 μg ml^{-1}, respectively).[565]

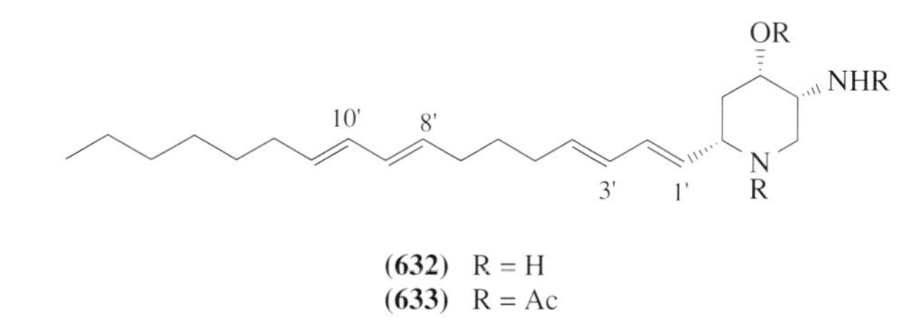

(632) R = H
(633) R = Ac

The ^{1}H NMR spectrum of the acetate (**633**) in $CDCl_3$ showed signals so broad that no signals were able to be assigned, and was quite different from the ^{1}H NMR spectrum of pseudodistomin B acetate (**624**). The 2D NMR experiments (^{1}H-^{1}H COSY, HSQC, and HMBC) of (**633**) were carried out in a CD_3OD solution, which showed relatively resolved signals. Since the ^{1}H NMR spectrum of the natural compound (**632**) in a C_5D_5N solution appeared better in resolution, the ^{1}H-^{1}H COSY and HSQC spectra of (**632**) were recorded in this solution. From these spectral data, pseudodistomin C (**632**) was suggested to consist of a piperidine moiety and an unsaturated side-chain; the piperidine ring has the same substituents (4-hydroxyl and 5-amino groups) as those of pseudodistomins A and B, (**621**) and (**622**), and the side-chain attached to C-2 contains two dienes. The positions of the two dienes were clarified unambiguously by the following degradations. Pseudodistomin C (**632**) was treated with ozone followed by $NaBH_4$ reduction and acetylation to give a crude product, from which the tetraacetate ((**634**), EIMS m/z 315 $(M+H)^+$ and 255 (M—$CH_3CONH_2)^+$) was obtained by HPLC purification, thus revealing one of the two dienes to be located at the 1′,3′-position. The second diene was deduced to be on the 8′,10′-position since 1,5-pentanediol diacetate (**635**) was detected by reversed-phase TLC and HPLC analyses from the crude mixture of the ozonolysis products. These alkenes were inferred to be all *E* from the coupling constants and the ^{13}C chemical shifts of the allylic methylenes.

D-**(634)**

(635)

Since the ^{1}H NMR spectrum of pseudodistomin C acetate (**633**) appeared quite different from that of pseudodistomin B acetate (**624**), stereochemical evidence of the piperidine ring portion of pseudodistomin C (**632**) was required. Thus, the tetraacetate (**634**), which was obtained by ozonolysis of (**632**), was prepared as an optically active form, as shown in Scheme 17. Oxazolidine aldehyde (**636**), prepared from L-serine,[566] was treated with allylmagnesium bromide to give a 1:1 diastereomeric mixture of allyl alcohols. After deprotection of the acetonide group and conversion into pivaloyl ester, the unnecessary *threo*-isomer was removed by silica gel column chromatography. The *erythro*-monopivaloate (**637**) was transformed via five steps into a benzyl carbamate (**638**), which was subjected to amide mercuration to give 2*R*- and 2*S*-piperidine derivatives (**639**) and (**640**) in the ratio of 54:46. Oxidative demercuration of (**639**) and (**640**) gave primary alcohols (**641**) and (**642**), respectively, which were deprotected and acetylated to afford tetraacetates L-(**634**) and (**643**), respectively. The ^{1}H NMR spectrum of the tetraacetate (**634**) obtained from a natural specimen of pseudodistomin C (**632**) was identical with that of the 2*R*,4*R*,5*S*-derivative L-(**634**). Since the tetraacetate (**643**) possesses the same relative configurations at the C-2, C-4, and C-5 positions on the piperidine ring as those of pseudodistomins A (**621**) and B (**622**), the relative configurations of pseudodistomin C (**632**) proved to be different from those of (**621**) and (**622**). The sign of optical rotation of synthetic L-(**634**) ($[\alpha]_D$ $-19\,^{\circ}$) was opposite to that of the tetraacetate (**634**) ($[\alpha]_D$ $+16\,^{\circ}$) derived from a natural specimen of (**632**). The absolute configuration of pseudodistomin C (**632**) was therefore revealed as 2*S*, 4*S*, and 5*R*. This result was, however, unexpected since the piperidine alkaloids isolated from the same tunicate possess different stereochemistries at the C-4 and C-5 positions. To obtain further unambiguous confirmation of this conclusion, the enantiomer D-(**634**) was prepared from D-serine by the same procedures as above, and subjected to chiral HPLC analysis (CHIRALPAK AD, Daicel Chemical Ind. Ltd.; 4.6 × 250 mm; flow rate: 0.5 ml min^{-1}; UV detection at 215 nm; eluent: hexane/2-propanol, 8:2), which established that the tetraacetate (**634**) (t_R 16.5 min) derived from natural specimen (**632**) showed the same retention time as the enantiomer D-(**634**) prepared from D-serine (D-(**634**): t_R 16.5 min; L-(**634**): t_R 15.1 min), thus firmly establishing the 2*S*, 4*S*, and 5*R*-configurations for pseudodistomin C (**632**).[565]

To provide further unambiguous evidence for the whole structure of pseudodistomin C (**632**), the total synthesis of (**632**) was investigated by the authors as follows.[567] The synthesis began with Garner's aldehyde, D-(**636**),[566] derived from D-serine as summarized in Scheme 18. The Grignard reaction of allylmagnesium bromide with D-(**636**) afforded a 1:1 mixture of *erythro*- and *threo*-homoallyl alcohols. To obtain the *erythro*-alcohol (**645**) practically, the diastereomeric mixture was oxidized with Dess–Martin periodinane in DMF to give the ketone (**644**), which was reduced with $Zn(BH_4)_2$ to give *erythro*-alcohol (**645**) in 96% *de* (60% *ee*, *vide infra*). The *erythro*-alcohol (**645**) was transformed into the *t*-butyl carbamate (**647**) via isomeric alcohol (**646**) in seven steps by the previous method;[565] the terminal amine was protected by the Boc group to simplify the deprotection. Amide mercuration of (**647**) with $Hg(OAc)_2$ in $ChCl_3$ afforded (2*S*)-piperidine (**648**) and its (2*S*)-isomer in a ratio of 1.5:1. The (2*S*)-piperidine derivative (**648**) was oxidatively demercurated to give alcohol, which was treated with diphenyl disulfide and tri-*n*-butylphosphine in pyridine followed by oxidation of the sulfide group with diphenyl diselenide and hydrogen peroxide to furnish phenylsulfone (**649**).

The side-chain moiety of (**632**) was prepared as shown in Scheme 19, starting from the known 1-bromo-3*E*,5*E*-decadienoyl bromide (**650**).[568] Condensation of (**650**) and sodium dimethyl malonate afforded dimethyl ester (**651**), which was heated with sodium chloride in wet DMSO at 190 °C to afford the corresponding ester (**652**). Reduction of the ester group of (**652**) with DIBAL in toluene and the Wittig reaction provided ethyl tetradecatrienoate (**653**) with all *E*-configurations revealed from coupling constants. The ester (**653**) was reduced with DIBAL and the resulting alcohol was oxidized with pyridinium chlorochromate (PCC) to afford the corresponding aldehyde, which was subjected to Julia alkenation with the phenylsulfone (**649**). The sulfone (**649**) was treated with *n*-butyllithium in THF in the presence of HMPA at −78 °C to produce the orange sulfone anion, which was allowed to react with the aldehyde obtained from (**653**), and then quenched with benzoyl chloride to afford a diastereomeric mixture of *β*-benzoyloxy sulfones. Treatment of the crude mixture with sodium amalgam resulted in formation of the tetraene (**654**) possessing the backbone skeleton of pseudodistomin C (**632**). The ^{1}H NMR of (**654**) revealed that the last generated $\Delta^{1',2'}$-double bond was *E* ($J_{1',2'}$ = 14.0 Hz), and the HPLC analysis of the tetraene (**654**) using a reversed-phase column showed a single peak predominantly, suggesting that the tetraene (**654**) possesses all *E*-configurations. Removal of the protective groups of (**654**) with 3N HCl afforded pseudodistomin C (**632**), whose ^{1}H NMR and EIMS spectra as well as R_f values on TLC were completely identical with those of the natural specimen.[565] The synthetic pseudodistomin C (**632**) was acetylated with acetic anhydride in pyridine to furnish pseudodistomin C triacetate (**633**), which was also identified

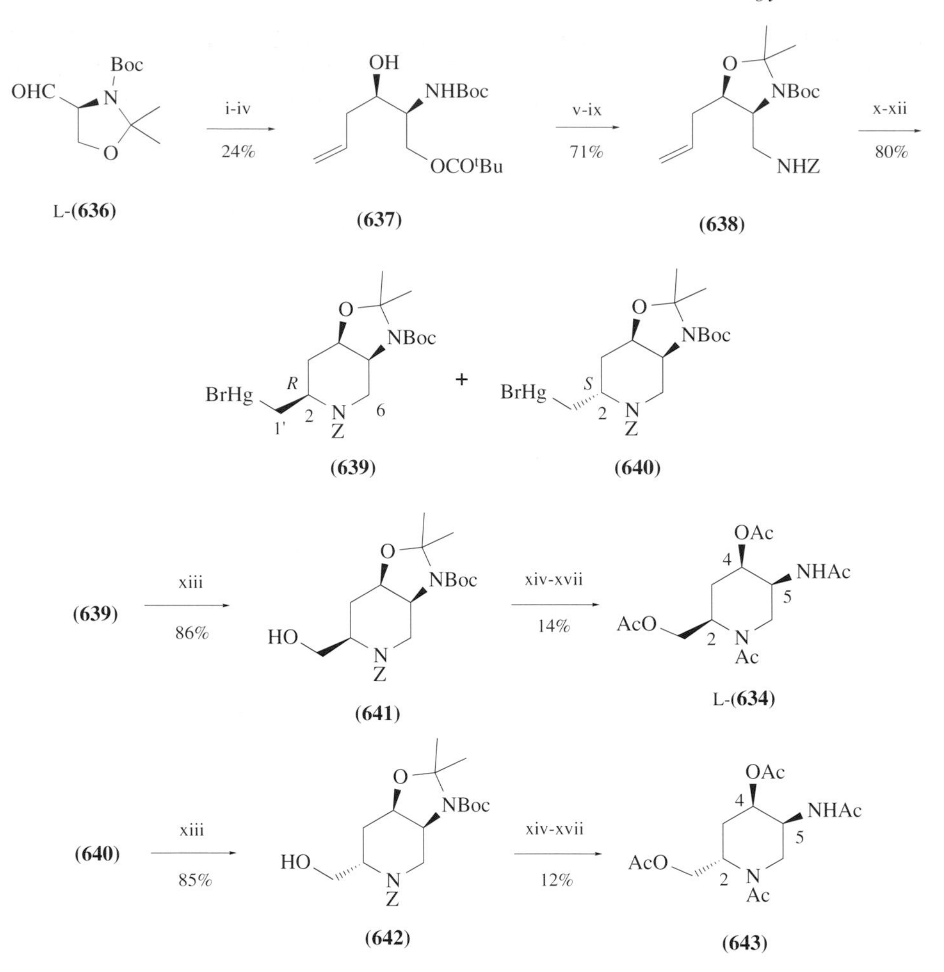

i, CH_2=$CHCH_2MgBr$; ii, *p*-TsOH, MeOH; iii, PivCl, pyridine; iv, SiO_2 column, hexane/EtOAc (3:1); v, DMP, BF_3·OEt; vi, 2.5 N KOH, MeOH; vii, Phthalimide, DIAD, PPh_3; viii, H_2NNH_2·H_2O, EtOH; ix, ZCl, 2N NaOH; x, $Hg(OCOCF_3)_2$, $CHCl_3$; xi, $NaHCO_3$; xii, NaBr; xiii, $NaBH_4$,O_2, DMF; xiv, TFA, CH_2Cl_2; xv, Ac_2O, pyridine; xvi, H_2, Pd/C, EtOH; xvii, Ac_2O, pyridine.

Scheme 17

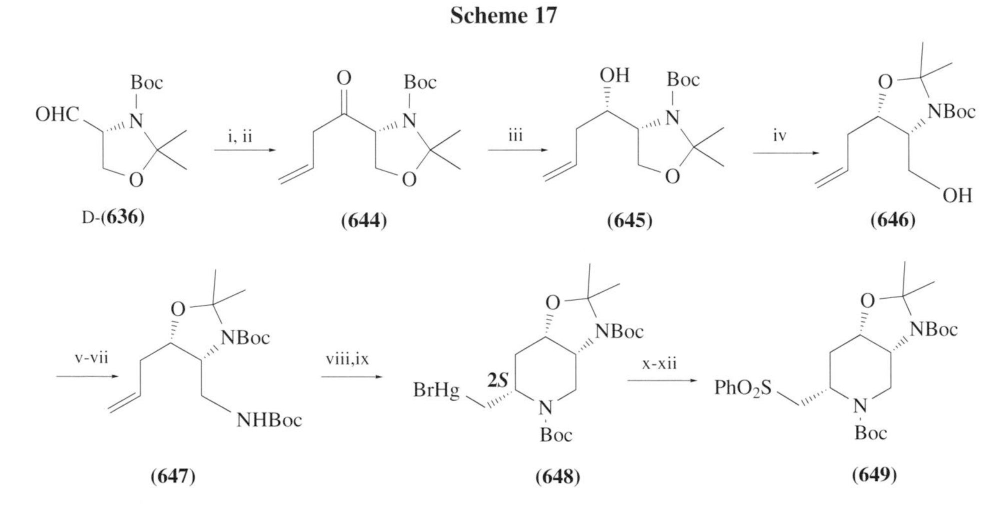

i, CH_2=$CHCH_2MgBr$ (quant.); ii, Dess–Martin periodinane, DMF (68%); iii, $Zn(BH_4)_2$, benzene-Et_2O (quant.); iv, ref. 2, (4 steps, 56%); v; Phthalimide, DIAD, PPh_3; vi, H_2NNH_2•H_2O, EtOH; vii, $(Boc)_2O$, 1N NaOH, dioxane (3 steps, 80%); viii, $Hg(OAc)_2$, $CHCl_3$; ix, NaBr, $NaHCO_3$ (2 steps, 56%; 2*R*-isomer, 28%); x, O_2, $NaBH_4$, DMF (90%); xi $(PhS)_2$, $Bu^n{}_3P$, pyridine (79%); xii, Ph_2Se_2, 30% H_2O_2, CH_2Cl_2-Et_2O (73%)

Scheme 18

with the triacetate (**633**)[565] derived from a natural specimen of (**632**) on the basis of ^{1}H NMR and EIMS spectra as well as TLC and HPLC examinations. The sign of the optical rotation of synthetic triacetate (**633**) ($[\alpha]_D$ +43° (*c* 1, $CHCl_3$)) was also the same as that of the natural one ($[\alpha]_D$ +85° (*c* 0.98, $CHCl_3$)).[565]

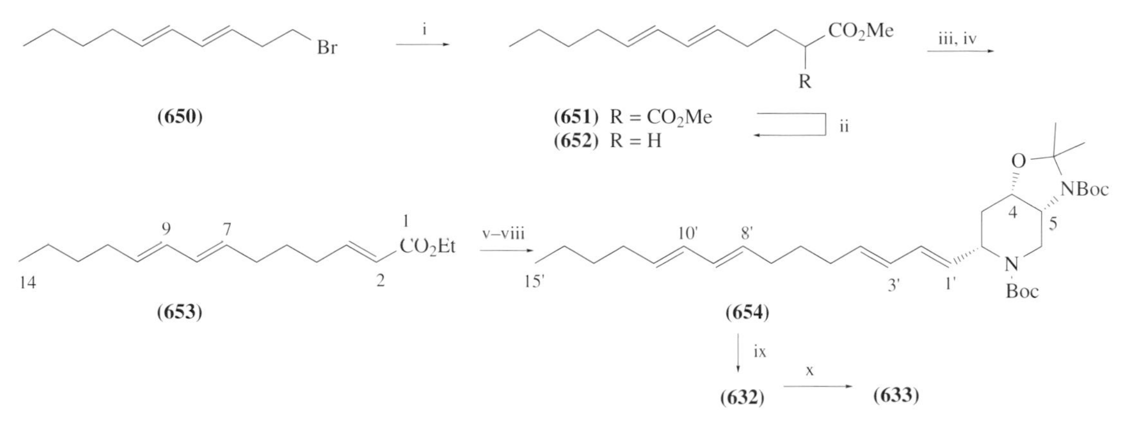

i, Na, MeOH, $CH_2(CO_2Me)_2$ (74%); ii, NaCl, DMSO, H_2O, 190 °C (74%); iii, DIBAL, toluene; iv, $Ph_3P{=}CHCO_2Et$, CH_2Cl_2 (2 steps, 68%); v, DIBAL, CH_2Cl_2 (80%); vi, PCC, CH_2Cl_2 (74%); vii, Bu^nLi, **649**, THF-HMPA; viii, Na-Hg, MeOH (13% from **649**); ix, 3N HCl, EtOAc (34%); x, Ac_2O, pyridine (56%)

Scheme 19

Thus the further structural confirmation of (**632**) by total synthesis was completed, although the absolute value of the optical rotation of the synthetic triacetate (**633**) was smaller than that of the natural specimen of (**633**). The optical purity of the synthetic compound was examined by means of chiral HPLC analysis after conversion of the corresponding alcohol derived from (**648**) into tetraacetate (**634**), which had been obtained by ozonolysis of (**632**),[565] to reveal that the synthetic tetraacetate (**634**) obtained in this study was 60% *ee*. The optical purity of synthetic (**633**) was estimated to be parallel to this result. This result may be attributable to partial racemization during the oxidation–reduction process to obtain the *erythro*-alcohol (**645**). In the authors' previous study,[565] chiral HPLC had shown that no crucial racemization occurred since the *erythro*-alcohol (**645**) and its *threo*-isomer were separated after conversion into monopivaloyl esters ((**637**) and its isomer) by four repeated silica gel chromatographies.

8.07.9.3 Sponge Metabolites

8.07.9.3.1 Manzamines and related alkaloids

In 1968 Higa and co-workers isolated a novel cytotoxic *β*-carboline alkaloid, named manzamine A (**655**), from a marine sponge *Haliclona* sp. collected off Manzamo, Okinawa, and the structure including absolute configuration was established by X-ray analysis.[569] In the following year the isolation of manzamines B (**656**), C (**657**), and D (**658**) from the *Haliclona* sponge was reported.[570,571]

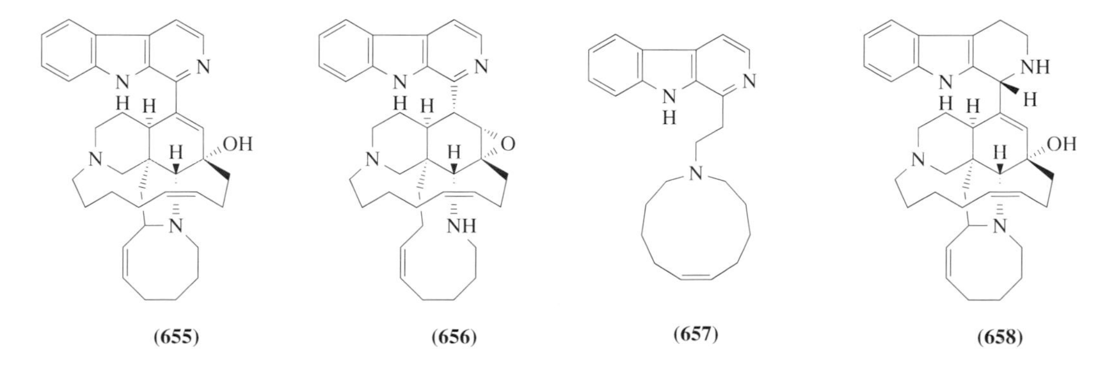

At almost the same time Kobayashi and co-workers independently isolated keramamines A and B (=manzamines A (**655**) and F (**659**), respectively) from an Okinawan marine sponge *Pellina* sp.[572] These unusual ring systems have attracted great interest as challenging targets for total synthesis and for their unprecedented biosynthetic path, to be resolved.

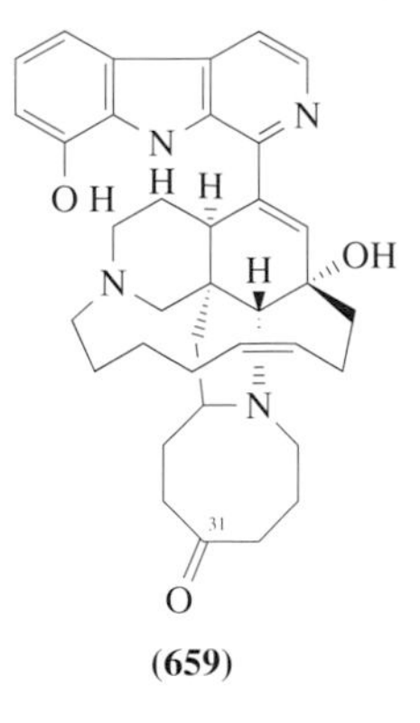

(659)

In 1992 Baldwin and Whitehead proposed a biogenetic path for manzamines A–C (**655**)–(**657**), as shown in Scheme 20.[573] The proposal suggested that *bis*-3-alkyldihydropyridine macrocycle (a), which can be derived from ammonia, a C_3 unit, and a C_{10} unit, might be converted through a Diels–Alder-type [4+2] intramolecular cycloaddition into a pentacyclic intermediate (b), which in turn led to manzamines A (**655**) and B (**656**) via a tetracyclic intermediate (c). Although Baldwin and Whitehead's biogenetic path was elegant and fascinated many chemists, this proposal was only a hypothesis without any experimental basis. In the same year Kobayashi and co-workers isolated two novel alkaloids, ircinals A (**660**) and B (**661**) from an *Ircinia* sponge,[574] which are very close to intermediate (c). Subsequently, from an *Amphimedon* sponge, they obtained some novel manzamine-related alkaloids, keramaphidin B[575] (**662**), which is quite similar to intermediate (b); keramaphidin C[576] (**663**) and keramamine C[576] (**664**), which seem to be biogenetic precursors of manzamine C (**657**); and ircinols A (**665**) and B[577] (**666**), which correspond to antipodes of the alcohol forms of ircinals A (**660**) and B (**661**).

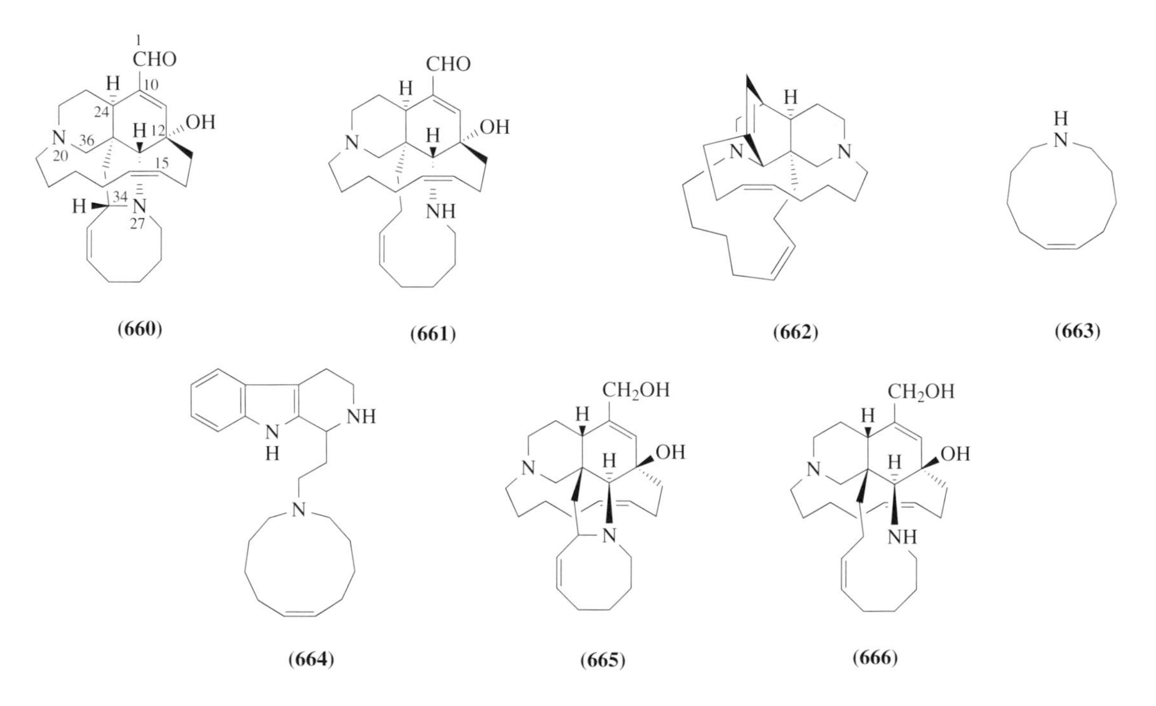

(660) **(661)** **(662)** **(663)**

(664) **(665)** **(666)**

The absolute configuration of ircinal A (**660**) was confirmed by the chemical correlation with manzamine A (**655**) (Scheme 21).[574] The Pictet–Spengler cyclization of (**660**) with tryptamine afforded manzamine D (**658**), which was then transformed into manzamine A (**655**) through ddq oxidation. The structure of ircinal B (**661**) was established by the chemical correlation with

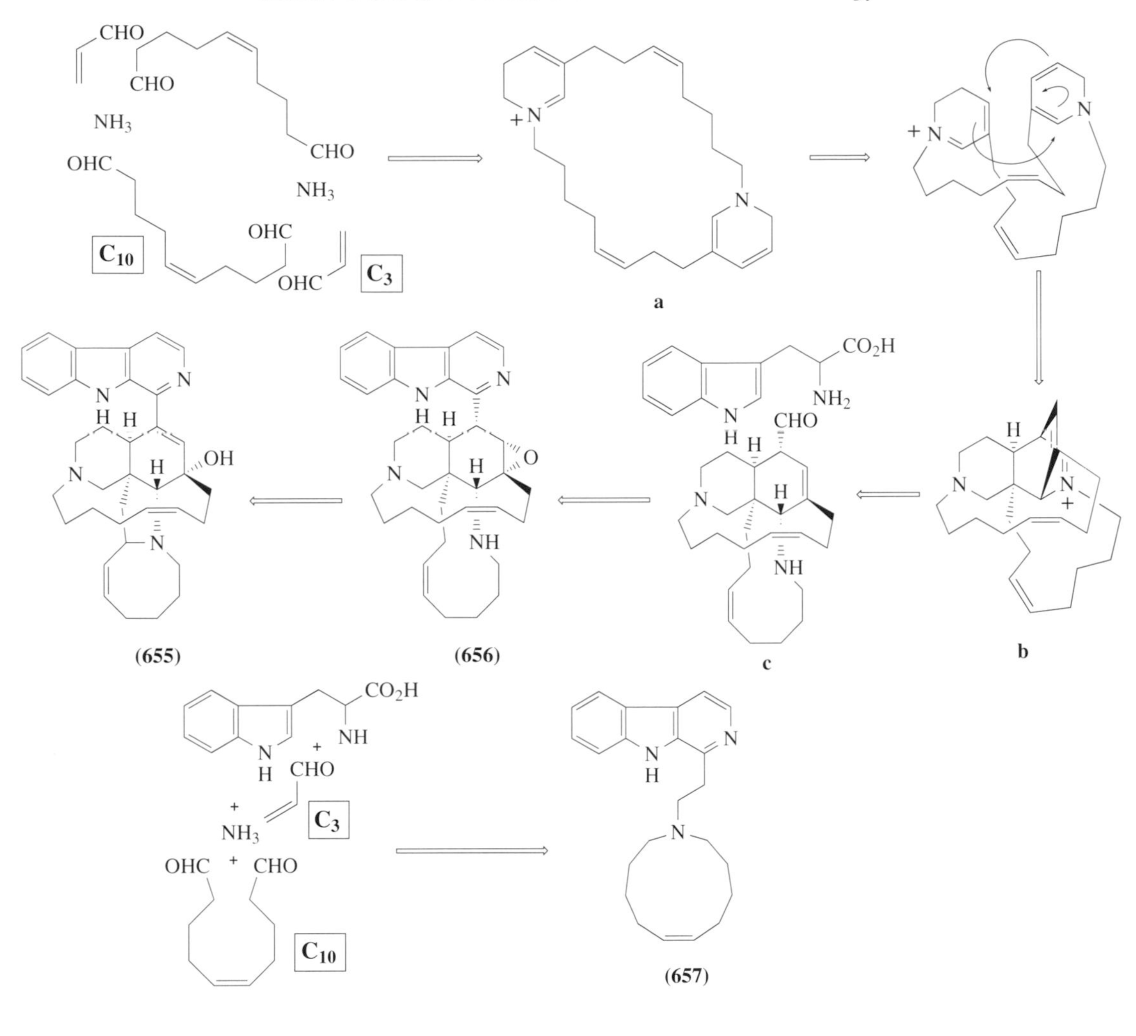

Scheme 20

manzamine B (**656**), as shown in Scheme 21. The Pictet–Spengler cyclization of (**656**) with tryptamine afforded manzamine H (**667**), which was then transformed into manzamine J (**668**) through ddq oxidation. On the other hand, treatment of manzamine B (**656**) with NaH also gave manzamine J (**668**). Manzamines A (**655**) and B (**656**) may be biogenetically generated from ircinals A (**660**) and B (**661**) through the Pictet–Spengler reaction with tryptamine or tryptophan followed by dehydrogenation, as shown in Scheme 21.

The planar structure of keramaphidin B (**662**), $C_{26}H_{40}N_2$, was elucidated by detailed analyses of ^{1}H and ^{13}C NMR data aided by 2D NMR experiments. The relative stereostructure was established by X-ray analysis using a suitable crystal of (**662**) grown in acetonitrile. Interestingly, the X-ray study revealed that the crystal of (**662**) was racemic though it possesses four asymmetric centers. The structure of keramaphidin B (**662**) corresponds to that of the pentacyclic intermediate (b) proposed by Baldwin and Whitehead.

Chiral HPLC analyses of the crystals and the filtrate of keramaphidin B (**662**) revealed that the ratio of (+)- and (−)-forms was ca. 1:1 for the crystals and 20:1 for the filtrate, respectively. On the other hand, the ratio of the crystal and the filtrate of keramaphidin B (**662**) was 5:95. These results suggested that (+)-keramaphidin B {(+)-(**662**)} was estimated to be ca. 97% of this sponge.[578]

(+)-Keramaphidin B ((+)-(**662**)) was subjected to reduction with Pd/C, and then oxidation with OsO_4 to give two dihydroxy products (**669**) and (**670**) in the ratio of ca. 9:1 (Scheme 22). The relative stereochemistry at C-9 and C-10 of (**669**) and (**670**) was elucidated from NOE data. Treatment of (**669**) with (*R*)-(−)- and (*S*)-(+)-α-methoxy-α-(trifluoromethyl)phenylacetyl chloride (MTPACl) gave the (*S*)- and (*R*)-MTPA esters, respectively. The ^{1}H chemical shift differences ($\Delta\delta = \delta_S - \delta_R$) indicated that the absolute configuration at C-10 of (**669**) was *R*. Thus the absolute

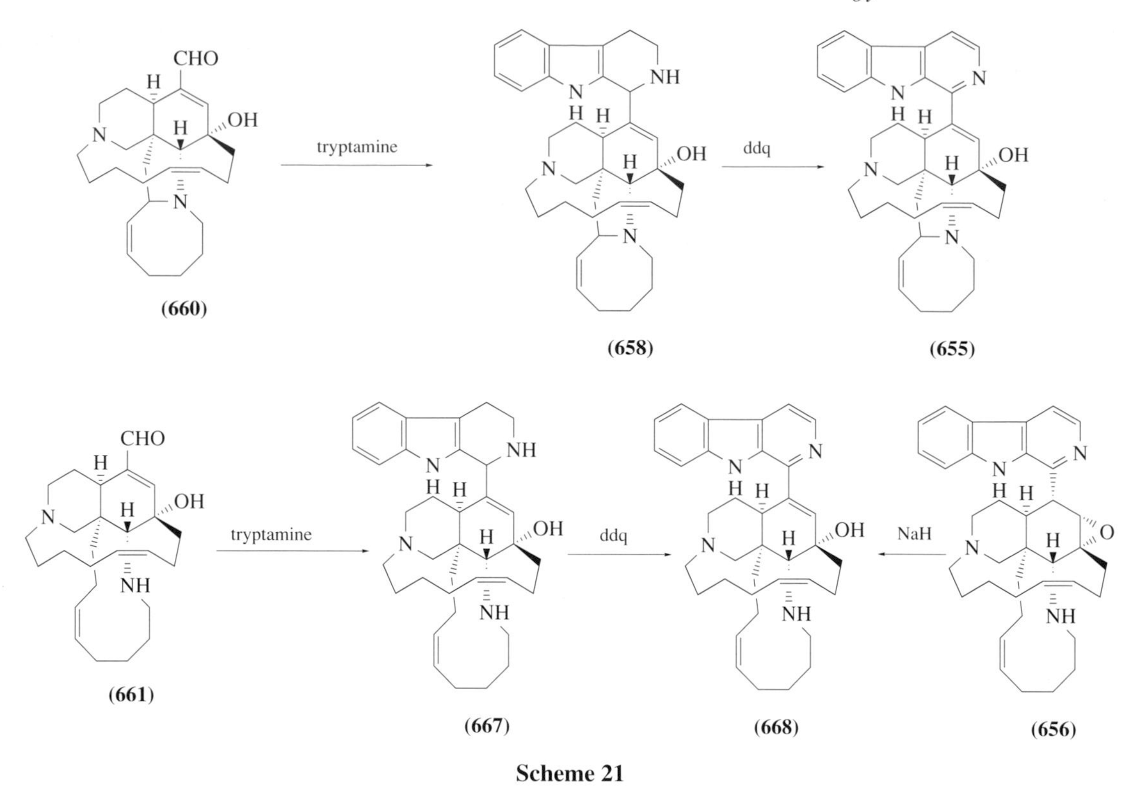

Scheme 21

configurations at C-1, C-4, C-4a, and C-8a of (+)-keramaphidin B ((+)-(**662**)) were concluded to be *R*, *S*, *R*, and *S*, respectively.[579]

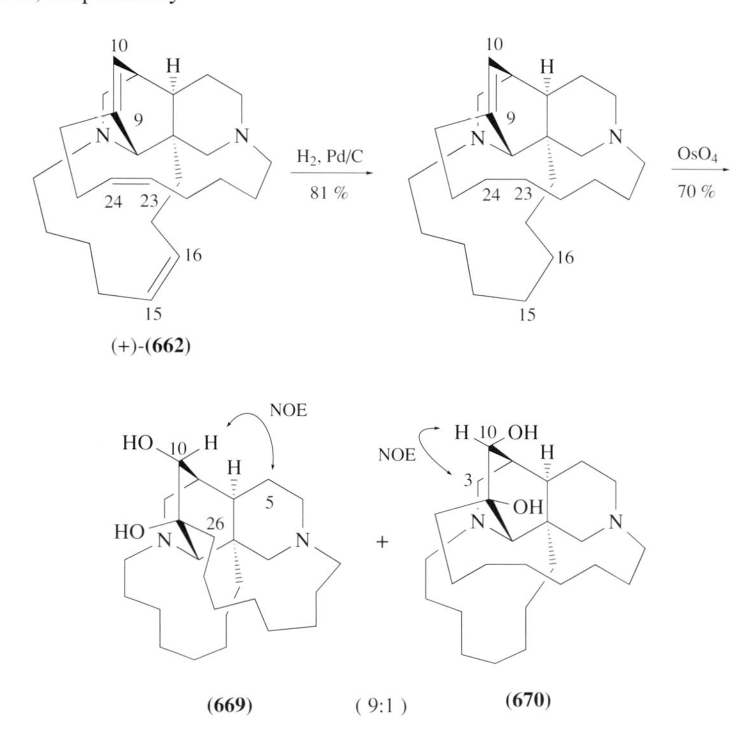

Scheme 22

Ircinol A (**665**) corresponded to the alcoholic form of (**660**). Treatment of ircinal A (**660**) with DIBAH afforded a reduction product (**671**) (Scheme 23), the spectral data of which were identical with those of ircinol A (**665**). However, the sign of the optical rotation was opposite ((**671**), $[\alpha]_D^{18}$

+20° (*c* 0.2, MeOH); (**664**), $[\alpha]_D^{18}$ −19° (*c* 0.5, MeOH)). This result revealed that ircinol A (**665**) was an enantiomer of the alcoholic form at C-1 of ircinal A (**660**), which has been shown to have the same absolute configuration as that of manzamine A (**655**).

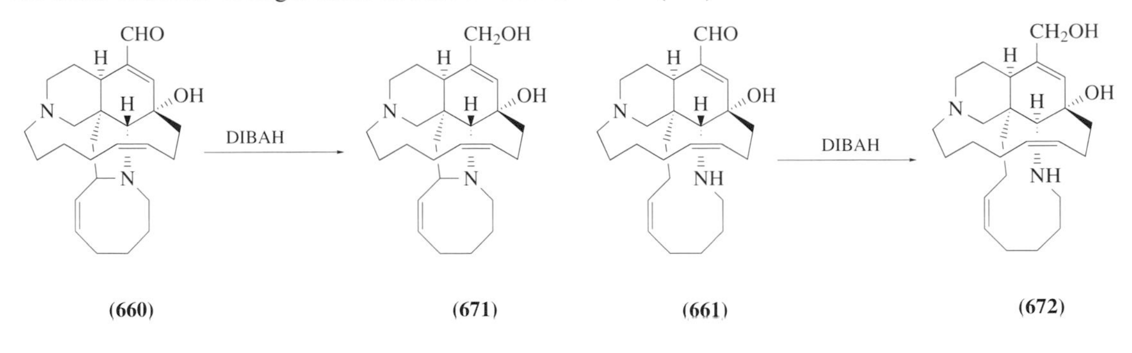

Scheme 23

Reduction of ircinal B (**661**) with DIBAH yielded a reduction product (**672**) (Scheme 23), which showed identical spectral data to those of ircinol B (**666**) but the opposite sign of optical rotation ((**672**), $[\alpha]_D^{18}$ +4.2° (*c* 0.2, MeOH); (**666**), $[\alpha]_D^{18}$ −2.8° (*c* 0.12, MeOH)). Thus the absolute stereochemistry of ircinol B was concluded to be as shown in Structure (**666**).

Manzamine L[578] (**673**) was isolated from the fractions containing manzamines D (**658**) and H (**667**). The spectral data of manzamine L (**673**) were similar to those of manzamine H (**667**), and the structure of (**673**) was elucidated to be a stereoisomer at C-1 of (**667**) on the basis of conversion of (**673**) with ddq into manzamine J (**668**). The absolute configurations at C-1 of manzamines D (**658**), H (**667**), and L (**673**) were deduced from the CD data and the molecular mechanics calculations using Macro Model version 5.0 to be *R*, *R*, and *S*, respectively. Interestingly, the 1*S*-isomer of manzamine D (**658**) has not been isolated from the *Amphimedon* or *Ircinia* sponge, but from the latter ircinals A (**660**) and B (**661**) and manzamines H (**667**) and J (**668**) were obtained. 6-Hydroxymanzamine A[580] (=manzamine Y[581] (**674**)) is the first manzamine congener with a hydroxyl group at C-6, although 8-hydroxyl analogues of manzamines such as manzamine F[582] (**659**) and 8-hydroxymanzamine A[583] (=manzamine G[584] (**675**)) have been reported previously. On the other hand, 3,4-dihydromanzamine A[580] (**676**) is the first 3,4-dihydro analogue of manzamines and is easily converted into manzamine A (**655**) by daylight.

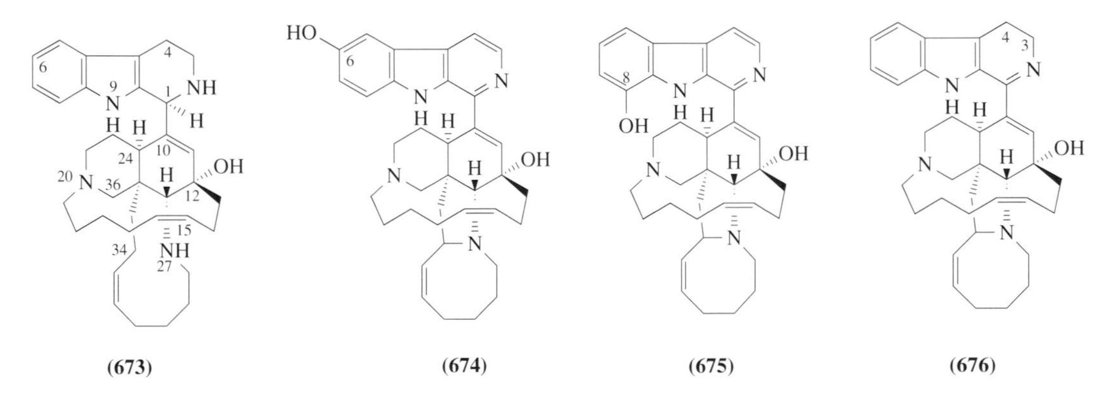

The *Amphimedon* sponge also contains several manzamine-related alkaloids with both dextrorotation and laevorotation. The minor enantiomer of keramaphidin B ((−)-(**662**)) is considered to possess the same configuration as most of the manzamine-related alkaloids with dextrorotation represented by manzamines A (**655**) and B (**656**), while the major enantiomer ((+)-(**662**)) may be correlated to those with laevorotation like ircinols A (**665**) and B (**666**). Altogether the ratio of dextrorotatory vs. laevorotatory alkaloids in this sponge was found to be about 15:1. On the other hand, the ratio of (−)- and (+)-form of keramaphidin B (**662**) was 3:97.

A plausible biogenetic path of manzamines A (**655**) and B (**656**) is shown in Scheme 24, which was elucidated on the basis of the structures and ratios of all the manzamine-related alkaloids isolated from this sponge. Baldwin and Whitehead's *bis*-3-alkyldihydropyridine (a) might be a biogenetic precursor of both enantiomers of keramaphidin B (**662**). The 2,3-iminium form of the

(−)-enantiomer ((−)-(**662**)) may be hydrolyzed to generate (+)-ircinals A (**660**) or B (**661**), which are probably converted through Pictet–Spengler cyclization with tryptamine into manzamines D (**658**), H (**667**), and L (**673**), respectively, and then dehydrogenated to manzamines A (**655**) and B (**656**), respectively. On the other hand, the (+)-enantiomer ((+)-(**662**)) may be associated with some antipodes of ircinals and manzamines such as ircinols A (**665**) and B (**666**).

Isolation of keramaphidin C (**663**) and keramamine C (**664**) together with manzamine C (**657**) and tryptamine seems to substantiate partly the biogenetic path of manzamine C (**657**), which may be derived from coupling of keramaphidin C (**663**) with tryptamine and a C_3 unit via keramamine C (**664**) (Scheme 25). On the other hand, keramaphidin C (**663**) is probably generated from a C_{10} unit and ammonia.

Many manzamines and related alkaloids have been reported from several marine sponges; manzamines E (**677**) and F (**659**) from an Okinawan sponge *Xespongia* sp.;[582] 8-hydroxymanzamine A (=manzamine G (**676**)) from an Indonesian sponge *Pachipellina* sp.;[583] xestomanzamines A (**678**) and B (**679**) and manzamines X (**680**) and Y (=6-hydroxymanzamine A) from Okinawan sponges *Xestospongia* and *Haliclona* sp.,[581] respectively; 1,2,3,4-tetrahydro-8-hydroxymanzamine A (**681**) and its 2-*N*-methyl derivative (**682**) from Papua New Guinean sponges *Petrosia* and *Cribochalina* sp.;[585] and 6-deoxymanzamine X (**683**) and the three 1-*n*-oxide derivatives (**684**)–(**686**) from a Philippine sponge *Xestospongia ashimorica*.[586] Unique *β*-carboline alkaloids, xestomanzamines A

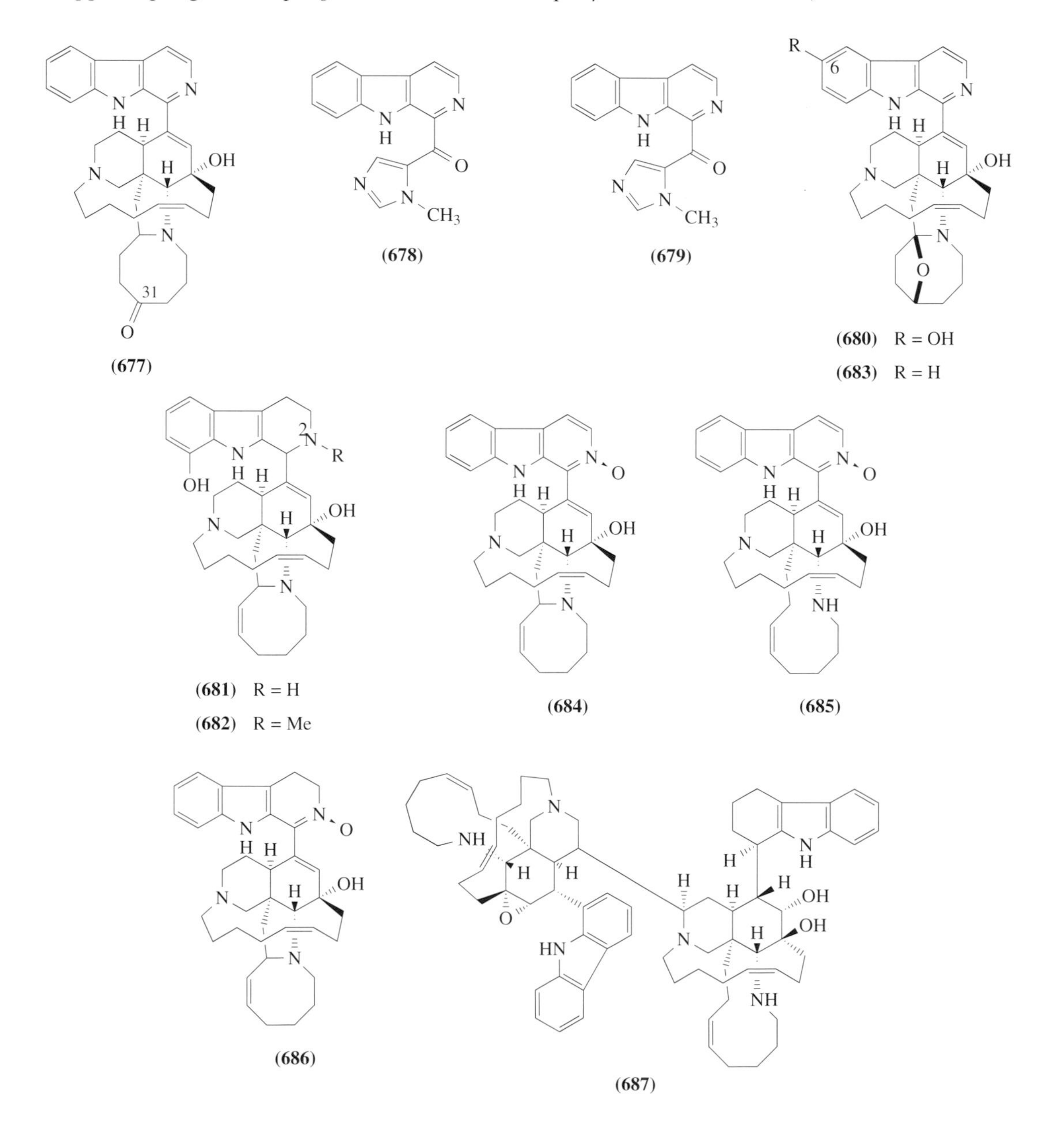

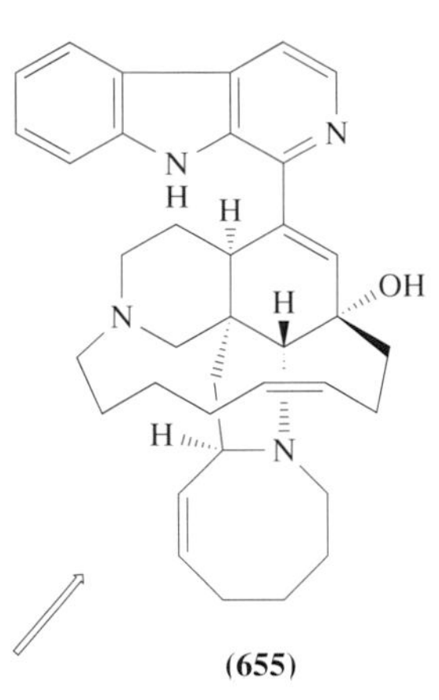

(655)

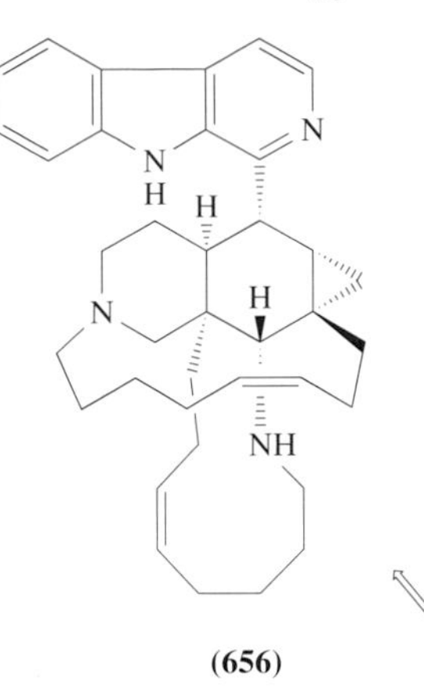

(656)

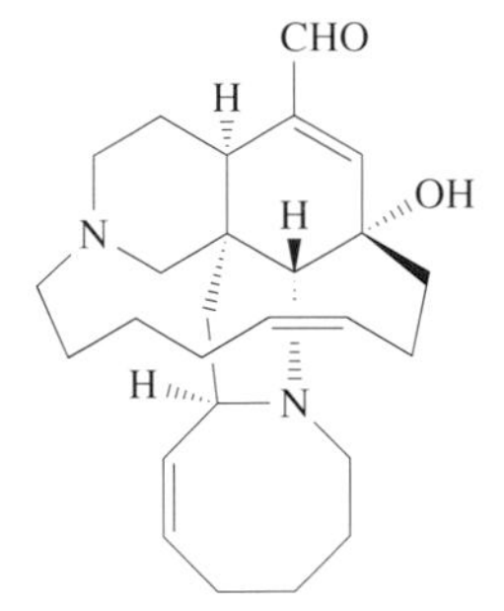

(660)

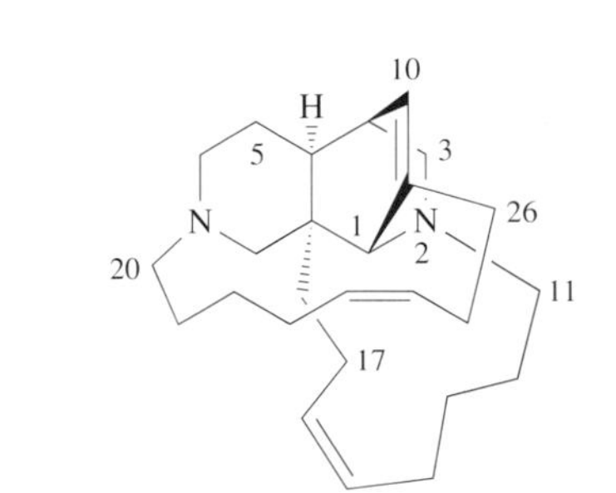

(–)-(662)

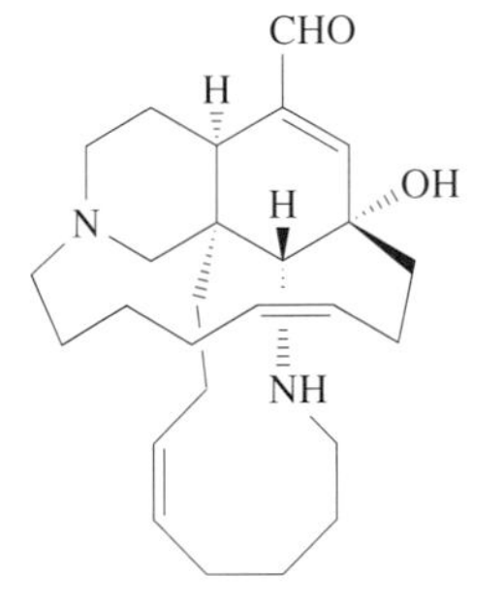

(661)

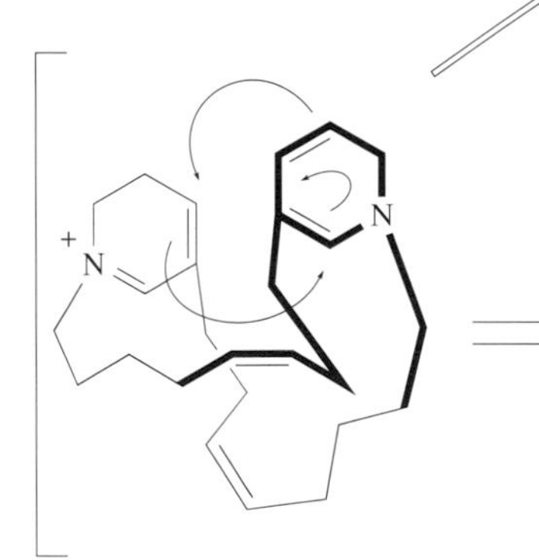

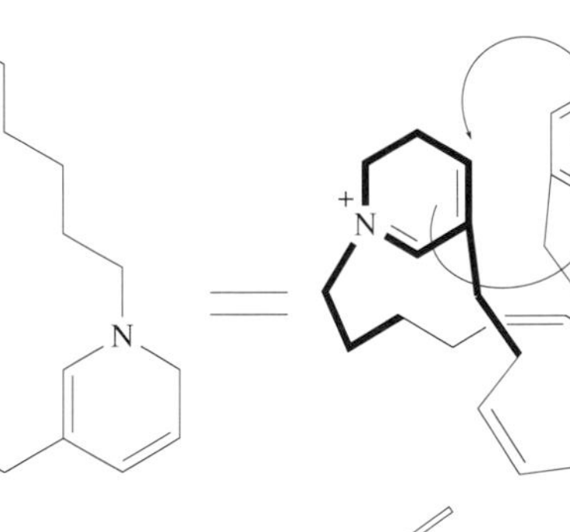

a

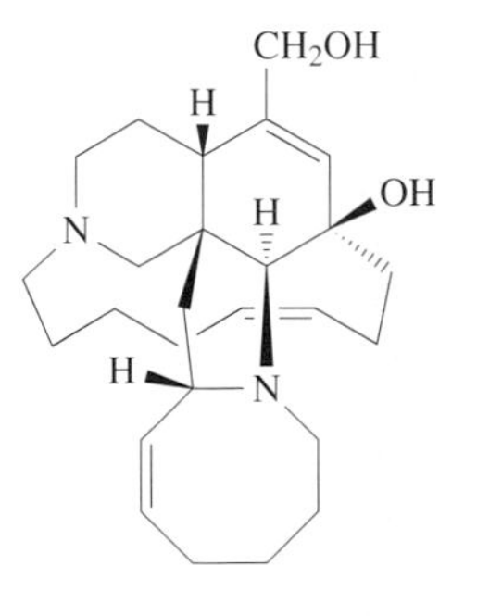

(665)

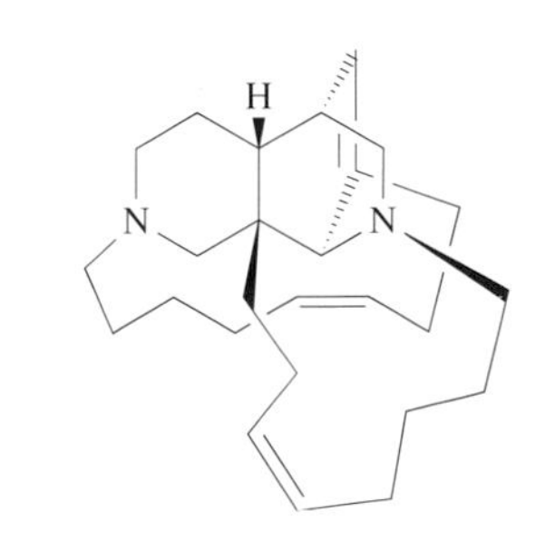

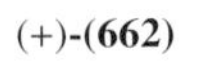

(+)-(662)

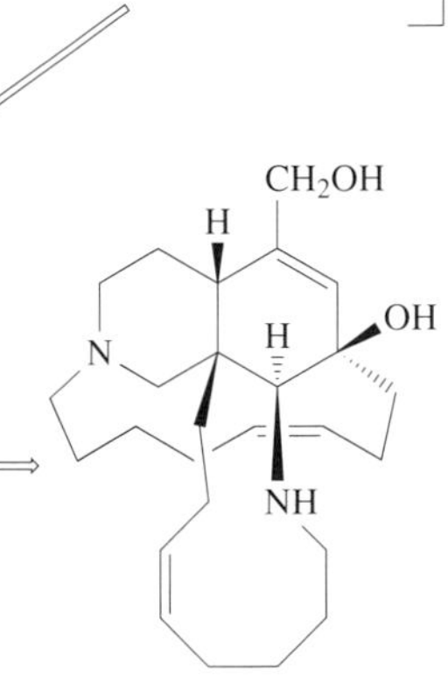

(666)

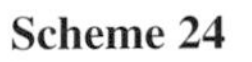

Scheme 24

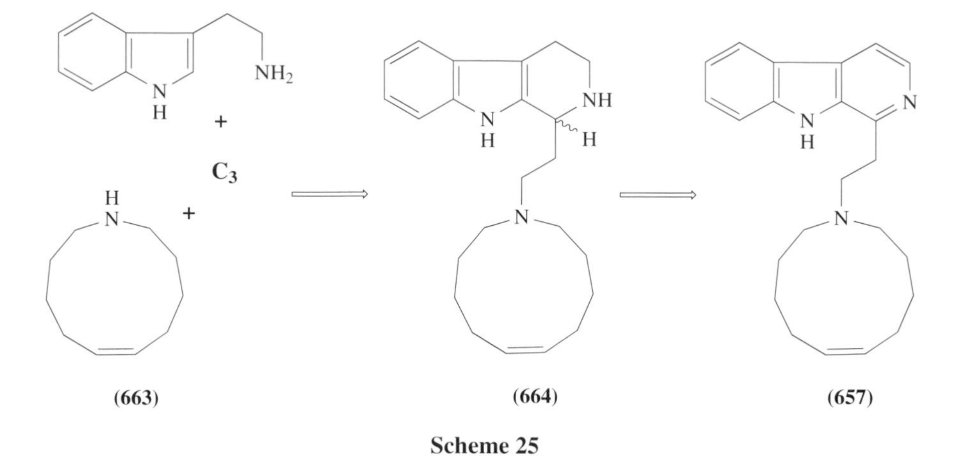

Scheme 25

(**678**) and B (**679**) may be biogenetically derived from *N*-methylhistidine and tryptamine. More recently, Scheuer and co-workers isolated the first manzamine dimer, named kauluamine (**687**), from an Indonesian sponge *Prianos* sp.[587]

Several alkaloids similar to keramaphidin B (**662**) were reported from two groups independently just before or after isolation of (**662**). Andersen and co-workers isolated eight pentacyclic alkaloids, named ingenamine (**688**),[588] ingamines A (**689**) and B (**690**),[589] and ingenamines B–F (**691**)–(**695**),[590] from the sponge *Xestospongia ingens* collected in Papua New Guinea. Crews and co-workers reported two new alkaloids, xestocyclamines A (**696**)[591] and B (**697**),[592] from a Papua New Guinean sponge *Xestospongia* sp. Ingenamine (**688**) corresponds to the 5-hydroxyl form of keramaphidin B (**662**). The skeletons of ingamines A (**689**) and B (**690**) differ from those of ingenamine (**688**) and keramaphidin B (**662**) in having a C_{12} alkyl chain between N-7 and C-9 in place of the C_8 alkyl chain in (**688**) and (**662**). Ingenamines B (**691**), C (**692**), and D (**693**) have a C_9 or C_{10} alkyl chain between N-7 and C-9, while ingenamines E (**694**) and F (**695**) possess a twelve-carbon alkyl chain between N-2 and C-8a. If keramaphidin B (**662**) and ingenamine (**688**) are generated from a pair of symmetrical dialdehydes (C_{10} unit), C_3 units, and ammonia as Baldwin suggested,[573] it could be that the macrocyclic rings of ingamines A (**689**) and B (**690**) and ingenamines B–F (**691**)–(**695**) are derived from a C_{11}, C_{12}, or C_{14} dialdehyde unit in addition to a C_{10} dialdehyde unit. The absolute configurations of ingenamine (**688**), ingamine A (**689**), and ingenamine E (**694**), which were determined by application of a modified Mosher method at the hydroxyl group on C-6,[590] were the same as those of keramaphidin B (**662**). Xestocyclamine A (**696**) was revised to be the Δ_{14} isomer of (**688**) by reassignment of its 2D NMR data, although the initial structure of xestocyclamine A was elucidated to be (**698**). The structure of xestocyclamine B (**697**) corresponds to the 15,16-dihydro form of ingenamine D (**693**).

Some alkaloids from marine sponges, which are reminiscent of Baldwin's bisdihydropyridine macrocycle (a), have been reported. In 1989 Fusetani *et al.* isolated new bistetrahydropyridine alkaloids, haliclamines A (**699**) and B (**700**) as cytotoxic constituents from a sponge *Haliclona* sp. collected from the Uwa Sea, Japan.[593] The structures consisted of two tetrahydropyridines linked through C_9 and C_{12} alkyl chains. Cyclostellettamines A–F[594] (**701**)–(**706**), isolated from the hydrophilic extracts of the sponge *Stelletta maxima* collected off the Sata Peninsula, Shikoku, Japan, were macrocyclic bispyridines linked through C_{12}–C_{14} alkyl chains.

Unique bis-quinolizidine alkaloids, petrosin (**707**)[595] and petrosins A (**708**)[596,597] and B (**709**)[596] were isolated from the Papua New Guinean sponge *Petrosia serita*, and the C_2-symmetrical structure of petrosin (**707**) was established by X-ray analysis, while petrosin A (**708**) was elucidated as mesomeric.[597] In 1984 Nakagawa *et al.* reported isolation of novel macrocyclic 1-oxa-quinolizidine alkaloids, xestospongins A–D (**710**)–(**713**), from the Australian sponge *Xestospongia exigua*, and the structures were established on the basis of X-ray analysis of xestospongin C (**712**).[598] In 1989 Kitagawa and co-workers isolated 10 new alkaloids, araguspongines B–H and J[599] (**714**)–(**721**) and aragupetrosine A (**722**),[600] together with petrosin (**709**) and petrosin A (**710**) from an Okinawan sponge *Xestospongia* sp. Araguspongines are bis-1-oxa-quinolizidine alkaloids, while aragupetrosine A (**721**), having a quinolizidine and a 1-oxa-quinolizidine ring is a hybrid of petrosin (**707**) and araguspongin F (**717**). Araguspongines F (**718**), G (**719**), H (**720**), and J (**721**) were obtained as optically pure compounds, while araguspongines B (**714**), D (**716**), and E (**717**) were isolated as enantiomeric mixtures or mesomeric compounds. In 1986 an interesting biogenetic path for petrosin

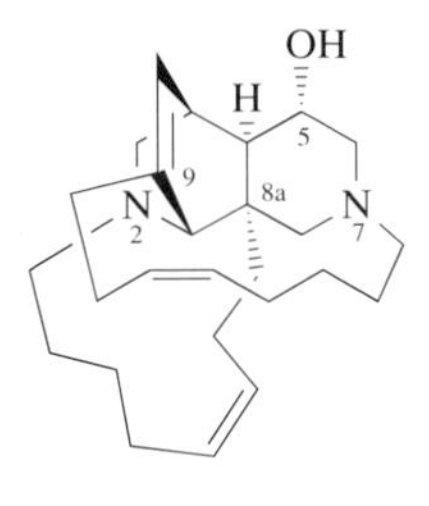

(688)

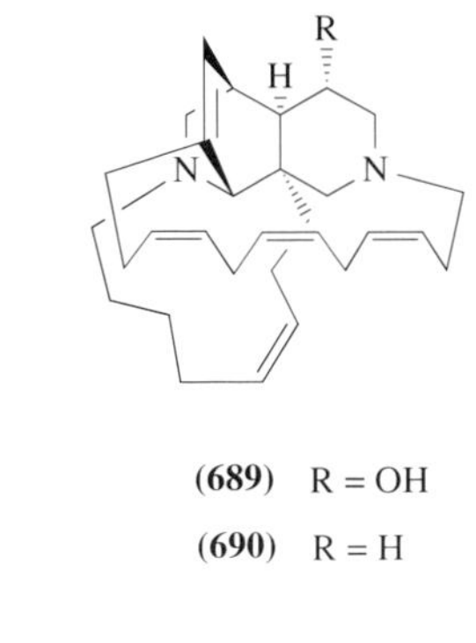

(689) R = OH

(690) R = H

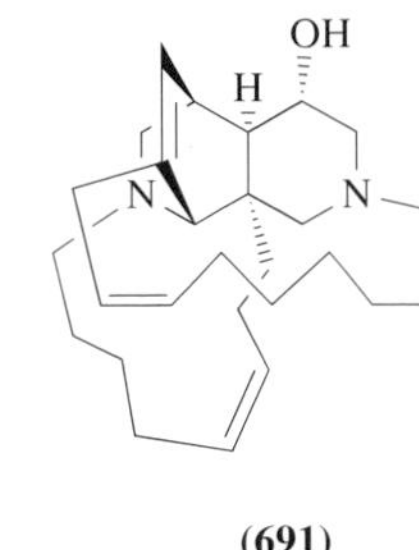

(691)

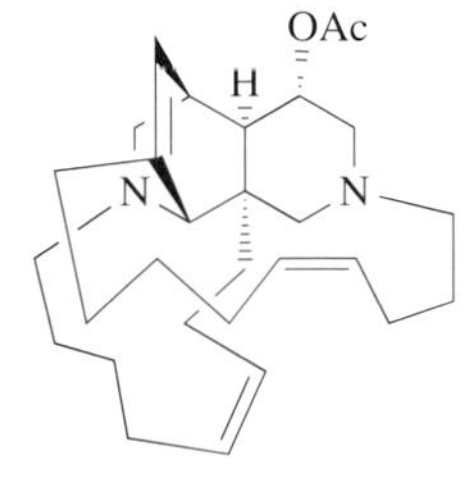

(692)

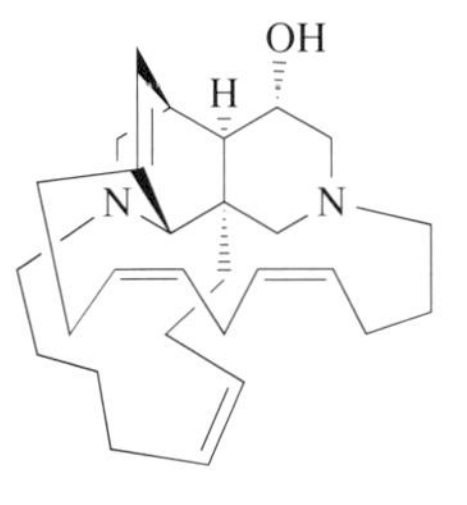

(693)

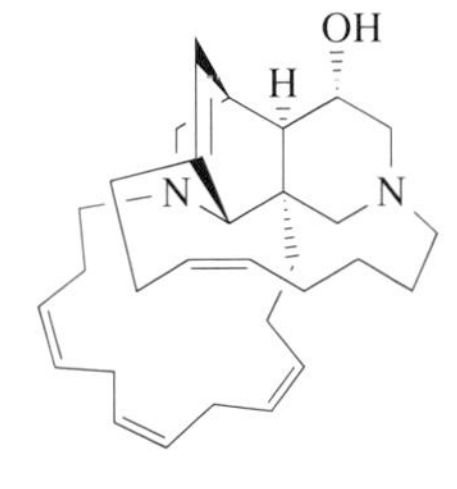

(694) R = OH

(695) R = H

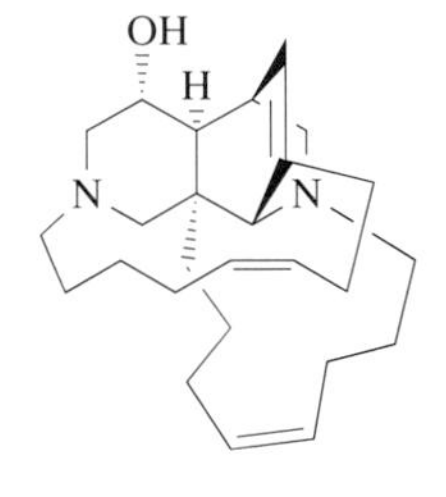

(696)

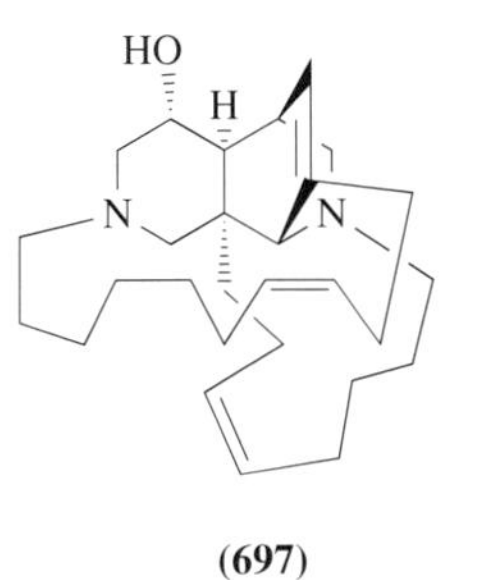

(697)

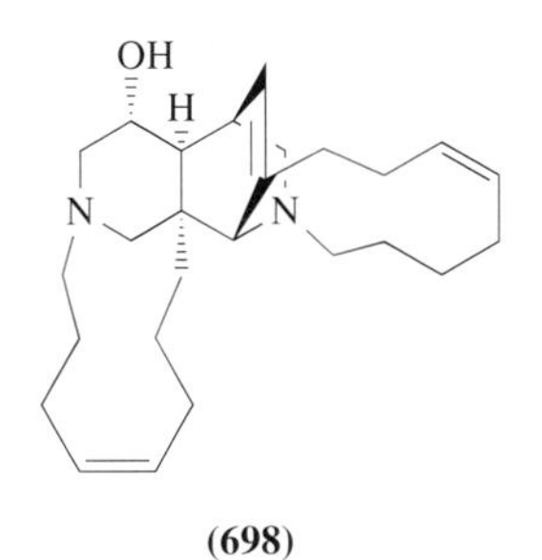

(698)

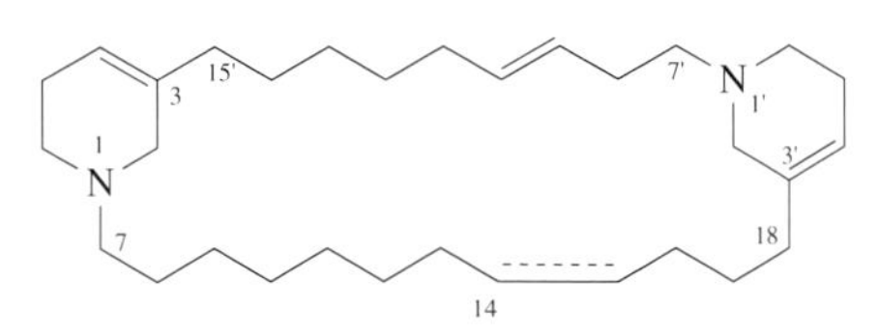

(699)

(700) :Δ^{14}

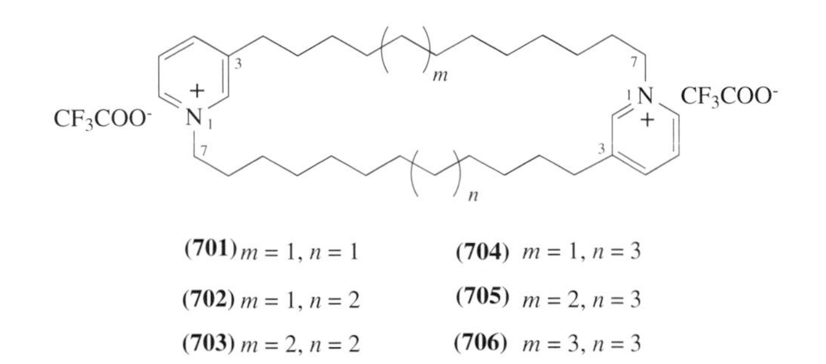

(701) $m = 1, n = 1$

(702) $m = 1, n = 2$

(703) $m = 2, n = 2$

(704) $m = 1, n = 3$

(705) $m = 2, n = 3$

(706) $m = 3, n = 3$

(707), petrosin A **(708)**, and xestospongin A–D **(710)**–**(713)** was proposed by Cimino *et al.*,[601] although those alkaloids were isolated from different genera of marine sponges. On the other hand, Kitagawa *et al.* indicated the biogenetic relationship between bisquinolizidine alkaloids and bis-1-oxa-quinolizidine alkaloids.[599,600] The biogenetic path is shown in Scheme 26; dimerization of two C_9-C_5N units (d) gives the macrocycle (e) followed by oxidation of C-12 and C-12′ to afford the diketone compound (f). Araguspongines B–H and J **(714)**–**(720)** may be generated from (e) via pathway A. On the other hand, aragupetrosine A **(722)** and petrosins **(707)** and **(708)** were probably generated via pathway B and C, respectively. Both precursors, (a) in Schemes 20 and 24 and (e) in Scheme 26 are similar to haliclamines A **(699)** and B **(700)** and cyclostellettamines A–F **(701)**–**(706)**. Demethylxestospongin B **(723)** has been isolated together with xestospongins B **(708)** and D **(710)** and araguspongin F **(718)** from a New Caledonian sponge *Xestospongia* sp.[602]

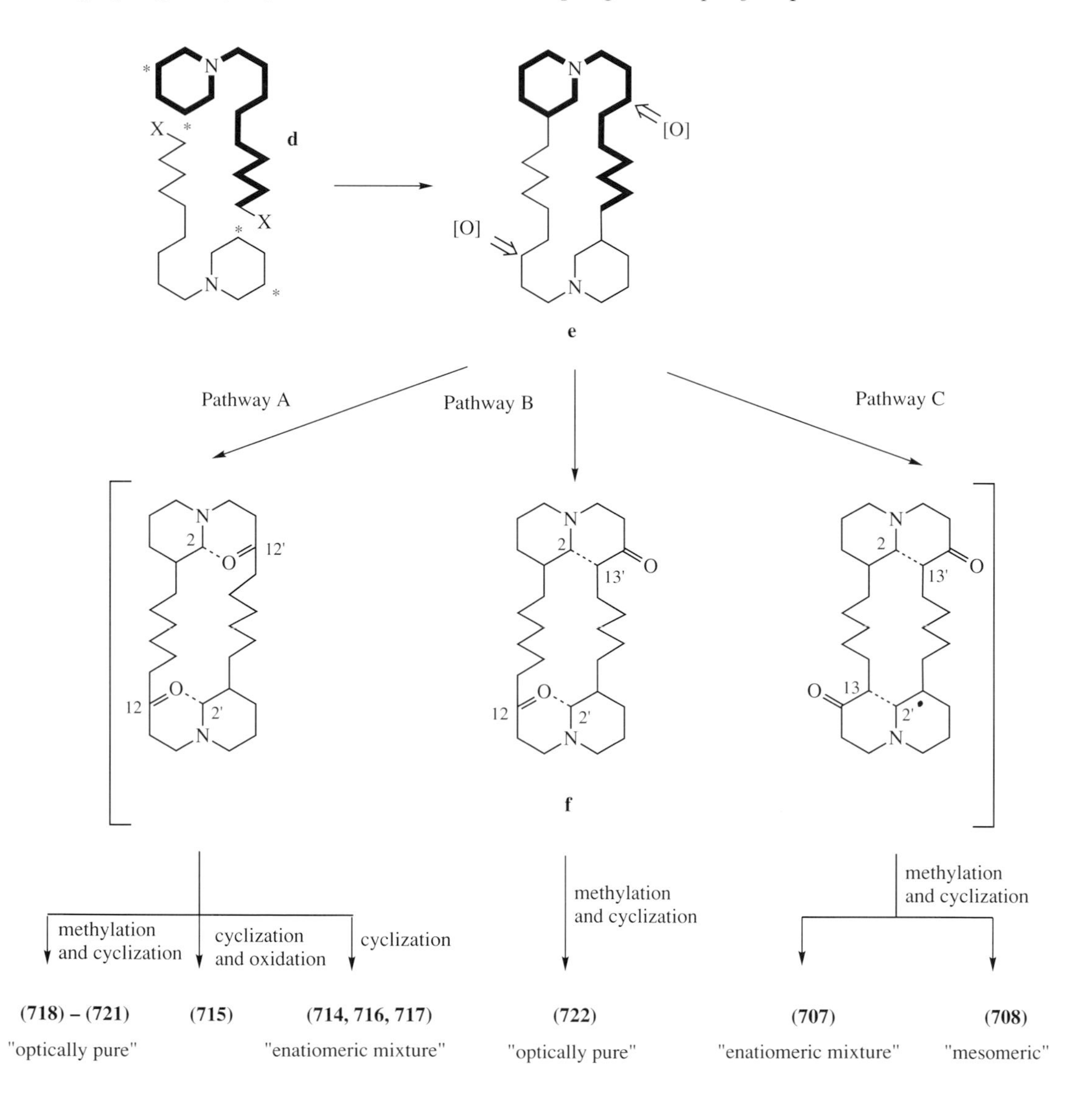

Scheme 26

The metabolic pattern of the Mediterranean sponge *Reniera sarai* is characterized by the presence of a series of unique complex polycyclic alkaloids. Cimino *et al.* reported the isolation of unprecedented alkaloids, sarains 1–3 **(724)**–**(726)**, isosarains 1–3 **(727)**–**(729)**, and saraines A–C **(730)**–**(732)**.[601,603–608] Sarains 1–3 **(724)**–**(726)**[601,608] possessed a *trans*-quinolizidine moiety linked to an unsaturated pyperidine ring directly and by two linear alkyl chains, while isosarains 1–3 **(727)**–**(729)**,[604,606,608] obtained as minor constituents of the sponge, were isomers at C-1, C-2, and C-9 of sarains 1–3 **(724)**–**(726)**, respectively. Saraine A **(730)**, the structure of which was assigned on the

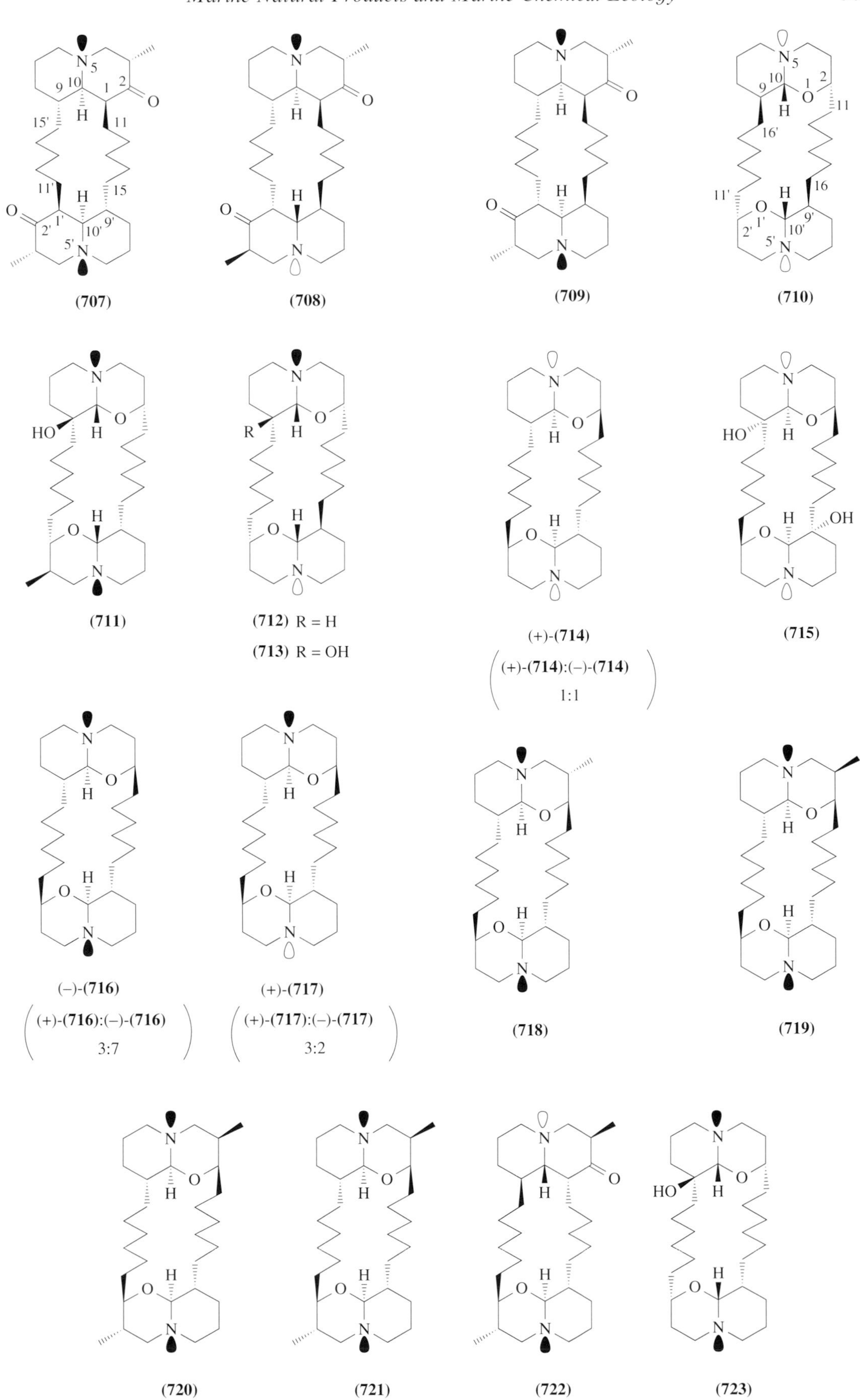
(707)
(708)
(709)
(710)
(711)
(712) R = H
(713) R = OH
(+)-(714)
(+)-(714):(–)-(714)
1:1
(715)
(–)-(716)
(+)-(716):(–)-(716)
3:7
(+)-(717)
(+)-(717):(–)-(717)
3:2
(718)
(719)
(720)
(721)
(722)
(723)

basis of combination of X-ray study of the acetyl derivative of **(730)**[603] and detailed spectral studies of **(730)** itself,[605] had a unique pentacyclic core with two macrocyclic rings. More recently structures of saraines B **(731)** and C **(732)** and the absolute stereochemistry of **(730)**–**(732)** were determined.[607] The biogenetic path of sarain-related alkaloids **(724)**–**(732)** proposed by Cimino *et al.* is analogous to Baldwin's proposal for manzamines (Scheme 21). The retro-biosynthesis of sarains 1–3 **(724)**–**(726)** and isosarains 1–3 **(727)**–**(729)** (Scheme 27(a)) proceeds through a partially reduced *bis*-3-alkylpyridine macrocycle which contains 10 carbons in an alkyl chain and 10, 11, or 12 carbons in another alkyl chain (χ). These macrocycles form sarains and isosarains through some intramolecular reaction.[608] On the other hand, macrocycles for saraines A–C **(730)**–**(732)** contain 12 carbons in an alkyl chain and 10, 11, or 12 carbons in another alkyl chain (χ) (Scheme 27(b)).[607]

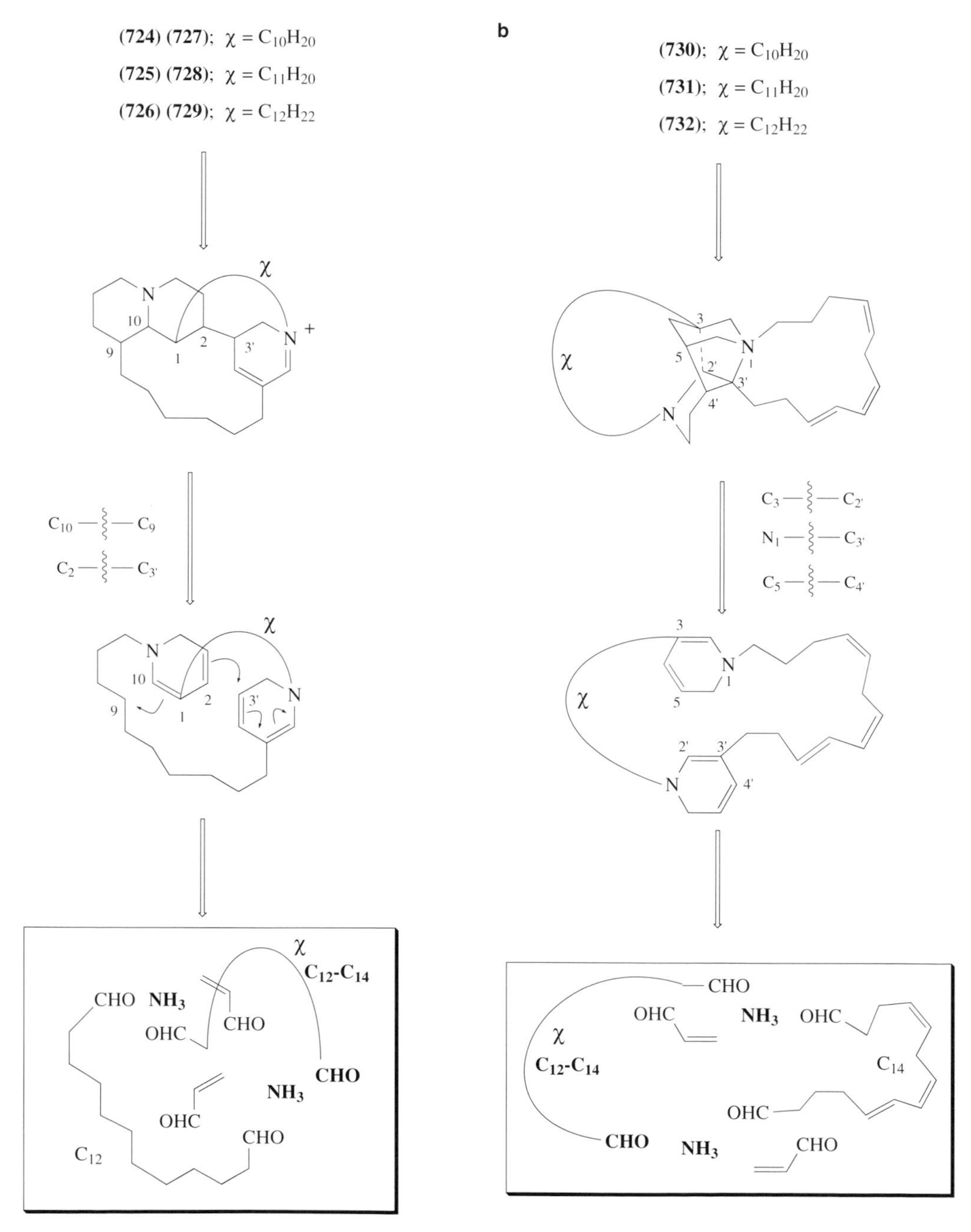

Scheme 27

Papuamine[609] **(733)** and haliclonadiamine[610] **(734)** were isolated from *Haliclona* sponges collected off Papua New Guinea and Palau, respectively. The structure of **(733)** was elucidated mainly by

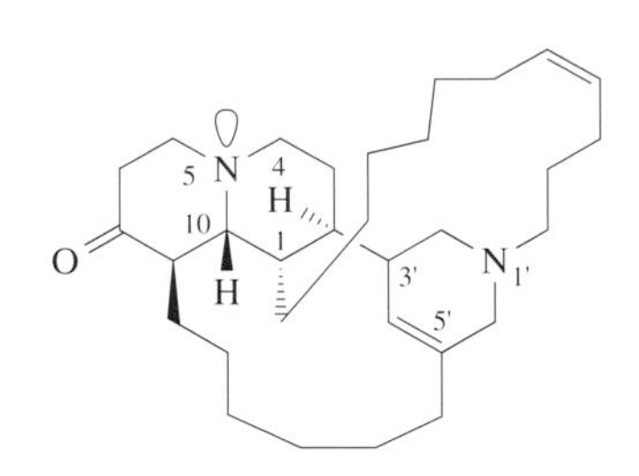

(724)

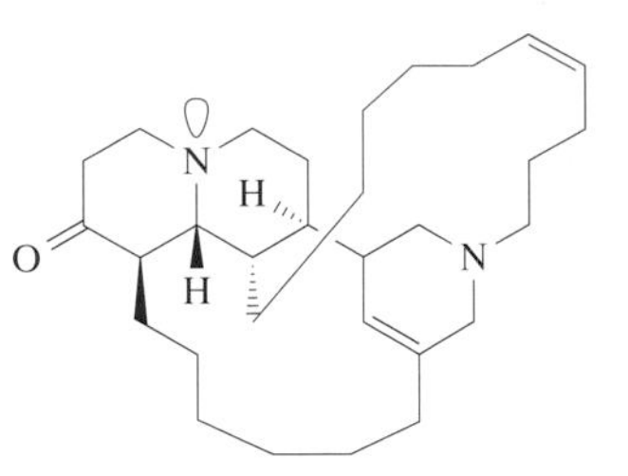

(725)

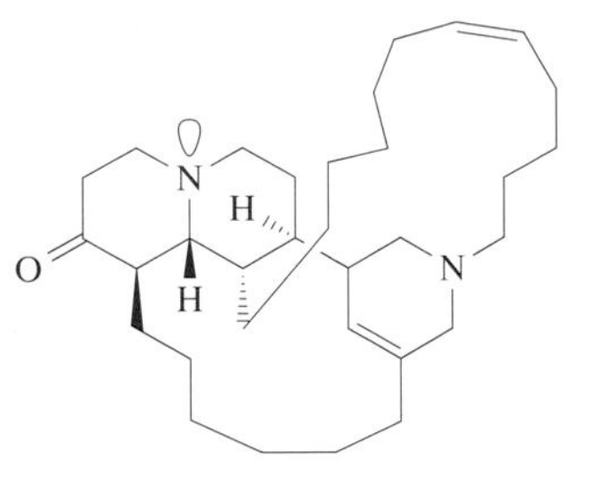

(726)

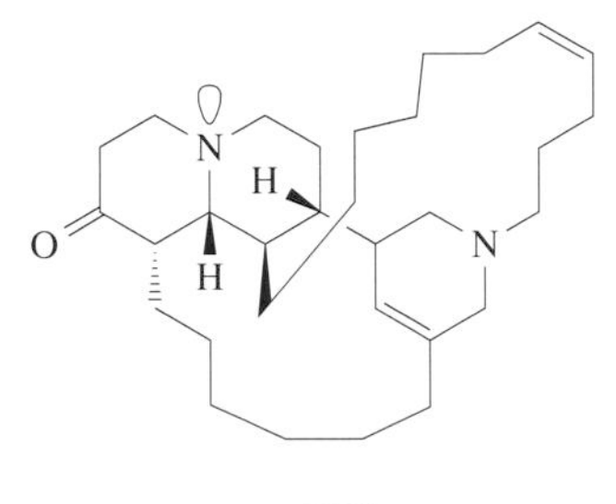

(727)

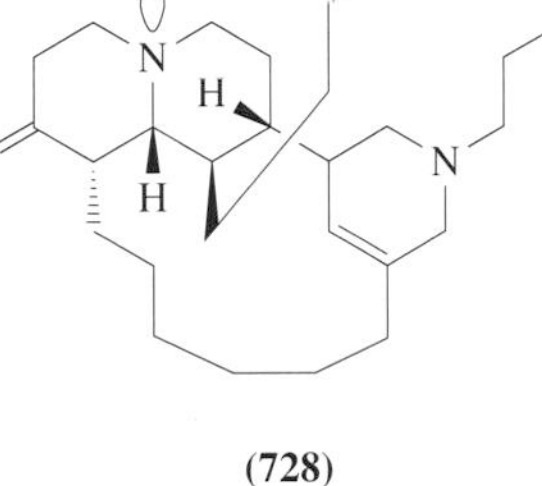

(728)

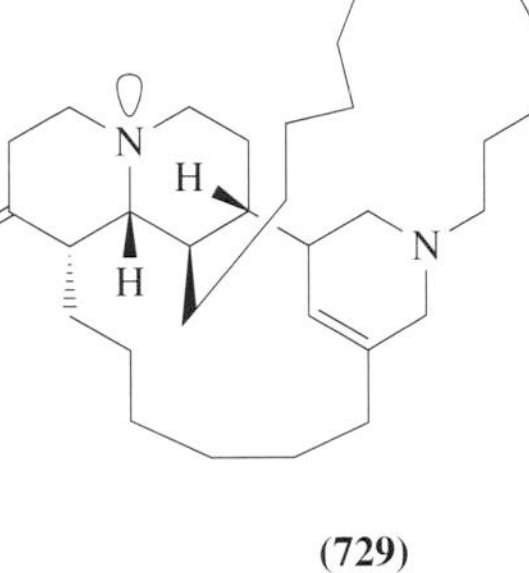

(729)

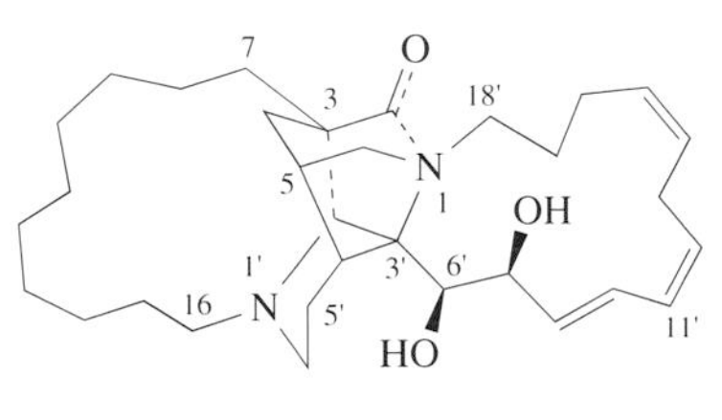

(730)

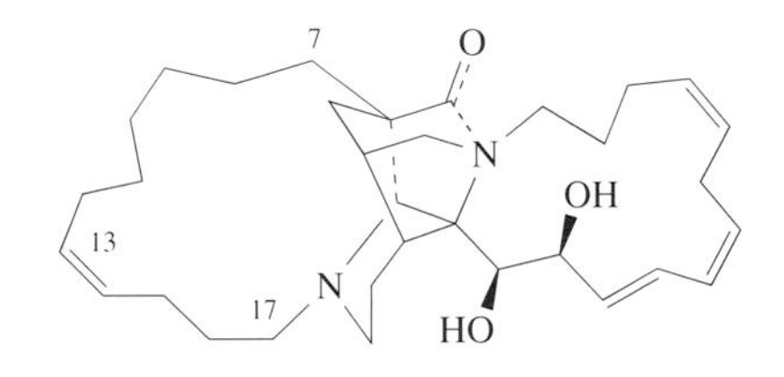

(731)

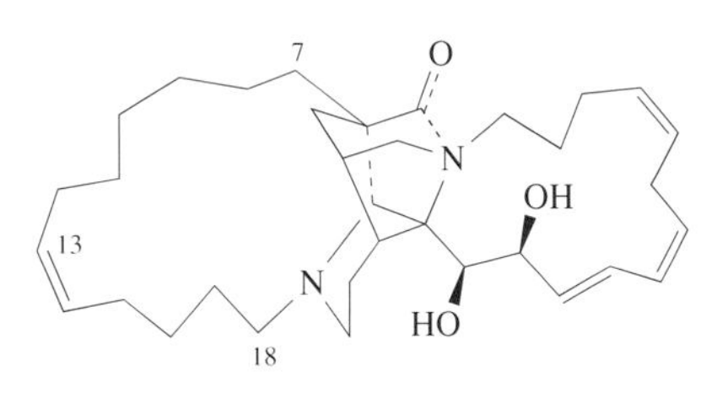

(732)

NMR data including INADEQUATE studies, while the structure of (**734**) was established by X-ray analysis of its diacetyl derivative. Haliclonadiamine (**734**) is an unsymmetrical diastereomer of papuamine (**733**). Crews *et al.* proposed the biogenetical and taxonomic consideration of marine diamine-containing alkaloids, in which they suggested that papuamine (**733**) and haliclonadiamine (**734**) may be formed by condensation of ammonia, acrolein, and acyclic aldehyde, which are Baldwin's biosynthetic building blocks of manzamines.[585]

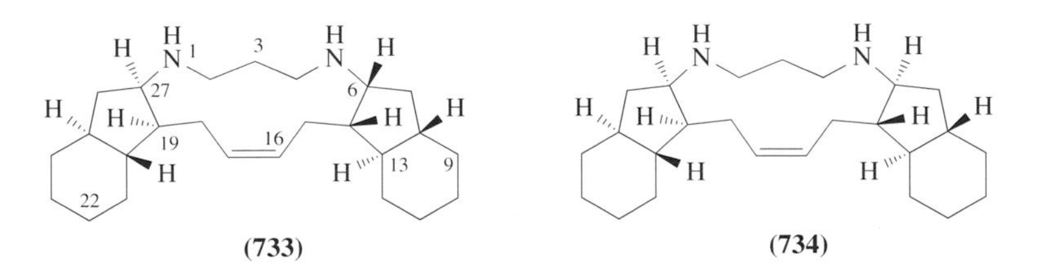

(733) **(734)**

An example of a novel class of pentacyclic alkaloid, madangamine A[611] (**735**), was isolated by Andersen and co-workers from the Papua New Guinean sponge *Xestospongia ingens*, from which ingenamine (**688**) and ingamines A (**689**) and B (**690**) were also isolated. The proposed biogenesis for madangamine A (**735**) is outlined in Scheme 28, which partly resembles Baldwin's proposal for the biogenesis of manzamines. The ingenamine-type intermediate [= ingenamine F (**694**)], which may be generated from a bis-3-alkylpyridine macrocycle (i) similar to (a) (Scheme 20) through a "4+2" cycloaddition reaction, can undergo rearrangement to generate the madangamine A skeleton.

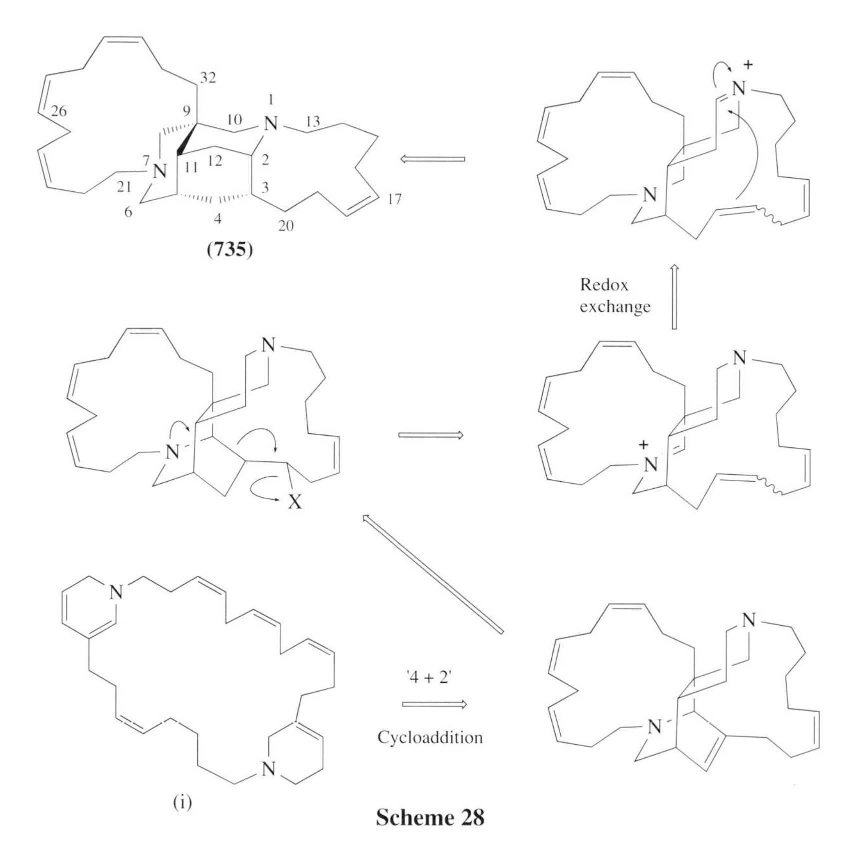

Scheme 28

Crews and co-workers isolated a novel tetracyclic diamine alkaloid, named halicyclamine A (**736**), from an Indonesian sponge *Haliclona* sp., and elucidated the structure on the basis of spectroscopic data.[612] More recently isolation of haliclonacyclamines A (**737**) and B (**738**) from a *Haliclona* sponge collected off the Great Barrier Reef was reported by Garson and co-workers, and the structure of (**739**) was assigned by X-ray analysis.[613] The relative stereochemistry at C-2 in halicyclamine A (**736**) was different from those of haliclonacyclamines A (**737**) and B (**738**). Crews and co-workers also reported the isolation of halicyclamine B (**739**) from an Indonesian *Xestospongia* sponge and its structural determination based on X-ray analysis.[614] The biogenetic path of these tetracyclic alkaloids is proposed by Garson and co-workers (Scheme 29); cleavage of the C-1—C-8a bond of Baldwin's pentacyclic intermediate (b) analogue followed by reduction may give halicyclamine A (**738**), while Crews and co-workers proposed that halicyclamine B (**739**) was probably derived from a homologue of (**736**) via 1,3-sigmatropic shift.

The taxonomy of sponge materials containing manzamine-related alkaloids is shown in Table 22. This result indicates that all the alkaloids described in this review are obtained mainly from sponges belonging to the two orders Haplosclerida and Nepheliospongida except for sponges of genera *Ircinia*, *Prianos*, and *Stelletta*. In particular, the *Xestospongia* and *Haliclona* sponges are rich sources of manzamine-related alkaloids.

Manzamines and related alkaloids are very interesting marine sponge-derived metabolites possessing unusual ring systems which are challenging targets for total synthesis and for unprecedented biogenesis studies. Further progress in extensive studies on isolation, structure elucidation, biogenesis, total synthesis, and bioactivity is expected for these miracle compounds.

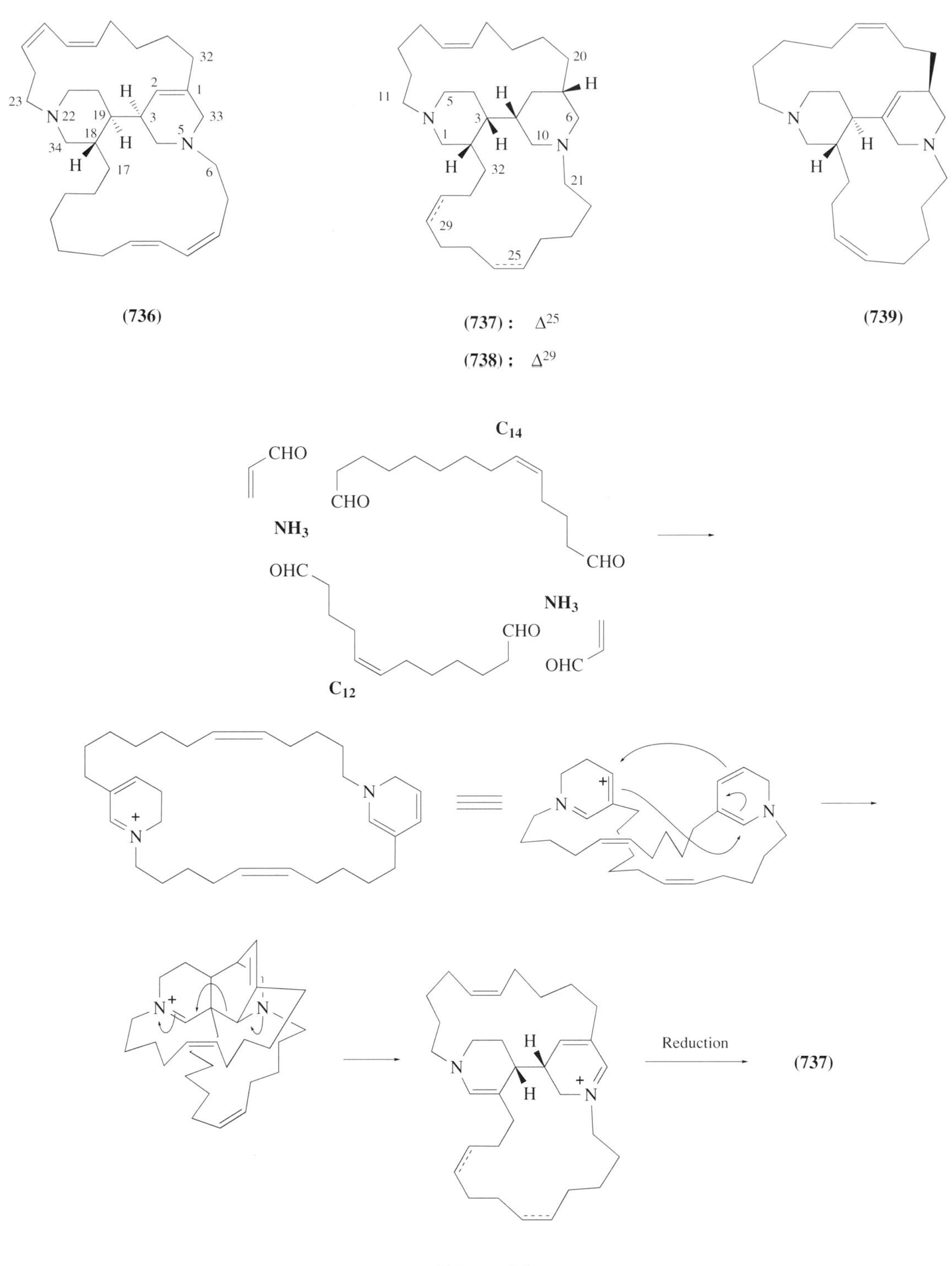

Scheme 29

8.07.9.3.2 *Metabolites of the genus* Theonella

Marine sponges of the genus *Theonella* frequently afford a variety of interesting secondary metabolites including unusual cyclic peptides[615] as well as polyoxygenated aliphatic compounds,[616] most of which exhibit significant biological activity. This section deals with several natural products isolated from marine sponges of the genus *Theonella*.

Table 22 Taxonomy of sponges containing manzamine-related alkaloids.

Order	*Family*	*Genus*	*Compounds*	*Ref.*
Subclass, Tetractinomorpha				
Chroristida	Stelletidae	*Stelletta*	**(701)**–**(706)**	593
Subclass, Ceractinomorpha				
Dictyoceratida	Thorectidae	*Ircinia*	**(655)**, **(656)**, **(658)**–**(660)**, **(666)**, **(667)**	574
Nepheliospongida	Oceanapiidae	*Pellina*	**(655)**, **(656)**	572
		Pachypellina	**(655)**, **(675)**	583
Nepheliospongiidae	*Petrosia*		**(707)**, **(708)**, **(709)**	595, 596
			(681), **(682)**	585
		Xestospongia	**(659)**, **(675)**, **(677)**	582, 584
			(655), **(675)**, **(677)**–**(680)**	581
			(710)–**(713)**	598
			(707), **(708)**, **(714)**–**(722)**	599, 600
			(706), **(708)**, **(716)**, **(723)**	601
			(696), **(697)**	591, 592
			(662), **(688)**–**(695)**, **(735)**	588–590, 611
			(739)	612
Haplosclerida	Niphatidae	*Cribochalina*	**(682)**	586
		Amphimedon	**(655)**–**(658)**, **(660)**–**(667)**, **(673)**–**(676)**	575–578, 580
	Chalinidae	*Haliclona*	**(655)**–**(658)**	569–571
			(699), **(700)**	593
			(655)–**(657)**, **(674)**	581
			(733)	609
			(733), **(734)**	610
			(736)	612
			(737), **(738)**	613
		Reneira	**(724)**–**(732)**	602–604, 607, 608
Poecilosclerida	Desmacidonidae	*Prianos*	**(687)**	587

Fusetani and his research group, during their screening program of Japanese marine invertebrates for potential biomedicals, found that a marine sponge of the genus *Theonella* inhibited various proteinases, particularly thrombin. The sponge was collected off Hachijo-jima Island, 300 km south of Tokyo, and was characterized by a brilliant yellow inner body which was different from that of another *Theonella* sponge containing theonellamides.[617] Two active substances, named cyclotheonamide A (**740**) and B (**741**) were isolated.[618] Cyclotheonamide A (**740**) was isolated from the 1987 collection by means of solvent partitioning, gel filtration, and three successive reverse-phase HPLCs, whereas cyclotheonamide B (**741**) was isolated from the 1989 collection. ^{1}H and ^{13}C NMR spectra indicated that cyclotheonamide A (**740**) contained proline, phenylalanine, 2,3-diaminoproprionic acid (Dpr), and two unusual amino acid residues: α-ketoarginine (K-Arg (**742**)) was an analogue of arginine with a carbonyl group between the carboxyl and methine carbons; the other one, V-Tyr (**743**), was a tyrosine analogue with an ethylene unit between the carboxyl and methine carbons. The α-amino group of Dpr was formylated. The sequence of the five residues was deduced by a combination of HMBC and NOESY experiments. Cyclotheonamide B (**741**) has an acetamide instead of the formylamide in (**740**). The stereochemistry of the amino acids except (**742**) was deduced by chiral GC analysis after degradation. Cyclotheonamide A (**740**) inhibited thrombin with IC_{50} 0.076 $\mu g\ ml^{-1}$. The stereochemistry of V-Tyr residue (**743**) was reassigned as having *S*-configuration during the study of total synthesis of (**741**) by Hagihara and Schreiber.[619] The K-Arg residue (**742**) was also revealed to possess *S*-configuration; this site was not defined in the original studies.

Further investigation of the extract of the marine sponges *Theonella swinhoei* led to isolation of three other analogues, cyclotheonamides C (**744**), D (**745**), and E (**746**).[620] Compounds (**744**) and (**745**) were isolated from *T. swinhoei* with a bright yellow interior from which (**740**) and (**741**) were obtained previously, while (**746**) was isolated from *T. swinhoei* with a white interior. Cyclotheonamides C (**744**), D (**745**), and E (**746**) inhibited thrombin with IC_{50} of 8.4, 8.2, and 28 nM, respectively, while they were inhibitory against trypsin with IC_{50} of 7.4, 63, and 370 mM, respectively. Cyclotheonamide C (**744**), in which the vinylogous tyrosine unit was replaced by a dehydrovinylogous tyrosine unit, showed thrombin-inhibiting activity comparable to that of cyclotheonamide A (**740**). Cyclotheonamide E (**746**), which included D-Ile and phenylacetyl-Ala amide

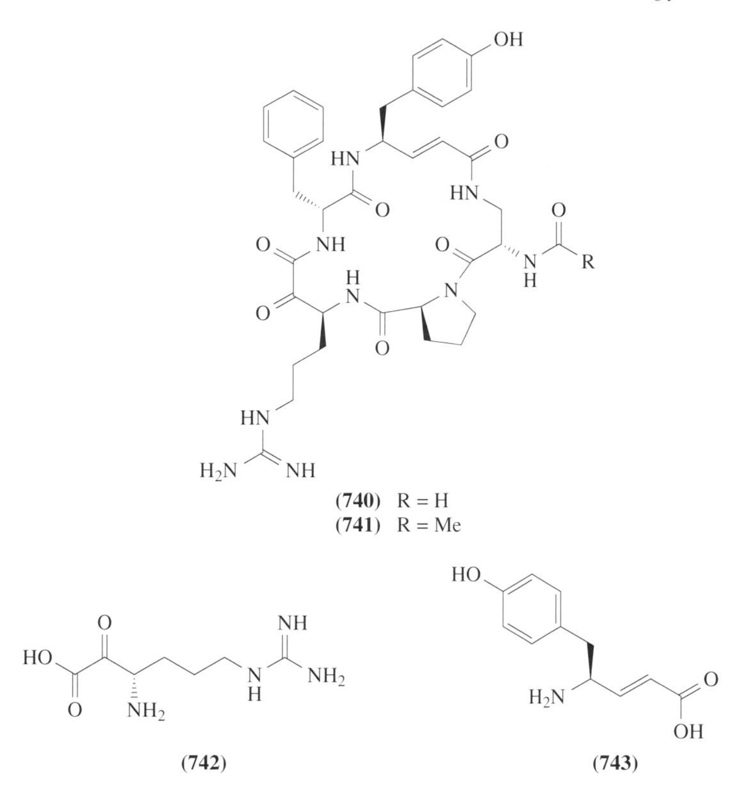

(740) R = H
(741) R = Me

(742) **(743)**

groups in place of D-Phe and formyl groups in **(740)**, was 20-fold less active. Interestingly, cyclotheonamide A **(740)** was not detected in the white variety of *T. swinhoei*. Cyclotheonamide B **(741)** was present only in the 1990 collection of the yellow variety and was never encountered thereafter. These observations may indicate the involvement of microbial symbionts in the synthesis of cyclotheonamides.

The unique structural features of cyclotheonamides in combination with their serine-protease inhibiting properties have motivated many researchers to study their total synthesis and mechanisms of actions. Total syntheses of cyclotheonamides have been accomplished by the research groups of Schreiber,[619] Maryanoff,[621] Wipf,[622] Shioiri,[623,624] and Ottenheijm.[625] Maryanoff *et al.* studied X-ray crystallography of a complex of cyclotheonamide A **(740)** with human α-thrombin, a protease central to the bioregulation of thrombosis and hemostasis.[621] This work (2.3 Å resolution) confirmed the structure of cyclotheonamide A **(740)** and revealed intimate details about its molecular recognition within the enzyme active site. Interactions due to the "Pro-Arg motif" (Arg occupancy of the S_1 specificity pocket; formation of a hydrogen-bonded two-strand antiparallel β-sheet with Ser^{214}-Gly^{216}) and the α-keto amide group of **(740)** were primarily responsible for binding to thrombin, with the α-keto amide serving as a transition-state analogue. A special interaction with the "insertion loop" of thrombin (Tyr^{60A}-Thr^{60I}) was manifested through engagement of the hydroxyphenyl group of **(740)** with Trp^{60D} as part of an "aromatic stacking chain." Biochemical inhibition data (K_i values at 37 °C) were obtained for **(740)** with thrombin and a diverse collection of serine proteases. Thus, **(740)** was just a moderate inhibitor of human α-thrombin ($K_i = 0.18\ \mu\ M^{-1}$) but a potent inhibitor of trypsin ($K_i = 0.023\ \mu M$) and streptokinase ($K_i = 0.035\ \mu M$). Maryanoff *et al.* also achieved total synthesis of cyclotheonamide A **(740)** and prepared several analogues **(747)**–**(751)** of **(740)** by a convergent synthetic protocol[626] in which a late-stage primary amine group was available for substitution. The ^{13}C NMR spectrum of (**740**) in D_2O showed virtually exclusive population by the hydrated form of the α-keto amide (*gem*-diol structure). They also prepared cyclotheonamide B **(741)** through an analogous transformation. The analogues **(747)**–**(751)**, as well as **(740)** and **(741)**, were examined for their ability to inhibit the serine protease α-thrombin, in comparison with suitable reference standards. They characterized Michaelis–Menten and slow-binding kinetics for the cyclotheonamide derivatives. An attempt was made to utilize the unoccupied hydrophobic S_3 subsite of thrombin via analogues **(747)** and **(748)**. Also, removal of the hydroxyphenyl group, which was thought to be involved in an aromatic stacking interaction with Trp^{60D} of thrombin, was explored via analogue **(751)**. The importance of the α-keto and alkene groups was examined via analogues (**749**) and (**750**), respectively. The relation of structure and function with the analogues proved to be less predictable than anticipated.[627]

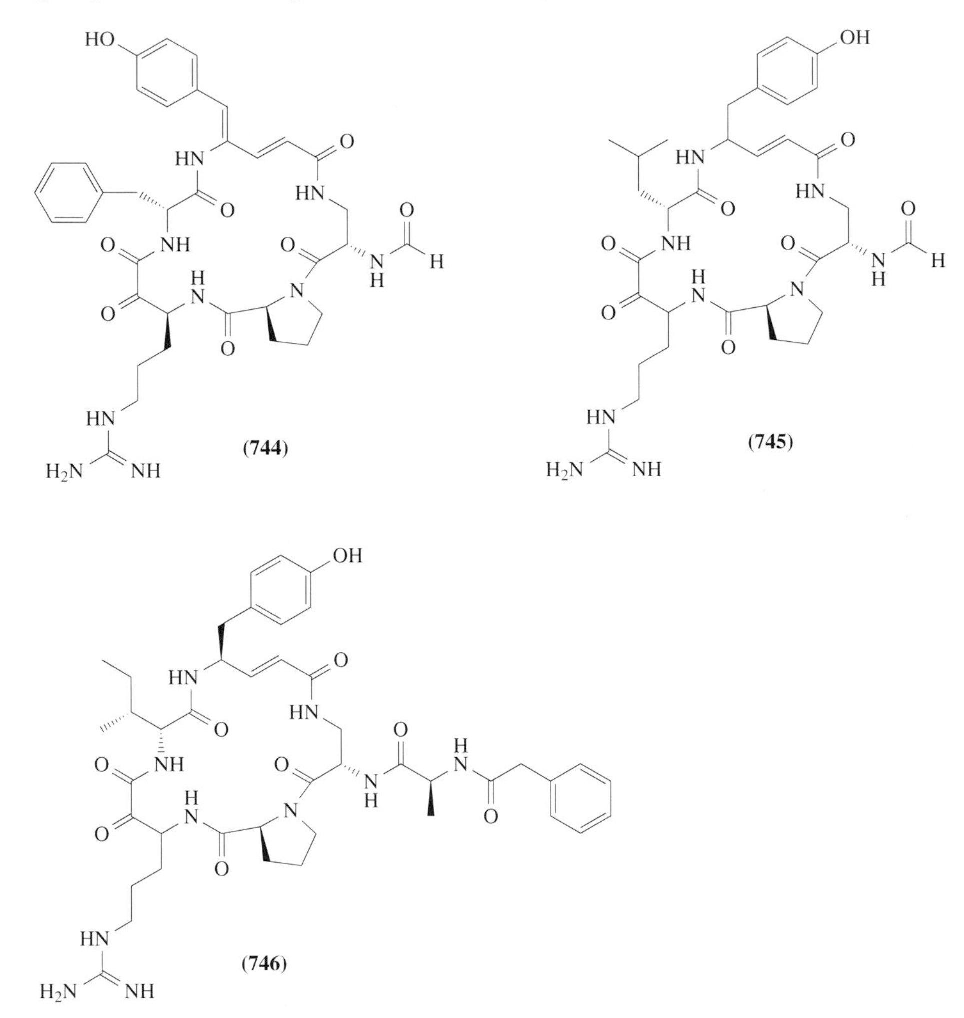

The treatment of cyclotheonamide A (**740**) with aqueous sodium carbonate or triethylamine at 23 °C generated two isomeric products. X-ray analysis of a complex with human α-thrombin indicated a ring-opened pentapeptide from cleavage at the α-keto amide bond. However, mass spectral data and a model study suggested a different product.[628] Clardy and co-workers independently studied a 2.0 Å resolution X-ray diffraction analysis of the cyclotheonamide A (**740**) complexed to bovine β-trypsin, which revealed the key interactions responsible for the slow and tight binding:[629] (**740**) formed a covalently linked hemiketal with Ser 195 Oγ as well as an extensive series of hydrogen bonds and hydrophobic interactions; in particular, the arginine side-chain used both direct and water-mediated hydrogen bonds to fill the S_1 pocket, the hemiketal oxygen utilized several main-chain hydrogen bonds to fill the oxyanion hole, and the side-chain of D-Phe used aromatic stacking to form a binding pocket.

From the same *Theonella* sponge collected at Hachijo-jima Island that contained cyclotheonamides, orbiculamide A (**752**), a cytotoxic cyclic peptide, was isolated.[630] It is a cyclic peptide with three unusual amino acid residues: 2-bromo-5-hydroxytryptophan (**753**); theoleucine (**754**), which is an α-keto homologue of leucine; and theoalanine (**755**), which may be formed by cyclization of a dipeptide, alanyl-vinylogous-serine. Standard amino acid residues were assigned by amino acid analysis and HPLC analysis of the acid hydrolyzate after derivatization with Marfey's reagent. The total structure was elucidated by interpretation of 2D NMR spectral data in CD_3OH and DMSO-d_6 and by chemical degradation. The absolute configurations of the unusual amino acid residues (**753**)–(**755**) were determined by chiral GC analysis after degradations (e.g., $NaIO_4/KMnO_4$. H_2O_2/aqueous NaOH). The 3-methylvaleric acid residue was revealed to have *S* stereochemistry by GC (OV-1) analysis after conversion into the (*S*)-1-naphthylethylamide. α-Keto-β-amino acids such as (**754**) appear to be a characteristic feature of peptides from sponges of the genus *Theonella*.

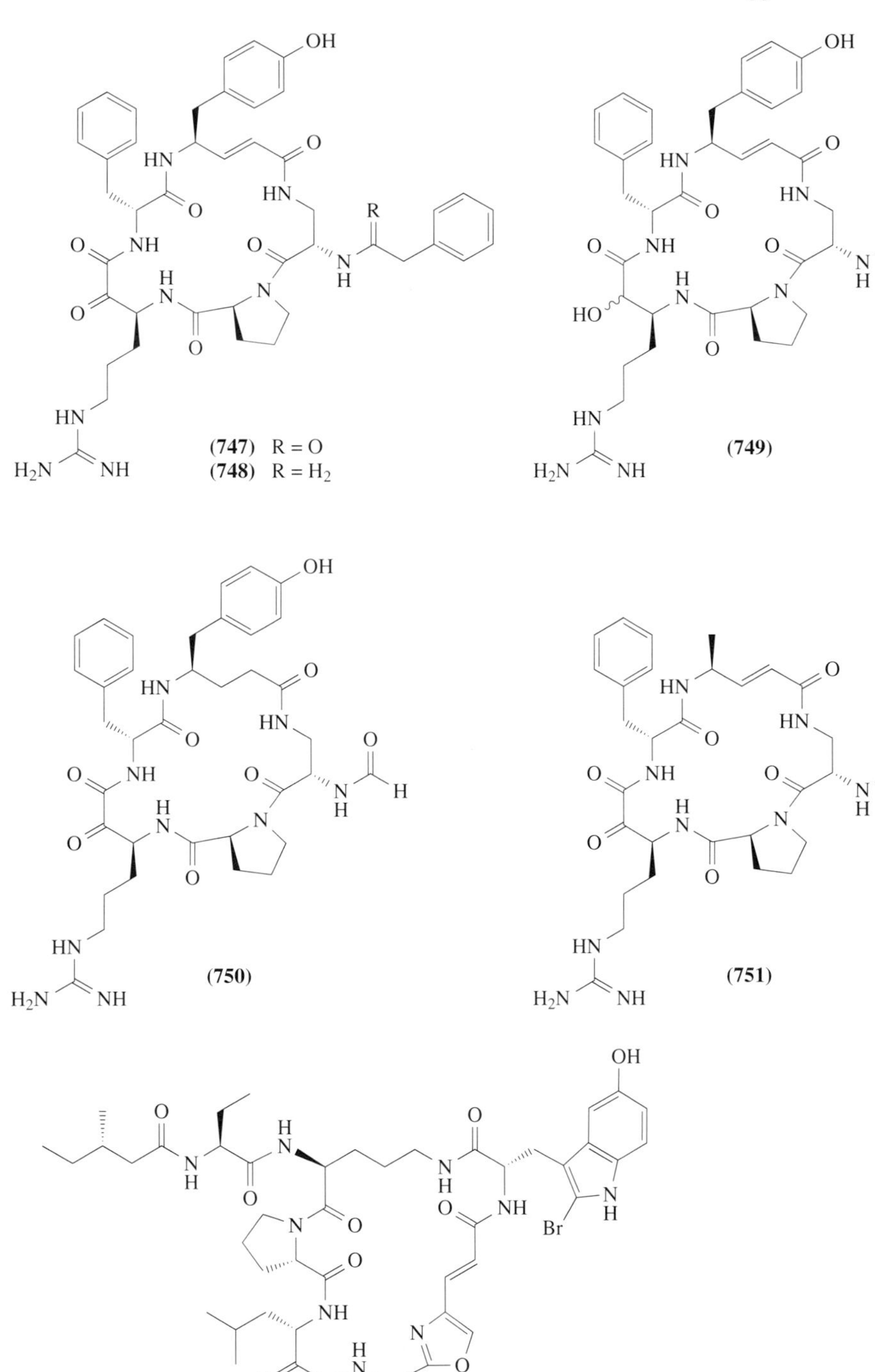

OH
HN
O
HN
R
O
NH
O
N
H
H
N
N
O
O
HN
H2N
NH
(747) R = O
(748) R = H2
HO
(749)
(750)
H
(751)
Br

(752)

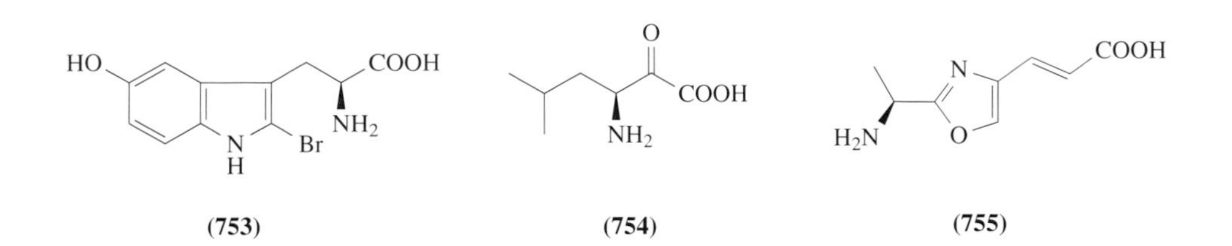

HO
COOH
NH2
N
H
Br
COOH
O
N
O
H2N
(753)
(754)
(755)

From a sponge of the genus *Theonella* collected off the Kerama Islands, Okinawa, Kobayashi *et al.* isolated keramamides B–D (**756**)–(**758**),[631] which were structurally related to orbiculamide A (**752**). Keramamides B–D (**756**)–(**758**) (5×10^{-8} M) inhibited the superoxide generation response of the human neutrophils elicited with a chemotactic peptide, *N*-formyl-Met-Leu-Phe (fMLP), but did not inhibit that induced by phorbol myristate, acetate, or immune complex. Another related peptide, keramamide F (**759**) was also obtained from the same sponge.[632] Keramamides B–D (**756**)–(**758**) and F (**759**) also contained α-keto β-amino acid residues. Keramamide F (**759**) was moderately cytotoxic. In addition, four related cyclic peptides, keramamides E (**760**), G (**761**), H (**762**), and J (**763**) were isolated from the sponge *Theonella* sp., and their structures were elucidated on the basis of spectral data, particularly 2D NMR and FAB MS/MS data, as well as by chemical means.[633] The 2-hydroxy-3-methylpentanoic acid which was generated from (**760**) by alkaline hydrolysis was revealed to have 2*S*,3*S*-configuration by comparison of retention times in chiral HPLC with those of 2*S*,3*S*-, 2*R*,3*R*-, 2*S*,3*R*-, and 2*R*,3*S*-isomers derived from L-Ile, D-Ile, L-allo-Ile, and D-allo-Ile through deamination with $NaNO_2$, respectively. Similarly the absolute configurations of the 2-hydroxy-3-methylpentanoic acid moiety in keramamides B–D (**756**)–(**758**) were determined to be 2*S*,3*S* by the same methods. Keramamide E (**760**) exhibited cytotoxicity against L1210 murine leukemia cells and KB human epidermoid carcinoma cells with IC_{50} values of 1.60 μg ml^{-1} and 1.55 μg ml^{-1}, respectively, whereas keramamides G (**761**), H (**762**), and J (**763**) showed no cytotoxicity ($IC_{50} > 10$ μg ml^{-1}).

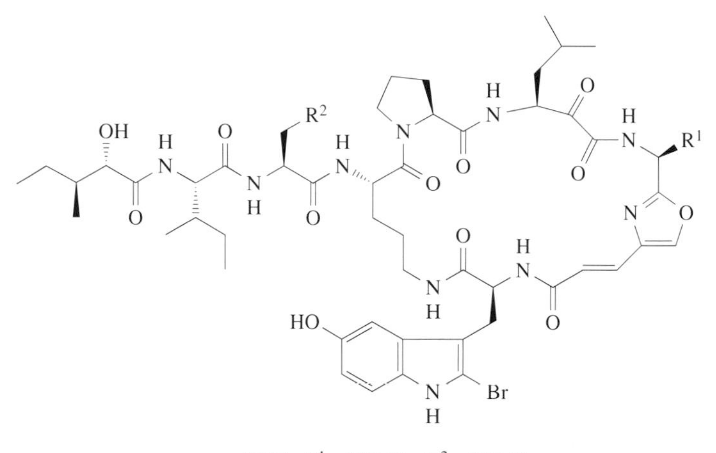

(**756**) $R^1 = CH_2Me$, $R^2 = CH_2Me$
(**757**) $R^1 = CH_2Me$, $R^2 = Me$
(**758**) $R^1 = Me$, $R^2 = Me$
(**760**) $R^1 = Me$, $R^2 = CH_2Me$

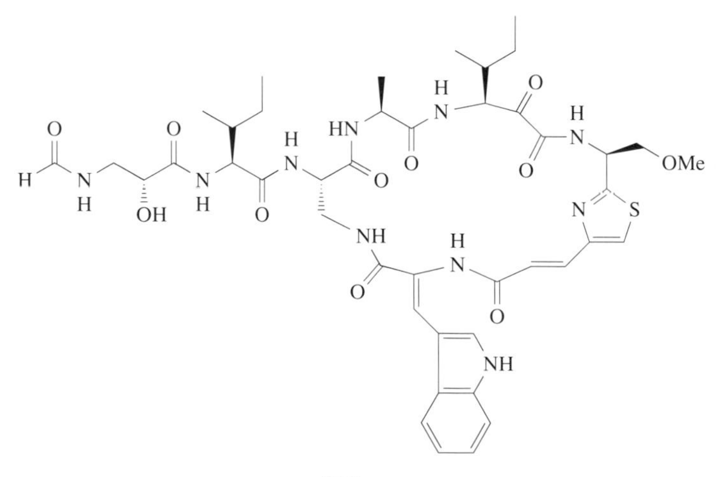

(**759**)

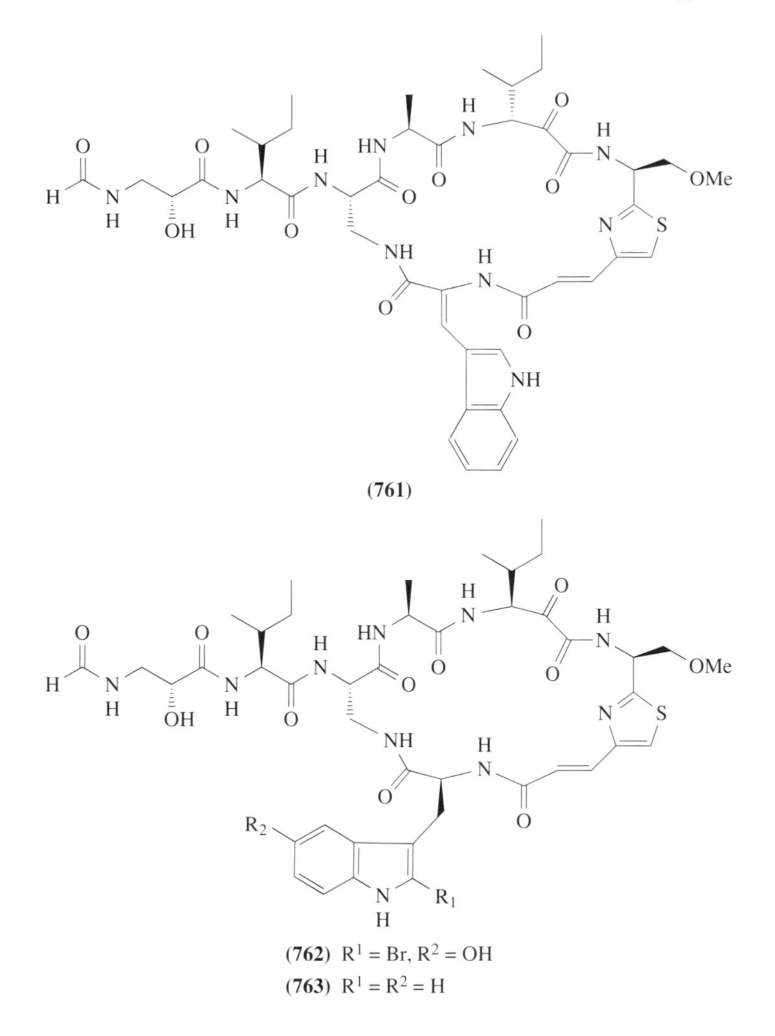

(761)

(762) $R^1 = Br, R^2 = OH$

(763) $R^1 = R^2 = H$

Kobayashi *et al.* also isolated a different class of cyclic peptides from *Theonella* sponges. From the same sponge that contained keramamides B–D **(756)**–**(758)**, which was brown-colored with a yellow inner body, keramamide A **(764)** was isolated,[634] and the structure was established as a unique hexapeptide containing a hitherto unknown amino acid, 6-chloro-5-hydroxy-*N*-methyltryptophan, and possessing an unusual ureido bond. The structural assignment was made on the basis of spectroscopic results. In particular, the proposed structure based on the NMR data was wholly supported by FABMS/MS evidence. Comparison of the daughter ions obtained by the CAD spectra of two precursor ions containing ^{35}Cl and ^{37}Cl, respectively, established the presence or not of a chlorine atom in a particular daughter ion. Keramamide A **(764)** exhibited inhibitory activity against sarcoplasmic reticulum Ca^{2+}-ATPase (IC_{50} 3×10^{-4} M). From another *Theonella* sponge with a white inner body, konbamide **(765)**, with calmodulin antagonistic activity, was isolated.[635] Konbamide **(765)** was elucidated to be a unique hexapeptide with an ureido bond and unusual amino acids such as 2-bromo-5-hydroxytryptophan and *N*-methylleucine. The absolute configurations of the unusual tryptophan-derived amino acid residues, 2-bromo-5-hydroxytryptophan and 6-chloro-5-hydroxy-*N*-methyltryptophan, contained in keramamide A **(764)** and konbamide **(765)**, respectively, were determined to be both L on the basis of chiral GC and HPLC analyses.[636] Schmidt and Weinbrenner synthesized both the D- and L-2-bromo-5-hydroxytryptophan containing forms of konbamide **(765)**.[637] The physical properties of the natural and the synthetic konbamide were compared; HPLC data, optical rotation data, and NMR spectra showed slight variations, whereas the mass spectra were the same. They suggested that the proposed structure of konbamide **(765)** cannot be corrected. Thus, the structure of konbamide **(765)** as well as keramamide A **(764)** is under reinvestigation by Kobayashi's group.

Five tridecapeptide lactones, named theonellapeptolides Ia **(766)**, Ib **(767)**, Ic **(768)**, Id **(769)**,[638], and Ie **(770)**, were isolated from the Okinawan marine sponge *Theonella swinhoei*.[639] A new HPLC

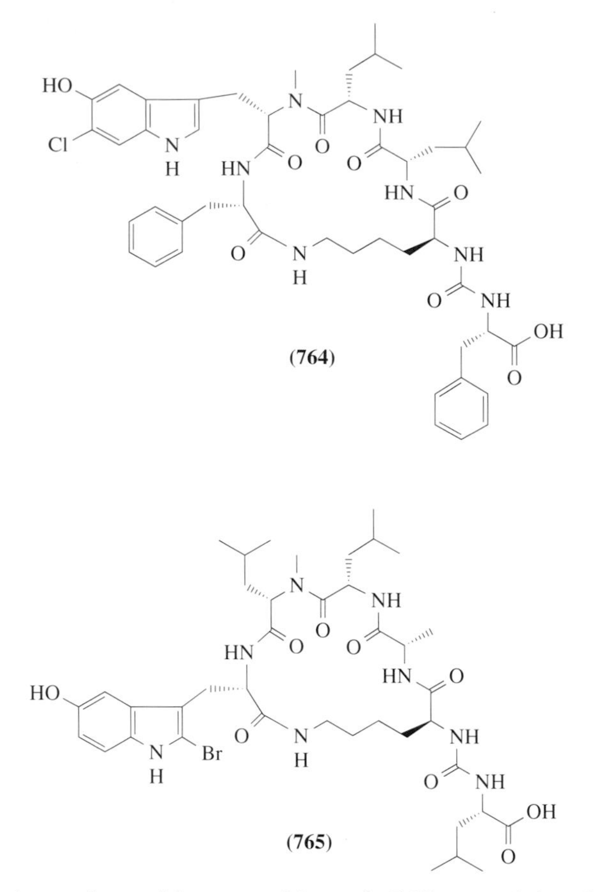

(764)

(765)

method for analyzing the amino acid composition of (**769**) containing *N*-Me amino acids was devised. The structures of these peptide lactones were determined on the basis of chemical and physicochemical examinations including HPLC–CD combined analysis of the amino acid compositions. Theonellapeptolides Ib (**767**), Ic (**768**), Id (**769**), and Ic (**770**) exhibited moderate cytotoxic activity towards for L1210 *in vitro* (IC_{50} 1.6, 1.3, 2.4, and 1.4 μg ml^{-1}, respectively), and theonellapeptolide Ie (**770**) exhibited ion-transport activities for Na^+ and K^+. Following the characterization of theonellapeptolides Ia–Ie (**766**)–(**770**), another new tridecapeptide lactone named

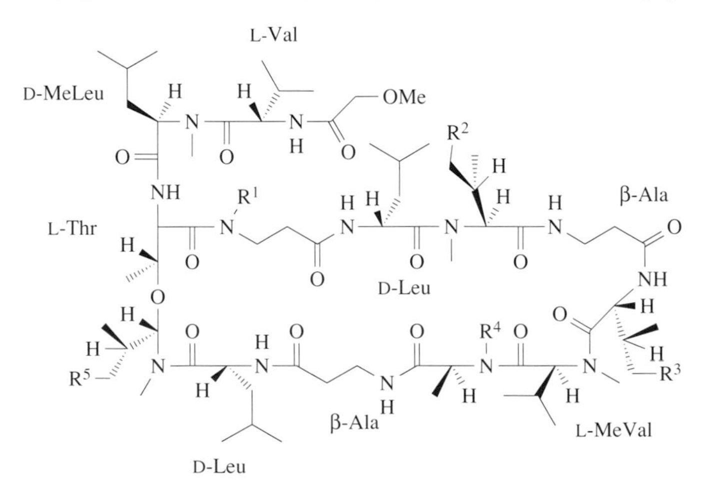

(**766**) $R^1 = H, R^2 = Me, R^3 = H, R^4 = Me, R^5 = Me$
(**767**) $R^1 = H, R^2 = H, R^3 = Me, R^4 = Me, R^5 = Me$
(**768**) $R^1 = H, R^2 = Me, R^3 = Me, R^4 = Me, R^5 = H$
(**769**) $R^1 = H, R^2 = Me, R^3 = Me, R^4 = Me, R^5 = Me$
(**770**) $R^1 = Me, R^2 = Me, R^3 = Me, R^4 = Me, R^5 = Me$
(**771**) $R^1 = H, R^2 = Me, R^3 = Me, R^4 = H, R^5 = Me$

theonellapeptolide IId (**771**) was isolated from the Okinawan marine sponge *Theonella swinhoei*.[640] Theonellapeptolide IId (**771**) prevented fertilization of the sea urchin *Hemicentrotus pulcherrimus* at concentrations of 25 μg ml^{-1} or greater but did not affect early embryonic development of fertilized eggs up to the gastrula stage.

Matsunaga and Fusetani isolated five cytotoxic bicyclic peptides, theonellamides A–E (**772**)–(**776**),[641] from the same *Theonella* sponge which contained theonellamide F (**777**),[617] an antifungal bicyclic peptide bridged by a histidinoalanine residue. Theonellamides A (**772**) and B (**773**) differed from theonellamide F (**777**) in three amino acid residues ((**772**), (**773**)/(**777**): 2*S*-isoserine (Iser)/*β*-alanine (*β*Ala), *β*-methyl-*p*-bromophenylalanine (MeBrPhe)/*p*-bromophenylalanine (BrPhe), and (5*E*,7*E*)-3-amino-4-hydroxy-6-methyl-8-phenyl-5,7-octadienoic acid (Apoa)/(5*E*,7*E*)-3-amino-4-hydroxy-6-methyl-8-(*p*-bromophenyl)-5,7-octadienoic acid (Aboa)). Additionally, theonellamide A (**772**) bore a *β*-D-galactose linked to the free imidazole nitrogen. Theonellamide C (**774**) was debromotheonellamide F. Theonellamide D (**775**) and E (**776**) were the *α*-L-arabinoside and *β*-D-galactoside of theonellamide F (**777**). Their structures were assigned on the basis of spectral data and chromatographic analyses of degradation products. The histidinoalanine residue (Hisala) in theonellamide F (**777**) was deduced to have L stereochemistry for the histidine portion and D for the alanine portion.[617] When the acid hydrolysate of theonellamide F (**777**) was prepared under standard conditions (6N HCl, 110 °C, 16 h), a 1:1 mixture of the LD- and the LL-isomers was obtained.[617] Under milder hydrolysis conditions (6N HCl, 107 °C, 8 h), the ratio of the LD- and the LL-isomers was 3:1.[617] In the mild acid hydrolysate (6N HCl, 110 °C, 3 h) of theonellamides A (**772**), the LD- and the LL-isomers were detected in a ratio of 5:1 by HPLC analysis of the Marfey derivative, suggesting the stereochemistry of the Hisala residue in (**772**) to be identical with that in (**777**).

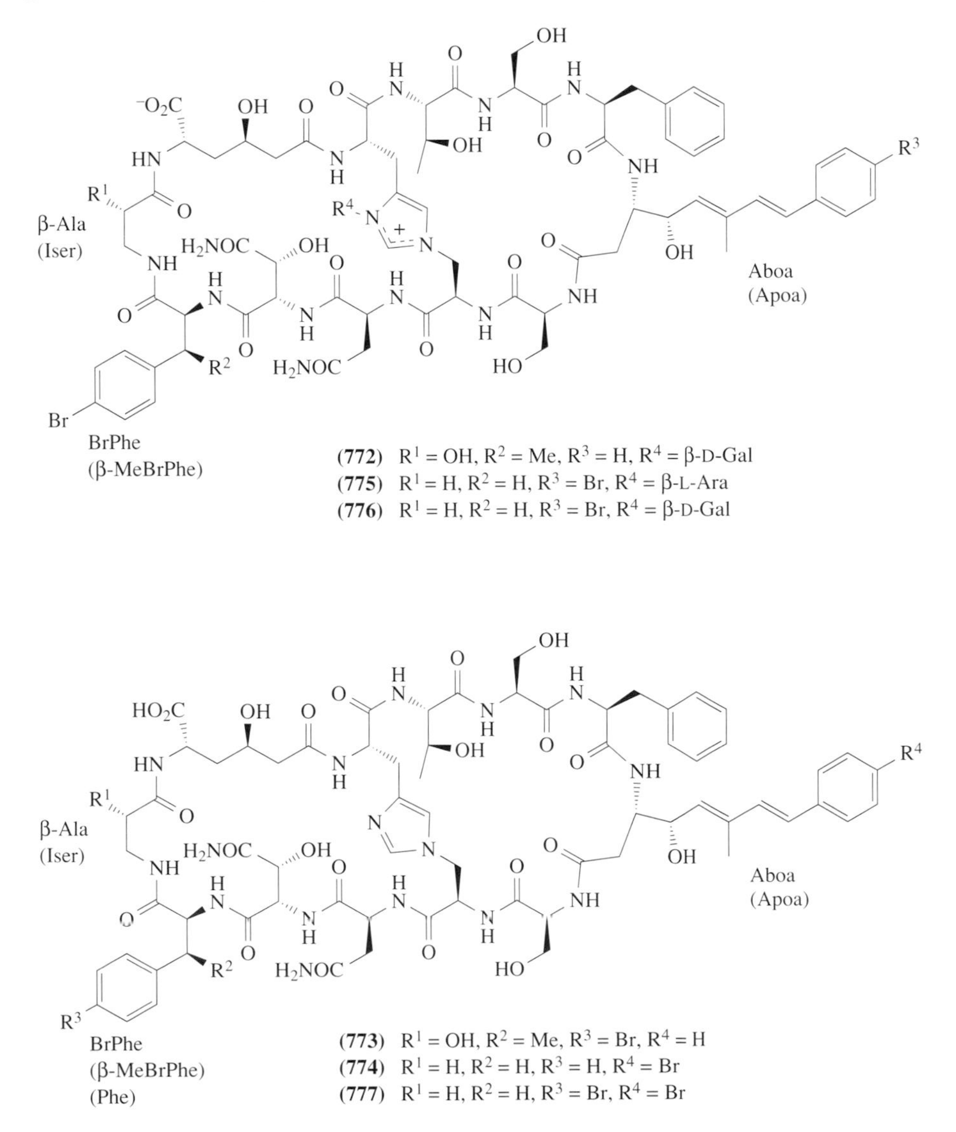

However, there was still a possibility of an overlap of the DL- and the DD-isomer peaks with the LL-isomer peak in the Marfey analysis. In order to exclude this possibility, the LD- and the LL-isomers were derivatized with Marfey's reagent prepared from D-Ala (D-Marfey's reagent), which introduced opposite chiral centers in the molecule. The LD-isomer and the LL-isomer derivatized with Marfey's reagent prepared from D-Marfey's reagent were enantiomeric to, and therefore chromatographically equivalent to, the DL-isomer and the DD-isomer derivatized with the conventional Marfey's reagent. All four peaks were well separated and the major Hisala residue liberated from **(772)** by mild acid hydrolysis was determined unambiguously as the LD-isomer. Theonellamides A–E **(772)**–**(776)** were moderately cytotoxic against P388 murine leukemia cells with IC_{50} values of 5.0, 1.7, 2.5, 1.7, and 0.9 μg ml^{-1}, respectively. Bewley and Faulkner also isolated theonegramide **(778)**, an antifungal glycopeptide consisting of D-arabinose joined to a bicycle dodecapeptide, which was structurally related to theonellamide F (777),[617] from the lithistid sponge *Theonella swinhoei* collected at Antolang, Negros Island, the Philippines.[642] Theonegramide **(778)** inhibited the growth of *Candida albicans* in the standard disk assay at a loading of 10 μg/disk.

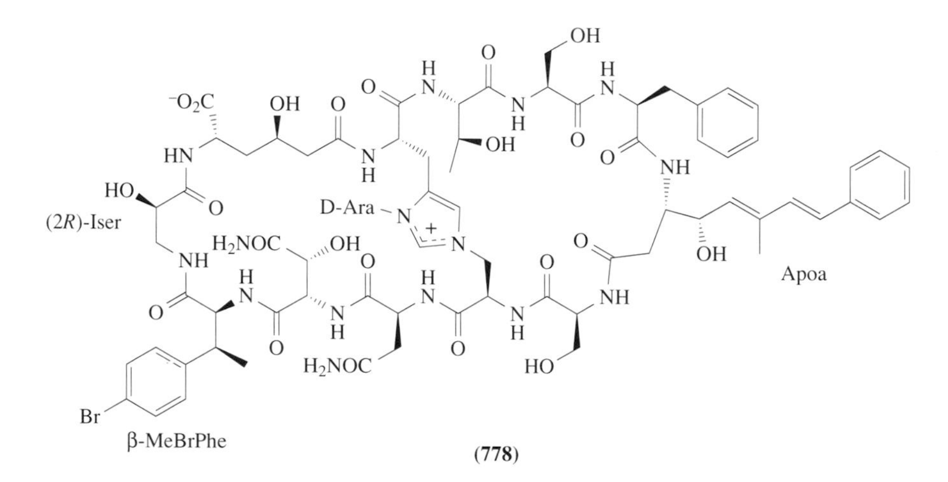

(778)

Nazumamide A **(779)**, a thrombin-inhibiting linear tetrapeptide, was isolated from a marine sponge, *Theonella* sp. which contained cyclotheonamides.[618] The structure of **(779)** was determined by interpretation of 2D NMR data and by chemical degradation as 2,5-dihydroxybenzoyl-L-arginyl-L-prolyl-L-isoleucyl-L-α-aminobutyric acid.[643] Nazumamide A **(779)** inhibited thrombin with an IC_{50} of 2.8 μg ml^{-1}, but not trypsin at a concentration of 100 μg ml^{-1}. Its total synthesis was achieved in an efficient manner using diethyl phosphorocyanidate (DEPC) as a coupling reagent.[644]

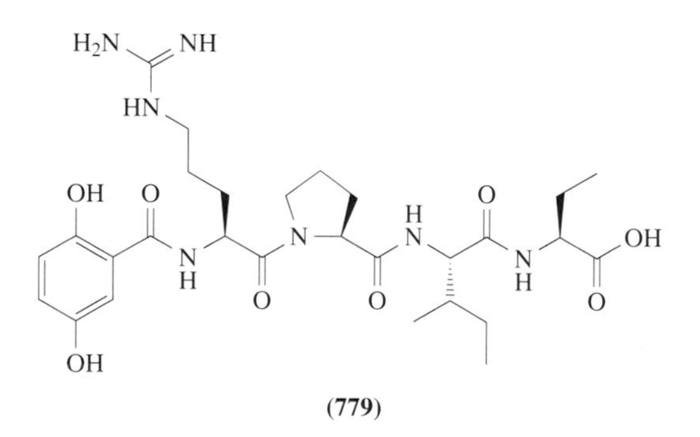
(779)

Motuporin **(780)**, a cyclic pentapeptide, was isolated from the marine sponge *Theonella swinhoei* collected off Motupore Island, Papua New Guinea.[645] The structure of motuporin **(780)** was elucidated by spectroscopic analysis and chemical degradation. Motuporin **(780)** inhibited protein phosphatase-1 in a standard phosphorylase phosphatase assay at a concentration of <1 nmol, making it one of the most potent PP1 inhibitors, and it also displayed considerable *in vitro* cytotoxicity against murine leukemia (P388: IC_{50} 6 μg ml^{-1}), human lung (A549: IC_{50} 2.4 μg ml^{-1}), ovarian (HEY: IC_{50} 2.8 μg ml^{-1}), colon (LoVo: IC_{50} 2.3 μg ml^{-1}), breast (MCF7: IC_{50} 12.4 μg ml^{-1}), and brain (U373MG: IC_{50} 2.4 μg ml^{-1}) cancer cell lines. Motuporin (**780**) was an analogue of nodularin, a hepatotoxic cyclic pentapeptide from the brackish-water blue-green alga *Nodularia spumigena*.[646] Motuporin (**780**) differed from nodularin by the replacement of a polar arginine residue with a nonpolar valine residue. The structural similarity between motuporin (**780**) and nodularin provided the suggestion that motuporin (**780**) was being produced by a blue-green alga.

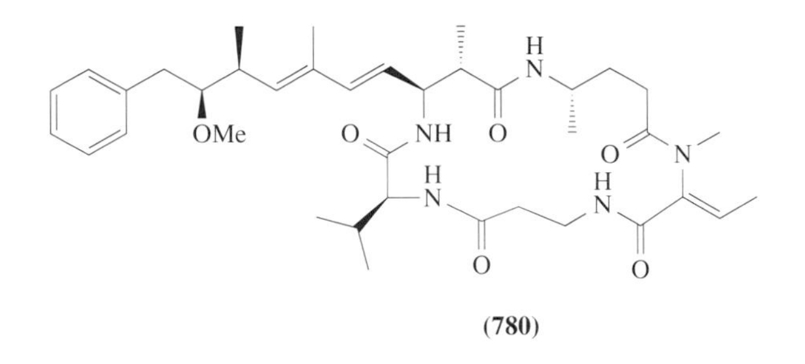

(780)

A sponge of the genus *Theonella*, collected near Perth, off Cape Vlamingh in Western Australia, contained a cyclic octapeptide, perthamide B (**781**).[647] The structure was elucidated by spectroscopic methods, revealing that (**781**) contained several unusual amino acid residues such as 3-amino-2-hydroxy-6-methyloctanoic acid, 2,3-dehydro-2-aminobutyric acid, *N*-methylglycine, *β*-hydroxy-asparagine, *O*-methylthreonine, *γ*-methylproline, and *m*-bromotyrosine. Perthamide B (**781**) weakly inhibited the binding of [^{125}I]interleukin-1.beta. to intact EL4.6.1 cells with an IC_{50} of 27.6 μM. However, the binding could not be differentiated from the toxicity of this compound.

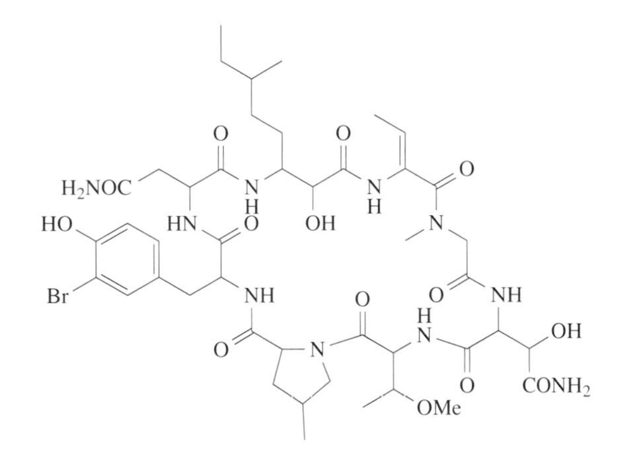

(781)

Fusetani and his research group isolated highly cytotoxic polypeptides, polytheonamides A–C, from the marine sponge *Theonella swinhoei*.[648] Interpretation of the 2D NMR data of the acid hydrolyzate of polytheonamide B (**782**) as well as amino acid analysis led to identification of Ala, Asp, *α*-Thr, Ser, Glu, Val, Gly, Ile, t-Leu, *β*-methylGlu, *β*-methylIle, *β*-hydroxyVal, and *β*-hydroxyAsp. The 2D NMR data of polytheonamide B (**782**) suggested the presence of *γ*-hydroxyl t-Leu, which was a new amino acid. The structure of polytheonamide B (**782**) was assigned to be linear 48-residue peptides with *N*-terminus blocked by a carbamoyl group, mainly by interpretation of spectral data.[649] Similarly polytheonamide A was analyzed by a spectroscopic method leading to the same amino acid sequence as (**782**). The chemical shift discrepancy of NH protons between polytheonamides A and B (**782**) was observed in residues 41–48, suggesting that the stereochemistry of amino acids in this region may differ between the two compounds. Polytheonamide A gave the same $(M+Pr_3N)^+$ ion in the FABMS at m/z 5033 as polytheonamide B (**782**). Polytheonamide C gave the $(M+Pr_3N)^+$ ion 14 mass units larger than (**782**). Interpretation of HOHAHA and NOESY data led to the amino acid sequence with Gln-46 residue in (**782**) being replaced by *β*-MeGln. Polytheonamides exhibited highly cytotoxic activity against L1210 with IC_{50} <4 ng ml^{-1}.

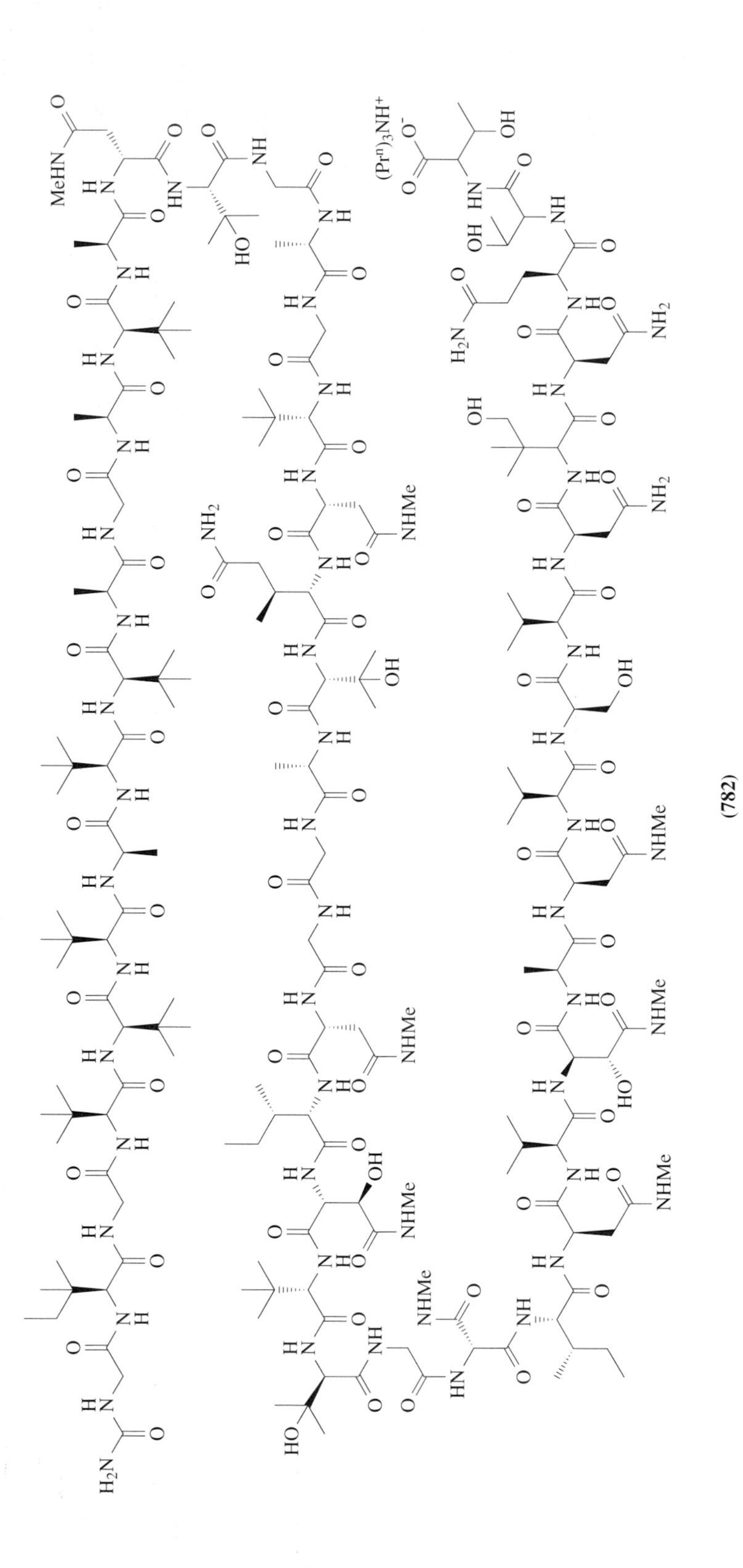

(782)

The absolute configuration of each amino acid residue of polytheonamide B (**782**) was determined by chiral chromatography of fragment peptides obtained by partial acid hydrolysis.[650] Amino acid residues in polytheonamide B (**782**) have alternating D- and L-configuration. The NMR spectra in $CD_3OH/CDCl_3$ (1:1) indicated that polytheonamide B (**782**) adopted a certain secondary structure. Conformational analysis was carried out by distance geometry calculations, using the DADAS90 program, with NMR parameters obtained in $CD_3OH/CDCl_3$ (1:1), i.e., a total of 378 distance constraints (160 intraresidue, 96 sequential, 80 long-range NOEs, and 42 hydrogen bonds) and 48 dihedral angle constraints. The calculated structure fitted a right-handed parallel β-helix structure, which was proposed for the structure of a pore-forming peptide, gramicidin A.

Swinholide A (**783**) is a potent cytotoxic macrolide, first isolated from the Red Sea sponge *Theonella swinhoei* by Carmely and Kashman in 1985, and its structure was initially described as a monomeric 22-membered macrolide.[651] Swinholide A (**783**) was revealed to have a 44-membered dimeric dilactone structure. The absolute stereostructure of (**783**) was obtained by means of X-ray crystallographic analysis of a dimeric diketone structurally related to (**783**) and by chemical derivatizations.[652] The configurations of each asymmetric carbon in (**783**) were similar to those of scytophycin C (**784**) previously isolated from the cultured blue-green alga *Scytonema pseudo-hofmanni*.[653] The crystal structure further revealed the conformational characteristic of the 44-membered ring structure, which was probably related to the cytotoxic activity.[654] Swinholide A (**783**) (21,21′-diolide) exhibited high cytotoxic activity against L1210 (IC_{50} 0.03 μg ml^{-1}) and KB (IC_{50} 0.04 μg ml^{-1}) tumor cells, while the activity of isoswinholide A (21,23′-diolide) was significantly lowered (IC_{50} values, 0.12 μg ml^{-1} and 1.4 μg ml^{-1} for L1210 and KB cells, respectively). Acidic treatment of swinholide A (**783**) provided several isomeric macrolides having different sizes of the dilactone ring structure. The *in vitro* cytotoxicities and *in vivo* antitumour activities of these dimeric macrolides, together with two monomeric macrolides which were synthesized from (**783**) were examined from the structure–activity correlation viewpoint. Two monomeric macrolides, 21-olide (**785**) and 23-olide (**786**), were obtained by treatment of the monomeric acid (**787**), which was prepared from (**783**) through alkaline hydrolysis, with 1-(3-dimethylaminopropyl)-3-ethyl-carbodiimide hydrochloride (EDCI) and 4-dimethylaminopyridine in 41% and 24% yields, respectively. These compounds, however, did not show promising antitumor activity.[655]

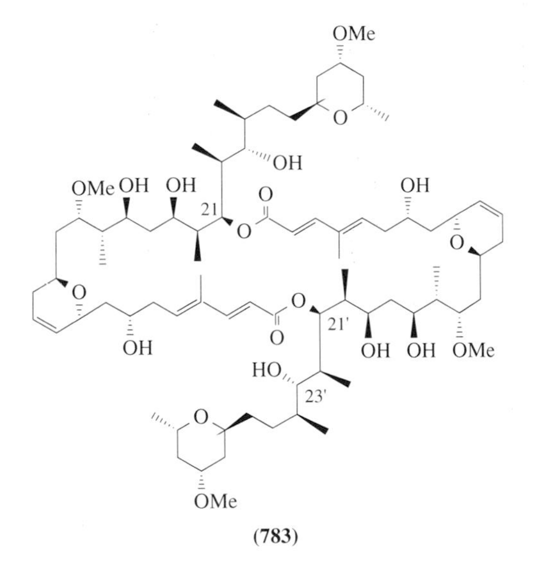

(783)

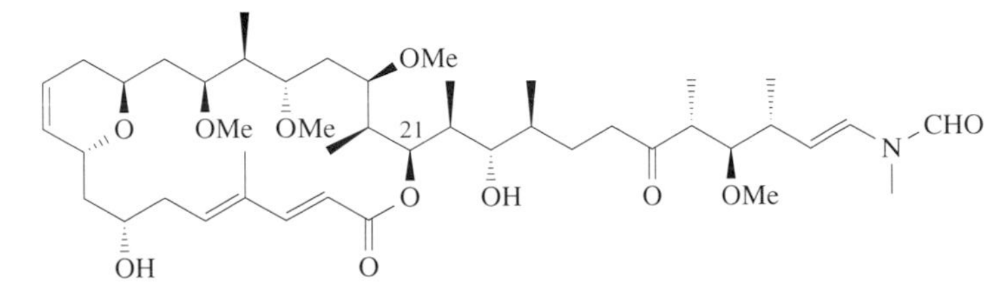

(784)

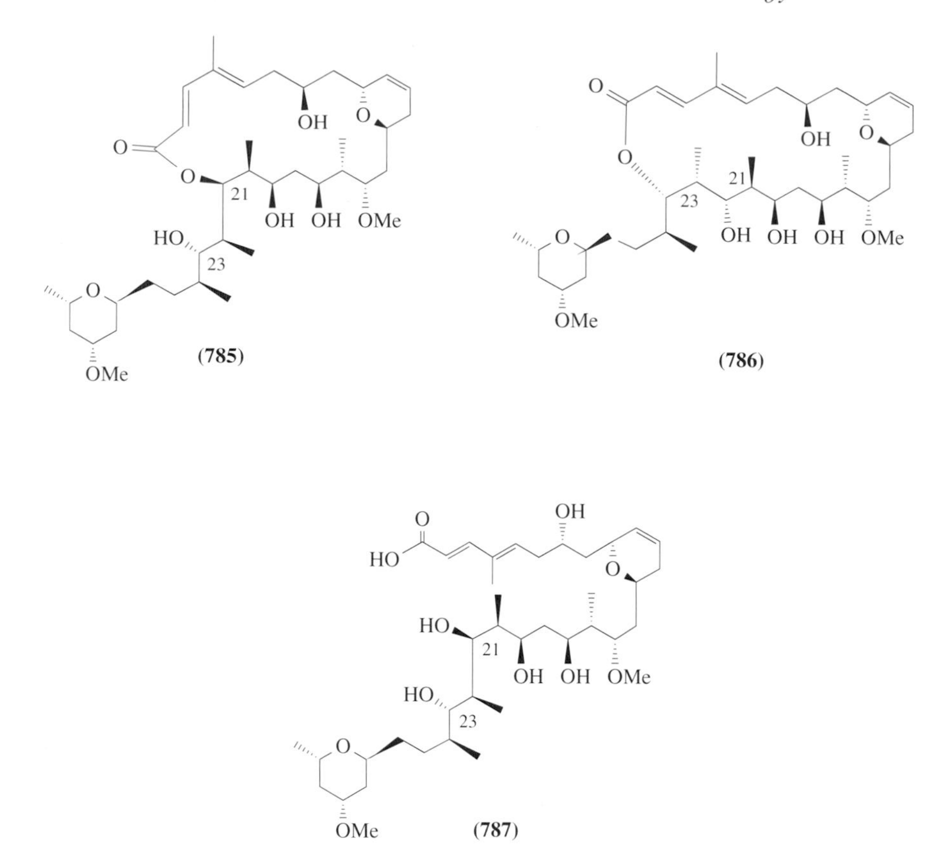

(785) **(786)** **(787)**

Thus, the marine sponge *Theonella swinhoei* yielded many important, bioactive natural products, most of which shared structural features with bacterial natural products. The presence of microbial symbionts in *T. swinhoei* was reported, and it was originally suggested that the cytotoxic macrolide swinholide A (**783**) and many of the bioactive cyclic peptides from *T. swinhoei* were all produced by symbiotic cyanobacteria. By transmission electron microscopy, four distinct cell populations were found to be consistently present in *T. swinhoei*: eukaryotic sponge cells, unicellular heterotrophic bacteria, unicellular cyanobacteria, and filamentous heterotrophic bacteria. Purification and chemical analyses of each cell type showed the macrolide swinholide A (**783**) to be limited to the mixed population of unicellular heterotrophic bacteria, and an antifungal cyclic peptide such as (**778**) occurred only in the filamentous heterotrophic bacteria. Contrary to prior speculation, no major metabolites were located in the cyanobacteria or sponge cells.[656]

The marine sponge *Theonella* sp., from which cyclotheonamides were isolated, contained orange pigments associated with cytotoxicity. The MeOH extract of the sponge was partitioned between water and ether, and the aqueous phase was extracted with Bu^nOH. The Bu^nOH phase was successively purified by Sephadex LH20 and reverse-phase chromatographies to yield aurantosides A and B (**788**) and (**789**), both as orange powders.[657] Aurantoside A (**788**) had a molecular formula of $C_{36}H_{46}N_2O_{15}Cl_2$ as revealed by FABMS, NMR spectra, and elemental analysis. Its gross structure was determined by the interpretation of 2D NMR spectra to be a tetramic acid glycoside: one side-chain was a dichlorinated conjugated heptaene, while the other was a methylenecarboxyamide; xylopyranose, arabinopyranose, and a 5-deoxypentofuranose, whose hydroxyl group on C-2 was methylated, were linked via nitrogen to the pyrrolidone ring, in order. Aurantoside B (**789**) gave almost superimposable NMR spectra to that of (**788**), except for the absence of the *O*-methyl group. Detailed analysis of spectral data indicated that the 2-*O*-methyl group of the 5-deoxypentose unit of aurantoside A (**788**) was replaced by a hydroxyl group in aurantoside B (**789**). Aurantosides A (**788**) and B (**789**) were cytotoxic against P388 and L1210 leukemia cells ((**788**), IC_{50} 1.8 μg ml^{-1} and 3.4 μg ml^{-1}, respectively; (**789**), IC_{50} 3.2 μg ml^{-1} and 3.3 μg ml^{-1}, respectively).

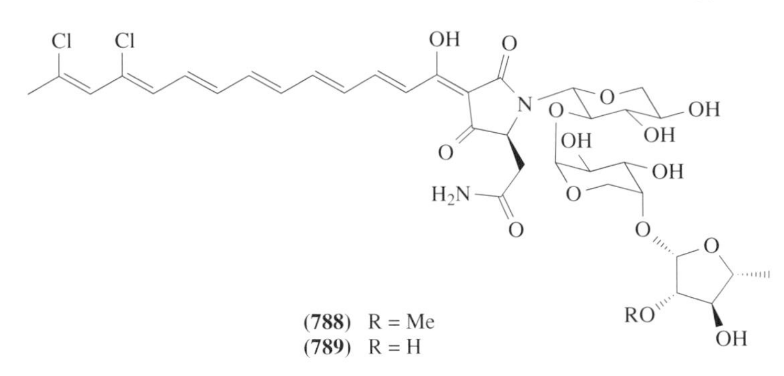

(**788**) R = Me
(**789**) R = H

Onnamide A (**790**) is a potent antiviral and highly cytotoxic compound, first isolated from a sponge of the genus *Theonella*.[658] Several other highly cytotoxic metabolites related to (**790**) were isolated from the sponges *Theonella*.[659,660] Theopederins A–E (**791**)–(**795**) were isolated from a marine sponge of the *Theonella* sp.[661] and their structures established mainly by extensive 2D NMR analyses as well a by comparison with spectral data of mycalamides A (**796**) and B (**797**).[662] Theopederins A–E (**791**)–(**795**) were highly cytotoxic against P388 murine leukemia cells with IC_{50} of 0.05, 0.1, 0.7, 1.0, and 9.0 ng ml^{-1}, respectively. Theopederins A and B (**791**) and (**792**) also showed promising antitumor activity against P388 (i.p.): T/C = 205% (0.1 mg kg^{-1} d^{-1}, treated on days 1, 2, and 4–6, i.p.), and T/C = 173% (0.4 mg kg^{-1} d^{-1}, treated on days 1, 2, and 4–6, i.p.), respectively. Theopederins A–C (**791**)–(**793**) might be generated by oxidative cleavage of a double bond of onnamide A (**790**),[658] the most abundant cytotoxic component of the water-soluble portion of the EtOH extract of the same sponge.

A novel tetrahydroprotoberberine alkaloid, theoneberine (**798**), was isolated from the Okinawan marine sponge of the species *Theonella*, collected off Ie Island, Okinawa, and its structure was elucidated on the basis of spectral and chemical means.[663] Theoneberine (**798**) was the first tetrahydroprotoberberine alkaloid isolated from marine organisms and also the first example as a naturally occurring tetrahydroprotoberberine alkaloid with substitution by bromine atoms. Biogenetically, theoneberine (**798**) seemed to be a unique isoquinoline alkaloid classified as a hybrid between 1-benzylisoquinolines and protoberberines, which was hitherto unknown from natural sources. Theoneberine (**798**) exhibited antimicrobial activity against Gram-positive bacteria (MIC values: *Staphylococcus aureus*, 16 μg ml^{-1}; *Sarcina lutea*, 2 μg ml^{-1}; *Bacillus subtilis*, 66 μg ml^{-1}; *Mycobacterium* sp. 607, 4 μg ml^{-1}), and was inactive against fungi or Gram-negative bacteria. Compound (**798**) also showed cytotoxicity against murine lymphoma L1210 and human epidermoid carcinoma KB cells *in vitro* with IC_{50} values of 2.9 μg ml^{-1} and ca. 10 μg ml^{-1}, respectively.

Theonezolide A (**799**), a novel macrolide, was isolated from the Okinawan marine sponge *Theonella* sp., collected off Ie Island, Okinawa, and the planar structure was elucidated on the basis of extensive spectroscopic analyses of (**799**) and its four ozonolysis products.[664] The 2D NMR techniques of gradient-enhanced HMBC and HSQC–HOHAHA along with the FABMS/MS experiment were applied and proved to be quite efficient for structural study of this long aliphatic molecule. Theonezolide A (**799**), $C_{79}H_{140}N_4O_{22}S_2$, was the first member of a new class of polyketide natural products consisting of two principal fatty acid chains with various functionalities such as sulfate ester, an oxazole, and a thiazole group, constituting a 37-membered macrocylic lactone ring bearing a long side-chain attached through an amide linkage. Theonezolides B (**800**) and C (**801**), two additional macrolides, were also isolated from the Okinawan marine sponge *Theonella* sp., collected of Ie Island, and their structures were elucidated on the basis of spectroscopic data as well as chemical degradation experiments. The absolute configurations of the terminal chiral center of theonezolides A (**799**), B (**800**), and C (**801**) were determined by synthesis of their ozonolysis products from L-alaninol.[665] Compounds (**799**)–(**801**) exhibited cytotoxicity against murine lymphoma L1210 and human epidermoid carcinoma KB cells *in vitro* (IC_{50} values for (**799**), both 0.75 μg ml^{-1}; for (**800**), 5.6 μg ml^{-1} and 11 μg ml^{-1}, respectively; and for (**801**), 0.3 μg ml^{-1} and 0.37 μg ml^{-1}, respectively).

Theonezolide A (**799**) caused a marked platelet shape change at low concentrations (0.2–0.6 μM).[666] Increasing concentrations of (**799**) to 6 μM or more caused shape change followed by a small but sustained aggregation. In a Ca^{2+}-free solution, theonezolide A-induced aggregation was markedly inhibited, although the marked shape change was still observed. Aggregation stimulated by (**799**) increased in a linear fashion with increasing Ca^{2+} concentrations from 0.1 mM to 3.0 mM. Furthermore, theonezolide A (**799**) markedly enhanced $^{45}Ca^{2+}$ uptake into platelets. Aggregation induced by theonezolide A (**799**) was inhibited by Arg-Gly-Asp-Ser, an inhibitor of fibrinogen

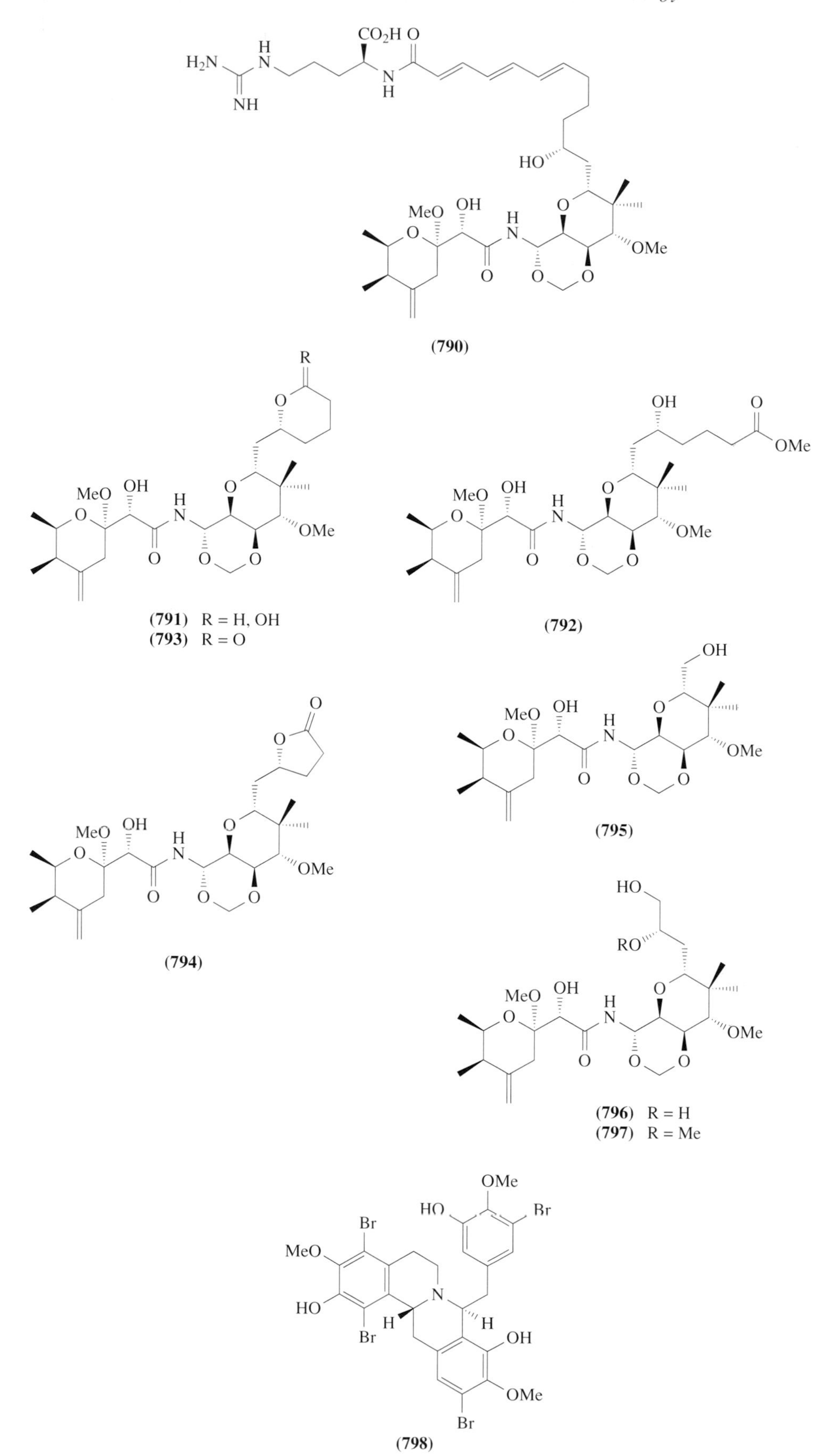

CO_2H O
H_2N
H
N
NH
N
H
HO
MeO
OH
H
N
O
O
OMe
O
O
O
O
(790)
R
O
MeO
OH
H
N
O
O
OMe
O
O
O
O
(791) R = H, OH
(793) R = O
OH
O
OMe
MeO
OH
H
N
O
O
OMe
O
O
O
O
(792)
OH
MeO
OH
H
N
O
O
OMe
O
O
O
O
(795)
O
O
MeO
OH
H
N
O
O
OMe
O
O
O
O
(794)
HO
RO
MeO
OH
H
N
O
O
OMe
O
O
O
O
(796) R = H
(797) R = Me
OMe
HO
Br
Br
MeO
HO
Br
N
H
H
OH
OMe
Br
(798)

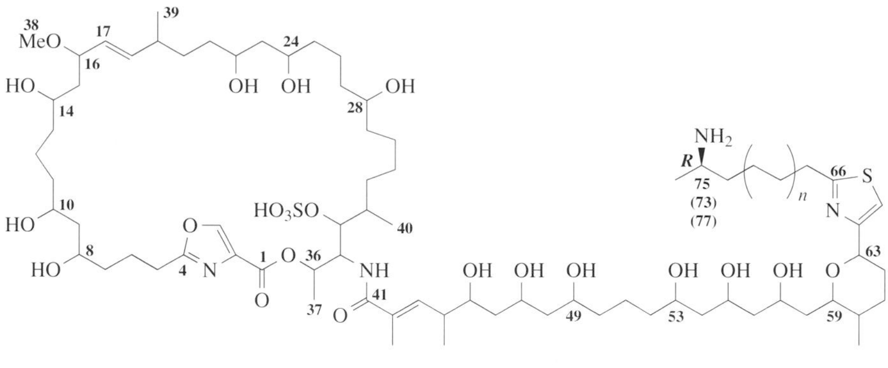

(799): $n = 3$, **(800)**: $n = 2$, **(801)**: $n = 4$

binding to glycoprotein IIb-IIa, H-7 and staurosporine, protein kinase C inhibitors, or genistein and tyrphostin A23, protein tyrosine kinase inhibitors, whereas shape change was blocked by genistein and tyrphostin A23. H-7 or staurosporine did not affect the theonezolide A-induced shape change. These results suggested that theonezolide A-induced platelet shape change was not dependent on external Ca^{2+}, whereas theonezolide A-induced aggregation was caused by an increase in Ca^{2+} permeability of the plasma membrane. It was also suggested that both aggregation and shape change induced by theonezolide A (**799**) were associated with protein phosphorylation by protein kinase C and tyrosine kinase.

8.07.9.3.3 Others

Four cyclic peptides, kapakahines A (**802**) and B (**803**).[667] and stereoisomers C (**804**) and D (**805**), were isolated from the marine sponge *Cribrochalina olemda*, collected at Pohnpei, Micronesia. Their structures including complete stereochemistry were elucidated by spectral analysis and chemical degradation.[668] Kapakahine B (**803**) is a cyclic hexapeptide, while kapakahines A (**802**), C (**804**), and D (**805**) are octapeptides. The unique structural feature of these peptides was the lack of an amide linkage between two tryptophan residues. Instead the ring was closed by a bond from the indole nitrogen of Trp-1 to the *β*-carbon of Trp-2, and all four peptides contained a tryptophan-derived α-carboline which was part of a fused tetracyclic system that also included an imidazolone derived from phenylalanine or tyrosine. Kapakahines A (**802**), B (**803**), and C (**804**) showed moderate cytotoxicity against P388 murine leukemia cells at IC_{50} values of 5.4, 5.0, and 5.0 μg ml^{-1}, respectively. Kapakahine D (**805**) did not show cytotoxicity at a concentration of 10 μg ml^{-1}. Kapakahine A (**802**) was tested for inhibitory activities against several enzymes [thrombin, trypsin, plasmin, elastase, papain, angiotensin converting enzyme, and protein phosphatase 2A (PP2A)], but showed only 15% inhibition against PP2A at a concentration of 30 μM (32 μg ml^{-1}). Against other enzymes, (**802**) did not show any activity at a concentration of 100 μg ml^{-1}.

Two antifungal and cytostatic bis-oxazole macrolides, phorboxazoles A (**806**) and B (**807**), were isolated from the marine sponge, *Phorbas* sp., collected near Muiron Island, Western Australia.[669] Structural assignment was accomplished through extensive 2D NMR spectroscopy including COSY, RCT, HMQC, and HMBC. The complete relative stereochemistry about the macrolide ring and the solution conformation of (**806**) were established from analysis of ROESY experiments. Phorboxazoles A (**806**) and B (**807**) exhibited *in vitro* antifungal activity against *Candida albicans* at 0.1 μg/disk and extraordinary cytostatic activity against a variety of human solid tumor cell lines (mean 50% Growth Inhibition (GI_{50}) $< 7.9 \times 10^{-10}$ M in the National Cancer Institute's 60 tumor cell line panel). For example, phorboxazoles A (**806**) inhibited growth of colon tumor cells HCT-116 (GI_{50} 4.36×10^{-10} M) and HT29 (3.31×10^{-10} M). Although all attempts to obtain crystals of either compound (**806**) or (**807**) suitable for X-ray diffraction studies unfortunately failed, the absolute configurations of 14 out of the 15 stereocenters in both (**806**) and (**807**) were elucidated by a combination of techniques including synthesis of a suitable model compound for NMR spectroscopic comparisons.[670] The absolute configuration of C-43 in phorboxazole A (**806**) was established as *R* by correlation with (*R*)-dimethyl methoxysuccinate, while 43*R* was also suggested for phorboxazole B (**807**) by CD comparison, thus completing the entire stereochemical determination of the phorboxazoles (**806**) and (**807**).[671]

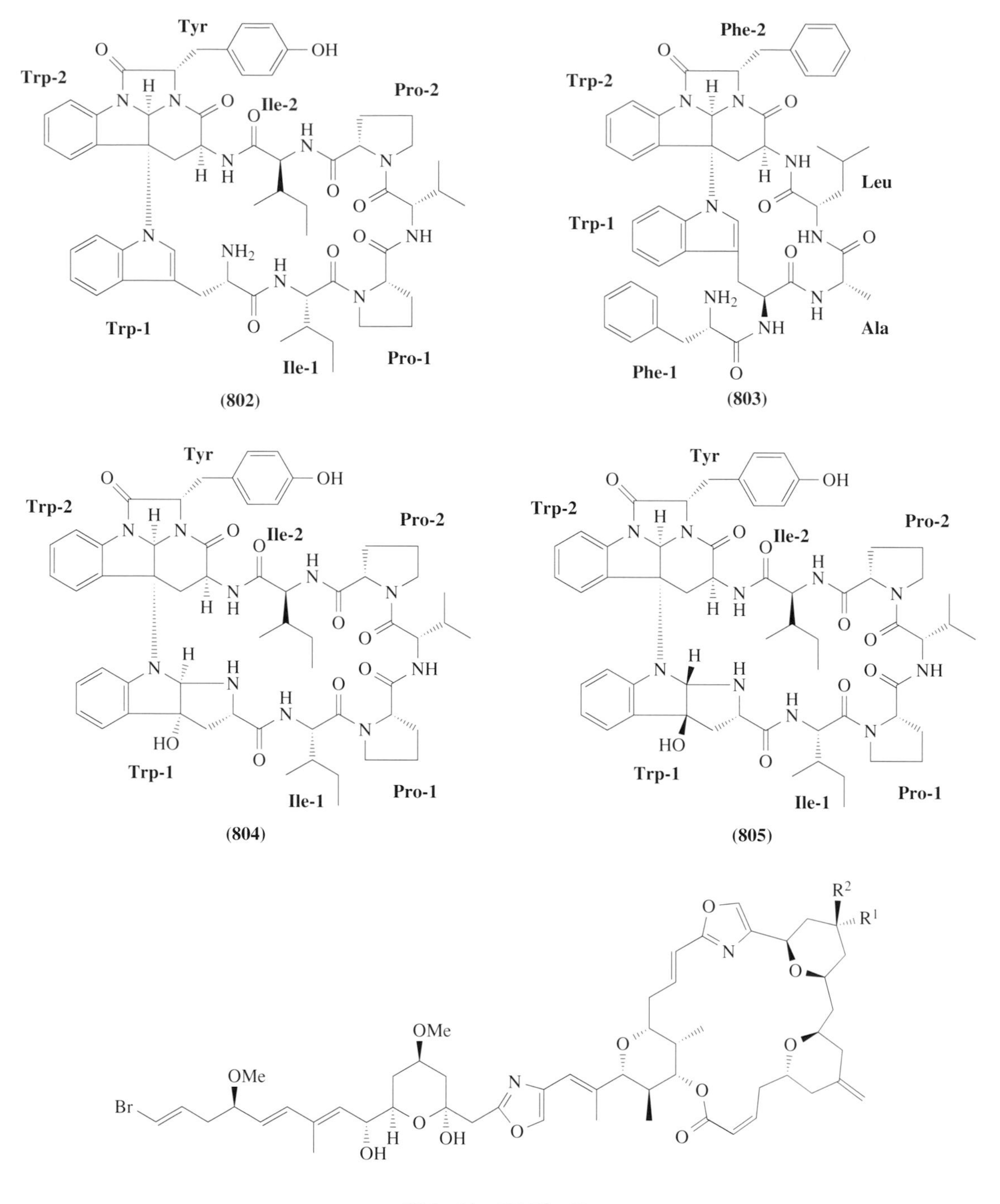

(802)

(803)

(804)

(805)

(806) $R^1 = OH, R^2 = H$
(807) $R^1 = H, R^2 = OH$

The lithistid sponge *Aciculites orientalis*, collected from the small island of Siquijor, located southeast of the large island of Negros, in the Philippines, contained three cyclic peptides, aciculitins A–C (**808**)–(**810**), that were identical except for homologous lipid residues.[672] The structure of the major peptide, aciculitin B (**809**), $C_{62}H_{88}N_{14}O_{21}$, was elucidated by interpretation of spectroscopic data. The absolute stereochemistries of the amino acid residues, the 2,3-dihydroxy-4,6-dienoic acid unit, and the sugar moiety were determined by combination of chiral GC analysis, chemical degradations, and synthesis of authentic samples. The aciculitins consisted of a bicyclic peptide that contained an unusual histidino-tyrosine bridge. Attached to the bicyclic peptide were C_{13}-C_{15} 2,3-dihydroxy-4,6-dienoic acids bearing D-lyxose at the 3-position. The aciculitins (**808**)–(**810**) inhibited the growth of *Candida albicans* at a loading of 2.5 μg/disk in the standard disk assay, and were cytotoxic toward the human colon tumor cell line HCT-116 with an IC_{50} of 0.5 μg ml^{-1}.

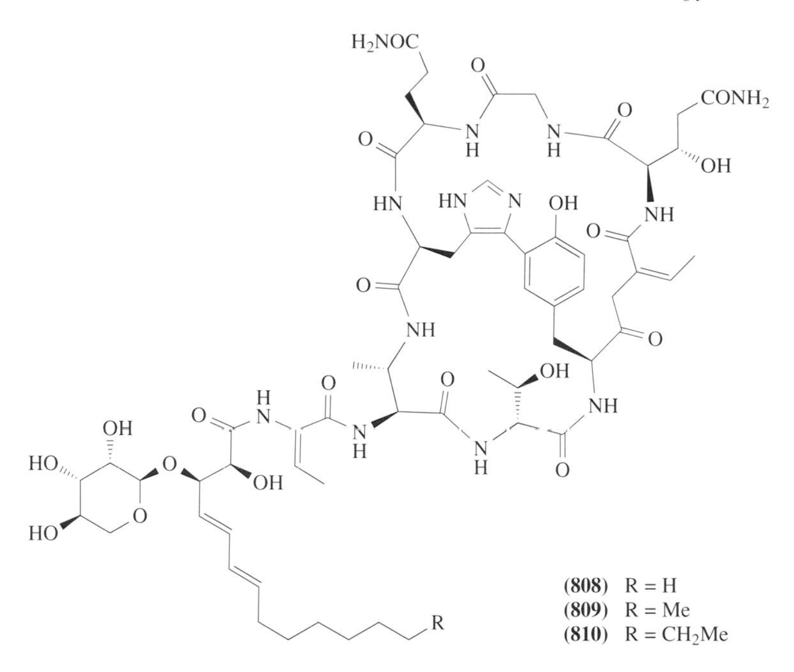

(808) R = H
(809) R = Me
(810) R = CH_2Me

Dysidiolide (**811**), a sesterterpene γ-hydroxybutenolide, was isolated from a sponge *Dysidea etheria* de Laubenfels, collected off Long Island, Bahamas at a depth of 18 m.[673] Its structure was determined by single-crystal X-ray diffraction. Dysidiolide (**811**) blocks the activity of a protein phosphatase called cdc25A. This enzyme catalyzes the removal of phosphate groups from a protein complex that signals cells to undergo mitotic division. Inhibitors of the enzyme arrest the cell cycle and are potentially useful in the treatment of cancer. The organization of the carbon skeleton of dysidiolide (**811**) was the first reported for the sesterterpene group of relatively rare natural products.

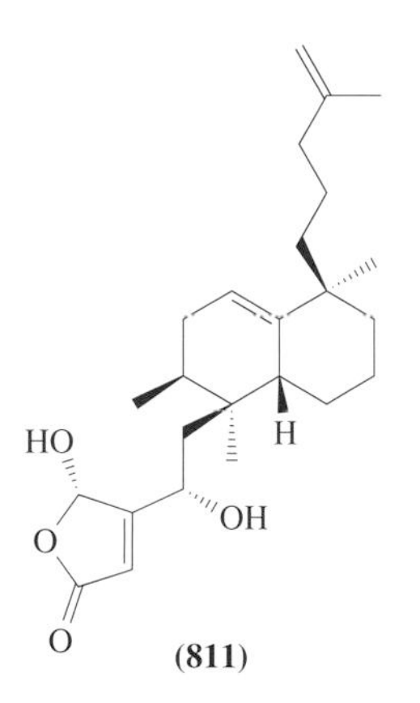

(811)

Callipeltin A (**812**) is a cyclic depsidecapeptide isolated from a shallow-water sponge of the genus *Callipelta* (order Lithistida), collected in the waters off the east coast of New Caledonia.[674] The structure of callipeltin A (**812**), which possesses the *N*-terminus blocked with a β-hydroxyl acid, and the *C*-terminus lactonized with a threonine residue, was determined by interpretation of spectral data, chemical degradation, and evaluation of the amino acids obtained by acid hydrolysis. Along with four common L-, one D-, and two *N*-methyl amino acids, it contained three new amino acid residues: β-methoxytyrosine (βOMeTyr), (2*R*,3*R*,4*S*)-4-amino-7-guanidino-2,3-dihydroxyheptanoic acid (AGDHE), and (3*S*,4*R*)-3,4-dimethyl-L-glutamine. Callipeltin A (**812**) was found to protect cells infected by human immunodeficiency (HIV) virus. The antiviral activity of callipeltin A (**812**) was measured on CEM4 lymphocytic cell lines infected with HIV-1 (Lai strain). In order to evaluate the antiviral activity of callipeltin A (**812**) the inhibition of cytopathic effects (CPE) induced by HIV-1 were studied, using MTT cell viability to determine the CD_{50} (50% cytotoxic dose) and ED_{50} (50% effective dose). At day six post-infection, callipeltin A (**812**) exhibited a CD_{50} of 0.29 $\mu g\ ml^{-1}$ and an ED_{50} of 0.01 $\mu g\ ml^{-1}$ giving a selectivity index (SI ratio CD_{50}/ED_{50}) of 29. The AZT reference had a CD_{50} of 50 μM and an ED_{50} of 30 nM.

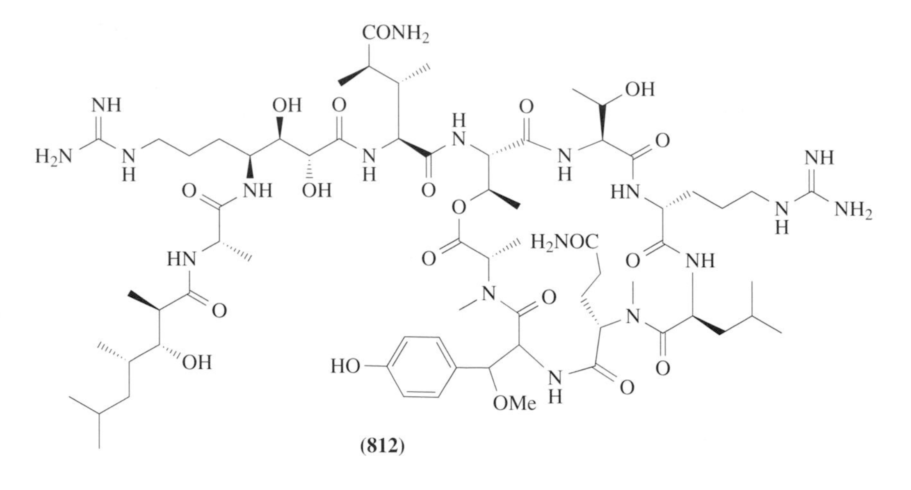

(812)

Following the characterization of callipeltin A (**812**), two other cytotoxic peptides, callipeltins B (**813**) and C (**814**), were isolated from the same New Caledonian sponge *Callipelta* sp.[675] Callipeltin B (**813**) possessed the same cyclic depsipeptidal structure as in callipeltin A (**812**) and differed from (**812**) by having the *N*-terminal 2,3-dimethylpyroglutamic acid unit instead of the tripeptide chain with the *N*-terminus blocked by a hydroxyl acid. Callipeltin C (**814**) was simply the acyclic callipeltin A. Callipeltins A–C (**812**)–(**814**) were cytotoxic against various human carcinoma cells *in vitro* and the activities of (**812**) and (**813**) exceeded significantly those of (**814**), thus suggesting the importance of the macrocycle for the bioactivity. Compounds (**812**) and (**813**) also exhibited antifungal activity against *Candida albicans* showing 30 mm and 9 mm inhibition at 100 μg/disk, respectively. Whereas callipeltin A (**812**) showed anti-HIV activity, callipeltins B (**813**) and C (**814**) proved to be inactive as antiviral compounds.

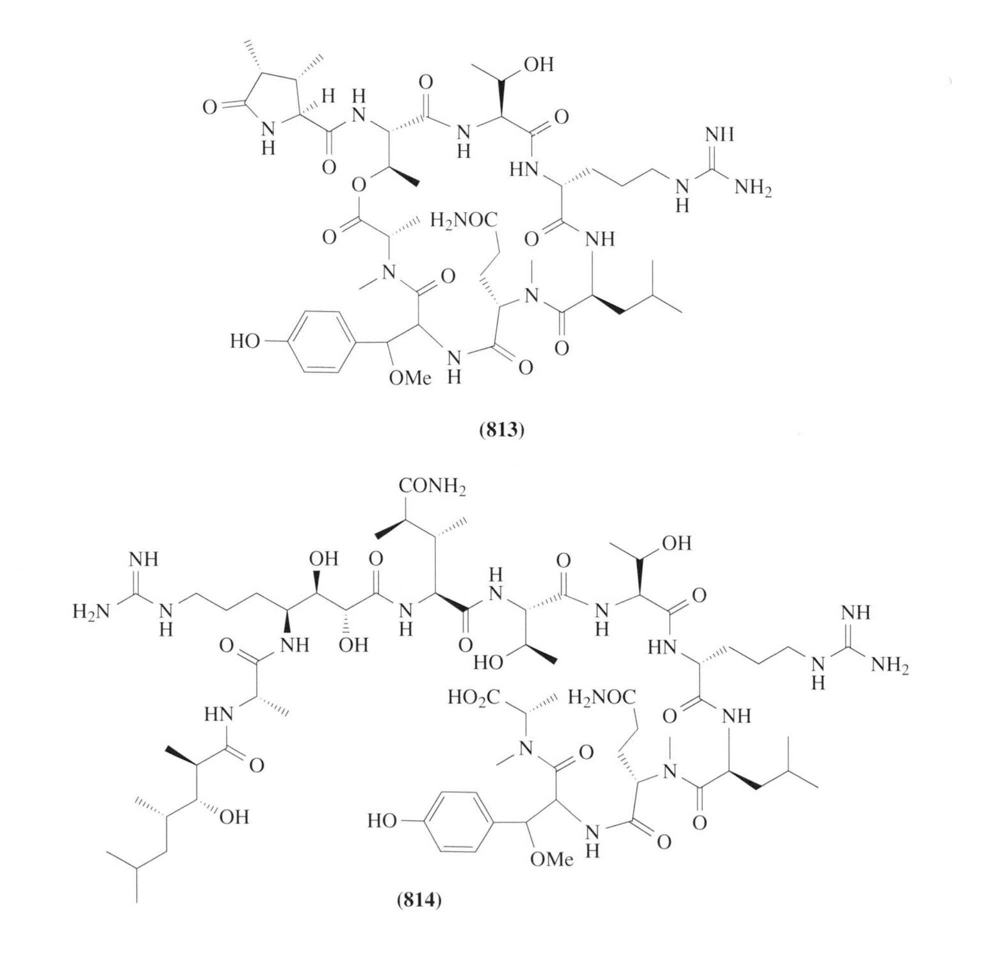

(813)

(814)

Further investigations on the dichloromethane extract from several collections of the same lithistid New Caledonian sponge *Callipelta* sp. (2.5 kg freeze-dried in total) resulted in the isolation of another cytotoxic component of a different class of natural products, named callipeltoside A (**815**) (3.5 mg yield; 1.4×10^{-4}%, dry weight).[676] Callipeltoside A (**815**) was a glycoside macrolide and its structural assignment was accomplished through extensive 2D NMR spectroscopy. The complete relative stereochemistry was proposed from the analysis of ROESY and NOE difference experiments. Callipeltoside A (**815**) represented the first member of a new class of marine-derived macrolides, containing unusual structural features including the previously unknown 4-amino-4,6-dideoxy-2-*O*-3-*C*-dimethyl-α-talopyranosyl-3,4-urethane unit (callipeltose), linked through an *a*-*O*-glycoside linkage to a hemiketal oxane ring, which was part of a 14-membered macrocycle lactone with a dienyne cyclopropane side-chain. Callipeltoside A (**815**) exhibited moderate cytotoxic activity with IC_{50} values against the NSCLC-N6 human bronchopulmonary non-small-cell lung carcinoma and P388 of 11.26 μg ml^{-1} and 15.26 μg ml^{-1}, respectively. Further, cell cycle analysis by flow cytometry assays of the NSCLC-N6 cell line treated with callipeltoside A (**815**) revealed a cell cycle-dependent effect, involving a dependent G1 blockage. These results were indicative of a blockage of NSCLC-N6 cell proliferation *in vitro* at the level of the G1 phase or by enzyme inhibition or inducing terminal cell differentiation. In the latter case callipeltoside A (**815**) would be an interesting mechanism-based lead.

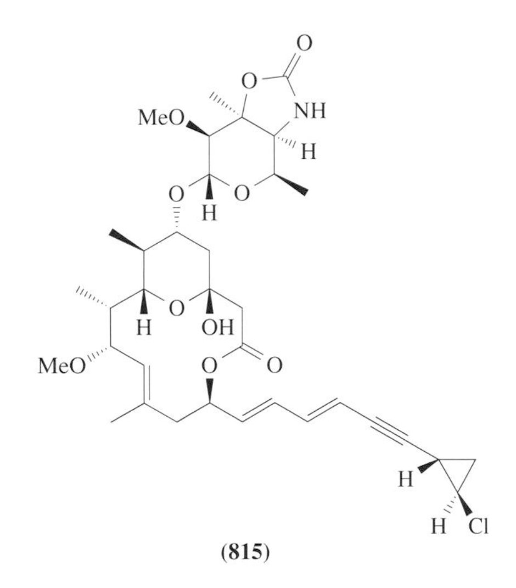

(**815**)

8.07.9.4 Microbial Metabolites

8.07.9.4.1 Bacteria

Following the discovery of over 6000 new compounds during investigations on marine natural products since the 1950s, marine microorganisms, including bacteria, fungi, and microalgae have come into focus as important new sources of biologically active compounds.[677,678] Bioactive metabolites of marine sponges have been extensively studied by marine natural products chemists (see Section 8.07.9.3). The unique metabolites of sponges are, however, often very minor constituents. It is often found that structurally similar compounds are obtained from sponges of different species and sometimes sponges of the same species contain quite different metabolites. These facts, as well as the structural characteristics of the compounds, strongly suggest that microorganisms living in or on sponges are responsible for the production of many bioactive compounds. Some of secondary metabolites which had been isolated from extracts of sponges were also obtained from the culture media of marine bacteria isolated from sponges. These are described below.

Cardellina and co-workers examined the microorganisms growing on or in sponges from Bermudan waters. From a specimen of the sponge *Tedania ignis* a bright orange-pigmented, Gram-positive bacterium was isolated and identified as a member of the genus *Micrococcus*. Culturing of the

Micrococcus sp. was successfully carried out and from the culture media three diketopiperazines (**816**)–(**818**) were isolated by chromatographic techniques, including centrifugal countercurrent chromatography.[679] Compounds (**816**)–(**818**) proved to be identical to those previously isolated from extracts of the sponge *Tedania ignis*.[680] From the fermentation culture extracts of the same *Micrococcus* sp. four benzothiazoles (**819**)–(**822**) were obtained.[681] These benzothiazoles were previously known as volatile constituents of cranberries, an aroma constituent of tea leaves, a fermentation product of yeast, or synthesized compounds.

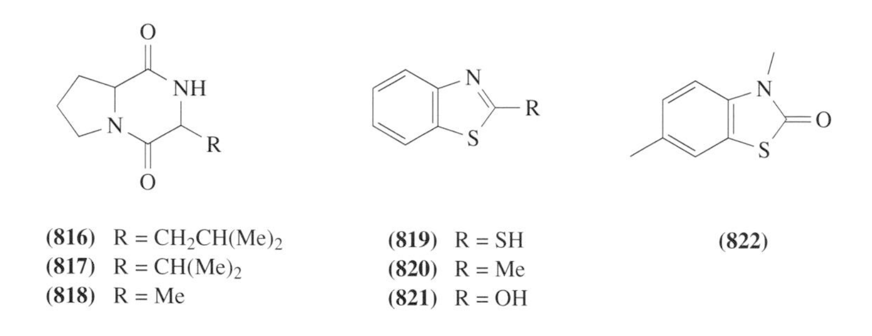

Elyakov *et al.* reported that a *Vibrio* sp. of bacteria isolated from the sponge *Dysidea* sp. collected at Eastern Samoa biosynthesized brominated diphenyl ethers on the basis of GC–MS analyses using 3,5-dibromo-2-(3′,5′-dibromo-2′-methoxyphenoxy)phenol (**823**) as a standard. Compound (**823**) was previously obtained from extracts of the sponge *Dysidea* sp.[682]

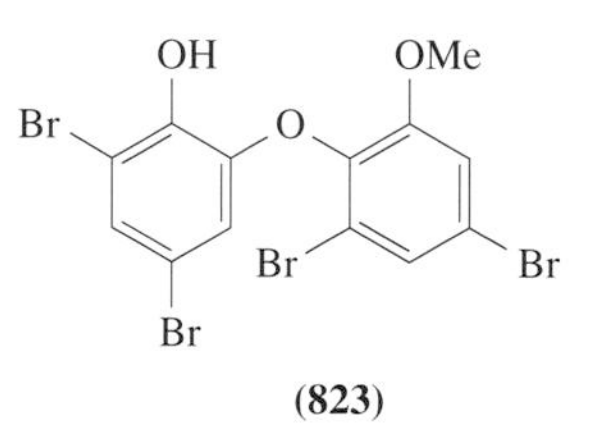

(**823**)

From a Japanese sponge *Halichondria okadai* a bacterium of a member of the genus *Alteromonas* was isolated and cultured in the laboratory. From the cultured mycelium of this bacterium Kobayashi and co-workers isolated a unique macrocylic lactam, alteramide A (**824**), possessing cytotoxic activity (IC_{50} values against P388, L1210, and KB cells, 0.1, 1.7, and 5.0 μg ml^{-1}, respectively).[683] The structure of compound (**824**) was elucidated by extensive analyses of the spectral data (e.g., ^1H-^1H COSY, HMQC, HMBC, NOESY, and ROESY) as well as chemical degradation into small fragments. Treatment of (**824**) with ozone followed by $NaBH_4$ reduction and acetylation afforded the bicyclo[3.3.0]octane derivative (**825**), while *β*-hydroxyornithine (**826**) was obtained from (**824**) through ozonolysis and oxidative workup with H_2O_2. The relative stereochemistry of the bicyclo[3.3.0]octane unit was established from NOESY data obtained for (**825**) and the absolute configurations of the C-23 and C-25 positions of (**824**) were defined through chiral HPLC examination of (**826**). Compound (**824**) possesses two diene groups which were vulnerable to an intramolecular photochemical [4+4] cycloaddition to give a hexacyclic compound (**827**). It should be noted that the structure of alteramide A (**824**) appears to be biogenetically related to those of an antibiotic ikarugamycin (**828**) produced by a terrestrial *Streptomyces* sp.[684] and an antifungal and cytotoxic metabolite discodermide (**829**) isolated from a Caribbean marine sponge *Discodermia dissoluta*.[685] Since compound (**824**) was obtained from a bacterium isolated from a marine sponge, discodermide (**829**) was strongly suggested to be of microbial origin.

An antibiotic indole trimer named trisindoline (**830**) was isolated from the culture of a bacterium of *Vibrio* sp. by Kitagawa and co-workers.[686] This bacterium was separated from the Okinawan marine sponge *Hyrtios altum*. The structure of trisindoline (**830**) was verified by synthesis of (**830**) from oxindole (**831**) through two steps, (i) copper(II) bromide in ethyl acetate under reflux for 3 h, and (ii) indole and silver carbonate in tetrahydrofuran at 25 °C for 1.5 h, giving 47% overall yield. Trisindoline (**830**) showed antibiotic activities (16, 17, and 10 mm diameter growth inhibitions for *E. coli*, *B. subtilis*, and *S. aureus*, respectively, at 10 μg/disk (i.e. = 8 mm)).

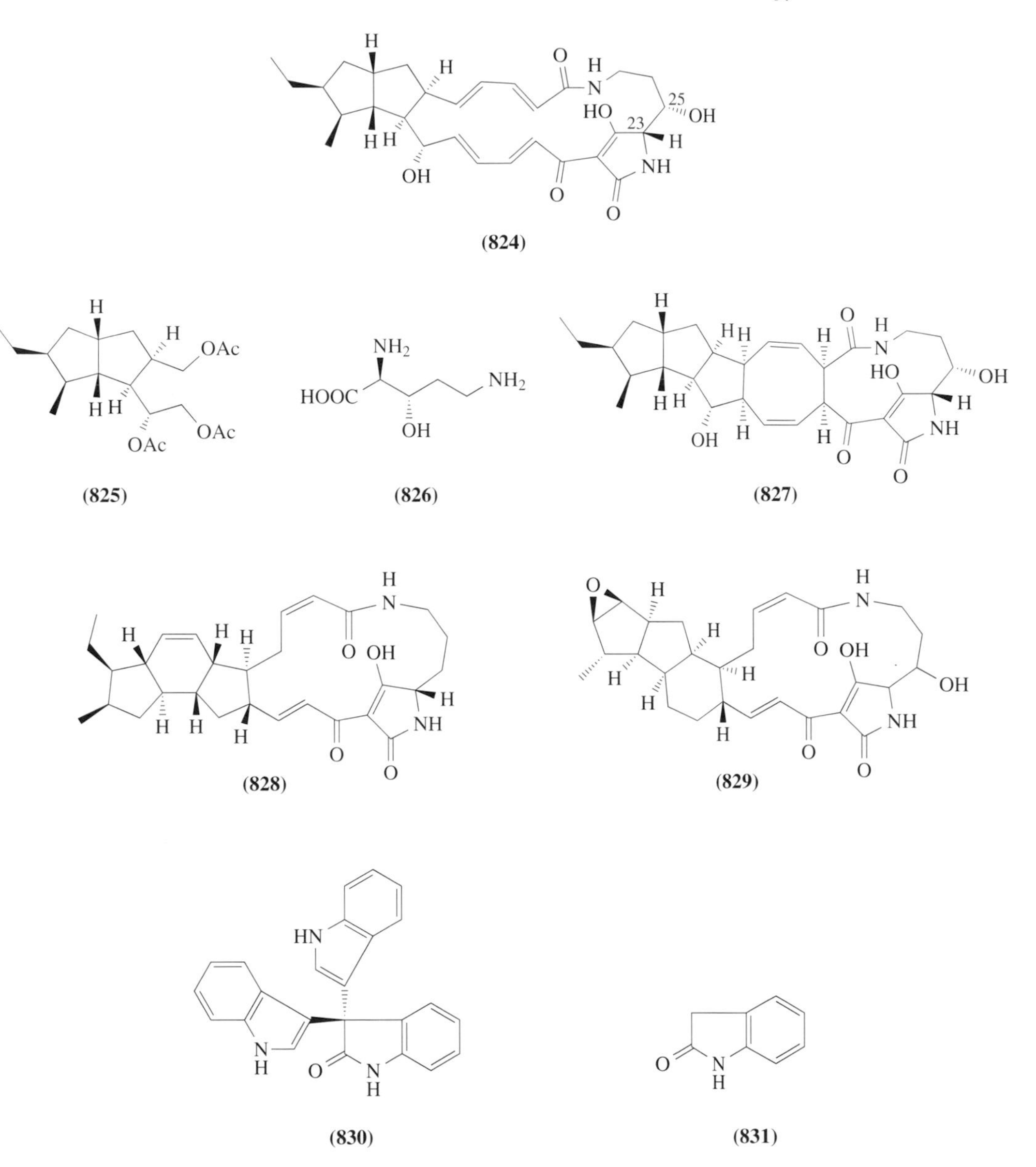

(824)

(825) (826) (827)

(828) (829)

(830) (831)

Imamura *et al.* isolated two new antimycin antibiotics, named urauchimycins A (**832**) and B (**833**), from a fermentation broth of a *Streptomyces* sp., which was obtained from an unidentified sponge collected at Urauchi-cove, Iriomote, Japan.[687] They were the first antimycin antibiotics which possess a branched side-chain moiety. They exhibited inhibitory activity against morphological differentiation of *Candida albicans*.

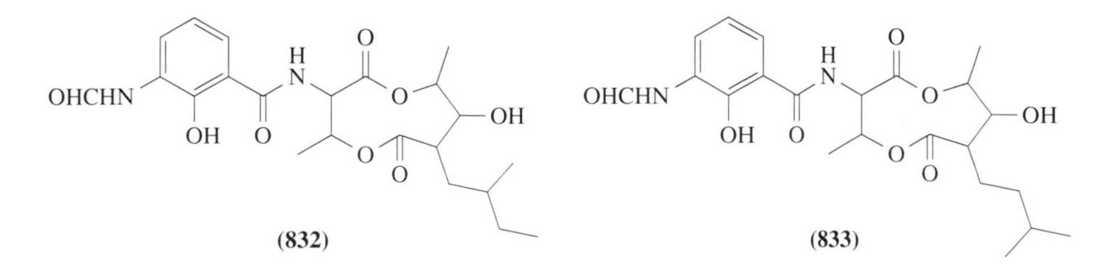

(832) (833)

A number of interesting carotenoids including aromatic carotenoids have been isolated from marine sponges. Carotenoids are noted as quenchers and/or scavengers of active oxygen species.

The origins of carotenoids remain unknown, especially whether sponges can modify dietary carotenoids into suitable structures through enzymatic bioconversion or whether the precursors and/or carotenoids themselves are biosynthesized by some symbiotic or co-existing microorganisms before being transferred to the host sponge and stored. To study the origins of the carotenoids in marine sponges, Miki *et al.* investigated a marine bacterium *Pseudomonas* sp. strain number KK10206C, which was associated with a marine sponge *Halichondria okadai*, to obtain a novel C_{50}-carotenoid, okadaxanthin (**834**), through a visible absorption spectrum-guided isolation procedure.[688] Its structure, 2,2′-bis(4-hydroxy-2-methyl-2-butenyl)-ε,ε-carotene, was elucidated mainly by spectroscopic methods. Okadaxanthin (**834**) turned out to be a potent singlet oxygen quencher, approximately 10 times as strong as α-tocopherol.

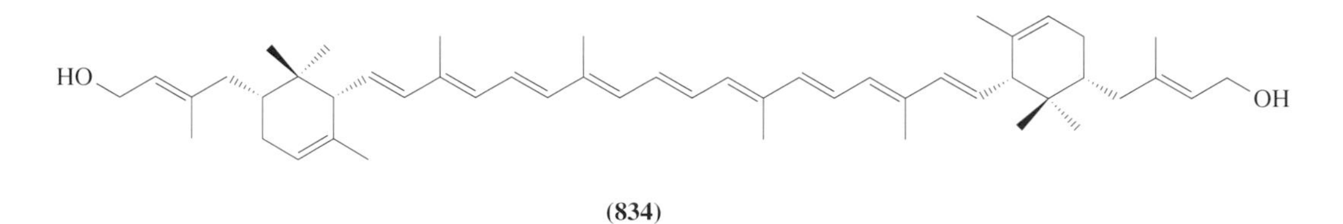

(834)

Baker and co-workers isolated a new diketopiperazine, cyclo-(L-proline-L-methionine) (**835**), along with five known diketopiperazines and two known phenazine alkaloid antibiotics from a strain of *Pseudomonas aeruginosa* obtained from a sponge *Isodictya setifera* Topsent (family Esperiopsidae), which was collected from Hut Point and Danger Slopes on Ross Island, Antarctica. Diketopiperazines including (**835**) were found to be inactive as antibiotics or cytotoxins.[689]

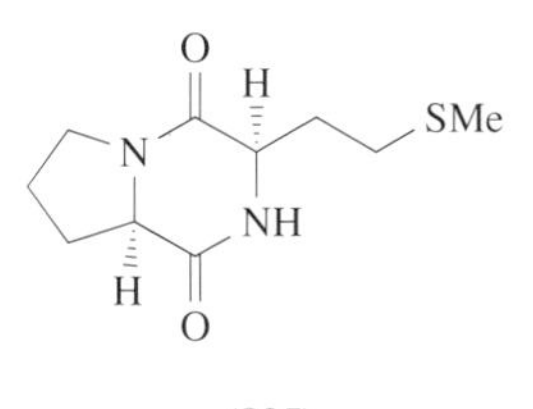

(835)

Marine bacteria obtained from other than sponges were also investigated as follows. Fenical and co-workers isolated eight new secondary metabolites, macrolactins A–F (**836**)–(**841**), and macrolactinic and isomacrolactinic acids (**842**) and (**843**), of an unprecedented C_{24} linear acetogenin origin, from the culture broth of an apparently taxonomically unclassifiable marine bacterium.[690] This bacterium was obtained from a deep-sea core, collected at a depth of 980 m in the North Pacific, and was a motile, Gram-positive, oxidase and catalase positive, unicellular bacterium with a strong salt requirement for growth. Macrolactin A (**836**), the parent aglycone, showed selective antibacterial activity against *Bacillus subtilis* and *Staphylococcus aureus* at concentrations of 5 μg/disk and 20 μg/disk, respectively, by using standard agar plate-assay disk methods. Macrolactin A (**836**) also inhibited B16-F10 murine melanoma cell replication with *in vitro* IC_{50} values of 3.5 μg ml^{-1}, and (**836**) was a potent inhibitor of *Herpes simplex* type I virus (strain LL), as well as type II virus (strain G) with IC_{50} values of 5.0 μg ml^{-1} and 8.3 μg ml^{-1}, respectively. Further macrolactin A (**836**) protected T-lymphoblast cells against human HIV viral replication. Unfortunately, fermentation of this deep-sea bacterium was unreliable, and macrolactin A (**836**) was no longer available in significant yield. The structures of macrolactins were determined by a combination of spectroscopic techniques that included extensive use of proton NMR spectroscopy. The complete relative and absolute stereochemistries of macrolactines were determined by a combination of ^{13}C-acetonide analysis, degradation, and chemical correlations in collaboration with Rychnovsky's research group.[691]

A marine-derived actinomycete of the genus *Streptomyces* was isolated from the surface of an unidentified gorgonian of the genus *Pacifigorgia* by Fenical and co-workers.[692] From the culture broth of the *Streptomyces* sp. two closely related novel compounds, octalactins A (**844**) and B (**845**), were isolated and their structures, including relative stereochemistries, were firmly established, with (**844**) being defined by single-crystal X-ray crystallographic analysis. These metabolites contain unusual saturated eight-membered lactone functionalities and the lactone ring has a boat-chair conformation with a *cis* lactone in the solid state, which was consistent with the result of the

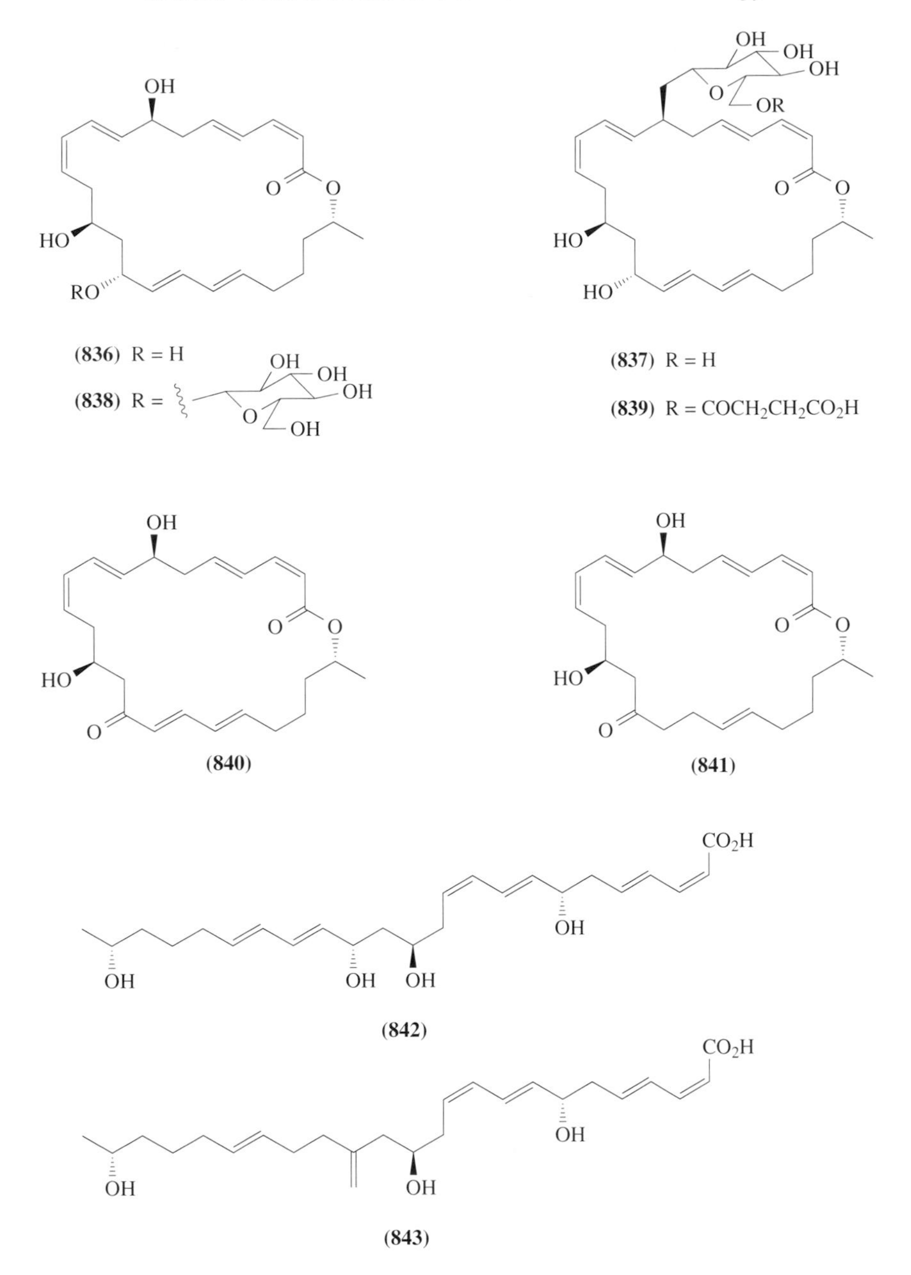

molecular mechanics calculation. Octalactin A (**844**) exhibited strong cytotoxic activity toward B16-F10 murine melanoma and HCT-116 human colon tumor cell lines with IC_{50} values of 0.0072 μg ml^{-1} and 0.5 μg ml^{-1}, respectively, while quite surprisingly octalactin B (**845**) was completely inactive in the cytotoxicity assay.

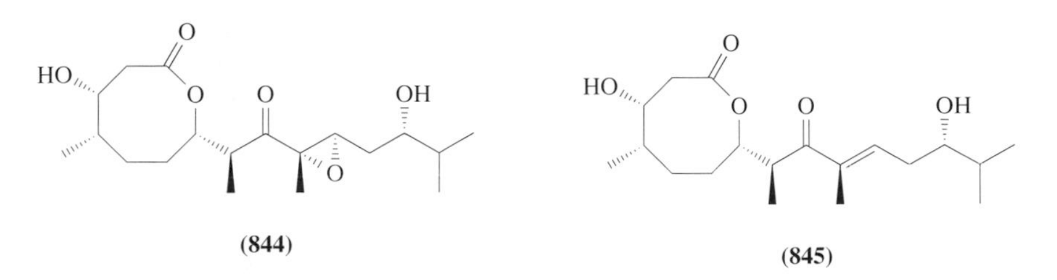

Fenical's research group reported the isolation of several more unique secondary metabolites from marine bacteria. From the culture of an unidentified marine bacterium of the order Actinomycetales, which was obtained by serial dilution and plating methods from a marine sediment sample collected in the shallow waters of Bodega Bay, California, a novel macrolide, maduralide (**846**), possessing a 24-membered lactone ring was isolated.[693] The 6-deoxy-3-*O*-methyl talopyranoside in maduralide (**846**) was unprecedented among macrolide glycosides. Maduralide (**846**) showed weak antibiotic activity against *Bacillus subtilis*. A new γ-lactone (**847**), (1′*R*,2*S*,4*S*)-2-(1-hydroxy-6-methylheptyl)-4-hydroxymethyl-butanolide, was isolated from the culture broth of a marine actinomycete, which was isolated from a sediment sample collected at 24 m depth in the Bahamas Islands.[694] Physiological functions for butanolide (**847**) were not demonstrated. Furthermore, four new alkaloid esters of the rare phenazine class (**848**)–(**851**) were isolated from the fermentation extract of a filamentous bacterium of an unknown *Streptomyces* sp., which was isolated from the shallow sediments in Bodega Bay.[695] The sugar ester functionalities of (**848**)–(**851**) were uncommon, and the natural quinovose in (**848**)–(**851**) was found to possess the rare L configuration from the sign of the Cotton effect in the CD spectrum, although D-quinovose was a common hexose found in the saponin glycosides from marine sea cucumbers. These phenazine esters were shown to exhibit modest broad-spectrum activity against numerous Gram-positive and Gram-negative bacteria. Phenazine (**848**) showed its most potent activity against *Hemophilus influenzae* with Minimum Inhibitory Concentration (MIC) values of 1 μg ml^{-1} and also inhibited *Clostridium perfringens* (MIC = 4 μg ml^{-1}). Phenazine (**851**) was more active overall, showing inhibitory activities against *E. coli* (4 μg ml^{-1}), *Salmonella enteritidis* (4 μg ml^{-1}), and *Clostridium perfringens* (4 μg ml^{-1}). These compounds were not appreciably cytotoxic against murine and human cancer cells tested *in vitro*.

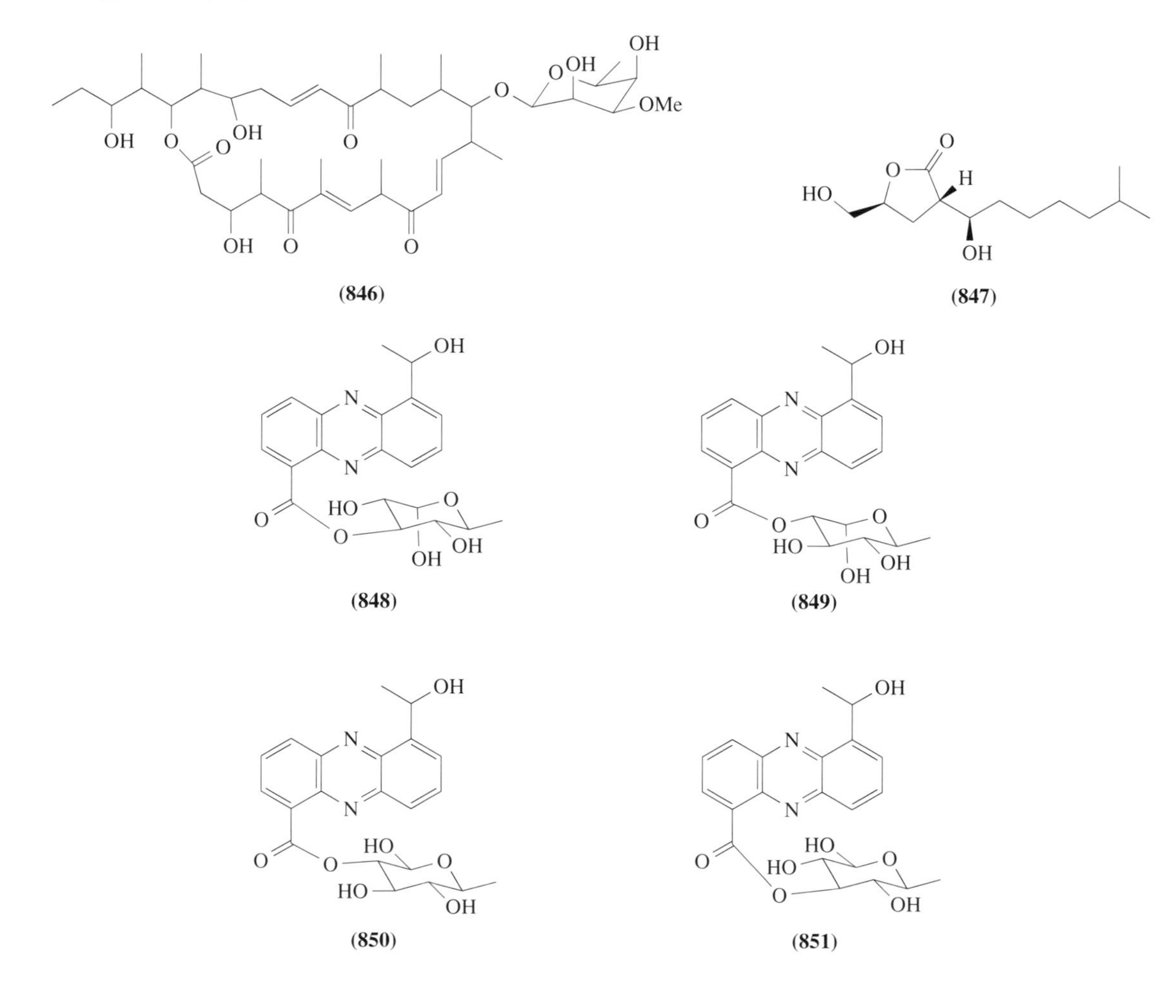

(**846**) (**847**) (**848**) (**849**) (**850**) (**851**)

Several bioactive peptides have been discovered from marine bacteria. Butler and co-workers isolated siderophores having peptide structures, named alterobactins A (**852**) and B (**853**), from an open-ocean bacterium *Alteromonas luteoviolacea*, which was found in oligotrophic and coastal waters.[696] A siderophore is a low-molecular-mass compound with a high affinity for ferric ion which is secreted by microorganisms in response to low-iron environments. Virtually all microorganisms require iron for growth. The paucity of iron in surface ocean water has spurred a lively debate concerning iron limitation of primary productivity, yet little is known about the molecular mechanisms used by marine microorganisms to sequester iron. Alterobactin A (**852**) had an exceptionally high affinity constant for ferric iron, suggesting that at least some marine microorganisms may have developed higher-affinity iron chelators as part of an efficient iron-uptake mechanism which is more effective than that of their terrestrial counterparts.

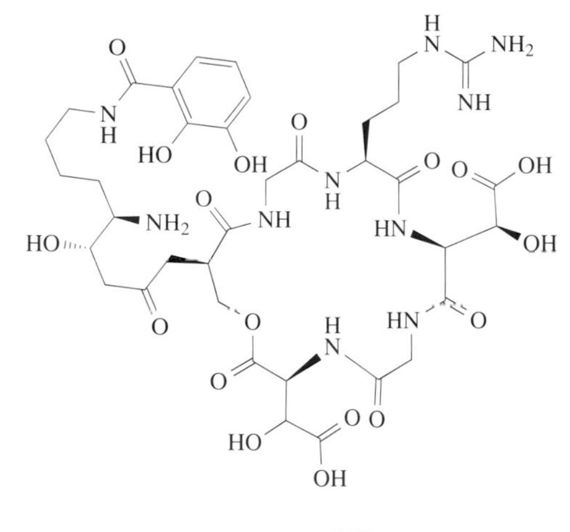

(852)

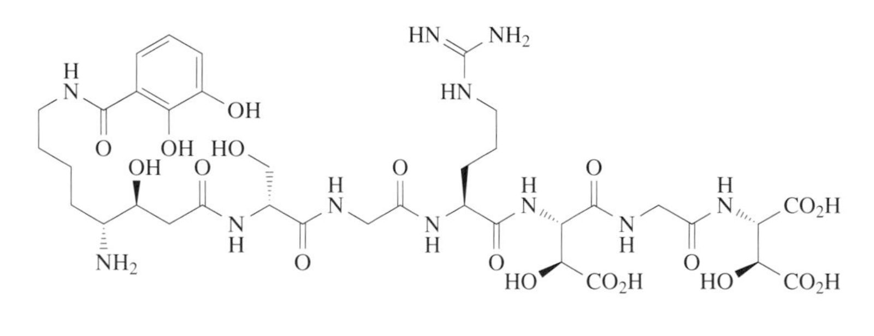

(853)

Fenical's research group isolated antiinflammatory depsipeptides, salinamides A (**854**) and B (**855**), from a marine Streptomycete, which was isolated from the surface of the jellyfish *Cassiopeia xamachana* collected in the Florida Keys.[697] Fermentation in seawater-based media, followed by vacuum-flash chromatography, and reversed-phase HPLC, yielded only salinamide A (**854**) as approximately 9% of the dry extract. A subsequent fermentation yielded almost equal portions of salinamides A (**854**) and B (**855**); the latter was a crystalline compound, and its structure was established by single-crystal X-ray diffraction analysis. The X-ray analysis determined only the relative stereostructure, and chiral GC analysis of the hydrolytic fragments revealed the absolute stereochemistry (D-Thr and L-allo-Thr). The core of salinamide B (**855**) was a bicyclic hexa-depsipeptide with two ester links involving serine and threonine residues as well as an aromatic ether link. The connectivity of salinamide B (**855**) severely limited the flexibility of the central core—a feature that complicated the NMR analysis by increasing relaxation times—and a CPK model showed a tightly packed molecular interior. Upon treatment with HCl, salinamide A (**854**) was converted into a chlorohydrin that was identical in all respects to salinamide B (**855**). Salinamides A (**854**) and B (**855**) exhibited moderate antibiotic activity against Gram-positive bacteria. The most potent *in vitro* activity was against *Streptococcus pneumoniae* and *Staphylococcus pyrogenes* with MIC values of 4 μg ml^{-1} for salinamide A (**854**) and 4 μg ml^{-1} and 2 μg ml^{-1} for salinamide B (**855**). More importantly, salinamides A (**854**) and B (**855**) showed potent topical antiinflammatory activity using the phorbol ester-induced mouse ear edema assay. Salinamide A (**854**) showed 84% inhibition of edema and salinamide B (**855**) showed 83% inhibition at the standard testing dose of 50 μg/ear.

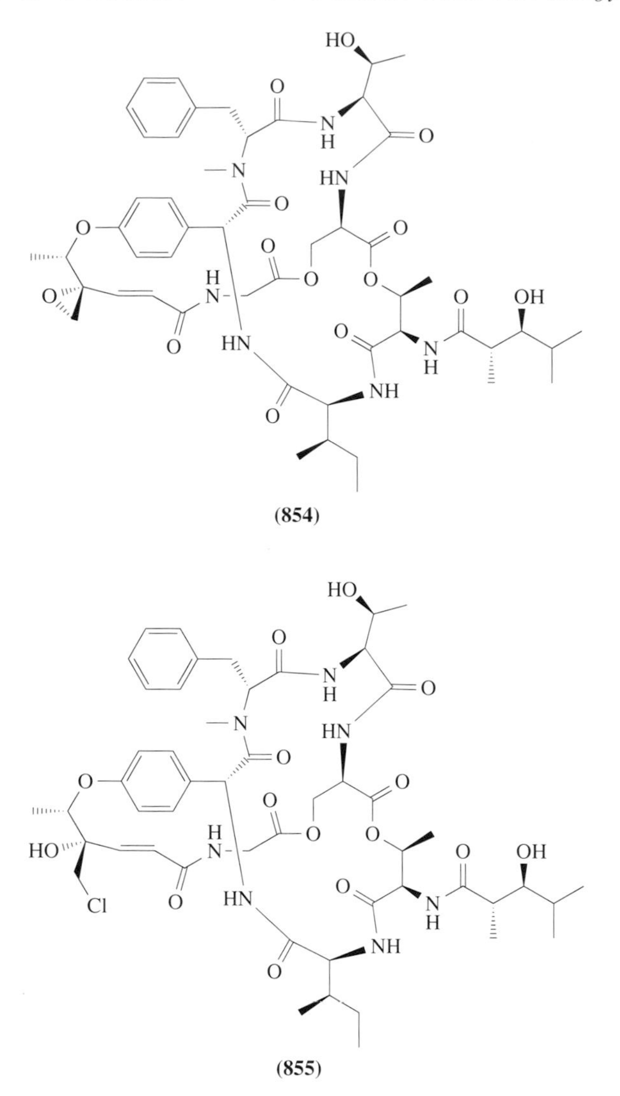

(854)

(855)

A cyclic acylpeptide, halobacillin (**856**), was isolated from cultures of a *Bacillus* species, obtained from a marine sediment core taken at 124 m near the Guaymas Basin, Mexico.[698] Halobacillin (**856**) was the first acylpeptide of the iturin class produced by a marine isolate. The iturins are a class of cyclic acylpeptides produced exclusively by several *Bacillus* species, and are characterized as polar cyclic heptapeptides with lipophilic β-acyloxy or β-amino fatty acid components. Halobacillin (**856**) exhibited moderate human cancer cell cytotoxicity, but in contrast to the iturins, no antifungal or antibiotic activity.

A known metabolite, andrimid (**857**), and three new related metabolites, moiramides A (**858**), B (**859**), and C (**860**), were isolated by Andersen and co-workers from solid agar cultures of a marine isolate of the bacterium *Pseudomonas fluorescens*, obtained from tissues of an unidentified tunicate collected at Prince of Wales Island in Moira Sound, Alaska.[699] Andrimid (**857**) was first isolated from cultures of an *Enterobacter* sp. that was an intracellular symbiont of the brown planthopper *Nilaparvata lugens*, and it was found to exhibit potent activity against *Xanthomonas campestris* pv. *oryzae*, the pathogen responsible for causing bacterial blight in rice plants.[700] Andrimid (**857**) and moiramide B (**859**) were found to exhibit potent *in vitro* inhibition of methicillin-resistant *Staphylococcus aureus* with MIC values of 2 $\mu g\ ml^{-1}$ and 0.5 $\mu g\ ml^{-1}$, respectively. Stable isotope incorporation experiments were carried out to elucidate the biogenesis of the acylsuccinimide fragment of andrimid (**857**) that was essential for antimicrobial activity, demonstrating that the acylsuccinimide fragment was derived from a combination of acetate and amino acid building blocks. It was proposed that the biosynthesis proceeded through a dipeptide-like intermediate formed from γ-amino-β-keto acids that were in turn formed from valine and glycine homologated with acetate, presumably via malonyl-CoA.

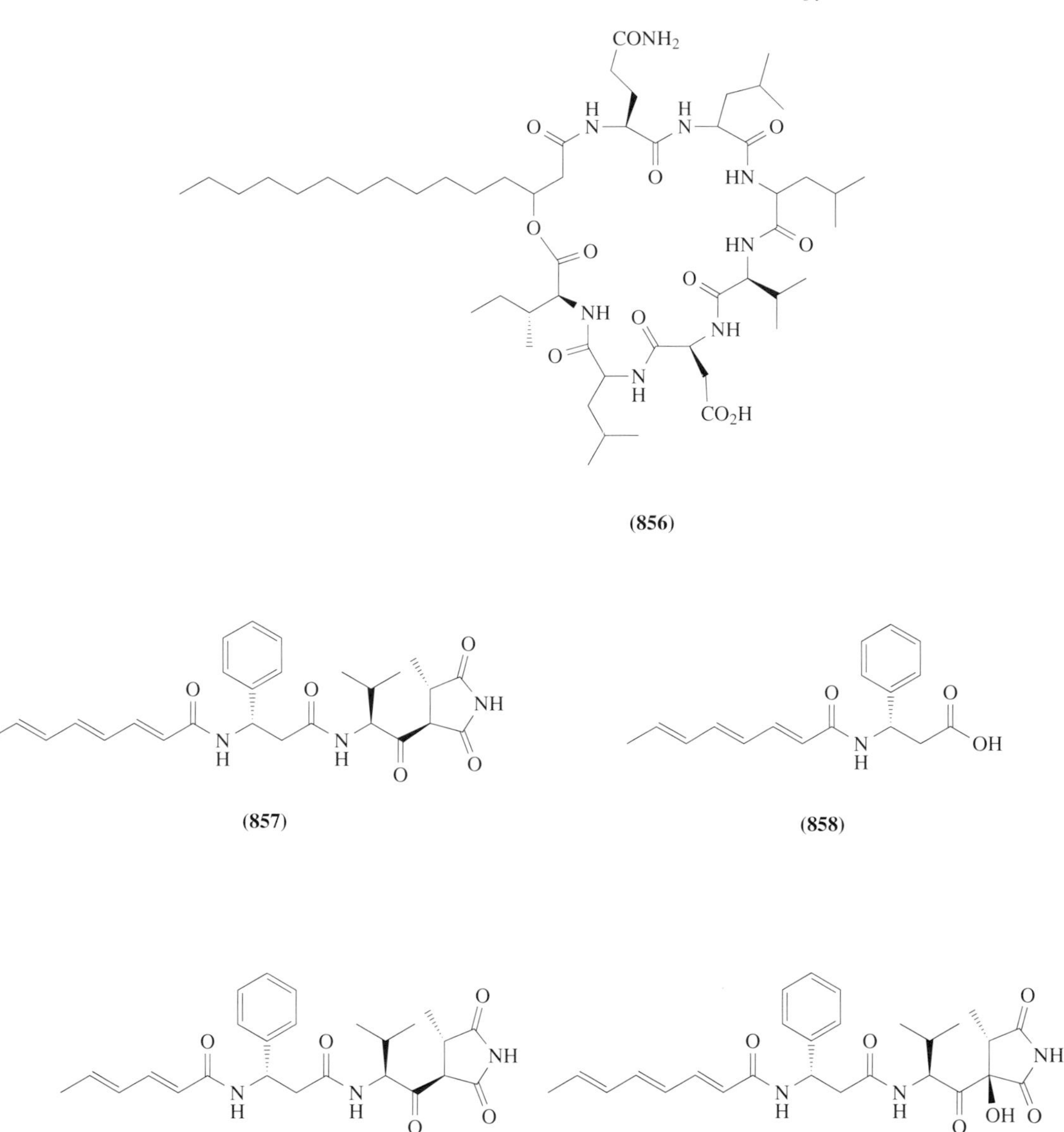

(856)

(857) **(858)**

(859) **(860)**

Andersen and co-workers also isolated a cyclic decapeptide antibiotic, loloatin B (**861**), from the culture of a *Bacillus* sp. obtained from the tissues of an unidentified tube worm collected at 15 m depth off of Loloata Island, Papua New Guinea.[701] Loloatin B (**861**) inhibited the growth of methicillin-resistant *Staphylococcus aureus* (MRSA), vancomycin-resistant *Enterococcus* sp. (VRE), and penicillin-resistant *Streptococcus pneumoniae* with MICs of 1–2 μg ml^{-1}. VRE first appeared in the early 1990s and they are rapidly spreading across North America. It was reported that an outbreak of VRE had a 73% mortality rate and there are no effective antibiotics currently available for such organisms.

Davidson and Schumacher isolated two new caprolactams, caprolactins A (**862**) and B (**863**), from the liquid culture of an unidentified Gram-positive bacterium obtained from a deep-ocean sediment sample.[702] Caprolactins A (**862**) and B (**863**), obtained as an inseparable mixture, were composed of cyclic-L-lysine linked to 7-methyloactanoic acid and 6-methyloctanoic acid, respectively. The structures were proposed using spectroscopic methods and confirmed by synthesis. Both caprolactins A (**862**) and B (**863**) were cytotoxic towards human epidermoid carcinoma KB cells and human colorectal adenocarcinoma LoVo cells, and exhibited antiviral activity towards *Herpes simplex* type II virus.

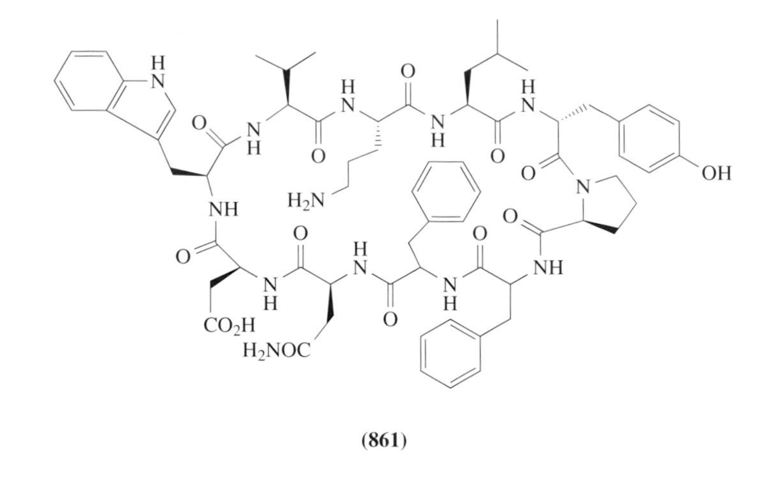

(861)

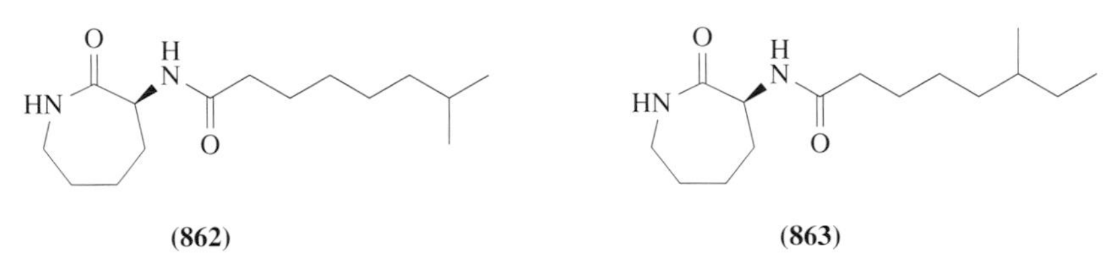

(862) **(863)**

Several pyrones and related compounds have been reported as marine bacterial metabolites. Oncorhyncolide (**864**) was isolated from liquid shake cultures of the bacterial isolate, which was obtained from a surface seawater sample taken near a chinook salmon (*Oncorhyncus tshawytscha*) net-pen farming operation.[703] This bacterial isolate was an oxidase-positive, Gram-negative bacillus that did not closely resemble any previously described species. Oncorhyncolide (**864**) was a biologically inactive but chemically novel metabolite. Although oncorhyncolide (**864**) had a relatively simple structure, it was not possible to predict the exact nature of its biogenetic origin with any degree of certainty. Stable isotope incorporation studies were therefore undertaken to show that all of the carbons, including the branching methyls at C-15 and C-16 in oncorhynolide (**864**) were derived from acetate.[704]

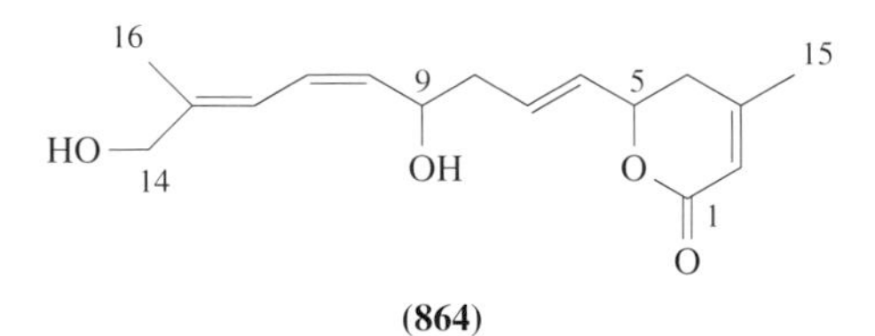

(864)

Four new α-pyrones, elijopyrones A–D (**865**)–(**868**), were isolated from a cultured marine actinomycete, which was obtained from a sediment collected from the San Elijo Lagoon, Cardiff, California.[705] The structures of elijopyrones A–D (**865**)–(**868**) were determined by comprehensive spectral analyses.

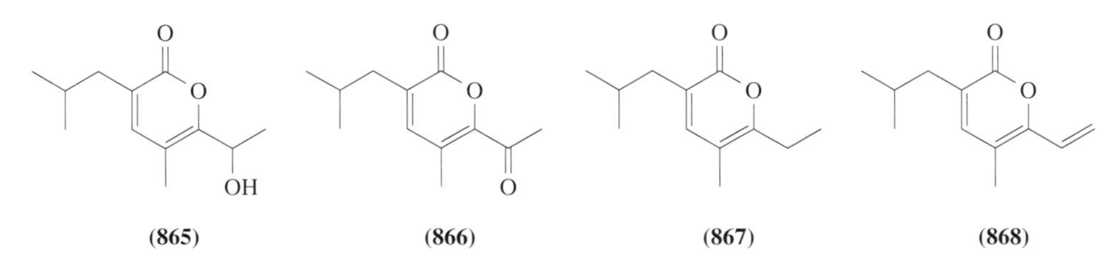

(865) **(866)** **(867)** **(868)**

Three cytotoxic α-pyrones, lagunapyrones A–C (**869**)–(**871**), were produced in fermentation by a marine bacterium, which was an unidentified actinomycete isolated from sediment collected in the Agua Hedionda Lagoon in Carlsbad, California.[706] The structure was assigned on the basis of comprehensive spectroscopic analyses, and transformation of lagunapyrone B (**870**) into its [1′,3′-$^{13}C_2$]-labeled acetonide allowed the relative stereochemistry of the flexible 1,3-diol moiety to be determined as 19R*, 20R*, and 21S*. Assignment of the relative stereochemistry at C-6 and C-7 was accomplished as 6R*, 7S* by computer analysis of the vicinal proton coupling constants, and by comparison of these values with two isomeric synthetic analogues. Since the C-6, C-7 stereochemistry could not be related to the relative stereochemistry assigned at C-19 through C-21, the alternative 6S*, 7R*, 19R*, 20R*, and 21S* configuration may also be assigned. The carbon skeleton of the lagunapyrones was not previously observed. Biosynthetically, these compounds could be derived by the condensation of acetate and/or propionate units, or by a combination including methylation of the carbon chain through the methionine pathway. Lagunapyrone B (**870**) showed modest *in vitro* cytotoxicity, $ED_{50} = 3.5\ \mu g\ ml^{-1}$, against the human colon cancer cell line HCT-116. A structurally similar metabolite, leptomycin, isolated from a *Streptomyces* sp. inhibits the proliferation of *Schizosaccharomyces pombe* in both the G1 and the G2 phases of the cell cycle.[707]

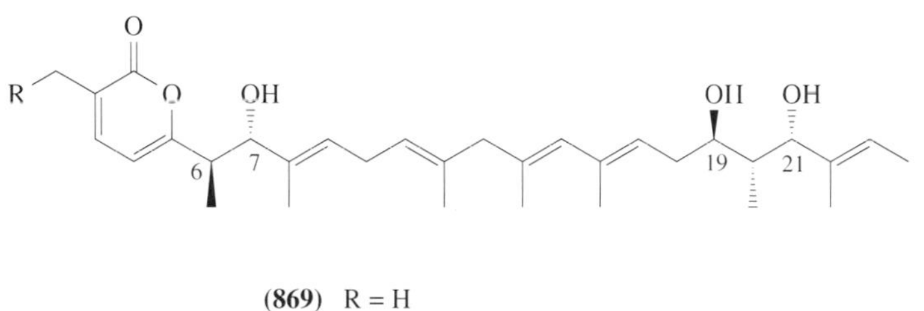

(**869**) R = H
(**870**) R = CH_2Me
(**871**) R = CH_2CH_2Me

Four new α-pyrone-containing metabolites, wailupemycins A–C (**872**)–(**874**) and 3-pi-5-deoxy-enterocin (**875**) were isolated together with known compounds enterocin (**876**) and 5-deoxyenterocin (**877**) from a *Streptomyces* sp., which was obtained from shallow water marine sediments collected at Wailupe beach park on the southeast shore of Oahu, Hawaii.[708] Compounds (**872**)–(**875**) were tested for antimicrobial activity against *Bacillus subtilis*, *Staphylococcus aureus*, and *Escherichia coli in vitro* using the paper disk diffusion method. Compound (**875**) exhibited activity against only *S. aureus* (18 mm zone of inhibition at 1 mg/6 mm disk), while compound (**872**) was inhibitory to only *E. coli* (15 mm zone at 0.1 mg/6 mm disk). Compounds (**873**) and (**874**) were inactive against all three test organisms at 0.1 mm/disk.

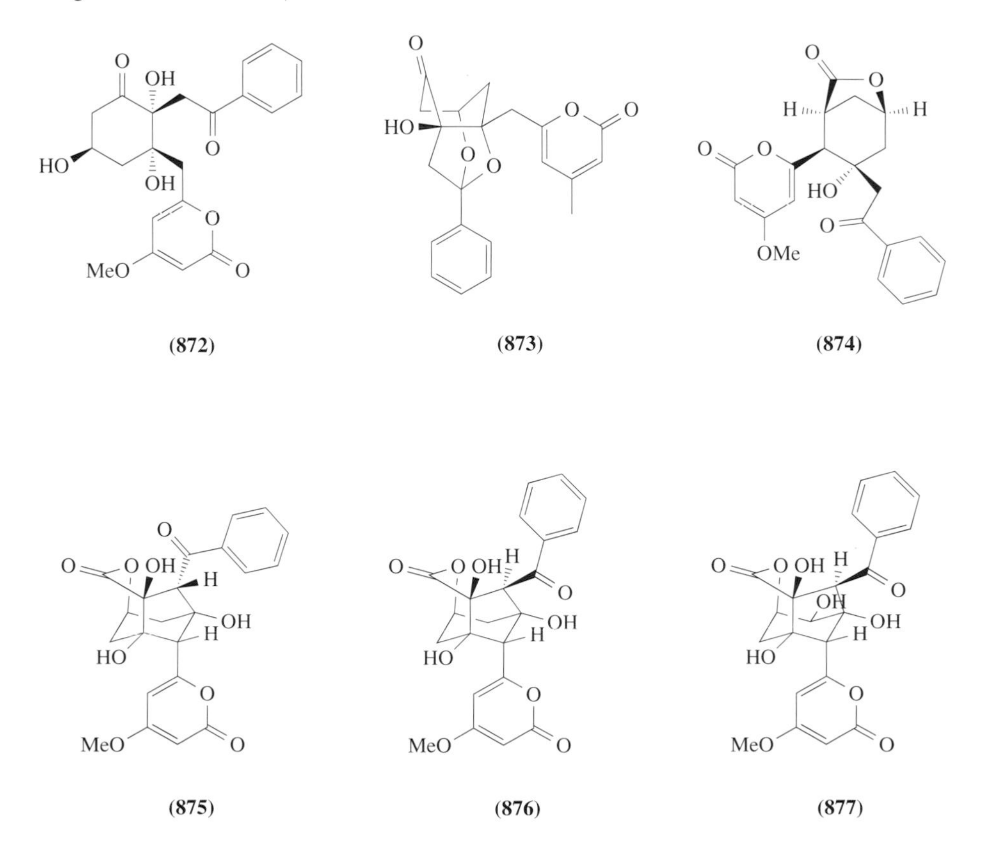

A new pluramycin-type metabolite, γ-indomycinone (**878**), was isolated together with some previously known metabolites from the culture broth of an actinomycete identified as *Streptomyces* sp. obtained from a deep-ocean sediment sample.[709] Pluramycin antibiotics are most commonly isolated from terrestrial *Streptomyces* sp., and contain an anthraquinone-γ-pyrone nucleus. γ-Indomycinone (**878**) bore a 1-hydroxy-1-methylpropyl side-chain attached to the anthraquinone-γ-pyrone nucleus. Although the crude EtOAc-extract of the entire culture broth of this bacterium was potently cytotoxic against the KB cell line with an MIC value of $\leqslant 0.001\ \mu g\ ml^{-1}$, the cytotoxicity of the new compound (**878**) was not described, but the rubiflavin antibiotic complex, which was related to (**878**), was shown to bind to DNA, inhibiting DNA synthesis.

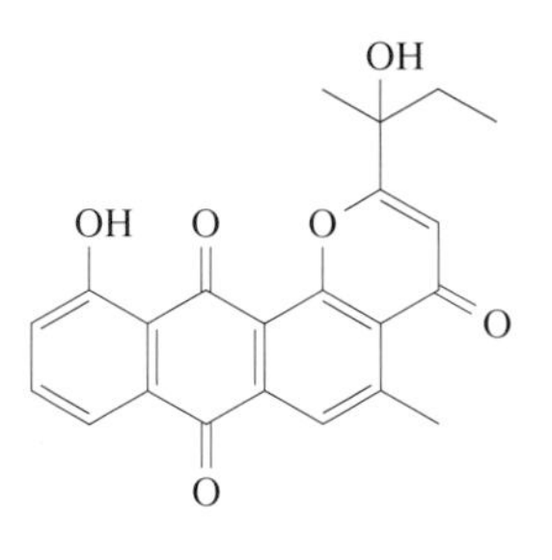

(878)

Magnesidin, a magnesium-containing antibiotic, was isolated from a marine bacterium *Pseudomonas magnesiorubra*, obtained from the surface of macro alga *Caulerpa peltate*, and was described to consist of two analogous components and magnesium.[710,711] Imamura *et al.* isolated magnesidin A (**879**), a component of magnesidin, from halophilic bacterium *Vibrio gazogenes*, obtained from marine mud, and its structure including stereochemistry was elucidated by mainly NMR data.[712] Magnesidin A (**879**) as well as its magnesium-free sample were active against microalga *Prorocentrum micans* at the concentration of $1\ \mu g\ ml^{-1}$ and weakly active against *Bacillus subtilis*.

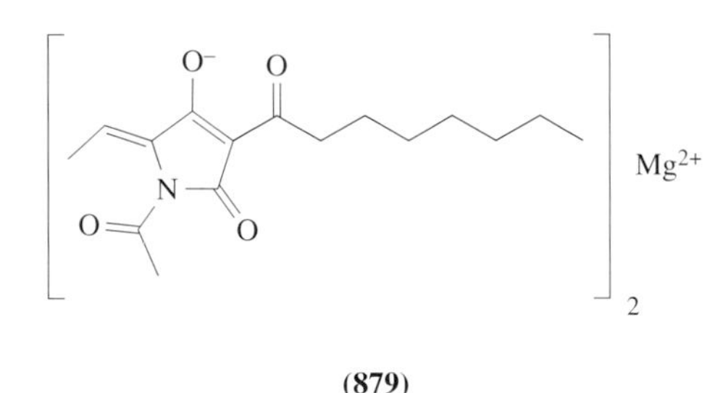

(879)

Kobayashi *et al.* investigated extracts of cultured marine bacterium *Flavobacterium* sp., which was isolated from the bivalve *Cristaria plicata* collected in Ishikari Bay, Hokkaido, and succeeded in isolating two new sphingolipids, flavocristamides A (**880**) and B (**881**),[713] possessing a sulfonic acid group. The structures of compounds (**880**) and (**881**) along with the stereochemistry of all chiral centers were elucidated on the basis of extensive spectroscopic analyses including FAB MS/MS measurements as well as degradation experiments such as acid hydrolysis and Lemieux oxidation. Both (**880**) and (**881**) exhibited marked inhibitory activity against DNA polymerase α. Ceramide-1-sulfonic acids were previously obtained from marine diatom *Nitzschia alba*,[714] and studies on the synthesis of ceramide-1-sulfonic acids were also reported.[715] An identical compound to flavocristamide B (**881**), named sulfobactin A, was isolated independently by Kamiyama *et al.* almost at the same time from the culture broth of terrestrial bacterium *Chryseobacterium* sp. (*Flavobacterium* sp.) as a von Willebrand factor (vWF) receptor antagonist.[716,717] Sulfobactin A (=flavocristamide B (**881**)) inhibited the binding of vWF to its receptor with an IC_{50} value of 0.47 μM, and it also inhibited ristocetin-induced agglutination in human platelets fixed with paraformaldehyde with an IC_{50} value of 0.58 μM.

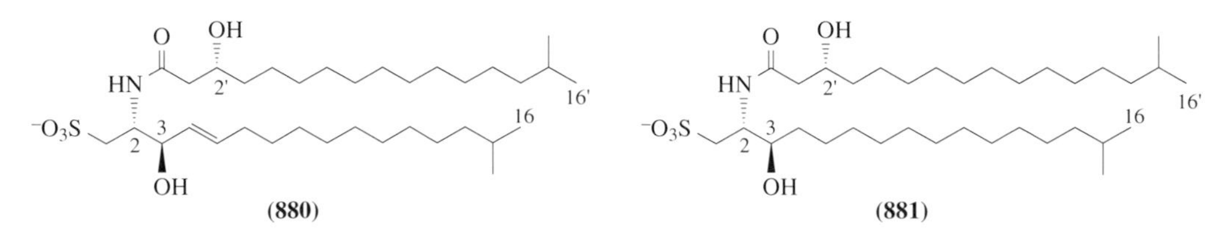

(880) **(881)**

8.07.9.4.2 Fungi

Chemical studies on the bioactive metabolites of fungal isolates from marine environments have been limited. Marine fungi, however, can be expected as a potential source of new bioactive substances from the following studies. Researchers of the Sankyo group isolated the novel platelet activating factor (PAF) antagonists, phomactins A (**882**),[718] B (**883**), B_1 (**884**), B_2 (**885**), C (**886**), D (**887**),[719] E (**888**), F (**889**), and G (**890**)[720] from the cultured broth of a marine deuteromycetes, *Phoma* sp., which lived in the shell of a crab *Chionoecetes opilio* collected off the coast of Fukui Prefecture, Japan. Phomactins are unique diterpenoids and their structures, including the absolute configurations, were established by the X-ray analyses performed on a mono-*p*-bromobenzoyl derivative of phomactin A (**882**) as well as a diketone derived from phomactin B (**883**). Phomactins B_1 (**884**) and B_2 (**885**) were chemically correlated with phomactin B (**883**). Compounds (**882**)–(**890**) exhibited PAF antagonistic activities. In particular, phomactin D (**887**) potently inhibited the binding of PAF to its receptors and PAF-induced platelet aggregation with IC_{50} values of 0.12 μM and 0.80 μM, respectively, while other compounds antagonized PAF action at higher concentrations. Phomactin A (**882**), however, exhibited no effect on adenosine diphosphate-, arachidonic acid-, and collagen-induced platelet aggregation. Phomactin A (**882**) was thus considered to be a new type of specific PAF antagonist. Chu *et al.* also reported several PAF antagonists with the same skeletal framework as phomactins.[721,722]

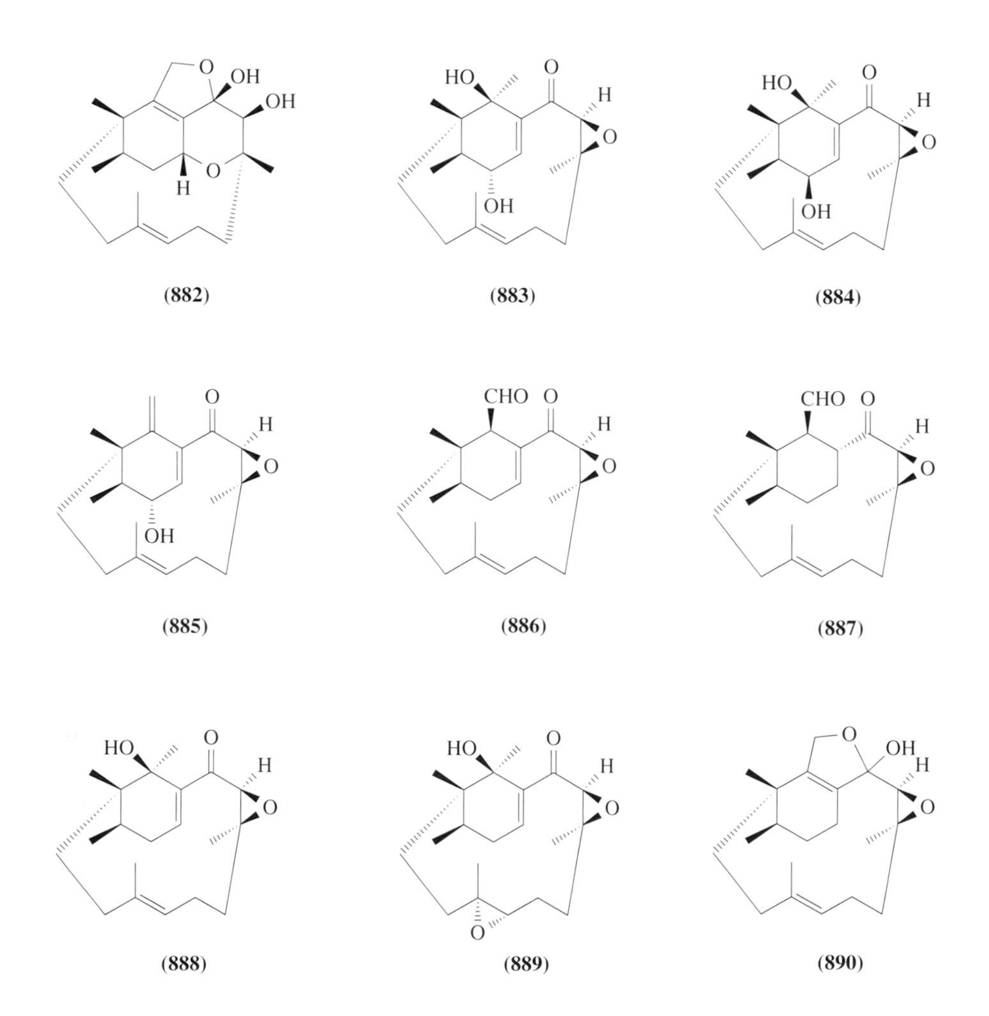

Two lipophilic tripeptides, fellutamides A (**891**) and B (**892**), have been isolated by Kobayashi's research group from the cultured mycelium of a fungus *Penicillium fellutanum* Biourge which was isolated from the gastrointestine of the marine fish *Apogon endekataenia* Bleeker, collected off Manazuru beach, Kanagawa Prefecture, Japan.[723] The structures of fellutamides were elucidated by applying several types of 2D NMR techniques in combination with FAB MS/MS data. The stereochemistries of amino acid residues were determined by chiral HPLC and GC analyses and the absolute configuration of 3-hydroxydodecanoic acid obtained through hydrolysis was assigned by comparison of the optical rotation with those of standard samples. Fellutamides A (**891**) and B (**892**) were potently cytotoxic against murine leukemia P388 (IC_{50} 0.2 μg ml^{-1} and 0.1 μg ml^{-1}, respectively) and L1210 (IC_{50} 0.8 μg ml^{-1} and 0.7 μg ml^{-1}, respectively) cells, and human epidermoid carcinoma KB (IC_{50} 0.5 μg ml^{-1} and 0.7 μg ml^{-1}, respectively) cells *in vitro*. The fungus *Penicillium fellutanum* is not a marine-specific one, since the same species have been isolated from terrestrial sources. The peptides (**891**) and (**892**) which are structurally related to the leupeptines[724] exhibited stimulating activity on nerve growth factor (NGF) synthesis.[725]

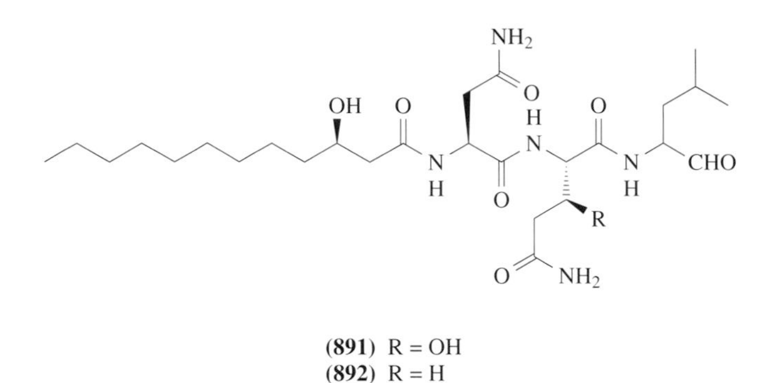

(**891**) R = OH
(**892**) R = H

Kitagawa and co-workers isolated a new polyketide compound, trichoharzin (**893**), from a culture of the imperfect fungus *Trichoderma harzianum*, which was separated from the marine sponge *Micale cecilia*, collected at Amami Island, Kagoshima Prefecture, Japan.[726] The absolute stereochemistry of trichoharzin (**893**) was determined by spectral data and CD spectrum of its dibenzoate derivative. Trichoharzin (**893**) was a new polyketide constructed with an alkylated decalin skeleton and esterified with 3-methylglutaconic acid, a rare acyl moiety. It exhibited inhibitory activity against morphological differentiation of *Candida albicans*.

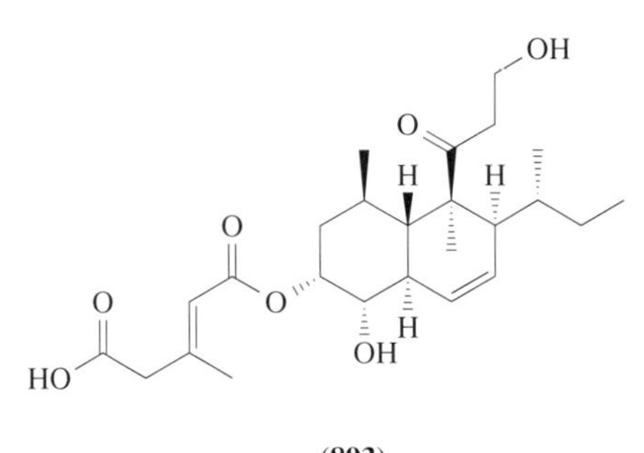

(**893**)

Crews and co-workers also reported studies on marine fungal metabolites. The salt-water culture of an unidentified fungus separated from an Indo-Pacific marine sponge of the genus *Stylotella* sp., collected in the Somosomo Strait near Taveuni, Fiji, was revealed to yield new tetraketide natural products, demethyl nectriapyrone A (**894**) and B (**895**).[727] These α-pyrone-containing compounds are analogous to nectriapyrone (**896**), previously reported from a terrestrial fungus, and also to the α-pyrones obtained from marine bacteria (see Section 8.07.9.4.1).

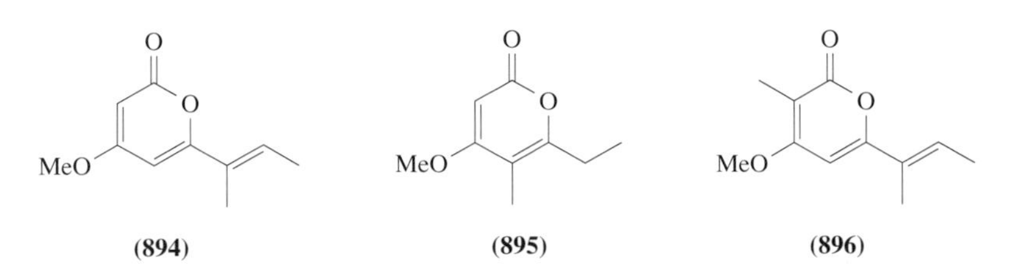

(**894**) (**895**) (**896**)

Three new chlorinated cyclic sesquiterpenes, chloriolins A–C (**897**)–(**899**), were isolated from an unidentified fungus which was initially separated from the Indo-Pacific marine sponge *Jaspis* aff. *johnstoni* and then cultured on marine media.[728] 2D NMR experiments, synthetic transformations, and X-ray crystallography were used to establish the structures of these compounds. The sesquiterpenes (**897**)–(**899**) were, however, inactive in the National Cancer Institute (NCI) disease-oriented screening program.

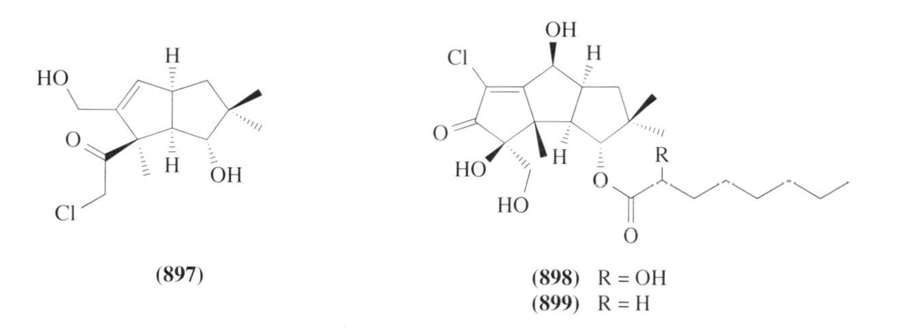

From the salt-water culture of an unidentified fungus separated from inside the encrusting sponge *Spirastrella vagabunda*, collected from the Togian Islands in central Sulawesi, Indonesia, a new polyketide natural product, 14,15-secocurvularin (**900**), was isolated by bioassay guided fractionation, and its structure was elucidated by spectral studies.[729] 14,15-Secocurvularin (**900**) exhibited mild antibiotic activity against *Bacillus subtilis*. Interestingly, curvularin (**901**) had been reported to be isolated from four different fungi (*Curvularia*, *Cochliobolus*, *Penicillium*, and *Alternaria*).

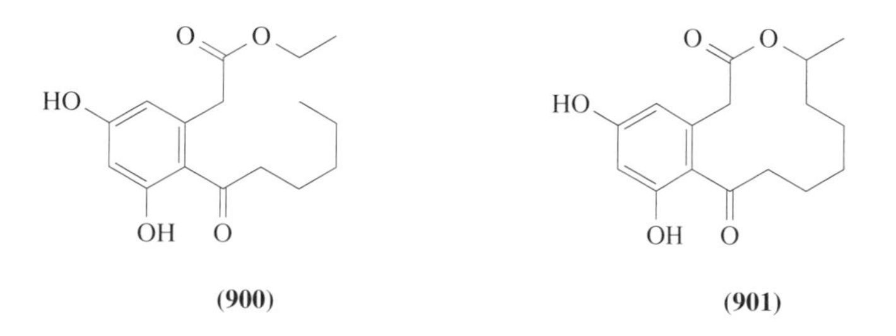

Chen *et al.* isolated three antimicroalgal substances, halymecins A–C (**902**)–(**904**), from the fungus, *Fusarium* sp. of an alga (*Halymenia dilatata*) isolate collected at Palau, by using centrifuged partition chromatography. Another fungus, *Acremonium* sp. of an alga isolate collected at Aburatsubo, Japan, contained two related metabolites, halymecins D (**905**) and E (**906**). Halymecins A–E (**902**)–(**906**) exhibited inhibition activity against diatom *Skeletonema costatum* at 10 $\mu g\ ml^{-1}$. Halymecins A (**902**) also showed antimicroalgal activity against blue-green alga *Oscillatoria amphibia* (500 $\mu g\ ml^{-1}$), green alga *Brachiomonas submarina* (67.5 $\mu g\ ml^{-1}$), and dinoflagellate *Prorocentrum micans* (67.5 $\mu g\ ml^{-1}$).[730] A structurally related antibacterial compound, exophilin A (**907**), was isolated by Nippon Suisan's research group in the culture of the marine microorganism *Exopiala pisciphila*, which was separated from a marine sponge *Mycale adhaerens*.[731] The absolute chemical structure of exophilin A (**907**) was elucidated as a trimer of (3*R*,5*R*)-3,5-dihydroxydecanoic acid by spectroscopic methods and analysis of a degradation product. Exophilin A (**907**) showed antimicrobial activity against Gram-positive bacteria. The marine microorganism *Exopiala pisciphila* is a member of the so-called "black yeasts," a group presenting taxonomic problems, because of their developmental plasticity and the limited number of morphological characteristics available for classification.

A new tricyclic sesquiterpene, isoculmorin (**908**), was isolated by Alam *et al.* from the marine fungus *Kallichroma tethys*, which was cultured in a seawater medium, and the structure was determined by X-ray crystallography.[732]

Numata *et al.* have reported a series of studies on bioactive metabolites of fungi separated from marine algae or fish. Communesins A (**909**) and B (**910**) were isolated from the mycelium of a strain of *Penicillium* sp. stuck on the marine alga *Enteromorpha intestinalis*, and (**909**) and (**910**) exhibited moderate to potent cytotoxic activity in the P-388 lymphocytic leukemia in cell culture with ED_{50} values of 3.5 $\mu g\ ml^{-1}$ and 0.45 $\mu g\ ml^{-1}$, respectively.[733]

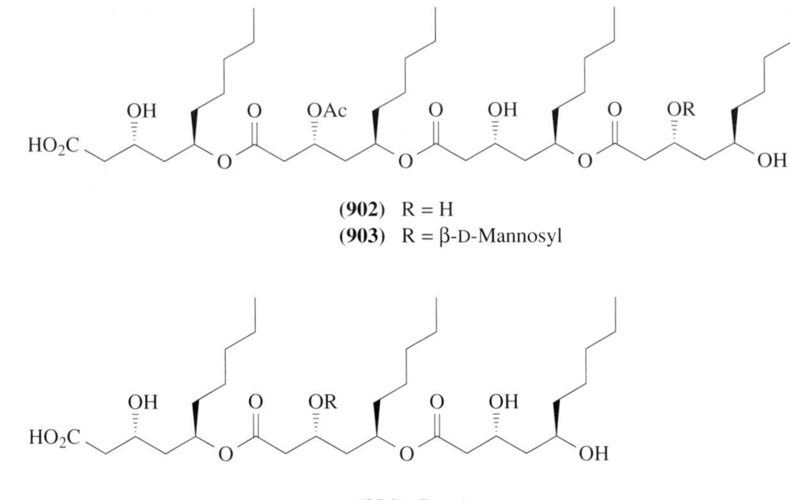

(902) R = H
(903) R = β-D-Mannosyl

(904) R = Ac
(907) R = H

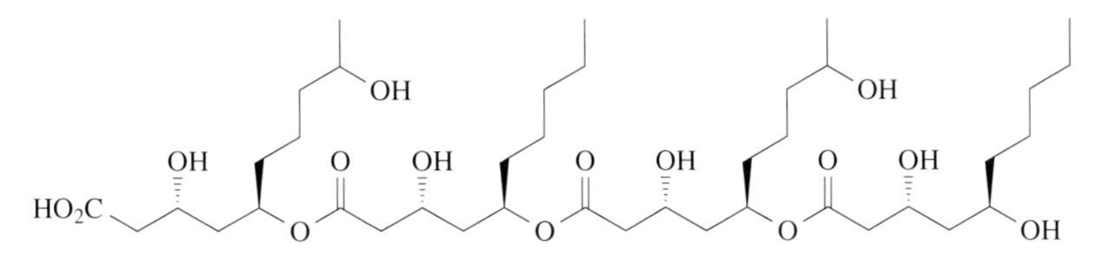

(905)

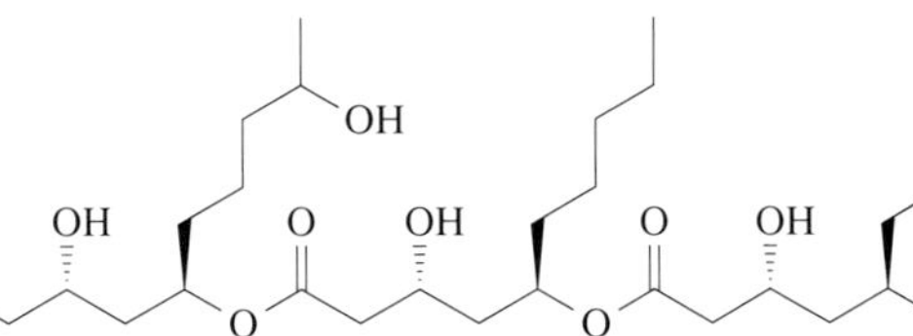
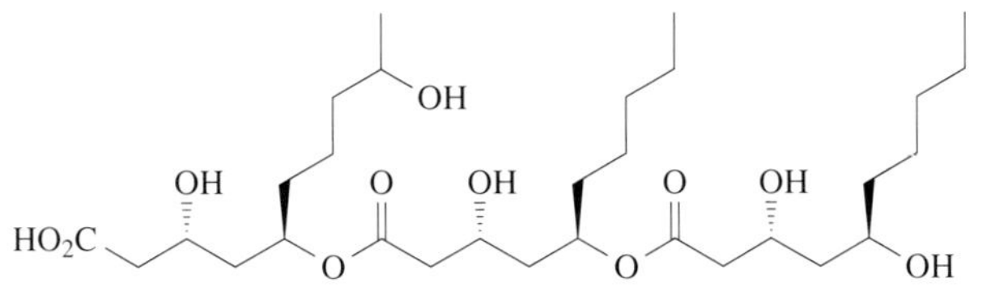

(906)

(908)

(909) **(910)**

Leptosin A (**911**) and its analogues leptosins B, C, D, E, F, G, G_1, G_2, H, I, J, K, K_1, and K_2 were isolated from a strain of *Leptosphaeria* sp. attached to the marine alga *Sargassum tortile*.[734–737] They were potently cytotoxic against cultured P388 cells ((**911**): ED_{50} 1.85 ng ml^{-1}) and leptosins A (**911**) and C (**912**) exhibited significant antitumor activity against Sarcoma 180 ascites (T/C 260% and 293%, respectively) at a dose of 0.5 mg kg^{-1} and 0.25 mg kg^{-1}, respectively. The structure of leptosin K (**913**) with a different configuration from that of leptosin A (**911**) or C (**912**) was determined by X-ray analysis. Leptosin K (**913**) was revealed to exist in a mixture of four conformers by X-ray and NOE studies.[737]

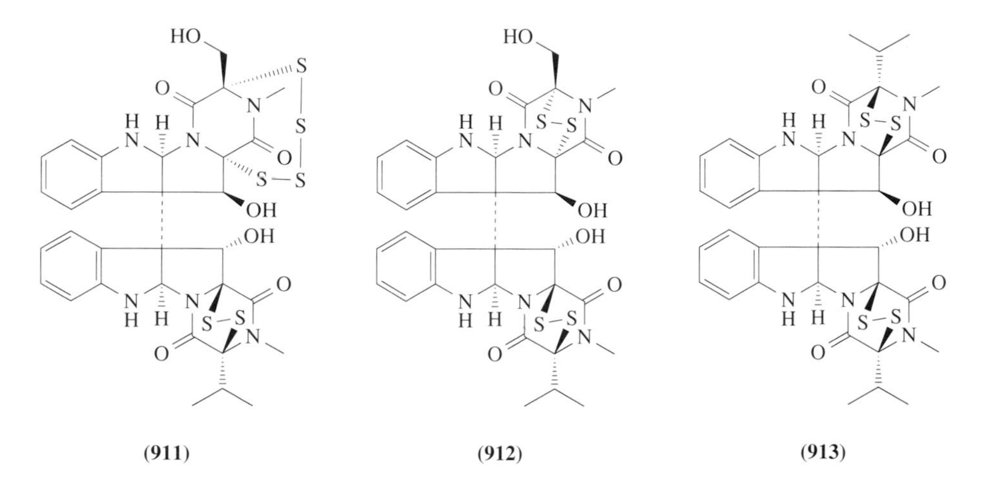

Seven quinazoline alkaloids, fumiquinazolines A–G (**914**)–(**920**) were isolated from the cultured mycelium of a strain of *Aspergillus fumigatus* which existed in the gastrointestinal tract of the salt-water fish *Pseudolabrus japonicus*.[738,739] Their structures were based on spectral and chemical evidences as well as X-ray diffraction analysis of (**919**). The absolute stereostructure of (**919**) was revealed by the production of L-(+)-alanine through its acid hydrolysis. All these compounds (**914**)–(**920**) exhibited moderate cytotoxic activity against cultured P388 lymphocytic leukemia cells.

Halichomycin (**921**), a new class of cytotoxic compounds, was isolated from a strain of *Streptomyces hygroscopicus* which was obtained from the gastrointestinal tract of the marine fish *Halichoeres bleekeri*.[740] Halichomycin (**921**) exhibited cytotoxicity (ED_{50} 0.13 μg ml^{-1}) in the P388 lymphocytic leukemia test system in cell culture. The relative stereochemistry of (**921**) was established by NOESY experiments. Although Numata and co-workers described this compound as a macrolide, the biosynthetic provenance and path of compound (**921**) appeared to be strange.

From the same fungal strain of *Penicillium* sp., which was originally separated from the marine alga *Enteromorpha intestinalis*, as that which produced communesins A (**909**) and B (**910**), a new class of cytochalasans, named penochalasins A (**922**), B (**923**), and C (**924**) were isolated and their stereostructures were established on the basis of NMR spectral analyses and chemical transformations.[741] Different conformations of penochalasin A (**922**) in $CDCl_3$ and pyridine-d_5 were determined by analysis of NMR data and molecular modeling. All three compounds exhibited cytotoxicity against cultured P388 cells (ED_{50} 0.3–0.5 μg ml^{-1}).

Furthermore, the same *Penicillium* sp. also produced four new cytotoxic metabolites, penostatins A–D (**925**)–(**928**), and their stereostructures were established on the basis of spectral analyses.[742] Compounds (**925**)–(**927**) exhibited significant cytotoxicity against cultured P388 cells (ED_{50} 0.8–1.2 μg ml^{-1}), while the cytotoxicity of (**928**) was weak (ED_{50} 11.5 μg ml^{-1}).

8.07.9.4.3 **Blue-green algae**

Blue-green algae or cyanobacteria (cyanophytes) have been demonstrated as a source of interesting natural products of significant bioactivities, in particular, by a series of leading studies by Moore *et al.*[743] His research group isolated debromoaplysiatoxin (**929**)[744] and lyngbyatoxin (**930**)[745] from the filamentous blue-green alga *Lyngbya majuscula* in the 1970s. These natural products had structural similarities to those isolated from sea hare *Aplysia* (aplysiatoxins) and antibiotics from *Streptomyces* (teleocidins), respectively. Moore further studied the metabolites of blue-green alga *Lyngbya*

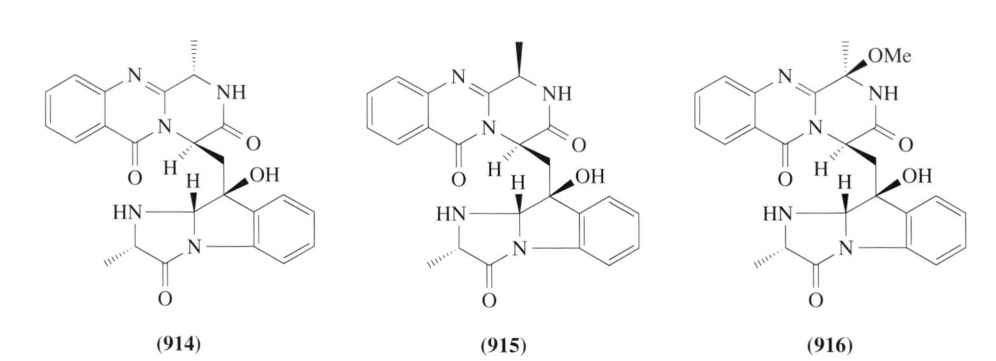

(914) (915) (916)

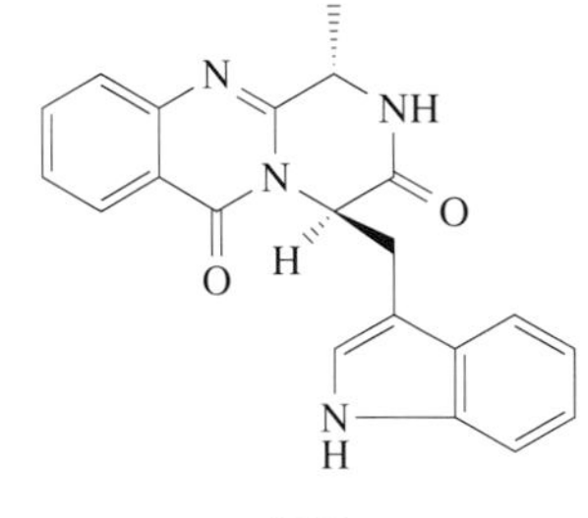

(917)

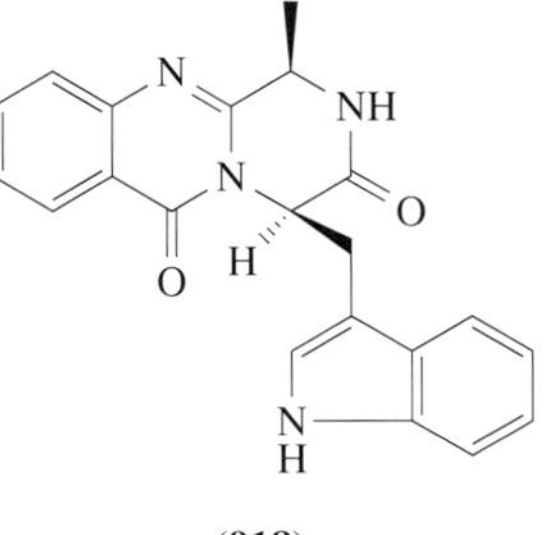

(918)

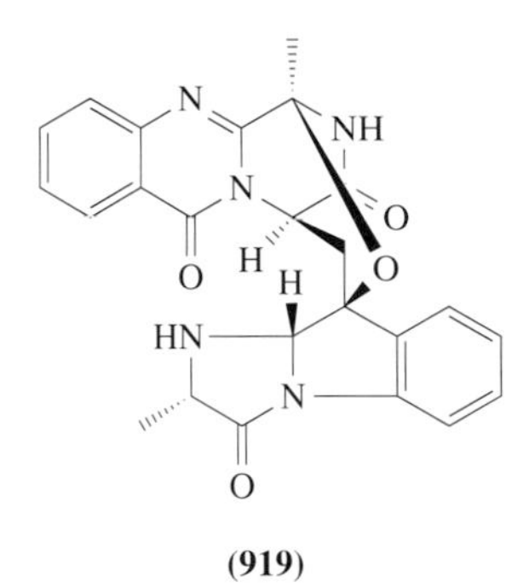

(919)

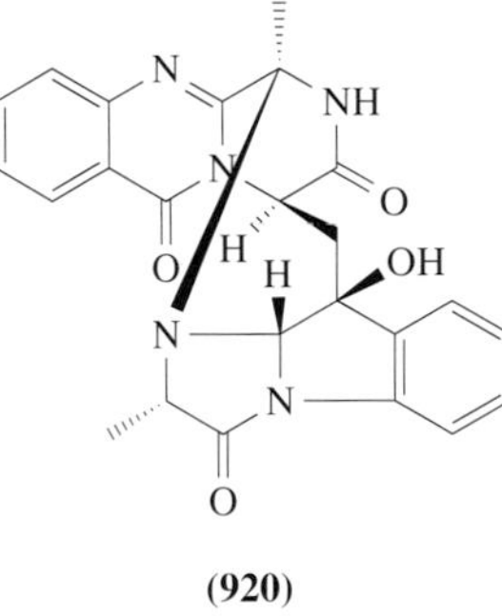

(920)

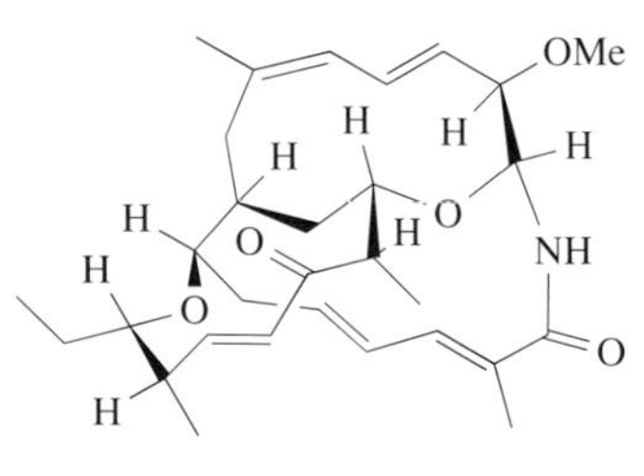

(921)

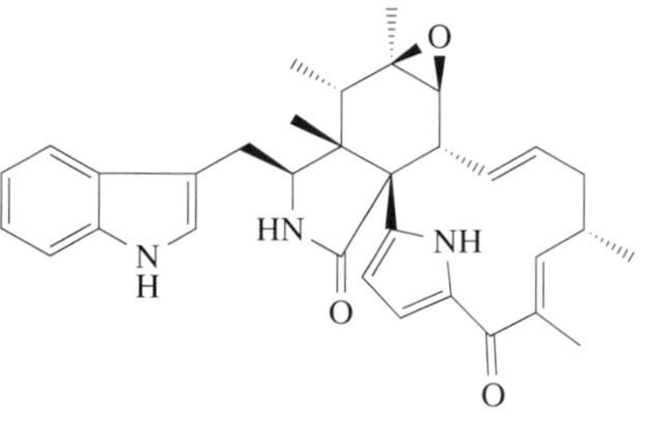

(922)

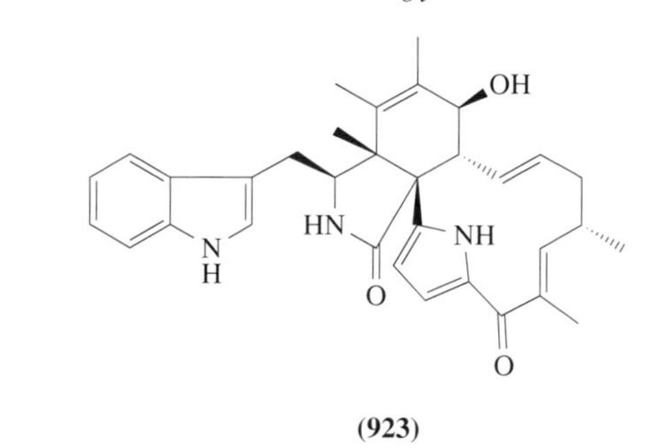

(923)

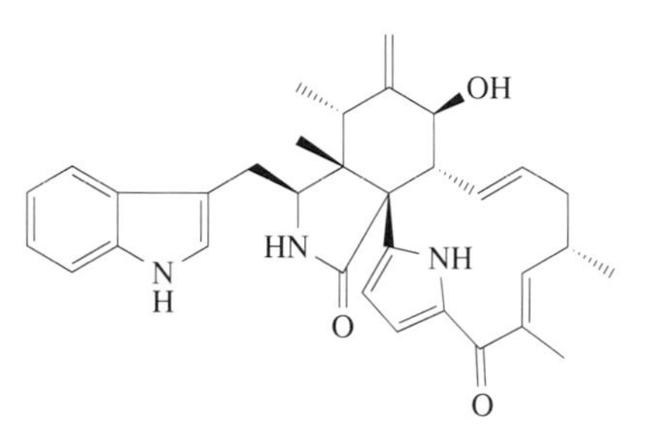

(924)

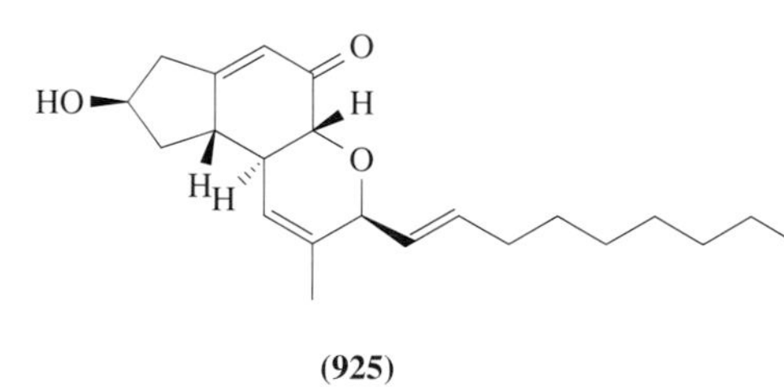

(925)

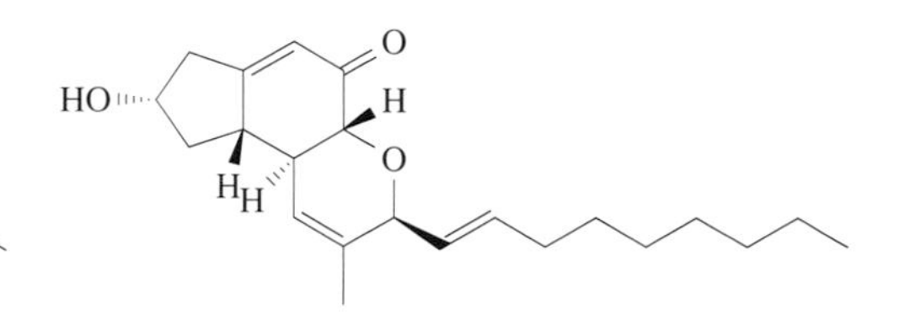

(926)

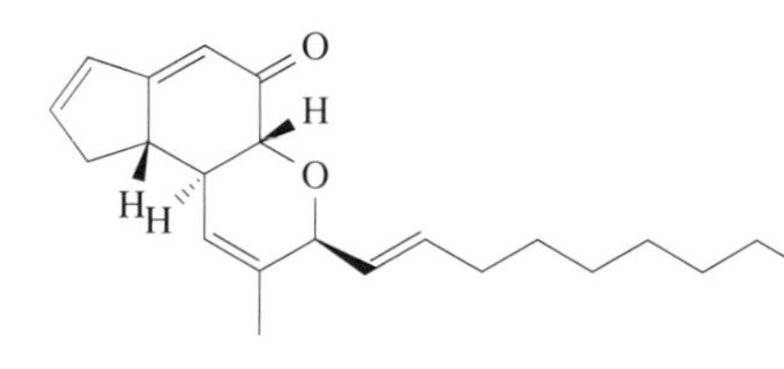

(927)

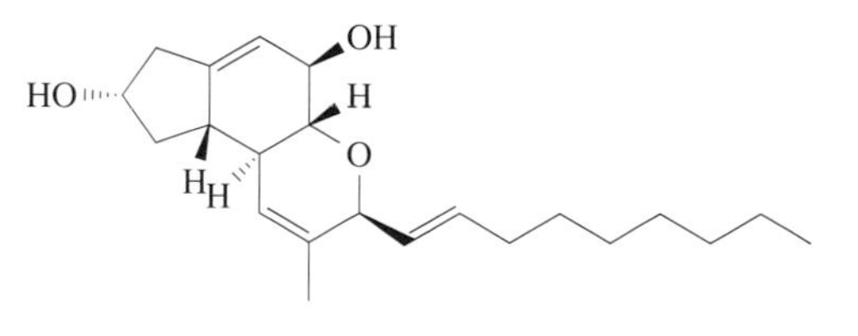

(928)

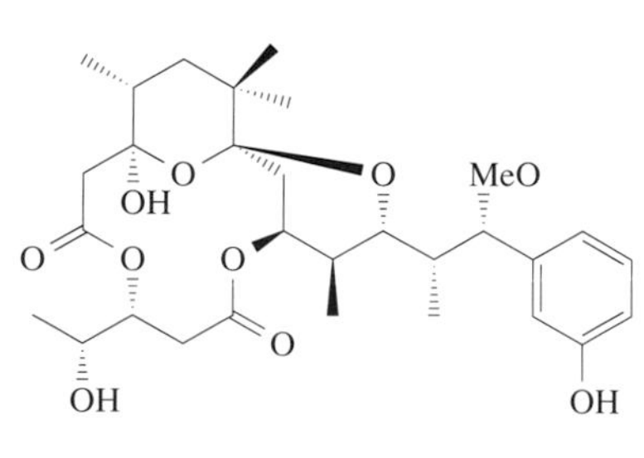

(929)

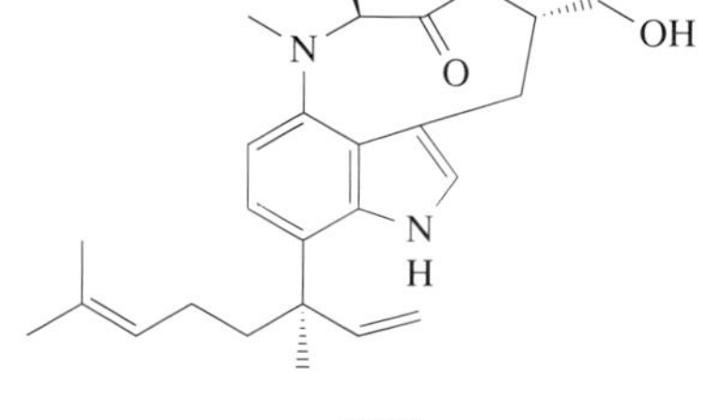

(930)

majuscula in the 1970s and 1980s and isolated a number of unique natural products such as majusculamides, malyngolide, malyngamides, malyngic acid, pukeleimides, and others.[746]

In 1994, Gerwick *et al.* studied the organic extract of *Lyngbya majuscula*, a Caribbean marine cyanobacterium collected from Curaçao, Netherland Antilles. The extract was found to be strongly cytotoxic against a Vero cell line and also highly toxic to brine shrimp ($LC_{50} = 25$ ng ml^{-1}). Using a brine shrimp assay to guide fractionation, a unique metabolite, named curacin A (**931**), was isolated as 8–10% of the crude extract and was found to be responsible for the potent brine shrimp toxicity ($LC_{50} = 3$ ng ml^{-1}) as well as mammalian cell antiproliferative activity ($IC_{50} = 6.8$ ng ml^{-1} in the Chinese hamster Aux B1 cell line).[747] The structure containing unique thiazoline and cyclopropane moieties was deduced by spectroscopic analyses, and the complete relative and absolute configuration of curacin A (**931**) was defined by comparison of products obtained from chemical degradation of the natural product with the same substance prepared by synthesis.[748] Curacin A (**931**) was partially hydrogenated with Wilkinson's catalyst to yield a mixture of 15,16-dihydro and 3,4,15,16-tetrahydro derivatives. This mixture was ozonized and the ozonide was reduced with excess dimethyl sulfide to yield 5-methoxyoctan-2-one (**932**). On the other hand, ozonolysis of curacin A (**931**) (−78 °C, $CHCl_3$, 2 min), followed by oxidative workup (H_2O_2, 45 °C, 16 h), and then reaction with excess CH_2N_2 in Et_2O gave a methyl sulfonate derivative (**933**). Optically active (**932**) and (**933**) were obtained by enantioselective syntheses. Thus, curacin A (**931**) was shown to have 2*R*, 13*R*, 19*R*, and 21*S* absolute configuration.

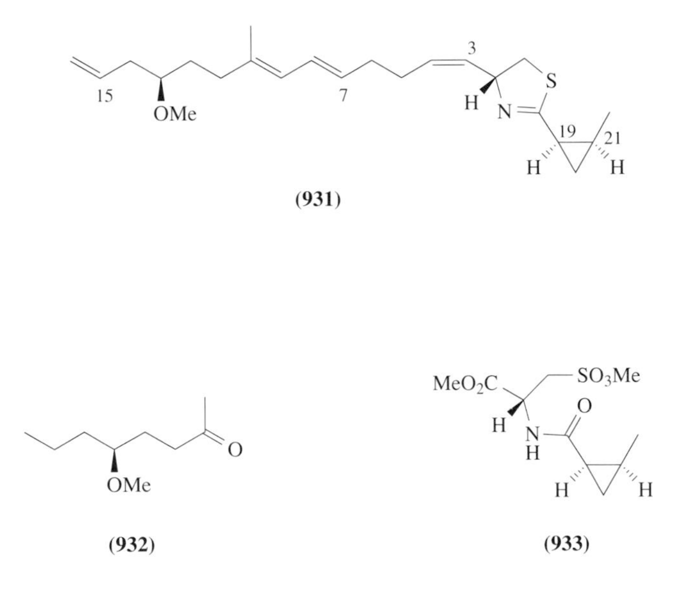

(**931**)

(**932**)

(**933**)

Curacin A (**931**) was examined in the NCI cell line screen, and its differential cytotoxicity pattern was evaluated by the COMPARE algorithm, indicating that curacin A (**931**) was an antitubulin agent. Pure curacin A (**931**) was revealed to be an antimitotic agent (IC_{50} values in three tumor cell lines, ranging from 7 nM to 200 nM) that inhibited microtubule assembly and the binding of colchicine to tubulin. Curacin A (**931**) probably binds in the colchicine site because it competitively inhibited the binding of [^{3}H]colchicine to tubulin with an apparent K_i value of 0.6 μM and stimulated tubulin-dependent GTP hydrolysis, as do most other colchicine-site agents. The binding of curacin A (**931**) to tubulin resembled the binding reactions of combretastatin A-4 and podophyllotoxin in contrast to that of colchicine in that it occurred as extensively on ice as at higher temperatures. However, once bound, the dissociation rate of curacin A (**931**) from tubulin was very slow, most closely resembling that observed with colchicinoids (thiocolchicine was the drug examined) than the faster dissociation that occurs with combretastatin A-4 and podophyllotoxin. Because the molecular structure of curacin A (**931**) is so different from that of previously described colchicine-site drugs (e.g., there is no aromatic moiety, and there are only two conjugated double bonds in its linear hydrocarbon chain), the activities of natural isomers, curacins B (**934**) and C (**935**),[749] and synthetic derivatives were also examined. Only modest enhancement or reduction of activity was observed with a variety of structural changes.[750]

Gerwick and co-workers further studied the metabolites of the blue-green alga *L. majuscula* and isolated several bioactive substances. From a Curaçao collection of *L. majuscula*, a new lipopeptide, malyngmide H (**936**), was isolated, guided by ichthyotoxic activity against goldfish, together with

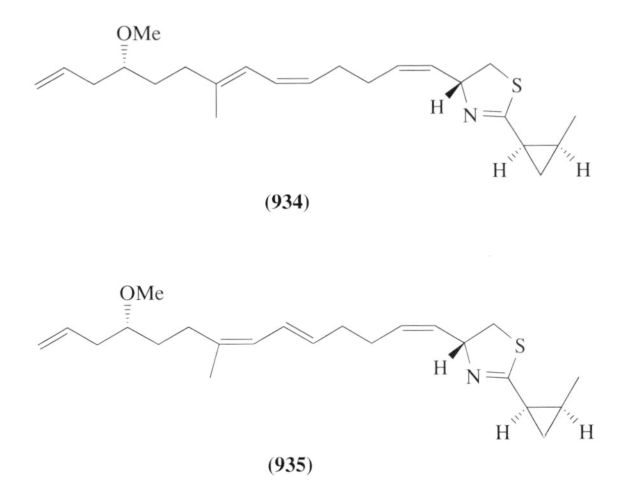

(934)

(935)

the corresponding free acid, 7-methoxytetradec-4*E*-enoic acid (**937**).[751] The absolute stereochemistry of the cyclohexenone moiety of malyngamide H (**936**) was deduced by a combination of 2D NOESY and exciton chirality CD spectroscopy. Malyngmide H (**936**) showed ichthyotoxic effect ($LC_{50} = 5$ μg ml^{-1}, $EC_{50} = 2$ μg ml^{-1}), but it was not active in brine shrimp lethality or molluskicidal assays. Malyngamide H (**936**) may be part of the natural defense of this cyanobacterium. An ichthyotoxic amide of 7(*S*)-methoxytetradec-4(*E*)-enoic acid, malyngamide I (**938**), was isolated from *L. majuscula* collected from shallow water at Uken, Okinawa.[752] Malyngamide I (**938**) was moderately toxic towards brine shrimp (*Artemia salina*, LD_{50} ca. 35 μg ml^{-1}) and goldfish (*Carassius auratus*, $LD_{50} < 10$ μg ml^{-1}). Data from the acetate derivative (**939**) of this amide were used for the probable structure

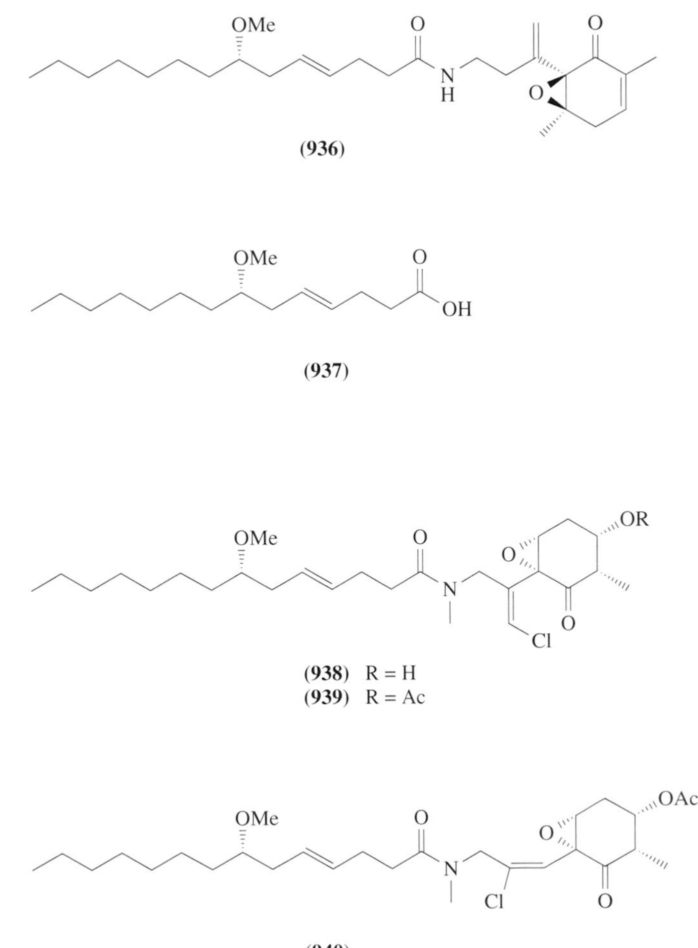

(936)

(937)

(938) R = H
(939) R = Ac

(940)

revision of stylocheilamide (**940**),[753] a previously reported metabolite from the sea hare *Stylocheilus longicauda*. The structure of stylocheilamide should likely be revised from (**940**) to (**939**), being in keeping with the established structures of the other chlorine-containing malyngamides and also consistent ecologically with the fact that *Stylocheilus longicauda* is believed to have a dietary origin from *L. majuscula*.

The lipid extract of a Curaçao collection of *L. majuscula*, which contained curacin A (**931**), exhibited several bioactivities including potent brine shrimp toxicity (*Artemia salina*, $LC_{50} = 25$ μg ml^{-1}), goldfish ichthyotoxicity (*Carassius auratus*, $LC_{50} = 25$ μg ml^{-1}), and molluskicidal activity (*Biomphalaria glabrata*, $LC_{100} < 100$ μg ml^{-1}). The brine shrimp toxicity was traced to curacin A (**931**), and part of the ichthyotoxicity was ascribed to malyngamide H (**936**). Fractionation using fish and snail bioassays led to the isolation of two other distinct classes of natural products from this organism, i.e., the most ichthyotoxic compound, antillatoxin (**941**) ($LD_{50} = 0.05$ μg ml^{-1}), and a strongly molluskicidal agent, barbaramide (**942**) ($LD_{100} = 10$ μg ml^{-1}). Antillatoxin (**941**)[754] was isolated in small yield as an amorphous powder (1.3 mg, 0.07% of extract) and its structure was elucidated on the basis of spectroscopic data to be composed of a lipid portion and three amino acid residues, glycine, *N*-methylvaline, and alanine. The stereochemistries of the *N*-methylvaline and alanine residues were both assigned to be the L configuration from chiral phase TLC analysis. The stereochemistry at the C-4 and C-5 positions in the lipid portion was examined using a combination of molecular modeling, NOESY data, *J* values, and CD spectroscopy. Modeling of antillatoxin's structure was accomplished using a dynamic simulated annealing protocol with the program XPLOR. The calculated structure of the 4*S*,5*R* stereoisomer was most consistently accounted for by the NMR and CD spectral data, thus assigning the absolute stereochemistry of antillatoxin (**941**) as 4*S*, 5*R*, 2′*S*, 5′*S*. On the other hand, barbaramide (**942**)[755] was a novel lipopeptide containing a trichloromethyl group and the methyl enol ether of a *β*-keto amide.

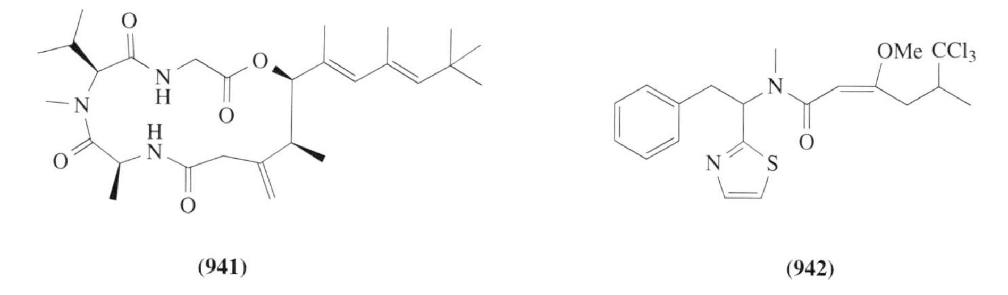

(941) **(942)**

The trichloromethyl portion of barbamide (**942**) closely resembled the trichloromethyl portion of dysidin, a polychlorinated amino acid derivative found in the sponge *Dysidea herbaceae*.[756] Microscopic investigations of *Dysidea* showed it to be rich in symbiotic filamentous cyanobacteria. A flow-cytometric separation of the symbiont *Oscillatoria spongeliae* from the sponge cells suggested that the polychlorinated amino acid derivatives were associated with the cyanobacterial filaments.[757] The finding of structurally similar components in the marine cyanobacterium *L. majuscula* provided further support for the cyanobacterial origin of these metabolites. It was interesting that the amine portion of barbamide (**942**) was the *N*-methyl equivalent of dolaphenine (*N*-methyldolaphenine), a structural component of the antineoplastic peptide dolastatin 10 isolated from the sea hare *Dolabella auricularia*.[758] Hence, it is conceivable that dolastatin 10, or at least a portion of it, also arises from cyanobacterial metabolism. Sea hare are well known to incorporate unique secondary metabolites from their algal diets. The crude lipid extract of a Curaçao collection of *L. majuscula* exhibited different biological properties including brine shrimp toxicity, goldfish ichthyotoxicity, and molluskicidal activity. Fractionation guided by these different biological assays led to isolation of four distinctly different classes of natural products, each of which was selective in its range of activity. The molluskicidal activity was due to barbamide (**942**), while the ichthyotoxicity was due to two components, antillatoxin (**941**) and malyngamide H (**936**). Finally, the brine shrimp toxicity was traced to a potent new antimitotic agent, curacin A (**931**). The apparently selective activity of each of these compounds to only one class of animal led to the speculation that these four kinds of compounds represent separate adaptations by this cyanobacterium to these different classes of predators.

In May 1994, a simultaneous blue-green algal bloom and a massive die-off of pelagic larval rabbitfishes (*Siganus argenteus* and *S. spinus*) occurred at Ypao beach, Guam. From this microbial assemblage which was composed of *Schizothrix calcicola* and *Lynbya majuscula*, a new herbivore antifeedant metabolite, ypaoamide (**943**), was isolated and its structure was determined spectroscopically by Paul and co-workers.[759] Isolated cells of the *L. majuscula* were found to produce ypaoamide (**943**) in laboratory culture based on GC–EIMS analysis. Ypaoamide (**943**) had structural similarities to malyngamides or majusculamides, suggesting that common biosynthetic pathways may be employed by different chemotypes of *L. majuscula*. The *t*-butyl lipid side-chain contained in ypaoamide (**943**) appeared unusual with little biosynthetic precedent other than in antillatoxin (**941**).

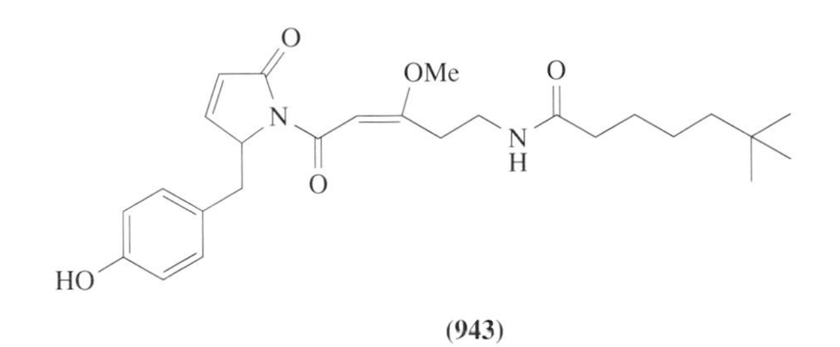

(**943**)

Gerwick *et al.* studied a suite of cytotoxic and antimicrobial cyclic peptides produced by the Caribbean cyanobacterium *Hormothamnion enteromorphoides*, which grows abundantly in the shallow coastal waters off northern Puerto Rico. The structure of the most lipophilic and abundant of these peptides, hormothamnin A (**944**), was determined by interpretation of physical data, principally high field NMR and FAB MS, in combination with chemical derivatization and degradations. Hormothamnin A (**944**) is a cyclic undecapeptide consisting of several unusual amino acid residues containing D-amino acids.[760]

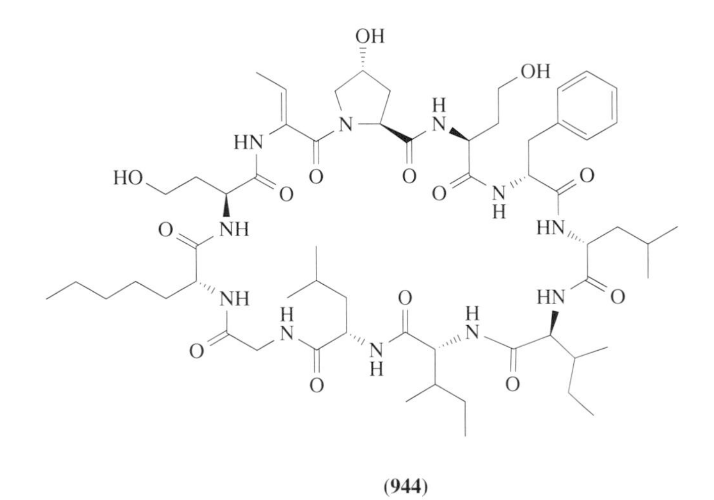

(**944**)

Nagle and Gerwick also isolated a series of cytotoxic metabolites, nakienones A–C (**945**)–(**947**) and nakitriol (**948**), from dead and necrotic branches of stony coral (*Acropora* sp.) which were completely covered with a gray-black mat of cyanobacteria (*Synechocystis* sp.) in the waters off Yonaine at Nakijin Village, Okinawa.[761] Nakienone A (**945**) was cytotoxic against KB and HCT 116 cell lines, with LD_{50} values of ca. 5 μg ml^{-1} and 20 μg ml^{-1}, respectively. Nakienone C (**947**) was generated from nakienone A (**945**) while obtaining the NMR spectra of (**126**) in $CDCl_3$ solution, presumably through an acid catalyzed rearrangement. Nakienone A (**945**) and C (**947**) were structurally similar to the didemnenones A–D,[762] isolated from *Didemnum* and *Trididemnum* sp., ascidians known to contain unicellular algal symbionts. The tunicate symbiont *Prochloron* is a prochlorophyte which is remarkably similar to the unicellular cyanobacterium *Synechocystis*.

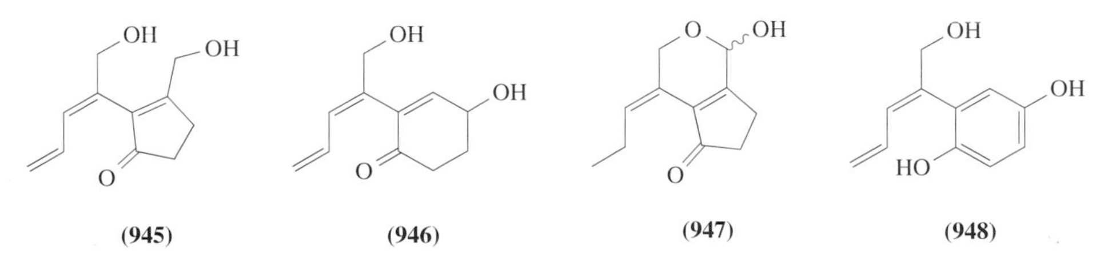

(**945**) (**946**) (**947**) (**948**)

From a marine reddish blue-green alga *Oscillatoria* sp., Murakami *et al.* isolated a macrolide, oscillariolide (**949**), possessing inhibitory activity of development of fertilized echinoderm eggs.[763] The blue-green alga was collected from Gokashowan-Bay, Mie Prefecture, Japan, and cultured in 1000 l tanks. Oscillariolide (**949**) had a unique 14-membered macrocyclic lactone with a side-chain bearing 1,3-polyol and brominated diene moieties.

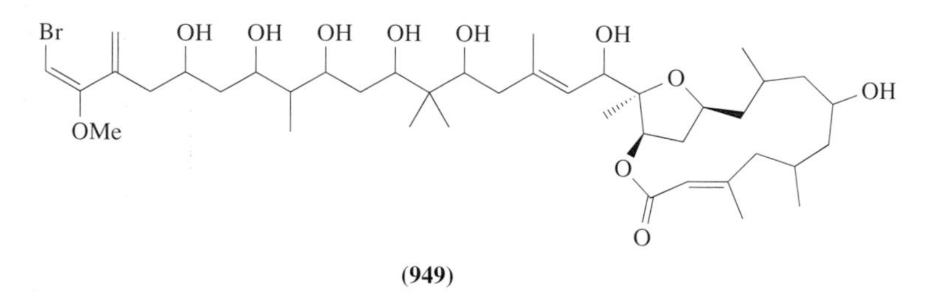

(949)

Ohizumi and co-workers isolated (2*S*)-1-*O*-palmitoyl-3-*O*-*β*-D-galactopyranosylglycerol (**950**) from the cultured cells of the marine Cyanophyceae *Oscillatoria rosea*, supplied by the NIES-collection (Microbial Culture Collection, the National Institute for Environmental Studies, Environmental Agency, Japan).[764] The monogalactosylacylglycerol (**950**) caused a concentration-dependent inhibition of platelet aggregation induced by U46619, a thromboxane A_2 analogue, with an IC_{50} value of 6.0×10^{-5} M. Compound (**950**) (100 μg ml^{-1}) only markedly inhibited platelet aggregation induced by U46619 and did not inhibit that induced by thrombin or ionomycin.

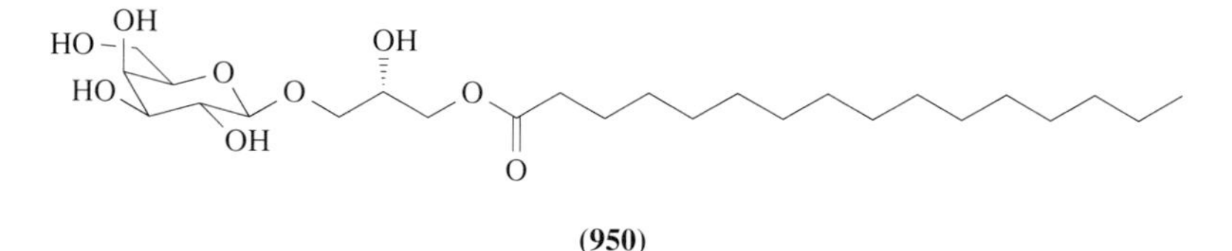

(950)

Although only a limited number of studies have been reported on the secondary metabolites of marine blue-green algae, those of freshwater or terrestrial origins have been extensively investigated by several research groups. In particular, potent cyclic heptapeptide hepatotoxins termed microcyctins and related peptides, produced by freshwater blue-green algae of the genera of *Microcystis*, *Anabaena*, *Nostoc*, and *Oscillatoria*, have been thoroughly studied, and more than 40 microcystins have been reported.[765] A number of cyclic or linear peptides related to microcystins have also been investigated from freshwater blue-green algae.[766–772] Microcystin LR (**951**) is found most often among the microcystins, and microcystins show strong inhibitory activity against protein phosphatases PP1 and PP2A and are also reported to be tumor promoters.[773–775]

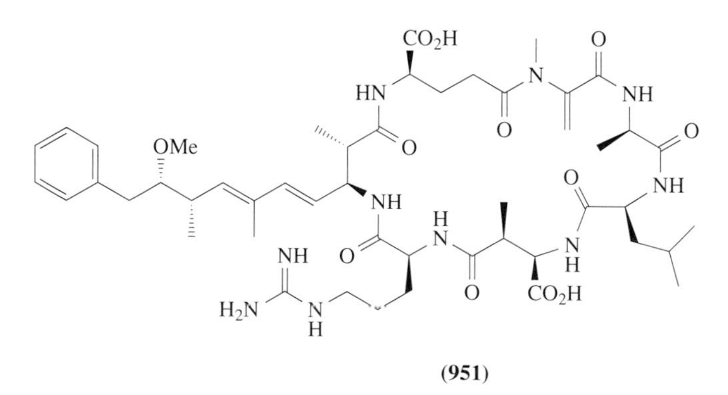

(951)

Murakami and co-workers described a series of studies on freshwater cyanobacterial peptides isolated as protease inhibitors including those against thrombin,[776], trypsin,[777] plasmin,[778] elastase,[779] chymotrypsin,[780] and angiotensin-converting enzyme.[781]

Since the early 1980s, Moore and co-workers have been continuously studying laboratory-cultured freshwater cyanobacterial natural products.[743,782] These cyanobacterial secondary metabolites have chemically unique structures, belonging to various biosynthetic classifications such as cyclic peptides (e.g. dendroamide A (**952**))[783] or depsipeptides (e.g., hapalosin (**953**)),[784] macrolides (e.g., tolytoxin

(**954**)),[785] isotactic polymethoxy-1-alkenes (e.g., mirabilene isonitrile A (**955**)),[786] cyclophanes (e.g., cylindrocyclophane A (**956**)),[787] porphyrins (e.g., tolyporphin (**957**)),[788] nucleosides (e.g., tubericidin (**958**)),[789] oxazole and/or thiazole-containing polythiazoline alkaloids (e.g., tantazole B (**959**)),[790,791] indole alkaloids with an isonitrile or isothiocyanate-containing isoprenoid moiety (e.g., welwitindolinone A isonitrile (**960**)),[792] chlorinated *N*-acylpyrrolinones (e.g., mirabimide E (**961**)),[793] and guanidium compounds (e.g., anatoxina(s) (**962**),[794] cylindrospermopsin (**963**)).[795] These unique compounds exhibit a wide variety of potentially useful bioactivities including cytotoxic, antifungal, antiviral, hepatotoxic, anticholinesterase, and multidrug resistance reversal effects.

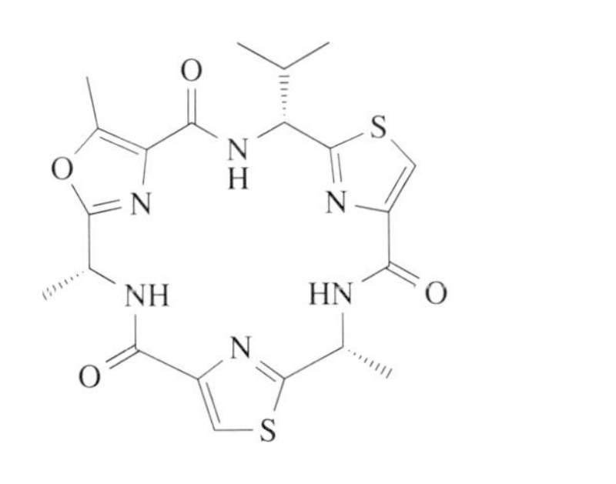

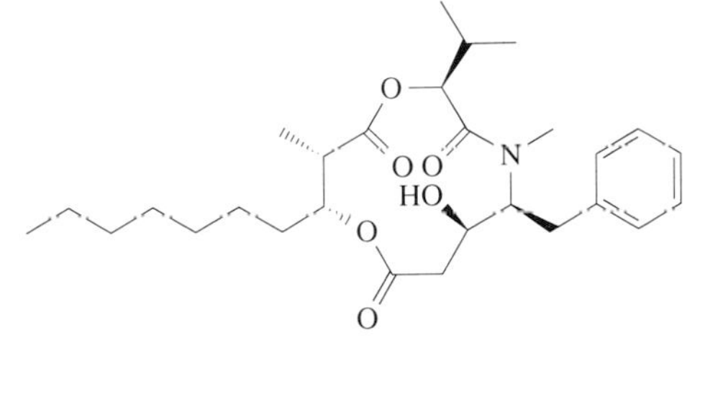

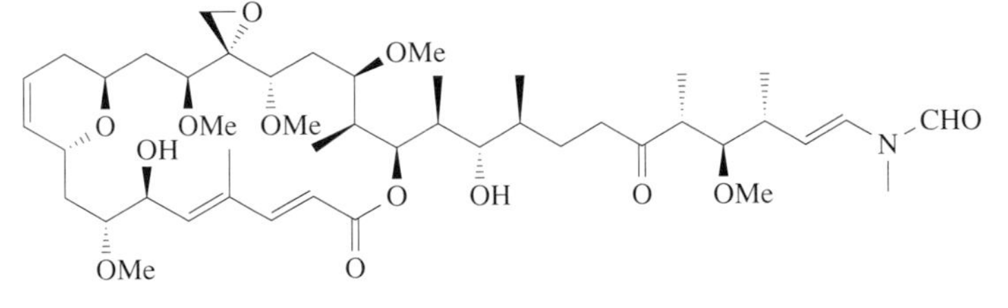

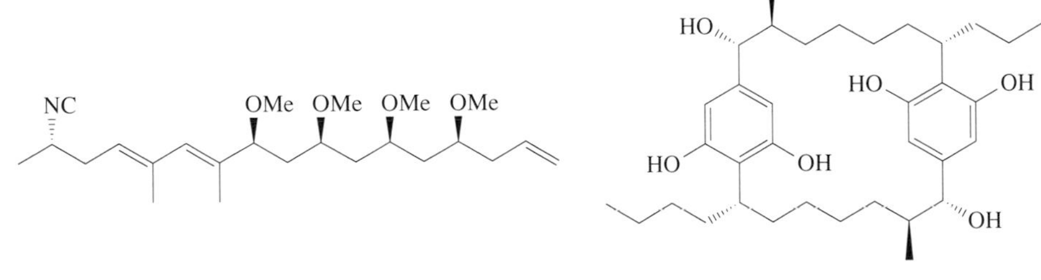

(955) **(956)**

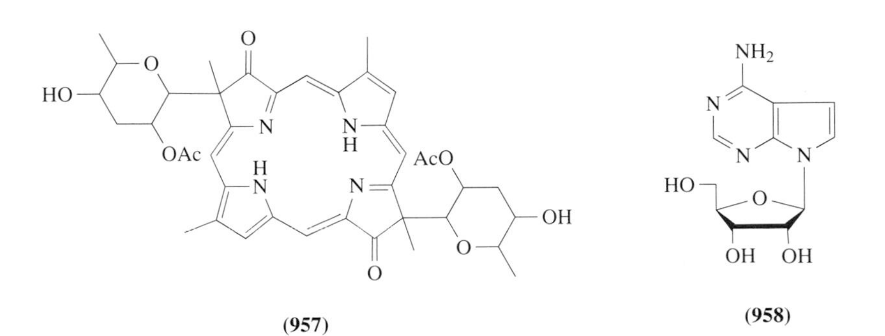

(957) **(958)**

Thus, blue-green algae (cyanobacteria) have been recognized as a rich source of structurally novel and biologically active natural products. It has been pointed out that most of these unique natural

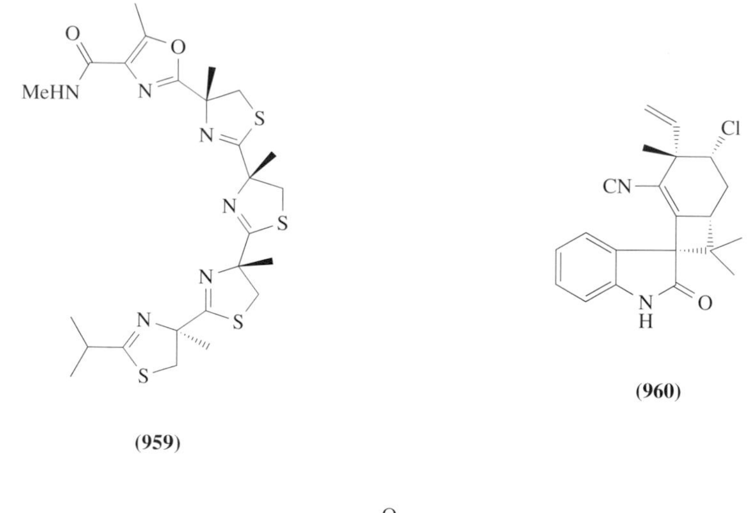

(959)

(960)

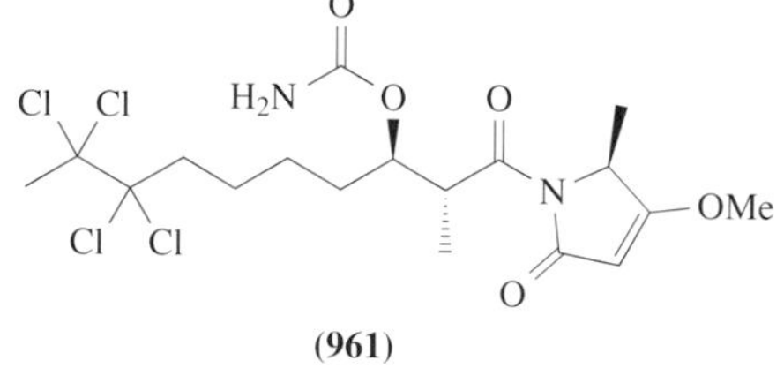

(961)

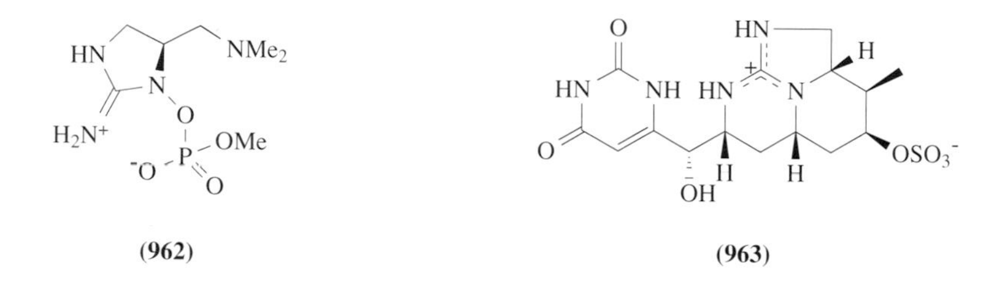

(962)

(963)

products will probably only be useful as biochemical research tools.[796] A few, however, have the potential for development into useful commercial products.[796] For example, cryptophycin-1 (**964**), a novel inhibitor of microtubule assembly from *Nostoc* sp. GSV 224, shows impressive activity against a broad spectrum of solid tumors implanted in mice, including multidrug-resistant ones. This cyclic depsipeptide (**964**) had previously been isolated from *Nostoc* sp. ATCC 53787 as an antifungal agent and its gross structure determined.[797] Moore and co-workers established the relative and absolute stereochemistry of cryptophycin-1 (**964**) using a combination of chemical and spectral techniques.[798] Several minor cryptophycins were also isolated from GSV 224 and their total structures were determined. The convergent total synthesis of cryptophycin-3 (**965**) was achieved and the chloro-*O*-methyltyrosine unit in cryptophycin-1 (**964**) was revised to have the D-configuration.[799] Using a modified isolation procedure devoid of MeOH, a further 18 cyclic cryptophycins were isolated from *Nostoc* sp. GSV 224 as minor constituents.[800] Acyclic cryptophycins were not found, indicating that acyclic cryptophycin analogues are artifacts of isolation as a consequence of using MeOH. The relative stereochemistry of cryptophycin-3 (**965**) was further rigorously established by X-ray crystallography. NOE studies show that the preferred conformations of most cryptophycins in solution differ from the conformation of cryptophycin-3 (**965**) in the crystal state. Although cryptophycin-1 (**964**) was relatively stable at pH 7, both in ionic and nonionic media, the ester bond linking L-leuicic acid and (*R*)-3-amino-2-methylpropanoic acid units was fairly labile to solvolysis and mild base hydrolysis. Structure–activity relationship studies indicated that the intact macrolide ring, the epoxide group, the chloro and *O*-methyl groups in the chloro-*O*-methyl-D-tyrosine unit, and the methyl group in the (*R*)-3-amino-2-methylpropanoic acid unit were needed for the *in vivo* activity of cryptophycin-1 (**964**).

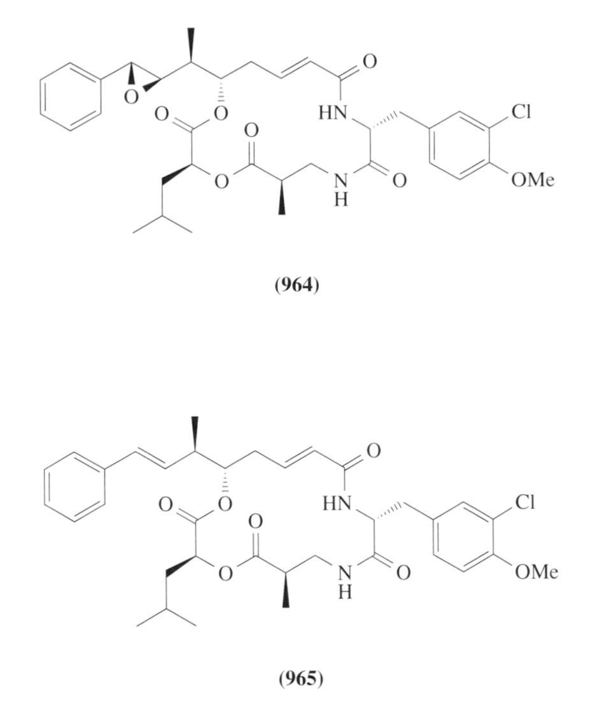

(964)

(965)

The mechanism of action of cryptophycin demonstrated that this compound potently disrupts the microtubule structure in cultured cells, and cryptophycin appeared to be a poor substrate for the drug-efflux pump P-glycoprotein in contrast with the *Vinca* alkaloids.[801] The site of cryptophycin interaction with tubulin was examined to reveal that crytophycin blocked the formation of vinblastine-tubulin paracrystals in intact cells and suppressed vinblastine-induced tubulin aggregation *in vitro*. Cryptophycin inhibited the binding of [^{3}H]vinblastine and the hydrolysis of [γ-^{32}P]GTP by isolated tubulin, but did not block the binding of colchicine. These results indicated that cryptophycin disrupts the *Vinca* alkaloid site of tubulin.[802,803]

Kitagawa and co-workers isolated a potent cytotoxic depsipeptide, arenastatin A (**966**) from the Okinawan marine sponge *Dysidea arenaria*,[804] and its structure including all absolute configurations was established by NMR and synthetic studies.[805] Interestingly, the chemical structure of arenastatin A (**966**) was very similar to that reported for cryptophycins, suggesting the participation of a presumably symbiotic cyanobacterium in the biosynthesis of arenastatin A (**966**) in the marine sponge. The total synthesis of arenastatin A (**966**) was achieved[806] and stereoisomers of (**966**) with different configurations at the epoxide moiety, a secondary methyl group on C-6, and/or the *O*-methyltyrosine unit were also prepared by applying improved methods to show that only (**966**) was potently cytotoxic (IC_{50} 5 pg ml^{-1} against KB cells) and other stereoisomers did not show cytotoxicity at concentrations below 0.1 μg ml^{-1}.[807]

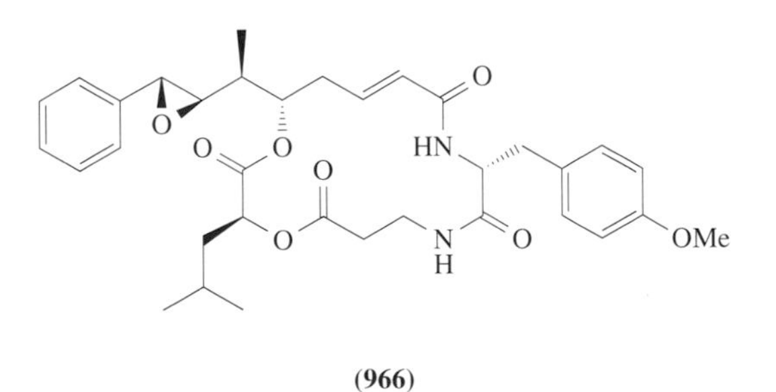

(966)

8.07.9.4.4 Dinoflagellates and other microalgae

In 1968 Sharma *et al*. isolated goniodomin from a marine dinoflagellate *Goniodoma* sp. and proposed that it contained five hydroxyl groups, a lactone ring, four ether linkages, and a dihydro-geranyl side-chain by functional group analysis.[808] In paper disk-agar plate tests, goniodomin (0.5–500 μg ml^{-1}) strongly inhibited *Cryptococcus neoformans*, *Trichophyton mentagrophytes*, and other fungi, but had little or no activity against bacteria. Twenty years later, Murakami *et al.* isolated an antifungal polyether macrolide, named goniodomin A (**967**), from *Goniodoma pseudogoniaulax* collected in a rock pool.[809] Its structure was elucidated by spectral data.

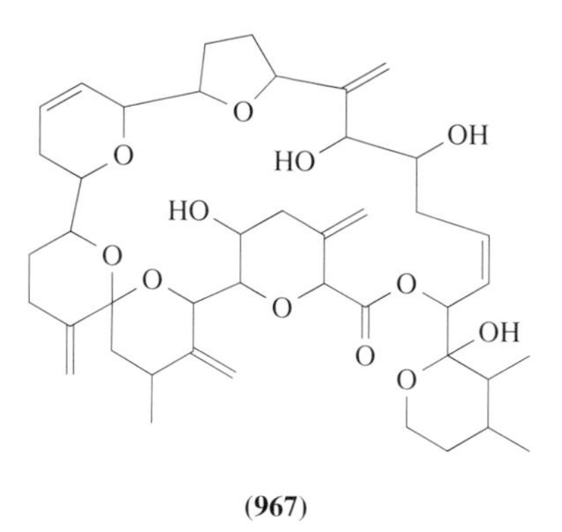

(**967**)

Goniodomin A (**967**) showed antifungal activity against *Mortierella ramannianus* at a concentration of 0.5 μg ml^{-1}, and inhibited the cell division of fertilized sea urchin eggs at 0.05 μg ml^{-1}. In addition, goniodomin A (**967**) administered i.p. at 0.6 mg kg^{-1} to P388 leukemia cell-inoculated mice prolonged the survival time. The i.p. LD_{50} values of (**967**) in male ICR mice were 1.2 mg kg^{-1} and 0.7 mg kg^{-1} at 24 h and 48 h, respectively. Morphological changes in the liver and thymus of male ICR mice were induced by (**967**).[810] Furthermore, Ohizumi and co-workers showed that goniodomin A (**967**) induces modulation of actomyosin ATPase activity mediated through conformational change of actin.[811] The effect of (**967**) was dependent on the concentration of actin, but not of myosin. The actomyosin ATPase activity was increased by pretreatment of actin (but not of myosin) with (**967**). Goniodomin A (**967**) induced a sustained and concentration-dependent increase in the fluorescence intensity (excitation wavelength, 277 nm; emission wavelength, 329 nm) of actin. The maximum response was obtained with concentrations of (**967**) in the 10^{-5} to 10^{-4} M range in the presence of 5 μM F-actin. However, the ATPase activity and fluorescence intensity of myosin were not changed by (**967**) at concentrations from 10^{-8} to 10^{-5} M. Interestingly, goniodomin A (**967**) induced a remarkable but transient increase in the fluorescence intensity of actomyosin in a concentration-dependent manner, with a peak at 3×10^{-7} M. This profile was quite similar to that found in the stimulation of the actomyosin ATPase activity induced by (**967**). To investigate further the effect of (**967**), actin was labeled with *N*-(1-pyrenyl)iodoacetamide. Goniodomin A (**967**) at 10^{-6} M had no effect on the fluorescence intensity of pyrenyl-actin (excitation wavelength, 365 nm; emission wavelength, 407 nm), but increasing concentrations of (**967**) to 3×10^{-6} M remarkably decreased its intensity. This effect was potentiated by heavy meromyosin. Actin molecules treated with (**967**) were completely sedimented by mild centrifugation (for 15 min at 12 000 *g*). Electron microscopic observations suggested that actin filaments associated with each other to form a gel in the presence of 3×10^{-6} M goniodomin A (**967**). The conformational change of actin molecules, resulting from stoichiometric binding of (**967**) to actin monomers in filaments, may modify the interaction between actin and myosin.

New polyhydroxypolyenes with potent vasoconstrictive activity, zooxanthellatoxin A (**968**) and B (**969**), were isolated from a cultured zooxanthella, *Symbiodinium* sp., isolated from the flatworm *Amphiscolops* sp. collected at Okinawa.[812] These compounds caused sustained contractions of isolated rabbit aorta at concentrations above 7×10^{-7} M; this effect was abolished in Ca^{2+}-free solution or in the presence of verapamil. Both compounds had relatively large molecular weights, containing a large number of oxygen atoms and alkenic carbons, thus differing from two other vasoconstrictive marine toxins, maitotoxin and palytoxin (see Sections 8.07.8.4.2 and 8.07.8.5.1), containing more alkenes than palytoxin, and fewer ethereal rings than maitotoxin. The structures of zooxanthellatoxin A (**968**) and B (**969**) were studied by chemical degradations (e.g., periodate oxidation and $NaBH_4$ reduction).[813,814] Zooxanthellatoxin A (**968**) gave a seco-acid upon treatment with a weak base and further hydrolyzed to afford a terminal carboxylic acid segment of the amide structure. Comparison of the DQF–COSY, TOCSY, HMQC, and HMBC spectra of (**968**) in CD_3OD or CD_3OD-C_5D_5N with those of its degradation products revealed a novel 62-membered lactone structure for (**968**).[815] Compound (**968**) contained several characteristic functionalities including a bisepoxide, a sulfate ester, an amide, two conjugated dienes, and many allylic alcohols. The structure of zooxanthellatoxin B (**969**) was also determined to be a 62-membered lactone by comparing spectral data and degradation products with those of zooxanthellatoxin A (**968**).[816]

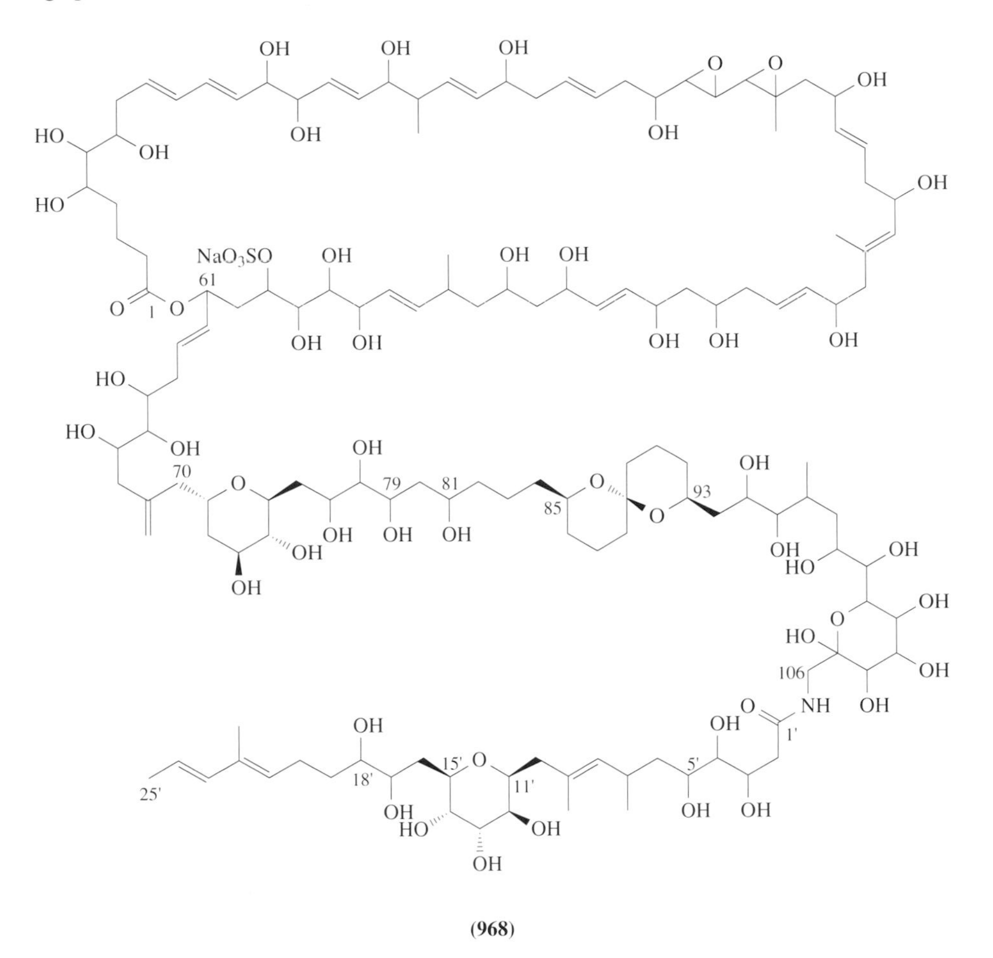

(**968**)

Two degradation products, (**970**) and (**971**), were synthesized and the absolute stereochemistry of the chiral centers contained in them (C-11′, C-12′, C-13′, C-14′, C-15′, C-17′, and C-18′; C-81, C-85, and C-93) were determined.[817,818] Ohizumi and co-workers revealed that zooxanthellatoxin A (**968**) caused aggregation in rabbit washed platelets in a concentration-dependent manner (1–4 μM), accompanied by an increase in cytosolic Ca^{2+} concentration. It was also suggested that (**968**) elicits Ca^{2+}-influx from platelet plasma membranes and the resulting increase in cytosolic Ca^{2+} concentration subsequently stimulates the secondary release of thromboxane A2 from platelets.[819]

Marine dinoflagellates have proven to be a rich source of unique natural products as described above (see also Sections 8.07.8.3–8.07.8.5). Among other dinoflagellates, the genus *Amphidinium* attracts much attention because of the production of bioactive substances such as amphidinolides (Section 8.07.9.4.5) and amphidinols. In 1991 Yasumoto and co-workers isolated a polyhydroxy-polyene antifungal substance, named amphidinol (**972**) (synonymous to amphidinol 1) from the cultures of the dinoflagellate *Amphidinium klebsii* mainly by gel permeation chromatography.[820] The dinoflagellate *Amphidinium* sp. was collected at Ishigaki Island, Japan. Antifungal activity (6 μg/disk) of amphidinol (**972**) against *Aspergillus niger* was three times more potent than amphotericin B; the hemolytic activity was 120 times that of standard saponin. UV and IR spectra, respectively, indicated the presence of a conjugated triene and a sulfate ester in the molecule. The latter moiety was further confirmed by determination of SO_4^{2-} after solvolysis. FAB-MS indicated a molecular weight of 1488 ($C_{73}H_{125}O_{27}SNa$). The major part of the structure was deduced by a combined use of 2D NMR, such as ^1H-^1H COSY, ^1H-^1H HOHAHA, phase-sensitive NOESY, and phase-sensitive C-H COSY. To enhance ^{13}C for NMR measurements, $NaH^{13}CO_3$ was added to the culture media, which yielded (**972**) with 3.5% isotope abundance. Hetero-COSY experiments, including phase-sensitive C-H

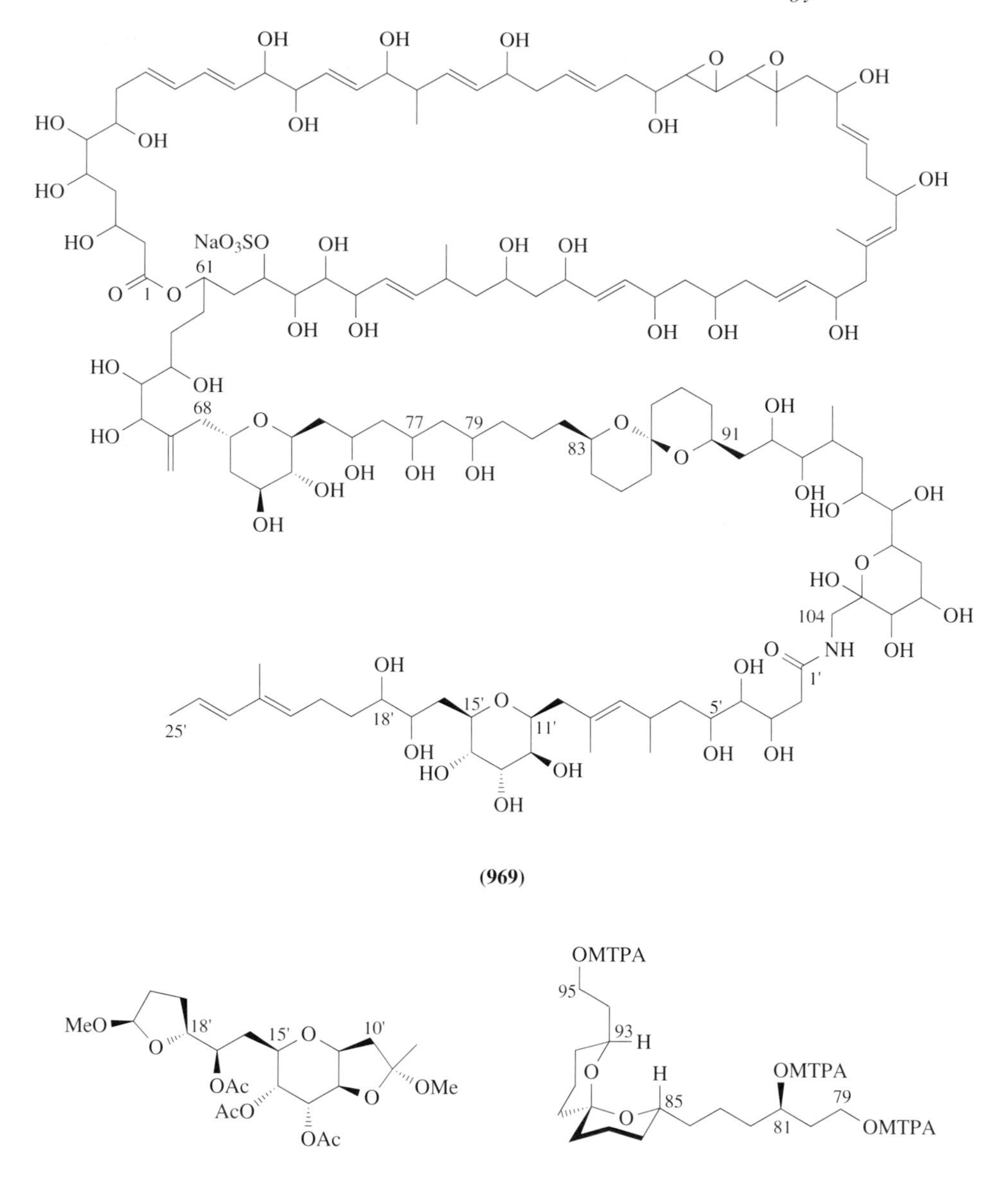

(969)

(970)

(971)

COSY and C-H HOHAHA, were successfully conducted with <3 mg of (**972**) (2 μmol). Tandem mass spectroscopy effectively clarified the structure in polyhydroxylated and polyene regions. Amphidinol (**972**) was the first member of a new class of polyhydroxy-polyenes. Notable structural features include a C_{69} chain with four C_1 branches, 21 hydroxyls, nine alkenes, and a sulfate ester.

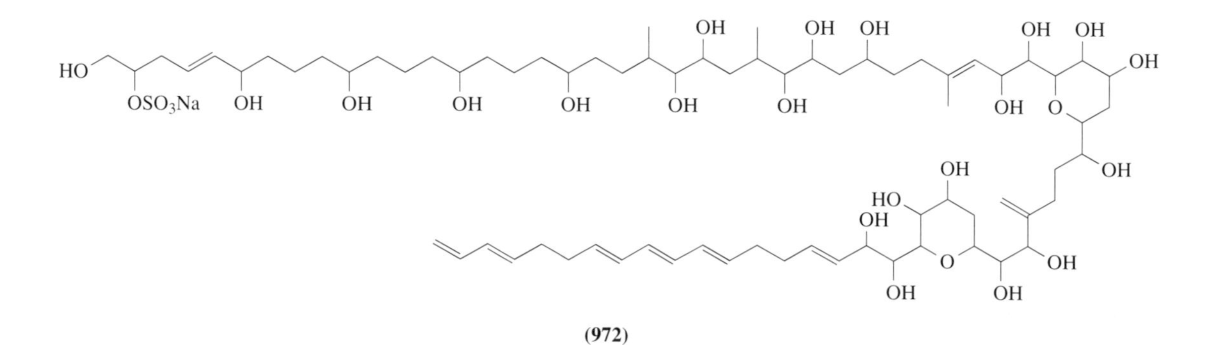

(972)

Another polyene-polyhydroxy compound, amphidinol 2 (**973**), was isolated as a potent hemolytic and antifungal agent from another cultured strain of dinoflagellate *Amphidinium klebsii* by Tachibana and co-workers.[821] The structure elucidated by spectroscopic methods turned out to be partly analogous to amphidinol 1 (**972**).

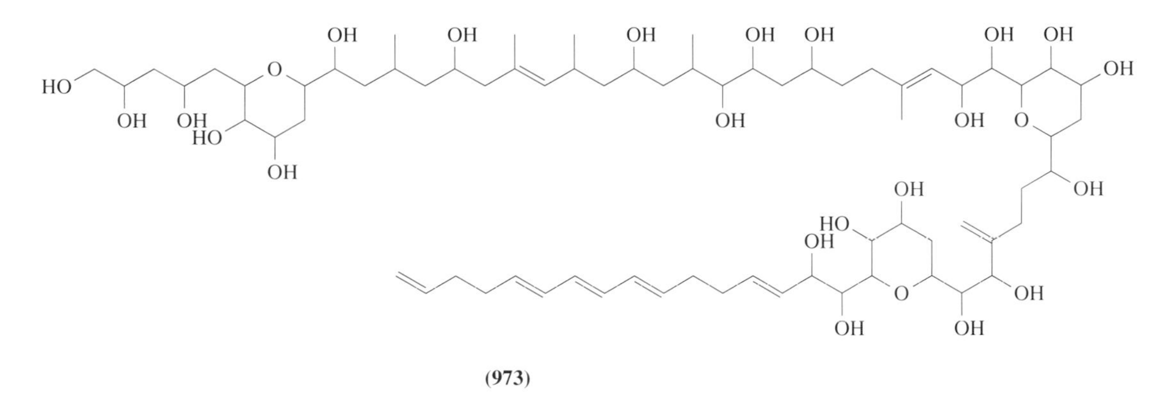

(973)

Kobayashi *et al.* studied a symbiotic dinoflagellate of the genus *Symbiodinium* isolated from the inside of gill cells of the Okinawan bivalve *Fragum* sp. From the toluene-soluble fraction of the extract of the harvested cells a sphingosine derivative, symbioramide (**974**),[822] was isolated by using silica gel column chromatographies. The structure of symbioramide (**974**) was elucidated on the basis of spectral and chemical means. The ^{1}H and ^{13}C NMR data suggested that compound (**974**) belongs to the ceramides. Acid hydrolysis of (**974**) afforded methyl 2-hydroxyoctadec-3*E*-enoate and 2*S*-amino-3*R*-hydroxyoctadecan-1-ol. The absolute configurations at the C-2 and C-3 positions of the sphingosine part of (**974**) were thus determined.

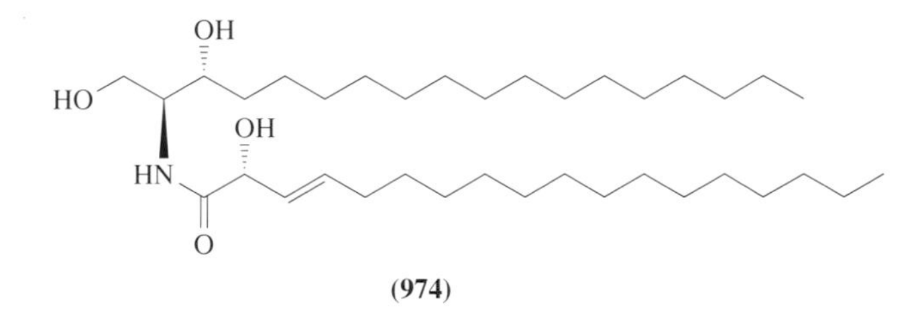

(974)

Symbioramide (**974**) exhibited weak cytotoxic activity against L1210 cells *in vitro* with an IC_{50} value of 9.5 μg ml^{-1}. Additionally, compound (**974**) was found to be a sarcoplasmic reticulum (SR) Ca^{2+}-ATPase activator. The Ca^{2+}-ATPase in the SR membrane plays a key role in muscle relaxation by energizing Ca^{2+}-pumping from the cytoplasm into the lumen of SR. Symbioramide (**974**) at 10^{-4} M activated SR Ca^{2+}-ATPase activity by 30%. This is the first example of an SR Ca^{2+}-ATPase activator of marine origin. Symbioramide (**974**) may serve as a valuable chemical tool for studying the regulatory mechanisms of SR Ca^{2+}-pumping systems. The α-hydroxy-β,γ-dehydro fatty acid contained in symbioramide (**974**) is rare from natural sources. Previously ceramides of α-hydroxyl fatty acids were obtained from the sponge *Dysidea etheria*.[823] The ceramides isolated from the sponge may be of microbial origin since sponges are known to possess symbiotic microorganisms. The absolute configuration of the C-2 position of the 2-hydroxyoctadec-3*E*-enoic acid moiety of symbioramide (**974**), which was previously unassigned, was established as *R* based on the total synthesis of symbioramide (**974**) achieved by Nakagawa *et al.*[824] Mori and Uenishi also studied the synthesis of the ceramide (**974**), and they found the interesting phenomenon that the optical rotation of (**974**) was influenced by the temperature of the sample solution in a cell for rotation measurements: $[\alpha]_D^{19} = +3.6°$, $[\alpha]_D^{23} = +0.76°$, $[\alpha]_D^{28} = -1.5°$, and $[\alpha]_D^{35} = -5.5°$ ($c = 0.31$ in $CHCl_3$).[825]

Soft corals are known to possess symbiotic microalgae. Kobayashi *et al.* isolated a diterpene, 16-deoxysarcophine (**975**),[826] with potent Ca-antagonistic activity from the Okinawan soft coral *Sarcophyton* sp. and established the structure by X-ray analysis. This was the first example of a Ca-antagonist from marine sources and of a nonalkaloid compound with such activity. Symbiotic microalgae of the soft coral *Sarcophyton* sp. were examined and a dinoflagellate belonging to the genus *Symbiodinium* was isolated from the polyps of the soft coral. 16-Deoxysarcophine (**975**) was detected by HPLC analysis of the extracts of the dinoflagellate *Symbiodinium* sp.

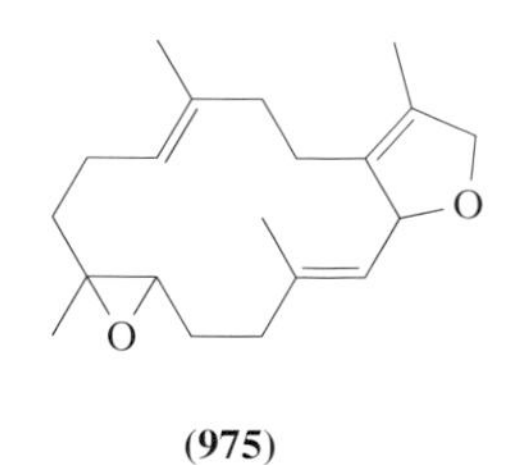

(**975**)

The Haptophyceae are microscopic, unicellular algae, which are widely distributed in the oceans and often constitute a major proportion of marine phytoplankton. Kobayashi *et al.* encountered a haptophyte of the genus *Hymenomonas*, which was isolated from an unidentified cylindrical stony coral collected at Sesoko Island, Okinawa. The extract of the harvested cells was partitioned between toluene and water and the aqueous phase was further extracted with chloroform. The toluene-soluble fraction was subjected to separation using Sephadex LH-20 and silica gel column chromatographies to give a sterol sulfate, hymenosulfate (**976**),[827] with potent Ca^{2+}-releasing activity in SR. The structure of (**976**) was elucidated by conversion of (**976**) into known (24*R*)-23,24-dimethylcholesta-(22*E*)-5,22-dien-3β-ol through acid hydrolysis. The presence of a sulfate group was confirmed by ion chromatography of sulfate ions liberated by hydrolysis. Hymenosulfate (**976**) is the first sterol sulfate isolated from marine microalgae. The major components of the toluene-soluble fraction of the extract of this haptophyte were glycolipids, mono- and digalactosyldiacylglycerols ((**977**) and (**978**), respectively), while the chloroform layer contained mainly octadecatetraenoic acid (**979**) along with a small amount of monogalactosylmonoacylglycerol (**980**). In the SR, the Ca^{2+}-releasing activity of (**976**) was 10 times more potent than that of caffeine, a well-known Ca^{2+}-releaser. The glycolipids (**977**), (**978**), and (**980**) exhibited inhibition of Na^{+}, K^{+}-ATPase activity with an IC_{50} value of 2×10^{-5} M each.

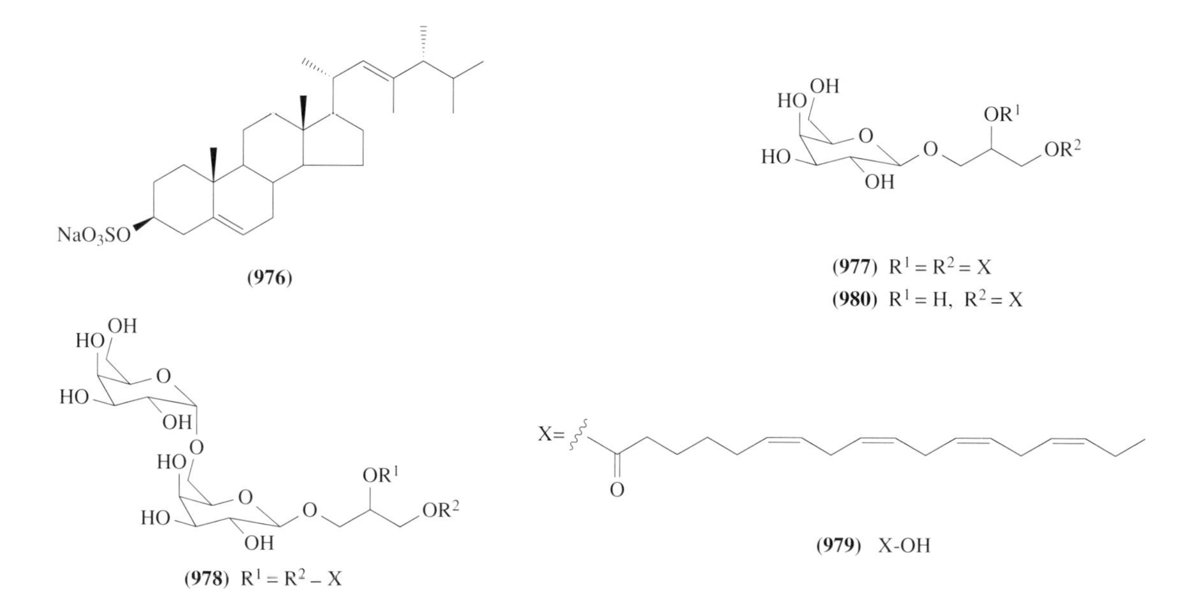

(**976**)

(**977**) $R^1 = R^2 = X$

(**980**) $R^1 = H$, $R^2 = X$

(**978**) $R^1 = R^2 - X$

(**979**) X-OH

From a cultured green colonial microalga, *Botryococcus braunii* Berkeley, Okada *et al.* isolated a new member of the natural carotenoids consisting of ketocarotenoid and its squalene analogue, botryoxanthin A (**981**).[828] The structure was determined by 2D NMR data. Botryoxanthin A (**981**) free from chlorophylls and intracellular carotenoids such as lutein could be extracted without breaking the cell wall. This implied that botryoxanthin A (**981**) may exist in the extracellular matrix like other secondary carotenoids and contribute to the color expression of the algal colonies. The presence of vicinally located C_1-branches (methyl and exomethylene) in the squalene moiety appears to be interesting in connection with the structures of microalgal metabolites like amphidinolides, as described in the following section.

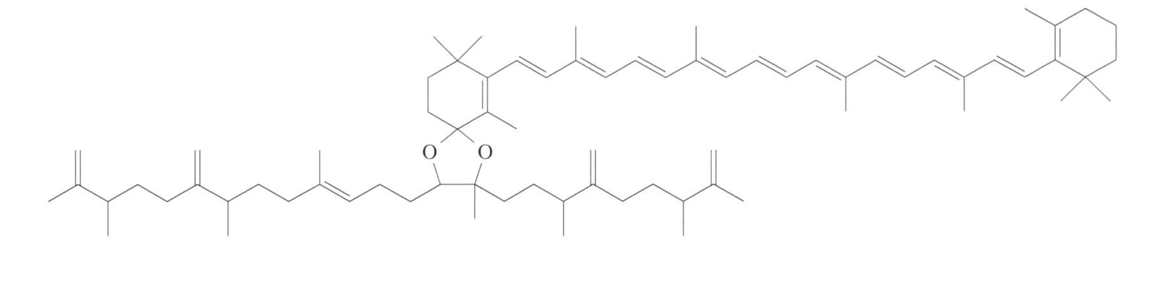

(981)

8.07.9.4.5 *Amphidinolides*

This section describes mainly the work carried out by the authors' research group on a series of unique macrolides, named amphidinolide, isolated from cultured marine dinoflagellate of the genus *Amphidinium*. During the course of studies on the search for bioactive substances from Okinawan marine organisms, the authors initiated a project on bioactive natural products of symbiotic marine microalgae in the mid 1980s. With help from previous studies by Yamasu who systematically collected symbiotic microalgae from marine invertebrates of Okinawan coastal waters,[829] large-scale cultures of the microalgae were carried out.

The first microalga utilized for research was a dinoflagellate belonging to the genus *Amphidinium* (strain number Y-5, 25 μm in length and 20 μm in width) isolated from the inner tissue of the host, a flatworm of the genus *Amphiscolops* Graff, 1905 (500 μm in length and 220 μm in width, green color), which was living on algae or seaweeds such as *Enteromorpha and Jania* spp. and collected at Chatan beach, Okinawa Island.[829] The unialgal cultures of *Amphidinium* sp. (Y-5) were grown in 3 l glass bottles containing 2 l of seawater medium enriched with Provasoli's ES supplement.[830,831] Cultures were incubated statically at 25 °C in an apparatus where illumination from a fluorescent light source was supplied in a cycle of 16 h light and 8 h darkness. After 2 weeks the culture was harvested by suction of the supernatant media with an aspirator, followed by centrifugation, to yield harvested cells ranging from 0.3 to 0.5 g l^{-1} of culture. The harvested cells were extracted with methanol/toluene (3:1) followed by partitioning between toluene and water. The toluene-soluble fraction was subjected to repeated silica-gel flash chromatography followed by reversed-phase HPLC resulting in isolation of four cytotoxic macrolides, amphidinolides A (**982**),[832] B (**983**),[833] C (**984**),[834] and D (**985**).[835]

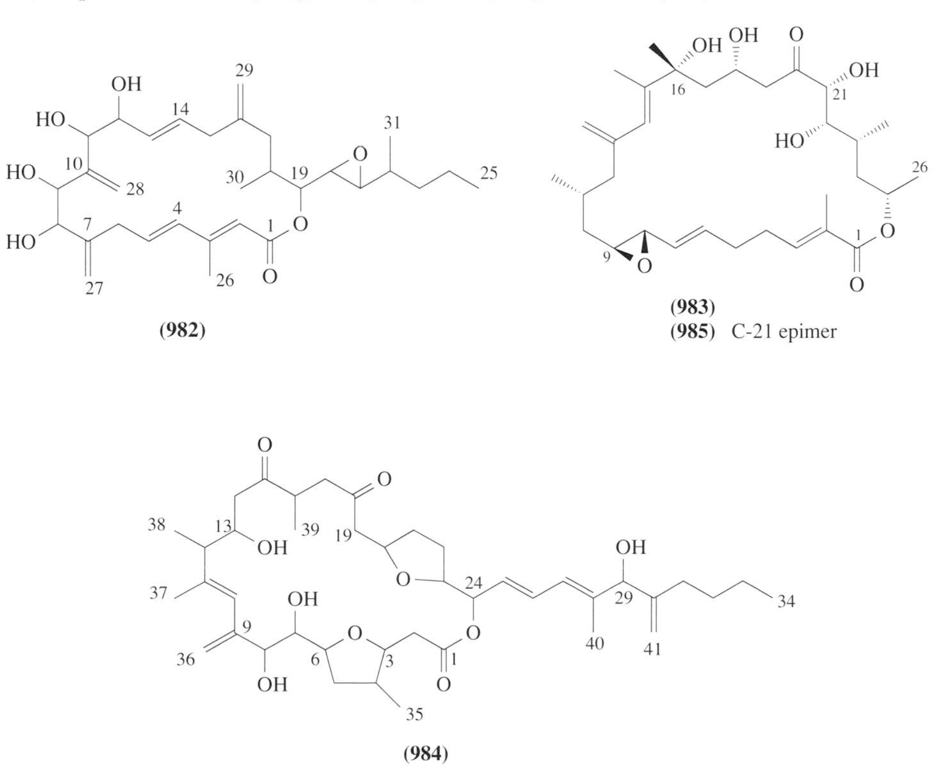

(982)

(983)
(985) C-21 epimer

(984)

Three other species of dinoflagellates of the genus *Amphidinium* (strain numbers Y-5′, Y-25, and Y-26) were also investigated. Their host animals were also Okinawan flatworms [Y-5′: *Amphiscolops* sp. (different from the host of Y-5); Y-25: *Amphiscolops breviviridis*; Y-26: *Amphiscolops magniviridis*]. From extracts of the cultured cells of these strains of *Amphidinium* spp. four other cytotoxic macrolides, amphidinolides E (**986**),[836] F (**987**),[837] G (**988**), and H (**989**),[838] were isolated. In addition to potent cytotoxic activity, amphidinolides B (**983**) and C (**984**) were shown to activate rabbit skeletal muscle actomyosin ATPase activity.[839] The chemical structures of these macrolides were elucidated mainly on the basis of extensive spectroscopic studies including several types of 2D NMR experiments (e.g., ^{1}H-^{1}H COSY, ^{1}H-^{13}C COSY, HMBC, HMQC, and NOESY spectra).

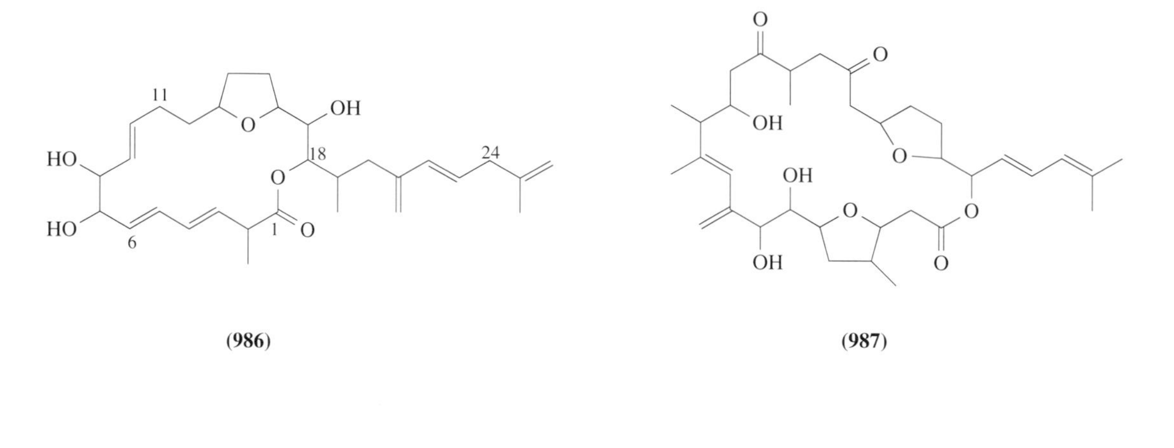

(986) **(987)**

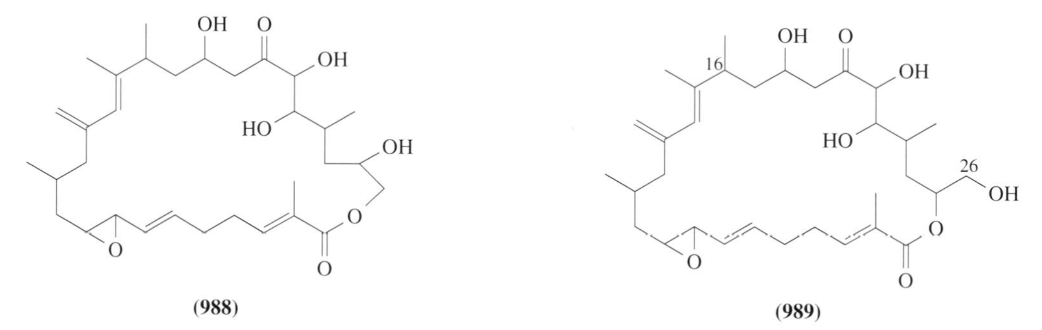

(988) **(989)**

Amphidinolide A (**982**), $C_{31}H_{46}O_7$, was suggested to possess an $\alpha,\beta,\gamma,\delta$-dienoate chromophore from its UV absorption spectrum (λ_{max} 265 nm). Its ^{1}H and ^{13}C NMR data revealed the presence of one epoxide and three exomethylene groups. Detailed analysis of the ^{1}H-^{1}H COSY spectrum allowed assignment of all proton signals and clearly established the proton-connectivities, starting with H-2 through to H-25, leading to a 20-membered macrolide structure. It was ambiguous whether a secondary methyl group was located at C-22 or C-23 due to overlapping of the signals for protons on C-22–C-24. It was, however, placed on C-22 because the ^{13}C chemical shifts of the C-23–C-25 signals suggested the presence of an *n*-propyl group instead of an ethyl group.[840] Efforts to elucidate the stereochemistry of amphidinolide A (**982**) were made by analyzing the phase-sensitive NOESY spectra of this 20-membered macrolide with nine chiral centers. The authors have proposed a possible stereostructure of amphidinolide A as (**990**) containing the relative configurations of the nine chiral centers of (**982**) and most sufficiently satisfying the NOESY data obtained in several solvent systems.

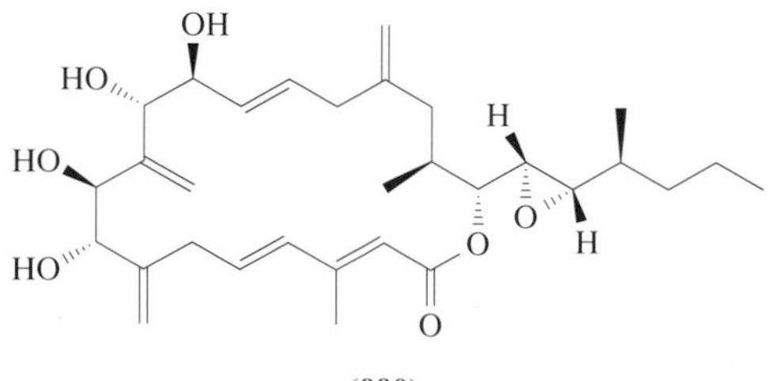

(990)

Amphidinolide B (**983**), $C_{32}H_{50}O_8$, possessed an α-methyl-α,β-unsaturated ester moiety, which was shown by the UV absorption (λ_{max} 222 nm) as well as the ^{13}C NMR chemical shifts. Its ^1H and ^{13}C NMR data also suggested the presence of one exomethylene, one epoxide, and one isolated ketone group. The ^1H-^1H COSY spectrum indicated the proton connectivities of three fragments (C-1–C-15, C-17–C-19, and C-21–C-26). Since the ^1H NMR chemical shifts for α positions to an sp^3 quaternary carbon (C-16) and ketone group (C-20) could be discriminated and NOE were observed between the methyl protons on C-15 and C-16, these three fragments were shown to be connected through C-16 and C-20, leading to a total structure consisting of a 26-membered macrocyclic lactone ring.

Amphidinolide C (**984**), $C_{41}H_{62}O_{10}$, exhibited a UV absorption maximum at λ_{max} 240 nm, implying the presence of a diene chromophore. Interpretation of the ^1H and ^{13}C NMR data suggested the presence of two tetrahydrofuran rings, two exomethylenes, and two isolated ketone groups. Extensive 2D NMR experiments were carried out on the tetraacetate (**991**) prepared from (**984**). Three partial structures (C-2–C-14, C-16–C-17, and C-19–C-34) were elucidated by analysis of the ^1H-^1H COSY and double relayed coherence transfer (RCT2) spectra. All protonated carbons were clearly assigned by ^1H-^{13}C COSY via one-bond couplings. The geometries of the double bonds were determined from NOE data. The connection of three partial structures through carbonyl groups (C-1, C-15, and C-18) was established by analyzing the ^1H-^{13}C long-range (two- and three-bond) couplings detected in the HMBC spectrum to give rise to a complete structure. The diene moiety (C-36, 9, 10, and 11) of the tetraacetate (**991**) was slowly oxidized by air to afford a [4+2] cycloaddition product (**992**), the spectral data of which provided additional proof for the structure of amphidinolide C (**984**).

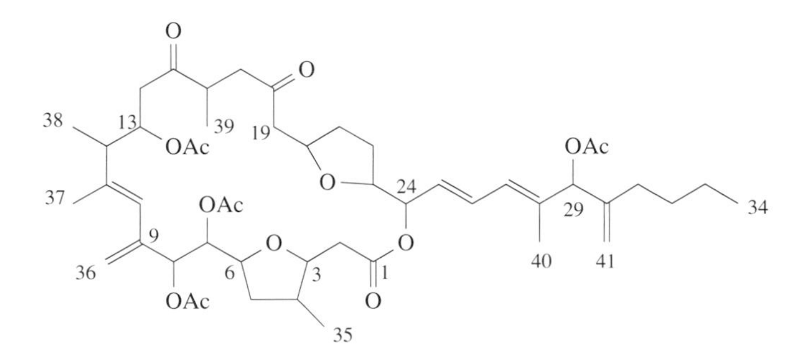

(991)

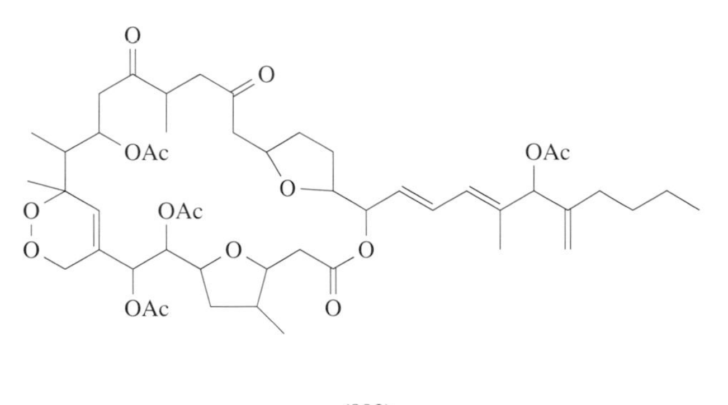

(992)

Amphidinolide E (**986**), $C_{30}H_{44}O_6$, showed a UV absorption maximum at λ_{max} 230 nm corresponding to a diene chromophore. The presence of two exomethylenes and one tetrahydrofuran ring was inferred by analyzing the ^1H and ^{13}C NMR data. Its ^1H-^1H COSY data afforded information on the proton-connectivities for three partial structures (C-2–C-14, C-15–C-19, and C-20–C-27). The geometries of the four disubstituted double bonds were deduced on the basis of the ^1H-^1H coupling constants determined by a *J*-resolved 2D NMR experiment. The partial structures were revealed to be linked to each other by ^{1}H chemical shift data together with the relayed-correlations observed through an RCT-COSY spectrum to construct a 19-membered macrocyclic lactone ring with an alkyl side-chain. Spectral investigations of amphidinolides D (**985**), G (**988**), and H (**989**), whose molecular formulas are all $C_{32}H_{50}O_8$, revealed that they are structurally related closely to amphidinolide B (**983**): amphidinolides B (**983**) and D (**985**) are stereoisomers at C-21; amphidinolides G (**988**) and H (**989**) are regioisomers of the lactone-terminal positions (C-25 or C-26); amphidinolides B (**983**) and H (**989**) are only different in the position of one hydroxyl group (C-16 or C-26).

Amphidinolide D (**985**), an epimer of amphidinolide B (**983**) at the C-21 position, was, however, about 100 times less cytotoxic than the latter. During the isolation process of the extract of the strain Y-5, 1 mol of MeOH was added on the C-8/C-9 epoxide of amphidinolide B (**983**) to generate compound (**993**), the IC_{50} value of which was 0.081 μg ml^{-1} against the L1210 cell, being considerably weaker than that of (**983**) (1/600). These results implied that the stereochemistry at C-21 and the presence of an epoxide at the C-8/C-9 position are quite important for the cytotoxic activity of these compounds, presumably due to the significant change of the molecular conformation.

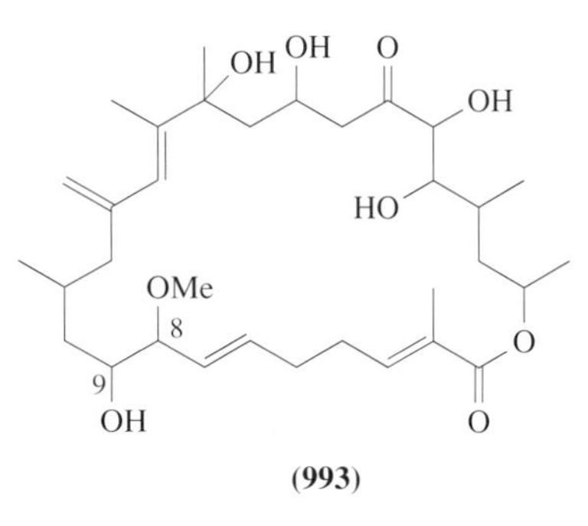

(**993**)

Amphidinolide F (**987**), $C_{35}H_{52}O_9$, proved to be analogous to amphidinolide C (**984**). The structural difference between amphidinolides F (**987**) and C (**984**) was found in the length of the alkyl side-chain, the former possibly being a biogenetic precursor of the latter. The cytotoxicities of amphidinolides C (**984**) and F (**987**) were also very different, the former being about 250 times as strong as the latter. The length of the side-chain may significantly affect the potency of cytotoxic activity.

After obtaining the series of macrolides, amphidinolides A–H (**982**)–(**989**), the mass-culturing of the dinoflagellates *Amphidinium* sp. (strain numbers Y-5 and Y-25) was continued.[841] Previous studies had revealed that fractionation by silica gel chromatography of the toluene-soluble portion of the extracts of these microalgae afforded several fractions exhibiting extremely potent cytotoxicity against murine lymphoma L1210 and human epidermoid carcinoma KB cells *in vitro*, with the inhibition values at 10 μg ml^{-1} being more than 90%. These inhibition values cannot be fully accounted for by estimating from the IC_{50} values of previously isolated amphidinolides. Thus, further investigations continued to search for other cytotoxic components of these dinoflagellates. As a result, several novel cytotoxic macrolides with various ring numbers, amphidinolides J (**994**), K (**995**), M (**996**), N (**997**), O (**998**), P (**999**), and Q (**1000**), were isolated, together with a new linear metabolite amphidinin A (**1001**) from strain Y-5, while a new 27-membered macrolide, amphidinolide L (**1002**), was isolated from strain Y-25.

In 1993 the authors succeeded in the isolation and structure elucidation of amphidinolide J (**994**),[842] a novel 15-membered macrolide, and determined its absolute stereochemistry by combination of degradation experiments and synthesis of optically active compounds. Further cultivation of the dinoflagellate *Amphidinium* sp. (strain Y-5) was carried out at 25 °C for 2 weeks in a seawater medium enriched with Erd-Schreiber (ES) nutrients. The harvested algal cells (920 g, wet weight, from 3300 l of culture) were extracted with MeOH/toluene (3:1) and the extracts were partitioned between toluene and water. The toluene-soluble fraction was subjected to a silica gel column ($CHCl_3$/MeOH, 95:5) followed by gel filtration on Sephadex LH-20 ($CHCl_3$/MeOH, 1:1). Subsequent separation by reversed-phase HPLC (ODS, 88% MeOH) afforded amphidinolide J (**994**) (0.0002% yield, wet weight) as a colorless oil.

The planar structure of amphidinolide J (**994**), $C_{24}H_{38}O_4$, was studied by detailed analyses of its ^{1}H and ^{13}C NMR data aided with 2D NMR experiments (^{1}H-^{1}H COSY, HSQC, HMBC, and

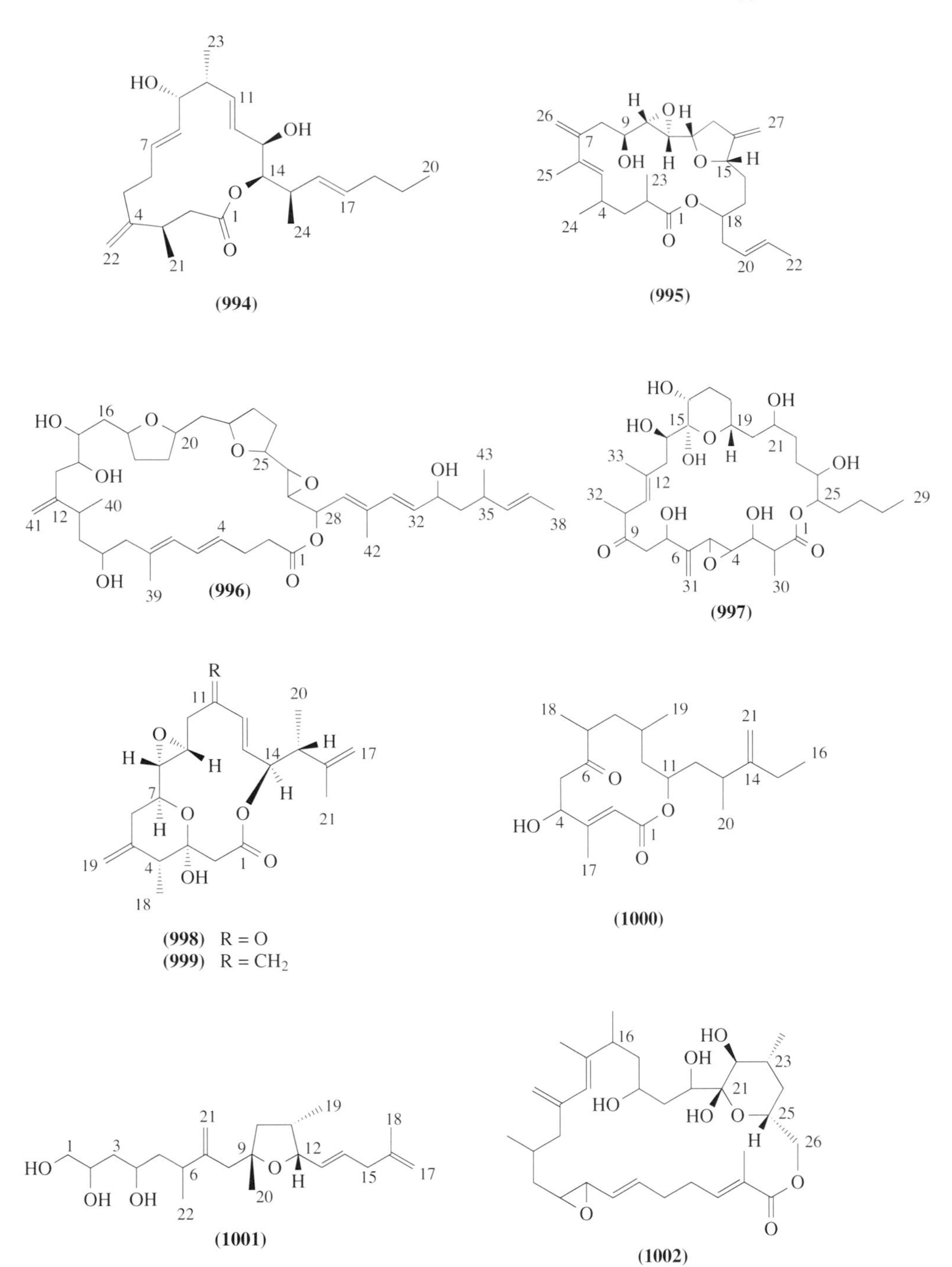

NOESY), thereby leading to a gross structure of **(994)** consisting of a 15-membered lactone ring with three disubstituted *E*-alkenes ($J_{7,8}$ = 15.0 Hz, $J_{11,12}$ = 15.8 Hz, and $J_{16,17}$ = 15.0 Hz). This gross structure was further confirmed by the structures of the degradation products **(1003)**–**(1008)** obtained by the following ozonolysis experiments. Treatment of **(994)** with ozone (−78 °C, 1 min) followed by $NaBH_4$ reduction and acetylation (Scheme 30) afforded a complex mixture, from which the normal and reverse-phase HPLC separations were carefully carried out to obtain degradation products **(1003)**–**(1005)**, corresponding to the C-1–C-7, C-8–C-11, and C-12–C-16 moieties of **(994)**, respectively. In addition, partial-degradation products **(1006)**–**(1008)** were also obtained and their structures provided further evidence for the proposed planar structure of **(994)**. For the unambiguous determination of the absolute configurations of the six chiral centers of **(994)**, the fragments **(1003)**–**(1005)** together with all their possible diastereomers were prepared in optically active forms.

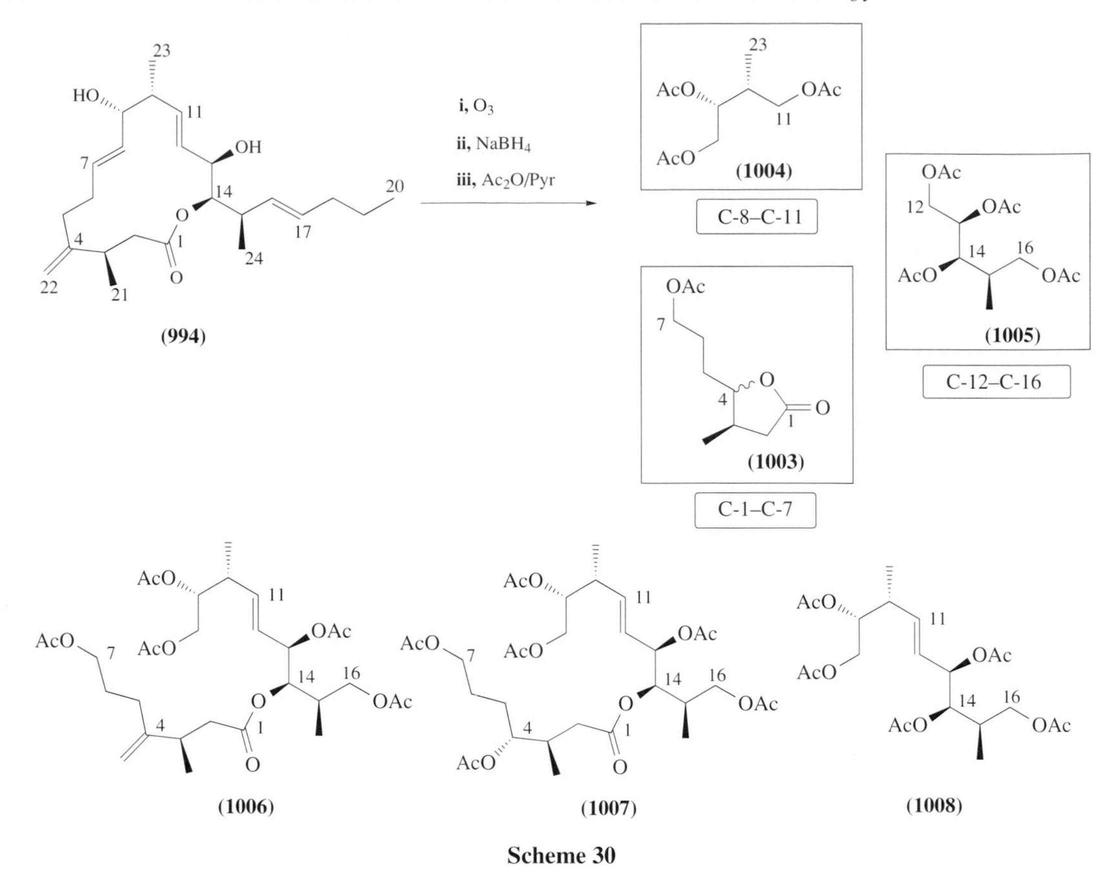

Scheme 30

The C-1–C-7 fragment (**1003**) was synthesized as shown in Scheme 31, starting from mono-protected 2(*S*)-methylpropane-1,3-diol (**1009**), which was readily supplied from (−)-methyl 3-hydroxy-2(*R*)-methylpropionate (**1010**). The Grignard addition to the corresponding aldehyde from (**1009**) afforded a diastereomeric mixture at C-4 in the ratio of 45:55, which was separated in the final step by silica gel HPLC. The 3,4-*syn* (**1003a**) and 3,4-*anti* (**1003b**) isomers thus obtained were completely identical with those from natural specimens including the signs of optical rotations [synthetic, (**1003a**): $[\alpha]_D$ +17° (*c* 1.0, $CHCl_3$); (**1003b**): $[\alpha]_D$ −22° (*c* 1.0, $CHCl_3$); natural, (**1003a**): $[\alpha]_D$ +17° (*c* 0.06, $CHCl_3$); (**1003b**): $[\alpha]_D$ −34° (*c* 0.2, $CHCl_3$)] to establish the 3*R*-configuration for (**994**).

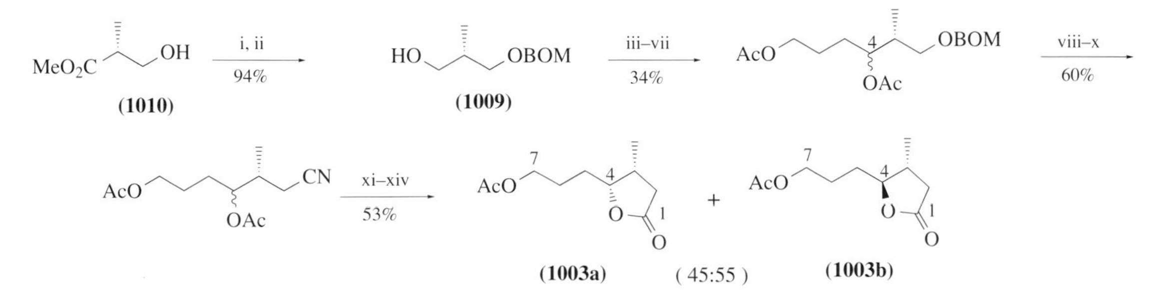

i, BOMCl, Pr^i_2NEt, CH_2Cl_2, room temperature, 44 h; ii, LAH, ether, room temperature, 30 min; iii, DMSO, $(COCl)_2$, CH_2Cl_2, –78 °C, 30 min, then Et_3N, 0 °C, 30 min; iv, $CH_2{=}CHCH_2CH_2MgBr$, ether, 50 °C, 40 min; v, O_3, MeOH, –78 °C, 2.5 h; vi, $NaBH_4$, MeOH, 0 °C, 1 h; vii, Ac_2O, pyridine, room temperature, 12 h; viii, H_2, Raney Ni (W-2), EtOH, room temperature, 48 h; ix, TsCl, pyridine, room temperature, 44 h; x, NaCN, DMSO, 85-90 °C, 2 h; xi, NaOH, H_2O_2, EtOH, 65 °C, 1.5 h, then 90 °C, 7 h; xii, 2M HCl, room temperature; xiii, Ac_2O, pyridine, room temperature, 11 h; xiv, HPLC separation

Scheme 31

The C-8–C-11 fragment (**1004**) and its *syn*-isomer (**1010**) were readily prepared ((i) reductive ozonolysis, (ii) deprotection, and (iii) acetylation; Scheme 32) from allyl alcohols ((**1012**) and (**1013**), respectively), which were obtained from (**1010**) via modifications of literature procedures.[843] The spectral data of the C-8–C-11 fragment obtained by degradation of (**994**) were identical with those of the *anti*-isomer (**1004**) and their optical data (synthetic, $[\alpha]_D$ +5.0° (*c* 1.0, $CHCl_3$); natural; $[\alpha]_D$ +2.8° (*c* 0.22, $CHCl_3$)) revealed the 9*R*,10*R*-configurations for (**994**).

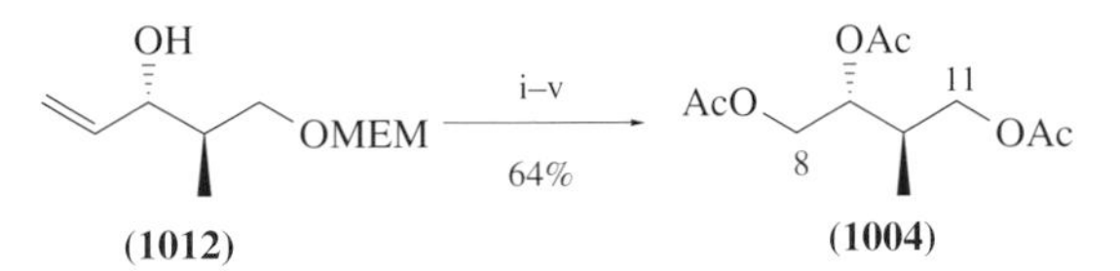

i, O_3, MeOH, –78 °C, 5 min; ii, $NaBH_4$, MeOH, 0 °C, 1 h; iii, Ac_2O, pyridine, room temperature, 38 h; iv, 4M HCl, THF, 50 °C, 2 h; v, Ac_2O, pyridine, room temperature, 20 h

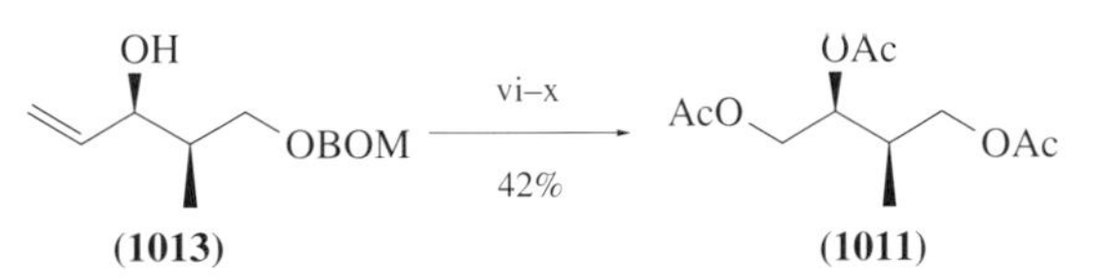

vi, O_3, MeOH, –78 °C, 1 min; vii, $NaBH_4$, MeOH, 0 °C, 45 min; viii, Ac_2O, pyridine, room temperature, 17 h; ix, H_2, 10% Pd-C, MeOH, room temperature, 11h; x, Ac_2O, pyridine, room temperature, 20 h

Scheme 32

Preparations of the C-12–C-16 fragment (**1005**) and its diastereomers (**1014**)–(**1016**) were achieved by applying Kishi's methods for pentose synthesis[844] (Scheme 33). The epoxy alcohol (**1017**), obtained from D-glyceraldehyde acetonide (**1018**), was treated with lithium dimethylcuprate to give 1,3-diol (**1019**) together with undesired 1,2-diol in the ratio of 1:1, which was separated in the final step by silica gel HPLC (hexane/EtOAc, 2:1). The diastereomers (**1014**)–(**1016**) were also obtained by similar procedures from the corresponding epoxy alcohols ((**1020**)–(**1022**), respectively). The C-12–C-16 fragment derived from (**994**) was identical with the *syn-anti* isomer (**1005**) including the sign of optical rotation (synthetic, $[\alpha]_D$ +41° (*c* 1.0, $CHCl_3$); natural $[\alpha]_D$ +44° (*c* 0.23, $CHCl_3$)),

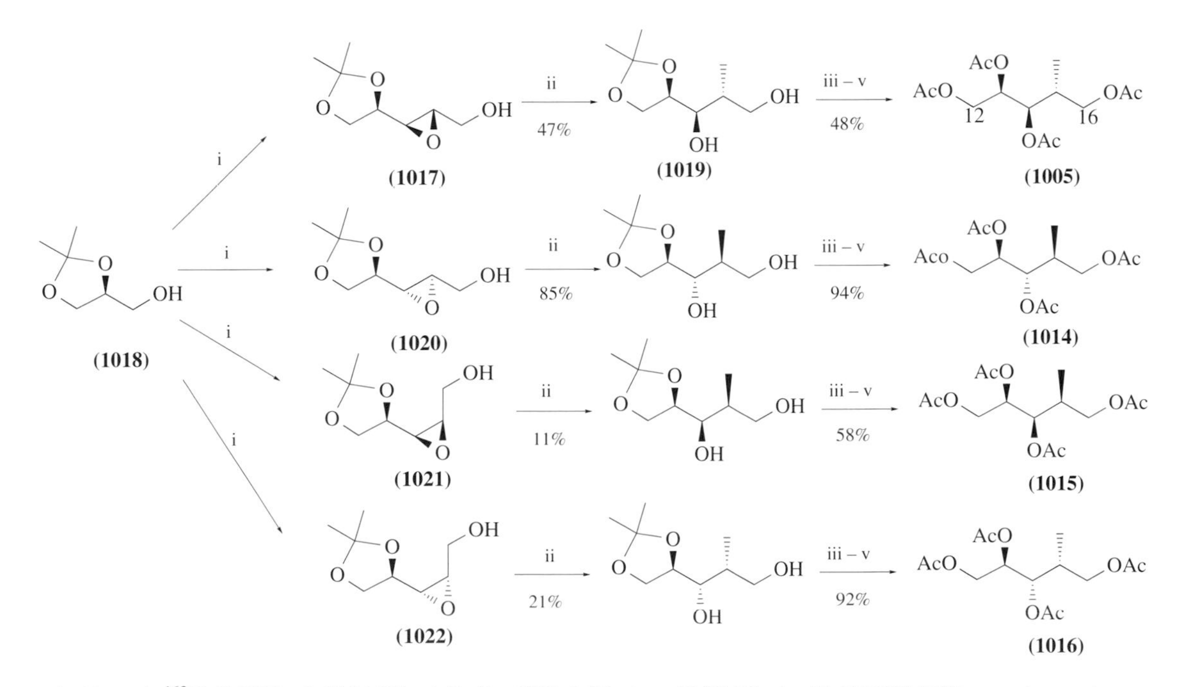

i, (4 steps);[163] ii, CuI (12 eq), MeLi (24 eq), Et_2O, –40°C, 4–9 h, then –23 °C, 30 min; iii, 1M HCl, THF, room temperature, 7–25 h or $AcOH/H_2O$ (4:1), 40 °C, 4 h; iv, Ac_2O, pyridine, room temperature, 11–20 h; v, HPLC separation

Scheme 33

thus determining the 13*R*,14*R*,15*R*-configurations for (**994**). From these results the structure of amphidinolide J was firmly established as (**994**) including the absolute stereochemistry of the six chiral centers.

In connection with our studies on cytotoxic macrolides from the dinoflagellate *Amphidinium* sp. (strain Y-5), we also continued examining the extract of the strain Y-25 *Amphidinium* sp., which was isolated from the Okinawan flatworm *Amphiscolops breviviridis* to result in the isolation of a new 27-membered macrolide, amphidinolide L (**1002**).[845] Amphidinolide L (**1002**), $C_{32}H_{50}0_8$, was isolated as a colorless oil in 0.0004% yield from ca. 1800 g (wet weight) of the harvested cells obtained from 17501 of culture of this alga. Detailed analysis of the 2D NMR data (^{1}H-^{1}H COSY, HOHAHA, HMQC, and HMBC) led to the planar structure of amphidinolide L (**1002**) as constructed from a 27-membered lactone ring with an epoxide and tetrahydropyran moiety, which corresponded to the 20-dihydro-21-dehydro derivative of amphidinolide G (**988**). The relative stereochemistry of the tetrahydropyran moiety (C-21, C-22, C-23, and C-25 positions) was elucidated on the basis of NOE and coupling constant data, and the absolute configurations of the C-23 and C-25 positions were established by synthesis of the tetraacetate (**1023**), corresponding to the C-21–C-26 fragment of (**1002**), starting from the optically active epoxy alcohol (**1024**)[846] (Scheme 34). The synthetic tetraacetate (**1023**) showed completely identical spectral data including the signs of the optical rotations (synthetic, $[\alpha]_D$ +64° (*c* 0.2, $CHCl_3$); natural, $[\alpha]_D$ +72±8° (*c* 0.01, $CHCl_3$)), with those of compound (**1023**), which was obtained from (**1002**) by treatment with $NaIO_4$ followed by $NaBH_4$ reduction and acetylation, indicating the 21*R*,22*S*,23*R*,25*S*-configurations. Investigations are continuing into the synthesis of the diastereomers of the C-15–C-26 fragment of (**102**) to determine the stereochemistry of that moiety; preparation of a diastereomer (**1025**) possessing 16*R*,20*R*-configuration from (−)-methyl 3-hydroxy-2(*R*)-methylpropionate (**1010**) (Scheme 31) has been achieved in an efficient manner with good yields.[847]

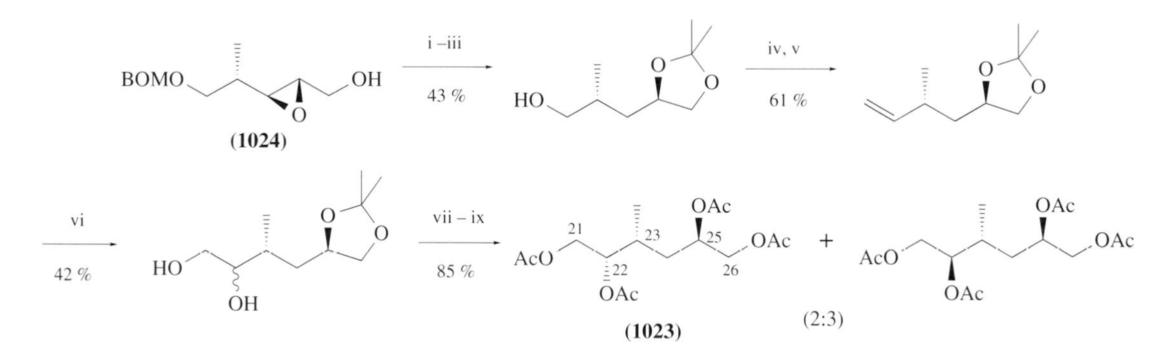

i, DIBAH, benzene, room temperature, 1 h; ii, $(MeO)_2CMe_2$, PPTS, CH_2Cl_2, room temperature, 5 h; iii, H_2, $Pd(OH)_2$, EtOH, room temperature, 18 h; iv, DMSO, $(COCl)_2$, CH_2Cl_2, -78 °C, 30 min, then Et_3N, –20 °C, 30 min; v, Ph_3PCH_3Br, Bu^nLi, THF, room temperature, 2 h; vi, OsO_4, pyridine, THF, room temperature, 4 h; vii, 1N HCl, THF, room temperature, 5 h; viii, Ac_2O, pyridine, room temperature, 18 h; ix, HPLC separation

Scheme 34

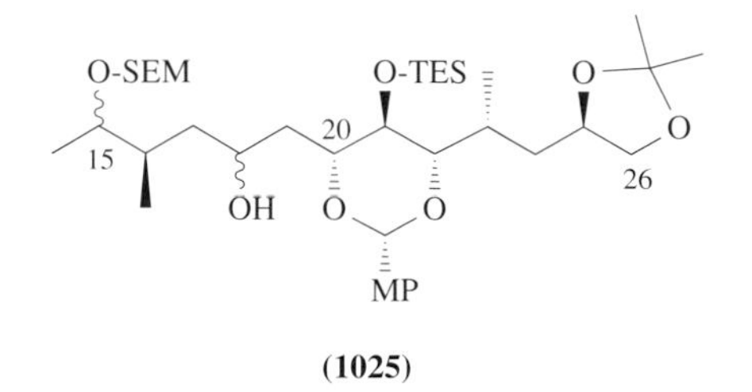

The cytotoxic fraction mainly containing amphidinolide J (**994**) was further examined carefully by separation using reversed-phase HPLC to give four new compounds, amphidinolides K (**995**), M (**996**), N (**997**), and amphidinin A (**1001**). Amphidinolide K (**995**),[848] $C_{27}H_{40}O_5$, was isolated in 0.0002% yield (wet weight), and its gross structure was elucidated by applying several types of 2D

NMR techniques using a 600 MHz spectrometer to deduce the planar structure as (**995**) containing a 19-membered macrocyclic lactone along with a diene, an epoxide, and a tetrahydrofuran moiety. We proposed the relative stereochemistry of the epoxide-tetrahydrofuran portion (C-9, C-10, C-11, C-12, and C-15) on the basis of the NOESY and coupling constant data, and the diene moiety was inferred to have *S-trans* conformation from the NOESY correlations (H-5 and one of H_2-8; H_3-25 and one of H_2-26).

Amphidinolide M (**996**),[849] $C_{43}H_{66}O_9$, was first isolated in 1986 from the dinoflagellate of this species (strain number Y-5) in ca. 0.0005% yield (wet weight). Unfortunately, the sample of (**996**) decomposed extensively during storage as a $CDCl_3$ solution, and the structural studies were interrupted. The quantity of these macrolides contained in the extracts of the cultured cells varied a little during the course of time and amphidinolide M (**996**) was not isolated again for several years. When amphidinolide M (**996**) was reisolated from the same strain of this cultured alga by careful HPLC examination, spectral studies of (**996**) and its tetraacetate were extensively carried out to suggest that compound (**996**) was a 29-membered macrolide with two dienes, two tetrahydrofuran (THF) rings, an exomethylene, and an epoxide. The stereochemistry of (**996**) remained undetermined; the NOESY data of (**996**), however, may have implied that the angular hydrogens of two THF portions were both *trans* since NOESY cross-peaks were significantly observed for H-15/H-20 and H-22/H-27 while no correlations between angular protons (H-17/H-20 and H-22/H-25) were visible.

Amphidinolide N (**997**)[850] was isolated from a relatively polar fraction by reversed-phase HPLC (Develosil ODS-5, 60% CH_3CN) separation. Compound (**997**) was extremely cytotoxic against murine lymphoma L1210 and human epidermoid carcinoma KB cells *in vitro* (*vide infra*); the cytotoxicity of (**997**) was the most potent of all amphidinolides that have ever been isolated. The structure of this macrolide, $C_{33}H_{54}O_{12}$, was interpreted by the extensive analysis of its spectroscopic data and was proposed to be (**997**), which was composed of a 26-membered macrolide containing a tetrahydropyran (THP) moiety with a hemiketal group, an epoxide, and an exomethylene group. The NOESY cross-peak observed between H-14 and H-19 might suggest that C-14 and H-19 were both axially oriented on the THP ring. The hydroxyl group on C-16 was deduced to be axial from the coupling constants ($J_{16,17a} = J_{16,17b} = 2.5$ Hz). After the isolation and gross structure of amphidinolide N (**997**) was published, the isolation and structure of caribenolide I (**1026**),[851] a related compound to (**997**), was described by Shimizu and co-workers; they isolated compound (**1026**) from a cultured free-swimming Caribbean dinoflagellate *Amphidinium* sp. Caribenolide I (**1026**) was reported to show strong cytotoxicity (IC_{50} 0.001 μg ml^{-1} or 1.6 nM) against both human colon tumor cell line (HCT 116 and its drug-resistant cell line, HCT 116/VM 46). This cytotoxicity was about 100 times higher than that of amphidinolide B (**983**) (IC_{50}, HCT 116, 0.122 μg ml^{-1}). Compound (**1026**) was also described to exhibit *in vivo* activity against murine tumor P388 (T/C: 150 at a dose of 0.03 mg kg^{-1}).

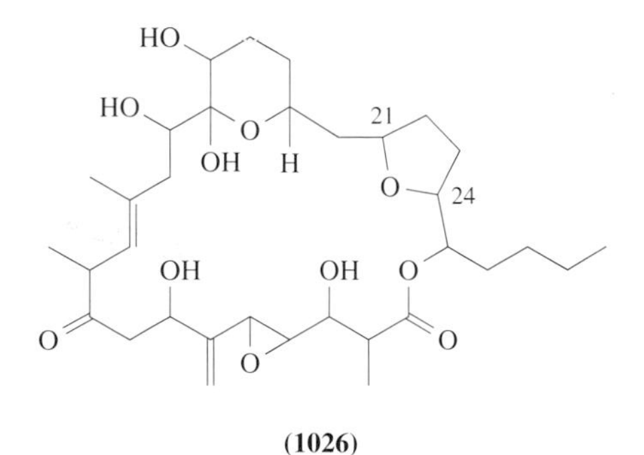

(**1026**)

Amphidinin A (**1001**), $C_{22}H_{38}O_4$,[852] was isolated from the macrolide-containing fraction by reversed-phase HPLC (Develosil ODS-5; 10 × 250 mm; 59% MeCN; flow rate: 2.5 ml min^{-1}; detection: RI and UV at 220 nm). Under this separation condition, compound (**1001**) had a very close retention time (t_R 27.6 min) to those of amphidinolides A (**981**) (t_R 29.1 min) and E (**986**) (t_R 26.0 min). Being different from all other cytotoxic metabolites isolated from this microalga, amphidinin A (**1001**) did not have the macrolide structure; the IR spectrum of (**1001**) showed no characteristic band due to a carbonyl group. Extensive NMR studies revealed that compound (**1001**) possessed a linear backbone skeleton with one THF moiety. Three hydroxyl groups are located on one end of the molecule, while there is a 2-methyl-1,4-pentadiene unit on the other end, constructing a hydrophilic and a hydrophobic moiety, respectively, in a linear molecule. The relative stereochemistry of the THF portion was suggested by the NOESY data. As a result, the gross structure of amphidinin A was deduced as (**1001**), but the relative and absolute configurations of the chiral centers of (**1001**) remained undefined. Although amphidinin A (**1001**) is a nonmacrolide, this compound has several structural relationships to previously isolated amphidinolides, implying that biogenesis of amphidinin A (**1001**) may be closely related to amphidinolides.

During examinations of the cytotoxic fraction of the extract of the Y-5 strain of this microalga, an allenic compound (**1027**) was also isolated;[853] this compound was identified including the CD spectral data as apo-9′-fucoxanthinone,[854] which was previously reported as a permanganate oxidation product of fucoxanthin. The deacetyl derivative of (**1027**) was known as a grasshopper ketone isolated from ant-repellent secretions of the large flightless grasshopper *Romalea microptera*[855] and also isolated from *Edgeworthia chrysantha*.[856]

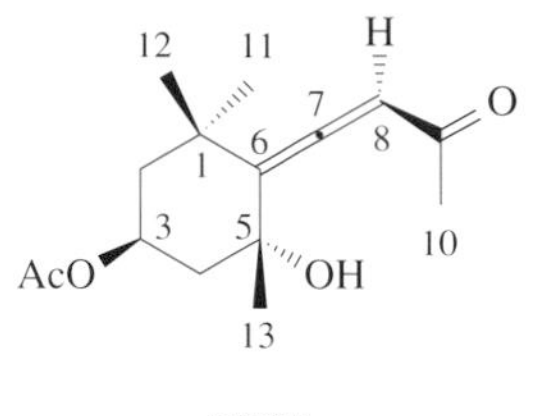

(**1027**)

During studies on the Y-5 strain of *Amphidinium* sp., the cytotoxic fractions which were less polar than amphidinolide J (**994**) were examined. The ^{1}H NMR spectra of these crude fractions exhibited significantly exomethylene signals. Previously isolated amphidinolides all contain exomethylene groups, which were detected as sharp singlets around 5 ppm in the ^{1}H NMR spectra of crude fractions in the latter stage of the isolation process. Thus, these fractions were further purified by reversed-phase HPLC to result in the isolation of three novel macrolides, amphidinolides O (**998**), P (**999**), and Q (**1000**). Amphidinolides O (**998**) ($C_{21}H_{28}O_6$) and P (**999**) ($C_{22}H_{30}O_5$)[857] were both novel 15-membered macrolides, and these two compounds are structurally related to each other. They both contain a THP moiety with a hemiketal group, an epoxide, and at least two exomethylene groups. The structural difference was found at the C-11 position; the C-11 ketone group of (**998**) was replaced by an exomethylene group for (**999**), which was indicated from the following observations: (i) two IR absorption bands due to carbonyl groups were observed for (**998**), but one (ν_{max} 1730 cm^{-1}) for (**999**); (ii) compound (**998**) showed a UV absorption maximum at 231 nm due to an enone moiety, while the UV absorption of (**999**) underwent a blue shift (λ_{max} 225 nm), which was assignable to a diene chromophore; (iii) the ^{13}C NMR of (**999**) showed no signal due to a conjugated ketone, instead of which NMR signals for another exomethylene group were observed (δ_C 118.1 (C-22) and 142.3 (C-11); δ_H 4.98 (1H, br s) and 4.85 (1H, br s) for H_2-22).

Amphidinolides O (**998**) and P (**999**) possess seven chiral centers; the relative configurations of five chiral centers contained in the THP and epoxide ring portion were both elucidated on the basis of the NMR data as 3*S**, 4*R**, 7*S**, 8*S**, and 9*S** for (**998**) and 3*S**, 4*R**, 7*S**, 8*R**, and 9*S** for (**999**). The relative stereochemistries of the remaining two chiral centers (C-14 and C-15) were investigated by the combination of the ^{1}H NMR data and molecular mechanics calculations. Four diastereomers were considered for each compound with (14*R**,15*R**)-, (14*R**,15*S**)-, (14*S**,15*R**)-, and (14*S**,15*S**)-configurations, and the Monte Carlo lowest-energy conformations were calculated with the MM2 force field. The NOESY spectrum of amphidinolide O (**998**) in C_6D_6 solution clearly revealed cross-peaks due to H-8/H-12 and H-10b/H-12 with no correlations for H-8/H-13 or H-10b/H-13 observed. On the other hand, the NOESY spectra of amphidinolide P (**999**) in C_6D_6 showed substantial correlations for H-8/H-13, H-10b/H-13, and H-12/H-22a, but no cross-peaks due to H-8/H-12, H-10b/H-12, or H-13/H-22a were visible. These observations implied that, in the C_6D_6 solution states, the C-11–C-13 enone moiety of (**998**) is abundantly *S-cis* while the *S-trans* conformation is predominant for the C-11–C-13 diene moiety of (**999**). This result was consistent with the calculation data of 14*R**-diastereomers for both (**998**) and (**999**). The relative stereochemistry of the remaining chiral center at C-15 was analyzed on the basis of comparison of the proton–proton coupling constant ($J_{14,15}$) between the observed and the calculated values. The observed $J_{14,15}$-values were 7.4 Hz and 9.3 Hz for (**998**) and (**999**), respectively. The calculated average values of $J_{14,15}$ for the 15*R**-diastereomers (9.2 Hz for (**998**) and 9.5 Hz for (**999**)) corresponded better than those of the 15*S**-diastereomers (2.5 Hz for (**998**) and 3.7 Hz for (**999**)). Thus, the relative configurations of amphidinolides O and P were deduced as 3*S**, 4*R**, 7*S**, 8*S**, 9*S**, 14*R**, and 15*R** for (**998**) and 3*S**, 4*R**, 7*S**, 8*R**, 9*S**, 14*R**, and 15*R** for (**999**). The structural difference between (**998**) and (**999**) is found only at the C-11 position, viz., the ketone group for (**998**) and the exomethylene group for (**999**). Natural product analogues with this type of structural difference (ketone/exomethylene) are believed to be quite rare. Compounds (**998**) and (**999**) are likely to be biogenetically related to each other; one may be a precursor of the other, but it is unknown which one preceded the other.

Amphidinolide Q (**1000**) ($C_{21}H_{34}O_4$)[858] was revealed to possess one ketone, one exomethylene, and four methyl groups by spectral data. The ^{13}C chemical shift of the C-17 methyl (δ_C 16.6) argued that the Δ^2-alkene was *E*, and this double bond was suggested to be conjugated with the C-1 ester carbonyl from the ^{13}C chemical shifts (C-2: δ_C 117.4; C-3: δ_C 155.4), which were also consistent with the UV absorption data of (**1000**) (MeOH, λ_{max} 222 nm, ε 10 300). Since the molecule of (**1000**) was inferred to contain one ring from the unsaturation degrees, the C-1 carbonyl had to be linked to the C-11 oxymethine to form a 12-membered lactone ring, which was coincident with the low-field resonance of H-11 (δ_H 5.28). The gross structure of amphidinolide Q was thus elucidated as (**1000**) having a novel backbone skeleton with a 12-membered macrocyclic lactone ring. Among the NOESY correlations observed for (**1000**), cross-peaks for H-2/H-8a, H-7/H-9, H-8a/H-10a, and H-9/H-11 were noteworthy, which may suggest that the H-7, H-9, and H-11 are oriented to the same side of the macrocycle plane, whereas the H-2, H-8a, and H-10a are directed otherwise. Further convincing evidence, however, has not been provided for the stereochemical assignment of the molecule of (**1000**).

In 1987 the authors first reported the isolation and basic structure of amphidinolide B (**983**) from the Y-5 strain of *Amphidinium* sp.,[833] and the basic structure was later revised partially.[835] In 1994 Shimizu and co-workers isolated three macrolides belonging to the amphidinolide B group (amphidinolides B_1 (**983**), B_2 (**1028**), and B_3 (**1029**)) from a free-swimming dinoflagellate *Amphidinium* sp. and reported their relative stereochemistry on the basis of the X-ray crystal structure of amphidinolide B_1.[859] The identity of amphidinolides B and B_1 was unambiguously established by direct comparison of HPLC and ^{1}H NMR data using each authentic sample.[860] The signs of the optical rotations of these two samples were the same. The absolute stereochemistry of amphidinolide B (**983**) was studied based on the synthesis of a degradation product (**1030**) and chiral HPLC analysis.

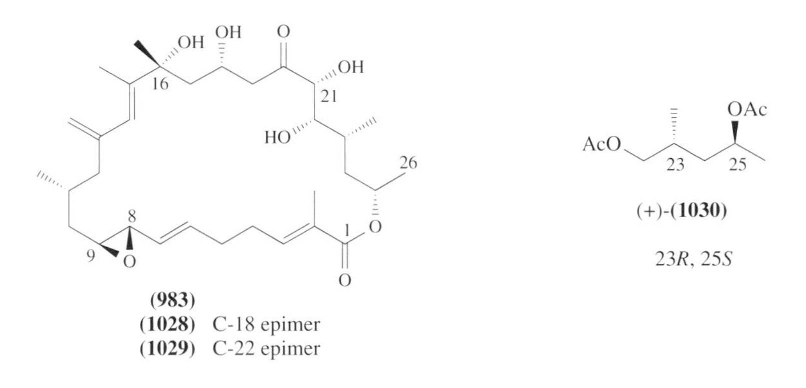

(**983**)
(**1028**) C-18 epimer
(**1029**) C-22 epimer

In advance of the degradation experiment both enantiomers of the C-22–C-26 fragment, (+)-(**1030**) and (−)-(**1030**) were prepared, as shown in Scheme 35, from (2*S*,4*S*)-(+)-pentanediol (**1031**) and (2*R*,4*R*)-(−)-pentanediol, respectively, both of which were available commercially. The chiral HPLC analysis (CHIRALCEL OD, Daicel Chemical Ind. Ltd.; 4.6 × 250 mm; flow rate: 1.0 ml min^{-1}; eluent: hexane/2-propanol (500:1); UV detection at 215 nm) of the enantiomers (+)- and (−)-(**1030**) showed that they were separable [(+)-(**1030**), t_R 23.2 min; (−)-(**1030**), t_R 22.3 min]. A MeOH adduct of amphidinolide B (**983**), which was obtained as an artifact of isolation and has a structure with a methoxyl and a hydroxyl group at the C-8 and C-9 positions,[835] was treated with $NaIO_4$ followed by $NaBH_4$ reduction and acetylation (Ac_2O/pyridine) to give the C-22–C-26 fragment (**1030**) after separation by normal-phase HPLC. The fragment (**1030**) thus obtained was subjected to chiral HPLC analysis as above and proved to be identical with (+)-(**1030**) (t_R 23.2 min), thus revealing that the C-22–C-26 fragment (**1030**) has (23*R*, 25*S*)-configurations. Since the relative stereochemistry of amphidinolide B_1 identical with (**983**) is known,[859] the absolute configurations of amphidinolide B (**983**) were concluded as 8*S*, 9*S*, 11*R*, 16*R*, 18*S*, 21*R*, 22*S*, 23*R*, and 25*S*, which was in agreement with results on the absolute configurations of amphidinolide L (**1002**).[845]

Shimizu and co-workers reported that amphidinolides B_2 (**1028**) and B_3 (**1029**), which they isolated concurrently with amphidinolide B_1 (**983**), were C-18 and C-22 epimers of (**983**), respectively.[859] The ^{1}H NMR spectra of amphidinolide B_2 (**1028**) and amphidinolide D (**985**) resembled each other very

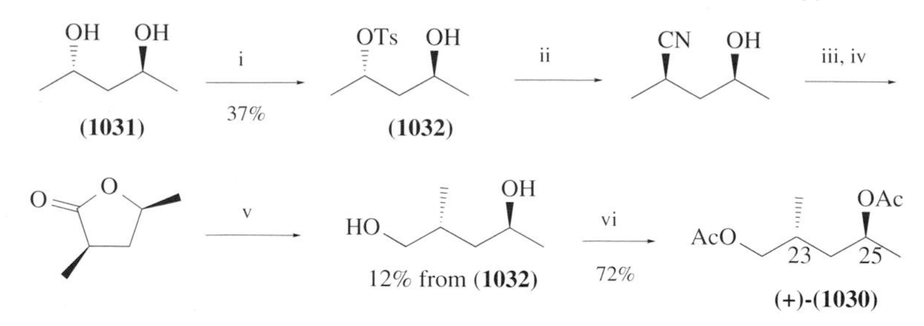

i, TsCl, pyridine; ii, NaCN, DMSO; iii, NaOH, H_2O_2, EtOH; iv, 2N HCl; v, LAH, Et_2O; vi, Ac_2O, pyridine

Scheme 35

well, indicating that these two compounds were identical. The authors had assigned the structure of amphidinolide D (**985**) to the C-21 epimer of (**983**),[835] and the conclusion of this different structural assignment ((**985**) or (**1027**)) was not provided.

The stereochemistry of amphidinolide C (**984**) has also been studied on the basis of NOESY experiments on (**984**) to propose the relative stereochemistry around a tetrahydrofuran ring moiety (C-1–C-9). A diastereomer of the C-1–C-9 fragment (**1033**) proposed from the NOESY data was synthesized as an optically active form, as shown in Scheme 36, to provide an authentic sample for use in studies on the degradation of amphidinolide C (**984**).[861]

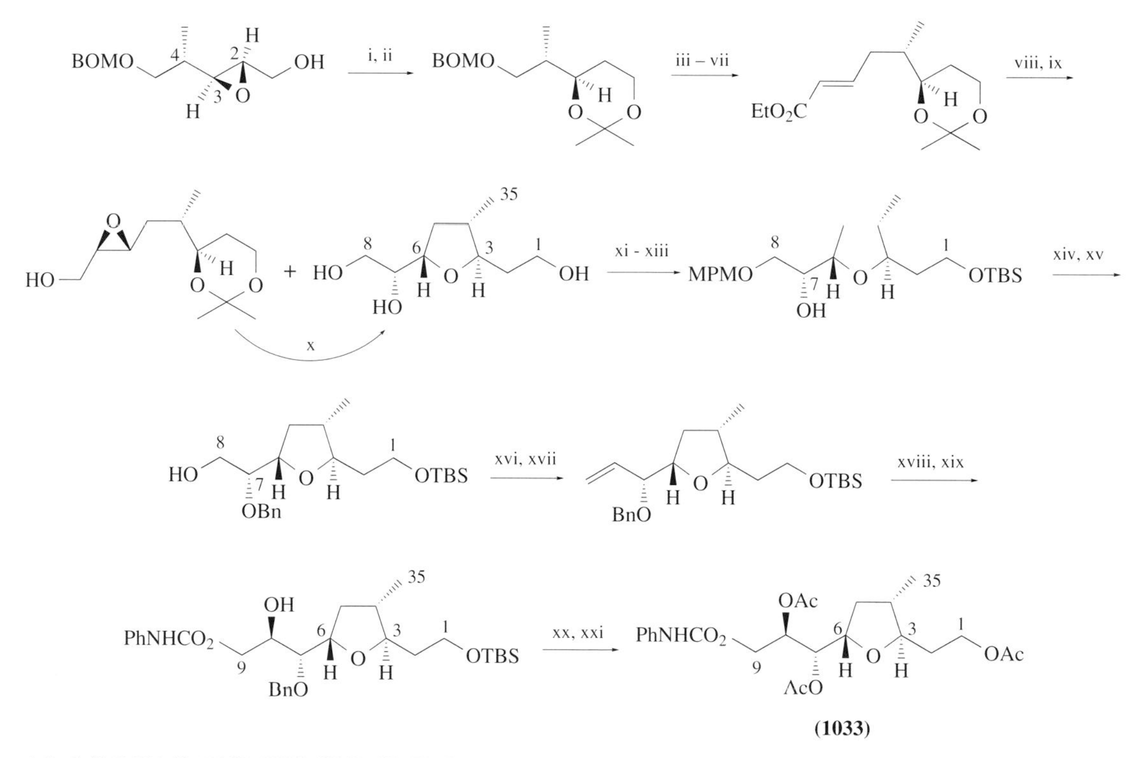

i, Red-Al (96%); ii, DMP, PPTS (99%); iii, H_2, Raney Ni (90%); iv, TsCl, Et_3N, DMAP (93%); v, NaCN, DMSO; vi, DIBAH; vii, $EtO_2CCH{=}PPh_3$ (3 steps, 79%); viii, DIBAH (86%); ix, (–)-DET, TBHP, $Ti(Pr^iO)_4$ (**7**, 62%; **8**, 16%); x, 1N H_2SO_4 (90%); xi, *p*-anisaldehyde dimethylacetal, TsOH (62%); xii, TBSCl, imidazole (86%); xiii, DIBAH (61%); xiv, BnBr, NaH (81%); xv, DDQ, phosphate buffer (81%); xvi, $(COCl)_2$, DMSO, Et_3N; xvii, Ph_3PCH_3Br, Bu^nLi (2 steps, 51%); xviii, AD-mix-α (56%); xix, PhNCO, DMAP, pyridine, then silica gel column (hexane/EtOAc, 5:2) (58%); xx, H_2, $Pd(OH)_2/C$; xxi, Ac_2O, pyridine (2 steps, 68%)

Scheme 36

Macrolide antibiotics from terrestrial microorganisms generally possess even-numbered macrocyclic lactones, which are reasonably derived from the polyketide biosynthesis. However, many amphidinolides comprise unusual odd-numbered macrocyclic lactone rings (amphidinolides C (**984**), 25-membered; E (**986**), 19-membered; F (**987**), 25-membered; G (**988**), 27-membered; J (**994**), 15-membered; K (**995**), 19-membered; L (**1002**), 27-membered; M (**996**), 29-membered; O (**998**), 15-membered; and P (**999**), 15-membered). The amphidinolides have some other unique structural features: (i) they have a variety of novel backbone-skeletons, isolated from one genus of microalga: (ii) all amphidinolides contain one or more exomethylene units; and (iii) vicinally located one-carbon branches (viz., methyl or exomethylene) are present in amphidinolides B (**983**), C (**984**), D (**985**), F (**987**), G (**988**), H (**989**), J (**994**), K (**995**), L (**1002**), M (**996**), O (**998**), P (**999**), and Q (**1000**), and amphidinin A (**1001**). In particular, the generation of an odd-numbered macrocyclic lactone ring, as well as structural feature (iii), was unable to be accounted for by the classical polyketide biosynthesis. The investigation of the biosynthesis of amphidinolides based on stable isotope incorporation experiments, although the sample size of the macrolides produced by the alga was not very high, requiring a large scale of culturing and a considerable number of ^{13}C-labeled precursors. The experimental results as well as the hypothesis on the biosynthesis of amphidinolide J (**994**), the most abundant macrolide in *Amphidinium* sp. (strain Y-5), are described below.[862]

The dinoflagellate *Amphidinium* sp. (strain Y-5) was cultured in 3 L glass bottles containing nutrient-enriched seawater medium, and feeding experiments were carried out with [1^{13}C], [2-^{13}C], and [1,2-$^{13}C_2$] sodium acetate and [methyl-^{13}C]-L-methionine. A summary of the conditions of the feeding experiments is shown in Table 23. The ^{13}C-labeled precursors were fed to the alga (610 μM for labeled sodium acetate and 93 μM for labeled methionine) in one portion 10–12 days after inoculation, and 2 days later the culture was harvested. The extract of the harvested cells was purified by improved procedures to afford ^{13}C-labeled amphidinolide J (**994**) (0.5–1 mg from 80–100 L of culture). The assignments of the ^{13}C NMR signals of (**994**) in C_6D_6 solution were fully established by HMQC and HMBC spectra and are presented in Table 24.

Table 23 Feeding experiments of ^{13}C-labeled precursors to *Amphidinium* sp. (Y-5).

				The day after inoculation	
Run	*^{13}C-Labeled precursors*	*Culture* (L)	*Concentration* (μM)	*Addition* (day)	*Harvest* (day)
1	1-^{13}C-NaOAc	200	610	10	14
2	1-^{13}C-NaOAc	100	610	10	12
3	2-^{13}C-NaOAc	100	610	12	14
4	1,2-$^{13}C_2$-NaOAc	100	610	10	12
5	methyl-^{13}C-(L)-methionine	80	93	10	12

The ^{13}C NMR spectrum of amphidinolide J (**984**) labeled from sodium [1-^{13}C] acetate showed significant enrichment of nine carbons (C-1, C-4, C-6, C-8, C-10, C-13, C-15, C-17, and C-19), while 15 carbons were enriched by sodium [2-^{13}C] acetate (C-2, C-3, C-5, C-7, C-9, C-11, C-12, C-14, C-16, C-18, C-20, C-21, C-22, C-23, and C-24). The ratios of the signal intensities over those of nonlabeled (**994**) are given in Table 24. Thus, all 24 carbons contained in amphidinolide J (**994**) were revealed to be derived from acetates. The ^{13}C NMR of (**994**) obtained from the feeding experiment of [methyl-^{13}C]-L-methionine did not show appreciable enrichment of any carbon. The ^{13}C-^{13}C coupling constants ($^1J_{CC}$) of (**994**) labeled with [1,2-$^{13}C_2$]-acetate (Table 24) indicated that the C_2 units for C-1/C-2, C-4/C-5, C-6/C-7, C-8/C-9, C-10/C-11, C-13/C-14, C-15/C-16, C-17/C-18, and C-19/C-20 originate from the same acetates. Interestingly, when the culture was harvested 4 days after feeding of sodium [1-^{13}C] acetate to the alga (run 1 of Table 23), the ^{13}C NMR of isolated (**994**) showed that all carbon atoms of (**994**) were enriched and almost all signals were observed with double satellite signals due to vicinal ^{13}C-^{13}C couplings. This phenomenon was considered to be observed probably because C-1 of acetate was cleaved via decarboxylation during passage through the TCA cycle and the released $^{13}CO_2$ was reincorporated during photosynthesis to give randomly labeled acetates, which led to all-carbon enriched (**994**). The labeling patterns of amphidinolide J (**994**) shown by the feeding experiments were quite unusual and are presented in Figure 18.

Table 24 Isotope incorporation results from the ^{13}C NMR data of amphidinolide J (**994**).[a]

Position	δ_C	*Intensity ratio (labeled/unlabeled)*[b]		*Assignment* "c" or "m"[c]	J_{CC} (Hz)
		[1-^{13}C]-*acetate*	[2-^{13}C]-*acetate*		[1,2-^{13}C$_2$]-*acetate*
1	171.6	1.41	1	c	57.8
2	39.9	1.01	1.72	m	57.8
3	34.6	0.88	1.59	m	
4	151.9	1.34	0.97	c	42.5
5	36.1	1.05	2.02	m	41.4
6	29.7	1.68	1.36	c	43.6
7	130.8	0.87	1.88	m	43.6
8	136.5	2.10	1.10	c	49.0
9	78.8	0.95	1.66	m	48.0
10	45.7	1.51	0.99	c	43.6
11	133.5	1.03	2.09	m	43.6
12	132.6	0.87	1.51	m	
13	72.6	1.46	1.05	c	42.5
14	79.9	1.15	2.33	m	41.4
15	39.5	1.52	1.16	c	42.5
16	133.6	0.76	1.61	m	43.6
17	131.5	1.53	0.95	c	42.5
18	35.3	0.94	2.50	m	42.5
19	23.4	1.58	1.01	c	34.9
20	14.2	1	1.98	m	34.9
21	22.2	0.96	1.81	m	
22	108.7	1.12	1.99	m	
23	19.0	1.03	1.78	m	
24	17.5	1.14	2.21	m	

[a] The ^{13}C NMR spectra were recorded in C_6D_6 solution on a Bruker ARX500 spectrometer at 125 MHz with sweep width of 35 700 Hz using Bruker's pulse program "zgpg30." Numbers of scans were ca. 13 000 and 25 728, for the samples from feedings of mono- and double-^{13}C labeled precursors, respectively. [b] Intensity of each peak in the labeled (**994**) divided by that of the corresponding signal in the unlabeled (**994**), normalized to give a ratio of 1 for an unenriched peak (C-20 for [1-^{13}C]-acetate labeling and C-1 for [2-^{13}C]-acetate labeling). [c] "c" denotes "carbon derived from C-1 of acetate," while "m" indicates carbon derived from C-2 of acetate."

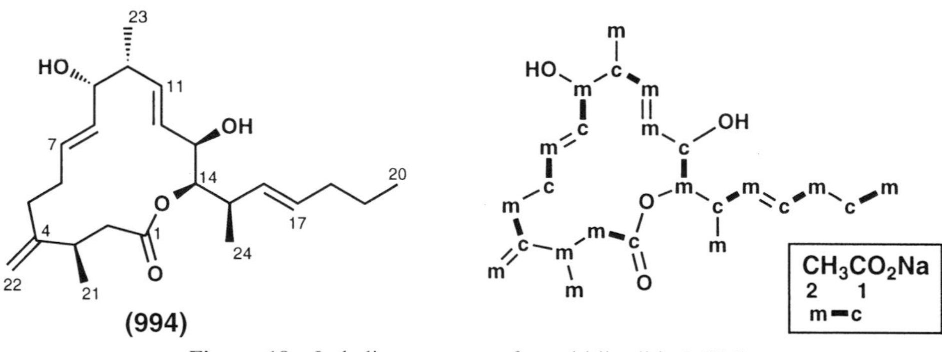

Figure 18 Labeling patterns of amphidinolide J (**994**).

Significantly, the C-3 and C-12 of (**994**) were derived from the methyl carbons of acetates, the carboxyl carbons of which were lost. Thus, the carbons constituting the 15-membered lactone ring were not constructed from the consecutive polyketide chain. This finding seems to justify the fact that the lactone ring size of (**994**) does not have to be even. The irregular labeling pattern of (**994**) could be interpreted as one possibility by assuming that the backbone carbons of (**994**) were biosynthetically derived from the precursors depicted in Figure 19. Units B (C-4 to C-9) and D (C-13 to C-20) are likely to be classical polyketides derived as a result of the condensation of three and four acetate units, respectively. Unit A (C-1/C-2/C-3/C-21) contains the "c-m-m-m" moiety and many come from a dicarboxylic acid like α-ketoglutarate after passage of acetate through the TCA cycle, which has been observed in the biosynthesis of brevetoxin B and okadaic acid (see Sections 8.07.8.3.1 and 8.07.8.5.2). Unit C (C-10/C-11/C-12) labeled "c-m-m" may be derived from succinate, corresponding to the six units in brevetoxin B. Units E, F, and G (C-22, C-23, and C-24) are one-carbon branches (an exomethylene and two secondary methyls), and they were demonstrated to be derived from the C-2 of acetates and attached to carbons in a linear chain derived from the C-1 of acetates (C-4, C-10, and C-15, respectively). One-carbon branching of this type is unusual in polyketide biosynthesis and has been previously reported only in few cases. Another one-carbon branch of C-21 also came from the C-2 of acetate. However, the condensation of this carbon to the linear chain occurred at the carbon (C-3) derived from the C-2 of acetate; thus, the participation of a dicarboxylic acid precursor was proposed for this moiety. How the vicinal locations of one-carbon branches are brought about in amphidinolides is an interesting question, and the results argued that two vicinal one-carbon branches (C-21 and C-22) of (**994**) were both derived from the C-2 of acetate but attached to the linear chain through different processes. It should be noted that the oxymethines at C-9 and C-14 of (**994**) are derived from the C-2 of acetate, and the origins of the oxygen atoms are unknown.

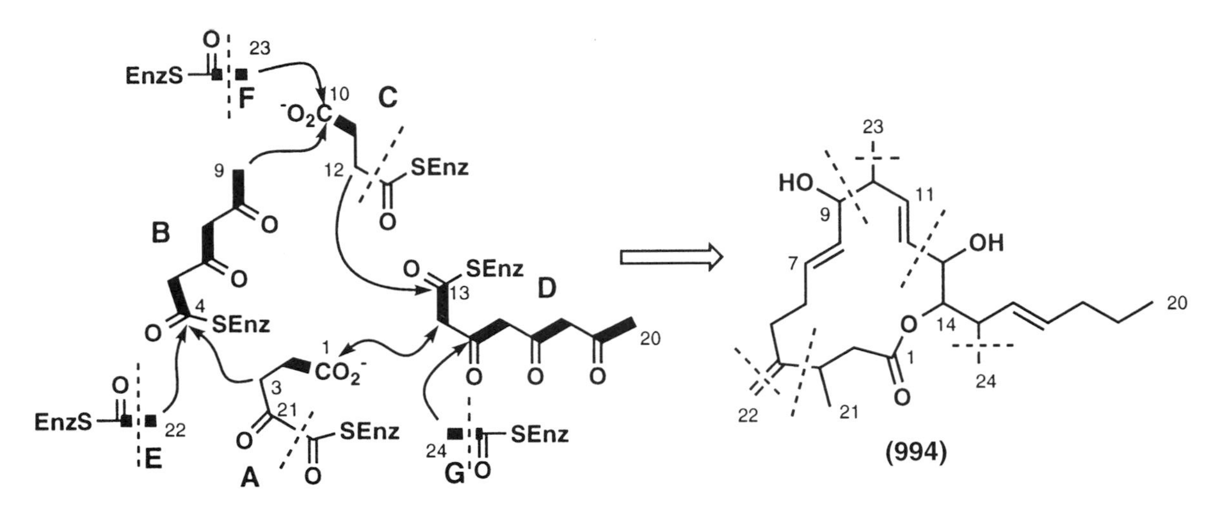

A: α-ketoglutarate; **B**, **D**: classical polyketide;
C: succinate; **E**, **F**, **G**: C-2 of acetate

Figure 19 Possible biosynthetic building blocks of amphidinolide J (**994**).

In connection with this work, the unusual 1,4-polyketides, amphidinoketides I (**1034**) and II (**1035**) were isolated from *Amphidinium* sp. by Shimizu's group[863] and a possible biosynthetic pathway involving the condensation of succinates was proposed. The 1,4-polyketide moieties of (**1034**) and (**1035**) may correspond to the C-15–C-18 portion of amphidinolides C (**984**) and F (**987**).

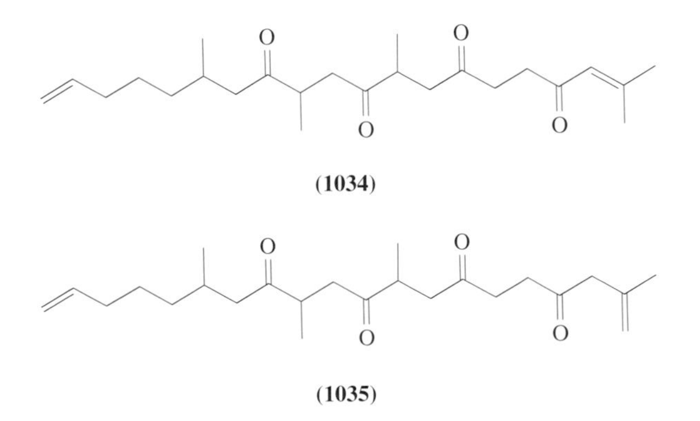

Wright and co-workers[329] proposed the biosynthetic path of amphidinolide J (**994**) through a carbon-deletion step involving oxidation of a methyl-derived carbon to yield an α diketide followed by a Favorski-type rearrangement, peroxide attack, and collapse of the cyclopropanone, which may account for the interrupted pattern of acetate units in the chain instead of the participation of dicarboxylic acid precursors (see Section 8.07.8.3.1).

The cytotoxicity data of amphidinolides against murine lymphoma L1210 and human epidermoid carcinoma KB cells *in vitro* are summarized in Table 25 together with their isolation yields from the four strains of the dinoflagellates *Amphidinium* sp. The level of the cytotoxic activity of amphidinolides B_1 (**983**), B_2 (**1028**), and B_3 (**1029**) against a different cell line (human colon tumor cell line HCT 116 (IC_{50} 0.122, 7.5, and 0.206 μg ml^{-1}, respectively), reported by Shimizu and co-workers,[859] were significantly lower than those of amphidinolides B (**983**) and D (**985**).

Table 25 Isolation yields and cytotoxicity data of amphidinolides.

Compounds		*Lactone ring size*	*Isolation yields* ($\times 10^{-4}$%)[a] *Strain number* Y-5[b]	Y-5′	Y-25	Y-26	*Cytotoxicity* (IC_{50}, μg ml^{-1}) L1210	KB
Amphidinolide A	(**982**)	(20)	20	4			2.0	5.7
Amphidinolide B	(**983**)	(26)	10			0.8	0.00014	0.0042
Amphidinolide C	(**984**)	(25)	15			0.3	0.0058	0.0046
Amphidinolide D	(**985**)	(26)	4				0.019	0.08
Amphidinolide E	(**986**)	(19)	4	3			2.0	10
Amphidinolide F	(**987**)	(25)				0.1	1.5	3.2
Amphidinolide G	(**988**)	(27)			20		0.0054	0.0059
Amphidinolide H	(**989**)	(26)			17		0.00048	0.00052
Amphidinolide J	(**994**)	(15)	60				2.7	3.9
Amphidinolide K	(**995**)	(19)	0.3				1.65	2.9
Amphidinolide L	(**1002**)	(27)			2		0.092	0.1
Amphidinolide M	(**996**)	(29)	4				1.1	0.44
Amphidinolide N	(**997**)	(26)	9				0.00005	0.00006
Amphidinolide O	(**998**)	(15)	1				1.7	3.6
Amphidinolide P	(**999**)	(15)	2				1.6	5.8
Amphidinolide Q	(**1000**)	(12)	0.5				6.4	> 10
Amphidinin A	(**1001**)		0.6				3.6	3.0

[a] Based on the wet weight of the harvested cells. [b] Isolation yields vary during the course of time.

8.07.10 REFERENCES

1. D. J. Faulkner, *Nat. Prod. Rep.*, 1996, **13**, 75.
2. D. J. Faulkner, *Nat. Prod. Rep.*, 1995, **12**, 223.
3. D. J. Faulkner, *Nat. Prod. Rep.*, 1994, **11**, 355.
4. D. J. Faulkner, *Nat. Prod. Rep.*, 1993, **10**, 497.
5. D. J. Faulkner, *Nat. Prod. Rep.*, 1992, **9**, 323.
6. D. J. Faulkner, *Nat. Prod. Rep.*, 1991, **8**, 97.
7. D. J. Faulkner, *Nat. Prod. Rep.*, 1990, **7**, 269.
8. D. J. Faulkner, *Nat. Prod. Rep.*, 1988, **5**, 613.
9. D. J. Faulkner, *Nat. Prod. Rep.*, 1987, **4**, 539.
10. D. J. Faulkner, *Nat. Prod. Rep.*, 1986, **3**, 1.
11. D. J. Faulkner, *Nat. Prod. Rep.*, 1984, **1**, 551.
12. D. J. Faulkner, *Nat. Prod. Rep.*, 1984, **1**, 251.
13. *Chem. Rev.* 1993, **93**, 1671 (special issue on "Marine Natural Products Chemistry").
14. J. Kobayashi and M. Ishibashi, in "The Alkaloids," eds. A. Brossi and G. A. Cordell. Academic Press, San Diego, CA, 1992, vol. 41, p. 41.
15. P. J. Scheuer (ed.), "Bioorganic Marine Chemistry," Springer-Verlag, Berlin, 1987–1992, vols. 1–6.
16. P. J. Scheuer (ed.), "Marine Natural Products—Diversity and Biosynthesis, Topics in Current Chemistry vol. 167," Springer-Verlag, Berlin, 1993.
17. G. R. Pettit, in "Progress in the Chemistry of Organic Natural Products," eds. W. Herz, G. W. Kirby, W. Steglich, and Ch. Tamm, Springer-Verlag, Wien, 1991, vol. 57, p. 158.
18. I. Wahlberg and A.-M. Eklund, in "Progress in the Chemistry of Organic Natural Products," eds. W. Herz, G. W. Kirby, R. E. Moore, W. Steglich, and Ch. Tamm, Springer-Verlag, Wien, 1992, vol. 59, p. 189.
19. I. Wahlberg and A.-M. Eklund, in "Progress in the Chemistry of Organic Natural Products," eds. W. Herz, G. W. Kirby, R. E. Moore, W. Steglich, and Ch. Tamm, Springer-Verlag, Wiene, 1992, vol. 60, p. 17.
20. L. Minale, R. Riccio, and F. Zollo, in "Progress in the Chemistry of Organic Natural Products," eds. W. Herz, G. W. Kirby, R. E. Moore, W. Steglich, and Ch. Tamm, Springer-Verlag, Wien, 1993, vol. 62, p. 75.
21. R. G. S. Berlinck, in "Progress in the Chemistry of Organic Natural Products," eds. W. Herz, G. W. Kirby, R. E. Moore, W. Steglich, and Ch. Tamm, Springer-Verlag, Wien, 1995, vol. 66, p. 178.
22. V. J. Paul (ed.), "Ecological Roles of Marine Natural Products," Cornell University Press, Ithaca, NY, 1992.
23. J. C. Coll, *Chem. Rev.*, 1992, **92**, 613.
24. I. Kitagawa and N. Fusetani (eds.), "Kaiyou Seibutsu no Kemikaru Shigunaru," Kodansha, Tokyo, 1989 (in Japanese).
25. T. Yasumoto (ed.), "Kagaku de Saguru Kaiyou Seibutsu no Nazo," Kagakudojin, Kyoto, 1992 (in Japanese).
26. K. Sakata, in "Bioorganic Marine Chemistry," ed. P. J. Scheuer, Springer-Verlag, Berlin, 1989, vol. 3, p. 115.
27. K. Sakata, in "Kaiyou Seibutsu no Kemikaru Shigunaru," eds. I. Kitagawa and N. Fusetani, Kodansha, Tokyo, 1989, p. 7 (in Japanese).

28. K. Sakata, in "Kagaku de Saguru Kaiyou Seibutsu no Nazo," ed. T. Yasumoto, Kagakudojin, Kyoto, 1992, p. 79 (in Japanese).
29. K. Ina and K. Sakata, in "Kaiyou Tennenbutsu Kaguku," ed. I. Kitagawa, Kagakudojin, Kyoto, 1987, p. 73 (in Japanese).
30. K. Ina, in "Seibutsu Kassei Tennen Busshitsu," ed. S. Shibata, Ishiyaku Shuppan, Yokyo, 1988, p. 130 (in Japanese).
31. Y. Hashimoto, S. Konosu, N. Fusetani, and K. Nose, *Bull. Jpn. Soc. Sci. Fish.*, 1968, **34**, 78.
32. A. W. Sangster, S. E. Thomas, and N. L. Tingling, *Tetrahedron*, 1975, **31**, 1135.
33. M. G. Pawson, *Comp. Biochem. Physiol.*, 1977, **56A**, 129.
34. K. Ina and H. Matsui, *Nippon Nôgeikagaku Kaishi*, 1980, **54**, 7 (in Japanese).
35. T. J. Hara, *Comp. Biochem. Physiol.*, 1977, **56A**, 559.
36. A. M. Mackie and J. W. Adron, *Comp. Biochem. Physiol.*, 1978, **60A**, 79.
37. A. M. Mackie, in "Chemoreception in Fishes," ed. T. J. Hara, Elsevier, Amsterdam, 1982, p. 275.
38. S. Murofushi, A. Sano, and K. Ina, *Nippon Nôgeikagaku Kaishi*, 1982, **56**, 255 (in Japanese).
39. K. Sakata, T. Itoh, and K. Ina, *Agric. Biol. Chem.*, 1984, **48**, 425.
40. K. Sakata and K. Ina, *Bull. Jpn. Soc. Sci. Fish*, 1985, **51**, 659.
41. K. Sakata, M. Tsuge, Y. Kamiya, and K. Ina, *Agric. Biol. Chem.*, 1985, **49**, 1905.
42. K. Sakata, T. Sakura, and K. Ina, *J. Chem. Ecol.*, 1988, **14**, 1405.
43. D. G. Müller, L. Jaenicke, M. Donike, and T. Akintobi, *Science*, 1971, **171**, 815.
44. T. Kajiwara, in "Kaiyou Tennenbutsu Kagaku," ed. I. Kitagawa, Kagakudojin, Kyoto, 1987, p. 53 (in Japanese).
45. T. Kajiwara, in "Kagaku de Saguru Kaiyou Seibutsu no Nazo," ed. T. Yasumoto, Kagakudojin, Kyoto, 1992, p. 71 (in Japanese).
46. L. Jaenicke and W. Boland, *Angew. Chem. Int. Ed. Eng.*, 1982, **21**, 643.
47. W. Boland, *Proc. Natl. Acad. Sci. USA*, 1995, **92**, 37.
48. A. Cook, J. A. Elvidge, I. Geukbribm, *Proc. Roy. Soc.*, 1948, **135**, 293.
49. D. G. Müller and L. Jaenicke, *FEBS Lett.*, 1973, **30**, 137.
50. D. G. Müller and C. E. Schmid, *Biol. Chem. Hoppe-Seyler*, 1988, **369**, 647.
51. D. G. Müller, G. Gassmann, and K. Lüning, *Nature*, 1979, **279**, 430.
52. K. Lüning and D. G. Müller, *Z. Pflanzenphysiol.*, 1978, **89**, 333.
53. T. Kajiwara, K. Kodama, and A. Hatanaka, *Tennen Yuki Kagoubutsu Toronkai Koen Yoshishu*, 1991, **33**, 337 (in Japanese).
54. W. Boland, W. A. König, R. Krebber, and D. G. Müller, *Helv. Chim. Acta*, 1989, **72**, 1288.
55. K. Stratmann, W. Boland, and D. G. Müller, *Angew. Chem. Int. Ed. Eng.*, 1992, **31**, 1246.
56. K. Stratmann, W. Boland, and D. G. Müller, *Tetrahedron*, 1993, **49**, 3755.
57. W. Boland, L. Jaenicke, D. G. Müller, and G. Gassmann, *Experientia*, 1987, **43**, 466.
58. G. Pohnert and W. Boland, *Tetrahedron*, 1994, **50**, 10 235.
59. J. Keitel, I. Fischer-Lui, W. Boland, and D. G. Müller, *Helv. Chim. Acta*, 1990, **73**, 2101.
60. W. Boland, G. Pohnert, and I. Maier, *Angew. Chem. Int. Ed. Eng.*, 1995, **34**, 1602.
61. G. Pohnert and W. Boland, *Tetrahedron*, 1996, **52**, 10 073.
62. N. Fusetani, in "Kaiyou Seibutsu no Kemikaru Shigunaru," eds. I. Kitagawa and N. Fusetani, Kodansha, Tokyo, 1989, p. 126 (in Japanese).
63. J. G. Dulka, N. E. Stacey, P. W. Sorensen, and G. J. Van Der Kraak, *Nature*, 1987, **325**, 251.
64. T. Yamasu, *Iden*, 1988, **42**, 12 (in Japanese).
65. C. R. Wilkinson and P. Fay, *Nature*, 1979, **279**, 527.
66. G. M. L. Patterson and N. W. Withers, *Science*, 1982, **217**, 1034.
67. R. D. Gates, O. Hoegh-Guldberg, M. J. McFall-Ngai, K. Y. Bil, and L. Muscatine, *Proc. Natl. Acad. Sci. USA*, 1995, **92**, 7430.
68. J. Kobayashi and M. Ishibashi, *Chem. Rev.*, 1993, **93**, 1753.
69. M. S. Gil-Turnes, M. E. Hay, and W. Fenical, *Science*, 1989, **246**, 116.
70. M. Murata, K. Miyagawa-Kohshima, K. Nakanishi, and Y. Naya, *Science*, 1986, **234**, 585.
71. Y. Naya, in "Kaiyou Tennenbutsu Kagaku," ed. I. Kitagawa, Kagakudojin, Kyoto, 1987, p. 87 (in Japanese).
72. K. Kondo, J. Nishi, M. Ishibashi, and J. Kobayashi, *J. Nat. Prod.*, 1994, **57**, 1008.
73. N. Fusetani, in "Kaiyou Seibutsu no Kemikaru Shigunaru," eds. I. Kitagawa and N. Fusetani, Kodansha, Tokyo, 1989, p. 146 (in Japanese).
74. K. Sakata, in "Kaiyou Seibutsu no Fuchaku Kikou," ed. T. Kajiwara, Koseisha-Kouseikaku, Tokyo, 1991, p. 113 (in Japanese).
75. Y. Shizuri, A. Isoai, and H. Kawahara, *Kagaku to Koygou*, 1996, **49**, 1021 (in Japanese).
76. H. Hirota, *Kagaku to Kogyou*, 1996, **49**, 1025 (in Japanese).
77. A. Yamada, K. Yamaguchi, H. Kitamura, K. Yazawa, S. Fukuzawa, G.-Y.-S. Wang, M. Kuramoto, and D. Uemura, *Tennen Yuki Kagoubutsu Toronkai Koen Yoshishu*, 1995, **37**, 224 (in Japanese).
78. W. A. Müller, *Differentiation*, 1985, **29**, 216.
79. N. C. Edwards, M. B. Thomas, B. A. Long, and S. J. Amyotte, *Roux's Arch. Dev. Biol.*, 1987, **196**, 381.
80. J. R. Pawlik and D. J. Faulkner, *J. Exp. Mar. Biol. Ecol.*, 1986, **102**, 301.
81. D. Kirchman, S. Graham, D. Reish, and R. Mitchell, *J. Exp. Mar. Biol. Ecol.*, 1982, **75**, 191.
82. R. M. Weiner, A. M. Segall, and R. R. Colwell, *Appl. Environ. Microbiol.*, 1985, **49**, 83.
83. S. L. Coon, D. B. Bonar, and R. M. Weiner, *J. Exp. Mar. Biol. Ecol.*, 1985, **94**, 211.
84. D. E. Morse, N. Hooker, H. Duncan, and L. Jensen, *Science*, 1979, **204**, 407.
85. A. N. C. Morse and D. E. Morse, *J. Exp. Mar. Biol. Ecol.*, 1984, **75**, 191.
86. H. G. Trapido-Rosenthal and D. E. Morse, *J. Comp. Physiol. B.*, 1985, **155**, 403.
87. A. J. Yool, S. M. Grau, M. G. Hadfield, R. A. Jensen, D. A. Markell, and D. E. Morse, *Biol. Bull.*, 1986, **170**, 255.
88. K. Kurata, M. Suzuki, K. Shiraishi, and K. Taniguchi, *Phytochemistry*, 1988, **27**, 1321.
89. J. C. Yvin, L. Chevolot, A. M. Chevolot-Magueur, and J. C. Cochard, *J. Nat. Prod.*, 1985, **48**, 814.
90. K. Tegtmeyer and D. Rittschof, *Peptides*, 1989, **9**, 1403.
91. D. J. Crisp, in "Marine Biodeterioration: An Interdisciplinary Study," eds. J. D. Costlow and R. C. Tipper, Naval Institute Press, 1984, p. 103.

92. K. Kon-ya, N. Shimidzu, N. Otaki, A. Yokoyama, K. Adachi, and W. Miki, *Experientia*, 1995, **51**, 153.
93. P. A. Kiefer, K. L. Rinehart, Jr., and I. R. Hooper, *J. Org. Chem.*, 1986, **51**, 4450.
94. P. A. Kiefer, R. E. Schwartz, M. E. S. Koker, R. G. Hughes, Jr., D. Rittschof, and K. L. Rinehart, *J. Org. Chem.*, 1991, **56**, 2965.
95. T. Okino, E. Yoshimura, H. Hirota, and N. Fusetani, *Tetrahedron Lett.*, 1995, **36**, 8637.
96. H. Hirota, Y. Tomono, and N. Fusetani, *Tetrahedron*, 1996, **52**, 2359.
97. T. Okino, E. Yoshimura, H. Hirota, and N. Fusetani, *Tetrahedron*, 1996, **52**, 9447.
98. S. Tsukamoto, H. Kato, H. Hirota, and N. Fusetani, *Tetrahedron*, 1996, **52**, 8181.
99. S. Tsukamoto, H. Kato, H. Hirota, and N. Fusetani, *J. Org. Chem.*, 1996, **61**, 2936.
100. S. Tsukamoto, H. Kato, H. Hirota, and N. Fusetani, *Tetrahedron Lett.*, 1996, **37**, 1439.
101. S. Tsukamoto, H. Hirota, H. Kato, and N. Fusetani, *Tetrahedron Lett.*, 1993, **34**, 4819.
102. S. Tsukamoto, H. Hirota, H. Kato, and N. Fusetani, *Experientia*, 1994, **50**, 680.
103. S. Tsukamoto, H. Hirota, H. Kato, and N. Fusetani, *Fisheries Sci.*, 1994, **60**, 319.
104. S. Tsukamoto, H. Kato, H. Hirota, and N. Fusetani, *Tetrahedron Lett.*, 1994, **35**, 5873.
105. S. Tsukamoto, H. Kato, H. Hirota, and N. Fusetani, *Tetrahedron*, 1994, **50**, 13 583.
106. T. Goto, in "Marine Natural Products: Chemical and Biological Perspectives," ed. P. J. Scheuer, Academic Press, New York, 1980, vol. 3, p. 179.
107. M. Ohashi, *Nippon Nôgeikagaku Kaishi*, 1992, **66**, 757 (in Japanese).
108. F. McCapra, *J. Chem. Soc., Chem. Comm.*, 1977, 946.
109. Y. Toya, *Nippon Nôgeikagaku Kaishi*, 1992, **66**, 742 (in Japanese).
110. Y. Toya, S. Nakatsuka, and T. Goto, *Tetrahedron Lett.*, 1983, **24**, 5753.
111. Y. Toya, S. Nakatsuka, and T. Goto, *Tetrahedron Lett.*, 1985, **26**, 239.
112. E. M. Thompson, S. Nagata, and F. I. Tsuji, *Proc. Natl. Acad. Sci. USA*, 1989, **86**, 6567.
113. M. Isobe, in "Kagaku de Saguru Kaiyou Seibutsu no Nazo," ed. T. Yasumoto, Kagakudojin, Kyoto, 1992, p. 87 (in Japanese).
114. Y. Ohmiya and F. I. Tsuji, *Nippon Nôgeikagaku Kaishi*, 1992, **66**, 752 (in Japanese).
115. S. Inouye, M. Noguchi, Y. Sakaki, Y. Takagi, T. Miyata, S. Iwanaga, T. Miyata, and F. I. Tsuji, *Proc. Natl. Acad. Sci. USA*, 1985, **82**, 3154.
116. H. Charbonneau, K. A. Walsh, R. O. McCann, F. G. Prendergast, M. J. Cormier, and T. C. Vanaman, *Biochemistry*, 1985, **24**, 6762.
117. D. Prasher, R. O. McCann, and M. J. Cormier, *Biochem. Biophys. Res. Commun.*, 1985, **126**, 1259.
118. S. Inouye, Y. Sakaki, T. Goto, and F. I. Tsuji, *Biochemistry*, 1986, **25**, 8425.
119. S. Inouye, S. Aoyama, T. Miyata, F. I. Tsuji, and Y. Sakaki, *J. Biochem.*, 1989, **105**, 473.
120. F. I. Tsuji, S. Inouye, T. Goto, and Y. Sakaki, *Proc. Natl. Acad. Sci. USA*, 1986, **83**, 8107.
121. K. Kurose, S. Inouye, Y. Sakaki, and F. I. Tsuji, *Proc. Natl. Acad. Sci. USA*, 1989, **86**, 80.
122. O. Shimomura, *Biochem. J.*, 1986, **234**, 271.
123. O. Shimomura, B. Musicki, and Y. Kishi, *Biochem. J.*, 1988, **251**, 405.
124. O. Shimomura, B. Musicki, and Y. Kishi, *Biochem. J.*, 1989, **261**, 913.
125. K. Teranishi and T. Goto, *Chem. Lett.*, 1989, 1423.
126. O. Shimomura, S. Inouye, B. Musicki, and Y. Kishi, *Biochem. J.*, 1990, **270**, 309.
127. O. Shimomura, *Biochem. J.*, 1995, **306**, 537
128. C. F. Qi, Y. Gomi, M. Ohashi, Y. Ohmiya, and F. I. Tsuji, *J. Chem. Soc., Chem. Commun.*, 1991, 1307.
129. T. Hirano, R. Negishi, M. Yamaguchi, F. Q. Chen, Y. Ohmiya, F. I. Tsuji, and M. Ohashi, *J. Chem. Soc., Chem. Commun.*, 1995, 1335.
130. O. Shimomura and F. H. Johnson, *Proc. Natl. Acad. Sci. USA*, 1978, **75**, 2611.
131. K. Hori, J. M. Anderson, W. W. Ward, and M. J. Cormier, *Bicohemistry*, 1975, **14**, 2371.
132. B. Musicki, Y. Kishi, and O. Shimomura, *J. Chem. Soc., Chem. Commun.*, 1986, 1566.
133. K. Teranishi, K. Ueda, M. Hisamatsu, and T. Yamada, *Biosci. Biotech. Biochem.*, 1995, **59**, 104.
134. K. Teranishi, M. Isobe, T. Yamada, and T. Goto, *Tetrahedron Lett.*, 1992, **33**, 1303.
135. K. Teranishi, M. Isobe, and T. Yamada, *Tetrahedron Lett.*, 1994, **35**, 2565.
136. T. Hirano and M. Ohashi, *Yuki Gosei Kagaku Kyokaishi*, 1996, **54**, 596 (in Japanese).
137. T. Hirano, I. Mozoguchi, M. Yamaguchi, F.-Q. Chen, M. Ohashi, Y. Ohmiya, and F. I. Tsuji, *J. Chem. Soc., Chem. Commun.*, 1994, 165.
138. F.-Q. Chen, T. Hirano, M. Ohashi, H. Nakayama, K. Oda, and M. Machida, *Chem. Lett.*, 1993, 287.
139. F.-Q. Chen, T. Hirano, Y. Hashizume, Y. Ohmiya, and M. Ohashi, *J. Chem. Soc., Chem. Commun.*, 1994, 2405.
140. F.-Q. Chen, J. L. Zheng, T. Hirano, H. Niwa, Y. Ohmiya, and M. Ohashi, *Chem. Chem. Soc., Perkin Trans. 1*, 1995, 2129.
141. O. Shimomura, *FEBS Lett.*, 1979, **104**, 220.
142. C. W. Cody, D. C. Prasher, W. M. Westler, F. G. Prendergast, and W. W. Ward, *Biochemistry*, 1993, **32**, 1212.
143. S. Inouye, T. Matsuno, S. Kojima, M. Kubota, T. Hirano, H. Niwa, M. Ohashi, and F. I. Tsuji, *Tennen Yuki Kagoubutsu Toronkai Koen Yoshishu*, 1996, **38**, 85 (in Japanese).
144. S. Inouye and F. I. Tsuji, *FEBS Lett.*, 1994, **341**, 277.
145. M. Chalfie, Y. Tu, G. Euskirchen, W. W. Ward, and D. C. Prasher, *Science*, 1994, **263**, 802.
146. T. Goto, H. Iio, S. Inoue, and H. Kakoi, *Tetrahedron Lett.*, 1974, 2321.
147. S. Inoue, S. Sugiura, H. Kakoi, K. Hasizume, T. Goto, and H. Iio, *Chem. Lett.*, 1975, 141.
148. M. Isobe, H. Takahashi, K. Usami, M. Hattori, and Y. Nishigohri, *Pure Appl. Chem.*, 1994, **66**, 765.
149. F. I. Tsuji and G. B. Leisman, *Proc. Natl. Acad. Sci. USA*, 1981, **78**, 6719.
150. H. Takahashi and M. Isobe, *Bioorg. Med. Chem. Lett.*, 1993, **3**, 2647.
151. H. Takahashi, Y. Yasuda, and M. Isobe, *Tennen Yuki Kagoubutsu Toronkai Koen Yoshishu*, 1994, **36**, 144 (in Japanese).
152. H. Takahashi and M. Isobe, *Chem. Lett.*, 1994, 843.
153. K. Uami and M. Isobe, *Tetrahedron Lett.*, 1995, **36**, 8613.
154. K. Usami and M. Isobe, *Tetrahedron*, 1996, **52**, 12 061.
155. K. Teranishi, K. Ueda, H. Nakao, M. Hisamatsu, and T. Yamada, *Tetrahedron Lett.*, 1994, **35**, 8181.

156. H. Nakamura, D. Takeuchi, and A. Murai, *Synlett*, 1995, 1227.
157. H. Nakamura, D. Takeuchi, M. Aizawa, C. Wu, and A. Murai, *Tennen Yuki Kagoubutsu Toronkai Koen Yoshishu*, 1996, **38**, 679 (in Japanese).
158. J. C. Dunlap, J. W. Hastings, and O. Shimomura, *Proc. Natl. Acad. Sci. USA*, 1980, **77**, 1394.
159. H. Nakamura, B. Musicki, Y. Kishi, and O. Shimomura, *J. Am. Chem. Soc.*, 1988, **110**, 2683.
160. H. Nakamura, Y. Kishi, O. Shimomura, D. Morse, and J. W. Hastings, *J. Am. Chem. Soc.*, 1989, **111**, 7607.
161. H. Nakamura, B. Musicki, Y. Kishi, D. Morse, J. W. Hastings, and O. Shimomura, *Tennen Yuki Kagoubutsu Toronkai Koen Yoshishu*, 1988, **30**, 276 (in Japanese).
162. H. Nakamura, Y. Oba, and A. Murai, *Tetrahedron Lett.*, 1993, **34**, 2779.
163. H. Nakamura, *Nippon Nôgeikagaku Kaishi*, 1992, **66**, 748 (in Japanese).
164. H. Nakamura, in "Kagaku de Saguru Kaiyou Seibutsu no Nazo," ed. T. Yasumoto, Kagakudojin, Kyoto, 1992, p. 97 (in Japanese).
165. T. Roenneberg, H. Nakamura, and J. W. Hastings, *Nature*, 1988, **334**, 432.
166. T. Roenneberg, H. Nakamura, L. C. Cranmer III, K. Ryan, Y. Kishi, and J. W. Hastings, *Experientia*, 1991, **47**, 103.
167. H. Nakamura, M. Takamatsu, Y. Kishi, L. D. Cranmer III, J. W. Hastings, and T. Roenneberg, *Tennen Yuki Kagoubutsu Toronkai Koen Yoshishu*, 1989, **31**, 268 (in Japanese).
168. H. Nakamura, M. Ohtoshi, O. Sampei, Y. Akashi, and A. Murai, *Tetrahedron Lett*, 1992, **33**, 2821.
169. H. Nakamura, K. Fujimaki, O. Sampei, and A. Murai, *Tetrahedron Lett.*, 1993, **34**, 8481.
170. H. Nakamura, Y. Oba, K. Fujimaki, M. Funahashi, and A. Murai, *Tennen Yuki Kagoubutsu Toronkai Koen Yoshishu*, 1993, **35**, 783 (in Japanese).
171. V. J. Paul and W. Fenical, in "Bioorganic Marine Chemistry," ed. P. J. Scheuer, Springer-Verlag, Berlin, 1987, vol. 1, p. 1.
172. P. Karuso, in "Bioorganic Marine Chemistry," ed. P. J. Scheuer, Springer-Verlag, Berlin, 1987, vol. 1, p. 31.
173. P. W. Sammarco and J. C. Coll, in "Bioorganic Marine Chemistry," ed. P. J. Scheuer, Springer-Verlag, Berlin, 1988, vol. 2, p. 87.
174. K. Tachibana, in "Bioorganic Marine Chemistry," ed. P. J. Scheuer, Springer-Verlag, Berlin, 1988, vol. 2, p. 117.
175. N. Fusestani, in "Kaiyou Seibutsu no Kemikaru Shigunaru," eds. I. Kitagawa and N. Fusetani, Kodansha, Tokyo, 1989, p. 7 (in Japanese).
176. N. Fusetani, *Kagaku to Seibutsu*, 1990, **28**, 728 (in Japanese).
177. J. R. Pawlik, *Chem. Rev.*, 1993, **93**, 1911.
178. K. Azumi and H. Yokosawa, *Seikagaku*, 1992, **64**, 237 (in Japanese).
179. K. Kurata, K. Shiraishi, T. Takato, K. Taniguchi, and M. Suzuki, *Chem. Lett.*, 1988, 1629.
180. K. Kurata, K. Taniguchi, K. Shiraishi, and M. Suzuki, *Tetrahedron Lett.*, 1989, **30**, 1567.
181. K. Kurata, K. Taniguchi, K. Shiraishi, and M. Suzuki, *Phytochemistry*, 1990, **29**, 3453.
182. K. Taniguchi, K. Kurata, and M. Suzuki, *Kagaku to Seibutsu*, 1994, **32**, 434 (in Japanese).
183. Y.-C. Shen, P. I. Tsai, W. Fenical, and M. E. Hay, *Phytochemistry*, 1993, **32**, 71.
184. V. J. Paul and K. L. Van Alstyne, *J. Exp. Mar. Biol. Ecol.*, 1992, **160**, 191.
185. I. Kubo, Y.-W. Lee, M. Pettei, F. Pilkiewicz, and K. Nakanishi, *J. Chem. Soc., Chem. Commun.*, 1976, 1013.
186. G. Cimino, S. De Rosa, S. De Stefano, R. Morrone, and G. Sodano, *Tetrahedron*, 1985, **41**, 1093.
187. V. Caprioli, G. Cimino, R. Colle, M. Gavagnin, G. Sodano, and A. Spinella, *J. Nat. Prod.*, 1987, **50**, 146.
188. G. Cimino, G. Sodano, and A. Spinella, *Tetrahedron*, 1987, **43**, 5401.
189. G. Cimino, G. Sodano, and A. Spinella, *J. Nat. Prod.*, 1988, **51**, 1010.
190. C. Avila, G. Cimino, A. Crispino, and A. Spinella, *Experientia*, 1991, **47**, 306.
191. R. R. Vardaro, V. Di Marzo, A. Crispino, and G. Cimino, *Tetrahedron*, 1991, **47**, 5569.
192. V. Di Marzo, R. R. Vardaro, L. De Petrocellis, G. Villani, R. Minei, and G. Cimono, *Experientia*, 1991, **47**, 1221.
193. V. Roussis, J. R. Pawlik, M. E. Hay, and W. Fenical, *Experientia*, 1990, **46**, 327.
194. G. Cimino, A. Spinella, and G. Sodano, *Tetrahedron Lett.*, 1989, **30**, 3589.
195. G. Cimino, A. Crispino, V. Di Marzo, A. Spinella, and G. Sodano, *J. Org. Chem.*, 1991, **56**, 2907.
196. G. Cimino, A. Crispino, V. Di Marzo, G. Sodano, A. Spinella, and G. Villani, *Experientia*, 1991, **47**, 56.
197. V. Di Marzo, G. Cimino, A. Crispino, C. Minardi, G. Sodano, and A. Spinella, *Biochem. J.*, 1991, **273**, 593.
198. G. Cimino, A. Passeggio, G. Sodano, A. Spinella, and G. Villani, *Experientia*, 1991, **47**, 61.
199. H. L. Sleeper and W. Fenical, *J. Am. Chem. Soc.*, 1977, **99**, 2367.
200. J. B. McClintock, B. J. Baker, M. T. Hamann, W. Yoshida, M. Slattery, J. N. Heine, P. J. Bryan, G. S. Jayatilake, and B. H. Moon, *J. Chem. Ecol.*, 1994, **20**, 2539.
201. P. Proksch, *Toxicon*, 1994, **32**, 639.
202. J. R. Pawlik, B. Chanas, R. J. Toonen, and W. Fenical, *Mar. Ecol. Prog. Ser.*, 1995, **127**, 183.
203. B. Chanas and J. R. Pawlik, *Mar. Ecol. Prog. Ser.*, 1995, **127**, 195.
204. A. Groweiss, U. Shmueli, and Y. Kashman, *J. Org. Chem.*, 1983, **48**, 3512.
205. M. S. Butler and R. J. Capon, *Aust. J. Chem.*, 1993, **46**, 1255.
206. S. De Rosa, R. Puliti, A. Crispino, A. De Giulio, C. A. Mattia, and L. Mazzarella, *J. Nat. Prod.*, 1994, **57**, 256.
207. M. R. Conte, E. Fattorusso, V. Lanzotti, S. Magno, and L. Mayol, *Tetrahedron*, 1994, **50**, 849.
208. A. Carotenuto, E. Fattorusso, V. Lanzotti, S. Magno, and L. Mayol, *Liebigs Ann.*, 1996, 77.
209. R. Teeyapant and P. Proksch, *Naturwissenschaften*, 1993, **80**, 369.
210. R. Teeyapant, P. Kreis, V. Wray, L. Witte, and P. Proksch, *Z. Naturforsch., C: Biosci.*, 1993, **48**, 640.
211. S. Albrizio, P. Ciminiello, E. Fattorusso, S. Magno, and J. R. Pawlik, *J. Nat. Prod.*, 1995, **58**, 647.
212. W. Miki, N. Otaki, A. Yokoyama, and T. Kusumi, *Experientia*, 1996, **52**, 93.
213. R. J. Capon, K. Elsbury, M. S. Butler, C. C. Lu, J. N. A. Hooper, J. A. P. Rostas, K. J. O'Brien, L.-M. Mudge, and A. T. R. Sim, *Experientia*, 1993, **49**, 263.
214. W. Fenical and J. R. Pawlik, *Mar. Ecol. Prog. Ser.*, 1991, **75**, 1.
215. M. Iorizzi, P. Bryan, J. McClintock, L. Minale, E. Palagiano, S. Maurelli, R. Riccio, and F. Zollo, *J. Nat. Prod.*, 1995, **58**, 653.
216. K. Azumi, H. Yokosawa, and S. Ishii, *Biochemistry*, 1990, **29**, 159.
217. K. Azumi, M. Yoshimizu, S. Suzuki, Y. Ezura, and H. Yokosawa, *Experientia*, 1990, **46**, 1066.

218. V. J. Paul, N. Lindquist, and W. Fenical, *Mar. Ecol. Prog. Ser.*, 1990, **59**, 109.
219. N. Lindquist and W. Fenical, *Experientia*, 1991, **47**, 504.
220. N. Lindquist, W. Fenical, L. Párkányi, and J. Clardy, *Experientia*, 1991, **47**, 503.
221. S. A. Thompson, K. Tachibana, K. Nakanishi, and I. Kubota, *Science*, 1986, **233**, 341.
222. M. G. Zagorski, D. G. Norman, C. J. Barrow, T. Iwashita, K. Tachibana, and D. J. Patel, *Biochemistry*, 1991, **30**, 8009.
223. C. J. Barrow, K. Nakanishi, and K. Tachibana, *Biochim. Biophys. Acta*, 1992, **1112**, 235.
224. A. Kawamura, K. Konoki, Y. Onishi, and K. Tachibana, *Tennen Yuki Kagoubutsu Toronkai Koen Yoshishu*, 1994, **36**, 152 (in Japanese).
225. G. Saberwal and R. Nagaraj, *J. Biol. Chem.*, 1993, **268**, 14 081.
226. K. S. Moore, S. Wehrli, H. Roder, M. Rogers, J. N. Forrest, Jr., D. McCrimmon, and M. Zasloff, *Proc. Natl. Acad. Sci. USA*, 1993, **90**, 1354.
227. R. M. Moriarty, L. A. Enache, W. A. Kinney, C. S. Allen, J. W. Canary, S. M. Tuladhar, and L. Guo, *Tetrahedron Lett.*, 1995, **36**, 5139.
228. A. Sadownik, G. Deng, V. Janout, and S. L. Regen, *J. Am. Chem. Soc.*, 1995, **117**, 6138.
229. G. Deng, T. Dewa, and S. L. Regen, *J. Am. Chem. Soc.*, 1996, **118**, 8975.
230. H. Onoki, K. Ito, N. Matsumori, K. Tachibana, and N. Fusetani, *Tennen Yuki Kagoubutsu Toronkai Koen Yoshishu*, 1993, **35**, 622 (in Japanese).
231. A.J. Kohn, *Ecol. Monogr.*, 1959, **29**, 47.
232. R. Endean, G. Parish, and P. Gyr, *Toxicon*, 1974, **12**, 131.
233. M. Kobayashi, J. Kobayashi, and Y. Ohizumi, in "Bioorganic Marine Chemistry," ed. P. J. Scheuer. Springer-Verlag, Berlin, 1989, vol. 3, p. 71.
234. H. Nakamura, J. Kobayashi, Y. Ohizumi, and Y. Hirata, *Experientia*, 1983, **39**, 590.
235. S. Sato, H. Nakamura, Y. Ohizumi, J. Kobayashi, and Y. Hirata, *FEBS Lett.*, 1983, **155**, 277.
236. Y. Ohizumi, S. Minoshima, M. Takahashi, A. Kajiwara, H. Nakamura, and J. Kobayashi, *J. Pharmacol. Exp. Ther.*, 1986, **239**, 243.
237. M. Kobayashi, C. H. Wu, M. Yoshii, T. Narahashi, H. Nakamura, J. Kobayashi, and Y. Ohizumi, *Pflügers Arch*, 1986, **407**, 241.
238. Y. Ohizumi, H. Nakamura, J. Kobayashi, and W. A. Catterall, *J. Biol. Chem.*, 1986, **261**, 6149.
239. T. Gonoi, Y. Ohizumi, H. Nakamura, J. Kobayashi, and W. A. Catterall, *J. Neurosci.*, 1987, **7**, 1728.
240. B. M. Olivera, W. R. Gray, R. Zeikus, J. M. McIntosh, J. Varga, V. de Santos, and L. J. Cruz, *Science*, 1985, **230**, 1338.
241. R. A. Myers, L. J. Cruz, J. E. Rivier, and B. M. Olivera, *Chem. Rev.*, 1993, **93**, 1923.
242. L. M. Kerr and D. Yoshikami, *Nature*, 1984, **308**, 282.
243. Y. Hatanaka, E. Yoshida, H. Nakayama, and Y. Kanaoka, *Chem. Pharm. Bull.*, 1990, **38**, 236.
244. Y. Hatanaka, E. Yoshida, H. Nakayama, T. Abe, M. Satake, and Y. Kanaoka, *FEBS Lett.*, 1990, **260**, 27.
245. Y. Hidaka, K. Sato, H. Nakamura, J. Kobayashi, Y. Ohizumi, and Y. Shimonishi, *FEBS Lett.*, 1990, **264**, 29.
246. J. M. Lancelin, D. Kohda, S. Tate, Y. Yanagawa, T. Abe, M. Satake, and F. Inagaki, *Biochemistry*, 1991, **30**, 6908.
247. K. Sato, Y. Ishida, K. Wakamatsu, R. Kato, H. Honda, Y. Ohizumi, H. Nakamura, M. Ohya, J. M. Lancelin, D. Kohda, and F. Inagaki, *J. Biol. Chem.*, 1991, **266**, 16 989.
248. Y. Kobayashi, T. Ohkubo, Y. Kyogoku, Y. Nishiuchi, S. Sakakibara, W. Braun, and N. Gô, *Biochemistry*, 1989, **28**, 4853.
249. A. Pardi, A. Galdes, J. Florance, and D. Maniconte, *Biochemistry*, 1989, **28**, 5494.
250. J. H. Davis, E. K. Bradley, G. P. Miljanich, L. Nadasdi, J. Ramachandran, and V. J. Basus, *Biochemistry*, 1993, **32**, 7396.
251. J. M. McIntosh, B. M. Olivera, L. J. Cruz, and W. R. Gray, *J. Biol. Chem.*, 1984, **259**, 14 343.
252. E. E. Mena, M. F. Gullak, M. J. Pagnozzi, K. E. Richter, J. Rivier, L. J. Cruz, and B. M. Olivera, *Neurosci. Lett.*, 1985, **260**, 9280.
253. J. Kobayashi, H. Nakamura, Y. Hirata, and Y. Ohizumi, *Biochem. Biophys. Res. Commun.*, 1982, **105**, 1389.
254. T. Gonoi, Y. Ohizumi, J. Kobayashi, H. Nakamura, and W. A. Catterall, *Mol. Pharmacol.*, 1987, **32**, 691.
255. J. Kobayashi, H. Nakamura, and Y. Ohizumi, *Br. J. Pharmacol.*, 1981, **73**, 583.
256. J. Kobayashi, H. Nakamura, and Y. Ohizumi, *Eur. J. Pharmacol.*, 1983, **86**, 283.
257. J. Kobayashi, H. Nakamura, and Y. Ohizumi, *Toxicon*, 1985, **23**, 783.
258. J. M. McIntosh, F. Ghomashchi, M. H. Gelb, D. J. Dooley, S. J. Stoehr, A. B. Giordani, S. R. Naisbitt, and B. M. Olivera, *J. Biol. Chem.*, 1995, **270**, 3518.
259. J. Kobayashi, H. Nakamura, Y. Hirata, and Y. Ohizumi, *Life Sci.*, 1982, **31**, 1085.
260. J. Kobayashi, H. Nakamura, Y. Hirata, and Y. Ohizumi, *Comp. Biochem. Physiol.*, 1983, **74B**, 381.
261. H. Schweitz, J. F. Renaud, N. Randimbivololona, C. Preau, A. Schmid, G. Romey, L. Rakatovao, and M. Lazdunski, *Eur. J. Biochem.*, 1986, **161**, 787.
262. J. Kobayashi, Y. Ohizumi, H. Nakamura, and Y. Hirata, *Toxicon*, 1981, **19**, 757.
263. H. Nakamura, J. Kobayashi, Y. Ohizumi, and Y. Hirata, *Experientia*, 1982, **38**, 897.
264. J. M. McIntosh, D. Yoshikami, E. Mahe, D. B. Nielsen, J. E. Rivier, W. R. Gray, and B. M. Olivera, *J. Biol. Chem.*, 1994, **269**, 16 733.
265. M. Fainzilber, J. C. Lodder, K. S. Kits, O. Kofman, I. Vinnitsky, J. V. Rietschoten, E. Zlotkin, and D. Gordon, *J. Biol. Chem.*, 1995, **270**, 1123.
266. M. Fainzilber, D. Gordon, A. Hasson, M. E. Spira, and E. Zlotkin, *Eur. J. Biochem.*, 1991, **202**, 589.
267. J. M. McIntosh, A. Hasson, M. E. Spira, W. R. Gray, W. Li, M. Marsh, D. R. Hillyard, and B. M. Olivera, *J. Biol. Chem.*, 1995, **270**, 16 796.
268. C. Hopkins, M. Grilley, C. Miller, K.-J. Shon, L. J. Cruz, W. R. Gray, J. Dykert, J. Rivier, D. Yoshikami, and B. M. Olivera, *J. Biol. Chem.*, 1995, **270**, 22 361.
269. H. Terlau, K.-J. Shon, M. Grilley, M. Stocker, W. Stühmer, and B. M. Olivera, *Nature*, 1996, **381**, 148.
270. T. Yasumoto and M. Yotsu-Yamashita, *J. Toxicol., Toxin Rev.*, 1996, **15**, 81.
271. T. Yasumoto and M. Murata, *Chem. Rev.*, 1993, **93**, 1897.

272. R. B. Woodward, *Pure Appl. Chem.*, 1964, **9**, 49.
273. T. Goto, Y. Kishi, S. Takahashi, and Y. Hirata, *Tetrahedron*, 1965, **21**, 2059.
274. K. Tsuda, S. Ikuma, M. Kawamura, K. Tachikawa, K. Sakai, C. Tamura, and O. Akamatsu, *Chem. Pharm. Bull.*, 1964, **12**, 1357.
275. Y. Kishi, M. Aratani, T. Fukuyama, F. Nakatsubo, T. Goto, S. Inoue, H. Tanino, S. Sugiura, and H. Kakoi, *J. Am. Chem. Soc.*, 1972, **94**, 9217.
276. Y. Kishi, T. Fukuyama, M. Aratani, F. Nakatsubo, T. Goto, S. Inoue, H. Tanino, S. Sugiura, and H. Kakoi, *J. Am. Chem. Soc.*, 1972, **94**, 9219.
277. K. Matsumura, *J. Agric. Food Chem.*, 1996, **44**, 1.
278. T. Yasumoto, D. Yasumura, M. Yotsu, T. Michishita, A. Endo, and Y. Kotaki, *Agric. Biol. Chem.*, 1986, **50**, 793.
279. T. Noguchi, J.-K. Jeon, O. Arakawa, H. Sugita, Y. Deguchi, Y. Shida, and K. Hashimoto, *J. Biochem.*, 1986, **99**, 311.
280. M. Yotsu, T. Yamazaki, Y. Meguro, A. Endo, M. Murata, H. Naoki, and T. Yasumoto, *Toxicon*, 1987, **25**, 225.
281. Y. Shimizu, in "Kaiyou Tennenbutsu Kagaku," ed. I. Kitagawa, Kagakudojin, Kyoto, 1987, p. 13 (in Japanese).
282. Y. Tasumoto and T. Michishita, *Agric. Biol. Chem.*, 1985, **49**, 3077.
283. M. Nakamura and T. Yasumoto, *Toxicon*, 1985, **23**, 271.
284. A. Endo, S. S. Khora, M. Murata, H. Naoki, and Y. Yasumoto, *Tetrahedron Lett.*, 1988, **29**, 4127.
285. S. S. Khora and T. Yasumoto, *Tetrahedron Lett.*, 1989, **30**, 4393.
286. T. Yasumoto, M. Yotsu, M. Murata, and H. Naoki, *J. Am. Chem. Soc.*, 1988, **110**, 2344.
287. M. Yotsu, T. Yasumoto, Y. H. Kim, H. Naoki, and C. Y. Kao, *Tetrahedron Lett.*, 1990, **31**, 3187.
288. M. Yotsu, Y. Hayashi, S. S. Khora, S. Sato, and T. Yasumoto, *Biosci. Biotech. Biochem.*, 1992, **56**, 370.
289. Y. Kotaki and Y. Shimizu, *J. Am. Chem. Soc.*, 1993, **115**, 827.
290. M. Yotsu-Yamashita, Y. Yamagishi, and T. Yasumoto, *Tetrahedron Lett.*, 1995, **36**, 9329.
291. M. Yotsu-Yamashita, A. Sugimoto, and T. Yasumoto, *Tennen Yuki Kagoubutsu Toronkai Koen Yoshishu*, 1993, **35**, 495 (in Japanese).
292. H. Terlau, S. H. Heinemann, W. Stühmer, M. Pusch, F. Conti, K. Imoto, and S. Numa, *FEBS Lett.*, 1991, **293**, 93.
293. H. Nakayama, Y. Hatanaka, E. Yoshida, K. Oka, M. Takanohashi, Y. Amano, and Y. Kanaoka, *Biochem. Biophys. Res. Commun.*, 1992, **184**, 900.
294. Y. Shimizu, in "Progress in the Chemistry of Organic Natural Products," eds. W. Herz, H. Grisebach, G. W. Kerby, and Ch. Tamm, Springer-Verlag, Wien, 1984, vol. 45, p. 235.
295. Y. Shimizu, *Chem. Rev.*, 1993, **93**, 1685.
296. Y. Shimizu, S. Gupta, K. Masuda, L. Maranda, C. K. Walker, and R. Wang, *Pure Appl. Chem.*, 1989, **61**, 513.
297. M. Kodama, T. Ogata, and S. Sato, *Agric. Biol. Chem.*, 1988, **52**, 1075.
298. K. Hashimoto and T. Noguchi, *Pure Appl. Chem.*, 1989, **61**, 7.
299. Y. Oshima, K. Sugino, and T. Yasumoto, in "Mycotoxins and Phycotoxins '88," eds. S. Natori, K. Hashimoto, and Y. Ueno, Elsevier, Amsterdam, 1989, p. 319.
300. H. Nakayama, *Kagaku to Seibutsu*, 1995, **33**, 585 (in Japanese).
301. M. Noda, S. Shimizu, T. Tanabe, T. Takai, T. Kayano, T. Ikeda, H. Takahashi, H. Nakayama, Y. Kanaoka, N. Minamino, K. Kangawa, H. Matsuo, M. A. Raftery, T. Horose, S. Inayama, H. Hayashida, T. Miyata, and S. Numa, *Nature*, 1984, **312**, 121.
302. M. Noda, H. Suzuki, S. Nima, and W. Stühmer, *FEBS Lett.*, 1989, **259**, 213.
303. E. Yoshida, H. Nakayama, Y. Hatanaka, and Y. Kanaoka, *Chem. Pharm. Bull.*, 1990, **38**, 982.
304. T. Yasumoto, M. Yotsu-Yamashita, A. Sugimoto, M. Isemura, and K. Nishimori, *Japan–U.S. Seminar on Bioorganic Marine Chemistry, Abstract*, 1994, p. 1.
305. T. Yasumoto, *Gendai Kagaku*, 1994, **282**, 47 (in Japanese).
306. Y. Li and E. Moczydlowski, *J. Biol. Chem.*, 1991, **266**, 15 481.
307. K. Tachibana, P. J. Scheuer, Y. Tsukitani, H. Kikuchi, D. Van Engen, J. Clardy, Y. Gopichand, and F. J. Schmitz, *J. Am. Chem. Soc.*, 1981, **103**, 2469.
308. M. Murata, M. Shimatani, H. Sugitani, Y. Oshima, and Y. Yasumoto, *Bull. Jpn. Soc. Sci. Fish.*, 1982, **48**, 549.
309. J. S. Lee, T. Yanagi, R. Kenma, and T. Yasumoto, *Agric. Biol. Chem.*, 1987, **51**, 877.
310. K. Akasaka, H. Ohrui, H. Meguro, and T. Yasumoto, *J. Chromatogr.*, 1996, **729**, 381.
311. A. Takai, C. Bialoja, M. Troschka, and J. C. Rüegg, *FEBS Lett.*, 1987, **217**, 81.
312. T. Yasumoto, M. Murata, Y. Oshima, M. Sano, G. K. Matsumoto, and J. Clardy, *Tetrahedron*, 1985, **41**, 1019.
313. M. Murata, Y. Murakami, M. Kumagai, T. Yanagi, T. Iwashita, and H. Naoki, *Tennen Yuki Kagoubutsu Toronkai Koen Yoshishu*, 1986, **28**, 192 (in Japanese).
314. T. Hu, J. Doyle, D. Jackson, J. Marr, E. Nixon, S. Pleasance, M. A. Quilliam, J. A. Walter, and J. L. C. Wright, *J. Chem. Soc., Chem. Commun.*, 1992, 39.
315. J. C. Marr, T. Hu, S. Pleasance, M. A. Quilliam, and J. L. C. Wright, *Toxicon*, 1992, **30**, 1621.
316. T. Hu, J. Marr, A. S. W. DeFreitas, M. A. Quilliam, J. A. Walter, J. L. C. Wright, and S. Pleasance, *J. Nat. Prod.*, 1992, **55**, 1631.
317. T. Hu, J. M. Curtis, J. A. Walter, and J. L. C. Wright, *J. Chem. Soc., Chem. Commun.*, 1995, 597.
318. T. Hu, J. M. Curtis, J. A. Walter, J. L. McLachlan, and J. L. C. Wright, *Tetrahedron Lett.*, 1995, **36**, 9273.
319. M. Suganuma, H. Fujiki, H. Suguri, S. Yoshizawa, M. Hirota, M. Nakayasu, M. Ojika, K. Wakamatsu, K. Yamada, and T. Sugimura, *Proc. Natl. Acad. Sci. USA*, 1988, **85**, 1768.
320. S. Nishiwaki, H. Fujiki, M. Suganuma, H. Furuya-Suguri, R. Matsushima, Y. Iida, M. Ojika, K. Yamada, D. Uemura, T. Yasumoto, F. J. Schmitz, and T. Sugimura, *Carcinogenesis*, 1990, **11**, 1837.
321. F. J. Schmitz, R. S. Prasad, Y. Gopichand, M. B. Hossain, D. van der Helm, and P. Schmidt, *J. Am. Chem. Soc.*, 1981, **103**, 2467.
322. M. Kuramoto, T. Ishida, N. Yamada, A. Yamada, D. Uemura, T. Haino, K. Yamada, Y. Ijuin, and K. Fujita, *Tennen Yuki Kagoubutsu Toronkai Koen Yoshishu*, 1993, **35**, 693 (in Japanese).
323. S. Matsunaga, *Gendai Kagaku*, 1992, **251**, 56 (in Japanese).
324. N. Matsumori, M. Murata, and K. Tachibana, *Tetrahedron*, 1995, **51**, 12 229.
325. K. Torigoe and T. Yasumoto, *Tennen Yuki Kagoubutsu Toronkai Koen Yoshishu*, 1991, **33**, 525 (in Japanese).
326. M. Norte, A. Padilla, and J. J. Fernández, *Tetrahedron Lett.*, 1994, **35**, 1441.

327. J. Needham, J. L. McLachlan, J. A. Walter, and J. L. C. Wright, *J. Chem. Soc., Chem. Commun.*, 1994, 2599.
328. J. Needham, T. Hu, J. L. McLachlan, J. A. Walter, and J. L. C. Wright, *J. Chem. Soc., Chem. Commun.*, 1995, 1623.
329. J. L. C. Wright, T. Hu, J. L. McLachlan, J. Needham, and J. A. Walter, *J. Am. Chem. Soc.*, 1996, **118**, 8757.
330. K. Torigoe, M. Murata, T. Yasumoto, and T. Iwashita, *J. Am. Chem. Soc.*, 1988, **110**, 7876.
331. T. Hu, A. S. W. deFreitas, J. M. Curtis, Y. Oshima, J. A. Walter, and J. L. C. Wright, *J. Nat. Prod.*, 1996, **59**, 1010.
332. M. Murata, M. Sano, T. Iwashita, H. Naoki, and T. Yasumoto, *Agric. Biol. Chem.*, 1986, **50**, 2693.
333. J. H. Jung, C. J. Sim, and C.-O. Lee, *J. Nat. Prod.*, 1995, **58**, 1722.
334. Z.-H. Zhou, M. Komiyama, K. Terao, and Y. Shimada, *Nat. Toxins*, 1994, **2**, 132.
335. K. Terao, E. Ito, T. Yanagi, and T. Yasumoto, *Toxicon*, 1986, **24**, 1141.
336. M. Murata, M. Kumagai, J. S. Lee, and T. Yasumoto, *Tetrahedron Lett.*, 1987, **28**, 5869.
337. H. Naoki, M. Murata, and T. Yasumoto, *Rapid Commun. Mass Spectrom.*, 1993, **7**, 179.
338. M. Satake, K. Terasawa, Y. Kadowaki, and T. Yasumoto, *Tetrahedron Lett.*, 1996, **37**, 5955.
339. T. Kusumi, H. Takahashi, P. Xu, T. Fukushima, Y. Asakawa, T. Hashimoto, Y. Kan, and Y. Inouye, *Tetrahedron Lett.*, 1994, **35**, 4397.
340. H. Takahashi, T. Kusumi, and Y. Kan, *Tennen Yuki Kagoubutsu Toronkai Koen Yoshishu*, 1996, **38**, 475 (in Japanese).
341. K. Terao, E. Ito, M. Oarada, M. Murata, and T. Yasumoto, *Toxicon*, 1990, **28**, 1095.
342. T. Yasumoto and M. Satake, *J. Toxicol., Toxin Rev.*, 1996, **15**, 91.
343. P. J. Scheuer, *Tetrahedron*, 1994, **50**, 3.
344. K. Tachibana, M. Nukina, Y.-G. Joh, and P. J. Scheuer, *Biol. Bull.*, 1987, **172**, 122.
345. M. Nukina, L. M. Koyanagi, and P. J. Scheuer, *Toxicon*, 1984, **22**, 169.
346. T. Yasumoto, I. Nakajima, R. Bagins, and R. Adachi, *Bull. Jpn. Soc. Sci. Fish.*, 1977, **43**, 1021.
347. M. Murata, A. M. Legrand, and T. Yasumoto, *Tetrahedron Lett.*, 1989, **30**, 3793.
348. M. Murata, A. M. Legrand, Y. Ishibashi, and T. Yasumoto, *J. Am. Chem. Soc.*, 1989, **111**, 8929.
349. M. Murata, A. M. Legrand, Y. Ishibashi, M. Fujui, and T. Yasumoto, *J. Am. Chem. Soc.*, 1990, **112**, 4380.
350. T. Suzuki, O. Sato, M. Hirama, Y. Yamamoto, M. Murata, T. Yasumoto, and N. Harada, *Tetrahedron Lett.*, 1991, **32**, 4505.
351. M. Sasaki, A. Hasegawa, and K. Tachibana, *Tetrahedron Lett.*, 1993, **34**, 8489.
352. M. Murata, A. M. Legrand, P. J. Scheuer, and T. Yasumoto, *Tetrahedron Lett.*, 1992, **33**, 525.
353. M. Satake, A. Morohashi, T. Yasumoto, and A. M. Legrand, *Tennen Yuki Kagoubutsu Toronkai Koen Yoshishu*, 1996, **38**, 481 (in Japanese).
354. M. Satake, M. Murata, and T. Yasumoto, *J. Am. Chem. Soc.*, 1993, **115**, 361.
355. M. Satake, M. Murata, and T. Yasumoto, *Tetrahedron Lett.*, 1993, **34**, 1975.
356. M. Satake, M. Murata, and T. Yasumoto, *Tennen Yuki Kagoubutsu Toronkai Koen Yoshishu*, 1992, **34**, 87 (in Japanese).
357. R. C. Crouch, G. E. Martin, S. M. Musser, H. R. Grenade, and R. W. Dickey, *Tetrahedron Lett.*, 1995, **36**, 6827.
358. Y. Ohizumi, *J. Toxicol., Toxin Rev.*, 1996, **15**, 109.
359. A. Yokoyama, M. Murata, Y. Oshima, T. Iwashita, and T. Yasumoto, *J. Biochem.*, 1988, **104**, 184.
360. M. Murata, T. Iwashita, A. Yokoyama, M. Sasaki, and T. Yasumoto, *J. Am. Chem. Soc.*, 1992, **114**, 6594.
361. M. Murata, H. Naoki, T. Iwashita, S. Matsunaga, M. Sasaki, A. Yokoyama, and T. Yasumoto, *J. Am. Chem. Soc.*, 1993, **115**, 2060.
362. M. Murata, H. Naoki, S. Matsunaga, M. Satake, and T. Yasumoto, *J. Am. Chem. Soc.*, 1994, **116**, 7098.
363. M. Satake, S. Ishida, T. Yasumoto, M. Murata, H. Utsumi, and T. Hinomoto, *J. Am. Chem. Soc.*, 1995, **117**, 7019.
364. M. Sasaki, T. Nonomura, M. Murata, and K. Tachibana, *Tetrahedron Lett.*, 1994, **35**, 5023.
365. M. Sasaki, T. Nonomura, M. Murata, and K. Tachibana, *Tetrahedron Lett.*, 1995, **36**, 9007.
366. M. Sasaki, N. Matsumori, M. Murata, and K. Tachibana, *Tetrahedron Lett.*, 1995, **36**, 9011.
367. N. Matsumori, T. Nonomura, M. Sasaki, M. Murata, K. Tachibana, M. Satake, and T. Yasumoto, *Tetrahedron Lett.*, 1996, **37**, 1269.
368. M. Sasaki, N. Matsumori, T. Maruyama, T. Nonomura, M. Murata, K. Tachibana, and T. Yasumoto, *Angew. Chem., Int. Ed. Engl.*, 1996, **35**, 1672.
369. T. Nonomura, M. Sasaki, N. Matsumori, M. Murata, K. Tachibana, and T. Yasumoto, *Angew. Chem., Int. Ed. Engl.*, 1996, **35**, 1675.
370. W. Zheng, J. A. DeMattei, J.-P. Wu, J. J.-W. Duan, L. R. Cook, H. Oinuma, and Y. Kishi, *J. Am. Chem. Soc.*, 1996, **118**, 7946.
371. M. Takahashi, Y. Ohizumi, and T. Yasumoto, *J. Biol. Chem.*, 1982, **257**, 7287.
372. J. W. Daly, J. Lueders, W. L. Padgett, Y. Shin, and F. Gusovsky, *Biochem. Pharmacol.*, 1995, **50**, 1187.
373. K. Konoki, M. Hashimoto, M. Murata, K. Tachibana, and T. Yasumoto, *J. Nat. Toxins*, 1996, **5**, 209.
374. H. Nagai, K. Torigoe, M. Satake, M. Murata, T. Yasumoto, and H. Hirota, *J. Am. Chem. Soc.*, 1992, **114**, 1102.
375. H. Nagai, M. Murata, K. Torigoe, M. Satake, and T. Yasumoto, *J. Org. Chem.*, 1992, **57**, 5448.
376. R. E. Moore, in "Progress in the Chemistry of Organic Natural Products," eds. W. Herz, H. Grisebach, G. W. Kirby, and Ch. Tamm, Springer-Verlag, Wien, 1985, vol. 48, p. 81.
377. R. E. Moore and G. Bartolini, *J. Am. Chem. Soc.*, 1981, **103**, 2491.
378. D. Uemura, K. Ueda, and Y. Hirata, *Tetrahedron Lett.*, 1981, **22**, 2781.
379. R. W. Armstrong, J.-M. Beau, S. H. Cheon, W. J. Christ, H. Fujioka, W.-H. Ham, L. D. Hawkins, H. Jin, S. H. Kang, Y. Kishi, M. J. Martinelli, W. W. McWhorter, Jr., M. Mizuno, M. Nakata, A. E. Stutz, F. X. Talamas, M. Taniguchi, J. A. Tino, K. Ueda, J. Uenishi, J. B. White, and M. Yonaga, *J. Am. Chem. Soc.*, 1989, **111**, 7530.
380. M. Usami, M. Satake, S. Ishida, A. Inoue, Y. Kan, and T. Yasumoto, *J. Am. Chem. Soc.*, 1995, **117**, 5389.
381. S. Gleibs, D. Mebs, and B. Werding, *Toxicon*, 1995, **33**, 1531.
382. H. Fujiki and M. Suganuma, *J. Toxicol., Toxin Rev.*, 1996, **15**, 129.
383. J. J. Monroe and A. H. Tashjian, Jr., *Am. J. Physiol.*, 1996, **270**, C1277.
384. Y. Shimizu, in "Marine Natural Products, Chemical and Biological Perspectives," ed. P. J. Scheuer, Academic Press, New York, 1978, p. 1.
385. Y. Y. Lin, M. Risk, S. M. Ray, D. Van Engen, J. Clardy, J. Golik, J. C. James, and K. Nakanishi, *J. Am. Chem. Soc.*, 1981, **103**, 6773.
386. Y. Shimizu, H.-N. Chou, H. Bando, G. Van Duyne, and J. C. Clardy, *J. Am. Chem. Soc.*, 1986, **108**, 514.

387. J. Pawlak, M. S. Tempesta, J. Golik, M. G. Zagorski, M. S. Lee, K. Nakanishi, T. Iwashita, M. L. Gross, and K. B. Tomer, *J. Am. Chem. Soc.*, 1987, **109**, 1144.
388. A. V. K. Prasad and Y. Shimizu, *J. Am. Chem. Soc.*, 1989, **111**, 6476.
389. M. S. Lee, D. J. Repeta, K. Nakanishi, and M. G. Zagorski, *J. Am. Chem. Soc.*, 1986, **108**, 7855.
390. H.-N. Chou and Y. Shimizu, *J. Am. Chem. Soc.*, 1987, **109**, 2184.
391. M. S. Lee, G.-W. Qin, K. Nakanishi, and M. G. Zagorski, *J. Am. Chem. Soc.*, 1989, **111**, 6234.
392. S. Cestele, F. Sampieri, H. Rochat, and D. Gordon, *J. Biol. Chem.*, 1996, **271**, 18 329.
393. K. C. Nicolaou, *Angew. Chem., Int. Ed. Engl.*, 1996, **35**, 589.
394. R. E. Gawley, K. S. Rein, G. Jeglitsch, D. J. Adams, E. A. Theodorakis, J. Tiebes, K. C. Nicolaou, and D. G. Baden, *Chem. Biol.*, 1995, **2**, 533.
395. S. Matile, N. Berova, and K. Nakanishi, *Chem. Biol.*, 1996, **3**, 379.
396. S. Matile and K. Nakanishi, *Angew. Chem., Int. Ed. Engl.*, 1996, **35**, 757.
397. S. Matile, N. Berova, K. Nakanishi, J. Fleischhauer, and R. W. Woody, *J. Am. Chem. Soc.*, 1996, **118**, 5198.
398. H. Ishida, A. Nozawa, K. Totoribe, N. Muramatsu, H. Nukaya, K. Tsuji, K. Tamaguchi, T. Yasumoto, H. Kaspar, N. Berkett, and T. Kosuge, *Tetrahedron Lett.*, 1995, **36**, 725.
399. A. Morohashi, M. Satake, K. Murata, H. Naoki, H. F. Kaspar, and T. Yasumoto, *Tetrahedron Lett.*, 1995, **36**, 8995.
400. R. C. Crouch, G. E. Martin, R. W. Dickey, D. G. Baden, R. E. Gawley, K. S. Rein, and E. P. Mazzola, *Tetrahedron Lett.*, 1995, **51**, 8409.
401. T. Seki, M. Satake, L. Mackenzie, H. F. Kaspar, and T. Yasumoto, *Tetrahedron Lett.*, 1995, **36**, 7093.
402. T. Kosuge, H. Zenda, A. Ochiai, N. Masaki, M. Noguchi, S. Kimura, and H. Narita, *Tetrahedron Lett.*, 1972, 2545.
403. T. Kosuge, K. Tsuji, K. Hirai, K. Yamaguchi, T. Okamoto, and Y. Iitaka, *Tetrahedron Lett.*, 1981, **22**, 3417.
404. S. Inoue, K. Okada, H. Tanino, K. Hashizume, and H. Kakoi, *Tetrahedron Lett.*, 1996, **50**, 2729.
405. T. Kosuge, K. Tsuji, K. Hirai, T. Fukuyama, H. Nukaya, and H. Ishida, *Chem. Pharm. Bull.*, 1985, **33**, 2890.
406. T. Kosuge, H. Zenda, and K. Tsuji, *Yakugaku Zasshi*, 1987, **107**, 665 (in Japanese).
407. T. Kosuge, K. Tsuji, K. Hirai, and T. Fukuyama, *Chem. Pharm. Bull.*, 1985, **33**, 3059.
408. M. Yotsu-Yamashita, R. L. Haddock, and T. Yasumoto, *J. Am. Chem. Soc.*, 1993, **115**, 1147.
409. M. Yotsu-Yamashita, T. Yasumoto, and R. L. Haddock, *Tennen Yuki Kagoubutsu Toronkai Koen Yoshishu*, 1992, **34**, 612 (in Japanese).
410. M. Yotsu-Yamashita, T. Seki, V. J. Paul, H. Naoki, and T. Yasumoto, *Tetrahedron Lett.*, 1995, **36**, 5563.
411. K. Fujiwara, S. Amano, and A. Murai, *Chem. Lett.*, 1995, 855.
412. T. Igarashi, M. Satake, and T. Yasumoto, *Tennen Yuki Kagoubutsu Toronkai Koen Yoshishu*, 1994, **36**, 89 (in Japanese).
413. T. Igarashi, M. Satake, and T. Yasumoto, *J. Am. Chem. Soc.*, 1996, **118**, 479.
414. L. Glendenning, T. Igarashi, and T. Yasumoto, *Bull. Chem. Soc. Jpn.*, 1996, **69**, 2253.
415. S. Zheng, F. Huang, S. Chen, X. Tan, J. Zuo, J. Peng, and R. Xie, *Zhongguo Haiyang Yaowu*, 1990, **9**, 33 (in Chinese).
416. T. Chou, T. Haino, A. Nagatsu, D. Uemura, S. Fukuzawa, S. Z. Zheng, and H. S. Chen, *Tennen Yuki Kagoubutsu Toronkai Koen Yoshishu*, 1994, **36**, 57 (in Japanese).
417. D. Uemura, T. Chou, T. Haino, A. Nagatsu, S. Fukuzawa, S.-Z. Zheng, and H. S. Chen, *J. Am. Chem. Soc.*, 1995, **117**, 1155.
418. T. Chou, T. Haino, M. Kuramoto, and D. Uemura, *Tetrahedron Lett.*, 1996, **37**, 4027.
419. T. Chou, O. Kamo, and D. Uemura, *Tetrahedron Lett.*, 1996, **37**, 4023.
420. T. Hu, J. M. Curtis, Y. Oshima, M. A. Quilliam, J. A. Walter, W. M. Watson-Wright, and J. L. C. Wright, *J. Chem. Soc., Chem. Commun.*, 1995, 2159.
421. K. L. Rinehart, V. Kishore, K. C. Bible, R. Sakai, D. W. Sullins, and K.-M. Li, *J. Nat. Prod.*, 1988, **51**, 1.
422. G. R. Pettit, C. L. Herald, D. L. Doubek, and D. L. Herald, *J. Am. Chem. Soc.*, 1982, **104**, 6846.
423. K. Yamada, M. Ojika, T. Ishigaki, Y. Yoshida, H. Ekimoto, and M. Arakawa, *J. Am. Chem. Soc.*, 1993, **115**, 11 020.
424. M. Ojika, H. Kigoshi, T. Ishigaki, I. Tsukada, T. Tsuboi, T. Ogawa, and K. Yamada, *J. Am. Chem. Soc.*, 1994, **116**, 7441.
425. H. Kigoshi, M. Ojika, T. Ishigaki, K. Suenaga, T. Mutou, A. Sakakura, T. Ogawa, and K. Yamada, *J. Am. Chem. Soc.*, 1994, **116**, 7443.
426. H. Kigoshi, K. Suenaga, T. Mutou, T. Ishigaki, T. Atsumi, H. Ishiwata, A. Sakakura, T. Ogawa, M. Ojika, and K. Yamada, *J. Org. Chem.*, 1996, **61**, 5326.
427. E. D. de Silva and P. J. Scheuer, *Tetrahedron Lett.*, 1980, **21**, 1611.
428. R. S. Jacobs, P. Culver, R. Langdon, T. O'Brien, and S. White, *Tetrahedron*, 1985, **41**, 981.
429. W. Fenical, *J. Nat. Prod.*, 1987, **50**, 1001.
430. H. Michibata, *Zool. Sci.*, 1996, **13**, 489.
431. M. Henze, *Hoppe-Seler's Z. Physiol. Chem.*, 1911, **72**, 494.
432. H. Michibata, *Comp. Biochem. Physiol.*, 1984, **78A**, 285.
433. H. Michibata, T. Terada, N. Anada, K. Yamakawa, and T. Numakunai, *Biol. Bull.*, 1986, **171**, 672.
434. H. Michibata, Y. Iwata, and J. Hirata, *J. Exp. Zool.*, 1991, **257**, 306.
435. R. W. Collier, *Nature*, 1984, **309**, 441.
436. R. K. Wright, in "Invertebrate Blood Cells," eds. N. A. Ratcliffe and A. F. Rowley, Academic Press, London, 1981, vol. 2, p. 565.
437. K. Kustin, D. S. Levine, G. C. McLeod, and W. A. Curby, *Biol. Bull.*, 1976, **150**, 426.
438. L. Botte, S. Scippa, and M. de Vincentiis, *Experientia*, 1979, **35**, 1228.
439. J. Hirata and H. Michibata, *J. Exp. Zool.*, 1991, **257**, 160.
440. L. V. Boas and J. C. Pessoa, in "Comprehensive Coordination Chemistry," eds. G. Wilkinson, R. D. Gillard, and J. A. McCleverty, Pergamon, Oxford, 1987, vol. 2, p. 453.
441. S. G. Brand, C. J. Hawkins, A. T. Marshall, G. W. Nette, and D. L. Parry, *Comp. Biochem. Physiol.*, 1989, **93B**, 425.
442. T. Ishii, I. Nakai, C. Numako, K. Okoshi, and T. Okake, *Naturwissenschaften*, 1993, **80**, 268.
443. D. A. Webb, *Publ. Staz. Zool. Napoli*, 1956, **28**, 273.
444. H.-J. Bielig, E. Bayer, H. D. Dell, G. Robins, H. Möllinger, and W. Rüdiger, *Protides Biol. Fluids*, 1996, **14**, 197.
445. I. G. Macara, G. C. McLeod, and K. Kustin, *Comp. Biochem. Physiol.*, 1979, **63B**, 299.
446. R. C. Bruening, E. M. Oltz, J. Furukawa, K. Nakanishi, and K. Kustin, *J. Am. Chem. Soc.*, 1985, **107**, 5298.

447. H. Michibata, T. Uyama, and J. Hirata, *Zool. Sci.*, 1990, **7**, 55.
448. D. E. Ryan, N. D. Ghatlia, A. E. McDermott, N. J. Turro, and K. Nakanishi, *J. Am. Chem. Soc.*, 1992, **114**, 9659.
449. D. H. Anderson and J. H. Swinehart, *Comp. Biochem. Physiol.*, 1991, **99A**, 585.
450. P. Frank, B. Hedman, R. K. Carlson, T. A. Tyson, A. L. Roe, and K. O. Hodgson, *Biochemistry*, 1987, **26**, 4975.
451. K. Kanamori and H. Michibata, *J. Mar. Biol. Ass. UK*, 1994, **74**, 279.
452. M. I. Agudelo, K. Kustin, and G. C. McLeod, *Comp. Biochem. Physiol.*, 1983, **75A**, 211.
453. C. J. Hawkins, P. Kott, D. L. Parry, and J. H. Swinehart, *Comp. Biochem. Physiol.*, 1983, **76B**, 555.
454. S. G. Brand, C. J. Hawkins, and D. L. Parry, *Inorg. Chem.*, 1987, **26**, 627.
455. P. Frank, R. M. K. Carlson, and K. O. Hodgson, *Inorg. Chem.*, 1986, **25**, 470.
456. T. Uyama, Y. Moriyama, M. Futai, and H. Michibata, *J. Exp. Zool.*, 1994, **270**, 148.
457. H. Michibata, J. Uchiyama, Y. Seki, T. Numakunai, and T. Uyama, *Biol. Trace Element Res.*, 1992, **34**, 219.
458. T. Uyama, J. Uchiyama, T. Nishikata, N. Satoh, and H. Michibata, *J. Exp. Zool.*, 1993, **265**, 29.
459. J. Wuchiyama and H. Michibata, *Acta Zool. (Stockholm)*, 1995, **76**, 51.
460. R. Endean, *Q. J. Microscop. Sci.*, 1960, **101**, 177.
461. M. J. Smith, *Biol. Bull.*, 1970, **138**, 379.
462. D. B. Carlisle, *Proc. Royal Soc. B.*, 1968, **171**, 31.
463. A. F. Rowley, *J. Exp. Zool.*, 1983, **227**, 319.
464. J. Kobayashi, G. C. Harbour, J. Gilmore, and K. L. Rinehart, Jr., *J. Am. Chem. Soc.*, 1984, **106**, 1526.
465. K. L. Rinehart, Jr., J. Kobayashi, G. C. Harbour, R. G. Hughes, Jr., S. A. Mizsak, and T. A. Scahill, *J. Am. Chem. Soc.*, 1984, **106**, 1524.
466. K. L. Rinehart, Jr., J. Kobayashi, G. C. Harbour, J. Gilmore, M. Mascal, T. G. Holt, L. S. Shield, and F. Lafargue, *J. Am. Chem. Soc.*, 1987, **109**, 3378.
467. Y. Nakamura, J. Kobayashi, J. Gilmore, M. Mascal, K. L. Rinehart, Jr., H. Nakamura, and Y. Ohizumi, *J. Biol. Chem.*, 1986, **261**, 4139.
468. J. Kobayashi, M. Ishibashi, U. Nagai, and Y. Ohizumi, *Experientia*, 1989, **45**, 782.
469. A. Seino, M. Kobayashi, J. Kobayashi, Y.-I. Fang, M. Ishibashi, H. Nakamura, K. Momose, and Y. Ohizumi, *J. Pharmacol. Exp. Ther.*, 1991, **256**, 861.
470. M. Adachi, M. Kakubari, and Y. Ohizumi, *Bio. Chem. Hoppe-Seyler*, 1994, **375**, 183.
471. Y.-I. Fang, M. Adachi, J. Kobayashi, and Y. Ohizumi, *J. Biol. Chem.*, 1993, **268**, 18 622.
472. M. Adachi, Y.-I. Fang, T. Yamakuni, J. Kobayashi, and Y. Ohizumi, *J. Pharm. Pharmacol.*, 1994, **46**, 771.
473. M. Adachi, M. Kakubari, and Y. Ohizumi, *J. Pharm. Pharmacol.*, 1994, **46**, 774.
474. Y. Takahashi, K. Furukawa, M. Ishibashi, D. Kozutsumi, H. Ishiyama, J. Kobayashi, and Y. Ohizumi, *Eur. J. Pharmacol.*, 1995, **288**, 285.
475. Y. Takahashi, K.-I. Furukawa, D. Kozutsumi, M. Ishibashi, J. Kobayashi, and Y. Ohizumi, *Br. J. Pharmacol.*, 1995, **114**, 941.
476. J. Kobayashi, M. Taniguchi, T. Hino, and Y. Ohizumi, *J. Pharmacol.*, 1988, **40**, 62.
477. J. Kobayashi, H. Nakamura, Y. Ohizumi, and Y. Hirata, *Tetrahedron Lett.*, 1986, **27**, 1191.
478. J. Kobayashi, J.-F. Cheng, T. Ohta, S. Nozoe, Y. Ohizumi, and T. Sasaki, *J. Org. Chem.*, 1990, **55**, 3666.
479. O. Murata, H. Shigemori, M. Ishibashi, K. Sugama, K. Hayashi, and J. Kobayashi, *Tetrahedron Lett.*, 1991, **32**, 3539.
480. J. W. Blunt, R. J. Lake, M. H. G. Munro, and T. Toyokuni, *Tetrahedron Lett.*, 1987, **28**, 1825.
481. R. J. Lake, M. M. Brennan, J. W. Blunt, M. H. G. Munro, and L. K. Pannell, *Tetrahedron Lett.*, 1988, **29**, 2255.
482. R. J. Lake, J. D. McCombs, J. W. Blunt, M. H. G. Munro, and W. T. Robinson, *Tetrahedron Lett.*, 1988, **29**, 4971.
483. K. F. Kinzer and J. H. Cardellina II, *Tetrahedron Lett.*, 1987, **28**, 925.
484. C. Debitus, D. Laurent, and M. Pais, *J. Nat. Prod.*, 1988, **51**, 799.
485. S. A. Adesanya, M. Chbani, and M. Pais, *J. Nat. Prod.*, 1992, **55**, 525.
486. M. Chbani, M. Pais, J. M. Delauneux, and C. Debitos, *J. Nat. Prod.*, 1993, **56**, 99.
487. A. Badre, A. Boulanger, E. Abou-Mansour, B. Banaigs, G. Combaut, and C. Francisco, *J. Nat. Prod.*, 1994, **57**, 528.
488. G. Q. Shen and B. J. Baker, *Tetrahedron Lett.*, 1994, **35**, 1141.
489. A. J. Blackman, D. J. Matthews, and C. K. Narkowicz, *J. Nat. Prod.*, 1987, **50**, 494.
490. A. Aiello, E. Fattorusso, S. Magno, and L. Mayol, *Tetrahedron*, 1987, **43**, 5929.
491. L. K. Larsen, R. E. Moore, and G. M. L. Patterson, *J. Nat. Prod.*, 1994, **57**, 419.
492. B. S. Davidson, *Chem. Rev.*, 1993, **93**, 1771.
493. K. L. Rinehart, Jr., J. B. Gloer, J. C. Cook, Jr., S. A. Mizsaki, and T. A. Scahill, *J. Am. Chem. Soc.*, 1981, **103**, 1857.
494. A. E. Wright, D. A. Forleo, G. P. Gunawardana, S. P. Gunasekera, F. E. Koehn, and O. J. McConnell, *J. Org. Chem.*, 1990, **55**, 4508.
495. K. L. Rinehart, T. G. Holt, N. L. Fregeau, J. G. Stroh, P. A. Keifer, F. Sun, L. H. Li, and D. G. Martin, *J. Org. Chem.*, 1990, **55**, 4512.
496. R. Sakai, J. G. Stroh, D. W. Sullins, and K. L. Rinehart, *J. Am. Chem. Soc.*, 1995, **117**, 3734.
497. R. Sakai, K. L. Rinehart, V. Kishore, B. Kundu, G. Faircloth, J. B. Gloer, J. R. Carney, M. Namikoshi, F. Sun, R. G. Hughes, Jr., D. G. Grávalos, T. G. de Quesada, G. R. Wilson, and R. M. Heid, *J. Med. Chem.*, 1996, **39**, 2819.
498. R. Sakai, K. L. Rinehart, Y. Guan, and A. H.-J. Wang, *Proc. Natl. Acad. Sci. USA*, 1992, **89**, 11 456.
499. T. Arai, K. Takahashi, A. Kubo, S. Nakagara, S. Sato, K. Aiba, and C. Tamura, *Tetrahedron Lett.*, 1979, 2355.
500. R. Sakai, E. A. Jares-Erijman, I. Manzanares, M. V. S. Elipe, and K. L. Rinehart, *J. Am. Chem. Soc.*, 1996, **118**, 9017.
501. J. Kobayashi, Y. Doi, and M. Ishibashi, *J. Org. Chem.*, 1994, **59**, 255.
502. Y. Doi, M. Ishibashi, and J. Kobayashi, *Tetrahedron*, 1994, **50**, 8651.
503. X. Fu, M. B. Hosssain, D. van der Helm, and F. J. Schmitz, *J. Am. Chem. Soc.*, 1994, **116**, 12 125.
504. S. Fukuzawa, S. Matsunaga, and N. Fusetani, *J. Org. Chem.*, 1994, **59**, 6164.
505. S. Fukuzawa, S. Matsunaga, and N. Fusetani, *J. Org. Chem.*, 1995, **60**, 608.
506. S. Fukuzawa, S. Matsunaga, and N. Fusetani, *Tetrahedron*, 1995, **51**, 6707.
507. S. Fukuzawa, S. Matsunaga, and N. Fusetani, *Tennen Yuki Kagoubutsu Toronkai Koen Yoshishu*, 1996, **38**, 73 (in Japanese).
508. G. R. Pettit, M. Inoue, Y. Kamano, D. L. Herald, C. Arm, C. Dufresne, N. D. Christie, J. M. Schmidt, D. L. Doubek, and T. S. Krupa, *J. Am. Chem. Soc.*, 1988, **110**, 2006.

509. S. Fukuzawa, S. Matsunaga, and N. Fusetani, *Tetrahedron Lett.*, 1996, **37**, 1447.
510. C. M. Ireland, T. F. Molinski, D. M. Roll, T. M. Zabriskie, T. C. McKee, J. C. Swersey, and M. P. Foster, in "Bioorganic Marine Chemistry," ed. P. J. Scheuer, Springer-Verlag, Berlin, 1989, vol. 3, p. 1.
511. M. P. Foster, G. P. Concepción, G. B. Caraan, and C. M. Ireland, *J. Org. Chem.*, 1992, **57**, 6671.
512. M. P. Foster and C. M. Ireland, *Tetrahedron Lett.*, 1993, **34**, 2871.
513. S. G. Toske and W. Fenical, *Tetrahedron Lett.*, 1995, **36**, 8355.
514. A. R. Carroll, Y. Feng, B. F. Bowden, and J. C. Coll, *J. Org. Chem.*, 1996, **61**, 4059.
515. P. Merfey, *Carlsberg Res. Commun.*, 1984, **49**, 591.
516. M. J. Vázquez, E. Quinoa, R. Riguera, A. Ocampo, T. Iglesias, and C. Debitus, *Tetrahedron Lett.*, 1995, **36**, 8853.
517. M. J. Smith, D. Kim, B. Horenstein, and K. Nakanishi, *Acc. Chem. Res.*, 1991, **24**, 117.
518. L. A. McDonald, J. C. Swersey, C. M. Ireland, A. R. Carroll, J. C. Coll, B. F. Bowden, C. R. Fairchild, and L. Cornell, *Tetrahedron*, 1995, **51**, 5237.
519. A. Rudi, I. Goldberg, Z. Stein, F. Frolow, Y. Benayahu, M. Schleyer, and Y. Kashman, *J. Org. Chem.*, 1994, **59**, 999.
520. R. J. Andersen, D. J. Faulkner, C.-H. He, G. D. Van Duyne, and J. Clardy, *J. Am. Chem. Soc.*, 1985, **107**, 5492.
521. N. Bontemps, I. Bonnard, B. Banaigs, G. Combaut, and C. Francisco, *Tetrahedron Lett.*, 1994, **35**, 7023.
522. L. A. McDonald, G. S. Eldredge, L. R. Barrows, and C. M. Ireland, *J. Med. Chem.*, 1994, **37**, 3819.
523. J. Kobayashi, J.-F. Cheng, M. R. Wälchli, H. Nakamura, Y. Hirata, T. Sasaki, and Y. Ohizumi, *J. Org. Chem.*, 1988, **53**, 1800.
524. A. R. Carroll and P. J. Scheuer, *J. Org. Chem.*, 1990, **55**, 4426.
525. A. R. Carroll, N. M. Cooray, A. Poiner, and P. J. Scheuer, *J. Org. Chem.*, 1989, **54**, 4231.
526. A. Rudi, Y. Benayahu, I. Goldberg, and Y. Kashman, *Tetrahedron Lett.*, 1988, **29**, 6655.
527. G. A. Charyulu, T. C. McKee, and C. M. Ireland, *Tetrahedron Lett.*, 1989, **30**, 4201.
528. P. A. Searle and T. F. Molinski, *J. Org. Chem.*, 1994, **59**, 6600.
529. B. S. Davidson, T. F. Molinski, L. R. Barrows, and C. M. Ireland, *J. Am. Chem. Soc.*, 1991, **113**, 4709.
530. P. W. Ford and B. S. Davidson, *J. Org. Chem.*, 1993, **58**, 4522.
531. V. Behar and S. J. Danishefsky, *J. Am. Chem. Soc.*, 1993, **115**, 7017.
532. P. W. Ford, M. R. Narbut, J. Belli, and B. S. Davidson, *J. Org. Chem.*, 1994, **59**, 5955.
533. M. Litaudon and M. Guyot, *Tetrahedron Lett.*, 1991, **32**, 911.
534. M. Litaudon, F. Trigalo, M.-T. Martin, F. Frappier, and M. Guyot, *Tetrahedron*, 1994, **50**, 5323.
535. B. S. Davidson, P. W. Ford, and M. Wahlman, *Tetrahedron Lett.*, 1994, **35**, 7185.
536. R. S. Compagnone, D. J. Faulkner, B. K. Carté, G. Chan, A. Freyer, M. E. Hemling, G. A. Hofmann, and M. R. Mattern, *Tetrahedron*, 1994, **50**, 12 785.
537. H. Kang and W. Fenical, *Tetrahedron Lett.*, 1996, **37**, 2369.
538. S. A. Abas, M. B. Hossain, D. van der Helm, F. J. Schmitz, M. Laney, R. Cabuslay, and R. C. Schatzman, *J. Org. Chem.*, 1996, **61**, 2709.
539. R. W. Schumacher and B. S. Davidson, *Tetrahedron*, 1995, **51**, 10 125.
540. H. Niwa, M. Watanabe, H. Inagaki, and K. Yamada, *Tetrahedron*, 1994, **50**, 7385.
541. H. Kang, P. R. Jensen, and W. Fenical, *J. Org. Chem.*, 1996, **61**, 1543.
542. N. Miyairi, H.-I. Sakai, T. Konomi, and H. Imanaka, *J. Antibiot.*, 1976, **29**, 227.
543. R. Riccio, R. B. Kinnel, G. Bifulco, and P. J. Scheuer, *Tetrahedron Lett.*, 1996, **37**, 1979.
544. L. A. McDonald, T. L. Capson, G. Krishnamurthy, W.-D. Ding, G. A. Ellestad, V. S. Bernan, W. M. Maiese, P. Lassota, C. Discafani, R. A. Kramer, and C. M. Ireland, *J. Am. Chem. Soc.*, 1996, **118**, 10 898.
545. N. B. Perry, J. W. Blunt, and M. H. G. Munro, *Aust. J. Chem.*, 1991, **44**, 627.
546. D. Enders and M. Finkam, *Liebigs Ann. Chem.*, 1993, 551.
547. E. A. Jares-Erijman, C. P. Bapat, A. Lithgow-Bertelloni, K. L. Rinehart, and R. Sakai, *J. Org. Chem.*, 1993, **58**, 5732.
548. P. A. Searle and T. F. Molinski, *J. Org. Chem.*, 1993, **58**, 7578.
549. M. F. Raub, J. H. Cardellina II, M. I. Choudhary, C. Z. Ni, J. Clardy, and M. C. Alley, *J. Am. Chem. Soc.*, 1991, **113**, 3178.
550. M. F. Raub and J. H. Cardellina II, *Tetrahedron Res.*, 1992, **33**, 2257.
551. F. Kong and D. J. Faulkner, *Tetrahedron Lett.*, 1991, **32**, 3667.
552. B. Steffan, *Tetrahedron*, 1991, **47**, 8729.
553. A. J. Blackman, C. Li, D. C. R. Hockless, B. W. Skelton, and A. H. White, *Tetrahedron*, 1993, **49**, 8645.
554. J. F. Biard, S. Guyot, C. Roussakis, J. F. Verbist, J. Vercauteren, J. F. Weber, and K. Boukef, *Tetrahedron Lett.*, 1994, **35**, 2691.
555. J. Kobayashi and M. Ishibashi, *Heterocycles*, 1996, **42**, 943.
556. M. Ishibashi, Y. Ohizumi, T. Sasaki, H. Nakamura, Y. Hirata, and J. Kobayashi, *J. Org. Chem.*, 1987, **52**, 450.
557. S. Knapp and J. J. Hale, *J. Org. Chem.*, 1993, **58**, 2650.
558. M. Nakagawa, A. Hasegawa, H. Kawamoto, R. Yamashita, and T. Hino, *110th Annual Meeting of the Pharmaceutical Society of Japan*, Sapporo, 1990, Abstract papers, p. 26 (in Japanese).
559. I. Utsunomiya, M. Ogawa, and M. Natsume, *Heterocycles*, 1992, **33**, 349.
560. T. Naito, Y. Yuumoto, I. Ninomiya, and T. Kiguchi, *Tetrahedron Lett.*, 1992, **33**, 4033.
561. T. Naito, M. Ikai, M. Shirakawa, K. Fujimoto, I. Ninomiya, and T. Kiguchi, *J. Chem. Soc., Perkin Trans. 1*, 1994, 773.
562. T. Kiguchi, Y. Yuumoto, I. Ninomiya, T. Naito, K. Deki, M. Ishibashi, and J. Kobayashi, *Tetrahedron Lett.*, 1992, **33**, 7389.
563. T. Kiguchi, Y. Yuumoto, I. Ninomiya, and T. Naito, *Tennen Yuki Kagoubutsu Toronkai Koen Yoshishu*, 1992, **34**, 392 (in Japanese).
564. M. Ishibashi, K. Deki, and J. Kobayashi, *J. Nat. Prod.*, 1995, **58**, 804.
565. J. Kobayashi, K. Naitoh, Y. Doi, K. Deki, and M. Ishibashi, *J. Org. Chem.*, 1995, **60**, 6941.
566. P. Garner and J. M. Park, *Org. Synth.*, 1992, **70**, 18.
567. Y. Doi, M. Ishibashi, and J. Kobayashi, *Tetrahedron*, 1996, **52**, 4573.
568. P. I. Svirskaya, S. N. Maiti, A. J. Jones, B. Khouw, and C. C. Leznoff, *J. Chem. Ecol.*, 1984, **10**, 795.
569. R. Sakai, T. Higa, C. W. Jefford, and G. Bernardinelli, *J. Am. Chem. Soc.*, 1986, **108**, 6404.

570. R. Sakai, S. Kohmoto, T. Higa, C. W. Jefford, and G. Bernardinelli, *Tetrahedron Lett.*, 1987, **28**, 5493.
571. T. Higa, in "Studies in Natural Product Chemistry B II," ed. Atta-ur-Rahman, Elsevier, Amsterdam, 1989, pp. 346–353.
572. H. Nakamura, S. Deng, J. Kobayashi, Y. Ohizumi, Y. Tomotake, T. Matsuzaki, and Y. Hirata, *Tetrahedron Lett.*, 1987, **28**, 621.
573. J. E. Baldwin and R. C. Whitehead, *Tetrahedron Lett.*, 1992, **33**, 2059.
574. K. Kondo, H. Shigemori, Y. Kikuchi, M. Ishibashi, T. Sasaki, and J. Kobayashi, *J. Org. Chem.*, 1992, **57**, 2480.
575. J. Kobayashi, M. Tsuda, N. Kawasaki, K. Matsumoto, and T. Adachi, *Tetrahedron Lett.*, 1994, **35**, 4383.
576. M. Tsuda, N. Kawasaki, and J. Kobayashi, *Tetrahedron Lett.*, 1994, **35**, 4387.
577. M. Tsuda, N. Kawasaki, and J. Kobayashi, *Tetrahedron*, 1994, **50**, 7957.
578. M. Tsuda, I. Inaba, N. Kawasaki, K. Honma, and J. Kobayashi, *Tetrahedron*, 1996, **52**, 2319.
579. J. Kobayashi, N. Kawasaki, and M. Tsuda, *Tetrahedron Lett.*, 1996, **37**, 8203.
580. J. Kobayashi, M. Tsuda, N. Kawasaki, T, Sasaki, and Y. Mikami, *J. Nat. Prod.*, 1994, **57**, 1737.
581. M. Kobayashi, Y.-J. Chen, S. Aoki, T. In, T. Ishida, and I. Kitagawa, *Tetrahedron*, 1995, **51**, 3727.
582. T. Ichiba, R. Sakai, S. Kohmoto, G. Saucy, and T. Higa, *Tetrahedron Lett.*, 1988, **29**, 3083.
583. T. Ichiba, J. M. Corgiat, P. J. Scheuer, and M. Kelly-Borges, *J. Nat. Prod.*, 1994, **57**, 168.
584. T. Higa, in "Proceedings of the First Princess Chulabhorn Science Congress 1987," 1987, pp. 450–459.
585. P. Crews, X.-C. Cheng, M. Adamczeski, J. Rodríguez, M. Jaspars, F. J. Schmitz, S. C. Traeger, and E. O. Pordesimo, *Tetrahedron*, 1994, **50**, 13 567.
586. R. A. Edrada, P. Proksch, V. Wary, L. Witte, W. E. G. Müller, and R. W. M. Van Soest, *J. Nat. Prod.*, 1996, **59**, 1056.
587. I. I. Ohtani, T. Ichiba, I. Isobe, M. Kelly-Borges, and P. J. Scheuer, *J. Am. Chem. Soc.*, 1995, **117**, 10 743.
588. F. Kong, R. J. Andersen, and T. M. Allen, *Tetrahedron Lett.*, 1994, **35**, 1643.
589. F. Kong, R. J. Andersen, and T. M. Allen, *Tetrahedron*, 1994, **50**, 6137.
590. F. Kong and R. J. Andersen, *Tetrahedron*, 1995, **51**, 2895.
591. J. Rodríguez, B. M. Peters, L. Kurz, R. C. Schatzman, D. McCarley, L. Lou, and P. Crews, *J. Am. Chem. Soc.*, 1993, **115**, 10 436.
592. J. Rodíguez and P. Crews, *Tetrahedron Lett.*, 1994, **35**, 4719.
593. N. Fusetani, K. Yasumuro, S. Matsunaga, and H. Hirota, *Tetrahedron Lett.*, 1989, **30**, 6891.
594. N. Fusetani, N. Asai, S. Matsunaga, K. Honda, and K. Yasumuro, *Tetrahedron Lett.*, 1994, **35**, 3967.
595. J. C. Braekman, D. Daloze, P. M. de Abreu, C. Piccinni-Leopardi, G. German, and M. Van Meerssche, *Tetrahedron Lett.*, 1982, **23**, 4277.
596. J. C. Braekman, D. Daloze, N. Defay, and D. Zimmermann, *Bull. Soc. Chim. Belg.*, 1984, **93**, 941.
597. J. C. Braekman, D. Daloze, G. Cimino, and E. Trivellone, *Bull. Soc. Chim. Belg.*, 1988, **97**, 519.
598. M. Nakagawa, M. Endo, N. Tanaka, and L. Gen-pei, *Tetrahedron Lett.*, 1984, **25**, 3227.
599. M. Kobayashi, K. Kawazoe, and I. Kitagawa, *Chem. Pharm. Bull.*, 1989, **37**, 1676.
600. M. Kobayashi, K. Kawazoe, and I. Kitagawa, *Tetrahedron Lett.*, 1989, **30**, 4149.
601. G. Cimino, S. De Stefano, G. Scognamiglio, G. Sodano, and E. Trivellone, *Bull. Soc. Chim. Belg.*, 1986, **95**, 783.
602. J.-C. Quirion, T. Sevenet, H.-P. Husson, B. Weniger, and C. Debitus, *J. Nat. Prod.*, 1992, **55**, 1505.
603. G. Cimino, C. A. Mattia, L. Mazzarella, R. Puliti, G. Scognamiglio, A. Spinella, and E. Trivellone, *Teterahedron*, 1989, **45**, 3863.
604. G. Cimino, A. Spinella, and E. Trivellone, *Tetrahedron Lett.*, 1989, **30**, 133.
605. G. Cimino, G. Scognamiglio, A. Spinella, and E. Trivellone, *J. Nat. Prod.*, 1990, **53**, 1519.
606. G. Cimino, A. Fantana, A. Madaio, G. Scognamiglio, and E. Trivellone, *Magn. Reson. Chem.*, 1991, **29**, 327.
607. Y. Guo, A. Madaio, E. Trivellone, G. Scognamiglio, and G. Cimino, *Tetrahedron*, 1996, **52**, 8341.
608. Y. Guo, A. Madaio, E. Trivellone, G. Scognamiglio, and G. Cimino, *Tetrahedron*, 1996, **52**, 14 961.
609. B. J. Baker, P. J. Scheuer, and J. N. Shoolery, *J. Am. Chem. Soc.*, 1988, **110**, 965.
610. E. Fahy, T. F. Molinski, M. K. Harper, B. W. Sullivan, D. J. Faulkner, L. Parkanyi, and J. Clardy, *Tetrahedron Lett.*, 1988, **29**, 3427.
611. F. Kong, R. J. Andersen, and T. M. Allen, *J. Am. Chem. Soc.*, 1994, **116**, 6007.
612. M. Jaspars, V. Pasupathy, and P. Crews, *J. Org. Chem.*, 1994, **59**, 3253.
613. R. D. Charan, M. J. Garson, I. M. Brereton, A. C. Willis, and J. N. A. Hooper, *Tetrahedron*, 1996, **52**, 9111.
614. B. Harrison, S. Talapatra, E. Lobkovsky, J. Clardy, and P. Crews, *Tetrahedron Lett.*, 1996, **37**, 9151.
615. N. Fusetani and S. Matsunaga, *Chem. Rev.*, 1993, **93**, 1793.
616. R. D. Norcross and I. Paterson, *Chem. Rev.*, 1995, **95**, 2041.
617. S. Matsunaga, N. Fusetani, K. Hashimoto, and M. Wälchli, *J. Am. Chem. Soc.*, 1989, **111**, 2582.
618. N. Fusetani, S. Matsunaga, H. Matsumoto, and Y. Takebayashi, *J. Am. Chem. Soc.*, 1990, **112**, 7053.
619. M. Hagihara and S. L. Schreiber, *J. Am. Chem. Soc.*, 1992, **114**, 6570.
620. Y. Nakao, S. Matsunaga, and N. Fusetani, *Bioorg. Med. Chem.*, 1995, **3**, 1115.
621. B. E. Maryanoff, X. Qiu, K. P. Padmanabhan, A. Tulinsky, H. R. Almond, Jr., P. Andrade-Gordon, M. N. Greco, J. A. Kauffman, K. C. Nicolaou, A. Liu, P. H. Brings, and N. Fusetani, *Proc. Natl. Acad. Sci. USA*, 1993, **90**, 8048.
622. P. Wipf and H. Kim, *J. Org. Chem.*, 1993, **58**, 5592.
623. J. Deng, Y. Hamada, T. Shioiri, S. Matsunaga, and N. Fusetani, *Angew. Chem. Int. Ed. Engl.*, 1994, **33**, 1729.
624. J. Deng, Y. Hamada, and T. Shioiri, *Tetrahedron Lett.*, 1996, **37**, 2261.
625. H. M. M. Bastiaans, J. L. van der Baan, and H. C. J. Ottenheijm, *Tetrahedron Lett.*, 1995, **36**, 5963.
626. B. E. Maryanoff, M. N. Greco, H.-C. Zhang, P. Andrade-Gordon, J. A. Kauffman, K. C. Nicolaou, A. Liu, and P. H. Brungs, *J. Am. Chem. Soc.*, 1995, **117**, 1225.
627. B. E. Maryanoff, H.-C. Zhang, M. N. Greco, K. A. Glover, J. A. Kauffman, and P. Andrade-Gordon, *Bioorg. Med. Chem.*, 1995, **3**, 1025.
628. B. E. Maryanoff, H.-C. Zhang, M. N. Greco, E. Zhang, P. Vanderhoff-Hanaver, and A. Tulinsky, *Tetrahedron Lett.*, 1996, **37**, 3667.
629. A. Y. Lee, M. Hagihara, R. Karmacharya, M. W. Albers, S. L. Schreiber, and J. Clardy, *J. Am. Chem. Soc.*, 1993, **115**, 12 619.
630. N. Fusetani, T. Sugawara, S. Matsunaga, and H. Hirota, *J. Am. Chem. Soc.*, 1991, **113**, 7811.

631. J. Kobayashi, F. Itagaki, H. Shigemori, M. Ishibashi, K. Takahashi, M. Ogura, S. Nagasawa, T. Nakamura, H. Hirota, T. Ohta, and S. Nozoe, *J. Am. Chem. Soc.*, 1991, **113**, 7812.
632. F. Itagaki, H. Shigemori, M. Ishibashi, T. Nakamura, T. Sasaki, and J. Kobayashi, *J. Org. Chem.*, 1992, **57**, 5540.
633. J. Kobayashi, F. Itagaki, H. Shigemori, T. Takao, and Y. Shimonishi, *Tetrahedron*, 1995, **51**, 2525.
634. J. Kobayashi, M. Sato, M. Ishibashi, H. Shigemori, T. Nakamura, and Y. Ohizumi, *J. Chem. Soc., Perkin Trans. 1*, 1991, 2609.
635. J. Kobayashi, M. Sato, T. Murayama, M. Ishibashi, M. R. Wälchli, M. Kanai, J. Shoji, and Y. Ohizumi, *J. Chem. Soc., Chem. Commun.*, 1991, 1050.
636. M. Ishibashi, Y. Li, M. Sato, and J. Kobayashi, *Nat. Prod. Lett.*, 1994, **4**, 293.
637. U. Schmidt and S. Weinbrenner, *Angew. Chem., Int. Ed. Engl.*, 1996, **35**, 1336.
638. I. Kitagawa, N. K. Lee, M. Kobayashi, and H. Shibuya, *Tetrahedron*, 1991, **47**, 2169.
639. M. Kobayashi, N. K. Lee, H. Shibuya, T. Momose, and I. Kitagawa, *Chem. Pharm. Bull.*, 1991, **39**, 1177.
640. M. Kobayashi, K. Kanzaki, S. Katayama, K. Ohashi, H. Okada, S. Ikegami, and I. Kitagawa, *Chem. Pharm. Bull.*, 1994, **42**, 1410.
641. S. Matsunaga and N. Fusetani, *J. Org. Chem.*, 1995, **60**, 1177.
642. C. A. Bewley and D. J. Faulkner, *J. Org. Chem.*, 1994, **59**, 4849.
643. N. Fusetani, Y. Nakao, and S. Matsunaga, *Tetrahedron Lett.*, 1991, **32**, 7073.
644. K. Hayashi, Y. Hamada, and T. Shioiri, *Tetrahedron Lett.*, 1992, **33**, 5075.
645. E. Dilip de Silva, D. E. Williams, R. J. Andersen, H. Klix, C. F. B. Holmes, and T. M. Allen, *Tetrahedron Lett.*, 1992, **33**, 1561.
646. K. L. Rinehart, K. Harada, M. Namikoshi, C. Chen, C. A. Harvis, M. H. G. Munro, J. W. Blunt, P. E. Mulligan, V. R. Beasley, A. M. Dahlem, and W. W. Carmichael, *J. Am. Chem. Soc.*, 1988, **110**, 8557.
647. N. K. Gulavita, S. A. Pomponi, A. E. Wright, D. Yarwood, and M. A. Sills, *Tetrahedron Lett.*, 1994, **35**, 6815.
648. T. Hamada, T. Sugawara, S. Matsunaga, and N. Fusetani, *Tetrahedron Lett.*, 1994, **35**, 719.
649. T. Hamada, T. Sugawara, S. Matsunaga, and N. Fusetani, *Tetrahedron Lett.*, 1994, **35**, 609.
650. T. Hamada, S. Matsunaga, N. Fusetani, M. Fujiwara, and K. Fujita, *Tennen Yuki Kagobutsu Toronkai Koen Yoshishu*, 1995, **37**, 695 (in Japanese).
651. S. Carmely and Y. Kashman, *Tetrahedron Lett.*, 1985, **26**, 511.
652. I. Kitagawa, M. Kobayashi, T. Katori, M. Yamashita, J. Tanaka, M. Doi, and T. Ishida, *J. Am. Chem. Soc.*, 1990, **112**, 3710.
653. M. Ishibashi, R. E. Moore, G. M. L. Patterson, C. Xu, and J. Clardy, *J. Org. Chem.*, 1986, **51**, 5300.
654. M. Doi, T. Ishida, M. Kobayashi, and I. Kitagawa, *J. Org. Chem.*, 1991, **56**, 3629.
655. M. Kobayashi, K. Kawazoe, T. Okamoto, T. Sasaki, and I. Kitagawa, *Chem. Pharm. Bull.*, 1994, **42**, 19.
656. C. A. Bewley, N. D. Holland, and D. J. Faulkner, *Experientia*, 1996, **52**, 716.
657. S. Matsunaga, N. Fusetani, Y. Kato, and H. Hirota, *J. Am. Chem. Soc.*, 1991, **113**, 9690.
658. S. Sakemi, T. Ichiba, S. Kohmoto, G. Saucy, and T. Higa, *J. Am. Chem. Soc.*, 1988, **110**, 4851.
659. S. Matsunaga, N. Fusetani, and Y. Nakao, *Tetrahedron*, 1992, **48**, 8369.
660. J. Kobayashi, F. Itagaki, H. Shigemori, and T. Sasaki, *J. Nat. Prod.*, 1993, **56**, 976.
661. N. Fusetani, T. Sugawara, and S. Matsunaga, *J. Org. Chem.*, 1992, **57**, 3828.
662. N. B. Perry, J. W. Blunt, M. H. G. Munro, and L. K. Pannell, *J. Am. Chem. Soc.*, 1988, **110**, 4850.
663. J. Kobayashi, K. Kondo, H. Shigemori, M. Ishibashi, T. Sasaki, and Y. Mikami, *J. Org. Chem.*, 1992, **57**, 6680.
664. J. Kobayashi, K. Kondo, M. Ishibashi, M. R. Wälchli, and T. Nakamura, *J. Am. Chem. Soc.*, 1993, **115**, 6661.
665. K. Kondo, M. Ishibashi, and J. Kobayashi, *Tetrahedron*, 1994, **50**, 8355.
666. M.-C. Rho, Y.-H. Park, S. Sasaki, M. Ishibashi, K. Kondo, J. Kobayashi, and Y. Ohizumi, *Can. J. Physiol. Pharmacol.*, 1996, **74**, 193.
667. Y. Nakao, B. K. S. Yeung, W. Y. Yoshida, P. J. Scheuer, and M. Kelly-Borges, *J. Am. Chem. Soc.*, 1995, **117**, 8271.
668. B. K. S. Yeung, Y. Nakao, R. B. Kinnel, J. R. Carney, W. Y. Yoshida, P. J. Scheuer, and M. Kelly-Borges, *J. Org. Chem.*, 1996, **61**, 7168.
669. P. A. Searle and T. F. Molinski, *J. Am. Chem. Soc.*, 1995, **117**, 8126.
670. P. A. Searle, T. F. Molinski, L. J. Brzezinski, and J. W. Leahy, *J. Am. Chem. Soc.*, 1996, **118**, 9422.
671. T. F. Molinski, *Tetrahedron Lett.*, 1996, **37**, 7879.
672. C. A. Bewley, H. He, D. H. Williams, and D. J. Faulkner, *J. Am. Chem. Soc.*, 1996, **118**, 4314.
673. S. P. Gunasekera, P. J. McCarthy, M. Kelly-Borges, E. Lobkovsky, and J. Clardy, *J. Am. Chem. Soc.*, 1996, **118**, 8759.
674. A. Zampella, M. V. D'Auria, L. G. Paloma, A. Casapullo, L. Minale, C. Debitus, and Y. Henin, *J. Am. Chem. Soc.*, 1996, **118**, 6202.
675. M. V. D'Auria, A. Zampella, L. G. Paloma, L. Minale, C. Debitus, C. Roussakis, and V. Le Bert, *Tetrahedron*, 1996, **52**, 9589.
676. A. Zampella, M. V. D'Auria, L. Minale, C. Debitus, and C. Roussakis, *J. Am. Chem. Soc.*, 1996, **118**, 11 085.
677. W. Fenical, *Chem. Rev.*, 1993, **93**, 1673.
678. B. S. Davidson, *Curr. Opin. Biotechnol.*, 1995, **6**, 284.
679. A. C. Stierle, J. H. Cardellina II, and F. L. Singleton, *Experientia*, 1988, **44**, 1021.
680. F. J. Schmitz, D. J. Vanderah, K. H. Hollenbeak, C. E. L. Enwall, Y. Gopichand, P. K. SenGupta, M. B. Hossain, and D. van der Helm, *J. Org. Chem.*, 1983, **48**, 3941.
681. A. A. Stierle, J. H. Cardellina II, and F. L. Singleton, *Tetrahedron Lett.*, 1991, **32**, 4847.
682. G. B. Elyakov, T. Kuznetsova, V. V. Mikhailov, I. I. Maltsev, V. G. Voinov, and S. A. Fedoreyev, *Experientia*, 1991, **47**, 632.
683. H. Shigemori, M.-A. Bae, K. Yazawa, T. Sasaki, and J. Kobayashi, *J. Org. Chem.*, 1992, **57**, 4317.
684. S. Ito and Y. Hirata, *Bull. Chem. Soc. Jpn.*, 1977, **50**, 1813.
685. S. P. Gunasekera, M. Gunasekera, and P. McCarthy, *J. Org. Chem.*, 1991, **56**, 4830.
686. M. Kobayashi, S. Aoki, K. Gato, K. Matsunami, M. Kurosu, and I. Kitagawa, *Chem. Pharm. Bull.*, 1994, **42**, 2449.
687. N. Imamura, M. Nishijima, K. Adachi, and H. Sano, *J. Antibiot.*, 1993, **46**, 241.
688. W. Miki, N. Otaki, A. Yokoyama, H. Izumida, and N. Shimidzu, *Experientia*, 1994, **50**, 684.
689. G. S. Jayatilake, M. P. Thornton, A. C. Leonard, J. E. Grimwade, and B. J. Baker, *J. Nat. Prod.*, 1996, **59**, 293.

690. K. Gustafson, M. Roman, and W. Fenical, *J. Am. Chem. Soc.*, 1989, **111**, 7519.
691. S. D. Rychnovsky, D. J. Skalitzky, C. Pathirana, P. R. Jensen, and W. Fenical, *J. Am. Chem. Soc.*, 1992, **114**, 671.
692. D. M. Tapiolas, M. Roman, W. Fenical, T. J. Stout, and J. Clardy, *J. Am. Chem. Soc.*, 1991, **113**, 4682.
693. C. Pathirana, D. Tapiolas, P. R. Jensen, R. Dwight, and W. Fenical, *Tetrahedron Lett.*, 1991, **32**, 2323.
694. C. Pathirana, R. Dwight, P. R. Jensen, W. Fenical, A. Delgado, L. S. Brinen, and J. Clardy, *Tetrahedron Lett.*, 1991, **32**, 7001.
695. C. Pathirana, P. R. Jensen, R. Dwight, and W. Fenical, *J. Org. Chem.*, 1992, **57**, 740.
696. R. T. Reid, D. H. Live, D. J. Faulkner, and A. Butler, *Nature*, 1993, **366**, 455.
697. J. A. Trischman, D. M. Tapiolas, P. R. Jensen, R. Dwight, W. Fenical, T. C. McKee, C. M. Ireland, T. J. Stout, and J. Clardy, *J. Am. Chem. Soc.*, 1994, **116**, 757.
698. J. A. Trischman, P. R. Jensen, and W. Fenical, *Tetrahedron Lett.*, 1994, **35**, 5571.
699. J. Needham, M. T. Kelly, M. Ishige, and R. J. Anderson, *J. Org. Chem.*, 1994, **59**, 2058.
700. A. Fredenhagen, S. Y. Tamura, P. T. M. Kenny, H. Komura, Y. Naya, K. Nakanishi, K. Nishiyama, M. Sugiura, and H. Kita, *J. Am. Chem. Soc.*, 1987, **109**, 4409.
701. J. Gerard, P. Haden, M. T. Kelly, and R. J. Andersen, *Tetrahedron Lett.*, 1996, **37**, 7201.
702. B. S. Davidson and R. W. Schumacher, *Tetrahedron*, 1993, **49**, 6569.
703. J. Needham, R. J. Andersen, and M. T. Kelly, *Tetrahedron Lett.*, 1991, **32**, 315.
704. J. Needham, R. J. Andersen, and M. T. Kelly, *J. Chem. Soc., Chem. Commun.*, 1992, 1367.
705. S. G. Toske, P. R. Jensen, C. A. Kauffman, and W. Fenical, *Nat. Prod. Lett.*, 1995, **6**, 303.
706. T. Lindel, P. R. Jensen, and W. Fenical, *Tetrahedron Lett.*, 1996, **37**, 1327.
707. M. Yoshida, M. Nishikawa, K. Nishi, K. Abe, S. Horinouchi, and T. Beppu, *Exp. Cell Res.*, 1990, **187**, 150.
708. N. Sitachitta, M. Gadepalli, and B. S. Davidson, *Tetrahedron*, 1996, **52**, 8073.
709. R. W. Schumacher, B. S. Davidson, D. A. Montenegro, and V. S. Bernan, *J. Nat. Prod.*, 1995, **58**, 613.
710. N. M. Gandhi, J. Nazareth, P. V. Divekar, H. Kohl, and N. J. de Souza, *J. Antibiot.*, 1973, **26**, 797.
711. H. Kohl, S. V. Bhat, J. R. Patell, N. M. Gandhi, J. Nazareth, P. V. Divekar, N. J. de Souza, H. G. Berscheid, and H.-W. Fehlhaber, *Tetrahedron Lett.*, 1974, 983.
712. N. Imamura, K. Adachi, and H. Sano, *J. Antibiot.*, 1994, **47**, 257.
713. J. Kobayashi, S. Mikami, H. Shigemori, T. Takao, Y. Shimonishi, S. Izuta, and S. Yoshida, *Tetrahedron*, 1995, **51**, 10487.
714. R. Anderson, M. Kates, and B. E. Volcani, *Biochim. Biophys. Acta*, 1978, **528**, 89.
715. K. Ohashi, S. Kosai, M. Arizuka, T. Watanabe, Y. Yamagiwa, T. Kamikawa, and M. Kates, *Tetrahedron*, 1989, **45**, 2557.
716. T. Kamiyama, T. Imuno, T. Satoh, S. Sawairi, M. Shirane, S. Ohshima, and K. Yokose, *J. Antiobiot.*, 1995, **48**, 924.
717. T. Kamiyama, T. Umino, Y. Itezono, Y. Nakamura, T. Satoh, and K. Yokose, *J. Antibiot.*, 1995, **48**, 929.
718. M. Sugano, A. Sato, Y. Iijima, T. Oshima, K. Furuya, H. Kuwano, T. Hata, and H. Hanzawa, *J. Am. Chem. Soc.*, 1991, **113**, 5463.
719. M. Sugano, A. Sato, Y. Iijima, K. Furuya, H. Haruyama, K. Yoda, and T. Hata, *J. Org. Chem.*, 1994, **59**, 564.
720. M. Sugano, A. Sato, Y. Iijima, K. Furuya, H. Kuwano, and T. Hata, *J. Antibiot.*, 1995, **48**, 1188.
721. M. Chu, M. G. Patel, V. P. Gullo, I. Truumess, M. S. Puar, and A. T. McPhail, *J. Org. Chem.*, 1992, **57**, 5817.
722. M. Chu, I. Truumees, I. Gunnarsson, W. R. Bishop, W. Kreutner, A. C. Horan, M. G. Patel, V. P. Gullo, and M. S. Puar, *J. Antibiot.*, 1993, **46**, 554.
723. H. Shigemori, S. Wakuri, K. Yazawa, T. Nakamura, T. Sasaki, and J. Kobayashi, *Tetrahedron*, 1991, **47**, 8529.
724. T. Aoyagi, T. Takeuchi, A. Matsuzaki, K. Kawamura, S. Kondo, K. Hamada, K. Maeda, and H. Umezawa, *J. Antibiot.*, 1969, **22**, 283.
725. K. Yamaguchi, T. Tsuji, S. Wakuri, K. Yazawa, K. Kondoh, H. Shigemori, and J. Kobayashi, *Biosci. Biotech. Biochem.*, 1993, **57**, 195.
726. K. Kobayashi, H. Uehara, K. Matsunami, S. Aoki, and I. Kitagawa, *Tetrahedron Lett.*, 1993, **34**, 7925.
727. L. M. Abrell, X.-C. Cheng, and P. Crews, *Tetrahedron Lett.*, 1994, **35**, 9159.
728. X.-C. Cheng, M. Varoglu, L. Abrell, P. Crews, E. Lobkovsky, and J. Clardy, *J. Org. Chem.*, 1994, **59**, 6344.
729. L. M. Abrell, B. Borgeson, and P. Crews, *Tetrahedron Lett.*, 1996, **37**, 8983.
730. C. Chen, N. Imamura, N. Nishijima, K. Adachi, M. Sakia, and H. Sano, *J. Antibiot.*, 1996, **49**, 998.
731. J. Doshida, H. Hasegawa, H. Onuki, and N. Shimidzu, *J. Antibiot.*, 1996, **49**, 1105.
732. M. Alam, E. B. G. Jones, M. B. Hossain, and D. van der Helm, *J. Nat. Prod.*, 1996, **59**, 454.
733. A. Numata, C. Takahashi, Y. Ito, T. Takada, K. Kawai, Y. Usami, E. Matsumura, M. Imachi, T. Ito, and T. Hasegawa, *Tetrahedron Lett.*, 1993, **34**, 2355.
734. C. Takahashi, A. Numata, Y. Ito, E. Matsumura, H. Araki, H. Iwaki, and K. Kushida, *J. Chem. Soc., Perkin Trans. 1*, 1994, 1859.
735. C. Takahashi, Y. Takai, Y. Kimura, A. Numata, N. Shigematsu, and H. Tanaka, *Phytochemistry*, 1995, **38**, 155.
736. C. Takahashi, A. Numata, E. Matsumura, K. Minoura, H. Eto, T. Shingu, T. Ito, and T. Hasegawa, *J. Antibiot.*, 1994, **47**, 1242.
737. C. Takahashi, K. Minoura, T. Yamada, A. Numata, K. Kushida, T. Shingu, S. Hagishita, H. Nakai, T. Sato, and H. Harada, *Tetrahedron Lett.*, 1995, **51**, 3483.
738. A. Numata, C. Takahashi, T. Matsushita, T. Miyamoto, K. Kawai, Y. Usami, E. Matsumura, M. Inoue, H. Ohishi, and T. Shingu, *Tetrahedron Lett.*, 1992, **33**, 1621.
739. C. Takahashi, T. Matsushita, M. Doi, K. Minoura, T. Shingu, Y. Kumeda, and A. Numata, *J. Chem. Soc., Perkin Trans. 1*, 1995, 2345.
730. C. Takahashi, T. Takada, T. Yamada, K. Minoura, K. Uchida, E. Matsumura, and A. Numata, *Tetrahedron Lett.*, 1994, **35**, 5013.
741. A. Numata, C. Takahashi, Y. Ito, K. Minoura, T. Yamada, C. Matsuda, and K. Nemoto, *J. Chem. Soc., Perkin Trans. 1*, 1996, 239.
742. C. Takahashi, A. Numata, T. Yamada, K. Minoura, S. Enomoto, K. Konishi, M. Nakai, C. Matsuda, and K. Nemoto, *Tetrahedron Lett.*, 1996, **37**, 655.
743. R. E. Moore, T. H. Corbett, G. M. L. Patterson, and F. A. Valeriote, *Curr. Pharm. Des.*, 1996, **2**, 317.

744. J. S. Mynderse, R. E. Moore, M. Kashiwagi, and T. R. Norton, *Science*, 1977, **196**, 538.
745. J. H. Cardellina II, F.-J. Marner, and R. E. Moore, *Science*, 1979, **204**, 193.
746. R. H. Moore, in "Marine Natural Products, Chemical and Biological Perspectives," ed. P. J. Scheuer, Academic Press, New York, 1981, vol. 4, p. 1.
747. W. H. Gerwick, P. J. Proteau, D. G. Nagle, E. Hamel, A. Blokhin, and D. L. Slate, *J. Org. Chem.*, 1994, **59**, 1243.
748. D. G. Nagle, R. S. Geralds, H.-D. Yoo, W. H. Gerwick, T.-S. Kim, M. Nambu, and J. D. White, *Tetrahedron Lett.*, 1995, **36**, 1189.
749. H.-D. Yoo and W. H. Gerwick, *J. Nat. Prod.*, 1995, **58**, 1961.
750. A. V. Blokhin, H.-D. Yoo, R. S. Geralds, D. G. Nagle, W. H. Gerwick, and E. Hamel, *Mol. Pharmacol.*, 1995, **48**, 523.
751. J. Orjala, D. Nagle, and W. H. Gerwick, *J. Nat. Prod.*, 1995, **58**, 764.
752. J. S. Todd and W. H. Gerwick, *Tetrahedron Lett.*, 1995, **36**, 7837.
753. A. F. Rose, P. J. Scheuer, J. P. Springer, and J. Clardy, *J. Am. Chem. Soc.*, 1978, **100**, 7665.
754. J. Orgala, D. G. Nagle, V. L. Hsu, and W. H. Gerwick, *J. Am. Chem. Soc.*, 1995, **117**, 8281.
755. J. Orjala and W. H. Gerwick, *J. Nat. Prod.*, 1996, **59**, 427.
756. R. Kazlauskas, R. O. Lidgard, R. J. Wells, and W. Vetter, *Tetrahedron Lett.*, 1977, 3183.
757. M. D. Unson and D. J. Faulkner, *Experientia*, 1993, **49**, 349.
758. G. R. Pettit, Y. Kamano, C. L. Herald, A. A. Tuinman, F. E. Boettner, H. Kizu, J. M. Schmidt, L. Baczynskyj, K. B. Tomer, and R. J. Bontems, *J. Am. Chem. Soc.*, 1987, **109**, 6883.
759. D. G. Nagle, V. J. Paul, and M. A. Roberts, *Tetrahedron Lett.*, 1996, **37**, 6263.
760. W. H. Gerwick, Z. D. Jiang, S. K. Agarwal, and B. T. Farmer, *Tetrahedron*, 1992, **48**, 2313.
761. D. G. Nagle and W. H. Gerwick, *Tetrahedron Lett.*, 1995, **36**, 849.
762. N. Lindquist, W. Fenical, D. F. Sesin, C. M. Ireland, G. D. Van Duyne, C. J. Forsyth, and J. Clardy, *J. Am. Chem. Soc.*, 1988, **110**, 1308.
763. M. Murakami, H. Matsuda, K. Makabe, and K. Yamaguchi, *Tetrahedron Lett.*, 1991, **32**, 2391.
764. M.-C. Rho, K. Matsunaga, K. Yasuda, and Y. Ohizumi, *J. Nat. Prod.*, 1996, **59**, 308.
765. M. Namikoshi, F. Sun, B. W. Choi, K. L. Rinehart, W. W. Carmichael, W. R. Evans, and V. R. Beasley, *J. Org. Chem.*, 1995, **60**, 3671.
766. R. E. Moore, J. L. Chen, B. S. Moore, G. M. L. Patterson, and W. W. Carmichael, *J. Am. Chem. Soc.*, 1991, **113**, 5083.
767. M. O. Ishitsuka, T. Kusumi, H. Kakisawa, K. Kaya, and M. M. Watanabe, *J. Am. Chem. Soc.*, 1990, **112**, 8180.
768. H. J. Shin, M. Murakami, H. Matsuda, and K. Yamaguchi, *Tetrahedron*, 1996, **52**, 8159.
769. S. Tsukamoto, P. Painuly, K. A. Young, X. Yang, Y. Shimizu, and L. Cornell, *J. Am. Chem. Soc.*, 1993, **115**, 11 046.
770. K.-I. Harada, T. Mayumi, T. Shimada, M. Suzuki, F. Kondo, and M. F. Watanabe, *Tetrahedron Lett.*, 1993, **34**, 6091.
771. T. Sano and K. Kaya, *Tetrahedron Lett.*, 1995, **36**, 5933.
772. D. E. Williams, M. Craig, C. F. B. Holms, and R. J. Andersen, *J. Nat. Prod.*, 1996, **59**, 570.
773. I. R. Falconer and T. H. Buckley, *Med. J. Aust.*, 1989, **150**, 351.
774. R. E. Honkanen, J. Zwiller, R. E. Moore, S. L. Daily, B. S. Khatra, M. Dukelow, and A. L. Boynton, *J. Biol. Chem.*, 1990, **265**, 19 401.
775. S. Yoshizawa, R. Matsushima, M. F. Watanabe, K.-I. Harada, A. Ichihara, W. W. Carmichael, and H. Fujiki, *J. Cancer Res. Clin. Oncol.*, 1990, **116**, 609.
776. H. Matsuda, T. Okino, M. Murakami, and K. Yamaguchi, *Tetrahedron*, 1996, **52**, 14 501.
777. M. Murakami, K. Ishida, T. Okino, Y. Okita, H. Matsuda, and K. Yamaguchi, *Tetrahedron Lett.*, 1995, **36**, 2785.
778. T. Okino, M. Murakami, R. Haraguchi, H. Munekata, H. Matsuda, and K. Yamaguchi, *Tetrahedron Lett.*, 1993, **34**, 8131.
779. T. Okino, H. Matsuda, M. Murakami, and K. Yamaguchi, *Tetrahedron*, 1995, **51**, 10 679.
780. H. J. Shin, M. Murakami, H. Matsuda, K. Ishida, and K. Yamaguchi, *Tetrahedron Lett.*, 1995, **36**, 5235.
781. T. Okino, H. Matsuda, M. Murakami, and K. Yamaguchi, *Tetrahedron Lett.*, 1993, **34**, 501.
782. G. M. L. Patterson, L. K. Larsen, and R. E. Moore, *J. Appl. Phycol.*, 1994, **6**, 151.
783. J. Ogino, R. E. Moore, G. M. L. Patterson, and C. D. Smith, *J. Nat. Prod.*, 1996, **59**, 581.
784. K. Stratmann, D. L. Burgoyne, R. E. Moore, G. M. L. Patterson, and C. D. Smith, *J. Org. Chem.*, 1994, **59**, 7219.
785. S. Carmeli, R. E. Moore, and G. M. L. Patterson, *J. Nat. Prod.*, 1990, **53**, 1533.
786. S. Carmeli, R. E. Moore, G. M. L. Patterson, Y. Mori, and M. Suzuki, *J. Org. Chem.*, 1990, **55**, 4431.
787. B. S. Moore, J. L. Chen, G. M. L. Patterson, R. E. Moore, L. S. Brinen, Y. Kato, and J. Clardy, *J. Am. Chem. Soc.*, 1990, **112**, 4061.
788. M. R. Prinsep, F. R. Caplan, R. E. Moore, G. M. L. Patterson, and C. D. Smith, *J. Am. Chem. Soc.*, 1992, **114**, 385.
789. S. L. Mooberry, K. Stratman, and R. E. Moore, *Cancer Lett.*, 1995, **96**, 261.
790. S. Carmeli, R. E. Moore, G. M. L. Patterson, T. H. Corbett, and F. A. Valeriote, *J. Am. Chem. Soc.*, 1990, **112**, 8195.
791. T. Fukuyama and L. Xu, *J. Am. Chem. Soc.*, 1993, **115**, 8449.
792. K. Stratmann, R. E. Moore, R. Bonjouklian, J. B. Deeter, G. M. L. Patterson, S. Schaffer, C. D. Smith, and T. A. Smitka, *J. Am. Chem. Soc.*, 1994, **116**, 9935.
793. S. Paik, S. Carmeli, J. Cullingham, R. E. Moore, G. M. L. Patterson, and M. A. Tius, *J. Am. Chem. Soc.*, 1994, **116**, 8116.
794. S. Matsunaga, R. E. Moore, W. P. Niemczura, and W. W. Carmichael, *J. Am. Chem. Soc.*, 1989, **111**, 8021.
795. I. Ohtani, R. E. Moore, and M. T. C. Runnegar, *J. Am. Chem. Soc.*, 1992, **114**, 7941.
796. R. E. Moore, *J. Ind. Microbiol.*, 1996, **16**, 134.
797. R. E. Schwartz, C. F. Hirsch, D. F. Sesin, J. E. Flor, M. Chartrain, R. E. Fromtling, G. H. Harris, M. J. Salvatore, J. M. Liesch, and K. Yudin, *J. Ind. Microbiol.*, 1990, **5**, 113.
798. T. Golakoti, I. Ohtani, G. M. L. Patterson, R. E. Moore, T. H. Corbett, F. A. Valeriote, and L. Demchik, *J. Am. Chem. Soc.*, 1994, **116**, 4729.
799. R. A. Barrow, T. Hemscheidt, J. Liang, S. Paik, R. E. Moore, and M. A. Tius, *J. Am. Chem. Soc.*, 1995, **117**, 2479.
800. T. Golakoti, J. Ogino, C. E. Heltzel, T. L. Husebo, C. M. Jensen, L. K. Larsen, G. M. L. Patterson, R. E. Moore, S. L. Mooberry, T. H. Corbett, and F. A. Valeriote, *J. Am. Chem. Soc.*, 1995, **117**, 12 030.
801. C. D. Smith, X. Zhang, S. L. Mooberry, G. M. L. Patterson, and R. E. Moore, *Cancer Res.*, 1994, **54**, 3779.

802. C. D. Smith and X. Zhang, *J. Biol. Chem.*, 1996, **271**, 6192.
803. R. Bai, R. E. Schwartz, J. A. Kepler, G. R. Pettit, and E. Hamel, *Cancer Res.*, 1996, **56**, 4398.
804. M. Kobayashi, S. Aoki, N. Ohyabu, M. Kurosu, W. Wang, and I. Kitagawa, *Tetrahedron Lett.*, 1994, **35**, 7969.
805. M. Kobayashi, M. Kurosu, N. Ohyabu, W. Wang, S. Fujii, and I. Kitagawa, *Chem. Pharm. Bull.*, 1994, **42**, 2196.
806. M. Kobayashi, M. Kurosu, W. Wang, and I. Kitagawa, *Chem. Pharm. Bull.*, 1994, **42**, 2394.
807. M. Kobayashi, W. Wang, N. Ohyabu, M. Kurosu, and I. Kitagawa, *Chem. Pharm. Bull.*, 1995, **43**, 1598.
808. G. M. Sharma, L. Michaels, and P. R. Burkholder, *J. Antibiot.*, 1968, **21**, 659.
809. M. Murakami, K. Makabe, K. Yamaguchi, S. Konosu, and M. R. Wälchli, *Tetrahedron Lett.*, 1988, **29**, 1149.
810. K. Terao, E. Ito, M. Murakami, and K. Yamaguchi, *Toxicon*, 1989, **27**, 269.
811. K. Furukawa, K. Sakai, S. Watanabe, K. Maruyama, M. Murakami, K. Yamaguchi, and Y. Ohizumi, *J. Biol. Chem.*, 1993, **268**, 26 026.
812. H. Nakamura, T. Asari, Y. Ohizumi, J. Kobayashi, T. Yamasu, and A. Murai, *Toxicon*, 1993, **31**, 371.
813. H. Nakamura, T. Asari, A. Murai, T. Kondo, K. Yoshida, and Y. Ohizumi, *J. Org. Chem.*, 1993, **58**, 313.
814. T. Asari, H. Nakamura, A. Murai, and Y. Kan, *Tetrahedron Lett.*, 1993, **34**, 4059.
815. H. Nakamura, T. Asari, A. Murai, Y. Kan, T. Kondo, K. Yoshida, and Y. Ohizumi, *J. Am. Chem. Soc.*, 1995, **117**, 550.
816. H. Nakamura, T. Asari, K. Fujimaki, K. Maruyama, A. Murai, Y. Ohizumi, and Y. Kan, *Tetrahedron Lett.*, 1995, **36**, 7255.
817. H. Nakamura, K. Fujimaki, and A. Murai, *Tetrahedron Lett.*, 1996, **37**, 3153.
818. H. Nakamura, K. Sato, and A. Murai, *Tetrahedron Lett.*, 1996, **37**, 7267.
819. M.-C. Rho, N. Nakahata, H. Nakamura, A. Murai, and Y. Ohizumi, *Br. J. Pharmacol.*, 1995, **115**, 433.
820. M. Satake, M. Murata, T. Yasumoto, T. Fujita, and H. Naoki, *J. Am. Chem. Soc.*, 1991, **113**, 9859.
821. G. K. Paul, N. Matsumori, M. Murata, and K. Tachibana, *Tetrahedron Lett.*, 1995, **36**, 6279.
822. J. Kobayashi, M. Ishibashi, H. Nakamura, Y. Hirata, T. Yamasu, T. Sasaki, and Y. Ohizumi, *Experientia*, 1988, **44**, 800.
823. S. H. Grode and J. H. Cardellina II, *Lipids*, 1983, **18**, 889.
824. M. Nakagawa, J. Yoshida, and T. Hino, *Chem. Lett.*, 1990, 1407.
825. K. Mori and K. Uenishi, *Liebigs Ann. Chem.*, 1994, 41.
826. J. Kobayashi, Y. Ohizumi, H. Nakamura, T. Yamakado, T. Matsuzaki, and Y. Hirata, *Experientia*, 1983, **39**, 67.
827. J. Kobayashi, M. Ishibashi, H. Nakamura, Y. Ohizumi, and Y. Hirata, *J. Chem. Soc., Perkin Trans. 1*, 1989, 101.
828. S. Okada, H. Matsuda, M. Murakami, and K. Yamaguchi, *Tetrahedron Lett.*, 1996, **37**, 1065.
829. T. Yamasu and A. Okazaki, *Galaxea*, 1987, **6**, 61.
830. L. Provasoli, in "Culture and Collection of Algae," eds. A. Watanabe and A. Hattori, Japanese Society of Plant Physiology, Tokyo, 1968, p. 63.
831. H. Iwasaki, in "Sourui Kenkyuhou," eds. K. Nishizawa and M. Chihara, Kyouritu-Shuppan, Tokyo, 1979, p. 281 (in Japanese).
832. J. Kobayashi, M. Ishibashi, H. Nakamura, Y. Ohizumi, T. Yamasu, T. Sasaki, and Y. Hirata, *Tetrahedron Lett.*, 1986, **27**, 5755.
833. M. Ishibashi, Y. Ohizumi, M. Hamashima, H. Nakamura, Y. Hirata, T. Sasaki, and J. Kobayashi, *J. Chem. Soc., Chem. Commun.*, 1987, 1127.
834. J. Kobayashi, M. Ishibashi, M. R. Wälchli, H. Nakamura, Y. Hirata, T. Sasaki, and Y. Ohizumi, *J. Am. Chem. Soc.*, 1988, **110**, 490.
835. J. Kobayashi, M. Ishibashi, H. Nakamura, Y. Ohizumi, T. Yamasu, Y. Hirata, T. Sasaki, T. Ohta, and S. Nozoe, *J. Nat. Prod.*, 1989, **52**, 1036.
836. J. Kobayashi, M. Ishibashi, T. Murayama, M. Takamatsu, M. Iwamura, Y. Ohizumi, and T. Sasaki, *J. Org. Chem.*, 1990, **55**, 3421.
837. J. Kobayashi, M. Tsuda, M. Ishibashi, H. Shigemori, T. Yamasu, H. Hirota, and T. Sasaki, *J. Antibiot.*, 1991, **44**, 1259.
838. J. Kobayashi, H. Shigemori, M. Ishibashi, T. Yamasu, H. Hirota, and T. Sasaki, *J. Org. Chem.*, 1991, **56**, 5221.
839. J. Kobayashi, *New J. Chem.*, 1990, **14**, 741.
840. R. J. Capon and D. J. Faulkner, *J. Org. Chem.*, 1984, **49**, 2506.
841. M. Ishibashi and J. Kobayashi, *Heterocycles*, 1997, **44**, 543.
842. J. Kobayashi, M. Sato, and M. Ishibashi, *J. Org. Chem.*, 1993, **58**, 2645.
843. D. R. Williams, P. A. Jass, H.-L. Allan Tse, and R. D. Gaston, *J. Am. Chem. Soc.*, 1990, **112**, 4552.
844. N. Minami, S. S. Ko, and Y. Kishi, *J. Am. Chem. Soc.*, 1982, **104**, 1109.
845. M. Tsuda, T. Sasaki, and J. Kobayashi, *J. Org. Chem.*, 1994, **59**, 3734.
846. K. Horita, K. Tanaka, and O. Yonemitsu, *Chem. Pharm. Bull.*, 1993, **41**, 2044.
847. M. Tsuda, H. Ishiyama, M. Sato, A. Hatakeyama, M. Ishibashi, and J. Kobayashi, *Yuki Gousei Symposium Koen Yoshishu*, 1994, **66**, 93.
848. M. Ishibashi, M. Sato, and J. Kobayashi, *J. Org. Chem.*, 1993, **58**, 6928.
849. J. Kobayashi, N. Yamaguchi, and M. Ishibashi, *J. Org. Chem.*, 1994, **59**, 4698.
850. M. Ishibashi, N. Yamaguchi, T. Sasaki, and J. Kobayashi, *J. Chem. Soc., Chem. Commun.*, 1994, 1455.
851. I. Bauer, L. Maranda, K. A. Young, Y. Shimizu, C. Fairchild, L. Cornell, J. MacBeth, and S. Huang, *J. Org. Chem.*, 1995, **60**, 1084.
852. J. Kobayashi, N. Yamaguchi, and M. Ishibashi, *Tetrahedron Lett.*, 1994, **35**, 7049.
853. Y. Doi, M. Ishibashi, N. Yamaguchi, and J. Kobayashi, *J. Nat. Prod.*, 1995, **58**, 1097.
854. R. Bonnet, A. K. Mallams, J. L. Tee, B. C. L. Weedon, and A. McCormick, *Chem. Comm.*, 1966, 515.
855. J. Meinwald, K. Erickson, M. Hartshorn, Y. C. Meinwald, and T. Eisner, *Tetrahedron Lett.*, 1968, 2959.
856. T. Hashimoto, M. Tori, and Y. Asakawa, *Phytochemistry*, 1991, **30**, 2927.
857. M. Ishibashi, M. Takahashi, and J. Kobayashi, *J. Org. Chem.*, 1995, **60**, 6062.
858. J. Kobayashi, M. Takahashi, and M. Ishibashi, *Tetrahedron Lett.*, 1996, **37**, 1449.
859. I. Bauer, L. Maranda, Y. Shimizu, R. W. Peterson, L. Cornell, J. R. Steiner, and J. Clardy, *J. Am. Chem. Soc.*, 1994, **116**, 2657.

860. M. Ishibashi, H. Ishiyama, and J. Kobayashi, *Tetrahedron Lett.*, 1994, **35**, 8241.
861. H. Ishiyama, M. Ishibashi, and J. Kobayashi, *Chem. Pharm. Bull.*, 1996, **44**, 1819.
862. J. Kobayashi, M. Takahashi, and M. Ishibashi, *J. Chem. Soc., Chem. Commun.*, 1995, 1639.
863. I. Bauer, L. Maranda, K. A. Young, Y. Shimizu, and S. Huang, *Tetrahedron Lett.*, 1995, **36**, 991.

Author Index

This Author Index comprises an alphabetical listing of the names of the authors cited in the text and the references listed at the end of each chapter in this volume.

Each entry consists of the author's name, followed by a list of numbers, for example

Templeton, J. L., 366, 385[233] (350, 366), 387[370] (363)

For each name, the page numbers for the citation in the reference list are given, followed by the reference number in superscript and the page number(s) in parentheses of where that reference is cited in the text. Where a name is referred to in text only, the page number of the citation appears with no superscript number. References cited in both the text and in the tables are included.

Although much effort has gone into eliminating inaccuracies resulting from the use of different combinations of initials by the same author, the use by some journals of only one initial, and different spellings of the same name as a result of the transliteration processes, the accuracy of some entries may have been affected by these factors.

Subject Index

PHILIP AND LESLEY ASLETT
Marlborough, Wiltshire, UK

Every effort has been made to index as comprehensively as possible, and to standardize the terms used in the index in line with the IUPAC Recommendations. In view of the diverse nature of the terminology employed by the different authors, the reader is advised to search for related entries under the appropriate headings.

The index entries are presented in letter-by-letter alphabetical sequence. Compounds are normally indexed under the parent compound name, with the substituent component separated by a comma of inversion. An entry with a prefix/locant is filed after the same entry without any attachments, and in alphanumerical sequence. For example, 'diazepines', '1,4-diazepines', and '2,3-dihydro-1,4-diazepines' will be filed as:-

Diazepines
1,4-Diazepines
1,4-Diazepines, 2,3-dihydro-

The Index is arranged in set-out style, with a maximum of three levels of heading. Location references refer to volume number (in bold) and page number (separated by a comma); major coverage of a subject is indicated by bold, elided page numbers; for example;

Furoxans, reactions, with electrophiles **2**, **1234–55**
  ring-cleavage **5**, 345

See cross-references direct the user to the preferred term; for example,

Mercaptans *See* Thiols

See also cross-references provide the user with guideposts to terms of related interest, from the broader term to the narrower term, and appear at the end of the main heading to which they refer, e.g.

Thiones
  See also
    Thioketones